THE LIBRARY
ST. MARY'S COLLEGE OF MARYLAND
ST. MARY'S CITY, MARYLAND 20686

081435

SPHECID WASPS OF THE WORLD

A GENERIC REVISION

ERRATA

SPHECID WASPS OF THE WORLD: A GENERIC REVISION

p. 44, left col., last line: indent *Gonius* Panzer entry as a synonym of *Palarus*.
p. 44, right col., line 18: insert = in front of *Morphota*.
p. 45, right col., line 31: Jurine should not be italicized.
p. 53, left col., lines 3 through 6: these should be indented several more spaces.
p. 61, right col., line 2: 469 should read 569.
p. 86, right col., line 21: Manke should read Menke.
p. 107, right col., line 23: Prionyxina should not be italicized.
p. 108, both cols.: page numbers need to be inserted in key in following sequence: 109, 119, 128, 124, 127.
p. 116, footnote 23 and right col., line 2: *triangulus* should read *triangulum*.
p. 117, right col., line 39: *triangulus* should read *triangulum*.
p. 184, left col., line 46: indent to align with line 45 (as a syn. of *dispar*).
p. 185, right col., line 37: *Passaboecus* should read *Passaloecus*.
p. 185, right col., line 40: no parens around Rohwer, delete *(Passaloecus)*.
p. 197, right col., line 49: Ammoplanima should read Ammoplanina.
p. 230, right col., line 34: 259 should read 250.
p. 259, left col., line 4: 1917 should read 1918.
p. 260, left col., line 12: *socias* should read *socia*.
p. 271, left col., line 2: *costale* should read *costae*.
p. 272, right col.: last 9 lines indent as synonyms of *brullii*.
p. 294, left col., line 51: 204 should read 304.
p. 295, left col., line 57: *Sadiostethus* should read *Saliostethus*.
p. 299, right col., line 30: *nufiventris* should read *rufiventris*.
p. 302, right col.: footnote 17 belongs at the bottom of page 301.
p. 330, left col, line 27: 329 should read 330.
p. 348, right col., line 15: *tuckumanum* should read *tucumanum*.
p. 360, left col., line 1: Oxybelus should be italic, *Latreille*, p. 364 should be in roman type.
p. 360, left col., line 4: Enchemicrum should be italic, *Pate*, p. 364 should be in roman type.
p. 374, right col., line 13: 000 should read 395.
p. 375, left col., line 6: *Pseudoturneira* should read *Pseudoturneria*.
p. 382, left col., line 35: Ribant should read Ribaut.
p. 406, right col., line 14: *advenas* should read *advena*.
p. 408, left col., line 10: *clariconis* should read *clarconis*.
p. 431, right col., line 24: (cly) should read (con).
p. 485, right col., line 58: Ethipian should read Ethiopian, *Orytus* should read *Orytthus*.
p. 572, right col., figure legend: *Odontsphex* should read *Odontosphex*.
p. 613, left col., line 5: 268 should read 286.
p. 615, right col., line 10: Petzer should read Peltzer.
p. 618, right col., lines 6 through 13 should read:
    1823. A description of some new species of hymenopterous insects. West. Quart. Rep. (Med., Surg., Nat. Sci.) 2:71-82.
    1837. Descriptions of new species of North American Hymenoptera and observations on some already described. Boston J. Nat. Hist. 1:361-416.
Schletterer, A.
    1887. Die Hymenopteren-Gattung *Cerceris* Latr. mit
p. 634, middle col., line 51: *aiurnnensis* should read *aiunensis*.
p. 648, middle col., line 57: Schluz should read Schulz.
p. 649, left col., line 21: 706 should read 506.
p. 649, left col., line 43: Shcaeffer should read Schaeffer.
p. 655, middle col., line 5: *froggati* should read *froggatti*.
p. 661, middle col., line 7: Cheessman should read Cheesman.
p. 661, middle col., line 57: *Jamacensis* should read *Jamaicensis*.
p. 666, middle col., line 42: *Rhompalum* should read *Rhopalum*.
p. 670, right col., line 57: Salsz should read Salz.
p. 676, middle col., line 68: Ribant should read Ribaut.
p. 676, middle col., line 71: *pernniger* should read *perniger*.
p. 677, middle col., line 41: 329 should read 330.
p. 691, left col., lines 31-32: *triangulus* should read *triangulum*.

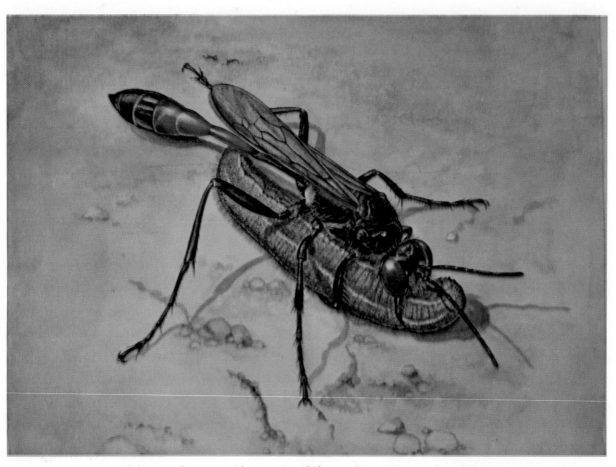
Ammophila azteca Cameron and its prey, the alfalfa caterpillar, *Colias eurytheme* Boisduval.

SPHECID WASPS OF THE WORLD
A GENERIC REVISION

R.M. BOHART and A.S. MENKE

in collaboration with
H.S. COURT, F.D. PARKER, E.E. GRISSELL,
and D.P. LEVIN

UNIVERSITY OF CALIFORNIA PRESS
BERKELEY · LOS ANGELES · LONDON

Issued June 25, 1976

University of California Press
Berkeley and Los Angeles, California

University of California Press, Ltd.
London, England

Copyright © 1976, by
The Regents of the University of California

ISBN 0-520-02318-8
Library of Congress Catalog Card Number: 72-87207
Printed in the United States of America

CONTENTS

INTRODUCTION, vii
BEHAVIOR, 1
ZOOGEOGRAPHY, 5
MORPHOLOGY, 7
    Head, 7
    Thorax, 10
    Gaster, 20
    Glossary of morphological terms, 21
SYSTEMATICS OF THE SPHECIDAE, 27
    Family characters, 27
    Systematic history and basic references, 28
    Phylogeny, 29
    Philosophy of supraspecific categories, 33
    Family statistics, 33
    Explanation of checklists, 34
SYNONYMIC GENERIC CATALOG OF THE SPHECIDAE, 37
KEYS TO SUBFAMILIES AND TRIBES, 57
    Key to subfamilies of Sphecidae, 57
    Key to tribes of Sphecidae, 58
    Key to tribes of Sphecidae in America north of Mexico, 60
SUBFAMILY AMPULICINAE, 63
    Key to genera of Ampulicinae, 66
SUBFAMILY SPHECINAE, 79
    Key to tribes of Sphecinae, 82
SUBFAMILY PEMPHREDONINAE, 155
    Key to tribes of Pemphredoninae, 158
SUBFAMILY ASTATINAE, 203
    Key to tribes and genera of Astatinae, 205
SUBFAMILY LAPHYRAGOGINAE, 217
SUBFAMILY LARRINAE, 221
    Key to tribes of Larrinae, 226

SUBFAMILY CRABRONINAE, 357
    Key to tribes of Crabroninae, 359
SUBFAMILY ENTOMOSERICINAE, 433
SUBFAMILY XENOSPHECINAE, 437
SUBFAMILY NYSSONINAE, 441
    Key to tribes of Nyssoninae, 445
SUBFAMILY PHILANTHINAE, 555
    Key to tribes of Philanthinae, 558
LITERATURE CITED, 593
ADDENDUM, 627
INDEX, 633

INTRODUCTION

When primitive man established his abode in caves, under cliff overhangs or in rudely constructed forest shelters, he came into close contact with wasps of the family Sphecidae. These insects had been making use of such protected places as nest sites for a long time, and there was probably an adverse reaction to the intrusion. The result may have been an uneasy truce and occasional "warfare." When man moved into more sophisticated dwellings, the wasps moved with him and proceeded to build nests under the eaves of his house, in his garden pathways, and in the twigs of his ornamental plants. Today, every farm boy and many city youngsters quickly learn to recognize the blue mud-dauber, the black and yellow mud-dauber, the cicada-killer, and the bembicin sand wasps. The relationship with these insects is generally unfriendly and results from a fear of their greatly overrated stinging powers. Actually, sphecids are mainly beneficial to man and are relatively harmless.

The curious and often elaborate actions of wasps have made them favorite subjects of studies on behavior. Without correct identifications, these studies are of little value. Yet determinations have not always been correct or easy to obtain. In the past century hundreds of scientists have contributed new information on the taxonomy of the several thousand kinds of sphecid wasps. This accumulation is now so large that there is a great need for its consideration from a worldwide standpoint and summarization in the form of revisions, keys, and catalogs.

Since there is no modern, world revision of Sphecidae, no continental revision, and very few comprehensive studies dealing with local faunae, many of the characteristics imputed to the genera and tribes are not reliable except at a regional level. This has resulted in widespread differences of opinion about use and composition of most genera, tribes, and subfamilies. We have taken a broad view and have created or retained categories that seem to convey the most information and that are at the same time morphologically defensible. On the whole this has led to a reduction of subgenera, recognition of more genera than previously, and a narrowing of morphological gaps as large, "catchall" genera have been partitioned in the interests of practicality. We hope that our classification will be viewed impartially and without traditional or conservative bias. Also, since so many genera are imperfectly known, it is to be expected that changes enlarging or decreasing the number will be made in the future as more information becomes available.

Our work then has been directed toward several different goals: (1) providing a worldwide perspective of the family including a reclassification of all categories down to the generic level, diagnoses, keys for all categories down through subgenus, illustrations, and distributional data; (2) the summarization of previously published data including biological information; (3) the notation of problem areas within various taxa; (4) the establishment of a standardized morphological nomenclature for the family; (5) provision of an up-to-date generic catalog; and (6) compilation of species and their synonyms.

We have tried to lay the groundwork for more detailed revisions at the generic or tribal level. Obviously, there is much undetected synonymy presently built into the Sphecidae. Likewise, there are myriads of species yet undiscovered. In a fast-moving science such as taxonomy it is the fate of any revision to be several years out-of-date by the time of its publication. This should in no way diminish its long-term value, however. Incorporation of new species and synonymic findings into a well-ordered revision is a relatively simple matter.

It is our hope that with the publishing of this book the state of knowledge in this family will have reached the point where most future generic revisions will not be limited to political areas. This is not to imply that studies of local faunae are not valuable; but what is needed are revisionary studies on a broader geographical basis, e.g., hemispheric or worldwide, if we are to unravel the many remaining problems.

Subjects that we have not studied or have touched upon only briefly in this book are: comparative anatomy of mouthparts; female terminalia; internal organs; eggs, larvae, and pupae; the metasternal area of the thorax; and leg details such as setation and coxal morphology. Studies in these areas may help to further clarify the classification of the family.

We have been able to study material of all but 5 of the 226 genera recognized in this book: *Anomiopteryx, Odontocrabro, Chimiloides, Leclercqia,* and *Towada.* One additional taxon, *Mellinusterius,* has never been satisfactorily identified, and we have been unable to locate type material. It may not be a sphecid.

Inevitably, our treatment of various sections has been somewhat uneven. The multiple authorship has been partly responsible, but differing extent of published knowledge has contributed, also. There are limitations in the present study, but we feel that we have done the best we can considering the time and material available. In this connection we are reminded of the subtly humorous statement by J. B. Parker (1929) with respect to his revision of the Stizini and Bembicini. He asked his readers for "a full measure of their generous sympathy," and to those who found his work "intolerably bad," he said, "I shall look with expectancy for a speedy publication of something better."

PROCEDURE

The present study was begun in 1964. Implementation of the work followed a series of steps: (1) a thorough search of literature and development of a working library of about 2,000 individual papers by some 400 authors; (2) assembling a collection representing nearly every genus in the Sphecidae with special emphasis on the type species of each genus; and (3) intensive study of wasps thus accumulated, comparison with descriptions in the literature, preparation of several thousand illustrations. Illustrations are often the most useful part of a taxonomic publication. We have tried to provide "habitus" or recognition drawings in lateral and face view as well as detailed figures to supplement the keys and descriptions.

As an important background for the present study, visits to a large number of museums were necessary. R. M. Bohart spent a total of seven months during two trips (1960, 1971) in western Europe, and A. S. Menke visited there for short periods in 1964 and 1965. Bohart, Menke, and F. D. Parker visited museums in the eastern United States at various times from 1960-1970, and Menke traveled to the Canadian National Collection at Ottawa and the Provancher Collection at Quebec in 1966.

We have had the helpful cooperation and support of museum curators and other scientists in all parts of the world, and many individual specialists gave us the benefit of their knowledge on systematic problems in addition to contribution of specimens and personal reprints. Without such help a project of this scope and magnitude would have been impossible.

Of the various collaborators on this book, R. M. Bohart and A. S. Menke have written the sections of general information and have acted as editors for the entire work. Individual authorship of taxonomic sections should be attributed as follows: Nyssoninae, Entomosericinae, and Xenosphecinae to Bohart; Sphecinae, Laphyragoginae, Larrinae, and Heliocausini to Menke; Ampulicinae and Philanthinae to Bohart and Menke; Crabroninae to Mrs. H. Court, except *Crossocerus* to D. Levin, *Crabro, Ectemnius,* and *Lestica* to Bohart; Pemphredoninae to Bohart and E. E. Grissell; and Astatinae to F. D. Parker.

ACKNOWLEDGMENTS

This project had the generous support of the National Science Foundation for three years under grants GB-3074 and GB-5839. The Guggenheim Foundation is thanked for awarding R. M. Bohart a grant which supported a six month visit to various museums in Europe. A. S. Menke would like to thank Sigma Xi - RESA (Scientific Research Society of America) for supporting a trip to Europe for the purpose of studying wasp types.

We have been fortunate to have the services of four fine artists: Judy Jay (Mrs. J. Skovlin), who drew the full figure profiles; Mrs. Karen Fulk, who rendered the many facial portraits; Mrs. Ellen Parker, who contributed several detailed drawings, and Mrs. Mary Benson, who painted the watercolor frontispiece. Most of the line drawings were made by the coeditors.

J. van der Vecht, Putten, Netherlands; W. J. Pulawski, Zoological Institute, Breslau (Wroclaw), Poland, and J. de Beaumont, Lausanne, Switzerland deserve special mention for taking the time to review certain sections of the book. Not only did they offer many useful criticisms, but they also freely gave us much new species synonymy and clarified the status of various names. In the checklists we have credited them in each specific case. Several people examined types, providing helpful notes on each, and assisted with the identification of various species: K. V. Krombein, D. Vincent, E. Riek, I. H. H. Yarrow, E. Rubio, J. Leclercq, K. Tsuneki, D. Levin, and M. Fritz. H. E. Evans and K. V. Krombein reviewed some of the biology sections. W. J. Pulawski was especially helpful in bringing to our attention obscure or unobtainable literature published in eastern Europe. R. Brumley performed extensive library research on wasp biology and contributed significantly to the section on Pemphredoninae. L. A. Stange established a background for our subsequent studies in the taxonomy and phylogeny of the same subfamily. G. S. Steyskal and R. Gagné translated a number of important foreign papers for us, and A. Gurney verified the use of many Orthoptera names. R. W. Matthews, J. Davidson, J. P. van Lith, O. W. Richards, and C. Sabrosky also assisted in various ways. To all of these people we are most grateful for their help.

A number of museums have lent or exchanged material specifically for this study. Including the names of the principal cooperators, the following is a list of these museums.

United States institutions: Academy of Natural Sciences at Philadelphia, Pennsylvania (S. Roback, the late H. Grant); American Museum of Natural History,

New York City (J. Rozen); Bishop Museum, Honolulu, Hawaii (C. Yoshimoto, L. Gressitt); California Academy of Sciences, San Francisco (C. MacNeill, P. Arnaud); California Insect Survey, University of California, Berkeley (J. Powell); Carnegie Museum, Pittsburgh, Pennsylvania (G. Wallace); Los Angeles County Museum of Natural History, California (C. Hogue, R. Snelling); Museum of Comparative Zoology, Harvard University, Cambridge, Massachusetts (H. Evans); U. S. National Museum, Washington, D. C. (K. Krombein, A. Menke); University of California, Davis (R. Schuster); University of California, Riverside (P. Timberlake, S. Frommer); University of Kansas, Lawrence (G. Byers); Utah State University, Logan (G. Bohart, W. Hanson).

Foreign institutions: Academia de Ciencias de Cuba, Havana (P. Alayo); Academy of Sciences, Moscow, USSR (Y. Popov); British Museum (Natural History), London (I. H. H. Yarrow, C. Vardy, R. W. Crosskey); Canadian National Collection, Ottawa (W. R. M. Mason); Commonwealth Scientific and Industrial Research Organization, Canberra, Australia (E. Riek); Department of Scientific and Industrial Research, Nelson, New Zealand (G. Kuschel); Fukui University Biological Laboratory, Japan (K. Tsuneki); Institut für Pflanzenschutzforschung Kleinmachnow (formerly Deutsches Entomologisches Institut), Eberswalde, German Democratic Republic (J. Oehlke); Institut Agronomique de l`Etat, Gembloux, Belgium (J. Leclercq); Instituto Miguel Lillo, Tucumán, Argentina (A. Willink, L. Stange); Kyushu University Entomological Laboratory, Fukuoka, Japan (K. Yasumatsu); Lunds Universitets Zoologiska Institution, Lund Sweden (H. Andersson); Musée Zoologique, Lausanne, Switzerland (J. de Beaumont, J. Aubert); Museo Argentina de Ciencias Naturales, Buenos Aires (M. Viana); Museo Civico di Storia Naturale, Genoa, Italy (D. Guiglia, E. Tortonese); Museo Civico di Storia Naturale, Venice, Italy (A. Giordani Soika); Museo ed Istituto di Zoologia Sistematica, Università di Torino, Italy (M. Zunino, U. Parenti, G. Bacci); Museo de Zoologia, Instituto Municipal de Ciencias Naturales, Barcelona, Spain (F. Español); Museu e Laboratorio Zoologico, Universidade de Coimbra, Portugal (M. de A. Diniz); Muséum d'Histoire Naturelle, Geneva, Switzerland (C. Besuchet); Muséum National d'Histoire Naturelle, Paris, France (S. Kelner-Pillault); Muzeul de Istorie Naturala "Grigore Antipa," Bucarest, Romania (X. Scobiola-Palade); National Museums of Rhodesia, Bulawayo (E. Pinhey); Naturhistorisches Museum, Vienna, Austria (M. Fischer); Rijksmuseum van Natuurlijke Historie, Leiden, Netherlands (J. van der Vecht); South Australian Museum, Adelaide (G. Gross, N. McFarland); Természettudományi Múzeum Allattára, Budapest, Hungary (L. Móczár, J. Papp); Transvaal Museum, Pretoria, South Africa (S. de Kock, the late G. van Son); University of Stellenbosch, South Africa (J. Theron); University Museum, Oxford, England (C. O'Toole, I. Lansbury); Zoological Institute, Leningrad, USSR (V. Tobias); Zoological Institute, Breslau (Wroclaw), Poland (W. Pulawski); Zoologische Sammlung des Bayerischen Staates, Munich, Federal Republic of Germany (F. Bachmaier); Zoologische Institut, Martin Luther Universität, Halle, German Democratic Republic (J. Hüsing); Zoologischen Museum der Humboldt Universität, Berlin, German Democratic Republic (E. Königsmann).

The laborious task of typing all of the entries for the index on 3 x 5 cards was done by Karen Menke. We thank her for handling this important and exacting job.

BEHAVIOR

The fascinating complexities of wasp behavior have been favorite subjects for biological studies. Although Kohl provided a firm foundation for systematics of the family Sphecidae and called attention to papers on biology published up to his time, much has transpired since. The number of species known in 1900 has more than doubled, and the amount of biological information has grown even more rapidly. Here, we have attempted to call attention to the most significant biological works, to present some broad generalities, and to give summaries of reported habits.

In an important volume recently published, "The comparative ethology and evolution of the sand wasps" by Howard E. Evans (1966a), there is considerable philosophical discussion about the concordance, or lack of it, between morphology and habits. Obviously, comparative ethology is a fruitful line of study, the results of which will sharpen our phylogenetic concepts. Structure and behavior patterns are interrelated components of adaptations, are subject to the same general rules of evolution, and can be used as circumstantial evidence in the construction of phylogenetic suppositions.

Evans (1966a, b) has stressed the fact that behavior patterns are plastic. Those that are fixed in some genera or species may be quite variable in others. The same observation can be made about morphological elements. Although the usual trend is toward complexity, it appears that behavior patterns have sometimes evolved from specific to general. Most sphecids are sufficiently advanced that they are fairly restricted as to prey, at least within a family or subfamily of arthropods. Yet, many of the most highly developed sphecids are again more general with respect to prey — *Lindenius, Crossocerus,* and especially *Microbembex,* which provisions with a great variety of prey, alive or long dead. This seeming trend toward simplicity is paralleled in morphology. Consider the evolutionary fate of midtibial spurs from the conventional two (many sphecids), to one (Larrinae, Bembicini and others), to none (males of *Dinetus* and a few *Didineis*). Another example involving structure concerns the dentition of the tarsal claws, which seems to have progressed from simple to one or two inner teeth (Sceliphronini, *Sphex, Parapsammophila*) and back to simple again as in most *Ammophila*. Dentition of the claws may be related to the type of prey or possibly to the type of nesting habitat substrate, but this has yet to be proven. A third example is the tendency toward elimination of mesopleural sulci and carinae. The more generalized sphecids, such as *Dolichurus* in the subfamily Ampulicinae, have an omaulus, sternaulus, and acetabular carina. At the other end of the spectrum none of these can be made out in the tribe Bembicini. Presumably, this modification has paralleled a tendency toward improved flight capability.

Adult sphecids feed on a variety of food. Since most species have short tongues they tend to frequent flowers with short corollas, such as Compositae, Euphorbiaceae, Polygonaceae, and Umbelliferae, in search of nectar. Certain wasps seem to prefer certain flowers, but the correlation is quite weak compared with bees or masarids. Other nectar sources are extrafloral nectaries and honeydew. Some sphecids, such as the Sphecinae and Philanthinae, are commonly collected on flowers, while others, such as the Crabroninae and Pemphredoninae, seek nourishment from honeydew. Thus one must know the habits of the wasps he wants to collect if he is to be successful. Some species feed on body fluids of their prey, as in *Mellinus* and *Philanthus*.

The sleeping habits of wasps have been the subject of numerous investigations. These have been summarized by Evans (1966b). We have observed large aggregations of *Ammophila* of several species on dry weed stems near Blythe, California, and of *Steniolia* near Truckee, California. At least in the latter case, the aggregations may serve a reproductive function (Evans and Gillaspy, 1964). Many wasps sleep singly on plants or in ground crevices or burrows. Males sometimes dig special burrows of their own for this purpose. A few sphecids remain active at night. Some *Liris*, for example, excavate very deep nests, and their digging goes on day and night. Some wasps such as *Sericophorus* start their activities very early in the morning.

Mating in Sphecidae is accomplished in a variety of ways. It may be preceded by prenuptial flights of males, as in *Stictia*, or the setting up of territories, as in *Astata* and *Sphecius*. Copulation may occur in the air, on plants, on the ground, or in nests. It is probable that copulation takes place only once in many species, but at least in *Oxybelus sericeus* the act is repeated many times in a single day (Bohart and Marsh, 1960). In the Trypoxylonini there are cases in which the male remains in the nest and guards it during the female's absence. Mating takes place on her return. The whole subject is a complex one, and reference should be made to Evans (1966b) for additional sources of information.

The majority of sphecids are predators, but some are cleptoparasitic; that is, their larvae develop on the provisions in the nest of some other wasp. The only known sphecid cleptoparasites are the Nyssonini and *Stizoides*. A very few sphecids behave almost like parasitoids (*Larra* and some *Chlorion*). In these two genera, prey paralysis is temporary allowing the wasp to deposit an egg; the host soon revives and regains its normal activities. The wasp egg hatches, and the larva begins feeding on the active host; but death of the latter is inevitable. The definition of parasitoid by Evans and Eberhard (1970) includes the absence in the wasp of a sting, otherwise *Larra* and *Chlorion* could be placed in this category. Table 1 gives types of prey used by the different tribes of the Sphecidae.

The development of prey-carrying mechanisms in wasps and their significance in evolution has been stressed by Evans (1962). This author has made the point that there is no important correlation between manner of prey carriage and type of prey. Evans has outlined several steps in development of prey carriage from primitive to advanced: (1) prey is seized with the mandibles and dragged backwards to the nest (as in Ampulicinae); (2) prey is straddled, seized in the mandibles, and dragged forward over the substrate (as in *Miscophus* of the subfamily Larrinae); (3) prey is carried in flight, held by the mandibles, often assisted by the legs (as in some Sphecinae and some Pemphredoninae); (4) prey is carried in flight, held by the midlegs, but unassisted by the mandibles (as in most Nyssoninae); (5) as in the third step, but prey is held by the hind legs (as in some *Oxybelus* of the subfamily Crabroninae); (6) prey is impaled on the sting and carried in flight (as in some *Oxybelus*); (7) prey is held in a clamp formed by the apical abdominal segment (as in *Listropygia* and *Clypeadon*). Evans has postulated that the more advanced types of prey carriage allow more rapid nest provisioning, tend to minimize attacks by predators and parasites during prey transport, enable the wasp after returning to reopen the nest without dropping the prey, and increase the prey searching distance.

Nests can be arbitrarily placed in two categories depending upon location: (1) underground (terricolus) nests and (2) aboveground nests. The second category includes twig (xylicolus) nests, those in crevices, and the familiar mud nests in barns and on houses. Details of nest construction vary widely. For instance, species with ground nests may regularly exhibit differences in depth of burrow, number of cells, and circumstances of entrance closure. Here again, patterns are usually constant in genera and may be relatively fixed among species, as in the tribe of Oxybelini. In an extreme case known to us, four species in the single genus *Oxybelus*, which may nest in the same area, are *sericeus*, *sparideus*, *emarginatus*, and *uniglumis*. Yet, they differ widely in many of their habits. Thus, the burrow may be constructed by simply digging a hole; or, as in one species (*sparideus*), excavation may be subterranean until the last moment. This unusual situation occurs when the wasp digs into dry sand, which closes behind her. Only after the burrow is finished does she return to the surface, reopen the entrance, and stabilize it with damp sand. The entrance may be closed when the wasp is away (*sparideus*) or open; the entrance may be marked by a carefully prepared circle of sand (*sparideus*) or plain. Prey may be brought to the nest impaled on the sting (*uniglumis*) or carried by the midlegs; prey may be a few muscoid flies (*sparideus*), many muscoid flies (*uniglumis*), salt marsh flies (*sericeus*), or many tiny midges (*emarginatus*). Males may be inconspicuous or, as in one species (*sericeus*), may stand guard near the nest entrance and mate repeatedly with the females. These differences in habits of four related species are further discussed under Oxybelini.

Not all sphecids construct an original burrow. For instance, the twig nesters generally take over the abandoned tunnels of bees. Similarly, a variety of wasps may adopt and remodel the nests of mud daubers. The most primitive type of nest occurs in *Larra* and some *Chlorion* where the wasp uses the burrow constructed by its victim. This type of biology is reminiscent of that in bethyloid and scolioid wasps.

The presence or absence of specialized rake setae on the front basitarsus is a taxonomic feature with important behavioral connotations. These comb-like modifications, which are almost universal in females of fossorial species of Sphecidae, are used in digging, cleaning, and closing the nest. In males with a well-developed rake it is presumed, and in many cases known, that they dig overnight shelters. It is no surprise that in most cases twig-nesting wasps are without a definitive rake. In ground-nesters the absence of a rake in females is relatively rare and no doubt bears behavioral implications. Thus, *Larra* has no rake, but neither does it construct an original nest. In the case of minute wasps, such as the pemphredonine *Pulverro*, the rake is hardly discernible. *Pulverro* has been stated by Bohart and Grissell (1972) to accomplish much of its transport of nest-excavated material in its mandibles and apparently omits the final closure of the burrow. Since the diameter of the entrance hole is only about 1.5 mm, most parasites or predators are eliminated by size, and the completed nest is soon closed by windblown dust and debris.

Predators and parasitic forms maintain a constant check on the populations of Sphecidae. Most of the destruction is accomplished in the cell shortly after its closure by the female wasp. The egg or young larva may

TABLE 1. Prey Records for the Tribes of Sphecidae

Tribe	Spiders	Collembola	Ephemeroptera	Odonata	Orthoptera	Psocoptera	Hemiptera	Homoptera	Thysanoptera	Neuroptera	Trichoptera	Lepidoptera	Mecoptera	Diptera	Coleoptera	Hymenoptera
Dolichurini					X											
Ampulicini					X											
Sceliphronini	X				X											
Sphecini					X											
Ammophilini					X							X			X	X
Psenini								X								
Pemphredonini		X					X	X								
Astatini							X									
Dinetini							X									
Laphyragogini	UNKNOWN															
Larrini					X		X	X				X				
Palarini																X
Miscophini	X				X	X	X	X				X	X			
Trypoxylonini	X															
Bothynostethini														X		
Scapheutini	UNKNOWN															
Oxybelini							X							X	X	
Crabronini		X	X	X	X	X			X	X	X	X	X	X	X	
Entomosericini	UNKNOWN															
Xenosphecini													X			
Mellinini														X		
Heliocausini							X									
Alyssonini								X								
Nyssonini*																X
Gorytini								X								
Stizini*					X											X
Bembicini**		X	X				X			X		X		X		X
Eremiaspheciini	UNKNOWN															
Philanthini																X
Aphilanthopsini																X
Odontosphecini	UNKNOWN															
Pseudoscoliini															X	
Cercerini															X	X

*The Nyssonini and *Stizoides* are cleptoparasites of Hymenoptera.
**Microbembex* uses almost any available prey, dead or moribund.

be killed directly, or the larva may starve to death. The "villains" are fleshflies (Sarcophagidae), beeflies (Bombyliidae), ruby wasps (Chrysididae), velvet ants (Mutillidae), and a variety of others. Many of the behavior patterns of wasps can be correlated with biological pressure from natural enemies. Nests may be located in rather concealed places; females and sometimes males (some *Oxybelus*) may chase intruders away from the nest entrance, or rarely, the male (some *Trypoxylon*) may block the entrance while the female is away. Some females are careful to close the burrow entrance whenever they depart. However, even when the female is most circumspect, parasites may gain entry. A beefly needs only an unguarded moment to pause in flight and cast its eggs into the burrow. Another remarkable situation is the sort of self-parasitism occurring in wasps that provision with fleshflies, which, although paralyzed, may give birth to living and voracious maggots (refer to biology under *Oxybelus*). Strepsiptera or "stylops" are probably the only true parasites of sphecids since they do not kill their host. Members of the Sphecinae are the most commonly attacked group of sphecids. Stylopized wasps often have abnormal morphology and color patterns. Beaumont (1955b) gave a good review of this subject.

In some of the more advanced sphecids the egg hatches soon after the first prey member is deposited, and the larva is fed progressively until it reaches maturity. This is taken a step further in a few *Ammophila*, which maintain several nests at once. One of these, *A. pubescens*, makes daily inspections of each nest to determine whether the egg laid on the initial provision has hatched. If so, additional prey are provided. This is termed delayed progressive provisioning. Evans (1966a) has pointed out that progressive provisioning may have a selective advantage through the protection afforded from natural enemies. When species such as certain bembicins oviposit in the empty cell, feed the larva progressively, and regularly clean the nests, it is nearly impossible for predators, parasites, or scavengers to interfere.

Sociality in Sphecidae is developed only slightly, except in a few cases. Evans (1964d) recorded the communal use of a burrow entrance by several females of the crabronine *Moniaecera asperata*. At the same time he mentioned similar observations by other workers, all relating to crabronine wasps. Other examples are known in the Larrinae: *Liris* and *Dalara*, Williams (1919, 1928a), and *Sericophorus*, Rayment (1955b); and in the Nyssoninae: *Sphecius*, Lin and Michener (1972). Another interesting situation is found in some *Isodontia* species where the nest has one large brood cell in which a number of wasp larvae develop amicably. Nests apparently are maintained over several generations in some *Dicranorhina* (Iwata and Yoshikawa, 1964), and the offspring of *Cerceris rubida* assist the mother in maintaining the nest (Grandi, 1961). A primitive form of sociality is now known in the sphecine genus *Trigonopsis* (Eberhard, 1972). Here several females may cooperate in building a mud nest, and the offspring tend to remain with the nest and mate. Thus, there is considerable inbreeding, and the nest may be maintained over several generations. Two examples of social behavior are known in the tribe Pemphredonini, subtribe Stigmina. Iwata (1964a) recorded three female and four male adults of *Carinostigmus* inhabiting a single burrow in a stem. One female was presumed to be the mother of the other six. Another nest contained a pupa, prepupa, and a medium-sized larva with 14 aphid prey. The evidence for sociality was indicative but not positive. However, in the related genus *Microstigmus* social behavior is well developed. Matthews (1968b) presented a detailed study based on *M. comes* in Costa Rica. Nests contained as many as 18 adults of both sexes and up to 18 cells with brood of all ages. Parental care and cooperation were observed. Since life histories in several related genera in the subtribe are imperfectly known, it is likely that further cases of sociality will be found.

A valuable summary of the behavior patterns of solitary wasps has been provided by Evans (1966b) and Evans and Eberhard (1970). In these articles the known information on such topics as adult feeding, sleeping aggregations, grooming, mating, nest construction, prey selection and paralyzation, prey carriage, orientation to the nest entrance, closing procedures, accessory burrow construction, and oviposition has been condensed. The very impressive survey of the behavior of the Hymenoptera by Kunio Iwata (1972) was received too late for the inclusion here of its pertinent details. Unfortunately for most specialists, this monumental book is entirely in Japanese; however, an English translation will soon be available through the National Science Foundation translation service.

The following workers have made extensive or important contributions to our knowledge of sphecid biology, and their pertinent publications are listed in the bibliography: H. E. Evans, C. Ferton, G. Grandi, K. Iwata, H. Janvier, K. V. Krombein, H. Maneval, E. T. Nielsen, R. W. Matthews, G. Olberg, G. W. and E. G. Peckham, J. Powell, P. Rau, K. Tsuneki, and F. X. Williams. A mention should be made also of a few outstanding monographs restricted to a single species or genus: G. P. Baerends (1941) on *Ammophila pubescens*, A. Steiner (1962) on *Liris nigra*, N. Tinbergen (1932, 1935), Tinbergen and Kruyt (1938) on *Philanthus triangulum*, E. T. Nielsen (1945) on *Bembix rostrata*, K. Tsuneki (1958) on *Bembix niponica*, H. Evans (1957b) on *Bembix*, and A. Huber (1961) on *Mellinus arvensis*.

ZOOGEOGRAPHY

Basic information on distribution is given under each genus, tribe, and subfamily. However, there are some patterns that warrant special mention.

Certain genera have disjunct or uneven ranges. Among these are *Odontosphex* with one species in northwest Africa and Saudi Arabia and three in southern South America; *Clitemnestra* with Australian and Chilean species; *Ochleroptera* with many New World species and one in eastern New Guinea; *Chalybion* with two New World species and many Old World forms; *Ancistromma* with many species in North America and several in the Mediterranean area; *Plenoculus* with many North American representatives, one on the Iberian Peninsula, and one in southwestern USSR; *Pisonopsis* with three species in western temperate North America and two in southern South America; *Prosopigastra* with many forms in Africa and Asia but only one in western North America; *Pseneo* with many species in the New World and two in the Philipine Islands; *Podagritus* with South American, Australian, and Nigerian species; *Neodasyproctus,* which is well distributed in the Ethiopian Region, but has only one species ranging from Australia to Fiji; *Xysma* with one species each in eastern United States and the Cape region of South Africa; and *Parapiagetia* with its species divided between the warmer parts of the Old World and southern South America. One tribe, the Bembicini, has an unbalanced generic distribution pattern. *Bembix* is cosmopolitan, but the 18 other genera of the tribe are restricted to the New World, especially in South America.

Continental areas with high endemism at the generic level are: southern Africa, Australia, and South America. Future collecting in these areas should yield the most fruitful results with respect to clarification of sphecid taxonomy. The deserts of the southwestern United States harbor a few endemic forms, such as *Xenosphex, Ammopsen,* and *Xerogorytes.* Xeric areas of the Palearctic Region, such as North Africa and southwestern USSR, show much endemism at both specific and generic levels. Island endemics are found in Hawaii (*Nesomimesa* and *Dienomimesa*) and Madagascar (*Hovanysson*). Island endemism at the specific level is common in *Psen, Pison,* and *Liris,* especially in the East Indies. This area serves as an illustration of the evolutionary route through isolation from subspecies to species and eventual generic distinction. The West Indies show the same trend to a lesser extent.

As might be expected, there is much evidence for traffic between North America and Asia via a Bering Straits passage. There is less evidence for a widespread exchange between North and South America; the direction seems to have been largely south to north. Examples are the limited North American distribution of the primarily South American genera *Bothynostethus, Ochleroptera, Podium, Eremnophila,* and *Microbembex.*

Several genera, which seem quite successful in North America, have not penetrated beyond Panama. Examples are *Pseudoplisus, Podalonia, Palmodes, Philanthus,* and *Eucerceris.* The large genus *Ammophila* is represented in North America by a number of species groups, but only one of these is found in South America.

The Sphecidae offers some evidence for a former connection between Australia and western South America. The genus *Clitemnestra* is the only taxon that is currently restricted to Australia and Chile, for example. Other genera suggest such a connection, but their current distribution is much broader.

Man has spread a few "tramp" species to many areas. *Sceliphron caementarium,* a native of North America, has been established on many Pacific islands and has reached Australia and Europe. Early traders transported *Sceliphron fuscum,* a Madagascan species, to New Caledonia. Several Old World *Trypoxylon* have been introduced into North America, and two North American species are found in Japan. The Old World *Pison argentatum* has been so widely distributed that its native land is in doubt. All of these species are readily subject to transportation as mature larvae in mud nests on machinery and the like or in all sorts of crevices aboard ship or plane.

Examples of individual species with broad ranges not influenced by man are: *Sphex ichneumoneus* and *Prionyx thomae,* northern United States to Argentina and Chile; *Sphex argentatus,* India to Japan and Australia; and *Prionyx viduatus,* South Africa to Taiwan.

Dryudella pinguis and *Pemphredon montanus* are examples of Holarctic species. Most Holarctic forms are twig-nesters. *Ammophila azteca* is an example of a species with very broad ecological tolerances. It is found in California from the beaches to high elevations in the mountains. It ranges across North America and northward into Canada where it crosses the Arctic Circle.

MORPHOLOGY

Over many years a confused and duplicative terminology has evolved to describe the parts of the sphecid body. We have adopted the most universal, most descriptive, or least ambiguous names. In certain cases we have coined new terms either because none were available or because those already in use were morphologically inappropriate. We have referred to basic works that do not deal specifically with the Sphecidae for some of our names and, also, in order to make comparisons of nomenclatures used in other families. Among these are Snodgrass (1910, 1935, 1956, 1963), Michener (1944, 1965), Duncan (1939), Ross (1936), Lanham (1952), Daly (1964), Richards (1956a, b, and 1962), Eickwort (1969), Stephen, Bohart and Torchio (1969), and Matsuda (1971).

Following the lead of Michener (1944) some authors have used the term mesosoma for the definitive thorax in Hymenoptera because it contains the first abdominal segment (propodeum). Under this system the definitive abdomen is called the metasoma. While this usage may be morphologically sound, we prefer to use the more conventional names, thorax and abdomen. We have followed Snodgrass (1960, 1963) in our use of terms such as flagellomere, tarsomere, tergum, sulcus, suture, and segment.

In the following discussion we have tried to explain our terminology, describe evolutionary morphological trends, and indicate the variety of peculiar structures in the family.

HEAD (FIG. 1)

The sphecid head is not unusual from a family standpoint. Yet, it is ordinarily distinctive at the generic and specific levels. In fact, most species can be identified by a careful study of the face. Among the vast variety of details, a few trends are apparent.

1. Mouthparts have become greatly elongated in a few generic groups. The lengthening may involve the cardo (as in *Laphyragogus*), the prementum, stipes, galea, palpi, and glossa, or all of these. In our descriptions mouthpart length usually refers to the relative length of the galea and glossa. In the Bembicini the prementum and stipes are not unusually developed but the galea and glossa are greatly lengthened. Even the labrum is long and acts as a sort of basal sheath. In some cases (such as *Bembix*), the overall mouthparts are long, but the palpi are short. In a related genus (*Stictiella*) the mouthparts, including the palpi, are long. Striking differences in length of palpi can occur within a genus. The relatively long mouthparts of some species of the sphecine genus *Prionyx* bear long maxillary palpi. In *parkeri* the labial palpus is equally long. However, in the closely related *thomae* the labial palpus is quite short. Mouthparts, whether long or short, are usually retracted and largely concealed behind the mandibles. Exceptions are the bembicin genera *Zyzzyx* and *Steniolia* in which the tongue is normally extended and may reach the midcoxae.

2. Irrespective of mouthpart length, there has been a tendency toward reduction in number of palpal segments. This occurs in somewhat scattered fashion in many of the most highly specialized sphecids. The standard palpal formula in generalized wasps is 6-4 (six maxillary palpal segments and four labial palpal segments). This becomes 5-3 in some *Tachytes*, *Eremiasphecium*, and some Crabronini. Intermediate combinations, such as 6-3, 6-1 and 5-4 also occur. In the Bembicini the most extreme reductions take place, culminating in *Bembix* with 4-2 and *Microbembex* with 3-1. The labial palpus in some *Prionyx* is also reduced to one segment. In our study we have not given the mouthparts the attention they deserve, and a thorough study of them throughout the family should provide additional phylogenetic information. The paper by Ulrich (1924) would probably make an excellent background for such a project.

3. Judging by some of the more primitive sphecids, the archetype mandible is relatively simple and probably bears a subapical tooth on the inner margin (fig. 1). This type is widespread throughout the family. Presumably in response to prey relationships, additional teeth may be present along the inner margin as in *Ammophila, Alysson, Mellinus, Pemphredon*. In Crabronini there is often a basal spine or process. In some Crabronini the mandible appears to be split at the apex (*Crossocerus, Moniaecera*), and if the subapical tooth is involved, the apex may be tridentate (*Ectemnius, Foxita*). A some-

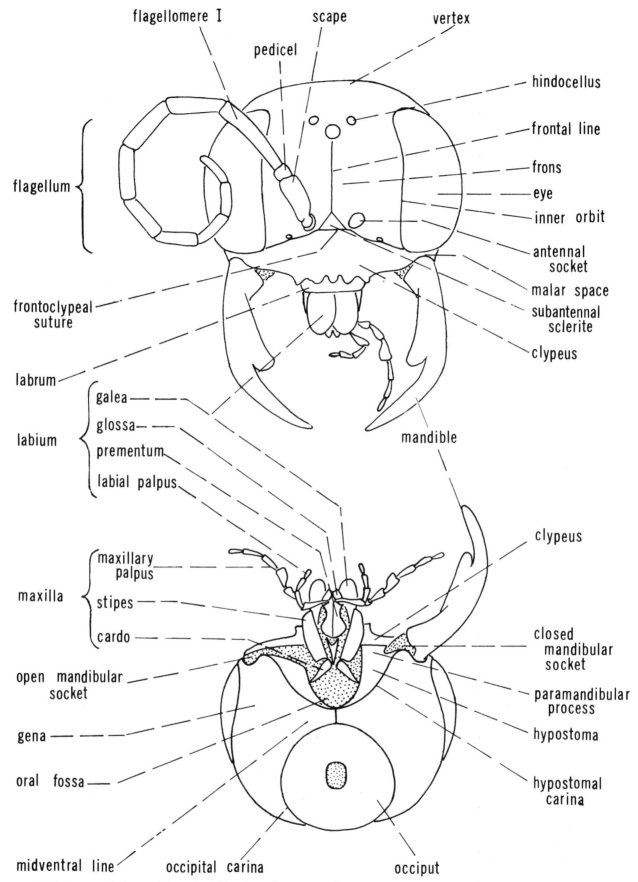

FIG. 1. Morphology of the sphecid head. A: front view of head. B: posteroventral view of a composite head.

what similar situation occurs in some Sphecini (*Isodontia*). In a few sphecids the mandible may have a cluster of teeth at the apex as in some Pemphredonini (*Paracrabro*). The completely simple mandible appears to be a derived condition and is found in parasitic forms (Nyssonini) and other specialized types, but the relatively simple, scolioid-like mandible of *Larra* may represent a generalized type.

One feature of special concern is the notch or step on the externoventral surface of the mandible between the middle and the basal third. This structure, for which no function is known, occurs in many places in the Larrinae and Crabroninae, and in a few related forms such as *Laphyragogus* and *Xenosphex*. However, it is not found elsewhere in the family.

4. The sphecid head is normally orthognathus, that is, oriented dorsoventrally, but in a few of the more specialized pemphredonines (Ammoplanina) and in some ampulicines and sceliphronins it is prognathus, directed forward, so that its long axis is nearly parallel with that of the rest of the body.

5. The occipital carina is generally distinct and when complete reaches the hypostomal carina or ends posterior to it at the midventral line. In many cases it is evanescent below. In some it is trenchant below or produced into a process. Exceptionally, in some ampulicines as well as in *Spilomena* and related pemphredonines it may be completely absent. In *Laphyragogus* the carina has an angular outline dorsad, and in *Pae* it bears dorsolateral expansions.

6. The hypostoma, in addition to bearing occasional processes (*Eremnophila, Dienomimesa*), may be extended forward to the clypeus, thus completing the sclerotization of the mandibular socket. This paramandibular process must be considered an advancement, which in some way strengthens the head or increases the efficiency of the mandible. It is not present in generalized aculeate wasps such as *Scolia*. Among sphecids the paramandibular process occurs sporadically in many of the subfamilies but is practically universal in the Philanthinae and Sphecinae. On the other hand it is absent in the Nyssoninae, Xenosphecinae, Laphyragoginae, Entomosericinae, and Astatinae. In the Larrinae closed sockets are found only in *Auchenophorus* and *Parapiagetia*. *Ampulex* is the only ampulicine with closed sockets. In the Pemphredoninae the sockets are open in all Psenini although often narrowly so. The sockets appear to be closed in all Pemphredonini. They are open in the crabronine tribe Oxybelini and also in the more generalized Crabronini (*Encopognathus, Entomocrabro, Anacrabro*).

7. Primitively the sphecid head is broad, that is, the eyes are widely separated and the inner orbits are essentially parallel (*Chlorion*). Narrowing of the frons and a marked convergence of the orbits above or below has taken place in many sphecids, and while the trends are not all in one direction within a subfamily, they must have considerable evolutionary significance. Examples of extreme convergence above are the holoptic males of some Astatinae, *Palarus* and *Prosopigastra;* but even in females the convergence may be marked in such diverse genera as *Tachysphex, Heliocausus,* and *Pseudoscolia.* Conversely, in a great many genera the inner eye margins converge below. When this results from a broadening of the lower part of the eyes, the facets in this area become enlarged. Extreme examples are found in many Crabronini, such Gorytini as *Pseudoplisus,* and the larrine genera, *Scapheutes* and *Bothynostethus.* The inner margins may also be notched (Trypoxylonini, Philanthini), incurved, outcurved, or sinuate.

The presence of extensive hair on the eyes is a specialization found in a few genera of Miscophini, Crabronini, Trypoxylonini, Gorytini, and Bembicini. Notable genera in which at least some species are concerned are *Nitela, Moniaecera, Huavea, Entomognathus, Pison, Megistommum,* and *Trichostictia.* The use of hair on the eyes as a diagnostic tool must be weighed carefully, however, since it can be detected on the eyes of most sphecids under the right light and magnification.

8. The ocelli are characteristically reduced in some genera or tribes. In the Larrini the hindocelli are represented only by oval to comma-shaped scars. In *Odontosphex* the hindocelli are nearly indistinguishable. All ocelli are somewhat deformed in *Palarus, Kohlia,* and *Heliocausus*. In Bembicini the ocelli are greatly reduced or absent, sometimes leaving a shiny scar. A feature found in Psenini and associated with the ocelli is the postocellar line, which connects the posterior margins of the hindocelli by a transverse groove. It occurs in *Pseneo* and related genera. This seems to be the same groove that is present in *Scolia* and *Holopyga* (Chrysididae).

9. The antenna consists of three true morphological segments; the scape that articulates with the antennal socket, the pedicel, and the terminal flagellum. The last is subdivided into segment-like units called flagellomeres. Typically, there are 10 flagellomeres in the female and 11 in the male; but in some genera two flagellomeres in the male are indistinguishably fused so that only 10 flagellomeres are discernable (*Ectemnius, Lestica, Sericophorus,* and some *Solierella* and *Trypoxylon,* for example). The female flagellum is rarely modified, but there are a variety of structures and deformities in the male. Presumably, these modifications of the male flagellum are sensory in function, and usually they are quite helpful in species discrimination. For example, flat or curved plate-like areas, here termed placoids, occur in varying numbers in species of *Sphex, Isodontia, Prionyx, Larra,* and *Liris*. Linear welts called tyli occur on the flagellum in males of *Passaloecus, Psenulus,* and *Astata,* among others. Sometimes a single row of setae is found along the length of the venter of the flagellum, as in some *Heliocausus* and some crabronines. The last flagellomere is often distinctively modified, as in the Alyssonini. A great variety of other types of swellings, projections, and deformities are found on the male flagellum throughout the family and especially in the Crabroninae. The twisted and deformed antenna of

Dinetus is noteworthy. The antenna is filiform in most sphecids, but the flagellomeres sometimes gradually thicken towards the apex forming a club. Occasionally the antenna is almost capitate (*Tanyoprymnus, Ammatomus*).

The scape is short and stout in primitive groups, but it becomes long and slender in more advanced forms such as the Crabroninae, which have a geniculate antenna. The pedicel sometimes offers valuable taxonomic characters, but in general it is a simple, short segment. It has been common practice, especially among European workers such as Kohl and Beaumont, to regard the pedicel as the first "segment" of the flagellum. This only leads to confusion when counting the flagellomeres in reference to some comparative character, and furthermore, it is morphologically incorrect.

The head has many other structural modifications that do not seem to illustrate trends but are peculiar enough to deserve mention. The frons may have a partial or complete frontal carina or groove from the midocellus to the interantennal area. When carinate, it may be developed into a median tubercle, an interantennal process, or lead to a transverse carina below the antennal sockets (some Pemphredoninae). In Trypoxylonini and Nyssonini the frontal carina may bifurcate above. Other modifications of the frons are a more or less pronounced scapal basin that is sometimes sharply margined above and laterally (*Foxita*), a median swelling as in *Prosopigastra* or lateral swellings as in *Liris* and *Larropsis,* a raised U-shaped plate that overlaps the antennal sockets as in *Dolichurus* and some *Trypoxylon,* and orbital foveae somewhat similar to those of Colletidae and Andrenidae, which are present in many Crabronini.

The vertex may be greatly raised as in *Ammoplanus*, or it may bear knobs or other projections as in some *Belomicrus*. The clypeus is particularly useful as a species character, but overall shape and dentition of the free margin may be of generic or tribal value. In nearly all of the Crabronini the clypeus is unusually transverse and is used as a tribal recognition character. The labrum is often concealed beneath the clypeus, but when exposed it may offer unusual features of generic rank (Pemphredonini, Sphecini, Larrini, Bembicini). The malar space is usually rather short in Sphecidae, but in a few cases, where the eyes have shortened, the space is considerably lengthened (some *Microstigmus*, some Miscophini). The gena sometimes has ventral prongs which may be associated with the occipital carina (*Gastrosericus, Solierella*).

In species that nest in sand a psammophore or "sand basket" is sometimes developed in females, and used to carry sand out of the nest. The psammophore consists mainly of long curving bristles (ammochaetae) from the genal area and the inferior margin of the mandible. These are used in conjunction with the foreleg psammophore. Examples are *Ammopsen* (Pemphredonini), *Belomicroides* (Oxybelini), and various Sphecini.

Position of the antennal sockets in relation to the clypeus would seem to have special taxonomic importance. Certainly in many more generalized sphecids, such as Ampulicinae, Astatinae, and Larrini, the sockets are contiguous with the clypeal margin or nearly so. Similar situations are found in many Nyssoninae and Pemphredonini. It could be argued that removal of the sockets by a socket diameter or more from the clypeus is a specialization, as in such advanced groups as Philanthinae, many Gorytini, Xenospheciae, *Laphyrogogus*, etc. However, there are so many exceptions and contradictions that it is unwise to draw sweeping conclusions on evolutionary trends. For example, the sockets are relatively high in *Tenthredo* (Tenthredinidae), low in *Scolia* (Scoliidae), moderately high in many Sphecini, high in Psenini, low in many Bembicini. Finally, in the specialized genus *Ammoplanops* (Pemphredonini) the sockets may be low in some species (as in other members of the tribe) but high in others.

THORAX (FIGS. 2-4, 29, 45, 94, 99A, 112, 158)

The most numerous and significant structural modifications are found on the thorax and associated propodeum. At least some of these are connected with flight, nest construction, prey seizure and carriage, and copulation. To fully appreciate the evolutionary implications it is well to start with a generalized or archetype sphecid thorax (fig. 2). Its main features, most of which are found in the tribe Sphecini, are given below:

1. Overall form elongated with the dorsal components distinct and not closely fitted to one another (as examples, Ampulicinae, most Sphecinae, *Lyroda, Larra,* and Alyssonini).

2. Scutum with complete notauli (for example, many Ampulicinae, some Pemphredonini, *Entomosericus*).

3. Mesopleuron with only two sulci: the episternal sulcus, which is complete ventrad to the anterior mesopleural margin, and the scrobal sulcus, which joins the episternal sulcus at approximately a right angle (for example, most Sphecinae, most Miscophini and Larrini, Astatinae, and Philanthini).

4. The metapleural sulcus or line extends dorsad from the hindcoxal cavity through the upper metapleural pit to the metanotum, so that the metapleuron is differentiated completely from the propodeum, and the pleuron is divided by the transmetapleural line into upper and lower areas as, for example, in most Ammophilini.

5. Wings are long (fig. 5): forewing with three submarginal cells, the first partly divided by a remnant of the first radial crossvein (1r) (for example, *Aphelotoma, Uniplectron,* and *Austrogorytes*); marginal cell long, apex acute (for example, many Pemphredoninae); two discoidal cells in addition to the subdiscoidal cell; hindwing with median and submedian cells large; jugal lobe constituting most of the anal area (Sphecinae, Larrini, Astatini, and Laphyragoginae) and retaining a remnant of A_3 (Sphecinae, some Larrini, and Laphyragoginae); costa and a remnant of subcosta present in hindwing (some Astatini and some Dolichurini).

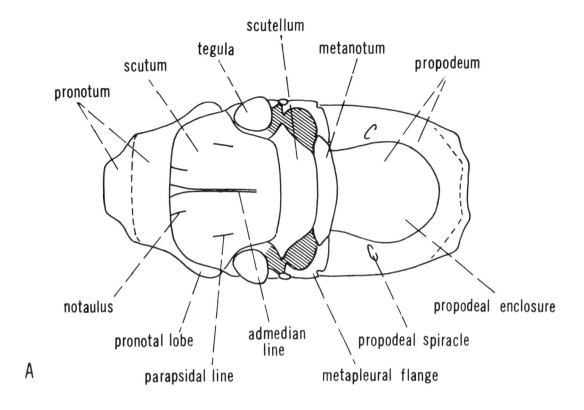
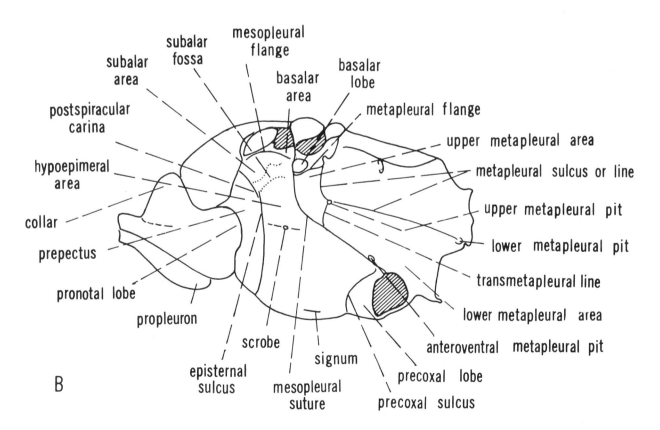

FIG. 2. Morphology of the generalized sphecid thorax, *Ammophila* (Ammophilini). A, dorsal; B, lateral.

6. Legs are short, stout, and relatively simple: midtibia with two apical spurs (Ampulicinae, Sphecinae, Astatinae, Xenospheciinae, and about two-thirds of the Nyssoninae); tarsus with five tarsomeres; tarsomeres with plantulae (some or all genera of Dolichurini, Sceliphronini, Ammophilini, Trypoxylonini, Pemphredoninae, Scapheutini, Nyssonini, Gorytini, Philanthini, Pseudoscoliini, Aphilanthopsini); claws simple.

From the basic thoracic pattern outlined above, sphecids have evolved many diverse specializations. There has been an obvious trend toward shortening and closer fitting of the thoracic components. The end result is a compact, streamlined, nearly spherical structure, which is characterized by the loss of most sulci and other coarse sculpture. This is best exemplified by the Bembicini. Simplification of the thorax is only one of the ways in which it has become specialized. The numerous carinae and sulci found on the thorax of many Gorytini and Crabronini must be considered as specializations also, although their significance is largely unknown. Likewise, the unusual lengthening of the thorax, as in some sceliphronins, is obviously a specialization that probably has to do with their nesting and prey searching behavior. Some wasps, such as *Penepodium complanatum* and a few of the Australian *Tachysphex*, are strongly flattened dorsoventrally, an adaptation to help in locating cockroaches in tight situations. Wing tendencies have been a reduction in wing length, loss of veins, and constriction or loss of the jugal lobe (Miscophini, Pemphredonini, Crabronini, for example). Reduction of venation is prevalent among smaller wasps, and apparently there is a direct relationship between body size and extent of venation (for example, Ammoplanina). Leg specializations include a variety of modifications in males correlated with copulation, and in females with nest construction and prey carriage. Also, in higher forms, the inner or both midtibial spurs are reduced or absent (Larrinae, Pemphredoninae, Bembecini). Plantulae are usually absent in the more specialized members of subfamilies and tribes.

Prothorax

The components of the prothorax are the notum and propleuron. The notum has a posterior collar and lateral lobes. The collar in its two extremes may be long and distinctly separated from the scutum (*Trigonopsis, Ampulex,* Alyssonini) or thin and closely appressed to the scutum (most Larrini, Stizini). A transverse carina (occasionally many) or sulcus may be found on the collar, and the humeral angles may be sharp (some *Nitela*, Nyssonini, and Crabronini), or the collar may be tuberculate (Ampulicinae and *Penepodium*). In all sphecids the posterior margin of the collar is essentially a straight line. This characteristic is one of the basic traits that separate the Sphecidae from other aculeate wasps. The pronotal lobes (= pronotal or humeral "tubercles" or lateral lobes of some authors) are protective shields covering the mesothoracic spiracles. Their presence is a prominent character of the Sphecidae. Matsuda (1971) used the term spiracular lobes for these structures, and while his term is perhaps more accurate, we prefer to retain the more commonly used name pronotal lobe. The lobe is in contact with the edge of the tegula or nearly so in a few sphecids (some Astatinae, Pemphredonini and Ampulicinae, *Holcorhopalum*), but usually the two are well separated.

The propleuron is ordinarily simple but it may bear a variety of tubercles, prongs, or carinae (*Gastrosericus, Trachypus, Trypoxylon*). Matsuda (1971) used the term cervicopleuron for this plate because it apparently contains the cervical sclerite.

Scutum (fig. 2A)

Three pairs of longitudinal grooves or lines are found on the scutum. The most median pair arises anteriorly and forms the admedian lines (anteroadmedian lines of Daly, 1964, 1965). They are often short and inconspicuous but may form a single line (Nyssonini) or define a median welt (some Bembicini and some *Ammophila*). Outside these lines are the notauli which may extend the entire length of the scutum in more generalized forms (most Ampulicini, some *Polemistus*, etc.), but most often they are evanescent or short. The outermost lines are the parapsidal lines which arise posteriorly and correspond to the mesopleural signa (Daly, 1964, 1965). They seem to have little taxonomic value. The paper of Tulloch (1929) should be consulted for further discussion of the notauli and parapsidal lines.

The lateral margin of the scutum is usually simple or somewhat reflexed. However, it projects over the tegulae to some extent in Nyssoninae, and all except the most primitive genera of that subfamily have a carina or line running obliquely backward and mesad from the margin near the rear of the tegula. This oblique carina delimits a small posterolateral area, which may be declivous.

Other scutal peculiarities are a transverse anterior carina in some *Polemistus*, the projection of the scutum over the pronotum as in *Pulverro*, and posterior, raised lobes that project over the scutellum in *Holcorhopalum*.

Scutellum

The scutellum is a relatively unmodified posterior sclerite of the mesonotum. In many genera there is a transverse, often pitted, anterior groove. In a few cases the scutellum is longitudinally ridged (*Austrogorytes, Oxybelus,* and some *Ammophila*). In *Zanysson* there are small winglike projections, and in females of most *Stizus* there is a posteromedian pubescent depression. In *Trirogma* the scutellum may be conically elevated.

Metanotum

This transverse sclerite is often erroneously called the postscutellum. Michener (1944) and Matsuda (1971) used the more appropriate term, metanotum. It is sometimes bilobed, trilobed, or serrate. In *Oxybelus* and related genera it bears striking wing-like expansions or squamae.

Mesopleuron

A complex system of secondary sulci and carinae has evolved on this sclerite in the Sphecidae. There is a clear pattern in the family, however, and it is usually possible to homologize the features found in different groups. Tracing the development of grooves and carinae is simplified by the presence of certain landmarks, which are found also in Scoliidae, Vespidae, Pompilidae, and in the bees. These marking points are the scrobe, subalar fossa, signum, and the pronotal lobe (fig. 2B). A peculiarity in Hymenoptera is the apparent extension of the mesothoracic pleuron to the midventral line and absence of a definable sternum. There have been two schools of thought on this question; whichever explanation is followed, the nomenclature of the parts is affected. Snodgrass (1910) and Richards (1956a) have suggested that the ventral region contains visible sternal elements. The other view, espoused by Michener (1944) and derived from various papers of Gordon Ferris, is that the sternum has infolded and is not visible externally. The recent viewpoint of Matsuda (1971) is that part of the sternum is visible but not differentiated. He used the term pleurosternum for the ventral region of the mesothorax. For taxonomic purposes it seems most practical to use the term mesopleuron for the entire mesothoracic pleuron, even though Matsuda has shown that the definitive mesopleuron consists almost entirely of the episternum. The ventral area can be called the pleurosternum (or more properly, mesopleurosternum) following Matsuda, or it can be called the ventral mesopleural area or ventral surface of the mesopleuron.

A diverse terminology for pleural grooves and plates has accumulated over the years. Michener (1944) gave a thorough treatment of pleural morphology in bees, and we have used his terms with some modification. Additional terms and their sources are: omaulus (Pate, 1936a), sternaulus (Viereck, 1916), prepectus (Snodgrass, 1910), hypersternaulus (Pate, 1940a), acetabular carina (Richards, 1956a), basalar area (Richards, 1972a), and verticaulus (Court and Bohart, 1966). Newly coined terms are subalar line or carina, basalar lobe, omaular area, subomaulus, subomaular area, mesopleuraulus, postspiracular carina, and postspiracular area.

The following list of some of the mesopleural terms that we employed includes the most significant synonyms. These were compiled from sphecid literature as well as from papers dealing with related families.

1. Episternal sulcus (figs. 2 B, 3 A)
    anterior oblique suture (Richards, 1956a, b)
    epicnemial sulcus or suture (Leclercq, 1954a, 1957e; Richards, 1956a, b)
    mesopleural sulcus (in part, Duncan, 1939)
    oblique groove (Evans, 1966a)
    pre-episternal suture (Michener, 1944)
    prepectal suture (in part, Rohwer, 1916)
    vertical median suture, a-b-c, and median episternal groove (Bequaert, 1918, p. 254)
2. Hypoepimeral area (fig. 2 B)
    epimeron (Kohl, 1896; Arnold, 1922)
    mesepimeron (Duncan, 1939)
    mesepimerum (Bequaert, 1918, p. 254)
    first rectangle (Menke, 1964a)
3. Midventral line
    discrimen (Michener, 1944)
    median sternal groove or mesolcus (Richards, 1956a)
4. Omaulus (figs. 3, 4, 45, 112, 158)
    epicnemial ridge or carina (Kohl, 1896; Beaumont, 1964b; Evans, 1966a)
    precoxal suture (in part, Richards, 1956a, b)
    prepectal suture (in part, Rohwer, 1916)
    prepectal carina (Leclercq, 1957e)
5. Postspiracular area
    postspiracular sclerite (Michener, 1944; Richards, 1956a)
6. Postspiracular carina (figs. 2 B, 3 A, 45)
    precoxal suture (in part, Richards, 1956a, b)
7. Precoxal lobe (figs. 2 B, 94 A)
    katepisternum (Michener, 1944)
    mesosternal lobe (Richards, 1956b)
    trochantin (Richards, 1956a, b)
8. Precoxal sulcus (figs. 2 B, 45)
    precoxal suture (Leclercq, 1957e)
    trochantinal suture (Richards, 1956a, b)
9. Preomaular area (fig. 112)
    anterior face of mesepisternum (Michener, 1944)
    epicnemium (Arnold, 1922; Pate, various papers; Evans, 1966a)
    prepectus (Leclercq, 1957e)
10. Prepectus (figs. 2 B, 112)
    anterior plate of mesepisternum (Richards, 1956a)
    upper plate of mesepisternum (Bequaert, 1918, p. 254)
    epicnemium (Richards, 1956a, b; Leclercq, 1957e)
    episternum (Handlirsch, 1887; Arnold, 1922)
    pre-episternal area (Michener, 1944)
    pre-episternum (Eickwort, 1969)
11. Scrobal sulcus (fig. 45)
    epimeral suture (in part, Beaumont, 1964b)
    episternaulus (Viereck, 1916)
    mesopleural sulcus (in part, Duncan, 1939)
    pleural suture (in part, Gillaspy, 1964)
    scrobal groove (in part, Evans, 1966a)
    b-d section of "mesopleural suture" (Bequaert, 1918, p. 254)
12. Scrobe (figs. 2 B, 112)
    episternal scrobe (Michener, 1944; Richards, 1956b)
    mesopleural pit (Pate, various papers)
    mesopleural stigma (Leclercq, 1954a)
13. Signum (fig. 2 B)
    precoxal suture (in part, Richards, 1956b)
    sternopleural suture (Bequaert, 1918, p. 254)
14. Sternaulus (fig. 158)
    precoxal carina or ridge (in part, Leclercq,

1954a; Evans, 1966a)
precoxal suture (in part, Richards, 1956a, b)
prepectal or epicnemial suture, f-g (Bequaert, 1918, p. 254)
15. Verticaulus (fig. 123)
precoxal carina (Leclercq, 1954a)

The general mesopleural pattern is outlined below. Further explanation can be obtained from the glossary and by consulting the figures and discussions under each subfamily, tribe, and genus.

The episternal sulcus is short in many sphecids, fading below its juncture with the scrobal sulcus, when the latter is present. In some groups it is entirely absent (fig. 4). Usually the length of the sulcus is a generic characteristic, but occasionally, as in *Ammophila,* it varies from long to very short. In the more primitive groups the episternal sulcus usually lies just behind the postspiracular carina (fig. 2 B, e.g., Sphecinae, Pemphredoninae). In more advanced genera the sulcus is widely separated from the carina (figs. 71 D, 112, e.g., *Tachytes, Trypoxylon,* Crabronini). In a few instances the shortened episternal sulcus is continuous with the scrobal sulcus, and the resultant episternal-scrobal sulcus arches from the subalar fossa to the mesopleural suture (e.g., Bembicini, some Gorytini, fig. 3 B). When an omaulus is present and the episternal sulcus is long, the latter extends across the preomaular area. In the nyssonine genus *Ochleroptera* the episternal sulcus curves forward around the pronotal lobe and crosses the omaulus near the latter's base, rather than extending ventrad in the usual fashion.

In many sphecids the anteroventral part of the mesopleuron is inflected, being set off by an oblique angulation. If the angulation is sharp-edged or cariniform, it is termed an omaulus (figs. 3, 4). When an omaulus is present the inflected region in front of it is called the preomaular area. However, because this area usually has a finer sculpture, it can often be identified as the preomaular area even though no omaulus is present.

The omaulus is reduced or lost in many specialized forms. Sometimes that part of the omaulus, between its origin beneath the pronotal lobe and its juncture with the lower end of the postspiracular carina, is absent or evanescent. In this case the omaulus forms a continuum with the postspiracular carina (common in some Gorytini). The omaulus is often joined below by either or both the sternaulus and acetabular carina.

Differentiating between the cariniform omaulus and the episternal sulcus is often difficult or impossible in some Pemphredoninae; especially in the Pemphredonini, for three reasons. First, the pleuron is often areolate or otherwise coarsely sculptured, and the episternal sulcus is thus obscured. Some *Passaloecus* exemplify this condition. A few *Passaloecus* such as *gracilis* have both an omaulus and an episternal sulcus, but most have only an episternal sulcus, and often it is a broadly areolate groove. In this case its rather sharp anterior edge may be misconstrued as an omaulus. Secondly, the episternal sulcus is far forward in a few genera, and thus is can be mistaken for the omaulus; but when present the latter is nearly always represented by an anterior carina as well as a sulcus. An example is found in *Pemphredon* and especially *Diodontus.* Here the episternal sulcus appears to be absent between the subalar fossa and the scrobal sulcus but present beyond, and its forward position makes it appear to be the omaulus. *Spilomena* and *Xysma* are particularly difficult cases. Here a vertical sulcus originates just beneath the apex of the pronotal lobe and this position makes it extremely difficult to decide whether it represents a displaced, weak omaulus, or a displaced episternal sulcus. It does not originate in the subalar fossa nor does it originate from the lower base of the pronotal lobe. Because it is basically a sulcus, we have arbitrarily called it the episternal sulcus. Another aspect of this particular case is the fact that the pronotal lobe touches the tegula, or the two are very narrowly separated. Apparently the anterior part of the mesothorax has become shortened in *Spilomena,* and its closest relatives, and as a consequence the episternal sulcus is not in its normal position with respect to the pronotal lobe. A third problem occurs in the Psenini, especially in genera such as *Deinomimesa* and *Nesomimesa,* where the episternal sulcus so closely parallels the omaulus that the two are essentially inseparable. The same is true in *Paracrabro* of the Pemphredonini.

The sternaulus is an essentially horizontal carina that runs along the lower lateral surface of the mesopleuron. It is located dorsad of the subpleural signum. When both an omaulus and sternaulus are present they usually form a continuous structure, and one must arbitrarily separate the two at the point where the obliquely vertical omaulus bends to a horizontal plane (fig. 158). The hypersternaulus (fig. 3 A) is a horizontal sulcus that extends posterad from the lower part of the episternal sulcus. Sometimes as in *Diodontus* and other Pemphredonini it extends obliquely upward from the episternal sulcus. The hypersternaulus and sternaulus are never found together on a thorax.

The scrobal sulcus is evanescent or absent in many higher sphecids, although in *Cerceris,* which has lost most of the mesopleural features, it is a broad, deep, valleylike impression. Usually the scrobal sulcus does not extend forward beyond its junction with the episternal sulcus, but in some Gorytini it may do so and attain the omaulus (fig. 158). Sometimes the episternal sulcus is weak or absent in gorytins with this arrangement.

The precoxal sulcus is present in most sphecids and delimits the precoxal lobes. Presumably, this is the generalized condition. The sulcus is absent in Ampulicinae and in certain of the more advanced sphecids (e.g., *Xenosphex,* some Stizini, and some Bembicini). Dorsad the precoxal sulcus may be associated closely with or appear to be a carina.

Other sulci and carinae occur sporadically in some sphecid groups: subomaulus, mesopleuraulus, and verticaulus. These can be identified by consulting the glossary and figures.

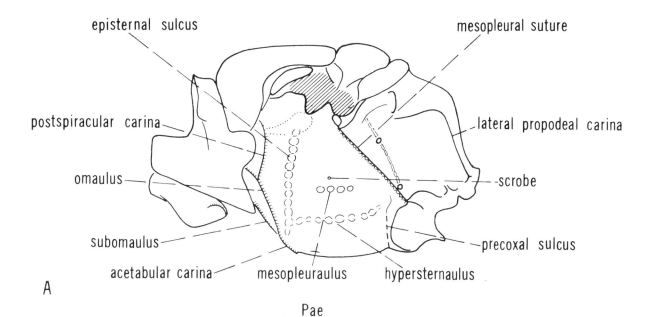

FIG. 3. Morphology of the sphecid thorax in two specialized types. A, *Pae* (Crabronini); B, *Sphecius* (Gorytini).

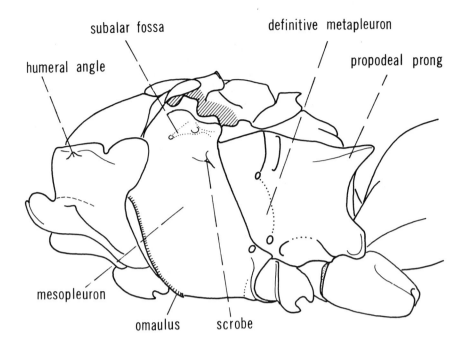
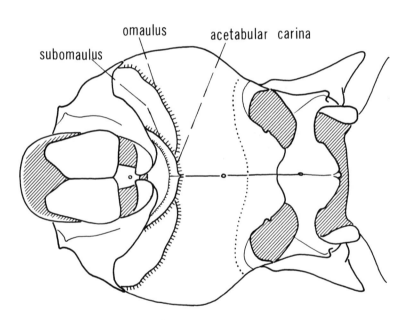

Zanysson texanus

FIG. 4. Morphology of the thorax in Nyssonini.

Between the episternal sulcus and mesopleural suture is an area known as the hypoepimeral area (fig. 2 B). It is delimited below by the scrobe and/or scrobal sulcus, and above by the subalar fossa. The sculpture of the hypoepimeral area is often an important diagnostic tool. Sometimes the area is gibbous.

The narrow zone above the subalar fossa is known as the basalar area (Richards, 1972a). Some authors are of the opinion that this area represents the epimeron. The basalar area is often margined below by a carina, the mesopleural flange (fig. 2B).

Some additional peculiarities of the mesopleuron are worth mentioning: a tooth-like process at the lower end of the omaulus (*Afrogorytes,* some *Hoplisoides*), a strong anteroventral tubercle in *Eremnophila opulenta,* and ventral tubercles in some *Prosopigastra.*

Metapleuron

The metapleuron is a rather simple plate primitively divided into upper and lower areas by the transmetapleural line (e.g., some Sphecinae, some Philanthinae, some Pemphredoninae, fig. 2 B). Landmarks are three pits, the upper pit, lower pit, and anteroventral pit. The first two lie on the metapleural sulcus, and the third is above the midcoxa just behind the mesopleural suture. The transmetapleural line extends between the upper and anteroventral pits. The posterior margin of the metapleuron, when complete, extends to the dorsal rim of the metacoxal cavity, but in about half the sphecids it is evanescent or absent below the upper pit (fig. 4, e.g., Larrinae, most Nyssoninae). In the latter situation the lower posterior boundary of the definitive metapleuron is formed by the transmetapleural line, and the resultant sclerite is wedge-shaped. In some cases where the metapleural sulcus is evanescent below, the lower area can still be recognized because its sculpture is different from that of the propodeum.

The metapleural flange (fig. 2 B, metanotal flange or lamina of authors) is an important structure, especially at the species level. In *Pison, Trypoxylon, Tachytes, Tachysphex, Chalybion, Podalonia, Ammophila,* and *Philanthus,* some species may have it expanded into a broad lamella. The metapleural flange may be continuous with the mesopleural flange. The metathoracic spiracle, which is located near the anterior end of the metapleural flange, is protected by an angular or circular setose plate, which according to Richards (1972a) is derived from the mesopleuron. Michener (1944) called it the peritreme, but Richards indicated that the true peritreme lies beneath it. Richards used the name epimeral lobe for this structure, but since it is derived from his basalar area (fig. 2 B) we think that basalar lobe is a more appropriate term. The basalar lobe is a conspicuous feature in the Sphecinae (fig. 2 B), but it occurs also in other groups, such as *Ampulex, Astata, Tachytes, Sphecius* and various Bembicini, where it seems to be a separate sclerite. In many other sphecids the basalar lobe is a simple, posterior extension, as in *Gorytes* (fig. 158).

The intercoxal carina or sulcus (fig. 112) is a conspicuous feature in certain genera, such as *Penepodium, Trigonopsis,* and *Trypoxylon.* In the last genus its shape is important as a group character.

In some groups the metapleuron may be grooved for reception of the hindfemur when the legs are retracted (some *Prionyx* for example).

We have not studied the metasternum in detail, but in *Ampulex* it is deeply notched apically and forms an inverted Y.

Wings

The terminology we use for the wings is taken largely from Haupt (1931) and Ross (1936) for veins and some cells, and from Lanham (1952) for some cells (fig. 5). The marginal cell has been called the radial cell by some workers, and the submarginal cells have sometimes been called cubital cells. The r-m crossveins are often referred to as the "transverse cubital veins" by various authors, but it is simpler to call them 1r-m and 2r-m. The m-cu crossveins of the forewing have usually been called the recurrent veins, and we have retained this usage in our keys and descriptions. Some authors have called cu-a in the hindwing the "transverse median nervure or vein"; the media has sometimes been called the "cubital vein" and the cubitus the "discoidal vein" (Evans, 1966a; Gillaspy, 1964). The anal area is often referred to as the "vannal" area, and the anal lobe as the "vannal" lobe.

Reduction of venation in the forewing has occurred, primarily, through the loss of one or more submarginal and discoidal cells and by a shortening of the marginal cell. The apex of the latter is often truncate, as in Larrinae and Astatinae, and in extreme cases it may be quite small. In a few genera, such as *Timberlakena, Miscophoides,* and *Saliostethus,* the apex of the marginal cell may be open. In *Microbembex* and *Ampulex* the cell is partially removed from the wing margin, and in *Anomiopteryx* it is completely so.

Reduction of submarginal cells occurs through the loss or joining of r-m crossveins. Petiolate submarginal cells illustrate an early step in the process (many Miscophini, *Kohliella,* some *Ammophila, Exeirus,* for example). Discoidal cells may be similarly lost. Genera that show extreme cell reduction are *Miscophoides, Belomicrus,* and *Ammoplanus.*

Hindwing venation is partially lost or evanescent in such genera as *Nitela, Spilomena, Ammoplanus,* and *Timberlakena,* but in most sphecids the reduction is brought about by a shortening of the median and submedian cells.

Important generic characteristics in the wings are the position of crossvein cu-a relative to the media, shape and number of various cells, termination of the recurrent veins on the media, size and shape of the stigma, whether or not the hamuli are divided into two groups, size of the jugal lobe, and amount of membrane beyond the cellular area of the forewing.

The placement of cu-a relative to divergence of the media from M+Cu would seem to have evolutionary significance. Considerable evidence points to a distal di-

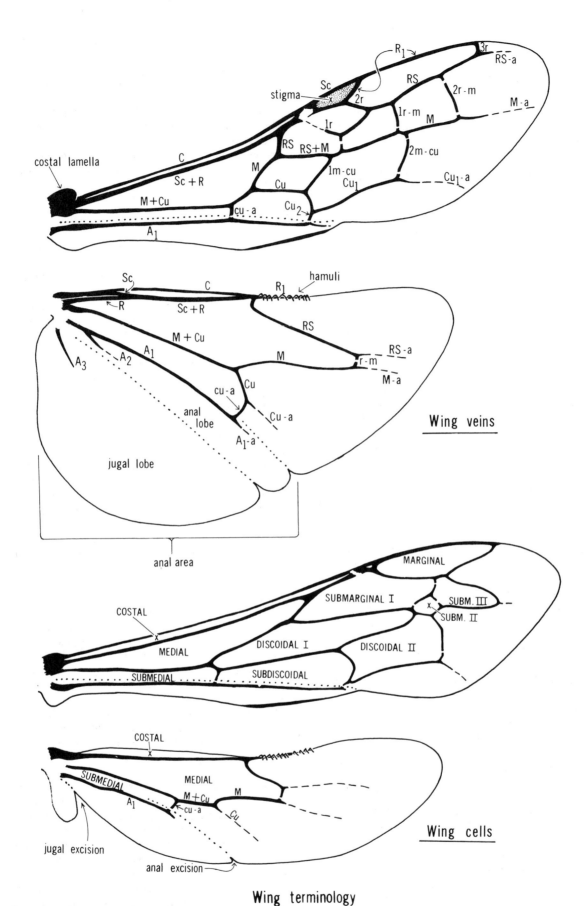

FIG. 5. Sphecid wing morphology.

vergence in the forewing as a primitive trait. In the hindwing, however, evidence is contradictory, and it may be that the divergence point has been influenced by movement of the crossvein. Concentration of cells toward the wing base with a shortening of the median cell in the forewing seems to have resulted in a movement of the media basad of cu-a. With a shortening of the submedian cell in the hindwing, the media seems to have been "stranded" distad of cu-a. However, in most cases it appears that the media rather than cu-a has changed position (*Philanthus,* Astatinae, *Laphyragogus,* some Bembicini).

Legs

The number of midtibial spurs is perhaps the most important phylogenetic feature of the legs in Sphecidae. The tendency for the inner spur to shorten or disappear is certainly an evolutionary advancement, although the reasons for it are not clear. The number of spurs is generally constant within a subfamily, but there are so many exceptions that this character cannot stand alone. The greatest discordance within a subfamily occurs in the Nyssoninae where two spurs is the rule; however, some males of *Oryttus,* some *Stizoides,* some *Alysson,* some *Heliocausus,* and all Bembicini have one, and there are none in males of *Didineis* and some *Bembix.*

Claw subteeth may occur in numbers from one to five on the inner claw margin. In the Sphecinae, Bohart and Menke (1963) postulated that one subtooth was primitive, and either two to five or a simple claw were derived situations. On this basis *Chlorion* (one subtooth) was generalized, but *Prionyx* (two to five subteeth) and *Ammophila* (often no subteeth) were specialized. If this theory is correct, the simple claws of most higher Sphecidae may represent a derived condition within the family. As an interesting corollary, in the family Chrysididae one subtooth (*Cleptes* and *Hedychridium*) seems to be generalized, whereas four to five subteeth (some *Omalus*) or no subteeth (*Chrysis*) is specialized.

Midcoxal separation is an important group character. They are widely separated in most Larrinae for example, but it is difficult to decide whether this is the generalized condition or a specialization. The midcoxa is a rather rounded, simple structure in primitive groups such as the Ampulicinae and Sphecinae, but in many sphecids there is a diagonal carina or ridgelike elevation on the dorsal surface that runs to the outer apex of the coxa (Larrinae, for example). Very often this carina is accompanied by the development of a transverse, subbasal impression on the outer lateral surface of the coxa. This depression appears to be a development that allows greater movement of the coxa.

Plantulae, which are small padlike organs that originate at the ventral apex of the tarsomeres, are found in various groups in the Sphecidae. Presumably they are a primitive holdover, and their presence is thus an indication of primitiveness.

A number of leg modifications are associated with ethology and as a result are primarily female features. Fossorial forms have evolved a foretarsal rake, which usually consists of stout and often bladelike setae. In some genera, such as *Tachysphex,* the rake may be composed of rather fine, hairlike setae. The rake is usually lateral, but in *Mesopalarus* the setae originate dorsally. Commonly, the tarsomeres of the foreleg are flattened asymmetrically in fossorial forms. This asymmetry is carried to an extreme in *Laphyragogus* and *Austrogorytes.* In what appears to be a specialization for prey seizure, the arolia are enlarged on the female front legs in several gorytin genera. Other notable female tarsal modifications are found in *Ampulex, Liris, Dicranorhina,* and some *Tachysphex.*

The hind leg may also be modified in fossorial wasps. The hindfemur is sometimes thickened toward the apex, and the latter may be a flat plate used in pushing soil from the burrow (Bothynostethini, Scapheutini, Odontosphecini, Cercerini, and Entomosericinae). Some *Tachytes* and all Alyssonini have a downward projecting lobe at the hindfemoral apex, which may function similarly. The hindtibia may bear rows of stout teeth, which can also function in nest excavation. There is a hindtibial rake in the Astatinae. In twig-nesting forms the foretarsal rake is usually lost and the legs in general are weakly setose.

The legs in males frequently have modifications, which presumably have evolved in connection with holding the female during copulation or as species recognition features for benefit of females. In the first category are the serrations, spines or excavations of the midfemora and midtibiae (some Bembicini, for example). A similar holding purpose may be served by the posteromedial fingerlike process on the forecoxa of some *Tachytes,* Bothynostethini, and *Solierella.* This process is often associated with an excavation on the inner face of the trochanter. The clublike hindfemora of *Dynatus* fall into this category also. Tarsomeres II and III of the midleg in male *Laphyragogus ajjer* are arcuate and each bears two stout, recurved apical spines. A posterobasal notch is found on the forefemur of some Larrini, such as *Tachysphex.* In males of some *Parapiagetia* the inner tarsal claw may be smaller than the outer claw. This may have a copulatory connotation. In the second category are the highly colorful shields of the foretibiae or foretarsi in some males of such genera as *Crabro* (fig. 120), *Crossocerus,* and *Didineis.* In males of Australian *Bembix* the foretarsi may be broadened and ornamented. Perhaps such shields have a sexual display function, although Evans and Matthews (1973b) suggest that they are used in digging.

Propodeum

Although technically of abdominal origin, the propodeum (epinotum or median segment of some authors) is discussed here because it is part of the definitive thorax. The generalized condition is an elongate, quadrangular structure with a dorsum, a sloping posterior face, and two vertical sides. These four faces are often separated by carinae. The evolutionary trend is toward a shortening of the propodeum, including a rounding of its boxlike contours (as in *Bembix,* for example), and the development of a dorsal enclosure. The U-shaped or V-shaped enclosure may extend onto the posterior face in forms with a shortened propodeum.

The propodeal spiracle is shielded in most sphecids by an anterior operculum, but this is absent in *Laphyragogus.* Below the spiracle there is a spiracular groove (stigmatal groove of authors) in some Sphecidae (some Sphecinae and some Nyssoninae, for example). A posterolateral propodeal tooth or prong occurs in *Larrisson, Heliocausus, Ampulex,* and most Nyssonini (fig. 4).

The sculpture on the propodeal dorsum is frequently of generic importance (*Astata* and *Dryudella,* for example). In Oxybelini the propodeum bears a process or mucro near the squamae of the metanotum (fig. 118). The form of the area surrounding the petiole socket may be of generic value, as in *Eremochares* and *Trypoxylon.*

In a few genera in which the gaster is petiolate or pedunculate, the petiole socket is isolated from the metacoxal sockets by a Y-shaped sclerite (figs. 32 E, F; 111 D) or a pair of sclerites (fig. 71 E) which Menke (1966a) called the propodeal sternite (Ampulicinae, Sphecinae, *Parapiagetia, Trypoxylon*) and Oeser (1971) named the "metapectus." This appears to be a secondary development associated with petiolation, and it is unlikely that it represents the original first abdominal sternum.

GASTER

Since the propodeum is the true first segment of the abdomen and is morphologically part of the thorax, it is convenient to call the definitive abdomen (true segments II through X) the gaster. The basal segment is then numbered I, and so on. Gastral segment I is narrowed in pedunculate and petiolate forms. Otherwise, gastral structure is fairly simple throughout the family, especially in the female. The various modifications of the male gaster relate mostly to copulation or may be recognition features for the female.

The following summary outlines the basic or generalized condition: gaster sessile; female with six and male with seven visible segments; neither sex with a pygidial plate; each tergum bearing a spiracle laterally but without a lateral carina and attendant laterotergite, nor with spiracular lobes; sterna without special pubescence, processes, or other peculiar structure; male with cerci that appear to originate on tergum VIII but belong to true segment X; volsella of male genitalia divided into an outer cuspis and an inner articulating digitus; and aedeagal penis valves with ventral teeth.

In most sphecids the gastral segments are simple, but in a number of forms their apices may be double edged (many Nyssonini, some Scapheutini) and sometimes spinose laterally (*Foxia*). In many larrines, as well as other sphecids, dark or more often silvery hair bands (fasciae) may be present.

There has been a general trend toward development of a peduncle or petiole, which shows up in all of the larger subfamilies except Astatinae. This condition is confined to the first segment, except in a few *Trypoxylon* and some *Cerceris.* There are several possible explanations for petiole development: to increase the maneuverability and flexibility of the gaster for stinging prey; to place more weight farther back to act as a balancing organ in flight, which might be especially useful during carriage of prey; to break up the body into two main areas in order to confuse predators; and to resemble some other insect for mutual protection. The fourth idea may well be tied in with the third, and differential clouding of the wings may give an appearance of petiolation when the wasp is not in flight (as in ant mimics, in some ampulicines and in the genus *Miscophus*).

In the Sphecinae and many Pemphredoninae the petiole consists of sternum I only. In petiolate forms of other sphecids both tergum and sternum are involved. The entire sternum may be long, and cylindrical, as in most Sphecinae, or it may be narrowed only basally, as in some Pemphredoninae. Within a single genus, *Diodontus* for instance, there may be a short but distinct petiole or no appreciable one.

The petiole of *Ammophila* is of special interest since at first glance it appears to be two segmented. Here, the narrowed tergum I is displaced to the tip of sternum I, and the rest of the gaster articulates with the tergal apex. The apex of sternum I is broadly separated from the base of sternum II. The space between them is membranous, but it also contains a ligament that connects the two sterna.

In some male Gorytini there may be only six visible segments, VII being weakly developed and concealed beneath VI. The second segment in *Ampulex* is much longer than those remaining, and in the male there are only three visible segments. In these wasps IV-VII are ordinarily withdrawn into III. In female *Ampulex* the apical segments are strongly compressed.

The first tergum often has a lateral carina (outside the spiracle) which delimits a lateral area, the laterotergite (epipleurite of some authors). Sometimes a carina and laterotergite are developed also on tergum II (fig. 71 G), and in *Anacrabro* (fig. 123) and *Belomicrus* each tergum has these developments. In *Anacrabro* the reflexed laterotergites and flattened sterna permit the wasp to roll into a ball. In genera such as *Palarus* and *Heliocausus* there is also a strong carina or lamella mesad of the lateral carina. In some petiolate forms the tergum is completely fused with the sternum (some Psenini, for example). *Deinomimesa* females have tergum I enlarged and greatly distorted in the form of an irregular crest. Males of the same species have an ordinary pemphredonine petiole.

Terga II to V in the female and II to VI in the male are most frequently simple and similar. A striking exception is in *Handlirschia* and most Bembicini where tergum VI has the spiracles on lateroventral spiracular lobes. Those of *Handlirschia* nearly meet ventrally (fig. 170).

The apical margins of terga II-IV of *Olgia spinulosa* bear fringes of short, stiff setae.

In many fossorial forms and especially in females, the last visible tergum, the pygidium, may bear a pygidial plate. Ordinarily this is a flattened area defined by lateral carinae, grooves or sharp angles. As a rule, these converge toward the apex to form a V or U-shaped plate, which may be used as a dirt-pusher during burrow construction. In some Crabronini and Psenini the pygidial plate is quite constricted or linear. At the other extreme is the greatly modified plate in some *Palarus, Heliocausus, Listropygia,* and *Clypeadon.* In the two last-named genera the pygidial plate combines with the legs to form an "ant-clamp," which is used to hold prey (Evans, 1962).

The generalized pygidial plate is presumably bare, flat, defined by grooves, and subtriangular. Setae, carinae, and other structures are considered to be specializations. The dense, appressed setae of *Tachytes, Alysson, Bothynostethus, Pterygorytes,* and others exemplify an advanced type of plate. The development of a pygidial plate in the male of such genera as *Tachytes, Liris,* and some *Dienoplus* is not common in the Sphecidae and must be considered as a specialized condition. The pygidial plate appears to have been secondarily lost in forms such as *Holotachysphex, Psenulus,* and *Trypoxylon,* whose habits have changed from fossorial to twig-nesting.

Terga VII to VIII of the female and VIII to IX of the male are concealed, reduced and/or fused with one another. The possession of cerci on apparent tergum VIII (fig. 9 A), a relatively primitive trait, is found only in Astatinae, *Eremiasphecium,* some Ampulicinae, and some Sphecinae.

Peculiarities of sternum I, in addition to its occasional role in petiolation, are: development of carinae from the basomedian area (single in Gorytini, Heliocausini, Stizini, and Bembicini; double in Alyssonini, Nyssonini, and Entomosericini, for example); and presence of a large spinelike process in the male of the *Trypoxylon* subgenus *Trypargilum.*

The most striking modifications of the sterna are found on sternum II. Modifications of this plate are rare in the female, but mention can be made of the pair of specialized basal areas in *Liris* and its relatives, as well as the unusually swollen and protuberant condition in *Sphodrotes, Pterygorytes,* and many Nyssonini. In the male a thick transverse ridge or flange occurs on sternum II in genera such as *Heliocausus, Larrisson,* and *Palarus.* In many male Bembicini there is a median longitudinal carina or hamate keel. In others of this tribe there is a double median process or submedian teeth.

Sterna II to V bear an assortment of apical, discal or basal tufts, fringes, setal rows, or large velvety patches, especially in males. The sterna sometimes have transverse grooves basad. These are deep, sharp-edged, and median in *Cerceris.* In bees Michener (1944) used the term graduli for these sulci (fig. 106 B).

Sternum VI in males and VII to a lesser extent may have prominent transverse or longitudinal processes, as in *Bembix.* Sternum VI in females may be somewhat keellike and strongly so in *Belomicroides* and *Metanysson.* In the latter genus the sternum is reduced to a sharp line.

The shape of sternum VIII in the male is often of generic or specific value. It may be concealed, or its apex may be visible. In many Psenini, *Sagenista,* and some *Sphecius* sternum VIII projects from the gastral apex as a pseudosting.

The female genitalia of aculeates have been discussed by Michener (1944) and Richards (1956b). We have not studied these organs intensively, and a careful study might reveal characteristics of suprageneric importance.

There are two basic patterns in the male genitalia. The primitive type has been outlined above (fig. 6 A). The other form in which the volsella is not differentiated into digitus and cuspis, presumably through the fusion or loss or one or the other or both, is the specialized condition (fig. 6 B). The primitive type is found in all Sphecinae except the subtribe Prionyxina; all Ampulicinae, Astatinae, Pemphredoninae, Laphyragoginae, Entomosericinae, Xenosphecinae, most Nyssoninae; and all Philanthinae except Cercerini. The Prionyxina volsella is of interest because the digitus is still identifiable, but it is immovably fused with the rest of the volsella, and thus it does not fit the definition. The volsella in this subtribe graphically illustrates the intermediate step in the evolution from a generalized volsella to the advanced simple type. The undifferentiated or simple volsella is characteristic of the Larrinae and Crabroninae, but it is also found in *Ammatomus,* Alyssonini, Nyssonini, and Cercerini.

The genitalia show an almost bewildering variety of form at both the generic and specific levels. In some genera, such as *Moniaecera* and *Pison,* the overall appearance of genitalia of different species, considered by themselves, would lead to formation of new genera. Some striking peculiarities of the genitalia are: the biramous gonostyle in *Entomosericus,* some *Pison,* and some *Trypoxylon;* the membranous sac at the aedeagal apex in some *Mellinus;* the articulating penis valve head in most *Eremnophila;* and the accessory lobes on the inner face of the gonostyle in *Larrisson.*

GLOSSARY OF MORPHOLOGICAL TERMS

acetabular carina: transverse carina on anterior part of mesothoracic venter, often connecting with lower end of omaulus (fig. 4)

admedian lines: most median pair of lines originating anteriorly on scutum (fig. 2 A)

aedeagus: sclerotized central organ of male genitalia composed of two penis valves (fig. 6).

ammochaetae: the long, fine setae that comprise the psammophore.

anal area of hindwing: posterobasal part of wing behind first or only anal vein and marked at wing margin by anal excision, usually divided into anal lobe and jugal

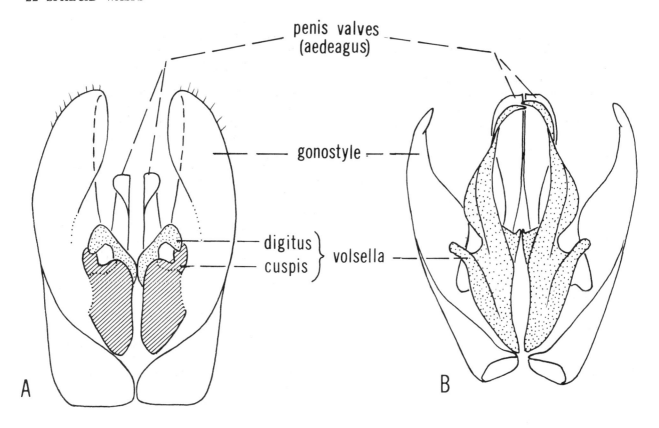

FIG. 6. Male genitalia of the two main types found in the family Sphecidae. A: volsella composed of a cuspis and an articulating digitus. B: volsella (stippled area) undifferentiated.

lobe, the two separated by a fold and the jugal excision at wing margin (fig. 5)

anal lobe of hindwing: see anal area

antenna: a basal scape, pedicel and terminal flagellum, the latter composed of a series of articles called flagellomeres (fig. 1)

anteroventral metapleural pit: see metapleural pits

areolate: relatively large, basinlike, irregular reticules (see reticulate (fig. 89 C)

arolium: a saclike organ between claws (pulvillus of authors)

axillae: lateral sclerites associated with scutellum but derived from scutum

basal vein: in forewing, that part of the media between its divergence from Cu and its fusion with radial sector (fig. 5)

basalar area: narrow area of mesopleuron dorsad of subalar fossa, often limited below by mesopleural flange (fig. 2 B)

basalar lobe: a posterior extension of basalar area that covers metathoracic spiracle (fig. 2 B) (= peritreme of authors)

basitarsus: the basalmost tarsomere

cerci: paired, one-segmented appendages of tergum VIII (true segment X) in some males, ordinarily concealed beneath tergum VII (fig. 9 A) (E. Smith, 1970) (= pygostyles of authors)

clavate: clublike, gradually thickening toward apex.

costal lamella: thin basal expansion of costal margin of forewing (fig. 5)

cu-a: cubito-anal crossvein of each wing, forming outer end of submedian cell (fig. 5)

cuspis: stationary outer apical extension of volsella in male genitalia (fig. 6)

digitus: articulating inner apical part of volsella in male genitalia (fig. 6)

episternal-scrobal sulcus: continuous arc originating in subalar fossa and passing through scrobe to mesopleural suture (fig. 3 B). It contains elements of both episternal and scrobal sulci

episternal sulcus: originating in subalar fossa and extending ventrally on mesopleuron; when complete, reaching anteroventral margin of mesothorax; often joined by scrobal sulcus (figs. 2 B, 3 B)

fimbriae: specialized rows of hairs on male sterna (fig. 132 D)

flagellomere: one of units or articles of flagellum (fig. 1)

frontal line or carina: median groove or carina leading, when complete, from midocellus to interantennal

area (fig. 1)
frontoclypeal suture: forms upper margin of clypeus (fig. 1) (= epistomal suture of Michener, 1944)
gaster: definitive abdomen composed of true second and following morphological segments, of which first consists of tergum I and sternum I in our terminology
gena: posterolateral part of head between outer orbit and occipital carina (temples or cheeks of some authors) (fig. 1)
gonostyle: outermost paired appendages of male genitalia, sometimes divided into basistyle and dististyle (fig. 6)
gradulus: transverse groove across tergum or sternum (fig. 106 B)
hamuli: hooks on margin of hindwing beyond origin of radial sector (RS) (fig. 5)
humeral angles: dorsolateral corners of pronotal collar (fig. 4)
hypersternaulus: groove originating anteriorly at lower part of episternal sulcus, usually horizontal (fig. 3 A)
hypoepimeral area: dorsoposterior area of mesopleuron defined by episternal sulcus and scrobal sulcus (fig. 2 B)
hypostoma: sclerotized area around oral fossa delimited outwardly by hypostomal carina (fig. 1)
hypostomal carina: U-shaped or broadly V-shaped carina on venter of head delimiting hypostoma (fig. 1)
humeral plate: small sclerite posterior to tegula (fig. 154)
inner orbit: inner margin of compound eye (fig. 1)
intercoxal carina: a ridge or carina extending from dorsal rim of mesocoxal cavity to same area of metacoxal cavity (fig. 112)
jugal lobe: posterobasal lobe of anal area on hindwing; when present, marked by jugal excision (fig. 5)
lateral carina: a carina or line usually found only on first tergum and positioned laterad of spiracle, delimiting laterotergite (fig. 71 G)
laterotergite: deflected lateral area of tergum, which is defined dorsally by lateral carina or line (epipleurite of authors)
lower metapleural area: that part of metapleuron beneath transmetapleural line, its definition dependent on presence of metapleural sulcus or line between upper and lower metapleural pits (fig. 2 B)
lower metapleural pit: see metapleural pits
malar space: area between compound eye and mandible socket, width measured as least distance between eye and socket (fig. 1) (= cheek of some authors)
mandibular notch: externoventral emargination or stepped angulation (fig. 87 B)
mesopleural flange: carina along lower margin of subalar area (fig. 2 B)
mesopleural pit: posterior pit on mesopleuron above midcoxa (fig. 3 B)
mesopleural suture: posterior margin of mesopleuron extending from midcoxal cavity to beneath wings (fig. 2 B)
mesopleuraulus: horizontal sulcus originating at episternal sulcus just below level of scrobe and extending posterad (fig. 3 A)
metapleural flange: carina or lamelliform extension of metapleuron surrounding hindwing base (fig. 2 B) (= metanotal flange of authors)
metapleural pits: three landmarks, upper metapleural pit on upper part of metapleural sulcus or line, anteroventral metapleural pit, and lower metapleural pit (or pocket) directly above hindcoxa (fig. 2 B)
metasternum: plate extending between mid and hindcoxae and identified basally by sternal apophyseal pit III
midtibial spurs: one or two movable, spinelike processes which are usually much larger than nearby setae, arising from rings set in membranous area at inner apex of midtibia; spur pectinate along its shaft (fig. 142 G, H) (calcars of some authors)
mucro: dorsobasal median projection of propodeum, usually spinelike and pointing obliquely upward (fig. 118)
notaulus (-i): paired lines or grooves on scutum, originating anteriorly and outside admedian lines (fig. 2 A). See Hopper, 1959, Proc. Ent. Soc. Wash. 61:155-171, and Forbes, 1940, Bull. Brooklyn Ent. Soc. 35:136-137 for derivation and spelling
oblique scutal carina: short line or carina originating at lateral edge of scutum usually opposite tegula and setting off posterolateral, often deflected corner of scutum (fig. 156)
ocellar scars: flattened opaque remnants of ocelli (fig. 61)
omaulus: ridge or carina originating at lower base of pronotal lobe and extending posteroventrally (fig. 3)
oral fossa: cavity within which mouthparts are attached (fig. 1)
orbital foveae: depressed, oval or elongate areas with a distinct rim, usually located along upper inner orbits (fig. 121 B, G) (sometimes called facial or supraorbital foveae)
palpal formula: number of segments in maxillary palpus compared with those in labial palpus as, for instance, 6-4
paramandibular process: forward extension of hypostoma, which on reaching underside of clypeus isolates mandibular socket from oral fossa; not clearly visible unless mandibles are spread (fig. 1)
parapsidal lines: short lateral scutal impressions or raised lines (fig. 2 A)
pedicel: second antennal segment, located between scape and flagellum (fig. 1)
peduncle: applied to basal segment of gaster; a narrowed, clavate stem attaching gaster to propodeum (fig. 126)
penis valve: paired, sclerotized elements of aedeagus (fig. 6)
petiole: slender, parallel-sided, or cylindrical stalk (fig. 22 N)
petiole socket: orifice on posterior end of propodeum in which gaster is inserted.
placoid (-s): special platelike, flat, or curved areas on

male flagellomeres that are bounded by ridges or depressed below level of surrounding integument (= fossula of authors) (fig. 28 D)

plantulae: small oval pads, which may be found apicomedially on underside of tarsomeres (fig. 19 A, B) (lamellate, oval, intersegmental pads of Bohart and Menke, 1963)

postocellar line: transverse impression that connects posterior margins of hindocelli

postspiracular carina: arcuate carina on anterodorsal part of mesopleuron associated with pronotal lobe (figs. 2 B, 3 A)

precoxal area of mesopleuron: area in front of midcoxa on lateral pleural surface

precoxal lobes: ventral mesopleural lobes between and in front of mesocoxae and set off by precoxal sulcus (figs. 2 B, 94 A)

precoxal sulcus: transverse groove found posteroventrally on mesopleuron (fig. 2 B)

preomaulal area: mesopleural area in front of omaulus or linear angulation that may replace omaulus; usually with finer texture than rest of mesopleuron (fig. 112)

prepectus: area of mesopleuron in front of episternal sulcus (fig. 2 B)

prescutellar sulcus: transverse, sometimes pitted groove at anterior margin of scutellum

pronotal collar: raised posterior part of pronotum (fig. 2 B)

pronotal lobe: posterolateral part of pronotum covering mesothoracic spiracle (fig. 2 B) (tubercles of authors)

propleuron: paired ventral plates of prothorax in front of forecoxae (fig. 2 B)

propodeal enclosure: area of propodeal dorsum usually delimited by grooves or carinae, sometimes extending onto posterior face of propodeum (fig. 2 A) (saeptum of Empey, 1969, 1971)

propodeal side: lateral, vertical face of propodeum

propodeal sternite: sclerite, which isolates petiole socket from hindcoxal cavities, usually single and Y-shaped, not clearly visible unless hindcoxae are separated (figs. 71 E, 111 D)

propodeum: true first abdominal segment that forms an integral part of thorax, delimited anteriorly by posterior margin of metanotum and by metapleural sulcus (fig. 2 A) (median segment or epinotum of authors)

psammophore: "sand basket" formed by hair fringes (ammochaetae) on lower surface of mandible, gena, propleuron, forefemur, and tibia (fig. 59); foretarsal rake also used as part of psammophore

pseudosting: sternum VIII of some males, which resembles a stinger

pygidial plate: specialized area of tergum VI (the pygidium) in female and VII in male, usually flattened and delimited by carinae or grooves (figs. 66, 74)

rake: linear series of setae on outer margin of foretarsus, which function as a rake; occurring in most females and some males (fig. 59)

recurrent veins: m-cu crossveins between media and cubitus of forewing, used with reference to their termination at submarginal cells (fig. 5)

reticulate: sculpture with appearance of relatively fine meshwork.

scapal basin: depression above antennal sockets within which scapes may rest (fig. 127 B)

scape: basal segment of antenna (fig. 1)

scrobal sulcus: a horizontal mesopleural groove that passes through scrobe (figs. 45, 158)

scrobe: pit or mark somewhat above and behind middle of mesopleuron (fig. 2 B)

scutellum: small posterior mesonotal plate (behind scutum) (fig. 2 A)

scutum: large anterior mesonotal plate (in front of scutellum) (fig. 2 A)

sessile gaster: one in which gaster swells uniformly and abruptly from point of insertion (fig. 65)

signum: short line or mark lateroventrally on mesopleuron (fig. 2 B), morphologically equivalent to parapsidal line of scutum.

spiracular groove: extending from propodeal spiracle toward metacoxa (stigmatal groove of authors) (figs. 29 D, 158)

spiracular lobe: laterobasal spiracle bearing lobes of tergum VII in males of most Bembicini and in *Handlirschia* (fig. 170 F)

sternaulus: horizontal lateroventral carina of mesopleuron above signum and extending from lower end of omaulus toward precoxal sulcus (fig. 158)

stigma: sclerotized area on leading edge of forewing basad of marginal cell and in front of first submarginal cell (fig. 5)

subalar fossa: depressed area of mesopleuron beneath forewing insertion, containing one to several pits (fig. 3 B)

subalar line: a horizontal line extending from subalar fossa to hindmargin of pronotum (fig. 3 B)

subantennal area: triangular or pentagonal sclerite of frons contiguous with clypeus and between antennal sockets (fig. 1)

subomaulus: carina originating from anterior margin of mesopleuron below omaulus and basically parallel to latter (fig. 3 A, 112)

sulcus: a secondary impression on a sclerite which does not represent a cleavage line between two sclerites (fig. 2)

suture: cleavage line between two sclerites (fig. 2)

tarsomere: one of tarsal units, of which first is often called basitarsus

tegula: ovoid plate over base of forewing (= squama of authors) (fig. 2 A)

transmetapleural line: runs between upper metapleural pit and anteroventral metapleural pit (fig. 2 B)

tylus (-i): linear welt or cariniform swelling on male flagellomere (tyloides of authors)

upper metapleural area: that part of metapleuron above transmetapleural line (fig. 2 B)

upper metapleural pit: see metapleural pits

vertex of head: top of head from ocellar triangle to top of occipital carina

verticaulus: mesopleural carina originating in front of midcoxa and extending dorsad vertically or obliquely (fig. 123); often continuous with sternaulus

volsella: paired structure associated with inner base of gonostyles (fig. 6)

SYSTEMATICS OF THE SPHECIDAE

FAMILY CHARACTERS

Our main purpose in this book has been to present a review of the infrafamilial structure of the Sphecidae rather than a comparison with related aculeate families. Although a thorough in-depth study of the Aculeata is yet to be made and is outside the scope of our book, we would like to put the Sphecidae into a frame of reference with respect to its closest relatives. For the following discussion of sphecid characters we have drawn upon the works of Börner (1919), Handlirsch (1925, 1933), Leclercq (1954a), Michener (1944), Richards (1956b, 1962), and Lanham (1960), as well as our own observations. At the family level diagnostic characters are few, often awkward to describe, and sometimes inconspicuous. We judge the following to be the most important distinctions between the Sphecidae and the bees, vespids, and pompilids.

Pronotum – Morphologically, as well as ethologically, the Sphecidae is the most diverse of the higher aculeate families. However, the pronotum offers several features that with rare exceptions, are constant and serve to differentiate the family.

The sphecid pronotum is essentially immovably joined with the mesothorax. Also, its posterior margin in dorsal view is nearly straight and usually separated from the scutum by a constriction that contributes to the formation of a characteristically raised collar. In a few sphecids the pronotal collar is sharp edged dorsally and very closely applied to the anterior part of the scutum. In *Liris, Bembix,* and *Stizus,* for example, there is usually no obvious constriction. Laterally the posterior margin of the notum always has a distinctive rounded prolongation or pronotal lobe that covers a spiracle. With few exceptions this lobe is well separated from the tegula, so that the scutum anterolaterally is in contact with the anterodorsal part of the mesopleuron. In some Ampulicinae, Pemphredonini, Astatinae, and the crabronine genus *Holcorhopalum*, the pronotal lobe is very close to the tegula, so that the scutum and mesopleuron are isolated or are in narrow contact.

The pronotum in bees is similar to sphecids, but it is always short and sometimes the hindmargin is concavely arcuate. The pronotal lobe is close to or touches the tegula, so that often the scutum and mesopleuron are isolated from each other.

The vespid pronotum is solidly joined with the mesothorax, and their dorsal surfaces are essentially continuous. The pronotum is prolonged posterolaterally and rather broadly touches the tegula, so that the scutum and mesopleuron are broadly isolated. This posterolateral pronotal extension gives the posterior margin of the pronotum a distinctive inverted U-shape. The pronotal lobe is weakly formed or absent in vespids.

The pompilid pronotum is not firmly joined with the mesothorax, nor is there any constriction between them. Instead, the pronotum is capable of considerable movement over the scutum. The thin, flat, hindmargin of the pronotum also makes the dorsal surfaces of the scutum and pronotum smoothly continuous. The hindmargin of the pronotum is either concavely arcuate or obtusely angled. Most pompilids have a moderately developed pronotal lobe. This lobe is very close to or touches the tegula so that the scutum is at most narrowly in contact with the mesopleuron.

Hind leg – One of the basic differences between sphecids and bees is the presence in the former and the absence in the latter of a cleaning pecten or brush in a slight depression basally on the inner side of the basitarsus (fig. 28P). This apparatus is opposed by the usually pectinate inner tibial spur. The hindbasitarsus also differs between these two groups in another way. In sphecids it is similar to the following tarsomeres with respect to width and crosssection. In bees the basitarsus is usually much broader than the succeeding tarsomeres and flattened, at least inside. There are marginal cases in the bees, and, unfortunately for the student, they occur in the most wasplike forms, the Nomadini, a parasitic group.

The hindleg in vespids and pompilids is similar to that of the Sphecidae.

Setae – Another basic difference between sphecids and bees is the presence of branched or plumose setae in the latter. This is the character most often referred to in texts, but it is difficult to use without recourse to high magnification. The setae of sphecids, vespids, and pompilids are simple, unbranched structures. Many sphecids have appressed metallic setae on the frons and clypeus. As pointed out by Stephen, Bohart, and Torchio (1969) nearly all bees that resemble sphecids lack metallic facial setae.

Wing folding — The sphecids, pompilids, and all bees except *Eulonchopria* differ from vespids in the absence of the wing folding mechanism found in that family.

Eyes — Except for the Trypoxylonini and Philanthini the inner eye margins are not notched or emarginate in the sphecids. Likewise, the inner margins are rarely emarginate in the bees or pompilids, but they are characteristically deeply notched in the vespids.

Cerci — These appendages of the male tenth tergum are absent in most higher sphecids but are present in the Dolichurini, some Sphecini, most Sceliphronini, Astatinae, and the Eremiaspheciini. Cerci are also absent in the bees and vespids, but they are present in the Pompilidae.

Thoracic pleura — Certain features of the pleura although not constant in the Sphecidae are still frequent enough to constitute subsidiary family traits. One of these is the presence in most sphecid genera of an episternal sulcus on the mesopleuron. This sulcus is notably absent only in the Cercerini and the Ampulicinae. However, in the latter there is usually a ventral remnant. In bees the episternal sulcus is common only in the primitive bees, Colletidae, Halictidae, and Andrenidae, although an episternal-scrobal sulcus is present in many other bees. The episternal sulcus is present in most vespids but appears to be absent in all pompilids.

SYSTEMATIC HISTORY AND BASIC REFERENCES

The first really modern classification of the Sphecidae was provided by Kohl (1896), although earlier authors such as Dahlbom (1843-1845) and Lepeletier (1845) made attempts at organization of the group. Kohl redescribed all the genera and arranged them first into "Gattungs-gruppen," which correspond roughly to current day subfamilies, then "Untergruppen," which are equivalent to tribes. He recognized 86 genera, and his overall treatment was quite conservative. Nevertheless, his classification was for the most part sound and it has been nearly universally followed to this day. A publication of Fox (1894d), a correspondent with Kohl, previewed Kohl's opus in English, but it was based on the New World fauna. Unfortunately, the contribution by Fox was overshadowed by the classification of Ashmead (1899), which was the exact opposite of Kohl's work in nearly every respect. Ashmead was an extreme splitter at the family level. He divided the Sphecidae into 12 families and recognized a total of 177 genera. His groupings were often heterogeneous, and his work was full of errors. In spite of this, his classification gained some support in North America, probably because Kohl's work was in German and not so accessible. In retrospect it is remarkable that, within three years, two such diametrically opposed phylogenetic schemes should have appeared. By way of comparison with Ashmead and after 70 years of additions, we recognize 226 genera, 33 tribes, and 11 subfamilies.

Kohl's "Gattungs-gruppen" were given subfamily names by Dalla Torre (1897) who made slight changes based on Handlirsch's (1887-95) monograph of the Nyssoninae. Börner (1919) and Handlirsch (1925) made further modifications. Ashmead's system was split even more by Brues and Melander (1932) who recognized 17 families, and by Essig (1942) who gave 21 families!

Other authors have made noteworthy contributions to the taxonomy of the Sphecidae. Radoszkowski (1892) pointed out the value of male genitalia in classification. He was one of the first to remark on the significance of cerci ("palpe genital"). Gutbier (1915) made one of the first attempts to summarize ethology of the Sphecidae, and Iwata (1942, 1972) has provided a more recent summary. Pate's (1937d) catalog of the genera of the Sphecidae and their type species cleared up and brought to light many nomenclatorial problems. Evans and Lin (1956a, b) and Evans (1957c, 1958a, 1959a, 1964a, c) studied the larvae of the Sphecidae and their implications in the phylogeny of the family. Iida (1967, 1969a, b) and Tsuneki and Iida (1969) are further publications on larvae.

At the generic and specific level a few authors have made great contributions toward establishing the identity of many species of older workers such as Linnaeus, De Geer, Fabricius, etc. Schulz (1905, 1906, 1911a, b, 1912) studied types of many early describers, but unfortunately his publications have often been ignored or overlooked. Richards (1935a) identified a number of Linnean species, and van der Vecht published several papers based on type studies, the most outstanding of which is his (1961a) treatise on Fabrician sphecid types. Beaumont has reported on the identity of many old European species in various papers.

Entomology of the 1800's consisted mainly of descriptive taxonomy with such workers as F. Smith, A. Costa, F. Morawitz, Cameron, Taschenberg, Radoszkowski, Saussure, A. Mocsáry, Cresson, and Spinola predominating.

Kohl, with his many monographic works in the late 1800's, ushered in a change from descriptive to revisionary taxonomy. Handlirsch and Fox were noteworthy contemporaries of Kohl. During the present century many individuals have made revisionary contributions. Of these, a few deserve special mention because of their broad contributions: Arnold, Beaumont, Evans, Gussakovskij, Krombein, Leclercq, Pate, Pulawski, Tsuneki, R. E. Turner, Williams, and Willink.

Listed below are important publications of a regional nature that contain keys to genera or species of certain geographic or political areas. Some of the older works cited have nearly outlived their usefulness, and there are no synopses for many regions. We have also included catalogs, and these are indicated by an asterisk.

British Isles — E. Saunders (1896)
France — Berland (1925b)
Switzerland — Beaumont (1964b)
Poland — Noskiewicz and Pulawski (1960)
Hungary — Bajári (1957), Móczár (1959)
Spain — Giner Marí (1943b), Ceballos* (1956, 1959, 1964)
Czechoslovakia — Zavadil and Snoflák (1948), Balthasar (1972)
German Democratic Republic — Oehlke (1970)
"Middle Europe" — Schmiedeknecht (1930)
Egypt — Honoré (1942, keys to genera only)
Ethiopian Region — Arnold (1922-1931, plus later supplements)
Madagascar — Arnold (1945)
India and Burma — Bingham (1897)
Hawaii — Yoshimoto (1958, 1960)
Indonesia — van der Vecht (1939, keys to genera only)
Philippine Islands — Baltazar* (1966)
Japan — Yasumatsu and Watanabe* (1964, 1965)
America, north of Mexico — Muesebeck et al.* (1951), Krombein* (1958d, 1967a)

Four volumes of a bibliographical nature are especially useful to students of sphecids:

Dalla Torre (1897) — "Catalogus Hymenopterorum" — Volume 8 is a synonymic catalog of Sphecidae.

Horn and Kahle (1935-1937) and Sachtleben (1961) — "Über entomologische Sammlungen" — Locations of important insect collections are given. Handwriting samples and labels of various workers are illustrated.

Maidl and Klima (1939) — This synonymic catalog was discontinued after completion of Nyssoninae and Astatinae. The generic assignments are faulty.

Zimsen (1964) — "The type material of J. C. Fabricius"

PHYLOGENY

It is often difficult in a group as large and diversified as the Sphecidae to grasp the significance or the role played by certain characters in the evolution of the family. It is necessary, however, to arrive at some conclusion regarding generalized or "primitive" traits and their specialized or "advanced" condition if one is to attempt to interpret phylogeny. It is obvious that evolution has often proceeded from the simple to the complex. Thus, a filiform antenna would ordinarily be considered generalized, whereas a pectinate or distorted one would be considered specialized. Unfortunately, there are so many exceptions to this rule that it must be used with extreme caution. Somewhat complex structures found in generalized Sphecidae (dentition of claws, numerous mesopleural sulci) may be secondarily lost so that in such cases simplification is the specialized condition. We have found that a study of features common in the more primitive hymenopterous families and preserved in some of the Sphecidae is the most productive way of making value judgements on evolutionary paths. On this basis we have assessed the subfamilies for their generalized versus specialized traits. At the same time we have realized that every existing subfamily has developed specializations of its own; and that different features have evolved at different rates in different groups. Some characters defy a logical evolutionary explanation; among these are the placement of antennae high or low on the face and the degree of separation of the midcoxae. The following list gives those features that appear to be of phylogenetic significance for the family as a whole.

An approximate idea of the evolutionary status of each subfamily was derived by a summation of specialized traits. Each character was given a range in value of 0 to 3. If a particular specialized character was found in about half of the genera in the subfamily, it was assigned a value of 2. If it was present in all or nearly all genera, it was given a value of 3. If the character was not represented or found only in a small number of genera, a value of 0 or 1 was assigned respectively. On this basis the subfamilies can be arranged in the following ascending order: Ampulicinae, 27; Sphecinae, 28; Astatinae, 31; Laphyragoginae, 36; Entomosericinae, 45; Xenosphecinae, 45; Pemphredoninae, 48; Philanthinae, 50; Nyssoninae, 57; Larrinae, 60; and Crabroninae, 70.

These results obscure the relatively high evolutionary level of some tribes, such as the Bembicini, which if compared with the list of specialized characters has a value of 64. The degree of evolutionary diversity is also not apparent. For example, the Larrinae and Nyssoninae include a wide array of types including some relatively unspecialized forms. By comparison, the Crabroninae is a rather uniformly specialized group in spite of the great number of genera.

Our ideas of the phylogenetic relationships of the subfamilies and their tribes are shown in the dendrogram (fig. 7). As this diagram implies, we think that the Ampulicinae, Sphecinae, and Pemphredoninae have few close ties with the rest of the sphecids, and they are shown as isolated branches. The remaining subfamilies are grouped along two main lines, one of which consists almost entirely of the Nyssoninae. The other branch, here called the larrine complex, includes several subfamilies, most of which can be arranged in a nearly linear fashion starting with the most primitive, Astatinae, going through the Laphyragoginae, Larrinae, and ending in the highly specialized Crabroninae. The Philanthinae seems to have been derived from the Larrine complex and perhaps is most closely allied with the Larrinae, but the group has diverged from the main larrine stem. The Entomosericinae and Xenosphecinae display characters intermediate between the larrine and nyssonine branches and are placed between them.

Larvae

The classification of the Sphecidae proposed by Evans (1959a, 1964a), based on larvae, is in general concordance with the one we have proposed based on adult morphology. For example, we agree with Evans in placing the Trypoxyloninae as a tribe in the Larrinae and in regarding the Ampulicidae as a subfamily of the Sphecidae. However, we have arrived at different conclusions in a few cases. He suggested that the Crabroninae should be included in the Larrinae. There is no doubt that such a move can be defended on adult morphology, because the crabronines are certainly closely linked to the larrines through the Bothynostethini and Scapheutini. However, for the sake of practicality we have kept the Crabroninae separate. Another point of difference with Evans' scheme is the status of the Mellinini. He would make this group a subfamily, but adult morphology does not offer the strong differences found in the larvae. This is discussed more fully in the systematics of the Mellinini. Based on larval characters, Evans associates the Astatinae and Philanthinae with the nyssonine branch in his dendrogram. However, adult morphology clearly relates these two groups more closely with the larrine stem.

Since larvae are known for only a relatively small number of genera, we have much more to learn about larval characters and their bearing on sphecid classification. The affinities of unusual forms such as *Laphyragogus, Entomosericus, Xenosphex, Odontosphex,* and *Auchenophorus,* for example, are still uncertain. Hopefully, the discovery and study of their larvae will shed further light on sphecid phylogeny.

Those who would make separate families of the sphecid subfamilies should consider Evans' (1964a) comments on larvae. He said that "the most striking discontinuity in larval structure [in the Sphecidae] occurs not between the Ampulicidae and the Sphecidae but between the Ampulicinae and Sphecinae on the one hand and all other sphecoids on the other." The same thing can be said on the basis of adult morphology. It would not be too unreasonable to separate the Ampulicinae and Sphecinae from the other subfamilies, as the family Sphecidae. The remaining groups would constitute the family Larridae. However, we do not feel that much would be gained by doing this, and furthermore it would impose a family name change that we feel is undesirable. The Sphecidae is perhaps unique in that there are an unusual number of aberrant relict genera. Among the more striking examples are *Entomosericus, Xenosphex, Laphyragogus,* and *Odontosphex.* Evolutionary extinction has taken care of such problem genera in most other Aculeate families. These genera possess traits that cut across tradi-

TABLE 2
Phylogenetic Characters in Sphecidae

	Generalized	*Specialized*
1.	Ocelli normal	Ocelli deformed
2.	Inner eye margins parallel	Eyes strongly converging or holoptic
3.	Mouthparts short	Mouthparts elongate or unusually modified
4.	Palpal formula 6-4	Palpi with fewer segments
5.	Mandibular socket open	Mandibular socket closed
6.	Pronotum long, not closely fitted to scutum	Pronotum short, closely appressed to scutum
7.	Pronotal lobe close to or touching tegula	Pronotal lobe and tegula well separated
8.	Notauli present, long	Notauli absent or short
9.	Propodeum elongate, dorsal surface horizontal	Propodeum short, dorsum sloping
10.	Episternal sulcus present, long	Episternal sulcus absent or short
11.	Omaulus absent	Omaulus present
12.	Lower metapleural area present	Lower metapleural area absent
13.	Marginal cell long, apex acuminate	Marginal cell short, apex truncate or open
14.	Three normal submarginal cells	Submarginal cells two or less, or one of them petiolate
15.	Outer veinlet of first submarginal cell angled and with a remnant of the first radial crossvein	Outer veinlet of first submarginal cell straight, not appendiculate
16.	Media diverging after cu-a in forewing	Media diverging before cu-a in forewing
17.	Forewing with three discoidal cells	Forewing with fewer than three discoidal cells
18.	Jugal lobe nearly as long as anal area	Jugal lobe small or absent
19.	Subcosta present in hindwing	Subcosta absent in hindwing
20.	Second and third anal veins present in hindwing	A_2 and A_3 absent
21.	Hindwing with medial and submedial cells	Medial and submedial cells incomplete
22.	Gaster sessile	Gaster petiolate
23.	Male with seven and female with six visible segments	Fewer segments visible
24.	Pygidial plate absent	Pygidial plate present
25.	Cerci present	Cerci absent
26.	Male sternum VIII broad, not appreciably narrowed apically	Sternum VIII narrowed or spinose apically, or modified into a false sting
27.	Volsella with digitus and cuspis	Volsella simple or absent
28.	Claws with inner teeth	Claws simple
29.	Midtibia with two apical spurs	Midtibia with one or no spurs
30.	Plantulae present	Plantulae absent

tional subfamily lines and in our view make impractical the establishment of several families from the subfamilies of the Sphecidae. *Entomosericus,* for example, links the Larrinae and Nyssoninae. Some would argue that these odd genera should be more or less arbitrarily placed in one or another subfamily (or family), but we feel that doing so would obscure an understanding of the relationships of the groups involved. Also, such arbitrary placement would lead to problems in defining groups. We feel that much more is gained by isolating these problem genera in monotypic subfamilies. This draws attention to them, and also makes it simpler to define the larger subfamilies, a task which is not easy even without them.

Sphecids and the Ancestry of Bees

Bees are certainly the closest relatives of the Sphecidae, and their differences are not always easily appreciated. In fact, we share the view of Michener (1944) and Bradley (1958) who suggested that bees should be included in the superfamily Sphecoidea.

A discussion of the ancestry of bees with respect to the Sphecidae would necessitate a study far larger than permitted within the scope of this book. After surveying the Sphecidae it is our impression that the subfamily Pemphredoninae contains some of the most beelike forms in the family with respect to general body form and wing venation. Some larrines, such as *Tachytes* and *Sericophorus,* are also beelike, but the resemblance is more superficial. Malyshev (1968) devoted considerable space to outline his hypothesis that bees probably arose from a pemphredonine-like ancestor. Similarities in nest construction between the more primitive bees and pemphredonines were important to his theory. He pointed out that among sphecids the bee habits (in *Hylaeus, Colletes*) of lining and partitioning nest cells with a membrane secreted by salivary glands is found only in pemphredonine wasps, notably *Psenulus.* The use of resin by the pemphredonine genus *Passaloecus* (Krombein, 1967b) to partition the nest is reminiscent of the same use of this substance by some megachilid, anthorphorid, and apid bees; and the discovery of sociality in the pemphredonine genus *Microstigmus* (Matthews, 1968a, b) adds further credence to Malyshev's hypothesis that bees and pemphredonines may have evolved from a common ancestor. Some species of the sceliphronin genus *Podium* use resin in nest construction too, and sociality is known in the related genus *Trigonopsis.* However, neither seems likely to be directly involved in the ancestry of bees.

Fossil Record

As far as sphecids are concerned the fossil record is quite fragmentary. The oldest sphecids are several forms described by Evans (1969b, 1973b) from Cretaceous amber of Canada and Siberia. The wings of these are very similar to living Pemphredoninae, and it seems likely that the Sphecidae must have been quite diversified before the end of the Mesozoic. Evans (1966a) remarked that he had seen Baltic Amber (Oligocene) in the collection of the Museum of Comparative Zoology containing sphecids belonging to, or close to, modern genera such as *Ampulex, Dolichurus, Pemphredon, Passaloecus,* and even the highly evolved Crabronini.

The descriptions of most supposed sphecid fossils do not include figures or photographs, and we have not felt obliged to evaluate the few taxa so far described. Evans (1966a) has examined the types of a few and reassigned some of these to other families. Listed below in decreasing chronological age are all of the fossils known to us that have been attributed to the Sphecidae. Annotations have been appended in a few cases.

Lower Cretaceous:
   Archisphex crowsoni Evans, 1969b (Similar to wing of *Psenulus,* a pemphredonine)

Upper Cretaceous:
   Lisponema singularis Evans, 1969b (Similar to wing of *Spilomena,* a pemphredonine)
   Taimyrisphex pristinus Evans, 1973b (Evans' placement of this taxon in Sphecidae only tentative. We concur because pronotum is unsphecidlike)
   Pittoecus pauper Evans, 1973b (Evans places this in Pemphredonini)

Eocene:
   Didineis solidescens Scudder, 1890 (Evans, 1966a placed this in the Eumenidae near *Alastor*)
   Hoplisus archoryctes Cockerell, 1922 (Evans, 1966a, suggested a relationship with *Harpactostigma* which now equals *Oryttus,* but it could just as easily be a *Psammaletes.*

Oligocene:
   Pison oligocenum Cockerell, 1908 (Redescribed as *Pison oligocaenum* by Cockerell in 1909)
   Tracheliodes succinalis (Cockerell), 1909 (*Crabro*)
   Tracheliodes tornquisti (Cockerell), 1909 (*Crabro*)
   Nysson rottensis Meunier, 1915. (Evans, 1966a, placed this in the *Eumenidae*; and Statz, 1936, identified it as *Alastor.*)
   Sceliphron tertiarium Meunier, 1915
   Philoponites clarus Cockerell, 1915
   Sceliphron brevior Cockerell, 1921
   †*Sphex obscurus* Statz, 1936, nec Fischer-Waldheim, 1843
   Philanthus annulatus Theobald, 1937
   †*Cerceris berlandi* Timon-David, 1944, nec Giner Marí, 1941

Miocene:
   Ammophila annosa Heer, 1865, nomen nudum?
   Ammophila inferna Heer, 1865
   Sphex giganteus Heer, 1867
   Ammophila gigantea Schöberlin, 1888, nomen nudum
   Ammophila minima Schöberlin, 1888, nomen nudum
   Tracheliodes mortuellus Cockerell, 1905
   Hoplisus sepultus Cockerell, 1906 (Evans, 1966a, said

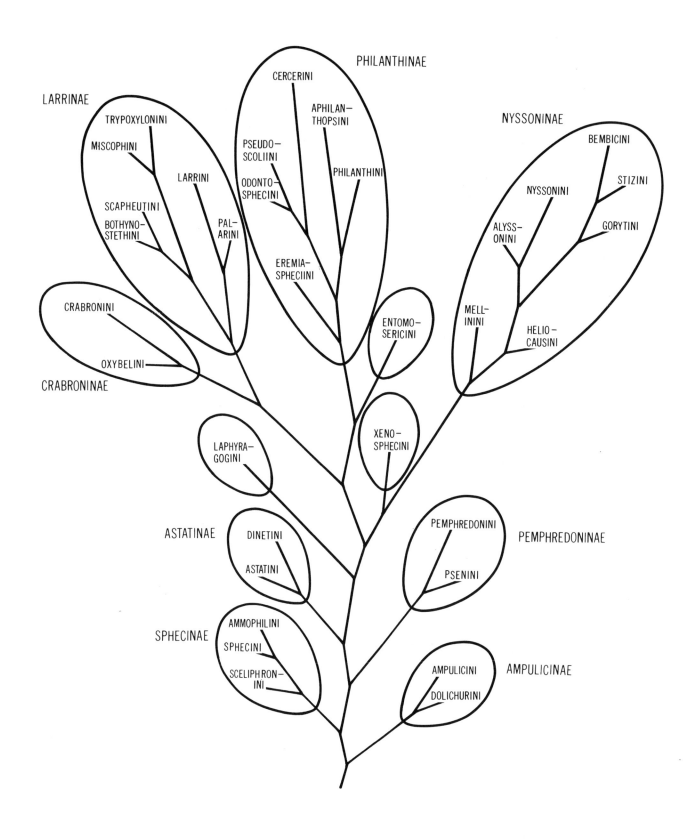

FIG. 7. Dendrogram showing presumed phylogeny of the family Sphecidae.

this is similar to *Gorytes*.)
Hoplisidea kohliana Cockerell, 1906 (Evans, 1966a, placed this in the Tiphiidae near *Anthobosca*)
Passaloecus scudderi Cockerell, 1906
Prophilanthus destructus Cockerell, 1906
Ammophila antiquella Cockerell, 1906
Larrophanes ophthalmicus Handlirsch, 1907
Chalybion mortuum Cockerell, 1907
Pison cockerellae Rohwer, 1908
Mellinus handlirschi Rohwer, 1908
Philanthus saxigenus Rohwer, 1909
Passaloecus fasciatus Rohwer, 1909
Crabro longoevus Cockerell, 1910 (= *Ectemnius*)
Sphex bischoffi Zeuner, 1931 (This appears to belong in the Ammophilini or Sceliphronini.)

PHILOSOPHY OF SUPRASPECIFIC CATEGORIES

One of the first taxonomic problems that we faced was the possibility that the Sphecidae might be divided into a number of families, following the lead of some previous workers. In support of this idea is the fact that sphecids are perhaps even more diversified than the bees that are customarily treated as five to a dozen or more families. A similar comparison can be made with the chalcidoid wasps which have been fragmented into many families. However, regardless of propriety, the Sphecidae cannot be divided satisfactorily into a series of families when considered from a world standpoint. Even subfamilies are difficult to delimit except on a local basis, and it is only when we reach the level of the tribe that the subdivisions become readily definable.

Our philosophy on the recognition of genera requires some explanation. The inadequacy of generic discretion and the wide disagreement among specialists as to generic limits have been major stumbling blocks in the systematics of aculeate wasps. In the present study we have tried to view the entire family objectively from a world standpoint. Ideally, assignment of an animal to a genus should apply a certain appearance, certain habits, and certain relationships with other genera. Obviously, the old genera of *Nysson, Gorytes, Sphex,* and *Crabro,* to name a few, imparted little of this sort of information. We believe that a genus should be that group of related species that can be separated structurally from other such groups with a reasonable degree of certainty, particularly if there is a biological connotation. From a practical point of view a genus of 25 species is more easily handled than one of 250 species. The degree of difference from other genera is a matter of less moment and can be considered in relation to the size of the genus. Therefore, not all genera are equivalent phylogenetically. For the most part we have avoided the use of subgenera and have treated the supraspecific categories either as "species groups" or as genera. The use of subgenera where they are defined only on characters of one sex seems particularly weak to us, and in most examples of this kind we have used species groups. Species group headings convey the same information that subgenera do, and they do not clutter our already overburdened zoological list of generic categories. Ordinarily, if we have found a subgenus that is distinct in both sexes, we have elevated it to genus. Exceptions to this rule are found in some of the larger genera such as *Liris, Pison, Sericophorus,* and *Crossocerus*. Our rather brief studies of such genera indicate that they contain distinct groups of a generic nature, but since we have not been in a position to make in-depth analyses of these taxa, and also because the fauna of many parts of the world such as Australia, New Guinea, and South America is not adequately known, we have decided to use subgenera. Further evidence may indicate that some of these are nothing more than species groups, while others are genera.

FAMILY STATISTICS

The number of species in our checklists totals 7,634 (not including subspecies, synonyms) that are distributed among 226 genera. This total includes 85 unknown or questionably assigned species, a few of which, especially in *Sphex* and *Crabro*, may not be sphecids. The number of species in each of the 11 subfamilies is shown in Table 3 below, which also indicates the largest genus in each group. Figures in parentheses give the additional number of species in each taxon that are unknown or doubtfully placed.

The Larrinae is the largest subfamily, but the largest genus, *Cerceris*, represents nearly 80 percent of the Philanthinae. For its size, the Larrinae contains comparatively few genera, and some of these rival the second largest sphecid genus *Trypoxylon*, in size: *Tachysphex* – 351, *Tachytes* – 268, and *Liris* – 263, *Tachysphex* and *Bembix* are the third and fourth largest sphecid genera respectively, followed by *Tachytes* and *Liris* in declining order. There is a considerable gap between these five and the largest of the remaining sphecid genera. The diversity of the Nyssoninae is clearly indicated by their having close to twice the number of genera of the Larrinae or the Crabroninae.

The body length of sphecids ranges from the diminutive 2 mm of some species of *Belomicrus, Miscophoides,* and *Timberlakena* to the 50 mm or more at-

TABLE 3.
Family Statistics

Subfamily	Species total	Number of genera	Largest genus
Larrinae	2,073+ (7)	39	*Trypoxylon*–359
Nyssoninae	1,473+ (3)	71	*Bembix* – 329
Crabroninae	1,211+(17)	44	*Oxybelus* – 217
Philanthinae	1,097+ (3)	11	*Cerceris* – 857+(2)
Pemphredoninae	714+ (1)	28	*Psenulus* – 121
Sphecinae	660+(52)	19	*Ammophila*–183+(4)
Ampulicinae	165+ (2)	6	*Ampulex*–116+(2)
Astatinae	145	5	*Astata*–76
Laphyragoginae	6	1	*Laphyragogus*–6
Xenosphecinae	3	1	*Xenosphex*–3
Entomosericinae	2	1	*Entomosericus*–2

tained by some specimens of *Dynatus nigripes, Sphex ingens,* and some *Parapsammophila.*

EXPLANATION OF CHECKLISTS

In preparing the species lists we have checked the literature through December 31, 1973 and in a few instances beyond this date. Undoubtedly, in spite of our efforts, a few names may have escaped detection.

For date of publication we have ordinarily used that indicated on the original paper. However, for papers the dates of which are unclear we have followed Sherborn and Woodward (1901, 1906), Sherborn (1922a, 1923, 1937), Stearn (1937), Griffin (1937), Blackwelder (1947, 1949), Bohart and Menke (1963), and Menke (1963a, 1974b).

In these lists we have given the species name, author, year, original generic name in parentheses if different from the present assignment, geographic distribution, subspecies if any, and critical synonymy. Synonyms and subspecies are listed in chronological order. Authors names of valid species and subspecies are contained in parentheses if the taxon was originally described in a different genus. This has not been done for authors of synonyms. The ending of each name has been changed to agree with the gender of the genus. An abbreviated example from *Sceliphron* illustrates the style outlined:

madraspatanum (Fabricius), 1781 (*Sphex*); Oriental
   Region
   ?*lugubre* Christ, 1791 (*Sphex*), nec Villers, 1789
   interruptum Palisot de Beauvois, 1806 (*Pelopaeus*)
   ssp. *tubifex* (Latreille), 1809 (*Pelopaeus*); Morocco to
      Tunisia, s. Europe to Iran
      pectorale Dahlbom, 1845 (*Pelopaeus*)
   ssp. *formosanum* van der Vecht, 1968; Taiwan

In this example we have indicated that *Sceliphron madraspatanum,* published in 1781 by Fabricius, is the valid name. Parentheses around Fabricius indicate that he described the species in a different genus, in this case *Sphex,* which is listed in parentheses behind the date of publication. This is followed by the distributional data for the species. Besides the nominate form *madraspatanum,* two subspecies are listed. Synonyms of *madraspatanum s.s.* are *lugubre* and *interruptum,* but the questionmark in front of the first of these indicates that its placement here needs confirmation, and "nec Villers, 1789" shows that the name is a junior homonym in *Sceliphron.* The name *pectorale* is listed as a synonym of the subspecies *tubifex.* All of the names with the exception of *formosanum* van der Vecht are indicated as having been described in genera other than *Sceliphron.*

Our recognition of subspecies is based on the latest available reference. There are obvious discrepancies in use of the trinomial in the literature. Some trinomials represent true geographical subspecies, others are simply color varieties or expressions of a cline. Still others appear to us to represent distinct species. For the most part we have left these problems to future workers because their interpretation is beyond the scope of this revision.

We have used a number of symbols in our lists. Questionmarks in front of species names not in synonymy indicate that the generic assignment is in doubt. A few of these taxa are even questionably placed in the Sphecidae. Questionmarks in front of names in synonymy and subspecies indicate that the status of the taxon within the genus is unclear. The dagger symbol (†) in front of a species' name indicates that it is a junior homonym currently without a replacement name. We have not renamed many of the homonyms discovered because we feel that in some cases revisionary work will indicate that replacement names are available through synonymy. We have examined a great number of holotypes in this study. We have placed asterisks in front of the names of those that we specifically want to indicate as having been seen by one of us.

Where one of us or one of our correspondents (J. van der Vecht, W. J. Pulawski, etc.) have made new synonymy, new combinations, or change of status, it is indicated at the end of the citation of the name: new synonymy by R. Bohart. In nearly every case, new synonymy is based on examination of holotypes. Lectotypes have also been designated, and these are appended at the end of the taxon in question.

We have based our spelling of authors names on the way they have most commonly appeared: De Geer, Vander Linden, Guérin-Méneville, du Buysson, Palisot de Beauvois, etc. An exception is Lepeletier de Saint-Fargeau. He usually referred to himself as St. Fargeau as attested by his signatures in our possession, but authors have traditionally called him Lepeletier.

Some of the works of Linnaeus such as "Pandora et Flora Rybyensis" also contain the name of one of his students. On the surface it would appear that taxa described in such papers should bear dual authorship. However, according to remarks on pages 83-84 in "A catalogue of the works of Linnaeus. . ." published in 1957 by Sandbergs Bokhandel, Stockholm, "the academic theses which have Linnaeus' name as Praeses contain nothing but the results of his own researches and investigations. At that time the professors at Upsala dictated most of the theses to their students and only rarely did a defendant write his thesis himself. One of the things the respondent had to do was to put the text into good or passable Latin. Sometimes, when the subject was complicated, this was not an easy task. Besides rendering the text into Latin, the defendant should pay for the printing of the theses. In this way, a professor, at least an industrious one like Linnaeus saved a lot of money." We assume that this discussion applies equally to Dahlbom's "Exercitationes Hymenopterologicae" in which a number of his students' names appear. We have attributed authorship of taxa in these cases solely to Linnaeus or Dahlbom. To do otherwise in the case of Dahlbom's work would pose problems because parts of some descriptions begin under one student's name and finish under another.

Sometimes a species in current literature has been attributed to an author who has given the best or most useful description rather than to the original describer. Most of these cases involve descriptions published before 1850. For example a few names attributed to Latreille, Dahlbom, or Vander Linden were described previously by Panzer or Fabricius. An example is *Bembix oculata* for which Latreille is usually given as the author, whereas actually it was described earlier by Panzer. In all such cases we have tried to ferret out the original author and date.

We have tried to follow the International Code of Zoological Nomenclature throughout, and this has resulted in a number of species name changes, especially for some rather well known Old World forms. It is our contention that rules are made to be followed, and that the only way to reach the goal of a stable nomenclature is to make the necessary changes when the need is discovered rather than to procrastinate for sentimental reasons. Failure of workers to follow Article 59 has created the largest source of name changing. Dalla Torre (1897) in his catalog of the Sphecidae used a very conservative generic concept, and his "lumping" of genera resulted in his proposing many new names for secondary homonyms. The same applies to the very early works of Villers (1789), Gmelin (1790), Turton (1802), and a few others who recognized only a few large wasp genera such as *Sphex* and *Vespa*. According to Article 59 (c) new names proposed before 1960 for secondary homonyms are valid even though homonymy no longer exists. The following example from *Cerceris* illustrates how we have indicated the application of article 59c in our checklists.

cornigera (Gmelin), 1790 (*Vespa*); India, new name for *Vespa cornuta* (Fabricius), (Art. 59c)

cornuta Fabricius, 1787 (*Crabro*); nec *Vespa cornuta* Linnaeus, 1758, now in *Synagris*.

This shows that Gmelin lumped *Crabro cornuta* Fabricius and *Vespa cornuta* Linnaeus under the genus *Vespa* and renamed Fabricius' junior homonym as *cornigera*.

Appended to the checklists of *Sphex* and *Bembix* species are lists of names published in auction catalogs. The rules of the International Commission on Zoological Nomenclature are not completely clear as to the status of such names, but under Article 8 of the Code it would appear that the paper in which they were published does not constitute a permanent, scientific record. Some names of Lichtenstein (1796) have been accepted for many years in other insect orders, and it would seem that the status of auction catalogs needs to be settled by the Commission.

We have treated generic names ending in ops as masculine (Art. 30a (i)(2)), and those ending in opsis as feminine (Art. 30a (i)).

Distribution as given in the species checklists often refers to areas rather than to political boundaries. For example, s. Africa refers to the southern one-fourth of that continent whereas S. Africa indicates the country of South Africa. The Ethiopian Region indicates the zoological region south of the Sahara desert. The use of China refers to the mainland and Taiwan to the island of Formosa. For island names in the East Indies we have followed the Atlas of the World published in 1970 by the National Geographic Society. The term "Middle East" refers particularly to Turkey, Syria, Israel, Jordan, Saudi Arabia, Iran, Iraq, and small associated countries. The term "sw. USSR" includes the areas or states of Transcaspia, Transcaucasia, Turkmen S.S.R., Uzbek S.S.R., Kazakh S.S.R., Tadzhik S.S.R., and Kirgiz S.S.R. We have used the current names of the various new African nations as well as others such as Bangladesh (East Pakistan) and Sri Lanka (Ceylon). Directional abbreviations used are e., s., n., w., se., nw., sw., ne., and centr. In the United States (U.S.), e. U.S. refers generally to the area east of the Mississippi River, w. U.S. to the area west of the Rocky Mountains crest, and centr. U.S. to the area in between. The abbreviation "sw. U.S." means the southern tier of states from west Texas to southern California. The term "U.S.: Transition zone" indicates a faunal zone as figured in the frontispiece of Muesebeck *et. al.*, (1951).

TABLE 4.
Family Arrangement

Ampulicinae	Crabroninae
Dolichurini	Oxybelini
Ampulicini	Crabronini
Sphecinae	Entomosericinae
Sceliphronini	Entomosericini
Sphecini	Xenospheciinae
Ammophilini	Xenospheciini
Pemphredoninae	Nyssoninae
Psenini	Mellinini
Pemphredonini	Heliocausini
Astatinae	Alyssonini
Astatini	Nyssonini
Dinetini	Gorytini
Laphyragoginae	Stizini
Laphyragogini	Bembicini
Larrinae	Philanthinae
Larrini	Eremiaspheciini
Palarini	Philanthini
Miscophini	Aphilanthopsini
Trypoxylonini	Odontospheciini
Bothynostethini	Pseudoscoliini
Scapheutini	Cercerini

SYNONYMIC GENERIC CATALOG OF THE SPHECIDAE

Listed here are all the recent genera of the family and their synonyms. Fossil genera are omitted but can be found in the section on fossil Sphecidae. The genera (and higher taxa) are arranged in what we consider the best phylogenetic order possible in a linear format. Subgenera are placed in parentheses, indented, and listed in chronological sequence within each genus. Synonyms are indented under each genus and subgenus. We have chosen to give only the name, author, year, and page of the original generic description. The complete citation can be found in the literature cited section at the end of the book.

The first entry after "type species" is always the binomen cited as the type by the designator. In most cases our wordage is an exact quotation, but we have spelled out the genus and author names in full and have added the proper date after the author's name if these were abbreviated or omitted by the designator. We have placed parentheses around the name of the author if the binomen designated as the type species is not the original combination for the species name. If authorship of the binomen designated as type species was attributed to someone other than the true author of the species, we have inserted "of" between the species name and the supposed author. We have also made certain that the species name agrees with the gender of the genus name with which it is combined. Quotes placed around the first entry mean that we have given the exact wordage used by the designator. This indicates that the identity of the type species is not clear without recourse to evidence external to the article in which the designation was made.

Our treatment of monotypic genera follows the above outline for designated type species, except that the first entry is the oldest reference to the species given in the original description of the genus.

Unless some clarification is needed, the first entry is followed by an author, year, and page reference to the person that designated the type species or by "monotypic" if the genus was based on one species and it was not specifically stated to be the type species by the author. We have used brackets [ ] after the first entry if clarification of the type designation is needed. Often they merely enclose the original binomen of the species or the original binomen of the species under which the type is now placed as a synonym. If the type species is now regarded as a subspecies, this information is given in brackets also. However, the current generic assignment of the subspecies is used if different from the original binomen. Brackets are also used to clarify the identity of the species cited in quotes. If more than one entry is enclosed in the brackets, they are listed in descending chronological order reading from left to right and ending with the original binomen (except where the type species is now a subspecies).

We have endeavored to follow the Code throughout, and this has resulted in some changes of authorship because a few names do not satisfy the provisions of Articles 11(f)ii, 13, or 42(d). Such cases are annotated. Most of these names have been validated by subsequent workers, but several [*Alpiothyreopus, Atrichothyreopus,* and *Trichothyreopus* Noskiewicz and Chudoba (1949), and *Euoxybelus* Noskiewicz and Chudoba (1950)] remain invalid. We have not followed the Code in a few cases where petitions are currently being considered by the Commission, since we anticipate favorable action on the appeals. These are discussed in footnotes. In accepting type species designations, we have followed the provisions of Opinions 11 and 136 for Latreille (1810) and Opinion 71 and Direction 32 for Westwood (1839-40). Another important Opinion is 135 which suppressed the "Erlangen List" published in 1801. There are other Opinions of the Commission that affect genera of the Sphecidae, but for the most part they deal with specific genera, and we have cited them under the genus to which they apply.

We have not repeated some of the nomina nuda and many of the misspellings noted by Pate (1937d) in his pioneer work on the nomenclature of the sphecid genera. His paper should be consulted for these as well as for the history of past nomenclatorial problems involving sphecid genera.

Only one generic name remains enigmatic as of this writing: *Mellinusterius* Meunier, 1889:24, type species:

Mellinusterius aphidium Meunier, 1889, monotypic. The color description of *Mellinusterius* is fairly detailed and should permit someone familiar with the Neotropical wasp fauna to identify *aphidium*. Perhaps it is a gorytin or scapheutin genus.

Pate (1937d) listed *Sericogaster* Westwood (1835) as an unknown sphecid genus, but our examination of its type species, *fasciatus*, lent through the courtesy of C. O'Toole of Oxford University, has revealed that this taxon belongs in the bee family Colletidae. See Menke and Michener (1973) for further details.

Derivations of many generic names are given by Kohl (1896) and Dalla Torre (1897).

SUBFAMILY AMPULICINAE
Tribe Dolichurini

Dolichurus Latreille, 1809:387. Type species: *Pompilus corniculus* Spinola, 1808, designated by Latreille, 1810:438.
> *Thyreosphex* Ashmead, 1904:282. Type species: *Thyreosphex stantoni* Ashmead, 1904, monotypic.

Paradolichurus Williams, 1960a:229. Type species: *Dolichurus californicus* Williams, 1960, original designation. New status by Bohart and Menke.

Aphelotoma Westwood, 1841a(April):152. Type species: *Aphelotoma tasmanica* Westwood, 1841, monotypic. Also described by Westwood, 1841b(June):16, and 1842:225.
> ? *Conocercus* Shuckard, 1840:180 (no species). Westwood, 1842:230 (footnote) tentatively synonymized *Conocercus* with *Aphelotoma*.

Austrotoma Riek, 1955:141. Type species: *Aphelotoma aterrima* Turner, 1907, monotypic.

Trirogma Westwood, 1841a(April):152. Type species: *Trirogma caerulea* Westwood, 1841, monotypic. Also described by Westwood, 1841b(June):16, and 1842:223.
> *Trirhogma* Agassiz, 1847:378. Emendation of *Trirogma* Westwood, 1841.

Tribe Ampulicini

Ampulex Jurine, 1807:132. Type species: "*Chlorion compressum* de Latreille et de Fabricius" [= *Chlorion compressum* (Fabricius), 1804, = *Sphex compressus* Fabricius, 1781], designated by Audouin, 1822:301.
> *Pronoeus* Latreille, 1809:56. Type species: *Dryinus aeneus* Fabricius, 1804, designated by Latreille, 1810:438.
> *Lorrheum* Shuckard, 1837:18. Type species: *Chlorion compressum* (Fabricius), 1804 [= *Sphex compressus* Fabricius, 1781], designated by Shuckard, 1837:18.
> *Rhinopsis* Westwood, 1844:68. Type species: *Rhinopsis abbottii* Westwood, 1844 [= *Ampulex canaliculata* Say, 1823], monotypic.
> *Waagenia* Kriechbaumer, 1874:55. Type species: *Waagenia sikkimensis* Kriechbaumer, 1874, monotypic.
> *Chlorampulex* Saussure, 1892:441. Type species: *Sphex compressus* Fabricius, 1781, designated by Pate, 1937d:18.

SUBFAMILY SPHECINAE
Tribe Sceliphronini
Subtribe Stangeellina

Stangeella Menke, 1962b:303. Type species: *Pelopoeus cyaniventris* Guérin-Méneville, 1831, original designation.

Subtribe Sceliphronina

Chlorion Latreille, 1802-1803:333. Type species: *Sphex lobatus* Fabricius, 1775, monotypic.
> *Chlorium* Schulz, 1906:193. Emendation of *Chlorion* Latreille, 1802-1803.

Penepodium Menke, new genus. Type species: *Pepsis luteipennis* Fabricius, 1804, present designation by Menke.

Dynatus Lepeletier, 1845:332. Type species: *Dynatus spinolae* Lepeletier, 1845, monotypic.
> *Stethorectus* F. Smith, 1847:349. Type species: *Stethorectus ingens* F. Smith, 1847 [= *Dynatus spinolae* Lepeletier, 1845], monotypic.

Podium Fabricius, 1804:183. Type species: *Podium rufipes* Fabricius, 1804, designated by Latreille, 1810:438.
> *Talthybius* Rafinesque-Schmaltz, 1815:125. New name for *Podium* Fabricius, 1804.
> *Ammophilus* Perty, 1833:141. Type species: *Ammophilus fumigatus* Perty 1833, designated by Pate, 1937d:7. Nec *Ammophilus* Latreille, 1829; possibly a lapsus for *Ammophila* W. Kirby, 1798.
> *Parapodium* Taschenberg, 1869:423. Type species: *Parapodium biguttatum* Taschenberg, 1869 [= *Podium rufipes* Fabricius, 1804], monotypic.

Trigonopsis Perty, 1833:141. Type species: *Trigonopsis abdominalis* Perty, 1833 [= *Podium rufiventre* Fabricius, 1804], monotypic.

Chalybion Dahlbom, 1843:21. Type species: "*Chalybion caerulem*"[1] [= *Sphex caeruleus* of Linnaeus, 1767:941, = *Sphex caeruleus* Linnaeus, 1763 (nec *Sphex caeruleus* Linnaeus, 1758), = *Sphex cyaneus* Fabricius, 1775 (nec *Sphex cyaneus* Linnaeus, 1758), = *Pelopeus* (sic) *californicus* Saussure, 1867], designated by Patton, 1881a:378.
> *Chalybium* Agassiz, 1847:77. Emendation of *Chalybion* Dahlbom, 1843.
> *Chalybium* Schulz, 1906:192. Emendation of *Chalybion* Dahlbom, 1843.

---

[1]The name *caeruleus* was not specifically mentioned by Dahlbom in his assignment of species to *Chalybion*. However, his third species *cyaneum* was obviously a lapsus for *caeruleus* of Linnaeus, 1767, since his bibliographic citation for *cyaneum* refers to *caeruleus* of Linnaeus, 1767. It is clear that Dahlbom simply made a mistake in substituting *cyaneum* (a Fabrician name) for *caeruleus* Linnaeus, 1767, and hence we accept Patton's designation.

(*Hemichalybion*) Kohl, 1918:79. Type species: *Pelopoeus eckloni* Dahlbom, 1845, designated by Pate, 1937d:30.
Sceliphron Klug, 1801:561. Type species: *Sphex spirifex* Linnaeus, 1758, designated by Bingham, 1897:235.
    Pelopoeus Latreille, 1802-1803:334. Type species: "*Pelopoeus spirifex*, Fab." [= *Pelopoeus spirifex* of Fabricius, 1804; = *Sphex spirifex* Linnaeus, 1758], designated by Latreille, 1810:438.
    Pelopaeus Latreille, 1804:180. Lapsus or emendation of *Pelopoeus* Latreille, 1802-1803.
    Sceliphrum Schulz, 1906:192. Emendation of *Sceliphron* Klug, 1801.
    (*Prosceliphron*) van der Vecht, 1968:192. Type species: *Sceliphron coromandelicum* (Lepeletier), 1845 [= *Pelopaeus coromandelicus* Lepeletier, 1845], original designation.

Tribe Sphecini

Subtribe Sphecina

Sphex Linnaeus, 1758:569. Type species: *Sphex flavipennis* Fabricius, 1793, designated by Internat. Comm. Zool. Nomencl., Opinion 180, 1946:571.
    Sphaex Scopoli, 1772:122. ? Lapsus or emendation of *Sphex* Linnaeus, 1758.
    Ammobia Billberg, 1820:105. Type species: *Pepsis argentata* (Fabricius), 1804 [= *Sphex argentatus* Fabricius, 1787], designated by Rohwer, 1911a:153.
    Proterosphex Fernald, 1905:165. Type species: *Sphex maxillosus* Fabricius, 1793 (nec *Sphex maxillosus* Poiret, 1787) [= *Sphex rufocinctus* Brullé, 1833], original designation.
    (*Fernaldina*) R. Bohart and Menke, 1963:130. Type species: *Sphex lucae* Saussure, 1867, original designation.
Isodontia Patton, 1881a:380. Type species: *Isodontia philadelphica* (Lepeletier), 1845 [= *Sphex philadelphicus* Lepeletier, 1845], original designation.
    Leontosphex Arnold, 1945:90. Type species: *Sphex leoninus* Saussure, 1890, original designation.
    Murrayella R. Bohart and Menke, 1963:137. Type species: *Sphex elegans* F. Smith, 1856, original designation.

Subtribe Prionyxina

Palmodes Kohl, 1890:112. Type species: *Chlorion occitanicum* (Lepeletier and Serville), 1828 [= *Sphex occitanicus* Lepeletier and Serville, 1828], designated by Fernald, 1906:318.
Chilosphex Menke, new genus. Type species: *Sphex argyrius* Brullé, 1833, present designation by Menke.
Prionyx Vander Linden, 1827:362. Type species: *Ammophila kirbii* Vander Linden, 1827 [recte *kirbyi*], monotypic.
    Priononyx Dahlbom, 1843:28. Type species: *Pepsis thomae* (Fabricius), 1804 [= *Sphex thomae* Fabricius, 1775], monotypic.
    Enodia Dahlbom, 1843:28. Type species: *Sphex albisectus* Lepeletier and Serville, 1828 [= *Ammophila kirbyi* Vander Linden, 1827], designated by Kohl, 1885b:164. Nec *Enodia* Hübner, 1819.
    Harpactopus F. Smith, 1856:264. Type species: *Harpactopus crudelis* F. Smith, 1856, designated by Patton, 1881a:384.
    Parasphex F. Smith, 1856:267. Type species: *Sphex albisectus* Lepeletier and Serville, 1828 [= *Ammophila kirbyi* Vander Linden, 1827], designated by Kohl, 1885b:164.
    Gastrosphaeria A. Costa, 1858:10. Type species: *Gastrosphaeria anthracina* A. Costa, 1858 [= *Sphex subfuscatus* Dahlbom, 1845], monotypic.
    Pseudosphex Taschenberg, 1869:420. Type species: *Pseudosphex pumilio* Taschenberg, 1869, monotypic. Nec *Pseudosphex* Hübner, 1818.
    Calosphex Kohl, 1890:113. Type species: *Sphex niveatus* Dufour, 1853, designated by Pate, 1937d:15.
    Neosphex Reed, 1894:627. Type species: *Neosphex albospiniferus* Reed, 1894, monotypic.

Tribe Ammophilini

Parapsammophila Taschenberg, 1869:469. Type species: *Parapsammophila miles* Taschenberg, 1869 [= *Ammophila cyanipennis* Lepeletier, 1845], designated by Pate, 1937d:48.
    Ceratosphex Rohwer, 1921b:671. Type species: *Sphex bakeri* Rohwer, 1921, original designation.
Hoplammophila Beaumont, 1960b:1. Type species: *Ammophila armata* (Illiger), 1807 [= *Sphex armatus* Illiger, 1807], original designation.
    Micadophila Tsuneki, 1962:28. Type species: *Ammophila aemulans* Kohl, 1901, original designation.
Podalonia Fernald, 1927:11[2]. Type species: *Ammophila violaceipennis* Lepeletier, 1845, designated by Internat. Comm. Zool. Nomencl., Opinion 857, 1968b:88. *Podalonia* Spinola, 1851, suppressed in same Opinion.
    Psammophila Dahlbom, 1842:2. Type species: *Psammophila affinis* (W. Kirby), 1798 [= *Ammophila affinis* W. Kirby, 1798], designated by Fernald, 1927:11. Nec *Psammophila* Brown, 1827.
Eremochares Gribodo, 1883:265. Type species: *Eremochares doriae* Gribodo, 1883 [= *Ammophila dives* Brullé, 1833], monotypic.
Eremnophila Menke, 1964c:875. Type species: *Ammophila opulenta* Guérin-Méneville, 1838, original designation.
Ammophila W. Kirby, 1798:199. Type species: *Sphex sabulosus* Linnaeus, 1758, designated by Internat.

---

[2]See systematics of *Podalonia* for discussion of *Podalonia* vs. *Pompilus*.

Comm. Zool. Nomencl., Opinion 180, 1946:571.

>*Ammophylus* Latreille, 1802-1803:332. Lapsus or emendation of *Ammophila* W. Kirby, 1798.

>*Miscus* Jurine, 1807:130 (no species). Type species: *Miscus campestris* (Latreille), 1809 [= *Ammophila campestris* Latreille, 1809], designated by Shuckard, 1837:79, (one of two species first included in *Miscus* by Latreille 1809:54)

>*Ammophilus* Latreille, 1829:322. Lapsus or emendation of *Ammophila* W. Kirby, 1798.

>*Coloptera* Lepeletier, 1845:387. Type species: *Coloptera barbara* Lepeletier, 1845, monotypic.

>*Argyrammophila* Gussakovskij, 1928a:7. Type species: *Ammophila induta* Kohl, 1901, original designation.

>*Apycnemia* Leclercq, 1961c:211. Type species: *Ammophila fallax* Kohl, 1884 [= *Ammophila hungarica* Mocsáry, 1883], original designation.

SUBFAMILY PEMPHREDONINAE

Tribe Psenini

Subtribe Psenina

Ammopsen Krombein, 1959a:18. Type species: *Ammopsen masoni* Krombein, 1959, original designation

Mimesa Shuckard, 1837:228. Type species: "*P. equestris* F." [= *Psen equestris* of Latreille, 1819, = *Trypoxylon equestre* Fabricius, 1804], original designation.

>*Aporia* Wesmael, 1852:272. Type species: *Mimesa equestris* (Fabricius), 1804 [= *Trypoxylon equestre* Fabricius, 1804], designated by Kohl, 1896:293. Nec *Aporia* Hübner, 1819.

>*Aporina* Gussakovskij, 1937:665. New name for *Aporia* Wesmael, 1852. Nec *Aporina* Fuhrmann, 1902.

Odontopsen Tsuneki, 1964b:12. Type species: *Psen hanedai* Tsuneki, 1964, original designation.

Mimumesa Malloch, 1933:16. Type species: *Psen niger* Packard, 1867, original designation.

Pseneo Malloch, 1933:7. Type species: *Psen kohlii* W. Fox, 1898, original designation.

Psen Latreille, 1796:122 (no species). Type species: *Sphex ater* Fabricius, 1794 [= *Crabro ater* Olivier, 1792], designated by Latreille, 1802-1803:338 (first included species).

>*Psenia* Stephens, 1829b:361. Type species: *Sphex ater* Fabricius, 1794 [= *Crabro ater* Olivier, 1792], designated by Pate, 1937d:54.

>*Dahlbomia* Wissman, 1849:9. Type species: *Sphex ater* Fabricius, 1794 [= *Crabro ater* Olivier, 1792], monotypic.

>*Mesopora* Wesmael, 1852:279. Type species: *Psen ater* of Vander Linden, 1829 [= *Sphex ater* of Panzer, 1799, = *Sphex ater* Fabricius, 1794, = *Crabro ater* Olivier, 1792], monotypic.

>*Caenopsen* Cameron, 1899:55. Type species: *Caenopsen fuscinervis* Cameron, 1899, monotypic.

>*Punctipsen* van Lith, 1968:125. Type species: *Mimesa exarata* Eversmann, 1849, original designation. New synonymy by R. Bohart.

Nesomimesa Perkins, 1899:8. Type species: *Nesomimesa hawaiiensis* Perkins, 1899, designated by Pate, 1937d:43.

Deinomimesa Perkins, 1899:11. Type species: *Deinomesa ferox* Perkins, 1899, designated by Pate 1937d:21.

Subtribe Psenulina

Pluto Pate, 1937d:51. New name for *Psenia* Malloch, 1933.

>*Psenia* Malloch, 1933:44. Type species: *Mimesa tibialis* Cresson, 1872, original designation. Nec *Psenia* Stephens, 1829.

Psenulus Kohl, 1896:293. Type species: "*M. fuscipennis* Dahlb." [= *Psen fuscipennis* Dahlbom, 1843], designated by Ashmead, 1899:224.

>*Neofoxia* Viereck, 1901:338. Type species: *Psen atratus* of Panzer, 1806 [= *Trypoxylon atratum* Fabricius, 1804, = *Sphex pallipes* Panzer, 1798], original designation.

>*Stenomellinus* Schulz, 1911b:142. Type species: *Psen dilectus* Saussure, 1892, monotypic.

>*Eopsenulus* Gussakovskij, 1934a:84. Type species: *Psenulus iwatai* Gussakovskij, 1934; original designation.

>*Nipponopsen* Yasumatsu, 1938:84. Type species: *Nipponopsen anomoneurae* Yasumatsu, 1938, original designation.

>*Diodontus* of authors (mainly American).

Tribe Pemphredonini

Subtribe Pemphredonina

Diodontus Curtis, 1834: text for pl. 496. Type species: *Pemphredon tristis* Vander Linden, 1829, designated by Internat. Comm. Zool. Nomencl., Opinion 844, 1968a:10.

>*Xylocelia* Rohwer, 1915:243. Type species: *Diodontus occidentalis*. W. Fox, 1892, original designation.

>*(Neodiodontus)* Tsuneki[3], 1972a:210. Type species: *Diodontus kohli* Tsuneki, 1972, original designation.

Pemphredon Latreille, 1796:128 (no species). Type species: *Pemphredon lugubris* (Fabricius), 1804 [= *Crabro lugubris* Fabricius, 1793], designated by Shuckard, 1837:193 (one of two species first included in *Pemphredon* by Latreille, 1802-1803:342).

---

[3]This taxon was described for a single male specimen. The subgenus is dubious without substantiation of the venational character on which it is based.

Cemonus Panzer, 1806:186. Type species: *Sphex unicolor* Panzer, 1798, (nec *Sphex unicolor* Fabricius, 1787) [= *Cemonus rugifer* Dahlbom, 1844], monotypic.

Cemonus Jurine, 1807:213. Type species: "*C. unicolor* F." [= *Pelopoeus unicolor* of Fabricius, 1804[4], = *Sphex unicolor* Panzer, 1798 (nec *Sphex unicolor* Fabricius, 1787), = *Cemonus rugifer* Dahlbom, 1844], designated by Shuckard, 1837:199.

Cenomus Gimmerthal, 1836:436. Lapsus or emendation of *Cemonus* Panzer or Jurine.

Dineurus Westwood, 1837:173. Type species: "*P. unicolor* Lat." [= *Pemphredon unicolor* of Latreille, 1809, = *Sphex unicolor* of Jurine, 1807:214, = *Pelopoeus unicolor* of Fabricius, 1804[4], = *Sphex unicolor* Panzer, 1798 (nec *Sphex unicolor* Fabricius, 1787), = *Cemonus rugifer* Dahlbom, 1844], original designation.

Ceratophorus Shuckard, 1837:195. Type species: *Pemphredon morio* Vander Linden, 1829, original designation.

Diphlebus Westwood, 1840:81. Type species: *Pelopoeus unicolor* of Fabricius, 1804[4], [= *Sphex unicolor* Panzer, 1798 (nec *Sphex unicolor* Fabricius, 1787), = *Cemonus rugifer* Dahlbom, 1844], original designation.

Chevrieria Kohl, 1883:658. Type species: *Pemphredon unicolor* (Fabricius), 1804 [= *Pelopoeus unicolor* of Fabricius, 1804[4], = *Sphex unicolor* Panzer, 1798 (nec *Sphex unicolor* Fabricius, 1787), = *Cemonus rugifer* Dahlbom, 1844], original designation.

Susanowo Tsuneki, 1972b:12[5]. Type species: *Pemphredon sudai* Tsuneki, 1972, original designation. New synonymy by R. Bohart.

Passaloecus Shuckard, 1837:188. Type species: *Pemphredon insignis* Vander Linden 1829, original designation. New name for *Xyloecus* Shuckard, 1837.

Xyloecus Shuckard, 1837: Conspectus of the genera, no. 25 (no species). Nec *Xyloecus* Serville, 1833.

Coeloecus C. Verhoeff, 1890:383. Type species: *Diodontus gracilis* Curtis, 1834, designated by Pate, 1937d:19

Heroecus C. Verhoeff, 1890:383 (no species). Type species: *Pemphredon insignis* Vander Linden, 1829, designated by Pate, 1937d:31 (first included species).

Polemistus Saussure, 1892:565. Type species: *Polemistus macilentus* Saussure, 1892, designated by Pate, 1937d:52. New status by Bohart and Menke.

Subtribe Stigmina

Arpactophilus F. Smith, 1864:36. Type species: *Arpactophilus bicolor* F. Smith, 1864, monotypic.

Harpactophilus Kohl, 1896:276. Emendation of *Arpactophilus* F. Smith, 1864.

Austrostigmus Turner, 1912c:55. Type species: *Austrostigmus queenslandensis* (Turner), 1908 [= *Stigmus queenslandensis* Turner, 1908], original designation. New synonymy by R. Bohart.

Paracrabro Turner, 1907:274. Type species: *Paracrabro froggatti* Turner, 1907, original designation.

Stigmus Panzer, 1804:Heft 86, pl. 7. Type species: *Stigmus pendulus* Panzer, 1804, monotypic.

Antronius Zetterstedt, 1838:442. Type species: *Stigmus pendulus* Panzer, 1804, monotypic. This name unavailable (Art. 11(d)).

Gonostigmus Rohwer, 1911c:559. Type species: *Gonostigmus typicus* Rohwer, 1911 (= *Stigmus temporalis* Kohl, 1892), original designation.

(*Atopostigmus*) Krombein, 1973:218. Type species: *Stigmus fulvipes* W. Fox, 1892, original designation.

Carinostigmus Tsuneki, 1954a:3. Type species: *Stigmus congruus* Walker, 1860, original designation. New status by Bohart and Menke.

Microstigmus Ducke, 1907a:28. Type species: *Microstigmus theridii* Ducke, 1907, monotypic.

Spilomena Shuckard, 1838:79. New name for *Celia* Shuckard, 1837.

Celia Shuckard, 1837:182. Type species: *Celia troglodytes* (Vander Linden), 1829 [= *Stigmus troglodytes* Vander Linden, 1829], original designation. Nec *Celia* Zimmermann, 1832.

Microglossa Rayment, 1930:212. Type species: *Microglossa longifrons* Rayment, 1930, original designation. Nec *Microglossa* Voight, 1831.

Microglossella Rayment, 1935:634. New name for *Microglossa* Rayment, 1930.

Taialia Tsuneki, 1971b:10. Type species: *Taialia formosana* Tsuneki, 1971, original designation. New synonymy by R. Bohart.

Xysma Pate, 1937e:94. Type species: *Ammoplanus ceanothae* Viereck, 1904, original designation.

Telexysma Leclercq, 1959:21. Type species: *Telexysma africana* Leclercq, 1959, original designation. New synonymy by R. Bohart.

---

[4] If Fabricius is considered to have described *unicolor* as a new species, as some authors contend ("at nostra minor"), then *Cemonus* Jurine, *Dineurus*, *Diphlebus*, and *Chevrieria* would be junior synonyms of *Psen* because the specimens Fabricius placed under the name *unicolor* were undoubtedly congeneric and probably conspecific with *Psen ater* (Fabricius), the type species of *Psen*. See Beaumont, 1937, pp. 35 and 42.

[5] Based on a single female from Japan with a frontal tooth, this subgeneric name does not seem to be adequately substantiated.

Subtribe Ammoplanina

Protostigmus Turner, 1918:356. Type species: *Protostigmus championi* Turner, 1918, monotypic.
> *Ammoplanopterus* Mochi, 1940:27. Type species: *Ammoplanopterus sinaiticus* Mochi, 1940, monotypic.

Anomiopteryx Gussakovskij, 1935:418. Type species: *Anomiopteryx paradoxa* Gussakovskij, 1935, original designation.

Pulverro Pate, 1937d:107. Type species: *Pulverro mescalero* Pate, 1937, original designation.

Ammoplanops Gussakovskij, 1931b:457. Type species: *Ammoplanops carinatus* Gussakovskij, 1931, original designation.

Ammoplanus Giraud, 1869:469. Type species: *Ammoplanus perrisi* Giraud, 1869 [? = *Ammoplanus wesmaeli* Giraud, 1869], designated by Pate, 1937d:7.
> *Hoplocrabron* De Stefani, 1887:60. Type species: *Hoplocrabron marathroicus* De Stefani, 1887, monotypic.
> *Ceballosia* Giner Marí, 1943:285. Type species: *Ammoplanus rjabovi* Gussakovskij, 1931, original designation. Nec *Ceballosia* Mercet, 1921.

Ammoplanellus Gussakovskij, 1931b:442. Type species: *Ammoplanus chorasmius* Gussakovskij, 1931, original designation. New status by R. Bohart.
> *(Parammoplanus)* Pate, 1939:391. Type species: *Ammoplanus apache* Pate, 1937, original designation. New status by R. Bohart.

Timberlakena Pate, 1939:374. Type species: *Timberlakena nolcha* Pate, 1939, original designation.
> *Mohavena* Pate, 1939:381. Type species: *Timberlakena yucaipa* Pate, 1939, original designation. New synonymy by R. Bohart.
> *Riparena* Pate, 1939:378. Type species: *Timberlakena cahuilla* Pate, 1939, original designation. New synonymy by R. Bohart.

SUBFAMILY ASTATINAE

Tribe Astatini

Uniplectron F. Parker, 1966a:766. Type species: *Uniplectron azureum* F. Parker, 1966, original designation.

Diploplectron W. Fox, 1893a:38. Type species: *Liris brunneipes* Cresson, 1881, monotypic.

Astata Latreille, 1796:xiii. Emendation of *Astatus* Latreille, 1796:114.
> *Astatus* Latreille, 1796:114 (no species). Type species: *Tiphia abdominalis* Panzer, 1798 [=*Sphex boops* Schrank, 1781], designated by Latreille, 1802-1803:337 (first included species). See also Opinion 139, Internat. Comm. Zool. Nomencl., 1943a.
> *Dimorpha* Panzer, 1806:126. Type species: *Tiphia abdominalis* Panzer 1798 [= *Sphex boops* Schrank, 1781], monotypic.

Dryudella Spinola, 1843:135. Type species: *Dryudella ghilianii* Spinola, 1843 [= *Astata tricolor* Vander Linden, 1829], designated by P. Verhoeff, 1951:152[6].

Tribe Dinetini

Dinetus Panzer, 1806:191. Type species: *Dinetus pictus* of Jurine, 1807 [= *Crabro pictus* of Panzer, 1794, = *Crabro pictus* Fabricius, 1793], designated by Latreille, 1810:438.
> *Dinetus* Jurine, 1807:207. Type species: *Crabro pictus* of Panzer, 1794 [= *Crabro pictus* Fabricius, 1793], monotypic. Nec *Dinetus* Panzer, 1806.

SUBFAMILY LAPHYRAGOGINAE

Laphyragogus Kohl, 1889:190. Type species: *Laphyragogus pictus* Kohl, 1889, monotypic.
> *Leianthrena* Bingham in Kohl, 1896:381. Type species: *Leianthrena Kohlii* Bingham, 1896, original designation.
> *Lianthrena* Bingham, 1897:212. Type species: *Lianthrena kohlii* Bingham, 1897 [= *Leianthrena kohlii* Bingham, 1896], original designation.

SUBFAMILY LARRINAE

Tribe Larrini

Subtribe Larrina

Larra Fabricius, 1793:220. Type species: *Larra ichneumoniformis* Fabricius, 1793 [= *Sphex anathema* Rossi, 1790], designated by Latreille, 1810:438.
> *Larrana* Rafinesque-Schmaltz, 1815:124. Emendation or new name for *Larra* Fabricius, 1793.
> *Lara* Drapiez, 1819:54. Lapsus or emendation of *Larra* Fabricius, 1793.
> *Monomatium* Shuckard, 1840:181 (no species). Type species: *Larraxena princeps* F. Smith, 1851, designated by Pate, 1935:246 (first included species).
> *Lyrops* Dahlbom, 1843:132. Type species: *Tachytes paganus* Dahlbom, 1843[7], monotypic. Nec *Lyrops* Illiger, 1807.
> *Larraxena* F. Smith, 1851:30. Type species: *Larraxena princeps* F. Smith, 1851, monotypic.
> *Larrada* F. Smith, 1856:273. Type species: *Larrada anathema* (Rossi), 1790 [= *Sphex anathema* Rossi, 1790], original designation.
> *(Cratolarra)* Cameron, 1900:34. Type species: *Cratolarra femorata* Cameron, 1900 (nec *Larra femorata* (Saussure), 1854)[8], monotypic. New status by Menke.

---

[6]Pate's (1937d:24) contention that Spinola's genus was monotypic is erroneous. See discussion by Verhoeff, 1951.

[7]Stadelmann (1897) placed *paganus* in *Larra* after studying the type.

[8]We have not renamed Cameron's species because it may be a junior synonym of some previously described form.

Liris Fabricius, 1804:227. Type species: *Sphex auratus* Fabricius, 1787 [= *Sphex aurulentus* Fabricius, 1787], designated by Patton 1881b:386.

    Lirisis Rafinesque-Schmaltz, 1815:124. Emendation or new name for *Liris* Fabricius, 1804.

    (*Motes*) Kohl, 1896:351. Type species: "*Notogonia odontophora* Kohl, 1892" [= *Larra odontophora* Kohl, 1894], designated by Pate, 1937d:41.

    (*Leptolarra*) Cameron, 1900:29. Type species: *Leptolarra reticulata* Cameron, 1900 (nec *Liris reticulata* (Saussure), 1892) [= *Leptolarra reticuloides* Richards, 1935, = *Liris nigricans* (Walker), 1871 subsp. *reticuloides* (Richards), 1935], designated by Richards, 1935a:164.

    Notogonia A. Costa, 1867:82. Type species: *Tachytes niger* of Vander Linden, 1829 [= *Larra nigra* of Latreille, 1804-1805, = *Pompilus niger* of Panzer, 1799, = *Sphex niger* Fabricius, 1775], monotypic. Nec *Notogonia* Perty, 1850.

    Caenolarra Cameron, 1900:28. Type species: *Caenolarra appendiculata* Cameron, 1900 (nec *Liris appendiculata* (Taschenberg), 1870) [= *Spanolarra rufitarsis* Cameron, 1900], monotypic.

    Spanolarra Cameron, 1900:32. Type species: *Spanolarra rufitarsis* Cameron, 1900, monotypic.

    Notogonius Howard, 1901(June):pl. 6, fig. 1. Lapsus for *Notogonia* A. Costa[9].

    Chrysolarra Cameron, 1901 (August):118. Type species: *Chrysolarra appendiculata* Cameron, 1901 (nec *Liris appendiculata* (Taschenberg), 1870)[10], designated by Pate, 1937d:18.

    Notogonidea Rohwer, 1911b:234. New name for *Notogonia* A. Costa, 1867.

    Dociliris Tsuneki, 1967a:26. Type species: *Larrada subtessellata* F. Smith, 1856, original designation. New synonymy by Menke.

    Nigliris Tsuneki, 1967a:27. Type species: *Notogonia japonica* Kohl, 1884 [= *Liris festinans* (F. Smith), 1858, subsp. *japonica* (Kohl)], original designation. New synonymy by Menke.

Dalara Ritsema, 1884b:560. New name for *Darala* Ritsema, 1884a.

    Darala Ritsema, 1884a:81. Type species: *Darala schlegelii* Ritsema, 1884, monotypic. Nec *Darala* Walker, 1855.

    Hyloliris Williams, 1919:49. Type species: *Hyloliris mandibularis* Williams, 1919, original designation.

Paraliris Kohl, 1884b:361. Type species: *Paraliris kriechbaumeri* Kohl, 1884 monotypic.

Dicranorhina Shuckard, 1840:181 (no species). Type species: *Piagetia intaminata* Turner, 1910, designated by Pate, 1937d:21 (one of four species first included in *Dicranorhina* by Turner, 1912b:199).

    Piagetia Ritsema, 1872:121. Type species: *Piagetia ritsemae* Ritsema, 1872, designated by Bingham, 1897:210.

Subtribe Tachytina

Gastrosericus Spinola, 1838:480. Type species: *Gastrosericus waltlii* Spinola, 1838, monotypic.

    Gasterosericus Dahlbom, 1845:467. Lapsus or emendation of *Gastrosericus* Spinola, 1838.

    Eparmostethus Kohl, 1907:167. Type species: *Eparmostethus madecassus* Kohl, 1907, monotypic.

    Paralellopsis Maidl, 1914:147. Type species: *Paralellopsis africanus* Maidl, 1914 [= *Gastrosericus neavei* Turner, 1913], original designation.

    Dinetomorpha Pate[11], 1937d:22. Type species: *Gastrosericus flavicornis* Gussakovskij, 1931, original designation.

    Gastrargyron Pate[11], 1937d:28. Type species: *Gastrosericus marginalis* Gussakovskij, 1931, original designation.

    Parallelopsis Pate, 1937d:47. Lapsus or emendation of *Paralellopsis* Maidl, 1914.

Tachytella Brauns, 1906:56. Type species: *Tachytella aureopilosa* Brauns, 1906, monotypic.

Larropsis Patton, 1892:90. Type species: *Larrada tenuicornis* F. Smith, 1856, original designation.

Ancistromma W. Fox, 1893b:487. Type species: *Larrada distincta* F. Smith, 1856, designated by Rohwer, 1911c:582.

Tachytes Panzer, 1806:129. Type species: *Pompilus tricolor* of Fabricius, 1798 [= *Sphex tricolor* Fabricius, 1793 (nec *Sphex tricolor* Schrank, 1781), = *Sphex tricoloratus* Turton, 1802, = *Tachytes obsoletus* (Rossi), 1792, subsp. *tricoloratus* (Turton), 1802], monotypic.

    Lyrops Illiger, 1807a:162. Type species: *Andrena etrusca* Rossi, 1790, monotypic. See also Illiger, 1807b:195.

    Tachyptera Dahlbom, 1843:133. Type species: *Apis obsoleta* Rossi, 1792, designated by Patton 1881b:391. Nec *Tachyptera* Berge, 1842.

    Holotachytes Turner, 1917:10. Type species: *Tachytes dichrous* F. Smith, 1856,

---

[9]There are numerous spelling errors in Howard's book, and we regard *Notogonius* as a lapsus for *Notogonia* Costa. This view was also shared by G. E. Bohart (1951:954) who nevertheless recognized *Notogonius nigripennis* (Fox) as type species, an unnecessary action. The specimen figured by Howard is in the U.S. National Museum and is *Liris argentata* (Palisot de Beauvois), not *nigripennis*.

[10]Cameron's species will probably prove to be a junior synonym of some other species, hence we have not renamed it.

[11]These names were first proposed by Gussakovskij (1931a), but since he did not designate type species [Art. 13(b)] they are not valid as of that date. However, they are available with Pate (1937d) as author.

original designation.
Calotachytes Turner, 1917:10. Type species: *Tachytes marshalli* Turner, 1912, original designation.
Tachyoides Banks, 1942:397. Type species: *Tachytes mergus* Fox, 1892, original designation.
Tachyplena Banks, 1942:397. New name for *Tachyptera* Dahlbom[12].
Tachynana Banks, 1942:398. Type species: *Tachytes obscurus* Cresson, 1872 [= *Tachytes chrysopyga* (Spinola), 1841, subsp. *obscurus* Cresson, 1872], original designation.
Tachysphex Kohl, 1883b:166. Type species: *Tachysphex filicornis* Kohl, 1883 [= *Tachytes fugax* Radoszkowski, 1877], designated by Bingham, 1897:192.
Schistosphex Arnold, 1922:137. Type species: *Schistosphex breijeri* Arnold, 1922, original designation. New synonymy by Menke.
Atelosphex Arnold, 1923a:177. Type species: *Atelosphex miscophoides* Arnold, 1923, original designation.
Parapiagetia Kohl, 1896:373. Type species: *Piagetia odontostoma* Kohl, 1884, original designation.
Lirosphex Brèthes, 1913:150. Type species: *Tachysphex subpetiolatus* Brèthes, 1909, original designation. New synonymy by Menke.
Psammosphex Gussakovskij, 1952:246. Type species: *Tachysphex genicularis* F. Morawitz, 1890, original designation.
Holotachysphex Beaumont, 1940:179. Type species: *Tachysphex holognathus* Morice, 1897, monotypic.
Phytosphex Arnold, 1951:153. Type species: *Tachysphex turneri* Arnold, 1923, original designation.
Haplognatha Gussakovskij, 1952:248. Type species: *Haplognatha prosopigastroides* Gussakovskij, 1952, original designation.
Prosopigastra A. Costa, 1867:88. Type species: *Prosopigastra punctatissima* A. Costa, 1867, monotypic.
Homogambrus Kohl, 1889:191. Type species: *Tachysphex globiceps* F. Morawitz, 1889, monotypic.
Hologambrus Morice, 1897:309. Lapsus or emendation of *Homogambrus* Kohl, 1889.
Kohliella Brauns, 1910:668. Type species: *Kohliella alaris* Brauns, 1910, monotypic.

Tribe Palarini

Palarus Latreille, 1802-1803:336. Type species: *Tiphia flavipes* Fabricius, 1793 (nec *Palarus flavipes* (Fabricius), 1781) [= *Palarus rufipes* Latreille, 1812], monotypic.
Gonius Panzer, 1806:176. Type species: *Philanthus flavipes* (Fabricius), 1804 [= *Crabro flavipes* Fabricius, 1781, = *Tiphia variegata* Fabricius, 1781], monotypic.
Gonius Jurine, 1807:203. Type species: *Philanthus flavipes* (Fabricius), 1793 [= *Crabro flavipes* Fabricius, 1781, = *Tiphia variegata* Fabricius, 1781], monotypic. Nec *Gonius* Panzer, 1806.

Tribe Miscophini

Lyroda Say, 1837:372. Type species: *Lyroda subita* Say, 1837, designated by Patton, 1881b:386.
Morphota F. Smith, 1856:293. Type species: *Morphota fasciata* F. Smith, 1856, designated by Pate, 1937d:41.
Odontolarra Cameron, 1900:35. Type species: *Odontolarra rufiventris* Cameron, 1900 [*Morphota formosa* F. Smith, 1859], monotypic.
Lyrodon Howard, 1901:pl. 6, fig. 5. Lapsus for *Lyroda* Say, 1837. Nec *Lyrodon* Goldfuss, 1837.
Sericophorus F. Smith, 1851:32. Type species: *Sericophorus chalybeus* F. Smith, 1851, monotypic.
Sericophorus Shuckard, 1840:181. Nomen nudum.
Tachyrrhostus Saussure, 1854:69. Type species: *Tachyrrhostus cyaneus* Saussure, 1854 [= *Sericophorus chalybeus* F. Smith, 1851], designated by Pate, 1937d:63.
(Zoyphidium) Pate, 1937d:68. New name for *Zoyphium* Kohl, 1893.
Zoyphium Kohl, 1893a:569. Type species: *Zoyphium sericeum* Kohl, 1893, monotypic. Nec *Zoyphium* Agassiz, 1847, which is an emendation of *Zuphium* Latreille, 1806.
Anacrucis Rayment, 1955:45. Type species: *Anacrucis laevigata* Rayment, 1955, original designation. New synonymy by Menke.
Sphodrotes Kohl, 1889:188. Type species: *Sphodrotes punctuosa* Kohl, 1889, monotypic.
Larrisson Menke, 1967b:29. Type species: *Sericophorus abnormis* Turner, 1914, original designation.
Mesopalarus Brauns, 1899:416. Type species: *Mesopalarus mayri* Brauns, 1899, monotypic.
Paranysson Guérin-Méneville, 1844:441. Type species: *Nysson abdominale* Guérin-Méneville, 1844, monotypic.
Helioryctes F. Smith, 1856:358. Type species: *Helioryctes melanopyrus* F. Smith, 1856, monotypic.
Pseudohelioryctes Ashmead, 1899:248. Type species: *Pseudohelioryctes foxii* Ashmead, 1899, original designation.
Plenoculus W. Fox, 1893c:554. Type species: *Plenoculus davisi* W. Fox, 1893, monotypic.
Ptygosphex[13] Gussakovskij, 1928a:18.

---

[12] Although Banks stated that *Tachyplena* was a new name for *Tachyptera*, he did designate *Tachytes mandibularis* Patton as type species, and therefore his generic name could be construed as a new genus.

[13] Spelled *Ptigosphex* on the generic heading, but obviously this is a printer's lapsus since the name is spelled *Ptygosphex* several times in the body of the description.

Type species: *Ptygosphex murgabensis* Gussakovskij, 1928, original designation.
  Pavlovskia Gussakovskij, 1935:424. Type species: *Pavlovskia tadzhika* Gussakovskij, 1935 [= *Ptygosphex murgabensis* Gussakovskij, 1928], original designation.
Solierella Spinola, 1851:349. Type species: *Solierella miscophoides* Spinola, 1851, monotypic.
  Silaon Piccioli, 1869:283. Type species: *Silaon compeditus* Piccioli, 1869, monotypic.
  Sylaon Piccioli, 1870:pl. 1. Lapsus or emendation of *Silaon* Piccioli.
  Niteliopsis S. Saunders, 1873:410. Type species: *Niteliopsis pisonoides* S. Saunders, 1873, monotypic.
  Ammosphecidium Kohl, 1878:701. Type species: *Ammosphecidium helleri* Kohl, 1878 [= *Silaon compeditus* Piccioli, 1869], monotypic.
  Sylaon Kohl, 1885a:209. Lapsus or emendation of *Silaon* Piccioli, 1869. Nec *Sylaon* Piccioli, 1870.
  Lautara Herbst, 1920:217. Type species: *Lautara jaffueli* Herbst, 1920, monotypic.
Miscophus Jurine, 1807:206. Type species: *Miscophus bicolor* Jurine, 1807, monotypic.
  (Nitelopterus) Ashmead in Kohl, 1896:497. Type species: *Nitelopterus slossonae* Ashmead, 1896, monotypic.
    Nitelopterus Ashmead, 1897:22. Type species: *Nitelopterus slossonae* Ashmead, 1897 [= *Nitelopterus slossonae* Ashmead, 1896], monotypic.
    Hypomiscophus Cockerell, 1898:321. Type species: *Miscophus arenarum* Cockerell, 1898, original designation.
    Miscophinus Ashmead, 1898:187. Type species: *Miscophinus laticeps* Ashmead, 1898, original designation.
Saliostethus Brauns in Kohl, 1896:448. Type species: *Saliostethus lentifrons* Brauns, 1896, monotypic.
  Mutillonitela Bridwell, 1920:396. Type species: *Mutillonitela mimica* Bridwell, 1920, original designation.
Saliostethoides Arnold, 1924:30. Type species: *Saliostethoides saltator* Arnold, 1924, original designation.
Miscophoides Brauns in Kohl, 1896:448. Type species: *Miscophoides handlirschii* Brauns, 1896, monotypic.
  (Miscophoidellus) Menke, new subgenus. Type species: *Miscophoides formosus* Arnold, 1924, present designation by Menke.
Nitela Latreille, 1809:77. Type species: *Nitela spinolae* Latreille, 1809, monotypic.
  Rhinonitela Williams, 1928a:97. Type species: *Rhinonitela domestica* Williams, 1928, original designation.
  (Tenila) Brèthes, 1913:153. Type species: *Nitela amazonica* Ducke, 1903, original designation.

Auchenophorus Turner, 1907:270. Type species: *Auchenophorus coruscans* Turner, 1907, original designation.

Tribe Trypoxylonini

Pisonopsis W. Fox, 1893:553. Type species: *Pisonopsis clypeata* W. Fox, 1893, monotypic.
Pison Jurine in Spinola, 1808:255. Type species: *Pison jurini* Spinola, 1808 [recte *jurinei*, = *Alyson ater* Spinola, 1808], monotypic.
  Tachybulus Latreille, 1809:75. Type species: *Tachybulus niger* Latreille, 1809 [= *Alyson ater* Spinola, 1808], monotypic.
  Nephridia Brullé, 1833:408. Type species: *Nephridia xanthopus* Brullé, 1833, monotypic.
  Pisonitus Shuckard, 1838:79. Type species: *Pison argentatum* Shuckard 1838, designated by Pate, 1937d:51.
  Pseudonysson Radoszkowski, 1876:104. Type species: *Pseudonysson fasciatus* Radoszkowski, 1876, monotypic.
  Taranga W. F. Kirby, 1883:201. Type species: *Taranga dubia* W. F. Kirby, 1883 [= *Pison spinolae* Shuckard, 1838], monotypic.
  Pisum Agassiz, 1847:293. Emendation of *Pison* Jurine, 1808. Nec *Pisum* Megerle, Megerle, 1811.
  Pisum Schultz, 1906:212. Emendation of *Pison* Jurine, 1808. Nec *Pisum* Megerle, 1811.
  (Pisonoides) F. Smith, 1858:104. Type species: *Pison obliteratum* F. Smith, 1858, monotypic.
    Parapison F. Smith, 1869b:298. Type species: *Pisonoides obliteratum* F. Smith, 1858, designated by Pate, 1937d:47.
  (Krombeiniellum) Richards, 1962:118. New name for *Paraceramius* Radoszkowski, 1887.
    Paraceramius Radoszkowski, 1887:432. Type species: *Paraceramius koreensis* Radoszkowski, 1887, monotypic. Nec *Paraceramius* Saussure, 1854.
  (Entomopison) Menke, 1968a:5. Type species: *Pison pilosum* F. Smith, 1873, original designation.
Aulacophilus F. Smith, 1869:305. Type species: *Aulacophilus vespoides* F. Smith, 1869, monotypic.
Pisoxylon Menke, 1968a:1. Type species: *Pisoxylon xanthosoma* Menke, 1968, original designation.
Trypoxylon Latreille, 1796:121 (no species). Type species: *Sphex figulus* Linnaeus, 1758, designated by Latreille, 1802-1803:339 (first included species).
  Tripoxilon Spinola, 1806:65. Lapsus or emendation of *Trypoxylon* Latreille, 1796.
  Apius Panzer, 1806:106. Type species: *Sphex figulus* of Panzer, 1801 [= *Sphex figulus* Linnaeus, 1758], monotypic.
  Apius Jurine, 1807:140. Type species:

Sphex figulus of Fabricius, 1775 [= *Sphex figulus* Linnaeus, 1758], designated by Morice and Durrant 1915: 394. Nec *Apius* Panzer, 1806.

Trypoxilon Jurine, 1807:141 and tableau comparatif, p. 2. Lapsus or emendation of *Trypoxylon* Latreille, 1796.

Trypoxylum Agassiz, 1847:380. Emendation of *Trypoxylon* Latreille, 1796.

Trypoxylum Schulz, 1906:212. Emendation of *Trypoxylon* Latreille, 1796. Nec *Trypoxylum* Agassiz, 1847.

Asaconoton Arnold, 1959:322. Type species: *Trypoxylon egregium* Arnold, 1959, original designation.

(Trypargilum) Richards, 1934:191. Type species: *Trypoxylon nitidum* F. Smith, 1856, original designation.

Tribe Bothynostethini

Bothynostethus Kohl, 1884b:344. Type species: *Bothynostethus saussurei* Kohl, 1884, monotypic.

Willinkiella Menke, 1968d:92. Type species: *Willinkiella argentina* Menke, 1968 [= *Pisonopsis argentinus* Schrottky, 1909], original designation.

Tribe Scapheutini

Scapheutes Handlirsch, 1887:278. Type species: *Scapheutes mocsaryi* Handlirsch, 1887, monotypic. See also Handlirsch, 1888:229.

Bohartella Menke, 1968d:95. Type species: *Bohartella scapheutoides* Menke, 1968, original designation.

SUBFAMILY CRABRONINAE

Tribe Oxybelini

Belomicroides Kohl, 1899:312. Type species: *Belomicroides schmiedeknechtii* Kohl, 1899, monotypic.

Brimocelus Arnold, 1927:62. Type species: *Belomicrus radiatus* Arnold, 1927, original designation.

Enchemicrum Pate, 1929:219. Type species: *Enchemicrum australe* Pate, 1929, original designation.

Belomicrus A. Costa, 1871:80. Type species: *Belomicrus italicus* A. Costa, 1871, monotypic.

Oxybeloides Radoszkowski, 1877:68. Type species: *Oxybeloides fasciatus* Radoszkowski, 1877 [= *Oxybelus radoszkowskyi* Dalla Torre, 1897], monotypic.

Oxybelomorpha Brauns in Kohl, 1896:475. Type species: *Oxybelomorpha kohlii* Brauns, 1896, monotypic.

Nototis Arnold, 1927:64. Type species: *Belomicrus bicornutus* Arnold, 1927, monotypic.

Pseudoxybelus Gussakovskij, 1933b:286. Type species: *Belomicrus persa* Gussakovskij, 1933, monotypic.

Oxybelus Latreille, 1796:129 (no species) Type species: *Crabro uniglumis* of Fabricius, 1775 [= *Vespa uniglumis* Linnaeus, 1758], designated by Latreille, 1802-1803:343 (first included species).

Notoglossa Dahlbom, 1845:514. Type species: *Notoglossa sagittata* Dahlbom, 1845 [= *Oxybelus lamellatus* Olivier, 1811], monotypic.

Alepidaspis A. Costa, 1882:35. Type species: *Alepidaspis diphyllus* A. Costa, 1882, monotypic.

Anoxybelus Kohl, 1923:274. Type species: *Oxybelus maidlii* Kohl, 1923, monotypic.

Gonioxybelus Pate[14], 1937d:28. Type species: *Oxybelus nigripes* Olivier, 1811 [= *Apis trispinosus* Fabricius, 1787], original designation.

Orthoxybelus Pate[14], 1937d:45. Type species: *Vespa uniglumis* Linnaeus, 1758, original designation.

Latroxybelus Noskiewicz and Chudoba, 1950:300. Type species: *Oxybelus latro* Olivier, 1811, monotypic.

Tribe Crabronini

Entomocrabro Kohl, 1905:356. Type species: *Crabro dukei* Kohl, 1905 [recte *duckei*], monotypic.

Anacrabro Packard, 1866:67. Type species: *Anacrabro ocellatus* Packard, 1866, monotypic.

Encopognathus Kohl, 1896:486. Type species: *Crabro braueri* Kohl, 1896, monotypic.

Karossia Arnold, 1929:409. Type species: *Karossia hessei* Arnold, 1929, original designation. New synonymy by H. Court.

Rhectognathus Pate, 1936b:147. Type species: *Encopognathus pectinatus* Pate, 1936, original designation.

Tsaisuma Pate, 1943b:57. Type species: *Lindenius wenonah* Banks, 1921, original designation.

Aryana Pate, 1943b:68. Type species: *Encopognathus oxybeloides* Pate, 1943 [= *Crabro bellulus* Schulz, 1906], original designation. Synonym of *Karossia*.

Entomognathus Dahlbom, 1844:295. Type species: *Crabro brevis* Vander Linden, 1829, monotypic.

(Toncahua) Pate, 1944b:341. Type species: *Entomognathus texanus* Cresson, 1887, original designation.

Florkinus Leclercq, 1956c: 2. Type species: *Encopognathus evolutionis* Leclercq, 1956, original designation.

(Koxinga) Pate, 1944b:341. Type species: *Entomognathus siraiya* Pate, 1944, original designation.

(Mashona) Pate, 1944b:341. Type species: *Thyreopus apiformis* Arnold, 1926, original designation.

Lindenius Lepeletier and Brullé, 1834:791. Type species: *Crabro albilabris* Fabricius, 1793, designated by Westwood, 1839:80.

Chalcolamprus Wesmael, 1852:590. Type

---

[14]These names were first proposed by Minkiewicz (1933), but since he did not designate type species [Art. 12(b)] they are not valid as of that date. However, the names are available with Pate (1937d) as author.

species: *Crabro albilabris* of Vander Linden, 1829 [= *Crabro albilabris* Fabricius, 1793], monotypic.

> *Trachelosimus* A. Morawitz, 1866:249. Type species: *Crabro armatus* Vander Linden, 1829 [= *Lindenius pygmaeus* (Rossi), 1794, subsp. *armatus* (Vander Linden), 1829], monotypic.

Holcorhopalum Cameron, 1904:264. Type species: *Holcorhopalum foveatum* Cameron, 1904, monotypic.

> *Amaripa* Pate, 1944b:344. Type species: *Amaripa thauma* Pate, 1944 [= *Holcorhopalum foveatum* Cameron, 1904], original designation.

Quexua Pate, 1942a:55. Type species: *Quexua cashibo* Pate, 1942, original designation.

> *Arecuna* Pate, 1942a:58. Type species: *Quexua essequibo* Pate, 1942, original designation.

Rhopalum Stephens, 1829a:34[15]. Type species: *Crabro rufiventris* Panzer, 1799 [= *Sphex clavipes* Linnaeus, 1758], designated by J. Curtis, 1837: text accompanying plate 656. See also Stephens, 1829b:366 and J. Curtis, 1829:123.

> *Euplilis* Risso, 1826:227. Type species: *Euplilis rufiventris* (Panzer), 1799 [= *Crabro rufiventris* Panzer, 1799, = *Sphex clavipes* Linnaeus, 1758], designated by Pate, 1935:246.
>
> *Physoscelus* Lepeletier and Brullé, 1834:804. Type species: *Crabro rufiventris* Panzer, 1799 [= *Sphex clavipes* Linnaeus, 1758], designated by Westwood, 1839:80.
>
> *Physoscelis* Westwood, 1839:80. Lapsus or emendation of *Physoscelus* Lepeletier and Brullé, 1834.
>
> (*Corynopus*) Lepeletier and Brullé, 1834:802. Type species: *Crabro tibialis* of Panzer, 1801 [= *Crabro tibialis* Fabricius, 1798, = *Sphex coarctatus* Scopoli, 1763], monotypic.
>
> *Dryphus* Herrich-Schaeffer, 1840:123. Type species: *Crabro tibialis* Fabricius, 1798 [= *Sphex coarctatus* Scopoli, 1763], monotypic.
>
> *Alliognathus* Ashmead, 1899:219. Type species: *Crabro occidentalis* W. Fox, 1895, original designation. New synonymy by Court and Bohart.
>
> (*Calceorhopalum*) Tsuneki, 1952b:111. Type species: *Rhopalum calceatum* (Tsuneki), 1947 [= *Crabro calceatus* Tsuneki, 1947, (nec *Crabro calceatus* Rossi, 1794), = *Rhopalum pygidiale* Bohart, new name herein], original designation.
>
> (*Latrorhopalum*) Tsuneki, 1952b:111. Type species: *Rhopalum latronum* (Kohl), 1915 [= *Crabro latronum* Kohl, 1915], original designation.
>
> (*Aporhopalum*) Leclercq, 1955d:3. Type species: *Rhopalum perforator* F. Smith, 1876, original designation.
>
> (*Zelorhopalum*) Leclercq, 1955d:4. Type species: *Rhopalum aucklandi* Leclercq, 1955, original designation.

Isorhopalum Leclercq, 1963:74. Type species: *Isorhopalum mayoni* Leclercq, 1963, original designation.

Podagritus Spinola, 1851:353. Type species: *Podagritus gayi* Spinola, 1851, monotypic.

> (*Echuca*) Pate, 1944b:354. Type species: *Crabro tricolor* F. Smith, 1856, original designation.
>
> (*Parechuca*) Leclercq, 1970:91. Type species: *Podagritus neuqueni* Leclercq, 1957, original designation.

Podagritoides Leclercq, 1957a:6. Type species: *Crabro oceanicus* Schulz, 1906, original designation.

Echucoides Leclercq, 1957a:7. Type species: *Podagritus piratus* Leclercq, 1957, original designation.

Notocrabro Leclercq, 1951a:47. Type species: *Notocrabro idoneus* (Turner), 1908 [= *Crabro idoneus* Turner, 1908], original designation.

> *Spinocrabro* Leclercq, 1954a:209. Unnecessary new name for *Notocrabro* Leclercq, 1951.

Moniaecera Ashmead, 1899:220. Type species: *Crabro abdominalis* W. Fox, 1895, original designation.

Huavea Pate, 1948a:58. Type species: *Moniaecera chontale* Pate, 1948, original designation.

Crossocerus Lepeletier and Brullé, 1834:763. Type species: *Crabro scutatus* Fabricius, 1787 [= *Sphex palmipes* Linnaeus, 1767], designated by Westwood, 1839:80.

> *Stenocrabro* Ashmead, 1899:216. Type species: *Crabro planipes* W. Fox, 1895, original designation.
>
> *Synorhopalum* Ashmead, 1899:218. Type species: *Crabro decorus* W. Fox, 1895, original designation.
>
> *Ischnolynthus* Holmberg, 1903:472. Type species: *Ischnolynthus foveolatus* Holmberg, 1903 [= *Crabro elongatulus* Vander Linden, 1829], monotypic.
>
> (*Blepharipus*) Lepeletier and Brullé, 1834:728. Type species: *Belpharipus nigritus* Lepeletier and Brullé, 1834, designated by Ashmead[16], 1899:215.
>
> *Coelocrabro* Thomson, 1874:262. Type species: *Crabro pubescens* Shuckard, 1837 [= *Blepharipus nigritus* Lepeletier and Brullé, 1834][17], designated by

---

[15]*Euplilis* Risso has priority over *Rhopalum*, and the name is generally used by North American authors; but Europeans, who have published most of the major works on the genus in recent years, still use *Rhopalum*. With this in view, the International Commission on Zoological Nomenclature has been asked (Menke, Bohart, and Richards, 1974a) to suppress *Euplilis* in favor of *Rhopalum* (see also Benson, Ferrière, and Richards, 1947b). In anticipation of favorable action by the Commission, we are using *Rhopalum*. We cannot accept the practice of some Europeans of attributing authorship of the name to Kirby. Stephens merely used Kirby's manuscript name when he published *Rhopalum*.

[16]Westwood's (1839:80) designation of *Crabro dimidiatus* Fabricius, 1781, as type species of *Blepharipus* is invalid since this species was only questionably included in the taxon by Lepeletier and Brullé (1834:731).

[17]Leclercq (1973a) has established the synonymy of *pubescens* with *nigritus*.

Richards, 1935:166.

Dolichocrabro Ashmead. 1899:216. Type species: *Dolichocrabro wickhamii* Ashmead, 1899, original designation.

(*Cuphopterus*) A. Morawitz, 1866:252. Type species: *Crabro subulatus* Dahlbom, 1845, monotypic.

Blepharipus of authors accepting Westwood's type species designation.

(*Hoplocrabro*) Thomson, 1874:262. Type species: *Crabro quadrimaculatus* Fabricius, 1793, monotypic.

(*Epicrossocerus*) Ashmead, 1899:215. Type species: *Crabro insolens* W. Fox, 1895, original designation.

(*Ablepharipus*) Perkins, 1913:390. Type species: *Ablepharipus podagricus* (Vander Linden), 1829 [= *Crabro podagricus* Vander Linden, 1829], monotypic.

(*Ornicrabro*) Leclercq, 1973b:291. Type species: *Crossocerus flavissimus* Leclercq, 1973, original designation.

(*Acanthocrabro*) Perkins, 1913:391. Type species: *Acanthocrabro vagabundus* (Panzer), 1798 [= *Crabro vagabundus* Panzer, 1798], monotypic.

Blepharipus of Leclercq, 1954.

(*Microcrabro*) Saussure, 1892:574. Type species: *Crabro micromegas* Saussure, 1892, monotypic.

Yuchiha Pate, 1944a:272. Type species: *Crossocerus xanthochilos* Pate, 1944, original designation.

(*Apocrabro*) Pate, 1944a:282. Type species: *Crossocerus aeta* Pate, 1944, original designation.

(*Nothocrabro*) Pate, 1944a:314. Type species: *Crabro nitidiventris* W. Fox, 1895, original designation.

(*Stictoptila*) Pate, 1944a:315. Type species: *Crabro confertus* W. Fox, 1895 [= *Crabro maculipennis* F. Smith, 1856], original designation.

(*Eupliloides*) Pate, 1946b:53. Type species: *Rhopalum albocollare* Ashmead, 1904, original designation.

(*Ainocrabro*) Tsuneki, 1954b:74. Type species: *Crossocerus aino* (Tsuneki), 1947 [= *Crabro aino* Tsuneki, 1947], original designation.

(*Pericrabro*) Leclercq, 1954a:219. Type species: *Thyreopus sociabilis* Arnold, 1932, original designation.

(*Oxycrabro*) Leclercq, 1961f:74. Type species: *Crabro acanthophorus* Kohl, 1892, original designation.

(*Paroxycrabro*) Leclercq, 1963:2. Type species: *Crossocerus sotirus* Leclercq, 1963, original designation.

(*Neoblepharipus*) Leclercq, 1968c:98. Type species: *Crossocerus potosus* Leclercq, 1968, original designation.

Fentis Tsuneki, 1971a:13. Type species: *Crossocerus quinquedentatus* Tsuneki, 1971, original designation.

(*Apoides*) Tsuneki, 1968c:5. Type species: *Crossocerus alticola* Tsuneki, 1968, original designation.

(*Alicrabro*) Tsuneki, 1968c:7. Type species: *Crossocerus rufiventris* Tsuneki, 1968, original designation.

(*Bnunius*) Tsuneki, 1971a:15. Type species: *Crossocerus domicola* Tsuneki, 1971, original designation.

Tracheliodes A. Morawitz, 1866:249. Type species: *Brachymerus megerlei* (Dahlbom), 1845 [= *Crabro megerlei* Dahlbom, 1845, = *Crossocerus curvitarsis* Herrich-Schaeffer, 1841], designed by Ashmead, 1899:219.

Brachymerus Dahlbom, 1845:519. Type species: *Crabro megerlei* Dahlbom, 1845 [= *Crossocerus curvitarsis* Herrich-Schaeffer, 1841], monotypic. Nec *Brachymerus* Chevrolet, 1835.

Fertonius Pérez in Ferton, 1892:341. Type species: *Crossocerus luteicollis* Lepeletier and Brullé, 1834 [= *Crabro quinquenotatus* Jurine, 1807], designated by Pate 1937d:27.

Crabro Fabricius, 1775:373. Type species: *Vespa cribraria* Linnaeus, 1758, designated by Internat. Comm. Zool. Nomencl., Opinion 144, 1943b:91. (*Crabro* Geoffroy, 1762 suppressed in same Opinion.)

Carabro Say, 1823:78. Lapsus for *Crabro*.

Thyreopus Lepeletier and Brullé, 1834:751. Type species: *Sphex cribrarius* (Linnaeus), 1767 [= *Vespa cribraria* Linnaeus, 1758], designated by Westwood, 1839:80. Strict synonym of *Crabro*.

Thyreocnemus A. Costa, 1871:64. Type species: *Crabro pugillator* A. Costa, 1871, monotypic. Strict synonym of *Crabro*.

Anothyreus Dahlbom, 1845:526. Type species: *Anothyreus lapponicus* of Dahlbom, 1845 [= *Crabro lapponicus* Zetterstedt, 1838], designated by Ashmead, 1899:214. Often treated as a subgenus of *Crabro*.

Paranothyreus Ashmead[18], 1899:213. Type species: *Crabro hilaris* F. Smith, 1856, original designation. Often treated as a subgenus of *Crabro*.

Synothyreopus Ashmead, 1899:213. Type species: *Crabro tumidus* (Packard), 1867 [= *Thyreopus tumidus* Packard, 1867], original designation. Often treated as a subgenus of *Crabro*.

Pemphilis Pate[19], 1944b:340. Type spe-

[18]Kohl (1896) coined this name, but because he used it as a secondary, not primary [Art. 42 (d)], division of the genus *Crabro*, the name is unavailable as of that date. Ashmead treated the taxon as a genus and thereby made it available.

[19]As far as we have been able to determine, *Pemphilis* Risso (1826:227) is a nomen nudum. There is no description, and the two species placed under it, "*Pemphilis palmata* Leach" and "*patellatus* Fab.," appear to be manuscript names. We have confirmed Pate's (1935:245) statement that Fabricius never used the latter name in his publications.

cies: *Vespa cribraria* Linnaeus, 1758, original designation. Strict synonym of *Crabro*.

Agnosicrabro Pate[20], 1944b:349. Type species: *Crabro occultus* Fabricius, 1804, original designation. Often treated as a subgenus of *Crabro*.

Dyscolocrabro Pate[20], 1944b:349. Type species: *Crabro chalybeus* Kohl, 1915, original designation. Often treated as a subgenus of *Crabro*.

Hemithyreopus Pate[20], 1944b:349. Type species: *Crabro loewi* Dahlbom, 1845, original designation. Often treated as a subgenus of *Crabro*.

Parathyreopus Pate[20], 1944:349. Type species: *Crabro filiformis* Radoszkowski, 1877, original designation. Often treated as a subgenus of *Crabro*.

Norumbega Pate, 1947d:12. Type species: *Thyreopus argus* Packard, 1867 (nec *Crabro argus* (Christ), 1791), [= *Crabro argusinus* R. Bohart, new name], original designation. Often treated as a subgenus of *Crabro*.

Pseudoturneria Leclercq, 1954a:208. New name for *Turneriola* Leclercq, 1951.

Turneriola Leclercq, 1951a:48. Type species: *Turneriola perlucida* (Turner), 1908 [= *Crabro perlucidus* Turner, 1908], original designation. Nec *Turneriola* China, 1933.

Piyuma Pate, 1944b:356. Type species: *Piyuma koxinga* Pate, 1944 [= *Crabro prosopoides* Turner, 1908], original designation.

Piyumoides Leclercq, 1963:60. Type species: *Crabro hewittii* Cameron, 1908, original designation.

Leclercqia Tsuneki, 1968c:14. Type species: *Leclercqia formosana* Tsuneki, 1968, original designation.

Towada Tsuneki, 1970d:1. Type species: *Crossocerus leclercqi* Tsuneki, 1959 [= *Crabro flavitarsus* Tsuneki, 1947], original designation.

Chimila Pate, 1944b:371. Type species: *Chimila pae* Pate, 1944, original designation.

Pae Pate, 1944b:364. Type species: *Pae paniquita* Pate, 1944, original designation.

Lamocrabro Leclercq, 1951a:48. Type species: *Lamocrabro nasicornis* (F. Smith), 1873 [= *Crabro nasicornis* F. Smith, 1873], original designation. (see addendum).

Chimiloides Leclercq, 1951a:50. Type species: *Chimiloides nigromaculatus* (F. Smith), 1868 [= *Crabro nigromaculatus* F. Smith, 1868], original designation.

Enoplolindenius Rohwer, 1911c:562. Type species: *Lindenius clypeatus* Rohwer, 1911, original designation.

(*Iskutana*) Pate, 1942c:390. Type species: *Enoplolindenius georgia* Pate, 1942 [= *Lindenius robertsoni* Rohwer, 1920], original designation.

Foxita Pate, 1942c:368. Type species: *Foxita atorai* Pate, 1942, original designation. (see addendum).

Taruma Pate, 1944b:360. Type species: *Taruma bara* Pate, 1944, original designation. (see addendum).

Vechtia Pate, 1944b:377. Type species *Crabro spinifrons* Bingham, 1897 [= *Crabro rugosus* F. Smith, 1857], original designation.

Hingstoniola Turner and Waterston, 1926:189. Type species: *Crabro duplicatus* Turner and Waterston, 1926, monotypic.

Arnoldita Pate, 1948b:156. Type species: *Thyreopus perarmatus* Arnold, 1926, original designation.

Odontocrabro Tsuneki, 1971a:22. Type species: *Odontocrabro abnormis* Tsuneki, 1971, original designation.

Neodasyproctus Arnold, 1926:373. Type species: *Thyreopus kohli* Arnold, 1926, monotypic.

Dasyproctus Lepeletier and Brullé, 1834:801. Type species: *Dasyproctus bipunctatus* Lepeletier and Brullé, 1834, monotypic.

Megapodium Dahlbom, 1844:295. Type species: *Megapodium westermanni* Dahlbom, 1844, designated by Pate, 1937d:37.

Megalopodium Schulz, 1906:202. Emendation of *Megapodium* Dahlbom, 1844.

Williamsita Pate, 1947c:107. Type species: *Crabro novocaledonicus* Williams, 1945, original designation.

(*Androcrabro*) Leclercq, 1950b:192. Type species: *Williamsita neglecta* (F. Smith), 1868 [= *Crabro neglectus* F. Smith, 1868], original designation.

Ectemnius Dahlbom, 1845:389. Type species: "*E. guttatus*, Dahlb." [= *Crabro guttatus* of Dahlbom, 1845, = *Crabro guttatus* Vander Linden, 1829], designated by Ashmead, 1899:172.

Clytochrysus A. Morawitz, 1864:453. Type species: *Crabro chrysostomus* Lepeletier and Brullé, 1834 [= *Crabro lapidarius* Panzer, 1804], designated by Richards, 1935a:168. Often treated as a subgenus of *Ectemnius*.

Thyreocerus A. Costa, 1871:65. Type species: *Crabro crassicornis* Spinola, 1808, monotypic. Often treated as a subgenus of *Ectemnius*.

Mesocrabro C. Verhoeff, 1892:70. Type species: *Crabro guttatus* Vander Linden, 1829, designated by Pate, 1937d:38. Strict synonym of *Ectemnius*.

Hypocrabro Ashmead, 1899:168. Type species: *Crabro decemmaculatus* Say, 1823, original designation. Often

---

We have searched through most of W. E. Leach's works for the name *palmata* but have not found it. In his foreword to volume 5, Risso implied that his friend Leach assisted with the insect portion, and it seems likely that Risso simply published some names that he copied from Leach determination labels. Since there is no evidence so far that *palmata* Leach is the same as *Crabro palmatus* Panzer or that *patellatus* Fab. is equivalent to *Crabro patellatus* Panzer, we cannot accept Pate's 1935 designation of *palmata* Leach as type species of *Pemphilis* Risso as valid. The first acceptable use of *Pemphilis* under the Code is that of Pate, 1944b.

[20]These names were first proposed by Kohl (1915), but since he used them as secondary, not primary [Arts. 11(f)ii and 42(d)], divisions of the genus *Crabro,* they are unavailable as of that date. Pate (1944b) treated them as subgenera thereby making them available. The names were not validated by Pate (1937d) because he maintained Kohl's usage of them as secondary divisions of *Crabro*.

treated as a subgenus of *Ectemnius.*

Pseudocrabro Ashmead, 1899:169. Type species: *Crabro chrysarginus* of Lepeletier, 1845 [= *Crabro chrysargyrus* Lepeletier and Brullé, 1834, = *Crabro decemmaculatus* Say, 1823], original designation. Synonym of *Hypocrabro.*

Xestocrabro Ashmead, 1899:169. Type species: *Crabro sexmaculatus* Say, 1823 [= *Crabro continuus* Fabricius, 1804], original designation. Synonym of *Hypocrabro.*

Xylocrabro Ashmead, 1899:169. Type species: *Crabro stirpicola* Packard, 1866, original designation. Synonym of *Hypocrabro.*

Metacrabro Ashmead, 1899:169. Type species: *Crabro kollari* Dahlbom, 1845 [= *Crabro lituratus* Panzer, 1804], original designation. Often treated as a subgenus of *Ectemnius.*

Protothyreopus Ashmead, 1899:170. Type species: *Crabro rufifemur* Packard, 1866, original designation. Often treated as a subgenus of *Ectemnius.*

Nesocrabro Perkins, 1899:25. Type species: *Crabro rubrocaudatus* Blackburn, 1887, designated by Pate, 1937d:42. Often treated as a subgenus of *Ectemnius.*

Oreocrabro Perkins, 1902:146. Type species: *Crabro abnormis* Blackburn, 1887, original designation. Often treated as a subgenus of *Ectemnius.*

Hylocrabro Perkins, 1902:147. Type species: *Crabro tumidoventris* Perkins, 1899, original designation. Synonym of *Oreocrabro.*

Melanocrabro Perkins, 1902:147. Type species: *Crabro curtipes* Perkins, 1899, original designation. Synonym of *Oreocrabro.*

Xenocrabro Perkins, 1902:148. Type species: *Crabro unicolor* F. Smith, 1856, original designation. Synonym of *Oreocrabro.*

Lophocrabro Rohwer, 1916:667. Type species: *Crabro singularis* F. Smith, 1856, [=*Vespa maculosa* Gmelin, 1790] original designation. Synonym of *Metacrabro.*

Merospis Pate, 1941b:121. Type species: *Ectemnius cyanauges* Pate, 1941, original designation. Often treated as a subgenus of *Ectemnius.*

Cameronitus Leclercq, 1950c:14. Type species: *Ectemnius menyllus* (Cameron), 1905 [= *Crabro menyllus* Cameron, 1905], original designation. Often treated as a subgenus of *Ectemnius.*

Apoctemnius Leclercq, 1950d:200. Type species: *Ectemnius domingensis* Leclercq, 1950, original designation. Often treated as a subgenus of *Ectemnius.*

Protoctemnius Leclercq, 1951d:105. Type species: *Crabro tabanicida* Fischer, 1929, original designation. Often treated as a subgenus of *Ectemnius.*

Yanonius Tsuneki, 1956c:129. Type species: *Crabro martjanowii* F. Morawitz, 1892, original designation. Often treated as a subgenus of *Ectemnius.*

Policrabro Leclercq, 1958c:106. Type species: *Crossocerus krusemani* Leclercq, 1950, original designation. Often treated as a subgenus of *Ectemnius.*

Iwataia Tsuneki, 1959b:8: *Crabro furuichii* Iwata, 1934, original designation. Often treated as a subgenus of *Ectemnius.*

Leocrabro Leclercq, 1968b:300. Type species: *Ectemnius leonesus* Leclercq, 1968, original designation. Treated as a subgenus of *Ectemnius* by Leclercq.

Ceratocrabro[21] Tsuneki, 1970d:1. Type species: *Ectemnius shimoyamai* Tsuneki, 1958, original designation. Treated as a subgenus of *Ectemnius* by Tsuneki.

Lestica Billberg, 1820:107. Type species: *Crabro subterraneus* Fabricius, 1775, designated by Rohwer, 1911a:154.

Solenius Lepeletier and Brullé, 1834:713. Type species: *Solenius interruptus* Lepeletier and Brullé, 1834, preocc. [= *Crabro confluentus* Say, 1837][22]. Often treated as a subgenus of *Lestica.*

Ceratoculus Lepeletier and Brullé, 1834:739. Type species: *Crabro alatus* Panzer, 1797, designated by Ashmead, 1899:170. Often treated as a subgenus of *Lestica.*

Thyreus Lepeletier and Brullé, 1834:761. Type species: *Crabro vexillatus* Panzer, 1797 [= *Apis clypeata* Schreber, 1759], monotypic. Nec *Thyreus* Panzer, 1806. See *Clypeocrabro* below.

Hypothyreus Ashmead, 1899:171. Type species: *Crabro subterraneus* Fabricius, 1775, original designation. Strict synonym of *Lestica.*

Clypeocrabro Richards, 1935a:167. New name for *Thyreus* Lepeletier and Brullé, 1834. Often treated as a subgenus of *Lestica.*

Ptyx Pate, 1947d:13. Type species: *Crabro pluschtschevskyi* F. Morawitz, 1891, original designation. Often treated as a subgenus of *Lestica.*

---

[21] We have reason to believe that this taxon may belong in *Lestica* or *Williamsita.* See systematics of *Ectemnius.*

[22] We anticipate a favorable vote on the petition of Court and Menke (1968), in which the International Commission on Zoological Nomenclature was requested to invalidate the type species designation of Westwood (1839). Westwood selected *Sphex vagus*, now known to be a *Mellinus,* as the type species of *Solenius.* (see addendum).

SUBFAMILY ENTOMOSERICINAE

Entomosericus Dahlbom, 1845:486. Type species: *Entomericus* (sic) *concinnus* Dahlbom, 1845, monotypic. Spelled *Enthomosericus* by Dahlbom, 1845, on Tabula Examinationis Synoptica Generis Nyssonidarum.

SUBFAMILY XENOSPHECINAE

Xenosphex Williams, 1954a:97. Type species: *Xenosphex xerophilus* Williams, 1954, original designation.

SUBFAMILY NYSSONINAE

Tribe Mellinini

Mellinus Fabricius, 1790:226. Type species: *Vespa arvensis* Linnaeus, 1758, designated by Curtis, 1836: text for pl. 580.
>*Millimus* Gimmerthal, 1836:449. Lapsus or emendation of *Mellinus* Fabricius, 1790.

Tribe Heliocausini

Heliocausus Kohl, 1892:210. Type species: *Heliocausus fairmairei* Kohl, 1892, monotypic.
>*Pseudolarra* Reed, 1894:638. Type species: *Arpactus larroides* Spinola, 1851, designated by Pate, 1937d:55.

Tribe Alyssonini

Alysson Panzer, 1806:169. Type species: *Pompilus spinosus* Panzer, 1801, designated by Morice and Durrant, 1915:406.
>*Alyson* Jurine, 1807:195. Type species: *Pompilus spinosus* Panzer, 1801, monotypic.

Didineis Wesmael, 1852:109. Type species: *Alyson lunicornis* of Vander Linden, 1829 [= *Pompilus lunicornis* Fabricius, 1798], monotypic.

Tribe Nyssonini

Nursea Cameron, 1902:275. Type species: *Nursea carinata* Cameron, 1902, monotypic.

Nippononysson Yasumatsu and Maidl, 1936:501. Type species: *Nippononysson rufopictus* Yasumatsu and Maidl, 1936, original designation.

Hyponysson Cresson, 1882:273. Type species: *Hyponysson bicolor* Cresson, 1882, monotypic.

Nysson Latreille, 1802-1803:340. Type species: *Crabro spinosus* Fabricius, 1775 [= *Sphex spinosus* J. Forster, 1771], designated by Shuckard, 1837:99. First use of this spelling by Latreille and evidently the one he originally intended since he used it exclusively in all subsequent works.
>*Nysso* Latreille, 1796:125 (no species)[23].

[23]*Nysso* has priority but has never been used. The International Commission on Zoological Nomenclature has been asked (Menke, Bohart, and Richards, 1974b) to suppress *Nysso* in favor of *Nysson* (see also Benson, Ferrière, and Richards, 1947a). We assume that the petition will receive a favorable vote and have therefore kept the popular spelling *Nysson*, rather than *Nysso*.

Apparently a printers lapsus for *Nysson*. Latreille, 1802-1803:140, placed two species in *Nysson*.
>*Nyssonus* Rafinesque-Schmaltz, 1815:124. Emendation of *Nysson* Latreille, 1802-1803.

Synnevrus A. Costa, 1859:16. Type species: *Synnevrus procerus* A. Costa, 1859 [= *Nysson epeoliformis* F. Smith, 1856], monotypic.
>*Synneurus* Gerstaecker, 1867:79. Emendation of *Synnevrus* A. Costa, 1859.

Neonysson R. Bohart, 1968c:228. Type species: *Nysson porteri* Ruiz, 1936, original designation.

Epinysson Pate, 1935:250. Type species: *Nysson basilaris* Cresson, 1882, original designation.

Hovanysson Arnold, 1945:52. Type species: *Brachystegus camelus* Arnold, 1945, original designation.

Brachystegus A. Costa, 1859:24. Type species: *Nysson dufourii* Lepeletier, 1845 [= *Nysson scalaris* Illiger, 1807], monotypic.

Acanthostethus F. Smith, 1869a:306. Type species: *Acanthostethus basalis* F. Smith, 1869 [= *Nysson mysticus* Gerstaecker, 1867], monotypic.
>? *Spalagia* Shuckard, 1840:181. Nomen nudum, synonymy teste Patton, 1909: 442.

Zanysson Rohwer, 1921a:404. Type species: *Nysson texanus* Cresson, 1872, original designation.

Perisson Pate, 1938b:156. Type species: *Nysson basirufus* Brèthes, 1913, original designation.

Cresson Pate, 1938b:153. Type species: *Nysson parvispinosus* Reed, 1894, original designation.
>*Cressonius* Bradley, 1956:257. Unnecessary emendation of *Cresson* Pate, 1938.

Antomartinezius Fritz, 1955:14. Type species: *Cresson patei* Fritz, 1955, original designation.

Foxia Ashmead, 1898:187. Type species: *Foxia pacifica* Ashmead, 1898, monotypic.

Losada Pate, 1940c:2. Type species: *Losada paria* Pate, 1940, original designation.

Idionysson Pate, 1940c:5. Type species: *Idionysson borero* Pate, 1940, original designation.

Metanysson Ashmead, 1899:326. Type species: *Nysson solani* Cockerell, 1895, original designation.
>*Huachuca* Pate, 1938b:185. Type species: *Metanysson arivaipa* Pate, 1938, original designation.

Tribe Gorytini

Clitemnestra Spinola, 1851:341. Type species: *Arpactus gayi* Spinola, 1851, original designation.
>? *Miscothyris* Shuckard, 1840:181. Nomen nudum.

>*Clytemnestra* Saussure, 1867:75. Emendation of *Clitemnestra* Spinola, 1851.

>*Miscothyris* F. Smith, 1869:307. Type species: *Miscothyris thoracicus* F. Smith, 1869, monotypic.

>*Clytaemnestra* Handlirsch, 1895:858. Emendation of *Clitemnestra* Spinola, 1851.

>*Astaurus* Rayment, 1955:55. Type species: *Astaurus hylaeoides* Rayment,

1955 [= *Arpactus plomleyi* Turner, 1940], original designation. New synonymy by R. Bohart.

Ochleroptera Holmberg, 1903:487. *Ochleroptera oblita* Holmberg, 1903, monotypic.

Paramellinus Rohwer, 1912:469. Type species: *Gorytes bipunctatus* Say, 1824, original designation.

Olgia Radoszkowski, 1877:33. Type species: *Olgia modesta* Radoszkowski, 1877, monotypic.

Kaufmannia Radoszkowski, 1877:43. Type species: *Kaufmannia maracandica* Radoszkowski, 1877, monotypic.

Argogorytes Ashmead, 1899:324. Type species: *Gorytes carbonarius* F. Smith, 1856, original designation.

Archarpactus Pate, 1937d:10. Proposed as a new name for *Arpactus* Jurine, 1801, which has no nomenclatural standing (see Opinion 135 of the Internat. Comm. Zool. Nomencl. 1939). Pate's name apparently has no legality since it does not meet any of the requirements of Article 13(a) of the Code.

Neogorytes R. Bohart, new genus. Type species: *Neogorytes ecuadorae* R. Bohart, present designation.

Exeirus Shuckard, 1838:71. Type species: *Exeirus lateritius* Shuckard, 1838, monotypic.

Exirus Schulz, 1906:199. Emendation of *Exeirus* Shuckard, 1838.

Dienoplus W. Fox, 1893b:548. Type species: *Dienoplus pictifrons* W. Fox, 1893, monotypic.

Arpactus Jurine, 1807:192. Type species: *Arpactus formosus* Jurine, 1807, designated by Shuckard, 1837:220. Nec *Arpactus* Panzer, 1805.

Harpactus Shuckard, 1837:221. Emendation of *Arpactus* Jurine, 1807.

Harpactes Dahlbom, 1843:147. Emendation of *Harpactus* Shuckard, 1837, nec *Harpactes* Swainson, 1833.

Hapalomellinus Ashmead, 1899:300. Type species: *Gorytes eximius* Provancher, 1888 (nec *Gorytes eximius* F. Smith, 1862) [= *Gorytes albitomentosus* Bradley, 1920], original designation.

Trichogorytes Rohwer, 1912:469. Type species: *Trichogorytes argenteopilosus* Rohwer, 1912, original designation.

Austrogorytes R. Bohart, 1967:155. Type species: *Gorytes bellicosus* F. Smith, 1862, original designation.

Gorytes Latreille, 1804:180. Type species: *Mellinus quinquecinctus* Fabricius, 1793, monotypic.

Arpactus Panzer, 1805:Heft 98, text for pl. 17. Type species: *Mellinus quadrifasciatus* Fabricius, 1804, monotypic.

Arpactus Panzer, 1806:164. Type species: *Mellinus quadrifasciatus* Fabricius, 1804, designated by Pate, 1937d:11. Nec *Arpactus* Panzer, 1805.

Euzonia Stephens, 1829b:363. Type species: *Mellinus quinquecinctus* Fabricius, 1793, designated by Pate, 1937d:27.

Hoplisus Lepeletier, 1832:61. Type species: "*H. 5-cinctus* St. Farg" [= *Hoplisus quinquecinctus* of Lepeletier, 1832, = *Mellinus quinquecinctus* Fabricius, 1793], designated by Westwood, 1839:80.

Euspongus Lepeletier, 1832:66. Type species: *Euspongus laticinctus* Lepeletier, 1832, designated by Westwood, 1839:80.

Pseudoplisus Ashmead, 1899:323. Type species: *Gorytes floridanus* W. Fox, 1891 [= *Pseudoplisus smithii* (Cresson), 1880, subsp. *floridanus* (W. Fox), 1891], original designation.

Laevigorytes Zavadil in Zavadil and Snoflak, 1948:66. Type species: *Gorytes kohlii* Handlirsch, 1888, monotypic.

Megistommum Schulz, 1906:200. New name for *Megalomma* F. Smith, 1873.

Megalommus Shuckard, 1840:181. Nomen nudum.

Megalomma F. Smith, 1873:405. Type species: "*Megalomma elegans* Smith" [= *Gorytes elegans* F. Smith, 1873 (nec *Gorytes elegans* Lepeletier, 1832), = *Gorytes procerus* Handlirsch, 1888], designated by Pate, 1937d:37. Nec *Megalomma* Westwood, 1841.

Stenogorytes Schrottky, 1911:28. Type species: *Megalomma melanogaster* Schrottky, 1911, original designation.

Neoplisus R. Bohart, 1967:159. Type species: *Gorytes notabilis* Handlirsch, 1888, original designation.

Eogorytes R. Bohart, new genus. Type species: *Gorytes fulvohirtus* Tsuneki, 1963, present designation by R. M. Bohart.

Lestiphorus Lepeletier, 1832:70. Type species: *Crabro bicinctus* Rossi, 1794, monotypic. See Internat. Comm. Zool. Nomencl., Opinion 675, 1963:331.

Lestophorus Agassiz, 1847:208. Emendation of *Lestiphorus* Lepeletier, 1832.

Hypomellinus Ashmead, 1899:299. Type species: *Gorytes rufocinctus* Fox, 1892 [= *Gorytes piceus* Handlirsch, 1888], original designation.

Mellinogastra Ashmead, 1899:300. Type species: *Gorytes mellinoides* W. Fox 1895, original designation.

Oryttus Spinola, 1836:xxiii. Type species: "*Arpactus concinnus*" [= *Sphex concinnus* Rossi, 1790], monotypic.

Agraptus Wesmael, 1852:108. Type species: *Gorytes concinnus* of Vander Linden, 1829 [= *Sphex concinnus* Rossi, 1790], monotypic.

Harpactostigma Ashmead, 1899:299. Type species: *Hoplisus velutinus* Spinola, 1851, original designation.

Arcesilas Pate, 1938d:60. Type species: *Gorytes mirandus* W. Fox, 1892, original designation.

Psammaletes Pate, 1936a:49. Type species: *Gorytes bigeloviae* Cockerell and Fox, 1897, original designation.

Handlirschia Kohl, 1896:425. Type species: *Sphecius aethiops* Handlirsch, 1889, monotypic.

Sphecius Dahlbom, 1843:154. Type species: *Sphecius speciosus* Dahlbom, 1843 (nec *Sphecius speciosus*

(Drury), 1773) [= *Sphex speciosus* Drury, 1773], monotypic.

    Hogardia Lepeletier, 1845:288. Type species: *Hogardia rufescens* Lepeletier, 1845 [= *Stizus hogardii* Latreille, 1806], by tautonymy [Art. 68(d)].

    (Sphecienus) Patton, 1879a:345. Type species: *Stizus nigricornis* Dufour, 1838, original designation.

    (Nothosphecius) Pate, 1937a:199. Type species: *Stizus grandidieri* Saussure, 1887, original designation.

Tanyoprymnus Cameron, 1905b:375. Type species: *Tanyoprymnus longitarsis* Cameron, 1905 [= *Gorytes moneduloides* Packard, 1867], monotypic.

    Ceratostizus Rohwer, 1921a:412. Type species: *Gorytes moneduloides* Packard, 1867, original designation.

Ammatomus A. Costa, 1859:36. Type species: *Gorytes coarctatus* Spinola, 1808, monotypic.

Kohlia Handlirsch, 1895:950. Type species: *Kohlia cephalotes* Handlirsch, 1895, monotypic.

    Stizobembex Gussakovskij, 1952:272. Type species: *Stizobembex pavlovskii* Gussakovskij, 1952, original designation.

Pterygorytes R. Bohart, 1967:157. Type species: *Gorytes valens* Fox, 1897, original designation.

Psammaecius Lepeletier, 1832:72. Type species: *Gorytes punctulatus* Vander Linden, 1829, monotypic.

Liogorytes R. Bohart, 1967:160. Type species: *Liogorytes catarinae* R. Bohart, 1967 [= *Gorytes polybia* Handlirsch, 1895], original designation.

Arigorytes Rohwer, 1912:469. Type species: *Gorytes coquillettii* Fox, 1895, original designation.

Xerogorytes R. Bohart, new genus. Type species: *Arigorytes anaetis* Pate, 1947, present designation by R. Bohart.

Hoplisoides Gribodo, 1884:276. Type species: *Hoplisoides intricans* Gribodo, 1884, monotypic.

    Icuma Cameron, 1905a:21. Type species: *Icuma sericea* Cameron, 1905, monotypic. New synonymy by R. Bohart.

Sagenista R. Bohart, 1967:157. Type species: *Hoplisus scutellaris* Spinola, 1841, original designation.

Afrogorytes Menke, 1967b:34. Type species: *Gorytes monstrosus* Handlirsch, 1894, original designation.

Tribe Stizini

Stizus Latreille, 1802-1803:344. Type species: "*Stizus ruficornis* Fabr." [= *Bembex ruficornis* Fabricius, 1787 (nec *Vespa ruficornis* J. Forster, 1771), = *Vespa ruficornis* J. Forster, 1771], designated by Blanchard, 1846:pl. 121.

    Larra Fabricius sensu Klug, 1845:pl. 46 and text (in part)[24].

    Larra Fabricius sensu F. Smith, 1856:337 (in part)[24].

    Megastizus Patton, 1879a:344. Type species: *Stizus brevipennis* Walsh, 1869, original designation.

    Stizolarra Saussure, 1887:9. Type species: *Sphex vespiformis* Fabricius, 1775, designated by Pate, 1937d:62.

    Megalostizus Schulz, 1906:199. Emendation of *Megastizus* Patton, 1879.

Stizoides Guérin-Méneville, 1844:438. Type species: *Larra fasciata* Fabricius, 1798 (nec *Stizus fasciatus* (Fabricius), 1781) [= *Sphex assimilis* Fabricius, 1787], designated by J. Parker, 1929:10.

    (Scotomphales) Vachal, 1900:233. New name for *Omphalius* Vachal, 1899 (see also Ann. Soc. Ent. France 68:843 (1900)).

    Omphalius Vachal, 1899:534. Type species: *Omphalius niger* Vachal, 1899 [= *Stizus niger* Radoszkowski, 1881], monotypic. Nec *Omphalius* Philippi, 1847.

    (Tachystizus) Pate[25], 1937d:63. Type species: *Crabro tridentatus* Fabricius, 1775, original designation.

Bembecinus A. Costa, 1859:4. Type species: *Bembecinus meridionalis* A. Costa, 1859, monotypic.

    Stizomorphus A. Costa, 1859:7. Type species: *Vespa tridens* Fabricius, 1781, monotypic.

    Gorystizus Pate[25], 1937d:29. Type species: *Vespa tridens* Fabricius, 1781, original designation.

    Lavia Rayment, 1953:123. Nomen nudum.

Tribe Bembicini

Bicyrtes Lepeletier, 1845:53. Type species: *Bicyrtes servillii* Lepeletier, 1845 [= *Monedula ventralis* Say, 1824], monotypic.

    Bembidula Burmeister, 1874:122. Type species: *Monedula discisa* Taschenberg, 1870, designated by J. Parker, 1917:56.

    Dumonela Reed, 1894:608. Type species: *Monedula sericea* Spinola, 1851 [= *Bembex variegata* Olivier, 1789], original designation.

Microbembex Patton, 1879b:364. Type species: *Bembex monodonta* Say, 1824, original designation.

Carlobembix Willink, 1958:47. Type species: *Carlobembix marthae* Willink, 1958, original designation.

Hemidula Burmeister, 1874:119. Type species: *Monedula singularis* Taschenberg, 1870, designated by J. Parker, 1917:2.

Rubrica J. Parker, 1929:53. Type species: *Monedula gravida* Handlirsch, 1890, original designation.

---

[24] *Larra* of Fabricius contained unrelated species of sphecids as well as a scoliid and a tiphiid. Latreille (1802-1803) recognized the conglomerate nature of the genus and restricted *Larra* to wasps now placed in the Larrinae. He established the new genus *Stizus* for the remaining sphecids of Fabricius' *Larra*. Klug and Smith both restricted *Larra* to species assigned by other authors to Latreille's *Stizus*. F. Smith (1856:273-74, 337) went so far as to designate a type species for *Larra* although his designation was antedated by that of Latreille, 1810, and others. Species placed in *Larra* by Klug and Smith are now assigned to *Stizus*, *Stizoides*, and *Bembecinus*.

[25] These names were first proposed by Minkiewicz (1933), but since he did not designate a type species [Art. 13 (b)], they are not valid as of that date. However, the names are available with Pate (1937d) as author.

Selman J. Parker, 1929:20. Type species: *Selman angustus* J. Parker, 1929 [= *Monedula notata* Taschenberg, 1870], original designation.
Stictia Illiger, 1807a:131. New name for *Monedula* Latreille, 1802-1803 (see also Illiger, 1807b:195).
    Monedula Latreille, 1802-1803:345. Type species: *Bembex signata* of Fabricius, 1781 [= *Vespa signata* Linnaeus, 1758], designated by Latreille, 1810:438. Nec *Monedula* Moehring, 1758.
Editha J. Parker, 1929:17. Type species: *Monedula magnifica* Perty, 1833, original designation.
Bembix[26] Fabricius, 1775: Characters Generum, p. [xxiii] (no species). Type species: *Bembex rostrata* of Fabricius, 1781 [= *Apis rostrata* Linnaeus, 1758], designated by Latreille, 1810:438[27] (one of three species included in *Bembyx* by Fabricius, 1775:361).
    Bembyx[26] Fabricius, 1775:361. Lapsus for *Bembix* Fabricius, 1775.
    Bembex[26] Fabricius, 1776:122. Emendation of *Bembix* Fabricius, 1775.
    Apobembex Pate[25], 1937d:9. Type species: *Bembex oculata* of Latreille, 1804-1805 [= *Bembex oculata* Panzer, 1801], original designation.
    Epibembex Pate[25], 1937d:26. Type species: *Apis rostrata* Linnaeus, 1758, original designation.
Trichostictia J. Parker, 1929:14. Type species: *Monedula vulpina* Handlirsch, 1890, original designation.
Zyzzyx Pate, 1937d:68. New name for *Therapon* J. Parker, 1929.
    Therapon J. Parker, 1929:12. Type species: *Stictia chilensis* Eschscholz, 1822, original designation. Nec *Therapon* Cloquet, 1819.
Stictiella J. Parker, 1917:21. Type species: *Monedula formosa* Cresson, 1872, original designation.
    Microstictia Gillaspy, 1963b:196. Type species: *Monedula femorata* W. Fox, 1895, original designation. New synonymy by R. Bohart.
Glenostictia Gillaspy in Gillaspy, Evans and Lin, 1962:563. Type species: *Monedula pulla* Handlirsch, 1890, original designation.
Xerostictia Gillaspy, 1963b:187. Type species: *Xerostictia longilabris* Gillaspy, 1963, original designation.

---

[26]According to Pate (1937:13, footnote 31) *Bembix* is the etymologically correct spelling. *Bembyx* appears to be a printer's lapsus, since Fabricius did not use or refer to this spelling in later works. However, *Bembex* is the spelling that he used in all of his later publications, and we agree with the evidence outlined by Pate (1937d:13, footnote 29) that it was an emendation of his earlier *Bembix*. If this reasoning is not accepted then the situation is more complicated. Latreille's 1810 type species designation is for the *Bembex* spelling, and since Fabricius (1776) included only *Vespa signata* Linnaeus, 1758 in *Bembex*, the taxon is monotypic if *Bembex* is not considered an emendation of *Bembix*. Latreille's designation would therefore be invalid, and *Bembex* would replace *Stictia*.

[27]Morice and Durrant (1915:400) gave the first type species designation for the spelling of *Bembix*: *Apis rostrata* Linnaeus, 1758.

Steniolia Say, 1837:367. Type species: *Bembex longirostra* Say, 1837, monotypic.

SUBFAMILY PHILANTHINAE
Tribe Eremiaspheciini

Eremiasphecium Kohl, 1897:67. Type species: *Eremiasphecium schmiedeknechtii* Kohl, 1897, monotypic.
    Shestakovia Gussakovskij, 1930a:275. Type species: *Shestakovia digitata* Gussakovskij, 1930, original designation.
    Mongolia Tsuneki, 1972a:230. Type species: *Mongolia steppicola* Tsuneki, 1972, original designation. New synonymy by Menke and Pulawski.

Tribe Philanthini

Philanthus Fabricius, 1790:224. Type species: *Philanthus coronatus* Fabricius, 1790 [= *Sphex coronatus* Thunberg, 1784], designated by Shuckard, 1837:246.
    Symblephilus Panzer, 1806:171. Type species: *Philanthus pictus* Panzer, 1797 [= *Vespa triangulum* Fabricius, 1775], designated by Pate, 1937d:62.
    Simblephilus Jurine, 1807:185. Type species: *Vespa triangulum* Fabricius, 1775, designated by Morice and Durrant, 1915:402.
    Cheilopogonus Westwood, 1834:441. Type species: *Cheilopogonus punctiger* Westwood, 1834 [= *Vespa gibbosa* Fabricius, 1775], monotypic.
    Philanthus Guérin-Méneville, 1835:pl. 71, fig. 8. Lapsus for *Philanthus* Fabricius, 1790.
    Anthophilus Dahlbom, 1844:190. Type species: *Philanthus politus* Say, 1824, designated by Ashmead, 1899:294.
    Chilopogon Kohl, 1896:329. Emendation of *Cheilopogonus* Westwood, 1834.
    Epiphilanthus Ashmead, 1899:294. Type species: *Philanthus solivagus* Say, 1837, original designation.
    Pseudanthophilus Ashmead, 1899:294. Type species: *Philanthus ventilabris* Fabricius, 1798, original designation.
    Oclocletes Banks, 1913:423. Type species: *Philanthus sanbornii* Cresson 1865, original designation.
    Ococletes Mickel, 1916a:407. Lapsus or emendation of *Oclocletes* Banks, 1913.
Trachypus Klug, 1810:41. Type species: *Trachypus gomesii* Klug, 1810 [= *Zethus elongatus* Fabricius, 1804], monotypic.
    Simblephilus Dahlbom, 1844:190. Type species: *Philanthus petiolatus* Spinola, 1841, monotypic. Nec *Simplephilus* Jurine, 1807.
    Philanthocephalus Cameron, 1890:86. Type species: *Philanthocephalus gracilis* Cameron, 1890, designated by Pate, 1937d:49.

TRIBE APHILANTHOPSINI
Subtribe Philanthinina

Philanthinus Beaumont, 1949a:194. Type species: *Philanthus integer* Beaumont, 1949, original designation.

Shestakoviella Gussakovskij, 1952:277. Type species: *Philanthus eximius* F. Morawitz, 1894 [= *Anthophilus quattuordecimpunctatus* F. Morawitz, 1888], original designation.

Subtribe Aphilanthopsina

Aphilanthops Patton, 1881c:401. Type species: *Philanthus frigidus* F. Smith, 1856, original designation.

Clypeadon Patton, 1897:13. Type species: *Aphilanthops quadrinotatus* Ashmead, 1890 [= *Philanthus laticinctus* Cresson, 1865], original designation.

Listropygia R. Bohart, 1959:106. Type species: *Aphilanthops bechteli* R. Bohart, 1959, original designation.

TRIBE ODONTOSPHECINI

Odontosphex Arnold, 1951:154. Type species: *Odontosphex bidens* Arnold, 1951, original designation.

TRIBE PSEUDOSCOLIINI

Pseudoscolia Radoszkowski, 1876:103. Type species: *Pseudoscolia maculata* Radoszkowski, 1876, monotypic.

Philoponus Kohl, 1889:193. Type species: *Philoponus dewitzii* Kohl, 1889, designated by Pate, 1937d:50. Nec *Philoponus* Thorell, 1887.

Acolpus Vachal, 1893:cclxv. Type species: *Aphilanthops theryi* Vachal, 1893, monotypic. Nec *Acolpus* Jayne, 1882.

Philoponidea Pate, 1937d:50. New name for *Philoponus* Kohl, 1889.

Philoponoides Giner Marí, 1945:372. Type species: *Philoponoides tricolor* Giner Marí, 1945, original designation.

TRIBE CERCERINI

Cerceris Latreille, 1802-1803:367. Type species: *Philanthus ornatus* Fabricius, 1790 [= *Sphex rybyensis* Linnaeus, 1771], designated by Latreille, 1810:438.

Nectanebus Spinola, 1838:489. Type species: *Nectanebus fischeri* Spinola, 1838 (nec *Cerceris fischeri* Spinola, 1838) [= *Nectanebus histerisnicus* Spinola, 1838], original designation. New synonymy by R. Bohart and Menke.

Diamma Dahlbom, 1844:225. Type species: *Diamma spinolae* Dahlbom, 1844 [= *Cerceris binodis* Spinola, 1841], monotypic. Nec *Diamma* Westwood, 1835.

Didesmus Dahlbom, 1845:502. New name for *Diamma* Dahlbom, 1844.

Apiraptrix Shestakov, 1923:101. Type species: *Cerceris rybyensis* (Linnaeus), 1771 [= *Sphex rybyensis* Linnaeus, 1771], original designation.

Paracerceris Brèthes, 1913:127. Type species: *Paracerceris tridentifera* Brèthes, 1913 [= *Cerceris pollens* Schletterer, 1887], monotypic.

Bucerceris Minkiewicz, 1933:253. Type species: *Cerceris bупresticida* Dufour, 1841, monotypic.

Stercobata Gussakovskij, 1935:445. Type species: *Cerceris bупresticida* Dufour, 1841, monotypic.

Apicerceris Pate[28], 1937d:8. Type species: *Sphex rybyensis* Linnaeus, 1771, original designation.

Apiratryx Balthasar, 1972:387, 397. Lapsus.

Eucerceris Cresson, 1865:104. Type species: *Eucerceris fulvipes* Cresson, 1865 [= *Cerceris cressoni* Schletterer, 1887], designated by Pate, 1937d:27.

---

[28]This name was first proposed by Minkiewicz (1933), but since he did not designate a type species [Art. 13(b)], it is not valid as of that date. However, the name is available with Pate (1937d) as author.

KEYS TO SUBFAMILIES AND TRIBES

Diversity within the Sphecidae is such that the lines of demarcation between subfamilies are not always definite, and the intermediate forms and exceptions to the rule in all of the larger groups make it nearly impossible to present a workable key to subfamilies on a worldwide basis. Although we have constructed such a key, we found it much easier to bypass the subfamilies by writing a key to all of the tribes. Even though exceptional forms sometimes require complicated couplets, most users will find it more reliable than the subfamily key. It is essential, however, that a user pay special attention to the alternatives in the longer couplets. Furthermore, specimens should be in good condition, with legs somewhat extended downward, and the wings positioned to reveal the jugal lobe of the hindwing. Sometimes it is useful or even necessary to have the mandibles spread or the mouthparts extended. High magnification (50X - 100X depending on wasp size) is a requirement for seeing certain leg details such as plantulae and claw teeth. The unusual terms in the keys are explained in the glossary, and most of the structures mentioned are illustrated in the general morphology section (figs. 1-6). The term "sternal petiole" refers to the rather cylindrical basal gastral segment of some wasps, which is composed of the sternum only. In these cases the corresponding tergum is not attached to the petiole socket but is found at the distal end of sternum I. The sternal petiole is most obvious in forms with a long petiole. For the benefit of the large North American audience that we anticipate will use this book, we have also included a tribal key based solely on the Nearctic fauna.

KEY TO SUBFAMILIES OF SPHECIDAE

1. Gaster with cylindrical petiole composed of sternum only unless it has two sections (as in *Ammophila*) (fig. 15) *and* jugal lobe of hindwing large, containing an anal vein (fig. 18) .... Sphecinae, p. 79
   Gaster variable, if petiolate and petiole composed of sternum only, then jugal lobe of hindwing very small 2
2. Midtibia with two apical spurs[1] or (in a few cases) none 3
   Midtibia with only one apical spur 6
3. Claws with one inner tooth (fig. 14 A-F), notauli usually present and very long .... Ampulicinae, p. 63
   Claws simple 4
4. Mandible toothed or notched externoventrally, hindwing jugal lobe much less than half length of anal area, mesopleuron without precoxal lobes (sw. U.S.) .... Xenosphecinae, p. 437
   Mandible simple externoventrally *or* other characters not as above 5
5. Hindwing jugal lobe less than half length of anal area; mandible not notched nor strongly angulate externoventrally ..... Nyssoninae, p. 441
   Hindwing jugal lobe usually much more than half length of anal area, but if not, then mandible notched externoventrally .... Astatinae, p. 203
6. Episternal sulcus curving forward and then downward parallel to front margin of mesopleuron (fig. 132 I), notauli extending over three-fourths of scutal length (partially obscured by punctation) (se. Europe and sw. Asia) . Entomosericinae, p. 433
   Episternal sulcus variable or absent, but not curving as above; notauli short or absent 7
7. Midcoxae contiguous (fig. 58 I), hindwing jugal lobe subequal to length of anal area (fig. 51 A), *and* ocelli normal; hindwing media diverging before cu-a (s. Palearctic) .... Laphyragoginae, p. 217
   Midcoxae variable but if contiguous, then hindwing jugal lobe one-half or less as long as anal area, *or* hindocelli deformed 8
8. Gaster with petiole composed of sternum only[2] .... some Pemphredoninae, p. 155

---

[1] The tibial spurs arise from the membranous area at the inner apex of the tibia. Under higher magnification they can be seen to be pectinate or serrate.

[2] Except in *Microstigmus* and a few related genera.

Gaster sessile or with petiole composed of both tergum and sternum 9

9. Gaster sessile, forewing with two or fewer submarginal cells and either (a) stigma as large as single discoidal cell; or (b) stigma nearly equal in area to first discoidal cell, two submarginal cells present (second never petiolate), mandible simple externoventrally, inner eye margins not angulate, and hindwing media arising at or before cu-a .... some Pemphredoninae, p. 155
   Gaster variable but if sessile, forewing with three submarginal cells *or* if fewer, then stigma much smaller than first discoidal cell (rarely no discoidal cell) *and* not agreeing with all other characters in alternative "b" above 10

10. Clypeus tripartite and completely divided by vertical sutures (fig. 80 I), hindfemur simple at apex, hindwing jugal lobe about half length of anal area, antennal sockets not contiguous with frontoclypeal suture (Palearctic and Ethiopian) (Palarini) .... Larrinae, p. 221
    Clypeus never completely divided by vertical sutures, hindfemur sometimes truncate at apex, socket placement variable 11

11. Hindocelli deformed or greatly reduced (figs. 61, 140, 182 M, 188) 12
    Hindocelli normal... 14

12. Hindwing media diverging before cu-a (n. Africa, Arabia, and s. S. America) (Odontosphecini) .... Philanthinae, p. 555
    Hindwing media diverging after cu-a 13

13. Hindwing jugal lobe subequal to length of anal area, rarely two-thirds as long as in some African forms (Larrini) .... Larrinae, p. 221
    Hindwing jugal lobe at most a little more than half length of anal area (Bembicini, Heliocausini)............ Nyssoninae, p. 441

14. With an oblique scutal carina (fig. 156), episternal-scrobal sulcus often present, metapleuron broad below, not tapering (Stizini) .... Nyssoninae, p. 441
    Without an oblique scutal carina 15

15. Propodeum with a sharp dorsal tooth or prong posterolaterally, omaulus present, but episternal sulcus absent except for a ventral trace in front of omaulus (Alyssonini) .... Nyssoninae, p. 441
    Propodeum not distinctly toothed or if so, episternal sulcus distinct 16

16. Antennal sockets contiguous with frontoclypeal suture, or if not contiguous, forewing with fewer than three submarginal cells, mandibular socket open (except for a few genera which have a single submarginal cell) .... Larrinae, p. 221, Crabroninae, p. 357
    Antennal sockets above frontoclypeal suture by at least one-third of a socket diameter, forewing with three submarginal cells, mandibular socket closed by a forward extension of the hypostoma .... Philanthinae, p. 555

KEY TO TRIBES OF SPHECIDAE

1. Tarsal claws toothed along inner margin or bifid, two midtibial spurs, jugal lobe of hindwing small or absent (Ampulicinae) 2
   Without all of above characteristics 3
2. Gaster sessile; forewing media diverging after cu-a, marginal cell apex ending at wing margin .... Dolichurini, p. 66
   Gaster petiolate; forewing media diverging somewhat before cu-a, marginal cell apex removed from wing margin .... Ampulicini, p. 74
3. Hindwing with jugal lobe large, comprising more than half length of anal area; gaster with long cylindrical petiole composed of sternum I[3], claws commonly toothed along inner margin............................... 4
   Hindwing with jugal lobe less than half length of anal area, *or* if not, gaster sessile *or* petiole made up of sternum and tergum, the latter not displaced to distal end of sternum; claws sometimes toothed... 6
4. Tarsi with plantulae (fig. 19 A,B) and/or claws of some legs with one mesal tooth on inner margin (if tooth basal or all claws simple then plantulae present and body metallic blue or black and yellow) .... Sceliphronini, p. 82
   Tarsi without plantulae (rare exceptions but body black or black and red); claw simple or with one or more basal teeth on inner margin.................................. 5
5. Claw with two or more teeth; second recurrent vein usually received by submarginal cell III, but if not then claw with at least three teeth *or* apicoventral setae of hindtarsomere V very broad, separated at base by no more than 1.5 setal widths (fig. 26B) ....Sphecini, p. 106
   Claw usually simple or with one tooth, but if with two teeth (Old World forms) then apicoventral setae of hindtarsomere V narrow, separated by three or more setal widths; second recurrent vein usually received by submarginal cell II .... Ammophilini, p. 134
6. Midtibia with a single spur, gaster with a sternal petiole *or* if not, forewing stigma at least nearly as large as discoidal cell I 7
   Without combination of single midtibial spur and sternal petiole *or* combination of single spur and unusually large forewing stigma in comparison with discoidal cell I.. 8
7. Forewing with three submarginal cells, antennal sockets placed well above frontoclypeal suture .... Psenini, p. 158
   Forewing with no more than two sub-

---

[3]The petiole appears two segmented in some Ammophilini due to the narrow tergum I that is attached to the apex of sternum I.

marginal cells, antennal sockets usually placed just above frontoclypeal suture. .... Pemphredonini, p. 174

8. Prestigmal length of first submarginal cell in forewing much more than half total cell length (fig. 180E)............................ 9
Prestigmal length of first submarginal cell in forewing not much more than half total cell length (fig. 5)....................... 10

9. Ocelli normal........................ Stizini, p. 523
Ocelli conspicuously reduced *or* vestigial and often leaving opaque scars .... Bembicini, p. 532

10. Midtibia with two apical spurs and/or scutum with an oblique scutal carina posterolaterally (fig. 156); collar short and transverse, hindwing with jugal lobe absent or occupying less than half length of anal area, mandible not notched or stepped externoventrally 11
Midtibia with one apical spur at most and no oblique scutal carina, *or* if with two spurs, collar long and rounded in profile, *or* jugal lobe occupying more than half length of anal area, *or* mandible stepped or notched externoventrally 13

11. Sternum I with a pair of carinae or ridges toward base but without a median ridge; scutum with a median furrow in place of admedian lines, oblique scutal carina present posterolaterally........................... Nyssonini, p. 461
Sternum I simple toward base *or* with a median ridge; scutum with admedian lines or rarely a median carina but no median furrow 12

12. Sternum I with a keellike basomedian ridge and sometimes other ridges, oblique scutal carina often present.Gorytini, p. 481
Sternum I simple toward base, oblique scutal carina absent, gaster pedunculate .... Mellinini, p. 445

13. Hindocelli deformed or represented by scars (figs. 61, 140, 188), clypeus not completely divided into three parts by longitudinal lines 14
Hindocelli normal *or* clypeus completely divided into three parts by longitudinal lines (fig. 80 I) 16

14. Hindwing media diverging before cu-a...................... Odontosphecini, p. 572
Hindwing media diverging after cu-a 15

15. Hindocellar scars oval, elliptic, or "tailed" (fig. 61) ....Larrini, p. 226
Hindocellar scars C-shaped (fig. 140); s. S. America ....Heliocausini, p. 449

16. Omaulus present, hindwing media diverging before cu-a, episternal sulcus not defined...................... Alyssonini, p. 453
Without all of above characteristics 17

17. Midtibia with two apical spurs, *or* if with none, jugal lobe of hindwing small, and mandible notched externo-ventrally (Astatinae) 18
Midtibia with one apical spur, *or* if with none, jugal lobe of hindwing more than half length of anal area, *or* mandible not notched externoventrally......... 19

18. Hindwing jugal lobe more than half length of anal area, mandible not notched externoventrally, midtibia with two apical spurs in both sexes .... Astatini, p. 205
Hindwing jugal lobe much less than half length of anal area, mandible notched externoventrally, females with two midtibial spurs but males with none .... Dinetini, p. 215

19. Hindwing media diverging before cu-a, mandible stepped or notched externoventrally, hindfemur not truncate distally 20
Without above combination of characters 21

20. Midtibia with two apical spurs, hindwing jugal lobe much less than half length of anal area .... Xenosphecini, p. 439
Midtibia with one apical spur, hindwing jugal lobe more than half length of anal area .... Laphyragogini, p. 219

21. Hindwing media diverging before cu-a, hindfemur truncate distally or with an apical spoon-like process 22
Hindwing media diverging after cu-a, *or* if not, hindfemur simple distally 24

22. Second submarginal cell of forewing petiolate, notauli short if present .... Scapheutini, p. 352
Second submarginal cell of forewing not petiolate............................. 23

23. Inner orbits converging below .... Entomosericini, p. 434
Inner orbits converging above .... Pseudoscoliini, p. 573

24. Inner orbits notched or angulate 25
Inner orbits entire 26

25. Hindwing jugal lobe small, much less than half length of anal area .... Trypoxylonini, p. 327
Hindwing jugal lobe large, at least half length of anal area .... Philanthini, p. 561

26. Hindfemur thickened toward apex which is blunt or truncate (figs. 90, 115F) 27
Hindfemur not blunt or truncate apically 28

27. Episternal sulcus well developed .... Bothynostethini, p. 349
Episternal sulcus absent ....Cercerini, p. 574

28. Forewing with one submarginal cell and scape as long as basal three flagellomeres together *or* forewing with submarginal cell confluent with first discoidal cell (fig. 116) 29
Forewing with more than one submarginal cell, *or* with one cell which is not confluent with first discoidal and scape shorter than basal three flagellomeres 30

29. Forewing with one submarginal cell and a separate first discoidal

cell (fig. 116 A-F) .... Crabronini, p. 370
Forewing with submarginal cell confluent with first discoidal cell (fig. 116 G-H) .... Oxybelini, p. 359

30. Clypeus tripartite, dividing lines distinct, complete and longitudinal (fig. 80 I), ocelli often deformed (fig. 80 A,B) .... Palarini, p. 286
Clypeus not or incompletely tripartite, ocelli normal 31

31. Hindwing jugal lobe much more than half length of anal area, antennal sockets well above clypeus .... Aphilanthopsini, p. 569
Hindwing jugal lobe not much more than half length of anal area, antennal sockets usually contiguous with clypeal margin 32

32. Hindwing media (if present) diverging beyond cu-a, palpal formula 6-4, mandibular sockets open except in *Auchenophorus*, which has only one forewing submarginal cell .... Miscophini, p. 291
Hindwing media diverging at cu-a, palpal formula 5-3, mandibular sockets closed (mandibles must be spread), forewing with three submarginal cells, small and rare wasps from Mediterranean and Transcaspian area, Mongolia .... Eremiaspheciini, p. 558

KEY TO TRIBES OF SPHECIDAE IN AMERICA NORTH OF MEXICO

1. Gaster with a cylindrical petiole composed of sternum only[4], tergum I removed to apex of sternum I (fig. 15), jugal lobe of hindwing comprising nearly all of anal area (Sphecinae) 2
Gaster without a petiole composed of sternum only, or if so, jugal lobe of hindwing comprising at most not much more than half of anal area 4

2. Tarsal claws of foreleg and midleg with two or more basal teeth along inner margin (fig. 26) .... Tribe Sphecini, p. 106
Tarsal claws of foreleg and midleg simple or each with a single tooth on inner margin 3

3. Tarsal claws simple or rarely with a basal tooth .... Tribe Ammophilini, p. 134
Tarsal claws with one mesal tooth (fig. 19D) .... Tribe Sceliphronini, p. 82

4. Midtibia with two apical spurs (rarely none) 5
Midtibia with one apical spur (rarely none) 12

5. Claws with one inner tooth or bifid (Fig. 14 C,E,) (Ampulicinae)
    a. gaster petiolate .... Tribe Ampulicini, p. 74
    b. gaster sessile .... Tribe Dolichurini, p. 66
Claws simple or with a basal lobe, not a tooth *(Pseudoplisus)* 6

6. Mandible toothed or notched externoventrally; hindwing jugal lobe small (fig. 134 A) (Xenospheciinae) .... Tribe Xenospheciini, p. 439
Mandible simple externoventrally 7

7. Hindwing jugal lobe more than half length of anal area (fig. 51 C-F) (Astatinae) .... Tribe Astatini, p. 205
Hindwing jugal lobe less than half length of anal area (fig. 134 C,E) (Nyssoninae) 8

8. Forewing with prestigmal length of first submarginal cell much more than length of wing beyond marginal cell .... Tribe Stizini (part), p. 523
Forewing with prestigmal length of first submarginal cell not distinctly more than length of wing beyond marginal cell or propodeum with a lateral tooth or prong 9

9. Pronotal collar low, longer than scutellum in dorsal view .... Tribe Alyssonini (part), p. 453
Pronotal collar shorter than scutellum in dorsal view 10

10. Admedian scutal lines fused to form a single furrow, propodeum with a lateral tooth or prong .... Tribe Nyssonini, p. 461
Admedian scutal lines well separated, propodeum not toothed nor pronged 11

11. Second submarginal cell of forewing receiving at least the second recurrent vein .... Tribe Gorytini, p. 481
Second submarginal cell of forewing receiving at most the first recurrent vein .... Tribe Mellinini, p. 445

12. Gaster with petiole composed of sternum only, tergum I removed to distal end of sternum I (Pemphredoninae) 13
Gaster not petiolate, or petiole composed of both tergum and sternum 14

13. Forewing with three submarginal cells .... Tribe Psenini, p. 158
Forewing with no more than two submarginal cells .... Tribe Pemphredonini (part), p. 174

14. Forewing either with (1) one discoidal cell, at most two submarginal cells, and a much enlarged stigma *or* (2) two to three submarginal cells, stigma nearly as large as first discoidal cell, two sessile submarginal cells, mandible simple externoventrally, inner orbits not angulate, and hindwing media diverging at or before cu-a .... Tribe Pemphredonini (part), p. 174
Without either combination above 15

15. Hindocelli deformed or greatly reduced (figs. 61, 182 M) 16
Hindocelli normal 17

16. Hindwing jugal lobe subequal to length of anal area (fig. 70 B,G) (Larrinae) .... Tribe Larrini, p. 226
Hindwing jugal lobe at most a

---
[4] Tergum I is often slender and petiolelike in Ammophilini but it is displaced to the apex of sternum I.

KEY TO SUBFAMILIES AND TRIBES 61

little more than half as long as anal area (Nyssoninae)....Tribe Bembicini, p. 532

17. With an oblique scutal carina posterolaterally (fig. 156 C); metapleuron broad below, not tapering (Nyssoninae).....Tribe Stizini (part), p. 523
    Without an oblique scutal carina 18
18. Propodeum with a small, sharp, dorsal tooth posterolaterally (Nyssoninae)........... Tribe Alyssonini (part), p. 453
    Propodeum not distinctly toothed 19
19. Antennal sockets placed above clypeus by at least one-third of a socket diameter, forewing with three submarginal cells (Philanthinae)................ 20
    Antennal sockets touching clypeus, or if not, then forewing with fewer than three submarginal cells (Larrinae) 22
20. Hindfemur ending in a flattened plate or truncation (fig. 190) ....Tribe Cercerini, p. 574
    Hindfemur ending normally 21
21. Inner orbits angulate (fig. 186 C) ....Tribe Philanthini, p. 561
    Inner orbits straight or smoothly curved (fig. 186 A,B)....................Tribe Aphilanthopsini, p. 469
22. Inner orbits angulate (fig. 104) often only one submarginal cell in forewing............................Tribe Trypoxylonini, p. 327
    Inner orbits not angulate.. 23
23. Forewing with more than one submarginal cell, or more than two discoidal cells, or both (subdiscoidal cell counted as discoidal cell) 24
    Forewing with one submarginal cell and one or two discoidal cells (fig. 116) 25
24. Inner orbits converging strongly below (fig. 114 A) ...Tribe Bothynostethini, p. 349
    Inner orbits not converging strongly below ....Tribe Miscophini, p. 291
25. Forewing with two well defined discoidal cells in addition to the submarginal cell (fig. 116 A-F) ....Tribe Crabronini, p. 370
    Forewing with only one discoidal cell (fig. 116 G-H) ....Tribe Oxybelini, p. 359

SUBFAMILY AMPULICINAE

The cockroach wasps, as these insects are called in reference to their prey, are a small, mainly tropical group of six genera. Most entomologists identify the subfamily by the rather bizarre genus *Ampulex,* which is the largest of the genera but by no means the most typical. The Ampulicinae is probably the most primitive subfamily in the Sphecidae, but many specializations are evident, especially in *Ampulex.* This subfamily has a rather isolated position in the family, but resemblances with the Sceliphronini of the Sphecinae indicate some relationship.

Diagnostic characters:

1. (a) Inner orbits entire, converging a little above or below or parallel (fig. 12); (b) ocelli normal.
2. (a) Antennae rather low on face, sockets contiguous with or slightly removed from frontoclypeal suture; (b) male with 13 and female with 12 an-

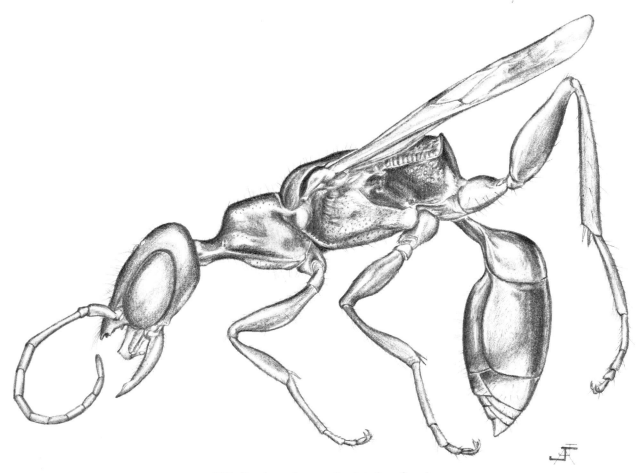

FIG. 8. *Ampulex canaliculata* Say, female

64 SPHECID WASPS

tennal articles.
3. Clypeus transverse or strongly nasiform.
4. (a) Mandible without a notch or angle on externoventral margin; (b) palpal formula 6-4; (c) mouthparts short although maxillary palpi often very long; (d) mandibular socket open except in *Ampulex;* (e) gular area of head usually broadly separating hypostoma from occipital area.
5. (a) Pronotum with a high collar which is moderately long to very long and often tuberculate; (b) pronotal lobe close to tegula, sometimes touching.
6. (a) Notauli usually present and extending length of scutum, usually deeply impressed; (b) no oblique scutal carina; (c) scutellum with pitted transverse basal sulcus.
7. (a) Episternal sulcus absent except sometimes for remnant between end of omaulus and anteroventral margin of pleuron; (b) omaulus usually present.
8. Lower metapleural area sometimes defined by differential sculpture, but definitive metapleuron often consisting of upper metapleural area only.
9. (a) Midtibia with two apical spurs; (b) midcoxal cavities widely separated but midcoxae approximate, sternal area between midcoxae narrow, often sharp-edged, midcoxae without dorsolateral carina or crest; (c) precoxal lobes absent; (d) hindfemur simple apically; (e) claws with a single tooth on inner margin or bifid; (f) foretarsal rake absent; (g) arolium minute.
10. (a) Propodeum moderately long to long; (b) enclosure usually present, U-shaped to nearly triangular; entirely dorsal; (c) propodeal sternite present.
11. (a) Forewing with two or three submarginal cells and two recurrent veins, first recurrent received by submarginal I or II, second recurrent received by II or III; (b) marginal cell acuminate apically, appendiculate, but apex approximating a right angle in some *Ampulex.*
12. (a) Jugal lobe absent or small; (b) hindwing media diverging before, at, or after cu-a; (c) hindwing sometimes with subcostal vein; (d) anal area without anal vein.
13. (a) Gaster sessile or with a petiole composed of sternum and tergum; (b) tergum I with lateral line or carina; (c) male gaster usually with three visible segments only, segments IV-VII in male capable of being withdrawn, usually only narrowly exposed if at all; (d) no pygidial plate; (e) sternum VI in female elongate, overlapping dorsally to form an exposed tapering tube through which the sting apparatus is exserted.
14. (a) Volsella of male genitalia with movable digitus and cuspis (fig. 9 C); (b) aedeagal head with teeth on ventral edge; (c) male sternum VIII as in fig. 9 B,D; (d) cerci present (fig. 9 A) except in Ampulicini.

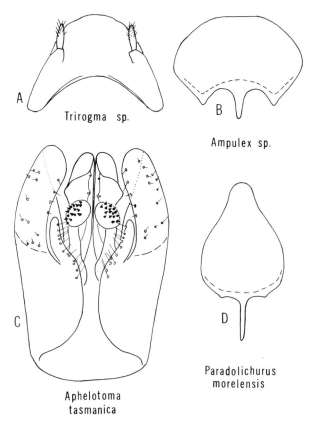

FIG. 9. Features of male ampulicines. A, tergum VIII; B, sternum VIII; C, ventral view of genitalia; D, sternum VIII.

Systematics: The most distinctive features of these wasps are: the generally elongate body with long, slender appendages; legs fitted for running with the femora distinctively drumstick-like, their distal third usually fusiform; collar angular and often with one to three tubercles or teeth; notauli long, usually deeply impressed; episternal sulcus absent on lateral surface of mesopleuron; propodeum often with posterolateral tooth-like projections; two midtibial spurs; gaster attenuate apically in female, but blunt in male, the latter with three normal-sized segments into which the remainder may be withdrawn.

Over the years opinions have differed as to the affinities and status of the ampulicines. This can be illustrated by some extreme examples: the two tribes recognized by us have sometimes been considered as separate families (see Handlirsch, 1925, p. 809), or they have been placed as tribes in separate subfamilies of the Sphecidae (Börner, 1919). Recent trends have been to regard this group as a family (Krombein, 1951; Rasnitsyn, 1966; Riek, 1970) or as a subfamily of the Sphecidae related to the Sphecinae (Leclercq, 1954a; Malyshev, 1968; Oehlke, 1969; Richards, 1972a). Pate (1938b) tentatively placed them as a tribe in the Nyssoninae, on the basis that the scutum overlies the tegula. Nagy (1969) even related the ampulicines to the "Dryinidae and Cleptidae." Evans (1959a) said that larval characters show that the "relationship between the ampulicids and the Sphecinae is very close indeed." He went on to say that the "forked process on the mesosternum" (i.e. metasternum) of the adult clearly separates them from the Sphecinae. Evans made the mistake of most authors in characterizing the ampulicines on the basis of characters found only in *Ampulex*. Although in this genus the metasternum is deeply divided posteriorly and forklike, this is not at all pronounced in the other genera. In fact the metasternal apex of the other genera is similar to that of many Sphecinae. Furthermore, the metasternal apex is deeply emarginate in some sphecids, although the general shape of the sternum is different (*Bembix*, for example).

The configuration of the pronotum and venation alone would seem to negate a possible relationship between ampulicines and bethyloids or other more primitive aculeates. At the same time these structures point to a relationship with the Sphecidae. In fact there appears to be no single character that will separate ampulicines from sphecids. Therefore, it seems logical to include the group as a subfamily in the Sphecidae. This arrangement is enhanced by the similarities between the ampulicines and certain genera in the sphecine tribe Sceliphronini. For example, the pronotum of *Trigonopsis, Dynatus,* and *Penepodium* is similar to that of *Ampulex,* and the rather prognathous head of some ampulicines is also found in some of the sceliphronin genera. It could be argued that these similarities are simply the result of convergence, but the propodeal sternite, the presence of plantulae between the tarsomeres, the tarsal claw teeth, the presence of cerci, and the division of the volsella into a digitus and cuspis are characteristics shared by both subfamilies. Also, the larval similarities cannot be ignored. The Ampulicinae differs from the Sphecinae in lacking precoxal lobes; in essentially lacking an episternal sulcus; in having the petiole, when present, composed of tergum and sternum; in having a lateral line or carina on tergum I; in having a small or no jugal lobe; and the absence of a third anal vein in the hindwing. An obvious objection to Pate's idea of including Ampulicinae in Nyssoninae is the toothed-claw condition in the former.

The Ampulicinae should be regarded as the most primitive sphecid subfamily. Generalized or unspecialized characters found throughout the subfamily are: inner orbits not strongly converging above or below, palpal formula 6-4, mandible without an externoventral notch, ocelli normal, body long and slender, pronotum not closely appressed to scutum, propodeal enclosure strictly dorsal, two midtibial spurs, no pygidial plate, and volsella divided into digitus and cuspis.

Other generalized characters of Ampulicinae that are not found universally in the subfamily are: mandibular socket open and continuous with the oral fossa; notauli complete; lower metapleural area well defined; tarsal plantulae present; forewing media diverging well after cu-a; forewing with three submarginal cells, the first partially divided by the remnant of a radial crossvein; forewing marginal cell acuminate apically and not diverging from wing margin; hindwing subcostal vein evident; and cerci present.

Obvious specializations of the subfamily are the reduction or absence of the hindwing jugal lobe and loss of the second anal vein, flattening of the head or posterior displacement of the mandibles in females, reduction or absence of the occipital carina, loss of the episternal sulcus except for a ventral remnant, and telescoping of the terminal male abdominal segments. Several other specializations occur in the genus *Ampulex*, which we put in a tribe by itself.

Sexual dimorphism is rather strong in the Ampulicinae. Females have mandibles specialized for gripping, the normal number of visible gastral segments, and generally fine punctation or other microsculpture. Males on the other hand have rather simple mandibles, usually only three visible gastral segments, and coarser microsculpture.

A dendrogram expressing the generic relationships as we see them is shown in fig. 10.

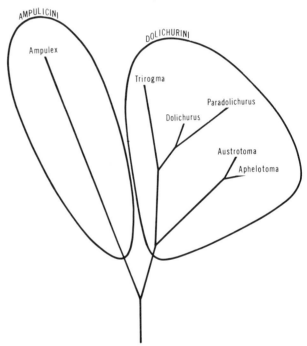

FIG. 10. Dendrogram of presumed relationships of ampulicine genera.

Biology: These wasps appear to provision only with cockroaches. Domestic roaches are included, and the wasps may enter houses in search of them. After a roach is captured and weakly paralyzed, the wasp grasps it by the antennal bases with her mandibles and walking backwards drags the prey to any suitable niche. Only one roach is provided per cell, and the egg is laid on the midcoxa. The nest is stoppered with bits of leaves, grass, and other debris.

The prey-niche sequence is a primitive type of biology, and the fact that the wasp walks backwards while dragging the prey is perhaps unique in the Sphecidae. This primitive type of prey carriage is found commonly in the Pompilidae.

KEY TO GENERA OF AMPULICINAE

1. Antennal sockets each with an overhanging frontal lobe (fig. 12 A); metasternum Y-shaped with arms directed posteriorly; petiole inserted between and on same level as hindcoxae; widespread (Tribe Ampulicini) .... *Ampulex* Jurine, p. 74
   Antennal sockets uncovered or both overlaid by a single median platform-like lobe; metasternum somewhat emarginate posteriorly but not Y-shaped; petiole inserted above and somewhat after hindcoxae (Tribe Dolichurini) ... 2
2. Antennal bases not covered by a median lobe 3
   Antennal bases covered by a median frontal platform (fig. 12 C,D) 4
3. Cubitus of hindwing extending beyond cu-a toward wing margin (fig. 11 F); propodeum without a sharp lateral spine on its posterior declivity; Australia ... *Aphelotoma* Westwood, p. 70
   Cubitus of hindwing not extending beyond cu-a (fig. 11 E); propodeum with a sharp posterolateral spine; Queensland, Australia ...... *Austrotoma* Riek, p. 70
4. Forewing media diverging at or before cu-a (fig. 11 D); hindwing media diverging after cu-a (fig. 11 D); abdomen distinctly petiolate; Oriental Region, Iraq .... *Trirogma* Westwood, p. 73
   Forewing media diverging after cu-a (fig. 11 C); hindwing media diverging before cu-a; abdomen practically sessile 5
5. Notauli well developed, complete to posterior scutal margin or nearly so; propodeal outline rather sharply bent in profile; hindwing jugal lobe present; widespread ...... *Dolichurus* Latreille, p. 66
   Notauli undeveloped; propodeal outline broadly rounded in profile; hindwing jugal lobe absent; California, Mexico .... *Paradolichurus* Williams, p. 69

Tribe Dolichurini

The wasps in this tribe are mostly small and inconspicuous However, a few *(Trirogma)* are moderately large and metallic greenish blue. They are generally found in wooded areas, running or "skipping" over leaf litter, on dead stumps, or on tree trunks. Many of the species have banded wings and a mimetic resemblance to ants.

Diagnostic characters:

1. Mandibular sockets open, continuous with oral fossa.
2. Clypeus broadly convex, sometimes with a median lobe, but never sharply rooflike.
3. Metasternum not Y-shaped nor deeply cleft posteriorly.
4. Hindcoxae subcontiguous.
5. Penultimate tarsomere similar in form to III, tarsomere I inserted toward end of IV.
6. Marginal cell of forewing not bending away from wing margin toward apex.
7. Gaster inserted after and above hindcoxae.
8. Swollen part of sternum II with a transverse sulcus or carina basally, or more rarely medially (sulcus at least present as lateral traces).
9. Apex of female gaster conical.
10. Cerci present.

Systematics: The tribe contains five of the six ampulicine genera, but the included 49 named species represent only a third of those contained in the subfamily. Relatively few are known from any one part of the world. Our ideas of relationships are expressed in the dendrogram (fig. 10). The presence of cerci is an indication of the primitive nature of the tribe. Most of its peculiarities, such as the flattened and broadened mandibles in *Aphelotoma* and *Austrotoma* (fig. 13 A,D), can be attributed to specialization for capture of prey. *Dolichurus* is the most generalized member of the tribe, the largest in number of species, and the only one that occurs on more than one continent.

Biology: All are thought to prey on cockroaches. This presumption seems likely but the only positive proof is in *Dolichurus* and *Trirogma*.

Genus Dolichurus Latreille

Generic description: Medium to small wasps, 5 to 13 mm long; mostly black, often with terminal abdominal segments red; sometimes with whitish marks on mandibles, clypeus, frontal platform, pronotum, and coxae; wings not banded, mandible rather short and stout, inner margin in female with two or three teeth (fig. 12 D), inner margin in male with a single subapical tooth; maxillary palpus usually long and slender, often as long as head height; labial palpus stouter and moderately short; male clypeus short, transverse, sometimes with a median carina, free edge usually denticulate; female clypeus with free edge simple and broadly rounded to somewhat produced and subtruncate (fig. 12 D); median carina often present, at least basally, a transverse discal row of four to eight stout bristles present; labrum concealed, apex rounded truncate in female; antennal sockets nearly contiguous, covered by a U-shaped and platformlike extension of frons; frontal carina or sulcus usually indicated but weak and intermittent; vertex raised above eyes, postocellar area longer than ocellar triangle in dorsal view of head; occipital carina U-shaped and ending opposite mandible base but short of hypostomal carina, or sometimes apparently absent; pronotal collar broader than long, somewhat broadly binodose laterally, shorter than scutum; scutum moderately convex to nearly flat, notauli distinct and nearly complete, admedian lines not evident but scutum often with a short anteromedian furrow, posterolateral area of scu-

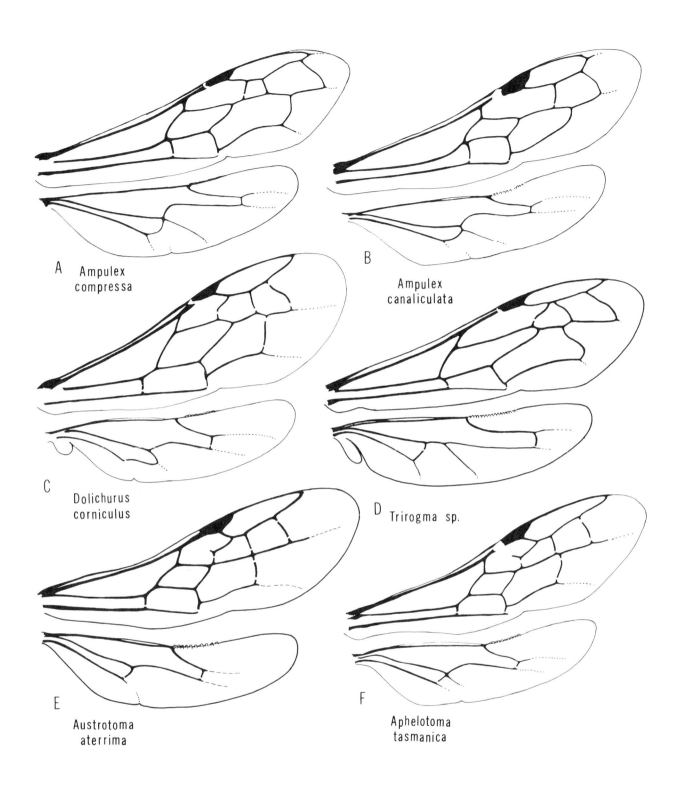

FIG. 11. Wings of Ampulicinae.

tum only slightly depressed; scutellum nearly flat; propodeum with enclosure broadly U-shaped, multicarinate to areolate, limited posteriorly by a strong and transverse carina that forms a decided angle in lateral view, lateral borders of enclosure irregular, posterior face of propodeum declivous, its sculpture finer than that of enclosure; mesopleuron with a distinct omaulus that joins an anteroventral episternal sulcus remnant to form an acute V, omaulus usually connecting with an acetabular carina which is sometimes incomplete, sternaulus often evanescent but usually at least indicated; scrobal sulcus present, often forming a broad groove which may continue forward to omaulus; hypoepimeral area bulging and often polished; intercoxal carina present but lower metapleural area not otherwise defined; tarsal plantulae present between flattened setae; forewing with media diverging well after cu-a, three submarginal cells of which I is much less than twice as long as II measured along posterior side, first intersubmarginal veinlet angled and often appendiculate, submarginal cell III trapezoidal to nearly triangular, first recurrent vein received by cell II, second by cell III; hindwing jugal lobe present but small, media diverging before cu-a that extends considerably distad of divergence point; gaster sessile; sternum II with a crescentic carina and/or sulcus across base of swollen part, faint or lateral only in some males.

Geographic range: The 34 listed species are divided by regions as follows: Oriental-17, Ethiopian-9, Palearctic-4, Neotropical-2, Nearctic-1, Australian-1. Several are found on large islands such as Taiwan-10, Celebes-2, and Madagascar-1. One species, *stantoni*, was introduced into the Hawaiian Islands by F. X. Williams (1919) where it has become established.

Systematics: These widespread wasps seem to qualify as the most generalized ampulicines. They share with *Aphelotoma* such presumably generalized features as forewing media diverging distad of cu-a, forewing with three submarginal cells of roughly similar length, angled and often appendiculate first intersubmarginal veinlet, tarsal plantulae, and well developed notauli. On the other hand, *Dolichurus* has the additional "primitive" characters of a distinct hindwing jugal lobe, well-marked scrobal sulcus, less elongate pronotal collar, and more ordinary mandibles. The principal specialization over *Aphelotoma* seems to be the presence of a frontal platform covering the nearly contiguous antennal sockets in *Dolichurus*.

The most closely related genus is *Paradolichurus*, which was treated by Williams (1960a, 1962) as a subgenus of *Dolichurus*. The absence of notauli, jugal lobe, and scrobal sulcus in *Paradolichurus* seems adequate for its separation.

The presence of a carina between the mid and hindcoxae, but the absence of a metapleural sulcus or carina (from hindcoxa to upper metapleural pit) allies *Dolichurus* with *Aphelotoma* and *Austrotoma* but distinguishes it from *Paradolichurus*, *Trirogma*, and *Ampulex*, all of which have the entire lower metapleural area somewhat distinguished.

The New World species of *Dolichurus* seem to differ from those of the Old World in having no occipital carina. Also, the clypeus is at least usually marked with white and not strongly produced toward the apex in the female. These differences may be of subgeneric value, but it should be pointed out that both *Trirogma* and *Ampulex* have some species without an occipital carina.

In most *Dolichurus* females the gaster is black with the apex red. In some cases, however, the gaster may be all black or all red. Species characters are found in markings, punctation, sculpture of the frons or propodeum, shape of the clypeus, and development of mesopleural and other sulci or carinae. Undoubtedly, the maxillary palpi, the carina across sternum II, and the male genitalia will offer distinctive features, also.

Of the 34 named species of *Dolichurus* we have studied six: *corniculus* (type of the genus) *haemorrhous*, *stantoni*, *ignitus*, *bipunctatus*, and *greenei*. In addition we have seen perhaps an equal number of unidentified species from the Ethiopian and Neotropical Regions.

There are no adequate keys. Most useful papers are those of Kohl (1893b), Yasumatsu (1936), Riek (1955), and Tsuneki (1967b). Tsuneki's paper contains exceptionally good structural drawings based on species from Taiwan.

Biology: The more important references on *Dolichurus* are: Williams (1919) for *stantoni*; Handlirsch (1889), Maneval (1932, 1939), Benoist (1927), Grandi (1954), Deleurance (1943), and Soyer (1947) for *corniculus*; and Ferton (1894) for *haemorrhous*. *Dolichurus stantoni* was described by Williams as being "perhaps the swiftest and most restless digger-wasp I have seen." Unlike *Ampulex*, this species apparently grabs the cockroach prey by any handy appendage such as a leg or cercus and then stings it. Amputation of the roach antennae by the wasp has been observed in *Dolichurus*. The prey is dragged by the wasp to a suitable nest site. The wasp walks backwards while holding the host's antennae with her mandibles.

Dolichurus nest in stems and other crevices just as do *Ampulex*. According to Ferton, *haemorrhous* was seen to bury its prey in a hole in the ground 7-8 cm deep, but it is not clear whether or not this was a pre-existing cavity. The cocoon of *Dolichurus* is tough and it differs from that of *Ampulex* by being lined inwardly with silk and covered on the outside with bits of earth and other debris. At least in *stantoni* the host is completely devoured, and the cocoon is therefore exposed, in contrast to the situation in *Ampulex*.

Prey records are listed below.

Dolichurus greenei		– *Parcoblatta* sp. (Krombein, 1955a)
"	*stantoni*	– *Blatella signata* (Brunner)
"	*haemorrhous*	– *Loboptera decipiens* (Germar)
"	*corniculus*	– *Phyllodromica megerlei* Fieber
		Ectobius lividus (Fabricius)
		Ectobius lapponicus Linnaeus
		Blatella germanica Linnaeus

Checklist of *Dolichurus*

abbreviatus Strand, 1913; Taiwan
abdominalis F. Smith, 1860; Celebes
alorus Nagy, 1971; Taiwan
amamiensis Tsuneki and Iida, 1964; Ryukyus, Taiwan
   puliensis Tsuneki, 1967
basuto Arnold, 1952; Lesotho
bicolor Lepeletier, 1845; centr. Europe
   ? *dahlbomi* Tischbein, 1852
bimaculatus Arnold, 1928; Rhodesia
bipunctatus Bingham, 1896; e. India, Burma, Sikkim
   reticulatus Cameron, 1899
carbonarius F. Smith, 1869; Australia: Queensland
cearensis Ducke, 1910; Brazil
†*clavipes* Cameron, 1897 (*Dolichusus*!); e. India, nec Dahlbom, 1829
corniculus (Spinola), 1808 (*Pompilus*); Europe
   clavipes Dahlbom, 1829 (*Pompilus*)
dromedarius Nagy, 1971; Taiwan
formosanus Tsuneki, 1967; Taiwan
gilberti Turner, 1912; e. India
greenei Rohwer, 1916; e. North America
guillarmodi Arnold, 1952; Lesotho
haemorrhous A. Costa, 1886; Italy, Portugal, Egypt
ignitus F. Smith, 1869; S. Africa, Rhodesia
   ssp. *contractus* Arnold, 1951; Ethiopia
kohli Arnold, 1928; S. Africa
laevis F. Smith, 1873; Brazil
   levis Dalla Torre, 1897, emendation
leioceps Strand, 1913; Taiwan
maculicollis Tsuneki, 1967; Taiwan
ombrodes Nagy, 1971; Taiwan
oxanus Nagy, 1971; Celebes
pempuchiensis Tsuneki, 1972; Taiwan
quadridentatus Arnold, 1940; Zaire
rubripyx Arnold, 1928; S. Africa
secundus Saussure, 1892; Madagascar
   tertius Saussure, 1892
shirozui Tsuneki, 1967; Taiwan
stantoni (Ashmead), 1904 (*Thyreosphex*); Philippines, Hawaii
taprobanae F. Smith, 1869; Sri Lanka, India, Burma, Sikkim
turanicus Gussakovskij, 1952; sw. USSR: Tadzhik S.S.R.
venator Arnold, 1928; Rhodesia

Genus Paradolichurus Williams

Generic description: Small wasps, 5 to 8 mm long, mostly black, gaster in female posteriorly red or brownish yellow, whitish markings may be present on mandibles, clypeus, frontal platform, pronotum, tegula, scutellum, coxae and tergum I; wings not banded; mandible rather short and stout, inner margin in female with three denticles, in male with a single subapical tooth; maxillary palpus slender, nearly as long as head height; labial palpus stouter and moderately short; clypeus in both sexes broadly convex, apex rounded truncate, not carinate, female clypeus with a transverse discal row of three slender white setae; antennal sockets nearly contiguous, covered by a U-shaped and platform-like extension of frons; frontal carina or sulcus hardly indicated; vertex raised above eyes, postocellar area more than 1.5 times as long as ocellar triangle in dorsal view of head; occipital carina U-shaped and ending well short of hypostomal carina; pronotal collar sloping rather evenly anteriorly, collar shorter than scutum, not binodose posterolaterally, no median sulcus; scutum moderately convex and with no evident notauli, parapsidal lines, admedian lines or median sulcus; scutellum simply convex; propodeal enclosure broadly U-shaped, carinate posteriorly, lateral margins irregular, surface mostly longitudinally ridged, posterior face somewhat irregularly cross ridged, no lateral teeth, propodeal profile rather evenly rounded; mesopleuron with omaulus but without evident sternaulus, acetabular carina, scrobal sulcus, hypoepimeral bulge, or anteroventral remnant of episternal sulcus; lower metapleural area defined by a carina from above hindcoxa to upper metapleural pit; tarsal plantulae present; wings not banded, forewing with media diverging after cu-a or sometimes interstitial with it, three submarginal cells of which I is about as long as II measured along posterior side, first recurrent vein received by cell I, second by cell II; hindwing without a jugal lobe, media diverging before cu-a that extends farther distad than point of divergence; gaster sessile; sternum II with a carina and associated sulcus extending across base of swollen part, sulcus curving posteriorly at sides.

Geographic range: *P. obidensis* was collected along the Amazon River at Obidos, Amazonas, Brazil. Its subspecies, *maranhensis*, was described from San Luis and Alcantara, Maranhão, Brazil. The single female of *californicus* was collected near Julian, San Diego County, California at an elevation of 4,000 feet in the Laguna Mountains. A third species, *morelensis* has been found in the Mexican states of Morelos: Cuernavaca, Alpuyeca, Lake Tequesquitengo, Huajintlan; Puebla: Petlalcingo; and Sinaloa: Chupaderos.

Systematics: There is an obvious relationship to *Dolichurus,* within which Williams (1960a) placed it as a subgenus. However, there are several points of apparent evolutionary advancement. These are the simple pronotum, absence of suctal sulci (especially the notauli), absence of most mesopleural sulci, and no hindwing jugal lobe.

The three known species of *Paradolichurus* differ in details of color and sculpture. *P. morelensis* and *obidensis* are the only ampulicines known to us with white spots on tergum I.

We have seen specimens of *californicus* and *morelensis* only. However, *obidensis* was rather completely described by Ducke (1903), and there seems little doubt that it belongs in this genus.

Biology: According to Ducke (1903) specimens of *obidensis* were collected as they were skipping over the ground along a partly sunlit wooded path. The holotype female of *californicus* was netted by F. X. Williams (1960a) "at the edge of a small clearing in a rather open oak and pine forest . . . ." Part of the type series of *morelensis* was taken by H. E. Evans "on dried leaves beneath large trees which were dripping honeydew".

Checklist of *Paradolichurus*

californicus (Williams), 1960 (*Dolichurus*); U.S.: s. California
obidensis (Ducke), 1903 (*Dolichurus*); Brazil: Amazonas
 ssp. *maranhensis* (Ducke), 1904 (*Dolichurus*); Brazil: Maranhão
morelensis (Williams), 1962 (*Dolichurus*); Mexico: Morelos, Puebla, Sinaloa

Genus Aphelotoma Westwood

Generic description: Mostly small wasps, 5 to 11 mm long, black and reddish brown, wings with a pale median band. Mandible of male with one subapical tooth on inner margin; female mandible very flat, expanded at outer base and strongly curved toward narrowly acute apex, inner margin with a subapical tooth followed by an irregularly dentate lobe (fig. 13 D); maxillary palpus small, much shorter than head height; clypeus short, apex in female truncate, lying over exposed and truncate or slightly emarginate labrum (fig. 13 D); antennal sockets separated by about a socket diameter or more, covered only at extreme base by frontal extensions; vertex not unusually raised, postocellar area in dorsal view about as long as ocellar triangle; occipital carina broadly U-shaped, not reaching hypostomal carina; malar space short in female but up to 2.5 ocellus diameters long in males; pronotal collar as long as scutum, with a median sulcus and a small anterolateral tooth; scutum with notauli complete, admedian lines and median sulcus evanescent, posterolateral area of scutum only slightly depressed; scutellum and metanotum simple, more finely sculptured than scutum; propodeum with enclosure broadly U-shaped, posterior margin carinate, lateral margins weaker and irregular, surface reticulate with reticules about as large as those on posterior face, no posterolateral tooth on declivity; mesopleuron with omaulus, sternaulus and anteroventral remnant of episternal sulcus but without acetabular carina, scrobal sulcus or bulging hypoepimeral area; intercoxal carina present but no other definition of a lower metapleural area; tarsal plantulae present; forewing with media diverging after cu-a, three submarginal cells, III quadrangular, first intersubmarginal veinlet angled and appendiculate, first recurrent vein received by cell I, second usually by III or interstitial (fig. 11 F); hindwing without a jugal lobe, media diverging at or slightly before or after cu-a, cubitus extending beyond cu-a (fig. 11 F); gaster with a very short petiole, sternum II with a sulcus across front of swollen part, sulcus curving backward at sides.

Geographic range: The eight named species are all from Australia.

Systematics: The relationship to *Austrotoma* is obvious. Together the two genera differ from *Dolichurus* by the well separated and weakly lidded antennal sockets, more expanded female mandibles, longer pronotal collar, absence of a scrobal sulcus or bulging hypoepimeral area, and absence of a hindwing jugal lobe. Except for the antennal socket character, the other differences seem to indicate a higher level of specialization for the endemic Australian genera. Riek (1955) pointed out that *Aphelotoma* differs from *Austrotoma* by having a lateral tooth on the pronotum but none on the propodeal declivity, and by having a short petiole. More significant in our opinion is the discrepancy in hindwing venation (fig. 11 E,F). The absence in *Austrotoma* of any trace of the cubitus beyond cu-a is rather remarkable.

Our view of *Aphelotoma* as a "primitive" but sidewise specialized offshoot is depicted in a dendrogram (fig. 10).

A review of the genus with keys to species was given by Riek (1955). Other papers of value are Westwood (1842) and Turner (1915a).

Through the kindness of E. F. Riek we have been able to study specimens of *affinis, nigricula,* and *tasmanica.*

Biology: Published information seems to be scanty, although it is assumed that cockroaches are the prey. Riek (1955) commented on their "habit of foraging for their prey in dead leaves and rubble and under bark. . . ." Riek said further that males prefer a smooth-barked living eucalyptus tree as an alighting point, whereas females use dead trees and stumps. He pointed out a resemblance in the jerky walking manner to ants of the genus *Myrmecia.* Turner (1915a) made a similar observation specifically on *A. tasmanica:* "Although of considerably smaller size, this wasp bears a considerable resemblance to ants of the genus *Myrmecia* . . . When alarmed the wasp often picks up a fragment of dead stick or leaf, which it carries in its mandibles, thus increasing the resemblance to the ant."

Checklist of *Aphelotoma* Westwood

affinis Turner, 1910; Australia: Queensland
auricula Riek, 1955; Australia: Queensland
fuscata Riek, 1955; Australia: New South Wales
melanogaster Riek, 1955; e. Australia
nigricula Riek, 1955; e. Australia
rufiventris Turner, 1914; Australia: Queensland
striaticollis Turner, 1910; Australia: Queensland
tasmanica Westwood, 1841 (Ann. Mag. Nat. Hist., p. 152); Tasmania
 ssp. *auriventris* Turner, 1907; s. Australia

Genus Austrotoma Riek

Generic diagnosis: Essentially as given for *Aphelotoma* except as follows: Small wasps, 6.0 to 6.5 mm long; clypeus of female rounding to a point in front, the projecting lobe overlaying an exposed and similarly shaped labrum (fig. 13 A); pronotal collar as long as scutum, with a fine median sulcus, without protuberances as seen in dorsal view; propodeal declivity with a sharp tooth near middle of lateral margin; forewing with media diverging after cu-a; hindwing with media diverging at cu-a but

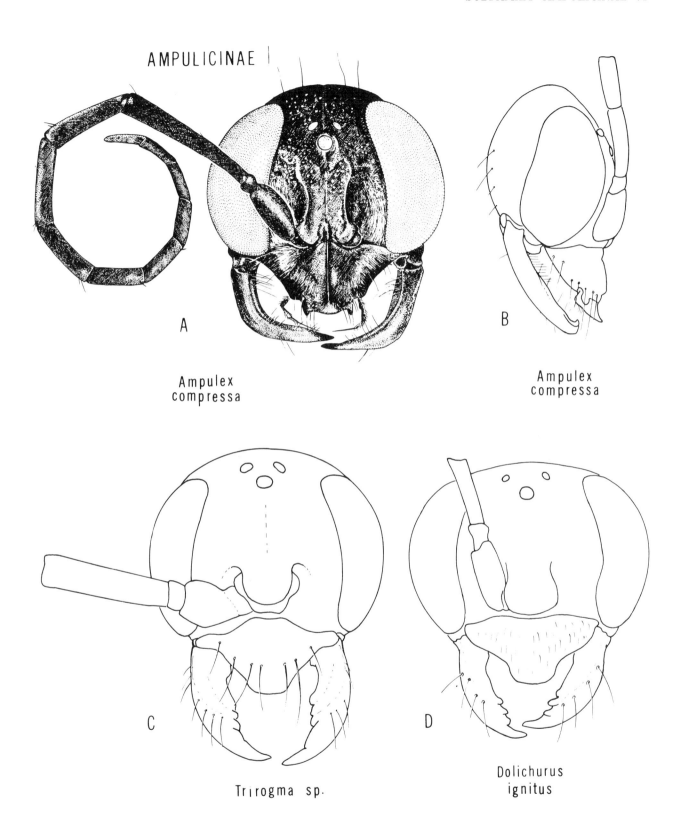

FIG. 12. Views of head of various female Ampulicinae.

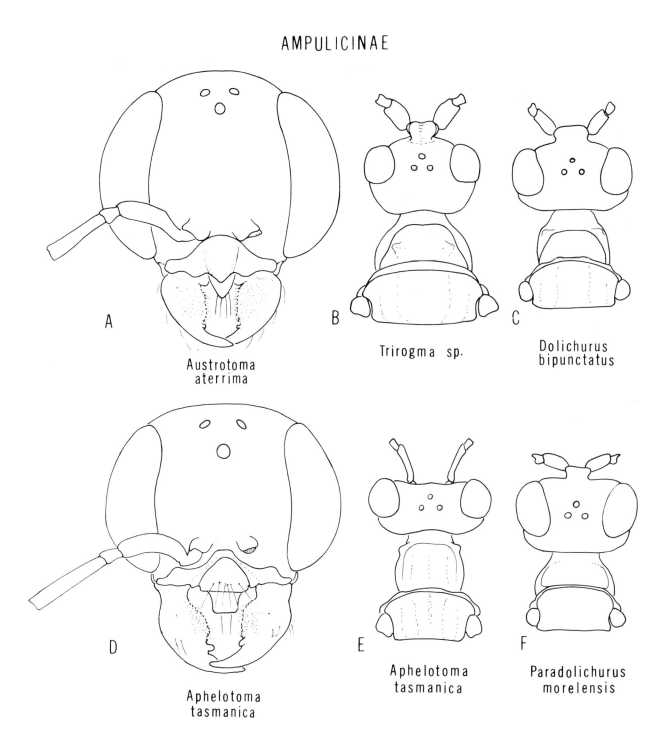

FIG. 13. Female ampulicines. A,D, face view; B,C,E,F, head and thorax as far as scutum, dorsal.

cubital vein not extending beyond cu-a (fig. 11 E); gaster sessile.

Geographic range: The single species has been reported from Mackay and Kuranda in Queensland, Australia, according to Riek (1955).

Systematics: The close relationship to *Aphelotoma* is discussed under that genus. The peculiar hindwing venation with a complete absence of the cubitus beyond cu-a is not found elsewhere in the Ampulicinae.

We have studied a female of *aterrima* sent to us by Riek. This specimen has a spot of white over each antennal socket and some white on the mandible base. We are not aware of any white markings in the species of *Aphelotoma*. *Austrotoma* was described and its affinities were discussed by Riek (1955).

Biology: Unknown.

Checklist of *Austrotoma*

aterrima (Turner), 1907 (*Aphelotoma*); Australia: Queensland

Genus Trirogma Westwood

Generic description: Medium to large wasps, 10 to 18 mm long, one species black, others greenish blue with mandibles sometimes white in males, wings not banded; mandible stout, inner margin in female with a weak subapical denticle and three more median ones (fig. 12 C), in male with one large median tooth; maxillary palpus lengthened but shorter than head height; male clypeus short with apex broadly concave and slightly emarginate; female clypeus with apex a reflexed lip with eight macrochaetae in a curved transverse row behind lip, no median clypeal carina; labrum concealed, its apex rounded truncate and multisetose in female, more pointed in male; antennal sockets nearly contiguous and covered by a U-shaped longitudinally grooved, platformlike extension of frons; frontal sulcus well developed or incomplete and evident only on upper part of frontal platform; vertex raised above eyes, postocellar area about twice as long as ocellar triangle in dorsal view of head; occipital carina a short and largely dorsal U, or absent; pronotal collar nearly as long as scutum, with a median furrow dividing two rounded or acute mammiform posterior lobes, pronotal lobe produced backward at inner posterior corner; scutum with notauli complete or only definite anteriorly, area between them raised and nearly even in females but sometimes furrowed between longitudinal swellings in males, parapsidal lines not evident but posterolateral corner of scutum strongly reflexed bordering a depressed area; scutellum simple in female but evenly convex to conically projecting upward in male; propodeal enclosure broadly U-shaped, margins irregular, lateral ones often indistinct, enclosure and posterior face with large areoles, posterior face declining abruptly or more gradually, dentiform or subdentiform angles often present posterolaterally on enclosure and laterally at middle of posterior surface; mesopleuron with omaulus, anteroventral remnant of episternal sulcus, and at least an anterior indication of sternaulus, scrobal sulcus continued to omaulus, hypoepimeral area bulging, no evident acetabular carina; lower metapleural area usually well defined; tarsal plantulae absent; forewing media diverging before or at cu-a, three submarginal cells of which I is about twice as long as II on posterior side, III is trapezoidal, first recurrent received by I or interstitial, second recurrent to II, first intersubmarginal veinlet curved but not strongly angled nor appendiculate, hindwing jugal lobe present but small, media diverging after cu-a; gaster with a short petiole; sternum II with a sulcus curving across front of swollen part of II or across middle of swollen part, curved forward laterally, sulcus sometimes interrupted narrowly or broadly toward middle.

Geographic range: The three named species are all in the Oriental Region. One of these, *caerulea*, extends as far west as Iraq. There are several undescribed species. We have studied two of these, one each from the Philippines and Celebes.

Systematics: We have studied about two dozen specimens of four species: *caerulea, prismatica,* and two undescribed forms, but none of *nigra*. The genus appears to be the most specialized of the Dolichurini, and its relative position is indicated in the dendrogram (fig. 10). The frontal platform is basically like that of *Dolichurus*, but there is a deep longitudinal groove. In some features it resembles *Ampulex*. These are the elongate first submarginal cell, relatively basal divergence of the forewing media, proximity of the third submarginal cell to the outer wing margin (contrast fig. 11 A,D), pronotal lobe angled posterodorsally, lower metapleural area defined, distinct gastral petiole, and absence of tarsal plantulae. These characters in common may indicate that *Ampulex* developed from a *Trirogma*-like ancestor. In any event, there are now many important differences of a tribal nature between the two. *Dolichurus* would seem to be the most closely related genus to *Trirogma* on the basis of general body form, clypeal and mandibular structure, presence of a frontal platform, and a hindwing jugal lobe. Primary differences are the valleylike shape of the frontal platform and the *Ampulex*-like features cited above.

Judging by the four species we have seen, specific characters are to be found in the shape of the male scutellum, sharpness of the pronotal tubercles, color of the mandibles, breadth of the clypeal "lip," length of the notauli, shape of the ocellar triangle, punctation of the gaster, and position of the transverse sulcus on sternum II.

Little has been written about the taxonomy, but reference should be made to Kohl (1893b) and Yasumatsu (1936).

Biology: Tsuneki (1972) captured a species identified as *caerulea* that was transporting a nymphal *Periplaneta australasiae* (Fabricius).

Checklist of *Trirogma*

caerulea Westwood, 1841 (Ann. Mag. Nat. Hist., p. 152); Iraq, Oriental Region
    caerulea Westwood, 1841 (J. Proc. Ent. Soc. Lond. 1:16).

caerulea Westwood, 1842 (Trans. Ent. Soc. Lond. 3:225)
nigra Cameron, 1903; Borneo
prismatica F. Smith, 1858; Borneo

Tribe Ampulicini

Members of this tribe are small to large wasps, many of which are metallic green to blue. Females are readily recognized by the laterally compressed terminal abdominal segments.

Diagnostic characters:

1. Mandibular sockets closed, separated from oral fossa by a paramandibular process of hypostoma that abutts back surface of clypeus.
2. Clypeus usually with a strongly elevated, median longitudinal ridge.
3. Metasternum Y-shaped, deeply cleft posteriorly.
4. Hindcoxae separated.
5. Penultimate tarsomere short, usually with a ventral mat of fine pubescence and without apical setae, tarsomere V usually inserted dorsally at base of IV (fig. 14 A-D).
6. Marginal cell of forewing curved away from wing margin towards apex.
7. Gaster inserted between and on same level as hindcoxae.
8. Swollen part of sternum II without a transverse sulcus or carina.
9. Apex of female gaster laterally compressed.
10. Cerci absent.

Systematics: This tribe contains the majority of the species in the subfamily, but all of these are in the genus *Ampulex*. Of the ten characters listed above, all but number eight probably represent specializations over the corresponding list for the tribe Dolichurini. Further details are given in the description and discussion of the genus *Ampulex*.

Biology: As far as known, all prey on cockroaches.

Genus Ampulex Jurine

Generic description: Small to large wasps, 5 to 33 mm long; black with thorax and rarely head partly to totally brown or reddish; or body metallic green or blue; legs frequently partly red; wings clear or infumate, commonly banded; inner eye margins straight or outcurved, usually converging above but sometimes parallel or converging below; female mandible bent upward at base so that its long axis forms an angle of 40° or less with a line through condyles (fig. 12 B), inner margin usually blade-like, without teeth except occasionally near apex; male mandible not bent upward, inner margin with a single tooth that varies from large to small; malar space absent except in some males; palpi short; clypeal elevation usually topped by a polished carina (fig. 12 A), free edge usually tridentate, especially in females; labrum thickened and often prominently exposed, usually truncate; antennal sockets usually separated by about a socket diameter or slightly less, each overlaid dorsally by a frontal extension; frontal carina or line present or absent; vertex sometimes raised above eyes, postocellar area at least twice as long as ocellar triangle or longer in dorsal view of head; occipital carina U-shaped, apparently always present in male but often evanescent or absent in female; pronotal lobe angulate dorsoposteriorly; pronotal collar broader than long to longer than broad, and slightly shorter than to longer than scutum; collar delimited laterally by a narrow, deeply impressed horizontal sulcus, which is infrequently evanescent posteriorly; dorsum of collar plain, or with a posteromedian conical elevation, often with a median longitudinal sulcus, which is usually most deeply impressed anteriorly, dividing collar into two lobes, collar sometimes bi- or trituberculate; scutum broadly convex in lateral view, notauli usually deeply impressed and with few exceptions extending to posterior margin, admedian lines usually absent but separate when present, posterolateral area of scutum deeply depressed, scutal margin slightly overhanging inner margin of tegula; scutellum and metanotum nearly flat; propodeum rectilinear, dorsum usually with a median longitudinal carina and two or more lateral carinae which converge posteriorly, surface with many transverse carinae, posterior face vertical, perpendicular to dorsal face, posterolateral angle usually produced into a tooth, propodeal side areolate; mesopleuron usually with an omaulus; sternaulus and anteroventral episternal sulcus remnant occasionally present and joining omaulus; hypoepimeral area sometimes bulging near scrobe; lower metapleural area defined by its smooth sculpture as contrasted with areolate propodeal side; tarsi without plantulae; forewing media diverging at or slightly before cu-a, two or three submarginal cells, the absence of first intersubmarginal veinlet accounting for two-celled wings, first submarginal cell two or more times as long as II in three-celled wings, first intersubmarginal veinlet straight, outer vein of submarginal cell III (or II) usually meeting marginal cell near apex of latter, first recurrent vein received by submarginal cell I, second recurrent received by III (II in two-celled wings); hindwing without jugal lobe, media diverging before, at, or after cu-a; petiole portion of sternum I often enveloping corresponding portion of tergum; sternum II with a strong, transverse, basal groove.

Geographic range: Ampulex is a widespread but largely tropical genus and most of its 118 species occur in the Ethiopian, Oriental and Neotropical Regions. *Ampulex compressa,* a native of the Orient, has been spread by man to many islands. It was intentionally introduced to Hawaii in 1940 as an aid in controlling domestic roaches, and it is now well established.

Systematics: Ampulex differs in a number of basic ways from all other ampulicines, as a glance at the systematics sections under the subfamily and tribes will show. As noted under the systematics of *Trirogma,* there are a number of similarities with *Ampulex,* and the two are doubtless related. Nevertheless, *Ampulex* has evolved a number of specializations such as the loss of cerci, the closed mandibular sockets, and the loss of the jugal lobe. These imply that the relationship is a rather distant one, and at the same time indicate that *Ampulex* is the most highly evolved member of the subfamily. The laterally compressed female gaster, the attachment of the petiole between the hindcoxae, the deeply cleft metasternum, the nasiform clypeus, and the peculiar female mandible, which is sickle-shaped and strongly bent upward near the base (as seen in lateral view), make *Ampulex* a very atypical sphecid. It is unfortunate that the subfamily has often been defined largely on some of these characters, inasmuch as they do not occur in the other genera.

The development of the strange clypeal form and female mandible in *Ampulex* is apparently related to prey capture. The accounts of Williams (1929, 1942) (see biology) indicate that the female always grasps the cockroach by the thin lateral margin of its pronotum. The mandibles of *Ampulex* females are bent up at their bases, so that when closed they form a clamp with the clypeus. When seizing prey the clypeus is thrust over the dorsum of the cockroach pronotum, and the mandibles are closed beneath it. The wasp thus has a firm grip on its prey. In some *Ampulex* the labrum projects beyond the clypeal apex and the mandible may have a dorsobasal notch. Probably these extreme specializations allow an even firmer grip on the prey.

The taxon *Rhinopsis*, which has been regarded both as a genus and subgenus, appears untenable. Originally it was proposed for those *Ampulex* with only two submarginal cells, but as Arnold (1928) correctly pointed out, the number of submarginals is sometimes not constant within a species. Bradley (1934) redefined *Rhinopsis* by using the nonappendiculate second recurrent vein (fig. 11 B) in conjunction with the two-celled condition. This restricts *Rhinopsis* to a few New World species and leaves many two-celled species in typical *Ampulex*. Differences in the head, collar, and gaster, as well as in the wings, show that a variety of groups are present within *Ampulex,* and it seems best to relegate *Rhinopsis* to the status of a species group.

Nagy (1971) recognized *Chlorampulex* as a separate genus, but it is isogenotypic with *Ampulex*. Furthermore, his statement that Jurine (1807) designated the type species of *Ampulex* is incorrect. Nagy also overlooked Pate's (1937d) type species designation for *Chlorampulex*.

At least two species of *Ampulex* are notable for their aberrant features. In *A. sikkimensis,* the type species of *Waagenia,* the fourth tarsomere is normal. It lacks a ventral hair mat, the apex bears setae just as the other articles do, and V is inserted near the apex of IV (fig. 14A, B). In all other *Ampulex* we have seen (about 30 species), tarsomere IV has a ventral hair mat and apical setae are absent. Usually the last tarsomere is inserted dorsally near the base of IV. According to the description and figures in Arnold (1928), the clypeus of *Ampulex arnoldi* has a broad, truncate median lobe, and the clypeal surface is only moderately convex. Also the sockets appear from the illustrations to be widely separated, and the male seems to have a malar space.

Species characters appear to be numerous. A few of the more useful ones, some of which have not been used before, are: sculpture including the development and form of facial and propodeal carinae; size and shape of the eyes, including the degree of convergence; details of the mandibles, clypeus, and labrum; the form of the palpi, especially the peculiar setiform development of the last labial palpus segment in some species; pronotal shape; completeness of the notauli, presence or absence of various mesopleural sulci; details and form of the basal gastral segments; tarsal characters including the degree of asymmetry of tarsomere III, the relative length and width of IV with respect to V; details of claw dentition; and wing venational differences.

The basic reference on *Ampulex* is that of Kohl (1893b), in which a key was provided for all of the species known to him, and the descriptions quoted for the remainder. Bradley (1934) differentiated the two Nearctic species. Other keys are: Schulz (1904) for some Neotropical species, Arnold (1928) for the Ethiopian Region, and Bingham (1897) and Yasumatsu (1936) for some of the Oriental species. *Ampulex* are rather rare in collections, and much work remains to be done on the systematics of the genus.

Biology: F. X. Williams (1929, 1942) summarized much of the early published accounts of *Ampulex* biology and included his own observations on *A. canaliculata* and *A. compressa*. Recently Krombein (1967b) gave a good account of *A. canaliculata*. *Ampulex* explore crevices and holes in tree bark and *A. compressa,* which provisions with *Periplaneta,* is often found in houses searching for prey. The gait of some of the smaller, somber colored *Ampulex* is rather slow, and *canaliculata,* for example, seems reluctant to fly even when disturbed by the net of a collector. These dark colored *Ampulex* are well camouflaged when walking on tree bark, and often resemble ants. This mimicry is enhanced by wing banding, which with the wings folded over the body improves the overall antlike appearance by suggesting a petiole. *Ampulex ruficornis* and *constanciae* closely resemble a stinging arboreal ant, *Sima rufonigra* Jerdon, with which they mingle according to Williams (1929). Turner (1926) noted mimicry between *A. formicoides* and *Sima nigra* Jerdon. The larger, metallic colored species are more active and do not mimic ants.

Ampulex females use their nasiform clypeus in conjunction with the mandibles during prey seizure. In those species for which prey capture has been described, the wasp grasps the lateral margin of the roach pronotum, the wasp's clypeus clamping down on the dorsal surface, while the mandibles are thrust under the pronotum. While holding the prey in this manner, the wasp curves her gaster under her body so that she can sting the prey in the ventral region of the thorax. Williams and others have noted rather frequently that the wasp, after paralyzing the prey, may amputate part of both antennae of

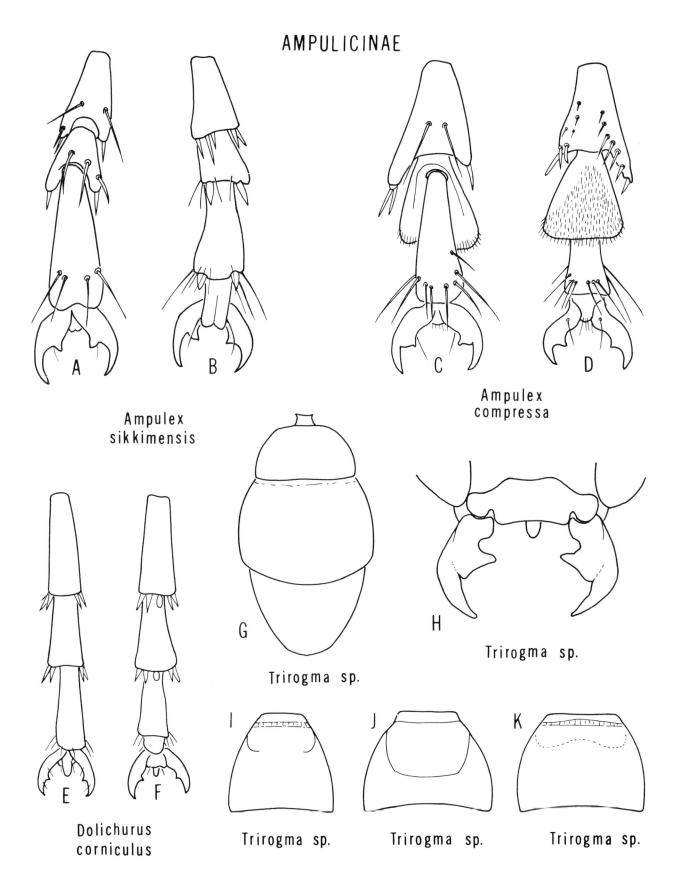

FIG. 14. Structural details of the Ampulicinae. A,C,E, and B,D,F, dorsal and ventral views, respectively, of distal female hindtarsomeres; G, dorsal aspect of male gaster, H, male clypeus and mandibles; I, male sternum II; J-K, female sternum II.

the host. The wasp then usually feeds on the exuding body fluid of the roach. When a suitable niche, such as a hollow plant stem or trap nest, is available, a female *Ampulex* may construct a multicellular nest. Each roach with its egg is isolated from those adjoining by a loose plug of debris.

Known prey of *Ampulex* species are tabulated below.

Ampulex compressa	— *Periplaneta americana* (Linnaeus)
	Periplaneta australasiae (Fabricius)
	Neostylopyga rhombifolia (Stoll)
Ampulex assimilis	— *Shelfordella tartara* Saussure
Ampulex fasciata	— *Ectobia livida* (Fabricius)
Ampulex canaliculata	— *Parcoblatta virginica* (Brunner)
Ampulex dissector	— *Periplaneta americana* (Linnaeus)
	Periplaneta australasiae (Fabricius)
	Periplaneta fuliginosa Serville
	Periplaneta picea Shiraki

The smooth, brown cocoon of *Ampulex* is drawn out nipple-like at both ends and enshrouded in a loose layer of silk. The cocoon is constructed within the remains of the host.

Checklist of *Ampulex*

aborensis Nurse, 1914; e. India
aenea (Fabricius), 1804 (*Dryinus*); Guinea
  ? *guerini* Dahlbom, 1843 (♂ only)
  dahlbomii Kohl, 1893
†*aenea* Spinola, 1841; s. India, nec Fabricius, 1804
aereola Bradley, 1934; Guyana
alisana Tsuneki, 1967; Taiwan
angusticollis Spinola, 1841; Costa Rica to Argentina
  dubia Kohl, 1893
?*annulipes* Motschulsky, 1863; Sri Lanka (possibly a bethyloid)
apicalis F. Smith, 1873; S. Africa
approximata Turner, 1912; w. India
arnoldi Brauns, 1928; Rhodesia
assamensis Cameron, 1903; e. India
assimilis Kohl, 1893; Guinea, Iraq
atrohirta Turner, 1915; Borneo
?*bispinosa* (Reich), 1795 (*Sphex*); French Guiana
bredoi Arnold, 1947; Zambia
brunneofasciata Giordani Soika, 1939; Ethiopia
bryanti Turner, 1914; Borneo
canaliculata Say, 1823; e. U.S.
  abbottii Westwood, 1844 (*Rhinopsis*)
  pensylvanica Haldeman, 1849
  melanognathus Rohwer, 1912 (*Rhinopsis*)
carinifrons Cameron, 1903; e. India
chalybea F. Smith, 1856; Africa
collator Bradley, 1934; Costa Rica
compressa (Fabricius), 1781 (*Sphex*); Ethiopian and Oriental Regions, Australia, East Indies, New Caledonia, Cook, Midway, Hawaii, St. Helena, Mauritius, Réunion, Chagos Archipelago, Seychelles
  sinensis Saussure, 1867
  striolata Saussure, 1892 (*Chlorampulex*)
compressiventris Guérin-Méneville, 1835 (pl. 70); s. Africa
  sibirica Fabricius, 1793 (*Sphex*), nec Christ, 1791 (Art. 58(12)).
  ? *guerini* Dahlbom, 1843 (♀ only)
conigera Kohl, 1893; Ethiopia, Somalia, Zambia
constanceae (Cameron), 1891 (*Rhinopsis*); w. India
crassicornis Kohl, 1898; centr. Africa
crawshayi Turner, 1917; Zaire to Malawi
crudelis Bingham, 1897; Sikkim, e. India
  trigona Cameron, 1902
cuprea F. Smith, 1856; China
cyanator (Thunberg), 1822 (*Ichneumon*); Guinea
  gratiosa Kohl, 1893 (synonymy from Roman, 1912)
cyanipes (Westwood), 1841 (Ann. Mag. Nat. Hist. (1)7: 152 (*Chlorion*); S. Africa
  cyanipes Westwood, 1841 (J. Proc. Ent. Soc. Lond. 1:16) (*Chlorion*)
  cyanipes Westwood, 1842 (Trans. Ent. Soc. Lond. 3:230) (*Chlorion*)
  cyaneipes Dalla Torre, 1897, emendation
cyanura Kohl, 1893; S. Africa
  africana Cameron, 1905
  capensis Cameron, 1905
  ssp. *monticola* Arnold, 1928; Rhodesia
  ssp. *rhodesiana* Arnold, 1928; Rhodesia
cyclostoma Gribodo, 1894; Mozambique
dentata Matsumura and Uchida, 1926; Ryukyus
denticollis (Cameron), 1910 (*Dolichurus*); Rhodesia, S. Africa
  ssp. *rufithorax* Arnold, 1931; S.-W. Africa
†*denticollis* Tsuneki, 1967; Taiwan, nec Cameron, 1910
difficilis Strand, 1913; Taiwan
dissector (Thunberg), 1822 (*Ichneumon*); Taiwan, China, Japan
  amoena Stål, 1857 (synonymy of this and following from Roman, 1912)
  novarae Saussure, 1867
  consimilis Kohl, 1893
  japonica Kohl, 1893
distinguenda Kohl, 1893; Celebes
dives Kohl, 1893; Gabon, Zaire, Fernando Póo
elegantula Kohl, 1893; Colombia
erythropus Kohl, 1893; Indonesia: Java, Borneo
esakii Yasumatsu, 1936; Taiwan
fasciata Jurine, 1807; Europe
  europaea Giraud, 1858
ferruginea Bradley, 1934; U.S.: Texas, Florida
formicoides Turner, 1926; Malaya
formosa Kohl, 1893; Senegal
fulgens Beaumont, 1970; Afghanistan
hellmayri Schulz, 1904; Brazil
himalayensis Cameron, 1903; e. India
honesta Kohl, 1893; Gabon, Zaire
hospes F. Smith, 1856; Borneo
  foveifrons Cameron, 1903
  ssp. *cognata* Kohl, 1893; Java

insularis F. Smith, 1857; Borneo, Malaya
interstitialis Cameron, 1903; e. India
javana Cameron, 1905; Java
khasiana Cameron, 1903; e. India
kristenseni Turner, 1917; Ethiopia; Tanzania: Tanganyika
kurarensis Yasumatsu, 1936; Taiwan
laevigata Kohl, 1893; Philippines
    levigata Dalla Torre, 1897, emendation
latifrons Kohl, 1893; Sikkim, e. India
    pulchriceps Cameron, 1900
    brevicornis Cameron, 1902
lazulina Kohl, 1893; Guinea, Senegal
longicollis Cameron, 1902; e. India
    trichiosoma Cameron, 1902
lugubris Arnold, 1947; Malawi, Mozambique
luluana Leclercq, 1954; Zaire
maculicornis (Cameron), 1889 (*Rhinopsis*); Mexico to Panama
    maculicollis Dalla Torre, 1897, lapsus
major Kohl, 1893; Malaya
melanocera Cameron, 1908; Tanzania: Tanganyika
metallica Kohl, 1893; Malaya
micado Cameron, 1905; ? Japan
micans Kohl, 1893; ? Mexico
minor Kohl, 1893; Brazil, Guyana
mirabilis Berland, 1935; China
mocsaryi Kohl, 1898; Java
moebii Kohl, 1893; Zaire, S. Africa
montana Cameron, 1903; e. India
moultoni Turner, 1915; Borneo
murotai Tsuneki, 1973; Taiwan
mutilloides Kohl, 1893; S. Africa
    sanguinicollis Brauns, 1898
nasuta Er. André, 1894; Tanzania: Zanzibar
nebulosa F. Smith, 1856; S. Africa
    cribrata Kohl, 1893
neotropica Kohl, 1893; Costa Rica, Panama
nigricans Cameron, 1899; e. India
nigrocaerulea Saussure, 1892; S. Africa, Rhodesia
    spiloptera Cameron, 1905
    jansei Cameron, 1910
nitidicollis Turner, 1919; Tanzania: Tanganyika
occipitalis Arnold, 1947; Malawi
overlaeti Leclercq, 1954; Zaire
pilipes Kohl, 1893; Guinea, Mozambique
pilosa Cameron, 1900; e. India, Borneo
psilopa Kohl, 1893; Guinea
purpurea (Westwood), 1844 (*Chlorion*); S. Africa
raptor F. Smith, 1856; Venezuela
regalis F. Smith, 1860; Celebes
rothneyi Cameron, 1902; e. India
ruficollis Cameron, 1888; Spain
ruficornis (Cameron), 1889; (*Rhinopsis*); e. India
ruficoxis Cameron, 1902; e. India
rufofemorata Cameron, 1903; Borneo
sagax Kohl, 1893; Guyana
satoi Yasumatsu, 1936; Korea
sciophanes (Nagy), 1971 (*Chlorampulex*); Taiwan
seitzii Kohl, 1893; China, Taiwan, Java
senex Bischoff, 1915; e. Africa
*sikkimensis (Kriechbaumer), 1874 (*Waagenia*); Sikkim
smaragdina F. Smith, 1858; Singapore
sodalicia Kohl, 1893; Malaya, Borneo, e. India
    striatifrons Cameron, 1902
    tricarinata Cameron, 1902
sonani Yasumatsu, 1936; Taiwan
sonnerati Kohl, 1893; Philippines
spectabilis Kohl, 1893; Guinea
splendidula Kohl, 1893; centr. Africa
surinamensis Saussure, 1867; Colombia, Guyana, Surinam, Brazil
sybarita Kohl, 1893; Java
takeuchii Yasumatsu, 1936; Taiwan
thoracica F. Smith, 1856; Brazil
toroensis Turner, 1919; Uganda, Senegal
trigonopsis F. Smith, 1873; Brazil
varicolor Turner, 1919; Vietnam
venusta Stål, 1857; s. Africa (? = *nebulosa*)
viridescens Arnold, 1947; Zambia, Zaire

SUBFAMILY SPHECINAE

Because of their generally large size and bright colors, these common inhabitants of fields, forests, and even vacant city lots are conspicuous. They are probably the best known members of the Sphecidae. Fossorial forms such as *Sphex, Ammophila,* and *Prionyx* are frequently called "digger wasps" or "sand wasps." The well known yellow and black mud-nest builders of the genus *Sceliphron,* the "mud daubers," also belong here. The sphecines are collectively called the "thread waisted wasps" in reference to their distinctive, cylindrical petiole. We recognize 19 genera, divided among three tribes. The subfamily is represented everywhere except the polar regions, although at least two species, *Ammophila azteca* and *mediata,* range north of the Arctic Circle.

The subfamily displays a variety of nesting habits. Some species of the fossorial genus *Chlorion* behave like parasitoids, while at the other extreme a primitive form of sociality is present in *Trigonopsis,* a genus of mud-nest builders. Some *Ammophila* maintain several nests simultaneously, and progressively provision each one, adding prey only as needed by the developing wasp larva. Some *Isodontia,* a genus that nests in pre-existing cavities in twigs and wood, rear a number of larvae in one large brood cell. Prey of sphecines range from spiders, cockroaches, crickets, grasshoppers, katydids, mantids, and walking sticks to lepidopterous and hymenopterous caterpillars. *Podium* is one of the few sphecids known to use resin in nest building.

Diagnostic characters:

1. (a) Inner orbits entire, converging or diverging below or parallel; (b) ocelli normal
2. (a) Antennae usually inserted near middle of face except in some Sceliphronini, sockets contiguous with to widely separated from frontoclypeal suture; (b) male with 13 and female with 12 antennal articles.
3. Clypeus transverse or higher than wide, rarely nasiform.
4. (a) Mandible without a notch or angle on externoventral margin; (b) palpal formula usually 6-4 (6-3 to 6-1 in some *Prionyx*); (c) mouthparts short to long; (d) mandibular socket closed except in *Stangeella* and males of some genera of Ammophilini; (e) occiput and hypostoma in contact or nearly so except in some Sceliphronini.
5. (a) Pronotum with a high, short collar or collar elongate, and somewhat depressed below level of scutum, rarely tuberculate; (b) pronotal lobe broadly separated from tegula.
6. (a) Notauli present or absent but weak and short when present; (b) no oblique scutal carina.
7. (a) Episternal sulcus usually present, usually extending to anteroventral region of pleuron; (b) omaulus absent.
8. Definitive metapleuron usually consisting of upper metapleural area only, except in Ammophilini.
9. (a) Midtibia with two apical spurs except in some Ammophilini; (b) midcoxal cavities and coxae usually approximate (broadly separated in a few *Penepodium* and *Isodontia*), midcoxae without dorsolateral carina or crest; (c) precoxal lobe present; (d) hindfemur simple apically except in *Dynatus;* (e) claws with one to five teeth on inner margin, or simple; (f) foretarsal rake present or absent; (g) arolium usually large.
10. (a) Propodeum moderately long to very long; (b) enclosure present, U-shaped, or absent; (c) propodeal sternite present.
11. (a) Forewing with three submarginal cells (rare exceptions in *Prionyx, Ammophila*) and two recurrent veins whose end point is variable; (b) marginal cell apex acuminate, rounded, or truncate.
12. (a) Jugal lobe large, occupying most of anal area (fig. 18); (b) hindwing media usually diverging at or after cu-a (slightly before in some Sceliphronini); (c) hindwing without subcostal vein but with third anal vein.
13. (a) Gaster with cylindrical petiole composed of sternum I except displaced tergum I also forms part of petiole in some Ammophilini; (b) tergum I without lateral carina or line; (c) male with six or seven visible segments; (d) no pygidial plate.
14. (a) Volsella of male genitalia usually with cuspis

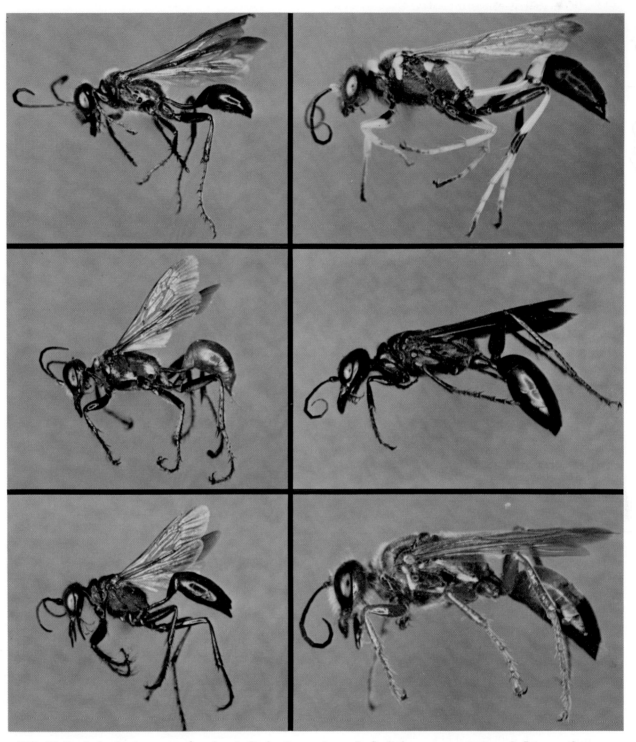

FIG. 15. Typical sphecine wasps, females. A, *Isodontia mexicana;* B, *Sceliphron caementarium;* C, *Prionyx thomae;* D, *Chlorion aerarium;* E, *Podalonia mexicana;* F, *Sphex ichneumoneus.*

and movable digitus, but digitus immovably fused in Prionyxina; (b) aedeagal head usually with teeth on ventral margin; (c) male sternum VIII variable; (d) cerci present or absent.

Systematics: The Sphecinae has a considerable number of universal diagnostic features, among the more important of which are: ocelli normal, mandible unnotched, jugal lobe large and containing an anal vein, gaster with a cylindrical sternal petiole, propodeal sternite present, tergum I without lateral line or carina, presence of two recurrent veins, and absence of an omaulus. Other important but nonuniversal diagnostic tools are: the closed mandibular sockets, long episternal sulcus, two midtibial spurs, toothed claws, hindwing media diverging after cu-a, and volsella with cuspis and movable digitus.

The subfamily classification we are using here is essentially the same as the one we proposed earlier (Bohart and Menke, 1963), but considerable refinements have been made after the study of much more material than was available when we made our initial investigation. Some of the features we used to separate the three tribes have proven to be unusable, and a completely new key to them is presented here. Even so, because of exceptional species in certain genera of each tribe it has been impossible to construct an absolutely "clean" key.

The following chart shows some of the characters that we feel are of significance in determining the relative evolutionary status of the tribes and their genera. A few features have not been listed. For example, we are not sure of the significance of the placoids. In our 1963 paper we regarded them as specialized structures, but their absence in many of the more highly evolved sphecines (*Trigonopsis, Sceliphron,* Ammophilini, for example) suggest that perhaps they are better regarded as primitive features. Some things such as the loss of the foretarsal rake in genera that have changed from fossorial to twig or mud-nesting habits, are obviously linked with these changes. The significance of open mandibular sockets is somewhat questionable in this subfamily. They are found only in the enigmatic sceliphronin genus *Stangeella* and some males of the tribe Ammophilini. We feel that open sockets in *Stangeella* are primitive, but in the Ammophilini they may have become open secondarily.

Of the three tribes, the Sceliphronini has retained the greatest number of generalized features, the most important and nearly universal being a single claw tooth, plantulae, an expanded third maxillary palpal segment and cerci. Specializations in this tribe include a general lengthening of the body and development of a semiprognathus head. Some species of *Podium* have the first recurrent vein ending on the first submarginal cell, but the corresponding hindwing has a mixture of primitive and specialized features. In this genus the jugal lobe has become reduced, so that the third anal vein lies along the wing margin, but the media often retains a primitive position with respect to cu-a. The partial loss of the hypostomal carina in females of some genera of the Sceliphronini is another interesting specialization in this tribe.

TABLE 5

Characters of phylogenetic significance in the Sphecinae

	Generalized	Specialized
1.	Sockets low on face	Sockets near middle of face
2.	Mandible simple	Mandible with inner teeth
3.	Mandible socket open	Mandible socket closed
4.	Mouthparts short	Mouthparts long
5.	Maxillary palpal segment III expanded on one side	Maxillary palpal segment III essentially symmetrical
6.	Propodeal enclosure absent	Propodeal enclosure present
7.	Episternal sulcus long	Episternal sulcus short, or absent
8.	Spiracular groove present	Spiracular groove absent
9.	Lower metapleural area defined	Lower metapleural area not defined
10.	Three submarginal cells	Two submarginal cells
11.	Submarginal cell II receiving only one recurrent vein	Submarginal cell II receiving both recurrent veins
12.	Hindwing media diverging at or slightly before cu-a	Hindwing media diverging well after cu-a
13.	Midtibia with two spurs	Midtibia with one spur
14.	Plantulae present	Plantulae absent
15.	Apicoventral setae of hindtarsomere V sctiform	Apicoventral setae of hindtarsomere V bladelike
16.	Claw with one tooth	Claw with two or more teeth, or simple
17.	Petiole short, composed of sternum I only	Petiole long, includes tergum I
18.	Cerci present	Cerci absent
19.	Volsella with movable digitus	Digitus immovable
20.	No yellow markings	Yellow pigment present

The Sphecini is obviously more specialized than the Sceliphronini. Among the important advancements here are the development of multitoothed claws, fusion of the digitus with the rest of the volsella, loss of the cerci, and development of broad, leaflike apicoventral setae on the last tarsomere. The first two of these are restricted to the subtribe Prionyxina. Morphologically this subtribe seems more advanced than the subtribe Sphecini, but biologically the Prionyxina is the more primitive of the two. Noteworthy specializations in the Prionyxina include: reduction in the number of labial palpal segments, reduction in the size of the stigma and narrowing of the submarginal cells, the tendency for loss of the third submarginal cell, and the notched labrum.

The Ammophilini is the most advanced tribe. Very few primitive features are retained, but among them the presence of plantulae in *Parapsammophila* and the widespread retention of a defined lower metapleural area are noteworthy. The petiole reaches its greatest length in this tribe, and in *Eremnophila* and *Ammophila* tergum I is very slender and displaced to the distal end of sternum I, so that the petiole is often said to be two segmented.

The development of long mouthparts reaches its zenith in *Ammophila* and *Podalonia*. The loss of one midtibial spur and one submarginal cell and the shortening of the episternal sulcus are interesting tendencies in the Ammophilini.

Sexual dimorphism is rather strong in the Sphecinae, especially in the head. Generally the male face is narrower than that of the female, and the orbits often converge below. The female orbits are often parallel. The clypeus usually differs markedly; that of the female presumably modified to suit nest building requirements or prey capture, whereas that of the male may be adapted for holding the female during copulation. A triangular to nasiform male clypeus is a specialized form. This sort of male clypeus is represented in all three tribes but is most common in the Ammophilini, notably *Ammophila* and *Eremnophila*. The male mandible is sometimes modified for use in copulation, and the hypostoma occasionally has prongs near the mandible base which may have a similar function. Shape of the female mandible reflects the nesting habits of the wasp. Fossorial forms usually have an arcuate mandible with one or more inner mesal teeth. In twignesters the mandible is often less arcuate, and the teeth are usually apical or subapical. Mudnesting forms generally have simpler mandibles.

The sexes often differ markedly in color, and in some cases sex associations have been confused because of this. Generally males are more extensively black and have paler wings than the females. Vestiture also may be differently colored in the two sexes, and in species with large geographic ranges, hair color as well as wing and body color patterns may vary considerably over the distribution of the species. This kind of variation has led to the naming of unnecessary new species in the past.

KEY TO TRIBES OF SPHECINAE[1]

1. Tarsi with plantulae (fig. 19 A,B)[2] and/or claws of some legs with one mesal[3] tooth on inner margin .... Sceliphronini, p. 82
   Tarsi without plantulae[4]; claw simple or with one or more basal teeth on inner margin 2
2. Claw with two or more teeth; second recurrent vein usually received by submarginal cell III (fig. 25 A-D), but if not then claw with at least three teeth *or* apicoventral setae of hindtarsomere V very broad, separated at base by no more than one and a half setal widths (fig. 26 B); tarsi without plantulae .... Sphecini, p. 106
   Claw usually simple or with one tooth, but if with two teeth (Old World forms) then apicoventral setae of hindtarsomere V narrow, separated by three or more setal widths; second recurrent vein usually received by submarginal cell II (fig. 35); tarsi sometimes with plantulae in one Old World genus .... Ammophilini, p. 134

Tribe Sceliphronini

This group consists mainly of slender wasps. The best known genera of the tribe are *Chlorion* and *Chalybion*, many species of which are metallic green or blue, and *Sceliphron* which contains the yellow and black "mud daubers." These three genera are cosmopolitan, but the remaining five genera of the Sceliphronini are restricted to the New World. The Sceliphronini probably has the broadest spectrum of ethological diversity of any sphecid tribe. Some species of the fossorial genus *Chlorion* have parasitoidlike habits, the most primitive type of biology in the Sphecidae. At the other extreme, the mud-nest building genus *Trigonopsis* displays a primitive form of sociality. Orthoptera, especially cockroaches, are the predominant prey of sceliphronin wasps, although spiders are provisioned by two genera.

Diagnostic characters:

1. (a) Antennal socket placement variable but usually somewhat below middle of face, separated from or contiguous with frontoclypeal suture; (b) male antenna usually with placoids (none in *Trigonopsis* and *Sceliphron*); (c) scape short, globose to elongate oval; (d) comparative lengths of flagellomeres I-II variable.
2. (a) Clypeus usually much broader than long but as long as broad in some *Chalybion;* (b) labrum broader than long except in *Dynatus* males, free margin usually straight.
3. (a) Mandible simple or with one or two inner teeth; (b) mandible socket closed by paramandibular process of hypostoma except in *Stangeella.*
4. (a) Maxillary palpal segment III usually expanded on one side; (b) hypostomal carina often incomplete in female; (c) occipital carina variable.
5. Collar short to long.
6. (a) Propodeal enclosure sometimes present; (b) spiracular groove present or absent; (c) episternal sulcus short or long; (d) lower metapleural area usually not defined, metapleural sulcus usually absent between upper and lower metapleural pits.
7. (a) Marginal cell apex variable; (b) three submarginal cells present; (c) both recurrent veins usually received by second submarginal cell but occasionally first ending on I or second on III
8. (a) Female with or without psammophore; (b) midcoxae usually nearly contiguous but very broadly

---

[1] A magnification of 25X or 50X is required to clearly see the claw and tarsal characters used here.

[2] Plantulae are very small in the southern South American sceliphronin genus *Stangeella* (fig. 19 B), and the claw tooth is basal; but this genus can be identified by its black body and wings and metallic blue gaster.

[3] The claw tooth is basal in a few species of the sceliphronin genera *Chlorion, Dynatus,* and *Sceliphron,* and all claws are simple in one yellow and black *Sceliphron* and two metallic blue *Chalybion;* but plantulae are present in all these cases.

[4] Some species of the Old World ammophilin genus *Parapsammophila* have small plantulae, but the one or two claw teeth are basal.

separated in a few *Penepodium;* (c) two apicoventral setae of hindtarsomere V setiform to moderately broad and bladelike, separated by two or more setal widths; (d) inner margin of claw with one tooth which is usually mesal, claw rarely simple; (e) tarsi with plantulae except in some *Podium* and most *Chalybion*

9. (a) Petiole consisting of sternum I only; (b) tergum VIII with cerci in males except in some *Chalybion* and many *Sceliphron;* (c) volsella with an articulating digitus.

Systematics: The most important unifying characteristics of the eight genera in the Sceliphronini are the single, usually mesal claw tooth, and the presence of plantulae. Unfortunately, neither feature is universal, and consequently the few exceptional species are technically difficult to isolate from the tribe Ammophilini. Except for their occurence in some species of the ammophilin genus *Parapsammophila,* plantulae are unique to the sceliphronins, but they are absent in a few *Podium* and many *Chalybion.* However, the claw teeth in these two genera are mesal, not basal, as in the few ammophilins with single-toothed claws. The claws of one or two pairs of legs are simple in one *Dynatus* and many *Chalybion,* and all the claws are simple in one *Sceliphron* and two *Chalybion.* However, these last three forms have plantulae. The claw of *Parapsammophila* usually has two teeth, although the more basal tooth is weakly formed or absent in a few species.

The distortion to one side of the third maxillary palpal segment occurs only in the Sceliphronini, but it is not universally found in the tribe. The two apicoventral setae of hindtarsomere V are generally narrower and hence more widely separated than in the Sphecini. Wing venation in the Sceliphronini is quite variable, but basically both recurrent veins end on the second submarginal cell.

The Sceliphronini is rather heterogeneous and eventually it may be desirable to recognize several subtribes. *Stangeella* with its open mandibular sockets is quite isolated from other genera in the tribe. This and other features of the morphology of *Stangeella* give the taxon the status of a linking form between the three tribes of the Sphecinae, and the classification of the subfamily might be improved if it were placed in a monotypic tribe, the Stangeellini. For the time being, however, we are only proposing placement of the genus in a subtribe, the Stangeellina. The remaining genera fall into the subtribe Sceliphronina. *Penepodium, Dynatus, Podium,* and *Trigonopsis* form a rather close knit group characterized by essentially simple mandibles and, with few exceptions, a small, oval patch of short pubescence at the base of tergum I. Each of the remaining genera, *Chlorion, Chalybion,* and *Sceliphron,* seem somewhat isolated. *Chlorion,* perhaps the most generalized genus in the tribe, is morphologically similar to the *Penepodium* group, and perhaps a *Chlorion*-like ancestor gave rise to this assemblage of four genera. *Chalybion* is an enigma. The subgenus *Chalybion* consists of forms that show affinities with the *Penepodium* group, but the subgenus *Hemichalybion,* especially in its yellow and black color pattern, but also in its general body form, seems closer to *Sceliphron.* Quite likely the explanation of the resemblance between *Hemichalybion* and *Sceliphron* is convergence of independently evolved traits, and *Sceliphron* may not be as closely related to *Chalybion* as has been thought by most previous workers. The face of *Sceliphron* is not too far removed from that of *Stangeella.* Perhaps the former evolved from an ancestor resembling the latter. It must also be pointed out, however, that there are resemblances between the faces of *Sceliphron* and some *Chalybion.* We have expressed our thoughts on the relationships of the sceliphronin genera in the dendrogram shown in figure 16.

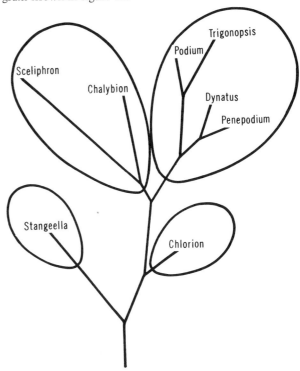

FIG. 16. Dendrogram showing suggested relationships of genera in the tribe Sceliphronini.

Specialization in this tribe has gone in the direction of elongating and slenderizing the head, thorax, and abdomen, and it is best exemplified by such species as *Podium plesiosaurus, Trigonopsis rufiventris, Penepodium complanatum, Chalybion frontale,* and *C. dolichothorax.* Yellow pigment is regarded as a specialization (Leclercq, 1954a), and this color is found in *Sceliphron, Chalybion,* subgenus *Hemichalybion,* and *Trigonopsis.* Loss of plantulae is a specialization in the more highly evolved forms of *Podium* and most *Chalybion.* Loss of the claw tooth occurs primarily in *Chalybion,* but some *Sceliphron* and *Dynatus* have simple claws. Apparently a basal claw tooth is the primitive condition. The claw tooth is basal in *Stangeella,* some *Chlorion,* some *Dynatus,* and some exceptional *Sceliphron.* The claw tooth has migrated to a subapical position in some of the more advanced *Podium* and a few *Sceliphron. Dynatus* has evolved peculiar hind legs, the clublike femur of the male being especially

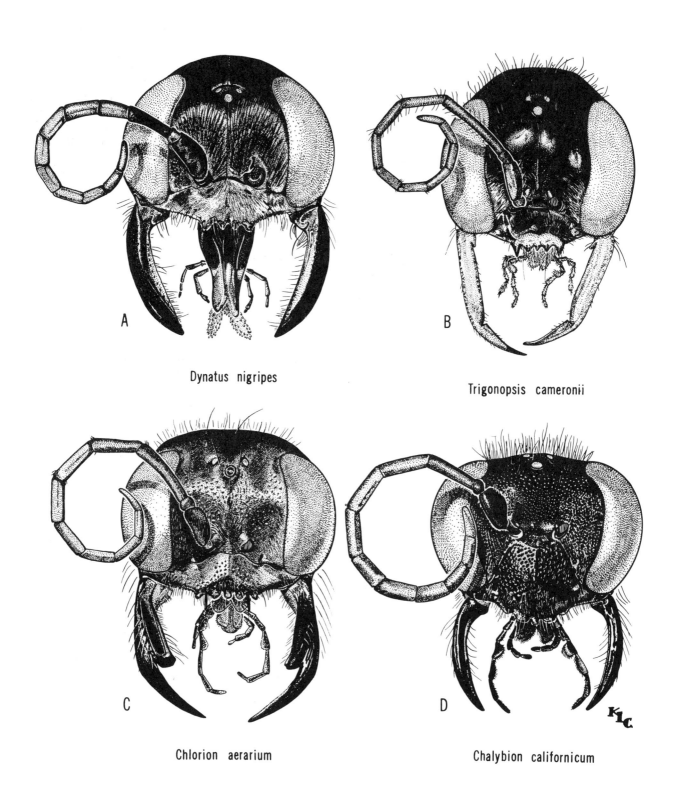

FIG. 17. Facial portraits of females in the tribe Sceliphronini.

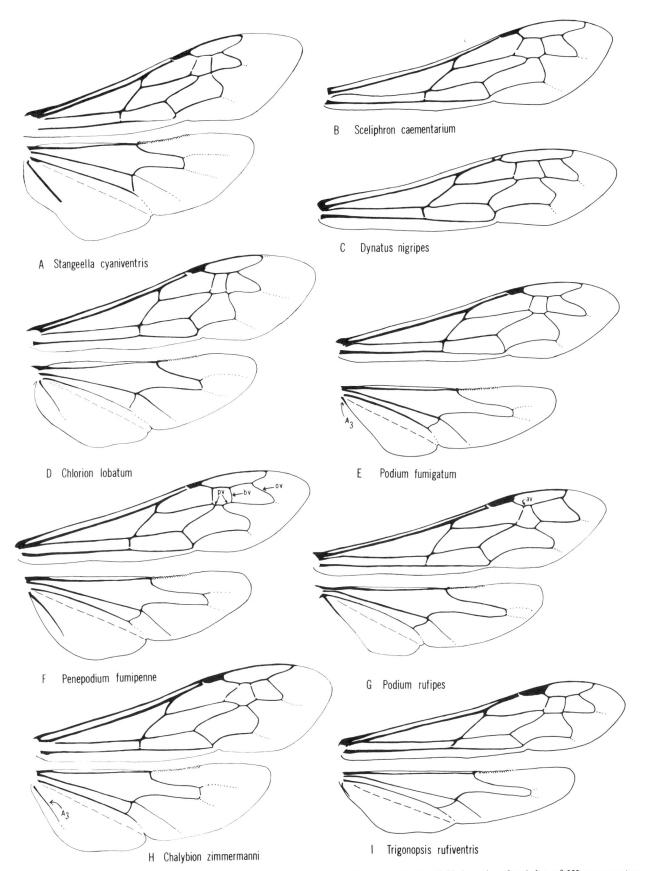

FIG. 18. Wings in the tribe Sceliphronini; av = anterior veinlet of submarginal cell II, bv = basal veinlet of III, ov = outer veinlet of III, pv = posterior veinlet of II, A_3 = third anal vein.

bizarre. Presumably, these leg modifications and the unusual nasiform male clypeus have a function during copulation. The offset flagellomeres of most male *Penepodium*, the notched female mandible in *Trigonopsis*, and the loss of the mesopleural suture in some *Chalybion* are other notable specializations in the Sceliphronini.

The loss of a foretarsal rake and development of a toothless mandible are obviously related to the change from fossorial nesting habits to nesting in pre-existing cavities and the building of free mudnests.

Biology: Chalybion and *Sceliphron* stock their nests with spiders, but all of the other genera provision with Orthoptera. *Stangeella* is one of the few sphecid genera to use Mantidae and walking sticks. Nests are made before the search for prey, except in those species of *Chlorion* that have a parasitoid-like ethology. Interestingly, in *Penepodium* and *Trigonopsis* the egg is laid on the prey before it is placed in the nest. A similar claim has been made for one *Chalybion* species. Some *Chlorion* do not construct a nest, but others of the genus as well as *Stangeella* and some *Penepodium* are fossorial. Mudnests are constructed by *Sceliphron* and *Trigonopsis*. The remaining genera including some *Penepodium* nest in preexisting cavities. The closing plug is partially made of resin in some *Podium*, and some *Chalybion* cover the closing plug with uric acid. The development of sociality in *Trigonopsis* makes it one of the most highly evolved sphecids. Nothing is known about the biology of *Dynatus*.

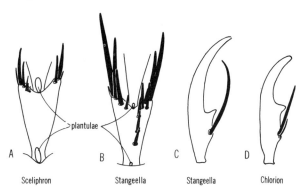

FIG. 19. Tarsal details in the tribe Sceliphronini; A,B, hindtarsomere IV, ventral, all setae not shown; C,D, hindtarsal claw, lateral.

Key to genera of Sceliphronini

1. Propodeum with U-shaped dorsal enclosure defined at least posterad by a semicircular sulcus or furrow 2
   Propodeal dorsum at most with a median, longitudinal sulcus, or furrow and/or posteromedian pit 4
2. Both recurrent veins received by second submarginal cell (fig. 18 B); spiracular groove absent; body usually with yellow areas; female without foretarsal rake ....*Sceliphron* Klug, p. 103
   Second recurrent vein received by third submarginal cell or interstitial between II and III (fig. 18 A,D); spiracular groove present; body without yellow but often metallic blue or green; female with a foretarsal rake 3
3. Claw tooth usually mesal (fig. 19 D); female clypeus with five large teeth (fig. 17 C), rarely only two or four teeth; antennal sockets contiguous with frontoclypeal suture in female, separated by less than a socket diameter in male; male sterna IV and V at most sparsely setose..... *Chlorion* Latreille, p. 88
   Claw tooth basal (fig. 19 C); female clypeus without teeth, margin trapezoidal (fig. 23 B); antennal sockets separated from frontoclypeal suture by a socket diameter or more; male sterna IV and V densely covered with velvety pile, southern South America.................. *Stangeella* Menke, p. 87
4. Collar with a posteromedian prominence or tubercle[5] *and* occipital carina ending close to hypostomal carina, the two separated by much less than length of oral cavity.. 5
   Collar with median, longitudinal sulcus or indentation[6], or if merely convex or with posteromedian prominence then occipital carina separated from hypostomal carina by at least length of oral cavity (fig. 20 B)........................... 6
5. Posterior veinlet of submarginal cell II no more than two-thirds length of posterior veinlet of submarginal cell III (fig. 18 F); spines of hindtibia not set in large, shallow, circular depressions; hindfemur fusiform, apex simple .... *Penepodium* Menke, p. 90
   Posterior veinlet of submarginal cell II at least as long as posterior veinlet of submarginal cell III (fig. 18 C); spines on outer part of closing surface of hindtibia set in large, shallow, circular depressions; female hindfemur of rather uniform thickness, apex with small ventral lobe (fig. 21 A), male hindfemur greatly enlarged apically, clublike (fig. 21 B-D) .... *Dynatus* Lepeletier, p. 92
6. Spiracular groove present but crossed by ridges; propodeal dorsum with median, longitudinal furrow crossed by many strong ridges and usually longitudinally bisected by a carina; female mandible with subapical, slitlike notch on inner margin (fig. 17 B)[7]........... *Trigonopsis* Perty, p. 96

---

[5] Collar simply convex in one *Penepodium* female but occipital and hypostomal carinae adjacent to one another, and body is black.

[6] Collar weakly or not indented in a few *Chalybion*, but body is metallic blue.

[7] One exception known.

Spiracular groove absent[8]; propodeal dorsum plain or with a simple median groove; mandible simple or with subapical tooth on inner margin (figs. 17D, 22 A,C,I,M)... 7

7. Third anal vein of hindwing broadly separated from wing margin, length at least one-third distance from wing base to jugal excision (fig. 18 H); episternal sulcus long, ending near anteroventral margin of pleuron; body usually metallic blue or black and yellow ...... *Chalybion* Dahlbom, p. 98

Third anal vein of hindwing contiguous with wing margin, short, extending no more than one-fourth distance from wing base to jugal excision, often obscure (fig. 18 E); episternal sulcus short, ending somewhat below level of pronotal lobe; head and thorax black, legs often partly red .... *Podium* Fabricius, p. 94

Subtribe Stangeellina

Diagnosis: Hypostoma not extending to back of clypeus, mandible socket open to oral fossa (fig. 20 A).

Discussion: The mandible form, basal placement of the claw tooth, presence of a spiracular groove, and the general body form of *Stangeella,* the only genus of the subtribe, are highly suggestive of the tribe Sphecini, but the presence of plantulae and the single claw tooth ally the genus with the Sceliphronini.

Genus Stangeella Menke

Generic description: Inner orbits straight for most of their length, converging below (fig. 23 B); vertex flat, not elevated above eyes; postocular area not elongate; face and vertex with long, dark, erect hair; frontal line impressed but weakly so above; flagellomere I longer than II; flagellomeres II-VI with placoids in male; sockets separated from frontoclypeal suture by about a socket diameter in female, slightly more than a socket diameter in male; face slightly depressed around tentorial pit in female; clypeus length about two-thirds width, disk convex, free margin projecting, trapezoidal, apex weakly emarginate (fig. 23 B); free margin of labrum shallowly, angularly emarginate, surface convex basally; female mandible long, strongly arcuate, apex sharp, inner margin with two mesal teeth (fig. 23 B), male mandible shorter, straight until near apex, inner margin with one subapical tooth; mouthparts moderately long, third maxillary palpal segment symmetrical; hypostomal carina ending near mandible socket; occipital carina incomplete below, ending just before reaching hypostomal carina; collar short, dorsomedially indented; scutellum strongly convex; propodeum moderately long, with U-shaped dorsal enclosure defined by sulcus, surface convex, finely, densely rugulose; propodeal side with spiracular groove; episternal sulcus long, ending at anteroventral margin of pleuron, scrobal sulcus weak or absent; lower metapleural area usually defined by a weak, arcuate sulcus; marginal cell rounded apically, submarginal cell II higher than wide, its basal veinlet longer than anterior veinlet (fig. 18 A), outer veinlet of submarginal cell III ending near middle of marginal cell; first recurrent vein received by submarginal cell II, second recurrent vein interstitial between II and III or rarely received by II or III; third anal vein of hindwing well separated from wing margin, long (fig. 18 A); midcoxae nearly contiguous, separated by less than petiole width; tibiae spinose; female foreleg with rake composed of long, bladelike spines; tarsi with very small, inconspicuous plantulae (fig. 19 B); apicoventral, terminal, bladelike setae of hindtarsomere V moderately broad, separated by three or more setal widths; claw tooth basal (fig. 19 C); petiole slightly longer than hindcoxa, about equal to length of hindbasitarsus; male sterna IV-VI with velvety pile; male sternum VIII triangular, apex narrowly rounded; head of digitus angular, penis valve head with teeth on ventral margin of basal lobe.

Geographic range: This monotypic genus is confined to southern South America.

Systematics: Except for appressed silver hair on the face and a blue-black gaster, *Stangeella cyaniventris* is a totally black wasp. The infumate wings usually have a greenish or bluish luster. This wasp ranges in length from 14 to 25 mm.

The basal claw tooth, clypeal shape, wide separation of the antennal sockets and frontoclypeal suture, and open mandibular sockets are the most diagnostic features of *Stangeella;* although the foretarsal rake, wing venation, and spiracular groove are also distinctive. The genus is unlike all other sceliphronin genera, but there is a vague resemblance to *Sceliphron* in the face and in some thoracic details such as the presence of a dorsal enclosure. The open mandibular socket is a primitive feature, and in view of the restricted range of this monotypic genus it is probably safe to assume that *Stangeella* is a relic.

Willink (1951) gave the most detailed description of *cyaniventris* (as a member of the genus *Chlorion*), but the papers of Menke (1962a) and Bohart and Menke (1963) contain figures and other data.

Biology: Janvier (1926, 1928) observed *Stangeella* in Chile. The colonial nests which are dug before prey search, consist of from 4 to 6 cells and are 10 to 15 cm deep. The burrow is closed by the wasp before she begins her search for prey. A large pebble is used to stopper the opening, and then fine particles are scraped over the opening to conceal it. Prey consist primarily of Mantidae, especially *Coptopteryx crenaticollis* (Blanchard) and occasionally *Coptopteryx gayi* (Blanchard), but Janvier noted a single instance where a *Stangeella* female brought in a walking stick of the genus *Bacteria.* Prey are flown to the nest site, the wasp grasping the mantid's prothorax with her mandibles. Paralysis is nearly complete. The egg is laid at the base of one of the forelegs of the first mantid provisioned. According to Janvier five to eight nymphal and adult mantids are placed in a cell. The cocoon exterior was described as white by Janvier.

Checklist of Stangeella

cyaniventris (Guérin-Méneville), 1831 (pl. 8) (*Pelopoeus*); Chile, Argentina, Uruguay, s. Brazil

---

[8] One *Podium* has a spiracular groove, but the propodeal dorsum has only a simple, shallow, median groove.

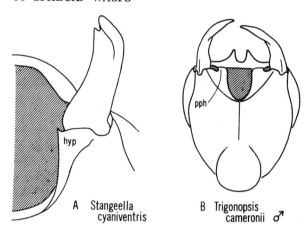

FIG. 20. Head details in the tribe Sceliphronini, ventral; A shows open mandibular socket; B shows closed mandibular sockets; hyp = hypostoma, pph = paramandibular process of hypostoma.

Subtribe Sceliphronina

Diagnosis: Hypostoma with paramandibular process that extends to back side of clypeus closing mandible socket (fig. 20 B).

Genus Chlorion Latreille

Generic description: Lower half of inner orbits straight, orbits parallel or converging above in female (fig. 17C), parallel or converging below in male; vertex usually elevated much above eyes especially in female; postocular area of head short (*boharti*) to long (females of *regale*, *hemiprasinum*); face often so broad in female that head is plainly broader than high; frons essentially glabrous to rather densely setose; frons in males often with a median, elevated, platelike area at upper margin of antennal sockets; male flagellomeres I or II through VII to X with special placoidlike areas; sockets contiguous with frontoclypeal suture in female, separated from suture by less than a socket diameter in male; face often sunken at tentorial pits especially in female; clypeus short in female, two to seven times as wide as high, disk moderately convex, somewhat ridge-like, free margin projecting, straight, or concave mesally and typically with five large teeth (exceptions: *splendidum*, *maxillosum*) lateral to and between which may be a number of much smaller teeth (fig. 17C); male clypeus about twice as wide as high, disk moderately to strongly convex, free margin with a weakly to strongly projecting median lobe which typically bears three teeth or lobes, center or lateral teeth or both sometimes absent or weak (*mirandum*, *maxillosum*, *viridicoeruleum*) margin simply arcuate and sinuate mesally in *lobatum*; free margin of labrum straight or concave; female mandible long and slender, apex acuminate, inner margin with a large mesal tooth (subapical in *maxillosum*) and often a second, smaller tooth; male mandible of normal length, inner margin with single, large, mesal, or subapical tooth; mouthparts short except in *maxillosum* which has elongate galea, third maxillary palpal segment flattened and expanded on one side (except in *lobatum*, *maxillosum*, and several Neotropical species); hypostomal carina ending at or near mandibular socket except in females of *cyaneum* and *boharti* where it is evanescent about half distance to socket; occipital carina complete, contiguous with apex of hypostomal carina or meeting latter before its apex; collar high, short to moderately long, dorsomedially indented except in some New World forms; notauli occasionally present but short; scutellum flat or convex, propodeum long, with U-shaped dorsal enclosure which is usually finely to coarsely, transversely ridged; propodeal side with spiracular groove (weak in *boharti*); episternal sulcus long, reaching anteroventral margin of pleuron, scrobal sulcus usually present but often weakly impressed; metapleural sulcus at most weakly indicated between upper and lower metapleural pits so that lower metapleural area not clearly defined; marginal cell apex narrowly rounded to truncate, submarginal cell II narrowed towards anterior margin of wing, its basal veinlet longer than anterior veinlet, outer veinlet of submarginal cell III ending at or slightly beyond middle of marginal cell; first recurrent vein received by submarginal II, second recurrent received by submarginal III (fig. 18D); third anal vein of hindwing well separated from wing margin (fig. 18D); midcoxae separated by a distance about equal to petiole width; hindfemur fusiform, thickest near base, apex simple; tibiae variably spinose; female foretarsus with rake composed of long spines which are sometimes bladelike; apicoventral terminal, bladelike setae of hindtarsomere V slender to moderately broad, separated by two or more times a setal width; claw tooth mesal or subbasal (fig. 19D); petiole usually about as long as hindcoxa, but up to two times as long as coxa in some males, petiole usually much less than length of hindbasitarsus but nearly equal in some males; sterna without special areas or vestiture except apices of VII-VIII in male often tufted; sternum VIII triangular, apex rounded or narrowly truncate; head of digitus rounded, ventral margin of penis valve head with a row of teeth.

Geographic range: The 18 species of *Chlorion* are scattered over the globe, but the genus is not represented in Australia or Europe. This pattern suggests an old, declining taxon. *Chlorion* is wide ranging in the New World, being found from Canada to Argentina. In the Old World most species occur from northeastern Africa to southwest Asia. The Ethiopian Region is penetrated deeply by only one species, *maxillosum*, which occurs over most of Africa. The only islands inhabited by *Chlorion* are Sri Lanka, Sumatra, and Java.

Systematics: *Chlorion* range in length from 16 to 37 mm. A third of the species are wholly metallic green or blue at least in one sex, and these are among the most handsome sphecids. The thorax is metallic green or blue in all New World species, but the legs and gaster of some South American forms are partly or all red. Except for the metallic green or blue *lobatum*, all Old World species are black or black and red insects, although the gaster often has a metallic tinge. *C. splendidum* has a red head, thorax, and legs. The wings in *Chlorion* vary from clear to yellow to darkly violaceous. The sexes are sometimes differently colored as in *viridicoeruleum,* and some species such as *hemipyrrhum* and *maxillosum* are polychromatic.

Chlorion is easily distinguished from other sceliphronins, except *Stangeella* and a few *Trigonopsis,* by the fact that submarginal cell II receives only the first recurrent vein. Subsidiary characters are the well developed female foretarsal rake and the spiracular groove. *Stangeella* shares these features, but recurrent vein II is either interstitial between submarginals II and III, or it is received by III. Unlike *Stangeella,* the antennal sockets touch the frontoclypeal suture in female *Chlorion,* and *Chlorion* males lack the velvety pubescence on sterna IV-VI of *Stangeella* males. The two genera are really not similar. For example, the head and thoracic structure is quite different, and the open mandibular socket of *Stangeella* indicates that there is no close relationship. *Trigonopsis* is not likely to be confused with *Chlorion,* but the presence of an intercoxal carina in the former is distinguishing.

Chlorion is a relatively homogeneous taxon, the most discordant element being *maxillosum* which has unusually long mouthparts and female mandibles, as well as rather short, fine, and uniform pectination of the inner hindtibial spur. Also, the long marginal setae of the clypeus are spread across the middle section of the clypeus in the female rather than divided into two distinct groups as in other species. Beaumont (1962) placed *maxillosum* in a species group by itself. This species is also notable for its color forms which have contributed to the many names in synonymy. Over a half dozen subspecies have been recognized (keyed in Arnold, 1928, and Berland, 1956) at various times in the past, but as Leclercq (1955c) has pointed out, many of them are not geographically isolated. We have retained only two subspecific forms here, our separation being based primarily on the color of the wings (yellow vs. black), since all other criteria seem unstable. Valdeyron-Fabre (1955) has found both of these wing color types in Tunisia. This suggests that either the two forms represent two species, or that these color differences have no real subspecific significance.

Most species of the genus can now be identified through the use of the modern keys of Beaumont (1962) for the western Palearctic forms, Menke (1961) for the Nearctic forms, and Menke and Willink (1964) for the Neotropical forms. Kohl (1890) can be used for identification of the Oriental and east Asian species.

Biology: Data is available for only three species of *Chlorion: maxillosum* (Valdeyron-Fabre, 1955, as *xanthocerus*), *lobatum* (Lefroy, 1904; Beadnell, 1906; Hingston, 1925-1926; Iwata, 1964a), and *aerarium* (Peckham and Peckham, 1898, 1900, as *coeruleum;* Krombein, 1953, 1958c, 1959b). Gryllidae are the prey of all three wasps, although Hungerford and Williams (1912) observed an *aerarium* female carrying a *Ceuthophilus* species (Gryllacrididae). An interesting transition from parasitoid-like to predatory behavior is displayed by these three species.

Chlorion maxillosum digs down to the burrow of its host, *Brachytrupes megacephalus* (Lefebvre), stings the gryllid either in the latter's burrow or on the soil surface should the cricket leave its burrow in an effort to escape. The wasp lays an egg near the first abdominal spiracle, then flies off. Paralysis is temporary, and the cricket revives in a few minutes. If outside its burrow, the cricket digs another. The wasp larva eventually kills its host and pupates in the cricket's burrow.

Chlorion lobatum behaves similarly to *maxillosum* up to the point of egg deposition. After chasing its host, *Brachytrupes achatinus* (Houttyn), from the burrow, the wasp stings the cricket, and then drags it back into its burrow holding the gryllid's antennae with her mandibles. The wasp then lays an egg behind one of the forecoxae of the host and closes the nest with soil. As in *maxillosum,* paralysis is temporary. Iwata noted that occasionally a *lobatum* female oviposits while its prey is still on the soil surface and then flies off leaving the cricket to revive and dig back into the soil.

Reports on *Chlorion aerarium* are rather fragmentary but nevertheless sufficient to indicate that this species has a more advanced type of nesting procedure. Nests are apparently excavated prior to searching for prey. *Gryllus rubens* Scudder (= *Acheta assimilis* in Krombein), *G. pennsylvanicus* (Burmeister), and *Anurogryllus muticus* (De Geer) seem to be common hosts. They are flown or carried on the ground to the nest. Up to seven crickets are placed in a cell, and temporary closures are made between forages for prey. The accounts of the Peckhams suggest that multicellular nests are constructed. The egg is laid "on the right side of the body" according to the Peckhams who also indicate that prey paralysis is weak.

Checklist of *Chlorion*

aerarium Patton, 1879; U.S.: transcontinental; n. Mexico
   nearcticum Kohl, 1890 (*Sphex*)
boharti Menke, 1961; Mexico: Baja California
consanguineum (Kohl), 1898 (*Sphex*); ? Ethiopia
cyaneum Dahlbom, 1843; U.S.: Arizona, Texas; n. Mexico
   occultum Kohl, 1890 (*Sphex*)
funereum Gribodo, 1879; Somalia to Chad, Algeria and Egypt; Aden; Iraq
   eximium Kohl, 1885 (*Sphex*), nec Lepeletier, 1845
   kohli Ed. André, 1888 (*Sphex*)
hemiprasinum (Sichel), 1863 (*Sphex*); Argentina, Uruguay
   bicolor Saussure, 1869
   pretiosum Taschenberg, 1869
hemipyrrhum (Sichel), 1863 (*Sphex*); Brazil, Argentina
   metallicum Taschenberg, 1869
   pallidipenne Taschenberg, 1869
   nobilitatum Taschenberg, 1869
   principale Strand, 1910 (*Sphex*)
   eximium Strand, 1910 (*Sphex*), nec Lepeletier, 1845

lepidum Strand, 1910 (*Sphex*)
hirtum (Kohl), 1885 (*Sphex*); Ethiopia to Egypt; Arabia; Israel
lobatum (Fabricius), 1775 (*Sphex*); Oriental Region
    viridis Barbut, 1781 (*Sphex*)
    ? *ferum* Drury, 1782 (*Sphex*)
    chrysis Christ, 1791 (*Sphex*)
    ? *smaragdinum* Christ, 1791 (*Sphex*)
    azureum Lepeletier & Serville, 1828, lectotype ♀, "Patrie inconnue" (Mus. Turin), present designation by Menke
    ? *ferox* Westwood, 1837
    azuzium Blanchard, 1840, lapsus
    ssp. *rugosum* F. Smith, 1856; Sumatra, Java
magnificum F. Morawitz, 1887; sw. USSR: Turkmen S.S.R.; Iran; Afghanistan; Pakistan
maxillosum (Poiret), 1787 (April) (*Sphex*); Palearctic Africa
    apicale Guérin-Méneville, 1849 (*Pronaeus*)
    rufipes Guérin-Méneville, 1849 (*Pronaeus*)
    nigripes Guérin-Méneville, 1849 (*Pronaeus*)
    ssp. *ciliatum* (Fabricius), 1787[9] (*Sphex*); Ethiopian Region, Tunisia
    xanthoceros Illiger, 1801 (*Sphex*)
    mandibulare Fabricius, 1804
    maxillare Palisot de Beauvois, 1805 (*Pepsis*)
    varipenne Reiche and Fairmaire, 1850
    instabile F. Smith, 1856 (*Pronaeus*)
    affine F. Smith, 1856 (*Pronaeus*)
    fulvipes Gerstaecker, 1857
    subcyaneum Gerstaecker, 1857
    zonatum Saussure, 1867
    unicolor Saussure, 1867
    ? *cyanescens* Radoszkowski, 1881 (*Sphex*)
    columbianum Gribodo, 1883, lectotype ♀, "Caracas, Columb." (Mus. Genoa), present designation by Menke
    massaicum Cameron, 1908 (*Sphex*)
    levilabris Cameron, 1910 (*Sphex*)
    kigonseranum Strand, 1916 (*Sphex*)
migiurtinicum (Giordani Soika), 1941 (*Sphex*); Somalia, Ethiopia
mirandum (Kohl), 1890 (*Sphex*); Colombia, Peru
regale F. Smith, 1873; sw. USSR: Turkmen S.S.R.; Iran, Afghanistan, Pakistan
    superbum Radoszkowski, 1887
semenowi F. Morawitz, 1890; sw. USSR, Iran
    ssp. *occidentale* (Beaumont), 1962 (*Sphex*); Egypt, Saudi Arabia, Israel
    bicolor Walker, 1871, nec Saussure, 1869
splendidum Fabricius, 1804; India
    campbellii W. Saunders, 1841 (*Pronaeus*)
    pulchrum Lepeletier, 1845 (*Sphex*)
    melanosoma F. Smith, 1856
    forficula Saussure, 1891
strandi Willink, 1951; s. S. America

---

[9] The preface to Fabricius' work was dated February 1787, and therefore his book must have appeared after Poiret's paper.

tibiale Strand, 1910 (*Sphex*), nec Fabricius, 1781
viridicoeruleum Lepeletier & Serville, 1828; Panama to Argentina

Penepodium Menke, new genus

Generic description: inner orbits arcuately bowed outward, slightly converging above in *goryanum* group and females of *luteipenne* group, slightly converging below in *foeniforme* group and males of *luteipenne* group; vertex much elevated above eyes; postocular area elongate, genae comprising up to one-half head length; clypeal disk and frons with long, erect, usually pale setae, similar vestiture often present on vertex; frons usually elevated at upper level of sockets, often dropping abruptly between sockets to clypeal margin; scape shorter than flagellomere I, which is longer than II except about equal in males of *luteipenne* group; middle flagellomeres of male offset from one another and usually bearing elongate, triangular placoids (fig. 22 H), except flagellum filiform and without placoids in *foeniforme* group; sockets contiguous with or narrowly separated from frontoclypeal suture; face usually not sunken at tentorial pits in male, but moderately depressed in female; clypeus short, two to nearly three times as wide as high, disk weakly to moderately convex in female, moderately to strongly convex in male, free margin with two sharp teeth separated by a semicircular or U-shaped emargination (fig. 22D), except females of *goryanum* group that have four teeth in two groups (fig. 22E); mandible simple, arcuate, apex usually acuminate; mouthparts short to moderately long, last two or three palpal segments usually short, third maxillary palpal segment similar to I and II, symmetrical, maxillary V sometimes broader than long; hypostomal carina ending near mandible socket except ending about half distance to socket in most females of *luteipenne* and *goryanum* groups; occipital carina ending just short of juncture with apex of hypostomal carina; pronotal length equal to or shorter than width, collar about as long as broad in *foeniforme* group, length equal to half to two-thirds width in other groups, surface sloping up to a polished, bare, posteromedian prominence or tubercle except merely convex in female of *complanatum*; scutellum flat; propodeum rather long, without dorsal enclosure but usually with a broad, shallow, median longitudinal furrow which may contain a weak carina, posterior face punctate or rugosopunctate; spiracular groove present in some *foeniforme* group species, but absent or weak in other groups; episternal sulcus ending a little beyond scrobal sulcus or extending nearly to anteroventral margin of pleuron; arcuate intercoxal sulcus and ridge present (weak in some males of *foeniforme* group); marginal cell rounded or narrowly truncate apically, stigma usually short, less than half length of marginal cell, shape of submarginal cell II variable, length of posterior veinlet of II at most two-thirds length of posterior veinlet of III, outer veinlet of III not parallel with basal veinlet of cell and ending at outer one-fourth of marginal cell; both recurrent veins usually received by submarginal cell II but occasionally first recurrent interstitial or received by first submarginal cell, second recurrent usually ending at middle of submarginal cell II (fig. 18F); third anal vein of hindwing separated from wing margin, long, extending at least one-third the distance to jugal excision (fig. 18F); midcoxae separated by one to four petiole widths, but usually about twice the petiole width, hindcoxa normal; hindfemur fusiform, thickest near base, apex simple; tibiae nearly spineless or with scattered stout spines near closing face; female foreleg without a rake, setae at apex of tarsomeres slender, sharp; apicoventral, terminal, bladelike setae of hindtarsomere V slender, nearly setiform, separated by about 10 setal widths; petiole 1.2-2.2 times length of hindcoxa, and about as long as hindbasitarsus; base of tergum I with an oval mat of short hairs except in most *foeniforme* group species and weak in some males of other groups; sterna III-IV in female with velvety micropubescence except in *foeniforme* group, sterna III-VI in male usually with similar vestiture; male sternum VII emarginate apically, VIII triangular, apex broadly to narrowly truncate or rounded; head of digitus angular, penis valve head with teeth along ventral margin and often with a basoventral lobe.

Type of genus: Pepsis luteipennis Fabricius, 1804.

Etymology: Pene (L. = almost) + Podium. Gender neuter.

Geographic range: Central Mexico to Argentina. Twenty-two species are currently recognized. An unsuccessful attempt was made to establish *P. haematogastrum* in Hawaii (Williams, 1928a).

Systematics: These are moderate to large wasps (16-39 mm). In a few species the abdomen is partly or all red, but the body except for the appendages is black in the vast majority. The legs are usually black but rarely may be partly or totally red. Wings may be clear with spots at the second submarginal cell and at the apex, or they may be amber or infuscate. In at least two species the propodeum is covered with dense, appressed, gold hair, but generally the body vestiture consists mainly of erect hair.

The essential difference between *Penepodium* and *Podium* is found in the hindwing. The third anal vein in *Penepodium* is long and well separated from the posterior margin of the wing (fig. 18 F). In *Podium* the third anal vein is short, often indistinct, and essentially contiguous with the wing margin (fig. 18 E). The collar is peaked in all but one *Penepodium*, whereas in *Podium* it is longitudinally impressed or at least flattened on the disk. Only one *Podium, agile,* is known with a posteromedian prominence, but other generic characters separate it from *Penepodium*. An intercoxal sulcus is present in all *Penepodium* but rarely is evident in *Podium*. The second recurrent vein usually ends at the middle of submarginal cell II in *Penepodium,* but in *Podium* it usually ends beyond the middle. In *Penepodium* the third maxillary palpal segment is similar to the first two, but in *Podium* the third is usually expanded on one side. The two or four toothed clypeus of female *Penepodium* contrasts with the five or more large teeth and many smaller teeth of the female clypeus in *Podium*. The offset flagellomeres found in most male *Penepodium* is a condition not found in *Podium*.

Differences between *Penepodium* and *Dynatus* are discussed under the systematics of the latter genus. Although *Trigonopsis* does not really resemble *Penepodium* because of its more elongate, slender form, especially the long head and prothorax, the basic differences should be mentioned. The long second anal vein of the hindwing (reaching at least one-third the distance to the jugal excision and usually halfway) of *Penepodium* contrasts sharply with the tapering, short (at best one-fourth the distance to the excision) vein of *Trigonopsis*. In *Penepodium* the occipital carina ends close to the hypostomal carina, but there is a broad gap between the two in *Trigonopsis*. Other differences between the two genera are found in the female mandible, female clypeus, stigma, venation, and with few exceptions in the male antenna.

Kohl's (1902) paper on *Podium s. l.* still provides the only keys for the identification of *Penepodium* species, and additional forms have been described since his work appeared. Some are known from one sex and a few of the "species" currently recognized may prove to be only geographic color forms. The species of *Penepodium,* of which we have seen examples of nearly all, conveniently fall into three species groups based on details of the collar length, female clypeus, and male antenna. The *luteipenne* (code letter L in accompanying checklist; = *flavipenne* group of Kohl) and the *goryanum* (code letter G) groups are closely allied, differing only in female clypeal details and other head features. The *foeniforme* group (F) is a small, discordant element in the genus. Its species share a tendency towards slenderness and dorsoventral flattening of the body. *P. complanatum,* of which we have seen the unique female type (Mus. Oxford), is extremely flattened, and as a consequence, the coxae, especially the middle pair, are widely separated. The flattening in this species is highly reminiscent of the strongly flattened Australian *Tachysphex*. The collar of *complanatum* is merely broadly convex, and there is no trace of a posteromedian prominence.

Biology: Williams (1928a) provided fairly detailed accounts of two species, and Richards (1937) added some fragmentary notes. According to Williams, *luteipenne* (as *flavipenne*) and *haematogastrum* nest in well drained, bare ground or, as in one *haematogastrum* site, a termite mound. Several wasps may nest in the same area. Excavation of the short burrow (up to 5 cm) is accomplished by biting out small lumps of soil with the mandibles. This is aided by moistening the soil with water obtained from some nearby source. The nest is left open until fully provisioned. Cockroaches of the genus *Epilampra* are the prey of *luteipenne* and *haematogastrum*. They are carried in flight to the nest dorsum up, the antennae of the roach grasped by the wasp's mandibles, and the body braced by the legs of the wasp. One to five roaches are provisioned per nest, and the egg is laid behind the forecoxa of the last roach collected. Interestingly, oviposition occurs on the soil surface near the burrow before the last roach is placed in the nest. Paralysis is evidently temporary or very weak, as Williams found that the prey often escaped from the nests that he opened. Nest closure is accomplished by moistening soil near the burrow and then biting out clumps with the mandibles which are applied to the burrow entrance. The cocoon is described as delicate, brown, and paperlike, similar to that of *Sceliphron*.

Both Williams and Richards recorded *goryanum* as gathering mud, and the latter gave a record of a female placing a roach in an abandoned *Passalus* hole in a stump. Richards also recorded *mocsaryi* as gathering mud.

The evidence is very fragmentary at this point, but what is available suggests that the species of the *luteipenne* group excavate nests in the ground and that species of the *goryanum* group use pre-existing cavities for nest sites.

Checklist of *Penepodium*

albovillosum (Cameron), 1888 (*Podium*); Panama, Brazil (G), lectotype ♂, Bugaba, Panama (Mus. London), present designation by Menke.
braziliense (Schrottky), 1903 (*Sceliphron*); Brazil (?L)
*complanatum (F. Smith), 1856 (*Podium*); "South America" (F)
distinguendum (Kohl), 1902 (*Podium*); Brazil (G)
egregium (Saussure), 1867 (*Podium*); Uruguay (L)
fallax (Kohl), 1902 (*Podium*); Brazil (L)
foeniforme (Perty), 1833 (*Ammophilus*); Mexico to Paraguay (F)
    nitidum Spinola, 1851 (*Podium*), lectotype ♂, Pará, Brazil (Mus. Turin), present designation by Menke
    bellum Cameron, 1888 (*Podium*)
fumipenne (Taschenberg), 1869 (*Podium*); Brazil, Paraguay, Argentina (L), lectotype ♂, "Parana" (Mus. Halle), present designation by Menke (? = *egregium*)
goryanum (Lepeletier), 1845 (*Podium*); Centr. and S. America (G)
*haematogastrum (Spinola), 1851 (*Podium*); Columbia, Brazil (L)
    *consanguineum (F. Smith), 1856 (*Podium*), new synonymy by Menke
hortivagans (Strand), 1910 (*Sceliphron*); Paraguay (L)
junonium (Schrottky), 1903 (*Sceliphron*); Brazil (L) (? = *haematogastrum*)
latro (Kohl), 1902 (*Podium*); Brazil (G)
luteipenne (Fabricius), 1804 (*Pepsis*); Panama to Brazil (L)
    flavipenne Latreille, 1809 (*Podium*)
    latreillei Spinola, 1851 (**Podium**)
    dubium Taschenberg, 1869 (*Podium*), lectotype ♀, "Nov. Frib." (Mus. Halle), present designation by Menke, new synonymy by Menke
mocsaryi (Kohl), 1902 (*Podium*); "Sudamerika" (G)
pauloense (Schrottky), 1903 (*Sceliphron*); Brazil (L)
princeps (Kohl), 1902 (*Podium*); Guyana, Surinam (L)
romandinum (Saussure), 1867 (*Podium*); Brazil (L)
spretum (Kohl), 1902 (*Podium*); Brazil (L)
taschenbergi (Kohl), 1902 (*Podium*); Brazil (L)
triste (Kohl), 1902 (*Podium*); Brazil (L)
viduatum (Kohl), 1902 (*Podium*); Brazil (L)

Genus Dynatus Lepeletier

Generic description: Inner orbits arcuately bowed out, converging above except parallel in some males (fig. 17 A); vertex elevated much above eyes; postocular area elongate, especially in males where genae usually are more than half head length; lower frons densely covered with long, semierect, upward directed and usually dark bristles, which contrast with essential absence of similar hair on clypeal disk and vertex; scape shorter than flagellomere I in male, more slender in female where it is equal to or longer than flagellomere I; flagellomere I longer than II; flagellomeres III-VI of male usually with very small, oval, subapical placoids; sockets touching frontoclypeal suture; face sunken at tentorial pits, the depression extending dorsad between socket and eye margin; clypeus short in female, about three times as wide as high, disk flat, free margin with a median lobe bearing four teeth, middle ones closer to each other than to lateral teeth; male clypeus about twice as wide as high, weakly to strongly projecting, often nasiform or hoodlike (fig. 21 E-J), free margin basically with four angles or teeth (fig. 21 F-J); labrum broader than long in female, as broad as long in male; female mandible scoop-like, apex broadly rounded, inner margin without teeth; male mandible acuminate apically, inner margin with a median tooth-like sinuation in *burmeisteri* (fig. 21 J), simpler in other species (fig. 21 F); mouthparts short to moderately long, last two or three palpal segments very short, sometimes broader than long, maxillary palpal segment III similar to I and II; hypostomal carina ending near mandible socket in male, disappearing about halfway to socket in female; occipital carina incomplete below; collar long, length equal to three-fourths width in male, somewhat shorter in female, surface sloping up to a polished, bare, posteromedian prominence; scutellum flat; propodeum very long, dorsum without enclosure but with broad, median, longitudinal furrow, posterior face coarsely, transversely ridged, impunctate; spiracular groove weakly indicated as a broad, shallow furrow traversed by ridges; episternal sulcus sinuate, long, but ending short of anteroventral margin of pleuron, scrobal sulcus present; arcuate intercoxal sulcus and ridge present; marginal cell very long, apex roundly truncate, stigma small, submarginal cell II broader than high, its basal veinlet equal to or shorter than anterior veinlet, posterior veinlet of submarginal II equal to or greater than length of posterior veinlet of submarginal III, outer veinlet of submarginal III sometimes parallel with basal veinlet of cell, ending at outer fifth of marginal cell; both recurrent veins received by submarginal cell II, the second vein ending at outer fourth of cell (fig. 18 C); third anal vein of hindwing separated from wing margin; midcoxae separated by 1.0-2.0 petiole widths, hindcoxa elongate, slightly more than twice width in female, about three times width in male and essentially cylindrical; hindfemur in female uniformly thick for most of length, apex with ventral lobe on both sides of socket

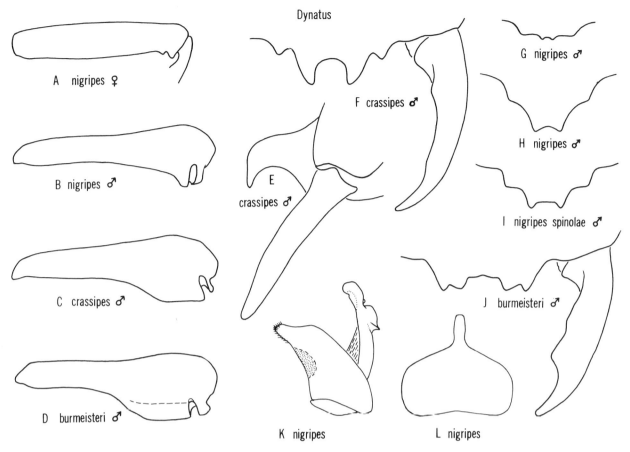

FIG. 21. Structural features of *Dynatus*; A-D, left hindfemur, lateral; E, lower part of head, lateral; F, outline of clypeus and mandible; G-I, outline of clypeus; J, outline of clypeus and mandible; K, male genitalia, lateral; L, male sternum VIII.

(fig. 21 A), male hindfemur club-like, apex greatly enlarged (fig. 21 B-D); tibiae with scattered spines, those near closing face of hindtibia set in large, circular depressions; female foreleg without a rake, setae at apex of tarsomeres stout, flattened, blunt; apicoventral, terminal, bladelike setae of hindtarsomere V moderately broad, separated by two to four setal widths; claw tooth mesal or subbasal (claw simple on foreleg and sometimes midleg of *burmeisteri*); petiole 1.2-1.5 times as long as hindcoxa in female, about equal to coxa in male, petiole slightly shorter than hindbasitarsus; base of tergum I with a shallow, oval depression which is covered with a dense mat of hair; sterna III or IV-VI in male, and II-IV or V in female covered with velvety micropubescence; male sternum VII emarginate apically, apex of VIII with a long, narrow, parallel-sided process (fig. 21 L); genitalia as in fig. 21 K, head of digitus angular, penis valve head complex, without ventral teeth.

Geographic range: This Neotropical genus contains three species. *Dynatus* ranges from central Mexico to Argentina.

Systematics: These black wasps are among the largest sphecids known, their length ranging from 31 to 52 mm. Usually the wings are blackish, but in *nigripes* they may be lightly infumate or amber. Pubescence is dark except for silvering of the clypeus, lower frons, upper pleuron, and sometimes metanotum and scutum laterally.

The bizarre, clublike hindfemora of the male, the usually nasiform clypeus in this sex, and the nonfusiform hindfemur of the female with its apicoventral lobes are quick sight identification features of *Dynatus*. The tuberculate, nonsulcate pronotum, the coarse ridges on the posterior face of the propodeum, the dimpled hindtibia, and the eighth male sternum are additional distinctions. *Dynatus* is most similar to *Penepodium*, since both genera have the same thoracic details, the form of the collar being especially notable. However, the forewing venation of *Dynatus* differs markedly from that of *Penepodium*. The width of submarginal cell II in *Dynatus* is about equal to or slightly greater than that of submarginal cell III. This is best expressed by measuring the lengths of the posterior veinlets (part of the media) of each cell. In *Penepodium* submarginal cell II is never more than five-eighths of the width of III, and very often it is only a third or less.

This genus is fascinating for several reasons. First, these large and presumably conspicuous wasps seem to be rarely taken as evidenced by their scarcity in collections. Secondly, nothing is known of their nesting habits, nor is there an explanation for the peculiar leg and clypeal developments in the male. Thirdly, the specific differences are few and confined almost entirely to the male sex, and strangely the male genitalia seem identical in the three apparent species. Another interesting aspect of *Dynatus* is the allometric development of the male clypeus and hindfemur. In the smallest male (31 mm) of *nigripes* that we have seen the clypeus is scarcely nasiform (fig. 21 G), nor is the hindfemoral club strongly developed. In contrast the largest males of this species (45 mm) have a strongly produced clypeus (fig. 21 H) and a very large hindfemoral club (fig. 21 B). Allometric growth is known elsewhere in the Sphecidae (Beaumont, 1943), but it is quite graphic in *Dynatus*.

Based on the material we have seen there seem to be three species, and possibly the subspecies *spinolae* represents a fourth. The male clypeus in this form is slightly different from typical *nigripes*.

Biology: Unknown. Menke collected a number of individuals of both sexes of *nigripes* on flowers of a single shrub adjacent to a gallery forest in savannah country in Venezuela. The scooplike mandibles of the female plus the absence of a foretarsal rake suggest that *Dynatus* is a mud user, and like the related genera *Penepodium* and *Podium* it may utilize pre-existing cavities for nests or perhaps make free mud nests. Fahringer's (1922) note that *Dynatus* was introduced to Turkey (but presumably never established) also suggests this kind of nesting behavior.

Checklist of *Dynatus*

burmeisteri (Burmeister), 1861 (*Podium*); Argentina
    lindneri Schulthess, 1937 (*Podium*), new synonymy
        by Menke
*crassipes (Cameron), 1897 (*Podium*); Mexico (Colima, Orizaba) to Costa Rica
nigripes (Westwood), 1832 (*Podium*); Peru, Colombia, Venezuela, Trinidad, Guyana, Brazil (Manacapuru)
    giganteus Erichson, 1848 (*Podium*)
**ssp. *spinolae* Lepeletier, 1845; French Guiana, Brazil (Santarém, Pernambuco, Rio de Janeiro)
*ingens F. Smith, 1847 (*Stethorectus*)

Genus Podium Fabricius

Generic description: Inner orbits arcuately bowed out, nearly always converging above, or in some males parallel (converging below in *friesei* female); vertex only slightly elevated above eyes except in *agile* group where head is nearly prognathus; postocular area not elongate, upper part of genae comprising less than half head length; face including vertex usually with long, erect hair which is mostly black in *fumigatum* group, mostly pale and sometimes sparse in *agile* and *rufipes* groups; inner surface of scape with dark bristles in *fumigatum* group, with fine, pale setae in *agile* and *rufipes* groups; flagellomere I longer than II in females and some males, I equal to II in most males, but rarely I shorter than II in this sex; male flagellum filiform, flagellomeres nearly always with flat or curved placoids; sockets touching frontoclypeal suture; face usually sunken at tentorial pits, the depression extending upward between socket and eye margin; clypeus short, two to three or more times as wide as high, disk flat to convex, free margin in female with five or seven large teeth, lateral to which are usually many small teeth (fig. 22 A,C), margin in male with two sharp, median teeth separated by a semicircular or U-shaped emargination (teeth truncate in large males of *tau*), margin occasionally also with a lateral notch; mandible long, slender in female and usually evenly arcuate, apex acuminate, inner margin simple in male but occasionally with a small, rounded tooth at extreme base in female (none in *agile* group); mouthparts short, third maxillary palpal segment slightly to broadly expanded on one side; hypostomal carina ending near mandible socket in male, extending about half distance to socket in female (except ends near socket in *plesiosaurus* female); occipital carina meeting apex of hypostomal carina in some *rufipes* group species but usually incomplete below, ending close to hypostomal carina, occipital, and hypostomal carinae separated by distance equal to or greater than length of oral cavity in *agile* group; collar short in *fumigatum* group, at most half as long as broad, indented medially, the indentation usually continued down nearly vertical front face of collar, collar rather long, disk with median longitudinal impression in *rufipes* and *agile* groups (except flat or with posteromedian indented prominence in some *agile* group females); scutellum flat; propodeum long, without dorsal enclosure but usually with weak, median, longitudinal groove, posterior face usually punctate or rugosopunctate; spiracular groove absent except in *batesianum*; episternal sulcus ending a short distance below level of pronotal lobe; intercoxal sulcus or carina usually absent (present in *agile* group, *fulvipes*, *tau*); marginal cell narrowly truncate to bluntly acuminate apically, stigma short to long, length (as measured on wing margin) up to two-thirds length of marginal cell, submarginal cell II strongly narrowed anteriorly in *agile* and *rufipes* groups, basal veinlet up to three times longer than anterior veinlet (fig. 18 C), submarginal cell II quadrate to moderately narrowed anteriorly in *fumigatum* group (fig. 18 E), length of posterior veinlet of submarginal cell II less than length of posterior veinlet of submarginal cell III (about half its length), outer veinlet of submarginal cell III parallel with basal veinlet of cell in *rufipes* group and ending close to apex of marginal cell in *agile* and *rufipes* groups (fig. 18 G), these veinlets nearly parallel in *agile* group, outer veinlet of submarginal cell III in *fumigatum* group not parallel with basal veinlet of cell and ending at outer third or fourth of marginal cell (fig. 18 E); first recurrent vein received by submarginal cell I or interstitial in *agile* and *rufipes* groups (fig. 18 G) (rarely received by submarginal II), first recurrent vein received by submarginal cell II in *fumigatum* group (fig. 18 E), second recurrent vein usually ending beyond middle of submarginal cell II (rarely received by submarginal III in *fumigatum* group); third anal vein of hindwing contiguous with wing margin, short, tapering (fig. 18 E,G); midcoxae usually separated by slightly less than petiole width, hindfemur fusiform, apex simple; tibiae spineless or with scattered spines or hairlike setae; female foreleg without rake; tarsi with plantulae except in *rufipes* and *agile* groups; apicoventral, terminal, bladelike setae of hindtarsomere V narrow to moderately broad, separated by two or more setal widths; claw tooth subapical in some species of *agile* group; petiole 1.1-1.5 times length of hindcoxa in female, 1.5-2.0 times length of hindcoxa in male, petiole length same as length of hindbasitarsus or longer; base of tergum I usually with a linear or oval patch of mixed short and long hairs which often arise from a shallow depression (absent in *agile* group); sterna IV-V of some females and most males of *fumigatum* group with velvety micropubescence; male sternum VII entire or shallowly and angularly emarginate apically; VIII triangular, apex rounded or truncate (fig. 22 F), sometimes weakly emarginate; head of digitus variable in shape, penis valve head with teeth along ventral margin, head usually with a basoventral lobe.

Geographic range: Two of the 20 species of this primarily Neotropical genus occur in the eastern half of North America, where one, *luctuosum*, ranges as far north as New York. Of the various cockroach-collecting sceliphronin genera, *Podium* appears to be the only one represented in the Greater Antilles. Cuba, Hispaniola, and Jamaica each have one apparently endemic species. Some species of *Podium* have a broad distribution; *rufipes*, for example, extends from the northeastern U.S. to Argentina.

Systematics: These primarily black wasps range in length from 11 to 25 mm. The legs are usually partly red, and the gaster is red in *plesiosaurus, batesianum, agile,* and a few other species. The antennae are black, but the mandibles are often reddish or yellowish. The wings are clear in most species but usually spotted or banded across the middle and at the submarginal cells. Yellow or darkly infuscate wings occur but are uncommon. The head and thorax of most *Podium* are rather densely covered with erect hair, and rather coarsely, closely punctate in contrast to *Trigonopsis*. Also, the integument of *Podium* is often dull or weakly shining.

As conceived by Kohl (1902) *Podium* was a broad genus that included *Dynatus* and *Trigonopsis* as subgenera, as well as the new genus *Penepodium* Menke. Richards (1937) was probably the first to recognize the validity and desirability of treating *Dynatus* and *Trigonopsis* as genera. Removal of the species placed in the new genus *Penepodium* makes *Podium* still more homogeneous. One could extend this splitting further by making subgenera or even genera out of the *fumigatum* group on one hand and the *agile* and *rufipes* groups on the other. Although we have seen nearly all the species, including many types, we feel that it is too early to take either of these courses. The fauna of South America is still little known, and species may yet be discovered that connect these species groups.

As constituted here, *Podium* is readily separated from all other sceliphronin genera by: 1) the short, tapering third anal vein of the hindwing that is contiguous with the wing margin, and 2) the rather short episternal sulcus that ends just below the level of the pronotal lobe. There are other important generic characters, but all are subject to rare exceptions. Among these are: pronotum indented and usually furrowed mesally, absence of a spiracular groove, absence of an intercoxal sulcus, and the small teeth on the female clypeus lateral to the large central ones.

We recognize three species groups in *Podium,* the *rufipes* group or *Podium s.s.* (code letter R in checklist), the *agile* group (A), and the *fumigatum* group (F). These are identical to groups recognized by Kohl except that he used the subgenus *Parapodium* for the *rufipes* group which included our *agile* group. The *fumigatum* group is the largest and presumably most generalized assemblage in the genus. The short collar, plantulae, reception by submarginal cell II of both recurrent veins and nonparallel basal and outer veinlets of submarginal cell III characterize this group. The *rufipes* group has a longer collar, the first recurrent usually either ends on submarginal I or is interstitial, plantulae are absent, and the basal and outer veinlets of submarginal cell III are parallel. The *agile* group has an even longer collar, plantulae are absent, and the occipital carina is broadly separated from the hypostomal carina. The arrangement of the recurrent veins is as in the *rufipes* group, but the veinlets of III are not quite parallel. *Podium plesiosaurus,* of which we have seen the unique female type (Mus. London), is a very slender species, and its very long collar (fig. 22 B) makes it appear at first glance to be a *Trigonopsis*. However, this species possesses all of the generic features of *Podium*. The face of *plesiosaurus* is unusually long and narrow (fig. 22 A).

Starting with the *fumigatum* group and ending with the *agile* group there is a nice evolutionary progression from relatively short, stocky forms to long, slender species. Other specializations such as absence of plantulae, migration of the recurrent veins basad, and wide separation of the hypostomal and occipital carinae coincide with this phylogenetic sequence of species groups. It is only a short step from the *agile* group to *Trigonopsis*, and this genus is closely allied to *Podium* as evidenced by their similar hindwing anal vein, the presence in some females of a small, basal mandibular tooth, the tendency for the stigma to be rather long, and the wide separation of the occipital and hypostomal carinae (in the *agile* group of *Podium*). Differences between the two genera are enumerated under the systematics of *Trigonopsis*.

The only comprehensive key is that of Kohl (1902), but Bohart and Menke (1963) keyed the North American species and Menke (1974a) treated the *agile* group. The genus is still in about the same state that Kohl left it. Some species are known from one sex and geographic color variation has yet to be investigated.

Biology: Podium generally inhabit woods, and females spend much of their time running over the surfaces of dead trees where they explore crevices and holes presumably in search of their cockroach prey or perhaps for nesting sites. Krombein's (1967b, 1970) papers include very detailed observations on *luctuosum* and especially *rufipes,* plus summaries of earlier work. Rau's (1937) paper is also important. *Podium luctuosum* and at least North American examples of *rufipes* apparently nest in any suitable cavity. Krombein has reared both species from tubular stick traps, and Rau found *rufipes* nesting in old nests of *Sceliphron caementarium* which had been superseded by *Osmia*. A Brazilian species identified by Williams (1928a) as *rufipes* nested in a plastered-over hollow in a termite mound, but he did not say that the wasp was the mason. However, Howes (1919) indicated that a Guyanese species he identified as *rufipes* makes tubular mud nests up to 10 cm long that contain as many as four cells. Howes also said that nests built on houses are covered with termite wings, spider skins, and bits of debris. His description of the long tubular nest agrees well with those made by some of the larger *Trypoxylon* subgenus *Trypargilum*, and in view of Krombein's and Rau's observations, one is inclined to either doubt the accuracy of Howes' account or to suspect that he misidentified his wasp.

In North America *rufipes* makes single-celled nests. The nest is left open until fully provisioned, and the closing plug is complex. Rau's wasps first made a mud plug and then covered it with pine resin. In the many nests observed by Krombein the entrance was first packed with a variety of objects (wood pulp, spider silk, cockroach feces, sand), and then resin was applied as a final touch. Neither Williams nor Howes mentioned the use of resin. Krombein examined two nests of *luctuosum*. One in a stick trap contained two cells separated by a partition of rotten wood and plastered mud, the other in a "boring" in a dead tree trunk was unicellular. The two-celled nest, which may not have been finished, was sealed with a plug of rotten wood, a layer of plastered mud, and then more rotten wood. Krombein noted only that the single-celled nest was sealed with mud. Menke observed a *luctuosum* female that made several trips to gather mud from a marshy area in woods before she was captured.

Prey used by *rufipes* are *Parcoblatta pensylvanica* (De Geer), *Chorisoneura texensis* Saussure and Zehntner, *Cariblatta lutea* (S. and Z.), *C. minima* Hebard, *Latiblatella rehni* Hebard, and *Eurycotis floridana* (Walker). *Parcoblatta uhleriana* (Saussure) and *P. virginica* (Brunner) are provisioned by *luctuosum*. Prey are flown to the nest and apparently held by the mandibles. The egg is laid behind the forecoxa of the first provision as in other roach-collecting sceliphronin genera, although Howes claimed that it was laid on the last provision. One to 14 roaches may be in a cell, but Krombein found the average to be six in the many *rufipes* nests he studied. The two-celled nest of *luctuosum* contained five and six prey respectively. Paralysis is essentially complete. The cocoon of *rufipes* was described by Krombein as being similar in shape and texture to that of *Sceliphron caementarium*.

The chalcid parasitoid *Melittobia chalybii* Ashmead is known to attack both *rufipes* and *luctuosum*. The chrysidid *Neochrysis panamensis* (Cameron) and the bombyliid *Lepidophora appendiculata* (Macquart) are cleptoparasites of *rufipes*.

Checklist of *Podium*

agile Kohl, 1902; French Guiana, Brazil (A)
*angustifrons Kohl, 1902; Panama, Guyana, French Guiana, Brazil, Argentina (F)
*batesianum Schulz, 1903; Brazil (R)
chalybaeum Kohl, 1902; Mexico (F)
denticulatum F. Smith, 1856; Mexico to Brazil (F), lectotype ♀, "Braz." (Mus. London), present designation by Menke, new status by Menke
    brevicolle Kohl, 1902, lectotype ♀, Bahia, Brazil (Mus. Paris); present designation by Menke, new synonymy by Menke
    *longipilosellum Cameron, 1912, new synonymy by Menke
foxii Kohl, 1902; Brazil (F)
*friesei Kohl, 1902; s. Mexico, Honduras, Ecuador, Paraguay (A)
fulvipes Cresson, 1865; Cuba (F)
fumigatum (Perty), 1833 (*Ammophilus*); Brazil (F)
  ssp. *bugabense* Cameron, 1888; Costa Rica, Panama, Guyana
  ssp. *aureosericeum* Kohl, 1902; Surinam
intermissum Kohl, 1902; Brazil (F)
iridescens Kohl, 1902; Mexico (F), lectotype ♀, Oaxaca, Mexico (Mus. Berlin); present designation by Menke
kohlii Zavattari, 1908; Peru (R) (? = *batesianum*)
krombeini Bohart and Menke, 1963; U.S.: Texas, California; Mexico (R)
luctuosum F. Smith, 1856; U.S.: Missouri to New York, Florida and Texas (F)
opalinum F. Smith, 1856; Jamaica (F), lectotype ♀, Jamaica (Mus. London), present designation by Menke
*plesiosaurus (F. Smith), 1873 (*Trigonopsis*); Brazil (A)
rufipes Fabricius, 1804; U.S.: Kansas, Illinois to Maryland and Florida; Centr. and S. America (R)
    biguttatum Taschenberg, 1869 (*Parapodium*), lectotype ♂, Venezuela (Mus. Halle), present designation by Menke
    carolina Rohwer, 1902
sexdentatum Taschenberg, 1869; Brazil, Argentina (F)
tau (Palisot de Beauvois), 1811 (*Pepsis*); Hispaniola (F)
trigonopsoides Menke, 1974; Brazil (A)

Genus Trigonopsis Perty

Generic description: inner orbits arcuately bowed out, usually slightly converging below or parallel (converging above in one species); vertex much elevated above eyes if head viewed with face vertical, but with head oriented in normal, semiprognathous condition vertex not raised much above eyes; postocular area elongate, genae about one-half length of head, and usually converging towards occiput so that head is triangular in dorsal view; erect setae of face and vertex pale, sparse, obscure; flagellomere I in female 1.4-1.5 times length of II, in male I is 1.3-1.5 times length of II; flagellum filiform in male and without placoids; sockets touching frontoclypeal suture; face sunken at tentorial pits, the depression extending upward between socket and eye margin; clypeus short, 2.0-3.0 times as wide as high, disk flat to slightly convex, free margin in female with five to seven teeth, outermost usually largest (fig. 17 B), margin in male usually with two sharp teeth or two bidentate lobes separated by a semicircular or U-shaped emargination, but triangularly produced in *violascens,* the apex truncate; free margin of labrum broadly, shallowly concave or straight; female mandible long, slender, straight, or slightly arcuate between base and subapical, inner notch beyond which it angles inward (fig. 17 B), inner margin sharp, with a subapical V-shaped notch (absent in one species) and usually a small rounded lobe at extreme base; male mandible simple or with a rounded, subbasal tooth on inner margin; mouthparts short, third maxillary palpal segment slightly to broadly expanded on one side; hypostomal carina ending near mandible socket in male, extending about half distance to socket or absent in female; occipital carina often lamelliform, incomplete below, horseshoe shaped, the ends separated from apex of hypostoma by length of oral fossa or more (fig. 20 B); pronotum and collar at least as long as broad, collar highly polished, surface convex or sloping upward to a posteromedian prominence or tubercle which may be indented; scutum strongly convex in lateral view, disk often with a deep, broad, median furrow; scutellum flat; propodeum moderately long, without dorsal enclosure but with median, longitudinal furrow which is pitted or crossed by many closely spaced, strong ridges, and usually bisected by a median carina (ridges or carina weak in occasional specimens), posterior face transversely ridged or rugosopunctate; spiracular groove present, crossed by ridges; episternal sulci long, both joining at midventral line near anteroventral margin of pleuron, scrobal sulcus present; arcuate intercoxal sulcus and ridge present; marginal cell narrowly rounded or bluntly acuminate apically, stigma very long, length equal to at least half length of marginal cell (as measured on wing margin), shape of submarginal cell II variable, length of posterior veinlet of submarginal cell II less than length of posterior veinlet of submarginal cell III (but equal to more than half its length), outer veinlet of submarginal cell III not parallel with basal veinlet of cell and ending at outer third or fourth of marginal cell; first recurrent vein received by submarginal cell II (rarely interstitial between I and II), second recurrent vein ending at outer fifth of submarginal II or interstitial between II and III (fig. 18 I), rarely received by submarginal cell III; third anal vein of hindwing narrowly separated from wing margin, thin, tapering, short, not exceeding one-fifth distance to jugal excision (fig. 18 I); midcoxae nearly contiguous, rarely separated by as much as petiole width; hindfemur fusiform, apex simple; tibiae with fine, hairlike setae only; female foreleg without rake; apicoventral, terminal, bladelike setae of hindtarsomere V narrow to moderately broad, separated by 2.5 times a setal width or more; petiole length usually about 3.0 times hindcoxal length, but sometimes less than twice hindcoxal length (*intermedia*), petiole length usually more than length of hindbasitarsus but occasionally slightly less; base of tergum I with oval mat of short hairs (weak in some males); sterna IV-VI in male with velvety micropubescence, male sternum VII angularly emarginate apically, VIII triangular, apex acuminate and deeply, narrowly emarginate (fig. 22 G) or apex broadly rounded; penis valve head with teeth along ventral margin, head with a basoventral lobe.

Geographic range: This Neotropical genus ranges from southern Mexico to Bolivia and southern Brazil. Ten species are currently recognized, but doubtless more of these rarely collected forest dwelling wasps await discovery.

Systematics: Trigonopsis are shiny, elongate (12-28 mm) wasps with long, slender legs and petiole, and usually a pronounced necklike pronotum. The body is black or blue black, but some species have a red gaster. Occasionally the appendages are partly reddish. The mandibles in the female and sometimes in the male are reddish to yellow. The clypeus is usually partly or completely yellow or cream colored. A pale narrow mark is usually found along the inner orbit on the extreme lower frons. The wings are clear or yellowish, and always have two dark bands, one near the middle of the wing and another through the submarginal cells. Appressed silver or gold vestiture is confined mainly to the lower frons and various small areas of the thorax, but especially notable is a broad band along the posterior margin of the mesopleuron. In contrast to *Podium* the head and thorax of *Trigonopsis* are generally sparsely, finely punctate, glossy, and with little erect hair.

Trigonopsis is usually quite easily identified simply by its slender, elongate body, necklike pronotum, and trigonal head, but a few species are proportionately shorter and more compact, and in some of these the head and pronotal characters break down. The strongly convex form of the scutum of *Trigonopsis* gives the thorax a distinctive appearance. When viewed laterally it tapers noticeably towards the petiole socket in most species. In *Podium* and *Penepodium* the scutum is much flatter, and the thorax viewed laterally has a rectangular, boxy appearance. The long forewing stigma is also distinctive, especially for separation from *Penepodium,* but it is also long in a number of *Podium*. The female mandible of *Trigonopsis* with its subapical notch on the inner margin is unique in the Sceliphronini, but one species lacks this notch. *Trigonopsis* has a distinctive median, longitudinal furrow on the propodeal dorsum which, except for occasional variant specimens, is crossed by many strong, transverse ridges that give the groove a pitted appearance. Additionally, this furrow is longitudinally bisected by a carina. This furrow plus the long episternal sulcus immediately separate *Trigonopsis* from *Podium*. Other features of *Trigonopsis* that will separate the genus from all but a few atypical *Podium* are: the broad space between the hypostoma and the ends of the occipital carina, the well developed spiracular groove (although crossed by ridges), the intercoxal sulcus, and the absence on the collar of a median, longitudinal impression. Differences from *Penepodium* are discussed under the latter.

Trigonopsis divides into two species groups on the form of the occipital carina, the sculpture of the propodeal dorsum, and the form of male sternum VIII. The *rufiventris* group (code letter R in checklist) contains the slender forms. The occipital carina is lamelliform, the propodeal dorsum is largely smooth, and sternum VIII has an apical slit (fig. 22 G). The collar in this group is convex except in *violascens,* in which there is a vaguely indented posteromedian prominence. The collar of *resplendens* is intermediate between that of *violascens* and the more typical species. Species of the *intermedia* group (code letter I) vary from long and slender to shorter and stockier. The occipital carina is low, the propodeal dorsum is transversely ridged (at least weakly), and sternum VIII is simply rounded apically. The collar is tuberculate in all species in this group, and it is less necklike. The petiole and stigma are shortest in this assemblage, and the head is not strongly trigonal. The mandible in one species studied lacks the subapical inner notch. The *intermedia* group is the more generalized of the two species groups.

Richards (1937), after a study of some of the types, produced a key to the species, but some of his nomenclature is out-of-date. Kohl's (1902) paper on *Podium s. l.* contains valuable data also. *Trigonopsis* are uncommon in collections, and some forms are still known from only one sex.

Biology: The papers of Williams (1928a), Rau (1933), Arlé (1933), Richards (1937), and the recent ones by Eberhard (1972, 1974) are the only references. Females construct free mudnests similar to those of *Sceliphron* on tree trunks, on the underside of leaves, on the lower surfaces of projecting rocks, and other sheltered situations. Wasps either gather mud (Williams) or use water (Richards, Eberhard) to soften soil for use in building their nests, and often the nests are built near streams. Sometimes water may be obtained from leaf axils or leaf surfaces. *Trigonopsis* nests consist of from one to 129 elongate cells which are cemented together and covered with mud. Blattidae are reported as prey of *cameronii, intermedia, moraballi,* and *rufiventris,* but according to Richards, *grylloctonus* provisions with Gryllidae. Prey are carried in flight dorsum up, head forward. Paralysis varies from partial to complete. The egg is laid on the first provision, and Eberhard says that *cameronii* lay the egg before the roach is placed in the cell. The egg is placed just behind one of the forecoxae of the host. A single cell may contain as many as 14 roaches.

The recent discovery by Eberhard of a primitive type of sociality in *cameronii* is the most interesting aspect of *Trigonopsis* biology. In a prelude to Eberhard's disclosure Richards (1937) reported that one six-celled nest (of which two were open) of *rufiventris* may have been the work of two female wasps. After removal of the nest from its substrate, two wasps were seen at the site, one with a roach. According to Eberhard, *cameronii* females may either build mudnests or renovate existing ones for re-use. After the first roach is placed in the nest, the opening is sealed with mud. One to three days passes before additional prey are placed in the cell. Up to seven days are required to complete the provisioning and a mud cap is placed on the cell at the end of each day. Eberhard found that a single nest may be maintained by one to four females, although each one tended to concentrate its building efforts on its own cell. Occasionally, however, a female would plaster or repair cells made by nestmates, and the wasps would act together to repel ants. Eberhard found that nestmates occasionally stole prey from open cells, but that this behavior was the exception rather than the rule. Also, females being robbed did not interfere. Eberhard stated that the offspring of a nest tended to return to the nest and provision cells there. Males remained on the nest and mated with emerging females, so that the wasps of any given nest tended to be closely related.

Males of *cameronii* are reported by Eberhard to guard nests.

Arlé reared the ichneumonid *Photocryptus apicalis* (Schmiedeknecht) from cells of *rufiventris.* Richards reported that sarcophagids of the genus *Amobia* were obtained from the nest of *violascens,* and that the bombyliid *Anthrax leucopyga* Marquart emerged from one cell of a nest of *moraballi.* Eberhard reared the rhipiphorid, *Macrosiagon lineare* (Le Conte), from a *cameronii* cell.

Checklist of *Trigonopsis*

cameronii (Kohl), 1902 (*Podium*); s. Mexico to Colombia (R), lectotype ♀, David, Chiriqui, Panama, Champion (Mus. Vienna), present designation by Menke

cyclocephalus F. Smith, 1873; Brazil, Bolivia (I)

grylloctonus Richards, 1937; Guyana (R)

intermedia Saussure, 1867; Brazil, Surinam, French Guiana, Guyana (I)

moraballi Richards, 1937; Guyana (I)

*resplendens (Kohl), 1902 (*Podium*); Brazil, Peru (R)

rufiventris (Fabricius), 1804 (*Podium*); Peru, Venezuela, Surinam, French Guiana, Guyana, Brazil (R)
  *abdominalis Perty, 1833
  haemorrhoidalis F. Smith, 1856

*soror Mocsáry, 1883; Brazil, Peru (R), new status by Menke
  *frivaldszkyi Mocsáry, 1883

vicina (Dalla Torre), 1897 (*Sceliphron*); Trinidad, Brazil, Guyana (I), new name for *Sceliphron affine* (Smith), (Art. 59c)
  affinis F. Smith, 1851, nec *Sceliphron affine* (Fabricius), 1793 (*Sphex*), now in *Sceliphron*

violascens (Dalla Torre), 1897 (*Sceliphron*); Trinidad, Guyana, Brazil, Peru (R), new name for *Sceliphron violaceum* Smith, (Art. 59c)
  violacea F. Smith, 1851, nec *Sceliphron violaceum* (Fabricius), 1775 (*Sphex*), now in *Chalybion*

Genus Chalybion Dahlbom

Generic description: Lower three-fifths of inner orbit usually straight, curving in towards midline of face above, orbits converging above or parallel except converging be-

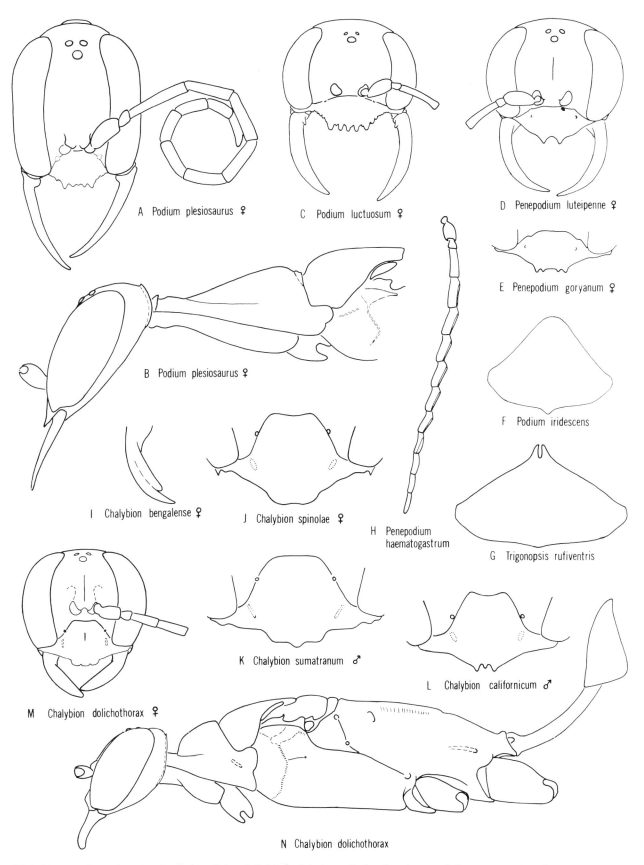

FIG. 22. Details in the tribe Sceliphronini A,C,D,M, facial views; B, head and part of thorax, lateral; E,J,L, clypeus; F,G, male sternum VIII; H, male antenna; I, right mandible; N, head, thorax and gastral segment I, lateral; A,B,M,N, drawn from holotypes.

low in many males; vertex not or only slightly elevated above eyes; postocular area not elongate; head with much erect hair; frons above sockets elevated as a flat or sometimes concave plate of which lower angles overhang sockets; flagellomere I about equal to length of II, male flagellomeres with placoids which may be irregular or weakly defined (none in *frontale*); sockets narrowly separated from or contiguous with frontoclypeal suture; clypeus long, width 1.0-1.33 times length, disk weakly to strongly convex, often with a median ridge, free margin in subgenus *Chalybion* with three teeth or lobes in males and some females, five in most females (fig. 17 D) (margin with median lobe and lateral sinuation in one African male and only with median excision in female of another African form), margin in subgenus *Hemichalybion* with a broad, median lobe or projection which is sometimes emarginate (fig. 22 J), margin trilobate in *sumatranum* male (fig. 22 K); mandible simple (fig. 17 D, 22 M) except with inner subapical tooth (fig. 22 I) in most females of subgenus *Chalybion;* mouthparts short, third maxillary palpal segment slightly to broadly expanded on one side; hypostomal carina ending near mandible socket; occipital carina usually a complete ring or nearly so, contiguous with or narrowly separated from hypostomal carina; collar usually about half as long as broad but occasionally shorter or longer, disk nearly always strongly indented mesally; propodeum long, without dorsal enclosure, posterior face often a gradual slope; propodeal side without spiracular groove; episternal sulcus extending to anteroventral margin of pleuron, scrobal sulcus usually present; upper metapleural pit small and mesopleural suture evanescent or absent for most of its length in *tibiale* group where pleura form a continuous, flat surface; marginal cell usually acuminate apically, submarginal cell II usually strongly narrowed anteriorly, rarely petiolate, receiving both recurrent veins; third anal vein of hindwing well separated from wing margin, long (fig. 18 H); midcoxae nearly contiguous; tibiae not spinose; female without a foretarsal rake; tarsi usually without plantulae; apicoventral bladelike setae of hindtarsomere V narrow, separated by four or more times a setal width; claw of hind leg often simple (all claws simple in *madecassum* group); petiole shorter or longer than hindcoxa or hindbasitarsus; male tergum VIII with cerci except in *madecassum* group; female sternum IV (and rarely III) sometimes with hair mat which may be in a depression, sterna IV-V of male with velvety micropubescence (II-VI with fimbriae in *saussurei* and *spinolae*); male sternum VIII triangular or with a slender to broad fingerlike apical process; head of digitus variable, penis valve head with teeth on ventral margin.

Geographic range: Two of the 31 species currently recognized occur in North and Central America; the remainder are Old World inhabitants. The genus is absent in South America, and occurs only on the Darwin Peninsula of Australia. The subgenus *Hemichalybion* is not represented in the New World. Through the activities of man, *C. bengalense,* an Oriental species, has been introduced to and become established in a number of areas, principally islands.

Systematics: Chalybion generally are metallic blue and range in length from 11 to 32 mm. The thorax is black in *fuscipenne* and some *Hemichalybion* species. Some of the *tibiale* group and the subgenus *Hemichalybion* have partially reddish brown mandibles, clypeus, and appendages. This color may even extend onto the thorax in *Hemichalybion*. Most *Hemichalybion* have partly yellow legs, especially the last pair, and a yellow petiole, a color pattern that is very similar to that found in most *Sceliphron*. Wings vary from clear to strongly infumate. Erect hair may be white or black, and the color sometimes differs between sexes, or geographically.

Most *Chalybion* are easily identified by their blue bodies, *Stangeella* and *Chlorion* being the only other sceliphronin genera that contain similarly colored species. The latter two genera differ from *Chalybion* in a number of characters among which the positions of the recurrent veins are immediately diagnostic. The following characters taken together separate *Chalybion* from all other sceliphronin genera: reception by submarginal cell I of both recurrent veins; presence (minor exceptions) of a median indentation or sulcus on the collar; long episternal sulcus; absence on the propodeum of a spiracular groove, a propodeal enclosure, or a median, longitudinal groove; absence of an intercoxal sulcus; and essentially equal lengths of flagellomeres I and II. Of these features, the absence of a propodeal enclosure is the most important character for separating *Chalybion* from *Sceliphron,* the genus with which it has traditionally been associated as a subgenus. Only one *Chalybion, californicum,* has a trace of an enclosure, and it is indicated only occasionally in this species and then usually by a slight crescentic elevation posterad. The equal lengths of flagellomeres I and II separates *Chalybion* from most *Sceliphron,* but these articles are also equal (males of *fasciatum*) or nearly so in a few species of the latter genus. The presence of placoids on the antennae of most male *Chalybion* is also a distinction from *Sceliphron,* as is clypeal dentition. However, van der Vecht (1961b) illustrated the clypeus of an unnamed female *Chalybion* which approaches *Sceliphron* in having a median notch. The usually strongly convex and ridged clypeus of *Chalybion* is distinctive, but the clypeus also bulges to some degree in all *Sceliphron*. The clypeus is only slightly convex in a few *Chalybion*. The petiole compared with the hindcoxa and hindbasitarsus is proportionately shorter in *Chalybion* than in *Sceliphron*. All *Sceliphron* have tarsal plantulae, but these structures are absent in most *Chalybion*.

Earlier, we (Bohart and Menke, 1963) recognized *Hemichalybion* as a genus, but we have since seen a better representation of species including the atypical *femoratum,* plus some of the more unusual *Chalybion s.s.* not previously available, and we now feel that the taxon should be included in *Chalybion* as a subgenus. The usually entire or somewhat bilobed clypeal margin of *Hemichalybion* species is about the only significant difference from *Chalybion;* and the trilobate margin in the male of *sumatranum* (fig. 22 K) shows that *Hemichalybion* is at best only weakly separable from *Chalybion*. Therefore, recognition of *Hemichalybion* even as a subgenus seems in doubt. Perhaps it could be recognized as a genus on the basis of yellow color, if the nonyellow marked *femoratum* were placed in *Chalybion s.s.* Hopefully, the biology of *femoratum* and other *Hemichalybion* species when known in more detail will help to clarify the status of these taxa. For example, *Chalybion s.s.* do not make mud nests, but some authors suggest that *Hemichalybion* species do.

Chalybion is generally regarded as a close relative of *Sceliphron*, but it may be more closely allied with *Podium* or *Chlorion*. The thorax of *Chalybion* tends towards elongation as in these two genera, unlike the shorter, stockier thorax of *Sceliphron*. The thorax of *Chalybion* is similar in most respects to that of *Podium*, and the clypeal dentition of *Chalybion s.s.* is not far removed from that of *Chlorion* and female *Podium*. On the other hand, the clypeus and more compact thorax of the subgenus *Hemichalybion* offer a transition between *Chalybion* and *Sceliphron*.

Morphologically *Chalybion s.s.* is diverse. The *tibiale* group (code letter T in checklist) consists of several African species in which the pleura are streamlined into a continuous smooth, flat surface, and the mesopleural suture is evanescent. This is also the only group in which reddish color is found on the appendages, etc. In the *madecassum* group (code letter M) we place *madecassum* and *frontale*, two forms with a very long petiole and long thorax. Most peculiarly, simple claws occur throughout, and there are no cerci in the male. The first tergum in this group is longer than high and almost forms part of the petiole. *Chalybion frontale* is one of the most unusual species in the genus. The plate above the antennal sockets is much more developed than in other species. Its lower margins are thickened, double edged, and reflexed upward so that overall the structure is very similar to the plate found in the dolichurin genera of the Ampulicinae. This species is also the only *Chalybion* we have seen that lacks placoids on the male antenna. Plantulae occur sporadically in the genus. The female mandible is simple in the subgenus *Hemichalybion* and in a few species of the subgenus *Chalybion*: *californicum*, *dolichothorax*, and the *madecassum* group. Of the 20 species we have seen, only a few have toothed hindclaws. The last female sternum in *accline*, *fabricator*, and *malignum* is broadly truncate. The clypeus is unusual in an unidentified Philippines female that belongs in this last assemblage. There is a median lobe weakly subdivided into three. On each side, some distance from this lobe is a small, acuminate tooth. The collar is long in *madecassum* and *dolichothorax* and is at most feebly, longitudinally sulcate. The thorax in the last species is very long in proportion to the rest of the body (fig. 22 N) and the face is rather narrow (fig. 22 M). The petiole is of normal length. This unusual wasp is known only from a female in the Paris Museum and the type.

In a few species, *laevigatum* and *japonicum* for example, wing venation is rather unstable, and various anomalies involving primarily the submarginal cells are not uncommon. Leclercq (1955c) has figured some of the variation encountered in *laevigatum*. According to van der Vecht (pers. comm.) the second submarginal cell is petiolate in one African species.

Species characters have been outlined by Kohl (1918), but we have noted a few things that seem important which he overlooked: the development of a lamelliform metapleural flange (as in *californicum*), the number and arrangement of placoids, male genitalia including the last sternum, the setation of the last tarsomere, and the degree of separation of the hypostomal and occipital carinae.

Kohl's (1918) revision of the world species of *Chalybion* (as a subgenus of *Sceliphron*) is still the basic work for identification. The discussion of van der Vecht (1961b) on the relationship between *Chalybion* and *Sceliphron* is also important. Arnold (1928) keyed some of the African species and Bohart and Menke (1963) keyed the New World forms. Beaumont (1965) and Leclercq (1966) provided useful notes and figures for discriminating certain Old World species.

Key to subgenera of *Chalybion*

1. Female clypeus with three or five lobes or teeth (with only a median notch in one African form), male clypeus with three lobes or teeth; legs and petiole without yellow; inner margin of female mandible usually with subapical tooth .... subgenus *Chalybion* Dahlbom
   Female clypeus with broad, projecting or shallowly emarginate median lobe, male clypeus similar or bilobate, if trilobate (fig. 22 K) then petiole yellow; inner margin of female mandible simple .... subgenus *Hemichalybion* Kohl

Biology: The basic references on *Chalybion s.s.* are: Rau (1915, 1928a,b, 1935b), Irving and Hinman (1935), and Muma and Jeffers (1945) on *californicum*; Rau (1940) and Ward (1970, 1971) on *zimmermanni*; Iwata (1939b,c, 1964a) and Yamamoto (1942, 1958) on *japonicum*; Dutt (1912), Bordage (1912), Friederichs (1918), Williams (1919), and Jayakar and Spurway (1963, 1965a,b) on *bengalense*.

All typical *Chalybion* species appear to mass provision with spiders and make their nests in preexisting cavities such as holes in wood, bamboo, and other plant stems, crevices in walls, and in abandoned mudnests of other wasps such as *Sceliphron* and *Trypoxylon*. Rau (1928a) noted a single instance where a *californicum* female opened a sealed nest of *Sceliphron caementarium*, removed the contents, and then introduced her own spider prey, but this appears to have been a very unusual case. *Chalybion* females use mud to partition and seal their nests, but instead of gathering the mud from damp areas as *Sceliphron* does, these wasps carry water to a source of mud near their nest, and then use the water to dampen the earth so that it can be plastered (Rau, 1928a,b; Ward, 1971). Old *Sceliphron* nests are a common mud source for *Chalybion californicum* and *zimmermanni*. Although technically *Chalybion* daub their nests with mud, they should probably be called water carrying wasps rather than mud daubers. The latter name is more appropriate for *Sceliphron* species. Rau (1928b) called *C. californicum* the "cow bird wasp," and Yamamoto (1958) used the name "blue burglar" in apparent reference to *C. bengalense,* both appelations referring to the appropriation of *Sceliphron* nests by these two species. *Chalybion bengalense* uses a variety of cavities for nesting sites, however, and although *californicum* has apparently only been observed in old *Sceliphron* nests, it seems likely that this species must also use other suitable cavities. With the exception of *californicum*, those *Chalybion* species that have been studied in detail, *japonicum, zimmermanni,* and *bengalense,* seem to prefer pre-existing cavities over empty mudnests for nest sites. For example, Ward (1971) found that only 15 percent of the 198 nests of *zimmermanni* that she studied were made in old *Sceliphron* nests, while 80 percent were made in pre-existing holes in wood.

One of the more interesting aspects of the nesting behavior of *Chalybion s.s.* is the fact that all of the species listed above except *californicum* seal their nests with an inner layer of mud and an outer layer of white material composed of either lime from walls or uric acid obtained from animal feces (Williams, 1919; Jayakar and Spurway, 1963; Yamamoto, 1942; Iwata, 1964a; and Ward, 1970, 1971). In contrast, *C. californicum* seals the "rented" *Sceliphron* nests with small clumps of mud (Rau 1928a; Muma and Jeffers, 1945). There is an interesting correlation between the type of nest closure and the form of the female mandible. *Chalybion californicum* has simple mandibles, but the females of species that use a white seal have a subapical tooth on the inner margin.

The wasp egg is laid on the abdomen of one of the spider prey, but there is disagreement as to when it is laid. For example, Rau (1928a) claimed that *californicum* females laid the egg on the last provision, but Muma and Jeffers (1945) stated that it was laid on the first provision. Bordage (1912) said that the egg was laid on the last spider in *bengalense*. Yamamoto (1942) claimed that *japonicum* laid the egg on the spider, before it was introduced to the cell. One of the common provisions of *C. californicum* is the black widow spider, *Latrodectus mactans* Fabricius (Rau, 1935b; Irving and Hinman, 1935; Muma and Jeffers, 1945). Some authors report that temporary mud closures are made on nests that take more than a day to provision.

The cocoon of *C. californicum* and *zimmermanni* is brown like that of *Sceliphron,* but it has a fragile, pale, outer cover of silk. It is not known whether this is true for all *Chalybion* because descriptions of cocoons of other species merely say that they are brown.

Data on *Hemichalybion* are fragmentary. According to van der Vecht (correspondence with Menke), the extensive papers by Verlaine (1925, 1926) on a species called *clypeatum* actually pertain to *Sceliphron spirifex*. Rudow (1912) found *femoratum* using a *Eumenes* mud nest, and Bonelli (1969) found the same species in an old mud nest which we presume was abandoned by a *Sceliphron*. Rudow described the cocoon of *femoratum* as light yellow. Brauns (1911a) said that *spinolae* nests in crevices just like *Chalybion*. Arnold (1928) said of *spinolae* (as *eckloni*), it is "frequently seen in houses, with an inconvenient partiality for building its mud-nests at the angles of walls . . .". Quite likely Arnold made the same mistake with *spinolae* that the Peckhams and Rau (early papers) made in their observations of *C. californicum* in not realizing that the wasp does not build its own mud nest, but takes empty *Sceliphron* nests for its own. According to van der Vecht (1961b) a specimen of *clypeatum* in the Paris Museum bears a label that says it is a common inhabitant of *Synagris* mud nests.

Chalybion is one of the sphecid genera that commonly forms "sleeping" aggregations in sheltered situations at night or during inclement weather (Bordage, 1912; Jayakar and Mangipudi, 1965; Williams, 1928a; Iwata, 1963; Bohart and Menke, 1963; Ward, 1972).

Checklist of *Chalybion*

Subgenus *Chalybion*

accline (Kohl), 1918 (*Sceliphron*); New Guinea
bengalense (Dahlbom), 1845 (*Pelopoeus*); Mozambique, Seychelles, Socotra, Mauritius; Oriental Region to ne. Australia, Johnston, Gilberts, Marianas, Solomons, Hawaii, Phoenix Is.
   violaceum Fabricius, 1775 (*Sphex*), nec Scopoli, 1763
   ? *ferum* Drury, 1782 (*Sphex*)
   ? *nitidulum* Christ, 1791 (*Sphex*)
   ? *smaragdinum* Christ, 1791 (*Sphex*)
   ? *ferox* Westwood, 1837 (*Chlorion*)
benoiti Leclercq, 1955; se. Africa: Mozambique (T)
?*bocandei* (Spinola), 1851 (*Ammophila*); Guinea
bonneti Leclercq, 1966; Madagascar
californicum (Saussure), 1867 (*Pelopeus*); N. America: transcontinental; Mexico, Hawaii, Bermuda
   caeruleum Linnaeus, 1763 (*Sphex*), nec Linnaeus, 1758
   cyaneum Fabricius, 1775 (*Sphex*), nec Linnaeus, 1758
convexum (F. Smith), 1876 (*Pelopoeus*); Rodriguez Is.
*dolichothorax (Kohl), 1918 (*Sceliphron*); China
fabricator (F. Smith), 1860 (*Pelopaeus*); Moluccas: Batjan
   gnavum Kohl, 1918 (*Sceliphron*), new synonymy by van der Vecht
frontale (Kohl), 1906 (*Sceliphron*); Celebes, Philippines (M)
fuscipenne (F. Smith), 1856 (*Pelopoeus*); Senegal to Gabon (T)
gredleri (Kohl), 1918 (*Sceliphron*); Zaire
heinii (Kohl), 1906 (*Sceliphron*); Sudan, Ethiopia, Saudi Arabia (T)
   planatum Arnold, 1951 (*Sceliphron*), new synonymy by van der Vecht
japonicum (Gribodo), 1883 (*Pelopoeus*); Japan[10]

---

[10] J. van der Vecht examined Gribodo's syntypes (Mus. Genoa) and notified us of this synonymy.

curvatum Ritsema, 1880, nec *Sceliphron curvatum* F. Smith, 1870, new synonymy by van der Vecht
  ritsemae Dalla Torre, 1897 (*Sceliphron*), new name for *curvatum* Ritsema
  japonicum Pérez, 1905, nec Gribodo, 1883, new synonymy by J. van der Vecht
  ssp. *punctatum* (Kohl), 1888 (*Pelopoeus*); China, Korea, Vietnam, Taiwan, Ryukyus, new status by J. van der Vecht
    inflexum Sickmann, 1894 (*Sceliphron*), new synonymy by van der Vecht
    degenerans Kohl, 1918 (*Sceliphron*), new synonymy by van der Vecht
klapperichi (Balthasar), 1957 (*Sceliphron*); Afghanistan
laevigatum (Kohl), 1888 (*Pelopoeus*); Ethiopian Region (T)
  chalybeum F. Smith, 1856 (*Pelopoeus*), nec Vander Linden, 1827
  levigatum Dalla Torre, 1897 (*Sceliphron*)
  ?*cubitaloide* Strand, 1910 (*Sceliphron*)
  nigrithorax Benoit, 1951 (*Sphex*)
  ssp. *sommereni* (R. Turner), 1920 (*Sceliphron*); Zaire, Kenya
    cyaneum Cameron, 1908 (*Ammophila*), nec Fabricius, 1775
    perpulchrum Arnold, 1934 (*Sceliphron*)
ommissum (Kohl), 1889 (*Pelopoeus*); ne. Mediterranean area
madecassum (Gribodo), 1883 (*Pelopoeus*) Madagascar, Seychelles, Isles Glorieuses, Comoros, Rodriguez (M)
  purpurescens Pérez, 1895 (*Sceliphron*)
malignum (Kohl), 1906 (*Sceliphron*); Sri Lanka, Sikkim, China
  horni Strand, 1915 (*Sceliphron*)
minos (Beaumont), 1965 (*Sceliphron*); Crete, Greece, Turkey
schulthessirechbergi (Kohl), 1918 (*Sceliphron*); Zaire (T)
targionii (Caruccio), 1872 (*Pelopoeus*); Mediterranean region, Iraq, Iran
  ?*flebile* Lepeletier, 1845 (*Pelopoeus*)
tibiale (Fabricius), 1781 (*Sphex*); Mozambique, S. Africa (T)
turanicum (Gussakovskij), 1935 (*Sceliphron*); sw. USSR
walteri (Kohl), 1889 (*Pelopoeus*); sw. USSR, Syria, Turkey
zimmermanni Dahlbom, 1843; U.S.: Texas to Indiana, N. Carolina; Hispaniola
  texanum Cresson, 1872 (*Pelopoeus*)
  ssp. *aztecum* (Saussure), 1867 (*Pelopoeus*); U.S.: Utah, Arizona to Texas; Mexico to Costa Rica; Revilla Gigedo Is.: Clarión
    monstrosum Kohl, 1918 (*Sceliphron*), new synonymy by Menke
  ssp. *peninsularum* Bohart and Menke, 1963; Mexico: Baja California

Subgenus *Hemichalybion*

clypeatum (Fairmaire), 1858 (*Pelopoeus*); Ethiopian Region
  brachystylus Kohl, 1888 (*Pelopoeus*)
  ssp. *lusingi* (Leclercq), 1955 (*Sceliphron*); Zaire
  ssp. *kiloense* (Leclercq), 1955 (*Sceliphron*); Zaire
femoratum (Fabricius), 1781 (*Sphex*); s. Europe, w. Asia, Afghanistan
  chalybeum Vander Linden, 1827 (*Pelopaeus*)
fuscum (Lepeletier), 1845 (*Pelopaeus*); Sri Lanka
  taprobanense Strand, 1915 (*Sceliphron*), new synonymy by Menke
spinolae (Lepeletier), 1845 (*Pelopaeus*); Ethiopian Region
  eckloni Dahlbom, 1845 (*Pelopoeus*); new synonymy by Menke
  ssp. *rufopictum* (Magretti), 1884 (*Pelopaeus*); Sudan, Ethiopia, Zaire
  ssp. *saussurei* (Kohl), 1918 (*Sceliphron*); S. Africa, new status by van der Vecht
sumatranum (Kohl), 1884 (*Pelopoeus*); Sumatra, Vietnam, Hainan

Fossil *Chalybion*

mortuum Cockerell, 1907

Genus Sceliphron Klug

Generic description: Inner orbits bowed out or lower three-fifths essentially straight, orbits converging above or parallel in female, variable in male, vertex not elevated above eyes, sometimes depressed below them; postocular area not elongate; head with much erect hair; flagellomere I longer than II except I slightly shorter than II (25:27) in male of *fasciatum;* male flagellum without placoids; sockets separated from frontoclypeal suture by at least 0.5 socket diameter; clypeus 0.5-0.75 as long as wide, disk slightly to strongly convex, free margin in female sharp edged and with a pair of broad, flat lobes separated by a slitlike excision, lobes often delimited laterally by a similar excision (fig. 23 A); male clypeus with two sinuations or lobes, or rarely with a single median lobe; mandible simple except with inner subapical tooth in some females; mouthparts short, third maxillary palpal segment expanded on one side; hypostomal carina ending near mandible socket in subgenus *Sceliphron,* evanescent about halfway to socket in *Prosceliphron;* occipital carina ending just short of hypostomal carina; collar short, disk usually indented mesally; propodeum moderately long, with U-shaped dorsal enclosure defined at least apically by a broad furrow; propodeal side without spiracular groove; episternal sulcus extending to anteroventral margin of pleuron, but often interrupted at level of pronotal lobe by an angular pleural bulge in subgenus *Sceliphron,* scrobal sulcus present or absent; metapleural sulcus sometimes present between upper and lower metapleural pits, lower metapleural area otherwise often differentiated from propodeum by sculpture; marginal cell rounded or acuminate apically, submarginal cell II broader than long and receiving both recurrent veins; third anal vein of hindwing well separated from wing margin; midcoxae essentially contiguous, outer surface of hindcoxa with angular bulge in *spirifex* group; tibiae not spinose; female without foretarsal rake; apicoventral, terminal,

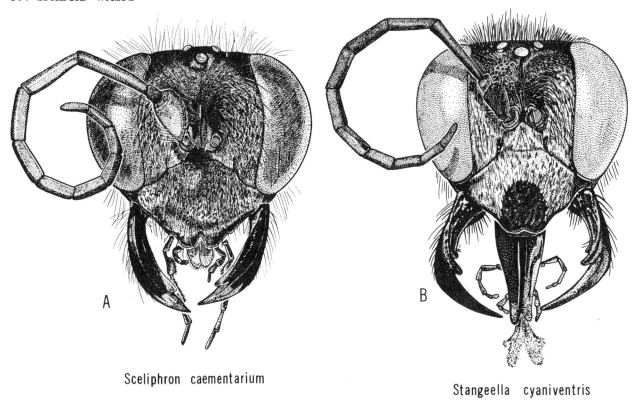

Sceliphron caementarium

Stangeella cyaniventris

FIG. 23. Facial portraits of females in the tribe Sceliphronini.

bladelike setae of hindtarsomere V narrow, separated by four or more times a setal width; claw tooth rarely subbasal or subapical, claws simple in *fistulare;* petiole length 2.5-3.6 times hindcoxal length, and 1.5-2.2 times length of hindbasitarsus; male tergum VIII with cerci in subgenus *Prosceliphron;* female sterna without hairmats; male sterna without or with weakly formed bands of velvety micropubescence; male sternum VIII broadly triangular; head of digitus variable, penis valve head with teeth on ventral margin, apex of head with lateral acuminate process in most species of subgenus *Sceliphron.*

Geographic range: Sceliphron is found in all of the temperate and tropical continental areas of the world and many islands. Thirty species are known, and most, including the subgenus *Prosceliphron,* are located in the Old World. Two species, *caementarium* and *fuscum,* have been introduced to a number of new areas through the activities of man.

Systematics: Sceliphron species range in length from 12 to 32 mm. The body is black, but it and/or the legs are conspicuously marked with yellow in most species. *S. fuscum* lacks yellow, but the head, thorax, and legs are partially reddish. The wings vary from clear to yellowish or amber.

The yellow maculations, very long petiole, and propodeal enclosure readily identify *Sceliphron.* The female clypeus is also distinctive. See systematics of *Chalybion* for other differences between that genus and *Sceliphron.*

J. van der Vecht (in van der Vecht and van Breugel, 1968) divided *Sceliphron* into two subgenera. The subgenus *Prosceliphron* differs from *Sceliphron s.s.* primarily in the presence of cerci, and the short hypostomal carina, although van der Vecht and van Bruegel used wing venation, eye, mandibular, and other differences. They divided the subgenus *Sceliphron* into two species groups, the *spirifex* group (code letter S) and the *madraspatanum* group (code letter M). The latter group contains species in which the female mandible has an inner subapical tooth, and the hindcoxa does not have a strong, angular bulge on the outer surface. In the *spirifex* group the female mandible with one exception (*arabs*) is simple and the hindcoxa has an angular bulge. The episternal sulcus is interrupted in all of the *spirifex* group, but it is also interrupted in a few of the *madraspatanum* group. *Sceliphron arabs* is peculiar in having a bituberculate scutellum, and *Sceliphron fistulare* has simple claw teeth. *Sceliphron deforme* has a strange humped first tergum.

Kohl's (1918) world revision of *Sceliphron* was updated by van der Vecht and van Breugel (1968). Van der Vecht's 1961 paper on *Sceliphron* also contains valuable data.

Biology: Much has been written about the mud daubers, as *Sceliphron* species are called, in reference to their habit of building free nests with mud collected from some moist spot. Kohl (1918) covered most of the early literature on these wasps. Other papers are those of Rau (1928b, 1946), Muma and Jeffers (1945), Shafer (1949), and Eberhard (1971) on *caementarium;* Krombein and

Walkley (1962) and White (1962) on *spirifex;* Janvier (1923, 1928), Orfila and Salellas (1929), and Bruch (1930) on *asiaticum;* Dow (1932) on *assimile;* Friederichs (1918) on *fuscum;* Katayama and Ikushima (1935), Iwata (1964a), and Spurway *et al.* (1964) on *madraspatanum;* Rau (1933) on *fistulare;* and Rotarides (1934), Mazek-Fialla (1936), and Myartseva (1968) on *destillatorium.*

Sceliphron females build their nests in a variety of sheltered situations, but some, such as *caementarium,* are commonly associated with human habitations. Each mudnest usually consists of several contiguous tubes or cells of mud. Each cell is mass provisioned with spiders, and the egg is laid on the first provision, although some American authors have stated that the egg is laid on later provisions in *caementarium.* Cells that have not been fully stocked in one day may be temporarily sealed at night. Usually the several cells of a nest are covered with a layer of mud, but White (1962) found that only 40 percent of the *spirifex* nests he studied were so covered. He postulated that this trait is being lost because the wasp is becoming more and more associated with the dwellings of man. Such sites offer more protection for the nest than the original ones (cliff overhangs, rock piles, etc.), and White supposed that the extra protection afforded by the outer mud cover is no longer necessary.

A variety of insects have been found parasitizing *Sceliphron* nests: Chrysididae of the genera *Chrysis, Trichrysis, Ceratochrysis,* and *Pyria;* Ichneumonidae of the genera *Acroricnus* and *Osprynchotus;* chalcids of the genus *Melittobia;* Mutillidae of the genus *Dolichomutilla;* bombyliid flies of the genera *Anthrax* and *Hyperalonia;* and Sarcophagidae of the genus *Amobia.*

Checklist of *Sceliphron*

Subgenus *Sceliphron*

annulatum (Cresson), 1865 (*Pelopoeus*); W. Indies: Greater Antilles, Bahamas (M)
    jamaicense Fabricius, 1775 (*Sphex*), nec *Sphex jamaicensis* (Drury), 1773
    ssp. *lucae* (Saussure), 1867 (*Pelopeus*); Mexico: Baja California, lectotype ♂, "Cap S. Lucas, Base California" (Mus. Geneva), present designation by Menke
arabs (Lepeletier), 1845 (*Pelopaeus*); sw. Asia (S)
    caucasicum Ed. André, 1888 (*Pelopoeus*)
argentifrons (Cresson), 1865 (*Pelopoeus*); Cuba (M)
assimile (Dahlbom), 1843 (*Pelopoeus*); U.S.: Texas; Mexico to Panama; W. Indies (M)
    nicaraguanum Kohl, 1918
asiaticum (Linnaeus), 1758 (*Sphex*); S. America (M)
    figulum Dahlbom, 1843 (*Pelopoeus*)
    vindex Lepeletier, 1845 (*Pelopaeus*)
    rufescens Strand, 1910
    ssp. *chilense* (Spinola), 1851 (*Pelopaeus*); Chile, lectotype ♀, Chile (Mus. Turin), present designation by Menke
caementarium (Drury), 1773 (*Sphex*); N. and Centr. America, W. Indies, Bermuda, Peru; Japan, Marianas, Marshalls, Hawaii, Australia, New Caledonia, Fiji, Samoa, Society Is., Marquesas Is., Gambier Is.; France, Germany, Madeira Is. (M)
    flavomaculatum De Geer, 1773 (*Sphex*)
    lunatum Fabricius, 1775 (*Sphex*)
    flavipes Fabricius, 1781 (*Sphex*)
    flavipunctatum Christ, 1791 (*Sphex*)
    affine Fabricius, 1793 (*Sphex*)
    architectum Lepeletier, 1845 (*Pelopaeus*)
    servillei Lepeletier, 1845 (*Pelopaeus*)
    solieri Lepeletier, 1845 (*Pelopaeus*)
    canadense F. Smith, 1856 (*Pelopaeus*)
    nigriventre A. Costa, 1864 (*Pelopoeus*)
    tahitense Saussure, 1867 (*Pelopeus*), lectotype ♂, Tahiti (Mus. Geneva), present designation by Menke
    economicum Curtiss, 1938 (*Sphex*), new synonymy by Menke
destillatorium (Illiger), 1807 (*Sphex*); s. Palearctic Region (S)
    flavipes Christ, 1791 (*Sphex*), ♀ only, nec Fabricius, 1781
    pensile Illiger, 1807 (*Sphex*)
    sardonium Lepeletier, 1845 (*Pelopaeus*)
    sardous Carrucio, 1872 (*Pelopoeus*)
    trinacriense De Stefani, 1889 (*Pelopoeus*)
fasciatum (Lepeletier), 1845 (*Pelopaeus*); W. Indies: Hispaniola, Guadeloupe, ? St. Vincent (M)
fistularium (Dahlbom), 1843 (*Pelopoeus*); Mexico to Argentina (S)
    bimaculatum Lepeletier, 1845 (*Pelopaeus*)[11]
    histrio Lepeletier, 1845 (*Pelopaeus*), lectotype ♂, Cayenne, French Guiana (Mus. Turin), present designation by Menke
fossuliferum (Gribodo), 1895 (*Pelopoeus*); Ethiopian Region (M)
    decipiens Arnold, 1952
    quartinae of authors, nec Gribodo
    ssp. *voeltzkowii* Kohl, 1909; Kenya to Malawi; Nigeria
    voeltzkovii Kohl, 1918
    masaicum Turner, 1919
    ssp. *complex* Kohl, 1918; Gabon, Guinea
fuscum Klug, 1801; Madagascar, Seychelles, Mauritius, Réunion, Aldabra, Comoros, New Caledonia (S)
    hemipterum Fabricius, 1798 (*Sphex*), nec Scopoli, 1772
    quodi Vachal, 1907
intrudens (F. Smith), 1859 (*Pelopaeus*); Celebes (M)
javanum (Lepeletier), 1845 (*Pelopaeus*); Java (S)
    ssp. *benignum* (F. Smith), 1859 (*Pelopaeus*); Borneo
    sintangense Strand, 1915
    ssp. *laboriosum* (F. Smith), 1859 (*Pelopoeus*); Indonesia; Aru, New Guinea
    lorentzi Cameron, 1911
    ssp. *nalandicum* Strand, 1915; Sri Lanka, India, Bangladesh

---

[11] J. van der Vecht and F. van Breugel (1968) placed this species in the synonymy of *asiaticum,* but Menke who has seen the supposed type in Turin identified it as *fistularium.*

ssp. *aemulum* Kohl, 1918; Philippines: Palawan
    luzonense Rohwer, 1922
ssp. *petiolare* Kohl, 1918; e. India to Vietnam and
    Sumatra
ssp. *tenggarae* van der Vecht, 1957; Indonesia: Komodo, Lombok, Sumbawa, Sumba, Flores
ssp. *timorense* van der Vecht, 1957; Indonesia: Timor
ssp. *chinense* van Breugel, 1968; se. China, Hainan to
    e. India
laetum (F. Smith), 1856 (*Pelopoeus*); Australia to Celebes, New Guinea, New Caledonia, Solomons, New Zealand, Guam, Carolines (S)
    cygnorum Turner, 1910
ssp. *maindroni* van der Vecht, 1968; Indonesia: Halmahera, Ternate, Batjan, Obi, Tanimbar
? *leptogaster* Cameron, 1905; S. Africa (may belong in Ammophilini)
madraspatanum (Fabricius), 1781 (*Sphex*); Maldives, Sri Lanka, India to Vietnam; Indonesia: Sumatra, Java, Borneo, Lesser Sundas, Buru, Ambon (M)
    maderospatanum Gmelin, 1790 (*Sphex*)
    ? *lugubre* Christ, 1791 (*Sphex*), nec Villers, 1789
    interruptum Palisot de Beauvois, 1806 (*Pelopaeus*)
    bilineatum F. Smith, 1852 (*Pelopaeus*)
    separatum F. Smith, 1852 (*Pelopaeus*)
ssp. *tubifex* (Latreille), 1809 (*Pelopaeus*); Morocco to Tunisia, s. Europe to sw. USSR and Iran
    pectorale Dahlbom, 1845 (*Pelopoeus*)
    transcaspicum Radoszkowski, 1886 (*Pelopoeus*)
ssp. *pictum* (F. Smith), 1856 (*Pelopoeus*); sw. USSR, Iran, to w. India, Oman
ssp. *conspicillatum* (A. Costa), 1864 (*Pelopoeus*); Philippines, New Guinea, Solomons, Carolines
ssp. *kohli* Sickmann, 1894; Vietnam, China, Japan, Ryukyus
ssp. *andamanicum* Kohl, 1918; Andaman Is., Nicobar Is.
ssp. *sutteri* van der Vecht, 1957; Indonesia: Sumbawa, Sumba
ssp. *formosanum* van der Vecht, 1968; Taiwan
pietschmanni Kohl, 1918; Iraq (S)
quartinae (Gribodo), 1884 (*Pelopoeus*); Ethiopian Region (M)
    fulvohirtum Arnold, 1928
spirifex (Linnaeus), 1758 (*Sphex*); s. Europe; sw. Asia; Africa, Canary Is., Cape Verde Is. (S)
    aegyptium Linnaeus 1758 (*Sphex*)
    atrum Scopoli, 1786 (*Sphex*)
    flavipes Christ, 1791 (*Sphex*), ♂ only, nec Fabricius, 1781
    aegyptiacum Klug, 1801
    spinifer Blanchard, 1840, lapsus

Subgenus *Prosceliphron*

bruinjnii (Maindron), 1878 (*Pelopaeus*); New Guinea
    bruynii Cameron, 1906, emendation
    fallax Kohl, 1918

coromandelicum (Lepeletier), 1845 (*Pelopaeus*); India, Sri Lanka, Burma
curvatum (F. Smith), 1870 (*Pelopoeus*); n. India, new status by van der Vecht
deforme (F. Smith), 1856 (*Pelopoeus*); India to China, sw. USSR: Kazakh S.S.R.[12]
? ssp. *rufopictum* (F. Smith), 1856 (*Pelopoeus*); Celebes
    flavofasciatum F. Smith, 1859 (*Pelopaeus*)
ssp. *unifasciatum* (F. Smith), 1860 (*Pelopoeus*); Indonesia: Ternate, Batjan
    affine Maindron, 1878 (*Pelopaeus*), nec Fabricius, 1793
? ssp. *atripes* (F. Morawitz), 1888 (*Pelopoeus*); sw USSR
? ssp. *tibiale* Cameron, 1899; e. India
    lineatipes Cameron, 1900
ssp. *koreanum* Uchida, 1925; Korea, Japan
ssp. *bicinctum* van der Vecht, 1957; Indonesia: Sumba
ssp. *taiwanum* Tsuneki, 1971; Taiwan; Japan: se. Honshu
ssp. *nipponicum* Tsuneki, 1972; Japan
fervens (F. Smith), 1858 (*Pelopoeus*); Borneo
? ssp. *murarium* F. Smith, 1864 (*Pelopoeus*); Indonesia: Ceram, Ambon
    rufipes Mocsáry, 1883 (*Pelopoeus*), nec *Sceliphron rufipes* (Fabricius), 1804
    mocsaryi Dalla Torre, 1897
formosum (F. Smith), 1856 (*Pelopoeus*); Australia, New Guinea; Indonesia: Ceram, Ternate
    papuanum Cameron, 1906, new synonymy by van der Vecht
funestum Kohl, 1918; Turkey
ocellare Kohl, 1918; New Britain
pulchellum Gussakovskij, 1933; Iran
rectum Kohl, 1918; Sikkim
shestakovi Gussakovskij, 1928; sw. USSR

Fossil *Sceliphron*

brevior Cockerell, 1921
tertiarium Meunier, 1915

Tribe Sphecini

This group consists mostly of large, robust wasps of which many species, especially in *Sphex,* are quite colorful. Some of the commoner ones have been the subjects of much study by ethologists. Because of this interest some of them now bear colloquial names such as the "Great Golden Digger" and the "Yellow Winged *Sphex*." Most wasps in this tribe are fossorial, but two of the five genera nest in preexisting cavities. However, it is presumed that they evolved from fossorial forms. Nests are provisioned mainly with Tettigoniidae and Acrididae.

Diagnostic characters:
1. (a) Sockets near middle of face and usually separated from frontoclypeal suture by less than a socket diameter; (b) male antenna with or without

---

[12] The arrangement of subspecies and synonymy under *deforme* was suggested by van der Vecht in correspondence.

placoids; (c) scape usually short, stout; (d) flagellomere I usually longer than II (exceptions in *Isodontia, Prionyx*).

2. (a) Clypeus large, about twice as wide as high to as high as wide; (b) labrum variable

3. (a) Inner margin of mandible with one or two teeth, or apex bifid or trifid; (b) mandibular socket closed by paramandibular process of hypostoma.

4. (a) Maxillary palpal segment III not asymmetrical; (b) hypostomal carina complete; (c) occipital carina incomplete below, disappearing just before meeting hypostomal carina.

5. Collar short, high.

6. (a) Propodeal enclosure rarely defined; (b) spiracular groove absent except in *Sphex;* (c) episternal sulcus long, extending ventrad to or nearly to anteroventral margin of pleuron (rare exceptions); (d) lower metapleural area usually not defined, metapleural sulcus usually absent between upper and lower metapleural pits.

7. (a) Marginal cell rounded or truncate apically; (b) three submarginal cells present (except two in one *Prionyx*); (c) first recurrent vein usually received by submarginal cell II, second recurrent usually received by submarginal III.

8. (a) Female with or without psammophore; (b) midcoxae nearly contiguous (except in one *Isodontia*); (c) two apicoventral setae of hindtarsomere V broad, bladelike, separated by less than twice a setal width (exceptions in *Sphex*); (d) inner margin of claw typically with two basal teeth, but up to five in some *Prionyx;* (e) tarsomeres without plantulae.

9. (a) Petiole consisting of sternum I only; (b) tergum VIII with or without cerci; (c) volsella with or without an articulating digitus.

Systematics: The five genera of the Sphecini form a very cohesive group. The basic complement of two claw teeth (up to five in some *Prionyx*), the reception by the second and third submarginal cells of one recurrent vein each (rare exceptions), the absence of plantulae, and the usually broad, bladelike, and narrowly separated apicoventral setae of hindtarsomere V separate the tribe from all other Sphecinae. Two-toothed claws are found elsewhere in the subfamily only in the ammophilin genera *Parapsammophila, Hoplammophila,* and *Eremochares;* but in these genera both recurrent veins usually end on the second submarginal cell, and the apicoventral tarsal setae are more broadly separated. Additionally, most *Parapsammophila* have plantulae. The few species of Sphecini with ammophilin type venation and only two claw teeth are excluded from the Ammophilini by their broad apicoventral tarsal setae.

We (Bohart and Menke, 1963) divided the Sphecini into two subtribes primarily on the basis of the shape of submarginal cell II. Our current studies confirm and strengthen the validity of these subtribes but indicate that the comparative lengths of the anterior and basal veinlets of submarginal cell II are a more reliable (though not infallible) diagnostic tool than the shape of the cell. In the Sphecina the basal veinlet is nearly always equal to or shorter than the anterior veinlet, while in the Prionyxina the basal veinlet is always longer than the anterior veinlet. Problems in the use of this subtribal difference arise in a few *Sphex* and one *Isodontia* where the basal veinlet may be as much as 20 percent longer than the anterior veinlet. Fortunately, this venational problem is overcome by the use of additional subtribal characters. Some of these can be found in the key to genera which follows, and others are noted in the subtribal diagnoses.

One distinction is worthy of note here, however. Generally it has been thought that in all Sphecinae the volsella has a movable digitus. In the Prionyxina, however, the digitus has apparently become fused with the rest of the volsella so that by definition there is no digitus. This evolutionary advancement in the Prionyxina emphasizes the phylogenetic "enigma" stated by Evans (1958a). He pointed out that biologically *Sphex* and *Isodontia* (= Sphecina) are farther advanced than *Prionyx* (= Prionyxina), but that morphologically *Prionyx* was more specialized than *Sphex* or *Isodontia*.

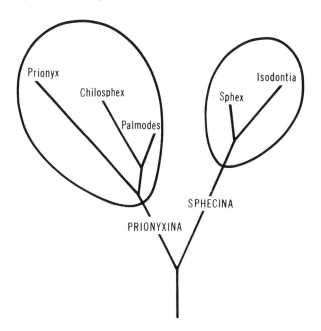

FIG. 24. Dendrogram showing suggested relationships of genera in the tribe Sphecini.

A suggested phylogeny of the Sphecini is shown in figure 24. *Palmodes* seems to be the most generalized form in the tribe, but it is perhaps meaningless to try to assess the relative evolutionary positions of the other four genera because of the discordance of various phylogenetic criteria. The greatest numbers of morphological specializations are found in certain groups of *Prionyx,* but its biology is relatively primitive. *Isodontia* and *Chilosphex,* which nest in preexisting cavities, seem to be derived from fossorial ancestors. The brood chamber in the nest of some species of *Isodontia* represents the most advanced type of nesting behavior in the tribe, although the nesting habits of some *Sphex* are also highly developed. The most complicated genitalia are found in the genus *Sphex*, but this may be a reflection of the age of the genus rather than advancement. The genitalia are fairly simple and rather uniform throughout the Prionyxina which, based on the absence of cerci and a digitus, is more specialized than the Sphecina.

Because the genera included here have until fairly recently been regarded by most authors as only subgenera of *Sphex*, much of the literature, including keys, has a rather broad scope. Some of the more important works of this type are: Kohl (1890), Dusmet and Mercet (1906), Roth (1925), Berland (1926-1929), Yasumatsu (1938a), Honoré (1944a), Berland and Bernard (1947), Willink (1951), and Berland (1956).

Biology: Various Orthoptera are the prey of this group. *Sphex, Prionyx,* and *Palmodes* are fossorial, but *Isodontia* and *Chilosphex* nest in preexisting cavities such as cracks in walls or hollow plant stems. The nest is prepared before obtaining prey except in some fossorial forms in the Prionyxina. Tsuneki (1963c) discussed the evolution of ethological traits within the east Asian members of the tribe.

Key to genera of Sphecini

1. Length of basal veinlet of submarginal cell II equal to or (more often) shorter than anterior veinlet (fig. 25 A-C)[13]; inner orbits straight below but curving in towards ocelli above (fig. 27 A, B); inner hindtibial spur closely, usually finely pectinate (fig. 28 N)[13] complete spiracular groove present or absent .... subtribe Sphecina ... 2
Length of basal veinlet of second submarginal cell greater than anterior veinlet (fig. 25 D, E); inner orbits straight or broadly bowed in towards midline of face (fig. 27 C, D); pectens of inner hindtibial spur usually coarse and well spaced at least near middle (fig. 28 L, P); spiracular groove absent .... subtribe Prionyxina ... 4
2. Propodeal side with complete spiracular groove (fig. 29 D), and length of petiole as measured along dorsum less than combined lengths of hindtarsomeres II-IV ....
.... *Sphex* subgenus *Sphex* Linnaeus, p. 00
Propodeal side without spiracular groove, or groove incomplete, not extending more than halfway from hindcoxa to spiracle; petiole length variable 3
3. Claw teeth perpendicular to inner margin (fig. 26 D); petiole measured along dorsum no more than two-thirds combined lengths of hindtarsomeres II-IV; anterior veinlet of submarginal cell III equal to or shorter than length of posterobasal veinlet (fig. 25 B)[14]; Holarctic ....
.... *Sphex* subgenus *Fernaldina* Bohart and Menke
Claw teeth obliquely oriented to inner margin (fig. 26 E); petiole equal to or longer than hindtarsomeres II-IV[15]; anterior veinlet of third submarginal cell usually conspciuously longer than length of posterobasal veinlet (fig. 25 C)[16]   *Isodontia* Patton, p. 00
4. Free margin of female clypeus entire or with a median notch (fig. 27 C); hindtarsal claw with two to five teeth on inner margin; male flagellum often with placoids on articles III-IV[17]; cosmopolitan ....
.... *Prionyx* Vander Linden, p. 00
Female clypeus with a truncate or slightly concave median lobe which is bounded laterally by a sinuation or notch (fig. 27 D); hindtarsal claw with two basal teeth on inner margin; male flagellum without placoids; Holarctic 5
5. Episternal sulcus extending ventrad nearly to anteroventral margin of pleuron; female with a well developed foretarsal rake ....
.... *Palmodes* Kohl, p. 00
Episternal sulcus ending at level of scrobe (fig. 29 B); female without a foretarsal rake (fig. 28 K); southern Europe, western Asia .... *Chilosphex* Menke, p. 00

Subtribe Sphecina

Diagnosis: Inner orbits straight below but curving in towards midline of face at level of ocelli, straight portions of inner orbits converging towards clypeus or parallel (fig. 27 A, B); scape short, globose or ovoid; male antenna usually with placoids; free margin of labrum not emarginate nor with long, fringing setae; malar space absent or very narrow; mouthparts moderately long to long; collar thick or sharp edged; admedian lines and parapsidal lines usually present, latter often long, notauli usually absent; mesopleural suture usually angulate below (exceptions mainly in *Isodontia*) so that upper meta-

---

[13] The basal veinlet is slightly longer than the anterior veinlet in a few Australian *Sphex* and in some males of *Isodontia edax,* but the hindtibial spur is finely pectinate. Some *Sphex* have rather coarsely pectinate spurs (fig. 28M), but in these the presence of a complete spiracular groove is diagnostic for the genus.

[14] In occasional variants of the Nearctic *S. (F.) lucae* the anterior veinlet is longer than the posterobasal veinlet, but the claw and petiole characters are diagnostic.

[15] The petiole is slightly shorter than the combined lengths of tarsomeres II-IV in some New World forms, but the claw and wing characters are diagnostic.

[16] Equal in one species.

[17] Males of a few Palearctic and South American *Prionyx* with only two claw teeth lack placoids and will key to *Palmodes*. However, the Palearctic males can be identified as *Prionyx* by the dense covering of appressed white vestiture on the head and thorax, and their generally red gaster with cream colored bands at the apex of each tergum. The South American males can be identified as *Prionyx* simply by the fact that *Palmodes* does not occur in the Neotropical Region.

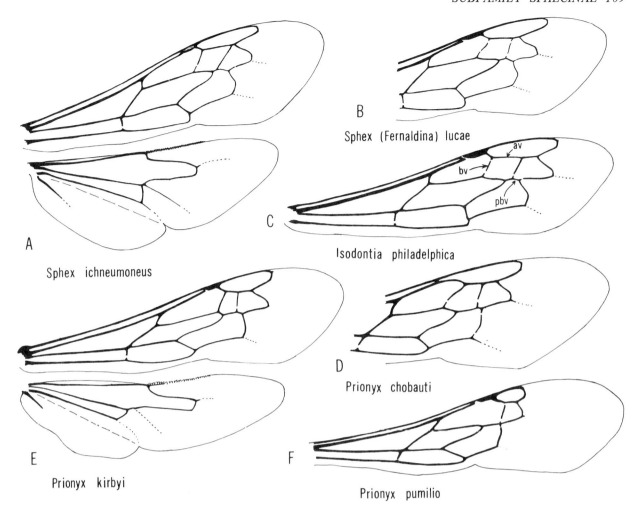

FIG. 25. Wings in the tribe Sphecini; bv = basal veinlet, av = anterior veinlet, pbv = posterobasal veinlet.

pleural area is not narrowly attenuate below (fig. 29 D); lower metapleural area not defined, metapleural sulcus not evident between upper and lower metapleural pits; spiracular groove present or absent; basal veinlet of second submarginal cell equal to or less than length of anterior veinlet (fig. 25 A-C) (rarely basal veinlet a little longer); first recurrent vein received by second submarginal cell, or (especially in *Sphex*) sometimes interstitial between II and III; tibiae and tarsi sparsely to densely spinose, without long spines in a few exceptional *Isodontia*; pectens of inner hindtibial spur closely spaced, usually fine (fig. 28 N) (coarse in some *Sphex*, fig. 28 M); tergum VIII usually with cerci (exceptions in *Isodontia*); male without scalelike micropubescence on sterna IV-V; volsella with an articulating digitus (fig. 30 E).

Discussion: The primary diagnostic features of the Sphecina are: 1) inner orbits incurving at level of ocelli; 2) labrum not fringed with long setae in female; 3) basal veinlet of submarginal cell II usually less than or equal to anterior veinlet; 4) inner hindtibial spur with close, fine pectination; 5) tergum VIII usually with cerci; 6) male sterna IV-V without micropubescence; and 7) volsella with an articulating digitus. As noted in the diagnosis there are exceptions to characters 3-5, but they are infrequent enough to be of minor importance.

This subtribe differs in a number of ethological traits from the Prionyxina. The nest is always prepared before collecting prey, and each cell is provided with several hosts which are usually flown to the nest. The egg is deposited on the venter of the pro- or mesothorax.

An interesting peculiarity in the Sphecina is the tendency for males to be larger than females in some of the bigger species such as *Sphex latreillei, ingens, permagnus,* and *Isodontia leonina.*

Genus Sphex Linnaeus

Generic description: Male antenna usually with one or more nonspiculate, flat (curved in subgenus *Fernaldina*) placoids which are most common on flagellomeres IV-VI (fig. 28 D); free margin of female clypeus usually arcuate, sharp edged, often with a small median notch, sometimes with a small median lobe which may be notched (fig. 27 A); clypeal margin usually broadly truncate in male, sometimes thickened, often broadly, shallowly emarginate (clypeus triangular in male of *argentina*); labrum in female usually with a median longitudinal carina

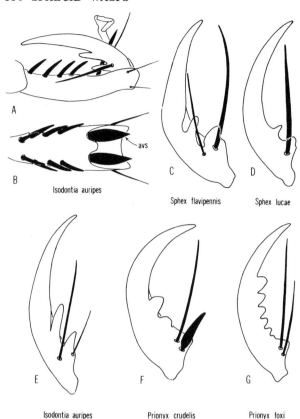

FIG. 26. Tarsal structure in the tribe Sphecini; A,B, hindtarsomere V, lateral and ventral, avs = apicoventral blade-like seta; C-G, hindtarsal claw, lateral, not all setae shown.

which may project apically (fig. 28 G), surface plain in male; female mandible long, broadest at middle, usually strongly arcuate, the axes of the base and apex approximating a right angle, apex usually attaining or surpassing base of opposite mandible when folded, inner margin with single, large tooth near middle, the upper surface of which is usually ridged and sulcate, proximal side of inner tooth often subtended by a small tooth (fig. 27 A) (female mandible in some Australasian species shorter, less arcuate, and inner tooth subapical); male mandible shorter and more slender than female, inner tooth sometimes subapical; malar space absent; collar with some thickness and usually standing apart from scutum; propodeal enclosure present or absent; propodeal side with a complete spiracular groove except absent in subgenus *Fernaldina;* basal veinlet of submarginal cell II usually less than or equal to length of anterior veinlet (fig. 25 A), but up to 1.2 times as long as anterior veinlet in some Australian forms; anterior veinlet of submarginal cell III (part of radial sector) usually shorter than posterobasal veinlet (part of media) (fig. 25 A); outer margin of female foreleg with well developed rake composed of long, often blade-like setae, tarsomere I usually asymmetrical and bearing rake setae on both outer and inner margins, II with two or more outer rake setae in addition to those at outer apex (fig. 28 I); apicoventral bladelike setae of last tarsomere of mid and hind legs narrow to broad, usually separated by about 1.5-2.0 times breadth of one seta, rarely separated by as much as 3.0-4.0 times a setal width (mainly Australasian forms); claws short to long, moderately to strongly arcuate in lateral profile, not prehensile, claw teeth perpendicularly (fig. 26 D) or obliquely oriented to inner margin (fig. 26 C); petiole usually straight, short, length measured along dorsum rarely more than two-thirds combined length of hindtarsomeres II-IV; tergum VIII of male with cerci; sterna of male with or without long erect setae, transverse bands of fimbriae absent, terminal sterna often with tufts or patches of short, dense setae; sternum VIII usually triangular, or rounded and with a small apicomedian projection, but unusually modified in a few species; ventral margin of penis valve head often with teeth which usually extend in a row around outer surface near apex, penis valve head sometimes inflated and/or bearing apicodorsal or lateral "wings" or other processes; digitus of volsella simple except in some species with highly modified aedeagus.

Geographic range: Sphex is a cosmopolitan genus and its 111 currently recognized species are distributed as follows (some species occur in more than one region): Palearctic – 15, Nearctic – 12, Neotropical – 28, Ethiopian – 22, Oriental – 19, and Australian – 21. Island endemics are found in the West Indies, Madagascar, the Seychelles, New Guinea, and on several of the islands of Melanesia and Indonesia. A few species have very extensive ranges. For example, *Sphex ichneumoneus* occurs from Canada to Argentina; *S. argentatus* ranges from India to Japan and Australia; and *S. fumicatus* is found over most of Africa and eastward to Sri Lanka.

Systematics: Sphex is used here in the strict sense meaning that the various "subgenera" such as *Isodontia, Prionyx* and *Palmodes* are excluded. Evans and Lin (1956a) were perhaps the first authors to elevate the various subgenera to genera using larval and ethological evidence to support their action. We (Bohart and Menke, 1963) found that adult morphology clearly justified raising the various subgenera. While their elevation has since gained full acceptance among New World authors, many Old World workers still cling to the obsolete view of including, under *Sphex,* a variety of morphologically and ethologically distinct taxa.

Our application of the name *Sphex* is in accordance with Opinion 180 of the International Commission on Zoological Nomenclature (1946), which states that *Sphex flavipennis* Fabricius is the type of the genus. Starting in 1905 and ending supposedly in 1946 with Opinion 180 the name *Sphex* has had a rather stormy history (see Bohart and Menke, 1963 for details). Among nearly all contemporary authors, Opinion 180 has gained complete acceptance.

The species of *Sphex* are moderate to very large wasps (11-47 mm). *Sphex ingens* is the giant of the genus and also one of the largest sphecids. Black is the basic color, and most of the Australian forms are entirely this shade, but elsewhere the gaster and legs may be partly or all red. In a few species the head and thorax are partly red also. Wings vary from clear to yellow, reddish, or black. The black wings of some species are iridescent green, blue or violet. Many species have extensive silvery, golden or coppery appressed hair on the head and thorax. The extensive metallic appressed hair of *cubensis, habenus* and some of the Australian species is quite showy. Erect hair varies from sparse to dense and wool-like, and the color may be white, yellow, red, or black. Males of *Sphex latreillei* that have dense, erect, red hair on the thorax and first tergum are among the gaudiest sphecids known.

Some species are dichromatic in one sex (red or black gaster in *habenus* females, for example). Furthermore, the two sexes are sometimes differently colored, as in *servillei* where the male has moderately infumate wings and the female has yellow wings, and in *sericeus* where males usually have pale hair and black legs but females have coppery hair and red legs. Wing, hair, and body color varies geographically in a number of species. This has contributed to superfluous names. Some of these polychromic forms have been discussed by Bohart and Menke (1963), Menke (1965a), and van der Vecht and Krombein (1955).

Except in the subgenus *Fernaldina*, which contains only two species, all *Sphex* possess a complete spiracular groove on the propodeum, and this character quickly separates the genus from *Isodontia*. The characters in the key to genera of the Sphecini readily separate *Fernaldina* from *Isodontia*. One of the primary features that we (Bohart and Menke, 1963) used to separate the two genera and one we particularly stressed in our key to genera was the proportions of the veinlets of submarginal cell III. Unfortunately this distinction between the two genera breaks down in occasional variants of some species, such as *lucae* and a number of Australasian forms. The latter area is particularly important to an understanding of the taxonomy of the genus *Sphex* because it contains a few perplexing forms which are atypical not only in wing venation, but also in the form of the mandible, the narrow ventral, terminal, blade-like setae of the last hindtarsomere (*darwiniensis* and *rugifer*, for example), and the absence of a median carina on the labrum. The last two features fortunately have no bearing on the distinctions between *Isodontia* and *Sphex*. For a fuller discussion of the differences between these two genera see systematics under *Isodontia*.

The absence of both a spiracular groove and cerci prompted us (Bohart and Menke, 1963) to describe *Fernaldina* as a genus for the New World species *lucae*. Beaumont (1967b) pointed out that the Old World species *melanocnemis* also belonged in *Fernaldina*, and he reduced it to a subgenus of *Sphex*. In most of its features (clypeus sharp edged, mandible shape, tarsal rake present, petiole length, wing venation, claw teeth perpendicular and biology) *Fernaldina* is *Sphex*-like, and it does seem more logical now to regard the taxon as a subgenus of *Sphex*. We have examined specimens of *melanocnemis*, a species not available to us in our 1963 study, and it is very similar to *lucae* in general body form, male genitalia, and in having placoids on flagellomeres III-VIII. The two species are obviously closely related. In addition to the characters mentioned above, *Fernaldina* differs from typical *Sphex* in having curved rather than flat placoids. The perpendicularly oriented claw teeth are not unique to *Fernaldina*; some *Sphex s.s.* have similar teeth. The male sterna are glabrous in *Fernaldina*, but this condition is also found in a few *Sphex s.s.* The disjunct distribution of *Fernaldina* is interesting. The group is perhaps a remnant of the specialized offshoot from which *Isodontia* evolved.

Species groups have yet to be established in *Sphex*, but a number of groups are obvious, based on the structure of the male genitalia, the presence or absence of a linear swelling in front of the upper part of the spiracular groove (present in *argentatus* and its relatives for example), and whether or not the metanotum is bituberculate. Doubtless other criteria are available.

Species characters are easy to find in most *Sphex*, especially in the male, but in some of the species complexes differences can be subtle. Male genitalia offer good characters in most forms, but they are sometimes identical or nearly so in closely related species. The number and arrangement of placoids on the male flagellum are very useful in species discrimination, but as in *Isodontia* there seems to be some variability within a few species. For example, *S. rufocinctus* males usually have placoids on flagellomeres III-VIII, but occasional specimens have them on III-VI, III-VII, or II-VIII. The same problem exists in *subtruncatus* (van der Vecht, 1957). Another placoid variable is the length and width of each one. A few species lack placoids (*schrottkyi, fumipennis, luctuosa, darwiniensis*, for example).

A number of *Sphex* in Australia and South America have unusual abdominal modifications in the male (*melanopus, cubensis, habenus, argentinus, schrottkyi, permagnus, ingens, mendozanus, latreillei, vestitus*, and *bilobatus*, for example). In these the last tergum and/or the last two or three sterna are modified in various ways, and the genitalia are usually bizarre and often large, so that they are partially exposed. In *argentinus* and *mendozanus* the last tergum is drawn out so that it projects far beyond the last sternum. In *argentinus* the surface has a large bowl-like depression, and the apex is truncate and emarginate. The last sternum of *schrottkyi, ingens*, and *mendozanus* is long and slender apically with the apex truncate or bilobed. In *vestitus* the last sternum is notched, and two carinae extend back from the sides of the notch forming a channel. In *bilobatus* the sternum ends in two long, fingerlike lobes.

Other notable peculiarities in some *Sphex* are: the triangular clypeus of *argentinus* males; the projecting, emarginate clypeus of *permagnus* and *schrottkyi* males; the apically expanded inner side of the hindtibia of *melanopus* and *cubensis;* the projections on the inner margin of the hindbasitarsis of *permagnus* males; the almost tubercular swelling at the anteroventral area of the mesopleuron in females of several Australian species; the rather blade-like projection at the ventral end of the petiole of one unidentified Australian female; and the coarse, rather rounded, ridges that traverse the propodeal dorsum in *sericeus, satanas,* and *rufiscutis.* Additional species characters can be found in leg setation, including the presence or absence of metallic hair; the form and orientation of the claw teeth; the shape and sculpture of the thorax; differences in the pectination of the inner hindtibial spur; and of course the shape and sculpture of the head, particularly the clypeus.

In spite of their large size and abundance, these showy wasps are still imperfectly known taxonomically in many parts of the world. Older keys are out-of-date nomenclatorially or taxonomically or both. In addition, all authors except Bohart and Menke (1963) have used *Sphex* in the broad sense so that the user often has to contend with species of *Prionyx, Palmodes, Isodontia,* etc., before coming to the *Sphex* portion of the key. Kohl (1890) is still a basic reference for the world fauna. Keys to the New World species are found in Bohart and Menke (1963) and Willink (1951) who reviewed the Nearctic and Argentine species, respectively. Menke (1964d, 1965a) published useful notes and figures for some of the New World species based on type studies. The outdated keys in Fernald (1906) are of some use in the New World. For the Old World the following keys are available: Berland and Bernard (1947), Giner Marí (1943b), and Dusmet and Mercet (1906) for Europe; Arnold (1928) for the Ethiopian Region; Roth (1925), Honoré (1944a), and Berland (1956) for north Africa; Bingham (1897) for India and environs; Yasumatsu (1938a) for eastern Asia; and Turner (1910) for Australia. The following works contain keys to closely related forms found in certain parts of the Old World: van der Vecht and Krombein (1955), van der Vecht (1957), Soika (1942), and Leclercq (1961e). The catalog of the Ethiopian species by Leclercq (1955c) is also useful.

Biology: These ground nesting wasps are usually gregarious. Most colonies seem to consist of fewer than 50 individuals, but occasionally groups of a hundred or more (Janvier, 1926; Bohart and Menke, 1963) or even thousands of wasps (Bingham, 1900) have been observed. *Sphex* colonies may be maintained for many years — 25 in the case of *Sphex ichneumoneus,* which is the longest on record (Fernald, 1945). The papers of Fabre (1856) Picard (1903) and Olberg (1966) on *S. rufocinctus*[18]; Rau and Rau (1918), Frisch (1937), and Ristich (1953) on *S. ichneumoneus;* Frisch (1938) on *S. pensylvanicus;* Cazier and Mortenson (1965d) on *S. lucae;* Piel (1935b) on *S. argentatus* (as *umbrosus*); Janvier (1926, 1928) on *S. latreillei;* Tsuneki and Iida (1969) on *S. subtruncatus* (as *haemorrhoidalis*); and Tsuneki (1963c) on *S. argentatus, diabolicus* ssp. *flammitrichus, rufocinctus* (as *maxillosus*), *inusitatus* ssp. *fukuiensis,* and *subtruncatus* (as *haemorrhoidalis*) are the most detailed of the many references on *Sphex* biology. Tsun-

[18] Fabre misidentified his species as *flavipennis* according to Berland (1923), who examined his material and identified the specimens as *maxillosus.* However, *S. maxillosus* is preoccupied and apparently must now be called *rufocinctus.*

eki's 1963 paper is by far the best. He discussed the pros and cons of several behavioral traits and included references to most of the papers on *Sphex* in Japanese that have appeared in the past. Olberg's 1966 paper contains excellent photographs.

Some of the species studied by the first students of wasp behavior were given common names which have become firmly established. "The Great Black Wasp" (*Sphex pensylvanicus*) of John Bartram, the early American naturalist, was perhaps the first fossorial sphecid reported upon in a scientific paper, according to Evans (1963b). The celebrated behaviorist Fabre gained some of his fame from his studies of the "Yellow winged *Sphex*" (*Sphex rufocinctus*), and the Peckhams' famous 1898 book on "The Solitary Wasps" introduced the well known "Great Golden Digger" (*Sphex ichneumoneus*).

Most species nest in rather open areas with little vegetation, but several, such as *pensylvanicus* (Reinhard, 1929b; Frisch, 1938; Rau, 1944b), *caliginosus* (Bequaert, 1937), and *sericeus* (Tsuneki and Iida, 1969; Piel, 1935b; Williams, 1919) seem to seek out dark, sheltered areas such as the floors of abandoned buildings for nest sites. Typically nests are multicellular and vary in depth from 3 to about 75 cm according to species, soil type, etc. Each cell is at the end of a side branch of the main tunnel. Species with single celled nests include *lucae* and *subtruncatus,* but the latter excavates several nests close together. Apparently some species close the nest while hunting for prey, while others leave it open. Prey are chiefly various kinds of Tettigoniidae, but Gryllacrididae and Gryllidae are occasionally used (see table 6). The prey collected by any particular wasp tend to belong to one species of katydid, but this varies with the season and locality. Prey are flown to the nest dorsum up, or if too large, the wasp may take short, hopping flights or even walk to the nest. The wasp grasps her prey by its antennae which are held by the mandibles. The middle legs of the wasp may assist in holding the prey during flight. Cazier and Mortenson found that females of *lucae* amputated the antennae of each katydid near the base. The egg is laid on the thoracic venter of the first provision. Mass provisioning is the rule, but progressive provisioning is typical of *subtruncatus. Sphex argentatus* has the interesting habit of digging accessory burrows, usually two, one on either side of the real burrow. These are dug after excavation of the nest burrow is completed and the entrance is temporarily closed. Tsuneki concludes at least tentatively that these accessory burrows, which are left open, serve to divert the attention of parasitic flies away from the real nest.

Sarcophagid and tachinid flies are known to parasitize *Sphex* nests, and some authors have noted that ants sometimes raid the nests of some species.

Checklist of *Sphex*

At the end of this list we have recorded 44 species that were originally described in *Sphex* but most probably do not belong in it. These are primarily names of Villers, Schrank, Müller, Gmelin and other contemporaries of Lin-

TABLE 6.
Summary of Prey Records of *Sphex*

Species	Gryllidae Gryllinae *Gryllus*	Oecanthinae *Oecanthus*	Gryllacrididae *Brachybaenus*	*Eremus*	*Gryllacris*	Tettigoniidae Conocephalinae *Conocephalus*	*Orchelimum*	Copiphorinae *Homorocoryphus*	*Neoconocephalus*	Decticinae *Atlanticus*	*Decticus*	*Neduba*	*Pholidoptera*	*Platycleis*	*Yersinella*	Listroscelinae *Hexacentrus*	Phaneropterinae *Amblycorypha*	*Burgilis*	*Cosmophyllum*	*Ducetia*	*Holochlora*	*Insara*	*Kuwayamaea*	*Microcentrum*	*Pachytrachelus*	*Phaneroptera*	*Scudderia*	*Tetana*	Pseudophyllinae *Acanthodis*	*Coccanotus*	*Idiarthron*
afer	X																														
argentatus			X			X		X			X			X		X				X	X		X			X					
argentinus				X																											
caliginosus					X				X									X													
cubensis						X																								X	X
diabolicus						X	X							X						X	X										
dorsalis						X																									
ichneumoneus			X		X	X	X			X	X	X					X										X		X		
inusitatus						X		X	X	X																					
latreillei																			X										X		
lucae																						X									
mendozanus																												X			
pensylvanicus	X									X	X		X	X	X									X	X	X					
rufocinctus														X											X	X					
sericeus								X								X				X											
subtruncatus						X																			X	X					

naeus. They represent species that have not been identified and some may not be sphecids.

Several common species of *Sphex* have been misidentified in the past (*haemorrhoidalis, fumicatus, argentatus, subtruncatus,* for example), and consequently published geographic ranges may differ from those given here. We have followed the most recent published discussion in these cases, and van der Vecht in correspondence with Menke has clarified many species.

Sphex subgenus *Sphex*

abyssinicus (Arnold), 1928 (*Chlorion*); Ethiopia
afer Lepeletier, 1845; nw. Africa, lectotype ♀, "*Sphex afra*" in Lepeletier's handwriting (Mus. Paris), present designation by Menke
   ssp. *sordidus* Dahlbom, 1845; Spain to sw. USSR and Afghanistan
     tristis Kohl, 1885
     ?*plumipes* Radoszkowski, 1886, nec Drury, 1773
     pachysoma Kohl, 1890
ahasverus Kohl, 1890; Australia
alacer Kohl, 1895; New Guinea
argentatus Fabricius, 1787; Oriental Region to ne. Australia
   ?*unicolor* Fabricius, 1787
   ?*umbrosus* Christ, 1791
   argenteus Turton, 1802
   argentifrons Lepeletier, 1845; lectotype ♂, Java (Mus. Turin), present designation by Menke
   plumiferus A. Costa, 1864
   nanulus Strand, 1913
   ssp. *fumosus* Kohl, 1890; Japan
argentinus Taschenberg, 1869; Argentina
ashmeadi (Fernald), 1906 (*Chlorion*); sw. U.S.; Mexico
atropilosus Kohl, 1885; Spain to sw. USSR
   atrohirtus Kohl, 1890
basilicus (R. Turner), 1915 (*Chlorion*); Australia
bilobatus Kohl, 1895; Australia
   canescens F. Smith, 1856; nec Scopoli, 1786
bohemanni Dahlbom, 1845; Uganda to S. Africa
   ?*abbottii* W. Fox, 1891
   kilimandjaroensis Cameron, 1908
   transvaalensis Cameron, 1910
brachystomus Kohl, 1890; New Britain
brasilianus Saussure, 1867; Brazil
caeruleanus Drury, 1773; Guinea to Zaire
   pulchripennis Mocsáry, 1883, new synonymy by Menke
caliginosus Erichson, 1848; Centr. and S. America; Lesser Antilles
   fuscus Lepeletier, 1845, nec Linnaeus, 1761
   erythropterus Cameron, 1888
camposi Campos, 1922 (pl. 1, fig. 4); Ecuador
carbonicolor van der Vecht, 1973; Australia
   carbonarius F. Smith, 1856, nec Scopoli, 1763
castaneipes Dahlbom, 1843; S. Africa
cognatus F. Smith, 1856; Indonesia: Batjan, Ceram, Ambon; New Guinea, Australia
   amator F. Smith, 1856
   formosus F. Smith, 1856
   opulentus F. Smith, 1856
confrater Kohl, 1890; New Britain, New Ireland, Solomons[19]
   ?*sieberti* Strand, 1910
cubensis (Fernald), 1906 (*Chlorion*); Cuba
   clavipes Kohl, 1890, nec Linnaeus, 1758
darwiniensis R. Turner, 1912; Australia
decipiens Kohl, 1895; S. Africa
   ssp. *meridionalis* (Arnold), 1947 (*Chlorion*); Zambia
decoratus F. Smith, 1873; Australia
deplanatus Kohl, 1895; Sri Lanka, Socotra Is.
diabolicus F. Smith, 1858; Sri Lanka, S. India to Sikkim, Borneo and Philippines
   ssp. *flammitrichus* Strand, 1913; Vietnam, China, Taiwan, Ryukyus, Japan
   aureopilosus Berland, 1928, lectotype ♂, Ba Cha, Tonkin (Mus. Paris), present designation by Menke
dorsalis Lepeletier, 1845; U.S.: California to Florida; Centr. and S. America; W. Indies
   singularis F. Smith, 1856
   chlorargyricus A. Costa, 1862
   micans Taschenberg, 1869, nec Eversmann, 1849
   dubitatus Cresson, 1872
   spiniger Kohl, 1890
dorycus Guérin-Méneville, 1838; New Guinea: Waigeo
   errabundus Kohl, 1898
ephippium F. Smith, 1856; Australia
ermineus Kohl, 1890; Australia
erythrinus Guiglia, 1938; Ethiopia
   erytrhinus Magretti, 1905, nomen nudum
ferrugineipes W. Fox, 1897; Brazil
finschii Kohl, 1890; Indonesia: Ewab, Aru; New Guinea, New Britain, Solomons, Australia
flavipennis Fabricius, 1793; Mediterranean region to Afghanistan
   bicolor Dahlbom, 1845; nec Fabricius, 1775
   cinereorufocinctus Dahlbom, 1845
   sellae Gribodo, 1873
   rufodorsatus De Stefani, 1887
flavovestitus F. Smith, 1856; U.S.: Virginia to Texas
   flavipes F. Smith, 1856; nec Fabricius, 1781
   flavitarsis Fernald, 1906 (*Chlorion*)
   ssp. *saussurei* (Fernald), 1906 (*Chlorion*); Mexico
   hirsutus Saussure, 1867, nec Scopoli, 1763
formosellus van der Vecht, 1957; Indonesia: Sumba, Timor; sw. Australia
fulvohirtus Bingham, 1890; Sri Lanka
fumicatus Christ, 1791; Africa: transcontinental, Cyprus, e. Mediterranean area, Aden, Socotra, Pembas, w. and s. India, Sri Lanka
   albifrons Fabricius, 1793, nec Villers, 1789
   metallicus Taschenberg, 1869
   argentiferus Walker, 1871
   magrettii Gribodo, 1894
   erebus W. F. Kirby, 1900
   davisi Fernald, 1907 (*Chlorion*)

---

[19] Java for *sieberti* is probably erroneus according to van der Vecht in litt.

umbrosus of authors, not Christ
    ssp. *voeltzkowii* Kohl, 1909; Madagascar
fumipennis F. Smith, 1856; Australia
    ssp. *antennatus* F. Smith, 1856; New Caledonia, New Hebrides
    ssp. *rouxi* Schulthess, 1915; Loyalty Is.
    splendidus Berland, 1928, nec Müller, 1776
funerarius Gussakovskij, 1934; China: Kansu Prov., Szechwan Prov.; Mongolia
gaullei Berland, 1927; Centr. African Rep., Ethiopia
?*gisteli* Strand, 1917; "India occidentalis"
    aurulentus Gistel, 1857, nec Fabricius, 1787
gratiosus F. Smith, 1856; Libya
guatemalensis Cameron, 1888; Mexico, Guatemala
habenus Say, 1832; U.S.: Virginia to Texas; Mexico
    lautus Cresson, 1872
    illustris Cresson, 1872
    princeps Kohl, 1890
    chrysophorus Kohl, 1890
    lanciger Kohl, 1895
haemorrhoidalis Fabricius, 1781; Liberia to Tanzania: Tanganyika[20]
    volubilis Kohl, 1895
    pachydermus Strand, 1916
    ssp. *umtalicus* Strand, 1916; Rhodesia
    ssp. *kobrowi* (Arnold), 1928 (*Chlorion*); S. Africa
    ssp. *mweruensis* (Arnold), 1947 (*Chlorion*); Zaire
    ssp. *basuto* (Arnold), 1947 (*Chlorion*); Lesotho
ichneumoneus (Linnaeus), 1758 (*Apis*); N., Centr., and S. America
    surinamensis Retzius, 1783 (*Nomada*)
    ?*aurifluus* Perty, 1833
    ?*aurocapillus* Templeton, 1841
    croesus Lepeletier, 1845
    dimidiatus Lepeletier, 1845, nec De Geer, 1773
    ?*sumptuosus* A. Costa, 1862
    ignotus Strand, 1916
incomptus Gerstaecker, 1871; Ethiopian Region
    nyanzae R. Turner, 1918 (*Chlorion*)
    ssp. *anonymus* Leclercq, 1955; Zaire, Zambia
ingens F. Smith, 1856; Brazil, Argentina, lectotype ♂, Brazil (Mus. London), present designation by Menke
inusitatus Yasumatsu, 1935; Japan: Kyushu; ne. China
    ssp. *fukuiensis* Tsuneki, 1957; Japan: Honshu
jamaicensis (Drury), 1773 (*Vespa*); U.S.: Florida; W. Indies
    jamaica Christ, 1791
    aurulentus Guérin-Méneville, 1835, nec Fabricius, 1787
    lanierii Guérin-Méneville, 1844
    ornatus Lepeletier, 1845
    fulviventris Kohl, 1890
jansei Cameron, 1910; S. Africa
kolthoffi Gussakovskij, 1938; China: Kansu Prov.
lanatus Mocsáry, 1883; Rhodesia, S. Africa, Zaire, Tanzania

[20] Records of *haemorrhoidalis* from the Oriental Region are erroneous and pertain to *subtruncatus* Dahlbom or some other species (see van der Vecht, 1957).

latreillei Lepeletier, 1831; Chile
    thunbergi Lepeletier, 1831
    chiliensis Lepeletier, 1845
latro Erichson, 1848; S. America
    clypeatus F. Smith, 1856
    roratus Kohl, 1890
libycus Beaumont, 1956; Libya
luctuosus F. Smith, 1856; Australia
madasummae van der Vecht, 1973; Burma to Philippines, New Guinea
    caerulescens Le Guillou, 1841 (Rev. Mag. Zool.), nec Reich, 1795
    caerulescens Le Guillou, 1841 (Ann. Soc. Ent. France), nec Reich, 1795
    maurus F. Smith, 1856, nec Fabricius, 1787
    nigerrimus A. Costa, 1864, nec Scopoli, 1763
malagassus Saussure, 1890 (pl. 18, Fig. 38); Madagascar
    malagassus Saussure, 1891
    malagassus Saussure, 1892
mandibularis Cresson, 1868; Cuba, Hispaniola
maximiliani Kohl, 1890; Mexico; lectotype ♂, Mexico (Mus. Vienna), present designation by Menke
melanopus Dahlbom, 1843; El Salvador to Argentina
    difficilis Spinola, 1851, new syn. by Menke
    proximus F. Smith, 1856, new syn. by Menke
    ruficaudus Taschenberg, 1869
    funestus Kohl, 1890, new syn. by Menke
melas Gussakovskij, 1930; sw. USSR: Turkmen S.S.R.
mendozanus Brèthes, 1909; Argentina
mimulus R. Turner, 1910; Australia
mochii Giordani Soika, 1942; Ethiopia (? = *gaullei*)
modestus F. Smith, 1856; Australia
    dolichocerus Kohl, 1890
    bannitus Kohl, 1895
muticus Kohl, 1885; Indonesia: Buru, Ambon (♂ only, see van der Vecht, 1973)
neavei (Arnold), 1928 (*Chlorion*); Zaire, Malawi
nigrohirtus Kohl, 1895; Guinea to Uganda, Angola, and Tanzania: Zanzibar; Fernando Póo
    camerunicus Strand, 1916
    conradti Berland, 1927, lectotype ♀, Fernando Póo (Mus. Paris), present designation by Menke
nitidiventris Spinola, 1851; Mexico to Argentina
    beatus Cameron, 1888
    neotropicus Kohl, 1890
nudus Fernald, 1903; U.S.: e. of Mississippi River
    bridwelli Fernald, 1903
obscurus (Fabricius), 1804 (*Pepsis*); Sri Lanka, India[21]
    ?*hirtipes* Fabricius, 1793
    cinerascens Dahlbom, 1843
    xanthopterus Cameron, 1889
observabilis (R. Turner), 1918 (*Chlorion*); Uganda, Zaire

[21] Kohl (1895) stated that Dahlbom's citation of "Guinea" was in error, and he suggested that *cinerascens* originated in the Oriental Region. Van der Vecht (1961) also gave Sri Lanka as the probable location. *S. hirtipes* was described from Guinea also, but van der Vecht (1961) said it may be synonymous with *obscurus*. Kohl (1890) tentatively considered *hirtipes* to be synonymous with *Prionyx crudelis*.

opacus Dahlbom, 1845; Mexico to Argentina
    iheringii Kohl, 1890
optimus F. Smith, 1856; Gambia
oxianus Gussakovskij, 1928; Turkey, Iran, Afghanistan, sw. USSR: Uzbek S.S.R.
   ssp. *nubilus* Beaumont, 1968; Israel
paulinierii Guérin-Méneville, 1843; Senegal to Zaire
    eximius Lepeletier, 1845
pensylvanicus Linnaeus, 1763 (Centuria Insectorum Rariorum); U.S.: transcontinental except in nw.; n. Mexico
    pensylvanicus Linnaeus, 1763 (Amoenitates Acad.)
    robustisoma Strand, 1916
permagnus (Willink); 1951 (*Chlorion*); Argentina
peruanus Kohl, 1890; Peru
praedator F. Smith, 1859; Celebes
    calopterus Kohl, 1890
    celebesianus Strand, 1913
    kohlianus Strand, 1913
   ssp. *tyrannicus* F. Smith, 1860; Indonesia: n. Moluccas: Batjan, Kajoa, Halmahera
   ssp. *luteipennis* Mocsáry, 1883; Indonesia: s. Moluccas: Ambon, Buru[22]
   ssp. *melanopoda* Strand, 1915; India, Sri Lanka
prosper Kohl, 1890; Venezuela
pruinosus Germar, 1817; Mediterranean area to Ethiopia, sw. USSR, and Burma
    vicinus Lepeletier, 1845
    scioensis Gribodo, 1879
    rothneyi Cameron, 1889
    retractus Nurse, 1903
resinipes (Fernald), 1906 (*Chlorion*); Hispaniola
    rufipes Lepeletier, 1845, lectotype ♀, Guadeloupe (Mus. Turin), present designation by Menke, nec Linnaeus, 1758
resplendens Kohl, 1885; Aru I., ? New Guinea
    nitidiventris F. Smith, 1859, nec Spinola, 1853
    gratiosus F. Smith, 1859, nec F. Smith, 1856
    gratiosissimus Dalla Torre, 1897
    lanceiventris Vachal, 1908
    wallacei R. Turner, 1908
    mertoni Strand, 1911
rhodosoma (R. Turner), 1915 (*Chlorion*); Australia
rufinervis Pérez, 1895; Seychelles, lectotype ♂, Mahé, Seychelles (Mus. Paris), present designation by Menke
rufiscutis (R. Turner), 1918 (*Chlorion*); Somalia to S. Africa
    haemorrhoidalis Magretti, 1898, nec Fabricius, 1781
   † ssp. *laevigatus* Arnold, 1951; Mali, nec Rossi, 1794
rufocinctus Brullé, 1833; s. Europe, s. Asia to China[23]

    maxillosus Fabricius, 1793, nec Poiret, 1787
    triangulus Brullé, 1833, nec Villers, 1789[23]
    ?*leuconotus* Brullé, 1833
    obscurus Fischer-Waldheim, 1843, nec Schrank, 1802, nec Fabricius, 1804
    pedibusnigris Zanon, 1925
   ssp. *mavromoustakisi* Beaumont, 1947; Cyprus
rugifer Kohl, 1890; Australia
satanas Kohl, 1898; Mali to Mozambique
    gorgon Kohl, 1913
schoutedeni Kohl, 1913; Zaire, Malawi
schrottkyi (Bertoni), 1918 (*Proterosphex*); Argentina
    luciati Brèthes, 1918, lectotype ♂, La Rioja, Argentina (Mus. Buenos Aires), present designation by Menke
semifossulatus van der Vecht, 1973; Australia
    argentifrons F. Smith, 1868, nec Lepeletier, 1845
sericeus (Fabricius), 1804 (*Pepsis*); Indonesia: Java, Bali, Sumba, Flores
   ssp. *fabricii* Dahlbom, 1843; India, Sri Lanka
    aurulentus Fabricius, 1793, nec Fabricius, 1787
    ferrugineus Lepeletier, 1845
   ssp. *lineolus* Lepeletier, 1845; Burma, Sumatra, China, Taiwan, Ryukyus
    lepeletierii Saussure, 1867
    rugosus Matsumura, 1912
   ssp. *godeffroyi* Saussure, 1869; s. New Guinea, nw. Australia
    aurifex F. Smith, 1873
    pallidehirtus Kohl, 1890
   ssp. *ferocior* van der Vecht and Krombein, 1955; Indonesia: Talaud, Sangi, Celebes, Batjan, Buru, Ambon
    ferox F. Smith, 1862, nec *Sphex* (*Chlorion*) *ferox* (Westwood), 1837[24]
   ssp. *nigrescens* van der Vecht and Krombein, 1955; Philippines
   ssp. *stueberi* van der Vecht and Krombein, 1955; New Guinea
   ssp. *wegneri* van der Vecht and Krombein, 1955; Borneo
servillei Lepeletier, 1845; s. Texas to Argentina
    fuliginosus Dahlbom, 1843; nec Scopoli, 1763
    chichimecus Saussure, 1867
    congener Kohl, 1890
    joergenseni Brèthes, 1913
stadelmanni Kohl, 1895; Rhodesia, Mozambique, S. Africa
   ssp. *integrus* (Arnold), 1928 (*Chlorion*); Mozambique, S. Africa
staudingeri Gribodo, 1894; New Guinea, Australia
subhyalinus W. Fox, 1899; Brazil
subtruncatus Dahlbom, 1843; India to Sumatra and

---

[22] African records of *luteipennis* pertain to a different and probably undescribed species.

[23] *Sphex rufocinctus* and *triangulus* are usually regarded as synonyms of *maxillosus* by European workers, but based on Menke's study of their types (Mus. Paris) they appear to be synonyms of *Sphex flavipennis* or *afer* in the case of *triangulus*. The condition of the types is poor and the final decision of their identity should be ascertained by some European specialist. Van der Vecht (1959) pointed out that *maxillosus* is preoccupied, but no action has been taken to conserve this well known name, and therefore we have used the next available name.

[24] *Chlorion* was still considered as a subgenus of *Sphex* when *ferocior* was proposed as a new name.

Japan[25]
   nigripes F. Smith, 1856, nec Fabricius, 1793
   erythropoda Cameron, 1889
   tsingtauensis Strand, 1916
   haemorrhoidalis of authors, in part, not Fabricius
  ssp. *siamensis* Taschenberg, 1869; Thailand
  ssp. *sulciscuta* Gribodo, 1894; Philippines
  ssp. *xuthus* van der Vecht, 1957; Indonesia: Sumba
  ssp. *coraxus* van der Vecht, 1957; Indonesia: Sumba
  ssp. *orius* van der Vecht, 1957; Indonesia: Flores, Sumba
taschenbergi Magretti, 1884; Chad, Sudan, Ethiopia, s. Arabian pen.
tepanecus Saussure, 1867; U.S.: Texas to Arizona; Mexico
   mexicanus Taschenberg, 1869, nec Saussure, 1867
texanus Cresson, 1872; U.S.: Kansas to Texas, Arizona
tinctipennis Cameron, 1888; Mexico to Brazil
tomentosus Fabricius, 1787; Guinea, Mali to Ethiopia and S. Africa
   tuberculatus F. Smith, 1873, nec Villers, 1789
   luteifrons Radoszkowski, 1881
torridus F. Smith, 1873; Madagascar, Europa, Comoros, Aldabra
vestitus F. Smith, 1856; Australia
   praetextus F. Smith, 1873
   imperialis Kohl, 1890

Subgenus *Fernaldina*

lucae Saussure, 1867; w. and se. U.S.; Mexico
   belfragei Cresson, 1872
melanocnemis Kohl, 1885; e. Mediterranean area

Fossil *Sphex*

bischoffi Zeuner, 1931
giganteus Heer, 1867
†*obscurus* Statz, 1936, nec Fischer-Waldheim, 1843, nec Fabricius, 1804 (*Pepsis*)

Nomina nuda in *Sphex*

formicaroides Gistel, 1837
synoecodes "Perez," Campos, 1922
tiphia Gistel, 1837

Species described in *Sphex* but probably belonging elsewhere, possibly not in the Sphecidae

albifrons Villers, 1789; "Europa, in montibus Gebennis capta"
annularis Poda, 1761; Italy
antarcticus Linnaeus, 1767; S. Africa
austriacus Schrank, 1781; Austria
bifasciatus O. Müller, 1776; Denmark
†*cinctus* Fabricius, 1793; Guinea, nec Scopoli, 1763
coccineus Gmelin, 1790; "Europam"
collaris Linnaeus, 1767; Spain
conicus Villers, 1789; "Europa, Gallia Aust."
†*conicus* Radoszkowski, 1877; sw. USSR, nec Villers, 1789
elongatus Villers, 1789; "Europa, Gallia Austr."
†*flavicornis* Villers, 1789; "Europa, Gallia Austr.", nec Fabricius, 1781
gibbosus Rossi, 1790; Italy
gregarius Scopoli, 1763; Austria
guttatus Gmelin, 1790; no locality
leucopterus Pallas, 1771; USSR
lineatus Villers, 1789; "Europa, Gallia Aust."
lucidus Villers, 1789; "Europa, Gallia Austr."
lugubris Villers, 1789; "Europa, Gallia Austr."
mauritanicus Linnaeus, 1767; Mauritania
   mauritianus Christ, 1791
melanochlorus Gmelin, 1790; "Europa"
militaris O. Müller, 1776; Denmark
mixtus Fabricius, 1794; "Americae Insulis"
naviculatus Villers, 1789; "Europa"
†*niger* O. Müller, 1775; Denmark, nec Fabricius, 1775
†*nigerrimus* Schrank, 1802; Germany, nec Scopoli, 1763
†*nobilis* O. Müller, 1776; Denmark, nec Scopoli, 1763
pygmaeus Schrank, 1785; Austria
quadrifasciatus Villers, 1789; "Europa, Gallia Austr."
†*quadrifasciatus* O. Müller, 1776; Denmark, nec Villers, 1789
semicinctus Villers, 1789; "Europa, Gallia Austr."
serotinus O. Müller, 1776; Denmark
spinipes Gmelin, 1790; no locality
splendidus O. Müller, 1776; Denmark
†*splendidus* Reich, 1795; French Guiana, nec Müller, 1776
stigma Linnaeus, 1767; S. Africa
tenthredinoides Scopoli, 1763; Austria
testaceus Gmelin, 1790; "Europa"
†*tomentosus* Gmelin, 1790; "Europa", nec Fabricius, 1787
triangulus Villers, 1789; "Europa"
†*tricolor* Reich, 1795; French Guiana, nec Schrank, 1781
trifasciatus O. Müller, 1776; Denmark
trimarginatus O. Müller, 1776; Denmark
†*vespiformis* Schrank, 1781; Austria, nec Fabricius, 1775

The following list contains *Sphex* names published in auction catalogs. Their status is unclear under the current rules of the I.C.Z.N. (article 8), and the descriptions are very brief. If these names are eventually declared valid, a number of names now in use will become junior homonyms. Sherborn (1899) has discussed the Lichtenstein catalog.

annulatus Lichtenstein, 1796; "Asia"
antaeus Lichtenstein, 1796; ?
antaeus Holthusius (Schneider), 1800; ?
argyrosticta Lichtenstein, 1796; ?
argyrostoma Lichtenstein, 1796; "India orientali"
aselenos Lichtenstein, 1796; "insulis maris pacifici"
aureonitens Lichtenstein, 1796; "Surinam"
brachyptera Lichtenstein, 1796; "India orientali"

---

[25] Dahlbom cited Africa as the locality for this species, but Schulz (1912) studied the type and surmised that it was an Oriental species, not African. Van der Vecht (1957) accepted Schulz' opinion.

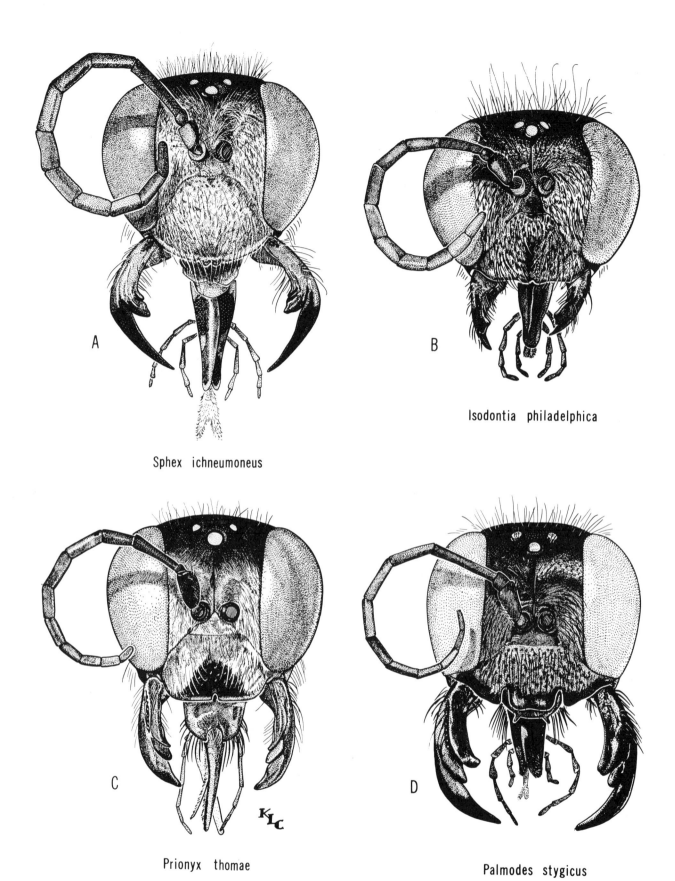

FIG. 27. Facial portraits of females in the tribe Sphecini.

chilon Lichtenstein, 1796; "India orientali"
chrysosticta Lichtenstein, 1796; "America"
chrysostoma Lichtenstein, 1796; "India orientali"
corynetes Lichtenstein, 1796; "Cap. bon. spei"
gigas Lichtenstein, 1796; "Surinam"
gnatho Lichtenstein, 1796; "Cap. bon. spei"
lineatus Lichtenstein, 1796; "India orientali"
manduco Lichtenstein, 1796; "Surinam"
manicatus Lichtenstein, 1796; ?
mesomelaena Lichtenstein, 1796; "India orientali"
nigellus Lichtenstein, 1796; "America"
nigripes Lichtenstein, 1796; ?
nigripes Holthusius (Schneider), 1800; ?
pumila Lichtenstein, 1796; ?
quinqueguttatus Lichtenstein, 1796; ?
ruficollis Lichtenstein, 1796; ?
rufilumbis Lichtenstein, 1796; "Surinam"
serpentinus Lichtenstein, 1796; ?
sinensis Lichtenstein, 1796; ?

Genus Isodontia Patton

Generic description: Two or more flagellomeres of male with flat or curved placoids which are sometimes spiculate; clypeal free margin sometimes more or less arcuate or obtusely angular but more often broadly trapezoidal (fig. 27 B), transverse part straight, sinuate, or concave (margin with a broad, truncate median lobe in *edax*), central part of margin often thickened or double edged, and often in females with a small median lobe or notch, latter sometimes bounded by two small lobes; surface of labrum plain, but in female sometimes with a median or lateral carina, or apex often with a pair of ridges or carinae which extend back along midline forming a short, narrow channel (fig. 28 H); mandible usually short, broadest at or near apex (very slender in *diodon*, fig. 28 T), apex not reaching base of opposite mandible when folded, apically bi- or tridentate (fig. 28 R, S), or with one or two (rarely three) subapical teeth on inner margin (fig. 28 Q) (in forms with single inner tooth it is occasionally located near middle), upper surface of teeth not carinate or ridged; malar space absent or very narrow; collar with some thickness and standing apart from scutum, or sharp edged and appressed to scutum; propodeum rarely with a defined dorsal enclosure, propodeal side without spiracular groove or groove indicated only near hindcoxa; basal veinlet of submarginal cell II sometimes equal to but more often shorter than length of anterior veinlet (fig. 25 C); anterior veinlet of submarginal cell III (part of radial sector) longer than posterobasal veinlet (part of media) or (rarely) both equal (variants of *chrysorrhoea*); midcoxae well separated in *edax*; outer surface of hindtibia spineless, or nearly so in *diodon*; outer margin of female foreleg usually without a tarsal rake (fig. 28 J), but when present rake spines are short and there is at most a single, slender seta on outer margin of tarsomere II (not including setae at outer apex), tarsomere I at most weakly asymmetrical; apicoventral bladelike setae of last tarsomere of mid and hind legs usually very broad (fig. 26 B), usually separated by a setal width or less, rarely separated by more than twice a setal breadth (*diodon, edax, chrysorrhoea*); claws usually long, moderately arcuate in lateral profile, prehensile, claw teeth obliquely oriented to inner margin (fig. 26 A); petiole often arcuate in lateral profile, length measured along dorsum usually more than (rarely slightly less than) combined lengths of hindtarsomeres II-IV; tergum VIII of male usually with cerci (absent in *elegans* groups and *paludosa*), sterna of male usually with some long, erect setae and/or transverse bands of appressed or suberect fimbriae, rarely without either; sternum VIII triangular, rounded or truncate apically and usually notched; ventral margin of penis valve head usually with teeth and a basal process; digitus of volsella usually simple.

Geographic range: Most of the 54 species of *Isodontia* are found in the Neotropical and Oriental Regions, each area having about 15 species. Five species are recognized in both the Ethiopian and Nearctic Regions and the Palearctic has four, but two of the latter are known only from Japan. Four species are known from Australia, but not all are endemic. Madagascar, New Britain, and New Guinea have one, two, and three species respectively. *Isodontia mexicana*, a North American species, has become established in Hawaii and France.

Systematics: These are moderate to large wasps (11-33 mm), with some males of *leonina* being the giants of the genus. Most species are black except that the gaster and legs may have varying amounts of red or yellow. Some species of the New World *apicalis* group are metallic blue or green. Wings vary from clear to yellowish or strongly infumate with bluish reflections. Erect body hair is usually black or white, but *leonina* is notable for the dense yellow hair on the legs and petiole of the male and on the propodeum.

The absence of a complete spiracular groove separates *Isodontia* from its close relative *Sphex* with the exception of the *Sphex* subgenus *Fernaldina*. *Fernaldina* is not closely allied with *Isodontia* as evidenced by *Sphex*-like details of its head and legs, but it presents problems in separating *Isodontia* and *Sphex* in a key. The wing characters used by us (Bohart and Menke, 1963) to separate *Isodontia* and *Sphex* in our key break down in the latter genus, especially in some Australian species of *Sphex s.s.* and in occasional variants of *Fernaldina* where the anterior veinlet of the third submarginal cell exceeds the length of the posterobasal veinlet. Nevertheless, the wing character is not without value because the majority of *Isodontia* species can be separated from most *Sphex* by the obviously greater length of the anterior veinlet as compared with the posterobasal veinlet. *Isodontia chrysorrhoea* is the only species of the genus that we have seen in which the two veinlets are occasionally equal. These veinlets are equal in a number of *Sphex* species also.

Unlike *Sphex*, *Isodontia* is not a fossorial genus and the females either lack a foretarsal rake or the rake is poorly developed. The tarsomeres are essentially bilaterally symmetrical in outline, and in those forms in which a weak rake is present (some species of the *apicalis* group) there is only a single lateral rake spine on the second tarsomere. Most *Isodontia* lack this single spine, although of course there are a few spines at the outer apex of the tarsomere. In contrast, *Sphex* females have a strongly developed rake, and the outline of tarsomere I is usually distinctly asymmetrical. Also, tarsomere II in *Sphex* usually has two or more lateral rake spines that are similar in length and form to those of tarsomere I.

There are also differences between the two genera in the form of the female mandible. In *Isodontia* the mandible is rather short so that the apex does not reach the base of the opposite mandible when closed, whereas in most *Sphex* the apex reaches or surpasses the base. The mandible apex is essentially bifid or trifid in most *Isodontia*, but in the least specialized species such as *edax, pelopoeiformis, longiventris,* and *simoni,* for example, there is a single subapical or even mesal tooth on the inner margin. The unusual mandible of *I. diodon* is a special case (fig. 28 T). In most *Sphex* the apical half of the unworn female mandible tapers gradually to a single sharp point. There is a single large inner tooth on the mandible in *Sphex,* and it is located near the middle of the margin in most species. A much smaller tooth is sometimes found at the proximal side of the base of the large tooth. Unlike *Isodontia,* the upper surface of the tooth in *Sphex* is usually ridged and sulcate. These formations extend back onto the mandible proper. The mandible of *Isodontia* is moderately arcuate from base to apex or even straight, but in *Sphex* the mandible is strongly arcuate so that the long axes of the base and apex approximate a right angle. The female mandible of a few Oriental and Australian *Sphex* is less strongly curved and the inner tooth is more subapical than mesal. Except for those male *Isodontia* with apically tridentate mandibles there are no strong differences between the mandibles of male *Sphex* and *Isodontia.*

The clypeus is usually thickened or double edged in *Isodontia,* but at least in female *Sphex* the free margin is sharp edged.

The long, arcuate petiole of most *Isodontia* contrasts rather sharply with the very short, straight petiole of most *Sphex.* Previously we (Bohart and Menke, 1963) compared the length of the petiole with the length of the hindcoxa in order to express the difference between the petioles of the two genera. In *Sphex* the petiole (measured along the dorsum) is less than or about equal to (or very slightly more than) the length of the hindcoxa. In *Isodontia* the petiole is usually 1.5 or more times the length of the hindcoxa. Unfortunately, the petiole of *Isodontia paludosa, edax, chrysorrhoea, dolosa,* and possibly others is closer to being equal to the length of the hindcoxa. It may be more practical to compare the petiole length with the combined lengths of hindtarsomeres II-IV, even though this method is not without similar drawbacks. In all *Sphex* studied the petiole is shorter than the combined length of the tarsomeres, whereas in *Isodontia* it is usually much longer than the tarsomeres, occasionally equal to them and shorter only in *dolosa* and some females of *ochroptera* and *harmandi.*

The use of subgenera in *Isodontia* was discussed by us in our paper on the Sphecinae (Bohart and Menke, 1963) in which we proposed the subgenus *Murrayella,* but we synonymized Arnold's (1945) subgenus *Leontosphex* under *Isodontia s.s.* We also established the *philadelphica* and *fuscipennis* species groups. After studying three-fourths of the species of *Isodontia* including some of the more unusual forms previously not available to us we have reached the conclusion that the existing subgeneric classification must be refined or discarded in favor of species groups. We feel that the latter course offers the best solution to infrageneric groupings in *Isodontia. Murrayella* was proposed for those species with apically tridentate mandibles in both sexes and no cerci in the male. The Old World species *paludosa* and *nigella* were cited as belonging in *Murrayella,* but after reappraisal only four New World species, i.e., the *elegans* group (*elegans, auripes, bruneri, mexicana*) seem assignable to the taxon as it was originally conceived. We have found that cerci are present in *nigella,* although they are very small. Cerci are absent in *paludosa,* but the mandible is more nearly bidentate, especially in the female. The large inner tooth is bifid, however, and when various species of *Isodontia* are studied, it becomes apparent that there is a progression from the bidentate (primitive) to the tridentate (specialized) condition, with the mandible of *paludosa* illustrating the intermediate state. This intermediate type is also found in females of *splendidula* and possibly other species. *Murrayella* could be redefined to include all species with fully tridentate mandibles without regard to the presence or absence of cerci, but this might result in a heterogeneous assemblage; and furthermore, in some species the mandibles are tridentate only in one sex (females of *chrysorrhoea* and *stanleyi* for example) or weakly tridentate in one sex (*ochroptera* males). A more comprehensive study will be required to solve this problem.

The *morosa* group (*morosa, harmandi, franzi, nigella,* etc.) contains a number of Oriental species that are similar to the *elegans* group except that cerci are present, and flagellomere I is sometimes equal to II. The labrum in both groups has two parallel carinae or ridges that extend back from the apex along the midline forming a groove. Krombein (1967b) thought this was possibly used in holding grass (for nest building) during flight to the nest. It is found also in the Old World *splendidula* group (*splendidula, stanleyi*), but in this assemblage the two carinae merge forming a Y. There is a long carina paralleling the lateral margin of the labrum also in the *splendidula* group. The isolated species *paludosa* has two weak carinae on the labrum. *Isodontia ochroptera* appears related to the *morosa* assemblage, but the labrum has only a single, short, median carina that does not reach the apex. Interestingly, the clypeal notch is strongly developed in *ochroptera* and bounded by two small lobes. This clypeal development probably serves the same purpose as the labral groove. Possibly the labral and clypeal notches have a function in prey transport.

There are a few unusual species in the genus. *Isodontia diodon,* for example, has peculiar, slender mandibles (fig. 28 T) and atypical genitalia. *Isodontia edax* has rather widely separated midcoxae, a striking truncate clypeal lobe, short, stout legs and unusual genitalia. Three other species have atypical genitalia: *guaranitica, leonina,* and *chrysorrhoea.* Before the species of the genus can be satisfactorily grouped, much more study will be required.

Species discrimination is not easy in some sections of the genus. The genitalia are often similar, and females may be difficult to associate with their respective males. The number of flagellomeres with placoids is generally considered to be constant for any given species, but it appears to us that at least in some forms there is variation. In *Isodontia fuscipennis* males, for example, placoids are on flagellomeres IV-VI, VII, or VIII, and in *I. philadelphica* placoids are found on flagellomeres V-VII in examples from the western United States and on flagellomeres V-VIII in eastern specimens. We have noted similar variation in some of the Old World species. The significance of this variation is unknown.

Bohart and Menke (1963) keyed the Nearctic forms and Willink (1951) keyed the species of southern South America. Arnold (1928) provided keys for the species of the Ethiopian Region, and Yasumatsu (1938a) keyed some of the Oriental forms. Kelner-Pillault (1962) separated the European species.

Biology: Isodontia are nonfossorial. They utilize preexisting cavities such as hollow plant stems, rolled leaves, abandoned bee burrows in logs or in the ground, or crevices between stones. The more important references on *Isodontia* nesting habits are those of Krombein (1967b, 1970) on *auripes, elegans,* and *mexicana;* C. Lin (1966) and Medler (1965) on *mexicana;* Piel (1933) on *nigella, harmandi, maidli;* Nicolas (1894) on *splendidula;* and De Stefani (1896, 1901), Rudow (1912), and Berland (1959), all of whom dealt with *paludosa.* Summaries and/or references of earlier authors can be found in most of these papers, and Tsuneki has listed the articles in Japanese that have appeared in the past. A good nontechnical account is that of Evans (1959c). Some of the above workers have reared *Isodontia* from stick traps.

Species such as *auripes, elegans, mexicana,* and *nigella* use grass stems or blades and similar plant and nonplant materials to divide their tubular nests into cells and to make the final closure. These species have earned the name "grass carrier (or carrying) wasps" because of their habit of flying to the nest carrying beneath them a long grass stem (up to 80 mm) held by the mandibles. An *Isodontia* flying with a long grass stem trailing off behind is an interesting sight (see figures in Evans or Piel). In some species grass stems are neatly coiled into wads which form the wall between two cells. The final closure of the nest in *auripes, mexicana,* and *nigella* takes place as the female fills the entrance hole with a broom-like tuft of grass stems which may protrude from the opening by as much as 50 mm.

Plant stems are not the only materials used to divide and close nests. Plant fibers (often derived from tree bark), moss, flower pappus, and bits of wood are used by some species. *Isodontia pelopoeiformis,* which is one of the more structurally primitive species, even supplements plant material with bits of soil or charcoal (Smithers, 1958; Heinrich, 1969).

In two species plant materials are used to construct or line the entire nest. In an *Isodontia* identified as *costipennis* (probably a misidentification) the nest is a tube woven of asclepiadaceous or apocynaceous seed hair (Mayer and Schulthess, 1923). Presumably this bag was located inside some kind of cavity, although the authors did not say so. The nest of *I. paludosa* is even more interesting. It closely resembles a small bird nest (see photo in Berland) with two concentric rings of plant material: an outer ring of rather coarse, coiled grass stems and an inner ring of thistle pappus that forms a cell. The overall diameter of the nest is 15 to 21 cm, but the depth is only 2 or 2.5 cm because the wasp builds the nest in narrow crevices between rocks, boards, or, as in one instance, the thin space between two beehives. This species could be called the "bird nest wasp."

Gryllidae and Tettigoniidae seem to be the normal prey of these wasps, but Iwata (1939c) found *Isodontia formosicola* provisioning with Blattidae. At least within a given area most *Isodontia* tend to provision with one species but occasional nests have mixed prey. Records for the genus are shown on table 7. *Oecanthus, Conocephalus,* and *Phaneroptera* seem to be favored.

The cells in the nests of *elegans, maidli, splendidula,* and most *nigella* contain a single wasp larva. However, *auripes* and *harmandi* females construct a single large brood cell (up to 10.5 cm in length) in which 2 to 12 larvae develop amicably. Apparently there is no cannibalism among the larvae unless insufficient food is provided by the mother wasp. Occasionally a female of *harmandi* may construct a nest with more than two brood cells. *Isodontia nigella* and *pelopoeiformis* sometimes construct nests that are intermediate between the unilarval, multicellular nests and the multilarval, unicellular type. In these species the cells are sometimes separated by flimsy partitions and/or some cells may contain two or more larvae. *Isodontia mexicana* constructs both multicellular and brood chamber nests. *I. paludosa* apparently constructs brood cells because Rudow (1912) showed seven cocoons in one nest of this species. Mass provisioning is the rule in *Isodontia,* and a temporary closure is made during the provisioning period. The number of prey per larva apparently varies according to the size of the former. Although De Stefani (1901) claimed that two eggs were invariably laid on each provision in *paludosa,* one egg per prey is typical for other species in the genus. The egg is laid on the thoracic venter of the first provision, and in brood cell nests additional eggs are laid in sequence on the next few prey brought to the nest. In brood cells containing large numbers of larvae, a few prey may be added without egg deposition

TABLE 7.
Summary of Prey Records of *Isodontia*

Species	Gryllidae			Tettigoniidae																						
	Gryllinae	Oecanthinae		Conocephalinae			Copiphorinae		Decticinae				Eneopterinae	Listroscelinae			Mecapodinae	Phaneropterinae							Pseudophyllinae	Tettigoniinae
	Gryllus	*Neoxabea*	*Oecanthus*	*Conocephalus*	*Odontoxiphidium*	*Orchelimum*	*Homorocoryphus*	*Neoconocephalus*	*Atlanticus*	*Eremopedes*	*Gampsocleis*	*Platycleis*	*Orocharis*	*Hexacentrus*	*Neobarrettia*	*Xiphidiopsis*	*Mecapoda*	*Dichopetala*	*Ducetia*	*Holochlora*	*Phaneroptera*	*Scudderia*	*Tylopsis*	*Amblycorypha*	*Pleminia*	*Tettigonia*
apicalis																								X		
auripes		X	X						X				X									X				
"costipennis"			X																							
elegans				X						X								X							X	
harmandi				X												X										
maidli				X																						
mexicana	X	X	X	X	X	X	X	X							X						X	X				
nigella											X		X							X	X					
paludosa																					X					
pelopoeiformis												X									X					
philadelphica			X																		X					X
splendidula																							X			

after the first four or five eggs have been laid. In brood cell nests some of the first eggs laid may hatch before the mother is finished with her provisioning, but there is no contact between them. Amputation of antennae and legs of the prey has been noted by several observers.

The brood cell with its multiple wasp larvae is the most interesting aspect of *Isodontia* biology. Such cells are unknown elsewhere in the Sphecidae, although cell walls may occasionally be absent in certain pemphredonines and crabronines (Ohgushi, 1945). Intriguing also is the fact that within the closely knit *elegans* species group, one species constructs the more primitive unilarval, multicellular nest (*elegans*), another makes the most advanced type of nest consisting of a multilarval brood chamber (*auripes*), and a third (*mexicana*) has both types of nests. Another interesting facet of *Isodontia* biology is the fact that the offspring of a single brood chamber are usually all of one sex (Tsuneki, 1963c, 1964d; Krombein, 1967b).

Flies of the families Tachinidae, Sarcophagidae (*Amobia*, *Senotainia*), Phoridae (*Megaselia*), Anthomyiidae (*Eustalomyia*), and Bombyliidae (*Anthrax*) have been found parasitizing nests of *Isodontia*. Chalcid wasps of the families Pteromalidae (*Epistenia coeruleata* Westwood) and Eulophidae (*Melittobia chalybii* Ashmead) have been reared from the nests of *I. elegans* (Parker and Bohart, 1966) and *auripes,* respectively. The mutillid *Sphaeropthalma* has been bred from the nest of *elegans.*

Checklist of *Isodontia*

abdita (Kohl), 1895 (*Sphex*); ? Sikkim (? = *aurifrons*)
    ssp. *nugenti* (Turner), 1910 (*Sphex*); Australia
albohirta (Turner), 1908 (*Sphex*); Australia
apicalis (F. Smith), 1856, p. 262 (*Sphex*); s. and se. U.S.
    cinerea Fernald, 1903
    harrisi Fernald, 1906 (*Chlorion*)
apicata (Bingham), 1897 (*Ammophila*); India
aurifrons (F. Smith), 1859 (*Sphex*); Sikkim, Indonesia, New Guinea
auripes (Fernald), 1906 (*Chlorion*); e. U.S.
    tibialis Lepeletier, 1845 (*Sphex*), nec Fabricius, 1781
auripygata (Strand), 1913 (*Sphex*); Taiwan
azteca (Saussure), 1867 (*Sphex*); Mexico, lectotype ♀, Cordova, Mexico (Mus. Geneva), present designation by Menke
    robusta Cameron, 1889 (*Sphex*)
bastiniana Richards, 1937; Guyana
boninensis (Tsuneki), 1973 (*Sphex*); Bonin Is.
bruneri (Fernald), 1943 (*Chlorion*); Cuba (? = *elegans*)
chrysorrhoea (Kohl), 1890 (*Sphex*); Sumatra, Borneo
    *apicalis F. Smith, 1856, p. 253 (*Sphex*), nec F. Smith, 1856, p. 262
    *hewitti Cameron, 1906 (*Sphex*)
costipennis (Spinola), 1851 (*Sphex*); Brazil
cyanipennis (Fabricius), 1793 (*Sphex*); Centr. and S. America
    nigrocoerulea Taschenberg, 1869 (*Sphex*)
    bipunctata Rohwer, 1913
diodon (Kohl), 1890 (*Sphex*); Burma, Malaya
    maia Bingham, 1894 (*Sphex*)
    ssp. *severini* (Kohl), 1898 (*Sphex*); Malaya, Java, Bali, Borneo
    malayana Cameron, 1902 (*Sphex*)
    ssp. *nigelloides* (Strand), 1915 (*Sphex*); Sri Lanka, India
    ssp. *philippensis* (Rohwer), 1922 (*Chlorion*); Philippines
    ssp. *alemon* van der Vecht, 1957; Indonesia: Sumba
dolosa (Kohl), 1895 (*Sphex*); Peru, French Guiana, Brazil
edax (Bingham), 1897 (*Sphex*); Sri Lanka, s. India to Sikkim, North Vietnam
egens (Kohl), 1898 (*Sphex*); New Britain
elegans (F. Smith), 1856 (*Sphex*); w. N. America
exornata Fernald, 1903; U.S.: Texas to Florida
    ?*instabilis* F. Smith, 1856 (*Sphex*)
formosicola (Strand), 1913 (*Sphex*); Taiwan
franzi (Cameron), 1902 (*Sphex*); Thailand, Borneo
fuscipennis (Fabricius), 1804 (*Pepsis*); Mexico to Argentina, ? Jamaica
    chrysobapta F. Smith, 1856 (*Sphex*)
    petiolata F. Smith, 1856 (*Sphex*), nec Drury, 1773
guaranitica Willink, 1951; Argentina, Brazil
harmandi (Pérez), 1905 (*Sphex*); Japan
insularis (Cameron), 1901 (*Sphex*); New Britain
jaculator (F. Smith), 1860 (*Sphex*); Indonesia: Batjan, new combination by J. van der Vecht
laevipes (W. Fox), 1897 (*Sphex*); Brazil
leonina (Saussure), 1890 (pl. 19, fig. 6) (*Sphex*); Madagascar
    leonina Saussure, 1891 (*Sphex*)
    leonina Saussure, 1892 (*Sphex*)
longiventris (Saussure), 1867 (*Sphex*); Guinea
    ?*meruensis* Cameron, 1908 (*Sphex*)
maidli (Yasumatsu), 1938 (*Sphex*); Japan
mexicana (Saussure), 1867 (*Sphex*); e. and s. U.S., Mexico, Centr. America, Hawaii, France, lectotype ♂, Orizaba, Mexico (Mus. Geneva), present designation by Menke
    apicalis Harris, 1835 (*Sphex*), nomen nudum
    apicalis Saussure, 1867 (*Sphex*), nec Smith, 1856
morosa (F. Smith), 1860 (*Sphex*); Indonesia, Batjan
    volatilis F. Smith, 1860 (*Sphex*), new synonymy by van der Vecht
    ? ssp. *triodon* (Kohl), 1890 (*Sphex*); s. China, Java, Borneo, Philippines
nigella (F. Smith), 1856 (*Sphex*); Oriental Region, Japan, e. USSR, Australia
    xanthognatha Pérez, 1905 (*Sphex*)
obscurella (F. Smith), 1856 (*Sphex*); Australia, Tasmania
ochroptera (Kohl), 1890 (*Sphex*); India to Celebes
paludosa (Rossi), 1790 (*Sphex*); n. Mediterranean region, Turkey, sw. USSR
    fuscata Dahlbom, 1843 (*Sphex*), nec Fabricius, 1793
    clavigera F. Smith, 1856 (*Sphex*)
    parthenia A. Costa, 1858 (*Sphex*)
    ?*eversmanni* Ed. André, 1888 (*Sphex*)

paranensis (Berland), 1927 (*Sphex*); Argentina, lectotype ♂, Bella Vista, Paraná, Corrientes Prov., Argentina (Mus. Paris), present designation by Menke

pelopoeiformis (Dahlbom), 1845 (*Sphex*); Ethiopian Region

pempuchi (Tsuneki), 1971 (*Sphex*); Taiwan

permutans (Turner), 1912 (*Sphex*); New Guinea

?*petiolata* (Drury), 1773 (*Sphex*); Jamaica (possibly a *Podium*)

   ?*vaga* Christ, 1791 (*Sphex*), nec Linnaeus, 1758

philadelphica (Lepeletier), 1845 (*Sphex*); U.S.: California to Florida and n. to Kansas and New York; Mexico

   macrocephala W. Fox, 1890 (*Sphex*)

   digueti Berland, 1927 (*Sphex*)

poeyi Pate, 1948; Cuba

praslinia (Guérin-Méneville), 1831 (pl. 8) (*Sphex*); New Ireland

sepicola (F. Smith), 1859 (*Sphex*); Indonesia: Aru, new combination by van der Vecht

simoni (du Buysson), 1897 (*Eremochares*); Rhodesia to South Africa, lectotype ♂, Hamman's Kraal, Transvaal, South Africa (Mus. Paris), present designation by Menke

   ?*trichionota* Cameron, 1910 (*Sphex*)

simplex (Kohl), 1898 (*Sphex*); New Guinea

sonani (Yasumatsu), 1938 (*Sphex*); Taiwan

splendidula (A. Costa), 1858 (*Sphex*); n. Mediterranean region; Turkey; Israel; ? Algeria

   ?*affinis* Lucas, 1848 (*Sphex*)

stanleyi (Kohl), 1890 (*Sphex*); Zaire to S. Africa

ustulata (Kohl), 1890 (*Sphex*); Indonesia: Timor

vidua (F. Smith), 1856 (*Sphex*); Australia

visseri Willink, 1951; Argentina

Subtribe Prionyxina

Diagnosis: Inner orbits essentially straight or broadly bowed in towards midline of face (fig. 27 C,D), straight portions converging below or, especially in females, parallel or slightly diverging below; scape short, ovoid, rarely moderately elongate; male antenna plain or with placoids, outline of labrum variable but fringed with long, conspicuous setae in female (fig. 28 E); malar space absent; mouthparts short to long; collar thick, separated from scutum by a constriction; admedian lines usually present, notauli usually absent; mesopleural suture not angulate below, upper metapleural area attenuate ventrad (fig. 29 C); lower metapleural area sometimes defined; spiracular groove absent; basal veinlet of submarginal cell II much longer than length of anterior veinlet (fig. 25 D); first recurrent vein received by submarginal cell II (exceptions in *Prionyx*), second recurrent received by submarginal cell III (exceptions in *Prionyx*); tibiae and tarsi moderately to densely spinose: pectens of inner hindtibial spur usually coarse and well spaced at middle (fig. 28 L,P); tergum VIII without cerci; male sterna IV-V with dull, scalelike or velvety micropubescence; volsella of one piece, without an articulating digitus (fig. 30 F-I).

Discussion: The primary characteristics of this subtribe are: 1) inner orbits straight or broadly bowed inward, 2) female labrum fringed with long setae, 3) basal veinlet of submarginal cell II much longer than anterior veinlet, 4) pectens of inner hindtibial spur coarse and well spaced at least at middle, 5) tergum VIII without cerci, 6) sterna IV-V usually with micropubescence, and 7) volsella solid, without a movable digitus. Minor exceptions to character (6) are found in *Prionyx,* and the pectens of the inner hindtibial spur are rather fine and close in *Chilosphex* (4), but otherwise these features are constant in the subtribe.

Ethologically the Prionyxina is more diverse than the Sphecina. The nest is dug after or before prey capture, and it is usually unicellular, although multicellular nests are known in a few species. Generally the cell is provisioned with a single large orthopteran, which with one known exception is dragged to the nest. The egg is laid on the upper base of the hindcoxa of the prey.

Genus Palmodes Kohl

Generic description: Male flagellum without placoids; female clypeus essentially flat, free margin with a truncate or slightly concave median lobe which is bounded laterally by a sinuation or notch (fig. 27 D); free margin of male clypeus sinuate or with an emarginate median lobe; female labrum with two widely separated fingerlike lobes on free margin (fig. 27 D); male labrum arcuate or truncate marginally, sometimes with two small triangular projections; female mandible with two teeth near middle of inner margin; male mandible with single, subapical inner tooth; mouthparts short; lower metapleural area at most weakly defined dorsally by an indistinct groove; metapleuron sometimes weakly depressed channellike; tibiae and tarsi moderately spinose; female foreleg with well developed foretarsal rake of long, often bladelike setae; apicoventral bladelike setae of last tarsomere of mid and hind legs very broad, usually separated by less than a setal breadth; petiole equal to or less than combined length of hindtarsomeres II-IV; last sternum of female conical or strongly keellike; male sternum VIII triangular or spatulate, apex rounded or truncate and often emarginate (fig. 30 A); penis valve head with teeth along ventral margin; apical lobe of volsella short and broad (fig. 30 F).

Geographic range: Palmodes is Holarctic, and its 20 species are equally divided between the Old and New World. Nine of the 10 North American species are restricted to the western part of the continent, and most of the Palearctic species are found around the Mediterranean.

Systematics: Palmodes are medium to large wasps (11-27 mm long) with largely black bodies. The gaster is partly or totally red in one or both sexes of some species, and a few are bichromatic in one sex. The wings are weakly to strongly infumate or sometimes yellowish. The legs of one Old World species are partly red.

Palmodes differs from *Prionyx* in very few characters. The median clypeal lobe of *Palmodes* females separates them from all *Prionyx* females. The presence of only two claw teeth in *Palmodes* is nearly diagnostic for the genus,

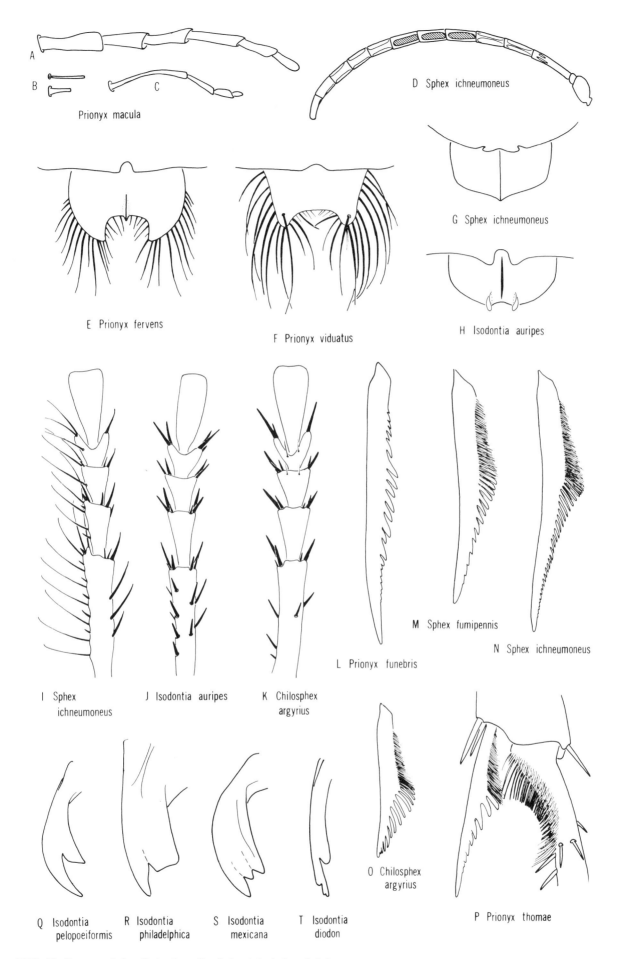

FIG. 28. Structural details in the tribe Sphecini; A-C, palpi drawn to same scale; A, maxillary palpus; B, labial palpus from same specimen as A; C, labial palpus from second specimen; D, male antenna with placoids on flagellomeres IV-VI; E-H, outline of labrum and clypeus; I-K, left female foretarsus, dorsal; L-O, inner hindtibial spur; P, inner hindtibial spur and basitarsal cleaning pecten; Q-T, right female mandible.

TABLE 8.
Summary of Prey Records for *Palmodes*

Prey	*californicus*	*carbo*	*dimidiatus*	*hesperus*	*laeviventris*	*praestans*	*occitanicus*
Gryllacrididae							
Cyphoderris		X					
Tettigoniidae							
Tettigoniinae							
Tettigonia							X
Decticinae							
Anabrus				X	X		
Atlanticus			X				X
Neduba	X			X			
Capnobates						X	
Pediodectes			X		X		
Ephipperinae							
Ephippiger							X
Phaneropterinae							
Platylyra	X						

but some Old World and Neotropical *Prionyx* also have two claw teeth. The absence of placoids on the male antenna of *Palmodes* is nearly diagnostic for the genus, but some Old World and a few Neotropical *Prionyx* males also have plain antennae. Unfortunately, some of these nonplacoid *Prionyx* have only two claw teeth, and the only positive means of identifying males to genus in these cases is by examination of the volsella. In *Palmodes* the apical volsellar process is short and broad (fig. 30 F), while in *Prionyx* males it is long and slender (fig. 30 I). Effective but nongeneric characters have been used in our generic key to divide these problem males.

Two species, *argyrius* and *pseudoargyrius*, have previously been assigned to *Palmodes*, but the ethology of *argyrius* (and presumbaly of *pseudoargyrius*) coupled with morphological differences has prompted us to place them in a separate genus.

The New World species were revised by Bohart and Menke (1961, 1963), and Roth (1963) revised the Old World forms. Species discrimination is often difficult in *Palmodes* depending in many forms on subtle differences in sculpture. It seems probable that some of the "subspecies" of the Palearctic *occitanicus* will prove to be species when sufficient material has been gathered for critical study. Roth divided the Palearctic species into several species groups and subgroups based on the shape of the last male sternum.

Biology: Three authors have published fairly detailed accounts of the nesting habits of several Nearctic *Palmodes:* La Rivers, 1945; Krombein, 1953, 1955b; and Evans, 1970. Biological data is available for only one Palearctic species, *occitanicus,* and the papers dealing with this species are mostly fragmentary (see Berland, 1928 and Tsuneki, 1963c for the more important references). *Palmodes* are mostly solitary ground nesters, although Evans (1970) found several *carbo* females nesting "somewhat gregariously," their nests separated by 30 to 50 cm. *P. occitanicus* nests have been found in cracks between the rocks of a wall, and in soil that accumulated under arched tiles of roofs as well as in the ground. The majority of the published accounts indicate that the nest is excavated before the prey is sought. Fabre (1915) stated that *occitanicus* females he observed always obtained their prey before digging their nests, but none of the subsequent observers of this species have corroborated his account. Similarly, Parker and Mabee (1928) said that *laeviventris* dug its nest after obtaining prey, but La Rivers (1945) found that this species invariably dug the nest first. In the case of *laeviventris* one possible explanation for this discrepancy is that several other black *Palmodes (stygicus, carbo, lissus)* may have been lumped together under this name, and one or more of them may dig the nest after catching prey. *Palmodes carbo* and *dimidiatus* make temporary nest closures, but *occitanicus* and *laeviventris* leave the nest open during the search for prey. *Palmodes carbo, hesperus, dimidiatus,* and *occitanicus* excavate single celled nests and use a single provision, but according to La Rivers (1945) *laeviventris* provisions one to four prey with two being the normal complement.

These are deposited in a linear fashion within the burrow and usually separated by a thin layer of soil. An egg is laid on each provision above the hindcoxa. Gryllacrididae and Tettigoniidae are used as prey by *Palmodes* (see table 8) and decticine katydids seem to be the most prevalent group. Generally the prey is much larger and heavier than the wasp, and it is dragged along the ground to the nest. The wasp grasps one antenna of the orthopteran with her mandibles, straddles the prey, and walks off with it. Amazingly, while bearing such loads, the wasp is able to scale perpendicular substrates to reach the nest.

Palmodes laeviventris is one of the most important natural predators of the Mormon Cricket (*Anabrus simplex*), and large outbreaks of the latter are often accompanied by increased populations of the wasp. La Rivers (1945) estimated that within a half square mile area in Nevada 30,000 wasps killed 500,000 Mormon Crickets.

The nyssonine wasp, *Stizoides unicinctus,* is a cleptoparasite of *Palmodes laeviventris* (La Rivers, 1945), and miltogrammine Sarcophagidae commonly infest the nest of *Palmodes*.

Checklist of *Palmodes*

californicus Bohart and Menke, 1961; N. America: California, Nevada to British Columbia
carbo Bohart and Menke, 1963; U.S.: Rocky Mtns. to West Coast
    morio Kohl, 1890 (*Sphex*), nec Fabricius, 1775
dimidiatus (De Geer), 1773 (*Sphex*); U.S.: transcontinental; n. Mexico
    violaceipennis Lepeletier, 1845 (*Sphex*)
    rufiventris Cresson, 1872 (*Sphex*)
    abdominalis Cresson, 1872 (*Sphex*), nec Drury, 1773
    opuntiae Rohwer, 1911 (*Chlorion*)
    daggyi Murray, 1951 (*Sphex*)
garamantis (Roth), 1959 (*Sphex*); Algeria
hesperus Bohart and Menke, 1961; U.S.: Great Basin to West Coast
insularis Bohart and Menke, 1961; U.S.: California Channel Islands
laeviventris (Cresson), 1865 (*Sphex*); U.S.: Great Basin
lissus Bohart and Menke, 1961; U.S.: California to Texas
melanarius (Mocsáry), 1883 (*Sphex*); Spain, n. Africa, Greece, Turkey, sw. USSR
    anatolicus Kohl, 1888 (*Sphex*)
    picicornis F. Morawitz, 1890 (*Sphex*)
minor (F. Morawitz),1890 (*Sphex*); Turkey, sw. USSR, Afghanistan
orientalis (Mocsáry), 1883 (*Sphex*); sw. USSR
occitanicus (Lepeletier and Serville), 1828 (*Sphex*); n. Mediterranian region
    proditor Lepeletier, 1845 (*Sphex*)
    confinis Dahlbom, 1845 (*Sphex*)
    montanus F. Morawitz, 1889 (*Sphex*)
    ssp. *perplexus* (F. Smith), 1856 (*Sphex*); China
    ?*solieri* Lepeletier, 1845 (*Sphex*)
    mandarina F. Smith, 1856 (*Sphex*)
? ssp. *australis* (Saussure), 1867 (*Harpactopus*); "Nova Hollandia" (erroneus)
    ssp. *syriacus* (Mocsáry), 1881 (*Sphex*); e. Mediterranean region, Afghanistan
    ssp. *puncticollis* (Kohl), 1888 (*Sphex*); sw. USSR
    ssp. *cyrenaicus* (Gribodo), 1924 (*Sphex*); Libya, lectotype ♂, Cirenaica (Mus. Genoa), present designation by Menke
    ssp. *barbarus* (Roth), 1963 (*Sphex*); Morocco to Algeria
    ssp. *gaetulus* (Roth), 1963 (*Sphex*); Algeria, Tunisia
    ssp. *ibericus* (Roth), 1963 (*Sphex*); Iberian Peninsula
pacificus Bohart and Menke, 1961; U.S.: coastal California
palmetorum (Roth), 1963 (*Sphex*); Algeria, Tunisia
parvulus (Roth), 1967 (*Sphex*); Turkey
praestans (Kohl), 1890 (*Sphex*); U.S.: Oregon to Texas
pusillus (Gussakovskij), 1931 (*Sphex*); sw. USSR, nw. China
sagax (Kohl), 1890 (*Sphex*); "New Holland"[26]
strigulosus (A. Costa), 1858 (*Sphex*); n. Mediterranean region; Turkey; sw. USSR
    ferus Dahlbom, 1843 (*Sphex*); nec Drury, 1782
    straboni Berland, 1927 (*Sphex*)
stygicus Bohart and Menke, 1961; U.S.: Great Basin

Chilosphex Menke, new genus

Generic description: male flagellum without placoids; female clypeus flat, free margin with a truncate median lobe; free margin of male clypeus sinuate; labrum broader than long, margin simple; inner margin of female mandible with two subapical teeth, apex thus appearing tridentate (fig. 29 A); male mandible with one subapical tooth on inner margin; mouthparts short; episternal sulcus short, continuous with scrobal sulcus, the two forming an arc (fig. 29 B); lower metapleural area defined dorsally by a weak ridge; metapleuron weakly depressed; tibiae and tarsi moderately spinose; female foreleg without a tarsal rake (fig. 28 K); pectens of inner hindtibial spur closely spaced and rather fine (fig. 28 O); apicoventral bladelike setae of last tarsomere of mid and hind legs separated by about a setal breadth; petiole about equal to combined lengths of hindtarsomeres II-IV; last sternum of female keellike; male sternum VIII spatulate, apex with a pair of short processes (fig. 30 B); penis valve head without teeth (fig. 30 C); apical lobe of volsella short and broad (fig. 30 G).

Type of genus: Sphex argyrius Brullé, 1833.

Etymology: Chilos = grass + Sphex meaning grass wasp in reference to the nesting habits of the genus. Gender masculine.

Geographic range: This genus occurs along the northern Mediterranean area from Spain to Israel and eastward to the southwestern USSR. Only two species are known.

Systematics: Medium to large wasps (16-22 mm long) with black bodies. The first two or three gastral segments are usually red, and the wings are clear or weakly infumate.

---

[26] According to Roth (1963) this is probably a Palearctic species, the type of which bears an erroneous locality label.

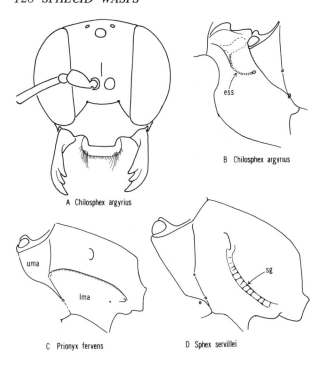

FIG. 29. Facial and Pleural views in the tribe Sphecini; A, female face; B, meso- and metapleuron, ess = episternal-scrobal sulcus; C,D, metapleuron and propodeum, uma = upper metapleural area, lma = lower metapleural area, sg = spiracular groove.

The absence of a foretarsal rake in the female and the short episternal sulcus separate *Chilosphex* from its close relative *Palmodes*. Additional differences are the simple labrum, the two short processes on the eighth sternum of the male, the absence of teeth on the aedeagal head, and a more finely pectinate inner hindtibial spur. The two species in *Chilosphex* have previously been assigned to *Palmodes,* but their nonfossorial habits and the morphological differences just enumerated argue for placement in a separate taxon.

Biology: Berland (1958) observed *argyrius* females nesting in the crevices between the stones of a wall in France. The nesting cavity was lined with stems of plants which were collected on the ground nearby. The stems were held in the mandibles and flown to the nest. Apparently each nest contains only one cell, and three or four prey are provisioned. Prey noted by Berland consisted of decticine Tettigoniidae of the genera *Pholidoptera* and *Metrioptera*. These were held by their antennae and dragged along the ground to the nest. The egg is laid just above the hindcoxa.

Checklist of *Chilosphex*

argyrius (Brullé), 1833 (*Sphex*); n. Mediterranean area to sw. USSR
  emarginatus Brullé, 1833 (*Sphex*), nec Villers, 1789
pseudoargyrius (Roth), 1967 (*Sphex*); Turkey

Genus Prionyx Vander Linden

Generic description: Flagellomere I sometimes equal to II or II longer than I in some males; male flagellomeres plain or with placoids on articles III-V or VI (rarely on II-VII or only III-IV), placoids usually flat and nonspiculate; female clypeus usually bulging, free margin arcuate to straight for most of its breadth, often with a median V or U-shaped notch (fig. 27 C); male clypeus arcuate to trapezoidal in outline, sometimes weakly emarginate; female labrum entire or broadly emarginate (fig. 28 F), or with a large U-shaped median notch in which case surface usually with a median carina which projects as a tooth into notch (fig. 28 E); male labrum entire or variably emarginate; female mandible with two teeth near middle of inner margin (one inner tooth weak or absent in *foxi, niveatus, trichargyrus*); male mandible with a single subapical inner tooth; mouthparts short to long, labial palpi sometimes very short and occasionally only with one to three segments (fig. 28 B); episternal sulcus extending ventrad to or nearly to anteroventral margin of pleuron except ending at level of scrobe in *pumilio* and occasional species variants in *kirbyi* group; lower metapleural area often defined in females and some males by a dorsal ridge or crest (fig. 29 C); metapleuron often depressed channellike for reception of hindfemur; three submarginal cells except two in some *pumilio*; first and second recurrent veins typically received by submarginal cells II and III respectively, but other combinations occur; tibiae and tarsi moderately to densely spinose; female foreleg with a well developed rake of long, usually bladelike setae; apicoventral bladelike setae of last tarsomere of mid and hind leg usually separated by not more than a setal breadth; claw with two to five basal teeth on inner margin (fig. 26 F,G); petiole varying from short, straight with length much less than combined length of hindtarsomeres II-IV, to elongate, arcuate, and slightly more than length of hindtarsomeres II-IV; last sternum of female varying from broadly, transversely arched to strongly keellike; male sterna occasionally with long, erect setae and IV-V usually covered with dull, scalelike, or velvety micropubescence which contrasts with adjacent sterna (III-VI densely fimbriate in *saevus*); sternum VIII triangular and often keellike, or spatulate, apex rounded or narrowly truncate and sometimes emarginate; penis valve head usually with teeth along ventral margin (fig. 30 D); apical lobe of volsella long and slender (fig. 30 H-I).

Geographic range: Prionyx is cosmopolitan, but 34 of the 56 species are Palearctic, and the majority of these occur in the Mediterranean area and sw. Asia. The Nearctic Region has seven species, and three of these are shared with the Neotropical Region which has 13 species. The Oriental Region has four species, but only one of these is endemic. The Ethiopian Region has seven species, but three of these extend beyond the area. Australia has two species, one of which ranges northward into the East Indies. The genus is notable for the number of wide-ranging species. *P. fervens* and *thomae* extend

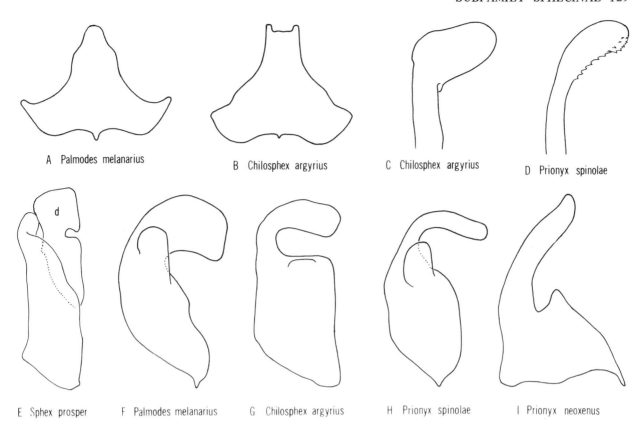

FIG. 30. Male terminalia in the tribe Sphecini; A,B, male sternum VIII; C,D, outer half of penis valve, lateral; E, left half of volsella, ventral, d = digitus; F-I, left half of volsella, lateral.

from the United States to Argentina. In the Old World *crudelis* ranges from South Africa to Turkey and eastward to sw. USSR and Sri Lanka; *P. viduatus* occurs from South Africa to the eastern Mediterranean and eastward to India, Taiwan, and the Ryukyus, and the distribution of *P. kirbyi* and *subfuscatus* extends from South Africa to Spain and eastward to China with *subfuscatus* reaching into India.

Systematics: Species of *Prionyx* range in length from 6.5 to 35 mm, *pumilio* and *macula* being the pygmy and giant of the genus, respectively. Black is the basic color, but the gaster is often totally or partly red. In some of the Old World species the terga are margined apically with white or cream bands. In a few species the head, thorax, and legs are partly or totally red. The wings may be clear, infumate, or yellowish. Erect vestiture may be black or pale and varies from dense to absent over most of the thorax. A number of species, especially in the Old World, have extensive areas of the head, thorax, and sometimes even the legs and gaster densely covered with appressed silver or gold hair (*trichargyrus, nigropectinatus,* and *macula,* for example). Many of the species have very bristly legs.

Because of its morphological diversity, *Prionyx* is difficult to characterize. About half of the species in the genus have three or more claw teeth, a feature that immediately separates them from related genera. Female *Prionyx* can be separated from *Palmodes* and *Chilosphex* by clypeal characters: margin entire or with a median notch. Males of about half of the species of *Prionyx* can be identified by the presence of placoids on the flagellum. About half of the remainder can be separated from *Palmodes* and *Chilosphex* by the claw tooth character mentioned above. On a purely structural basis the few remaining males, which have only two claw teeth and no placoids, apparently can be identified as *Prionyx* only by examining the apical process of the volsella. In *Prionyx* it is always long and slender (though arcuate), whereas in *Palmodes* and *Chilosphex* it is short and broad. Since this genitalic feature is not generally accessible we have used geographic and color differences in our generic key to simplify identification of these problem cases.

Now, as in 1963, our research indicates that the species group provides the best way to handle infrageneric groupings in *Prionyx*. We have been able to study many of the Old World forms that were previously unavailable to us, and it is obvious, as Beaumont (1968b) recently indicated, that some additional groups are warranted. Species groups are based primarily on the number of claw teeth, clypeal and labral details, absence or presence of placoids and whether or not they are flat or curved, the form of the last female sternum, and markings.

Identical numbers of claw teeth in different groups of *Prionyx* do not necessarily indicate a close relationship. For example, the *niveatus* and *crudelis* groups both have two claw teeth but in its gross morphology the latter group is obviously more closely allied to the *thomae* group in which the claw has three to five teeth. The *niveatus* group is much more similar to the *kirbyi* group that has three or four claw teeth. The *kirbyi* and *foxi* groups are closely related, both having similar morphology even down to the sharing of a keellike last sternum in the female. The *foxi* group has five claw teeth, however. The *pumilio* group includes *pumilio, spinolae, erythrogaster* and *herrerai. P. pumilio* differs from the other three species in having a short episternal sulcus, a thicker collar, a more elongate thorax, and often possessing only two submarginal cells; in its general facies *pumilio* is similar to the *niveatus, kirbyi,* and *foxi* groups, but this may be a matter of convergence rather than an indication of close affinity. Certainly the other three *pumilio* group species do not immediately reflect such a relationship. Rather, they seem to have closer ties with the *thomae* group.

There are more species groups among the *Prionyx* with placoids than we envisioned in our 1963 paper, and some of our characterizations were not accurate. The *crudelis* group, for example, should be restricted to those species with two claw teeth (three in occasional males), a median U-shaped notch in the female clypeus and labrum, and flat, nonspiculate placoids. In this group the female claw often has a short, stout, inner basal seta (fig. 26 F), the stigma is frequently open apically, and the metanotum is usually tuberculate, especially in the female. The metapleuron is depressed channel-like in some species of the *crudelis* group. Beaumont (1968b) used the name *subfuscatus* group for this assemblage, but we prefer to retain our 1963 appellation, primarily because *crudelis* is the type species of *Harpactopus,* a generic name that may some day be used as a subgenus. The *globosus* group needs little updating except to indicate that the female labrum has a U-shaped notch and the metanotum is not tuberculate. Beaumont (1968b) divided his *macula* group into two subgroups which in our opinion should be full groups. We restrict the *macula* group to his second sub-group, which consists of species with two claw teeth, curved placoids, placoid-bearing flagellomeres thicker and much longer than other flagellomeres, female clypeal margin entire or with a weakly obtuse and V-shaped emargination, labial palpi often with fewer than four segments, stigma open apically, inner hindtibial spur with coarse pectens only, and male sternum VI with a reflexed rounded lobe at outer apical angle. Beaumont gave additional distinctions. We propose that the first subgroup under Beaumont's *macula* group be called the *stschurowskii* group. It is much closer to the *crudelis* group than to the *macula* group, but it differs from the former in lacking the U-shaped clypeal and labral notches, and the stigma is closed apically. We have placed all New World placoid-bearing species in the *thomae* group, but possibly several groups will eventually be recognized. In the *thomae* group the claw has three to five teeth; flagellomeres I and II of the male are usually conspicuously shorter than the remaining articles; the clypeus and labrum have a U-shaped notch in the female; and with one exception, all species have flat, nonspiculate placoids. *P. simillimus* has curved, spiculate placoids. The *thomae* group is closely allied with the *crudelis* group. The preceding discussion about species groups must be considered tentative at this time. Many other features, such as mouthparts, leg setation, genitalia, and wing venation need to be studied before we know how best to group the species of *Prionyx.*

Prior to our work on the Sphecinae (Bohart and Menke, 1963) most of the species groups in the genus *Prionyx* were regarded as subgenera in the old, broadly conceived genus *Sphex.* In our 1963 paper we brought these "subgenera" together under the oldest generic name, *Prionyx.* Our current research further substantiates our belief that this was a phylogenetically sound move, but even today some authors cling to the older, clearly outmoded system. At this time, we do not feel that subgenera are useful in *Prionyx,* but when the morphology and ethology of this protean genus are better known it may be desirable to have them. For example, the genus is divisible into two nearly equal sections based on the presence or absence of placoids. Placoids are absent in the *niveatus, pumilio, kirbyi,* and *foxi* species groups. With the possible exception of the *pumilio* group, which as constituted here may not be homogeneous, these species' groups seem to represent a natural assemblage. The second section (placoids present) includes the *crudelis* (= *subfuscatus* group of Beaumont, 1968b), *globosus, stschurowskii, macula* (Beaumont, 1968b), and *thomae* species groups. This section is not as homogeneous as the first and would be even less so if all or part of the *pumilio* group was included.

Intriguing is the fact that in the two sections outlined above, morphological specializations have evolved at different rates. For example, the number of claw teeth has increased in both sections, but placoids have developed in only one of them. The clypeal and labral notches, which are presumed to be specializations, have evolved primarily in the section with placoid bearing species. This section also shows a reduction of labial palpal segments and the open stigma. Yet, the petiole has remained comparatively short and the gaster short and robust. In contrast the following specializations have developed in the nonplacoid section: long petiole and slender gaster, keellike last sternum, extensive appressed pubescence, cream colored tergal bands, and loss of the second submarginal cell. The *macula* group has the most phylogenetic discordance. This group, which retains the primitive two toothed claw, relatively simple labrum and clypeus, short petiole and stout gaster, and the nonparallelogram shape of submarginal cell II would seem to be the most generalized in the subtribe. At the opposite extreme, the same group has evolved placoids, an open stigma, reduction of labial palpal segments, and lobes at the outer corner of male sternum VI.

Venation is variable in *Prionyx* especially in the *foxi, kirbyi, niveatus,* and *stschurowskii* groups. In some of their species the first recurrent vein is often interstitial between submarginal cells I and II (*viduatus, niveatus, stschurowskii,* for example), or even received by I (*vidu-*

atus, niveatus). The second recurrent vein is more stable, but in *funebris* (*macula* group), *foxi,* and *stschurowskii,* for example, it may be interstitial between submarginal II and III or even be received by II in some *stschurowskii. Prionyx pumilio* typically has only two submarginal cells, the true second submarginal cell missing through the loss of the first r-m crossvein (fig. 25 F). Here the first recurrent vein is received by the first submarginal cell and the second recurrent is interstitial or occasionally (especially in females) also received by the first submarginal cell. Some *pumilio* have three submarginals, in which case the second recurrent vein is received by the second submarginal cell. Some individuals of this species have one wing with two and one wing with three cells. Another wing peculiarity is the open stigma (fig. 25 D) found in the *macula* and *crudelis* groups. Here that portion of vein R_1 which normally forms the outer end of the stigma is absent or evanescent. Typically in *Prionyx,* and for the subtribe for that matter, the second submarginal cell is a narrow parallelogram, and it is often described as being "higher than wide." Exceptions to this form are found in some *Prionyx* (fig. 25 D), in which the posterior veinlet (part of the Media) of the cell is much longer than the anterior veinlet (part of the Radial Sector). This presumably primitive condition is most common in the *macula* group.

The number of claw teeth varies from specimen to specimen in some species. For example, *saevus* may have two or three claw teeth, *neoxenus* three to four teeth, and *thomae* four to five teeth. The size of claw teeth generally increases distad, and apparently the basalmost tooth is sometimes undeveloped. The arrangement of placoids on the male antenna seems much more constant in *Prionyx* than in *Sphex* and *Isodontia,* but there is some variation in the *crudelis* group. For example, *crudelis* and *saevus* have placoids on flagellomeres 3-4 or 5.

Appreciation of species characters can be gained by consulting the works cited below. Past authors have not given enough attention to the labrum, the mouthparts including the comparative lengths of the palpi, the shape of the placoids, various wing details, and the last sternum of the male.

The North American *Prionyx* were keyed by Bohart and Menke (1963), but F. Parker's (1960) revision should be consulted. Willink (1951) keyed most of the South American species, but some of his nomenclature must be corrected by consulting our checklist. No comprehensive keys to the Old World *Prionyx* have been published since the time of Kohl (1890), but the following are useful: Arnold (1928), Ethiopian Region; Roth (1925), Honoré (1944a), Berland (1956), Palearctic Africa; Yasumatsu (1938a), eastern Asia; Berland and Bernard (1947), France.

Biology: Evans (1958c) produced an excellent review of the habits of the New World species and included data on one Old World form. Krombein (1964b) increased our knowledge of the Nearctic species *parkeri.* Important Old World references are Ferton (1902, 1912a), Berland (1925a), Chandler (1928), Grandi (1934), Piel (1935b), Benz (1959), Tsuneki (1963c), Tsuneki and Iida (1969), Iwata (1964a), and Kazenas (1968). The ethology of *Prionyx* is nearly as diverse and perplexing with respect to phylogeny as the morphology of the genus.

Most accounts indicate that *atratus, crudelis, fervens, globosus, parkeri, subfuscatus, thomae,* and *viduatus* obtain prey before digging a nest, but the common Old World species *kirbyi* digs its nest first and closes it temporarily. Three South American *Prionyx* also are reported to dig the nest first: *spinolae* (Janvier, 1926, 1928), *bifoveolatus* (Liebermann, 1931), and *fervens* (Conil, 1878). Evans (1958c) related an observation made on *fervens* in Mexico, contrary to Conil's report, that the wasp dug the nest after catching prey.

Prey are exclusively Acrididae (see table 9) that are transported over the ground except in *spinolae* which carries its prey in flight. The female straddles its victim and pulls the grasshopper forward by holding its antennae with the mandibles. The wasp's front legs often assist in holding the grasshopper. The prey is often left on some support such as a tuft of grass while the wasp digs her burrow. *Prionyx* are solitary nesters except for *spinolae* which nest in colonies of up to 100 individuals. Typically the nest is short, single celled and provisioned with only one grasshopper. However, *Prionyx spinolae* makes multicellular nests. In this species the cells are made in series, each separated by a layer of soil, and each cell is provisioned with five to ten acridids. Like *kirbyi, Prionyx spinolae* makes a temporary closure. Liebermann (1931) found that nests of *bifoveolatus* may contain up to three cells, and Ferton (1902) found one nest of *kirbyi* with two cells. The egg is laid on the upper edge of the hindcoxal membrane of the grasshopper, and in the case of *spinolae* it is laid on the first provision.

Based on the few species thus far studied, the sequence prey-nest or nest-prey correlates fairly well with the species groups outlined earlier. The *crudelis* and *thomae* groups obtain prey before digging the nest. Exceptions apparently are found in *fervens* and *bifoveolatus,* both of the *thomae* group. Erroneous observations may explain this discrepancy. More disconcerting is the fact that two rather closely allied species, *kirbyi* and *viduatus,* are diametrically opposed in this behavior sequence.

Some *Prionyx* species are regarded as important enemies of certain migratory grasshoppers. C. Williams (1933) and Haskell (1955) observed large numbers of *crudelis* accompanying swarms of the Desert Locust, *Schistocera gregaria,* in east Africa. These authors indicate that the wasps follow the swarms, but actual migration of the wasps remains to be proven, as Haskell himself pointed out.

Miltogrammine Sarcophagidae are common cleptoparasites of *Prionyx* and the sphecid *Stizoides* has been found to usurp the provisions of *P. atratus.* Liebermann (1931) found mutillids in the nests of *bifoveolatus.*

Checklist of Prionyx

afghaniensis (Beaumont), 1970 (*Sphex*); Afghanistan
atratus (Lepeletier), 1845 (*Sphex*); N. America
   labrosus Harris, 1835 (*Sphex*); nomen nudum

TABLE 9.
Summary of Prey Records for *Prionyx*

Species	Acrididae	Ageneotettix	Amphitornus	Aulocara	Mermiria	Orphulella	Cantantopinae	Calliptamus	Catantops	Scotussa	Cyrtacanthacridinae	Melanoplus	Paraidemona	Schistocerca	Trigonophymus	Gomphocerinae	Chorthippus	Dasyhippus	Dociostaurus	Euchorthippus	Omocestus	Oedipodinae	Arphia	Celes	Dissosteira	Encoptolophus	Gastrimargus	Locusta	Oedipoda	Paradalophora	Scirtetica	Sphragemon	Sphingonotus	Trilophidia	Trimerotropis	Xyleus
atratus		X										X											X		X					X		X			X	
bifoveolatus			X	X	X																															
crudelis										X				X																						
fervens								X						X			X	X	X	X																
kirbyi								X	X									X	X	X	X															
nigropectinatus												X		X							X															
niveatus																																	X			
parkeri												X					X						X		X						X					
spinolae?													X	X?	X																					
subfuscatus																			X								X	X	X							
thomae			X			X			X								X						X		X		X	X	X					X	X	
viduatus																			X																X	X

brunnipes Cresson, 1872 (*Priononyx*)
bifoveolatus (Taschenberg), 1869 (*Priononyx*); s. S. America
    striatulus Brèthes, 1909 (*Sphex*), lectotype ♂, "Buen. Ayres" (Mus. Buenos Aires), present designation by Menke, new synonymy by Menke
    subexcisus Brèthes, 1909 (*Sphex*), lectotype ♀, "Rep. Arg.?" (Mus. Buenos Aires), present designation by Menke, new synonymy by Menke
    wagneri Berland, 1927 (*Sphex*), lectotype ♂, Icano, Prov. Santiago del Estero, Argentina (Mus. Paris), present designation by Menke, new synonymy by Menke
    caridei Lieberman, 1931 (*Sphex*), new synonymy by Menke
canadensis (Provancher), 1887 (*Priononyx*); Canada, n. U.S.
    excisus Kohl, 1890 (*Sphex*)
chobauti (Roth), 1925 (*Sphex*); Morocco, Algeria
crudelis (F. Smith), 1856 (*Harpactopus*); Zambia to Libya; Mauritius; e. Mediterranean area; sw. USSR; Saudi Arabia to India and Sri Lanka
    ?*hirtipes* Fabricius, 1793 (*Sphex*) (possibly a *Sphex*, teste van der Vecht, 1961a)
    rufipennis Fabricius, 1793 (*Sphex*), nec De Geer, 1778
    aegyptius Lepeletier, 1845 (*Sphex*), nec Linnaeus, 1758, lectotype ♀, "Egypte" (Mus. Turin), present designation by Menke
    grandis Radoszkowski, 1876 (*Sphex*)
    ?*turcomanicus* Radoszkowski, 1893 (*Sphex*)
damascenus (Beaumont), 1968 (*Sphex*); Syria
elegantulus (Turner), 1912 (*Sphex*); China
erythrogaster (Rohwer), 1913 (*Callosphex*); Peru
fervens (Linnaeus), 1758 (*Sphex*); U.S.: California to Texas; Centr. and S. America
    johannis Fabricius, 1804 (*Pepsis*)
    doumerci Lepeletier, 1845 (*Sphex*)
    striatus F. Smith, 1856 (*Priononyx*)
    laerma Cameron, 1897 (*Sphex*)
foxi Bohart and Menke, 1963; U.S.: California to Texas
    ferrugineus W. Fox, 1892 (*Sphex*), nec Lepeletier, 1845
fragilis (Nurse), 1903 (*Sphex*); Pakistan, India
funebris (Berland), 1927 (*Sphex*); Ethiopia to S. Africa
globosus (F. Smith), 1856 (*Sphex*); Australia
gobiensis (Tsuneki), 1971 (*Sphex*); Mongolia
guichardi (Beaumont), 1967 (*Sphex*); Turkey
haberhaueri (Radoszkowski), 1872 (*Sphex*); Israel, sw. USSR, Iran, Afghanistan
herrerai (Brèthes), 1926 (*Sphex*); Peru
    ?*vilarrubiai* Giner Marí, 1944 (*Sphex*)
inda (Linnaeus), 1758 (*Sphex*); Ethiopian Region
    indostana Linnaeus, 1764 (*Sphex*)
    tyrannus F. Smith, 1856 (*Harpactopus*)
    vagus Radoszkowski, 1881 (*Sphex*), nec Linnaeus, 1758
    englebergi Brauns, 1889 (*Sphex*)
insignis (Kohl), 1885 (*Sphex*); Syria

judaeus (Beaumont), 1968 (*Sphex*); Israel
kirbyi (Vander Linden), 1827 (*Ammophila*); s. Europe, n. Africa, s. Palearctic Region in Asia
    albisectus Lepeletier and Serville, 1828 (*Sphex*), lectotype ♀, "S. albisecta" in Lepeletier's handwriting (Mus. Paris), present designation by Menke
    ssp. *marginatus* (F. Smith), 1856 (*Parasphex*); Ethiopian Region
    sjoestedti Cameron, 1908 (*Sphex*)
    curvilineatus Cameron, 1912 (*Sphex*)
    congoensis Berland, 1927 (*Sphex*), lectotype ♀, Libreville, Gabon (Mus. Paris), present designation by Menke
    alluaudi (Berland, 1927 (*Sphex*), lectotype ♀, "Assinie, Cote occid. Afrique" (Mus. Paris), present designation by Menke
    chudeaui Berland, 1927 (*Sphex*)
kurdistanicus (Balthasar), 1953 (*Sphex*); Iraq
†*leuconotus* (F. Morawitz), 1890 (*Sphex*); sw. USSR; nec Brullé, 1833 (? = *viduatus*)
lividocinctus (A. Costa), 1858 (*Enodia*); Mediterranean area, Turkey, Iran, sw. USSR
    isselii Gribodo, 1880 (*Priononyx*)
    obliquestriatus Mocsáry, 1883 (*Enodia*)
    graecus Mocsáry, 1883 (*Enodia*)
    ssp. *oasis* (Tsuneki), 1971 (*Sphex*); Mongolia
    ssp. *apakensis* (Tsuneki), 1971 (*Sphex*); n. China
macula (Fabricius), 1804 (*Pepsis*); n. Africa, Israel, Kuwait, Saudi Arabia, Iran, Iraq
    eatoni E. Saunders, 1910 (*Sphex*)
    ssp. *lugens* (Kohl), 1889 (*Sphex*); sw. USSR: Armenian S.S.R.; Iran; Afghanistan
melanotus (F. Morawitz), 1890 (*Sphex*); sw. USSR (? = *radoszkowskyi*)
neoxenus (Kohl), 1890 (*Sphex*); Argentina, Chile
    ?*melaenus* Spinola, 1851 (*Sphex*)
    ommissus Kohl, 1890 (*Sphex*)
    melanogaster Brèthes, 1910 (*Sphex*)
    nigricapillus Berland, 1927 (*Sphex*), lectotype ♂, Arequipa, Peru (Mus. Paris), present designation by Menke
    gayi Berland, 1927 (*Sphex*), lectotype ♂, "Chili" (Mus. Paris), present designation by Menke
nigropectinatus (Taschenberg), 1869 (*Sphex*); Egypt, Sudan, Iran, ? India
    ?*dives* Lepeletier, 1845 (*Sphex*)
    ?*nivosus* F. Smith, 1856 (*Sphex*)
    maracandicus Radoszkowski, 1877 (*Podium*)
niveatus (Dufour), 1853 (*Sphex*); Egypt to Morocco; Israel; Saudi Arabia; Iran; sw. USSR, lectotype ♀, Ponteba (Mus. Paris), present designation by Menke
    albopectinatus Taschenberg, 1869 (*Enodia*)
    ?*suavis* F. Morawitz, 1893 (*Sphex*)
    ssp. *ettingol* (Tsuneki), 1971 (*Sphex*); Mongolia
notonitidus (Willink), 1951 (*Chlorion*); Argentina, Chile
nudatus (Kohl), 1885 (*Sphex*); nw. Africa, sw. Europe, e. Mediterranean area to Iran and Afghanistan
parkeri Bohart and Menke, 1963; transcontinental in N. America, s. to s. Mexico

perezi (Berland), 1927 (*Sphex*); Senegal, Algeria, lectotype ♀, Senegal (Mus. Paris), present designation by Menke (? = *viduatus*)
persicus (Mocsáry), 1883 (*Sphex*); Iran, sw. USSR
?*hispidus* F. Morawitz, 1890 (*Sphex*)
*pseudostriatus (Giner Marí), 1944 (*Sphex*); Peru (? = *fervens*)
 *vaqueroi Giner Marí, 1944 (*Sphex*), new synonymy by Menke
pumilio (Taschenberg), 1869 (*Pseudosphex*); Argentina, Chile
 ?*dolichoderus* Kohl, 1890 (*Sphex*)
 ?*albospiniferus* Reed, 1894 (*Neosphex*)
radoszkowskyi (Kohl), 1888 (*Sphex*); sw. USSR, Uzbeck S.S.R.
reymondi (Roth), 1954 (*Sphex*); Algeria
saevus (F. Smith), 1856 (*Harpactopus*); Australia
 ssp. *harpax* (Kohl), 1898 (*Sphex*); Indonesia: Sumba, Komodo, Flores, Timor, Wetar
senegalensis (Arnold), 1951 (*Sphex*); Senegal
senilis (Morice), 1911 (*Sphex*); n. Africa, Saudi Arabia
 biskrensis Roth, 1925 (*Sphex*)
sennae (Mantero), 1901 (*Sphex*); Argentina
simillimus (Fernald), 1907 (*Chlorion*); Argentina
 tucumanensis Strand, 1910 (*Sphex*)
sirdariensis (Radoszkowski), 1877 (*Sphex*); Iran, sw. USSR
songaricus (Eversmann), 1849 (*Sphex*); Turkey, Israel; sw. USSR; Iran; Iraq, Afghanistan
 tenuicornis F. Morawitz, 1890 (*Sphex*)
spinolae (F. Smith), 1856 (*Sphex*); Chile
 chilensis Spinola, 1851 (*Sphex*), nec Lepeletier, 1845, lectotype ♀, "Chili", (Mus Turin), present designation by Menke
stschurowskii (Radoszkowski), 1877 (*Sphex*); sw. USSR, Iran, Afghanistan
 ssp. *hyalipennis* (Kohl), 1895 (*Sphex*); n. Africa, Israel, Saudi Arabia, Iraq
subatratus (R. Bohart), 1958 (*Priononyx*); w. U.S.; Mexico
subfuscatus (Dahlbom), 1845 (*Sphex*); Mediterranean area; sw. Asia; India; China
 soror Dahlbom, 1845 (*Sphex*)
 nigritus Lucas, 1848 (*Sphex*), nec Fabricius, 1781
 desertorum Eversmann, 1849 (*Sphex*)
 subfuscatus Eversmann, 1849 (*Sphex*), nec Dahlbom, 1845
 chrysopterus Ruthe and Stein, 1857 (*Enodia*)
 anthracinus A. Costa, 1867 (*Gastrosphaeria*)
 ? ssp. *namkumiensis* (Laidlaw), 1929 (*Sphex*); s. and e. India
 ? ssp. *rhodesianus* (Arnold), 1936 (*Chlorion*); Rhodesia, Tanzania
 ssp. *albovillosulus* (Giordani Soika), 1942 (*Sphex*); Somalia
 ? ssp. *rukwaensis* (Arnold), 1959 (*Sphex*); Tanzania
sundewalli (Dahlbom), 1845 (*Enodia*); S. Africa
thomae (Fabricius), 1775 (*Sphex*); U.S.: w. 100th merid., Mississippi to Georgia; Centr. and S. America, W. Indies
 crucis Fabricius, 1804 (*Pepsis*)
 pubidorsum A. Costa, 1862 (*Enodia*)
 antillarum Saussure, 1867 (*Priononyx*)
 mexicanus Saussure, 1867 (*Priononyx*)
 edwardsi Cameron, 1903 (*Sphex*), lectotype ♂, Ambato, Ecuador (Mus. London), present designation by Menke
 platensis Brèthes, 1908 (*Sphex*), lectotype ♂, "Nov. Frib." (Mus. Buenos Aires), present designation by Menke
 ?*altibia* Strand, 1911 (*Sphex*)
trichargyrus (Spinola), 1838 (*Sphex*); Senegal to Egypt
 leucosoma Kohl, 1890 (*Sphex*)
viduatus (Christ), 1791 (*Sphex*); Africa; e. Mediterranean area; Saudi Arabia to India; Socotra; China; Taiwan; Ryukyus
 pubescens Fabricius, 1793 (*Sphex*)
 canescens Dahlbom, 1843 (*Enodia*)
 micans Eversmann, 1849 (*Sphex*)
 granti W. F. Kirby, 1900 (*Sphex*)
 platynotus Matsumura, 1912 (*Sphex*)
 ?*zanoni* Gribodo, 1925 (*Priononyx*)
 ssp. *argentatus* (Mocsáry), 1883 (*Enodia*); sw. USSR, new status by Beaumont
 mocsaryi Kohl, 1885 (*Sphex*)
 ssp. *pollens* (Kohl), 1885 (*Sphex*); e. Mediterranean area
vittatus (Kohl), 1884 (*Enodia*); sw. USSR; Israel; Afghanistan (? = *haberhaueri*)
zarudnyi (Gussakovskij), 1933 (*Sphex*); Iran

Nomen nudum in *Prionyx*

noverca Kaye, 1910 (*Pseudosphex*)

Tribe Ammophilini

These rather large wasps have a distinctive, long, slender gaster and usually a long petiole. One genus, *Hoplammophila*, nests in preexisting cavities in wood, but the other five genera of the tribe are fossorial. *Ammophila*, the largest genus of the tribe, is its best known member thanks largely to the statements of early students of wasp behavior, who claimed that some species are tool users and hence possess intelligence. A few *Ammophila* have been shown to construct and maintain several nests simultaneously, a rather amazing feat. Lepidopterous and hymenopterous caterpillars are provisioned by most genera, but *Eremochares* stores its nests with grasshoppers.

Diagnostic characters:

1. (a) Sockets near middle of face and usually separated from frontoclypeal suture; (b) male antenna without placoids; (c) scape ovate to slender and parallel sided; (d) flagellomere I nearly always longer than II (equal in a few *Podalonia*).
2. (a) Clypeus large, broader than long to longer than broad; (b) labrum variable but usually broader than long.
3. (a) Mandible with one to three inner teeth, rarely simple; (b) mandible socket closed by paraman-

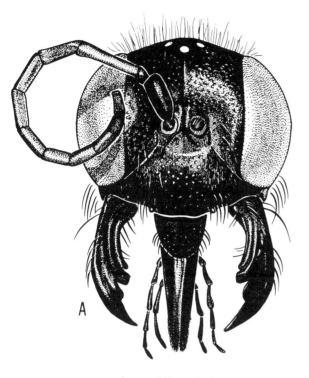
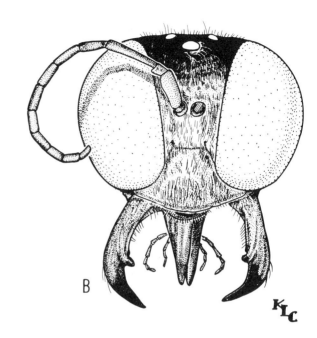

Ammophila sabulosa   Eremochares mirabilis

FIG. 31. Facial portraits of females in the tribe Ammophilini.

dibular process of hypostoma except in males of four genera.
4. (a) Maxillary palpal segment III symmetrical; (b) hypostomal carina complete; (c) occipital carina incomplete below, ending just short of hypostomal carina, or meeting hypostomal carina at or near its apex.
5. Collar short to long.
6. (a) Propodeal enclosure usually present and defined by differential sculpture rather than a sulcus; (b) spiracular groove absent; (c) episternal sulcus usually long, extending to or nearly to anteroventral margin of pleuron (short or absent in some *Ammophila*); (d) lower metapleural area usually present, defined by ridge or sulcus between upper and lower metapleural pits.
7. (a) Marginal cell apex variable; (b) three submarginal cells present (except one or two in some *Ammophila*); (c) both recurrent veins usually received by submarginal cell II (fig. 35 A-B).
8. (a) Female with psammophore; (b) midcoxae contiguous or nearly so; (c) two apicoventral setae of hindtarsomere V setiform to moderately broad, blade-like, separated by three or more setal widths (only two widths in some *Podalonia*); (d) inner margin of claw simple or with one or two basal teeth; (e) tarsi without plantulae except in most *Parapsammophila.*
9. (a) Petiole usually consisting of sternum I only (tergum I part of petiole in *Ammophila* and some *Eremnophila*); (b) tergum VIII without cerci; (c) volsella with articulating digitus.

Systematics: It is difficult to give a clear-cut diagnosis for the Ammophilini. The absence of cerci is a consistent feature, and the reception by the second submarginal cell of both recurrent veins is a nearly universal character. Exceptions to this wing pattern are found in *Parapsammophila, Eremochares,* and *Ammophila.* This wing feature is shared with many Sceliphronini, but the absence of plantulae in all ammophilins with the exception of some species of *Parapsammophila* is a more basic distinction. Claw form in the Ammophilini is not diagnostic because the claws may be simple, single, or double toothed. However, when present, the teeth are basal, unlike the mesal tooth of the vast majority of sceliphronins. The ammophilin genera with two toothed claws are most easily separated from the Sphecini by the narrower apicoventral setae of hindtarsomere V and/ or the reception of both recurrent veins on submarginal cell II. The petiole and gaster characters employed earlier by us (Bohart and Menke, 1963) in separating ammophilin genera with two claw teeth from similarly dentate sphecin genera do not hold up.

We are following Menke (1966a) in our recognition of genera. Some Old World authors still follow Kohl (1906) in treating these taxa as subgenera of one large, portmanteau genus, *Ammophila;* but because these entities are easily distinguished from one another morphologically, there is no valid reason for continuing this practice. It should also be noted that biology, so far as known, supports the recognition of these taxa as genera. Generic differences are based mainly on claw dentition, characters of the petiole (sternum I), and the length of the mouthparts.

The Ammophilini is a rather homogeneous group, and it is dominated by one evolutionary trend: the tendency towards lengthening of the petiole and slenderization of the gaster. These specializations are least developed in *Podalonia,* especially in the female. In this genus the petiole is composed only of sternum I, which is often no longer (measured along the dorsum) than the hindcoxa. Tergum I, at least in most females, has the normal bell shape common to most nonpetiolate wasps. It is usually higher than long or as long as high in lateral profile. *Parapsammophila, Eremochares,* and *Hoplammophila* have a longer petiole, and tergum I is usually drawn out so that it is longer than high. Some females and many males of *Podalonia* have this type of tergum. The highest degree of abdominal elongation is found in *Ammophila* and some *Eremnophila,* where tergum I may be very long and slender, often equaling the length of sternum I, and is usually displaced to near the apex of sternum I. In these genera the petiole is clearly "two segmented" and the gaster articulates with the apex of tergum I. Sternum I is widely separated from sternum II, but the two are connected by a ligament. Coordinated with the lengthening of tergum I is the displacement posterad of the spiracle. The reason for abdominal lengthening in the Ammophilini is not clear, but two hypotheses can be made. Perhaps the stalklike abdominal arrangement facilitates the stinging of caterpillar prey, which have long bodies and a major ganglion in each segment. The orthopterous prey of *Eremochares* does not support this idea, however, since the wasp does not have to contend with an organism possessing many consecutive nerve ganglia; but this genus may have switched from caterpillars to grasshoppers. Another theory is that the long abdomen serves as a balancing organ during flight, especially while carrying prey.

Lengthening of the mouthparts is another interesting evolutionary trend in this tribe. The galea is short in *Eremochares* and most *Parapsammophila,* moderately long in *Hoplammophila,* some *Eremnophila,* and one or two species of *Parapsammophila.* The galea is long in most species of *Podalonia, Ammophila,* and *Eremnophila.*

We regard open mandibular sockets as a generalized feature in the Sphecidae. Therefore, our discovery that the sockets are open in males of *Parapsammophila, Eremochares,* and some *Podalonia* and *Ammophila* came as a surprise, because we regard the Ammophilini as the most specialized of the three sphecine tribes. It could be argued that the sockets have become secondarily open, but evidence in *Parapsammophila,* certainly one of the most primitive ammophilin genera, contradicts this theory. In this genus the sockets are usually rather broadly open in the male, but in the species with the longest mouthparts the sockets are narrowly open. However, open sockets in the highly specialized genus *Ammophila* tend to support the premise that they have become secondarily open.

The phylogenetic significance of claw teeth or the lack of them is difficult to assess. If the Sceliphronini are to be regarded as the most primitive sphecines, then perhaps a single claw tooth should be considered the generalized claw condition in the Ammophilini. If this is true then specialization in the Ammophilini has gone in two directions: loss of the tooth on the one hand, and the development of a second tooth on the other. However, another factor is involved. It seems likely that the ammophilins arose from an ancestor more like the Sphecini than the Sceliphronini, and two toothed claws are the generalized condition in the Sphecini. Hence, the most probable evolutionary trend in the Ammophilini has been from two toothed claws to simple claws. On this basis *Parapsammophila, Eremochares,* and *Hoplammophila* possess the most generalized claws, and *Podalonia, Eremnophila,* and *Ammophila* the most specialized. In the genera with two claw teeth, the more basal tooth is associated with the long setae that project from the base of the claw towards the apex. In some cases, this basal tooth is more like an angular swelling than a tooth, and its origin may not be the same as that of the distal tooth.

The arolium is very small in the female of some species of *Podalonia* and *Ammophila,* but it is nearly always large in the corresponding male, leading us to assume that this reduction in size is a specialization. In *Ammophila formicoides,* a highly specialized species, the arolium is small or nonexistent in both sexes.

The loss of the outer midtibial spur is a widespread tendency in the tribe, but it does not occur in two rather specialized genera, *Hoplammophila* and *Eremnophila.* The presence of a single spur is nearly constant in *Eremochares,* but in the other genera the loss of one spur is usually in the more specialized groups only.

The presence of plantulae in most species of *Parapsammophila* is another surprising discovery, but one that substantiates our feeling that the genus is the most primitive of the tribe. The most generalized venation found in the Ammophilini occurs in some species of *Parapsammophila.* In *P. algira* for example, submarginal cell III receives the second recurrent vein. It is less easy to assess the relative evolutionary status of the remaining genera because of nonconcordance of many features. For example, *Podalonia* has the most generalized abdomen in the tribe, but the claws are usually simple and the mouthparts are long. The rather universal habit of excavating a nest after obtaining prey is a primitive trait of *Podalonia,* but the use of lepidopterous larvae for provisioning probably should be regarded as an advancement. The narrow face of *Eremochares* coupled with the loss of one midtibial spur, long petiole, and modified petiole socket indicate that this is a specialized genus, and yet it has claw teeth and short mouthparts. *Eremochares dives* stocks its nests with relatively primitive prey, grasshoppers; but the species is a progressive provisioner, an advanced type of nest stocking. Available evidence suggests that this species may even maintain several nests simultaneously. *Hoplammophila* has changed from fossorial to xylicolous nesting, but morphological specializations are mainly the closed mandibular sockets, triangular male clypeus, and unusual genitalia. *Eremnophila* and *Ammophila* are obviously the most specialized genera in the tribe because of their "two-segmented" petiole and long mouthparts, among other things. Some *Ammophila* exhibit the most advanced biology known in the tribe: delayed progressive provisioning and maintenance of several nests at one time.

Because the genera we recognize were until fairly recently included as subgenera under *Ammophila,* many of the keys to species are broad in scope. A few such papers are: Roth (1928, 1929), Kohl (1906), Alfieri (1946), Arnold (1928), and Tsuneki (1967d).

Biology: Except for *Hoplammophila,* which has changed to a xylicolous nesting habit, all ammophilins are fossorial. As far as known, most *Podalonia* obtain prey before excavating a nest, but the reverse is true in the other genera, although a few *Ammophila* have also been reported to obtain prey first. *Eremochares* provisions with grasshoppers, but lepidopterous and sometimes hymenopterous caterpillars are stored by the other genera of the tribe. F. D. Parker (personal communication) has observed *Ammophila* carrying weevil larvae. Mass provisioning is employed by many species, but progressive provisioning is known in *Eremochares* and *Ammophila.*

Key to genera of Ammophilini

1. Episternal sulcus curving back to scrobe from subalar fossa, then extending obliquely ventrad to anteroventral area of mesopleuron (fig. 32 G); New World forms .... *Eremnophila* Menke, p. 146
   Episternal sulcus extending straight down from subalar fossa (sometimes absent), not passing through scrobe (fig. 32 H) 2
2. Claw usually simple but if with single basal tooth on inner margin then mouthparts very long, galea attaining base of stipes when folded 3
   Claw with one or two basal teeth on inner margin (fig. 33 A-C); galea never extending beyond middle of stipes when folded and usually shorter; Old World forms 4
3. Apex of sternum I (petiole) meeting and often overlapping base of II (fig. 32 B); spiracle of tergum I located before apex of sternum I (lateral profile) (fig. 32 A); petiole often bent upward at level of tergum I base.. *Podalonia* Fernald, p. 141
   Apex of sternum I not reaching base of II, intervening space usually long and consisting of membrane and a ligament (fig. 32 C-D); spiracle of tergum I located at or beyond level of sternum I apex (fig. 32 D); sternum I usually bent downward or straight at level of tergum I base ...... *Ammophila* W. Kirby, p. 147
4. Petiole socket nearly completely surrounded by propodeal tergum (fig. 32 F); inner orbits of female strongly converging below (fig. 31 B) .... *Eremochares* Gribodo, p. 145
   Petiole socket broadly bounded ventrally by T-shaped propodeal sternite (fig. 32 E); inner orbits of female only slightly converging below or parallel or diverging below 5
5. Tarsomere I of female foreleg strongly asymmetrical, its outer apex prolonged (fig. 32 I), free margin of male clypeus rounded or truncate; tarsi usually with plantulae .... *Parapsammophila* Taschenberg, p. 137
   Tarsomere I of female foreleg nearly symmetrical, outer apex only slightly produced (fig. 32 J); male clypeus triangular; plantulae absent............... *Hoplammophila* Beaumont, p. 140

Genus Parapsammophila Taschenberg

Generic description: Inner orbits straight, converging below in male, usually parallel in female or nearly so; frontal line weak to absent; scape globose to slender, terminal male flagellomeres sometimes with tyli; sockets separated from frontoclypeal suture (or line tangential to lower margins of tentorial pits) by a socket diameter in female, more in male; free margin of female clypeus with a truncate median lobe or merely straight for most of its width, margin of male clypeus rounded or truncate, labrum usually broader than long, truncate to arcuate; mandible with one or two mesal teeth on inner margin in female, one to four teeth of variable size and placement in male; mouthparts usually short, but galea long and attaining middle of stipes when folded in one or two species; hypostoma closing mandible socket in female, but socket broadly to narrowly open in male; collar short, anterior surface vertical or nearly so; petiole socket broadly bounded ventrally by propodeal sternite (fig. 32 E); mesosternal area sometimes with anterior concavity (in lateral profile) to accomodate forecoxae; episternal sulcus running straight down from subalar fossa, not passing through scrobe; inferior metapleural area often defined; marginal cell rounded apically; shape of submarginal cell II highly variable; second recurrent vein rarely received by submarginal III (*algira*), first or second recurrent vein occasionally interstitial; female foretarsus with rake of long, sometimes blade-like spines, outer apex of tarsomeres, especially I, prolonged (fig. 32 I); posterior surface of forecoxa sometimes with spur or angulation, midcoxae nearly contiguous; midtibia usually with two spurs (outer spur absent or very short in most *algira, foleyi,* and *turanica*); tarsi usually with plantulae but sometimes very small (fig. 34 D); apicoventral setae of

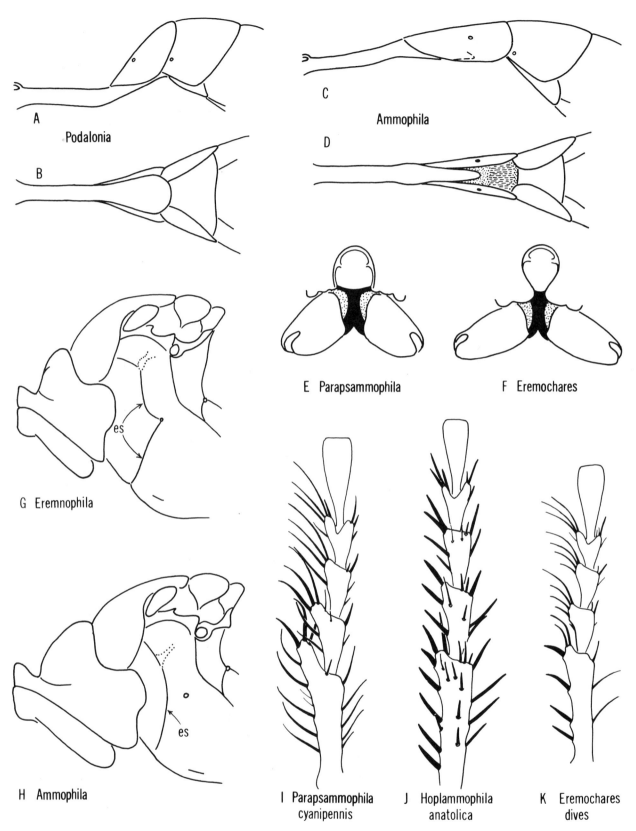

FIG. 32. Structural details in the tribe Ammophilini; A-D, gastral segments I-II; E,F, posterior view of petiole socket, hindcoxae, and propdeal sternite (solid black); G,H, lateral view of anterior half of thorax (es = episternal sulcus); I-K, female left foretarsus, dorsal.

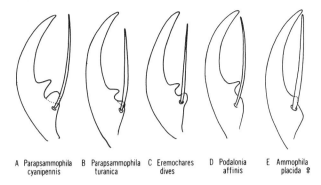

FIG. 33. Lateral view of hindtarsal claws in the tribe Ammophilini.

A Parapsammophila cyanipennis B Parapsammophila turanica C Eremochares dives D Podalonia affinis E Ammophila placida ♀

hindtarsomere V moderately broad to slender; claw usually with two basal teeth, distal tooth acute, proximal tooth blunt or rounded and bearing one or more long setae at base (fig. 33 A), proximal tooth sometimes poorly formed or absent (fig. 33 B); arolium moderate to large; petiole consisting of sternum I only, moderately long to long, apex reaching base of sternum II, tergum I longer than high (lateral profile) and in some species rather slender, especially in males, but still dilated distally; spiracle of tergum I located at middle or slightly beyond; sides of male sternum VIII converging towards truncate or sinuate apex; penis valve head with or without teeth on ventral margin, gonostyle not biramous apically.

Geographic range: Parapsammophila is an Old World genus, and most of its 19 species are restricted to Africa. However, several African forms range eastward to the deserts of the southern Palearctic Region. Two species are known in India, and one each in the Philippines and Australia.

Systematics: These wasps range in length from 17 to 52 mm, the biggest species ranking among the largest known Sphecidae. Some forms are entirely black, including the violaceous wings, but in a few the head, thorax, abdomen, and legs are partially or totally red. The sexes of some species are differently colored. The wings are often clear. Erect body hair may be scanty or dense and varies from black to silver. The head and thorax are densely clothed with appressed silver hair in a few species.

Parapsammophila is the only genus in the Ammophilini in which tarsal plantulae occur, but they are absent in *bakeri* and *eremophila,* very small in some females, and absent in some males. The configuration of the petiole socket (compare figs. 32 E and 32 F) immediately separates *Parapsammophila* from *Eremochares.* Unlike *Eremochares* the inner orbits of the female do not converge strongly below in *Parapsammophila,* and the foretarsal rake spines of tarsomere I are mostly long and fine in the latter genus rather than short and stout as in *Eremochares.* Generally there are two midtibial spurs in *Parapsammophila* and usually only one in *Eremochares.*

The outer apical prolongation of the first female foretarsomere and the strongly developed rake of *Parapsammophila* contrast sharply with the weak rake and only slight asymmetry of foretarsomere I of *Hoplammophila*

females (compare figs 32 I and 32 J). The males of these two genera are easily differentiated by the clypeal shape, but the simple gonostyle of *Parapsammophila* is also diagnostic. The gonostyle of *Hoplammophila* is biramous.

The presence in *Parapsammophila* of one or two claw teeth distinguishes the genus from *Eremnophila,* and all but a very few *Podalonia* and *Ammophila.* Species of the last two that have a single claw tooth have very long mouthparts. In contrast, the few *Parapsammophila* with only one well developed claw tooth have very short mouthparts.

Parapsammophila of Bohart and Menke (1963) and Menke (1966a) have never been subjected to a thorough, modern review. Kohl (1906) keyed the Palearctic species, but he placed some in *Eremochares.* The figures and notes provided by Beaumont (1968b) supplement Kohl's work. Roth (1928) and Alfieri (1946) keyed some of the Palearctic African forms, and Arnold (1928) diagnosed the species of the Ethiopian Region, but these keys are mostly inadequate.

Biology: Unknown, but presumably these wasps are fossorial.

Checklist of *Parapsammophila*

algira (Kohl), 1901 (*Ammophila*); n. Africa to Afghanistan
  *caelebs Kohl, 1901 (*Ammophila*), new synonymy by Menke
  bituberculata Bytinski-Salz, 1955 (*Ammophila*)
  gibba Alfieri, 1961 (*Ammophila*)
bakeri (Rohwer), 1921 (*Sphex*); Philippines
caspica (Gussakovskij), 1930 (*Ammophila*); e. Mediterranean area, sw. USSR
  macularis Gussakovskij, 1930 (*Ammophila*), new synonymy by Pulawski
  sacra Bytinski-Salz, 1955 (*Ammophila*), new synonymy by Pulawski
consobrina (Arnold), 1928 (*Sphex*); S. Africa (? = *testaceipes*)
cyanipennis (Lepeletier), 1845 (*Ammophila*); n. Africa
  reticollis A. Costa, 1864 (*Ammophila*)
  miles Taschenberg, 1869, lectotype ♂, "Chartum" (Mus. Halle), present designation by Menke
dolichostoma (Kohl), 1901 (*Ammophila*); Saudi Arabia, Egypt
errabunda (Kohl), 1901 (*Ammophila*); e. centr. Africa
eremophila (Turner), 1910 (*Ammophila*); Australia
erythrocephalus (Fabricius), 1781 (*Sphex*) India
  fuscipennis F. Smith, 1870 (*Ammophila*)
  ?*violaceipennis* Cameron, 1889 (*Ammophila*), nec Lepeletier, 1845
  ?*indica* Dalla Torre, 1897 (*Ammophila*)
foleyi (Beaumont), 1956 (*Ammophila*); n. Africa
funerea (Nurse), 1903 (*Ammophila*); w. India
herero (Arnold), 1928 (*Sphex*); Rhodesia, S.-W. Africa, Botswana
lateritia Taschenberg, 1869; n. Africa, lectotype ♀, "Chartum" (Mus. Halle), present designation by Menke
  monilicornis Morice, 1900

litigiosa (Kohl), 1901 (*Ammophila*); "Africa?"
ludovicus (F. Smith), 1856 (*Ammophila*); Ethiopian Region
ponderosa (Gerstaecker), 1871 (*Ammophila*); Ethiopian Region
  ?*gigantea* Kohl, 1901 (*Ammophila*)
testaceipes (R. Turner), 1918 (*Sphex*); Tanzania
turanica F. Morawitz, 1890; n. Africa, s. Asia
  lutea of authors, not Taschenberg, 1869
unguicularis (Kohl), 1901 (*Ammophila*); Mozambique

Genus Hoplammophila Beaumont

Generic description: Inner orbits straight for most of their length, diverging below or parallel; frontal line impressed; scape ovoid; sockets separated from frontoclypeal suture by a socket diameter in male, somewhat less in female; female clypeus convex, free margin with broad, truncate, median lobe delimited laterally by a reflexed tooth, male clypeus triangularly produced, tip reflexed, clypeal disk usually with a short, median ridge or elevated process; labrum rounded, as long as broad in male, somewhat shorter in female; mandible with one or two mesal or subapical teeth on inner margin in female, male with one mesal or subapical tooth; mouthparts moderately long, when folded galea sometimes reaching middle of stipes; hypostoma closing mandible socket; collar moderately long, anterior surface sloping gradually to neck; dorsal enclosure of propodeum defined by differential sculpture or a ridge; petiole socket broadly bordered below by propodeal sternite; episternal sulcus running straight down from subalar fossa, not passing through scrobe; marginal cell rounded apically; submarginal cell II strongly narrowed anteriorly, posterior veinlet of cell two or more times as long as anterior veinlet; psammophore hairs of femur weak apically and sparse or absent on tibia, foretarsal rake present, composed of short, stout spines, outer apex of tarsomere I weakly produced, tarsomeres II-IV essentially symmetrical (fig. 32 J); forecoxa sometimes with spur at inner apex, midcoxae essentially contiguous; midtibia with two spurs; apicoventral setae of hindtarsomere V slender to moderately broad; claw with one or two basal teeth, if two, proximal tooth blunt and bearing one or more long setae at base; arolium moderate to large; petiole consisting of sternum I only, moderately long to long, apex reaching base of sternum II, tergum I longer than high (lateral profile) but dilated distally, its spiracle located about at middle; male sternum VIII emarginate apically or variably truncate; penis valve head of unusual form, without ventral row of teeth, gonostyle biramous apically (fig. 34 A).

Geographic range: Hoplammophila occurs from southern Europe to Japan and Taiwan. With the exception of Taiwan, it is not known to occur in the Oriental Region. Three of the four species occur in Europe and western Asia.

Systematics: These wasps range in length from 19 to 36 mm and are totally black except that gastral segment II and usually III are red. Erect thoracic hair is usually

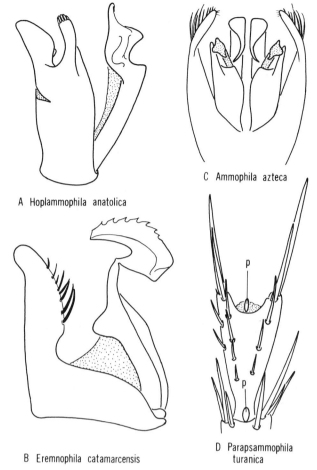

FIG. 34. A,B, male genitalia, lateral; C, male genitalia, ventral (digitus stippled); D, hindtarsomere IV and apex of III, ventral, p = plantula.

pale and there is a patch of appressed silver hair on the mesopleuron near the pleural suture. Wings are clear or moderately infumate.

The absence of a long projection at the outer apex of foretarsomere I of the female, the short rake spines, and the triangular male clypeus immediately separate *Hoplammophila* from *Parapsammophila*, the only genus with which it might be confused. The absence of plantulae and the complex male genitalia, including the biramous gonostyle, are additional differences from *Parapsammophila*. The presence of one or two claw teeth combined with the rather short mouthparts separate *Hoplammophila* from *Podalonia* and *Ammophila*.

The genus was originally proposed as a subgenus of *Ammophila s.l.* by Beaumont (1960b), but Menke (1966a) elevated it to genus. This is supported by biological as well as morphological evidence. Of the various characteristics cited by Beaumont as diagnostic we have found that the hindwing venation has no generic value. We have noted a number of specific differences in the male that have not been stressed before: genal punctation, form of the mandibular teeth, and the presence or absence of an angular lobe at the mandibular socket margin of the hypostoma. Good differences also exist in the female mandible. Beaumont's paper contains excellent figures of the unusual genitalia among species of *Hoplammophila* and other specific details. We have studied all four species of the genus which Menke (1974c) has keyed.

Biology: The habits of only one species, *aemulans*, are known (Iwata, 1938c; Tsuneki, 1963b, 1968d). This species nests in abandoned burrows in tree trunks and presumably other preexisting cavities. Tsuneki obtained a number of nests from bamboo stick traps placed horizontally in piles of logs. Prey are Notodontidae of the genus *Gonoclostera*. These are carried on foot, venter up, and are held by the mandibles of the wasp. Nests contain one to four cells in linear sequence. One caterpillar is placed in a cell and the egg is laid on its side. Cells are separated by a compound plug. First is a layer of pebbles and bits of wood and other debris. This is followed by a thick layer of dried soil, which is made by bringing in wet earth which is then compacted into a mass that dries hard. The closing plug of the nest is similar to the cell partitions. Tsuneki (1968d) hypothesized that the loose inner part of the partition and closing plug is an atavistic trait which reflects the fossorial ancestry of the genus. The presence of a foretarsal rake in the female also suggests fossorial ancestry, although presumably these wasps use the rake to gather materials for the partitions and closing plug. It will be interesting to see if other species of the genus have nesting habits similar to *aemulans*.

Checklist of *Hoplammophila*

aemulans (Kohl), 1901 (*Ammophila*); e. Asia, Japan, Taiwan
    rhinoceros Strand, 1913 (*Ammophila*)
anatolica (Beaumont), 1960 (*Ammophila*); Turkey
armata (Illiger), 1807 (*Sphex*); n. Mediterranean region[27]
clypeata (Mocsáry), 1883 (*Ammophila*); n. Mediterranean region

Genus Podalonia Fernald

Generic description: Inner orbits straight, parallel or more rarely converging below in female, converging below in male; frontal line impressed or absent; scape oval or elongate oval, flagellomere I longer than II or only approximately so in males, terminal flagellomeres often with tyli in male; sockets in female nearly contiguous with frontoclypeal suture or line tangential to lower margins of tentorial pits, sockets separated by about a socket diameter in male; free margin of female clypeus either with broad, truncate, median lobe which is usually sharply angulate laterally (median lobe subdivided into several teeth in three North American forms), or margin merely arcuate or sinuate; margin of male clypeus straight or rounded, sometimes broadly shallowly emarginate (with median tooth in *minax*); labrum usually broader than long, apex rounded, truncate or emarginate; female mandible with one large mesal tooth on inner margin which usually is subtended basad and sometimes distad by a smaller tooth; male mandible with one mesal or nearly subapical tooth on inner margin; mouthparts long, galea reaching base of stipes when folded except ending at basal one-fourth in *dispar*; hypostoma closing mandible socket in female and a few males; collar short, anterior surface sloping steeply to neck; propodeal enclosure present or absent; petiole socket broadly bounded ventrally by propodeal sternite; episternal sulcus running straight down from subalar fossa, not passing through scrobe; lower metapleural area often defined; marginal cell apex acuminate to bluntly rounded; shape of submarginal cell II variable; female foretarsus with rake of moderate to long spines which are stout, stiff, or flexible and bladelike, outer apex of tarsomeres usually prolonged, but tarsomeres only slightly asymmetrical in a few species; inner apex of forecoxa occasionally with an obscure tooth or prominent spine in one or both sexes, midcoxae narrowly separated; midtibia in both sexes sometimes with outer spur very short or absent, especially in New World forms; pectens of inner hindtibial spur usually coarser and widely spaced on distal half, but sometimes fine and closeset throughout; apicoventral setae of hindtarsomere V moderately broad to slender, separated by two or more setal widths; claw usually simple, but with one sharp basal tooth on inner margin in a few species (fig. 33 D); arolium moderate to large in male, moderate to small in female; petiole consisting of sternum I only, as short as hindcoxa to moderately long, usually longer in male than female, often bent upward at base of tergum I, apex reaching base of sternum II (fig. 32 B); tergum I in female strongly dilated posterad, usually shorter than to as long as high, occasionally longer than high, surface usually sloping up from petiole at an angle ranging between 45-90° (lateral profile) (fig. 32 A), tergum in male less dilated than female, usually longer than high, surface more gradually sloping up from petiole; spiracle of tergum I at or before middle (fig. 32 A), rarely slightly beyond middle in male; sides of male sternum VIII converging towards apex which is usually emarginate; penis valve head usually with teeth on ventral margin and often a basal spinelike process, gonostyle not biramous.

Geographic range: With the exception of South America, *Podalonia* is represented in all of the major temperate and tropical land areas of the world. Of the 66 species recognized, 20 are in the New World. The greatest number of species occur in the Mediterranean area and southwestern Asia. One species, *tydei*, has a very broad range. It is found on the Canary and Madeira Islands, through the Mediterranean area, and eastward to China and the Oriental Region. The only representative of *Podalonia* in Australia is a subspecies of *tydei*. There are no records of the genus from the islands between Australia and mainland Asia.

---

[27] Illiger (1807a) proposed this name in the genitive case, *"Sph. armatae"*, on page 91. On plate VI, figs. A and B, the species is identified as *"Sph. sabulosa major etrusca"* and *"S. sabulosa major"* respectively.

Systematics: Species of *Podalonia* range in length from 10-27 mm, the African *sheffieldi* being the largest. They are black wasps, but the gaster is commonly partially to completely red. Partially red legs are only found in a few species: *minax, argentipilis,* and *erythropus,* for example. *Podalonia caerulea* and *parallela* are metallic blue black. The sexes are sometimes differently colored. For example, in *luctuosa* and *communis* the female is completely black, but the male has a partially red gaster. The head and thorax are densely clothed with black or white erect hair in most species, and hair color may vary between sexes and/or geographically. Males and some females have appressed silver hair on the face, but with few exceptions the rest of the head and thorax are devoid of such vestiture. Wings vary from clear to darkly violaceous.

Despite the recent generic diagnoses for *Podalonia* by Bohart and Menke (1963), Pulawski (1965b), and Menke (1966a), a number of Old World authors continue to recognize this taxon as a subgenus of *Ammophila*. However, *Podalonia* is easily distinguished morphologically from *Ammophila,* and the fairly constant biological difference between the two is additional evidence that they deserve recognition as separate genera. In *Podalonia* the first sternum (petiole) reaches the base of the second sternum, and the spiracle of tergum I is located basad of the apex of sternum I as seen in lateral view (fig. 32 A, B). In *Ammophila* sternum I ends before reaching the base of sternum II, and the intervening space is membranous or sometimes partially filled by a small ligament or plate, which Oeser (1971) regards as part of sternum I (fig. 32 D). The spiracle of tergum I is at the level of the apex of sternum I or beyond it in *Ammophila* (see fig. 32 C, D). The strongly dilated (or bell shaped) tergum I of female *Podalonia* contrasts rather sharply with the slender tergum I of female *Ammophila,* but this difference is less appreciable in the male. The bending upward of the petiole at the base of tergum I in most *Podalonia* is another distinction from *Ammophila*. Some authors have used the coarse pectination of the inner hindtibial spur common to many *Podalonia* species as a generic distinction from *Ammophila* but there are exceptions. The long mouthparts and straight episternal sulcus separate *Podalonia* from other genera in the tribe. Some species of *Podalonia* have the shortest petiole in the tribe. Sternum I as measured on the dorsal surface is often no longer than the hindcoxa.

The name *Podalonia* is now attributed to H. T. Fernald (1927) rather than to Spinola (1853) because of the recent ruling of the International Commission on Zoological Nomenclature (Opinion 857, 1968) made on behalf of the petition of Menke, Bohart, and van der Vecht (1966). Townes (1973) declared that the "sphecid genus *Podalonia* . . . should be called *Pompilus* . . .", but he neglected to note that the International Commission on Zoological Nomenclature issued Opinion 166 in 1945 in which *Pompilus pulcher* Fabricius was designated as type species, thus placing the genus in the spider wasp family Pompilidae. This Opinion seems to have been generally accepted.

Species discrimination in *Podalonia* is difficult. Male genitalia and clypeal shape generally offer good differences, but females of some species are especially hard to separate. Fine details of thoracic and facial sculpture are used to identify many species. The metapleural flange is broadly lamelliform in some species and is a useful diagnostic tool. The setation of the foretarsal rake, size of the arolia, and shape of the foretarsomeres in the female may offer differences in this sex.

Wing venation anomalies are fairly common in this genus, and they usually involve partial or total loss of one submarginal cell or the subdivision of some cell by a spurious vein. Occasionally, the second or third submarginal cell is nearly petiolate. The incompletely closed mandibular socket of most males, the loss in some species of one midtibial spur, and the presence in a few forms of a claw tooth are interesting morphological facets. Species with claw teeth that we have seen are *affinis, argentipilis* (♀ only), *caucasica,* and *ebenina*. Interestingly, the arolium is larger than usual in the females of these species. Sometimes a very small tooth is discernable on some of the claws of species such as *hirsuta* and *alpina*.

The work of Kohl (1906) contains a key to the Palearctic species known at that time. Beaumont in a number of papers has cleared up the identity of or diagnosed a number of western Palearctic forms. Tsuneki (1971d) updated Kohl's key for the eastern Palearctic species and included many fine figures. Arnold (1928) keyed the species of the Ethiopian Region. Murray (1940) revised the New World species.

Townes (1973) has revived the matter of the identity of *Sphex viaticus* Linnaeus, 1758, a species that has enjoyed recognition as a pompilid by most wasp taxonomists for the past 25 years. He would place *viaticus* in *Podalonia* where it would replace the well known Old World species *hirsuta* (Scopoli). Linnaeus' description clearly refers to a pompilid as van der Vecht (1958) has shown, but the biological annotation applies to a sphecid. Townes infers that Linnaeus in his "Fauna Svecica", 1761, recognized the dual nature of *viaticus* because he described an undisputed pompilid, *fuscus,* and placed it right after *viaticus,* and also emended the description of the latter by eliminating reference to biology and by mentioning the presence of black hair. Because the red segments of the gaster are still said to be black banded the description of *viaticus* clearly is still muddled. Townes' argument that Linnaeus or one of his contemporaries such as De Geer or Villers restricted the species to the Sphecidae is not substantiated when these authors' works are carefully examined. For example, De Geer in volume 2 of his "Memoires pour servir a l'Histoire des Insectes", 1771, gave the first clearly sphecid description for *viaticus,* including a figure, but he failed to say that he was revising the interpretation of Linnaeus' species or otherwise acting as "first revisor". Furthermore, one infers from recommendation 74A of the Code that "first revisors" in the sense of Townes cannot be regarded as having legally restricted *viaticus* to the

Sphecidae. This is a point that could stand clarification by the Commission. Richards (1935a) recognized a sphecid specimen in the Linnaean Collection in London as "the type of *Sphex viatica* Linnaeus ...", but Linnaeus based his description on a mixed series as Townes and others have demonstrated; therefore, the London specimen can at best be a syntype. Townes regards Richards' 1935 statement as a lectotype designation, but we cannot agree since Richards did not indicate that *viaticus* had to have been based on more than one specimen or that he was selecting the London one as the type. According to van der Vecht (letter to Menke) Richards told him in a recent conversation that it was not his intention to designate the Linnaean Collection specimen as lectotype, and he thought that this was made clear in his introductory remarks. Whether or not to construe Richards' published statement as lectotypy is another situation that needs clarification by the Commission. The first unquestioned lectotype designation for *viaticus* was made by van der Vecht (1958) who restricted the species to the Pompilidae by selecting a figure mentioned by Linnaeus as the lectotype. Townes argues that van der Vecht's lectotype designation is invalid because it was not in agreement with "the first revisor (Linnaeus, 1761, or Villers, 1789)", but the Code makes such agreement a recommendation only (Art. 74A). At this point we cannot regard *viaticus* as a sphecid, and we can see no benefit in Townes' attempt to change a situation that has stabilized in recent years. Some of the issues raised by Towne's paper have been submitted to the International Commission on Zoological Nomenclature by Sabrosky (1974) for clarification.

Biology: Murray (1940) summarized most of the papers published on New World species but did not include those of J. Parker (1915) and Williams (1928b). Additional references are those of Evans (1963a, 1970). Unfortunately, the identity of the wasps observed in the earlier accounts is uncertain in most cases. A number of Old World species have been studied: *affinis* by Marchal (1892), Palmer and Stelfox (1931), Olberg (1952, 1959), Diederichs (1953), and Teschner (1959); *atrocyanea* by Tsuneki (1968c); *ebenina* by Myartseva (1963); *hirsuta* by Roth (1928), Bougy (1935), Deleurance (1941), Grandi (1957), Fulcrand and Gervet (1968), Gervet and Fulcrand (1970), and Myartseva (1969); *tydei* by Roth (1928), Soika (1933), Deleurance (1941), Hemmingsen (1960), and Myartseva (1969); and *tydei* ssp. *suspiciosa* by Chandler (1925) and Bristowe (1971). Earlier references can be found in some of these papers, and excellent photographs are contained in Olberg's and Diederichs' works.

Species of *Podalonia* generally capture their prey before excavating a nest, but Myartseva (1963) found that *ebenina* sometimes digs the burrow before seeking prey, in which case the burrow is usually temporarily stoppered. Tsuneki (1968c) noted the same habit in an unidentified east Asian species. Lepidopterous larvae are provisioned by *Podalonia,* and most reports indicate that hairless, nocturnal-feeding caterpillars of the moth family Noctuidae are preferred. These cutworms burrow into the soil in the daytime. *Podalonia* females walk over the soil surface searching for a buried cutworm, and when one is located they dig down to it and pull it out. Other types of caterpillars have been reported as prey, however. Williams (1928b) and Murray (1940) recorded hairy, nonburrowing caterpillars of the genus *Malacosoma* (tent caterpillars, family Lasiocampidae) were used by *sericea* (identity not certain) and *occidentalis*. Balduf (1936) found a *Podalonia* species using diurnal notodontid caterpillars; and the Old World species *hirsuta* sometimes provisions with diurnal noctuid and notodontid larvae, some of which are rather hairy. According to Roth (1928) this species has been observed using the very hairy larvae of the Gypsy Moth (*Lymantria,* Lymantriidae), and Ferton (1914) saw a *hirsuta* female excavating a nest after she had hung a hairy nymphalid butterfly caterpillar on a nearby twig.

A paralyzed caterpillar is carried venter up and is held by the wasp's mandibles. Transport is on the ground. Usually the wasp hangs the caterpillar on some support such as a leaf axil, twig, or lays it on some grass off the ground. Then a search is made for a spot in which to excavate the rather shallow, single-celled nest. This often is a considerable distance (up to 20 feet) from the place where the prey is left. One caterpillar is used per cell, and the egg is laid on its side.

An interesting aspect of *Podalonia* behavior is the tendency for females to assemble in tight clusters of up to several hundred individuals in protected situations, such as in crevices in rocks or under tree bark during the late summer and early fall (Grandi, 1925, 1929; Roth, 1928; Maneval, 1939). By marking individuals in the fall, Maneval established that at least some of these aggregations remain through the winter, the females reappearing during favorable weather the following year. Hicks (1931) and Tsuneki (1968c) found that females occasionally excavate special burrows in which they spend the night or day, in the case of inclement weather. Hicks found four females in one burrow. One of us (Menke) found a large number of female *Podalonia communis* clustering under the loose bark of a fallen tree during a thunderstorm in the mountains of southern California.

Miltogrammine Sarcophagidae have been reported infesting the nests of various species of *Podalonia*. Sometimes the caterpillar provisioned by a wasp has already been parasitized by a braconid (*Meteorus*) or ichneumonid (*Netelia*). Hicks (1933) found that the *Netelia* larva attacked the *Podalonia* larva after the original host was consumed.

Checklist of *Podalonia*

affinis (W. Kirby), 1798 (*Ammophila*); Palearctic Region
   ariasi Mercet, 1906 (*Ammophila*)
   lutaria of authors, not Fabricius, 1787
  ssp. *concolor* (Brullé), 1839 (*Ammophila*); Canary Is.
   nigra Brullé, 1839 (*Ammophila*)
  ssp. *ulanbaatorensis* (Tsuneki), 1971 (*Ammophila*); Mongolia, n. China
afghanica Balthasar, 1957; Afghanistan

albohirsuta (Tsuneki), 1971 (*Ammophila*); Mongolia
alpina (Kohl), 1888 (*Ammophila*); Europe, sw. Asia
altaiensis (Tsuneki), 1971 (*Ammophila*); Mongolia
andrei (F. Morawitz), 1889 (*Ammophila*); China, Mongolia
argentifrons (Cresson), 1865 (*Ammophila*); w. U.S.
argentipilis (Provancher), 1887 (*Pelopoeus*); sw. U.S.
    morrisoni Cameron, 1888 (*Ammophila*)
    nicholi Carter, 1924 (*Psammophila*)
aspera (Christ), 1791 (*Sphex*); distribution unknown
atrocyanea (Eversmann), 1849 (*Psammophila*); e. Europe, centr. Asia, n. China
    ?*psilocera* Kohl, 1888 (*Ammophila*)
bolanica (Nurse), 1903 (*Ammophila*); Pakistan (? = *hirsuta mervensis*)
caerulea Murray, 1940; U.S.: California
canescens (Dahlbom), 1843 (*Psammophila*); Ethiopian Region
    incana Dahlbom, 1843 (*Psammophila*)
    capensis Lepeletier, 1845 (*Ammophila*)
    longipilosella Cameron, 1910 (*Ammophila*)
  ssp. *madecassa* (Kohl), 1909 (*Ammophila*); Madagascar
caroli (Pérez), 1907 (*Ammophila*); Trucial Coast, Saudi Arabia
caucasica (Mocsáry), 1883 (*Psammophila*); sw. USSR, Mongolia, n. China
chalybea (Kohl), 1906 (*Ammophila*); Mongolia
clypeata Murray, 1940; w. U.S.
communis (Cresson), 1865 (*Ammophila*); w. U.S., Mexico
  ssp. *intermedia* Murray, 1940; Mexico
  ssp. *atriceps* (F. Smith), 1856 (*Ammophila*); Mexico to Panama, lectotype ♂, Mexico (Mus. London), present designation by Menke
    alpestris (Cameron), 1888 (*Ammophila*)
compacta Fernald, 1927; U.S.: California, Oregon
dispar (Taschenberg), 1869 (*Psammophila*) Egypt, lectotype ♂, "Chartum" (Mus. Halle), present designation by Menke
    ?*strenua* Walker, 1871 (*Ammophila*), nec Cresson, 1865
    ?*walkeri* Dalla Torre, 1897 (*Ammophila*)
ebenina (Spinola), 1838 (*Ammophila*); n. Africa, Middle East
    micipsa Morice, 1900 (*Psammophila*)
erythropus (F. Smith), 1856 (*Ammophila*); Senegal, Gambia
    *rufipes Lepeletier, 1845 (*Ammophila*), nec Guérin-Méneville, 1831, new synonymy by Menke
fera (Lepeletier), 1845 (*Ammophila*); Europe, Middle East
    abeillei Marquet, 1881 (*Ammophila*)
    polita Mocsáry, 1883 (*Psammophila*)
    morawitzi Ed. André, 1886 (*Ammophila*)
flavida (Kohl), 1901 (*Ammophila*); Mongolia, n. China
gobiensis (Tsuneki), 1971 (*Ammophila*); Mongolia, n. China
  ssp. *chahariana* (Tsuneki), 1971 (*Ammophila*); n. China

gulussa (Morice), 1900 (*Psammophila*); Algeria[28]
harveyi (Beaumont), 1967 (*Ammophila*); Turkey
hirsuta (Scopoli), 1763 (*Sphex*); Europe, Mongolia, n. China
    arenaria Fabricius, 1787 (*Sphex*); nec Linnaeus, 1758
    arenosa Gmelin, 1790 (*Sphex*)
    argentea W. Kirby, 1798 (*Ammophila*)
    macrogaster Dahlbom, 1831 (*Sphex*)
    viatica of authors, not Linnaeus, 1758[29]
  ssp. *mervensis* (Radoszkowski), 1887 (*Ammophila*) Mediterranean Islands, centr. Asia
    ebenina Lepeletier, 1845 (*Ammophila*), nec Spinola, 1838
  ssp. *nepalensis* (Zavattari), 1909 (*Ammophila*); Nepal
hirsutaffinis (Tsuneki), 1971 (*Ammophila*); Mongolia
hirticeps (Cameron), 1889 (*Ammophila*); India
kaszabi (Tsuneki), 1971 (*Ammophila*); Mongolia
kozlovii (Kohl), 1906 (*Ammophila*); Mongolia, Tibet
luctuosa (F. Smith), 1856 (*Ammophila*); N. America: transcontinental, lectotype ♀, Rocky Mtns. (Mus. London), present designation by Menke
    pacifica Melander and Brues, 1902 (*Psammophila*)
luffi (E. Saunders), 1903 (*Ammophila*); Europe
    arenaria Luderwaldt, 1897 (*Ammophila*), nec Fabricius, 1787
mahatma (Turner), 1918 (*Sphex*); Tibet
mandibulata (W. F. Kirby), 1889 (*Ammophila*); Afghanistan
marismortui (Bytinski-Salz), 1955 (*Ammophila*); Mediterranean region, Iraq
masinissa (Morice), 1900 (*Psammophila*); n. Africa, Israel
    massinissa of auctt.
mauritanica (Mercet), 1906 (*Ammophila*); n. Africa, Canary Is.
melaena Murray, 1940; w. U.S.
merceti (Kohl), 1906 (*Ammophila*); Spain
mexicana (Saussure), 1867 (*Ammophila*); w. U.S., Mexico
mickeli Murray, 1940; w. U.S.
minax (Kohl), 1901 (*Ammophila*); Egypt, Libya
    confalonierii Guiglia, 1932 (*Ammophila*)
moczari (Tsuneki), 1971 (*Ammophila*); Mongolia
montana (Cameron), 1888 (*Ammophila*); Mexico, Guatemala
    jason Cameron, 1888 (*Ammophila*)
    quadridentata Cameron, 1888 (*Ammophila*)
nigriventris (Gussakovskij), 1935 (*Ammophila*); China
nigrohirta (Kohl), 1888 (*Ammophila*); sw. USSR, Mongolia
obo (Tsuneki), 1971 (*Ammophila*); Mongolia
occidentalis Murray, 1940; w. U.S.
parallela Murray, 1940; U.S.: California
piceiventris (Cameron), 1888 (*Ammophila*); Guatemala
pseudocaucasica Balthasar, 1957; Afghanistan

[28] Morice (1911) designated the ♂ as type of *gulussa* thus restricting the name to *Podalonia*. Morice's "♀" of *gulussa* is *Parapsammophila algira*.

[29] See systematics for discussion of this name.

pubescens Murray, 1940; sw. U.S., Mexico
puncta Murray, 1940; centr. U.S.
pungens (Kohl), 1901 (*Ammophila*); sw. USSR: Kazakh S.S.R.; Mongolia
robusta (Cresson), 1865 (*Ammophila*); N. America: transcontinental; Mexico to Costa Rica
rothi (Beaumont), 1951 (*Ammophila*); Morocco, ? Turkey
schmiedeknechtii (Kohl), 1898 (*Ammophila*); Egypt, Libya
    saharae Giner Marí, 1945 (*Ammophila*)
sericea Murray, 1940; w. U.S.
sheffieldi (R. Turner), 1918 (*Sphex*); Malawi, S. Africa
sonorensis (Cameron), 1888 (*Ammophila*); w. U.S., Mexico, lectotype ♀, n. Sonora, Mexico (Mus. London), present designation by Menke
    differentia Murray, 1940, new synonymy by Bohart
turcestanica (Dalla Torre), 1897 (*Ammophila*); sw. USSR
    hirticeps F. Morawitz, 1893 (*Ammophila*), nec Cameron, 1889
*tydei (Le Guillou), 1841 (*Ammophila*); Canary Is., Madeira Is.
    madeirae Dahlbom, 1843 (*Psammophila*)
    ssp. *senilis* (Dahlbom), 1843 (*Psammophila*); s. Europe, Asia
    klugii Lepeletier, 1845 (*Ammophila*)
    psammodes Lepeletier, 1845 (*Ammophila*)
    capuccina A. Costa, 1858 (*Ammophila*)
    ?*spinipes* F. Smith, 1878 (*Ammophila*)
    lanuginosa Marquet, 1881 (*Ammophila*)
    ?*laeta* Bingham, 1897 (*Ammophila*)
    ?*durga* Nurse, 1903 (*Ammophila*)
    errabunda Mercet, 1906 (*Ammophila*), nec Kohl, 1901
    homogenea Mercet, 1906 (*Ammophila*)
    ssp. *argentata* (Lepeletier), 1845 (*Ammophila*); n. Africa
    ssp. *suspiciosa* (F. Smith), 1856 (*Ammophila*); Australia
    ssp. *apakensis* (Tsuneki), 1971 (*Ammophila*); n. China
valida (Cresson), 1865 (*Ammophila*); w. U.S.
    grossa Cresson, 1872 (*Ammophila*)
violaceipennis (Lepeletier), 1845 (*Ammophila*); e. U.S.
    cementaria F. Smith, 1856 (*Ammophila*)

Genus Eremochares Gribodo

Generic description: Inner orbits straight or slightly bowed inward, strongly converging below (fig. 31 B); frontal line absent; scape slender, sockets separated from frontoclypeal suture by about two socket diameters in female, three or more in male; free margin of female clypeus arcuate or with broad median lobe which is notched laterally and sometimes mesally, male clypeus slightly to moderately projecting, narrowly to broadly truncate; labrum broader than long, convex in male, margin arcuate or roundly triangular; inner margin of female mandible with one large mesal tooth which may be subtended by a smaller tooth, male mandible simple or with one small mesal tooth on inner margin; mouthparts short; hypostoma closing mandible socket in female, but socket broadly to narrowly open in male; collar moderately long, anterior surface sloping gradually to neck; petiole socket nearly surrounded by propodeal tergite (fig. 32 F); mesosternal area with anteromedian process directed towards forecoxae; episternal sulcus running straight down from subalar fossa, not passing through scrobe; lower metapleural area poorly or not defined; marginal cell apex rounded; submarginal cell II variable in shape, but posterior veinlet rarely twice as long as anterior veinlet; second recurrent vein usually interstitial between II and III; female foretarsus with rake of rather short, stout spines along outer margin of tarsomere I, apex of I and margins of II-IV usually with longer, finer spines, tarsomeres asymmetrical but I with outer apex only moderately produced (fig. 32 K); posterior surface of forecoxa sometimes with an angulation, midcoxae nearly contiguous; midtibia usually with one spur (outer spur sometimes present but short); apicoventral setae of hindtarsomere V moderately broad; claw with two basal teeth, proximal tooth bearing one or more long setae at base (fig. 33 C); arolium large; petiole consisting of sternum I only, very long, apex reaching base of sternum II, tergum I longer than high (lateral profile), but dilated distally, tergum I spiracle located at about middle; sides of male sternum VIII converging apically, apex roundly truncate; penis valve head with row of teeth on ventral margin, gonostyle not biramous apically.

Geographic range: Eremochares occurs in the Mediterranean area, ranging southward to Sudan and eastward to southwestern and southcentral Asia. Five species are known.

Systematics: These wasps are 17 to 26 mm long. *Eremochares* are slender, black, or black and red wasps with clear wings. Females have dense appressed silver hair on the face and pleura, and it often extends onto the gena and thoracic dorsum. Males have less appressed hair but are generally covered with much erect pale hair, which is absent or much sparser in the female. *E. mirabilis* females are red except for areas of the head, but the color is obscured by the extensive appressed silver hair.

The configuration of the petiole socket (fig. 32 F) immediately identifies *Eremochares,* although some *Ammophila* species approach it. The mesosternal process is also distinctive, but a similar development occurs in a few *Ammophila.* There is usually only one midtibial spur in *Eremochares,* but there are two in some *lutea.* One leg of one male of over 50 specimens of *dives* examined had 2 spurs, one being very short. In the female the large eyes, strongly converging inner orbits, and resultant narrow frons and clypeus are distinctive, as are the stout rake spines of the first foretarsomere.

Three species have been studied by us, and in *mirabilis* and *lutea* both recurrent veins end on submarginal cell II. However, the second recurrent generally is interstitial in *dives.* In occasional specimens of *dives* the second recurrent ends just inside the second or third submarginal cell.

There is no key for all of the species of the genus. Useful references are Menke (1966b) and Kohl (1906).

Biology: Eremochares dives, the only species so far studied, is the only ammophilin known to provision with Orthoptera (Smirnov, 1915; Beaumont, 1951a; Myartseva 1969; Kazenas, 1970). Myartseva was able to rear larvae on noctuid caterpillars. *E. dives* usually nests in colonies. Nests are dug in barren compact, damp soil and average about 8.5 cm deep. The burrow is nearly vertical and ends in a single cell. After completion of the excavation the nest is usually temporarily stoppered. One or two large soil lumps are placed in the entrance, and soil obtained by excavating an accessory burrow nearby is spread on top. Prey consists of immature Acrididae representing a wide variety of genera. The hopper is flown to the nest right side up and is held by the wasp's mandibles and forelegs. Upon arrival at the nest the grasshopper is left on the surface while the wasp opens and inspects the burrow. At this time the prey is vulnerable to thievery by ants and commonly by other female *dives.* After the first hopper is placed in the cell an egg is laid on the thoracic venter near the front legs. According to Kazenas, the wasp again temporarily seals the burrow and waits about four days before adding more prey, although daily inspections are made each morning. The progressive provisioning of this species takes seven or eight days and four to six grasshoppers are supplied the developing wasp larva. The daily inspections by the wasp led Kazenas to suggest that *dives* may maintain several nests at once, as in some *Ammophila.* Final closure of the nest includes a cover of moist clay that makes a hard seal upon drying. The accessory burrow dug by these wasps is the source of most of the soil used for the temporary and final closures and accordingly it becomes deeper as the days go by. The wasp closes the accessory burrow when the real nest is sealed for the last time.

Miltogrammine sarcophagids and mutillids of the genus *Dasylabris* infest the nests of *dives.*

Checklist of *Eremochares*

dives (Brullé), 1833 (*Ammophila*); Mediterranean region sw. USSR, Afghanistan, lectotype ♀, "Moree" (Mus. Paris), present designation by Menke
   melanopus Lucas, 1848 (*Ammophila*), lectotype ♂, "Algerie" (Mus. Paris), present designation by Menke
   festiva F. Smith, 1856 (*Ammophila*)
   elegans F. Smith, 1856, (*Ammophila*)
   limbata Kriechbaumer, 1869 (*Ammophila*)
   nigritaria Walker, 1871 (*Ammophila*)
   doriae Gribodo, 1882, lectotype ♀, "Tunisi" (Mus. Genoa), present designation by Menke
   retowskii Konow, 1887 (*Parapsammophila*)
   orichalceomicans Strand, 1915 (*Ammophila*)
ferghanica (Gussakovskij), 1930 (*Ammophila*); sw. USSR: Uzbek S.S.R.
kohlii (Gussakovskij), 1928 (*Ammophila*); sw. USSR: Turkmen S.S.R.
lutea (Taschenberg), 1869 (*Parapsammophila*); Sudan
mirabilis (Gussakovskij), 1928 (*Ammophila*); sw. USSR: Turkmen S.S.R.

Genus Eremnophila Menke

Generic description: Inner orbits straight for most of their length or weakly bowed inward, converging below in male, parallel or diverging below in female; frontal line impressed or absent; scape globose to slender; sockets separated from frontoclypeal suture (or line tangential to lower margins of tentorial pits) by a socket diameter or more in male and some females, by slightly less than a socket diameter in some females; free margin of female clypeus usually with a broad, truncate or sinuate median lobe which is delimited laterally by an angle or tooth, but clypeal margin without lobe in *melanaria,* male clypeus triangular or with midapical tooth or truncate; labrum longer than broad, apex broadly to narrowly rounded and reaching middle of stipes or beyond when appressed; mandible with one large mesal tooth on inner margin in female which is subtended on each side by a smaller tooth, inner margin also with basal angulation; inner margin in male with one mesal or nearly subapical tooth except two small teeth in *melanaria;* mouthparts moderately long to long, galea reaching middle of stipes or beyond when folded, but completely or largely concealed by labrum; hypostoma closing mandibular socket; collar short, anterior surface vertical or nearly so, dorsum with two depressions in female so that collar appears trilobed (except simple in *eximia* group); propodeal socket broadly bounded ventrally by propodeal sternite; episternal sulcus angulate, curving posterad from subalar fossa to scrobe and then extending obliquely ventrad, becoming evanescent on venter of pleuron (fig. 32 G); marginal cell apex rounded or roundly truncate; submarginal cell II strongly narrowed anteriorly, posterior veinlet two or more times length of anterior veinlet; psammophore hairs often weak on tibia and distal half of femur, female foretarsus with rake of rather long spines, outer apex of tarsomeres, especially I, prolonged; midcoxae separated by half petiole diameter or less; midtibia with two spurs; apicoventral setae of hindtarsomere V slender, sometimes nearly setiform; claw usually simple, rarely with one or two very weak basal teeth on inner margin; arolium large; petiole usually consisting of sternum and tergum I, the latter elongate, only slightly dilated distad except in *eximia* group, sternum I long to very long, apex reaching base of sternum II in *eximia* group but ending at level of spiracle in other groups, the space between sterna I and II membranous or covered by a small plate, spiracle of tergum I located at about apical third except at middle or slightly beyond in *eximia* group; sternum VIII rounded, truncate or triangular, male genitalia of unusual form, base of gonostyle prolonged ventrad, L-shaped, not biramous apically, head of aedeagus flexibly joined with stalk (fig. 34 B).

Geographic range: This is primarily a Neotropical genus although one of its nine species occurs in North America, where it gets as far north as southern Canada.

Systematics: Species of *Eremnophila* range in length from 15 to 35 mm, and nearly all are completely black. One Argentine species, *auromaculata,* has partly reddish legs, and the gaster is red except for sternum I and tergum III. *E. eximia* is black bodied over most of its range, but specimens from Bolivia, southern Brazil, and Argentina are colored like *auromaculata.* The wings are clear or weakly infumate in this genus. With rare exceptions the thorax has distinctive patches of appressed silver hair on the mesopleuron, next to the propodeal spiracle and petiole socket, and on the pronotal lobe. In addition, the scutum often has appressed silver hair mesally.

Menke (1964c) proposed *Eremnophila* as a subgenus of *Ammophila* but later (1966a) elevated it to a genus. The angular form of the episternal sulcus (fig. 32 G) immediately separates *Eremnophila* from all other ammophilin genera. The unusually long labrum and bizarre male genitalia are also distinctive. The articulating aedeagal head is unique.

In addition to their unusual genitalia males display most of the other specializations found in *Eremnophila.* Sternum II has a peculiar angular profile in *melanaria* males. The hypostoma has a tonguelike projection in the *opulenta* and *binodis* groups, and sternum VIII in the latter assemblage has a subapical spinelike process. The clypeus in most males is triangular or bears an apical tooth.

Menke (1964c) divided the genus into several groups and keyed the species. The *eximia* group (code letter E in checklist) seems to represent the most primitive section of the genus. Tergum I is usually dilated distally, and its spiracle is mesal or nearly so. Sternum I reaches the base of II and the male genitalia are the simplest in the genus. Unlike the other groups the collar is not trilobate in the female. The monotypic *melanaria* (M) and *opulenta* (O) groups are interesting because of the angular bulge and nipplelike projection, respectively, on the mesopleuron of both sexes. The *binodis* group is clearly the most advanced. Besides the male features enumerated above, the petiole includes tergum I, and sternum I does not reach the base of II.

Biology: Evans (1959b) summarized information on *aureonotata,* one of two species which has been studied. It excavates short, single celled burrows and makes a temporary closure before searching for prey. A single large lepidopterous larva (Notodontidae) is provisioned, and the egg is laid on the side of the caterpillar. Rau's (1922) hesperiid prey record needs to be confirmed. Prey are transported venter up across the ground. The wasp holds the larva with her mandibles. The completed nest is closed with soil particles and bits of debris. Richards' (1937) observations of an *opulenta* female agree with those made on *aureonotata,* except that the prey consisted of a sphingid larva.

Checklist of *Eremnophila*

asperata (W. Fox), 1897 (*Ammophila*); Brazil (E)

aureonotata (Cameron), 1888 (*Ammophila*); Canada and U.S. e. 100th merid. to El Salvador (B).

auromaculata (Pérez), 1891 (*Ammophila*); Argentina, Uruguay (E)
 giacomellii Schrottky, 1910 (*Ammophila*)

binodis (Fabricius), 1798 (*Sphex*); Panama to Argentina (B)
 abbreviata Fabricius, 1804 (*Pelopoeus*)
 *brasiliana Brèthes, 1909 (*Ammophila*), new synonymy by Menke
 guiana Cameron, 1912 (*Ammophila*)
 oxystoma Cameron, 1912 (*Ammophila*)

catamarcensis (Schrottky), 1910 (*Ammophila*); Argentina (B)

eximia (Lepeletier), 1845 (*Ammophila*); S. America (E)
 eugenia F. Smith, 1856 (*Ammophila*)
 nigrocincta Fernald, 1907 (*Sphex*)
 friedrichi Schrottky, 1909 (*Ammophila*)
 trimaculigera Strand, 1910 (*Ammophila*)

melanaria (Dahlbom), 1843 (*Ammophila*); Mexico to Argentina (M)
 lobicollis Cameron, 1912 (*Ammophila*)
 ?*miliaris* Cameron, 1888 (*Ammophila*)
 ?*iridipennis* Cameron, 1888 (*Ammophila*)
 ?*velutina* Schrottky, 1910 (*Ammophila*)

opulenta (Guérin-Méneville), 1838 (*Ammophila*); Mexico to Argentina (O)
 bimaculigera Strand, 1910 (*Ammophila*)

willinki (Menke), 1964 (*Ammophila*); Brazil to Argentina (B)

Genus Ammophila W. Kirby

Generic description: Inner orbits straight or slightly bowed, converging below in both sexes, or sometimes parallel or diverging below in female (fig. 31 A); frontal line variable; scape oval to elongate and parallel sided, terminal flagellomeres often with tyli in male; sockets in female nearly contiguous with or separated by as much as a socket diameter from frontoclypeal suture, in male by one to two diameters; free margin of female clypeus typically with a sinuate or truncate median lobe which is usually delimited laterally by a small tooth or angle (fig. 31 A), lobe sometimes with median emargination, clypeal margin sometimes sinuately transverse or broadly arcuate and without teeth; clypeal margin of male variable; labrum broader than long to twice as long as broad, truncate, rounded or emarginate, sometimes with apicomedian process; inner margin of female mandible with one to three mesal teeth (sometimes subapical), inner margin in male with one or two subapical teeth (sometimes mesal), rarely with ventrobasal tooth; mouthparts usually long to very long, galea in most species reaching or exceeding base of stipes when folded; hypostoma closing mandible socket (narrowly open in some males), hypostoma with a projecting lobe or spine near mandible in a few males; collar short to long; propodeal enclosure defined by differential sculpture or vestiture (sometimes obscured by vestiture); petiole socket usually broadly or narrowly bounded ventrally by propodeal sternite; anteroventral area of mesopleuron sometimes concave for reception of forecoxae, depression sometimes margined by carina which may form two lobes or depression accompanied by mesal process directed towards forecoxae; episternal sulcus usually present, running straight down from subalar fossa, not passing through scrobe, length variable, ending at level of scrobe or extending to anteroventral margin of pleuron; lower metapleural area usually defined; marginal cell apex acuminate to rounded; shape of submarginal cell II variable, submarginal cell III occasionally absent (fig. 35 C) or petiolate on radius (fig. 35 B); first or second recurrent vein sometimes interstitial, or rarely first received by submarginal I; female foretarsus with rake composed of short, stiff spines or long, slender, flexible, blade-like spines, outer apex of tarsomeres, especially I, usually prolonged; posterior face of forecoxa sometimes angular or armed with a spine, midcoxae contiguous or somewhat separated; midtibia usually with two spurs, but outer spur sometimes small or absent, inner hindtibial spur finely, closely pectinate; apicoventral setae of hindtarsomere V setiform to blade-like, claw usually simple but with one sharp, basal tooth on inner margin in *hungarica* group and some females of *nigricans* group (fig. 33 E); arolium small to large; petiole very long, consisting of cylindrical sternum I and slender, weakly dilated tergum I, thus appearing two segmented, gaster articulating with apex of tergum I (fig. 32 C); sternum I usually bent downward at base of tergum I or straight (fig. 32 C), rarely bent upward at this point, apex of sternum I not reaching base of II, both connected by a ligament visible in intervening space which is often very long (fig. 32 D); tergum I often as long as sternum I, its dorsal surface usually forming a fairly continuous plane with sternum I (lateral profile), thickest at level of spiracle (lateral profile) and usually parallel sided or constricting posterad of spiracle (continuously dilating from base to apex in some species of *nigricans* and *hungarica* groups); spiracle of tergum I located beyond middle; sternum VIII of male variable but usually truncate apically and often emarginate; penis valve head usually with teeth on ventral margin and often a basal spine-like process, gonostyle not biramous (fig. 34 C).

Geographic range: With 187 species, this cosmopolitan genus is the largest in the Ammophilini. The Palearctic Region has 70 species, but many of them are restricted to the xeric areas of central Asia and the Mediterranean area. Sixty species are found in America north of Mexico, and two-thirds of these are in the ecologically diverse western half of the continent. Mesoamerica has 23 species, some of which are shared with North America, but this area still has a distinctive fauna. The Ethiopian Region has 25 species, only a few of which are shared with the Palearctic fauna. *Ammophila* is poorly represented in the Oriental Region (10 species), South America (7 species), and Australia (5 species). The representatives in the last two regions represent intrusions by single species groups. Island endemics are few: West Indies – 2, Philippines – 1, Celebes – 1.

Systematics: Ammophila range in length from 8 to 37 mm. Although a very few species are entirely black, most forms have a partly or all red gaster, and the legs are commonly partly red. In a few species the head and/or thorax are red. Males usually are less extensively red than the female, and sometimes the male is quite differently colored. The wings are generally clear, but less often they are yellowish or infumate. The head and thorax of some species, especially those from xeric habitats, are densely covered with appressed silver hair which obscures most of the underlying sculpture. Most species, however, are less densely clothed; the silvering consists of pleural bands or spots and a general covering on the frons and clypeus. Some species do not have appressed silver hair. Erect hair when present may be pale or dark, and sometimes it varies geographically.

The gap between the apex of sternum I and the base of sternum II separates *Ammophila* from all other genera in the tribe, with the exception of *Eremnophila*. In *Ammophila* the episternal sulcus does not curve back to the scrobe before passing on down to the ventral region of

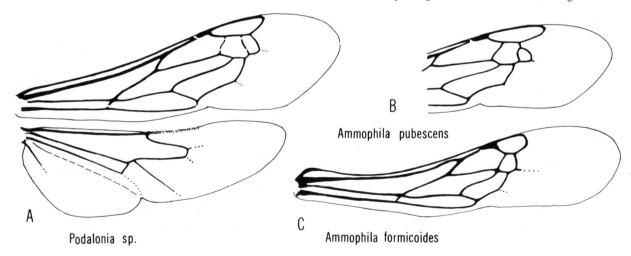

FIG. 35. Wings in the tribe Ammophilini.

the pleuron, as it does in *Eremnophila*. Most people have difficulty separating *Ammophila* and *Podalonia*, but the petiole characters, once appreciated, make separation easy (see discussion under *Podalonia*).

In the past a number of subgenera have been recognized in *Ammophila*: *Apycnemia* Leclercq, *Miscus* Jurine, *Argyrammophila* Gussakovskij, and *Coloptera* Lepeletier. It is the opinion of Menke, who has a revision of the New World *Ammophila* in progress, that species groups are a better solution to infrageneric groupings in this genus. This opinion is based on a worldwide study of the genus. Species groups are based primarily on length of the episternal sulcus, sculpture of the thoracic dorsum, male genitalia, and number of midtibial spurs.

Apycnemia was proposed for Kohl's (1896) *fallax* group, which is distinguished by the presence in both sexes of a basal claw tooth and a short episternal sulcus. The Old World *fallax* group, which should be called the *hungarica* group because of synonymy, is related to the New World *nigricans* group (Menke, 1970), as shown by similar male genitalia and the fact that females of some species have a claw tooth. The episternal sulcus is usually long in the *nigricans* group and the propodeal sculpture differs between the two groups. Morphologically, the *hungarica* and *nigricans* groups may represent the most primitive section of the genus. In addition to the claw tooth, the separation between sternum I and II is rather short in some species in these groups.

The New World *azteca* group (Menke, 1967a) and the Old World *sabulosa* group are characterized by having a long episternal sulcus, a usually smooth pronotum and scutum, and by the aedeagus. These two groups could be united, since they are really only geographically distinct. Together they contain the majority of the species of *Ammophila*, and a number of subgroups could be recognized. The species usually assigned to the subgenus *Miscus*, a taxon proposed for forms with a petiolate third submarginal cell (fig. 35 B), belong in the *sabulosa* group. Menke's (1966b) New World *urnaria* group is allied to the *sabulosa* and *azteca* groups, but it is distinguished from them by the fact that the episternal sulcus ends at about the level of the scrobe. The loss of one midtibial spur is a tendency in the *urnaria* group. Species of these three groups seem to be rapidly evolving.

Argyrammophila was established for a few unusual Old World species of *Ammophila* that have an elongate collar, a humped scutum, a long episternal sulcus, usually only one midtibial spur, and extensive appressed silver vestiture. We prefer to call this assemblage the *induta* group. *Ammophila induta* has an unusually long labrum, but the galea is shorter than in other species of the group that we have studied. The form of the thorax, especially the configuration of the petiole socket, is suggestive of the genus *Eremochares*. Wing venation in the *induta* group is different from the norm in *Ammophila*. The recurrent veins tend to be interstitial, and in some examples the first recurrent ends on submarginal cell I.

The *nasuta* group of Roth (1928) is an Old World assemblage, in which there is usually only one midtibial spur, the episternal sulcus is long, and the male has a narrow face and elongate, nasiform clypeus. The male genitalia are usually bizarre. The New World counterpart of this group is the California coast inhabitant *nasalis*. The *nasuta* and *induta* groups contain some of the most specialized *Ammophila*. Both groups have very long mouthparts.

The species of the New World *procera* group (Menke, 1964a) and the Old World *clavus* group (Kohl, 1906) are only geographically distinct. They have a short or no episternal sulcus, the collar often elongate, both it and the scutum usually transversely ridged, and the aedeagus has a distinctive form. There is a tendency for reduction in size of the arolium in females. We consider Kohl's *gracillima* group as part of the *clavus* group. Most of the Old World and all of the New World forms placed in *Coloptera,* a taxon proposed for species with only two submarginal cells, are assignable to the *clavus* and *procera* groups. Two noteworthy species of the latter group are the ant mimics, *wrightii* and *formicoides*. Small size, flattened pronotal collar, no episternal sulcus, very small arolia in the female of *wrightii* and both sexes of *formicoides,* as well as a tendency for small wings and reduced venation are interesting morphological aspects of these two forms. *Ammophila wrightii* displays an interesting west to east clinal change in the number of submarginal cells. Specimens from California and Arizona usually have three submarginal cells, and those from about the 104th meridian eastward have two. Wings with partial third submarginals or specimens with two cells in one wing and three in the other are found in the zone between these areas. Occasional examples of *wrightii* have only one submarginal cell, and the two recurrent veins simply end on the media beyond the cell. In *formicoides,* which always has two submarginals, the first recurrent vein is often interstitial or even received by the first submarginal.

Ammophila coronata has a pair of semicircular swellings on the upper frons and no episternal or scrobal sulci, but this species evidently should be assigned to the *clavus* group based on male genitalia. *Ammophila striata* has a long episternal sulcus but otherwise is obviously a member of the *clavus* group.

Species characters can be appreciated by studying the papers of Menke, Beaumont, Tsuneki, and Kohl. Kohl's (1906) work is still a valuable reference for the identification of Palearctic species. Roth (1928) and Alfieri (1946) provided keys for the identification of north African forms; Arnold (1928) keyed the species of the Ethiopian Region, although these are outdated. Beaumont (1963) keyed the species of the *nasuta* group. Tsuneki (1967d, 1971d) provided keys to east Asian species. Fernald's (1934) revision of the North American species still contains the only keys for identification of our species, but his work is full of errors. Menke's (1964a,b, 1966b, 1967a, 1970) papers contain descriptions of new species and notes and figures useful for distinguishing them from previously named forms.

Biology: The early disclosure of the Peckhams (1898), that some members of *Ammophila* used a "tool" during nest closure and therefore might possess intelligence, made this genus a popular object of study for students of behavior, and a considerable literature has built up since then. H. E. Evans (1959b) and others have since shown that the "tool using" habit is the result of a gradual succession of instinctive behavioral elements. Evans (1959b) and Powell (1964) have reviewed most of the literature on North American species and included their own extensive observations. Since then two more important papers have been published by Evans (1965, 1970), and Menke cleared up the identity of most of the species observed by Hicks (Menke, 1965b). Baerends' (1941) classic study of *Ammophila pubescens* (as *campestris*) contains references to most of the earlier literature on Old World species. His work was summarized in English by Tinbergen (1958). Important Old World papers published since Baerends' paper are: Adriaanse (1948), Teschner (1959), Olberg (1952, 1959), Grandi (1957, 1961, 1962), Myartseva (1969), Bonelli (1966, 1967, 1969), and Tsuneki (1968c). Olberg's works contain many fine photographs which show various nesting activities.

Ammophila are generally solitary nesters, but a few are reported to be gregarious. Nests are dug prior to searching for prey according to most reports, but Hicks (1934) and Roth (1928) claimed the reverse in *wrightii* and *haimatosoma*, respectively. Interestingly, these two species belong to the *procera* and *clavus* groups, respectively, which are closely related. Also, these two species have very small arolia, a characteristic of many *Podalonia*, most of which excavate the nest after prey has been obtained.

Ammophila nests are generally simple, short burrows ending in a single cell, but Janvier (1928) and Frisch (1940) have described two-celled nests. When the excavation is completed, the wasp usually makes a temporary closure. Lepidopterous and hymenopterous (sawfly) caterpillars are provisioned by *Ammophila*. Recently, Frank Parker (personal communication) discovered *A. azteca* provisioning with larvae of the weevil *Hypera postica* in Utah. For most species, prey selection seems dependent upon the species available. Hairless larvae are most commonly used, but moderately to densely hairy caterpillars are sometimes provisioned. Depending on the size of the larva, prey are either flown to the nest or dragged over the ground. The caterpillar is carried head foremost and upside down and is held by the wasp's mandibles, which are often assisted by the front pair of legs. One to 11 prey may be placed in a cell. The egg is laid on the first provision near the cephalic end of the caterpillar.

Mass provisioning is the most common type in *Ammophila*, but Baerends (1941), Powell (1964), Evans (1965), and Tsuneki (1968c) documented the occurrence of progressive provisioning in *pubescens, azteca, pruinosa, harti,* and *sabulosa*, although this trait is apparently not necessarily constant within a species. All of these species belong in the closely related *azteca* and *sabulosa* species groups.

Baerends (1941), Evans (1965), and Tsuneki (1968c) found that *pubescens, azteca,* and *sabulosa,* respectively, maintain several nests at one time. After the initial caterpillar is provisioned and an egg has been laid, the nest is temporarily sealed. At least in *pubescens,* periodic visits are made by the wasp to ascertain whether the egg has hatched. Prior to this event no additional caterpillars are provisioned. The ability of these wasps to remember the location of several nests and also their changing requirements is remarkable.

Tsuneki (1968c) reported several cases of a peculiar behavior that he observed in *sabulosa*, which he interpreted as a primitive type of parasitism. He saw two females that dug up completed nests presumably made by some other *sabulosa* female. The wasp then pulled out the caterpillar, ate the attached egg, and then stung the already paralyzed host. The burrow was then cleaned out, the caterpillar reintroduced and an egg laid. Nest closure followed.

Baerends (1941) and Tsuneki (1968c) both devoted considerable attention to the mechanisms used by *Ammophila* in orientation and nest finding. Teschner's paper (1959) included a discussion of male recognition of females by abdominal color pattern. *Ammophila* commonly form loose sleeping aggregations on scrub and tall grasses on which they spend the night (Evans and Linsley, 1960). Two North American species, *wrightii* and *formicoides*, are interesting because of their close resemblance to red or brown myrmicine ants. These small wasps have rather short wings, and they are often found running about on the ground in association with *Pogonomyrmex* for which they are easily mistaken.

Miltogrammine sarcophagids commonly infest *Ammophila* nests.

Checklist of *Ammophila*

aberti Haldeman, 1852; w. U.S., Mexico
    urnaria Lepeletier, 1845, nec Dahlbom, 1843
    tarsata F. Smith, 1856
    yarrowi Cresson, 1875
    transversus Fernald, 1934 (*Sphex*)
acuta (Fernald), 1934 (*Sphex*); w. U.S.
adelpha Kohl, 1901; sw. USSR
aellos Menke, 1966; Mexico
afghanica Balthasar, 1957; Afghanistan
airensis Berland, 1950; Niger (? = *theryi*)
albotomentosa Morice, 1900; Algeria
altigena Gussakovskij, 1930; centr. Asia
aphrodite Menke, 1964; sw. U.S.
apicalis Guérin-Méneville, 1835 (pl. 70); W. Indies
    guerinii Dalla Torre, 1897
arabica W. F. Kirby, 1900; Saudi Arabia
ardens F. Smith, 1868; Australia
?*areolata* Walker, 1871; Saudi Arabia
argyrocephala Arnold, 1951; Ethiopia
†?*arvensis* Lepeletier, 1845; "Amerique Septentrionale", nec Dahlbom, 1843 (? = *procera* or *Podalonia*)
asiatica Tsuneki, 1971; Mongolia
assimilis Kohl, 1901; e. Mediterranean region, Afghanistan

atlantica Roth, 1928; n. Africa
atripes F. Smith, 1852; Oriental Region: mainland Asia
    ?*dimidiata* F. Smith, 1856, nec Christ, 1791
    pulchella F. Smith, 1856
    simillima F. Smith, 1856
    humbertiana Saussure, 1867, lectotype ♀, "Trincom., Ceylon" (Mus. Geneva), present designation by Menke
    longiventris Saussure, 1867, lectotype ♂, "Trincom., Ceylon" (Mus. Geneva), present designation by Menke
    spinosa F. Smith, 1873
    buddha Cameron, 1889
  ssp. *taschenbergi* Cameron, 1890; Java
    erythropus Taschenberg, 1869, nec Smith, 1856
  ssp. *japonica* Kohl, 1906; Japan, Korea, s. Manchuria, lectotype ♂, Kofou, Japan (Mus. Paris), present designation by Menke
  ssp. *formosana* Strand, 1913; Taiwan, Ryukyus
    taiwana Tsuneki, 1967
aucella Menke, 1966; Mexico, Guatemala
aurifera R. Turner, 1908; Australia
azteca Cameron, 1888; Canada and U.S.: transcontinental
    pilosa Fernald, 1934 (*Sphex*)
    aculeata Fernald, 1934 (*Sphex*)
    nuda Murray, 1938 (*Sphex*), nec Fernald, 1903
    brevisericea Murray, 1951
  ssp. *clemente* Menke, 1967; California: San Clemente Island
barbara (Lepeletier), 1845 (*Coloptera*); Algeria, Morocco
  ssp. *semota* Beaumont, 1967; Turkey
barbarorum Arnold, 1951; Ethiopia
basalis F. Smith, 1856; India, Sri Lanka
    nigripes F. Smith, 1856
    orientalis Cameron, 1889
bechuana (Turner), 1929 (*Sphex*); s. Africa
bella Menke, 1966; U.S.: Arizona, Mexico
bellula Menke, 1964; U.S.: Arizona, New Mexico
beniniensis (Palisot de Beauvois), 1806 (*Sphex*); Africa
    tenuis Palisot de Beauvois, 1806 (*Sphex*)
    cyaniventris Guérin-Méneville, 1843
    ?*rugicollis* Lepeletier, 1845
    lugubris Gerstaecker, 1857
    guineensis Ritsema, 1874
    coeruleornata Cameron, 1910
    maculifrons Cameron, 1910
    massaica Cameron, 1910
    sjoestedti Cameron, 1910
  †ssp. *tomentosa* (Arnold), 1920 (*Sphex*); s. Africa, nec Fabricius, 1787
  ssp. *imerinae* Saussure, 1892; Madagascar
boharti Menke, 1964; U.S.: California, Nevada
bonaespei Lepeletier, 1845; s. Africa, lectotype ♀, "cap de Bonne Esperance" (Mus. Turin), present designation by Menke
    *rubriceps Taschenberg, 1869, new synonymy by Menke
braunsi (Turner), 1919 (*Sphex*); s. Africa

breviceps F. Smith, 1856; w. U.S., Mexico
brevipennis Bingham, 1897; India
californica Menke, 1964; w. U.S.
calva (Arnold), 1920 (*Sphex*); s. Africa, new status by Menke
campestris Latreille, 1809; Europe, Asia
    retusa Gistel, 1848
    ?*neoxena* F. Smith, 1856
    ? *striaticollis* F. Morawitz, 1888
    ? *nigrina* F. Morawitz, 1888
    ? *separanda* F. Morawitz, 1890
caprella Arnold, 1951; Ethiopia
cellularis Gussakovskij, 1930; Siberia
centralis Cameron, 1888; Mexico, Centr. America, lectotype ♂, "El Reposo" (Mus. London), present designation by Menke
    consors Cameron, 1888, lectotype ♂, "N. Yucatan" (Mus. London), present designation by Menke
    nigrocaerulea Cameron, 1888
clavus (Fabricius), 1775 (*Sphex*); Australia
cleopatra Menke, 1964; U.S.: transcontinental; Mexico
    juncea of authors, not Cresson, 1865
coachella Menke, 1966; sw. U.S.
conditor F. Smith, 1856; se. U.S.
confusa A. Costa, 1864; Senegal
conifera (Arnold), 1920 (*Sphex*); s. Africa
cora Cameron, 1888; Mexico, lectotype ♂, "San Geronimo, Guat." (Mus. London), present designation by Menke
cybele Menke, 1970; Cuba
coronata A. Costa, 1864; Philippines
    superciliaris Saussure, 1867
crassifemoralis (Turner), 1919 (*Sphex*); Africa
dantani Roth, 1933; n. Africa
dejecta Cameron, 1888; U.S.: Arizona; Mexico
dentigera Gussakovskij, 1928; sw. USSR, Iran
deserticola Tsuneki, 1971; Mongolia, ne. China
djaouak Beaumont, 1956; Libya
dolichocephala Cameron, 1910; s. Africa
dolichodera Kohl, 1884; S. Africa
    macrocola Kohl, 1884
    pulchricollis Cameron, 1910
    lukombensis Cameron, 1912
dubia Kohl, 1901; Egypt
dysmica Menke, 1966; w. U.S.
elongata Fischer-Waldheim, 1843; s. Russia
evansi Menke, 1964; e. U.S.
exsecta Kohl, 1906; Syria
extremitata Cresson, 1865; w. U.S.
eyrensis R. Turner, 1908; Australia
femurrubra W. Fox, 1894; sw. U.S.
fernaldi (Murray), 1938 (*Sphex*) e. U.S. to Arizona, Mexico
*ferrugineipes Lepeletier, 1845; s. Africa, new status by Menke
    erythrospila Cameron, 1905
    dunbrodyensis Cameron, 1905
    meruensis Cameron, 1910
    ? *curvistriata* Cameron, 1910

ferruginosa Cresson, 1865; w. U.S. except coastal states
    collaris Cresson, 1865
    cressoni H. Smith, 1908 (*Sphex*)
?*filata* Walker, 1871; Egypt
formicoides Menke, 1964; U.S.: Arizona, New Mexico; Mexico
formosensis Tsuneki, 1971; Taiwan
    formosana Tsuneki, 1967, nec Strand, 1913
gaumeri Cameron, 1888; Mexico, centr. America, lectotype ♀, "N. Yucatan" (Mus. London), present designation by Menke
    micans Cameron, 1888
gegen Tsuneki, 1971; Mongolia
gracilis Lepeletier, 1845; S. America
    mutica Dahlbom, 1845
    moneta F. Smith, 1856
    fragilis F. Smith, 1856
    ? *sauvis* Burmeister, 1872
    ? *arechavaletai* Brèthes, 1909
gracillima Taschenberg, 1869; Egypt, Sudan, centr. Asia, lectotype ♀, "Chartum" (Mus. Halle), present designation by Menke
    longicollis Kohl, 1884
    debilis F. Morawitz, 1889
    producticollis Morice, 1900 (♂ only)
grandis Gistel, 1857; Italy
guichardi Beaumont, 1956; Libya
haimatosoma Kohl, 1884, Mediterranean region
  ssp. *sinaitica* Alfieri, 1946; Egypt
harti (Fernald), 1931 (*Sphex*); centr. U.S.
    argentata Hart, 1907, nec Lepeletier, 1845
hemilauta Kohl, 1906; n. Africa
hermosa Menke, 1966; w. U.S.
heydeni Dahlbom, 1845; Mediterranean region, Asia
    ? *attenuata* Christ, 1791 (*Sphex*)
    ? *rubra* Radoszkowski, 1876 (? = *quadraticollis*)
    rubra Radoszkowski, 1877, nec Radoszkowski, 1876
    iberica Ed. André, 1886
  ssp. *rubriventris* A. Costa, 1864; Sicily, Corsica
  ssp. *sarda* Kohl, 1906; Sardinia, Cyprus
holosericea (Fabricius), 1793 (*Sphex*); "Barbarie"
    sericea Lepeletier and Serville, 1828
honorei Alfieri, 1946; Egypt
horni Schulthess, 1927; Sudan
hungarica Mocsáry, 1883; Iberian peninsula to Hungary and sw. USSR, Iran, Cyprus, new status by Menke
  * *turcica* Mocsáry, 1883
  * *hispanica* Mocsáry, 1883
    fallax Kohl, 1884, lectotype ♀, Amasia, Turkey (Mus. Vienna), present designation by Menke
hurdi Menke, 1964; sw. U.S.
imitator Menke, 1966; U.S.: Arizona; Mexico
induta Kohl, 1901; centr. Asia
insignis F. Smith, 1856; Africa
    transvaalensis Cameron, 1910
    promontorii Arnold, 1920 (*Sphex*)
  ssp. *egregia* Mocsáry, 1881; Syria, Israel
  ssp. *litoralis* (Arnold), 1920 (*Sphex*); s. Africa
    nigricollis Arnold, 1960

insolita F. Smith, 1858; Celebes
  ssp. *argyropleura* van der Vecht, 1957; Indonesia: Sumba
  ssp. *auricollis* van der Vecht, 1957; Indonesia: Timor
  ssp. *ruficoxa* van der Vecht, 1957; Indonesia: Wetar, Kisar
instabilis F. Smith, 1856; Australia
    impatiens F. Smith, 1868
juncea Cresson, 1865; U.S.: transcontinental; Mexico
    montezuma Cameron, 1888
judaeorum Kohl, 1901; Jordan, Israel (? = *theryi*)
kalaharica (Arnold), 1935 (*Sphex*); S. Africa
karenae Menke, 1964; w. U.S.
kennedyi (Murray), 1938 (*Sphex*); U.S.: transcontinental
    vulgaris Cresson, 1865, nec W. Kirby, 1798
koppenfelsii Taschenberg, 1880; w. Africa
laeviceps F. Smith, 1873; Chile
  ? *chilensis* Reed, 1894
  ? *ruficollis* Reed, 1894
laevicollis Ed. André, 1886; Spain, s. France
laevigata F. Smith, 1856; India, Sri Lanka, Vietnam, Thailand
    bicellulalis Strand, 1915
lampei Strand, 1910; Peru, Chile
    nigripes Reed, 1894, nec Smith, 1856
    peruviana Rohwer, 1913 (*Sphex*)
leoparda (Fernald), 1934 (*Sphex*); e. U.S.
laticeps (Arnold), 1928 (*Sphex*); S. Africa
lativalvis Gussakovskij, 1928; sw. USSR
leclercqi Menke, 1964; Spain
    yarrowi Leclercq, 1961, nec Cresson, 1875
macra Cresson, 1865; w. U.S.
marshi Menke, 1964; U.S.: California, Nevada
mcclayi Menke, 1964; U.S.: California, Nevada
mediata Cresson, 1865; w. U.S.
mescalero Menke, 1966; sw. U.S., Mexico
mimica Menke, 1966; U.S.: California, Arizona
minor Lepeletier, 1845; Algeria
mitlaensis Alfieri, 1961; Egypt
modesta Mocsáry, 1883; Spain
moenkopi Menke, 1967; U.S.: Arizona
monachi Menke, 1966; U.S.: Nevada
mongolensis Tsuneki, 1971; Mongolia
murrayi Menke, 1964; U.S.: California
nasalis Provancher, 1895; U.S.: California
    craspedota Fernald, 1934 (*Sphex*)
nasuta Lepeletier, 1845; Algeria, lectotype ♂, Oran, Algeria (Mus. Turin), present designation by Menke
nearctica Kohl, 1889; w. U.S.
nefertiti Menke, 1964; U.S.: Great Basin area
nigricans Dahlbom, 1843; e. U.S.
    intercepta Lepeletier, 1845
nitida Fischer-Waldheim, 1843; USSR
novita (Fernald), 1934 (*Sphex*); sw. U.S., Mexico
?*obscura* Bischoff, 1911; Africa: Zaire[30]

---

[30] Bischoff described *obscura* as a variety of *rufipes* Lepeletier which he misidentified as a species of *Ammophila*. True *rufipes* is a *Podalonia* and is now known as *erythropus*.

occipitalis F. Morawitz, 1890; sw. USSR
parapolita (Fernald), 1934 (*Sphex*); w. U.S.
parkeri Menke, 1964; w. U.S.
peckhami (Fernald), 1934 (*Sphex*); U.S.: Colorado to Mexico
    willistoni Fernald, 1934 (*Sphex*)
peringueyi (Arnold), 1928 (*Sphex*); S. Africa
philomela Nurse, 1903; India
picipes Cameron, 1888; U.S.: Arizona; Centr. America
    alticola Cameron, 1888
    volcanica Cameron, 1888
    chiriquensis Cameron, 1888
    fragilis of authors, not F. Smith, 1856
pictipennis Walsh, 1869; e. U.S.; Mexico
    anomala Taschenberg, 1869, lectotype ♀, Illinois (Mus. Halle), present designation by Menke
    ? *nigropilosa* Rohwer, 1912 (*Sphex*)
pilimarginata Cameron, 1912; Guyana
placida F. Smith, 1856; w. U.S., Mexico
platensis Brèthes, 1909; Argentina
poecilocnemis Morice, 1900; Algeria
polita Cresson, 1865; w. U.S.
procera Dahlbom, 1843; U.S.: transcontinental; Mexico; Guatemala
    procera Lepeletier, 1845, nec Dahlbom, 1843
    saeva F. Smith, 1856; lectotype ♀, California (Mus. London), present designation by Menke
    gryphus F. Smith, 1856, lectotype ♀, Charleston, Fla. (Mus. London), present designation by Menke
    barbata F. Smith, 1856
    ceres Cameron, 1888, lectotype ♂, "San Geronimo, Guat." (Mus. London), present designation by Menke
    championi Cameron, 1888, lectotype ♀, "San Geronimo, Guat." (Mus. London), present designation by Menke
    striolata Cameron, 1888
producticollis Morice, 1900 (♀ only); Algeria, sw. USSR
    divina Kohl, 1901
    argentina Gussakovskij, 1930, new synonymy by Pulawski
proxima (F. Smith), 1856 (*Coloptera*); Guinea
pruinosa Cresson, 1865; w. U.S.
pseudonasuta Bytinski-Salz, 1955; Israel
pubescens Curtis, 1836; Europe
    pubescens Curtis, 1828-1829, nomen nudum
    arvensis Dahlbom, 1843 (*Miscus*)
    susterai Snoflak, 1943
    alpicola Beaumont, 1945
    adriaansei Wilcke, 1945
pulawskii Tsuneki, 1971; Mongolia
punctata F. Smith, 1856; n. India
punctaticeps (Arnold), 1920 (*Sphex*); s. Africa
quadraticollis A. Costa, 1893; Tunisia
regina Menke, 1964; w. U.S.
roborovskyi Kohl, 1906; Mongolia, Iran
rubiginosa Lepeletier, 1845; S. Africa
rubripes Spinola, 1838; Mediterranean area
    propinqua Taschenberg, 1869

syriaca Mocsáry, 1883
    erminea Kohl, 1901
ruficollis A. Morawitz, 1890; sw. USSR
ruficosta Spinola, 1851; Argentina, Chile
rufipes Guérin-Méneville, 1831 (pl. 9) (*Ammophilus*); Peru
    ? *variolosa* Giner Marí, 1944
†*rugicollis* Gussakovskij, 1930; sw. USSR, Iran, nec Lepeletier, 1845
sabulosa (Linnaeus), 1758 (*Sphex*); Europe, Asia
    hortensis Poda, 1761 (*Sphex*)
    frischii Fourcroy, 1785 (*Ichneumon*)
    dimidiata Christ, 1791 (*Sphex*), nec De Geer, 1773
    vulgaris W. Kirby, 1798
    pulvillata Sowerby, 1805, new synonymy by Menke
    mucronata Jurine, 1807 (*Sphex*)
    cyanescens Dahlbom, 1845
    ? *vischu* Cameron, 1889
    ssp. *vagabunda* F. Smith, 1856; China, Taiwan
    ssp. *infesta* F. Smith, 1873; Japan (montane), Korea
        marginalis Pérez, 1905, new synonymy by Menke
    ssp. *touareg* Ed. André, 1886; Algeria
        oraniensis Roth, 1928
    ssp. *kamtschatica* Gussakovskij, 1933; e. USSR: Kamchatka Penin.
    ssp. *nipponica* Tsuneki, 1967; Japan (lowland), ne. China, Korea
sarekandana Balthasar, 1957; Afghanistan
sareptana Kohl, 1884; sw. USSR, Turkey
saussurei (du Buysson), 1897 (*Coloptera*); S. Africa
shoshone Menke, 1967; U.S.: Wyoming, Utah
sickmanni Kohl, 1901; China
    ssp. *wusheensis* Tsuneki, 1967; Taiwan (? = *subassimilis*)
sinensis Sickmann, 1894; China
†*sjoestedti* Gussakovskij, 1935; China, nec Cameron, 1910
slovaca Zavadil, 1937; Czechoslovakia (? = *campestris*)
smithii F. Smith, 1856; India
solowiyofkae Matsumura, 1911; USSR: Sakkalin (? = *sabulosa nipponica*)
stangei Menke, 1964; w. U.S.
strenua Cresson, 1865; w. U.S., Mexico
    dubia Fernald, 1934 (*Sphex*), nec Kohl, 1901
    denningi Murray, 1951
striata Mocsáry, 1878; Siberia
    ssp. *nadigi* Roth, 1932; n. Africa
strumosa Kohl, 1906; n. Africa, Israel
subassimilis Strand, 1913; Taiwan
tekkensis Gussakovskij, 1930; sw. USSR
terminata F. Smith, 1856; Canary Islands
    apicalis Brullé, 1839, nec Guérin-Méneville, 1835
    ssp. *mocsaryi* Frivaldsky, 1876; s. Europe, w. Asia
        minuta Frivaldsky, 1876
        rhaetica Kohl, 1879
        julii Fabre, 1879
        moksari Marquet, 1881, lapsus or emendation
        ? *kirgisica* F. Morawitz, 1890
    ssp. *electa* Kohl, 1901, n. Africa, lectotype ♀, Sidi

Maklouf, Algeria (Mus. Paris), present designation by Menke
   ssp. *turkestana* Kohl, 1906, sw. USSR
      turcestanica Kohl, 1906, nec Dalla Torre, 1897
theryi (Gribodo), 1894 (*Coloptera*); Algeria, new status
tsunekii Menke, new name for *tomentosa* Tsuneki; Mongolia
    tomentosa Tsuneki, 1971, nec Arnold, 1920
tuberculiscutis (Turner), 1919 (*Sphex*); e. Africa
tyrannica Cameron, 1890; ? India
unita Menke, 1966; U.S.: Great Basin area
urnaria Dahlbom, 1843; e. U.S.
   inepta Cresson, 1872
   floridensis Fernald, 1934 (*Sphex*)

varipes Cresson, 1865; centr. U.S., Mexico
   comanche Cameron, 1888
vulcania du Buysson, 1897; s. Africa
wahlbergi Dahlbom, 1845; s. Africa
wrightii (Cresson), 1868 (*Coloptera*); w. U.S.
zanthoptera Cameron, 1888; U.S.: Arizona; Mexico
   trichiosoma Cameron, 1888
   xanthoptera Cameron, 1900

Fossil *Ammophila*

annosa Heer, 1865; Miocene. Nomen nudum?
antiquella Cockerell, 1906; Florissant, Colorado
inferna Heer, 1865; Miocene

SUBFAMILY PEMPHREDONINAE

The pemphredonines are medium small to tiny sphecids, many of which are recognized by one or more of the following characters: a stemlike sternal petiole, a cuboidal head, and a disproportionately enlarged forewing stigma. These criteria are useful in a general way only, since convergence with sphecids in other subfamilies creates certain technical difficulties, discussed below in the section on systematics.

Pemphredonines are often abundant, but the smaller forms are usually overlooked by collectors. The females provision their nests with Homoptera or in a few cases with Thysanoptera or Collembola.

Diagnostic characters:

1. (a) Inner orbits essentially parallel, converging below or somewhat above, eyes usually rather widely separated; (b) ocelli normal.
2. (a) Antennae low on face or near middle, sockets contiguous with or well separated from frontoclypeal suture; (b) male with 13 and female with 12 antennal articles.
3. Clypeus transverse, often very short.
4. (a) mandible without notch or step on externoventral margin, apex sometimes bi- to pentadentate, inner margin simple or with teeth; (b) palpal formula 6-4; (c) mouthparts short; (d) mandible socket usually open in Psenini, usually closed in Pemphredonini.
5. (a) Pronotal collar short, usually high, often transversely carinate; (b) pronotal lobe and tegula well separated in Psenini, but the two touching or narrowly separated in most Pemphredonini.
6. (a) Notauli usually present, sometimes very long; (b) no oblique scutal carina; (c) admedian lines separated.
7. (a) Episternal sulcus present or absent; (b) omaulus present or absent; (c) acetabular carina and hypersternaulus sometimes present.
8. Definitive metapleuron consisting of upper metapleural area (except lower area also present in *Odontopsen*).
9. (a) Midtibia with one apical spur; (b) midcoxae nearly contiguous or broadly separated, with dorsolateral crest or carina except in some Pemphredonini; (c) precoxal lobes present but very short in some Pemphredonini; (d) hindfemoral apex simple; (e) claw simple; (f) plantulae present or absent.
10. (a) Propodeum short to moderately long; (b) dorsal enclosure triangular when present; (c) propodeal sternite absent.
11. (a) Forewing with three, two, one, or no submarginal cells and one or two recurrent veins whose end points are variable; (b) marginal cell apex usually acute, rarely truncate or open; (c) three or fewer discoidal cells present; (d) forewing media diverging after cu-a.
12. (a) Jugal lobe small, no more than half length of anal area; (b) hindwing media diverging before, at, or after cu-a; (c) subcostal vein absent; (d) jugal lobe without an anal vein.
13. (a) Gaster sessile or more often petiolate, the petiole usually composed of sternum I only (tergum included in *Microstigmus*); (b) tergum I without lateral carina except in some Pemphredonini; (c) male with seven and female with six visible gastral segments; (d) pygidial plate present or absent.
14. (a) Volsella with digitus and cuspis; (b) cerci absent.

Systematics: The Pemphredoninae as treated here is a subfamily containing two tribes of nearly equal size, 28 genera, and 715 species. Most of the genera have a distinct petiole made up entirely from the first sternum. Petiolate forms are readily identified to subfamily by this characteristic in addition to the single midtibial spur, moderate jugal lobe of the hindwing, and the absence of a spoon-shaped distal truncation on the hindfemur. However, *Microstigmus*, at least, seems to have both tergum and sternum in the petiole. Nonpetiolate forms, in addition to the above characteristics, have no more than two submarginal cells in the forewing and a much enlarged stigma (as large as the only discoidal cell); or if there are two or three discoidal cells, the stigma is nearly as large as the first discoidal; two sessile submarginal cells are present, the man-

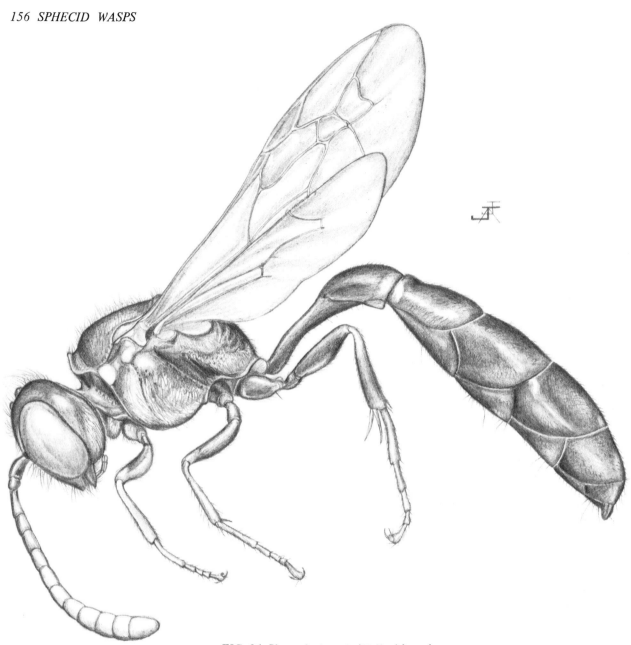

FIG. 36. *Pluto clavicornis* (Malloch), male.

dible is simple externoventrally, the inner eye margins are not angulate, and the hindwing media diverges at or before cu-a. The complications in the nonpetiolate forms are a reflection of the parallel reduction in wing venation and cells in Pemphredonini and certain of the Larrinae and do not indicate a close relationship between the two groups.

We have only made a cursory study in the Pemphredoninae of a number of morphological features that have important diagnostic or phylogenetic value in other subfamilies: 1) mandible socket open or closed, 2) plantulae present or absent, 3) lateral tergal carina present or absent, 4) midcoxa simple or with oblique, dorsal crest or carina, and 5) whether or not the tegula and pronotal lobe are in contact. The generic and phylogenetic relevance of these features should be investigated.

The position of the Pemphredoninae with respect to other subfamilies is expressed in the dendogram, fig. 7. The two tribes contain many highly specialized forms. On the basis of reduction in size and wing veins the Pemphredonini seem to have reached a more advanced stage of evolution than the Psenini. The supposed relationships within the tribes are indicated in figs. 37 and 38. A matter of some interest in the phylogeny of the subfamily is the recent finding of a fossil in Upper Cretaceous amber from Cedar Lake, Manitoba, Canada. Evans (1969b) described the genus as *Lisponema* and pointed out the apparent close relationship to *Spilomena*. He further noted that the occurrence of such a specialized wasp in the Upper Cretaceous indicates that the twig-nesting habit was developed very early in the aculeates. A second Upper

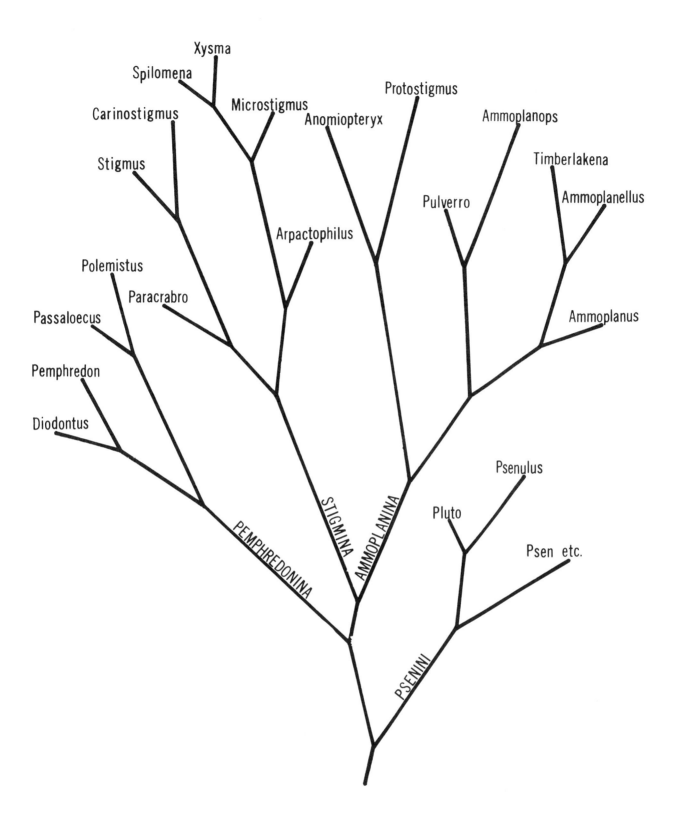

FIG. 37. Dendrogram suggesting relationships in the Pemphredonini.

Cretaceous pemphredonine genus, *Pittoecus,* has just been described by Evans (1973b).

KEY TO TRIBES OF PEMPHREDONINAE

Forewing with three submarginal cells; antennal sockets placed well above clypeal margin, usually near middle of face...*Psenini,* p. 158
Forewing with no more than two submarginal cells; antennal sockets usually placed just above clypeal margin..............................*Pemphredonini,* p. 174

Tribe Psenini

The psenins are medium small wasps ranging in length from 4 to 15 mm. All are distinctly petiolate although the petiole in *Ammopsen* is short. This genus is peculiar in other respects. It contains the smallest wasps in the tribe and the only ones without an omaulus. Also, there is no transverse pronotal carina, and a "sandbasket" is formed by the foretarsal rake of the female together with the genal ammochaetae. It is the only genus without a trace of a frontal carina. *Ammopsen* may be the most generalized psenin genus.

The distal divergence of the hindwing media in *Psenulus* and *Pluto* has led us to divide the tribe into two subtribes, Psenulina and Psenina (fig. 41). These subtribes correspond to the suprageneric groups labeled as "Psenuli and Pseni" by Gittins (1969).

The episternal sulcus is well developed throughout the Psenini, although it is somewhat abbreviated in *Nesomimesa.* Typically it starts out at the forward part of the subalar fossa close behind the postspiracular carina. It descends almost vertically and ends below against the omaulus with which it forms a long acute angle. It may continue past its juncture with the omaulus as far as the anteroventral margin of the pleuron. Our brief studies of one or two species from each genus indicate that the mandibular sockets are open in the Psenini, which bolsters our contention that the tribe is more generalized than the Pemphredonini. Also, the pronotal lobe appears to be well separated from the tegula throughout the Psenini but the converse is true in many members of the other tribe. One interesting morphological feature of Psenini, except in *Ammopsen* and *Pluto,* is the projection of male sternum VIII, which is carried as an upturned pseudosting.

The recurrent veins seem to be unusually plastic in relation to their endpoints on the submarginal cells. The common situation is for both recurrents to end on submarginal cell II, but in *Ammopsen* the usual endpoints are on I and III. Similarly, in *Psen* and *Pluto* the recurrents are usually to II and III. Exceptions occur in some species or specimens of all of these and occasionally on the two forewings of a single specimen.

Our ideas of generic relationships are expressed in the dendrogram, fig. 38. The recognition of *Mimesa, Mimumesa, Odontopsen,* and *Pseneo* as distinct genera is a more liberal view than that of previous workers, except

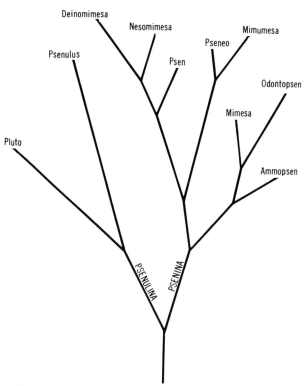

FIG. 38. Dendrogram suggesting relationships in the Psenini.

Gittins (1969) who raised *Mimumesa* to generic status. Gittins based his decision primarily on mesopleural differences as well as differing biologies. In most respects we agree with Gittins as discussed below and in the systematics section under *Mimumesa.*

Differing views on generic status in the Psenini are illustrated by *Mimesa,* which was proposed by Malloch (1933) as a subgenus of *Psen.* Krombein (1951) elevated it to genus but included *Mimumesa.* Van Lith (1959), Beaumont (1937, 1964b), and Oehlke (1965) have continued to treat *Mimesa* and *Mimumesa* as subgenera of *Psen.* It was pointed out by van Lith that the differences separating *Psen* and *Mimumesa* are less striking than those separating the latter from *Mimesa.* Therefore, rather than treating *Mimesa* as a separate genus and *Mimumesa* as a subgenus, he preferred to consider all of the categories and *Pseneo* as subgenera of *Psen.* While agreeing with the facts presented by van Lith, we feel that since the categories can be separated morphologically, and a considerable number of species is involved, it is more practical to treat the above mentioned groups as genera.

The tribe contains 10 genera and some 332 species distributed in all the faunal regions but especially in the Holarctic and Oriental ones.

The biology differs from that of the Pemphredonini particularly in the prey. Whereas the above tribe provisions primarily with aphis and to a lesser extent with immature scales, thrips and Collembola, the Psenini use mostly cicadellid, fulgoroid, membracid, cercopid, and psyllid "plant-hoppers". Most useful publications dealing generally with the tribe are Malloch (1933), Beaumont (1937), Gussakovskij (1937), Spooner (1948), Leclercq (1961a), Mingo (1964), van Lith (1959, 1962, 1965), Oehlke (1965), Gittins (1969), and Danks (1971).

Key to genera of Psenini

1. Hindwing media diverging at or beyond cu-a (fig. 41 A,B) ..... subtribe Psenulina 2
Hindwing media diverging well before cu-a (fig. 41 C,D) ..... subtribe Psenina 3
2. Hindcoxa with a downwardly directed bristle, most conspicuous in female (fig. 39 L); frontal carina low and simple; female foretarsus with a short rake (Neotropical and Nearctic) .... *Pluto* Pate, p. 169
Hindcoxa without a special bristle; frontal carina unusually raised between antennal bases and connected below with a cross carina, at least in males, female foretarsus without a rake (widespread except S. America) .... *Psenulus* Kohl, p. 171
3. Omaulus absent or strong and curving semicircularly forward toward prothorax, never curving posteriorly nor joining an acetabular carina; scrobal sulcus not deeply impressed, hypoepimeral area not well defined, usually more strongly punctate or ridged than median area of mesopleuron; frontal carina usually incomplete (fig. 40 A) 4
Omaulus strong, continued by acetabular carina to midventral line (fig. 43 E), or ending just as it becomes ventral (fig. 43 L) or as it turns posteriorly; scrobal sulcus deeply impressed, hypoepimeral area usually smooth and bulging; frontal carina usually complete from midocellus to interantennal area (fig. 40 B,C,D). 6
4. Omaulus absent, occipital carina not extending halfway to hypostomal carina (fig. 39 E) pronotal collar rounded beneath dense pubescence; female pygidial plate weakly differentiated laterally; male sternum VI with subapical row of long, flattened setae (w. U.S.) .... *Ammopsen* Krombein, p. 159
Omaulus present (fig. 43 H), occipital carina reaching to or almost to hypostomal carina (fig. 39 D); pronotal collar with a transverse carina, at least laterally; female pygidial plate with strong lateral carinae; male sternum VI without long flattened setae.... 5
5. Occipital carina becoming higher as it approaches hypostomal area and ending in a tooth; pronotal collar with carina at sides only (Japan) .... *Odontopsen* Tsuneki, p. 162
Occipital carina of even height; pronotal collar with carina extending all across (essentially Holarctic) ..... *Mimesa* Shuckard, p. 161
6. Omaulus continued to midventer by an acetabular carina (fig. 43 E) 7
Omaulus not continued to midventer by an acetabular carina (fig. 43 L) 8
7. Petiole polished above and without carinae or obvious grooves, outwardly directed hairs along inside of laterodorsal carina inconspicuous or absent; apex of clypeus nearly always with a broad bevel broken by several vertical carinae (New World and Philippine Is.) .... *Pseneo* Malloch, p. 164
Petiole carinate or posteriorly sulcate above, with conspicuous outwardly directed hairs along inside of laterodorsal carina; apex of clypeus thin, not broadly beveled (widespread, except Australia and S. America) .... *Mimumesa,* Shuckard, p. 163
8. Petiole with a distinct, although sometimes rather weak (rooflike), mediodorsal, longitudinal carina (fig. 39 G); female with clypeus directed forward, exposing labrum (fig. 40 C); male without special hair tufts on gastral sterna III and IV (Hawaiian Islands) .... *Deinomimesa* Perkins, p. 167
Petiole without a mediodorsal longitudinal carina, at least in males, female with clypeus directed downward, partially covering labrum 9
9. Petiole laterally and ventrally with fine, inconspicuous hair; female with bladelike or sometimes toothlike genal process (fig. 40 D); male without sternal fimbriae (Hawaiian Islands) ...... *Nesomimesa* Perkins, p. 167
Petiole with conspicuous outstanding hair laterally and ventrally; female without a genal process; male nearly always with apical fimbriae on sternum IV and often on III (widespread) ..... *Psen* Latreille, p. 165

Subtribe Psenina

Genus Ammopsen Krombein

Generic diagnosis: Clypeal apex slightly concave in male, a little convex in female, edge thin; mandible weakly bidentate at apex in male, simple in female; frontal carina undeveloped; occipital carina present dorsally but absent ventrally; no genal process but a row of long ammochaetae on female gena; pronotal carina not developed as a sharp ridge, collar covered with appressed hair; omaulus absent, this area and mesopleuron generally covered with dense silvery hair; partially obscured scrobal sulcus present, hypoepimeral area not strongly bulging; female foretarsus with a rake; plantulae absent; forewing with recurrent veins usually going to submarginal cells I and III (fig. 41 C), submarginal III nearly square; hindwing media diverging before cu-a; propodeum granulate rather than reticulate; petiole short, less than twice as long as broad in dorsal view; male sternum VI with a subapical row of long flattened setae, VIII straplike and apically rounded; female pygidial plate broadly triangular, lateral carinae indistinct, surface with short dense silvery setae.

Geographic range: This monotypic genus occurs sparingly in sandy areas of southern California, Arizona, New Mexico, Nevada, and Utah.

Systematics: Ammopsen is one of the most distinctive genera of the Psenini. The single species is one of the

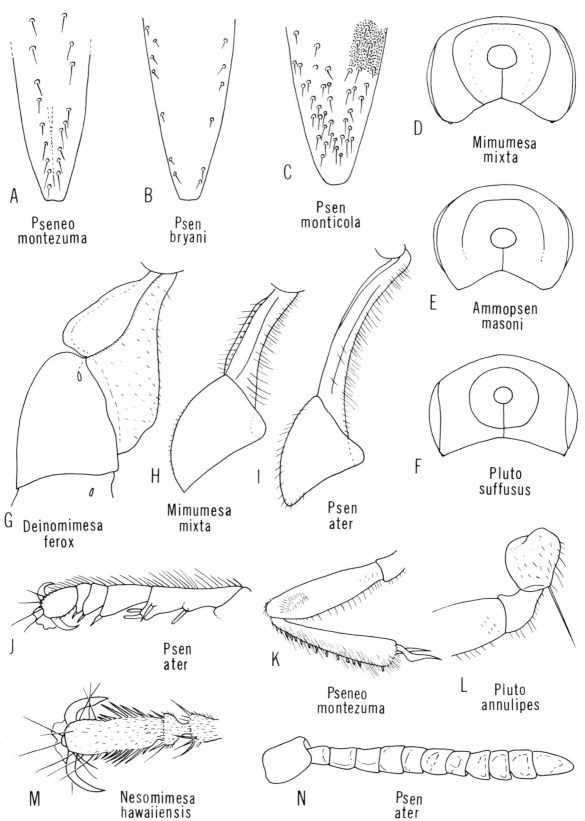

FIG. 39. Structural details in the tribe Psenini; A-C, pygidium, female; D-F, head, ventral; G-I, base of abdomen, lateral view, female; J, midtarsus, semilateral; K, femur and tibia; L, hindcoxa and trochanter, female; M, hindtarsal apex, female; N, male antenna, ventral.

smallest of the tribe, averaging 4 mm in length. Also unusual is the dense silvery pile which covers much of the body. Peculiar structural features are the rounded pronotal collar, absence of a frontal carina, absence of an omaulus, presence of genal ammochaetae, unusually short petiole for members of this tribe, unusual male sternal features, and large but weakly defined female pygidial plate. All of these indicate that *Ammopsen* is generalized. Although the genus is unique, there may be a distant relationship to *Mimesa*. A rather complete description is given by Krombein (1959a).

Biology: Unknown. The presence of a foretarsal rake, genal ammochaetae ("sand basket") and pygidial plate suggest a ground-nesting habit. Adults are most commonly collected on *Euphorbia* mats.

Checklist of *Ammopsen*

masoni Krombein, 1959; U.S.: s. California, Arizona, New Mexico, Utah, Nevada

Genus Mimesa Shuckard

Generic diagnosis: Clypeal apex simple to quadridentate, edge rather thin, female sometimes with a preapical transverse or nasiform elevation; mandible simple or bidentate apically; frontal carina not distinct for its entire length; occipital carina joining hypostomal carina well before midventral line of head; no genal process; pronotal carina complete; omaulus weak or moderate, curving semicircularly forward toward prothorax (fig. 43 H), no acetabular carina; scrobal sulcus weakly impressed; hypoepimeral area moderately convex, striate or well punctured; female foretarsus with a rake; plantulae present; recurrent veins of forewing usually ending at submarginal cell II; hindwing media diverging before cu-a; propodeum posteriorly rough, multicarinate or reticulate, dorsomedial area of petiole smooth or rough, sometimes with a carinule or a low longitudinal swelling; a row of fine laterodorsal hair; male sterna without fimbriae, VIII an upturned pseudosting; female pygidial plate often broadly triangular, densely pubescent.

Geographic range: Of the 53 listed species about two-thirds are Palearctic, one-third Nearctic, one Oriental. The Palearctic forms are rather generally distributed, including one species in Egypt. The Nearctic fauna is also widespread. One species, *mexicana* Cameron, was described from the mountains of Durango, Mexico, at an elevation of nearly 6,000 feet.

Systematics: Mimesa are medium-small wasps which almost always have one or more of the postpetiole segments red. However, there are a few species of *Mimumesa* and *Pluto* with similar general appearance and markings. Among the genera in which the hindwing media diverges before cu-a, *Mimesa* and *Odontopsen* are the only ones in which the omaulus curves semicircularly forward to end before reaching the midline or ends at the midline, just behind the prothorax and hidden under the front coxae (fig. 43 H). The omaulus is sometimes indistinct toward its ventral terminus, but it can usually be discerned as part of a curving ridge. Differentiation from the monotypic *Odontopsen* is based on the simple occipital carina of *Mimesa* and the complete transverse carina on the pronotal collar.

The genus is rather homogeneous, and the species are often difficult to distinguish. Useful characters are the degree and disposition of pilosity, form of the clypeus including the preapical tubercles which are sometimes present, sculpture of the propodeum and petiole, and punctation, especially of the mesopleuron. The relative length of the petiole is useful, also. For instance, in *beckeri* it is about as broad as long in dorsal view, about twice as long as broad in *crassipes,* and nearly five times as long as broad in *grandii*. Malloch (1933) keyed the North American species, but the genus needs revising. European species were keyed most recently by Oehlke (1965).

Biology: Incomplete biologies are known for several species of *Mimesa.* Spooner (1948) summarized the known records for *equestris* and *lutaria.* Williams (1914b), Grandi (1954), and Krombein (1961a) gave brief accounts of the nest habits and prey of *ezra, grandii,* and *basirufa* respectively. Tsuneki (1970a) described the nests of *bidentata.*

Nesting sites are excavated primarily in level to slightly sloping firm soil or sandy clay, but Spooner mentioned one specimen of *lutaria* nesting in the vertical slope of a dune "blow-out." The main burrow appears to descend vertically 15-24 cm and then may branch in several directions horizontally for up to 35 cm in some species. Each nest apparently contains several cells, up to three being recorded by Spooner (1948). The burrows are left open during provisioning. Often a slight craterlike depression may be evident at the opening produced by the circle of excavated sand. Williams suggested that *ezra* may form a cone of agglutinized sand grains about the nest entrance. So far as is known all prey consists of members of the family Cicadellidae, with up to 20 being placed in each cell.

Several species of the family Chrysididae are known to parasitize *Mimesa.* Móczár (1967) listed the following: *Elampus panzeri* (F.) on *M. equestris* (as *Psen bicolor* Shuckard), *E. constrictus* (Foerster) on *M. bicolor,* and *M. lutaria* (as *Psen shuckardi* Wesmael). In addition, Spooner (1948) listed *Elampus spina* (Lepeletier) as a parasite of *M. lutaria* in Sweden.

Checklist of *Mimesa*

aegyptiaca Radoszkowski, 1876; Egypt
agalena Gittins, 1966; U.S.: California
albopilosa (Tsuneki), 1972 (*Psen*); Mongolia
angulicollis (Tsuneki), 1972 (*Psen*); Mongolia
arizonensis (Malloch), 1933 (*Psen*); U.S.: Arizona
barri Gittins, 1966; U.S.: Idaho, Oregon, California
basirufa Packard, 1867; U.S.: transcontinental
   nebrascensis H. Smith, 1908
beckeri Tournier, 1889; s. USSR to Mongolia
bicolor (Jurine), 1807 (*Psen*); n. and centr. Europe
   basalis Stephens, 1829 (*Trypoxylon*), new name for *equestris* of Curtis, 1824

equestris of authors, nec Fabricius, 1804
bidentata (Gussakovskij), 1937 (*Psen*); Siberia, Mongolia
brevis Maidl, 1914; USSR
breviventris F. Morawitz, 1891; s. USSR
bruxellensis Bondroit, 1934; Europe: Belgium, France
    rossica Gussakovskij, 1937 (*Psen*)
caucasica Maidl, 1914; Eurasia
    tenuis Oehlke, 1965 (*Psen*)
chinensis (Gussakovskij), 1937 (*Psen*); China
concors (Gussakovskij), 1937 (*Psen*); Mongolia
coquilletti (Rohwer), 1910 (*Psen*); U.S.: California, Nevada
crassipes A. Costa, 1871; s. and centr. Europe
    ochroptera A. Costa, 1871
    carbonaria A. Costa, 1871, nec Smith 1865
    costae Ed. André, 1888
    helvetica Tournier, 1889
    lixivia Tournier, 1889
    pannonica Maidl, 1914
cressonii Packard, 1867; U.S.
    denticulata Packard, 1867
    conica H. Smith, 1908
    ssp. *atriventris* (Malloch), 1933 (*Psen*); Canada: Ontario
dawsoni Mickel, 1916; U.S.: Nebraska
dzingis (Tsuneki), 1972 (*Psen*); Mongolia
edentata (Malloch), 1933 (*Psen*); U.S.: California
equestris (Fabricius), 1804 (*Trypoxylon*); Eurasia
    rufa Panzer, 1805 (*Psen*)
    bicolor of authors, nec Jurine, 1807
ezra (Pate), 1944 (*Psen*); U.S.: Kansas, Colorado, Pennsylvania, Maryland
    argentifrons Cresson, 1865 (June, p. 487) (*Mimesa*), nec Cresson, 1865 (Jan. p. 152) (*Psen*)
fallax F. Morawitz, 1893; sw. USSR: Kazakh S.S.R.
filippovi (Gussakovskij), 1937 (*Psen*); China
grandii Maidl, 1933; Mediterranean; sw. USSR; Iran
    carinata Gussakovskij, 1937 (*Psen*)
granulosa (W. Fox), 1898 (*Psen*); U.S.: South Dakota, Montana, Idaho, New Mexico
gregaria (W. Fox), 1898 (*Psen*); U.S.: Wyoming, Colorado, New Mexico
    ssp. *simplex* (Malloch), 1933 (*Psen*); U.S.: Idaho, Utah, Colorado
gussakovskiji (Beaumont), 1941 (*Psen*); se. Europe
hissarica (Gussakovskij), 1935 (*Psen*); sw. USSR: Kazakh S.S.R.
impressifrons (Malloch), 1933 (*Psen*); U.S.: Washington
inflata (van Lith), 1968 (*Psen*); centr. New Guinea
jacobsoni (Gussakovskij), 1937 (*Psen*); sw. USSR: Tadzhik S.S.R.
kaszabi (Tsuneki), 1972 (*Psen*); Mongolia
lutaria (Fabricius), 1787 (*Sphex*); Eurasia
    shuckardi Wesmael, 1852
    japonica Pérez, 1905
    dispar Gussakovskij, 1937
maculipes (W. Fox), 1893 (*Psen*); n. America: Ontario to Florida
    nigrescens Rohwer, 1910 (*Psen*)
    perplexa Rohwer, 1910 (*Psen*)
mexicana Cameron, 1891; Mexico
mongolica F. Morawitz, 1889; Mongolia, centr. China
nigrita Eversmann, 1849; Europe to Mongolia
    sibirica Beaumont, 1937 (*Psen*)
pauper Packard, 1867; U.S.: e. of Rocky Mts.; Canada
    cingulata Packard, 1867
pekingensis (Tsuneki), 1971 (*Psen*); China
polita (Malloch), 1933 (*Psen*); U.S.: Nebraska, New Mexico
proxima Cresson, 1865; U.S.: Nebraska, Colorado, New Mexico, Washington
punctifrons (Malloch), 1933 (*Psen*); U.S.: California
punctipleuris (Gussakovskij), 1937 (*Psen*); e. Mongolia
pygidialis (Malloch), 1933 (*Psen*); Canada: Alberta; U.S.: Michigan, Colorado
sabina Gittins, 1966; U.S.: California
sciaphila (Arcidiacono), 1965 (*Psen*), Sicily, new status by Bohart
shestakovi (Gussakovskij), 1937 (*Psen*); sw. USSR: Kazakh S.S.R.
sublaevis (Beaumont), 1954 (*Psen*); n. Africa
unicincta Cresson, 1865; Canada: British Columbia; U.S.: California, Colorado
vindobonensis Maidl, 1914; Austria, Mongolia

Genus Odontopsen Tsuneki

Generic diagnosis: Clypeal apex emarginate in male, bluntly quadridentate in female, edge thin; mandible bidentate at apex; frontal carina incomplete below; occipital carina reaching hypostomal carina far from midventral line of head, lower part of carina projecting strongly in male, developed into a large and forward-directed process in female; pronotum with several oblique lateral carinules but without a complete ridgelike carina; omaulus curving toward prothorax (as in fig. 43 H), no acetabular carina; scrobal sulcus distinct but not deeply impressed, hypoepimeral area not strongly bulging, weakly reticulate; plantulae apparently absent; forewing recurrent veins ending at submarginal cells I and II or both at II; hindwing media diverging before cu-a; propodeum extensively smooth, some areas striate; petiole dorsally with a median longitudinal carina and a series of outward-pointing hairs along laterodorsal margin; male sterna without frimbriae, VIII an upturned pseudosting; female pygidial plate subtriangular, closely punctured and with abundant short stiff hair.

Geographic range: This monotypic genus is known only from Japan in the mountains of central Honshu Island.

Systematics: Our knowledge of this genus is based on the excellent description and figures of Tsuneki (1964b) and examination of a male paratype of *hanedai* borrowed from the U.S. National Museum. Tsuneki stated that the genal process of the female creates a superficial similarity to *Nesomimesa*, but in *Odontopsen* the process is developed from the occipital carina, and there are many other points of dissimilarity. The hindwing venation, forward curving omaulus, and well developed occipital carina ally this genus with *Mimesa*, as suggested by Tsuneki. However, the prolongation of the occipital carina into a ventral projection and the absence of a complete transverse pronotal carina have led us to consider *Odontopsen* as a genus.

Biology: Unknown

Checklist of *Odontopsen*

hanedai (Tsuneki), 1964 (*Psen*); Japan: Honshu Island

Genus Mimumesa Malloch

Generic diagnosis: Clypeal apex emarginate, bidentate to quadridentate, edge rather thin; mandible bidentate apically; frontal carina complete, occipital carina directed toward hypostomal carina and usually reaching it before midventral line of head (fig. 39 D); no genal process; pronotal carina complete, omaulus continuous with acetabular carina, usually angled backward at the juncture (fig. 43 E); scrobal sulcus deeply impressed, hypoepimeral area bulging and smooth or finely punctate; female foretarsus with at least a short rake; plantulae present; submarginal cell II receiving both recurrent veins, hindwing media diverging before cu-a; propodeum posteriorly rough, usually coarsely reticulate; dorsomedial area of petiole usually bicarinate, the two carinae fusing posterad, a few Oriental species posteriorly sulcate, laterodorsal hair row prominent; male sterna without fimbriae, VIII an upturned pseudosting; female pygidial plate broadly to narrowly triangular, densely bristled or nearly smooth with a row of bristles on each side.

Geographic range: Of the 29 listed species about half (15) are Nearctic, three are Neotropical, 10 are Palearctic; and possibly one is Oriental. Several of the Nearctic forms occur in Canada and Alaska. The Neotropical species are West Indian or from the Caribbean coast of Central America. One of the West Indian forms is represented by a subspecies on the island of Trinidad, the only known penetration of South America. The Old World fauna extends into the Oriental Region on the large islands of Taiwan and Celebes. Apparently, the African continent is outside the distribution pattern.

Systematics: The decision to treat *Mimumesa* as a genus is discussed in the systematic section of the tribe Psenini. In this respect we are in agreement with Gittins (1969), although his reference to differences in biology between *Mimesa* and *Mimumesa* is only partially correct (see Tsuneki, 1970a). The continuous omaular-acetabular carina (fig. 43 E) relates *Mimumesa* to *Pseneo* and differentiates it from *Psen* as well as *Mimesa*. The distinction from *Pseneo* is discussed under that genus. Most of the species have the abdomen entirely black, but in the red-marked exceptions, such as *clypeatus* and *petiolatus,* there is a strong superficial resemblance to *Mimesa,* the species of which almost always have a red band before the middle of the gaster. In any case, attention should be paid to the connection of the omaulus with the acetabular carina, the bulging hypoepimeron, and the complete frontal carina, all of which readily distinguish *Mimumesa* from *Mimesa.*

The Palearctic species have been reviewed by Beaumont (1937, 1964b), the Oriental and east Asian ones by van Lith (1965), and the Nearctic ones by Malloch (1933).

Biology: Tsuneki (1959a) provided important and complete nesting observations for four species of *Mimumesa* in Japan. Yasumatsu and Watanabe (1964) reviewed host records for the same area. Spooner (1948) provided a review of nesting and prey information for several European forms. Gurney (1951) gave nesting information for an American species.

Adult *Mimumesa* appear to excavate nests either in clay banks or in decaying wood. According to Tsuneki, those species that excavate burrows in clay banks dig 2-5 cm vertical to the surface of the bank and then dig about 10 cm straight down. As many as 28 cells have been counted from a single nest. In *littoralis* each cell is temporarily closed during provisioning, and upon completion of one or two cells the branch tunnel leading to them is tightly packed with earth. Prey for all known species of *Mimumesa* are various species of delphacids and cicadellids. Approximately five to 10 immobilized leafhoppers are placed in each cell. In *littoralis* the female wasp oviposits on the last prey member, attaching the egg at its cephalic pole to the outer base of the hindcoxa. Tsuneki suggested that the orientation of the egg to the hindcoxa "may be the rule of oviposition among the members of *Mimumesa.*"

Mimumesa dahlbomi is the only Old World species known to nest in decaying wood. However, Spooner reviewed several references which suggest that *unicolor* may also nest in wood or plant stems. The following summary of nesting habits of *dahlbomi* (including the subspecies *pacificus*) is taken from Tsuneki (1954c, 1959a). Specimens of *dahlbomi* were found nesting in the moist wood of a dead tree 70 cm in diameter. Abandoned beetle burrows were used as entrance holes. The wasp excavated vertically for 1-2 cm, at which point the main burrow turned at a right angle parallel to the surface of the wood. In each nest one to three larval cells were found, each provisioned with 10 to 15 delphacids and cicadellids. The egg was placed on the last prey member as in the clay-nesting species of *Mimumesa* except that the cephalic pole was directed anteriorly.

A nesting site of *M. nigra* about 30 cm above ground in a decaying fence post was described by Gurney (1951) in Virginia. Cells contained adult female leafhoppers of the genus *Agallia*. Tightly woven silk cocoons were about 7 mm long, pale brown, and contained a dried dark meconium at one end. Adult wasps chewed an opening through the other end when emerging.

Three species of Chrysididae are known to parasitize *Mimumesa* (Móczár, 1967); *Chrysis succincta* L., *Hedychridium ardens* Cocquebert, and *Omalus auratus* (L.). All three species prey upon *M. unicolor.*

Checklist of *Mimumesa*

atratina (F. Morawitz), 1891 (*Mimesa*); Europe
  ?*carbonaria* Tournier, 1889 (*Mimesa*), nec F. Smith, 1865
  belgica Bondroit, 1931 (*Mimesa*)
    ssp. *longula* (Gussakovskij), 1937 (*Psen*); Japan
    sameshimai Yasumatsu, 1937 (*Psen*)
beaumonti (van Lith), 1949 (*Psen*); Belgium
bermudensis (Malloch), 1933 (*Psen*); Bermuda
canadensis (Malloch), 1933 (*Psen*); Canada: Ontario; U.S.: Alaska, New York, Michigan, Colorado, North Dakota
clypeata (W. Fox), 1898 (*Psen*); w. and nw. U.S.; N.W. Territory, Alaska
coloradoensis (Cameron), 1908 (*Psen*); U.S.: Colorado
cylindrica (W. Fox), 1898 (*Psen*); U.S.: Arizona
dahlbomi (Wesmael), 1852 (*Mimesa*); Europe to Mongolia
  concolor Radoszkowski, 1891 (*Psen*)
  ssp. *pacifica* (Tsuneki), 1959 (*Psen*); Japan
fuscipes (Packard), 1867 (*Psen*); U.S.: Massachusetts
interstitialis (Cameron), 1908 (*Psen*); U.S.: New Mexico
johnsoni (Viereck), 1901 (*Mimesa*); U.S.: New York to Virginia
kashmirensis (Nurse), 1903 (*Psen*); n. India
leucopus (Say), 1837 (*Psen*); Canada: Alberta; e. U.S.
  elongata Packard, 1867 (*Psen*)
littoralis (Bondroit), 1934 (*Mimesa*); Eurasia to Japan
  borealis Dahlbom, 1842 (*Mimesa*), ♂ only
  ?*fulvitarsis* Gussakovskij, 1937 (*Psen*)
  ?*celtica* Spooner, 1948 (*Mimesa*)
longicornis (W. Fox), 1898 (*Psen*); U.S., Cuba, Centr. America
  striata Viereck, 1901 (*Mimesa*)
  floridana Rohwer, 1910 (*Psen*)
mandibularis (H. Smith), 1908 (*Mimesa*); U.S.: Nebraska
mellipes (Say), 1837 (*Psen*); centr. and e. U.S.
  chalcifrons Packard, 1867 (*Psen*)
mixta (W. Fox), 1898 (*Psen*); N. America: Transition Zone
  alticola Viereck, 1903 (*Psen*)
  similis Rohwer, 1910 (*Psen*)
modesta (Rohwer), 1915 (*Psen*); Puerto Rico
  ssp. *recticulata* (Malloch), 1933 (*Psen*); Trinidad
nigra (Packard), 1867 (*Psen*); N. America: Transition Zone
?*petiolata* (F. Smith), 1864 (*Psen*); Celebes (? = *Psen*)
propinqua (Kincaid), 1900 (*Psen*); U.S.: Alaska
psychra (Pate), 1944 (*Psen*); Canada: Hudson Bay
  ?*borealis* F. Smith, 1856 (*Mimesa*), nec Dahlbom, 1842
regularis (W. Fox), 1898 (*Psen*); U.S.: New Jersey, Pennsylvania
sibiricana R. Bohart, n. name for *sibirica* Gussakovskij; Siberia, Netherlands
  sibirica Gussakovskij, 1937 (Oct.) (*Psen*); nec Beaumont, 1937 (March)
spooneri (Richards), 1948 (*Mimesa*); British Isles
unicolor (Vander Linden), 1829 (*Psen*); Europe to Afghanistan
  borealis Dahlbom, 1842 (*Mimesa*), ♀ only
  fuscipennis Radoszkowski, 1891 (*Psen*)
  palliditarsis E. Saunders, 1904 (*Mimesa*)
vanlithi (Tsuneki), 1959 (*Psen*); Japan
  ssp. *meridionalis* (Tsuneki), 1972 (*Psen*); Taiwan
wuestneii (Faester), 1951 (*Psen*); Austria

Genus Pseneo Malloch

Generic diagnosis: Clypeus bidentate or tridentate, nearly always with apex thickened, grooved transversely and with two or three longitudinal carinae; mandible simple or bidentate apically; frontal carina complete (rarely indistinct near midocellus); occipital carina strong and complete to hypostomal carina well before midventral line of head; no genal process; pronotal carina complete, humeral angles often sharp; omaulus continuous with acetabular carina; scrobal sulcus deeply impressed, hypoepimeral area bulging and polished, punctate or carinulate; female foretarsus with a rather weakly defined rake; hindfemur with a discrete inner distal hair patch or sometimes with inner pubescence in a longitudinal strip; plantulae apparently absent; forewing recurrent veins usually both ending at submarginal cell II; hindwing media diverging before cu-a; propodeum usually reticulate posteriorly, rarely granulate; metasternum usually truncate or weakly emarginate posteriorly and not much narrowed anteriorly; petiole smooth above, not carinate or sulcate, laterodorsal hair practically absent except rarely posteriorly; male sterna without fimbriae, VIII an upturned pseudosting; female pygidial plate somewhat narrowly triangular and with a row of bristles on each side, or more generally bristly, apex of plate rarely notched.

Geographic range: This genus is the only one of the tribe that is predominantly Neotropical with eight species ranging from Argentina to Mexico. The Nearctic fauna includes four species. Four are known from the Philippines and Malaya.

Systematics: The hindwing venation and continuous omaular-acetabular carina relate this genus to *Mimumesa*. Typical *Pseneo* are readily separated from all other psenins by the transverse groove or bevel of the clypeal apex, this groove crossed by longitudinal carinae, which form two or three teeth in front view of the face. In addition, there is a peculiar discrete patch of inner distal hair on the hindfemur. Unfortunately, neither of these characters is absolute. *Pseneo leytensis* has no discrete patch of femoral hair, although the inner surface of the hindfemur is unusually hairy over its entire length. *Pseneo irwini* has the hair patch but it extends basad beyond the middle of the femur. Furthermore, *irwini* has the clypeus thinly edged and simply bilobed. To differentiate such deviants from *Mimumesa* the petiole can be used. In *Pseneo* it is smoothly rounded dorsally rather than carinate or sulcate, polished, and without conspicuous hair. In *Mimumesa* it is carinate or longitudinally sulcate dorsally, often roughened, and with conspicuous rows of hair.

The Nearctic species were treated by Malloch (1933) and considerably revised by Krombein (1950c).

Biology: Only scattered notes have been published on species of this genus. The weakly defined tarsal rake of the female suggests a groundnesting habit. This was partially confirmed by Krombein (1950c), who recorded the burrow of *P. punctatus carolina* in the soil of a flowerpot. Each cell contained four or five leafhoppers of *Homalodisca triquetra* (F.) and *Graphocephala coccinea* (Foerster). Prey of other species were *Graphocephala* species for *P. simplicicornis* and "leafhoppers" for *P. Kohlii,* according to Krombein (1963a).

Checklist of *Pseneo*

argentina (Brèthes), 1910 (*Mimesa*); Argentina, new combination by Bohart
auratus (van Lith),1959 (*Psen*); Indonesia: Java, Bali, Sumatra; new combination by Bohart
    ssp. *mindoroensis* (van Lith), 1959 (*Psen*); Philippines: Mindoro
    ssp. *multipunctatus* (van Lith), 1959 (*Psen*); Philippines: Negros, Luzon
    ssp. *miltoni* (van Lith), 1965 (*Psen*); Malaya
aurifrons (Taschenberg), 1875 (*Mimesa*); Brazil, new combination by Bohart
claviventris (Cameron), 1891 (*Mimesa*); Mexico: Guerrero, new combination by Bohart
irwini R. Bohart and Grissell, 1969; El Salvador
joergenseni (Brèthes), 1913 (*Psen*); Argentina, new combination by Bohart
kohlii (W. Fox), 1898 (*Psen*); U.S.: New York to Kansas
    angulatus Malloch, 1933 (*Psen*)
    fulvipes Malloch, 1933 (*Psen*)
leytensis R. Bohart and Grissell, 1969; Philippines: Leyte
montezuma (Cameron), 1891 (*Mimesa*); e. Mexico to Costa Rica
    longiventris Cameron, 1891 (*Mimesa*), new synonymy by Bohart
montivagus (Dalla Torre), 1897 (*Mimesa*); Mexico: Guerrero
    monticola Cameron, 1891 (*Mimesa*), nec Packard, 1867
pulcher (Cameron), 1891 (*Mimesa*); Mexico: Guerrero, new combination by Bohart
punctatus (W. Fox), 1898 (*Psen*); U.S.: South Dakota, Nebraska, Colorado
    ssp. *ferrugineus* (Viereck), 1901 (*Mimesa*); U.S.: s. Florida
    ssp. *carolina* (Rohwer), 1910 (*Psen*); e. and se. U.S.
simplicicornis (W. Fox), 1898 (*Psen*); e. U.S.
spicatus (Malloch), 1933 (*Psen*); U.S.: New Mexico, Arizona
townesi (van Lith), 1959 (*Psen*); Philippines: Luzon
tridentatus (van Lith),1959 (*Psen*); Malay Penin., new combination by Bohart
    ssp. *chrysomaila* (van Lith), 1965 (*Psen*); Burma

Genus Psen Latreille

Generic diagnosis: Clypeal apex emarginate to bidentate, edge rather thin or at least not transversely grooved; mandible bidentate apically; frontal carina complete; occipital carina joining hypostomal carina before midventral line of head; no genal process; pronotal carina complete; omaulus ending as it becomes ventral and turning a little posteriorly, acetabular carina sometimes partially developed but not continuous with omaulus; scrobal sulcus deeply impressed, hypoepimeral area bulging, smooth or weakly punctate; female foretarsus with a weak rake; plantulae present (always?); second recurrent vein of forewing commonly ending at submarginal cell III; hindwing media diverging before cu-a; propodeum usually coarsely reticulate posteriorly, sometimes multicarinate; dorsomedian area of petiole usually smooth, rarely with coarse punctures, without carinae but rarely with a posterior longitudinal groove, no conspicuous laterodorsal hair but abundant strong hair lateroventrally; male sternum IV nearly always with apical frimbriae, III usually fimbriate also, VIII an upturned pseudosting; female pygidial plate subtriangular, narrow or broad, sparsely or densely bristled.

Geographic range: Over half of the 78 species of *Psen* are Oriental. The genus is not found in South America or Australia. Representatives occur on many large islands such as the Ryukyus, Taiwan, Philippines, Moluccas, Malay Archipelago, New Ireland, New Hebrides, Sri Lanka, Madagascar, and Cuba.

Systematics: The hindwing venation and backward-curving omaulus, which is not connected to the acetabular carina, ally *Psen* with *Deinomimesa* and *Nesomimesa.* The smooth dorsal surface of the petiole (at least no dorsomedian carina) and the conspicuous lateroventral hair of the petiole distinguish *Psen.* The males of all but one known species, *exaratus,* have apical fimbriae on sternum IV or III and IV. Chiefly for this reason van Lith (1968) established the subgenus *Punctipsen* to include *exaratus* and its subspecies.

Psen is the second largest genus in the tribe and there is considerable variation, as might be expected. This has been well illustrated by van Lith (1959). For instance, the posterior surface of the propodeum may be striate or more often coarsely reticulate; the female pygidial plate is ordinarily microreticulate and with two rows of bristles but it may occasionally be smooth or have numerous bristles; an acetabular carina may be well developed, short or absent; the petiole is often very long and slender in Oriental forms; males of several Palearctic species, such as *ater, exaratus* and *aurifrons* have the basitarsus of the midlegs spinose or otherwise modified; and most species are all black but many have the petiole or additional parts of the gaster red. An unusual number of subspecies have been described in *Psen*. Many of these are simply varieties but some have small structural distinctions and should probably be classed as species.

The Oriental and eastern Asiatic species have been treated by van Lith (1959, 1965, 1968); the Palearctic ones by Beaumont (1937, 1964b) and Gussakovskij (1937); and the Nearctic ones by Malloch (1933).

Biology: Information on habits has been published on about 14 species. These make burrows in stumps or dead trees (Barth, 1907b; Iwata, 1938b; Tsuneki (1959a), in hard sand (Sickmann, 1893), and in earthern (or clay) banks (Tsuneki, 1959a, van Lith, 1959). Four of the species, *affinis, aurifrons, betremi,* and *exaratus,* provision with Cicadellidae (Tsuneki, 1959a; van Lith, 1949). Six species, *ater, coriaceus, curvipilosus, erythopoda, richardsi,* and *vechti,* provision with Cercopidae (Iwata, 1938b; van Lith, 1959; Malloch, 1933; Tsuneki, 1959a). One species, *emarginatus,* provisions with *Membracidae* (Barth, 1907b) and one, *emarginatus,* with Fulgoridae (van Lith, 1959).

Evidence on burrow construction is limited. In rotten wood the nests may be multicellular with the burrows branched. Tsuneki (1959a) has recorded observations on such nests of *richardsi* from Honshu, Japan. Several to 10 wasps inhabited each decaying stump. "The nest belonged to a multicellular type with branched tunnels and end cells." Each cell contained four to six cercopids, *Eoscarta assimilis* (Uhler). Tsuneki also reported a nest of *aurifrons* in a clay bank. The burrow was vertical to the cliff face and ended in a single cell containing two rice plant leafhoppers, *Nephotettix apicalis* Motschulsky.

Checklist of *Psen*

affinis Gussakovskij, 1937; Russia, Japan, Korea
   ssp. *grahami* van Lith, 1965; centr. China
alishanus Tsuneki, 1967; Taiwan
amboinensis van Lith, 1965; Indonesia: Moluccas, Ambon
angulifrons van Lith, 1965; Philippines: Mindanao
assamensis van Lith, 1965; e. India
ater (Olivier), 1792 (*Crabro*); Europe, sw. USSR, Mongolia, Korea, Japan
    ater Fabricius, 1794 (*Sphex*)
    unicolor Fabricius of Panzer, 1804 (*Pelopoeus*)
    compressicornis Fabricius, 1804 (*Pelopoeus*)
    serraticornis Jurine, 1807
    atratus Jurine, 1807
aureohirtus Rohwer, 1921; Philippines: Luzon
   ssp. *rufopetiolatus* van Lith, 1959; Philippines: Negros
aurifrons Tsuneki, 1959; Japan
    orientalis Gussakovskij, 1934, nec Cameron, 1890
    caocinnus Tsuneki, 1973, unnecessary new name for *aurifrons* Tsuneki
bakeri Rohwer, 1923; Philippines: Luzon
barthi Viereck, 1907; U.S.: Connecticut, Pennsylvania, Maryland, Wisconsin
    myersianus Rohwer, 1909 (*Mimesa*)
betremi van Lith, 1959; Java
bishopi van Lith, 1968; Solomons
bnun Tsuneki, 1971; Taiwan
brinchangensis van Lith, 1965; Malaya
bryani Perkins and Cheesman, 1928; Samoa
carbonarius (F. Smith), 1865 (*Mimesa*); Indonesia: Moluccas (Morotai)
cheesmanae Krombein, 1949; New Hebrides
clavatus Cameron, 1890; India
congolus Leclercq, 1961; Zaire
coriaceus van Lith, 1959; Philippines: Mindoro, Negros, Mindanao, Bohol

curvipilosus van Lith, 1959; Java
dzimm Tsuneki, 1959; Japan
elisabethae van Lith, 1959; Java, Sumatra; Vietnam
   ssp. *auricomus* van Lith, 1965; Malaya, South China Sea islands
   ssp. *madrasiensis* van Lith, 1968; s. India
emarginatus van Lith, 1959; Java, Borneo
erraticus F. Smith, 1860; Celebes
erythopoda Rohwer, 1910; U.S.: New York to N.C.
   erythropoda Malloch, 1933, emend.
eurypygus van Lith, 1965; n. India
exaratus (Eversmann), 1849 (*Mimesa*); Eurasia
    superbus Tournier, 1889 (*Mimesa*)
    picicornis F. Morawitz, 1892 (*Mimesa*)
   ssp. *santoro* Yasumatsu, 1942; Ryukyus: Amami
   ssp. *intermedius* Tsuneki, 1966; Ryukyus: Okinawa
   ssp. *taiwanus* Tsuneki, 1966; Taiwan
   ssp. *indicus* van Lith, 1968; India
fuscinervis (Cameron), 1899 (*Caenopsen*); e. India
   ?*nigrinervis* Cameron, 1902
hakusanus Tsuneki, 1959; Japan
   ssp. *seminitidus* van Lith, 1965; China; Taiwan
    kohli Gussakovskij, 1934, nec W. Fox, 1898
heinrichi van Lith, 1968; Celebes
hirashimai Tsuneki, 1966; Ryukyus: Amami
koreanus Tsuneki, 1959; Korea
krombeini van Lith, 1965; s. India
kulingensis van Lith, 1965; s. China, Japan
lieftincki van Lith, 1959; w. Malaya, Sumatra, Taiwan
   ssp. *minor* van Lith, 1965; e. Malaya
   ssp. *nigripennis* Tsuneki, 1971; Taiwan
lobicornis van Lith, 1973; Nepal
madecassus Arnold, 1945; Madagascar
marjoriae van Lith, 1968; Philippines: Luzon
metalensis Turner, 1912; Sri Lanka
melanosoma Rohwer, 1921; Philippines: Negros, Luzon, Mindanao
monticola (Packard), 1867 (*Mimesa*); e. U.S.
nepalensis van Lith, 1968; n. India
nigriventris van Lith, 1965; Philippines: Luzon
nitidus van Lith, 1959; Indonesia, s. India
   ssp. *takasago* Tsuneki, 1967; Taiwan
   ssp. *himalayensis* van Lith, 1973; Nepal
novahibernicus van Lith, 1965; Melanesia: New Ireland
ohnonis Tsuneki, 1973; Japan
opacus van Lith, 1959; Philippines: Luzon
   ssp. *gressitti* Tsuneki, 1966; Ryukyus: Amami
orientalis Cameron, 1890; n. India
   reticulatus Cameron, 1902
patellatus Arnold, 1924; S. Africa
paulus van Lith, 1968; New Guinea
   ssp. *baduriensis* van Lith, 1968; West New Guinea: Japen
   ssp. *subtilis* van Lith, 1968; Solomons
pilosus van Lith, 1965; Malaya
politiventris Rohwer, 1921; Philippines: Luzon
   ssp. *bellus* van Lith, 1965; Philippines: Mindanao
   ssp. *pahangensis* van Lith, 1965; Malaya
refractus Nurse, 1902; India

meridianus van Lith, 1965
regalis van Lith, 1968; Solomons
richardsi Tsuneki, 1959; Japan
rubicundus van Lith, 1959; w. Java
    ssp. *lawuensis* van Lith, 1959; e. Java
ruficrus van Lith, 1965; New Guinea
rufiventris Cameron, 1890; s. India
rufoannulatus Cameron, 1907; n. India
sauteri van Lith, 1968; Taiwan
sedlaceki van Lith, 1968; New Guinea
shirozui Tsuneki, 1966; Taiwan
shukuzanus Tsuneki, 1972; Taiwan
    longicornis Tsuneki, 1967, nec W. Fox, 1898
silvaticus Arnold, 1924; Rhodesia
simlensis van Lith, 1968; n. India
striolatus (Cameron), 1892(*Mimesa*); Mexico: Guerrero, new combination by Bohart
tanoi Tsuneki, 1967; Taiwan
terrigenus van Lith, 1959; Java
toxopeusi van Lith, 1959; Celebes
triangulatus van Lith, 1959; Java, Sumatra
unifasciculatus Malloch, 1933; U.S.: New Mexico
ussuriensis van Lith, 1959; se. USSR: Ussuri
    orientalis Gussakovskij, 1932 (*Mimesa*), nec Cameron, 1890
    ssp. *tsunekii* van Lith, 1965; Australasia
    mandibularis Tsuneki, 1959; nec H. Smith, 1906
vadosus van Lith, 1968; Solomons
vechti van Lith, 1959; Java
    ssp. *birmanicus* van Lith; 1965; Burma
venetus Pate, 1946; Cuba
yasumatsui Gussakovskij, 1934; Japan
yomasanus van Lith, 1965, Burma

Genus Nesomimesa Perkins

Generic diagnosis: Clypeal apex rounded or obtusely pointed to triangled or trilobed, rather thinly edged; mandible bidentate apically; flagellum slender, swollen apically; frontal carina incomplete below; a lower genal process present in female, sometimes very strong (fig. 40 D), not developed from occipital carina which is complete to or almost to hypostomal carina before midventral line of head; pronotal carina complete; omaulus ending just as it becomes ventral, sometimes turning slightly posteriorly, no acetabular carina; episternal sulcus disappearing below juncture with scrobal sulcus which is moderately impressed; hypoepimeral area bulging strongly and finely reticulate; female foretarsus without a rake, last tarsomere on each leg of female with strong short basolateral bristles (see fig. 39 M); plantulae present; forewing recurrent veins both ending at submarginal cell II; hindwing media diverging before cu-a; propodeum extensively smooth, traversed by a few carinae; petiole without carinae on dorsal surface in male, sometimes with a dull carina in female, dorsolateral carinae weak or absent, an inconspicuous row of laterodorsal hair, lateral and ventral hair inconspicuous; male sterna without fimbriae, VIII an upturned pseudosting; female pygidial plate subtriangular, densely bristled.

Geographic range: The six described species of *Nesomimesa* are all from the Hawaiian Islands, two from Kauai, two from Oahu, one from Hawaii, and one from Maui, Lanai, and Molokai.

Systematics: Except for the original descriptions and figures by Perkins (1899), the only paper of consequence is that of Yoshimoto (1958), which presents a key, numerous illustrations, and a discussion of taxonomic characters, some not previously considered. Yoshimoto separated *Nesomimesa* from *Deinomimesa* on the basis of the weaker propodeal carinae, more slender and smooth petiole, largely concealed labrum, unswollen forefemur, and the slightly pointed or rounded clypeal apex in the male. The species of *Nesomimesa* were differentiated by Yoshimoto by the shape of the genal process of the female when well developed; by the structure of the ventral apex of the petiole; wing color; and microsculpture of the integument. Male genital differences were figured.

Nesomimesa and *Deinomimesa* are easily distinguished in the female, but similarity of the males indicates the close relationship.

Biology: Biological data have been contributed by Williams (1927) and Perkins (1899). *Nesomimesa* are generally found in upland areas from 1,400 to 4,000 feet. Perkins observed adults of *N. kauaiensis* "burrowing in the hard ground in bare spots in the forest." Williams reported *N. antennata* burrowing in rich soil in fern-covered areas at an elevation of 1,400 feet.

According to Williams, *N. antennata* and *N. hawaiiensis* each nest in small colonies. The burrows are several inches deep and probably have several cells. *N. antennata* burrows are marked by "little heaps of earth" often concealed by dead or living fern fronds. Prey for this species consists of various species of plant hoppers (Cixiidae, Flatidae) which are partially paralyzed by the adult wasp. Of special interest is the use of the sugarcane leafhopper (*Perkinsiella saccharicida* Kirk.) in provisioning for larvae of *N. hawaiiensis*. Bridwell (1919) reported that he had found *N. antennata* "in very large numbers flying about in drizzly rain over staghorn ferns (*Gleichenia dichotoma*). A great majority were males, but an occasional female was found, and when one of them settled on a leaf the males would swarm about her ..." Bridwell did not observe copulation. One female was carrying an adult leafhopper, *Oliarus* species (probably *kaonohi* Kirkaldy). Perkins' record of Tipulidae used as prey is questionable, since all other recorded prey are Homoptera.

Checklist of Nesomimesa

antennata (F. Smith), 1856 (*Mimesa*); Hawaiian Is.: Oahu
hawaiiensis Perkins, 1899; Hawaiian Is.: Hawaii
kauaiensis Perkins, 1899; Hawaiian Is.: Kauai
nitida Perkins, 1899; Hawaiian Is.: Lanai, Molokai, Maui
perkinsi Yoshimoto, 1959; Hawaiian Is.: Oahu
sciopteryx Perkins, 1899; Hawaiian Is.: Kauai

Genus Deinomimesa Perkins

Generic diagnosis: Clypeus greatly raised, directed forward, exposing all or most of large labrum (fig. 40 C),

PSENINI

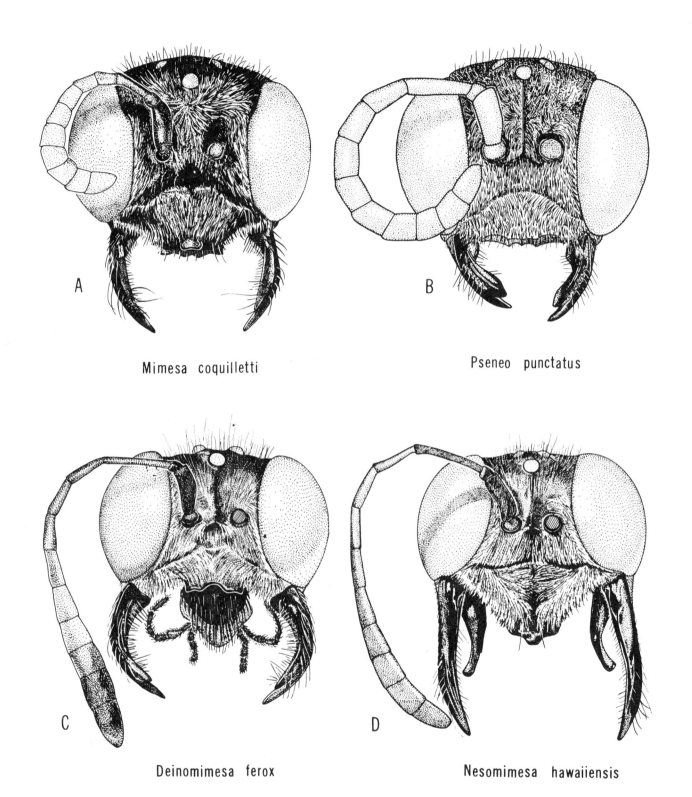

FIG. 40. Facial portraits of females in the tribe Psenini.

apex weakly bilobed, rather thinly edged; mandible bidentate apically; flagellum slender for basal five or six articles, then expanded (fig. 40 C); frontal carina incomplete below; occipital carina complete to hypostomal carina before midventral line of head; no genal process; pronotal carina complete; omaulus turning posteriorly and ending just as it becomes ventral, no acetabular carina; episternal sulcus disappearing below juncture with scrobal sulcus which is moderately impressed; hypoepimeral area bulging strongly, finely punctate; female foretarsus without a rake, last tarsomere on each leg of female with strong, short basolateral bristles (fig. 39 M); plantulae apparently absent; forewing recurrent veins ending at submarginal cell II; hindwing media diverging before cu-a; propodeum extensively granulate but also with well developed carinae; petiole with a dorsomedian longitudinal carina which in females may be extraordinarily raised crestlike so that petiole appears deformed (fig. 39 G); an inconspicuous row of laterodorsal hair; male sterna without fimbriae, VIII an upturned pseudosting; female pygidial plate subtriangular, densely bristled.

Geographic range: Each of the five known species is confined to a single island of the Hawaiian group.

Systematics: *Deinomimesa* are black wasps about 8 to 13 mm in length, the females with the clypeus raised and directed forward so as to largely expose the labrum (fig. 40 C). Another unique feature in females is the greatly raised dorsal keel of the petiole (fig. 39 G), far exceeding anything of this nature elsewhere in the entire family. As in *Nesomimesa* the antennae are markedly swollen on the distal four or five articles (fig. 40 C). The peculiar bristling of the last tarsomere in females (fig. 39 M), presumably an aid in handling prey, occurs also in *Nesomimesa* and strengthens the idea that the two genera are closely related. *Deinomimesa* also has obvious affinities with *Psen*, and the three genera have similar hindwing venation as well as mesopleural structure. The five species of *Deinomimesa* have been reviewed by Yoshimoto (1958), who gave redescriptions, a key to species, and illustrations. The species are differentiated by shape and carination of the petiole in the male and details of the clypeus in the female. Male genitalia provide further points of separation.

Biology: The only biological record is that of Timberlake (1920), who observed *D. haleakalae* nesting in horizontal burrows dug in banks along a trail at an elevation of 5,000 feet on Maui. Provisions were immature and adult leafhoppers of the genus *Nesophrosyne*.

Checklist of *Deinomimesa*

cognata Perkins, 1899; Hawaiian Is.: Kauai
ferox Perkins, 1899; Hawaiian Is.: Kauai
haleakalae Perkins, 1899; Hawaiian Is.: Maui
hawaiiensis Perkins, 1899; Hawaiian Is.: Hawaii
punae Perkins, 1899; Hawaiian Is.: Hawaii

Subtribe Psenulina
Genus Pluto Pate

Generic diagnosis: Clypeal apex relatively simple, sometimes nearly truncate, edge rather thin; mandible simple at apex; frontal carina complete but not connected below with a transverse carina; occipital carina generally complete to midventral line, sometimes nearly touching hypostomal carina; no genal process; pronotal carina complete and sharp; notauli faint to moderate, sometimes complete; omaulus ending as it becomes ventral, but its terminus curved backward toward signum (fig. 36); acetabular carina short when present; scrobal sulcus not deeply impressed; hypoepimeral area bulging and rarely with coarse sculpture; female foretarsus with a short rake; plantulae apparently absent; forewing recurrent veins usually ending at submarginal cells II and III; hindwing media diverging after cu-a; hindcoxa with a distinctive downward-pointing bristle, most conspicuous in female (fig. 39 L); dorsal surface of petiole smooth, lateral hair often prominent, but laterodorsal hair practically absent; male sterna without fimbriae, VIII often concealed, somewhat flattened dorsoventrally, pubescent at apex, not carried as a pseudosting; female pygidial plate subtriangular, relatively broad, densely bristled.

Geographic range: This New World genus contains 22 species. About a third of these are Neotropical. One species is found in Cuba and another in Cuba and Puerto Rico. The Nearctic range essentially covers the southern half of the United States from California to Maryland.

Systematics: The distal divergence of the hindwing media (fig. 41 B) indicates a relationship to *Psenulus*. Other characteristics in common are the complete frontal carina; bulging hypoepimeral area, rather well developed notauli, and generally smooth dorsal surface of the petiole. On the other hand there are many differences that argue against an especially close relationship. These are the backward curving omaulus of *Pluto*, more posteroventral placement of the occipital carina, foretarsal rake in the female, distinctive hindcoxal bristle, less stinglike male sternum VIII, and better developed female pygidial plate. The most diagnostic and unique feature of *Pluto*, particularly in the female, but present also in the male, is the downwardly directed bristle of the hindcoxa.

The most recent comprehensive work on *Pluto* (as *Psenia*) was published by Malloch (1933). The Neotropical species, many of which are here assigned to *Pluto* for the first time, are in need of revision.

Biology: Notes on the habits of *Pluto albifacies* (Malloch) were given by Evans (1968). About 200 females were nesting in a sandy clay bank above the Red River in Wilbarger Co., Texas during July. The burrows were essentially vertical and surrounded with a circular tumulus. Each burrow terminated in several cells. Prey were nymphs and adults of a small green leafhopper, *Opsius stactogalus* Fieber.

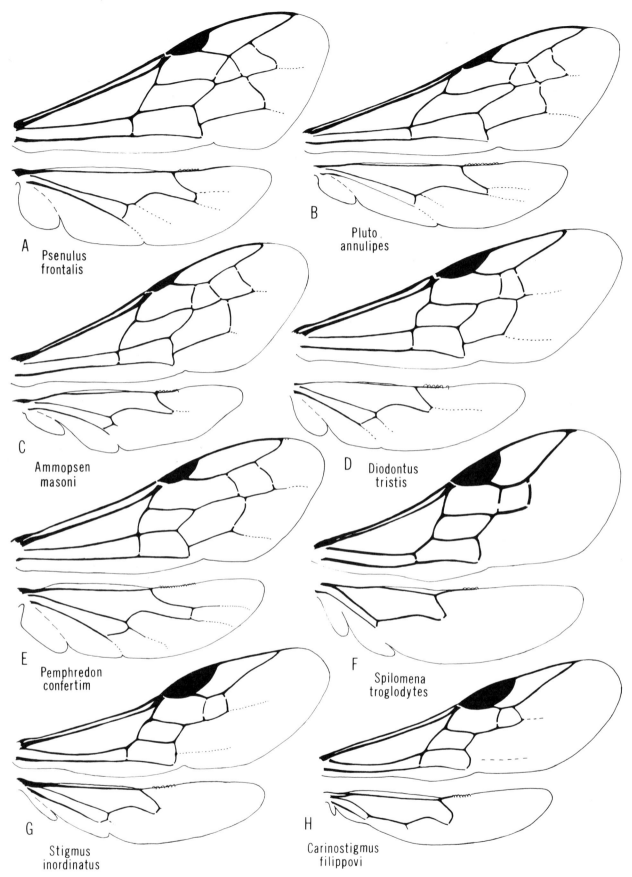

FIG. 41. Wings in the subfamily Pemphredoninae.

Checklist of *Pluto*

aerofacies (Malloch), 1933 (*Psenia*); U.S.: Texas, Mexico
albifacies (Malloch), 1933 (*Psenia*); U.S.: Iowa
angulicornis (Malloch), 1933 (*Psenia*); U.S.: Texas
annulipes (Cameron) 1891 (*Psen*); Mexico, new combination by Bohart
arenivagus Krombein, 1950; U.S.: North Carolina, Georgia
argentifrons (Cresson), 1865 (Jan., p. 152) (*Psen*); Cuba
atricornis (Malloch), 1933 (*Psenia*); Cuba; Puerto Rico
brevipetiolatus (Rohwer), 1910 (*Psenulus*); U.S.: California
clavicornis (Malloch), 1933 (*Psenia*); U.S.: Arizona; Mexico
littoralis (Malloch), 1933 (*Psenia*); U.S.: Maryland
longiventris (Malloch), 1933 (*Psenia*); U.S.: Arizona, California
marginatus (Malloch), 1933 (*Psenia*); U.S.: South Carolina, Louisiana
medius (F. Smith), 1856 (*Psen*); Brazil, new combination by Bohart
minutus (Malloch), 1933 (*Psenia*); U.S.: Texas
pallidistigma (Malloch), 1933 (*Psenia*); U.S. Arizona
rufibasis (Malloch), 1933 (*Psenia*); U.S.: Georgia
sayi (Rohwer), 1910 (*Psenulus*); U.S.: Austral zone e. of California
smithii (W. Fox), 1897 (*Psen*); Brazil, new combination by Bohart
suffusus (W. Fox), 1898 (*Psen*); U.S.: New Mexico, California, Nevada
texanus (Malloch), 1933 (*Psenia*); U.S.: Texas
tibialis (Cresson), 1872 (*Mimesa*); U.S.: Texas to District of Columbia and Virginia
townsendi (Cockerell), 1911 (*Psenulus*); Peru, new combination by Bohart

Genus Psenulus Kohl

Generic diagnosis: Clypeal apex nearly simple to quadridentate, edge rather thin or at least not beveled; mandible bidentate to tridentate at apex and often with a more basal inner tooth; frontal carina complete to interantennal area where it often projects strongly and where it is connected in all males and most females to a prominent transverse carina; occipital carina ending in hypostomal carina before midventral line of head (except one Javan species, see systematics); no genal process (except in above mentioned Javan species); pronotal carina prominent and complete; notauli well impressed and sometimes complete to scutellum; omaulus curving toward prothorax where it may become faint; acetabular carina sometimes present but not joining omaulus; scrobal sulcus weakly or not at all impressed, hypoepimeral area somewhat bulging, not coarsely sculptured; female foretarsus without a well defined rake; plantulae present; forewing recurrent veins usually ending at submarginal cell II; hindwing media diverging after cu-a (fig. 41 A); dorsomedian area of petiole generally smooth, sometimes with a median sulcus or a low longitudinal swelling, laterodorsal hair usually sparse or inconspicuous; male sterna without fimbriae, VIII an upturned pseudosting; female pygidial plate elongate triangular or nearly linear, usually reduced and sometimes absent.

Geographic range: The 121 listed species are widespread but particularly numerous (68) in the Oriental Region. Representation in other regions are as follows: Palearctic – 26; Ethiopian – 21, Nearctic – 4, Neotropical – 1, Australian – 3. Areas which seem to be devoid of *Psenulus* are notably South America, Melanesia, and Polynesia. Island inhabitants, in addition to those of the Oriental Region, are found on Japan and Madagascar.

Systematics: American authors have used the name *Diodontus* for *Psenulus,* but because of Opinion 844 (International Commission on Zoological Nomenclature, 1968), which was passed in favor of the petition of Bohart and Menke (1965), *Diodontus* is now applicable to a genus in the Pemphredonini.

Psenulus is presumably related to *Pluto* as discussed under the latter genus. The 121 known species of *Psenulus* make it the largest genus in the tribe Psenini. Palearctic species have been keyed by Beaumont (1937, 1964b), Blüthgen (1949), and Schmidt (1971); Ethiopian ones by Leclercq (1961a), and Nearctic ones, rather inadequately, by Malloch (1933). Krombein (1950b) greatly improved Malloch's presentation. The Oriental species have been fully treated by van Lith (1962, 1972). Thirteen species groups were separated morphologically by van Lith (1962), and this was expanded to 26 some ten years later (1972).

The variation within the genus is extensive and the more notable points are as follows. There is an unusual amount of sexual dimorphism as exemplified by the different shape of the lower frontal process in many species. Holarctic species are mostly black and generally have a rather short, dorsally sulcate, somewhat quadrate petiole. Many Oriental species are extensively yellow and the petiole is rather long, not dorsally sulcate, and more rounded in cross section. Several groups of Oriental species have the interantennal carina broadened and excavate. One Javan species, *dentatus,* has the occipital carina incomplete below and ending in a large process. Mandibles have two or three teeth apically. The second submarginal cell of the forewing is usually subtriangular but may be subquadrate, and the recurrent veins end at the submarginal cells in various combinations. The notauli may be fairly short but often extend to the scutellum. The prescutellar sulcus may be foveolate or rarely simple. The propodeum may be extensively smooth or coarsely reticulate. Females of the Holarctic *fuscipennis-frontalis* group have specialized apical hair fringes on sterna IV and V. The female pygidial plate is generally reduced but may form only a line.

Biology: Information on biology of less than a dozen species has been published. All of these nest in cavities such as hollow stems of elderberry, sumac, ash, raspberry, bamboo, or grass. Abandoned beetle borings are utilized also. An interesting "semiaquatic" nesting site described by Pagden (1933) for *Psenulus sogatophagus* was floating grass stems of *Sacciolepis myosuroides* Ridley. Prey falls in three general categories. 1) Aphididae of many genera are stored by *pallipes* and its subspecies according to Freeman (1938), Krombein (1951, 1963a), Yasumatsu and Watanabe (1964), and other authors. *P. fuscipennis* also stores aphids (Janvier, 1962). 2) Psyllidae are stored by *alienus* according to Parker and Bohart (1966), by *anomoneurae* (Yasumatsu and Watanabe, 1964), and by *concolor* (Spooner, 1948 and Danks, 1971). 3) Leafhoppers and delphacids are used as prey by three yellow-marked species of the Far East. *P. iwatae* stores a delphacid (Yasumatsu and Watanabe, 1964), *pulcherrimus* uses cicadellids (van Lith, 1962), and *sogatophagus* stores two sugarcane pests, *Sogata furcifera* Horv. (Delphacidae) and the leafhopper *Nephotettix bipunctata* F. (Pagden, 1933). In nearly all cases the prey consisted of both nymphs and adults. Additions to the biologies of *iwatae, pallipes,* and *anomoneurae* were given by Tsuneki (1970a).

Rather detailed observations were made by Krombein (1955a, 1958a, 1967b) on *pallipes* ssp. *parenosas* in eastern United States. This species stores as many as 27 paralyzed aphids (*Macrosiphum*) in a single cell. The cells are lined with a subopaque membranous coating, and the partitions are made of cemented bits of wood fiber up to 2 mm in thickness. Single burrows (abandoned beetle borings) contain up to ten cells. The egg is laid on one of the last-provisioned aphids. The wasp larva spins a delicate, subopaque, white cocoon with a dark and tough cap. There are at least two generations a year and probably three or more in some localities.

Mating in *Psenulus* takes place on the ground with the pair lying end to end (van Lith, 1951; Danks, 1971).

Benno (1957) has reared a single chysidid (*Trichrysis cyanea* Linnaeus) from a collection of nests of *Psenulus concolor* and *P. schencki*. The latter species is also attacked by the ichneumonid *Perithous divinator* (Rossi) in England according to Danks (1971).

Checklist of *Psenulus*

ajaxellus (Rohwer), 1923 (*Diodontus*); Philippines: Mindanao
alboscutellatus Arnold, 1945; Madagascar
alienus (Krombein), 1950 (*Diodontus*); U.S.: California
anomoneurae (Yasumatsu), 1938 (*Nipponopsen*); Japan
    mandibularis Tsuneki, 1959
annamensis van Lith, 1972; Vietnam
araucarius van Lith, 1972; New Guinea
avernus Leclercq, 1961; Zaire
bakeri (Rohwer), 1921 (*Diodontus*); Philippines: Luzon, Mindanao, Samar
  ssp. *boholensis* van Lith, 1962; Philippines: Bohol
  ssp. *canlaoensis* van Lith, 1962; Philippines: Negros
baltazarae van Lith, 1962; Philippines: Luzon
  ssp. *luteus* van Lith, 1962; Philippines: Sibuyan
bengalensis van Lith, 1972; India
benoiti Leclercq, 1961; Zaire
berlandi Beaumont, 1937; France
    haemorrhoidalis Berland, 1925 (*Psen*), nec A. Costa, 1871
bicinctus Turner, 1912; India: Assam
bidentatus (Cameron), 1910 (*Psen*); Ethiopia, Zaire
    rubrocaudatus Turner, 1912
birganjensis van Lith, 1973; Nepal
bisicatus van Lith, 1972; Pakistan
brevitaris Merisuo, 1937; n. Europe: Finland
capensis Brauns, 1899; Zaire, Rhodesia, Ethiopia
    pauxillus Arnold, 1947
    laevior Arnold, 1951
  ssp. *latiannulatus* Cameron, 1910; Nigeria, Zaire
  ssp. *basilewskyi* Leclercq, 1955; Zaire
carinifrons (Cameron), 1902 (*Psen*); nw. India
  ssp. *scutellatus* Turner, 1912; Australia, New Guinea, E. Indies; Philippines: Mindanao
    extremus van Lith, 1966
  ssp. *iwatai* Gussakovskij, 1934; Japan
  ssp. *rohweri* van Lith, 1962; Java; Taiwan; Philippines: Luzon
  ssp. *malayanus* van Lith, 1969; Malaya, Borneo, Sumatra, Vietnam
  ssp. *bismarkensis* van Lith, 1970; New Britain, Lavongai
cavifrons van Lith, 1962; Philippines: Samar
ceylonicus van Lith, 1972; Sri Lanka
chariis van Lith, 1972; Solomons
chillcotti van Lith, 1973; Nepal
compactus van Lith, 1962; Sumatra
concolor (Dahlbom), 1843 (*Psen*); n. and e. Europe, England
    intermedius Schenck, 1857 (*Psen*)
    ambiguus Schenck, 1857 (*Psen*)
continentis van Lith, 1962: Malaya
corporali van Lith, 1962: Sumatra
crabroniformis (F. Smith), 1858 (*Mellinus*); Borneo
  ssp. *sumatranus* (Ritsema), 1880 (*Psen*); Sumatra, Java
cypriacus van Lith, 1973; Cyprus
dentatus van Lith, 1962; Java
dilectus (Saussure), 1892 (*Psen*); Madagascar
diversus van Lith, 1962; Singapore
ealae Leclercq, 1961; Zaire
elegans van Lith, 1962; Java
erraticus (F. Smith), 1861 (*Psen*); Celebes
  ssp. *basilanensis* (Rohwer), 1921 (*Diodontus*); Philippines: Basilan; Borneo, Singapore
  ssp. *butuanensis* van Lith, 1962; Philippines: Mindanao
erusus Leclercq, 1961; Zaire
esuchus (Rohwer), 1923 (*Diodontus*); Borneo, Philippines
filicornis (Rohwer), 1923 (*Diodontus*); Philippines: Basilan
formosicola Strand, 1915; Taiwan
frontalis (W. Fox), 1898 (*Psen*); w. U.S.
    occidentalis Malloch, 1933 (*Diodontus*), nec W. Fox 1892
    hesperus Pate, 1944 (*Diodontus*)
fulgidus Arnold, 1945; Madagascar
fulvicornis (Schenck), 1857 (*Psen*); Germany
fuscipennis (Dahlbom), 1843 (*Psen*); Europe to sw. USSR
    nylanderi Dahlbom, 1845 (*Psen*)
    dufouri Dahlbom, 1845 (*Psen*)

?*nigratus* "Dahlbom" Brischke, 1862 (*Psen*)
  procerus A. Costa, 1871 (*Psen*)
  ssp. *japonicus* Tsuneki, 1959; Japan, Korea
fuscipes Tsuneki, 1959; Japan
garambae Leclercq, 1961; Zaire
ghesquierei Leclercq, 1961; Zaire
godavariensis van Lith, 1973; Nepal
hemicyclius van Lith, 1962; Philippines: Palawan
hoozanius van Lith, 1972; Taiwan
interstitialis Cameron, 1906; New Guinea; Australia: Queensland
  lutescens Turner, 1907 (*Psen*)
  ssp. *luzonensis* (Rohwer), 1921 (*Diodontus*); Philippines: Luzon, Biliran, Negros, Tarawakan
  ssp. *davanus* (Rohwer), 1923 (*Diodontus*); Philippines: Mindanao, Mindoro
  ssp. *pseudolineatus* van Lith, 1962; Philippines: Palawan, Balabac
  ssp. *baliensis* van Lith, 1962; Indonesia: Bali
  ssp. *solomonensis* van Lith, 1972; Solomons
jalapensis R. Bohart and Grissell, 1969; Mexico
kohli Arnold, 1923; Rhodesia
laevigatus (Schenck), 1857 (*Psen*); Europe
  distinctus Chevrier, 1870 (*Psen*)
laevis Gussakovskij, 1928; sw USSR: Tadzhik S.S.R.
lamprus van Lith, 1972; Celebes
lubricus (Pérez), 1902 (*Psen*); Japan
luctuosus Arnold, 1929; Rhodesia
lusingae Leclercq, 1961; Zaire
luteopictus (Rohwer), 1921 (*Diodontus*); Philippines: Luzon, Negros
  ssp. *calapanensis* van Lith, 1962; Philippines: Mindoro
maai van Lith, 1967; Borneo
macrodentatus van Lith, 1962; Sumatra
maculatus van Lith, 1962; Malaya
  ssp. *javanensis* van Lith, 1962: Java
maculipes Tsuneki, 1959; Japan
mauritii van Lith, 1969; Malaya
maurus (Rohwer), 1921 (*Diodontus*): Philippines: Luzon
melanonotus van Lith, 1969; Indonesia: Sumbawa
meridionalis Beaumont; 1937; Europe: s. France
†*montanus* (Cameron), 1907 (*Psen*); n. India, nec A. Costa, 1868
multipictus (Rohwer), 1921 (*Diodontus*); Philippines: Luzon
nasicornis van Lith, 1972; Celebes
neptunus van Lith, 1972; Celebes
nietneri van Lith, 1972; Sri Lanka
nigeriae Leclercq, 1961; Sierra Leone, Nigeria
nigrolineatus (Cameron), 1907 (*Mellinus*); Sri Lanka, Malaya, Borneo
  ssp. *ajax* (Rohwer), 1921 (*Diodontus*); Philippines: Luzon
  ssp. *flavicornis* van Lith, 1962; Philippines: Sibuyan
  ssp. *dubius* van Lith, 1962; Philippines: Mindanao
  ssp. *sulphureus* van Lith, 1962; Sumatra
nigromaculatus (Cameron), 1907 (*Mellinus*); Borneo
nikkoensis Tsuneki, 1959; Japan
nipponensis Yasumatsu, 1942; Korea
noonadanius van Lith, 1970; Philippines: Balabac, Palawan
orinus van Lith, 1973; Nepal
ornatus (Ritsema), 1876 (*Psen*); Japan
  ssp. *kankauensis* Strand, 1915; Taiwan
  ssp. *tritis* van Lith, 1962; Philippines: Palawan
  ssp. *pempuchiensis* Tsuneki, 1971; Taiwan
pagdeni van Lith, 1962; Malaya
pallidicollis van Lith, 1972; Borneo
pallipes (Panzer), 1798 (*Sphex*); Europe, n. Africa, Syria, Siberia
  atratus Fabricius, 1804 (*Trypoxylon*)
  montanus A. Costa, 1868 (*Psen*)
  haemorrhoidalis A. Costa, 1871 (*Psen*)
  minutus Tournier, 1889 (*Psen*)
  nigricornis Tournier, 1889 (*Psen*)
  rubicola Harttig, 1931
  puncticeps Gussakovskij, 1933, nec Cameron, 1906
  gussakovskij van Lith, 1973, new name for *puncticeps* Gussakovskij
  ssp. *chevrieri* (Tournier), 1889 (*Psen*); Switzerland
  ssp. *pygmaeus* (Tournier), 1889 (*Psen*); Switzerland
  ssp. *parenosas* (Pate), 1944 (*Diodontus*); U.S.
  ssp. *yamatonis* Tsuneki, 1959; Japan
pan Beaumont, 1967; Turkey
parvidentatus van Lith, 1972; Taiwan
paulisae Leclercq, 1961; Zaire
penangensis (Rohwer), 1923 (*Diodontus*); Malaya
pendleburyi van Lith, 1962; Borneo
peterseni van Lith, 1970; Philippines: Tawi Tawi
philippinensis (Rohwer), 1921 (*Diodontus*); Philippines: Luzon
  ssp. *dapitanensis* (Rohwer), 1923 (*Diodontus*); Philippines: Mindanao
pseudajax van Lith, 1962; Java
  ssp. *holtmanni* van Lith, 1972; Philippines: Culion, Busuanga
pulcherrimus (Bingham), 1896 (*Psen*); Burma, Vietnam
  ssp. *projectus* van Lith, 1962; Indonesia: Java, Krakatau, Kangean Is.
  ssp. *eburneus* van Lith, 1969; India
puncticeps (Cameron), 1907 (*Psen*); Oriental Region
  antennatus Rohwer, 1923 (*Diodontus*)
quadridentatus van Lith, 1962; Malaya
  ssp. *formosanus* Tsuneki, 1966; Taiwan
reticulosus Arnold, 1945; Madagascar
rufobalteatus (Cameron), 1904 (*Psen*); e. India
rugosus van Lith, 1962; Philippines: Mindanao
saltitans Arnold, 1958; Rhodesia
sandakaensis (Rohwer), 1923 (*Diodontus*); Borneo, Sumatra
schencki (Tournier), 1889 (*Psen*); n. and centr. Europe, England
  simplex Tournier, 1889 (*Psen*)
  longulus Tournier, 1889 (*Psen*)
scutatus (Rohwer), 1921 (*Diodontus*); Philippines: Luzon, Samar, Negros
  ssp. *sibuyanensis* van Lith, 1962; Philippines: Sibuyan
  ssp. *mindanaoensis* (Rohwer), 1923 (*Diodontus*); Philip-

pines: Mindanao
  ssp. *borneensis* (Rohwer), 1923 (*Diodontus*); Borneo
scutellatus Turner, 1912; Australia: Queensland
segrex van Lith, 1972; Celebes
separatus van Lith, 1962; Java
sinclairi Lal, 1939; India
singularis van Lith, 1962; Philippines: Luzon
sogatophagus Pagden, 1933; Malaya
stuckenbergi Arnold, 1962; e. Africa
suifuensis van Lith, 1972; China
suluensis van Lith, 1970; Philippines: Tawi Tawi
taihorinus Strand, 1915; Taiwan
tanakai Tsuneki, 1959; Japan
trevirus Leclercq, 1961; Zaire
trimaculatus van Lith, 1962; Malaya
trisulcus (W. Fox), 1898 (*Psen*); e. and centr. U.S.
    corusanigrens Rohwer, 1920 (*Diodontus*)
    sulcatus Malloch, 1933 (*Diodontus*)
tuberculifrons (Rohwer), 1921 (*Diodontus*); Philippines: Luzon
  ssp. *decoratus* van Lith, 1962; Philippines: Sibuyan
turneri Arnold, 1927; S. Africa
varius van Lith, 1962; Malaya
xanthognathus Rohwer, 1910; Philippines: Luzon
  ssp. *centralis* van Lith, 1969; Philippines: se. Luzon to Mindanao, Palawan
xanthonotus van Lith, 1969; Taiwan
yoshimotoi van Lith, 1969; North Borneo
  ssp. *pontilis* van Lith, 1970; Philippines: Tawi Tawi

Tribe Pemphredonini

A member of this tribe is difficult to characterize in simple terms since there is so much diversity in size and structure among the three subtribes. The basic difference from Psenini is the presence of no more than two submarginal cells. This seems to make a natural division. The low placement of the antennal sockets is generally a useful distinguishing character, but their location is variable in the pemphredonin genus *Ammoplanops*. Within

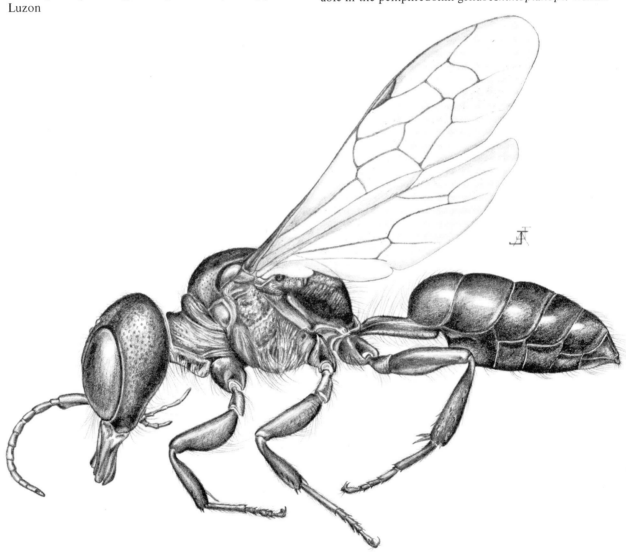

FIG. 42. *Pemphredon confertim* W. Fox, female.

certain genera of this tribe some species are petiolate and others are nonpetiolate, so the presence or absence of a petiole cannot even be used as a generic character.

The episternal sulcus is well developed in some members of the tribe and weakly so or not at all in others. In the subtribe Pemphredonina its presence or development is sometimes questionable. In *Pemphredon* and *Diodontus* it is difficult to decide whether the carina forming the anterior margin of a vertical broad pitted groove is the omaulus or simply the rather sharp margin of the episternal sulcus. The problem is further compounded by the fact that the pitted groove is broken by cross carinules dorsad so that it does not clearly arise from the subalar fossa. We are assuming in most cases (see *Diodontus*) that this pitted groove is the episternal sulcus because the lower end continues to the anteroventral margin of the mesopleuron. Furthermore, a comparison with the related genus *Passaloecus* strengthens this hypothesis, because in the latter genus the corresponding groove is connected to the subalar fossa. Based on the material we have seen, the presence of an omaulus in these three genera cannot be verified except for *Passaloecus gracilis* and related species. The allied genus *Polemistus* has the omaulus well developed. In the subtribe Stigmina there is an omaulus but no definitive episternal sulcus (except for a remnant below the omaulus), whereas in Ammoplanina the reverse is true.

The Pemphredonini is more specialized than the Psenini as evidenced by the reduced venation, trend towards an enlarged stigma, frequent touching of the pronotal lobe and tegula, closed mandibular sockets of most genera, and development of a lateral carina on tergum I in some genera. *Timberlakena* and its relatives exhibit some of the most reduced venation in the Sphecidae. Our ideas of the generic relationships are expressed in fig. 37. The tribe contains about half of the species in the subfamily.

There is a dearth of comprehensive treatments of the tribe. Most useful are those of Pate (1937c) and Leclercq (1959). Also of special value are papers by W. Fox (1892c) and Beaumont (1964b).

Key to subtribes of Pemphredonini

1. Forewing with three discoidal cells; stigma small to moderate, in any case smaller than marginal cell.............. Pemphredonina, p. 176
Forewing with one or two discoidal cells; stigma large, often approaching or surpassing marginal cell in size................. 2
2. Omaulus present or mesopleuron with a sulcus which descends from beneath posterior apex of pronotal lobe ... Stigmina, p. 185
Omaulus absent but episternal sulcus present and descending from subalar fossa as usual...................... Ammoplanina, p. 194

For convenience of the user all of the genera in the tribe are combined in a single key with an indication of the subtribal affiliations.

Key to genera of Pemphredonini

1. Forewing with two recurrent veins and three discoidal cells; stigma small or moderate in size (fig. 41 E) (subtribe Pemphredonina)........... 2
Forewing with one recurrent vein and two discoidal cells, rarely one; stigma large (figs. 41 F, 47 A)................................. 5
2. Episternal sulcus well developed, extending from subalar fossa to hypersternaulus and beyond; hypersternaulus horizontal; labrum with apex entire, usually roundly produced (fig. 44 B,D); mandible with two or three teeth; female without pygidial plate; hindtibia without a series of spines along posterior margin 3
Episternal sulcus incomplete, not evident between subalar fossa and hypersternaulus; hypersternaulus rising obliquely posterad (fig. 43 D); labrum emarginate or entire; mandible with two to six teeth; female with pygidial plate; hindtibia often with a series of spines along posterior margin........................ 4
3. Inner orbits nearly parallel or at least not converging strongly below, interocular distance at midocellus not more than one-third greater than least interocular distance (fig. 44 B); gena without long, erect setae ventrally; omaulus rarely present; midflagellar articles longer than broad (Holarctic and Oriental Regions)...... *Passaloecus* Shuckard, p. 182
Inner orbits converging strongly below, interocular distance at midocellus more than one-third greater than least interocular distance (fig. 44 D); gena with scattered long, erect setae, ventrally; omaulus present; midflagellar articles broader than long (widespread) .... *Polemistus* Saussure, p. 184
4. Abdomen in dorsal view with petiole longer than wide; labrum with apex entire (sometimes weakly notched) (Holarctic Region) ... *Pemphredon* Latreille, p. 179
Abdomen in dorsal view with petiole wider than long; labrum emarginate (fig. 43 G) (widespread except Neotropical and Australian Regions .... *Diodontus* Curtis, p. 176
5. Forewing with elongate marginal cell which is larger than stigma and closed apically (fig. 41 H); pronotal collar with complete transverse carina; omaulus present or mesopleuron with a sulcus behind a shelflike edge descending below posterior apex of pronotal lobe (subtribe Stigmina) 6
Forewing with short marginal cell, at most subequal in area to stigma, often open (fig. 47 G); pronotal collar without a complete transverse carina; omaulus absent, episternal sulcus descending from subalar fossa (Subtribe Ammoplanina) 12
6. Gaster in dorsal view with petiole much longer than wide.. 7
Gaster in dorsal view with petiole indistinct or no longer than wide 9
7. Acetabular carina absent; omaulus continued ventrally by a sulcus which ends

near anteroventral margin of pleuron or in a large, anterior, circular depression; each eye margined with a rather broad, pitted groove and carina; a median longitudinal ridge present on lower frons (Old World except Australia) .... *Carinostigmus* Tsuneki, p. 189
Acetabular carina present and continuous with omaulus; margins of eyes simple or narrowly sulcate; lower frons without a ridgelike process 8

8. Frons below simple, without median spinelike process; subomaulus present (fig. 45); female mandible with three teeth; male clypeus densely covered with appressed silvery pubescence; dorsal face of hindtibia without a row of spines (widespread except Australia) ..... *Stigmus* Panzer, p. 188
Frons below with a median hooklike projection (fig. 44 C); subomaulus absent; female mandible with five teeth (fig. 44 C); male clypeus without silvery pubescence; dorsal face of hindtibia with a row of weak spines (Australia) .... *Paracrabro* Turner, p. 186

9. With no more than one submarginal cell; occipital carina not evident 10
With two closed submarginal cells 11

10. With one large, closed submarginal cell (fig. 47 B) (S. America) .... *Microstigmus* Ducke, p. 191
With submarginal area of forewing completely open (fig. 47 A); radius and media of hindwing reduced to basal stubs (eastern U.S., South Africa) .... *Xysma* Pate, p. 193

11. Occipital carina absent; hindwing media not diverging before cu-a, not separated from Cu (fig. 41 F) (widespread) .... *Spilomena* Shuckard[1], p. 192
Occipital carina well developed; hindwing media diverging before cu-a (as in fig. 41 D); frons between scapal grooves often with a laminate carina (Australia to East Indies)................... *Arpactophilus* F. Smith, p. 186

12. Forewing marginal cell complete but not reaching anterior margin (fig. 47 C,H), two closed submarginal cells 13
Forewing marginal cell incomplete or complete and reaching anterior margin (fig. 47 E,G) 14

13. Second submarginal cell of forewing petiolate in front (fig. 47 H) (Mediterranean area).. *Protostigmus* Turner, p. 194
Second submarginal cell of forewing sessile in front (fig. 47 C) (sw. USSR) .... *Anomiopteryx* Gussakovskij, p. 196

14. Stigma broadly lenticular, tapering to a point distally (fig. 47 E); marginal cell of forewing complete 15
Stigma subglobose (fig. 49 A) 16

15. Submarginal cell in line distad with marginal and first discoidal cells (fig. 47 F), no backward projecting spur on submarginal cell (w. U.S. and sw. USSR)............. *Ammoplanops* Gussakovskij, p. 197
Submarginal cell angled out of line with marginal and first discoidal cells (fig. 47 E), and with a spur which projects obliquely backward (w. U.S., w. Canada)............................*Pulverro* Pate, p. 196

16. Hindwing without closed cells except for costal cell; forewing with one or two complete or more often incomplete submarginal cells (fig. 49 A-C) (w. U.S.) .... *Timberlakena* Pate, p. 200
Hindwing with two closed cells in addition to costal cell; forewing with one submarginal cell which is complete................................ 17

17. Forewing marginal cell complete (fig. 47 D) (Holarctic, Ethiopian).*Ammoplanus* Giraud, p. 197
Forewing marginal cell incomplete, R_1 not extending beyond stigma (fig. 47 G)............. 18

18. Hindtibia relatively straight in profile, clypeus emarginate at apex; terminal male sterna modified; female pygidial plate present (U.S.) .. *Ammoplanellus* subg.*Parammoplanus* Pate, p. 198
Hindtibia bowed in profile; clypeus nearly truncate apically; terminal male sterna simple; female pygidial plate absent (Holarctic, Ethiopian).*Ammoplanellus* Gussakovskij *s.s.*, p. 198

Subtribe Pemphredonina

The four genera of the subtribe all have two submarginal cells, three discoidal cells, two recurrent veins, and a relatively small stigma (fig. 41 D,E). The hindwing media diverges before cu-a. The hypersternaulus is nearly always well developed, and the hypoepimeral area is poorly to moderately well defined. The mandible sockets are closed and plantulae are present in the few species of each genus that we have checked. *Pemphredon* contains some of the largest wasps in the subfamily, some species attaining a modest 12 mm in length. Perhaps the smallest in the subtribe are those of *Polemistus* which may be as short as 3 mm. *Passaloecus* is notable for its distinctive mandibles (fig. 44 B). Likewise, *Polemistus* has the inner orbits converging rather strongly below (fig. 44 D), contrary to the rule in the subfamily. *Diodontus* has the labrum distinctly notched and *Pemphredon* has a rather long petiole.

Prey are almost exclusively aphids. Both ground-nesting and twig-nesting forms are found in the subtribe although *Diodontus* seems to nest only in the ground.

Genus Diodontus Curtis

Generic diagnosis: Mandible bidentate at most; labrum emarginate apically; face without a marked scapal basin, eyes not much, if any, closer below than above (fig. 43 G); occipital carina forming a complete circle or evanescent near oral fossa; pronotum transversely carinate or often rounded; omaulus sometimes present and then continuous with postspiracular carina (absent between latter and pleural margin); episternal sulcus present, represented as a broad, pitted groove separated from subalar fossa by cross ridges; hypersternaulus present, extending obliquely upward posterad (fig. 43 D);

---
[1] The fossil genus *Lisponema* Evans also keys out here.

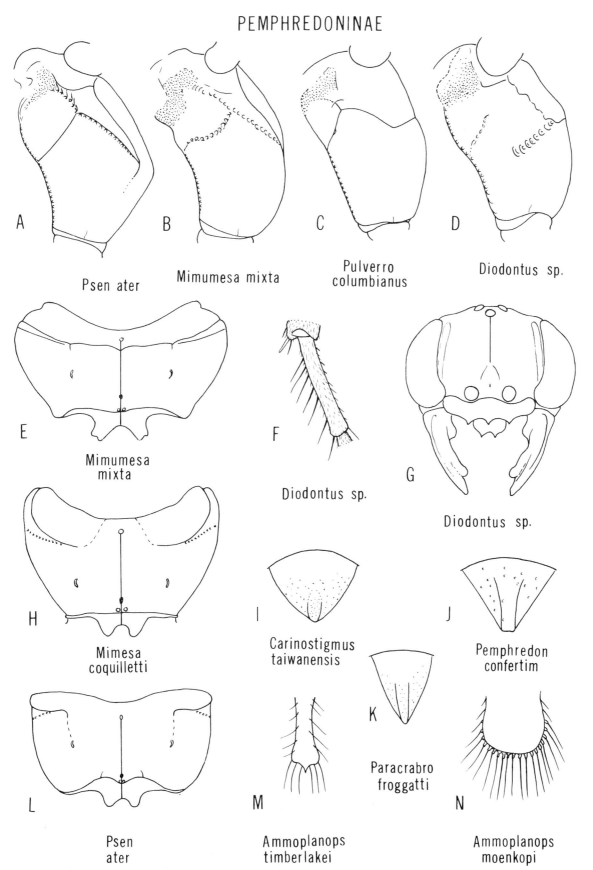

FIG. 43. Structural details in the subfamily Pemphredoninae; A-D, mesopleuron, lateral; E,H.L. mesopleuron, ventral; F, forebasitarsus, female; G, face view, female; I-K, pygidium, female; M,N, apex of male sternum VIII.

female foretarsus with a rake composed of short setae (fig. 43 F); hindtibia with series of spines along posterior face; submarginal cells each receiving a recurrent vein (fig. 41 D); gaster sessile or subsessile; female pygidial plate well defined.

Geographic range: Diodontus is known from all major continents except Australia and South America. Seventy-six species are recognized. Insular representatives include two species on the Canary Islands and one on Corsica. The Palearctic fauna has about half of the species. Seven poorly known forms occur in India, some of which have Palearctic affinities. The Ethiopian Region has only three species, all from the southern part. In the Nearctic Region 27 species have been described.

Systematics: Diodontus is most similar to *Pemphredon* in that females have a well defined pygidial plate, and a tarsal rake is usually present. In addition to the key characters (couplet 4) *Diodontus* and *Pemphredon* differ as follows: female *Pemphredon* have a narrowed, spoon-like pygidial plate, while that of *Diodontus* is broadly triangular; development of the vertex is much more pronounced in *Pemphredon;* the hindtibia of *Diodontus* is usually spinose; the mandibles of *Pemphredon* usually have three to six teeth, while *Diodontus* has two.

The identification of the pleural sulci and carinae is particularly difficult in *Diodontus.* For example, *Diodontus ater, vallicolae,* and *nigritus* seem to have a cariniform omaulus as well as a broad, foveate episternal sulcus, the two being contiguous. Perhaps the rather sharp anterior edge of the latter is misconstrued as an omaulus. In these species the "omaulus" is continuous with the post-spiracular carina, but there is still a weakly formed omaulal ridge originating under the pronotal lobe, which joins the lower end of the postspiracular carina and is continued ventrad as the "omaulus". In *D. leguminifera,* on the other hand, the omaulus is indistinct and only the episternal sulcus seems defined.

Tsuneki (1972a) proposed a new subgenus and species, *D. (Neodiodontus) kohli,* for a male from Mongolia in which there is only a single submarginal cell. The validity of this subgenus should be confirmed by additional material attesting to the wing condition.

Keys have been given by Kohl (1901) for the Palearctic Region; Tsuneki (1972a) for central Asia; Leclercq (1959) for Africa; Beaumont (1964b) for Switzerland; Fox (1892c) and Mickel (1916b) for North America; and Krombein (1939) for New York State. The American species are much in need of revisionary work, since the keys of Fox and Mickel are out-of-date and incomplete.

Through a mixup in generotypes the species of *Diodontus* have generally been placed in *Xylocelia* Rohwer by American workers. The International Commission has recently ruled in favor of *Diodontus* with *tristis* Vander Linden as the type. For details of the *Diodontus-Xylocelia* controversy see Bohart and Menke (1965) and Opinion 844 of the International Commission on Zoological Nomenclature (1968).

Biology: Powell (1963) studied *P. occidentalis* (W. Fox) in detail and reviewed the North American literature. Krombein (1958a, 1963a) studied *D. virginianus* Rohwer. Evans (1970) presented information on *D. argentinae* and *ater.* Biology of Old World species was supplied by Janvier (1962), Ferton (1908), Arnold (1923b), and Grandi (1961).

In contrast to related pemphredonine genera, which employ twigs or beetle borings, *Diodontus* nests in soil. Nests occur in flat ground or vertical banks, and both sand and clay soils are utilized. Krombein (1958a) reported *virginianus* as nesting in clumps of soil between roots of uprooted tree stumps in New York. Burrow construction is variable. One nest of *virginianus* consisted of a straight vertical shaft 32 mm in depth (Krombein, 1958a). Powell (1963) reported that most nests entered the soil at an angle, and almost all included a more or less vertical part in their lower section. Janvier (1962) showed the nest of the European *tristis* Vander Linden and *minutus* Fabricius as consisting of five branches arising from a main burrow. With one exception aphid prey have been reported. *D. virginianus* in New York captured the leafhopper *Typhlocyba* sp. In Maryland the same species was observed storing nymphs of the woolly alder aphid, *Proceiphilus tesselatus* (Fitch) (Krombein, 1958a). *D. occidentalis* in California provisioned with four different aphids. According to Powell (1963) a prey-selection sequence developed as follows: *Myzus* (probably *persicae*) was first taken, probably since it occurred nearest the nest site. A few days later females had expanded their area of of searching to include outlying zones where cream cups (*Platystemon*) were common, and *Aphis* became the dominant prey. Later, as the *Aphis* supply diminished, *Acrythosiphon* was taken from peripheral areas. A single immature *Rhopalosiphum* sp. was recovered from one of the cells. The prey are transported in the wasp's mandibles (usually venter up) to a previously prepared cell, and the nest entrance is left open while the wasp is away on a provisioning flight. Upon returning to the nest with prey the wasp closes the burrow with a loose plug of soil from within. Powell (1963) stated that completed cells of *occidentalis* contained 23-30 aphids. Completed cells of *virginianus* contained five woolly alder aphids, and the wasp egg was attached to the second sternum along the midline and extended forward between the coxae to the hindmargin of the anterior coxae (Krombein, 1958a).

Hymenopterous parasites are commonly associated with nesting *Diodontus.* Powell (1963) listed the following parasites of *D. occidentalis: Omalus cressoni* (Aaron) (Chrysididae), *Morsyma ashmeadii* W. Fox, and *Smicromutilla pomelli* Mickel (Mutillidae). In addition, the anthomyiid *Leucophora innupta* Huckett and the beefly *Lepidanthrax inauratus* (Coq.) were taken at the *occidentalis* nest site, but their relationships with the wasp are not clearly understood. Krombein (1963a) recorded *Omalus intermedius* (Aaron) as a parasite of *virginianus,* and (1958a) the mutillid *Ephuta scrupea* (Say) was taken at the nest site. *Leucophora sociata* (Meigen) was associated with *virginianus,* but as with *occidentalis,* the relationships are not clear. Grandi

(1961) recorded *Chrysis leachii* (Shuckard) as a parasite of the European *D. minutus*.

Checklist of *Diodontus*

adamsi Titus, 1909; U.S.: Michigan
afer Morice, 1911; Algeria
americanus Packard, 1867; U.S.: Maine, Wisconsin
antennatus (Mickel), 1916 (*Xylocelia*); U.S.: Nebraska
argentinae Rohwer, 1909, U.S.: Colorado, Wyoming
arnoldi Leclercq, 1959; Rhodesia, Lesotho
asiaticus Tsuneki, 1972; Mongolia
ater (Mickel), 1916 (*Xylocelia*); U.S.: Nebraska, Kansas
atratulus Taschenberg, 1875; S. Africa
  nitidus Arnold, 1923
beulahensis (Rohwer), 1917 (*Xylocelia*); U.S.: New Mexico
bidentatus Rohwer, 1911; Canada: New Brunswick; U.S.: New York, Michigan
brachycerus Kohl, 1898; Tunisia
  ssp. *oasicola* Tsuneki, 1972; Mongolia
brevilabris Beaumont, 1967; Turkey
brunneicornis Viereck, 1906; U.S.: Kansas
changaiensis Tsuneki, 1972; Mongolia
clarus Pulawski, 1964; Egypt
cockerelli Rohwer, 1909; U.S.: Colorado
collaris Tsuneki, 1972; Mongolia
crassicornis Gribodo, 1894; Algeria
crassicornus Viereck, 1904; U.S.: Oregon
denticollis Tsuneki, 1972; Mongolia
dziuroo Tsuneki, 1972; Mongolia
flavitarsis W. Fox, 1892; U.S.: Colorado
fletcheri Turner, 1917; India
florissantensis Rohwer, 1909; U.S.: Colorado
franclemonti (Krombein), 1939 (*Xylocelia*); ne. U.S.
fraternus Rohwer, 1909; U.S.: Colorado
freyi Bischoff, 1937, Canary Is.
friesei Kohl, 1901; Israel, Tunisia, Algeria, Spain
gegen Tsuneki, 1972; Mongolia
geniculatus Cameron, 1898; n. India
gillettei W. Fox, 1892; U.S.: Idaho, Colorado, Nebraska
handlirschii Kohl, 1888; Austria to Mongolia
hyalipennis Kohl, 1892; sw. USSR
insidiosus Spooner, 1938; British Is., Switzerland
kaszabi Tsuneki, 1972; Mongolia
kohli Tsuneki, 1972; Mongolia
kuroo Tsuneki, 1972; Mongolia
leguminiferus Cockerell and W. Fox, 1897; U.S.: New Mexico
longicornis Beaumont, 1960; Libya
luperus Shuckard, 1837; Palearctic Region
maestus (Mickel), 1916 (*Xylocelia*); U.S.: Colorado, Nebraska
major Kohl, 1901; Austria
  ssp. *gobiensis* Tsuneki, 1972; Mongolia
medius Dahlbom, 1845; Germany, Scandinavia, USSR
  tristis Dahlbom, 1842, nec Vander Linden, 1829
  dahlbomi A. Morawitz, 1864
metathoracicus (Mickel), 1916 (*Xylocelia*); U.S.: Nebraska, Missouri
minutus (Fabricius), 1793 (*Crabro*); Palearctic Region
  parvulus Radoszkowski, 1877 (*Passaloecus*)
mongolicus Tsuneki, 1972; Mongolia
monticola Tsuneki, 1972; Mongolia
mookoensis Tsuneki, 1972; Mongolia
  ssp. *cervinicornis* Tsuneki, 1972; Mongolia
moricei Kohl, 1901; Egypt
neomexicanus Rohwer, 1909; U.S.: New Mexico
nigritus W. Fox, 1892; U.S.: Colorado
obo Tsuneki, 1972; Mongolia
occidentalis W. Fox, 1892; U.S.: California, Arizona, Nebraska
oraniensis (Lepeletier), 1845 (*Pemphredon*); Algeria
  punicus Ed. André, 1888.
  ssp. *gracilipes* E. Saunders, 1904; Canary Is. (Tenerife)
puncticeps Gussakovskij, 1935; sw. USSR: Uzbek S.S.R.
†*punicus* Gribodo, 1894; Algeria, Tunisia, nec Ed. André, 1888
pygmaeus Tsuneki, 1972; Mongolia
reticulatus Cameron, 1905; w. India
ruficornis F. Morawitz, 1890; Austria
rugosus W. Fox, 1892; U.S.: Colorado, Montana, Nebraska, Illinois
rusticus Nurse, 1903; India
saegeri Leclercq, 1959; S. Africa
schmiedeknechti Kohl, 1898; Algeria
selectus Nurse, 1903; India
siouxensis (Mickel), 1916 (*Xylocelia*); U.S.: Nebraska
spinicollis Gussakovskij, 1933; Iran
spiniferus (Mickel), 1916 (*Xylocelia*); U.S.: Nebraska
striatus (Mickel), 1916 (*Xylocelia*); U.S.: Nebraska
striolatus Cameron, 1897; India
temporalis Kohl, 1901; Israel
tenuis Nurse, 1903; India
tiemudzin Tsuneki, 1972; Mongolia
tristis (Vander Linden), 1829 (*Pemphredon*); Palearctic Region
  dahlbomi Thomson, 1870, nec A. Morawitz, 1864
vallicollae Rohwer, 1909; U.S.: Colorado
  salicis Rohwer, 1909, new synonymy by R. Bohart
virginianus (Rohwer), 1917 (*Xylocelia*); U.S.: New York, Virginia

Genus Pemphredon Latreille

Generic diagnosis: Mandible with three to six stout teeth or rarely with only two; labrum rounded apically (fig. 44 A); face without a marked scapal basin; eyes broadly separated, inner orbits often nearly parallel but sometimes converging a little below, especially in males; head usually prolonged behind eyes (fig. 42); occipital carina joining hypostomal carina before midventral line of head or evanescent toward ventral terminus; pronotum transversely rounded, occasionally somewhat ridged; omaulus absent; episternal sulcus a broad, foveate groove which does not clearly arise from subalar fossa, sometimes weakly defined or absent above hypersternaulus, latter extending obliquely upward posterad (fig. 42); female foretarsus occasionally with a short rake; hindtibia with or without a series of spines along posterior face; submarginal cells each receiving a recurrent vein or 1 receiving both; gaster with a long petiole, longer than hindcoxa (fig. 42); female pygidial plate generally narrow and somewhat spoonlike (fig. 43 J), rarely reduced to a carina.

Geographic range: *Pemphredon* is almost exclusively a Holarctic genus. Of the 53 listed species 9 are Nearctic (in addition to the 3 Holarctic forms, *lethifer, inornatus,* and *montanus*); 40 are Palearctic; and 1 is presumably Oriental (Taiwan). Insular representatives include several species from Sakhalin (USSR), three from the Kuriles (USSR), one from Taiwan and two from Japan.

Systematics: The similarities and differences between *Pemphredon* and *Diodontus* were discussed under the latter. This is the second largest genus in the Pemphredonina. The American forms have received little attention and are poorly known. The studies of Rohwer (1917) and W. Fox (1892c) are badly outdated. A. Wagner (1931), Beaumont (1964b), Merisuo and Valkeila (1972), and Valkeila and Leclercq (1972) have studied the European species, and Tsuneki (1951) revised the species of Japan and adjacent regions.

The mesopleural details are somewhat more clear than in *Diodontus*. We have seen no omaulus in the species studied, but there is a foveate sulcus that seems to be interpretable as the episternal sulcus even though it does not clearly arise from the subalar fossa (*morio, lugens,* for example). In species such as *grinnelli* there may be only a vague suggestion of an episternal sulcus above the hypersternaulus.

There appear to be three principal species groups (sometimes considered subgenera) based on facial structure and wing venation. In the *lugubris* group (= *Pemphredon* s.s.) the outer vein of submarginal cell II meets the marginal cell well beyond its middle, the second recurrent ends well onto submarginal cell II which is generally longer than high, and the interantennal area is simple. The *morio* group (= *Ceratophorus*) differs by having the outer vein of submarginal cell II meeting the marginal cell near its middle, by having a blunt interantennal projection, and submarginal cell II usually higher than long. The *lethifer* group (= *Cemonus,* = *Dineurus*) has the outer vein of submarginal cell II meeting the marginal cell near its middle, the second recurrent ending before or slightly on submarginal cell II, no interantennal tubercle, and submarginal cell II usually as long or longer than high. In the checklist the apparent species group designation is indicated by a final symbol in parentheses: *lugubris* group (P), *morio* group (Cer), and *lethifer* group (C).

Considerable new synonymy has been suggested, but it is likely that much more will be discovered in the future.

Biology: Tsuneki (1952a) studied the habits of the Japanese species of *Pemphredon* and gave a general account of the biology of the genus. Janvier (1960, 1961a) studied in detail the life histories of several European species. Additional studies of Old World species are those of: Grandi (1961, 1962), Yasumatsu and Watanabe (1964), Leclercq (1953), Müller (1911), and Danks (1971). The North American species have been studied by Krombein (1960a, 1963a, 1964b), Rau (1928b, 1946, 1948), Rau and Rau (1918), Reinhard (1929a), and Parker and Bohart (1966).

Pemphredon commonly nest in ready-made tubes such as beetle borings, dried reeds, canes, grass culms, and stems of several plants (*Sambucus, Rubus, Hibiscus, Ailanthus, Artemisia,* and others). Nests are also excavated in decayed wood of fallen trees, logs, or telegraph poles. In addition, galls of *Cynips* and *Lipara* are utilized as nesting sites. Nests that are not placed in twigs or stems are often many branched and complete (see Tsuneki, 1952a and Janvier, 1960, 1961a for illustrations of nests.) Partitions and plugs are made of pith scraped from the sides of the burrow. When old nests are reutilized the partitions often contain dried aphids from cells in which the wasp larvae died. The wasps usually construct one or more spare tunnels in their burrows. According to Tsuneki (1952a) these short, blind tunnels are filled with pith or sawdust until they are later used as brood chambers. Grandi (1961) stated, however, that the spare tunnels are never filled with pith.

Pemphredon preys only on aphids. Prey records suggest the wasps are not particularly specific in host selection, and they probably take what is most readily available. The aphids may be stung to paralysis or pressed to death between the wasp's mandibles. Adults often puncture and feed on aphids without transporting them to a nest. The aphids are carried in the wasp's mandibles to a cell that has been previously prepared. During provisionary flights the nest entrance remains open. The wasp lays its egg after the cell has been completely provisioned (10 to 60 or more aphids), but the egg is put on an aphid near the bottom or middle of the cell. This requires the wasp to move through the mass of prey until she can reach an aphid that lies behind the middle of the cell. The position of the egg on an individual aphid is variable.

Ohgushi (1954) suggested the rudiments of social behavior in his study of *lethifer fabricii.* He pointed out that pithy partitions are sometimes omitted, so that more than one larva may occupy and mature in a cell. Also, the female may live in the open end of a twig or even in the open last cell after her larvae have matured.

Several species of Hymenoptera and a few Diptera are known to parasitize *Pemphredon.* The host-parasite associations are given below in tabular form.

Checklist of *Pemphredon*

austriacus (Kohl), 1888 (*Diphlebus*); Austria, Algeria (C)
balticus Merisuo, 1972; Finland to Germany (P)
beaumonti Hellén, 1955; Finland (P)
bipartior W. Fox, 1892; s. and e. U.S. (C)
 harbecki Rohwer, 1910, new synonymy by Bohart
bucharicus Gussakovskij, 1952; sw. USSR: Tadzhik S.S.R. (C)
clypealis Thomson, 1870; Europe (Cer)

TABLE 10
Parasites Associated with *Pemphredon*

Host	Parasite	Reference
Pemphredon confertim	Chrysididae: *Omalus janus* (Haldeman) *Omaulus purpuratus* (Provancher) Ichneumonidae: *Perithous mediator neomexicanus* Viereck Bombyliidae: *Anthrax irroratus* Say Sarcophagidae: *Senotainia trilineata* Vanderwump	Parker and Bohart, 1966 " " " "
P. concolor	Chrysididae: *Omalus janus* (Haldeman)	Reinhard, 1929a
P. flavistigma	Chrysididae: *Omalus* sp.	Tsuneki, 1952a
P. grinnelli	Chrysididae: *Omalus trilobatus* Bohart and Campos *Omalus cressoni* (Aaron) Pteromalidae: *Habrocytus analis* (Ashmead) Eurytomidae: *Eurytoma stigma* Ashmead Bombyliidae: *Anthrax irroratus* Say	Parker and Bohart, 1966 " " " "
P. lethifer	Chrysididae: *Omalus auratus* (L.)	Grandi, 1961; Leclercq, 1953; Benno, 1957; Tsuneki, 1952a
	Omalus sinuosus Auctt. *Omalus pusillus* F. Ichneumonidae: *Perithous divinator* Rossi Gasteruptionidae: *Gasteruption variolosum* Abeille Eurytomidae: *Eurytoma nodularis* Boheman Torymidae: *Diomorus calcaratus* Nees	Krombein, 1960a Benno, 1957 Grandi, 1961; Leclercq, 1953; Thomas, 1964 Grandi, 1961 " "
P. lugubris	Chrysididae: *Omalus violaceus* Scopoli *Omalus constrictus* (Förster) *Omalus auratus* (L.) *Omalus punctatus* (Uchida) Ichneumonidae: *Perithous divinator* Rossi	Móczár, 1967 " " Tsuneki, 1952a Grandi, 1961
P. rugifer	Chrysididae: *Omalus auratus* (L.) *Omalus aeneus* (F.) *Omalus auratus* (L.) *Omalus coeruleus* (Dahlbom) *Omalus constrictus* (Förster) Ichneumonidae: *Perithous* sp. *Hoplocryptus* sp. *Caenocryptus* sp.	Tsuneki, 1952a Janvier, 1960; Móczár, 1967 Janvier, 1960 Janvier, 1960 Móczár, 1967 Janvier, 1960 " "

concolor Say, 1824; U.S.: transcontinental in Transition Zone (P)
   morio Cresson, 1865, nec Vander Linden, 1829
   concolor Say of Provancher, 1882
   provancheri Dalla Torre, 1897
   cressonii Dalla Torre, 1897
   shawii Rohwer, 1917, new synonymy by Bohart
confertim W. Fox, 1892; U.S.: Washington, Oregon, California (P)
   errans Rohwer, 1917, new synonymy by Bohart
coracinus Valkeila, 1972; n. Mediterranean area (C)
diervillae Iwata, 1933; Japan (C)
dispar Valkeila, 1972; Morocco (C)
enslini Wagner, 1931; Europe (C)
fennicus Merisuo, 1972; Finland (P)
flavistigma Thomson, 1874; n. Europe, s. Russia, Korea, Japan (P)
foxii Rohwer, 1917; U.S.: Pennsylvania, New Jersey (P)
fuscipennis (Cameron), 1897 (*Cemonus*); India (C)
geminus Valkeila, 1972; sw. USSR (C)
grinnelli (Rohwer), 1910 (*Ceratophorus*); w. U.S. (C)
   gennelli Rohwer, 1910 (*Ceratophorus*)
   utahensis Rohwer, 1911 (*Ceratophorus*)
   giffardi Rohwer, 1917 (*Cemonus*), new synonymy by Bohart
inornatus Say, 1824; Holarctic (C)
   tenax W. Fox, 1892, new synonymy by Bohart
   shuckardi A. Morawitz (*Cemonus*), 1864, new synonymy by Bohart
   dentatus Puton, 1871, new synonymy by Bohart
japonicus Matsumura, 1912; Japan: Kurile Is. (P)
koreanus Tsuneki, 1951; Korea (P)
krombeini Tsuneki, 1960; Japan: Hokkaido (P)
   mandibularis Tsuneki, 1951, nec Cresson, 1865
lethifer (Shuckard), 1837 (*Cemonus*); Holarctic (C)
   strigatus Chevrier, 1870 (*Cemonus*)
   littoralis Wagner, 1918 (*Diphlebus*)
   fuscatus Wagner, 1918 (*Diphlebus*)
   neglectus Wagner, 1918 (*Diphlebus*)

minutus Wagner, 1918 (*Diphlebus*)
confusus Wagner, 1931
brevipetiolatus Wagner, 1931
fabricii Müller, 1911 (*Cemonus*)
levinotus Merisuo, 1972; USSR: Tadzhik S.S.R. (C)
lugens Dahlbom, 1842; n. and centr. Europe, USSR (P)
lugubris (Fabricius), 1793 (*Crabro*); n. Eurasia (P)
   luctuosus Shuckard, 1837
   pacificus Gussakovskij, 1933
maurusius Valkeila, 1972; Morocco (C)
minor Gussakovskij, 1952; sw. USSR: Tadzhik S.S.R. (C)
montanus Dahlbom, 1845; Holarctic (P)
   angularis W. Fox, 1892, new synonymy by Bohart
morio Vander Linden, 1829; Eurasia (Cer)
   anthracinus F. Smith, 1851
   carinatus Thomson, 1870
   intermedius Tsuneki, 1951, new synonymy by Bohart
mortifer Valkeila, 1972; Europe, Turkey, w. USSR, Sakhalin, Korea, Japan (C)
nannophyes Merisuo, 1972; USSR: Kazakh S.S.R. (C)
nearcticus Kohl, 1890; w. U.S. (P)
   cockerelli Rohwer, 1909, new synonymy by Bohart
nescius Merisuo, 1972; Italy (C)
ocellaris Gimmerthal, 1836; USSR? (Liefland)
oreades Valkeila, 1972; Pakistan (P)
orientalis Valkeila, 1972; China (C)
pilosus (Gimmerthal), 1836 (*Cenomus*); USSR? (Liefland)
platyurus Gussakovskij, 1952; sw. USSR: Tadzhik S.S.R. (C)
podagricus Chevrier, 1870; Austria, e. USSR: Ussuri; Japan (P)
   ?*laeviceps* Gussakovskij, 1933
punctifer Merisuo, 1972; USSR: Sakhalin Penin. (C)
rileyi W. Fox, 1892; U.S.: California (P)
rugifer Dahlbom, 1844; Eurasia (C)
   unicolor Panzer, 1798 (*Sphex,* text only), nec Fabricius, 1787
   unicolor Panzer, 1798 (*Crabro,* in error, plate only)
   ?*jurinii* Stephens, 1829, new name for *unicolor* of Jurine, 1807, plate 11
   lethifer Thomson, 1870, nec Shuckard, 1837
   solivagus Bondroit, 1932 (*Cemonus*)
sedulus Merisuo, 1972; USSR: Tadzhik S.S.R. (C)
scythicus Valkeila, 1972; USSR: Russian S.F.S.R. (C)
shirozui Tsuneki, 1966; e. Asia: Taiwan (P)
sudai Tsuneki, 1972; Japan (Cer)
tener Valkeila, 1972; Algeria (C)
tinctipennis Cameron, 1908; U.S.: Arizona (P)
trichogastor Valkeila, 1972; sw. USSR (C)
tridentatus Gussakovskij, 1952; sw. USSR: Tadzhik S.S.R.; Uzbek S.S.R. (C)
virginianus Rohwer, 1917; U.S.: New York, Pennsylvania, Virginia (P)
wesmaeli (A. Morawitz), 1864 (*Cemonus*); Europe (C)
   scoticus Perkins, 1929

Genus Passaloecus Shuckard

Generic diagnosis: Mandible with two or three week teeth at most, female mandible characteristically shaped (fig. 44 B); labrum nearly triangular, rounded at apex (fig. 44 B); face with weak scapal basin; male antenna usually with tyli; eyes broadly separated, sometimes converging a little below; head moderately developed behind eyes; occipital carina complete to midventral line of head, sometimes contiguous with hypostomal carina at this point; pronotum with a transverse carina; omaulus rarely present; episternal sulcus well defined, extending from subalar fossa to anteroventral area of pleuron; hypersternaulus essentially horizontal; scrobal sulcus present or absent; hypoepimeral area sometimes well defined; female foretarsus without a well developed rake; hindtibia without a series of posterior spines; each submarginal cell receiving a recurrent vein; gaster with a short petiole which is not longer than broad, female pygidial plate absent.

Geographic range: Passaloecus is primarily a Holarctic genus, but one of its 21 species reaches Taiwan in the Oriental Region. Ten species are Palearctic, seven Nearctic, and four are Holarctic. Some of the latter are probably adventive in the New World.

Systematics: Passaloecus is most closely related to *Polemistus.* Although clearly referable to the Pemphredonina these genera differ markedly from *Diodontus* and *Pemphredon.* The following characters are shared by *Passaloecus* and *Polemistus* but distinguish the two from *Diodontus* and *Pemphredon*: episternal sulcus strongly developed and clearly arising from subalar fossa, pygidial plate absent, mandibles with weak teeth and obliquely subtruncate, hypoepimeral area usually more pronounced, pronotum with a distinct transverse carina. In addition to the characters given in couplet 3 (see page 175) of the key to genera, *Passaloecus* differs from *Polemistus* as follows: the omaulus is absent (except in *Passaloecus gracilis, vandeli, borealis*), the petiole is broader than long in *Passaloecus,* and the scapal basin is deeper in *Polemistus.* The presence or absence of the scrobal sulcus and omaulus may prove to offer good group characters in *Passaloecus.*

A list of species including those of *Polemistus* was given by Yasumatsu (1934). Fox (1892c) and Krombein (1938) keyed some of the American species, and a review of the Nearctic species is now in progress. Beaumont (1964b) keyed the species of Switzerland, Ribaut (1952) keyed the species of France, Merisuo (1973b) keyed the Finnish forms, Leclercq (1959) keyed the African species, and Tsuneki (1955, 1971c) revised the species of Japan. Taxonomic notes were given by Valkeila (1961) and Merisuo (1972). Yarrow (1970) has given an important clarification of synonymy among European species and established some species groups.

Biology: The most comprehensive studies of *Passaloecus* biology are those of Janvier (1961b) and Tsuneki (1955). Other works include Grandi (1954, 1961), Leclercq (1939, 1940, 1941), Krombein (1955a, 1956a, 1958a, 1958b, 1960a, 1961a, 1961b, 1963a, 1967b), Thomas (1962), Fye (1965), Yasumatsu and Watanabe (1964), Iwata (1938b), Bonelli (1969), Danks (1971), and Merisuo (1973a).

PEMPHREDONINI

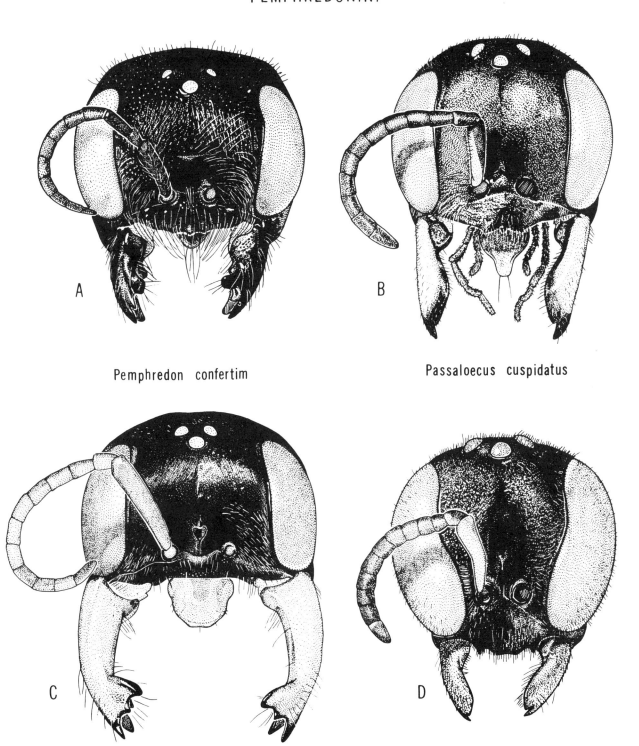

FIG. 44. Facial portraits of females in the tribe Pemphredonini.

With one exception the species of *Passaloecus* nest in wood. Nest sites include abandoned beetle borings; stems of *Rhus, Rosa, Sambucus, Cornus, Symphoricarpus, Eunonymus, Arundo;* bamboo and others; bark of pine, cedar, and oak; decayed wood; and *Cynips* galls. In France *pictus* nests in sandy soil (Janvier, 1961b). Nest architecture is usually a simple, linear arrangement, but Tsuneki (1955) illustrated a complex nest of *insignis* (as *monilicornis*) in which the entrance burrow branched into 22 brood cells. The nests are partitioned and plugged with coniferous resin or small grains of earth or pebbles. All species provision their cells with aphids; host data are too insufficient to generalize about specificity, but related genera are not very specific and provision with the nearest aphid source. The aphids are either stung to paralysis or pressed to death in the wasp's mandibles and are carried in the mandibles to a previously prepared cell. The number of aphids stored per cell varies from about six to over 60. The position in the cell of the egg-bearing aphid is not constant, and the egg is placed on the venter or the side of one of the aphids.

A number of species of Chrysididae (*Omalus*) and Ichneumonidae (*Poemenia* and *Lochetica*) have been reported as parasites of *Passaloecus*.

Checklist of *Passaloecus*

annulatus (Say), 1837 (*Pemphredon*); U.S.: transcontinental; Canada: Ontario
    rivertonensis Viereck, 1904
    equalis Viereck, 1906, new synonymy by D. Vincent
angustus Gussakovskij, 1952; sw. USSR: Tadzhik S.S.R. Uzbek S.S.R.
armeniacae Cockerell and W. Fox, 1897; w. N. America
*borealis Dahlbom, 1844[2], n. Sweden, s. Norway, Alps, Pyrenees; Alaska, w. Canada; U.S.: Rocky Mtns. s. to Utah, Colorado
brevilabris Wolf, 1958, n. & centr. Europe
clypealis Faester, 1947; Europe, Japan
    yamato Tsuneki, 1955
corniger Shuckard, 1837; Europe
    ssp. *hakusanicus* Tsuneki, 1955; Japan
    ssp. *anthrisci* Wolf, 1958; Germany
cuspidatus F. Smith, 1856; N. America: transcontinental
    mandibularis Cresson, 1865 (*Pemphredon*)
    distinctus W. Fox, 1892
dispar W. Fox, 1892; new syn. by D. Vincent
dubius Tsuneki, 1955; Japan
eremita Kohl, 1893; Austria
gracilis (Curtis), 1834 (*Diodontus*); w. Europe; U.S.: Indiana, Pennsylvania, New Jersey to Texas along coast
    insignis Vander Linden, 1829 (*Pemphredon*)(♂ only)
    *turionum Dahlbom, 1844, new synonymy by D. Vincent
    brevicornis A. Morawitz, 1864
    insignis of authors

*insignis (Vander Linden), 1829 (*Pemphredon*)(♀ only); Palearctic Region; Taiwan; Canada: Alberta; ne. U.S. to Virginia
    *monilicornis Dahlbom, 1842, new synonymy by D. Vincent
    roettgeni C. Verhoeff, 1890
    dahlbomi Sparre-Schneider, 1905, in Kohl
    dahlbomi Sparre-Schneider, 1909
    shuckardi Yasumatsu, 1934
    ithacae Krombein, 1938, new synonymy by D. Vincent
    taiwanus Tsuneki, 1967
longipes Merisuo, 1973; USSR
marginatus (Say), 1837 (*Pemphredon*); U.S.: Pennsylvania
melanocrus Rohwer, 1911; w. U.S.
melanognathus Rohwer, 1910; U.S.: Oregon, California
nipponicola Tsuneki, 1955; Japan, new status by R. Bohart
pictus Ribaut, 1952; s. Europe, Cyprus
relativus W. Fox, 1892; U.S.: Utah, Colorado, Nevada Arizona
*singularis Dahlbom, 1844; Europe, ne. U.S., and Utah, Colorado
    tenuis A. Morawitz, 1864
    gertrudis Krombein, 1938, new synonymy by D. Vincent
    gracilis of authors, *insignis* of authors
    ssp. *mongolicus* Tsuneki, 1972; Mongolia
vandeli Ribaut, 1952; France

Fossil *Passaloecus*

fasciatus Rohwer, 1909; U.S.: Colorado (Tertiary)
scudderi Cockerell, 1906; U.S.: Colorado (Florissant)

Genus Polemistus Saussure

Generic diagnosis: Mandible with two weak teeth at most, female mandible obliquely subtruncate; labrum nearly triangular, rounded at apex; face with a definite scapal basin; eyes strongly converging below (fig. 44 D), inner margins pitted a little and bowed out toward middle of frons; flagellar articles mostly short and broader than long; head moderately developed behind eyes; occipital carina complete to midventral line of head; scutum curving rather sharply down to pronotum; omaulus present; episternal sulcus extending from subalar fossa to anteroventral region of pleuron, hypersternaulus usually present, essentially horizontal, scrobal sulcus usually present; hypoepimeral area weakly or strongly defined; female foretarsus without a rake; hindtibia without a series of posterior spines; each submarginal cell receiving a recurrent vein; gaster with a short petiole which is a little longer than broad; female pygidial plate absent.

Geographic range: Occurring in all zoogeographical regions but often represented by only one or a few species. Present on several islands, such as Madagascar, Réunion, Rodriquez, Taiwan, Luzon, and Tokuno (Ryukyus). Not previously reported from the United States but now known to be present in Arizona and Utah.

---

[2] We are following Yarrow, 1970 in the use of this name, as well as *clypealis, corniger, gracilis,* and *insignis*.

Systematics: We have seen specimens of *pusillus, braunsii, abnormis, luzonensis,* and several unidentified species, most of which have been lent to us from the U.S. National Museum. In all of these the omaulus is distinct and the inner orbits converge noticeably below. In all species there is a tendency for the thorax to be lengthwise rectangular with a sharp slope across the front one-fifth or one-sixth of the scutum. In one undetermined species from the Philippine island of Mindanao this slope is accentuated by a strong cross-carina at the summit. Several Old World species, such as *braunsii,* have the notauli prolonged to the posterior margin of the scutum to form a slender U. In *luzonensis* and other Oriental species the U may also be present but more squarely cornered posteriorly. In these species the admedian lines and parapsidal lines are rather well developed so that the scutum may bear six longitudinal grooves or even eight (in *barabbas*) in varying degrees of completeness. Since the grooves are generally pitted, the resulting scutum may be intensively sculptured. In *pusillus* and other unnamed American species the scutum is plain over most of its surface. In all species there is a tendency for the inner orbital grooves to be pitted. However, this condition, which occurs in a number of other pemphredonines, is weak in some species and prominent in others. Several African species (*braunsii, bequaerti, schoutedeni*) and Asian species (*formosus, sumatrensis, siamensis, abnormis*) have a prominent median tooth or spine just above the antennal sockets on the lower frons. A discernible and sometimes deeply pitted hypersternaulus is almost a universal feature of the genus. *P. abnormis* has no trace of it, however. Most species have a sharp sublateral corner or tooth on the free edge of the clypeus. In *abnormis* this structure seems to be represented by a large lateral lobe. The propodeum may be finely to coarsely reticulate. The area between the upper edge of the scapal basin and the midocellus is frequently simple, but in *luzonensis* and other Oriental species it is set off with a strong transverse carina with associated transverse wrinkles. In addition there are median and submedian carinae which nearly outline large oblong areolae.

A close relationship with *Passaloecus* is indicated. Both genera have a well developed episternal sulcus, a similar and peculiar mandible, entire labrum, and no female pygidial plate. Separation points are the narrow face, the convergence below of the inner eye margins, and the short midflagellar articles of *Polemistus.* Also useful is the presence of an omaulus in *Polemistus* and its usual absence in *Passaloecus.*

The African species have been keyed by Leclercq (1959) and some of the Oriental species were keyed by Tsuneki (1971c).

Biology: There have been only a few accounts of the biology of *Polemistus.* Rau (1943) observed *pusillus* in Mexico where it utilized old *Trypoxylon* nests and provisioned them with aphids. The nests were sealed with a transparent, glasslike substance (perhaps a resin ?) and some of the cell partitions were made of the same material. Parasites associated with *pusillus* were a species of *Omalus* (Chrysididae) and one of *Monodontomerus* (Torymidae).

Williams (1928a) studied *luzonensis* in the Philippines. The nests were made in beetle borings in wood posts and stored with aphids taken from the fig, *Ficus nota.* As in *pusillus,* the nests were sealed with a translucent, resinlike substance. The chalcid parasite, *Ecdamua* sp., was associated with *luzonensis.*

Pagden (1933) reported on the habits of *barabbas* in Malaya. Nests were made in holes in wood over the heads of sunken nails and in old scolytid burrows in the walls of a house and in a garden gate. Partitions between cells and the closing plug were made with a waxy material pilfered from nests of bees of the genus *Trigona.* Prey were wingless adults and mature nymphs of the aphid "*Aphis laburni,* Kalt., the carrier of bunchy top in ground-nuts."

Checklist of *Polemistus*

abnormis (Kohl), 1888 (*Passaloecus*); Europe, Japan
alishanus (Tsuneki), 1971 (*Passaloecus*); Taiwan
annulicornis (Tsuneki), 1966 (*Passaloecus*); e. Asia, Ryukyus
bandraensis (Giner Marí), 1945 (*Passaloecus*); w. centr. India
barabbas (Pagden), 1933 (*Passaloecus*); Malaya
bequaerti (Arnold), 1929 (*Pemphredon*); Zaire
braunsii (Kohl), 1905 (*Passaloecus*); Zaire and Ethiopia to S. Africa
    striatifrons Cameron, 1910 (*Passaloecus*)
    dorsalis Kohl, 1912 (*Passaloecus*)
    ghesquieri Leclercq, 1955 (*Passaloecus*)
  ssp. *ferrugineipes* (Arnold), 1929 (*Pemphredon*); Zaire
  ssp. *apterinus* (Leclercq), 1959 (*Passaloecus*); Zaire
dudgeoni (Nurse), 1903 (*Passaloecus*); India, new combination by I. Yarrow
exul (Turner), 1907 (*Passaloecus*); Australia
formosus (Tsuneki), 1967 (*Passaboecus*); e. Asia: Taiwan
levipes (Bingham), 1897 (*Passaloecus*); Burma, new combination by I. Yarrow
luzonensis (Rohwer), 1919 (*Passaloecus*); Philippines: Luzon; Hawaii
macilentis Saussure, 1892; Madagascar
*pusillus Saussure, 1892; U.S.: Utah, Arizona; Mexico (type is a male)
reticulatus (Cameron), 1898 (*Passaloecus*); India
schoutedeni (Leclercq), 1959 (*Passaloecus*); Zaire
siamensis (Cockerell), 1931 (*Passaloecus*); Thailand
stieglmayri (Kohl), 1905 (*Passaloecus*); Brazil
sumatrensis (Maidl), 1925 (*Passaloecus*); Sumatra, Taiwan
  ssp. *yoshikawai* (Tsuneki), 1961 (*Passaloecus*); Laos

Subtribe Stigmina

The Stigmina are well represented in all faunal regions of the world. Seven genera, plus one fossil, are included, the individuals ranging in length from 7 mm *(Arpactophilus)* to 2 mm (*Spilomena*). In this subtribe the forewing stigma is enlarged but still covers less area than the more elongate marginal cell. Also, the pronotal collar has a complete transverse carina, and the omaulus is present, at least in *Arpactophilus, Carinostigmus, Paracrabro,* and *Stigmus*. The groove along the posterior edge of the omaulus may represent the episternal sulcus, but, if so, it has lost its connection with the subalar fossa above and has retained its identity only below where it curves forward (*Carinostigmus* for example). In *Spilomena, Xysma,* and *Microstigmus* the situation is somewhat different. Here, the groove does not so obviously lie along a carina, and it originates dorsally under the posterior apex of the pronotal lobe. Thus it might be an episternal sulcus that has been displaced forward. The situation here is similar to that in *Diodontus* and *Pemphredon*. In most of the genera there are two submarginal and two discoidal cells in the forewing (fig. 41 G), but *Microstigmus* has only one submarginal (fig. 47 B), and *Xysma* is without closed submarginal cells and has only one discoidal cell (fig. 47 A). The single recurrent vein is received by the first submarginal cell in all genera of the subtribe, except it occasionally is interstitial or received by II in *Arpactophilus*. *Carinostigmus* is the only genus in which the hindwing media diverges after cu-a (fig. 41 H), although it diverges at cu-a in *Microstigmus* and *Spilomena*. Most veins of the hindwing of *Xysma* are evanescent beyond basal stubs. The petiole is moderately long in *Stigmus, Carinostigmus,* and *Paracrabro* but short or absent in the other genera. The mandible sockets are closed in *Stigmus, Carinostigmus, Arpactophilus,* and *Paracrabro,* but we have not checked the other genera for this character. Plantulae are present in *Stigmus,* but they appear to be absent in the remaining genera. Further investigation of these last two features is needed.

Habits of this group are unusually interesting since subsociality has been claimed for *Carinostigmus* and advanced social behavior for *Microstigmus*. Prey provisions for *Carinostigmus* are aphids, for *Microstigmus* are Collembola and thrips, for *Stigmus* are aphids, for *Spilomena* are thrips and small Homoptera, and for *Xysma* are thrips.

Genus Arpactophilus F. Smith

Generic diagnosis: Mandible with two apical teeth; sometimes with malar space; labrum apically truncate; scapal basin well developed on either side of middle ridge which may be raised into a high laminate carina, sometimes nearly reaching midocellus; clypeus not densely silvery; eyes broadly separated, inner margins converging somewhat above; pitted grooves sometimes present along orbits; occipital carina complete to midventral line of head which it joins far behind oral fossa; pronotum with a transverse carina; notauli faint; omaulus sometimes continued by an acetabular carina, hypersternaulus present, essentially horizontal, scrobal sulcus present; female foretarsus without a rake; hindtibia with many short bristles; stigma large, two to three times as long as high, covering about half as much area as marginal cell; R_1 extending to apex of marginal cell (fig. 41 F); two submarginal cells; hindwing media diverging before cu-a; abdomen sessile or nearly so; female pygidial plate absent, last tergum sometimes transversely indented (as in *steindachneri*).

Geographic range: The genus is restricted to the Australasian Region, 11 species from Australia and one (*bicolor* F. Smith) from Misool Island in the Moluccas.

Systematics: Characteristic but not universal in the genus is the rough integument of the head and thorax often expressed in the form of numerous parallel ridges (fig. 46 D). Key characters are the complete occipital carina, forewing with two submarginal cells and a large marginal cell, transverse pronotal carina, presence sometimes of a distinct acetabular carina, and essentially sessile abdomen. We have synonymized *Austrostigmus* Turner for the following reasons: *Arpactophilus* had supposedly differed by the nearly triangular second submarginal cell. This is subject to great variation between species; and in any case *queenslandensis,* the type of *Austrostigmus,* has this cell nearly triangular. Turner (1916a) associated a transverse pitted groove at the base of the scutellum with all species of *Austrostigmus*. However, this groove, the prescutellar sulcus, is found also in some typical *Arpactophilus*. Both sorts have a carina along the inner orbits. Finally, the high interantennal carina of typical *Arpactophilus* is not generically diagnostic, since intergrades occur in species originally assigned to both names.

The genus does not seem to have any close relatives but may be closest to *Paracrabro*. However, the latter has the gaster petiolate, and it has more complex dentition on the mandibles.

Biology: Unknown.

Checklist of *Arpactophilus*

approximatus (Turner), 1916 (*Austrostigmus*); Australia: Queensland
arator Turner, 1908; Australia: Queensland
bicolor F. Smith, 1864; E. Indies: Misool
dubius (Turner), 1916 (*Austrostigmus*); Australia: Queensland
glabrellus (Turner), 1916 (*Austrostigmus*); Australia: W. Australia
kohlii Turner, 1908; Australia: Queensland
queenslandensis (Turner), 1908 (*Stigmus*); Australia: Queensland
reticulatus (Turner), 1912 (*Austrostigmus*); Australia: Queensland
ruficollis (Turner), 1916 (*Austrostigmus*); Australia: Queensland
steindachneri Kohl, 1883; Australia
 ssp. *deserticolus* Turner, 1936; Australia: W. Australia
sulcatus Turner, 1908; Australia: Queensland
tricolor Turner, 1908; Australia: Queensland

Genus Paracrabro Turner

Generic diagnosis: Mandible with five apical teeth, at least in female; labrum trapezoidal, broadly and shallowly emarginate at apex (fig. 44 C); face with a shallow

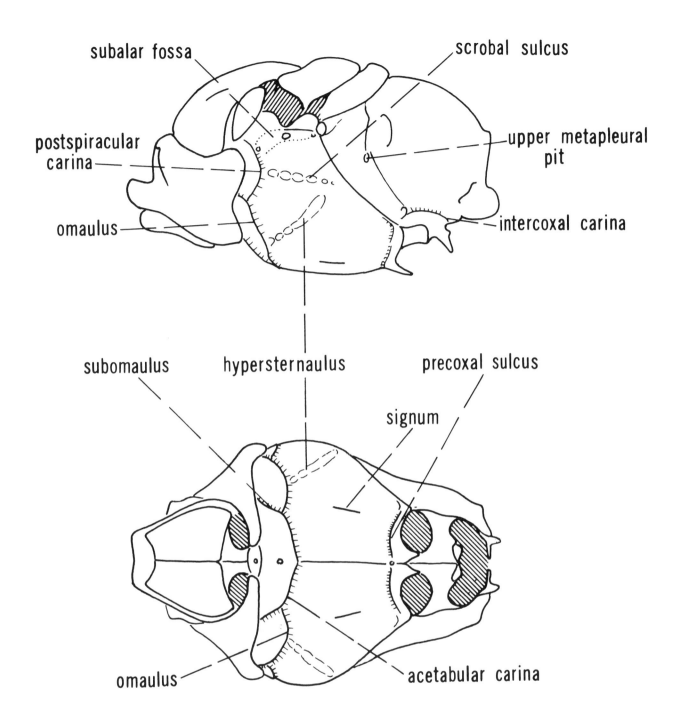

FIG. 45. Morphology of the thorax in *Stigmus*, a specialized genus in the tribe Pemphredonini.

scapal basin, lower frons with a hooked process; eyes broadly separated, converging a little above in female; crenate grooves along orbits narrow; head well developed and broad behind eyes; occipital carina present, complete to midventral line of head; pronotum with a transverse carina; notauli present but short; omaulus connecting with an acetabular carina; no definitive episternal sulcus, except below omaulus, hypersternaulus present, extending obliquely upward; scrobal sulcus absent; female foretarsus without a rake; hindtibia with a series of posterior spines; stigma large, about twice as long as high, covering slightly less area than marginal cell; two submarginal cells; hindwing media diverging before cu-a; petiole more than twice its diameter, a little longer than hindcoxa; pygidial plate forming a narrow angle.

Geographic range: This monotypic genus is restricted to Australia, where it occurs in the states of Victoria and New South Wales.

Systematics: Related to *Stigmus* and *Carinostigmus* but differing from the former by the frontal process, the emarginate labrum, and the spinose hindtibia. From *Carinostigmus* it differs mostly by the narrower orbital grooves, presence of an acetabular carina, spinose hindtibia, stouter forewing stigma, and the media diverging before cu-a. We have seen only a female of *Paracrabro*, furnished us by E. F. Riek.

Biology: Unknown.

Checklist of *Paracrabro*

froggatti Turner, 1907; Australia: Victoria

Genus Stigmus Panzer

Generic diagnosis: Mandible with two, or more often, three apical teeth; labrum subtriangular, rather broadly rounded at apex; face with a shallow scapal basin, lower frons without a tooth or longitudinal median carina; clypeus with dense, silvery hair, at least in male; eyes broadly separated, sometimes converging below, especially in males; pitted grooves along orbits narrow or absent; head moderately developed behind eyes; occipital carina present, complete to midventral line of head; pronotum with a transverse carina; notauli indicated or developed; omaulus and subomaulus well developed and connecting with acetabular carina; no definitive episternal sulcus but oblique hypersternaulus present and often joining above with pitted scrobal sulcus to describe a triangle or quadrangle (fig. 45); female foretarsus without a rake; hindtibia without a series of posterior spines; stigma large, about twice as long as high, covering nearly as much area as marginal cell; two submarginal cells (fig. 41 G); hindwing media diverging before cu-a (about as in fig. 41 G); petiole at least twice its diameter, longer than and sometimes twice as long as hindcoxa; female pygidial plate present but sometimes reduced to a groove.

Geographic range: The 30 species of *Stigmus* are divided as follows: Nearctic – 8, Neotropical – 10, Palearctic – 8 and Oriental – 4. Among the Neotropical forms are eight species from Mexico, the Lesser Antilles, and Central America. The three Oriental species are from the islands of Amami (northern Ryukyus), Taiwan, and Borneo.

Systematics: Generally small but not minute, mostly black pemphredonines with a large stout stigma and two discoidal cells in the forewing. The genus is related to *Carinostigmus* and *Paracrabro*, but it has a subomaulus as well as an omaulus (fig. 45). *Stigmus* also differs from the former by the reduced orbital grooves, simple lower frons, presence of an acetabular carina, stouter forewing stigma, and the hindwing media diverging before cu-a. The differences from *Paracrabro* are given under that genus. Tsuneki (1954a) gave keys, descriptions, and many figures of the Eurasian species. Valkeila (1956) and Beaumont (1964b) have discussed the European species. The American forms have been keyed by Krombein (1973) who placed *fulvipes* in a new subgenus, *Atopostigmus*.

Biology: American species have been studied by Krombein (1956a, 1958a, b, 1961a, 1963a), Wasbauer and Simonds (1964), Eickwort (1967), Rau (1928b), M. Smith (1923b), and Richardson (1915). Old World species have been studied by Janvier (1962) and Tsuneki (1970a). *Stigmus* nests in twigs of *Erythrina, Paeonia, Sambucus, Chionanthus, Syringa, Prunus, Polyphorus, Sassafras,* and other plants, structural timber, or galls. They frequently use preexisting cavities but may excavate their own nests. The cells are placed in linear series and are separated by masticated wood taken from scrapings of the bore. Cell length and partition length are highly variable (Eickwort, 1967 and Wasbauer and Simonds, 1964). In one case (Krombein 1961a) no partitions divided the prey into separate cells. *Stigmus* provisions its cells with aphids, and the prey is carried to the nest in the wasp's mandibles. Krombein (1961a) found that *americanus* provisioned paralyzed aphids (related genera often squeeze the prey between their mandibles). Janvier (1962) removed 23 aphids from one cell of the European *pendulus* and 27 from another cell. Wasbauer and Simonds (1964) found cells of *inordinatus* with a range of 12 to 30 aphids. *S. americanus* places its egg on the thoracic venter and abdomen of its aphid prey.

Omalus (Chrysididae) has frequently been recorded as a parasite of *Stigmus*. *S. inordinatus* is parasitized by *Omalus variatus* (Aaron), *O. glomeratus* (Buysson), *O. cressoni* (Aaron) (Parker and Bohart, 1966) and *O. iridescens* (Norton) (Davidson, 1895). *S. americanus* is parasitized by *O. iridescens* (Krombein, 1958d) and *O. purpuratus* (Provancher) (Bohart and Campos, 1960). Invrea (1941) and Móczár (1967) reported *O. aeneus* (F.) as a parasite of an undetermined Old World species of *Stigmus*. Benoist (1942) recorded *O. interandinus* Benoist as a parasite of *S. rumipambensis* in Ecuador.

Checklist of *Stigmus*

americanus Packard, 1867; e. North America; Canada: Brit. Columbia, Northwest Territory; U.S.: Washington
   lucidus Rohwer, 1909
   coloradensis Rohwer, 1911

aphidiperda Rohwer, 1911; U.S.: Pennsylvania to North Carolina
convergens Tsuneki, 1954; Japan
    ssp. *ami* Tsuneki, 1971; Taiwan
cuculus Dudgeon in Nurse, 1903; n. India
flavicornis Tsuneki, 1954; Japan
fraternus Say, 1824; e. U.S. to Missouri
    conestogorum Rohwer, 1911
    raui Rohwer, 1923
fulvicornis Rohwer, 1923; U.S.: Mississippi
fulvipes W. Fox, 1892; U.S.: transcontinental; Mexico
glabratus Kohl, 1905; Chile
hexagonalis W. Fox, 1897; Brazil, lectotype ♀, Chapada, Brazil (Carnegie Mus.), present designation by Bohart
hubbardi Rohwer, 1911; w. U.S.
inordinatus W. Fox, 1892; w. U.S.; Canada: Brit. Columbia
    coquilletti Rohwer, 1911
    reticulatus Mickel, 1918
    ssp. *universitatis* Rohwer, 1909; Colorado to Connecticut
japonicus Tsuneki, 1954; Japan
kansitakuanus Tsuneki, 1971; Taiwan
marginicollis (Cameron), 1908 (*Psen*); Borneo
montivagus Cameron, 1891; Mexico
munakatai Tsuneki, 1954; Japan
neotropicus Kohl, 1890; Brazil
nigricoxis Strand, 1911; Ecuador
parallelus Say, 1837; Mexico
patagonicus Mantero, 1901; s. Argentina
pendulus Panzer, 1804; Europe
    ater Jurine, 1807
podagricus Kohl, 1890; U.S.: Texas, Arizona; Mexico
    ssp. *tarsalis* Krombein, 1973; U.S.: centr. and s. Texas to Florida
quadriceps Tsuneki, 1954; Japan
rumipambensis Benoist, 1942; Ecuador
shirozui Tsuneki, 1964; e. Asia: n. Ryukyus
    ssp. *alishanus* Tsuneki, 1971; Taiwan
smithii Ashmead, 1900; Lesser Antilles (St. Vincent)
solskyi A. Morawitz, 1864; n. and w. Europe
    europaeus Tsuneki, 1954
    verhoeffi Tsuneki, 1954
temporalis Kohl, 1892; s. Mexico to Panama
    typicus Rohwer, 1911 (*Gonostigmus*)
thoracicus Ashmead, 1900; Lesser Antilles (St. Vincent)

Genus Carinostigmus Tsuneki

Generic diagnosis: Mandible with two (male) or three (female) apical teeth; labrum subtriangular, rounded at apex; face with a shallow scapal basin, lower frons with a median carina or ridge which may be produced into a spine or T-shaped hook; clypeus without dense silvery hair; eyes broadly separated, sometimes converging a little below, especially in males; pitted and rather broad grooves along orbits (fig. 46 C); head well developed behind eyes; occipital carina present, complete to midventral line of head and separated from hypostomal carina; notauli developed but sometimes short; omaulus present; no definitive episternal sulcus except below omaulus; acetabular carina and subomaulus absent; pitted hypersternaulus present, extending obliquely posterad, sometimes joining pitted scrobal sulcus, latter usually present (fig. 45); female foretarsus without a rake; hindtibia without a series of posterior spines; stigma large, about three times as long as high, covering considerably less area than marginal cell; two submarginal cells; hindwing media diverging well beyond cu-a (fig. 41 H), petiole much longer than twice its diameter, in any case longer than hindcoxa; female pygidial plate present, oval to teardrop-shaped in outline (fig. 43 I).

Geographic range: The 21 species of *Carinostigmus* are restricted to the Old World. A majority of the species (11) occur in the Ethiopian Region including two (*nubilipennis* Arnold and *tenellus* Arnold) from Madagascar. The Oriental fauna has most of the remaining species (8) ranging from Taiwan to India. Island representatives are found on Taiwan, Hainan, and Sri Lanka. Only two species occupy the fringes of the Palearctic Region: *marocensis* Tsuneki from Morocco and *filippovi* Gussakovskij from Japan.

Systematics: The most obvious recognition character is the presence of broad and areolate grooves along the eye margins (fig. 46 C). Since the width and areolation of these grooves is a matter of degree, other features discussed under *Stigmus* and *Paracrabro* are more diagnostic. The Asian species have been keyed, described, and several of them figured by Tsuneki (1954a, 1963d, 1966b, 1971c). In proposing his new subgenus, Tsuneki emphasized the armed frons, crenate grooving on the head, long petiole, long forewing stigma and nonsilvery clypeus. However, he neglected to describe the distal origin of the hindwing media (although he figured it). This characteristic differentiates the genus from both *Stigmus* and *Paracrabro* (fig. 41 G, H) as well as other members of the subtribe. We have seen no Ethiopian material, but descriptions and figures of Arnold (1945) covering the species of Madagascar and those of Leclercq (1959) treating the species of central Africa seem fairly typical of *Carinostigmus,* as we know it.

Biology: Arnold (1924), Yasumatsu and Watanabe (1964), Iwata (1964a), and Tsuneki (1970a) have reported on *Carinostigmus* biology, the most detailed account being that of Iwata.

The African *C. gueinzius* Turner nests in the straws of thatch (Arnold, 1924, under the name *rugosifrons* Arnold). Two Asian species, *iwatai* Tsuneki (from Hainan I.) and *monstrosus* Tsuneki (from Thailand) nest in stems of the composite *Eupatorium elephantus*. The nests of *monstrosus* are about 1.65 mm in diameter and 8 to 14 cm in length; cells are approximately 10 mm long and separated by plugs of pith 3 to 6 mm long (Iwata, 1964a). Aphids are the only known prey of *Carinostigmus* and are probably transported in the wasp's mandibles, as in *Stigmus.* Yasumatsu and Watanabe (1964) provided the only record of determined prey; *C. filippovi* Gussakovskij stores *Agrioaphis kuricola* Matsu-

PEMPHREDONINI

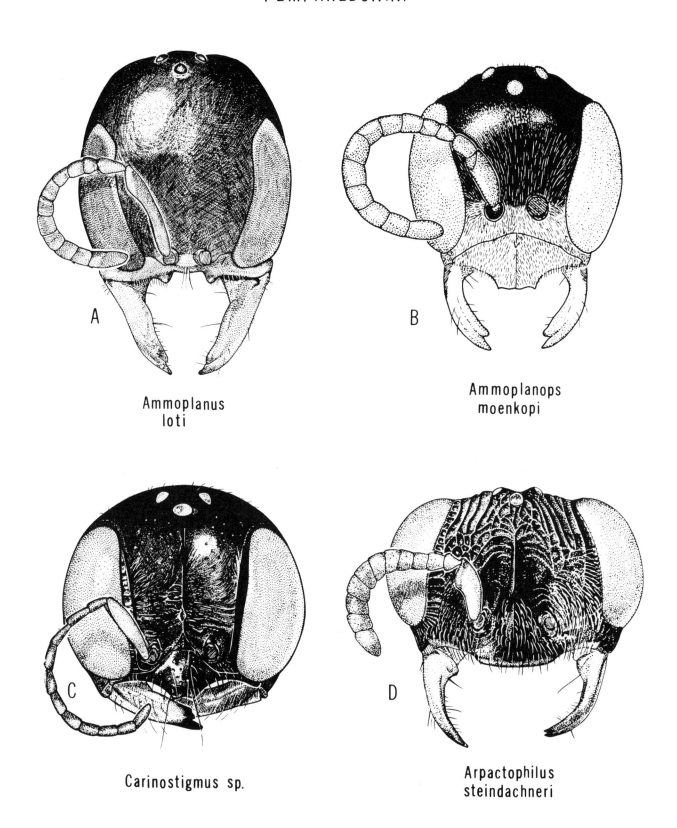

FIG. 46. Facial portraits of females in the tribe Pemphredonini.

mura in Japan. According to Iwata the egg of *iwatai* is placed at the bottom of the burrow, and aphids are provisioned progressively, creating a subsocial situation. He also considered *monstrosus* to be subsocial, since in one nest he found three female and four male adults. One adult female had worn wings and was assumed to be the mother of the remaining six adults. In addition, analysis of another nest in which cell 1 contained a pupa, cell 2 a prepupa, and cell 3 a medium sized larva with 14 aphids, prompted Iwata to claim the cells of *monstrosus* are progressively provisioned. However, the evidence for subsociality in species of *Carinostigmus* still needs confirmation.

Parasities of *Carinostigmus* are as yet unknown.

Checklist of *Carinostigmus*

aterrimus Turner, 1917; s. India
barbatus (Arnold), 1960 (*Stigmus*); Mozambique
congruus (Walker), 1860 (*Stigmus*); s. India; Sri Lanka
   niger Motschulsky, 1863 (*Stigmus*)
emirus (Leclercq), 1959 (*Stigmus*); Rhodesia, S. Africa
filippovi (Gussakovskij), 1934 (*Stigmus*); Japan
formosanus (Tsuneki), 1954 (*Stigmus*); Taiwan
   taiwanus Tsuneki, 1966 (*Stigmus*)
gueinzius (Turner), 1912 (*Stigmus*); Zaire, Rhodesia, S. Africa
   rugosifrons Arnold, 1923 (*Stigmus*)
guillarmodi (Arnold), 1955 (*Stigmus*); Lesotho
harudus (Leclercq), 1961 (*Stigmus*); Zaire
iwatai (Tsuneki), 1954 (*Stigmus*); Hong Kong, Taiwan; China: Hainan
johannis (Arnold), 1927 (*Stigmus*); Zaire, Rhodesia, S. Africa
   ssp. *hybridus* (Leclercq), 1959 (*Stigmus*); centr. and s. Africa
levifrons (Arnold) 1947 (*Stigmus*); Rhodesia
maior (Maidl), 1925 (*Stigmus*); Sumatra
marocensis (Tsuneki), 1956 (*Stigmus*); Morocco
monstrosus (Tsuneki), 1963 (*Stigmus*); Thailand
nubilipennis (Arnold), 1944 (*Stigmus*); Madagascar
pseudoscutus (Leclercq), 1959 (*Stigmus*); Zaire
saigusei (Tsuneki), 1966 (*Stigmus*); Taiwan
tenellus (Arnold), 1944 (*Stigmus*); Madagascar
thailandinus (Tsuneki), 1963 (*Stigmus*); Thailand
ugandicus (Leclercq), 1959 (*Stigmus*); Uganda

Genus Microstigmus Ducke

Generic diagnosis: Mandible with two apical teeth; malar space often very long; scapal basin shallow, lower frons with a weak longitudinal ridge; clypeus sparsely haired; eyes broadly separated, inner margins essentially parallel, no broad orbital grooves; occipital carina absent; pronotum with a transverse carina; notauli at most weakly developed; mesopleuron with sulcus originating below posterior apex of pronotal lobe and extending ventrad to anterior margin of pleuron (? = episternal sulcus); oblique hypersternaulus usually present; acetabular carina and scrobal sulcus absent; female foretarsus without a rake; hindtibia without posterior spines; stigma large, about twice as long as high and covering nearly half as much area as marginal cell; R_1 complete to end of marginal cell (fig. 47 B); one long closed submarginal cell; hindwing media diverging at cu-a; petiole short, hardly longer than wide, composed of both tergum and sternum; female pygidial plate absent.

Geographic range: The species are all Neotropical and have been reported from Paraguay to Costa Rica. Seventeen species are known.

Systematics: The three genera, *Microstigmus, Spilomena,* and *Xysma* share the following characters: no occipital carina, a shallow scapal basin, bidentate mandibles, a weak ridge on the lower frons, no apparent omaulus, no scrobal sulcus, similar hindwing venation, abdomen sessile or nearly so, notauli weak or absent, hindtibia not bristly or spinose, and no pygidial plate. The two most diagnostic features of the group are the absence of the occipital carina and the indistinguishable point of divergence of the media from Cu in the hindwing. In most pemphredonines the media diverges before the cubito-anal crossvein (fig. 47 E) or beyond it (fig. 41 H). In these cases the cubitus proper is represented by a short posteriorly directed branch. In the present group, however, the combined media and cubitus continue to the radiomedial crossvein without a posterior branch (fig. 47 B), and for practical purposes the media diverges at cu-a.

The three genera are nicely separated by the forewing venation: *Spilomena* has two closed submarginal cells, *Microstigmus* has one, and *Xysma* has none. In addition, the forewing radius is complete to the end of the marginal cell in *Microstigmus*.

Several interesting but variable features are found in *Microstigmus*. *M. myersi* and *wagneri* have the malar space short as usual in pemphredonines but *comes, eberhardi, bicolor, nigripes, lobifex, thripoctenus, theridii,* and *guianensis* have it relatively long. Most species have the ocellar triangle higher than wide. There appears to be a strong tendency in the genus toward pale (straw-colored) integument. *M. theridii, thripoctenus, nigripes, lobifex, guianensis,* and *comes* are extensively pale, and *brunniventris* and *eberhardi* are mostly pale except for the abdomen. Areolation of the thorax is often carried to an extreme in *Microstigmus*. All species examined have the propodeum and metanotum broadly areolate. In addition, such species as *nigripes, thripoctenus,* and *lobifex* have the scutellum and mesopleuron divided into a network by secondary carinae. A few of these ridges can be discerned on *bicolor*. In *lobifex* the scutum is similarly covered with subdivisions. The scape is also unusually long in *Microstigmus*.

Other variable characteristics that are useful at the species level are the stoutness of the flagellomeres which are nearly quadrate in *wagneri, guianensis, myersi, thripoctenus,* and *brunniventris* but mostly more slender in *theridii, comes, eberhardi, bicolor, nigripes,* and *lobifex*.

The close relationship of *Microstigmus* and *Spilomena* is supported by larval studies of Evans and Matthews (1968a) on *M. comes*. These authors pointed out critical similarities in the absence of setae, presence of antennal papillae, a semicircular labrum, and a tridentate mandible.

Richards (1972b) revised the genus.

Biology: Microstigmus nests in dense tropical forests of American jungles. The nests are suspended beneath leaves of *Coccoloba pubescens* (Myers, 1934) or *Crysophila guagara* Allen (Matthews, 1968a) and are made of fibers or waxy bloom of the supporting plant. Nests may be incorporated with bits of rotten wood (Richards, 1932) or earth pellets (Turner, 1929). The plant fibers, formed into a conical, baglike structure about 12 mm long, are not woven but held loosely by fine threads of a silklike material. The silk becomes more prominent near the apex of the bag and wholly predominates in a coiled thin pedicel of about 12 mm (longer if straightened). Cells are pocketlike cavities (from one to 18) in the lower half of the nest. Adults reside in the upper, hollow part just below the entrance (near the base of the pedicel). The inside of this part of the nest is covered with a smooth, rigid, translucent coating (Matthews, 1968b). Prey consists of Collembola (Iwata, 1942; Myers, 1934; Matthews, 1968b) or of thrips (Matthews, 1970). Ducke (1907a) originally suggested *Microstigmus* as a parasite of theridiid spiders, since the wasp's nests superficially resemble those of Theridiidae. Myers' (1934) study of a Trinidad *Microstigmus* (probably *myersi* Turner) showed that the food mass, consisting almost entirely of entomobryids (one sminthurid) with the legs and antennae removed, was secured to the cell wall with silk. Matthews (1968b) recorded about 1,200 unmutilated Collembola as prey of *M. comes*. About 85 percent were entomobryids, and 15 percent were sminthurids. The number of prey provisioned varies from about 13 to more than 50.

Myers (1934) found one nest contained an unidentified pteromalid, and a different unidentified chalcidoid emerged from a *Microstigmus* pupa. Richards (1935b) reared a braconid parasite, *Heterospilus microstigmi* Richards from *M. theridii*. The same parasite was found by Matthews (1968b) on *M. comes*.

Matthews (1968a, b) gave the first detailed description of sociality among sphecid wasps, when he described the habits of *Microstigmus comes*. However, Iwata (1964a) had previously claimed subsociality for the pemphredonine genus *Carinostigmus*. Matthews' studies were made in Costa Rica and his evidence can be outlined as follows: Most nests contained two or more females and ovarian dissection indicated reproductive dominance (one always had a larger oocyte than its nest mates) without apparent external morphological differences. No two cells in a nest are ever at the same stage of development, further substantiating that only one egg is available per colony at any time regardless of the number of females present. In a nest containing two females both were observed to carry prey to the same cell; thus females cooperate in provisioning one cell at a time. Cooperation among females in nest defense was observed. All nests with more than one adult always had at least one adult present — a possible division of labor between foragers and those left to protect the nest. Larval and pupal cells in active nests have no fecal pellets or meconial remains, while those kept without adults (in covered petri dishes) accumulate fecal matter, thus indicating parental care. Although many new adults emerged during the study, no new nests were found; apparently the offspring associate with previously established nests. Larvae spin no cocoons and appear to have no spinnerets (Evans and Matthews, 1968a).

Checklist of *Microstigmus*

adelphus Richards, 1972; Panama
arlei Richards, 1972; Brazil
bicolor Richards, 1971; Ecuador, Guyana
brunniventris Rohwer, 1923; Paraguay
comes Krombein, 1967; Costa Rica
eberhardi Richards, 1971; Colombia
guianensis Rohwer, 1923; Guyana
  hingstoni Richards, 1932
lobifex Richards, 1972; Guyana
luederwaldti Richards, 1972; Brazil
myersi Turner, 1929; Trinidad
miconiae Richards, 1972; Panama, Venezuela
nigrifex Richards, 1972; Guyana
pallidus Richards, 1972; Colombia
soror Richards, 1972; Colombia
theridii Ducke, 1907; Brazil, Guyana
thripoctenus Richards, 1970; Costa Rica
wagneri du Buysson, 1907; Brazil

Genus Spilomena Shuckard

Generic diagnosis: Mandible with two apical teeth; malar space short to moderate; labrum apically truncate; scapal basin barely indicated; lower frons with a short longitudinal ridge; clypeus not densely silvered; eyes broadly separated, inner margins essentially parallel or a little converging above; simple grooves sometimes developed along orbits; occipital carina absent; pronotum with a transverse carina; notauli weakly indicated; apparent episternal sulcus displaced forward and arising beneath posterior apex of pronotal lobe; no definite omaulus; no acetabular carina or scrobal sulcus, a weak horizontal hypersternaulus sometimes indicated; female foretarsus without a rake; hindtibia without posterior spines; stigma large, covering about half as much area as marginal cell; R_1 extending to end of marginal cell (fig. 41 F); two submarginal cells; hindwing media diverging at cu-a (not separated from Cu — see *Microstigmus*) (fig. 47 B); petiole absent or short and no longer than broad as seen from above; female pygidial plate absent or narrow.

Geographic range: Spilomena is recorded from all zoogeographical realms. There is only a single South American representative, *chilensis* Herbst from Chile. However, we have examined material of several undescribed species from Ecuador, Brazil, and Argentina. The number of described species from each region is as follows: Nearctic — 5; Palearctic — 22; Ethiopian — 9; Oriental — 4; Australian — 9; Neotropical — 1. Insular forms include *canariensis* Bischoff from the Canary Islands, *seyrigi* Arnold from Madagascar, and *hobartia* Turner from Tasmania.

Systematics: Related to *Xysma* and *Microstigmus,* as discussed under the latter genus but differing from both in having two complete submarginal cells and from *Xysma* in having R_1 extending to the marginal cell apex. *Spilomena* are small wasps, the males of which have yellow facial markings. Generally, the sculpture of the head is very fine, but the Australian *longiceps* has the face longitudinally striate. Most of the species are black, but a few have extensive reddish or straw-colored areas.

We have synonymized *Taialia* Tsuneki (1971c) under *Spilomena* because it seems to agree with all of the important generic characters. Tsuneki's type species, *formosana,* is known only from the female, and it appears to differ from other *Spilomena* mainly in its longer malar space (a variable character in this subtribe – see *Microstigmus*).

The Palearctic species were treated in some detail by Blüthgen (1953). Yarrow (1969) has given a key to the five British species. Tsuneki (1971b) revised the Japanese species. Hellén (1954), Van Lith (1955) Beaumont (1956b), and Valkeila (1957) treated various European forms. Leclercq (1959) keyed the Ethiopian species.

Biology: American species of *Spilomena* have been studied by Krombein (1956b, 1958b, 1963a). Blüthgen (1953, 1960), Valkeila (1957, 1961), Beaumont (1964b), Arnold (1923b), Tsuneki (1956b, 1970a), and Danks (1971) recorded biological information on Old World species.

The species nest in twigs, structural timber, and decayed wood. The primary prey appears to be immature thrips, but Krombein (1956b) recovered adult thrips from nests of *pusilla* in West Virginia. Also, Blüthgen (1953) stated that immature Psyllidae are provisioned by *troglodytes.* Arnold (1923b) recorded immature Coccidae as prey, and Beaumont (1964b) said that aphids and coccids are stored by *Spilomena.* The species carry the prey in their mandibles (Tsuneki, 1956b, 1970a).

Parasites of *Spilomena* (several British species) were recorded by Danks (1971) as the ichneumonid *Neorhachodes enslini* (Rushka), the chalcidoids *Diomerus armatus* Boheman and *Lonchetron fennicum* Graham, and the chloropid *Oscinella nigerrima* (Macquart).

Checklist of *Spilomena*

alboclypeata Bradley, 1906; U.S.: transcontinental; Canada: Brit. Columbia
ampliceps Krombein, 1952; U.S.: W. Virginia
arania Leclercq, 1961; Zaire
ausiana Leclercq, 1959; sw. Africa
australis Turner, 1910; Australia: Queensland
barberi Krombein, 1962; U.S.: transcontinental; Canada: Ontario
beata Blüthgen, 1953; England, France, Switzerland
bimaculata (Rayment) 1930 (*Microglossa*); Australia: Victoria
canariensis Bischoff, 1937; Canary Is.
capitata Gussakovskij, 1931; sw. USSR: Kazakh S.S.R.
chilensis Herbst, 1920; Chile
curruca (Dahlbom), 1843; (*Celia*); n. Europe
dedzcli Tsuneki, 1971; Japan
differens Blüthgen, 1953; England, centr. and n. Europe
elegantula Turner, 1916; Australia
enslini Blüthgen, 1953; England, centr. and n. Europe
expectata Valkeila, 1957; Finland
formosana (Tsuneki), 1971 (*Taialia*); Taiwan
foxii Cockerell, 1897; U.S.: California, Arizona, New Mexico, Colorado
fulvicornis Gussakovskij, 1931; sw. USSR: Uzbek S.S.R.
hobartia Turner, 1914; Australia: Tasmania
indostana Turner, 1918; w. centr. India
iridescens Turner, 1916; Australia
jacobsoni Maidl, 1925; Sumatra
japonica Tsuneki, 1956; Japan
kaszabi Tsuneki, 1972; Mongolia
koikensis Tsuneki, 1971; Japan
laeviceps Tsuneki, 1956; Japan
longiceps Turner, 1916; Australia
longifrons (Rayment), 1930 (*Microglossa*); Australia: Victoria
luteiventris Turner, 1936; Australia: Queensland
merceti Arnold, 1923; S. Africa
mocsaryi Kohl, 1898; Hungary, Czechoslovakia
   zavadili Šnoflak, 1942
mongolica Tsuneki, 1972; Mongolia
nikkoensis Tsuneki, 1971; Japan
obliterata Turner, 1914; Malaya
obscurior Gussakovskij, 1952; sw. USSR: Tadzhik S.S.R.
pondola Leclercq, 1959; Zaire, S. Africa
punctatissima Blüthgen, 1953; centr. and s. Europe
pusilla (Say), 1837 (*Stigmus*); centr. and e. U.S.
robusta Arnold, 1927; S. Africa
roshanica Gussakovskij, 1952; sw. USSR: Tadzhik S.S.R.
rudesculpta Gussakovskij, 1952; sw. USSR: Tadzhik S.S.R.
rufitarsus (Rayment), 1930 (*Microglossa*); Australia: Victoria
seyrigi Arnold, 1945; Madagascar
stevensoni Arnold, 1924; Rhodesia
troglodytes (Vander Linden), 1829 (*Stigmus*); England, widespread in Europe, centr. Russia, Azores
   minutissimus Radoszkowski, 1877 (*Stigmus*)
turneri Arnold, 1927; S. Africa
vagans Blüthgen, 1953; England, n. and centr. Europe
wittei Leclercq, 1959; Zaire

Genus Xysma Pate

Generic diagnosis: Mandible with two apical teeth; scapal basin shallow, lower frons with a weak longitudinal ridge; clypeus sparsely haired; eyes broadly separated, inner margins undulate but essentially parallel; no grooves along orbits; occipital carina absent; pronotum with a transverse carina; notauli absent; apparent episternal sulcus displaced forward and arising beneath posterior apex of pronotal lobe; no definite omaulus; no acetabular carina, scrobal sulcus or hypersternaulus; female foretarsus with a rake; hindtibia without posterior spines; stigma large, less than twice as long as high, smaller than marginal cell; R_1 absent beyond stigma, marginal cell thus "open" anteriorly (fig. 47 A); no closed submarginal cells; first discoidal cell complete, second absent and subdiscoidal cell weakly sclerotized posteriorly, often appearing open behind (fig. 47 A); radial sector, M + Cu, and anal vein of hindwing not well sclerotized beyond basal stubs, costal cell only closed cell of wing; gaster sessile; female pygidial plate absent.

Geographic range: This genus is known from eastern United States (Pennsylvania to Georgia) and South Africa.

Systematics: Xysma are small black wasps about 2 mm in length. The genus is known from a few specimens of two species. We have studied two females collected by Krombein. *Xysma* seems to be a highly specialized derivative from *Spilomena*-like stock. The absence of an occipital carina and pygidial plate and the essentially sessile gaster, as well as pleural features, relate *Xysma* to *Microstigmus* and *Spilomena*. However, it is easily separated by the absence of R_1 along the marginal cell, the absence of submarginal cells in the forewing (fig. 47 A), and the reduced hindwing venation. In *Xysma* the tendency of the group toward weak notauli is carried to completion, since they are no longer visible.

Telexysma Leclercq was described from a single female from the Cape region of South Africa. The single discoidal cell was made the principal basis of separation. Leclercq was misled by the *Xysma ceanothi* wing drawing of Pate (1937c) which depicted two discoidal cells. Specimens of *ceanothi* that we have studied show the second discoidal cell only partly sclerotized. In a slide-mounted wing the second discoidal appears broadly open posteriorly. This agrees closely with the figure of *Telexysma africana* given by Leclercq (1959). The principal difference between the wing of *ceanothi* and *africana* is the somewhat smaller marginal cell of the latter.

A complete generic description, a detailed description of the only known American species, and a discussion of affinities was given by Pate (1937c).

Biology: Krombein (1958a) studied *X. ceanothae* (Viereck) in Arlington, Virginia and on Plummers Island, Maryland. At Arlington, female wasps entered anobiid borings in a cowshed wall. Krombein presumed that *Xysma* excavated galleries in the anobiid frass as do some species of *Spilomena*. On Plummers Island, the wasps nested in rafters of a cabin porch. Both colonies provisioned their nests with thrips. The prey are transported in the wasp's mandibles and are held venter up.

Checklist of *Xysma*

africana (Leclercq), 1959 (*Telexysma*); S. Africa (Cape Province)
ceanothae (Viereck), 1904 (*Ammoplanus*); e. U.S.

Subtribe Ammoplanina

The Ammoplanina are tiny or very small wasps 2 to 3 mm long. The head is prognathus (directed forward) rather than orthognathus (directed downward). The seven genera appear to be restricted to the Holarctic and Ethiopian regions. The wings display a great amount of venational reduction in this group. Most genera have one submarginal cell and two discoidal cells (including the subdiscoidal cell). *Anomiopteryx* and *Protostigmus* have two submarginals, but their marginal cell is so reduced that its apex is far removed from the wing margin (fig. 47 C,H). The marginal cell is open in *Ammoplanellus* and *Timberlakena,* and sometimes R_1 is absent or envanescent beyond the stigma (figs. 47 G, 49 A). Some *Timberlakena* have two submarginals while others have one or no closed cells. The stigma is broadly lenticular in *Protostigmus, Anomiopteryx, Ammoplanops,* and *Pulverro* (fig. 47 A,E,F,H) but subglobose in *Ammoplanus, Ammoplanellus,* and some *Timberlakena* (figs. 47 D,G, 49 A). *Pulverro* and *Ammoplanops* have the normal complement of hindwing cells, and the divergence of the media before cu-a in these two genera is the most generalized of the subtribe (fig. 47 E). *Anomiopteryx* and *Protostigmus* also have all cells intact, but the media appears to diverge at cu-a (fig. 47 H). In *Ammoplanellus* and *Ammoplanus* hindwings the cells are still recognizable, but the veins are weakly sclerotized, and the media diverges well beyond cu-a (fig. 47 D,G). In *Timberlakena* the veins of the hindwing are at most represented only by basal stubs except that the costal cell is usually formed (fig. 49 A-C). The pronotal collar is not transversely carinate, and the mesopleuron lacks an omaulus and acetabular carina but the episternal sulcus is well developed and normally placed. The petiole when present is very short. We have only checked a few examples of some genera, but the mandible sockets appear to be closed and plantulae absent. These need to be further studied. The most useful publications on the group are those of Gussakovskij (1931b), Pate (1937c), and Leclercq (1959).

Too little is known about the habits of the Ammoplanina to permit generalization. Thrips are the only reported prey.

Genus Protostigmus Turner

Generic diagnosis (based on male): Mandible with two apical teeth; scapal basin shallow; eyes broadly separated, inner margins slightly converging above; occipital carina absent, scrobal sulcus very weak; hindtibia posteriorly spinulose; stigma large, broadly lenticular, acute at apex, more than twice as large as marginal cell which is short, truncate and removed from wing margin (fig. 47 H); two submarginal cells of which second is petiolate in front; hindwing media diverging at cu-a; propodeum from above shorter than breadth at middle; petiole very short.

Geographic range: Protostigmus contains two species, both from the Mediterranean area: *championi* Turner from Tunisia and *sinaiticus* Mochi from the Sinai Peninsula.

Systematics: As discussed under *Anomiopteryx*, these two genera are very close and possibly should be merged. Since only a few male specimens are known in the two genera, it is difficult to form proper generic concepts. For-

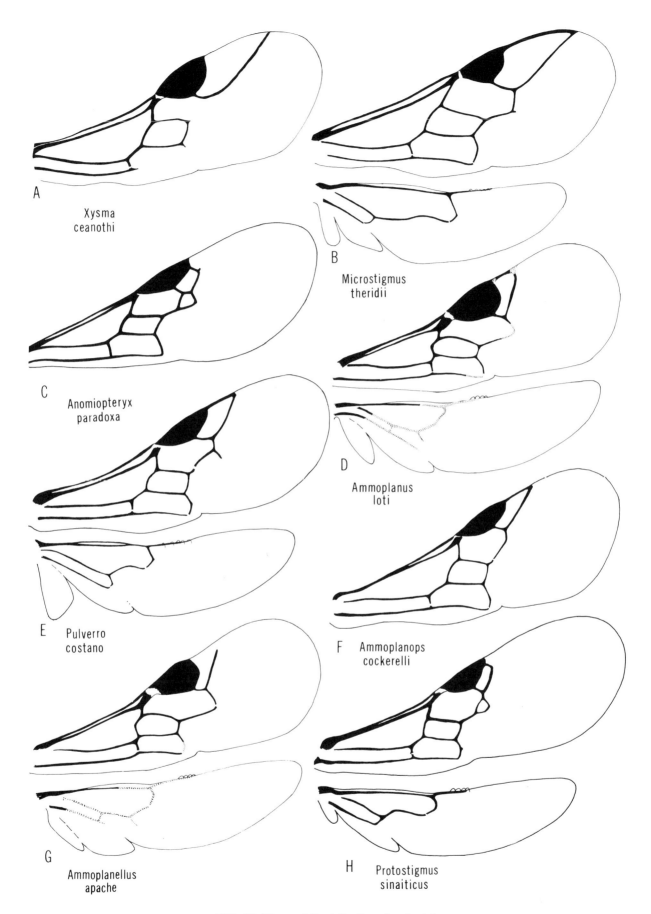

FIG. 47. Wings of the tribe Pemphredonini.

tunately, the species involved have been well described and illustrated, *A. paradoxa* by Gussakovskij (1935), *P. championi* by Turner (1918), and *P. sinaiticus* by Mochi (1940). We have studied the holotype male of *sinaiticus* lent by the United States National Museum. In this specimen R_1 ends just beyond the stigma. Also, sterna III and IV each have a patch of median, posteriorly directed fimbriae. The clypeus, lower frons laterally and mandibles are ivory-yellow.

Biology: Unknown.

Checklist of *Protostigmus*

championi Turner, 1918; Tunisia
sinaiticus (Mochi), 1940 (*Ammoplanopterus*); e. Mediterranean Region: Sinai

Genus Anomiopteryx Gussakovskij

Generic diagnosis (based on male): Mandible with two apical teeth; scapal basin shallow; eyes broadly separated, inner margins slightly converging above, occipital carina absent (presumably); pronotum elongate; scrobal sulcus very weak; hindtibia posteriorly spinulose; stigma large, broadly lenticular but distally blunt, covering more area than marginal cell which is short, truncate, and removed from wing margin (fig. 47 C); two submarginal cells; hindwing media diverging at cu-a; propodeum from above longer than breadth at middle; petiole short, about as long as broad seen from above; male pygidial plate present.

Geographic range: This monotypic genus is known only from southwestern USSR (Uzbek S.S.R.)

Systematics: We have seen no specimens of this genus, and the above generic diagnosis is based on the single known male described and figured by Gussakovskij (1935). Within the subtribe Ammoplanina its only close relative is *Protostigmus,* which seems to differ chiefly in having the second submarginal cell petiolate (fig. 47 H) and the propodeum shorter. These seem to be rather weak generic characters, and collection of more material as well as additional species may result in reduction of *Anomiopteryx* to a junior synonym. Both genera have the marginal cell entirely removed from the anterior wing margin.

Biology: Unknown.

Checklist of *Anomiopteryx*

paradoxa Gussakovskij, 1935; sw. USSR: Uzbek S.S.R.

Genus Pulverro Pate

Generic diagnosis: Mandible with two distal teeth, sometimes with an inferior median tooth; labrum entire; female scape about one-fourth as long as entire antenna, scapal basin barely indicated; lower frons with at most an indication of a median longitudinal ridge, except in some males; eyes broadly separated, inner margins essentially parallel, vertex not unusually raised above compound eyes; occipital carina absent dorsally but present ventrally; notauli evident anteriorly, scutum raised above pronotum and sometimes overhanging it; scrobal sulcus distinct; female foretarsus with a short rake; hindtibia with some short posterior bristles; stigma broadly lenticular, pointed apically, at least as large as marginal cell; R_1 extending to end of marginal cell; one submarginal and two discoidal cells; submarginal cell out of line with marginal and first discoidal cells, and with a spur which projects obliquely backward (fig. 47 E); hindwing with two closed cells in addition to the costal cell, media diverging before cu-a; petiole absent; female pygidial plate present, subtriangular, and somewhat shiny; male sternum VIII fan shaped to subtriangular, apical margin serrulate.

Geographic range: The seven described species of *Pulverro* are essentially Nearctic, although one species ranges well south into Mexico.

Systematics: The distinctive forewing venation (fig. 47 E) readily separates *Pulverro* from *Ammoplanops* as well as from all other genera of Pemphredoninae. The submarginal cell has an obliquely backward pointing distal spur which seems to be a remnant of a second submarginal cell. Judging from wing folds, the spur represents the distal vein of a petiolate cell which is in the process of disappearing. Except for venation, *Pulverro* agrees closely with *Ammoplanops* in important structural details. The species of *Pulverro* have been discussed and keyed by Pate (1937c). Useful specific characters are the convexity of the scutum, formation and spinulation of the clypeus, and in males the dentition of the mandibles and the shape of sternum VIII.

Biology: Bohart and Grissell (1972) have published the only information. They studied a loosely knit colony of *Pulverro monticola* nesting in the side banks of a dirt road in the California Sierra. Burrows, which had a diameter of about 2 mm, entered the sloping banks almost horizontally then angled downward to a depth of about 85 mm. Short side burrows near the bottom led to cells provisioned with up to 21 thrips, mostly *Frankliniella moultoni* Hood but partly *Aeolothrips fasciatus* (Linnaeus). The thrips were adults for the most part and were carried in the mandibles of the wasps. Burrow entrances were marked by a mound of tumulus on the downhill side, and they were not closed during provisioning or after completion of the nest.

Checklist of *Pulverro*

californicus Eighme, 1973; U.S.: California
chumashano Pate, 1937; U.S.: w. California
   costano Pate, 1937, new synonymy by Bohart
columbianus (Kohl), 1890 (*Ammoplanus*); w. U.S. and sw. Canada
   eriogoni Rohwer, 1909 (*Ammoplanus*)
   colorado Pate, 1937
constrictus (Provancher), 1895 (*Anacrabro*); U.S.: s. California
   serrano Pate, 1937
laevis (Provancher), 1895 (*Anacrabro*); U.S.: s. California

mescalero Pate, 1937; sw. U.S., n. and centr. Mexico
monticola Eighme, 1969; n. California

Genus Ammoplanops Gussakovskij

Generic diagnosis: Mandible with two distal teeth, sometimes with other denticles; labrum entire; female scape about one-fourth as long as entire antenna, scapal basin barely indicated; antennal sockets sometimes placed well above clypeal margin; lower frons without a longitudinal median ridge; clypeus well silvered in males, less so in females; eyes broadly separated, inner margins essentially parallel, vertex not unusually raised above compound eyes; occipital carina absent or faintly indicated ventrally; notauli very faint; scutum in profile humped above pronotum; scrobal sulcus present; female foretarsus with a short rake; hindtibia with many posterior bristles; stigma broadly lenticular, pointed apically, at least as large as marginal cell; R_1 extending to end of marginal cell; one submarginal and two discoidal cells; submarginal cell in line distad with marginal and first discoidal cells (fig. 47 F); hindwing with two closed cells in addition to costal cell, media diverging before cu-a; petiole absent; female pygidial plate present, subtriangular and somewhat shiny; male sternum VIII straplike or fan shaped, and apical margin minutely serrulate (fig. 43 M,N).

Geographic range: Ammoplanops occurs only in the Nearctic and Palearctic Regions and the 11 described species are confined to Asia and western United States.

Systematics: Ammoplanops and *Pulverro* are close and are best separated by the distinctive wing venation (fig. 47 E,F). *Ammoplanops* is the only genus of the subtribe in which the marginal, submarginal, and discoidal cells have their distal veins nearly in a line.

Discussion of systematic position and differentiation of species are to be found principally in Gussakovskij (1931b) and Pate (1939).

The position of the antennal sockets varies considerably, and they may be placed well above the clypeal margin (*cockerelli* and *moenkopi*), close to the margin as usual in Pemphredonini (*vierecki*), or intermediate (*cressoni*). This variation contravenes an otherwise useful way of separating the two tribes of the subfamily. Most of the 11 described Nearctic species can be separated in the male on the basis of the clypeus, which is frequently snoutlike and with spines or other projections. Male sternum VIII is also of considerable value (fig. 43 M,N) and its narrow or spatulate form divides two groups of Nearctic species. Palearctic species seem to be more generalized. Females have relatively few specific characters.

Biology: No published accounts of nesting are available. The presence of a female tarsal rake and pygidial plate suggests that the species nest in soil. Adults are attracted to flowers of *Eriogonum, Croton, Guterrezia, Baileya, Dalea, Lepidium, Larrea, Prosopis, Malvastrum, Tamarix, Helianthus, Chrysothamnus, Heliotropium, Anethum, Cleomella,* and *Baccharis*.

Checklist of *Ammoplanops*

ashmeadi Pate, 1939; U.S.: s. California
carinatus Gussakovskij, 1931; sw. USSR: Kazakh S.S.R., Transcaspia
cockerelli (Ashmead), 1903 (*Ammoplanus*); U.S.: Texas to Utah and California
cressoni Pate, 1939; U.S.: s. California, Arizona, Nevada
foxi Pate, 1939; U.S.: s. California
laticeps Gussakovskij, 1931; sw. USSR
moenkopi Pate, 1939; U.S.: Arizona, New Mexico, Utah
mongolicus Tsuneki, 1972; Mongolia
timberlakei Pate, 1939; U.S.: s. California
tuberculifer Gussakovskij, 1931; sw. USSR: Kazakh S.S.R.
vierecki Pate, 1939; U.S.: New Mexico

Genus Ammoplanus Gussakovskij

Generic diagnosis: Mandible with two distal teeth; labrum bilobed; scapal basin shallow, somewhat grooved medially; lower frons without a longitudinal median ridge; clypeus not well silvered, usually emarginate medially, particularly in females, labrum filling cut out area; eyes broadly separated, inner margins usually converging below, sometimes nearly parallel; vertex broadly rounded, usually raised well above compound eyes; occipital carina present dorsally, disappearing ventrally; notauli present anteriorly but faint; scutum convex but not much raised above pronotum; scrobe usually small, scrobal sulcus hardly indicated; female foretarsus without a rake; hindtibia gently curved in profile, without obvious bristles, stigma subglobose, considerably larger than marginal cell; R_1 extending to end of complete marginal cell; one submarginal and two discoidal cells, submarginal cell far out of line with marginal and first discoidal cells (fig. 47 D); hindwing with two closed cells in addition to costal cell, media diverging far beyond cu-a (fig. 47 D); petiole practically absent; female pygidial plate present; male sternum VIII broadly straplike.

Geographic range: The 41 known species occur in the Nearctic (7), Palearctic (28), and Ethiopian (6) Regions. The Nearctic forms are mainly western and southwestern with only one species (*unami*) from eastern United States (Pennsylvania). The Palearctic species are roughly from four subregions: southwestern USSR, Mongolia, central and eastern Europe, and the Mediterranean.

Insular representatives include *marathroicus* from Sicily and *insularis* from the Balearics. The Ethiopian species are divided between the southern part of the continent and Madagascar.

Systematics: In the subtribe Ammoplanima only *Ammoplanus* and *Ammoplanellus* have in combination the globose stigma, one closed submarginal cell, and in the hindwing two closed discal cells. The two genera are separated primarily by the forewing marginal cell, which is complete in *Ammoplanus* and open anteriorly and distally in *Ammoplanellus* (compare fig. 47 D, G). Information on American species has been furnished largely by Pate (1937c, 1939, 1943a, 1945); on Palearctic species by Gussakovskij (1931b), Maréchal (1938), Giner Marí (1943a), and Blüthgen (1954, which includes a key to central European species); and on African species by Arnold (1945, 1959) and Leclercq (1959).

Species of *Ammoplanus* generally have the clypeal apex moderately to strongly incised. Sometimes the cut out area has a median projection, which may vary from a tooth (as in *wesmaeli*) to a spatulate process (as in *rjabovi, ceballosi* and an undescribed species from California). Giner Marí (1943a) attached subgeneric value to this characteristic as well as to the color of the stigma and modifications of the hindtarsus of the male. We agree with Pate (1945) that *Ceballosia* Giner Marí is better considered as a species group.

Biology: Adult wasps are attracted to such flowers as *Eriogonum, Rhamnus, Eriodictyon, Lepidium, Holodiscus,* and *Prunus.*

Checklist of *Ammoplanus*

angularis Gussakovskij, 1952; sw. USSR: Tadzhik S.S.R.
bischoffi Maréchal, 1938; Italy
brevicornis Arnold, 1945; Madagascar
ceballosi Giner Marí; 1943; Spain
chemehuevi Pate, 1943; U.S.: s. California
cradockensis Arnold, 1959; S. Africa
curvidens Tsuneki, 1972; Mongolia
diversipes Gussakovskij, 1931; sw. USSR
dusmeti Giner Marí, 1943; Spain
egregius Arnold, 1945; Madagascar
gegen Tsuneki, 1972; Mongolia
handlirschi Gussakovskij, 1931; centr. and e. Europe; sw. and s. USSR
   sibiricus Gussakovskij, 1931
   metatarsalis Gussakovskij, 1931
hissaricus Gussakovskij, 1952; sw. USSR: Tadzhik S.S.R.
hofferi Šnoflak, 1943; Czechoslovakia
insularis Giner Marí, 1945; s. Europe: Balearic Is.
jucundus Arnold, 1945; Madagascar
kaszabi Tsuneki, 1972; Mongolia
kohlii Schmiedeknecht, 1898; Algeria
latiscapus Leclercq, 1959; S. Africa
loti Pate, 1943; U.S.: s. California
maidli Gussakovskij, 1931; Tunisia, Algeria
mandibularis Cameron, 1903; S. Africa
   capensis Arnold, 1923
marathroicus (De Stefani), 1887 (*Hoplocrabron*); Sicily
mongolensis Tsuneki, 1972; Mongolia
monticola Gussakovskij, 1952; sw. USSR: Tadzhik S.S.R.
nasutus Tsuneki, 1972; Mongolia
perrisi Giraud, 1869; Spain (? = *wesmaeli*)
platytarsus Gussakovskij, 1931; sw. USSR
pragensis Šnoflak, 1946; Czechoslovakia
quabajai Pate, 1943; U.S.: California
rjabovi Gussakovskij, 1931; sw. USSR: Caucasus Mts.
sechi Pate, 1943; U.S.: s. California
serratus Tsuneki, 1972; Mongolia
shestakovi Gussakovskij, 1931; sw. USSR
simplex Gussakovskij, 1952; sw. USSR: Tadzhik S.S.R.
tetli Pate, 1943; U.S.: s. California
transcaspicus Gussakovskij, 1931; sw. USSR
unami Pate, 1937; U.S.: Pennsylvania
vanyumi Pate, 1943; U.S.: California, Idaho
wesmaeli Giraud, 1869; Europe
   pulchrior Maréchal, 1938
   minor Maréchal, 1938
   dentatus Šnoflak, 1946
zarcoi Giner Marí, 1943; Spain

Genus Ammoplanellus Gussakovskij

Generic diagnosis: Mandible with two distal teeth; female labrum deeply bilobed, at least in female; scapal basin shallow, somewhat grooved medially; lower frons without a longitudinal median ridge; clypeus not well silvered, apex not deeply but sometimes distinctly emarginate; eyes broadly separated, inner margins usually converging below, sometimes nearly parallel; vertex narrowly to broadly rounded, raised well above compound eyes; occipital carina present dorsally but usually evanescent ventrally; notauli present anteriorly but faint; scutum convex but not much raised above pronotum; scrobe moderate, sulcus absent; female foretarsus without a rake, hindtibia gently curved in profile or somewhat distorted (*Ammoplanellus s.s.*); stigma subglobose, considerably larger than marginal cell; R_1 not extending beyond stigma, marginal cell thus open anteriorly as well as at apex; submarginal cell far out of line with marginal and first discoidal cells (fig. 47 G); hindwing with two closed cells in addition to costal cell, media diverging well beyond cu-a; petiole practically absent; female pygidial plate present (subgenus *Parammoplanus*) or absent (typical subgenus); male sternum VIII broadly straplike.

Geographic range: Of the 14 described species of *Ammoplanellus* seven are from the southern part of Africa including three from Madagascar, four from United States, and three from Asia. Two of the four Nearctic species are assigned to the subgenus *Parammoplanus.*

Systematics: The close relationship between *Ammoplanellus* and *Ammoplanus* has been discussed under the latter genus. Previous authors have treated *Ammoplanellus* as a subgenus, but the open forewing marginal cell seems sufficient to split off the 14 species involved from the 40 species of *Ammoplanus.* A discussion of affinities within the genus and with related genera was given by Gussakovskij (1931b), Pate (1939, 1945), and Leclercq (1959). We have followed Pate and Leclercq in recognizing *Parammoplanus* as a subgenus of *Ammoplanellus* based on two Nearctic species. These have the following characteristics differentiating them from the typical subgenus: a pygidial plate present in the female, terminal sterna modified in the male, hindtibiae gently curved in profile, mesopleural scrobe relatively small, pronotal lobes dark or at least concolorous with the rest of the pronotum, and propodeum not dorsally areolate. Conversely, *Ammoplanellus s.s.* has no pygidial plate in the female, terminal male sterna are simple, the hindtibia is bowed in profile, the scrobe is relatively large, the pronotal lobes are contrastingly light, and the propodeum is dorsally areolate. We have been able to study relatively few of the species, so the above subgeneric differences may not be absolute.

Biology: Arnold (1924) observed *rhodesianus* Arnold nesting in sandy banks and cracks of an old mud wall in southern Africa. Pate (1945) reported many specimens of *xila* Pate "flying rapidly about holes in a chair from which had previously emerged small powder-post flat-headed borers." Pate assumed that *xila* was nesting in the abandoned buprestid borings.

Adults frequent flowers of *Eriogonum, Ceanothus, Dalea, Apocynum, Baccharis, Acacia, Sapindus, Adenostoma, Heteromeles, Rhamnus, Chrysothamnus, Prosopis* and *Tetradymia.*

Checklist of *Ammoplanellus*
(P = subgenus *Parammoplanus* Pate)

apache (Pate), 1937 (*Ammoplanus*); U.S.: California, Arizona, New Mexico (P)
aussi (Leclercq), 1959 (*Ammoplanus*); S. Africa
bezicus (Leclercq), 1959 (*Ammoplanus*); S. Africa
chorasmius (Gussakovskij), 1931 (*Ammoplanus*); sw. USSR
claripennis (Arnold), 1945 (*Ammoplanus*); Madagascar
clypealis (Tsuneki), 1972 (*Ammoplanus*); Mongolia
consobrinus (Arnold), 1927 (*Ammoplanus*); Rhodesia, S. Africa
lenape (Pate), 1937 (*Ammoplanus*); U.S.: Pennsylvania

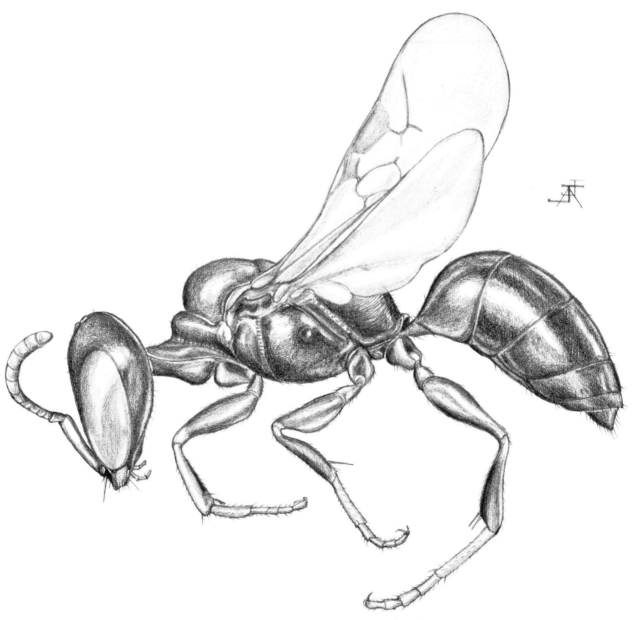

FIG. 48. *Timberlakena yucaipa* Pate, female.

(P)

ssp. *olamentke* (Pate), 1943 (*Ammoplanus*); U.S.: s. California
madecassus (Kohl), 1909 (*Ammoplanus*); Madagascar
orbiculatus (Tsuneki), 1972 (*Ammoplanus*); Mongolia
orientalis (Arnold), 1944 (*Ammoplanus*); Madagascar, new status by Bohart
rhodesianus (Arnold), 1924 (*Ammoplanus*); Rhodesia
umatilla (Pate), 1945 (*Ammoplanus*); U.S.: Washington
xila (Pate), 1945 (*Ammoplanus*); U.S.: Arizona

Genus Timberlakena Pate

Generic diagnosis: Mandible relatively simple, with a small subapical tooth; labrum bilobed in female, broadly emarginate in male, female scape about a third as long as entire antenna, scapal basin barely indicated; lower frons without a longitudinal median ridge; clypeus appearing bare in most views, not densely silvered; eyes broadly separated, inner margins essentially parallel, vertex not unusually raised above compound eyes; occipital carina present dorsally, but faint; notauli present anteriorly but often weakly impressed; scutum convex, raised above level of pronotum and sometimes overhanging it slightly; no scrobal sulcus but scrobe unusually prominent; female foretarsus without an obvious rake; hindtibia minutely bristled; stigma subglobose, equal to or larger than marginal cell; R_1 extending a short distance above marginal cell or not present beyond stigma, marginal cell thus incomplete in front as well as at apex (fig. 48); two discoidal cells and one or two submarginal cells which are often partly open (fig. 49B); hindwing without closed cells except for costal cell, vestigial media diverging after cu-a; petiole absent; female pygidial plate absent; male sternum VIII straplike and minutely serrate apically.

Geographic range: Timberlakena is endemic in western United States, occurring in Idaho, Arizona, and California.

Systematics: The five described species of *Timberlakena* are among the smallest sphecids, averaging about 2 mm in length. Within the subtribe Ammoplanina the subglobose stigma and absence of discal cells in the hindwing set the genus apart. Its nearest relatives seem to be *Ammoplanus* and *Ammoplanellus,* both of which have two discal cells in the hindwing.

The species of *Timberlakena* differ among themselves in minor details of wing venation. On this basis Pate (1939) described three subgenera: *Mohavena* for *yucaipa, Riparena* for *cahuilla,* and *Timberlakena* for the other three forms. We have studied all of these as well as two undescribed species, and it seems to us that the degree of sclerotization of R_1 beyond the stigma, of the veins bounding the submarginal cells, and of the hindwing mediocubitus is not sufficient to separate subgenera. The gradual disappearance of veins and cells seems to have occurred in an almost linear series: *cahuilla-yucaipa-ocha-hualga-nolcha.* In this series *cahuilla* has two essentially complete submarginal cells (fig. 49 A) and a fairly complete hindwing M + Cu, whereas *nolcha*

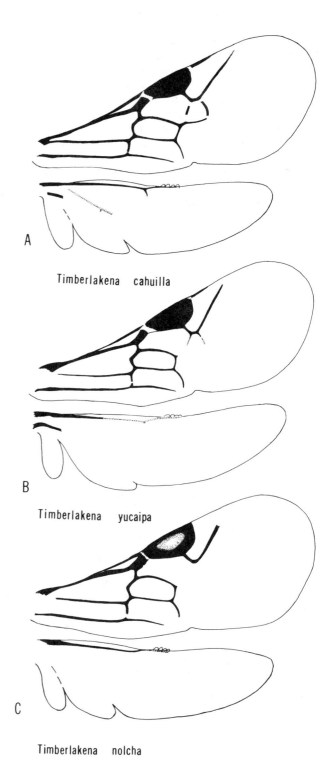

FIG. 49. Wings in the pemphredonin genus *Timberlakena.*

has only one submarginal cell which is incomplete (fig. 49 C), and the hindwing M + Cu is merely a basal stub.

Biology: Adults frequent flowers such as *Eriogonum, Acacia, Prosopis, Cleomella, Salsola, Larrea, Olneya, Lepidospartum, Eriodictyon,* and *Phacelia.*

Checklist of *Timberlakena*

cahuilla Pate, 1939; U.S.: s. California

hualga Pate, 1939; U.S.: California, new status by Bohart
nolcha Pate, 1939; U.S.: California, Arizona, Idaho
ocha Pate, 1939; U.S.: s. California, Arizona
yucaipa Pate, 1939; U.S.: s. California

SUBFAMILY ASTATINAE

Members of the subfamily range in size from about 4 to 16 mm, with *Diploplectron* the smallest and *Astata* the largest; they occur in most life zones. The Astatinae is a small group which retains many characters of the archaic sphecid stock. This does not mean that the group should be considered primitive, in fact the contrary is evident in many of their specialized features. In the two largest genera, *Astata* and *Dryudella,* the males have developed holoptic eyes and are often observed perched on some object, like a clod, pebble, or twig. From this vantage point they make swift, darting flights and generally return to the same perch; their flights are faster than the eye can follow. As far as is known, all species make nests in the soil and provision them exclusively with Hemiptera.

Diagnostic characters:

1. (a) Inner orbits entire, converging above (fig. 53); (b) ocelli normal.
2. (a) Antennae low on face, sockets contiguous with frontoclypeal suture; (b) female with 12 and male with 13 antennal articles.
3. Clypeus transverse, often very short.
4. (a) mandible without notch or angle on externoventral margin except in Dinetini, inner margin with subapical tooth; (b) palpal formula 6-4; (c) mouthparts short; (d) mandible socket open.
5. (a) Pronotum with short collar; (b) pronotal lobe and tegula separated but often very narrowly.
6. (a) Scutum with or without notauli; (b) no oblique scutal carina; (c) admedian lines well separated, area between flat or rarely grooved.
7. (a) Mesopleuron with episternal sulcus; (b) omaulus absent; (c) scrobal sulcus present except in some Dinetini.
8. Definitive metapleuron consisting of upper metapleural area only.
9. (a) Midtibia with two apical spurs (none in *Dinetus* males); (b) midcoxae narrowly separated or contiguous, with dorsolateral carina or crest; (c) hindcoxae subcontiguous and metasternum on same plane as mesosternum; (d) precoxal lobes present but very short; (e) hindfemur simple apically; (f) claw simple; (g) plantulae present.
10. (a) Propodeum moderately long; (b) dorsal enclosure U-shaped when present; (c) no propodeal sternite.
11. (a) Forewing with two or three submarginal cells, II usually receiving both recurrent veins but first sometimes received by I (fig. 51 B, E); (b) marginal cell truncate apically; (c) outer veinlet of submarginal I sometimes with R_1 stub (fig. 51 F).
12. (a) Hindwing jugal lobe subequal to length of anal area except about half its length in Dinetini (fig. 51 B, C); (b) media diverging before cu-a; (c) anal lobe with remnant of second anal vein except in Dinetini (fig. 51 B); (d) subcosta usually present.
13. (a) Gaster sessile; (b) tergum I with lateral carina (weak in *Diploplectron*; (c) male with seven visible terga; (d) pygidial plate present
14. (a) Male genitalia with volsella differentiated into digitus and cuspis; (b) cerci present

Systematics: This group has been treated variously as a subfamily or as a tribe, usually in the Larrinae. We believe that the Astatinae are distinct from the Larrinae because of the following characters that do not appear in that subfamily: two midtibial spurs (or none), cerci present, volsella with digitus and cuspis. The divergence of the hindwing media before cu-a distinguishes the Astatinae from all Larrinae except the tribe Scapheutini.

There appears to be some relationship with the Laphyragoginae. The wings of both groups are very similar (see fig. 51 A, C). Other shared characteristics are the bulging frons, small ocelli, tendency toward holoptic eyes, differentiated volsella, and similar makeup of thoracic sulci, propodeum, and gaster. However, the Laphyragoginae differ in having one midtibial spur, antennal sockets higher on the face, no cerci, and the hindwing jugal lobe equally broad in the two sexes. It is likely that the Astatinae and the Laphyragoginae diverged from the ancestral larrine stock at an early period. Both have retained more primitive characteristics than present-day Larrinae.

Inclusion of the Dinetini in the subfamily Astatinae is based on the agreement of midtibial spurs in the female, hindwing venation, cerci and genitalic structures. Yet, there are several features that could be used to justify subfamily status for the seven known species of *Dinetus*. These are moderate hindwing jugal lobe, no trace of A_2 in the hindwing, notauli absent, antennae usually rolled or twisted, antennal sockets distinctive, and only two submarginal cells in the forewing. In spite of these differences, it seems best to treat the group as a tribe in the Astatinae, especially in view of the similar habits of using Hemiptera as prey and the "perching" characteristic of the males.

The wings of astatines exhibit three unusual and primitive features: (1) a stub of r_1 in submarginal cell I of the forewing (fig. 51 E), (2) an A_2 remnant in the hindwing of Astatini (fig. 51 C), and (3) a differentiated subcostal vein in the hindwing of *Astata* (fig. 51 C), *Dryudella* and *Diploplectron*. The A_2 remnant is absent in Dinetini as well as the subcostal vein, and the latter is only weakly indicated in *Uniplectron*. Another generalized feature is the divergence of the media before cu-a in the hindwing. The divergence of the forewing media after cu-a as a rule in Astatini and usually before in Dinetini is the expected evolutionary progression. This can be said also for the reduction from three submarginal cells in the forewing (Astatini) to two such cells (Dinetini). Another interesting wing feature is the expanded jugal lobe in *Astata* and *Dryudella* males associated with holoptic eyes and presumably correlated with their "perching" activities. Finally, the large area of the forewing beyond the cellular portion is a characteristic of the Astatinae. In all cases the length of the membranous part beyond the cells is greater than any dimension of the distal submarginal cell.

Evolutionary trends in the Astatinae are enumerated on table 11 and the relationships of the genera are shown on fig. 50.

TABLE 11.
Evolutionary Trends Within the Astatinae

	Generalized	*Specialized*
1.	Without white pigment	White pigment present
2.	Compound eyes dichoptic	Compound eyes holoptic
3.	Frons descending smoothly from vertex	Frons with shelflike ridge or swellings
4.	Flagellomeres without tyli	Flagellomeres with tyli, articles twisted
5.	Scape shorter than flagellomere I	Scape longer than flagellomere I
6.	Flagellomere I longer than II	Flagellomere I as long as II
7.	Malar space very short, mandible nearly touching compound eye	Malar space long, mandible well separated from compound eye
8.	Clypeus with truncate median lobe	Clypeus with toothed median lobe
9.	Setae on labrum simple	Setae on labrum forked
10.	Mandible entire externoventrally	Mandible notched externoventrally
11.	Labial palpal segment II symmetrical	Labial palpal segment II asymmetrical
12.	Psammophore weak, on foretarsus only	Psammophore strong, and present on coxa and gena
13.	Thorax not bulbous	Thorax bulbous
14.	Scutum with notauli	Scutum without notauli
15.	Episternal sulcus complete, scrobal sulcus present	Episternal sulcus partly evanescent, scrobal sulcus absent or very weak
16.	Forewing media diverging after cu-a	Forewing media diverging before cu-a
17.	Marginal cell elongate	Marginal cell shortened
18.	Forewing with three submarginal cells	Forewing with two submarginal cells
19.	Forewing with r_1 stub	Forewing without r_1 stub
20.	Second submarginal cell longer than first	Second submarginal cell shorter than first
21.	Jugal lobe of hindwing as long as anal lobe, jugal excision shallow	Jugal lobe of hindwing short, jugal excision deep
22.	A_2 remnant and subcosta present in hindwing	A_2 remnant and subcosta absent
23.	Two midtibial spurs	No midtibial spurs
24.	Tergum III narrower than II	Tergum III as wide as II
25.	Sterna with simple apical margins	Sternal margins modified
26.	Sterna flat, without median hump	Sterna with median hump
27.	Sterna without hairbrush	Sterna with hairbrush
28.	Pygidial plate well developed	Pygidial plate weak or absent

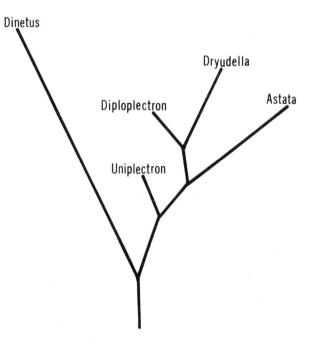

FIG. 50. Dendrogram suggesting relationships in the Astatinae.

Regional works on the Astatinae are those of W. Fox (1892b) and F. D. Parker (1962c, 1964, 1968a, 1969, 1972) for North America; F. D. Parker (1968b) for South America; Gussakovskij (1927), P. M. F. Verhoeff (1951), and Pulawski (1955, 1959) for Eurasia; and Arnold (1924) for Africa.

KEY TO TRIBES AND GENERA OF ASTATINAE

1. Forewing with two submarginal cells (fig. 51 B), vein A_2 of hindwing absent; jugal lobe of hindwing moderate, length not more than three-fourths of anal area (fig. 51 B) (Dinetini) .... *Dinetus* Panzer, p. 216
   Forewing with three submarginal cells[1] (fig. 51 D), vein A_2 of hindwing present; jugal lobe of hindwing large, nearly as long as anal area (fig. 51 C) (Astatini) 2
2. Antenna with 13 articles, abdomen with seven visible terga, males 3
   Antenna with 12 articles, abdomen with six visible terga, females 6
3. Compound eyes dichoptic (fig. 53 D) 4
   Compound eyes holoptic (fig. 53 C) 5
4. Submarginal cell II receiving both recurrent veins toward middle of cell (fig. 51 F), flagellomere V as long as VI, labial segment II asymmetrical, hindwing without a distinct jugal excision (fig. 51 F) .... *Uniplectron* F. Parker, p. 208
   Submarginal cell II receiving second recurrent vein only (first recurrent vein may be interstitial or nearly so) (fig. 51 E), flagellomere V shorter than VI, labial segment II symmetrical, hindwing with deep jugal excision (fig. 51 E) ..... *Diploplectron* W. Fox, p. 208
5. Malar space longer than midocellus diameter (fig. 53 C), submarginal cell II shorter than I on media, tergum I narrower than II .... *Dryudella* Spinola, p. 213
   Malar space short, base of mandible nearly touching lower eye margin (about as in fig. 53 B), submarginal cell II longer than I on media, tergum I as wide as II .... *Astata* Latreille, p. 211
6. Pygidial plate bordered with stout, recurved marginal spines, submarginal cell II as long as or often longer than I as measured on media (fig. 51 C) .... *Astata* Latreille, p. 211
   Pygidial plate not bordered by stout spines (fig. 51 E); submarginal cell II short, not more than 0.75 length of I (fig. 51 E) 7
7. Old World species 8
   New World species (including Hawaiian Islands) 9
8. With combination of (1) first recurrent vein interstitial or received by submarginal cell I, median lobe of clypeus truncate, and propodeal enclosure coarsely sculptured, *or* (2) first recurrent vein interstitial or received by submarginal cell I, clypeus tridentate, and propodeal enclosure granulate .... *Diploplectron* W. Fox, p. 208
   Without either of above combinations .... *Dryudella* Spinola, p. 213
9. With combination of (1) first recurrent vein interstitial or received by submarginal cell I, and median lobe of clypeus tridentate; *or* (2) first recurrent vein received toward base of submarginal cell II, clypeus bidentate, and scutum partly polished .... *Diploplectron* W. Fox, p. 208
   Without either of above combinations 10
10. Flagellomere I as long as II; metallic blue, median clypeal lobe rounded .... *Uniplectron* F. Parker, p. 208
    Flagellomere I longer than II, if metallic blue, then median clypeal lobe 3-toothed .... *Dryudella* Spinola, p. 213

Tribe Astatini

These alert and often brightly colored wasps vary in appearance from the beelike *Astata* (fig. 52) to the antlike *Diploplectron* (fig. 54). One can expect to find examples of this tribe in all major life zones, although the range of most genera is relatively restricted. Only *Astata* has numerous species in most zoogeographic regions.

Diagnostic characters:

1. Eyes often holoptic in males.
2. (a) Scape shorter than flagellomere I; (b) antennal socket rim oblong; antennae not twisted or rolled; (c) frons with median sulcus, at least below.
3. (a) Mandible entire externoventrally; (b) some palpal segments asymmetrical; (c) cardo short and stout; (d) malar space variable.
4. (a) Pronotal lobe almost touching tegula; (b) psammophore short, not present on forecoxa or gena.
5. (a) Scrobal sulcus well defined; (b) episternal sulcus complete to forecoxal cavity (fig. 56); (c) notauli present anteriorly or practically absent.
6. Two midtibial spurs in both sexes.
7. (a) Forewing with three submarginal cells; (b) veinlet forming outer side of submarginal cell I angulate and appendiculate with r_1 (fig. 51 C); (c) forewing media diverging beyond cu-a; (d) jugal lobe of hindwing nearly as long as anal area; (e) vein A_2 of hindwing present basally (fig. 51 C).

Systematics: The tribe is composed of four genera and contains 137 described species. It is rather remotely related to the Dinetini, and they are easily separated by characters given in the key.

The wing venation is rather consistent among the genera except for *Dryudella*. In this genus the first recurrent vein ends either in submarginal cell I or II, or is interstitial. This leads to confusion with a small percentage of *Diploplectron* females where clypeal, head, and pronotal characters must be considered. The development of the expanded jugal lobe of the male hindwing is peculiar to *Astata* and *Dryudella*, as is the holoptic condition in males.

Most species are quite distinctive and are widely distributed. The subspecies concept is widely used for the

---

[1] Occasional mutant forms may have fewer than three cells, in which case the presence of A_2 in the hindwing is diagnostic.

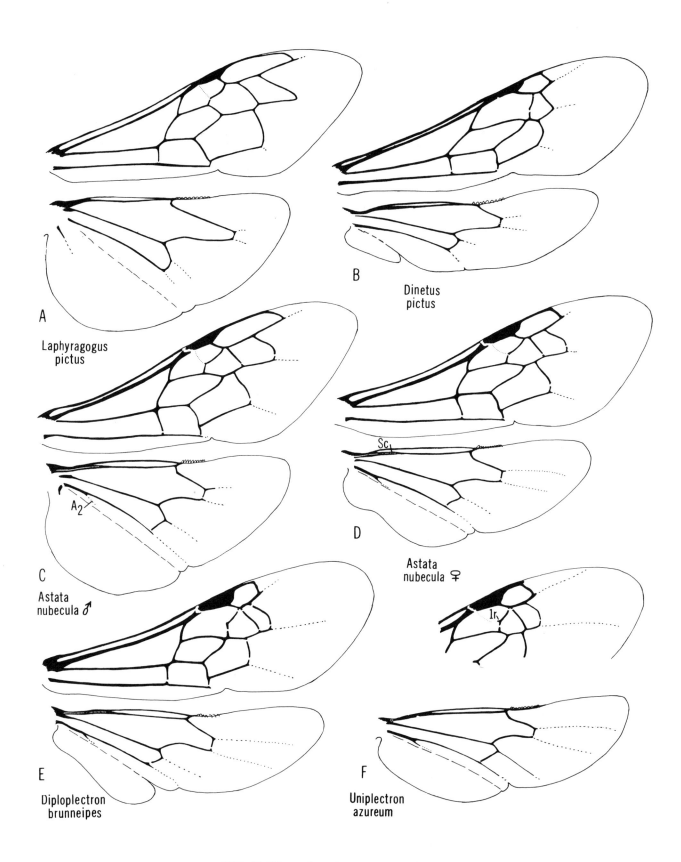

FIG. 51. Wings of Laphyragoginae and Astatinae.

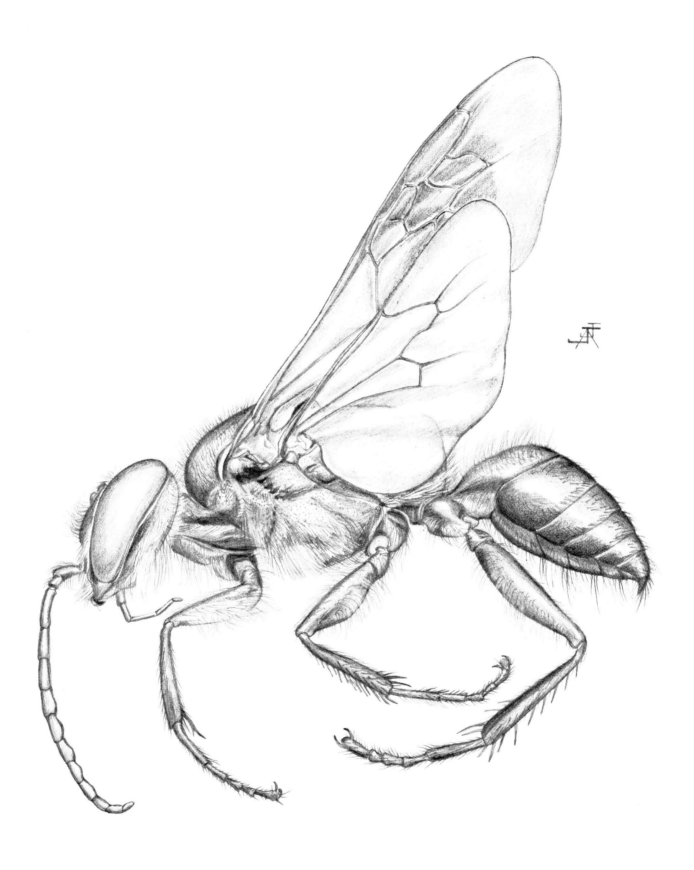

FIG. 52. *Astata occidentalis* Cresson, male.

Old World fauna, and names appear to be applied more to nationalistic boundaries rather than to geographic ones, (examples: *Astata m. miegii* Dufour—France, *m. scapularis* Kohl—Italy, *m. escalerai* Giner Marí—Morocco) On the other hand, several subspecies are described from the same country (*Astata f. fuscistigma* Cameron, *f. albopilosella* Cameron and *f. nana* Arnold are all from South Africa). Perhaps when the Old World fauna is revised, many of the existing subspecific names will be synonymized or, more likely, raised to specific rank.

Biology: So far as is known, the Astatini are ground nesting and provision nests with adults and nymphs of the heteropteran families Lygaeidae, Pentatomidae, Scutelleridae, Cydnidae, and Reduviidae. The nests are made in a variety of soil types and are either single or multicellular.

Genus Uniplectron F. Parker

Generic diagnosis: Second labial palpal segment asymmetrical, setae on labrum simple, clypeus with rounded truncate median lobe, flagellomere I as long as II, length of malar space less than midocellus, eyes dichoptic (fig. 53), body without white pigmentation, thorax relatively long, marginal cell of forewing short, both recurrent veins ending at submarginal cell II which is shorter than I, jugal lobe of hindwing not broadened in male nor excised, gastral segments not unusually modified, female pygidial plate weak and without strong setae.

Geographic range: Known only from Mexico.

Systematics: The single species shares characters with both *Dryudella* and *Diploplectron,* the wings like *Dryudella* and body otherwise like *Diploplectron*. The generic characters of *Uniplectron* most nearly approach those of the subfamily archetype (Table 11) resulting in a comparative value of 6. This suggests a relict status.

Biology: The type series was collected in an arid tropical habitat where the wasps were crawling among the thorny branches of an *Acacia* (F. Parker, 1966a).

Checklist of *Uniplectron*

azureum F. Parker, 1966; Mexico

Genus Diploplectron W. Fox

Generic diagnosis: Second labial palpal segment symmetrical, setae on labrum simple, malar space at least as long as midocellus, eyes dichoptic, flagellomeres with or without tyli; yellow to white pigments often present on face and thorax of male; coxae normal, not flanged; notum flattened, not bulbous; notauli often evanescent, dorsal surface of pronotum curved posterad in profile, especially in male; marginal cell of forewing reduced, not longer than stigma; submarginal cell I widest, first recurrent vein received by submarginal cell I or interstitial or nearly so; submarginal cell II shorter than I, rarely absent in mutant specimens; hindwing jugal and anal excisions deep; gastral segments not unusually modified, sterna not emarginate nor with hairbrushes in male; female pygidial plate weak, without strong spines.

Geographic range: The zoogeographic regions in which the genus is represented and the number of species listed for each are: Nearctic — 13, Neotropical — 2, Palearctic — 2, and Ethiopian — 1.

Systematics: Diploplectron is related to *Dryudella,* and this relationship is apparent when females of the *Dryudella tricolor* group are compared with those of *Diploplectron*. The males, however, are not at all alike since in *Dryudella* they have holoptic eyes and an enlarged jugal lobe, whereas *Diploplectron* males lack these structures.

The Nearctic species can be split into five groups based on the body color pattern and configuration of the male clypeus. In the *kantsi, viereckii,* and *orizabense* groups the gaster is red, and the male clypeus is produced into a pointed or narrowly truncate snout. In the *brunneipes* and *beccum* groups the abdomen is usually black (red in one form), and the male clypeus is bifurcate. The clypeus is bidentate in all Nearctic females.

The three Old World species can be separated from the New World forms by the presence of a psammophore on the hind leg of the former group. This brush is found in other genera of the New World, but not in *Diploplectron*. *D. kriegeri* Brauns is a peculiar species in that the integument is rippled by close, thin ridges. No other known species even approaches this type of sculpture, and *kriegeri* seems to be a relic. The other two species, *palearcticum* Pulawski and *asiaticum* Pulawski, are quite small (4 mm) and nearly impunctate. They are easily separated by clypeal characters. The males of these two species have the ocelli lower on the frons than other species groups.

A male we have seen of *palearcticum* Pulawski has one forewing with one submarginal cell and the other with one submarginal cell and part of a second cell. Pulawski's drawing of the holotype shows three submarginal cells. The placement of the recurrent veins in the forewing is another variable character, not only among individuals, but among species. The first two submarginal cells each receive a recurrent vein in most *Diploplectron*. However, a Mexican species, *D. neotropicum* Parker, has both recurrent veins ending at submarginal cell II or interstitial.

Krombein (1939) gave a key to *Diploplectron* and F. Parker (1972) has since revised this genus. Pulawski (1965a) described the third Palearctic form and listed differences among species known from that region.

Diploplectron has a value of 9 when its characters are compared with those of the archetype, indicating that it is among the less specialized genera of the Astatini (fig. 54). Pulawski (1965a) suggested that the genus may be highly evolved and that the dichoptic eyes of *Diploplectron* and *Uniplectron* may have been secondarily derived from a holoptic ancestor. However, it seems more logical to assume that the archetype and these genera have always been dichoptic, rather than to speculate that the holoptic condition was developed, then reversed. The Holoptic condition and the broad jugal lobe appear to be secondary modifications relating to the perching and mating activities of the males.

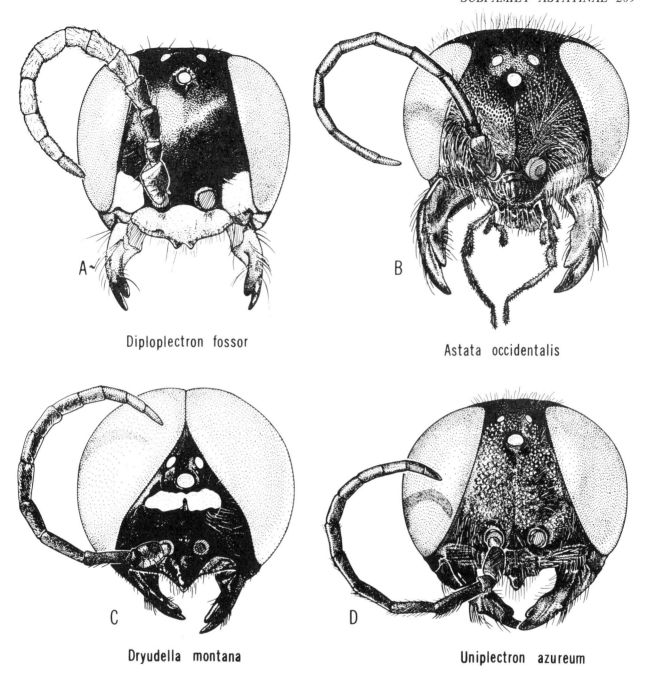

FIG. 53. Facial portraits in the tribe Astatini; A,C,D, males; B, female.

The name *Diploplectron* stems from two words: *diplos* = double and *plectron* = a tool for plucking a stringed instrument. The name is neuter and probably refers to the forked clypeal lobe of the type species.

Biology: F. X. Williams (1946) gave observations of nesting in two unidentified species. One of these (probably *peglowi*) used small, black lygaeid nymphs, *Sphagisticus nebrilosus* (Fallén), as prey, and the other, *californicum*, used lygaeids, *Rhyparochromus californicus* Van Duzee, and *Emblethis vicarius* Horvath. Parker (1972) reported *peglowi* nesting in a sandbank at Davis, California and provisioning its nests with a coreid, *Aufeius impressicollis* Stål. Kurczewski (1972) has given the most detailed account of the biology of *peglowi*. He found that nests contained three or four cells and the entrance was left open during the hunt for prey. The cells were stocked with three to six lygaeid bugs, especially *Sphaerobius insignis* (Uhler), but also *Lygaeus* sp. Prey were usually flown to the nest. The bugs were transported venter to venter and were grasped by the wasps legs as well as her mandibles. The egg is laid longitudinally on the thoracic venter of one of the prey so that it lies between the coxae.

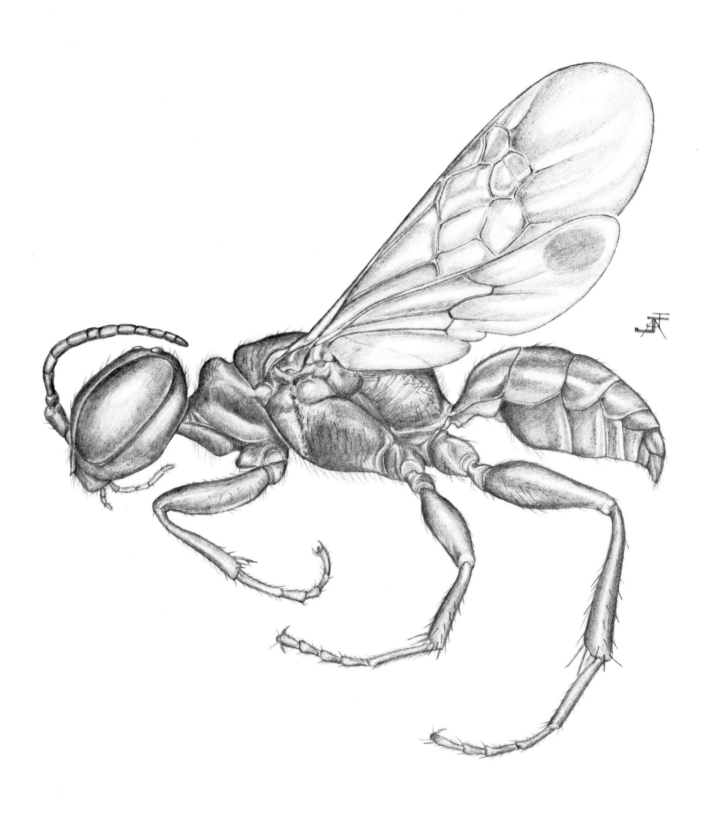

FIG. 54. *Diploplectron brunneipes* (Cresson), male.

Chrysidids of the genus *Hedychridium* were observed inspecting *peglowi* nests by Kurczewski.

In 1964 Bohart and Parker discovered a small colony of *fossor* on Mt. Rose, Nevada at an elevation of 8,000 feet. The site was a sloping, gravelly area with small clumps or mats of grass. In one nest they found the wasp and its prey which were lygaeid nymphs. A female of *vierecki* from New Mexico was captured and pinned with a nymphal cydnid bug, *Microporus obliquus* Uhler.

Because of their small size and inconspicuous habits, *Diploplectron* are rarely collected. Yet, by concentrating attention on mats of *Euphorbia* and Bermuda grass we have been able to amass over 100 in a day. Other likely places to search are among stems of low-growing weeds and under clumps of *Verbena*.

Checklist of *Diploplectron*

asiaticum Pulawski, 1965; sw. USSR: Turkmen S.S.R.; Mongolia
beccum F. Parker, 1972; w. U.S.
brunneipes (Cresson), 1881 (*Liris*); w. U.S.
    bidentatum Ashmead, 1899
    foxii Ashmead, 1899
    bidentatiforme Rohwer, 1909
californicum F. Parker, 1972; w. U.S.
diablense Williams, 1951; U.S.: California
ferrugineum Ashmead, 1899; w. U.S.
    ashmeadi Rohwer, 1909
    cressoni Rohwer, 1909
    relativum Rohwer, 1909
fossor Rohwer, 1909; w. N. America: nw. Canada to s. Calif.
    rufoantennatum Rohwer, 1909
irwini F. Parker, 1972; U.S.: Arizona
kantsi Pate, 1941; sw. U.S., Mexico
kriegeri Brauns, 1898; s. Africa
neotropicum F. Parker, 1972; Mexico
orizabense F. Parker, 1972; s. Mexico
palearcticum Pulawski, 1958; Egypt
peglowi Krombein, 1939; boreal N. America, Mexico: Baja California
reticulatum Williams, 1946; sw. U.S.
secoense F. Parker, 1972; U.S.: California
sierrense F. Parker, 1972; U.S.: California, Nevada
vierecki Pate, 1941; sw. U.S.

Genus Astata Latreille

Generic diagnosis: Second labial palpal segment asymmetrical, setae on labrum forked, malar space less than midocellus diameter, frons without shelflike ridge, male flagellomeres with tyli, thorax bulbous, marginal cell of forewing long, second submarginal cell II as long as I, recurrent veins ending at submarginal cell II, hindwing jugal lobe enlarged in males and not excised, coxae often flanged or flattened, gastral segment II as wide as I in males, sterna modified or simple, but often with hairbrush in males, pygidial plate with strong setae in females.

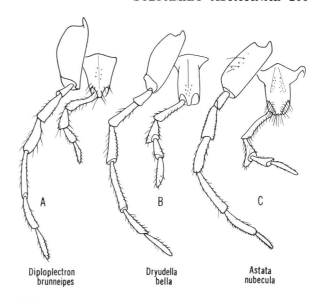

FIG. 55. Mouthparts of the Astatinae; left, maxillary palpus; right, labial palpus.

Geographic range: Astata is represented on all continents except Australia. About 30 of the 76 species occur in the Palearctic Region. The remainder are about equally divided among the Nearctic, Oriental, Ethiopian, and Neotropical Regions.

Systematics: In the past Astata has generally been lumped with *Dryudella*, but F. Parker (1962c) recently distinguished the two. *Astata* is easily separated from *Dryudella* as indicated in the key.

Most species are separated from one another by the shape of the clypeus and differences in form of the eyes and antennae, but some females can be distinguished only by punctation. In most species the length of the marginal cell is a sex-linked character — long in males and short in females (see fig. 51 D). In two species, *australasiae* and *nevadica*, the marginal cell is quite short.

The Nearctic and Neotropical species have been revised by F. Parker (1962c, 1964, 1968a,b). Kohl (1885a), Gussakovskij (1927), and Pulawski (1955, 1959) have published keys to most of the Palearctic species. Arnold (1924) keyed the Ethiopian ones, but only a partial key exists for the Oriental forms (Bingham, 1897).

Biology: Evans (1957d) made extensive biological studies on three species of Nearctic *Astata* and summarized the existing literature. Powell and Burdick (1960) observed the nesting habits of *occidentalis* in California and compared their observations with those of Evans. Other noteworthy biological references are those of Ferton (1901, 1908), Tsuneki (1947a, 1969b), and Grandi (1961).

Male *Astata* are often observed perched on some object, such as a clod, stone, or twig, whereas the females are generally associated with nesting activities. We are not able to explain the male's behavior, but it may have something to do with mating habits. In flight they are swifter than the eye can follow, which seriously handicaps observation. Their holoptic eyes and broad hindwing jugal lobe apparently are correlated with their "perching" and rapid flight, respectively, and an investigation of these habits would make an excellent evolutionary study.

More data are available concerning the female's activities. Although they are speedy fliers, they are no match for the males. Their nests are found in a variety of situations: beneath overhanging vegetation; in bare, open, hard packed soil; or on sandy slopes. Evans (1957d) has summarized the seven important aspects of *Astata* nesting behavior as follows: (1) The nest is complex and multicellular with smooth-walled cells, which are often in short series separated by earthen partitions. (2) Prey accumulates in the burrow, and then a cell is prepared to receive it. (3) Prey is completely paralyzed. (4) Provisions are adult and nymphal bugs, especially Pentatomidae. (5) The prey are packed venter-down in the cell with the egg beneath the bottom bug; the larva is in an inverted position during feeding. (6) The egg is attached to the bug's prosternum and extends backward along the midventer. (7) Prey is carried venter-up and held in the mandibles during flight. Points 1 and 5 are distinctive, 2 is reminiscent of the Philanthinae, and the others are found in several other subfamilies.

Checklist of *Astata*

albovillosa Cameron, 1890; Mexico, Guatemala
alpaca F. Parker, 1968; Peru
alpestris Cameron, 1890; Mexico
affinis Vander Linden, 1829; Spain to Italy; ne. Africa
    ssp. *ariadne* Pulawski, 1959; Crete
    ssp. *jerichoensis* Pulawski, 1957; e. Mediterranean region
    ssp. *radoszkowskii* Pulawski, 1957; sw. USSR
aschabadensis Radoszkowski, 1893; sw. USSR: Turkmen S.S.R.
australasiae Shuckard, 1838; Chile, Peru (New Holland is erroneous)
    chilensis Saussure, 1854
    dimidiata Taschenberg, 1880
    dispar Reed, 1894
apostata Mercet, 1910; Spain
bakeri F. Parker, 1962; w. U.S.; Mexico
bechteli F. Parker, 1962; w. U.S.
bicolor Say, 1823; e. U.S.; Mexico
    terminata Cresson, 1872
    pygidialis W. Fox, 1892 (*Astatus*)
bigeloviae Cockerell and W. Fox, 1897 (*Astatus*); sw. U.S.
blanda Saussure, 1892; Madagascar
boharti F. Parker, 1962; sw. U.S.
bola F. Parker, 1968; Argentina
boops (Schrank), 1781 (*Sphex*); Europe, w. Asia, n. Africa, India
    abdominalis Panzer, 1798 (*Tiphia*)
    pompiliformis Panzer, 1804 (*Larra*)
    oculata Jurine, 1807 (*Dimorpha*)
    victor Curtis, 1829
    agilis F. Smith, 1875
    sicula Kohl, 1884 (*Astatus*)
    cobosi Giner Marí, 1946
    ssp. *picea* A. Costa, 1867; Sardinia, Corsica
    carbonaria Kohl, 1885 (*Astatus*)
    ssp. *canariensis* Pulawski, 1974; Canary Is.
brevitarsis Pulawski, 1958; Hungary
cleopatra Pulawski, 1959; Egypt
clypeata F. Parker, 1962; centr. U.S.: east slope Rocky Mts.; Mexico
compta Nurse, 1909; w. India
    absoluta Nurse, 1909
cosquin F. Parker, 1968; Argentina
costae A. Costa, 1867; s. Europe, n. Africa
    costae Piccioli, 1869
    gracilis Gussakovskij, 1927 (*Astatus*)
    ssp. *parvula* Gussakovskij, 1927 (*Astatus*); sw. USSR: Kazakh S.S.R.
diabolica Balthasar, 1957; Afghanistan
diversipes Pulawski, 1955; Turkey
dominica Pate, 1948; Hispaniola
enslini Maidl, 1926; Sudan
femorata F. Parker, 1963; U.S.; Arizona
flavipennis (Turner), 1917 (*Dimorpha*); Kenya
fumipennis E. Saunders, 1910; Algeria
fuscistigma Cameron, 1905; s. Africa
    ssp. *albopilosella* Cameron, 1910; s. Africa
    ssp. *nana* Arnold, 1946; s. Africa
gallica Beaumont, 1942; France, Morocco
gigas Taschenberg, 1870; Mexico to Brazil
    strigosa Kohl, 1888 (*Astatus*)
glabra Kohl, 1892 (*Astatus*); Brazil
gracilicornis Arnold, 1924; Rhodesia
graeca Beaumont, 1965; Greece, Bulgaria, Turkey, Syria, Israel
histrio Lepeletier, 1845; Algeria
jucunda Pulawski, 1959; Bulgaria, Hungary
kashmirensis Nurse, 1909; nw. India
    stecki Beaumont, 1942
    ssp. *melanotica* Pulawski, 1973; Israel
laeta E. Saunders, 1910; Algeria
leila Pulawski, 1967; Turkey
leuthstromi Ashmead, 1897; N. America
lubricata Nurse, 1903; Egypt, w. India
    eremita Pulawski, 1959
lucinda Nurse, 1904, w. India
lugens Taschenberg, 1870; Brazil
lusitanica Pulawski, 1974; Portugal
maculata Radoszkowski, 1877; sw. USSR: Turkmen S.S.R.
melanaria Cameron, 1905; s. Africa
miegii Dufour, 1861; Iberian Penin., s. France
    provincialis Richards, 1928
    ssp. *scapularis* Kohl, 1889 (*Astatus*); e. Mediterranean region, sw. USSR
    pelops Morice, 1902 (*Astatus*)
    ssp. *escalerai* Giner Marí, 1946; Morocco, Algeria
mexicana Cresson, 1881; sw. U.S., Mexico
minax Arnold, 1945; Madagascar
minor Kohl, 1885 (*Astatus*); Europe, n. Africa, sw. USSR

?*vanderlindenii* Robert, 1833
moralesi Giner Marí, 1945; Spanish Sahara
nevadica Cresson, 1881; w. U.S., Mexico
nigra F. Smith, 1856; Algeria
    unicolor Lepeletier, 1845, nec Say, 1824
nigricans Cameron, 1889; India
nubecula Cresson, 1865; N. America
    nigropilosa Cresson, 1881
obscura Arnold, 1932; Rhodesia
occidentalis Cresson, 1881; N. America
    apicipennis Cameron, 1890
    tinctipennis Cameron, 1890
    sayi W. Fox, 1893 (*Astatus*)
pontica Pulawski, 1958; Bulgaria
quettae Nurse, 1903; Hungary to Mongolia, India
    fletcheri (Turner), 1917 (*Dimorpha*)
    hirsutula Gussakovskij, 1927 (*Astatus*)
    tibialis Gussakovskij, 1927 (*Astatus*)
    hungarica Pulawski, 1958
radialis E. Saunders, 1910; Algeria
resoluta Nurse, 1909; Pakistan
rufipes Mocsáry, 1883; s. Europe, sw. USSR, w. Siberia, n. Africa
    massiliensis Morice, 1902 (*Astatus*)
    sareptana Gussakovskij, 1927 (*Astatus*)
    ssp. *echingol* Tsuneki, 1971; Mongolia
ruficaudata (Turner), 1917 (*Dimorpha*); Malawi
    ssp. *hova* Arnold, 1945; Madagascar
rufitarsis F. Smith, 1856; s. Africa
rufofemorata Arnold, 1945; Madagascar
rugifrons Arnold, 1946; Zambia
sabulosa Gussakovskij, 1927 (*Astatus*); sw. USSR
selecta Nurse, 1909; w. India
spinolae Saussure, 1854; Chile
stevensoni Arnold, 1924; Rhodesia
tarda Cameron, 1896; Sri Lanka
trochanterica Beaumont, 1953; Morocco
tropicalis Arnold, 1924; Zambia
unicolor Say, 1824; N. America
    insularis Cresson, 1865
    rufiventris Cresson, 1872
vaquero F. Parker, 1967; Argentina
williamsi F. Parker, 1962; w. U.S.

Genus Dryudella Spinola

Generic diagnosis: Second labial palpal segment variable, setae on labrum simple, malar space longer than midocellus, frons with a shelflike ridge, eyes of male holoptic, thorax bulbous, pronotum in profile flat and abruptly declivous posteriorly, marginal cell of forewing short, submarginal cell II shorter than I, recurrent veins ending at submarginal cells I and II or both at II, hindwing jugal lobe enlarged in males and not excised, gastral segments not unusually modified, female pygidial plate variable but without strong setae.

Geographic range: This genus is mostly Holarctic, but some species extend into the Ethiopian and Neotropical Regions. The 42 species are distributed as follows: Nearctic – 9, Palearctic – 29, Ethiopian – 1, Oriental – 2. One additional species (*pinguis*) is Holarctic in its distribution.

Systematics: The relationship of several *Dryudella* species groups to other astatine genera is easily seen when specimens of each are compared (fig. 56). The Holarctic *tricolor* group shares many characters with *Diploplectron*, and the females are very hard to separate. The *stigma* group, also Holarctic, is like some *Astata,* and the metallic blue *caerulea* group closely resembles *Uniplectron.*

The structural variation among species of *Dryudella* is surprisingly great. For instance the median lobe of the female clypeus may be undeveloped, truncate, rounded, subserrate, bilobed or trilobed. The first recurrent vein usually terminates at submarginal cell II, but in some species of the *stigma* group and most of the *tricolor* group, the first recurrent is interstitial or terminates at submarginal cell I. In some females (*caerulea*) the anal lobe of the hindwing is enlarged but with the anal and jugal excisions weak. *D. caerulea* is peculiar also in having a basally directed spur on the hindwing radial sector. The propodeal enclosure is always granulate but may have some longitudinal striae. Females in which venation approaches that of *Diploplectron* can be distinguished by the weak curvature of the postocular area as seen in dorsal view of the head. These forms also have the pronotal profile strongly flattened.

When the characters of *Dryudella* are rated against the archetype (table 11) a value of 13 is obtained. This indicates that among the astatine genera, *Dryudella* is the most specialized.

The Nearctic species have been revised by F. Parker (1969). Palearctic species have been treated by Pulawski (1955, 1959, 1967), who recorded 17 species, but there are 13 additional names in the literature. P. Verhoeff's (1951) paper has valuable data on types. The Ethiopian fauna is limited, with only a single species known.

Biology: Life history studies are lacking for most *Dryudella*. However, our knowledge of some species has been enhanced by the works of several authors: Ferton (1901, 1908) recorded the prey of the European *tricolor* and *stigma*. F. Williams (1946) presented unequaled observations after rearing successive generations of *immigrans* in glass jars. P. Verhoeff (1951) cited prey records for three European species, *stigma, pinguis* and *freygessneri*. Evans (1963a) described a nest and prey of *montana*. F. Parker (1969) listed new prey records for the following species: *rhimpa, pernix, picta,* and *montana.* The following hemipteran families have been recorded as prey of *Dryudella*: Pentatomidae, Scutelleridae, Lygaeidae, Reduviidae, Cydnidae, Alydidae, and Rhopalidae.

Checklist of *Dryudella*

amenartais (Pulawski), 1959 (*Astata*); Egypt, Iraq
ammochtona (Pulawski), 1959 (*Astata*); Egypt
anatolica (Pulawski), 1967 (*Astata*); Turkey
aquitana (Pulawski), 1970 (*Astata*); France

214 SPHECID WASPS

FIG. 56. *Dryudella bella* (Cresson), male.

beaumonti (Pulawski), 1959 (*Astata*); n. Africa
bella (Cresson), 1881 (*Astata*); w. U.S.; Mexico
bifasciata (Schulthess), 1926 (*Astata*); Morocco to Egypt
caerulea (Cresson), 1881 (*Astata*); w. U.S.; Mexico
elbae (Pulawski), 1959 (*Astata*); Egypt
elegans (Cresson), 1881 (*Astata*); w. N. America
erythrosoma (Pulawski), 1959 (*Astata*); Morocco, Spain
eurygnatha (Pulawski), 1967 (*Astata*); Turkey
femoralis (Mocsáry), 1877 (*Astata*); centr. Europe
flavoundata (Arnold), 1923 (*Astata*); s. Africa

africana Arnold, 1923 (*Diploplectron*)
    quadrisignata Arnold, 1924 (*Astata*)
freygessneri (Carl), 1920 (*Astata*); centr. Europe
frontalis (Radoszkowski), 1877 (*Astata*); sw. USSR: Kazakh S.S.R.
immigrans (Williams), 1946 (*Astata*); Hawaii, w. U.S., Mexico
kaszabi (Tsuneki), 1971 (*Astata*); Mongolia
maculifrons (Cameron), 1889 (*Astata*); n. India
maroccana (Giner Marí), 1946 (*Astata*); Morocco
    ?*melanica* Giner Marí, 1946 (*Astata*)

millsi Cockerell, 1914; U.S.: Colorado
modesta Mocsáry, 1879; Hungary
mongolica (Tsuneki), 1971 (*Astata*); Mongolia
montana (Cresson), 1881 (*Astata*); w. U.S.
    florissantensis Rohwer, 1909 (*Diploplectron*)
nephertiti (Pulawski), 1959 (*Astata*); Egypt
nuristanica (Balthasar), 1957 (*Astata*); Afghanistan
obo (Tsuneki), 1971 (*Astata*); Mongolia
opaca (Pulawski), 1959 (*Astata*); Egypt
orientalis (F. Smith), 1856 (*Astata*); India
    interstitialis Cameron, 1907 (*Astata*)
osiriaca (Pulawski), 1959 (*Astata*); Egypt
pernix F. Parker, 1969; s. U.S.
picta (Kohl), 1888 (*Astatus*); Mexico, w. N. America
    kohli Cameron, 1890 (*Astata*)
    asper W. Fox, 1893 (*Astatus*)
    aspera Dalla Torre, 1897 (*Astata*)
    asperiformis Rohwer, 1909 (*Astata*)
picticornis (Gussakovskij), 1927 (*Astatus*); sw. USSR:
  Moldavian S.S.R.
pinguis (Dahlbom), 1832 (*Larra*); Europe; USSR; U.S.:
  Alaska, Colorado
    pinguis Zetterstedt, 1838 (*Larra*); nec Dahlbom, 1832
      (*Larra*)
    jaculator F. Smith, 1845 (*Astata*)
rhimpa F. Parker, 1969; w. U.S., Mexico
quadripunctata (Radoszkowski), 1877 (*Astata*); sw.
  USSR: Kazakh S.S.R.
sepulchralis (Beaumont), 1968 (*Astata*); Canary Is.
similis (Gussakovskij), 1927 (*Astatus*); Mongolia
stigma (Panzer), 1806-1809 (*Dimorpha*); Europe to
  Siberia
    intermedia Dahlbom, 1843 (*Astata*)
    major Grönblom, 1946 (*Astata*)
  ssp. *bajanica* (Tsuneki), 1971 (*Astata*); Mongolia
tegularis (F. Morawitz), 1889 (*Astatus*); Mongolia
tricolor (Vander Linden), 1829 (*Astata*); Europe, w. Asia
    ghilianii Spinola, 1843
    cincta Spinola, 1843 (*Dimorpha*)
    spinolae A. Costa, 1867
    dimidiata A. Costa, 1867
    emeryana A. Costa, 1867
    lineata Mocsáry, 1879
    calopteryx Gussakovskij, 1952 (*Astatus*)
    monticola Giner Marí, 1945 (*Astata*)
  ssp. *monochroma* (Pulawski), 1973 (*Astata*); Israel,
    Syria
xanthocera (Pulawski), 1961 (*Astata*); Lebanon

Tribe Dinetini

The seven species of the monogeneric tribe Dinetini are small, colorful wasps with a larroid appearance. The peculiar twisted and rolled antennae of the males are a striking detail. The group is entirely Old World and restricted to the Palearctic of Africa and Eurasia.

Diagnostic characters:

1. Eyes dichoptic.
2. (a) Scape longer than flagellomere I; (b) rim of antennal socket commashaped with a basal projection (fig. 57 A); (c) antenna rolled or twisted toward apex; (d) frons with median sulcus in lower part.
3. (a) Mandible notched externoventrally; (b) palpal segments essentially symmetrical; (c) cardo long and slender; (d) no malar space.
4. (a) Pronotal lobe well separated from tegula; (b) psammophore extending from gena to base of forecoxa to end of foretarsus.

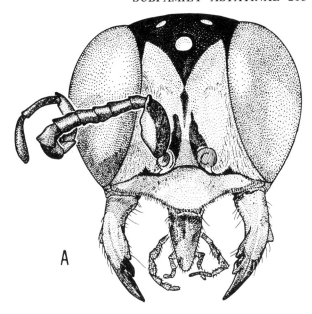
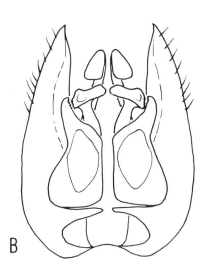

Dinetus pictus

FIG. 57. *Dinetus pictus* male; A, facial portrait; B, genitalia.

5. (a) Scrobal sulcus weak or evanescent; (b) episternal sulcus ending opposite forecoxal cavity, but not turning forward; (c) notauli absent.
6. Female with two midtibial spurs, male with none.
7. (a) Forewing with two submarginal cells; (b) veinlet forming outer side of submarginal cell I not angulate or appendiculate; (c) forewing media diverging close to cu-a but sometimes on either side of it; (d) hindwing jugal lobe 0.5 to 0.6 as long as anal area; (e) vein A_2 of hindwing absent.

Systematics: Kohl (1896) and Beaumont (1960c) placed this tribe in the subfamily Larrinae because of the following characters: notched mandibles, the convergence of the eyes towards the vertex, and the appendiculate marginal cell. Although phylogenetically important, these characters by themselves are insufficient evidence for including the Dinetini in the Larrinae. We believe the tribe should be placed with the Astatinae because of the following shared characteristics: females with two midtibial spurs, volsella composed of digitus and cuspis, cerci present, media of hindwing diverging before cu-a, and midcoxae subcontiguous. The first three of these characters differentiate from Larrinae as we have limited the subfamily. The last two characters are rarely found in larrines. Bolstering the relationship with astatines are the habits in *Dinetus* of provisioning with bugs and the "perching" of the males.

On the other hand, the relationship of *Dinetus* with other Astatinae is not close. Probably, *Dinetus* is a relict genus which split early from astatine stock. It has become rather specialized as evidenced by the notched mandibles, long bulbous scape, rolled or twisted flagellum, elimination of some thoracic sulci, reduced wing venation, and shorter hindwing jugal lobe.

Genus Dinetus Panzer

Generic diagnosis: As given for tribe. See fig. 57 A for a front view of the head of a representative species; male genitalia, fig. 57 B.

Geographic range: Palearctic Region. Eight species are known.

Systematics: Beaumont (1960c) revised the genus and recognized seven species distributed in three species groups. All except *pictus* are uncommon in collections. We have seen examples of *pictus, cereolus,* and *venustus*.

Biology: Beaumont (1960c) has summarized the observations of previous workers, particularly Ferton (1914) and Grandi (1931, 1961). Ethological data have been recorded only for *pictus*.

As in *Astata,* male *Dinetus* perch on pebbles, clods, or twigs. However, *Dinetus* perch close to the nesting site and copulate with the females at every opportunity.

A variety of nesting sites is utilized, on slopes, river banks, and bare, open forest floors; the soil type is usually sandy. The burrow is crooked and varies in depth from 6-10 cm, ending in a series of cells. Unlike the Astatini, these wasps make pellets of soil which they pick up in their mandibles and remove from the burrow by flying backwards out of the entrance and dropping the cargo a few feet away. Whether the wasp is at home or away, the nest entrance is always closed.

As in the Astatini, nests are provisioned with Heteroptera. In the case of *pictus,* at least, the preferred prey are nymphs of Nabidae: *Nabus myrmicoides* Costa and *N. rugosus* Linnaeus. Nymphs of the lygaeid, *Aphanus pineti* Herrich-Schaefer, are provisioned, also. The bugs are placed venter-up (contrary to *Astata*) four to seven per cell, with the egg glued on the sternum of the bottom bug. Obviously, the habits of species other than *pictus* need to be known before a worthwhile comparison can be made with other astatines.

Checklist of *Dinetus*

arenarius Kazenas, 1973; sw. USSR: Kazakh S.S.R.
cereolus Morice, 1897; Egypt
   ssp. *politus* Turner, 1917; ne. India
dentipes E. Saunders, 1910; Algeria, Tunisia, Egypt
   gracilis Giner Marí, 1945
nabataeus Beaumont, 1960; Egypt
pictus (Fabricius), 1793 (*Crabro*); Europe
   guttatus Fabricius, 1793 (*Sphex*); nec Gmelin, 1790
   ceraunius Rossi, 1794 (*Crabro*)
simplicipes E. Saunders, 1910; Algeria
   perezi Ferton, 1914
pulawskii Beaumont, 1960; Egypt
venustus Beaumont, 1956; Morocco

SUBFAMILY LAPHYRAGOGINAE

This monogeneric subfamily contains a few species that are distributed across North Africa and eastward in southern Asia. *Laphyragogus* species resemble the Philanthinae because of their red and yellow banded abdomens, but the true affinities of the group appear to be with the Astatinae on one hand and the Larrinae on the other. Apparently, the ethology of *Laphyragogus* is unknown.

Diagnostic characters (based largely on *L. pictus,* the only species available for study):

1. (a) Inner orbits broadly emarginate and converging above (fig. 58 D); (b) ocelli normal.
2. (a) Antennae rather low on face but sockets not contiguous with frontoclypeal suture; (b) female with 12 and male with 13 antennal articles.
3. Clypeus transverse but not narrowly so.
4. (a) Mandible with a large angular tooth on externoventral margin (fig. 58 C), and one or two subbasal teeth on inner margin; (b) palpal formula 6:4; (c) mouthparts short, but cardo elongate and broad, the stipes short, broad, triangular; (d) mandible socket open.
5. (a) Pronotum with a low collar; (b) pronotal lobe separated from tegula.
6. (a) Scutum without notauli (sometimes faintly visible); (b) no oblique scutal carina.
7. (a) Mesopleuron with episternal sulcus; (b) omaulus absent.
8. Definitive metapleuron consisting of upper metapleural area only.
9. (a) Midtibia with one apical spur; (b) midcoxae contiguous and with dorsolateral carina; (c) hindcoxae separated and metasternum elevated, not on same plane as mesosternal area; (d) precoxal lobes present but very short; (e) hindfemur simple apically; (f) claw simple; (g) plantulae absent.
10. (a) Propodeum long; (b) dorsal enclosure rather weakly defined, triangular, the apex extending onto vertical posterior face; (c) no propodeal sternite.
11. (a) Forewing with three submarginal cells and two recurrent veins which end at second cell; (b) marginal cell obliquely truncate, appendiculate (fig. 51 A).
12. (a) Hindwing with large, broad jugal lobe (nearly equal to length of anal area), base of lobe with remnant of an anal vein (fig. 51 A); (b) media diverging before cu-a.
13. (a) Gaster sessile; (b) tergum I with lateral carina; (c) male with seven visible terga, sternum VII present, longer than preceding sterna; (d) pygidial plate present.
14. (a) Male genitalia with volsella differentiated into a digitus and cuspis; (b) cerci absent.

Systematics: Laphyrogogus has been regarded as an isolated genus of the Larrinae following Kohl (1896), who placed it in the Larrinae because of the tooth on the outer margin of the mandible, the single long midtibial spur, and the appendiculate marginal cell. However, he did point out the similarity between the wings of *Laphyrogogus* and *Astata* (Astatinae) on the one hand, and the general resemblance to the philanthines in the form of the eyes and abdomen on the other hand. Ashmead (1899) placed *Laphyragogus* in his "Lyrodinae," a heterogeneous group of genera. Beaumont (1959), in his review of the genus, accepted Kohl's placement of *Laphyragogus* in the Larrinae.

We believe there are good reasons for removing *Laphyragogus* from the Larrinae and placing it in a subfamily by itself. First, the wing venation is not typical of the Larrinae (fig. 51 A). The primary differences from the larrine wing are: first, the media of the hindwing diverges from Cu + M before cu-a, and there is a very broad, large jugal lobe. A few larrines do possess the hindwing features of *Laphyragogus:* large jugal lobe in Larrini (but not as broad as in *Laphyragogus*), and the media diverges before cu-a in the Scapheutini. Second, the volsella of the male genitalia in *Laphyragogus* is divided into a digitus and cuspis (fig. 58 A). In all Larrinae the volsella is simple and often reduced or absent. Thirdly, the makeup of the mouthparts is peculiar. The stipes are short and broad (triangular), with the consequence that the mouthparts cannot be folded flat into the oral fossa.

218 SPHECID WASPS

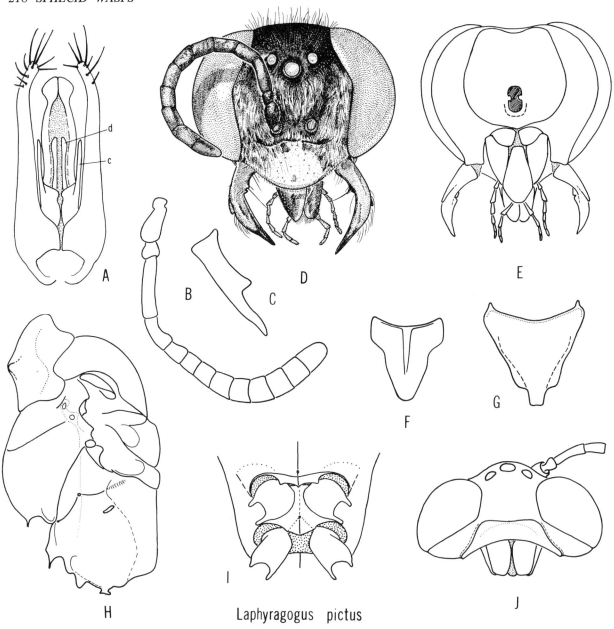

FIG. 58. *Laphyragogus pictus;* A, male genitalia, d = digitus, c = cuspis; B, male antenna; C, male mandible; D-E, facial and ventral views of female head, resp.; F, male sternum VIII; G, female tergum VI; H, thorax, lateral; I, mid and hindcoxae; J, dorsoposterior view of male head.

Instead they protrude (fig. 58 E). Fourthly, the midcoxae are essentially contiguous in *Laphyragogus,* and the hindcoxae are separated. Contiguous midcoxae are known in only a few larrine genera such as *Miscophus.* The metasternal area is elevated above the mesosternal area rather than being on the same plane as is the rule in most Sphecidae.

The striking similarity of the wings of *Laphyragogus* and *Astata* (compare figs. 51 A and 51 C) and the presence of a digitus and cuspis in the male suggest that the genus is more closely allied to the Astatinae than the Larrinae. The contiguous midcoxae lend some weight to this assumption. However, unlike the Astatinae, *Laphyragogus* lacks cerci and has only one midtibial spur. To place *Laphyragogus* in the Astatinae would, we feel, stretch the limits of the subfamily to the point where it would be difficult to defend its status as a subfamily. *Laphyragogus* is suggestive also of the Philanthinae, but the latter have the mandibular sockets closed by a forward extension of the hypostoma, and thus a relationship between the two groups must be rather remote. Under the circumstances it seems best to set *Laphyragogus* aside in its own subfamily. Its position relative to the other subfamilies must lie between the Astatinae and

SUBFAMILY LAPHRAGOGINAE

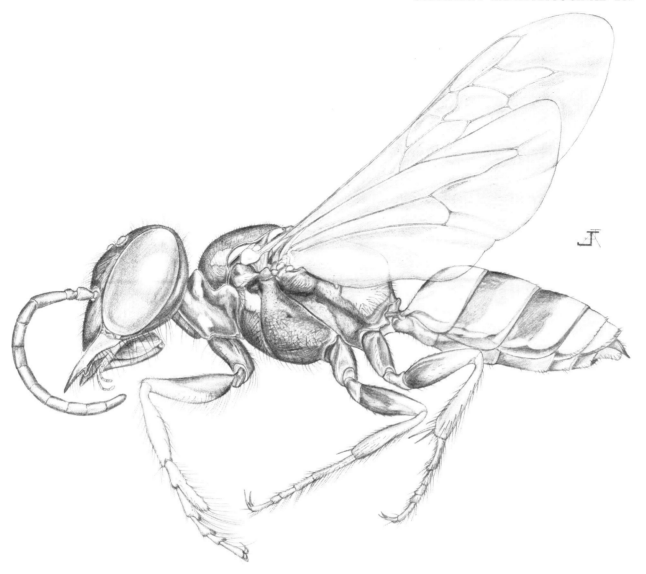

FIG. 59. *Laphyragogus pictus* Kohl, female.

Larrinae with a possible distant relationship with Philanthinae.

Tribe Laphragogini
Genus Laphyragogus Kohl

Generic diagnosis: Inner orbits more strongly converging in male; face swollen between ocelli and antennal sockets; antenna moderately long, simple in female, flagellomere II and sometimes III arcuate or strongly elbowed in male (except in *ajjer*), remaining articles gradually thickening towards apex (fig. 58 B); clypeal margin entire, arcuate (female) or with a truncate median lobe (male); occipital carina trapeziform above, the angles nearly touching eyes (fig. 58 E, J) and forming a complete circle below, occiput strongly depressed dorsally (fig. 58 E); no malar space; mandible with one or two teeth on inner margin (fig. 58 D); labrum small, bilobed, hidden beneath clypeus, pronotal collar narrow, much below level of scutum (fig. 58 H); scutum strongly convex anteriorly; mesosternal area depressed anteriorly (except in *ajjer*), depression bounded by a carina which is suggestive of an omaulus but which is more ventral in location; metapleuron and part of hypoepimeral area depressed, forming a channel for reception of hindfemur; propodeal spiracle without an operculum; legs moderately long, slender in male; a well-developed foretarsal rake in female, tarsomeres strongly asymmetrical (fig. 59), male foretarsus nearly symmetrical but with a weak rake; arolium very small in female, moderate to large in male; last tergum with a flat pygidial plate in both sexes, plate projecting laterally over the sides of tergum (fig. 58 G); sternum VII of male with special patches of long hair and a variably shaped median apical projection, sternum VIII roughly triangular, apex rounded (fig. 58 F); digitus and cuspis slender, elongate (fig. 58 A).

Geographic range: The genus is known from North Africa, southwest and central Asia, and northwest India. Six species are currently recognized.

Systematics: The large jugal lobe of the hindwing, the angular form of the mouthparts, and the trapeziform course of the occipital carina dorsally are distinctive features of *Laphyragogus.*

A combination of generalized and specialized features are displayed. Among the former are: crossvein between submarginals I and II of forewing angled; media of hindwing diverging before cu-a; jugal lobe of hindwing very large and possessing a remnant of an anal vein; midcoxae contiguous; propodeal spiracle without an operculum; and volsella with a digitus and cuspis. Specialized features are: mandible with an external tooth; angular form of the mouthparts; trapeziform occipital carina; and single midtibial spur. It is probable that *Laphyragogus* is an archaic element in the Sphecidae.

The reason for the peculiar development of the mouthparts and occipital carina (with the attendant depression of the occiput dorsally) may be connected with the angle of the head in relation to the body. Apparently, the head is normally tilted, so that the mandibles and mouthparts are directed anteriorly instead of downward as in most sphecids. The depressed occiput permits clearance of the pronotum, which otherwise would touch the back of the head and limit a backward tilt. The angular mouthparts may have developed in conjunction with this almost prognathous head posture. Although the habits of these wasps are unknown, it is tempting to postulate that they sit on the ground with legs spread (in the manner of bembicines) waiting for passing prey. It is not hard to imagine that the semiprognathous head is held close to the surface. This posture would give the wasp a broad field of vision.

The nonoperculate propodeal spiracle is unusual and perhaps unique to *Laphyragogus.* The operculum presumably has developed in other wasps to shield the spiracles from flying sand kicked up by the wasp during nest excavation. Considering that *Laphyragogus* is a desert dweller, the absence of the operculum is even more peculiar.

The very small female arolia are peculiar also. In most wasps, the female arolia are as large or larger than those of the male.

We have studied only *Laphyragogus pictus* Kohl and have relied on Beaumont's (1959) review of the genus for details of other species. Color is quite variable. Several species are extensively yellow, and some have red markings also. One species, *Laphyragogus ajjer,* differs in several respects from all other *Laphyragogus.* Flagellomere II is not arched in the male, the anterior mesosternal area is not depressed, and the last tergum bears a basal, strongly elevated, triangular pygidial plate. Tarsomeres II and III of the midleg are arcuate, and each bears two heavy, recurved apical spurs. The terminal gastral sterna do not follow the structural pattern of the other species of *Laphyragogus.*

Biology: Unknown.

Checklist of *Laphyragogus*

ajjer Beaumont, 1958; Algeria
   ssp. *orientalis* Beaumont, 1970; Iran
kohlii (Bingham) in Kohl, 1896 (*Leianthrena*); n. India
   kohlii Bingham, 1897 (*Lianthrena*)
pectinatus Beaumont, 1959; n. Africa, Israel
pictus Kohl, 1889; n. Africa, Israel
turanicus Gussakovskij, 1952; sw. USSR: Tadzhik S.S.R.
visnagae Beaumont, 1959: Algeria

SUBFAMILY LARRINAE

With over 2,000 species, this is by far the largest subfamily in the Sphecidae. Its morphological and biological diversity rivals that of the Nyssoninae, and it would be difficult to describe a "typical" larrine. For the most part they are fairly compact, often dark colored wasps. However, the gaster is sometimes slender or petiolate as in the well known genus *Trypoxylon;* and a few larrines, such as *Palarus,* are brightly colored. Size ranges from some of the tiniest sphecids known (2 mm) to large wasps that approach a length of 30 mm. Most species are fossorial, and the common term "digger wasps" probably is the best general appelation for the subfamily. A few genera are twig nesters, and some *Trypoxylon* make free mud nests. Prey range from spiders, Orthoptera and Hemiptera to coleopterous, dipterous and hymenopterous adults, and lepidopterous larvae, but spiders and Orthoptera are the predominant groups provisioned.

Diagnostic characters:

1. (a) Inner orbits usually converging above, sometimes parallel, rarely converging below except in Bothynostethini and Scapheutini; (b) hindocelli normal except scarlike in Larrini and deformed in Palarini.
2. (a) Antennae low on face except near middle in *Trypoxylon* and *Pisoxylon* (Trypoxylonini), sockets contiguous with frontoclypeal suture or separated from it; (b) female with 12 and most males with 13 antennal articles (exceptions in Miscophini).
3. Clypeus transverse except about as high as broad in some Trypoxylonini.
4. (a) Externoventral margin of mandible entire, stepped or notched, inner margin usually with one or two subbasal teeth, less commonly simple or subapically bidentate; (b) palpal formula 6-4 except 5-3 in some *Tachytes* (Larrini); (c) mouthparts usually short but long in some Larrini; (d) mandible socket open except in *Auchenophorus* (Miscophini) and some *Parapiagetia* (Larrini).
5. Pronotal lobe separated from tegula.
6. (a) Scutum with or without notauli; (b) no oblique scutal carina opposite tegula.
7. (a) Mesopleuron usually with an episternal sulcus; (b) omaulus present only in *Bohartella* (Scapheutini) and some *Trypoxylon;* (c) acetabular carina present only in some Bothynostethini and Scapheutini; (d) sternaulus absent.
8. Metapleuron consisting only of upper metapleural area, definitive pleuron tapering ventrad.
9. (a) Midtibia with one apical spur; (b) midcoxae usually separated, each with a dorsolateral carina or prominent crestlike elevation (except simple in *Auchenophorus*); (c) precoxal lobe present, large (rare exceptions); (d) hindfemur simple apically except truncate and otherwise modified in Scapheutini, Bothynostethini, most *Tachytes,* and a few *Palarus;* (d) claws simple except toothed within in some female Larrini; (e) plantulae absent except in some Trypoxylonini and Scapheutini.
10. (a) Propodeum very short to long; (b) dorsal enclosure present or absent; (c) propodeal sternite absent except in *Parapiagetia* (Larrini) and some *Trypoxylon.*
11. (a) Number of submarginal cells varying from three to none, but three the normal complement; (b) usually two but sometimes one or no recurrent veins present, second received by submarginal II in wings with three submarginal cells; (c) marginal cell truncate or acuminate apically, apex rarely open in some Miscophini.
12. (a) Hindwing jugal lobe varying from as long as anal area to very small or absent; (b) media diverging after cu-a except before cu-a in Scapheutini; (c) second anal vein absent except in some Larrini and Palarini.
13. (a) Gaster usually sessile, but sometimes pedunculate or petiolate in which case petiole consists of tergum and sternum; (b) tergum I and occasionally II with lateral carina (rare exceptions); (c) male with seven visible terga, sternum VII present and similar to preceding sterna; (d) pygidial plate present or absent.
14. (a) Volsella simple, without digitus, rarely absent (Palarini, some Miscophini); (b) cerci absent.

Systematics: As constituted here, the Larrinae consists of 39 genera which are distributed among six tribes; Larrini, Palarini, Miscophini, Trypoxylonini, Bothynostethini, and Scapheutini. The inclusion of the Trypoxylonini is the primary divergence from most current concepts of the subfamily, but our studies of adult morphology and Evans' work (1964a) on larvae indicate that the recognition of a subfamily Trypoxyloninae is unwarranted. Interestingly, Kohl's concept (1885a) of the "Larriden" included most of the forms contained in our Larrinae plus *Astata* in the old broad sense of this genus. As more genera became known (some of them obviously presenting difficulties for the classifier), Kohl (1896) found it necessary to divide the group in 4 "Gattungsgruppe": *Pison, Miscophus, Larra,* and *Astata.* The last is equivalent to the Astatinae, tribe Astatini, while his Gattungsgruppe *Pison* is identical with our concept of the tribe Trypoxylonini. His *Miscophus* group corresponds with our Miscophini as far as it goes. Kohl's *Larra* group is equivalent to our Larrini, but he associated as satellites a number of odd genera such as *Laphyragogus, Dinetus,* and *Palarus* and also several miscophine elements, so that overall the assemblage was quite heterogeneous.

We have placed elsewhere several discordant elements, *Laphyragogus, Dinetus, Eremiasphecium,* and *Odontosphex,* which have at one time or another been included in the Larrinae. On the other hand, the bothynostethin and scapheutin genera, which have often been associated with the Nyssoninae following the lead of Handlirsch (1887-1895) and Kohl (1896), are properly associated with the Larrinae as shown by Menke (1968d). Unfortunately, the end result of this subtraction and addition is not a more easily definable subfamily, but we believe that a more natural assemblage has been achieved. There is strong evidence based on larvae (Evans, 1964a) and adult morphology that the Crabroninae should also be included in the Larrinae. If enlarged to that extent it would be impossible to define the subfamily without many exceptions. This, plus the fact that the Crabroninae is a large, rather homogeneous and thus easily defined group, has caused us to exclude it from the Larrinae. As it is, there are no characters in the subfamily that will, by themselves, separate the Larrinae from all other subfamilies. In fact, there are only two features in the group that are universal and significant at the subfamily level: 1) midtibia with one apical spur, and 2) volsella simple, not differentiated into cuspis and movable digitus. These two characters in combination are found throughout the Crabroninae, but they occur elsewhere only in the tribe Cercerini of the Philanthinae and in some males of two nyssonine genera (*Didineis, Cresson*). These facts emphasize the close relationship between the Larrinae and the Crabroninae.

Most Larrinae have two or three submarginal cells in contrast to the single cell found in the Crabroninae. The distinctive fusion of the single submarginal and discoidal cells in the crabronine tribe Oxybelini is not found in any of the single-celled larrine genera. The emarginate inner orbits in the larrine tribe Trypoxylonini separates its single submarginal cell members (*Trypoxylon, Pisoxylon*) from the Crabronini, and in addition there is little general resemblance between the two groups. A few larrine genera in the tribe Miscophini also have only a single submarginal cell in the forewing, but the length of the scape is usually diagnostic. In the tribe Crabronini the scape is usually about half the length of the flagellum. In the Miscophini the scape is much shorter. Of course, the venational patterns of these odd miscophins differ considerably from the Crabronini.

Another basic difference between the Larrinae and at least the crabronine tribe Crabronini is that the mandibular socket is open in the former and closed in most genera of the latter. There are a few exceptional genera in both groups that contradict this distinction, but the difference is important. The sockets are apparently open in all crabronines of the tribe Oxybelini.

Unfortunately, the volsella, by virtue of being concealed and confined to one sex, is not a practical tool for casual identification of taxa under a microscope. Therefore, it is necessary to employ a number of nonuniversal characters in various combinations to isolate the Larrinae from other subfamilies possessing a single midtibial spur, e.g.: Crabroninae; Laphyragoginae; Philanthinae, Entomosericinae; the following Nyssoninae; Bembicini, some Heliocausini and Stizini, some males of *Didineis* (Alyssonini) and *Cresson* (Nyssonini); and some Ammophilini (Sphecinae). Some of these characters can be found in the keys to the tribes of the Sphecidae provided at the end of the systematics section of the family. One very useful feature for identifying a wasp as a larrine is the presence, on the externoventral margin of the mandible, of a notch or step. This character is found elsewhere only in many Crabroninae, the Laphyragoginae, the Xenosphecinae, and the Dinetini of the Astatinae, but the last two possess the nonlarrine characteristics of two midtibial spurs (none in some *Dinetini*). Unfortunately the notch or step is not universal in the Larrinae. Another valuable diagnostic character in the Larrinae is the divergence of the hindwing media after cu-a. Only in the Scapheutini does the media diverge before cu-a.

Points of difference between the Larrinae and Crabroninae have been adequately detailed above. Larrines can be separated from other groups with a single midtibial spur as follows. The combination in *Laphyragogus* of a large jugal lobe and the media diverging before cu-a in the hindwing is not found in any larrine. It is not easy to separate larrines that do not have notched or stepped mandibles from the Philanthinae, but in most cases one or more of the following differences will distinguish the two groups. The second recurrent vein is received by the second submarginal cell in larrines that possess three submarginal cells. In the Philanthinae there are always three submarginal cells, and the second recurrent vein is received by the third submarginal except in the isolated tribes Eremiaspheciini and Odontosphecini. In *Odontosphex* the divergence of the hindwing media before cu-a and the large jugal lobe separates the genus from all larrines. The combination in *Eremiasphecium* of normal ocelli and a 5:3 palpal formula separates the genus from all larrines. A second and major difference between the two subfamilies is found in the mandibular sockets. In the Larrinae the sockets are open except in *Auchenophorus* and some *Parapiagetia*. In all Philanthinae, except *Odontosphex,* the sockets are closed by a forward extension of the hypostoma, the paramandibular process. Unlike all philanthines, *Auchenophorus* has only one submarginal cell, and *Parapiagetia* has notched mandibles. Although the socket difference between the two subfamilies is a primary diagnostic feature, it is difficult to use in ordinary sorting of specimens because the mandibles usually have to be spread to see the character. A third feature that separates many larrines from many philanthines and one that has already been alluded to above is the point of divergence of the hindwing media. It diverges after cu-a in all larrines except the Scapheutini, but it diverges before cu-a in several philanthine groups including the perplexing tribe Odontosphecini.

The entomosericine characteristic of the inner orbits converging strongly below is found only in the larrine tribes Scapheutini, Bothynostethini, and some Trypoxylonini, but the media of the hindwing diverges after cu-a in the last two instead of before cu-a as in *Entomosericus.* The Scapheutini share with *Entomosericus* the same hindwing venation, but the second submarginal cell is petiolate in the former and also the notauli are absent or very weak and short.

The absence in all larrines of the oblique scutal carina, a structure common to most nyssonines including the Bembicini, Stizini, and Nyssonini, separates the subfamily from these three tribes. The deformed, scarlike hindocelli of the Heliocausini separates the group from all Larrinae except the Larrini, but the oval, oblong, elliptic, or "tailed" scars of the latter contrast with the C-shaped scars of *Heliocausus.*

Most of the foregoing has been aimed at the novice reader, since most taxonomists familiar with sphecids can separate larrines from the groups with which we have compared them, simply on the general habitus of the wasps. The above discussion also emphasizes the problems encountered when trying to find a morphological basis for separating large, diverse but natural groups from others.

It is guesswork to say where the Larrinae developed, but the Old World has twice the number of endemic genera found in the New World (16 vs. 8), and the preponderance of these are located in Africa. The tribes Larrini, Miscophini, and Trypoxylonini have cosmopolitan genera, but the Palarini is confined to the Old World, and the Bothynostethini and Scapheutini are essentially Neotropical elements of the New World. The Larrini and Miscophini reach their greatest diversity in the Old World, and their origins are presumed to be in that area. The Trypoxylonini is probably a New World development, judging by the three endemic genera and the tremendous diversity in Neotropical *Trypoxylon.* Several genera are worthy of mention because of their disjunct and/or uneven distribution patterns. *Ancistromma* and *Plenoculus* each contain many species in North America, but both possess two or three species in southern Europe and southwestern Asia. The widespread Old World genus *Prosopigastra* has a single representative in the southwestern United States. *Parapiagetia* occurs all over Africa and eastward to India and southwestern USSR, but there are quite a few species of the genus also in southern South America. Australian endemism is found only in the tribe Miscophini, which is represented there by four unique genera.

The probable evolutionary relationships of the six larrine tribes are depicted in the dendogram that accompanies the section on the phylogeny of the Sphecidae at the beginning of the book. The Miscophini is perhaps the most generalized group, but reduction of wing venation and the small size of the volsella are evolutionary advancements in the tribe. The Trypoxylonini, Bothynostethini, and Scapheutini are rather easily derived from a premiscophine stock. The Trypoxylonini is notable for the development of emarginate eyes, reduction of wing venation, and abdominal petiolation. The Bothynostethini and Scapheutini seem to be closely allied. This observation is based on the general similarity of the face, thorax, and legs of their genera; but dissimilarity between the two groups in wing venational details and the male genitalia, including the eighth sternum, also suggest the possibility of convergence or parallel evolution of unrelated groups. It is for this reason that we have elevated the subtribes recognized by Menke (1968d) to tribes. Hopefully, larval studies will shed further light on their affinities. The major trend in these two tribes has been the development of sinuate inner orbits, which strongly converge below, and the truncate hindfemoral apex. There is scarcely any doubt that the Bothynostethini and Scapheutini are closely allied with the Crabroninae.

A second major evolutionary branch in the Larrinae is formed by the tribe Larrini. This group exhibits the most generalized wing venation in the subfamily, and the tribe has retained the primitive large jugal lobe and a remnant of a second anal vein. On the other hand, the scarlike hindocelli represent a high degree of specialization. The tendency for strong convergence of the eyes above which culminates in the holoptic condition in *Prosopigastra,* is another noteworthy feature of the Larrini. The affinities of the tribe Palarini are not obvious, and it is associated with the Larrini only tentatively. The deformed hindocelli, convergence of the eyes dorsad, and the tripartite clypeus are the main criteria for this association. Species in some genera of the Larrini have the clypeus partially divided. In addition to the three specializations listed above for the Palarini, the peculiar abdominal modifications are noteworthy.

In the following table some of the features that appear to be useful in determining the phylogeny and classification of the Larrinae are listed.

Other interpretations of some of the characters in this list could be given. For example, the emarginate or notched inner orbits may be a primitive feature rather than a specialization. Similar eyes are found in scolioid

Phylogenetic Characters in Larrinae

	Unspecialized or primitive	*Specialized or advanced*
1.	Male antenna with 13 articles	Male antenna with 12 articles
2.	Male antenna simple	Male antenna with sensory areas or otherwise modified
3.	Antennal sockets contiguous with frontoclypeal suture	Sockets separated from frontoclypeal suture
4.	Inner orbits parallel	Inner orbits strongly converging above or below
5.	Inner orbits straight or arcuate	Inner orbits emarginate or sinuate
6.	Eyes bare	Eyes setose
7.	Eye facets of uniform size	Facets of lower part of eye enlarged
8.	Ocelli round and with transparent convex lens	Ocelli deformed, lens flattened or opaque
9.	Frontal surface flat or broadly convex	Frons variably swollen, tuberculate, abruptly elevated, or carinate
10.	Clypeus transverse, simple	Clypeus as high as wide, or tripartite
11.	Labrum short, hidden behind clypeus	Labrum long, prominently exposed
12.	Mandible without inner teeth	Mandible toothed within
13.	Mandible socket open to oral fossa	Mandible socket closed by paramandibular process of hypostoma
14.	Mouthparts short	Mouthparts long
15.	Palpal formula 6:4	Palpal formula 5:3
16.	Gena simple	Gena tuberculate or spinose below, or carinate
17.	Occipital carina attaining hypostomal carina, or forming a complete circle	Occipital carina disappearing below
18.	Pronotal collar elongate, as high as and not closely appressed to scutum	Collar short, sharpedged, lower than and closely appressed to scutum
19.	Propleuron simple	Propleuron with tubercles, projections or carinae
20.	Notauli present, long	Notauli short or absent
21.	Propodeum long, smooth, without dorsal enclosure or special sculpture	Propodeum short, with dorsal enclosure or special sculpture
22.	Episternal sulcus present, extending to anteroventral margin of pleuron, and originating at anterior end of subalar fossa	Episternal sulcus absent, or present but originating near middle of subalar fossa and not reaching anteroventral margin of pleuron
23.	Omaulus absent	Omaulus present
24.	Mesopleuron without hypersternaulus, acetabular carina or ventral tubercles	Mesopleuron with hypersternaulus, acetabular carina, or ventral tubercles
25.	Propodeal sternite absent	Propodeal sternite present
26.	Male foreleg simple	Male forecoxa, trochanter and/or femur modified
27.	Midcoxae separated	Midcoxae approximate
28.	Hindcoxa simple	Hindcoxa with sensory pit or spinose
29.	Hindfemur simple apically	Hindfemur obliquely truncate apically
30.	Male hindfemur simple	Male hindfemur angulate or spinose basally
31.	Hindtibia simple	Hindtibia carinate
32.	Female foretarsal rake present	Female foretarsal rake absent
33.	Tarsomere V symmetrical, short	Tarsomere V asymmetrical, or long
34.	Claw simple, short	Claw with inner teeth and/or elongate and prehensile
35.	Media of forewing diverging before or at cu-a	Media diverging after cu-a
36.	Marginal cell long, acuminate apically	Marginal cell short, truncate or open apically
37.	Three submarginal cells	Fewer than three submarginal cells
38.	Submarginal cells four sided, not petiolate	Second or third submarginal cell petiolate
39.	Two discoidal cells	One or no discoidal cells
40.	Both recurrent veins independently received by submarginal cell II	Recurrent veins joining before reaching submarginal cell II
41.	Subdiscoidal cell present	Subdiscoidal cell absent
42.	Hindwing with medial and submedial cells	Hindwing without medial and submedial cells
43.	Jugal lobe large	Jugal lobe small or absent
44.	Gaster sessile	Gaster pedunculate or petiolate
45.	Tergum I without lateral carina	Tergum I and II with lateral carina
46.	Tergal apices single edged	Tergal apices double edged
47.	Pygidial plate asetose	Pygidial plate setose
48.	Sterna simple	Sterna with various prominences or special setae
49.	Male sternum VIII rounded or truncate apically	Male sternum VIII emarginate or bispinose apically
50.	Volsella present	Volsella absent

and vespoid wasps, and the same condition in the Trypoxylonini (and Philanthini) may just be a holdover of a primitive trait. However, we regard the eye notch as a special development.

Several characters have been omitted from the foregoing list because their evolutionary significance, at least in the Larrinae, is uncertain. The presence of an externoventral notch or step on the mandible could be a specialization over a simple mandible, or it could represent a primitive character. There is better evidence for the latter. For example, the development of a pygidial plate in males is certainly an advanced condition. Within *Liris* the presence of a plate in males is found primarily in the subgenus *Liris*, a group characterized by the absence of a mandibular notch or its weak development. One infers from this that the advanced mandibular condition is the loss of the notch. The function of the notch is unknown. Notched mandibles seem to be universally present in the more primitive genera of each larrine tribe such as *Larra, Larropsis, Lyroda, Pisonopsis, Willinkiella,* and *Scapheutes.* However, a notch may or may not be present in more advanced genera, and it is variably developed or in the process of being "phased out" in a few (*Palarus, Liris, Gastrosericus, Holotachysphex, Tachytes, Solierella, Pison,* and *Bothynostethus*). Similar mandibles are found in male Mutillidae, but the fact that the presence of a notch is sex linked in this family offers little support for assuming that the same condition in larrines is a primitive feature.

The significance of the presence or absence of a pygidial plate is likewise in doubt. For example, it is uncertain whether or not twig nesting forms such as *Nitela* and *Trypoxylon*, which lack a pygidial plate, ever had one, although in the latter there is evidence that the pygidial plate has been lost. Some species of *Pisonopsis*, a more primitive relative of *Trypoxylon*, have pygidial plates. It may be that some genera have never developed a pygidial plate, while others have developed one and secondarily lost it when presumably changing from a fossorial habit to a twig-nesting one.

It is also difficult to say whether the divergence of the hindwing media before cu-a in the Scapheutini is a generalized or specialized feature, but a glance at the wings of other groups such as the Dolichurini, Astatinae, and Psenini, which are among the more generalized groups of the Sphecidae, suggests that the point of divergence of the media in the Scapheutini is a primitive trait. The presence of plantulae in some scapheutins supports this view, since they are regarded as primitive structures.

The following summarizes the main evolutionary trends in the Larrinae:

Head: The ocelli become flattened and somewhat deformed in *Palarus*, but the lens remains transparent. Nearly complete obliteration of the lens takes place in all Larrini. Strong convergence of the eyes above occurs in most Larrini. Strong convergence below is a specialization of the Bothynostethini, Scapheutini, and some Trypoxylonini, where it is usually accompanied by enlargement of the lower facets of the eyes. The frons has been modified by a variety of carinae in some *Trypoxylon, Nitela,* and *Solierella,* and it has become strongly elevated in *Kohliella,* most *Prosopigastra,* and some *Tachysphex.* The linear frontal swellings of the Larrina are also extreme developments. Closure of the mandibular sockets has occurred in *Auchenophorus* and some *Parapiagetia.* The mouthparts have lengthened in some *Tachytes* and *Tachysphex,* and palpal reduction has taken place in some *Tachytes.* The reduction in the number of male antennal flagellomeres has occurred in *Sericophorus* and some *Solierella* and *Trypoxylon.*

Thorax: The tendencies are towards shortening and rounding of the propodeum, shortening and closer fitting of the pronotal collar with the scutum, and reduction of pleural sulci as in the Miscophini and Larrini, or development of the omaulus and proliferation of other grooves and carinae as in *Trypoxylon* and *Bohartella.* The development of a propodeal sternite occurs in certain petiolate forms (*Trypoxylon* and *Parapiagetia*).

Gaster: The beginnings of the petiolate condition are found in genera such as *Lyroda, Pison,* and *Dicranorhina* which have pedunculate gasters. Petiolation is found most notably in *Trypoxylon*, but some *Parapiagetia* have a short petiole. *Kohliella* lacks lateral tergal carinae, but most larrines have the carina on tergum I. *Prosopigastra, Holotachysphex, Pisonopsis,* and *Pison* have developed the carina on tergum II as well. The volsella is reduced or lost in *Palarus* and in many Miscophini. The eighth sternum in the male has developed a pair of long processes in the Scapheutini and some *Trypoxylon.*

Wings: Overall reduction of venation and size of the jugal lobe has taken place in the Miscophini and Trypoxylonini. In both tribes reduction starts with the petiolation of submarginal cell II, and is followed by the disappearance of it or of submarginal cell III. This is accompanied by the disappearance of the outer discoidal cell. Concordant with this in the Miscophini is the shortening of the submarginal cell and reduction and loss of the jugal lobe. The marginal cell is very small in *Saliostethus* and some *Miscophus* and is open in *Saliostethoides* and *Miscophoides.* In the last genus the submarginal and discoidal cells and the jugal lobe have disappeared. There are no closed cells in the hindwing of *Nitela*. An interesting development in the Trypoxylonini is the division of the hamuli into two groups. Coalescence of the recurrent veins in certain genera of the Larrini is another interesting trend.

Legs: The most striking developments are the modifications of the male front and hindlegs (fingerlike projection on inner posterior angle of forecoxa, concavity of foretrochanter and base of forefemur, and angle or prong near base of hindfemur), which in the case of the foreleg occur sporadically throughout the subfamily and presumably function during copulation. Interesting modifications occur in the female as well but are mostly confined to genera of the Larrini. Among these should be mentioned the prehensile claws of most *Liris* and related genera and the attendant lengthening of the last tarsomere. The last tarsomere is asymmetrical in females of *Dicranorhina* and some *Tachysphex,* and in the latter, one claw is often shorter than its mate. Unequal claws are also found in males of some *Parapiagetia.* Claw teeth are developed in females of *Kohliella* and some *Liris.*

KEY TO TRIBES OF LARRINAE

1. Hindocelli reduced to flat, opaque scars of various shapes (fig. 61 A-M); jugal lobe of hindwing large, nearly as long as anal area (fig. 70) .... tribe Larrini, p. 226
   Hindocellus with a convex or somewhat flattened, complete, transparent lens; jugal lobe of hindwing small or absent, never more than one-half length of anal area (figs. 102 G, 113) 2
2. Inner orbits emarginate[1] (fig. 104) .... tribe Trypoxylonini, p. 327
   Inner orbits straight, bowed or sinuate but not emarginate (figs. 80, 114) 3
3. Clypeus divided into three parts by two vertical suturelike lines (fig. 80 I); antennal sockets separated from frontoclypeal suture by one-third or more of a socket diameter; hindocelli usually flattened and elliptical; Old World .... tribe Palarini, p. 286
   Clypeus not divided into three parts; antennal sockets essentially contiguous with frontoclypeal suture[2]; hindocelli round and convex 4
4. Hindfemoral apex simple, femur thickest near middle, gradually tapering toward apex (fig. 87); inner orbits parallel or converging above[3]; cosmopolitan .... tribe Miscophini, p. 291
   Hindfemoral apex with an outer, oblique truncation which often terminates ventrally as a lamella or projection (fig. 115 F); apical half of femur of equal thickness or increasing in thickness towards apex; inner orbits strongly converging below[4]; New World ... 5
5. Hindwing media diverging after crossvein cu-a (fig. 113 B); veinlet forming outer margin of submarginal cell III meeting marginal cell near middle of latter .... tribe Bothynostethini, p. 349
   Hindwing media diverging before crossvein cu-a (fig. 113 A); veinlet forming outer margin of submarginal cell III meeting marginal cell near apex of latter .... tribe Scapheutini, p. 352

Tribe Larrini

Morphologically the Larrini is considered to be the most primitive tribe in the Larrinae, and yet its chief characteristic, the hindocellar deformation, represents a high degree of specialization. These wasps are usually of medium size. Most species of the subtribe Larrina are somber colored, but many members of the Tachytina are partially red, and the gaster is often banded with appressed silver or golden vestiture. The more robust and hairy species of the genus *Tachytes* are rather beelike.

With over 1,100 species the Larrini ranks as the largest tribe of the subfamily, having more than twice the number of species found in the Trypoxylonini. It is also the largest tribe in the Sphecidae. For its size the Larrini has comparatively few genera (15), but some of these are among the largest in the family.

Nearly all of the Larrini are fossorial and can truly be called "digger wasps" or "sand loving wasps," but one genus, *Holotachysphex,* is a twig nester. Although Orthoptera are the predominant prey, Hemiptera, Homoptera, and lepidopterous larvae are used in two genera. There is some evidence for sociality in *Dalara.*

The larger genera of the tribe are cosmopolitan, but two small ones, *Tachytella* and *Kohliella,* are restricted to southern Africa and are presumed to be relics. *Prosopigastra, Ancistromma,* and *Parapiagetia* have peculiar disjunct ranges.

Diagnostic characters:

1. (a) Inner orbits usually converging above (sometimes parallel in *Parapiagetia* and *Gastrosericus,* sometimes converging below in *Gastrosericus*); (b) midocellus often reduced in size, hindocelli reduced to elliptic, oval, accent marklike, or commalike scars which are not translucent.
2. (a) Antennal sockets contiguous with frontoclypeal suture or narrowly separated from it (a little more removed in a few *Tachytes*); (b) male with 11 flagellomeres.
3. (a) Clypeus transverse, usually with a median lobe of various shapes; (b) labrum usually hidden or barely projecting beyond clypeal margin (prominently exposed only in some *Tachysphex*).
4. (a) Mandible notched or angulate externoventrally or entire, inner margin with or without teeth; (b) mouthparts usually short (long in some *Tachytes* and *Tachysphex*); (c) mandibular socket open except in some *Parapiagetia.*
5. Pronotal collar short.
6. Scutum with or without notauli.
7. (a) Episternal sulcus present (except in some *Tachysphex*); (b) hypersternaulus absent.
8. (a) Female foretarsus with rake (except in *Paraliris, Holotachysphex,* and some *Liris*); (b) hindfemur simple apically (except in most *Tachytes*); (c) midcoxae widely separated (except in some *Gastrosericus* and *Tachysphex*); (d) arolia equal on all legs, variable in size; (e) male foreleg unmodified in Larrina, variable in the Tachytina.
9. (a) Propodeum short to long; (b) enclosure absent (except in some *Prosopigastra*); (c) propodeal sternite absent (except in *Parapiagetia*).

---

[1] Weakly angulate in some males of *Pisonopsis* (fig. 104 B).

[2] Except in *Mesopalarus* but clypeal character is diagnostic (fig. 87 D).

[3] Weakly converging below in exceptional forms but femoral characters are diagnostic.

[4] Inner orbits parallel in *Willinkiella* from Argentina but femoral characters are diagnostic.

10. (a) Three submarginal cells (except two in *Gastrosericus*), the second occasionally petiolate in *Larra*, third petiolate in *Kohliella;* (b) two recurrent veins present, both received by submarginal cell II; (c) marginal cell apex truncate (roundly acuminate in many *Tachytes*).
11. (a) Jugal lobe large, nearly equal to length of anal area (somewhat shorter in *Tachytella* and some *Gastrosericus*); (b) hamuli not divided into two groups.
12. (a) Gaster sessile (pedunculate in *Dicranorhina* and some *Parapiagetia,* petiolate in some *Parapiagetia*); (b) pygidial plate present or absent.
13. (a) Volsella present; (b) gonostyle simple (except in some Larrina).

Systematics: Two features, the scarlike hindocelli and the large jugal lobe, distinguish the Larrini from other tribes of the Larrinae. The hindocelli in the Palarini are deformed, but the lens is still complete and translucent. In the Larrini the hindocelli are essentially represented by opaque scars, although in every genus there appears to be a remnant of the lens. This remnant is in the form of a narrow, somewhat translucent band which borders the anterior or mesal edge of the scar. It usually extends around the inner end of the scar. The rest of the scar is dark or opaque. The surface may be somewhat elevated and shining, but the rear or lateral edge is usually difficult to define, and the lens remnant usually provides the easiest means of determining its shape and length.

As constituted here the Larrini contains the Tachytini of G. E. Bohart (1951) which is reduced to a subtribe. Maintenance of the Tachytini would tend to obscure the rather close relationship with the Larrini *s.s.* and furthermore, on a worldwide basis it is difficult to separate the two taxa. The characteristic linear facial swellings of the Larrina are not found in the Tachytina, but the weak frontal swellings of *Larropsis* and *Ancistromma* and the raised frons of *Tachytella* are suggestive of the frons in the Larrina.

There are differences between the two subtribes in the ocelli, but they are not completely reliable. The midocellus is situated in a rather deep, more or less transverse depression in all of the genera of the Larrina except *Dicranorhina*. In the Tachytina the midocellus is on the same general level as the surrounding integument, or at most set in a very small, shallow, circular depression. The hindocellar scars are very small and elliptical in the Larrina, and their long axes are oriented on a line drawn between the eyes. In the Tachytina the scars are larger, often commalike, and their long axes usually form a definite obtuse or acute angle. However, in some *Parapiagetia* the long axes of the scars are nearly on a line drawn between the eyes. In the Tachytina there is a more or less circular swelling between the mid and hindocelli which may have a linear impression. However, in the Larrina the area between the ocelli varies from flat to swollen.

The episternal sulcus originates at the anterior end of the subalar fossa in all Larrina, but in most Tachytina it originates at the middle of the fossa. Exceptions in the Tachytina are found primarily in *Parapiagetia* and *Gastrosericus*. Evans' (1964c) revised generic key to the larvae of the Larrinae indicates that there is a basic divergence of larval characters between the Larrina and Tachytina.

The genus *Odontosphex,* originally described as a larrine genus, was placed in the Philanthinae by Menke (1967d). There is an additional point of difference between this genus and the larrines that has not been stressed. The hindocellar scars of *Odontosphex* are quite close to the midocellus, the distance between them being somewhat less than a midocellus diameter. In the Larrini there is a much greater distance between the mid and hindocelli. This is further evidence that *Odontosphex* does not belong in the Larrini.

A dendrogram that depicts our current thoughts on the relationships of the 15 genera of the Larrini is shown in fig. 60. While the phylogeny of the Larrina appears fairly straightforward, the relationships within the Tachytina are more obscure. *Tachytella* and *Gastrosericus* appear somewhat isolated. *Larropsis, Ancistromma,* and *Tachytes* seem to be assignable to one phyletic line, and the remaining genera cluster around *Tachysphex*.

Larra and *Larropsis* are probably the most generalized members of the Larrina and Tachytina, respectively; and *Larropsis* would probably have to be considered as most closely approximating the archetype from which the tribe evolved, even though its elongate ocellar scars conflict somewhat with this hypothesis. The oval scars common to most *Tachysphex* probably typify the most generalized ocellar condition. Most of the genera in the Tachytina are highly specialized in different ways. It would be difficult to single out any one as the most advanced, but certainly *Tachytes* ranks near the top. In the Larrina, *Dicranorhina* is probably the most advanced genus although some groups within *Liris* are probably equally specialized. Biologically *Tachytes* and *Dicranorhina* rank near the top of their respective subtribes. *Larropsis* and *Larra* on the other hand have primitive nesting behaviors. In fact *Larra,* with parasitoidlike habits, exhibits the most primitive type of biology known in the Sphecidae.

Some of the variables within the tribe are noteworthy. The mandibles are notched externoventrally in most of the genera, but they are entire in *Dalara, Paraliris,* and some species of *Liris, Gastrosericus, Tachytes,* and *Holotachysphex*. The ocellar scars of *Parapiagetia* and most *Tachysphex* are fairly large and more or less oval, and this condition is probably the most generalized one. From this pattern, the scars have evolved to the very small, elliptical marks of the Larrina on one hand and to the elongate, "tailed" types found in various genera of the Tachytina. The very long scars of *Tachytes* are the most distorted in the Larrini.

A pygidial plate is present in most of the genera, but it is absent in *Holotachysphex,* a twig nester, and weakly defined in a few *Tachysphex* and *Prosopigastra*. *Tachysphex mendozanus,* an aberrant species, also lacks a pygidial plate. The apex of the last male sternum is rounded

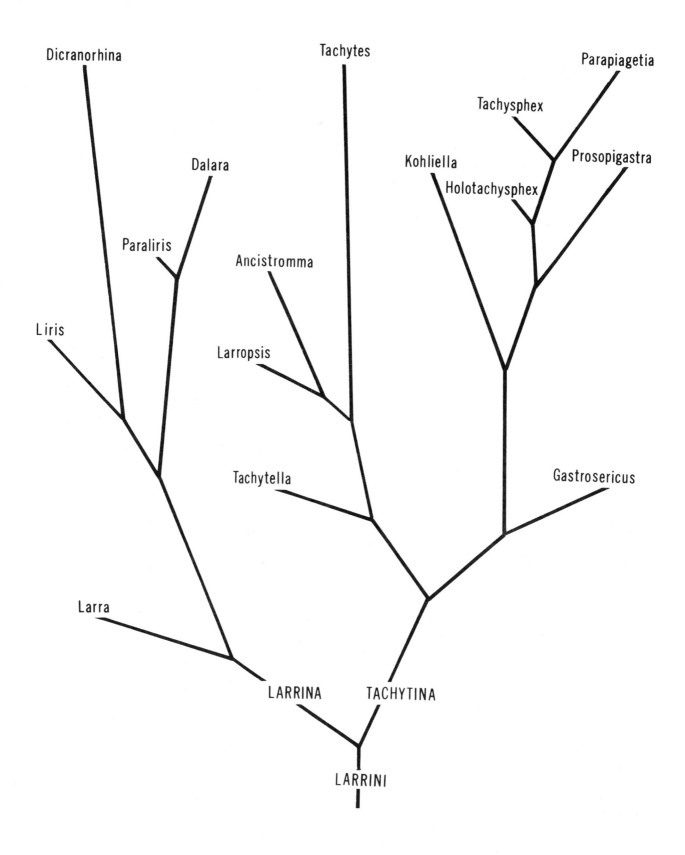

FIG. 60. Dendrogram suggesting relationships of genera in the tribe Larrini.

or truncate in *Larropsis, Ancistromma, Tachytella, Kohliella,* and some species of *Gastrosericus, Parapiagetia,* and *Tachytes.* This is presumed to be generalized. In the Larrina the last sternum is rounded or truncate, but it is often emarginate in *Larra* and *Liris. Tachysphex, Holotachysphex, Prosopigastra,* and some *Parapiagetia, Gastrosericus,* and *Tachytes* have an emarginate sternal apex. This is the more advanced condition.

The claws are simple in most larrines, but there is a single tooth on the inner margin in *Kohliella* females and one or two teeth on the inner margin in some *Liris* females. The last tarsomere is modified in various ways. In most female Larrina except *Larra* the tarsomere is elongate, somewhat elbowed in lateral profile, and has a ventral mat of pubescence. The claws are often long and prehensile as well. The last female tarsomere of the foreleg of *Dicranorhina* is asymmetrical. The claws are unequal in length in a few *Gastrosericus* and some *Parapiagetia.* In the latter genus the last tarsomere is usually quite long. In *Tachysphex* females of species that capture cockroaches for prey the last tarsomere is asymmetrical, and one claw is usually shorter than its mate. The various tarsal modifications that are restricted to the female are probably associated in some way with nesting behavior.

Wing venation is fairly constant in the Larrini. Reduction occurs through loss of the third submarginal cell in *Gastrosericus,* or by shortening of the marginal cell, as in *Kohliella* and *Gastrosericus.* Coalescense of veins contributes to the upside down petiolation of the third submarginal cell in *Kohliella,* an oddity in the Sphecidae, and the occasional petiolation of the second submarginal cell in *Larra.* The recurrent veins join before meeting the media in some species of *Liris, Dicranorhina, Gastrosericus,* and *Tachytella.* The bananalike shape of the third submarginal cell in many *Tachytes* and *Tachysphex* is an interesting development in this tribe.

Other specializations and peculiarities of the Larrini include: various modifications of the male foreleg, including a posterobasal notch on the femur in various genera of the Tachytina; modifications of the male hindfemur in some genera of the Larrina; loss of the episternal sulcus in one group of *Tachysphex;* closed mandibular sockets in some *Parapiagetia;* the modified hindfemoral apex in most *Tachytes;* the propodeal sternite of *Parapiagetia;* the holoptic eyes in some male *Prosopigastra;* and the ventral mesopleural tubercles of some *Prosopigastra.*

Biology: Nothing is known about the biology of *Paraliris, Tachytella,* or *Kohliella,* and all that is known about *Holotachysphex* is that it nests in twigs. The other genera provision with Orthoptera of various families except for *Prosopigastra,* which collects Hemiptera and Homoptera, and a few *Tachytes,* which catch lepidopterous larvae. Evidence suggests that *Parapiagetia* may also provision with lepidopterous larvae. *Larra* behaves like a parasitoid in that it does not excavate a nest. Some larrines start their nests in preexisting holes and burrows while other initiate excavation at the soil surface. The nests of some *Liris* are extremely deep with lengths exceeding 1.5 meters. Sociality is a possibility in *Dalara,* and one nest may continue to be used over several generations in *Dicranorhina.*

Key to New World genera of Larrini

1. Frons just below midocellus with a transverse swelling extending from eye to eye (interrupted medially) and joining a linear swelling along inner orbit to form an M or inverted U; midocellus in a broad depression; ocellar scars very small, narrow, elliptical, their long axes on a straight line drawn between eyes (fig. 61 A) subtribe Larrina 2
   Frons variable but without swellings as above; midocellus not in a broad depression; ocellar scars oval, commalike or golf club shaped, not unusually small, their long axes obliquely oriented (fig. 61 B, F-K) subtribe Tachytina 3
2. *Female:* last tarsomere evenly arcuate in lateral view, sides diverging most of way to apex (dorsal view) (fig. 64 P), ventral surface without hairmat; claw not prehensile and without an inner tooth (fig. 64 P); pronotal collar flat or slightly arcuate in front view (fig. 63 C); outer surface of foretibia with a row of three or more stout bristles on apical half (fig. 64 A); pygidial plate glabrous and shiny, no transverse row of setae apically (fig. 66 A); *Male:* propodeal side densely punctate, shiny (fig. 62); apical half of outer surface of foretibia usually with one or more erect bristles; mandible without inner teeth ....
   .... *Larra* Fabricius, p. 233
   Female: last tarsomere angled in lateral view, sides parallel on apical half (fig. 64 O); ventral surface usually with dense hairmat; claw prehensile (often folded against last tarsomere), sometimes with an inner tooth (fig. 64 Q); pronotal collar wedgelike in front view (fig. 63 D); outer surface of foretibia at most with one preapical bristle (fig. 64 C); pygidial plate usually extensively setose and usually with a transverse row of stout apical setae (fig. 66 D-H); *Male:* propodeal side dull and impunctate *or* if shiny then at most with sparse pinprick punctures[5]; outer surface of foretibia without bristles; mandible often with one or two inner basal teeth ....
   .... *Liris* Fabricius, p. 238
3. Ocellar scars very long, golf club or comma-shaped, long axes of scars subparallel, not exceeding an angle of 70°, distance between midocellus and end of tail less than length of scar (fig. 61 G); pygidial plate present in both sexes and usually clothed

---

[5] *Liris tenebrosa* has a highly polished thorax, and the propodeal side is rather closely punctate, but the mandible has a subapical inner tooth.

with dense setae which obscure integument (fig. 74 B) .... *Tachytes* Panzer, p. 260
Ocellar scars oval or oblong, or if elongate (accent-mark-shaped) then long axes of scars forming an angle of 80° or more (fig. 61 F, I, J); distance between midocellus and lower end of scar equal to or greater than length of scar; pygidial plate usually present in female but bare or sparsely setose (fig. 74 A, E–G), male usually without a pygidial plate 4

4. Petiole socket isolated from metacoxal cavities by a pair of dark propodeal sternites (fig. 71 E); gaster slender, often petiolate (fig. 75); inner tarsal claw of male much shorter than outer claw (fig. 73 P), s. S. America .... *Parapiagetia* Kohl, p. 277
Petiole-metacoxal cavity completely membranous, gaster sessile, usually robust; male inner tarsal claw not greatly reduced 5

5. Female foretarsomere II with three or more rake spines which are long and fine[6] (fig. 73 D, H); male sternum VIII emarginate apically (bispinose) (fig. 76 H) 6
Female foretarsomere II with no more than two rake spines which are usually bladelike or thornlike (fig. 73 A); male sternum VIII rounded apically (fig. 76 E) (occasionally weakly emarginate) 7

6. Gastral tergum II without a lateral carina, male forefemur with a basoventral notch or depression (fig. 73 K) *Tachysphex* Kohl, p. 267
Gastral tergum II with a lateral carina (fig. 71 G); male forefemur simple, one North American species ....
.... *Prosopigastra* A. Costa, p. 282

7. Subalar fossa bordered below by a sharp longitudinal carina (fig. 71 B); least interocular distance greater than combined length of pedicel and flagellomere I (except equal in one species) .... *Larropsis* Patton, p. 257
Subalar fossa not bordered below by a sharp carina, mesopleural surface sloping uninterrupted into fossa; least interocular distance usually equal to or less than combined length of pedicel and flagellomere I .... *Ancistromma* W. Fox, p. 259

Key to Old World genera of Larrini

1. Forewing with two submarginal cells; Africa, s. Asia .... *Gastrosericus* Spinola, p. 252
Forewing with three submarginal cells 2

2. Third submarginal cell petiolate, the stalk ending on the media (fig. 70 F); face with a large V-shaped swelling below midocellus; S. Africa .... *Kohliella* Brauns, p. 286
Third submarginal cell not petiolate 3

3. Petiole socket isolated from metacoxal cavities by a pair of dark propodeal sternites (fig. 71 E); male inner tarsal claw often greatly reduced (fig. 73 P); mandibular sockets closed (fig. 79 B); Africa, s. Palearctic Region, Oriental Region ....
.... *Parapiagetia* Kohl, p. 277
Petiole-metacoxal cavity completely membranous; male inner tarsal claw not greatly reduced; mandibular sockets open (fig. 79 D) ... 4

4. Ocellar scars very small, narrow, elliptical, their long axes on a straight line drawn between eyes (fig. 61 A); frons just below midocellus with a transverse swelling extending from eye to eye and interrupted by median frontal line, and a linear swelling (weak or absent in *Dicranorhina*) along inner orbit which joins transverse swelling to form an M or inverted U (fig. 67 A–D) 5
Ocellar scars oval, commalike, or golf club-shaped, their long axes obliquely oriented or subparallel, not on a line drawn across head perpendicular to median frontal line (fig. 61 B–M); frons variable but without swellings as above 9

5. Mandible bidentate apically (fig. 67 C–D), or mandible very long (fig. 67 D), mandible with inner basal tooth and at most a weak notch on outer margin (fig. 67 C–D); se. Asia 6
Mandible simple, not unusually long, often with a strong notch or step on outer margin, sometimes with an inner basal tooth (figs. 67 B, 72 A–C) 7

6. Body with long, erect hair; female hindtibia without a strong carina on apical half; male hindfemur without a large basal process .... *Paraliris* Kohl, p. 259
Body without erect hair, at most with thin pruinose pubescence; female hindtibia with a strong carina on apical half; male hindfemur with a large basal process (fig. 64 L) .... *Dalara* Ritsema, p. 248

7. Gaster pedunculate, first segment longer than broad in dorsal view (fig. 63 A); ocelli on a nearly flat surface (fig. 67 B); Old World tropics .... *Dicranorhina* Shuckard, p. 250
Gaster not pedunculate, first segment about as broad as long in dorsal view (fig. 65); midocellus usually in a broad depression, area between ocelli often elevated (fig. 61 A, 67 A); cosmopolitan 8

8. *Female*: last tarsomere evenly arcuate in lateral view, sides diverging most of way to apex (dorsal view) (fig. 64 P), ventral surface without dense hairmat; claw not prehensile and without an inner tooth (fig. 64 P); propodeal side densely punctate, shiny and not strongly ridged (fig. 62); pronotal collar flat or arcuate in front view (fig. 63 C); pygidial plate glabrous and shiny, apex without a transverse row of stout setae (fig. 66 A–B); *male*: propodeal side densely punctate, shiny, and with a few short anterior ridges at most[7]; mandible without

---

[6] Some roach collecting *Tachysphex* have only two rather short rake spines, but their peculiar asymmetrical last tarsomere (fig. 73 E,N) is diagnostic.

[7] The propodeal side in a *Larra* from Guadalcanal is largely covered with coarse diagonal ridges which give way to close punctation dorsad.

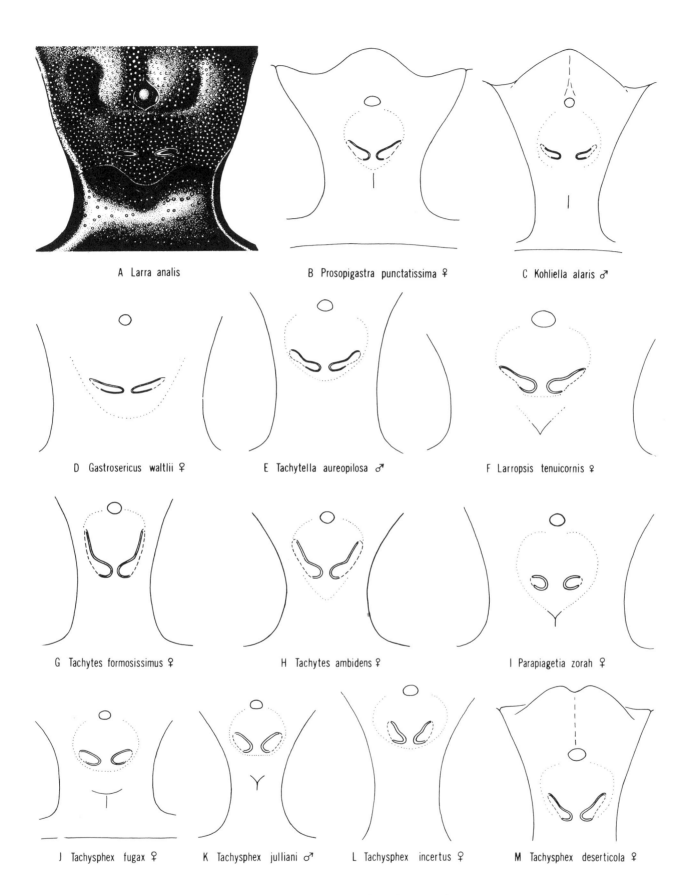

FIG. 61. Ocelli in the tribe Larrini.

inner teeth (fig. 72 A) ...... *Larra* Fabricius, p. 233
Female: last tarsomere angled in lateral view, sides parallel on apical half (fig. 64 O), ventrally with a dense hairmat[8]; claw prehensile (often folded against last tarsomere), sometimes dentate (fig. 64 Q); propodeal side usually dull and impunctate or occasionally shiny but then not densely punctate, sometimes with extensive diagonal ridging; pronotal collar wedgelike in front view (fig. 63 D); pygidial plate usually setose and usually with a transverse row of stout apical setae (fig. 66 C–E); *male:* propodeal side impunctate and dull, or if shiny, then at most with sparse pinprick punctures, mandible often dentate (fig. 72 B–C) ....
.... *Liris* Fabricius, p. 238

9. Ocellar scars very long, golf club or commashaped, long axes of scars forming an angle of 70° or less, often subparallel, distance between midocellus and end of tail less than length of scar (fig. 61 G–H); both sexes with pygidial plate which is usually densely setose, the setae usually silver or golden (fig. 74 B–C) .... *Tachytes* Panzer, p. 260
Ocellar scars oval or oblong (fig. 61 B, J–L), or if elongate (commalike or accent mark shaped fig. 61 E-F,M) then long axes of scars forming an angle of 80° or more; distance between midocellus and lower end of scar equal to or greater than length of scar; pygidial plate usually present in female but bare or sparsely setose (fig. 74 A, E–H), male usually without a pygidial plate 10

10. Tergum II without lateral carina[9] 11
Tergum II with lateral carina at least on basal half (fig. 71 G) 13

11. Female foretarsomeres I-II with numerous, usually closeset, long, fine flexible rake spines[10] (fig. 73 D); apex of male sternum VIII emarginate (fig. 76 H); frons just above antennal sockets with a pair of tubercles which are usually polished (fig. 69 E); ocellar scars oval or oblong (fig. 61 J-L), rarely accent mark shaped (fig. 61 M) ....
.... *Tachysphex* Kohl, p. 267
Female foretarsomere I with four to five widely spaced, stiff rake spines which are about as long as tarsomere II, tarsomere II with only two rake spines which are located at outer apex (fig. 73 A–B); apex of male sternum VIII rounded or truncate (fig. 76 C–E); frons without a pair of prominences just above antennal sockets; ocellar scars commalike (fig. 61 E–F) 12

12. Pronotal collar sharp edged, fitting closely with scutum; third submarginal cell elon-

---

[8] Tarsomere characters not valid for some African and Australian species, but propodeal side and collar are diagnostic.

[9] A trace of a carina is sometimes present basally in *Ancistromma* males, but the rounded apex of sternum VIII is diagnostic.

[10] A few *Tachysphex* females have only two rake spines on tarsomere II, but the facial tubercles above the antennal sockets are diagnostic.

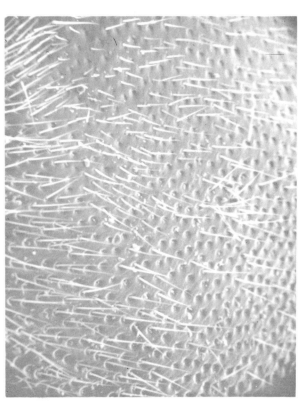

FIG. 62. Section of propodeal side of *Larra analis* showing punctation (100 X).

gate (fig. 71 C); scape short, stout (fig. 79 F); occipital carina meeting hypostomal carina; Mediterranean area, s.w. USSR, s. Africa .... *Ancistromma* W. Fox, p. 259
Pronotal collar with some thickness, standing apart from scutum (fig. 71 F); third submarginal cell short (fig. 70 H); scape elongate, gradually broadening distad (fig. 79 E); occipital carina ending short of hypostomal carina; s. Africa ..*Tachytella* Brauns, p. 256

13. Female with foretarsal rake composed of long, fine, flexible spines (fig. 73 H); male forefemur simple; both sexes with a prominent frontal swelling which is usually glabrous, polished and often moundlike (fig. 78 D); eyes sometimes holoptic in male (fig. 78 C) .... *Prosopigastra* A. Costa, p. 282
Female without a foretarsal rake (fig. 73 G); male forefemur notched posterobasally; frons without a prominent swelling; male eyes dichoptic .*Holotachysphex* Beaumont, p. 282

Subtribe Larrina

Diagnosis: Midocellus in a broad, deep, transverse depression except in *Dicranorhina*; hindocellar scars usually not at rear of a distinctly circular swelling, scars elliptic, their long axes on a line drawn between eyes or nearly so (fig. 61 A); frons with several linear swellings which form an M or H figure: a transverse swelling beneath midocellus which is usually partially interrupted by a frontal line, and a swelling which closely parallels inner orbit, latter swelling delimited mesad by a linear depression; antenna moderately long to long; malar space undeveloped except in *Dalara* and *Paraliris;* occipital carina joining hypostomal carina at or near apex of latter; collar wedge shaped in front view except in female and some male *Larra;* admedian lines present except in some *Liris,* and separated; episternal sulcus originating at anterior end of subalar fossa (fig. 63 B), extending to anteroventral margin of pleuron except in *Dicranorhina;* scrobal sulcus present; legs moderately long to long, arolia small to moderate; genitalia not laterally compressed, head of aedeagus with teeth on ventral margin of penis valves.

Discussion: This is a rather homogeneous assemblage of five genera that is chiefly characterized by the small, elliptic, ocellar scars, whose long axes are more or less on a line drawn between the eyes, the linear frontal swellings, and the origin of the episternal sulcus at the anterior end of the subalar fossa. The Larrina is primarily a tropical group of genera, the few temperate species belonging to *Larra* and *Liris.* Orthoptera are the exclusive prey.

Genus Larra Fabricius

Generic description: Upper inner angle of eye often bordered by a deep sulcus; midocellar lens usually convex; hindocellar scars very small, often obscure; frontal line a deep impression which is interrupted at transverse swelling; frons and vertex moderately to strongly shining, the former often impunctate, vertex usually densely and rather coarsely punctate; scape long but stout, scape, and pedicel often conspicuously shining in contrast to flagellum, especially in females of subgenus *Cratolarra,* flagellomeres usually longer than broad and bearing placoids in male; clypeus simple, nearly flat to moderately and broadly swollen, free margin usually broadly arcuate and sharp edged, although a truncate or rounded median lobe present in some males (especially in *Cratolarra*); labrum quadrangular or angular, often projecting beyond clypeus, rarely with median notch; inner margin of mandible simple (sometimes with a small, weak subbasal cleft or suggestion of a tooth) (fig. 72 A), externoventral margin with a subbasal notch; dorsum of collar in female as high as scutum (rare exceptions) and usually with some thickness, straight or slightly arcuate when viewed anteriorly (fig. 63 C); male collar usually lower than scutum, dorsum rather sharp edged and often forming a wedge fitting closely with scutum (similar to fig. 63 D); notauli usually absent, or weakly indicated; propodeum rather long, rectangular, sides parallel in dorsal view or constricted at level of spiracles (fig. 63 E–F); propodeal surface moderately to densely punctate, propodeal side usually moderately to strongly shining and usually smooth, punctures on side sometimes finer and sparser anteriorly (fig. 62), (in exceptional forms[11] a few short ridges present anteriorly or punctation grading to weak and fine rugosopunctation dorsally); terga often polished in female, transverse apical fasciae often present in male, absent or weak in female, tergum I with lateral carina, II with a weak carina basally; female sternum II simple, convex basally; female with a pygidial plate defined by grooves or carinae, surface glabrous, usually shining, impunctate to moderately punctate, rarely densely punctate, apex without a transverse row of stout setae (fig. 66 A, B); male pygidial plate at most poorly defined, surface punctate and setose; femora and tibiae short and stout in female, closing face of hindfemur not excavate nor with a basal process; female tibiae with rows of stout spines on outer surface ranked as follows: one or two rows on I except none in *Cratolarra* (fig. 64 A, J), three on II, and two on III, male tibiae more weakly, sparsely spined especially in *Cratolarra,* spine rows of male foretibia in subgenus *Larra* sometimes represented by a single spine (fig. 64 I); hindtibia often angular in cross section but without polished, sharpedged longitudinal carina; female with foretarsal rake of stout spines, males with or without a weak rake; last tarsomere normal, evenly arcuate in lateral view and gradually broadening from base to apex in dorsal view (parallel sided toward apex in *L. dux*), venter without dense hairmat, claws short and without inner teeth (fig. 64 P); submarginal cell II infrequently petiolate (fig. 70 A) or triangular; apex of male sternum VIII variable (fig. 63 J); volsella slender, elongate especially in subgenus *Larra,* base of gonostyle with a pair of lobes which form a circular enclosure (fig. 63 K).

Geographic range: 65 species are listed in *Larra,* and the majority are inhabitants of the tropics. One species is known from the Nearctic Region, and only three or four are found in the Palearctic Region. The subgenus *Cratolarra* is restricted to the Old World tropics. Island endemics are found in Madagascar and in Australia.

At least four species of *Larra* have been introduced to new areas in an effort to control mole crickets (Gryllotalpidae). The Oriental *polita luzonensis* is established in the Hawaiian Islands (F. Williams, 1928a); and *bicolor,* a South American wasp, is now common in Puerto Rico (Wolcott, 1938). *Larra amplipennis,* another Oriental species, was introduced to Hawaii, but it did not become established (Williams, 1928a). The South American species, *scapteriscica* and *bicolor,* were also unsuccessfully introduced to Hawaii.

Systematics: These are small to large (6-25 mm long), black wasps which often have a partially or totally red gaster. The legs are frequently partly red in the Old World subgenus *Cratolarra,* but except for *dux,* the species of the subgenus *Larra* have black legs. The body, especially the gaster, is usually shiny, and the head and thorax are usually rather densely, finely to coarsely punctate. The legs, especially in the female, are short,

---

[11] The propodeal side in a male *Larra (Cratolarra)* from Guadalcanal is entirely covered by diagonal ridges, but there are punctures between the ridges dorsally.

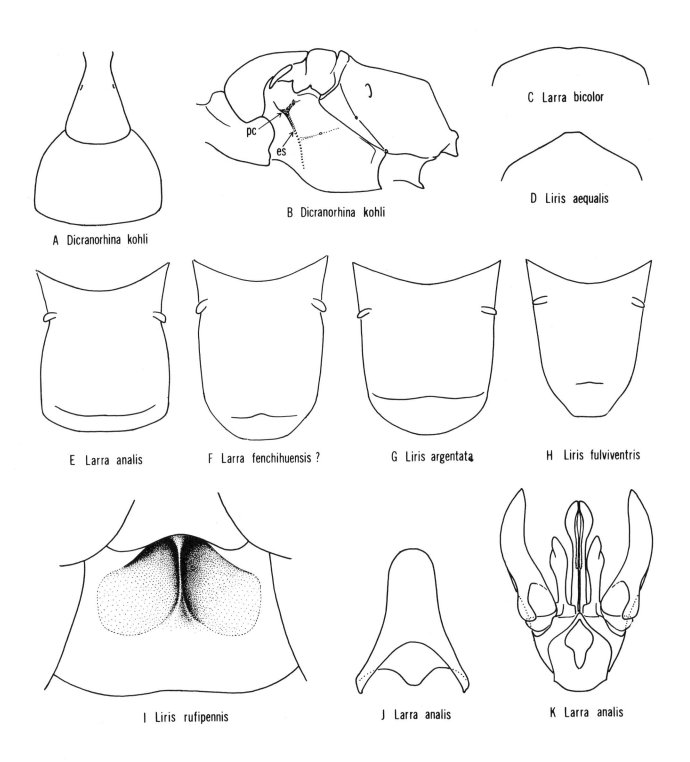

FIG. 63. Morphology in the subtribe Larrina; A, terga I-II; B, pleuron, pc = postspiracular carina, es = episternal sulcus; C,D, outline of female pronotal collar, front view; E-H, female propodeum, dorsal; I, female sternum II; J, male sternum VIII, K, male genitalia, ventral.

stout, and spiny. Morphology and ethology suggest that *Larra* is the most primitive genus in the subtribe Larrina, and the genus vies with *Larropsis* as the most primitive genus of the Larrini.

Structurally *Larra* and the genus *Liris s.l.* are very similar, and most if not all of the features used by previous workers to separate the former from the latter are unreliable. In part this is because certain characters apply only to females, a fact which has not always been made clear by some authors. The first three of the following characters fall into this category: 1) pronotal collar thick, straight on top, and as high as scutum; 2) sternum II without a pair of special flattened oval areas basally; 3) pygidial plate bare and without a transverse row of stout setae apically; 4) claws short and without inner teeth; 5) foretibia with spine rows; 6) no teeth on inner margin of mandible; 7) propodeum constricted at level of spiracles; and 8) integument shining. *Larra* itself is rather constant in its morphology, and its biology is different from that of *Liris*. The biological differences offer a fairly strong argument for maintaining the two taxa as separate genera. The problem of separation of *Larra* and *Liris* centers essentially on the broad spectrum of structural diversity in the latter. Practically every distinctive morphological feature of *Larra* (and there are not many) has a counterpart in some section of *Liris s.l.*

We have found only one characteristic common to both sexes of *Larra* which will, in the majority of cases, separate the genus from *Liris*. This is the sculpture of the propodeal side. Usually it is smooth, shining, and densely punctate (fig. 62). The punctures tend to become finer and sparser anteriorly, sometimes resembling pinpricks. The sculpture of the propodeal side in *Liris* is variable, but most often it is rather smooth and dull. It is never densely punctate. The propodeal side in a few exceptional *Liris* is smooth, shiny, and punctate, but the punctures are very fine, resembling pinpricks. Frequently, there are strong diagonal ridges on the side in *Liris,* but there are no punctures between them. In addition to the normal punctation some male *Larra* of the subgenus *Cratolarra* have a few short diagonal ridges or striatopunctation anterad. We have also seen a *Cratolarra* male from Guadalcanal in which the propodeal side is extensively ridged, but there are many punctures between the ridges.

The lack of teeth on the inner margin of the mandible is a positive characteristic of both sexes of *Larra,* but unfortunately the same condition is universal in *Liris,* subgenus *Motes,* and is not uncommon among males of the subgenus *Leptolarra*. Furthermore, the mandibles must be spread (not an easy task) to ascertain the presence or absence of teeth. When used in conjunction with the propodeal side sculpture the mandibular character should enable anyone to separate *Larra* and *Liris.*

In addition to the female generic characters listed at the beginning of this discussion (1-3), we have found with one exception a generally excellent difference between the two genera in the form of the last tarsomere of the female. This tarsomere is normally formed in *Larra* females (and males), i.e., gradually broadening from base to apex in dorsal view and evenly arcuate in lateral view (fig. 64 P). The venter of the tarsomere is not covered by a mat of short, dense setae, although there are some spines and setae scattered over the surface. The last female tarsomere in *Liris*, especially on the mid and hindlegs, is specially modified in all but a few species. Instead of broadening gradually to the apex, the tarsomere is parallel sided from about its middle to the apex. In lateral view the tarsomere is not evenly arcuate but instead is angled most sharply at or about the midpoint, resembling a banana (fig. 64 O). The ventral surface is covered by a mat of short, dense velvety hairs. This tarsal character is an excellent one for separating most female *Larra* and *Liris*. *Larra dux* is the only species that we have seen with a *Liris*-like tarsus, but there is no ventral hairmat.

Larra divides into two groups which may be regarded as subgenera. The name *Cratolarra* Cameron, which has been misapplied in *Liris* by Williams (1928a) and Tsuneki (1967a), is available for one of the subgenera. Richards (1935a, p. 163) established that *Cratolarra* was a true *Larra* and not a *Liris s.l.*

Key to subgenera of *Larra*

Outer face of foretibia with one or two spine rows (fig. 64 A), or with at least one subapical spine in male (fig. 64 I); cosmopolitan .... subgenus *Larra* Fabricius
Outer face of foretibia without spines (fig. 64 J), Old World Tropics .... subgenus *Cratolarra* Cameron

A number of features suggest that *Cratolarra* is more advanced than *Larra s.s.* The inner orbits converge more strongly above, especially in males; spines are absent on the outer face of the foretibia; the pygidial plate usually is bordered laterally by a carina rather than by just a sulcus; and the volsella tends to be shorter.

When one studies the phylogenetic relationships between *Larra s.l.* and *Liris,* it is immediately apparent that *Cratolarra* represents an evolutionary stepping stone from *Larra s.s.* to *Liris s.l.* The advanced features of *Cratolarra* enumerated above, with the possible exception of the volsellar character, are shared with *Liris*. Also, it is in *Cratolarra* that propodeal sculpture sometimes verges on that found in some *Liris*. Furthermore, the more slender body of *Cratolarra* is suggestive of *Liris*. For further data see the discussion under *Liris*.

Taxonomic characters used in *Larra* consist mainly of details of the head, although the form of the collar, thoracic sculpture, body color, and male genitalia offer specific differences. From our observations it would appear that the color of the gaster may vary within a species and hence its reliability as a species character is in doubt. Two atypical species are worth mentioning. The second submarginal cell is petiolate in the unrelated species, *dux* Kohl and *princeps* F. Smith. Material we have studied of the latter species suggests that the character is not con-

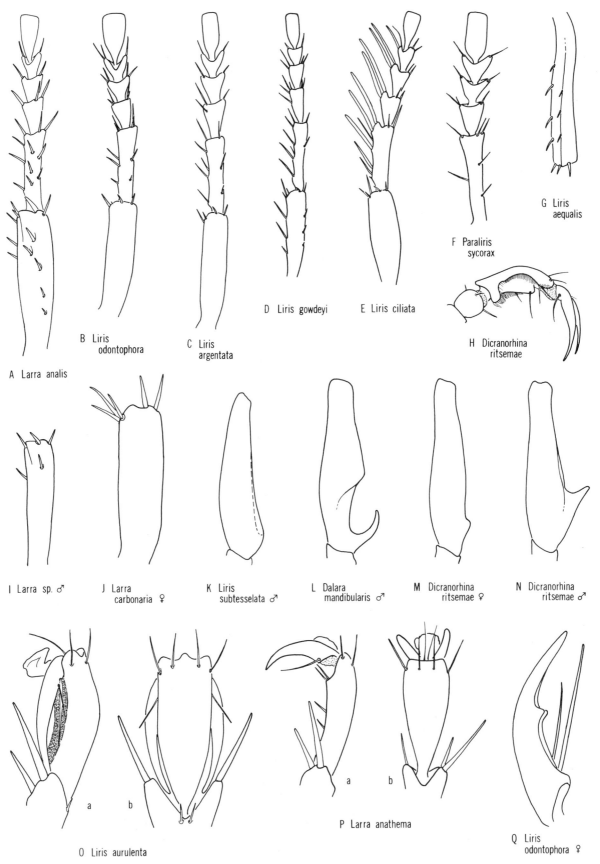

FIG. 64. Leg details in the subtribe Larrina; A-E, female foretibia and tarsus, left dorsal; F, female foretarsus, left, dorsal; G, female hindtibia; H, female foretarsomeres IV-V, inner, lateral; I-J, foretibia, left dorsal; K-N, hindfemur, left, lateral; O-P, female hindtarsomere V, lateral (a) and dorsal (b); Q, bidentate hindtarsal claw.

stant. The taxa *Monomatium* and *Larraxena* were proposed for *princeps*, but this species is a *Larra s.s.*, and we have placed these two generic names in synonymy. *Larra dux*, of which we have seen only the female type (Mus. Vienna), is odd for other reasons. It is the only species in the subgenus *Larra* with red legs. More importantly the last tarsomere is parallel sided as in *Liris*, and the pronotal collar is sharp edged, closely applied to and somewhat below the level of the scutum.

Larra is in need of revision. Existing keys are regional and sometimes incomplete or otherwise outdated. F. Williams (1928a) keyed the South American and Philippine species, Tsuneki (1967a) keyed the Taiwanese species, Turner (1916b) keyed the Australian species, and Arnold (1923a, 1945) keyed the species of southern Africa and Madagascar.

Biology: *Larra* preys upon mole crickets (Gryllotalpidae). Records of *Larra* preying on crickets of the family Gryllidae are probably erroneus. *Larra* females do not construct their own nest and prey paralysis is temporary, the host reviving soon after egg deposition. Reminiscent of parasitoids, the egg of *Larra* is small, and females have a high egg output, each female probably laying more than 30 eggs during her lifespan. Because *Larra* has a sting it does not fit the definition of a parasitoid, although in most other respects it behaves like one. *Chlorion* is the only other pseudoparasitoid known in the Sphecidae. This primitive biology is more characteristic of bethylid and tiphiid wasps.

The basic references of *Larra* ethology are those of Williams (1928a), C. E. Smith (1935), Iwata and Tanihata (1963), Tsuneki (1969a), and Nambu (1970). Most of these writers have successfully reared *Larra* in glass containers in the laboratory, which has permitted close observation of stinging, egg deposition, etc. There are no apparent biological differences between the two subgenera. Females search the soil surface for mole cricket burrows. When one is located they enter either by using the entrance or by digging down to one of the passageways. The occupant, which is much larger than the wasp, usually leaves the burrow when disturbed by the wasp, the latter in hot pursuit. Stinging usually occurs after the cricket is on the soil surface, but females sometimes paralyze the cricket while it is still in the burrow. Apparently, in the latter case the victim is dragged to the surface for oviposition. Although the egg is ordinarily laid while the cricket is immobile on the soil surface, some *Larra* have been observed to drag the cricket into the burrow before oviposition. The egg is laid on the thorax, usually on the venter between the legs. Seemingly the deposition site is different from species to species. Also, each species of *Larra* tends to be host specific. Paralysis of the cricket lasts only a few minutes and the host soon burrows into the ground. Pupation occurs in the host's burrow. Occasionally the cricket already bears the egg of a previous *Larra* female. In this case the egg may be removed by the second *Larra* female. C. E. Smith (1935) noted a few instances where a *Larra analis* female deposited two eggs on a cricket. Some crickets accidently escape their fate by molting soon after they have been victimized. Smith recorded one case of parthenogenesis in *L. analis*. The unfertilized larva developed and pupated normally, but no adult emerged. Smith also found that *L. analis* was parasitized by a nematode. *Larra* females spend the night in burrows in the soil.

Checklist of *Larra*

Species preceded by a ? (not including synonyms) may belong in *Liris*, *Tachytes*, or *Tachysphex*. Subgeneric assignment, when known, is indicated by (L) for *Larra* and (C) for *Cratolarra*.

†*abdominalis* Guérin-Méneville, 1849; Ethiopia; nec Say, 1823
†*aethiops* (F. Smith), 1873 (*Larrada*); Brazil, Ecuador, nec Cresson, 1865, a *Tachysphex* (L)
alecto (F. Smith), 1858 (*Larrada*); Singapore
altamazonica Williams, 1928; Ecuador, Brazil (L)
amplipennis (F. Smith), 1873 (*Larrada*); Japan, Ryukyus, Taiwan, Thailand, Philippine Is., (L)
   sanguinea Williams, 1928
   aeripilosa Tsuneki, 1963
analis Fabricius, 1804; e. U.S. (L)
   canescens F. Smith, 1856 (*Larrada*), new synonymy by Bohart
   americana Cresson, 1872 (*Larrada*), nec Saussure, 1867
   cressonii W. Fox, 1893
anathema (Rossi), 1790 (*Sphex*); s. Europe, n. Africa, Iraq (L)
   ichneumoniformis Fabricius, 1793
   ?*teutona* Fabricius, 1804 (*Pompilus*)
   grandis Chevrier, 1872 (*Tachytes*)
   ssp. *melanaria* Kohl, 1880; Italy, Morocco
   ssp. *nudiventris* A. Costa, 1893; Tunisia
angustifrons Kohl, 1892; Java
apicepennis Cameron, 1904; e. India
betsilea Saussure, 1887; Madagascar (L)
   ssp. *proditor* Kohl, 1891; Morocco to S. Africa
      pseudanathema Kohl, 1894
      semirubra Bischoff, 1913
   ssp. *rufa* Arnold, 1929; Zaire
bicolor Fabricius, 1804; Puerto Rico, Venezuela, Guyana, Surinam, Brazil (L)
   americana Saussure, 1867 (*Larrada*)
bicolorata Cameron, 1904; e. India
braunsii Kohl, 1898; Brazil, Peru (L)
bulawayoensis Bischoff, 1913; Rhodesia (C)
   ssp. *occidentalis* Arnold, 1923; Cameroon
burmeisterii (E. Lynch Arribalzaga), 1884 (*Larrada*); Argentina (L)
carbonaria (F. Smith), 1858 (*Larrada*); Singapore, ? Burma, Taiwan, Japan, Philippine Is., Sumatra, Java (C)
   docilis F. Smith, 1873 (*Larrada*) (♂ only, "♀" = ♂ *Liris*)
   erebus F. Smith, 1873 (*Larrada*)
   sparsa Rohwer, 1911
cassandra Schrottky, 1902; Argentina (L)
coelestina (F. Smith), 1873 (*Larrada*); China
   caelestina Dalla Torre, 1897

corrugata Turner, 1912; New Guinea (C)
?*diversa* (Walker), 1871 (*Larrada*); Egypt
*dux (Kohl), 1892 (*Larraxena*); Zaire, Rhodesia, Zanzibar, Senegal (L)
erratica Bingham, 1897; Sri Lanka, Burma
erythropyga Turner, 1916; Malawi, Lesotho (L)
extrema Dahlbom, 1845; S. Africa
femorata (Saussure), 1854 (*Tachytes*); Australia, Tasmania (C)
   ?*scelesta* Turner, 1908
†*femorata* (Cameron), 1900 (*Cratolarra*); e. India, nec Saussure, 1854 (C)
fenchihuensis Tsuneki, 1967; Taiwan (C)
fuscinerva Cameron, 1900; n. India
*gastrica (Taschenberg), 1870 (*Larrada*); Argentina, Venezuela, lectotype ♂, Parana, Dec. (Mus. Halle), present designation by Menke
godmani Cameron, 1889; Mexico (L)
glabrata (F. Smith), 1856 (*Larrada*); Celebes
guiana Cameron, 1912; Guyana (L)
heydenii Saussure, 1890 (pl. 12, fig. 18); Madagascar (C)
   heidenii Saussure, 1891 (*Notogonia*), emendation
impressifrons Arnold, 1923; Rhodesia, S. Africa
madecassa Saussure, 1887; Madagascar (L)
mansueta (F. Smith), 1865 (*Larrada*); New Guinea
maura (Fabricius), 1787 (*Sphex*); India, ? China, ? Thailand (C)
   ? ssp. *rechingeri* Kohl, 1907; New Britain
?*mediterranea* Gistl, 1857; "Jonicae Insulae"
melanocnemis Turner, 1916; Australia (L)
melanoptera Kohl, 1884; Mauritius
mendax (F. Smith), 1865 (*Larrada*); Indonesia: Moluccas: Halmahera (C)
mundula Kohl, 1894; Sierra Leone
neaera Nurse, 1903; Pakistan
nigripes (Fabricius), 1793 (*Sphex*); S. Africa
obscurior Dalla Torre, 1897; China
   obscura Sickmann, 1894, nec Magretti, 1884
outeniqua Arnold, 1923; S. Africa (C)
pacifica Williams, 1928; Ecuador (L)
pagana (Dahlbom), 1843 (*Tachytes*); West Indies: Virgin Is. (L) (see Stadelmann, 1897)
paraguayana Strand, 1910; Paraguay (L)
parvula Schrottky, 1903; Brazil (L)
polita (F. Smith), 1857 (*Larrada*); Java, Borneo (C)
   ssp. *rufipes* (F. Smith), 1859 (*Larrada*); Celebes, new combination by van der Vecht
   ssp. *luzonenensis* Rohwer, 1919; Philippines, Taiwan, Ryukyus, Hawaii, new status by van der Vecht Vecht
princeps (F. Smith), 1851 (*Larraxena*); S. America (L)
psilocera Kohl, 1884; Tasmania
pusilla Arnold, 1932; Rhodesia
pygidialis Cameron, 1904; e. India
saussurei Kohl, 1892; Madagascar (C)
   prismatica Saussure, 1887; nec F. Smith, 1858
scapteriscica Williams, 1928; Brazil (L)
similis (Mocsáry), 1891 (*Larrada*); China
simillima (F. Smith), 1856 (*Larrada*); India (L)

fuscipennis Cameron, 1889, nec F. Smith, 1856
   ssp. *personata* (F. Smith), 1859 (*Larrada*); Celebes
   ssp. *sumatrana* Kohl, 1884; Sumatra
sinensis (Mocsáry), 1891 (*Larrada*); China
tarsata (F. Smith), 1860 (*Larrada*); Indonesia: Moluccas: Batjan
transcaspica F. Morawitz, 1894; sw. USSR, Iran
   lativalvis Kohl, 1899
transandina Williams, 1928; Ecuador, Brazil, Trinidad (L)
variipes Saussure, 1892; Madagascar (C)
   carbonaria Saussure, 1892, nec F. Smith, 1858
   carbunculus Dalla Torre, 1897
zarudniana Gussakovskij, 1933; Iran

Genus Liris Fabricius

Generic description: Upper inner angle of eye not bordered by a deep sulcus; midocellar lens usually convex, hindocellar scars as in *Larra;* frons and vertex dull to moderately shining (rarely highly polished), sometimes densely micropunctate (especially in subgenus *Motes*); frontal line as in *Larra;* scape usually long, slender (shorter, more robust in *Motes*), scape and pedicel surfaces similar to that of flagellum, flagellomeres usually longer than broad and with placoids in male; clypeus often with a median longitudinal ridge, free margin usually with a rounded or truncate median lobe (triangular in some males) which often has a median notch, free margin of lobe often with a polished, asetose bevel, clypeal margin usually concave laterally; labrum transverse or quadrangular, hidden or occasionally projecting slightly beyond clypeus, free margin usually with a median notch (exceptions mainly in *Motes*); inner margin of mandible with one or two subbasal teeth (fig. 72 C) except in subgenus *Motes* (fig. 72 B), and some males of subgenus *Leptolarra* (occasionally weak in some *Leptolarra* females), or rarely mandible apex with subapical tooth (*tenebrosa*), externoventral margin at most with a weak subbasal V-shaped notch (subgenus *Liris,* fig. 72 C), or with a large semicircular notch or strong angulation (*Motes* and *Leptolarra,* fig. 72 B); dorsum of collar in both sexes usually below level of scutum, and usually thin (sharp edged), wedge-shaped (fig. 63 D) and usually fitting closely with scutum (collar thicker and/or as high as scutum in some *Motes* and *Leptolarra*); notauli usually present; propodeum rather long, rectangular or somewhat rounded posteriorly, propodeal surface dull to strongly shining, dorsum and posterior face variably sculptured (rugosopunctate, ridged, reticulate, smooth, etc.), sides usually smooth or diagonally ridged, rarely with very fine pinprick punctures; terga usually dull and often with transverse apical fasciae, tergum I with lateral carina, II sometimes with a weak carina basally; female sternum II usually with a pair of flattened, oval areas basally, which are separated by a basomedian longitudinal ridge or bulge (fig. 66 I) (similar areas weakly indicated in exceptional males); male sterna sometimes with special erect or decumbent vestiture arranged in apical rows or tufts, sternum II sometimes strongly bulging basally, hindmargin of sternum II sometimes with a median pro-

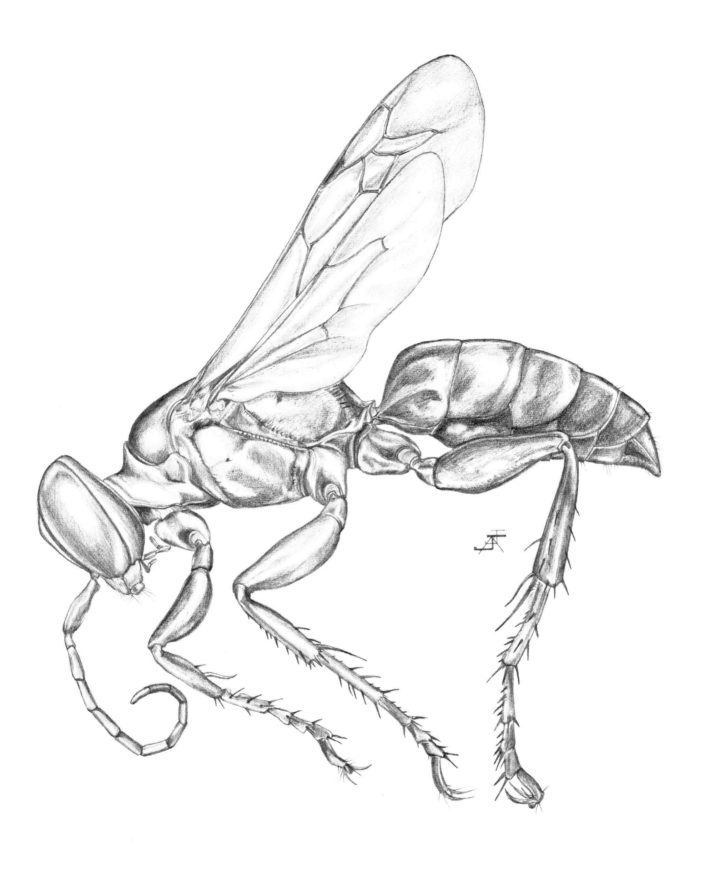

FIG. 65. *Liris argentata* (Palisot de Beauvois), female

cess, sternum VII sometimes modified, sterna III-VII sometimes depressed; female with a pygidial plate (fig. 66 C-H) which is defined by carinae, surface glabrous, pruinose, or partially to fully covered with appressed setae, a few scattered erect hairs sometimes intermixed in latter case, surface dull or shining, punctate or impunctate, apex with a transverse row of stout setae (fig. 66 D-H) except in subgenus *Motes* (fig. 66 C) and *tenebrosa,* apex sometimes emarginate; male sometimes with a pygidial plate (especially in subgenus *Liris*), surface covered with fine decumbent setae and apex broadly emarginate; femora and tibiae usually slender in both sexes, closing face of male hindfemur sometimes flattened or excavate (fig. 64 K), foretibia in both sexes without spine rows on outer face except in a few exceptional species in which foretarsal rake is strongly developed (fig. 64 D), midtibia with two or three and hindtibia with two spine rows, but each often composed of two or three slender widespaced spines, hindtibia usually at least angulate in cross section or with a longitudinal carina the edge of which is polished (fig. 64 G) (*Motes* apparently always without a polished carina); female foretarsus in *Motes* without a rake or rake very weak (fig. 64 B), rake weak (one or two spines per tarsomere) to strong (numerous bladelike spines) in *Leptolarra* and *Liris* (fig. 64 C-E), male with or without a rake; last female tarsomere of mid and hind legs angled in profile (fig. 64 O), sides parallel on apical half in dorsal view (fig. 64 O), and venter with dense hairmat (a few exceptional species from Africa and Australia have normally shaped tarsomere V and/or no ventral hairmat), last male tarsomere normal, female claws usually long and prehensile, inner margin of claw usually simple (except two teeth set side by side in female *Motes* (fig. 64 Q) and two *Leptolarra*, and one tooth in a few females of other *Leptolarra*); recurrent veins of forewing sometimes joining before reaching submarginal cell; apex of male sternum VIII rounded or truncate, often emarginate; volsella long and slender, sometimes exceeding length of aedeagus; gonostyle similar to *Larra* or becoming very complex in many *Leptolarra* and *Liris s.s.* where inner face may bear special setae and entire structure may have one or two points of articulation which allow complex folding of gonostyle.

Geographic range: This is a large cosmopolitan genus containing over 260 species, most of which inhabit the tropics. This number will doubtless increase considerably when the Neotropical fauna has been studied. The subgenus *Liris* is confined to the Old World. A few species have been accidentally introduced by man to new areas such as Hawaii, which has one immigrant from the New World and two from the Oriental Region.

Systematics: We are using *Liris* in the broad sense; that is, it includes, besides *Liris s.s.*, the taxa that have been known under the following names: *Motes, Notogonidea, Notogonia,* and *Leptolarra.* This is not new, because Arnold (1945) and more recently Beaumont (1961c) have used this concept. Our studies indicate that there is indeed little morphological evidence to support the retention of these taxa as separate genera. In fact, one could go a step farther and unite *Liris* with *Larra,* because the characters which separate the two are rather weak, especially in the male.

Liris is a morphologically diverse group, but in general the species are dull black. Red when present is usually confined to the legs, although the propodeum is red in *fulvipes;* and at least five species, *braunsi, conspicua, fulviventris, rubella,* and *rubricata,* have a completely red gaster as in *Larra.* Silvery or brassy appressed vestiture as fasciae on the gastral terga, or as patches elsewhere, is common in *Liris.* The tergal vestiture reaches its greatest development in *L. haemorrhoidalis.* Several Neotropical *Liris* have striking patterns of appressed vestiture on the thorax and gaster. Unlike *Larra,* the integument is usually dull in *Liris,* but there are sufficient exceptions to make this only a generalization. The abdominal dullness is often due to a fine, appressed vestiture. Species range in length from 5 to 30 mm.

Some of the difficulties of separating *Liris* from *Larra* have been discussed under *Larra.* Most of the differences between the taxa are based on the female as can be seen in the keys to genera. It is the male that presents problems. The only feature common to both sexes, which can be used to isolate *Liris* from *Larra,* is the sculpture of the propodeal side, and even this has its drawbacks. The propodeal side is dull, smooth or minutely granulate in most *Liris,* and is always impunctate except in a handful of cases. In these latter (*tenebrosa* and *lutusator* are examples) the side is shiny, as in *Larra,* but the punctation is sparse and very fine, resembling pinpricks. In other *Liris* the propodeal side may be rugose or diagonally ridged and sometimes shiny. We have seen no *Liris* with the shiny, smooth and densely punctate propodeal side found in *Larra,* but it is quite possible that such species exist. As presently understood, the biological differences between the two taxa supports their maintenance as discreet genera.

Some discussion of a few of the features that have been used to separate *Liris* from *Larra* is necessary in order to emphasize the problems involved. In both genera the collar is wedge shaped except in *Larra* females. In *Liris bembesiana* and *ciliata* the collar is weakly wedge shaped in the female. The mandible in *Liris* has one or two subbasal teeth on the inner margin except in the subgenus *Motes* and in the males of some species in the subgenus *Leptolarra (agilis, bembesiana, ciliata, flavipennis, lutusator, miscophoides,* and *simulatrix* are examples). In *Larra,* subgenus *Cratolarra,* and in most *Liris* there are no spines on the outer face of the foretibia. However, in females (and some males) of a few *Liris* in which the foretarsal rake is well developed, there is a row of spines on the foretibia (*aurulenta, cowani, croesus, gowdeyi,* and *haemorrhoidalis,* for example). This is a characteristic of *Larra s.s.* In *Liris* the sides of the propodeum generally converge posteriorly when viewed from above (fig. 63 H). In *Larra* there is usually a constriction at about the level of the spiracles so that the sides diverge posteriorly. However, in both genera there are some marginal or intermediate cases (fig. 63

F,G). The hindtibia in many *Liris* has a sharp, polished carina, but in all *Motes* and a fair number of *Leptolarra* this carina is absent, as in *Larra*. The second sternum in female *Liris* nearly always has a pair of specialized, flattened, oval areas basally, which are set apart by a linear or tubercular swelling. *L. tenebrosa* is one notable exception. *Larra* has a simple sternum. There are a few *Liris* that nearly approximate *Larra* in this character (*bembesiana*, for example). The pygidial plate in *Larra* is always glabrous and without a transverse row of setae apically. Unfortunately, all *Liris* of the subgenus *Motes* have a similar pygidial plate, although in these species there is usually a fine bloom on the surface. In the rest of *Liris,* with the exception of *tenebrosa*, the pygidial plate bears an apical row of stout setae, and the plate surface is usually partially or totally covered with decumbent setae. However, in some *Leptolarra* the plate surface is glabrous or sparsely setose. The last tarsomere of the mid and hind legs in nearly all female *Liris* is modified and distinct from the normal *Larra* tarsomere. It is parallel sided on the apical half to two-thirds in dorsal view, angled most strongly at about the midpoint in lateral view, and bears a dense mat of short setae ventrally. In addition, the claws are long and prehensile. Exceptions are *Liris australis*, which has a *Larra*-like tarsomere in all respects, and *odontophora, bembesiana*, and *ciliata*, which have the tarsomere formed nearly like *Larra*, but it has a ventral hairmat. At least one *Larra (dux)* has the tarsomere formed as in *Liris*, but the claws are short, not prehensile, and there is no ventral hairmat.

In his 1961 paper on the *Liris* of the Mediterranean Region, Beaumont (1961c) proposed a species group classification for the taxa of that area and indicated the homologies between his groups and the various generic names that have been proposed over the years. He did not use any of these generic names as subgenera. In a later paper (Beaumont, 1965) he suggested that his *odontophora* group could be recognized as a subgenus, namely *Motes*. Arnold (1945) used *Motes* as a subgenus but in a much broader sense than that proposed by Beaumont. We have found it convenient to recognize three subgenera, *Liris s.s., Motes,* and *Leptolarra*. The first two are small, rather homogeneous, and correspond to Beaumont's *aurulenta* and *odontophora* groups. *Leptolarra* contains the vast majority of *Liris* species and is a very diverse assemblage. The following key illustrates the differences among the three subgenera. It can be seen that males offer little in the way of subgeneric characters (genitalia have not been studied in this regard).

Key to subgenera of *Liris*

1. Mandible entire or with a small V-shaped notch on externoventral margin (fig. 72 C); Old World .... *Liris* Fabricius
Mandible with a broad, deep, emargination (fig. 72 B) on externoventral margin 2
2. Apex of female pygidial plate with transverse row of stout setae (fig. 66 D), surface usually at least partially covered with appressed setae or hairs; female (and usually male) mandible with one or two teeth on inner margin near base (fig. 72 C); female tarsal claw usually simple but when toothed within, the pygidial plate is not completely glabrous; hindtibia often with a sharp, polished carina (fig. 64 G)........... *Leptolarra* Cameron
Apex of female pygidial plate without transverse row of stout setae (fig. 66 C) surface glabrous or at most with a fine dustlike bloom; mandible of both sexes without inner teeth (fig. 72 B); inner margin of female tarsal claw with two teeth set side by side (fig. 64 Q); hindtibia without a sharp, polished carina .... (some males of *Leptolarra* will key out here) .... *Motes* Kohl

The subgenus *Motes* is the most *Larra*-like group within *Liris*. Males of *Motes* are so similar to those of *Larra* that the propodeal side sculpture discussed earlier is the only feasible distinguishing feature, except association with females. The only biological information for *Motes* is that prey consists of Gryllidae, and this supports the placement of *Motes* in *Liris* rather than *Larra*.

The mandibular character used to separate *Leptolarra* and *Liris s.s.* is perhaps not phylogenetically sound. Beaumont (1961c), for example, indicated a close relationship between his *aurulenta* group (equivalent to *Liris s.s.*) and his *memnonia* group which is a section of the subgenus *Leptolarra*. Furthermore, the presence or absence of a mandibular notch is subject to some variation. For example, the mandible of *L. (Liris) mandibularis* has a rather large V-shaped notch in the female and an angulation or step in the male. However, it is more convenient to recognize *Leptolarra* rather than to merge it with *Liris s.s.* The subgenus *Leptolarra* is such a diverse group that it may eventually benefit from division into several subgenera. Since a number of generic names are available for this purpose, we have listed below some of Beaumont's species groups followed by their equivalent generic names. These are listed in order of their priority.

memnonia group:	*Chrysolarra* Cameron, 1901
	Dociliris Tsuneki, 1967
nigricans group:	*Leptolarra* Cameron, 1900
	Nigliris Tsuneki, 1967
nigra group:	*Spanolarra* Cameron, 1900
	Notogonia Costa, 1867 (preoccupied)
	Caenolarra Cameron, 1900
	Notogonidea Rohwer, 1911

According to Beaumont (1961c:239) and van der Vecht (correspondence), the holotypes of the type species of *Spanolarra* and *Caenolarra* are probably the opposite sexes of a single species. When Tsuneki (1967a) proposed *Dociliris* and *Nigliris* he was apparently unaware that earlier names of Cameron were available. The *nigra* group is the most likely contender for subgeneric status of those recognized by Beaumont and placed by us in *Leptolarra*. F. Williams (1928a) and Tsuneki (1967a) im-

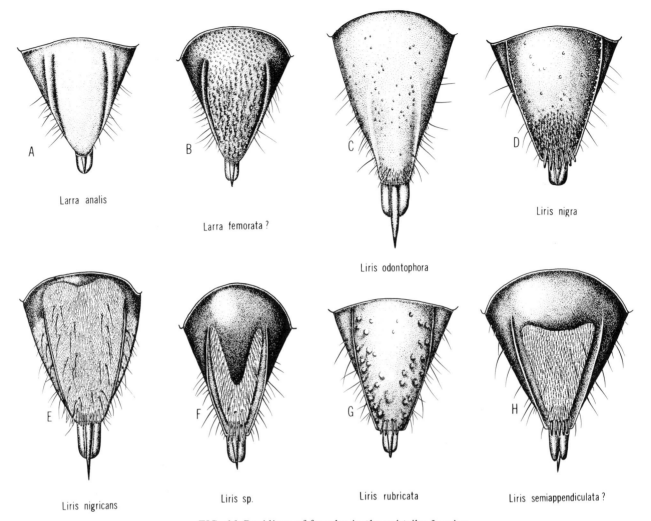

FIG. 66. Pygidium of females in the subtribe Larrina.

properly referred to *Cratolarra* some Oriental species that belong in this group. Richards (1935a) pointed out that *Cratolarra* is a group within *Larra*. The essential difference between *Spanolarra*, as the *nigra* group should be called if it is eventually regarded as a subgenus, and the similar subgenus *Motes* is the presence in the former of basal teeth on the inner margin of the female mandible, the presence of a transverse row of setae at the pygidial plate apex, and the thick collar. Males of the two groups are assigned to their proper taxon only with difficulty.

Beaumont's groups do not encompass all of the possible divisions within *Leptolarra*. There are elements in Africa, Australia, and the Neotropical Region that are equally distinctive, and a few species are very unusual. In the Neotropical Region there is great variation in pygidial morphology (fig. 66 F-H). Here, also, the pygidial plate apex is emarginate in some species. The mandible apex is bidentate in a few Neotropical species, with *tenebrosa* especially notable in this regard. Females of a few *Leptolarra* have claw teeth: *croesus*, *rufipennis*, (two teeth side by side as in *Motes*), *denticulata*, *dentipes*, and *neavei*, for example. The majority of females of *subtessellata* studied by us have simple claws, but a few have toothed claws. Some *Leptolarra* have a well developed foretarsal rake with spines long and bladelike (fig. 64 E): *ciliata*, *bembesiana*, *pictipennis*, and *sericosoma*, for example.

Liris s.s. contains two distinctive groups. The typical *aurulenta* group consists of species confined primarily to mainland areas. These forms have rather elongate bodies, and erect pubescence is sparse or absent, but appressed hair is abundant and frequently brassy. The foretibia often has a row of spines in the female. The second or *melania* group consists mostly of shorter, black wasps which have much erect hair. These predominate in the islands of the western Pacific.

The subgeneric components of *Larra* and *Liris* can be arranged phylogenetically in an almost linear fashion: *Larra* s.s. – *Cratolarra* – *Motes* – *Spanolarra* – *Leptolarra* – *Liris* s.s. Some of the criteria used in arriving at this scheme are tabulated below

Species characters are numerous in *Liris*, and for these details the reader should consult the excellent illustrations and discussions found in Beaumont (1961c), Tsuneki

TABLE 13
Phylogenetic Characters in *Larra* and *Liris*

Generalized	Specialized
1. Mandible without inner teeth	Mandible with inner teeth
2. Mandible notched externally	Mandible entire (notch presumed to be secondarily lost in *Liris s.s.*)
3. Pronotum not wedge shaped, thick, not closely fitting with scutum	Pronotum wedge shaped, thin, appressed to scutum
4. Foretibia with spines on outer surface	Foretibia without spines (presumably secondarily redeveloped in some *Leptolarra* and *Liris s.s.*)
5. Pygidial plate bare, apex without transverse row of setae	Pygidial plate setose, apex with transverse row of setae
6. Hindtibia not carinate	Hindtibia with a sharp, polished carina

(1967a), and Williams (1928a).

A number of regional keys to species exist, but some of the older ones are outdated. The important references are Krombein (1954b) for North America and (1949, 1950a) for Micronesia, Beaumont (1961c) for the Mediterranean area, Tsuneki (1967a) for Taiwan, F. X. Williams (1928a) for the Philippines, Turner (1916a, b) for Australia, and Arnold (1923a, 1945) for the Ethiopian Region and Madagascar, respectively.

The Neotropical *Liris* fauna is unworked, and because it is large and diverse no satisfactory subgeneric or other supraspecific classification can be attained without a thorough knowledge of it. Clearly, there is a need for a taxonomic study of the *Liris* of this region.

Biology: Liris has a more advanced biology than *Larra*. Prey is captured, stung, but only weakly paralyzed and brought to a nest constructed by the wasp or more commonly to a preexisting cavity such as a bee burrow that is located beforehand. The weak paralysis of prey is important because it indicates that in this aspect of behavior *Liris* has not progressed much over *Larra*. Steiner's (1962) paper is a classic study of one species, *L. nigra*, but the main emphasis of his work is hunting and stinging of prey. He has reared *L. nigra* in the laboratory for many years (Steiner, 1965). Notable publications on *Liris* behavior are: F. Williams (1914a, 1919, 1928a), Rau and Rau (1918), Piel (1935a) Krombein and Evans (1955), Iwata (1964a), Steiner (1968, 1971), Okumura (1971), and Tsuneki (1969a). Other references can be found in these papers.

Crickets of the family Gryllidae may be the exclusive prey of all subgenera of *Liris*, but Arnold (1945) stated that camel crickets (Gryllacrididae) are used by *L. brunnipennis*. In the same paper Arnold said that a mole cricket (Gryllotalpidae) was associated with one specimen of *L. incerta*, but that gryllids were the common prey of this species. The gryllotalpid record seems questionable. After paralysis the prey is either flown or dragged over the ground in an upright position to the nest site. The wasp grasps the antennae of the cricket with her mandibles during transport. One can speculate that the development of inner teeth on the mandible in *Liris* is correlated with prey carriage. *Larra*, which does not transport prey, does not have mandibular teeth. The clypeal notch found in some *Liris* females may have a similar correlation.

Published accounts indicate that *Liris* females apparently prefer to use preexisting burrows or galleries in the soil for nesting purposes. These cavities may be modified by the wasp. In some species, *nigra* for example, the female will excavate its own burrow if no preexisting cavities can be found. This species has a weakly developed foretarsal rake. Some species of the subgenus *Liris* excavate very deep nests, usually using some preexisting burrow as a starting place. F. Williams (1928a) said that nests of *L. (Liris) haemorrhoidalis* subsp. *magnifica* are as deep as 1.3 m and as long as 2.2 m. This species has a moderately developed foretarsal rake. Excavation of this nest may take several weeks according to Williams. *Liris haemorrhoidalis* works on the nest both day and night. The nests of *Leptolarra* such as *nigra*, *festinans* subsp. *japonica*, and *argentata* are much shorter and shallower (9 cm long, 6.5-12.5 cm deep). The nest of *haemorrhoidalis* terminates in a number of branches, each with two to nine cells. On the other hand, *Leptolarra* usually have simple, one-celled nests. Williams (1914a), Ferton (1901), and Tsuneki (1969a) found, however, that *argentata*, *nigra*, and *japonica* sometimes have multicellular nests.

The available evidence, though fragmentary, suggests that the biology of the subgenus *Liris* is more highly developed than that of *Leptolarra*. Unfortunately, nothing is known about the nesting habits of *Motes*, except that prey consists of Gryllidae. Williams (1919, 1928a) noted that several females of the subgenus *Liris* shared a single nest entrance, but each female had her own nest within, so that sociality is apparently absent.

Cells of *Liris* are usually stocked with two to seven adult and immature crickets. Occasionally only one cricket is provisioned per cell. Apparently the egg is laid on the last provision and attached to the thoracic venter as in *Larra* (for exceptions see Okumura, 1971). Nests are left open until provisioning is completed, and closure is accomplished by filling the upper part of the burrow with loose bits of plant material, small pebbles, etc. Williams (1928a) referred to *L. lutusator* as a mud dauber, but probably it merely uses mud to close the nest or partition cells within it.

Checklist of *Liris*

Species preceded by ? (not including names in synonymy) may belong in *Tachytes*, *Tachysphex*, or *Larra*. Subgeneric assignment is indicated by (L) for subgenus *Liris* and (M) for subgenus *Motes*. Species without either appellation belong in *Leptolarra*, or else their subgeneric assignment is unknown.

abbreviata (Turner), 1908 (*Notogonia*); Australia
abyssinica (Arnold), 1933 (*Notogonidea*); Ethiopia

aciculata (Cameron), 1905 (*Notogonia*); e. India
aedilis (F. Smith), 1869 (*Larrada*); Celebes
aequalis (W. Fox), 1893 (*Notogonia*); N. America, Mexico
   ?*montezuma* Cameron, 1889 (*Notogonia*)
    nigripennis W. Fox, 1893 (*Notogonia*), nec Cameron, 1889
    nigripennata Dalla Torre, 1897 (*Larra*)
    occidentalis Viereck, 1903 (*Notogonia*)
    subaequalis Rohwer, 1909 (*Notogonia*)
agilis (F. Smith), 1856 (*Larrada*); n. Africa, Canary Is., Cape Verde Is., Jordan
    costae Magretti, 1884 (*Notogonia*)
    cooperi Beaumont, 1950
agitata (Turner), 1908 (*Notogonia*); Australia
alaris (Saussure), 1892 (*Notogonia*); Madagascar
  ssp. *felina* (Arnold), 1923 (*Notogonidea*); Rhodesia, S. Africa
alberti (Arnold), 1943 (*Leptolarra*); Zaire
    alberti Arnold, 1944
albopilosa Tsuneki, 1967; Taiwan
angustiventris (Arnold), 1923 (*Notogonidea*); Rhodesia
antaka (Saussure), 1890 (pl. 19, fig. 7) (*Notogonia*); Madagascar
    ancara Saussure, 1891 (*Notogonia*)
    antaca Saussure, 1892 (*Notogonia*), emendation
  ssp. *transvaalensis* (Cameron), 1910 (*Notogonia*); Ethiopian Region
    ?*massaica* Cameron, 1908 (*Notogonia*)
    brevicarinata Cameron, 1910 (*Notogonia*)
    pretoriaensis Cameron, 1910 (*Notogonia*)
    griseola Arnold, 1923 (*Notogonidea*)
anthracina Kohl, 1892; ? Sikkim (L)
†*anthracina* (Cameron), 1903 (*Notogonia*); e. India, nec Kohl, 1892
antica (F. Smith), 1856 (*Larrada*); Brazil, Trinidad (M)
   ?*rufogeniculata* Cameron, 1912 (*Tachysphex*)
antilles (Krombein), 1953 (*Motes*); Bahamas, Cuba
apicalis (W. Fox), 1896 (*Notogonia*); e. Africa
apicipennis (Cameron), 1889 (*Notogonia*); Panama to Texas
*appendiculata (Taschenberg), 1870 (*Larrada*); "Congon." (? = Brazil), new combination by Menke
†*appendiculata* (Cameron), 1901 (*Chrysolarra*); e. India, nec Taschenberg, 1870, nec Cameron, 1900 (? = *deplanata* Kohl)
argentata (Palisot de Beauvois), 1811 (*Larra*); N. America, Cuba, Bahamas, Hawaii
    pensylvanica Palisot de Beauvois, 1811 (*Larra*), new synonymy by Bohart and Menke
    pennsylvanica F. Smith, 1856 (*Larrada*), emendation
    *nuda Taschenberg, 1870 (*Larrada*), new synonymy by K. V. Krombein
argenticauda (Cameron), 1889 (*Notogonia*); Mexico
    chrysura Cameron, 1889 (*Notogonia*), new synonymy by Bohart
argentifrons (Cameron), 1889 (*Notogonia*); Guatemala
aterrima (F. Smith), 1856 (*Larrada*); S. Africa
atrata (Spinola), 1805 (*Larra*); s. Europe, n. Africa, e. Mediterranean region, Iraq, Iran, Canary Is., Cape Verde Is.
    micans Spinola, 1806 (*Larra*)
    nigrita Lepeletier, 1845 (*Tachytes*)
    pharaonum Kohl, 1906 (*Notogonia*)
atropos Gribodo, 1894; Rhodesia, Mozambique, S. Africa (L)
    africana Turner, 1917
atrox Arnold, 1959; Zambia
aureosericea (Cameron), 1901 (*Chrysolarra*); e. India
aurifrons (F. Smith), 1859 (*Larrada*); n. India, Celebes (L)
    nitida Cameron, 1913
aurulenta (Fabricius), 1787 (*Sphex*); India to Japan, Indonesia, Marianas, Hawaii, Carolines, Marshalls (L)
    aurata Fabricius, 1787 (*Sphex*), nec Linnaeus, 1758, a chrysidid
    opulenta Lepeletier, 1845 (*Tachytes*)
    auropilosa Rohwer, 1911 (*Tachytes*)
    purpureipennis Matsumura and Uchida, 1926 (*Tachytes*)
?*australis* (Saussure), 1854 (*Tachytes*); Australia
avellanipes (Saussure), 1890 (pl. 19, fig. 8) (*Notogonia*); Madagascar
    avellanipes Saussure, 1891 (*Notogonia*)
  ssp. *flavivena* Arnold, 1947; Rhodesia
bakeri (Williams), 1928 (*Notogonidea*); Philippines
basilissa (Turner), 1908 (*Notogonia*); Australia
beata (Cameron), 1889 (*Notogonia*); Mexico (? = *aequalis*)
bella (Lepeletier), 1845 (*Tachytes*); Brazil
†*bella* (Rohwer), 1911 (*Notogonia*); Panama, Costa Rica, nec Lepeletier, 1845
bembesiana (Bischoff), 1913 (*Notogonia*); Rhodesia
  ssp. *pseudocroesus* (Arnold), 1940 (*Motes*); Rhodesia
bengalensis (Cameron), 1903 (*Notogonia*); e. India
bequaerti (Arnold), 1929 (*Notogonidea*); Zaire
bicolor (W. F. Kirby), 1900 (*Notogonia*); Socotra I.
bidentata (Arnold), 1923 (*Notogonidea*); Rhodesia (M)
bradleyi (Maidl), 1925 (*Notogonia*); Sumatra
braueri Kohl, 1884; Egypt, Jordan, Saudi Arabia, Iran, Sri Lanka (L)
braunsi (Arnold), 1923 (*Notogonidea*); Rhodesia
bredoi Benoit, 1951; Zaire
brunnipennis Arnold, 1945; Madagascar (? = *radialis*)
caeruleipennis (Maidl), 1925 (*Notogonia*); Sumatra
campestris (F. Smith), 1856 (*Larrada*); Brazil, Paraguay
carolinensis Yasumatsu, 1941; Carolines (L)
championi (Cameron), 1889 (*Notogonia*); Guatemala
cheesmanae Menke, new name; New Hebrides
    nitida Cheesman, 1937 (*Notogonidea*), nec Cameron 1913
chrysobapta (F. Smith), 1862 (*Larrada*); Celebes
chrysonota (F. Smith), 1869 (*Larrada*); Australia
    crassipes F. Smith, 1873 (*Larrada*)
ciliata (F. Smith), 1856 (*Larrada*); Rhodesia, S. Africa
cleopatra Beaumont, 1961; Egypt, Sudan, Israel
clypeata (F. Smith), 1873 (*Larrada*); New Caledonia, New Hebrides (L)
commixta (Turner), 1908 (*Notogonia*); Australia
?*conjungens* (Walker), 1871 (*Larrada*); ne. Africa:

Dahlak Archipelago
consobrina Arnold, 1959; Tanzania: Tanganyika
conspicua (F. Smith), 1856 (*Larrada*); India
    pulchripennis Cameron, 1889 (*Notogonia*)
    luteipennis Cameron, 1890 (pl. 9, fig. 2) (*Notogonia*), nec Cresson, 1869
corniger Williams, 1936; Solomons (L)
coronalis (F. Smith), 1856 (*Larrada*); Brazil
cowani (W. F. Kirby), 1883 (*Larrada*); Madagascar (L)
    pedestris Saussure, 1892
crassicornis (Maidl), 1925 (*Notogonia*); Sumatra
crepitans Leclerq, 1967; Madagascar
croesus (F. Smith), 1856 (*Larrada*); Ethiopian Region
    liroides Turner, 1913 (*Motes*)
    pseudoliris Turner, 1913 (*Notogonia*)
    deceptor Turner, 1916 (*Motes*)
  ssp. *congoensis* (Arnold), 1932 (*Notogonidea*); Zaire
cubitalis (Saussure), 1887 (*Notogonia*); Mauritius
decorata (F. Smith), 1856 (*Larrada*); Brazil
dejecta Arnold, 1945; Madagascar
denticulata (Turner), 1920 (*Notogonia*); S. Africa, Ghana
dentipes (Turner), 1917 (*Notogonia*); Ghana
  ssp. *tanganyikae* Arnold, 1947; Tanzania: Tanganyika
deplanata (Kohl), 1884 (*Notogonia*); Sri Lanka, ne. India
  ssp. *binghami* Tsuneki, 1967; n. India, Taiwan, Ryukyus
diabolica (F. Smith), 1873 (*Larrada*); Ethiopian Region (L)
    capitalis Radoszkowski, 1881 (*Tachytes*)
    opipara Kohl, 1894 (*Larra*)
    violaceipennis Cameron, 1908, nec Cameron, 1889
*distinguenda (Spinola), 1841 (*Larra*); French Guiana
?*dives* (Lepeletier), 1845 (*Tachytes*); "Caroline"
docilis (F. Smith), 1873 (*Larrada*) (♀ only which was cited erroneously as a ♂, see *Larra carbonaria*); Japan, Ryukyus, Taiwan, Philippines, Marianas, Hawaii
    tisiphone F. Smith, 1873 (*Larrada*), nec F. Smith, 1858
    tisiphonoides Dalla Torre, 1897 (*Larra*)
    manilensis Rohwer, 1910 (*Notogonia*)
    vortex Tsuneki, 1966
domingana (Strand), 1911 (*Notogonia*); Ecuador
dominicana Evans, 1972; W. Indies: Dominica
ducalis (F. Smith), 1860 (*Larrada*); Sri Lanka, India, Burma, Celebes (? = *aurifrons*) (L)
    nigripennis Cameron, 1889
dyscheira (Saussure), 1892 (*Notogonia*); Madagascar
    dasycheira Dalla Torre, 1897 (*Larra*)
elegans (Bingham), 1897 (*Larra*); Burma
erythropoda (Cameron), 1890 (pl. IX, fig. 5) (*Notogonia*); ? India
erythropyga (Arnold), 1940 (*Motes*); Rhodesia
erythrotoma (Cameron), 1908 (*Notogonia*); Tanzania: Tanganyika
esakii Yasumatsu, 1941; Caroline Is. (L)
extensa (Walker), 1860 (*Larrada*); Sri Lanka
*facilis (F. Smith), 1873 (*Larrada*); Brazil

*fasciata (F. Smith), 1873 (*Larrada*); Brazil
ferrugineipes (Lepeletier), 1845 (*Tachytes*); Senegal
  ?*pallipes* F. Smith, 1856 (*Larrada*)
  ?*pallidipes* Dalla Torre, 1897 (*Larra*)
festinans (F. Smith), 1859 (*Larrada*); Celebes, Philippines, Marianas, Carolines, Australia, Fiji, Samoa, New Caledonia, Solomons (status of *festinans* and subspecies according to van der Vecht in correspondence)
    manilae Ashmead, 1904 (*Notogonia*)
    retiaria Turner, 1908 (*Notogonia*)
    williamsi Rohwer, 1919 (*Notogonidea*)
  ssp. *japonica* (Kohl), 1884 (*Notogonia*); Japan, Taiwan
flavinerva (Cameron), 1900 (*Leptolarra*); e. India (M)
  ?*larroides* Williams, 1928 (*Motes*)
flavipennis (Williams), 1928 (*Notogonidea*); Philippines (? = *rufitarsis* Cam.)
flavitincta (Arnold), 1940 (*Motes*); Rhodesia, Ghana
formosana Tsuneki, 1973; Taiwan
foveiscutis (Cameron), 1913 (*Notogonia*); New Guinea: Waigeo
fuliginosa (Dahlbom), 1843 (*Larra*); West Indies: Cuba, Puerto Rico, Dominica; U.S.: Florida
    dahlbomi Cresson, 1865 (*Larrada*)
fulvipes Fabricius, 1804; Brazil, Guyana, Paraguay, Argentina (M)
    *angustata Taschenberg, 1870 (*Larrada*), new synonymy by Menke
    rufithorax Ducke, 1908 (*Motes*)
    longiventris Cameron, 1912 (*Tachysphex*), new synonymy by Menke
fulviventris (Guérin-Méneville), 1844 (*Lyrops*); Cuba, Jamaica, St. Vincent (M)
funerea (F. Smith), 1864 (*Larrada*); New Guinea: Waigeo
fuscata Tsuneki, 1971; Taiwan
fuscinerva (Cameron), 1905 (*Notogonia*); e. India (? = *subtessellata*)
fuscistigma (Cameron), 1903 (*Notogonia*); e. India
ganahlii (Dalla Torre), 1897 (*Larra*); Madagascar
    rufipes Saussure, 1892 (*Notogonia*); secondary homonym of *Larra rufipes* F. Smith
gastrifera (Strand), 1910 (*Notogonia*); Paraguay
gibbosa Kohl, 1892; "Arabia" (L)
gowdeyi (Turner), 1913 (*Notogonia*); Uganda
gracilicornis (Arnold), 1923 (*Notogonidea*); Ghana, Rhodesia, Madagascar
gryllicida Evans, 1972; W. Indies: Dominica, St. Vincent, Barbados
haemorrhoidalis (Fabricius), 1804 (*Pompilus*); Mediterranean region, Africa, Middle East, w. India, Sri Lanka
    auriventris Guérin-Méneville, 1835 (*Lyrops*)
    savignyi Spinola, 1838 (*Lyrops*)
    orichalcea Dahlbom, 1843
    illudens Lepeletier, 1845 (*Tachytes*)
    rubricans Pérez, 1895
  ?ssp. *magnifica* Kohl, 1884; Australia
  ?ssp. *jocositarsa* Saussure, 1887; Madagascar (? = *cowani*)
hanedai Tsuneki, 1971; Taiwan

incerta Arnold, 1945; Madagascar
indica (Cameron), 1903 (*Notogonia*); e. India
inopinata Beaumont, 1961; Cyprus, e. Mediterranean region (M)
insularis (Saussure), 1867 (*Larrada*); Nicobar Is.
intermedia (Cameron), 1903 (*Notogonia*); e. India
†*intermedia* Tsuneki, 1972; Taiwan, nec Cameron, 1903
†*iridipennis* (Maidl), 1925 (*Notogonia*); Sumatra, nec Cameron, 1900
iriomotoensis Tsuneki, 1972; Ryukyus
irrorata (F. Smith), 1856 (*Larrada*); Senegal to Uganda
    fraudulenta Kohl, 1894 (*Larra*)
    ssp. *tibialis* (Arnold), 1932 (*Notogonidea*); Zaire
ituriensis Benoit, 1951; Zaire
jaculator (F. Smith), 1856 (*Larrada*); e. India
    chapmani Cameron, 1900 (*Notogonia*)
†*japonica* (Cameron), 1901 (*Chrysolarra*); Japan, nec Kohl, 1884
jucunda Arnold, 1959; Tanzania: Tanganyika
karnyi (Maidl), 1927 (*Notogonidea*); Sumatra
keiseri (Leclercq), 1962 (*Motes*); Madagascar
khasiana (Cameron), 1905 (*Notogonia*); e. India
krombeini Menke, new name; Palau Is. (L)
    williamsi Krombein, 1949, nec Rohwer, 1919
kuchingensis (Cameron), 1909 (*Notogonia*); Borneo
labiata (Fabricius), 1793 (*Sphex*); W. Indies
    ignipennis F. Smith, 1856 (*Larrada*)
laboriosa (F. Smith), 1856 (*Larrada*); Philippines, Marianas, Burma, Taiwan, Hawaii
    crawfordi Rohwer, 1910 (*Notogonia*)
larriformis (Williams), 1928 (*Notogonidea*); Philippines, Taiwan
larroides (Williams), 1928 (*Motes*); Philippines, Singapore (M)
    ssp. *taiwanus* (Tsuneki), 1967 (*Motes*); Taiwan
*laterisetosa (Spinola), 1851 (*Larra*); Brazil, new combination by Menke
ligulata (Williams), 1928 (*Notogonidea*); Philippines
*limpidipennis (F. Smith), 1873 (*Larrada*); Brazil
liriformis (Williams), 1947 (*Notogonidea*); Fiji
litoralis Tsuneki, 1963; Thailand
longicornis (Cameron), 1900 (*Larra*); e. India
longitarsis (Cameron), 1900 (*Leptolarra*); e. India
luctuosa (F. Smith), 1856 (*Larrada*); Hispaniola, Brazil
luteipennis (Cresson), 1869 (*Larrada*); Cuba
lutusator (Williams), 1928 (*Notogonidea*); Costa Rica to Bolivia
maidli (Arnold), 1929 (*Notogonidea*); Sudan
    minima Maidl, 1924 (*Notogonia*), nec Arnold, 1923
mandibularis Menke, new name; Philippines (L)
    intermedia Williams, 1928, nec Cameron, 1903, nec Arnold, 1923
melania Turner, 1916; Australia, Guadalcanal (L)
memnonia (F. Smith), 1856 (*Larrada*); Egypt, Ethiopian Region
    * *semiargentia* Taschenberg, 1870 (*Larrada*), new synonymy by Menke ("Amer. Merid." an error)
    funebris Radoszkowski, 1876 (*Tachytes*)
    obscura Magretti, 1884 (*Larrada*)
    setigera Arnold, 1940 (*Motes*)
    sepulchralis of Arnold, 1923 (*Notogonidea*)
    radialis of Arnold, 1929
    ssp. *coloripes* (Arnold), 1940 (*Motes*); Madagascar
mescalero (Pate), 1943 (*Motes*); U.S.: California to Texas; n. Mexico
mindanao Menke, new name; Philippines, Borneo
    mindanaoensis Williams, 1928, (p. 79) (*Notogonidea*), nec Williams, 1928 (p. 83) (*Liris*)
mindanaoensis Williams, 1928 (p. 83); Philippines (L)
minima (Arnold), 1923 (*Notogonidea*); Rhodesia, Madagascar
    ssp. *lacustris* Arnold, 1947; Zambia
miscophoides (Arnold), 1923 (*Notogonidea*); Rhodesia
    ssp. *aegyptiaca* Beaumont, 1961; Egypt
modesta (F. Smith), 1859 (*Larrada*); Indonesia: Aru, Batjan (? L)
    vindex F. Smith, 1860 (*Larrada*), new synonymy by van der Vecht
montivaga (Cameron), 1908 (*Notogonia*); Tanzania: Tanganyika
mordax Kohl, 1892; Borneo (L)
morio Kohl, 1892; ? Sikkim
morrae (Strand), 1910 (*Notogonia*); Paraguay
muesebecki (Krombein), 1954 (*Motes*); U.S.: Florida, Texas
murina (Dahlbom), 1843 (*Tachytes*); "America boreali" (Stadelmann, 1897, is the authority for placement in *Liris*)
neavei (Turner), 1917 (*Notogonia*); Sierra Leone, Nigeria, Uganda, Malawi
negrosensis (Williams), 1928 (*Notogonidea*); Philippines, Singapore (? = *funerea*)
nigra (Fabricius), 1775 (*Sphex*); s. Europe, n. Africa, sw. USSR, Iraq, Afghanistan, India
    pompiliformis Panzer, [1806-1809] (*Larra*), nec Panzer, 1804
    nigra of Vander Linden, 1829 (*Tachytes*)
    uniocellatus Dufour, 1834 (*Anoplius*)
    ? *confusa* Radoszkowski, 1887 (*Larra*)
    ? *nigriventris* Cameron, 1889 (*Larra*)
    ? *germabensis* Radoszkowski, 1893 (*Tachytes*)
    ? *nana* Bingham, 1897 (*Larra*)
    ? *iridipennis* Cameron, 1900 (*Larra*)
    baguenai Giner Marí, 1934 (*Notogonia*)
    ssp. *namana* (Bischoff), 1913 (*Notogonia*); S. Africa
nigricans (Walker), 1871 (*Larrada*); Africa, Cape Verde Is., Canary Is., e. Mediterranean region, Madagascar, Seychelles, Iraq
    argyropyga A. Costa, 1875 (*Notogonia*)
    sculpturata Kohl, 1892 (*Notogonia*)
    reticulata Saussure, 1892 (*Notogonia*)
    palumbula Kohl, 1894 (*Larra*)
    jugurthae Gribodo, 1894 (*Larra*)
    mahensis Cameron, 1908 (*Notogonia*)
    punctipleura Cameron, 1908 (*Notogonia*)
    ssp. *reticuloides* (Richards), 1935 (*Leptolarra*); India
    reticulata Cameron, 1900 (*Leptolarra*), nec Saussure, 1892

indica Arnold, 1945, nec Cameron, 1903
?*nigripes* (Saussure), 1867 (*Larrada*); Australia (? = *Tachytes*, see Turner, 1916a:250)
  nitens (Arnold), 1929 (*Notogonidea*); Zaire
  nugax (Kohl), 1894 (*Larra*); S. Africa, Malawi, Ethiopia
    dixeyi Bingham, 1912 (*Notogonia*)
    pilosifrons Turner, 1916 (*Notogonia*)
  obliquetruncata (Turner), 1908 (*Notogonia*); Australia
  obtusedentata (Maidl), 1925 (*Notogonia*); Sumatra
 odontophora (Kohl), 1894 (*Larra*); Guinea, Liberia, Zaire (M)
  opalipennis (Kohl), 1898 (*Notogonia*); n. Africa
    ssp. *cypriaca* Beaumont, 1961; Cyprus
  ordinaria (Arnold), 1932 (*Notogonidea*); Zaire
  ornatitarsis (Cameron), 1911 (*Notogonia*); New Guinea
  pacificatrix (Turner), 1908 (*Larra*); New Hebrides
  panamensis (Cameron), 1889 (*Notogonia*); Panama (? = *apicipennis*)
  parva (Cameron), 1903 (*Notogonia*); e. India
  peruana (Brèthes), 1924 (*Notogonia*); Peru (M)
†*peruana* (Brèthes), 1926 (*Tachytes*); Peru, ? nec Brèthes, 1924, new combination by Bohart
  picipes (Cameron), 1903 (*Notogonia*); e. India
  pictipennis (Maidl), 1924 (*Notogonia*); Sudan
  piliventris (Cameron), 1903 (*Notogonia*); e. India
  pilosa (Cameron), 1903 (*Notogonia*); e. India
  pitamawa (Rohwer), 1919 (*Cratolarra*); Philippines, Taiwan, Borneo, Malay Penin., India (? = *fuscinerva* Cam.)
 plebeja (Taschenberg), 1870 (*Larrada*); Brazil, Peru, new combination by Menke, lectotype ♂, "Lagoa Santa", (Mus. Halle), present designation by Menke
 pluto (F. Smith), 1856 (*Tachytes*), Brazil
  politica (Dalla Torre), 1897 (*Larra*); "Congon." (? = Brazil), Peru, new combination by Menke
   * *polita* Taschenberg, 1870 (*Larrada*), nec *Larra polita* Smith, 1857
  praedatrix (Strand), 1910 (*Notogonia*); Paraguay
  praetermissa (Richards), 1928 (*Notogonia*); s. Europe, n. Africa, e. Mediterranean region, possibly a subspecies of *japonica* Kohl (i.e. *festinans*) according to Tsuneki (1967)
    schulthessi Giner Marí, 1942 (*Leptolarra*)
  primania (Kohl), 1894 (*Larra*); Liberia
  pruinosa (F. Smith), 1873 (*Larrada*); Brazil
†*pruinosa* (Cameron), 1901 (*Chrysolarra*); Borneo, nec F. Smith, 1873
  pulcherrima (Cameron), 1902 (*Notogonia*); w. India
  pygmaea (Cameron), 1903 (*Notogonia*); e. India
  quadrifasciata (F. Smith), 1856 (*Larrada*); Brazil
  radialis (Saussure), 1887[12] (*Notogonia*); Madagascar
    radamae Saussure, 1891 (*Notogonia*)
  radulina Evans, 1972; W. Indies: Dominica, Barbados, Mustique

  recondita (Turner), 1916 (*Notogonia*); Australia
  regina (Turner), 1908 (*Notogonia*); Australia
  robusta (Williams), 1928 (*Notogonidea*); Philippines, Borneo, Malay Penin.
    ssp. *planata* Tsuneki, 1963; Thailand
  robustoides (Williams), 1928 (*Notogonidea*); Philippines, Borneo, Malay Peninsula
  rohweri (Williams), 1928 (*Notogonidea*): Philippines, Malay Penin., Taiwan
 rubella (F. Smith), 1856 (*Larrada*); Ethiopian Region, new combination by Menke (M)
    cyphononyx Kohl, 1894 (*Larra*), new synonymy by Menke
    arnoldi Benoit, 1951
  rubricata (F. Smith), 1856 (*Larrada*); Brazil, Trinidad
  rufipennis Fabricius, 1804; Brazil, Ecuador, Peru; W. Indies: St. Vincent
    compressifemur Giner Marí, 1944 (*Notogonia*), new synonymy by Menke
  rufitarsis (Cameron), 1900 (*Spanolarra*); e. India, ? Philippines
    appendiculata Cameron, 1900 (*Caenolarra*), nec Taschenberg, 1870
    ? *flavipennis* Williams, 1928 (*Notogonidea*)
  rufoscapa (Cameron), 1905 (*Notogonia*); S. Africa, Botswana
  rugifera (Turner), 1918 (*Motes*); Uganda, Ghana (M)
  sabrina (Leclercq), 1961 (*Motes*); Madagascar
  sabulosa (F. Smith), 1864 (*Larrada*); Indonesia: Ceram
  sagax Kohl, 1892; ? Sikkim (L)
  samoa Menke, new name; Samoa
    samoensis Williams, 1928 (p. 34), (*Notogonidea*), nec Williams, 1928:36
  samoensis Williams, 1928 (p. 36); Samoa, Carolines (L)
  scabriuscula Arnold, 1945; Madagascar
  semiappendiculata (Cameron), 1912 (*Tachysphex*); Guyana
  sepulchralis (Gerstaecker), 1857 (*Lyrops*); Mozambique (see Arnold, 1940:123, 142; 1945:132)
    sepulchralis Gerstaecker, 1862 (*Lyrops*)
    intermedia Arnold, 1923 (*Notogonidea*), nec Cameron, 1903
    ssp. *hova* Arnold, 1945; Madagascar
  serena (Turner), 1908 (*Notogonia*); Australia
  sericosoma (Turner), 1913 (*Notogonia*); Kenya
  seychellensis (Cameron), 1907 (*Notogonia*); Seychelles
    rufofemorata Cameron, 1907 (*Notogonia*)
  silvicola (Williams), 1928 (*Notogonidea*); Philippines, Borneo
  simulatrix (Arnold), 1923 (*Notogonidea*); Rhodesia
  solstitialis (F. Smith), 1856 (*Larrada*); Ethiopian Region, Madagascar
    femoralis Saussure, 1887 (*Notogonia*)
    cnemophila Cameron, 1908 (*Notogonia*)
    ssp. *anubis* Beaumont, 1970; Egypt
    tibialis Beaumont, 1961; nec Arnold, 1923
  sophiae Evans, 1972; W. Indies: Dominica
  spathulifera (Turner), 1916 (*Notogonia*); Australia
  splendens (Ashmead), 1900 (*Motes*); W. Indies: St.

[12] The history of this name is involved. See Arnold (1940, p. 123) for details. Leclercq (1961d, p. 114) designated a lectotype for *radialis* apparently unaware that Arnold had already done so.

Vincent (? M)
sternalis (Rohwer), 1914 (*Notogonidea*); Guatemala, Honduras, Costa Rica
strenua (Cameron), 1905 (*Notogonia*); e. India
striaticollis (Cameron), 1903 (*Notogonia*); e. India
subfasciata (Walker), 1871 (*Larrada*); e. Mediterranean region to Zaire
    egregia Arnold, 1929 (*Notogonidea*)
subpetiolata (F. Smith), 1856 (*Larrada*); Brazil
subtessellata (F. Smith), 1856 (*Larrada*); e. Mediterraneanean region, sw. Asia, Oriental Region, Ryukyus, Marshalls, Hawaii, Fiji
    exilipes F. Smith, 1856 (*Larrada*)
    insularis Cameron, 1913 (*Notogonia*), nec Saussure, 1867, new synonymy by van der Vecht
    luzonensis Rohwer, 1919 (*Notogonidea*)
sulcifrons (Cameron), 1905 (*Notogonia*); e. India
sumatrensis (Maidl), 1927 (*Notogonidea*); Sumatra
surusumi Tsuneki, 1966; Ryukyus, Taiwan
    shirozui Tsuneki, 1966
tachytoides Tsuneki, 1963; Thailand
tegularis (Cameron), 1902 (*Notogonia*); Sarawak
†*tegularis* (Cameron), 1905 (*Notogonia*); e. India, nec Cameron, 1902
*tenebrosa (F. Smith), 1873 (*Larrada*); Brazil, Panama
testaceicornis (Cameron), 1905 (*Notogonia*); e. India
thaiana Tsuneki, 1963; Thailand, ? Philippines
?† *liroides* Williams, 1928 (*Notogonidea*), nec Turner, 1913
thysanomera (Kohl), 1894 (*Larra*); Gabon, Ghana
    ssp. *usambaraensis* (Cameron), 1908 (*Notogonia*); Burundi
tinctipennis (Cameron), 1889 (*Notogonia*); Panama
tisiphone (F. Smith), 1858 (*Larrada*); Borneo
townesi (Krombein), 1949 (*Motes*); Palau Is.
transversa Cheesman, 1955; New Caledonia
trifasciata (F. Smith), 1856 (*Larrada*); Hispaniola, Puerto Rico, Virgin Is.
tristis (F. Smith), 1856 (*Larrada*); Borneo
trivittata (W. F. Kirby), 1900 (*Tachytes*); Socotra
    expedita Kohl, 1906 (*Notogonia*)
tropicalis Arnold, 1960; Uganda, Zaire, Liberia
truncata (F. Smith), 1856 (*Larrada*); Brazil
truncatula (Dalla Torre), 1897 (*Notogonia*); Costa Rica (? = *beata*)
    truncata Cameron, 1889 (*Notogonia*), nec F. Smith, 1856
uelensis Benoit, 1951; Zaire
umbripennis (Cameron), 1902 (*Notogonia*); Borneo
utopica (Leclercq), 1961 (*Motes*); Madagascar
vagans (Arnold), 1940 (*Motes*); Liberia
varipilosa (Cameron), 1903 (*Notogonia*); e. India
vigilans (F. Smith), 1856 (*Larrada*); China
vinulenta (Cresson), 1865 (*Larrada*); Cuba, Puerto Rico
    ssp. *muspa* (Pate), 1943 (*Motes*); U.S.: Florida
violaceipennis (Cameron), 1889 (*Notogonia*); Panama
†*violaceipennis* Cameron, 1904; e. India, nec Cameron, 1889
vivax (Cameron), 1905 (*Notogonia*); e. India
voeltzkowii (Kohl), 1909 (*Notogonia*); Madagascar
vollenhovia (Ritsema), 1874 (*Larrada*); Guinea, Nigeria
    vollenhovenia Schulz, 1904 (*Notogonia*), emendation
wheeleri Arnold, 1960; Rhodesia
xanthoptera Arnold, 1945; Madagascar
    ssp. *nyasae* Arnold, 1947; Malawi

Genus Dalara Ritsema

Generic description: Upper inner angle of eye not bordered by a deep sulcus; midocellar lens weakly convex; hindocellar scars small; frontal line impressed, interrupted at transverse swelling; frons and vertex dull, impunctate; scape long, slender, surface of scape and pedicel not contrasting with that of flagellum, flagellomeres about twice as long as broad and bearing weak placoids; clypeus with a prominent, truncate, median lobe (fig. 67 D); labrum quadrangular, hidden; inner margin of mandible with a large subbasal tooth, externoventral margin entire, mandible apex bidentate in female (fig. 67 D), male mandible greatly elongate and bearing a small tooth on inner margin near apex (fig. 67 D); malar space 1.0-2.0 midocellar diameters long in male; collar as in *Liris,* dorsum sharp edged, lower than and closely fitting with scutum; notauli extending nearly to level of tegulae; propodeum moderately long, quadrangular, sides converging posteriorly in dorsal view, propodeum dull, impunctate, dorsum finely rugose; gaster dull or weakly shining, impunctate, tergal fasciae poorly developed, tergum I with lateral carina; female sternum II with a pair of special oval areas basally as in *Liris* but without a basomedian ridge or prominence; female with a pygidial plate defined by carinae, shining, surface rugosopunctate and with median ridge (fig. 68 A) apex with a transverse row of broad flattened setae (four in *mandibularis,* the only species in which the female is known); male without a pygidial plate; femora and tibiae long and slender, closing face of male hindfemur with a prominent bulge basally, or with a hooklike process and associated depression (fig. 64 L); mid and hindtibiae of both sexes with rows of moderately stout spines arranged as follows: three on II, two on III; hindtibia with a longitudinal, sharp edged carina in female, that of male merely angulate in cross section; female with a rather weak foretarsal rake; female last tarsomere of mid and hind leg angled at midpoint in lateral view and bearing a dense hairmat ventrally, claws long, prehensile and without inner teeth; male tarsomere similar but claws shorter and ventral hairmat absent; apex of male sternum VIII roundly truncate and with a small median notch; genitalia about as in *Liris,* volsella long.

Geographic range: This small genus is recorded only from the East Indies.

Systematics: These wasps range in size from 9 to 12 mm. The body is largely dull black and without erect hair. The mandible may be yellowish, and the legs may be partly reddish. Except for a cloudy marginal cell the wings are clear.

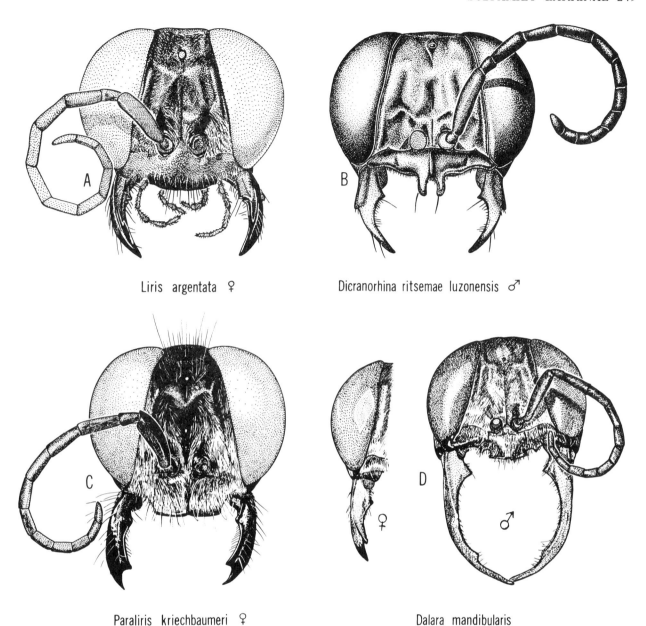

FIG. 67. Facial portraits in the subtribe Larrina; C, holotype.

The characteristics of the mandible, especially the long, slender male mandible, provide the only means of separating *Dalara* from its very close relative *Liris*. Possibly *Dalara* should be considered as a subgenus of *Liris*, although its biology argues otherwise. Van der Vecht (1950) discussed differences between the two species. F. Williams (1928a) provided excellent figures of various features of *D. mandibularis*, the only species known from both sexes.

Biology: F. X. Williams (1919) has provided the only account of nesting behavior. He found three *Dalara mandibularis* females nesting in a fallen, decayed tree trunk. It is not entirely clear from Williams' statements whether these females utilized one nest entrance or were merely nesting in close proximity within the log. He said that "two specimens were seen to enter a hole" and also that he dug into the "colony, in which there appeared to be three actively working females." From this it can be inferred that the species is communal, but sociality remains to be proven. In a later paper Williams (1928a) suggested that *Dalara* uses preexisting cavities in wood. The burrows are irregular, and he noted that two of them seemed to have a common entrance. In the nest studied the cells were scattered along the burrow courses in a haphazard manner. Some stored cells did not contain eggs or larvae of the wasp. Prey, which are flown to the nest, consist of short-winged wood crickets of the genus *Calyptotrypus*.

Checklist of *Dalara*

mandibularis (Williams), 1919 (*Hyloliris*); Philippines
schlegelii (Ritsema), 1884 (*Darala*); Java, Sumatra

Genus Paraliris Kohl

Generic description: Upper inner angle of eye not bordered by a deep sulcus; midocellar lens weakly convex; hindocellar scars very small, obscure; orbital swellings of frons much more prominent than transverse swelling; frontal line a deep impression which is interrupted at transverse swelling; frons below transverse swelling shining, impunctate, ocellar area of vertex duller and weakly rugose or shallowly, coarsely punctate, vertex behind ocelli shining and punctate; scape long, slender, surface not contrasting with that of flagellum, flagellomeres two or more times as long as wide and without placoids; clypeus with a prominent, broad, truncate, median lobe (fig. 67 C); labrum transverse, rectangular, hidden; inner margin of mandible with a prominent tooth, mandible apex broadly bidentate (fig. 67 C), externoventral margin not notched (fig. 67 C); malar space about as long as midocellus; collar as in *Liris,* dorsum sharp edged, lower than and closely fitting with scutum, latter anteromedially depressed for reception of collar; notauli present, extending nearly to level of tegulae; propodeum moderately long, quadrangular, rounded posteriorly, sides slightly converging posteriorly in dorsal view, propodeal dorsum finely rugose, side shining, smooth, punctate; disk of terga shining, terga and sterna punctate, punctures becoming denser and coarser laterally, tergal fasciae absent, terga I and II with lateral carina, that of II incomplete, female sternum II with a basomedian prominence but without special oval areas; female with pygidial plate defined by carinae, surface shining, slightly raised along midline, moderately covered by coarse setigerous punctures (fig. 68 B), male without a pygidial plate; femora and tibiae long and slender, closing face of hindfemur not excavate nor with a basal process; female mid and hindtibiae with two rows

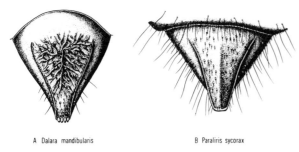

A Dalara mandibularis   B Paraliris sycorax

FIG. 68. Pygidium of females in the subtribe Larrina.

of long, slender spines on outer surface, male with a single row; hindtibia weakly angular in cross section, without a sharp carina; female without a foretarsal rake (fig. 64 F); female last tarsomere of mid and hind leg angled at midpoint in lateral view, parallel sided on apical half in dorsal view and bearing a dense hairmat ventrally; claws long, prehensile, without inner teeth; male last tarsomere similar but claws shorter and ventral hairmat absent; recurrent veins sometimes joining at submarginal cell; apex of male sternum VIII truncate; genitalia about as in *Liris,* volsella long.

Geographic range: This small genus is restricted to the Oriental Region. The type species of the genus *kriechbaumeri* was purported by Kohl to be an African wasp when he described it, but van der Vecht (correspondence) after studying the type found it to be a Javanese species.

Systematics: These are moderately large (17mm), black wasps, the body and legs distinctively covered with rather dense, long, pale, erect hair. A close relationship with *Liris* is obvious, but the apically bidentate mandible without an externoventral notch is distinctive. Some *Liris* have apically bidentate mandibles, but the mandible in these is also notched externally. The dense, coarse punctation laterally on the terga and sterna, and the smooth shiny, punctate propodeal side are also diagnostic features of *Paraliris.* The bare pygidial plate without an apical row of setae and the absence of a foretarsal rake are additional characters in the female.

There is no key for the three species, but van der Vecht is currently revising the genus. We have seen examples of all three including the type of *kriechbaumeri* (Mus. Munich), and they are very similar.

Biology: Unknown. The absence of a tarsal rake suggests that *Paraliris* may be a twig nester.

Checklist of *Paraliris*

faceta Bingham, 1897; e. India, Burma (? = *sycorax*)
kriechbaumeri Kohl, 1883; Java (S. Africa is erroneous)
sycorax (F. Smith), 1857 (*Larrada*); Borneo

Genus Dicranorhina Shuckard

Generic description: Upper inner angle of eye not bordered by a deep sulcus; midocellar lens weakly convex; hindocellar scars very small; orbital swellings of frons and associated linear depressions very weak, area above transverse swelling nearly flat; frontal line often weakly impressed; frons and vertex dull, impunctate; scape long, slender, surface not contrasting with that of flagellum, flagellomeres two or three times as long as broad and without placoids; clypeus in female with a truncate, median lobe, free margin of male clypeus sometimes like female (*kohli, palawanensis, nigra*), but more commonly with two fingerlike projections (fig. 67 B); labrum rectangular, hidden; inner margin of mandible with a tooth near base, externoventral margin notched or with a steplike angulation; dorsum of collar thickened (fig. 63 B), lower than and closely fitting with scutum, latter depressed mesally for reception of collar; notauli present, extending nearly to level of tegulae; propodeum moderately long, rectangular, sides converging posteriorly in dorsal view, propodeum dull, impunctate, dorsum finely rugose or ridged; episternal sulcus disappearing before reaching anterior margin of pleuron; gaster pedunculate, segment I longer than broad in dorsal view (fig. 63 A), terga dull or weakly shining, impunctate and usually without apical fasciae, terga I and II with lateral carina, that of II weak and incomplete; female sternum II usually with a pair of rather weak special oval areas basally as in *Liris,* but these at most weakly separated by a median ridge or prominence; female with a pygidial plate defined by carinae, shining, punctate, each puncture containing a seta, but apex usually with a transverse row of stout setae, but this row often difficult to distinguish because of other subapical setae; male tergum VII often flattened but without a distinct pygidial plate; femora, tibiae and female forecoxa long, slender, closing face of hindfemur usually with a basal tubercle or clawlike process (fig. 64 M-N), and often flattened or excavate especially in males (femur simple in females of *kohli, intaminata, wollastoni,* and possibly others); mid and hindtibiae of both sexes with rows of long, fine spines arranged as follows: two or three on II, two on III; hindtibia rather sharply angulate in cross section but without a polished, sharp edged carina; female with a rather weak foretarsal rake composed of a few long, fine spines; female foretarsomere V, including claws, asymmetrical and venter hollowed out, the claws long and prehensile (fig. 64 H); other last tarsomeres of female and all in male normal, without ventral hairmat, claws short; recurrent veins sometimes joining at or just before meeting submarginal cell; apex of male sternum VIII truncate or rounded, often weakly emarginate; genitalia about as in *Larra,* volsella short or long.

Geographic range: Dicranorhina, in which 12 species are currently recognized, occurs only in the Old World tropics. Two species are found in Africa, four in the continental part of the Oriental Region, and one in Australia. The remainder are known from various islands in Indonesia and the western Pacific. *D. ritsemae luzonensis* was apparently introduced to the Palau Islands (Krombein, 1949, 1950a) and Hawaii (Weber, 1950).

Systematics: These are medium sized (7-12 mm long), black wasps which characteristically have a dark spot on the forewing in the area of the marginal and submarginal cells. The body does not have erect hair, and the legs and thorax sometimes are partly red.

Among the several genera closely allied with *Liris, Dicranorhina* is the most distinctive. The pedunculate gaster immediately identifies the genus; and the low, rather thick collar, incomplete episternal sulcus, weak orbital swellings of the frons, and flat ocellar substrate are also diagnostic. When present the hindfemoral process or tubercle and the male clypeal projections are good recognition features. The scooplike last tarsomere of the female foreleg is unusual and probably is correlated in some way with nesting or prey capture. All of these features suggest that *Dicranorhina* is the most specialized genus in the subtribe.

There is no key for the genus. Arnold (1923b) contrasted the two African species, van der Vecht (1937) discussed some of the Indonesian forms, and F. X. Williams (1928a) figured and described the Philippine species. Some are known by one sex, and it seems likely that a few of the Cameron names will prove to be synonyms.

Biology: Williams (1928a), Dupont and Franssen (1937), and Iwata and Yoshikawa (1964) have published ethological data for *ritsemae* ssp. *luzonensis, cavernicola,* and *ruficornis,* respectively. *Dicranorhina cavernicola* nested in dry soil deep in a cave. The other two species were observed nesting in the ground beneath stilt houses, and *luzonensis* was also seen nesting in steep banks. The prey of *ruficornis* was a cricket of the genus *Ornebius,* but the other species provisioned with wood crickets of the genus *Cycloptilum.*

The account of Iwata and Yoshikawa is the most detailed. Prey were transported on the wing, venter up. The wasp used her mandibles and forelegs to hold the prey. The nest was left open during provisioning. The most elaborate of the two nests excavated was more than 27 cm long, about 2.5 cm deep, and it had a sinuate course. Cells were scattered along the edge of the tunnel, and those nearest the nest entrance were the oldest, containing the pupal remains of hatched wasps. The cells that were currently being provisioned were at the end of the burrow. These contained two to five crickets, and the egg was laid on the thoracic venter of one. Iwata and Yoshikawa found two freshly emerged females in the second nest, which, when viewed in the light of the arrangement of old and new cells in both nests studied, suggested to the authors that the offspring from each nest may continue using the nest as their own by extending the burrow made by their mother. Thus, a nest may be maintained over several generations in *Dicranorhina.* This theory was also suggested by Dupont and Franssen who found many empty pupal cases on the soil surface near the nest openings of *cavernicola.*

Two colonies of *cavernicola* were found 45 and 60 feet from the cave entrance, respectively. The deeper of the two colonies was in total darkness. The 15 *cavernicola* nests dug up were 15-20 cm long and 7-12 cm deep. The nest entrances were ringed with a little wall of excavated earth, and burrows were unbranched, except in one nest where one passage ended blindly at 5 cm and the other in a brood chamber with prey. The cells of *cavernicola* contained one to five crickets. The egg and young larvae of *cavernicola* are found just below the mouthparts of the host. Larger larvae are found on the side of the cricket thorax or abdomen. Prey are flown to the nest in this species, but the method of navigation in the dark recesses of the cave is unknown.

Checklist of *Dicranorhina*

cavernicola (van der Vecht), 1937 (*Piagetia*); Java
fasciatiipennis (Cameron), 1889 (*Piagetia*); Sri Lanka, s. India
intaminata (Turner), 1910 (*Piagetia*); Australia
kohli (Brauns), 1898 (*Piagetia*); S. Africa, Malawi,

Rhodesia
   striata Cameron, 1905 (*Piagetia*)
     ssp. *montana* (Leclercq), 1955 (*Piagetia*); Zaire
nigra (Maidl), 1925 (*Piagetia*); Sumatra, Solomons
palawanensis Williams, 1928; Philippines: Palawan
ritsemae (Ritsema), 1872 (*Piagetia*); Java
     ssp. *luzonensis* Rohwer, 1919; Philippines, Palau Is., Hawaii
ruficollis (Cameron), 1904 (*Piagetia*); Singapore
ruficornis (Cameron), 1889 (*Piagetia*); India, Thailand
woerdeni (Ritsema), 1872 (*Piagetia*); w. centr. Africa
wollastoni Turner, 1912; New Guinea
varicornis (Cameron), 1904 (*Piagetia*); India, Sikkim

Subtribe Tachytina

Diagnosis: Midocellus not in a broad, transverse depression; hindocellar scars at rear of a roughly circular swelling, their long axes usually forming an angle of less than 140° (up to 175° in some *Parapiagetia*); frons flat to variably convex, without transverse linear swelling beneath midocellus; antenna short to moderately long, flagellomeres without placoids in male, usually longer than broad; malar space undeveloped except in some *Ancistromma*, *Gastrosericus*, and *Tachysphex;* occipital carina joining apex of hypostomal carina except in *Tachytella* and some *Gastrosericus;* collar arcuate or flat in front view; admedian lines separated when present, notauli sometimes present; episternal sulcus originating near middle of subalar fossa (fig. 71 A,D) except origin is anterior in *Parapiagetia* and some *Gastrosericus;* legs short to moderately long, arolia moderate; genitalia laterally compressed, volsella and gonostyle fringed ventrally with setae, except gonostyle bare in some *Tachytes*.

Discussion: The genera of the Tachytina are much less homogeneous than those of the Larrina. *Gastrosericus*, *Tachytella*, and *Parapiagetia* are the most discordant genera. The subtribe is distinguished primarily by the obtuse to acute angle formed by the ocellar scars (exceptions in *Parapiagetia*), the absence of linear swellings on the frons as described for the Larrina, and the origin of the episternal sulcus near the middle of the subalar fossa (exceptions in *Parapiagetia* and *Gastrosericus*). Two of the 10 genera in the Tachytina, *Tachytella* and *Kohliella*, are endemic to southern Africa. The largest genera, *Tachytes* and *Tachysphex*, are cosmopolitan, and they have many species in temperate regions as well as the tropics. Prey consists mainly of Orthoptera, but Hemiptera and Homoptera are used by *Prosopigastra*, and lepidopterous larvae are known to be used by a few *Tachytes* and perhaps also *Parapiagetia*.

Genus Gastrosericus Spinola

Generic description: Inner orbits usually converging above or parallel, rarely slightly convergent below (fig. 69 A,B); hindocellar scars long, narrow, their long axes forming an angle of 130° to 145° (fig. 61 D); frons weakly to strongly raised, elevation usually somewhat flattened, often reflexed downward near sockets, and set off from inner orbit by a fine linear impression; frons without median prominences just above sockets; frontal line fine, most strongly impressed near sockets; scape moderately long to long and slender, flagellomeres as long as broad in some males; clypeus rather flat or with a central swelling, disk of clypeus occasionally with tubercles or a large, emarginate lamella (fig. 67 B), clypeal free margin sinuate or broadly arcuate, or more commonly with a median lobe of variable shape (triangular, fig. 69 A, truncate or rounded), and sometimes deeply, broadly, emarginate laterally (fig. 69 B), margin often with lateral lobes or teeth; labrum rectangular to deeply bilobate, hidden, sometimes very small; inner margin of mandible simple or with one or two subbasal teeth, or with a subapical tooth in some females (*capensis*, *waltlii*), externoventral margin usually notched but entire in *marginalis* and *mongolicus*, or merely angulate as in *swalei* and *turneri;* malar space developed in some males; gena in female sometimes with one or two tubercles ventrally, gena with a lamella and associated carina in both sexes of *braunsi;* occipital carina usually disappearing just before reaching hypostomal carina; dorsum of collar usually with some thickness, usually much lower than and sometimes closely appressed to scutum; propodeum short to moderately long, female propleuron sometimes with prominent posterolateral process (*madacassus*, *swalei*) (fig. 73 R); episternal sulcus sometimes originating at anterior end of subalar fossa (*waltlii* group, *apostoli*, *neavei*, *turneri*), sulcus ending just as it reaches venter of pleuron, scrobal sulcus absent; terga impunctate; tergum I with lateral carina, II rarely with trace of carina basally; both sexes with a pygidial plate margined by carinae laterally, surface in female sparsely to densely covered with decumbent setae (*waltlii* and *marginalis* groups, *neavei*), or with a few setae apically, or glabrous and punctate (impunctate in *silverlocki*); male sterna III-IV in *waltlii* group each with a band of long fimbriae which are often set in large shallow depressions; female forecoxa with a tubercle or large angular process ventrally in *braunsi*, *lamellatus*, *madacassus*, *neavei*, *pulchellus*, and *temporalis*, male forecoxa similarly armed in *lamellatus* and *bidentatus;* male foretrochanter often excavate posterobasally but forefemur simple; midcoxae sometimes narrowly separated; mid and hindtibiae sparsely to moderately set with short, stout or long, slender spines which are arranged in rows, two or three on II, two on III, hindtibia not ridged; female with a foretarsal rake composed of stiff, short or long spines which are usually well spaced, tarsomere II with only two rake spines; male usually with a rake; inner claw of mid and hindtarsi occasionally shorter than outer claw; marginal cell broadly truncate apically, sometimes very short (*marginalis* group, *moricei*); only two submarginal cells present, second receiving both recurrent veins which may join before reaching cell (fig. 70 C,D); jugal lobe sometimes about three-fourths length of anal area; apex of male sternum VIII variable (fig. 76 A,B); volsella flat, apex acutely triangular or with a slender, spinose process which is often hooklike; head of aedeagus without teeth.

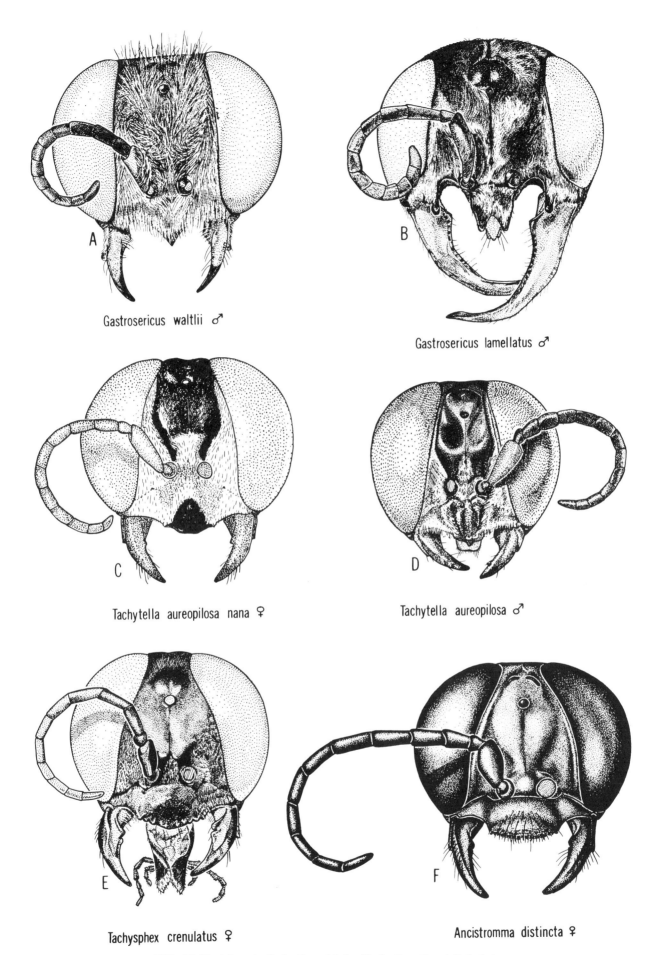

FIG. 69. Facial portraits in the subtribe Tachytina; C and D, holotypes.

Geographic range: Gastrosericus is an Old World genus in which 42 species are currently recognized. They are distributed as follows: Ethiopian Region — 22 plus one Madagascan endemic, Palearctic Region — 15, and Oriental Region — 4.

Systematics: These wasps are 4 to 10 mm long and black, but sometimes the gaster is partly or all red. In *marginalis* the red terga are bordered by pale yellow. Often the legs and antennae are extensively red or yellow. The wings are infuscate in *neavei*, but they are clear in most other species. Tergal fasciae are common, and appressed silver vestiture is usually found on the head and thorax. Dense suberect hair occurs in species of the *waltlii* group.

Gastrosericus is easily identified by the presence of only two submarginal cells (fig. 70 C,D), the shape of the ocellar scars (fig. 61 D), and the incomplete episternal sulcus (not reaching anteroventral margin of pleuron). Within the Larrini, *Gastrosericus* appears to have no close relatives. The morphological diversity in this rather small genus is remarkable, and may indicate that *Gastrosericus* is an old, declining group which nonetheless shows considerable specialization.

Casual examination of a few of the more unusual species prompts one to think initially that more than a single genus is involved. However, intensive study shows that there is no practical way to divide *Gastrosericus* because of transitional forms. The same applies to the recognition of subgenera, as Arnold (1922:114, 1927:116) correctly pointed out. Because some species, especially those of the Oriental Region, are inadequately described and have not been studied since their description, and because many are known by only one sex, it is impossible here to give more than a preliminary outline of some possible species groups in *Gastrosericus*. We have studied 18 species.

The *waltlii* group contains *capensis, drewseni, guigliae, lanuginosus, moricei, rufitarsis,* and *waltlii*, and possibly *aiunensis, shestakovi,* and *wroughtoni*. The female pygidial plate is covered with setae, sterna III-IV have frimbriae in the male, the apex of male sternum VIII is rounded, the recurrent veins are not petiolate on submarginal cell II, and the inner margin of the male mandible is simple. The female mandible of *capensis, rufitarsis,* and *waltlii* has a subapical tooth on the inner margin but no basal teeth (fig. 72 K); and although we have not seen females of *drewseni* and *lanuginosus,* we presume they have the same type of mandible because they are similar to *waltlii*. In *moricei,* a transitional species, the female mandible has a basal but no subapical tooth, and furthermore the head and thorax are clothed with appressed silver hair instead of the suberect vestiture characteristic of the other five species in the group. The volsella of *moricei* has apical spinose processes instead of the triangular apex found in *waltlii* and its closest allies.

The *marginalis* group consists of *marginalis* and *mongolicus*, species in which the mandibles have no teeth or notch (fig. 72 I); the recurrent veins are not petiolate on the submarginal cell; the female pygidial plate is densely covered with a fine, very short, silver pubescence; male sterna II-IV have long, decumbent setae mesally; and male sternum VIII is emarginate apically. The marginal cell is very short, an unusual condition, but one found also in *moricei*. Gussakovskij (1931a) proposed the subgeneric name *Gastrargyron*[13] for this assemblage, and it could be defended as a subgenus based on the mandible, marginal cell and pygidial plate. However, the absence of a mandibular notch is a rather weak character, inasmuch as some unrelated species such as *swalei* have angulate rather than notched mandibles.

The remaining species, most of them from the Ethiopian Region, cannot be satisfactorily grouped at this time, but most appear to be more or less closely related to one another. The pygidial plate is glabrous in many of these forms, but there are setae apically in *swalei, apostoli, braunsi, chalcithorax,* and *turneri,* and the entire surface is setose in *neavei, rothneyi,* and *fluviatilis*. The male sterna lack fimbriae or special setae. The apex of sternum VIII is usually rounded (fig. 76 A) but it is truncate in *apostoli* and emarginate in *divergens* (fig. 76 B). The recurrent veins are often petiolate on the submarginal cell, but this sometimes varies within a species. The jugal lobe is shorter in this group than it is in the other species of the genus. The unusual clypeal, genal, pronotal, propleural and coxal developments in *Gastrosericus* are confined to members of this assemblage, but many of the included species lack these specializations. Inner subbasal mandibular teeth are present (fig. 72 L) or absent, and there is either a notch or step on the externoventral margin (fig. 72 J). The mandible is long and slender in *silverlocki* and *lamellatus*. In the latter species the mandible does not taper gradually to a point but remains broad nearly to the apex, where it is obliquely angled. In a species we have determined as *rothneyi,* the inner claw of the mid and hindtarsi is shorter than its mate, especially in the male. The same condition exists in the male of *neavei*. The volsella varies from the type found in *waltlii* to one in which there are two darkly sclerotized hooklike processes apically (fig. 77 A), as in *divergens* and *neavei*.

We have noted that in species diagnoses no one has used the distinctive depressions found on the side and top of pronotum in many of the African species. Also, the collar may be transversely carinate anteriorly as in *lamellatus*.

The only keys are those of Arnold (1922) for the Ethiopian Region and Gussakovskij (1931a) for some of the Palearctic species. Both are outdated.

Biology: Iwata and Yoshikawa (1964) observed a colony of *Gastrosericus rothneyi* in Thailand. Nests were excavated in bare, sandy soil to depths ranging from 5.5 to 10.7 cm, and each ended in a single cell. The burrows were closed during search for prey which were *Tridactylus*. Arnold (1922) cited pygmy mole crickets as the prey of *Gastrosericus simplex*. Arnold (1945) also said that two specimens of *madecassus* were pinned with a "larval grasshopper" and a "small cercopid." The latter

---

[13] Gussakovskij's name must be attributed to Pate, 1937d. See generic catalog.

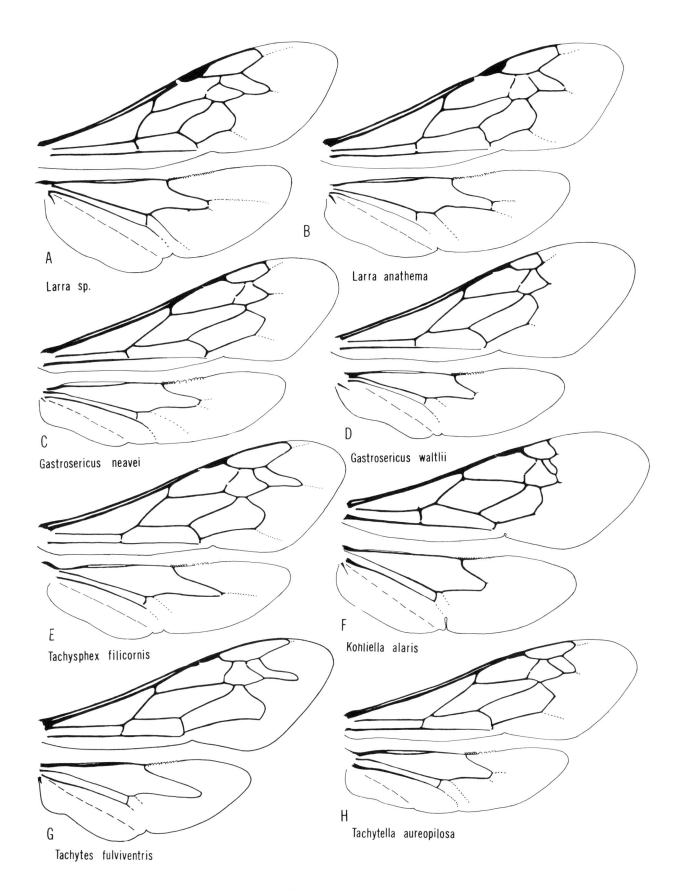

FIG. 70. Wings in the tribe Larrini.

seems an unlikely prey record. *Gastrosericus rothneyi* flies with its prey venter up but the manner of holding the prey is not clear. The number of prey per cell is unknown because Iwata and Yoshikawa did not study completed nests, but it is clear that the egg is not laid on the first provision.

Checklist of *Gastrosericus*[14]

aiunensis Giner Marí, 1945; Spanish Sahara
apostoli Beaumont, 1967; Turkey, Egypt
attenuatus Turner, 1912; Ghana
bidentatus Arnold, 1922; Rhodesia
braunsi Arnold, 1922; Rhodesia
    ssp. *unicolor* Arnold, 1929; Rhodesia
capensis (Brauns), 1906 (*Gasterosericus*); S. Africa
chalcithorax Arnold, 1922; S. Africa
decipiens Arnold, 1955; Lesotho
divergens Arnold, 1922; Rhodesia
drewseni (Dahlbom), 1845 (*Gasterosericus*); n. Africa
electus Nurse, 1903; w. India
eremorum Beaumont, 1955; Morocco
flavicornis Gussakovskij, 1931; sw. USSR
fluviatilis Arnold, 1951; Mali
funereus Gussakovskij, 1931; sw. USSR
guigliae Beaumont, 1956; Libya, Egypt
karooensis (Brauns), 1906 (*Gasterosericus*); S. Africa
lamellatus Turner, 1912; Rhodesia
lanuginosus Arnold, 1922; Rhodesia
laticeps Arnold, 1922; Rhodesia
madecassus (Kohl), 1907 (*Eparmostethus*); Madagascar
marginalis Gussakovskij, 1931, sw. USSR, Egypt
modestus Arnold, 1922; Rhodesia
mongolicus Gussakovskij, 1931; Mongolia
moricei E. Saunders, 1910; n. Africa
neavei Turner, 1913; Kenya, Rhodesia
    africanus Maidl, 1914 (*Paralellopsis*)
  ssp. *reversus* Arnold, 1951; Mali
niger (Dufour), 1853 (*Dinetus*); Algeria
oraniensis (Brauns), 1906 (*Gasterosericus*); S. Africa
pratensis Arnold, 1929; Rhodesia
pulchellus Arnold, 1929; Rhodesia
rothneyi Cameron, 1889; Sri Lanka, e. India, Burma, Thailand
    binghami Cameron, 1897
rufitarsis Cameron, 1902; w. India, Pakistan
sanctus Pulawski, 1973; Israel
senegalensis Arnold, 1951; Senegal
shestakovi Gussakovskij, 1931; sw. USSR
silverlocki Turner, 1912; Zambia, Rhodesia
simplex Arnold, 1922; Rhodesia
swalei Turner, 1916; Rhodesia
temporalis Beaumont, 1955; Morocco
turneri Arnold, 1922; Rhodesia
waltlii Spinola, 1838; n. Africa, Cyprus, Rhodes, Turkey, USSR, Mongolia

maracandicus Radoszkowski, 1877
rufiventris F. Morawitz, 1889
  ssp. *dubius* Gussakovskij, 1931; sw. USSR; Iran
wroughtonii Cameron, 1889; w. India

Genus Tachytella Brauns

Generic description: Inner orbits converging above; hindocellar scars elongate, somewhat constricted near middle, their long axes forming an obtuse angle (fig. 61 E); frons with a glabrous, broad, raised, flattened or weakly concave area (fig. 69 C,D) and without single or paired tubercles above antennal sockets; frontal line very fine, weakly impressed; scape long and broadening distad (fig. 79 E); clypeus with central swelling which in male has a weak to strong mediolongitudinal furrow, free margin with a median lobe which in male has a median tooth (fig. 69 D); labrum truncate, hidden or narrowly exposed; inner margin of mandible with a weak subbasal tooth in female, simple in male, externoventral margin with a notch; occipital carina ending just short of hypostomal carina; dorsum of collar with some thickness, below level of and not closely appressed to scutum (fig. 71 F); propodeum short, dorsum depressed in front of spiracles; episternal sulcus long but ending well before reaching anteroventral margin of pleuron, scrobal sulcus absent, subalar fossa not margined below by a carina; terga impunctate, tergum I with lateral carina; both sexes with a pygidial plate margined by carinae, that of female glabrous, sparsely punctate, narrowly triangular and with truncate apex, that of male setose, moderately punctate, broadly triangular, slightly convex; male sterna VI-VIII with dense rows or tufts of erect fimbriae; male foreleg unmodified; midtibia with two and hindtibia with three rows of moderately long spines, hindtibia angular in cross section, not ridged; female with foretarsal rake composed of short, widely spaced spines of which there are four on tarsomere I and two on II (fig. 73 B); male with a feeble rake, marginal cell narrowly truncate apically, submarginal cell III short, recurrent veins sometimes joining before attaining media (petiolate); jugal lobe of hindwing about two-thirds length of anal area (fig. 70 H); apex of male sternum VIII truncate (fig. 76 C), the truncation sometimes feebly concave (fig. 76 D); volsella long, slender, head of aedeagus with teeth (fig. 77 B).

Geographic range: This relict genus is represented by a single species which is known only from three localities in the Republic of South Africa: Willowmore, Worcester, and Resolution, the last located in the Albany District that surrounds Grahamstown. These areas are about 100 miles from the southern seacoast.

Systematics: Tachytella aureopilosa is a small (7-8 mm long), black wasp in which gastral segments I, I-II or I-III, the forefemur, the midfemur, and hindfemur apically, the tibiae and tarsi are red. The head and thorax have much appressed silver vestiture, but erect hair is absent. Gastral terga I-IV or V have rather weak marginal fasciae, and the wings are clear. We have seen the types of *aureopilosa* (a male, Mus. Pretoria) and its subspecies *nana*

---

[14] *Gastrosericus errans* Turner, 1936, described from Australia, represents a new genus in the Miscophini.

(a female, Mus. Bulawayo), as well as a paratype male of the latter. Two males have recently been found in the British Museum, one of which has been deposited in the U.S. National Museum. The length of the type of *aureopilosa* is 8 mm instead of 9 mm as given by Brauns (1906). Arnold's (1923b) measurement of 13 mm, which appears to be based on the same specimen, is presumed to be erroneus. Arnold (1936) stated that the male of *nana* is 5.8 mm long, but the paratype male seen by us is 7.5 mm. Arnold's *nana* appears to be no more than a subspecies of *aureopilosa,* and it is quite likely that the supposed subspecific differences merely represent infraspecific variation. Certainly the genitalia of *nana* and *aureopilosa* are quite similar. The aedeagal heads differ only in minor details of shape and number of teeth. Also, the propodeal and eighth sternum differences cited by Arnold are not nearly as great as he suggested.

Because of the paucity of *Tachytella* material (apparently only six specimens are known), it is difficult to assess the characters of the genus with respect to their value as generic criteria and their significance in relating the genus to other taxa in the subtribe. The relatively short jugal lobe (for the Larrini), the flat, raised frons, the absence of suprantennal tubercles, the elongate, constricted, commalike ocellar scars, and incomplete occipital carina are the most distinctive generic features of *Tachytella.* The female tarsal rake and the simple male foreleg are important also. At first glance *Tachytella* looks much like *Tachysphex,* and the female pygidial plate is quite like those found in most *Tachysphex.* The male genitalia are built on the same patterns as *Tachysphex,* but the *Tachytella* subgenital plate is more like that of *Larropsis* and *Ancistroma.* The mandible, ocellar scar, and female rake of *Tachytella* also resemble *Larropsis* and *Ancistromma,* and it would appear that the genus is more closely allied with these two genera than with *Tachysphex.* Probably *Tachytella* is an offshoot of the ancestral stock that gave rise to both the *Tachysphex* line and the *Larropsis-Tachytes* line.

The only papers dealing with the genus are those by Brauns and Arnold cited above.

Biology: Unknown.

Checklist of *Tachytella*

aureopilosa Brauns, 1906; S. Africa: Willowmore, Worcester
  ssp. *nana* Arnold, 1936; S. Africa: Resolution

Genus Larropsis Patton

Generic description: Inner orbits converging above; least interocular distance greater than combined length of pedicel and flagellomere I (except about equal in *interocularis*); hindocellar scars commalike or accent-mark-shaped, their long axes forming an obtuse angle (fig. 61 F); frons weakly to moderately convex discally, usually raised into a weak linear swelling along inner orbit, depressed above antennal socket and usually with a small central tumescence just above sockets; frontal line evanescent to deeply impressed; scape short, stout; clypeus sometimes partly divided into three sections by a pair of vertical impressions, clypeal free margin with a prominent, broad, median lobe which is sharply angulate laterally in female, narrow and rounded in male; labrum transverse, rarely notched, hidden; inner margin of mandible usually without subbasal teeth except in some females, externoventral margin with a deep notch; collar with some thickness or sometimes sharpedged, below or as high as and not closely appressed to scutum; propodeum moderately long; episternal sulcus extending to anteroventral margin of pleuron, scrobal sulcus sometimes present; subalar fossa margined below by one or more horizontal carinae (fig. 71 B); terga impunctate to finely, densely punctate, tergum I and sometimes II with lateral carina; female with pygidial plate, surface shining, sparsely to densely punctate, sometimes completely covered with appressed setae but these usually restricted to apex; male tergum VII flattened and punctate but without pygidial carinae; male forecoxa sometimes with a posteromedian fingerlike process; inner face of male foretrochanter usually excavate basally; male forefemur usually with a basal projection and associated weak depression (fig. 73 L); mid and hindtibiae moderately to thickly set with long spines which are arranged in rows, hindtibia sometimes with a longitudinal ridge or carina; both sexes with a foretarsal rake, spines stiff, moderately long to long, sometimes bladelike, tarsomere II with only one or two spines (fig. 73 A); marginal cell obliquely truncate apically; apex of male sternum VIII rounded (fig. 76 E), occasionally weakly emarginate (*deserta, arizonensis, vegeta*); volsella long, flat; head of aedeagus with teeth.

Geographic range: This is a Nearctic genus with 25 species. Most of these occur in the west, and none are recorded from the Mississippi basin.

Systematics: Larropsis range in length from 6 to 14 mm. The body is black, but the abdomen is partly or all red in some species. In *chilopsidis* the female is red except for parts of the head, and the male has a red gaster and partly red legs. The wings are frequently infumate in this genus.

In 1966 Bohart and Bohart elevated *Ancistromma* to a genus, thus making *Larropsis* a more homogeneous group. *Larropsis* is identified by the presence of a horizontal carina on the lower side of the subalar fossa (fig. 71 B). In some species this carina is paralleled by one or more additional carinae on the hypoepimeral area.

Although the weak linear swellings along the inner orbit of *Larropsis* and *Ancistromma* suggest a relationship with *Larra* and its relatives, these two genera are certainly more closely allied with *Tachytes* and *Tachysphex.* The close alliance with *Tachytes* is shown by the form of the ocellar scars, the foretarsal rake, the process of the male forecoxa, and the setose pygidial plate. In the past the facial swellings along the orbits have been given considerable weight in separating *Larropsis* (and *Ancistromma*) from *Tachysphex,* but they are frequently very weak. We have found other features that are generally better

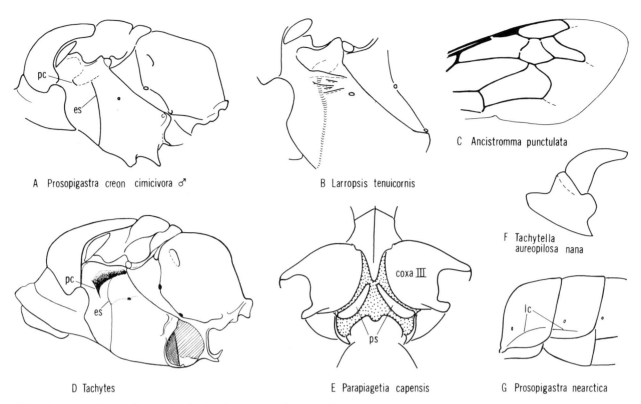

FIG. 71. Morphology of the subtribe Tachytina; A, thorax, left lateral, note tubercles on mesopleuron below; B, meso- and metapleuron, left; C, forewing, distal part; D, thorax, left lateral; E, hindcoxae and base of gaster; F, pronotum and scutum, left lateral; G, gastral segments I-II, lateral (pc = postspiracular carina, es = episternal sulcus, ps = propodeal sternites, lc = lateral carina).

for separation of these taxa. The frons in *Larropsis* and *Ancistromma* usually has a small moundlike swelling medially just above the antennal sockets (fig. 69 F). *Tachysphex* usually have two shiny swellings in this location (fig. 69 E). Females of *Larropsis* and *Ancistromma* have only two rake spines on the second foretarsomere, and these are generally dark colored and stiff (fig. 73 A). In *Tachysphex* this tarsomere usually bears numerous, closeset, long, fine, flexible, hairlike spines (fig. 73 D). Often the spines are pale in *Tachysphex*. There are some species of *Tachysphex* that have only two spines on tarsomere II, but even so they are still usually long, hairlike and flexible. The pygidial plate in female *Larropsis* and *Ancistromma* is setose at least apically (fig. 74 A), but in *Tachysphex* it is usually asetose. Those species of *Tachysphex* that have setose pygidial plates are separable from *Larropsis* and *Ancistromma* by the frontal or tarsal rake characters. Males of these genera are more easily separated. The apex of sternum VIII is rounded or at most very shallowly emarginate (fig. 76 E) in *Ancistromma* and *Larropsis,* while in *Tachysphex* the apex is bispinose or weakly trispinose (fig. 76 H). The male forefemoral differences between these two genera and *Tachysphex* are fairly constant, also, the problem cases occurring mainly in *Tachysphex* (compare figs. 73 K, 73 L).

Larropsis probably can be regarded as the most generalized member of the Larrini. Certainly it competes with *Larra* as the most primitive genus in the tribe.

The species of *Larropsis* were recently revised by Bohart and Bohart (1966).

Biology: Data for *Larropsis* are fragmentary but presumably its biology is similar to *Ancistromma*. Prey consists of gryllacridid crickets of the genus *Ceuthophilus* in *divisa* (Williams, 1914a), and *Ammobaenetes* in *filicornis* (Bohart and Bohart, 1966).

Checklist of *Larropsis*

arizonensis G. and R. Bohart, 1966; sw. U.S.
atra Williams, 1914; centr. and sw. U.S.
chilopsidis (Cockerell and W. Fox), 1897 (*Ancistromma*); sw. U.S.; n. Mexico
    zerbeii Viereck, 1906 (*Ancistromma*)
    tachysphecoides Viereck, 1906 (*Ancistromma*)
conferta (W. Fox), 1893 (*Ancistromma*); centr. U.S.; Mexico
    paenerugosa Viereck, 1906 (*Ancistromma*)
    bruneri H. Smith, 1906 (*Ancistromma*)
    minor Williams, 1914
    gracilis Rohwer, 1915
consimilis (W. Fox), 1893 (*Ancistromma*); sw. and centr. U.S.
    vegetoides Viereck, 1906 (*Ancistromma*)
deserta G. and R. Bohart, 1966; U.S.: California
discreta (W. Fox), 1893 (*Ancistromma*); se. U.S.

divisa (Patton), 1879 (*Larra*); centr. U.S.
elegans G. and R. Bohart, 1966; U.S.: Texas, New Mexico
filicornis Rohwer, 1911; centr. U.S.
    yatesi Mickel, 1917
greenei Rohwer, 1917; centr. and e. U.S.
interocularis G. and R. Bohart, 1966; U.S.: Kansas, Arizona
lucida G. and R. Bohart, 1966; U.S.: Arizona
rugosa (W. Fox), 1893 (*Ancistromma*); centr. and sw. U.S.; n. Mexico
sericea G. and R. Bohart, 1966; centr. U.S.
snowi G. and R. Bohart, 1966; U.S.: Arizona
sonora G. and R. Bohart, 1966; n. Mexico: Sonora
sparsa G. and R. Bohart, 1966; sw. U.S.; n. Mexico
striata G. and R. Bohart, 1966; U.S.: California
tenuicornis (F. Smith), 1856 (*Larrada*); w. U.S.; Baja California
testacea G. and R. Bohart, 1966; U.S.: Kansas
texensis G. and R. Bohart, 1966; U.S.: Texas
uniformis G. and R. Bohart, 1966; w. U.S.; n. Mexico
vegeta (W. Fox), 1893 (*Ancistromma*); centr. and sw. U.S.
washoensis G. and R. Bohart, 1966; U.S.: Nevada

Genus Ancistromma W. Fox

Generic diagnosis: Details as described for *Larropsis* except for the following: least interocular distance equal to or less than (slightly greater than in *portiana* and *granulosa*) combined length of pedicel and flagellomere I; long axes of ocellar scars forming an obtuse or right angle; clypeal free margin with a prominent median lobe which is angulate laterally (fig. 69 F) (lobe less prominent in some males), lobe sometimes with a median notch; labrum usually with median notch in female; inner margin of mandible with one or two subbasal teeth except in some males; malar space absent except in *shappirioi* and *sericifrons* males; collar usually sharp edged, usually below and closely appressed to scutum; subalar fossa not margined below by a carina; female with pygidial plate surface shining, punctate, with appressed setae apically (fig. 74 A); male tergum VII flattened, punctate, sometimes with lateral pygidial carina.

Geographic range: This is primarily a Nearctic genus, but five of the 16 species occur in the Old World. Three of the Old World forms are found in the Mediterranean Region, the fourth is known from the southwestern USSR, and the fifth from South Africa. Synonymy of some of the Palearctic species is likely.

Systematics: These wasps are very similar to *Larropsis* in size and form, but the gaster is more frequently red in *Ancistromma*. Bohart and Bohart (1966) elevated *Ancistromma* to genus after having used the taxon as a subgenus of *Larropsis* (Bohart and Bohart, 1962). The two taxa are so similar that either arrangement could be followed. Certainly they are no more distinct than some groups within *Liris, Tachytes,* and *Tachysphex*. The absence of a horizontal carina below the subalar fossa in *Ancistromma* is the principal feature for separation from *Larropsis*. The stronger convergence of the inner orbits in *Ancistromma* is diagnostic in all but two or three species. See discussion under *Larropsis* for relationships with other genera.

Ancistromma is a specialized offshoot of *Larropsis* as evidenced by the narrower least interocular distance, better developed inner mandibular teeth, notched labrum, and development in several species of a malar space.

The Nearctic forms have been recently revised by Bohart and Bohart (1962).

Biology: Evans (1958b) published detailed data on *Ancistromma distincta*. Ferton (1912a) gave some fragmentary notes on the Old World species *punctulata* under the manuscript name of *laevidorsis* Ferton. Females of both species flush gryllid crickets from under stones, leaves, and other shelter. Evans noted that prey consisted only of adult *Nemobius fasciatus* (De Geer) in the New York population of *distincta* that he studied. We have seen specimens of this species pinned with a nymph of *Gryllus* and an adult female of *Nemobius allardi* Alexander and Thomas. Bohart and Bohart (1962) recorded gryllacridid crickets as prey of *A. aurantia* (*Ceuthophilus fusiformis* Scudder), and *capax* (*Ceuthophilus* sp.). *Ancistromma distincta* carries its lightly paralyzed prey over the ground or in short hopping flights. The cricket is held at the base of its antennae by the wasps' mandibles and carried dorsum up. Nests are made in the bottom of pre-existing cavities such as caved-in parts of mole burrows. Evans hypothesized that once a *distincta* female locates a suitable cavity she continues to use it till she dies. The five nests that he dug up contained from one to nine cells. They were 10 to 20.5 cm beneath the soil surface. One to three crickets were provisioned per cell, and the egg was laid beneath the fore and midcoxae of the first provision.

Evans found that about half of the *distincta* cells were parasitized by miltogrammine Sarcophagidae.

Checklist of *Ancistromma*

asiatica Gussakovskij, 1935; sw. USSR
    shestakovi Gussakovskij, 1935
aurantia (W. Fox), 1891 (*Larra*); centr. U.S.
    aurulenta W. Fox, 1893
bradleyi (G. and R. Bohart), 1962 (*Larropsis*); U.S.: California, Oregon
capax W. Fox, 1893; w. U.S.
    dolosa W. Fox, 1893
    dolosana Rohwer, 1915 (*Larropsis*)
    picina Mickel, 1916 (*Larropsis*)
corrugata (G. and R. Bohart), 1962 (*Larropsis*); w. N. America
distincta (F. Smith), 1856 (*Larrada*); n. N. America
    semirufa Banks, 1921 (*Larropsis*)
europaea Mercet, 1910; Spain
granulosa (G. and R. Bohart), 1962 (*Larropsis*); w. U.S.
hurdi (G. and R. Bohart), 1962 (*Larropsis*); U.S.: California
maligna Mercet, 1910; Spain (? = *europaea*)

obliqua (F. Smith), 1856 (*Larrada*); S. Africa
platynota (G. and R. Bohart), 1962 (*Larropsis*); U.S.: Arizona
portiana (Rohwer), 1911 (*Larropsis*); U.S.: New Mexico, Texas
punctulata (Kohl), 1884 (*Tachysphex*); Morocco, Algeria, Cyprus
    laevidorsis Pérez, Ferton, 1912, 1923 (*Tachysphex*), nomen nudum, placement according to W. J. Pulawski in correspondence
    ssp. *melanaria* (Kohl), 1888 (*Tachysphex*); sw. USSR
    socias Kohl, 1888 (*Tachysphex*)
sericifrons H. Smith, 1906; centr. and sw. U.S.
    rubens Mickel, 1918 (*Larropsis*)
shappirioi (G. and R. Bohart), 1962 (*Larropsis*); e. U.S.

Genus Tachytes Panzer

Generic description: Inner orbits converging above, sometimes narrowly separated at vertex; hindocellar scars golf club shaped, (i.e. with a long tail), long axes of tails subparallel or forming an angle up to 70°, length of scar longer than distance between midocellus and lower end of scar (fig. 61 G,H); frons weakly to moderately convex and just above antennal sockets with a median swelling which is sometimes binodose as in *Tachysphex;* frontal line present or absent; scape usually moderately long and rather stout, but especially long and slender in *mergus* group, flagellomeres usually longer than broad; sockets contiguous with frontoclypeal suture or separated from it by up to one-third a socket diameter; clypeus usually convex, sometimes partially divided by a pair of vertical suturelike lines into three sections, free margin usually with a prominent median lobe of various shapes, lobe often bounded by one or more small teeth, clypeal surface with prominent subapical ridges, tubercles, or teeth in some females of *mergus* and *ambidens* groups; labrum arcuate to truncate, sometimes with median notch, hidden or narrowly exposed; inner margin of mandible with one or two subbasal teeth except in some males, externoventral margin usually with a strong angle or notch, but notch occasionally small or absent (fig. 72 D-H); mouthparts usually short to moderately long but long in *basilicus* and *marshalli* groups, palpal formula 5-3 in latter group; dorsum of collar narrow, much lower than and closely appressed to scutum (fig. 71 D); propodeum short to long; episternal sulcus extending to anteroventral margin of pleuron, scrobal sulcus absent or weakly impressed; subalar fossa not margined below; terga impunctate to very finely punctate, tergum I with lateral carina; both sexes with pygidial plate, surface densely covered with appressed metallic (rarely black or brown) setae, except setae sparse in some Australian forms and *mergus* group (surface nearly asetose in *nitidiusculus*) (fig. 74 B-D); female pygidial plate flat, triangular, margined by carinae laterally, apex usually rounded, male pygidial plate sometimes without lateral carinae and apex often more broadly rounded or truncate than in female; male sterna sometimes with special hair

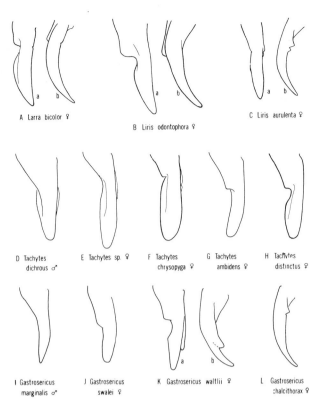

FIG. 72. Right mandible in the tribe Larrini; A-C, lateral (a) and anterior (b) aspects; D-J, lateral; K, lateral (a) and anterior (b); L, anterior.

tufts or marginal fringes; male forecoxa with a triangular or fingerlike process posteriorly in *distinctus, chilensis, abdominalis, mergus, ambidens,* and *pygmaeus* groups (fig. 73 M); male foretrochanter and forefemur with basal depression in first five groups (fig. 73 M), male forefemur with semicircular basal notch in *dichrous* group; apex of hindfemur in most females with a dorsal oblique truncation which is covered with appressed setae, and an outer ventral platelike area or thin, downward directed process (fig. 73 J) (absent or weak in some males and in *mergus* group); foretibia usually with an outer row of long rake spines, mid and hindtibiae with three rows of spines (sometimes only two on hindtibia), spines of hindfemur sometimes very short, stout, upper row sometimes associated with a longitudinal ridge or carina; female with a well developed foretarsal rake which is composed of widely spaced (on tarsomere I), short (usually no longer than tarsomere II), stiff spines, tarsomere II usually with only two spines (fig. 73 C); male with or without a rake; apex of marginal cell narrowly rounded (fig. 70 G) or truncate (fig. 76 F-G); apex of male sternum VIII bispinose (emarginate) or truncate; volsella usually long and slender; ventral surface of gonostyle usually fringed with setae; head of aedeagus rarely with teeth (fig. 77 C).

Geographic range: Tachytes is fairly evenly represented throughout the temperate and tropical world. The 268 species are distributed among the Zoological Regions as follows: 28 Nearctic, 64 Palearctic, 88 Ethiopian, 40 Oriental, 12 Australasian, and 27 Neotropical. In addition four species traverse the Nearctic and Neotropical Regions. Madagascar has seven endemics, two species are known in the New Hebrides, and one occurs in New Guinea.

Systematics: Tachytes are 4 to 24 mm long, but average about 12 mm. They are mostly stout-bodied, hairy, beelike wasps. Many have appressed silver or gold vestiture on the head, thorax and abdomen, the hair usually forming transverse bands or fasciae on the last. Most are mainly black, but the gaster and legs are sometimes partly or all red. Wings are often clear or yellowish, but occasionally they are infuscate or black.

The long "tail" of the ocellar scar quickly separates *Tachytes* from other genera of the subtribe Tachytina. In most *Tachytes* the tails are subparallel, and the distance between the midocellus and the end of the tail is always less than the length of the scar. Usually in both sexes of *Tachytes* the pygidial plate is densely covered with appressed silver, gold, or brassy setae. Within the subtribe, the truncate hindfemoral apex and associated platelike area (fig. 73 J) are peculiar to *Tachytes*. The genus is certainly one of the most advanced in the Larrini as evidenced by the long mouthparts and 5-3 palpal formula of some species, the hindfemoral truncation just mentioned, the strongly deformed ocelli, the long male coxal process in some groups, the setose pygidial plate, and the generally hairy body. Although *Tachytes* is often confused with *Tachysphex* (see systematics of the latter for differences), the two are probably not closely related. An association with *Larropsis* and *Ancistromma* seems to be more logical.

Turner (1917) and Banks (1942) proposed a number of subgenera for some of the groups within *Tachytes*. Pulawski (1962) in a masterful work on the Palearctic species recognized only one of these, *Holotachytes* Turner, which contains a single, aberrant, Old World species. He demonstrated that the characters used by Banks for *Tachynana* Banks are variable, and he also suggested that *Calotachytes* Turner was of questionable value. Our study of *Tachytes* indicates that Pulawski's use of the species group for infrageneric divisions of *Tachytes* is the most realistic approach in this large genus, especially considering that the species in some parts of the world are not satisfactorily known (Australia, India). We are not recognizing the subgenus *Holotachytes*, preferring instead to call this taxon the *dichrous* group. *Tachytes dichrous* does not have an externoventral notch on the mandible, the propodeum is long and without vestiture above, the outer apicoventral plate of the hindfemur is margined anteriorly by a row of setae, the hindtibia has a serrate longitudinal carina, the male forefemur has a semicircular basal notch similar to that found in *Tachysphex*, and the head of the aedeagus has teeth on the ventral margin of the penis valves.

Similarly, we are using the name *marshalli* group for *Calotachytes*. This assemblage consists of Ethiopian species in which the palpal formula is 5-3, the basal labial segment and galea are very long, the mandibular notch is small, and the outer apicoventral plate of the hindfemur is weak or absent.

The Old World *basilicus* group first proposed by Arnold (1923a) and later redefined by Pulawski (1962) has the same characteristics as the *marshalli* group, except the palpal formula is 6-4. The two groups are obviously closely allied, indicating that if *Calotachytes* is to be recognized it should also include the *basilicus* group. Arnold's concept of the *basilicus* group was based on pubescence rather than mouthparts, and some of the species he assigned to it, *rhodesianus* for example, probably do not belong in Pulawski's updated interpretation of the assemblage. Likewise, Arnold's (1923a) *praestabilis* group, in which the galea is very long, appears to be allied or synonymous with the *basilicus* group *sensu* Pulawski.

The majority of the Palearctic species fall into two large assemblages, the *obsoletus* and *ambidens* groups. The first of these is equivalent to *Tachytes* in the strictest sense; i.e., mandible with an externoventral notch, galea short, palpi 6-4, marginal cell apex narrowly rounded, hindfemoral apex usually with an apicoventral plate, mid and hindbasitarsus usually with one or more spines on outer margin, and male foreleg unmodified. Pulawski also recognized three small groups which are closely allied with the *obsoletus* group: the *maculicornis, etruscus,* and *patrizii* groups. Because of synonymy the last should be called the *comberi* group.

The *ambidens* group contains species in which the male forecoxa has an inner posterior fingerlike process, and the forefemur and foretrochanter have basal depressions or excavations. The female clypeus in this group usually has special ridges, tubercles, or teeth. The mandibular notch is variable, sometimes being rather weak. The marginal cell is truncate apically, and the mid and hindbasitarsus have no outer spines. The tails of the ocellar scars display the greatest divergence found in the genus, their long axes approaching an angle of 70° in some species (fig. 61 H). The plate at the hindfemoral apex is usually quite broad in this group, being oval instead of linear. Pulawski's small *aeneus* group apparently is related to the *ambidens* group, and his isolated, monotypic, *pygmaeus* group has a forecoxal process in the male and a glabrous propodeal dorsum.

The New World species have been studied by several workers of which Banks (1942) published the most comprehensive paper. Although he illuminated many useful characters for our *Tachytes,* he nevertheless produced an unsound classification. Banks tried to group our species on, among other things, the presence or absence of erect hair ventrally on the hindfemur, but as Pulawski has correctly pointed out, there are intergrades between the two conditions. Nevertheless, the use of hindfemoral vestiture in grouping species is useful when other characters are used in conjunction with it. Recent extensive investigations of New World *Tachytes* by R. M. Bohart (unpublished) make it possible to improve synonymy and to assign our species to groups.

The *aurulentus* group is equivalent to *Tachytes* in the strict sense and to the Old World *etruscus* group. Here the male antennae, forecoxae, and foretrochanters are simple: the female hindtibia has slender spines, the female pygidial plate is densely setose, and the hindfemur bears at least a few, often many, long hairs beneath. A subgroup with red hindtibia includes *aurulentus, auricomans, columbiae, crassus, exornatus, harpax, praedator, validus,* and *varians.* A second subgroup with dark hindtibia includes *badius, ermineus, floridanus, hades, leprieurii,* and *sayi.*

The *pepticus* group agrees with the preceding subgroup except that male flagellomeres VII-IX are broadened asymmetrically or sometimes IX-X or IX-XI are enlarged. Also, there is practically no outstanding hair beneath the hindfemur. In this group are *californicus, chelatus, fulviventris, nevadensis, pennsylvanicus, pepticus, sculleni,* and *spatulatus.*

The *distinctus* group is composed of medium-sized to large wasps with practically no hair beneath the hindfemur but with a rather slender prolongation on the inner side of the male forecoxa (fig. 73 M). Associated with the prolongation are a posterobasal depression on the trochanter and a basoventral groove on the forefemur. The female has dense pygidial setae (fig. 74 B) and bristles on the hindtibia. Included are *amazonus, argyrofacies, aureovestitus, chrysocercus, costalis, distinctus, fraternus, frontalis, guatemalensis, jucundus,* and *tricinctus.*

The *chilensis* group of small to medium-sized wasps bridges the gap between the preceding and following ones. It differs from the *distinctus* group mainly in having abundant hair, or at least a basal patch, on the underside of the hindfemur. Also, the prolongations of the male forecoxae are shorter and more triangular, and the groove on the forefemur is usually weak to absent. In the *chilensis* group are *andreniformis, chilensis, concinnus, imperialis, pretiosus,* and *setosus.*

The *abdominalis* group are mostly small *Tachytes,* similar to the above but with at most a few scattered long hairs toward the base of the hindfemur beneath. Also, the females have dentiform setae on the hindtibia, and male flagellomere I is shorter than II (except in *birkmanni*). Included are *abdominalis, birkmanni, chrysopyga, excellens, intermedius, micantipygus, obductus, parvus, simulans,* and *staegeri.*

The *mergus* group, which Banks considered as a separate genus *Tachyoides,* is quite similar to the *abdominalis* group, exhibiting the short forecoxal prolongation and short flagellomere I in the male. However, both sexes have the setae of the pygidial plate sparse (sometimes nearly absent in females, fig. 74 D), and the hindtibial setae are bristlelike in females rather than dentiform. Additional features are the long female scape, sicklelike female mandibles, unusual teeth or other modifications on the female clypeus, and a rather simple apex to the hindfemur rather than a platelike truncation. Although *mergus* and its relatives form a very distinctive group, we prefer not to recognize the subgenus *Tachyoides.* Sparse pygidial setation is found in a number of Australian *Tachytes,* which are otherwise typical members of the genus (fig. 74 C). The female clypeal traits of the *mergus* group have their counterpart in the Old World *ambidens* group. The simple hindfemoral apex appears not to be found elsewhere in *Tachytes,* but the condition is approached in the *marshalli* and *basilicus* groups. Included in the *mergus* group are *mergus, nitidiusculus, ornatipes,* and *rhododactylus.*

Species discrimination in *Tachytes* is difficult in some sections of the genus, particularly with respect to associating females with males. In a few cases several species have been recognized from one geographic area on the basis of distinctive male differences, but the females cannot be reliably associated with their respective males.

A species character of special value is the number of silvery or golden tergal hair bands. This usually involves the first four terga but sometimes only two or three and occasionally the first five. The number may be the same in the two sexes or different. The number of bands seems to be quite constant in most species, but in a few it seems to be undergoing evolutionary change. A good example is *chrysopyga.* In this species males always have the basal four terga with bands. More northern females (ssp. *obscurus*) have three bands, Caribbean females (ssp. *argentipes*) have a trance of silver on tergum IV, Brazilian females (*chrysopyga* s.s.) have four bands, and the Argentinean race (ssp. *staegeri*) has three bands.

An interesting character of specific value is found in *intermedius,* where the mesopleural and metapleural flanges are broadly lamellate, and the metapleural flange overlies a slender vertical process on the basal margin of the propodeum. It may be significant that this otherwise ordinary member of the *abdominalis* group has unusual prey (pygmy mole crickets).

The Palearctic Region is the only area for which a thorough, modern treatment of *Tachytes* is available (Pulawski, 1962), and his paper should be consulted for an insight into the structures used in species definition. Keys to species for other areas are: North America north of Mexico, Banks (1942) and Bohart (1962b); Ethiopian Region, Turner (1917) and Arnold (1923b); Madagascar, Arnold (1945); India, Bingham (1897); Taiwan, Tsuneki, (1967a) and Japan, Tsuneki (1964c); the Philippines, F. X. Williams (1928a); Australia, Turner (1916c). Some of the foregoing are obviously out-of-date.

Biology: The best references on *Tachytes* are those of F. X. Williams (1914a, 1928a), J. B. Parker (1921), Krombein and Kurczewski (1963), Kurczewski (1966a), Kurczewski and Ginsburg (1971), Evans and Kurczewski (1966), C. S. Lin (1967), Pulawski (1962), and Tsuneki (1969a). Pulawski's paper summarizes most of the Old World observations.

Nests are dug in a variety of situations. Some species start their tunnels in preexisting holes such as rodent burrows. N. Lin (1966) reported that *distinctus* used Cicada Killer (*Sphecius speciosus*) emergence holes as starting places for nests. This agrees with observations by Rau (1923) that *distinctus* used a variety of preexisting holes in which to start a nest. Burrows range in length from 7 cm (*intermedius*) to nearly 1 m (*praedator*), and 7.5

to 70 cm below the soil surface. Nests are sometimes excavated at night in *praedator* (C. S. Lin, 1967), *distinctus* (F. Williams, 1914a), and *nipponicus* (Tsuneki, 1969a). Most species form a mound of excavated soil (tumulus) at or around the nest entrance. The nests are usually multicellular, the cells either being placed at the ends of branches that radiate from the main burrow or along the burrow. The main entrance is left open during provisioning except in *mergus.*

Orthoptera of the families Acrididae, Tettigoniidae, Tetrigidae, and Tridactylidae are provisioned by most groups of *Tachytes.* Gussakovskij (1952) declared that *T. bidens (ambidens* group) uses lepidopterous larvae of the family Geometridae. This aberrant behavior has been confirmed by Pulawski (in letter to Menke) who observed "several females of *T. ambidens* carrying small caterpillars in Tedshen" (Turkmen S.S.R.). Prey specificity is fairly constant at the family level, and it coincides fairly well with the species groups used by us. For example, *aurulentus, crassus, exornatus, harpax, praedator,* and *validus,* all members of the *aurulentus* group, use katydids exclusively. The Old World species *sinensis* and *nipponicus,* which also use tettigoniids, appear to belong to the *etruscus* group which is equivalent to our *aurulentus* group. Prey records for species of the *pepticus, distinctus,* and *obsoletus* groups consist of Acrididae. *Tachytes abdominalis* and *chrysopyga (abdominalis* group) also use acridids, but *abdominalis* has been seen using Tetrigidae as well. *Tachytes obductus,* another *abdominalis* group species, is recorded using tetrigids. Finally, two species, *mergus (mergus* group*)* and *intermedius* (*abdominalis* group) use pygmy mole crickets (Tridactylinae), exclusively.

Prey are usually flown to the nest. The wasp holds onto the orthopteran's antennae with her mandibles and usually supports the prey with the legs during flight. Prey are held dorsum or venter up during transport. One to 13 prey are placed in the cell, and the egg is laid after provisioning of the cell is completed. Paralysis appears to be complete or nearly so in all of the species reported upon, with the exception of *mergus* and *intermedius* which use tridactylids. F. Williams (1928a), Kurczewski (1966a), and Krombein and Kurczewski (1963) noted that the prey of these species were quite mobile and could still jump.

Both *mergus* and *intermedius* hunt their prey in the manner of *Larra,* i.e., they walk over the soil surface tapping the substrate with their antennae in their search for the burrows of the tridactylid prey. When a pygmy mole cricket is located, the wasp digs down to its burrow. If on target the wasp grasps the cricket's head with her mandibles (very long and slender in the case of *mergus*) and removes it from its burrow. Sometimes the digging of the wasp causes the cricket to flee its burrow by leaping from its entrance. Observers have reported that the wasp is very quick and usually succeeds in catching the cricket in midair in this circumstance.

Sarcophagid flies have been found as parasites in *Tachytes* nests by several observers.

Checklist of *Tachytes*

abdominalis (Say), 1823 (*Larra*); U.S.: Texas to Arizona, Utah; Mexico
abercornensis Arnold, 1959; Zambia
admirabilis Turner, 1916; Uganda
aeneus E. Saunders, 1910; Algeria
aestuans Turner, 1916; Australia
agadiriensis Nadig, 1933; Morocco
alacris Arnold, 1960; Rhodesia
albonotatus Walker, 1871; ? Egypt
alfierii Pulawski, 1962; Egypt
amazonus F. Smith, 1856; N. and S. America
   clypeatus Taschenberg, 1870; new synonymy by Bohart
   scalaris Taschenberg, 1870; new synonymy by Bohart
   rufofasciatus Cresson, 1872; new synonymy by Bohart
   fervens F. Smith, 1873; new synonymy by Bohart
   dives Holmberg, 1884, nec Lepeletier, 1845; new synonymy by Bohart
   rufomaculatus Cameron, 1889 (*Tachysphex*); new synonymy by Bohart
   holmbergii Dalla Torre, 1897, given as a n. name for *dives*
   fiebrigi Brèthes, 1909; new synonymy by Bohart
   nigricaudus Brèthes, 1909; new synonymy by Bohart
   mimeticus Schrottky, 1909; new synonymy by Bohart
   anisitsi Strand, 1910; new synonymy by Bohart
   rufoannulatus Strand, 1910; new synonymy by Bohart
ambidens Kohl, 1884; sw. USSR
   ambidens Kohl, 1898
   caucasicus Radoszkowski, 1886 (♀ only)
andreniformis Cameron, 1889; Mexico
   anthreniformis Dalla Torre, 1897
antillarum Cameron, 1906; Cuba
antillarum Cmaeron, 1906; Cuba
apiformis F. Smith, 1856; Brazil, Argentina
approximatus Turner, 1908; Australia
archaeophilus Pulawski, 1962; Egypt
argenteovestitus Cameron, 1910; S. Africa
argenteus Gussakovskij, 1932; e. Mediterranean region, sw. USSR, Iran
   radoszkowskii Beaumont, 1936
argyreus (F. Smith), 1856 (*Larrada*); n. Africa to sw. USSR, n. India
   melanopygus A. Costa, 1893 (Rend. Accad. Sci. Fis. Mat. Naples, also Atti Accad. Sci. Fis. Mat. Naples)
   pygmaeus Kohl, 1888 (♂ only)
   debilis Pérez, 1907 (*Tachysphex*)
   beludzhistanicus Gussakovskij, 1932 (♂ only)
   ssp. *thalassinus* Pulawski, 1962; Cyprus
argyrofacies Strand, 1910; Paraguay, Argentina
argyropis Saussure, 1887; Madagascar
asiagenes Pulawski, 1962; sw. USSR
assamensis Cameron, 1904; n. India
associatus Turner, 1917; Mozambique
astatiformis Tsuneki, 1963; Thailand
astutus Nurse, 1909; India, Burma, Thailand, Taiwan
   shirozui Tsuneki, 1966

aureocinctus Cameron, 1905; Borneo
aurichalceus Kohl, 1891; w. and centr. Africa
auricomans Bradley, 1919; U.S.: Georgia, new status by Bohart
    distinctus of authors, nec F. Smith, 1856
aurifex F. Smith, 1858; Borneo
auripes Berland, 1942; Zaire
aurovestitus F. Smith, 1873; Brazil
aurulentus (Fabricius), 1804 (*Larra*); e. U.S. and Canada
    mandibularis Patton, 1881
    propinquus Rohwer, 1909
    duplicatus Rohwer, 1920
badius Banks, 1942; U.S.: Texas to Arizona
bakeri Williams, 1928; Philippines
basilicus (Guérin-Méneville), 1844 (*Lyrops*); Senegal to Algeria and Egypt
    superbiens Morice, 1911
    guichardi Arnold, 1951
beludzhistanicus Gussakovskij, 1932; Iran (♀ only)
bidens Gussakovskij, 1952; sw. USSR, Iran
bimetallicus Turner, 1917; S. Africa
birkmanni Rohwer, 1909; U.S.: Texas
    obscuranus Rohwer, 1909 (♂ only)
    atomus Banks, 1942; new synonymy by Bohart
biskrensis E. Saunders, 1910; Algeria
borneanus Cameron, 1902; Borneo, Philippines
    banoensis Rohwer, 1919
braunsi Turner, 1917; S. Africa
bredoi Arnold, 1947; Zambia
brevipennis Cameron, 1900; e. India
    maculitarsis Cameron, 1900
brevis Walker, 1871; Somalia
brunneomarginatus Arnold, 1947; Zambia
brunneus Pulawski, 1962; Sudan
bulawayoensis Bischoff, 1913; Rhodesia
californicus R. Bohart, 1962; U.S.: California to Washington
cameronianus Morice, 1897; Egypt
carinatus Berland, 1945; Mali
cataractae Arnold, 1923; Rhodesia
celsissimus Turner, 1917; e. India
cephalotes Walker, 1871; Egypt
ceratophorus Pulawski, 1962; Iran
chelatus R. Bohart, 1962; U.S.: Arizona
chilensis (Spinola), 1851 (*Larra*); Chile, Argentina; lectotype ♀, "Chili, D. Gay", Mus. Turin, present designation by Bohart
    gayi Spinola, 1851 (*Larra*); lectotype ♂, "Chili, D. Gay", Mus. Turin, present designation by Bohart
    spinolae Reed, 1894
    dichrous Herbst, 1921, nec F. Smith, 1856
chivensis Pulawski, 1962; sw. USSR
chrysocercus Rohwer, 1911; U.S.: Arizona, Texas; Mexico
    apache Banks, 1942, new synonymy by R. Bohart
chrysopyga (Spinola), 1841 (*Lyrops*); s. Mexico to s. Brasil, lectotype ♀, "Cayenne, D. Le Prieur", Mus. Turin, present designation by Bohart
    columbianus Saussure, 1867; new synonymy by Bohart
    argentipes Cameron, 1889; nec F. Smith, 1856; new synonymy by Bohart
    argentricus Dalla Torre, 1897; n. name for *argentipes* Cameron
    ssp. *staegeri* Dahlbom, 1845; Argentina
    ssp. *argentipes* F. Smith, 1856; Caribbean Islands, new status by Bohart
        insularis Cresson, 1865; new synonymy by Bohart
    ssp. *obscurus* Cresson, 1872; U.S., n. Mexico; new status by Bohart
        texanus Cresson, 1872; new synonymy by Bohart
        hirsutifrons Banks, 1942; new synonymy by Bohart
chudeaui Berland, 1942; Mali
    ssp. *niger* Berland, 1942; Niger
cinerascens Arnold, 1935; S. Africa
columbiae W. Fox, 1892; e. U.S.
comberi Turner, 1917; n. Africa, Arabian penin. Pakistan
    tricinctus Pérez, 1907, nec Fabricius, 1804
    patrizii Guiglia, 1932
    atlanticus Berland, 1942
compactus Arnold, 1951; Senegal
concinnus F. Smith, 1856; Brazil
confusus Arnold, 1923; Rhodesia
contractus Walker, 1871; ? Dahlak Archipelago, Ethiopia
copiosus Arnold, 1945; Madagascar
corniger Gussakovskij, 1952; sw. USSR
    cornigera Gussakovskij, 1952 (orig. orthography)
    trichopus Pulawski, 1962
costalis Taschenberg, 1870; Brazil
crassus Patton, 1881; e. U.S.
danae Arnold, 1923; Malawi
decoratus Walker, 1871; Egypt;? Arabian penin.
decorsei Berland, 1942; Mali
dichrous F. Smith, 1856; s. and n. centr. Africa
    ferox F. Smith, 1873 (*Larrada*)
    ssp. *hospes* Bingham, 1898; Iraq, India, Sri Lanka
dilaticornis Turner, 1916; Ethiopia, Kenya
diodontus Pulawski, 1962; Egypt
discrepans Arnold, 1951; Senegal
dispersus Turner, 1916; Australia
disputabilis Turner, 1917; Malawi, Rhodesia
distanti Turner, 1917; S. Africa
distinctus F. Smith, 1856; N. & S. America
    fulvipes F. Smith, 1856 (*Larrada*)
    elongatus Cresson, 1872, new synonymy by Bohart
    yucatanensis Cameron, 1889, new synonymy by Bohart
    contractus W. Fox, 1892, nec Walker, 1871, new synonymy by Bohart
    seminole Banks, 1942
    austrinus Banks, 1942
    ssp. *bimini* Krombein, 1953; Bahamas
diversicornis Turner, 1918; Ghana, Ethiopia, Mali, Pakistan

griseolus Arnold, 1951
rufitibialis Arnold, 1951
ermineus Banks, 1942; sw. U.S.; Mexico
erynnis Turner, 1917; Zaire to S. Africa
etruscus (Rossi), 1790 (*Andrena*); s. Europe; sw. Asia
    argentatus Brullé, 1833
europaeus Kohl, 1884; nw. Africa, Europe, Asia
    tricolor Fabricius of Panzer, 1801 and 1806
  ssp. *orientis* Pulawski, 1962; China
eurous Pulawski, 1962; Israel
excellens Cameron, 1912; Guyana
exclusus Turner, 1917; Ghana
exornatus W. Fox, 1893; sw. U.S.; Mexico
falciger Arnold, 1951; Mauritania
famelicus Pulawski, 1962; sw. USSR
fatalis Turner, 1916; Australia
fervens F. Smith, 1873; Brazil
fervidus F. Smith, 1856; India
fidelis Pulawski, 1962; Arabian penin.
flagellatus Nurse, 1903; w. India
flavocinereus Arnold, 1945; Madagascar
floridanus Rohwer, 1920; U.S.: Florida to Arizona
    foxi Banks, 1942; new synonymy by Bohart
    comanche Banks, 1942; new synonymy by Bohart
formosissimus Turner, 1908; Australia
fraternus Taschenberg, 1870; Centr. and S. America
    ferrugineipes Cameron, 1889; nec Lepeletier, 1845
      (in *Larra*); new synonymy by Bohart
    asuncionis Strand, 1910; new synonymy by Bohart
    ametinus Cameron, 1912; new synonymy by Bohart
freygessneri Kohl, 1881; Mediterranean region to sw. USSR and Iran
frontalis F. Smith, 1873; Brazil
fruticis Tsuneki, 1964; Japan, Korea, Taiwan (? = *magellanicus,* fide Tsuneki '67)
fucatus Arnold, 1951; Egypt, Mauritania
    serapis Pulawski, 1962
†*fulvipes* Schrottky, 1903; Brazil, nec F. Smith, 1856
fulviventris Cresson, 1865; U.S.: N. Dakota to N. Mexico and Texas; Mexico
    caelebs Patton, 1879
    coelebs Dalla Torre, 1897
  ssp. *rossi* R. Bohart, 1962; U.S.: s. California; Mexico: Baja California
fulvopilosus Cameron, 1904; n. India, China
    andreniformis Cameron, 1902, nec Cameron, 1889
    fulvovestitus Cameron, 1904
gondarabai (Guiglia), 1943 (*Tachysphex*); Ethiopia, new combination by Pulawski
    gondarabai Guiglia, 1950 (*Tachysphex*)
gracilicornis Arnold, 1944; Mozambique
guatemalensis Cameron, 1889; e. U.S. to Guatemala
    coxalis Patton, 1892; new synonymy by Bohart
habilis Turner, 1917; Malawi, Mali
hades Schrottky, 1903; Brazil
    minos Schrottky, 1903; new synonymy by Bohart
hamiltoni Turner, 1917; Zaire, Kenya
    harpax Patton, 1881; e. U.S.

    dubitatus Rohwer, 1909
hengchunensis Tsuneki, 1967; Taiwan
hirsutus F. Smith, 1856; S. Africa
illabefactus Pulawski, 1962; Iraq
imperialis Saussure, 1867; Chile, Argentina
    saussurei Reed, 1894
indicus Dalla Torre, 1897; India
    ornatipes Cameron, 1889 (Mem. Proc. Manch. Lit. Philo. Soc.), nec Cameron, 1889 (Biologia Centrali-Amer.).
indifferens Arnold, 1945; Madagascar
indulgens Cheesman, 1937; New Hebrides
inexorabilis Turner, 1917; Uganda to S. Africa
instabilis Turner, 1917; Malawi
integer Gussakovskij, 1932; Iran, sw. USSR
intermedius (Viereck), 1906 (*Tachysphex*); e. U.S. to Texas
    minutus Rohwer, 1909, new synonymy by Bohart
    maestus Mickel, 1916, new synonymy by Bohart
    austerus Mickel, 1916, new synonymy by Bohart
    amiculus Banks, 1942, new synonymy by Bohart
interstitialis Cameron, 1900; Sri Lanka
irreverens Cheesman, 1937; New Hebrides
irritabilis Turner, 1917; Malawi, Rhodesia
jucundus F. Smith, 1856; Brazil
kristenseni Turner, 1917; Ethiopia, S. Africa
labilis Turner, 1917; Malawi
lachesis Turner, 1917; Malawi
lamborni Arnold, 1934; Dahomey
lamentabilis Arnold, 1951; Ghana
lanuginosus Pulawski, 1962; sw. USSR, Afghanistan
latifrons Tsuneki, 1964; Japan
lepidus Arnold, 1929; Rhodesia
leprieurii (Spinola), 1841 (*Lyrops*); French Guiana
    apiformis F. Smith, 1856; new synonymy by Bohart
levantinus Pulawski, 1962; Syria
lingnaui Arnold, 1933; S. Africa
longirostris Arnold, 1947; Zambia
lugubris Walker, 1871; Egypt
maculicornis E. Saunders, 1910; n. Africa
maerens Turner, 1917; Malawi
    moerens Arnold, 1923
magellanicus Williams, 1928; Philippines
maindroni Berland, 1942; Somalia
manjikuli Tsuneki, 1963; Thailand
maroccanus Pulawski, 1962; Morocco
    aegyptiacus Pulawski, 1962, new synonymy by W. Pulawski
marshalli Turner, 1912; Rhodesia, Malawi
matronalis Dahlbom, 1845; Iberian Penin., n. Africa; e. Mediterranean region; sw. USSR
    obesus Kohl, 1884
megaerus Turner, 1917; Ghana
melancholicus Arnold, 1923; Rhodesia
memnon Turner, 1916; Malawi
meraukensis Cameron, 1911; New Guinea
mergus W. Fox, 1892; U.S.: Arizona to New Jersey and Florida; Mexico

minor Rohwer, 1909
obscuranus Rohwer, 1909; ♀ only
ariellus Banks, 1942 (*Tachyoides*), new synonymy by R. Bohart
micantipygus Strand, 1910; Paraguay
midas Arnold, 1929; Rhodesia
mirus Kohl, 1894; Mozambique, Burundi
  pulchrivestitus Cameron, 1908
mitis Turner, 1916; Australia
modestus F. Smith, 1856; India, Burma, China, Japan, Taiwan
  ?*dilwara* Nurse, 1903
    maculipennis Cameron, 1904
  ssp. *neglectus* Turner, 1917; Malawi, Mali
monetarius F. Smith, 1856; n. India
mongolicus Tsuneki, 1972; Mongolia
mutilloides Walker, 1871; Africa
nasicornis Gussakovskij, 1952; sw. USSR; ne. China
  rhinoceros Gussakovskij, 1952
natalensis Saussure, 1854; S. Africa
nevadensis R. Bohart, 1962; U.S.: Nevada, Washington to California
nigrescens Berland, 1942; Zaire
nigroannulatus Bischoff, 1913; Rhodesia
  glabriusculus Arnold, 1923
  ssp. *lichtenburgensis* Arnold, 1923; S. Africa
nigropilosellus (Cameron), 1910 (*Liris*); Rhodesia, Malawi, S. Africa
  gigas Bischoff, 1913
niloticus Turner, 1918; n. Africa; sw. USSR
  curiosus Gussakovskij, 1952
nipponicus Tsuneki, 1964; Japan, Korea
nitidiusculus (F. Smith), 1856 (*Larrada*); Brazil, Surinam
nitidulus (Fabricius), 1793 (*Crabro*); India, Burma
nomarches Pulawski, 1962; Egypt
notabilis Turner, 1917; Malawi, Rhodesia, Uganda
nudiventris Turner, 1917; Uganda, Zaire
obductus W. Fox, 1892; e. U.S.
observabilis Kohl, 1894; Zanzibar, Malawi, Kenya, Rhodesia, Uganda, Angola, Zaire
obsoletus (Rossi), 1792 (*Apis*); s. and centr. Europe, sw. USSR
  ssp. *tricoloratus* (Turton), 1802 (*Sphex*); nw. Africa
    tricolor Fabricius, 1793 (*Sphex*); nec Schrank, 1781
    tingitanus Pate, 1937
  ssp. *occidentalis* Pulawski, 1962; s. France, Spain
oppositus Turner, 1917; Zambia
†*opulentus* Nurse, 1909; India, nec Lepeletier, 1845
ornatipes Cameron, 1889; Guatemala, Panama
oviventris Saussure, 1891, Madagascar
pallidiventris Arnold, 1947; Zambia
parvus W. Fox, 1892; U.S.: transcontinental
  arizonicus Banks, 1942; new synonymy by Bohart
  pattoni Banks, 1942; new synonymy by Bohart
pennsylvanicus Banks, 1921; U.S.: transcontinental
pepticus (Say), 1837 (*Lyrops*); e. U.S. to Texas
  sericatus Cresson, 1872; new synonymy by Bohart
  cressoni Banks, 1942; new synonymy by Bohart
  inferioris Banks, 1942; new synonymy by Bohart
perornatus Turner, 1917; Uganda
picticornis Arnold, 1945; Madagascar
plagiatus Walker, 1871; ? Egypt
plutocraticus Turner, 1910; Australia
popovi Pulawski, 1962; sw. USSR
praedator W. Fox, 1892; e. U.S.
praestabilis Turner, 1917; Uganda
pretiosus Cameron, 1912; Guyana
priesneri Pulawski, 1962; Sudan
procerus A. Costa, 1882; Sardinia, n. Africa
  procerus A. Costa, 1883
  incognitus Pulawski, 1962
pulchricornis Turner, 1917; Malawi, Rhodesia, Mozambique
  kolaensis Turner, 1917
punctuosus Arnold, 1923; S. Africa
pygmaeus Kohl, 1888 (♀ only) Africa, India, Sri Lanka
  basalis Cameron, 1889
  calvus Turner, 1929
  seminudus Arnold, 1951
rarus Arnold, 1960; Rhodesia
relucens Turner, 1916; Australia
repandus (Fabricius), 1787 (*Crabro*); s. India
rhodesianus Bischoff, 1913; Mali, Rhodesia, Botswana
rhododactylus Taschenberg, 1870; Argentina
  ?*callosus* Kohl, 1892; tentative new synonymy by Bohart
rostratus Berland, 1942; Chad
rubellus Turner, 1908; Australia
rufipalpis Cameron, 1904; n. India
†*rufipes* Berland, 1942 (♂ only); Niger, nec Aichinger, 1870
rufiscutis Turner, 1918; Malawi
rufomarginatus Arnold, 1945; Madagascar
sacricola Pulawski, 1962; Israel
sagani Guiglia, 1943; Ethiopia
  sagani Guiglia, 1950
saharicus Pulawski, 1962; Algeria, Mali
  rufipes Berland, 1942 (♀ only), nec Aichinger, 1870
salvus Kohl, 1906; Socotra
saundersii Bingham, 1897; w. India; Burma, Borneo, Taiwan
  shiva Nurse, 1903
  varipilosus Cameron, 1905
  formosanus Tsuneki, 1966
  ssp. *suluensis* Williams, 1928; Philippines
sayi Banks, 1942; w. U.S.
  brevipilis Banks, 1942; new synonymy by Bohart
  hesperus Banks, 1942; new synonymy by Bohart
sculleni R. Bohart, 1962; sw. U.S.
sedulus F. Smith, 1860; E. Indies: Kajoa (Halmahera)
senegalensis Berland, 1942; Senegal
separabilis Turner, 1917; S. Africa
setigera Kohl, 1898; Brazil
setosus Taschenberg, 1870; Brazil
sheppardi Arnold, 1940; Rhodesia

sibiricus Gussakovskij, 1932; se. USSR: Ussuri
silverlocki Turner, 1917; Zambia
silvicola Williams, 1928; Philippines (? = *sinensis*)
silvicoloides Williams, 1928; Philippines
simillimus Schulthess, 1926; nw. Africa
simulans F. Smith, 1873; Brazil
simulatrix Turner, 1917; Nigeria
sinensis F. Smith, 1856; India, Thailand, China, Korea, Japan
    japonicus Kohl, 1888
    rothneyi Cameron, 1889
  ssp. *fundatus* Rohwer, 1911; Taiwan
  ssp. *yaeyamanus* Tsuneki, 1972; Ryukyus
sinuatus Pulawski, 1962; Israel
sjoestedti Cameron, 1908; Kenya
  ssp. *formosus* Arnold, 1923; Rhodesia
  ssp. *karrooensis* Arnold, 1923; S. Africa
spatulatus W. Fox, 1892; w. U.S.
  basirufus Rohwer, 1909, new syonymy by Bohart
  utahensis Banks, 1942
sulcatus Turner, 1916; Australia
surigensis Williams, 1928; Philippines (? = *modestus*)
tabrobane Cameron, 1900; w. India, Sri Lanka
  proximus Nurse, 1903
tachyrrhostus Saussure, 1854; Australia
tarsalis (Spinola), 1838; (*Lyrops*); Egypt, Iran, sw. USSR
  tarsatus F. Smith, 1856, p. 281 (*Larrada*)
  turcomanica Radoszkowski, 1893
†*tarsalis* (Dalla Torre), 1897 (*Tachysphex*); India, new name for *tarsatus* F. Smith (Art. 59c), nec Spinola, 1838 (? = *vestitus*)
  tarsatus F. Smith, 1856, p. 297, nec *Tachysphex tarsatus* (Say), 1823 (*Larra*), now in *Tachysphex*
tassilicus Pulawski, 1962; Algeria
testaceinerva Cameron, 1908; Kenya
tomentosus Kohl, 1891; Ghana, Malawi, Uganda, S. Africa
toyensis Tsuneki, 1971; Taiwan
transvaalensis Cameron, 1905; S. Africa
trichopygus Pulawski, 1962; Sinai
tricinctus (Fabricius), 1804; (*Liris*); W. Indies
  tricinctus Dahlbom, 1843
  cubensis Cresson, 1865
trigonalis Saussure, 1867; Java
tuberculatus Giner Marí, 1944; Peru
turneri Arnold, 1923; Rhodesia
ugandensis Turner, 1917; Uganda
vagus Radoszkowski, 1877; sw. USSR, China
validus Cresson, 1872; e. U.S. to Texas
  breviventris W. Fox, 1892 (*breviventris* Cresson, 1872, was a manuscript name only)
  calcaratus W. Fox, 1892, new synonymy by Bohart
  calcaratiformis Rohwer, 1909, new synonymy by Bohart
  belfragei Banks, 1942, new synonymy by Bohart
  coloradensis Banks, 1942, new synonymy by Bohart
  quadrifasciatus Dreisbach, 1948, new synonymy by Bohart
varians (Fabricius), 1804 (*Liris*); Surinam
velox F. Smith, 1856; Gambia, Rhodesia, Zaire (? = *monetarius*)
  neavei Turner, 1917
versatilis Turner, 1917; Malawi, Uganda
vestitus (F. Smith), 1873 (*Larrada*); n. India
  tarsatus F. Smith, 1856, p. 297, nec *Tachysphex tarsatus* (Say), 1823
  tarsalis Dalla Torre, 1897 (*Tachysphex*), nec Spinola, 1838
vicinus Cameron, 1889; India, Burma
vischnu Cameron, 1889 (errata, p. viii); India
  virchu Cameron, 1889 (orig. orthography)
  vischu Dalla Torre, 1897, emendation
  vishnu Bingham, 1897, emendation
  vischnu Schulz, 1906, emendation
volubilis Turner, 1917; Mozambique, Tanzania: Tanganyika
xenoferus Rohwer, 1911; w. India
yerburyi Bingham, 1897; Sri Lanka

Genus Tachysphex Kohl

Generic description: Inner orbits converging above, often narrowly separated at vertex; hindocellar scars usually oval or oblong (fig. 61 J-L), their long axes forming an angle of 80° to 130°, scars sometimes with a "tail" shaped like an accent mark (fig. 61 M), but length of scar shorter than or at most equal to distance between midocellus and lower end of scar; frons weakly to strongly convex, and just above antennal sockets with a pair of prominences which are usually polished, a pair of diverging linear swellings or transverse polished bands sometimes associated with these (fig. 69 E); frontal line present or absent; scape moderately long, slender to stout, flagellomeres usually longer than broad; clypeus usually strongly convex, sometimes partially divided by a pair of vertical suturelike lines into 3 sections, free margin with a prominent median lobe which may be arcuate, truncate, or angular and variably toothed; labrum sometimes with a median notch and usually hidden, or sometimes triangular, convex, and prominently exposed (fig. 79 C); inner margin of mandible with one or two subbasal teeth except in female of *ramses* and some males, externoventral margin with a deep notch; malar space absent except in some species of *brevipennis* group; mouthparts short to moderately long; dorsum of collar usually thin, usually much lower than and closely appressed to scutum; propodeum short to moderately long, episternal sulcus usually extending ventrad nearly to anteroventral margin of pleuron but absent in *brevipennis* group, scrobal sulcus usually absent; subalar fossa not margined below by carinae (except partially in *punctifrons*); terga impunctate to finely, sparsely punctate (densely punctate in exceptional forms – *brevipennis* group, *mendozanus*), tergum I with lateral carina; female with a flat, triangular pygidial plate margined by carinae laterally (fig. 74

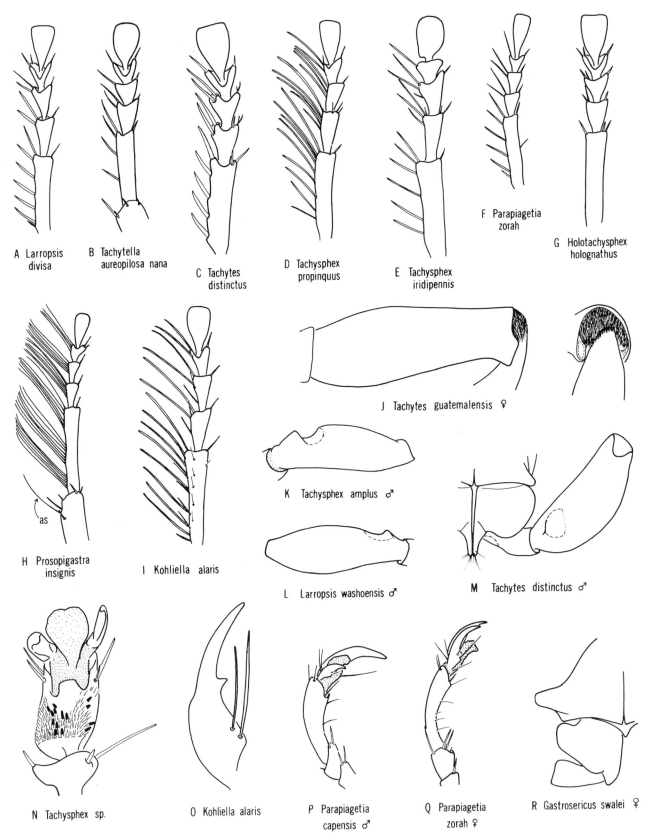

FIG. 73. Leg details in the subtribe Tachytina; A-I, left female foretarsus, dorsal, as = apicoventral tibial seta; J, hindfemur, lateral and end-on views; K-L, forefemur, lateral; M, forecoxa to femur, ventral; N, foretarsomeres IV-V of female, ventral; O, toothed hindclaw of female; P, hindtarsomeres IV-V, inner; Q, hindtarsomeres IV-V, outer; R, propleuron to trochanter, right, ventral.

E-H) (carinae weak in exceptional forms) apex often narrowly to broadly truncate, rarely emarginate (fig. 74 H), surface shining or dull, impunctate to densely punctate, usually asetose or with a few scattered inconspicuous setae (sparsely to moderately setose in some exceptional species, fig. 74 G), pygidial plate absent in *mendozanus*; male tergum VII usually flattened but rarely with lateral pygidial carinae (*brullii*), metallic setae absent; male sterna sometimes with marginal hair fringes, male sterna III-IV sometimes with oblique grooves (graduli); male forecoxa and trochanter unmodified; male forefemur with a shallow depression or semicircular notch on ventral side near base (fig. 73 K) (near middle in a few species) except simple in *geniculatus* group and a few other species; midcoxae sometimes narrowly separated; outer side of foretibia sometimes with a few spines mid and hindtibiae sparsely to moderately set with short or long spines which are arranged in rows, hindtibia not ridged; female with a well developed foretarsal rake which is usually composed of long, fine, closeset, flexible spines, tarsomere II usually with numerous rake spines but sometimes as few as two (fig. 73 D-E); male with or without a foretarsal rake; last tarsomere in some females short, stout, and asymmetrical, with unequal or deformed prehensile claws (fig. 73 N); marginal cell apex narrowly truncate; apex of male sternum VIII bispinose (fig. 76 H), sometimes weakly trispinose; volsella long, slender, often bearing a mediodorsal lobe, head of aedeagus nearly always with teeth on ventral margin (fig. 77 D).

Geographic range: Tachysphex, the largest genus in the Larrini with 351 listed species, is well represented on all continents. The Nearctic Region has 62 and the Neotropical Region 29. The Old World species are divided thusly; 138 in the Palearctic Region, 78 in the Ethiopian Region, 17 in the Oriental Region, and 19 in Australia. Madagascar has 14 endemic species, New Zealand 1, and several small oceanic islands have single representatives. Two species have been introduced to Hawaii, one from North America, and one from the Orient.

Systematics: Tachysphex are 4 to 18 mm long but mostly 6 to 10 mm. They are usually black with the abdomen and legs often partly or all red. Vestiture varies from sparse to dense, obscuring sculpture. The gastral terga often bear apical fasciae, but they are seldom as conspicuous as in *Tachytes*.

The principal diagnostic characteristics of *Tachysphex* are: ocellar scars oval, oblong, or sometimes accent-mark-shaped (with a straight "tail") in which case the length of the scar is less than or at most equal to the distance between the midocellus and the lower end of the tail, and the long axes of the scars form an angle ranging between 80° and 130°; frons with a pair of small, usually glabrous, polished tubercles just above antennal sockets; externoventral margin of mandible with a notch; only gastral tergum I with lateral carina; pygidial plate glabrous or with a few nonmetallic setae; female foretarsal rake with closeset, long, flexible spines of which there are nearly always three or more on tarsomere II; male forecoxa and trochanter simple, but forefemur nearly always notched or with a basoventral depression; apex of male sternum VIII bispinose or weakly trispinose.

Tachysphex is best contrasted with *Tachytes,* inasmuch as they are often associated in collections, and together they constitute the major part of the subtribe Tachytina. *Tachytes* differs from *Tachysphex* in having elongate golf-club shaped ocellar scars, the long tails of which are nearly parallel. The distance between the end of the tail and the midocellus is less than the length of the scar. Both sexes of *Tachytes* have a pygidial plate and the surface is nearly always densely covered with metallic setae. Male *Tachytes* often have the forecoxa and trochanter modified, but the forefemur usually is simple or at most "dented" basally. The *Tachysphex* type of forefemoral notch is found only in *Tachytes dichrous*. The female foretarsal rake in *Tachytes* consists of widely separated, stiff spines. *Tachysphex* never has the apically truncated hindfemur and associated platelike area found in most *Tachytes*.

Tachysphex is one of the dominant sphecid genera and also one of the most highly evolved members of the Larrini as shown by the development in some groups of long mouthparts, a projecting labrum, a specialized asymmetrical last female tarsomere, and the loss of the episternal sulcus in one group. The genus forms an evolutionary branch within the Tachytina around which a number of genera (*Parapiagetia, Holotachysphex, Prosopigastra,* and *Kohliella*) can be grouped. The taxon *Holotachysphex,* which Beaumont (1940) proposed as a subgenus of *Tachysphex*, is sufficiently distinct to be recognized as a genus in our opinion.

The following discussion outlines some of the more distinctive groups within *Tachysphex* and also attempts to show the broad spectrum of morphological diversity in this taxonomically difficult genus. Beaumont (1936a, b, 1947a, b) established a group classification for the Palearctic fauna. It has recently been updated by Pulawski (1971), who consolidated some of Beaumont's groups and changed the names in some cases. Pulawski's names are used in our outline.

The labrum in the *panzeri* and *geniculatus* (= *luxuriosus* group of Beaumont) groups is prominently exposed (fig. 79 C), and the galea and glossa are longer than in other *Tachysphex*. The male forefemur is simple in the *geniculatus* group, an atypical condition in the genus, and one which is known elsewhere only in *dimidiatus* Saussure and *sericeus* ssp. *flavofimbriatus* according to Pulawski (in correspondence with Menke) who has seen the types. We have also seen one unidentified Australian male with a simple forefemur.

The last tarsomere is asymmetrical (fig. 73 N) in females of the *obscuripennis* (= *lativalvis* group of Beaumont) and *brullii* (= *bicolor* group of Pulawski) groups[15] and in a number of other *Tachysphex* from various parts

---

[15] Pulawski (1971) included the *obscuripennis* group in his *bicolor* (recte *brullii*) group because of similar tarsal modifications, among other things. However, *brullii* uses tettigoniids for prey unlike the blattid prey of *obscuripennis* and its relatives. For this reason we have kept the group separate.

of the world whose affinities are unknown (*inconspicuus, depressiventris, fanuiensis, halictiformis, harpax, hippolytus, insulsus, nigerrimus, scaurus, stevensoni, suavis, subcoriaceus, theseus,* and *venator,* for example). Instead of expanding gradually distad, the tarsomere in these forms is usually short, stout, and one claw is either shorter than its mate or distorted and thickened. The ventral surface of the tarsomere usually has special matlike vestiture. With the exception of the *brullii* group in which Tettigoniidae are the only known prey, these species, where prey records are known, provision with cockroaches (*obscuripennis, suavis, blattivorus, inconspicuus, fanuiensis, nigerrimus*), and it seems logical to assume that this tarsal modification is related to prey preference. Some of the Australian species with this type of tarsomere have strongly dorsoventrally flattened bodies (*depressiventris* for example), and it is tempting to speculate that this latter condition has evolved in response to their presumed habit of searching for cockroaches in tight crevices such as beneath tree bark. Some of the roach-collecting species have only two or three rake spines on foretarsomere II, which is atypical for the majority of *Tachysphex,* although some species with normal last tarsomeres such as *beaumonti* and *brevipecten* display the same condition. Likewise some roach collectors have only a few, rather short, well spaced rake spines on tarsomere I. Similar rakes occur in a few species in other sections of the genus.

The episternal sulcus is absent or very short in species of the *brevipennis* group that includes, in addition to those species listed by Pulawski (1971), *lugubris, miscophoides,* and *minutus.* Arnold (1923b) proposed the generic name *Atelosphex* for *miscophoides,* an African representative of the group, but Beaumont (1940, 1947a) and Pulawski (1971) consider this taxon invalid because the assemblage is closely related to species having a normally developed sulcus (various forms in the *pompiliformis* group). The marginal and third submarginal cells tend to be shorter than normal in the *brevipennis* group, and some species have a short malar space and tergal punctation. Perhaps *Atelosphex* could be used as a subgenus, but until the supraspecific classification of *Tachysphex* is well established it seems best to keep the name in synonymy.

The frons of several Old World species (*isis, osiris, deserticola,* fig. 61 M) has a strong central bulge reminiscent of *Prosopigastra,* but it is covered with appressed pubescence unlike the polished, glabrous bulge common to most species in the latter genus. The two frontal tubercles are quite prominent in some species such as *bipustulosus* and *pusulosus.* In *minutulus* a pair of broad, polished bands are associated with these prominences. The ocellar scars attain their greatest elongation in *deserticola* (fig. 61 M) and various species of the *panzeri, geniculatus, erythropus* (= *fluctuatus* group of Beaumont), and *isis* groups. Generally, the more elongate the scars, the more acute is the angle formed by their long axes (see fig. 61 L,M). The pronotal collar is nearly as high as the scutum in *carli.*

Tachysphex breijeri has a rather broadly emarginate pygidial apex (fig. 74 H), and this prompted Arnold (1922) to describe the species in a new genus, *Schistosphex.* We have seen the type, and it is simply a rather large *Tachysphex* with a peculiar pygidial plate. We have seen other African *Tachysphex* with emarginate pygidial plates although the notch is narrower (*asinus, barkeri, saevus*). The pygidial plate is sparsely to moderately setose (fig. 74 G) in a very few *Tachysphex* (*brullii, inconspicuus,* and *vitiensis,* for example). The lateral pygidial carinae are weak in some species (*deserticola, erythropus,* and *miscophoides,* for example), and the last tergum has no trace of a pygidial plate in *mendozanus.* This last species, of which we have seen only a female, is rather aberrant. It could be placed in *Holotachysphex,* except that it has a well developed foretarsal rake and does not have a lateral carina on tergum II. According to Pulawski (correspondence with Menke) sterna III-IV of the male have oblique grooves similar to those found in some male *Tachysphex,* such as *terminatus* and *ashmeadi.* Some idea of the pygidial variation in *Tachysphex* can be gained by looking at fig. 74 E-H. Male *Tachysphex* rarely have lateral pygidial carinae, but *brullii* is an exception. The male gastral sterna of *euxinus* and some species in the *panzeri* and *erythropus* groups bear apical rows of fimbriae or patches of setae.

For discussions and illustrations of species characters the papers of Beaumont (1936b, 1947a), R. Bohart (1962a), and especially Pulawski (1971) should be consulted. A useful character that appears to have been little used so far in species discrimination is the degree of development of the metapleural flange. For example, it is broad in *plicosus, fusus,* and *terminatus* and narrow in *nigrior* and *williamsi.* With the exception of the western Palearctic species, which have recently been monographed by Pulawski (1971), *Tachysphex* in most areas of the world need revision. Existing keys to species of other areas are outdated for the most part: North America, W. Fox (1893b) and F. Williams (1914a); Ethiopian Region, Arnold (1924) and Madagascar, Arnold (1945); France, Beaumont (1936b); Egypt, Beaumont (1947a); India, Bingham (1897); Australia, Turner (1916b). Kurczewski (1971) keyed the species of Florida, and Pulawski (1974) keyed the Neotropical forms.

Biology: Important papers on New World *Tachysphex* are those of F. Williams (1914a), Rau and Rau (1918), Janvier (1928), Callan (1942), Strandtmann (1953), Newton (1956), Krombein (1963c, 1964b,) Kurczewski (1966b, c, 1967b), and Kurczewski and Harris (1968). The most extensive observations on Old World *Tachysphex* have been published by Grandi (1926a, 1928a, b, 1930, 1934, 1954, and 1961). Adlerz (1904), Cheesman (1928), Cros (1936), Deleurance (1945a), Williams (1945), Olberg (1959), Bonelli (1966, 1969), and Tsuneki (1969a) have also written interesting accounts. Pulawski (1971) has summarized much of the biology of Old World species in his monograph and has also discussed the relationship between biology and morphology of the adults.

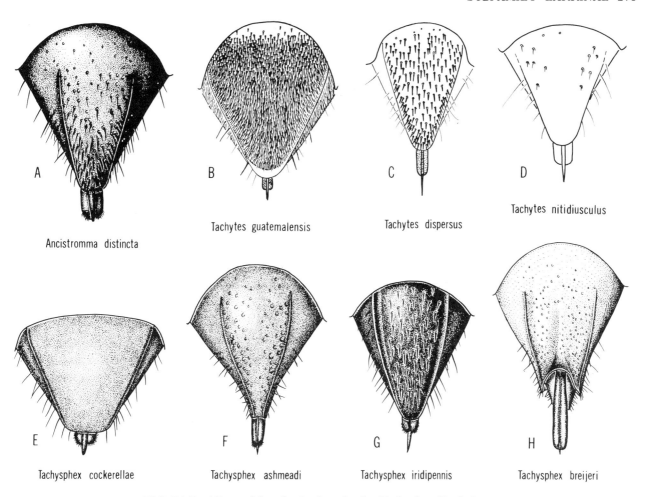

FIG. 74. Pygidium of females in the subtribe Tachytina, H = holotype.

The nest is dug in a variety of situations: open sandy areas, vegetated areas, bluffs, etc. *Tachysphex costale* subsp. *fertoni* uses abandoned burrows of *Philanthus* or ant tunnels for its nest. *Tachysphex* species excavate the burrow before searching for prey, and the burrow may or may not be temporarily closed during prey search. The cells are 3 to 6 cm beneath the soil surface. Prey consists of various families of Orthoptera, and according to Beaumont (1954c), Grandi (1961), and Pulawski, (1971) prey preferences of different species groups tend to be fairly specific at the family level: Acrididae and sometimes Tettigoniidae (*panzeri* and *pompiliformis* groups), Mantidae (*erythropus, albocinctus, julliani, schmiedeknechti* groups), Gryllidae (*plicosus* group), Blattidae (*obscuripennis* group), and Tettigoniidae (*brullii* group). On the other hand, *Tachysphex terminatus* is known to provision with Acrididae, Tettigoniidae, and Tetrigidae, the first two sometimes together in a cell. Hemipterous and homopterous prey have been recorded by two authors, but both records are questionable. Ferton (1905) observed *Tachysphex rufiventralis* (a subspecies of *incertus*) with a fulgorid, *Hysteropterum grylloides* Fabricius, but the correctness of this report is doubtful. Janvier (1928) reported that *T. pisonoides* stored its nest with bugs of the genera *Phymata* and *Nabis,* but this prey record plus his description of the nest sounds more like *Astata.*

Prey are flown or carried over the ground to the nest venter up. The wasp grasps the antenna of the host with her mandibles, and her legs assist in holding the prey in flight. Nests are usually multicellular and one to eight or more prey are placed in a cell. The egg is laid after the last provision is stored, and it is often placed on the largest orthopteran. The egg is fastened either to the venter of the "neck" of the host or somewhere on the thoracic venter.

Chrysidids, mutillids, bombyliids, and miltogrammine sarcophagids have been recorded as parasites of *Tachysphex* by various authors. See Pulawski (1971) for more specific data.

Kurczewski (1966c) published an interesting paper on male *Tachysphex* behavior. Males of *T. similis* and *terminatus,* which have a foretarsal rake, dig short resting burrows that they occupy at night and during periods of cloudy or very hot weather. *Tachysphex apicalis* males do not have a tarsal rake and apparently cannot dig. They seek shelter in preexisting holes.

The varied ethology of the genus suggests that *Tachysphex* would be a very rewarding group for a behaviorist to study. For instance, it would be interesting to learn the function of the asymmetrical last tarsomere in the roach-collecting *Tachysphex*.

Checklist of *Tachysphex*

acutemarginatus Strand, 1910; Paraguay, Brazil, Argentina
acutus (Patton), 1880 (*Larra*); e. U.S.
    bruesi Rohwer, 1911
adelaidae Turner, 1910; Australia
adjunctus Kohl, 1885; Europe, n. Africa
advenus Pulawski, 1974; Brazil
aemulus Kohl, 1906; Aden, Abd al Kuri Island
aequalis W. Fox, 1893; w. U.S.
    rufitarsis Cameron, 1889 (*Larra*), nec Spinola, 1851
    johnsoni Rohwer, 1911
    opwanus Rohwer, 1911
    washingtoni Rohwer, 1917
aethiopicus Arnold, 1923; Rhodesia, new status by W. Pulawski
aethiops (Cresson), 1865 (*Larrada*); w. U.S.
agilis (F. Smith), 1856 (*Tachytes*); s. Africa
agnus Pulawski, 1971; Tunisia, Israel, Turkey
alayoi Pulawski, 1974; Cuba, Virgin Is., Jamaica, Puerto Rico, U.S.: Florida
albocinctus (Lucas), 1848 (*Tachytes*); Iberian Penin., Africa, e. Mediterranean region, Pakistan, w. India
    ruficrus Dufour, 1853 (*Tachytes*)
    syriacus Kohl, 1888
    heliopolites Morice, 1897, (♂ only)
    mantiraptor Ferton, 1912
    dusmeti Giner Marí, 1934
    ssp. *peculator* Nurse, 1909; Iran, sw. USSR, w. India
    argyrius Gussakovskij, 1933
alpestris Rohwer, 1908; U.S.: Colorado
ambiguus Arnold, 1923; Rhodesia
    ssp. *congoensis* Arnold, 1924; Zaire
ambositrae Leclercq, 1967; Madagascar
amplus W. Fox, 1893; sw. U.S.
    gillettei Rohwer, 1911
    neomexicanus Rohwer, 1911
anceps Arnold, 1945; Madagascar
angularis Mickel, 1916; U.S.: Nebraska (? = *sonorensis* Cameron)
angustatus Pulawski, 1967; Greece, Turkey, sw. USSR, Iran
antennatus W. Fox, 1893; U.S.: Texas
antillarum Pulawski, 1974; Cuba, Puerto Rico
anubis Pulawski, 1964; Egypt
apakaensis Tsuneki, 1971; China
apicalis W. Fox, 1893; U.S.: Florida; Cuba
    fumipennis W. Fox, 1893
apoctenus Pulawski, 1974; Brazil, Argentina
araucanus Pulawski, 1974; Chile
arenarius Arnold, 1947; Rhodesia
argentatus Gussakovskij, 1952; Egypt, Israel, Syria, Turkey, sw. USSR
argenticeps Arnold, 1959; Tanzania: Tanganyika

argentifrons Arnold, 1924; S. Africa
ashmeadii W. Fox, 1893; w. U.S.
    posterus W. Fox, 1893, new synonymy by R. Bohart
    spinosus W. Fox, 1893
    spissatus W. Fox, 1893
    rufipes Provancher, 1895 (*Larra*), nec F. Smith, 1858, nec Aichinger, 1870
asinus Arnold, 1923; Rhodesia
asperatus W. Fox, 1893; U.S.: California
aterrimus Arnold, 1924; Rhodesia
atlanteus Beaumont, 1955; Morocco
?*atratus* (Lepeletier), 1845 (*Tachytes*); s. Europe (may be a *Liris*)
aureopilosus Tsuneki, 1972; Mongolia
auriceps Cameron, 1889; w. India, Sri Lanka
auropilosus Turner, 1917; e. Africa
†*australis* (Saussure), 1867 (*Tachytes*); Australia, nec Saussure, 1854
barkeri Arnold, 1923; S. Africa
beaumonti Pulawski, 1971; sw. USSR: Kazakh S.S.R., Turkmen S.S.R.
beidzmiao Tsuneki, 1971; China
belfragei (Cresson), 1872 (*Larrada*); centr. and e. U.S.
    minimus W. Fox, 1892 (*Tachytes*), new synonymy by R. Bohart
bengalensis Cameron, 1889; Sri Lanka, India, Burma, Thailand, Philippines, Marianas, Carolines, Hawaii
    brevitarsis Kohl, 1901
    ssp. *yaeyamanus* Tsuneki, 1971; Ryukyus
bipustulosus Arnold, 1949; Rhodesia
    bituberculatus Arnold, 1923, nec Cameron, 1905
bituberculatus Cameron, 1905; e. India
blattivorus Gussakovskij, 1952; sw. USSR
boharti Krombein, 1963; e. U.S.
brachycerus Arnold, 1940; S. Africa
brasilianus Pulawski, 1974; Brazil
braunsi Arnold, 1923; S. Africa
    ssp. *boer* Arnold, 1929; S. Africa
    debilis Arnold, 1924; nec Pérez, 1907
    ssp. *rufopictus* Arnold, 1929; S. Africa
*breijeri (Arnold), 1922 (*Schistosphex*); S. Africa, new combination by Menke
breviceps Pulawski, 1974; Argentina
brevipecten Beaumont, 1955; n. Africa
brevipennis Mercet, 1909; Iberian Penin.
brevipes Pulawski, 1971; sw. USSR: Uzbek S.S.R.
brinckerae Turner, 1917; S. Africa
brullii (F. Smith), 1856 (*Tachytes*); s. Europe, sw. USSR
    bicolor Brullé, 1856 (*Tachytes*), nec *Tachytes bicolor* (Fabricius), 1804 (*Larra*), now in *Larra*
    brullii F. Smith, 1856 (*Tachytes*), original orthography new name for *bicolor* Brullé. (Art. 59c)
    spoliatus Girard, 1863 (*Tachytes*)
    rufipes Aichinger, 1870 (*Tachytes*)
    montanus Radoszkowski, 1886 (*Tachytes*), nec Cresson, 1865
    monticola Dalla Torre, 1897

ssp. *galileus* Beaumont, 1947; Israel, Turkey
bruneiceps Arnold, 1923; Rhodesia, Madagascar
buyssoni Morice, 1897; Chad, Egypt, Iraq
capensis (Saussure), 1867 (*Tachytes*); S. Africa
carli Beaumont, 1947; Morocco, Gibraltar
changi Tsuneki, 1967; Taiwan, Ryukyus, Sri Lanka
cheops Beaumont, 1940; Mauritania, Libya, Egypt, Israel
chephren Beaumont, 1940; Egypt, Israel
clarconis Viereck, 1906; U.S.: Kansas
clypeatus Arnold, 1947; Zambia
cockerellae Rohwer, 1917; U. S.: California to Colombia
conceptus Pulawski, 1974; Argentina, Chile
conclusus Nurse, 1903; w. India
confrater Pulawski, 1971; sw. USSR: Turkmen S.S.R.
consanguineus Arnold, 1924; Rhodesia, S. Africa
consocius Kohl, 1892; Mediterranean region, sw. USSR, Africa
    cabrerai Mercet, 1909
    minutulus Arnold, 1923, new synonymy by Pulawski
    ssp. *mookonis* Tsuneki, 1972; Mongolia
convexus Pulawski, 1971; sw. USSR: Turkmen S.S.R., Kazakh S.S.R.
coquilletti Rohwer, 1911; w. U.S.
    dentatus Williams, 1914
coriaceus (A. Costa), 1867 (*Tachytes*); Europe
    reiseri Kohl, 1901
costae (De Stefani), 1881 (*Tachytes*)[16]; Mediterranean region, sw. USSR, Iran
    zarudnyi Gussakovskij, 1933
    tadzhicus Gussakovskij, 1952
    ssp. *canariensis* Beaumont, 1968; Canary Is.
    ssp. *fertoni* Pulawski, 1971; Algeria
crassicornis Arnold, 1945; Madagascar
crassiformis Viereck, 1906; U.S.: Kansas
crassipes Arnold, 1923; Rhodesia
    ssp. *claripennis* Arnold, 1924; Rhodesia
crenulatus W. Fox, 1893; w. U.S.
crocodilus Pulawski, 1971; Egypt
ctenophorus Pulawski, 1971; sw. USSR: Kazakh S.S.R. Siberia
cubanus Pulawski, 1974; Cuba, **Jamaica**
dakotensis Rohwer, 1923; U.S.: N. Dakota
decorus W. Fox, 1893; U.S.: N. Dakota
denisi Beaumont, 1936 (♂ only); sw. Europe, n. Africa
depilosellus Turner, 1917; Zambia
    ssp. *fallax* Arnold, 1924; S. Africa
depressiventris Turner, 1916; Australia
descendentis Mercet, 1909; Iberian Penin.
deserticola Beaumont, 1940; Egypt, Libya, Sudan
desertorum F. Morawitz, 1894; n. Africa, Israel, sw. USSR, Iraq, Iran, w. China
    abjectus Kohl, 1901
diabolicus Arnold, 1923; Rhodesia
    ssp. *analis* Arnold, 1924; S. Africa
    ssp. *claripes* Arnold, 1924; Rhodesia
    ssp. *trifasciatus* Arnold, 1924; S. Africa

dicksoni Arnold, 1962; S. Africa
dignus Kohl, 1889; Cyprus, Turkey, Israel, Syria, sw. USSR, Iran
    digenus Kohl of Dalla Torre, 1897
discrepans Turner, 1915; Australia: Tasmania
diversilabris Arnold, 1960; S. Africa
dzinghis Tsuneki, 1972; Mongolia
ebeninus Arnold, 1929; Rhodesia
egregius Arnold, 1924; Rhodesia
eldoradensis Rohwer, 1917; U.S.: California
erythraeus Mickel, 1916; U.S.: Nebraska
erythrophorus Dalla Torre, 1897; Egypt, sw. USSR, Pakistan, nw. India
    erythrogaster Cameron, 1889; nec Costa, 1882
    latissimus Turner, 1917
    pectoralis Pulawski, 1964
erythropus (Spinola), 1838 (*Lyrops*); s. Europe, n. Africa Turkey, sw. USSR, Arabian penin., Iran, w. India
    maracandicus Radoszkowski, 1877 (*Tachytes*)
    fluctuatus Gerstaecker of Kohl, 1885
    heliopolites Morice, 1897 (♀ only)
    inventus Nurse, 1903
    mantivorus Beaumont, 1940
euxinus Pulawski, 1958; Bulgaria, Turkey, Syria, Lebanon
excelsus Turner, 1917; n. Spain, e. France, Pakistan, Tibet, Mongolia
    mysticus Pulawski, 1971
excisus Arnold, 1945; Madagascar
?† ssp. *dimidiatus* Saussure, 1892, Madagascar, nec Panzer, 1806-1809
eximius Pulawski, 1971; sw. USSR: Turkmen S.S.R.
exsectus W. Fox, 1893; w. U.S.
fanuiensis Cheesman, 1928; Tuamotu Archipelago, Society Is., New Caledonia
fasciatus Morice, 1897; Egypt, Algeria
ferrugineus Pulawski, 1967; Turkey, sw. USSR
flavogeniculatus (Taschenberg), 1880 (*Tachytes*); Ethiopia
formosanus Tsuneki, 1971; Taiwan, Ryukyus
fortior Turner, 1908; Australia
fugax (Radoszkowski), 1877 (*Tachytes*); Mediterranean area, Canary Is., Ethiopian Region, sw. USSR
    filicornis Kohl, 1883
    ssp. *excerptus* Turner, 1917; Rhodesia
fulgidus Arnold, 1924; S. Africa
fulvicornis Turner, 1918; n. Africa, Arabian penin., e. India
    imperfectus Beaumont, 1940
†*fulvicornis* Tsuneki, 1972; Mongolia, nec Turner, 1918
fulvitarsis (A. Costa), 1867 (*Tachytes*); Europe, nw. Africa, w. Asia, Iran, Siberia
    acrobates Kohl, 1878 (*Tachytes*)
    strigosus Mocsáry, 1879 (*Tachytes*)
    dubius Radoszkowski, 1886 (*Tachytes*) (♂ only)
    caucasicus Radoszkowski, 1886 (*Tachytes*) (♀ only, ♂ = *Tachytes*)
    bipunctatus F. Morawitz, 1891
    ssp. *erythrogaster* (A. Costa), 1882 (*Tachytes*); Sardinia, Corsica

[16] See addendum.

erythrogaster A. Costa, 1883 (*Tachytes*)
fuscispina Pulawski, 1971; sw. USSR: Turkmen S.S.R.
fusus W. Fox, 1893; Nearctic Region, Hawaii
   foxii Rohwer, 1908
gagates Arnold, 1940; Rhodesia
galapagensis Rohwer, 1924; Galapagos Is.
gegen Tsuneki, 1971; China
geniculatus (Spinola), 1838 (*Lyrops*); Algeria, Egypt, Arabian penin., Syria
georgii Arnold, 1940; S. Africa
  ssp. *montivagus* Arnold, 1944; Lesotho
gibbus Pulawski, 1974; El Salvador to Colombia
glaber Kohl, 1906; Abd al Kuri Island
gracilicornis Mercet, 1909; n. Africa
   eduardi Morice, 1910
  ssp. *baal* Pulawski, 1971; Israel
gracilitarsis Morice, 1910; Morocco, Algeria, s. Spain
graecus Kohl, 1883; se. Europe: Greece, Yugoslavia, Bulgaria; Turkey, Lebanon
grandii Beaumont, 1965; s. Europe, Egypt, w. Asia, sw. USSR
grandissimus Gussakovskij, 1933; n. Africa, Israel, Arabian penin., sw. USSR, Iran
gujaraticus Nurse, 1909; Egypt, Israel, Saudi Arabia, sw. USSR, w. India
   laniger Pulawski, 1964
gussakovskii Pulawski, 1971; sw. USSR: Kazakh S.S.R.
halictiformis Arnold, 1945; Madagascar
harpax Arnold, 1923; Rhodesia, S. Africa
helianthi Rohwer, 1911; w. U.S.
helveticus Kohl, 1885; Europe, sw. USSR, Mongolia
  ssp. *aegyptiacus* Morice, 1897; Egypt
  ssp. *quadrifasciatus* Pulawski, 1971; Cyprus, Jordan, sw. USSR
hermia Arnold, 1924; Rhodesia
  ssp. *angustus* Arnold, 1924; Rhodesia
heterochromus Beaumont, 1955; Morocco
hippolyta Arnold, 1924; Rhodesia, S. Africa
horus Beaumont, 1940; Egypt, Israel
hostilis Kohl, 1901; sw. USSR
hurdi R. Bohart, 1962; U.S.: California
*hypoleius (F. Smith), 1856 (*Tachytes*); Australia
idzekii Tsuneki, 1971; China
imbellis Turner, 1908; Australia
incanus Beaumont, 1940; Mauritania
incertus (Radoszkowski), 1877 (*Tachytes*); s. Europe, nw. Africa, sw. Asia
   pygidialis Kohl, 1883
  ssp. *nattereri* Kohl, 1888; Egypt, Sudan
  ssp. *rufiventralis* Ferton, 1905; Corsica, Sardinia
  ssp. *kallipygus* Pulawski, 1971; Algeria
inconspicuus (W. F. Kirby), 1890 (*Tachytes*); Mexico to Argentina, Fernando de Noronha
   blatticidus Williams, 1941
inextricabilis Pulawski, 1971; Egypt, Israel, Jordan, Syria
instructus Nurse, 1909; w. India
   striolatus Cameron, 1908, nec Cameron, 1903
insulsus Arnold, 1945; Madagascar

*iridipennis F. Smith), 1873 (*Tachytes*); Mexico to Argentina
isis Beaumont, 1940; n. Africa, Syria
jujuyensis Brèthes, 1913; Surinam, Brazil, Argentina
julliani Kohl, 1883; Mediterranean region, sw. USSR, Iran
   semenovi Gussakovskij, 1933
  ssp. *africanus* Pulawski, 1971; Morocco, Algeria, Libya
  ssp. *nigripes* Tsuneki, 1972; Mongolia
karrooensis Arnold, 1923; S. Africa
kaszabi Tsuneki, 1972; China
kodairai Tsuneki, 1972; China
krombeini Kurczewski, 1971; U.S.: Florida
lacertosus Arnold, 1944; Rhodesia
laevifrons (F. Smith), 1856 (*Larrada*); e. U.S.
   levifrons Dalla Torre, 1897
lanatus Arnold, 1947; Zambia
laticauda Gussakovskij, 1933; e. Mediterranean area, sw. USSR, Iran
laticeps Arnold, 1924; Rhodesia
latifrons Kohl, 1884; n. Africa, Greece, Turkey, sw. USSR, Iran
   cyrenaicus Beaumont, 1947
leensis Rohwer, 1911; U.S.: Kansas, Texas
   consimiloides Williams, 1913
lihyuetanus Tsuneki, 1971; Taiwan
limatus Arnold, 1924; S. Africa
lindbergi Beaumont, 1956; Cape Verde Is.
linsleyi R. Bohart, 1962; w. U.S.
liriformis Pulawski, 1967; Turkey, Syria, Lebanon, sw. USSR
  ssp. *tenax* Pulawski, 1971; Israel
longipalpis Beaumont, 1940; ne. Africa
  ssp. *simplex* Pulawski, 1971; Israel, Iran, sw. USSR: Turkmen S.S.R.
lucillus Pulawski, 1971; sw. USSR: Turkmen S.S.R.
lugubris (Arnold), 1924 (*Atelosphex*); Rhodesia
luxuriosus Morice, 1897; Libya, Egypt, Sudan
   seth Pulawski, 1964
mackayensis Turner, 1908; Australia
maidli Beaumont, 1940; n. Africa, sw. USSR
   simillimus Gussakovskij, 1952
malkovskii Pulawski, 1971; sw. USSR: Kazakh S.S.R.
marshalli Turner, 1917; Rhodesia, Malawi
  ssp. *terrificus* Arnold, 1935; Botswana
mauretanus Pulawski, 1971; Morocco, Algeria
maurus Rohwer, 1911; U.S.: Texas
mediterraneus Kohl, 1883; Mediterranean region, sw. USSR, Iran
  ssp. *collaris* Kohl, 1898; e. Africa
melas Kohl, 1898; Syria to e. USSR: Irkutsk
  ssp. *eatoni* E. Saunders, 1910; Algeria, Egypt
mendozanus Brèthes, 1913; Argentina
micans (Radoszkowski), 1877 (*Tachytes*); n. Africa, sw. USSR, Iran
micromegas Saussure, 1892; Madagascar, Seychelles
   sikorae Saussure, 1892 (pl. 27, fig. 2)

miniatulus Arnold, 1924; Rhodesia
minutulus Arnold, 1923; Rhodesia
minutus Nurse, 1909; nw. India
   lilliputianus Turner, 1917
miscophoides (Arnold), 1923 (*Atelosphex*); S. Africa
mocsaryi Kohl, 1884; Iberian Penin., se. Europe, sw.
  USSR, Afghanistan, Libya
  ssp. *algirus* Kohl, 1892; Algeria, Morocco
    maroccanus Beaumont, 1947
moczari Tsuneki, 1972; Mongolia
modestus Arnold, 1924; S. Africa
montanus (Cresson), 1865 (*Larrada*); w. U.S.
    compactus W. Fox, 1893
    inusitatus W. Fox, 1893
morawitzi Pulawski, 1971; sw. USSR
morosus (F. Smith), 1859 (*Tachytes*); Celebes
mundus W. Fox, 1893; centr. U.S.
    glabrior Williams, 1914, new synonymy by R. Bohart
mycerinus Beaumont, 1940; n. Africa
nambui Tsuneki, 1973; Ryukyus
naranhun Tsuneki, 1971; China
nasalis F. Morawitz, 1893; sw. USSR, Mongolia, Turkey
    abditus Kohl, 1898
    mongolicus Kohl, 1898
nigerrimus (F. Smith), 1856 (*Tachytes*); New Zealand
    sericops F. Smith, 1856 (*Tachytes*)
    depressus Saussure, 1867 (*Tachytes*)
    nigerrimus "White", Butler, 1874 (pl. 7, fig. 14)
     (*Astata*)
    helmsii Cameron, 1888 (*Tachytes*)
nigrescens Rohwer, 1908; U.S.: Colorado
nigricolor (Dalla Torre), 1897 (*Larra*); Japan, Korea,
  Taiwan
    *nigricans F. Smith, 1873 (*Larrada*), nec Walker, 1871,
    japonicus Iwata, 1933
nigrior W. Fox, 1893; w. U.S.
nigrocaudatus Williams, 1914; centr. U.S.
niloticus Pulawski, 1964; Egypt
nitelopteroides Williams, 1958; Mexico: Baja California
nitidior Beaumont, 1940; n. Africa, s. Europe, w. Asia
nitidissimus Beaumont, 1952; n. Africa, se. Europe, w.
  Asia
nitidiusculus (F. Smith), 1856 (*Larrada*); Brazil
nitidus (Spinola), 1805 (*Astata*); n. and centr. Europe
    borealis Pulawski, 1971
  ssp. *ibericus* (Saussure), 1867 (*Tachytes*); s. Spain,
    Canary Is., n. Africa, Syria, Israel
nonakai Tsuneki, 1971; China
notogoniaeformis Nadig, 1933; n. Africa
novarae (Saussure), 1867 (*Tachytes*); Nicobar Is.
nubilipennis Beaumont, 1950; n. Africa
oberon Arnold, 1923; Rhodesia
  ssp. *mashona* Arnold, 1929; Rhodesia
  ssp. *halophilus* Arnold, 1940; Africa
obscuripennis (Schenck), 1857 (*Tachytes*); Europe,
  Turkey
    lativalvis Thomson, 1870 (*Tachytes*)
  ssp. *gibbus* Kohl, 1885; s. Europe, nw. Africa

philippi E. Saunders, 1910
obscurus Pulawski, 1971; Canary Is.
octodentatus Arnold, 1924; Rhodesia
  ssp. *inermis* Arnold, 1924; Rhodesia
omoi Guiglia, 1943; Ethiopia
    omoi Guiglia, 1950
opacus F. Morawitz, 1893; Lebanon, Syria, sw. USSR,
  Iran, w. China
osiris Beaumont, 1940; n. Africa
pacificus Turner, 1908; Australia, Tasmania
palopterus (Dahlbom), 1845 (*Tachytes*); n. Africa, Israel
  Arabian penin.
panzeri (Vander Linden), 1829 (*Tachytes*); Europe, nw.
  Africa, w. Asia, Sri Lanka
    aurifrons Lucas, 1848 (*Tachytes*)
    discolor Frivaldsky, 1876 (*Tachytes*)
    aurifrons Cameron, 1900 (*Tachytes*), nec Lucas, 1848
    ceylonicus Cameron, 1900 (*Tachytes*)
  ssp. *rufiventris* (Spinola), 1838 (*Lyrops*); Corsica,
    Sardinia
  ssp. *oraniensis* (Lepeletier), 1845 (*Tachytes*); nw. Africa,
    Israel
  ssp. *pulverosus* (Radoszkowski), 1886 (*Tachytes*); n.
    Africa, Israel, sw. USSR, Iran, w. India
    ablatus Nurse, 1909
  ssp. *pentheri* Cameron, 1905; Rhodesia, S. Africa
    caliban Arnold, 1923
  ssp. *sycorax* Arnold, 1923; Rhodesia
  ssp. *dolosus* Arnold, 1923; Rhodesia
  ssp. *nanus* Arnold, 1924; S. Africa
  ssp. *zavattarii* Guiglia, 1939; Ethiopia
  ssp. *fortunatus* Beaumont, 1968; Canary Is.
  ssp. *cyprius* Pulawski, 1971; Cyprus
  ssp. *sareptanus* Pulawski, 1971; sw. USSR: Kazakh
    S.S.R.
parvulus (Cresson), 1865 (*Larrada*); w. U.S.
    consimilis W. Fox, 1893, new syn. by Bohart
    argyrotrichus Rohwer, 1911, new syn. by Bohart
pauxillus W. Fox, 1893; California
pechumani Krombein, 1938; e. U.S.
pectinatus Pulawski, 1974; Argentina
pekingensis Tsuneki, 1971; China
perniger Arnold, 1947; Madagascar
persa Gussakovskij, 1933; e. Mediterranean area, Iran
  ssp. *confinis* Gussakovskij, 1952; sw. USSR
  ssp. *catharinae* Pulawski, 1964; Sinai Penin., Israel,
    Syria
  ssp. *nigripes* Pulawski, 1967; sw. USSR: Kazakh S.S.R.
    Bulgaria; Greece; Turkey; Iran
persistans Turner, 1916; Australia
pilosellus Pulawski, 1971; sw. USSR: Kazakh S.S.R.,
  Turkmen S.S.R.
pilosulus Turner, 1908; Australia
pisonoides (Reed), 1894 (*Larrada*); Chile
pisonopsis Pulawski, 1974; Chile
plenoculiformis Williams, 1914; U.S.: Kansas
plesius Rohwer, 1917; w. U.S.

plicosus (A. Costa), 1867 (*Tachytes*); Mediterranean region, w. Palearctic, e. India
    gallicus Kohl, 1883
    striolatus Cameron, 1903
pompiliformis (Panzer), 1804 (*Larra*); Europe, nw. Africa, s. Palearctic Region to Siberia and nw. India
    dimidiatus Panzer, 1806-1809 (*Larra*)
    jokischianus Panzer, 1806-1809 (*Larra*)
    nigripennis Spinola, 1808 (*Tachytes*)
    austriacus Kohl, 1892
    rufoniger Bingham, 1897
    projectus Nurse, 1903
      pectinipes of authors, not Linnaeus, 1758 (which is a pompilid)
powelli R. Bohart, 1962; U.S.: California
priesneri Beaumont, 1940; n. Africa
propinquus Viereck, 1904; w. U.S.
prosopigastroides Bischoff, 1913; Rhodesia
    sipapomae Arnold, 1923
psammobius (Kohl), 1880 (*Tachytes*); Europe, sw. USSR, Siberia
pseudopanzeri Beaumont, 1955; sw. Europe, Morocco
psilocerus Kohl, 1884; Mexico
ptah Pulawski, 1964; Egypt, Israel, Aden
pugnator Turner, 1908; Australia
pulcher Pulawski, 1967; Turkey, Syria, Lebanon, Israel, sw. USSR
punctatus (F. Smith), 1856 (*Larrada*); S. Africa
punctatiformis Arnold, 1923; S. Africa
puncticeps Cameron, 1903; e. India, Taiwan, Philippines, Tasmania
    varihirtus Cameron, 1903
    rugidorsatus Turner, 1915
    mindorensis Williams, 1928
punctifrons (W. Fox), 1891 (*Larra*); e. U.S.
    fedorensis Rohwer, 1911
punctipes Pulawski, 1967; Turkey
punctiventris Arnold, 1924; Rhodesia, Zambia
pusulosus Beaumont, 1955; n. Africa, Syria, Turkey
quadricolor (Gerstaecker), 1857 (*Lyrops*); Mozambique
quadrifurci Pulawski, 1971; Egypt; sw. USSR: Turkmen S.S.R.
quebecensis (Provancher), 1882 (*Larra*); ne. N. America
radiatus Gussakovskij, 1952; sw. USSR
radoszkowskyi F. Morawitz, 1893; sw. USSR
ramses Pulawski, 1971; Egypt
reedi Menke in Pulawski, 1974; Chile
    gayi Spinola, 1851 (*Larra*) (♀ only. ♂ lectotype = *Tachytes chilensis*)
    erythropus Herbst, 1921, nec Spinola, 1838
remotus Pulawski, 1974; Colombia
rhodesianus Bischoff, 1913; Rhodesia
robustior Williams, 1914; U.S.: Texas, Kansas
    crenuloides Williams, 1914
rubicundus Pulawski, 1971; sw. USSR
*ruficaudis (Taschenberg), 1870 (*Tachytes*); Mexico to Paraguay
    *pullulus Strand, 1910 (*Tachytes*),

rufitarsis (Spinola), 1851 (*Larra*); Chile
  ?*rufipes* Reed, 1894 (*Tachytes*), nec Aichinger, 1870
    rufiventris Reed, 1892 (*Larrada*), nec Spinola, 1838
rugosus Gussakovskij, 1952; se. Europe, Turkey, Syria, sw. USSR
    rhodius Beaumont, 1960
saevus Arnold, 1924; S. Africa
sanguinosus Mickel, 1916; U.S.: Kansas, Nebraska
saturnus Arnold, 1924; S. Africa
saundersi Mercet, 1909; Iberian penin.
scabrosus Arnold, 1929; Rhodesia
scaurus Arnold, 1945; Madagascar
schlingeri R. Bohart, 1962; w. U.S.
schmiedeknechti Kohl, 1883; n. Africa, e. Mediterranean region, sw. USSR, Iran, w. India
    psilopus Kohl, 1884
    heliophilus Nurse, 1909
    calopteryx Gussakovskij, 1933
    fasciipennis Gussakovskij, 1933
    ornatipennis Gussakovskij, 1933
  ssp. *satanas* Pulawski, 1971; Syria
schoenlandi Cameron, 1905; S. Africa
  ssp. *detritus* Arnold, 1924; S. Africa
  ssp. *luctuosus* Arnold, 1924; S. Africa
sculptilis W. Fox, 1893; U.S.: Colorado
    sphecodoides Rohwer, 1911
sculptiloides Williams, 1914; U.S.: Kansas
selectus Nurse, 1909; Libya, e. Mediterranean region, w. India
    actaeon Beaumont, 1960
semirufus (Cresson), 1865 (*Larrada*); w. U.S.
    punctulatus H. Smith, 1906, nec Kohl, 1884
    puncticeps H. Smith, 1908, nec Cameron, 1903
    giffardi Rohwer, 1917
sepulcralis Williams, 1914; e. U.S.
    maneei Banks, 1921
sericans Gussakovskij, 1952; Cyprus, Syria, sw. USSR
  ssp. *gracilis* Pulawski, 1971; Egypt, Algeria
sericeus (F. Smith), 1856 (*Larrada*); Ethiopian region
    fluctuatus Gerstaecker, 1857 (*Lyrops*)
    fluctuatus Gerstaecker, 1862 (*Lyrops*)
  ssp. *kalaharicus* Arnold, 1924; Botswana
  ssp. *flavofimbriatus* Arnold, 1945; Madagascar
sexinus Leclercq, 1961; Madagascar
seyrigi Arnold, 1945; Madagascar
siitanus Tsuneki, 1971; China
similis Rohwer, 1910; e. U.S.
    similans Rohwer, 1910
sinaiticus Pulawski, 1964; Sinai Penin.
sonorensis (Cameron), 1889 (*Larra*); Mexico: Sonora
sordidus Dahlbom, 1845; sw. USSR, Iran, Rhodes, Cyprus, Israel, Turkey
    lebedevi Gussakovskij, 1952
speciosissimus Morice, 1897; n. Africa, Israel, Syria, Iran
    redivivus Kohl, 1901
spinulosus Pulawski, 1975; Surinam, Brazil
    spinosus Pulawski, 1974, nec Fox, 1893
splendidulus F. Morawitz, 1893; sw. USSR

spretus Kohl, 1901; sw. USSR
stachi Beaumont, 1936; Israel, sw. USSR
stevensoni Arnold, 1924; Rhodesia
stimulator Turner, 1916; Australia
suavis Arnold, 1929; Rhodesia, Madagascar
subandinus Pulawski, 1974; Brazil, Argentina
subcoriaceus Arnold, 1945; Madagascar
subdentatus F. Morawitz, 1893; Turkey, sw. Europe, sw. USSR
subeditus Leclercq, 1961; Madagascar
subfimbriatus Arnold, 1924; S. Africa
subfuscatus Turner, 1917, Malawi, Zambia
   strigatus Turner, 1917
subopacus Turner, 1910; Australia
   debilis Turner, 1908, nec Pérez, 1907
sulcidorsum Beaumont, 1950; Algeria, Morocco
svetlanae Pulawski, 1971; sw. USSR
tarsatus (Say), 1823 (*Larra*); Nearctic Region
   *dubius W. Fox, 1893, nec Radoszkowski, 1886, new synonymy by Bohart
   dubiosus Dalla Torre, 1897
    hitei Rohwer, 1908
    zimmeri Mickel, 1916
tarsinus (Lepeletier), 1845 (*Tachytes*); n. Africa, s. Palearctic Region
tenuicornis Bischoff, 1913; S. Africa
tenuipunctus W. Fox, 1893; w. N. America
   granulosus Mickel, 1916
tenuis Turner, 1908; Australia
terminatus (F. Smith), 1856 (*Larrada*); Nearctic Region, Bahamas, Colombia, Brazil
   minor Provancher, 1887 (*Larra*)
tessellatus Dahlbom, 1845; Greece, Turkey
testaceipes Bingham, 1897; Burma
texanus (Cresson), 1872 (*Larrada*); sw. U.S.; Mexico
theseus Arnold, 1951; Ghana
tinctipennis Cameron, 1904; n. India, China
titania Arnold, 1923; Rhodesia
   ssp. *willomorensis* Arnold, 1924; S. Africa
tridentatus Arnold, 1924; Rhodesia
triquetrus W. Fox, 1893; w. U.S.
truncatifrons Turner, 1908; ? Australia
tuckeri Arnold, 1923; S. Africa
undatus (F. Smith), 1856 (*Tachytes*); Brazil, Surinam, Argentina, Chile
   herbstii Kohl, 1905
unguiculatus Arnold, 1924; S. Africa
unicolor (Panzer), 1806-1809 (*Larra*); Palearctic Region
   jurinii Drapiez, 1819 (*Lara*)
    jurinei Vander Linden, 1829 (*Tachytes*), emendation
   nitidus of authors, nec Spinola
   ssp. *simonyi* Kohl, 1892; w. Canary Is.
vanrhynsi Arnold, 1940; S. Africa
venator Arnold, 1960; Zambia
verhoeffi Pulawski, 1971; Syria, Israel
vestitus Kohl, 1892; n. Africa, Israel
viarius Arnold, 1947; Rhodesia
villosus Arnold, 1947; Madagascar

vitiensis Williams, 1928; Fiji Is.
vulneratus Turner, 1917; Zambia
   ssp. *foucauldi* Beaumont, 1952; Algeria, Chad
walkeri Turner, 1908; Australia
waltoni Arnold, 1940; S. Africa
wheeleri Rohwer, 1911; U.S.: Kansas, Texas
williamsi R. Bohart, 1962; U.S.: California
yarrowi Beaumont, 1960; n. Africa, Israel

Nomina nuda in *Tachysphex*

antigae "Tourn." in Antiga and Bofill, 1904
reticulatus "Pérez" in Antiga and Bofill, 1904

Genus Parapiagetia Kohl

Generic description: Inner orbits converging above, or parallel but bowed outward; hindocellar scars oval, their long axes forming an angle of 140° - 175° (fig. 61 I); frons rather flat to moderately, broadly convex, sometimes with a linear depression adjacent to inner orbit at level of midocellus, frons with a pair of polished prominences just above sockets (fig. 78 B); frontal line weakly to deeply impressed; scape moderately long, slender, flagellomeres usually longer than broad; clypeus rather flat or with a central convexity which in females may bear a transverse dentate crest, the area between the crest and the clypeal free margin polished, glabrous, clypeal free margin in male with a triangular or rounded median lobe which may be notched, margin in female with a small median lobe and blunt lateral teeth, or variably dentate; labrum transverse or quadrangular, sometimes emarginate, hidden or narrowly exposed; inner margin of mandible simple or with a mesal subbasal cleft; externoventral margin with a notch or step, paramandibular process of hypostoma well developed in Old World species, usually touching back of clypeus (fig. 79 B); collar with some thickness, lower than and usually not closely appressed to scutum; propodeum short to moderately long; origin of episternal sulcus usually at or close to anterior end of subalar fossa, sulcus ending abruptly just as it reaches ventral surface of pleuron, or occasionally attaining anteroventral margin of pleuron (*odontostoma* group, one Chilean sp.), scrobal sulcus absent, petiole socket isolated from metacoxal sockets by a pair of dark sclerites (fig. 71 E); gaster sessile, or pedunculate, or segment I with a short petiole, gaster compact or long and slender, often clavate, terga impunctate, tergum I with lateral carina which is sometimes distinct only basally; female with a triangular pygidial plate margined by carinae laterally, apex rounded or truncate, surface dull to shiny, punctate, sparsely to densely covered with setae in *erythropoda* group; male usually with lateral pygidial carinae, surface similar to female, apex emarginate in *vernalis*; male sterna II-III or IV in *erythropoda* group each with a broad, semicircular, apical depression set with long, sparse, appressed hair; male foreleg unmodified; tibiae set with spines which are arranged in rows: one on I

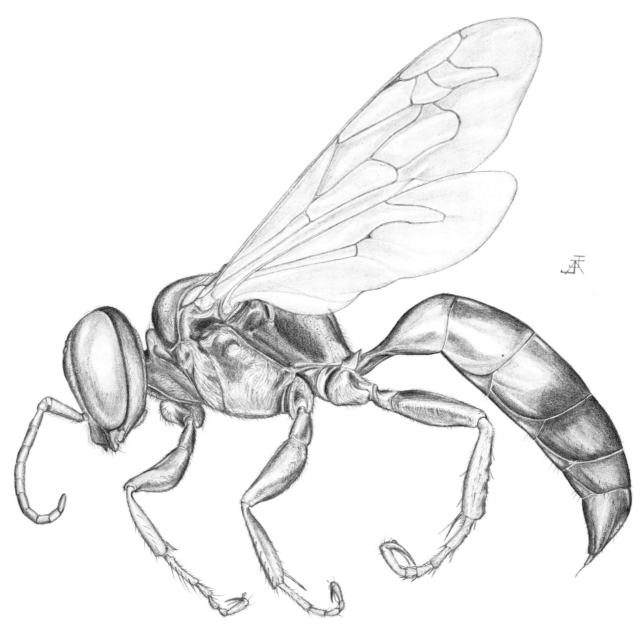

FIG. 75. *Parapiagetia* species in *subpetiolata* group, female.

(often very fine), three on II, and two or three on III, spines of uppermost row in the last sometimes short, stout, and set on tubercles, hindtibia sometimes angular in cross section, not ridged; female with a foretarsal rake composed of rather short to moderately long spines of which there are only two on tarsomere II (fig. 73 F); male with a similar but less well developed rake; last tarsomere of mid and hind legs usually long and arcuate in side view, claws of mid and hind legs of female long, slender, prehensile (figs. 73 Q, 75), outer claw slightly shorter than its mate in females of *odontostoma* group, inner claw of mid and hind legs of male much shorter than outer claw in *erythropoda* and *subpetiolata* groups (fig. 73 P); marginal cell truncate apically, short in *odontostoma* group; apex of male sternum VIII rounded (fig. 76 J, *erythropoda* and *subpetiolata* groups, or emarginate (fig. 76 I, *odontostoma* group); volsella long, slender; head of aedeagus with teeth (fig. 77 F, *odontostoma* and *subpetiolata* groups), or with only a basoventral toothlike lobe (fig. 77 E, *erythropoda* group).

Geographic range: Nineteen species are currently recognized in *Parapiagetia*, and they are distributed as follows: two in southern South America, two in the Ethiopian Region plus two on Madagascar, three in the

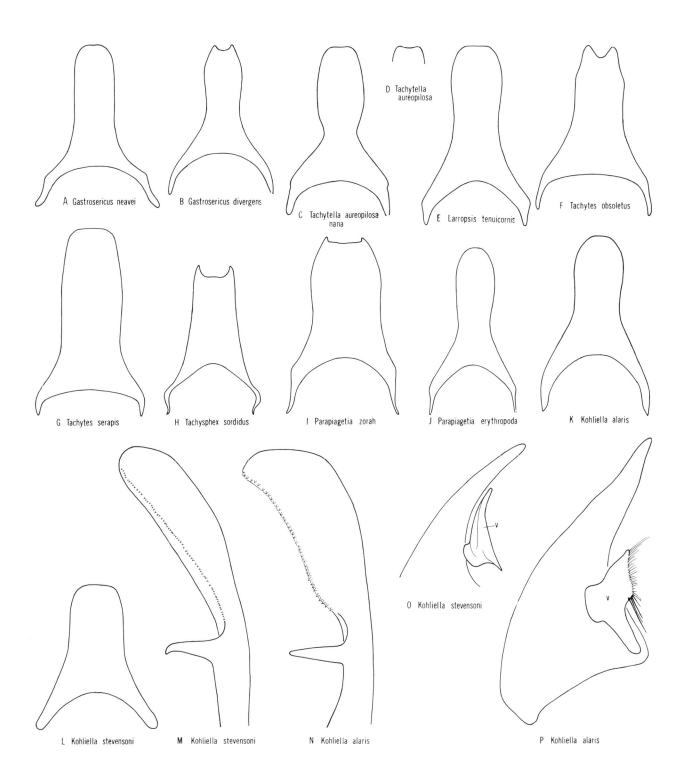

FIG. 76. Genitalic morphology in the subtribe Tachytina; A-L, sternum VIII (D, apex only, holotype); M-N, penis valve, lateral; O,P, volsella and gonostyle, right, inner (setation omitted), v = volsella, only apical half of gonostyle in O.

Oriental Region, and the remainder in the Palearctic Region. The discovery that *Lirosphex* Brèthes is a synonym of *Parapiagetia* indicates for the first time that this genus occurs outside the Old World. This peculiar disjunct pattern is reminiscent of the philanthine genus *Odontosphex*.

Systematics: Parapiagetia species range in length from 4 to 11.5 mm. They are black, the legs partly red or yellow, and the gaster sometimes partly or all red. The wings are clear or slightly infumate. Vestiture is variable, but usually there are tergal fasciae.

The character which immediately sets *Parapiagetia* apart from other Larrini is the presence of a pair of sclerites in the petiole-metacoxal cavity which isolate the petiole socket from the coxal sockets. Similar developments are found only in the Sphecinae, some Ampulicinae, and some *Trypoxylon*, and it seems clear that this structure, here termed the propodeal sternite, is directly related to the development of abdominal petiolation. *Parapiagetia* is probably most closely allied with *Tachysphex* as shown by the similar ocellar scars, supra-antennal tubercles, and wing venation; but unlike most *Tachysphex*, *Parapiagetia* has simple inner mandibular margins, only two rake spines on foretarsomere II, and a simple male forefemur. Also, in most *Parapiagetia* the episternal sulcus does not reach the anteroventral margin of the pleuron; the gaster is usually long and slender, similar to that of some *Trypoxylon;* sternum VIII of the male is usually round apically; and tarsomere V of the mid and hind legs is usually as long as or longer than tarsomere I. In *Tachysphex* the episternal sulcus is usually complete, or nearly

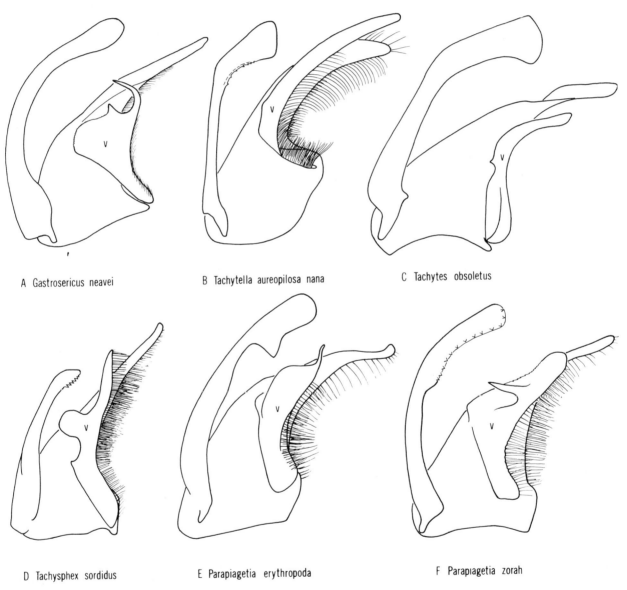

A Gastrosericus neavei   B Tachytella aureopilosa nana   C Tachytes obsoletus

D Tachysphex sordidus   E Parapiagetia erythropoda   F Parapiagetia zorah

FIG. 77. Male genitalia in the subtribe Tachytina, right inner half, setation not shown on C and somewhat schematic on others, v = volsella.

so, the gaster is compact, sternum VIII is bispinose, and tarsomere I is longer than V. The extension of the hypostoma to or nearly to the back of the clypeus in the Old World species of *Parapiagetia* is distinctive, and this condition is found elsewhere in the Larrinae only in *Auchenophorus*. The shortening of the inner claw on the mid and hind legs of all *Parapiagetia* males except those of the *odontostoma* group is a peculiar feature of the genus. The unusual morphology of *Parapiagetia*, its relatively few species, and its disjunct distribution suggests that this is a relict genus but one which has evolved a number of specializations.

Beaumont (1960a) outlined a tentative species group classification for *Parapiagetia,* which was refined by Pulawski (1961). The *odontostoma* group (*Parapiagetia s.s.*) contains only three species that have in common the following: marginal cell short; episternal sulcus reaching anteroventral margin of pleuron or nearly so; gaster pedunculate; pygidial plate glabrous; male claws equal in length or nearly so; male sterna without setal depressions, sternum VIII emarginate; and aedeagus with teeth along ventral edge (fig. 77 F). *P. odontostoma, mongolica,* and *zorah* are placed in this group. Dr. Beaumont also recognized another assemblage, here called the *genicularis* group, which he associated closely with the *odontostoma* group. The *genicularis* group is equivalent to *Psammosphex* Gussakovskji, and it contains *genicularis,* and *piagetiodes,* species which are the most *Tachysphex*-like of all *Parapiagetia*. According to Beaumont (we have not seen examples of this group) these species differ from those of the *odontostoma* group in having a more compact gaster, a longer marginal cell, and in female clypeal details. Presumably this group represents the most generalized *Parapiagetia*. The third group recognized by Beaumont contains, as far as we can determine, the remaining Old World species with the exception of *richteri* and *vernalis*. Pulawski termed this the *denticulata* group, which, because of recent synonymy, should be called the *erythropoda* group. It is characterized as follows: marginal cell long, episternal sulcus ending as it reaches venter and pleuron, gaster more compact and elliptical in shape, pygidial plate setose, inner claw of male mid and hind legs much shorter than its mate (fig. 73 P), male sterna with setal depressions, sternum VIII rounded, and aedeagus without teeth (fig. 77 E). We can add a fourth assemblage, the *subpetiolata* group, which contains the New World species of *Parapiagetia,* most of which are undescribed. Brèthes (1913) proposed a new genus, *Lirosphex,* for these forms, but clearly they are congeneric with *Parapiagetia*. The *subpetiolata* group has characteristics of both the *odontostoma* and *erythropoda* groups: marginal cell long; episternal sulcus ending as it reaches venter of pleuron; gaster long, slender and often petiolate; pygidial plate asetose; inner claw of male mid and hind legs much shorter than its mate; male sterna simple, sternum VIII rounded, aedeagus with teeth. One undescribed Chilean species has a complete episternal sulcus and may not belong in this group. In South American species, where petiolation is best developed, the first sternum is usually cylindrical basally and the first tergum is represented basally by a narrow dorsal strip. The tergum, and to a lesser extent the sternum, expand at the level of the spiracle to form the bell shape of a normal first gastral segment (fig. 75). The open mandible socket (fig. 79 D) is unique to the New World forms.

Aside from Arnold's (1922) key to the two species of southern Africa there are no keys available for identifying species of this genus. The papers of Beaumont (1955a, 1956a, 1960a) and Pulawski (1961) contain figures and descriptions of a few species that will give some indication of the specific characters used in the genus. We have noted that the form of the carina found on the side of the pronotum along the lower margin is sometimes distinctive. For example, it is lamelliform in a few South American species.

Biology: Arnold (1945) listed immature Tetrigidae as prey of *Parapiagetia longicornis*. Beaumont (1952b) reported that an unknown *Parapiagetia* species was seen by a colleague in the act of transporting a caterpillar. Nothing else is known about the genus.

Checklist of *Parapiagetia*

capensis Brauns, 1910; S. Africa
   ssp. *ferox* Arnold, 1922; Rhodesia, Ethiopia, Senegal, Mauritania
   ssp. *rhodesiana* Arnold, 1922; Rhodesia
capitalis (E. Saunders), 1910 (*Tachysphex*); Algeria
erythropoda (Cameron), 1889 (*Tachytes*); Libya, Egypt, India
   erythropa Dalla Torre, 1897 (*Tachytes*)
   denticulata Morice, 1897 (*Tachytes*)
   saharica Beaumont, 1956
genicularis (F. Morawitz), 1890 (*Tachysphex*); Egypt, sw. USSR, Pakistan
   integra Kohl, 1892 (*Tachysphex*)
joergenseni (Brèthes), 1913 (*Lirosphex*); Argentina
kaszabi Tsuneki, 1972; Mongolia
longicornis Arnold, 1945; Madagascar
mongolica (F. Morawitz), 1889 (*Piagetia*); Mongolia
odontostoma (Kohl), 1884 (*Piagetia*); Sinai Penin., Egypt, Libya
   saussurei Kohl, 1894 (*Piagetia*), lapsus for *odontostoma*
piagetiodes (E. Saunders), 1910 (*Tachysphex*); n. Africa
pluridentata Arnold, 1945; Madagascar
richteri Beaumont, 1970; Iran
rufescens (Gussakovskij), 1952 (*Psammosphex*); sw. USSR
subpetiolata (Brèthes), 1909 (*Tachysphex*); Paraguay
substriatula (Turner), 1917 (*Tachysphex*); n. India
tridentata Tsuneki, 1972; Mongolia
vernalis Brauns, 1910; S. Africa
wickwari Turner, 1914; Sri Lanka
zorah Beaumont, 1955; n. Africa (? = *mongolica*)

Genus Holotachysphex Beaumont

Generic description: Inner orbits converging above, bowed outward; hindocellar scars elongate, oval, their long axes approximating a right angle; frons broadly, weakly convex, densely punctate and with a pair of polished tubercles above antennal sockets; frontal line absent or weakly impressed; scape short, expanded apically; clypeal free margin with a triangular or narrow, rounded or truncate median lobe; labrum rounded or truncate, sometimes emarginate, hidden; inner margin of mandible with a single large tooth near middle, externoventral margin notched or entire; dorsum of collar thin, much lower than and closely appressed to scutum; propodeum moderately long; episternal sulcus extending to ventral region of mesopleuron where it ends before reaching anteroventral margin of pleuron, scrobal sulcus weak or absent; terga densely punctate, I-II with lateral carina; pygidial plate absent, tergum VI convex, acuminate, sparsely punctate; male sterna II-III or IV with large patches of velvety pubescence; male forecoxa and trochanter unmodified; male forefemur with semicircular posterobasal notch; tibiae with only a few inconspicuous spines, hindtibia not ridged; foretarsal rake undeveloped, tarsomere I with a few short lateral spines, II with a single apicolateral seta (fig. 73 G); wings similar to *Tachysphex,* marginal cell truncate apically; apex of male sternum VIII emarginate; volsella long, slender, with a mediodorsal lobe; head of aedeagus with teeth on ventral margin.

Geographic range: The six species in this small genus are found in Africa, Madagascar, the eastern Mediterranean area, southwestern USSR, and the Oriental Region.

Systematics: Holotachysphex range in length from 6.5 to 11 mm. The body is black, but the tergal margins are pale. The gaster is largely or totally red, at least in females of species from the Ethiopian Region. The legs are sometimes partially reddish. The wings are clear.

Holotachysphex was described as a subgenus of *Tachysphex* by Beaumont (1940), and although the two taxa are similar in general facies there are a number of basic differences that warrant the recognition of *Holotachysphex* as a genus. Among these are: presence of lateral carinae on terga I and II, absence of a foretarsal rake, absence of a pygidial plate, the generally dense punctation of the body, and the velvety sternal patches of the male. The genitalia, male subgenital plate, pair of supraantennal tubercles (fig. 78 A), and notched male forefemur ally *Holotachysphex* most closely with *Tachysphex,* but the lateral carina on tergum II and the dense body punctation point to a relationship with *Prosopigastra,* also.

Following Pulawski (1971) we are including *Phytosphex* Arnold in *Holotachysphex.* The only distinction between the two is the presence of an externoventral notch on the mandible of the former. We are not recognizing *Phytosphex* as a subgenus because the small number of species in *Holotachysphex* does not seem to warrant such action. Also, the presence or absence of a notch is a weak character in other genera such as *Tachytes, Liris,* and *Bothynostethus,* where it varies by degrees from one extreme to the other. Admittedly, intermediate mandibles are not known in *Holotachysphex.*

The species of the genus are poorly known and only one sex has been described for some. Species characters are found in the configuration of the clypeus and degree of punctation of various areas of the body. No keys are available but the papers of Beaumont (1947a, 1960a), Gussakovskij (1952), Arnold (1923a, 1945), and Pulawski (1967) contain figures. *H. sacalava* was described as a subspecies of *tuneri* by Arnold but is clearly specifically distinct, based on material identified by Arnold and seen by us.

Biology: Arnold (1923a) said that *turneri* nests in hollow stems of Aloe and *Datura.* The cells are separated by partitions of earth and little pebbles. The type of prey and other details are unknown.

Checklist of *Holotachysphex*

holognathus (Morice), 1897 (*Tachysphex*); Egypt, Crete, w. India, Sri Lanka
   integer Morice, 1897 (*Tachysphex*), nec Kohl, 1892
   pollux Nurse, 1903 (*Tachysphex*)
mochii (Beaumont), 1947 (*Tachysphex*); Rhodes, Turkey, Cyprus
   schwarzi Pulawski, 1967 (*Tachysphex*)
pentapolitanus (Beaumont), 1960 (*Tachysphex*); Libya
prosopigastroides (Gussakovskij), 1952 (*Haplognatha*); sw. USSR: Tadzhik, S.S.R.
sacalava (Arnold), 1945 (*Tachysphex*); Madagascar, new status by A. S. Menke
turneri (Arnold), 1923 (*Tachysphex*); S. Africa, Ethiopia
   ssp. *transvaalensis* (Arnold), 1924 (*Tachysphex*); S. Africa

Genus Prosopigastra A. Costa

Generic description: Inner orbits converging above, most strongly so in male in which eyes are sometimes holoptic; hindocellar scars accent-mark shaped, sometimes constricted near middle, their long axes usually approximating a right angle (fig. 61 B), but sometimes as broad as 135° in females; frons in Old World species with a prominent, central swelling which is usually moundlike (fig. 78 D), swelling glabrous, impunctate and shining (punctate, setose, and dull in *nubigera* male); frons in New World species moderately, broadly swollen, punctate, disk of frons glabrous and shining in female (fig. 78 E) but covered by appressed setae in male; frons without a pair of polished tubercles above antennal sockets; frontal line often present on lower surface of frontal swelling but rarely above it, sometimes absent; scape moderately long, slender, or expanded apically, basal flagellomeres often rounded out beneath in males; clypeal free margin with a median lobe, clypeus often bulging at base of lobe; labrum truncate, hidden; inner margin of mandible with a single large tooth near middle, externoventral margin with a deep notch; dorsum of collar thin, much lower than and closely appressed to scutum; propodeum

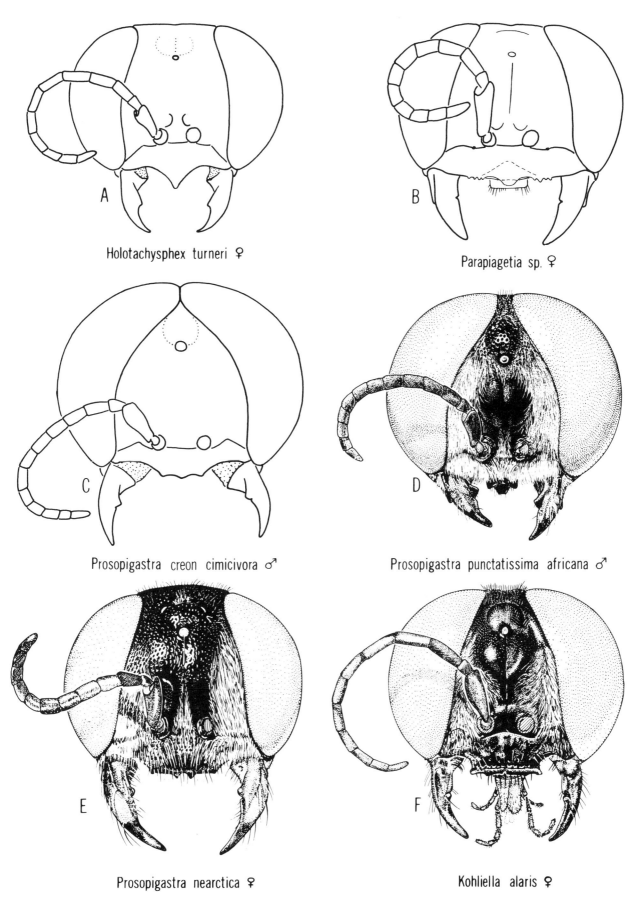

FIG. 78. Facial outlines and portraits in the subtribe Tachytina; B, Chilean species.

short, sometimes with semicircular dorsal enclosure; episternal sulcus extending ventrad to or nearly to anteroventral margin of pleuron; scrobal sulcus weak or absent; males frequently and females occasionally with a ventral mesopleural tubercle, pleuron often with a tubercle or transverse (or vertical) carina just before midcoxa (fig. 71 A); terga moderately to densely punctate, terga I-II with lateral carina (fig. 71 G); pygidial plate present in female but usually bordered only apically by carinae or angles and these sometimes very weak, surface shining, punctate, flat or convex, usually asetose or with a few scattered inconspicuous setae apically; pygidial plate absent in male but tergal apex often produced and truncate or emarginate; male sterna II-V sometimes with transverse welts or carinae; inner posterior angle of forecoxa in both sexes sometimes with dorsal, triangular lamella; male foretrochanter and femur unmodified; female foretibia with a long, hairlike seta apicoventrally, mid and hindtibiae moderately set with spines arranged in rows, uppermost row on hindtibia usually set on tubercles, hindtibia not ridged; female with well developed foretarsal rake composed of very fine, long, spines which are closeset at least at the apex of each tarsomere (fig. 73 H), tarsomere II with more than three spines; male with a weak foretarsal rake; marginal cell short, apex broadly truncate, apex of male sternum VIII emarginate; volsella long, slender, sometimes bearing a mediodorsal lobe; head of aedeagus with teeth on ventral margin.

Geographic range: Of the 36 species currently recognized in *Prosopigastra* all but one are found in the Old World. The single New World form, *nearctica,* occurs over a wide range of altitudes in California but always in a dry, sandy environment. The species is also found in Arizona.

Four of the Old World species are restricted to the Ethiopian Region, and three are known from the western fringe of the Oriental Region. The remaining species are about equally divided between the Mediterranean area and the deserts of southwestern USSR with a few of these being found in both regions.

Systematics: These stocky wasps are 4 to 12 mm long, the head and thorax black, but the gaster usually partly or all red in one or both sexes. In species with a black gaster the tergal margins are often pale. The legs are often red and sometimes marked with yellow. The wings are clear, but the forewing is occasionally faintly infumate across the discoidal-submarginal area. Pubescence is silvery and is dense mainly on the face and thoracic sides. Tergum I sometimes has a conspicuous cover of appressed silver hair.

The principal diagnostic features of *Prosopigastra* are: lateral carina on terga I-II, dense tergal punctation, and the facial prominence. Additional characters are the closeset, long, fine rake spines and glabrous pygidial plate of the female, the absence of a posterobasal notch on the male forefemur, the generally shiny and often densely punctate head and thorax, the short propodeum, and the short, truncate marginal cell. When present, the holoptic condition and ventral mesopleural tubercles are also diagnostic.

Prosopigastra probably is most similar to *Holotachysphex*. Both have shiny, densely punctate bodies and lateral carinae on terga I and II, but *Holotachysphex* does not have a facial prominence, a tarsal rake, or a pygidial plate. The male forefemur is notched in *Holotachysphex*. *Prosopigastra* may also be related to *Tachysphex,* but the latter has a carina only on tergum I, the terga are at most sparsely punctate, and the male forefemur is usually notched. A few *Tachysphex* have a strong facial swelling (*isis, osiris, deserticola,* for example), but it is covered with appressed setae. *Prosopigastra* lacks the two tubercles found just above the antennal sockets in *Tachysphex* (and *Holotachysphex*), although some *Prosopigastra* have a pair of slightly raised, polished bands in this area (*nearctica,* for example). Generally in *Tachysphex,* hindtarsomere I is shorter than the combined length of the following three tarsomeres, but in *Prosopigastra* these measurements are about equal. This generic distinction breaks down in some of the roach-collecting *Tachysphex* where tarsomere I is frequently equal to the following three tarsomeres due to the shortness of tarsomere IV. Probably *Prosopigastra* should be regarded as a somewhat isolated and highly evolved member of the subtribe Tachytina.

The eyes are holoptic in the males of about one-third of the species. Kohl (1889) described *Homogambrus* for the male of *globiceps,* and various authors have recognized this taxon as a subgenus of *Prosopigastra.* Gussakovskij (1933a) appears to have identified *Homogambrus* solely by the holoptic condition, thus placing females in the taxon only through association with known males. On the other hand, Beaumont (1954d) tried to differentiate *Homogambrus* using characters of both sexes, but these break down in the Ethiopian fauna. Also, the male of *gaetula,* which is holoptic according to Beaumont's (1950) figure, is placed by him in *Prosopigastra.* It appears to us that the recognition of *Homogambrus* is untenable. The single New World species, *nearctica,* does not fit conveniently into either subgenus because it lacks a facial prominence even though it is typically *Prosopigastra* in all other respects. It does not seem useful to establish a subgenus for this odd species.

Species characters are found in the configuration of the clypeus, the degree of eye convergence, the development of the frontal prominence, the degree of punctation of various parts of the body, wing venation, presence or absence of forecoxal lamellae, presence or absence and degree of development of ventral mesopleural tubercles, male sternal features, and male genitalia, among others.

Keys to Palearctic species can be found in Mercet (1907) and Gussakovskij (1933a), of which the latter is more complete but still outdated. Arnold (1922) keyed three of the four Ethiopian species. Beaumont (1947c, 1954d, 1955a, 1956a) and Pulawski (1958b, 1965c) offered useful data on various Palearctic species, and R. Bohart (1958b) gave the pertinent details of the New World species. Many species are poorly known and a revision is needed.

Biology: The only references to *Prosopigastra* biology appear to be those of Ferton (1912a,b) on *P. creon* ssp. *cimicivora* and *punctatissima* and Arnold (1922) on *P. neavei.* Ferton's papers are the most detailed, but unfortunately he lumped the two species in describing their habits. His observations were made on a small colony of wasps that nested in a barren, sandy area along the ocean shore. Plantain grew nearby. According to Ferton, *cimicivora* and *punctatissima* utilized preexisting burrows of other Hymenoptera and cicindelid beetles for their nests. They preferred vertical burrows. Side tunnels that led to a number of cells were excavated in these at depths ranging from 2.5 to 20.5 cm. During provisioning the nest entrance was left open in both species. *P. cimicivora* stored only lygaeid bugs of one species, *Apterola pedestris* (Stål), but *punctatissima* used adult fulgoroids of the family Tropiduchidae (*Ommatissis binotatus* Fieber) and lygaeids of the genus *Nysius.* Prey were collected on the nearby plantain and paralyzed on the plant or on the ground. The mode of transport to the nest is presumed to be on the wing, but this is not stated by Ferton. However, he indicated that the wasps do not deposit their prey on the ground near the nest prior to entry but go directly into the nest with prey. In three cells examined by Ferton there were 13 to 19 prey. The egg probably is laid on the first provision because Ferton stated that it is always on one of the prey at the back of the cell. The egg is affixed to the venter of the host between the first and second pair of legs. Paralysis of the prey in *cimicivora* is not total. The nest is closed with debris.

Arnold's brief notes on *neavei* indicate that the species nests in hard, claylike soil. The burrow is about 7.5 cm long, 2.5 cm beneath the surface, and ends in three to six cells. An average of six pentatomid bugs were found in each cell. The egg was glued to the base of the abdomen.

Checklist of *Prosopigastra*

boops Gussakovskij, 1933; sw. USSR: Turkmen S.S.R.
bulgarica Pulawski, 1958; Bulgaria, Turkey
capensis Brauns, 1906; S. Africa
creon (Nurse), 1903 (*Homogambrus*); Cyprus, sw. USSR: Turkmen S.S.R., Pakistan, w. India
  acanthophora Gussakovskij, 1933
  cypriaca Beaumont, 1954
  ssp. *cimicivora* (Ferton), 1912 (*Homogambrus*) (Bull. Soc. Ent. Fr.); n. Africa
    cimicivora Ferton, 1912 (Ann. Soc. Ent. Fr.), nomen nudum
  ssp. *carinata* Arnold, 1922; Rhodesia
desertorum Gussakovskij, 1933; sw. USSR: Turkmen S.S.R.
falsa (F. Morawitz), 1893 (*Tachysphex*); sw. USSR: Turkmen S.S.R.
fumipennis Gussakovskij, 1952; sw. USSR: Tadzhik S.S.R.
gaetula Beaumont, 1950; Algeria
gigantea Gussakovskij, 1935; sw. USSR: Uzbek S.S.R.
globiceps (F. Morawitz), 1889 (*Tachysphex*); Mongolia
handlirschi Morice, 1897; n. Africa; e. Mediterranean region; Spain
  crosi Ferton, 1912
  tunetana Gussakovskij, 1933
insignis E. Saunders, 1910; n. Africa, Syria, Iraq
  angustifrons Schulthess, 1928
kizilkumii (Radoszkowski), 1877 (*Tachytes*); sw. USSR: Uzbek S.S.R., Turkmen S.S.R.
  roseiventris F. Morawitz, 1894 (*Homogambrus*)
kohli Mercet, 1907; Iberian Penin.
laevior Morice, 1897; Morocco, Egypt, Sudan
  sericans Morice, 1897 (*Hologambrus*)
latifrons Gussakovskij, 1933; sw. USSR: Turkmen S.S.R., Uzbek S.S.R.
lissipes Pulawski, 1973; Israel
major (F. Morawitz), 1890 (*Homogambrus*); sw. USSR: Turkmen S.S.R.
menelaus (Nurse), 1903 (*Homogambrus*); Pakistan, w. India
minima Beaumont, 1956; n. Africa
mocsaryi Brauns, 1906; S. Africa
moricei Mercet, 1907; Algeria
nearctica R. Bohart, 1958; U.S.: California, Arizona
neavei Turner, 1917; Malawi, Zambia, Rhodesia
nubigera Gussakovskij, 1933; sw. USSR: Turkmen S.S.R.
nuda (Nurse), 1903 (*Tachysphex*); w. India

A Kohliella stevensoni ♂

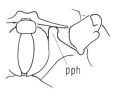
B Parapiagetia capensis ♀

C Tachysphex panzeri ♂

D Parapiagetia sp. ♀

E Tachytella aureopilosa ♂

F Ancistromma punctulata ♀

FIG. 79. Head morphology in the subtribe Tachytina; A, clypeus and mandibles, paratype; B, oral area, pph = paramandibular process of hypostoma; C, clypeus, labrum and mandibles; D, oral area with open mandibular socket, Chilean species; E-F, left scape to flagellomere I, E = holotype.

oasicola Tsuneki, 1972; Mongolia
orientalis Beaumont, 1947; Crete, Cyprus, e. Mediterranean region, sw. USSR
pilosa Arnold, 1959; Tanzania: Tanganyika
punctatissima A. Costa, 1867; sw. Europe
  ssp. *africana* Beaumont, 1955; n. Africa
riparia Gussakovskij, 1952; sw. USSR: Tadzhik S.S.R.
rufiventris Gussakovskij, 1933; sw. USSR: Turkmen S.S.R., Uzbek S.S.R.
thalassina Gussakovskij, 1933; sw. USSR
turcomanica Gussakovskij, 1933; sw. USSR: Turkmen S.S.R.
werneri (Maidl), 1924 (*Homogambrus*); Egypt, Sudan
zalinda Beaumont, 1955; n. Africa, e. Mediterranean region

Genus Kohliella Brauns

Generic description: Inner orbits converging above; hindocellar scars oblong, somewhat comma-like, their long axes forming an angle approximating 130° (fig. 61 C); frons with a large, prominent, V-shaped swelling and a small, central convexity just above sockets which is bisected by frontal line, latter impressed (*stevensoni*) or absent (*alaris*) as it passes over V-shaped prominence (fig. 78 F); scape moderately long; clypeus with a central convexity, clypeal free margin with a prominent median lobe which is truncate (*alaris*, fig 78 F); or angular and notched (*stevensoni*, fig. 79 A); labrum transverse, hidden; inner margin of mandible with a median cleft in female, with (*alaris*) or without (*stevensoni*) a tooth in male, externoventral margin with a notch; mouthparts including palpi laterally compressed, dorsum of collar thin, lower than and closely appressed to scutum; propodeum moderately long; episternal sulcus extending to anteroventral margin of pleuron, scrobal sulcus absent; terga impunctate or with faint punctation, tergum I without lateral carina except at extreme base; female with a triangular, glabrous, sparsely punctate, shiny pygidial plate which is defined laterally by angles rather than carinae; male tergum VII flattened and punctate; male forecoxa unmodified; male forotrochanter weakly depressed posterobasally; male forefemur with a posterobasal notch which is lined with a mat of short setae; tibiae set with long, fine spines arranged in rows: two on I, three on II and III, hindtibia angular in cross section, not carinate; both sexes with well developed foretarsal rake composed of long, slender spines which are bladelike in female (fig. 73 I), tarsomere II with one rake spine in male, two in female, foretarsomeres with shorter subsidiary rake spines dorsally and ventrally; inner margin of female claw with a tooth (fig. 73 O); marginal cell short, truncate apically; third submarginal cell petiolate on the media (fig. 70 F); apex of male sternum VIII rounded (fig. 76 K-L); volsella about half as long as gonostyle or shorter, setose ventrally (fig. 76 O-P); head of aedeagus with very fine teeth along ventral margin, a long ventrally directed spine present near base of aedeagal head (fig. 76 M-N).

Geographic range: The two species of *Kohliella* occur in southern Africa.

Systematics: Kohliella alaris is black and ranges from 6 to 9 mm in length, while *K. stevensoni*, of which only the male is known, is black with gastral segments I-III and the legs red. *Kohliella stevensoni* is 9 mm long. The wings are clear in both species.

The petiolate third submarginal cell, laterally compressed mouthparts, absence of a lateral carina on tergum I, V-shaped facial swelling, toothed female claw, aedeagal spine and short volsella are unique features of *Kohliella*. This is the only genus in the subtribe Tachytina known to have claw teeth. Although *Kohliella* is similar to *Tachysphex* in general facies only a few features, such as the form of the collar, the male forefemoral notch, and the bare pygidial plate, are common to both. *Kohliella* is probably best regarded as a specialized relic.

Biology: Unknown. The highly developed female foretarsal rake, which is unusually strong in the male as well, the compressed mouthparts, and the toothed female claw, suggest that the behavior and ethology of this genus should be very interesting.

Checklist of *Kohliella*

alaris Brauns, 1910; S. Africa, Rhodesia
stevensoni Arnold, 1924; Rhodesia

Tribe Palarini

This Old World tribe contains a single genus *Palarus*. The bee and wasp killers, as the species of *Palarus* might be called in reference to their hymenopterous prey, resemble wasps of the subfamily Philanthinae because of their similar color patterns. *Palarus* are medium to large, rather stout-bodied wasps. The abdomen is banded or spotted with yellow, the legs are yellowish or reddish, and there are areas of red or yellow on other parts of the body. Only a few species are largely black. The thorax is short and stout, and the sessile gaster is broad at the base, but it usually tapers uniformly distad giving it a distinctive triangular shape. At least one species is occasionally an important pest of honeybees.

Diagnostic characters:

1. (a) inner orbits strongly converging above, eyes usually separated at vertex by approximately a midocellus diameter (fig. 80 I); (b) midocellus of normal size but commonly glazed and somewhat depressed, hindocelli smaller, usually flattened, and oval or linear rather than round (fig. 80 A,B) but still with a transparent lens.
2. (a) Antennal sockets rather low on face but not contiguous with frontoclypeal suture, separated from it by one-third to one-half a socket diameter; (b) male antenna with 13 articles.
3. Clypeus tripartite, divided by distinct sutures (fig. 80 I).
4. (a) Mandible usually notched or angulate on externoventral margin, inner margin with one or two small teeth near middle; (b) mouth parts short;

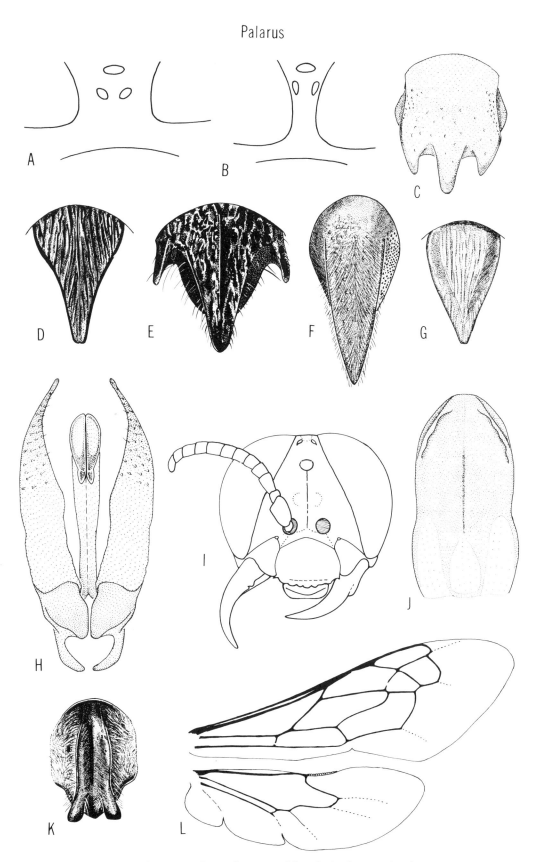

FIG. 80. Structural details in the genus *Palarus;* A-B, ocellar area of female *latifrons* and *rufipes,* resp.; C,E,K, pygidium of male *ambustus, latifrons,* and *rufipes,* resp.; D,F,G, female pygidial plate of *latifrons, rufipes,* and *handlirschi* resp.; H, male genitalia of *variegatus;* I, face of female *rufipes;* J, sternum VIII of male *variegatus;* L, wings of *rufipes.*

(c) mandible socket open.
5. Pronotal collar short, high, often closely appressed to scutum.
6. Scutum with notauli absent or only weakly developed anteriorly.
7. Episternal and scrobal sulci present.
8. (a) Foreleg with well developed tarsal rake in both sexes; (b) hind femur simple apically; (c) midcoxae widely separated.
9. Propodeum short; (b) enclosure usually well defined, triangular, apex extending onto posterior vertical face; (c) no propodeal sternite.
10. (a) Three submarginal cells; (b) two recurrent veins, both received by submarginal II, or first ends on I; (c) marginal cell truncate, appendiculate (fig. 80 L).
11. (a) Jugal lobe about half length of anal area; (b) hamuli not divided into two groups.
12. (a) Gaster sessile; (b) pygidial plate present in both sexes.
13. (a) Volsella apparently absent (fig. 80 H); (b) gonostyle simple.

Systematics: The affinities of *Palarus* are not clear. Börner (1919) established the tribe Palarini for *Palarus* and *Dinetus,* but these two genera are not allied. Handlirsch (1925) removed *Dinetus* and placed it in a new tribe, the Dinetini, which we have transferred to the Astatinae. Handlirsch (1925, 1933) placed *Palarus* in the Larrinae, and Beaumont (1949b) followed this arrangement. This placement can be defended on the basis of several larroid features found in *Palarus:* single midtibial spur, strong convergence of the eyes towards the vertex, notched mandible (variable in both *Palarus* and the Larrinae) with one or two inner basal teeth, abnormal hindocelli, and similar genitalia. However, none of these characters are unique to the Larrinae and most are subject to variation within the group. Thus, the argument could just as easily be made that their presence in *Palarus* is a matter of parallel evolution and not indicative of a close relationship. For example, deformed ocelli have evolved in the Nyssoninae (Bembicini, Heliocausini) and also in the Philanthinae (Odontospfhecini). The *Palarus* habit of using Hymenoptera as prey is not known in any larrid, but is common in some Philanthinae. Indeed, there is rather striking resemblance between *Palarus* and the philanthines as far as color pattern is concerned, and one could probably defend (albeit weakly) placing *Palarus* in this subfamily on morphological grounds. Both have one midtibial spur and the genitalia and wings of *Palarus* are quite similar to those of the Cercerini, the wings being particularly like *Eucerceris* (compare figs. 80 L and 184 F). The male hindfemur of *P. handlirschi* has an apicolateral process similar to *Cerceris,* but analogous structures are also found in the Larrinae.

There are a number of peculiar or unique features in *Palarus* that cast doubt on the placement of this genus in the Larrinae (or Philanthinae). The tripartite clypeus (fig. 80 I) of *Palarus* is unique, although it is true that the clypeus in some Larrinae (*Tachytes* and a few exceptional *Liris,* for example) has a suggestion of such division, but the clear suturelike lines found in *Palarus* are absent or incomplete in most of these examples. The same applies to the philanthine clypeus. The abdominal modifications in *Palarus* are peculiar, and similar structures such as the transverse keel on sternum II are found only in a few other Sphecidae: *Larrisson* (Larrinae), *Heliocausus* (Nyssoninae), and several other nyssonine groups. The combination of wing features found in *Palarus* is also distinctive: the rather small hindwing jugal lobe, and the shape of the third submarginal cell including the proximity of its outer side to the apex of the marginal cell.

The larva of *Palarus* has been discussed by Evans (1958a). Although he retained Palarini as a tribe in the Larrinae, it is clear from his description that the larva of *Palarus* differs in a number of ways from all other known larrine larvae.

For the present it seems wisest to maintain the Palarini as a larrine tribe. Only after more has been learned about sphecid biology and larvae can the relationships of *Palarus* be properly evaluated.

Genus Palarus Latreille

Generic diagnosis: Eyes nearly holoptic (except in *P. latifrons* group), hindocelli flattened and deformed (except in *P. latifrons* and *handlirschi* groups); face above antennal sockets swollen; antenna short, scape short, stout, flagellomeres about as long as wide, basal male flagellomeres often with shiny angular areas ventrally and terminal article sometimes truncate; occipital carina joining hypostomal carina about midway between apex of oral fossa and mandible base; mandible notched or angular externoventrally (except some species of *histrio* group); labrum short, broadly triangular; legs short and stout, mid and hindlegs with numerous stout spines, arolium large; submarginal cell II triangular, sometimes briefly petiolate, crossvein $r-m_3$ of forewing meeting marginal cell near its apex; media of hindwing diverging after crossvein cu-a; vertical face of gastral tergum I concave, tergum I with lateral cariniform or lamelliform prominences mesad of the spiracle in the *variegatus* and *histrio* groups (these are mesad of lateral tergal carina); sterna, especially II, variously modified into thick transverse keels, or tubercles, or special platelike areas in both sexes (except *latifrons* group, females of *handlirschi* group, and some females of *histrio* group); male sternum VIII strongly sclerotized, apex broadly rounded or bluntly acuminate, sides usually thickened or upturned towards apex (fig. 80 J).

Geographic range: Palarus extends through all of Africa, southern Europe, and across southern Asia as far as China and Korea. It is not recorded from Japan or southeast Asia. Thirty-two species are currently recognized; they are nearly equally divided between Africa and Asia. However, the range of a few extends over both areas.

Systematics: The most distinctive features of *Palarus* are the tripartite clypeus, the deformed ocelli, the moderate jugal lobe, and the abdominal modifications. There

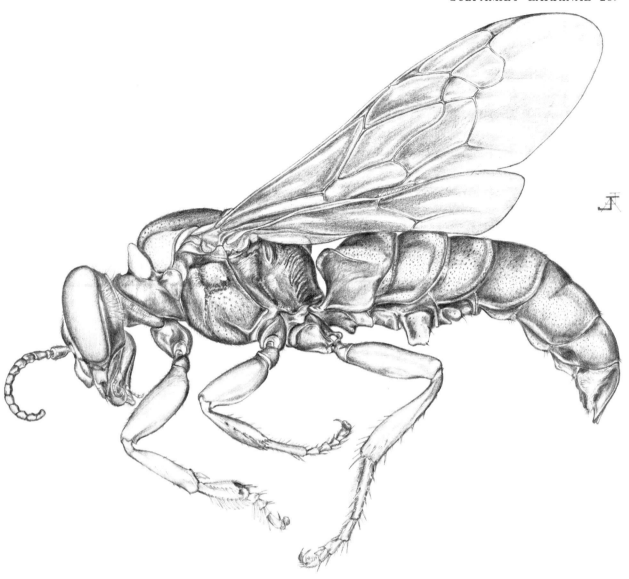

FIG. 81. *Palarus rufipes* Latreille, male.

are several distinct groups within *Palarus,* and Turner (1911) divided the genus into five or six species groups based primarily on the structure of the last tergum of the male. Beaumont (1949b) refined Turner's groupings by employing more structural features of both sexes. He reduced the number of species groups to four.

The *latifrons* group includes only one African and one Asian species. Unlike all other *Palarus* the eyes of these species are broadly separated at the vertex, and the hindocelli are the most ordinary in the genus (fig. 80 A). Also, the abdominal venter is little modified and tergum I lacks the laterobasal carinae. All of these features suggest that the *latifrons* group is the most generalized in the genus.

The *handlirschi* group has one Asian and three African species. In these the eyes are holoptic or nearly so, and unlike other *Palarus* the first recurrent vein is received by submarginal cell I. The male has a strong transverse keel on sternum II, but the female sternum is simple.

The *variegatus* and *histrio* groups contain the majority of *Palarus,* which are about equally divided between them. These two groups are separated from each other primarily by pygidial differences. Features they hold in common and that set them apart from the *latifrons* and *handlirschi* groups are: eyes nearly holoptic (fig. 80 B, I), recurrent veins received by submarginal II or interstitial, tergum I nearly always with a strong cariniform or lamelliform ridge laterally (mesad of the lateral carina) (fig. 81), and sterna variously modified in both sexes. Some species in the *histrio* group lack a notch on the mandible (*histrio, parvulus, hastatifrons*), and the first male tarsomere of certain species in the *variegatus* group is modified into a large flattened plate (*rufipes, dongalensis*).

The pygidial plate displays important differences at both the specific and group levels. The female plate is as follows in each group: *latifrons* group — surface covered with heavy longitudinal ridges (fig. 80 D); *handlirschi* group with a V-shaped carina on the pygidial surface which parallels the pygidial carinae, the pygidial surface is also longitudinally ridged (fig. 80 G); *variegatus* group — surface covered with very fine longitudinal ridges (fig. 80 F); *histrio* group — surface smooth but punctate. The male pygidial plate differs as follows: *latifrons* group — acuminate or roundly truncate apically, a spine on the tergum lateral to base of plate (fig. 80 E); *handlirschi* group — pygidial carinae weak, apex subtruncate; *variegatus* group — plate rather narrow, strongly elevated, usually bifurcate apically or acute (fig. 80 K); *histrio* group — plate broad, trifurcate apically (fig. 80 C).

Allometric growth poses an interesting problem for students of *Palarus*. This applies primarily to the various abdominal features, such as the transverse keel on the second sternum. Within one species the smaller individuals may have a small or indistinct keel. Larger individuals have a strong keel. Similar variation is found in the width of the interocular space.

Biology: The activities of *Palarus* have been fairly well documented by Brauns (1911a), Ferton (1912a), Ahrens (1925), Clausen, Gardner, and Sato (1932), Móczár (1952), Grandi (1961), and Tsuneki (1969a). Observations of earlier workers have been summarized by one or more of these authors. Most studies have been of *P. variegatus*, and the following notes pertain to this species unless otherwise stated. Ahren's paper (in Russian) deals with *variegatus* also, but because his observations differ in some respects from those of other workers his paper will be discussed separately.

The nest is excavated in packed sandy soil or heavy clay and usually in ground that is rather barren of plant life. Such areas often support large numbers of nesting females. Several authors have observed that the female (males also?) spends the night inside the nest, sealing the entrance behind her with sand. Brauns stated that preexisting holes may be utilized for this purpose in south African species. Both males and females of African *Palarus* frequent flowers according to Brauns. The male seeks out nesting females, and when one is located he circles above her and finally lands upon her.

The nest is excavated before prey capture and contains a single cell. The burrows penetrate the soil at an angle and average about 30.5 cm in length. The cell is a few centimeters below the soil surface. Details of the burrow have been described only for *P. variegatus*. In this species the burrow has several vertical undulations before the cell is reached. During construction of the nest the female may continue her excavations for a long time (20-30 minutes) without coming to the surface. Thus, as she digs, soil is pushed back behind her blocking the burrow entrance. Apparently nest closure during the hunt for prey varies from wasp to wasp. Several authors have noted that the entrance is often left open during the hunt.

Palarus mass provision with hymenopterous adults of a great many families (Ichneumonidae, Tiphiidae, Scoliidae, Mutillidae, Sphecidae, Pompilidae, Vespidae, and various bees), and no prey specificity is apparent. The female may stay near her nest waiting for some wasp to fly by. In this case the *Palarus* flies up and catches the prey in mid air, then falls or flies to the ground in order to sting and paralyze the victim. In other cases the *Palarus* female may fly in search of prey. The female flies to the nest with her prey and if the nest entrance is blocked she leaves the victim temporarily on the ground, opens the nest, goes inside, comes back out head first and grasps the prey, and then pulls it inside the nest after her (walking backwards). If the nest is already open on arrival, the *Palarus* goes directly into the nest with her prey. Five to 12 prey are stocked in a cell, and the egg is laid on the venter of the first provision. During final nest closure the pygidial plate of the female is used to tamp the soil.

Several authors have noted that the heads of the prey have been twisted, presumably by the *Palarus* (Dufour, 1841; Móczár, 1952; Grandi 1961). In some cases the head has been rotated so many times that it is easily separated from the body. Móczár was of the opinion that this strange behavior was a substitute method of paralyzing the prey. This explanation seems unlikely in view of the well-developed sting of female *Palarus*, and Grandi has apparently observed stinging of prey in *Palarus*. No one has observed the act of "head twisting," and its significance awaits explanation.

Ahren's (1925) report on *P. variegatus* contains a few questionable observations. He stated that some nests had up to four cells, although he was unable to follow the burrow to each cell in these supposed multicellular nests. This suggests that he may have confused several nests as one. However, several times he observed two females exiting from a single burrow. If these latter observations are correct, they give more credence to his records of multicellular nests. Common use of a single burrow by more than one female is rare in the Sphecidae. Another interesting aspect of Ahren's paper is the statement that egg deposition was delayed until several provisions had been made. This is supported by Tsuneki (1969a), although Grandi (1961) stated that the egg is laid on the first provision.

In South Africa, *Palarus* can be a nuisance for beekeepers. Brauns (1911a) reported that *Palarus latifrons*, which often utilizes honey bees as prey, is commonly found in great numbers near beehives and can exact a considerable toll. This species has earned the name, "Banded Bee Pirate."

Checklist of *Palarus*

The parentheses at the end of each citation contain the code letters that denote the species group to which the taxon belongs — (L) = *latifrons* group; (HA) = *handlirschi* group; (V) = *variegatus* group; (HI) = *histrio* group.

ambustus Klug, 1845; n. Africa (HI)
  ssp. *confusus* Turner, 1911; nw. Africa
    histrio Lepeletier, 1845 (*Astata*), nec Spinola, 1838
aurantiacus Radoszkowski, 1893; sw. USSR: Turkmen S.S.R. (V)
beaumonti Bytinski-Salz, 1957; Turkey (HI)
bernardi Beaumont, 1949; Sudan (V)
bisignatus F. Morawitz, 1890; sw. USSR (HI)

comberi Turner, 1911; ne. India (HA)
dongalensis Klug, 1845; n. Africa, Arabian penin. (V)
  rufipes Spinola, 1838; nec Latreille, 1812
  decipiens Honoré, 1941
fortistriolatus Cameron, 1907; India (V)
funerarius F. Morawitz, 1889; Mongolia, India, sw. USSR (HI)
  gracilis Kohl, 1889
  quiescens Nurse, 1903
handlirschi Brauns, 1912; s. Africa (HA)
  ssp. *nigrior* Arnold, 1923; Rhodesia
  ssp. *occidentalis* Arnold, 1929; S.-W. Africa
hastatifrons Turner, 1919; Israel, Jordan (HI)
  ssp. *africanus* Beaumont, 1949; nw. Africa
  ssp. *oceanicus* Beaumont, 1949; Morocco
histrio Spinola, 1838; n. Africa (HI)
  histrio Dahlbom, 1845, nec Spinola, 1838
  lepidus Klug, 1845
incertus Radoszkowski, 1893; sw. USSR: Turkmen S.S.R. (HI) (? = *bisignatus*)
interruptus (Fabricius), 1787 (*Crabro*); Sri Lanka, India (L)
  interruptus Fabricius, 1787 (*Bembex*)
  indicus Gmelin, 1790 (*Vespa*), new name for *Vespa interrupta* (Fabricius), 1787 (*Bembex*)
  orientalis Kohl, 1884
klugi Menke, new name for *indicus* Nurse; India
  indicus Nurse, 1903, nec Gmelin, 1790
laetus Klug, 1845; Egypt (HI)
  annulatus Walker, 1871 (*Larra*), nec Klug, 1845
  walkeri Handlirsch, 1892 (*Stizus*)
  eximius Honoré, 1941
  ssp. *fabius* Nurse, 1903; India, Iraq, Iran
  ssp. *disputabilis* Morice, 1911; Algeria
latifrons Kohl, 1884; s. Africa (L)
  curvilineatus Cameron, 1905
  lineatifrons Cameron, 1905
maculatus Dahlbom, 1845; s. Africa (?)
multiguttatus Arnold, 1960; Kenya (V)
nursei Turner, 1911; India (HI)
obesus Arnold, 1951; Mauritania (HA)
oneili Brauns, 1898; s. Africa (V)
parvulus Beaumont, 1949; n. Africa (HI)
pentheri Brauns, 1898; s. Africa (V)
pictiventris F. Morawitz, 1890; sw. USSR: "Transcaspia" (HI)
rothschildi Magretti, 1908; e. Africa (V)
rufipes Latreille, 1812; nw. Africa (V)
  flavipes Fabricius, 1793 (*Tiphia*), nec Fabricius, 1781
  humeralis Dufour, 1853
saundersi Morice, 1897; n. Africa (V)
seraxensis Radoszkowski, 1893; sw. USSR: Turkmen S.S.R. (HI)
spinolae Saussure, 1854; Egypt, Arabian Penin. (V)
  fulviventris Latreille, 1812
  ssp. *niger* Beaumont, 1949; n. Africa
turneri Brauns, 1912; s. Africa (HA)
variegatus (Fabricius), 1781 (*Tiphia*); Europe, w. Asia, China (V)

flavipes Fabricius, 1781 (*Crabro*)
auriginosus Eversmann, 1849
  ssp. *affinis* F. Morawitz, 1893; sw. USSR: Tadzhik S.S.R.
  ssp. *varius* Sickmann, 1894; China, Korea, Quelpart Island
saishiuensis Okamoto, 1924

Tribe Miscophini

This is a highly diversified group of wasps with respect to both morphology and biology. It contains some of the most primitive larrine genera, as well as some highly specialized types. Miscophins are moderate to very small wasps, a few of which rank among the smallest sphecids known. Coinciding with the strong tendency for reduction in body size in this tribe is the trend toward reduction in wing venation.

Spiders, Orthoptera, Hemiptera, Homoptera, Psocoptera, lepidopterous larvae, and Diptera are used in nest provisioning by various miscophin genera. The habits of six of the 14 genera are still unknown.

Endemism is high in the Miscophini. A number of genera are known only from Australia or southern Africa, for example, which may be an indication that the Miscophini is an old group. Only two genera, *Lyroda* and *Nitela,* are cosmopolitan.

Diagnostic characters:

1. (a) Inner orbits usually either parallel or converging above (figs. 87, 91); (b) ocelli normal.
2. (a) Antennal sockets low on face, contiguous with frontoclypeal suture (except in *Mesopalarus*); (b) male with 11 flagellomeres except only 10 in *Sericophorus* and some *Solierella.*
3. (a) Clypeus transverse, variable, usually with a median lobe and/or teeth; (b) labrum small, usually hidden.
4. (a) Mandible usually notched or angulate externoventrally (exceptions: *Mesopalarus, Nitela, Auchenophorus,* and most *Solierella*), inner mandibular margin with or without teeth; (b) mouthparts usually short; (c) mandibular socket open except in *Auchenophorus.*
5. Pronotal collar variable.
6. Scutum usually without notauli.
7. (a) Episternal sulcus present except in *Saliostethoides* and some *Saliostethus* and *Miscophoides;* (b) hypersternaulus present in *Nitela* and *Auchenophorus.*
8. (a) Female foreleg with a tarsal rake except in *Nitela, Auchenophorus,* most *Solierella,* and some *Plenoculus* and *Miscophus;* (b) hindfemur simple apically, usually thickest towards base; (c) midcoxae widely separated to contiguous; (d) arolium variable; (e) male forecoxa and trochanter unmodified except in *Solierella,* male forefemur simple.
9. (a) Propodeum short to long; (b) enclosure present or absent; (c) propodeal sternite absent.
10. (a) Number of submarginal cells ranging from three

to none, second often petiolate; (b) usually two recurrent veins but none in *Miscophoides,* and submarginal II absent in *Saliostethus, Saliostethoides, Nitela,* and *Auchenophorus,* and sometimes absent or evanescent in a few *Miscophus,* termination point of recurrent veins variable; (c) marginal cell apex variable (open in *Saliostethiodes* and *Miscophoides,* and some *Saliostethus* and *Miscophus*).

11. (a) Jugal lobe small to moderate, never more than half length of anal area (absent in *Saliostethiodes, Miscophoides,* and *Auchenophorus*); (b) hamuli not divided into two groups.
12. (a) Gaster sessile (pedunculate in some *Lyroda* and *Miscophus*), (b) pygidial plate present or absent.
13. (a) Volsella present or absent, (b) gonostyle usually simple.

Systematics: The Miscophini is separated from the other larrine tribes by the following combination of characters: 1) ocelli normal, 2) inner orbits not emarginate and usually converging above or parallel, 3) jugal lobe of hindwing never more than half as long as anal area and usually less, sometimes absent, 4) hindfemur simple apically, 5) clypeus not divided by vertical sutures into three parts.

The five characteristics above serve to separate the tribe from other larrines but not from the Crabroninae. The typical crabronine characteristic of one submarginal and one discoidal cell in the forewing is also found in a few miscophins, *Nitela, Auchenophorus, Saliostethus,* and a few atypical *Miscophus.* The length of the scape is diagnostic in these cases, however. The crabronine scape is about half the length of the flagellum, whereas in the Miscophini it is much less than half the flagellar length. The inner orbits in the Crabronini usually converge below, and correspondingly the eye facets are larger towards the clypeus. In the Miscophini the inner orbits very rarely converge below and then only slightly. The lower eye facets are not conspicuously enlarged in the Miscophini.

None of the miscophin genera have the oxybelin characteristic of fusion of the single submarginal and discoidal cells. Neither do the miscophins possess the metanotal and propodeal processes which are nearly universal in the Oxybelini. Only one oxybelin genus, *Belomicroides,* lacks these, and one species in particular, *santschii,* is very similar to *Plenoculus* except for the fused forewing cells.

Various sections of the Miscophini as here constituted have been recognized as separate tribes at different times by different authors. *Lyroda* has usually been associated with the Larrini (or Tachytini), but the normal ocelli preclude this association; and there are certainly many more points of similarity between *Lyroda* and some miscophin genera, such as *Paranysson, Sericophorus,* and *Plenoculus,* than there are to any of the genera in the Larrini. Furthermore, Evans (1964c) indicated that the larva of *Lyroda* is similar to that of *Plenoculus,* and he placed the genus in the Miscophini.

Dalla Torre (1897) proposed the subfamily Sericophorinae for *Sericophorus, Sphodrotes,* and *Paranysson* (Turner, 1914, used the name Paranyssoninae for the same group), but there is little to justify the retention of this assemblage.

The tribe (or subfamily) Nitelini has been used in the past for *Nitela, Miscophus,* and related genera or simply for *Nitela.* It would not be unreasonable to recognize this tribe if it contained only *Nitela* and the seemingly related *Auchenophorus,* but until the male of the latter is better known we feel it is best to remain conservative on this point. These two genera are the only miscophins with a hypersternaulus. Both have similar venation, simple mandibles, and no pygidial plate or tarsal rake. The closed mandibular sockets of *Auchenophorus* is a discordant feature in the Miscophini.

The possible phylogenetic relationships of the miscophin genera are indicated by the dendrogram in fig. 82. One should view this diagram with an open mind since many of the genera are poorly known. The males of some await discovery, and much is still unknown about ethology as well as the larvae. When more information is available in these areas the dendrogram will no doubt need to be modified

The Miscophini is the least homogeneous tribe in the Larrinae. For example, it contains several genera (*Larrisson, Mesopalarus,* and *Auchenophorus*), the affinities of which are not entirely clear. On the other hand, some genera are obviously closely allied (*Lyroda, Paranysson, Plenoculus, Solierella,* and *Miscophus*). These are the most typical genera of the tribe, and they form a core around which the remaining genera can be grouped.

Lyroda probably is the most generalized member of the tribe. Its wing venation is not reduced, and the body is elongate. Prey consists of Orthoptera. The greatly reduced venation of *Miscophoides* and *Saliostethoides* ranks these genera among the most specialized in the tribe. *Sericophorus,* which provisions with flies, appears to be the most biologically advanced genus. Although the miscophin ancestors probably did not have notched mandibles, most of the genera in the Miscophini do. *Auchenophorus* and *Nitela* do not have notched mandibles and possibly never did have. The notch is absent in most *Solierella,* and presumably in this genus the notch is in the process of being "phased out" probably in response to the switch from ground nesting to twig nesting. The pygidial plate is present in most miscophin genera, but is absent in *Nitela, Auchenophorus, Sphodrotes, Miscophus, Saliostethus, Miscophoides,* and all but a few *Solierella.* In *Solierella* this loss may be due to a change to twig nesting, but some *Solierella* still nest in the ground although these use abandoned burrows. *Nitela* and *Auchenophorus* may never have had a pygidial plate. *Miscophus* nest in loose sandy soil and possibly have lost the need for a pygidial plate. Perhaps, however, *Miscophus* and most of the members of its phylogenetic branch have never had one. This hypothesis is supported by the fact that *Saliostethoides,* presumably one of the most advanced genera on this line has a pygidial plate. The habits of *Saliostethus* and *Miscophoides* are unknown.

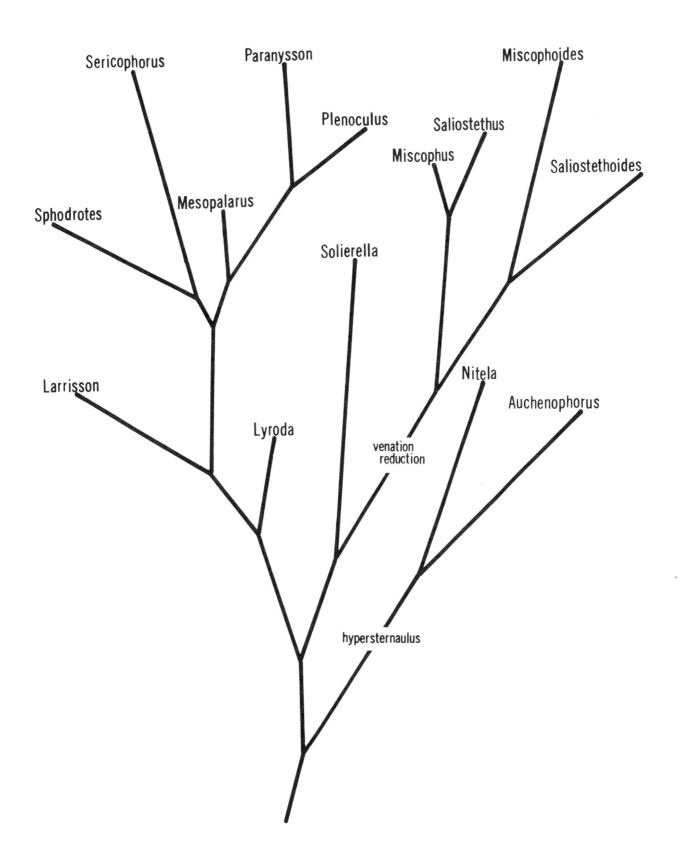

FIG. 82. Dendrogram suggesting relationships in the Miscophini.

Reduction of wing venation is confined primarily to two phyletic lines: *Miscophus* and its relatives, and *Nitela* and *Auchenophorus*. Reduction in body size seems to be correlated with loss of venation. Other specializations or peculiarities in the Miscophini are: 1) development of a long malar space (*Nitela*, some *Solierella*, some *Miscophoides*, one *Plenoculus*); 2) loss of hindwing jugal lobe (*Auchenophorus*, *Saliostethoides*, *Miscophoides*); 3) modification of male forecoxae and trochanter (*Solierella*); 4) loss of episternal sulcus (*Saliostethoides*, some *Saliostethus*, and *Miscophoides*); 5) hindcoxa with prong or tubercle (*Paranysson*); 6) modification of male sterna (*Larrisson*); 7) reduction in number of male flagellomeres (*Sericophorus*, some *Solierella*); 8) hypostoma with paramandibular process (*Auchenophorus*).

Biology: There are data for only eight of the genera in the Miscophini. *Lyroda, Sericophorus, Paranysson, Plenoculus, Miscophus, Sphodrotes* and some *Solierella* are ground nesters. *Nitela* and some *Solierella* are twig nesters. Prey ranges from spiders to Diptera. The most interesting biology is found in *Sericophorus* which utilizes flies. In one instance, Rayment (1955b) reported the use of one burrow by two females.

Key to genera of Miscophini

1. Forewing with three submarginal cells (figs. 83 A-E, 84 A-C) 2
   Forewing with two, one, or no submarginal cells (figs. 83 F, 84 D-K, 85 A-H) 8
2. Pronotal collar with three dorsal prominences (fig. 86 E); submarginal cell II trapezoidal in shape, four to six sided, not petiolate (fig. 83 A) (rarely three sided or triangular), and receiving both recurrent veins or first interstitial .... *Lyroda,* Say p. 295
   Pronotal collar arcuate or flat dorsally; second submarginal cell three sided, triangular, often petiolate (fig. 83 B-D) (rarely four sided, but if so still basically triangular and first recurrent vein received by first submarginal cell (fig. 83 E) 3
3. Genera from Australia and New Guinea 4
   From the New World and from the Palearctic, Oriental, and Ethiopian Regions in the Old World 6
4. Second submarginal cell petiolate (fig. 83 D); both recurrent veins received by second submarginal or first interstitial; second gastral sternum elongate, angulate anteriorly in profile (fig. 88 B) .... *Sphodrotes* Kohl, p. 204
   Second submarginal cell not petiolate (fig. 83 B,E); first recurrent received by first submarginal cell 5
5. Occipital carina ending at hypostomal carina; male (seven gastral terga) antenna with 12 articles; male sternum II simple, without a transverse flange ....
   .... *Sericophorus* F. Smith, p. 299
   Occipital carina disappearing before reaching hypostomal carina; male antenna with 13 articles (fig. 87 B); male sternum II with a thick transverse flange (fig. 88 D) (female unknown) ....
   .... *Larrisson* Menke, p. 304
6. Externoventral margin of mandible usually simple (fig. 92 E) (weakly angulate in a few species, fig. 92 F); frons with a V-shaped swelling which often bears a V-shaped carina (fig. 91 A,B), swelling when viewed from above accented by arrangement of facial pubescence (fig. 98 F); pygidial plate absent in both sexes (fig. 92 H) (feebly indicated in a South American species); male foretrochanter concave or hollowed posterobasally (figs. 92 A-C) and forecoxa often with a posterior process (fig. 92 B); outer side of hindtibia spineless or with three or four widely spaced spines (fig. 92 J) ....
   .... *Solierella* Spinola, p. 311
   Externoventral margin of mandible notched or angulate (fig. 92 G); frons broadly swollen or flat, without a V-shaped swelling (fig. 91 C); pygidial plate present in female and most males, clearly defined by carinae in female (figs. 89 H, 92 I) (except in *Paranysson inermis*); male foretrochanter and coxa not modified (fig. 92 D); hindtibia usually with numerous spines in one or more rows (fig. 92 K) 7
7. Propodeal dorsum finely granulate, occasionally finely ridged; female hindcoxa without a ventral spine or tubercle; wings usually clear; occipital carina incomplete below; North and Centr. America, Mediterranean area, Transcapian area ....
   .... *Plenoculus* W. Fox, p. 308
   Propodeal dorsum coarsely areolate or reticulate (*Paranysson inermis* only has widely spaced longitudinal ridges) (fig. 89 C); female hindcoxa with a ventral tubercle or spinelike process (except in *P. inermis*) (fig. 86 F); wings usually strongly infumate; occipital carina meeting hypostomal carina; Ethiopian Region, Saudi Arabia to Burma ....
   .... *Paranysson* Guérin-Méneville, p. 306
8. Externoventral margin of mandible entire (fig. 98 E); forewing with only one submarginal and one discoidal cell (fig. 85 F-H); mesopleuron with hypersternaulus (fig. 99 A) 9
   Externoventral margin of mandible notched or angulate (fig. 92 G) *or* if not then forewing with two submarginal and two discoidal cells (fig. 83 F, 84 D); mesopleuron without a hypersternaulus 10
9. Hindwing with jugal lobe but without closed cells (fig. 85 F,G); mandibular socket not separated from oral fossa by forward extension of hypostoma (fig. 99 B); cosmopolitan ...... *Nitela* Latreille, p. 322

Hindwing without a jugal lobe but with medial and submedial cells (fig. 85 H); subdiscoidal cell of forewing closed behind (fig. 85 H); mandibular socket isolated from oral fossa by forward extension of hypostoma (fig. 99 C); Australia ....*Auchenophorus* Turner, p. 325

10. Forewing with two discoidal and two submarginal cells, the second submarginal four sided, not petiolate (figs. 83 F; 84 D) 11

    Forewing with two, one, or no discoidal and submarginal cells, second submarginal (if present) three sided and petiolate (figs. 84 E-K; 85 A-E) 13

11. Second submarginal cell receiving both recurrent veins or first interstitial (fig. 83 F); antennal sockets not contiguous with frontoclypeal suture (fig. 87 D); female with a pygidial plate (fig. 89 F); (male unknown); S. Africa .... *Mesopalarus* Brauns, p. 305

    Second submarginal cell receiving only second recurrent vein, first received by first submarginal (fig. 83 C)........................... 12

12. Female with a pygidial plate; male (seven gastral terga) antenna with 12 articles, Australia ....
    .... *Sericophorus frontale* (Turner)

    Female without a pygidial plate; male antenna with 13 articles; America and Mediterranean area ....
    .... some *Solierella* Spinola, p. 311

13. Forewing with one or two submarginal cells, and marginal cell usually closed apically (figs. 84 E-K; 85 A-C); hindwing with a jugal lobe (fig. 84 F-I, 85 A-B); Old World and New World................................. 14

    Forewing without submarginal cells and marginal cell open apically (fig. 85 D-E); hindwing without a jugal lobe (fig. 85 D-E); s. African 15

14. Midcoxae contiguous behind or nearly so (fig. 94 A); precoxal sulcus strong (fig. 94 A); free clypeal margin usually excised laterally (fig. 91 D); episternal sulcus present; forewing variable (figs. 84 E-K, 85 B) but commonly with two submarginal cells and one or two discoidals; Old and New World......... *Miscophus* Jurine, p. 314

    Midcoxae usually widely separated (fig. 94 B); precoxal sulcus usually absent or weak (fig. 94 B); free clypeal margin entire (fig. 97 C); episternal sulcus sometimes absent; forewing with only one submarginal and one discoidal cell (fig. 85 A-C); s. Africa .... *Sadiostethus* Brauns, p. 319

15. Forewing without discoidal cells (fig. 85 E); female without a pygidial plate ....
    .... *Miscophoides* Brauns, p. 322

    Forewing with one discoidal cell (fig. 85 D); female with a triangular pygidial plate bounded by carinae (fig. 89 E) ....
    .... *Saliostethoides* Arnold, p. 320

Genus Lyroda Say

Generic description: Inner orbits essentially parallel or converging above (fig. 86 D); antenna elongate, scape usually short and stout, flagellomeres longer than broad, those of male unmodified; clypeal free margin with several teeth laterally (and sometimes mesally) in female, margin with a median truncate or weakly toothed lobe in male; labrum quadrangular, apex sometimes visible; inner margin of mandible with two subbasal teeth in female, none in male, externoventral margin notched or angulate in both sexes; mouthparts short; occipital carina incomplete, disappearing short of hypostomal carina; collar moderately long, usually trituberculate dorsally (fig. 86 E); admedian lines of scutum nearly contiguous or widely separated; propodeum long, dorsal enclosure not defined, dorsum usually with a median longitudinal carina, lateral carina sometimes present; mesopleuron with scrobal sulcus; gaster sessile except pedunculate in two species, both sexes with a pygidial plate defined by carinae, surface densely or sparsely covered with short setae; legs long, slender, foreleg with a weak tarsal rake in both sexes, spines few; midcoxae separated, hindcoxae contiguous; hindtibia with two parallel rows of spines, the uppermost borne on a weak ridge, arolia large in both sexes; marginal cell of forewing truncate, appendiculate (fig. 90 A); three submarginal cells, second submarginal cell four to six sided (rarely three sided), not petiolate, and receiving both recurrent veins or first recurrent interstitial; length of hindwing jugal lobe equal to or less than half of anal area; male sternum VIII as in fig. 90 L; volsella well differentiated (fig. 90 A).

Geographic range: Represented in all regions, but in the Palearctic *Lyroda* is known only from Japan. Eighteen species have been described.

Systematics: These are elongate wasps with slender legs. The usually four-sided, non-petiolate second submarginal cell (fig. 83 A), the rather long pronotal collar with its three dorsal prominences (fig. 86 E) and the long propodeum separate *Lyroda* from other miscophins.

The rather unspecialized wing venation, elongate body, trace of a spiracular groove in some species, and primitive biology (orthopteran prey and ground nests) suggest that *Lyroda* is perhaps the least specialized of the miscophin genera and possibly the most primitive genus in the Larrinae. In any case, *Lyroda* probably comes closest to typifying the progenitor of the various miscophin genera, and likewise suggests an excellent archetype for the subfamily. The few widely distributed species offers further evidence that the genus is old and on the decline. Phylogenetically, *Lyroda* leads most conveniently to the *Paranysson-Plenoculus-Solierella-Miscophus* line, but the large arolium, form of the clypeus and non-petiolate second submarginal cell indicate a relationship with *Sericophorus* also.

The tribal placement of *Lyroda* has a varied history. Some authors (Dalla Torre, 1897; Handlirsch, 1925; Arnold, 1936) have associated it with the Larrini, although the normal ocelli and small hindwing jugal lobe

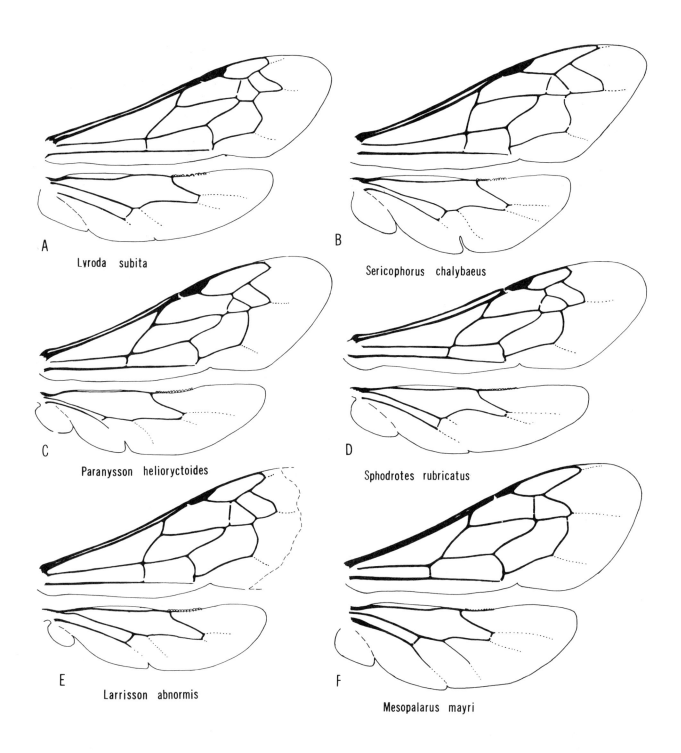

FIG. 83. Wings in the tribe Miscophini.

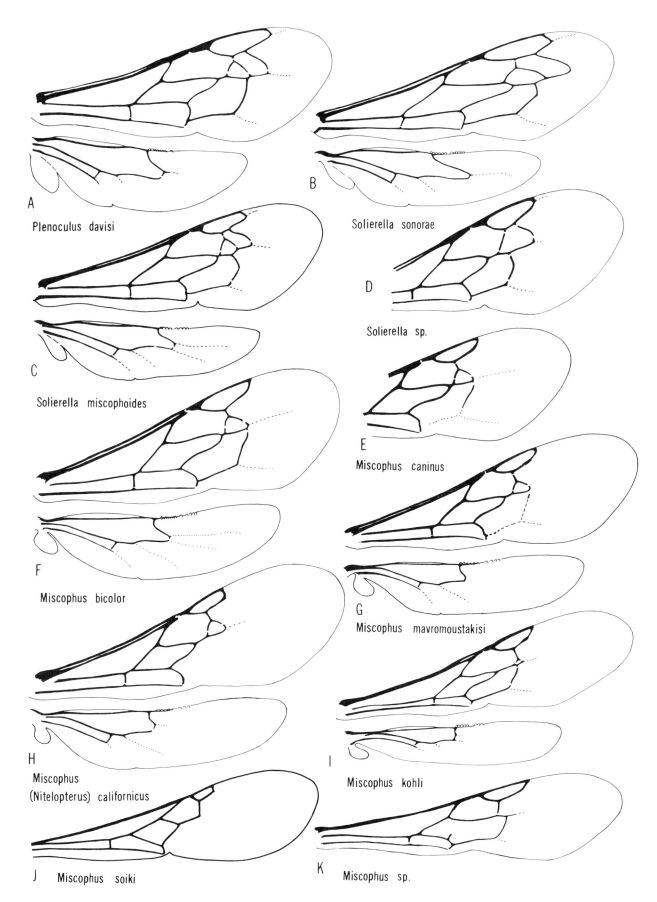

FIG. 84. Wings in the tribe Miscophini.

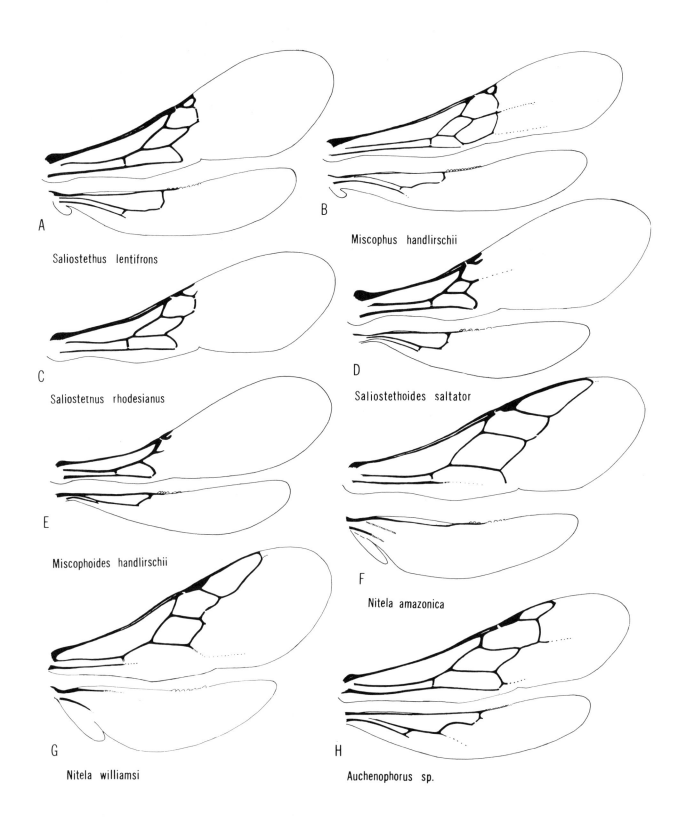

FIG. 85. Wings in the tribe Miscophini. E = holotype.

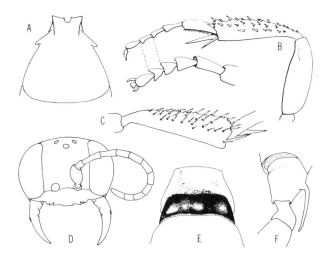

FIG. 86. Structural details in the tribe Miscophini; A, *Sericophorus (Zoyphidium) sericeus* holotype, tergum I; B, *Larrisson abnormis* male, right hindleg, lateral and dorsal; C, *Paranysson melanopyrus* female, left hindtibia, lateral; D, *Lyroda subita,* face; E, *Lyroda subita,* pronotal collar; F, *Paranysson melanopyrus* female, right hindcoxa, ventral.

clearly negate this assignment. Ashmead (1899) put *Lyroda* in a heterogeneous group he termed the Lyrodinae. Fox (1894d) placed *Lyroda* in a monotypic tribe, the Lyrodini, in his broad subfamily Bembicinae. Krombein (1951) assigned the taxon to the Miscophini. This last placement is clearly the most logical and is well supported morphologically.

A few structural variables are worth discussion. In *venusta* the second submarginal cell is (always?) triangular but not petiolate. In at least two Australian species, *minima* and *queenslandensis,* the first recurrent vein is interstitial. In another Australian species (apparently undescribed) the marginal cell is very short, its outer veinlet being interstitial with the outer veinlet of the third submarginal cell. The marginal cell has a very long appendiculation. In this species and *minima,* the inner orbits definitely converge towards the vertex. The first gastral segment broadens gradually in some species (*aethiopica, venusta*) giving the gaster a pedunculate appearance. Tergum I has laterobasal carinae (not to be confused with lateral carina delimiting laterotergite), which may converge towards the posterior margin of the tergum or extend transversely across its base. These carinae appear to offer good specific differences.

There is no key that includes all species. The North American forms were keyed by Fox (1893b) and Williams (1914a).

Biology: These wasps are ground nesters and provision with Orthoptera of the families Gryllidae (used by *L. subita*) and Tetrigidae (used by *L. japonica, formosa,* and *madecassa*). The wasp grasps the antennae of the prey with her mandibles (Patton, 1892; Iwata, 1963, 1964a) and flies to the nesting site either directly (Iwata, 1963)

or in short, gliding flights (Iwata, 1964a). Tsuneki and Iida (1969) stated that *L. japonica* carries prey with the midlegs and does not use the antennae of the prey. According to Evans (1964a), *L. subita* walks or flies with prey. Evans was of the opinion that *L. subita* always nests in preexisting burrows or cavities. The burrow of *Lyroda* is rather long and deep, with the cell 15-30 cm below the soil surface (Iwata, 1938a, 1964a; Peckham and Peckham, 1898; Evans, 1964e; Tsuneki and Iida, 1969). The entrance is left open while the wasp is away (Tsuneki, 1969a). Mass provisioning is the rule (Iwata, 1964a), although the Peckhams account (1898) suggests progressive provisioning in *L. subita.* They dug up the nest of one female just after she had dragged in a cricket. In the cell they found the wasp, three crickets, and a wasp larva "a day or two old." Tsuneki and Iida (1969) found that the egg was laid on the last provision in *L. japonica.* Evans (1964e) found one nest of *L. subita* with two cells, and Tsuneki and Iida (1969) noted the same for *L. japonica.* Evans also found that *Lyroda* nests were parasitized by miltogrammine flies

Checklist of *Lyroda*

aethiopica Kohl, 1894; Guinea, Rhodesia
argenteofacialis (Cameron), 1889 (*Astata*); India (? = *formosa*)
concinna (F. Smith), 1856 (*Morphota*); Brazil
fasciata (F. Smith), 1856 (*Morphota*); Brazil
formosa (F. Smith), 1859 (*Morphota*); India, Sikkim, Burma, Philippines, Celebes, Solomons
    nufiventris Cameron, 1900 (*Odontolarra*)
harpactoides (F. Smith), 1856 (*Morphota*); Brazil
japonica Iwata, 1933; Japan
    ssp. *takasago* Tsuneki, 1967; Taiwan
madecassa Arnold, 1945; Madagascar
michaelsensi Schulz, 1908; Australia
    ssp. *tasmanica* Turner, 1914; Australia: Tasmania
minima Turner, 1936; Australia
nigra (Cameron), 1904 (*Odontolarra*); Sikkim
queenslandensis Turner, 1916; Australia
salai Giner Marí, 1945; India
subita (Say), 1837 (*Lyrops*); N. America
    arcuata F. Smith, 1856 (*Larrada*)
    cockerelli Rohwer, 1909
taiwana Tsuneki, 1967; Taiwan
tridens (Taschenberg), 1870 (*Morphota*); Brazil
triloba (Say), 1837 (*Lyrops*); N. America
    caliptera Say, 1837 (*Lyrops*), nomen nudum or lapsus for *triloba*
venusta Bingham, 1897; Burma, Borneo, ? Philippines

Genus Sericophorus F. Smith

Generic description: Inner orbits slightly converging above or parallel (fig. 87 A); antenna short, stout, clavate, terminal flagellomeres usually broader than long, both sexes with 12 articles, those of male sometimes modified; clypeal free margin with broad median lobe which is bounded laterally by two teeth in female (deeply emarginate in *clypeatus*), male without lateral teeth; labrum short, broad,

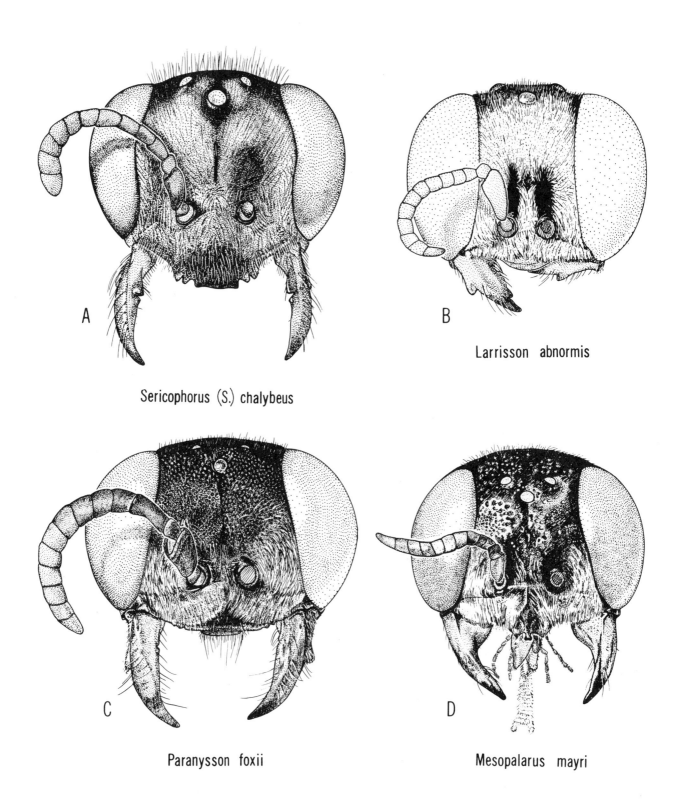

FIG. 87. Facial portraits in the tribe Miscophini; A,D, females; B,C, males. C = holotype. D = paratype

usually hidden; inner margin of mandible with two subbasal teeth, externoventral margin with a subbasal notch; mouthparts short; occipital carina joining hypostomal carina just before the latter's apex; collar short; scutum with or without notauli, admedian lines well separated; propodeum short, dorsal enclosure not defined, dorsum with a median longitudinal foveolate sulcus (subgenus *Sericophorus*, fig. 89 B), or a carina (*Zoyphidium*, fig. 89 D), propodeum with a lateral carina (*Sericophorus*), or not (*Zoyphidium*); posterolateral corner of propleuron sometimes tuberculate, margin sometimes lamellate; mesopleuron sometimes with scrobal sulcus; female with a pygidial plate which is bounded by a carina at least apically, surface densely clothed with setae, male tergum VII flattened and setose; legs short, stout, foreleg with a tarsal rake composed of short spines in both sexes (absent or weak in some males); midcoxae separated, hindcoxae contiguous; hindtibia with long spines in rows, or spines scattered and sometimes few; tarsomere V usually inflated (fig. 93 A), arolium moderate to large, that of foreleg often larger than those on other legs; marginal cell of forewing narrowly truncate and appendiculate apically (*Sericophorus*, fig. 83 B), or acuminate (*Zoyphidium*); three submarginal cells (two in *frontale*, the second four sided), II three sided (sometimes four sided but still basically triangular) and receiving second recurrent vein, first recurrent received by I; length of hindwing jugal lobe equal to or less than half length of anal area; male sternum VIII as in fig. 90 J, genitalia with very small volsella (fig. 90 B).

Geographic range: Restricted to Australia, Tasmania, and New Guinea. Sixty-seven species are currently recognized. No species are recorded from New Guinea in the literature, but we have seen one male from Nadzab (U.S. National Museum).

Systematics: Sericophorus generally are robust, beelike wasps although some species are elongate. Some are metallic blue or green with red legs, while others are black with red legs and red areas on the thorax and abdomen. The nonpetiolate second submarginal cell (fig. 84 B), reception of the first recurrent vein by submarginal I, and form of the occipital carina (joins hypostomal carina) separate *Sericophorus* from its closest relatives, *Lyroda, Paranysson,* and *Sphodrotes.* The ten flagellomeres of the male antenna are found also in some *Solierella* and *Trypoxylon.* This feature plus the fact that some species have the beginnings of what appear to be orbital foveae (*bicolor, relucens*) suggests that *Sericophorus* is a very advanced group. The biology supports this view.

Rayment (1955a) suggested that *Sericophorus* was the closest sphecid link with the bees, and some species are quite beelike in appearance. According to Michener (1944) some colletid bees (the most primitive bee family) have orbital foveae. Thus, *Sericophorus* may be close to the sphecid prototype which gave rise to the bees.

In the past *Zoyphidium* has been recognized as a genus. Turner (1914) and Cockerell (1932) declared that it could be separated from *Sericophorus* because the marginal cell was not truncate and appendiculate. Cockerell discussed subsidiary characters but pointed out that they were variable. We have come to the conclusion that *Zoyphidium* is at most a subgenus of *Sericophorus.* The marginal cell differences are slight, and we have been unable to find other constant characters to support the separation of *Zoyphidium* as a genus. The following key to the subgenera of *Sericophorus* will illustrate the rather tenuous differences.

Key to subgenera of *Sericophorus*

Marginal cell narrowly truncate, appendiculate (fig. 83 B); propodeum usually with lateral carina (fig. 89 B); propodeal dorsum usually with a broad foveolate median sulcus (fig. 89 B); media of forewing often in same plane as cu-a, usually diverging from M+Cu at or beyond cu-a (fig. 83 B) .... subgenus *Sericophorus*

Marginal cell acuminate; propodeum usually without lateral carina; propodeal dorsum usually with a narrow median sulcus containing a longitudinal carina (fig. 89 D); media of forewing usually not on same plane as cu-a and usually diverging from M+Cu before cu-a ....
.... subgenus *Zoyphidium*

Sericophorus sericeus, the type species of *Zoyphidium,* has gone unrecognized since its description by Kohl. We have borrowed it from the Humboldt University Museum in Berlin. The spiracles of tergum I are borne on peculiar backward projecting thornlike processes (fig. 86 A), but *sericeus* is congeneric with the remaining species assigned to *Zoyphidium. Sericophorus sericeus* and several other species have two basal ridges on tergum I as well as prominences of one kind or another on sternum I. Modifications of the posterolateral area of the propleuron have been noted in *sericeus* and a few other *Sericophorus,* also. Other useful specific differences are more obvious: clypeal details, presence or absence of a median scutellar prominence, sculpture of the propodeum, and the form of the antenna. We have not seen specimens of *frontalis,* the only species of *Sericophorus* reported to have two submarginal cells. Turner's description indicates other differences from the rest of *Sericophorus,* and *frontalis* may not be congeneric.

Turner (1914) provided the first key to *Sericophorus.* Hacker and Cockerell (1922) described new species of *Zoyphidium* and gave a key. Rayment (1955a) published an extensive paper on *Sericophorus* in which he described a great many new species and two new genera, one of which, *Anacrucis,* is a synonym of *Zoyphidium.* The other genus, *Astaurus,* which he considered a link between bees and *Sericophorus,* is a synonym of the gorytine genus *Clitemnestra*[17]. Unfortunately, Rayment's keys to species are almost useless, and furthermore, he was apparently unaware of Hacker's and Cockerell's papers and some of Turner's species.

Biology: Rayment (1953, 1955a, b, 1959) reported on a number of species but his dialogues are sometimes confusing. The best account of *Sericophorus* biology is by

Matthews and Evans (1971) who summarized some of Rayment's work and corrected the names of his prey. *Sericophorus* species provision with muscoid flies and this has earned them the name "Policemen Flies."

Some *Sericophorus* have unusual behavioral traits. *S. viridis* and *teliferopodus* excavate their nests during the predawn hours and hunt prey during the first hour after sunrise. More intriguing is the fact that these two species catch only male flies. A possible explanation of this was offered by Matthews and Evans. The most commonly caught prey of *viridis, Calliphora tibialis* Macquart, is related to calliphorids that are very active during the early morning. Males perch and challenge passing insects, but females are usually crawling on the ground or moving from place to place. Thus a *Sericophorus* female is apt to capture male flies, especially because they engage in aerial combat with intruders. Rayment (1955b) reported that two females of *S. victoriensis* utilized a single nest, and that *viridis roddi* mated within the nest, but Matthews and Evans were unable to confirm this behavior in their field work. Rayment also noted territorial behavior in *sydneyi* females.

Sericophorus viridis and *teliferopodus* nest in colonies but *relucens* is essentially a solitary nester. The main shaft of the burrow is vertical and cells are placed in a radiating pattern at the ends of lateral, horizontal tunnels. Nests of *relucens* are 8 to 13 cm deep and have only one or two cells. *S. viridis* nests have up to 23 cells, and the main tunnel may reach a depth of 50 cm. The nest entrance apparently is never closed but individual cells are sealed when completely stocked. The nest opening is usually surrounded by a ring of excavated soil, and according to Rayment the large female arolia are used as "scoops" during excavation.

Prey are flown to the nest venter up and only the hindlegs of the wasp hold the prey. The flies are dead or deeply paralized. Three to 12 flies are placed venter up in a cell. The egg is laid on the prothoracic venter of the first provision, but only after mass provisioning is completed. *S. relucens,* unlike other species studied, hunts prey throughout the day and both sexes of flies are provisioned. There is also less prey specificity in this species. Thirteen species of Diptera representing four families were utilized by *relucens* according to Matthews and Evans.

The nyssonin *Acanthostethus portlandensis* (Rayment) is apparently a cleptoparasite of *Sericophorus*. Miltogrammine sarcophagids, chloropids, and mutillids have also been implicated as parasites of *Sericophorus* nests.

Checklist of *Sericophorus*

Subgenus *Sericophorus*

aliceae Turner, 1936; W. Australia
bicolor F. Smith, 1873; sw. and centr. Australia
brisbanensis Rayment, 1955; e. Australia
carinatus Rayment, 1955; se. Australia
castaneus Rayment, 1955; e. Australia
chalybaeus F. Smith, 1851; sw. and se. Australia, Tasmania
    cyaneus Saussure, 1854 (*Tachyrrhostus*)
  ssp. *fulleri* Rayment, 1955; Canberra
claviger (Kohl), 1892 (*Tachyrrhostus*); se. Australia
  ssp. *burnsiellus* Rayment, 1955; Victoria
cliffordi Rayment, 1955; Australia: Victoria
cockerelli Menke, new name; se. Australia
    hackeri Rayment, 1955, nec *hackeri* Cockerell, 1932
cyanophilus Rayment, 1955; e. Australia
elegantior Rayment, 1955; W. Australia
froggatti Rayment, 1955; se. Australia
funebris Turner, 1907; ne. Australia
gracilis Rayment, 1955; e. Australia
inornatus Rayment, 1955; e. Australia
lilacinus Rayment, 1955; e. Australia
littoralis Rayment, 1955; S. Australia
metallescens Rayment, 1955; se. Australia
minutus Rayment, 1955; se. Australia
nigror Rayment, 1955; S. Australia
niveifrons Rayment, 1955; e. Australia
occidentalis Rayment, 1955; sw. Australia
patongensis Rayment, 1955; se. Australia
pescotti Rayment, 1955; se. Australia
raymenti Menke, new name; Australia: Tasmania
    rufipes Rayment, 1955, nec *rufipes* Rohwer, 1911
relucens F. Smith, 1856; Australia: widespread
  ?*rufipes* Rohwer, 1911 (*Zoyphium*)
  ssp. *nigricornis* Rayment, 1955; Australia: scattered records across continent
  ssp. *ruficornis* Rayment, 1955; Australia: scattered records across continent
rufobasalis Rayment, 1955; e. and se. Australia
rufotibialis Rayment, 1955; Australia: Canberra
rugosus Rayment, 1955; e. Australia
sculpturatus Rayment, 1955; W. Australia
spryi Rayment, 1955; se. Australia
subviridis Rayment, 1955; se. Australia
sydneyi Rayment, 1955; W. Australia
tallongensis Rayment, 1955; se. Australia
teliferopodus Rayment, 1955; se. Australia
  ssp. *okiellus* Rayment, 1955; Victoria: Melton
victoriensis Rayment, 1955; se. Australia
violaceus Rayment, 1955; S. Australia
viridis (Saussure), 1854 (*Tachyrrhostus*); sw. and se. Australia
  ssp. *roddi* Rayment, 1955; New S. Wales

Subgenus *Zoyphidium*

affinis (Hacker & Cockerell), 1922 (*Zoyphium*); e. Australia
argyreus (Hacker & Cockerell), 1922 (*Zoyphium*); e. Australia
asperithorax (Rayment), 1955 (*Anacrucis*); W. Australia
cingulatus (Rayment), 1955 (*Anacrucis*); se. Australia
clypeatus (Rayment), 1955 (*Anacrucis*); se. Australia
collaris (Hacker & Cockerell), 1922 (*Zoyphium*); e. Australia
crassicornis (Cockerell), 1914 (*Zoyphium*); e. Australia
dipteroides Turner, 1912; ne. Australia
doddi (Turner), 1912 (*Zoyphium*); ne. Australia

[17] Edgar Riek verified this synonymy for us after examining Rayment's types.

emarginatus (Hacker & Cockerell), 1922 (*Zoyphium*); e. Australia
erythrosoma (Turner), 1908 (*Zoyphium*); ne. and e. Australia
ferrugineus (Rayment), 1955 (*Anacrucis*); se. Australia
flavofasciatus (Turner), 1916 (*Zoyphium*); e. Australia
frontalis (Turner), 1908 (*Zoyphium*); e. Australia
fuscipennis (Hacker & Cockerell), 1922 (*Zoyphium*); Australia: Tasmania
hackeri (Cockerell), 1932 (*Zoyphium*); Australia: Queensland
humilis (Cockerell), 1932 (*Zoyphium*); Australia: Queensland: Bribie I.
iridipennis (Turner), 1914 (*Zoyphium*); Australia: Tasmania
kohlii (Turner), 1908 (*Zoyphium*); e. Australia
laevigatus (Rayment), 1955 (*Anacrucis*); se. Australia
niger (Hacker & Cockerell), 1922 (*Zoyphium*); e. Australia
ornatus (Hacker & Cockerell), 1922 (*Zoyphium*); e. Australia
punctuosus (Rayment), 1955 (*Anacrucis*); w. Australia
pusillus (Hacker & Cockerell), 1922 (*Zoyphium*); e. Australia
rufonigrus (Turner), 1908 (*Zoyphium*); n. centr. Australia
sericeus (Kohl), 1893 (*Zoyphium*); S. Australia
splendidus (Hacker & Cockerell), 1922 (*Zoyphium*); e. Australia
striatulus (Rayment), 1955 (*Anacrucis*); se. Australia
tuberculatus (Turner), 1936 (*Zoyphium*); W. Australia

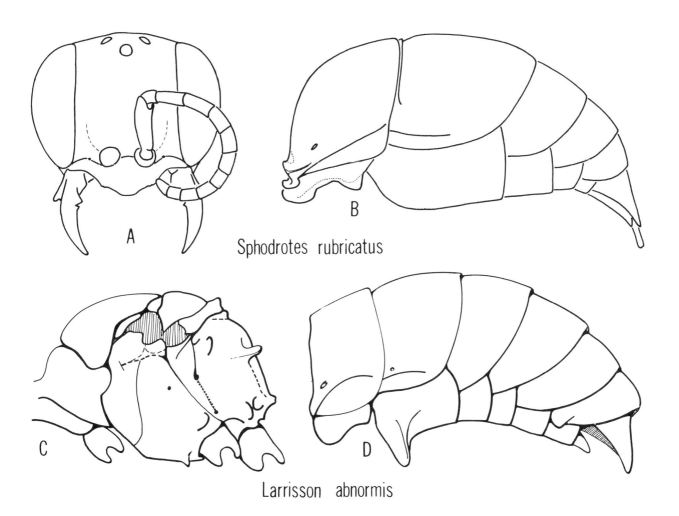

FIG. 88. Structural features in the tribe Miscophini; A, face; B,D, gaster, lateral; C, thorax, lateral.

Genus Sphodrotes Kohl

Generic description: Inner orbits parallel or very slightly converging below (fig. 88 A); antenna moderately long, scape long, flagellomeres longer than broad, those of male unmodified; frons broadly, shallowly depressed above antennal sockets forming a scapal basin; clypeal free margin with a broad median lobe which is often bounded laterally by one or more small teeth; labrum rounded or quadrangular, hidden; inner margin of mandible with two subbasal teeth, externoventral margin notched or angulate; mouthparts short; occipital carina incomplete, disappearing before reaching hypostomal carina; collar short; admedian lines of scutum nearly contiguous; propodeum moderately long, dorsal enclosure not defined but dorsum with a median longitudinal carina and surface foveolate, propodeum usually with a posterolateral tubercle; prepectus with a preomaulal area although omaulus absent; tergum I with a strongly raised, angulate, laterobasal carina mesad of lateral carina, sternum I with a posteromedian prominence, sternum II usually angular in profile basally (fig. 88 B); neither sex with a pygidial plate although female tergum VI is triangular and somewhat flattened; legs slender, moderately long, female with a foretarsal rake, spines widely spaced, male with a weak rake; midcoxae separated, hindcoxae contiguous; hindtibia with one or more rows of spines; tarsi somewhat flattened, tarsomere V not swollen, arolia moderately large; marginal cell of forewing truncate, but not or only weakly appendiculate (fig. 83 D); three submarginal cells, second petiolate and receiving both recurrent veins (first recurrent interstitial in *pilosellus*); length of hindwing jugal lobe equal to or less than half length of anal area; male sternum VIII as in fig. 90 M; volsella not clearly differentiated from gonostyle (fig. 90 C).

Geographic range: Restricted to Australia and Tasmania. Only six poorly known species have been described.

Systematics: These wasps have partially or totally red legs and the gaster is often extensively red. The very coarse, dense punctation of the entire body, the presence of a scapal basin, and the long angular form of the second sternum (fig. 88 B) are distinctive features. The parallel inner orbits (fig. 88 A) and the petiolate second submarginal cell which receives both recurrents (fig. 83 D) are additional recognition features. Probably the genus was derived from a *Sericophorus*-like ancestor, but *Sphrodrotes* does not appear to be closely allied with any miscophin genus. Specific differences are fairly numerous. We have noted a few that have not been previously utilized: the structure of sternum I, the form of the laterobasal carina of tergum I, and the form of the gena. *S. cygnorum* appears to be the most distinctive species. The gena is broad giving the head a "square" appearance dorsally, the dorsal surface of the pronotal collar is flat, and the mesopleuron is sharply angulate anteriorly, delimiting a preomaulal area. Other species have such an angulation, but it is not as strong.

Turner (1914) provided a key for the species, but some are known from only one sex.

Biology: Evans (1973c) reported on *nemoralis*. Nests are multicellular, 45 to 52 cm deep, and the entrances are left open at all times. The pentatomoid bug prey were flown to the nest. The bugs were stored at the end of the main burrow of the nest and then a cell was excavated. Six to 15 bugs were placed in a cell.

Checklist of *Sphodrotes*

cygnorum Turner, 1910; Australia
marginalis Turner, 1914; Australia
nemoralis Evans, 1973; Australia
pilosellus Turner, 1910; Australia
punctuosus Kohl, 1889; Australia, Tasmania
rubricatus Turner, 1910; Australia

Genus Larrisson Menke

Generic diagnosis: (Based on male only) Inner orbits diverging above and below (fig. 87 B); scape elongate but flagellomeres about as long as wide except terminal article longer than wide; clypeus with a prominent truncate median lobe (fig. 87 B); labrum short, rounded, hidden; mandible with an inner subapical tooth and a mesal angulation, externoventral margin with a mesal notch; mouthparts short; occipital carina incomplete, disappearing just before joining hypostomal carina; collar short; admedian lines of scutum well separated; propodeum very short, dorsum without enclosure, propodeum with a posterolateral prong (fig. 88 C); mesopleuron with a sharp tubercle in front of midcoxae (fig. 88 C); tergum I concavely truncate basally, tergum VII with a flat pygidial plate which is punctate, hairless and truncate apically but not margined by carinae; sternum II with a thick transverse flange (fig. 88 D); legs short and stout, foreleg with a tarsal rake composed of moderately long, widely spaced spines; midcoxae widely separated, hindcoxae essentially contiguous; hindfemur with an outer subapical setose ridge (fig. 86 B); hindtibia with three parallel rows of stout spines (fig. 86 B); tarsomeres II-V of all legs strongly dorsoventrally compressed (fig. 86 B); tarsomere V not enlarged, arolia moderate, marginal cell of forewing apically truncate, appendiculate, three submarginal cells, II triangular but not petiolate and receiving second recurrent vein, first recurrent received by submarginal I (fig. 83 E); length of hindwing jugal lobe equal to or less than half length of anal area (fig. 83 E); sternum VIII thickened, spatulate (fig. 90 G); volsella elongate, gonostyle with inner accessory lobes (fig. 90 F).

Geographic range: The single species, *abnormis*, is known only from western Australia.

Systematics: This poorly known genus appears to have no close relatives within the Miscophini. The restricted range of the genus coupled with certain morphological features considered as indicators of primitiveness in the Miscophini (nonpetiolate second submarginal cell, large volsella, subapical mandibular teeth, male tarsal rake and pygidial plate) suggest that *Larrisson* is a relic. Nonetheless, the genus displays a number of specializations, some of which are unique or of rare occurrence in the Miscophini: propodeum very short and bearing a posterolateral process, transverse flange on sternum II of male, tarsomeres flattened, gonostyle with inner appendages, mandible notched.

Turner (1914) described *abnormis* in *Sericophorus* but noted that "a new genus will probably have to be erected when more material is available." The general build of *Larrisson* is unlike *Sericophorus* or *Sphodrotes*, its only possible Australian miscophin relatives. The form of the inner orbits, the subapical mandibular tooth, the incomplete occipital carina, the triangular second submarginal cell, the very short propodeum and its posterolateral process, and the flattened tarsi, when taken together, separate *Larrisson* from *Sericophorus* and *Sphodrotes*. Sternum VIII and the genitalia of male *Larrisson* are also quite different from these two genera.

Except for the head *Larrisson* is similar in most respects to the genus *Palarus* as pointed out by Menke (1967b), but these similarities are doubtless only examples of convergent evolution. When the female of *Larrisson* has been discovered, and some insight into the biology of the genus has been gained, perhaps it will be possible to more accurately determine the affinities of this interesting taxon. Likewise, when the female is known, certain features such as the subapical mandibular tooth, the mesopleural tubercle, the outer subapical setose ridge on femur III, and the flange on sternum II, may prove to be restricted to the male. Also, a few of these characters may be specific differences only.

Biology: Unknown.

Checklist of *Larrisson*

abnormis (Turner), 1914 (*Sericophorus*); w. Australia

Genus Mesopalarus Brauns

Generic description: (Based on female only) Inner orbits converging above, antenna short, scape short, stout, flagellomeres about as long as wide (fig. 87 D); sockets low on face but separated from frontoclypeal suture by about one-third of a socket diameter (fig. 87 D); clypeus with a strongly raised and inverted Y-shaped carina in middle, free margin with a blunt central tooth and an angle laterally (fig. 87 D); labrum transverse, narrow, hidden; mandible without teeth, externoventral margin weakly and obtusely angulate subbasally; mouthparts moderately long, especially the glossa, but palpi short and slender; occipital carina incomplete disappearing before reaching hypostomal carina; collar very short, closely appressed to scutum; admedian lines of scutum obscured by punctures; propodeum short, dorsal enclosure V-shaped, its apex extending onto posterior face, its smooth surface contrasting with surrounding coarse sculpture, enclosure with a broad median sulcus and a transverse basal sulcus crossed by many ridges; margin of posterolateral angle of propleuron lamellate, projecting in a gibbous angle over ventral margin of pronotum; mesopleuron with scrobal sulcus, prepectus with preomaular area although omaulus absent; tergum VI with a triangular, hairless pygidial plate which is bounded by strongly raised carinae, surface with a median longitudinal ridge (fig. 89 F); legs short and stout, female foreleg with a well developed tarsal rake composed of long, fine, closeset setae, those of tarsomeres II-III originating on dorsum (fig. 93 B); midcoxae widely separated, hindcoxae contiguous or nearly so, hindtibia with two parallel rows of short, stout spines, the uppermost row borne on prominent tubercles, tarsomere V not enlarged, arolia moderate sized; marginal cell of forewing truncate, appendiculate, two submarginal cells, II four sided and receiving both recurrent veins or first recurrent interstitial (fig. 83 F); length of hindwing jugal lobe equal to or less than half of anal area (fig. 83 F).

Geographic range: Known only from southern Africa. Two species have been described in this presumably relict genus.

Systematics: Mesopalarus is superficially similar to *Paranysson* in color, size, and general build, but there are many basic differences: inner margin of mandible without teeth (fig. 87 D), antennal sockets not contiguous with frontoclypeal suture (fig. 87 D), mouthparts longer, occipital carina not reaching hypostomal carina, gena inflated, giving head a globose appearance, tarsal rake composed of silky, closeset hairs and those of distal tarsomeres originating dorsally (fig. 93 B), propodeal enclo-

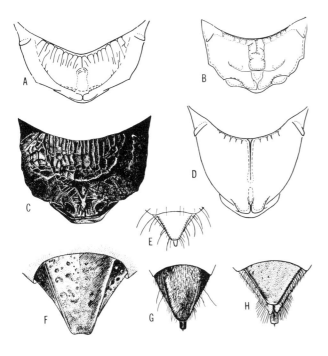

FIG. 89. Propodeum (A-D) and female pygidium (E-H) in the tribe Miscophini; A, *Mesopalarus mayri* paratype; B, *Sericophorus rufescens;* C, *Paranysson foxii* paratype; D, *Sericophorus (Zoyphidium) sericeus* holotype; E, *Saliostethoides saltator;* F, *Mesopalarus turneri;* G, *Lyroda subita;* H, *Paranysson melanopyrus.*

sure largely smooth, and forewing with only two submarginal cells, the second nonpetiolate (fig. 83 F).

The affinities of *Mesopalarus* are problematical, especially since the male is unknown. Certainly the mandible is not typically larroid. Rather, it resembles the mandible of some philanthine genera, and the overall structure of the head and its appendages is similar to some species of the philanthine genus *Pseudoscolia*. On the other hand, these two genera differ widely in some respects, and, unlike the Philanthinae, the mandible sockets are open in *Mesopalarus*. Arnold (1923b) regarded the genus as a close ally of *Paranysson,* but in view of the many differences this does not seem likely. Hopefully, the discovery of the male will shed more light on the relationships of *Mesopalarus*. Until then it seems best to consider the genus as an aberrant but specialized miscophin possibly distantly allied with *Paranysson*.

We have seen an example of each species and the two are very close structurally. *M. turneri* may prove to be only a color form of *mayri*.

Biology: Unknown.

Checklist of *Mesopalarus*

mayri Brauns, 1889; S. Africa
turneri Arnold, 1931; S. Africa

Genus Paranysson Guérin-Menéville

Generic description: Inner orbits converging above (fig. 87 C); antenna moderately long, scape short, stout, flagellomeres about as long as wide, those of male unmodified; clypeal free margin variable but usually with small sublateral teeth; labrum transverse, narrow, free margin sometimes visible; inner margin of mandible with two subbasal teeth, externoventral margin notched or angulate; mouthparts short; occipital carina joining hypostomal carina just before the latter's apex; collar short; scutum with notauli absent or poorly defined, admedian lines nearly contiguous or separated; propodeum short, dorsal enclosure usually well defined (fig. 89 C), roughly triangular, its apex extending onto posterior face, lateral propodeal carina present (except in *oscari* and *inermis*) although irregular, and often forming a tooth or angle, dorsal and posterior faces of propodeum areolate or reticulate (except in *inermis*); mesopleuron with scrobal sulcus, prepectus with a preomaular area although omaulus absent; both sexes with a hairless pygidial plate margined by carinae (except in *inermis*), that of male margined by carinae only at apex; legs short, stout, female foreleg with tarsal rake composed of long, stout, widely spaced spines, male with a weak rake; midcoxae widely separated, hindcoxae contiguous or nearly so; inner ventral side of hindcoxae with a long spinelike process in female (fig. 86 F) (except *helioryctoides* which has a small tubercle, and *inermis* which is plain), hindtibia with several parallel rows of short stout spines, the uppermost row borne on prominent tubercles (fig. 86 C); tarsomere V not swollen, arolia small to moderate; marginal cell of forewing truncate or rounded apically, appendiculate (fig. 83 C); three submarginal cells, II usually petiolate and receiving second recurrent vein, first recurrent received by I or interstitial; length of hindwing jugal lobe equal to, or less than half length of anal area; male sternum VIII as in fig. 90 I; volsella apparently absent, apex of gonostyle with an articulating setose process (fig. 90 E).

Geographic range: Africa, Aden, and Burma. More collecting in southern Asia will probably narrow the gaps in the present distribution pattern. Ten species are currently recognized, and all but one are from Africa.

Systematics: These compact, rather stout, beelike wasps are small to medium (5-15 mm long). *Paranysson* is easily recognized because of the distinctive and nearly uniform color pattern exhibited by the species. The head and thorax are black, the gaster and legs are red, and the wings are infumate. The single exception to this pattern is the only non-African species, *assimilis,* which differs in having clear wings and largely black legs. The areolate or reticulate propodeal enclosure (fig. 89 C) and the very uniform, close punctation of the head, scutum and mesopleuron (punctures nearly contiguous) are sculptural details which will usually distinguish *Paranysson* from its closest allies, *Plenoculus* and *Solierella*. These three genera have the same venational pattern but the hindtibia in *Paranysson* bears several parallel rows of short stout spines and the uppermost row is borne on prominent tubercles. The hindtibia of *Solierella* is nearly spineless. The hindcoxal process in female *Paranysson* is unique in the Miscophini (fig. 86 F). The reported absence of such a process in females of *assimilis* and *foxii* is unsubstantiated since examination of the types of both species has proven that they are males not females. The female of *Paranysson inermis* does lack a coxal process, however, and furthermore the propodeal dorsum is merely longitudinally ridged. The weakly defined pygidial plate of *inermis* is an additional aberrant feature of this species. The pygidial apex in *inermis* has a small V-shaped emargination in contrast to the acuminate or narrowly truncate pygidial apex of the other species. The articulating apical process on the gonostyle (fig. 90 E) is an interesting generic feature of *Paranysson*. The setose nature of this process may indicate that it is actually the volsella which has migrated distad.

Most keys to species are of limited usefulness because they are based on females only (Turner, 1914; Arnold, 1923b). However, Leclercq (1968a) has recently produced a more workable but still incomplete key. The sexes of some species have not been associated or described, and the types of at least two have been wrongly sexed, thus compounding this problem. Guiglia (1948) redescribed the type of *abdominalis*. Specific characters appear to be numerous, and the reviser of this genus should not have a difficult task. Beside the sculptural differences between species the clypeus, antenna, hindcoxa, and genitalia offer good diagnostic characters. The clypeal margin may be single or double edged (thickened) and the outline is distinctive. The first flagellomere may be elongate as in *ab-*

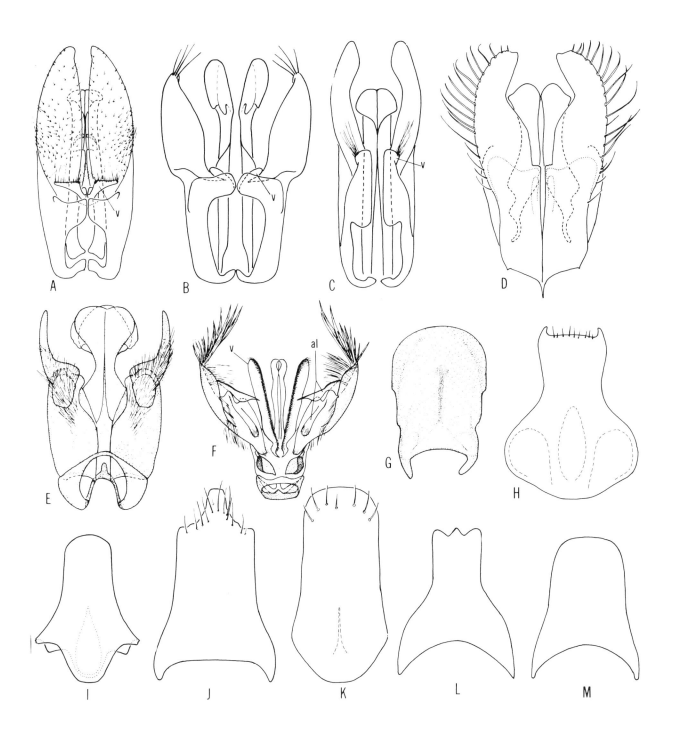

FIG. 90. Male genitalia (A-F) and sternum VIII (G-M) in the tribe Miscophini; A,L, *Lyroda subita;* B,J, *Sericophorus (Zoyphidium)* sp.; C,M. *Sphodrotes cygnorum;* D,K, *Plenoculus davisi;* E,I, *Paranysson foxii* holotype; F,G, *Larrisson abnormis;* H, *Solierella jaffueli;* v = volsella, al = accessory lobes.

dominalis and *foxii*. In the latter, this article is broadly shallowly depressed ventrally (fig. 87 C). The form of the coxal spine is distinctive in females and the hindcoxa in males of some species we have seen is distinctively modified. In *P. oscari* the palpi are very densely hairy. Arnold (1951) pointed out that the apical margin of the first sternum is truncate in *oscari* and *helioryctoides,* but the visibility of this character depends on whether or not segment II is fully extended. The mesosternal areas of the thorax appear to differ in some species we have studied. In *foxii,* for example, the midventral line is crossed by a V-shaped carina anteriorly (acetabular carina remnant?). The surface is scooped out on both sides of the V.

Paranysson was originally described as a subgenus of *Nysson*. This error was compounded by Cresson (1882), who assigned several North American *Zanysson* species to the genus; and was continued by Handlirsch (1887), Fox (1894d), Kohl (1896), and Ashmead (1899). Both Kohl and Ashmead recognized the genus as a larrine but under generic names that were then not known to be synonyms of *Paranysson,* namely *Helioryctes* and *Pseudohelioryctes*. Turner (1914) clarified the matter by synonymizing *Helioryctes* and *Pseudohelioryctes* with *Paranysson* and removing the genus from the Nyssoninae. He allied the genus with *Sericophorus* and *Sphodrotes,* which Kohl considered as isolated larroid genera. Dalla Torre (1897) proposed the subfamily Sericophorinae for these genera, but Turner suggested the name Paranyssoninae after the addition of *Paranysson*. In view of the morphological similarities between the sericophorine and miscophine genera, it is surprising that neither these nor later workers proposed combining the groups under one tribe.

Biology: Bequaert (1933) observed a colony of *Paranysson melanopyrus* in the Congo. Nests were dug in sandy soil and were very deep (75 cm). The main shaft terminated in several horizontal branches, each with a cell. Nymphal pentatomid bugs were provisioned, six or seven to a cell.

Checklist of *Paranysson*
(Gender masculine)

abdominalis (Guérin-Méneville), 1844 (*Nysson*); Senegal
*assimilis (Bingham), 1897 (*Helioryctes*) (syntypes = ♂♂); Aden, s. India, Burma
brevispinosus Arnold, 1929; Rhodesia, Zaire
bumbanus Leclercq, 1968; Zaire
*foxii (Ashmead), 1899 (*Pseudohelioryctes*); (type = ♂) Ethiopia
helioryctoides (Turner), 1912 (*Nysson*); Zambia
inermis Leclercq, 1968; Zaire
melanopyrus (F. Smith), 1856 (*Helioryctes*); Gambia, Sierra Leone, Liberia, Nigeria, Zaire, Uganda
   congoensis Arnold, 1929
oscari Turner, 1914; Liberia, Zaire, Zambia, S. Africa
   servus Arnold, 1929
quadridentatus (Cameron), 1910 (*Helioryctes*); Zaire, Zambia, Rhodesia, S. Africa

Genus Plenoculus W. Fox

Generic description: Inner orbits converging above (fig. 91 C); antenna short to moderately long, scape short, flagellomeres usually about as wide as long or wider than long, terminal article often more elongate, male flagellum sometimes modified and often enlarged beyond first two or three articles; frons broadly and evenly convex, without V or Y-shaped carinae, facial vestiture uniform, not reflecting a V-shaped pattern from above under various lights; clypeus often very narrow, free margin sinuate or with a truncate or angular median lobe which sometimes has a median notch, lateral teeth often present, clypeus sometimes dimorphic, males often with a lateral "hairbrush" (fig. 91 C); labrum transverse, sometimes bilobed; inner margin of mandible simple or with one or two subbasal teeth, externoventral margin notched or angulate; malar space absent except in *platycerus;* mouthparts short; occipital carina incomplete below and sometimes above; collar short; admedian lines of scutum separated, sometimes widely; propodeum short, dorsal enclosure defined by its lack of pubescence, no delimiting carinae except apically, enclosure triangular or U-shaped, apex extending onto posterior face, surface finely granulate, sometimes weakly ridged; propleuron usually simple, female and usually male with a flat hairless pygidial plate which is usually punctate, its apex rounded or truncate (fig. 92 I); legs moderately long, foretarsal rake present in both sexes although sometimes weak in both/or absent in male, composed of widely spaced stout or bladelike spines (fig. 93 F); midcoxae separated, hindcoxae contiguous; hindtibia variably spinose, spines usually long and widely spaced although in some species densely spined, in others feebly so (fig. 92 K); tarsomere V not swollen, arolia small to moderate; marginal cell of forewing truncate or rounded apically, appendiculate (fig. 84 A), three submarginal cells, II petiolate and receiving second recurrent vein, first recurrent received by submarginal I; length of hindwing jugal lobe equal to or less than half length of anal area; male sternum VIII as in fig. 90 K; volsella greatly reduced or absent, apex of aedeagus without teeth, apical half of gonostyle sometimes articulating with basal half (fig. 90 D).

Geographic range: North America to southern Mexico in the New World, and the Iberian Peninsula and southwest USSR in the Old World. Eighteen species are recognized, only two of which occur in the Old World.

Systematics: This is a genus of small black wasps (3.0 to 6.5 mm) in which the gaster is often partly or all red. The clypeus is frequently yellow in the male, and the legs often have yellow stripes or spots in both sexes.

Plenoculus is probably most closely related to *Paranysson,* but the genus is more easily confused with *Solierella*. The presence of a broadly triangular, carina-delimited pygidial plate which is blunt apically identifies female *Plenoculus*. Nearly all *Solierella* females have a conical, distally attenuate tergum VI, with no pygidial plate (compare fig. 92 H and I). A few exceptional *Solierella*

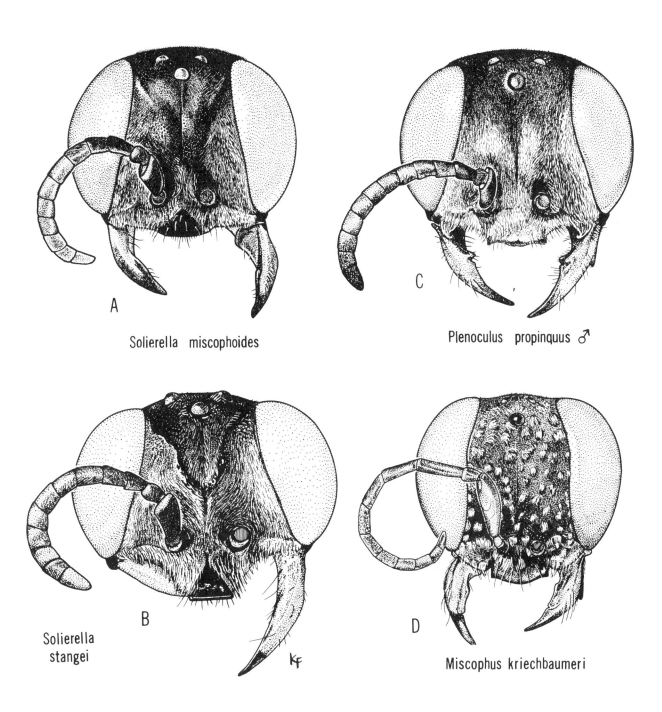

FIG. 91. Facial portraits in the tribe Miscophini; A,B,D, females; C, male. B = holotype.

females have a narrow pygidial plate which is delimited by carinae only apically. What appear to be weak lateral pygidial carinae in the other *Solierella* are actually rows of short setae. Male *Plenoculus* and *Solierella* have been difficult to separate until now, but we have found a character which makes separation fairly easy. In *Plenoculus* the male foretrochanter is simple (fig. 92 D), but the trochanter of *Solierella* males is nearly always concave or scooped out posterobasally (fig. 92 B). In addition, *Solierella* males do not have lateral clypeal hairbrushes whereas many *Plenoculus* males do. Furthermore, *Plenoculus* males, with the exception of *platycerus*, have no malar space, whereas most males of *Solierella* do. There are other differences between the two genera which are common to both sexes. The frons in *Plenoculus* is broadly convex and unadorned with carinae or humps, and the facial hair is uniform and unidirectionally oriented so that no special patterns are visible in different lights when the head is viewed from above. In *Solierella* the frons may bear a V-shaped carina, or at least it has a Y or V-shaped swelling that usually is much less pubescent than the rest of the frons. This sparsely haired swelling usually stands out as black V-shaped area when the head is viewed from above. We have seen only one or two species of *Solierella* in which the frons is uniformly pubescent. These seem to undescribed. In *Plenoculus* the frons appears uniformly silver in dorsal view. The frontal carina sometimes extends onto the clypeus in *Solierella*. The mandibles of *Plenoculus* are either notched or angulate externoventrally (fig. 92 G), but in *Solierella* they are usually simple (fig. 92 E) although a few species have an angulation (fig. 92 F). The hindtibia in *Plenoculus* (fig. 92 K) is more strongly spined than in *Solierella* (fig. 92 J).

Plenoculus is quite variable in certain head features, most notably the clypeus. Three groups can be defined on the basis of the female clypeus. The free margin is broadly arched or triangularly produced and is edentate in the first group: *boregensis, cockerelli, cuneatus, mexicanus,* and *timberlakei*. In the second group the free margin is sinuate and has one or two lateral teeth: *gillaspyi, hurdi, sinuatus*. The females of the remaining species (*beaumonti, boharti, davisi, deserti, murgabensis, palmarum, parvus, propinquus, stygius*) have a truncate median lobe which is bounded laterally by several teeth. The lobe may be medially emarginate. Within these three groups the male clypeus varies considerably and is often quite different from that of the female. In the commonest male clypeal form the free margin has a median truncate lobe. Group III males of certain species possess a lateral clypeal hair brush, and the sterna bear transverse welts or tubercles: *boharti, davisi, deserti, propinquus,* and *stygius*. *Plenoculus beaumonti* and *playtcerus* males have hairbrushes, but the sterna are simple.

At the specific level, **distinguishing** features are found in the form of the clypeus, the shape of the ocellar triangle, the form of the flagellum especially in the male, the occipital carina, propodeal and pygidial sculpture, and male genitalia. The males of *davisi* and *sinuatus* have a

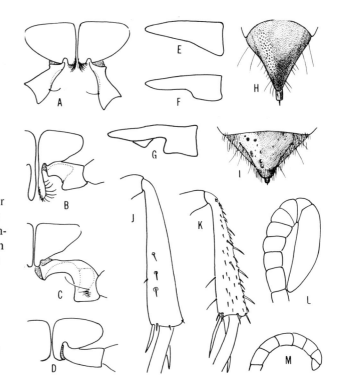

FIG. 92. Structural details in the tribe Miscophini; A-D, male forecoxa and trochanter; E-G, left mandible, lateral; H-I, female tergum VI; J-K, left hindtibia; lateral; L-M, male antenna, distal. A, E, *Solierella plenoculoides;* B, *S. boharti;* C, *S. mirifica;* D, *Plenoculus propinquus;* E,J, *Solierella miscophoides;* F, *S. albipes;* G,I,K, *Plenoculus davisi;* H, *Solierella jaffueli;* L, *S. semirugosa;* M, *S. californica*.

tubercle on the ventral side of the mesopleuron. The propleuron bears a strong transverse posterior ridge in the female of *hurdi*. The male of *platycerus* is quite striking in two respects: flagellum broad and strongly flattened, and a long malar space.

The New World species were recently treated by F. Williams (1960b) who discussed species groups, provided a key, and gave specific characters. However, additional information is still needed. Several species are known only by the types, and only one sex has been described in some.

Biology: Williams (1960b) summarized earlier reports and provided additional data. Evans (1961) and Kurczewski (1968) have given the most detailed account of nest architecture and wasp behavior. Most of our information pertains to *Plenoculus davisi*, which nests in sandy soil and provisions with bugs of the family Miridae. Williams reported also that immature aphids were recorded as prey by one observer. Nests are uni or multicellular, and the entrance is left open during search for prey which are flown to the nest venter up. Two to 24 bugs are placed in each cell, and the egg is laid on the last provision. Males dig shelter burrows in which they spend the night.

Williams observed *cockerelli* and found that this species provisions with moth larvae of the family Pyralidae. As in *davisi* the prey are carried to the nest in flight. There are biological data for only one other species, *stygius,* in which females have been captured carrying mirid bugs.

Checklist of *Plenoculus*

beaumonti Andrade, 1957; Spain
boharti Williams, 1960; U.S.: California
boregensis Williams, 1960; U.S.: California
   ssp. *perniger* Williams, 1960; U.S.: California
cockerellii W. Fox, 1893; U.S.: Texas to California
cuneatus Williams, 1960; U.S.: California, Nevada
davisi W. Fox, 1893; Canada to Mexico
   abdominalis Ashmead, 1899
   apicalis Williams, 1914
  ssp. *atlanticus* Viereck, 1902; U.S.: Eastern Seaboard Texas
  ssp. *gracilis* Williams, 1960; U.S.: California
  ssp. *mojavensis* Williams, 1960; U.S.: New Mexico to California
  ssp. *transversus* Williams, 1960; U.S.: California
deserti Williams, 1960; U.S.: California
gillaspyi Krombein, 1938; U.S.: Texas
hurdi Williams, 1960; Mexico
mexicanus Williams, 1960; Mexico
murgabensis (Gussakovskij), 1928 (*Ptigosphex, Ptygosphex*); sw. USSR
   tadzhikus Gussakovskij, 1935 (*Pavlovskia*)
palmarum Williams, 1960; U.S.: California
parvus W. Fox, 1897; U.S.: New Mexico
platycerus Menke, 1968; Mexico
propinquus W. Fox, 1893; w. U.S.
   rufescens Cockerell, 1898
sinuatus Williams, 1960; U.S.: California
stygius Williams, 1960; U.S.: California, Arizona
timberlakei Williams, 1960; U.S.: California, Arizona

Genus Solierella Spinola

Generic description: Inner orbits converging above (fig. 91 A-B); antenna short to moderately long, male usually with 13 articles (12 in some), scape short or elongate, flagellum variable, terminal article in male often much longer than those preceding it, or short and conical (fig. 92 L-M); frons with a V-shaped tumescence (rare exceptions) which may bear a V or Y-shaped carina, the stem of the latter sometimes extending onto the clypeus, facial vestiture nearly always accentuating the V-shaped tumescence when head viewed from above; clypeal free margin with a rounded or truncate median lobe in female, that of male similar or with one to three median teeth; labrum spatulate or transverse; inner margin of mandible simple, externoventral margin usually simple but angulate in a few species; malar space often present, especially in male; mouthparts short; occipital carina incomplete below or meeting hypostomal carina; collar short, admedian lines of scutum separated when present; propodeum short, dorsal enclosure sometimes delimited by carinae although always distinct because of its lack of vestiture, V- or U-shaped surface variable, granulate to foveolate; tergum VI conical or weakly flattened in female, without pygidial plate but often with two apically converging rows of fine short setae which superficially appear as pygidial carinae, (apex of tergum weakly carinate in one Neotropical species), male tergum VII often flattened but without a pygidial plate; legs moderately long, foretarsal rake absent in most species (fig. 93 G), but present in some females (fig. 93 H-I); inner posterior angle of male forecoxa sometimes with a process (fig. 92 B), male foretrochanter excavated posterobasally (weakly so in *plenoculoides*) (fig. 92 A-C); midcoxae narrowly separated, hindcoxae contiguous, hindtibia spineless or with at most three or four widely spaced, thin spines (fig. 92 J): tarsomere V not swollen, arolia small to moderate; marginal cell of forewing truncate, appendiculate (narrowly so or acuminate in *dispar*) (fig. 84 B-D), usually 3 submarginal cells, II petiolate (except in wings without submarginal III, fig. 84 D, and receiving both recurrent veins, or first recurrent received by submarginal I or interstitial (fig. 84 B, C); length of hindwing jugal lobe equal to or less than half length of anal area; male sternum VIII as in fig. 90 H; volsella greatly reduced or absent, head of aedeagus with teeth.

Geographic range: The Nearctic Region with 36 species and the Palearctic Region with 31 species are the main centers of distribution for *Solierella*. Most of the Nearctic species are found in the drier regions of western North America. The Neotropical Region is represented by 10 species, but many more await description so that the New World probably will prove to be the stronghold of *Solierella*. Four species are known from southern Africa and Madagascar, one from India, and one occurs in the Philippine Islands. One Nearctic form, *peckhami,* has been introduced to Hawaii and the Marshall Islands.

Systematics: Solierella are small to medium (2.5-11 mm) black wasps. The gaster and last pair of legs are often partly or all red, and yellow markings are common on the thorax and legs. The genus is closely allied with *Plenoculus* and *Paranysson*. For differences between *Solierella* and these two genera the reader is referred to the discussions in the systematics section of each of these.

There have been attempts in the past to split *Solierella* into two or more genera or subgenera. Wing venation and imagined mandibular differences have been the basis for such separation. We have studied examples of the type species of *Solierella (miscophoides)* and those of its generic synonyms *(compedita, pisonoides, jaffueli)*. The mandible in all four species is simple externally, that is, not notched nor angulate; but Rohwer (1922c) stated that *Solierella* was generically distinct from *Silaon* because of the notched or angulate mandibles in the former. The wing characteristics employed by Ashmead (1899) in separating *Niteliopsis, Silaon,* and *Solierella* are too variable to be useful. Actually, the type species of these three taxa appear to be more closely allied to each other than they are to many of the other groups within *Solierella*. *Lautara* is the most distinctive of the generic synonyms, because of the presence of a female foretarsal rake (fig. 93 I) and a short forecoxal process in the male. F. Williams (1950) set up rather loosely defined species groups for the Nearctic *Solierella*, but he did not recognize subgenera. Beaumont (1964b) did not use species groups or subgenera, and he pointed out that *pisonoides*, the type of *Niteliopsis*, has essentially the same type of mandible (simple) as *compedita*, the type of *Silaon*. Earlier workers had stated that *pisonoides* had a weakly angulate mandible. Some authors (Krombein, 1951, for example) have suggested that subgenera probably should be recognized because of biological differences between the various groups within *Solierella*. We have not studied this in sufficient detail to make a judgement in this matter, but it is clear in light of the structural variables discussed below that a meaningful infrageneric classification can be made only after a worldwide study of the genus. Also, it is obvious that knowledge of *Solierella* biology is still fragmentary in some groups, so that a subgeneric division at this time based on biology would be premature.

Specific characters, some of which may have group significance, are numerous on the head. The majority of the species have simple mandibles (fig. 92 E), but a few are roundly to sharply angled externoventrally: *clypeata*, *albipes*, *mandibularis*, *stangei*, *prosopidis*, *pectinata*, and *dispar* (fig. 92 F). The frontal swelling bears a V or Y-shaped carina in some species. The antennae vary considerably. The flagellomeres may be several times longer than wide as in *striatipes* and its relatives, or shorter than wide at the other extreme. The male flagellum offers good specific characters in many species. The length of the last article may be very short and conical, or quite elongate and sometimes equal to the length of five or six of the preceding articles (figs. 92 L,M). The male antenna usually has 13 articles, but in a few Old World species there are only 12: *andradei*, *babaulti*, *chivensis*, *guichardi*, *madagascariensis*, *nitida*, *pallidipes*, *pisonoides*, and *verhoeffi*. Details of the clypeus are usually distinctive. In most species the male has a well developed malar space. *Solierella miscophoides* and *jaffueli* are notable exceptions. A malar space is also present in a few females (*prosopidis*, for example). The gena in *stangei* bears a large downward projecting prong.

The foreleg in the female is usually devoid of a tarsal rake (fig. 93 G), but *jaffueli*, *pectinata*, *boharti*, *sonorae*, *major*, *masoni*, and *fossor* have a rake composed of one or two long spines at the apex of each tarsomere (fig. 93 H-I). The rake is best developed in *jaffueli* in which tarsomere I bears several long spines (fig. 93 I). The foretrochanter is concave posterobasally in all male *Solierella* studied by us. However, the concavity is shallow in at least one species, *plenoculoides* (fig. 92 A), but at the other extreme it is sometimes very deep, as in *mirifica* (fig. 92 C). The outer limit of the concavity often bears one or more long, incurved bristles (fig. 92 A, C). In some species the male forecoxa bears a long fingerlike projection at its inner posterior angle (fig. 92 B): *jaffueli*, *boharti*, *major*, *son-*

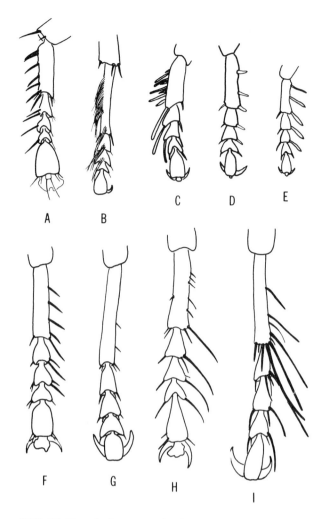

FIG. 93. Female foretarsus in the tribe Miscophini; A-C, right leg, D-I, left leg; A, *Sericophorus chalybeus*; B, *Mesopalarus mayri* paratype; C, *Saliostethus mimicus* paratype; D, *Miscophoides handlirschii* holotype; E, *Saliostethoides saltator*; F, *Plenoculus davisi*; G, *Solierella miscophoides*; H, *Solierella striatipes*; I, *Solierella jaffueli*.

orae, and *striatipes*. In others the projection is replaced by long bristles (fig. 92 A). The hindcoxa often bears an inner dorsal carina which may form a tooth basally. The inner margin of the hindbasitarsus is sometimes obtusely angulate in the male.

The forewings vary considerably in *Solierella* (see fig. 84 B-D). The media may diverge before or after crossvein cu-a, or it is interstitial. The shape of the submarginal cell is often distinctive. There is sometimes a lack of constancy in these characters even within species, and in fact the two forewings of a specimen may differ (see figures in Beaumont, 1964c). Furthermore, in *australis* (according to Williams 1950, plate 17), *syriaca*, and *seabrai* there may be only two submarginal cells, the second of which is not petiolate. Occasionally, specimens of these latter species are found in which the second submarginal cell and second discoidal cell are open apically. These are merely genetic freaks. For further remarks on specific characters in *Solierella* the reader is referred to Williams (1950) and Beaumont (1964c).

There is no key which includes all the North American species of *Solierella*. Williams' (1950) key pertains to the Californian species, although it does include 27 of the 35 North American species as well. He has described several species since, and there has been some name changing, so the usefulness of this key is limited. Ducke (1907b) presented a key for some of the Brazilian species. Gussakovskij (1928b, 1930c), Beaumont (1964c), and Arnold (1923b, 1945) have treated most of the Old World species.

Biology: Our knowledge of *Solierella* biology is fragmentary for the most part. However, the papers by F. X. Williams (1927, 1950), Newton (1960), Kurczewski (1967a), Krombein (1967b), Carrillo and Caltagirone (1970), and Beaumont (1964c) contain detailed accounts and/or summaries of previous work. *Solierella* nest in preexisting cavities in stems, twigs, almond hulls, and galls or in abandoned burrows in the ground. Most twig nesters provision with Hemiptera, but small grasshoppers are utilized by *plenoculoides*, and *sayi* stores Psocoptera. Ground nesters provision with one or two large grasshoppers (*fossor, striatipes*) or numerous Hemiptera of various families (*inermis, corizi, nitens, seabrai*). The prey is flown to the nest or is carried along the ground venter up in the case of *striatipes*, which provisions with grasshoppers larger than itself. The egg is laid before or behind the forecoxae in *inermis, sayi,* and *peckhami* or behind the hindcoxae in *striatipes*. Nests often contain several loosely packed cells each containing up to 13 prey in the species which provision with bugs. The nests are loosely closed with pebbles and pieces of grass.

Carrillo (1967) and Carrillo and Caltagirone (1970) described the larval development in *peckhami* and *blaisdelli*. There are only two larval instars, and the first instar contains a preeclosion and posteclosion phase.

Chrysididae of the genera *Hedychridium* and *Pseudolopyga* have been reported parasitizing *Solierella* by Bohart and Brumley (1967) and Carrillo and Caltagirone (1970).

Checklist of *Solierella*

abdominalis Williams, 1950; U.S.: California
aegyptia Kohl, 1898; Egypt
affinis (Rohwer), 1909 (*Niteliopsis*); w. U.S.
albipes (Ashmead), 1899 (*Pleneculus!*); w. U.S.
amazonica Ducke, 1904; Brazil
andradei Beaumont, 1957; Morocco
antennalis Beaumont, 1956; Libya
antennata Ducke, 1907; Brazil
arcuata Williams, 1950; U.S.: California
atra Reed, 1894; Chile
australis Williams, 1950; U.S.: California
babaulti Beaumont, 1956; Libya
bactriana Gussakovskij, 1930; sw. USSR
bicolor Williams, 1950; w. U.S.: California
blaisdelli (Bridwell), 1920 (*Silaon*); w. U.S.

boharti Williams, 1950; U.S.: California
    lasseni Williams, 1950
boregensis Williams, 1958; U.S.: California
bridwelli Williams, 1950; U.S.: California
californica Williams, 1950; U.S.: California
canariensis E. Saunders, 1904; Canary Is.
    ssp. *heterocera* Beaumont, 1968; Canary Is.
capparidis Gussakovskij, 1928; sw. USSR
chilensis Kohl, 1892; Chile
chivensis Gussakovskij, 1928; sw. USSR
clypeata Williams, 1950; U.S.: California
compedita (Piccioli), 1869 (*Silaon*); Mediterranean area
    helleri Kohl, 1878 (*Ammosphecidium*)
    xambeui Er. André, 1896 (*Sylaon*)
    ssp. *cretica* Beaumont, 1964; Crete
    ssp. *cypriaca* Beaumont, 1964; Cyprus
corizi Williams, 1950; U.S.: California to Texas
dispar Pulawski, 1964; Egypt, Canary Is.
flavicornis Gussakovskij, 1928; sw. USSR
fossor (Rohwer), 1909 (*Niteliopsis*); w. U.S.
    foxii Viereck, 1906 (*Niteliopsis*); nec Viereck, 1902
foxii (Viereck), 1902 (*Plenoculus*); U.S.: New Jersey
fusciventris Gussakovskij, 1930; sw. USSR
guichardi Beaumont, 1956; Libya
gussakovskiji Menke, new name; sw. USSR
    affinis Gussakovskij, 1928, nec *affinis* Rohwer, 1909
inermis (Cresson), 1872 (*Nysson*); U.S.: Florida
insidiosa Beaumont, 1964; Syria, Cyprus
iresinides (Rohwer), 1914 (*Silaon*); Guatemala
jaffueli (Herbst), 1920 (*Lautara*); Chile
kansensis (Williams), 1914 (*Niteliopsis*); U.S.: Kansas
lagunensis (Williams), 1928 (*Silaon*); Philippines
levis Williams, 1950; U.S.: California
longicornis Pulawski, 1964; Egypt
lucida (Rohwer), 1909 (*Niteliopsis*); U.S.: Colorado
    licida Rohwer, 1909 (*Niteliopsis*), misspelling
madagascariensis Arnold, 1945; Madagascar
major (Rohwer), 1917 (*Silaon*); U.S.: Washington to California
mandibularis Beaumont, 1957; Morocco
masoni Williams, 1959; U.S.: California
mexicana (Rohwer), 1911 (*Silaon*); Mexico
minarum Ducke, 1907; Brazil
mirifica Pate, 1934; U.S.: Arizona
miscophoides Spinola, 1851; Chile
    spinolae Kohl, 1892, new synonymy by Menke
modesta (Rohwer), 1909 (*Niteliopsis*); U.S.: Colorado
nigrans Krombein, 1951; U.S.: transcontinental
    nigra Rohwer, 1909 (*Niteliopsis*), nec Ashmead, 1899
nigridorsum Pulawski, 1964; Egypt
nitens Williams, 1950; U.S.: California
nitida Gussakovskij, 1928; sw. USSR
nitraria Pulawski, 1967; n. Africa, Egypt
obscura Beaumont, 1956; Libya
pallidipes Arnold, 1945; Madagascar
peckhami (Ashmead), 1897 (*Plenoculus*); U.S.: transcontinental, Hawaii, Marshalls
    nigra Ashmead, 1899 (*Plenoculus*)
    rohweri Bridwell, 1920 (*Silaon*)

arenaria Krombein, 1939
pectinata Pulawski, 1964; Egypt, Canary Is.
pisonoides (S. Saunders), 1873 (*Niteliopsis*); ne. Mediterranean area
　paradoxa Gussakovskij, 1930
platensis Brèthes, 1913; Argentina
plenoculoides (W. Fox), 1893 (*Niteliopsis*); e. U.S. to Rocky Mts.
　ssp. *similis* (Bridwell), 1920 (*Silaon*); U.S.: California, Oregon
prosopidis Williams, 1959: U.S.: California
　mandibularis Williams, 1958, nec Beaumont, 1957
quitensis (Benoist), 1942 (*Sylaon*); Ecuador
rhodesiana Arnold, 1923; S. Africa
sayi (Rohwer), 1909 (*Niteliopsis*); w. U.S.
scrobiculata Arnold, 1923; s. Africa
seabrai Andrade, 1950; n. Mediterranean area
　ssp. *corsa* Beaumont, 1964; Corsica
semirugosa Williams, 1958; U.S.: California
sonorae Williams, 1950; U.S.: California
stangei Menke, 1968; Argentina
striatipes (Ashmead), 1899 (*Niteliopsis*); U.S.: California
syriaca Beaumont, 1964; Syria
timberlakei Williams, 1950; U.S.: California
turneri Dutt, 1917; e. India
vandykei Williams, 1950; U.S.: California
verhoeffi Beaumont, 1964; ne. Mediterranean area
vierecki (Rohwer), 1909 (*Niteliopsis*); w. U.S.
　parva Rohwer, 1909 (*Niteliopsis*)
weberi Williams, 1955; U.S.: California
xanthocera Gussakovskij, 1930; sw. USSR
zimini Gussakovskij, 1928; sw. USSR

Genus Miscophus Jurine

Generic diagnosis: Inner orbits variable, usually converging above although often sinuate below (fig. 91 D), orbits sometimes bowed outward; antenna moderately long, scape short or elongate, flagellomeres usually longer than wide although shorter in male; frons usually simple, but with a pair of broad shallow depressions in the *kohli* group and two divergent carinae in *niloticus,* frontal line weakly to strongly impressed; clypeal free margin usually with a lateral notch or excision which delimits a truncate or angulate median lobe, or free margin weakly sinuate, without a definite median lobe, clypeus often dissimilar between sexes; labrum hidden, quadrangular or rectangular; inner margin of mandible simple, externoventral margin notched; no malar space, mouthparts short; occipital carina ending at hypostomal carina or just before it; pronotal collar usually short, dorsum often with a median prominence, collar longer (length about half width) in *handlirschii, kohli,* and *soikai* groups; entire pronotum about as long as scutum except longer in *handlirschii, kohli,* and *soikai* groups; admedian lines of scutum narrowly separated, obliterated in *handlirschii* group; propodeum short to moderately long, dorsum with no enclosure but a median longitudinal carina usually present except in *handlirschii* group in which several longitudinal

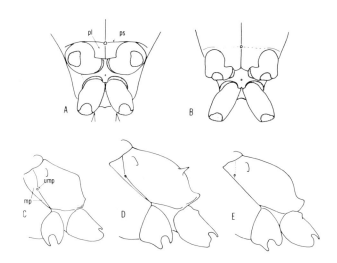

FIG. 94. Structural details in the thorax of Miscophini; A-B, venter of meso- and metathorax; C-E, lateral view of metapleuron and propodeum; A, *Miscophus* sp., pl = precoxal lobe, ps = precoxal sulcus; B, *Saliostethus lentifrons;* C, *Miscophus (Nitelopterus)* sp., mp = metapleuron, ump = upper metapleural pit; D, *Miscophus kohli;* E, *Miscophus handlirschii.*

parallel ridges or carinae are present; apex of propodeal dorsum with two spinelike projections in *kohli* group (fig. 94 D); mesopleuron sometimes with scrobal sulcus, hypoepimeral area strongly gibbous in *handlirschii* group; precoxal sulcus strong, precoxal lobes well defined (fig. 94 A); metapleuron usually well defined (fig. 94 C), but narrow and parallel sided or absent below upper metapleural pit in *soikai, kohli,* and *handlirschii* groups (upper metapleural pit often contiguous with mesopleural suture in these groups) (fig. 94 D-E); gaster sessile (pedunculate in *ichneumonoides*), tergum VI conical or weakly flattened in female, without a pygidial plate; legs moderately long to long and slender, foretarsal rake present in most females although weak in some or represented by a single spine per tarsomere, rake usually absent in male; midcoxae narrowly separated to subcontiguous (fig. 94 A), dorsal carina of midcoxa often forming a tubercle basally, hindcoxae contiguous; hindtibia sometimes spineless, but usually with two to six widely spaced, long spines on outer side; tarsomere V not swollen, arolia small; marginal cell of forewing truncate or acuminate apically (fig. 84 E-K), sometimes greatly reduced or open (*handlirschii* group, fig. 85 B); two submarginal cells usually present, II petiolate (fig. 84 E-I), or greatly reduced or absent in the *handlirschii* and *kohli* groups (figs. 84 I,K; 85 B), always absent in the *soikai* group (fig. 84 J); two discoidal cells usually present (figs. 84 F,I; 85 B), but second absent due to loss of second recurrent vein in *soikai* group (fig. 84 J), and subgenus *Nitelopterus* (fig. 84 H), second recurrent sometimes faint in *bicolor* group (fig. 84 E, G), discoidals I and II fused in one species of the *kohli* group due to loss of first recurrent (fig. 84 K); first recurrent vein received by submarginal I, second recurrent received by submarginal II when latter is present; forewing media usually diverging after cu-a, but occasionally interstitial; hindwing radial sector short in *handlirschii* and *kohli* groups; hindwing jugal lobe small, much less than half length of anal area (fig. 84 F); male sternum VIII variable (fig. 96 E-J); volsella weakly differentiated, but its free margin densely setose (fig. 96 A-B).

Geographic range: Miscophus is represented in all regions except Australasia. About 150 species are known, but most of these are found in the Old World, especially the Mediterranean area. About a dozen species are known from North America. The genus is most diverse in the Mediterranean area and Ethiopian Region, but the subgenus *Nitelopterus* is endemic to North America. The Oriental Region has four Indian species, and one has been described from the Neotropical Region. The latter species, *exoticus*, needs to be verified as a Neotropical element.

Systematics: These are small to medium (2.8-10 mm), mostly somber colored wasps with rather short wings that are usually infuscate apically. The genus is allied with *Plenoculus* and *Solierella* on one hand, and with the south African endemics, *Saliostethus*, *Saliostethoides*, and *Miscophoides*, on the other. The absence of submarginal cell III in *Miscophus* separates it from *Plenoculus* and nearly all *Solierella*. The few *Solierella* with only two submarginal cells have a nonpetiolate second submarginal, whereas in *Miscophus* submarginal cell II is always petiolate when present. Unlike the majority of *Solierella*, the mandible in *Miscophus* is always externoventrally notched. The midcoxae are closer together in *Miscophus* than in any other of the miscophine genera, except *Miscophoides* subgenus *Miscophoidellus*, some *Nitela*, and one *Saliostethus*. For differences between *Miscophus* and its African relatives see the discussion under each genus.

The most striking and at the same time most perplexing feature of *Miscophus* is the tendency toward reduction or loss of certain wing veins, reduction in cell size, and overall shortening of the wings. These tendencies probably are correlated with the habit of walking during prey search rather than flying. Even with prey they usually move in short hopping flights (see biology). Most species have venation as shown in fig. 84 F, but the second recurrent vein is completely absent in two groups (figs. 84 H, J, 95), and is evanescent or incomplete in a few species (fig. 84 E, G). The group in which there are two submarginal cells but in which the second recurrent is absent is found only in North America, and because of this venational peculiarity these species have been regarded as representing a genus, *Nitelopterus*. However, the presence of an incomplete second recurrent vein in several Old World species (*chrysis, caninus, mavromoustakisi, niloticus*) and one undescribed North American species indicates to us that *Nitelopterus* is better treated as a subgenus of *Miscophus*, as Andrade (1956b) already has suggested. A second wing variable is the reduction or loss of submarginal cell II (fig. 84 I-K, 85 B). This occurs in various species groups. In the Old World species *venusta*, submarginal cell II is present in the female but not in the male. The second recurrent vein is also absent in this form. The marginal cell is sometimes incomplete (open apically) in some species of the *handlirschii* group (*portoi* for example); and in one undescribed species of the *kohli* group the second recurrent vein is interrupted, thus fusing the two discoidal cells (fig. 84 K). The latter species also has lost the second submarginal cell.

Another important variable is the narrowing of the metapleuron below the upper metapleural pit (fig. 94 D-E). The normal condition is shown by fig. 94 C. In the *kohli* and *soikai* groups and a few isolated species (*bytinskii, cyanescens,* and *coerulescens*), the metapleuron is very narrow and nearly parallel sided below the pit (fig. 94 D). In the *handlirschii* group the metapleuron is not discernable below the pit (fig. 94 E). Other thoracic details of importance at both the group and specific level are the form of the propodeum (rounded or angulate in profile, with two spines in the *kohli* group), the length and shape of the pronotum, and the form of the hypoepimeral area (gibbous in the *handlirschii* group and a few other species). The form of sternum VIII in the male is of group importance (fig. 96 E-J). In the *helveticus* group (= *gallicus* group) the plate terminates in two spinelike processes (fig. 96 I). In the *bicolor* group and *Nitelopterus* the sternum is truncate apically and usually bears four short teeth (fig. 96 E), and in the *handlirschii* group the sternum is rounded apically (fig. 96 J). The sternum in the *kohli* group (fig. 96 F) and some of the other African species (fig. 96 G-H) is somewhat similar to the *bicolor* type.

The most important works on *Miscophus* are those of the late N. F. de Andrade (1952, 1953, 1954, 1956a, b, 1960). He formed a species group classification based on the Palearctic species of *Miscophus*, although in his 1956b paper he did discuss some of the extralimital groups and related genera. He presented a key to the species groups in his last paper (1960). P. M. F. Verhoeff (1955), and Beaumont (1952b) have also made a contribution to the species group classification of *Miscophus*.

There is no key to the New World *Miscophus*. Andrade's papers contain keys and descriptions of all the Palearctic species with the exception of those described by Pulawski (1964) and some of those described by Balthasar (1953). Arnold (1923b) presented a key to the species of the Ethiopian Region, but it is outdated because of later descriptions of new species. The Oriental species have not been treated.

As Andrade has indicated in his papers, the supraspecific classification of *Miscophus* cannot be considered final in its present state due to the paucity of material of certain peculiar forms, and also because the faunas of some areas are still poorly known. The Ethiopian Region, for example, contains a number of peculiar *Miscophus* that do not fit conveniently into any of the existing species groups. *Miscophus kohli, ichneumonoides,* and one undescribed species form an assemblage, here termed the *kohli* group,

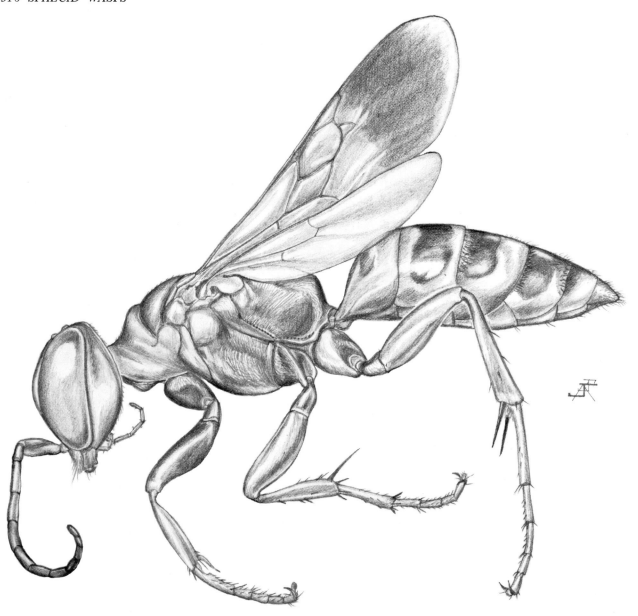

FIG. 95. *Miscophus (Nitelopterus) laticeps* (Ashmead), female.

which is distinguished by the presence of two propodeal spines (fig. 94 D), a narrow metapleuron (fig. 94 D), two shallow facial depressions, and an elongate body. However two other African species seen by us (*cyanescens, coerulescens*) weakly display one or more of these features. Three other south African species, *kriechbaumeri, bellulus,* and *crispus,* have distinctive scalelike tufts of setae on the head (fig. 91 D) and thorax. We have called this assemblage, the *kriechbaumeri* group. *Miscophus stevensoni* has subapical tergal swellings. The descriptions of other African species indicate other peculiarities. In North America the endemic subgenus *Nitelopterus* predominates, but there is one described typical *Miscophus* of the *bicolor* group (*americanus*), and several others of uncertain affinities await description.

Biology: Most of what is known about these wasps concerns only the subgenus *Nitelopterus* (Krombein and Evans, 1954, 1955; Krombein and Kurczewski, 1963; Krombein, 1964b; Evans, 1963a, Cazier and Mortenson, 1965b; and Powell, 1967). Data on typical *Miscophus* are for the most part fragmentary (Ferton, 1896, 1914; Hartman, 1905; Crevecoeur, 1930; Krombein, 1963a). Kurczewski (1969) has given new data for both subgenera and summarized much of the previous work. The various reports indicate that in general the nesting activities of the species are similar. Nests are dug in loose, sandy soil and are short,

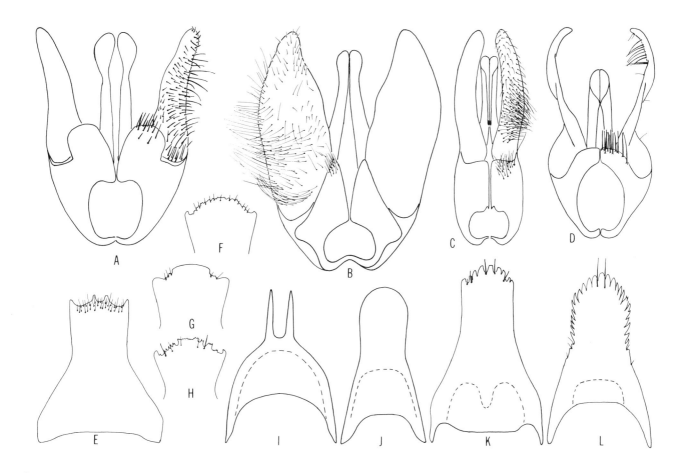

FIG. 96. Male genitalia (A-D) and sternum VIII (E-L) in the tribe Miscophini; A,E, *Miscophus bicolor;* B, *Miscophus (Nitelopterus)* sp.; C,K, *Saliostethus capicola;* D,L, *Saliostethoides saltator;* F, *Miscophus* sp. near *kohli;* G, *Miscophus cyanescens;* H, *Miscophus coerulescens;* I, *Miscophus gallicus;* J, *Miscophus handlirschii.*

varying from 2-5 cm in length. Nests usually have only one cell, but Evans (1963a) reported as many as six cells per nest of *evansi,* and Crevecoeur (1930) noted three to six in *spurius.* Small, usually immature spiders serve as larval food, and mass provisioning is the rule. The available data suggest that each species of *Miscophus,* at least in the subgenus *Nitelopterus,* may collect only certain kinds of spiders. The egg is laid on one of the last provisions. Some species leave the nest open between provisions (Kurczewski, 1969).

Females search for prey on the ground, but there is evidence that plant dwelling spiders are also collected. The female grasps the front legs of the spider with her mandibles and usually walks or makes short hopping flights to the nest. The prey is held venter up. Although progression when walking is usually forward, Cazier and Mortenson noted that females of *laticeps* and a species near *texanus* walked backwards dragging the prey. Hartman (1905) stated that after the female (*Miscophus americanus*) grasps the spider, she "slings it on her back and marches off with it".

Some authors have noted several legs missing on the prey, and the significance of this has various interpretations (Kurczewski, 1969).

Miscophus seem to spend most of their time walking rather than flying, and the wings are short and narrow in some groups. Presumably, the shortening of the wings and reduction of venation is a selective response to their preference for walking. They resemble ants when on the ground and are often seen running about in their company. They hold their wings flat over the gaster, and the clear wing base and smoky tip accentuate mimicry with the ant petiole and gaster respectively.

Crevecoeur (1930) reported Eulophidae of the genus *Melittobia* parasitizing nests of *Miscophus spurius.*

Checklist of *Miscophus*

abscondicus Andrade, 1960; Balearics
aegyptius Morice, 1897; Egypt to Syria
    politus Honoré, 1944
aenescens (Bridwell), 1920 (*Hypomiscophus*); U.S.:

Oregon
aenigma Honoré, 1944; Egypt
affinis Pulawski, 1964; Egypt
agadiriensis Andrade, 1954; Morocco, Spanish Sahara
akrofisianus Balthasar, 1953; Cyprus
albomaculatus Andrade, 1960; Israel, Canary Is.
albufeirae Andrade, 1952; Portugal, ? France
    ssp. *anatolicus* Beaumont, 1967; Turkey
alfierii Honoré, 1944; Egypt, Israel
americanus W. Fox, 1890; e. U.S.
andradei P. Verhoeff, 1955; Portugal
antares Andrade, 1956; Algeria
arenarum Cockerell, 1898; U.S.: New Mexico
arnoldi Turner, 1929; S. Africa
ater Lepeletier, 1845; Europe
    maritimus F. Smith, 1858
atlanteus Andrade, 1956; Morocco
bellulus Arnold, 1925; Rhodesia
belveriensis Andrade, 1960; Balearics
benidormicus P. Verhoeff, 1955; Spain
berlandi Andrade, 1956; Spanish Sahara
bicolor Jurine, 1807; Europe, sw. Asia
    dubius Panzer, 1806-1809 (***Larra***)
    metallicus C. Verhoeff, 1890
  ssp. *guigliaae* Andrade, 1953; Cyprus
  ssp. *bulganicus* Tsuneki, 1972; Mongolia
bonifaciensis Ferton, 1896; Corsica
bytinskii P. Verhoeff, 1955; Israel
californicus (Ashmead), 1898 (*Miscophinus*); U.S.:
    California
canariensis Beaumont, 1968; Canary Is.
  ssp. *nigrifemur* Beaumont, 1968; Canary Is.
caninus Andrade, 1953; Cyprus, Turkey
ceballosi Andrade, 1954; Algeria
chrysis Kohl, 1894; Somalia
clypearis Honoré, 1944; Egypt
coerulescens Arnold, 1923; Rhodesia
collaris Honoré, 1944; Egypt
concolor Dahlbom, 1844; Europe
    insubricus A. Costa, 1867
    moravicus Balthasar, 1957
corsicus Andrade, 1960; Corsica
crispus Arnold, 1925; S. Africa
ctenopus Kohl, 1884 (March); n. Africa, Saudi Arabia
    manzonii Gribodo, 1884 (May)
    rubriventris Honoré, 1944, nec Ferton, 1896
    gigas Giner Marí, 1945
    honorei Balthasar, 1953
cyanescens Turner, 1917; Rhodesia
cyanurus (Rohwer), 1909 (*Miscophinus*); U.S.: Colorado
cypriacus Andrade, 1953; Cyprus
  ssp. *obscurus* Andrade, 1954; ne. Mediterranean area
deserti Berland, 1943; n. Africa
deserticola Turner, 1929; S. Africa
difficilis Nurse, 1903; India
dispersus Andrade, 1954; n. Africa
eatoni E. Saunders, 1903; Canary Is., Mediterranean area
elegans Andrade, 1960; Syria
evansi (Krombein), 1963 (*Nitelopterus*); U.S.: Wyoming

eximius Gussakovskij, 1936; Mongolia
exoticus Taschenberg, 1870; Brazil
flavopictus Pulawski, 1964; Egypt
funebris Honoré, 1944; Egypt
galei (Rohwer), 1909 (*Miscophinus*); U.S.: Colorado
garianensis Andrade, 1956; Libya
gegensumus Tsuneki, 1971; China
gibbicollis Giner Marí, 1945; Spanish Sahara, Morocco
gineri P. Verhoeff, 1955; Spain
gobiensis Tsuneki, 1972; Mongolia
grangeri Beaumont, 1968; Algeria
gratiosus Andrade, 1960; Syria
gratuitus Andrade, 1954; sw. USSR
guichardi Beaumont, 1968; Canary Is.
gussakovskiji Andrade, 1954; sw. USSR
handlirschii Kohl, 1892; Algeria, Tunisia
hebraeus Andrade, 1954; Israel
heliophilus Pulawski, 1968; Sudan
helveticus Kohl, 1883; s. Europe (? = *italicus* A. Costa)
    gallicus Kohl, 1884
    rufus Giner Marí, 1943
  ssp. *rubriventris* Ferton, 1896; Corsica, Spain to Italy
  ssp. *viator* Andrade, 1954; Algeria
hissaricus Gussakovskij, 1935; sw. USSR
histronicus Balthasar, 1953; Cyprus
ichneumonoides Arnold, 1929; Rhodesia
imitans Giner Marí, 1945; n. Africa, Israel
impudens Andrade, 1960; Morocco
inconspicuus Andrade, 1960; Egypt
infernalis Arnold, 1929; Rhodesia
insolitus Andrade, 1953; Cyprus, Turkey
    beaumonti Balthasar, 1953
insulicola Balthasar, 1953; Cyprus
italicus A. Costa, 1867; Italy
kansensis (Slansky), 1969 (*Nitelopterus*); U.S.: Kansas
karrooensis Arnold, 1923; S. Africa
kohli Brauns, 1899; S. Africa
kriechbaumeri Brauns, 1899; S. Africa
laticeps (Ashmead), 1898; (*Miscophinus*); U.S.: California
levantinus Balthasar, 1953; Cyprus
littoreus Andrade, 1960; Spanish Sahara, Morocco
luctuosus Andrade, 1960; Corfu, Cyprus
lugubris Arnold, 1929; Rhodesia
lusitanicus Andrade, 1952; n. Mediterranean area
  ssp. *nomadus* Andrade, 1953; Cyprus
  ssp. *thracius* Pulawski, 1962; Bulgaria
maculipes Arnold, 1945; Madagascar
maurus (Rohwer), 1909 (*Miscophinus*); U.S.: Colorado
mavromoustakisi Andrade, 1953; Cyprus
  ssp. *cappadocicus* Beaumont, 1967; Turkey
merceti Andrade, 1952; Portugal to France
    hispanicus "Mercet" Ceballos, 1943, nomen nudum
  ssp. *orientalis* Beaumont, 1967; Turkey
mimeticus Honoré, 1944; Egypt to sw. USSR
minutus Andrade, 1953; Cyprus, Turkey
mochii Arnold, 1940; n. Africa, Israel
modestus Arnold, 1929; Rhodesia
mongolicus Tsuneki, 1972; Mongolia
montanus Gussakovskij, 1935; sw. USSR

nevesi Andrade, 1952; Portugal
nicolai Ferton, 1896; w. Mediterranean area
   ssp. *rufescens* Andrade, 1960; Portugal, Spain
niger Dahlbom, 1844; Europe
nigricans Cameron, 1907; "Matheran" (? India)
nigrescens (Rohwer), 1909 (*Miscophinus*); U.S.: Colorado
nigriceps (Rohwer), 1911 (*Miscophinus*); U.S.: California
nigripes Honoré, 1944 (p. 140); Egypt, Sudan
    nigripes Honoré, 1944 (p. 142), nec Honoré, 1944, p. 140
    lotus Andrade, 1954
niloticus Honoré, 1944; Egypt
nitidior Beaumont, 1968; Canary Is.
nobilis Andrade, 1960; Morocco
numidus Beaumont, 1968; Algeria
obscuritarsis Pulawski, 1964; Egypt
occidentalis Andrade, 1960; Morocco
oraniensis Brauns, 1906; S. Africa
othello Balthasar, 1953; Cyprus
papyrus Andrade, 1954; n. Africa
pardoi Andrade, 1954; Morocco
pharaonis Arnold, 1940; Libya to Israel
    frater Honoré, 1944
portoi Andrade, 1956; Portugal
postumus Bischoff, 1922; Europe
pretiosus Kohl, 1884; n. and e. Mediterranean area
   ssp. *brunnescens* Honoré, 1944; Egypt
primogeniti Andrade, 1954; Canary Is.
pseudomimeticus Andrade, 1960; n. Africa, Israel
pseudonotogonia Brauns, 1899; S. Africa
pulcher Andrade, 1953; Cyprus
punicus Andrade, 1954; Algeria, Tunisia
quettaensis Nurse, 1903; India
reptans Arnold, 1962; Rhodesia
rhodesianus Turner, 1917; Rhodesia
rothneyi Bingham, 1897; India
rufiventris Tsuenki, 1972; Mongolia
sallitus Andrade, 1960; Cabo Verde Is.
scintillans Andrade, 1956; n. Morocco
sericeus Radoszkowski, 1876; n. Africa, sw. USSR
seyrigi Arnold, 1945; Madagascar
similis F. Morawitz, 1896; sw. USSR
sirius Andrade, 1956; n. Africa
slossonae (Ashmead) in Kohl, 1896 (*Nitelopterus*); U.S.: Florida
    slossonae Ashmead, 1897 (*Nitelopterus*)
   ssp. *barberi* (Krombein), 1954 (*Nitelopterus*); U.S.: Florida
soikai Beaumont, 1952; Algeria
sordidatus Arnold, 1945; Madagascar
specularis Andrade, 1960; sw. USSR
spurius (Dahlbom), 1832 (*Larra*); Europe
   ?*unicolor* Schummel, 1831, nomen nudum
stevensoni Arnold, 1923; Rhodesia
susterai Balthasar, 1953; Cyprus
syriacus Andrade, 1960; Syria, Lebanon
tagiurae Andrade, 1954; Libya
temperatus Balthasar, 1953; Cyprus
texanus (Ashmead), 1898 (*Miscophinus*); U.S.: Texas
timberlakei (Bridwell), 1920 (*Hypomiscophus*); U.S.: California
tinctus Andrade, 1956; Morocco
transcaspicus Andrade, 1960; sw. USSR
tsunekii Andrade, 1960, Korea
unigena Balthasar, 1953; Cyprus
venustus Beaumont, 1969; Turkey
verecundus Arnold, 1925; Rhodesia
verhoeffi Andrade, 1952; Portugal to France
   ssp. *nitidus* Andrade, 1960; Morocco
yermasoyensis Balthasar, 1953; Cyprus
zakakiensis Balthasar, 1953; Cyprus

Nomina nuda in *Miscophus*
guigliae Gussakovskij in Soika, 1939
venetianus Gussakovskij in Soika, 1939

Genus Saliostethus Brauns

Generic description: Inner orbits usually convergent above (fig. 97 C) (sinuate but essentially parallel in *rhodesianus*); antenna moderately long, scape stout or slender and elongate, flagellomeres longer than broad or quadrate; frons simple, frontal carina or line absent; clypeal free margin entire, sinuate or straight; labrum transverse, hidden but free margin with about six stout, flat setae which in females project beyond clypeal margin (fig. 97 C); inner margin of mandible simple, externoventral margin with a notch or angle; no malar space; mouthparts short; occipital carina incomplete below, disappearing well before reaching hypostomal carina; pronotal collar 0.25-0.5 times as long as wide, pronotum as long as or longer than scutum; scutum without admedian lines; propodeum short, dorsum with no enclosure and without a well defined median carina except in *rhodesianus;* mesopleuron without episternal sulcus (except in *lounsburyi* and *rhodesianus*) or other sulci; precoxal sulcus usually weak or absent, precoxal lobes thus not clearly defined (fig. 94 B) (sulcus strong in *rhodesianus*); metapleuron not defined below upper metapleural pit which is nearly contiguous with mesopleural suture; tergum VI conical in female, without a pygidial plate; legs moderately long, foretarsal rake present in female and composed of stout or bladelike spines (fig. 93 C), male without a rake; midcoxae widely separated (except in *rhodesianus*), hindcoxae contiguous; hindtibia with three to 12 long spines on outer side; tarsomere V not swollen, arolium small; forewing with marginal cell greatly reduced (open apically in *rhodesianus*, fig. 85 C), and only one submarginal and discoidal cell present (second recurrent vein completely absent) (fig. 85 A,C); hindwing radial sector short; hindwing jugal lobe very small (fig. 85 A); male sternum VIII as in fig 96 K; volsella poorly differentiated but its free margin setose (fig. 96 C).

Geographic range: Five species are known in this endemic south African genus.

Systematics: Saliostethus are small (3-6.5 mm), brown, or brown and black wasps with short, apically infumate wings. The legs are usually pale brown or yellowish, and the metanotum and sometimes the scutellum have yellow markings.

This poorly known genus is closely allied to *Miscophus*, and the two are difficult to separate. The most distinctive features of *Saliostethus* are: presence of only one submarginal and discoidal cell, metapleuron undefined below upper metapleural pit, free clypeal margin entire, labrum with stout flat marginal setae, midcoxae usually widely spaced, and absence (usually) of the episternal sulcus. This last feature would be an excellent diagnostic character except that the sulcus is weakly present in *Saliostethus lounsburyi* and strongly indicated in *rhodesianus*. The widely spaced midcoxae would also offer an excellent means of separation from *Miscophus* if it were not for the narrowly separated coxae of *rhodesianus*. The wing venation is a good character, but in the *Miscophus soikai* group, which contains but two poorly known species, there is only one submarginal and discoidal cell. However, unlike *Saliostethus,* the *soikai* group does have a narrow metapleuron, and furthermore the clypeus is notched laterally. In *Saliostethus* the clypeal margin is entire although sometimes sinuate or with a median angular lobe. None of the *Miscophus* studied have stout flat setae on the labrum, but there are many *Miscophus* that we have not seen, including the *soikai* group. Male sternum VIII and the genitalia of *Saliostethus* are much as in *Miscophus* (fig. 96 C, K). Arnold (1923b, 1945) stated that *Saliostethus* females did not have a tarsal rake, but this is an error (see fig. 93 C).

We have seen examples of all five species. *Saliostethus lounsburyi* and *mimicus,* described originally in a new genus *Mutillonitela,* are certainly congeneric with *lentifrons,* the type of *Saliostethus;* but both species possess a distinctive, short, velvety, purple pubescence on the head, whereas the head of *lentifrons* is nearly glabrous and shining. *Saliostethus mimicus* has a deep narrow sulcus on the vertex of the head between the lateral ocellus and the inner eye margin. This same area in *lentifrons* has a slightly raised oval scar instead. *Saliostethus rhodesianus* and *capicola* lack both features. The open marginal cell of *rhodesianus* (fig. 85 C), episternal sulcus, and narrowly separated midcoxae make it distinctive. *S. rhodesianus* has scalelike tufts of hair on the frons and to some extent also on the pleura. Similar tufts are found in the *kriechbaumeri* group of the genus *Miscophus*. Perhaps *rhodesianus* is an aberrant *Miscophus* and as such should be removed to that genus. *Saliostethus* is a heterogeneous assemblage, and more collecting and study will be necessary to determine the proper status of the taxon and its species. There is no key available.

Biology: Little is known. Bridwell (1920), stated that "both species were taken running along the bare sand and resemble closely the small Mutillidae which are found there, until disturbed when they escape by flying." As in *Miscophus* the wing pattern doubtless enhances their resemblance to ants or mutillids

Checklist of *Saliostethus*

lentifrons Brauns, 1899; S. Africa
capicola Arnold, 1923; S. Africa
*mimicus (Bridwell), 1920 (*Mutillonitela*); S. Africa (type in Mus. Washington)
*lounsburyi (Bridwell), 1920 (*Mutillonitela*); S. Africa (type in Mus. Washington)
rhodesianus Arnold, 1929; Rhodesia

Genus Saliostethoides Arnold

Generic description: Inner orbits converging above or subparallel (fig. 97 D); antenna short, scape elongate, flagellomeres longer than broad; frons simple, without frontal carina or line; clypeal free margin entire, but with median lobe in male; labrum transverse, hidden, free margin with long setae, some of which are flattened; inner margin of mandible with suggestion of a subbasal tooth, externoventral margin notched (fig. 97 D); no malar space; mouthparts short; occipital carina incomplete below, ending well before reaching hypostomal carina; pronotal collar short, pronotum about as long as scutum; scutum without admedian lines; propodeum short, dorsum with no enclosure but a median longitudinal carina; mesopleuron without episternal sulcus or other sulci; metapleuron not defined, upper metapleural pit indistinct; female tergum VI with a glabrous, triangular pygidial plate which is margined by carinae (fig. 89 E), male without a pygidial plate; legs moderately long; female with a foretarsal rake (fig. 93 E), male with a very weak one; midcoxae widely separated, hindcoxae contiguous; hindtibia with six to nine widely spaced, long slender spines on outer side; tarsomere V not swollen, arolium small; forewing venation greatly reduced, only medial, submedial, discoidal I and subdiscoidal cells present (fig. 85 D); hindwing radial sector very short; hindwing without a jugal lobe (fig. 85 D); male sternum VIII as in fig. 96 L, volsella poorly differentiated, but its free margin setose (fig. 96 D).

Geographic range: Saliostethoides contains a single species that is endemic to southern Africa.

Systematics: Saliostethoides saltator is one of the smallest sphecids known (2-4 mm). The legs, gaster, and pronotum are reddish, and the short wings are infumate except for a pale band across the middle. The head, thorax, gaster, and legs are covered with long, bristly setae.

Saliostethoides is easily recognized by its wing venation (fig. 85 D). The marginal cell is open, there are no submarginals and only one discoidal cell. The hindwing has lost the jugal lobe. Other distinctive features are: the absence of the episternal sulcus, metapleuron undefined below upper metapleural pit, serrate male eighth sternum, and presence of a pygidial plate in the female.

This genus is intermediate between *Saliostethus* and *Miscophoides* as far as venation is concerned, but the presence of a pygidial plate in the female indicates that *Saliostethoides* is probably not too close to either genus. However, the serrately margined eighth sternum is derivable from that of *Miscophus* or *Saliostethus*.

Biology: Unknown, although Arnold (1924) said that "in its movements this insect resembles *Miscophoides*."

Checklist of *Saliostethoides*

saltator Arnold, 1924; Rhodesia

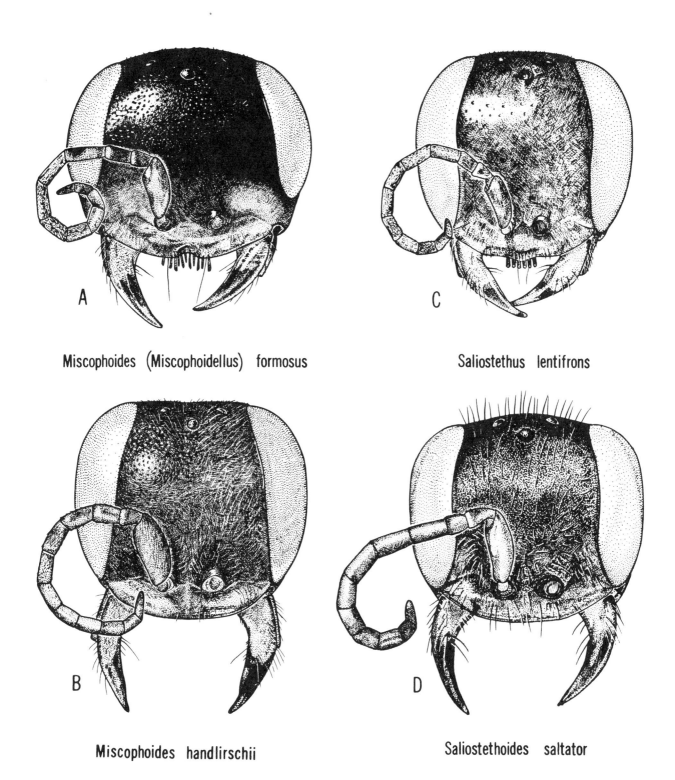

FIG. 97. Facial portraits of females in the tribe Miscophini. B is based on the holotype, C on a paratype.

Genus Miscophoides Brauns[18]

Generic description: Inner orbits converging above (fig. 97 A, B); antenna short, scape moderately long but very stout, flagellomeres longer than broad; frons simple, usually without frontal carina or line; clypeal free margin entire (fig. 97 B) or with a median emargination (fig. 97 A) (subgenus *Miscophoidellus*); labrum hidden; transverse, free margin with about six long flattened setae which project beyond edge of clypeus (fig. 97 A); inner margin of mandible simple, externoventral margin notched; a long malar space in subgenus *Miscophoidellus* (fig. 97 A); mouthparts short although palpi long; occipital carina incomplete below, ending well before reaching hypostomal carina; pronotal collar moderately long, with a posterolateral depression next to scutum, or short and without depression (*Miscophoidellus*), pronotum as long as or longer than scutum; admedian lines of scutum very short and separated; propodeum short, dorsum without enclosure or median longitudinal carina; mesopleuron with an episternal sulcus (absent in *Miscophoidellus*), other sulci and carinae lacking; metapleuron not defined below upper metapleural pit, which is nearly contiguous with mesopleural suture; tergum VI conical in female (apex rounded in *formosus*), without a pygidial plate; legs moderately long; a weak foretarsal rake of short spines in female (fig. 93 D); midcoxae widely separated (narrowly separated in *Miscophoidellus*), hindcoxae contiguous; hindtibia with few to many slender spines on outer side; tarsomere V not swollen, arolium small; forewing venation greatly reduced, only medial, submedial and subdiscoidal cells present (fig. 85 E); hindwing with very short radial sector, submedial cell short and narrow; hindwing without a jugal lobe (fig. 85 E); form of male sternum VIII and genitalia unknown.

Geographic range: Miscophoides is endemic to southern Africa and contains three species.

Systematics: These are small (2 to 5 mm) brown wasps with abbreviated, banded wings. The open marginal cell and absence of both submarginal and discoidal cells is the most striking feature of the genus and probably represents the highest degree of venational reduction in the Sphecidae. This remarkable wing pattern is about the only feature which unites the three described forms. These species divide readily into two groups which appear to warrant generic or at least subgeneric status. Because of the paucity of material and resultant poor knowledge of the two groups it seems best for the time being to recognize them as subgenera. A new subgenus *Miscophoidellus* Menke, type species *Miscophoides formosus* Arnold, is proposed for two of the species. The following key will demonstrate the salient features.

Key to subgenera of *Miscophoides*

Episternal sulcus present; malar space absent free margin of clypeus entire (fig. 97 B); dorsum of pronotal collar with a posterolateral depression; midcoxae separated .... subgenus *Miscophoides* Brauns

Episternal sulcus absent; a long malar space present (fig. 97 A); free margin of clypeus with a median emargination (fig. 97 A); pronotal collar without depressions; midcoxae subcontiguous .... subgenus *Miscophoidellus* Menke, new subgenus

The subgenus *Miscophoidellus* contains *minutus* and *formosus*. We have not studied *minutus*, but Arnold's (1952) description and figure leave little doubt that it is a *Miscophoidellus*. The general build of *handlirschii* is different from *formosus* and *minutus*. The former has a more elongate body. The collar and gaster are especially elongate. Both *formosus* and *minutus* have partly yellow legs and thorax, whereas *handlirschii* is all brown except for the tegula and pronotal lobe.

Biology: Little is known. Arnold (1924) states that "this pretty wasp (*formosus*) hops over the sand, keeping its wings in constant motion; when disturbed it takes short jerky flights."

Checklist of *Miscophoides*

Subgenus *Miscophoides*

*handlirschii Brauns, 1896; S. Africa

Subgenus *Miscophoidellus*

formosus Arnold, 1924; Rhodesia
minutus Arnold, 1952; S. Africa

Genus Nitela Latreille

Generic description: Inner orbits converging above (fig. 98 B-D); inner and/or outer orbits sometimes margined by a carina; eyes sometimes with a fine pubescence; antenna variable, pedicel, and flagellomeres usually longer than broad; antennal socket very low on face (fig. 98 A-D), surrounded by a basin which may be carinate above (fig. 98 A-C); frontal carina usually weak or absent, sometimes bifurcate dorsad (fig. 98 B), or lamelliform as in subgenus *Tenila* (fig. 98 D, E); clypeus usually very narrow, mesally tumid and usually bearing a vertical carina which may be continuous with frontal carina, clypeal margin usually with a median truncate or sinuate lobe; labrum quadrate or subtriangular, its free margin sometimes visible beyond clypeal margin; inner margin of mandible usually (always?) with a subapical tooth in female, apex blunt or weakly bifid in some females, inner margin usually toothless in male (subapically dentate in *carinifrons*, with a large subbasal tooth in *darwini*), externoventral margin without a notch; malar space usually present in both sexes; mouthparts short, labial palpus very short, first segment much longer than remaining three; occipital carina incomplete below, usually ending far short of hypostomal carina; mandibular socket open, (fig. 99 B); pronotal collar moderately long and often with a transverse sulcus which is usually interrupted at midline by a carina or triangular prominence, front margin of collar sometimes carinate, humeral angles rounded or angulate; admedian lines of scutum short

---

[18] We have studied only females of this genus and have relied on published descriptions for male characters.

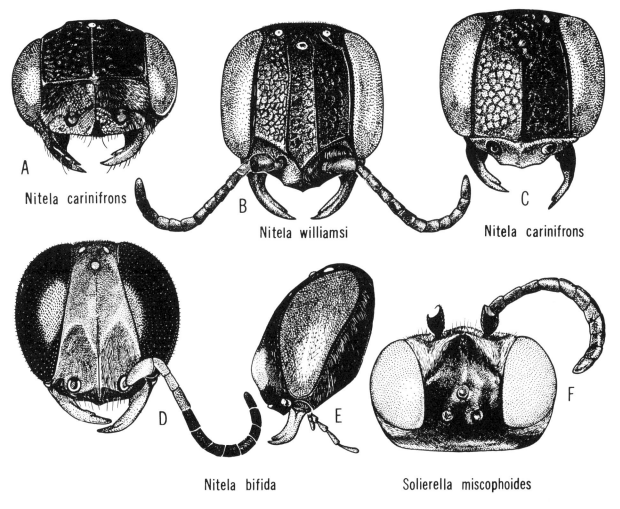

FIG. 98. Facial portraits of females in the tribe Miscophini.

and widely separated when present; scutellum simple or with a narrow sulcus along anterior margin; propodeum long, dorsum without enclosure but usually longitudinally ridged, coarsely reticulate or areolate, posterior face nearly vertical, flat, and sometimes margined by a carina; propleuron usually simple; mesopleuron with an episternal sulcus and a hypersternaulus (fig. 99 A); metapleuron sometimes poorly defined beneath upper metapleural pit; pygidial plate absent, segment VI conical in female and sometimes laterally compressed; legs moderately long, foretarsal rake absent; midcoxae separated, occasionally narrowly so, hindcoxae contiguous; hindtibia spineless; tarsomere V no swollen, arolia small; forewing marginal cell truncate apically (sometimes narrowly so) and appendiculate, forewing with only one submarginal and discoidal cell, and subdiscoidal cell open posteriorly because A_1 is evanescent beyond cu-a, recurrent vein usually received by submarginal I (fig. 85 F), or received beyond in several exceptional species (*rugosa, williamsi, yasumatsui*) (fig. 85 G); hindwing venation greatly reduced and jugal lobe small (fig. 85 F,G); male sternum VIII variable (fig. 99 E, G); volsella at most weakly differentiated from gonostyle (fig. 99 D, F).

Geographic range: Nitela is on every continent. The Ethiopian Region with 11 has the largest share of the 43 species. The rest are about equally divided among the other regions. The genus attains its greatest morphological diversity in the Oriental and Neotropical Regions. Two species, *darwini* and *austrocaledonica,* are known only from oceanic islands, the Galapagos and New Caledonia, respectively.

Systematics: These are small (2.5-6.5 mm), largely black, and often coarsely sculptured wasps. The legs and other appendages may be pale, and in several Ethiopian species the gaster is all or partly reddish. The thorax and gaster are partly red in the Neotropical *guiana.*

Nitela is readily separated from all other miscophin genera by the reduced wing venation. The hindwing in *Nitela* has few veins, no closed cells (fig. 85 F, G), and the subdiscoidal cell of the forewing is open behind. Other features of *Nitela* are: the subapically toothed female mandible and absence of an externoventral mandibular notch or angle in both sexes, long pedicel, labial palpus, longitudinal ridging or coarse reticulation of propodeal dorsum, basins around antennal sockets, and presence of a hypersternaulus on mesopleuron.

The Australian genus *Auchenophorus* may be closely allied with *Nitela*, but the former has more normal hindwing venation with closed cells, and the subdiscoidal cell of the forewing is completely closed (fig. 85 H). However, there is no jugal lobe in *Auchenophorus*, and the closed mandibular socket of this genus is unique in the Miscophini. Probably *Nitela* is most closely allied (albeit distantly) with *Solierella* or *Miscophus*.

There is considerable variation in the configuration of carinae and other surface structures on the head and thorax of *Nitela*. Althouth this variation does not reach the complexity found in the genus *Trypoxylon*, still it poses similar problems at the generic level. Two taxa, *Tenila* and *Rhinonitela*, heretofore considered as genera, have been included by Menke (1968e) in *Nitela*, the first as a subgenus. Both of these "genera" were distinguished from *Nitela* primarily by strange carinal developments on the head. *Tenila* is distinctive by its lamelliform frontoclypeal carina (fig. 98 D, E) and the finely but rather densely and uniformly pubescent eyes. The hairy eyes seem rather unimportant since some typical *Nitela* have sparsely pubescent eyes; but with no intermediates between *Tenila* and typical *Nitela* in the form of the frontoclypeal lamella, we have retained *Tenila* as a subgenus.

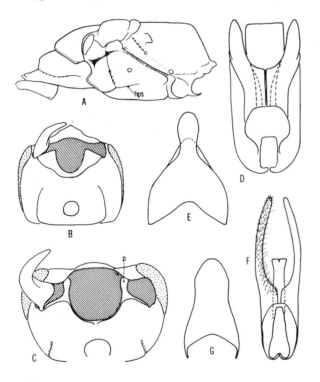

FIG. 99. Structural details in Miscophini; A, *Nitela townesorum* thorax, lateral, hps = hypersternaulus; B, *Nitela williamsi* head, ventral; C, *Auchenophorus* sp. head, ventral, p = paramandibular process of hypostoma; D, *Nitela darwini*, male genitalia; E, *Nitela darwini*, male sternum VIII; F, *Nitela (Tenila) amazonica*, male genitalia, ventral; G, *Nitela amazonica*, male sternum VIII

The type species of *Rhinonitela*, *domestica*, and a recently described relative, *williamsi*, differ rather markedly from other *Nitela*. The frontal carina is continuous with the clypeal carina and bifurcates dorsad (fig. 98 B). The antennal socket basins are margined above by a transverse V-shaped carina, and the inner orbits are bordered by a carina (fig. 98 B). The outer orbit of *williamsi* is margined by a weak sulcus which is bounded by a ridgelike swelling. *Nitela domestica* has a very fine, weak carina but no marginal sulcus. The posterior face of the propodeum is bounded by a carina in both species, but this carina is also found in some typical *Nitela*. Based on *domestica* and *williamsi*, *Rhinonitela* would seem to warrant at least subgeneric status simply because of the facial carinae. However, there are at least two other species of *Nitela* that complicate such a distinction: *carinifrons* and *rugosa*. The face of *carinifrons* is similar to the *Rhinonitela* type except that the frontal carina does not bifurcate (fig. 98 A, C). The outer orbit of *carinifrons* is margined by a coarsely pitted sulcus which is bounded by a carina. The posterior face of the propodeum is margined by a carina. Unlike typical *Nitela* males, the mandible of *carinifrons* is subapically dentate, but since males of *domestica* and *williamsi* are undescribed, the significance of this feature is unknown. The pronotal humeri of *carinifrons* are angulate instead of the rounded *Rhinonitela* type. The humeri of typical *Nitela* are rounded or angulate. *Nitela rugosa* lacks the facial carinae of *domestica*, *williamsi*, and *carinifrons*, but it does have a pitted sulcus and carina along the outer orbit and a carina around the posterior face of the propodeum. The pronotal collar is most similar to *carinifrons* but the humeri are more sharply angulate and there is a strong transverse carina on the collar.

Clearly it is too early to make a final decision as to the status of *Rhinonitela*, but for the time being it seems best to follow Menke's (1968e) suggestion of not recognizing it even as a subgenus because of the existence of intermediate or linking forms like *rugosa* and *carinifrons*.

Species of *Nitela* are separated mainly by head and thoracic sculptural differences, form of the pronotum, details of the head, and wing venation. The pronotum is noteworthy because it varies from smooth with rounded humeri to carinate with sharply pointed humeri. The genitalia and male eighth sternum also offer good distinctions, but so far they have been little used by taxonomists. The female mandible of some species has been reported to be simple, but the absence of a subapical inner tooth needs to be verified in these cases. All females studied by us have subapically dentate mandibles which can be seen only when the mandibles are spread.

Pate (1937b) presented a key to three of the five North American *Nitela*, and Krombein (1950d, 1968) presented notes for distinguishing the others. Arnold (1940) keyed the species of the Ethiopian Region, Tsuneki (1956a) keyed those of Japan, Turner (1916d) keyed those of Australia, and F. X. Williams (1928a) keyed those of the Philippines. Menke (1969) provided a key for the species of the subgenus *Tenila*.

Key to subgenera of Nitela

Face with a continuous lamelliform fronto-clypeal carina (fig. 98 D, E); eyes uniformly pubescent; Neotropical Region .... subgenus *Tenila* Brèthes

Frontoclypeal carina if present not lamelliform; eyes bare or weakly and unevenly pubescent .... subgenus *Nitela* Latreille

Biology: The papers of Janvier (1962), Valkeila (1955), Ahrens (1949), Maneval (1929), and Vincens (1910) on *Nitela spinolae,* and Iwata (1939a) on *N. domestica* are the only extensive sources.

The species nest in the ends of twigs, in cynipid galls, or in beetle burrows in wood. The nest of *spinolae* contains two to six cells which are arranged in linear fashion, and each cell is separated by a plug composed of small wood chips and other materials (see fig. 13 in Janvier). The nests of *domestica* are single celled. Psocoptera appear to be the predominant prey, although *spinolae* is known to use Aphididae (Vincens, 1910; Janvier, 1962) and Psyllidae (Maneval, 1929). Janvier (1962) found aphids in 31 *spinolae* nests. The number of prey per cell varies from six to 38, and the egg is laid transversely between the fore and midcoxae. Valkeila (1955) recorded chalcidoid parasites in some *spinolae* nests.

Checklist of Nitela
Subgenus Nitela

australiensis Schulz, 1908; Australia, Tasmania
    nigricans Turner, 1910
austrocaledonica Williams, 1945; New Caledonia
bicornis Williams, 1928; Philippines
braunsii Arnold, 1940; S. Africa
capicola Brauns, 1911; S. Africa
carinifrons Menke, 1968; Costa Rica
cerasicola Pate, 1937; U.S.: New York
collaris Turner, 1926; Sri Lanka, India, Malaya
costaricensis Brauns, 1911; Costa Rica
domestica (Williams), 1928 (*Rhinonitela*); Philippines, Japan, ? Taiwan
darwini Turner, 1916; Galapagos
fallax Kohl, 1884; Europe
floridana Pate, 1934; U.S.: Florida
henrici Turner, 1926; Sri Lanka
kurandae Turner, 1908; Australia
leoni Krombein, 1968; U.S.: Florida
lubutuana Arnold, 1929; Zaire
    ssp. *nyasae* Arnold, 1947; Malawi
luzonensis Williams, 1928; Philippines, Borneo
maxima Maidl, 1925; Sumatra
merceti Brauns, 1911; S. Africa
    transvaalensis Brauns, 1911
mochii Arnold, 1940; Egypt
ochripes Arnold, 1940; Rhodesia
ohgushii Tsuneki, 1956; Japan
oxiana Gussakovskij, 1945; sw. USSR
parallela Arnold, 1929; Rhodesia
pendleburyi Turner, 1926; Malaya
promontorii Brauns, 1911; S. Africa
reticulata Ducke, 1908; Brazil
retifera Arnold, 1940; Rhodesia
rhodesiae Arnold, 1940; Rhodesia
rufiventris Turner, 1916; Malawi
rugosa Williams, 1928; Philippines, ? Malaya
rugosissima Arnold, 1940; Rhodesia
schmidti Brauns, 1911; Costa Rica
sculpturata Turner, 1916; Australia
    reticulata Turner, 1908, nec Ducke, 1908
spinolae Latreille, 1809; Europe
townesorum Krombein, 1950; U.S.: California
virginiensis Rohwer, 1923; U.S.: e. of Mississippi River
williamsi Menke, 1968; Malaya
yasumatsui Tsuneki, 1956; Japan

Subgenus Tenila

amazonica Ducke, 1903; Brazil, Trinidad
bifida Menke, 1969; Costa Rica
guiana (Williams), 1928 (*Rhinonitela*); Guyana

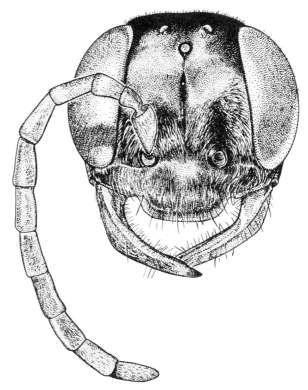

FIG. 100. Facial portrait of *Auchenophorus* sp., female.

Genus Auchenophorus Turner

Generic description: (Based on a female of *coruscans* and a female tentatively identified as *aeneus*. No males have been studied). Inner orbits converging above (fig. 100); orbits not margined by sulci; eyes bare; antenna long, scape short and stout, flagellomeres longer than broad; frons simple but with an incomplete frontal carina which is tuberculate below; clypeal free margin with a broad median lobe rounded at corners, clypeal disk swollen; labrum transverse, hidden; inner and externoventral margins of mandible simple; no malar space; occipital carina incomplete below, ending far short of hypostomal carina; hypostoma with paramandibular process, mandibular socket isolated from oral fossa (fig. 99 C); pronotal collar rather long, obliquely angulate in profile, sometimes transversely carinate; scutum without admedian lines; scutellum with a broad sulcus along anterior margin; propodeum long, dorsum with a V-shaped enclosure which is coarsely ridged or areolate; mesopleuron with an episternal sulcus and hypersternaulus; metapleuron sometimes poorly defined beneath upper metapleural pit; female tergum VI flattened and apex rounded but pygidial carinae absent, legs moderately long, foretarsal rake absent; mid and hindcoxae separated; hindtibia with a single row of spines; tarsomere V not swollen, arolium small; forewing marginal cell rounded, not appendiculate, only one submarginal and discoidal cell, subdiscoidal cell closed behind, recurrent vein received by submarginal cell (fig. 85 H); hindwing with median and submedial cells, but radial sector very short (fig. 85 H); no hindwing jugal lobe; male genitalia undescribed.

Geographic range: The three species are endemic to Australia.

Systematics: These are small to medium sized wasps (5-10 mm) with a distinctive, elongate body and banded wings (fig. 101). Although the ground color is black or metallic blue, the thorax, abdomen and appendages may be partly or extensively red. Turner's (1916d) key to species is misleading with respect to color, as a glance at his species descriptions (Turner, 1907) and our material indicates.

Auchenophorus is a poorly known but very interesting genus of uncertain affinities. When Turner (1907) described the taxon he thought that it was an ampulicine genus, but later (1910) he revised his opinion and placed it near *Nitela*. A relationship with the nyssonine tribe Alyssonini is suggested by the general body form of *Auchenophorus,* but there are a number of basic differences between this genus and the alyssonins. *Auchenophorus* would seem to be related to *Nitela* because of the hypersternaulus, fairly similar forewing venation and mandibles and a ridged or foveate propodeal dorsum.

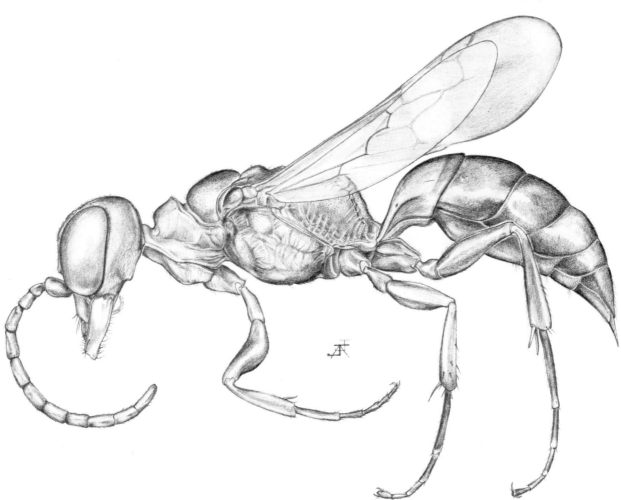

FIG. 101. *Auchenophorus* sp., female.

The hypersternaulus offers the best evidence for a close relationship between the two genera because no other taxa in the Miscophini have this structure. However, unlike *Nitela*, *Auchenophorus* has closed cells in the hindwing and no jugal lobe (fig. 85 H). The swollen clypeus of *Auchenophorus* and its broad median lobe are also distinctive. The closed mandibular socket, the dorsally flattened and apically rounded female tergum VI, the row of spines on the hindtibia, and the noncontiguous hindcoxae are other differences from *Nitela*. These points throw doubt on the supposed relationship. The closed mandibular socket is especially striking because it is known elsewhere in the Larrinae only in *Parapiagetia*. Furthermore, *Auchenophorus* is the only larrine with simple midcoxae, that is, there is no dorsolateral carina or crest. These facts in addition to other peculiarities of the genus suggest that *Auchenophorus* has no close relatives. The genus could be placed in a monotypic tribe, but until more is known about it (knowledge of the male genitalia would be helpful) it seems best to regard *Auchenophorus* as an aberrant miscophin genus.

Notable specific differences are found in the length and form of the pronotum, thoracic sculpture, and details of the frons and vertex. In the two species we have studied the mesosternum has a deep median groove posterad. Anterad this groove terminates in a large pit. The depth of this groove appears to be a specific character. The maxillary palpi have six segments but the first is short and difficult to see. This probably explains Turner's (1907) statement that the palpi are five-segmented. All the segments are short in *coruscans,* but in the species we have tentatively identified as *aeneus* the last four are elongate and the first segment is extremely short.

Turner (1916) provided a key to the species, but because it is based largely on color it is difficult to use.

Biology: Unknown except that Turner (1916d) said of *fulvicornis* "when the wings are closed this species closely resembles *Ephutomorpha impressiventris* André and other similarly coloured Mutillidae, with which it is found running on the ground."

Checklist of *Auchenophorus*

aeneus Turner, 1907; Australia
coruscans Turner, 1907; Australia
fulvicornis Turner, 1907; Australia

Tribe Trypoxylonini

Wasps of this tribe are particularly interesting because they are a highly developed branch of the subfamily Larrinae both in structure and biology. The distinctive angularly emarginate inner orbits, and the tendency towards reduced forewing venation are characteristics that isolate the group from other larrines. Most species are black although red and yellow are often present in tropical forms and Australian representatives. The common slender-bodied *Trypoxylon* wasps are well known, and the more than 350 species make the genus one of the largest in the family. Nests are in preexisting cavities such as plant stems and beetle burrows in wood, or free mud nests are made on sheltered surfaces. Only a few retain the more primitive habit of excavating nests in the ground. The close association of some *Trypoxylon* and *Pison* with human habitations and the resultant common habit of nesting in various household niches has earned them a variety of names such as "keyhole" wasps. All genera provision with spiders, and mass provisioning is the rule. Perhaps the most interesting feature in trypoxylonin biology is the participation of the male in nest building in some *Trypoxylon* and *Pison*. This is an unusual trait in solitary wasps, and much is still to be learned about its significance in the Trypoxylonini. Since the nests are made in almost any cavity, a number of *Trypoxylon* and *Pison* have been distributed through commerce and have become established in areas far removed from their native lands. In fact, the origins of some are unknown due to their widespread distribution and long association with man. The tribe is cosmopolitan, but it is most diversified in the Neotropical Region where several endemic generic elements are found.

Diagnostic characters:

1. (a) Inner orbits angularly emarginate (notched) (fig. 110 A, B), emargination weak in some male *Pisonopsis*, fig. 104 B; (b) ocelli normal.
2. (a) Antennal sockets low on face, contiguous with frontoclypeal suture (fig. 104 A-C, F) (except in *Trypoxylon* and *Pisoxylon*, fig. 104 D, G-L); (b) male with 11 flagellomeres except 10 in one *Trypoxylon*.
3. (a) Clypeus variable but usually with a median projection or lobe; (b) labrum small.
4. (a) Externoventral margin of mandible entire (except notched in *Pisonopsis* and *Pison*, subgenus *Entomopison*), inner margin simple (a single median tooth in some *Pison* and *Trypoxylon s.s.*, apically bidentate in some female *Trypoxylon*); (b) mouthparts short; (c) mandibular socket open.
5. (a) Pronotal collar high, usually narrow and not closely appressed to scutum; (b) posterolateral margin of propleuron usually lamellate, posterolateral angle usually declivous and set off from rest of propleuron by an inner ridge or hump.
6. Scutum with or without notauli.
7. (a) Episternal sulcus present; (b) omaulus, and subomaulus sometimes present in *Trypoxylon*.
8. (a) Female foreleg without a tarsal rake; (b) hindfemur simple apically; (c) midcoxae widely separated to contiguous, (d) arolia moderate to large, equal on all legs; (e) male forecoxa and trochanter unmodified.
9. (a) Propodeum short to long; (b) enclosure often undefined; (c) propodeal sternite present in many *Trypoxylon s.s.*
10. (a) One, two, or three submarginal cells, II petiolate when three are present; (b) two recurrent veins except one in wings with only one submarginal

cell (*Trypoxylon* and *Pisoxylon*), end point of veins variable; (c) marginal cell acuminate except in *Pisonopsis* and some *Pison*.
11. (a) Jugal lobe small, less than half length of anal area; (b) hamuli usually divided into two groups except in most *Pisonopsis* and some *Pison*.
12. (a) Gaster sessile or petiolate; (b) pygidial plate not defined except in some *Pisonopsis* females.
13. (a) Volsella present; (b) gonostyle simple or bifurcate apically.

Systematics: W. Fox (1894d) first proposed the Trypoxylonini as a tribe in his heterogeneous Bembicinae. Later workers, including the most recent, have recognized the group as a subfamily, although it has sometimes been given family status. Evans (1964a), in a summation of his sphecid larval studies and their taxonomic implications, expressed the opinion that this group should be reduced to a tribe in the Larrinae. Our studies of the adults have led us to concur with Evans' suggestion. *Pisonopsis* links the Trypoxylonini with the rest of the Larrinae. Its wing venation, emarginate eyes, biology, and larval morphology (see Evans, 1959a, for discussion of larva) definitely ally *Pisonopsis* with *Pison* and *Trypoxylon,* and in fact, it is separated only with difficulty from *Pison*. On the other hand, the weakly emarginate eyes of some males, the presence of a pygidial plate in some females, and the externoventral notch of the mandible relate *Pisonopsis* to the Miscophini (this is where *Pisonopsis* has recently been placed; see Krombein, 1951). Furthermore, Evans (1959a) stated that the larva of *Pisonopsis* is intermediate in some respects between the trypoxylonin wasps and the larrines. The discovery by Menke (1968c) of notched mandibles in some neotropical *Pison* is perhaps even better evidence that the Trypoxylonini should be regarded as a larrine tribe. There is no doubt that *Pison* and *Trypoxylon* are closely related, but notched mandibles are a larrine character. Some authors (Arnold, 1955, for example) have proposed that *Pison* should be placed in its own subfamily, the Pisoninae. Such a move cannot be substantiated.

In the last analysis, the Trypoxylonini can be isolated only on the basis of having notched eyes and a tendency for reduced wing venation. There is also a trend towards division of the hindwing hamuli into two groups (fig. 102 D, F-J). This division is present in only one species of *Pisonopsis,* but it occurs in about half the *Pison* species studied, and the hamuli always appear to be divided in the remaining genera.

Presumably, the Trypoxylonini have evolved from some ancestral stock of ground-nesting miscophin wasps with externoventrally notched mandibles, nonemarginate inner orbits, three submarginal cells (second petiolate) and a sessile gaster. From this it must be inferred that originally all trypoxylonins had notched mandibles, and females had a pygidial plate. The discovery that some species of *Pisonopsis, Pison,* and *Aulacophilus* still possess tarsal plantulae is intriguing in view of the overall specialized nature of the Trypoxylonini. *Pisonopsis* is the most generalized genus of the tribe, and *Trypoxylon* the most specialized. The loss of the external mandibular notch in most members of the tribe may be related to the change from ground nesting to twig or crevice nesting. The most advanced type of nest is the free mud cell or cells constructed by some *Pison* and *Trypoxylon.* The subsociality in some *Trypoxylon* is another advancement. The reduction of wing venation and lengthening of the gaster are specializations, the significance of which is not clear. The loss of one or more submarginal cells has occurred in *Aulacophilus, Pisoxylon,* and *Trypoxylon*. This loss is accompanied by a lengthening and petiolation of the gaster, except in *Pisoxylon*.

Some specializations in the tribe are restricted mainly to *Trypoxylon*. Among these should be mentioned the presence of an omaulus (fig. 112); extension of the precoxal carina dorsally where it parallels the mesopleural suture (fig. 112); various frontal carinae (fig. 104 G-L); the presence of a supposed sense organ on the ventral surface of the hindcoxa (fig. 111 D); and the propodeal sternite (fig. 111 D). The propodeal sternite in *Trypoxylon* may be regarded as a specialization connected with lengthening and petiolation of the abdomen. This sternite is found in all members of the subfamily Sphecinae and some Ampulicinae. It occurs also in the genus *Parapiagetia,* a rather specialized larrin genus that sometimes has a petiolate gaster.

In the Trypoxylonini the pronotum and propleuron often bear peculiar carinae and/or sulci, which are most evident in *Pison* and *Trypoxylon*. In most of the genera the posterolateral angle of the propleuron is set off inwardly by a ridge or swelling, and the margin of the angle is lamellate, projecting over the ventral edge of the pronotum (fig. 112). This structure is most pronounced in *Pison* and *Trypoxylon*. Richards (1934) termed this area the "rounded tubercle of the propleuron." We suggest that posterolateral declivity of the propleuron is a better term. The various pronotal and propleural structures are usually excellent specific characters.

Biology: The prey are spiders and nests are mostly in preexisting cavities (bee and beetle burrows, *Sceliphron* and vespid mud nests, plant stems, etc.). At least one *Pison* has been reported to excavate burrows in the soil. Some *Trypoxylon* and *Pison* make free mud nests, and some of the former are subsocial. Males are known to stay in the nest during its construction and provisioning by the female in some *Trypoxylon* (*Trypargilum*), and there is a similar report for one *Pison*.

Key to genera of Trypoxylonini

1. Forewing with one submarginal cell (fig. 102 F-J); antennal sockets not contiguous with frontoclypeal suture (fig. 104 D, G-L) 2
   Forewing with two or three submarginal cells (fig. 102 A-E); antennal sockets contiguous with frontoclypeal suture (fig. 104 A-C, F) 3

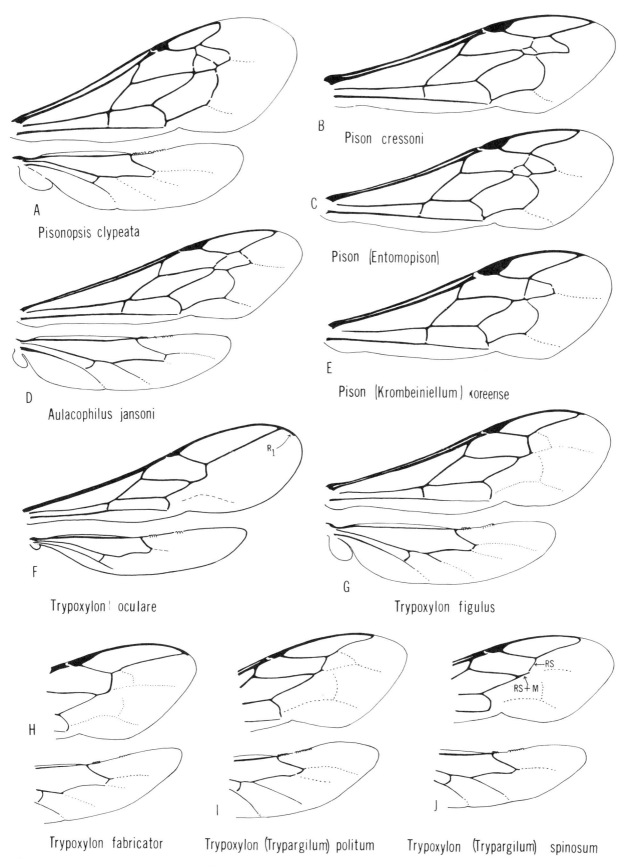

FIG. 102. Wings in the tribe Trypoxylonini.

2. Gaster long, clavate, often petiolate, segment I usually slender, at least two times as long as wide (fig. 109); Old and New World .... *Trypoxylon* Latreille, p. 339
Gaster compact, sessile, segment I not elongate (fig. 106 A); S. America .... *Pisoxylon* Menke, p. 338
3. Gaster petiolate, segment I rodlike (tergum nodose at apex) and nearly as long as remaining segments combined (fig. 107); mesopleuron with many coarse horizontal ridges; Neotropical Region .... *Aulacophilus* F. Smith, p. 337
Gaster compact, sessile, segment I not petiolate or at most pedunculate in dorsal view; mesopleuron without coarse horizontal ridges; Old and New World 4
4. Marginal cell of forewing rounded or truncate distally, the apex not or only slightly extending beyond outer veinlet of submarginal cell III (fig. 102 A); externoventral margin of mandible notched or strongly angulate (fig. 104 A); sterna III-IV with a lateral oblique groove (fig. 106 B); female tergum VI usually flattened or with a distinct pygidial plate bounded by carinae; New World .... *Pisonopsis* W. Fox, p. 329
Marginal cell of forewing acute distally, apex extending well beyond outer veinlet of submarginal cell III, (fig. 102 B-C, E); *or if apex rounded and/or not extending much beyond submarginal cell III (exceptional Old World species), then externoventral margin of mandible not notched; externoventral margin of mandible entire (except in some Neotropical forms but wing characteristics typical); sterna without oblique grooves; female tergum VI conical, sometimes weakly keeled along midline; cosmopolitan .... *Pison* Jurine, p. 332

Genus Pisonopsis W. Fox

Generic description: Inner orbits converging above or parallel, emargination weak in male (fig. 104 A, B); frons simple or with a faint median longitudinal line or carina; basal flagellomeres of male excavate beneath; clypeus transverse (width three or more times height), free margin with a median triangular lobe (lobe truncate in *australis*); labrum small, hidden, quadrangular or longer than wide; externoventral margin of mandible with a strong angle (notched); occipital carina incomplete, disappearing just before hypostomal carina; scutum without notauli, propodeum short, enclosure not or poorly defined, dorsum with a median longitudinal sulcus; gaster sessile, terga I-II with lateral carina, each side of sterna III-IV usually with a transverse oblique groove (gradulus) which rarely extends to middle of plate; female tergum VI with a triangular carina-delimited, hairless pygidial plate in *clypeata* and *australis*, tergum VI weakly flattened, or conical in both sexes of other species; midcoxae widely separated, hindcoxae contiguous; arolia moderate, forewing with three submarginal cells, recurrent veins usually received by II or interstitial, sometimes first recurrent received by submarginal I (fig. 102 A); marginal cell rounded apically, apex not or only slightly surpassing outer veinlet of submarginal III (fig. 102 A); hindwing hamuli not divided into two groups (except in *birkmanni*); male sternum VIII as in fig. 103 C; gonostyle simple (fig. 105 D).

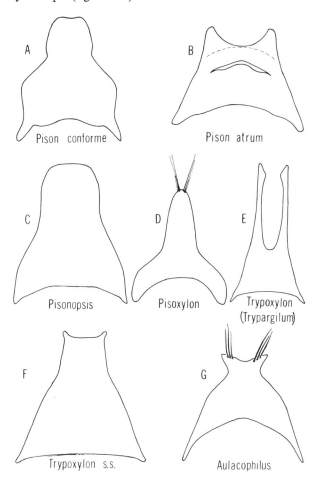

FIG. 103. Male sternum VIII in genera of the tribe Trypoxylonini.

Geographic range: The five species currently recognized in *Pisonopsis* occur in two widely separate areas: western North America and southern South America. This disjunct pattern suggests that *Pisonopsis* is a relict group.

Systematics: Pisonopsis is not easily distinguished from *Pison*, except by the combination of externoventrally notched mandibles, distally rounded marginal cell (fig. 102 A), and obliquely grooved sterna III-IV (fig. 106 B). Of these three features only the sternal grooves (graduli) are unique to *Pisonopsis*, but they are short and weak in some species, such as *areolata*. We have studied nearly 40 species of *Pison* and have not seen graduli; but there are 100 or so species we have not seen, and it is possible that the grooves are present in some of these. Also, graduli are often difficult to see in *Pisonop-*

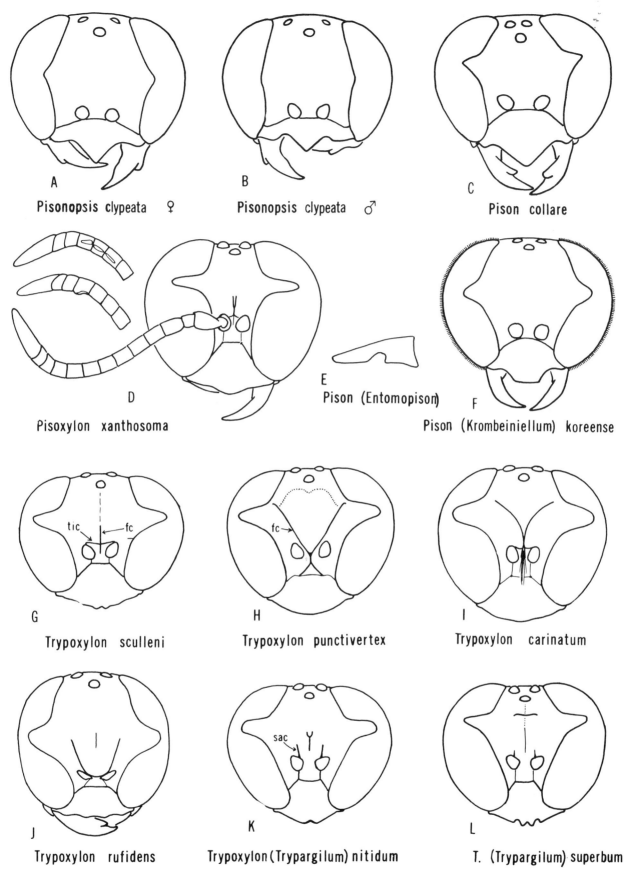

FIG. 104. Facial details in the tribe Trypoxylonini; E, mandible, lateral; tic = transverse interantennal carina; fc = frontal carina; sac = supra-antennal carina.

sis because of the telescoping of the segments. At least two *Pison* (*sericeum* and *eremnon*) have short, distally rounded marginal cells like those of *Pisonopsis*. The presence of a pygidial plate in two *Pisonopsis* (*clypeata* and *australis*) is a feature not found in any *Pison,* but the absence of a pygidial plate in the remaining three *Pisonopsis* is disconcerting. The genitalia of *Pisonopsis* do not differ significantly from the pattern found in *Pison,* although they are more simplified than those of many *Pison* (compare fig. 105 A-C, D).

F. Williams (1954b) revised the North American species, but there is some question about the status of his subspecies. They may represent species; further study of the situation is needed. The most distinctive form is the Argentinean *australis* which has a strong nasiform projection on the clypeal disk of the female. Menke (1968c) discussed the status of some of the South American *Pisonopsis*.

Biology: H. Janvier (1928) gave a good account of the nest of the Chilean species, *areolata*. Bamboo shoots were utilized, and the nests contained three to seven cells, each separated by a thin wall of clay. The North American species, *P. birkmanni*, the female of which has no pygidial plate, has been observed nesting in various plant stems (F. Williams, 1954b; Parker and Bohart, 1966, 1968). Small lumps of soil were used by this species to separate the cells. *Pisonopsis clypeata,* a species with a pygidial plate, has been reported to nest in bee burrows in the ground (Linsley, MacSwain, and Smith, 1952). Small pebbles and wood fragments were used to separate the cells. Evans (1969a) observed *clypeata* nesting in what appeared to be abandoned burrows of *Philanthus* and a tiger beetle. The burrow is left open until provisioning is complete (Evans, 1970). Parker and Bohart (1968) found *clypeata* nesting in *Sambucus* trap stems, indicating that species with a pygidial plate do not always nest in the soil.

Checklist of *Pisonopsis*

australis Fritz, 1965; Argentina
areolata (Spinola), 1851 (*Pison*); Chile, ? Argentina
　　variicornis Reed, 1894 (*Pison*)
　? *anomala* Mantero, 1901
birkmanni Rohwer, 1909; U.S.: Texas, California; Mexico
clypeata W. Fox, 1893; U.S.: Nevada, California
　　ssp. *occidentalis* Williams, 1954; U.S.: California
triangularis Ashmead, 1899; U.S.: Colorado to California
　　ssp. *californica* Williams, 1954; U.S.: California

Genus Pison Jurine

Generic description: Inner orbits usually converging above, infrequently parallel (converging below in some *Krombeiniellum* and in *Pison melanocephalum*); eyes covered with short dense pile in subgenus *Krombeiniellum* (fig. 104 F); frons simple or with a short median longitudinal line or carina; basal flagellomeres of male sometimes excavate or otherwise modified; clypeus variable but usually about twice as wide as high, free margin usually with a truncate or V-shaped median lobe; labrum quadrangular, apex sometimes bilobed; mandible usually simple but with an inner subapical or mesal tooth in some species, externoventral margin notched only in subgenus *Entomopison;* occipital carina rarely a complete circle, usually disappearing just before reaching midline of hypostomal carina, if a complete circle, it may or may not be contiguous with apex of hypostomal carina; scutum without or with weak notauli, propodeum short to moderately long, dorsum with a median longitudinal sulcus and/or carina but no enclosure; propodeum often with lateral carina or areolate sulcus running between spiracle and petiole socket; gaster sessile (rarely pedunculate, subgenus *Pisonoides,* fig. 106 C), terga I-II with lateral carina, sterna without graduli, male sterna sometimes with central tubercles or transverse welts; female tergum VI conical, without a pygidial plate; midcoxae widely separated to subcontiguous, hindcoxae contiguous; arolia moderate and equal; forewing usually with three submarginal cells (fig. 102 B-C), II always petiolate and sometimes greatly reduced, if II is lacking (wing with only two submarginals, fig. 102 E), then III is definitive submarginal II and is not petiolate, end point of recurrent veins in three-celled wings highly variable, in two-celled wings recurrent I received by submarginal I and second recurrent usually interstitial; marginal cell usually acuminate distally (rounded in *eremnon* and *sericeum*) and apex usually well beyond outer veinlet of last submarginal cell; hindwing hamuli sometimes divided into two groups; male sternum VIII variable (fig. 103 A-B); gonostyle simple or biramous, volsella often greatly enlarged and partially fused with gonostyle (fig. 105 A-C).

Geographic range: Cosmopolitan but most of the 145 species occur in the Southern Hemisphere and a third of the species are found in Australia. A large number of endemic forms have evolved on islands in the Pacific Ocean (Yasumatsu, 1953) and to a lesser extent on islands in other areas. The small number of species in the Northern Hemisphere in constrast to the large number in isolated Southern Hemisphere areas such as Australia and South America (mostly undescribed in the latter), plus the many oceanic species, suggests that the genus was at one time much bigger but is now restricted to isolated areas where the species are protected from competition with more advanced wasps such as *Trypoxylon*.

Since many species nest in cavities in lumber or other items commonly transported by ships, a number of *Pison* have been successfully introduced to different parts of the world. *Pison koreense,* an Oriental species, was introduced into the United States after World War II (Krombein, 1958e). The geographic origins of a few species are unknown (*argentatum, hospes, iridipenne*) because they were apparently distributed long ago. *Pison argentatum,* with the greatest range of any *Pison,* is known from Madagascar to Hawaii.

The subgenus *Pison* occurs throughout the world except for North America north of Mexico. *Pisonoides* contains two Australian and one Oriental species, and *Ento-*

mopison is a small Neotropical subgenus. *Krombeiniellum* has representatives in the Oriental and Neotropical Regions, and one species, *koreense*, has been introduced into North America.

Systematics: The presence of two or three submarginal cells in the forewing (fig. 102 B-C, E) and the sessile (rarely pedunculate) abdomen distinguish *Pison* from all other trypoxylonine genera, except *Pisonopsis*. *Pison* apparently lack the graduli found on the sterna of *Pisonopsis,* and, except for the subgenus *Entomopison, Pison* have the externoventral margin of the mandible entire. In the subgenus *Entomopison* the marginal cell is acuminate distally, and the apex extends well beyond the outer veinlet of submarginal cell III. The marginal cell is rounded or truncate distally in *Pisonopsis*, and the apex is about even with the outer veinlet of submarginal cell III.

Most *Pison* are black although the legs are occasionally red, and a number of Australian species are extensively red. Length varies from 5 to 21 mm. The largest species is probably the east Asian *regale*.

Of the many variables in *Pison* none is more striking than the forewing venation. Wings have three or two submarginal cells, and the two-celled condition is clearly the result of complete reduction of the second cell. This is easily seen in long series of specimens of species in which submarginal II varies all the way down to pinhole size. Examples of such species are the African *xanthopum, denticeps,* and *inaequale*. In the last-named species one wing may have two cells and the other three. The name *Pisonoides* was proposed for two-celled *Pison*, but this condition occurs in unrelated groups of species. Menke (1968a) restricted *Pisonoides* to three Australasian species with a pedunculate gaster (fig. 106 C) and two submarginal cells (*obliteratum, icariodes,* and *difficile*). It remains to be seen whether or not *Pisonoides* can be defended even with these two characters. *Pison lobiferum,* a species from Madagascar with three submarginal cells, has a rather narrow gastral segment I (see Arnold, 1945).

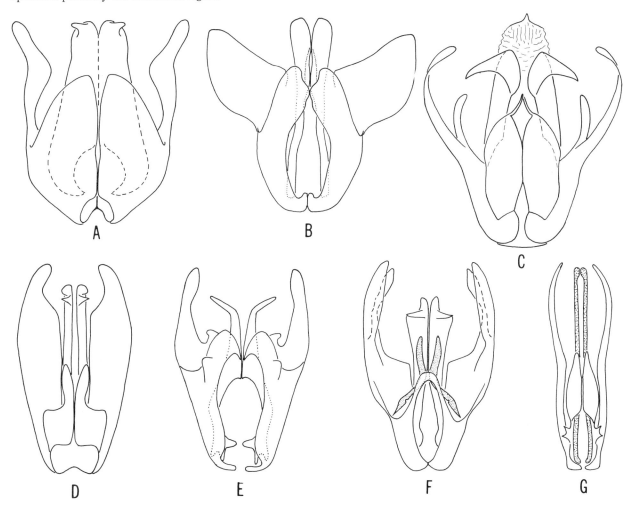

FIG. 105. Male genitalia of various trypoxylonin genera; A, *Pison conforme;* B, *Pison spinolae;* C, *Pison aureosericeum;* D, *Pisonopsis clypeata;* E, *Pisoxylon xanthosoma;* F, *Trypoxylon (Trypoxylon) sculleni* (Stippled structure is volsella); G, *Trypoxylon (Trypargilum)* sp.

The subgenus *Krombeiniellum* originally contained species with two submarginal cells and hairy eyes. Recently *Krombeiniellum* with three submarginal cells have been noted in Central and South America (Krombein, 1958e; Menke, 1968c). This subgenus can still be defined on the basis of densely hairy eyes, but some Neotropical *Pison (Pison)* have sparsely hairy eyes. Thus, there is some doubt as to the validity of *Krombeiniellum*, also.

The most variable wing pattern is produced by the recurrent veins. In wings with three submarginal cells the first recurrent is received by I or II or is interstitial. The second recurrent is received by II or III or is interstitial. Of nine possible arrangements of recurrent veins in three-celled *Pison*, we have seen seven. The subgenus *Pisonitus* has been used for *Pison* in which the second recurrent vein is received at the middle of the second submarginal cell. The value of this taxon is dubious in view of the above facts, and we have not recognized it. In two-celled wings there is less variability. The first recurrent is always received by I. The second recurrent is usually interstitial in two-celled *Krombeiniellum* and received just inside submarginal II in *Pisonoides* and two-celled species of *Pison (Pison)*. In *Pison (Krombeiniellum) stangei*, both recurrents are received by submarginal I.

The apex of the marginal cell is usually acuminate in *Pison*, but in *P. eremnon*, *sericeum*, and probably a few other species it is rounded as in *Pisonopsis*.

The mandible is usually simple, but some species (*spinolae* and *eremnon*, for example) have the beginnings of an inner mesal tooth, and in *regale* and *collare* there is a large inner tooth (fig. 104 C). There is a subapical inner tooth in some Neotropical species such as *cameronii*. The Neotropical subgenus *Entomopison* is characterized by a notch on the externoventral margin of the mandible (fig. 104 E). The basal flagellomeres are excavate and/or tuberculate in males of *algericum*, *fenestratum*, and *chilense*.

The sterna of males in some species (*insigne*, *impunctatum*, *regale*, *tuberculatum*, *nigellum*, *korrorense*, *iridipenne*, *westwoodii*, and *oakleyi*) bear polished tubercles or transverse welts. Male sternum VIII and genitalia have not been thoroughly investigated by us, but it is obvious that there is considerable variation in these structures (figs. 105 A-C; 103 A, B). The hindwing hamuli need to be studied. In some *Pison* there are two groups, as in *Trypoxylon*.

In addition to the features discussed above some other useful species characters are: sculpture of frons, thorax and gaster; shape of clypeus; placement of ocelli; details of hypostomal carina, prementum and stipes; details of propleura and pronotum; and appearance of the metapleural flange.

With the exception of the Neotropical members, there are keys to nearly all species in the genus. Turner (1916e) gave keys for the Ethiopian and Australian Regions as well as India and environs. He listed species known up to 1916, and his paper is a good taxonomic starting point. Gussakovskij (1937) covered the Palearctic forms, and those of the

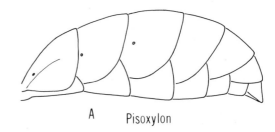

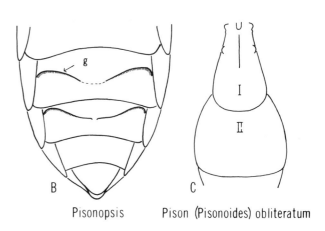

FIG. 106. Gastral features of Trypoxylonini; A, male, lateral; B, venter of *P. triangularis* female, g = gradulus; C, terga I-II, dorsal.

Ethiopian Region have been treated recently by Leclercq (1965). Beaumont (1961b) revised the species inhabiting the Mediterranean area. Yasumatsu (1935b, 1939) revised the *Pison* of Japan and China. Krombein (1949, 1950a) keyed the *Pison* of Micronesia, and Yasumatsu (1953) published a useful supplement to this work. Menke (1968c) described some new Neotropical species, discussed the subgenera of *Pison*, and keyed the species of *Krombeiniellum*. There are many undescribed *Pison* in the Neotropics, and a revision is needed.

Key to subgenera of *Pison*

1. Externoventral margin of mandible notched or strongly angulate (fig. 104 E); S. America .... subgenus *Entomopison* Menke, p. 337
   Mandible entire 2
2. Eyes densely covered with short hair (fig. 104 F), Oriental Region, New World .... subgenus *Krombeiniellum* Richards, p. 337
   Eyes bare 3
3. Gaster sessile, not pedunculate, segment I bellshaped in dorsal view, width of segment I much more than two-thirds greatest width of II and usually subequal to II; forewing with two or three submarginal cells; Old and New World subgenus *Pison* Jurine, p. 335
   Gaster pedunculate, sides of segment I subparallel in dorsal view (fig. 106 C), apical width of segment I about half to two-thirds greatest width of II; forewing with

two submarginal cells; Australasian
Region .... subgenus *Pisonoides* F. Smith, p. 337

Biology: The more important papers on *Pison* biology are those of Bordage (1912) on *argentatum;* Micheli (1933) on *atrum;* Pagden (1934) on *argentatum, obliteratum, punctifrons,* and *erythropum;* H. Masuda (1939) and Tsuneki (1970b) on *strandi;* Sheldon (1968) on *koreense;* Iwata (1964a, b) on *argentatum, koreense, obliteratum,* and *punctifrons;* Yoshimoto (1965) on *argentatum;* Horne (1870) on *erythropum* (as *rufipes* F. Smith) and *rugosum;* Janvier (1928) on *chilense;* and Cowley (1962) on *spinolae.*

Some species nest in plant stems or beetle burrows in wood and partition the cells with mud or cemented pellets (*atrum, strandi, punctifrons, rugosum*) (good photos in Masuda and in Tsuneki). Oniki (1970) found a Brazilian *Pison* building mud cells in the nests of a hummingbird. The observations by Masuda and Tsuneki of a male guarding the nest while the female is provisioning (*P. strandi*) is quite interesting. Similar behavior has not been reported for any other *Pison* but has been noted in some *Trypoxylon* subgenus *Trypargilum.*

Pison argentatum, ignavum, koreense, erythropum, and *obliteratum* make free mud nests in sheltered situations, although *argentatum* may make its mud cells inside cavities. These nests are usually made up of separate cells laid in rows or simply in irregular groups (see figures in Horne, 1870; Iwata, 1964b; and photos in Yoshikawa, 1964). Sometimes an aggregate of cells may be covered with a layer of mud. The mud nests of other wasps such as *Eumenes* and *Sceliphron* are sometimes used by *Pison.* The expropriated cells are subdivided by walls of mud.

Janvier's observations on *Pison chilense* are of particular interest. According to him, this species excavates its own burrows in the banks of stream beds. During nest excavation the female continually wets the soil with water obtained at the stream. The cells are arranged in a linear fashion, as in twig-nesting species, and each is separated by a wall of mud. Krombein's (1950a) notation of *Pison nigellum* Krombein nesting in galleries in the soil is more understandable in the light of Janvier's observations.

Pison cells are stocked with a variety of spiders, although some species seem to show preference for certain spider families. The egg is laid on the last provision according to Masuda, Sheldon, and Yoshimoto.

Sheldon (1968) summarized the known parasites of *Pison.* These include Eulophidae, Chrysididae, and Diptera.

Checklist of *Pison*
Subgenus *Pison*

aberrans Turner, 1908; Australia
agile (F. Smith), 1869 (*Parapison*); Sri Lanka
algiricum Kohl, 1898; nw. Africa
allonymum Schulz, 1906; Ethiopian Region
    iridipenne Cameron, 1905, nec F. Smith, 1879
    rhodesianum Bischoff, 1913
    karrooense Arnold, 1924
ashmeadi Turner, 1916; Philippines
    punctulatum Ashmead, 1905, nec Kohl, 1883
assimile Sickmann, 1894; China
argentatum Shuckard, 1838; Mauritius, Madagascar, India, Malaysia, Borneo, Philippines, Hawaii, Marianas, Samoa, Fiji, Carolines, Solomons
    fuscipalpe Cameron, 1901
    argenteum Ashmead, 1904 (*Pisonitus*)
atripenne Gussakovskij, 1938; China
atrum (Spinola), 1808 (*Alyson*); Europe
    jurini Spinola, 1808
    nigrum Latreille, 1809 (*Tachybulus*)
auratum Shuckard, 1838; Australia
aureosericeum Rohwer, 1915; Australia
aurifex F. Smith, 1869; Australia
auriventre Turner, 1908; Australia
basale F. Smith, 1869; Australia
caliginosum Turner, 1908; Australia
cameronii Kohl, 1893; Mexico or Peru
    fasciatum Kohl, 1883, nec Radoszkowski, 1876
carinatum Turner, 1917; centr. Africa, Egypt, Cyprus
    cyprium Gussakovskij, 1937
    xanthopus of Turner, 1916
chilense Spinola, 1851; Chile
collare Kohl, 1887; Bismarck Archipelago, Solomons
conforme F. Smith, 1869; Mexico
congenerum Turner, 1916; Australia
cressoni Rohwer, 1911; Nicaragua
decipiens F. Smith, 1869; Australia
denticeps Cameron, 1910; S. Africa
deperditum Turner, 1917; Australia
dimidiatum F. Smith, 1869; Australia
dives Turner, 1916; Australia
eremnon Menke, 1968; Guyana, Brazil
erythrocerum Kohl, 1884; Australia
    ruficorne F. Smith, 1869 (*Parapison*), nec F. Smith, 1856
erythrogastrum Rohwer, 1915; Australia
erythropus Kohl, 1884; India
    rufipes F. Smith, 1869 (*Parapison*), nec Shuckard, 1838
    rufipes F. Smith, 1870 (*Parapison*), nec Shuckard, 1838
esakii Yasumatsu, 1937; Marianas
exclusum Turner, 1916; Australia
exornatum Turner, 1916; Australia
exultans Turner, 1916; Australia
fasciatum (Radoszkowski), 1876 (*Pseudonysson*), sw. Asia, ? India
fenestratum F. Smith, 1869; Australia
    nitidum F. Smith, 1868, nec F. Smith, 1858
festivum F. Smith, 1869; Australia
flavolimbatum Turner, 1917; Guyana
fraterculum Turner, 1916; Australia
fuscipenne F. Smith, 1869; Australia
glabrum Kohl, 1908; Samoa
hanedai Tsuneki, 1973; Ogasawara Is.

hospes F. Smith, 1879; Singapore, Cocos Is., Marshalls, Marianas, Carolines, Samoa, Fiji, Tonga, Marquesas, Society Is., Ellice Is., Hawaii
    palauense Yasumatsu, 1937, nomen nudum
hissaricum Gussakovskij, 1937; sw. USSR: Tadzhik S.S.R.
humile Arnold, 1945; Madagascar
ignavum Turner, 1908; Australia, Fiji, New Caledonia, Society Is., Samoa, Marquesas, Caroline
impunctatum Turner, 1912; New Guinea, Society Is., Marquesas
inaequale Turner, 1916; Malawi
inconspicuum Turner, 1916; Australia
infumatum Turner, 1908; Australia
insigne Sickmann, 1894; China
insulare F. Smith, 1869; New Hebrides Is.; Hawaii; Banks Is.
iridipenne F. Smith, 1879; Hawaii, Samoa, Marshalls, Marianas, Bolabola Is., Carolines, Fiji, Marquesas, Tuamotu Archipelago, Society Is., Bismarck Archipelago, ? Australia
isolitum Turner, 1911; Seychelles
kohlii Bingham, 1897; Burma, Borneo
    aureopilosum Cameron, 1909
korrorense Yasumatsu, 1937; Palau Is. (? = *iridipenne*)
laeve F. Smith, 1956; U.S.: Georgia? (USSR: Georgian S.S.F.? or New Georgia?)
lobiferum Arnold, 1945; Madagascar
lutescens Turner, 1916; Australia
maculipenne F. Smith, 1860; Brazil
mandibulatum Turner, 1916; Australia
marginatum F. Smith, 1856; Australia
mariannense Yasumatsu, 1953; Marianas
melanocephalum Turner, 1908; Australia
meridionale Turner, 1916; Australia
minicum Arnold, 1945; Madagascar
montanum Cameron, 1910; Ethiopian Region
morosum F. Smith, 1856; New Zealand
multistrigatum Turner, 1917; Malawi
nigellum Krombein, 1949; Carolines
nitidum F. Smith, 1858; Indonesia: Ewab, Aru
noctulum Turner, 1908; Australia
novocaledonicum Krombein, 1949; New Caledonia
oakleyi Krombein, 1949; Guam
  ssp. *rotaense* Tsuneki, 1968; Marianas: Rota
  ssp. *boninense* Tsuneki, 1973; Ogasawara Is., Bonin Is.
obesum Arnold, 1958; centr. Africa
orientale Cameron, 1897; India
pallidipalpe F. Smith, 1863; Indonesia: Ceram
papuanum Schulz, 1904; New Guinea (? = *collare* Kohl)
    morosum F. Smith, 1864, nec F. Smith, 1856
    constrictum Turner, 1912
pasteelsi Leclercq, 1965; centr. Africa
peletieri Le Guillou, 1841 (Rev. Mag. Zool.); Australia
    pelletieri Le Guillou, 1841 (Ann. Soc. Ent. France)
perplexum F. Smith, 1856; Australia
pertinax Turner, 1908; Australia
petularum Leclercq, 1965; centr. Africa
ponape Krombein, 1949; Carolines
pregustum Leclercq, 1965; centr. Africa
premunitum Leclercq, 1965; centr. Africa
priscum Turner, 1908; Australia
pulchrinum Turner, 1916; Australia
punctifrons Shuckard, 1838; Oriental Region, Japan, Bonins, Marianas, Carolines, Marshalls, Hawaii; ? St. Helena
    suspiciosum F. Smith, 1858
    fabricator F. Smith, 1869
    striolatum Cameron, 1896
    lagunae Ashmead, 1904
    javanum Cameron, 1905
    japonicum Gussakovskij, 1937
punctulatum Kohl, 1884; Australia
regale F. Smith, 1852; China
repentinum Arnold, 1940; Rhodesia, Malawi
rothneyi Cameron, 1896; India
    crassicorne Cameron, 1897
ruficorne F. Smith, 1856; Australia
rufipes Shuckard, 1838; Australia, Tasmania
rufitarse Arnold, 1944; Rhodesia
rugosum F. Smith, 1856; India, Pakistan
    appendiculatum Cameron, 1897
sarawakense Cameron, 1903; Sarawak
scabrum R. Turner, 1908; Australia
scruposum Arnold, 1955; Rhodesia
separatum F. Smith, 1869; Australia
sericeum Kohl, 1888; s. Europe
seyrigi Arnold, 1945; Madagascar
simillimum F. Smith, 1869; Australia
simulans Turner, 1915; Tasmania
sogdianum Gussakovskij, 1937; sw. USSR: Uzbek S.S.R.
speculare Turner, 1911; Seychelles
spinolae Shuckard, 1838; Australia, Tasmania, New Zealand
    australe Saussure, 1853
    tasmanicum F. Smith, 1856
    dubium W. F. Kirby, 1883 (*Taranga*)
    pruinosum Cameron, 1898
strandi Yasumatsu, 1935; Japan
    iwatai Yasumatsu, 1935
strenuum Turner, 1916; Australia
strictifrons Vachal, 1907; New Caledonia
strigulosum Turner, 1917; Ghana
susanae Cheesman, 1955; New Caledonia
suspicax Kokujev, 1912; sw. USSR: Caucausus
tahitense Saussure, 1867; Society Is., Samoa, Cook Is., Fiji, Ellice Is., Marquesas, Marshalls, Tabuai Is.
    rechingeri Kohl, 1908
tenebrosum Turner, 1908; Australia
testaceipes Turner, 1916; Nigeria
tibiale F. Smith, 1869; Australia
tosawai Yasumatsu, 1935; Bonins
transvaalense Cameron, 1910; s. Africa
    clypeatum Cameron, 1910
trukense Yasumatsu, 1953; Carolines
tuberculatum F. Smith, 1869; New Zealand
ugandense Arnold, 1955; Uganda
ussuriense Gussakovskij, 1937; se. USSR: Maritime Terr.

vestitum F. Smith, 1856; Australia
virosum Turner, 1908; Australia
wagneri Arnold, 1932; e. Africa
westwoodii Shuckard, 1838; Australia, Tasmania
  ? *obliquum* F. Smith, 1856
wollastoni Turner, 1916; St. Helena
xanthopus (Brullé), 1833 (*Nephridia*) centr. and w. Africa
  ? *obscurum* Shuckard, 1838

Subgenus *Pisonoides*

difficile Turner, 1908; Australia
icarioides Turner, 1908; Australia
obliteratum F. Smith, 1858; India, Malaysia, Indonesia

Subgenus *Krombeiniellum*

browni (Ashmead), 1905 (*Pisonoides*); Philippines
differens Turner, 1916; India, Java
duckei Menke, 1968; Brazil, Trinidad
koreense (Radoszkowski), 1887 (*Paraceramius*); Korea, China, Japan, e. U.S.
krombeini Menke, 1968; centr. America
neotropicum Menke, 1968; Brazil
plaumanni Menke, 1968; Brazil
stangei Menke, 1968; Argentina

Subgenus *Entomopison*

aureofaciale Strand, 1910; Paraguay, Brazil
convexifrons Taschenberg, 1870; Brazil
pilosum F. Smith, 1873; Brazil

Fossil *Pison*

cockerellae Rohwer, 1908
oligocenum Cockerell, 1908
  oligocaenum Cockerell, 1909

Genus Aulacophilus F. Smith

Generic description: Inner orbits converging above or below; frons simple or with a faint frontal line or carina; flagellomeres shorter and thicker towards apex in both sexes; clypeus about twice as wide as high, swollen, margin variable; labrum small, hidden; mandible simple; occipital carina incomplete, disappearing just short of hypostomal carina; collar moderately long to short, scutum without notauli; propodeum moderately long, with no enclosure, dorsum with or without longitudinal sul-

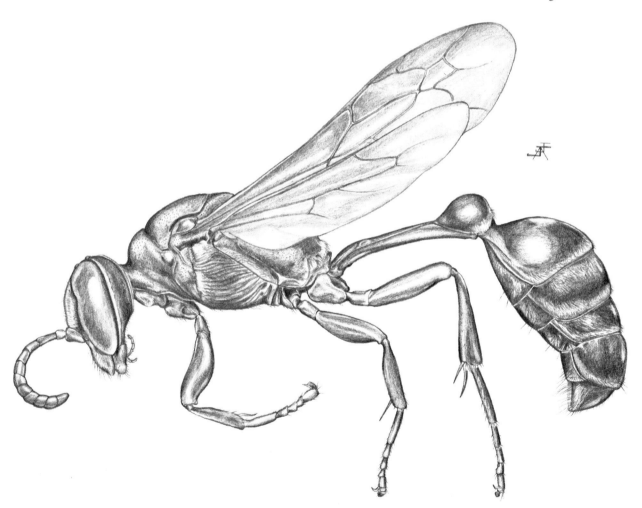

FIG. 107. *Aulacophilus jansoni* Turner, female.

cus and/or carina, no lateral carina separating propodeal side from dorsum; mesopleuron with coarse longitudinal ridging between episternal sulcus and mesopleural suture (fig. 107); metapleuron parallel sided for most of its length; gaster petiolate, segment I rodlike (except for bulblike apical swelling of tergum) and nearly as long as remaining segments combined; tergum I and sometimes II with lateral carina; sterna without graduli; female tergum VI conical; midcoxae separated, hindcoxae contiguous; arolium moderately large on all legs in both sexes; forewing with two submarginal cells, II receiving both recurrent veins or first recurrent interstitial; marginal cell acuminate distally, apex well beyond outer vein of second submarginal cell (fig. 102 D); male sternum VIII as in fig. 103 G; gonostyle slender, volsella large and fused basally with gonostyle.

Geographic range: Neotropical Region: southern Mexico to Brazil and Peru. Three species are known.

Systematics: This is a very distinctive genus (fig. 107). The slender, apically nodose petiole and longitudinally ridged mesopleuron are characteristics that make identification of *Aulacophilus* very easy. These features indicate that *Aulacophilus* is not closely allied to the remaining trypoxylonin genera. The petiole is composed of both tergum and sternum, not the sternum alone as stated by Turner (1916e).

A key to the three species of *Aulacophilus* can be found in Turner's (1916e) paper on *Pison*.

Biology: Unknown.

Checklist of *Aulacophilus*

eumenoides Ducke, 1904; Trinidad, Brazil
jansoni Turner, 1916; Mexico to Nicaragua
vespoides F. Smith, 1869; Brazil, Peru

Genus Pisoxylon Menke

Generic description: Inner orbits slightly converging below (fig. 104 D); frontal carina narrowly Y-shaped, base of Y with a short perpendicular branch (transverse interantennal carina) at upper level of antennal sockets which are not contiguous with frontoclypeal suture; clypeus nearly as high as wide, trapezoidal; labrum small, with two narrow fingerlike processes, the tips of which are barely visible and bear long setae; mandible simple, occipital carina meeting hypostomal carina just before apex of latter; notauli absent, propodeum moderately long, dorsum with a median, longitudinal sulcus, but no enclosure, no lateral carina separating propodeal side from dorsum; gaster sessile, tergum I with lateral carina, sterna without graduli; midcoxae essentially contiguous, hindcoxae contiguous; arolia moderately large, forewing with one submarginal cell and one recurrent vein, marginal cell acuminate, its anterior veinlet extending beyond apex of cell (similar to *Trypoxylon fabricator* fig. 102 H); sternum VIII as in fig. 103 D; genitalia as in fig. 105 E.

Geographic range: South America. Only one species known.

Systematics: *Pisoxylon* is very similar to *Trypoxylon* but differs in having a compact, sessile gaster (fig. 106 A). The gaster is always slender and clavate in *Trypoxylon*, and segment I is often so narrow that it forms a petiole. Some *Trypoxylon* have a short first segment (*rufidens* group for example), but the gaster is long and narrow, not compact as in *Pisoxylon*. Menke (1968a) described *Pisoxylon* from a male. There is a female from Petropolis, Brazil in the U.S. National Museum that may be *xanthosoma*, but the body is largely brown and black. *Pisoxylon* would appear to exemplify the probable ancestor of *Trypoxylon*.

Biology: Unknown.

Checklist of *Pisoxylon*

xanthosoma Menke, 1968; Peru

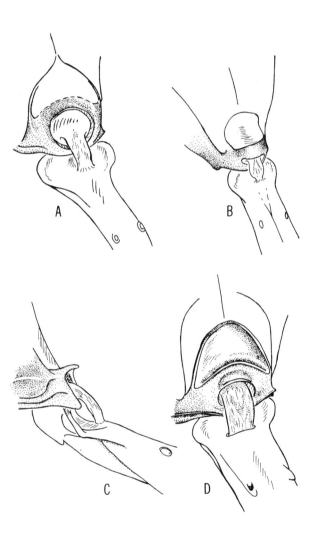

FIG. 108. Petiole socket details in *Trypoxylon*; A, *bicolor (figulus* group*)*; B,C, species in *fabricator* group; D, *T. (Trypargilum) nitidum*.

Genus Trypoxylon Latreille

Generic description: Inner orbits converging above or below, or parallel; frons usually variably adorned with carinae which originate at dorsal margin of antennal sockets (supraantennal carina), or one that runs between them (transverse interantennal carina), frontal carina often bifurcating above, its arms sometimes enclosing a heart or shield-shaped area (fig. 104 H); male antenna with 13 articles (12 in *mandibulatum*) and often some flagellomeres are modified; antennal sockets not contiguous with frontoclypeal suture (nearly contiguous in a few exceptional species); clypeus variable but usually height is equal to at least half of width, sometimes as high or higher than wide; labrum small to moderately large, apex sometimes visible when unfolded, free margin entire or deeply and narrowly notched, or with two narrow fingerlike processes; mandible simple (with an inner mesal tooth in some species of *fabricator* group, apically bidentate in females of *rufidens, mandibulatum,* and *oculare* groups, these three groups in subgenus *Trypoxylon*); occipital carina variable, incomplete below, or meeting hypostomal carina before its apex, or a complete circle which may or may not be contiguous with hypostomal carina, scutum with or without notauli; propodeum short to long, dorsum sometimes with a central longitudinal sulcus and sometimes a U-shaped en-

FIG. 109. *Trypoxylon arizonense* W. Fox, female.

closure, sides of propodeum often delimited dorsally by a carina (lateral propodeal carina) which runs between spiracle and petiole socket; dorsal margin of petiole socket a broad, flat or concave band (subgenus *Trypargilum,* fig. 108 D), or a narrow, convex band which may bear a backward projecting lamella, or simply sharply edged and/or lamellate (subgenus *Trypoxylon,* fig. 108 A-C); propodeal sternite present in many species of subgenus *Trypoxylon* (fig. 111 D); mesopleuron with an omaulus in some *Trypargilum;* gaster elongate, clavate, segment I usually two or more times as long as apical width and often forming a petiole (shorter in *rufidens* group); only tergum I with lateral carina which is sometimes feeble; sterna without graduli, male sterna sometimes modified in *Trypargilum;* female tergum VI conical, without a pygidial plate, midcoxae separated to subcontiguous, hindcoxae subcontiguous or contiguous; female hindcoxae with a small, ventral organ in the form of a perforated tubercle (fig. 111 D) or semitransparent spot in most of subgenus *Trypoxylon*; arolia moderate to large, equal on all legs in both sexes; forewing with one submarginal cell and one recurrent vein (fig. 102 F-J), marginal cell usually acuminate apically, veinlet forming anterior margin of cell (R_1) ending at cell apex (*Trypargilum*) (fig. 102 I-J) or extending beyond (*Trypoxylon* and *superbum* group of *Trypargilum*) (fig. 102 F-H); male sternum VIII with two long apical process in *Trypargilum* (fig. 103 E), variable in subgenus *Trypoxylon* but if with two apical processes they are short (fig. 103 F); gonostyle simple in *Trypargilum* (fig. 105 G), but often biramous in *Trypoxylon* (fig. 105 F); volsella simple.

Geographic range: Cosmopolitan but poorly represented in Australia and the various Pacific island groups. 359 species have been described, but there are undoubtedly many more, especially in South and Central America where the genus attains its greatest diversity. The subgenus *Trypargilum* is endemic to this area and nearly one-half of the known species of the genus occur in the Western Hemisphere. The Ethiopian Region contains the largest number of species in the Old World, with approximately 75. The Palearctic and Oriental Regions each have about half this number, and Australia has only one or two. A handful of species are island endemics. The Old World is represented only by the subgenus *Trypoxylon* in which we include *Asaconoton.* The *figulus* and *scutatum* groups predominate, but three smaller species groups are present (and endemic); the *brevipenne, carpenteri,* and *mandibulatum* groups.

Considering the number of species of *Trypoxylon* and *Pison*, their distribution, and similar biologies it can be speculated that *Trypoxylon* is a more recent and successful group than *Pison*. Perhaps as *Trypoxylon* invaded areas inhabited by *Pison*, competition forced *Pison* into refuges such as Australia. That island continent is the stronghold of *Pison* and presumably its isolation has prevented *Trypoxylon* from gaining a foothold there. This may explain why there are many more *Pison* species in Australia than elsewhere on the globe.

Three *Trypoxylon* are Holarctic in their distribution. *Trypoxylon figulum* occurs across Europe and Asia and apparently was introduced to the northeastern United States during colonial times. *Trypoxylon frigidum* and *pennsylvanicum* are two North American species recent-

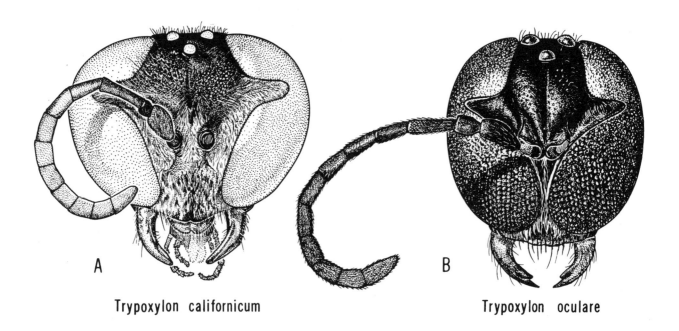

FIG. 110. Facial portraits in the tribe Trypoxylonini.

ly reported in eastern Asia by Tsuneki (1956c). When and in which direction the spread of these two species occurred is not known. At least two other species have been successfully transported by commerce (*philippinense* and *bicolor* to Hawaii), but in general, introduction of *Trypoxylon* to new areas has not been as great as in *Pison*.

Systematics: The single submarginal cell, together with the slender, clavate abdomen distinguish *Trypoxylon* from its relatives (fig. 109). Length varies from 4 or 5 mm (*rufidens, clavicerum*) to 23 or 24 mm (*politum*). Species of the subgenus *Trypoxylon* are smaller on the average than those of *Trypargilum*. Most *Trypoxylon* are black but in *Trypargilum* many species have extensively red or yellow gasters. The entire body of a few species in both subgenera is largely yellow. The degree of abdominal petiolation is most variable in the subgenus *Trypoxylon* where it varies from practically none in the *rufidens* group where the gaster is best described as clavate and some species of the *figulus* group, to extreme petiolation in some species of the *figulus* and *fabricator* groups. The species of *Trypargilum* are fairly uniform in their general appearance, but the wide array of morphological types within the subgenus *Trypoxylon* is impressive. The most striking differences in the latter are found in the various carinae of the frons (see fig. 104 G-J). The general structure of the head and thorax varies considerably in this subgenus also.

The genus contains a number of distinct groups; Richards (1934) split *Trypoxylon* into two subgenera, *Trypoxylon* and *Trypargilum*. Also, he proposed a world species group classification. Richards used a number of features in diagnosing his subgenera, all of which are subject to some variation, making it difficult to clearly separate the two taxa. Krombein (1967a, b), on the strength of biological differences and unspecified morphological criteria, recognized *Trypargilum* and *Trypoxylon* as genera. We feel that this is not sufficiently supported by morphology. Probably the best character for separation of *Trypargilum* from *Trypoxylon* is the appearance of the part of the propodeum that forms the dorsal margin of the petiole socket. Unfortunately, this is a difficult area to see unless the gaster is depressed, the legs are out of the way, and a good light source is available. In *Trypargilum* the socket is bounded dorsally by a broad bandlike structure, the surface of which is transversely flat or concave (fig. 108 D) but never convex. This appearance is quite constant in *Trypargilum* and is easily recognized after some experience with these wasps. On the other hand in the subgenus *Trypoxylon*, there is considerable variation in the form of the dorsal margin of the socket. The essential difference from *Trypargilum* is that the margin is convex rather than flat or concave. Some *Trypoxylon* have a narrow convex band, while in others the margin is simply a sharp edge that may or may not bear a median backward projecting lamella (fig. 108 A-C).

Various wing characters were used at the subgeneric level by Richards, also. In *Trypargilum* R_1 of the forewing (costa of Richards) does not extend beyond the apex of the marginal cell (fig. 102 I-J) except in the *superbum* group. In *Trypoxylon* R_1 apparently always projects beyond the apex (fig. 102 F-H). This character is awkward because the anterior margin of the forewing is usually narrowly infumate along R_1 and beyond the marginal cell apex. It is often difficult, therefore, to tell whether or not R_1 exceeds the apex. Another forewing character is the angle formed by the juncture of the outer veinlet of the first submarginal cell (the radial sector) with RS + M (fig. 102 J). According to Richards this angle is more obtuse in *Trypargilum* (110-135°) (fig. 102 I-J) and nearly a right angle in *Trypoxylon* (80-100°) (fig. 102 F-H). We have seen exceptions, however, and perhaps this is why Richards did not use this feature in his key to subgenera. A third wing character is the spacing between the two groups of hindwing hamuli. In *Trypargilum* the space is less than or at most equal to the length of the outer row of hamuli (fig. 102 I-J). In *Trypoxylon* the length of the space is usually much greater than the length of the outer row (fig. 102 F, H), but this is not the case in *T. figulus,* the type species of *Trypoxylon* (fig. 102 G); and we have noted other exceptions.

In *Trypargilum*, according to Richards, the second gastral and following gastral sterna are usually devoid of pubescence (except marginally), whereas *Trypoxylon* species are supposed to have broadly pubescent sterna. We have found exceptions in both subgenera, and the usefulness of this character is doubtful (Richards also noted exceptions in his subgeneric key).

In his key to subgenera Richards stated that the intercoxal carina ("metapleural keel") is "usually straight" in the subgenus *Trypoxylon*. This statement is subject to interpretation, because the carina is often arcuate to some degree in *Trypoxylon* and is rather strongly so in the *brevipenne* and *rufidens* groups. In *Trypargilum* it is usually strongly arcuate (fig. 111 B) except in the *albitarse* group (fig. 111 A, C).

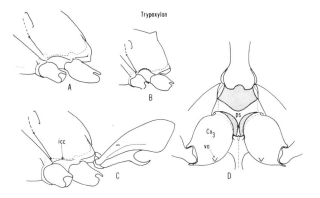

FIG. 111. Structural details of *Trypoxylon;* A, *(Trypargilum)* sp.; B, *tridentatum* male; C, *politum* male; D, *bicolor*. (vo = ventral organ, icc = intercoxal carina, ps = propodeal sternite)

The carinae of the face are quite variable, but nevertheless there is a fairly constant difference between the two subgenera. In *Trypargilum* there is never a transverse interantennal carina ("transverse or limiting keel" of Richards), and the surface of the frons between the antennal sockets and above them is essentially on one plane, except for the frontal carina and the prominence on which it is sometimes situated (figs. 104 K-L, 110 A). Most *Trypoxylon* have a transverse interantennal carina (fig. 104 G, I), or else the frons is transversely elevated near the dorsal margin of the sockets, so that the frontal surface between and above the sockets is not on one continuous plane (fig. 104 J). If neither of these situations are evident in *Trypoxylon*, then usually the frontal carina is strongly elevated between the sockets (fig. 104 H). However, we have noted one exception to this in the *fabricator* group, and a thorough study of the *figulus* group might reveal exceptions to all three of these possibilities.

The presence of a propodeal sternite (fig. 111 D) in many species of the subgenus *Trypoxylon* (*brevipenne, rufidens, scutatum, fabricator,* and *figulus* groups is a feature heretofore unrecorded in the genus. This sternite is absent in the *marginatum* group, the *oculare* group, the New World representatives of the *scutatum* group, some species of the *fabricator* group, and is weak or absent in many species of the *figulus* group. The only other sphecids with this sternite are the Sphecinae, some Ampulicinae, and the genus *Parapiagetia* (Larrinae).

There appears to be a subgeneric difference between *Trypargilum* and *Trypoxylon* in the male genitalia and sternum VIII as pointed out by Richards. The gonostyle ("stipes" of Richards) is simple, and sternum VIII bears two long parallel processes in *Trypargilum*. In *Trypoxylon* the gonostyle is usually divided apically into a dorsal and ventral lobe ("squama"), and sternum VIII is quite variable (if apical processes are present they are short).

Richards' subgeneric division is undoubtedly a basic natural cleavage, but as can be seen from the above discussion the two groups are not easy to characterize because of the species group diversity within each subgenus. Some of the species groups proposed by Richards might be given subgeneric status, and doing so would ease the current difficulty in diagnosing subgenera.

In *Trypargilum*, for example, both the *albitarse* group (=*politum* group of Sandhouse, 1940) and the *superbum* group might be split off from the remaining species groups as subgenera. *Trypargilum* would thus be restricted to three closely allied species groups: the *nitidum, spinosum,* and *punctulatum* groups. These three groups are distinguished from each other with difficulty, and perhaps they should be united as a single species group. Color is the main criterion for separation of the *punctulatum* group. The *spinosum* group is chiefly identified by having a broadly lamellate metapleural flange ("earlike", Richards), but this character is variable and we have seen many examples that could be placed arbitrarily in either the *spinosum* group or *nitidum* group (the latter having a narrower flange). Other possible diagnostic features for separation of these three groups do not seem to be constant: presence or absence of omaulus, form of the inner dorsal "keel" of hindcoxa, and the length of precoxal carina ("pleural suture").

Some of the species groups in the subgenus *Trypoxylon* are likely candidates for subgeneric status also: the *rufidens* group, *scutatum* group, *marginatum* group, and perhaps the *brevipenne* group. On the other hand, the distinctions between the *figulus* and *fabricator* groups seem particularly weak. We have not seen examples of the *fiebrigi, mandibulatum,* or *carpenteri* groups and therefore cannot evaluate them. A thorough world study of the subgenus *Trypoxylon* should be made before any groups are elevated to subgenus, because the limits of some groups are not clear. Furthermore, we have seen species from the Neotropics (undescribed) that would either require new species groups or the redefinition of existing ones. Menke (1968a) described one of these, *oculare,* and suggested that it be placed in a separate group.

Arnold's (1959) subgenus *Asaconoton* was proposed for a species with a nasiform frontal carina. Although we have not seen *egregium*, the type species, it seems to us that this subgenus is no more distinct than some of the species groups of the subgenus *Trypoxylon,* and we have synonymized *Asaconoton*. There are a number of Oriental *Trypoxylon* with frontal carinae similar to *egregium*, such as *T. pacificum*.

A number of keys for *Trypoxylon* have been published. Richards (1934) is the basic work for the New World, especially the Neotropics; and of course, his species group classification applies to the world fauna. His treatment of the North American species was incomplete, but Sandhouse (1940) gave a good account of our species. Krombein (1962) modified Sandhouse's key to include the new species, *clarkei* Krombein. Other useful papers on the Neotropical species are those of Richards (1936, 1969) and Gemignani (1933, 1940, 1941). Richards' statement that many more Neotropical species await discovery is doubtless correct as evidenced by the many unnameable species in the collection of the University of California, Davis. The Ethiopian species have recently been treated by Leclercq (1965). Gussakovskij (1936) provided a key to the Palearctic forms, and Tsuneki (1956c) gave an updated treatment of the eastern Palearctic fauna. Tsuneki (1966a, 1967c, 1971f) revised the species of Taiwan and surrounding islands. The Indoaustralian area has not received a modern revision, but Bingham (1897) provided a key to the species of India.

Richards (1934) gave an excellent basic account of the morphology of *Trypoxylon*, and the reader is referred to his paper for specific details. Some of his terms are not in current vogue, however, and for clarity we have relabelled the thorax (fig. 112). The list below compares Richards' terminology with the structural names used by us.

TABLE 14.
Comparative Terminology in *Trypoxylon*

Terminology of Richards, 1934	Terms used here and/or suggested changes
1. vertex	frons
2. dorsal part of occiput	vertex
3. limiting keels of antennal scrobes	supraantennal carina (fig. 104 K)
4. supraantennal keel	frontal carina or line (fig. 104 G-H)
5. transverse or limiting keel	transverse interantennal carina (fig. 104 G)
6. occipital keel	occipital carina
7. keel surrounding buccal cavity	hypostomal carina
8. tempora	gena
9. pronotal tubercle or posterior lobe of pronotum	pronotal lobe
10. proepisternal keel	transverse carina of propleuron
11. "rounded tubercle" of propleuron	posterolateral declivity of propleuron (fig. 112)
12. pleural suture	precoxal carina (fig. 112)
13. "strong furrow" delimiting "prepectus," or prepectal suture	episternal sulcus (fig. 112)
14. dorsal keel of prepectus	omaulus (fig. 112)
15. ventral keel of prepectus	subomaulus (fig. 112)
16. lateral lobe of the metanotum or metapostnotum	metapleural flange
17. metapleural keel	intercoxal carina (fig. 112)
18. mesopleural-metapleural suture	mesopleural suture (fig. 112)
19. metapleural-propodeal suture	metapleural sulcus or line (fig. 112)
20. "keel defining sides of propodeum above"	lateral propodeal carina (fig. 112)
21. stipes (of genitalia)	gonostyle
22. squama (of genitalia)	gonostyle (ventral lobe of)
23. sagitta	aedeagus or penis valves

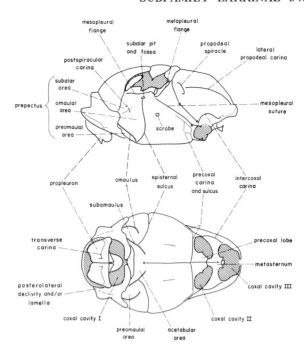

FIG. 112. Hypothetical *Trypoxylon* thorax.

The following provisional keys to the New World subgenera and species groups are offered as a supplement to Richards' key of 1934.

Provisional key to New World subgenera of *Trypoxylon*[19]

Transverse interantennal carina absent, frontal surface between and above sockets continuously flat (except for frontal carina and prominence on which it is sometimes borne), not interrupted by transverse sculpture (fig. 104 K-L, 110 A); frontal carina weak if extending between sockets but often strong near middle of frons, if bifurcating above, arms short and parallel forming a small U (fig. 104 K); dorsal margin of petiole socket a broad, flat or concave band (fig. 108 D); omaulus sometimes present (fig. 112); propodeal side not limited dorsally by a carina (fig. 111) (in *albitarse* group a carina originates at petiole socket and disappears before reaching halfway to spiracle, fig. 111 A-C); propodeal sternite absent[20]; two groups of hindwing hamuli usually separated by less than length of apical (outer) row (fig. 102 I-J) .... subgenus *Trypargilum* Richards, p. 348
Transverse (may be angulate) interantennal carina present (fig. 104 G-I), or if not, then frons just above or between sockets

[19] Certain facial carinae and propodeal socket details are difficult to see without good light and high magnification. It is essential that all characters in each couplet be used to assure correct subgeneric placement.
[20] Hindcoxae must be spread apart to see this structure.

strongly or abruptly elevated (fig. 104 H, J) (rare exceptions noted in *fabricator* group); frontal carina weak or strong, often bifurcating above, the arms sometimes rejoining to enclose a shield-shaped area on frons (fig. 104 H-I); dorsal margin of petiole socket convex, narrowly bandlike or merely sharp edged, often with a backward projecting lamella (fig. 108 A-C); omaulus absent; propodeal side often limited dorsally by a carina or foveate sulcus between spiracle and petiole socket (fig. 112); propodeal sternite often present (fig. 111 D); two groups of hindwing hamuli usually separated by more than length of apical row (fig. 102 F-H) ....
.... subgenus *Trypoxylon* Latreille, p. 345

Provisional key to species groups of subgenus *Trypargilum*

1. Frons with a short, strongly elevated, transverse carina just beneath median ocellus (fig. 104 L); R_1 usually extending beyond apex of marginal cell .... *superbum* group
   Frons just beneath median ocellus without a short transverse carina; R_1 not extending beyond apex of marginal cell 2
2. Propodeum with a strong carina at laterodorsal margin of petiole socket, carina directed towards spiracle but extending only one-fifth to half the distance[21] (fig. 111 A, C); omaulus absent (or at most barely discernible beneath pronotal lobe); intercoxal carina weakly arcuate or nearly straight (fig. 111 A, C); male sternum I often with a large spinelike process (fig. 111 C); large black wasps with white hindtarsi; wings usually infumate .... *albitarse* group
   Propodeum usually without a posterolateral carina but if present then very short and omaulus well developed; intercoxal carina strongly arcuate (fig. 111 B); male sternum I without a process; color variable
   nitidum, spinosum, and *punctulatum* groups

Provisional key to New World species groups of subgenus *Trypoxylon*

1. Metapleural flange broadly lamellate, the lamella clear or pale 2
   Metapleural flange narrowly or not lamellate, lamella black or nearly so 3
2. Frontal carina simple, not bifurcating above (faintly bifurcate above in a few species) ....
   .... *fabricator* group
   Frontal carina bifurcating above, the arms long and divergent (one species from southern S. America) .... *fiebrigi* group

3. Frontal carina bifurcating above, the arms diverging, often rejoining to form a shield-shaped frontal area (fig. 104 H, I) 4
   Frontal carina absent or simple, or if bifurcate then arms are short and parallel or subparallel (fig. 104 G, J) 5
4. Frontal carina below bifurcation simple between sockets (may bifurcate below sockets) (fig. 104 H); precoxal carina long, extending dorsad parallel to mesopleural suture (fig. 112); arms of frontal carina not enclosing median ocellus if they rejoin .... *marginatum* group
   Frontal carina between sockets split longitudinally by a slitlike depression from which project several long pale bristles (carina not bifurcating below sockets) (fig. 104 I); precoxal carina short, not extending dorsally parallel to mesopleural suture; arms of frontal carina enclosing median ocellus if they rejoin[22]; one Mexican and one North American species .... *scutatum* group
5. Frons above antennal sockets with a broad flat, raised horseshoe or U-shaped area (fig. 104 J); frontal carina faint or represented by an impressed line 6
   Frons above antennal sockets without a flat U-shaped elevation but often prominent just above or between sockets (fig. 104 G); frontal carina and interantennal carina weak or strong .... *figulus* group
6. Propodeal sternite present; R_1 of forewing only slightly extending beyond apex of marginal cell; terga I-III strongly convex in lateral profile .... *rufidens* group
   Propodeal sternite absent; R_1 of forewing extending well beyond apex of marginal cell (fig. 102 F); terga essentially flat in profile .... *oculare* group

Biology: Richards (1934) summarized most of the observations published up to that time. Since then a number of other important papers have appeared: Yoshimoto (1964), Krombein (1954a, 1956b, 1959c, 1967b, 1970), Fattig (1936), Vesey-Fitzgerald (1936, 1950), Maréchal (1936), Muma and Jeffers (1945), C. Lin (1969), Freeman (1938), Kurczewski (1963), Medler (1967), Rau (1943, 1944a), H. Wagner (1955), Nambu (1966-1967, 1971a, b), Tsuneki (1970b), Matthews and Matthews (1968), and Paetzel (1973). Rau (1935a) corrected the names of the species on which he had published earlier (Rau, 1933). Krombein's (1967b) recent book includes some of the most detailed nesting studies to date.

Most *Trypoxylon* nest in preexisting cavities, such as in the ends of plant stems and in beetle or bee burrows in wood or in the ground. Mud nests of other wasps are frequently used also. Around dwellings they may nest in nail holes, keyholes, folded newspapers, pipestems, and so on. *Trypoxylon* are often called "keyhole" wasps.

---

[21] In *T. laeve* the carina is very short and feeble, but the intercoxal carina is weakly arched.

[22] Common in Old World species of *scutatum* group, but not known in any New World species.

The cavity used for the nest is divided into cells with mud.

All species of the *albitarse* group (*Trypargilum*) and most of those in the *fabricator* group (*Trypoxylon*) make free mud nests in sheltered places. These mud cell builders are commonly called "mud daubing wasps." The large nests of the *albitarse* group are long and tubular, some up to 63 cm in length. The wasp biologist, Phil Rau, coined the name "pipe-organ wasps" for these species. An especially interesting behavior in the subgenus *Trypargilum* is the guarding of the nest by the male. During the female's absence the male often sits inside the nest with his head in the entrance. Rau called such species "Patriarchate wasps". In the interesting account by Paetzel (1973) on a species probably misidentified as "*rubrocinctum*" the male assisted the female in nest preparation and cell closure. Often the male took prey brought to the nest by the female and placed it in the cell so that she was free to continue the hunt (see also Rau, 1928b on *T. clavatum*). According to Fattig (1936), the male of *politum* (as *politiforme*) helped in nest construction by bringing in mud. Rau (1933) stated that at one nest of *T.* (*Trypoxylon*) *fabricator* (as *rugifrons*) "the male was almost always present", but the presence of a male during nest building needs to be confirmed in the subgenus *Trypoxylon*.

Although most reports indicate that species of the *fabricator* group (subgenus *Trypoxylon*) make mud nests, at least two species, *T. johnsoni* and *clarkei*, nest in preexisting cavities (Richards, 1934; Krombein, 1962, 1967b). According to Richards there is one dubious report that *T. rejector*, a member of the *figulus* group, makes free mud nests.

Richards (1934) and Vesey-Fitzgerald (1936) reported that females of *fabricator*, *maidli*, and *manni* (all *fabricator* group) were gregarious, building contiguous nests or at least in close proximity to one another.

There appears to be no prey specificity in *Trypoxylon*, but Krombein (1967b) found that in the subgenus *Trypargilum* each species collected certain types of spiders, e.g., wandering spiders or snare builders, suggesting that each species hunts in a different manner. Matthews and Matthews (1968) have enlarged on this aspect of *Trypargilum* biology.

The cocoons of *Trypargilum* species are usually dark brown and have a varnished appearance because of the incorporation of larval salivary secretions. Those of the subgenus *Trypoxylon* are often more delicate, being spun of white subopaque silk. Krombein (1967b) and Matthews and Matthews (1968) found that the cocoons of *Trypargilum* were specifically different.

Checklist of *Trypoxylon*
Subgenus *Trypoxylon*

abditum Arnold, 1924; centr. and s. Africa
accumulator F. Smith, 1875; India
acutangulum Arnold, 1951; Niger
aegyptium Kohl, 1906; Egypt
    aegyptiacum "Kohl" of Gussakovskij, 1936
aeneipenne Arnold, 1943; centr. Africa
albipes F. Smith, 1856; Albania
aldrichi Sandhouse, 1940; w. N. America
algoense Arnold, 1924; S. Africa
ambiguum Tsuneki, 1956; Japan
ammophiloides A. Costa, 1871; Sardinia
arabicum Gussakovskij, 1936; Arabian penin.
arnoldi Menke, new name; centr. and e. Africa
    cognatum Arnold, 1924, nec Cameron, 1897
arroyense Richards, 1934; Mexico to Panama
arudum Leclercq, 1965; centr. Africa
ashmeadi Baltazar, 1966; Philippine Is. (? = *obsonator*)
    elongatum Ashmead, 1905, nec F. Smith, 1856
asinum Leclercq, 1965; centr. Africa
asuncicola Strand, 1910; Brazil, Paraguay, Argentina
    ssp. *levius* Richards, 1934; Brazil
attenuatum F. Smith, 1851; Europe, Asia
    ssp. *hannibalis* Gribodo, 1896; Algeria
backi Sandhouse, 1940; e. U.S.
balteatum Cameron, 1889; Mexico
barberi Richards, 1934; Guatemala, Panama
bellardi Richards, 1934; Venezuela
biarti Richards, 1934; Mexico
bicolor F. Smith, 1856; Malaysia, Java, Hawaii
bidentatum W. Fox, 1891; w. U.S.
    morrisoni Richards, 1934
bilobatum Tsuneki, 1961; Thailand
bourgeoisi Strand, 1911; Ecuador
brasilianum Saussure, 1867; Brazil
braunsi Arnold, 1924; S. Africa
brevipenne Saussure, 1867; Guinea to Rhodesia
breviventre Arnold, 1959; e. Africa
bridwelli Sandhouse, 1940; U.S.: Texas
buddha Cameron, 1889; India
caldesianum Richards, 1934; Costa Rica, Colombia
cameronii Dalla Torre, 1897; Mexico
    palliditarse Cameron, 1889, nec Saussure, 1867
    tabascense Strand, 1910
canaliculatum Cameron, 1889; India
capense Cameron, 1905; S. Africa
capitale Richards, 1934; Trinidad
carcinum Leclercq, 1965; centr. Africa
carinatum Say, 1837; e. N. America
cariosum Arnold, 1959; Rhodesia
carpenteri Richards, 1933; Uganda
castoris Leclercq, 1965; centr. Africa
cataractae Arnold, 1924; centr. and s. Africa
    ssp. *madecassum* Arnold, 1945; Madagascar
catinum Leclercq, 1965; centr. Africa
cavallum Leclercq, 1965; centr. Africa
centrale W. Fox, 1895; Baja California
chichidzimaense Tsuneki, 1973; Bonin Is.
chichimecum Saussure, 1867; Mexico
chihpense Tsuneki, 1971; Taiwan
chinense Gussakovskij, 1936; China
chingi Tsuneki, 1971; Taiwan
chirindense Arnold, 1936; Zaire, Uganda, Rhodesia
chosenense Tsuneki, 1956; Korea
cinctellum Richards, 1934; Panama

clarkei Krombein, 1962; e. U.S.
clavicerum Lepeletier and Serville, 1828; Palearctic Region
    tibiale Zetterstedt, 1840
    ssp. *gussakovskiji* Tsuneki, 1974; se. USSR: Maritime Prov.
        pygmaeum Gussakovskij, 1933, nec Cameron, 1900
    ssp. *exiguum* Tsuneki, 1956; Japan
cocorite Richards, 1934; Trinidad
cognatum Cameron, 1897; Himalaya Mts.
coloratum F. Smith, 1857; Borneo
confrater Kohl, 1894; Liberia, Zaire
connexum Turner, 1908; Australia
cornigerum Cameron, 1889; Mexico to Argentina
crassifrons Tsuneki, 1963; Thailand
crassipunctatum Arnold, 1946; Malawi
cricetum Leclercq, 1965; Zaire
crudele Richards, 1934; Panama
darium Leclercq, 1965; Zaire
dendrophilum Arnold, 1959; Zaire; Tanzania: Tanganyika
deuterium Leclercq, 1965; Zaire
dubiosum Tsuneki, 1964; Ryukyus
duckei Richards, 1934; Brazil
dyeri Richards, 1934; Honduras
ebneri Maidl, 1924; Sudan
ecuadorium Richards, 1934; Colombia, Ecuador (? = *bourgeoisi*)
    ssp. *meridionale* Richards, 1934; Colombia
egregium Arnold, 1959; Zaire; Tanzania: Tanganyika
elegantulum F. Smith, 1860; Celebes (? = *gracilescens*)
elgonense Arnold, 1956; Kenya
elongatum F. Smith, 1856; Sierra Leone
emeritum Leclercq, 1965; Burundi
errans Saussure, 1867; Reunion, Mauritius, Seychelles
    gardneri Cameron, 1907
erythrozonatum Cameron, 1902; Maldives
eugeniae Gussakovskij, 1936; sw. USSR: Tadzhik S.S.R.
excellens Strand, 1910; Paraguay
eximium F. Smith, 1859; Indonesia: Ewab, Aru
eyeni Leclercq, 1965; Zaire
fabricator F. Smith, 1873; Panama to Brazil
    gracile Taschenberg, 1875
    brevicarinatum Cameron, 1912
fastigium W. Fox, 1893; N. America
    carinifrons W. Fox, 1891, nec Cameron, 1889
    subfrigidum Rohwer, 1909
    nigrellum Rohwer, 1909
fenchihuense Tsuneki, 1967; Taiwan
ferox F. Smith, 1860; Celebes (? = *gracilescens*)
fiebrigi Richards, 1934; Paraguay
figulus (Linnaeus), 1758 (*Sphex*); Holarctic Region
    fuliginosum Scopoli, 1763 (*Sphex*)
    apicale W. Fox, 1891
    ssp. *major* Kohl, 1883; Europe
    ssp. *medium* Beaumont, 1945; Europe
        rubi Wolf, 1959
    ssp. *minor* Beaumont, 1945; Europe
    ssp. *koma* Tsuneki, 1956; Korea
    ssp. *yezo* Tsuneki, 1956; Japan
    ssp. *barbarum* Beaumont, 1957; Morocco
        fieuzeti Giner Marí, 1959
fitzgeraldi Richards, 1934; Trinidad
flavimanum Arnold, 1946; Malawi
fletcheri Turner, 1918; India
florale Richards, 1934; Brazil
formosicola Strand, 1922; Taiwan, Ryukyus
    ssp. *kankauense* Strand, 1922; Taiwan
    ssp. *calcarale* Strand, 1922; Taiwan
    ssp. *inornatum* Matsumura, 1926; Ryukyus
        amamiense Tsuneki, 1964
foveatum Cameron, 1904; S. Africa
frigidum F. Smith, 1856; N. America
    plesium Rohwer, 1920
    ssp. *cornutum* Gussakovskij, 1933; se. USSR: Maritime Prov.; Japan
    ssp. *chongar* Tsuneki, 1956; Korea
frioense Richards, 1934; Colombia
frontale F. Smith, 1856; "Africa"
fronticorne Gussakovskij, 1936; Palearctic Region
fulvocollare Cameron, 1904; India
funatui Tsuneki, 1963; Thailand
geniculatum Cameron, 1902; India
gounellei Richards, 1934; Brazil
gracilescens F. Smith, 1860; Taiwan, Celebes, India (?= subsp. of *eximium*)
    petioloides Strand, 1922
gracilicorne Arnold, 1946; Malawi
gracillimum F. Smith, 1863; Indonesia: Misool
grenadense Richards, 1934; W. Indies
gussachaos Menke, new name; sw. USSR: Kazakh S.S.R.
    carinifrons Gussakovskij, 1936, nec Cameron, 1889
gustatum Leclercq, 1965; Zaire
hova Saussure, 1892; Madagascar
imayoshii Yasumatsu, 1938; Japan
inconstans Arnold, 1946; Rhodesia
indicum Menke, new name; n. India
    ornatipes Cameron, 1913, nec W. Fox, 1891
infimum Arnold, 1940; Rhodesia
insolitum W. Fox, 1897; Brazil; lectotype ♀, Santarém, Brazil; Carnegie Mus., present designation by Menke
intrudens F. Smith, 1870; India
isigakiense Tsuneki, 1973; Ryukyus
javanum Taschenberg, 1875; Java
johnsoni W. Fox, 1891; e. U.S.
    ornatipes W. Fox, 1891
    adelphiae Sandhouse, 1940
judicum Leclercq, 1965; Zaire
kabeyae Leclercq, 1965; Zaire
kalimantan Menke, new name, Borneo
    annulipes Cameron, 1903; nec Taschenberg, 1875
kansitakum Tsuneki, 1971; Taiwan
kapiricum Leclercq, 1965; Zaire
kaszabi Tsuneki, 1971; Mongolia
katangae Leclercq, 1965; Zaire
khasiae Cameron, 1904; India
klapperichi Balthasar, 1957; Afghanistan

kodamanum Tsuneki, 1972; Japan
kohli Arnold, 1924; Zaire, Rhodesia, S. Africa
koikense Tsuneki, 1956; Japan
kolazyi Kohl, 1893; Europe
kolthoffi Gussakovskij, 1938; China
konosuense Tsuneki, 1968; Japan
koreanum Tsuneki, 1956; Korea
koshunicon Strand, 1922; Taiwan
kumaso Tsuneki, 1966; Ryukyus
kyotoense Tsuneki, 1966; Japan
lacustre Arnold, 1956; Zaire
　? *insulum* Arnold, 1959
latiscutatum Arnold, 1946; Zaire, Malawi
　ssp. *rectirugosum* Arnold, 1959; Tanzania: Tanganyika
layouanum Evans, 1972; Dominica
leptogaster Kohl, 1894; centr. Africa
letiferum Arnold, 1946; Rhodesia
leucorthrum Richards, 1934; Trinidad
lissonotum Cameron, 1910; S. Africa
lucidum Arnold, 1959; Zaire, Rhodesia
lusingum Leclercq, 1965; Zaire
luteosignatum Arnold, 1945; Madagascar
mabwense Leclercq, 1965; Zaire
magrettii Gribodo, 1884; Ethiopia
maidli Richards, 1934; Brazil (? = *parvum*)
　ssp. *bodkini* Richards, 1934; Trinidad, Guyana
malaisei Gussakovskij, 1933; se. USSR: Maritime Prov.; Korea, Japan
mandibulatum Richards, 1933; Sri Lanka, India
manni Richards, 1934; Trinidad, Brazil
marginatum Cameron, 1912; Panama, Guyana
marginifrons Cameron, 1912; Zaire
massaicum Cameron, 1910; Tanzania: Tanganyika
mazaruni Richards, 1934; Guyana
mediator Nurse, 1903; India
melanocorne Strand, 1922; Taiwan
melanurum Cameron, 1902; Maldives, Laccadives
montanum Schulz, 1906; India
　　placidum Cameron, 1904, nec F. Smith, 1863
monticola Tsuneki, 1956; Japan
montivagum Arnold, 1940; Zaire, Rhodesia
moraballi Richards, 1934; Guyana, Brazil
　ssp. *boliviense* Richards, 1934; Bolivia
murotai Tsuneki, 1973; Taiwan
mutilatum Richards, 1934; Bolivia
nagamasae Tsuneki, 1963; Thailand
nambui Tsuneki, 1966; Japan
　　kinkadzanense Tsuneki, 1971
nigrispine Cameron, 1905; Mexico
nipponicum Tsuneki, 1956; Japan
nitidissimum Richards, 1934; Brazil, Bolivia
nodosicorne Turner, 1917; India
nodosum Arnold, 1944; Rhodesia
obsonator F. Smith, 1873; India, Taiwan, Japan
　　hyperorientale Strand, 1922
　ssp. *tropicale* Tsuneki, 1961; Thailand, s. Taiwan
oculare Menke, 1968; Brazil, Peru
okinawanum Tsuneki, 1966; Ryukyus: Okinawa
ordinarium Richards, 1934; Brazil

orientale Cameron, 1904; India
orientinum Richards, 1969; Cuba
pachygaster Richards, 1934; Trinidad
pacificum Gussakovskij, 1933; se. USSR: Maritime Prov.; Korea, Japan
pan Arnold, 1956; Rhodesia
parvum Schrottky, 1902; Argentina
patruele Arnold, 1924; S. Africa
paulisum Leclercq, 1965; Zaire
peltopse Kohl, 1906; Brazil
pennsylvanicum Saussure, 1867; e. N. America
　ssp. *japonense* Tsuneki, 1956; Japan
pentheri Richards, 1934; Brazil
petiolatum F. Smith, 1857; Borneo
philippinense Ashmead, 1904; Philippines, Hawaii
pileatum F. Smith, 1856; India
　　nigricans Cameron, 1889, nomen nudum
placidum F. Smith, 1863; Indonesia: Misool, Australia
posterorubrum Richards, 1934; Mexico
providum F. Smith, 1860; Indonesia: Batjan; Solomons
pulawskii Tsuneki, 1956; Japan
puliense Tsuneki, 1967; Taiwan
pumilio Arnold, 1959; Lesotho
punctatissimum Arnold, 1924; centr. and s. Africa, Madagascar
punctivertex Richards, 1934; Texas to Brazil
puttalamum Strand, 1922; Sri Lanka
pygmaeum Cameron, 1900; India
quadriceps Tsuneki, 1971; Taiwan
rejector F. Smith, 1870; India
regulare Viereck, 1906; centr. U.S.
responsum Nurse, 1903; India
　ssp. *regium* Gussakovskij, 1933; se. USSR: Maritime Prov.
　ssp. *hatogayuum* Tsuneki, 1956; Japan
　ssp. *ryukyuense* Tsuneki, 1966; Ryukyus
　ssp. *taiwanum* Tsuneki, 1967; Taiwan
richardsi Sandhouse, 1940; e. U.S.
rubiginosum Gussakovskij, 1936; sw. USSR: "Caucaso"
rubrifemoratum Richards, 1934; Panama, Trinidad
　ssp. *rubellum* Richards, 1934; Argentina
rufidens Cameron, 1905; Mexico to Panama
　ssp. *trinidadianum* Richards, 1934; Trinidad
rufimanum Spinola, 1851; Brazil
rugiceps Dalla Torre, 1897; Mexico
　　rugifrons Cameron, 1889, nec F. Smith, 1873
rugifrons F. Smith, 1873; Brazil
saitamaense Tsuneki, 1973; Japan
sanctum Richards, 1934; Brazil
sapporoense Tsuneki, 1960; Japan
schmidti Richards, 1936; Costa Rica
schmiedeknechtii Kohl, 1906; Java
sculleni Sandhouse, 1940; w. U.S.
scutatum Chevrier, 1867; s. Europe
　　scutigerum Taschenberg, 1881
　　quartinae Gribodo, 1884
scutiferum Taschenberg, 1875; Brazil
scutifrons Saussure, 1892; Madagascar
segregatum Richards, 1934; Paraguay, Argentina

senegambicum Kohl, 1906; Senegal
  palaeotropicum Schulz, 1906
seyrigi Arnold, 1945; Madagascar
shannoni Richards, 1934; Peru
shimoyamai Tsuneki, 1958; Japan
shirozui Tsuneki, 1966; Ryukyus
silvestre Richards, 1934; Guyana
sinuosiscutis Arnold, 1945; Madagascar
sogdianum Gussakovskij, 1952; sw. USSR: Tadzhik S.S.R.
solivagum Arnold, 1946; Zaire, Malawi
staudingeri Richards, 1934; Peru
stevensoni Arnold, 1924; Zaire, Rhodesia
stieglmayri Richards, 1934; Brazil
stroudi Gribodo, 1884; centr. and s. Africa
subpileatum Strand, 1922; Taiwan
succinctum Cresson, 1865; Cuba
sulcatoides Richards, 1934; Argentina
sulcifrons Gussakovskij, 1936; sw. USSR: Turkman S.S.R., Uzbek S.S.R.
syriacum Mercet, 1906; Syria
tainanense Strand, 1922; Taiwan
  taiwanense of Tsuneki, 1966
takasago Tsuneki, 1966; Ryukyus
tanoi Tsuneki, 1967; Taiwan
testaceicorne Cameron, 1907; India
testaceipes Arnold, 1924; Zaire, Rhodesia
thaianum Tsuneki, 1961; Thailand
timberlakei Sandhouse, 1940; sw. U.S.
tinctipenne Cameron, 1889; India
toltecum Saussure, 1867; Mexico
tremulum Arnold, 1946; Malawi
tricolor Sickmann, 1894; China
trigeminum Richards, 1934; Guyana, Brazil
trinidadense Richards, 1934; Trinidad, Paraguay
triste Arnold, 1924; Rhodesia
trochanteratum Cameron, 1902; India
tuberculifrons Arnold, 1945; Madagascar
tubulentum Arnold, 1946; Zaire, Malawi
turkestanicum Gassakovskij, 1936; sw. USSR: Uzbek S.S.R.
unguicorne Richards, 1934; Bolivia
urichi Richards, 1934; Panama, Trinidad
vallicola Tsuneki, 1971; Taiwan
varipes Pérez, 1905; se. USSR: Maritime Prov.; China, Japan
varipilosum Cameron, 1901; Singapore
viduum Arnold, 1951; Ethiopia
volitans Arnold, 1956; Uganda
wheeleri Richards, 1936; Panama
winthemi Richards, 1934; Brazil
xenophon Richards, 1934; Panama
zikae Arnold, 1956; Uganda
zurki Leclercq, 1965; Zaire

Subgenus *Trypargilum*

aestivale Richards, 1934; Costa Rica to Paraguay
agamemnon Richards, 1934; Panama
albitarse Fabricius, 1804; S. America
  palliditarse Saussure, 1867
  rostratum Taschenberg, 1875
  leucotrichium Rohwer, 1912
  palliditarse of Richards, 1934
albonigrum Richards, 1934; Brazil
alleni Richards, 1934; Colombia
annulatum Brèthes, 1913; Argentina
apicipenne Cameron, 1889; Mexico
arizonense W. Fox, 1891; w. U.S. (? = *fusciventre* and/or *sonorense*)
  rufozonale W. Fox, 1891
armatum Taschenberg, 1875; Venezuela, Colombia
atkinsoni Richards, 1934; Panama, Colombia
aureovestitum Taschenberg, 1875; Argentina
  ? *tuckumanum* Brèthes, 1913
aurifrons Shuckard, 1837; Colombia to Paraguay
  ornatum F. Smith, 1856
  latiornatum Cameron, 1912
barticense Richards, 1934; Guyana, French Guiana, Surinam
bensoni Richards, 1934; Nicaragua
bicalcaratum Richards, 1934; Colombia, Peru (? = *apicipenne*)
bogotense Richards, 1934; Colombia, Peru
brethesi Gemignani, 1941; Argentina
bryanti Richards, 1934; Brazil, Argentina
buchwaldi Richards, 1934; Ecuador
busckii Richards, 1934; Panama
californicum Saussure, 1867; U.S.: California
campaspe Richards, 1934; Mexico
cariniceps Richards, 1934; Bolivia
carinifrons Cameron, 1889; Mexico
clavatum Say, 1837; e. N. America
  annulare Dahlbom, 1844
  cockerellae Rohwer, 1909
  quintile Viereck, 1906
collinum F. Smith, 1856; se. U.S.
  ssp. *rubrocinctum* Packard, 1867; e. U.S.
correntium Brèthes, 1909; Argentina
cubense Richards, 1934; Cuba
emdeni Richards, 1934; Peru
excavatum F. Smith, 1856; W. Indies
fractum Richards, 1934; Brazil
fugax Fabricius, 1804; Brazil to Colombia
  ? *colombianum* Saussure, 1867
  medianum W. Fox, 1897, new synonymy by Menke; lectotype ♂, Corumba, Brazil, Carnegie Mus., present designation by Menke
  ssp. *santamartae* Richards, 1934; Colombia
fuscipenne Fabricius, 1804; S. America
fusciventre Cameron, 1889; Mexico
gandarai Rohwer, 1912; Mexico
johannis Richards, 1934; se. U.S.
joergenseni Brèthes, 1910; Argentina
  ? *melanopteron* Richards, 1934
lactitarse Saussure, 1867; Mexico
laeve W. Fox, 1897; Brazil; lectotype ♂, "Maran", Brazil, Carnegie Mus., present designation by Menke
laevifrons F. Smith, 1873; Brazil, Guyana

luteitarse Saussure, 1867; Mexico
lynchi Brèthes, 1913; Argentina
  ? *gaucho* Richards, 1934
majus Richards, 1934; Brazil
mayri Richards, 1934; Cuba
melanoleucum Richards, 1934; Brazil
mexicanum Saussure, 1867 (p. 77); Mexico
monstrificum Kohl, 1905; Bolivia
mutatum Kohl, 1885; Argentina, Paraguay
    coloratum Taschenberg, 1875, nec F. Smith, 1857
  ssp. *confluens* Richards, 1934; Uruguay
nattereri Richards, 1934; Brazil, Peru
nitidum F. Smith, 1856; Mexico to Argentina
    bahiae Saussure, 1867
    annulipes Taschenberg, 1875
    rufosignatum Taschenberg, 1875
    fallax W. Fox, 1897, new synonymy by Menke; lectotype ♂, Maruru, Brazil, Carnegie Mus., present designation by Menke
  ssp. *aztecum* Saussure, 1867; Mexico
    fulvispine Cameron, 1889
  ssp. *schulthessi* Richards, 1936; Costa Rica
niveitarse Saussure, 1867; Brazil
obidense Richards, 1934; Brazil
olfersi Richards, 1934; Argentina
opacum Brèthes, 1913; Argentina
  ? *alutaceum* Richards, 1934
optimum Richards, 1934; Brazil, Argentina
  ssp. *rugosum* Menke, new name; Brazil
    regium Richards, 1934, nec Gussakovskij, 1933
orizabense Richards, 1934; Mexico
pectorale Richards, 1934; Mexico, Trinidad
platense Brèthes, 1913; Argentina
  ? *mendozanum* Richards, 1934
politum Say, 1837; e. U.S.
    neglectum Kohl, 1884
    basale Rohwer, 1912
    politiforme Rohwer, 1912
    albitarse of Richards, 1934
poultoni Richards, 1934; S. America
punctulatum Taschenberg, 1875; Brazil, Paraguay
rogenhoferi Kohl, 1884; Brazil to Argentina
  ? *holoneurum* Schrottky, 1910
  ? *argentinum* Brèthes, 1910
  ? *festivum* Brèthes, 1913
  ? *incognitum* Brèthes, 1913
rohweri Richards, 1934; Hispaniola
salti Richards, 1934; U.S.: Arizona, Mexico to Ecuador
saundersi Richards, 1934; ? Colombia
saussurei Rohwer, 1912; Mexico
    mexicana Saussure, 1867 (p. 78), nec Saussure, 1867, p. 77
schnusei Richards, 1934; Peru
scrobiferum Richards, 1934; Guyana
semiflavum Richards, 1934; Bolivia
sonorense Cameron, 1889; Mexico
spatulatum Richards, 1934; Mexico
spinosum Cameron, 1889; Texas to Panama
    cinereohirtum Cameron, 1889
striatum Provancher, 1888; e. N. America to Argentina (? = *lactitarse*)
    cinereum Cameron, 1889
    albopilosum W. Fox, 1891
    planoense Rohwer, 1909
subimpressum F. Smith, 1856; W. Indies
superbum F. Smith, 1873; Brazil
  ssp. *fulvipes* Cameron, 1889; Panama
  ssp. *superciliosum* Richards, 1934; Bolivia
surinamense Richards, 1934; Guyana, French Guiana, Surinam, Brazil, Peru
tenoctitlan Richards, 1934; Mexico, Guatemala
texense Saussure, 1867; centr. to se. U.S.
    sulcus La Munyon, 1877
    aureolum Rohwer, 1909
    relativum Rohwer, 1909
transversistriatum Strand, 1910; Paraguay
tridentatum Packard, 1867; N. America
    projectum W. Fox, 1891
  ssp. *archboldi* Krombein, 1959; Florida
triodon Richards, 1934; Argentina
vagulum Richards, 1934; Colombia
vagum F. Smith, 1873; Panama to Brazil
xanthandrum Richards, 1934; Panama
xantianum Saussure, 1867; Mexico: Baja California

Nomina nuda in *Trypoxylon*

annulatum "Cameron" Richards, 1934
aureosericeum Schrottky, 1913
hispanicum "Mercet" Richards, 1934
meridionale Gussakovskij, 1936 (p. 663), nec Richards, 1934
simlaense "Cameron" Richards, 1934
geniculatum Cameron, 1889 (may = *geniculatum* Cameron, 1902)

Tribe Bothynostethini

Small to medium-sized black wasps with yellow markings in some forms. The "square" head, compact thorax, and color pattern give them a crabronine appearance, and this look is enhanced in *Bothynostethus* where the inner orbits converge strongly below. The clypeus is often covered with a silver "moustache" similar to that in crabronines. Two genera are included, *Bothynostethus* and *Willinkiella*.

Diagnostic characters:
1. (a) Inner orbits converging below or nearly parallel; (b) ocelli normal.
2. (a) Antennal sockets low on face, contiguous with frontoclypeal suture; (b) male with 11 unmodified flagellomeres; (c) occipital carina strong but disappearing well before reaching hypostomal carina.
3. (a) Clypeus transverse; (b) labrum small, hidden.
4. (a) Externoventral margin of mandible entire or angled, toothed or notched, inner margin with one or two teeth except simple in *Bothynostethus*

males; (b) mouthparts short to moderately long; (c) mandibular socket open.
5. Pronotal collar narrow, about as high as scutum.
6. Notauli absent or weak.
7. (a) Episternal sulcus present; (b) omaulus and postspiracular carina absent; (c) acetabular carina sometimes present but weak.
8. (a) Female foretarsus with a weakly developed rake, tarsomeres essentially symmetrical; (b) hindfemur clublike, obliquely truncate distally (fig. 115 F); (c) midcoxae separated, hindcoxae contiguous; (d) arolia moderate in both sexes; (e) male forecoxa and trochanter usually modified; (f) plantulae apparently absent.
9. (a) Propodeum short, dorsally with a transverse basal or subbasal foveate sulcus; (b) enclosure nearly triangular when defined; (c) no propodeal sternite
10. (a) Three submarginal cells present, II petiolate; (b) two recurrent veins, both received by submarginal II (first recurrent may end on I in *Bothynostethus*); (c) outer veinlet of submarginal III ending near middle of marginal cell, apex of latter usually acuminate.
11. (a) Jugal lobe small, at most half as long as anal area; (b) hamuli not divided into two groups; (c) media diverging after cu-a.
12. (a) Gaster sessile, tergum I with lateral carina; (b) pygidial plate well defined in both sexes, covered with short bristles or hair as in *Tachytes*; (c) male sternum VIII tongue shaped.
13. (a) Volsella a large rolled plate (C-shaped in profile); (b) gonostyle simple.

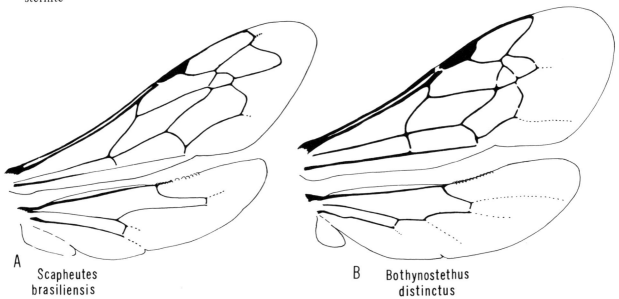

FIG. 113. Wings in the tribes Scapheutini (A) and Bothynostethini (B).

Systematics: This tribe was originally proposed by Fox (1894d), although he included *Plenoculus*, which is not closely related. Both Kohl (1896) and Handlirsch (1925) pointed out the similarity of *Bothynostethus* to *Scapheutes*. Menke (1968d) divided the Bothynostethini into two subtribes, the Bothynostethina and Scapheutina, each of which included two genera. These two groups share the clublike and distally truncate hindfemur (fig. 115 F), petiolate submarginal cell II (fig. 113), and (in most species) the strong convergence below of the inner orbits (fig. 114 A, C, D). This combination sets them apart from other Larrinae and indicates strong specialization. Truncation of the hindfemur is weak in some *Bothynostethus* males, however. In spite of the considerable similarities between Menke's two subtribes their evolutionary divergence is obvious, and we have decided to accord tribal status to both. The Bothynostethini has the more specialized wing venation with the hindwing media diverging after cu-a (fig. 113 B), longer mouthparts (in *Willinkiella*), unnotched mandibles (in some *Bothynostethus*), and no omaulus. The Scapheutini, on the other hand, are less specialized, as shown by the divergence of the hindwing media before cu-a and other wing details (fig. 113 A), short mouthparts, comparatively weak modification of the male forecoxae, and the presence of plantulae. Some advancements in the Scapheutini include the bispinose male sternum VIII (fig. 115 B), convergence below of the inner orbits, double-edged tergal margins, and the presence in *Bohartella* of an omaulus (fig. 115 G). The phyletic inconsistencies plus the rather restricted distributions of the genera making up the two tribes suggest that they are relics.

Besides the features enumerated above, the Bothynostethini differs from the Scapheutini in that the outer veinlet of submarginal cell III is received near the middle of the marginal cell (fig. 113 B), by the presence of a basal or subbasal transverse foveate groove on the propodeal dorsum (fig. 115 H), and by the form of the volsella (fig. 115 D).

There are some reasons for allying the Bothynostethini with the Miscophini, and Krombein (1951) placed *Bothynostethus* in that tribe. The parallel inner orbits of *Willinkiella*, petiolate submarginal cell II, sessile gaster, and normal ocelli lend credence to the idea. However, the genitalia and hindfemur of the Bothynostethini are distinguishing.

Some striking crabronine features are exhibited which suggest that the tribe may represent a remnant of the ancestral stock that gave rise to the Crabroninae. The similarities are: eyes converging strongly below with associated enlargement of lower facets, transverse clypeus bearing a "moustache," rather "square" head, weak female tarsal rake with little tarsal asymmetry, and short propodeum. The crabronine genera *Entomocrabro* and *Entomognathus* are like *Bothynostethus* in many ways. This is further discussed under systematics of the subfamily Crabroninae.

Biology: All are presumably ground nesting. *Bothynostethus* is known to provision with Coleoptera (Chrysomelidae), and this is interesting since *Entomognathus* also uses beetles of this family.

Key to genera of Bothynostethini

Inner orbits moderately to strongly converging below (fig. 114 A), propodeum coarsely areolate posteriorly and posterolaterally (fig. 115 I), externoventral margin of mandible usually entire (N. and S. America) .... *Bothynostethus* Kohl, p. 351

Inner orbits essentially parallel (fig. 114 B), propodeum smooth posteriorly (fig. 115 H), mandible toothed externoventrally (Argentina) .... *Willinkiella* Menke, p. 352

Genus Bothynostethus Kohl

Generic diagnosis: Inner orbits converging below, eye facets largest opposite antennal sockets (fig. 114 A); antenna short; externoventral margin of mandible usually entire (toothed or notched in *nitens, kohlii, duckei* and *dubius*), inner margin with two subbasal teeth in female, none in male; mouthparts short; pronotal collar without transverse carina (except in *collaris*); propodeal surface coarsely areolate posteriorly (weakly so in one North American species), dorsal enclosure evident only by its lack of areolation, but dorsum with a median longitudinal sulcus or fovea and transverse basal sulcus crossed by ridges; acetabular carina short when present, extending only a little on either side of midventral line; gastral terga and sterna simple, except sternum I usually with two basal, parallel ridges which delimit a short channel; male forecoxa longitudinally channeled mesad and usually with an apical process, trochanter concave basally (except in *duckei*); marginal cell of forewing usually acute apically (truncate and appendiculate in *duckei*); first recurrent vein received by submarginal cell II, or vein interstitial or rarely received by I; forewing media commonly interstitial with cu-a, but sometimes diverging beyond it in forms with externoventral margin of mandible entire, or media sometimes diverging before cu-a in forms with externoventral tooth or notch on mandible.

Geographic range: New World, although primarily Neotropical. Ten species are known.

Systematics: The strongly converging eyes and areolate propodeum separate *Bothynostethus* from *Willinkiella*. In addition, the lack of inner basal mandibular teeth in male *Bothynostethus* is distinctive.

There are several variables in the genus. One is the presence or absence of an externoventral mandibular tooth or notch. Based on our material it appears that a notched mandible is found in about half the species, all of which are South American. The male of *duckei*, which has recently been discovered, has simple forecoxae and trochanters, unlike males in the other species we have studied. The end point of the first recurrent vein also varies. In most species it is interstitial between submarginal cells I and II, or it is received by II. In *aberrans* and possibly a few other species, it is received by I. The forewing media usually diverges before cu-a in forms with notched mandibles but after cu-a in species with entire mandibles. However, the media is often interstitial in species with either mandibular type, and establishment of subgenera does not appear to be feasible or warranted. In one species, *duckei*, the marginal cell is truncate and appendiculate.

Some species are known from one sex only. From the material on hand it is obvious that there are some undescribed forms, and *Bothynostethus* would be an excellent genus for a taxonomic study. The keys and discussions of Ducke (1902, 1904), Handlirsch (1888), and Schulz (1904) are of limited usefulness.

Biology: Two records of nesting habits have been made on *distinctus*. Cazier and Mortenson (1965c), working in Arizona, noted that the entrance of a nest was in a rodent burrow. Prey was a chrysomelid beetle of the genus *Monoxia* (Galerucinae). Kurczewski and Evans (1972) made more extensive observations on a small colony in New York. The female wasps remodeled preexisting tunnels leading into vertical sandbanks beneath protecting overhangs. The entrances were not closed during the wasp's absence to search for galerucine chrysomelid beetles of two genera. The beetles were carried in flight right side up and head forward, the wasp's mandibles holding the beetle's antennae. Nests had up to eight branches ending in cells that were 15 to 45 cm from the surface.

The use of beetles by *Bothynostethus* is most interesting since no other Larrinae are known to provision with Coleoptera. The possible evolutionary significance of this is discussed under the crabronine genus *Entomognathus*.

Checklist of *Bothynostethus*
(N = notched mandible)

aberrans Ducke, 1902; Brazil
clypearis Ducke, 1904; Brazil
collaris Ducke, 1904; Brazil, Argentina
distinctus W. Fox, 1891; U.S.: transcontinental
dubius Ducke, 1902; Brazil (N)
duckei Menke, 1968; Brazil (N)
kohlii Ducke, 1902; Brazil (N)
nitens Handlirsch, 1888; Brazil (N)
*paraensis (Spinola), 1851 (*Pison*); Brazil, new combination by Menke
saussurei Kohl, 1884; Mexico

Genus Willinkiella Menke

Generic diagnosis: Inner orbits parallel, eye facets only slightly larger below (fig. 114 B); antenna short; mandible with a prominent externoventral angle or tooth (notched), inner margin with two subbasal teeth in female, one in male; mouthparts moderately long; pronotal collar without a transverse carina; propodeal surface smooth posteriorly, not areolate, dorsal enclosure not defined but dorsum with a median fovea and a transverse basal sulcus which is crossed by ridges; gastral terga and sterna simple; male forecoxa deeply, longitudinally channeled mesad and bearing an apical process, trochanter concave basally (fig. 115 E); both recurrent veins of forewing received by submarginal II; forewing media diverging before cu-a.

Geographic range: The single species is known only from Argentina.

Systematics: The smooth propodeum (fig. 115 H) and parallel inner orbits are diagnostic (fig. 114 B). Unlike some *Bothynostethus,* the mandibles in *Willinkiella* are notched, and the forewing media diverges before cu-a. Modifications of the male forecoxa have reached a high state in *Willinkiella* (fig. 115 E). Yellow markings are more extensive in *Willinkiella* than in *Bothynostethus.* As an example, the gaster is yellow banded in the former but not in the latter.

Biology: Unknown.

Checklist of *Willinkiella*

argentina (Schrottky), 1909 (*Pisonopsis*); Argentina
  argentina Menke, 1968, new synonymy by M. Fritz who has studied material identified by Schrottky as his species

Tribe Scapheutini

Small to medium-sized black wasps similar to those of the Bothynostethini but with different wing venation and male abdominal structure. Two genera are included, *Scapheutes* and *Bohartella.*

Diagnostic characters:

1. (a) Inner orbits converging below, eye facets larger below; (b) ocelli normal.
2. (a) Antennal sockets low on face, contiguous with frontoclypeal suture; (b) male with 11 flagellomeres.
3. (a) Clypeus transverse; (b) labrum small, hidden.
4. (a) Externoventral margin of mandible entire or with an angle or tooth, inner margin with one or two teeth; (b) mouthparts short; (c) mandible socket open.
5. Pronotal collar narrow, about as high as scutum.
6. Notauli absent or weak.
7. (a) Episternal sulcus present; (b) omaulus and postspiracular carina present or absent; (c) acetabular carina present but interrupted near midline of venter.
8. (a) Female foretarsus with a weakly developed rake, tarsomeres essentially symmetrical; (b) hindfemur clublike, obliquely truncate distally; (c) midcoxae separated, hindcoxae contiguous; (d) arolia moderate in both sexes; (e) male forecoxa and trochanter usually modified; (f) plantula usually present on tarsomere IV.
9. (a) Propodeum short, dorsally without a transverse basal sulcus; (b) enclosure roughly triangular, defined by sculpture or ridges; (c) no propodeal sternite.
10. (a) Three submarginal cells present, II petiolate; (b) two recurrent veins, both received by submarginal II; (c) outer veinlet of submarginal cell III ending near apex of marginal cell, apex of latter truncate and usually appendiculate.
11. (a) Jugal lobe small, at most half as long as anal area; (b) hamuli not divided into two groups; (c) media diverging before cu-a;
12. (a) Gaster sessile, tergum I with lateral carina; (b) pygidial plate well defined in both sexes, covered with short bristles or hair as in *Tachytes;* (c) male sternum VIII bispinose apically.
13. (a) Volsella greatly reduced, forming a small plate at base of gonostyle; (b) gonostyle simple.

Systematics: Relationships with Bothynostethini have been discussed under that tribe. The diagnostic differences in Scapheutini are the divergence of the hindwing media before cu-a, termination of the outer veinlet of submarginal cell III near the apex of the marginal cell, the absence of a transverse basal propodeal groove, the bispinose male sternum VIII, and the greatly reduced volsella. The presence of a plantula on tarsomere IV in both scapheutin genera (always?) is another distinction. Plantulae are known elsewhere in the Larrinae only in the Trypoxylonini.

Kohl (1896) and Handlirsch (1925) regarded *Scapheutes* as an alyssonin genus, and the genus does have some of the characteristics of this nyssonine tribe: hindfemoral apex with process, bispinose male sternum VIII, and the hindwing venation, but the single long midtibial spur and the inner subbasal tooth or teeth on the mandible of *Scapheutes* indicate that it belongs in the Larrinae.

BOTHYNOSTETHINI

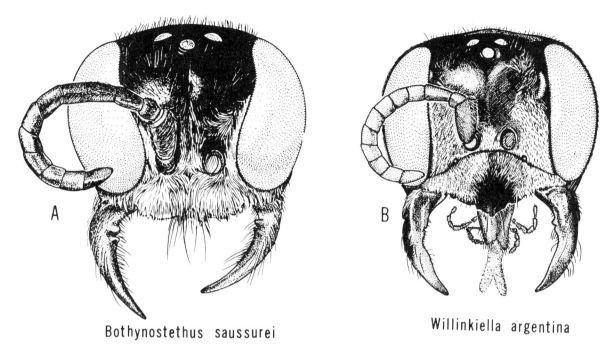

Bothynostethus saussurei

Willinkiella argentina

SCAPHEUTINI

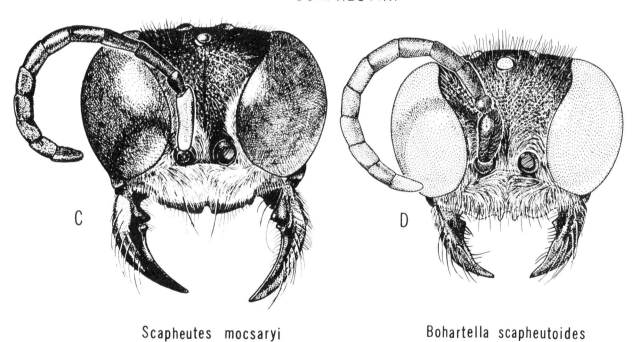

Scapheutes mocsaryi

Bohartella scapheutoides

FIG. 114. Facial portraits of females in the tribes Bothynostethini and Scapheutini, D = holotype.

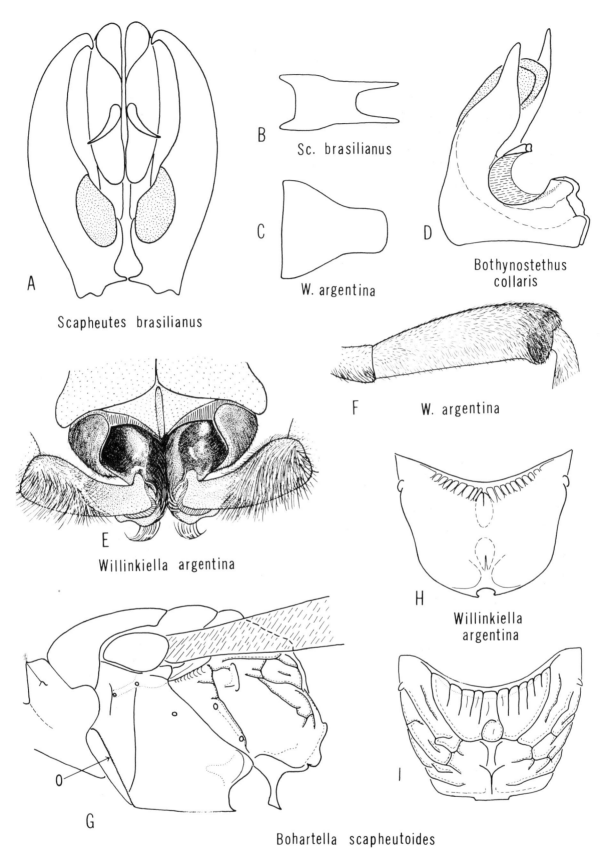

FIG. 115. Structural details in the tribes Scapheutini and Bothynostethini; A, male genitalia, ventral; B,C, male sternum VIII; D, male genitalia, lateral; E, base of forelegs in male; F, hindfemur, female; G, thorax, lateral, o = omaulus; H,I, propodeum, dorsal.

The Scapheutini may be closer to the protocrabronin stock than the Bothynostethini because the omaulus, carinate pronotal collar and lobe and postocular carina of *Bohartella* are features found in one or more crabronine genera, such as *Anacrabro, Encopognathus, Vechtia, Hingstoniola, Quexua*, and *Enoplolindenius.* The long scape of *Scapheutes* is also crabroninlike (fig. 114 C).

Biology: Unknown.

Key to genera of Scapheutini

Mesopleuron without an omaulus; mandible with an external tooth; propodeum smooth posteriorly (similar to fig. 115 H); occipital carina ending before reaching hypostomal carina; each gastral tergum bearing a transverse pit row subapically (S. America) .... *Scapheutes* Handlirsch, p. 355

Mesopleuron with an omaulus (fig. 115 G); externoventral margin of mandible entire; propodeum coarsely areolate posteriorly (fig. 115 I); occipital carina meeting hypostomal carina just before apex of latter; gastral terga without transverse pit rows (Brazil) .... *Bohartella* Menke, p. 355

Genus Scapheutes Handlirsch

Generic diagnosis: Antenna short to moderately long, scape rather long (fig. 114 C), middle male flagellomeres longer and broader than other articles, gena without vertical carina along outer orbit; occipital carina strong but ending abruptly below before reaching hypostomal carina; externoventral margin of mandible with a prominent angle or tooth, inner margin in female with two subbasal teeth, one in male; pronotal collar without a transverse carina; propodeum smooth posteriorly, not areolate, dorsal enclosure present but defined by differential sculpture rather than by carinae or sulci, enclosure with a fine median longitudinal carina; omaulus and postspiracular carina absent, acetabular carina present but interrupted at midline of venter, episternal sulcus meeting and crossing acetabular carina but very weak between it and anterior margin of pleuron; each gastral tergum with a subapical, obtusely V-shaped transverse pit row, apical margin of basal terga sometimes thickened and double edged; male forecoxa and trochanter weakly modified, slightly concave, not channeled.

Geographic range: South America.

Systematics: The notched mandible, incomplete occipital carina, ecarinate collar, smooth propodeum, and tergal pit rows are diagnostic. *Scapheutes* also differs from *Bohartella* in the absence of both an omaulus and a genal carina. In *Bohartella* the propodeal enclosure is defined by carinae, but in *Scapheutes* it is evident only because of its distinctive sculpture. The inner margin of the female mandible has two teeth in *Scapheutes,* but there is only one in *Bohartella.*

Five species have been described but each is known only from one sex. The type of the genus, *mocsaryi,* was described from a male. The taxonomy of the group is fragmentary, and Schulz's (1904) key dealt with only three species.

Biology: Unknown.

Checklist of *Scapheutes*

brasilianus Handlirsch, 1895; Brazil
flavopictus (F. Smith), 1860 (*Pison*); Brazil
friburgensis Brèthes, 1913; Brazil
laetus (F. Smith), 1860 (*Pison*); Brazil
mocsaryi Handlirsch, 1887; Brazil

Genus Bohartella Menke
(Based on female only, male unknown)

Generic diagnosis: Antenna moderately long but scape short and bulbous (fig. 114 D); gena with a vertical carina just anterior to the latter's apex; outer margin of mandible entire, inner margin with one subbasal tooth; pronotal collar with a transverse carina; pronotal lobe traversed by a carina; propodeum areolate posteriorly, dorsal enclosure defined by ridges (fig. 115 I); omaulus present, postspiracular carina present but not reaching omaulus, acetabular carina present but broadly interrupted at midline of venter, continuous with omaulus laterally, episternal sulcus crossing omaulus and only slightly weaker beyond (fig. 115 G); apical margins of basal gastral terga thickened and double edged, terga without subapical transverse pit rows.

Geographic range: This monotypic genus is known only from Brazil.

Systematics: The simple external margin of the mandible, the omaulus (fig. 115 G), genal and pronotal carinae, the complete occipital carina, and the areolate propodeum (fig. 115 I) are distinctive features of *Bohartella.* The acetabular carina is much weaker in *Bohartella* than in *Scapheutes,* being only a short continuation of the omaulus.

Biology: Unknown.

Checklist of *Bohartella*

scapheutoides Menke, 1968; Brazil

SUBFAMILY CRABRONINAE

The crabronines include a great variety of forms which are characterized especially by the cuboidal head and the single discrete submarginal cell of the forewing. In addition there is one recognition character of less than absolute value that is useful to the collector in the field. It is the frequent presence of a transverse silvery clypeus which, together with the vertical pale stripes on the scapes, forms an inverted pi-like mark (∥). This trait which is most common in the Crabronini has earned them the name "silver mouth (or mustache) wasps". Many crabronines are fossorial, but some genera in the Crabronini nest in plant stems and various pre-existing cavities. Twelve orders of insects serve as prey for these wasps, but Diptera predominate.

Diagnostic characters:

1. (a) Eyes moderately separated or more often converging strongly below, inner orbits entire, facets often larger below; (b) ocelli normal
2. (a) Antennal sockets close to frontoclypeal suture; (b) male with 11 to 13, female with 12 antennal articles.
3. Clypeus transverse as a rule.
4. (a) Mandible sometimes with externoventral notch, inner margin simple or dentate; (b) palpal formula 6-4 to 5-3; (c) mouthparts short; (d) mandible socket open or closed.
5. (a) Pronotal collar short; (b) pronotal lobe and tegula separated.
6. (a) Notauli short, usually indistinct; (b) admedian lines narrowly separated or practically contiguous; (c) scutum without oblique carina but with lateral flange which partly overlaps tegula.
7. (a) Episternal sulcus present, scrobal sulcus usually absent; (b) omaulus, sternaulus, hypersternaulus, verticaulus, mesopleuraulus, and acetabular carina present or absent.
8. Definitive metapleuron usually consisting of upper area only.
9. (a) Midtibia with one apical spur, sometimes none in male, rarely none in females; (b) midcoxae slightly to broadly separated, without dorsolateral carina; (c) precoxal sulcus and lobes present; (d) hindfemoral apex sometimes thickened and truncate; claw simple; foretarsus with or without a rake; (e) plantulae absent.
10. (a) Propodeum relatively short; (b) dorsal enclosure sometimes present; (c) no propodeal sternite.
11. (a) Forewing with one submarginal cell or this fused with first discoidal cell, third discoidal cell absent; (b) marginal cell apex truncate or rounded; (c) media diverging beyond cu-a; (d) stigma tapering to a point well beyond base of marginal cell.
12. (a) Hindwing jugal lobe small, shorter to longer than submedian cell; (b) media diverging beyond cu-a; (c) second anal vein and subcosta absent.
13. (a) Gaster sessile to pedunculate, peduncle clearly composed of tergum and sternum; (b) tergum I with lateral carina, one also usually on II and rarely on following terga; (c) male with seven exposed terga; (d) pygidial plate present in females and some males.
14. (a) Volsella simple; (b) cerci absent.

Systematics: Technically, the Crabroninae can be separated from the Larrinae by the following combination of characters: a single submarginal cell, scape nearly to fully half as long as flagellum, inner orbits not strongly nor angularly emarginate. Further discussion is given under the systematics of Miscophini and Trypoxylonini, neither of which seem closely related to crabronines, but both of which have species with a single submarginal cell in the forewing.

The subfamily contains two tribes, Crabronini and Oxybelini, that are readily divided by the fusion of the forewing submarginal cell with the first discoidal cell in the latter (fig. 116 G). In spite of this specialization, the Oxybelini seem to be more generalized than all but a few Crabronini such as *Entomocrabro* and *Encopognathus*. The two evolutionary lines shown in fig. 7 seemingly diverged from a *Bothynostethus*-like form in the Larrinae. Points of similarity between *Bothynostethus* and the more generalized crabronines are the rather

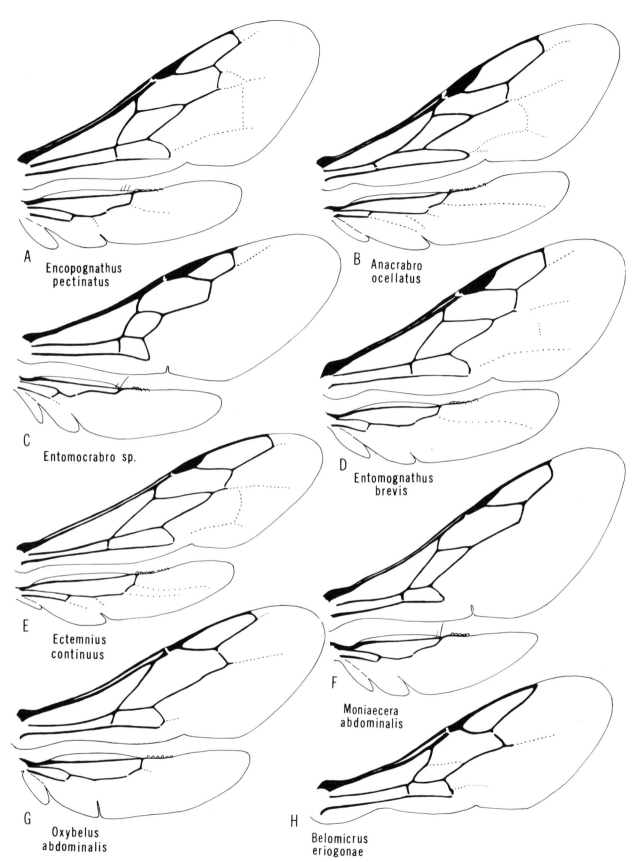

FIG. 116. Wings in the subfamily Crabroninae.

cuboidal head, inner orbits converging below, mandible externoventrally notched (in some *Bothynostethus*), clypeus rather transverse and silvery, mandible socket open, hindwing media diverging after cu-a, jugal lobe small, a single midtibial spur, propodeum short. The similarity is further appreciated by comparing *Bothynostethus* with the list of 19 presumably generalized characters of Crabronini in Table 11. Only three of these are not in agreement. On the other hand there are some obvious differences between *Bothynostethus* and Crabronini like *Entomocrabro, Entomognathus,* and *Encopognathus*. These are the absence of an omaulus or of orbital foveae in *Bothynostethus* and the presence of three submarginal cells (II petiolate) as well as an obliquely truncate hindfemur. Possibly of evolutionary relevance is the fact that in wings of some *Encopognathus* faint remnants of a petiolate submarginal cell II can be discerned (fig. 116 A). Also, some *Oxybelus* have an obliquely truncate hindfemur. A similar argument of relationship could be made for the larrine tribe Scapheutini with the more generalized Oxybelini and Crabronini. The presence of an omaulus in the scapheutin genus *Bohartella*, and the rather long scapes in *Scapheutes* may point toward Crabroninae, also.

Nesting behavior similarities and differences between *Bothynostethus* and *Entomognathus* are discussed under "Biology" of the latter. Concerning larvae, Kurczewski and Evans (1972) have pointed out many features that the two genera have in common. The main difference is the presence of accessory fingerlike lobes on the prothorax of *Bothynostethus,* as in many Larrinae but not in Crabroninae.

The similarities in adult morphology, nesting behavior, and larval structure among several larrine and crabronine genera mentioned above strengthen our belief that the two subfamilies are closely related, but they do not alter our conclusion that they should be treated separately for practical considerations.

Overall, the evidence suggests that the Crabroninae arose from a larrine ancestor before the divergence of Bothynostethini from Scapheutini. The separation of Oxybelini from Crabronini must have taken place soon afterward (fig. 7).

Key to tribes of Crabroninae

Submarginal and discoidal cells fused (fig. 116 G-H) .... Oxybelini, p. 359
Submarginal and discoidal cells separate (fig. 116 A-F) .... Crabronini, p. 370

Tribe Oxybelini

The oxybelin wasps are small and rather bee-like forms with quick motions and rapid flight. They are most often seen as they light on sand and other dry ground, or as they fly around the tops of flowering bushes such as *Baccharis*. These wasps are fossorial and prey of most species are flies including some injurious forms, such as blackflies, tsetse flies and houseflies (R. Bohart and Schlinger, 1957). The prey of *Belomicrus* consists of Hemiptera and Coleoptera. The habit in some *Oxybelus* of carrying the prey impaled on the sting represents one of the most advanced prey carriage mechanisms known in the Sphecidae.

Tribal diagnosis: Small forms, most less than 6 mm long; eyes naked, inner orbits parallel to arcuate, and divergent above and below; frons without a distinct scapal basin, at most with a polished area behind each scape; ocellar triangle broader than long; occipital carina incomplete, separated from hypostomal carina; antennal sockets well separated from each other and from inner orbits; clypeus transverse; male antenna with 13 articles, female with 12; scape ecarinate; male flagellum simple, without ventral hair fringe; palpal formula 6:4; mandible apically simple and acuminate, externoventral margin notched or entire, inner margin simple or dentate, mandibular socket open; metanotum often with squamae; propodeum often with a mucro (fig. 117); mesopleuraulus absent; midtibia with one apical spur; submarginal cell confluent with discoidal cells I and II (fig. 116 G); jugal lobe shorter than submedian cell; gaster sessile; male pygidial plate trapezoidal to rectangular; female pygidial plate triangular and flat.

Systematics: The peculiar venation of the forewing is practically diagnostic (fig. 116 G). The great majority of species are in the genera *Oxybelus* and *Belomicrus,* and these are characterized by the metanotal squamae opposing the propodeal mucro (fig. 117). *Enchemicrum* has these structures, also, but *Brimocelus* has only the mucro, and *Belomicroides* has no mucro and sometimes no squamae. Although the squamae-mucro combination is unique to Oxybelini, well developed squamae are found in some species of *Encopognathus* (Crabronini). In addition to the wing venation, oxybelins have two characters that differ from the situation found in most crabronins. These are the nearly equal interocular distance above and below, and the absence of a distinct scapal basin.

Key to genera of Oxybelini

1. Terga I-III and sometimes IV and V with lateral carina (fig. 117); Holarctic, Ethiopian.................. *Belomicrus* A. Costa, p. 360
   Only terga I-II with lateral carina 2
2. Metanotal squamae small or absent, propodeal mucro absent; sternum VI in female with a high keel; Palearctic and Ethiopian .... *Belomicroides* Kohl, p. 360
   Metanotal squamae well developed, propodeal mucro present (fig. 118 A-F); sternum VI in female not keel shaped.............. 3
3. Scutum strongly arched, much higher in profile than rather reduced pronotal collar; S. Africa .... *Brimocelus* Arnold, p. 364
   Scutum not strongly arched and not rising higher in profile than pronotal collar 4
4. Scutellum, at least posteriorly, and metanotum with a median longitudinal carina (fig. 118 F); verticaulus present; world-

wide except Australia . Oxybelus *Latreille, p. 364*
Scutellum and metanotum without a median carina (fig. 118 D); verticaulus absent; U.S. .... Enchemicrum *Pate, p. 364*

Genus Belomicroides Kohl

Generic diagnosis: Frons simple; mandible with externoventral margin usually notched, sometimes entire; psammophore present in both sexes, weak in male; pronotal collar ecarinate, at nearly same level as scutum; scutum simple, not highly arched; scutellum carinate laterally, with or without weak median carina; metanotum with lateral carinae sometimes lamellate; postspiracular carina, omaulus, acetabular carina, verticaulus, sternaulus, and hypersternaulus absent; propodeum without mucro, lateral propodeal carina well developed to absent; male forecoxa simple or with oblique sulcus on anterior face; female foretarsus with rake; gaster with terga rounded laterally except I-II with lateral carina; venter convex; terga II-IV in males sometimes with transverse basal pubescent grooves; male pygidial plate trapezoidal, female sternum VI with lamellate median keel.

Geographic range: Ethiopian and Palearctic Regions. Three species have been described from southern Africa, four from the Mediterranean area, and one from southwestern USSR.

Systematics: Belomicroides is distinguished by the absence of a mucro on the propodeum, the weak development of the lateral carinae of the metanotum, and the lamellate median keel on sternum VI of the female. The metanotal carinae vary from simple to strongly lamellate and posteriorly angulate but are not differentiated into discrete squamae. This genus, the most generalized in the Oxybelini, offers one of the closest links between the Crabroninae and Larrinae.

Biology: Unknown.

Checklist of *Belomicroides*

arenarius Arnold, 1960; Rhodesia
fergusoni Beaumont, 1960; Greece: Rhodes
marleyi Arnold, 1927; S. Africa
maurusius Pate, 1931; Algeria
pictus Arnold, 1927; Rhodesia, S. Africa
santschii (Schulthess), 1925 (*Oxybelus*); Tunisia
schmiedeknechtii Kohl, 1899; Tunisia
zimini (Gussakovskij), 1952 (*Belomicrus*); USSR: Tadzhik S.S.R., Ukrainian S.S.R.

Genus Belomicrus A. Costa

Generic diagnosis: Frons simple, vertex with (fig. 119 A) or without postocular tubercles; gena with or without carina; mandible with apex simple or with preapical flange (fig. 119 A), externoventral margin notched or entire, internal margin simple to notched or dentate medially; female with psammophore (fig. 117); pronotal collar ecarinate and rounded laterally to carinate and/or angulate laterally; scutum simple, not highly arched; scutellum carinate or simple medially, carinate laterally; metanotum with discrete squamae; postspiracular carina, omaulus, hypersternaulus and sternaulus present or absent; acetabular carina absent; precoxal area with or without tubercle; propodeal mucro usually present, lateral propodeal carina present (fig. 117); tarsomere V not enlarged, female foretarsus with strong rake, legs sometimes modified; gaster with terga I-III and sometimes IV and V with lateral carina and usually sharply bent under laterally, venter flat or slightly concave (fig. 117); male with or without subquadrate pygidial plate; sternum VI of female simple.

Geographic range: Palearctic, Ethiopian, and Nearctic Regions. 63 species are known. This genus is most abundant in warm, arid situations. Twenty-five species are found in North America. Thirteen of these are endemic to California, 10 are distributed in various parts of the western United States with some extending into Canada or Mexico, one is recorded from the Midwest and one from the southeastern states. Twelve species are found in the Ethiopian Region, one in Ethiopia and the remainder in southern Africa. Of the 26 Palearctic species, eight are Mediterranean, eight are described from southwestern USSR, two are recorded from Iran and Pakistan, two are found in Europe, and one, *italicus*, is widespread.

Systematics: Belomicrus is the only genus in the Oxybelini in which the first three terga are sharply bent under laterally (fig. 117). In this character the genus is similar to the crabronin genus *Anacrabro* (fig. 123) from which it may be easily distinguished by the venation of the forewing and the presence of the metanotal squamae. The distinct squamae also differentiate *Belomicrus* from *Belomicroides* and *Brimocelus* which at most have lamellate lateral carinae on the metanotum. *Oxybelus* and *Enchemicrum* are further distinguished by the apically truncate forewing marginal cell (fig. 116 G) and the lack of a genal psammophore in the female. In *Belomicrus* the apex of the marginal cell is acute (fig. 116 H) and the genal psammophore is present in the female (fig. 117).

In North America there are two easily differentiated species groups: the *forbesii* group in which the squamae are joined to form a posteriorly notched triangle (fig. 118 B, C) and the *cladothricis* group in which the squamae are well separated and not part of a triangle (fig. 118 A, E). Examples of the first group above are *forbesii, penuti, franciscus, maricopa,* and *serrano;* of the second group are *cladothricis, eriogonae, sechi, viereki,* and *pachappa.* Preliminary study of Old World species indicates that *italicus*, the genotype, might fit into the *forbesii* group. South African species exhibit a variety of squamal forms that have been assigned to the subgeneric names *Nototis* and *Oxybelomorpha* by Arnold (1927). Obviously, a worldwide study of subgenera or groups is needed.

Important taxonomic references are Kohl (1923), Arnold (1927), and Pate (1940a, b).

Biology: Information has been published by Williams (1936) on *franciscus,* by Valkeila (1963) on *borealis,* and by Evans (1969a, 1970) on *forbesii.* Williams re-

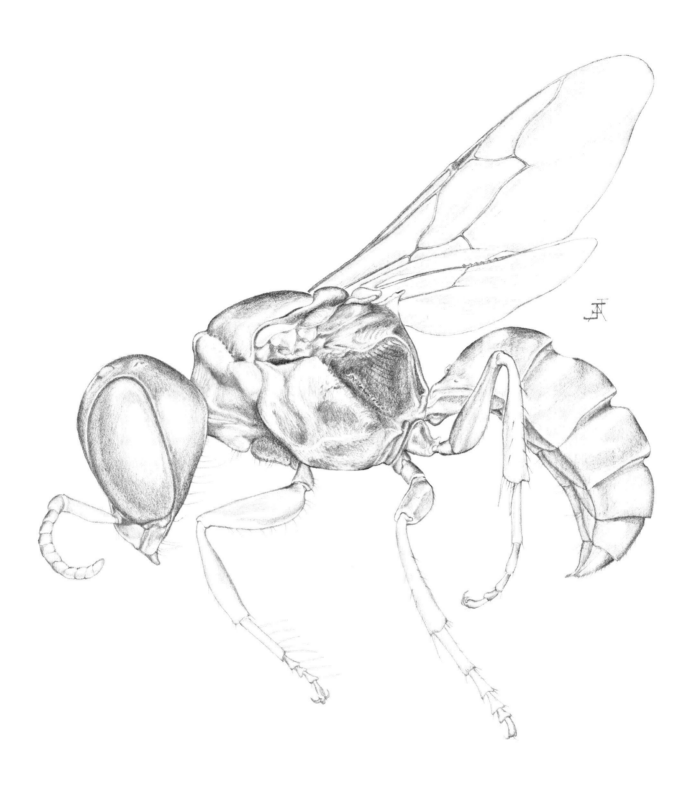

FIG. 117. *Belomicrus eriogonae* Pate, female.

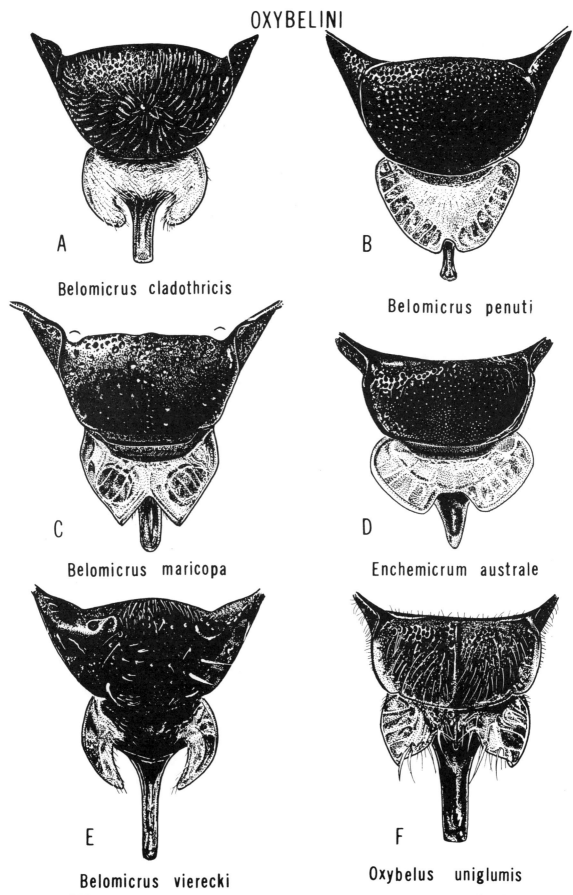

FIG. 118. Scutellum, metanotum and propodeal mucro in the tribe Oxybelini, females.

corded his species as nesting in sand in San Francisco, California. The female flew out of the burrow backward and obliquely upward as she distributed excavated material. No tumulus accumulated, and the nest entrance was left open during provisioning. Prey were beetles of the genus *Trichochrous* (Melyridae: Dasytinae). Valkeila's description of nesting was similar to that of Williams, and the prey was *Dasytis niger* (L.), which is a dasytine beetle also. Valkeila observed three nests that entered the ground at an angle of 30 degrees and then leveled off near the cluster of four cells at the terminus. The nests were about 4-5 cm long, and one of them contained 34 beetle prey. The egg was laid between the coxae of the beetle. Evans' observations differed in several respects, although the details of construction were similar. Working in Wyoming, he found burrows in compact, sandy-clay, pebbly soil. Burrow length varied from 6 to 10 cm, and horizontal depth was 4 to 7 cm. Up to four cells were closely clustered near the terminus. Prey were immature or adult mirids, *Orectoderus obliquus* Uhler, five to nine per cell. A parasite was *Senotainia trilimeata* Wulp (Sarcophagidae) and a predator, *Philanthus pulcher*.

There are several instances where species in different genera of the same tribe may take prey of different families (in Bembicini, for example). Also, some species of sphecids may have several families and orders of prey taken by a single individual (*Lindenius, Microbembex*). However, it is quite unusual for species in the same genus to select prey of different families on a seemingly fixed basis as in *Belomicrus* and *Ectemnius*. Substantiation for this situation in the former genus is given by unpublished observations of R. M. Bohart. In 1956, Bohart and MacSwain collected numerous *Belomicrus franciscus* above Placerville, California. The female wasps were carrying melyrid beetles into their nests. These were *Amecocerus cervicalis* Blaisdell (Melyridae: Dasytinae) and an unnamed species of *Amecocerus* (H. B. Leech det.). In the same year, Bohart observed nests of *B. penuti* and *B. coloratus* in the mountains of Nevada County, California. In these cases, the provisions were all immature bugs of the family Miridae. It would seem to be significant that *penuti, coloratus,* and *forbesii* are all in the same species subgroup and different from the subgroup to which *franciscus* belongs.

The host and other nesting information is unknown for *B. cladothricis* and the many other species in its group. These tiny wasps occur in western United States and are among the smallest sphecids. It is interesting to speculate that their prey may be something quite different from either Melyridae or Miridae.

Checklist of *Belomicrus*

affinis Gussakovskij, 1952; sw. USSR: Tadzhik S.S.R.
antennalis Kohl, 1899; Russia, Austria
apache Pate, 1940; U.S.: New Mexico, Arizona
bicornutus Arnold, 1927; S. Africa
borealis Forsius, 1923; Finland
braunsii Kohl, 1923; S. Africa
bridwelli Pate, 1940; U.S.: Virginia, Tennessee, Florida
caesariensis Pate, 1931; Algeria, Spain, Portugal
   ssp. *maurusius* Beaumont, 1957; Morocco
   ssp. *oceanicus* Beaumont, 1957; Morocco
   ssp. *saharicus* Beaumont, 1958; Algeria
cahuilla Pate, 1940; U.S.: California
cladothricis (Cockerell), 1895 (*Oxybelus*); w. U.S., Mexico: Baja California
   minidoka Pate, 1940; new synonymy by R. Bohart
   prosopidis Pate, 1940; new synonymy by R. Bohart
coloratus Baker, 1909; U.S.: Idaho, Nevada, California
cookii Baker, 1909; U.S.: California
crassus Arnold, 1936; S. Africa
cucamonga Pate, 1940; U.S.: California
dunensis Beaumont, 1957; Morocco
eriogoni Pate, 1940; U.S.: California, Oregon, Nevada, Utah; new status by R. Bohart
excisus Gussakovskij, 1952; sw. USSR: Tadzhik S.S.R.
ferrieri Kohl, 1923; S. Africa
forbesii (Robertson), 1889 (*Oxybelus*); w. U.S. and Canada
   columbianus Kohl, 1892 (*Oxybelus*)
   larimerensis Rohwer, 1908
franciscus Pate, 1931; U.S.: California
   quemaya Pate, 1940; new synonymy by R. Bohart
funestus Arnold, 1929; S. Africa
istam Pate, 1940; U.S.: California
italicus A. Costa, 1871; centr. and s. Europe, centr. Asia; n. Africa
   obscurus Kohl, 1892 (*Oxybelus*)
jurumpa Pate, 1940; U.S.: California
kaszabi Tsuneki, 1972; Mongolia
kohlianus Schulthess, 1926; Tunisia
kohlii (Brauns), 1896 (*Oxybelomorpha*); S. Africa
kuznetzovi Gussakovskij, 1952; sw. USSR: Tadzhik S.S.R.
maricopa Pate, 1947; U.S.: Arizona, New Mexico
mescalero Pate, 1940; U.S.: New Mexico, California
meyeri Kohl, 1923; Iran; Pakistan
minimus Gussakovskij, 1952; sw. USSR: Tadzhik S.S.R.
minutissimus Arnold, 1936; S.-W. Africa
mirificus Kohl, 1905; Ethopia, Egypt
modestus (Kohl), 1892 (*Oxybelus*); sw. USSR
mongolicus Tsuneki, 1972; Mongolia
mono Pate, 1940; U.S.: California; new status by R. Bohart
moricei Kohl, 1923; Israel
multifasciatus Tsuneki, 1972; Mongolia
nirginus Kazenas, 1971; sw. USSR: Kazakh S.S.R.
odontophorus (Kohl), 1892 (*Oxybelus*); Algeria, sw. USSR
pachappa Pate, 1940; U.S.: California
parvulus (Radoszkowski), 1877 (*Oxybelus*); sw. USSR: Turkmen S.S.R.
   femoralis Kohl, 1900
patei Beaumont, 1950; Algeria, Morocco
penuti Pate, 1940; U.S.: California, Oregon, Nevada
persa Gussakovskij, 1933; Iran
potawatomi Pate, 1947; U.S.: Iowa

querecho Pate, 1940; U.S.: New Mexico
radoszkowskyi (Dalla Torre), 1897 (*Oxybelus*); sw USSR: Turkmen S.S.R.
    fasciatus Radoszkowski, 1877 (*Oxybeloides*), nec Dahlbom, 1845 (*Oxybelus*)
rhodesianus Arnold, 1927; Rhodesia
rufiventris Arnold, 1936; S. Africa
schulthessii Kohl, 1923; sw. USSR: Turkmen S.S.R.
sechi Pate, 1940; U.S.: s. California
serrano Pate, 1940; U.S.: California
sordidus Arnold, 1927; S. Africa
steckii Kohl, 1923; Europe: France, Spain, Portugal
  ssp. *maroccanus* Beaumont, 1956; Morocco
timberlakei Pate, 1940; U.S.: California
tuktum Pate, 1940; U.S.: California
turneri Arnold, 1927; S. Africa
vanyume Pate, 1940; U.S.: California
viereckii Pate, 1940; sw. U.S.; Mexico: Baja California, Puebla
waterstonii Kohl, 1923; Israel
wouroukatte Beaumont, 1967; Turkey

Genus Brimocelus Arnold

Generic diagnosis: Frons simple; mandible with externoventral margin notched; weak psammophore present in female; pronotal collar low, narrow, ecarinate; scutum strongly arched above level of pronotum; scutellum longitudinally carinate medially and laterally, dentate posterolaterally; metanotum concave; lateral margins strongly lamellate, bent upward, and angulate posteriorly, posterior margin elongate medially, forming short trough with notched apex; postspiracular carina, omaulus, acetabular carina, sternaulus, and hypersternaulus absent; precoxal area simple; propodeum with mucro and lateral propodeal carina present; tarsomere V normal, female foretarsus with weak rake; gaster with terga rounded laterally, venter convex; terga of male without transverse pubescent grooves; pygidial plate of both sexes triangular, sternum VI of female simple.

Geographic range: Ethiopian Region. One species known.

Systematics: Brimocelus is distinguished from other oxybelins by the reduced pronotal collar, the highly arched scutum, and the development of the metanotal margins. These are more strongly lamellate than in *Belomicroides* but not differentiated into discrete squamae as in *Belomicrus*, *Oxybelus*, and *Enchemicrum*.

Biology: Unknown.

Checklist of *Brimocelus*

radiatus (Arnold), 1927 (*Belomicrus*); S. Africa

Genus Enchemicrum Pate

Generic diagnosis: Clypeus with a weak mediodiscal tubercle; frons with short median longitudinal carina; mandible with externoventral margin notched in male, sinuate in female; female without psammophore, at most with a few ammochaetae on mandible; pronotal collar carinate; scutum with longitudinal median depression; scutellum carinate laterally, ecarinate medially (fig. 118 D), edentate posterolaterally; metanotum with discrete squamae; postspiracular carina, omaulus, and hypersternaulus present; acetabular carina and sternaulus absent; precoxal area simple; propodeum with mucro and lateral propodeal carinae present; tarsomere V enlarged, female foretarsus with few short rake setae; gaster with terga rounded laterally except I and II with lateral carina, venter convex; pygidial plate of male trapezoidal; sternum VI of female simple.

Geographic range: Central and southern U.S. Only one species is known.

Systematics: Enchemicrum and *Oxybelus* are closely related genera, which may be distinguished from other oxybelins by the following combination of characters: metanotal squamae discrete, marginal cell truncate apically, abdominal venter convex, and psammophore absent. *Enchemicrum* differs from *Oxybelus* as follows: verticaulus absent, scutum strongly arched, scutellum and metanotum ecarinate medially, mandible externoventrally notched in male, sinuate in female.

Biology: The nesting habits of *Enchemicrum australe* were observed near Willis, Oklahoma by R. M. Bohart and Holland (1966). The two burrows were located 150 feet apart on opposite sides of a borrow-pit pond. They were dug vertically in firmly-packed sand, had one completed cell about 7.5 cm down and a second one started. The cells contained between 12 and 16 flies each. The prey were of uniform size, averaging three mm in length and included two species of Ephydridae: *Paralimna texana* Cresson and *Zeros flavipes* Cresson, and one species of Dolichopodidae: *Medetera californiensis* Wheeler. When hunting prey the females left the burrow entrance open, and on returning they flew directly in. No males or parasites were observed near the nests, but one burrow was invaded by ants, *Iridomyrmex pruinosus analis* (E. André) and *Dorymyrmex pyramicus* (Roger), which were eating the prey in the unfinished cell. Both sexes of *E. australe* were seen on flowers of *Tamarix*.

Checklist of *Enchemicrum*

australe Pate, 1929; U.S.: Georgia to Arizona, n. to Illinois

Genus Oxybelus Latreille

Generic diagnosis: Clypeus usually with a prominent mediodiscal tooth (fig. 119 B) or longitudinal carina; frons simple; mandible with externoventral margin sinuate; female without genal ammochaetae; pronotal collar usually carinate; scutellum carinate medially, at least posteriorly, laterally with a thin flange (fig. 118 F); metanotum usually with a median carina and with squamae which usually appear as two lateral leaflike expansions or lobes (fig. 118 F); postspiracular carina, omaulus and acetabular carina present; hypersternaulus present or evanescent, sternaulus absent or incomplete, precoxal carina strong; propodeum with mucro and lateral propodeal carinae present; tarsomere V enlarged, female foretarsus with a short rake, a less well developed rake

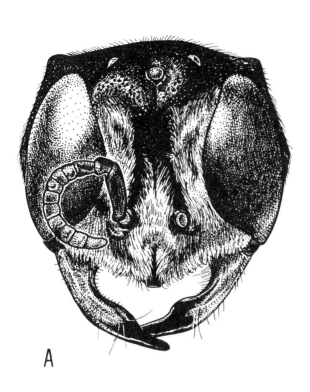
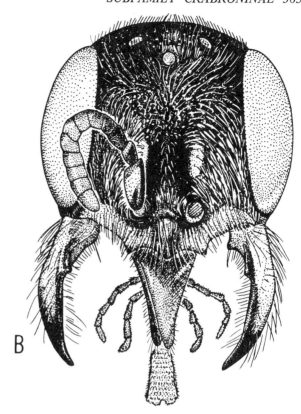

Belomicrus maricopa **Oxybelus ventralis**

FIG. 119. Facial portraits of females in the tribe Oxybelini.

usually present in male; hindfemur simple distally or keeled; only terga I-II of gaster with lateral carina or line, venter convex; pygidial plate of male trapezoidal, lateral margins usually distinct; sternum VI of female simple.

Geographic range: Oxybelus is the largest genus in the subfamily Crabroninae, followed by *Crossocerus* and *Ectemnius.* The genus is worldwide except for the Australian area. The more than 215 listed species are well represented in all other Regions but especially the Palearctic. An unidentified species occurs on the Galapagos Islands.

Systematics: Oxybelus are generally recognized as small (3-10 mm long), stout wasps with spiny legs and coarse sculpture, at least on the thorax. As would be expected, the morphological differences are considerable. The presence of both squamae and mucro, a strong precoxal carina and acetabular carina, the laterally ecarinate terga III-V, the enlarged last tarsomere, and the median carina of the scutellum are adequate for separation from other Oxybelini. The closest relative seems to be *Enchemicrum,* which has no precoxal, acetabular, or median scutellar carinae.

Specific and group characters are especially to be found in the form of the clypeus, ocellar area, pronotal collar, scutum, squamae, mucro, hindfemur, and last tergum. Markings, including eye color, are also useful but are subject to variation. Pubescence, particularly of the forewing median cell and of the propodeal dorsum, offers good characters for some species.

Groups can be based on the form of the mucro and squamae. Without accounting for many intermediate and individual sorts, four general types and a few examples are as follows: (1) mucro lamelliform, squamae with inner margins evenly incurved — *lamellatus, andalusiacus,* and *lingula;* (2) mucro spinose, squamae with inner margins evenly incurved — *major, ventralis,* and *argentatus;* (3) mucro gutterlike, squamae with inner margins evenly incurved or at least without a strong inner lobe (fig. 118 F) — *uniglumis, packardii,* and *aurantiacus;* (4) mucro flared distally and deeply notched, each squama with an inner lobe which is often strong — *emarginatus, similis,* and *cordatus.* Group (1) is Old World and best represented in the Ethiopian Region. Groups (2) and (3) have many species and are typically Holarctic. Group (4) is best represented in the New World, both North and South America.

More comprehensive and useful papers are those of Kohl (1884a), Brèthes (1913), Arnold (1927), Yasumatsu (1935a), Pate (1937e), P. Verhoeff (1948), Faester (1949), Bohart and Schlinger (1956, 1957), Blüthgen (1954), Guiglia (1953), Móczár (1958), and Beaumont (1964b). Guiglia's reference contains a good bibliography.

Biology: Oxybelus burrows often occur in loose aggregations, are usually short (6-10 cm) and most often are located in sand or sandy soil. Banks along a watercourse, a lake margin, or protected areas behind an ocean beach are favored sites for the nests. The burrows may end in one or a few cells that are stocked with a great variety of flies, but specially Muscoidea. During provisioning, the nest entrance may be left open or closed, depending upon the species. Prey are carried to the nest impaled on the sting (*uniglumis, subulatus, bipunctatus*) or held with the midlegs beneath the body of the wasp (*emarginatus, subcornutus, sericeus*). The wasp often alights with its prey near the nest, then flies up and directly down into the entrance. In *bipunctatus* the wasp frequently impales the prey on the sting only after reaching the nest. An egg is laid on the first prey in a cell. The prey may be numerous if they are small midges or as few as three if a medium-sized fly is provisioned. Such diverse families as Ephydridae, Asilidae, Syrphidae, Dolichopodidae, Tachinidae, Calliphoridae, Therevidae, Chironomidae, Sarcophagidae, and Muscidae have figured in prey records; but the last four families have appeared most often.

The most frequently observed species in Europe as well as America is *uniglumis.* Hamm and Richards (1930) have given a summary of notes by other workers and added some of their own. This species is one that carries its prey on the sting. The fly is captured in the air or as it alights on grass stems or tree stumps. Paralysis follows a sting in the neck region. The fly is impaled on the sting so that it trails "upside down behind the wasp's tail." Although the *uniglumis* female closes its burrow whenever it is prey gathering, several sarcophagid flies (*Sphecapata conica* Fallén, *Metopia leucocephala* Rossi) are common parasites.

Several unusual observations have been made on species of *Oxybelus.* Bohart and Marsh (1960) reported that males of *sericeus* in California acted as guards near the nest and struck at intruders such as chrysidids, miltogrammine sarcophagids, or other male *Oxybelus*, driving them away. Copulation took place whenever a female emerged from her burrow or returned to it with prey. A matched pair remained faithful during construction of a single nest. Frequent matings were also seen by Krombein (1964b) with respect to *emarginatus* in Florida. This author also noted the close proximity of ant nests (*Dorymyrmex*) to burrows of *emarginatus* without apparent conflict. Bohart, Lin, and Holland (1966) gave details of nesting habits of *sparideus* on the sandy shore of Lake Texoma, Oklahoma. Here the burrows were nearly vertical, about 9 cm deep and ended in two or three cells. They were stocked with three to six flies which averaged about the same size as the wasp. An entrance-closing ritual was performed whenever the wasp left the nest. This involved the bringing up of a plug of damp sand and tamping of the closure in circular fashion by the wasp before it flew away. The resulting "crater" seemed to serve as a landmark for the returning wasp laden with prey. A sort of "self-parasitism" was also reported in which the *O. sparideus* females included paralyzed sarcophagid females in some of their cells. The maggots emerged from the mother's body and destroyed the cell contents.

An excellent paper that gives much new information and summarizes previous reports of *Oxybelus* nesting behavior is that of Peckham *et al.* (1973). These authors found that association of the sexes at the nest entrance in *subulatus* was much as described above for *sericeus.* Also, they grouped 8 species with respect to digging behavior: *bipunctatus* and *uniglumis* are "rakers" and "pushers"; *subulatus* specializes in "carrying"; *exclamans* is a "raker", and *sericeus, emarginatus, sparideus* and *subcornutus* are "pushers". Species that close the nest during provisioning are *bipunctatus, uniglumis, exclamans,* and *sparideus.* Those that leave the nest open are *sericeus, subulatus, cressonii, emarginatus* and *subcornutus.*

Some observations imply great constancy in the family, species, or even sex of prey. For instance, cells of *linsleyi* excavated by Bohart (unpublished notes) frequently contained only male Therevidae of a single species. *O. sericeus* prey reported by Bohart and Marsh (1960) were all one species of Ephydridae (*Ephydra riparia* Fallén). Prey of *O. subcornutus* were reported by Peckham *et al.* as male Syrphidae exclusively. Conclusions based on these data might be misleading. All evidence suggests that *Oxybelus* females may show some preferences but essentially they are opportunists and capture flies that are abundant and locally available.

Checklist of *Oxybelus*

abdominalis Baker, 1896; U.S. w. of 100th mer., n. Mexico
   calligaster Viereck, 1906 (*Notoglossa*)
acutissimus Bischoff, 1911; Zaire
   ssp. *propinquus* Arnold, 1927; Ethiopia, Rhodesia, S. Africa
aestuosus Bingham, 1897; Burma
aethiopicus Cameron, 1906; S. Africa
agilis F. Smith, 1856; India
   sabulosus F. Smith, 1856
agnitus Brèthes, 1913; Argentina
†*albimaculatus* Kazenas, 1972; sw. USSR: Kazakh S.S.R., nec Mickel, 1918
albipes F. Morawitz, 1894; sw. USSR: Turkmen S.S.R.
albopictus Radoszkowski, 1877; sw. USSR: Turkmen S.S.R.
americanus Spinola, 1841; French Guiana
analis Cresson, 1865; Cuba, Nicaragua, Bermuda
   ssp. *bimini* Krombein. 1953; Bahamas
andalusiacus Spinola, 1843; nw. Africa, Balearics, Sardinia, Corsica, Spain, s. France
   arabs Lepeletier, 1845
   frondigera A. Costa, 1883 (*Notoglossa*)
   andalusiaticus Móczár, 1958
   andalsiacus Tsuneki, 1961
andinus Brèthes, 1913; Argentina
angustus Saussure, 1892; S. Africa
argentatus Curtis, 1833; England, Channel Is., Belgium,

Netherlands
? *decimmaculatus* Donovan, 1806 (*Vespa*)
   ferox Shuckard, 1837
   nigricornis Shuckard, 1837
ssp. *aculeatus* Thomson, 1870; Sweden, Denmark
   quattuordecimnotatus Dahlbom, 1845, nec Jurine, 1807
   dahlbomi Schulz, 1906
ssp. *treforti* Sajó, 1884; Austrai, Hungary, Czechoslovakia
ssp. *bouwmani* P. Verhoeff, 1948; Netherlands
ssp. *debeaumonti* P. Verhoeff, 1948; centr. Europe
ssp. *gerstaeckeri* P. Verhoeff, 1948; centr. and s. Europe
ssp. *mongolicus* Tsuneki, 1972; Mongolia
argenteopilosus Cameron, 1891; w. U.S., w. Mexico
argentinus Brèthes, 1913; Argentina
argypheus R. Bohart and Schlinger, 1956; w. U.S., n. Mexico
arnoldi Benoit, 1951; Zaire
aurantiacus Mocsáry, 1883; Europe, n. Africa
   rufitarsis F. Morawitz, 1894
aurifrons F. Smith, 1856; Brazil
aztecus Cameron, 1891; Mexico, centr. America
bareii Radoszkowski, 1893; sw. USSR: Turkmen S.S.R.
bechuanae Arnold, 1936; Botswana
bicornutus Arnold, 1929; S.-W. Africa
bipunctatus Olivier, 1811; Europe, Japan, e. U.S., e. Canada
   nigroaeneus Shuckard, 1837
   laevigatus Schilling, 1848
   ssp. *thermophilus* Beaumont, 1950; Algeria, Morocco
braunsi Arnold, 1927; S. Africa
bugabensis Cameron, 1891; Mexico, Panama
   longispina Cameron, 1891; new synonymy by R. Bohart
   bugobensis Dalla Torre, 1897
californicus R. Bohart and Schlinger, 1956; w. U.S., Mexico
callani Pate, 1943; Trinidad, Brazil
canaliculatus Radoszkowski, 1877; sw. USSR: Turkmen S.S.R.
   ssp. *seraksensis* Radoszkowski, 1893; sw. USSR: Turkmen S.S.R., Spain
canalis R. Bohart and Schlinger, 1956; sw. U.S., n. Mexico
canescens Cameron, 1890; India
carinatus Gussakovskij, 1933; Iran
catamarcensis (Schrottky), 1909 (*Notoglossa*); Paraguay
caucasicus Radoszkowski, 1893; USSR: Georgia
ceylonicus Cameron, 1896; Sri Lanka
chilensis Reed, 1894; Chile
citrinulus Gussakovskij, 1952; sw. USSR: Tadzhik S.S.R.
citrinus Radoszkowski, 1893; sw. USSR: Turkmen S.S.R.
clandestinus Kohl, 1905; Chile
cocacolae P. Verhoeff, 1968; Morocco, Canary Is.
cochise Pate, 1943; sw. U.S., n. Mexico: Chihuahua, Baja California
cocopa Pate, 1943; U.S.: California
collaris Kohl, 1884; Arabian penin.

comatus Reed, 1894; Chile
congophilus Benoit, 1951; Zaire
coniferus Arnold, 1951; Ethiopia
continuus Dahlbom, 1845; Egypt
cordatus Spinola, 1851; Chile
cordiformis Gussakovskij, 1952; sw. USSR: Tadzhik S.S.R.
cornutus Robertson, 1889; U.S. w. of 100 mer., Mexico: Jalisco, Durango
   quadricolor Cockerell and Baker, 1896
   polygoni Rohwer, 1909
crandalli R. Bohart and Schlinger, 1956; U.S.: Arizona, Arkansas
cressonii Robertson, 1889; U.S.: Utah and Texas, e. to Virginia, n. to Michigan
cristatus Saussure, 1892; Madagascar
curviscutis Arnold, 1927; Ethiopian Region
decipiens Brèthes, 1913; Argentina
decorosus (Mickel), 1916 (*Notoglossa*); U.S.: New York, Michigan, Florida
†*deserticola* Tsuneki, 1972; Mongolia, nec Beaumont, 1950
diphyllus (A. Costa), 1882 (*Alepidaspis*); Sardinia
   ssp. *pharao* Kohl, 1884; n. Africa
dissectus Dahlbom, 1845; centr. Europe, n. Africa
   monachus Gerstaecker, 1867
   morosus Chevrier, 1868
   ssp. *eburneofasciatus* Dahlbom, 1845; France, Spain
      dufouri Marquet, 1896
   ssp. *elegans* Mocsáry, 1879; Hungary, Czechoslovakia, Mongolia
   ssp. *tingitanus* Beaumont, 1950; Morocco
dissimilis Arnold, 1934; Rhodesia
dusmeti E. Pérez, 1966; Spain
eburneoguttatus Arnold, 1952; Rhodesia
eburneus Radoszkowski, 1877; sw. USSR: Turkmen S.S.R.
   turkestanicus Mocsáry, 1879
elongatus Radoszkowski, 1877; sw. USSR: Turkmen S.S.R.
emarginatus Say, 1837; s. Canada, U.S., Mexico
   dilutus Baker, 1896
   trifidus Cockerell and Baker, 1896
   americanus Robertson, 1901 (*Notoglossa*)
   pacificus Rohwer, 1909 (*Notoglossa*)
   minor Mickel, 1916 (*Notoglossa*)
exclamans Viereck, 1906; U.S., n. Mexico
   townsendi Rohwer and Cockerell, 1908
   argentarius Mickel, 1916
   pectorosus Mickel, 1918
eximius Sickmann, 1894; China
fedtschenkoi Radoszkowski, 1877; sw. USSR: Turkmen S.S.R., Tadzhik S.S.R.
fischeri Spinola, 1838; Egypt
   africanus Kohl, 1884
   ssp. *tegularis* E. Saunders, 1903; Canary Is.
fissus Lepeletier, 1845; France
flavicornis Arnold, 1927; Botswana, Rhodesia, S. Africa

ssp. *nyassae* Arnold, 1927; Tanzania
flavipes Cameron, 1890; India
†*flavipes* Kazenas, 1972; sw. USSR: Kazakh S.S.R., nec Cameron, 1890
flaviventris Arnold, 1927; Rhodesia
fossor Rohwer and Cockerell, 1908; U.S. w. of 100th mer., Mexico: Baja California
    umbrosus Mickel, 1916
    puente Pate, 1943
fraudulentus Arnold, 1940; S.-W. Africa
    decipiens Arnold, 1929, nec Brèthes, 1913
frontalis Robertson, 1889; U.S.: e. of Rocky Mts.; Mexico: Veracruz
fulvicaudis Cameron, 1908; India
fulvopilosus Cameron, 1890; India
furcifer Turner, 1917; India
fuscohirtus Gussakovskij, 1930; sw. USSR: Tadzhik S.S.R.
glasunowi F. Morawitz, 1894; sw. USSR: Turkmen S.S.R.
gobiensis Tsuneki, 1972; Mongolia
guichardi Beaumont, 1950; Algeria
haemorrhoidalis Olivier, 1811; Europe
    ssp. *bicolor* Schilling, 1848; Germany
harraricus Arnold, 1927; Ethiopia
hastatus Fabricius, 1804; Spain, Morocco, Tunisia
    lancifer Olivier, 1811
hessei Arnold, 1929; S.-W. Africa
hurdi R. Bohart and Schlinger, 1956; Mexico
imperialis Gerstaecker, 1867; S. Africa
inornatus (Robertson), 1901 (*Notoglossa*); e. U.S.
insularis Kohl, 1884; Sri Lanka
†*interruptus* Brèthes, 1913; Argentina, nec Cresson, 1865
joergenseni Brèthes, 1913; Argentina
kirgisicus Radoszkowski, 1893; USSR: Kirghiz
kizilkumii Radoszkowski, 1877; sw. USSR: Turkmen S.S.R.
krombeini R. Bohart and Schlinger, 1956; U.S.: California
laetus Say, 1837; e. U.S.
    ssp. *fulvipes* Robertson, 1889; se. U.S. coast
    floridanus Robertson, 1901
lamellatus Olivier, 1811; Cyprus, e. Mediterranean area, n. Africa, w. India
    savignyi Spinola, 1838
    sagittatus Dahlbom, 1845 (*Notoglossa*)
    squamosus F. Smith, 1875
    forticarinatus Cameron, 1908
    ssp. *bicolorisquama* Strand, 1922; Taiwan
    kankauensis Strand, 1922
    ssp. *banksi* Ashmead, 1905; Philippines
    ssp. *divaricatus* Pate, 1938; Philippines: Mindanao
    ssp. *aequipunctatus* Pate, 1938; Philippines: Luzon
lanceolatus Gerstaecker, 1867; Saudi Arabia
latidens Gerstaecker, 1867; centr. and e. Europe, Middle East
    psammobius Kohl, 1884
    ssp. *flavitibialis* Tsuneki, 1971; Mongolia
latifrons Kohl, 1892; USSR: Caucasus

latilineatus Cameron, 1908; India
latro Olivier, 1811; Europe to Mongolia
    armiger Olivier, 1811
    affinis Marquet, 1881
    opacus Tournier, 1901
    ssp. *rugulosus* Móczár, 1957; Hungary
lepturus Arnold, 1927; Rhodesia
lewisi Cameron, 1890; Sri Lanka, India, Japan, Taiwan
    sakuranus Tsuneki, 1966
limatus Arnold, 1927; Rhodesia
lineatus (Fabricius), 1787 (*Nomada*); centr. and s. Europe
    bellicosus Olivier, 1811
linguiferus Turner, 1917; India
lingula Gerstaecker, 1867; Rhodesia
    pinnatus Saussure, 1892
    spiniferus Cameron, 1905
    striatiscutis Cameron, 1905
    kalaharicus Bischoff, 1913
linsleyi R. Bohart and Schlinger, 1956; U.S.: California
lubricus Beaumont, 1950; Algeria
macswaini R. Bohart and Schlinger, 1956; U.S.: California, Arizona
maculipes F. Smith, 1856; centr. Europe, Middle East, sw. USSR
    solskii Radoszkowski, 1877
    acuticornis F. Morawitz, 1891
    pictipes F. Morawitz, 1891
maidlii Kohl, 1923; Pakistan
major Mickel, 1916; e. U.S.
mandibularis Dahlbom, 1845; Europe, sw. USSR
    sericatus Gerstaecker, 1867
maracandicus Radoszkowski, 1877; sw. USSR: Turkmen S.S.R.
marginatus F. Smith, 1856; Brazil
marginellus Spinola, 1851; Chile
marginicollis Gussakovskij, 1933; Iran
matabele Arnold, 1927; Rhodesia
merwensis Radoszkowski, 1893; sw. USSR: Turkmen S.S.R.
    mervensis Dalla Torre, 1897
metopias Kohl, 1894; Mozambique
mexicanus Robertson, 1889; Mexico, Belize, Nicaragua
minutissimus F. Morawitz, 1892; sw. USSR
moczari Tsuneki, 1972; Mongolia
†*modestus* Brèthes, 1913; Argentina, nec Kohl, 1892
mucronatus (Fabricius), 1793 (*Crabro*); centr. and s. Europe
    pugnax Olivier, 1811
    bellus Dahlbom, 1844
    ambiguus Gerstaecker, 1867
    scutellaris A. Costa, 1871
    meridionalis Mocsáry, 1879
    nigriventris Tournier, 1901
    maculiventris Tournier, 1901
    immaculatus Guiglia, 1944
    bistillatus Balthasar, 1972
    ssp. *moricei* Beaumont, 1950; Canary Is., Algeria, Libya

nanus Bingham, 1897; Burma
nasutus Bischoff, 1913; S.-W. Africa
natalensis Arnold, 1927; S. Africa
neuvillei Magretti, 1908; Kenya
niger Robertson, 1889; se. Canada, e. U.S.
nigritulus Turner, 1917; India: Assam
†*nigriventris* Tsuneki, 1963; Thailand, nec Tournier, 1901
nipponicus Tsuneki, 1966; Japan
    ssp. *formosus* Tsuneki, 1968; Taiwan
oasicola Tsuneki, 1972; Mongolia
occitanicus Marquet, 1896; France
packardii Robertson, 1889; U.S.
    texanus Robertson, 1889
    heterolepis Cockerell and Baker, 1896
    defectus Cockerell and Baker, 1896
    unicus Mickel, 1918
    carolinus Banks, 1921
paenemarginatus (Viereck), 1906 (*Notoglossa*); U.S.: Kansas
pallidus Arnold, 1927; Ethiopia, Tanzania
palmetorum Beaumont, 1950; Egypt, Mauritania
    deserticola Beaumont, 1950
pamparum Brèthes, 1913; Argentina
paracochise R. Bohart and Schlinger, 1956; U.S.: Arizona, Texas; n. Mexico
paraguayensis Brèthes, 1909; Paraguay
parisinus (Kittel), 1828 (*Crabro*); France
parvus Cresson, 1865; U.S. w. of 100th mer., n. Mexico
    intermedius Baker, 1896
    coloradensis Baker, 1896
    incisura Mickel, 1916 (*Notoglossa*)
paucipunctatus Arnold, 1927; S. Africa, Rhodesia
pectoralis F. Morawitz, 1893; sw. USSR: Turkmen S.S.R., Tadzhik S.S.R.
peringueyi Saussure, 1892; S. Africa
perornatus Arnold, 1944; Madagascar
philippinensis Pate, 1938; Philippines
    ssp. *reticulatus* Pate, 1938; Philippines: Mindanao, Samar
phyllophorus Kohl, 1898; Egypt, Algeria
pictisentis Cameron, 1908; India to Thailand
    aurifrons Cameron, 1902; nec F. Smith, 1856
        pictiscutis "Cameron" Turner, 1917, in error
pictus Arnold, 1927; Rhodesia, Mali
pilosus Arnold, 1927; Ethiopia
pitanta Pate, 1943; sw. U.S.
platensis Brèthes, 1901; Argentina
polyacanthus A. Costa, 1882; Sardinia
polyceros Pate, 1943; Venezuela, French Guiana
pulawskii Tsuneki, 1972; Mongolia
pygidialis Gussakovskij, 1952; sw. USSR: Tadzhik S.S.R.
pyrura (Rohwer), 1914 (*Notoglossa*); Guatemala
quattuordecimnotatus Jurine, 1807; centr. and s. Europe, n. Africa, sw. USSR
    crassipes Walckenaer, 1817 (*Crabro*), nec Fabricius, 1798, new synonymy by J. Leclercq
    punctatus Walckenaer, 1817 (*Crabro*), new synonymy by J. Leclercq
    quattuordecimguttatus Shuckard, 1837, lapsus or emendation
    furcatus Lepeletier, 1845
    quadrinotatus A. Costa, 1883, nec Say, 1824
    maritimus Marquet, 1896
rancocas Pate, 1943; U.S.: New Jersey
raptor Lepeletier, 1845; France
rejectus Baker, 1896; U.S.: Colorado
robertsonii Baker, 1896; w. U.S.
    varicoloratus Baker, 1896
    hirsutus Baker, 1896
    apicatus H. Smith, 1908
    glenensis H. Smith, 1908
robustus Cameron, 1890; India
rubrocaudatus Arnold, 1927; Rhodesia, Tanzania, S. Africa
ruficaudis Cameron, 1905; Rhodesia, S. Africa
    capensis Cameron, 1905
    ssp. *melanarius* Arnold, 1927; S. Africa
ruficornis F. Smith, 1856; India
rufipes Taschenberg, 1880; Ethiopia
rufopictus F. Morawitz, 1892; sw. USSR
sarafschani Radoszkowski, 1877; sw. USSR: Turkmen S.S.R.
sericeus Robertson, 1889; U.S.; Mexico: Baja California, Nayarit
    delicatus Mickel, 1918
    crocatus Krombein, 1955; new synonymy by R. Bohart
similis Cresson, 1865; w. U.S.
    striatifrons Mickel, 1916 (*Notoglossa*)
solitarius Arnold, 1927; Rhodesia
sparideus Cockerell, 1895; sw. U.S., n. Mexico
spectabilis Gerstaecker, 1867; sw. Europe, nw. Africa
spinulosus Gussakovskij, 1935; sw. USSR: Turkmen S.S.R.
stevensoni Arnold, 1927; Rhodesia
strandi Yasumatsu, 1935; Japan
subcornutus Cockerell, 1895; U.S., n. Mexico
    cockerellii Baker, 1896
    punctatus Baker, 1896
    striatus Baker, 1896
    denverensis Rohwer, 1909
subcristatus Saussure, 1892; Madagascar
subspinosus Klug, 1835; s. Europe, Middle East, n. Africa
subtilis Gussakovskij, 1935; sw. USSR: Tadzhik S.S.R.
subulatus Robertson, 1889; U.S.
    mucronatus Packard, 1867, nec Fabricius, 1793
    packardi Dalla Torre, 1890, nec Robertson, 1889
    acutus Baker, 1896
    albosignatus H. Smith, 1908
    mottensis Mickel, 1918
suluensis Pate, 1938; Borneo; Philippines: Palawan
taenigaster (Viereck), 1906 (*Notoglossa*); w. U.S., Mexico
    fastigatus Mickel, 1916
    albomaculatus Mickel, 1918 (*Notoglossa*)
taprobanensis Pate, 1930; Sri Lanka
    ceylonicus Cameron, 1900, nec Cameron, 1896

tarijensis Brèthes, 1913; Argentina
taschenbergi Kohl, 1884; S. Africa
   fasciatus Taschenberg, 1875, nec Dahlbom, 1845
tengu Tsuneki, 1972; Mongolia
timberlakei R. Bohart and Schlinger, 1956; U.S.: California
timidus Chevrier, 1868; centr. Europe
transcaspicus Radoszkowski, 1888; sw. USSR
transiens Turner, 1917; India
† ssp. *thaianus* Tsuneki, 1963; Thailand, nec Tsuneki, 1961
tricolor Gussakovskij, 1952; sw. USSR: Tadzhik S.S.R.
tridentatus F. Smith, 1856; India, Burma
trispinosus (Fabricius), 1787 (*Apis*); continental Europe
   nigripes Olivier, 1811
uniglumis (Linnaeus), 1758 (*Vespa*); Europe to Mongolia, N. America
   uniglummis Christ, 1791 (*Vespa*), lapsus
   punctatus; Fabricius, 1793 (*Nomada*)
   tridens Fabricius, 1793 (*Crabro*)
   ? *decimmaculatus* Donovan, 1806 (*Vespa*)
   pygmaeus Olivier, 1811
   quadrinotatus Say, 1824
   impatiens F. Smith, 1856
   interruptus Cresson, 1865
   fallax Gerstaecker, 1867
   brodiei Provancher, 1883
   montanus Robertson, 1889
   hispanicus Giner Marí, 1943
uturoae Cheesman, 1928; Society Is.
varians F. Morawitz, 1891; USSR
variegatus Wesmael, 1852, centr. and s. Europe
   pulchellus Gerstaecker, 1867
ventralis Fox, 1894; U.S.: Pacific coast states; Mexico: Baja California
   manni Rohwer, 1909
venustus Sickmann, 1894; China
verhoeffi Beaumont, 1950; Algeria
victor Lepeletier, 1845; centr. and s. Europe, Mongolia, n. Africa
   dubius Dahlbom, 1845
   fasciatus Dahlbom, 1845
   simplex Dahlbom, 1845
   analis Gerstaecker, 1867, nec Cresson, 1865
   elegantulus Gerstaecker, 1867
   incomptus Gerstaecker, 1867
   melancholicus Chevrier, 1868
willowmorensis Arnold, 1927; S. Africa
woosnami Arnold, 1927; S. Africa
xanthogaster Pate, 1938; Borneo
xerophilus R. Bohart and Schlinger, 1956; sw. U.S.
zavattari Guiglia, 1943; Somalia

Nomina nuda in *Oxybelus*

antigae "Tournier", Antiga and Bofill, 1904
eburneus "L. Dufour", Dours, 1874

Tribe Crabronini

Crabronins are a large and varied group of wasps currently divided among 39 genera. As a rule, they can be recognized by the large, cuboidal head, relatively stout thorax, and stout legs. The gaster is sometimes pedunculate, and then it characteristically reaches its greatest size behind the middle. Entomologists often tend to equate the crabronins with the relatively large and distinctive species in the genus *Crabro* (fig. 120). However, many of the species are small, some as short as 2 mm. The considerable variability in evolutionary development is indicated by table 15.

Tribal diagnosis: Small to medium large forms, 2 to 14 mm long; eyes naked or rarely hairy, inner orbits nearly always converging below, often strongly so; eye facets usually enlarging ventrad; frons often with a definite scapal basin which may be margined laterally, dorsally, or both; ocellar triangle various; occipital carina complete or incomplete; antennal sockets usually close to each other and to inner orbits; clypeus usually transverse; male antenna with 12 or 13 articles, female with 12; scape often with one or two carinae; male flagellum sometimes broadened and/or fringed beneath; palpal formula 6-4, 6-3 or 5-3; mandible apex simple or bidentate, rarely tridentate, inner margin simple or dentate, inner base sometimes with a large tooth or prong; mandible socket closed except in *Entomocrabro, Anacrabro,* and *Encopognathus;* metanotum and propodeum without mucro; metanotum usually without squamae; submarginal cell separated by a vein from discoidal cell I (fig. 116 D); gaster often pedunculate; male pygidial plate present or absent, female pygidial plate triangular and flat or narrowed and guttered.

Systematics: Nearly all Crabronini have the eyes expanded below and inwardly, crowding the antennal bases. This is accomplished in part by an enlargment of the facets in the affected area. In most Oxybelini the eyes are simply bowed inward medially and the facets are only a little enlarged. An exceptional crabronin is *Anacrabro* in which the eyes are bowed in similar to those of many Oxybelini although the facets are still larger mesad.

The most generalized Crabronini are sometimes called the "*Encopognathus* series". Included genera are *Encopognathus, Anacrabro, Entomocrabro,* and *Entomognathus.* All of these have the mandible notched externoventrally and simple apically. The female pygidium bears a flat, triangular plate used as a "dirt pusher" during ground nest construction. The palpal formula is 6-4. All but *Entomognathus* have the mandibular socket open as in the Oxybelini.

In more advanced genera of the Crabronini a number of evolutionary trends are discernible: (1) the mandibular socket becomes closed; (2) the mandible loses the externoventral notch and becomes divided apically; (3) the palpal formula is reduced to 6-3 as in *Odontocrabro* and *Tracheliodes* or to 5-3 as in *Rhopalum* and related genera; (4) the male flagellum may become expanded, fringed, or shortened to 10 articles; (5) shield-like structures may develop in males on the forefemur as in one *Ectemnius,* foretibia as in many *Crabro,* or foretarsus as

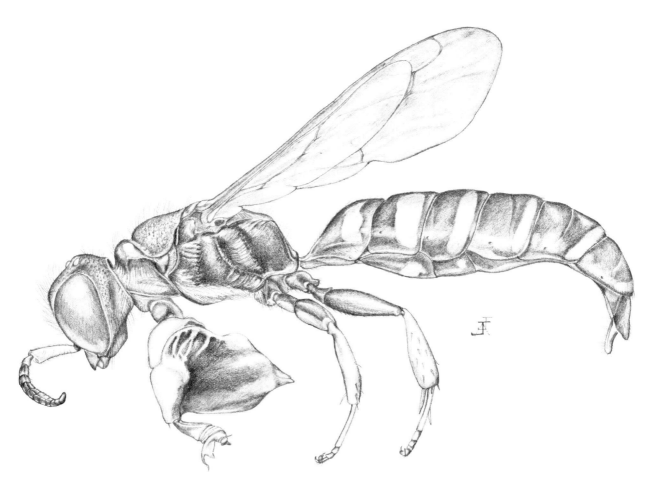

FIG. 120. *Crabro latipes* F. Smith, male.

in some *Crossocerus* and *Lestica*; (6) the gaster may become pedunculate as in *Rhopalum, Moniaecera,* and *Arnoldita*; (7) and the pygidium may become greatly narrowed and gutter-like or even spiniform. These and other trends are depicted in Table 15.

The 5-3 palpal formula referred to in number 3 of the above list can be used to partially characterize eight genera which can be considered the *Rhopalum* series. Included are *Rhopalum, Podagritus, Podagritoides, Echucoides, Notocrabro, Isorhopalum, Moniaecera* and *Huavea*. In these genera the body form is usually slender and elongate, the gaster is pedunculate, and the hindtibia is often clavate. *Podagritus, Echucoides* and *Podagritoides* are so closely related to *Rhopalum* that they could be considered as a group of subgenera of the latter. However, if the four genera were combined, the resulting entity would be large and heterogeneous. Therefore, as long as they can be separated satisfactorily, it is more convenient to consider them as distinct.

In some situations it is difficult to guess the evolutionary significance of various characters. An example is the disposition of the ocelli in a broad triangle as in *Crabro* or an equilateral one as in *Crossocerus*. Also with doubtful implications is the apparent movement of the recurrent vein terminus on the submarginal cell. It may be near the middle as in *Lindenius* (fig. 122 E), beyond the middle as in *Crabro* (fig. 122 F), and near the distal end as in *Ectemnius* (fig. 122 G).

Of the many references on Crabronini two book-length works are outstanding. These are Kohl (1915) which dealt with the species of the Palearctic Region and Leclercq (1954a) which had a much broader geographical coverage of the tribe. Pate's (1944b) paper on the genera is also important.

Biology: The habits of more than half of the genera are unknown. Eight genera are fossorial and presumably this is the primitive nesting habit in the Crabronini. The more advanced genera like *Ectemnius* and *Crossocerus* contain twig and pre-existing cavity nesting forms in addition to ground nesting species. A few genera such as *Dasyproctus* have apparently completely changed over to twig nesting. There is such diversity of prey (12 orders of insects) in the Crabronini that it is difficult to say what the original food preference was in the group, but Evans (1969a) has suggested that most likely it was Hemiptera. Today only four genera are known to use true bugs and only two of these do so commonly. Flies are most commonly provisioned, but there is con-

372 SPHECID WASPS

siderable prey diversity in some of the large genera (*Lindenius, Rhopalum, Crossocerus, Ectemnius*) with 8 different orders taken by *Crossocerus*. Some genera apparently use only one order of insects and prey in some of these is unusual: Coleoptera (*Entomognathus*), ants (*Encopognathus, Tracheliodes*), and Lepidoptera (*Lestica*).

Presocial behavior is known in *Moniaecera, Dasyproctus, Rhopalum, Crossocerus* and *Ectemnius* (see Evans, 1964d, Bowden, 1964, Peters, 1973). Here several females may share one burrow entrance, but presumably each wasp has its own nest. Division of labor among several females has been implied by Peters in one species of *Crossocerus*.

Literature on crabronin biology is extensive but several authors have provided summaries or reviews for species from certain regions: Kohl (1915), Hamm and Richards (1926), Leclercq (1954a), and Tsuneki (1960). Fain (1973) has listed the saproglyphid mites associated with various crabronins.

Key to genera of Crabronini

1. Mandible notched externoventrally, simple apically (fig. 122 J); if notch is indistinct (some females of *Anacrabro*), then terga I-IV are laterally carinate and folded under 2
   Mandible not notched externoventrally, often dentate apically (fig. 121 I) 5
2. Ocellar triangle equilateral (as in fig. 121 B), distal angle of discoidal cell I about 60 degrees (fig. 116 C); Mexico to S. America .... *Entomocrabro* Kohl, p. 375
   Ocellar triangle broader than long (as in fig. 121 G), distal angle of discoidal cell I 45 degrees or less (fig. 116 A, B, D) 3
3. Terga I-IV laterally carinate and sharply folded under, sterna flat or concave (fig. 123), New World .... *Anacrabro* Packard, p. 377
   Terga III-IV without lateral carina, rounded laterally, sterna convex 4
4. Eyes obviously hairy; N. America to Panama, Old World except Australia .... *Entomognathus* Dahlbom, p. 381
   Eyes bare or with indistinct and sparse hairs; w. U.S., Mexico, Old World except Australia .... *Encopognathus* Kohl, p. 379
5. Vertex with sharp tubercle behind compound eye (fig. 125), Neotropical.................... 6
   Vertex without sharp tubercle behind compound eye 7
6. Gaster sessile to pedunculate, but first segment never nodose; acetabular carina present; gena carinate from postocular tubercle to posterior mandibular condyle; (fig. 125) .... *Quexua* Pate, p. 385
   Gaster with first segment pedunculate, nodose at apex; acetabular carina absent; gena ecarinate .... *Holcorhopalum* Cameron, p. 385
7. Scapal basin with lateral carinae which may be continuous dorsally (figs. 121 A, 127 B) 8
   Scapal basin ecarinate laterally (figs. 121 D,

TABLE 15.
Evolutionary Assessment of Selected Genera in the Crabronini

Arbitrary scores: + (always) = 0, † (sometimes) = 1, * (never) = 2.

Presumed generalized characters of Crabronini	Entomocrabro	Anacrabro	Lindenius	Encopognathus	Entomognathus	Crabro	Moniaecera	Foxita	Ectemnius	Lestica	Arnoldita
1 Mandibular socket open	+	+	*	+	*	*	*	*	*	*	*
2 Mandible with exterventral notch	+	+	*	+	*	*	*	*	*	*	*
3 Mandible apically simple and acuminate	+	+	+	+	+	†	*	*	*	*	†
4 Palpal formula 6-4	+	+	+	+	+	+	*	+	+	+	*
5 Male antenna with 13 articles	+	+	+	+	†	†	+	+	+	†	*
6 Male flagellum not unusually expanded nor fringed	+	+	†	†	†	†	†	*	†	†	*
7 Scape ecarinate	+	+	+	†	+	†	+	*	†	†	*
8 No projection between antennal sockets	+	+	†	†	+	+	†	+	+	+	*
9 Scapal basin ecarinate	+	+	+	+	+	+	*	†	+	*
10 Gena simple, not constricted nor toothed	+	*	†	†	†	†	†	+	+	†	†
11 Recurrent vein joining submarginal cell near its middle	+	+	+	+	†	*	+	†	*	†	+
12 Male femora, tibiae, or tarsi not shield-like	+	+	+	+	+	†	+	+	†	†	+
13 Omaulus and postspiracular carina continuous	+	+	†	*	*	+	*	+	+	+	+
14 Verticaulus absent	+	*	+	†	*	+	+	*	*	*	*
15 Acetabular carina absent	+	+	+	+	†	+	+	*	*	†	+
16 Midtibia of male spurred	+	+	+	+	+	†	+	+	*	+	+
17 Gaster sessile or nearly so	+	+	+	+	+	+	*	+	+	+	*
18 Pygidial plate of female of the flat-triangular type	+	+	+	+	+	+	*	†	†	†	*
	0	4	8	9	11	12	15	18	18	19	24

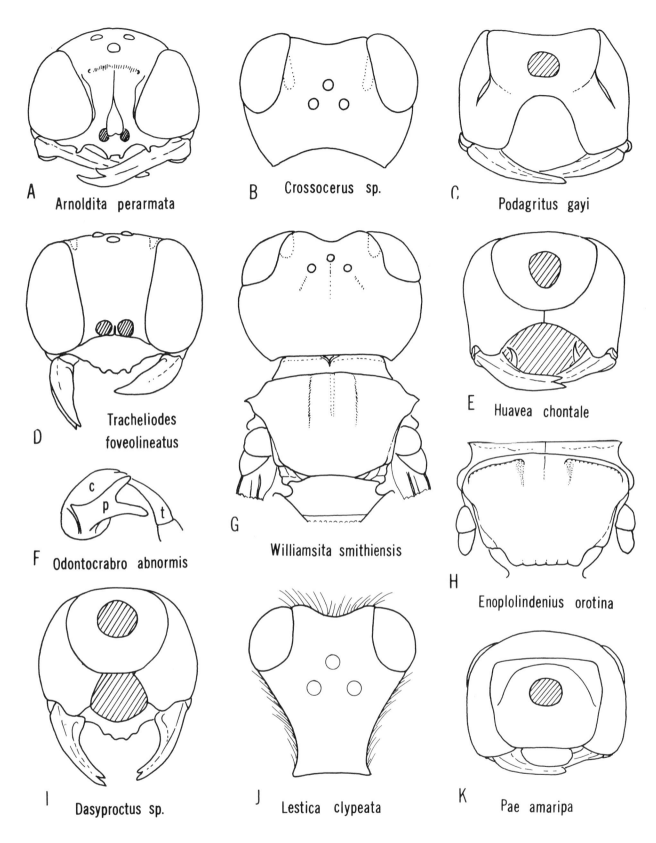

FIG. 121. Structural details in the tribe Crabronini; all figures except J are based on females; figs. A and D, facial views; B, G,H,J, dorsal views; C,E,I,K, ventral view of head; F, ventral view of base of foreleg, c = coxa, p = projection, t = trochanter (redrawn from Tsuenki, 1971a).

127 A) 13
8. Scapal basin not enclosed on three sides by a continuous carina (fig. 121 A) 9
Scapal basin enclosed laterally and dorsally by a strong continuous carina (fig. 127 B) 10
9. Scapal basin weakly carinate laterally, ecarinate dorsally, bisected by longitudinal carina bearing one or two processes (fig. 121 A), gaster pedunculate; Ethiopian .... *Arnoldita* Pate, p. 418
Scapal basin strongly carinate laterally, ecarinate but well defined dorsally, simple medially; gaster sessile; Neotropical................ *Taruma* Pate, p. 416
10. Scapal basin bisected by longitudinal carina extending downward from dorsal carina; recurrent vein joining submarginal cell beyond middle (as in fig. 116 D); Oriental .... *Hingstoniola* Turner and Waterston, p. 417
Scapal basin ecarinate medially; recurrent vein joining submarginal cell at middle (as in fig. 116 B) 11
11. Scutum with transverse anterolateral carinae (fig. 121 H); mandible apically simple; female pygidial plate broad, flat, coarsely punctate; New World .... *Enoplolindenius* Rohwer, p. 414
Scutum without transverse anterolateral carinae; mandible apically bidentate in male, tridentate in the female; female pygidial plate strongly narrowed and concave 12
12. Dorsal carina of scapal basin developed into a downcurved lamina (fig. 122 I); sternaulus present; Oriental .... *Vechtia* Pate, p. 417
Dorsal carina of scapal basin simple; sternaulus absent; Neotropical (fig. 127 B) .... *Foxita* Pate, p. 415
13. Palpal formula 6-3; foretrochanter unusually slender and elongate (fig. 129) or forecoxa with bifurcate appendage anteriorly (fig. 121 F); gaster sessile or subsessile 14
Palpal formula 6-4 or 5-3; foretrochanter not unusually slender and forecoxa without a bifurcate process; gaster sessile to pedunculate . 15
14. Inner orbits strongly converging below; forecoxa with a bifurcate process; Taiwan (fig. 121 F) .... *Odontocrabro* Tsuneki, p. 418
Inner orbits nearly parallel to moderately converging below (fig. 121 D); forecoxa without a bifurcate process; Holarctic .... *Tracheliodes* A. Morawitz, p. 404
15. Palpal formula 5-3; gaster pedunculate (fig. 128), slender, elongate; omaulus sometimes absent; first tergum often nodose at apex; hindfemur often swollen 16
Palpal formula 6-4; gaster usually sessile or subsessile, omaulus present (figs. 122 H, 126, 128) 23
16. Occipital carina forming a complete circle separated from hypostomal carina (fig. 121 E), mandible with inner basal spinous tooth (fig. 121 E); omaulus present (fig. 128) 17
Occipital carina incomplete or joining hypostomal carina (fig. 121 C); mandible without inner basal spinous tooth; omaulus present or absent 18
17. Hypersternaulus, subomaulus, postspiracular and acetabular carina absent; precoxal area simple or with a tooth; female pygidial plate broad, flat, coarsely punctate; N. American .... *Moniaecera* Ashmead, p. 394
Hypersternaulus, subomaulus, postspiracular, and acetabular carina present; precoxal area with verticaulus extending to scrobe; female pygidial plate punctured basally, elongate, and smooth apically; N. American .... *Huavea* Pate, p. 000
18. Omaulus consisting of a short carina extending posteroventrally from pronotal lobe; mesopleuron with a deep depression behind pronotal lobe; Oriental .... *Isorhopalum* Leclercq, p. 390
Mesopleuron without omaulus, or postspiracular carina and/or vertical part of omaulus present; area behind pronotal lobe without enlarged depression 19
19. Omaulus and postspiracular carina absent; or if omaular area is angulate, recurrent vein ending at middle of submarginal cell (as in fig. 122 E) 20
Omaulus and postspiracular carina present; or if not, recurrent vein ending well beyond middle of submarginal cell (as in fig. 122 F) ... 21
20. Tergum I with a median, dorsoposterior spine (fig. 122 H); Australia .... *Notocrabro* Leclercq, p. 394
Tergum I without a spine; widespread (fig. 126) .... *Rhopalum* Stephens, p. 387
21. Recurrent vein joining submarginal cell near its middle; omaulus strong but short; propleuron with a strong, forward directed tooth; female pygidial plate gutterlike; Fiji .... *Podagritoides* Leclercq, p. 393
Recurrent vein joining submarginal cell well beyond its middle; or if not, omaulus absent; propleuron not toothed; female pygidial plate essentially flat 22
22. Scapal basin with a median tubercle; hindtibia not claviform; most male flagellomeres with a dorsal, subapical hair; female pygidial plate delimited only toward apex; S. America .... *Echucoides* Leclercq, p. 394
Scapal basin without a median tubercle (fig. 127A), hindtibia often claviform; male flagellomeres without dorsal hair; female pygidial plate well defined; Australasian and Neotropical .... *Podagritus* Spinola, p. 390
23. Pronotal collar unnotched medially 24
Pronotal collar with a median notch (fig. 121 G) 28
24. Transverse frontal carina absent[1] 25
Transverse frontal carina present, broken medially (fig. 130 B) 26
25. Verticaulus present (see fig. 123), female pygidial plate pebbled and with a median longi-

[1] Some *Crossocerus* may run to this point but are distinguished by the absence of a verticaulus and by their finely sculptured or shiny propodeal enclosure.

tudinal carina, propodeum finely sculptured; Oriental and Australasian .... *Piyuma* Pate, p. 409
Verticaulus absent, female pygidial plate smooth and slightly concave, propodeum coarsely areolate; Australia .... *Pseudoturneira* Leclercq, p. 409
26. Scape bicarinate longitudinally; occipital carina incomplete, separated from hypostomal carina (as in fig. 121 E); Oriental .... *Piyumoides* Leclercq, p. 410
Scape ecarinate; occipital carina joining hypostomal carina (as in fig. 121 I) 27
27. Propodeum finely sculptured, enclosure defined; Oriental .... *Leclercqia* Tsuneki, p. 411
Propodeum smooth, shining, enclosure not defined; Japan .... *Towada* Tsuneki, p. 411
28. Ocelli in an equilateral or subequilateral triangle (fig. 121 B); propodeum smooth or finely sculptured; no verticaulus; widely distributed .... *Crossocerus* Lepeletier and Brullé, p. 397
Ocelli in a low triangle (fig. 121 G) except in a few *Lestica* which have the gena greatly compressed toward the vertex (fig. 121 J) and a few *Ectemnius,* which have a verticaulus; propodeum various 29
29. Gaster with a slender peduncle which is nodose toward the apex (as in fig. 126) male antenna with 13 articles, female pygidial plate gutterlike (as in fig. 122 D) 30
Gaster sessile to pedunculate but not nodose toward the apex 31
30. Body mostly dull, mat; verticaulus weak but unusually elongate above, approaching scrobe; scape bicarinate; Oriental, Australasian, Ethiopian .... *Dasyproctus* Lepeletier and Brullé, p. 419
Body shiny; verticaulus strong but short, hardly extending above coxal cavity; scape ecarinate; Ethiopian, Australasian .... *Neodasyproctus* Arnold, p. 418
31. Occipital carina in dorsoposterior view outlining a polished rectangular area rather than a simple arc (fig. 121 K), mesopleuraulus present (fig. 3 A) 32
Occipital carina in dorsoposterior view forming a simple arc, mesopleuraulus absent 33
32. Occipital carina forming a complete circle, strongly or weakly flanged below; Mexico to S. America .... *Pae* Pate, p. 412
Occipital carina incomplete, ending below in a dentiform process; Neotropical .... *Lamocrabro* Leclercq, p. 413
33. Verticaulus absent but sometimes replaced by an angle or sharp tooth; female pygidial plate always flat (fig. 122 C) 34
Verticaulus present (very short in some *Ectemnius*); female pygidial plate flat or more commonly gutterlike (fig. 122 D) 35
34. Jugal lobe longer than hindwing submedian cell (fig. 122 A); mandible simple apically; pygidial plate present in male but no tibial shield; Holarctic .... *Lindenius* Lepeletier and Brullé, p. 382
Jugal lobe shorter than hindwing submedian cell (fig. 122 B); mandible usually bidentate apically, pygidial plate usually absent in male but tibial shield often present (fig. 120); Holarctic, Centr. America .... *Crabro* Fabricius, p. 406
35. Submarginal cell with posterobasal veinlet no more than twice as long as distoposterior veinlet, usually less; male antenna with 13 articles........................... 36
Submarginal cell with posterobasal veinlet plainly more than twice as long as distoposterior veinlet[2], latter usually shorter than end vein of cell (fig. 122 G); male antenna with 12 or 13 articles 37
36. Mandible simple or truncate apically, pronotal collar ecarinate, gaster black and yellow, female pygidial plate narrowly gutterlike; S. America .... *Chimila* Pate, p. 412
Mandible bidentate apically, pronotal collar transversely carinate, gaster red, female pygidial plate flat; Australia .... *Chimiloides* Leclercq, p. 413
37. Orbital foveae absent or shallow and evanescent, if limited by a fine inner ridge then upper frons without close moderate to coarse punctation; male with 12 antennal articles or rarely 13; widespread .... *Ectemnius* Dahlbom, p. 422
Orbital foveae distinct (fig. 121 G); upper frons with coarse or moderate and close punctation 38
38. Male with 13 antennal articles, mesopleuron, terga, and vertex with moderate to fine punctation, female pygidial plate narrowly gutterlike; Australia, New Caledonia, New Hebrides .... *Williamsita* Pate, p. 421
Male with 12 antennal articles; mesopleuron (fig. 131), terga, and vertex usually with coarse punctation, female pygidial plate sometimes flat; widespread .... *Lestica* Billberg, p. 428

Genus Entomocrabro Kohl

Generic diagnosis: Eyes with few scattered fine, erect hairs, inner orbits convergent below; scapal basin simple or with median tubercle above antennal sockets, ecarinate above; orbital foveae present in female, indistinct in male; ocelli large, ocellar triangle equilateral; gena simple; occipital carina incomplete, separated from hypostomal carina; antennal sockets contiguous to each other and to inner orbits; male antenna with 13 articles; scape ecarinate; male flagellum not modified, without ventral hair fringe; palpal formula 6-4; mandible apically simple, externoventral margin notched beneath, inner margin notched and bidentate medially; pronotal collar low, narrow, rounded or carinate medially, carinate but not angulate laterally; scutum slightly or strongly arched above level of pronotum, with deep fovea anteriorly on notauli; scutal flange terminating on lateral margin of scutum

---

[2] Exceptions are a few species of *Ectemnius* from the Oriental Region. Males of these have antennae with 12 articles. For location of veinlets, see fig. 122 E.

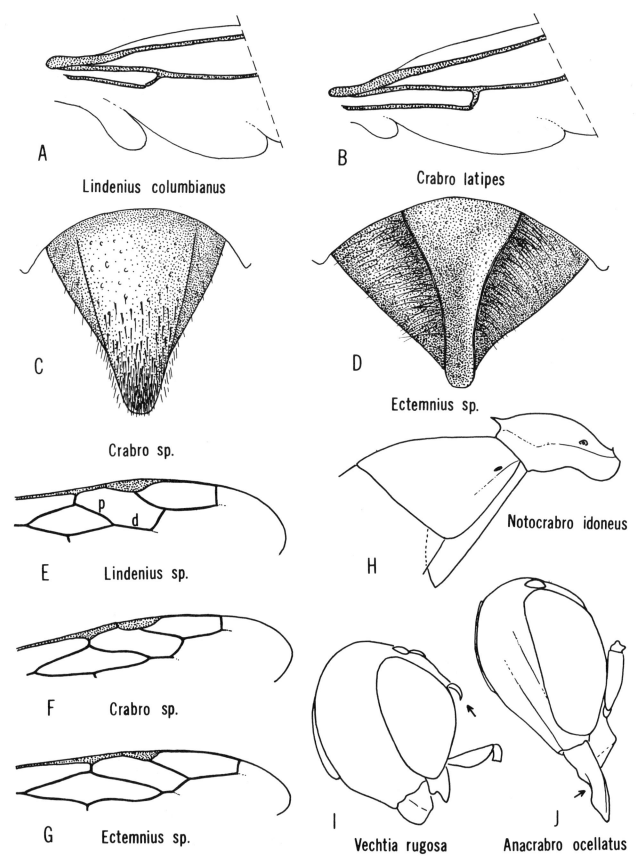

FIG. 122. Structural details in the tribe Crabronini; A-B, basal venation of hindwing; C-D, female pygidium; E-G, cells near stigma of forewing, p = posterobasal veinlet, d = distoposterior veinlet; H, lateral view of terga I-II; I,J, female head, lateral, arrow indicates frontal lamella on *Vechtia* and mandibular notch on *Anacrabro*, resp.

above tegula; prescutellar sulcus foveate; scutellum simple or with trough on anterior third, metanotum simple; postspiracular carina continuous with omaulus, sternaulus complete from omaulus to midcoxa, or incomplete or absent, hypsternaulus present; acetabular carina, verticaulus, and mesopleuraulus absent; propodeum moderately sculptured, enclosure well defined or areolate, lateral propodeal carina present; legs simple or forebasitarsus flattened; recurrent vein joining submarginal cell before its middle, discoidal cell I rhomboidal, discoidal cell II shortened (fig. 116 C); hindwing jugal lobe slightly shorter than submedian cell; gaster subsessile; pygidial plate of male trapezoidal-quadrate, that of female broad, flat coarsely punctate, triangular.

Geographic range: Neotropical Region. Five species have been described from South America, one from Guatemala and Mexico.

Systematics: From other members of the *Encopognathus* series, *Entomocrabro* is easily distinguished by the rhomboidal discoidal cell I (fig. 116 C). On the basis of its numerous generalized characters as shown in table 15, *Entomocrabro* may be the most primitive of the crabronin genera. Pate reviewed the genus (1941a).

Biology: Leclercq (1950a) reported that *terricola* nested in the ground and preyed on cicadellids.

Checklist of *Entomocrabro*

amahuaca Pate, 1941; Peru
bequaerti Pate, 1941; Guatemala, Mexico
duckei (Kohl), 1905 (*Crabro*); Brazil
richardsi Pate, 1941; Guyana
sacuya Pate, 1941; Peru
terricola Leclercq, 1950; Ecuador

Genus Anacrabro Packard

Generic diagnosis: Eyes naked, inner orbits arcuate, divergent above and below (fig. 130 A); frons without concave scapal basin, at most with polished area behind each scape; orbital foveae present in female, present or indistinct in male; ocellar triangle broader than long; gena narrowed ventrally, flattened posteroventrally, with carina delimiting angle between lateral and posterior planes; postocular carina present; occipital carina incomplete, separated from hypostomal carina; antennal sockets well separated from each other and from inner orbits; male antenna with 13 articles; scape ecarinate; male flagellum without ventral hair fringe; palpal formula 6-4; mandible apically simple, acuminate, externoventral margin distinctly to indistinctly notched, inner margin bidentate at basal one-third, mandibular socket open; pronotal collar and lobe broad, flat, margined anteriorly by continuous carina; collar angulate laterally, lobes with translucent spot; scutum with hind angles expanded, angulate; scutal flange terminating on lateral margin above tegula; axillae broad flattened, carinate; and lamellate laterally, scutellum and metanotum simple; postspiracular carina continuous with omaulus, verticaulus extending to scrobe and joined ventrally by precoxal carina (fig. 123); sternaulus present as continuation of omaulus; acetabular carina, hypersternaulus, and mesopleuraulus absent; metapleuron with diagonal carina originating above upper metapleural pit and ending above metacoxa; propodeum short, coarsely sculptured, dorsal face with horizontal row of large foveae, posterior face coarsely areolate; lateral propodeal carina well developed; legs with femora, especially forefemur, expanded and compressed; forebasitarsus of male distorted basally; recurrent vein joining submarginal cell at or before middle (fig. 116); median vein arising well after cu-a, discoidal cell II elongate, equal in length to discoidal cell I, apically acute; jugal lobe longer than hindwing submedian cell; gaster sessile, terga I-IV or VI with lateral carina, laterotergites bent under so that venter is flat or concave (fig. 123); male pygidial plate quadrate, that of female broad, flat, coarsely punctate, triangular.

Geographic range: New World. Eight of the 12 species are South American; the remainder are found in the U.S., Mexico and Central America.

Systematics: *Anacrabro* is a highly evolved member of the *Encopognathus* series. It is the only genus of crabronins in which the laterotergites are sharply bent under and the venter of the gaster is flat or concave. Other distinctive features include the verticaulus, the carina on the metapleuron, and the elongate discoidal cell II. In spite of these specializations it is a relatively primitive genus as indicated by its position in table 15.

Leclercq (1951b, 1973c) keyed the Neotropical members of the genus.

Biology: Barth (1909) found *Anacrabro ocellatus* nesting in a bank near the shore of Cedar Lake, Wisconsin. There the adults were commonly found on the flowers of Umbelliferae and Compositae from July through September. The female excavated her nest first by biting off a piece of soil and flying backward 15 to 45 cm to drop it. After penetrating some distance she carried out the loads of soil between her forelegs with her head flexed over them to hold them in place. A single cell was constructed at the end of the main tunnel, provisioned and sealed off; then additional short tunnels excavated off the main one, each with a single terminal cell. The female presumably captured the prey, *Lygus pratensis* (L.), on nearby flowers and carried the bug pressed against her thorax by the mid and hind legs and with the abdomen curved over it. The bugs were stored with the head toward the end of the cell, and the egg was attached to the thorax at the neck. The prey did not respond to stimulation and were presumed dead. The entrance to the nest was never closed, even at night or when the wasp was absent. On returning to the nest she always flew directly into it. On occasion, she made brief halts on the way to the entrance; but in no case did she alight on the ground and crawl in, even when blown off course by the wind.

Parasitic flies were observed around the nests. In one case a wasp attacked a tachinid she found inside her nest but did not seem to harm it. In one cell the *Lygus* bugs were being eaten by ants (*Monomorium*).

Krombein (1963a) also found *ocellatus* nesting in the ground and preying on *Lygus*. For further detail on this

FIG. 123. *Anacrabro ocellatus* Packard, female.

species see Kurczewski and Peckham (1970).

F. X. Williams (1928a) reported that *cimiciraptor* stored Miridae.

Checklist of *Anacrabro*

argentinus Brèthes, 1913; Argentina
benoistianus Leclercq, 1951; Ecuador
boerhaviae Cockerell, 1895; n. Mexico, sw. U.S., new status by R. Bohart
cimiciraptor Williams, 1928; Brazil
coruleter Pate, 1947; Colombia
eganus Leclercq, 1973; Brazil
fritzi Leclercq, 1973; Bolivia
golbachi Leclercq, 1973; Argentina
meridionalis Ducke, 1908; Brazil, Peru
mocanus Leclercq, 1973; Guatemala, Costa Rica
ocellatus Packard, 1866; e. and centr. U.S.
   rugosopunctatus Provancher, 1883 (*Thyreopus*), nec Taschenberg, 1875
   rugosulopunctatus Dalla Torre, 1897 (*Crabro*), new name for *rugosopunctatus*
   robertsoni Rohwer, 1920, new synonymy by Bohart
  ssp. *micheneri* Leclercq, 1973; centr. and s. Mexico
salvadorius Leclercq, 1973; Mexico, El Salvador

Genus Encopognathus Kohl

Generic diagnosis: Eyes naked, inner orbits converging below or arcuate; scapal basin simple, ecarinate above, with or without lamelliform tooth between antennal sockets; orbital foveae distinct to evanescent; ocellar triangle broader than high; vertex simple or with transverse carina behind ocelli; gena simple or with foveate postocular sulcus; occipital carina complete or incomplete, sometimes joining hypostomal carina; antennal sockets separated from each other, contiguous to or separated from inner orbits; male antenna with 13 or 12 articles; scape simple or partly unicarinate and lamellate; male flagellum simple or modified, without ventral hair fringe; palpal formula 6-4; mandible apically simple, acuminate; externoventral margin notched subbasally, inner margin subbasally notched and bidentate or edentate; mandibular socket open; pronotal collar reduced and lower than scutum to well developed at level of scutum, ecarinate and rounded laterally to carinate and angulate laterally; scutum simple; scutal flange terminating on lateral margin above tegula; prescutellar sulcus weakly foveate to narrowed and efoveate; scutellum and axillae simple or carinate laterally and posteriorly; metanotum simple or with posterior lamellae; postspiracular carina absent on specimens examined; omaulus, verticaulus, hypersternaulus, sternaulus and acetabular carina present or absent, mesopleuraulus absent; propodeum moderately to coarsely sculptured, enclosure well defined or areolate, lateral propodeal carina present or absent; legs simple or modified; recurrent vein joining submarginal cell near middle, jugal lobe shorter than or equal to submedian cell (fig. 116 A); gaster sessile, terga mostly rounded laterally, venter convex; pygidial plate of male trapezoidal-quadrate, of female broad, flat, coarsely punctate, triangular.

Geographic range: Widespread except in the Australasian Region. Twenty-three species are known.

Systematics: *Encopognathus* is distinguished from *Anacrabro* by the convex venter of the gaster and from *Entomognathus* by the absence of erect hair on the eyes. Some species of the *braueri* group have the metanotum bearing a well developed squama like that in many Oxybelini. In *chirindensis* the squama is *Oxybelus*-like, and in *brownei* it is *Belomicrus*-like. In many respects the genus is rather primitive (Table 15).

Karossia was described from a single female by Arnold (1929). He considered it an annectant genus having affinities with *Thyreopus s.l.* and *Oxybelus*. Arnold stated that the midtibia had two apical spurs, but specimens sent by Arnold to the British Museum have only a single spur. Misled by this error Pate (1936b), who had seen no Old World *Encopognathus*, described his subgenus *Rhectognathus* and placed *Karossia* in a separate tribe, the Karossiini. The subgenera *Aryana* and *Tsaisuma* were added by Pate (1943b). The former is a synonym of *Karossia*. Also, there are two rather isolated species, *braunsi* and *isolatus*, which do not fit into any of the subgenera. It seems to us that this complicated situation is best handled by means of species groups which can be separated according to the following key:

1. Sculpture coarse; verticaulus present; Old World 2
Sculpture fine; verticaulus absent 3
2. Gena with foveate postocular sulcus; hypersternaulus absent; male antenna with 12 articles (*Encopognathus* s.s.); sixteen species .... *E. braueri* group
Gena without foveate postocular sulcus; hypersternaulus present; male antenna with 13 articles (*Karossia*); two species .... *E. hessei* group
3. Propodeum areolate overall; male antenna with 12 articles; New World .... *E. pectinatus* group
Propodeal enclosure defined and smooth or ridged 4
4. Pronotal collar carinate laterally (*Tsaisuma*); sw. U.S.; one species .... *E. wenonah* group
Pronotal collar rounded, ecarinate laterally; Old World 5
5. Hypersternaulus present; India; one species .... *E. isolatus* group
Hypersternaulus absent; Spain; one species .... *E. braunsi* group

Pate (1943b) reviewed the western American forms and presented a key to the African species in the subgenus *Encopognathus* (1948b). Leclercq (1958b, 1963) described new species and reviewed the Asian forms.

Biology: The members of this genus nest in sandy soil. Arnold (1932) reported *E. chirindensis* attacking an ant *Tetramorium setuliferum* Emery and storing ants of additional species.

Checklist of *Encopognathus*

[Species group assignment is indicated by code letters

at the end of citations: *braueri* group (*Encopognathus s.s.*) (E), *braunsi* group (B), *hessei* group (K), *isolatus* group (I), *pectinatus* group (R), *wenonah* group (T)]

acanthomerus Pate, 1948; Nigeria (E)
africanus Leclercq, 1955; Tanzania (E)
alcatae Leclercq, 1963; Philippines: Mindoro (E)
argentatus (Lepeletier and Brullé), 1834 (*Lindenius*); India, China (K)
    argenteolineatus Cameron, 1890 (*Oxybelus*)
bellulus (Schulz), 1906 (*Crabro*); India, Sri Lanka (K)
    bellus Cameron, 1890 (*Oxybelus*), nec Dahlbom, 1844 (in *Oxybelus*)
    oxybeloides Pate, 1943
braueri (Kohl), 1896 (*Crabro*); Senegal, Gambia (E)
braunsi Mercet, 1915; Spain (B)
bridwelli Pate, 1948; Nigeria (E)
brownei Turner, 1917; Ethiopian Region (E)
egregius Arnold, 1926 (*Thyreopus*)
chapraensis (Turner), 1917 (*Entomognathus*); Oriental Region (E)
chirindensis (Arnold), 1932 (*Thyreopus*); Rhodesia (E)
esoterus Leclercq, 1963; Philippines: Luzon (E)
gombaki Leclercq, 1963; Malaya (E)
granulatus (Arnold), 1926 (*Thyreopus*); Zaire (E)
hessei (Arnold), 1929 (*Karossia*); S. Africa (K)
isolatus (Turner), 1917 (*Entomognathus*); India (I)
lumpuri Leclercq, 1958; Malaya (E)
pectinatus Pate, 1936; U.S.: California (R)
rhodesianus (Arnold), 1932 (*Thyreopus*); Rhodesia, S.-W. Africa (E)
rufiventris Timberlake, 1940; U.S.: California (R)
rugosopunctatus Turner, 1912; S. Africa (E)
thaianus Tsuneki, 1963; Thailand (E)
wenonah (Banks), 1921 (*Lindenius*); U.S.: California (T)

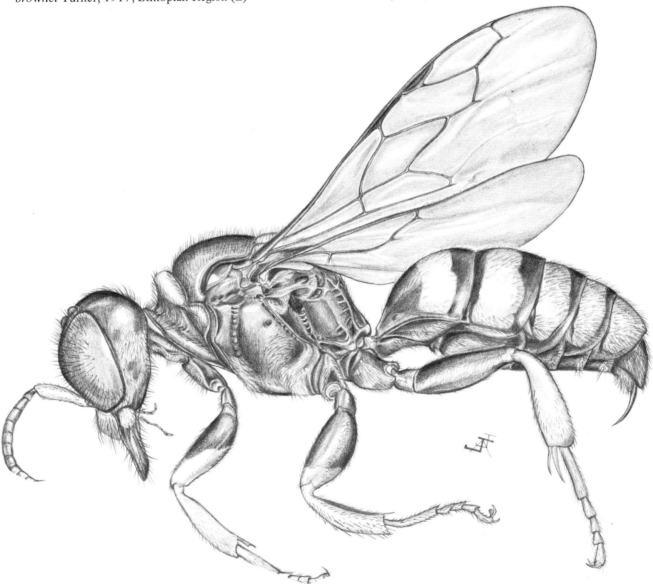

FIG. 124. *Entomognathus texanus* (Cresson), female.

Genus Entomognathus Dahlbom

Generic diagnosis: Eyes covered by erect hair, (fig. 124), inner orbits convergent below; scapal basin simple, ecarinate above; orbital foveae distinct to indistinct; ocellar triangle broader than high; vertex simple or with median sulcus either extending backward from anterior ocellus or reduced to short impression between hindocelli; gena simple or with foveate postocular sulcus; occipital carina complete or incomplete, separated from or joining hypostomal carina; with or without spiniform process beneath genae, with or without median carina extending upward towards vertex; antennal sockets separated from each other and from inner orbits; male antenna with 13 or 12 articles; scape ecarinate; male flagellum simple or modified, without ventral hair fringe; palpal formula 6-4; mandible apically simple, acuminate, externoventral margin notched medially, inner margin dentate or notched and bidentate; pronotal collar rounded, ecarinate medially, carinate or ecarinate laterally; scutum rounded or slightly depressed medially, simple or with deep foveae laterally at parapsidal lines and anteriorly at notauli; scutal flange terminating on lateral margin above tegula; prescutellar sulcus simple or foveate; scutellum simple or with posterior foveae; metanotum simple; postspiracular carina, omaulus, and acetabular carina present or absent; sternaulus absent or present as continuation of omaulus; hypersternaulus and mesopleuraulus absent, verticaulus present and continuous with sternaulus or absent; propodeum moderately sculptured, enclosure well defined or areolate, lateral propodeal carina present or absent; legs simple or modified; recurrent vein joining submarginal cell at or beyond middle; marginal cell obliquely to squarely truncate apically, with R_1 terminating at or extending beyond apex (fig. 116 D); hindwing jugal lobe shorter to longer than submedian cell; gaster sessile, terga III-V or III-VI with posterior margins emarginate to straight; pygidial plate of male trapezoidal-quadrate, of female broad, flat, coarsely punctate, triangular.

Geographic range: Nearctic, Neotropical, Palearctic, Ethiopian, and Oriental Regions. 42 species are listed. The subgenus *Toncahua* is confined to the New World, *Koxinga* to the Orient, and *Entomognathus s.s.* to the Palearctic Region. *Mashona* is found in both the Ethiopian and the Oriental Regions.

Systematics: Entomognathus is distinguished from other genera in the *Encopognathus* series by the hairy eyes (fig. 124).

Possible evolutionary connections with Bothynostethini are discussed under systematics of the subfamily. Some behavioral similarities are given under "Biology" below.

Pate (1944b) divided *Entomognathus* into four subgenera on the basis of wing venation, carinae of the pronotum and mesopleuron, number of articles of the male antenna, presence or absence of a postocular sulcus, shape of the posterior margins of abdominal terga, and geographical distribution. Leclercq (1963), after studying Indian forms, found variation in some of the characters. The following key is modified from Pate following the suggestions of Leclercq.

Key to subgenera of *Entomognathus*

1. Mesopleuron with sternaulus and verticaulus; male antenna with 13 articles 2
   Mesopleuron without verticaulus; sternaulus absent or present as incomplete backward extension of omaulus; male antenna with 12 or 13 articles 3
2. Forewing with R_1 extending beyond apex of marginal cell (Oriental Region) ...... *Koxinga* Pate
   Forewing with R_1 not extending beyond apex of marginal cell (Ethiopian and Oriental Regions) .... *Mashona* Pate
3. Male antenna with 12 articles; abdominal terga III-V or III-VI with posterior margins angularly emarginate medially (New World) .... *Toncahua* Pate
   Male antenna with 13 articles; abdominal terga with posterior margins straight medially (Palearctic Region).. *Entomognathus* Dahlbom

Three regional keys have been published: Arnold (1945) on the species of the subgenus *Mashona* found on Madagascar, Leclercq (1961b) on the species of *Mashona* found in continental Africa, and Krombein (1963d) on the species of *Toncahua* found in the eastern United States.

Biology: The members of this genus nest in the soil and prey on chrysomelid beetles. Most of the published observations are on *brevis,* the most widespread Palearctic species. The following information is taken from Grandi (1961) unless otherwise noted.

The burrows are difficult to find because the wasps construct chimneys of soil 10-25 mm high above the entrances. These were observed also by Benoist (1915). The burrows extend as far as 12 cm into the soil, their conformation being determined by the soil characteristics. The cells are provisioned with 14-24 chrysomelid beetles, the number and species apparently depending on what is available. The prey are permanently paralyzed but respond to stimuli. After provisioning is complete the egg is laid transversely on the venter of a beetle at the junction of the prothorax and mesothorax. The larva consumes only the soft parts of the beetle. Benoist (1915) reported that the chitinous remains were not incorporated into the cocoon as in *Lindenius.* Instead, soil and quartz grains are mixed with the larval secretions.

Most of the prey species listed for *brevis* belong to the subfamily Halticinae with one record of Cryptocephalinae. Arnold (1932, 1945) found Ethiopian species preying on chrysomelid beetles also, one possibly belonging to the subfamily Galerucinae. Most recently, Cazier and Mortenson (1965c) captured a female *E. texanus* carrying a *Diabrotica tricincta* Say (Galerucinae).

Miller and Kurczewski (1972) reviewed information on nesting behavior of *Entomognathus* and gave details of their own studies of *memorialis* in New York and

Pennsylvania. Nest entrances were in the vertical faces of sand banks and between dense rootlets. Differences from *brevis* were: 3-9 prey per cell rather than 14-24 (the beetles were larger), no entrance turrets, and many cells were in tandem and the maximum per nest was 20 rather than 10. Miller and Kurczewski commented on similarities and differences between nesting behavior of *Entomognathus* and *Bothynostethus* (Larrinae). Both use chrysomelid beetles and the egg is laid "larrinelike" transversely on the venter of the prey. Differences, at least between *memorialis* and *Bothynosthethus distinctus* are that in the former original galleries are made, the prey is held during transport venter up and without the aid of the mandibles, and the cocoon is entirely free of prey remains.

Checklist of *Entomognathus*

[Subgeneric assignment is indicated by capital letters following citation: *Entomognathus* (E), *Koxinga* (K), *Mashona* (M), and *Toncahua* (T)]

apiformis (Arnold), 1926 (*Thyreopus*); Rhodesia (M)
arenivagus Krombein, 1963; U.S.: N. Carolina, Florida (T)
arnoldi R. Bohart and Menke; Madagascar, new name for *tridens* Arnold
    tridens Arnold, 1944 (*Crabro*), nec Fabricius, 1798
bidentatus (Arnold), 1927 (*Thyreopus*); Rhodesia (M)
brevis (Vander Linden), 1829 (*Crabro*); Palearctic Region (E)
    apicalis Lepeletier and Brullé, 1834 (*Lindenius*)
breviusculus (Gussakovskij), 1952 (*Crabro*); sw. USSR (E)
collarti Leclercq, 1961; Zaire (M)
dentifer (Noskiewicz), 1929 (*Crabro*); Europe to e. Mediterranean (E)
    permixtus Nouvel and Ribant, 1956
diversicornis (Arnold), 1944 (*Crabro*); Madagascar (M)
euryops (Kohl), 1899 (*Crabro*); Tunisia (E)
evolutionis (Leclercq), 1956 (*Encopognathus*); Mexico (T), new combination by J. Leclercq
faunus (Arnold), 1944 (*Crabro*); Madagascar (M)
fortuitus (Kohl), 1915 (*Crabro*); Spain (E)
geometricus Leclercq, 1955; Mexico, Panama (T)
ignavus (Arnold), 1927 (*Thyreopus*); Rhodesia (M)
lenapeorum Viereck, 1904; e. U.S. (T)
libanonis (Kohl), 1905 (*Crabro*); Cyprus, Syria, Iran (E)
memorialis Banks, 1921; e. U.S. (T)
mexicanus Cameron, 1904; Mexico (T)
midus (Arnold), 1944 (*Crabro*); Madagascar (M)
mimicus (Arnold), 1944 (*Crabro*); Madagascar (M)
?*nanus* (Cameron), 1890 (*Crabro*); India (M)
narratus Leclercq, 1963; India (M)
nathani Leclercq, 1963; India (M)
nitidus (Cameron), 1890 (*Oxybelus*); India (K)
patricius (Arnold), 1932 (*Thyreopus*); Rhodesia (M)
pulicus Leclercq, 1963; India (M)
ruficaudatus (Arnold), 1944 (*Crabro*); Madagascar (M)
rugosissimus Turner, 1917; Malawi, Rhodesia (M)
sahlbergi (A. Morawitz), 1866 (*Crabro*); USSR: e. Siberia, Mongolia (E)
schmidti Beaumont, 1967; Turkey (E)
schmiedeknechtii (Kohl), 1905 (*Crabro*); e. Mediterranean (E)
singarae Leclercq, 1963; India (M)
siraiya Pate, 1944; N. Viet Nam, Malaya: Selangor; Taiwan (K)
stevensoni (Arnold), 1926 (*Thyreopus*); Rhodesia, S. Africa (M)
    ssp. *fraternus* (Arnold), 1936 (*Thyreopus*); Rhodesia
subnasutus (Arnold), 1927 (*Thyreopus*); Rhodesia (M)
surgicus Leclercq, 1961; Zaire (M)
swellendamensis (Arnold), 1934 (*Thyreopus*); S. Africa (M)
syrittus Leclercq, 1961; Zaire (M)
texanus (Cresson), 1887 (*Crabro*); e. U.S. and s. to Mexico (T)
    panurgoides Viereck, 1904 (*Anothyreus*)
tricoloripes (Arnold), 1934 (*Thyreopus*); S. Africa (M)
verecundus (Arnold), 1932 (*Thyreopus*); Rhodesia (M)

Genus Lindenius Lepeletier and Brullé

Generic diagnosis: Eyes naked or sparsely covered with very short, erect hair, inner orbits convergent below; scapal basin ecarinate, with or without median tubercle, orbital foveae distinct, ocellar triangle broader than high, gena simple or with carina joining occipital carina and base of mandible; occipital carina incomplete and separated from hypostomal carina, sometimes terminating ventrally in a tooth or lobe; antennal sockets contiguous or separated, separated from inner orbits, scape ecarinate; male antenna with 13 articles, male flagellum simple or modified, without ventral hair fringe; palpal formula 6-4; mandible apically simple and acuminate or with subapical lobe on inner margin in some males, inner margin with median tooth, externoventral margin entire, internoventral carina dilated to form median lobe in some males; pronotal collar rounded to transversely carinate and angulate laterally; scutum simple or with median longitudinal depression or trough; axillae, scutellum, and metanotum simple; scutal flange terminating on lateral margin of scutum above tegula; prescutellar sulcus foveate; postspiracular carina and omaulus present except in *luteiventris,* continuous; acetabular carina, verticaulus, sternaulus, and mesopleuraulus absent; hypersternaulus present or absent, propodeum moderately to finely sculptured, enclosure defined, lateral propodeal carina well developed to absent; legs with tarsi modified in some males, otherwise simple; recurrent vein joining submarginal cell near its middle (fig. 122 E); hindwing jugal lobe longer than submedian cell (fig. 122 A); gaster sessile with quadrate to triangular pygidial plate in male; female with broad, flat, triangular pygidial plate which may be elongated or narrowed apically.

Geographic range: Holarctic Region. The majority of the 58 species are Palearctic (47) and of these 29 occur in countries bordering the Mediterranean. Ten species occur in the Nearctic Region.

Systematics: The members of the genus *Lindenius* are small, dark, stout-bodied wasps with the following combination of characters: mandible with apex simple, externoventral margin entire; ocellar triangle broad and low, scapal basin ecarinate; and jugal lobe of hindwing longer than submedian cell. This last characteristic is the only one that separates *Lindenius* from *Crabro*, but the two genera do not seem to be closely related. The genus has been variously associated with *Entomognathus, Encopognathus,* and *Oxybelus*. Its division into subgenera and groups has been similarly controversial. The most logical arrangement is that of Beaumont (1956c), who outlined five species groups and appended a number of "isolated" species. The following summary stresses the most important characters of each group.

1. *L. melinopus* group. Mandible with inner basal tooth small, in male internal carina lobate; clypeal free edge with a small submedian notch; female pygidial plate with sides converging very little, apex truncate. Two eastern Palearctic species.

2. *L. albilabris* group. Mandible with inner basal tooth small, inner lobe absent in male; clypeal free edge with a small submedian notch; female pygidial plate narrowed and hairy toward apex. Nine Palearctic species (= *Lindenius s.s.*).

3. *L. ibericus* group. Mandible with inner basal tooth rather strong, inner lobe present in male; clypeal free edge with a deep submedian sinus; female pygidial plate narrowed and hairy toward apex. Six Mediterranean area species.

4. *L. pygmaeus* group. Mandible with inner basal tooth very strong, inner lobe absent in male; clypeal free edge with a deep submedian sinus; female pygidial plate narrowed and hairy toward apex. Nine Palearctic and nine Nearctic species (= *Trachelosimus*).

5. *L. mesopleuralis* group. Mandible with inner basal tooth small, inner lobe absent in male; male clypeus as in *pygmaeus* group, female clypeal free edge with broad median lobe that is denticulate; hypersternaulus present; pygidial plate of female unusually broad, broadly rounded at apex, entirely covered by flattened pilosity. Two Mediterranean and two Transcaspian species.

Biology: Information on nesting site and prey of three Old World species, *albilabris, panzeri,* and *pygmaeus,* was summarized by Leclercq (1954a). In addition, the habits are at least partly known for three New World species, *columbianus, tecuya,* and *tylotis*. On the basis of these studies, some generalizations can be made. Nests are constructed in sandy or hard-packed soil. The openings, which are marked by a ring of tumulus, are not closed during provisioning. The burrows are rather shallow (6-10 cm) and end in one to nine cells. Prey are usually small flies, frequently Chloropidae, or Hemiptera, particularly Miridae. Hymenoptera may be taken also, expecially in the Chalcidoidea, Braconidae, and Formicidae. Parasites are miltogrammine Sarcophagidae, *Myrmosa* (Tiphiidae), and *Hedychridium* (Chrysididae).

Some detail on the Old World species was given by Hamm and Richards (1926). A published American record is that of Evans (1970) on *columbianus*. His findings were fragmentary but duplicated those reported from Europe. The most complete study on *Lindenius* biology was made by H. K Court (1961) in an unpublished master's thesis. Mrs. Court detailed findings on *tylotis* and *columbianus*. She also reported on further data obtained by R. Bohart on *columbianus* and *tecuya*. In the case of *tylotis,* the females were nesting in flat, bare, silty ground situated above a creek in Davis, California. Many new burrows were present in a restricted area, as well as many from a previous year. The tunnels were vertical for 2 to 5 cm and led rather horizontally into several side branches. These terminated in one to three cells each. The completed nests were 7-15 cm long and contained 4-24 cells. Prey were in the three orders, Diptera, Hemiptera, and Hymenoptera, in that order of frequency. The average number of prey per cell was about eight. The larvae constructed silken cocoons into which the remains of the prey were incorporated. The most frequently observed parasite was *Myrmosa bradleyi* Roberts.

Observations on *columbianus* are much the same except that the nests are usually in gravelly soil and may be rather solitary. Court observed prey capture in this species at Soda Springs, California. Females circled umbels of *Carum* and struck at suitable insects on the flowers. Then, the prey were apparently stung in flight. They were carried back to the nest secured by the hind legs of the wasp and held upside down with the head forward. The wasp generally plunged directly into the entrance but occasionally alighted momentarily first. If the prey was taken away at this time and placed near the entrance, the wasp would go into the nest as usual and reappear in a minute or two. Then she would sometimes ignore the prey or sometimes would retrieve it and take it into the nest.

Nest closure has not been observed. Since the foretarsal rake of *Lindenius* is rather short, there may be no closing ritual in this genus. In the case of *tylotis* completed nests were not closed at the surface until rains plugged some of them with mud. Males of *Lindenius* are often collected in abundance as they circle bushes. At least in the case of *tylotis,* they may spend considerable time within old or active burrows, often to be chased out of the latter by females. It was assumed by Court that mating took place within the nests, but there is a bit of evidence to the contrary. Bohart (unpublished observation) collected a pair of *columbianus* in flight early in the season at Sagehen Creek, Nevada County, California. The pair remained in copula in the killing bottle and were mounted on a pin, still attached.

Bohart's notes on *tecuya* were made at Foster Park, Ventura County, California in 1959. Burrows were scattered in sand and gravel at the side of a dry riverbed. Contents of one cell 5 cm below the surface were 47 insects in six families of flies, three of chalcidoids, and one of Hemiptera (Anthocoridae).

Checklist of *Lindenius*

[Assignment to species groups indicated by letters as follows: *albilabris* group (alb), *ibericus* group (ibe), isolated species (i.sp.), *melinopus* group (mel), *mesopleuralis* group (mes), *pygmaeus* group (pyg).]

abditus (Kohl), 1898 (*Crabro*); Greece (alb)
aegyptius (Kohl), 1888 (*Crabro*); Egypt, Morocco, Libya, s. Spain (mes)
affinis Kazenas, 1973; sw. USSR: Kazakh S.S.R.
albilabris (Fabricius), 1793 (*Crabro*); Europe to Mongolia and N. Korea (alb)
    aenescens Dahlbom, 1838 (*Crabro*)
  ssp. *manchurianus* Tsuneki, 1967; Manchuria
anatolicus Beaumont, 1967; Turkey (i. sp.)
aptus Marshakov, 1973; sw. USSR: Turkmen S.S.R.
armaticeps (W. Fox), 1895 (*Crabro*); s. Canada, U.S. ne. from Colorado and Texas (pyg)
    flaviclypeus W. Fox, 1895 (*Crabro*)
    zellus Rohwer, 1909 (*Crabro*)
atlanteus Beaumont, 1956; Morocco (pyg)
buccadentis Mickel, 1916; U.S. e. from Wyoming, n. from Arkansas (pyg)
cabrerae Leclercq, 1960; Spain (i. sp.)
californicus Court and R. Bohart, 1958; U.S.: California (pyg)
ceballosi Leclercq, 1959; Spain (i. sp.)
columbianus (Kohl), 1892 (*Crabro*); U.S., s. Canada (pyg)
    errans W. Fox, 1895 (*Crabro*)
    pinguis W. Fox, 1895 (*Crabro*)
    salicis Cockerell, 1897 (*Ammoplanus*)
crenicornis Marshakov, 1973; sw. USSR: Tadzhik S.S.R.
crenulifer (Kohl), 1905 (*Crabro*); Syria, Israel, Lebanon (alb)
difficillimus (Kohl), 1915 (*Crabro*); Egypt, Libya (ibe)
effrenus (Kohl), 1915 (*Crabro*); Algeria, Tunisia (ibe)
fastidiosus Beaumont, 1967; Turkey (i. sp.)
gobiensis Tsuneki, 1972; Mongolia (pyg)
gussakovskii Marshakov, 1973; sw. USSR: Kazakh S.S.R.
haemodes (Kohl), 1905 (*Crabro*); Egypt, Ethiopia (pyg)
hamiger (Kohl), 1915 (*Crabro*); sw. USSR: Transcaspia (?alb)
hamilcar (Kohl), 1899 (*Crabro*); Mediterranean area, Canary Is. (alb)
    cogens Kohl, 1915 (*Crabro*)
hannibal (Kohl), 1898 (*Crabro*); Tunisia, Algeria, Morocco, Portugal (ibe)
harbinensis Tsuneki, 1967; Manchuria (pyg)
hasdrubal Beaumont, 1956; Algeria (i. sp.)
helleri (Kohl), 1915 (*Crabro*); Israel, Turkey, Greece (alb)
ibericus (Kohl), 1905 (*Crabro*); s. France, Spain, Portugal (ibe)
  ssp. *alticollis* Beaumont, 1956; Morocco, Algeria
  ssp. *humilicollis* Beaumont, 1956; France, Spain, Portugal
ibex Kohl, 1883; Turkey, Greece: Corfu, Siros, Peloponnesos; Algeria (?) (alb)

  ssp. *syriacus* (Kohl), 1905 (*Crabro*); Israel
inyoensis Court and R. Bohart, 1958, U.S.: California, Nevada (pyg)
irrequietus (Kohl), 1915 (*Crabro*); centr. Asia (? pyg)
kaszabi Tsuneki, 1972; Mongolia (i. sp.)
laevis A. Costa, 1871; Italy, Romania, Yugoslavia, Czechoslovakia, Hungary (alb)
    levis Dalla Torre, 1897 (*Crabro*)
    rhaibopus Kohl, 1915
latebrosus (Kohl), 1905 (*Crabro*); USSR: Siberia (pyg)
latifrons (W. Fox), 1895 (*Crabro*); U.S.: Texas (pyg)
latitarsis Marshakov, 1973; sw. USSR: Kazakh S.S.R. (alb)
leclercqi Beaumont, 1956; Algeria, Tunisia (i. sp.)
luteiventris (A. Morawitz), 1866 (*Crabro*); n. Spain (i. sp.)
    fulviventris "Pérez" Antiga and Bofill, 1904 (*Crabro*), nomen nudum
  ssp. *tenebrosus* (Kohl), 1915 (*Crabro*); s. Spain, Morocco
major Beaumont, 1956; France, Spain, Morocco (mel)
melinopus (Kohl), 1915 (*Crabro*); France, Spain, Morocco, Algeria, Tunisia (mel)
merceti (Kohl), 1915 (*Crabro*); Spain (i. sp.)
mesopleuralis (F. Morawitz), 1890 (*Crabro*); centr. Asia, Turkey (mes)
  ssp. *mediterraneus* (Kohl), 1915 (*Crabro*); Italy, France, Spain
montezuma (Cameron), 1891 (*Crabro*); sw. U.S., Mexico (pyg)
    dugesianus Leclercq, 1950
nasutus Gribodo, 1884; Italy
neomexicanus Court and R. Bohart, 1958; U.S.: New Mexico, Colorado (i. sp.)
nitidus Beaumont, 1967; Turkey (i. sp.)
ocliferius (F. Morawitz), 1896 (*Crabro*); sw. USSR: Transcaspia (mes)
pallidicornis (F. Morawitz), 1890 (*Crabro*); sw. USSR: Transcaspia (? mes)
panzeri (Vander Linden), 1829 (*Crabro*); Europe, w. Asia, n. Africa (pyg)
    venustus Lepeletier and Brullé, 1834
  ssp. *mongolicus* Tsuneki, 1972; Mongolia
parkanensis Zavadil in Zavadil and Šnoflak, 1948; Czechoslovakia, Romania, Hungary (ibe)
    ponticus Beaumont, 1956
peninsularis (Kohl), 1915 (*Crabro*); Spain (ibe)
prosopiformis (Nurse), 1903 (*Crabro*); India (pyg)
pygmaeus (Rossi), 1794 (*Crabro*); s. France, Italy, Spain, Portugal (pyg)
  ssp. *armatus* (Vander Linden), 1829 (*Crabro*); Europe, Middle East
    curtus Lepeletier and Brullé, 1834
    kratochvili Šnoflak, 1948
    mixtus Šnoflak in Zavadil and Šnoflak, 1948
  ssp. *algirus* (Kohl), 1892 (*Crabro*); Algeria, Morocco
satschouanus (Kohl), 1915 (*Crabro*); centr. Asia (? pyg)
spilostomus (Kohl), 1899 (*Crabro*); Tunisia, Algeria,

Morocco (alb)
subaeneus Lepeletier and Brullé, 1834; Europe (i. sp.)
tecuya Pate, 1947; U.S.: California (pyg)
tylotis Court and R. Bohart, 1958; U.S.: California (pyg)

Genus Holcorhopalum Cameron

Generic diagnosis (male unknown): Eyes apparently bare; inner orbits arcuate below, strongly divergent above; scapal basin shallowly concave, simple, ecarinate; upper frons bisected by a furrow extending from anterior ocellus to scapal basin; orbital foveae absent; ocellar triangle equilateral, ocelli large; vertex with sharp tubercle behind compound eye; gena simple, ecarinate; occipital carina incomplete, separated from hypostomal carina; antennal sockets contiguous or separated from each other, contiguous with inner orbits; scape ecarinate; palpal formula 6-4; mandible apically simple, acuminate, inner margin weakly bidentate before middle, externoventral margin entire; pronotal collar ecarinate, rounded anteriorly and laterally, pronotal lobe flat and attaining the tegula; scutum with admedian lines present as longitudinal welt, notauli with deep, elongate subcuneate fovea, hind angle produced as a thick, conical, backward-projecting, spinoid tubercle; axillae with lateral margins sharply reflexed upward into laminate plates; scutellum may be modified, metanotum simple; postspiracular carina continuous with omaulus, hypersternaulus present; acetabular carina, sternaulus, and mesopleuraulus absent; propodeum coarsely sculptured, dorsal face areolate, lateral propodeal carina present; legs simple; recurrent vein joining submarginal cell before middle, discoidal cell I rhomboidal; hindwing jugal lobe slightly longer than submedian cell; gaster pedunculate, first segment nodose at apex; female pygidial plate broad, flat, triangular.

Geographic range: Neotropical Region. One species is known from Mexico and Guyana.

Systematics: Holcorhopalum like *Quexua* has postocular tubercles on the vertex and the mandible is simple apically and externoventrally. However, the first segment of the gaster is pedunculate and nodose as in *Rhopalum*, the discoidal cell I of the forewing is rhomboidal as in *Entomocrabro* (see fig. 116 C), and the gena is ecarinate. In addition *Holcorhopalum* is the only genus of crabronins in which the pronotal lobe reaches the tegula.

Biology: Unknown.

Checklist of *Holcorhopalum*

foveatum Cameron, 1904; Mexico, Guyana
    thauma Pate, 1944 (*Amaripa*), new synonymy by J. Leclercq

Genus Quexua Pate

Generic diagnosis: Eyes bare or sparsely covered with erect hair (fig. 125), inner orbits convergent below; scapal basin ecarinate above, without median tubercle, with or without lateral carinae paralleling inner orbits and extending dorsally to supraorbital areas; orbital foveae represented by ovate, shiny, flat or callused areas; ocellar triangle equilateral, with deep median fovea or sulcus between hindocelli; vertex with sharp tubercle behind eye and a carina along gena from tubercle to posterior mandibular condyle; occipital carina incomplete, sometimes bifurcate below, mandibular branch present or absent, hypostomal branch complete or incomplete (may be reduced to a short spur on hypostomal carina); antennal sockets contiguous to or slightly separated from each other and inner orbits; male antenna with 13 articles; scape ecarinate and cylindrical or longitudinally bicarinate and flattened anteriorly; male flagellum not modified, without ventral hair fringe; palpal formula 6-4; mandible with apex simple, acute, externoventral and inner margins simple, pronotal collar rounded or sharply carinate anteriorly, rounded laterally, scutum simple or with admedian lines deeply impressed and terminating posteriorly in foveae, lateral flange terminating on lateral margin of scutum above tegula; scutellum simple or with tubercles, axillae and metanotum simple; postspiracular carina, omaulus, and acetabular carina continuous; sternaulus, hypersternaulus, verticaulus, and mesopleuraulus absent; propodeum moderately sculptured, dorsal face areolate or with defined enclosures; lateral propodeal carina well developed, bifurcate below; legs simple; recurrent vein joining submarginal cell near middle (fig. 125); hindwing jugal lobe shorter than or as long as submedian cell; gaster subpedunculate to pedunculate, not nodose at apex (fig. 125); pygidial plate of male quadrate, that of female broad, flat, coarsely punctate, triangular.

Geographic range: Neotropical Region. Six species are found in South America, and a seventh extends north to Costa Rica.

Systematics: Members of the genus *Quexua* are easily recognized by the postocular tubercles with the genal carina extending downward from them to the mandible base. In addition the mandibles are simple apically and externoventrally, the ocellar triangle is nearly equilateral and with a fovea midway between the hindocelli, the gaster is pedunculate or subpedunculate but not nodose, and both sexes have a pygidial plate.

When Pate described the genus (1942a), he divided it into two subgenera, *Quexua s.s.* and *Arecuna. Quexua s.s.* contained five species distinguished by the following combination of characters: frons with lateral carina paralleling inner orbits, antennal scape usually longitudinally carinate, genal carina more widely separated from posterior orbit than in *Arecuna*, occipital carina terminating at posterior mandibular condyle, clypeal lobe quadridentate, forewing with marginal cell squarely truncate apically and R_1 not extending beyond apex; longer apical spur of tibia III "cultriform with apex acuminate," female pygidial plate shagreened, coarsely punctate, and usually bisected by a longitudinal carinule. The monotypic *Arecuna* was characterized as follows: front and scape ecarinate, genal carina paralleling posterior orbit more closely than in *Quexua s.s.*, occipital carina joining hypostomal carina, clypeal lobe truncate, edentate, marginal cell obliquely truncate apically, with R_1 extending

FIG. 125. *Quexua verticalis* (F. Smith), female.

beyond apex, longer apical spur of tibia III "elongate and aciculate," and female pygidial plate shining, with few scattered punctures, ecarinate medially. Leclercq (1955a) pointed out that contrary to Pate's diagnosis, in all *Quexua s.s.* he had examined both the mandibular and hypostomal branches of the occipital carina were present. In the course of this study a male *Arecuna* has been found in which the wing venation and genal carina are typical; but the occipital carina has both branches present, the clypeus is quadridentate, frons and scape are carinate, and the tibial spur III is "cultriform." Under the circumstances, it appears that the recognition of subgenera is not warranted.

Pate (1942a) further divided *Quexua s.s.* into two species groups: *llameo* group (monotypic) and *cashibo* group (*cashibo, witoto,* and *pano*). Leclercq (1955a) synonymized *llameo* with *verticalis* and added two new species to the *cashibo* group. Pate's original group diagnoses may be modified as follows:

1. In the *verticalis* group (monotypic) the scape is ecarinate, the admedian lines and notauli are deeply impressed, and the female pygidial plate is ecarinate medially.

2. In the *cashibo* group (*cashibo, witoto, pano, ricata,* and *inca*) the scape is longitudinally bicarinate, the admedian lines and notauli are weakly impressed, and the female pygidial plate is bisected by a longitudinal carina.

Biology: Unknown.

Checklist of *Quexua*

cashibo Pate, 1942; Peru, Bolivia
essequibo Pate, 1942; Guyana
inca Leclercq, 1955; Peru, Bolivia
pano Pate, 1942; Peru, Bolivia
ricata Leclercq, 1955; Costa Rica, Bolivia
verticalis (F. Smith), 1873 (*Crabro*); Colombia, Brazil, Peru, Ecuador, Bolivia
    llameo Pate, 1942
witoto Pate, 1942; Colombia

Genus Rhopalum Stephens

Generic diagnosis: Small to moderate sized, elongate forms; eyes bare, inner orbits convergent below; scapal basin ecarinate dorsally, with or without supraantennal projection; orbital foveae distinct to indistinct; ocellar triangle equilateral or slightly broader than long; gena simple or depressed ventrally, lower border ridged; occipital carina not forming a complete circle, separated from hypostomal carina, sometimes terminating ventrally in a tooth or angle; antennal sockets contiguous to or slightly separated from each other and inner orbits; male antenna with 13 articles; scape ecarinate; male flagellum simple or modified, without ventral hair fringe; palpal formula 5-3; mandible apically bidentate, sometimes enlarged and lamellate in male, externoventral margin entire, inner margin edentate; pronotal collar ecarinate anteriorly, rounded, angulate or spiniform laterally, median notch deep to obsolete; propleura simple or dentate anteriorly; axillae, scutellum and metanotum simple; mesopleuron with postspiracular carina and omaulus usually absent, omaular area usually rounded, hypersternaulus, sternaulus, and mesopleuraulus absent, precoxal area simple or with a short verticaulus (*longinodum*); propodeum finely sculptured, enclosure well defined to absent, lateral propodeal carina present or absent, midtibia with an apical spur, hindtibia weakly to strongly swollen; forewing recurrent vein joining submarginal cell near middle or sometimes beyond; gaster usually pedunculate, slender, first segment often nodose; male usually

FIG. 126. *Rhopalum coarctatum* (Scopoli), male.

without pygidial plate, female pygidial plate variable, concave and shining to flat and dull with sides incompletely carinate (see also key to subgenera).

Geographic range: Cosmopolitan. 109 species are listed. Subgenera *Aporhopalum* and *Zelorhopalum* are found in New Zealand; *Latrorhopalum* in central Asia and the Far East; and *Calceorhopalum* in Japan, the north Pacific islands, and South America. The nominate subgenus and *Corynopus* are widely distributed.

Systematics: Most *Rhopalum* have the omaular area rounded and ecarinate, and the recurrent vein joins the submarginal cell near its middle. Intermediate forms between *Rhopalum* and *Podagritus* vary in either or both of these characters. In the most recent revision of the *Rhopalum* series Leclercq (1970) noted that all incontestable *Podagritus* could be separated from all Northern Hemisphere *Rhopalum* by the conformation of the female pygidial plate (broad, flat, triangular, and distinctly punctate). Unfortunately, a few Australian *Rhopalum* have the female pygidial plate nearly flat and somewhat punctate. Similarly, most *Podagritus* males have a punctate pygidial plate, whereas *Rhopalum* males do not. Yet, there are a few species in both genera in which this difference is doubtful, and if a female were not available generic assignment might be difficult.

American workers have used the name *Euplilus* but in most of the world literature *Rhopalum* has been employed. Menke, Bohart and Richards (1974b) have appealed to the International Commission on Zoological Nomenclature for conservation of *Rhopalum,* and we are using the name in anticipation of a favorable outcome.

Rhopalum has been variously divided into subgenera on the basis of modifications of the female pygidium and male genitalic, foreleg, and antennal characters. The number of intermediate species indicates that species groups may eventually be the most practical method of division. For the present we are using the current subgeneric categories with the exception of *Alliognathus* which we are synonymizing under *Corynopus*. *R. occidentale,* the single species assigned to *Alliognathus,* has the female pygidial plate a little broader apically than in typical *Corynopus,* and the basal male flagellomeres are only slightly dentate beneath instead of strongly so or excavate. These differences do not appear to be subgenerically significant.

The following key is modified from Tsuneki (1952b) and Leclercq (1955e, 1957c, 1970). These papers should be consulted for additional details on the genus.

Key to subgenera of *Rhopalum*
(based mainly on females)

1. Head and thorax dull, orbital foveae distinct; gonobase of male genitalia Y-shaped in dorsal view; female pygidial plate greatly narrowed toward apex, bisected above by a median longitudinal carina; central and eastern Asia .... *Latrorhopalum* Tsuneki
Head and thorax shining, orbital foveae indistinct; gonobase of male genitalia U-shaped 2
2. Pygidial area of female dull, punctate 3
Pygidial area of female shiny 4
3. Pygidial area of female bisected by a longitudinal carina; basal segment of gaster without an apical node; New Zealand .... *Zelorhopalum* Leclercq
Pygidial area of female not longitudinally carinate; basal segment of gaster usually nodose apically; widely distributed .... *Rhopalum* Stephens
4. Pygidial area of female at most gibbous basally, without a distinct longitudinal carina; gonostyle of male genitalia usually more than twice length of aedeagus; widely distributed .... *Corynopus* Lepeletier and Brullé
Pygidial area of female with a distinct longitudinal carina; gonostyle of male genitalia less than twice length of aedeagus 5
5. Pygidial plate of female greatly narrowed and concave apically; basal flagellomeres of male somewhat notched beneath; Japan, n. Pacific Is., S. America .... *Calceorhopalum* Tsuneki
Pygidial plate of female often narrow but rather broadly rounded or truncate toward apex which is nearly flat; basal flagellomeres of male simple; New Zealand .... *Aporhopalum* Leclercq

Biology: A summary of older papers was given by Leclercq (1954a). As might be expected from the customarily narrowed female pygidial plate most species nest in twigs or reeds. Provisions are commonly small flies of great variety but especially Chironomidae. Psocoptera and Aphididae (winged and wingless) are frequently used, also. Exceptional prey reported by Leclercq (1954a) are winged ants, psyllids, and Microlepidoptera. Tsuneki (1960) summarized biology of Japanese forms and later (1973b) gave additional information. Observations on different species indicated that the egg was laid on the first, second, and last prey member, and Tsuneki suggested that progressive provisioning might sometimes occur.

Species in which the female pygidial plate is relatively broad and nearly flat might be expected to nest in the ground. This is partly confirmed by observations on *Rhopalum variitarse* in Canberra, Australia by Evans and Matthews (1971a). This species was observed nesting in the sloping side of a furrow along a fence in sandy loam. The nest entrance was open when the wasp returned with prey. At the bottom of an oblique, straight tunnel of 10 cm length were two short closed off galleries ending in spherical cells of about 4 mm diameter. These were packed tightly with five small flies each. Families represented were Stratiomyidae, Dolichopodidae, Lauxaniidae, and Tachinidae.

Janvier (1928) found several females of *longinodum* sharing a burrow entrance.

Checklist of *Rhopalum*

[Subgeneric assignments indicated at end of citations by code letters as follows: *Aporhopalum* (A), *Calceorhopalum* (Ca), *Corynopus* (Co), *Latrorhopalum* (L), *Rhopalum*

(R), *Zelorhopalum* (Z)].

ammatticum Leclercq, 1963; India (Ca)
angulicolle Cameron, 1904, Mexico (R)
angustipetiolatum Tsuneki, 1971; Taiwan (L)
anteum Leclercq, 1957; Australia: Victoria (R)
antillarum Leclercq, 1957; Cuba (R)
aucklandi Leclercq, 1955; N. Zealand (Z)
australiae Leclercq, 1957; Australia: A.C.T. (R)
austriacum (Kohl), 1899 (*Crabro*); centr. Europe (R)
avexum Leclercq, 1963; India, Philippines (R)
bamendae (Leclercq), 1961 (*Podagritus*); Nigeria
beaumonti Móczár. 1957: Hungary (R)
bohartum Tsuneki, 1966; Taiwan, Ryuku Is., (CA)
    bohartorum Tsuneki, 1968; (emendation)
brevinodum (Spinola), 1851 (*Physoscelis*); Chile, Argentina
    pucarense Leclercq, 1970, new synonymy by J. Leclercq
?*bruchi* Schrottky, 1909; Argentina (? = *Podagritus*)
calderoni Leclercq, 1970; Ecuador, Bolivia (Ca)
calixtum Leclercq, 1957; Australia: Queensland, N.S. Wales (R)
calverti (Pate), 1947 (*Euplilis*); Costa Rica (R)
canlaoni Leclercq, 1963; Philippines (Ca)
changi Tsuneki, 1968; Taiwan (L)
claudii (Janvier) 1928 (*Crabro*); Chile (name validated by illustration of larva)
clavipes (Linnacus), 1758 (*Sphex*); centr., s. Europe; U. S. (R)
    rufiventre Panzer, 1799 (*Crabro*)
    ssp. *jessonicum* Bischoff, 1922 (*Crabro*); Japan
claviventre (Cresson), 1865 (*Crabro*); Cuba, Jamaica (R)
coarctatum (Scopoli), 1763 (*Sphex*); Holarctic (Co)
    crassipes Fabricius, 1798 (*Crabro*)
    tibialis Fabricius, 1798 (*Crabro*), nec Olivier, 1792
    modestum Rohwer, 1908, new synonymy by R. Bohart
coriolum Leclercq, 1957; Australia: N.S. Wales (R)
crassinodum (Spinola), 1851 (*Physoscelis*); Chile (Ca)
    herbstii Kohl, 1905 (*Crabro*), new synonymy by J. Leclercq
cruentatum (Arnold), 1944 (*Crabro*); Rhodesia
    ssp. *belgarum* Leclercq, 1955; Ruanda
decavum Leclercq, 1963; Philippines: Sibuyan Is. (R)
dedarum Leclercq, 1957; Australia: W. Australia, N.S. Wales, A.C.T., Queensland (R)
dineurum Leclercq, 1957; Tasmania (R)
diopura (Pate), 1947 (*Euplilis*); Venezuela (R)
domesticum Williams, 1928; Philippines (R)
ebetsuense Tsuneki, 1952; Japan (R)
erraticum Tsuneki, 1968; Taiwan (L)
eucalypti Turner, 1915; Tasmania (R)
expeditionis Leclercq, 1955; Tibet (L)
exultatum Leclercq, 1970; Ecuador (Ca)
fenimorum Leclercq, 1970; Mexico
formosanum Tsuneki, 1966; Taiwan (Ca)
frenchii (Turner), 1908 (*Crabro*); Australia (R)

gorongozae (Arnold), 1960 (*Crabro*); Madagascar
gracile Wesmael, 1852; Europe, Turkmen S.S.R., Japan (R)
    nigrinum Kiesenwetter, 1849; nec *Crabro nigrinus* H.-S.
    kiesenwetteri A. Morawitz, 1866 (*Crabro*), new name for *Crabro nigrinus* (Kiesenwetter), (Art. 59c)
    simplicipes F. Morawitz, 1888 (*Corynopus*)
grahami Leclercq, 1957; Australia: A.C.T., Victoria
grenadinum (Pate), 1947 (*Euplilis*); W. Indies: Grenada (R)
guttatum Tsuneki, 1955; Japan
hakodatense Tsuneki, 1960; Japan (R)
hanedai Tsuneki, 1973; Japan
harpax Leclercq, 1957; Australia: Victoria (R)
heterocerum (Mantero), 1901 (*Crabro*); Argentina
hillorum Leclercq, 1963; India (Ca)
hombceanum Tsuneki, 1973; Taiwan (L)
ichneumoniforme (Arnold), 1927 (*Thyreopus*); S. Africa, Madagascar
    ssp. *stramineipes* (Arnold), 1932 (*Thyreopus*); Rhodesia, Zaire
iridescens Turner, 1917; India: Kashmir (L)
kedahense Leclercq, 1957; Malaya (R)
kerangi Leclercq, 1957; Australia: Victoria
kuehlhorni Leclercq, 1957; Australia: S. Australia, A.C.T.
kuwayamai Tsuneki, 1952; Kurile Is. (Ca)
    ssp. *nikkoense* Tsuneki, 1956; Japan
laticorne (Tsuneki), 1947 (*Crabro*); Korea, USSR: Ussuri (L)
latronum (Kohl), 1915 (*Crabro*); Japan (L)
littorale Turner, 1915; Tasmania, Australia: A.C.T., Victoria, W. Australia (R)
longinodum (Spinola), 1851 (*Physoscelis*); Chile, Argentina (Ca)
    chilensis Reed, 1894 (*Crabro*)
    droserum Leclercq, 1957 (*Podagritus*), new synonymy by J. Leclercq
macrocephalum Turner, 1915; Australia: Queensland, Victoria, N.S. Wales (R)
magellanum (Leclercq), 1957 (*Podagritus*); Chile: Tierra del Fuego (Ca), new combination by J. Leclercq
minusculum Leclercq, 1963; Basilan Is. (R)
murotai Tsuneki, 1973; Taiwan (L)
mushaense Tsuneki, 1971; Taiwan (R)
neboissi Leclercq, 1957; Australia: Victoria (R)
nicaraguaense Cameron, 1904; Guatemala to Costa Rica (R)
    opacum Rohwer, 1914
nipponicum (Kohl), 1915 (*Crabro*); Japan (Co)
    ssp. *hokkaidense* Tsuneki, 1952; Japan
notogeum Leclercq, 1957; Australia: W. Australia
occidentale (W. Fox), 1895 (*Crabro*); U.S. (Co)
    carolina Banks, 1921
oriolum Leclercq, 1957; Malaya (R)
pallipes (Lepeletier and Brullé), 1834 (*Physoscelis*); Brazil to Argentina
    pallidipes Dalla Torre, 1897 (*Crabro*)
parcimonium Leclercq, 1963; Philippines (R?)
pedicellatum Packard, 1867; U.S. w. of 100th mer. (Co)
    rubrocinctum Peckham and Peckham, 1898

arapaho Pate, 1947 (*Euplilus*)
perforator F. Smith, 1876; N. Zealand (A)
petiolatum (Nurse), 1912 (*Crabro*); India
plaumanni Leclercq, 1970; Brazil (R)
potosium Leclercq, 1970; Bolivia (R)
prisonium Leclercq, 1970; Paraguay (R)
pygidiale R. Bohart; Japan, new name for *calceatum* (Ca)
    calceatum Tsuneki, 1947 (*Crabro*), nec Rossi, 1794 (now in *Gorytes*)
quitense (Benoist), 1942 (*Crabro*); Ecuador (R)
rotolum Leclercq, 1970; Mexico (R)
rufigaster Packard, 1867; U.S. e. of 100th mer. (Co)
    lucidum Rohwer, 1909
rumipambae Leclercq, 1970; Ecuador, Bolivia (Co)
seychellense Turner, 1912; Seychelles
    oceanicum Turner, 1911 (*Crabro*), nec Schulz, 1906
shirozui Tsuneki, 1965; Taiwan (L)
simalurense (Maidl), 1925 (*Crabro*); Malaysia, Philippines (Ca)
spinicollum Tsuneki, 1968; Taiwan (Ca)
succineicollarum Tsuneki, 1952; Japan (R)
    ssp. *taiwanum* Tsuneki, 1971; Taiwan
sumatrae Leclercq, 1950; Sumatra (R)
taeniatum Leclercq, 1957; Australia: Queensland (R)
taipingshanum Tsuneki, 1968; Taiwan (L)
tayalum Tsuneki, 1966; Taiwan (R?)
tenuiventre (Turner), 1908 (*Crabro*); Australia: Queensland, Australian Capital Territory
tepicum Leclercq, 1957; Australia: Tasmania (R)
testaceum Turner, 1917; Australia: Queensland
tongyaii Tsuneki, 1963; Thailand (R)
transiens (Turner), 1908 (*Crabro*); Australia: Victoria, N.S. Wales
tristani (Pate), 1947 (*Euplilis*); Costa Rica (R)
tsunekiense Leclercq, 1957; Malaya (R)
tubarum Leclercq, 1957; Australia: Queensland (R)
tuberculicorne Turner, 1917; Australia: Queensland
variitarse Turner, 1915; Australia: Australian Capital Territory, Victoria, Tasmania
venustum Tsuneki, 1955; Japan (R)
watanabei Tsuneki, 1952; Japan (Ca)
    arasianum Tsuneki, 1972
    ssp. *tsuifenicum* Tsuneki, 1972; Taiwan
wusheense Tsuneki, 1973; Taiwan (L)
xenum Leclercq, 1957; Australia: W. Australia, N.S. Wales
yercaudi Leclercq, 1963; India (R)
zealandum Leclercq, 1955; N. Zealand (Z)

Genus Isorhopalum Leclercq

Generic diagnosis: Eyes bare, inner orbits convergent below; scapal basin simple or with minute tubercle above sockets, ecarinate; orbital foveae distinct, linear; ocellar triangle equilateral; postocular sulcus present, weak; gena broad, sometimes flattened posteroventrally; occipital carina incomplete, joining hypostomal carina; antennal sockets contiguous to each other and to inner orbits; male antenna with 13 articles; scape ecarinate; male flagellum simple or at most flattened beneath, without ventral hair fringe; palpal formula 5-3; mandible apically bidentate, externoventral margin entire; pronotal collar ecarinate, rounded anteriorly, spiniform laterally, without median notch; scutum, axillae, scutellum, and metanotum simple; scutal flange terminating on lateral margin of scutum above tegula; mesopleuron with deep depression behind pronotal lobe, omaulus represented by short (basal) remnant; postspiracular carina, acetabular carina, sternaulus, hypersternaulus, verticaulus, and mesopleuraulus absent, subalar area elongate; propodeum finely sculptured, enclosure not delimited, lateral propodeal carina absent; legs with male basitarsus I simple or modified, hindtibia strongly swollen; recurrent vein joining submarginal cell at or slightly before its middle; hindwing jugal lobe equal in length to submedian cell; gaster pedunculate, slender, elongate; male without pygidial plate, that of female short, concave, polished.

Geographic range: Oriental Region. Two species are known.

Systematics: Isorhopalum is very closely related to *Rhopalum* and *Podagritus* and shares with them the slender, elongate body form, pedunculate gaster, clavate hindtibia, and reduced number of palpal segments. However, in *Isorhopalum* the omaulus is only a short, basal remnant that originates beneath the pronotal lobe. Also, the postspiracular carina is absent. In *Rhopalum* both the omaulus and postspiracular carina are absent. In *Podagritus* the postspiracular carina and omaulus may be present, in which case they are continuous. However, the omaulus is absent between the pronotal margin and the point of juncture with the postspiracular carina. In addition, *Isorhopalum* has a deep depression behind the pronotal lobe and an elongate subalar area.

Biology: Unknown.

Checklist of *Isorhopalum*

marunum Leclercq, 1963; India
mayoni Leclercq, 1963; Philippines

Genus Podagritus Spinola

Generic diagnosis: Moderate sized, elongate forms; eyes bare, inner orbits convergent below (fig. 127); scapal basin ecarinate dorsally, sometimes with a low projection between antennal sockets; orbital foveae distinct or evanescent; ocellar triangle equilateral or broader than long; gena simple or depressed posteroventrally, sometimes with a tubercle or carina; occipital carina joining hypostomal carina or incomplete ventrally and often ending in a tooth or angle (fig. 121 C); antennal sockets close to each other and to orbits; male antenna with 13 articles, scape ecarinate; male flagellomeres I and/or IV usually concave beneath, a short and partial fringe sometimes present ventrally; palpal formula 5-3; mandible bidentate or simple at apex, externoventral margin entire, inner margin simple or with a slight median angle; pronotal collar variable; propleura simple; scutum, axillae, scutellum and metanotum simple; postspiracular

CRABRONINI

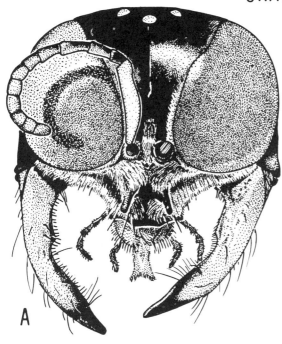

Podagritus gayi

Foxita sp.

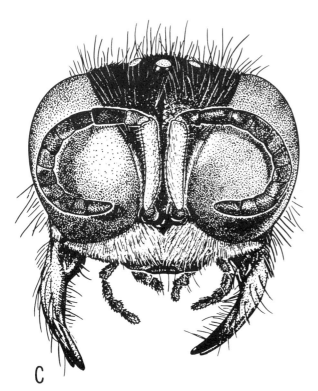

Crabro latipes

Moniaecera asperata

FIG. 127. Facial portraits of some females in the tribe Crabronini.

carina and omaulus present or absent, when absent omaular area is usually sharply angulate forming two perpendicular planes; acetabular carina, mesopleuraulus and sternaulus absent; hypersternaulus usually absent, verticaulus short or absent; metapleuron plainly depressed below level of mesopleuron and propodeum; propodeum often well sculptured, enclosure not or weakly delimited, lateral propodeal carina present or absent; male foretarsomere I often somewhat irregular; hindtibia often claviform; recurrent vein usually joining submarginal cell well beyond middle; hindwing jugal lobe about as long as submedian cell; gaster pedunculate; male usually with a rather well defined pygidial plate; that of female well defined, nearly flat, triangular, and punctate.

Geographic range: South America, Australia and New Zealand. 50 species are listed.

Systematics: Most *Podagritus* have the forewing recurrent vein joining the submarginal cell well beyond its middle, and the postspiracular carina and omaulus are usually present. However, some species have either the recurrent vein joining the submarginal cell just beyond the middle, or the omaular area is angulate but ecarinate. In either of these cases the alternate character is always typical for the genus.

Pate (1944b) noted that the Australian forms then known differed from those found in South America and suggested that it might be desirable to place them in a separate subgenus, *Echuca.* Leclercq, in his revisional studies (1955d, 1957a, d, f) found South American species that showed a mosaic of the characters of the two subgenera. He therefore modified Pate's diagnoses and described two additional subgenera, *Podagritoides* and *Echucoides,* which he later (1970) raised to genus. A third subgenus, *Parechuca,* was also described by Leclercq (1970). The three subgenera now in *Podagritus* can be separated as follows:

1. Metapleuron distinctly ridged above and somewhat gibbose; omaulus well developed and reaching episternal sulcus below; prepectus (area in front of episternal sulcus) flat; South America .... *Podagritus* Spinola s.s.
   Metapleuron smooth above or with very superficial sculpture, not gibbose 2
2. Prepectus rather convex, omaulus absent or incomplete below; area behind episternal sulcus not completely flat; South America and New Zealand .... *Parechuca* Leclercq
   Prepectus flat, separated from pre-omaular area by an omaulus or by a difference in plane; area behind episternal sulcus flat; Australia .... *Echuca* Pate

Four New Zealand species previously in *Rhopalum* are now in *Parechuca.* Six Neotropical species originally described as *Echuca* (*almagrus, geraesae, getricus, magellanus, sorbicus,* and *valdiviae*) have not been reassigned to subgenera. In addition, Leclercq (1970) suggested that examination of the type of *bruchi,* now placed in *Rhopalum,* may prove it to be a *Podagritus.*

Biology: Janvier (1928) found *P. gayi* nesting in sand on the banks of the Cautin River in Chile. The female loosened the sand with her mandibles and ejected it from the burrow with her legs. The main tunnel ran horizontally into the bank for 40-50 cm and then descended vertically. Single cells were excavated at the end of short horizontal galleries which branched off the vertical tunnel. The first cell was uppermost, and each one was provisioned and then closed with a plug of sand before the next gallery was started farther down and on the opposite side of the main tunnel. The nests studied contained an average of 10-12 cells.

The wasp hunted her dipteran prey on flowers, especially those of *Foeniculum officionale.* The species captured (*Toxomerus, Melanostoma, Hylemya, Sarcophaga, Heliomyza,* and *Acinia*) depended on which was most abundant at the time of provisioning. After seizing the fly by its neck or thorax, she curled her abdomen under it and stung it between the legs. Then she grasped it in her mandibles and flew back to the nest. The total number of prey per cell depended on their size; one cell contained 17 *Toxomerus,* another seven *Sarcophaga flavifrons.* The egg was laid between the legs of one of the first prey. The larva hatched in two to three days and spent the next five days feeding. When the supply of flies was devoured, the larva rested four to five days and then spun its cocoon, incorporating the remains of the prey into the walls. Metamorphosis started in November and was completed towards the end of December.

Evans and Matthews (1971a) found a number of *Podagritus leptospermi* nesting in a sloping gravel bank near Canberra, Australia. The burrows, dug in sandy loam containing pebbles, were vertical, relatively straight, and 21 to 29 cm long. The entrances were surrounded by flat mounds of soil 6 to 10 cm in diameter and were left open when the female was away. As in *gayi,* single cells were excavated at the end of lateral galleries. One nest had two cells, the other three. The paralyzed dipteran prey were carried dangling from the middle of the wasps' bodies, apparently grasped by the midlegs. On returning to the nest, the female first descended slowly towards the entrance, then plunged directly in. The flies were first stored in the bottom of the burrow, then 4 to 6 were placed in a cell, head in, venter up. The egg was fastened between the head and prothorax. Twenty-seven tachinid and therevid flies were recovered from nests of provisioning females.

Checklist of *Podagritus*

[Assignment to subgenus is indicated by code letters at end of citations: *Podagritus* (Po), *Echuca* (E), *Parechuca* (Pa).]

acollae Leclercq, 1970; Peru (Pa)
aemulans (Kohl), 1905 (*Crabro*); Chile, Ecuador, Peru (Po)
  aequadoricus Strand, 1911 (*Crabro*)
albipes (F. Smith), 1878 (*Rhopalum*); New Zealand (Pa)
alevinus Leclercq, 1957; Australia: Victoria (E)
aliciae (Turner), 1915 (*Rhopalum*); Australia: W.

Australia (E)
alutaceus Leclercq, 1951; Ecuador (Pa)
anerus Leclercq, 1955; Australia: Victoria (E)
arechavaletai (Brèthes), 1909 (*Crabro*); Uruguay (?), new combination by J. Leclercq
　ssp. *joergenseni* (Brèthes), 1913 (*Crabro*); Argentina, new status by Leclercq
aricae Leclercq, 1957; Chile, Peru, Argentina (Po)
　ssp. *carrascoi* Fritz, 1971; Peru, new status by J. Leclercq
bordai Fritz, 1971; Bolivia (Po)
brethesi Leclercq, 1951; Brazil (Po)
burnsi Leclercq, 1955; Australia: Victoria (E)
carolus Leclercq, 1955; Australia: S. Australia (E)
catherinae Fritz, 1971; Brazil (Po)
concordius Leclercq, 1970; Argentina, Brazil (Po)
cora (Cameron), 1888 (*Crabro*); New Zealand (Pa)
　　carbonarius F. Smith, 1856 (*Crabro*), nec Dahlbom, 1838
　　carbonicolor Dalla Torre, 1897 (*Crabro*)
　　jocosum Cameron, 1898 (*Crabro*)
cygnorum (Turner), 1915 (*Rhopalum*); Australia: W. Australia (E)
doreeni Leclercq, 1955; Australia: A.C.T., Victoria (E)
edgarus Leclercq, 1957; Australia: N.S. Wales (E)
erythropus (Brèthes), 1913 (*Crabro*); Argentina (Patagonia) (Po)
exegetus Leclercq, 1970; Peru (Pa)
?*fulvohirtus* (Cameron), 1891 (*Crabro*); Mexico (Po)
gayi Spinola, 1851; Chile, Argentina (Po)
　patagonicus Holmberg, 1903 (*Rhopalum*), new synonymy by J. Leclercq
imbelle (Turner), 1915 (*Rhopalum*); Australia: W. Australia (E)
kiatae Leclercq, 1955; Australia: Victoria, Tasmania (E)
krombeini Leclercq, 1955; Australia: N.S. Wales (E)
leptospermi (Turner), 1915 (*Rhopalum*); widespread in Australia (E)
lynchii (Holmberg), 1903 (*Rhopalum*); Argentina (Po), new combination by J. Leclercq
　pamparum Brèthes, 1913 (*Crabro*); new synonymy by J. Leclercq
marcellus Leclercq, 1955; Australia: W. Australia (E)
mullewanus Leclercq, 1970; Australia: W. Australia, S. Australia (E)
neuqueni Leclercq, 1957; Argentina (Pa)
nigriventris (Brèthes), 1913 (*Crabro*); Argentina, Paraguay, Brazil (Po)
　sombratus Leclercq, 1951
parrotti (Leclercq), 1955 (*Rhopalum*); New Zealand (Pa)
peratus Leclercq, 1955; Australia: W. Australia (E)
pius (Strand), 1910 (*Crabro*); Brazil, Peru, Uruguay, Argentina (?), new combination by J. Leclercq
　geraesae Leclercq, 1951, new synonymy by J. Leclercq
pizarrus Leclercq, 1957; Argentina (Po)
polybia Schrottky, 1909; Paraguay (Po)
rhopaloides Leclercq, 1951; Ecuador (Pa)
rieki Leclercq, 1957; Australia: A.C.T., N.S. Wales, Victoria (E)
riveti (Strand), 1911 (*Crabro*); Ecuador (Pa)
rufotaeniatus (Kohl), 1905 (*Crabro*); Chile (E), new combination by J. Leclercq
　getricus Leclercq, 1957, new synonymy by J. Leclercq
sorbicus Leclercq, 1957; Chile, Argentina (?)
swalei (Leclercq), 1955 (*Rhopalum*); New Zealand (Pa)
terpenus Leclercq, 1957; Argentina (Po)
tricolor (F. Smith), 1856 (*Crabro*); widespread in Australia (E)
　militaris Turner, 1908 (*Crabro*)
valdiviae Leclercq, 1951; Chile, Argentina (E)
　almagrus Leclercq, 1957, new synonymy by J. Leclercq
valenciai Fritz, 1971; Chile (Po)
venturii (Schrottky), 1902 (*Crabro*); Argentina, Paraguay (Po)
willinki Leclercq, 1957; Argentina (Po)
yarrowi Leclercq, 1955; Australia: W. Australia (E)

Genus Podagritoides Leclercq

Generic diagnosis: Moderate sized, very slender, elongate forms which are mostly shiny black; eyes bare, inner orbits convergent below; scapal basin ecarinate dorsally, area above antennal sockets simple; orbital foveae evanescent; ocellar triangle equilateral; gena rounded; occipital carina ending ventrally before reaching hypostomal carina, terminating in a denticle; antennal sockets close to each other and to orbits; male antenna with 13 articles; scape ecarinate; male flagellomere I deeply excavate, no ventral hair fringe; palpal formula 5-3; mandible falciform but enlarged and bidentate at apex; pronotal collar transversely rugulose, rounded laterally; propleura with a strong, forward projecting tooth; scutum, axillae, scutellum and metanotum simple; postspiracular carina and omaulus present, latter strong but short, ending in a protuberance below before reaching episternal sulcus; acetabular carina, sternaulus and mesopleuraulus absent; metapleuron without inferior spiracular carina but superior one distinct; propodeum mostly smooth, enclosure not defined, lateral propodeal carina present; legs rather simple; hindtibia moderately claviform; recurrent vein joining submarginal vein at distal third; jugal lobe about as long as submedian cell; gaster elongate-pedunculate; male with a distinct, punctate pygidial plate, that of female narrowed and concave apically.

Geographic range: Australasian Region where its single species is known only from the island of Fiji.

Systematics: Recently, Leclercq (1970) removed subgenus *Podagritoides* from *Podagritus* and elevated it to generic rank. Its single unique character is the presence of a strong, forward directed tooth on the propleuron. Less exclusive features are the weakly sculptured integument, reception of the recurrent vein at the middle of the submarginal cell, and the strong omaulus which ends in a prominent angle below before reaching the episternal sulcus. Its close relatives are *Podagritus* and *Echucoides* which differ as indicated in the key to genera and in the key provided by Leclercq (1957d).

Biology: Unknown.

Checklist of *Podagritoides*

oceanicus (Schulz), 1906 (*Crabro*); Fiji

Genus Echucoides Leclercq

Generic diagnosis: Moderate sized, slender, elongate forms; eyes bare, inner orbits convergent below; scapal basin ecarinate dorsally, with a small tooth above antennal sockets; orbital foveae evanescent; ocellar triangle equilateral; gena rounded; occipital carina joining hypostomal carina; antennal sockets close to each other and to orbits; male antenna with 13 articles, scape ecarinate; flagellomeres simple except for a dorsal hair on II and following; palpal formula 5-3; mandible stout, short, apically bidentate; pronotal collar rounded, not angled laterally, median incision not sharp; propleura simple; scutum, axillae, scutellum and metanotum simple; postspiracular carina and a fine omaulus present, latter reaching episternal sulcus below; acetabular carina, sternaulus and hypersternaulus absent; metapleuron on same plane as mesopleuron and propodeum, latter without spiracular carina; precoxal suture distinct; propodeum generally smooth, enclosure not defined, lateral propodeal carina absent; legs rather simple, hindtibia not claviform; recurrent vein joining submarginal cell near its distal third; jugal lobe about as long as submedian cell; gaster pedunculate, tergum I rather nodose posteriorly; male without a pygidial plate, that of female distinct toward apex only, where it is somewhat depressed.

Geographic range: Neotropical Region in Peru and Bolivia. Two species are known.

Systematics: Echucoides is distinctive in having a small median tubercle at about the middle of the scapal basin. Other characters of importance are the relatively flat pleural area, the single dorsal hair near the apex of most male flagellomeres, the nonclaviform hindtibia, the fine but distinct omaulus that reaches the episternal sulcus below, and the incomplete female pygidial plate that is delimited only toward the apex. Key characters and a discussion of relationships with *Podagritus* and *Podagritoides* were given by Leclercq (1957d, 1970).

Biology: Unknown.

Checklist of *Echucoides*

cercericus (Leclercq), 1957 (*Podagritus*); Peru, Bolivia
piratus (Leclercq), 1957 (*Podagritus*); Peru

Genus Notocrabro Leclercq

Generic diagnosis: Moderately small forms marked with yellow; eyes bare, inner orbits convergent below; scapal basin simple, ecarinate above; orbital foveae distinct, elongate-teardrop shaped in both sexes; ocellar triangle slightly broader than long, anterior ocellus smaller than hindocelli; gena narrowed ventrally; occipital carina incomplete ventrally; antennal sockets contiguous with or slightly separated from inner orbits, separated from each other by a narrow raised area covered with hair; male antenna with 13 articles; scape ecarinate, outer edge dilated (especially in male); male flagellum with basal articles modified, with ventral hair fringe; palpal formula 5-3; mandible apically bidentate, externoventral margin entire; pronotal collar ecarinate, rounded anteriorly and laterally, notched medially; scutum, axillae, scutellum, and metanotum simple. prescutellar sulcus very narrow, efoveate; mesopleuron sharply angulate anteriorly but without omaulus; postspiracular carina, acetabular carina, sternaulus, hypersternaulus, mesopleuraulus and verticaulus absent; propodeum finely sculptured, enclosure not defined by a furrow or carina, posterior face covered with silvery pubescence, lateral propodeal carina present posteriorly; legs slightly flattened, hindtibia swollen, recurrent vein joining submarginal cell at its middle; jugal lobe longer than submedian cell; gaster pedunculate, first segment slightly nodose at apex, with median, backward-projecting spine on dorsoposterior face of node (fig. 122 H); male with quadrate, shining pygidial plate; female with narrowed, elongate, slightly concave, shining, triangular pygidial plate.

Geographic range: Australia. Only one species is known.

Systematics: Notocrabro is easily recognized by the spine on tergum I (fig. 122 H). Its other characters fall within the range of the *Rhopalum* series. Leclercq (1954a) gave the palpal formula as 5-4, but on the single pair we have seen it is 5-3.

Leclercq (1954a) proposed a new name, *Spinocrabro*, for his genus *Notocrabro* (1951a) under the impression that the latter was homonymous with *Nothocrabro* Pate, 1944. However, since *noto* and *notho* have entirely different derivations, and there is a one letter difference in the generic names, Leclercq's new name is unnecessary.

Biology: Unknown.

Checklist of *Notocrabro*

idoneus (Turner), 1908 (*Crabro*); Australia: Queensland
spinulifer Turner, 1918 (*Rhopalum*)

Genus Moniaecera Ashmead

Generic diagnosis: Eyes bare or densely covered with short, erect hair, inner orbits convergent below; scapal basin ecarinate above, simple or with small to large and nasiform interantennal tubercle (fig. 127 D); orbital foveae present in female, absent or indistinct in male; ocellar triangle equilateral or broader than long; gena simple or with shallow postocular sulcus which may be foveate; occipital carina forming a complete circle separated from hypostomal carina; antennal sockets contiguous but separated from inner orbits; male antenna with 13 articles; scape ecarinate; male flagellum simple or modified, with or without ventral hair fringe; palpal formula 5-3; mandible apically bidentate, lower tooth sometimes diverging downward; externoventral margin entire, with basal tooth in some females; inner margin with large, spinelike, backward-projecting tooth at base (fig. 127 D); pronotal collar rounded medially, sharply angulate or carinate laterally; scutum, axillae, scutellum and metanotum simple, omaulus present, precoxal area simple or with sharp tubercle; postspiracular and acetabular carinae, sternaulus, hypersternaulus, verticaulus, and mesopleuraulus absent; propodeum finely sculptured, enclosure absent or weakly defined, lateral propodeal carina absent or weak; male fore and midbasitarsi simple or modified; hindtibia normal, not swollen; recurrent vein joining submarginal cell at or before its middle; jugal lobe slightly longer than submedian cell; gaster pedunculate to subpedunculate, first segment 1.0 to 1.5 times length of second, slightly nodose at apex; male without pygidial plate, that of female broad, flat, coarsely punctate, triangular.

Geographic range: North America to southern Mexico. Four described species are restricted to the southwestern U.S., and a fifth, *abdominalis,* extends east to Georgia and north to Kansas. Additional undescribed species are known from the southwestern states and Mexico.

Systematics: Moniaecera and *Huavea* are closely related genera in the *Rhopalum* series. They may be distinguished from other members of the series by the following characters: mandible with inner basal spinous tooth (fig. 127 D), occipital carina complete and separated from hypostomal carina (fig. 121 E), pronotal collar angulate laterally, omaulus present, forewing recurrent vein joining submarginal cell at or before its middle, and hindtibia not strongly clavate (fig. 128). *Moniaecera* differs from *Huavea* as follows: postspiracular and acetabular carinae and hypersternaulus absent, precoxal area simple or with tubercle, pygidial plate of female broad, flat, coarsely punctate, triangular. Pate (1948a) gave a review of the genus. Court and Bohart (1966) compared the genus with *Huavea.*

Biology: Observations have been made on two species. Hartman (1905) found *Moniaecera abdominalis* nesting on the bank of the Colorado River near Austin, Texas during the months of August and September. The burrow, excavated in the middle of a small elevation of sand, ran horizontally for 7.5 cm and then nearly vertically for 11 cm. No cells had been constructed, but the female was storing cicadellids, *Tylozygus bifidus* (Say), in a pocket at the bottom of the tunnel. When hunting she did not close the entrance of the nest, and on returning to it she hovered briefly several centimeters above it before plunging in.

Cazier and Mortenson (1964) studied *M. asperata* at a site near Portal, Arizona during late May and early June. Although this species also was found in a sandy area the nest was excavated in rocky, hard-packed soil in a shallow borrow pit. The burrow was 4 mm in diameter and descended vertically for 5 cm, then at a steep angle for another 6.5 cm. The single completed cell branched off the main tunnel 10 mm from the bottom. It contained a nearly mature larva, 13 cicadellids (*Circulifer tenellus* (Baker), *Empoasca abrupta* Delong), two psyllids (*Paratrioza* sp., *Aphalaroida* sp.), and the remains of a number of other leafhoppers. The female was storing additional prey at the bottom of the tunnel prior to constructing the second cell. On leaving the nest she did not close the entrance, and on returning she flew directly in without first hovering outside.

Evans (1964d) found *M. asperata* nesting communally at a site on the banks of the Rio Grande near Lajitas, Brewster County, Texas. The nests were dug in powdery clay-sand which had a hard crust. They were spaced 1.5 to 3 m apart and were shared by two or three females. The burrows descended vertically and then branched, presumably with each female nesting in her own tunnel. When digging, a female would fly out head first with a load of soil held between her head and thorax and drop it about 0.5 m from the entrance. In the nest that was studied there were four completed cells 7-10 mm apart in a cluster at the bottom of the tunnel which was 14 cm long. The cells contained an average of 20 prey of the following families: Diptera: Chironomidae; Homoptera: Psyllidae, Cicadellidae, and Hemiptera: Miridae. The female carried her prey near the middle of her body, probably holding it with her mid and hind legs. On returning to the nest, which was never closed, she hovered outside briefly before entering.

Checklist of *Moniaecera*

abdominalis (W. Fox), 1895 (*Crabro*); U.S.: Arizona to Georgia, n. to Kansas
asperata (W. Fox), 1895 (*Crabro*); sw. U.S.
evansi Pate, 1947; U.S.: Arizona
foxiana Pate, 1948; U.S.: California
pinal Pate, 1947; U.S.: California

Genus Huavea Pate

Generic diagnosis: Eyes sparsely or densely covered with short, erect hair, inner orbits convergent below; scapal basin with interantennal tubercle which is not nasiform, ecarinate above; orbital foveae present in female, absent or indistinct in male; ocellar triangle broader than long; gena with shallow postocular sulcus, otherwise simple; occipital carina forming a complete circle, separated from hypostomal carina; antennal sockets contiguous but separated from inner orbits; male antenna with 13 articles; scape ecarinate; male flagellum not modified, without ventral hair fringe; palpal formula 5-3; mandible apically bidentate, externoventral margin simple, inner margin with large, spinelike backward-projecting tooth at base; pronotal collar ecarinate, rounded anteriorly, sharply angulate laterally, notched medially; scutum, axillae, scutellum, and metanotum simple; prescutellar sulcus foveate; postspiracular carina continuous with omaulus, acetabular carina continuous with subomaulus but separated from omaulus, verticaulus present and extending to scrobe, hypersternaulus present (fig. 128), sternaulus and mesopleuraulus absent; propodeum moderately sculptured, enclosure well defined; lateral propodeal carina present; legs simple; recurrent vein joining submarginal cell at its middle (fig. 128); jugal lobe slightly longer than submedian cell; gaster pedunculate, first segment approximately 1.5 times length of second, slightly nodose at apex; last tergum of male simple or medially concave, pygidial plate of female an elongate triangle, apical half narrowed, shining, impunctate.

Geographic range: Southwestern United States and Mexico. Two species known.

Systematics: Huavea was originally described as a subgenus of *Moniaecera* by Pate (1948a), who had seen only the holotype male specimen of *chontale*. During the course of this study additional material of both sexes of *chontale* and a second species, *pima,* were examined; and in 1966, Court and Bohart raised *Huavea* to full generic rank.

The characters that are common to both *Moniaecera* and *Huavea* are given in the discussion of the former. Those that distinguish *Huavea* are the following: postspiracular carina, acetabular carina, subomaulus, verticaulus, and hypersternaulus present; pygidial plate of female elongate, punctate basally, narrowed and smooth apically.

Biology: Unknown.

Checklist of *Huavea*

chontale (Pate), 1948 (*Moniaecera*); Mexico
pima Court and R. Bohart, 1966; U.S.: Arizona; Mexico: Baja California

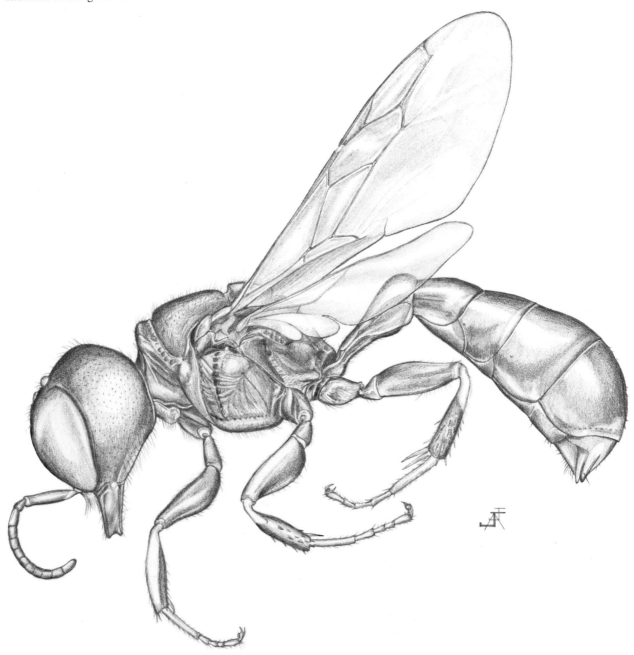

FIG. 128. *Huavea chontale* (Pate), female.

Genus Crossocerus Lepeletier and Brullé

Generic diagnosis: Small to medium sized wasps, length 3-13 mm; finely punctate and sculptured; mostly black or brown with varying yellow or white maculation, occasionally with immaculate red abdomen. Head subquadrate to rectangular in dorsal view; eyes naked, facets larger anteroventrally than posterodorsally, inner orbits convergent below; scapal basin glabrous, ecarinate dorsally; orbital foveae distinct to indistinct; ocelli forming an equilateral triangle or nearly so (fig. 121 B); frons usually with a median impressed frontal line; ocellocular distance subequal to interocellar distance; occipital carina usually obsolescent ventrally, very rarely a complete ring, occasionally terminating in a tooth or stout spine; scape ecarinate, flagellum usually simple, in male with 11 articles, sometimes with tyli and usually with a ventral hair fringe; antennal sockets contiguous with each other, with inner orbit and with frontoclypeal suture; palpal formula 6-4; mandible apically acuminate, bidentate, or tridentate, sometimes a medial tooth on inner margin, externoventral margin entire; malar space very small or absent; pronotal collar rounded to sharply ridged dorsally, frequently with anterolateral carina or angle; scutum finely punctate; omaulus and episternal sulcus present; scrobal sulcus, hypersternaulus, mesopleuraulus, verticaulus, and sternaulus absent; precoxal area sometimes with a tubercle; propodeum finely sculptured, rarely smooth or coarsely sculptured, usually with some indication of a dorsal enclosure which is bisected by a median furrow, lateral carina generally present but seldom attaining spiracles; wings hyaline, setulose, sometimes with dark apical spots; marginal cell elongate and apically truncate; recurrent vein received near middle of submarginal cell; jugal lobe present, variable in length; legs simple or modified; abdomen sessile to pedunculate, peduncle often nodose apically; segments II-V or VI subcylindrical; pygidial plate of female almost always present, and either flat, triangular and punctate, or excavate, more or less narrowed apically, and usually polished; males usually without pygidial plate, sternum VIII modified or simple.

Geographic range: Widely distributed, but only one species known from the Australasian Region (Solomon Is.). *Crossocerus* is the largest genus in the Crabronini and of the approximately 200 species listed below, more than half occur in the Holarctic Region, and more than half of these are Palearctic. One species, *C. annulipes,* is truly Holarctic. The Oriental Region has about 47 species; the Ethiopian, 24; and the poorly known Neotropical fauna is represented by about 10 species, one of which is the adventive European entity, *C. elongatulus* Vander Linden. Two *Crossocerus* are recorded only from Madagascar.

The Oriental fauna is the most diverse, having 14 subgenera, half of which are endemic. In contrast, the Palearctic, with about twice as many species, has nine subgenera but only one endemic subgenus.

Systematics: The component forms of *Crossocerus* are quite varied and have strong habitus similarities to many other crabronin genera. However, they can always be diagnosed by the following combination of characters: ocellar triangle equilateral or nearly so, palpal formula 6-4, omaulus present, verticaulus absent.

Crossocerus is related to *Crabro,* the latter having the same palpal formula and similar wing venation, thoracic structure, and ethology. Among the diverse exponents of these two genera, the ocellar triangle is the only fully distinguishing feature. *Crabro* has the ocelli arranged in a low triangle or curved line. Additionally, the propodeum of that genus is usually areolate or coarsely ridged, and the enclosure, when present, is not polished. In *Crossocerus* the propodeum usually has fine sculpture and a polished dorsal enclosure.

Some pedunculate *Crossocerus* strongly resemble *Rhopalum,* but the latter has no omaulus, and its palpal formula is 5-3. Other crabronins that have the ocelli in an equilateral triangle may also be confused with *Crossocerus,* but these have a palpal formula of less than 6-4.

There are about 20 subgenera currently recognized. It is likely that a careful study will result in the raising of some of these to generic status and the downgrading of others to species groups. Pate (1944a) presented a key to the subgenera of which the following is an updated and expanded version.

Key to subgenera of *Crossocerus* Lepeletier and Brullé (does not include *Ornicrabro*)

1. Female with broad, flat, coarsely punctate, triangular pygidial plate; male with last tergum more coarsely punctate than penultimate tergum; inner margin of mandible edentate 2
   Female with pygidial plate usually narrowed and excavate apically; rarely flat and with lateral margins only weakly incurved, but in such cases the disk is polished; males with last abdominal tergum not more coarsely punctate than penultimate tergum 5
2. Mandible of female acuminate or blunt at apex; male with tergum VII elongate triangular, not transverse; gaster sessile 3
   Mandible of female bidentate or tridentate at apex; male with tergum VII short, transverse 4
3. Occipital carina terminating in stout spine; clypeal lobe quadridentate; male antenna ventrally fringed with white hair ....
   .... *Hoplocrabro* Thomson
   Occipital carina ventrally obsolescent; clypeal lobe truncate; male antenna unfringed ....
   .... *Microcrabro* Saussure
4. Mandible of female tridentate at apex; upper frons with lateral glabrous swelling; gaster subsessile, immaculate red; mesopleuron without precoxal tubercle; male unknown .... *Alicrabro* Tsuneki
   Both sexes with mandible bidentate; upper frons normal; gaster sessile to pedunculate, immaculate black; mesopleural tubercle

present or absent ....
.... *Crossocerus* s.s. Lepeletier and Brullé
5. Mandible simple, acuminate at apex, with a strong premedian tooth on inner margin; mesopleuron with a foveate ridge reaching coxal insertion; gaster subsessile ....
.... *Pericrabro* Leclercq
Mandible bidentate or tridentate at apex; other characters variable 6
6. Occipital carina forming a complete circle; mandible bidentate at apex in both sexes; tergum VII of male very small; female mandible with weak medial tooth on inner margin; gaster sessile; very small wasps, ± 4 mm ....
.... *Oxycrabro* Leclercq
Occipital carina not a complete circle, mandible and gaster variable 7
7. Mandible with broad, weak medial tooth on inner margin; gaster immaculate black, pedunculate, first segment at least twice as long as broad at apex; occipital carina terminating in a weak tooth .... *Apoides* Tsuneki
Mandible evenly curved on inner margin or with sharp medial tooth; gaster with first segment longer than broad at apex 8
8. Mandible with medial tooth on inner margin 9
Mandible edentate on inner margin 16
9. Mesopleuron with precoxal tubercle.................. 10
Mesopleuron without precoxal tubercle 14
10. Mandible tridentate at apex; gaster immaculate black and sessile, tergum I somewhat nodose at apex; medium small wasps, ± 6 mm; male unknown .... *Bnunius* Tsuneki
Mandible tridentate or bidentate at apex, if the former, then gaster yellow maculate, tergum I sessile to pedunculate but not nodose at apex; size variable 11
11. Forewing with anterodistal fuscous spots; hindtibia slender; female mandibular apex tridentate; male with forebasitarsus sinuate or twisted spirally; gaster maculate, sessile 12
Forewing without spots; hindtibia variable; mandible bidentate in both sexes; forebasitarsus of male simple; gaster usually immaculate, sessile to pedunculate 13
12. Female with pygidial plate flat, subtriangular, lateral margins weakly narrowed; male with forefemur posteriorly dentate, mandible bidentate at apex, orbital foveae distinct ....
.... *Acanthocrabro* Perkins
Female with pygidial plate strongly narrowed and apically excavate; male with forefemur edentate, mandible tridentate at apex, orbital foveae indistinct ...... *Stictoptila* Pate
13. Pronotal collar without lateral carina; female pygidial plate equilateral, with trilobate depression, not sharply margined; small wasps, less than 7 mm .. *Ablepharipus* Perkins
Pronotal collar with anterolateral humeral carina; female pygidial plate attenuate apically, with V-shaped elevation at base, sharply margined; medium-sized wasps, 7-9 mm .... *Neoblepharipus* Leclercq
14. Mandible bidentate at apex in both sexes; pygidial plate very strongly narrowed, lateral margins approximate in apical half; propodeum with neither lateral carina nor dorsal enclosure; sessile, immaculate black; very small wasps, 5 mm or less ....
.... *Epicrossocerus* Ashmead
Mandible tridentate at apex in females; pygidial plate with lateral margins not approximate; gaster usually maculate, wasps longer than 5 mm 15
15. Gaster subsessile, first tergum longer than broad at apex, black marked with yellow or white, rarely immaculate black or red; forewing without dark spots; female with flat, subtriangular pygidial plate, lateral margin feebly sinuate; male with metacoxa basoventrally dentate, forefemur posteriorly carinate, midtibia carinate on outer surface, orbital foveae indistinct ....
.... *Cuphopterus* A. Morawitz
Gaster sessile, first segment shorter than apical width, black marked with yellow; forewing with anterodistal dark spots; female pygidial plate narrowed and excavate; males with metacoxa edentate, forefemur and midtibia ecarinate, orbital foveae distinct .... *Nothocrabro* Pate
16. Head large, transverse; mandible bidentate at apex in both sexes, pronotal collar rounded; female with pygidial plate feebly excavate; small wasps, less than 5 mm ....
.... *Paroxycrabro* Leclercq
Head subequal in width to thorax, subquadrate; other characters variable; longer than 5 mm 17
17. Gaster sessile or subsessile 18
Gaster pedunculate 19
18. Gaster with lateral maculation on tergum III, subsessile, tergum I a little longer than apical width; mesopleuron without precoxal tubercle; propodeum with lateral carina not developed; pronotal collar somewhat pointed at humerus, mandible bidentate at apex in both sexes; medium large, robust wasps, male about 10 mm, female about 13 mm ....
.... *Ainocrabro* Tsuneki
Gaster immaculate black, sessile, tergum I as long as broad at apex; mesopleuron with or without precoxal tubercle; propodeum usually with lateral carina on posterior face; pronotal collar rounded laterally or with anterior carina at humerus; mandible tridentate or rarely bidentate in females; small to medium wasps, usually 10 mm or less ....
.... *Blepharipus* Lepeletier and Brullé
19. Peduncle nodose at apex, separated from gaster by constriction; hindwing with jugal lobe longer than submedian cell; with acetabular carina; antennal scape clavate; mesopleuron with or without precoxal tubercle; pygidial plate of female absent .... *Eupliloides* Pate
Peduncle gradually widened apically and broadly joined to gaster, hindwing with jugal lobe shorter than submedian cell; with-

out acetabular carina; antennal scape slender, mesopleuron with sharp precoxal tubercle; pygidial plate present in female .... *Apocrabro* Pate

There are many references on systematics of *Crossocerus*. Especially useful are those of Kohl (1915), Leclercq (1954a, 1956d, 1958a, 1968c), Pate (1944a), and Tsuneki (1954b).

Biology: The species of *Crossocerus*, on the basis of nesting substrate, fall into two groups. One, which contains the subgenera *Crossocerus, Hoplocrabro, Alicrabro,* and *Pericrabro,* is, with a couple of exceptions, fossorial, nesting in sandy or clay soil. This habit is correlated with the female pygidial plate — a broad, flat triangular sand-tamper. The other group, in which *Blepharipus* and its allies belong, have the pygidial plate at least partially narrowed and excavate; these wasps are xylicolous or rubicolous. They construct burrows or renovate existing galleries, burrows, and cavities in old logs, stumps, branches, and posts.

Various small Diptera are the principal prey for this genus, but a few *Crossocerus* provision with Homoptera, and some take Trichoptera, Microlepidoptera, Hemiptera, and even Psocoptera, Mecoptera and Ephemeroptera (Tsuneki, 1960).

Davidson and Landis (1938) reported on the common Holarctic crabronin, *C. (Blepharipus) annulipes* in Ohio. They reared this wasp from some of its usual haunts, an old peach tree stump, a rotting fence post, and a porch. The stored prey were leafhoppers of the genera *Empoasca, Typhlocyba,* and *Erythroneura,* averaging 20 per cell. There were many cells per gallery which were "long and winding and penetrated both hard and rotting wood." Each cell was sealed by a plug of chewed wood. Hamm and Richards (1926) reported similar findings for this species in England, with the addition of the prey genus *Alebra*. Tsuneki (1960) listed jassid leafhoppers as fodder for *annulipes* in Japan, and he further noted that prey carriage was venter-to-venter, the wasp using her middle legs, her mandibles grasping the leafhopper's antennae. Female *annulipes* usually flew directly to the burrow entrance and entered with the paralyzed prey head first and venter up. The egg was white, 1.2 by 0.3 mm, and was laid with its anterior end on the leafhopper's neck.

Tsuneki offered observations on many of the Japanese *Crossocerus*. Prey carriage of the venter-to-venter, midleg clasp style was usual. *C. (Blepharipus) nigritus* used her hind legs to capture a fly that was sitting on a leaf. She then slowly moved forward over the prey until the sting was over the gula. Then she pushed up the head of the fly and stung it in the neck. The nest of *nigritus* was compoundly branched, with linearly arranged cells within some branches. Oviposition occurred after each cell was completely stocked. In Europe, Hamm and Richards (1926) recorded this wasp nesting in *Sambucus* and *Typha* and provisioning with a wide assortment of small flies. The nests had four to six cells each, and the males hatched out a few days before the females. According to Tsuneki (1960) *C. (B.) walkeri* stores its nests with various Ephemeroptera, and *C. (B.) cinxius* provisions with empidid flies, pentatomid bugs, and Psyllidae.

Among the fossorial *Crossocerus, C. (C.) maculiclypeus* is rather common and was the subject of a study by Kurczewski et al. (1969) in New York. They noted that the females apparently remodeled old wasp and bee burrows, selecting those on slopes or cliffs rather than on flat sand. They dug shallow burrows with the cells less than 1 cm beneath the surface. The cell number ranged from four to nine, and each cell averaged 13 flies when completely provisioned. *Platypalpus holoserica* Melander (Empididae) was the commonest prey; others included Dolichopodidae (*Thrypticus*), Agromyzidae, Psilidae, and Chamaemyiidae. Storage was ventral side up, head inward, and the egg was attached with the anterior end on the fly's neck, perpendicular to the fly's longitudinal axis. The finished cells were partially or completely closed with sand.

Krombein (1964a) found *C. (C.) planipes* nesting in a shaded area of bare forest soil. The burrow descended at a 20° - 30° angle ending in a cell about 3 cm below the surface. The cell, partially provisioned, contained five *Drapetis* sp. (Empididae). Another cell, about 2 cm deep, contained 16 of the same flies and one egg. The cells were ovoid and 0.6 cm long. A nearby burrow had a tumulus ring around the entrance mostly on the downhill side. This burrow angled down 3 cm to a depth of only 0.8 cm, then dropped of at a 60° angle to a point 2 cm below the surface. This cell was also stocked with empidids, 13 *Chersodromia* sp.

Of the 13 species of *Crossocerus* studied by Krombein (1963a) at Plummers Island, Maryland, all but one, *planifemur*, were multivoltine. *C. (C.) planifemur* is one of the very few wasps in the typical subgenus that is recorded as nesting in wood.

Biological data for the smaller subgenera are rather few. Tsuneki (1960) recorded *C. (Ainocrabro) aino* as a predator of *Panorpa bicornuta* MacLachlan. He also observed *C. (Ablepharipus) fukuiensis* nesting in old canes and similar tubes. Provisions were mainly Mycetophilidae, and the egg was laid as in *maculiclypeus*. Tsuneki records *C. (Neoblepharipus) amurensis* storing Trichoptera, Psocoptera, and Diptera. Also in Japan, Tsuneki noted *C. (Acanthocrabro) vagabundus* taking tipulids and other flies, and even tortricid moths for prey. The nests are constructed in old wood, are usually linear, and have few branches. The female bites off the fly's legs at the trochanter before dragging the prey, head first, into the burrow. The antennae and body remain paralyzed and quivering inside. Hamm and Richards (1926) reported the same behavior in Europe for this species.

Crossocerus (Cuphopterus) dimidiatus stores flies in preexisting cavities according to Peters (1973). He observed several females sharing one nest entrance and suggested that there was some division of labor. Evans (1964d) reported burrow sharing by two other species, *C. (Crossocerus) elongatulus* and *C. (Blepharipus) megacephalus* (as *leucostomoides*).

Crossocerus as a group seem to prefer arboreal habitats and avoid desert, range, and other xeric areas. They are seldom taken on flowers. Krombein (1951) and Leclercq (1961b) reported several species feeding on honeydew.

Checklist of *Crossocerus*

[Subgeneric assignment is indicated by code letters in parentheses at end of each citation as follows: *Ablepharipus* (Abl), *Acanthocrabro* (Aca), *Ainocrabro* (Ain), *Alicrabro* (Ali), *Apocrabro* (Apoc), *Apoides* (Apoi), *Blepharipus* (Ble), *Bnunius* (Bnu), *Crossocerus* (Cro), *Cuphopterus* (Cup), *Epicrossocerus* (Epi), *Eupliloides* (Eup), *Hoplocrabro* (Hop), *Microcrabro* (Mic), *Neoblepharipus* (Neo), *Nothocrabro* (Not), *Ornicrabro* (Orn), *Oxycrabro* (Oxy), *Paroxycrabro* (Par), *Pericrabro* (Per), *Stictoptila* (Sti).]

acanthophorus (Kohl), 1892 (*Crabro*); Europe (Oxy)
acephalus Leclercq, 1958; Zaire (Mic)
adhaesus (Kohl), 1915 (*Crabro*); sw. Iran, Cyprus (Cro)
aeta Pate, 1944; s. India, Malaysia, Java, Borneo, Philippines (Apoc)
    ssp. *loa* Pate, 1944; Sumatra, Taiwan
aino (Tsuneki), 1947 (*Crabro*); Japan (Ain)
albocollaris (Ashmead), 1904 (*Rhopalum*); s. India, Malaysia, Borneo, Philippines (Eup)
    elongatus Dudgeon in Nurse, 1903 (*Crabro*), nec Lepeletier and Brullé, 1834
    princesa Pate, 1948
    holtensis Leclercq, 1950
alticola Tsuneki, 1968; Taiwan (Apoi)
amurensis (Kohl), 1892 (*Crabro*); Japan; USSR: Siberia (Neo)
angelicus (Kincaid), 1900 (*Crabro*); U.S. and Canada w. of 100th meridian (Hop)
    vierecki H. Smith, 1908 (*Crabro*)
    boulderensis Rohwer, 1909 (*Crabro*)
    spinibuccus Viereck, 1909 (*Crabro*)
annandali (Bingham), 1908 (*Crabro*); India: Himalayas (Aca)
annulipes (Lepeletier and Brullé), 1834 (*Blepharipus*); Holarctic (Ble)
    gonager Lepeletier and Brulle, 1834
    nigritus Gimmerthal, 1836 (*Crabro*)
    ambiguus Dahlbom, 1842 (*Crabro*)
    capito Dahlbom, 1845 (*Crabro*), new synonymy by J. Leclercq
    parkeri Banks, 1921 (*Blepharipus*)
    davidsoni Sandhouse, 1938 (*Crabro*)
    ssp. *hokkaidensis* Tsuneki, 1954; Japan: Hokkaido
ardens (Cameron), 1890 (*Crabro*); e. India (Cro)
asiaticus Tsuneki, 1967; Manchuria (Hop)
assamensis (Cameron), 1902 (*Crabro*); e. India (Cup)
assimilis (F. Smith), 1856 (*Crabro*); Palearctic Region (Abl)
    affinis Wesmael, 1852 (*Crabro*), nec Rossi, 1792
    socius Thompson, 1870
    tirolensis Kohl, 1878
aswad (Nurse), 1902 (*Crabro*); s. India (Cro ?)
barbipes (Dahlbom), 1845 (*Crabro*); Palearctic Region (Ble)
    hirtipes A. Morawitz, 1866 (*Crabro*)
binotatus Lepeletier and Brullé, 1834; w. Palearctic Region (Cup)
    signatus Panzer, 1798 (*Crabro*), nec Olivier, 1792, ?*monstrosus* Dahlbom, 1845 (*Crabro*), tentative new synonymy by J. Leclercq
    confusus Schulz, 1906, new name for *signatus* Panzer
bispinosus Beaumont, 1967; Turkey (Cro)
bnun Tsuneki, 1971; Taiwan (Abl)
bougainvillae Pate, 1946; Solomon Is.: Bougainville, Russell (Eup)
brahmanus Leclercq, 1956; s. India (Cro)
brunniventris (Arnold), 1932 (*Thyreopus*); Rhodesia (Mic)
    ssp. *bifidus* Leclercq, 1958; Zaire
    ssp. *wittei* Leclercq, 1958; Zaire
    ssp. *bekiliensis* Arnold, 1944 (*Crabro*); Madagascar
bulawayoensis (Arnold), 1932 (*Thyreopus*); Zaire, Rhodesia (Ble)
burungaensis (Arnold), 1934 (*Thyreopus*); Zaire (Ble ?)
callani Pate, 1941; Trinidad (Ble)
capitalis Leclercq, 1956; Zaire (Mic)
capitosus (Shuckard), 1837 (*Crabro*); Palearctic Region (Ble)
    annulus Dahlbom, 1838 (*Crabro*)
    ssp. *yamato* Tsuneki, 1960; Japan: Hokkaido, Honshu
    ssp. *yeto* Tsuneki, 1960; Japan: Hokkaido
cetratus (Shuckard), 1837 (*Crabro*); Palearctic Region (Ble)
    vanderlindenii Dahlbom, 1838 (*Crabro*)
    dilatatus Herrich-Schaeffer, 1841
    inornatus Matsumura, 1911 (*Crabro*)
cheesmani Leclercq, 1955; New Guinea (Eup)
chromatipus Pate, 1944; w. U.S. (Cro)
    pictipes W. Fox, 1895 (*Crabro*), nec Herrich-Schaeffer, 1841
cinctipes (Provancher), 1882 (*Blepharipus*); n. U.S. and Canada, transcontinental (Ble)
    niger Provancher, 1888 (*Crabro*), nec Lepeletier and Brullé, 1834
    nigror W. Fox, 1895 (*Crabro*)
    nigrior W. Fox, 1896 (*Crabro*), emendation
    servus Dalla Torre, 1897
    cinctitarsis Ashmead, 1901 (*Stenocrabro*)
    columbiae Bradley, 1906 (*Blepharipus*)
    stygius Mickel, 1916 (*Thyreopus*)
    utensis Mickel, 1916 (*Thyreopus*)
cinxius (Dahlbom), 1838 (*Crabro*); n. Europe, Japan (Ble)
congener (Dahlbom), 1844 (*Crabro*); Europe (Abl)
decorus (W. Fox), 1895 (*Crabro*); sw. U.S., Mexico (Cro)
denticoxa (Bischoff), 1932 (*Crabro*); Germany, Czecho-

slovakia (Cro)
denticrus Herrich-Schaeffer, 1841; Europe, Algeria, Manchuria, Japan, Taiwan (Cro)
derivus Leclercq, 1968; Mexico: Guerrero (Cro)
diacanthus (Gussakovskij), 1930 (*Crabro*); India: Kashmir (Cro)
dimidiatus (Fabricius), 1781 (*Crabro*); Europe (Cup)
   subpunctatus Rossi, 1790 (*Crabro*)
   sexmaculatus Olivier, 1792 (*Crabro*)
   signatus Olivier, 1792 (*Crabro*)
   serripes Panzer, 1797 (*Crabro*)
   subpunctatus Rossi of Illiger, 1807 (*Crabro*)[3]
   notatus Illiger, 1807 (*Crabro*), new name for *Crabro maculatus* of Rossi, 1790
   pauperatus Lepeletier and Brullé, 1834 (*Blepharipus*)
   armipes Siebold, 1844 (*Crabro*)
  ssp. *sapporoensis* (Kohl), 1915 (*Crabro*); Japan
distinguendus (A. Morawitz), 1866 (*Crabro*); Europe, Algeria (Cro)
   mucronatus Thomson, 1870 (*Crabro*), nec Fabricius, 1793
distortus Leclercq, 1955; India (Ble)
domicola Tsuneki, 1971; Taiwan (Bnu)
elongatulus (Vander Linden), 1829 (*Crabro*); w. Palearctic s. of Arctic Circle, n. Africa, Argentina (Cro)
   affinis Lepeletier and Brullé, 1834
   annulatus Lepeletier and Brullé, 1834
   luteipalpis Lepeletier and Brullé, 1834
   morio Lepeletier and Brullé, 1834
   pallidipalpis Lepeletier and Brullé, 1834
   varipes Lepeletier and Brullé, 1834
   hyalinus Shuckard, 1837 (*Crabro*)
   obliquus Shuckard, 1837 (*Crabro*)
   propinquus Shuckard, 1837 (*Crabro*)
   proximus Shuckard, 1837 (*Crabro*)
   transversalis Shuckard, 1837 (*Crabro*)
   elongatus Lepeletier, 1845, lapsus for *elongatulus*
   brevis Eversmann, 1849 (*Crabro*)
   scutellaris F. Smith, 1851 (*Crabro*), nec Gimmerthal, 1836
   foveolatus Holmberg, 1903 (*Ischnolynthus*)
   berlandi Richards, 1928 (*Crabro*)
  ssp. *trinacrius* Beaumont, 1964; Sicily
emarginatus (Kohl), 1898 (*Crabro*); s. USSR: Ussuri to Japan (Cro)
   pacificus Gussakovskij, 1933 (*Crabro*)
eques (Nurse), 1902 (*Crabro*); India: Simla (Abl)
esau Beaumont, 1967; Turkey (Cro)
exiguus (Vander Linden), 1829 (*Crabro*); Palearctic Region (Cro)
   aphidum Lepeletier and Brullé, 1834
ezrae (Cameron), 1891 (*Crabro*); Mexico: Chilpancingo (Cro)
federationis Leclercq, 1961; Malaysia (Oxy)
fergusoni Pate, 1944; w. U.S. (Ble)
flavissimus Leclercq, 1973; Taiwan (Orn)
flavopictus (F. Smith), 1856 (*Crabro*); n. India, China, Japan, Sumatra, Java (Cup)
  ssp. *kansitakuanus* Tsuneki, 1971; Taiwan
fossuleus Leclercq, 1958; Zaire (Mic)
fukuiensis Tsuneki, 1960; Japan (Abl)
  ssp. *bambosicola* Tsuneki, 1970; Taiwan
gemblacencis Leclercq, 1968; Costa Rica (Mic)
gerardi Leclercq, 1956; India (Cro)
glabricornis (Arnold), 1926 (*Thyreopus*); S. Africa (Cup)
guerrerensis (Cameron), 1891 (*Crabro*); Mexico: Guerrero (Epi)
guichardi Leclercq, 1972; France (Neo)
hakusanus Tsuneki, 1954; Japan; Korea; Kurile Is., USSR: Sakhalin; Ryukyus (Ble)
harringtonii (W. Fox), 1895 (*Crabro*); s. centr. Canada, U.S. w. to New Mexico (Ble)
hedgreni (Kjellander), 1954 (*Crabro*); Sweden (Ble)
hewittii (Cameron), 1908 (*Crabro*); Borneo (Cro)
heydeni Kohl, 1880; Europe (Ble)
  ssp. *nipponius* Tsuneki, 1966; Japan
hingstoni Leclercq, 1950; Tibet (Cro)
hirashimai Tsuneki, 1966; Ryukyus (Ble)
hirtitibia (Arnold), 1944 (*Crabro*); Madagascar (Ble)
hiurai Tsuneki, 1966; Japan (Ble)
hospitalis Leclercq, 1961; Nigeria, Cameroon (Mic)
imitans (Kohl), 1915 (*Crabro*); Baltic and North Sea Is. and coasts (Cro)
impressifrons (F. Smith), 1856 (*Crabro*); U.S. and Canada e. of 100th mer. (Ble)
   tibialis Say, 1824 (*Crabro*), nec Oliver, 1792
   pusillus Harris, 1835 (*Crabro*), nomen nudum
   scutellatus Packard, 1867 (*Blepharipus*)
   harrisii Packard, 1867 (*Blepharipus*)
   tridentatus Rohwer, 1909 (*Crabro*), nec Fabricius, 1775
indonesiae Leclercq, 1961; Java (Abl)
insolens (W. Fox), 1895 (*Crabro*); U.S.: Colorado (Epi)
italicus Beaumont, 1964; Italy (subgenus ?)
jason (Cameron), 1891 (*Crabro*); Mexico: Guerrero
jubilans (Kohl), 1915 (*Crabro*); Cyprus, Asia (Cro)
   majuscula Kohl, 1915 (*Crabro*)
kamateensis Tsuneki, 1971; Taiwan (Abl)
kockensis Leclercq, 1950; Sumatra (Mic)
kohli (Bischoff), 1922 (*Crabro*); India: Kashmir; Mongolia; sw. USSR (Cro)
klapperichi Beaumont, 1963; Afghanistan (Cro)
krusemani Leclercq, 1950; Sumatra (Aca)
larutae Leclercq, 1961; Malaysia (Abl ?)
lentus (W. Fox), 1895 (*Crabro*); Canada and e. U.S. (Cro)
leontopolites Pate, 1946; Malaysia (Eup)
leucostoma (Linnaeus), 1758 (*Sphex*); n. Europe, Japan (Ble)

---

[3] *C. subpunctatus* has generally been regarded as authored by Illiger, 1807, but the name is attributable to Rossi.

carbonarius Dahlbom, 1838 (*Crabro*)
rugosus Herrich-Schaeffer, 1841
podagricus Vander Linden of Dahlbom, 1842 (*Crabro*)
melanarius Wesmael, 1852 (*Crabro*), new name for *podagricus*
lindbergi (Beaumont), 1954 (*Crabro*); Canary Is. (Cro)
lipatus Leclercq, 1961; Zaire (Mic)
lundbladi (Kjellander), 1954 (*Crabro*); Sweden (Cro)
maculiclypeus (W. Fox), 1895 (*Crabro*); Canada, w. and centr. U.S. (Cro)
    daeckei Rohwer, 1910 (*Thyreopus*)
maculipennis F. Smith, 1856; U.S.: Transition and Upper Austral zones (Sti)
    maculatus Lepeletier and Brullé, 1834 (*Blepharipus*), nec Fabricius, 1781 (*Crabro*)
    pictus F. Smith, 1856 (*Crabro*), nec Fabricius, 1793, new name for *maculatus*
    ventralis W. Fox, 1895 (*Crabro*)
    confertus W. Fox, 1895 (*Crabro*), new synonymy by Levin and Bohart
    canonicola Viereck, 1908 (*Crabro*), new synonymy by Levin and Bohart
    albertus Carter, 1925 (*Crabro*), new synonymy by Levin
maculitarsis (Cameron), 1891 (*Crabro*); Mexico: Guerrero; U.S.: Arizona, Texas (Neo)
malaisei (Gussakovskij), 1933 (*Crabro*); e. USSR: Ussuri (Cup)
megacephalus (Rossi), 1790 (*Crabro*); Palearctic Region (Ble)
    ? *bidens* Haliday, 1833 (*Crabro*), nec Schrank, 1802
    niger Lepeletier and Brullé, 1834
    rufipes Lepeletier and Brullé, 1834
    laeviceps F. Smith, 1856 (*Crabro*), new name for *rufipes*
    bison A. Costa, 1884 (*Crabro*), new synonymy by J. Leclercq
    ? *zaidamensis* Radoszkowski, 1887 (*Crabro*)
    leviceps Dalla Torre, 1897 (*Crabro*), emendation
    leucostomoides Richards, 1935 (*Crabro*)
    leucostomus of authors, not Linnaeus, 1758
melanius (Rohwer), 1911 (*Thyreopus*); N. America: Boreal Zone, s. Colorado to n. Mexico (Ble)
melanochilos Pate, 1944; Taiwan (Mic)
micromegas (Saussure), 1892 (*Crabro*); Madagascar (Mic)
mielatti Leclercq, 1961; Zaire (Mic)
minamikawai Tsuneki, 1966; Japan (Ble)
minimus (Packard), 1867 (*Blepharipus*); e. Canada, transcontinental in U.S. (Cro)
    propinquus W. Fox, 1895 (*Crabro*); nec Shuckard, 1837
    pelas Pate, 1944
minitulus (Arnold), 1944 (*Crabro*); Rhodesia (Ble ?)
morawitzi (Gussakovskij), 1952 (*Crabro*); sw. USSR: Tadzhik S.S.R. (Cro)
nelli (Viereck), 1904 (*Stenocrabro*); U.S.: New York, Pennsylvania (Abl)
    unicus Patton, 1879 (*Blepharipus*)

nigricornis (Provancher), 1888 (*Blepharipus*); U.S.: transcontinental (Ble)
nigritus (Lepeletier and Brullé), 1834 (*Blepharipus*); Europe, Algeria, Russia, Japan (Ble)
    pubescens Shuckard, 1837 (*Crabro*)
    diversipes Herrich-Schaeffer, 1841
    tischbeinii Dahlbom in Tischbein, 1850 (*Crabro*), nomen nudum
    inermis Thomson, 1870 (*Crabro*)
    melanogaster Kohl, 1879 (*Crabro*)
    sambucicola C. Verhoeff, 1891 (*Crabro*)
nikkoensis Tsuneki and Tanaka, 1955; Japan (Ble)
nitidicorpus Tsuneki, 1968, Taiwan (Ble)
nitidiventris (W. Fox), 1892 (*Crabro*); e. U.S. (Not)
odontophorus (Cameron), 1890 (*Crabro*); e. India (Cup)
onoi (Yasumatsu), 1939 (*Crabro*); s. Manchuria (Cro)
opacifrons (Tsuneki), 1947 (*Crabro*); Japan: Hokkaido (Cro)
ornatipes (Turner), 1918 (*Rhopalum*); n. Nigeria (Mic)
ovalis Lepeletier and Brullé, 1834; w. Palearctic Region (Cro)
    exiguus Vander Linden of Shuckard, 1837 (*Crabro*)
    punctus Zetterstedt, 1838 (*Crabro*)
    anxius Wesmael, 1852 (*Crabro*)
    shuckardi F. Smith, 1856 (*Crabro*), nec Dahlbom, 1838; new name for *exiguus* of Shuckard
    ovatus Schulz, 1906 (*Crabro*)
palmipes (Linnaeus), 1767 (*Sphex*); Europe and British Is. (Cro)
    palmarius Schreber, 1784 (*Sphex*)
    scutatus Fabricius, 1787 (*Crabro*)
    ornatus Lepeletier and Brullé, 1834
    scutellaris Gimmerthal, 1836 (*Crabro*)
    gracilis Eversmann, 1849 (*Crabro*)
    decoratus F. Smith, 1856 (*Crabro*), new name for *ornatus*
    ssp. *chosenensis* Tsuneki, 1957; Korea
parcorum Leclercq, 1958; Zaire (Ble)
pauxillus (Gussakovskij), 1933 (*Crabro*); e. USSR: Ussuri; Japan (Ble)
pavlovskii (Gussakovskij), 1952 (*Crabro*); sw. USSR, Cyprus (Cro)
?*perpusillus* (Walker), 1871 (*Crabro*); Egypt
phaeochilos Pate, 1944; Mexico: Jalisco (Mic)
pignatus Leclercq, 1968; Peru, Bolivia (Neo)
planifemur Krombein, 1952; U.S.: Maryland, W. Virginia (Cro)
planipes (W. Fox), 1895 (*Crabro*); transcontinental in U.S. and Canada (Cro)
    incavus W. Fox, 1895 (*Crabro*)
    cockerelli Rohwer, 1908 (*Crabro*)
pleuracutus Leclercq, 1973; Taiwan (Cro)
podagricus (Vander Linden), 1829 (*Crabro*); Palearctic Region (Abl)
    vicinus Dahlbom, 1842
    snoflaki Zavadil in Zavadil and Šnoflak, 1948 (*Crabro*)

punctatus Šnoflak in Zavadil and Šnoflak, 1948 (*Crabro*)
porexus Leclercq, 1968; Brazil: Santa Catarina (Neo)
potosus Leclercq, 1968; Bolivia (Neo)
pseudopalmarius (Gussakovskij), 1933 (*Crabro*); e. Palearctic Region (Cro)
pullulus (A. Morawitz), 1866 (*Crabro*); Siberia (Cro)
pusanoides Leclercq, 1963; India (Cro)
pusanus Leclercq, 1956; India (Cro)
pyrrhus Leclercq, 1956; India (Apoc)
quadrimaculatus (Fabricius), 1793 (*Crabro*); Palearctic e. to Siberia, Mongolia (Hop)
    quadripunctatus Fabricius, 1793 (*Crabro*)
    murorum Latreille, 1804-1805 (*Crabro*)
    levipes Vander Linden, 1829 (*Crabro*)
    bimaculatus Lepeletier and Brullé, 1834
    laevipes Lepeletier and Brullé, 1834, emendation
    quinquemacutatus Dahlbom, 1838 (*Crabro*)
    rotundarius Dahlbom, 1838 (*Crabro*)
quinquedentatus Tsuneki, 1971; Taiwan (Neo)
raui Rohwer, 1923; U.S.: Missouri (Epi)
rectangularis (Gussakovskij), 1952 (*Crabro*); sw. USSR: Tadzhik S.S.R. (Cro)
repositus (Arnold), 1944 (*Crabro*); Rhodesia (Mic)
republicus Leclercq, 1963; s. India (Mic)
    denticornis Gussakovskij, 1933 (*Crabro*), nec F. Smith, 1879
rimatus Leclercq, 1963; s. India (Mic)
riparium (Arnold), 1926 (*Thyreopus*); s. Rhodesia (Mic)
    nemoralis Arnold, 1926 (*Thyreopus*)
ruandensis (Arnold), 1932 (*Rhopalum*); Zaire (subgenus ?)
rufiventris Tsuneki, 1968; Taiwan (Ali)
ruwenzoriensis (Arnold), 1926 (*Thryeopus*); Uganda (Cup)
sciaphillus Leclercq, 1961; Zaire (Ble)
scutellifer (Dalla Torre), 1897 (*Crabro*); w. U.S. (Cro)
    scutellatus Say, 1824 (*Crabro*), nec Scheven, 1781
    eriogoni Rohwer, 1908 (*Crabro*)
segregatus Leclercq, 1958; Zaire (Cro)
senonus Leclercq, 1961; Zaire (Mic)
shibuyai (Iwata), 1934 (*Crabro*); Japan (Abl)
similis (W. Fox), 1895 (*Crabro*); e. U.S. and e. Canada (Cro)
    flavitrochantericus Viereck, 1904 (*Stenocrabro*)
simlaensis (Nurse), 1902 (*Crabro*); India, Java (Cro)
sinicus Leclercq, 1954; China (Ble)
    chinensis Gussakovskij, 1936 (*Crabro*), nec Sickmann 1894
slimmatus Leclercq, 1963; Philippines: Luzon (Cro)
sociabilis (Arnold), 1932 (*Thyreopus*); Rhodesia, Madagascar (Per)
sotirus Leclercq, 1963; Philippines: Mindanao (Par)
spangleri Krombein, 1962; U.S.: Maryland, N. Carolina (Cro)
spilaspis (Cameron), 1907 (*Dasyproctus*); Borneo (Eup)
spinigerus (Cameron), 1904 (*Rhopalum*); Mexico (Cro)
stictochilos Pate, 1944; ne. U.S. (Ble)
strangulatus (Bischoff), 1930 (*Crabro*); sw. USSR, Mongolia, n. India (Cro)
stricklandi Pate, 1944; N. America: Boreal Zone of Alberta to centr. Colorado (Ble)
styrius (Kohl), 1892 (*Crabro*); Great Britain, Germany, Switzerland (Ble)
subulatus (Dahlbom), 1845 (*Crabro*); e. Palearctic, Ural Mts. to Japan (Cup)
    suzukii Matsumura, 1912 (*Crabro*)
    monstrosus of authors
sugihari (Iwata), 1938 (*Crabro*); Kurile Is. and Sakhalin Is. (Ble)
sulcus (W. Fox), 1895 (*Crabro*); e. U.S. (Cro)
    plesius Rohwer, 1912 (*Stenocrabro*)
surusumi Tsuneki, 1971; Taiwan (Cup)
sutshanicus (Gussakovskij), 1933 (*Crabro*); e. USSR: Ussuri (Ble)
taiwanus Tsuneki, 1968; Taiwan (Abl)
takasago Tsuneki, 1957; Korea, Taiwan (Ble)
tanakai Tsuneki, 1954; Japan, Kurile Is., Sakhalin Is. (Ble)
tanoi Tsuneki, 1968; Taiwan (Ble)
tara Beaumont, 1967; Turkey (Ble)
tarsalis (W. Fox), 1895 (*Crabro*); U.S.: transcontinental (Ble)
tarsatus (Shuckard), 1837 (*Crabro*); Europe (Cro)
    palmatus De Stefani, 1884 (*Crabro*), nec Panzer, 1797
    palmipes auctt., nec Linnaeus, 1767
    ssp. *richardsi* (Beaumont), 1950 (*Crabro*); Algeria
taxus Leclercq, 1956; India (Cro)
temporalis (Gussakovskij), 1952 (*Crabro*); sw. USSR: Tadzhik S.S.R. (Ble)
tersus Kazenas, 1971; sw. USSR: Kazakh S.S.R. (Cro)
toledensis Leclercq, 1971; Spain (Cro)
traductor (Nurse), 1902 (*Crabro*); e. India (Cup)
tropicalis (Arnold), 1947 (*Crabro*); Tanzania (Cup)
tsuifengensis Tsuneki, 1968; Taiwan (Abl)
turneri (Arnold), 1927 (*Thyreopus*); S. Africa (Ble)
tyunzendzianus Tsuneki, 1955; Japan (Ble)
uchidai (Tsuneki), 1947 (*Crabro*); Japan: Hokkaido (Cro)
universitatis (Rohwer), 1909 (*Crabro*); U.S.: Colorado (Epi)
ursidus Leclercq, 1956; India (Apoc)
vagabundus (Panzer), 1798 (*Crabro*); Palearctic Region (Aca)
    J. Leclercq
    quinquemaculatus Lepeletier and Brullé, 1834 (*Blepharipus*)
    lefebvrei Lepeletier and Brullé, 1834
    fasciatus A. Costa, 1871 (*Crabro*)
    esakii Yasumatsu, 1942 (*Crabro*)
    ssp. *yamatonicus* (Tsuneki), 1948 (*Crabro*); Japan
    ssp. *koreanus* Tsuneki, 1957; Korea
varius Lepeletier and Brullé, 1834; Palearctic Region (Cro)
    varus Lepeletier and Brullé, 1834; original orthography, nec (Panzer), 1799

pusillus Lepeletier and Brullé, 1834
striatulus Lepeletier and Brullé, 1834, p. 775
spinipectus Shuckard, 1837 (*Crabro*), in part
varius Lepeletier, 1845, emendation
striatus Lepeletier, 1845, lapsus or emendation of *striatulus*
intricatus F. Smith, 1856 (*Crabro*), new name for *striatulus* (as *striatus*)
lepeletieri F. Smith, 1856 (*Crabro*), new name for *striatulus* (as *striatus*)
?*varus* (Panzer), 1799 (*Crabro*); Austria
verhoeffi Tsuneki, 1967; Manchuria (Ble)
viennensis Leclercq, 1968; Costa Rica (Mic)
walkeri (Shuckard), 1837 (*Crabro*); Palearctic Region (Ble)
   geniculatus Shuckard, 1837 (*Crabro*), nec Olivier, 1792
   scaposus "Zetterstedt", in Dahlbom, 1844, nomen nudum
   clypearis Schenck, 1857 (*Crabro*)
   cloevorax Nielson, 1900 (*Coelocrabro*)
wesmaeli (Vander Linden), 1829 (*Crabro*); Palearctic Region (Cro)
   ziegleri Lepeletier and Brullé, 1834
   maurus Lepeletier and Brullé, 1834 (*Ceratocolus*),
wickhamii (Ashmead), 1899 (*Dolichocrabro*); U.S.: transcontinental (Ble)
   ater Cresson, 1865 (*Crabro*), nec Olivier, 1792
   wickhami Ashmead, 1902 (*Dolichocrabro*)
   pammelas Pate, 1944
xanthochilos Pate, 1944; se. U.S. (Mic)
xanthognathus (Rohwer), 1911 (*Thyreopus*); Mexico (Cro)
yanoi (Tsuneki), 1947 (*Crabro*); Japan (Cup)
yasumatsui (Tsuneki), 1947 (*Crabro*); Japan (Cro)
   ssp. *mongolensis* Tsuneki, 1972; Mongolia
yerburii (Cameron), 1898 (*Crabro*); Sri Lanka (Cro)

Genus Tracheliodes A. Morawitz

Generic diagnosis: Eyes sparsely covered with very short, erect hair, inner orbits nearly parallel or convergent below; frons slightly concave but without distinct scapal basin, simple or with small spinoid tubercle above antennal sockets, ecarinate above; orbital foveae present in female, present or absent in male; ocellar triangle equilateral or broader than long; gena simple; occipital carina incomplete, separated from hypostomal carina; antennal sockets subcontiguous or separated from each other, well separated from inner orbits; male antenna with 13 articles; scape ecarinate; male flagellum simple or with tyli beneath, with or without ventral hair fringe; palpal formula 6-3; mandible apically bidentate (fig. 129), externoventral and inner margins simple; pronotal collar ecarinate, rounded anteriorly, rounded or sharply angulate laterally, notched medially; scutum, axillae, scutellum, and metanotum simple; postspiracular carina continuous with omaulus; acetabular carina, sternaulus, verticaulus, hypersternaulus, and mesopleuraulus absent; propodeum finely sculptured, enclosure absent or very weakly indicated, lateral propodeal carina absent; foreleg with trochanter slender, elongate; hindtibia clavate; recurrent vein joining submarginal cell at or near middle; jugal lobe as long as submedian cell; gaster sessile or subpedunculate (fig. 129), male without pygidial plate, that of female elongate, narrowed, either flat or concave apically.

Geographic range: Holarctic Region. Two species are found in central and southern Europe and the Mediterranean region and three in the western United States. In addition, three fossil species, two in Baltic amber, and one in Miocene shales from Colorado have been referred to this genus.

Systematics: Tracheliodes may be distinguished from other crabronins by the elongate foretrochanter (fig. 129) and the 6-3 palpal formula. Furthermore, the broad frons has the antennal sockets well separated from the inner orbits which are only moderately convergent below.

Kohl (1915) reviewed the systematics, and this was amplified by Pate (1942b). Pate's hypothesis of a close geographic-systematic relationship on a species and subspecies basis between *Tracheliodes* and their ant prey is doubtful. More evidence is obviously needed.

Biology: Reports on habits of *Tracheliodes* are available for two European species (*curvitarsis, quinquenotatus*) and two North American forms (*foveolineatus, hicksi*). Kohl (1915) reviewed the European studies, and a summary of present knowledge except for *foveolineatus* was given by Pate (1942b). *T. quinquenotatus* nests in various ground situations as well as in old walls. In *curvitarsis* the nests are in abandoned wood-boring beetle holes in trees. *T. foveolineatus* has been found to nest in plant stems (Parker and Bohart, 1966). So far as known, the prey in California and Nevada are always worker ants as follows: *quinquenotatus* takes *Tapinoma erraticum* Latreille and *Tapinoma nigerrimum* Nylander, *curvitarsis* uses *Liometopum microcephalum* Panzer, *hicksi* captures *Liometopum* sp., and *foveolineatus* provisions with *Liometopum occidentale luctuosum* Wheeler.

Grandi (1961) summarized his previous findings on *Tracheliodes quinquenotatus*. Burrows in hard sand may extend about 4 cm vertically but up to 30 cm horizontally. There is some evidence that females may use previously existing burrows, since with their strongly narrowed pygidium they are not well adapted for digging. Each cell is provisioned with a tightly packed mass of up to 94 paralyzed *Tapinoma* ant workers. The egg is attached behind the forecoxae of an ant positioned about a third of the way from the bottom of the cell. The mature larva forms a silken cocoon to which ant fragments are attached. The species is multivoltine, at least in warmer climates. Grandi recorded parasites in Italy as *Miltogramma punctatum* Meigen (Sarcophagidae), *Hammomyia sociata* Meigen (Anthomyiidae), *Smicromyrme rufipes* Fabricius (Mutillidae), and *Chrysis*

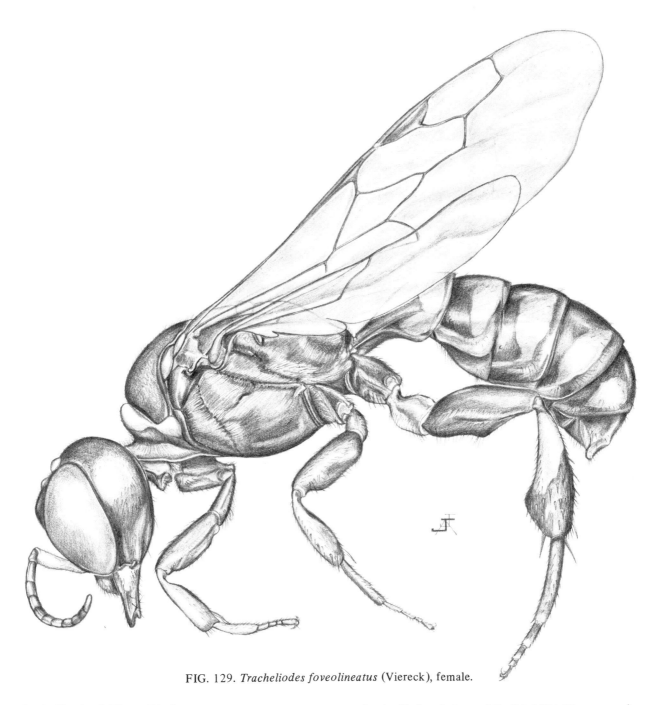

FIG. 129. *Tracheliodes foveolineatus* (Viereck), female.

leachi Shuckard (Chrysididae).

Checklist of *Tracheliodes*

amu Pate, 1942; U.S.: New Mexico, Arizona
curvitarsis (Herrich-Schaeffer), 1841(*Crossocerus*);
  Austria, Italy, Germany
    megerlei Dahlbom, 1845 (*Crabro*)
    filigranus A. Costa, 1871 (*Crabro*)
    rondani "Spinola" A. Costa, 1871 (*Blepharipus*),
      nomen nudum
foveolineatus (Viereck), 1909 (*Crabro*); U.S.: California
  and Oregon to Colorado
hicksi Sandhouse, 1936; U.S.: Colorado, Arizona
quinquenotatus (Jurine), 1807 (*Crabro*); Mediterranean
  area
    luteicollis Lepeletier and Brullé, 1834 (*Crossocerus*)
    trochantericus Herrich-Schaeffer, 1841 (*Ceratocolus*)
    bucephalus F. Smith, 1856 (*Crabro*)
    politus A. Costa, 1871 (*Blepharipus*), nomen nudum
    festivus Marquet, 1881 (*Crossocerus*)
    herinaceus Gribodo, 1894 (*Crabro*)
    formicarius Ferton, 1896 (*Fertonius*)
    quinquesignatus "Jurine" Bignell, 1900 (*Crabro*)

Fossil *Tracheliodes*

mortuellus Cockerell, 1905; Florissant, Colorado Shale
  (Miocene)
succinalis (Cockerell), 1909 (*Crabro*); Baltic Amber
tornquisti (Cockerell), 1909 (*Crabro*); Baltic Amber

Genus Crabro Fabricius

Generic description: Eyes bare, inner orbits convergent below (fig. 127 C); scapal basin simple; orbital foveae usually distinct, sometimes definitely margined, ocellar triangle broad; vertex simple or with a fine longitudinal line, its punctation similar to that in front of ocellar triangle; gena simple or extensively depressed behind mandible; occipital carina incomplete ventrally, sometimes nearly reaching hypostomal carina; antennal sockets usually close together and to inner orbits, rarely separated from each other by a socket diameter (*thyreophorus*); male antenna with 13 articles; scape usually ecarinate; male flagellum simple to greatly broadened, with or without a ventral hair fringe; palpal formula 6-4; mandible apically simple or bidentate, externo-ventral margin unnotched, inner margin sometimes with a small tooth; pronotal collar usually ecarinate except laterally and toward humeral angles; scutum simple, notauli short but often somewhat ridged anteriorly; prescutellar sulcus foveate; metanotum simple; omaulus present, joining a strong postspiracular carina above and usually a weak carina extending beneath anterior part of pronotal lobe; acetabular carina absent, sternaulus sometimes marked by a foveate groove; scrobal sulcus, hypersternaulus and mesopleuraulus absent; verticaulus absent or present only as a tubercle connected to a ridge from midcoxal cavity; propodeal sculpture often coarse, enclosure not polished, usually not well defined; lateral propodeal carina present, at least near petiole socket; legs simple or more often highly modified; male foretibia frequently shieldlike (fig. 120), male forefemur often with an assortment of projections, male midtibia rarely without a spur (*chalybeus*); recurrent vein ending beyond middle but not beyond distal third of submarginal cell whose apicoventral veinlet is at least as long as distal veinlet (fig. 122 F); marginal cell truncate distally; jugal lobe much shorter than hindwing submedian cell (fig. 122 B); gaster sessile; pygidial plate of male sometimes defined, that of female flat and subtriangular.

Geographic range: The genus is essentially Holarctic. However, of the nearly 70 listed species a few in the *tumidus* group are found in Costa Rica and subtropical parts of Mexico.

Systematics: As indicated in the key to genera, *Crabro* is distinguished from *Lindenius* chiefly by the short jugal lobe (see fig. 122 E, F). Actually, the two genera are not very similar and would rarely be confused, since *Crabro* is nearly always much larger, and most species are extensively pale marked. A relationship to *Ectemnius* and *Lestica* seems evident, but the more basal termination of the recurrent vein in *Crabro* along with the absence of an acetabular carina will permit separation. *Crossocerus* may also be confused with *Crabro*, but the nearly equilateral ocellar triangle of the former distinguishes it. In the majority of its species the males of *Crabro* are unique in the possession of a tibial shield (fig. 120). Similar shields occur in males of other crabronins, but they are either extensions of the femur (*Ectemnius cyanauges*) or the tarsus (some *Crossocerus*, *Lestica clypeata*). The ornamentation of the tibial shield, together with its shape, seem to be specific.

Kohl (1915) presented a key to species groups or "Untergruppen" based on males. Eight of these groups were identified by Latin names, and most subsequent authors have treated these as subgenera. Kohl's key contains a number of errors, and in the light of present-day knowledge some of his categories are not phylogenetically sound. In any case we prefer not to recognize subgenera. The following is a modified key to species groups in which *Synothyreopus* in the usual sense has been divided into three groups, represented by *tumidus*, *advenas*, and *thyreopus*. Since no specimens have been studied in the *occultus*, *chalybeus*, or *filiformis* groups, they are included on the basis of literature.

Key to groups of *Crabro* based on males

1. Forefemur and foretibia not unusually broadened; mandible bifid at apex 2
   Forefemur and/or foretibia unusually broadened, or if not, mandible is simple at apex 5
2. Pygidial plate well developed; flagellum without a noticeable ventral hair fringe 3
   Pygidial plate absent; flagellum with or without a noticeable ventral hair fringe 4
3. Midtibial spur present; foretarsus normal; abdomen not metallic; five species (*Agnosicrabro* Pate) .... *C. occultus* group
   Midtibial spur absent; foretarsus somewhat broadened; abdomen with a metallic sheen; one species (*Dyscolocrabro* Pate) .... *C. chalybeus* group
4. Flagellum with a noticeable ventral hair fringe; six species (*Paranothyreus* Ashmead) .... *C. hilaris* group
   Flagellum without a noticeable ventral hair fringe; four species (*Anothyreus* Dahlbom) .... *C. lapponicus* group
5. Mandible simple at apex 6
   Mandible bifid at apex 7
6. Flagellum fringed beneath; two species (*Parathyreopus* Pate) .... *C. filiformis* group
   Flagellum not fringed beneath; eight species (*Synothyreopus* Ashmead) .... *C. tumidus* group
7. Forefemur curved, strongly produced basally, ending in one or two points; foretibia at most with a very small shield which is not or hardly broader than tibia proper; four species (*Hemithyreopus* Pate) ...... *C. loewi* group
   Forefemur without a long basal projection ending in one or two points; foretibia with a rather large shield, or at least one which is considerably broader than tibia proper (fig. 120) 8
8. Tarsomere V with an inner lateral process which may sometimes be small (fig. 120); flagellum usually with prominent curled hairs beneath; pygidial plate absent; 31 species (*Crabro* Fabricius) .... *C. cribrarius* group
   Tarsomere V essentially symmetrical; flagellum with at most short, straight hairs beneath 9

9. Pygidial plate present; sculpture of upper frons coriaceous but not microridged; one named and one unnamed species .... *C. thyreophorus* group
Pygidial plate absent; upper frons often striatopunctate; nine species .... *C. advena* group

We have not seen specimens of a number of Eurasian species assigned above to the *cribrarius* group (cr). Furthermore, some of these were based on females, and others were quite inadequately described. Therefore, some eventually may be relocated in the *advenus* group (ad).

In the second part of couplet 1 in the above key there is an indication that simple mandibles may be associated with unbroadened legs. This refers to the male of an undescribed species of the *tumidus* group from Mexico in the University of California at Davis collection. It has relatively ordinary leg structure without a tibial shield.

At the species rather than group level, characters of importance, in addition to the obvious markings, punctation, and pubescence, are antennal shape in males; microridging of the frons, scutum, and mesopleuron; details of male leg structure, development of the sternaulus as well as the tubercle in front of the midcoxa; and the degree of propodeal ridging or reticulation.

Systematic works of special value in the past 80 years have been those of W. Fox (1895) on American species, Dalla Torre (1897) on overall synonymy, Kohl (1915) on *Crabro* in general, and Beaumont (1964b) on central European species.

Biology: Information is available for more than a dozen species including representatives of six species groups: *cribrarius* group (*Crabro* s.s.), *tumidus* group (*Synothyreopus*), *lapponicus* group (*Anothyreus*), *advena* group, *loewi* group (*Hemithyreopus*), and *hilaris* group (*Paranothyreus*). Behavioral differences among them appear to be more specific than supraspecific. Leclercq (1954a) gave a summary of habitat and prey preferences reported up to that time. Subsequent findings have all been in close agreement.

As a general statement it can be said that *Crabro* nest in the ground and provision with various species of flies. The prey are carried to the nest venter up and secured by the midlegs of the wasp. The burrows extend into soil, usually of a sandy nature, for 6 to 45 cm and end in a few to as many as 15 cells at the ends of terminal branches. Tumulus normally accumulates around the burrow entrance which is not closed while the wasp is away. The prey are killed or completely paralyzed, and an egg is laid on the throat of the first fly in the cell. Prey accumulate at the bottom of the burrow until enough are present to fill a cell, which is then constructed. Prey are placed in the cell venter up at the beginning, but this may vary later on. Common parasites are miltogrammine Sarcophagidae.

Differences among species have been discussed by Evans (1960b). These are: Gregarious tendencies — some species may be solitary but others, such as *monticola*, may have aggregations of up to 100 nests. The nesting site may be in sand or hard-packed dirt on the level or, as in *argusinus*, in sloping sand banks. The tunnel may average 8 cm in length (*argusinus*) or up to about 30 cm (*monticola*). Cells per nest may be few (*argusinus*: 1-5), several (*advenus*: 2-8), or many (*monticola*: 11-15). Cell size varies with the size of the wasp, with a length of 9 mm in *argusinus* to 17 mm in *monticola*. Type of prey seems to be of some specific value. *C. argusinus* took Dolichopodidae and Ephydridae; *advenus* collected mainly Muscoidea; and *monticola* provisioned chiefly with Tabanidae. These groups of flies appear to be favored ones for other species, judging by the literature survey of Leclercq (1954a). Other groups which are used frequently are Rhagionidae (by *C. lapponicus*), Therevidae, Stratiomyidae, and Syrphidae.

In addition to the papers cited above, others of special interest are Hamm and Richards (1926) on *scutellatus*, *peltarius*, and *cribrarius;* Rau and Rau (1918) on *cingulatus;* Adlerz (1903) on *lapponicus;* Merisuo (1932) on *loewi;* and Evans (1964e) on *venator*.

Checklist of *Crabro*

[Species groups are indicated in parentheses at the end of each citation; *C. advena* group (ad), *C. chalybeus* group (ch), *C. cribrarius* group (cr), *C. filiformis* group (fi), *C. hilaris* group (hi), *C. lapponicus* group (la), *C. loewi* group (lo), *C. occultus* group (oc), *C. thyreophorus* group (th), *C. tumidus* group (tu).]

advena F. Smith, 1856; e. U.S. and Canada (ad)
    pegasus Harris, 1835, nomen nudum
    succinctus Cresson, 1865
    pegasus Packard, 1867 (*Thyreopus*)
    signifer Packard, 1867 (*Thyreopus*), new synonymy by R. Bohart
    elegans Provancher, 1883 (*Thyreopus*)
    discretus W. Fox, 1895, new synonymy by R. Bohart
aequalis W. Fox, 1895; U.S. e. of 100th merid. (hi)
    rugicollis Viereck, 1904 (*Paranothyreus*)
    knoxensis Mickel, 1916 (*Thyreopus*), new synonymy by R. Bohart
alpestris Cameron, 1891; Mexico (tu)
alpinus Imhoff, 1863; Europe (cr)
    interruptus Lepeletier and Brullé, 1834, p. 755 (*Thyreopus*), nec *Crabro interruptus* Fabricius, 1787 (now in *Palarus*)
    lactarius Chevrier, 1867 (*Thyreopus*)
    interruptulus Dalla Torre, 1897, new name for *interruptus* L. & B.
altaicus F. Morawitz, 1892; Mongolia, mts. of centr. Asia (cr)
alticola Cameron, 1891; Mexico (tu)
altigena Dalla Torre, 1897; sw. USSR: Tadzhik S.S.R. (cr)
    alticola F. Morawitz, 1893, nec Cameron, 1891
argusinus R. Bohart; U.S. and Canada, new name for *argus* Packard (cr)
    argus Harris, 1835, nomen nudum
    argus Packard, 1867 (*Thyreopus*), nec *Sphex argus*

Christ, 1791 which = *Crabro cribrarius*
biguttatus F. Morawitz, 1892; Siberia (la)
brachycarpae Rohwer, 1908; U.S.: Colorado (ad)
   gillettei Rohwer, 1908, new synonymy by R. Bohart
caspicus (F. Morawitz), 1888 (*Blepharipus*); sw. USSR: Transcaspia (lo)
chalybeus Kohl, 1915; Japan (ch)
cingulatus (Packard), 1867 (*Thyreopus*); centr. and e. U.S., Mexico (hi)
   clariconis Viereck, 1906, new synonymy by R. Miller
cognatus W. Fox, 1895; U.S. except Pacific states (hi)
comberi Leclercq, 1950; n. India (oc)
conspicuus Cresson, 1865; w. U.S. (cr)
   medius W. Fox, 1895
costaricensis Cameron, 1891; Costa Rica (tu)
cribrarius (Linnaeus), 1758 (*Vespa*); Palearctic Region (cr)
   patellarius Schreber, 1784 (*Sphex*)
   argus Christ, 1791 (*Sphex*)
   longus Christ, 1791 (*Sphex*)
   lunatus Christ, 1791 (*Sphex*), nec Fabricius, 1775
   palmatus Panzer, 1797
   cribratus Eversmann, 1849, lapsus or emendation
   inornatus Mocsáry, 1901
   hypotheticus Kokujev, 1927
cribrellifer (Packard), 1867 (*Thyreopus*); e. U.S. (cr)
   sinuatus Provancher, 1883 (*Thyreopus*), nec Fabricius, 1804
   provancheri W. Fox, 1895; new name for *sinuatus*
femoralis F. Morawitz, 1891; Siberia, Mongolia (lo)
filiformis Radoszkowski, 1877; sw. USSR (fi)
flavoniger Dutt, 1921; India (la)
florissantensis Rohwer, 1909; U.S.: Colorado and Wyoming to Oregon and Washington (ad)
fratellus Kohl, 1915; Mongolia (oc)
funestus Kohl, 1915; Sikkim, Tibet (cr)
gulmargensis Nurse, 1903; n. India (cr)
gussakovskiji R. Bohart; Mongolia, new name for *tricolor* Gussakovskij (fi)
   tricolor Gussakovskij, 1938, nec F. Smith, 1856
henrici Krombein, 1951; e. U.S. (ad)
   vierecki Rohwer, 1910 (*Thyreopus*), nec H. Smith, 1908
hilaris F. Smith, 1856; centr. and e. U.S. (hi)
  ssp. *rufibasis* (Banks), 1921 (*Thyreopus*); U.S.: Florida, new status by R. Bohart
hispidus W. Fox, 1895; U.S.: Pacific Coast States; British Colombia (ad)
ingricus (F. Morawitz), 1888 (*Thyreopus*); Europe, ne. USSR (th)
juniatae Krombein, 1938; e. U.S. (cr)
korbi (Kohl), 1883 (*Thyreopus*); sw. Europe (ad)
koreanus Tsuneki, 1947; Korea (cr)
lacteipennis Rohwer, 1909; U.S.: Colorado, Arizona, New Mexico (tu)
lapponicus Zetterstedt, 1838; e. Europe (la)
largior W. Fox, 1895; U.S.: transcontinental (cr)
latipes F. Smith, 1856; U.S., Canada, Alaska (cr)
   gryphus Harris, 1835, nomen nudum
   vicinus Cresson, 1865
   coloradensis Packard, 1867 (*Thyreopus*)
   elongatus Provancher, 1888 (*Thyreopus*), new synonymy by R. Bohart
   canadensis Dalla Torre, 1897, new name for *elongatus*
   viciniformis Viereck, 1907, new synonymy by R. Bohart
   pratus Carter, 1925, new synonymy by R. Bohart
loewi Dahlbom, 1845; Europe, sw. USSR (lo)
   loeuvi Thomson, 1870, lapsus
   lowei Dalla Torre, 1897, emendation
   jaroschewskyi F. Morawitz, 1892
maeklini A. Morawitz, 1866; Siberia, Mongolia, Lappland (la)
malyshevi Ahrens, 1933; USSR (lo)
mocsaryi Kohl, 1915; sw. & se. USSR (cr)
mongolicus Tsuneki, 1958; Mongolia (oc)
monticola (Packard), 1867 (*Thyreopus*); e. U.S., e. Canada (cr)
occultus Fabricius, 1804; Algeria, Tunisia (oc)
   numidicus Gribodo, 1896
okabei Yasumatsu, 1944; China (cr)
pallidus W. Fox, 1895; w. U.S. (cr)
peltarius (Schreber), 1784 (*Sphex*); Palearctic Region (cr)
   patellatus Panzer, 1797
   dentipes Panzer, 1797
   mediatus Fabricius, 1798
  ssp. *bilbaoensis* Leclercq, 1960; Spain
peltatus Fabricius, 1793; Europe (cr)
   flavipes Lepeletier and Brullé, 1834 (*Blepharipus*), nec Lepeletier and Brullé, 1834, p. 699
   luteipes F. Smith, 1856, new name for *flavipes*
   rhaeticus Aichinger and Kriechbaumer, 1870
peltista Kohl, 1888; U.S.: Arizona; Mexico to Nicaragua (tu)
   incertus W. Fox, 1895
pleuralis W. Fox, 1895; w. U.S. (cr)
?*pubescens* (Gmelin), 1790 (*Vespa*); Europe
pugillator A. Costa, 1871; s. Europe, sw. Asia (cr)
scutellatus (Scheven), 1781 (*Sphex*); Palearctic Region (cr)
   scutularius Schreber, 1784 (*Sphex*)
  ? *quatuormaculatus* Christ, 1791 (*Sphex*)
   pterotus Panzer, 1801
  ? *reticulatus* Lepeletier and Brullé, 1834 (*Ceratoculus*) (sic)
   petrosus Eversmann, 1849, lapsus or emendation of *pterotus*
sibiricus A. Morawitz, 1866; Siberia, Mongolia, China (cr)
signaticrus F. Morawitz, 1893; sw. USSR: Turkmen S.S.R. (cr)
snowii W. Fox, 1896; e. U.S. (hi)
tenuiglossa Packard, 1866; centr. and e. U.S.: Canada: Ontario, Alberta (tu)
   discifer Packard, 1867 (*Thyreopus*)
tenuis W. Fox, 1895; U.S. from Michigan to Colorado

and Washington; Canada: Alberta (cr)
thyreophorus Kohl, 1888; U.S.: Nevada, California, Oregon (th)
tuberculiger Kohl, 1915; sw. USSR: Turkmen S.S.R. (cr)
tumidus (Packard), 1867 (*Thyreopus*); centr. and e. U.S. (tu)
uljanini Radoszkowski, 1897; sw. USSR (cr)
ussuriensis Gussakovskij, 1933; USSR: Ussuri (cr)
venator (Rohwer), 1911 (*Thyreopus*); Mexico (tu)
vernalis (Packard), 1867 (*Thyreopus*); N. Amer.: transcontinental (ad)
  bruneri Mickel, 1916; new synonymy by R. Bohart
villosus W. Fox, 1895; U.S.: California (cr)
virgatus W. Fox, 1895; w. U.S.; Canada: Alberta (ad)
  veles Carter, 1925, new synonymy by R. Bohart
werestchagini Gussakovskij, 1933; USSR (cr)

Unknown species described in *Crabro*

caramuru Holmberg, 1884; Uruguay
geniculatus Fabricius, 1793; Germany
grassator Bingham, 1898; n. India
†*lunatus* Schrank, 1802; Germany: Bavaria, nec Christ, 1791 (now in *Crabro*)
luxuriosus A. Costa, 1871; Italy (? = *Ectemnius*)
†*quinquecinctus* Schrank, 1802; Germany: Bavaria, nec Fabricius, 1787
tricuspis Schrank, 1796; Germany: Bavaria (not a crabronine teste Leclercq in litt.)
urophori Radoszkowski, 1877; sw. USSR
varicornis Fabricius, 1798; Germany (not a crabronine teste Leclercq in litt.)

Genus Pseudoturneria Leclercq

Generic diagnosis (male unknown): Eyes bare, inner orbits convergent below; scapal basin simple, ecarinate above; orbital foveae indistinct; ocellar triangle slightly broader than high, anterior ocellus smaller than hindocelli; gena simple, postocular sulcus absent; occipital carina incomplete, separated from hypostomal carina; antennal sockets slightly separated from each other and from inner orbits; scape ecarinate; palpal formula probably 6-4; mandible apically bidentate, lower tooth shorter than upper, externoventral margin entire; pronotal collar low, rounded anteriorly and laterally, without median notch; scutum with notauli and admedian lines; axillae ecarinate laterally; prescutellar sulcus narrow, efoveate; postspiracular carina continuous with omaulus, precoxal area simple; acetabular carina, sternaulus, hypersternaulus, and mesopleuraulus absent; propodeum coarsely areolate, dorsal enclosure not defined; lateral propodeal carina present; foretarsus without rake; recurrent vein joining submarginal cell slightly beyond its middle; jugal lobe longer than submedian cell; gaster subpedunculate, pygidial plate elongate-triangular, smooth and slightly concave apically.

Geographic range: One Australian species known.

Systematics: This genus is probably related to *Piyuma,* although it also resembles *Crossocerus.* It is distinguished from both by the rugose-areolate propodeum. Like *Piyuma* the ocellar triangle is slightly broader than high, the pronotal collar is without a median notch, the prescutellar sulcus is narrow and efoveate, and the dull, reddish-orange gaster is subpedunculate. However, the pygidial plate is elongate, shining, and ecarinate medially, and the precoxal area is simple. Only the holotype female is known. An important reference is Leclercq (1951a) where the genus was described (as *Turneriola,* nec China) and incorporated in a key.

Biology: Unknown.

Checklist of *Pseudoturneria*

perlucida (Turner), 1908 (*Crabro*); Australia: Queensland

Genus Piyuma Pate

Generic diagnosis: Eyes naked, inner orbits convergent below; scapal basin simple, ecarinate above; orbital foveae absent; ocellar triangle broader than long; gena simple; occipital carina incomplete, separated from hypostomal carina; antennal sockets contiguous, slightly separated from or contiguous to inner orbits; male antenna with 13 articles; scape longitudinally bicarinate; male flagellum simple, with ventral hair fringe; palpal formula 6-4; mandible apically bidentate, externoventral margin entire, inner margin simple; pronotal collar ecarinate, rounded anteriorly and laterally, without median notch; scutum, scutellum, and metanotum simple; scutal flange terminating on anterior margin of scutum at pronotal lobe; axillae broadened; prescutellar sulcus efoveate; postspiracular carina continuous with omaulus, verticaulus present; acetabular carina, sternaulus, hypersternaulus and mesopleuraulus absent; propodeum finely sculptured, enclosure weakly defined, lateral propodeal carina well developed; legs simple; recurrent vein joining submarginal cell at its middle; jugal lobe slightly shorter than submedian cell; gaster subsessile; male without pygidial plate, that of female flat, coriaceous, narrowed apically, bisected by longitudinal carina.

Geographic range: Oriental and Australasian Regions. One species is widely distributed throughout Southeast Asia to northern Australia, and four have been described from Malaya, Sarawak, and the Philippines. The zoogeography was discussed by Leclercq (1956a).

Systematics: Piyuma, like *Piyumoides,* has the pronotal collar ecarinate and without a median notch, the scutal flange terminates on the anterior margin of the scutum, the prescutellar sulcus is narrow and efoveate, and the female pygidial plate is broad and coriaceous. However, unlike *Piyumoides* the upper frons is simple, and the pygidial plate of the female is bisected by a longitudinal carina. Important references are Pate (1944b) and Leclercq (1951a, 1956a).

Biology: On Taiwan Iwata (1941) found *Piyuma prosopoides* nesting in abandoned beetle burrows in the rotten wood of a dead tree. The nests were linear and contained 2 to 4 larval cells and a few empty cells. The larval cells were stocked mainly with small Diptera: Tephritidae, Drosophilidae, and Stratiomyiidae. However, one cell included a winged psocopteran. The empty cells were nearest the entrance, and the outermost one was always closed by a hard, transparent membrane made from a water soluable, gummy substance. The source of this gum was unknown, but in Bangkok, Thailand, Iwata (1964a) found this species collecting a brown gummy secretion exuded from a wound on the trunk of *Samanea saman.*

In the Philippines, Williams (1928a) found the same species nesting in an abandoned termite tunnel in a bamboo upright in a nipa house. Small Diptera were being stored.

Checklist of *Piyuma*

accepta Leclercq, 1963; Malaysia: Singapore
butuana Leclercq, 1963; Philippine Is: Sibuyan
familiaris (F. Smith), 1858 (*Crabro*); Sarawak, new combination by J. Leclercq
prosopoides (Turner), 1908 (*Crabro*); Philippines, Taiwan, Borneo; Australia: Queensland
   dentipleuris Cameron, 1908 (*Crabro*)
   makilingi Williams, 1928 (*Crabro*)
   iwatai Yasumatsu, 1942 (*Crabro*)
   koxinga Pate, 1944
selangori Leclercq, 1957; Malaya

Genus Piyumoides Leclercq

Generic diagnosis (male not seen, descriptions of male characters taken from Leclercq, 1963): small forms having reddish or dark gaster. Eyes bare, inner orbits convergent below, anterior facets greatly enlarged, at least in female; scapal basin simple; frons just above scapal basin with broad, arcuate, carinate ridge broken medially by impressed frontal line (fig. 126 B) which may be nearly obsolete in some males; orbital foveae small; ocellar triangle equilateral; postocular sulcus present, narrow, efoveate; gena simple; occipital carina incomplete, separated from hypostomal carina; antennal sockets contiguous to each other and to inner orbits; male antenna with 13 articles; scape longitudinally bicarinate; male flagellum with only last article modified; palpal formula 6-4; mandible apically bidentate, externoventral margin entire; pronotal collar ecarinate, rounded anteriorly and laterally, without median notch; scutum with admedian lines and notauli; axillae, scutellum, and metanotum simple; prescutellar sulcus narrow, without foveae; mesopleuron with postspiracular carina continuous with omaulus, verticaulus present; acetabular carina, sternaulus, hypersternaulus, and mesopleuraulus absent; propodeum finely sculptured, enclosure not or very weakly defined, lateral propodeal carina present posteriorly only; forebasitarsus of female flattened; recurrent vein joining submarginal cell slightly before its middle; jugal lobe slightly longer than submedian cell; gaster subsessile, reddish in female, darker in male, tergum VII of male subtruncate, without pygidial plate; female pygidial plate coriaceous, slightly convex basally, narrowed and flattened posteriorly, lateral carinae present only on apical half.

Geographic range: Oriental Region. One species has been described from North Borneo and Sarawak and one from the Philippine Islands.

Systematics: Piyumoides, Towada, and *Leclercqia* are

FIG. 130. Facial portraits of females in the tribe Crabronini.

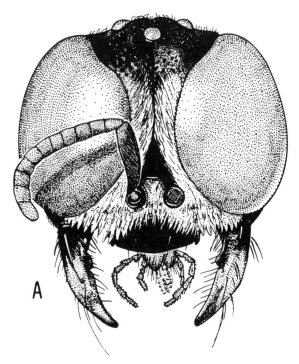

Anacrabro boerhaviae

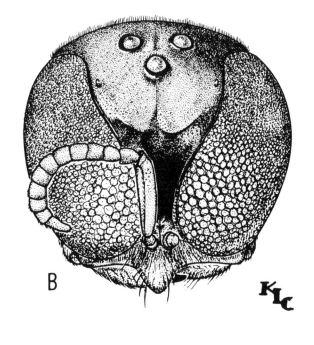

Piyumoides sp.

three closely related and as yet rarely collected genera from the Orient. In each the upper frons has an arcuate, carinate ridge broken medially (fig. 130 B), the anterior facets of the eyes are greatly enlarged at least in the female, the ocelli are arranged in an equilateral triangle, the pronotal collar is not notched medially, and the female pygidial plate is granular as in *Piyuma*. In addition *Piyumoides* and *Towada* have the prescutellar sulcus narrow and without foveae as well as the female pygidial plate ecarinate medially and with lateral carinae present only on the apical half. *Piyumoides* is distinguished from both *Towada* and *Leclercqia* by having the scape bicarinate and the occipital carina incomplete and separated from the hypostomal carina. In addition there is a verticaulus in *Piyumoides*.

Biology: Unknown.

Checklist of *Piyumoides*

hewitti (Cameron), 1908 (*Crabro*); Sarawak, N. Borneo
narcissus Leclercq, 1963; Philippines

Genus Leclercqia Tsuneki

Generic diagnosis (No specimens seen, description taken from Tsuneki, 1968c): Eyes bare, inner orbits convergent below, scapal basin glabrous, with median impressed line, frons just above scapal basin with broad, arcuate, carinate ridge which extends backward along inner orbits, broken medially by frontal line; ocellar triangle equilateral; postocellar sulcus absent, gena simple; occipital carina joining hypostomal carina; antennal sockets contiguous, separated from inner orbits; male antenna with 13 articles; scape ecarinate but with two spines at apex in both sexes; male flagellum with tyli and fringe of hair beneath, articles IX-XI especially modified; palpal formula probably 6-4; mandible apically bidentate, externoventral margin entire, inner margin with weak tooth; pronotal collar ecarinate, rounded anteriorly and laterally, probably without median notch; scutum with admedian lines and notauli present, lateral flange probably continuous to pronotal lobe; axillae and scutellum simple, prescutellar sulcus foveate, metanotum with deep, anterior lunate impression; postspiracular carina continuous with omaulus, precoxal area with tubercle (stronger in male); acetabular carina, sternaulus, hypersternaulus, and mesopleuraulus absent; mesosternum deeply excavate in male; propodeum finely sculptured, enclosure defined; legs of male modified, fore and midfemora expanded, all basitarsi modified; recurrent vein joining submarginal cell at its middle; gaster subsessile, sternum II of male with base "transversely furrowed and more or less roundly raised medianly just behind the furrow," male with tergum rounded, its surface with circular impression; female pygidial plate coriaceous, slightly convex basally, narrowed apically, carinate medially and laterally.

Geographic range: Oriental Region. One species known.

Systematics: Leclercqia is closely related to both *Piyuma* and *Piyumoides*. As in *Piyumoides*, the upper frons has an arcuate, carinate ridge that is broken medially, the ocellar triangle is equilateral, the occipital carina joins the hypostomal carina, and the antennal scape is ecarinate (some of these characters are variable in *Piyumoides*). However, all the articles of the male flagellum are modified to some degree, the prescutellar sulcus is foveate, the precoxal area has an obtuse tooth, sternum II of the male is modified, the apical tergum of the male has a broadly rounded disk, and the female pygidial plate is carinate medially and laterally.

Biology: The type series was collected while the specimens were licking honeydew on bamboo leaves.

Checklist of *Leclercqia*

formosana Tsuneki, 1968; Taiwan

Genus Towada Tsuneki

Generic diagnosis (no specimens seen, description taken from Tsuneki, 1947b, 1959b, 1970d): Small forms marked with yellow. Eyes apilose, inner orbits convergent below, anterior facets greatly enlarged, scapal basin simple; frontal line anteriorly bifurcate with an oblique carina on either side not reaching eye; orbital foveae indistinct, fine, elongate, ocellar triangle equilateral; gena simple; occipital carina nearly forming a complete circle, joining or very slightly separated from hypostomal carina; antennal sockets contiguous to each other and to inner orbits; male antenna with 13 articles, without erect hair beneath, flagellomere II relatively long, III short; scape ecarinate; palpal formula 6-4; mandible apically bidentate, externoventral margin entire; pronotal collar subcarinate (in lateral view, narrowed dorsally), rounded anteriorly and laterally, without median notch; scutellum and metanotum simple, prescutellar sulcus not foveate; mesopleuron with postspiracular carina continuous with omaulus, precoxal area roundly swollen, without verticaulus; acetabular carina, sternaulus, hypersternaulus, and mesopleuraulus absent; propodeum smooth and shining, dorsal enclosure not defined; legs simple; forewing with recurrent vein joining submarginal cell near middle; gaster subsessile, black, not spotted with yellow; first segment only slightly longer than wide at apex; female pygidial plate finely granulate, dull, apical half carinate laterally, bluntly raised medially.

Geographic range: Japan. One species known.

Systematics: Towada is very closely related to *Piyumoides*, and Tsuneki (1963e) placed the female in that genus soon after it was described by Leclercq. Later, Tsuneki (1970d) removed it to a new genus. An association of the female with the male of *flavitarsus* was made by Tsuneki (1973a), at which time he decided to reduce *Towada* to a subgenus of *Crossocerus*. Since Tsuneki (1959c) said that the pronotum had no trace of a median notch or groove, and this was confirmed by a figure, we think this character may be sufficient to separate the genus from *Crossocerus*. Consequently, we have tentatively retained it as distinct.

The characteristics shared by *Towada* and *Piyumoides* have been discussed in the systematics section of the latter genus. *Towada* is distinguished from *Piyumoides* principally by having the scape ecarinate and the occipital carina forming a nearly complete circle which touches or joins the hypostomal carina. In addition, the precoxal area is roundly swollen, but this character is probably of value only at the specific level. *Towada* is most easily separated from *Leclercqia* by the female pygidial plate which is ecarinate medially and by the male flagellum which is not fringed beneath.

Biology: Undescribed.

Checklist of *Towada*

flavitarsus (Tsuneki), 1947 (*Crabro*); Japan
leclercqi Tsuneki, 1959 (*Crossocerus*)

Genus Chimila Pate

Generic diagnosis: Eyes naked, inner orbits convergent below; scapal basin simple, without median tubercle, ecarinate; orbital foveae weak or absent; ocellar triangle broader than long; gena simple; occipital carina joining hypostomal carina; antennal sockets contiguous to each other and to inner orbits; male antenna with 13 articles; scape bicarinate; male flagellum without ventral hair fringe; palpal formula 6-4; mandible apically simple, inner margin with preapical lobe and large, acute basal tooth, externoventral margin entire; pronotal collar ecarinate, rounded anteriorly and laterally; scutum transversely microridged on anterior half, obliquely microridged posterolaterally, admedian lines and notauli present, lateral flange broad, overlying tegula, continuous anteriorly to pronotal lobe; axillae broad, flattened; scutellum and metanotum simple, prescutellar sulcus narrow, efoveate; postspiracular carina joining omaulus, verticaulus present, acetabular carina, sternaulus, hypersternaulus, and mesopleuraulus absent; propodeum finely sculptured, enclosure delimited, lateral propodeal carina present; legs stout, simple; recurrent vein joining submarginal cell beyond its middle; jugal lobe half length of hindwing submedian cell; gaster sessile, male without pygidial plate, that of female very narrow, concave.

Geographic range: Neotropical Region. This genus is known only from one species collected in Colombia.

Systematics: Chimila appears to be most closely related to *Pae*. The ocelli are arranged in a very broad triangle, the mandible is apically simple but with a preapical angle in both sexes of *Chimila* and in the male of *Pae*. Also, the clypeus in *Chimila* and male *Pae* is similar in shape, the scutum is microridged, the scutal lamina is broad, overlapping the tegula and terminating at the anterior margin of the scutum, the recurrent vein joins the submarginal cell beyond the middle, and the jugal lobe is shorter than the submedian cell. However, in *Chimila* the occipital carina is simple posteriorly and joins the hypostomal carina, the pronotal collar is ecarinate, the mesopleuraulus and hypersternaulus are absent, the prescutellar suture is narrow and efoveate, and the pygidial plate is narrowed and elongate apically. A complete description was given by Pate (1944b).

Biology: Unknown.

Checklist of *Chimila*

pae Pate, 1944; Colombia

Genus Pae Pate

Generic diagnosis: Medium to small forms having gaster marked with yellow. Eyes bare, inner orbits convergent below, anterior facets greatly enlarged in female; scapal basin simple, ecarinate orbital foveae indistinct, linear; ocellar triangle low, broad; gena swollen; occipital carina forming a complete circle, separated from hypostomal carina, flanged and foveate ventrally, dilated outward posterolaterally so that in posterior view it outlines a polished, rectangular area (fig. 121 K); hypostomal carina simple or expanded laterally to form lobe at hind angle; antennal sockets contiguous, touching or slightly separated from inner orbits; male antenna with 13 articles; scape longitudinally bicarinate; male flagellum not modified and without ventral hair fringe; palpal formula 6-4; mandible of male apically acute, with a very small preapical tooth below and at most a slight subapical expansion on inner margin; mandible of female apically bidentate with subapical tooth on inner margin, stout, extending only to edge of clypeal lobe, not bent downward; both sexes with strong basal tooth on inner margin, externoventral margin entire; clypeal lobe prolonged medially into a nasiform process in female, but simple, flat in male; pronotal collar transversely carinate, angulate to subangulate laterally, notched medially; lateral angle extending outward at least to line even with anterior apex of tegula; scutum with strong admedian lines and notauli, sculpture sometimes rugose anteriorly or longitudinally ridged, lateral lamina well developed, overlapping tegula; prescutellar sulcus foveate; scutellum longitudinally ridged, metanotum simple or longitudinally ridged; axillae broadened, carinate and bent upward laterally; postspiracular carina, omaulus, and acetabular carina continuous, hypersternaulus and mesopleuraulus present (fig. 3 A), sternaulus absent, verticaulus strong to barely indicated or absent; propodeum moderately sculptured, enclosure well defined, lateral propodeal carina present; legs simple; forebasitarsus without rake; recurrent vein joining submarginal cell at about its distal two-fifths; jugal lobe half to two-thirds length of submedian cell; gaster subsessile to sessile, marked with yellow; male without pygidial plate, that of female narrowed apically, concave, shining, spout shaped.

Geographic range: Neotropical Region. One species has been found only in South America, the other extends north to Mexico.

Systematics: The main features of the pleuron are shown in fig. 3 A. In *Pae* and the closely related genus *Lamocrabro* both a hypersternaulus and mesopleuraulus are present on the mesopleuron, and the occipital carina is dilated outward posterolaterally so that in posterior view it outlines a polished, rectangular area (fig. 121 K). In addition, the female clypeus has a characteristic median nasiform process. *Pae* and *Lamocrabro* have been compared by Leclercq (1951a).

The most important character separating the two genera is the development of the occipital carina. In *Pae* (six specimens seen), it forms a complete circle (but may be weak medially below in female *amaripa*) and is strongly or weakly flanged and foveate. In *Lamocrabro,* according to Leclercq's description (1951a) and the one specimen seen, the occipital carina is incomplete and terminates in a dentiform process. No males of *Lamocrabro* are known and the additional distinguishing characteristics are based on the females. In *Pae* the apex of the mandible is relatively shorter, extending only to the edge of the clypeal lobe and stout, not bent downward; the pronotal collar is broader, its lateral angle extending outward at least to the line even with the anterior apex of the tegula; and the first segment of the gaster is stout and sessile to slightly subsessile. In *Lamocrabro* the apex of the mandible is elongate when closed extending well beyond the edge of the median clypeal lobe and bent downward; the pronotal collar is narrow, its lateral angle being on a line with the hind angle of the scutum. Furthermore, the first segment of the gaster is more slender and subpedunculate (see Addendum).

No *Pae* male has been described previously. We have studied two males of which one from Cordoba, Mexico is probably *paniquita,* and the other from Brazil is probably *amaripa.* On the generic level they differ from the females primarily by the characters given in the generic diagnosis. In addition, the general sculpture is more coarse; the occipital carina is more strongly flanged and foveate; the transverse carina of the pronotal collar is stronger, expanding into a broader lamella; the scutum is coarsely rugose anteriorly, strongly longitudinally striate posteriorly and on the scutellum and metanotum; and the precoxal carina is well developed, bounded anteriorly by foveae. An exception to the tendency for coarse sculpture in the male is the development of the hypostomal carina. In the presumed *amaripa* it is simple, and in the presumed *paniquita* it is expanded laterally but not produced into a strong dentoid process at the hind angle.

Biology: Unknown.

Checklist of *Pae*

amaripa Pate, 1944; Guyana, Ecuador, Brazil, Bolivia
paniquita Pate, 1944; Mexico, Colombia

Genus Lamocrabro Leclercq

Generic diagnosis (male unknown): Medium-sized forms with yellow-marked gaster. Eyes bare, inner orbits convergent below; scapal basin simple, ecarinate; orbital foveae small, oval; ocellar triangle low, broad; gena swollen; occipital carina not forming a complete circle, separated from hypostomal carina, dilated outward posterolaterally so that in posterior view it outlines a polished, rectangular area, terminating ventrally in a dentiform projection; hypostomal carina sometimes expanded laterally to form a lobe at hind angle; antennal sockets contiguous with each other and with inner orbits; scape unicarinate (Leclercq, 1951a) or bicarinate; palpal formula 6-4; mandible of female with apex bidentate, elongate, when closed extending well beyond edge of median clypeal lobe and bent downward, inner margin with strong basal tooth and small subapical tooth, externoventral margin entire; female clypeal lobe prolonged medially into a nasiform process; pronotal collar transversely carinate, rounded to subangulate laterally, notched medially, lateral angle on line with hind angle of scutum; scutum with strong admedian lines and notauli, lateral lamina well developed, overlapping tegula; prescutellar sulcus foveate, axillae broadened, carinate and bent upward laterally; scutellum and metanotum simple; postspiracular carina, omaulus and acetabular carina continuous, hypersternaulus and mesopleuraulus present, sternaulus absent, precoxal area simple (Leclercq, 1951a) or with swollen, vertical ridge; propodeum moderately sculptured, enclosure well defined, lateral propodeal carina present; legs simple, front basitarsus without rake; recurrent vein joining submarginal cell at about its distal one-third; jugal lobe shorter than submedian cell; gaster subpedunculate, female pygidial plate narrowed apically, concave, shining, spout-like.

Geographic range: Neotropical Region. The single species is Brazilian.

Systematics: The characteristic development of the occipital carina posteriorly and the presence of the mesopleuraulus differentiate this rarely collected genus from all crabronins except its closest relative *Pae.* The characters that separate these two entities are discussed in the systematics section of the latter (see addendum).

In his original generic description Leclercq (1951a) stated that the scape was unicarinate and the precoxal area simple. However, on our only specimen the scape is bicarinate and the precoxal area has a vertical, slightly swollen ridge. Evidently these characters are variable at the generic level.

Biology: Unknown.

Checklist of *Lamocrabro*

nasicornis (F. Smith), 1873 (*Crabro*); Brazil
　　dentatus F. Smith, 1873 (*Crabro*), new synonymy by
　　　J. Leclercq

Genus Chimiloides Leclercq

Generic diagnosis (no specimens seen, description taken from Leclercq (1951a, 1954a): Eyes bare, inner orbits convergent below; scapal basin simple, upper and lower frons sometimes separated by a densely sculptured zone which is almost carinate; orbital foveae present but sometimes obscured by coarse sculpture; ocellar triangle broader than long, nearly flat to subequilateral; gena simple or with posterior orbits paralleled by a narrow foveate sulcus; occipital carina joining posterior edge of hypostomal carina; antennal sockets slightly separated from each other and from inner orbits; male antenna with 13 articles; scape bicarinate, male flagellum simple or notched and dentate; pedicel with or without short brush of hair beneath apex; palpal formula 6-4; mandible apically bidentate, externoventral margin entire; inner margin with obtuse premedian tooth; pronotal collar with transverse lamelliform carina which is translucent, angulate laterally, notched medially; scutum and scutellum coarsely sculptured; omaulus and acetabular carina present, sternaulus, hypersternaulus, and mesopleuraulus absent; verticaulus present, sometimes prolonged at base by a projecting tubercle bearing a strong tuft of silver hair, propodeum coarsely sculptured, sometimes reticulate dorsally; lateral propodeal carina present or absent; legs stout, simple or modified; midtibia without an apical spur in some males; female with rake on tarsus and tibia; recurrent vein joining submarginal cell at its outer third; jugal lobe shorter than submedian cell; gaster sessile, segments telescoped so sexes are difficult to distinguish; male without pygidial plate, that of female flat, triangular.

Geographic range: Australia. Three species are known.

Systematics: Members of this genus are stout-bodied, coarsely sculptured wasps with a sessile gaster which is at least partly brick red. They may be distinguished from other crabronins by the following combination of characters: scape bicarinate, mandible apically bidentate, occipital carina joining hypostomal carina, pronotal collar transversely carinate, male without pygidial plate, that of female broad, flat, not bisected by a median carina. Descriptions of the three known species were given by Leclercq (1954a).

Biology: Unknown.

Checklist of *Chimiloides*

doddii (Turner), 1908 (*Crabro*); Australia: Queensland; ♂ lectotype (described as ♀)

erythrogaster Turner, 1910 (*Crabro*)

nigromaculatus (F. Smith), 1868 (*Crabro*); Australia: Queensland

piliferus Leclercq, 1954; Australia; Queensland, Northern Territory

    doddii Turner, 1908 (*Crabro*), part (orig. described as ♂)

Genus Enoplolindenius Rohwer

Generic diagnosis: Eyes naked or sparsely covered with very short, erect hair, inner orbits convergent below; scapal basin enclosed laterally and dorsally by a strong carina, median tubercle absent or very weak; front bisected by carina extending from anterior ocellus to transverse carina; orbital foveae distinct or indistinct; ocellar triangle broader than high; gena simple or with posterior orbits paralleled by strong foveate sulcus, weak carina, or weak and simple sulcus; occipital carina flanged and foveate, forming a complete circle, separated from or joining hypostomal carina; antennal sockets contiguous to each other, contiguous to or slightly separated from inner orbits; male antenna with 13 articles; scape uni or bicarinate; male flagellum simple or modified, with or without ventral hair fringe; palpal formula 6-3; mandible apically simple, acuminate (or with small preapical tooth in some males); externoventral margin entire, inner margin simple or with denticle on basal third; pronotal collar transversely carinate anteriorly, sharply angulate laterally; scutum with transverse submarginal carina extending outward from notauli (fig. 121 H), admedian lines present; axillae simple or enlarged and carinate laterally; scutellum and metanotum simple; postspiracular carina, omaulus, and acetabular carina present, continuous, verticaulus present, sometimes angled forward as a partial sternaulus, hypersternaulus absent, mesopleuraulus absent or present as short, foveate sulcus just below or confluent with scrobe; propodeum moderately sculptured, dorsal face coarsely areolate, lateral propodeal carina well developed; legs with foretarsus of male modified, otherwise simple; male midtibia with or without apical spur; recurrent vein joining submarginal cell near its middle; jugal lobe slightly shorter to slightly longer than submedian cell; gaster sessile or subsessile; male with or without quadrate pygidial plate, that of female broad, flat, coarsely punctate, triangular.

Geographic range: New World. *Enoplolindenius* occurs from Nebraska (42°N.) to Argentina (30°S.) with the greatest number of species in northern South America. Eighteen species have been described.

Pate (1942c) noted that, according to the locality records then available, the subgenus *Iskutana* was peripheral in distribution while *Enoplolindenius s.s.* was central. He hypothesized that *Iskutana*, the more primitive of the two, arose first in northern or central South America and spread outward as this area was invaded by *Enoplolindenius s.s.* Leclercq (1951a) presented additional data which refuted this hypothesis. He pointed out that while *Iskutana* has been found farther north and south both subgenera are distributed over the central part of the range.

Systematics: Enoplolindenius, like *Foxita*, has the scapal basin completely enclosed by a strong carina. It is easily distinguished from the latter genus by the simple and acuminate apex of the mandible, the transverse submarginal carina extending laterad from the notauli on the scutum (fig. 121 H), and the broad, flat, triangular pygidial plate of the female.

Pate (1942c) proposed the subgenus *Iskutana* for those *Enoplolindenius* that have the posterior orbits simple, the male antenna modified, and the male pygidial plate present. However, the study of material now available indicates that this diagnosis must be revised.

All *Enoplolindenius s.s.* have the posterior orbits margined by a strongly foveate, carinate sulcus. In *Iskutana* the posterior orbits may be simple or bordered by a weak simple carina or a weakly foveate, carinate sulcus. The carina always parallels the eye more closely than in *Enoplolindenius s.s.*

In *Enoplolindenius s.s.* the male flagellum is simple or at most has the basal articles flattened beneath. In *Iskutana* it may be fringed with hair beneath and is modified in one of two ways: the basal articles are expanded or notched beneath, or the apical article is truncate and expanded beneath into an acute process. The scape in *Enoplolindenius s.s.* is always bicarinate and in the male bears a tuft of strong, curved setae at the apex. In *Iskutana* it may be either uni or bicarinate and bears at most only weak setae.

Enoplolindenius s.s. males lack a pygidial plate. In *Iskutana* it may be distinct, but in some specimens the lateral carinae become obsolete, and the pygidial area is difficult to distinguish.

Key to subgenera of *Enoplolindenius*

Posterior orbit simple or bordered by a weak, simple carina or a weakly foveate, marginate sulcus; male flagellum with or without ventral hair fringe and either basal or apical articles modified .... *Iskutana* Pate

Posterior orbit bordered by a strongly foveate, carinate sulcus; male flagellum simple or at most with basal articles flattened beneath .... *Enoplolindenius* Rohwer

Pate (1942c) also divided each subgenus into three species groups. Since then ten species have been described or assigned to the genus by Leclercq (1951a, 1968c). These break down some of the distinctions between the groups, and their redefinition is required.

Most of the characters used by Pate to distinguish his *Enoplolindenius s.s.* groups are specific: i.e. the pronotal collar simple or foveate, the axillae rounded or carinate laterally, and the male foretarsus patellate and fringed or merely flattened. The remaining character, the postocular carina incomplete and directed to the anterior mandibular condyle, can still be used to set off *boyaca* and *serrei*. However, these two species are known only from their types (both female).

Iskutana contains two fairly well defined entities. The *robertsoni* group combines Pate's *georgia* and *orotina* groups. It is composed of species in which the occipital carina forms a complete circle separated from the hypostomal carina and the basal articles of the male flagellum are modified. The three included species (*robertsoni, yucatanensis,* and *orotina*) are found in North America. Pate's *humahuaca* group may be retained for those species in which the occipital carina joins the hypostomal carina and the last article of the male flagellum is modified. The eight included species (*humahuaca, callangae, sucrensis, fiebrigi, hilota, benoisti, jaragua,* and *nisera*) are all found in South America. However, according to Leclercq's key (1968c), the occipital carina is separated from the hypostomal carina in the female of *nisera*. This provides an intermediate form joining the two groups.

Characters, which vary independently of the groups, include the presence or absence of the postocular carina, the modification of the axillae, and the presence or absence of foveae on the pronotal collar. The mesopleuraulus forms a short foveate sulcus (*humahuaca*), is reduced to a single fovea below the scrobe (*fiebrigi*), or is enlarged so that the foveae are confluent with each other and the scrobe (*robertsoni*).

Biology: Unknown.

Checklist of *Enoplolindenius*

[Subgeneric assignment is indicated by letter symbols after each citation: *Enoplolindenius* (E), *Iskutana* (I).]

benoisti Leclercq, 1951; Ecuador (I)
boyaca Pate, 1942; Colombia (E)
callangae Leclercq, 1968; Peru (I)
chibcha Pate, 1942; Mexico to Venezuela and Ecuador (E)
    paria Pate, 1942
clypeatus (Rohwer), 1911 (*Lindenius*); U.S.: Texas (E)
chrysis (Lepeletier and Brullé) 1834 (*Crossocerus*); Brazil (I), new combination by J. Leclercq
fiebrigi Leclercq, 1968; Paraguay (I)
hilota Leclercq, 1968; Bolivia (I)
humahuaca Pate, 1942; Argentina (I)
jaragua Pate, 1942; Brazil (I)
nisera Leclercq, 1968; Bolivia (I)
orotina Pate, 1942; Costa Rica (I)
partamona Pate, 1942; Guyana (E)
pugnans (F. Smith), 1873 (*Crabro*); Mexico, Brazil (E)
    circumscriptus Kohl, 1892 (*Crabro*), new synonymy by J. Leclercq
    mexicanus Cameron, 1904 (*Crabro*), new synonymy by J. Leclercq
    stirocephalus Cameron, 1912 (*Crabro*)
    aymara Pate, 1942
robertsoni (Rohwer), 1920 (*Lindenius*); U.S.: e. of 100th meridian (I)
    georgia Pate, 1942
    ponca Pate, 1942
serrei Leclercq, 1951; Costa Rica (E)
sucrensis Leclercq, 1968; Bolivia (I)
yucatanensis (Cameron), 1891 (*Crabro*); Mexico (I)

Genus Foxita Pate

Generic diagnosis: Eyes sparsely covered with very short, erect hair, inner orbits convergent below; scapal basin horizontally ridged, without median tubercle, enclosed laterally and dorsally by strong carina (fig. 127 B); front bisected by carina extending from anterior ocellus to transverse carina; orbital foveae distinct or indistinct; ocellar triangle equilateral; postocular sulcus distinct in males; gena simple; occipital carina complete to incomplete, flanged and foveate, separate from or contiguous with hypostomal carina; antennal sockets contiguous to each other and to inner orbits; male antenna with 13 articles; scape unicarinate; male flagellum with or without ventral hair fringe, last article modified; palpal formula 6-4; mandible apically bidentate in male, tridentate in female, externoventral margin entire, inner margin with tooth or angle before middle (fig. 127 B); pronotal collar carinate anteriorly, rounded or angulate laterally; scutum without anterior transverse carinae, with or without notauli and admedian lines; axillae simple or modified; scutellum with or without deep prescutellar sulcus across anterior third; metanotum simple; postspiracular carina, omaulus and acetabular carina continuous, verticaulus present, hypersternaulus present or absent, mesopleuraulus indicated by large fovea just below and confluent with scrobe or absent; sternaulus absent; propodeum moderately sculptured, dorsal face coarsely areolate or with poorly defined enclosure, lateral propodeal carina well developed; legs simple; recurrent vein joining submarginal cell at its middle; jugal lobe equal in length to submedian cell; gaster sessile or subsessile; male without pygidial plate, that of female very narrow, dorsal surface concave.

Geographic range: Neotropical Region. One species

has been found in Mexico, the remaining eight were described from South America.

Systematics: Foxita is easily distinguished from all New World genera, except *Enoplolindenius,* by the deeply concave scapal basin which is completely margined by a sharp carina (fig. 127 B). It differs from *Enoplolindenius* in the bi or tridentate mandible, the absence of an anterolateral transverse carina on the scutum, and the narrowed, concave pygidial plate of the female. In these characters it more closely resembles the Old World *Vechtia.* Without doubt it is a highly evolved member of the Crabronini (table 15).

Pate (1942c) divided the five species then known into three groups. The discovery of additional species, some of which are intermediate, makes it advisable to define only two groups as follows:

1. The *senci* group (*senci, galibi, patei, beieri,* and *boliviae*) is the more generalized. It is distinguished by the following characters: hypostoma rounded apically and distinctly separated from the occipital carina which is moderately flanged and foveate, but weak or absent medially below; orbital foveae indistinct; transverse pronotal carina distinct but not broadened into a translucent lamina; foveate prescutellar sulcus of normal size; both axillar margins rounded; hypersternaulus absent. Notauli and admedian lines are absent in *senci* (Pate, 1942c) but present in all other species. In *boliviae* the abdomen is black, in other *Foxita* it is marked with yellow.

2. The *atorai* group (*atorai, acavai, asuncionis*) is the more modified of the two and is distinguished by the following characters: apex of hypostoma acute and subcontiguous with the occipital carina which is strongly flanged, foveate and complete below; orbital foveae distinct at least in females; transverse pronotoal carina broadened into a translucent lamina; prescutellar sulcus greatly broadened and excavate to form a trough which is about two-fifths the length of the scutellum (in the first three species), or well developed (one-fourth to one-third the length of the scutellum) but not excavate behind the foveae (*asuncionis*); axillae simple or with one tooth or both edges sharply margined; hypersternaulus present or absent (within a series it may vary from an impressed line to a strongly foveate furrow).

Several characters vary independently of the groups, i.e., the last male flagellar article may be simple as in *acavai* and *boliviae* or modified as in *curvicollis* and *atorai;* the pronotal collar may be rounded or sharply angulate laterally.

The basic work on *Foxita* is that of Pate (1942c). Leclercq (1955b) gave a key to the species of the genus.
Biology: Unknown.

Checklist of *Foxita*

acavai Pate, 1942; Guyana
asuncionis (Strand), 1910 (p. 141, *Cerceris*); Paraguay, Guyana
    curvicollis Cameron, 1912 (*Crabro*), new synonymy by J. Leclercq
    ssp. *woyowai* Pate, 1942; Guyana, new status by J. Leclercq
atorai Pate, 1942; Guyana
beieri Leclercq, 1955; Bolivia
boliviae Leclercq, 1955; Bolivia
galibi Pate, 1942; Guyana
patei Leclercq, 1951; Mexico
senci Pate, 1942; Peru
?*tarumoides* Leclercq, 1954; Brazil (? = *Taruma*)
    megacephala F. Smith, 1873 (*Crabro*), nec Rossi, 1790

Genus Taruma Pate

Generic diagnosis (male unknown, no specimens seen, description taken from Pate (1944a) and notes made from holotype by J. Wasbauer): Eyes bare, inner orbits convergent below; scapal basin concave, smooth, partly delimited above by a lateral carina which disappears medially; upper frons bisected by a shallow impression in which lies a very fine longitudinal carinule; orbital foveae absent; ocellar triangle broader than high; gena simple; occipital carina incomplete, separated from hypostomal carina, weakly flanged and foveate; antennal sockets contiguous to each other and to inner orbits; scape longitudinally unicarinate; palpal formula 6-4; mandible of female apically tridentate, externoventral margin entire; pronotal collar transversely carinate anteriorly, angulate laterally, weakly notched medially; scutum, axillae, scutellum, and metanotum simple; prescutellar sulcus foveate; postspiracular carina, omaulus, and acetabular carina continuous, verticaulus present; sternaulus, hypersternaulus, and mesopleuraulus absent; propodeum finely sculptured; dorsal enclosure defined by foveate sulcus; lateral propodeal carina present; legs simple; foretarsus (of female) slightly flattened and with a weak rake; recurrent vein joining submarginal cell at its middle; jugal lobe slightly longer than submedian cell; gaster sessile, female with partly polished, elongate, triangular, pygidial plate which is somewhat narrowed and concave apically.

Geographic range: Neotropical Region. One species is known from Guyana.

Systematics: Judging from Pate's description, *Taruma* is very similar to *Foxita* and not to *Piyuma* as he suggested. Like *Foxita,* and unlike *Piyuma,* the upper frons in *Taruma* is bisected by a carina (weak), the scape is unicarinate, the mandible of the female is apically tridentate, the pronotal collar is carinate, the acetabular carina is present, the prescutellar suture is distinctly foveate, the abdomen is sessile, and the pygidial plate is narrowed and concave apically. Apparently *Taruma* is separated from *Foxita* as follows: carina delimiting scapal basin incomplete dorsomedially, scapal basin smooth, not horizontally ridged, foretarsus of female slightly flattened, and pygidial plate possibly less strongly narrowed apically (abdomen now missing on type). Pate probably considered the incomplete occipital carina to be of primary importance in diagnosing *Taruma,* but members of the *senci* group of *Foxita* show a range of variation of the occipital carina from weak to completely absent beneath (see Addendum).

Biology: Unknown.

Checklist of *Taruma*

bara Pate, 1944; Guyana

Genus Vechtia Pate

Generic diagnosis: Eyes sparsely covered with very short, erect hair, inner orbits convergent below; scapal basin concave, smooth, delimited laterally by a carina and dorsally by a transverse lamella which is downcurved medially (fig. 122 I); front bisected by a carina extending from anterior ocellus to apex of lamella; orbital foveae linear, adjacent to eye margin in female, obscured by furrow bordering eye in male; ocellar triangle broader than high; gena well developed, with weakly foveate sulcus bordering eye margin from orbital fovea downward across gena to posterior mandibular condyle; occipital carina complete, flanged, foveate, separated from hypostomal carina; antennal sockets contiguous, separated from inner orbits; male antenna with 13 articles; scape carinate on outer margin and basal half of inner margin; male flagellum simple, without ventral hair fringe; palpal formula 6-4; mandible apically bidentate in male, tridentate in female, externoventral margin entire; pronotal collar and lobes strongly carinate anteriorly, angulate laterally; scutum without anterior transverse carina, with strong notauli and admedian lines; prescutellar sulcus well developed and foveate; axillae moderately broadened, inner margin sometimes carinate; scutellum with longitudinal median carina and transverse carina at posterior third; metanotum simple; postspiracular carina, omaulus, and acetabular carina continuous; verticaulus and sternaulus present; hypersternaulus and mesopleuraulus absent; propodeum moderately sculptured, dorsal face delimited posteriorly by carinae; lateral propodeal carina well developed; midtibia of male short, slightly more than half length of midfemur, without apical spur but prolonged into sharp spinelike projection on inner side; recurrent vein joining submarginal cell at its middle; jugal lobe equal to or slightly longer than submedian cell; gaster sessile; male without pygidial plate, that of female very narrow and concave.

Geographic range: Oriental Region. One species is widely distributed throughout Southeast Asia east to New Guinea. Another is found in southern India.

Systematics: Vechtia may be easily distinguished from all other genera by the downcurved lamina limiting the scapal basin (fig. 122 I). It most closely resembles the Neotropical genus *Foxita*. The systematics have been discussed by Pate (1944a) and Leclercq (1951a, 1957b).

Biology: Unknown.

Checklist of *Vechtia*

prerugosa Leclercq, 1963; India
rugosa (F. Smith), 1857 (*Crabro*); se. Asia
    bucephalus F. Smith, 1864 (*Crabro*), nec F. Smith, 1856

spinifrons Bingham, 1897 (*Crabro*)
    ssp. *forticarinata* Leclercq, 1951; Borneo

Genus Hingstoniola Turner and Waterston

Generic diagnosis (female unknown[4]): Eyes naked, inner orbits convergent below; scapal basin "delimited laterally and dorsally by an extremely strong carina," bisected by carina extending downward from dorsal carina; upper front with "neither carina nor furrow extending from the anterior ocellus to the middle of the transverse carina delimiting the scapal basin"; orbital foveae absent; ocellar triangle broader than high; postocular sulcus foveate, delimited by carina; gena simple; occipital carina complete, flanged and foveate, separate from hypostomal carina; antennal sockets separated from each other, contiguous to or separated from inner orbits; male antenna with 13 articles; scape bicarinate, carinae becoming indistinct proximally; male flagellum modified, with or without fringe of hair beneath; palpal formula 6-4; mandible apically bidentate in male, externoventral margin entire, inner margin with tooth on basal half; pronotal collar carinate anteriorly, notched medially, angulate laterally; scutum without anterior transverse carinae; notauli and admedian lines present or obscured by coarse sculpture; prescutellar sulcus well developed and foveate; axillae moderately broadened; scutellum margined laterally and posteriorly by carina; metanotum coarsely sculptured; mesopleuron with postspiracular carina, omaulus, and acetabular carina continuous; verticaulus present; sternaulus, hypersternaulus, and mesopleuraulus absent; propodeum coarsely sculptured, enclosure areolate; lateral propodeal carina well developed; legs with fore and midtarsi of male modified, forefemur with spine or carina, midtibia without apical spur (one specimen seen); recurrent vein joining submarginal cell beyond its middle; jugal lobe slightly shorter than submedian cell; gaster sessile; male without pygidial plate.

Geographic range: Oriental Region. One species has been described from Sikkim and a second from Malaysia and Borneo.

Systematics: Hingstoniola may be distinguished from other crabronins by the marginate scapal basin bisected longitudinally by a simple carina. It appears to be most closely related to *Foxita* and relatives by the complete, flanged, and foveate occipital carina, the carinate scape and pronotum, the presence of the acetabular carina and verticaulus, and the tendency toward loss of the midtibial spur in the male. However, unlike *Foxita* and *Vechtia,* the upper frons is not bisected by a carina extending forward from the anterior ocellus, and the sculpture is dull and coarse.

Biology: Unknown.

Checklist of *Hingstoniola*

duplicata (Turner and Waterston), 1926 (*Crabro*); Sikkim

---

[4] Only one headless male seen, description of head characters taken from Turner and Waterston (1926), Pagden (1934), and notes made by C. R. Vardy.

pagdeni Leclercq, 1954; Malaya, Borneo
   fimbriata Pagden, 1934 (*Crabro*), nec Rossi, 1790
   parviornata "Cameron" Leclercq, 1951 (*Crabro*)
      nomen nudum

Genus Arnoldita Pate

Generic diagnosis (no male seen); Eyes bare, inner orbits arcuate or convergent below; scapal basin concave, weakly carinate laterally, ecarinate dorsally, bisected by longitudinal carina bearing one or two median processes (fig. 121 A); front bisected by furrow extending from anterior ocellus to scapal basin; orbital foveae present; ocellar triangle equilateral; postocular sulcus distinct; gena with or without tooth behind mandible; occipital carina incomplete, flanged and foveate, separate from hypostomal carina; antennal sockets contiguous and contiguous with or separate from inner orbits; male antenna (known for one species only) with 11 articles (first two or three flagellar articles fused according to Arnold, 1940); scape unicarinate; male flagellum with ventral hair fringe; palpal formula 6-3; mandible apically simple and acuminate or truncate, or bidentate (fig. 121 I), inner margin simple or dentate, externoventral margin entire; pronotal collar carinate anteriorly, angulate laterally; scutum with admedian lines and notauli present, latter foveate anteriorly; scutal flange terminating on lateral margin above tegula; axillae, scutellum, and metanotum simple; postspiracular carina, omaulus, verticaulus, hypersternaulus, and sternaulus present; acetabular carina and mesopleuraulus absent; propodeum moderately sculptured, enclosure well defined, lateral propodeal carina well developed; legs simple or modified; recurrent vein joining submarginal cell before its middle; jugal lobe shorter than submedian cell; gaster pedunculate, elongate, and slender; female pygidial plate narrowed apically, dorsal surface shallowly concave; structure of tergum VII of male unknown because this was not mentioned by Arnold in his descriptions.

Geographic range: Ethiopian Region. One species has been found in Zaire and Nigeria, the other two in Rhodesia.

Systematics: Members of this African genus are most quickly recognized by the incompletely margined scapal basin bisected by the longitudinal carina bearing at least one process (fig. 121 I). Pate (1948b) correctly considered *Arnoldita* to be a highly specialized and aberrant genus related to *Foxita*. The relationship was based on the strongly flanged and foveate occipital carina, the transversely carinate pronotal collar, and the presence of the postspiracular carina, omaulus, precoxal carina, sternaulus, hypersternaulus, and lateral propodeal carina. However, the scapal basin is ecarinate laterally and dorsally, the front is bisected by a furrow extending forward from the anterior ocellus, the occipital carina is incomplete, and the acetabular carina is absent. Without question it is one of the most advanced genera in the Crabronini (table 15).

Biology: Arnold (1926) reported that a male *senex* was reared from abandoned bostrichid burrows in a wooden post.

Checklist of *Arnoldita*

canalifera (Arnold), 1944 (*Crabro*); Rhodesia
perarmata (Arnold), 1926 (*Thyreopus*); Zaire, Nigeria
senex (Arnold), 1926 (*Thyreopus*); Rhodesia

Genus Odontocrabro Tsuneki

Generic diagnosis (male unknown, no specimens seen, description taken from Tsuneki, 1971a): Small, dark, shining forms with black gaster. Eyes naked, inner orbits convergent below; scapal basin simple, ecarinate above; frontal furrow extending from frons backwards between hindocelli to occiput; orbital foveae slightly impressed; ocellar triangle broader than high; gena swollen laterally; occipital carina not forming a complete circle, separated from hypostomal carina, not terminating in a tooth; antennal sockets separated from each other and from inner orbits; scape ecarinate; palpal formula 6-3; mandible of female apically bidentate; inner margin with tooth on basal half, externoventral margin entire; pronotal collar "flattened above, tuberculate on both sides of medial furrow and at the sides strongly toothed"; scutum with admedian lines and notauli present; prescutellar sulcus foveate; scutellum raised; mesopleuron with "postspiracular sclerite 2 (of Richards) markedly broad, not separated from the epicnemial area by a carina and provided on outer margin with a short triangular protuberance," omaulus present, acetabular carina apparently absent, sternaulus, hypersternaulus and mesopleuraulus absent, precoxal area simple, venter with "deep elliptic hollow on posterior portion in middle which is filled with comparatively long hairs"; propodeum with microreticulate dorsal enclosure, lateral propodeal carina present on posterior half; legs modified, forecoxa with long, bifurcate, large appendage in front (fig. 121 F), tarsi modified; recurrent vein joining submarginal cell at about its middle; gaster sessile, first segment as long as wide; female pygidial plate surface dull, divided into three areas by a median ridge which forks above.

Geographic range: Oriental Region. Only one species from Taiwan is known.

Systematics: This unusual genus, known only from the female (Tsuneki, 1971a), is recognized by the presence of a long, bifurcate appendage on the forecoxa (fig. 121 F), absence of the postspiracular carina, and a palpal formula of 6-3. In addition, the pronotal collar is tuberculate medially and strongly toothed laterally. Also, the pygidial plate is dull and divided into three areas by a median ridge which forks above.

Biology: Unknown.

Checklist of *Odontocrabro*

abnormis Tsuneki, 1971; Taiwan

Genus Neodasyproctus Arnold

Generic diagnosis: Eyes naked, inner orbits convergent

below; scapal basin without median tubercle, ecarinate; orbital foveae distinct, linear; ocellar triangle broader than high; gena simple; occipital carina incomplete, separate from or joining hypostomal carina; antennal sockets contiguous to each other and to inner orbits; male antenna with 13 articles; scape ecarinate; male flagellum modified, without ventral hair fringe; palpal formula 6-4; mandible apically bidentate in male, tridentate in female, inner and externoventral margins entire; pronotal collar carinate or ecarinate anteriorly, rounded laterally, notched medially; scutum, axillae, scutellum, and metanotum simple; scutal flange terminating on lateral margin of scutum above tegula; prescutellar sulcus foveate; postspiracular carina and omaulus present, acetabular carina, sternaulus, hypersternaulus, and mesopleuraulus absent, verticaulus strong but short; propodeum moderately sculptured, enclosure defined or not defined, lateral propodeal carina present or absent; legs simple; recurrent vein joining submarginal cell beyond its middle; jugal lobe shorter than submedian cell; gaster with first segment elongate-pedunculate; male with pygidial plate at most weakly delimited, female pygidial plate apically narrowed and concave.

Geographic range: Ethiopian and Australasian Regions. Nine species have been described from Africa and one from the Fiji Islands and Australia.

Systematics: Neodasyproctus is very similar to *Dasyproctus*. However, the integument is shining in the former and may be coarsely punctate. Also, the verticaulus is shorter in *Neodasyproctus*. Important references are those of Arnold (1945) and Leclercq (1951c).

Biology: Arnold (1926) reported that H. Brauns found *Neodasyproctus kohli* nesting in hollow stems.

Checklist of *Neodasyproctus*

basutorum (Turner), 1929 (*Thyreopus*); S. Africa
densepunctatus (Arnold), 1945 (*Crabro*); Madagascar
ealaenis Leclercq, 1951; Zaire
eburneopictus (Arnold), 1945 (*Crabro*); Madagascar
ferrierei Leclercq, 1951; Madagascar
kohli (Arnold), 1926 (*Thyreopus*); S. Africa
libertinus (Arnold), 1929 (*Thyreopus*); Liberia
protensus (Arnold), 1945 (*Crabro*); Madagascar
striolatus (Arnold), 1945 (*Crabro*); Madagascar
veitchi (Turner), 1917 (*Crabro*); Fiji, Australia

Genus Dasyproctus Lepeletier and Brullé

Generic diagnosis: Wasps with integument opaque or dull; eyes bare, inner orbits convergent below; scapal basin concave, simple or delimited dorsally by a carina which may be lamellate; orbital foveae distinct to evanescent; ocellar triangle broader than high; gena simple, broadened ventrolaterally and with a ventral tubercle (*javanus*), or bisected by a carina paralleling posterior orbit and attaining posterior mandibular condyle (*venans*); occipital carina a complete circle tangent to hypostomal carina or joining hypostomal carina (fig. 121 I); antennal sockets contiguous with each other and with inner orbits; male antenna with 13 articles; scape bicarinate; male flagellum simple or modified, without ventral hair fringe; palpal formula 6-4; mandible apically bidentate in male, tridentate in female, externoventral margin entire, inner margin usually edentate; pronotal collar with anterior carina which reaches pronotal lobes in males and most females, or in females of *javanus* group curves anteroventrally before reaching lobes; scutum simple, its lateral flange terminating above tegula; scutellum and metanotum simple; postspiracular carina, omaulus, and acetabular carina present, continuous; verticaulus elongate, sometimes weak; sternaulus, hypersternaulus, and mesopleuraulus absent; propodeum moderately sculptured, dorsal face microridged, rugose, or areolate, enclosure not or weakly defined, lateral propodeal carina well developed; legs simple or with hindfemur modified; recurrent vein joining submarginal cell beyond its middle; jugal lobe shorter than submedian cell; gaster with first segment elongate-pedunculate; male without pygidial plate, female pygidial plate strongly narrowed, concave.

Geographic range: Ethiopian, Oriental and Australasian Regions. 67 species are known. The zoogeography of *Dasyproctus* has been discussed in detail by Leclercq (1958a). This genus is widely distributed throughout the intertropical areas bounding the Indian Ocean. The Ethiopian Region is richest in both variety and number of species. The range of one of these, *arabs*, extends east to Pakistan via the desert bounding the Palearctic Region.

Systematics: Dasyproctus and *Neodasyproctus,* like *Williamsita,* are evidently descendants of the ancestral forms of *Ectemnius.* In all three and most *Ectemnius* the recurrent vein reaches the submarginal cell beyond its middle, the verticaulus is present, the ocelli are arranged in a broad triangle, the female pygidial plate is narrowed and concave apically, and the male antenna has 13 articles. However, in *Dasyproctus* and *Neodasyproctus* the first segment of the gaster is elongate-pedunculate, and the midtibial spur is present in both sexes. *Dasyproctus* is distinguished by its dull integument and elongate verticaulus.

The genus has been revised by Leclercq who first published on Oriental forms (1956b, 1963) and then covered the African fauna, gave a key to species, discussed species groups, etc. (1958a). The Asian forms were keyed by Leclercq (1972a).

Biology: Dasyproctus species make their nests in plant stems and provision with Diptera. The African species *bipunctatus* is considered a pest because it makes nests in live, flowering stems of ornamental plants. In Malaya *D. buddha* makes nests in solid dead grass stems (*Coelorrachis glandulosa* Staph.) and provisions with chloropids and otitids (Pagden, 1934). The uncompleted nest examined by Pagden contained four cells separated by pith. J. van der Vecht (1951) gave details of another Oriental form, *agilis* (as *ceylonicus*). Nests were made in sorghum stems and prey consisted mainly of *Rivellia basilaris* Wiedemann (Otitidae), although *Musca* and *Antherigona* (Muscidae), and *Eumerus* (Syrphidae) were also provisioned. Eggs were laid on the venter of the fly in the neck region. Carpenter (1942) found that the otitid fly *Rivellia trigona* Hendel was the predominant prey of the African species *bipunctatus lichtenbergensis*, and that up to 13 flies were placed in a cell. Michener (1971) examined a nest of another African form, *stevensoni*. The flies contained in three cells represented two genera of Milichiidae.

The most detailed study is that of Bowden (1964) who dealt with two African species, *bipunctatus* and *kibonotensis*. The latter has been regarded as a subspecies of *bipunctatus*, but Bowden found biological as well as morphological differences and elevated *kibonotensis* to species. *D. bipunctatus* burrows are made in green stems of various monocots. Flowering spikes of *Gladiolus* often collapse and die as a result of the tunneling. On the other hand, *kibonotensis* makes its nests in solid, live stems of *Rubus* species. From the entrance on the side of a stem, the nest proceeds in both directions through the stem. The section directed towards the apical meristem is completed first in both species. Nests of *bipunctatus* contain between six and 19 cells, but those of *kibonotensis* apparently never exceed seven or eight cells. Cells are separated by pith. Prey used by *kibonotensis* represented five families, but *Trirhithrum coffeae* Bezzi (Tephritidae) and *Lonchaea laevis* Bezzi (Lonchaeidae) were the most common. *Dasyproctus bipunctatus* also used five families of Diptera but most prey were *Antherigona* (Muscidae). Bowden theorized on the basis of prey that *bipunctatus* searched in grassy areas, while *kibonotensis* hunted in understory shrubs. Ten to 21 flies are stored per cell in the two species. The egg is usually laid on the first provision on the venter of the neck.

Bowden found one nest of *kibonotensis* that appeared to be the work of two females using a common entrance. A small *Rubus* bush may contain a dozen or more active nests and the perennial nature of the plant is responsible for a kind of presocial behavior in this species. Parental nests are available for habitation by progeny for considerable periods, and Bowden feels that some nests are continuously occupied for more than two months. He suggests that there is a possibility that some progeny coexist with their female parent and that it is conceivable that there is a continuity of experience within a *Rubus* colony in which generations may overlap.

Carpenter (1942) reared the mutillid *Promecilla unicingulata* Bischoff from cells of *bipunctatus*, and Bowden (1964) found miltogrammine Sarcophagidae in cells of *kibonotensis*.

Checklist of *Dasyproctus*

abax Leclercq, 1958; Zaire
agilis (F. Smith), 1858 (*Crabro*); Oriental Region incl. Taiwan, Philippines, Borneo, Celebes
   ceylonicus Saussure, 1867
   orientalis Cameron, 1890 (*Crabro*)
   indicus Saussure, 1892 (*Crabro*)
   infantulus Kohl, 1894 (*Crabro*)
   ? *revelatus* Cameron, 1898 (*Crabro*)
   impetuosus Cameron, 1901 (*Crabro*)
   philippinensis Ashmead, 1904
   funestus Turner, 1917
   palmerii "Cameron" Leclercq, 1950, nomen nudum
angusticollis (Arnold), 1926 (*Thyreopus*); Rhodesia, Zaire, Malawi
angustifrons (Arnold), 1927 (*Thyreopus*); S. Africa
araboides Leclercq, 1958; Mali
arabs (Kohl), 1894 (*Crabro*); Pakistan, Iran, Syria, Israel, Egypt, Algeria, Uganda, Somalia
   obockensis Leclercq, 1949
artisanus Leclercq, 1972; Philippines
aurovestitus Turner, 1912; Sierra Leone, Uganda, Malawi, S. Africa
australis Leclercq, 1972; Australia
barkeri (Arnold), 1927 (*Thyreopus*); Zaire, S. Africa, Uganda
   nyholmi Arnold, 1952 (*Crabro*)
basifasciatus (Arnold), 1951 (*Crabro*); Ethiopia
benoiti Leclercq, 1958; Zaire
bipunctatus Lepeletier and Brullé, 1834; Ethiopian Region
   massaicus Cameron, 1908
   jucundus Arnold, 1926 (*Thyreopus*)
   ssp. *rabiosus* (Kohl), 1894 (*Crabro*); Guinea, Nigeria
   ssp. *funereus* (Arnold), 1926 (*Thyreopus*); Rhodesia
   ssp. *lichtenburgensis* (Arnold), 1926 (*Crabro*); S. Africa
   † ssp. *lugubris* (Arnold), 1927 (*Crabro*); Kenya to S. Africa, nec Fabricius, 1793
   ssp. *avius* (Arnold), 1951 (*Crabro*); Ethiopia, Uganda
   ssp. *tervureni* Leclercq, 1958; Zaire
bredoi (Arnold), 1947 (*Crabro*); Zaire, Burundi, Zambia
buddha (Cameron), 1889 (*Rhopalum*); Oriental Region
   brookii Bingham, 1896 (*Crabro*)
   taprobane Cameron, 1898 (*Crabro*)
   idrieus Cameron, 1903 (*Crabro*)
   musaeus Cameron, 1903 (*Crabro*)
   testaceipalpis Cameron, 1908
burnettianus Turner, 1912; Australia: Queensland
cevirus Leclercq, 1963; Philippines: Negros, Mindanao, Sumbawa
collaris (Arnold), 1932 (*Thyreopus*); Zaire, Burundi, Ruanda, Rhodesia
   vumbuiensis Arnold, 1940 (*Crabro*)
conator (Turner), 1908 (*Crabro*); Australia: Queensland
croceosignatus (Arnold), 1940 (*Crabro*); Ethiopian Region
   bicuspidatus Arnold, 1944 (*Crabro*)
crudelis (Saussure), 1892 (*Crabro*); Madagascar
dubiosus (Arnold), 1926 (*Thyreopus*); Ethiopian Region
erythrotoma (Cameron), 1905 (*Crabro*); S. Africa
expectatus Turner, 1912; Australia: N.S. Wales, Victoria
ferox (Saussure), 1892 (*Crabro*); Madagascar
   immanis Saussure, 1892 (*Crabro*)
fortunatus Beaumont, 1968; Canary Is.
frater (Dahlbom), 1844 (*Megapodium*); S. Africa
immitis (Saussure), 1892 (*Crabro*); Ethiopian Region, Cape Verde Is.
   braunsii Kohl, 1894 (*Crabro*)
   sjoestedti Cameron, 1908

jacobsoni (Kohl), 1908 (*Crabro*); Indonesia
  muiri Turner, 1912
javanus Leclercq, 1956; Java, Borneo
jungi (Maa), 1936 (*Crabro*); China: Szechwan; Taiwan, new combination by J. Leclercq
  quinquemaculatus Maa, 1936 (*Crabro*), nec Lepeletier and Brullé, 1834
  formosanus Tsuneki, 1968, new synonymy by J. Leclercq
kibonotensis Cameron, 1908; Ethiopia, Tanzania, Rhodesia, Zaire, Uganda
  pullatus Arnold, 1944 (*Crabro*)
  uniguttatus Arnold, 1951 (*Crabro*)
kutui Leclercq, 1958; Zaire
lambertoni Leclercq, 1958; Madagascar
liberiae Leclercq, 1958; Liberia
lignarius (F. Smith), 1864 (*Crabro*); Moluccas: Morotai
localis Leclercq, 1958; Rhodesia, S.-W. Africa
medicus Leclercq, 1958; Nigeria, Zaire
oedignathus (Arnold), 1933 (*Thyreopus*); Ethiopia, Kenya
opifex (Bingham), 1897 (*Crabro*); Burma, Malaya
oppidanus Leclercq, 1972; Sikkim
pentheri Leclercq, 1956; Sri Lanka, Java, Philippines, Flores
pulveris (Nurse), 1902 (*Crabro*); India
quadricolor (W. F. Kirby), 1900 (*Rhopalum*); Socotra
ralumus Leclercq, 1972; New Britain
rebellus Leclercq, 1958; Ruanda
ruficaudis (Arnold), 1926 (*Thyreopus*); Ethiopian Region
saevus (Saussure), 1892 (*Crabro*); Madagascar
sakalavus Leclercq, 1967; Madagascar
sandakanus Leclercq, 1972; North Borneo
saussurei (Kohl), 1894 (*Crabro*); Madagascar
  infrarugosus Arnold, 1944 (*Crabro*)
scotti (Turner), 1911 (*Crabro*); Seychelles
simillimus (F. Smith), 1856 (*Crabro*); Ethiopia, Burundi, Kenya, S. Africa
solitarius (F. Smith), 1859 (*Crabro*); Indonesia: Aru
stevensoni (Arnold), 1926 (*Thyreopus*); Ethiopian Region
  stevensonianus Arnold, 1940 (*Crabro*)
  occidentalis Arnold, 1951 (*Crabro*)
temporalis Leclercq, 1963; North Borneo
townesi Leclercq, 1963; Philippines: Negros, Mindanao
toxopterus Leclercq, 1963; Philippines: Mindanao
tyronus Leclercq, 1963; North Borneo
uruensis Leclercq, 1972; Celebes
vaporus Leclercq, 1963; Philippines
vechtinus Leclercq, 1957; Celebes
venans (Kohl), 1894 (*Crabro*); Carolines: Palau
  immaculatus Krombein, 1949
verutus Rayment, 1932; Australia: Victoria
westermanni (Dahlbom), 1844 (*Megapodium*); Zaire, Rhodesia, S. Africa, S.-W. Africa
  schoenlandi Cameron, 1905
  rhodesiensis Arnold, 1926 (*Thyreopus*)
yorki Leclercq, 1956; Australia: Queensland, W. Australia
yorkoides Leclercq, 1972; Philippines

Genus Williamsita Pate

Generic diagnosis: Eyes naked, inner orbits convergent below; scapal basin simple, without median tubercle, ecarinate; orbital foveae distinct or indistinct; ocellar triangle broader than high (fig. 121 G); gena simple; occipital carina incomplete, flanged and foveate or simple, separated from hypostomal carina, curving anteriorly beneath head, sometimes ending in a tubercle; antennal sockets contiguous or slightly separated by a low ridge, contiguous to or slightly separated from inner orbits; male antenna with 13 articles; scape ecarinate or unicarinate; flagellum with first article elongate in both sexes, flagellum of male simple or modified, without ventral hair fringe; palpal formula 6-4; mandible apically bidentate in both sexes, female with preapical tooth on inner margin; externoventral margin entire, inner margin without basal tooth but sometimes expanded medially into an acute angle; pronotal collar carinate anteriorly, rounded or angulate laterally; scutum with strong admedian lines and notauli (fig. 121 G), scutal flange terminating on lateral margin above tegula, prescutellar sulcus foveate; axillae, scutellum, and metanotum simple; postspiracular carina continuous with omaulus, acetabular carina present or absent, verticaulus present; sternaulus, hypersternaulus, and mesopleuraulus absent; propodeum weakly to moderately sculptured, dorsal face with enclosure weakly or not defined, or coarsely rugose; lateral propodeal carina absent; legs of male sometimes modified by having foretarsus flattened, foretibia expanded apically, forefemur with spine beneath, midbasitarsus distorted, or midfemur flattened; male midtibia without apical spur or at most with spur very small; recurrent vein joining submarginal cell at its outer third; hindwing normal or very narrow, jugal lobe a third to a half length of submedian cell, gaster sessile; male with quadrate pygidial plate, that of female very narrow, concave.

Geographic range: Australasian Region. The eight members of this genus are known only from Australia and neighboring islands.

Systematics: Williamsita resembles *Ectemnius* very closely and is evidently a relic of the common ancestor of *Ectemnius* and *Lestica* (Pate, 1947c; Leclercq 1950b, 1954a). It is distinguished by the following combination of characters: male antenna with 13 articles; orbital foveae distinct in females (fig. 121 G), distinct or indistinct in males; pygidial plate of female strongly narrowed and concave, that of male quadrate; midtibial spur absent or very small in male.

Williamsita was originally based on a single species. Leclercq (1950b) transferred seven species described by Turner as *Crabro* (*Solenius*) into *Williamsita* and divided it into two subgenera: *Williamsita s.s.* and *Androcrabro*.

Key to subgenera of *Williamsita*

1. Male flagellum simple; occipital carina angling forward to parallel hypostomal carina for

a third to half its length, terminating abruptly or in a tubercle; hindwing very narrow .... *Williamsita* s.s.
Male flagellum with at least one article modified; occipital carina short and without tubercle; hindwing normal ... *Androcrabro* Leclercq

Biology: Evans and Matthews (1971a) reported three females of *W. bivittata* in the National Museum of Victoria pinned with prey identified as *Calliphora tibialis* Macquart (Calliphoridae). A label indicated they had been taken from nests in a willow log at Pascoe Vale, Victoria.

Checklist of *Williamsita*

(All except *W.* (*Williamsita*) *novocaledonica* are in subgenus *Androcrabro*)

bivittata (Turner), 1908 (*Crabro*); Australia: Victoria
manifestata (Turner), 1915 (*Crabro*); sw. Australia
neglecta (F. Smith), 1868 (*Crabro*); s. Australia
novocaledonica (Williams), 1945 (*Crabro*); New Caledonia
ordinaria (Turner), 1908 (*Crabro*); Australia: Queensland
serena (Turner), 1915 (*Crabro*); New Hebrides
smithiensis Leclercq, 1954; Australia: Queensland, Australian Capital Territory
   tridentata F. Smith, 1868 (*Crabro*), nec Fabricius, 1775
tasmanica (F. Smith), 1856 (*Crabro*); Australia: Tasmania

Genus Ectemnius Dahlbom

Generic diagnosis: Eyes bare, inner orbits convergent below; scapal basin simple or with a weak carina medially above; orbital foveae indistinct (rarely rather distinct in *krusemani* group); ocellar triangle broad (nearly equilateral in a few species of *krusemani* group); vertex simple or with a weak sulcus extending backward from anterior ocellus, its punctation usually similar to that in front of ocellar triangle; gena simple; occipital carina complete to hypostomal carina or nearly so; antennal sockets close together and to inner orbits (a little removed from latter in *rubrocaudatus* group); male antenna with 12 or rarely 13 articles; scape ecarinate, unicarinate or bicarinate; male flagellum simple or modified, without ventral hair fringe; palpal formula 6-4, mandible apically bidentate or tridentate, externoventral margin unnotched, inner margin often with a large tooth, rarely edentate; pronotal collar carinate or sometimes ecarinate; scutum simple. lateral flange extending in front of tegula, notauli sometimes ridged; prescutellar sulcus foveate; metanotum simple; postspiracular carina continuous with omaulus and acetabular carina, sternaulus sometimes present as forward extension of verticaulus which is always at least L-shaped, hypersternaulus and mesopleuraulus absent; propodeum coarsely or sometimes finely sculptured, enclosure usually well defined, rarely (in *krusemani* group) polished; lateral propodeal carina present or absent; legs simple or modified, male midtibia sometimes without an apical spur; recurrent vein joining submarginal cell beyond its distal third (fig. 122 G) (rare exceptions), marginal cell rather truncate distally; jugal lobe shorter than submedian cell; gaster sessile or pedunculate, pygidial plate of male sometimes defined, that of female strongly narrowed apically and spout shaped (fig. 122 D), rarely flat and triangular.

Geographic range: Worldwide but only two species in Australia. About 160 species have been described. Numerous closely related species are a feature of the fauna in the Hawaiian Islands.

Systematics: Most *Ectemnius* are easily recognizable by their rather coarse sculpture, low ocellar triangle, well developed and L-shaped verticaulus, gutter-shaped pygidial area in the female (fig. 122 D), and the 12 articles of the male antenna. However, a number of small species groups or "subgenera" are not in agreement with one or more of the more common characters. To understand the odd groups it is necessary to have a picture of the most commonly encountered situation, agreeing for the most part with the "subgenus *Hypocrabro*": yellowish or whitish markings rather extensive; mandible with a strong inner tooth; male antenna with 12 articles, flagellomere I in both sexes longer than II but not twice as long; antennal sockets contiguous with ocular margin or almost so; clypeal pubescence appressed and silvery or golden; scapal basin not margined above by even a faint carina; pronotal collar with a dorsal carina which does not extend onto pronotal lobes; forefemur of male not expanded shieldlike; thoracic sculpture rather coarse and close, often ridged but scutum without microridges anteriorly and longitudinal ones medially; scutellum rather simply convex; verticaulus not extending far forward; pygidial plate of male undeveloped; pygidial plate of female not flat but gutter shaped and greatly narrowed posteriorly (fig. 122 D). The 16 species groups (or subgenera) in current use are best identified by their exceptions to the above list. We prefer to consider them as species groups rather than as subgenera, but for the benefit of those who take a different view, we have used the equivalent subgeneric names along with a letter designation which follows each species name in the checklist.

E. domingensis group (A, = *Apoctemnius* Leclercq). Verticaulus (precoxal carina of Leclercq) connected to omaulus by sternaulus or nearly so; tergum I coarsely punctate and rather flattened; male pygidial plate sometimes developed but rounded.

E. menyllus group (Ca, = *Cameronitus* Leclercq). Scutum punctate-shagreened or punctate, not ridged; male pygidial plate developed and subquadrate.

E. lapidarius group (Cl, = *Clytochrysis* A. Morawitz). Flagellomere I in both sexes at least twice as long as II; pronotal collar frequently ecarinate; mandible without an inner tooth; thoracic sculpture mostly not coarse.

E. guttatus group (E, = *Ectemnius* Dahlbom s.s.). Scapal basin at least faintly margined by a carina at upper middle; female clypeus often strongly produced and truncate medially.

E. decemmaculatus group (H, = *Hypocrabro* Ashmead). Generally agreeing with basic list of characters in the *domingensis* group.

E. furuichii group (I, = *Iwataia* Tsuneki). Scutum with sparse and somewhat longitudinally joined punctures; male pygidial plate developed and subquadrate; male flagellomere IV excavated beneath and produced apically (contrary to *menyllus* group); one species from Madagascar and another from Japan and Taiwan.

E. leonesus group (L, = *Leocrabro* Leclercq). Male antenna with 13 articles, flagellomere I shorter than II; scutellum at posterior fourth with a transverse ridge preceded by a row of pits; verticaulus prolonged anteriorly and curving ventrally; one species from South America.

E. cyanauges group (Mer, = *Merospis* Pate). Integument distinctly metallic blue; male forefemur shieldlike; one species from Cuba.

E. lituratus group (Met, = *Metacrabro* Ashmead). Scutum with transverse microridges anteriorly and longitudinal ones behind them.

E. rubrocaudatus group (N, = *Nesocrabro* Perkins). Body often all dark or almost so; mandible with a weak inner tooth; clypeal pubescence reddish gold to silvery buff, semierect; antennal sockets well removed from ocular margins; thoracic sculpture mostly fine; six species from Hawaiian Islands.

E. abnormis group (O, = *Oreocrabro* Perkins). Body often all dark or nearly so; mandible with a weak internal tooth; thoracic sculpture mostly fine; female pygidial plate more nearly flat and *Crabro*-like than usual; many species from Hawaiian Islands.

E. krusemani group (Pol, = *Policrabro* Leclercq). Integument extensively smooth or finely sculptured, propodeal enclosure polished; pronotal collar ecarinate; flagellomere I a little longer or shorter than II; much like *Crossocerus* but verticaulus present; forewing different, and male antennae with 12 articles; four species from India to East Indies.

E. tabanicida group (P-1, = *Protoctemnius* Leclercq). Male antenna with 13 articles, flagellomere I shorter than II; one species from Brazil.

E. rufifemur group (P-2, = *Protothyreopus* Ashmead). Female pygidial plate a flat triangle as in *Crabro;* scape ecarinate; two North American species.

E. crassicornis group (T, = *Thyreocerus* A. Costa). Male flagellomeres II-IV flattened and dilated, excavate beneath; flagellomere I shorter than II; carina of pronotal collar extending onto pronotal lobes; eight species from southern Europe to southwest Asia.

E. martjanowii group (Y, = *Yanonius* Tsuneki). Female flagellomere I at least twice as long as II; mesopleuron leathery or dull, without regular punctation; propodeal enclosure nearly all smooth; three species from Siberia and Sikkim to mountains of Japan and China.

Ectemnius seems most closely related on the one hand to *Lestica* and on the other to *Crossocerus*. Only the more coarsely sculptured *Ectemnius* would be confused with *Lestica,* and the well defined orbital foveae of the latter are sufficient for separation. The *krusemani* group of species seems to lean toward *Crossocerus* as indicated by Leclercq (1958c). We have seen examples of *forestus* and *krusemani.* The somewhat high ocellar triangle, rather distinct orbital foveae, and polished integument of its species places them outside the usual concept of *Ectemnius.* The presence of a verticaulus and the 12 articles of the male antenna differentiate them from *Crossocerus.*

Characters useful at the specific level, in addition to the group distinctions, are found in the color pattern, sculpture, relative length of flagellomere I, nature of the excavations commonly seen on male flagellomeres, clypeal shape, presence or absence of a basoventral tooth or lamina on the male forefemur, shape of the basal midtarsomeres of males, and details of the female pygidial plate.

Ectemnius was considered as a part of *Crabro* by most authors before Pate's classic "Genera of pemphilidine wasps" (1944b). It is the second largest genus in the tribe, exceeded in number of species only by *Crossocerus*. It is certainly one of the more highly evolved crabronin genera as indicated by its position in table 15. In the past 20 years, most contributions to our knowledge of *Ectemnius* have come from Jean Leclercq and Katsuji Tsuneki but particularly from the former. Partial keys to subgenera were given by Leclercq (1954a, 1958c, 1968b) and keys to species by Leclercq (1958c, d) in southeast Asia and (1968b) in Latin America. Beaumont (1964b) gave a useful illustrated key to the species of central Europe. An important recent paper on Latin American species is that of Leclercq (1972b).

Biology: As might be expected in such a large genus considerable biological information has accumulated involving more than 30 species. There does not seem to be much correlation between habits and species groups (or subgenera) in choice of nesting site or of prey. However, no data are available for the *rufifemur, crassicornis, krusemani, martjanowii, cyanauges,* or *leonesus* groups.

Ectemnius burrows are often made in decaying wood of logs or stumps but sometimes in rather sound lumber (Krombein, 1963a: *cephalotes*). A considerable number of smaller species nest in pithy twigs (Parker and Bohart, 1966: *spiniferus;* Krombein, 1963a: *paucimaculatus, stirpicola;* Tsuneki, 1960: *rubicola*). Ground nests have been reported mostly for Hawaiian species. Some of these have the female pygidial plate relatively broad and nearly flat (Yoshimoto, 1960: *rubrocaudatus, compactus, atripennis, unicolor*). We are not aware of published information on the *rufifemur* group of North America (*dilectus, rufifemur*), but the flat pygidial plates of females are often observed to be worn in museum specimens and bear globules of soil. In addition there is a short rake on the female front basitarsus. Consequently, they are almost certainly ground-nesting. Burrows in decayed wood are frequently multiple and complicated, with as many as 24 branches, each ending in a cell (Tsuneki, 1960: *ruficornis,* as *nigrifrons*). Occassionally, a number of branches may contain two cells so that the total may reach 40 (Tsuneki, 1960: *spinipes*). During provisioning the burrow entrances may be left open (Tsuneki, 1960: *nigritarsus*) or closed (Tsuneki,

1960: *konowi*). Many species are multivoltine (Krombein, 1963a), and it is common for them to re-use old nests in logs, remodeling and extending them (Tsuneki, 1960). Tsuneki reported also that the egg is laid on the venter of the neck of the first prey in the cell but is not deposited until the cell is fully provisioned.

Adult flies are the favored prey for *Ectemnius*. Literature reports ordinarily list Syrphidae, Stratiomyidae, and Muscoidea; less often Tabanidae, Dolichopodidae, and Therevidae; and rarely Tipulidae, Ephydridae, Acroceridae, Pipunculidae, and a variety of small flies. More unusual prey are Lepidoptera, Ephemeroptera, Neuroptera, and Orthoptera. According to Tsuneki (1960), *E. spinipes* provisions with a large variety of adult moths but especially Pyralidae. One lycaenid butterfly was also given as prey. The same author also confirmed the use of long-horned grasshoppers (*Conocephalus*) by *E. furuichii* as previously reported by Iwata (1941). The records of Chrysopidae (Yasumatsu and Watanabe, 1964: *schlettereri*) and mayflies (Tsuneki, 1960: *rubicola*) appear to be "mistakes" by the wasp or merely supplements to the normal fly provisions.

According to Evans (1964d), the Old World species *sexcinctus* (as *quadricinctus*) nest communally, several females sharing a common burrow entrance.

A variety of parasites were recorded during an in-depth study by Krombein (1964c) of the twig-nesting *E. paucimaculatus*. They were in the families Phoridae, Anthomyiidae, Perilampidae, Torymidae, and Platygasteridae.

The most detailed work on biology is that of Tsuneki (1960), who included 12 Japanese species. Other extensive reports are those of Yoshimoto (1960), who summarized previous findings of 13 Hawaiian species, and Krombein (1963a), who presented data on eight species in eastern United States. Older references are given by Kohl (1915) and Leclercq (1954a).

Checklist of *Ectemnius*

abnormis (Blackburn), 1887 (*Crabro*); Hawaiian Is.: Oahu (O)
abyssinicus (Arnold), 1947 (*Crabro*); Ethiopia (Met)
adspectans (Blackburn), 1887 (*Crabro*); Hawaiian Is.: Hawaii, Molokai, Maui (N)
    daemonius Perkins, 1899 (*Nesocrabro*)
agycus (Cameron), 1904 (*Crabro*); Himalaya Mts., Java, Sumatra, Malaya, Philippines (Pol), new combination by J. Leclercq
    forestus Leclercq, 1958, new synonymy by J. Leclercq
albomaculatus Tsuneki, 1966; Ryukyus (Ca)
alishanus Tsuneki, 1968; Taiwan (Ca)
alpheus Pate, 1946; w. U.S. (H)
ammanitus Leclercq, 1958; India: Assam; China: Tonkin (Ca)
aprunatus Leclercq, 1968; Mexico: Orizaba (H)
arcuatus (Say), 1837 (*Crabro*); Nearctic Region (H), new status by R. Bohart
    packardii Cresson, 1865 (*Crabro*)
    honestus Cresson, 1865 (*Crabro*)
    villosifrons Packard, 1866 (*Crabro*)
    nokomis Rohwer, 1908 (*Crabro*)
    chrysargynus Mickel, 1917 (*Solenius*), lapsus and misidentification of *chrysargyrus*
asiaticus Leclercq, 1950; Sikkim (Y)
atriceps (Cresson), 1865 (*Crabro*); Nearctic Region (E)
    brunneipes Packard, 1866 (*Crabro*), new synonymy by R. Bohart
    foxii Kincaid, 1900 (*Crabro*)
atripennis (Perkins), 1899 (*Crabro*); Hawaii (O)
auriceps (Cresson), 1865 (*Crabro*); Cuba, Bahamas (H)
aztecus Leclercq, 1951; Mexico: Orizaba; Brazil (H)
basiflavus (Brèthes), 1910 (*Crabro*); Mexico to Argentina (H)
    umbrosus Schrottky, 1914 (*Xylocrabro*)
    cassus Leclercq, 1956
berissus Leclercq, 1972; Argentina (H)
besseyae (Rohwer), 1908 (*Xylocrabro*); w. U.S. (H)
bidecoratus (Perkins), 1899 (*Nesocrabro*); Hawaii (N)
bogorensis Leclercq, 1958; S. India, Java, Singapore, Philippines (Ca)
boletus Leclercq, 1958; Malaya (Ca)
    ssp. *gedehensis* Leclercq, 1958; Java
borealis (Zetterstedt), 1838 (*Crabro*); Palearctic Region (E)
    bipunctatus Zetterstedt, 1838 (*Crabro*), nec Fabricius, 1787
    nigrinus Herrich-Schaeffer, 1841 (*Crabro*)
    gredleri Kohl, 1887 (*Lindenius*), new synonymy by J. Leclercq
burgdorfi Leclercq, 1968; Costa Rica (Ce)
carinatus (F. Smith), 1873 (*Crabro*); S. America, incl. Trinidad and Curacao (A)
    maculicornis Taschenberg, 1875 (*Crabro*)
    bahiacus Leclercq, 1950
    capricornicus Leclercq, 1950
    riojacus Leclercq, 1950
    esterensis Leclercq, 1950
cavifrons (Thomson), 1870 (*Crabro*); Palearctic Region incl. British Isles (Ce)
    ?*interruptefasciatus* Retzius, 1783 (*Sphex*)
    aurarius Matsumura, 1912 (*Crabro*)
    ssp. *kizanensis* Tsuneki, 1972; Taiwan
    ssp. *nipponensis* Tsuneki, 1972; Japan
centralis (Cameron), 1891 (*Crabro*); sw. U.S., Mexico to Colombia and Trinidad (A)
cephalotes (Olivier), 1792 (*Crabro*); Europe, e. U.S. (Met)
    floralis Olivier, 1792 (*Crabro*)
    geniculatus Olivier, 1792 (*Crabro*)
    tibialis Olivier, 1792 (*Crabro*)
    ?*cephalotes* Panzer, 1799 (*Crabro*), nec Olivier, 1792
    striatus Lepeletier and Brullé, 1834 (*Crabro*), p. 707
    ornatus Lepeletier and Brullé, 1834 (*Crabro*)
    striatus Lepeletier and Brullé, 1834, p. 744 (*Cerato-*

colus), nec Lepeletier and Brullé, 1834, p. 707
striatulus Lepeletier and Brullé, 1834, p. 737 (*Blepharipus*)
lindenius Shuckard, 1837 (*Crabro*)
shuckardi Dahlbom, 1838 (*Crabro*)
interruptus Dahlbom, 1845 (*Crabro*), unnecessary new name for *shuckardi,* nec Fabricius, 1787
fargeii F. Smith, 1856 (*Crabro*), new name for *Ceratocolus striatus*
lindensis Inchbald, 1859 (*Crabro*), emendation or lapsus
aciculatus Provancher, 1882 (*Crabro*)
ruthenicus F. Morawitz, 1892 (*Crabro*)
lindenii "Inchbald" Dalla Torre, 1897 (*Crabro*)
chagrinatus Leclercq, 1950; Costa Rica, Ecuador (Ce)
ssp. *cayerae* Leclercq, 1968; Argentina
chrysites (Kohl), 1892 (*Crabro*); India, Japan, Korea, Taiwan, Ryukyus, Philippines, e. U.S.S.R. (Met)
auricomus Bingham, 1897 (*Crabro*)
khasianus Cameron, 1902 (*Crabro*)
butuani Leclercq, 1963, new synonymy by J. Leclercq
clearei Leclercq, 1964; Guyana (H)
compactus (Perkins), 1899 (*Nesocrabro*); Hawaiian Is.: Kauai, Lanai, Oahu (N)
lanaiensis Perkins, 1899 (*Nesocrabro*)
confinis (Walker), 1871 (*Crabro*); Mediterranean area, sw. USSR, n. India (H)
laevigatus De Stefani, 1884
pedicellaris F. Morawitz, 1889 (*Crabro*)
flavicollis F. Morawitz, 1892 (*Crabro*)
hannonis Gribodo, 1896 (*Crabro*)
balucha Nurse, 1903 (*Crabro*), ♂ only
subtilus "Perez", Antiga and Bofill, 1904 (*Crabro*), nomen nudum
conglobatus (Turner), 1908 (*Crabro*); Australia: Queensland (Ca)
continuus (Fabricius), 1804 (*Crabro*); Holarctic Region (H)
sexmaculatus Say, 1824 (*Crabro*), nec Olivier, 1792
punctatus Lepeletier and Brullé, 1834 (*Solenius*), p. 720
punctatus Lepeletier and Brullé, 1834 (*Ceratocolus*), p. 749, nec Lepeletier and Brullé, 1834, p. 720
fuscitarsis Herrich-Schaeffer, 1841 (*Crabro*)
sulphureipes F. Smith, 1856 (*Crabro*)
impressus F. Smith, 1856 (*Crabro*), new name for *punctatus* Lepeletier and Brullé, p. 749
fuscitarsus "Herrich-Schaeffer" Schenck, 1857 (*Crabro*), lapsus
vagatus F. Smith, 1869 (*Crabro*)
granulatus Walker, 1871 (*Crabro*)
rugosopunctatus Taschenberg, 1875 (*Crabro*)
validus De Stefani, 1884 (*Crabro*)
vagans Fokker, 1887 (*Crabro*)
slossonae Ashmead, 1902 (*Xylocrabro*)
bisexmaculatus Viereck, 1910 (*Crabro*), new name for *sexmaculatus* Say

sayi Cockerell, 1910 (*Crabro*), new name for *sexmaculatus* Say
giffardi Rohwer, 1917 (*Solenius*)
vagus of authors, not Linnaeus
ssp. *rufitarsis* (Dalla Torre), 1897 (*Crabro*); Canary Is., new name for *rufipes* (Art. 59c)
rufipes Brullé, 1839 (*Crabro*), nec *Crabro rufipes* (Lepeletier and Brullé), 1834 (*Ceratocolus*)
corporaali Leclercq, 1950; Sumatra (Ce)
corrugatus (Packard), 1866 (*Crabro*); Nearctic Region (E)
pauper Packard, 1866 (*Crabro*)
operus Rohwer, 1908 (*Crabro*)
drymocallidis Rohwer, 1908 (*Crabro*)
corvidus Leclercq, 1961; Malaya; India: Assam (H)
craesus (Lepeletier and Brullé), 1834 (*Solenius*); W. Indies (A)
crassicornis (Spinola), 1808 (*Crabro*); Europe, sw. USSR (T)
hybridus Eversmann, 1849 (*Crabro*)
punctulatus De Stefani, 1884
siculus De Stefani, 1884
crippsi (Arnold), 1927 (*Thyreopus*); Rhodesia (Met)
ssp. *mozambicus* (Arnold), 1960 (*Crabro*); Mozambique
crudator Leclercq, 1968; Brazil, Argentina (H)
cuernosi Leclercq, 1963; Philippines: Negros, Panay (T)
curictensis (Mader), 1939 (*Crabro*); e. Europe (T)
curtipes (Perkins), 1899 (*Crabro*); Hawaii (O)
cyanauges Pate, 1936; Cuba (Mer)
dartanus Leclercq, 1968; Brazil, Argentina, Paraguay (H)
decemmaculatus (Say), 1823 (*Crabro*); e. of Rocky Mts. in U.S., Mexico (H)
chrysargyrus Lepeletier and Brullé, 1834 (*Crabro*), new synonymy by Bohart
chrysarginus Lepeletier, 1845 (*Crabro*), lapsus
chrysargurus Dahlbom, 1845 (*Crabro*), lapsus
collinus F. Smith, 1856 (*Crabro*)
aurifrons F. Smith, 1856 (*Crabro*)
novanus Rohwer, 1911 (*Crabro*)
tequesta Pate, 1946, new status by Bohart
defiguratus Leclercq, 1972; Bolivia (H)
dilaticornis (F. Morawitz), 1893 (*Crabro*); sw. USSR: Turkmen S.S.R. (T)
dilectus (Cresson), 1865 (*Crabro*); Nearctic Region (P-2)
bigeminus Patton, 1879 (*Crabro*)
megacephalus Rohwer, 1908 (*Crabro*), nec Rossi, 1790
dilectiformis Rohwer, 1909 (*Crabro*)
crassiceps Mickel, 1916 (*Crabro*)
discrepans (Giffard), 1915 (*Melanocrabro*); Hawaiian Is.: Kauai I. (O)
distinctus (F. Smith), 1856 (*Crabro*); Hawaiian Is.: Oahu I. (O)
notostictus Perkins, 1899 (*Crabro*)
dives (Lepeletier and Brullé), 1834 (*Solenius*); Holarctic Region (E)
octonotatus Lepeletier and Brullé, 1834 (*Solenius*)

alatulus Dahlbom, 1838 (*Crabro*)
pictipes Herrich-Schaeffer, 1841 (*Crabro*)
octavonotatus Lepeletier, 1845 (*Solenius*), emendation
auratus F. Smith, 1856 (*Crabro*)
montanus Cresson, 1865 (*Crabro*), nec Gistel, 1857
cristatus Packard, 1866 (*Crabro*)
cubiceps Packard, 1866 (*Crabro*)
heraclei Rohwer, 1908 (*Crabro*)
montivagans Strand, 1917 (*Crabro*), new name for *montanus*
dizoster Pate, 1947; Jamaica (A)
domingensis Leclercq, 1950; Ecuador (A)
dominicanus Evans, 1972; W. Indies: Dominica (H)
dungensis Leclercq, 1958; Java (T)
ssp. *wattanapongsiri* Tsuneki, 1963; Thailand
embeliae Leclercq, 1958; Malaya, Java, Thailand, Philippines (Ca)
erebus Leclercq, 1958; Malaya (Ca)
esterensis Leclercq, 1950; Argentina (A)
excavatus (W. Fox), 1892 (*Crabro*); U.S.: Florida; Mexico; Costa Rica (H)
banksi Rohwer, 1909 (*Crabro*), new status by Bohart
ravinus Leclercq, 1968
flagellarius (F. Morawitz), 1892 (*Crabro*); sw. USSR, India (T)
balucha Nurse, 1903, in part
flavipennis (Lepeletier and Brullé), 1834 (*Ceratocolus*); Ecuador to Argentina (H)
cordillierae Leclercq, 1950
flavohirtus Tsuneki, 1954; Japan (Ca)
fliranus Leclercq, 1968; Paraguay, Brazil (A)
fossorius (Linnaeus), 1758 (*Sphex*); Europe, USSR, Turkey (Met)
bucephalus Christ, 1791 (*Sphex*)
aspidiphorus Schrank, 1802 (*Crabro*)
synonymy by J. Leclercq
fuscipennis Lepeletier and Brullé, 1834 (*Solenius*), p. 714, nec Lepeletier and Brullé, 1834 (*Crabro*), p. 710
grandis Lepeletier and Brulié, 1834 (*Solenius*)
tetraedrus Dahlbom, 1845 (*Solenius*)
fumipennis F. Smith, 1856 (*Crabro*), new name for *fuscipennis*
montanus Gistl, 1857 (*Crabro*)
my by J. Leclercq
fredericismithi (Schulz), 1906 (*Crabro*); Hawaiian Is.: Kauai (O)
affinis F. Smith, 1879 (*Crabro*), nec Rossi, 1792
fulvicrus (Perkins), 1899 (*Crabro*); Hawaii (O)
fulvopilosellus (Cameron), 1902 (*Crabro*); Sikkim; India: Assam (Met)
ctenopus Cameron, 1907 (*Crabro*)
fulvopilosus Meade-Waldo, 1915 (*Crabro*), emendation or lapsus
furuichii (Iwata), 1934 (*Crabro*); Japan (I)
ssp. *formosanus* Tsuneki, 1960; Taiwan
fuscipennis (Lepeletier and Brullé), 1834 (*Crabro*); India: Bengal (Ce)

guadalupensis Leclercq, 1972; "Guadalupa" (H)
guttatus (Vander Linden), 1829 (*Crabro*); Europe (E)
laportei Lepeletier and Brullé, 1834 (*Crabro*)
borealis Dahlbom, 1838 (*Crabro*), nec Zetterstedt, 1838
spinicollis Herrich-Schaeffer, 1841 (*Crabro*)
parvulus Herrich-Schaeffer, 1841 (*Crabro*)
laportaei Lepeletier, 1845 (*Crabro*), lapsus
pictus Schenck, 1857 (*Crabro*), nec Fabricius, 1793
divitoides C. Verhoeff, 1892 (*Crabro*)
parvulus Herrich-Schaeffer of C. Verhoeff, 1892 (*Crabro*)
verhoeffii Dalla Torre, 1897 (*Crabro*), new name for *parvulus*
schenckii Cockerell, 1907 (*Crabro*), new name for *pictus*
haleakalae (Perkins), 1899 (*Crabro*); Hawaiian Is.: Maui I. (O)
hawaiiensis (Perkins), 1899 (*Crabro*); Hawaii I. (O)
hebetescens (Turner), 1908 (*Crabro*); Australia: Queensland (H)
hispanicus (Kohl), 1915 (*Crabro*); Spain (H)
hypsae (De Stefani), 1894 (*Crabro*); Mediterranean area (H)
serotinus De Stefani, 1894
laetus "Pérez" Antiga and Bofill, 1904 (*Crabro*), nomen nudum
insignis (F. Smith), 1856 (*Crabro*); India (Met)
invalidus Leclercq, 1958; Malaya: Selangor (Pol)
iridifrons (Pérez), 1905 (*Crabro*); Japan (Met)
konowi (Kohl), 1905 (*Crabro*); e. Palearctic Region (Met)
rubropictus Matsumura, 1911 (*Crabro*)
kriechbaumeri (Kohl), 1879 (*Crabro*); Europe (Met)
krusemani (Leclercq), 1950 (*Crossocerus*); Indonesia (Pol)
lapidarius (Panzer), 1804 (*Crabro*); Holarctic Region (Cl)
sinuatus Fabricius, 1804 (*Crabro*)
?*cinctus* Spinola, 1806 (*Crabro*), nec Rossi, 1790
chrysostomus Lepeletier and Brullé, 1834 (*Crabro*), nec Gmelin, 1790
comptus Lepeletier and Brullé, 1834 (*Crabro*)
xylurgus Shuckard, 1837 (*Crabro*)
interstinctus F. Smith, 1851 (*Crabro*)
obscurus F. Smith, 1856 (*Crabro*)
gracilissimus Packard, 1866 (*Crabro*)
denticulatus Packard, 1866 (*Crabro*)
effossus Packard, 1866 (*Crabro*)
papagorum Viereck, 1908 (*Crabro*), new synonymy by R. Bohart
leonesus Leclercq, 1968; Argentina, Brazil, Surinam (L)
lesticoides Leclercq, 1950; French Guiana (H)
lituratus (Panzer), 1804 (*Crabro*); Europe incl. British Isles (Met)
petiolatus Lepeletier and Brullé, 1834 (*Solenius*)
fasciatus Lepeletier and Brullé, 1834 (*Ceratocolus*)
reticulatus Lepeletier and Brullé, 1834 (*Ceratoculus*)
kollari Dahlbom, 1845 (*Crabro*)

argenteus Schenck, 1857 (*Crabro*)
vestitus F. Smith, 1858 (*Crabro*)
intermedius A. Morawitz, 1866
dallatorreanus Kohl, 1880 (*Crabro*)
transiens Kohl, 1915 (*Crabro*)
lysias (Cameron), 1905 (*Crabro*); n. India (Ce)
mackayensis (Turner), 1908 (*Crabro*); Australia: Queensland (H)
maculosus (Gmelin), 1790, p. 2761 (*Vespa*); U.S. e. of 100th meridian; new name for *Vespa maculatus* (Fabricius) (Art. 59c) (Met)
    maculatus Fabricius, 1781 (*Crabro*), nec *Vespa maculata* Linnaeus, 1763, now in *Vespula*
    singularis F. Smith, 1856 (*Crabro*), new synonymy by Bohart
    frigidus F. Smith, 1856 (*Crabro*)
    quadrangularis Packard, 1866 (*Crabro*)
    quatuordecimmaculatus Packard, 1866 (*Crabro*)
    oblongus Packard, 1866 (*Crabro*)
    trapezoideus Packard, 1866 (*Crabro*)
    quadrangulus E. T. Cresson, Jr., 1928, lapsus
mandibularis (F. Smith), 1879 (*Crabro*); Hawaiian Is. Oahu, Molokai, Maui, Lanai (O)
    denticornis F. Smith, 1879 (*Crabro*)
    mauiensis Blackburn, 1887 (*Crabro*)
martjanowii (F. Morawitz), 1892 (*Crabro*); s. centr. USSR, Sikkim, China, Japan, Korea (Y)
    dubiosus Ashmead, 1904 (*Clytochrysus*)
    arreptus Kohl, 1915 (*Crabro*), new synonymy by J. Leclercq
    ssp. *tibeticus* Leclercq, 1950; Tibet
    ssp. *insulicola* Tsuneki, 1971; Taiwan
massiliensis (Kohl), 1883 (*Thyreocerus*); France, Spain, Algeria (T)
    bulgaricus Balthasar and Hrubant, 1967
mayeri (DeWitz), 1881 (*Crabro*); Puerto Rico (H)
melanotarsis (Cameron), 1902 (*Crabro*); India: Mts. of Assam (Ca)
    elvinus Cameron, 1905 (*Crabro*)
    monozonus Cameron, 1905 (*Crabro*)
    ssp. *changi* Tsuneki, 1971; Taiwan
menyllus (Cameron), 1905 (*Crabro*); n. India (Ca)
meridionalis (A. Costa), 1871 (*Crabro*); e. Europe, Mediterranean area (H)
    finitimus F. Morawitz, 1894 (*Crabro*)
    impressus Smith of authors
molokaiensis (Perkins), 1899 (*Crabro*); Hawaiian Is.: Molokai, Maui (O)
monticola (Perkins), 1899 (*Crabro*); Hawaiian Is.: Oahu, Kauai (O)
neptunus Leclercq, 1958; Borneo, Singapore (Met)
nielseni (Kohl), 1915 (*Crabro*); China, Korea (H)
nigritarsus (Herrich-Schaeffer), 1841 (*Crabro*); Palearctic Region, Indochina (Ca)
    trinotatus A. Costa, 1871 (*Crabro*)
    munakatai Tsuneki, 1947 (*Crabro*)
odyneroides (Cresson), 1865 (*Crabro*); U.S. w. of 100th meridian; Mexico (H)
    ariel Cameron, 1891 (*Crabro*)

pacuarus Leclercq, 1972; Costa Rica, Colombia (H)
pahangi Leclercq, 1958; Malaya (Ca)
palamosi Leclercq, 1964; Spain (E)
palitans (Bingham), 1896 (*Crabro*); India, Sri Lanka, Borneo, Philippines, Taiwan (Ca)
    orius Leclercq, 1958
    bornicus Leclercq, 1958
    cetonicus Leclercq, 1958
    palitoides Leclercq, 1963
    paxinus Leclercq, 1963
paucimaculatus (Packard), 1866 (*Crabro*); U.S.: Texas to New York (H)
pelotarum Leclercq, 1968; Colombia, Brazil, Argentina (H)
    cubiceps Taschenberg, 1875 (*Crabro*), nec Packard, 1866
pempuchi Tsuneki, 1971; Taiwan (Ca)
pendleburyi Leclercq, 1958; Malaya (Ca)
persicus (Kohl), 1888 (*Crabro*); Iran (H)
plutonius Leclercq, 1958; Celebes (Met)
polynesialis (Cameron), 1881 (*Crabro*); Hawaii I. (O)
praeclarus (Arnold), 1944 (*Crabro*); Madagascar (H)
praevius (Kohl), 1915 (*Crabro*); sw. USSR (E)
productus (W. Fox), 1897 (*Crabro*); Brazil (A)
proletarius (Mickel), 1916 (*Crabro*); Nearctic Region (E)
    parvulus Packard, 1866, nec Herrich-Schaeffer, 1841, new synonymy by R. Bohart
radiatus (Pérez), 1905 (*Crabro*); Japan (Met)
    mizuho Tsuneki, 1948 (*Crabro*)
    mizuko (and *mizubo*) Leclercq, 1954, lapsus
recuperatus Leclercq, 1972; Panama (H)
reginellus Leclercq, 1954; Australia: Queensland (H)
    cinctus Turner, 1908 (*Crabro*), nec Rossi, 1790
riosorum Leclercq, 1968; Argentina (H)
rubicola (Dufour and Perris), 1840 (*Solenius*); Europe (H)
    microstictus Herrich-Schaeffer, 1841 (*Crabro*)
    larvatus Wesmael, 1852 (*Crabro*)
    pumilus A. Costa, 1871
    ssp. *nipponis* Tsuneki, 1960; Japan
rubrocaudatus (Blackburn), 1887 (*Crabro*); Hawaii I. (N)
ruficornis (Zetterstedt), 1838 (*Crabro*); Holarctic Region, Mexico: Guerrero (Cl), new status by J. Leclercq
    aurilabris Herrich-Schaeffer, 1841 (*Crabro*)
    nigrifrons Cresson, 1865 (*Crabro*)
    contiguus Cresson, 1865 (*Crabro*)
    septentrionalis Packard, 1866 (*Crabro*)
    planifrons Thomson, 1870 (*Crabro*)
    hector Cameron, 1891 (*Crabro*)
    longipalpis C. Verhoeff, 1892 (*Crabro*)
    vestor Ashmead, 1899 (*Crabro*)
    lineatotarsis Matsumura, 1911 (*Crabro*)
    chipsanii Matsumura, 1911 (*Crabro*)
    ssp. *taiwanus* Tsuneki, 1968; Taiwan
rufifemur (Packard), 1866 (*Crabro*); Nearctic Region (P-2)
    ssp. *orizabinus* Leclercq, 1968; Mexico
rufipes (Lepeletier and Brullé), 1834 (*Ceratocolus*); U.S.

e. of 100th meridian (H), new status by R. Bohart (see Addendum)
  texanus Cresson, 1872 (*Crabro*), new synonymy by R. Bohart
  ssp. *ais* Pate, 1946; U.S.: Florida
rugifer (Dahlbom), 1845 (*Crabro*); Europe (H)
satan Pate, 1946; U.S.: California to New Mexico (H)
saxatilis (Cameron), 1891 (*Crabro*); Mexico to Colombia (A)
scaber (Lepeletier and Brullé), 1834 (*Solenius*); U.S. e of 100th meridian (H)
  ssp. *rufescens* Krombein, 1954; Fla.
schlettereri (Kohl), 1888 (*Crabro*); Palearctic Region (H)
  jakowlewi F. Morawitz, 1892 (*Crabro*)
  chinensis Sickmann, 1894 (*Crabro*)
  ?*nursei* Kohl, 1915 (*Crabro*)
  ?*obstrictus* Gussakovskij, 1933
  ssp. *sakaguchii* (Matsumura and Uchida), 1926 (*Crabro*); Taiwan, Ryukyus
    sagakuchii Leclercq, 1954, lapsus
  ssp. *ishigakiensis* Tsuneki, 1972; Ryukyus: Ishigaki I.
schwarzi (Rohwer), 1911 (*Crabro*); Guatemala (H)
semipunctatus (Lepeletier and Brullé), 1834 (*Crabro*); Brazil to Guatemala (H)
  opulentus F. Smith, 1856 (*Crabro*)
  championi Cameron, 1891 (*Crabro*)
  atitlanae Cameron, 1891 (*Crabro*)
  fumosus Brèthes, 1910 (*Crabro*)
sennacus Leclercq, 1968; Brazil (H)
servitorius Leclercq, 1972; Brazil, Peru, Argentina (H)
sexcinctus (Fabricius), 1775 (*Crabro*); sw. USSR, w. China (Ce)
  quadricinctus Fabricius, 1787 (*Crabro*)
  quatuorcinctus Christ, 1791 (*Sphex*)
  tibialis Olivier, 1792 (*Crabro*)
  octomaculatus Preyssler, 1793 (*Crabro*)
  zonatus Panzer, 1797 (*Crabro*)
  vespiformis Panzer, 1798 (*Crabro*)
  octomaculatus Schrank, 1802 (*Crabro*)
  flavipes Lepeletier and Brullé, 1834 (*Crabro*), nec Fabricius, 1781
  tetraedus Blanchard, 1840 (*Crabro*)
  saundersi Perkins, 1899 (*Crabro*)
seyrigi (Arnold), 1944 (*Crabro*); Madagascar (I)
?*shimoyamai* Tsuneki, 1958; Japan, (subgenus *Ceratocrabro* Tsuneki, may belong in *Williamsita*)
slateri (Arnold), 1926 (*Thyreopus*); centr. Africa (H)
  ssp. *nigrescens* (Arnold), 1944 (*Crabro*); Madagascar
sodalis (Bingham), 1897 (*Crabro*); n. India (Pol)
sonorensis (Cameron), 1891 (*Crabro*); sw. U.S., w. Mexico (H)
  montivagus Cameron, 1891 (*Crabro*)
  imbutus W. Fox, 1894 (*Crabro*)
  ferrugineipes Rohwer, 1908 (*Crabro*)
spiniferus (W. Fox), 1895 (*Crabro*); w. U.S. (H)
  conspiciendus Mickel, 1918 (*Solenius*)
spinipes (A. Morawitz), 1866 (*Crabro*); Palearctic Region (Met)
  bulsanensis Kohl, 1879 (*Crabro*)
  tetracanthus Pérez, 1905 (*Crabro*)
  jozankeanus Matsumura, 1912 (*Crabro*)
stirpicola (Packard), 1866 (*Crabro*); U.S. e. of 100th meridian (H)
stygius (Blackburn and Kirby), 1880 (*Crabro*); Hawaiian Is.: Oahu (N)
tabanicida (Fisher), 1929 (*Crabro*); Brazil (P-1)
taino Pate, 1947; Hispaniola (H)
teleges Pate, 1946; Galapagos (H)
trichiosomus (Cameron), 1904 (*Crabro*); India: Himalayas, Assam (Ca)
  himalayensis Cameron, 1905 (*Crabro*)
trifasciatus (Say), 1824 (*Crabro*), U.S., s. Canada (H)
tsuifenicus Tsuneki, 1971; Taiwan (Ca)
tumidoventris (Perkins), 1899 (*Crabro*); Hawaiian Is. (O)
  leucognathus Perkins, 1899 (*Crabro*)
unicolor (F. Smith), 1856 (*Crabro*); Hawaiian Is. (O)
  nesiotes Pate, 1937 (*Crabro*), unnecessary new name for *unicolor* F. Smith
varentzowi (F. Morawitz), 1894 (*Crabro*); sw. USSR: Transcaspia (H)
violaceipennis (Cameron), 1907 (*Crabro*); Sikkim (Ca)
walteri (Kohl), 1899 (*Crabro*); sw. USSR (H)
weberi Yoshimoto, 1960; Hawaii (O)
wickwari (Turner), 1920 (*Crabro*); Sri Lanka (Met)
yosemite Pate, 1946; U.S.: California (Cl)
yoshimotoi R. Bohart, new name; Hawaii (N)
  giffardi Yoshimoto, 1960, nec Rohwer, 1917

Fossil *Ectemnius*

longoevus (Cockerell), 1910 (*Crabro*); Florissant, Colo.

Genus Lestica Billberg

Generic diagnosis: Eyes bare, inner orbits convergent below; scapal basin simple; orbital foveae distinct, long oval, margined; ocellar triangle broad except in males with constricted genae; vertex faintly grooved behind midocellus, its punctation usually noticeably finer and more regular than that in front of ocellar triangle; gena sometimes much constricted in males, rarely with a genal carina; occipital carina reaching hypostomal carina or nearly so; antennal sockets close together and to inner orbits; male antenna with 12 articles; scape ecarinate or carinate; male flagellum slender and simple, or stout and with articles somewhat swollen toward base, rarely with a ventral hair fringe; palpal formula 6-4; mandible apically bidentate or tridentate, inner margin with or without a low tooth; clypeus usually with a sharp median carina, somewhat truncate and produced apically, flanked by a sharp submedian or sublateral tooth; scutum simple, lateral flange ending opposite front of tegula; notauli weak or apparently absent; prescutellar sulcus narrowly but deeply foveate; metanotum simple; postspiracular carina continuous with omaulus; acetabular carina present, absent, or

FIG. 131. *Lestica cinctella* (W. Fox), male.

incomplete; sternaulus absent or faintly indicated; verticaulus present, nearly straight or L-shaped; enclosure moderately to weakly defined, rough; lateral propodeal carina present or absent; legs often greatly modified, males and many females without a midtibial spur, male tarsi frequently lamellate, male forefemur often with a ventral spine (fig. 131) or deformed and unusually hairy; recurrent vein joining submarginal cell at its distal fifth or beyond, marginal cell truncate distally (fig. 131); jugal lobe shorter than submedian cell; gaster sessile, pygidial plate of female strongly narrowed and gutterlike or rather flat and subtriangular.

Geographic range: The genus occurs on all continents and some large islands, particularly those in the Oriental Region. In the New World only the *confluenta* group is represented. There are 38 listed species.

Systematics: Lestica is obviously close to *Ectemnius* and is perhaps an evolutionary extension of it. The combination of coarse thoracic sculpture (fig. 131) and definite orbital foveae with raised edges will distinguish *Lestica*. In most cases the rather coarse and uneven punctation in front of the ocellar triangle of *Lestica* will distinguish the two genera, also. Many *Lestica* males have a ventral spine on the forefemur, but this occurs also in *Ectemnius maculosus*. Another common but not exclusive character is the frequent dilation of the tarsi, especially the front basitarsus of male *Lestica*. In *clypeata* this produces an analogous situation to that found in *Ectemnius cyanauges* where the femur is dilated, and in many species of *Crabro* in which the tibiae are shieldlike.

The use of subgenera in *Lestica* has resulted from the formalizing of the groups established by Kohl (1915) and the addition of *Ptyx* by Pate (1947d). Thus, Leclercq (1954a) recognized five subgenera: *Lestica, Ceratocolus, Ptyx, Clypeocrabro,* and *Solenius.* Several subsequent authors have remarked on exceptions and have thrown doubt on the validity of subgenera. Material available to us includes only one species in each of the "subgenera" except *Solenius,* where we have eight species. Subgeneric characters in previous use that seem to be obviously variable above the species level are (1) the tooth on the inner margin of the mandible, (2) the termination point of the occipital carina beneath the head, (3) constriction of the male head in the genal area with consequent narrowing of the ocellar triangle, and (4) presence or absence of a scapal carina. Because of the variability in these characters we prefer to use species groups, but for those who wish to use subgenera we suggest the following arrangement based on females:

Lestica Billberg *s.s.* (= *subterraneus* group), including *Ptyx* (= *pluschtschevskyi* group) and *Ceratocolus* (= *alatus* group): female mandible bidentate at apex and relatively long, pygidial plate rather flat and subtriangular, terricolous forms.

Solenius Lepeletier and Brullé (= *confluenta* group), with *Clypeocrabro* (= *clypeata* group) as a synonym: female mandible tridentate at apex and relatively short; pygidial plate scooplike and strongly narrowed on distal half, flanked by hair tufts, lignicolous forms.

On a group basis the separation within *Lestica s.s.* seems to be based primarily on males: head constricted behind in *pluschtschevskyi* group and a malar space present; head constricted behind in *alatus* group, but practically no malar space; and head not constricted behind in *subterraneus* group. The *clypeata* group and *confluenta* group differ as indicated in the paragraphs above; in addition the former has the male head constricted behind (fig. 121 J), whereas it is more normal in the latter. Also, in American material available to us, males in the *confluenta* group have the forefemur spinose beneath. There have been no adequate keys published since Kohl (1915) in which the Palearctic fauna was treated.

Biology: Records are available for eight of the 37 listed species. In general it can be said that adult Lepidoptera are used as prey, especially small Noctuidae and such Microlepidoptera as pyralids and tortricids. There are records of two species using an occasional butterfly (Lycaenidae). Observations made prior to 1850 on *clypeata* and reported by Kohl (1915) give flies as prey. Since more modern records summarized by Leclercq (1954a) give moths as prey for *clypeata,* the earlier identifications may well have been incorrect.

Present evidence supports the ideas of Pate (1947d) that those females with a rather flat pygidial plate (*subterranea* and *alata* groups) make ground nests. The *pluschtschevskyi* group would presumably fit here too, but nothing is known of its habits. On the other hand, females that have the pygidial plate greatly narrowed on the distal half, concave and flanked by hair tufts, (*confluenta* and *clypeata* groups) are lignicolous, that is, nest in old wood or occasionally in plant stems.

The most extensive data on nesting habits are those of Tsuneki (1960) on *alata, reiteri, camelus, heros,* and *collaris* in Japan. In the case of *alata* the burrows were 3 to 15 cm below the surface of the ground and the tunnels were strongly curved or sinuate. A mound of tumulus marked the opening which was not closed in the absence of the wasp. Presumably, there were a number of small branches at the end of the main tunnel and each branch ended in a cell that contained about seven prey (2-9). The egg was attached to the first moth prey on the underside of the neck, but it was not laid until provisioning was complete. The cell wall was slightly smoothed, and the moth prey were packed into a "sausage, with their wings forming the outer epidermis and with their bodies constituting the inside meat." The wasp larvae used the moth wings as an outer layer on their cocoons.

Lestica nesting in wood frequently gain access through cracks or old beetle galleries. Also, they frequently re-use and remodel old nests of their own species. Their burrows are typified by those of *reiteri* on which Tsuneki (1960) made detailed observations. Moths were seized by the neck from above by the midlegs of the wasp and flown to the nest. A typical burrow was in a large decayed tree stump. It was highly multiple and extended about 16 cm into the wood. A few side branches diverged near the entrance but most of them, perhaps a dozen, were more nearly terminal. Each side branch ended in one to three cells in a linear arrangement. As many as 27 cells were found in a single nest. The number of prey per cell averaged about seven, and these were fashioned into a "sausage" as described for *alata.*

Checklist of *Lestica*

[Species groups and the corresponding subgeneric names now in use are indicated by letter symbols at the end of each citation: *alata* group (*Ceratocolus*) (ala), *clypeata* group (*Clypeocrabro*) (cly), *confluenta* group (*Solenius*) (con), *pluschtschevskyi* group (*Ptyx*) (plu), *subterranea* group (*Lestica*) (sub).]

alacer (Bingham), 1896 (*Crabro*); Sumatra, Sikkim, Taiwan (con)
alata (Panzer), 1797 (*Crabro*); Palearctic Region (ala)
    basalis F. Smith, 1856 (*Crabro*)
    japonica Schulz, 1904 (*Crabro*))
aurantiaca (Kohl), 1915 (*Crabro*); sw. USSR: Turkmen S.S.R. (ala)
bibundica Leclercq, 1972; Cameroun (con)
camelus (Eversmann), 1849 (*Crabro*); Japan, USSR (cly)
    sapporensis Matsumura, 1911 (*Crabro*)
cinctella (Fox), 1895 (*Crabro*); w. U.S.
clypeata (Schreber), 1759 (*Apis*); Europe, Middle East (cly)
    clypearia Schreber, 1784 (*Sphex*)
    ovata Christ, 1791 (*Sphex*)
    vexillata Panzer, 1797 (*Crabro*)
    lapidaria Fabricius, 1804 (*Crabro*), nec Panzer, 1804
    nigridens Herrich-Schaeffer, 1841 (*Crabro*)

collaris (Matsumura), 1912 (*Crabro*); Japan (con)
    aberrans Gussakovskij, 1933 (*Crabro*)
combinata Leclercq, 1963; Philippines: Mindanao (con)
compacta (Kohl), 1915 (*Crabro*); Iran (sub)
confluenta (Say), 1837 (*Crabro*); U.S., s. Canada (con)
    interrupta Lepeletier and Brullé, 1834 (*Solenius*), nec *Crabro interruptus* (Lepeletier and Brullé), 1834 (*Thyreopus*)
    dubia F. Smith, 1856 (*Crabro*), new name for *Crabro interruptus* (Lepeletier and Brullé), 1834 (*Solenius*) (Art. 59c). Smith's new name precludes use of *interruptus*, and we have used the next available name, *confluenta*.
    confluens LeConte, 1859 (*Crabro*), lapsus
    bella Cresson, 1865 (*Crabro*), new synonymy by R. Bohart
    atrifrons Cresson, 1865 (*Crabro*)
    eburnea Taschenberg, 1875 (*Crabro*)
    cinctibella Viereck, 1908 (*Crabro*), new synonymy by R. Bohart
    opwana Rohwer, 1908 (*Crabro*)
    townsendi Rohwer, 1911 (*Crabro*)
    planaris Mickel, 1916 (*Crabro*), new synonymy by R. Bohart
    *seamansi Carter, 1925 (*Solenius*), new synonymy by R. Bohart
consolator Leclercq, 1963; Borneo (con)
constanceae (Cameron), 1891 (*Crabro*); s. Mexico to Argentina (con)
constricta Krombein, 1949; Carolines: Palau; Philippines; China; Taiwan (con)
cubensis (Cresson), 1865 (*Crabro*); Cuba (con)
dasymera Pate, 1948; Nigeria (ala)
eurypus (Kohl), 1898 (*Crabro*); e. Europe (plu)
florkini Leclercq, 1956; Mexico: Orizaba; Costa Rica (con)
    colorata Leclercq, 1956
heros (Kohl), 1915 (*Crabro*); Japan (con)
indonesica Leclercq, 1958; Sumatra (con)
lieftincki Leclercq, 1958; Indonesia: Borneo, Java, Malaysia (con)
    peraki Leclercq, 1958
luzonia Leclercq, 1963; Philippines: Luzon (con)
molucca Leclercq, 1956; Indonesia: Ambon (con)
nitobei (Matsumura), 1912 (*Crabro*); Japan (sub)
ochotica (A. Morawitz), 1866 (*Crabro*); Siberia, China (sub)
plumata Leclercq, 1963; Borneo (con)
pluschtschevskyi (F. Morawitz), 1891 (*Crabro*); S. Europe, w. USSR (plu)
primitiva Leclercq, 1958; Indonesia (con)
producticollis (Packard), 1866 (*Crabro*); U.S. (con)
    quadrimaculata Provancher, 1882 (*Crabro*), also 1883, nec Fabricius, 1793
    quadripunctata Provancher, 1883 (*Crabro*), nec Fabricius, 1793
pygialis (Pérez), 1903 (*Ceratocolus*); Japan (ala)
quadriceps (Bingham), 1897 (*Crabro*); nw. India (con)
reiteri (Kohl), 1915 (*Crabro*); Japan (cly)
    kuramaensis Iwata, 1938 (*Crabro*)
    kuramensis Tsuneki, 1948 (*Crabro*), lapsus
relicta Leclercq, 1951; Australia (con)
sculpturata (F. Smith), 1873 (*Crabro*); Columbia, Peru, Brazil, Guyana, Surinam (con)
    sulphurata Dalla Torre, 1897 (*Crabro*), lapsus
siblina Leclercq, 1972; Celebes (con)
subterranea (Fabricius), 1775 (*Crabro*); Europe (sub)
    philanthoides Panzer, 1801 (*Crabro*)
sylvatica (Arnold), 1932 (*Thyreopus*); Rhodesia (con)
wollmanni (Kohl), 1915 (*Crabro*); sw. USSR (cly)

SUBFAMILY ENTOMOSERICINAE

Only two species are known in this small and relict subfamily. Both of these are Palearctic and have a general resemblance to small species of *Tachytes*. Actually, they seem to occupy an intermediate position between such larrine wasps as *Bothynostethus* and such nyssonine wasps as *Alysson*. This idea is illustrated graphically in fig. 7. Nesting habits and prey are unknown.

Diagnostic characters:
1. (a) Inner orbits converging below (fig. 132 A); (b) ocelli normal.
2. (a) Antenna somewhat below middle of face, sockets slightly above frontoclypeal suture; (b) male with 13, female with 12 antennal articles.
3. Clypeus somewhat broader than long.
4. (a) Mandible simple externoventrally, inner margin obtusely angulate before middle (fig. 132 A); (b) palpal formula 6-4; (c) mouthparts moderately long; (d) mandibular socket open.
5. (a) Collar as high as scutum; (b) pronotal lobe and tegula separated.
6. (a) Notauli essentially complete, fading shortly in front of scutellum; (b) no oblique scutal carina.
7. (a) Mesopleuron with episternal sulcus which curves forward around pronotal lobe and then straight down, ending as it reaches venter and turns posteriorly (fig. 132 I); (b) scrobal sulcus present; (c) omaulus, sternaulus and acetabular carina absent.
8. Definitive metapleuron consisting of upper metapleural area only (fig. 132 I).
9. (a) Midtibia with one long, apical spur; (b) midcoxae essentially contiguous, and with oblique dorsolateral crest; (c) precoxal lobes present; (d) hindfemur truncate apically, with apicoventral process on outer side (fig. 132 B); (e) claw simple; (f) plantulae absent.
10. (a) Propodeum short; (b) dorsal enclosure present, triangular (fig. 132 J); (c) propodeal sternite absent.
11. (a) Forewing with three submarginal cells, II receiving both recurrent veins, distance from submarginal III to wing apex less than greatest dimension of cell; (b) marginal cell acute apically (fig. 134 B).
12. (a) Jugal lobe small; (b) hindwing media diverging before cu-a; (c) subcostal vein and second anal vein absent (fig. 134 B).
13. (a) Gaster sessile; (b) tergum I with lateral carina; (c) male with seven visible segments, sternum VII exposed, VIII almost if not entirely hidden; (d) pygidial plate present.
14. (a) Volsella with digitus and cuspis; (b) gonostyle biramous (fig. 132 H); (c) cerci absent.

Systematics: The two known species form a single tribe and genus. As in other relict forms, relationships are difficult to determine, and morphological evidence points in different directions. Similarities and differences with respect to the Nyssoninae and particularly the Alyssonini have been pointed out under systematics of the latter. The preponderance of evidence argues against a close relationship. A somewhat stronger case can be made for association with the larrine tribe, Bothynostethini and particularly with *Bothynostethus*. Characters shared by the two genera are: a single strong midtibial spur; eyes converging below; hindfemur distally truncate; female foretarsal rake with many short spines; marginal cell ending acutely at wing margin; mandible entire and angled before middle of inner edge; jugal lobe rather small. On the other hand there are numerous dissimilarities which lead us to place *Entomosericus* in a separate subfamily that presumably followed an evolutionary course parallel to that of *Bothynostethus*. The differences in *Entomosericus* are: differentiation of cuspis and digitus in the male genitalia; more rounded pronotal collar; elongate notauli; small propodeal enclosure; elongate discoidal cell I of forewing; submarginal cell II not petiolate; hindwing media diverging well before cu-a; episternal sulcus curving forward and then downward parallel to front margin of mesopleuron (contrary to all Larrinae and Philanthinae).

Certainly, *Entomosericus* has morphological similarities to several rather distantly related genera. Among these is the philanthine *Odontosphex* in which the mandible and legs are strikingly like those of *Entomosericus*. In fact, if the legs alone are considered — abbreviated female foretarsal rake, single strong midtibial spur, hindfemoral truncation, dentate hindtibia — it would be difficult to distinguish generically among *Bothynostethus, Entomosericus,* and *Odontosphex.*

The male genitalia seem unusually distinctive. Yet, the biramous gonostyle is found also in the larrine genera *Trypoxylon* and *Pison* (compare figs. 132 H and 105 F). In the latter two genera the volsella is simple, whereas it is differentiated in *Entomosericus.*

Biology: Unknown.

Tribe Entomosericini

Genus Entomosericus Dahlbom

Generic diagnosis: Clypeus more than half as long as broad, free margin weakly quinquedentate; male antennal flagellum with tyli, last article incurved (fig. 132 E); antennal sockets slightly above clypeus, no process between or above them; inner eye margins undulate; ocelli in a broad triangle; pronotum broadly rounded in lateral view; scutum with notauli nearly complete and admedian lines converging posterad (fig. 132 J); scutellum with prescutellar sulcus, media diverging well before cu-a (fig. 134 B), ends of recurrent veins rather close together on submarginal cell II; hindwing media diverging well before cu-a; foretarsal rake in both sexes with many short setae (fig. 134 B); hindfemur truncate distally and with a prominent outer apical lobe, hindtibia with a double outer row of toothlike spines (fig. 132 B); propodeum short, not areolate, enclosure a small and essentially dorsal triangle (fig. 132 J); no spiracular groove; sternum I with a longitudinal groove beyond a projecting basal triangle; terga I and II with large laterotergites; gaster well punctured, terga with transverse subapical depressions which are best developed on II to IV; male sternum III with a prominent apical brush, VI with exposed area longer than broad and with an outcurving double carina (fig. 132 D); male sternum VIII plate shaped (fig. 132 G); female pygidial plate flat with raised edges, closely micropunctate, fringed with abundant hair from lateral area of tergum (fig. 132 F); male genitalia with digitus and cuspis differentiated, gonostyle biramous (fig. 132 H).

Geographic range: Palearctic: eastern Mediterranean area and southwest Asia.

Systematics: Good descriptions and a few figures were given by Handlirsch (1888a). A short discussion of phylogeny was offered by Beaumont (1954a).

Biology: Unknown.

Checklist of *Entomosericus*

concinnus Dahlbom, 1845 (*Entomericus!*); Greece, sw. USSR

kaufmanni Radoszkowski, 1877; Greece, Iran, sw. USSR

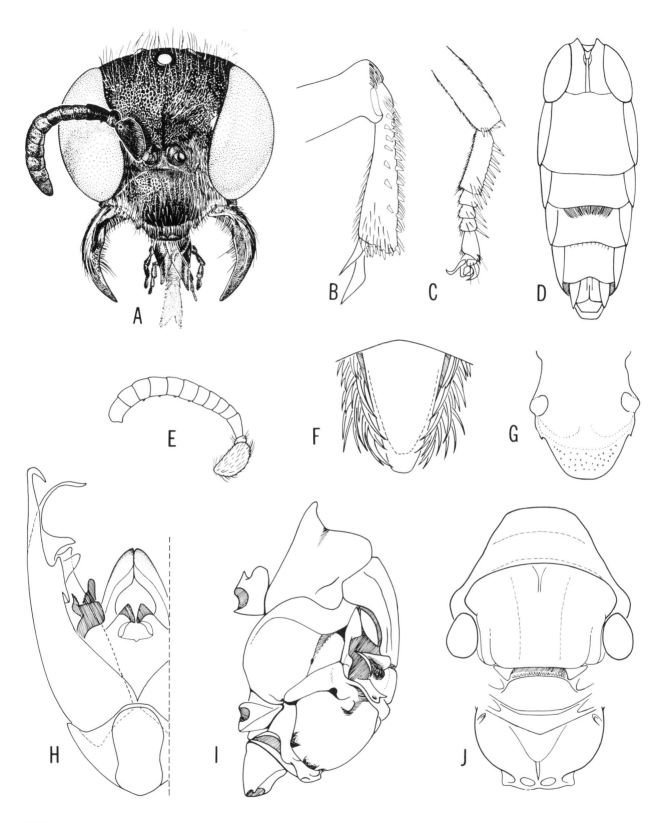

FIG. 132. Structural details in the subfamily Entomosericinae, *Entomosericus concinnus;* A, facial portrait, female; B, hindfemoral apex and tibia, female; C, foreleg, female; D, male gaster, ventral; E, male antenna; F, pygidium, female; G, sternum VIII, male; H, male genitalia, ventral; I,J, lateral and dorsal views of thorax.

SUBFAMILY XENOSPHECINAE

The three known species of this subfamily occur in a relatively small area of the southwestern United States. The adults are medium-sized to small, black and white wasps (fig. 133) occurring in sandy areas where they may alight on the sand or frequent flowers, such as *Eriogonum* or *Pectis*.

The phylogeny of the group has not been firmly established. Some features point toward Nyssoninae, others toward Larrinae or Philanthinae. Since there are two well developed midtibial spurs and an externoventral mandibular tooth, we have placed the group as a subfamily between Nyssoninae and Larrinae but diverging very early fron the nyssonine stem (fig. 7).

Diagnostic characters:

1. (a) Eyes converging below (fig. 141 B), inner orbits somewhat angularly emarginate above middle; (b) ocelli normal.
2. (a) Antennae below middle of face, sockets slightly removed from frontoclypeal suture; (b) male with 13, female with 12 articles.
3. Clypeus transverse but not always strongly so.
4. (a) Externoventral margin of mandible notched or toothed, inner margin with one or two subapical teeth (fig. 141 B); (b) palpal formula 6-4; (c) mouthparts short; (d) mandible socket open.
5. (a) Collar short, low, below scutum; (b) pronotal lobe and tegula separated.
6. (a) Notauli short, not exceeding half length of scutum; (b) no oblique scutal carina.
7. (a) Episternal sulcus absent except for a short remnant at subalar fossa; (b) scrobal sulcus weak or absent; (c) omaulus, sternaulus, and acetabular carina absent.
8. Definitive metapleuron consisting of upper metapleural area only.
9. (a) Midtibia with two apical spurs; (b) midcoxae contiguous, with oblique dorsolateral crest and notched posteriorly; (c) precoxal sulcus and lobes absent; (d) hindfemoral apex simple; (e) claws simple; (f) plantulae absent.
10. (a) Propodeum short; (b) dorsal enclosure present, U-shaped; (c) propodeal sternite absent.
11. Forewing with three submarginal cells, I and II each receiving a recurrent vein or first recurrent interstitial (fig. 134 A).
12. (a) Jugal lobe small; (b) hindwing media diverging before cu-a; (c) subcosta and second anal vein absent (fig. 134 A).
13. (a) Gaster sessile; (b) tergum I with lateral carina absent or present only on basal half; (c) male with seven terga and eight sterna exposed; (d) female with weakly defined pygidial plate.
14. (a) Volsella with digitus and cuspis (fig. 137 A); (b) cerci absent.

Systematics: The three known species form a single tribe and genus. The most closely related group of wasps appears to be the tribe Mellinini, a peripheral element of the subfamily Nyssoninae (fig. 7). *Xenosphex* and *Mellinus* have a surprisingly large number of similarities which are nearly matched, however, by differences. The principal similarities are: mandible toothed toward inner apex; no malar space nor genal carina, occipital carina incomplete; antennal sockets close to but not touching clypeus; inner orbits outcurved; notauli and admedian lines distinct anteriorly; scutellum and metanotum raised and relatively simple; no omaulus, sternaulus nor acetabular carina; forewing media diverging after cu-a or very close to it; recurrent veins widely separated, normally with not more than one received by submarginal cell II; hindwing media diverging before cu-a; jugal lobe small; arolia alike on all legs; midcoxae approximate; midtibia with two apical spurs; metapleuron simple, narrowed gradually below; no spiracular groove on propodeum; propodeal enclosure U-shaped, essentially dorsal; sternum I simple toward base, not obviously carinate; males of some species with eight sterna visible and VIII plate-like; female with pygidial plate; volsella with differentiated digitus and cuspis.

On the other hand, the differences between *Xenosphex* and *Mellinus* are substantial: mandible of *Xenosphex* with externoventral tooth or notch, *Mellinus* without; antennal sockets approximate in *Xenosphex*, well separated in *Mellinus;* inner orbits strongly converging below in *Xenosphex*, not so in *Mellinus;* collar depressed and appressed to scutum in *Xenosphex*, but ridgelike and distinct in *Mellinus*; episternal sulcus extremely short in *Xenosphex*, but extending far ventrally in *Mellinus*; submarginal cell II of *Xenosphex* petiolate or triangular and receiving second recurrent vein, but submarginal cell II in *Mellinus* trapezoidal and receiving first recurrent at most; marginal cell truncate distally in *Xenosphex*, acute in *Mellinus*; hindwing submedian cell obtuse distally in *Xenosphex*, acute in *Mellinus*; *Xenosphex* female with a foretarsal rake, *Mellinus* without one; midcoxa acutely notched on inner posterior surface in *Xenosphex*, stepped down in *Mellinus*; gaster sessile in *Xenosphex*, pedunculate in *Mellinus*; no precoxal sulcus or lobes in *Xenosphex*, present in *Mellinus*; tergum I in *Xenosphex* with laterotergite absent or only weakly differentiated at base, present in *Mellinus*; tergum I curving around and under in *Xenosphex*, its lateral margins nearly in contact ventrally thus broadly concealing sternum I, sternum broadly exposed in *Mellinus*; outer apex of hindfemur simple in *Xenosphex*, narrowly hamate in *Mellinus*; female sternum VI smoothly incurved in *Xenosphex*, carinate and angled out toward apex in *Mellinus*.

The evidence points to a fairly close relationship to *Mellinus* based on retention of generalized characteristics but a somewhat relict status following rather early separation. In addition to structural similarities, our knowledge

FIG. 133. *Xenosphex xerophilus* Williams, female.

of the biology of *Xenosphex,* although insufficient, strengthens the connection between the two genera. Both use flies as prey, and both have unusually powerful femora which are utilized in springing toward the prey from a perch.

Tribe Xenosphecini

Genus Xenosphex Williams

Generic diagnosis: Clypeal apex simply curved or with a median truncate lobe (fig. 141 B); labrum concealed; mandible with one or two preapical inner teeth (in addition to externoventral tooth); scape flattened laterally, more than twice as long as flagellomere I, last article incurved in male; no process between antennal bases which are closer together than to clypeus (fig. 141 B); inner eye margins a little angulate at upper third of eye, converging strongly toward clypeus; ocelli in a broad triangle; frons with a pair of elongate silvery hair patches (fig. 141 B); forewing media interstitial with cu-a or nearly so; submarginal cell II triangular or with a short petiole, receiving second recurrent vein (first recurrent sometimes interstitial but more often received by cell I, fig. 134 A); hindwing media diverging a short distance before cu-a; pronotal collar depressed beneath somewhat overhanging scutum; both notauli and admedian lines evident, former faint; scutellum and metanotum raised and relatively simple; propodeum without dorsolateral carinae or teeth, enclosure essentially dorsal, not divided medially, no spiracular groove, foretarsal rake well developed in female, weak in male; basal midtarsomere curved; hindfemur simple at apex; sternum I simple rather than carinate toward base; terga I and II curving under laterally, covering nearly all of sternum I and two-thirds of sternum II; gaster nearly impunctate except for pygidium which has a weakly margined plate in both sexes; male sterna III to V with outstanding hairbrushes, sternum VIII platelike or greatly narrowed; female sternum VI simple, incurved in profile; male genitalia with digitus and cuspis differentiated (fig. 137 A).

Geographic range: Southwestern United States. Fewer than 100 specimens of the subfamily are known in collections.

Systematics: The general habitus is shown in fig. 133, the face in fig. 141 B. The genus was reviewed by F. Parker (1966b), who gave a key to the three known species based on the shape of the clypeus, flagellomere I, mandibular dentition, and markings.

Biology: The adults are usually collected as they alight on the surface of the sand in dry desert washes. Occasionally, they are taken on adjacent flowers, such as *Eriogonum* and *Pectis.* The prey appears to be Diptera based on the single record by Parker (1966b), who observed a wasp carrying a bombyliid (*Lordotus miscellus* Coquillett). Two of the species, *xerophilus* and *boharti,* have been collected in the spring only, whereas *timberlakei* has been taken in the spring and in October subsequent to the passage of floodwaters.

Checklist of *Xenosphex*

boharti F. Parker, 1966; U.S.: se. California
timberlakei Williams, 1955; U.S.: se. California, nw. Arizona, s. Nevada
xerophilus Williams, 1954; U.S.: s. California, nw. and s. Arizona, s. Nevada

SUBFAMILY NYSSONINAE

In this large and diversified subfamily there are more than 1400 known species. The majority are medium-small to medium-large wasps, but a few, such as the well-known cicada killers are nearly 40 mm long. The most appropriate name is probably the "sand wasps" as pointed out by Evans (1966a). Certainly all species nest in the ground, and the great majority are found in sandy habitats. Most nyssonids are predaceous, and nests are provisioned with a variety of other insects from grasshoppers to flies. Individual species are fairly selective as a rule. Certain groups, including the genus *Nysson* from which the subfamily gets its name, are cleptoparasites.

The variety of shapes, sizes and markings within the subfamily make it impossible to visualize a typical nyssonine. However, to some extent this can be done at the tribal level, and the seven tribes taken together seem to form a natural group.

Diagnostic characters:

1. (a) Eyes with inner margins essentially parallel or converging below, if converging above (Heliocausini and some Bembicini), then (b) at least hindocelli somewhat distorted and scar-like.
2. (a) Mandible not notched nor toothed externoventrally, inner subteeth, when present, associated with apical tooth; (b) mandible socket open; (c) palpal formula 6-4 except in many Bembicini.
3. (a) Scutum without complete notauli; (b) an oblique scutal carina (see glossary and fig. 156) often present.
4. (a) Episternal sulcus present or absent, sometimes forming a continuous arc with scrobal sulcus (episternal-scrobal sulcus, fig. 3 B); (b) omaulus usually present (absent in Mellinini, Stizini, Bembicini and in some genera or species of other tribes).
5. (a) Midtibia with two apical spurs (only one in some males, in certain Heliocausini and Stizini, and in all Bembicini and rarely none in some Alyssonini, Bembicini), (b) claws simple, (c) midcoxae essentially contiguous except in Nyssonini, (d) plantulae present or absent.
6. No propodeal sternite.
7. (a) Forewing with two or three submarginal cells of which II usually receives at least one recurrent vein, (b) three discoidal cells present.
8. (a) Hindwing media diverging before, at or beyond cu-a; (b) jugal lobe usually small, rarely absent, at most half as long as anal area (in Heliocausini).
9. (a) Gaster usually sessile but peduncle when present, made up of both tergum and sternum; (b) male gaster with six or seven exposed terga; (c) sternum VIII often modified and largely or entirely concealed; (d) male genitalia usually with volsella differentiated into cuspis and digitus.

Systematics: We have included seven tribes and 71 genera in the subfamily. This is a departure from accepted modern practice, as exemplified by Pate (1938b), Krombein (1958d, 1967a) and Beaumont (1964b) in two important particulars. First, we have assigned ampulicines, *Xenosphex* and *Entomosericus* to other subfamilies, and we have added *Heliocausus* and reassigned *Mellinus* to the Nyssoninae. Second, we have raised a number of subgenera to generic rank and have created genera from certain species groups where it seemed expedient and phylogenetically sound. This is in accord with our philosophy expressed in the Introduction.

Six of the seven tribes are cosmopolitan. However, the subfamily has had its greatest development in the Western Hemisphere where all but 14 of the 71 genera are represented. Furthermore, 38 are restricted to the New World, 12 are endemic to North America, and 15 to South America. Handlirsch (1887-1895) is an important source of information for species named before 1895.

As herein defined the Nyssoninae encompasses a large and diversified group of genera. Unfortunately, there is no single characteristic that unites them. The oblique scutal carina is a unique structure in the subfamily. Yet three of the tribes and 10 of the 71 genera do not have it. These are presumably ancestral or relict types: *Alysson, Didineis, Heliocausus, Mellinus, Exeirus, Argogorytes, Neogorytes, Olgia, Clitemnestra,* and *Ochleroptera*. Two midtibial spurs are certainly a subfamily feature. Yet there has been a reduction to one spur in scattered instances and universally in the Bembicini. In *Oryttus* and *Cresson* the reduction is in the male sex only, but in *Stizoides* and

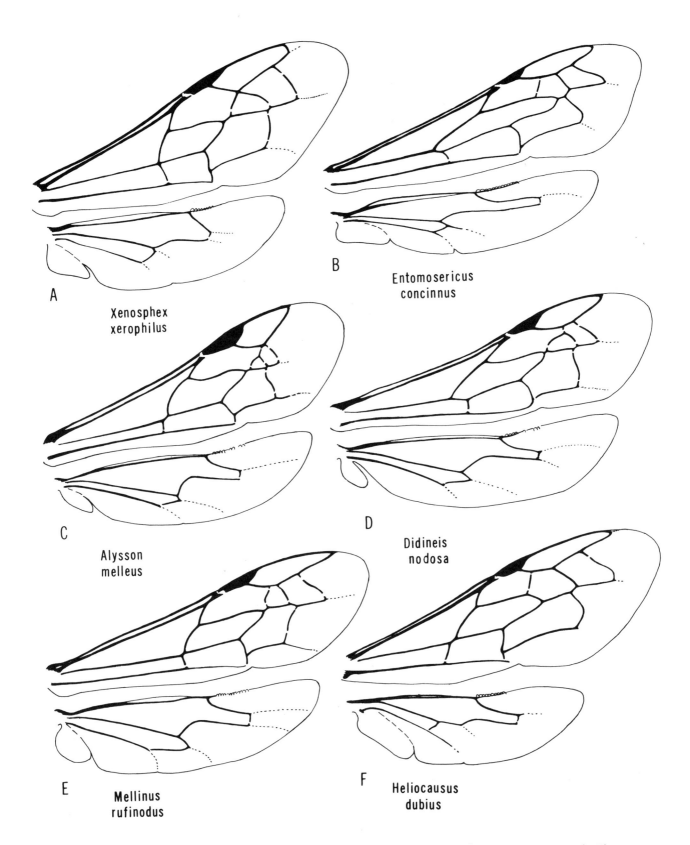

FIG. 134. Wings in the subfamilies Xenosphecinae (A), Entomosericinae (B), and Nyssoninae (C-F).

Heliocausus it may appear in both sexes. Other characteristics similarly suffer from exceptions, or are negative ones, such as the simple externoventral edge of the mandible.

Therefore, a morphological definition of the subfamily must be a somewhat negative and nonexclusive one: those forms having an oblique scutal carina plus 10 other genera, at least some of whose species have two midtibial spurs, and no species with an externoventral notch or tooth on the mandible. In addition, the mandible is either simple apically, or subteeth are associated with the apical one, there are two to three submarginal cells of which the second receives at least one of the two recurrents (not in *Mellinus*), the jugal lobe is no longer than half the entire anal area, the marginal cell of the forewing is not broadly truncate distally, and there is no propodeal sternite.

Origins of the Nyssoninae are obscure. It must be supposed that the ancestral stock was a prelarrine form with two midtibial spurs, mandible unnotched externoventrally, and the volsella with a conventional digitus and cuspis. A somewhat related group with a relict status is the Xenosphecinae. The single genus in this subfamily has the midtibia two spurred and the jugal lobe moderately small. However, there is no omaulus, the mandible is notched or dentate externoventrally, and the forewing marginal cell is broadly truncate distally.

Within the subfamily Nyssoninae evolutionary progress is obvious even though the exact routes and underlying reasons can only be surmised. The cactus diagram in fig. 135 is an attempt to show the relationships of the seven tribes. We place the Gorytini in a central position because of its large size and because all seven modern tribes could have developed from a gorytine ancestor similar to *Clitemnestra* or *Argogorytes*. The Heliocausini may be an exception. *Heliocausus* is the only genus, and it appears to be a small, relict offshoot without close modern relatives. Since it has a number of larroid features a divergence from the main nyssonine stem could have taken place earlier than fig. 7 indicates. The characters shared by at least some members of both *Heliocausus* and Larrinae are absence of an omaulus, abortive ocelli, single midtibial spur, forewing and hindwing medial veins arising beyond cu-a, appendiculate marginal cell of the forewing, and convergence of the eyes toward the vertex. On the other hand, the two miditibial spurs, the C-like form of the ocellar scars, and well-formed omaulus of some less specialized *Heliocausus,* as well as the moderate jugal lobe and the divided volsella, are convincing arguments against a close association with the Larrinae. Finally, Evans (1972) has recently described the larva as nyssoninelike.

Placed somewhat doubtfully is the tribe Mellinini with the single genus *Mellinus*. Alliance with the Nyssoninae is urged by the absence of any strong reasons for separating it. Evans (1959a, 1964a, 1966a) has taken the

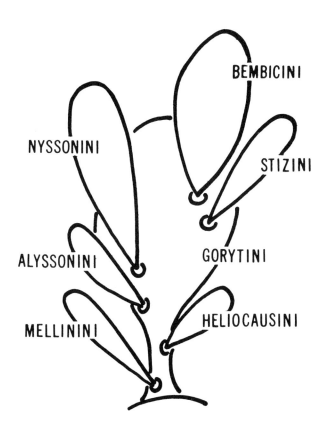

FIG. 135. Cactus-like diagram showing possible relationships of the seven tribes in the subfamily Nyssoninae.

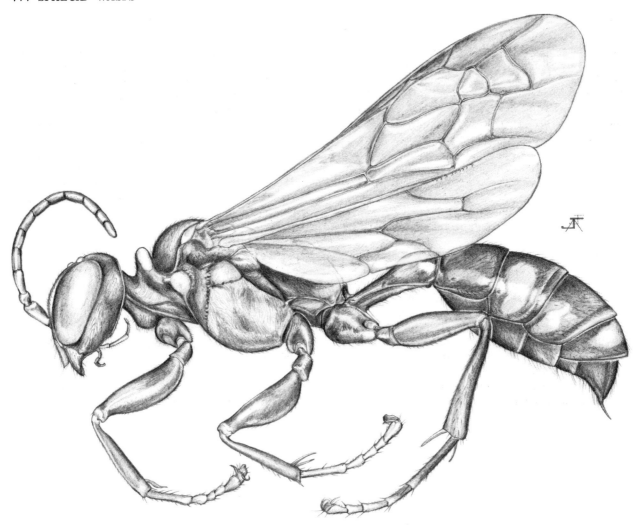

FIG. 136. *Mellinus bimaculatus* Packard, female.

opposite view and would maintain the genus as a separate subfamily on the basis of larval structure and on ethological grounds. Undoubtedly, *Mellinus* has a rather early origin but exhibits some specialized traits such as the nodose form of the first gastral segment. This feature is certainly not exclusive, and since an early divergence from the Nyssonine stem seems reasonable we prefer to include it.

Among the other nyssonine tribes there are obvious close ties between Alyssonini and Nyssonini and similarly, between Stizini and Bembicini. Characters shared by at least some genera of Alyssonini and Nyssonini are the double carina of tergum I, petiolate submarginal cell II, distally truncate hindfemur, and dorsolaterally toothed propodeum. The long misunderstood genus, *Nursea,* has many alyssonin features, most notable of which are the slender body form and relatively smooth integument. Yet, it is undoubtedly in the Nyssonini and closely related to *Nippononysson.* Therefore, it can be considered a sort of connecting link as discussed in the phylogeny sections of the two tribes.

The Stizini and Bembicini are likewise closely related and share a compact thorax, no omaulus, and a long first submarginal cell in the forewing. Ties to the Gorytini are suggested by several structural peculiarities shared with such genera as *Sphecius, Handlirschia,* and *Kohlia.*

In order to gain some concept of relative evolutionary development, a list has been made of 50 presumably unspecialized or primitive structures within the subfamily as compared with the specialized or derived condition (table 16). Fourteen species, representing all seven tribes, have been rated according to this list. Rather low on the scale (0-11) are *Clitemnestra chilensis* (0), *Mellinus rufinodus* (5), *Gorytes angustatus* (7), *Heliocausus fairmairei* (10), and *Alysson melleus* (11). A moderately specialized group of species (13-17) are *Lestiphorus piceus* (13), *Nippononysson rufopictus* (14), *Didineis nodosa* (15), and *Stizus occidentalis* (17). Highly advanced forms (19-26) are *Bicyrtes ventralis* (19), *Foxia pacifica* (21), *Bembecinus quinquespinosus* (22), *Metanysson arivaipa* (24), and *Bembix rostrata* (26).

A summary of the main evolutionary structural trends

within the subfamily Nyssoninae follows:

Head: The normal ocelli become slightly deformed in *Kohlia*, progressively more distorted in bembicin genera. Mandibles become simple in some Gorytini and in Nyssonini, other mouthparts become longer in Stizini and especially Bembicini, but palpal reduction takes place. The face changes from concave laterally in *Clitemnestra* and *Exeirus* to rather convex, as in Nyssonini, to greatly narrowed below as in *Afrogorytes*.

Thorax: There is a strong tendency toward consolidation, with elimination of sulci, overlapping and streamlining of notal elements, posterior extension of the propodeal enclosure, development of the oblique scutal carina, narrowing or welting of the area between the scutal admedian lines, movement of the origin of the medial vein basad in the forewing, reduction of wing membrane beyond the cellular area, and a tendency toward reduction to one midtibial spur as in some Gorytini and Stizini, and in all Bembicini.

Gaster: There is occasional development of pedunculation of the basal segment, as in *Mellinus* and in the *Lestiphorus* group. Sternum I may be simple as in *Mellinus*, with one carina as in Alyssonini, with two carinae as in Nyssonini, with three as in Gorytini or again with one as in Stizini and Bembicini. Male tergum VII may be simple, with ventral spiracle-bearing lobes, or tripartite to even quadripartite in some *Stictia*. Male sternum VIII may be simple as in *Clitemnestra*, sting shaped as in some *Sphecius*, bispinose as in *Oryttus*, sword shaped as in *Hoplisoides*, three pronged as in *Stizus*, or four pronged as in *Glenostictia*.

The episternal and scrobal sulci vary in form and development in the Nyssoninae. In the typical or generalized condition the episternal sulcus runs vertically ventrad and the scrobal sulcus joins it at approximately a right angle as in *Mellinus, Heliocausus, Argogorytes* and *Neogorytes* (figs. 136, 162 A). In some highly evolved forms the episternal sulcus is absent (Nyssonini, fig. 4, *Afrogorytes, Bembecinus*), or it may curve posterad, forming a continuous arc with the scrobal sulcus as in *Sphecius* (fig. 3 B), some Stizini and Bembicini. In some instances the scrobal sulcus extends forward horizontally beyond its juncture with the episternal sulcus and may meet the omaulus as in *Gorytes* and related genera (fig. 158). If the forward branch is obliquely directed towards the omaulus as in *Ochleroptera* and *Arigorytes,* it may be interpreted as a continuation of the episternal sulcus.

The metapleuron has extensive variation also. It usually consists of the upper metapleural area only and this may narrowly taper below (Alyssonini) or may be quite broad below (Heliocausini). The metapleural pits vary in size, even within a genus (*Hoplisoides,* fig. 178) and rarely the anteroventral pit may be confluent with the posteroventral mesopleural pit (some *Pseudoplisus*).

References of special importance are Handlirsch (1887-1895), Pate (1938b), Beaumont (1954a, 1964b), and Evans (1966a).

KEY TO TRIBES OF NYSSONINAE

1. Sternum I with two ridges diverging posteriorly from between hindcoxae or a single ridge which forks toward middle of sternum; submarginal cell II petiolate or forewing with only two submarginal cells; admedian lines essentially fused into a single median groove 2
   Sternum I basomedially simple or with a single ridge which does not bifurcate posteriorly 3
2. Oblique scutal carina present (see fig. 156); median groove of scutum strong; body sculpture usually rather coarse; pronotal collar ridgelike .... Nyssonini, p. 461
   Oblique scutal carina absent; median groove of scutum faint and only present anteriorly; body sculpture rather fine except for propodeum; pronotal collar broadly rounded .... Alyssonini, p. 453
3. Sternum I simple toward base; oblique scutal carina absent; gaster pedunculate; hindwing media diverging well before cu-a .... Mellinini, p. 445
   Sternum I with a ridge basally (between hindcoxae) 4
4. Oblique scutal carina absent; hindocelli deformed, c-like; eyes converging above, hindwing media diverging after cu-a .... Heliocausini, p. 449
   Oblique scutal carina present or, if not, then hindocelli are normal 5
5. Labrum exserted and at least as long as broad, omaulus absent, hindocelli deformed and often scarlike, only one midtibial spur, prestigmal length of submarginal cell I more than twice height of cell .... Bembicini, p. 532
   Labrum, if exserted, not as long as broad; hindocelli rarely deformed and if so, omaulus present 6
6. Prestigmal length of submarginal cell I more than twice height of cell, omaulus absent, propodeal enclosure extending far onto vertical slope, scutellum with lamelliform edge overlapping metanotum .... Stizini, p. 523
   Without above combination of characters .... Gorytini, p. 481

Tribe Mellinini

The single genus *Mellinus* contains moderately small to moderately large wasps occurring mostly in the Holarctic Region. They are rather uniform in general structure, and fig. 136 is typical. In the New World they are uncommon as attested by the relatively few preserved in museums. The two European species are more abundant, particularly *Mellinus arvensis* which frequents livestock range and collects muscoid flies in the vicinity of animal feces.

The status of the tribe, with its nine current species, is an interesting and controversial subject which is dealt with under the section Systematics.

TABLE 16.
Phylogenetic Characters in the Nyssoninae

Unspecialized or primitive	*Specialized or derived*
1. Male antenna relatively simple, last article neither very long nor very short	Male antenna with tyli, deformities, projections, or last article very long or short
2. Inner eye margins about as far apart above as below, bowed inward	Inner eye margins nearly straight, converging slightly or diverging below
3. Eyes essentially bare	Eyes hairy
4. Ocelli normal, rounded	Ocelli reduced or distorted
5. Frons without a projecting lobe between antennal bases	Frons with a projecting lobe between antennal bases
6. Frons unarmed above antennae and with a long median sulcus	Frons toothed or crested above antennae or without a definite median line or sulcus
7. Labrum rather short and inconspicuous	Labrum exserted, longer than broad
8. Mandible toothed subapically	Mandible simple
9. Maxillary palpus with 6 segments	Maxillary palpus with 5 to 3 segments
10. Labial palpus with 4 segments	Labial palpus with 3 to 1 segments
11. Pronotal collar well separated from scutum	Pronotal collar closely appressed to scutum
12. Scutal area between admedian lines broad and flat	Scutal area between admedian lines a narrow welt or a single line
13. No oblique scutal carina	Oblique scutal carina present
14. Scutellum not overlapping metanotum	Scutellum overlapping metanotum
15. Omaulus present	Omaulus absent
16. Episternal sulcus extending ventrad, scrobal sulcus joining it at an angle	Episternal and scrobal sulci joining into a simple arc, or obscure
17. Sternaulus absent	Sternaulus present
18. Acetabular carina absent	Acetabular carina present
19. Male midcoxa simple	Male midcoxa apically toothed or spined
20. Hindfemur simple at apex	Hindfemur with an apical process
21. Hindtibia and male midfemur simple posteriorly	Hindtibia or male midfemur serrate or dentate posteriorly
22. Two midtibial spurs	One midtibial spur or none, at least in male
23. Female foretarsal rake present	Female foretarsal rake absent
24. Male foretarsus simple	Male foretarsus unusually expanded
25. Female fore arolium not enlarged	Female fore arolium enlarged
26. Media of forewing arising after cu-a	Media of forewing arising before cu-a
27. Forewing with 3 submarginal cells	Forewing with 2 submarginal cells
28. Marginal cell long and tapering distally	Marginal cell rounded distally, subtruncate or pulled away from margin
29. Prestigmal area of submarginal cell I short	Prestigmal area of submarginal cell I long (as in Stizini-Bembicini)
30. Stigma as large as outline of scape	Stigma smaller than outline of scape
31. Submarginal cell II sessile	Submarginal cell II petiolate
32. Hindwing media diverging at or before cu-a	Hindwing media diverging well after cu-a
33. Hindwing median cell with 2 appendices	Hindwing median cell with 1 appendix
34. Metanotum low and simple, not overlapping propodeum	Metanotum ridge-like, dentate or overlapping propodeum
35. Propodeal enclosure essentially dorsal and horizontal	Propodeal enclosure with posterior third or more essentially vertical
36. Propodeum not spined dorsolaterally	Propodeum spined or toothed dorsolaterally.
37. Gaster sessile, basal segment broad	Gaster pedunculate
38. Tergum I with only one lateral carina extending beyond spiracle	Tergum I with 2 lateral carinae extending beyond spiracle
39. Sternum I simple, or with a median carina and sometimes submedian carinae	Sternum I without a median carina but with submedian carinae
40. Tergal and sternal apices single edged	Tergal and sternal apices double edged
41. Tergal apices not bordered with flattened setae	Tergal apices bordered with flattened setae
42. Males without conspicuous fimbriae or hair mats on sterna II-V	Males with conspicuous fimbriae or hair mats on sterna II-V, II-III or III-V
43. Sterna simple laterally	Sterna II-V dentate laterally
44. Female pygidial plate well developed	Female pygidial plate obscure
45. Female sternum VI simple	Female sternum VI longitudinal and linear
46. Male sternum II relatively simple	Male sternum II with a definite process
47. Male tergum VII simple basally	Male tergum VII with a median section and laterobasal spiracular lobes
48. Male tergum VII simple apically	Male tergum VII apically with 2 or more strong lobes or spines
49. Male sternum VIII simple apically	Male sternum VIII with 1 to 4 spines apically
50. Seven visible male terga	Six visible male terga

Diagnostic characters:

1. (a) Inner eye margins outcurved, thus converging both above and below, sometimes one more than the other (fig. 141 D); (b) ocelli with normal lenses.
2. (a) Labrum not prominent; (b) palpal formula 6-4; (c) mandible socket open.
3. Scutum (a) with admedian lines separated, sometimes enclosing a welt; (b) notauli moderately developed anteriorly; (c) no oblique scutal carina; (d) scutellum convex, not at all lamelliform.
4. (a) Omaulus absent; (b) acetabular carina, sternaulus and other mesopleural landmarks absent except for long, strong episternal sulcus and a scrobal sulcus which may be weak (fig. 136).
5. (a) Metapleuron separated by a sulcus from propodeum, consisting of upper area only, attenuate below; (b) no spiracular groove; (c) no spines or teeth on propodeum dorsolaterally.
6. (a) Midcoxae approximate, precoxal sulcus complete; (b) midtibia with two apical spurs; (c) female without a foretarsal rake; (d) plantulae present.
7. Forewing (a) with media commonly diverging beyond cu-a, sometimes interstitial; (b) three submarginal cells of which II is trapezoidal and receives neither recurrent vein (first recurrent may be interstitial) (fig. 134 E).
8. (a) Hindwing media diverging before cu-a; (b) jugal lobe moderately small (fig. 134 E).
9. (a) First gastral segment a narrow peduncle; (b) sternum I not carinate toward base; (c) male with seven visible terga and sternum VIII broad and plate-like; (d) male genitalia with volsella differentiated into cuspis and digitus (fig. 137 B); (e) female pygidial plate distinct and bluntly wedge shaped.

Systematics: Mellinus has been treated variously as an isolated subfamily or as an aberrant member of the nyssonine complex. Most modern workers have leaned toward the latter placement. However, Evans (1959a, 1964a) has described larval characters that have convinced him that *Mellinus* should be treated as a separate subfamily with larrine relationships. The principal characters in question concern the position of the anus and the development of antennal papillae. Evans hypothesizes that a terminal anus and no papillae are the generalized larval condition. Nyssoninae is one of several subfamilies in which there are antennal papillae, yet *Mellinus* and the Larrinae have no papillae. Furthermore, both have the anus preapical and directed ventrad, rather than terminal as in other sphecids.

We agree that *Mellinus* branched off at a rather early date, but the two midtibial spurs of the adult and the absence of any one adult character or even a combination of two that will rule the genus out of the Nyssoninae persuades us to treat the category as an early diverging tribe on the nyssonine stem (figs. 7, 135, 155).

The larval evidence presented by Evans, while of obvious phylogenetic value, can be explained in another way. The development in the larva of antennal papillae seems to be an inherent tendency in the Sphecidae since it has taken place in the Philanthinae, Nyssoninae, Astatinae, and some Pemphredoninae (Evans, 1964a). Since the presence of antennal papillae among so many subfamilies can hardly be evidence of close relationship but rather of parallel development, the absence of papillae can be viewed simply as retention of the generalized condition and proof of an early branching on the evolutionary "tree." The preapical anus in the larvae of Larrinae and *Mellinus* has a more positive basis, but this may be parallel development also, similar to such adult Hymenoptera trends as cerci loss, reduction in midtibial spurs, and changes in tarsal claw dentition.

As suggested by F. D. Parker (1966b), *Mellinus* may well be related to *Xenosphex,* another presumably relict genus. The long list of similarities is given in the systematic discussion of *Xenosphex*. Primary differences in *Mellinus* are the absence of an outer mandibular tooth, a simple midcoxa, a trapezoidal submarginal cell II which does not receive the second recurrent vein, and a pedunculate first gastral segment.

As a tribe, the Mellinini differs from other Nyssoninae by a combination of: a pedunculate gaster, no omaulus, and no oblique scutal carina. Others features of note that occur in various other nyssonines are the distinct and narrowly raised pronotal collar, the rather simple, U-shaped dorsal propodeal enclosure, and the anteriorly distinct notauli. To these adult differences should be added the larval characters of the preapical rather than apical anus and the nonpapillate rather than papillate antennal area.

Biology: Refer to the discussion under the genus.

Genus Mellinus Fabricius

Generic diagnosis: Male antennal flagellum relatively simple but tyli may be present ventrally on several articles, antennal sockets well separated, situated a little above clypeus; eyes broadly separated by rather flat frons, converging slightly below or occasionally above, inner orbits outcurved; labrum concealed, tongue short, maxillary palpus moderately long (fig. 141 D), mandible with two subteeth (female) or one subtooth (male) toward apex of inner margin; free clypeal margin with three medial denticles (fig. 141 D); female without a foretarsal rake (fig. 136); arolia alike on all legs; pronotal collar rather thin, raised, distinct from scutum; outer apex of hindfemur hamate, forming a small plate, especially in female metapleuron smooth, gradually narrowed below; propodeal enclosure large, U-shaped, dorsal, in American species smooth except for shagreening, medially roughened in European species; stigma of forewing rather long, its area about twice that of tegula, marginal cell long and slender, ending acutely at wing margin (fig. 134 E); submedial cell of hindwing ending acutely, media diverging well before cu-a, jugal lobe considerably larger in outline than tegula; gastral segment I a peduncle, laterotergite of tergum I abbreviated to a narrow strip; female sternum VI with a median ridge or carina toward the truncate apex; male genitalia with differentiated cuspis and digitus (fig. 137 B), aedeagal apices often enclosing

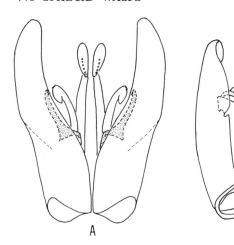

FIG. 137. Male genitalia, ventral, in Xenosphecinae and Nyssoninae; A, *Xenosphex xerophilus;* B, *Mellinus rufinodus.*

a large membranous sac.

Geographic range: Of the nine definite species of *Mellinus,* three are Eurasian (Palearctic), three are essentially Nearctic, and three are New World Tropical. One of the latter extends north into southeastern California, and another extends south into Brazil.

Systematics: As discussed under the tribe and in the systematics section under Xenosphecinae, *Mellinus* is a small, relict genus which presumably diverged rather early from the nyssonine stem (fig. 7). It has retained such generalized features as: two midtibial spurs; short tongue; no omaulus; no oblique scutal carina; notauli evident; simple propodeum with dorsal enclosure; raised and well separated pronotal collar, scutellum and metanotum; recurrent veins widely separated at their forward ends; submarginal cell II trapezoidal, submarginal cell III long and distally acute; and larval antennal areas not papillate.

Principal specializations seem to be the pedunculate gaster, the slightly hamate outer apex of the hindfemur, and the preapical position of the anus in the larva.

Although the species of *Mellinus* are remarkably uniform in most morphological details and fig. 136 is rather typical, the species differ widely in markings from the nearly all black *pygmaeus* to the extensively yellow *imperialis* or the yellow and red marked *rufinodus.* However, markings are subject to considerable variation within a species. This is especially true of the face which may be nearly all black or largely yellow. The flavid form of *rufinodus* has not been named, but that of *bimaculatus* was called *wolcotti* by H. S. Smith (1908a), and that of *abdominalis* was called *personatus* by Fox (1894c). The yellow face seems to represent individual variation rather than racial difference. More reliable characters are the markings of the antennae and legs, shape of the clypeus, details of the male antennae, shape of the peduncle, pubescence of the male sterna, shape of male sternum VIII, and the male genitalia.

The species of *Mellinus* seem to fall into two groups based on the genitalia and sternum VIII in the male (we have seen no males of *alpestris* and no material of *obscurus*). In the first group, which includes *rufinodus, bimaculatus,* and *imperialis,* the volsella is long and slender, the aedeagus consists of two curved and flattened pieces enclosing a huge membranous sac (fig. 137 B), and male sternum VIII is notched apically. In the second group, which includes *arvensis* (type of the genus), *crabroneus, abdominalis,* and *pygmaeus,* the volsella is stout, the aedeagus is not unusual, and male sternum VIII is entire. Useful references are Handlirsch (1888a), Fox (1894b), and Beaumont (1964b).

Biology: Much information on biology was given in the review of the biology of British fossorial wasps by Hamm and Richards (1930). The following is a condensation based on their notes concerning *M. arvensis.*

The species is gregarious, and the nests are generally in exposed sandy areas. The entrance is left open during construction and provisioning. The tunnel may be 30 to 50 cm deep with one to 10 cells on branches near the bottom. The cells are blocked off with earthen plugs when they are complete. The prey seems to be exclusively in the Diptera and generally in the Muscoidea.

The female wasp hunts mainly about dung but also on the leaves of trees where flies may be attracted. She often selects flies seated on the dung and stalks them in a "catlike" manner. "Her antennae are stretched straight forwards, and when she is within two centimeters of the fly she springs from her crouching position to alight on the back of the victim." The wasp then holds the fly by grasping a wing between her mandibles, the thorax held in the powerful front legs, and bends her abdomen around to sting the fly beneath its thorax. Then, with sting still in place, the fly is turned over, its proboscis gripped by the mandibles of the wasp, the sting withdrawn, and the fly carried off to the nest. Both the method of capture and the gripping of the fly by its proboscis are unusual features.

Each cell is stocked with four to nine flies, and an egg is laid on the last prey member, fixed diagonally across the sternum between the forelegs and midlegs. The fullgrown larva is purplish and spins a light yellow, soft, parchmentlike cocoon surrounded by a layer of loosely agglutinized sand and prey remains.

The muscoid prey fall into the following recorded genera: (Muscidae) *Musca, Muscina, Stomoxys, Orthellia, Helina, Hydrotaea, Myospila, Fannia, Phaonia, Pyrellia;* (Anthomyiidae) *Anthomyia, Scatophaga, Hylemya, Hydrophoria;* (Calliphoridae) *Calliphora, Pollenia, Onesia, Lucilia;* (Tachinidae) *Ceromasia, Carcelia, Zenillia, Phyrxe, Lydina;* (Sarcophagidae) *Sarcophaga, Paramacronychia.*

Other families of flies include Tabanidae: *Haematopota;* Syrphidae: *Syrphus* and *Spilogaster;* Tephritidae: *Tephritis;* Dryomyzidae: *Dryomyza;* Stratiomyiidae: *Sargus;* and Rhagionidae: *Rhagio.*

Parasites given by Hamm and Richards were the sarcophagids, *Macronychia griseola* Fallén and *Sphecapata conica* Fallén. Grandi (1954) added a third sarcophagid, *Metopia argyrocephala* Meigen (as *M. leucocephala* Rossi).

Additional information on the habits of *M. arvensis* was given by Bristowe (1948). His observations were generally similar to those of Hamm and Richards. However, he noted that the fly-laden wasp alights close to the burrow, "clambers off the fly without losing its grip of the proboscis, turns around and walks backwards into the burrow dragging the fly by its proboscis." Up to seven branches were found near the bottom end of the burrows. "The flies are packed tightly together with their heads pointing towards the entrance. The wasp pulls them into the burrow head first, so on reaching the cell it must turn them round in order to place them in their final resting position." Male wasps were observed by Bristowe to frequent the burrow entrances and to stalk and spring on one another as well as upon females. "When a male succeeds in mounting a female's back he grips her with his legs and tibial spurs, facing in the same direction as the female, then he edges backwards, tapping her abdomen with his own, until in a position to mate."

The important paper on *arvensis* by Huber (1961) provided additional data and presented a detailed exposition of known biology.

Hamm and Richards noted also that *Mellinus crabroneus* (as *sabulosa*) apparently hunts on flowers, and its prey differs somewhat on that account. These are mainly syrphids and muscids, captured and carried in the same manner described for *arvensis*. The burrows are 5 to 8 cm deep, curved medially, and the cells contain 12 to 15 flies each.

Yasumatsu and Watanabe (1964) recorded prey of *Mellinus obscurus* in Japan. These were all in muscoid genera: *Lucilia, Mesembrina, Sarcophaga,* and *Scatophaga*.

Checklist of *Mellinus*

abdominalis Cresson, 1882; w. U.S.
  personatus W. Fox, 1894, new synonymy by R. Bohart
alpestris Cameron, 1890; Mexico
arvensis (Linnaeus) 1758, (*Vespa*); Europe
  vagus Linnaeus, 1758 (*Sphex*)
  superbus Harris, 1776 (*Vespa*)
  tricinctus Schrank, 1781 (*Vespa*)
  clavatus Retzius, 1783 (*Sphex*)
  ?*infundibuliformis* Fourcroy, 1785 (*Vespa*), may be a *Cerceris*
  ?*petiolatus* Fourcroy, 1785 (*Vespa*)
  bipunctatus Fabricius, 1787 (*Crabro*)
  ?*gibbus* Villers, 1789 (p. 228) (*Sphex*), nec Linnaeus, 1758
  melanosticta Gmelin, 1790 (*Vespa*)
  ?*arthriticus* Rossi, 1790 (*Crabro*) (possibly Scoliidae)
  ?*rachiticus* Rossi, 1790 (*Crabro*) (possibly Scoliidae)
  vacus Rossi, 1790 (*Crabro*), lapsus for *vagus*
  annularis Christ, 1791 (*Sphex*)
  succinctus Olivier, 1792 (*Vespa*)
  diversus Olivier, 1792 (*Vespa*)
  labiatus Olivier, 1792 (*Crabro*)
  quinquemaculatus Fabricius, 1793 (*Philanthus*)
  u-flavum Panzer, 1794 (*Crabro*)
  capistratus Schrank, 1796 (*Crabro*), new synonymy by J. Leclercq
  pratensis Jurine, 1807
  annulatus Gimmerthal, 1836 (*Millimus!*)
  compactus Handlirsch, 1888
  alpinus Handlirsch, 1888
  ibericus Dusmet and Alonso, 1931
bimaculatus Packard, 1867; e. U.S.
  wolcotti H. S. Smith, 1908, new synonymy by R. Bohart
crabroneus (Thunberg), 1791 (*Sphex*); Eurasia, new name for *Sphex sabulosus* (Fabricius), (Art. 59c)
  sabulosus Fabricius, 1787 (*Crabro*), nec *Sphex sabulosus* Linnaeus, 1758, now in *Ammophila*
  ruficornis Villers, 1789 (*Sphex*), new name for *Sphex sabulosa* (Fabricius), 1787, (Art. 59c), nec *Sphex ruficornis* Fabricius, 1775 (a pompilid)
  sabulosus Olivier, 1792 (*Crabro*), nec Fabricius, 1787
  ruficornis Fabricius, 1793, nec Villers, 1789
  frontalis Panzer, 1797 (*Crabro*)
  petiolatus Panzer, 1797 (*Crabro*)
  fulvicornis Fabricius, 1804
?*globulosus* (Fourcroy), 1785 (*Vespa*), may be a *Cerceris*
imperialis R. Bohart, 1968; U.S.: s. California; Mexico: Sonora
obscurus Handlirsch, 1888; Korea, Japan
  tristis Pérez, 1905
pygmaeus Handlirsch, 1888; Mexico; Brazil
rufinodus Cresson, 1865; w. N. America

Fossil *Mellinus*

handlirschi Rohwer, 1908; U.S.: Colorado (Florissant Shales)

Tribe Heliocausini

This tribe is restricted to southern South America and contains only nine species. The "suntails" (approximate translation of *Heliocausus*) are wasps of small to medium size and are rather compactly built. The abdominal ground color is black or red or a combination of the two, and the terga are usually yellow banded. Yellow spots often occur elsewhere on the body. The strong convergence of the eyes towards the vertex, the deformed ocelli, and the strongly convex scutum are distinctive features.

Diagnostic characters:

1. (a) Inner orbits converging above (fig. 141 C) (sometimes slightly diverging beyond ocellar triangle in some females); (b) all ocelli deformed, scarlike (except midocellus with transparent lens in one group).
2. (a) Labrum not prominent; (b) palpal formula 6-4; mouthparts short.
3. (a) Scutum with admedian lines well separated; (b) notauli absent; (c) no oblique scutal carina; (d) scutellum without posterior overlapping edge.
4. (a) Omaulus sometimes present; (b) episternal

450 SPHECID WASPS

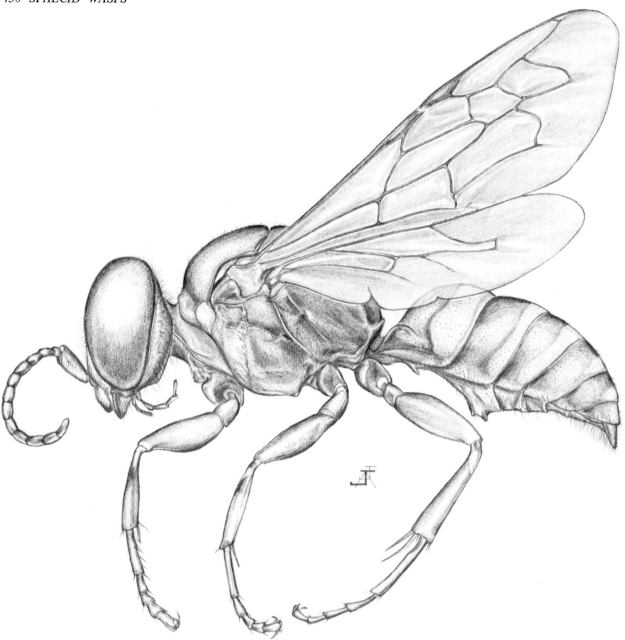

FIG. 138. *Heliocausus dubius* Kohl, male.

sulcus present, long; (c) scrobal sulcus present (fig. 138).

5. (a) Metapleuron consisting of upper area only, not tapering below; (b) spines or prongs sometimes on propodeum dorsolaterally.

6. (a) Midcoxae approximate; (b) midtibia with one apical spur (except two in some females); (c) foretarsal rake well developed in females; (d) plantulae present.

7. (a) Forewing media diverging after cu-a; (b) stigma moderate in size; (c) three submarginal cells of which II is not petiolate and receives both recurrent veins (fig. 134 F).

8. (a) Hindwing media diverging well beyond cu-a;

(b) jugal lobe a little more than half entire anal area (fig. 134 F).

9. (a) Gaster sessile; (b) sternum I with median ridge toward base; (c) male with seven visible terga, sternum VII apparently absent; (d) male genitalia with volsella differentiated into cuspis and digitus (fig. 6A); (e) female pygidial plate not or only slightly delimited.

Systematics: Ashmead (1899) placed *Heliocausus* in the "Larridae" (essentially equivalent to our Larrinae), subfamily "Lyrodinae"; the latter an ill defined ("....the ocelli are always distinct, normal, never aborted...."), heterogeneous assemblage of genera. Handlirsch (1925,

1933) showed more logic by establishing the tribe Heliocausini for the genus. He placed this tribe in the Nyssoninae despite the fact that *Heliocausus* did not fit his subfamily diagnosis. Subsequent authors of papers on the classification of the Nyssoninae have either tentatively placed *Heliocausus* in the subfamily (Beaumont, 1954a) or have omitted mention of the genus (Pate, 1938b; Evans, 1966a).

There are several features of *Heliocausus* that suggest a relationship with the Nyssoninae: simple mandible with a subapical tooth, approximate midcoxae, reduction or loss of sternum VII, volsella with digitus and cuspis (fig. 6 A), and compatible wing venation. The presence of an omaulus in most species, and the tendency for the development of posterolateral spines or prongs on the propodeum also suggest the Nyssoninae, as does the fact that some female *Heliocausus* have two midtibial spurs. The broadening of the metapleuron below also occurs in a number of nyssonine genera.

Heliocausus does have some larroid characteristics, however: strong convergence of the eyes towards the vertex, deformed ocelli, and a single midtibial spur in most species; but these features are found also in some Nyssoninae (Bembicini). Moreover, the ocellar deformation is of the bembicine type; that is, all three ocelli are scarlike, and the transparent part of the lens is C-shaped. The development of a transverse keel on male sternum II in *Heliocausus* is found in the Bembicini but also in the larrine genera, *Larrisson,* and *Palarus.* Morphological evidence suggests, then, that *Heliocausus* is a nyssonine. Evans (1972) found the larva to be somewhat gorytinlike, which adds additional support to this view.

Herbst (1921a) said that *Heliocausus* provisions with immature Orthoptera ("Heuschrecken"). This needs confirmation in view of Evans' statement that Fritz observed *Heliocausus larroides* preying on leafhoppers, a more typical nyssonine prey.

Everything considered, we feel justified in placing *Heliocausus* in its own tribe and associating it with the Nyssoninae as a relatively primitive, relict group (fig. 7).

Genus Heliocausus Kohl

Generic diagnosis: Convergence of eyes above strongest in males, inner margins slightly bowed out toward middle; ocelli flattened, deformed, lens C-shaped, midocellus wholly transparent in *fiebrigi* group and smaller than hindocelli (fig. 140 A); hindocellus in male contiguous with inner eye margin or nearly so; face essentially flat but with broad shallow sulci adjacent to inner orbits near vertex in female. these sulci converging behind ocelli to form a V; antenna short, articles variously modified in male, sockets contiguous with frontoclypeal suture; clypeus transverse, apical margin without teeth, disk usually swollen; head behind ocelli broad in female, depressed in male; occipital carina usually incomplete, disappearing just before hypostomal carina; mandible with a single subapical tooth on inner margin (fig. 141 C); pronotum with a high, narrow collar, thinner mesad; scutum strongly convex anteriorly in male; mesopleuron with a horizontal carina which forms ventral limit of subalar pit; metapleuron broad below; legs short, stout, arolia moderate to large; marginal cell narrowly rounded at apex and weakly appendiculate; hindwing media diverging well beyond cu-a; jugal lobe well developed and slightly more than half length of anal area (fig. 134 F); propodeum short and broad, triangular enclosure well defined with apex partly on vertical face, propodeal side usually set off from posterior face by a carina which forms an angle or sharp process dorsally (stronger in males) (fig. 139 A); tergum I with lateral carina, II with a transverse groove and often a carina at or near normal posterior limit of I; male sternum II with a transverse, keellike flange, male sternum VIII with three basal projections and either attenuate or rounded apcially (fig. 139 F, H); last male tergum usually fingerlike, arclike in cross section (fig. 139 G) (trispinose in *tridens,* fig. 139 E); female tergum VI conical, no pygidial plate except slightly developed at apex in some species.

Geographic range: Heliocausus is known only from Chile and Argentina. Of the nine species, three are apparently endemic to Chile, and the remainder are from Argentina.

Systematics: Heliocausus is easily recognized by the combination of deformed ocelli, eyes converging above, usual presence of a lateral propodeal angle or prong, and a broad metapleuron. The keel on male sternum II is also distinctive.

For a genus of such limited distribution and relatively few species, *Heliocausus* displays considerable morphological diversity in a number of features, which in most sphecid genera are normally constant. This suggests a relict status. For example, females of most (perhaps all) species in the *fiebrigi* group have two midtibial spurs. A second variable is the omaulus which is very strong in the *fiebrigi* group but is weakly indicated or absent in the *larroides* group. Certain variables are restricted to males. For example, the antenna of *larroides* is simple, but in the related species, *fairmairei* and *dubius,* the basal flagellomeres are slightly offset from one another and somewhat swollen ventrally (fig. 139 C). This is more evident in *fraternus.* In *fiebrigi, argentinus,* and *joergenseni* the flagellomeres are rather ordinary in appearance, but ventrally on each article (except the last one or two) there is a narrow placoid which forms with its counterparts a continuous channel. This channel bears a complete single row of short, curved setae (fig. 139 B), a feature reminiscent of some male Crabronini such as *Crossocerus nitidiventris.* Sternum VIII is variable, also. In some species it is blunt apically (fig. 139 F), while in others the apex bears a long narrow process (fig. 139 H). The last male tergum is usually fingerlike and rounded apically (fig. 139 G), but in *tridens* it is a broad plate with a sharp median apical projection and a shorter blunter projection apicolaterally (fig. 139 E).

We have seen all but one of the species (*joergenseni*) but only one sex in a few. Nevertheless, the material on hand suggests at least two species groups, one Chilean and the other Argentinean.

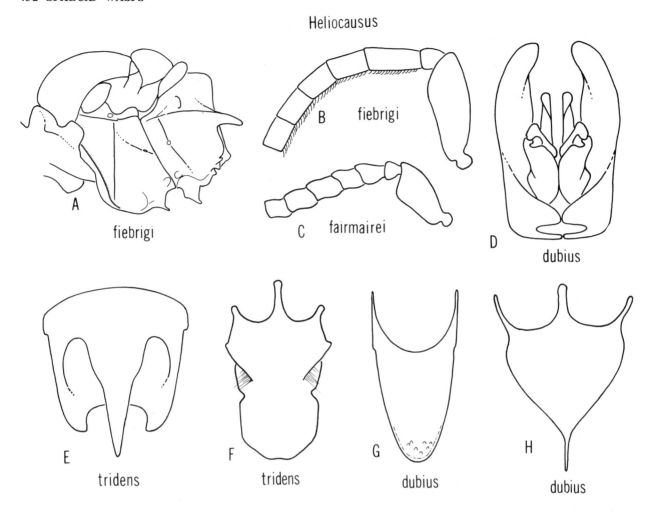

FIG. 139. Structural details in *Heliocausus*; A, thorax, lateral; B,C, male antenna; D, male genitalia; E,G, male tergum VII; F,H, male sternum VIII.

H. larroides group: one midtibial spur in both sexes, omaulus weak or absent, tergum I with lateral carina (mesad of spiracle, not to be confused with lateral carina delimiting laterotergite) (fig. 138), midocellus as large as and of same configuration as hindocelli (i.e. all ocelli with C-shaped rim) and all ocelli with dark pigment (i.e. not transparent except at C-shaped rim) (fig. 140 B), male antenna without ventral row of setae, Chilean species.

H. fiebrigi group: male with one midtibial spur, female with two, omaulus strong, tergum I without lateral carina, midocellus smaller than hindocelli and largely or wholly translucent (fig. 140 A) (hindocelli as in *larroides* group), male antenna usually with ventral row of setae (fig. 139 B), Argentinean species.

Heliocausus tridens does not fit either group satisfactorily. It seems to belong to the *fiebrigi* group; but the omaulus is weak, the male antenna lacks a setal row, and the last tergum is tridentate, so perhaps it should constitute a third group. We have seen only a male of *fraternus* which is an Argentine species that agrees with all of the features of the *larroides* group. Until the female is studied its affinities remain in doubt.

Evolutionary trends within the genus include the suppression of one midtibial spur and the omaulus, reduction of the pygidial plate in the female, and development of a carina mesad of the spiracle on tergum I. All of these trends are found in the *larroides* group.

Based on our material it is apparent that at least in *fairmairei* there is considerable variation in abdominal color pattern, and this may pose a problem for future workers in the genus. There is no key to the species of *Heliocausus*.

Biology: Very little is known about the ethology of these wasps. Herbst (1921a) said that the Chilean species nest in fine sand in sunny locations, they are gregarious, and they prey on immature Orthoptera. Jaffuel and Pirion (1926) said that *dubius* collected small bugs, "pequenos hemipteros," and *larroides* collected small white spiders, "pequenas aranas blancas." These observations are surprising and need confirmation. Manfredo Fritz (Evans, 1972) found *Heliocausus larroides* preying on leafhoppers.

Checklist of *Heliocausus*
(brackets enclose letter denoting species group)

argentinus Brèthes, 1913; Argentina (F)
dubius Kohl, 1905; Chile (L)
fairmairei Kohl, 1892; Chile (L)
   maculatus Reed, 1894
   ? *obscurus* Reed, 1894

fiebrigi Brèthes, 1909; Argentina (F)
fraternus Brèthes, 1913; Argentina (?L)
joergenseni Brèthes, 1913; Argentina (F)
larroides (Spinola), 1851 (*Arpactus*?); Chile (L)
mendozanus Brèthes, 1913; Argentina (F)
tridens Brèthes, 1913; Argentina (F)

Tribe Alyssonini

The alyssonins are small, slender wasps, most of which are taken only infrequently by the collector. The two well defined genera have a presumably relict status and come close to the theoretical ancestors of the Nyssonini. Figs. 141 A and 144 illustrate the alyssonin face and body profile. However, only a few species of *Didineis* have the striking foreleg expansion shown in fig. 144. All of the Alyssonini, as far as is known, stock their ground nests with saltatory Homoptera.

Diagnostic characters:

1. (a) Eyes broadly separated, inner orbits essentially parallel or converging slightly below (fig. 141 A); (b) ocelli well developed.
2. (a) Labrum small, usually concealed; (b) palpal formula 6-4.
3. (a) Scutum with admedian lines close together, sometimes short or faint; (b) notauli absent; (c) no oblique scutal carina but tegula partly covered by lateral scutal flange; (d) scutellum normally convex, not overlapping metanotum.
4. (a) Omaulus present; (b) no acetabular carina nor sternaulus; (c) episternal sulcus undefined above, represented by a pitted remnant between lower end of omaulus and anteroventral pleural margin; scrobal sulcus extending to omaulus, extension horizontal in *Alysson* but obliquely downward in *Didineis* (fig. 142 L, M).
5. (a) Metapleuron consisting of upper area only, tapering below; (b) propodeum rather long, depressed, rectilinear, areolate, bearing small dorsolateral teeth or angles (fig. 144).
6. (a) Midcoxae approximate, precoxal sulcus complete; (b) midtibia with two apical spurs, one or none (in some male *Didineis*) (fig. 142 G-I); (c) foretarsal rake of female weakly developed; (d) plantulae present.
7. (a) Forewing media diverging before or after cu-a (fig. 134 C, D); (b) stigma rather large (area greater than that of tegula); (c) three submarginal cells, II petiolate and receiving second or both recurrent veins.
8. (a) hindwing media diverging before cu-a; (b) jugal lobe small (fig. 134 C, D).
9. (a) Gaster sessile; (b) sternum I with a double carina toward base; (c) male with seven visible sterna in addition to protruding pair of spines from sternum VIII (fig. 144); (d) male genitalia with long gonobase, volsella rather simple but elongate, cuspis and digitus not differentiated (fig. 143); (e) female pygidial plate distinct, sharply edged, setose

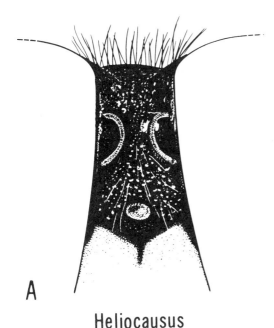

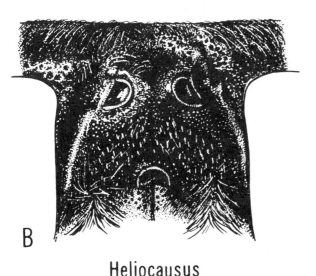

FIG. 140. Vertex area in females of two species of *Heliocausus*.

454 SPHECID WASPS

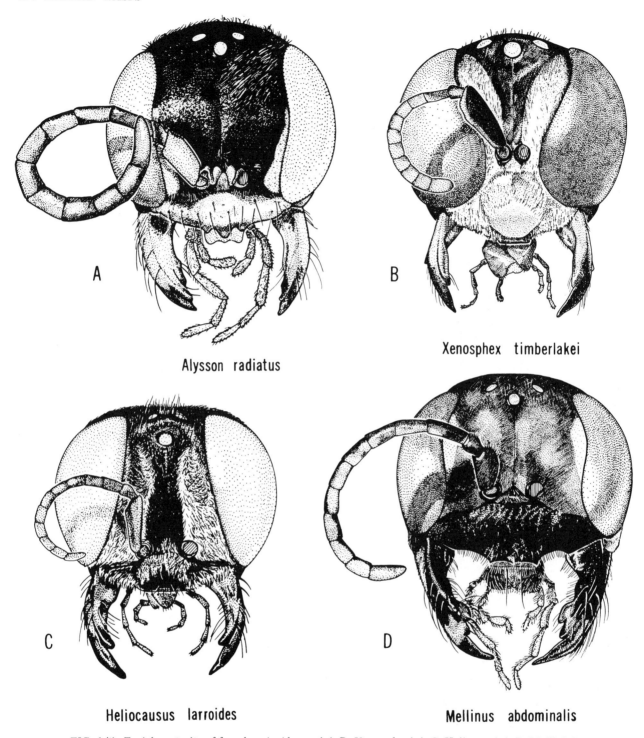

FIG. 141. Facial portraits of females; A, Alyssonini; B, Xenosphecini; C, Heliocausini; D, Mellinini.

(fig. 142 A, B).

Systematics: The alyssonins nearly fit the evolutionary concept of the nyssonine ancestral stock for the tribe Nyssonini. This is indicated on cactoid fig. 135 and the subfamily dendrogram (fig. 7). The phylogenetic association of the two tribes is reinforced by the discovery that the genus *Nursea* with its many alyssonin features is a generalized member of the Nyssonini. At first glance *Nursea* is strikingly like *Alysson*. The slender form, extensively polished integument, rectilinear and areolate propodeum with dorsolateral angles, broad frons, evanescent foretarsal rake, divergence of forewing media beyond cu-a, large stigma, and bristly female pygidial plate are all characteristics in common. Further study, however,

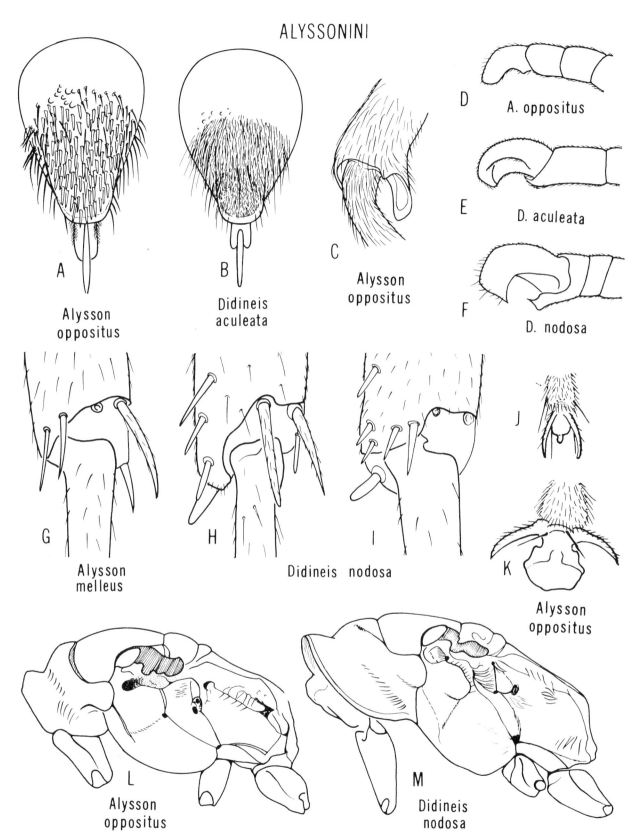

FIG. 142. Structural details in the tribe Alyssonini; A,B, female pygidium; C, apex of hindfemur, female; D-F, apex of male antenna, lateral; G-I, apex of midtibia; G,H, females; I, male; J,K, arolia and claws at apex of female midleg and foreleg, respectively; L,M, thorax, lateral.

reveals many differences, some of which are tribal in nature. In the latter category are (in *Nursea*) the short and appressed pronotal collar, widely separated midcoxae, and oblique scutal carina. Other differences shown by *Nursea* are the slightly separated admedian scutal lines, small female front arolium, two rather than three submarginal cells, absence of a jugal lobe, distal divergence of the hindwing media, and simple hindfemoral apex. Obviously, the progression of *Alysson-Nursea-Nippononysson-Nysson* is not a straight-line affair. However, a similar sort of evolutionary path may well have been followed.

Alysson and *Didineis* have been considered as a separate subfamily (Ashmead, 1899) or more frequently as a tribe in the Nyssoninae (Krombein, 1951). Sometimes the two genera have been associated with *Entomosericus* (Beaumont, 1954a), but we feel that there is stronger evidence against than for this idea. Points in common with *Entomosericus* are long ponotal "collar," hindwing media diverging before cu-a, jugal lobe moderately small, hindfemur with a spoonlike outer apical lobe, no spiracular groove on propodeum, and last male antennal article incurved (fig. 142 D-F).

Differences between *Entomosericus* and alyssonins make up a longer and in our opinion a more significant list. They are presented below in table 17.

TABLE 17.
Comparison of Structural Characters in Entomosericini and Alyssonini.

	Entomosericus	*Alysson* and *Didineis*
1.	Mandible simple	Mandible bidentate or rarely tridentate
2.	Clypeus quinquedentate, about two thirds as long as broad	Clypeus tridentate, about one third as long as broad
3.	Eyes converging below	Eyes nearly parallel within
4.	Female with foretarsal rake short but strong, fore arolium small	Female with weak foretarsal rake, fore arolium much enlarged
5.	Notauli present	Notauli absent
6.	Omaulus absent	Omaulus present
7.	Episternal sulcus well developed, scrobal sulcus meeting it at a right angle	Episternal sulcus evanescent above, scrobal sulcus horizontal or inclined downward anteriorly
8.	Forewing submarginal cell II not petiolate, recurrent veins ending close together, stigma small, submarginal cell I about half as long as discoidal I	Forewing submarginal cell II petiolate, recurrent veins ending far apart, stigma large, submarginal cell I nearly as long as discoidal I
9.	Propodeum short, sloping in profile, not dentate nor areolate, posterior face with a median groove	Propodeum long, rectilinear, dentate and areolate, posterior face without a median groove
10.	Sternum I with a single carina basally	Sternum I with a double carina basally
11.	Terga with transverse subapical depressions	Terga not depressed subapically
12.	Female pygidial plate bare, sides strongly raised	Female pygidial plate with broadened setae and bristles
13.	Male sternum VIII a broadly rounded lobe	Male sternum VIII deeply incised and bispinose
14.	Male gonostyle biramous	Male gonostyle simple
15.	Volsella differentiated into digitus and cuspis	Volsella undifferentiated

Taking the entire subfamily Nyssoninae into account, the Alyssonini differ by the rounded pronotal "collar" which is about as long as the scutellum or even longer. Furthermore, the tribe is distinguished by a combination of characters: petiolate submarginal cell II, no oblique scutal carina, episternal sulcus evanescent above, and omaulus present.

The arrangement of the midtibial spurs is unusual and deserving of special note. After clearing and slide-mounting midlegs from representatives species it was found that in *Alysson* the anterior spur is present, but the posterior one is marked only by a basal ring (fig. 142 G-I). In some specimens there may appear to be two spurs, but under high magnification one of these is merely a bristle arising on the edge surrounding the spur-bearing membranous socket. The true spur can be distinguished by the many secondary micropectinations along its shaft. In *Didineis* most species have two apical spurs (fig. 142 H), but in males of the *nodosa* group only the two basal rings are left (fig. 142 I). Thus, in females and most males of *Didineis* there are two spurs, in *Alysson* there is one, and in some males of *Didineis* there are none.

In the specialization table the Alyssonini rate from "11" to "15" which places the tribe well below the Nyssonini ("14" to "24").

The 55 listed species are primarily Holarctic, but a few must be considered Oriental and Ethiopian. The two genera, *Didineis* and *Alysson,* are rather similar, yet they are easily separated by the characters given in the key to genera and in the pertinent systematics sections.

Biology: The two genera seem to have similar habits. Refer to the generic biology sections.

Key to genera of Alyssonini

Forewing media diverging beyond cu-a or very near it (fig. 134 C), male last antennal article incurved but not strongly opposed by a projection from article XII (fig. 142 D), metapleuron about half as long as high (fig. 142 L), a pair of pale spots almost invariably present on tergum II (N. America, Eurasia, S. Africa) .... *Alysson* Panzer, p. 457

Forewing media diverging before cu-a by at least the latter's length (fig. 134 D), last male antennal article strongly incurved and opposed by a projection from article XII (fig. 142 E,F), metapleuron much less than half as long as high (fig. 142 M), tergum II without pale spots (Holarctic,

Bangladesh) .... *Didineis* Wesmael, p. 458

Genus Alysson Panzer

Generic diagnosis: Clypeus broad, apically thin and tridentate, labrum sometimes protruding slightly as a nearly rectangular lobe (fig. 141 A), mandible with a small subapical denticle and in some females with a toothlike angle near middle of inner edge, antenna relatively simple and last article in male incurved (fig. 142 D); frons simple and with median groove rather faint; forewing with media generally diverging beyond cu-a, rarely interstitial or slightly before it; stigma large; marginal cell stout and ending acutely at wing margin; submarginal cells II-III small, former always petiolate and latter occasionally; hindwing media diverging strongly toward front wing margin in basal part (fig. 134 C); jugal lobe small but much larger than tegula: arolium of female foreleg enlarged, rake feebly developed, midtibia with one apical spur; outer spex of hindfemur with a prominent spoon shaped lobe (fig. 142 C); pronotum with a strong transverse depression just in front of scutum; scutellum with a transverse anterior row of pits; propodeal enclosure entirely dorsal, triangular or U-shaped, areolate; pleuron (fig. 142 L) with scrobal sulcus extending almost horizontally forward to omaulus; metapleuron unusually stout, about half as long as high (fig. 142 L); propodeum somewhat areolate, with a dorsolateral angle or tooth and a second angle below it; gaster slender, relatively simple; female pygidial plate well formed (fig. 142 A), covered with scalelike setae and bristles; male sternum VII weakly sclerotized and more or less hidden beneath VI, VIII deeply emarginate and usually visible as two protruding spines; male genitalia with cuspis possibly indicated by a subbasal angle (fig. 143 B).

Geographic range: Of the 30 species listed, eight occur in America north of Mexico. In addition we have seen several undescribed Mexican species from Guadalajara, Nayarit, and Puebla. Of the remaining 22 species, 15 are Palearctic including one from Algeria and two fron northern India, three are Oriental, and four are Ethiopian including two from Madagascar.

Systematics: In addition to the characters given in the generic key, *Alysson* differs from *Didineis* by the more abrupt divergence of the hindwing media, single midtibial spur, more prominent cross groove on the pronotum, more nearly horizontal scrobal sulcus, and less prominent or entirely hidden male sternum VII.

Useful keys to species have been given by Handlirsch (1888a) for the world fauna, by Kokujev (1906) for Russia, and by Fox (1894a) for American representatives.

Biology: Information on the habits of *Alysson* has been summarized by Evans (1966a) who obtained his data from personal observations on *A. melleus;* and from records of Hartman (1905) on *melleus,* Rau and Rau (1918) on *melleus,* Kohl (1880) on *spinosus,* Olberg (1959) on *spinosus,* Ferton (1901) on *ratzeburgi,* and Yasumatsu and Masuda (1932) on *cameroni.*

According to Evans' summary, nesting is gregarious in cool, moist situations, often in sandy areas. The female

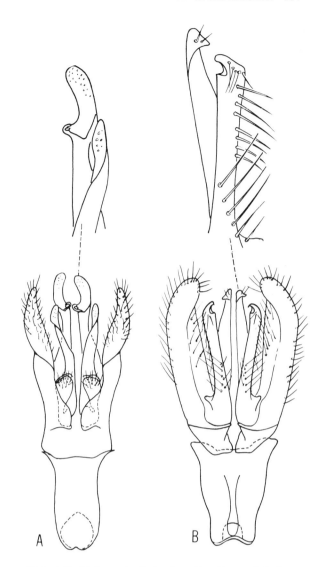

FIG. 143. Male genitalia in the tribe Alyssonini; A, *Didineis nodosa;* B, *Alysson spinosus.*

wasp digs a vertical burrow and pushes the excavated earth into a pile around the entrance. The short length of the tunnel through this tumulus assumes a slant of about 45 degrees with the surface. One to five cells per nest have been recorded, and the entrance is left open during provisioning. Prey are a variety of nymphal and adult leafhoppers and occasionally Cercopidae or Fulgoridae. The prey is carried in flight to the vicinity of the nest, then the last few centimeters on foot. As many as two dozen prey may be placed in a single cell. An egg is then laid on one of the last "hoppers" longitudinally on the side of the venter of the thorax, and the cell is closed. Final closure of the burrow is rather thorough but may leave a small mound.

Adults are not collected on flowers, but males are attracted to honeydew on leaves. Evans (1966a) reported an instance of copulation in *melleus* which took place in a nesting area. The male circled the female several times as she was walking on the sand. The male then landed on top of her, thrust his wings far forward, and affected coitus by twisting the tip of his abdomen sidewise. The period of contact was 15 seconds, after which the male flew off and the female resumed her walking over the sand.

Checklist of *Alysson*

(variant but improper spellings are *Alyson* and *Allyson*)

annulipes Cameron, 1897; e. India
cameroni Yasumatsu and Masuda, 1933; Japan
conicus Provancher, 1889; N. America
costai Beaumont, 1953; Italy
erythrothorax Cameron, 1902; n. India
flavomaculatus Cameron, 1901 (*Alyson*); U.S.: New Mexico
formosanus Tsuneki, 1968; Taiwan
guichardi Arnold, 1951; Ethiopia
guignardi Provancher, 1887; N. America; transcontinental
  *interstitialis Cameron, 1902, new synonymy by Bohart
  *petiolatus Cameron, 1902, new synonymy by Bohart
guillarmodi Arnold, 1944; s. Africa
harbinensis Tsuneki, 1967; Manchuria
katkovi Kokujev, 1906; Poland
madecassus Arnold, 1945; Madagascar
maracandensis Radoszkowski, 1877; sw. USSR: Kazakh S.S.R.
  incertus Radoszkowski, 1877
melleus Say, 1837; N. America e. of 100th mer.
ocellatus Beaumont, 1967; Turkey
oppositus Say, 1837; N. America e. of Rocky Mts.
pertheesi Gorski, 1852; Eurasia
  festivus Mocsáry, 1879
pictetii Handlirsch, 1895; Algeria
radiatus W. Fox, 1894; w. U.S.
ratzeburgi Dahlbom, 1843; Eurasia
ruficollis Cameron, 1898; Sri Lanka
seyrigi Arnold, 1945; Madagascar
spinosus (Panzer), 1801 (*Pompilus*); Europe
  bimaculatus Panzer, 1798 (*Sphex*), nec Fuesslin, 1775 (a scoliid)
  fuscatus Fabricius, of Panzer, 1798 (*Sphex*) (misidentification)
  ssp. *jaroslavensis* Kokujev, 1906; Russia
striatus W. Fox, 1894; U.S.: Washington, D.C., Long Island, New York
taiwanus Sonan, 1947; Taiwan
testaceitarsis Cameron, 1902; n. India
triangulifer Provancher, 1887; N. America e. of 100th mer.
  ssp. *shawi* Bradley, 1920; w. U.S., w. Canada
tricolor Lepeletier and Serville, 1825; Europe, sw. USSR
verhoeffi Tsuneki, 1967; Manchuria

Genus Didineis Wesmael

Generic diagnosis: Clypeus broad, apically thin and tridentate, labrum concealed, mandible with a small subapical tooth and in some females with an angle near middle of inner edge, antennal bases closer to clypeus than to each other, antenna nodose beneath in some males (*nodosa* group), last two articles modified so as to oppose each other (fig. 142 E,F), frons simple and with median groove rather faint; forewing with media diverging before cu-a by length of latter or more; stigma large; marginal cell ending acutely at wing margin; submarginal cells II-III small, II petiolate; hindwing media diverging well before cu-a but nearly parallel to it (fig. 134 D); jugal lobe small but much larger than tegula; arolium of female foreleg enlarged, rake feebly developed; midtibia with two apical spurs except for males of *nodosa* group, which have none; outer apex of hindfemur with a prominent spoon shaped lobe; pronotum with a transverse posterior impression which is largely hidden by scutum; propodeal enclosure dorsal, triangular or U-shaped; pleuron (fig. 142 M) with scrobal sulcus extending forwards and downwards toward end of omaulus; metapleuron about one-third as long as high; propodeum areolate in enclosure and to lesser extent elsewhere, with a dorsolateral angle or tooth (fig. 144); gaster slender, relatively simple, female pygidial plate margined posteriorly (fig. 142 B); male sternum VII well developed, VIII bispinose, its spines usually protruding (fig. 144); male genitalita as in fig. 143 A.

Geographic range: We have listed 26 species, all from the Holarctic Region except *orientalis* from Bangladesh. Nine species are known from North America including one from Cuba. On the whole, *Didineis* are uncommon, but like other such groups they may be locally abundant.

Systematics: Members of this genus are small, slender wasps, similar to those of the gorytin genus, *Dienoplus*. They are closely related to *Alysson* but are distinguished at a glance because they have no pale spots on the second tergum.

One distinctive American group represented by *nodosa* has the male antenna strongly nodose beneath and the male foreleg expanded (fig. 144). Other species in this group are *dilata, latimana, peculiaris,* and *viereckii.*

Fox (1894b) gave a key to the three American species known at that time. A much more up-to-date synopsis was that of Malloch and Rohwer (1930). In Europe the most useful keys are those of Handlirsch (1888a) and Gussakovskij (1937). The latter included 14 Palearctic species.

Biology: Not well known but apparently similar to that of *Alysson*. Strandtmann (1945) found a female *texana* carrying a fulgorid. Ferton (1912a) recorded nesting habits of *lunicornis* in Europe. Its prey were leafhoppers and fulgorids. Published information has been summarized by Evans (1966a).

Checklist of *Didineis*

aculeata (Cresson), 1865 (*Alyson*); Cuba
bactriana Gussakovskij, 1937; s. centr. USSR
barbieri Beaumont, 1968; Algeria
botscharnikovi Gussakovskij, 1937; sw. USSR
bucharica Gussakovskij, 1937; s. centr. USSR
clavimana Gussakovskij, 1937; USSR
crassicornis Handlirsch, 1888; Hungary

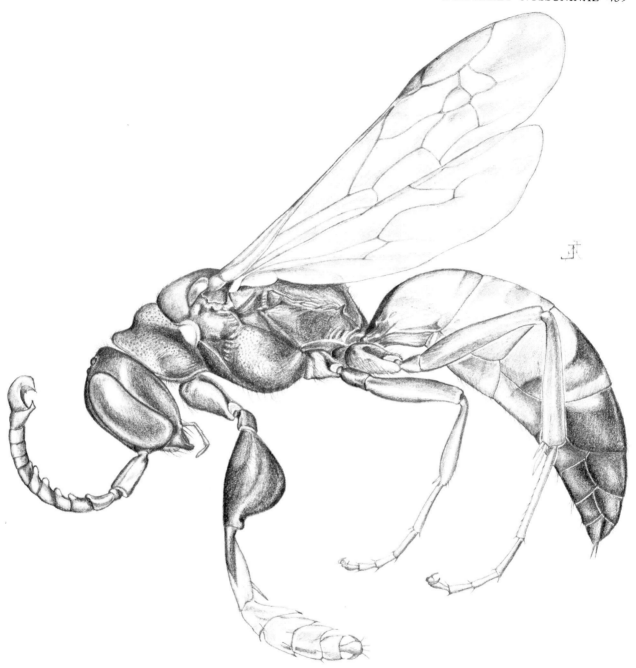

FIG. 144. *Didineis nodosa* W. Fox, male.

dilata Malloch and Rohwer, 1930; U.S.: Wisconsin, Nebraska
koshevnikovi Kokujev, 1906; w. USSR
latimana Malloch and Rohwer, 1930; U.S.
latro (Beaumont), 1967 (*Alysson*); Turkey
lunicornis (Fabricius), 1798 (*Pompilus*); Europe
    kennedii Curtis, 1836
nigricans Morice, 1911; Algeria
nodosa W. Fox, 1894; w. U.S.
    santacrucae Bradley, 1920
      clypeata Malloch and Rohwer, 1930
ogloblini Gussakovskij, 1937; sw. USSR

orientalis Cameron, 1897; Bangladesh
pannonica Handlirsch, 1888; Hungary, Turkey
peculiaris W. Fox, 1895; U.S.: Iowa to California; Mexico: Baja California
ruthenica Gussakovskij, 1937; w. USSR
sibirica Gussakovskij, 1937; USSR: Siberia
    ssp. *nipponica* Tsuneki, 1968; Japan
stevensi Rohwer, 1923; U.S.: N. Dakota
texana (Cresson), 1872 (*Alyson*); N. America e. of 100th mer., e. Arizona
turanica Gussakovskij, 1937; sw. USSR
viereckii Rohwer, 1911; U.S.: Kansas

NYSSONINI

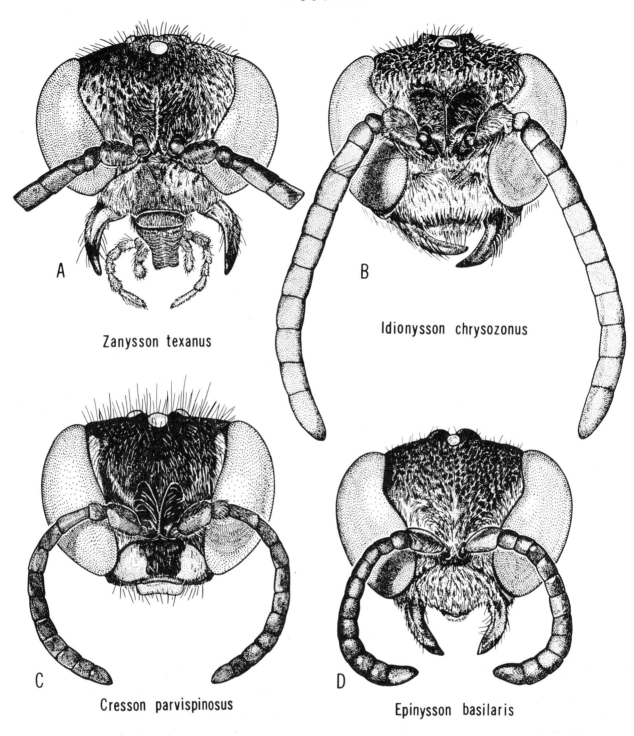

FIG. 145. Facial portraits of females in the tribe Nyssonini.

crassicornis Viereck, 1906, nec Handlirsch, 1888
wuestneii Handlirsch, 1888; Europe, Middle East
zimini Gussakovskij, 1937; w. USSR

Tribe Nyssonini

The Nyssonini are medium to small wasps, generally black or reddish, and usually with a limited amount of yellow maculation. All except one species have a heavily reinforced or "armored" integument, presumably as defense against the stings or bites of host wasps (fig. 151). The propodeal spines, reminiscent of those in the parasitic wasp family Chrysididae, may have a protective function also. They occur in all but one of the 18 genera.

Insofar as known, all Nyssonini are cleptoparasites of gorytin and larrin wasps. However, the paucity of data must be emphasized. Of the 216 species listed there is definite or presumptive proof of a parasite-host relationship for less than 10 percent.

Little is known of the habits of these wasps away from nesting areas of the hosts. They are occasionally collected on flowers, and some species are attracted to trees or bushes supplied with aphid honeydew. Most specimens are females collected as they fly close to the ground in rather slow and searching fashion.

Diagnostic characters:

1. (a) Eyes converging below, often strongly, inner margins slightly to moderately emarginate above (fig. 145); (b) ocelli well developed.
2. (a) Labrum small to rather large, in latter case broadly subrectangular; (b) palpal formula 6-4; (c) pronotum closely appressed to scutum.
3. (a) Scutum with admedian lines fused into a single line at or very near anterior edge; (b) without notauli, (c) with oblique scutal carina (fig. 156 D); (d) scutellum without posterior overlapping edge.
4. (a) Omaulus usually present, sometimes indistinct; (b) acetabular carina present, sometimes faint; (c) subomaulus present (fig. 4); (d) episternal and scrobal sulci absent or evanescent; (e) sternaulus absent; (f) verticaulus sometimes present (some *Nysson*).
5. (a) Metapleuron not always separated from propodeum by a distinct sulcus or line beneath upper metapleural pit; (b) propodeum spined dorsolaterally (fig. 4) except in *Nippononysson*.
6. (a) Midcoxae widely separated, precoxal sulcus complete or sometimes obscure; (b) midtibia with two apical spurs (except one in male *Cresson*); (c) foretarsal rake weak or absent; (d) arolia small or in one genus absent (*Antomartinezius*); (e) plantulae present but usually very small.
7. (a) Forewing media diverging at or before cu-a (except *Nursea*); (b) stigma rather small (except *Nursea* and *Nippononysson*); (c) two or three submarginal cells, true second cell petiolate (when present) and receiving at least one recurrent vein (fig. 148 A-C).
8. (a) Hindwing media diverging before to after cu-a; jugal lobe small or absent.
9. (a) Gaster with tergum I nearly as broad as II and with lateral line or carina; (b) sternum I with a double ridge converging basally; (c) male with seven visible terga; (d) male sternum VIII simple; (e) male genitalia with volsella undifferentiated into cuspis and digitus (fig. 150 F); (f) female pygidial plate usually well developed but sometimes indistinct, and with entire pygidium serrate or multidentate (fig. 153 D, J).

Systematics: Cactoid fig. 135 and the subfamily dendogram (fig. 7) illustrate the supposed origin and position of the tribe. We visualize it as arising rather early from an alyssonin or prealyssonin ancestor. Judging from *Nursea*, the body was not especially armored at first, the forewing media diverged beyond cu-a, and the stigma was large. Although the earliest nyssonins probably had submarginal cell II of the forewing petiolate, the number of cells was soon reduced in several phylogenetic lines. This usually took place by the loss of the basal vein of submarginal cell II, but in one instance (*Hyponysson*) it was the distal vein of III that was lost. A variety of dentition on the hindtibia developed along the evolutionary lines that culminated in *Zanysson* and *Metanysson* (fig. 146). The terga and sterna became double edged in many genera and in some a regular fringe of flattened setae was added. Males of several genera developed white hair brushes beneath sterna II to V (fig. 154). The essentially platelike female pygidium (fig. 153 D) became serrate and multidentate (fig. 153 F, J) in various species of the lines ending in *Metanysson* and *Foxia*. Arolia, which were originally small, became smaller yet and finally in *Antomartinezius* were lost altogether. The two midtibial spurs have been a remarkably constant feature, but one of these was finally lost in males of *Cresson*. Sternum VI in the female has become narrower and more sharply rounded or ridged in cross-section in many advanced genera. The ultimate in this was reached by *Metanysson* where the sternum has become saber shaped.

In summary the main morphological characters by which most Nyssonini may be recognized are the armored integument (except *Nursea*), spined propodeum (except *Nippononysson*), widely separated midcoxae, single median scutal line, double ridge on sternum I, petiolate submarginal cell II (when three cells are present), and possession of a strong oblique scutal carina.

Taking all of its features into account the tribe must be considered as highly advanced on the evolutionary scale although on a line separate from the Stizini-Bembicini. In the specialization table the Nyssonini rate about "14" to "24" which places the tribe a little above Stizini and a little below Bembicini.

In an effort to reduce the more than 200 listed species to manageable proportions, we have recognized 18 genera, thus abandoning the conservative stand that has kept most of the species in *Nysson*. Of the 18 genera, 13 occur in the Western Hemisphere, and of these 11 are found only there. Five are endemic to South America and only one to North America.

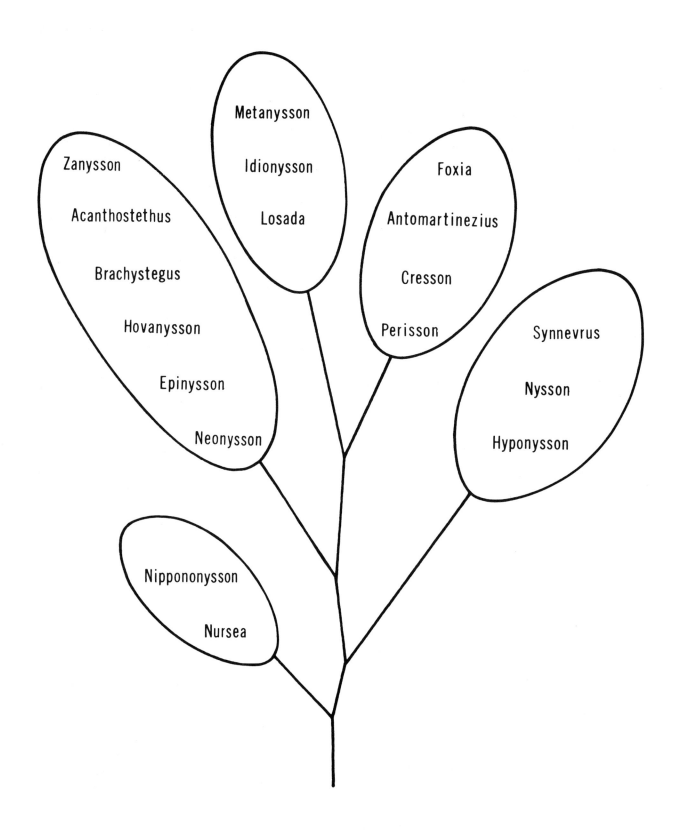

FIG. 146. Dendrogram suggesting relationships in the Nyssonini.

Literature on the group is quite scattered, but the papers of Handlirsch (1887, 1895) and Pate (1938b) are important landmarks.

Biology: In the Nyssonini the cleptoparasitic habit depends on several circumstances. The parasites must search out burrows of suitable hosts. They enter and quickly deposit an egg in a concealed place in one of the first of the host's stored provisions. The parasite closes the burrow upon leaving. The parasite egg hatches first; the young larva, armed with sharp mandibles, seeks out the host egg, punctures it, and assumes the role of the host larva.

An overall view of the host relationships of the Nyssonini is given in table 18. The literature sources are under the respective genera. Here the nyssonins are on the left, and their reputed host genera are on the right.

TABLE 18.
Host-parasite Associations in Nyssonini

Nysson	Host	Tribe
Nysson daeckei	*Hoplisoides*	
Nysson dimidiatus	*Dienoplus, Hoplisoides*	
Nysson fidelis	*Gorytes*	
Nysson interruptus	*Gorytes, Argogorytes*	
Nysson lateralis	*Gorytes*	Gorytini
Nysson maculatus	*Dienoplus*	
Nysson spinosus	*Gorytes, Argogorytes*	
Nysson trimaculatus	*Gorytes, Oryttus*	
Epinysson guatemalensis	*Hoplisoides*	
Epinysson bellus	*Hoplisoides*	
Epinysson moestus	*Hoplisoides*	
Epinysson basilaris	*Hoplisoides*	
Acanthostethus hentyi	*Sericophorus*	Miscophini
Acanthostethus portlandensis	*Sericophorus*	
Brachystegus scalaris	*Tachytes*	Larrini
Zanysson plesius	*Tachytes*	
Metanysson arivaipa	*Cerceris*	Cercerini
Metanysson coahuila	*Cerceris*	

In order to appreciate the trend of host-parasite relationships it is interesting to speculate on table 18 in comparison with the evolutionary dendrogram (fig. 145). Since the biologies of *Nursea* and *Nippononysson* are unknown it can only be guessed that the earliest hosts of the tribe were in the Gorytini. Most of the known associations are still of this sort, but there is strong indication that the more advanced members of the *Zanysson* line have adapted to hosts in the Larrinae, whereas in the *Metanysson* line at least *Metanysson* has become parasitic on *Cerceris*. Unfortunately, nothing is known of the four genera in the *Foxia* line. Possibly, *Foxia* may attack some of the gorytin genera about whose biology so little is known. On the other hand, some of the smaller *Tachytes*, such as *abdominalis*, occupy the same range as *Foxia* and could be hosts. Perhaps the peculiar South American genera, *Cresson, Antomartinezius,* and *Perisson,* may be associated with the similarly endemic Heliocausini. Nys-

sonin biology will be a fertile field for future biological studies.

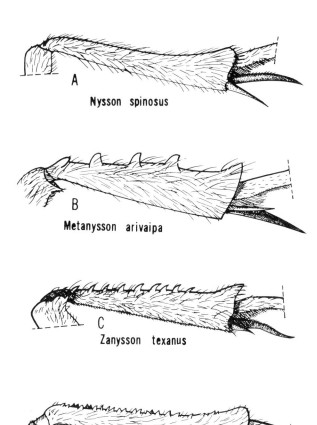

FIG. 147. Profiles of representative tibiae in the tribe Nyssonini.

Key to New World genera of Nyssonini

1. Hindtibia with teeth or stout spines along posterior surface (fig. 147 B-D) 2
   Hindtibia with hair and bristles only on posterior surface (fig. 147 A) 5
2. Forewing with two submarginal cells (fig. 148 H); female sternum VI reduced to a linear ridge; terga edged with close-set, flattened setae (N. and S. America) .... *Metanysson* Ashmead, p. 480
   Forewing with three submarginal cells; female sternum VI not unusually narrowed 3
3. Hindtibia with many small teeth along posterior edge (as in fig. 147 D); lower frons without a prominent crest; male with apical fimbriae on sterna II to V; (Chile) .... *Neonysson* R. Bohart, p. 470
   Hindtibia with four to 15 stout teeth in a single row (fig. 147 B, C); lower frons with a prominent crest; male without

sternal brushes 4
4. Hindtibia with a long single row of stout teeth (fig. 147 C); inner orbit rather evenly concave (fig. 145 A); margins of terga thickened but without flattened setae (N. and S. America)............ *Zanysson* Rohwer, p. 475
  Hindtibia with four stout teeth (about as in fig. 147 B); inner orbit angularly emarginate (fig. 145 B); margins of terga not thickened but edged with close-set, flattened setae (S. America) .. *Idionysson* Pate, p. 480
5. Sterna III-IV, and sometimes others, lobate, dentate or spinose laterally 6
  Sterna III-IV not obviously lobate, dentate or spinose laterally 8
6. Frons without a Y-shaped crest, sometimes with a simple raised ridge or point; second recurrent vein ending well beyond submarginal cell II (fig. 148 B) (N. and S. America) .... *Foxia* Ashmead, p. 478
  Frons with a sharply Y-shaped crest above antennal insertions (S. America) 7
7. Hindtarsus with a small but evident arolium; female foretarsus without a well developed comb (Argentina) .... *Perisson* Pate, p. 475
  Hindtarsus without an arolium; female foretarsus with a well developed comb (Argentina) .... *Antomartinezius* Fritz, p. 476
8. Posterior margins of terga simple, not thickened nor double edged 9
  Posterior margins of terga thickened and double edged, at least dorsolaterally 11
9. Forewing with only two submarginal cells (fig. 148 I), II petiolate (Nearctic) .... *Hyponysson* Cresson, p. 467
  Forewing with three submarginal cells, II petiolate 10
10. Posterior margins of terga without a regular fringe of flattened setae, sometimes with shaggy, pointed setae; second recurrent vein ending at or only slightly beyond submarginal cell II (fig. 148 A); female pygidial plate well developed (worldwide except S. America and Australia) .... *Nysson* Latreille, p. 467
  Posterior margins of terga with prominent flat setae which are apically notched forming a complete, regular fringe on I-V; second recurrent vein ending well beyond submarginal cell II; female pygidial plate hardly developed (Venezuela) .... *Losada* Pate, p. 479
11. Media of hindwing diverging at cu-a (fig. 148 E); posterior tergal margins gradually thicker toward dorsolateral edges (Holarctic) .... *Synnevrus* A. Costa, p. 470
  Media of hindwing diverging well beyond cu-a

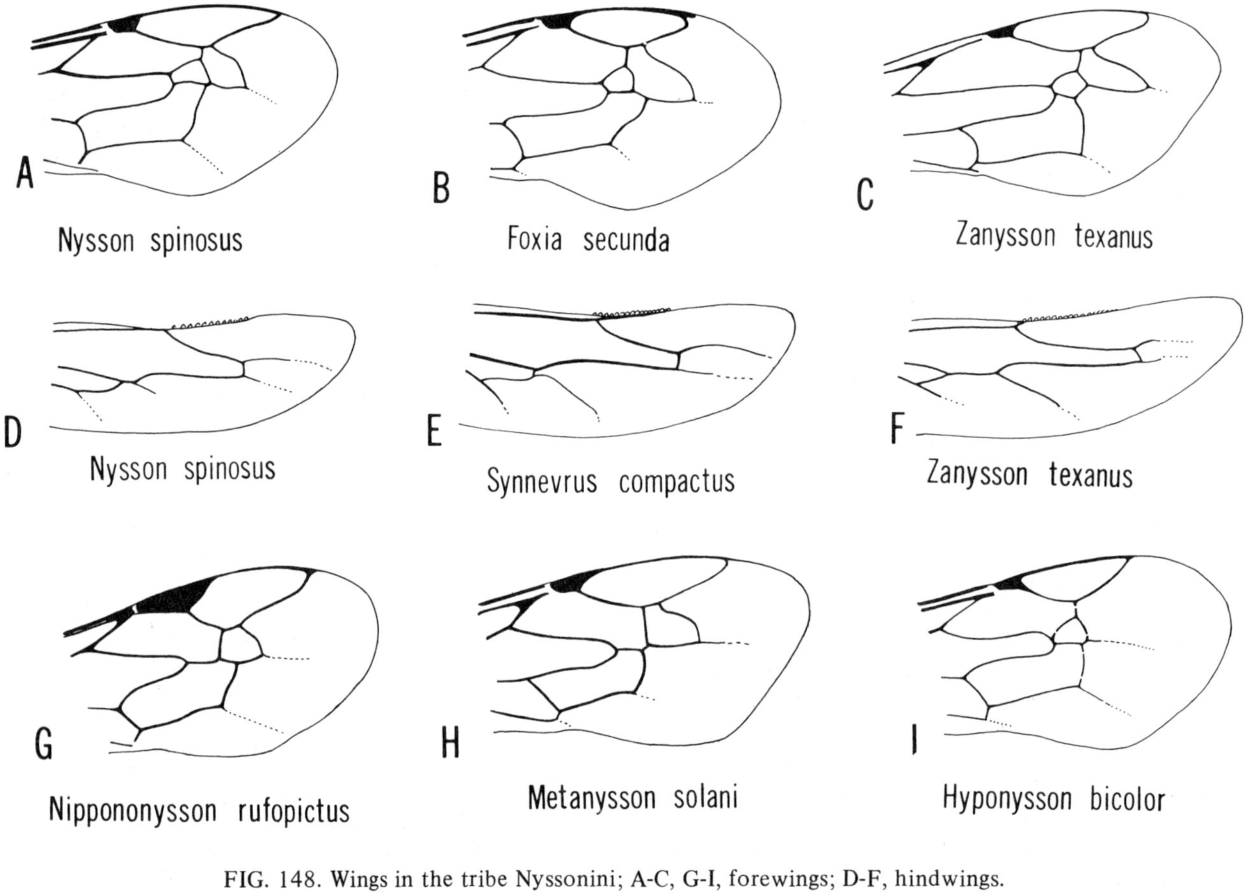

A Nysson spinosus
B Foxia secunda
C Zanysson texanus
D Nysson spinosus
E Synnevrus compactus
F Zanysson texanus
G Nippononysson rufopictus
H Metanysson solani
I Hyponysson bicolor

FIG. 148. Wings in the tribe Nyssonini; A-C, G-I, forewings; D-F, hindwings.

(as in fig. 148 F) 12
12. Tergal apices rather abruptly thicker dorsolaterally; frons without a Y-shaped crest
(fig. 145 D); male midtibia two spurred
(New World) .... *Epinysson* Pate, p. 471
Tergal apices not abruptly thicker dorsolaterally; frons with a sharply Y-shaped
crest extending upward from between antennal bases (fig. 145 C); male midtibia one
spurred (Chile) .... *Cresson* Pate, p. 476

Key to Old World genera of Nyssonini

1. Forewing with only two submarginal cells,
II not always petiolate (figs. 148 G, 149 D) 2
Forewing with three submarginal cells, II
petiolate (fig. 148 A) 4
2. Stigma of forewing much smaller than submarginal cell II; midtibia with a longitudinal, posterior, serrulate ridge (as in fig.
147 D); hindwing with a distinct jugal
lobe (Australia)... *Acanthostethus* F. Smith, p. 473
Stigma of forewing about as large as submarginal cell II; hindtibia with hair and bristles
only; hindwing without a distinct jugal lobe 3
3. Frons and scutum polished and practically
impunctate; inner eye margins hardly
emarginate (fig. 149 B); metanotum a
low convexity (n. India)... *Nursea* Cameron, p. 465
Frons and scutum with micro and macropunc-
tures; inner eye margin a little emarginate
above middle of face; metanotum trans-
versely ridged (e. coast of Asia) ....
.... *Nippononysson* Yasumatsu and Maidl, p. 465
4. Hindtibia irregularly serrate or serrulate
posteriorly (fig. 147 D); males with
prominent hair tufts on sterna II-V
(Eurasia, Africa)...... *Brachystegus* A. Costa, p. 472
Hindtibia with hairs and bristles
only (fig. 147 A) 5
5. Posterior margins of terga not thickened
nor double edged (worldwide except
S. America and Australia.. *Nysson* Latreille, p. 467
Posterior margins of terga thickened
and double edged 6
6. Media of hindwing diverging from cu-a
or very near it (fig. 148 E); male with-
out sternal hair tufts (Holarctic) ....
.... *Synnevrus* A. Costa, p. 470
Media of hindwing diverging well beyond
cu-a; male with prominent sternal hair
tufts (at least in one species)
(Madagascar)................ *Hovanysson* Arnold, p. 472

Genus Nursea Cameron

Generic diagnosis: Only one female known; front view of head (fig. 149 B); clypeal apex thin, convex; basal flagellar articles longer than broad, XI twice as long and large as X; face not ridged, a fine carina from antennal bases halfway to median ocellus; inner eye margins slightly converging below, hardly emarginate above; frons polished and practically impunctate; forewing with two submarginal cells (fig. 149 D) (more basal vein of usual second cell missing); marginal cell only twice as long as broad, distally pointed at wing margin; stigma large, covering as much area as distal (apparent second) submarginal cell; media diverging after cu-a; hindwing media diverging far beyond cu-a, jugal lobe atrophied (fig. 149 E); mesonotum polished and practically impunctate, scutellum with a deep median groove and sublobate (fig. 149 C); arolia present, small; foretarsus without a rake; posterior surface of hindtibia with hairs and small bristles only; outer apex of hindfemur without a spoon-like lobe; metanotum low, longitudinally ridged, ridges continued onto propodeal enclosure (fig. 149 C); propodeal teeth present as slightly acute angles among areolae; gaster (fig. 149 A) moderately slender, segments single edged posteriorly, not margined with close-set, flattened setae; sternum II moderately convex; sternum VI compressed but not saber shaped; pygidial plate short, hairy and bristly (fig. 149 F).

Geographic range: The genus is based on the holotype female of *carinata* collected by C. S. Nurse at Simla, Himachal Pradesh State, India (Mus. London). This area lies in the Himalayan foothills of northern India.

Systematics: The phylogenetic position of this genus has been a mystery since its original description in 1902 by Cameron, who put it near *Tiphia* (Tiphidae) and *Scolia* (Scoliidae). In correspondence with I.H.H. Yarrow at the British Museum, Dr. Yarrow pointed out his conclusion that *Nursea* belonged to the Nyssoninae on the basis of the oblique scutal carina and the two midtibial spurs. Our subsequent examination of the specimen showed it to be in the tribe Nyssonini and close to *Nippononysson*.

The extensively polished integument of *Nursea* differentiates this genus from all others in the tribe and provides a possible link with an alyssonin ancestor. Other characters reinforcing an association with Alyssonini are the general makeup of the body, quite similar propodeum with lateral spines derived from carinae, relatively large forewing stigma, origin of forewing media beyond cu-a (as in *Alysson*), and a bristly female pygidial plate (as in *Didineis*).

The relationship with *Nippononysson* is discussed under that genus. The main differences are (in *Nursea*) the smoother integument of the head and thorax, the distal derivation of the forewing media, and the presence of propodeal teeth.

Biology: Unknown.

Checklist of *Nursea*

carinata Cameron, 1902; n. India

Genus Nippononysson Yasumatsu and Maidl

Generic diagnosis: Clypeal apex thin, narrowly flanged and convex; labrum small, reddish; basal flagellar articles rather slender, longer than broad; last male flagellomere rather simple, obconic, larger than preceding article, not seamed; a small, short longitudinal ridge on frons above antennal bases; inner eye margins converging gently below, a little emarginate above; frons with abundant large and small punctures as usual in tribe; forewing with two submarginal cells (more basal vein of usual second cell

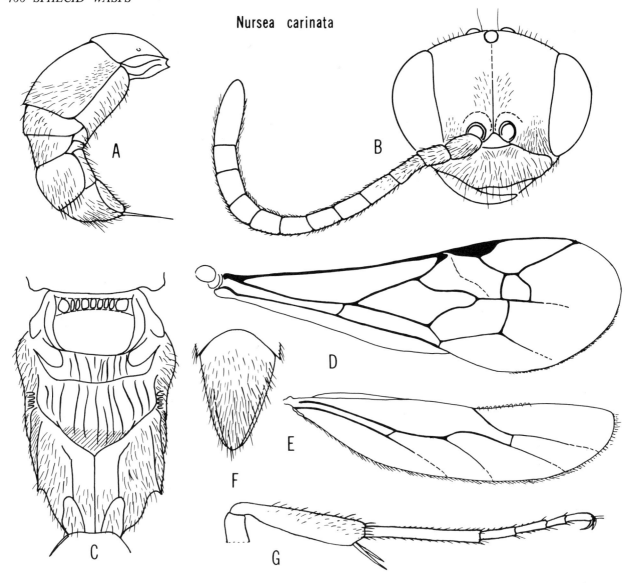

FIG. 149. *Nursea carinata* female holotype; A, profile of gaster; B, face; C, Dorsal view of scutellum, metanotum and propodeum; D-E, wings; F, pygidium; G, tibia and tarsus of midleg.

missing); marginal cell only twice as long as broad, distally pointed at wing margin; stigma large, covering more area than distal (apparent second) submarginal cell (wing venation about as in fig. 149 D, except for origin of forewing media which is interstitial with cu-a); hindwing media diverging far beyond cu-a; jugal lobe present but extremely small, much smaller in outline than tegula; scutum coarsely punctate as usual in tribe; scutellum weakly bilobed; arolia present; female foretarsus without a rake, male midtibia with two strong spurs; posterior surface of hindtibia with hairs and small, rather stout bristles only, apex of hindtibia a little reflexed and projecting, but rounded; outer apex of hindfemur without a spoon-like lobe; metanotum raised into a sharp transverse ridge; no propodeal spines but lateral area quite deeply areolate; gaster moderately slender, segments single edged posteriorly, not margined with close-set, flattened setae; sternum II bulging abruptly and strongly; female sternum VI convex; female pygidial plate small, nearly triangular; male sterna II to V with white-tufted apices (at least in *rufopictus*), but tufts not very discrete, tending toward a double row; male tergum VII with two strong teeth.

Geographic range: Oriental Region. The three species are from Hokkaido Island to Amami Island (Japan), the Philippine Islands, and Turkey.

Systematics: This genus is very close to *Nursea* and can be considered a derivative of it. The principal differences in *Nippononysson* are the coarser sculpture (more like others in the tribe), weak frontal keel, more convergent as well as less emarginate eyes, origin of the forewing media at cu-a rather than beyond it, sharply transverse metanotum, and absence of propodeal teeth. Male characteristics are not comparable since *Nursea* is known from the female only.

A comparison of *adiaphilus* with *rufopictus* and comments on phylogeny were made by Krombein (1943). Except for *Nursea,* we agree with his statement that "The thinner, more finely sculptured integument, lack of propodeal spines or teeth, and very large stigma seem to indicate that *Nippononysson* belongs much closer to the primitive nyssonine stock than any of the existing genera."

Biology: Unknown.

Checklist of *Nippononysson*

adiaphilus Krombein, 1943; Philippines
rufopictus Yasumatsu and Maidl, 1936; Japan, Ryukyus
inexspectatus Beaumont, 1967; Turkey

Genus Hyponysson Cresson

Generic diagnosis: Clypeal apex broadly truncate, not beveled; labrum small and reddish, basal flagellar article a little longer than broad, others shorter except conical last one, flagellum relatively slender and simple, frons simple, rather finely punctate; inner eye margins converging below but broadly separated, gently emarginate above; forewing with two submarginal cells (third missing), II petiolate and receiving both recurrent veins (fig. 148 I); marginal cell a little stubby distally but ending at wing edge; stigma smaller than submarginal cell II; hindwing media diverging well beyond cu-a; jugal lobe larger in outline than tegula; arolia present; female foretarsus without a comb; male midtibia two spurred; posterior surface of hindtibia with hairs and small bristles only, outer apex of hindtibia roundly projecting; outer apex of hindfemur without a spoon shaped lobe; metanotum roughened but essentially simple; metapleuron rather broad below; propodeal spines unusually small, no projecting lower tooth above hindcoxa; gaster moderately stout, sternum II not bulging as in many nyssonins, segmental margins thin edged and not fringed; female sternum VI strongly convex but not saber shaped; female pygidial plate rounded apically; male sterna II to V without hairbrushes; male tergum VII narrowly tridentate.

Geographic range: The two species together cover the southern and western parts of the United States.

Systematics: Although originally proposed as a genus by Cresson (1882), *Hyponysson* has usually been considered as a group or a subgenus of *Nysson.* We prefer to treat it as a genus on the basis of the constant and peculiar wing venation (fig. 148 I). At the same time it must be said that there is a strong resemblance to certain species of *Nysson,* notably *gagates.* The two species of *Hyponysson* differ from each other in several minor details. *H. bicolor* is larger, the male tergum VII is equidentate, and the female pygidial plate is broader and not so well defined.

Except for the forewing cell reduction, *Hyponysson* appears to be a rather generalized member of the tribe.

Biology: The only biological observation was made by Rau (1922) with respect to *raui* entering burrows of the panurgid bee, *Calliopsis nebraskensis* Crawford. Since there are no authenticated cases of association between bees and nyssonines, the habits of *raui* need much further investigation.

Checklist of *Hyponysson*

bicolor Cresson, 1882; w. U.S.
raui (Rohwer), 1917 (*Nysson*); s. and se. U.S.

Genus Nysson Latreille

Generic diagnosis: Clypeal apex thin, labrum not much exserted; flagellum various (fig. 152 D,G); frons plain above antennal bases or rarely with a raised point, inner eye margins strongly or moderately converging below, moderately and evenly emarginate above middle; forewing with three submarginal cells, recurrent veins both received by petiolate cell II (fig. 148 A), or exceptionally second recurrent interstitial or slightly on III; marginal cell tapering or nearly rounded distally but ending at wing margin, stigma smaller than submarginal cell II; hindwing media diverging near cu-a (fig. 148 D) or well beyond it; jugal lobe larger in outline than tegula; arolia present; female foretarsus without a comb; male midtibia with two spurs; posterior surface of hindtibia with hairs and small bristles only (fig. 147 A); outer apex of hindtibia projecting roundly or rather sharply; hindfemur without a spoon shaped outer distal truncation; metanotum relatively simple, often longitudinally carinate or areolate; lower metapleural area sometimes well defined (*N. tristis,* for example); propodeal spines large, slender, or small, no secondary tooth below; gaster stout, segments single edged without a regular setal fringe but sometimes with shaggy pubescence; female sternum VI convex, sometimes narrowed and projecting beyond tergum VI but not saber-like; female pygidial plate rounded or nearly pointed apically; male sterna II to V without white hair tufts or rarely with some evidence of them (*argenticus, fulvipes*); male tergum VII usually bidentate but sometimes tridentate or quinquedentate.

Geographic range: Widespread in continental areas of the world but apparently poorly represented in the Ethiopian Region, and absent from Australasia and South America. Over 80 species are known.

Systematics: Nysson is a variable genus as might be expected, since with 83 listed species it is also the largest in its tribe. Its differentiating morphological features are the absence of a Y-shaped frontal projection, three submarginal cells in the forewing with the recurrents ending on or very close to II (fig. 148 A), simple hindtibia, thin-edged abdominal segments without a regular setal fringe, and a flat female pygidial plate. One or more of the foregoing are found in other genera, but not in this combination.

Closest relatives are probably *Hyponysson,* which has only two submarginal cells, and *Synnevrus* which has the segmental margins double.

In most literature the divergence of the hindwing media at cu-a is considered of prime importance. As Beaumont (1954a) and others have pointed out, the type of the genus, *spinosus*, is variable in this respect. In the United States a similar situation occurs in *simplicicornis*. This has confused the limits of the genus as discussed under *Epinysson*. According to our arrangement, most *Nysson* have the hindwing media diverging at or near cu-a but in a number of American species the media diverges well beyond it (R. Bohart, 1968a). These are *trichrus, schlingeri, gagates, timberlakei, argenticus, euphorbiae, pumilus, bakeri, aridulus,* and *rufoflavus*. Species with a raised point above the antennal insertions are *timberlakei, euphorbiae* and *argenticus*. These three species are remarkable also for the shaggy silver pubescence margining the abdominal segments and for the greatly enlarged apex of the male antenna. In *timberlakei*, male tergum VII is quinquedentate, whereas in the other two it is tridentate. The tridentate condition of male tergum VII is not uncommon. In the American fauna it occurs in *simplicicornis, chumash, hesperus, gagates, trichrus, argenticus, euphorbiae,* and *bakeri*. An interesting modification of sternum VI in the female of certain American species probably indicates group status. Here VI narrows and is prolonged to a pointed or very narrow apex which is well beyond the tip of the pygidium, and it has a median longitudinal carina. This occurs in *chumash, rufiventris, simplicicornis, hesperus, trichrus* and *schlingeri*. It should be noted that the last two of these have the hindwing media diverging far beyond cu-a. In many cases male genitalia show diagnostic differences (fig. 150 A, B, F, G). Important references are Handlirsch (1887, 1895), Beaumont (1964b), Cresson (1882) and R. Bohart (1968a).

Biology: The association of *Nysson* and its hosts has been summarized by Evans (1966a). In his table 9 the following *Epinysson* should be eliminated: *hoplisivora, bellus, moestus,* and *tuberculatus*. The remainder show the following positive associations: *daeckei* on *Hoplisoides nebulosus* and *Gorytes canaliculatus*; *fidelis* on *Gorytes canaliculatus*; *dimidiatus* on *Hoplisoides latifrons*; and *maculatus* on *Dienoplus tumidus*. Other species with presumptive evidence of association are *lateralis* on *Gorytes canaliculatus*; *dimidiatus* on *Dienoplus elegans* and *Dieno-*

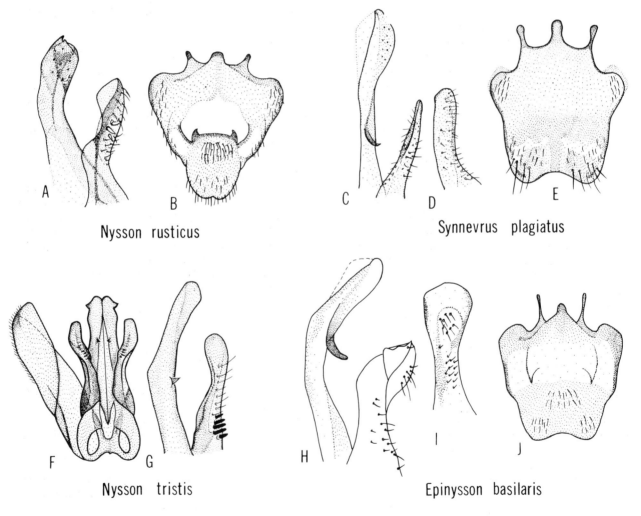

FIG. 150. Structural details in males of the tribe Nyssonini; A,C,G,H, apex of volsella and right side of aedeagus in side view; B,E,J, sternum VIII; D,I, ventral view of volsellar apex; F, ventral view of genitalia, gonostyle removed on right.

plus tumidus; maculosus on *Gorytes laticinctus, Lestiphorus bicinctus,* and *Oryttus concinnus; maculatus* on *Dienoplus laevis; interruptus* on *Gorytes quadrifasciatus* and *Argogorytes fargeii;* and *spinosus* on *Gorytes quadrifasciatus, Argogorytes fargeii,* and *Argogorytes mystaceus.* A questionable association is that of *dimidiatus* on *Bembecinus tridens.* In summary it appears that *Nysson* is a cleptoparasite of various gorytin genera including *Gorytes, Argogorytes, Oryttus, Dienoplus,* and *Hoplisoides.*

Checklist of *Nysson*

alicantinus Mercet, 1909; Spain
argenteofasciatus Radoszkowski, 1879; sw. USSR: Kazakh S.S.R.
argenticus R. Bohart, 1968; U.S.: California, Arizona
aridulus R. Bohart, 1968; U.S.: California
bakeri R. Bohart, 1968; U.S.: California
barrei Radoszkowski, 1893; sw. USSR
basalis F. Smith, 1856; India
    ssp. *taiwanus* Tsuneki, 1968; Taiwan
bohemicus Zavadil, 1951; e. Europe
braunsii Handlirsch, 1900; s. Africa
cardinalis Gussakovskij, 1929; sw. USSR: Transcaspia
castaneus Radoszkowski, 1877; sw. USSR: Kazakh S.S.R.
castellanus Mercet, 1909; Spain
chevrieri Kohl, 1879; Eurasia
chiengmaiensis Tsuneki, 1963; Thailand
chumash Pate, 1940; U.S.: California
compactus Cresson, 1882; w. U.S.
curtulus F. Morawitz, 1892; sw. USSR: Kazakh S.S.R.
daeckei Viereck, 1904; e. U.S.
dimidiatus Jurine, 1807; Eurasia
    wesmaeli Lepeletier, 1845
    distinguendus Chevrier, 1867
    decemnotatus A. Costa, 1869
doriae Gribodo, 1884; Borneo
?*dubius* Olivier, 1811; Arabian penin.
dusmeti Mercet, 1909; Spain
dutti Turner, 1921; India
erubescens Morice, 1911; Algeria
erythropoda Cameron, 1890; India
euphorbiae R. Bohart, 1968; sw. U.S.; Mexico: Sinaloa
fidelis Cresson, 1882; U.S.
fraternus Mercet, 1909; Spain
fulvipes A. Costa, 1859; Europe, n. Africa, Middle East
fulviventris Tsuneki, 1970; Mongolia
gagates Bradley, 1920; U.S.
ganglebaueri Kohl, 1912; Europe
gerstaeckeri Handlirsch, 1887; Rhodes, Middle East
hesperus R. Bohart, 1968; w. U.S.
horni Strand, 1913; Sri Lanka
hrubanti Balthasar, 1972; Czechoslovakia
humilis Handlirsch, 1895; USSR: Armenian S.S.R.
ibericus Handlirsch, 1895; Spain
inornatus Beaumont, 1967; Turkey
interruptus (Fabricius), 1798 (*Mellinus*); Europe
    spinosus Fabricius, 1804 (*Ceropales*), nec J. Forster, 1771
    panzeri Lepeletier, 1845
    shuckardi Wesmael, 1852
kolazyi Handlirsch, 1887; "Palearctic"
konowi Mercet, 1909; Spain
laevis Pulawski, 1964; Egypt
lapillus Beaumont, 1965; Lebanon
lateralis Packard, 1867; U.S.
laufferi Mercet, 1904; Spain
maculosus (Gmelin), 1790 (*Sphex*); Palearctic, new name for *maculatus* Fabricius, (Art. 59c)
    maculatus Fabricius, 1787 (*Sphex*), nec Drury, 1775, nec *Sphex maculata* (Fabricius), 1775 (*Evania*), now in Pompilidae
    dissectus Panzer, 1800 (*Mellinus*)
    guttatus Olivier, 1811
    omissus Dahlbom, 1845
    lineolatus Schenck, 1857
    dubius A. Costa, 1859, nec Olivier, 1811
miegi Mercet, 1904; Spain
mimulus Valkeila, 1964; Europe
    handlirschi Hellén, 1955, nec Schmiedeknecht, 1896
minutus Arnold, 1929; s. Africa
monachus Mercet, 1909; Spain
nanus Handlirsch, 1898; Algeria
neorusticus R. Bohart, 1968; U.S.: w. of 100th mer.
niger Chevrier, 1868; Europe
    ssp. *pekingensis* Tsuneki, 1971; China
parietalis Mercet, 1909; Spain
pratensis Mercet, 1909; Spain
pumilus Cresson, 1882; w. U.S.
pusillus Beaumont, 1953; Morocco
quadriguttatus Spinola, 1808; Europe
    ?*quadriguttatus* Olivier, 1811, nec Spinola, 1808
recticornis Bradley, 1920; w. U.S.
roubali Zavadil, 1937; e. Europe
rufiventris Cresson, 1882; w. U.S.
    punctatus Ashmead, 1897, new synonymy by R. Bohart
rufoflavus R. Bohart, 1968; U.S.: California
rufus Handlirsch, 1895; Egypt
rugosus Cameron, 1890; India
ruspolii Schulthess, 1893; Somalia
rusticus Cresson, 1882; w. U.S.
    ssp. *sphecodoides* Bradley, 1920; U.S.: California, new status by R. Bohart
ruthenicus Birula, 1912; w. USSR
schlingeri R. Bohart, 1968; U.S.: California
schmiedeknechtii Handlirsch, 1900; n. Africa
sexguttatus Gussakovskij, 1953; sw. USSR: Tadzhik S.S.R.
simplicicornis W. Fox, 1896; N. America
    maculipes Mickel, 1916 (*Brachystegus*)
    minimus Rohwer, 1921
    kaskaskia Pate, 1938
spinosus (J. Forster), 1771 (*Sphex*); Europe
    ?*bidens* Linnaeus, 1767 (*Vespa*)
    spinosus Fabricius, 1775 (*Crabro*), nec J. Forster, 1771

tricinctus Fabricius, 1793 (*Mellinus*)
trilineatus Turton, 1802 (*Vespa*)
geniculatus Lepeletier, 1845, nec Olivier, 1811
   ssp. *malaisei* Gussakovskij, 1933; Asia
subtilis W. Fox, 1896; N. America
susterai Zavadil, 1951; e. Europe
timberlakei R. Bohart, 1968; sw. U.S.
trichrus (Mickel), 1916 (*Brachystegus*); U.S. e. of 100th mer.
   nigripes Provancher, 1887, nec Spinola, 1808
   melanopus Pate, 1938, new name for *nigripes*
tridens Gerstaecker, 1867; Europe, Algeria, Mongolia
   ssp. *melas* M. Müller, 1920; Europe
trimaculatus (Rossi), 1790 (*Crabro*); Europe
   nigripes Spinola, 1808
   geniculatus Olivier, 1811
   ssp. *japonicus* Tsuneki, 1964; Japan
tristis Cresson, 1882; w. U.S.
varelai Mercet, 1909; Spain, Algeria
variabilis Chevrier, 1867; Europe
   friesei Handlirsch, 1887
willowmorensis Brauns, 1911; s. Africa

Genus Synnevrus A. Costa
(variant spelling is *Synneurus*)

Generic diagnosis: Clypeal apex not sharply beveled, labrum reddish to dark; basal flagellar articles stout, last article in male flattened or incurved beneath and with a longitudinal seam (fig. 152 B); no pronounced frontal process; inner eye margins moderately converging below, slightly emarginate above middle; forewing with three submarginal cells, III greatly narrowed in front and sometimes joining petiole of II which normally receives both recurrent veins; stigma smaller than II; marginal cell ending rather bluntly at wing margin; hindwing media diverging at or very near cu-a (fig. 148 E); jugal lobe larger in outline than tegula; male midtibia two spurred; arolia present; female foretarsus without a recognizable rake; posterior surface of hindtibia with hairs and bristles only; outer apex of hindfemur without a spoon shaped truncation; metanotum roughened; lower metapleural area usually well defined; propodeal spines strong, no secondary teeth between them and gastral insertion; gaster stout, segments double edged but with a regular fringe of flattened setae; female sternum VI convex; female with pygidial plate which is rounded or subtruncate apically (fig. 153 L); male sterna II to V without white hair brushes or lateral teeth; male tergum VII essentially bidentate (fig. 153 I).

Geographic range: The genus appears to be primarily Holarctic including north Africa. Of the 20 listed species, five are Nearctic.

Systematics: Synnevrus has been downgraded by most authors and treated as the "*Nysson epeoliformis* group." We feel that it is sufficiently large and distinctive to warrant generic status. The primary characteristics are the three submarginal cells, simple hindtibia, no male sternal brushes, double-edged abdominal segments, and hindwing media diverging from or very close to cu-a (fig. 148 E). The last character suggests that *Synnevrus* may be derived from some species of *Nysson*. We have not seen several Old World species, notably *militaris* and *decemmaculatus*. From the descriptions these fit the main characters of *Synnevrus* as we have outlined them. Literature is widely scattered.

Biology: Unknown.

Checklist of *Synnevrus*
(usually cited as *Synneurus* but original orthography and derivation indicates a "v" is correct)

aequalis (Patton), 1879 (*Nysson*); U.S.
aurinotus (Say), 1837 (*Nysson*); U.S. e. of 100th mer.
   freygessneri Handlirsch, 1887 (*Nysson*)
   angularis H. Smith, 1908 (*Nysson*)
   marlatti Rohwer, 1921 (*Nysson*), new synonymy by R. Bohart
costae (Handlirsch), 1901 (*Nysson*); Algeria
decemmaculatus (Spinola), 1808 (*Nysson*); Europe, Middle East
   decemguttatus Dahlbom, 1845 (*Nysson*), lapsus
   variolatus A. Costa, 1869 (*Nysson*)
   mopsus Handlirsch, 1898 (*Nysson*)
epeoliformis (F. Smith), 1856 (*Nysson*); Eurasia
   procerus A. Costa, 1859
   ssp. *ditior* (Morice), 1911 (*Nysson*); Algeria
excavatus (Turner), 1914 (*Nysson*); India
grandissimus (Radoszkowski), 1877 (*Nysson*); sw. USSR
   ssp. *mongolensis* (Tsuneki), 1970 (*Nysson*); Mongolia
guichardi (Beaumont), 1967 (*Nysson*); Turkey
handlirschi (Schmiedeknecht), 1896 (*Nysson*); Algeria
harveyi (Beaumont), 1967 (*Nysson*); Turkey
intermedius (Viereck), 1907 (*Nysson*); U.S.: Texas to California
   coyotero Pate, 1940 (*Nysson*), new synonymy by R. Bohart
maderae R. Bohart, 1968; U.S.: Arizona
militaris (Gerstaecker), 1867 (*Nysson*); Europe, Middle East
notabilis (Handlirsch), 1895 (*Nysson*); sw. USSR: Armenia
plagiatus (Cresson), 1882 (*Nysson*); U.S.
quadricolor (Arnold), 1951 (*Nysson*); ne. Africa
rufoniger (Turner), 1912 (*Nysson*); Rhodesia
rufopictus (F. Smith), 1856 (*Nysson*); Asia (?)
trichopygus (Beaumont), 1967 (*Nysson*); Turkey
uniformis (Pérez), 1895 (*Nysson*); Algeria

Genus Neonysson R. Bohart

Generic diagnosis: Clypeus concave apically and thin edged, ridged above to make a bevel; labrum dark, rounded, showing only a little beneath clypeus, basal flagellar articles slender in female, stout in male except for I; last male flagellomere incurved beneath, longitudinally seamed, larger than preceding article; front a little swollen above antennal bases and with a short raised median line; inner eye margins converging moderately below, evenly emarginate above; forewing with three submarginal cells, II petiolate and receiving both recurrent veins; marginal cell pointed distally and ending at wing margin; submarginal cells II and III unusually small, former sometimes smaller than stigma; hindwing media diverging far beyond cu-a; jugal lobe larger in outline than tegula; arolia present; female foretarsus without a recognizable rake; male midtibia two spurred; posterior surface of hindtibia with many small teeth rather generally distributed; outer apex of hindtibia rounded and spinose; outer apex of hindfemur not spoon shaped; metanotum simple; propodeal spines small but prominent, flattened dorsoventrally; a stout tooth well above hindcoxa, projecting outward and backward; gaster moderately stout, segments single-edged, simple, not fringed with flattened setae; female sternum VI sharply convex and with a weak median carina; female with a narrow pygidial plate, rounded or rather blunt apically; male sterna II to V with a double row of narrow hair tufts which divide sterna in three nearly equal parts; male tergum VII bidentate and with a strong median lobe.

Geographic range: Known only from Chile.

Systematics: As discussed by R. Bohart (1968c), in most respects *Neonysson* is in morphological agreement with *Brachystegus* which occurs in Africa and Eurasia. The two important characters in common are the multidentate posterior surface of the hindtibia (as in fig. 147 D) in combination with the three submarginal cells. The only other genera approaching this condition are *Acanthostethus*, which has only two submarginal cells, and *Zanysson*, which has the hindtibial teeth large and arranged in a single regular row (fig. 147 C). *Neonysson* differs from all of these by its simply edged abdominal segments and by the projecting tooth below each propodeal spine. Other differences of possible generic significance are the double row of hair tufts on the sterna of male *porteri*, and the rather "duck-billed" propodeal spines. It probably can be regarded as close to the ancestral type for *Brachystegus* and *Acanthostethus*.

Biology: Unknown.

Checklist of *Neonysson*

herbsti R. Bohart, 1968; Chile
porteri (Ruiz), 1936 (*Nysson*); Chile

Genus Epinysson Pate

Generic diagnosis: Face as in fig. 145 D; clypeus ending in a small, rounded, upturned point or lobe, thin edged; labrum inconspicuous; basal flagellar articles stout (fig. 145 D), last male flagellomere larger than preceding article, concave beneath, seamed, obliquely truncate distally; mandible without a basoventral, laminate projection; frons sulcate above antennal bases but not ridged or crested; inner eye margins converging strongly below, emarginate above (fig. 145 D); forewing with three submarginal cells, II petiolate and receiving both recurrent veins (as in fig. 148 A); marginal cell rather pointed and terminating at wing margin; stigma much smaller than submarginal cell II; hindwing media diverging far beyond cu-a (as in fig. 148 F); jugal lobe larger in outline than tegula; arolia present; female foretarsus without a rake; male midtibia two spurred; posterior surface of hindtibia with hairs and small bristles only; outer apex of hindtibia a rounded projection; outer apex of hindfemur not spoon shaped; metanotum simple to somewhat transversely ridged; propodeal spines short, stout, outwardly directed; no projecting lower tooth above hindcoxa; gaster stout, segments with a narrow double edge which swells abruptly laterad; no prominent flattened setae on segmental apices; female sternum VI convex; female pygidial plate rounded apically; male sterna II to V without hairbrushes; male tergum VII bidentate, the points usually long and slender.

Geographic range: New World. Twenty-two species are known.

Systematics: Epinysson has been a much misunderstood category since its establishment by Pate as a subgenus in 1935 to include many of the American nyssonins previously placed in *Brachystegus*. The confusion was implemented by Pate himself who placed prime significance on the position of the hindwing media. His doubts on the subject were expressed in the following statement (1938b): "A more natural division of *Nysson* might be affected by basing the separation into subgenera upon the structure of the apical margins of the abdominal tergites and sternites, i.e., whether they are simple as in *N. (Nysson) lateralis* Packard and *Nysson (Hyponysson) bicolor* Cresson, or double as in *Nysson (Epinysson) basilaris* Cresson." The unreliability of the position of the hindwing media has been pointed out by Beaumont (1954a) as quoted under his discussion of *Nysson* (p. 318). Certainly, the point of divergence of the hindwing media has great phylogenetic significance, but like most other morphological characters it is subject to variation within some genera (such as *Nysson*) or even within a single species. In *Epinysson* the hindwing media always diverges well beyond cu-a, but the distinctive double-edged, laterally swollen terga and sterna are much more diagnostic. In addition the simple frons, three submarginal cells, and simple hindtibia distinguish it from other genera in the tribe.

The genus, with 22 species, is a relatively large one but morphologically a close-knit unit. Along with *Neonysson* it can be considered close to the progenitor of *Brachystegus*, *Acanthostethus*, and *Zanysson*.

Principal references are Cresson (1882), Handlirsch (1887), Pate (1938b), and R. M. Bohart (1968c).

Biology: Species of this genus appear to be parasites of *Hoplisoides* (Gorytini). Records show *Epinysson guatemalensis hoplisivora* on *Hoplisoides costalis* (Reinhard, 1925a,b, 1929b), *E. bellus* on *H. tricolor* (Evans, et al., 1954), *E. moestus* on *H. hamatus* (Powell and Chemsak, 1959b), and *E. basilaris tuberculatus* on *H. nebulosus* (Evans, 1966a). It is interesting to note that the distribution of *Hoplisoides* and *Epinysson* in the New World coincide closely. The former is found also in Eurasia and Africa where it is preyed upon, at least in Europe, by species of *Nysson*.

The references of Reinhard, cited above, give us the clearest picture of *Epinysson* habits. *E. hoplisivora* were seen flying about or resting near the *H. costalis* burrows. The female nyssonin waited until the host wasp departed, then broke through the temporary nest closure with her front legs and remained inside for a few seconds. The egg was laid during this period, placed dorsally beneath the wings of one of the first provisioned treehoppers in the cell. The adult *hoplisivora* closed the nest upon leaving. The well concealed parasite egg was 1.34 mm long or about half as long as that of the host. It hatched early and, the young larva immediately sought out and destroyed the egg of its host, after which it consumed the provisions.

Checklist of *Epinysson*

albomarginatus (Cresson), 1882 (*Nysson*); U.S.: Nevada
arentis R. Bohart, 1968; U.S.: California
aztecus (Cresson), 1882 (*Nysson*); Mexico
basilaris (Cresson), 1882 (*Nysson*); se. U.S.
    ssp. *tuberculatus* (Handlirsch), 1887 (*Nysson*); U.S. e. of 100th mer., new status by R. Bohart
    tramosericus Viereck, 1904 (*Nysson*), new synonymy by R. Bohart
    dakotensis Rohwer, 1921 (*Nysson*), new synonymy by R. Bohart
bellus (Cresson), 1882 (*Nysson*); N. America e. to Oklahoma
    clarconis Viereck, 1906 (*Nysson*), new synonymy by R. Bohart
bifasciatus (Brèthes), 1913 (*Nysson*); Argentina
borinquinensis (Pate), 1937 (*Nysson*); Puerto Rico
    basirufus Rohwer, 1915 (*Nysson*), nec Brèthes, 1913
casali (Fritz), 1970 (*Nysson*); Argentina
desertus R. Bohart, 1968; sw. U.S.
guatemalensis (Rohwer), 1914 (*Nysson*); Centr. America
    ssp. *hoplisivora* (Rohwer), 1923 (*Nysson*); U.S.: Washington D.C. to Florida, new status by R. Bohart
inconspicuus (Ducke), 1910 (*Nysson*); Brazil
mellipes (Cresson), 1882 (*Nysson*); U.S.
    submellipes Viereck, 1904 (*Nysson*), new synonymy by R. Bohart
metathoracicus (H. Smith), 1908 (*Brachystegus*); centr. to sw. U.S.
moestus (Cresson), 1882 (*Nysson*); U.S.: Pacific states
    barberi Rohwer, 1921 (*Nysson*), new synonymy by R. Bohart
opulentus (Gerstaecker), 1867 (*Nysson*); U.S.
    seminole Bradley, 1920 (*Nysson*), new synonymy by R. Bohart
    foxii Rohwer, 1921 (*Nysson*), new synonymy by R. Bohart
    maiae Pate, 1938 (*Nysson*), new synonymy by R. Bohart
orientalis (Alayo), 1969 (*Nysson*); Cuba
pacificus (Rohwer), 1917 (*Nysson*); w. U.S.
partamona (Pate), 1938 (*Nysson*); Guyana
sigua (Pate), 1938 (*Nysson*); Panama
tomentosus (Handlirsch), 1887 (*Nysson*); Brazil
torridus R. Bohart, 1968; w. U.S.
zapotecus (Cresson), 1882 (*Nysson*); Mexico

Genus Hovanysson Arnold

Generic diagnosis: (Based primarily on *H. camelus*) Clypeal apex beveled, truncate; basal flagellar articles moderately slender, flagellomeres IV and following longer than broad; last flagellomere nearly three times as long as broad; frons sometimes with a weakly Y-shaped crest above antennal bases; inner eye margins broadly separated, converging a little below, slightly emarginate above; forewing with three submarginal cells, II small, petiolate and receiving first recurrent, III receiving second recurrent; hindwing media diverging far beyond cu-a; male midtibia with two spurs; posterior surface of hindtibia with hairs and small bristles only; outer apex of hindfemur with a truncate spoon-like lobe; metanotum relatively simple; propodeal spines stout, short; gaster moderately stout, tergum I with a discontinuous transverse carina across summit, segments double edged but not obviously fringed nor toothed; female sternum VI convex; female pygidial plate broadly rounded posteriorly; male sterna II to V with white hair tufts, at least in *camelus;* male tergum VII with apex bidentate or bilobed.

Geographic range: Known only from two Madagascan species.

Systematics: A paratype female and another pair of *camelus* were examined briefly by R. Bohart in 1960 at the British Museum. In the diagnosis above, partial notes have been amplified from Arnold's (1945) description. The category was considered by Arnold as a subgenus of *Brachystegus* to which it seems to be closely related. The principal points of difference in *Hovanysson* are the simple hindtibia, moderately slender antennal flagellum, and reception of the second recurrent vein of the forewing by submarginal cell III. We have not seen *albibarbis,* but presume that it is in this genus.

Biology: Unknown.

Checklist of *Hovanysson*

albibarbis (Arnold), 1945 (*Nysson*); Madagascar
camelus (Arnold), 1945 (*Brachystegus*); Madagascar

Genus Brachystegus A. Costa

Generic diagnosis: Clypeus concave apically and thin edged, sometimes ridged above to give the appearance of a bevel; labrum rounded and usually not protruding far beyond clypeus; basal flagellar articles usually stout; last male flagellomere incurved beneath, longitudinally seamed, larger than preceding article; a median longitudinal ridge present just above antennal bases, sharp and high, or low and rather dull; inner eye margins converging moderately below, evenly emarginate above; forewing with three submarginal cells, II petiolate and receiving both recurrent veins; marginal cell rather rounded distally but ending at wing margin; stigma much smaller than submarginal cell II; hindwing media diverging far beyond cu-a; jugal lobe larger in outline than tegula; arolia present; female foretarsus without a recognizable rake; male midtibia two spurred; posterior surface of hindtibia with many small teeth usually associated with a longitudinal carina (fig. 147 D); outer apex of hindtibia projecting bluntly, spinose; outer apex of hindfemur narrowly spoon shaped; metanotum rough to transversely ridged; propodeal spines strong, no projecting lower tooth well above hindcoxa; gaster stout, segments double edged, thicker laterally and somewhat lobate, especially on sterna II to V or some of them; no prominent flattened setae on segmental apices; female sternum VI convex; female pygidial plate rounded or truncate apically; male sterna II to V with white median hair brushes; male tergum VII bidentate to quinquedentate.

Geographic range: Essentially Palearctic and Ethiopian. Eighteen species known.

Systematics: Since we have seen authoritative material of only 10 species of the 18 listed, we have had to rely on the comments of others, especially Beaumont (1954a) for the characters of importance. Most previous authors have considered the genus in a subordinate position as the "*Nysson scalaris* group" (Handlirsch, 1895) or as a subgenus (Beaumont, 1954a). Considering its size and relative stability we prefer to treat it as a genus. The rather close relationship with the Chilean *Neonysson* is discussed under that genus. Also, the possibility that *Nysson fulvipes* and *N. rufus* are annectant species is discussed under *Nysson*. Among the Old World genera, *Brachystegus* seems closest to *Acanthostethus*. Both have a ridged lower frons, multidentate hindtibia (fig. 147 D), tufted male sterna, double edged abdominal segments, and the hindwing media arising far beyond cu-a. The important difference is the three submarginal cells in *Brachystegus* instead of two in *Acanthostethus*.

Biology: Deleurance (1943) presented evidence that *Brachystegus scalaris* is a parasite of *Tachytes europaeus* in southern France. The two species were frequently found in nesting areas of the latter, and the *Tachytes* was observed to attack the nyssonin, carry it off, and deposit it at a distance from the nest. The *Brachystegus* were seen to test the soil with their antennae in and about the open entrances of the burrows and to disappear within. In one such case an excavated nest revealed two paralyzed grasshoppers one of which had an egg attached to the side near a hindleg. This was presumably a *Brachystegus* egg since the *Tachytes* egg would have been farther forward and on the venter. A rough closing of a burrow by a *Brachystegus* on leaving was observed by Deleurance. Since the nyssonin has no comb its "raking" efforts would naturally be inefficient.

Checklist of *Brachystegus*

arnoldi Benoit, 1951; Zaire
auranticus Arnold, 1951; Mali
braueri (Handlirsch), 1887 (*Nysson*); Algeria
capensis (Handlirsch), 1887 (*Nysson*); S. Africa
cataractae Arnold, 1940; s. Africa
decipiens (Arnold), 1929 (*Nysson*); s. Africa
decoratus (Turner), 1914 (*Nysson*); India
   dubitatus Turner, 1914 (*Nysson*)
fraterculus (Gussakovskij), 1933 (*Nysson*); sw. USSR
gregoryi (Turner), 1912 (*Nysson*); e. Africa
incertus (Radoszkowski), 1877 (*Nysson*); Turkey, sw. USSR: Kazakh S.S.R.
ludovici (Turner), 1920 (*Nysson*); s. Africa
magrettii (Mantero), 1917 (*Nysson*); Ethiopia
nasutus (Cameron), 1910 (*Nysson*); Kenya
pieli (Yasumatsu), 1943 (*Nysson*); China
rhodesiae (Arnold), 1929 (*Nysson*); s. Africa
scalaris (Illiger), 1807 (*Nysson*); Eurasia, new name for *interruptus* (Art. 59c)
   interruptus Latreille, 1803 (*Nysson*), nec *Nysson interruptus* (Fabricius), 1798 (*Mellinus*), now in *Nysson*
   rufipes Olivier, 1811 (*Nysson*)
   auratus Schummel, 1834 (*Nysson*)
   dufourii Lepeletier, 1845 (*Nysson*)
   dufouri Dahlbom, 1845 (*Nysson*), nec Lepeletier, 1845
   dufourianum Blanchard, 1849 (*Nysson*)
   decemnotatus F. Morawitz, 1890 (*Nysson*), nec A. Costa, 1869
   transcaspicus Radoszkowski, 1893 (*Nysson*)
senegalensis Arnold, 1951; Mali
violaceipennis (Cameron), 1904 (*Nysson*); Himalayas: Sikkim

Genus Acanthostethus F. Smith

Generic diagnosis: Clypeal apex concave, thin or beveled; labrum broad, nearly rectangular and reddish; basal flagellar articles rather slender; last male flagellomere obconic or incurved beneath, seamed longitudinally; frontal process above antennal bases bluntly Y-shaped, sometimes identified as two parallel carinae which diverge above (*punctatissimus*); inner eye margins moderately converging below and emarginate above; forewing with two submarginal cells (basalmost vein of usual second cell missing); apparent cell II petiolate in front or sometimes sessile to truncate; recurrent veins both received by I, or second recurrent interstitial; marginal cell bluntly rounded to wing margin distally; stigma short, much smaller than distal submarginal cell; hindwing media diverging far beyond cu-a; jugal lobe a little smaller or larger in outline than tegula; arolia present; female foretarsus without a rake; male midtibia with two spurs; posterior surface of hindtibia with at least traces of a serrate longitudinal ridge, outer apex of hindtibia usually rather sharply and obliquely truncate; outer apex of hindfemur without a spoon shaped lobe; metanotum a little irregular but essentially simple; propodeal spines moderate but distinct; gaster stout; segments double edged and broader laterally, at least on sterna, not fringed with flattened setae; many species with angles or teeth laterally on sterna; female sternum VI broadly convex to compressed and medially carinate but not saber shaped; pygidial plate rounded apically or serrately edged (*nudiventris*); male sterna II to V with prominent hair tufts (*mysticus*) or with tufts barely indicated (*punctatissimus*); male tergum VII usually ending in about four teeth in addition to a median lobe, but sometimes bidentate.

Geographic range: Australia including Tasmania. Fifteen species are known.

474 SPHECID WASPS

Systematics: Acanthostethus is the dominant member of the tribe in Australia, and its morphological diversity may indicate the need of future subgeneric or generic division. It is probably closest to *Brachystegus* from which it differs primarily in the reduction of the submarginal cells. The principal reference, including a key to species, is that of Turner (1915a).

Biology: Rayment (1953) reported an association of *A. portlandensis,* and *A. hentyi* with *Sericophorus victoriensis* Rayment in Victoria, Australia. The wasps were taken in and about the burrows of the larrine wasp. Matthews and Evans (1971) similarly reported *portlandensis* as a presumptive parasite of *Sericophorus viridis.* Rayment reported an observation of copulation in *hentyi:* "When I visited the site of the nests of *Sericophorus victoriensis,* Raym. along the Cape Nelson-road, on a beautiful morning in March, 1953 . . . . three of the wasps were flying in and out among the stems in a cluster of braken ferns and other plants, and were only a few inches above the sandy ground of the road-side . . . . Suddenly, one of them, the female 'landed' on a small stick and was immediately followed by a male, and mating took place while the female was still resting on the stick."

Checklist of *Acanthostethus*

brisbanensis (Turner), 1915 (*Nysson*); Australia: Queensland
confertus (Turner), 1915 (*Nysson*); Australia: Queensland
gilberti (Turner), 1915 (*Nysson*); Australia: Queensland
hentyi (Rayment), 1953 (*Nysson*); Australia: Queensland
minimus (Turner), 1915 (*Nysson*); Australia: Queensland
moerens (Turner), 1915 (*Nysson*); sw. Australia
mysticus (Gerstaecker), 1867 (*Nysson*); S. and W. Australia
   basalis F. Smith, 1869
nudiventris (Turner), 1915 (*Nysson*); sw. Australia
obliteratus (Turner), 1910 (*Nysson*); W. Australia
portlandensis (Rayment), 1953 (*Nysson*); Australia: Victoria
punctatissimus (Turner), 1908 (*Nysson*); Australia: Queensland
saussurei (Handlirsch), 1887 (*Nysson*); S. Australia
spiniger (Turner), 1908 (*Nysson*); Australia: Queensland
tasmanicus (Turner), 1915 (*Nysson*); Australia: Tasmania
triangularis (Turner), 1940 (*Nysson*); W. Australia

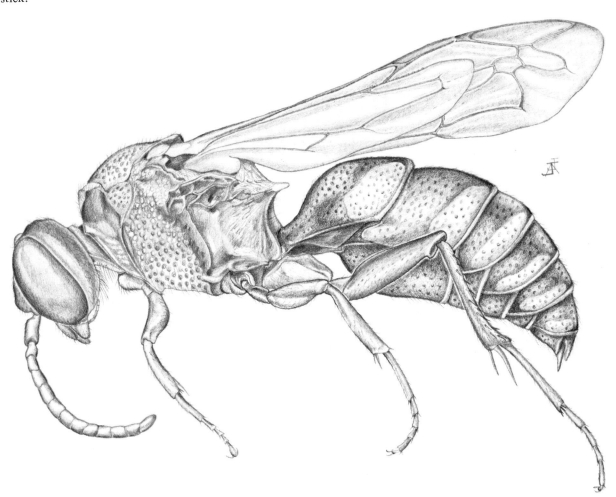

FIG. 151. *Zanysson texanus fuscipes* (Cresson), male.

Genus Zanysson Rohwer

Generic diagnosis: Face represented by fig. 145 A, side view of entire body by fig. 151; clypeal apex concave, beveled; labrum rather broad, reddish or dark, basal flagellar articles stout, last male flagellomere somewhat compressed beneath and with a longitudinal seam; an often serrate, sharply projecting frontal ridge just above antennal bases; inner eye margins moderately converging below, gently emarginate above middle (fig. 145 A), thoracic structure as in fig. 4; forewing with three submarginal cells, II rather diamond shaped, petiolate and receiving both recurrent veins (fig. 148 C); marginal cell ending somewhat bluntly at wing margin; stigma smaller than submarginal cell II; hindwing media diverging far beyond cu-a (fig. 148 F); jugal lobe larger in outline than tegula; arolia present; male midtibia with two spurs; posterior surface of hindtibia with a long single row of stout teeth (fig. 147 C); outer apex of hindtibia sharply pointed; hindfemur distally with an outer spoon shaped truncation (somewhat as in *Cerceris*, much as in *Alysson*); metanotum bilobed; propodeal spines strong (fig. 151), no secondary teeth between them and gastral insertion; gaster stout, most segments double edged posteriorly but without regular, prominent, close-set, flattened setae; female sternum VI convex; female pygidial plate rounded apically (fig. 153 D); no hair brushes on male sterna II to V; male tergum VII ending in three teeth or in several pairs (fig. 153 A).

Geographic range: New World from Canada to Argentina. Seventeen species known.

Systematics: The 17 listed species are remarkably similar in most respects. The primary generic characters are the long row of distinct teeth on the hindtibia (fig. 147 C), crested frons (fig. 145 A), three submarginal cells (fig. 148 C), absence of sternal brushes in the male (fig. 151), double edged abdominal segments which are not regularly fringed, and simple female pygidial plate (fig. 153 D).

Principal references are Cresson (1882), Handlirsch (1887, 1895), and Pate (1938b).

Biology: Cockerell (1903) first suggested that *Zanysson* parasitized *Tachytes*. He reported *Z. texanus* entering burrows of *Tachytes exornatus* at Las Cruces, New Mexico. Evans (1966a) has published observations by M. A. Cazier at Portal, Arizona, that tend to confirm the relationship. Cazier found a close association between *Zanysson plesius* (given as *tonto*) and *Tachytes distinctus*. Female *Zanysson* entered nests of the *Tachytes* several times and stayed up to 10 seconds. Considering the rather large size of most *Zanysson*, it is certainly possible that they are parasites of some of the larger species of *Tachytes*.

Checklist of *Zanysson*

argentinus (Brèthes), 1913 (*Nysson*); Argentina
armatus (Cresson), 1865 (*Nysson*); Cuba
changuina Pate, 1938; Panama
croesus (Handlirsch), 1895 (*Nysson*); Brazil
dives (Handlirsch), 1887 (*Nysson*); Mexico
    chichimeca Pate, 1938, new synonymy by R. Bohart
fasciatus (Olivier), 1811 (*Nysson*); S. America
foveiscutis (Gerstaecker), 1867 (*Nysson*); Brazil
gayi (Spinola), 1851 (*Nysson*); Chile, Argentina
luteipennis (Gerstaecker), 1867 (*Nysson*); Brazil
luxuriosus (Schrottky), 1910 (*Paranysson*); Argentina
macuxi Pate, 1938; Brazil: Amazonas
marginatus (Spinola), 1841 (*Nysson*); S. America
mexicanus (Cresson), 1882 (*Paranysson*); Mexico
    longispinis Cameron, 1905 (*Nysson*), new synonymy by R. Bohart
pilosus (F. Smith), 1873 (*Nysson*); Brazil
plesius (Rohwer), 1921 (*Nysson*); U.S.
    matinecoc Pate, 1938
    tonto Pate, 1940, new synonymy by R. Bohart
texanus (Cresson), 1872 (*Nysson*); U.S.
    ssp. *fuscipes* (Cresson), 1882 (*Paranysson*); w. U.S., new status by R. Bohart
    aureobalteatus Cameron, 1901 (*Nysson*), new synonymy by R. Bohart
varipilosellus (Cameron), 1912 (*Nysson*); Guyana

Genus Perisson Pate

Generic diagnosis: Clypeal apex concave, beveled, labrum pale orange, broad, subrectangular; basal flagellomeres moderately slender, first one or two somewhat longer than broad, last male flagellomere stout, a little incurved beneath, subtruncate at apex, not seamed (fig. 152 A); mandible with an acutely rounded basoventral, laminate projection; a strong, Y-shaped frontal keel above antennal bases; inner eye margins converging below, slightly emarginate above, width of frons at midocellus nearly twice that across antennal bases; forewing with three submarginal cells, II petiolate and receiving first recurrent vein, second recurrent received at basal third to fifth of submarginal cell III; marginal cell a little stubby at apex but ending on wing margin; stigma smaller than submarginal cell II; hindwing media diverging far beyond cu-a; jugal lobe larger in outline than tegula; arolia present although small; foretarsus with a very short and weak rake in both sexes; male midtibia two spurred; posterior surface of hindtibia with hairs and small bristles only; outer apex of hindtibia bluntly pointed, spinose; outer apex of hindfemur without a spoon shaped truncation; metanotum somewhat uneven but essentially simple; propodeal spines rather slender; gaster stout, segments double edged posteriorly, a little thicker laterally and with posteriorly projecting flat lobes on terga and sterna II to V, marginal silvery fringe on segments present but small and inconspicuous; female sternum VI broadly convex; female pygidial plate not well defined laterally, quinquedentate but more evenly so than in fig. 153 J; male sterna II to V with white median hair brushes; male tergum VII rather evenly quinquedentate (about as in fig. 153 G).

Geographic range: Argentina.

Systematics: The confusion among the identities of *Cresson*, *Antomartinezius*, and *Perisson* is explained in the discussion under *Antomartinezius*. To further complicate the problem, the male syntype of *Perisson basirufus* was misassociated by Brèthes and is actually *Antomartinezius*

476 SPHECID WASPS

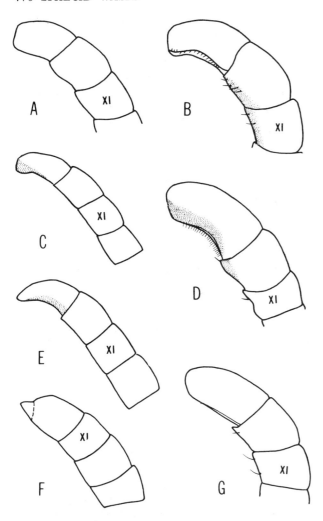

FIG. 152. Profile of male antennal apex in the tribe Nyssonini; A, *Perisson basirufus*; B, *Synnevrus plagiatus*; C, *Antomartinezius patei*; D, *Nysson tristis*; E, *Cresson parvispinosus*; F, *Foxia navajo*; G, *Nysson spinosus*.

patei. Pate (1938b) considered the female as the holotype but actually it was only a syntype. This female, which was lent to us from the National Museum of Argentina, is hereby designated the lectotype.

Part of the difficulty is that the two species, *A. patei* and *P. basirufus,* are superficially similar although they differ in markings, wing venation, absence or presence of arolia, and other details.

From *Cresson* the important differences in *Perisson* are the basoventral mandibular projection, the reception of the second recurrent vein by submarginal cell III at some distance from its base, the laterally lobed terga and sterna, and the quinquedentate rather than serrate last tergum in both sexes. From *Antomartinezius* the generic differences are presence of arolia, distal position of the second recurrent vein, and acutely rather than broadly rounded basoventral mandibular projection.

Biology: Unknown.

Checklist of *Perisson*

basirufus (Brèthes), 1913 (*Nysson*); Argentina

Genus Cresson Pate

Generic diagnosis: Face represented by fig. 145 C; clypeal apex concave, with a rounded bevel; labrum white, broad, subrectangular; basal flagellar articles rather slender (fig. 145 C); last male flagellomere slender, incurved beneath, tapering nearly to a point, not seamed (fig. 152 E); a small but definite, Y-shaped frontal keel present above antennal bases; inner eye margins converging strongly below, gently emarginate above (fig. 145 C); forewing with three submarginal cells, II petiolate and receiving both rather widely spaced recurrent veins; marginal cell rounded distally, faintly appendiculate, somewhat removed from wing margin; stigma smaller than submarginal cell II; hindwing media diverging far beyond cu-a; jugal lobe larger in outline than tegula; arolia present but small; female foretarsus without a recognizable rake; male midtibia with a single strong spur; posterior surface of hindtibia with hairs and small bristles only, outer apex of hindtibia pointed and bearing a small spine; outer apex of hindfemur without a spoon like lobe; metanotum transversely ridged; propodeal spines small but distinct; gaster stout, segments evenly double edged posteriorly but without prominent, close-set, flattened setae; female sternum VI convex; female pygidium broad, serrate laterally, bidentate apically (fig. 153 K), no definite plate; male sterna II to V with white median hair tufts; male tergum VII tridentate to quinquedentate apically (fig. 153 H).

Geographic range: This monotypic genus occurs in Chile.

Systematics: The phylogenetic position of *Cresson* is indicated in the dendrogram (fig. 146). The close relationship with *Antomartinezius* has been discussed under that genus. Briefly, the principal differentiating characters of *Antomartinezius* from *Foxia* and related genera are: no lateral lobes on terga or sterna, arolia present, male midtibia one spurred; female without a recognizable foretarsal comb, and female pygidium multiserrate.

Checklist of *Cresson*

parvispinosus (Reed), 1894 (*Nysson*); Chile

Genus Antomartinezius Fritz

Generic diagnosis: Clypeal apex concave, beveled, labrum white, broad, subrectangular; basal flagellomeres slender, longer than broad; last male flagellomere slender, incurved beneath, not seamed (fig. 152 C); mandible with a broadly rounded, basoventral, laminate projection; a small but definite, Y-shaped frontal keel present above antennal bases; inner eye margins converging strongly below, slightly emarginate above; forewing with three submarginal cells, II petiolate and receiving one or both recurrent veins, second recurrent sometimes interstitial or received basally by cell III, marginal cell rounded distally, faintly appendiculate, somewhat removed from wing margin; stigma smaller than submarginal cell II; hindwing

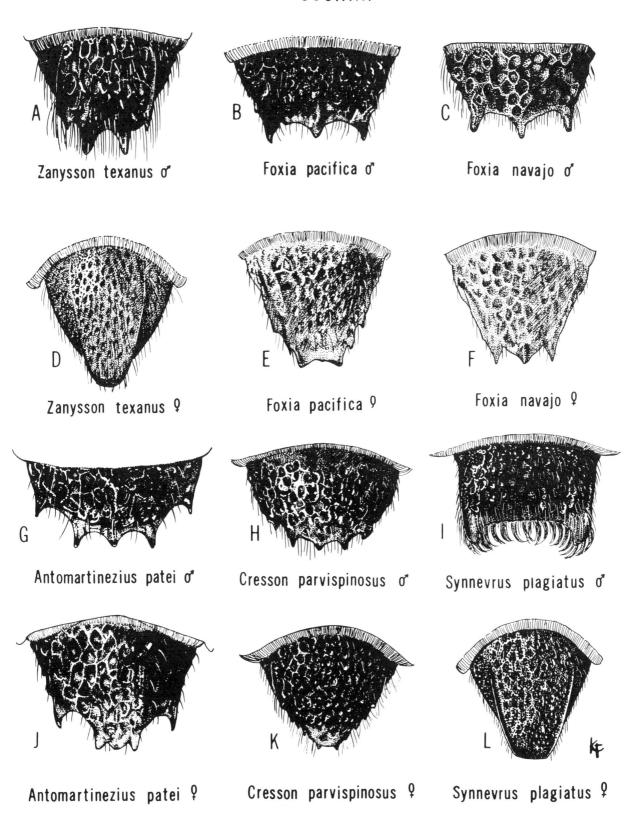

FIG. 153. Pygidium of various genera in the tribe Nyssonini.

media diverging far beyond cu-a; jugal lobe larger in outline than tegula; arolia absent; female foretarsus with a moderately developed rake, male foretarsus with weakly indicated comb; male midtibia with two spurs; posterior surface of hindtibia with hairs and small bristles only, outer apex of hindtibia pointed and bearing two small spines; outer apex of hindfemur without a spoon shaped truncation; metanotum somewhat uneven but essentially simple; propodeal spines small but distinct, gaster stout, segments double edged posteriorly, a little thicker laterally and with posteriorly projecting flat lobes on terga and sterna II to V, no prominent marginal fringe of flattened setae; female sternum VI broadly convex; female pygidium broad, quinquedentate (fig. 153 J), no definite plate; male sterna II to V with white median hair tufts; male tergum VII rather regularly quinquedentate (fig. 153 G).

Geographic range: Argentina. Two species are known.

Systematics: Antomartinezius patei was originally "described" by Pate (1938b), who misidentified two females in the Cornell University collection as *Nysson (Perisson) basirufus* Brèthes. After examination of the type material of *Perisson,* Fritz (1955) corrected the error and named Pate's species *Cresson (Antomartinezius) patei.* The male of *patei* has not been described previously, but we have seen several Argentine males and also one of *fritzi.*

The three genera, *Cresson, Perisson,* and *Antomartinezius* are remarkably alike in general appearance and even in many morphological details. For instance, all have a small, Y-shaped frontal prominence, similar wing venation, double-edged abdominal segments, white-tufted male sterna, and male sternum VII ordinarily five spined. However, they differ rather fundamentally. *Cresson* has no lateral lobes on terga or sterna but does have arolia, the male midtibia is one spurred, and the female pygidium is multidentate (serrate). *Perisson* has lateral lobes on terga and sterna II to V, arolia are distinct, the male midtibia is two spurred, and the female pygidium is quinquedentate. *Antomartinezius* has lateral lobes on terga and sterna II to V but has no trace of arolia, the male midtibia is two spurred, and the female pygidium is quinquedentate. Additionally, the fairly well developed foretarsal rake in females of *Antomartinezius* probably has generic value in this group.

A relationship with *Foxia* is apparent as shown in fig. 146. The different male antennae of *Foxia,* as well as the absence of a Y-shaped frontal crest, place this genus somewhat apart.

Biology: Unknown.

Checklist of *Antomartinezius*

patei (Fritz), 1955 (*Cresson*); Argentina
fritzi R. Bohart, 1968; Argentina

Genus Foxia Ashmead

Generic diagnosis: Side view of male exemplified by fig. 154; clypeal apex pointed or bidentate, thin, not beveled; labrum reddish or dark, protruding only slightly beyond clypeus; basal flagellar articles moderately stout, usually longer than broad; last male flagellomere very small (fig. 152 F) or in one case about half as large as preceding article (*pacifica*); frons without a definite keel but with a sharp point in one species (*pacifica*); inner eye margins strongly converging below, gently emarginate above; forewing with three submarginal cells, II small, petiolate and receiving first recurrent, III receiving second recurrent; marginal cell rather rounded distally but appressed to wing margin (fig. 148 B); stigma moderately small but sometimes larger than reduced submarginal cell II; hindwing media diverging far beyond cu-a (as in fig. 148 F); jugal lobe larger in outline than tegula; arolia present; female foretarsus with a barely perceptible rake; male midtibia two spurred; posterior surface of hindtibia with hairs and small bristles only, outer apex of hindtibia pointed; outer apex of hindfemur without a spoon shaped lobe; metanotum simple to transversely ridged; propodeal spines well developed; gaster stout, segments doubled edged and a little thicker laterally, most sterna toothed or sharply angled laterally, posterior margins of segments with a thick covering of fine silvery setae which may be short or may form a fringe as long as an ocellus diameter (fig. 154); female sternum VI convex; female pygidium serrate laterally and bidentate apically (*pacifica,* fig 153 E), tridentate (*navajo,* fig 153 F), or plate shaped (*deserticola*); male sterna II to V with white median hair tufts (fig. 154); male tergum VII tridentate (fig. 153 B,C).

Geographic range: The 10 described species occur in the southwestern and far western United States south to Argentina.

Systematics: The relationships of *Foxia* with *Cresson, Antomartinezius,* and *Perisson* have been discussed especially under *Antomartinezius. Foxia* seems to occupy a position a little apart from the other three as indicated by the more simple frons, sharp lateral angles of most sterna, far distal position of the second recurrent vein, and especially the greatly reduced last flagellomere in the male. The simple hindtibia and hindfemur, far distal hindwing media, and three submarginal cells distinguish it from other Nyssonini with which it might be confused.

Within the genus the greatest divergence seems to be in the form of the female pygidium. This is posteriorly rounded and plate shaped in *divergens* and *deserticola* but dentate to serrate in *navajo, pacifica,* and *secunda.* The female of *cuna* is unknown. In a recent paper Fritz (1972) described four South American species and gave a key to the genus.

Biology: The hosts are unknown. In southern California and New Mexico *Foxia* are often collected at flowers of desert willow, *Chilopsis linearis* Covanilles (Sweet).

Checklist of *Foxia*

cuna Pate, 1938; Panama
deserticola Fritz, 1959; Argentina, Bolivia
divergens (Ducke), 1903 (*Nysson*); Brazil
garciai Fritz, 1972; Argentina

FIG. 154. *Foxia secunda* Rohwer, male.

martinezi Fritz, 1972; Argentina
navajo Pate, 1938; sw. U.S.
pacifica Ashmead, 1898; U.S.: California
pirita Fritz, 1972; Argentina
secunda (Rohwer), 1921 (*Nysson*); sw. U.S.
tercera Fritz, 1972; Argentina

Genus Losada Pate

Generic diagnosis: Female only known. Clypeal apex broadly concave, beveled, labrum reddish, not much exposed; antenna rather simple, basal flagellomeres a little longer than broad; frontal ridge prominent, Y-shaped; inner eye margins converging strongly below, gently emarginate above; three submarginal cells, II petiolate and receiving first recurrent vein, second recurrent received at basal third to fourth of submarginal cell III; marginal cell rather broad, ending at wing margin, stigma smaller than submarginal cell II; hindwing media ending well beyond cu-a (as in fig. 148 F); jugal lobe larger in outline than tegula; scutal punctation large and shallow, obscuring median line; arolia present; female foretarsus without a rake; posterior surface of hindtibia unusually hairy but not dentate; apex of hindtibia projecting acutely; hindfemur without a distal spoon shaped truncation; metanotum with a median carina so that from an oblique anterior view it is weakly tridentate; propodeal spines small, no secondary tooth below; gaster stout, sternum II moderately convex, segments single edged and fringed with regular, close-set, flattened setae; female sternum VI convex, tapering distally almost to a point; female pygidial plate posterior only, narrow, occupying only a small part of tergum VI.

Geographic range: We have seen only two female specimens of this genus. They are *paria* (holotype at Mus. Philadelphia, and topotype at Mus. Washington) from San

Esteban, Venezuela. Other recorded specimens are a short series of syntypes of *mutilloides* collected at Belem and Itaituba in the state of Pará, Brazil by A. Ducke.

Systematics: The above generic description has been based on *paria* but Ducke's description of *mutilloides* agrees in all important particulars. Specifically, the latter differs in markings, the basal four tarsomeres and the propodeal spines being white, whereas in *paria* they are pale reddish. Pate (1938b) mistakenly but doubtfully assigned *mutilloides* to *Foxia*, then did not connect it with his new genus, *Losada*.

Specimens of *Losada* are unusually hairy on the wings, legs, and body proper. This led Ducke to compare them with a male mutillid. Other special features are the nearly conical sixth gastral segment, the beautifully fringed gastral segments, the simple hindtibia, the Y-shaped frontal process, and the termination of the recurrent veins at submarginal cells II-III. This last situation is otherwise found in the tribe only in *Foxia, Perisson,* and *Hovanysson* as a regular occurrence. In some other genera, such as *Antomartinezius*, the second recurrent may be interstitial or barely onto the third submarginal cell.

Relationships with *Metanysson* and *Idionysson* are discussed under the latter genus.

Biology: Unknown, all recorded specimens taken sweeping vegetation.

Checklist of *Losada*

mutilloides (Ducke), 1903 (*Nysson*); Brazil: Amazon Basin
paria Pate, 1940; Venezuela

Genus Idionysson Pate

Generic diagnosis: Face represented by fig. 145 B; clypeal apex concave, beveled; labrum dark, basal flagellar articles stout; last male flagellomere obconic, incurved beneath, longitudinally seamed; frontal ridge strongly projecting, Y-shaped; inner eye margin strongly and a little angularly emarginate above middle (fig. 145 B); forewing with three submarginal cells, II petiolate and receiving both recurrent veins; marginal cell blunt or rounded distally; stigma smaller than submarginal cell II; hindwing media diverging far beyond cu-a (as in fig. 148 F); jugal lobe larger in outline than tegula; arolia present; female foretarsus without a rake; male midtibia with two spurs; posterior surface of hindtibia with a row of four teeth (as in fig. 147 B), basal two sometimes weak; outer apex of hindtibia sharply pointed; hindfemur without a distal spoon shaped truncation; median scutal line indistinct; metanotum bilobed or bidentate; propodeal spines stout, a small secondary tooth partway between each spine and insertion of gaster; gaster stout, segments single edged but fringed with close-set, flattened setae, female sternum VI nearly flat or weakly convex, truncate or a little emarginate apically; female pygidial plate rather convex in profile, somewhat "dented" and broadly U-shaped posteriorly, weakly margined; male sterna II to V without special hair tufts; male tergum VII ending in three sharp teeth.

Geographic range: South America including Brazil, Bolivia, Paraguay, Uruguay, and Argentina. Three species are known.

Systematics: The single edged abdominal segments prominently fringed with flattened setae ally this genus with *Metanysson* and *Losada*. *Idionysson* shares with *Metanysson* the distinctive four-toothed hindtibia but differs from it in the complete and petiolate second submarginal cell, the rather broad female sternum VI, and the absence of male sternal tufts. With *Losada* the Y-shaped frontal crest is possessed in common, as well as the three submarginal cells. However, *Idionysson* has the hindtibia four toothed. From both of the other two genera, *Idionysson* differs by the rather angular emargination of the upper inner orbit (fig. 145 B) and by the secondary tooth below the propodeal spine similar to that in *Neonysson* but smaller.

Specific characters are the degree of bulge of sternum II, the shape of female sternum VI, the size of the hindtibial teeth, the development of a genal carina, the interocellar tuberculation, and details of the propodeum.

Biology: Unknown.

Checklist of *Idionysson*

borero Pate, 1940; Paraguay
chrysozonus (Gerstaecker), 1867 (*Nysson*); S. America
cordialis Fritz, 1970; Bolivia; Brazil: São Paulo

Genus Metanysson Ashmead

Generic diagnosis: Clypeal apex concave, beveled; labrum reddish or dark; basal flagellar articles stout; last male flagellomere obconic, simple, not seamed; frontal keel sometimes present just above antennal bases; inner eye margin shallowly and faintly emarginate above middle; forewing with two submarginal cells (more basal vein of usual second cell missing), recurrent veins received on I and II or both on I (fig. 148 H); marginal cell rounded distally and a little pulled away from margin; stigma smaller than submarginal cell II; hindwing media diverging far beyond cu-a; jugal lobe larger in outline than tegula; arolia present; female foretarsus without a rake; male midtibia with two spurs; posterior surface of hindtibia with a row of four stout teeth (fig. 147 B), outer apex of hindtibia sharply pointed; hindfemur with a small, outer, distal, spoon shaped lobe; median line of scutum impressed for entire length; metapleuron very broad below and ill defined posteriorly; metanotum simple or bidentate; propodeal spines strong, no secondary tooth below; gaster stout, segments single edged but fringed with regular, close-set, flattened setae; female sternum VI reduced to a sharp line; female pygidium plate shaped, the lateral margins usually serrulate; male sterna II to V with white median hair tufts; male tergum VI dentate dorsolaterally; male tergum VII bidentate to multidentate.

Geographic range: The 12 known species are found from southwestern United States to Argentina.

Systematics: Metanysson seems to be the most specialized genus of the Nyssonini. In support of this idea are the two submarginal cells, distally rounded marginal cell (fig. 148 H), lobed hindfemur, four toothed hindtibia (fig. 147 B), tufted male sterna; and especially the linear sternum VI of the female. This unique feature seems to have resulted from a sidewise envelopment of the sterna by the strongly developed tergum VI. In some specimens the sternum is pulled down slightly and looks like a saber set on edge in the sheath formed by tergum VI.

A comparison with its relatives, *Losada* and *Idionysson*, is given in the discussion of the latter. A point not previously stressed is that a median line extends the entire length of the scutum in species of *Metanysson* that we have seen. This line is faint in *Idionysson* and imperceptible in *Losada*.

In most species of *Metanysson* the female pygidial plate has serrate edges and a posterior point. However, in *alfkeni* the pygidium has smoothly carinate edges. The wing venation is somewhat variable. Pate (1938b) established the subgenus *Huachuca* for *arivaipa* because the second recurrent vein was received well over on apparent submarginal cell II. Fritz (1957) has shown that this characteristic is variable in some species of *Metanysson* and has synonynized the subgenus. Although position of the second recurrent vein in *arivaipa* seems quite constant, all other characteristics are as expected, and we agree with the conclusions of Fritz.

A key to *Metanysson* was given by Pate (1938b). Additional information was given in a partial key (only one sex of each form) to four Argentinean species by Fritz (1959) and a more complete key by Fritz some years later (1970a).

Based on *arivaipa*, the genus rates about "24" in the specialization table topped in the subfamily only by the bembicins, *Bembix* (26), *Rubrica* (26), and *Zyzzyx* (25).

Biology: Evans (1966a) reported findings of M. A. Cazier at Portal, Arizona. A female *M. arivaipa* was seen to enter a burrow of *Cerceris graphica* and remain for seven minutes while the *Cerceris* was away. Also, on two occasions females of *Metanysson coahuila* were seen in and about the burrows of *Cerceris conifrons*. On this evidence a biological association seems very likely although not proven.

Checklist of *Metanysson*

alfkeni (Ducke), 1904 (*Nysson*); Brazil, Argentina
    foersteri Fritz, 1958
arivaipa Pate, 1938; sw. U.S.
carcavalloi Fritz, 1959; Argentina
catamarcensis (Schrottky), 1910 (*Paranysson*); Argentina
coahuila Pate, 1938; sw. U.S.
diezguitas Fritz, 1958; Argentina
fraternus Fritz, 1970; Argentina
layano Pate, 1938; Brazil
lipan Pate, 1938; U.S.: Texas
solani (Cockerell), 1895 (*Nysson*); sw. U.S.
tropicalis Fritz, 1970; Boliva
yavapai Pate, 1938; sw. U.S.

Tribe Gorytini

Wasps of this tribe are best known to the public through the single genus *Sphecius* which contains the large and showy "cicada killers." The remainder of the tribe are mostly medium to small forms which are rather unobtrusive. Many of the gorytins have a rapid and seemingly erratic flight, with frequent deceleration while circling bushes. Attraction to flowers seems less than in many other wasps.

From the standpoint of the systematist the tribe is most interesting because of the great variety in form despite the rather homogeneous nesting habits. There are over 400 known species arranged in 31 genera, which makes the Gorytini the most diversified tribe in the subfamily, but only two-fifths as large as the Bembicini in number of species. There has been a slightly greater evolutionary development in the New World than in the Old. Yet, the tribe is well represented on all continents and many islands.

Diagnostic characters:
1. (a) Inner eye margins converging below or sometimes as close above as below, (b) ocelli well developed (slightly reduced in *Kohlia*).
2. (a) Labrum usually not prominent but in *Sphecius* and *Kohlia* with exposed part respectively half to two-thirds as long as broad, (b) palpal formula 6-4.
3. Scutum (a) with admedian lines usually well separated, (b) notauli sometimes moderately developed, (c) with an oblique scutal carina (except *Exeirus, Argogorytes, Ochleroptera* and related genera, all of which are positive for characters 4a, 6a, and 9b), (d) scutellum without a posterior overlapping lamelliform edge (except slightly in *Sphecius* and *Kohlia*).
4. (a) Omaulus usually present (not in evolved forms like *Ammatomus, Pterygorytes,* and *Handlirschia*), (b) acetabular carina and mesopleural sulci variable among genera.
5. (a) Metapleuron consisting of upper area only, and separated by a sulcus from propodeum; (b) no spines or teeth on propodeum dorsolaterally.
6. (a) Midcoxae approximate, precoxal sulcus complete or evanescent; (b) midtibia with two apical spurs (except some males of *Austrogorytes* and *Oryttus,* (c) most females with a foretarsal rake; (d) plantulae present or absent.
7. Forewing (a) with media diverging before or after cu-a, (b) prestigmal length of submarginal cell I usually about equal to wing length beyond marginal cell, (c) stigma medium to small, (d) three submarginal cells of which II is not petiolate (except *Exeirus,* fig 157 E) and receives at least one recurrent vein.
8. (a) Hindwing media diverging before, at or after cu-a; (b) jugal lobe small or rarely absent (*Pterygorytes*).
9. (a) Gaster with segment I sometimes constricted or pedunculate, (b) sternum I with a median ridge

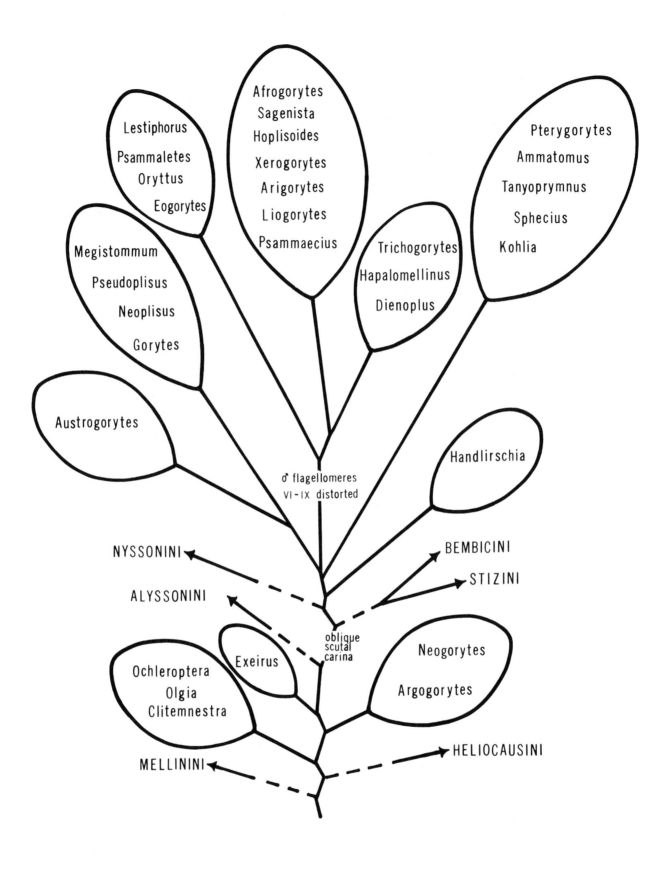

FIG. 155. Dendrogram illustrating suggested relationships of genera in the tribe Gorytini and points of departure for other tribes in the Nyssoninae.

toward base, (c) male with six or seven visible terga and with sternum VIII often modified, (d) male genitalia differentiated into cuspis and digitus (except *Ammatomus*), (e) female pygidial plate usually distinct and wedge shaped.

Systematics: The position of the tribe is best illustrated by cactoid fig. 135 in which it is shown to contain the presumed points of origin for most of the other tribes. The supposed relationships are given in different perspective and greater detail in dendrogrammatic fig. 155. Here, the six lowermost branches are characterized by the absence of the oblique scutal carina (fig. 156), which is obviously of great evolutionary significance. Beyond this important development there seem to be six principal branches, two of which involve the Nyssonini and the Bembicini-Stizini.

Directing attention to the least specialized genera, *Clitemnestra* and its relatives, *Exeirus* and *Argogorytes*, these are small and probably relict elements that are apparently not closely related despite the sharing of some primitive features. It may be significant that all three occur in Australia, and this circumstance supports that continent's claim to the birthplace of the tribe. *Clitemnestra* is generally conceded to be the most generalized member of the Gorytini and perhaps of the entire Nyssoninae. It is practically confined to Australia and Chile, in each of which there are a number of distinctive species. *Clitemnestra* is one of the many close faunal ties between the two countries.

Among the more advanced Nyssoninae, all of which have an oblique scutal carina (fig. 156), are the three large tribes, Nyssonini, Stizini, and Bembicini, as well as the bulk of the Gorytini. As outlined on fig. 170 the genus *Handlirschia* is a small relict offshoot, known only from the male holotype from South Africa. It combines gorytin and stizin features, but its wing venation definitely places it in the Gorytini.

The remaining genera can be visualized on three main branches of the dendrogram: the *Sphecius* group with strong suggestions of stizin relationships in the compact makeup of the thorax, the *Gorytes* branch with an Australian *Austrogorytes*-like ancestor as a possibility, and a tripartite branch in which male flagellomeres VIII to XI are specially distorted. The only suggestion of this antennal character among the other genera is the presence of ventral pits on flagellomeres VIII to XI of *Neoplisus* or the distortion of many of the more distal articles starting with flagellomere V or VI in a variety of genera such as *Sphecius* and *Handlirschia*.

Among the three upper central elements on the dendrogram, the *Dienoplus* group is separated by hindwing venation, the other two by structure of the mesopleuron. Of these three the *Hoplisoides* group is the least cohesive (fig. 178).

One puzzling feature is the occurrence of plantulae in most of the gorytin genera but their apparent absence in *Exeirus, Neoplisus, Eogorytes, Sphecius, Kohlia, Xerogorytes, Sagenista,* and *Afrogorytes*. For the most part we have considered these structures to be an indication of primitiveness. However, *Exeirus* seems to be only a sidewise development from primitive stock, whereas *Hoplisoides,* which has plantulae, seems rather highly evolved. In some genera, such as *Oryttus,* the plantulae are quite small and inconspicuous, but in others, such as *Austrogorytes,* they are prominent.

With respect to endemicity 10 genera occur in both the Old and the New World, nine are confined to the Old World, and 12 to the New World. Of the New World genera five (*Arigorytes, Xerogorytes, Trichogorytes, Hapalomellinus,* and *Tanyoprymnus*) are North American only, and only three (*Pterygorytes, Neogorytes,* and *Liogorytes*) are restricted to South America. As might be expected, the two most generalized types, *Clitemnestra* and *Argogorytes,* are found in both Hemispheres.

Among more general references on the tribe are those of Tsuneki (1963a), Beaumont (1952b, 1964b), Evans (1966a), and Handlirsch (1888a, 1888b, 1889, 1895).

Biology: All species of gorytins are predaceous and nest in the ground according to published information. The prey are generally nymphs or adults of one of the "hoppers" in the homopterous families Cicadellidae, Membracidae, Psyllidae, Fulgoridae, and Cercopidae. The genera *Exeirus* and *Sphecius* provision with cicadas (Cicadidae). In some cases only one or a few paralyzed individuals are stored in each cell, but as many as 100 leafhoppers may be furnished for a larva, as in *Hoplisoides glabratus* (see under *Hoplisoides* biology). Parasites commonly reported are in the tribe Nyssonini and miltogrammine Sarcophagidae. Chrysididae, such as *Hedychridium,* are involved, also, but, as in most other groundnesting forms, little is known of these.

Key to genera of Gorytini

1. Hindwing media diverging more than 1.0 midocellus diameter beyond cu-a (fig. 157 A).... 2
   Hindwing media diverging before cu-a or not more than 1.0 midocellus diameter beyond it (fig. 157 D,G).................................. 9
2. Reflexed lateral margin of scutum interrupted opposite posterior part of tegula by an oblique carina (sometimes faint) which delimits a small or nearly truncate lateroposterior, often declivous area (fig. 156); posterior veinlet of submarginal cell II (between recurrent veins) not more than one-fourth as long as posterior length of submarginal cell I (fig. 157 B) 3
   Lateral margin of scutum reflexed upward and uninterrupted opposite tegula; posterior veinlet of submarginal cell II more than one-fourth as long as posterior length of submarginal cell I (figs. 157 A, 162 C) 5
3. Inner orbits distinctly converging below (fig. 159 H), least interocular distance a little more than half that measured at middle of ocellar triangle; gaster pedunculate; w. N. America........ *Hapalomellinus* Ashmead, p. 496
   Inner orbits essentially parallel or converging only slightly below (fig. 159 G); least interocular distance not less than two-

484 *SPHECID WASPS*

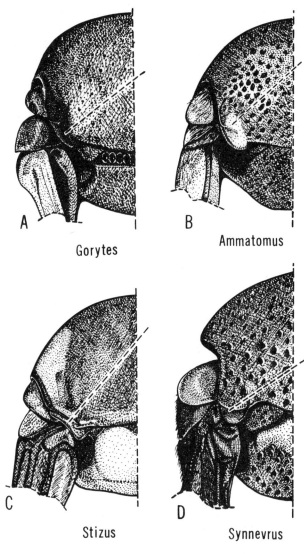

FIG. 156. Oblique scutal carina at posterolateral apex of scutum in four genera of Nyssoninae.

thirds that measured at middle of ocellar triangle; gaster not pedunculate 4
4. Pleuron completely and rest of body mostly covered with dense silvery appressed pubescence which hides sculpture; marginal cell of forewing rounded or blunt distally (fig. 157 B); w. N. America............. *Trichogorytes* Rohwer, p. 497
Pleuron and rest of body not so completely covered with pubescence; marginal cell of forewing acute distally; widespread except Australia .... *Dienoplus* W. Fox, p. 495
5. Frons narrower at level of midocellus than shortly below it (fig. 159 B), omaulus continued ventrally as an acetabular carina, male sternum VIII sword shaped, often exserted 6
Frons broader at level of midocellus than below it (fig. 159 A); omaulus continued ventrally only as a fine seam, ending ventrally before midline; male sternum VIII normally concealed, broadly rounded at apex 7
6. Gaster with segment I forming a slender peduncle (fig. 162 A); omaulus disappearing above as it approaches pronotal lobe; hindwing media diverging from cu-a at practically a right angle (fig. 162 C); Peru, Ecuador .... *Neogorytes* R. Bohart, p. 492
Gaster with segment I not unusually narrow; omaulus strong throughout; hindwing media diverging from cu-a at an obviously obtuse angle; worldwide .... *Argogorytes* Ashmead, p. 491
7. Frons without a long median sulcus, male sterna III to V fimbriate, hypoepimeral area hardly indicated; Palearctic.................... *Olgia* Radoszkowski, p. 491
Frons with a long median sulcus, male sterna III to V without special hairs, hypoepimeral area well marked.......................... 8
8. Gaster somewhat pedunculate, tergum I not more than half as broad posteriorly as II; N. and S. America, New Guinea .... *Ochleroptera* Holmberg, p. 489
Gaster not pedunculate, tergum I considerably more than half as broad at apex as II; S. America and Australia .... *Clitemnestra* Spinola, p. 485
9. Submarginal cell II petiolate (fig. 157 E); large wasps; Australia .... *Exeirus* Shuckard, p. 494
Submarginal cell II not petiolate 10
10. Hindwing jugal lobe practically absent or extremely small (fig. 157 C) 11
Hindwing jugal lobe well developed, usually larger than tegula (fig. 157 D)................ 12
11. Spiracular groove of propodeum present; propodeal enclosure essentially dorsal; omaulus present; Australia .... *Austrogorytes* R. Bohart, p. 498
Spiracular groove of propodeum absent; propodeal enclosure extending well onto posterior face of propodeum; omaulus absent; Brazil .... *Pterygorytes* R. Bohart, p. 513
12. Scutal admedian lines fused and replaced by a median longitudinal carina; s. Africa .... *Afrogorytes* Menke, p. 522
Scutal admedian lines present, not replaced by a median carina 13
13. Female with two rake setae on fore basitarsus before apex; male without special modifications on last four flagellomeres (fig. 167); spiracular groove present 14
Without above combination of characters 16
14. Propodeal enclosure with sculpture or longitudinal ridging, at least along most of anterior sulcus; Holarctic and Ethiopian .... *Gorytes* Latreille, p. 500
Propodeal enclosure without longitudinal ridging or general sculpture except sometimes in anterolateral corners, bounding sulci simple or appearing "pitted"...................... 15
15. Gaster pedunculate, tergum I somewhat nodose; male sterna III-V with erect

apical fimbriae; Centr. and S.
America .... *Megistommum* Schulz, p. 503
Gaster with tergum I usually somewhat constricted, but not at all nodose, sterna without obvious fimbriae; Nearctic and Ethiopian .... *Pseudoplisus* Ashmead, p. 502
16. Mesopleuron with no trace of a sternaulus 17
Mesopleuron with a complete or incomplete sternaulus 21
17. Omaulus absent 18
Omaulus present 19
18. Inner orbits nearly parallel, widely separated (fig. 174 B); male sterna III-V with short and erect apical fimbriae (fig. 170 C), tergum I stout; S. Africa. *Handlirschia* Kohl, p. 508
Inner orbits strongly converging below (as in Fig. 165 D); male sterna without fimbriae, segment I sometimes pedunculate; Old World .... *Ammatomus* A. Costa, p. 512
19. Ocellar lenses somewhat reduced so that each member of ocellar triangle is flattened externally; episternal sulcus continued downward almost vertically to omaulus; Old World .... *Kohlia* Handlirsch, p. 513
Ocellar lenses normal; episternal sulcus curving backward and joining scrobal sulcus (fig. 3 B) 20
20. Clypeus laterally almost a right angle or more often obtuse; mandible bidentate; arolia of female nearly equal on all legs; widespread except S. America ....
.... *Sphecius* Dahlbom, p. 509
Clypeus laterally acute; mandible simple; female foreleg with arolium much larger than on other legs; U.S. and Centr. America .... *Tanyoprymnus* Cameron, p. 511
21. Segment I pedunculate, tergum strongly humped towards apex 22
Segment I sometimes narrowed but tergum evenly curved, not strongly humped towards apex 23
22. Hindwing cu-a gently curved (fig. 157 F), mesopleuron often sparsely and rather finely punctate; male sternum VIII bispinose at apex (fig. 168 A); widespread except Australasia ... *Lestiphorus* Lepeletier, p. 505
Hindwing cu-a rather strongly curved or bent near forward end (as in fig. 157 H); mesopleuron coarsely punctate; male sternum VIII sword shaped (fig. 168 D); North America .... *Psammaletes* Pate, p. 508
23. Acetabular carina present, distinct and complete, or subomaulus continued and somewhat projecting ventrad; male sternum VIII sword shaped (fig. 175 B,C), male sterna without apical fimbriae 24
Acetabular carina absent, incomplete or indistinct; subomaulus short; male sternum VIII notched or bifid at tip (figs. 166 B, 176 A) (except in *Eogorytes* which has apical fimbriae on sterna III-V) 26
24. Propodeum coarsely areolate dorsolaterally, scutum essentially smooth; acetabular carina present; male without concealed and basal hairbrushes on sterna V and VI; Mexico to S. America......... *Sagenista* R. Bohart, p. 522
Propodeum usually punctate or rarely smooth dorsolaterally, not coarsely areolate; scutum usually well punctured; male with concealed hairbrushes on sterna V and VI 25
25. Acetabular carina present, subomaulus not continued ventrad; widespread except Australasia .... *Hoplisoides* Gribodo, p. 517
Acetabular carina absent, subomaulus continued ventrad; Palearctic ....
.... *Psammaecius* Lepeletier, p. 515
26. Spiracular groove absent or very weak 27
Spiracular groove distinct (rarely obscured by very coarse sculpture) 29
27. Propodeal enclosure smooth except for pitted boundary sulci; forewing not pictured; male sternum VIII bispinose apically (fig. 166 B); Mexico to Argentina .... *Neoplisus* R. Bohart, p. 504
Propodeal enclosure at least partly sculptured; male sternum VIII with a shallow apical notch (fig. 176 A,B) 28
28. Forewing pictured; frontal sulcus sharp; male sterna without velvety hair mats but V and VI with concealed basal hairbrushes; sw. U.S. ... *Xerogorytes* R. Bohart, p. 517
Forewing not pictured; frontal sulcus indistinct; male sterna with velvety hair mats on III-V but no concealed basal brushes on V and VI; w. U.S. and w. Mexico .... *Arigorytes* Rohwer, p. 516
29. Propodeal enclosure smooth except for median groove and pitted boundary sulci; episternal sulcus ending on scrobal sulcus which continues forward to omaulus (as in fig. 158); arolium on foreleg of female not unusually large .... *Liogorytes* R. Bohart, p. 516
Propodeal enclosure longitudinally ridged; episternal and scrobal sulci (when well marked) forming a continuous, arc-like or angled groove (sometimes obscured by rough sculpture) (as in fig. 3 B); arolium on foreleg of female much larger than those on other legs 30
30. Hindwing media diverging well before cu-a (fig. 157 G), male sterna III-V with apical fimbriae, male sternum VIII sword shaped; e. Asia ....
.... *Eogorytes* R. Bohart, p. 505
Hindwing media diverging at or very near cu-a (fig. 157 H) male sterna III-V without apical fimbriae, male sternum VIII bispinose apically (fig. 168 B,C); Holarctic, Ethipian, Chile . *Orytus* Spinola, p. 506

Genus Clitemnestra Spinola

Generic diagnosis: Medium small to small wasps; frons broad to narrow, sometimes broader below than above, least interocular distance usually a little above antennal sockets, width of frons as great or greater at level of mid-

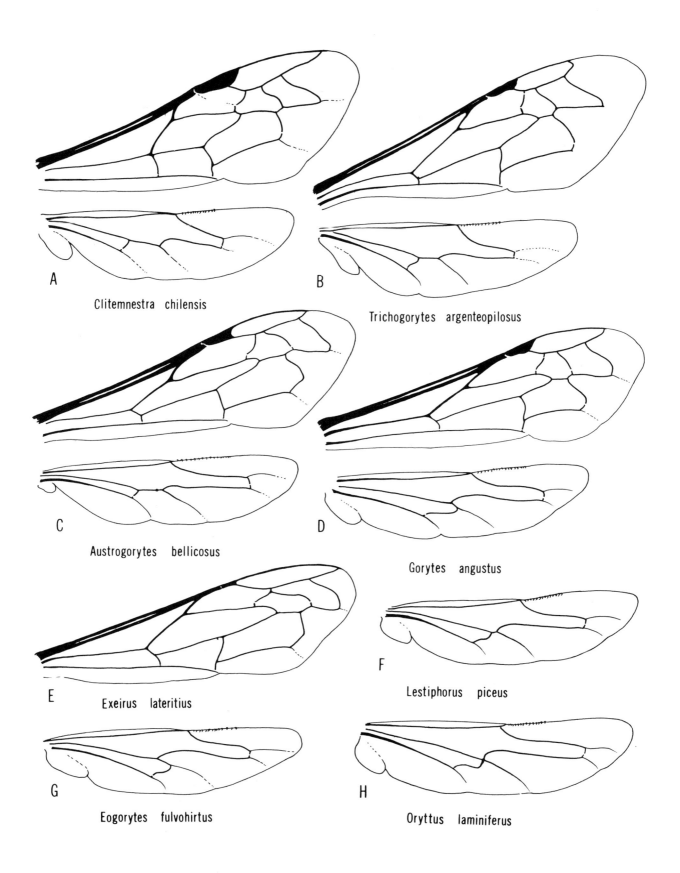

FIG. 157. Wings in the Gorytini.

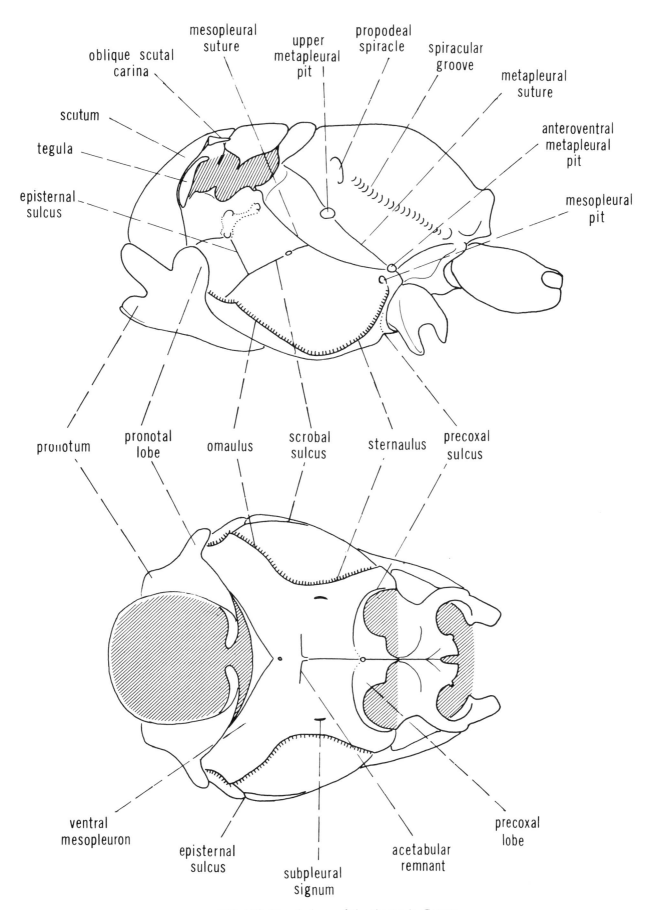

FIG. 158. Morphology of the thorax in *Gorytes*.

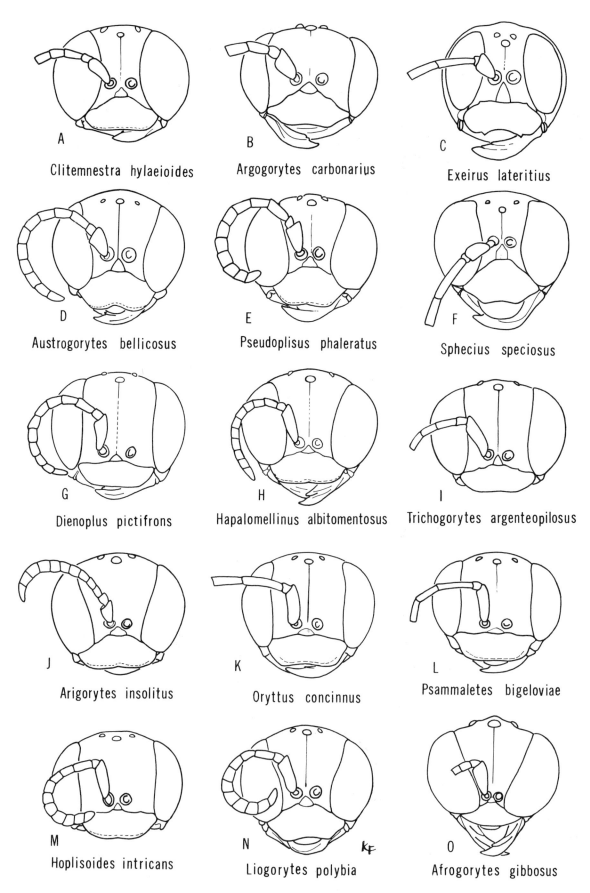

FIG 159. Facial outlines of females in the tribe Gorytini.

ocellus as a little below (fig. 158 A), inner eye margins rather strongly convex; last four male flagellomeres not specially modified; mandible with an inner subtooth; labrum normally visible; frons with a distinctly impressed line from midocellus to interantennal area; pronotal collar elevated, distinct from scutum; prescutellar sulcus efoveate; females of some species without a foretarsal rake; female arolia nearly equal in size; no posterolateral oblique scutal carina; episternal sulcus curving forward to upper end of omaulus, scrobal sulcus ending on episternal sulcus at a right angle; no acetabular carina nor sternaulus; midtibia with two apical spurs; tarsomere V curved beneath and more than twice as long as IV; forewing margin nearly straight near base of costa, media diverging at or slightly beyond cu-a, stigma moderate, one recurrent vein ending at submarginal cell I and one at cell II or both at II, veinlet between ends of recurrent veins about two-sevenths as long as posterior veinlet of submarginal cell I (fig. 157 A); jugal lobe somewhat larger than tegula but less than twice as long, hindwing media diverging well beyond cu-a (fig. 157 A); no spiracular groove; gaster not pedunculate; male with seven visible terga, sterna without fimbriae, sternum VIII simple and not unusually narrowed distally (fig. 160 A); female pygidial plate densely bristly; typical male genitalia as in fig. 161 B.

Geographic range: Three species have been described from Chile and nine from Australia.

Systematics: Three closely related genera, *Clitemnestra, Ochleroptera* and *Olgia* can be considered as relatively unspecialized Gorytini. Among the member species they qualify for practically every character previously listed as unspecialized in the Nyssoninae. Along with *Exeirus, Argogorytes,* and *Neogorytes* they lack even a trace of the oblique scutal carina which is found in all other gorytins (see fig. 156).

Clitemnestra qualifies as the least specialized and hence the most "primitive" in the subfamily. Its position is indicated by a "0" rating obtained from table 16. The features that separate it from its close relatives are the unconstricted basal gastral segment and the nonfimbriate sterna in the male. We have seen examples of all described species except *mimetica* and *megalophthalma*. One of the most distinctive forms is the Australian *thoracica* in which the claws are blunt and the male has a shovel like projection on sternum II. The claws are slightly to markedly unequal in most species of the genus. This is best observed on the female hind leg and involves the enlargement of the inner claw. The condition is well developed in *plomleyi, lucidula,* and *chilensis,* among others, but not in *thoracica* and *guttatula.* In a few species the first recurrent vein may be interstitial or may be received by submarginal cell I. This latter condition seems to be the rule in *plomleyi, perlucida,* and *lucidula.*

Biology: Some information comes from Janvier (1928), who observed a colony of *C. chilensis* near Santiago, Chile, and both *chilensis* and *gayi* near Temuco in central south Chile. The nests of *chilensis* were in clay hills and in a soft riverbed. The burrow entrances were close together in the Santiago colony but more isolated at Tenuco. The entrances were open during provisioning. Both mandibles and the weakly raked front legs were used in digging. The burrows were multicellular, and the prey was a fulgorid (*Dictyophora*), about 10 to a cell. The egg was attached to a midleg of the first of the almost completely paralyzed prey, and it hatched after five or six days. The larval feeding period was about a week, and then a stout cocoon was formed of silk and earth.

The habits of *C. gayi* were somewhat similar, but this wasp makes its populous nests in clay banks of forested areas, the entrances often concealed by mosses and lichens. Its provisions were unidentified homopterous nymphs and adult Membracidae.

Evans (1966a) has questioned several unusual features of *Clitemnestra* biology as reported by Janvier. First, the prey is carried in flight between the mandibles, aided by the front legs instead of with the middle legs as usual in Gorytini. Furthermore, the nest is provisioned over a period of several days, the cell nearest the entrance is the oldest, the egg is laid on a midleg of the first prey in the cell, and the period before eclosion is five or six days. Evans has suggested the need for confirmation by other workers. With respect to oviposition on a midleg of the first prey member of a cell, compare the similar situation in *Argogorytes mystaceus* discussed under the biology section of that genus.

Evans and Matthews (1971b) gave cicadellids and cixiids as presumptive prey of *plomleyi* in Australia.

Checklist of *Clitemnestra*
(variant but improper spelling is *Clytemnestra*)

chilensis (Saussure), 1867 (*Harpactus*); Chile
duboulayi (Turner), 1908 (*Gorytes*); W. Australia
gayi (Spinola), 1851 (*Arpactus*); Chile
guttatula (Turner), 1936 (*Arpactus*); W. Australia
lucidula (Turner), 1908 (*Gorytes*); Australia: Queensland
megalophthalma (Handlirsch), 1895 (*Gorytes*); Australia
mimetica (Cockerell), 1915 (*Gorytes*); Australia: N.S. Wales
multistrigosa Reed, 1894; Chile
perlucida (Turner), 1916 (*Gorytes*); Australia: Queensland
plomleyi (Turner), 1940 (*Arpactus*); se. Australia
   hylaeioides Rayment, 1955 (*Astaurus*)
sanguinolenta (Turner), 1908 (*Gorytes*); Australia: Queensland
thoracica (F. Smith), 1869 (*Miscothyris*); W. Australia

Genus Ochleroptera Holmberg

Generic diagnosis: Medium small to small wasps; inner eye margins strongly curved and closest near middle of eyes in front view, width of frons at level of midocellus as great or greater than a little below (as in fig. 159 A); last four male flagellomeres not specially modified; mandible with an inner subtooth; labrum not normally visible; frons with a longitudinal groove below midocellus; pronotal collar elevated, distinct from scutum; prescutellar sulcus efoveate; females with at most a weak foretarsal rake; female arolia nearly equal in size; no posterolateral oblique scutal carina; episternal sulcus curving forward to meet upper end of omaulus, scrobal sulcus ending on episternal sulcus at a right angle; no acetabular carina nor sternaulus; midtibia with two apical spurs; tarsomere V curved beneath and more than twice as long as IV; forewing margin nearly straight near base of costa, media diverging at or slightly beyond cu-a, stigma moderate, one recurrent vein ending at submarginal cell I and one at II or both at II, veinlet between ends of recurrent veins about one-fourth as long as posterior veinlet of submarginal cell I; jugal lobe somewhat larger than tegula but less than twice as long, hindwing media diverging well beyond cu-a; no spiracular groove; gaster pedunculate to subpedunculate, tergum I a third to half as broad as II; male with six visible terga, sterna without fimbriae, sternum VIII simple and not unusually narrowed distally (fig. 160 C); female pygidial plate densely bristly.

Geographic range: The genus is especially developed in South America where eight named species are known to occur and several more are undescribed. Three additional ones are found in North America: *bipunctata* from the United States, *jamaica* from the West Indies, and *championi* from Guatemala. One species, *novaguineensis*, occurs in eastern New Guinea (R. M. Bohart, 1970a).

Systematics: The constricted, subpedunculate to pedunculate first gastral segment differentiates this genus from both *Olgia* and *Clitemnestra,* its close relatives. Considering that *Clitemnestra* in the New World is practically limited to Chile, *Ochleroptera* can be considered as an evolutionary derivative of the former, which has spread to many parts of South and North America. Since *Clitemnestra* probably developed in Australia it is presumed that *Ochleroptera* was an offshoot there, having migrated along with the parent stock to Chile, and one species found its way north to New Guinea.

Many species have blue or green reflections from the vertex and notum. Tergum I may be slightly to strongly constricted and evenly confluent with tergum II in profile to distinctly nodose.

Biology: Strandtmann (1945) noted a few individuals of *Ochleroptera bipunctata* nesting in flower boxes in Texas, and Pate (1946a) added a prey record. The most complete observations, also on *bipunctata,* were published by Evans (1966a). The nests were located in sloping or vertical banks in coarse sandy soil of medium moisture content. There was no tumulus around the entrance holes, which were closed in the absence of the wasp. The prey were in four families of saltatory Homoptera: Cicadellidae, Cercopidae, Fulgoridae, and Psyllidae, which were in addition to treehoppers (Membracidae) as given by Strandtmann. The manner of prey carriage, stocking of the cell and oviposition were all usual as outlined under biology for the tribe. This contrasts with the oviposition behavior reported for the related genera, *Clitemnestra* and *Argogorytes,* and may represent an evolutionary advancement. Evans observed as many as three and possibly seven cells per burrow. The number of prey per cell was six to 18. Additional prey species were given by Evans (1968).

Checklist of *Ochleroptera*

aenea (Handlirsch), 1888 (*Gorytes*); Brazil, Argentina, Venezuela
    parvula Handlirsch, 1888 (*Gorytes*), new synonymy by R. Bohart
    subtilis Handlirsch, 1895 (*Gorytes*), new synonymy by R. Bohart
bipunctata (Say), 1824 (*Gorytes*); U.S.
championi (Cameron), 1890 (*Gorytes*); Guatemala
colorata (W. Fox), 1897 (*Gorytes*); Brazil
hirta (Handlirsch), 1888 (*Gorytes*); Brazil
jamaica Pate, 1947; Jamaica, Cuba
novaguineensis R. Bohart, 1969; e. New Guinea
oblita Holmberg, 1903; Argentina
pygmaea (Brèthes), 1913 (*Gorytes*); Argentina
sphaerosoma (Handlirsch), 1895 (*Gorytes*); Brazil
tenera (Handlirsch), 1895 (*Gorytes*); Venezuela
violacea (Handlirsch), 1888 (*Gorytes*); Brazil

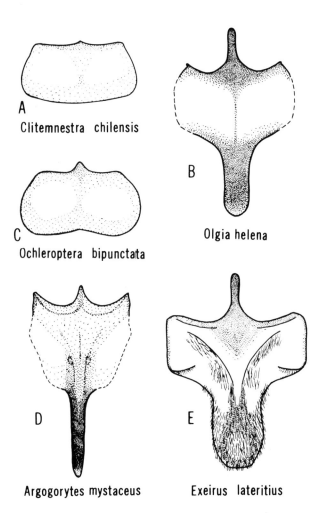

FIG. 160. Male sternum VIII in the tribe Gorytini.

Genus Olgia Radoszkowski

Generic diagnosis: Small wasps; inner eye margins strongly curved and closest near middle of eyes in front view; last four male flagellomeres not specially modified; mandible simple; labrum not normally visible; frons without a definite longitudinal median groove; pronotal collar elevated, distinct from scutum; female foretarsal rake moderately developed; female arolia nearly equal in size; no posterolateral oblique scutal carina; epistemal sulcus vertical, ending on omaulus (absent in *spinulosa*), scrobal sulcus evanescent or absent; no acetabular carina nor sternaulus; midtibia with two apical spurs; tarsomere V nearly straight and sometimes less than twice as long as IV; forewing margin nearly straight near base of costa, media diverging at cu-a; stigma moderate, recurrent veins ending at submarginal cell II, veinlet between recurrents about one-fourth as long as posterior veinlet of submarginal cell I; jugal lobe somewhat larger than tegula, hindwing media diverging well beyond cu-a; no spiracular groove; gaster not pedunculate; male with seven visible terga but last one very small, sterna III and IV with short but dense apical fimbriae, sternum VIII somewhat narrowed distally (fig. 160 B); female pygidial plate densely bristly.

Geographic range: According to Beaumont (1953a), who has published the only review of the genus, there are five species in southeastern Europe, central Asia, and north Africa.

Systematics: The two species we have seen are *helena* and *spinulosa,* kindly furnished by Dr. Jacques de Beaumont. Judging from this material as well as from Beaumont's excellent figures, key, and descriptions, it appears that *Olgia* is close to *Clitemnestra* but somewhat more advanced. This is indicated by the simple mandibles, ungrooved frons, narrowed sternum VIII, and fimbriate male sterna of *Olgia*. We view this genus as a Palearctic evolutionary extension of the Australian *Clitemnestra.*

Biology: Unknown.

Checklist of *Olgia*

bensoni (Beaumont), 1950 (*Gorytes*); nw. Africa
helena Beaumont, 1953; Greece, Middle East
maracandica (Radoszkowski), 1877 (*Kaufmannia*); sw. USSR
modesta Radoszkowski, 1877; sw. USSR: Turkmen S.S.R.
spinulosa Beaumont, 1953; sw. USSR: Caucasus; s. Yugoslavia; Middle East

Genus Argogorytes Ashmead

Generic diagnosis: Male antenna simple, no depressions or polished areas beneath last four articles, basal flagellomeres subequal in length; no sharply impressed line from midocellus to antennal socket area; inner eye margins sinuate, frons broad but a little narrowed toward clypeus, breadth at level of midocellus less than that a short distance below (fig. 159 B); labrum usually slightly visible; female foretarsus without a rake; pronotal collar elevated, not appressed to scutum; lateral margin of scutum a thin flange which has no oblique carina op-

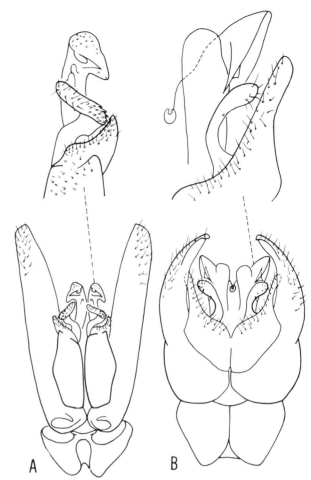

FIG. 161. Male genitalia in the tribe Gorytini; A, *Clitemnestra plomleyi*, B, *Argogorytes mystaceus*.

ate, female scutellum usually (always ?) with a posteromedian pubescent depression; omaulus strongly developed throughout, epistemal sulcus a vertical pitted groove that ends ventrad on omaulus, scrobal sulcus ending on epistemal sulcus at a right angle; acetabular carina present, sternaulus absent; forewing costa bowed out a little near base, media diverging before, at or beyond cu-a, stigma relatively large, marginal cell tapering to a moderate point distally, recurrents both received by submarginal cell II, distance between their ends more than one-third posterior length of submarginal cell I; hindwing media diverging at an obtuse angle well beyond cu-a, jugal lobe larger in outline than tegula and about twice as long; propodeum without a spiracular groove or with a weakly defined one, enclosure with a deep median groove; gaster sessile or slightly and broadly pedunculate; sternum II humped; male with seven visible terga; male sternum VIII slender apically, sword shaped, upturned (fig. 160 D); female pygidial plate with dense, fine bristles toward apex; male genitalia as in fig. 161 A).

Geographic range: Occurring in all continental Faunal Regions except the Ethiopian. We have listed 24 species.

Species are known from many of the islands of the western Pacific.

Systematics: Two closely related genera, *Argogorytes* and *Neogorytes,* have the undulate inner margin of the eyes arranged so that the frons is narrower at the level of the midocellus than it is slightly below it (fig. 159 B). Also, there is a well developed acetabular carina on the venter of the mesopleuron, and the prescutellar sulcus is foveate in contrast to the condition in *Clitemnestra* and its relatives.

The species of *Argogorytes* are remarkably uniform in structure, and there seems little justification for subgeneric divisions. We have studied types or named material of about two-thirds of the species including a series of the New Zealand generotype, *carbonarius*. In species we have seen, the scutellum of the female bears a posteromedian pubescent depression or roughened area as in some *Stizus*. The downwardly bulging second sternum is especially pronounced in *carbonarius* but is nearly as strong in the North American *nigrifrons*. The characteristic coarse channel which divides the propodeal enclosure longitudinally is rather small and basal in *carbonarius*. The American species, *nigrifrons, sapellonis, similicolor,* and *areatus,* have terga II to V in the male and II to IV in the female with a dense apical fringe of slightly broadened hairs. We have not noted this condition in the Old World species.

Argogorytes is certainly a "primitive" gorytine genus, but it is more specialized than *Clitemnestra* as indicated by the undulate inner eye margins (fig. 159 B) and the sword shaped end of male sternum VIII (fig. 160 D). The Australian *Argogorytes* have a superficial resemblance to species of *Austrogorytes* and may well be close to the ancestral stock of that genus.

Biology: Nesting habits are known for two Palearctic species of *Argogorytes*. Hamm and Richards (1930) summarized information on *mystaceus* and on *fargeii* (as *campestris*). Tsuneki (1965c) added information on the subspecies, *mystaceus grandis,* in Japan. In all cases the prey were spittlebug nymphs of the genus *Aphrophora* (Cercopidae). The wasps obtained these by plucking the nymphal bugs from their protective foam. In the case of *fargeii,* as reported by Adlerz (1903) in south Sweden, the prey was carried by the midlegs of the wasp, sternum to sternum. Burrows, which were always left open while the wasps were away, penetrated 9 cm down and 9 cm horizontally into the bare clay or gravel slopes. Each nest had six to nine cells with 19 to 27 prey deposited with heads pointing in. The egg was not seen. In Tsuneki's observations on *mystaceus grandis,* the burrows were closed in the absence of the wasp. In this case the egg was found attached to the outside of one of the hindcoxae of the first prey member of the cell. This is contrary to the usual fashion in Gorytini but similar to that given by Janvier for *Clitemnestra*. A complete cell held four nymphs. *Nysson spinosus* was given by Hamm and Richards as a parasite of both *fargeii* (as *campestris*) and *mystaceus*.

Evans and Eberhard (1970) discussed a sort of "parasitism" of *Argogorytes mystaceus* in which flowers of the orchid *Ophrys insectifera* produce an odor resembling the female sex attractant pheromone. After the male has been drawn to the flower, the color and shape of the blossom stimulates an attempted copulation and effects pollination.

Checklist of *Argogorytes*

areatus (Taschenberg), 1875 (*Gorytes*); Brazil
basalis (F. Smith), 1860 (*Gorytes*); Indonesia: Ambon
caerulescens (Turner), 1914 (*Gorytes*); Sri Lanka
carbonarius (F. Smith), 1856 (*Gorytes*); New Zealand
   trichiosoma Cameron, 1888 (*Gorytes*)
constrictus (F. Smith), 1859 (*Gorytes*); Indonesia: Aru
crucigera (Hacker and Cockerell), 1922 (*Arpactus*); e. Australia
fairmairei (Handlirsch), 1893 (*Gorytes*); Algeria
fargeii (Shuckard), 1837 (*Gorytes*); Eurasia
fuliginosus Tsuneki, 1968; Taiwan
hispanicus (Mercet), 1906 (*Gorytes*); Europe
matangensis (Turner), 1914 (*Gorytes*); Borneo
mongolensis Tsuneki, 1970; Mongolia
mystaceus (Linnaeus), 1761 (*Sphex*); Eurasia
   campestris Linnaeus, 1761 (*Vespa*)
   inimicus M. Harris, 1776 (*Vespa*)
   longicornis Rossi, 1790 (*Sphex*)
   bicinctus Fabricius, 1793 (*Crabro*), suppressed by I.C.Z.N., 1967, Opinion 675
   arpactus Fabricius, 1804 (*Mellinus*)
   flavicinctus Donovan, 1808 (*Vespa*)
   croceipes Eversmann, 1849 (*Gorytes*)
   tonus Bondroit, 1933 (*Gorytes*)
   ssp. *grandis* (Gussakovskij), 1933 (*Gorytes*); USSR, Japan
nigrifrons (F. Smith), 1856 (*Gorytes*); N. America
   bollii Cresson, 1872 (*Gorytes*)
   neglectus Rohwer, 1911 (*Gorytes*)
nipponis Tsuneki, 1963; Japan
przewalskyi Kazenas, 1971; sw. USSR: Kazakh S.S.R.
   przewalskii Kazenas, 1972
rubrosignatus (Turner), 1915 (*Arpactus*); W. Australia
rufomixtus (Turner), 1914 (*Gorytes*); Australia: N. S. Wales
sapellonis (Baker), 1907 (*Gorytes*); sw. U.S.
secernendus (Turner), 1915 (*Arpactus*); se. Australia
similicolor (Dow), 1933 (*Gorytes*); Paraguay
stenopygus (Handlirsch), 1895 (*Gorytes*); Celebes
tonkinensis (Yasumatsu), 1943 (*Gorytes*); China
vagus (F. Smith), 1859 (*Gorytes*); Indonesia: Aru

Neogorytes R. Bohart, new genus

Generic diagnosis: No impressed line from midocellus to antennal socket area; inner eye margins sinuate, frons broad but a little narrowed toward clypeus, breadth at level of midocellus less than that a short distance below (fig. 162 D); labrum visible beneath clypeus; foretarsus without a rake; pronotal collar elevated, not appressed to scutum; lateral margin of scutum a thin flange which has no oblique carina opposite posterior edge of tegula;

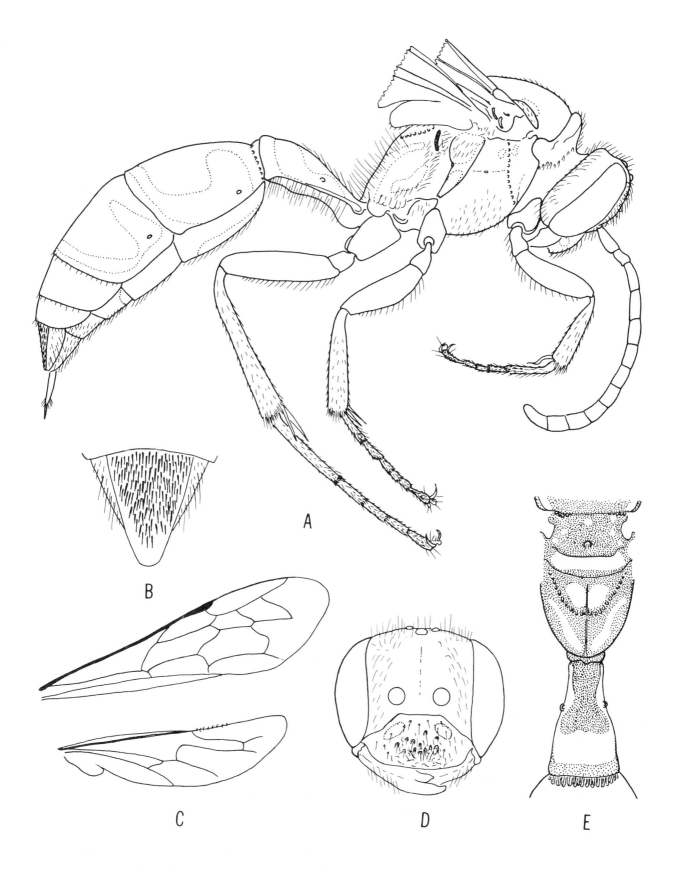

FIG. 162. *Neogorytes ecuadorae* female holotype; A, lateral, yellowish markings enclosed in dotted lines; B, pygidium; C, wings; D, facial view; E, scutellum to tergum I to show yellowish markings, dorsal.

prescutellar sulcus foveate, scutellum with a small medioposterior pubescent depression; omaulus fading out above before reaching pronotal lobe, episternal sulcus a vertical pitted groove that ends ventrad on omaulus, scrobal sulcus ending on episternal sulcus at a right angle; acetabular carina present, sternaulus absent; forewing costa bowed out a little near base, media arising before cu-a, stigma relatively large, marginal cell tapering to a moderate point distally, recurrents both received by submarginal cell II, distance between their ends more than one-third posterior length of submarginal cell I; media of hindwing diverging at nearly a right angle well beyond cu-a (fig. 162 C), jugal lobe about twice as long as tegula; propodeum with a spiracular groove, enclosure and propodeum below it with a deep median groove; gaster with segment I in the form of a slender peduncle, sternum II a little humped (fig. 162 A); female pygidial plate densely covered with moderately fine short bristles (fig. 162 B); male unknown.

Type species: Neogorytes ecuadorae R. Bohart

Geographic range: Known only from the highlands of Peru and Ecuador.

Systematics: The type species and an undescribed species from Peru bear a close relationship to *Argogorytes,* and they can be considered as a pedunculate extension of that genus. In addition to the pedunculate gaster the genus differs from *Argogorytes* in having the omaulus fading out completely before it reaches the pronotal lobe, in the more distinct spiracular groove, and in the remarkably abrupt divergence of the hindwing media (fig. 162 C). The general body structure is like that of the more slender species of *Ochleroptera,* but the different frons, larger jugal lobe, spiracular groove, pitted scutellum, and several other features readily distinguish it.

The somewhat advanced status of *Neogorytes* and the close ties with *Argogorytes* are indicated in the dendrogram (fig. 155).

Biology: Unknown.

Checklist of *Neogorytes*

ecuadorae R. Bohart (described below); Ecuador

Neogorytes ecuadorae R. Bohart, new species

Female: Length 11 mm; black with extensive yellow to orange markings, antennae black, clypeus with two basal yellow spots, narrow yellow stripes across pronotal collar and on scutum opposite tegula, legs mostly orange, other pale markings indicated in fig. 162 E. Pubescence fulvous and mostly erect on head and thorax; macropubescence less abundant on abdomen; darker areas of terga with thick, brownish, appressed micropubescence; pygidial plate thickly covered, except near apex, with brown setae which have coppery glints; wing membrane with close, dark microsetae. Structural features described for genus and illustrated in fig. 162.

Holotype, female (Mus. Harvard: MCZ), Banos, Tungurahua, Ecuador, 1900 m (W. C. MacIntyre).

Genus Exeirus Shuckard

Generic diagnosis: Large wasps; antenna simple, no special modifications beneath last four articles; basal flagellomeres longer than scape; mandible with an inner subtooth; a sharply impressed line from midocellus to antennal socket area; inner eye margins convex, about as close above as below, least interocular distance about equal to eye breadth (fig. 159 C); labrum usually hidden; female foretarsus with a well developed rake, basitarsus with three long spines before apex, arolia subequal on all legs; midtibia with two strong apical spurs; pronotal collar elevated, not appressed to scutum; lateral margin of scutum a thin flange which has no oblique carina opposite posterior edge of tegula, posterolateral corner raised somewhat tooth shaped; prescutellar sulcus simple; episternal sulcus curving forward, meeting omaulus nears its upper end, scrobal sulcus ending on episternal sulcus at a right angle; no acetabular carina nor sternaulus; forewing costa nearly straight near base, media arising before cu-a, stigma very small, marginal cell tapering to a moderate point distally; submarginal cell II petiolate and receiving both recurrent veins, distance between their ends about one-third posterior length of submarginal cell I, cells unusually close to apex of wing (fig. 157 E); media of hindwing interstitial (or nearly so) with cu-a, which is gently curved and almost longitudinal; jugal lobe much larger than tegula; propodeum with a strong spiracular groove, enclosure with a deep median groove; gaster sessile; tergum II with a transverse, shelflike, basal groove; female pygidial plate hardly indicated posteriorly, densely bristly; male with seven visible terga; male sternum VIII broadly rounded apically (fig. 160 E).

Geographic range: Australia (eastern part). One species known.

Systematics: The petiolate submarginal cell II (fig. 157 B) makes *Exeirus* unique among the Gorytini. Its tribal assignment has always been questionable, and it appears to be a rather isolated, relict genus. However, practically all of its features are gorytin. In size, general shape, and prey (cicadas) the single species is reminiscent of *Sphecius.* However, the relationship is not close, and *Exeirus* must have diverged very early, as indicated in fig. 155.

Biology: The habits have been recorded by Froggatt (1903), McCulloch (1923), and Musgrave (1925). Their observations have been summarized by Evans (1966a). Female wasps feed on tree sap exuding from punctures made by cicadas. Burrows are about the size of "mouseholes" and extend nearly vertically 23 to 47 cm into sandy soil. The entrance is left open while the wasp is seeking the cicada prey, which may be any of half a dozen large species. The wasp, after stinging its prey, drops with it to the ground and then drags or half flies with it "through grass and weeds, over sticks or anything else that may be between it and the burrow" (McCulloch, 1923). A nest may contain several cells, each with a single cicada prey to which an egg is attached on the venter near the base of a leg.

Checklist of *Exeirus*

lateritius Shuckard, 1837; Australia
lanio Stål, 1857 (*Sphecius*)

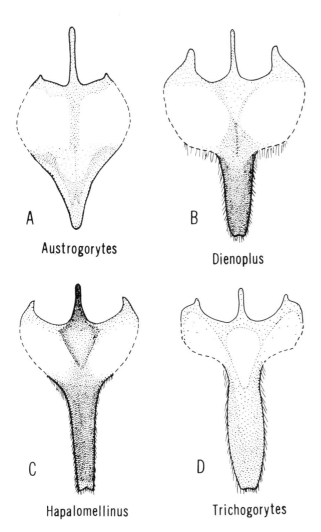

FIG. 163. Male sternum VIII in the tribe Gorytini; A, *Austrogorytes* sp.; B, *Dienoplus mendicus;* C, *Hapalomellinus albitomenosus;* D, *Trichogorytes argenteopilosus.*

Genus Dienoplus W. Fox

Generic diagnosis: Small wasps; frons broad, flat, sides nearly parallel (fig. 157 G), median sulcus usually weak; clypeus short and broad; labrum inconspicuous; last four male flagellomeres specially modified, flattened or concave beneath; mandible with an inner subtooth; pronotal collar thin, nearly even all across, rather closely appressed to scutum; female foretarsal rake well developed; female arolia nearly equal in size; posterolateral oblique scutal carina present; prescutellar sulcus foveate; omaulus and sternaulus present; acetabular carina absent; episternal sulcus when present running ventrad to omaulus or ending at anterior end of scrobal sulcus which joins it at about a right angle; metapleuron gradually narrowed to a point below; forewing not spotted, media diverging before cu-a, stigma moderate, veinlet of submarginal cell II between recurrents hardly one-fourth as long as posterior length of submarginal cell I; jugal lobe considerably larger than tegula, hindwing media diverging more than 1.0 midocellus diameter beyond cu-a; midtibia with two apical spurs; propodeal enclosure almost wholly dorsal; spiracular groove present; gaster not pedunculate; male with seven terga normally exposed, sterna without fimbriae, sternum VIII greatly narrowed and blade shaped distally but not bispinose (fig. 163 B); female pygidial plate relatively bare.

Geographic range: Primarily Holarctic, but a few species occur in southern Africa, India, and Burma. We have listed 54 species.

Systematics: The three genera, *Dienoplus, Hapalomellinus,* and *Trichogorytes,* appear to be related on the basis of similarities in sternum VIII of the male (fig. 163), but more especially on hindwing venation, since this is the main character separating them from *Hoplisoides* and its relatives. Although least developed in *Dienoplus,* there seems to be a strong tendency in the group toward fine, dense, silvery pubescence. This is especially true on the pleuron where the thick pile may hide the carinae and sutures ordinarily used in classification.

Dienoplus are small, active wasps which are often mistaken for *Tachysphex* by collectors until their pale markings and broad frons are observed. Most species have considerable amounts of rust-red coloration on the abdomen, as in *pictifrons* and *mendicus;* at the base of the gaster, as in *elegans* and *exiguus;* on the thorax, as in *laevis;* or over most of the body, as in *gyponae* and *citipes.* Pale abdominal markings may be present as subapical tergal bands, as in *elegans* and *laevis;* or as lateral spots on tergum II and a median one on V, as in *gyponae* and *citipes.*

We have been able to study about one-half of the 54 known species. For the remainder, reliance has been placed on the original descriptions and on the remarks of Beaumont (1954a) and Handlirsch (1888b, 1895). In much of the older literature the group has been considered as *Gorytes* or under the subgeneric name of *Harpactus.*

Biology: The nests are simple ones in sandy soil. There may be as few as two or as many as 15 cells per burrow in some species. Prey consists of nymphs and adults of Cicadellidae and Cercopidae. The beautiful red North American species, *gyponae,* was observed by F. X. Williams (1914b) to provision with the leafhopper, *Gyponana cinerea* Uhler. The hoppers were both stung and malaxated before being carried to the nest. As usual, the egg is laid on one side of the venter of the paralyzed leafhopper. The ethology for the genus has been briefly summarized by Evans (1966a).

A number of parasites have been reported. As given by Hamm and Richards (1930) these are *Nysson macu-*

losus and *N. dimidiatus* on *Dienoplus tumidus,* as well as the chrysidid, *Hedychridium rosae* Rossi. *Nysson dimidiatus* is also a parasite of *Dienoplus elegans* (Ferton, 1901). *Dienoplus lunatus* is attacked by the chrysidid, *Hedychridium integrum* Dahlbom (Olberg, 1959).

Checklist of *Dienoplus*

adventicus (Beaumont), 1967 (*Gorytes*); Turkey
affinis (Spinola), 1808 (*Gorytes*); Europe, Middle East
    carceli Lepeletier, 1832 (*Arpactus*)
annulatus (Eversmann), 1849 (*Harpactus*); USSR
arenarum (Beaumont), 1953 (*Gorytes*); Morocco
aureus (Beaumont), 1959 (*Gorytes*); Israel
castor (Handlirsch), 1898 (*Gorytes*); n. Africa
citipes Krombein, 1954; U.S.: Florida
coccineus (Balthasar), 1954 (*Gorytes*); Israel
consanguineus (Handlirsch), 1888 (*Gorytes*); Eurasia
    transiens A. Costa, 1888 (*Harpactus*)
creticus (Beaumont), 1965 (*Gorytes*) Greece
croaticus (Vogrin), 1954 (*Harpactus*); Yugoslavia
cyrenaicus (Beaumont), 1956 (*Gorytes*); Libya
decipiens (Arnold), 1936 (*Arpactus*); s. Africa
delicatulus (Morice), 1911 (*Gorytes*); Algeria
elegans (Lepeletier), 1832 (*Gorytes*); Europe, Middle East
  ssp. *siculus* Beaumont, 1968 (*Gorytes*); Sicily
escalerae (Turner), 1912 (*Gorytes*); n. Africa
exiguus (Handlirsch), 1888 (*Gorytes*); Europe
ferrugatus (Gussakovskij), 1928 (*Gorytes*); sw. USSR
fertoni (Handlirsch), 1910 (*Gorytes*); Corsica
formosus (Jurine), 1807 (*Arpactus*); ? Europe, Turkey
funereus (Giner Marí), 1945 (*Gorytes*); w. Africa
gyponae (Williams), 1914 (*Harpactus*); U.S.: Kansas
hissaricus (Gussakovskij), 1953 (*Gorytes*); sw. USSR
histrio (Saussure), 1892 (*Harpactus*); Madagascar
impudens (Nurse), 1903 (*Gorytes*); India
laevis (Latreille), 1792 (*Mutilla*); Europe, Middle East, Manchuria
    cruentus Fabricius, 1798 (*Sphex*)
    ruficollis Fabricius, 1798 (*Evania*)
    cruentatus Latreille, 1805 (*Mellinus*), lapsus
    caucasicus Radoszkowski, 1884 (*Harpactus*)
    morawitzi Radoszkowski, 1888 (*Harpactus*)
    levis Dalla Torre, 1897 (*Gorytes*)
  ssp. *aegyptiacus* (Schulz), 1904 (*Gorytes*); Egypt
  ssp. *alicantina* (Mercet), 1906 (*Gorytes*); sw. Europe
  ssp. *saharae* (Giner Marí), 1945 (*Gorytes*); n. Africa
  ssp. *dzinghis* Tsuneki, 1971; Mongolia
lateritius (Handlirsch), 1888 (*Gorytes*); Mexico
lepidus (Arnold), 1944 (*Gorytes*); Ethiopia
leucurus (A. Costa), 1884 (*Harpactes*); Sardinia
lunatus (Dahlbom), 1832 (*Larra*); Eurasia
    belgicus Wesmael, 1839 (*Gorytes*)
mendicus (Handlirsch), 1893 (*Gorytes*); w. N. America
moravicus (Šnoflak), 1946 (*Gorytes*); e. and s. Europe, Middle East
mundus (Beaumont), 1950 (*Gorytes*); Algeria
niger (A. Costa), 1858 (*Harpactes*); Europe
obscurus (Beaumont), 1969 (*Gorytes*); Turkey
ornatus (F. Smith), 1856 (*Harpactus*); Burma
osdroene (Beaumont), 1969 (*Gorytes*); Turkey
picticornis (Vogrin), 1954 (*Harpactus*); Yugoslavia
pictifrons W. Fox, 1893; w. N. America
    howardi Ashmead, 1899 (*Harpactus*)
pollux (Handlirsch), 1898 (*Gorytes*); n. Africa
pulchellus (A. Costa), 1859 (*Harpactes*); Europe
pyrrhobasis (Morice), 1911 (*Gorytes*); n. Africa
quadrisignatus (Palma), 1869 (*Harpactus*); s. Europe: Sicily
  ssp. *ifranensis* (Nadig), 1934 (*Gorytes*); Morocco
  ssp. *lugubris* (Beaumont), 1960 (*Gorytes*); Libya
rufithorax (Brauns), 1911 (*Gorytes*); s. Africa
sareptanus (Handlirsch), 1888 (*Gorytes*); USSR
schwarzi (Beaumont), 1965 (*Gorytes*); Greece
tauricus (Radoszkowski), 1884 (*Harpactus*); se. Europe; sw. USSR
transcaspicus (Kokujev), 1910 (*Gorytes*); sw. USSR
tumidus (Panzer), 1801 (*Pompilus*); Europe
  ssp. *japonensis* Tsuneki, 1963; Japan
turcmenicus (Radoszkowski), 1893 (*Harpactus*); sw. USSR
varipes Tsuneki, 1971; Mongolia
vicarius (Handlirsch), 1895 (*Gorytes*); s. Africa
  ssp. *karooensis* (Brauns), 1911 (*Gorytes*); s. Africa
vividus (Turner), 1921 (*Arpactus*); India
walteri (Handlirsch), 1888 (*Gorytes*); sw. USSR

Genus Hapalomellinus Ashmead

Generic diagnosis: Small wasps, frons narrowed toward clypeus (fig. 159 H); median frontal sulcus distinct; labrum inconspicuous, last four male flagellomeres specially modified, flattened or concave beneath; mandible with an inner subtooth; pronotal collar thin, nearly even all across, rather closely appressed to scutum; female foretarsal rake well developed; female arolia nearly equal in size; posterolateral oblique scutal carina present; prescutellar sulcus foveate; omaulus usually present as a short line of pits, sternaulus as a very short and isolated indication or absent; episternal sulcus running ventrad to omaulus, scrobal sulcus ending on episternal sulcus at a right angle; no acetabular carina; all mesopleural grooves usually obscured by silvery pubescence; metapleuron slender but not much narrowed below; forewing not spotted, media diverging before cu-a, stigma moderate, veinlet of submarginal cell II between recurrents hardly one-fourth posterior length of submarginal cell I; jugal lobe considerably larger than tegula, hindwing media diverging more than 1.0 midocellus diameter beyond cu-a; midtibia with two apical spurs; propodeal enclosure nearly smooth except for "stitching" associated with median groove; spiracular groove present, sometimes incomplete; gaster pedunculate, tergum I about twice as long as broad; male with seven terga normally exposed, sterna without fimbriae, sternum VIII distally narrowed and sword shaped (fig. 163 C); female pygidial plate broadly triangular, sparsely punctate, nearly bare.

Geographic range: Southwestern United States in desert areas. Three species are known.

Systematics: The pedunculate condition immediately

separates wasps of this genus from *Dienoplus* and *Trichogorytes* which are closely related. A remnant of the omaulus may be represented by a pitted line (*albitomentosus* and *pulvis*) or a seam (*timberlakei*). In *teren* no omaulus is apparent. In any case the scrobal sulcus is not extended to the omaulus. Dense white patches or bands of appressed hair are characteristic. On the terga the hair covers yellow integumental bands. *H. teren* is an unusual, almost all red, species.

Biology: The three species of *Hapalomellinus* are inconspicuous wasps that are seen uncommonly, yet may occur in considerable numbers in the desert during certain years. They are attracted to flowers of *Euphorbia*, *Croton*, *Chilopsis*, *Salsola*, and *Pectis*. Two rarely met species, *teren* and *pulvis*, can be collected near Blythe, California in October at flowers of *Pectis papposa* during years when summer floods come down from the mountains and bring out the flowers.

Information on nesting is drawn exclusively from Cazier and Mortenson (1965a), who observed *H. albitomentosus* in Cochise County, Arizona. The burrows were made in bare sandy soil and ended in one or two cells provisioned with adults and some nymphs of the cicadellid, *Stragania robusta* (Uhler). The prey are carried by the midlegs, and 14 to 15 are placed in a cell with the egg attached to the last one. Parasites are miltogrammine sarcophagids of the genus *Senotainia*, which follow the *Hapalomellinus* closely and even land on the prey while the wasp is in flight. Digging behavior and orientation were discussed at length by Cazier and Mortenson. The wasp exits from the burrow head first or in reverse while pushing out dirt. The tumulus is spread evenly from time to time. When leaving the burrow in search of provisions or when sleeping, the wasp plugs the entrance.

Checklist of *Hapalomellinus*

albitomentosus (Bradley), 1920 (*Gorytes*); sw. U.S.
   eximius Provancher, 1888 (*Gorytes*), nec F. Smith, 1862
pulvis R. Bohart, 1971; U.S.: New Mexico to s. California
teren Pate, 1946; U.S.: Arizona, s. California

Genus Trichogorytes Rohwer

Generic diagnosis: Small wasps, densely clothed with appressed pubescence; frons broad, nearly flat, slightly narrowed below (fig. 159 I); clypeus short and broad, all yellow in males; labrum inconspicuous; last four male flagellomeres specially modified, flattened or concave beneath; mandible with an inner subtooth; pronotal collar thin, nearly even all across, rather closely appressed to scutum; female foretarsal rake well developed; female arolia nearly equal in size; posterolateral oblique scutal carina present; prescutellar sulcus foveate; with or without an omaulus; episternal sulcus (as seen in denuded specimens) descending vertically to middle of mesopleuron, scrobal sulcus joining it at a right angle; no acetabular carina nor sternaulus; metapleuron gradually narrowed to a point below; forewing not spotted, media diverging before cu-a, stigma moderate, marginal cell short and

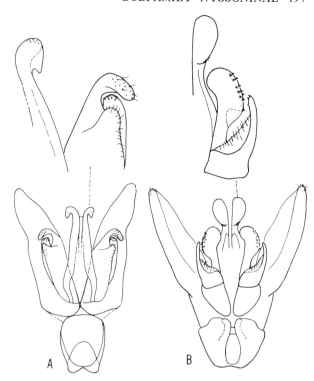

FIG. 164. Male genitalia in the tribe Gorytini; A, *Xerogorytes anaetis*; B, *Trichogorytes argenteopilosus*.

blunt (fig. 157 B), veinlet of submarginal cell II between recurrents hardly one-fourth posterior length of submarginal cell I (fig. 157 B); jugal lobe considerably larger than tegula, hindwing media diverging more than 1.0 midocellus diameter beyond cu-a; midtibia with two apical spurs; propodeal enclosure almost wholly dorsal; spiracular groove present but not normally visible; gaster not pedunculate; male with seven terga normally exposed, sterna without fimbriae, sternum VIII narrowed distally but bladelike and blunt (fig. 163 D); female pygidial plate broadly triangular, nearly bare or with appressed silvery pubescence; typical male genitalia: fig. 164 B.

Geographic range: Western North America from western Texas to New Mexico, Arizona, Utah, Nevada, southern California, and Sonora, Mexico. Two species known.

Systematics: Perhaps two dozen specimens are known among the species of this little-known genus. The most obvious morphological feature is the dense mat of silvery or greyish-silvery pubescence over much of the body. This completely hides the sculpture of the pleuron and partially obscures many other areas. The genus is closely related to *Dienoplus* but is somewhat higher on the evolutionary scale (fig. 155). The species differ from each other in details of pubescence, markings, shape of the frons and propodeal enclosure, and punctation of the pygidial plate. In *cockerelli* there are traces of an omaulus, but apparently there is none in *argenteopilosus*.

Biology: Unknown.

Checklist of *Trichogorytes*

argenteopilosus Rohwer, 1912; sw. U.S.
cockerelli (Ashmead), 1899 (*Harpactus*); sw. U.S.

Genus Austrogorytes R. Bohart

Generic diagnosis: Male antennal flagellum with tyli beneath most articles, terminal one usually incurved beneath; eyes converging below, often strongly, frons narrowest just below antennal sockets (fig. 157 D); labrum protruding slightly beneath clypeus, mandible with two subteeth on inner margin; female foretarsal rake present, two or three rake setae on basitarsus before apex; female with arolium of foreleg larger than those of other legs; pronotal collar rather thin, rounded, nearly even all across, not adhering closely to scutum; scutum distinctly and often closely punctate, lateral margin of scutum with an oblique carina opposite posterior edge of tegula; prescutellar sulcus foveate; episternal sulcus distinct in upper part (weakly in *spinicornis*), lower part continued forward almost horizontally to omaulus but forming an obtuse angle with scrobal sulcus; acetabular carina present (in *bellicosus*), weak or well developed; sternaulus absent or present; metapleuron broad above, abruptly narrowed in lower one-half; spiracular groove of propodeum present; propodeal enclosure essentially dorsal; metanotum longitudinally ridged; forewing media arising before cu-a, stigma small, veinlet of submarginal cell II between recurrent veins much longer than opposite anterior parallel veinlet and about one-third posterior length of submarginal cell I (fig. 157 C); jugal lobe smaller than outline of tegula, hindwing media diverging beyond cu-a, usually well beyond (fig. 157 C); female usually with a fairly well developed pygidial plate; male sterna IV to VI usually fimbriate; male sternum VII normally visible; male sternum VIII blade shaped to spatulate (fig. 163 A).

Geographic range: Australia. Fourteen species are known.

Systematics: The genus appears most closely related to *Gorytes* from which it differs especially by the long midposterior veinlet of submarginal cell II of the forewing, by the usually distal position of the media in the hindwing, and by the jugal lobe which is no larger than the tegula (fig. 157 C).

According to Evans and Matthews (1971b), the larva of *bellicosus* is somewhat *Gorytes*-like, but it has a three-toothed mandible. They characterized it as "a very generalized nyssonine larva." This fits our concept of the phylogenetic position based on adult structure.

Original types of all but three of the described species were examined by R. M. Bohart at the British Museum in 1960 and 1971. Two others, *A. spinicornis* and *aurantiacus*, were kindly lent by the South Australian Museum. Also, determined specimens of *ciliatus* have been seen.

Austrogorytes species have much in common, yet the genus shows a great deal of diversity. In the male there are two well developed midtibial spurs in *aurantiacus, chrysozonus, consuetipes, frenchii,* and *obesus*. The other 10 species have only one such spur. Females of some species, such as *browni*, have three comb bladelets on the basal article of the foretarsus before the apex; others, such as *bellicosus*, may have two or exceptionally three. The male foretarsus often has the terminal article swollen and articles II to IV variously modified. In *browni* this tendency culminates in long lobes on II to IV. Many *Austrogorytes* have the plantulae unusually large, especially on the foretarsus. Most species have the metanotum and propodeal enclosure coarsely striate and the remainder of the propodeum coarsely punctate or areolate. However, in *browni* the metanotum has large, separated punctures, and the propodeal enclosure is smooth except for the pitted median groove and lateral margin. The hindwing media commonly diverges well beyond cu-a. However, in *ciliatus* and *browni* it is somewhat closer, and in *aurantiacus* the media is nearly interstitial with cu-a. The sternaulus may be absent, as in *aurantiacus* and *browni;* indicated, as in *bellicosus;* or well marked, as in *tarsatus* and *spinicornis.* In most species tergum I is fairly broad, but in *consuetipes* and *spinicornis* it is definitely narrowed. Male sterna IV to VI generally have obvious fimbriae, but *obesus* and *spryi* are exceptions. The much reduced jugal lobe is characteristic of the genus. This condition in Gorytini is found elsewhere only in the unrelated genus, *Pterygorytes*. R. E. Turner (1915a) gave the most complete treatment of the species and offered a partial key. Species characters were discussed by R. M. Bohart (1967).

Biology: Evans and Matthews (1971b) described the nesting behavior of *bellicosus.* This species makes multicellular nests in coarse sandy soil and uses adults and nymphal bugs of the family Eurymelidae as provisions. During construction of the burrow, which measures 15 to 30 cm in length, a mound of tumulus accumulates at one side of the entrance. Soil is pushed out with aid of the hind legs. Depending upon their size, 5 to 27 bugs are provisioned per cell. The egg is positioned longitudinally on the side of the last prey. A chloropid parasite, *Lasiopleura* sp., was reared from the contents of one cell.

Checklist of *Austrogorytes*

aurantiacus (Turner), 1915 (*Arpactus*); W. Australia
bellicosus (F. Smith), 1862 (*Gorytes*); S. Australia
    dizonus Handlirsch, 1895 (*Gorytes*)
browni (Turner), 1936 (*Arpactus*); W. Australia
chrysozonus (Turner), 1915 (*Arpactus*); Australia: Queensland
ciliatus (Handlirsch), 1895 (*Gorytes*); S. Australia
consuetipes (Turner), 1915 (*Arpactus*); Australia: N. S. Wales
cygnorum (Turner), 1908 (*Gorytes*); W. Australia
frenchii (Turner), 1908 (*Gorytes*); se. Australia
obesus (Turner), 1915 (*Arpactus*); W. Australia
perkinsi (Turner), 1912 (*Gorytes*); Australia: Queensland
pretiosus (Turner), 1915 (*Arpactus*); W. Australia
spinicornis (Turner), 1915 (*Arpactus*); W. Australia
spryi (Turner), 1915 (*Arpactus*); Australia: Victoria
tarsatus (F. Smith), 1856 (*Gorytes*); S. Australia

GORYTINI

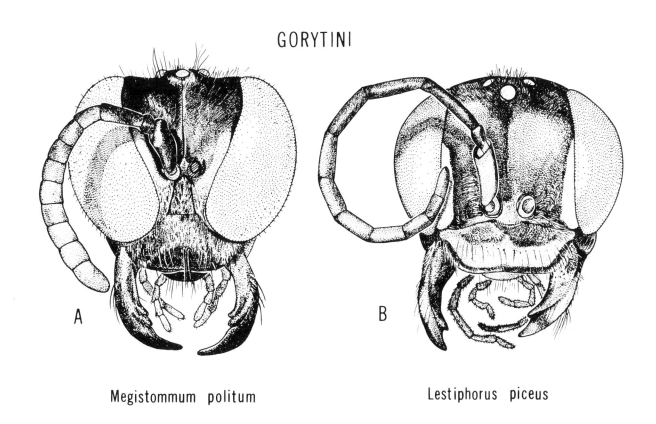

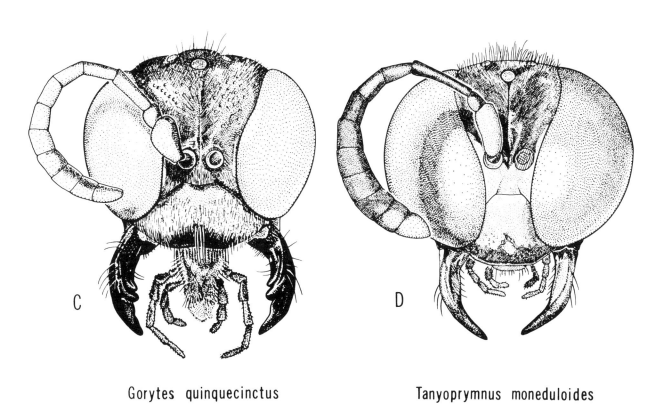

FIG. 165. Facial portraits of females in the tribe Gorytini.

eximius F. Smith, 1862 (*Gorytes*), new synonymy by R. Bohart

Genus Gorytes Latreille

Generic diagnosis: Medium-sized to fairly large wasps; inner eye margins converging below, often strongly (fig. 165 C); mandible usually with two inner subteeth; labrum inconspicuous; male flagellomeres often with tyli but last four articles not specially modified; pronotal collar rather thin, rounded, nearly even all across, not closely adherent to scutum; female foretarsal rake well developed, fore basitarsus with two long setae before apex; female arolia usually nearly equal in size; posterolateral oblique scutal carina present (fig. 156 A); prescutellar sulcus foveate; episternal sulcus when well impressed ending on scrobal sulcus which is continued forward in nearly a straight line to omaulus (fig. 158); acetabular carina incomplete or obscure; sternaulus complete or essentially so; forewing costal margin gently or sometimes strongly bowed out near base, media diverging before cu-a, stigma moderate, veinlet of submarginal cell II between recurrents about one-fourth posterior length of submarginal cell I (fig. 157 D); jugal lobe considerably larger than tegula (fig. 157 D), hindwing media generally diverging before cu-a but in some species at or a little beyond it; midtibia with two apical spurs; propodeal enclosure either extensively sculptured or longitudinally ridged, ridges sometimes confined to a frontal strip; spiracular groove present, usually distinct; gaster tapering toward base, sometimes nearly pedunculate; male with seven terga normally exposed, sterna without fimbriae, sternum VIII narrowly blade shaped, often minutely notched at apex (fig. 166 D); female pygidial plate variable, flat, usually long; typical male genitalia: fig. 169 B.

Geographic range: Holarctic and Ethiopian Regions. A majority of the 55 listed species occur in the Palearctic Region.

Systematics: In three closely related genera, *Gorytes*, *Pseudoplisus*, and *Megistommum*, the scrobal sulcus appears to be continued forward to the omaulus in nearly a straight line, and females have two long setae before the apex of the fore basitarsus. The possession of an oblique scutal carina places them well above such genera as *Exeirus*, *Argogorytes*, or *Clitemnestra* on the evolutionary scale, but the relatively simple male antenna puts them below *Neoplisus*, *Eogorytes*, and *Hoplisoides*.

The generic name *Gorytes* has contained many diverse elements in the past which have been gradually separated out. As now treated, it contains the species originally placed under "*Euspongus*" or "*Hoplisus*" and their relatives. There are considerable variations in structure as would be expected in a large genus. Many species have the propodeum very rough. Examples are *quinquecinctus*, *quadrifasciatus*, *laticinctus*, and *canaliculatus*. Most *Gorytes* have the hindwing media diverging well before cu-a, but a few, such as *atrifrons*, have it interstitial or slightly beyond. Generally, the convergence of the eyes toward the clypeus is rather pronounced, although less extreme than in the related genus *Pseudoplisus*. Some species, such as *laticinctus*, *quadrifasciatus*, *atrifrons*, and *canaliculatus*, have the least interocular distance of the female much greater than one eye breadth; others, such as *quinquecinctus*, *provancheri*, and *simillimus*, have the eyes separated by about one eye breadth. The European species have been considered by Beaumont (1953b).

The nearest relative of *Gorytes* is *Pseudoplisus*, which has been most frequently treated as a species group or as a subgenus. Since over 30 species are assignable to *Pseudoplisus*, it is convenient to treat it as a genus while admitting the similarity. As far as we have been able to determine, the ridged or sculptured propodeal enclosure of *Gorytes* is a constant separational character.

Biology: A superficial perusal of the literature gives the impression that a great deal of biological information has been published on this genus. However, most of the wasps quoted are now considered to be in other genera. The nesting behavior has been summarized by Evans (1966a), who recounted rather complete to partial information on *canaliculatus*, *simillimus*, *atricornis*, and *deceptor* from North America, as well as *pleuripunctatus*, *planifrons*, *laticinctus*, and *sulcifrons* from Europe. Recorded prey are in the families Cidadellidae, Fulgoridae, Cercopidae, and Membracidae. In general a species of *Gorytes* selects its prey from a single family of Homoptera and often a single species or genus. Prey of *canaliculatus* has most frequently been recorded as the leafhopper, *Macropis viridis* (Fitch) (Krombein, 1964a; R. Bohart and Holland, 1965; Evans, 1966a). Leafhoppers of the genus *Idiocerus* were given by both Krombein (1963a) and Evans (1966a). On the other hand, leafhopper prey in the genera *Norvellina*, *Oncopsis*, *Orientis*, *Paraphlepsius* and *Stragania* along with the fulgorid, *Haplaxius*, were recorded by Evans (1966a). Evans questioned the inclusion of the membracids, *Cyrtolobus fenestratus* Fitch and *Atymna inornata* Say as reported by Barth (1907a), but they certainly cannot be ruled out. Prey records gleaned from the summary by Evans (1966a) are as follows:

G. simillimus:	*Gyponana flavolineata* (Fitch), *Gyponana octolineata* (Say), *Scaphoideus productus* Osborne (all Cicadellidae)
G. atricornis:	*Aphrophora parallela* (Say) (Cercopidae) *Cyrtolobus tuberosus* (Fairmaire) (Membracidae)
G. deceptor:	*Spissistylus constans* (Walker) (Membracidae)
G. laticinctus:	*Philaenus spumarius* Linnaeus (Cercopidae)
G. planifrons:	*Issus coleoptratus* Fabricius (Fulgoridae)
G. pleuripunctatus:	*Dictyophora* sp. (Fulgoridae)
G. sulcifrons:	*Philaenus spumarius* Linnaeus (Cercopidae)

It is probably fair to say of *Gorytes*, as of many other

sphecids, that prey selection is a mixture of inherited preference (in this case for Homoptera and perhaps leafhoppers) and opportunism, when the favored prey is rare or when some other homopteran within the "hopper" category is abundant.

Most detailed reports of nesting activities are those of Barth (1907a), Krombein (1964a), and Evans (1966a) on *canaliculatus;* Maneval (1939) on *planifrons;* and Krombein (1936, 1952) on *simillimus.* Nests may be far apart or rather close together. The burrows are mostly about 12 cm deep and may contain as many as four cells. The number of provisions varies according to the size of both host and prey, but as few as four and as many as 19 have been recorded per completed cell. The nest entrance is closed while the wasp is away, and provisioning may extend over several days for a single burrow. The prey is carried in flight, mostly by the middle legs. The egg is laid on the last prey in the cell and is attached longitudinally to the venter of the thorax at one side. Eclosion from the egg to larval maturity takes about four days.

Parasites are miltogrammine sarcophagids; *Nysson daecki, N. lateralis,* and *N. fidelis;* and the mutillid, *Timulla leona* Blake; all listed for *G. canaliculatus* by Evans (1966a). Hamm and Richards (1930) gave *Nysson interruptus* and *N. trimaculatus* as parasites of *G. quadrifasciatus* in England.

Checklist of *Gorytes*

africanus Mercet, 1905; Morocco
aino Tsuneki, 1963; Japan
albidulus (Lepeletier), 1832 (*Hoplisus*); Europe, Middle East, Mongolia
    dissectus (Panzer), 1801 (*Mellinus*); nec Panzer, 1800
    albilabris Lepeletier, 1832 (*Euspongus*)
    elegans F. Smith, 1856
albosignatus W. Fox, 1892; w. N. America
ambiguus Handlirsch, 1888; centr. Asia
angustus (Provancher), 1895 (*Hoplisus*); w. N. America
atricornis Packard, 1867; N. America
    rugosus Packard, 1867
    decorus W. Fox, 1895, new synonymy by R. Bohart
    elegantulus H. Smith, 1908 (*Hoplisus*)
atrifrons W. Fox, 1892; w. U.S.
canaliculatus Packard, 1867; U.S.: transcontinental
    geminus Handlirsch, 1888, new synonymy by R. Bohart
    asperatus W. Fox, 1895, new synonymy by R. Bohart
    corrugis Mickel, 1918 (*Hoplisus*)
cochisensis R. Bohart, 1971; U.S.: Arizona
cribratus F. Morawitz, 1892; sw. USSR
deceptor Krombein, 1958; e. U.S.
dorothyae Krombein, 1950; e. U.S.
    ssp. *russeolus* Krombein, 1954; se. U.S.
eous Gussakovskij, 1933; e. Asia
fallax Handlirsch, 1888; Eurasia
flagellatus R. Bohart, 1971; w. U.S.
flaviventris F. Morawitz, 1894; sw. USSR
foveolatus Handlirsch, 1888; Europe and n. Africa, Middle East
    longicornis Handlirsch, 1898, nec Rossi, 1790
    usurpator Schulz, 1906
    dichrous Mercet, 1906
    rubrocinctulus Strand, 1910
hakutozanus Tsuneki, 1963; Korea
harbinensis Tsuneki, 1967; Manchuria
hebraeus Beaumont, 1953; Europe, Middle East
heptapotamiensis Kazenas, 1972; USSR: Kazakh S.S.R.
intrudens Nurse, 1903; India
jonesi (Turner), 1920 (*Arpactus*); s. Africa
koreanus Handlirsch, 1888; e. Asia
kulingensis (Yasumatsu), 1943 (*Hoplisus*); China
laticinctus (Lepeletier), 1832 (*Euspongus*); Europe, Middle East
limbellus R. Bohart, 1971; U.S.: California
maculicornis (F. Morawitz), 1889 (*Hoplisus*); China
mcateei Krombein and R. Bohart, 1962; e. U.S.
melpomene (Arnold), 1936 (*Arpactus*); s. Africa
mongolicus Tsuneki, 1970; Mongolia
neglectus Handlirsch, 1895; n. Europe, Siberia
nevadensis W. Fox, 1892; w. U.S.
nigrifacies (Mocsáry), 1879 (*Hoplisus*); e. Europe, Middle East
pieli Yasumatsu, 1943; Mongolia
planifrons (Wesmael), 1852 (*Hoplisus*); Europe
pleuripunctatus (A. Costa), 1859 (*Hoplisus*); e. Europe, Middle East
    tirolensis Kohl, 1880 (*Hoplisus*)
    fraternus Mercet, 1906
    ssp. *barbarus* Beaumont, 1953; Morocco
procrustes Handlirsch, 1888; s. Europe
prosopis R. Bohart, 1971; U.S.: California, Oregon
provancheri Handlirsch, 1895; w. N. America, new status by R. Bohart
    laticinctus Provancher, 1888, nec Lepeletier, 1832
proximus Handlirsch, 1893; e. Europe
quadrifasciatus (Fabricius), 1804 (*Mellinus*); Eurasia
    vicinus Lepeletier, 1832 (*Euspongus*)
    montivagus Mocsáry, 1878 (*Hoplisus*)
quinquecinctus (Fabricius), 1793 (*Mellinus*); Europe, Middle East
    ?*calceatus* Rossi, 1794 (*Crabro*)
    ?*arenarius* Panzer, 1798 (*Mellinus*)
    cinctus Latreille, 1805
    ?*ruficornis* Latreille, 1805
    sinuatus A. Costa, 1869 (*Hoplisus*)
quinquefasciatus (Panzer), 1798 (*Mellinus*); Eurasia
    ?*lacordairei* Lepeletier, 1845 (*Hoplisus*)
    eburneus Chevrier, 1870 (*Hoplisus*)
    geminatus A. Costa, 1869 (*Hoplisus*)
    intercedens Handlirsch, 1893;
    anceps Mocsáry, 1879 (*Hoplisus*)
    mauretanicus Handlirsch, 1898
    ssp. *levantinus* Pulawski, 1961; e. Mediterranean
radoszkowskyi Handlirsch, 1888; Korea
schlettereri Handlirsch, 1893; Europe
    ssp. *ponticus* Beaumont, 1967; Turkey
schmiedeknechtii Handlirsch, 1888; se. Europe, Middle East
simillimus F. Smith, 1856; N. America
    ephippiatus Packard, 1867
    gyponacinus Rohwer, 1911

sogdianus Gussakovskij, 1952; sw. USSR
sulcifrons (A. Costa), 1869 (*Hoplisus*); e. and s. Europe, sw. USSR
  laevigatus Kohl, 1880 (*Hoplisus*)
takeuchii Tsuneki, 1963; Korea
tricinctus (Pérez), 1905 (*Hoplisus*); Japan, Korea
umatillae R. Bohart, 1971; w. U.S.
  vicinus Handlirsch, 1893, nec Lepeletier, 1832
verhoeffi Tsuneki, 1967; Manchuria

Fossil *Gorytes*[1]

sepultus (Cockerell), 1906 (*Hoplisus*); Miocene shale, U.S.: Florissant, Colorado

Genus Pseudoplisus Ashmead

Generic diagnosis: Medium-sized to fairly large wasps; inner eye margins converging below, often strongly (fig. 159 E); female mandible with two inner subteeth; labrum inconspicuous; last four male flagellomeres not specially modified; pronotal collar rather thin, rounded, even all across, not closely adherent to scutum; female foretarsal rake well developed, fore basitarsus with two long setae before apex; female arolia nearly equal in size (rare exceptions in African species); posterolateral oblique scutal carina present; prescutellar sulcus foveate or efoveate; episternal sulcus when present ending on scrobal sulcus which is continued forward in a nearly straight line to omaulus; acetabular carina incomplete or obscure; sternaulus complete or in Ethiopian species usually incomplete; forewing costal margin usually bulged out abruptly toward base, media diverging before cu-a, stigma moderate, veinlet of submarginal cell II between recurrents about one-fourth posterior length of submarginal cell I; jugal lobe considerably larger than tegula, hindwing media diverging before cu-a; midtibia with two apical spurs; propodeal enclosure smooth except sometimes for pitted marginal lines and oblique rugulae at anterior corners; spiracular groove distinct, least so in Ethiopian species, gaster tapering toward base, tergum I usually somewhat constricted but not at all nodose; male with seven terga normally exposed; sterna without obvious fimbriae, sternum VIII narrowly bladelike distally (fig. 166 C); female pygidial plate variable, flat, usually long.

Geographic range: Most of the 33 species are confined to North America, but one is European, and four have been described from central and southern Africa.

Systematics: There is a close relationship with *Gorytes* as discussed under that genus. In addition to the difference in the propodeal enclosure, *Pseudoplisus* tends to have the entire body more smooth, the eyes more strongly converging, and the gaster somewhat pedunculate. The largely Ethiopian group containing *natalensis, kohlii, effugiens, nyasicus* and *ranosahae* have the gaster subpedunculate. Also, the metanotum is flattened and on the same plane as the propodeal enclosure. Arnold (1929) included *Gorytes jonesi* in this group, which he called "Section I." However, *jonesi* has the propodeal enclosure longitudinally rugose, so we have placed it in *Gorytes* rather than *Pseudoplisus*. Zavadil in Zavadil and Snoflák (1948) gave the group subgeneric status under the name *Laevigorytes* with *G. kohlii* as the type. In our opinion *Laevigorytes* is a synonym of *Pseudoplisus*. The subpedunculate characteristic is present in many species, for instance the American *tanythrix* and *nigricomus*. Also, *tritospilus* has the flattened notal characteristic.

The species are subject to considerable color variation; particularly if the range is broad. The commonest American species is *phaleratus,* which is mostly dark in New England, black with red and yellow in southern United States, and mostly red to yellow in the southwest U.S. and northern Mexico. It belongs to a group in which the prescutellar sulcus is efoveate. The most extensive papers on the genus are those of R. Bohart (1968e, 1969).

Biology: Very little is known of the habits. Arnold (1929) reported the cercopid, "*Ptyelus grossus*" as prey for *natalensis*.

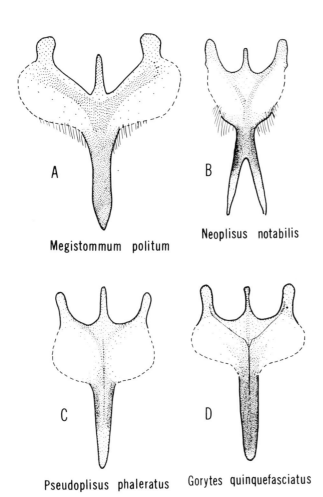

FIG. 166. Male sternum VIII in the tribe Gorytini.

[1] We have accepted the opinion of Evans (1966a) that this fossil represents *Gorytes* or a related genus.

Checklist of *Pseudoplisus*

abdominalis (Cresson), 1865 (*Gorytes*); w. U.S.
aequalis (Handlirsch), 1888 (*Gorytes*); Mexico
    handlirschi Cameron, 1890 (*Gorytes*), nec Morawitz, 1890, new synonymy by R. Bohart
        cameronis Handlirsch, 1895 (*Gorytes*), new name for *handlirschi* Cameron
apicalis (F. Smith), 1856 (*Gorytes*), U.S.
    propinquus Cresson, 1868 (*Gorytes*), new synonymy by R. Bohart
butleri R. Bohart, 1969; U.S.: Arizona
californicus R. Bohart, 1969; U.S.: California; Mexico: Baja California
catalinae R. Bohart, 1969; U.S.: Ariz.
claripennis R. Bohart, 1969; U.S.: Arizona
divisus (F. Smith), 1856 (*Gorytes*); U.S. e. of 100th mer.
    bipartitus Handlirsch, 1888 (*Gorytes*), new synonymy by R. Bohart
    varipunctus H. Smith, 1908, new synonymy by R. Bohart
effugiens (Brauns), 1911 (*Gorytes*); s. Africa
    fugax Turner, 1915 (*Arpactus*)
erugatus R. Bohart, 1969; Mexico: Baja California
fasciatus (W. Fox), 1895 (*Gorytes*); w. U.S.
flavidulus R. Bohart, 1969; U.S.: Arizona
guadalajarae R. Bohart, 1969; Mexico
hadrus R. Bohart, 1968; Mexico
imperialis R. Bohart, 1969; U.S.: California
kohlii (Handlirsch), 1888 (*Gorytes*); Europe
montanus (Cameron), 1890 (*Gorytes*); Mexico
natalensis (F. Smith), 1856 (*Gorytes*); s. Africa
    africanus Radoszkowski, 1881 (*Lestiphorus*)
nigricomus R. Bohart, 1969; U.S.: Arizona
notipilis R. Bohart, 1969; Mexico
nyasicus (Turner), 1915 (*Arpactus*); s. Africa
ocellatus R. Bohart, 1969; U.S.: California; Mexico: Baja California
oraclensis R. Bohart, 1969; U.S.: Arizona
phaleratus (Say), 1837 (*Gorytes*); N. America
    flavicornis Harris, 1835 (*Odynerus*), nomen nudum
    fulvipennis F. Smith, 1856 (*Gorytes*)
    modestus Cresson, 1865 (*Gorytes*)
    flavicornis Packard, 1867 (*Gorytes*)
    rufoluteus Packard, 1867 (*Gorytes*)
    alpestris Cameron, 1890 (*Gorytes*), new synonymy by R. Bohart
    alticola Cameron, 1890 (*Gorytes*), new synonymy by R. Bohart
    papagorum Viereck, 1908 (*Gorytes*), new synonymy by R. Bohart
    subaustralis Viereck, 1908 (*Gorytes*), new synonymy by R. Bohart
ranosahae (Arnold), 1945 (*Gorytes*); Madagascar
rubiginosus (Handlirsch), 1888 (*Gorytes*); Mexico
rufomaculatus (W. Fox), 1895 (*Gorytes*); centr. and w. U.S.
samiatus R. Bohart, 1969; Mexico
smithii (Cresson), 1880 (*Gorytes*); centr. U.S.
    infumatus Mickel, 1916
    ssp. *floridanus* (W. Fox), 1891 (*Gorytes*); U.S.: Florida, Oklahoma
    foveolatus W. Fox, 1890 (*Hoplisus*), nec Handlirsch, 1888
tanythrix R. Bohart, 1969; U.S.: Texas, Oklahoma
tritospilus R. Bohart, 1969; Mexico, El Salvador
venustus (Cresson), 1865 (*Gorytes*); centr. U.S., Mexico
    venustiformis Rohwer, 1911 (*Gorytes*), new synonymy by R. Bohart
werneri R. Bohart, 1969; U.S.: Arizona

Genus Megistommum Schulz

Generic diagnosis: Medium-sized wasps; inner eye margins converging strongly toward clypeus (fig. 165 A); mandible of female with two inner subteeth; labrum inconspicuous; male flagellum sometimes with tyli but last four articles not specially modified; pronotal collar rather thin, rounded, even all across, not closely adherent to scutum, female foretarsal rake well developed, fore basitarsus with two long setae before apex; female arolia nearly equal in size; posterolateral oblique scutal carina present; prescutellar sulcus nearly always efoveate; scrobal sulcus continued forward in nearly a straight line to omaulus (in some species it fades anteriorly into a shallow depression), episternal sulcus curving backward toward scrobe; acetabular carina incomplete or obscure; sternaulus complete; forewing costal margin bulging strongly toward base, media diverging before cu-a, stigma moderate, veinlet of submarginal cell II between recurrents about one-fourth posterior length of submarginal cell I; jugal lobe considerably larger than tegula, hindwing media diverging before cu-a; midtibia with two apical spurs; propodeal enclosure smooth except rarely for faint pitting in median groove; spiracular groove distinct, propodeum generally smooth; gaster pedunculate, tergum I somewhat nodose (fig. 167); male with seven terga normally exposed, sterna III to V with erect apical fimbriae, sternum VIII protruding as a narrow blade (fig. 166 A); female pygidial plate smooth or nearly so.

Geographic range: All of the eight named species are Central and South American.

Systematics: The definitely pedunculate abdomen and the fimbriate male sterna distinguish *Megistommum* from its close relatives, *Gorytes* and *Pseudoplisus*. *Megistommum* appears to be the most advanced of the three. In *M. politum* the eyes have short, erect, scattered hair. The most distinctive species is *splendidum* (fig. 167), which is liberally adorned with long, straggly, golden hair and has the prescutellar sulcus foveate. Also in this species, the swelling of the base of the forewing is incised distally to produce a definitive pubescent lobe.

Biology: Unknown.

Checklist of *Megistommum*

megalommiforme (Strand), 1910 (*Gorytes*); Paraguay
melanogaster (Schrottky), 1911 (*Megalomma*); Peru
mimetes (Handlirsch), 1901 (*Gorytes*); Brazil
nigriceps (F. Smith), 1873 (*Megalomma*); Brazil
politum (F. Smith), 1873 (*Megalomma*); Brazil
    petiolatum Taschenberg, 1875 (*Hoplisus*)
procerulides (Strand), 1910 (*Gorytes*); Paraguay

504 SPHECID WASPS

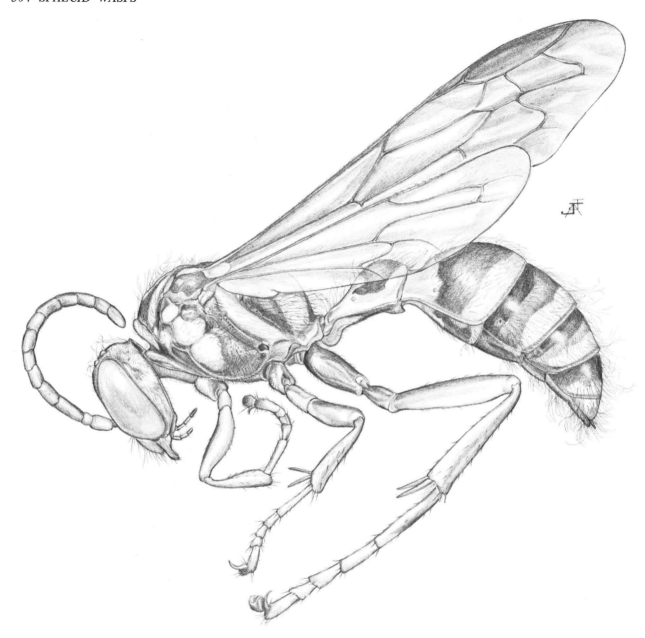

FIG. 167. *Megistommum splendidum* (Handlirsch), male.

?*melanoxanthum* Schrottky, 1911 (*Megalomma*)
procerus (Handlirsch), 1888 (*Gorytes*); Brazil
 elegans F. Smith, 1873 (*Megalomma*), nec Lepeletier, 1832
splendidum (Handlirsch), 1888 (*Gorytes*); Mexico, Centr. America
 centrale Cameron, 1890 (*Gorytes*)

Genus Neoplisus R. Bohart

Generic diagnosis: Medium-sized wasps; inner eye margins converging strongly toward clypeus; mandible usually with two inner subteeth; labrum inconspicuous; last four or five male flagellomeres with shiny spots or tyli on under surface; pronotal collar rather thin, rounded, even all across, not closely adherent to scutum; female foretarsal rake well developed, fore basitarsus with three or four long setae before apex; female foreleg arolium slightly larger than arolia of other legs; posterolateral oblique scutal carina present; prescutellar sulcus foveate; episternal sulcus angled at juncture with scrobal sulcus and continued forward and obliquely downward to or nearly to omaulus; acetabular carina incomplete or obscure; sternaulus complete; forewing costal margin not strongly bulging toward base, media diverging before cu-a, stigma moderate, veinlet of submarginal cell II between recurrents about one-fourth posterior length of submarginal cell I; jugal lobe considerably larger than tegula, hindwing media diverging before cu-a; midtibia with two apical spurs; propodeal enclosure smooth except for anterior pitted margin; spiracular groove essentially absent; gaster tapering toward base, tergum I a little constricted; male with seven terga normally exposed; sterna without fimbriae, sternum VIII deeply bifurcate, usually visible as two projecting spines (fig. 166 B); female pygidial plate striatopunctate.

Geographic range: Mexico to Argentina. Nine species are known.

Systematics: Neoplisus is related to *Gorytes* and *Pseudoplisus*, neither of which occurs in South America. Males are easily distinguished by the form of sternum VIII (fig. 166 B) which usually appears as two protruding spines. Females have more than two long setae on the fore basitarsus before the apex. Also, both sexes have only a faint indication of the spiracular groove. Males of known species have small platelets or depressions beneath the last four or five flagellomeres. Submarginal cell III forms a more nearly regular trapezoid than in related genera. The nine described species differ widely in markings but little in structure. Hence, a thorough study may reveal considerable synonymy.

Biology: Unknown.

Checklist of *Neoplisus*

balteatus (Cameron), 1890 (*Gorytes*); Guatemala
bruchi (Schrottky), 1909 (*Hoplisus*); Argentina, Brazil
    cearensis Ducke, 1910 (*Gorytes*), new synonymy by R. Bohart
facilis (F. Smith), 1873 (*Gorytes*); Brazil
foxii (Handlirsch), 1901 (*Gorytes*); Brazil
fumipennis (F. Smith), 1856 (*Gorytes*); Brazil
imitator (Handlirsch), 1901 (*Gorytes*); Brazil
notabilis (Handlirsch), 1888 (*Gorytes*); Brazil to Mexico
    fuscipennis Cameron, 1890 (*Gorytes*)
schrottkyi (Fritz), 1964 (*Gorytes*); Argentina
specialis (F. Smith), 1873 (*Gorytes*); Brazil
    partitus W. Fox, 1897 (*Gorytes*), new synonymy by R. Bohart
    bergii Handlirsch, 1901 (*Gorytes*), new synonymy by R. Bohart

Genus Eogorytes R. Bohart, new genus

Generic diagnosis: Medium-sized to fairly large wasps; eyes converging a little below; antennae slender, last four male flagellomeres concave and polished beneath; mandible with two subapical teeth; labrum normally concealed; pronotal collar rounded, distinct from scutum; foretarsal rake of female moderate, basitarsus with three spines before apex; foretarsal arolium of female much larger than those on other legs; posterolateral oblique scutal carina present; prescutellar sulcus foveate; omaulus present; acetabular carina undeveloped; episternal and scrobal sulci forming a continuous arc, the episternal-scrobal sulcus; sternaulus broadly interrupted medially; two midtibial spurs; forewing slightly curved out toward base of costa, media diverging before cu-a, stigma moderate, veinlet of submarginal cell II between recurrent veins about one-fourth posterior length of submarginal cell I, second recurrent ending just before end of submarginal cell II; jugal lobe much larger than tegula, hindwing media diverging well before strongly curved cu-a (fig. 157 G); propodeal enclosure coarsely and longitudinally ridged; spiracular groove distinct; tergum I moderately slender; male with seven visible terga, sterna III-V with prominent apical fimbriae, sternum VIII distally slender and swordlike; female pygidial plate well developed, striatopunctate.

Type of genus: Gorytes fulvohirtus Tsuneki.

Geographic range: Palearctic. Japan (central Honshu), Korea, China (Kiangsi and Szechwan Provinces), and Taiwan. Two species are known.

Systematics: Through the kindness of Professor K. Tsuneki, we have studied a male of *Eogorytes fulvohirtus* from Japan. We have also seen a pair from Kiangsi Province, China as well as a female of an allied species (possibly *taiwanus*) from Szechwan Province, China. *Eogorytes* seems to be intermediate in some respects between *Gorytes* and *Oryttus*. The relationships of the scrobal and episternal sulci to the omaulus are much as in *Gorytes* but the enlarged arolium on the foreleg of the female and the modified male flagellomeres VIII to XI are as in *Oryttus*. Also, the female has more than two long setae on the fore basitarsus before the apex. Peculiar to *Eogorytes* are the broadly interrupted sternaulus and the fimbriate male sterna. In the two species seen the antennae are unusually long and slender, tergum II is unusually stout, and the body is mostly covered with long hair.

Biology: Unknown.

Checklist of *Eogorytes*

fulvohirtus (Tsuneki), 1963 (*Gorytes*); Japan, Korea, ne. China
taiwanus (Tsuneki), 1971 (*Gorytes*); Taiwan

Genus Lestiphorus Lepeletier

Generic diagnosis: Moderate-sized wasps; inner eye margins essentially parallel, least interocular distance two to three times median clypeal length (fig. 165 B); mandible with an inner subtooth; labrum inconspicuous, flagellum long and slender, article I longer than scape, last four articles in male distinctively flattened or concave beneath; pronotal collar thin and rather closely appressed to scutum; female foretarsal rake well developed, fore basitarsus with three long setae before apex; female foreleg arolium much larger than other arolia, male arolia about equal on all legs; posterolateral oblique scutal carina present; prescutellar sulcus foveate; episternal and scrobal sulci faint or absent, scrobal sulcus not continued forward to omaulus, omaulus and sternaulus present, latter somewhat discontinuous; no acetabular carina; forewing usually spotted, media diverging before cu-a, stigma moderate, veinlet of submarginal cell II between recurrents about one-fourth posterior length of submarginal cell I; jugal lobe considerably larger than tegula, hindwing media diverging at or very near cu-a which is nearly straight (fig. 157 F); midtibia with two apical

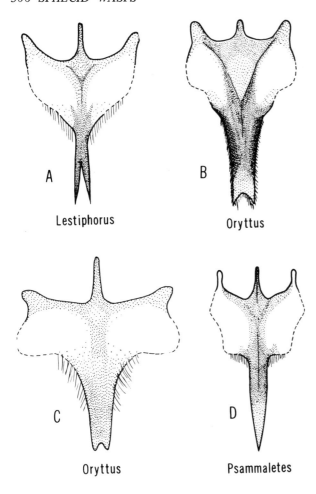

FIG. 168. Male sternum VIII in the tribe Gorytini; A, *Lestiphorus piceus;* B, *Oryttus laminiferus;* C, *Oryttus velutinus;* D, *Psammaletes crucis.*

spurs; propodeal enclosure usually longitudinally striate, unusually large and occupying more than half of horizontal surface of propodeum; spiracular groove indistinct except near spiracle; gaster pedunculate, tergum I strongly curved toward apex in profile; male with seven terga normally visible, sterna without fimbriae; sternum VIII deeply emarginate apically and bispinose (fig. 168 A); female pygidial plate variable, punctate to striate, sometimes short.

Geographic range: All of the zoogeographical regions are included except the Australasian. Of the 17 named species most occur in the Holarctic Region. Only one is known from South America and one from Ethiopian Africa; it is curious that there are no records from oceanic islands, even one as large as Australia. Most species are relatively rare in collections.

Systematics: Lestiphorus, Oryttus and *Psammaletes* form a related group of medium-sized wasps, rather slender-bodied, and often with long antennae. The faint scrobal sulcus is not continued forward anteriorly. The female has the foreleg arolium enlarged. Also, the last four flagellomeres of the male are distinctively flattened or concave beneath, and the hindwing media diverges at or very close to cu-a. These last two features occur also in *Hoplisoides* and its relatives.

Species of *Lestiphorus* are readily recognized by the usually fine sculpture, pedunculate abdomen, long antennae, generally spotted wings, and broad frons. In the only Ethiopian Region species, *mimicus,* the sculpture is unusually coarse on the scutum, mesopleuron and propodeal enclosure. Also, the first tergum is not much humped and forms only a weak, though definite, peduncle. *Lestiphorus* has a superficial similarity to *Mellinus* which differs in many important structural details such as the simple sternum I and the absence of the oblique scutal carina. As indicated in fig. 155, the genus is moderately advanced.

Biology: Bernard (1934) reported the prey of *L. bicinctus* as the spittlebug, *Philaenus spumarius* Linnaeus (Cercopidae). Benno (1966) reported *Nysson maculosus* (Gmelin) (as *trimaculatus* Rossi) was a not uncommon parasite of *L. bicinctus* in the Netherlands.

Checklist of *Lestiphorus*

becquarti (Yasumatsu), 1943 (*Gorytes*); China
bicinctus (Rossi), 1794 (*Crabro*); Europe (validated in Opinion 675, I.C.Z.N., 1963)
bilunulatus A. Costa, 1869; Europe
    semistriatus Schmiedeknecht, 1881
    ssp. *yamatonis* Tsuneki, 1963; Japan
cockerelli (Rohwer), 1909 (*Gorytes*); U.S.: Colorado, Nebraska
    williamsi Mickel, 1916 (*Mellinogastra*)
densipunctatus (Yasumatsu), 1943 (*Gorytes*); China
egregius (Handlirsch), 1893 (*Gorytes*); w. USSR
greenii (Bingham), 1896 (*Gorytes*); Sri Lanka
icariiformis (Bingham), 1908 (*Gorytes*); India
mellinoides (W. Fox), 1895 (*Gorytes*); U.S.: Texas
mimicus (Arnold), 1931 (*Arpactus*); sw. Africa
mitjaevi (Kazenas), 1972 (*Gorytes*); USSR: Kazakh S.S.R.
oreophilus (Kuznetzov-Ugamskij), 1927 (*Gorytes*); centr. Asia
pacificus (Gussakovskij), 1933 (*Gorytes*); e. Asia
peregrinus (Yasumatsu), 1943 (*Gorytes*); China
persimilis (Turner), 1926 (*Gorytes*); Thailand
piceus (Handlirsch), 1888 (*Gorytes*); w. U.S.; Canada
    rufocinctus W. Fox, 1892 (*Gorytes*)
sericatus (F. Smith), 1856 (*Gorytes*); Brazil

Genus Oryttus Spinola

Generic diagnosis: Moderate-sized wasps; inner eye margins essentially parallel or converging somewhat toward clypeus (fig. 169 K), especially in males; mandible with an inner subtooth; labrum inconspicuous; flagellum generally long and slender with first article at least three-fourths as long as scape, last four articles in male distinctively flattened or concave beneath; pronotal collar appressed to scutum or rather distinct; female foretarsal rake well developed, fore basitarsus with three long setae before apex; female foreleg arolium much larger than other arolia; male hind leg with arolium considerably

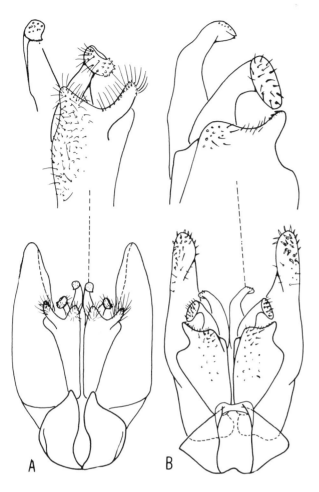

FIG. 169. Male genitalia in the tribe Gorytini; A, *Oryttus laminiferus;* B, *Gorytes quinquefasciatus.*

larger than on other legs; posterolateral oblique scutal carina present; prescutellar sulcus foveate; episternal and scrobal sulci often evanescent or obscured by coarse sculpture but the two joining at about a right angle when well impressed, forming an angular episternal-scrobal sulcus; omaulus present; sternaulus complete or nearly so; no acetabular carina; forewing spotted or extensively clouded, media diverging before cu-a, stigma moderate, veinlet of submarginal cell II between recurrents about one-fourth posterior length of submarginal cell I; jugal lobe considerably larger than tegula, hindwing media diverging at or very near cu-a which is sometimes strongly curved anterodistally (fig. 157 H); male midtibia often with only one midtibial spur; propodeum with longitudinal ridges which are coarse, sometimes irregular and usually complete; spiracular groove well defined if not obscured by coarse sculpture; gaster a little constricted at base of tergum II (most definitely in *velutinus*), tergum I gently curved toward apex in profile; male with seven terga normally visible, sterna without erect fimbriae (but see discussion of *velutinus* below), sternum VIII emarginate apically and bispinose (fig. 168 B,C); typical male genitalia: fig. 169 A; female pygidial plate punctate to striate.

Geographic range: Found widely in the Holarctic Region, two species in the Ethiopian Region, and one in Chile We have placed 13 species in this genus.

Systematics: The American forms have been treated by most previous authors as *Harpactostigma* of which the type species, *velutinus,* is Chilean. Most characters of *velutinus* are typical of *Oryttus* but the subpedunculate abdomen in the Chilean species and the subbasal patches of soft hair on sterna IV to VI of the male are distinguishing features that may warrant recognition of a monotypic subgenus.

In addition to at least some constriction of the basal gastral segment, other characteristics of *Oryttus* are the rough sculpture of the thorax on at least some sclerites, the long antennae with the last four articles specialized in the male, the hindwing venation near cu-a (fig. 157 H), the enlarged arolium on the foretarsus of the female, and the enlarged arolium on the hindtarsus of the male. The North American species (except *umbonatus*) have the forefemur expanded and concave beneath, the details of which furnish taxonomic characters. The pronotum shows an interesting and presumably evolutionary transition from rounded and not appressed to the scutum (*velutinus*), rounded and appressed (*mirandus*), to appressed and transversely grooved (*infernalis, arnoldi, laminiferus, lapazae*). According to our observations, species with a single apical midtibial spur in the male are *gracilis, infernalis, laminiferus, lapazae,* and *yumae;* with two such spurs are *concinnus, mirandus, velutinus,* and *umbonatus.* Important papers on the genus are those of Pate (1938a) and R. M. Bohart (1968d). The latter contains up-to-date keys and synonymy of American species.

Biology: Ferton (1901, 1910) listed the cicadellid, *Solenocephalus obsoletus* Germar, and the fulgorid, *Hysteropterum grylloides* (Fabricius), as prey of *O. concinnus* in western Europe. Deleurance (1945b) gave the prey of *concinnus* as the fulgorid, *Issus coleoptratus* (Geoffroy). He described the nesting site of slightly moist, heavy, hard-topped soil where burrows were excavated year after year. Provisioning took place in the afternoon. The egg was laid, as usual, laterally on the venter above the midleg. The various stages were described along with those of a parasite, *Nysson trimaculatus.* An American species, *O. laminiferus,* was observed by Gittins (1958) to nest in the vertical face of a clay bank in Idaho. The female wasp entered a small crack which terminated in a 16 cm tunnel, extending downward and curving to one side. Provisions were nymphal and adult *Scolops hesperius* Uhler (Fulgoridae).

Checklist of *Oryttus*

arnoldi (Benoit), 1951 (*Gorytes*); Zaire
concinnus (Rossi), 1790 (*Sphex*); Europe
  ssp. *paradisiacus* (Beaumont), 1967 (*Gorytes*); Turkey
gracilis (Patton), 1879 (*Hoplisus*); e. U.S.
  ssp. *arapaho* (Pate), 1938 (*Harpactostigma*); U.S.: Colorado, Texas, Kansas
    rutilus Pate, 1938 (*Harpactostigma*); U.S.

houskai (Balthasar), 1954 (*Gorytes*); Middle East
infernalis (Handlirsch), 1888 (*Gorytes*); Greece, Middle East
kaszabi Tsuneki, 1971; Mongolia
kraepelini (Brauns), 1899 (*Gorytes*); se. Africa
laminiferus (W. Fox), 1895 (*Gorytes*); w. U.S.
    ruficornis Provancher, 1888 (*Gorytes*), nec Latreille, 1805
    flavicornis Baker, 1907 (*Hypomellinus*)
    rufulicornis Maidl and Klima, 1939 (*Harpactostigma*)
lapazae R. Bohart, 1968; Mexico: Baja California
mirandus (W. Fox), 1892 (*Gorytes*); U.S.: Nevada, California
umbonatus (Baker), 1907 (*Hoplisoides*); U.S.: California
    femoratus Bradley, 1920 (*Gorytes*)
velutinus (Spinola), 1851 (*Hoplisus*); Chile
yumae R. Bohart, 1968; U.S.: Arizona, se. California

Genus Psammaletes Pate

Generic diagnosis: Medium-sized wasps; inner eye margins essentially parallel in female (fig. 157 L), slightly convergent below in male, least interocular distance two or more times median clypeal length, mandible with an inner subtooth; labrum inconspicuous; flagellum long and slender, first article at least as long as scape, last four articles in male distinctively flattened or concave beneath; pronotal collar somewhat appressed to scutum; female foretarsal rake well developed, fore basitarsus with three long setae before apex; female foreleg arolium much larger than other arolia, male hindleg arolium a little larger than other aolia; thorax coarsely sculptured (rugae and pitting); posterolateral oblique scutal carina present; prescutellar sulcus foveate; episternal and scrobal sulci evanescent but forming an angular episternal-scrobal sulcus as in *Oryttus* when well impressed; omaulus and sternaulus complete; no acetabular carina; forewing spotted, media diverging before cu-a, stigma moderate, veinlet of submarginal cell II between recurrents about one-fourth posterior length of submarginal cell I; jugal lobe considerably larger than tegula, hindwing media diverging at or very near cu-a which is strongly curved anterodistally (as in fig. 157 H); midtibia with two apical spurs; propodeal enclosure moderate in size, occupying not more than one-third of horizontal surface of propodeum, coarsely striate; spiracular groove distinct above, fading out below; gaster pedunculate, tergum I strongly curved toward apex in profile; male with seven terga normally visible, sterna without fimbriae; sternum VIII swordlike, not emarginate nor bispinose (fig. 168 D); female pygidial plate punctate to striate.

Geographic range: North America. Three species are known.

Systematics: Pate (1936a) reviewed the genus. A fairly close relationship to *Oryttus* seems indicated, but the strongly pedunculate and coarctate gaster of *Psammaletes* provides easy separation. Furthermore, male sternum VIII is quite different. The three described species are rare in collections.

Biology: Pate (1946a) recorded the prey of *mexicanus* (as *pechumani*) as the fulgorid, *Ormenoides venusta* (Melichar).

Checklist of Psammaletes

bigeloviae (Cockerell and W. Fox), 1897 (*Gorytes*); sw. U.S.; Mexico
crucis (Cockerell and W. Fox), 1897 (*Gorytes*); centr. and sw. U.S.; n. Mexico
    tricinctus Mickel, 1916 (*Hypomellinus*)
    venustus Mickel, 1916 (*Hypomellinus*)
mexicanus (Cameron), 1890 (*Gorytes*); e. U.S.; Mexico
    pechumani Pate, 1936; new synonymy by R. Bohart

Genus Handlirschia Kohl

Generic diagnosis (based on male holotype): Facial conformation as in fig. 174 B; last eight flagellomeres somewhat distorted, frontal groove mostly effaced, inner eye margins nearly straight, widely separated, converging slightly below; labrum plainly evident; mandible with strong inner subtooth; pronotum rather depressed, separated from scutum by a vertical transverse groove topped by a sharp edge; scutum with lateral oblique carina well

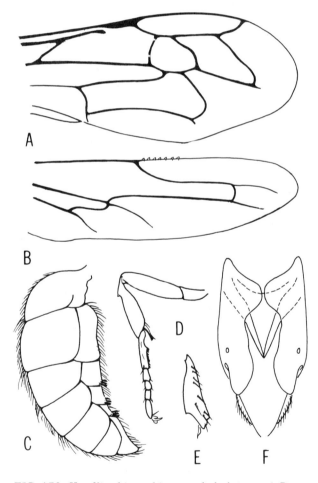

FIG. 170. *Handlirschia aethiops*, male holotype; A,B, wings; C, abdomen, lateral; D, foreleg; E, midtibia; F, gastral segment VII, ventral.

developed; omaulus, episternal sulcus, sternaulus and acetabular carina absent, scrobal sulcus weakly evident posteriorly; midtibia with two strong apical spurs; forewing about as in *Gorytes* (fig. 170 A), margin nearly straight toward base, media arising before cu-a, recurrents both received by submarginal cell II, veinlet between them short; hindwing media diverging slightly before cu-a (fig. 170 B), jugal lobe larger in outline than tegula; propodeum without a spiracular groove, somewhat projecting posteriorly, much as in *Bembecinus;* propodeal enclosure smooth, mostly dorsal, ending posteriorly in a greatly enlarged polished pit; propodeum otherwise with large separated punctures and medioposterior carinulae; gaster sessile; sternum II humped; seven visible terga of which VII has a large ventral spiracular lobe (fig. 170 F); sterna III to V with short, erect apical fimbriae (fig. 170 C); sternum VIII narrowed to a sharp spine; genitalia (according to Handlirsch, 1889) similar to those of *Sphecius*.

Geographic range: South Africa.

Systematics: This genus is similar in many ways to *Sphecius* and *Kohlia*. However, *Handlirschia* is the only known gorytin with a spiracular lobe on tergum VII, and neither *Sphecius* nor *Kohlia* have fimbriae on the sterna. For these reasons we have put it on a separate line of evolutionary development in the general direction of the Bembecini and Stizini, both of which commonly exhibit spiracular lobes on tergum VII as well as the reduced sectioning of the mesopleuron. In fact *Handlirschia* would probably be put in Stizini except for the rather typical gorytin wing venation.

We have studied the male holotype of *aethiops* in Mus. Vienna. Some of its features may be of specific rather than generic value. The legs are peculiar, the foretibia being toothed externobasally, the midtibia at the anterior two-thirds (figs. 170 D,E). The spiracular lobes of tergum VII actually close over the reduced sternum VII (fig. 170 F). The body is all black except for the yellowish-red legs, reddish antennae, and yellow facial markings. Gess (1973) recently described both sexes of a second species.

Biology: Unknown.

Checklist of *Handlirschia*

aethiops (Handlirsch), 1889 (*Sphecius*); S. Africa (presumably Zululand)
tricolor Gess, 1973; S. Africa: Transvaal.

Genus Sphecius Dahlbom

Generic diagnosis: Large wasps, some attaining a length of more than 35 mm; inner eye margins converging slightly to strongly below (fig. 157 F); mandible with an inner subtooth; labrum prominent, exposed part often half as long as broad; antenna long, gradually enlarged toward apex but not obviously clubbed, last four articles in male not specially modified in contrast to other articles; impressed median line on frons often weak or incomplete; pronotal collar thin and closely appressed to scutum; female foretarsal rake weakly developed, fore basitarsus with three or more stout setae before apex; female arolia about equal in size; posterolateral oblique scutal carina present; prescutellar sulcus foveate; episternal-scrobal sulcus curved backward around hypoepimeron, scrobal sulcus not continued forward toward omaulus which is distinct (fig. 3 B); sternaulus absent; acetabular carina present in Palearctic species; metapleuron gradually or hardly narrowed below; scutellum overlapping metanotum; forewing with media diverging before cu-a; stigma unusually small; veinlet of submarginal cell II between recurrents short; jugal lobe considerably larger than tegula, hindwing media diverging well before cu-a; midtibia with two apical spurs; propodeal enclosure large, extensively vertical as well as horizontal; no spiracular groove; abdomen stout, tergum I broad; male with seven terga normally visible, sterna without fimbriae, sternum VIII stinglike or sword shaped (fig. 172 A); female pygidial plate flat, sharply margined, closely punctate.

Geographic range: Of the 22 named species, three are Nearctic, two are Neotropical, three are Ethiopian, one is Australasian, one is Oriental, and 12 are Palearctic including north Africa.

Systematics: Five seemingly related genera are *Sphecius, Kohlia, Ammatomus, Tanyoprymnus,* and *Pterygorytes*. They are medium to large wasps with considerable resemblance to stizins. All have the thorax unusually compact, a tendency towards streamlining of the notum, and a stout propodeum. There is no spiracular groove and no sternaulus. In neither *Ammatomus* nor *Pterygorytes* is there an omaulus. These are advanced features pointing toward the Stizini and Bembicini. The rather large and usually exposed labrum as well as the lack of distinctive modifications embracing just the last four male flagellomeres are also notable.

The large size and robust shape of *Sphecius* will readily distinguish them from other gorytins. However, they have many close counterparts in the Stizini from which the presence of an omaulus in *Sphecius* is the quickest means of separation. The entirely different wing venation indicates that much of the similarity is superficial, yet *Sphecius* obviously heads an evolutionary line in the direction of *Stizus*. This explains why so many of the species were originally described in *Stizus*.

Pate (1937a) suggested the division of *Sphecius* into three subgenera, which appears to be quite logical according to the material we have studied. The addition of the acetabular carina as a characteristic of the Palearctic forms was proposed by Beaumont (1954a). Roth (1951) revised the Palearctic species.

Key to subgenera of *Sphecius*

1. Acetabular carina present; male mid basitarsus greatly swollen, excavate beneath, forming with end of tibia a sort of antennal cleaner (fig. 171 F,G); female with second article of midtarsus strongly asymmetrical (Palearctic).................................*Sphecienus* Patton
   Acetabular carina absent; male mid basitarsus simple (fig. 171 E); female with second article of midtarsus strongly asym-

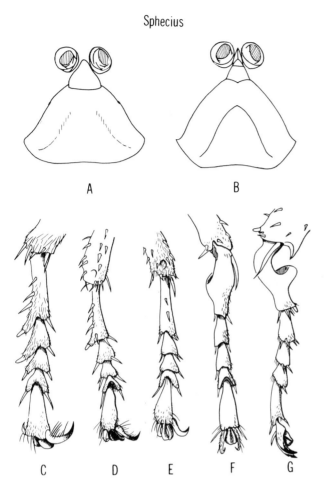

FIG. 171. Anatomical details in the genus *Sphecius;* A, clypeus of female *speciosus;* B, clypeus of female *pectoralis;* C, female midtarsus of *speciosus;* D, female midtarsus of *pectoralis;* E, male midtarsus of *speciosus;* F-G, two views of male midtarsus of *antennatus*.

metrical, or hardly so 2
2. Male with last flagellomere not strongly incurved beneath; female with second article of midtarsus quite asymmetrical, produced outwardly and spined at apex (fig. 171 C); female clypeus without well defined basal and discal planes (fig. 171 A) (New World) .... *Sphecius* Dahlbom
Male with last flagellomere strongly incurved beneath; female with second article of midtarsus hardly asymmetrical (fig. 171 D); female clypeus with distinct basal and discal planes (fig. 171 B) (Ethiopian, Australasian, Oriental).................................. *Nothosphecius* Pate

Biology: Members of this genus are well known as "cicada killers," but detailed accounts of habits are mostly restricted to the North American *speciosus*. Another common name for this species is "the ground hornet." A review of the published information has been given by Evans (1966a). According to Evans, the most important papers are those by Riley (1892), Howes (1919), Davis (1920), Reinhard (1929b), Dow (1942), Dambach and Good (1943), and N. Lin (1963). The species is active during July and August. Both sexes are encountered on flowers and in nesting areas. Males exhibit territoriality and investigate all intruders in a vigorous manner, often butting them. If the invader is another male, a grappling encounter may take place. These territories range in size up to about 100 square feet in the vicinity of an emergence hole, not necessarily that of the male in question. Mating occurs when a newly emerged female flies through a territory. The male clasps the female from behind, and the pair alight in a tree, on a clump of weeds, or on the ground where copulation takes place. As in some other Nyssoninae, the copulating pair initially have the male superimposed, but shortly form a straight line with their heads pointing in opposite directions. Copulation may occupy 45 minutes or more.

The burrows angle into the soil for a distance of 0.3 to 1.2 m, ending in a number of branches and cells. Each branch may terminate in two or three loosely grouped cells. Digging of the burrows may be accomplished during the day or overnight. The dirt is raked with the forelegs, then pushed out behind the female wasp with her hind legs, resulting in a large tumulus. Burrow entrances are left open during provisioning and may never be closed. However, cells are rather carefully closed when complete. Depending upon their size, one to four adult cicadas of the genus *Tibicen* are provisioned per cell. They are usually paralyzed by a sting in the membrane at the base of a front leg. The prey is carried venter to venter and is dragged or flown to the nest where it is stored, venter up in a cell. The egg is laid on the last cicada in a cell, and it is placed longitudinally on the venter alongside a midcoxa. There is some evidence that female wasps govern the sex of the offspring by depositing unfertilized (male) or fertilized eggs (female). In each case the female young have the more abundant provisions. Lin and Michener (1972) reported that as many as four females of *speciosus* may provision a single nest simultaneously, although only one of them made the initial excavation. The significance of this, and other behavior in *speciosus,* to the development of sociality in wasps is discussed at length by Lin and Michener.

The principal natural enemies are miltogrammine flies, *Senotainia trilineata* (Wulp) and (presumably) *Metopia argyrocephala* (Meigen). The *Senotainia* have been observed by Reinhard (1929b) to larviposit on cicadas as they are being dragged into the burrow. Maggots may eliminate 30 percent (Evans, 1966a) to 50 percent (Reinhard, 1929b) of the wasp brood.

Information on other *Sphecius* from various parts of the world primarily involve identification of the prey. Genera of cicadas reported as provisions are *Tibicen, Diceroprocta, Tettigades, Cicada, Playpleura, Poecilopsaltria,* and *Tamasa*.

Checklist of *Sphecius*

antennatus (Klug), 1845 (*Larra*); Eurasia

aberrans Eversmann, 1849 (*Stizus*)
  ssp. *impressus* Kokujev, 1910; USSR: Tadzhik S.S.R.
citrinus Arnold, 1929; s. Africa
claripennis Morice, 1911; Algeria
conicus (Germar), 1817 (*Stizus*); Europe, Middle East (see Bischoff, 1937)
    luniger Eversmann, 1849 (*Stizus*)
  ssp. *syriacus* (Klug), 1845 (*Larra*); Middle East
  ssp. *creticus* Beaumont, 1965; Crete
convallis Patton, 1879; U.S.: w. of 100th mer., Mexico: Baja California
    raptor Handlirsch, 1889
grandidieri (Saussure), 1887 (*Stizus*); Madagascar
    freyi Handlirsch, 1892
grandis (Say), 1823 (*Stizus*); U.S. to Centr. America
    fervidus Cresson, 1872 (*Stizus*)
    nevadensis Cresson, 1874 (*Stizus*)
hemixanthopterus Morice, 1911; Algeria
hogardii (Latreille), 1809 (*Stizus*); W. Indies; U.S.: se Florida
    rufescens Lepeletier, 1845 (*Hogardia*)
  ssp. *bahamas* Krombein, 1953; W. Indies: Bimini
intermedius Handlirsch, 1895; Algeria
malayanus Handlirsch, 1895; Indonesia
milleri Turner, 1915; s. Africa
  ssp. *aurantiacus* Arnold, 1940; ne. Africa
nigricornis (Dufour), 1838 (*Stizus*); Mediterranean area
pectoralis (F. Smith), 1856 (*Stizus*); Australia
percussor Handlirsch, 1889; se. Europe, sw. USSR
persa Gussakovskij, 1934; Iran
quartinae (Gribodo), 1884 (*Stizus*); nw. Africa
schulthessi Roth, 1951; n. Africa
speciosus (Drury), 1773 (*Sphex*); N. America
    tricinctus Fabricius, 1775 (*Vespa*)
    vespiformis Latreille, 1818 (*Stizus*)
    speciosus Dahlbom, 1843, nec Drury, 1773
spectabilis (Taschenberg), 1875 (*Stizus*); Brazil
  ssp. *nobilis* Brèthes, 1910; Argentina
turanicus Roth, 1959; Iran
uljanini (Radoszkowski), 1877 (*Stizus*); sw. USSR
    lutescens Radoszkowski, 1877 (*Stizus*)
    terminus W. F. Kirby, 1889 (*Stizus*)

Genus Tanyoprymnus Cameron

Generic diagnosis: Medium-sized wasps; eyes large, inner margins converging strongly below (fig. 165 D); mandible not dentate; labrum usually visible; antennae of the two sexes similar in shape, clublike (fig. 165 D), last four articles in male not specially modified; median line on frons distinctly impressed; pronotal collar thin and closely appressed to scutum; female foretarsal rake weak, fore basitarsus with four setae before apex; female foreleg arolium larger than other arolia; female forefemur sharply carinate beneath; posterolateral oblique scutal carina present; prescutellar sulcus efoveate; episternal-scrobal sulcus curving around hypoepimeron; no sternaulus nor acetabular carina; metapleuron greatly narrowed below; forewing with media diverging before cu-a, stigma small, veinlet of submarginal cell II between recurrents short; jugal lobe larger than tegula, hindwing media diverging well before cu-a; midtibia with two apical spurs; hindtarsus nearly twice as long as midtarsus; propodeal enclosure largely dorsal, smooth except for microsculpture; no spiracular groove; abdomen not pedunculate; male with seven terga normally visible, sterna without fimbriae, sternum VIII broad, deeply emarginate, usually protruding as two spines (fig. 172 B); male genitalia with digitus of volsella differentiated (fig. 173 B); female pygidial plate densely setose as in many *Tachytes* and *Alysson*.

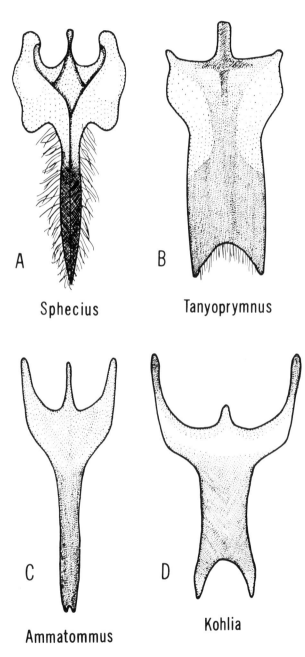

FIG. 172. Male sternum VIII in the tribe Gorytini; A, *Sphecius speciosus;* B, *Tanyoprymnus moneduloides;* C, *Ammatommus coarctatus;* D, *Kohlia cephalotes.*

512 SPHECID WASPS

Geographic range: United States to Central America.

Systematics: The close relationship with *Ammatomus* has been discussed under that genus. The nondentate mandible, presence of an omaulus, bispinose male sternum VIII (fig. 172 B), and differentiated digitus in *Tanyoprymnus* are important separating characters. The genus seems to be about on a par with *Kohlia, Sphecius,* and *Ammatomus* in evolutionary advancement.

Biology: The principal recorded evidence on this monotypic genus is that of Krombein (1959b), who found two females nesting in the vertical surface of a sand bank in coastal North Carolina. The single burrow examined was found to angle in and downward for about 13 cm. An incomplete cell contained four nymphs of the dictyopharine fulgorid, *Rhynchomitra microrhina* (Walker). The other wasp was captured as it was carrying an adult fulgorid of the same species.

Evans (1966a) reported a female collected in Florida by F. E. Kurczewski and pinned with a nymphal dictyopharine fulgorid.

R. M. Bohart and Holland (1965) found a nest near Lake Texoma, Oklahoma. Their unpublished observations are as follows: "We caught a female on July 13 emerging from a nest in hard-packed red sand near a borrow-pit pond. The burrow entered the nearly vertical surface of a small gully and proceeded slightly downward for 9 cm, then vertically for 14 cm and finally backward almost horizontally for 7 cm, terminating in a cell. Provisions in the cell were 16 adult dictyopharine fulgorids of *Scolops sulcipes* (Say) (J. P. Kramer det.)."

Checklist of *Tanyoprymnus*

moneduloides (Packard), 1867 (*Gorytes*); N. America
  belfragei Cresson, 1872 (*Gorytes*)
  longitarsis Cameron, 1905

Genus Ammatomus A. Costa

Generic diagnosis: Medium-sized wasps; eyes large, inner margins converging strongly below; mandible with an inner subtooth; labrum usually visible; antennae of sexes similar in shape, clublike, last four articles in male not specially modified; median line on frons distinctly impressed; pronotal collar thin and closely appressed to scutum; female foretarsal rake weak, with more than two setae in basitarsal comb before apex; female foreleg arolium larger than other arolia; female forefemur often sharply carinate beneath; posterolateral oblique scutal carina present but faint (fig. 156 B) or merely an angle in the margin; prescutellar sulcus simple; episternal-scrobal sulcus usually evanescent but curving around hypoepimeron when well impressed; omaulus, sternaulus and acetabular carina absent; metapleuron greatly narrowed below; forewing with media diverging before cu-a, stigma small, veinlet of submarginal cell II between recurrents short; jugal lobe larger than tegula, hindwing media diverging well before cu-a; midtibia with two apical spurs; hindtarsus nearly twice as long as midtarsus; propodeal enclosure convex, largely dorsal; no spiracular groove; abdomen not pedunculate or somewhat so; male with seven terga normally visible; sterna without fimbriae, sternum VIII swordshaped with a small apical emargination (fig. 172 C); male genitalia with digitus of volsella undifferentiated (fig. 173 A); female pygidial plate densely setose as in many *Tachytes* and *Alysson*.

Geographic range: Old World. Twenty-seven species are known.

Systematics: In many respects *Ammatomus* closely resembles the New World *Tanyoprymnus*. Important differences in the former are the dentate mandible, absence of an omaulus, male sternum VIII much narrower, and volsella undifferentiated. *Pterygorytes* also lacks an omaulus and has the mandible dentate, but otherwise it differs in many other ways.

Of the 26 known species, we have studied only 11: *africanus, alipes, coarctatus, decoratus, elongatulus, icarioides, lenis, mesostenus, rogenhoferi, rufonodis,* and *spiniferus latus.* However, these are remarkably alike in essential characters. Observed variations are in the constriction of tergum I: weak in *rogenhoferi* but pronounced in *coarctatus* and most of the other spe-

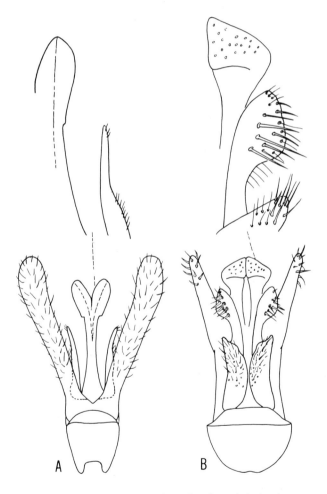

FIG. 173. Male genitalia in the tribe Gorytini; A, *Ammatomus coarctatus;* B, *Tanyoprymnus moneduloides.*

cies. In *rogenhoferi* females the outer corners of the hind-femur is lobate somewhat as in *Cerceris*. In females of many species the forefemur is carinate beneath. This same characteristic is found in *Tanyoprymnus* but not in *Kohlia, Pterygorytes,* or *Sphecius*. In some species, such as *africanus*, the scutum may be rather smooth, whereas in others, such as *spiniferus*, the scutum is coarsely punctate. Pulawski (1973) reviewed the Palearctic species.

Biology: Practically unknown. A specimen of *icarioides* from Townsville, Queensland, collected by G. F. Hill, is pinned with the presumed prey, an adult fulgorid (Flatinae).

Checklist of *Ammatomus*

africanus Turner, 1912; S. Africa
alipes (Bingham), 1897 (*Gorytes*); India
asiaticus (Radoszkowski), 1886 (*Lestiphorus*); sw. USSR
austrinus (Bingham), 1912 (*Gorytes*); Australia
biguttatus Arnold, 1945; Madagascar
coarctatus (Spinola), 1808 (*Gorytes*); Algeria, Cyprus, s. Europe to sw. USSR, Iran
    handlirschii F. Morawitz, 1890 (*Gorytes*)
    mavromoustakisi Balthasar, 1954 (*Gorytes*)
crassicornis (Mantero), 1917 (*Gorytes*); ne. Africa
decoratus (Handlirsch), 1888 (*Gorytes*); W. Australia
    ornatus F. Smith, 1868, nec F. Smith, 1856
elongatulus Turner, 1920; se. Africa
fallax Arnold, 1945; Madagascar
fuscipes (Arnold), 1929 (*Arpactus*); Rhodesia
gemellus (Arnold), 1929 (*Arpactus*); s. Africa
icarioides (Turner), 1908 (*Gorytes*); Australia: Queensland
lenis (Nurse), 1903 (*Gorytes*); India
madecassus (Schulthess), 1918 (*Gorytes*); Madagascar
mesostenus (Handlirsch), 1888 (*Gorytes*); Egypt, Israel to Iraq, Iran
  ssp. *rhopalocerus* (Handlirsch), 1895 (*Gorytes*); Algeria Morocco
  ssp. *nikolajevskii* (Gussakovskij), 1928 (*Gorytes*); sw. USSR to Afghanistan
pictipes (Arnold), 1929 (*Arpactus*); s. Africa
pretoriensis (Arnold), 1936 (*Arpactus*); se. Africa
rogenhoferi (Handlirsch), 1888 (*Gorytes*); Eurasia
rubicundus Arnold, 1945; Madagascar
rufonodis (Radoszkowski), 1877 (*Hoplisus*); Israel, Turkey, sw. USSR, Iran
  ssp. *saharae* (Handlirsch), 1895 (*Gorytes*); Algeria
seyrigi Arnold, 1945; Madagascar
sinensis (Yasumatsu), 1943 (*Gorytes*); China
spiniferus (du Buysson), 1897 (*Gorytes*); se. Africa
  ssp. *latus* Arnold, 1951; Ethiopia
stevensoni (Arnold), 1929 (*Arpactus*); s. Africa
thaianus Tsuneki, 1963; Thailand
yoshikawai Tsuneki, 1963; Cambodia

Genus Kohlia Handlirsch

Generic diagnosis: Medium-sized wasps, inner eye margins essentially parallel or a little diverging above (fig. 174 C), anterior clypeal margin arcuate and laterally sharp, clypeus with basal and apical planes, mandible simple or with an inner subtooth; labrum exposed, usually visible, last four male flagellomeres not specially modified; median line on frons incomplete; ocelli obovate, flattened externally; pronotal collar depressed, thin and closely appressed to scutum; tarsal rake well developed in both sexes, fore basitarsal comb with numerous setae before apex; female arolia small and about equal in size; notum somewhat flattened overall, scutellum overhanging metanotum; posterolateral oblique scutal carina present; prescutellar sulcus efoveate; episternal sulcus descending vertically to omaulus, scrobal sulcus ending on it at a right angle; no sternaulus nor acetabular carina; metapleuron gradually narrowed but broadly truncate below; forewing with media diverging after cu-a, stigma moderate, marginal cell bent away from wing margin, veinlet of submarginal cell II between recurrents short; jugal lobe much larger than tegula, hindwing media diverging well before cu-a; midtibia of male with one or two apical spurs; propodeum bent sharply in profile, enclosure extending well onto vertical face; no spiracular groove; abdomen not pedunculate; male with seven terga normally visible, sterna without fimbriae, sternum VIII broadly bladelike and distally excavated (fig. 172 D); female pygidial plate distinct, not densely setose.

Geographic range: Southern Asia (Tadzhik), eastern and southern Mediterranean area, and southern Africa. Three species are known.

Systematics: Kohlia is a peculiar genus which is most probably a relic. It combines such primitive characters as the distal origin of the forewing media and simple male antenna with many advanced features reminiscent of the Stizini and Bembicini. Some of these are the deformed ocelli (as in Bembicini), forewing marginal cell bent away from the margin (as in *Microbembex*), compact thorax with flattened notum and overlapping scutellum, and bent clypeus (as in *Rubrica, Editha,* etc.).

Two variable characters of importance are the midtibial spurs of the male, which are essentially single in *Kohlia coxalis;* and the mandible, which is simple in *coxalis* and *cephalotes* but dentate subapically in *pavlovskii*.

Biology: Unknown.

Checklist of *Kohlia*

cephalotes Handlirsch, 1895; s. Africa
coxalis Morice, 1897; n. Africa, Israel
pavlovskii (Gussakovskij), 1952 (*Stizobembex*); sw. USSR: Tadzhik S.S.R.

Genus Pterygorytes R. Bohart

Generic diagnosis: Medium large wasps, inner eye margins converging strongly toward clypeus; mandible with an inner subtooth; labrum exposed; male flagellum concave and polished beneath last three articles; median impressed line on frons complete; pronotal collar thin and rather closely appressed to scutum; female tarsal rake moderate, three setae before apex on fore basitarsus; foreleg arolium of female much larger than others; scutum, scutellum, metanotum and propodeum divided by simple sutures, fitting

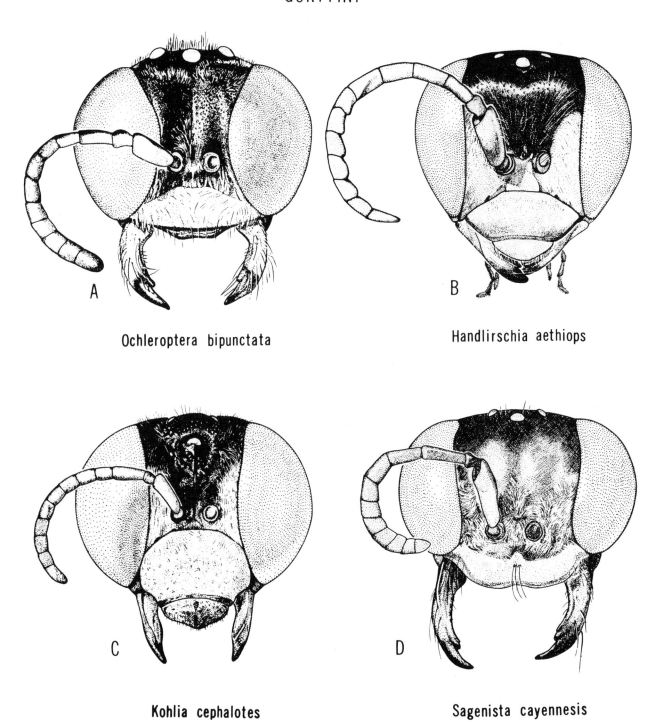

FIG. 174. Facial portraits in the tribe Gorytini, A,C,D, females; B, male.

smoothly and flatly (as in *Stizus*); episternal-scrobal sulcus curving around hypoepimeron; no omaulus, sternaulus nor acetabular carina; metapleuron greatly narrowed and linear below; forewing with media diverging before cu-a, stigma small, veinlet of submarginal cell II between recurrents short; jugal lobe undeveloped, hindwing media diverging a little or considerably (*triangularis*) before cu-a which is nearly straight; midtibia with two apical spurs; propodeal enclosure extending well onto vertical face of propodeum; no spiracular groove; abdomen not pedunculate; sternum II swollen subbasally and topped by a median shiny tubercle; tergum III with a polished mediobasal triangle; male with seven terga normally visible; sternum VIII narrowed on apical half (about as in fig. 175 A); female pygidial plate flat, densely covered with stout setae (as in many *Tachytes*).

Geographic range: The two known species and a third undescribed one are from Brazil.

Systematics: Features of the genus were illustrated and its relationships were discussed by R. M. Bohart (1967). The thoracic structure is remarkably stizin, but the wings with their normal-sized submarginal cell I are clearly gorytin. In other Gorytini without an omaulus, such as *Handlirschia*, *Ammatomus* and *Trichogorytes* in part, the jugal lobe is well developed.

We consider *Pterygorytes* to be among the most advanced members of the tribe and on a par with *Lestiphorus* and *Afrogorytes*. The species are moderately large, stout wasps known from few specimens.

Checklist of *Pterygorytes*

triangularis (F. Smith), 1873 (*Gorytes*); Brazil: Pará
valens (W. Fox), 1897 (*Gorytes*); Brazil: Mato Grosso and São Paulo

Genus Psammaecius Lepeletier

Generic diagnosis: Medium small wasps; inner eye margins converging strongly below; median frontal sulcus distinct; labrum inconspicuous; last four male flagellomeres specially modified, flattened or concave beneath; first flagellomere less than three-fourths as long as scape; mandible with an inner subtooth; pronotal collar simple, thin at middle, rather closely appressed to scutum; female foretarsal rake well developed; with three setae on basitarsus before apex; female arolia nearly equal in size; posterolateral oblique scutal carina present; episternal and scrobal sulci evanescent but forming an angular episternal-scrobal sulcus as in *Oryttus* when well impressed; omaulus and sternaulus present; no acetabular carina but subomaulus (anteriormost carina of mesopleuron) continued below in nearly a straight line to ventral midcarina, just before which it is somewhat bent and projecting almost tooth shaped; forewing somewhat pictured, media diverging before cu-a, stigma moderate, veinlet of submarginal cell II between recurrents short; jugal lobe larger than tegula, hindwing media diverging at or near cu-a; midtibia with two apical spurs; propodeal enclosure almost wholly dorsal; spiracular groove present; gaster subpedunculate at most; male with seven terga normally visible, sterna V and VI with basal and ordinarily concealed hairbrushes, sternum VIII sharply pointed (fig. 175 A); female pygidial plate well developed, subtriangular, not densely setose.

Geographic range: Palearctic. Five species are known.

Systematics: The five known species have been treated by Beaumont (1952a, 1954a). Some are black and yellow, others have varying amounts of red on the abdomen. The relationship with *Hoplisoides* and *Xerogorytes* is close. Male sterna V and VI in all three genera have basal (concealed) hairbrushes which become erect when exposed (see fig. 178 B). Also, the wings are ordinarily pictured. However, the peculiar subomaulus of *Psammaecius* and the rather distinct spiracular groove of the propodeum differentiate it from the other two genera.

Biology: Ferton (1905) reported briefly on the nesting behavior of *punctulatus* in France. The burrows were in compact sand and about 8.5 cm deep. A leafhopper, *Sol-*

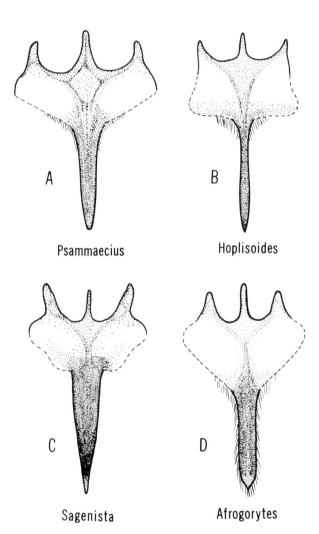

FIG. 175. Male sternum VIII in the tribe Gorytini; A, *Psammaecius punctulatus;* B, *Hoplisoides spilopterus;* C, *Sagenista cayennensis;* D, *Afrogorytes gibbosus.*

enocephalus obsoletus Germar, was provisioned at the rate of four or five per cell. These were carried against the venter of the wasp by means of the midlegs. The egg was laid on the side of the venter of the leafhopper's thorax.

Checklist of *Psammaecius*

austeni Turner, 1919; Israel
eremorum Beaumont, 1952; Algeria
luxuriosus (Radoszkowski), 1877 (*Hoplisus*); sw. USSR
punctulatus (Vander Linden), 1829 (*Gorytes*); Eurasia
versicolor Beaumont, 1959; Israel

Genus Liogorytes R. Bohart

Generic diagnosis: Medium small to medium large wasps; inner eye margins converging moderately towards clypeus (fig. 159 N); median frontal sulcus distinct; labrum inconspicuous; last four male flagellomeres specially modified, flattened or concave beneath; first flagellomere less than three-fourths as long as scape; mandible with an inner subtooth; pronotal collar simple, not closely appressed to scutum; female foretarsal rake well developed, with three setae on basitarsus before apex; female arolia nearly equal in size; posterolateral oblique scutal carina present; episternal sulcus ending on scrobal sulcus which is continued forward in a nearly straight line to omaulus (similar to fig. 158); no acetabular carina; sternaulus well developed; forewing not pictured, media diverging before cu-a; stigma moderate, veinlet of submarginal cell II between recurrents short; jugal lobe larger than tegula, hindwing media diverging slightly to considerably before cu-a which is rather strongly curved or angled anteriorly; midtibia with two apical spurs, male mid basitarsus usually curved and spinose in front; propodeal enclosure almost wholly dorsal, smooth, lateral sulci pitted or "stitched," median groove well developed; spiracular groove present; gaster pedunculate or subpedunculate; male with seven terga normally visible, sterna without basal or apical hairbrushes, sternum VIII plainly notched at apex (fig. 176 C), male genitalia: fig. 179 A; female pygidial plate flat and longitudinally striate, moderately setose.

Geographic range: The six described species are all South American.

Systematics: Species of *Liogorytes* superficially resemble *Gorytes* and could be considered to replace the latter in South America. The incurved male mid basitarsus is a striking feature. We have not seen the male of *joergenseni*, so its placement is a little doubtful.

A close relationship with *Arigorytes* seems evident. In addition to the midleg character of the male, differences are the curved or bent cu-a of the hindwing in *Liogorytes*, the plain rather than woolly male sterna, and the distinctly grooved frons. There are some similarities to *Xerogorytes*, but the latter has tergum I only slightly constricted, male sterna V and VI with concealed basal hairbrushes, and the forewing spotted.

Biology: Unknown.

Checklist of *Liogorytes*

cordobensis (Fritz), 1964 (*Gorytes*); Argentina
joergenseni (Brèthes), 1910 (*Gorytes*); Argentina
llanoi (Fritz), 1964 (*Gorytes*); Argentina
patagonicus (Fritz), 1959 (*Harpactostigma*); Argentina
polybia (Handlirsch), 1895 (*Gorytes*); Brazil
  catarinae R. Bohart, 1967, new synonymy by R. Bohart
unicinctus (Brèthes), 1913 (*Gorytes*); Argentina

Genus Arigorytes Rohwer

Generic diagnosis: Small wasps; inner eye margins converging below but least interocular distance 1.5 to 2.0 times length of scape (fig. 159 J); frons with merely traces of a groove below midocellus; labrum inconspicuous; last four male flagellomeres specially modified, flattened or concave beneath; first flagellomere less than three-fourths as long as scape; mandible with an inner subtooth; pronotal collar narrow, raised a little medially; not closely appressed to scutum; female foretarsal rake well developed, basitarsus with three blade shaped setae before apex; female arolia nearly equal; posterolateral oblique scutal carina present; episternal sulcus meeting scrobal sulcus at nearly a right angle, then directed downward and forward to omaulus; no acetabular carina; sternaulus usually complete; metapleuron gradually narrowed below; forewing not pictured, media diverging before cu-a; veinlet between recurrents short; jugal lobe larger than tegula, hindwing media diverging at or slightly before cu-a which is nearly straight; midtibia with two apical spurs; male hindtarsus often inflated; propodeal enclosure broadly triangular, well sculptured; spiracular groove indistinct; gaster not pedunculate; male with seven terga normally visible, sterna III to V with velvety hair mats, no concealed fimbriae on sterna V and VI, sternum VIII narrowed to a slightly notched tip (fig. 176 A); female pygidial plate broadly triangular, punctate, nearly bare.

Geographic range: The five known species are all from southwestern United States with one species occurring also in the mountains of Baja California, Mexico.

Systematics: The species form two groups in one of which the male hindtarsus is much larger than the midtarsus. Species in this group are *insolitus, coquillettii, ruficrus,* and *coachellae*. *Xerogorytes anaetis* was placed in *Arigorytes* by Pate (1947b), but this species has a furrowed frons, incomplete scrobal sulcus, medially depressed pronotal collar, pictured forewing, bent hindwing cu-a, and other features which rule it out (R. M. Bohart, 1967). The male of *anaetis* was unknown to Pate, but it was subsequently discovered by R. M. Bohart in 1958. The absence of sternal hair mats and the peculiar sternum VIII furnished additional evidence of its distinction. *Arigorytes* is obviously related to *Hoplisoides* and represents a small Nearctic offshoot from this evolutionary branch. The most recent paper on the group is that of R. Bohart (1971).

Biology: Unknown.

Checklist of *Arigorytes*

coachellae R. Bohart, 1971; sw. U.S.; Mexico: Baja California
coquillettii (W. Fox), 1895 (*Gorytes*); U.S.: California

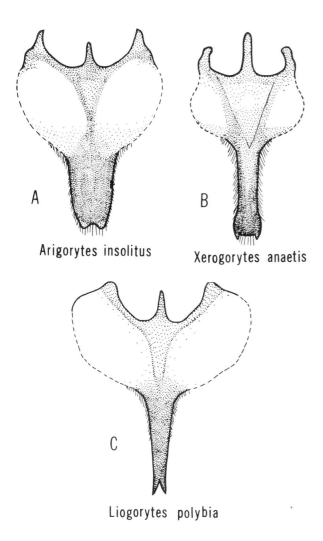

FIG. 176. Male sternum VIII in the tribe Gorytini.

insolitus (W. Fox), 1895 (*Gorytes*); w. U.S.
    clavatus Baker, 1907 (*Hoplisoides*)
ruficrus R. Bohart, 1971; U.S.: California
smohalla Pate, 1947; w. U.S.

Genus Xerogorytes R. Bohart, new genus

Generic diagnosis: Small wasps; inner eye margins converging rather strongly below; median frontal groove sharp; labrum inconspicuous; last four male flagellomeres specially modified, flattened or concave beneath; first flagellomere less than two-thirds as long as scape; mandible with an inner subtooth; pronotal collar somewhat thickened laterally, depressed medially, rather appressed to scutum; female foretarsal rake well developed, basitarsus with three bladelike setae before apex; female arolia nearly equal; posterolateral oblique scutal carina present; episternal and scrobal sulci evanescent but arrangement similar to *Arigorytes;* sternaulus complete; no acetabular carina or subomaulus; metapleuron somewhat constricted below, metapleural groove coarsely pitted; scutum moderately punctate; forewing pictured, media diverging before cu-a; stigma moderate, veinlet of submarginal cell II between recurrents short; jugal lobe larger than tegula, hindwing media diverging at or very near cu-a which is bent distally; midtibia with two apical spurs; propodeal enclosure triangular, boundaries pitted, surface at least partly and obliquely carinulate; spiracular groove indistinct; gaster with tergum I stout, only slightly set apart from II; male with seven visible terga, sterna V and VI with basal and normally concealed hairbrushes; VIII broadly bitoothed (fig. 176 B); typical male genitalia: fig. 164 A; female pygidial plate sharply margined, coarsely and longitudinally striatopunctate.

Type of genus: Arigorytes anaetis Pate.
Geographic range: Southwestern United States.
Systematics: This monotypic genus is known from only a few collections in southern Arizona and southwestern New Mexico. Pate (1947b) based his original description of *anaetis* on the female and assigned it with some misgivings to *Arigorytes*. R. Bohart collected a series of both sexes near Rodeo, New Mexico in 1958 and recognized the many differences between the male and those of *Arigorytes*. The species was placed by Bohart (1967) in *Liogorytes*, but several features of *anaetis* argue against this course. These are the absence of a spiracular groove on the propodeum, the simple male mid basitarsus, a spotted forewing, a stout tergum I, concealed basal hairbrushes on male sterna V and VI (see fig. 178 B), and the rather distinctively shaped male sternum VIII (fig. 176 B). In many respects it is closer to *Hoplisoides* or *Psammaecius,* but the absence in *anaetis* of an acetabular carina or of a subomaulus as well as the forward extension of the scrobal sulcus are differential characters of importance. Therefore, it seems best to consider it as a separate genus.
Biology: Unknown.

Checklist of *Xerogorytes*

anaetis (Pate), 1947 (*Arigorytes*); sw. U.S.

Genus Hoplisoides Gribodo

Generic diagnosis: Medium to small wasps; inner eye margins often nearly parallel and widely separated (fig. 159 M), sometimes converging below, especially in males (fig. 178 D); median frontal groove often indistinct; labrum inconspicuous; last four male flagellomeres specially modified, flattened or concave beneath (fig. 178 A); first flagellomere less than three-fourths as long as scape; mandible with an inner subtooth; pronotal collar a little thinner medially, rather closely appressed to scutum; female foretarsal rake well developed, basitarsus with three bladelike setae before apex (fig. 177); female arolia usually equal; posterolateral oblique scutal carina present; episternal sulcus, when present, ending at level of scrobe, scrobal sulcus sometimes weak and ending at episternal sulcus at nearly a right angle; sternaulus and acetabular carina present but sometimes weak and briefly interrupted; no subomaulus; scutum usually coarsely punctate; forewing usually pictured (fig. 177), media diverging before cu-a, stigma moderate, veinlet of submarginal cell II between recurrents short; jugal lobe larger than tegula, hindwing

FIG. 177. *Hoplisoides spilopterus* (Handlirsch), female.

media diverging at or very near cu-a (fig. 177); midtibia with two apical spurs; propodeal enclosure usually with longitudinal carinulae, lateral boundaries sometimes indistinct; spiracular groove present but not well impressed; gaster not pedunculate, male with six or sometimes seven normally visible terga, sterna V and VI with basal and concealed hairbrushes (fig. 178 B), sternum VIII sword shaped and pointed apically (fig. 175 B); female pygidial plate distinct, often long and ovoid-triangular, sides sometimes bent.

Geographic range: The 68 listed species are distributed over all continents except Australia.

Systematics: Closely related genera are *Psammaecius* and *Afrogorytes*. Beaumont (1952a) pointed out the structural peculiarity of the pleuron in the former, and Menke (1967b) did the same for the latter. Even with these two categories removed, the genus is still the largest one in the tribe and has considerable variation. Punctation is generally rather coarse, especially on the scutum. Yet, *semipunctatus* and *glabratus* are obvious exceptions. The acetabular carina is generally strong, but in *glabratus* it is rather weak; both *dentatus* and *diversus* have a prominent tooth marking the forward end of the sternaulus in males; pictured wings are practically universal, but the dark spots are sometimes quite discrete (as in *spilopterus*, fig. 177) and sometimes rather diffuse. In many species the males have a group of long hairs laterally on the free edge of the clypeus. These hairs often clump together to form a sort of "whisker," which may be very strong as in *hamatus* (fig. 178 H) and *denticulatus* or moderate as in *dentatus* and *placidus*. In *tricolor* and *glabratus* (fig. 178 G) the "whisker" is absent.

In North America three main groups may be distinguished: (1) the *denticulatus* group in which males have thick woolly hair on sterna IV to V (fig. 178 E) and cu-a is sharply bent in both sexes; (2) the *placidus* group in which the anteroventral metapleural pit is larger than the midocellus (fig. 178 C); and (3) the *costalis* group in which the above mentioned pit is not larger than the midocellus (fig. 178 F).

Distinctive South American forms are the vespid-like *jordani* and *vespoides*, the extensively polished *semipunctatus*, and the peculiar *bifasciatus*. The last named has the eyes of the male converging strongly below (fig. 178 D), much as in *Afrogorytes*.

The African species, including the generotype, *intricans*, are very similar to *placidus* from eastern United States. The small anteroventral metapleural pit of the *intricans* group is a minor separational point. The African group was called "*Arpactus*. Section II" by Arnold (1929).

Generally speaking, *Hoplisoides* appears to be moderately advanced on the evolutionary scale. It appears to be related to *Sagenista* and *Arigorytes*, differing from

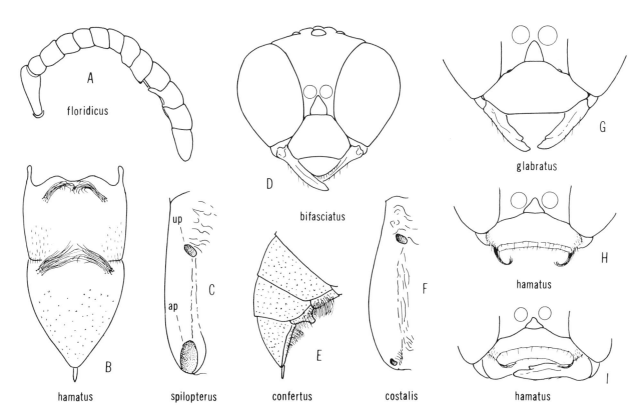

FIG. 178. Male structural details in *Hoplisoides;* A, antenna, lateral; B, sterna V-VI; C, metapleural area, left, up = upper metapleural pit, ap = anteroventral metapleural pit; D, facial outline; E, metapleural area, left; F, gastral segments IV-VI, lateral; G-I, lower part of face; G-H, front view; I, end-on view of clypeal truncation.

the former by a less sculptured propodeum and from the latter by the presence of a distinct acetabular carina. Furthermore, neither of the above named genera has concealed basal hairbrushes on sterna V and VI of the male.

The most useful publications on the genus to date are those of Beaumont (1952a), which treats the Palearctic species and distinguishes them from *Psammaecius*, and R. M. Bohart (1968b).

Biology: Considerable information is available on the nesting habits of *Hoplisoides*, but only a few complete studies have been made. These concern *costalis* (Reinhard, 1925a, b, 1929b), *tricolor* (Evans, *et al.*, 1954), *hamatus* (Powell and Chemsak, 1959b), and *placidus nebulosus* (Evans, 1966a). A summary of these and other reports has been given by Evans (1966a).

Prey seems to be practically confined to a single family of Homoptera for a given wasp species. Thus, Membracidae are used by *hamatus, costalis, placidus nebulosus, spilopterus, umbonicida,* and *manjikuli*; Cicadellidae are used by *denticulatus, tricolor,* and *glabratus*; and Fulgoridae (in the broad sense) are used by *latifrons,* and *punctuosus*.

Observations of Evans and others on *placidus nebulosus* are fairly typical for the genus. Female wasps select rather open sandy areas, although the nest entrance may be concealed beneath a leaf or overhanging rock. The nests are generally solitary, but a single female may construct several nests near each other. Burrows slant into the soil for about 9 cm and end in one to three cells. The burrows are almost invariably closed while the wasp is away and partially so when she sleeps overnight in a nearly completed nest. Prey is carried head forward, venter to venter, and is held by the midlegs of the wasp in flight. It is then shifted to the hind legs as the wasp enters its burrow. Treehoppers are stacked with head pointed in, and an egg is laid on the top (last) one alongside the coxae. Sand is scraped into the completed burrow and is pounded into place with strong blows of the tip of the abdomen. Prey consists of both adults and nymphs, about 10 to 15 per cell, and belonging to the following treehopper genera as summarized by Evans (1966a): *Campylenchia, Enchenopa, Entylia, Microcentrus, Palonica, Publilia, Spissistylus, Telamona,* and *Vanduzea.*

Very similar observations on *Hoplisoides hamatus* (as *Psammaecius adornatus*) were reported by Powell and Chemsak (1959b). A point of difference was that some 80 nests were loosely grouped along a roadway. Secondly, females apparently spent the night away from the nesting area. Thirdly, the prey were nymphs only of a membracid, presumably of the genus *Stictocephala.*

The following three previously unpublished notes are added to known information. *Hoplisoides placidus nebulosus:* several nests were observed during June and July 1965 at Lake Texoma, Oklahoma by R. M. Bohart and J. F. Holland. The burrows were in white sand near the crest of a dune and slanted some 9 cm into the ground. One cell contained 11 nymphs of the membracid, *Entylia concisa* Walker, and another cell yielded nymphs and adults of the membracids, *Entylia concisa, Tylopelta brevis* Van Duzee, and *Cerasa* sp. (determinations by R. C. Froeschner).

Hoplisoides hamatus: F. D. Parker collected specimens of *Epinysson moestus* entering burrows of *hamatus* on 12 May 1961 near Lake Berryessa, Napa Co., California.

Hoplisoides glabratus: R. Bohart and D. Linsdale observed nesting of *glabratus* near Rodeo, New Mexico on 28 August 1958. Burrows were about 4.5 cm long and slanted into the sandy soil. One oval cell was packed with 90 nymphal and eight adult leafhoppers, identified by J. P. Kramer as *Aceratagallia uhleri* (Van Duzee). An egg measuring 2.2 by 0.66 mm was attached to the last prey member. The leafhoppers were apparently being gathered among nearby flowers of a yellow composite, *Baileya pleniradiata.* Many female wasps were collected in partial concealment in the petals beneath the flower heads.

A variety of parasites have been reported to prey upon species of *Hoplisoides.* Briefly, these are as follows. *Hoplisoides costalis:* a nyssonin, *Epinysson guatemalensis hoplisivora* (Reinhard, 1925b); a chrysidid, *Elampus viridicyaneus* Norton (Krombein, 1963a); a sarcophagid, *Amobia aurifrons* Townsend (Reinhard, 1929b). *Hoplisoides hamatus:* a nyssonin, *Epinysson moestus* (Powell and Chemsak, 1959b and F. D. Parker, above record). *Hoplisoides latifrons:* a nyssonin, *Nysson dimidiatus* (Maneval, 1939). *Hoplisoides placidus nebulosus:* nyssonins, *Nysson daeckei* (Evans, 1966a), *Epinysson basilaris tuberculatus* (Evans, 1966a). *Hoplisoides tricolor:* a nyssonin, *Epinysson bellus* (Evans, *et al.*, 1954).

Checklist of *Hoplisoides*

aglaia (Handlirsch), 1895 (*Gorytes*); s. Africa
   euphrosyne Handlirsch, 1895 (*Gorytes*)
alaya (Pate), 1947 (*Psammaecius*); Dominican Republic
asuncionis (Strand), 1910 (*Gorytes*); Paraguay
ater (Gmelin), 1790 (*Vespa*); W. Indies, new name for *Vespa tricincta* (Fabricius), (Art. 59c)
   tricinctus Fabricius, 1775 (*Crabro*), nec *Vespa tricincta* Fabricius, 1775, now in *Sphecius*
   tristrigatus Fabricius, 1798 (*Mellinus*)
   behni Dahlbom, 1842 (*Lestiphorus*)
   scitulus Cresson, 1865 (*Harpactus*)
bandraensis (Giner Marí), 1954 (*Gorytes*); India
basutorum (Arnold), 1958 (*Gorytes*); s. Africa
bifasciatus (Brèthes), 1909 (*Gorytes*); Paraguay
bipustulatus (Arnold), 1945 (*Gorytes*); Madagascar
boranensis (Guiglia), 1940 (*Arpactus*); Ethiopia
braunsii (Handlirsch), 1901 (*Gorytes*); se. Africa
calliope (Arnold), 1936 (*Arpactus*); Zaire
carinatus R. Bohart, 1968; U.S.: Ariz.; Mexico: Son.
cazieri R. Bohart, 1968; U.S.: Arizona
confertus (W. Fox), 1895 (*Gorytes*); N. America
   imperialensis Bradley, 1920 (*Gorytes*), new synonymy by R. Bohart
confusus (Dutt), 1922 (*Gorytes*); India
†*confusus* (Alayo), 1969 (*Psammaecius*); Cuba, nec Dutt, 1922
costalis (Cresson), 1872 (*Gorytes*); N. and S. America
   ssp. *pygidialis* (W. Fox), 1895 (*Gorytes*); n. centr.

U.S., new status by R. Bohart
craverii (A. Costa), 1869 (*Hoplisus*); Europe, sw. USSR, Middle East, Mongolia
    ottomanus Mocsáry, 1879 (*Hoplisus*)
    ssp. *merceti* (Beaumont), 1950 (*Gorytes*); Spain
dentatus (W. Fox), 1893 (*Gorytes*); sw. U.S.
denticulatus (Packard), 1867 (*Gorytes*); N. America
    barbatulus Handlirsch, 1888 (*Gorytes*), new synonymy by R. Bohart
    ssp. *hypenetes* (Handlirsch), 1895 (*Gorytes*); Centr. and S. America, new status by R. Bohart
diversus (W. Fox), 1895 (*Gorytes*); U.S.: California
emeryi (Gribodo), 1894 (*Hoplisus*); se. Africa
eurynome (Arnold), 1956 (*Gorytes*); s. Africa
feae (Handlirsch), 1895 (*Gorytes*); Peru
ferrugineus (Spinola), 1838 (*Hoplisus*); n. Africa
    imsganensis Nadig, 1934 (*Gorytes*)
floridicus R. Bohart, 1968; U.S.: Florida
fuscus (Taschenberg), 1875 (*Hoplisus*); Brazil, raised from synonymy by R. Bohart
gazagnairei (Handlirsch), 1893 (*Gorytes*); Algeria
    ssp. *maroccanus* (Dusmet), 1925 (*Gorytes*); Morocco
    ssp. *distinguendus* (Yasumatsu), 1939 (*Gorytes*); China, e. USSR
glabratus R. Bohart, 1968; sw. U.S.
hamatus (Handlirsch), 1888 (*Gorytes*); w. N. America
    spilographus Handlirsch, 1895 (*Gorytes*), new synonymy by R. Bohart
    arizonensis Baker, 1907, new synonymy by R. Bohart
    adornatus Bradley, 1920 (*Gorytes*), new synonymy by R. Bohart
homonymus (Schulz), 1906 (*Gorytes*); India
    politus Bingham, 1897 (*Gorytes*), nec F. Smith, 1873
iheringii (Handlirsch), 1893 (*Gorytes*); Brazil
impiger (Bingham), 1897 (*Gorytes*); Burma, lectotype ♂, Tenasserim, Burma (Mus. London), present designation by R. Bohart
insularis (Cresson), 1865 (*Harpactus*); Cuba
intricans Gribodo, 1884; se. Africa
iridipennis (F. Smith), 1856 (*Gorytes*); Brazil to Mexico
    fasciatipennis Cameron, 1890 (*Gorytes*), new synonymy by R. Bohart
jaumei (Alayo), 1969 (*Psammaecius*); Cuba
jentinki (Handlirsch), 1895 (*Gorytes*); Indonesia: Timor
jibacoa (Alayo), 1969 (*Psammaecius*); Cuba
jordani (Handlirsch), 1895 (*Gorytes*); Paraguay
knabi (Rohwer), 1911 (*Gorytes*); Mexico
latifrons (Spinola), 1808 (*Gorytes*); Eurasia
    pulchellus Wesmael, 1851 (*Hoplisus*)
    minutus Mocsáry, 1879 (*Hoplisus*)
liberiensis (Arnold), 1936 (*Arpactus*); Liberia
manjikuli Tsuneki, 1963; Thailand
marshalli (Turner), 1915 (*Arpactus*); s. Africa
mendozanus (Brèthes), 1913 (*Gorytes*); Argentina, Brazil, Venezuela
montivagus (Arnold), 1951 (*Gorytes*); Ethiopia
morrensis (Strand), 1910 (*Gorytes*); Paraguay
mweruensis (Arnold), 1952 (*Gorytes*); s. Africa
orientalis (Handlirsch), 1888 (*Gorytes*); India
    tricolor F. Smith, 1875 (*Gorytes*), nec Cresson, 1868

panamensis (Maidl and Klima), 1939 (*Gorytes*); Panama
    maculipennis Cameron, 1890 (*Hoplisus*), nec Giraud, 1861
pictus (F. Smith), 1856 (*Gorytes*); India
    capitatus Nurse, 1902 (*Gorytes*), new synonymy by R. Bohart
placidus F. Smith, 1856; se. U.S.
    rufipes F. Smith, 1856 (*Gorytes*)
    ssp. *nebulosus* (Packard), 1867 (*Gorytes*); e. N. America, new status by R. Bohart
    armatus Provancher, 1887 (*Gorytes*), new synonymy by R. Bohart
    microcephalus Handlirsch, 1888 (*Gorytes*), new synonymy by R. Bohart
    pergandei Handlirsch, 1888 (*Gorytes*)
    harringtonii Provancher, 1888 (*Philanthus*), new synonymy by R. Bohart
    ssp. *birkmanni* Baker, 1907; sw. U.S., new status by R. Bohart
    pruinosus Baker, 1907, new synonymy by R. Bohart
projectus R. Bohart, 1968; U.S.: California
punctifrons (Cameron), 1890 (*Gorytes*); w. N. America incl. Mexico
    gulielmi Viereck, 1907 (*Gorytes*), new synonymy by R. Bohart
punctuosus (Eversmann), 1849 (*Hoplisus*); Eurasia
    punctatus Kirschbaum, 1853 (*Hoplisus*)
    crassicornis A. Costa, 1859 (*Hoplisus*)
    maculipennis Giraud, 1861 (*Hoplisus*)
    ibericus Mercet, 1906 (*Gorytes*)
    ssp. *curtulus* (A. Costa), 1893 (*Gorytes*); n. Africa
quedenfeldti (Handlirsch), 1895 (*Gorytes*); Algeria
remotus (Turner), 1921 (*Arpactus*); India
schubotzii (Arnold), 1932 (*Arpactus*); Zaire
semipunctatus (Taschenberg), 1875 (*Hoplisus*); Argentina; s. U.S. (probably introduced)
spilopterus (Handlirsch), 1888 (*Gorytes*); w. U.S.
    maculatus Handlirsch, 1895 (*Gorytes*), new synonymy by R. Bohart
    pogonodes Bradley, 1920 (*Gorytes*), new synonymy by R. Bohart
splendidulus (Bradley), 1920 (*Gorytes*); sw. U.S. to California
thalia (Handlirsch), 1895 (*Gorytes*); s. Africa
    transvaalensis Cameron, 1910 (*Gorytes*)
tricolor (Cresson), 1868 (*Gorytes*); U.S., Mexico: Sinaloa
    helianthi Rohwer, 1911 (*Gorytes*)
    rufocaudatus Mickel, 1916 (*Hoplisus*)
umbonicida Pate, 1941; S. America: Trinidad
umtalicus (Arnold), 1929 (*Arpactus*); s. Africa
vespoides (F. Smith), 1873 (*Gorytes*); centr. and S. America
    robustus Handlirsch, 1888 (*Gorytes*), new synonymy by R. Bohart
    sericeus Cameron, 1905 (*Icuma*), new synonymy by R. Bohart
    auropilosellus Cameron, 1912 (*Gorytes*), new synonymy by R. Bohart
whitei (Cameron), 1905 (*Hoplisus*); s. Africa

Genus Sagenista R. Bohart

Generic diagnosis: Medium to small, mostly black wasps; inner eye margins nearly parallel and widely separated or converging somewhat below, head broader than long in front view; median frontal groove weak or indistinct; labrum inconspicuous; last four male flagellomeres specially modified, flattened or concave beneath; first flagellomere two-thirds to nine-tenths as long as scape; mandible with an inner subtooth; pronotal collar rather even all across and closely appressed to scutum; female foretarsal rake well developed, basitarsus with three bladelike setae before apex; foreleg arolium of female much larger than other arolia; female midtibia stoutly produced at upper apex and bearing a finger shaped spine; posterolateral oblique scutal carina present; episternal sulcus short when present, ending at level of scrobe, scrobal sulcus when present joining episternal sulcus at about a right angle; acetabular carina present, sternaulus complete or partial; metapleuron gradually narrowing below; scutum impunctate; forewing spotted, banded or strongly darkened toward base, media diverging before cu-a, stigma moderate, veinlet of submarginal cell II between recurrents short; jugal lobe larger than tegula, hindwing media diverging at or very near cu-a; propodeal enclosure striate and/or areolate, propodeum otherwise coarsely areolate; spiracular groove not defined; gastral segment I narrow to stout but swelling evenly into II; male with seven visible terga, no concealed or visible sternal hairbrushes, sternum VIII usually protruding and sting shaped (fig. 175 C); female pygidial plate with disk slightly raised and punctate as opposed to terminal smooth part.

Geographic range: South America, especially Brazil. One of the six species ranges into Central America and Mexico.

Systematics: Sagenista is related to *Hoplisoides* as evidenced by general appearance, the presence of an acetabular carina, and the indistinct spiracular groove on the propodeum. However, the coarsely areolate propodeum, smooth scutum, enlarged female foreleg arolium, oddly spined female midtibia, distinctive female pygidial plate, and absence of concealed basal brushes on male sterna V to VI indicate a genus somewhat removed from others in the group. The most striking species is the type of genus, *scutellaris*. It has beautifully banded wings, a bright orange scutellum, and silvery micropubescence on the propodeum and tergum I. Other species are more sombre, but all have pictured wings and the elaborate propodeal network which is the trademark of the genus. We have seen a female of an Argentine species which is unusual in having flagellomere I about as long as the scape. A generic description and illustrations were given by R. M. Bohart (1967).

Biology: F. X. Williams (1928a) has published the only report of which we are aware. He recorded *Sagenista brasiliensis* (as *Gorytes*) nesting near Belem, Brazil in banks of rich soil along the jungle margins. The short burrows were located in masses of earth exposed by the roots of trees blown down by wind. Cells were provisioned with nymphal and adult fulgorids of several genera including *Dictyophora* and presumably *Thionia*. One cell contained six immature fulgorids, and another had five adults and one nymph. The wasp egg was attached to the side of the thorax. The unearthed and unhatched cocoons were "more or less enveloped by the remains of the homopterous victims and consisted of a soil cask that was gently rounded at the fore end and more narrowed and drawn out a little, nipplelike at the base." We have seen one female *Sagenista* from Rio de Janeiro (at Mus. Washington) pinned with a large adult membracid, presumably its prey.

Checklist of *Sagenista*

austerus (Handlirsch), 1893 (*Gorytes*); Brazil
brasiliensis (Shuckard), 1838 (*Gorytes*); Brazil
cayennensis (Spinola), 1841 (*Hoplisus*); Centr. and S. America
   anthracipenellus Taschenberg, 1875 (*Hoplisus*)
scutellaris (Spinola), 1841 (*Hoplisus*); S. America
   sanguinans Dominique, 1901 (*Harpactes*), new synonymy by R. Bohart
   fiebrigi Brèthes, 1909 (*Gorytes*), new synonymy by R. Bohart
seminiger (Dahlbom), 1843 (*Hoplisus*); Brazil
sepulchralis (Handlirsch), 1888 (*Gorytes*); Brazil

Genus Afrogorytes Menke

Generic diagnosis: Medium small wasps; inner eye margins converging strongly below (fig. 157 O); median frontal groove present but somewhat irregular; labrum incon-

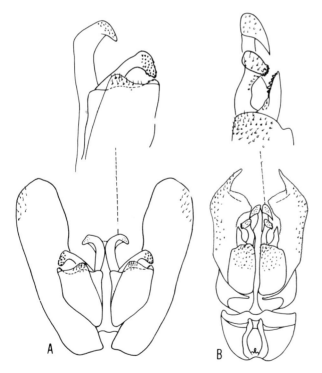

FIG. 179. Male genitalia in the tribe Gorytini; A, *Liogorytes polybia;* B, *Afrogorytes gibbosus.*

spicuous; vertex moundlike; last four male flagellomeres specially modified, flattened or concave beneath, last one elongate and arcuate; first flagellomere very short in male, less than half as long as scape in female; mandible with an inner subtooth; pronotal collar rather closely appressed to scutum, sharply edged in front near middle; female foretarsal rake moderately developed, basitarsus with two long blades before apex; female foreleg arolium not much larger than others; posterolateral oblique scutal carina present; episternal sulcus absent, scrobal sulcus foreshortened and slanting backward and downward; omaulus, sternaulus and acetabular carinae present, omaulus toothed below; metapleuron broadened below; scutum coarsely punctate, admedian lines joining to form a median carina; forewing lightly pictured, media diverging before cu-a, stigma moderate, veinlet of submarginal cell II between recurrents short; jugal lobe larger than tegula; hindwing media diverging slightly beyond cu-a which is strongly curved near distal end; midtibia with two apical spurs; propodeal enclosure small, longitudinally striate; no spiracular groove; gaster stoutly pedunculate; male with seven terga normally visible, sterna without concealed or visible hairbrushes, sternum VIII sword shaped distally, typical male genitalia: fig. 179 B; female pygidial plate well developed, smooth medially.

Geographic range: Southern Africa.

Systematics: A close relationship with *Hoplisoides* is evident. Principal distinguishing features of *Afrogorytes* are the single carina of the admedian scutal area, the pedunculate abdomen, and the unusually stout legs. Among the other structures figured and discussed briefly by Menke (1967b), the ventral tooth shaped process at the lower end of the omaulus is particularly striking. In some respects this resembles the tooth present in *Hoplisoides dentatus* and may indicate a mutant tendency in this group of wasps. A narrowing of the frons below is seen in varying degrees in *Hoplisoides,* particularly in males. *Afrogorytes* carries it to an extreme (fig. 157 O). There appear to be three closely related species in the genus, but this has not been thoroughly explored.

Biology: Unknown.

Checklist of *Afrogorytes*

gibbosus (Arnold), 1936 (*Arpactus*); s. Africa
monstrosus (Handlirsch), 1894 (*Gorytes*); s. Africa
silverlocki (Turner), 1913 (*Gorytes*); s. Africa

Tribe Stizini

Stizins are medium-small to large, rather stout wasps which are often yellow and black, red and black, or more rarely, all black. Several species of the genus *Stizus* superficially resemble certain cicada killers of the gorytin genus *Sphecius*, whereas others are remarkably similar to the Bembicini.

The largest members of the tribe are in the genus *Stizus,* the Nearctic species of which are rather rare in spite of the abundance of their presumed Orthopteran hosts. On the other hand the medium to small species of *Bembecinus* are among the commonest bush visiting wasps in relatively dry regions.

Diagnostic characters:

1. (a) Eyes converging below (sometimes only slightly), (b) ocelli well developed.
2. (a) Labrum exserted but exposed area broader than long (fig. 180 A-C), (b) palpal formula 6-4.
3. (a) Scutum with admedian lines well separated, (b) notauli absent or faint, (c) oblique scutal carina present (fig. 156 C), (d) scutellum with lamelliform edge overlapping metanotum (fig. 180 I).
4. (a) Omaulus absent, (b) acetabular carina and sternaulus absent, (c) episternal and scrobal sulci forming a continuous arc, the episternal-scrobal sulcus (absent in *Bembecinus*), thus defining a hypoepimeral area.
5. (a) Metapleuron consisting of upper metapleural area only, separated by a sulcus from propodeum; (b) upper lateral surface of propodeum not spined nor dentate, (c) propodeal enclosure extending far onto vertical slope.
6. (a) Midcoxae slightly separated, (b) midtibia with two apical spurs (except a few *Stizoides*), (c) arolia equal on all legs (except some female *Bembecinus*), (d) female with a foretarsal rake, (e) plantulae absent.
7. (a) Forewing with media diverging close to or before cu-a, (b) prestigmal length of submarginal cell I much more than length of wing beyond marginal cell, (c) stigma small (fig. 180 D,E), (d) three submarginal cells, II not petiolate (except some *Bembecinus*), receiving both recurrent veins (fig. 180 D,E).
8. (a) Hindwing costa well developed, (b) media diverging before cu-a (fig. 180 G,H), jugal lobe about half as long as anal area.
9. (a) Gaster sessile, tergum I nearly as broad as II; (b) sternum I with a median ridge toward base, (c) male with seven visible terga, VII with laterobasal spiracular lobes (fig. 180 M), (d) male sternum VIII with three distal prongs (fig. 180 J,K), (e) male genitalia with volsella differentiated into cuspis and digitus, (f) female pygidial plate often absent or weakly indicated (fig. 180 F).

Systematics: As discussed under the systematics of the subfamily, the Stizini occupy an intermediate position between gorytins and bembicins (fig. 7).

Within the Nyssoninae, the Stizini exhibit such generalized characters as two midtibial spurs, palpi of usual type, admedian scutal lines separated and the area between them flat, normal ocelli; sessile abdomen, and three submarginal cells. Some of the specializations that occur throughout the Stizini are: labrum exserted, oblique scutal carina present, scutellum overlapping metanotum with a thin lamina, principal landmark sulci on mesopleuron reduced to an episternal-scrobal sulcus that defines hypoepimeral area or sulcus absent, propodeal enclosure extending far onto vertical slope, submarginal cell I greatly elongate, sternum I with a median ridge

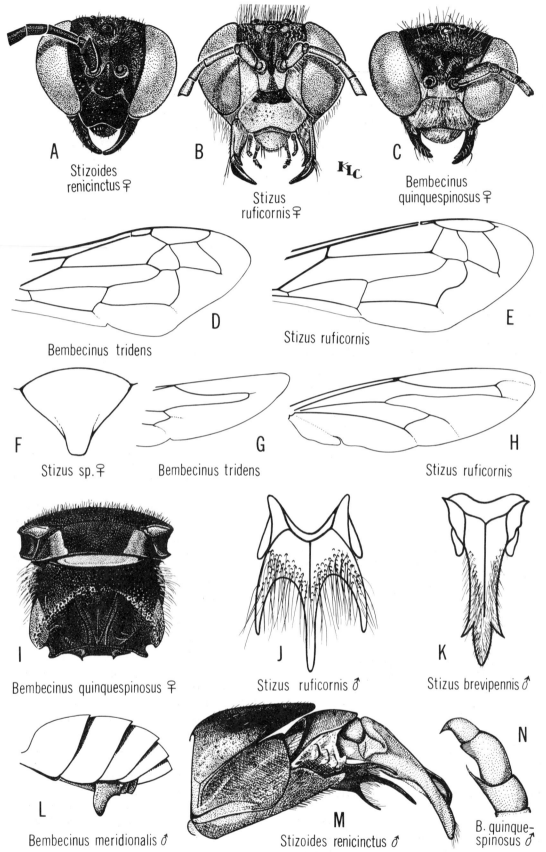

FIG. 180. Morphology in the tribe Stizini; A-C, facial portraits, D,E, forewing venation; F, pygidium; G,H, hindwing venation; I, scutellum, metanotum and propodeum, dorsal; J,K, sternum VIII; L, gastral segments I-V, lateral; M, terminalia, lateral; N, last three flagellomeres, lateral.

toward base, male tergum VII with laterobasal spiracular lobes, and male sternum VIII with three prongs. Also occurring in some Stizini are simple mandibles, petiolate submarginal cell II, enlarged arolium of foreleg in female, reduction of female pygidial plate, and only one midtibial spur.

The many structural similarities between *Sphecius,* a gorytin genus, and *Stizus* led earlier workers to place them in the same tribe. Undoubtedly many of the similarities can be explained by parallel development, but to do this entirely, strains the imagination. However, if it were not for the omaulus and shorter first submarginal cell, *Sphecius* would be placed in the *Stizini.* In any case it appears that *Stizus* developed from a *Sphecius*-like form. *Stizoides* can be viewed as an evolutionary excursion into cleptoparasitism, and *Bembecinus* as a sideline which took advantage of progressive nest provisions using a different type of prey.

Biology: This is discussed under the genera, but in summary it can be said that all members of the tribe are ground nesting; prey is Orthoptera or Mantodea in the case of *Stizus* and *Stizoides,* although the latter is a cleptoparasite; and *Bembecinus* utilizes a variety of homopterous "hoppers." As pointed out by Evans (1955b), the evolution in nesting habits starts with *Stizus,* which mass provisions, but it is already specialized since it lays its egg on the first grasshopper prey rather than the last. *Bembecinus* has advanced much farther, since it lays an egg on a special mound in the empty cell then provisions progressively as the larva grows.

Key to genera of Stizini

1. Propodeum compressed into a ridge posterlaterally, the ridge usually notched (fig. 180 I); episternal-scrobal sulcus absent; hindwing media diverging close to cu-a; median cell with only one distal appendix (fig. 180 G); male flagellomere IX with a spinelike apical projection, at least in Western Hemisphere forms (fig. 180 N); mandible usually dentate subapically (fig. 180 C); cosmopolitan............ *Bembecinus* A. Costa, p. 529
   Propodeum not ridgelike posterolaterally; episternal-scrobal sulcus present; hindwing media diverging far before cu-a (fig. 180 H); median cell with two distal appendices (fig. 180 H); male flagellomere IX without a spinelike projection 2
2. Eyes only slightly closer below than above (fig. 180 B); mandible dentate subapically (fig. 180 B); a low and somewhat double convexity present below midocellus; widespread but not Neotropical nor Australian .... *Sitzus* Latreille, p. 525
   Eyes converging strongly below (fig. 180 A); mandible simple (fig. 180 A); a prominent hump below midocellus; widespread but not Neotropical nor Australian ....
   .... *Stizoides* Guérin-Méneville, p. 527

Genus Stizus Latreille

Generic diagnosis: Eyes converging slightly below, at least narrower below than above (fig. 180 B); a low and somewhat double convexity below midocellus; mandible dentate subapically; exposed part of labrum moderately convex, usually a fourth to a third as long as broad; male antenna without a projection on flagellomere IX; scutellum typically (but not always) with a medial, pubescent depression in the female; propodeum not compressed into a ridge posterolaterally; midtibia with two apical spurs; forewing with submarginal cell II open in front (fig. 180 E); hindwing media diverging far before cu-a (fig. 180 E), median cell with two distal appendices; precoxal sulcus usually present.

Geographic range: The third largest genus in the subfamily behind *Bembix* and *Bembecinus.* We have listed 127 species, and these range over much of the temperate and tropical areas of the world with the apparent exception of southeast Asia, Australia, and South America. In North America there are four described species: *brevipennis* from Maryland to South Dakota and Texas, *iridis* from Utah and California, *occidentalis* from California, and *texanus* from southwestern U.S.

Systematics: Wasps of this genus are mostly medium sized to large (up to 34 mm long), often black and yellow with some red markings (as in *brevipennis, texanus*), some mostly yellow (as in *iridis, occidentalis*), a few largely reddish (as in *maronis* from Africa), and even essentially black forms (as in *fuscatus* from Egypt).

Stizus differs from the other two genera of the tribe especially by its subparallel inner eye margins. On the whole the genus is remarkably uniform in structure, most of the species being stout and resembling those of the gorytin genus *Sphecius.* Several authors have distinguished two main groups based on the presence or absence in the female of a pubescent depression on the scutellum. We have not studied enough material to rule on the validity of subgenera, but if these were recognized, about two-thirds of the species would fall in *Stizolarra* Saussure, without the scutellar depression, and the remainder would be in the typical subgenus.

All of the American species are in *Stizus s.s.,* and they exhibit several structural differences. Males of *brevipennis* and *texanus* have sternum VIII narrow, the median prong stout and turned down stinglike (fig. 180 K), as in many males of *Sphecius.* In the other two species, sternum VIII has three rather straight and slender prongs (as in fig. 180 J). The females of *brevipennis* and *texanus* differ from the other two by having a rather well formed pygidial plate. Thus, we have two Nearctic groups, one of which (*occidentalis* and *iridis*) is very close to *ruficornis,* type of the genus.

Among the exotic species, one is deserving of particular note. *S. scoliaeformis* from southwest Africa has the scutellar depression but differs from all other *Stizus* in the oblique rather than vertical slope of the pronotum. Furthermore, sternum II is carinate medially, and in the female the pygidial plate is well developed.

Particularly illuminating papers on *Stizus* are those of Handlirsch (1892), Mercet (1906), Mochi (1939c), Dow (1941), and Lohrmann (1943).

Biology: Nests are provisioned with grasshoppers, katydids, or mantids. In a brief note, Williams (1914b) recorded the habits of *brevipennis* in Kansas. The female wasp flies about the stems of *Helianthus* until it finds a tettigoniid (*Conocephalus*), which it deposits in a ground nest. Deleurance (1941) gave *Mantis religiosa* as prey of *ruficornis* in France.

A rather complete study is that reported by Tsuneki (1965b) on *Stizus pulcherrimus*. The burrows were dug in gravelly soil near an old riverbed in Seoul, Korea. The nests were communal and interspersed among those of *Cerceris* and *Lestica*. Some 20 *Stizus* nests were marked with heaps of debris at the entrances. Prey were nymphs and adults of Acrididae and, less frequently, Tettigoniidae. As many as eight hoppers were placed in each cell, the first one supplied with an egg attached near the wing base. Nest entrances were closed while the wasp was foraging. Usually two side holes or accessory burrows were constructed near each nest. The fully-grown larva constructed a tightly woven silken cocoon festooned with sand grains. Hosts recorded by Tsuneki were *Chorthippus dubius* (Zubovsky), *Parapleurus alliaeus* Germar, *Aiolopus tamulus* Fabricius, *Stauroderus schmidti* Ikonnikov, *Acrida lata* (Motschulsky), *Trilophida annulata* Thunberg, and *Conocephalus maculatus* (Le Guillou).

Checklist of *Stizus*

adelphus W. F. Kirby, 1900; Aden
aegyptius Lepeletier, 1845; Egypt
aestivalis Mercet, 1906; Spain
   ssp. *merceti* Beaumont, 1951; Morocco
anchoratus Mantero, 1917; ne. Africa
anchorites Turner, 1920; Egypt, Israel
annulatus (Klug), 1845 (*Larra*); Eurasia, n. Africa, Middle East
   subapicalis Walker, 1871 (*Larra*)
   eversmanni Radoszkowski, 1877
   kisilkumii Radoszkowski, 1877
   picticornis F. Morawitz, 1891
apicalis Guérin-Méneville, 1844; Senegal, Ethiopia
arenarum Handlirsch, 1892; Algeria
arnoldi Mochi, 1939; Egypt
atrox (F. Smith), 1856 (*Larra*); centr. and s. Africa
   flavomaculatus F. Smith, 1873 (*Larra*)
   pentheres Handlirsch, 1895
aurifluus Turner, 1916; Ghana
basalis Guérin-Méneville, 1844; n. and nw. Africa
   infuscatus Klug, 1845 (*Larra*)
baumanni Handlirsch, 1901; e. and se. Africa
beccarii Mantero, 1917; Ethiopia
bensoni Arnold, 1951; Mali
berlandi Arnold, 1945; Madagascar
biclipeatus (Christ), 1791 (*Vespa*); India
   ruficornis Fabricius, 1787 (*Tiphia*), nec J. Forster, 1771
   cingulatus Fabricius, 1798 (*Larra*)
   nubilipennis F. Smith, 1856 (*Larra*)
bipunctatus (F. Smith), 1856 (*Larra*); Eurasia
bizonatus Spinola, 1838; Egypt
   sphegiformis Klug, 1845 (*Larra*)
brevipennis Walsh, 1869; U.S. e of 100th mer.
   brendeli Taschenberg, 1875 (*Larra*)
breyeri Arnold, 1936; s. Africa
chrysorrhoeus Handlirsch, 1892; s. Africa
coloratus Nurse, 1903; India
combustus (F. Smith), 1856 (*Larra*); n. and centr. Africa; Israel
   fuliginosus Klug, 1845 (*Larra*), nec Dahlbom, 1843
continuus (Klug), 1835 (*Larra*); Europe
delessertii Guérin-Méneville, 1844; India
deserticus Giner Marí, 1945; n. Africa
dewitzii Handlirsch, 1892; s. Africa
   erythraspis Cameron, 1910
   propodealis Bischoff, 1915
dispar F. Morawitz, 1888; sw. USSR: Turkmen S.S.R.; Iran
elegans Dahlbom, 1845; Egypt
ellenbergeri Arnold, 1929; s. Africa
emir Handlirsch, 1901; sw. USSR: Transcaspia
erythraeensis Mantero, 1917; Ethiopia
?*erythrocephala* (Taschenberg), 1880 (*Larra*); Ethiopia, possibly not a *Stizus*
euchromus Handlirsch, 1892; sw. USSR
eugeniae Gussakovskij, 1935; centr. and sw. USSR
excellens Lohrmann, 1943; s. Africa
   eximius Arnold, 1936, nec F. Morawitz, 1894
eximius F. Morawitz, 1894; sw. USSR: Turkmen S.S.R.
fasciatus (Fabricius), 1781 (*Bembex*); s. Europe, Middle East, Mongolia
   integer Fabricius, 1804 (*Bembex*)
   terminalis Dahlbom, 1845
fedtschenkoi Radoszkowski, 1877; USSR: Kazakh S.S.R.
ferrandii Magretti, 1898; Somalia
ferrugineus (F. Smith), 1856 (*Larra*); equatorial Africa
   ?*ritsemae* Handlirsch, 1895
franzi Turner, 1916; Zaire
   neavei Kohl, 1913, nec Turner, 1912
   congoensis Mantero, 1917
fuscatus Bingham, 1897; Egypt
fuscipennis (F. Smith), 1856 (*Larra*); s. Africa
gobiensis Tsuneki, 1970; Mongolia
gracilipes Handlirsch, 1892; sw. USSR: Caucasus
hamatus Arnold, 1929; s. Africa
handlirschi Radoszkowski, 1893; s. Asia, Middle East
hebraeus Balthasar, 1954; Israel
hispanicus Mocsáry, 1883; Mediterranean area
   villosus A. Costa, 1887
histrio F. Morawitz, 1888; sw. USSR: Turkmen S.S.R.
hugelii Handlirsch, 1892; India
hyalipennis Handlirsch, 1892; Mediterranean area
imperator Nurse, 1903; India
imperialis Handlirsch, 1892; s. Africa
   tulbaghensis Arnold, 1936
   ssp. *conspicuus* Arnold, 1952; Kenya
iridis Dow, 1942; U.S.: Utah, California

jordanicus Lohrmann, 1942; Jordan
kaszabi Tsuneki, 1971; Mongolia
kiseritzkii Gussakovskij, 1928; sw. USSR
koenigi F. Morawitz, 1888; sw. USSR
   tages W. F. Kirby, 1889
kohlii Mocsáry, 1883; Syria
lacteipennis Mocsáry, 1883; sw. USSR
lepidus (Klug), 1845 (*Larra*); Egypt
lineatus (Fabricius), 1793 (*Bembex*); loc. unknown
lohrmanni R. Bohart; s. Africa, new name for *arnoldi* Lohrmann, nec Mochi, 1939
   pubescens Arnold, 1929, nec Klug, 1835
   arnoldi Lohrmann, 1943, nec Mochi, 1939
lughensis Magretti, 1898; Somalia
luteotaeniatus Gussakovskij, 1933; Iran
lutescens Radoszkowski, 1877; USSR: Kazakh S.S.R.
marnonis Handlirsch, 1892; Sudan
marshalli Turner, 1912; s. Africa
marthae Handlirsch, 1892; Egypt, Iran, Israel
   cheops Morice, 1897
   cheops Bingham, 1897, nec Morice, 1897
melanoxanthus (F. Smith), 1856 (*Larra*); India
melanurus Handlirsch, 1892; s. Africa
melleus (F. Smith), 1856 (*Larra*); India
mocsaryi Handlirsch, 1895; n. Africa
mongolicus Tsuneki, 1971; Mongolia
multicolor Turner, 1916; e. and s. Africa
nadigi Roth, 1933; Morocco
neavei Turner, 1912; s. Africa
nigriventris Arnold, 1951; Mali
niloticus Handlirsch, 1892; Egypt to Israel
occidentalis J. Parker, 1929; U.S.: California, Arizona
ocellatus Gistel, 1857; loc. unknown
orientalis Cameron, 1890; India
pauli Mantero, 1917; Ethiopia
perrisii Dufour, 1838; Europe
   ssp. *ibericus* Beaumont, 1962; Spain
pictus Dahlbom, 1845; Egypt
pluschtschevskii Radoszkowski, 1888; Eurasia
praestans F. Morawitz, 1893; sw. USSR: Turkmen S.S.R.
pulcherrimus (F. Smith), 1856 (*Larra*); China
pygidialis Handlirsch, 1892; se. Europe
quartinae (Gribodo), 1884 (*Sphecius*); Guinea
raddei Handlirsch, 1889; sw. USSR: Turkmen S.S.R.
rapax Handlirsch, 1892; Egypt
richardsi Arnold, 1959; s. Africa
rubellus Turner, 1912; s. Africa
rubroflavus Turner, 1916; Gambia
rufescens (F. Smith), 1856 (*Larra*); India, Sri Lanka, China
ruficornis (J. Forster), 1771 (*Vespa*); Europe, w. USSR, Middle East
   ruficornis Fabricius, 1787 (*Bembex*), nec J. Forster, 1771
   gaditana Gmelin, 1790 (*Vespa*)
   pubescens Klug, 1835 (*Larra*)
   ornatus Dahlbom, 1845
   komarovi Radoszkowski, 1888
   scutellaris W. F. Kirby, 1900
   distinguendus Handlirsch, 1901
   ssp. *strigatus* Mochi, 1939; Egypt
   ssp. *eremicus* Bytinski-Salz, 1955; Middle East
rufipes (Fabricius), 1804 (*Liris*); Algeria
   grandis Lepeletier, 1845
   gigas Walsh and Riley, 1868
rufiventris Radoszkowski, 1877; USSR: Kazakh S.S.R.
   sarmaticus F. Morawitz, 1891
   ssp. *compar* Handlirsch, 1892; Egypt, Iran
rufocinctus Dahlbom, 1845; Egypt
rufoniger Mochi, 1939; Egypt
saharae Roth, 1934; Libya
saussurei Handlirsch, 1895; s. Africa
savignyi Spinola, 1838; ne. Africa, Israel
   succineus Klug, 1845 (*Larra*)
schmiedeknechti Handlirsch, 1898; Egypt
scoliaeformis Arnold, 1929; S.W. Africa
scolioides Kokujev, 1902; sw. USSR
sexfasciatus (Fabricius), 1793 (*Bembex*); India
spectrum Handlirsch, 1901; Afghanistan
spinulosus Radoszkowski, 1876, Egypt
stevensoni Arnold, 1936; s. Africa
storeyi Turner, 1920; Egypt
tenuicornis (F. Smith), 1856 (*Larra*); Gambia
texanus Cresson, 1872; U.S.: Texas, Arizona; Mexico: Chihuahua
transcaspicus Radoszkowski, 1893; sw. USSR, Iraq
tricolor Handlirsch, 1892; Middle East
tunetanus A. Costa, 1893; Tunisia
vespiformis (Fabricius), 1775 (*Sphex*); India, Mauritius, Sri Lanka
vespoides (Walker), 1871 (*Larra*); ne. Africa, Israel
   magnificus F. Smith, 1873 (*Larra*)
   argenteus Taschenberg, 1875 (*Larra*)
wheeleri Arnold, 1959; s. Africa
zhelochovtzevi Gussakovskij, 1952; sw. USSR
zimini Gussakovskij, 1928; Iran, sw. USSR
zonatus (Klug), 1845 (*Larra*); n. Africa
   dimidiatus Taschenberg, 1875 (*Larra*)
zonosoma Handlirsch, 1895; Egypt

Genus Stizoides Guérin-Méneville

Generic diagnosis: Eyes converging strongly below (fig. 180 A); a rather prominent frontal hump below midocellus; mandible simple; labrum strongly arched, exposed part about two-thirds as long as broad; male antenna without a projection on flagellomere IX; scutellum of female not depressed medially; propodeum rounded rather than ridgelike posterolaterally; midtibia with two apical spurs (except in subgenus *Scotomphales* in which there is a distinctive mediobasal pubescent spot on sternum II); forewing with submarginal cell II open in front or rarely sessile, not petiolate; hindwing media diverging far before cu-a, median cell with two distal appendices; precoxal sulcus present or absent; terminal gastral segments of male as in fig. 180 M.

Geographic range: We have listed 28 species: 26 from the Old World and two from North America. Apparently, only Australia and southeast Asia, among the warmer parts of the world, have no recorded species. Since there has been no revision of the genus in this century, it is likely that synonymy will alter the figure of 26 species for the Old World, and new finds will undoubtedly swell the number. Of the two North American forms, *foxi* is a rare species in southern Arizona and Baja California, whereas *unicinctus* is more abundant and widespread in the United States. The latter species is most common west of the Mississippi River, penetrates southwestern Canada, and is known from Zacatecas, Mexico.

Systematics: These wasps are about the size of a small *Polistes*. A black and red color pattern with blackish wings is dominant in the genus and may be an example of warning coloration. The red markings occur in a variety of body areas, depending on the species. One of the North American species (*renicinctus*) has tergum II marked with red. The other (*foxi*) is all black.

The important characteristics are the converging eyes, simple mandible, frontal hump, and strongly arched labrum. The Midtibial spurs are typically two, but in some species, such as the Palearctic *tridentatus*, the posterior spur is small. In still others, such as the African species *niger* and *fenestratus*, there is only a single spur in both sexes. This condition controverts an important tribal distinction between Stizini on the one hand and Bembicini on the other. However, the oblique scutal carina and normal ocelli of *Stizoides* are immediate points of separation. The single-spurred species have been placed in the subgenus *Scotomphales* Vachal by Gillaspy (1963a) who gave similar rank to *Tachystizus* Pate and *Stizoides s.s.* On the basis of the tentative characters given by Gillaspy and on an examination of more than half of the known species in the genus, the following key to subgenera is offered:

Key to subgenera of *Stizoides*

1. Without a distinctive, mediobasal, micropunctate and pubescent spot on sternum II (Holarctic; Ethiopian .... *Tachystizus* Pate
   With a distinctive, mediobasal, micropunctate and pubescent spot on sternum II 2
2. Midtibia with one apical spur (Ethiopian) .... *Scotomphales* Vachal
   Midtibia with two apical spurs (Palearctic; Ethiopian) ....
   .... *Stizoides* Guérin-Méneville, *s.s.*

Biology: Based on the few observations that have been made on the nesting habits of *Stizoides*, it seems reasonably certain that they are cleptoparasites of Sphecinae and possibly also of Bembicini.

In this country, Williams (1914b) reported *renicinctus* victimizing *Prionyx atratus* in Kansas by ovipositing on its grasshopper provisions. Ahrens and Ahrens (1953) made a similar report on the European *tridentatus* which parasitizes *Sphex maxillosus*.

Rau and Rau (1918) gave a description of the manner in which *Prionyx thomae* is parasitized by *Stizoides renicinctus* in Missouri. The grasshopper-provisioned and closed *thomae* nest was excavated by the *renicinctus*, which destroyed the *thomae* egg, laid one of its own, and then refilled the hole.

The most complete report has been that of La Rivers (1945), who observed the habits and parasites of *Palmodes laeviventris* in Elko County, Nevada. One of the most destructive parasites was *Stizoides renicinctus*, which decreased the effectiveness of the *Palmodes* as a predator of the Mormon Cricket, *Anabrus simplex* Haldeman. The *renicinctus* female dug into the completed nest, destroyed the *Palmodes* egg, replaced it on the *Anabrus*, and rather carelessly refilled the hole, leaving an evident depression. La Rivers reported that if flesh flies had previously attacked the *Anabrus*, the invading *Stizoides* did not deposit an egg of her own and did not reclose the nest. A 7.5 percent parasitization, which amounted to about 10,000 destroyed *Palmodes* in a 350 square yard plot, was estimated.

Ahrens and Ahrens (1953) mentioned a persistent, low, searching flight of female *tridentatus* over the host *Sphex* colonies. In 1958 on the western slope of the Santa Rita Mountains of Arizona, R. Bohart observed many females of *foxi* exhibiting the same sort of searching flight. In this case, no host was identified. A possible instance of *Stizoides* cleptoparasitism involving *Bembix* in the Lake Victoria area of east Africa, as observed by G. D. Carpenter, was reported by Gillaspy (1963a). Sleeping aggregations of *renicinctus* have been noted by several authors, among them, Rau (1938) and Gillaspy (1963a).

Checklist of *Stizoides*

abdominalis (Dahlbom), 1845 (*Stizus*); Egypt
*amoenus (F. Smith), 1856 (*Larra*); Mali
  bicolor Taschenberg, 1875 (*Larra*), nec Fabricius, 1804
assimilis (Fabricius), 1787 (*Sphex*); India
  fasciatus Fabricius, 1798 (*Larra*), nec *Stizus fasciatus* (Fabricius), 1781 (*Bembex*)
  calopteryx Handlirsch, 1892 (*Stizus*), new name for *fasciatus* Fabricius, 1798
*blandinus (F. Smith), 1856 (*Larra*); India
citrinus (Klug), 1845 (*Larra*); Egypt
*conscriptus (Nurse), 1903 (*Stizus*); Pakistan
*cornutus (F. Smith), 1873 (*Larra*); India
crassicornis (Fabricius), 1787 (*Tiphia*); Mediterranean, sw. USSR
  rufipes Olivier, 1789 (*Bembex*)
  fulvipes Eversmann, 1849 (*Stizus*)
ctenopus (Arnold), 1929 (*Stizus*); s. Africa
cyanipennis (Saussure), 1887 (*Stizus*); Madagascar
cyanopterus (Gussakovskij), 1928 (*Stizus*); sw. USSR
egregius (Gussakovskij), 1928 (*Stizus*); sw. USSR
*erythrogaster (Turner), 1917 (*Stizus*); India
*fenestratus (F. Smith), 1856 (*Larra*); centr. and s. Africa
foxi Gillaspy, 1963; U.S.: Arizona; Mexico: Baja California
funebris (Handlirsch), 1900 (*Stizus*); s. Africa
klugii (F. Smith), 1856 (*Larra*); Egypt
  apicalis Klug, 1845 (*Larra*), nec Guérin-Méneville,

1844 (*Stizus*)
ssp. *numidus* (Schulz), 1905 (*Stizus*); Algeria
melanopterus (Dahlbom), 1845 (*Stizus*); w. Asia, se. Europe
  concolor Eversmann, 1849 (*Stizus*)
mionii Guérin-Méneville, 1844; Senegal, Mali
niger (Radoszkowski), 1881 (*Stizus*); equatorial Africa
  niger Vachal, 1899 (*Omphalius*)
*persimilis (Turner), 1918 (*Stizus*); e. Africa
poecilopterus (Handlirsch), 1892 (*Stizus*); ne. Africa, Iran
renicinctus (Say), 1823 (*Stizus*); N. America
  unicinctus Say, 1824 (*Stizus*), emendation
*simpsoni (Turner), 1916 (*Stizus*); Ghana
stenopus (Arnold), 1929 (*Stizus*); s. Africa
tridentatus (Fabricius), 1775 (*Crabro*); Mediterranean area to centr. Asia
  bifasciatus Fabricius, 1798 (*Larra*)
  unifasciatus Radoszkowski, 1877 (*Stizus*)
tuberculiventris (Turner), 1912 (*Stizus*); e. and s. Africa
verhoeffi Bytinski-Salz, 1955; Israel

Genus Bembecinus A. Costa

Generic diagnosis: Eyes converging below, often strongly (fig. 180 C); a moderate convexity below midocellus, no special hump; mandible dentate subapically (fig. 180 C); exposed part of labrum moderately convex, about half as long as broad; male antenna usually with a spinelike prolongation of flagellomere IX (fig. 180 N) (present in all American forms); scutellum of female not depressed medially; propodeum compressed and sharply edged posterolaterally, the edge usually notched below (fig. 180 I) (notched in all American forms); midtibia with two apical spurs; forewing with submarginal cell II sometimes petiolate; hindwing media diverging close to cu-a, median cell with more forward distal appendage only (fig. 180 G); precoxal and episternal-scrobal sulci absent.

Geographic range: Cosmopolitan and the second largest genus in the subfamily. *Bembecinus* occurs on all the continents and in many island groups. We have listed 150 species, with the largest share in the Ethiopian Region. Arnold (1929) considered some 30 species on the African continent, and in 1945 he treated five additional ones from Madagascar. Willink (1949) listed eight Neotropical species, and Krombein and Willink (1951) recorded five from North America, one of which extended across the Isthmus of Panama. Beaumont (1954b) gave 26 species for the Palearctic Region. J. van der Vecht (1949) keyed the Indo-Australian forms.

Systematics: These medium-sized wasps are readily identified by most collectors of Hymenoptera. Yet, it remained for J. Parker (1929) to recognize the validity of Costa's genus a full 70 years after it was proposed. In the meantime, Handlirsch, Kohl and Arnold, among others had considered all stizins as *Stizus* Latreille, The American fauna is abundantly distinct morphologically, but many of the characteristics break down when the Ethiopian species groups are considered. Of special value are the compressed lateroposterior angles of the propodeum (fig. 180 I), which are subject to considerable variation but are still a major generic separation point. Even more important features are the single appendix of the median cell in the hindwing (fig. 180 G) and the absence of the episternal-scrobal sulcus. On the other hand the spine of the eleventh male antennal article, which is characteristic of so many species (fig. 180 N), is absent in a number of African forms, such as *oxydorcus, cinguliger, haplocerus, laterimacula, rhopaloceroides, mutabilis, rhopalocerus, mirus,* and *assentator.*

Several authors, such as Handlirsch (1908), Arnold (1929), Lohrmann (1943), and Beaumont (1954b) have contributed to the group concept in *Bembecinus* in order to break up the large genus into more useful fractions. Exact agreement on group limits has not been achieved. In the suggested outline below, we have largely followed Beaumont (1954b) with respect to the Palearctic fauna and have condensed Arnold's arrangement of the Ethiopian forms.

Male antennal article XI simple.
1. *B. inermis* group (including *caffer* and related species). Antenna relatively slender; sterna simple; female front arolium not unusually enlarged. Ethiopian and Australian Regions.
2. *B. cinguliger* group (including *oxydorcus, rhopalocerus* and related species). Antenna clavate; sterna simple or with a prong on male sternum II; female front arolium much larger than other arolia. Ethiopian Region.

Male antennal article XI produced into a spine beneath (fig. 180 N).
3. *B. discolor* group. Fore basitarsus of female about 1.5 times as long as broad; male sternum VI with a small tooth, II and III unarmed. Palearctic Region (n. Africa).
4. *B. tridens* group (including *loriculatus*). Fore basitarsus of female at least 2.0 times as long as broad; male sterna usually without teeth or prongs; male hindfemur tapering gradually toward apex, not spined posteriorly; female hindfemur at least 3.0 times as long as broad; punctation of female sterna usually well spaced. Worldwide.
5. *B. peregrinus* group (includes *meridionalis, crassipes, gynandromorphus,* and related species). Fore basitarsus of female at least 2.0 times as long as broad; male sterna II, III and VI often armed (fig. 180 L); male hindfemur rather truncate distally or deformed, posterior (inner) aspect almost always spined; female hindfemur less than 3.0 times as long as broad, female sterna finely and densely punctate. Palearctic, Ethiopian, and possibly Neotropical Regions.

American species fall into the *tridens* group, although the South American *berlandi, bridarollii,* and *kuehlhorni* may have ridgelike projections on male sternum III. At least in *kuelhorni,* these projections may vary greatly in size. Most of the American species have the posterolateral area of the propodeum incurved or incised. Species with little or no excision of the propodeum are the North American *wheeleri* (*tridens* group), the Palestinean *bytinskii* (*tridens* group), and the South African *rhopalocerus* and *rhopaloceroides* (*cinguliger* group). Except for *inermis*, most Australian species seem to belong to the *tridens* group.

An interesting and difficult problem facing students of *Bembecinus* is the phenomenon of xanthochroism in certain species. Arnold (1945) illustrated the range in males of the Madagascan *mirus* from nearly all black to nearly all yellow, the female being rather consistently on the darker side of intermediate. Tsuneki (1965a) discussed a similar case in the Asian *hungaricus*. In North America similar situations have led to confusion and synonymy, as pointed out by Krombein and Willink (1951). Our species, which exhibit xanthochroism, are *quinquespinosus*, *neglectus*, and *wheeleri*. The capture on a single bush of structurally identical "yellow" and "black" males of *Bembecinus* is a striking and puzzling experience. Krombein and Willink offered arguments against humidity, distribution, or seasonal influences as responsible factors. They suggested the possibility of food involvement, since increased size seems to be correlated with increased amount of yellow.

Biology: The species nest in hard dirt or more frequently in sandy situations, sometimes in large colonies. For this reason they may appear locally in great numbers. The larvae are fed progressively on a variety of leafhoppers and other Homoptera in the "hopper" category. An excellent treatise on the ethology of *Bembecinus* is that of Evans (1955b). Additional information was given by Evans (1966a) and Evans and Matthews (1971b). In the 1955 paper a report of the nesting activities of *neglectus* in Kansas is followed by a comparison of prey-gathering and nesting habits of gorytins, stizins, and bembicins with a suggested evolution of behavioral types. Some of the features of *Bembecinus*, which are unique in the Stizini and which place the genus relatively high on the evolutionary scale, are change in prey from Orthoptera to Homoptera, change from mass provisioning to progressive provisioning, change from laying the egg on the first prey in the cell to laying it in the empty cell, preparation of a small mound of earth by gluing sand grains together and then attaching an egg at a 45° angle on top of it. The progressive provisioning and deposition of the egg in an empty cell are traits like those of the more advanced Bembicini, such as *Stictia*. Evans noted that *Bembecinus* always closes the nest when leaving and spends little time there when not engaged in provisioning. Thirteen papers dealing with biology were cited by Evans (1955b).

Natural enemies listed by Evans (1966a) were the mutillid, *Smicromyrme viduata* Pallas, on *B. tridens* in Italy; *Nysson dimidiatus* as a possible cleptoparasite of *B. tridens* in France; and the chrysidids, *Hedychrum chalybeum* Dahlbom and *Holopyga chrysonota* Forster, both on *B. tridens* in Europe.

Checklist of *Bembecinus*

acanthomerus (Morice), 1911 (*Stizus*); n. Africa
aemulus (Handlirsch), 1895 (*Stizus*); s. Africa
agilis (F. Smith), 1873 (*Larra*); Argentina to Guatemala
    cingulatus F. Smith, 1856 (*Larra*); nec Fabricius, 1798
    ornaticauda Cameron, 1912 (*Bembidula*)
alternatus van der Vecht, 1949; Java
angustifrons (Arnold), 1940 (*Stizus*); s. Africa
anthracinus (Handlirsch), 1892 (*Stizus*); New Guinea; Indonesia: Timor
  ssp. *ogasawaraensis* Tsuneki, 1970; Bonins
  ssp. *mukodzimaensis* Tsuneki, 1970; Bonins
antipodum (Handlirsch), 1892 (*Stizus*); Australia
argentifrons (F. Smith), 1856 (*Larra*); se. Africa
asiaticus Gussakovskij, 1936; centr. Asia, Middle East
asphaltites Beaumont, 1968; Israel
assentator (Arnold), 1945 (*Stizus*); Madagascar
asuncionis (Strand), 1910 (*Stizus*); Paraguay, Argentina
atratus (Arnold), 1936 (*Stizus*); s. Africa
barbarus (Beaumont), 1950 (*Stizus*); Algeria
barkeri (Arnold), 1940 (*Stizus*); s. Africa
berlandi Willink, 1952; Brazil
bernardi Beaumont, 1954; Egypt
bidens (Arnold), 1933 (*Stizus*); Ethiopia
bicinctus (Taschenberg), 1875 (*Larra*); Argentina
bimaculatus (Matsumura and Uchida), 1926 (*Stizus*); Ryukyus
    okinawanus Sonan, 1928 (*Stizus*)
    ryukyuensis Tsuneki, 1968
bishoppi Krombein and Willink, 1951; U.S.: Texas
boer (Handlirsch), 1900 (*Stizus*); s. Africa
    rufipes Arnold, 1929 (*Stizus*), nec Fabricius, 1804
    willowmorensis Arnold, 1936 (*Stizus*)
bolivari (Handlirsch), 1892 (*Stizus*); Argentina to Honduras
    dubius F. Smith, 1856 (*Larra*); nec Panzer, 1806-1809
    excisus Handlirsch, 1892 (*Stizus*)
    arechavaletai Brèthes, 1909 (*Stizus*)
    spegazzinii Brèthes, 1909 (*Stizus*)
    nectarinoides Ducke, 1910 (*Stizus*)
borneanus (Cameron), 1903 (*Stizus*); Borneo
braunsii (Handlirsch), 1894 (*Stizus*); e. and s. Africa
    pruinosus Cameron, 1910 (*Stizus*)
    pulchritinctus Cameron, 1912 (*Stizus*)
bredoi (Arnold), 1940 (*Stizus*); s. Africa
bridarollii Willink, 1949; Argentina
buyssoni (Arnold), 1929 (*Stizus*); s. Africa
    flavitarsis Arnold, 1960 (*Stizus*)
bytinskii Beaumont, 1954; Israel
caffer (Saussure), 1854 (*Stizus*); s. Africa
carinatus Lohrmann, 1942; s. Europe
carpetanus (Mercet), 1906 (*Stizus*); sw. Europe
cinguliger (F. Smith), 1856 (*Larra*); s. Africa
    clavicornis Handlirsch, 1892 (*Stizus*)
comechingon Willink, 1949; Argentina
comberi (Turner), 1912 (*Stizus*); Sri Lanka
consobrinus (Handlirsch), 1892 (*Stizus*); Brazil, Argentina
corpulentus (Arnold), 1929 (*Stizus*); e. and s. Africa
crassipes (Handlirsch), 1895 (*Stizus*); Spain

cyanescens (Radoszkowski), 1887 (*Stizus*); Iran, sw. USSR, China
cyprius Beaumont, 1954; Cyprus, Turkey
   ssp. *rhodius* Beaumont, 1960; Rhodes
   ssp. *creticus* Beaumont, 1961; Crete
dentipes (Gussakovskij), 1933 (*Stizus*); Iran
dentiventris (Handlirsch), 1895 (*Stizus*); s. Africa
diacanthus Beaumont, 1967; Turkey
discolor (Handlirsch), 1892 (*Stizus*); n. Africa
distinctus (Arnold), 1955 (*Stizus*); sw. Africa
egens (Handlirsch), 1892 (*Stizus*); Australia
escalerae (Turner), 1912 (*Stizus*); s. Africa
facialis (Handlirsch), 1892 (*Stizus*); Indonesia: Aru
fertoni (Handlirsch), 1908 (*Stizus*); Algeria
flavipes (F. Smith), 1856 (*Larra*); nw. Africa
flavopictus (Arnold), 1936 (*Stizus*); sw. Africa
flexuosefasciatus (Mantero), 1917 (*Stizus*); Ethiopia
fraterculus (Arnold), 1929 (*Stizus*); s. Africa
gazagnairei (Handlirsch), 1892 (*Stizus*); n. Africa
gorytoides (Handlirsch), 1895 (*Stizus*); Australia
gracilicornis (Handlirsch), 1892 (*Stizus*); sw. USSR
   barrei Radoszkowski, 1893 (*Stizus*)
gracilis (Schulthess), 1893 (*Stizus*); e. Africa
gusenleitneri Beaumont, 1967; Turkey
gynandromorphus (Handlirsch), 1892 (*Stizus*); Middle East
haemorrhoidalis (Handlirsch), 1900 (*Stizus*); Ethiopian Region
haplocerus (Handlirsch), 1895 (*Stizus*); s. Africa
hebraeus Beaumont, 1968; Israel
herbsti (Arnold), 1929 (*Stizus*); s. Africa
hirtiusculus (Arnold), 1945 (*Stizus*); Madagascar
hirtulus (F. Smith), 1856 (*Larra*); Australia
hoplites (Handlirsch), 1892 (*Stizus*); s. Africa
   ssp. *minor* (Arnold), 1929 (*Stizus*); s. Africa
hungaricus (Frivaldsky), 1876 (*Larra*); Europe
   ssp. *sibiricus* (Mocsáry), 1901 (*Stizus*); Siberia, Mongolia
   ssp. *formosanus* (Sonan), 1928; Taiwan, Ryukyus, Korea
     quadrimaculatus Sonan, 1928 (*Stizus*)
     quinquemaculatus Sonan, 1928 (*Stizus*)
     kotoshonus Sonan, 1934 (*Stizus*)
   ssp. *japonicus* (Sonan), 1934 (*Stizus*); Japan
     hirsutus Sonan, 1934 (*Stizus*)
   ssp. *amamiensis* Tsuneki, 1965; n. Ryukyus
   ssp. *verhoeffi* Tsuneki, 1965; Manchuria
hyperocrus (Arnold), 1929 (*Stizus*); s. Africa
inermis (Handlirsch), 1892 (*Stizus*); Fiji
   pacificus Turner, 1917 (*Stizus*)
innocens Beaumont, 1967; Turkey
insularis (Handlirsch), 1892 (*Stizus*); Indonesia
   socius Handlirsch, 1892 (*Stizus*)
   sarawakensis J. Parker, 1937
   stenaspis J. Parker, 1937
jacksoni (Arnold), 1955 (*Stizus*); Kenya
javanus (Handlirsch), 1892 (*Stizus*); Java
kachelibae (Arnold), 1960 (*Stizus*); Kenya
karooensis (Arnold), 1936 (*Stizus*); s. Africa
kobrowi (Arnold), 1936 (*Stizus*); s. Africa
kotschyi (Handlirsch), 1892 (*Stizus*); Sudan
kuehlhorni Willink, 1953; Argentina
laterimacula (Handlirsch), 1895 (*Stizus*); s. Africa
   ssp. *euteles* (Handlirsch), 1895 (*Stizus*); s. Africa
   ssp. *fraudulentus* (Arnold), 1951 (*Stizus*); Senegal
lateralis (Bingham), 1897 (*Stizus*); India
latericaudatus (Arnold), 1929 (*Stizus*); s. Africa
laticinctus (Arnold), 1929 (*Stizus*); s. Africa
latifascia (Walker), 1871 (*Larra*); e. Africa: Dahlak Archipelago
littoralis van der Vecht, 1949; Indonesia
lomii (Guiglia), 1945 (*Stizus*); Ethiopia
loriculatus (F. Smith), 1856 (*Larra*); Ethiopian Region
maior (Handlirsch), 1895 (*Stizus*); s. Africa
   major Dalla Torre, 1897 (*Stizus*), emendation
mattheyi (Beaumont), 1951 (*Stizus*); n. Africa
mayri (Handlirsch), 1892 (*Stizus*); n. Africa
meridionalis A. Costa, 1859; s. Europe, Middle East
mexicanus (Handlirsch), 1892 (*Stizus*); Mexico, Centr. America
   guttulatus Handlirsch, 1892 (*Stizus*)
   chichimecus Saussure, 1892 (*Stizolarra*)
mirus (Arnold), 1945 (*Stizus*); Madagascar
mitulus (Arnold), 1929 (*Stizus*); s. Africa
modestus (F. Smith), 1860 (*Larra*); Moluccas: Batjan
moneduloides (F. Smith), 1856 (*Larra*); U.S.: Florida
monodi (Berland), 1950 (*Stizus*); Mali
monodon (Handlirsch), 1895 (*Stizus*); Africa, Australia
multiguttatus (Arnold), 1951 (*Stizus*); Mali
mutabilis (Arnold), 1929 (*Stizus*); s. Africa
naefi (Beaumont), 1951 (*Stizus*); Morocco
nambui Tsuneki, 1973; Ryukyus
nanus (Handlirsch), 1892 (*Stizus*); e. and centr. U.S.
   ssp. *strenuus* Mickel, 1918; U.S.: Nebraska to Texas
   ssp. *floridanus* Krombein and Willink, 1951; U.S.: s. Florida
neglectus (Cresson), 1872 (*Monedula*); centr. and s. centr. U.S.
   xanthochrous Handlirsch, 1892 (*Stizus*)
nemoralis (Arnold), 1960 (*Stizus*); s. Africa
nigriclypeus (Sonan), 1928 (*Stizus*); Taiwan
nyasae (Turner), 1912 (*Stizus*); s. Africa
   ssp. *robustus* (Arnold), 1951 (*Stizus*); Mali
oxydorcus (Handlirsch), 1900 (*Stizus*); s. Africa
   johannis Cameron, 1905 (*Stizus*)
†*pacificus* Tsuneki, 1968; Taiwan, Korea, nec Turner, 1917
pallidicinctus van der Vecht, 1949; Indonesia
papuanus (Cameron), 1906 (*Stizus*); New Guinea
penpuchiensis Tsuneki, 1968; Taiwan
peregrinus (F. Smith), 1856 (*Larra*); e. Europe, Mediterranean area, Middle East
   erberi Mocsáry, 1881
   ssp. *biarmatus* Mocsáry, 1883; Turkey
podager (Beaumont), 1949 (*Stizus*); Morocco
polychromus (Handlirsch), 1895 (*Stizus*); s. Africa
posterus (Sonan), 1928 (*Stizus*); Taiwan
prismaticus (F. Smith), 1858 (*Larra*); Borneo, New

Guinea
proteus (Arnold), 1929 (*Stizus*); s. Africa
proximus (Handlirsch), 1892 (*Stizus*); India
pulchellus (Mercet), 1906 (*Stizus*); Spain
pusillus (Handlirsch), 1892 (*Stizus*); Indonesia
quadristrigatus (Arnold), 1929 (*Stizus*); S.-W. Africa
    ssp. *somalicus* (Arnold), 1940 (*Stizus*); Somalia
    ssp. *dubiosus* (Guiglia), 1945 (*Stizus*); Ethiopia
quinquespinosus (Say), 1823 (*Nysson*); w. U.S. to Panama
    godmani Cameron, 1890 (*Stizus*), new synonymy by R. Bohart
    lineatus Cameron, 1890 (*Stizus*), new synonymy by R. Bohart
    flavus Cameron, 1890 (*Stizus*), new synonymy by R. Bohart
    subalpinus Cockerell, 1898 (*Stizus*), new synonymy by R. Bohart
    cressoni Cameron, 1904 (*Nysson*), new synonymy by R. Bohart
rectilateralis (Arnold), 1945 (*Stizus*); Madagascar
reversus (F. Smith), 1856 (*Larra*); Sumatra
revindicatus (Schulz), 1906 (*Stizus*); Middle East
    schmiedeknechti Handlirsch, 1900 (*Stizus*), nec Handlirsch, 1898
    houskai Balthasar, 1954 (*Stizus*)
    ssp. *anatolicus* Beaumont, 1968; sw. Turkey
rhopaloceroides (Arnold), 1929.(*Stizus*); s. Africa
rhopalocerus (Handlirsch), 1895 (*Stizus*); s. Africa
schwarzi Beaumont, 1967; Turkey
semperi (Handlirsch), 1892 (*Stizus*); Philippines
signatus (Handlirsch), 1892 (*Stizus*); Australia
simillimus (F. Smith), 1859 (*Larra*); Indonesia: Aru, Waigeo
    magrettii Handlirsch, 1892 (*Stizus*)
sipapomae (Arnold), 1929 (*Stizus*); s. Africa
solitarius (Arnold), 1936 (*Stizus*); s. Africa
spinicornis (Saussure), 1887 (*Stizolarra*); Madagascar
spinifemur (Beaumont), 1949 (*Stizus*); Morocco
suada (F. Smith), 1864 (*Larra*); Indonesia: Halmahera
sudanensis (Arnold), 1951 (*Stizus*); Mali
tanoi Tsuneki, 1971; s. Ryukyus
tenellus (Klug), 1845 (*Larra*); ne. Africa
touareg Beaumont, 1954; Egypt
trichionotus (Cameron), 1913 (*Stizus*); Indonesia: Waigeo
tridens (Fabricius), 1781 (*Vespa*); Eurasia
    cinctus Rossi, 1790 (*Crabro*)
    repandus Panzer, 1800 (*Mellinus*)
    sinuatus Latreille, 1804-1805 (*Stizus*)
    satsumanus Sonan, 1934 (*Stizus*)
    ssp. *errans* Beaumont, 1950; Corsica
    ssp. *insulanus* Beaumont, 1954; Corsica
    ssp. *caesius* (Compte Sart), 1959 (*Stizus*); Majorca
    ssp. *mongolicus* Tsuneki, 1970; Mongolia
turneri (Froggatt), 1917 (*Stizus*); Australia
validior Gussakovskij, 1952; sw. USSR: Tadzhik S.S.R.
varians (Arnold), 1945 (*Stizus*); Madagascar
veniperdus (Lohrmann), 1942 (*Stizus*); India
versicolor (Handlirsch), 1892 (*Stizus*); e. Indonesia, New Britain
wheeleri Krombein and Willink, 1951; U.S.: Arizona
witzenbergensis (Arnold), 1929 (*Stizus*); s. Africa
zibanensis (Morice), 1911 (*Stizus*); n. Africa

Tribe Bembicini

The sand wasps are a familiar sight around beaches or sand dunes where they may be seen cruising or darting, chasing prey or each other. They are medium to large, stout bodied, and nearly always elaborately marked with whitish or yellow. Their flight is swift, and they are capable of hovering, darting, or traveling at an incredible speed in a straight line. They are probably the best fliers of the Nyssoninae, if not the entire family. If females are captured and handled carelessly, they are able to sting, but otherwise they are not at all dangerous even though their actions sometimes appear threatening.

Diagnostic characters:

1. (a) Eyes converging below inner margins nearly parallel (fig. 182 C) or most often converging slightly above (fig. 182 B); (b) ocelli always at least slightly deformed, sometimes reduced to scars.
2. (a) Labrum exserted and with exposed area at least as long as broad (fig. 182 A-C, I, J), (b) palpal formula 6-4 to 3-1.
3. (a) Scutum with admedian lines separated, (b) notauli absent, (c) with an oblique scutal carina, (d) scutellum with a lamelliform edge which overlaps metanotum.
4. (a) Omaulus, acetabular carina, sternaulus, and precoxal sulcus absent; (b) episternal and scrobal sulci usually present, forming a continuous arc or episternal-scrobal sulcus which defines hypoepimeral area.
5. (a) Metapleuron separated by a sulcus from propodeum, lower metapleural area not defined except minimally in a few cases; (b) upper lateral surface of propodeum not spined nor dentate, (c) propodeal enclosure extending far onto vertical slope.
6. (a) Midcoxae slightly separated, (b) midtibia with one apical spur, (rarely none in male *Bembix*) (c) arolia equal on all legs, (d) female with a foretarsal rake.
7. (a) Forewing media diverging before cu-a, (b) prestigmal length of submarginal cell I much more than length of wing beyond marginal cell, (c) stigma small, (d) three submarginal cells, II not petiolate and receiving both recurrent veins.
8. (a) Hindwing costa well developed, media diverging after cu-a, jugal lobe less than half as long as anal area.
9. (a) Gaster sessile, tergum I nearly as broad as II; (b) sternum I with a median ridge toward base, (c) male with seven visible terga, VII with laterobasal spiracular lobes (except in *Bicyrtes* and *Microbembex*); (d) male sternum VIII ending in one to four prongs (fig. 182 G); (e) male genitalia with volsella differentia-

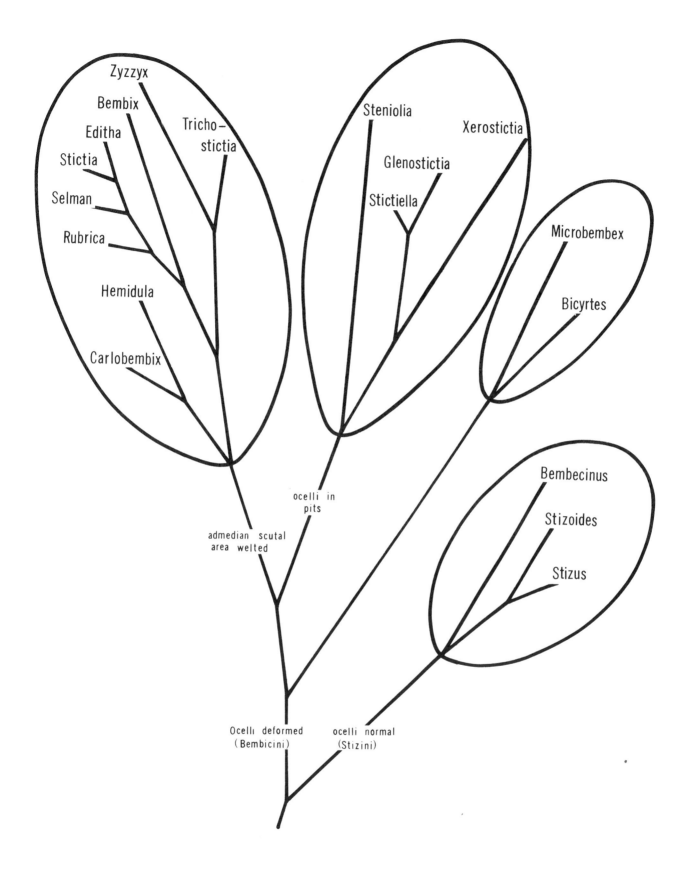

FIG. 181. Dendrogram suggesting relationships in the Bembicini and Stizini.

ted into cuspis and digitus, (f) female pygidial plate usually very weakly indicated posteriorly, sometimes rather well defined.

Systematics: The close relationship of the Bembicini and Stizini is indicated by the many tribal characters common to both. The principal points of departure of the bembicins are the deformed ocelli, the longer labrum, and the more distal position of the media in the hindwing. Also, the single-spurred midtibiae are differentiating from most of the stizins as well as the great majority of other Nyssoninae.

The reduction of pleural sulci, streamlining of the notum, and rearrangement of wing veins to reduce the distal membrane and increase the percentage of vein-supported forewing area, all seem to have a relationship to flight characteristics. These features are shared with the Stizini and to a lesser extend with such gorytins as *Sphecius* and *Kohlia*.

Within the Bembicini there seem to be three rather well defined evolutionary lines as indicated in fig. 181. The branch that culminates in *Microbembex* presumably developed the earliest and was probably "prestizin" since there are no spiracular lobes on male tergum VII. Other distinctions are the rather concave posterior slope of the propodeum, and the arcuate midocellar scar. To this may be added the prey association with Hemiptera (in *Bicyrtes*) or a wide variety of live and dead arthropods (in *Microbembex*). This branch is well represented in North and South America. The second branch is identified with its best known genus, *Stictiella*. Its outstanding morphological feature is the placement of the ocellar elements in somewhat raised depressions (fig. 182 M). The ocelli are not completely scarlike, having reduced lenses. Spiracular lobes are present on male tergum VIII, and the propodeum is flat or somewhat rounded posteriorly. The prey are flies, butterflies (in *Stictiella*), or rarely, wild bees. This evolutionary complex is centered in North America with very limited extension to the south. The third and largest group of genera can be termed the *Bembix* branch. Here the scutum is raised into an obvious welt anteromedially. Spiracular lobes are present on male tergum VIII, the posterior propodeal face is flat or rounded, and the ocelli are reduced in varying amount. Flies are the dominant prey, but *Editha* and *Zyzzyx* (rarely) use butterflies, and *Bembix* have been known to use damselflies, bees, wasps, and antlions (Evans and Matthews, 1973b). *Bembix* is worldwide, but the other genera are restricted to the western Hemisphere, particularly South America.

In many respects, such as wing venation and pleural structure, the Bembicini are remarkably uniform. The most obvious evolutionary tendencies are in ocelli, where there is a progressive reduction in size; in palpi, where there is a decline in number of segments; and in the tongue, which tends toward greater length. Each of these tendencies is manifested to some degree in each of the three evolutionary lines.

Of the many papers dealing with the taxonomy of the tribe, the most useful ones for an understanding of the genera are Handlirsch (1889, 1890, 1893), J. Parker (1929), Willink (1947), Gillaspy (1963b, 1964), and Bohart and Horning (1971). The last named contains an up-to-date key to the genera.

In addition to morphological evolutionary changes there are progressive behavioral tendencies that can be discerned within the Bembicini. Some of the more obvious of these are: the change from solitary nests to large, high density colonies; change from short corolla nectar sources to deep corolla types associated with tongue development and supposedly greater need for liquid in xeric areas; clustering in small or large numbers overnight, as reported for *Glenostictia, Steniolia, Zyzzyx,* and *Rubrica,* which may be involved with mating efficiency; construction of false burrows by some *Bembix* and *Rubrica* which may reduce pressure of parasitism; an inner closure of the nest just before the cell in some *Bembix* and *Stictia* which may also reduce parasitism; a general progression in prey composition from Hemiptera to Diptera to Lepidoptera, with occasional use of Hymenoptera (by *Bembix citripes* and *Glenostictia scitula*), Homoptera (by *G. scitula*), Odonata and Neuroptera (by *Bembix*) and, in the case of *Microbembex,* the utilization of a wide variety of live and dead arthropods (in the matter of food selection we can dimly see the evolutionary advantages of food specialization up to a certain point, but under some conditions there may be a further advantage in a return to a broad range of prey); the highly important trend from mass provisioning to progressive provisioning with certain intermediate steps, and along with this trend a change from lightly paralyzed to completely paralyzed or dead prey, as well as deposition of the egg on the last prey member instead of the first, or even the deposition in the newly constructed empty cell; a refinement of progressive provisioning in the cell cleaning procedure practiced by some *Bembix* and *Rubrica.*

Each of the above behavioral changes presumably has a selective advantage to the species in question. Sometimes pressure of parasitism is involved, and in other cases it may be a matter of reproductive efficiency or food economy. An extensive discussion of behavioral evolution has been given by Evans (1966a).

Natural enemies of Bembicini are most commonly the miltogrammine Sarcophagidae. However, Bombyliidae and Conopidae attack *Bembix* (see G. Bohart and MacSwain, 1939); *Dasymutilla* are common parasites; and Chrysididae are constant drains on populations of *Steniolia* (by *Parnopes edwardsii* Cresson), of *Bembix* (by *Parnopes chrysoprasinus* H. Smith), and of *Microbembex* (by *Parnopes fulvicornis* Cameron). The beetle family Rhipiphoridae (*Macrosiagon*) parasitizes *Bembix,* at least in some localities (Barber, 1915), and Strepsiptera (*Pseudoxenos*) are occasional parasites of *Bembix* and *Microbembex* (R. Bohart, 1941; Evans and Matthews, 1973b).

Key to genera of Bembicini

1. Ocelli depressed (fig. 182 C,J,M); inner

orbits essentially parallel as a rule, usually about as broad above as below (fig. 182 C,J) 2
Ocelli or ocellar scars not depressed below surrounding surface of face, inner orbits essentially parallel or diverging below (fig. 182 I) 5
2. Palpal formula 4-2, 3-2 or 3-1; midocellar plane longer than broad 3
Palpal formula 6-4; N. America 4
3. Labrum evenly convex toward base and about twice as long as its basal breadth, palpal formula 4-2; sw. U.S. and Baja California ...... *Xerostictia* Gillaspy, p. 552
Labrum swollen toward base so that a cross section would be irregularly convex, palpal formula 3-2 or 3-1; N. America and Ecuador ...... *Steniolia* Say, p. 552
4. Clypeus moderately to strongly and evenly convex, midocellar plane usually broader than long, male midfemur with inferior edge irregularly carinate to serrate (fig. 182 F) to emarginate or (in small species) rarely entire ...... *Stictiella* J. Parker, p. 550
Clypeus rather low, convexity often irregular and with beveled area near tentorial pits; male midfemur entire or with a large subapical notch; midocellar plane not broader than long ...... *Glenostictia* Gillaspy, p. 551
5. Palpal formula 5-3, 4-2, or 3-1 or in any case less than 6-4 6
Palpal formula 6-4 8
6. Marginal cell in distal half slightly bent away from wing margin, tergum VII of male without a spiracular lobe; New World ...... *Microbembex* Patton, p. 538
Marginal cell in distal half adhering to wing margin, tergum VII of male with a spiracular lobe 7
7. Palpal formula 4-2 (fig. 182 H) or 3-1; midocellus usually reduced to slender transverse scar, rarely with traces of a lens or with a nearly semicircular lens; worldwide ...... *Bembix* Fabricius, p. 543
Palpal formula 5-3, ocelli with an approximately semicircular lens; Chile and Argentina ...... *Zyzzyx* Pate, p. 549
8. Midocellus reduced to a scar, rarely with traces of a lens; New World 9
Midocellus approximating a semicircle with a distinct and somewhat translucent lens; South America 11
9. Lateral angles of propodeum projecting backward, tergum VII of male without spiracular lobe ...... *Bicyrtes* Lepeletier, p. 535
Lateral angles of propodeum not projecting backward, tergum VII of male with a spiracular lobe 10
10. Midocellar scar forming part of a circle, not flattened or indented medially; vertex in front view considerably depressed below level of eyes ...... *Stictia* Illiger, p. 541
Midocellar scar somewhat transverse, flattened or indented posteromedially; vertex in front view not depressed below level of eyes (fig. 182 A) ......
...... *Rubrica* J. Parker, p. 540
11. Mandible not toothed along inner margin; Argentina 12
Mandible subapically dentate 13
12. Midocellus a half circle with front edge concave; lateral ocellus flattened externally, male femora relatively simple ...... *Hemidula* Burmeister, p. 539
Midocellus an ellipse of about three-fifths of a circle, lateral ocellus circular, male forefemur greatly flattened and expanded, male midfemur very irregularly dentate ...... *Carlobembix* Willink, p. 539
13. Eyes with long dense hair, male tergum VII with median lobe rounded, male midcoxa with a long tooth or spine ...... *Trichostictia* J. Parker, p. 549
Eyes without hair, male tergum VII without a rounded median lobe 14
14. Inner eye margins diverging rather strongly below (fig. 182 B), male tergum VIII truncate medially and acute laterally ...... *Editha* J. Parker, p. 542
Inner eye margins slightly diverging below, male tergum VII with median and lateral notched projections ...... *Selman* J. Parker, p. 541

Genus Bicyrtes Lepeletier

Generic diagnosis: Flagellomere I often shorter than scape; eyes with inner margins somewhat converging below; least interocular distance usually a little less than twice length of flagellomere I; midocellus a transverse scar, indented medioposteriorly; vertex a little above top of eyes; clypeus convex laterally but rather flattened medially, not ridged anteriorly and ending below antennal sockets; labrum about as long as broad, usually simply convex but sometimes (*cingulata*) with a discal projecting ridge; tongue of moderate length, palpal formula 6-4; mandible subapically dentate; area of scutum between admedian lines nearly flat; many species with two distal appendices on discoidal cell II; hindwing media diverging well beyond cu-a; male midcoxa sometimes toothed; male midfemur sometimes with a basoventral process but without a row of teeth; propodeum with posterior surface somewhat concave, lateral angles compressed and wedge like (except two South American species: *anisitsi, cingulata*); female tergum VI sometimes with a well defined pygidial plate (*capnoptera, fodiens*); male tergum VII without spiracular lobes, sometimes with a lateral angle or tooth; male sternum VIII with three prongs.

Geographic range: Of the 23 listed species of *Bicyrtes,* 12 are restricted to South America, eight to North America, and the remaining extend on both sides of Panama. One of these, *variegata,* has been found from Texas to Argentina.

Systematics: In some respects, such as the eyes converging below, submarginal cell II narrowed in front, and wedgelike lateral propodeal angles, *Bicyrtes* is like the stizin genus *Bembecinus*. These similarities led Lohrmann (1948) to derive *Bicyrtes* and *Bembecinus* along a separate branch from all other stizins and bembicins. This arrange-

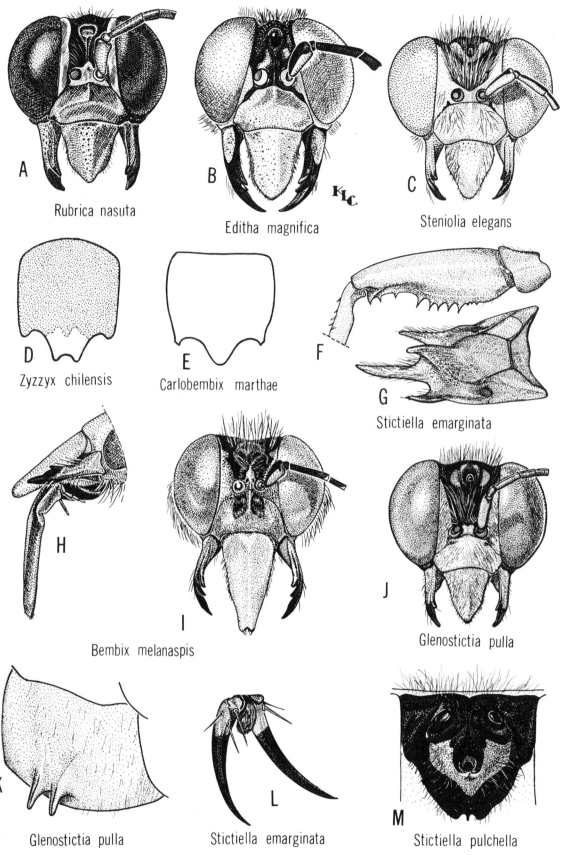

FIG. 182. Morphology in the tribe Bembicini; A-C,I,J, female facial portraits; D,E, male tergum VII; F, male midfemur; G, male sternum VIII; H, mouthparts, lateral; K, male sternum II; L, female hindtarsal claws; M, female ocellar area, dorsal.

ment, which appears to be based on parallel development, placed *Bicyrtes* and *Microbembex* at the opposite ends of the system. We feel that the evidence is strong for a close association of these two genera, with *Microbembex* being obviously the more highly developed. Lohrmann's scheme does not explain the absence of spiracular lobes on male tergum VII in *Bicyrtes* and *Microbembex,* or their full development in *Bembecinus.* The possibility that these lobes developed along separate lines, then completely atrophied, is rather unlikely.

It seems to us that *Bicyrtes* is an offshoot arising rather low on the evolutionary branch which derived the bembicins from a prestizin ancestor (fig. 181). The completely reduced midocellus argues against considering *Bicyrtes* as "primitive." Characters associating it with *Microbembex,* in addition to the absence of spiracular lobes, are the somewhat concave vertical face of the propodeum, the arcuate midocellar scar, the rather broad and flat scutal area between the admedian lines, the nearly parallel or anteriorly converging inner eye margins, the rather simple nature of the abdominal markings, the rather short labrum, and the rather simple midlegs of the male. The fact that neither genus is dependent upon flies for prey is of possible significance. The important distinctions of *Microbembex* are the reduced palpi and the edentate mandibles. South American *Bicyrtes* were treated by Willink (1947). The species of North America were keyed by Bohart and Horning (1971).

Biology: Bicyrtes is one of four genera in the tribe that regularly provision with insects other than Diptera. In the case of *Bicyrtes,* the prey are generally nymphal bugs, but adults may be used, and such families as Pentatomidae, Coreidae, and Reduviidae are involved most frequently. A fairly complete study of habits was made by M. Smith (1923a) on *quadrifasciata* in Mississippi. Nests were located in a strip of sand along a road. Burrows were 15 to 20 cm in depth and sloped at a 45 degree angle with the surface. The single terminal cells were stored with nymphs of nine species of bugs in the three families cited above. The egg was attached near the beak of the prey.

Krombein (1955b) also reported on the habits of *quadrifasciata,* but in much greater detail than did M. Smith. He found the burrows to be 6 to 11 cm long, and the cells were stocked with Coreidae and Pentatomidae, the former predominating. Mass provisioning was practiced, and the number of nymphs in completely stocked cells was nine to 14. The prey was carried beneath the wasp, head forward and venter up, clasped by the middle and posterior legs. At the moment of entrance into the burrow, the forelegs of the wasp were digging, the midlegs clasping, and the hind legs bracing. The egg was laid upright between the coxae of the first prey in the cell. The burrow was always closed when the wasp was away, but the egg did not hatch until after final closure. Krombein noted that the miltogrammine fly, *Senotainia rubriventris* Macquart was a persistent parasite.

As an addendum to the discrepancy in burrow length as reported by M. Smith and Krombein, a *quadrifasciata* female was discovered by R. Bohart and Holland (1965) near Lake Texoma, Oklahoma stocking a 30 cm deep burrow in hard red sand with nymphs of *Leptoglossus oppositus* (Say).

Among others who have recorded observations on habits of B. *quadrifasciata* are J. Parker (1917), Rau and Rau (1918), and Evans (1966a).

References to biology of other species contain information similar to that above, except that two or more terminal cells are sometimes noted, one closed off before the other is provisioned. Brèthes (1918) and Vesey-Fitzgerald (1940) studied *discisa,* J. Parker (1917) studied *ventralis,* Janvier (1928) reported on *variegata,* and Evans (1966a) studied *fodiens.* In the case of *variegata,* the nest terminated in as many as three cells, exclusively provisioned with a Chilean leaf-footed plant bug.

Evans (1966a) reported natural enemies as predominantly miltogrammine Sarcophagidae. However, he reared the chrysidid, *Holopyga ventralis* (Say), from nests of *Bicyrtes quadrifasciata* and B. *fodiens.* Evans' book should be consulted for many additional details and a summary of ethology.

Checklist of *Bicyrtes*

affinis (Cameron), 1897 (*Bembidula*); U.S.: s. Arizona; Mexico

angulata (F. Smith), 1856 (*Monedula*); French Guiana, Brazil, Paraguay, Argentina

angulifera (Strand), 1910 (*Bembidula*); Paraguay

anisitsi (Strand), 1910 (*Bembidula*); Paraguay, Argentina
   tridentata Strand, 1910 (*Bembidula*)
   bradleyi J. Parker, 1929

capnoptera (Handlirsch), 1889 (*Bembidula*); U.S.; Mexico: Baja California
   mesillensis Cockerell, 1898
   annulata J. Parker, 1917
   tristis C. Fox, 1923

cingulata (Burmeister), 1874 (*Bembidula*); Argentina
   cisandina Pate, 1936
   ssp. *micans* (Handlirsch), 1889 (*Bembidula*); Brazil to Argentina

diodonta (Handlirsch), 1889 (*Bembidula*); Mexico: Orizaba, Nayarit
   oribatea Pate, 1936

discisa (Taschenberg), 1870 (*Monedula*); Mexico to S. America

fodiens (Handlirsch), 1889 (*Bembidula*); e. to sw. U.S., Mexico, Centr. America
   burmeisteri Handlirsch, 1889 (*Bembidula*), new synonymy by R. Bohart

insidiatrix (Handlirsch), 1889 (*Bembidula*); e. and s. U.S.

lilloi Willink, 1947; Argentina, Bolivia

mendica (Handlirsch), 1889 (*Bembidula*); Argentina

odontophora (Handlirsch), 1889 (*Bembidula*); Venezuela to Bolivia, Argentina

ornaticauda (Cameron), 1912 (*Bembidula*); Guyana

paraguayana (Strand), 1910 (*Bembidula*); Paraguay, Argentina
   patei Willink, 1947

pexa J. Parker, 1929; French Guiana
quadrifasciata (Say), 1824 (*Monedula*); U.S. e. of Rocky Mts.
  sallei Guérin-Méneville, 1844
simillima (F. Smith), 1856 (*Monedula*); Brazil to Argentina
  defecta Brèthes, 1909 (*Bembex*)
  quinquemaculata J. Parker, 1929
  pullata J. Parker, 1929
  erfilai Willink, 1952, lapsus
  orfilai Willink, 1952
spinosa (Fabricius), 1794 (*Bembex*); Ecuador, Panama, W. Indies
tricolorata J. Parker, 1929; Argentina, Brazil
  sola J. Parker, 1929
variegata (Olivier), 1789 (*Bembex*); U.S.: Texas; Mexico to Argentina
  sericea Spinola, 1851 (*Monedula*)
  guiana Cameron, 1912 (*Bembex*), new synonymy by R. Bohart
ventralis (Say), 1824 (*Monedula*); U.S., s. Canada, n. Mexico
  servillii Lepeletier, 1845
  parata Provancher, 1888 (*Monedula*)
  meliloti Johnson and Rohwer, 1908 (*Bembidula*)
viduata (Handlirsch), 1889 (*Bembidula*); sw. U.S.; Mexico
  gracilis J. Parker, 1917

Genus Microbembex Patton

Generic diagnosis: Flagellomere I usually a little shorter than scape; eyes with inner margins essentially parallel, separated usually by more than an eye breadth; ocelli completely reduced, midocellar scar linear, transverse, arcuate, not in a pit nor on an elevation; vertex a little depressed below top of eyes; clypeus rather unevenly convex, strongly bulging, sometimes nasiform; labrum somewhat longer than broad, flattened toward base; palpal formula varying from 4-2 to 3-1, more commonly the latter; mandible edentate; area of scutum between admedian lines rather broad, flat or a little depressed; male midfemur simple; forewing marginal cell bent distinctly away from wing margin in distal third to half; hindwing media diverging after cu-a, distance not more than half length of cu-a; propodeum bulging at median summit and somewhat overhanging; male sternum II often with an apicomedian keellike process; male tergum VII without spiracular lobes, nearly simple to quadripartite; male sternum VIII ending in a downcurved spine; female tergum VI nearly always notched and bidentate apically.

Geographic range: Twenty-one species are currently recognized, all from the New World. Ten of these occur in North America and the remaining 11 in South America. Seven species are known from the United States but all except the widespread *monodonta* occur west of the 100th meridian.

Systematics: The partial removal of the marginal cell from the edge of the forewing readily distinguishes this genus. The 3-1 palpal formula is quite distinctive, but one species, *bidens*, has a 4-2 formula as in most *Bembix*. Likewise, the apically notched female pygidium is unique, but an Amazonian species, *pygidialis*, has a rounded pygidium. The completely reduced midocellus is a character shared with several other genera, but in *Microbembex* even the scar may be almost obliterated. Lohrmann (1948) divided the genus into two groups of approximately equal size. One of these was the "*Difformis*-Gruppe" which contained species between 13 and 17 mm in length, the males with lateral angles on tergum VII, females with tergum VI always ending in two points, and males generally with a transverse swelling on sternum II bearing one or two teeth. Typical examples cited by Lohrmann were *difformis, argentina, bidens, gratiosa,* and *uruguayensis.* The second or "*Monodonta*-Gruppe" was characterized by smaller size, 10 to 13 mm in length, the male usually with a keel-like process on sternum II, and male tergum VII simple laterally. Typical examples were *mondonta, anilis, hirsuta,* and *ciliata.*

Microbembex appears to be related to *Bicyrtes*, although it is considerably more specialized. The relationship is discussed under *Bicyrtes* and is illustrated graphically (fig. 181). Characteristics of some of the North American species have been discussed by R. Bohart (1970b). A key to species of the United States was given by Bohart and Horning (1971).

Biology: Four bembicin genera, *Microbembex, Bicyrtes, Editha,* and *Stictiella,* regularly provision their nests with prey other than Diptera. However, *Microbembex* differs from the other three in that a great variety of insects are taken either alive or dead. The most complete report is that of J. Parker (1917) on *M. monodonta* in Ohio. Nests were very abundant in lake sand dunes. Copulation took place soon after the females emerged and often before they had made their first flight. Males appeared to be polygamous. The burrows were 20 to 30 cm long and ended in a single cell, at the far end of which an egg was placed upright before the first prey was brought. The newly hatched larva was fed progressively on dead and disabled insects picked from the sand surface by the female *Microbembex*. Parker described the competition for prey that was frequently observed. "The struggles at the mouth of the burrow for the possession of a dead insect are frequent and furious, the contestants grappling and rolling over and over on the sand. Frequently it happens that the prey is dropped in the struggle, and while the pair of contestants are rolling on the sand a third wasp comes along and settles the quarrel by quietly carrying off the coveted treasure." Prey mentioned specifically by Parker were houseflies, stableflies, mayflies, and midges.

Evans (1966a) summarized biological information on the genus as follows: large colonies may occur in sandy areas; adults spend periods of inactivity in short burrows in the sand; nests are usually less than a foot deep, unicellular, and not marked by tumulus at the entrance; the female closes the nest when she is away from it; provisioning is progressive; the egg is laid erect in the cell, and first provisions are brought to the newly hatched larva; prey consists of an amazing conglomeration of dead and moribund arthropods.

Alcock and Ryan (1973) studied a nesting site of *nigrifrons* in central Washington. Behavior was similar to that reported for other species. Alcock and Ryan altered details in the vicinity of the nests during provisioning by *nigrifrons,* and concluded that when landmarks close to the nest were moved, a disorientation of the wasp resulted. However, the effect was only temporary and a reliance by the wasp on alternate and more distinct landmarks was suggested.

Natural enemies noted by Parker were *Exoprosopa* (Bombylliidae) and *Dasymutilla* (Mutillidae), as well as some small red ants. Also important are chrysidid wasps of the genus *Parnopes.* G. Bohart and MacSwain (1940) reported about 20 percent parasitism of *M. californica* (given as *aurata*) cocoons at Antioch, California, by *Parnopes fulvicornis* Cameron (given as *westcottii* Melander and Brues, a synonym). An odd parasite of *M. nigrifrons* is the Strepsipteran *Pseudoxenos lugubris* (Pierce), a single female of which had matured in a wasp collected by G. Bohart at Antioch, California. This specimen was taken in an area where the natural host, *Ammophila,* was abundant and commonly parasitized (R. Bohart, 1941).

Checklist of *Microbembex*

anilis (Handlirsch), 1893 (*Bembex*); Venezuela
argentifrons (Cresson), 1865 (*Bembex*); W. Indies
argentina Brèthes, 1913; Argentina
    nasuta J. Parker, 1929
argyropleura R. Bohart, 1970; U.S.: California, Nevada, Arizona, Utah
aurata J. Parker, 1917; U.S.: w. Texas to s. California
bidens J. Parker, 1929; Argentina
californica R. Bohart, 1970; U.S. and n. Mexico w. of Continental Divide
ciliata (Fabricius), 1804 (*Bembex*); S. America
    sulphurea Spinola, 1851 (*Bembex*)
cubana R. Bohart, new name; Cuba
    armata Cresson, 1865 (*Bembex*), nec Sulzer, 1776
difformis (Handlirsch), 1893 (*Bembex*); Brazil
equalis J. Parker, 1929; Peru
gratiosa (F. Smith), 1856 (*Bembex*); Brazil
    natalis J. Parker, 1929
hirsuta J. Parker, 1917; U.S.: Texas, New Mexico
monodonta (Say), 1824 (*Bembex*); N. and Centr. America e. of Continental Divide
    occidentalis Johnson and Rohwer, 1908
    tarsalis Rohwer, 1914
nigrifrons (Provancher), 1888 (*Bembex*); U.S. and Mexico w. of 100th mer.
    neomexicana Johnson and Rohwer, 1908
    deltaensis Johnson and Rohwer, 1908
patagonica (Brèthes), 1913 (*Bembex*); Argentina
pygidialis (Handlirsch), 1893 (*Bembex*); Brazil
rufiventris R. Bohart, 1970; U.S.: s. centr. California
subgratiosa (Strand), 1910 (*Bembex*); Paraguay
tricosa J. Parker, 1929; Jamaica
uruguayensis (Holmberg), 1884 (*Bembex*); Uruguay, Paraguay, Argentina
    schrottkyi Willink, 1947

Genus Carlobembix Willink

Generic diagnosis: Flagellomere I longer than scape; inner eye margins nearly straight, moderately diverging below; midocellus about a semicircle in extent but transversely elliptical, not on an elevation; lateral ocellus convex and circular; vertex not lower than top of eyes; clypeus rather evenly contoured, dorsal point about even with lower edge of antennal sockets; labrum about 1.3 times as long as broad, cross section near base a rather smooth curve; palpal formula 6-4; mandible edentate; area of scutum between admedian lines a distinct welt; mesopleuron anteroventrally with two sharp teeth; male midcoxa with a short apical tooth; male midfemur with a tooth beyond middle, followed by an irregular excavation; hindwing media diverging shortly after cu-a; male tergum VII with large spiracular lobes which nearly meet ventrally, apex of VII with three narrow lobes (fig. 182 E); male sternum VIII ending in a downcurved spine, and with a transverse basal keel.

Geographic range: The two males and two females of the type series of *Carlobembix marthae* are all of the recorded material of the genus (Willink, 1958). The type locality is Santiago del Estero, Argentina.

Systematics: As pointed out by Willink (1958), this genus is similar to *Hemidula,* especially considering the simple mandibles. Since only three species are involved in the two genera, there is some difficulty in sorting generic from specific characters. The practically normal rather than laterally flattened lateral ocelli, the elliptical rather than arclike midocellus, and the more nearly parallel inner orbits of *Carlobembix* are distinguishing. In the male the peculiar midfemur, the mesopleural ventral teeth, and the transverse basal keel of sternum VIII are distinctive.

Willink (1958) illustrated the generic characters. We have been able through Dr. Willink's cooperation to study a paratype male. It agrees well with the description and figures, except that the midocellus is elliptical and 1.5 times as broad as long. In the original description the statement was made, "con ocelos normales," which applies only to the lateral ocelli.

Biology: No information is available.

Checklist of *Carlobembix*

marthae Willink, 1958; Argentina

Genus Hemidula Burmeister

Generic diagnosis: Flagellomere I longer than scape; inner eye margins nearly straight but converging above so that least interocular distance is about equal to clypeal length; midocellus semicircular posteriorly and incurved anteriorly; lateral ocellus a little flattened externally; ocellar lenses convex and apparently functional; vertex not lower than top of eyes; clypeus rather flattened on lower three-fifths, then sloping abruptly toward antennal sockets;

labrum slightly longer than broad, cross section near base a rather smooth curve; palpal formula 6-4; mandible edentate; area of scutum between admedian lines a distinct welt; male midcoxa and midfemur not toothed; hindwing media diverging shortly beyond cu-a; male tergum VII with a rounded median apical lobe and acute lateral lobes; male sternum VIII with short lateral teeth and a stout median spine.

Geographic range: Argentina in Mendoza, San Juan, and Santiago del Estero.

Systematics: Through the kindness of Dr. Abraham Willink we have been able to study one female of *burmeisteri* and two females of *singularis*. Male characters have been taken from Willink (1947, 1958).

Many features are shared by *Hemidula* and *Carlobembix*. Both have functional ocelli which are not on a prominence or in a depression, frons broadening somewhat below, mandible simple, median scutal welt well formed, and similar wing venation. *Hemidula*, however, has the ocelli more extensively deformed and lacks many of the male peculiarities as discussed under *Carlobembix*. Both genera have the marginal cell of the forewing tight against R_1 (without a separate apical appendix). However, in material we have seen the marginal cell is more broadly rounded toward the wing apex in *Hemidula*. One presumably generic difference lies in sternum VIII of the male. In *Hemidula* there is a laterobasal tooth, whereas in *Carlobembix* there is a transverse basal ridge which expands medially.

Biology: Willink (1947) reported finding a large colony of *burmeisteri* nesting in a field covered with saltpetre. Entrance holes were made by preference in small depressions. The nests were 10 to 12 cm deep and penetrated the ground at an angle. Prey was not observed.

Checklist of *Hemidula*

burmeisteri Willink, 1947; Argentina
singularis (Taschenberg), 1870 (*Monedula*); Argentina

Genus Rubrica J. Parker

Generic diagnosis: Flagellomere I longer than scape; eyes with inner margins slightly diverging below (fig. 182 A); ocelli reduced to scars without lenses, anterior ocellar scar somewhat crescent shaped but flattened posteromedially, not in a pit nor elevated; vertex about on a level with top of eyes; clypeus with a longitudinal carina above, lower three-fifths of clypeus somewhat flattened, then sloping rather abruptly toward antennal sockets; labrum a little longer than wide (fig. 182 A), cross section near base almost a smooth curve; palpal formula 6-4; mandible subapically dentate; area of scutum between admedian lines a distinct welt; male midcoxa with a small, apical tooth; male midfemur with a subapical excavation and tooth; hindwing media diverging a little beyond cu-a; male sternum II with a median process; male tergum VII with large spiracular lobes, middle section acute laterally and with a median projection which is somewhat double; male sternum VIII ending in a stout, downcurved spine.

Geographic range: Five species are known, ranging from southern Mexico to Argentina.

Systematics: Rubrica is related to *Stictia, Selman,* and *Editha,* all associated at one time under the unavailable name, *Monedula* Handlirsch. These four genera have in common: toothed mandibles, inner eye margins somewhat diverging below (fig. 182 A,B), palpal formula 6-4, scutum welted anteromedially, midocellus or its scar not depressed nor on an elevation, male midfemur with a subapical notch and stout tooth, marginal cell of forewing not pulled away from R_1 but broadly rounded distally, male tergum VII with spiracular lobes, main section acute laterally (sometimes notched) and truncate or notched at median apex; male sternum VIII ending in a stout, downcurved spine. From the other genera in the group, *Rubrica* can be distinguished by the somewhat flattened shape of the midocellus scar (fig. 182 A). Additionally, it differs from *Editha* by the more distal origin of the forewing media, the notched rather than truncate median lobe of male tergum VII, and the nondepressed vertex. Males of *Rubrica* have the front basitarsus somewhat ornamented and, in the case of *gravida,* expanded. The usually reddish gena and reddish areas on the gaster are distinctive and prompted the name *Rubrica.*

Biology: According to Willink (1947), *R. nasuta* (as *surinamensis*) prefers rather compact soil for nesting purposes but will tolerate arid or humid climatic zones. Brèthes (1902), Bodkin (1917), Vessey-Fitzgerald (1940), and Llano (1959) have furnished many details of nesting habits. Large colonies may occur in areas devoid of vegetation. The burrows consist of a gently sloping first part of about 15 cm and a terminal horizontal part of about 13 cm ending in a cell. In most cases the nest entrance is closed while the wasp is foraging for prey, but no inner closure has been reported. According to Vesey-Fitzgerald, Diptera are the exclusive provisions, especially Syrphidae (*Volucella obesa* Fabricius) and Tabanidae (*Tabanus occidentalis* Linnaeus, and others). Other families used are Stratiomyidae, Asilidae, Nemestrinidae, Muscidae, and Tachinidae (Llano, 1959). Two species of flies which presumably live in nests of *Rubrica* in Trinidad (Vesey-Fitzgerald, 1940) are a miltogrammine sarcophagid, *Pachygraphomyia spinosa* Malloch, and a chloropid scavenger, *Hippelates pusio* Loew. A bombyliid parasite of *R. nasuta* in Argentina is *Hyperalonia morio* (Fabricius) according to Copello (1933). Brèthes (1902) reported that the egg of *nasuta* is attached in erect position to the wing base of the fly prey. He indicated, also, that provisioning is progressive and that cell cleaning may be regularly performed by the parent wasp. In *Rubrica gravida* the construction of false burrows has been recorded, reminiscent of the actions of some *Bembix* (Llano, 1959). In Guyana *denticornis* is one of the "cowfly tigers," so-called from its habit of taking Tabanidae in the vicinity of cattle (Bodkin, 1917). The known information on habits of *Rubrica* has been summarized by Evans (1966a).

Checklist of *Rubrica*

adumbrata (Handlirsch), 1890 (*Monedula*); Brazil

caesarea (Handlirsch), 1890 (*Monedula*); Brazil, Colombia
denticornis (Handlirsch), 1890 (*Monedula*); Mexico to Peru
gravida (Handlirsch), 1890 (*Monedula*); Argentina, Brazil, Bolivia
nasuta (Christ), 1791 (*Vespa*); Argentina to Colombia
    surinamensis De Geer, 1778 (*Apis*), nec Linnaeus, 1758
    striata Fabricius, 1793 (*Bembex*)
    continua Fabricius, 1804 (*Bembex*)

Genus Selman J. Parker

Generic diagnosis: Flagellomere I longer than scape; eyes with inner margins diverging slightly below; ocelli reduced externally, midocellus a little more than a semicircle, not on a mound, with a lens which is most obvious posteriorly; vertex about level with top of eyes; clypeus with a longitudinal ridge above, lower two-thirds of clypeus rather evenly convex, then sloping abruptly toward antennal sockets; labrum a little longer than wide, cross section near base a rather smooth curve; palpal formula 6-4; mandible subapically dentate; area of scutum between admedian lines a strong welt; male midfemur with an apicoposterior tooth; hindwing media diverging slightly beyond cu-a; male tergum VII with spiracular lobes, middle section of VII with median and lateral notched projections; male sternum VIII ending in a stout, downcurved spine.

Geographic range: Selman notatus is the only known species, and it occurs in Argentina.

Systematics: The name *Selman* seems to have been derived from the Greek selma, a neuter noun meaning deck, floor, or seat. However, in its present form *Selman* is a coined name; it was obviously considered as masculine by J. Parker, who named his new type of the genus *angustus* (a synonym of *notatus*). Consequently, we are treating it as masculine, and we are using *notatus* instead of the original *notata*.

The close resemblance to *Stictia* and the single known species make *Selman* a rather weak entity. However, the case is strengthened by the fact that *Stictia* is a large genus, well represented in Argentina, yet all of its species differ from *Selman* in a few significant points. These are the lens-bearing ocelli and nondepressed vertex of *Selman*.

Biology: Nothing is known.

Checklist of *Selman*

notatus (Taschenberg), 1870 (*Monedula*); Brazil to Argentina
    angustus J. Parker, 1929

Genus Stictia Illiger

Generic diagnosis: Flagellomere I about as long as scape or a little shorter; eyes with inner margins slightly wider below than above; ocelli completely reduced, midocellar scar with outline approaching a circle or somewhat more than a semicircle, not in a pit nor on an elevation; vertex depressed below top of eyes; clypeus usually rather evenly convex; labrum a little longer than broad, a little flattened above, cross section near base almost a smooth curve; palpal formula 6-4; mandible subapically dentate; area of scutum between admedian lines a distinct welt; male midfemur with a strong subapical notch and tooth; hindwing media diverging a little beyond cu-a; male tergum VII with large spiracular lobes, middle section acute to narrowly truncate laterally, rather narrowly produced and notched medially; male sternum VIII ending in a stout, downcurved spine.

Geographic range: Most of the 26 listed species occur in South America, but three reach the United States: *carolina* from southern U.S. east of the 100th meridian and north to Pennsylvania; *signata* from Florida; and *vivida* from southern Texas (Evans, 1957a).

Systematics: Stictia are mostly large, robust, and handsomely marked wasps related to *Rubrica, Selman,* and *Editha*. The characters of this group are discussed under *Rubrica*. In the relatively depressed vertex, *Stictia* differs from the other three genera. It is distinguished, also, by the large, nearly circular, opaque midocellus.

Biology: The predilection of members of this genus for Tabanidae as prey gives them local economic importance. The well known eastern United States species, *carolina*, has been dubbed the "horse guard," and it fully lives up to its name. This large and handsome wasp sometimes nests in tremendous colonies numbering thousands of individuals. Its local value can be seen by the fact that, as reported by Hine (1906), 30 to 60 horseflies may be fed progressively to a single larva. Hine's observations were made on several colonies in southwestern Louisiana in sandy areas near the Gulf of Mexico. He pointed out the interesting fact that horses and cattle show no anxiety at the approach of the rather noisily flying *carolina* but register great concern toward botflies and some tabanids. Furthermore, practically all of the prey were female horseflies of several species. Exceptional records were a stratiomyid (*Odontomyia*), the screwworm fly (*Cochliomyia*), one specimen each of a common large mosquito (*Psorophora ciliata*), and a muscid (*Orthellia*). Hine reported that the larval chambers were 15 cm or more underground. According to Evans (1957b), the egg of *Stictia carolina* is laid in the empty cell. In *signata, vivida,* and *heros* the egg is laid on the side of the first fly prey (Evans, 1966a). Also, *carolina* does not clean out the cell before final closing of the burrow. This probably accounts for scavengers of the family Phoridae reared by Evans (1957b) from the nest of *carolina*.

C. Lin (1971) studied three colonies of *Stictia carolina* in southern Oklahoma. He recorded various male activities: sun dance, courtship, and nuptial flight. Horseflies were the common prey, but skipper butterflies and small cicadas were occasionally provisioned during periods of "competition for food."

Bates (1876) described the "guarding" activities of *Stictia signata* on the Amazon River, the wasps even collecting specimens of horseflies on the observer. Richards (1937) reviewed previous biological records pertaining to *signata* and added some of his own. In Guyana, *sig-*

nata is a riverbank species, nesting in small sandy islands in the riverbed below the first falls of the Essequibo River. An unusual feeding record for *signata* was that of Howard, Dyar, and Knab (1912). In this case, mosquitoes, *Aedes aegypti* (Linnaeus) were captured on the wing and fed upon by the adult wasps.

Evans (1957a) reported the nesting and prey of *Stictia vivida* in Cameron County, Texas. The nests were in firm sand, well back from the Gulf shore. They were simple burrows, about 65 cm long, ending in a single cell in which the larva was being fed specimens of *Tabanus texanus* Hine. The flies were carried in the usual manner of bembicins, that is, held tightly against the underside of the thorax by the front and middle legs. The nest was closed upon leaving, both with an inner plug just in front of the cell and with an outer one. Closures may be omitted during progressive provisioning of relatively large larvae (Evans, 1966a).

Checklist of *Stictia*

andrei (Handlirsch), 1890 (*Monedula*); Peru
antiopa (Handlirsch), 1890 (*Monedula*); Surinam, Venezuela
arcuata (Taschenberg), 1870 (*Monedula*); Brazil to Argentina
belizensis (Cameron), 1907 (*Monedula*); Centr. America
caribana Fritz, 1972; Colombia, Venezuela
carolina (Fabricius), 1793 (*Bembix*); se. U.S. to New Mexico
croceata (Lepeletier), 1845 (*Monedula*); Peru
   proserpina Handlirsch, 1890 (*Monedula*)
decemmaculata (Packard), 1869 (*Monedula*); Brazil
decorata (Taschenberg), 1870 (*Monedula*); Chile, Peru, Argentina
dives (Handlirsch), 1890 (*Monedula*); Mexico
flexuosa (Taschenberg), 1870 (*Monedula*); Brazil, Argentina
   carbonaria Burmeister, 1874 (*Monedula*)
heros (Fabricius), 1804 (*Bembex*); centr. America to Brazil
infracta J. Parker, 1929; Peru
insulana Lohrmann, 1942; Venezuela
maccus (Handlirsch), 1895 (*Monedula*); French Guiana to Argentina
   lineata Fabricius of authors (see van der Vecht, 1961)
maculata (Fabricius), 1804 (*Bembex*); Mexico to Brazil, Peru
maculitarsis Schrottky, 1913; Brazil, Surinam
medea (Handlirsch), 1890 (*Monedula*); Surinam to Brazil
mexicana (Handlirsch), 1890 (*Monedula*); Mexico
pantherina (Handlirsch), 1890 (*Monedula*); Colombia to Brazil
punctata (Fabricius), 1775 (*Bembyx*); Mexico to Argentina
   ?*pictithorax* Strand, 1910 (*Monedula*)
signata (Linnaeus), 1758 (*Vespa*); Florida to Argentina
   vespiformis De Geer, 1773 (*Apis*)
   insularis Dahlbom, 1845 (*Monedula*)
  ssp. *aricana* Lohrmann, 1948; Peru: Arica
sombrana J. Parker, 1929; Peru
trifasciata J. Parker, 1929; Paraguay
vivida (Handlirsch), 1890 (*Monedula*); Mexico; U.S.: Texas
volucris (Handlirsch), 1890 (*Monedula*); Colombia

Genus Editha J. Parker

Generic diagnosis: Flagellomere I longer than scape; eyes with inner margins diverging toward clypeus, least interocular distance (near vertex) less than length of flagellomere I (fig. 182 B); ocelli not depressed, lenses evident but reduced externally, midocellus a little more than a semicircle; vertex about level with top of eyes; clypeus with a longitudinal carina above, continuous with interantennal ridge, lower half of clypeus flattened and on a different plane from upper part; labrum somewhat longer than wide, cross section near base a rather smooth curve; palpal formula 6-4; mandible subapically dentate; area of scutum between admedian lines a distinct welt; male midcoxa with a small, apical tooth; male midfemur with a stout, apicoposterior tooth; forewing media diverging well in front of cu-a; hindwing media diverging slightly beyond cu-a; male tergum VII with spiracular lobes small, apex of VII truncate medially and acute laterally; male sternum VIII ending in a stout, downcurved spine.

Geographic range: The six known species are all in the Neotropical Region.

Systematics: These are the largest and most handsome of the sand wasps. They range in size from about 25 to 45 mm in length and are strikingly marked. The genus is based on relatively few specimens, many in the original type material. J. Parker (1929) gave a key to the species which was modified with one added species by Araujo (1939). Finally, Willink (1947) called attention to a sixth species from Argentina and redescribed it.

Editha has much in common with *Stictia*, but the former is distinctive in having the eyes more narrowed above, vertex not depressed below eye level, forewing media diverging before rather than beyond cu-a, clypeus more unevenly contoured (fig. 182 B), and middle section of tergum VII in male truncate rather than emarginate apically.

Biology: Bondar (1930) reported *adonis* as preying upon Lepidoptera of several families in Brazil. Specimens of *adonis* sent to the Hope Museum at Oxford University were associated with the presumed prey which were six species in four genera of skipper butterflies (Hesperiidae) according to Poulton (1917). Bondar described the burrows as being about 22 cm long and dug in areas of sparse vegetation. Richards (1971) confirmed the use of butterflies as prey (by *magnifica*), but these seemed to be primarily Pieridae of the genus *Phoebis*.

Checklist of *Editha*

adonis (Handlirsch), 1890 (*Monedula*); Brazil, Paraguay, Bolivia
   stridulans Strand, 1910 (*Monedula*)

diana (Handlirsch), 1890 (*Monedula*); Brazil
fuscipennis (Lepeletier), 1845 (*Monedula*); Brazil, Paraguay
    zetterstedti Dahlbom, 1845 (*Monedula*)
integra (Burmeister), 1874 (*Monedula*); Argentina, Brazil
magnifica (Perty), 1833 (*Monedula*); Brazil, lectotype ♂, "Brasil", (Mus. Munich), designated by A. Menke
pulcherrima J. Parker, 1929; Uruguay

Genus Bembix Fabricius

Generic diagnosis: Eyes usually diverging slightly below, least interocular distance generally greater than length of flagellomere I (fig. 182 I); midocellus reduced to an arcuate line (except in a very few species which have traces of a lens); vertex slightly below top of eyes; clypeus relatively short, broad, arched, apically concave; labrum sometimes more than twice as long as broad, often notched apically, usually flattened toward base; palpal formula 4-2 (fig. 182 H), rarely 3-1; mandible subapically dentate; area of scutum between admedian lines a moderate but distinct welt; legs variable but many species with midfemur serrate below and/or dentate subapically; forewing with veinlet between first two submarginal cells usually markedly crooked; hindwing media diverging beyond cu-a by a length usually about equal to cu-a; male tergum VII with spiracular lobes and sometimes with a laterobasal tooth on main section; male sternum II often with a keel, VI often with a process of varying form; male sternum VIII ending in a downcurved spine.

Geographic range: Cosmopolitan and by far the largest genus in the subfamily. We have listed 329 species which are concentrated in Australia (80) and the Ethiopian Region (about 90); the Holarctic and Oriental Regions are well represented however. Twenty three species are known from North America and 11 occur in South America.

A measure of the distribution within the United States, based on Evans and Matthews (1968b), is that two species are found only east of the 100th meridian, eight species on both sides of the meridian, and nine species only to the west of it. The only polytypic species in North America is *americana*, which has distinctive subspecies on islands of the West Indies as well as Californian offshore islands.

Systematics: In addition to the large number of species of *Bembix*, there is such diversity in structure that it is difficult to characterize the genus. The most reliable characters are the greatly or totally reduced anterior ocellus, with only an arcuate scar or narrow lunule remaining, and the remarkable reduction in palpal segments. To these should be added the less unique features of dentate mandibles, apex of marginal cell close to R_1, median scutal welt, and propodeum not compressed laterally nor prominent medially. Other useful, but not absolute, characters are the irregular veinlet between submarginal cells I and II, and the rather long veinlet between cu-a, and the forking of M + Cu in the hindwing.

Among the many unusual features occurring in some species are the long labrum (more than twice as long as broad) in *rugosa* and *magdalena*, traces of a lens in the anterior ocellus of the *dentilabris* group, and maculate wings in *melanaspis* and *nubilipennis*. All of the above species are North American. Even more striking peculiarities are shown by some of the Australian species: the stout transverse and shovel like process on male sternum II in *lamellata*, the forked process on the same sternum of *furcata*, the great reduction (*uloola*) or loss (*munta* and others) of the midtibial spur in males, and the shield-like front basitarsus of some males which may be accompanied by the development of a lamelliform foretibial spur (see figures in Evans and Matthews, 1973b). Representatives of the Australian fauna have been studied through the courtesy of E. F. Riek.

Lohrmann (1948) divided the North American fauna into six species groups. Rearranging the grouping somewhat, Evans and Matthews (1968b) recognized five of the former groups and added a new one. The groups as exemplified by typical species were: *belfragei, amoena, cinerea, americana, texana,* and *pallidipicta*. Evans and Matthews (1973b) proposed 12 species groups for the Australian fauna.

Useful keys can be found in Handlirsch (1893) and J. Parker (1929) for the world, Turner (1915b) for Australia which is superseded by Evans and Matthews (1973b), Evans and Matthews (1968b) for North America, Bohart and Horning (1971) for California, Leclercq (1961d) for Madagascar and Priesner (1958) for Egypt.

The relationships of *Bembix* with other genera in the tribe are indicated in fig. 181. Its rather specialized nature is indicated by an evolutionary index of "27" based on table 16. Because of the diverse nature of the genus, the type species, *rostrata*, was selected to obtain the index number.

Within the genus *Bembix* it is possible that *dentilabris* is one of the least advanced species. Some of the characters that lead to this conclusion as pointed out by Evans (1960), are: the nearly normal lateral ocelli and the lens-bearing midocellus, midlegs with serrate midfemora and other structures used during copulation, male sterna without processes, and female pygidial plate somewhat developed.

On the other hand, both structural characters and behavioral traits as outlined by Evans (1960) put *occidentalis* high on the evolutionary tree and, therefore, among the most advanced of all wasps.

Biology: The sand wasps are well known residents of sandy beach areas. They can lay claim to beneficial status since they provision their nests with flies which are often noxious species. *Bembix hinei*, whose prey is largely *Tabanus*, has been called the "small horse guard" to differentiate it from the "horse guard," *Stictia carolina*. Information on ethology of *Bembix* is much more complete than that on any other large genus of Nyssoninae. Comprehensive works which have appeared in recent years are those of Nielsen (1945), Tsuneki (1958), and Evans (1957b, 1966a). Nielsen's study was concentrated on the type species of the genus, *rostrata*, in Denmark. Tsuneki dealt especially with *niponica* in Japan, and Evans compared the behavior of many

North American species. Incorporated with Evans' original studies were observations previously published by Peckham and Peckham (1898), Rohwer (1909), J. Parker (1910, 1917, 1925), Rau and Rau (1918), G. Bohart and MacSwain (1939, 1940), and Strandtmann (1953). The following summary of *Bembix* habits in and about the nesting area is taken largely from Evans (1957b, 1966a).

Bembix species nest gregariously and maintain their colonies from year to year. They are not social in the usual sense but may take part in mass "flights of intimidation" against intruders in the colony. They readily become accustomed to human presence and are not truly aggressive or dangerous. Males take part in prolonged flight rituals termed "sun dances" by the Raus (1918). The sexes normally meet in the air, and copulation takes place on a plant or on the ground in the ensuing 30 seconds to two minutes. Mating in both sexes may occur more than once. Polyandry and polygamy have been noted especially for *occidentalis*. The serrate and toothed midfemur in males of many species is used to keep the wings of the female immobile during the copulatory act. Apparently, in those species without the serrations, such as *occidentalis*, the copulatory period is a rather "rough and tumble" affair. Wasps of either sex may occasionally utter a simple chirp, but a loud, shrill chirping is produced by the male when mating.

Nests are built in sandy or rather hard soil and consist typically of a mound of excavated earth which is often dispersed by the wind, a sloping tunnel of several feet in depth, a rather horizontal terminal branch or cell burrow, and a more vertical spur in which the female may rest or into which she may pull the plug of earth closing off the cell burrow. Males dig shallow burrows in which they pass the night. The broadened front tarsi found in many Australian forms may assist in "digging through coarse, stony soil" (Evans and Matthews, 1973b).

Cell provisioning is almost invariably progressive, and prey usually consists of a variety of Diptera, but especially Muscoidea, Tabanidae, Syrphidae, Dolichopodidae, and Therevidae. The Nematocera are conspicuously absent in prey records. The placement of the *Bembix* egg seems to have evolutionary significance. It may be attached to the first fly prey, as in *rostrata*, *niponica*, and most of the American species considered by Evans (1957b); placed erect and attached to the floor of the empty cell, as in *troglodytes* and *texana;* leaned obliquely against the end of the cell, as in *pallidipicta;* or laid flat on the bottom of the cell near the middle, as in *occidentalis*.

Flies are seized and paralyzed either in flight or at rest and are then transported to the nest, held by the middle legs of the wasp. The fly is released at the burrow entrance, grasped with the hind legs and dragged inside. The prey are not as carefully paralyzed as in cases of mass provisioning by other types of wasps, but since they are ordinarily consumed in a day or so, a fresh condition over a considerable period is not essential. At or near larval maturity the wasp closes the cell, and its occupant spins a cocoon which incorporates sand grains. Then, the female may construct a second or third cell, or the nest may be filled, and the entrance concealed. Periods between summer broods are spent as prepupa and pupa. Overwintering takes place in the prepupal form.

Some exceptional habits have been noted by Evans (1966a). In *hinei* there are mass provisioning and multicellular nests, as in many less specialized nyssonine genera. Accessory burrows are constructed near the real ones by *troglodytes* and *pallidipicta*. Presumably, this may confuse potential parasites. Females of *cinerea* try to steal flies from one another in flight or at the burrow entrance. In the nest of *belfragei* the entrance is not closed during excursions for prey, but there is an inner closure of the cell gallery. *B. dentilabris* is crepuscular and Evans (1960a) related this to the fact that this is one of the few *Bembix* with apparently functional ocelli. In the case of *texana* only enough food is brought for current use of the larva, and fly remains are removed outside the nest by the wasp.

A recent and important study of Australian *Bembix* is that of Evans and Matthews (1973b). Their findings of behavior in 20 species largely parallel those reported elsewhere. However, these authors have considerably amplified our knowledge of prey selection in the genus. The use of damselflies by Australian *Bembix* was reported first by Wheeler and Dow (1933) but this was not verified until the recent work of Evans and Matthews. They found that about one-third of the species they observed preyed on insects of orders other than Diptera. Several preyed on bees; one on wasps, bees and flies; two species attacked damselflies; and one preyed upon antlions.

Both Nielson (1945) and Tsuneki (1958) reported observations made within nests fitted with panes of glass (Nielsen) or glass tubes (Tsuneki). The latter author made some experiments on learning with simple mazes. Also, in slightly quaint English, he reported some fascinating maternal care behavior of the wasp in the cell: "1. At the time of inspection of the brood-cell the wasp touches the larva with the antennae, usually holding it between them. 2. When the larva has been too near the accumulation of the food at the entrance the wasp catches it with her jaws and drags it toward the middle of the cell. 3. When the prey is scarce within reach of the larva she carries a prey or two to it from the pile of food at the entrance. Thus she appears to care for the larva not directly to take food from the provision. 4. Under the experimental condition when the larva is long exposed to light or some unfavourable event occurred upon it, she carries it to the dark portion of the tunnel. 5. Rarely she cleans the chamber carrying the remains of food out of the nest."

Definite or presumed parasites are Sarcophagidae (*Opsidea, Phrosinella, Senotainia, Amobiopsis*), Bombyliidae (*Anthrax, Villa, Exoprosopa*), Conopidae (*Physocephala*), Rhipiphoridae (*Macrosiagon*), Stylopidae (*Pseudoxenos*), Mutillidae (*Dasymutilla, Timulla*), and Chrysididae (*Parnopes*). A predator recorded by G. Bohart and MacSwain (1939) on *occidentalis* was *Proctacanthus occi-*

dentalis Hine (Asilidae).

Although flesh flies are frequently used as prey, progressive provisioning apparently prevents the one-sided competition provided by flesh fly larvae emerging from the paralyzed maternal fly, as reported for *Oxybelus sparideus* by R. Bohart, *et. al.,* (1966).

Checklist of *Bembix*
(variant but improper spellings are *Bembex* and *Bembyx*)

abercornensis Arnold, 1960; s. Africa
abragensis Priesner, 1958; Egypt
admirabilis Radoszkowski, 1893; sw. USSR: Transcaspia
affinis Dahlbom, 1844; loc. unknown
afra Handlirsch, 1893; s. Africa
agrestis J. Parker, 1929; Ethiopia
alacris J. Parker, 1929; Somalia
abata J. Parker, 1929; e. centr. Africa
albicapilla Arnold, 1946; s. Africa
albidula Turner, 1917; w. Africa
albofasciata F. Smith, 1873; s. Africa
    karschii Handlirsch, 1893
albopilosa Arnold, 1929; s. Africa
aldabra J. Parker, 1929; Indian Ocean: Aldabra
alfierii Priesner, 1958; Egypt
allunga Evans and Matthews, 1973; n. Australia
americana Fabricius, 1793; Virgin Is., Puerto Rico
    muscicapa Handlirsch, 1893
    separanda Handlirsch, 1893
    foxi J. Parker, 1917
  ssp. *antilleana* Evans and Matthews, 1968; Cuba
  ssp. *comata* J. Parker, 1917; Pacific Coast of N. America to Mexico
    nevadensis Rodeck, 1934
  ssp. *hamata* C. Fox, 1923; U.S.: California islands: Santa Cruz to San Miguel
    lucida C. Fox, 1923
    sanctae-rosae Cockerell, 1940
  ssp. *nicolai* Cockerell, 1938; U.S.: California islands: San Nicolas
  ssp. *spinolae* Lepeletier, 1845; N. America except Pacific Coast
    similans W. Fox, 1895
    connexa W. Fox, 1895
    primaaestate Johnson and Rohwer, 1908
amoena Handlirsch, 1893; w. U.S., w. Canada
anomalipes Arnold, 1929; s. Africa
arenaria Handlirsch, 1893; Syria
arlettae Beaumont, 1970; Iran
atrifrons F. Smith, 1856; Australia: s. Tropic of Capricorn
    flavilabris F. Smith, 1873
    funebris Turner, 1910
atrospinosa Turner, 1917; s. Africa
aureofasciata Turner, 1910; w. Australia
barbara Handlirsch, 1893; Algeria, Morocco
baringa Evans and Matthews, 1973; w. Australia
bataviana Strand, 1910; Java
baumanni Handlirsch, 1893; s. Africa
bazilanensis Yasumatsu, 1933; Philippines
belfragei Cresson, 1873; s. centr. U.S.
    cressonis Handlirsch, 1893
    insignis Handlirsch, 1893
    cressonii Dalla Torre, 1897, lapsus
bellatrix J. Parker, 1929; Brazil
bequaerti Arnold, 1929; s. Africa
  ssp. *dira* Arnold, 1929; s. Africa
berontha Evans and Matthews, 1973; w. Australia
bicolor Radoszkowski, 1877; Iran, Middle East to Afghanistan, Mongolia
    femoralis Radoszkowski, 1877
    bipunctata Radoszkowski, 1877, nec Dufour, 1861
    barbiventris F. Morawitz, 1889
bidentata Vander Linden, 1829; Eurasia
    dalmatica Kriechbaumer, 1869
boarliri Evans and Matthews, 1973; centr. Australia
boonamin Evans and Matthews, 1973; w. Australia
borneana Cameron, 1901; Borneo
borrei Handlirsch, 1893; Indonesia: Timor
  ssp. *thaiana* Tsuneki, 1963; Thailand
brachyptera Arnold, 1952; s. Africa
braunsii Handlirsch, 1893; w. centr. Africa
brullei Guérin-Méneville, 1831 (*Bember!*, pl. 9); Chile
    ventralis Dahlbom, 1844
    ventralis Lepeletier, 1845, nec Dahlbom, 1844
    emarginata Sichel, 1867
brunneri Handlirsch, 1893; Algeria
bubalus Handlirsch, 1893; s. Africa
budha Handlirsch, 1893; India
    buddha Dalla Torre, 1897, emendation
bulloni Giner Marí, 1945; n. Africa
buntor Evans and Matthews, 1973; sw. Australia
burando Evans and Matthews, 1973; ne. Australia
burraburra Evans and Matthews, 1973; w. Australia
cameroni Rohwer, 1912; centr. Mexico, sw. U.S.
    festiva J. Parker, 1929
    rohweri Lohrmann, 1948
cameronis Handlirsch, 1893; s. Africa
    cameronii Dalla Torre, 1897, lapsus
canescens (Gmelin), 1790 (*Apis*); sw. USSR: Caspian Sea area
capensis Lepeletier, 1845; s. Africa
    natalis Dahlbom, 1845
    miscella Mercet, 1904
capicola Handlirsch, 1893; s. Africa
carinata F. Smith, 1856; Ethiopia
    undulata Dahlbom, 1845, nec Spinola, 1838
    venator F. Smith, 1856
carripan Evans and Matthews, 1973; w. Australia
chlorotica Spinola, 1838; Egypt
chopardi Berland, 1950; nw. Africa
ciliciensis Beaumont, 1967; Turkey
cinctella Handlirsch, 1893; s. Europe, Turkey
  ssp. *enslini* Bytinski-Salz, 1955; Israel
cinerea Handlirsch, 1893; e. U.S. w. to Texas
citripes Taschenberg, 1870; S. America
    inops Handlirsch, 1893
    inopides Strand, 1910
    gradilis J. Parker, 1929
    bahiae J. Parker, 1929

subcitripes Willink, 1947
compedita Turner, 1913; s. Africa
    kohlii Turner, 1912, nec Morice, 1897
cooba Evans and Matthews, 1973; se. Australia
coonundura Evans and Matthews, 1973; w. Australia
cultrifera Arnold, 1929; s. Africa
    ssp. *ypsilon* Arnold, 1951; Ethiopia
cursitans Handlirsch, 1893; w. Australia
dahlbomii Handlirsch, 1893; n. Africa
    ?*glauca* Dahlbom, 1845, nec Fabricius, 1787
    ssp. *sabulosa* Bytinski-Salz, 1955; Israel
denticauda Arnold, 1946; s. Africa
dentilabris Handlirsch, 1893; U.S.: Texas to s. California; n. Mexico: Morelos to Baja California
    u-scripta W. Fox, 1895
    arcuata J. Parker, 1917
difformis Handlirsch, 1893; Brazil
dilatata Radoszkowski, 1877; sw. USSR: Turkmen S.S.R.
diversidens Arnold, 1946; s. Africa
diversipennis F. Smith, 1873; Ethiopian Region
diversipes F. Morawitz, 1889; centr. Asia
doriae Magretti, 1884; n. Africa
    arabica Lohrmann, 1942
dubia Gussakovskij, 1934; Iran
eburnea Radoszkowski, 1877; sw. USSR to Afghanistan
    kirgisica F. Morawitz, 1891
egens Handlirsch, 1893; w. and centr. Australia
eleebana Evans and Matthews, 1973; Australia
expansa Gribodo, 1894; Ethiopia
fantiorum Arnold, 1951; w. Africa
filipina Lohrmann, 1942; Philippines
finschii Handlirsch, 1893; New Guinea
fischeri Spinola, 1838; Egypt
    ssp. *tibesti* Beaumont, 1956; Libya
fischeroides Magretti, 1892; Somalia
flavescens F. Smith[2], 1856; Canary Is.
    ssp. *bolivari* Handlirsch, 1893; s. Europe, n. Africa
        algeriensis Lohrmann, 1942, nec Schulz, 1905
    ssp. *fonti* Mercet, 1905; Spanish Sahara: Rio de Oro
    ssp. *citrina* Mercet, 1905; Morocco
    ssp. *picturata* Bytinski-Salz, 1955; Mediterranean area
    ssp. *inimica* Beaumont, 1957; Tunisia
    ssp. *kittyae* Beaumont, 1957; Libya, Egypt
flavocincta Turner, 1912; s. Africa
flavifrons F. Smith, 1856; e. Australia
    saussurei Handlirsch, 1893
flavipes F. Smith, 1856; n. centr. and ne. Australia
    brevis Lohrmann, 1942
flaviventris F. Smith, 1873; s. Australia
    calcarina Handlirsch, 1893
forcipata Handlirsch, 1893; Tanzania
    massaica Cameron, 1910
formosana Bischoff, 1913; Taiwan
    metamelanica Strand, 1923
fossoria F. Smith, 1878; Burma

fraudulenta Arnold, 1929; s. Africa
freygessneri Morice, 1897; Egypt
    dissimilis W. F. Kirby, 1900; Egypt
    decipiens Priesner, 1958; Egypt
frommeri R. Bohart, 1970; U.S.: e. California
frontalis Olivier, 1789; India
fucosa J. Parker, 1929; Burma
fumida J. Parker, 1929; Japan
furcata Erichson, 1842; s. Australia incl. Tasmania
fuscipennis Lepeletier, 1845; s. Africa
    ssp. *centralis* Berland, 1950; Mali
galactina Dufour, 1853; nw. Africa
    fallax Mercet, 1905
    parkeri Lohrmann, 1942
ganglbaueri Handlirsch, 1893; sw. USSR: Caspian region
gelane Evans and Matthews, 1973; e. centr. Australia
geneana A. Costa, 1867; Sardinia
    melanostoma A. Costa, 1867
generosa J. Parker, 1929; Somalia
gillaspyi Evans and Matthews, 1968; U.S.: s. California
ginjulla Evans and Matthews, 1973; se. Australia
glauca Fabricius, 1787; India
    indica Handlirsch, 1893
gobiensis Tsuneki, 1971; Mongolia
goyarra Evans and Matthews, 1973; w. Australia
gracilens J. Parker, 1929; sw. Africa
gracilis Handlirsch, 1893; sw. USSR
grisescens Dahlbom, 1845; s. Africa
guigliae Arnold, 1951; Senegal
gunamarra Evans and Matthews, 1973; n. centr. and ne. Australia
handlirschella Ferton, 1912; n. Africa
    handlirschi Ferton, 1912, nec Cameron, 1902
handlirschi Cameron, 1902; Maldive Is. (Indian Ocean)
harenarum Arnold, 1929; s. Africa
hedickei Giner Marí, 1945; n. Africa
hesione Bingham, 1893; India
heteracantha Gussakovskij, 1934; Iran
hexaspila J. Parker, 1929; India
hinei J. Parker, 1917; s. centr. U.S.
hokana Evans and Matthews, 1973; w. Australia
holoni Bytinski-Salz, 1955; Israel
houskai Balthasar, 1953; Israel
hova Saussure, 1892; Madagascar
incognita J. Parker, 1929; Asia ?
infumata Handlirsch, 1893; centr. Mexico
    nubilipennis Cameron, 1890, nec Cresson, 1872
inscripta Dahlbom, 1844; loc. unknown
irritata Nurse, 1903; India
isabellae Beaumont, 1970; Iran
joeli Bytinski-Salz, 1955; Israel
johnstoni Turner, 1912; Uganda
    tenebrosa J. Parker, 1929
jordanicus Lohrmann, 1942; Israel
julii Fabre, 1879; s. Europe, n. Africa
    occitanica Mocsáry, 1883
junodi Arnold, 1929; s. Africa
kamulla Evans and Matthews, 1973; Australia

---

[2] *Bembex flavescens* Lichenstein, 1796 is a prior name, but it was apparently offered in error as a new generic assignment for the Indian wasp, *Vespa flavescens* Fabricius, 1775. Also it is an auction catalog name and may not be valid.

khasiana Cameron, 1904; India: Assam
kohlii Morice, 1897; Egypt
kora Evans and Matthews, 1973; e. Australia
kriechbaumeri Handlirsch, 1893; w. centr. Africa
  ssp. *scitula* Arnold, 1929; s. Africa
kununurra Evans and Matthews, 1973; n. centr. Australia
labiata Fabricius, 1798; Europe
labidura Handlirsch, 1893; centr. Africa
lactea Cameron, 1901; India
laeta J. Parker, 1929; e. Africa
lamellata Handlirsch, 1893; Australia
latebrosa Kohl, 1909; Madagascar
latifasciata Turner, 1915; Australia
latigenata Willink, 1947; Argentina
latitarsis Handlirsch, 1893; n. India
leeuwinensis Turner, 1915; se. Australia
levis J. Parker, 1929; e. centr. Africa
liberiensis J. Parker, 1929; Liberia
lineatifrons Cameron, 1910; Tanzania: Tanganyika
littoralis Turner, 1908; Australia
    flavolatera J. Parker, 1929
liturata Turner, 1917; s. Africa
  ssp. *flavopicta* Arnold, 1929; s. Africa
liventis J. Parker, 1929; e. centr. Africa
lobatifrons Turner, 1913; Kenya
lobimana Handlirsch, 1893; w. Australia
longipennis J. Parker, 1929; e. Africa
loorea Evans and Matthews, 1973; centr. Australia
loupata J. Parker, 1929; e. Africa
lunata Fabricius, 1793; India
lusca Spinola, 1838; Egypt; Sudan
    pectoralis Dahlbom, 1845
lutescens Radoszkowski, 1879; sw. USSR
luzonensis J. Parker, 1929; Philippines
madecassa Saussure, 1890 (pl. 11, fig. 2); Madagascar
    militaris Saussure, 1890 (pl. 11, fig. 3)
    crinita Saussure, 1891
magarra Evans and Matthews, 1973; nw. Australia
magdalena C. Fox, 1926; Mexico: Baja California
maidli Schulthess, 1927; Iran
maldiviensis Cameron, 1902; Maldive Is. (Indian Ocean)
maliki Evans and Matthews, 1973; s. centr. Australia
mareeba Evans and Matthews, 1973; Australia: coastal Queensland
marhra Evans and Matthews, 1973; w. Australia
marsupiata Handlirsch, 1893; sw. Australia
megadonta Cameron, 1904; India
megerlei Dahlbom, 1845; e. Europe, w. Asia
    ?*undata* Dahlbom, 1845
    sarafschani Radoszkowski, 1877
melanaspis J. Parker, 1917; sw. U.S.; Mexico: Baja California
melancholica F. Smith, 1856; India
melanopa Handlirsch, 1893; s. Africa
  ssp. *algoensis* Arnold, 1946; s. Africa
    litoralis Arnold, 1929, nec Turner, 1908
melanosoma Gribodo, 1894; se. Africa
melanura F. Morawitz, 1889; centr. Asia
    asiatica Radoszkowski, 1893

merceti J. Parker, 1929; Spain
    handlirschi Mercet, 1904, nec Cameron, 1902
mianga Evans and Matthews, 1973; Australia: s. Tropic of Capricorn
mildei Dahlbom, 1845; s. Europe
mima Handlirsch, 1893; centr. Mexico
minya Evans and Matthews, 1973; se. Australia
modesta Handlirsch, 1893; w. centr. Africa
moebii Handlirsch, 1893; Ethiopia
    testaceicauda Handlirsch, 1893
mokari Evans and Matthews, 1973; s. centr. Australia
moma Evans and Matthews, 1973; Australia
monedula Handlirsch, 1893; Ethiopia
moonga Evans and Matthews, 1973; se. Australia
multipicta F. Smith, 1873; Mexico, Centr. America, n. S. America
    frioensis J. Parker, 1929
mundurra Evans and Matthews, 1973; sw. and se. Australia
munta Evans and Matthews, 1973; w. Australia
musca Handlirsch, 1893; Australia
    mackayensis Turner, 1910
nasuta Morice, 1897; Egypt
nigrocornuta J. Parker, 1929; India, Burma
nigropectinata Turner, 1936; w. Australia
nilotica Priesner, 1958; Egypt
niponica F. Smith, 1873; Japan
    miserabilis J. Parker, 1922
  ssp. *picticollis* F. Morawitz, 1889; China to Mongolia
notabilis Arnold, 1952; Kenya
nubilipennis Cresson, 1872; centr. and sw. U.S., centr. Mexico
    nubilosa J. Parker, 1929
nupera J. Parker, 1929; Kenya
obtusa Turner, 1917; s. Africa
occidentalis W. Fox, 1893; w. and sw. U.S., n. Mexico
    beutenmuelleri W. Fox, 1901
    obsoleta Howard, 1901
ochracea Handlirsch, 1893; s. Africa
octosetosa Lohrmann, 1942; e. and se. Australia
oculata Panzer, 1801; s. Europe, Middle East, Iran, Afghanistan
    ?*pallescens* Giorna, 1791 (*Vespa*), see Schulz, 1906, Berliner Ent. Zeit. 51:303
    oculata Panzer of Latreille, 1804-1805
    latreillei Lepeletier, 1845
    neglecta Dahlbom, 1845
    ?*rossii* Dahlbom, 1845
    panzeri Handlirsch, 1893
    hispanica Mercet, 1904
    ebusiana Giner Marí, 1934
  ssp. *basalis* Dahlbom, 1845; w. Asia, e. Europe
  ssp. *soror* Dahlbom, 1845; Egypt
  ssp. *fuscilabris* Mocsáry, 1883; s. and e. Europe
  ssp. *pannonica* Mocsáry, 1883; Hungary
  ssp. *candiotes* Schulz, 1906; Crete
  ssp. *ceballosi* Giner Marí, 1945; Morocco
  ssp. *gegen* Tsuneki, 1971; Mongolia
  ssp. *mongolica* Tsuneki, 1971; Mongolia

odontopyga Turner, 1917; s. Africa
olba Evans and Matthews, 1973; centr. Australia
olivacea Fabricius, 1787; s. Europe, n. Africa, Turkey
    olivacea Cyrillo, 1787, nec Fabricius, 1787
    ?*clypeata* Christ, 1791 (*Vespa*)
    ?*apilinguaria* Christ, 1791 (*Vespa*)
    senilis Fabricius, 1804
    ?*octopunctata* Donovan, 1810 (*Bembex*)
    notata Dahlbom, 1845
    mediterranea Handlirsch, 1893
    maroccana Mercet, 1905
    pardoi Giner Marí, 1945
    saharae Giner Marí, 1945
olivata Dahlbom, 1845; s. Africa
    ?*intermedia* Dahlbom, 1845
oomborra Evans and Matthews, 1973; ne. Australia
opima Turner, 1917; s. Africa
opinabilis J. Parker, 1929; e. Africa
orientalis Handlirsch, 1893; India, Burma
ornatilabiata Cameron, 1910; Tanzania
ourapilla Evans and Matthews, 1973; w. Australia
ovans Bingham, 1893; Burma
palaestinensis Lohrmann, 1942; Israel
†*pallescens* Priesner, 1958; Egypt, nec Giorna, 1791
pallida Radoszkowski, 1877; sw. USSR: Kazakh S.S.R.
pallidipicta F. Smith, 1873; Mexico, U.S.
    pruinosa W. Fox, 1895
palmata F. Smith, 1856; e. and n. Australia
    tridentifera F. Smith, 1873
palona Evans and Matthews, 1973; Australia
papua Handlirsch, 1893; New Guinea area
parkeri Lohrmann, 1942; Algeria
parvula F. Morawitz, 1896; sw. USSR: Transcaspia
pectinipes Handlirsch, 1893; Australia
    palmata F. Smith, 1873, nec F. Smith, 1856
persa Schulthess, 1927; Iran
persimilis Turner, 1917; India
physopoda Handlirsch, 1893; Brazil
pikati Evans and Matthews, 1973; centr. Australia
pillara Evans and Matthews, 1973; w. Australia
pinguis Handlirsch, 1893; India to Java
piraporae J. Parker, 1929; Brazil
placida F. Smith, 1856; Colombia
planifrons F. Morawitz, 1891; sw. USSR, Iran
    mervensis Radoszkowski, 1893
portschinskii Radoszkowski, 1884; sw. USSR, Middle East, Mongolia
    seminigra F. Morawitz, 1889
priesneri Beaumont, 1965; Egypt
promontorii Lohrmann, 1942; e. Australia
pugillatrix Handlirsch, 1893; Philippines to New Guinea
pulka Evans and Matthews, 1973; centr. Australia
quadrimaculata Taschenberg, 1870; loc. unknown
quinquespinosa J. Parker, 1929; Philippines
radoszkowskyi Handlirsch, 1893; n. and e. Africa
    radoszkowskii Dalla Torre, 1897, emendation
rava Arnold, 1952; Kenya
recurva J. Parker, 1929; Cameroon
refuscata J. Parker, 1929; e. Africa
regia J. Parker, 1929; equatorial Africa
regnata J. Parker, 1929; centr. e. Africa
relegata Turner, 1917; India
residua J. Parker, 1929; China
robusta Lohrmann, 1942; Philippines
rostrata (Linnaeus), 1758 (*Apis*); Palearctic Region
    armata Sulzer, 1776 (*Vespa*)
    rostrata Gmelin, 1790 (*Apis*), nec Linnaeus, 1758
    vidua Lepeletier, 1845
    dissecta Dahlbom, 1845
    gallica Mocsáry, 1883
    paradoxa Giner Marí, 1943
  ssp. *algeriensis* Schulz, 1905; n. Africa
rufiventris Priesner, 1958; Egypt
rugosa J. Parker, 1917; U.S.: Arizona
salina Lohrmann, 1942; Israel
sayi Cresson, 1865; U.S., n. Mexico
    latifrons J. Parker, 1917
scaura Arnold, 1929; s. Africa
scotti Turner, 1912; Nigeria
seculata J. Parker, 1929; India
semoni Cameron, 1905; New Guinea
severa F. Smith, 1873; Australia
sibilans Handlirsch, 1893; s. Africa
silvestrii Maidl, 1914; Nigeria
sinuata Panzer, 1804; s. Europe
    sinuata Panzer of Latreille, 1804-1805
  ssp. *mauretanica* Beaumont, 1951; nw. Africa
smithii Handlirsch, 1893; Brazil
spatulata J. Parker, 1929; India
speciosa Arnold, 1929; s. Africa
spiritalis J. Parker, 1929; Brazil
splendida Arnold, 1951; w. Africa
stadelmanni Handlirsch, 1893; Kenya
    stadelmannii Dalla Torre, 1897, emendation
stenobdoma J. Parker, 1917; U.S.: w. Texas to s. California
stevensoni J. Parker, 1929; centr. and s. Africa
subeburnea Tsuneki, 1971; Mongolia
sulphurescens Dahlbom, 1845; India
taiwana Bischoff, 1913; Taiwan
    kosemponis Strand, 1922
    lutea Sonan, 1927
tarsata Latreille, 1809; s. Europe, sw. USSR: Turkmen S.S.R. (usually given as a synonym of *Bembix integra* Panzer, 1805, which was a misidentification of *integra* Fabricius, 1793, which is synonymous with *Stizus fasciatus* (Fabricius)
    cristatus Mocsáry, 1883
taschenbergii Handlirsch, 1893; India
tenuifasciata J. Parker, 1929; Nigeria
texana Cresson, 1872; se. and s. U.S.
    ?*fasciata* Fabricius, 1804, nec Fabricius, 1781
thooma Evans and Matthews, 1973; Australia
tibooburra Evans and Matthews, 1973; w. and centr. Australia
torosa J. Parker, 1929; New Guinea
tranquebarica (Gmelin), 1790 (*Vespa*); India, new name for *Vespa repanda* (Fabricius), (Art. 59c)

repanda Fabricius, 1787, nec *Vespa repanda* (Fabricius), 1787 (*Crabro*), now in *Tachytes*
    trepanda Dahlbom, 1844
transcaspica Radoszkowski, 1893; sw. USSR, Iraq
trepida Handlirsch, 1893; se. Australia
    victoriensis Lohrmann, 1942
triangulifera Arnold, 1944; s. Africa
    stevensoni Arnold, 1929, nec J. Parker, 1929
tricolor Dahlbom, 1845; Guinea, Senegal
troglodytes Handlirsch, 1893; n. Mexico; sw. U.S. to Kansas
    helianthopolis J. Parker, 1917
truncata Handlirsch, 1893; centr. Mexico
tuberculiventris Turner, 1908; Australia
turca Dahlbom, 1845; e. Europe, n. Africa, Turkey
    melaena F. Smith, 1856
ugandensis Turner, 1913; Uganda
uloola Evans and Matthews, 1973; w. Australia
ulula Arnold, 1929; s. Africa
undeneya Evans and Matthews, 1973; Australia
undulata Spinola, 1838; S. Africa
usheri Arnold, 1960; e. Africa
variabilis F. Smith, 1856; Australia
    raptor F. Smith, 1856
    crabroniformis F. Smith, 1873
vasta Lohrmann, 1942; Malaya
velox Handlirsch, 1893; Tanzania: Zanzibar
venusta Arnold, 1929; s. Africa
versuta Arnold, 1946; s. Africa
vespiformis F. Smith, 1856; Australia
wadamiri Evans and Matthews, 1973; n. centr. Australia
wagleri Gistel, 1857; Portugal
wangoola Evans and Matthews, 1973; s. Australia
wanna Evans and Matthews, 1973; se. Australia
warawara Evans and Matthews, 1973; se. Australia
weberi Handlirsch, 1893; China
    ssp. *lama* Tsuneki, 1971; Mongolia
weema Evans and Matthews, 1973; w. Australia
westermanni Spinola, 1838; Guinea
westoni Bingham, 1893; Burma
wilcannia Evans and Matthews, 1973; Australia
wiluna Evans and Matthews, 1973; Australia
wollowra Evans and Matthews, 1973; w. Australia
wolpa Evans and Matthews, 1973; Australia
wowine Evans and Matthews, 1973; w. Australia
yalta Evans and Matthews, 1973; w. Australia
yunkara Evans and Matthews, 1973; w. Australia
zarudnyi Gussakovskij, 1934; Iran
zonata Klug, 1835; s. Europe
    bipunctata Dufour, 1861
    lichtensteini Mocsáry, 1883

Lichtenstein auction catalog names
(See discussion at end of *Sphex* checklist for status of these).

cingulata Lichtenstein, 1796; "India orientali"
guttata Lichtenstein, 1796; "cap. bon. Spei"

Genus Trichostictia J. Parker

Generic diagnosis: Flagellomere I longer than scape; eyes hairy, inner margins nearly straight but obviously diverging below; midocellus semicircular, on a slight elevation; vertex a little lower than top of eyes; clypeus rather evenly contoured but with a basomedial ridge, dorsal point of clypeus reaching well between antennal sockets; labrum fully 1.5 times as long as broad, cross section near base a rather smooth curve; palpal formula 6-4; mandible subapically dentate; area of scutum between admedian lines a distinct welt; male midcoxa with a prominent apical tooth; male midfemur sometimes toothed toward apex; hindwing media diverging slightly beyond cu-a; male tergum VII with spiracular lobes small; apex of VII rounded medially and acute laterally; male sternum VIII ending in a stout, downcurved spine.

Geographic range: The three known species are all in the Neotropical Region.

Systematics: The hairy eyes distinguish these species at once from all other bembicins. A few other sphecids have this condition, such as *Entomognathus,* some *Huavea,* some *Pison,* and some *Megistommum.* In other respects *Trichostictia* has much in common with *Editha* and *Zyzzyx,* agreeing for the most part in venational, ocellar, and male abdominal features. Generic characters have been illustrated by J. Parker (1929) and Willink (1947).

Biology: No information is available.

Checklist of *Trichostictia*

brunneri J. Parker, 1929; Peru, Chile
guttata (Taschenberg), 1870 (*Monedula*); Brazil to Argentina
vulpina (Handlirsch), 1890 (*Monedula*); Chile, Peru, Bolivia

Genus Zyzzyx Pate

Generic diagnosis: Flagellomere I longer than scape; inner eye margins nearly straight, slightly diverging below; midocellus semicircular, on a low elevation; vertex a little lower than top of eyes; clypeus rather evenly contoured but with a basomedial ridge; dorsal point of clypeus reaching just above lower edge of antennal sockets; labrum nearly twice as long as broad, cross section near base a rather smooth curve; palpal formula 5-3 (terminal articles tiny); mandible subapically dentate; tongue long, reaching to midcoxae; area of scutum between admedian lines a distinct welt; male midcoxa with a long apical tooth; male midfemur serrate below and toothed subapically; hindwing media diverging a short way beyond cu-a; male tergum VII with spiracular lobes small and incompletely distinguished; apex of VII emarginate medially and acute laterally (fig. 182 D); male sternum VIII ending in a stout, downcurved spine.

Geographic range: The single species, *chilensis* is South American.

Systematics: Zyzzyx chilensis has been placed in *Stictia, Bembix, Monedula, Therapon* (a homonym), and *Zyzzyx.* It is obviously related to *Trichostictia* and somewhat less so to *Editha.* The 5-3 palpal formula of *Zyzzyx* is distinctive. The unusually long tongue, which is nearly as long as that of *Steniolia,* and the quadridentate male tergum VII (fig. 182 D) are noteworthy, also. Generic characters have been illustrated by J. Parker (1929) and Willink (1947).

Biology: Janvier (1928) has reported in detail on the nesting habits of *Z. chilensis* in colonies located over most of the length of Chile. Nesting sites were in sandy or gravelly areas of river banks or similar areas. The wasps did not spend the night in burrows but arrived between 8:00 and 10:00 a.m. and departed between 5:00 and 6:00 p.m. Females began excavation of galleries about the middle of December. Each tunnel angled into the soil for a short distance and ended in an ovoid cell in which an egg was deposited on end. On hatching, the larva was fed progressively by the mother on a diet of paralyzed flies, starting with small ones and ending with the largest species. The burrows were left open most of the time, but the entrance to the cell was closed whenever the wasp was away. Prey consisted of Therevidae, Asilidae, Stratiomyidae, Mydaidae, Tabanidae, Syrphidae, Sarcophagidae, and Tachinidae. After a larva was nearly mature and the cell was closed, another was started, so that as many as six cells might share a single burrow. Thus, one multicelled nest could account for about 150 flies.

During the nesting season, Janvier observed a number of overnight aggregations on branches of pine, locust, and other plants. These sleeping groups were composed of both sexes, the percentages of which varied with the season, males predominating in the earlier periods. One such aggregation, noted on January 4, contained 168 males and 23 females. Some wasps were marked with red paint, and in others the antennae were removed. Presumably, these traveled about 0.5 to 1.0 mile to the nesting site and returned to the same area. Aggregations formed on the same branches night after night, but the same wasps were not always present on a particular tree. Since wasps with amputated antennae flew in the direction of the nesting site and returned in the evening to the overnight area, it can be assumed that orientation is largely a matter of sight.

Checklist of *Zyzzyx*

chilensis (Eschscholz), 1822 (*Stictia*); Chile, Peru, Argentina
    peruvianus Guérin-Méneville, 1835 (pl. 70) (*Monedula*)
    orbignyi Guérin-Méneville, 1844 (*Monedula*)
    ?*odontomerus* Handlirsch, 1890 (*Monedula*)

Genus Stictiella J. Parker

Generic diagnosis: Flagellomere I often shorter than scape; eyes with inner margins a little converging to a little diverging below, least interocular distance about twice as long as flagellomere I; ocelli placed in elevated pits. floor of midocellar pit slightly broader than long (fig. 182 M) to nearly twice as broad; vertex about as high on top of eyes or considerably higher, usually depressed between eyes and vertex hump. clypeus smoothly convex, not ridged anteriorly, slightly to strongly protuberant in profile; labrum simple, somewhat longer than broad, cross section near base a rather smooth curve; tongue, when exserted, less than 1.5 times vertical eye length; palpal formula 6-4; mandible subapically dentate; area of scutum between admedian lines nearly flat; hindwing media diverging somewhat beyond cu-a; male midfemur nearly always with a row of spinelike teeth (fig. 182 F), a somewhat serrate ridge, or a definite inferior emargination; male midbasitarsus often curved and with a few large subbasal setae; claws long and arolium small to nearly absent (fig. 182 L); male tergum VII with small to large spiracular lobes, midsection simple or slightly emarginate distally; male sternum VIII with three prongs and sometimes a discal tooth (fig. 182 G).

Geographic range: The 16 known species are United States forms, and one extends into Mexico.

Systematics: Stictiella, as originally considered by J. Parker (1929) contained all species with ocelli in pits except those assigned to *Steniolia.* Gillaspy in Gillaspy, *et al.* (1962) and Gillaspy (1963b) divided *Stictiella* into four genera: *Glenostictia* Gillaspy with 11 species, *Microstictia* Gillaspy with four species, *Xerostictia* Gillaspy with one species, and the remaining 12 species in *Stictiella* proper.

Xerostictia is certainly distinct on the basis of the mouthparts. The other three categories are closely allied and not separated by characters of obvious generic value. However, the considerable number of species involved, apparent differences in biology, and consideration of the shape of the midocellar plane, in combination with other characteristics make it desirable to distinguish *Stictiella* and *Glenostictia.* On the other hand, *Microstictia* seems to be just a rather well defined species group of *Stictiella.*

To summarize, there are four genera with ocelli in pits (fig. 181). *Xerostictia* and *Steniolia* have reduced palpi and elongate tongue. The latter genus also has the labrum unevenly convex in cross section. In the two genera with 6-4 palpal formula, *Stictiella* has the midocellar plane slightly to considerably broader than long, whereas it is slightly to considerably longer than broad in *Glenostictia.* In species of each genus, which are borderline with respect to the midocellar plane, *Stictiella* has a more strongly convex clypeus, and the male midfemur is irregularly carinate to serrate rather than simple or subapically notched (fig. 182 F). Our present partial knowledge of biology indicates that *Stictiella* utilizes Lepidoptera for prey, whereas *Glenostictia* provisions with Diptera. If this should prove to be the rule, it constitutes a character of systematic importance.

The bowed male midbasitarsus is an interesting feature of many *Stictiella.* Males of most species have a process on sternum II which assumes a variety of forms from a simple ridge (*callista*) to a partially double and hairy tubercle (*tuberculata*) to a pair of teeth (*emarginata*). A flattening and dilation of the front distitarsus of the male is an interesting structural peculiarity in four species: *formosa, speciosa, spinifera,* and *tuberculata.*

The species assigned to *Microstictia* by Gillaspy (1963b) are *divergens, exigua, femorata,* and *minutula.*

They form a group with the midocellus nearly semicircular, sometimes nearly twice as broad as long. The clypeus is moderately convex, and the male midfemur is simple or broadly emarginate. The antennal sockets are removed from the clypeal margin by about the length of the midocellar plane (considerably less in other *Stictiella*). Also, male sternum VIII has a stout discal process in addition to three prongs. However, this occurs also in *emarginata*, a species outside the group (fig. 182 G).

Biology: The ethology of *Stictiella* was discussed in detail by Gillaspy, *et al.* (1962) and added to by Krombein (1964b) with respect to *serrata*. A summary was given by Evans (1966a). The following points are based on Evans' summary: Biological information is available on *Stictiella formosa, serrata, evansi, pulchella, callista,* and *emarginata*. Nests are in dry sandy soil, mostly near water, and are solitary or in small aggregations. Adults often visit flowers and pass inactive periods outside the burrows. Mounds created in nest excavation may be leveled (*serrata, pulchella, evansi*) or left largely untouched (*formosa*). A typical nest is 15 to 30 cm long, oblique, ending in a horizontal terminal cell or in several cells (up to 17 in *formosa*). An outer closure is maintained but no inner closure. Prey are small Lepidoptera, the butterfly families being Hesperiidae, Lycaenidae, and Nymphalidae; the moths being Noctuidae, Pyralidae, and others. A cell may contain all of one species or may have two or three families mixed. Numbers of individual prey have been recorded as seven to 19 per cell. Mass provisioning is the rule, but the egg is laid on the first prey in the cell, thus setting the stage for progressive provisioning which may occur in part if the egg hatches early. Principal natural enemies are miltogrammine Sarcophagidae.

Checklist of *Stictiella*

callista J. Parker, 1917; sw. U.S.
corniculata Mickel, 1918; U.S.: Wyoming to California
divergens J. Parker, 1917; U.S.: Kansas
emarginata (Cresson), 1865 (*Monedula*); U.S., s. Canada
    mamillata Handlirsch, 1890 (*Monedula*)
evansi Gillaspy, 1961; Mexico
exigua (W. Fox), 1895 (*Monedula*); U.S.: Montana
femorata (W. Fox), 1895 (*Monedula*); U.S.: Florida, Texas
formosa (Cresson), 1872 (*Monedula*); centr. U.S.
minutula (Handlirsch), 1890 (*Monedula*); U.S.: Texas
nubilosa Gillaspy, 1963; U.S.: western states
plana (W. Fox), 1895 (*Monedula*); centr. U.S.
pulchella (Cresson), 1865 (*Monedula*); U.S.: Colorado to California
    melanosterna J. Parker, 1917
serrata (Handlirsch), 1890 (*Monedula*); e. U.S., Mexico
speciosa (Cresson), 1865 (*Monedula*); centr. and w. U.S. and Canada
spinifera (Mickel), 1916 (*Stictia*); centr. U.S. and Canada
    melampous J. Parker, 1917
tuberculata (W. Fox), 1895 (*Monedula*); w. U.S.

Genus Glenostictia Gillaspy

Generic diagnosis: Flagellomere I about equal to or a little shorter than scape; eyes with inner margins very slightly converging below, or essentially parallel (fig. 182 J), least interocular distance usually about 1.5 times length of flagellomere I; ocelli in elevated pits; midocellar plane slightly to considerably longer than broad (fig. 182 J); vertex about level with top of eyes or a little lower, area between vertex hump and eye depressed; clypeus rounded but not at all protuberant, not ridged toward base which extends about as far dorsally as lower level of antennal sockets; labrum simple, a little longer than broad, tongue less than 1.5 times vertical eye height; palpal formula 6-4; mandible subapically dentate; area of scutum between admedian lines nearly flat; hindwing media diverging shortly to well beyond cu-a; male midfemur simple or broadly and deeply notched near apex (*tenuicornis*), not serrate; male tergum VII with rounded spiracular lobes, middle section simple; male sternum II often with projections (fig. 182 K); male sternum VIII with three prongs and a slender discal tooth (only a low carina in *clypeata* Gillaspy).

Geographic range: According to Gillaspy (1959, 1963c) and Gillaspy, *et al.* (1962), there are 11 species of *Glenostictia,* one of which, *pictifrons,* is rather widely distributed over the United States. The others are confined largely to areas west of the 100th meridian and many of these to the southwestern states and western Mexico.

Systematics: Glenostictia is probably most closely related to *Stictiella* in which the palpal formula is also 6-4. From *Stictiella* the flatter clypeus and nonserrate male midfemur are differences in addition to the narrower midocellar plane. Many species of *Glenostictia* have male sternum II with a pair of spinelike teeth or a bispinose process (fig. 182 K). One species without the characteristic is *clypeata.*

Biology: Published reports have been summarized and new evidence offered by Gillaspy, *et al.* (1962). *Glenostictia* are progressive provisioners and lay an egg on the first prey in the cell. The burrows are 10 to 25 cm deep and closed when the female is absent. Prey of *pulla* were given by La Rivers (1942) as *Helophila* (Syrphidae) and *Sarcophaga* (Sarcophagidae). The same species, according to Gillaspy, *et al.* (1962) used *Hylemya* (Muscidae), *Senotainia* and *Sarcophaga* (Sarcophagidae), *Stomatomyia* (Tachinidae), *Eupeodes* (Syrphidae), *Psilocephala* (Therevidae), and several genera of Bombyliidae. Gillaspy, *et al.* (1962) noted prey associated with museum specimens: Apioceratidae and Syrphidae for *clypeata* and Bombyliidae for *pictifrons*. Finally, *scitula* was found to use Scenopinidae and small Asilidae but more frequently used several species of bees (*Perdita*). One cocoon was discovered with 32 forewings of *Perdita* tacked to it. *G. scitula* females were found to spend the night in burrows. On the other hand, overnight clusters of both sexes of *pulla* were discovered with as many as 20 wasps participating.

Checklist of *Glenostictia*

argentata (C. Fox), 1923 (*Stictiella*); Mexico; U.S.: California
bifurcata (C. Fox), 1923 (*Stictiella*); Mexico; U.S.: s. California
   directa C. Fox, 1923 (*Stictiella*)
   albicera C. Fox, 1923 (*Stictiella*)
bituberculata (J. Parker), 1917 (*Stictiella*); sw. U.S.
clypeata (Gillaspy), 1959 (*Stictiella*); sw. U.S., Mexico
gilva Gillaspy, 1963; U.S.: Arizona, California
megacera (J. Parker), 1917 (*Stictiella*); w. U.S.; Mexico
pictifrons (F. Smith), 1856 (*Monedula*); e. and s. U.S. to California
   inermis Handlirsch, 1890 (*Monedula*)
   denverensis Cameron, 1908 (*Monedula*)
pulla (Handlirsch), 1890 (*Monedula*); w. U.S.
   usitata W. Fox, 1895 (*Monedula*)
scitula (W. Fox), 1895 (*Monedula*); sw. U.S.
   villosa W. Fox, 1895 (*Monedula*)
tenuicornis (W. Fox), 1895 (*Monedula*); sw. U.S.
terlinguae (C. Fox), 1928 (*Stictiella*); U.S.: Texas

Genus Xerostictia Gillaspy

Generic diagnosis: Flagellomere I a little shorter than scape in male and a little longer in female; eyes with inner margins diverging about equally above and below, least interocular distance (at middle of frons) about 1.5 times length of flagellomere I; ocelli placed in elevated pits, midocellar plane nearly oval, a little longer than broad; vertex rising slightly above top of eyes; clypeus rather strongly and smoothly convex, not carinate toward base above; labrum nearly twice as long as broad, cross section near base forming a rather smooth curve; tongue exserted, reaching about to midcoxae when extended; palpal formula 4-2; mandible subapically dentate; area of scutum between admedian lines nearly flat; hindwing media arising well beyond cu-a; male midfemur with a row of spines beneath; male tergum VII with stout and apically truncate spiracular lobes, middle section simple; male sternum VIII with three prongs and a slender, discal, downwardly directed tooth.

Geographic range: The one known species, *longilabris*, occurs in southern California and Arizona. A subspecies, *longilabris boharti* was collected in Baja California, Mexico.

Systematics: Among the genera with recessed ocelli, to which *Xerostictia* is related, it can be distinguished by the 4-2 palpal formula, the unusually long labrum, and the short and truncate spiracular lobes of male tergum VII. It shares with *Stictiella* the serrate male midfemur and protuberant clypeus. On the other hand it resembles *Steniolia* in the long tongue, slightly elongate midocellar plane, and long ventral tooth on sternum VIII of the male.

Biology: Unknown.

Checklist of *Xerostictia*

longilabris Gillaspy, 1963; U.S.: Arizona, California
   ssp. *boharti* Gillaspy, 1963; Mexico

Genus Steniolia Say

Generic diagnosis: Flagellomere I and pedicel together longer than scape; eyes with inner margins essentially parallel, slightly broader to slightly narrower below; ocelli placed in elevated pits, midocellar plane oval, longer than broad, lens present in part; vertex depressed below top of eyes; clypeus with a longitudinal carina or angle above, which is continuous with a ridge between antennae; labrum a little longer than its basal width, strongly convex and a little bilobate toward base so that a cross section does not form a smooth curve; tongue exserted, reaching at least to midcoxae; palpal formula 3-1 or 3-2; mandible subapically dentate; area of scutum between admedian lines nearly flat; hindwing media diverging slightly beyond cu-a; male with a tooth near median apex of sternum II; male tergum VII with spiracular lobes large, sometimes nearly meeting ventrally, midsection nearly truncate distally; male sternum VIII with three stout prongs and a discal, downwardly directed tooth; body elaborately marked with yellow or whitish spots and stripes; male usually larger than female and with longer wings.

Geographic range: The 15 known species are distributed rather equally among the life zones from Lower Sonoran to Canadian. All of the United States species occur west of the 100th meridian. Several species are Central American and two of these have been collected in northwestern South America.

Systematics: Species of the genus *Steniolia* are moderate-sized, long-tongued sand wasps belonging to a group of four genera characterized by having ocelli in somewhat elevated pits. Of these genera, only *Steniolia* and *Xerostictia* have the palpi reduced. *Steniolia* differs from other genera of the group by its very long tongue, the irregular cross section of the labrum, and the three maxillary segments.

A detailed study of the genus has been made by Gillaspy (1964). He concluded that it is a moderately advanced bembicin with specializations centered in mouthparts.

Biology: Evans and Gillaspy (1964) discussed known information on *Steniolia*, most of it based on *S. obliqua*. The main points emphasized were as follows: the long tongue enables *Steniolia* to take nectar from flowers with deep corollas; mating apparently takes place in large clusters which last overnight; nests are usually in powdery earth rather than sand, are shallow and ordinarily unicellular; the nest entrance, but not the cell gallery, is closed when the female is away; hunting females hover with a shrill wing buzz before darting after prey – this is usually Bombyliidae but sometimes may be Syrphidae or other flies; the egg is laid erect on the first fly in the cell and provisioning is progressive; parasites are miltogrammine Sarcophagidae, *Parnopes edwardsii* (Cresson) (Chrysididae) and *Dasymutilla* (Mutillidae).

Checklist of *Steniolia*

californiensis Gillaspy, 1964; U.S.: California; Mexico: Baja California

dissimilis C. Fox, 1923; U.S.: Arizona; Mexico: Sonora

duplicata Provancher, 1888; sw. U.S.; Mexico: Chihuahua, Sonora, Baja California
    edwardsii Patton, 1894
    meridionalis C. Fox, 1923

elegans J. Parker, 1929; Mexico, w. and sw. U.S.

eremica Gillaspy, 1964; w. U.S.

guatemalensis (Rohwer), 1914 (*Stictia*); s. Mexico to Ecuador

longirostra (Say), 1837 (*Bembex*); Mexico,
    montezuma F. Smith, 1856
    longirostris Handlirsch, 1889, emendation

mexicana Gillaspy, 1964; Mexico: Jalisco, Baja California

nigripes J. Parker, 1917; U.S.: California; Mexico: Baja California

obliqua (Cresson), 1865 (*Monedula*); U.S.: Colorado to Pacific states

powelli Gillaspy, 1964; Mexico: Baja California

scolopacea Handlirsch, 1889; U.S.: California, w. Nevada; Mexico: Baja California
    ssp. *albicantia* J. Parker, 1917; nw. California to British Columbia

sulfurea W. Fox, 1901; U.S.: California

tibialis Handlirsch, 1889; w. U.S.

vanduzeei Gillaspy, 1964; U.S.: California, Nevada

SUBFAMILY PHILANTHINAE

With about 1,100 described species the Philanthinae qualifies as one of the largest subfamilies of Sphecidae. Over three-fourths of these are contained in *Cerceris*, the largest sphecid genus. These wasps are mostly medium-sized, and practically all are colorfully ornamented with bands and spots of yellow or red. As the subfamily name implies, the philanthines are commonly found on or about flowers, and considerable quantities of pollen may adhere to them. As far as is known, all species are ground nesting, and the provisions are either Hymenoptera, particularly aculeates, or Coleoptera. The largest species of the subfamily is *Cerceris synagroides* which reaches 30 mm in length.

Diagnostic characters:
1. (a) Eyes rather widely separated, inner orbits entire (fig. 186 A) or emarginate (fig. 186 C), slightly to strongly converging above; (b) ocelli normal except in Odontosphecini.
2. (a) Antennal placement variable, sockets separated from frontoclypeal suture (fig. 186) except in Eremiaspheciini (fig. 185); (b) male with 13, female with 12 antennal articles.
3. Clypeus transverse but sometimes with an upper median lobe.
4. (a) Mandible without externoventral notch, inner margin simple or dentate; (b) palpal formula 6-4 except 5-3 in Eremiaspheciini; (c) mouthparts short but stipes and mentum often elongate; (d) mandible socket closed by paramandibular process of hypostoma except narrowly open in Odontosphecini.
5. (a) Pronotal collar short; (b) pronotal lobe and tegula separated.
6. (a) Notauli usually distinct; (b) admedian lines separated, at least basally; (c) no oblique scutal carina.
7. (a) Episternal and scrobal sulci sometimes present; (b) omaulus, sternaulus, and acetabular carina absent.
8. Definitive metapleuron usually consisting of upper metapleural area only, but lower area sometimes defined.
9. (a) Midtibia with one apical spur (fig. 190); (b) midcoxae nearly contiguous or well separated, without dorsolateral carina except in Odontosphecini, Eremiaspheciini, and some Pseudoscoliini; (c) precoxal sulcus and lobes usually present, latter sometimes very short, sulcus and lobes weak or absent in Pseudoscoliini; (d) hindfemoral apex sometimes thickened and truncate; (e) claw simple; (f) foretarsus with a rake, sometimes weak especially in males; (g) plantulae absent except in Philanthini, Pseudoscoliini, and most Aphilanthopsini.
10. (a) Propodeum short to moderately long; (b) dorsal enclosure present; (c) no propodeal sternite.
11. (a) Forewing with three submarginal cells, II and III each usually receiving one recurrent vein (fig. 184 B) (except in Eremiaspheciini and Odontosphecini, fig. 184 A,G); (b) marginal cell apex usually truncate or rounded; (c) media diverging before, at, or most often beyond cu-a; (d) stigma tapering to a point well beyond base of marginal cell.
12. (a) Jugal lobe small to large; (b) hindwing media diverging before, at, or after cu-a; (c) second anal vein and subcosta absent.
13. (a) Gaster sessile, pedunculate, or occasionally petiolate, petiole clearly composed of tergum and sternum; (b) tergum I with lateral carina (weak in Eremiaspheciini); (c) male with seven exposed terga; (d) pygidial plate present except in Philanthini.
14. (a) Volsella simple (fig. 6 B) or with digitus and cuspis; (b) cerci absent except in Eremiaspheciini.

Systematics: The Philanthinae contains six tribes and 11 genera. The huge genus *Cerceris* contains a disproportionate number of species in relation to the other genera. In three tribes there is only a single genus, and in each of these cases the species are rather few (4, 6, and 19). There are no simple recognition characters for the subfamily. The nearly universal closed mandibular socket is diagnostic but not readily seen unless the mandibles are spread. Most males have a tuft of incurving hairs from the lower lateral area of the clypeus, and these often become agglutinated with nectar (fig. 186, A,D). Similar brushes

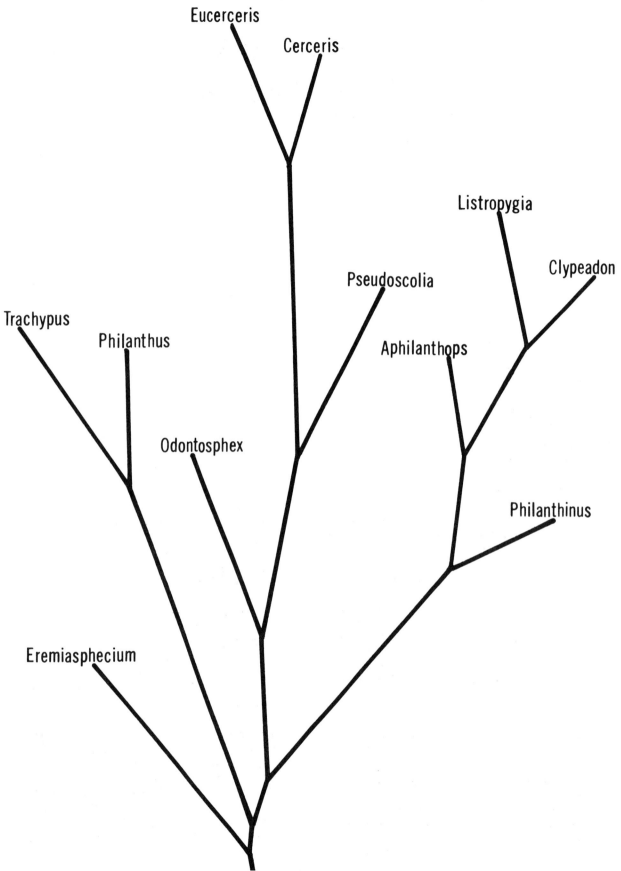

FIG. 183. Dendrogram suggesting relationships in the Philanthinae.

occur in a few other wasps, such as the nyssonine genus *Hoplisoides*. The head is large, broad, and usually maculate, and the clypeus usually has a dorsomedian lobe. The gaster is often pedunculate, sometimes petiolate, and nearly always extensively maculate. There is a full complement of wing cells that vary in details of shape and relative position of veins. The subfamily is certainly not homogeneous, the two most discordant elements being *Odontosphex* and *Eremiasphecium*, each comprising a tribe. Menke (1967d) has discussed the position of *Odontosphex* and has clarified its relationships. The principal deviation from more typical philanthines is the presence in *Odontosphex* of vestigial hindocelli, a characteristic of less than subfamily value in other branches of the Sphecidae. The incompletely closed mandibular sockets constitute another unique circumstance in the subfamily. Placement of *Eremiasphecium* is more controversial since there are several features not found in other Philanthinae. These are the reduced palpi, antennal sockets placed contiguous with the clypeal margin, cerci in the male, and unusual terminations for the recurrent veins. Again, it appears that while these characters are indicative of an isolated and relict status, there are many philanthine features, and inclusion within Philanthinae as the least specialized genus seems warranted.

The Philanthinae seem to occupy a position intermediate between Larrinae and Nyssoninae but probably closer to the former. The single midtibial spur, transverse clypeus, eyes converging above, complete wing venation, frequently large jugal lobe, absence of an omaulus or an oblique scutal carina, and the short tongue, all would point to a derivation from the larrine stock as suggested in fig. 7.

Our ideas of the interrelationships among tribes and genera are given in fig. 183. In this dendrogram the three main divisions, identified by the dominant genus in each, are the *Philanthus* branch with simple hindfemur, episternal sulcus, notched eyes; the *Cerceris* branch with truncate hindfemur, normal (6-4) palpi, no episternal sulcus; and the *Clypeadon* branch with simple hindfemur, episternal sulcus, entire eyes, and normal (6-4) palpi. In addition, *Eremiasphecium* is shown as a basal offshoot with reduced (5-3) palpi. Several of the tribes have unique characters that identify them. Thus, the Philanthini have no pygidial plate, and the inner eye margins are notched. Cercerini have an undifferentiated volsella and a deeply impressed and extended scrobal sulcus; Eremiaspheciini have a 5-3 palpal formula; and Odontosphecini have vestigial hindocelli.

There are two sorts of structural combinations among the genera, the first occurring within the branches and the second involving several branches. In the first category are the reduced jugal lobe and widely separated midcoxal cavities (*Pseudoscolia, Cerceris, Eucerceris*), undifferentiated volsella and petiolate submarginal cell II (*Cerceris, Eucerceris*), female with a pygidial ant clamp (*Clypeadon, Listropygia*), and episternal sulcus undeveloped, together with truncate hindfemur (entire *Cerceris* branch). In the second category are several combinations that might be considered as primitive or generalized in the subfamily. These are the absence of a delimited subantennal sclerite (*Odontosphex, Pseudoscolia, Eremiasphecium, Philanthinus*), no lateral brush on the male clypeus (*Odontosphex, Eremiasphecium*), and the hindwing media always diverging at or before cu-a (*Eremiasphecium, Odontosphex, Pseudoscolia, Philanthinus*). In the second category, but presumably specialized, are the cleft female sternum VI (*Philanthinus, Clypeadon, Listropygia, Cerceris, Eucerceris*) and distal divergence of the hindwing media (tribe Cercerini and subtribe Aphilanthopsina). Expression of such characters which involve more than one phylogenetic branch can be explained as latent potentials in the subfamily.

In an effort to assess evolutionary levels of advancement among the 11 genera of Philanthinae, we have made a list of presumably generalized characters within the subfamily and an opposing list of presumably advanced characters. On the basis of the number of advanced characters appearing in each genus the ratings are: *Eremiasphecium*, 5; *Philanthinus*, 10; *Odontosphex* and *Aphilanthops*, 11; *Clypeadon*, 13; *Philanthus, Trachypus,* and *Pseudoscolia*, 14; *Listropygia*, 15; *Cerceris*, 25; *and Eucerceris*, 27. The list is given below:

TABLE 19.
List of Characters of Presumably Phylogenetic Value in Philanthinae

Presumably generalized in	*Presumably specialized*
1. Antennal sockets contiguous with clypeus	Sockets well above clypeus
2. Antennal sockets more than 2.0 diameters apart	Sockets less than 2.0 diameters apart
3. No sharp interantennal carina	Sharp interantennal carina
4. Hindocelli normal	Hindocelli vestigial
5. Ocellocular distance not much reduced	Ocellocular distance reduced
6. Inner eye margins bowed out (away from each other)	Inner eye margins not bowed out
7. Inner eye margins simple	Inner eye margins notched
8. Face not unusually swollen	Face unusually swollen
9. No lateral clypeal brushes in male	Clypeal brushes present
10. No brushes on mandible	Mandibular brushes present
11. Mandible dentate	Mandible edentate
12. Mandible socket open	Mandible socket closed
13. Palpal formula 6-4	Palpal formula 5-3

14. No delimited subantennal sclerite	Subantennal sclerite delimited
15. Pronotal collar raised and distinct	Collar depressed or appressed
16. Episternal sulcus well developed	Episternal sulcus short or absent
17. Scrobal sulcus normal	Scrobal sulcus valley-like
18. Midcoxal cavities not widely separated	Midcoxal cavities well separated
19. Hindfemur simple apically	Hindfemur truncate apically
20. Plantulae present	Plantulae absent
21. Forewing marginal cell acute apically	Marginal cell rounded apically
22. Outer vein of submarginal cell III ending at or before apical third of marginal cell	Outer vein ending beyond apical third of marginal cell
23. First recurrent ending on submarginal cell I or or interstitial between I and II	First recurrent ending on submarginal cell II
24. Second recurrent ending on submarginal cell II or interstitial between II and III	Second recurrent ending on submarginal cell III
25. Second submarginal cell sessile in front	Second cell petiolate
26. Hindwing media diverging near or before cu-a	Hindwing media well after cu-a
27. Jugal lobe at least half as long as anal area	Jugal lobe less than half anal area
28. Gaster without a petiole or peduncle	Gaster with a petiole or peduncle
29. Female with a pygidial plate	No pygidial plate
30. Female without an ant-clamp apparatus	Pygidium part of an ant clamp
31. Female sternum VI simple	Female sternum VI cleft
32. Male sternum VIII simple and thinly edged in posterior view	Sternum VIII broadly triangular in posterior view
33. Male with cerci	Male without cerci
34. Volsella with digitus and cuspis	Volsella undifferentiated
35. Terga without median or submedian transverse grooves	Terga with median or submedian transverse grooves
36. No white or yellow markings	White or yellow markings present

Valuable references at the suprageneric level are Kohl (1891), Pate (1947a), Beaumont (1949a), R. Bohart (1966), and Menke (1967d).

As might be expected in large genera, there are a number of uncorrected cases of homonymy that have turned up in *Cerceris* and *Philanthus*. Where possible, these have been referred to the authors for suitable action. We are reluctant to propose new names because the categories in question are unknown to us and many may be synonyms. Future revisers should make the decisions.

KEY TO TRIBES OF PHILANTHINAE

1. Apex of hindfemur simple; episternal sulcus present and usually reaching ventral area of mesopleuron; volsella with a digitus and cuspis 2
   Apex of hindfemur somewhat truncate, flattened area sometimes kidney shaped or forming an apicoventral process; episternal sulcus absent or very short; volsella simple or divided into a digitus and cuspis 4
2. Inner orbit of eye sharply angled or notched (fig. 186 C, D) (weak in some *Philanthus* males whose eyes converge strongly towards vertex)... tribe Philanthini, p. 561
   Inner orbit not interrupted by an angle or notch (fig. 186 A,B) 3
3. First recurrent vein received by submarginal cell II, second recurrent by submarginal III; palpal formula 6-4; antennal sockets removed from frontoclypeal suture by at least half a socket diameter (fig. 186 A,B); male without cerci .... tribe Aphilanthopsini, p. 569
   First recurrent vein received by submarginal cell I or interstitial between cells I and II, second recurrent by II or interstitial between II and III; palpal formula 5-3; antennal sockets practically touching frontoclypeal suture (fig. 185); male with cerci, small wasps .... tribe Eremiaspheciini, p. 558
4. Mesopleuron with a broad, deep scrobal sulcus (fig. 190); hindwing media diverging well beyond cu-a (fig. 184 B); volsella simple (fig. 6 B) ...... tribe Cercerini, p. 574
   Scrobal sulcus absent or weak; hindwing media diverging before cu-a (fig. 184 A); volsella with digitus and cuspis 5
5. Hindocelli normal; second recurrent vein received by submarginal cell III; jugal lobe about half as long as anal area or less (fig. 184 C); midcoxae widely separated .... tribe Pseudoscoliini, p. 573
   Hindocelli vestigial; second recurrent vein received by submarginal cell II; jugal lobe nearly as long as anal area (fig. 184 A); midcoxae almost contiguous .... tribe Odontospheciini, p. 572

Tribe Eremiaspheciini

These are small and rather localized Old World wasps which appear to be in a relict status. The single genus *Eremiasphecium* has seven known species. The typical facial characteristics are shown in fig. 185; and the distinctive wing venation is shown in fig. 184 G.

Systematics: Tribal characteristics are: white or yel-

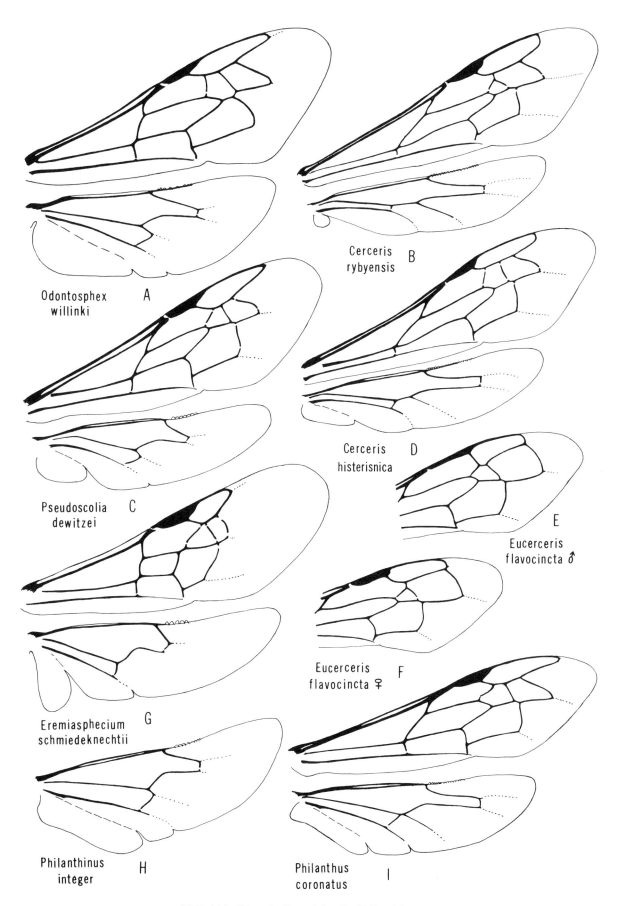

FIG. 184. Wings in the subfamily Philanthinae.

low markings present, mandible edentate, mandible socket closed, antennal sockets apparently touching clypeus and more than 2.0 diameters apart, no sharp interantennal carina, hindocelli normal, face not swollen above antennae, no lateral clypeal brushes in male, palpal formula 5-3, episternal sulcus well developed, scrobal sulcus weak or absent, midcoxal cavities close together, hindfemur simple apically, marginal cell of forewing narrowed toward apex but not markedly so, submarginal cell II narrowed in front, sometimes petiolate, first recurrent terminating on submarginal cell I or interstitial, second recurrent on cell II or interstitial, hindwing media diverging at cu-a, jugal lobe occupying more than half length of anal area, female with pygidial plate, cerci present, volsella differentiated into digitus and cuspis.

Relationships of the tribe are discussed under Systematics of the genus.

Genus Eremiasphecium Kohl

Generic diagnosis: Ocellocular distance not reduced, inner eye margins simple and bowed away from each other (fig. 185), no delimited subantennal sclerite, clypeus usually with a pair of rounded teeth, pronotal collar high and distinct from scutum, propodeal enclosure relatively large when defined, and almost wholly dorsal, submarginal cell III much reduced and no larger than II (fig. 184 G), jugal excision deep (fig. 184 G), female sternum VI simple.

Geographic range: Six of the seven known species are found in the southwest USSR. One of them (*schmiedeknechtii*) occurs also in Egypt and the Canary Islands. The seventh species was described from Mongolia.

Systematics: Relatively little is known about these small and inconspicuous wasps. Kohl (1897) based the genus on *schmiedeknechtii,* one of the two species that we have had available for study. Gussakovskij (1930a) described five more species, placing them in his new genus *Shestakovia,* that Pate (1935) synonymized with *Eremiasphecium.* Beaumont (1968a) gave additional description and figures of *schmiedeknechtii.* Tsuneki (1972a) recently described another species, *steppicola,* that he placed in a new genus *Mongolia.* Menke has studied the unique female type, and it agrees perfectly with all the characters of *Eremiasphecium.* The mandible of *steppicola* was figured by Tsuneki as being dentate within, but it is simple (fig. 185). The labrum of *steppicola* differs from *schmiedeknechtii* in having two notches (fig. 185 A).

The relationships of *Eremiasphecium* have been in doubt, although Kohl originally considered it to be a philanthine genus related to *Pseudoscolia.* Gussakovskij placed it close to the larrine genus *Miscophus.* Tsuneki considered the genus to be allied with the pemphredonine genus *Spilomena.* Menke (1967d) included *Eremiasphecium* as a tribe in his tribal key of the Philanthinae, and Beaumont (1968a) concurred with this assignment. Most of the features of *Eremiasphecium* are compatible with Philanthinae, particularly if it is considered as a rather generalized and relict type. The presence of a well developed episternal sulcus, rather basal termination of

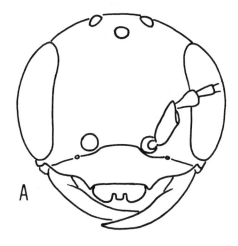

Eremiasphecium
steppicola

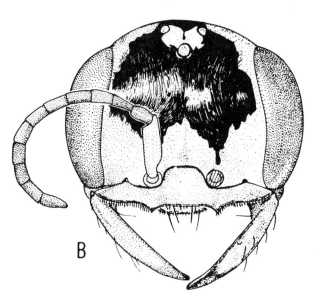

Eremiasphecium
schmiedeknechtii

FIG. 185. Facial views of female *Eremiasphecium,* A = holotype.

the recurrent veins (fig. 184 G), large jugal lobe (fig. 184 G), and well developed pygidial plate mark it as generalized but do not rule it out of either Larrinae or Philanthinae. The occurrence of cerci in the male is further evidence of its primitive status, and these appendages are not found elsewhere in Philanthinae or Larrinae. The closed mandible socket is definitely nonlarrine, and the reduced (5-3) palpi are an additional oddity. Characters which seem to point toward placement low on the philanthine line and away from Larrinae are: mandibles edentate, midcoxal cavities close together, hindwing media diverging at cu-a (fig. 184 G), and volsella differentiated into digitus and cuspis. The contracted venation, closed mandible socket, and reduced palpi are obvious specializations.

Biology: Unknown.

Checklist of *Eremiasphecium*

crassicorne (Gussakovskij), 1930 (*Shestakovia*); sw. USSR: Turkmen S.S.R.
desertorum (Gussakovskij), 1930 (*Shestakovia*); sw. USSR: Turkmen S.S.R.
digitatum (Gussakovskij), 1930 (*Shestakovia*); sw. USSR: Turkmen S.S.R.
longiceps (Gussakovskij), 1930 (*Shestakovia*); sw. USSR: Turkmen S.S.R.
ornatum (Gussakovskij), 1930 (*Shestakovia*); sw. USSR: Turkmen S.S.R.
schmiedeknechtii Kohl, 1897; Canary Is.; Egypt; sw. USSR: Turkmen S.S.R., Kazakh S.S.R.
   bicolor Gussakovskij, 1930 (*Shestakovia*)
steppicola (Tsuneki), 1972 (*Mongolia*); Mongolia

Tribe Philanthini

The colorful wasps in this tribe are relatively abundant and familiar sights on and around flowers. In general they are less heavily sculptured than their equally common counterparts in the Cercerini. Even when the integument bears large pits, the background in Philanthini is usually smooth or finely textured. Both pedunculate and nonpedunculate forms occur, the former mostly in the genus *Trachypus,* which represents the tribe in the New World tropics, the latter in the widespread genus *Philanthus.*

Nests are provisioned with Hymenoptera and especially with bees. This has given rise to the common name of "bee-wolf" for species of *Philanthus.* Honey bees have been identified several times as prey, but Halictidae and Anthophoridae are more commonly attacked.

Systematics: The following are tribal characteristics: mandible usually edentate, antennal sockets located above clypeus and more than two diameters apart, no sharp interantennal carina, lateral ocelli normal, inner eye margins notched (fig. 186 C, D), face rather swollen, lateral clypeal brushes nearly always present in male, palpal formula 6-4, episternal sulcus well developed, scrobal sulcus usually present but only a simple groove, midcoxal cavities rather narrowly separated, hindfemur simple at apex, plantulae present, marginal cell of forewing acutely narrowed toward apex, submarginal cell II not petiolate in front, first recurrent terminating on submarginal cell II and second recurrent on cell III, jugal lobe occupying more than half length of entire anal area (figs. 184 I, 187), female tergum VI simple and without a pygidial plate, male without cerci, volsella differentiated into digitus and cuspis, white or yellow markings generally present.

Key to genera of Philanthini

Last antennal article somewhat rounded apically and with a partly ventral, oval polished spot; first gastral segment usually broader than long; Old and New Worlds but not South America ... *Philanthus* Fabricius, p. 561
Last antennal article obliquely and sharply truncate, the end flat and polished; first gastral segment more than twice as long as broad in both sexes (fig. 187); Neotropical Region north to Brownsville, Texas .... *Trachypus* Klug, p. 568

Genus Philanthus Fabricius

Generic diagnosis: Last antennal article rounded and a little flattened at apex which bears an oval polished area extending onto ventral surface, ocellocular distance greatly reduced in some males, a subantennal sclerite present and defined by lines from antennal sockets through tentorial pits to clypeus, pronotum raised and not closely appressed to scutum, hindwing media diverging near or in front of cu-a, gaster sessile or if pedunculate or petiolate (a few Old World species, especially males), segment I no more than twice as long as broad, female sternum VI not cleft apically.

Geographic range: The genus contains about 135 species and is well represented in the Ethiopian, Palearctic, and Nearctic Regions. Small numbers are also Oriental and Neotropical. Species in the last category are from Cuba and Central America. Australia and South America are the only continents from which the genus is not known.

Systematics: About 45 species have been studied critically. The genus is closely related to *Trachypus,* and although there is a clear separation among New World species on the basis of the sessile versus pedunculate or petiolate gaster, the same cannot be said for the Old World. In *Philanthus notatulus* from the Oriental Region, the male has a clearly petiolate abdomen, the basal segment of the gaster being about twice as long as broad. However, segment I in the female is a peduncle that is only about as long as broad and this sex qualifies as a more ordinary *Philanthus.* Madagascan species, *madagascariensis* and *radamae,* have segment I longer than broad in both sexes. The antennal difference given in the key and generic diagnoses seem to be adequate for generic separation in any case.

A number of subgenera have been proposed, but few authors have used them consistently. Many subgeneric names are applicable to only one or a few distinctive species. In others the characters are not sufficiently discrete. For example, the *ventilabris* group, with two cari-

PHILANTHINAE

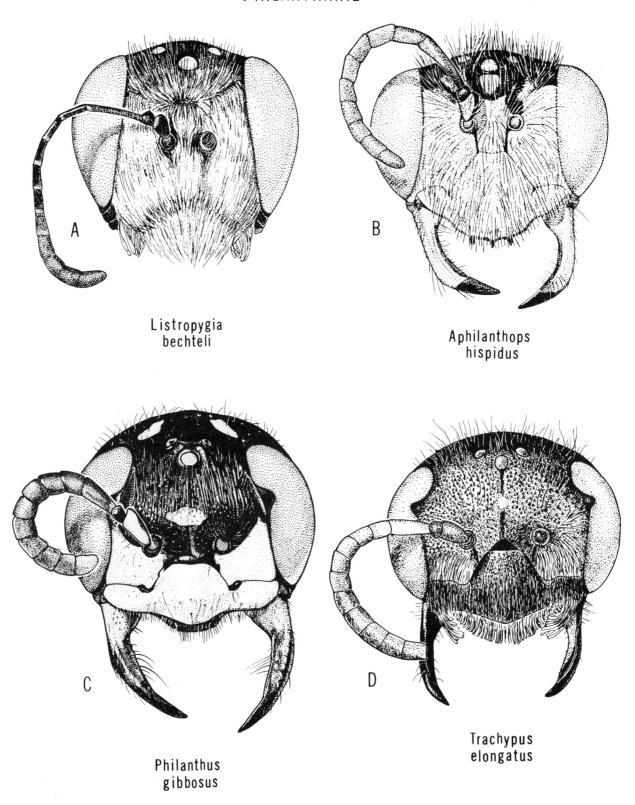

FIG. 186. Facial portraits in the subfamily Philanthinae; A,D, males; B,C, females.

nae across the pronotal collar (*Pseudoanthophilus*) is monotypic. The *politus* (*Anthophilus*) and *sanbornii* (*Oclocletes*) groups are based primarily on punctation peculiarities and strong convergence of the eyes above, respectively. Unfortunately, there are species in these groups, which have one or more of the critical characters weakly developed. In our opinion, species groups are preferable to subgenera, even though no comprehensive study has been made on this basis.

Species groups have been utilized by various workers. For instance, Strandtmann (1946) treated the North American *Philanthus* in the *sanbornii*, *gibbosus*, *politus*, and *bilunatus* groups in addition to four rather isolated species. Also, Beaumont (1961a) remarked on the *rutilus*, *genalis*, *coronatus*, and *coarctatus* groups of the Palearctic Region.

As in most other Philanthinae, male genitalia show only slight differences between species. Most useful characters are found in the dentition and other details of the clypeus, punctation, shape of the frons, carinae and form of the pronotal collar, and shape of gastral segment I. Other features useful in identification are: overall pubescence, color of the clypeal brush in males, fimbriae on male sterna (*crabroniformis*), acute rather than obtuse ocellar triangle (*albopilosus*), relative height of the eye emargination, dentate rather than edentate mandibles (*notatulus*), presence of a malar space (*barbiger* males), absence of the male clypeal brushes (*albopilosus*), and presence of a metapleural flange at the base of the hindwing (*politus*).

Some of the more useful papers and the areas covered are those of Cresson (1865b) for the United States, Kohl (1891) for the world, Arnold (1925) for southern Africa, Mochi (1939b) for Egypt, Arnold (1945) for Madagascar, Strandtmann (1946) and R. Bohart (1972) for the United States, Beaumont (1949a) for north Africa, Beaumont (1951b) for Europe, Beaumont (1961a) for the Palearctic Region, and van der Vecht (1966) for the Oriental Region.

Biology: The nesting habits of about 10 species of North American *Philanthus* have been studied in some detail: Reinhard (1924), Powell and Chemsak (1959a), Evans and Lin (1959), Evans (1964b, 1966b, 1970, 1973a) and Armitage (1965). The last author gave a short review of works on the biology of North American *Philanthus*. In Europe, *P. triangulum* has been the object of many studies associated primarily with its hunting activity involving the honeybee. Hamm and Richards (1930) gave a general review of biology for *triangulum*. Pelzer (1936) discussed aspects of prey capture and stinging as well as parasitism in *triangulum*, and Rathmayer (1962) gave a detailed account of the sting and poison of this species. Tinbergen (1932, 1935), Tinbergen and Kruyt (1938), and Tinbergen and Van der Linde (1938) discussed experimental work on stimuli leading to the hunting response in *triangulum* as well as the ability of this species to orient both to the nesting site and to the area where prey is captured. Female wasps normally located their nests by landmarks but could occasionally find nests up to 1,000 m distant when released in unfamiliar territory, by first a process of random searching, followed by directed searching when known landmarks were found. Olberg (1953) has written a short book on the predatory nature of *triangulum* with methods for its control (cultural and chemical). Included are many excellent photographs which illustrate various aspects of behavior.

The nesting behavior of *Philanthus* varies somewhat in detail among the species studied. All species nest in small to large aggregations in bare, sandy soil. Nests are excavated in level to slightly sloping areas (*bicinctus*, *crabroniformis*, *lepidus*, *pacificus*, *politus*, *pulcher*, *zebratus*) or steep slopes or vertical banks (*bilunatus*, *gibbosus*, *solivagus*), but the latter group of species occasionally nest in flat areas. In such situations where tumulus accumulates during excavation, only *pacificus*, *politus*, and *pulcher* level the mound. *P. lepidus* constructs one or two, rarely up to five, blind, "accessory burrows" (Evans, 1966c) at each nest site. These burrows remain open at all times and are never provisioned (Evans, 1964b). *P. zebratus* and *crabroniformis* also construct accessory burrows (Evans, 1970), and he suggests that they play a role in diverting parasites from the true nest in the first species, and serve as quarries for soil used in nest closure in the second species. In general, burrow length depends primarily upon wasp size although soil texture may dictate otherwise. A small wasp (e.g. *politus*) may excavate only to about 10 cm, whereas a larger species (e.g. *solivagus*) may dig to 85 cm (Evans and Lin, 1959). Prey consists of numerous species of Halictidae, but other bees in the Andrenidae, Anthophoridae, Apidae, Colletidae, and Megachilidae may be taken as well. *P. bicinctus* uses *Bombus* as prey, and *crabroniformis* and *sanbornii* as well as *triangulum* are known to take *Apis mellifera* as prey (G. Bohart, 1954; Evans, 1955a; Evans and Lin, 1959). More rarely Vespidae, Sphecidae, Ichneumonidae, Braconidae, Scelionidae, or Chrysididae may be used (Powell and Chemsak, 1959a; Evans, 1966b, 1970). *Philanthus* do not appear to be prey specific. For instance, *solivagus* nests have been found to contain 27 species of bees and wasps in five families. Single nests may have individual cells filled with a mixture of bees or with the same bee species (Evans and Lin, 1959). As in many other sphecids, prey selection seems to depend on the abundance of prey in an area and its size in relation to the provisioning wasp.

A captured bee is stung immediately between the forecoxae and then is carried to the nest held in the wasp's middle legs. The wasp alights at the nest entrance and digs the nest open with its forelegs (except *sanbornii* and *solivagus* which do not close their nests). Several bees (up to about 10) may be captured and stored in turn, midway in the burrow or at its terminus, until the wasp begins to excavate cells. *P. solivagus*, according to Evans and Lin (1959), may excavate its brood cells first and then provision them directly instead of storing bees in the burrow for future use. In general, cells may be excavated at the end of the burrow (*politus*, *lepidus*) or along the main burrow trunk (*solivagus, gibbosus, bilunatus*). Most nests contain from four to seven or more cells, and each cell may be provisioned with six to 18 bees or wasps. Complete nests take from five to 10 days, and one to three nests may be constructed during the early or late summer months depending on the species of *Philanthus* (Evans and Lin, 1959).

A presumptive stage in presociality involving temporary communal nesting has been described for *gibbosus* by Evans (1973a). Apparently several emerging wasps of both sexes may live in the parental nest for several days, the males sometimes for life. The females soon leave to establish new nests, but one may remain to occupy and expand the original quarters. Rarely, a single nest may be provisioned for a short time by two females. Communal use of burrows has also been reported for the closely related genus *Trachypus* (Evans and Matthews, 1973a).

Numerous parasites have been recorded about the nesting sites of *Philanthus,* but few have actually been reared from the host. The sarcophagid parasite *Senotainia trilineata* Wulp was reared from cells of *gibbosus* (Reinhard, 1924), *crabroniformis* (Evans, 1970), and *zebratus* (Evans, 1966b, 1970). Another sarcophagid, *Phrosinella pilosifrons* Allen, was reared from *crabroniformis, pulcher,* and *zebratus* by Evans (1970). *Physocephala chrysorrhoea* Meigen and *P. vittata* Fabricius were reared from adult *Philanthus* independently by Kroeber and Bouwman (in Hamm and Richards, 1930). Adult *Philanthus* have also been attacked and killed by the asilid *Deromyia discolor* Loew (Reinhard, 1924).

Several species of Hymenoptera are known to parasitize or destroy *Philanthus*. Móczár (1967) listed the following chrysidids attacking *Philanthus*: *Chrysis ignita ignita* Linnaeus and *Omalus auratus* Linnaeus (on *P. triangulum*); *Hedychrum intermedium* Dahlbom (on *P. coronatus* and *triangulum*); *Hedychrum gerstaeckeri* Chevrier (on *Philanthus* sp.). Evans (1964b) suggested that *Dasymutilla nigripes* (Fabricius) (Mutillidae) is a probable parasite of *Philanthus lepidus;* Shappirio (in Evans, 1964b) observed the same species of mutillid entering nests of *P. gibbosus*. Rathmayer (1962) observed the sphecid *Palarus variegatus* capture and kill *Philanthus* adults for its own nest provisions.

Checklist of *Philanthus*

adamsoni Arnold, 1952; Kenya
albopictus Taschenberg, 1880; Ethiopia
albopilosus Cresson, 1865; centr. U.S. s. to Chihuahua, Mexico
    simillimus Cresson, 1865
amabilis Arnold, 1946; Tanzania
ammochrysus Schulz, 1905; n. Africa
    krugeri Schulthess, 1926
  ssp. *psammophilus* Bytinski-Salz, 1959; Israel
angustifrons Kohl, 1891; sw. USSR
arizonicus R. Bohart, 1972; U.S.: Arizona
arnoldi Berland, 1936; Nigeria
asmarensis Giordani Soika, 1939; Ethiopia
    asmarensis Arnold, 1940
avidus Bingham, 1896; Burma
banabacoa Alayo, 1968; Cuba
barbatus F. Smith, 1856; U.S.: Idaho & Oregon to Colorado & N. Mexico; Mexico: Chihuahua to Puebla
    albifrons Cresson, 1865, new synonymy by R. Bohart
    henricus Dunning, 1898, new synonymy by R. Bohart
barbiger Mickel, 1916; centr. U.S., new status by R. Bohart
basalis F. Smith, 1856; Sri Lanka, India
  ssp. *clypeatus* Tsuneki, 1963; Thailand
basilewskyi Leclercq, 1955; Ruanda
bicinctus (Mickel), 1916 (*Ococletes*); U.S.: Colorado, Utah
    hirticulus Mickel, 1918 (*Ococletes*)
bilineatus Gravenhorst, 1807; locality unknown
bilunatus Cresson, 1865; U.S. e. of Rocky Mtns.; Canada: Ontario
    scelestus Cresson, 1879
    assimilis Banks, 1915, nec Banks, 1913
    consimilis Banks, 1923, nec Kohl, 1891
bimaculus Saussure, 1891; Madagascar (see footnote under *madagascariensis*)
    immitis Saussure, 1892
bredoi Arnold, 1946, Tanzania
bucephalus F. Smith, 1856; Gambia to S. Africa
    spilaspis Cameron, 1910
  ssp. *rotundicollis* Arnold, 1924; Nigeria
camerunensis Tullgren, 1904; Cameroon
capensis Dahlbom, 1845; S. Africa
  ssp. *nordicus* Leclercq, 1962; Tanzania
coarctatus Spinola, 1838; n. Africa
    niloticus F. Smith, 1873
coronatus (Thunberg), 1784 (*Sphex*); Europe
    coronatus Fabricius, 1790, nec Thunberg, 1784
  ssp. *ibericus* Beaumont, 1970; Spain
    occidentalis Beaumont, 1951, nec Strandtmann, 1946
  ssp. *orientalis* Bytinski-Salz, 1959; Israel
crabroniformis F. Smith, 1856; w. U.S.
    flavifrons Cresson, 1865, new synonymy by R. Bohart
    californicus Cresson, 1879, new synonymy by R. Bohart
    sublimis Cresson, 1879, new synonymy by R. Bohart
    magnificus Provancher, 1895 (*Liris*), nec Kohl, 1883, new synonymy by R. Bohart
crotoniphilus Viereck and Cockerell, 1904; w. U.S.; Mexico: Coahuila
curvimaculatus (Cameron), 1910 (*Trachypus*); Ethiopia
decemmaculatus Eversmann, 1849; USSR
dentatus Cameron, 1902; e. India
depredator F. Smith, 1856; India
desertorum F. Morawitz, 1890; sw. USSR
dichrous Kohl, 1894; Ghana, Cameroon, Zaire
  ssp. *dolosus* Kohl, 1894; Gambia to Zaire
  ssp. *abyssinicus* Arnold, 1925; Ethiopia
dimidiatus Klug, 1845; Ethiopia
dufourii Lucas, 1848; nw. Africa, Spain

bolivari Mercet, 1914
elegantissimus Dalla Torre, 1897; n. India
    elegans F. Smith, 1873, nec F. Smith, 1856, which is a *Trachypus*
femoralis Arnold, 1946; Zambia
flagellarius Turner, 1918; Tanzania
flavipes Arnold, 1949; Zambia
formosanus Tsuneki, 1973; Taiwan
fossulatus Turner, 1918; Somalia
foveatus Arnold, 1933; Zaire
fuscipennis Guérin-Méneville, 1844; w. & centr. Africa
    temerarius Kohl, 1894
    ? *usambarensis* Stadelmann, 1897 (pl., fig. 11)
    ? *undussumae* Stadelmann, 1897 (expl. of fig. 11, p. 55)
    ssp. *consimilis* Kohl, 1891; s. & e. Africa
        reticulatus Cameron, 1908
    † ssp. *laetus* Gribodo, 1895; Mozambique, nec Fabricius, 1793, now in *Cerceris*
genalis Kohl, 1891; Libya, Egypt
gibbosus (Fabricius), 1775 (*Vespa*); transcontinental in N. America, s. to El Salvador
    punctatus Say, 1824
    punctiger Westwood, 1835 (*Cheilopogonus*)
    gibbosus Dahlbom, 1844 (*Anthophilus*), nec Fabricius, 1775
    nodosus Klug, 1846 (lapsus for *gibbosus* Fabricius)
    xanthostigma Cameron, 1891, new synonymy by R. Bohart
    maculifrons Cameron, 1891, new synonymy by R. Bohart
    cockerelli Dunning, 1897
    chilopsidis Cockerell, 1898, new synonymy by R. Bohart
    maculiventris Cameron, 1905 (*Anthophilus*), new synonymy by R. Bohart
    melanaspis Cameron, 1905 (*Anthophilus*), new synonymy by R. Bohart
glaber Kohl, 1891; Siberia
gloriosus Cresson, 1865; sw. & centr. U.S.; Mexico: Chihuahua
    insignatus Banks, 1913
gwaaiensis Arnold, 1929; s. Africa
hellmanni (Eversmann), 1849 (*Anthophilus*); e. USSR, n. China, Mongolia
    ssp. *sibiricus* Radoszkowski, 1888; Siberia
histrio Fabricius, 1804; Ethiopian Region
    schoenherri Dahlbom, 1845
    formosus F. Smith, 1856
    flavolineatus Cameron, 1908
    trichiocephalus Cameron, 1910
    ssp. *eritreanus* Bischoff, 1915; Ethiopia
    ssp. *distinctus* Arnold, 1951; Ethiopia
impatiens Kohl, 1891; S. Africa
inversus Patton, 1879; U.S.: Iowa to California; Canada: Alberta
kaszabi Tsuneki, 1971; Mongolia
kizilkumii Radoszkowski, 1877; sw. USSR
kohlii F. Morawitz, 1890; sw. USSR

kokandicus Radoszkowski, 1877; sw. USSR
komarowi F. Morawitz, 1890; sw. USSR
laticeps Arnold, 1925; Nigeria
lepidus Cresson, 1865; e. U.S. w. to Texas and Colorado
    carolinensis Banks, 1913
    reductus Banks, 1921
levini R. Bohart, 1972; U.S.: California
limatus Bingham, 1909; Uganda, Ruanda
    ruandanus Bischoff, 1915
lingyuanensis Yasumatsu, 1935; n. China
loeflingi Dahlbom, 1845; S. Africa
    innominatus Bingham, 1902
    ssp. *meneliki* Arnold, 1925; Ethiopia
madagascariensis[1] Brèthes, 1910; Madagascar
    petiolatus Saussure, 1892, nec Spinola, 1841
    ranavalonae Arnold, 1945
major Kohl, 1891; S. Africa
melanderi Arnold, 1925; s. Africa
michelbacheri R. Bohart, 1972; Mexico: Baja California
minor Kohl, 1891; n. Africa
mongolicus F. Morawitz, 1889; Mongolia
multimaculatus Cameron, 1891; w. U.S.; Mexico
    anna Dunning, 1897, new synonymy by R. Bohart
    multiannulatus Dalla Torre, 1897, lapsus
    annae Dunning, 1898, emendation
    cleomae Dunning, 1898, new synonymy by R. Bohart
    subversus Banks, 1915, new synonymy by R. Bohart
    yakima Banks, 1919, new synonymy by R. Bohart
namaqua Arnold, 1925; s. Africa
nasalis R. Bohart, 1972; U.S.: California
natalensis Arnold, 1925; s. Africa
neomexicanus Strandtmann, 1946; sw. U.S. to California
nepalensis Bingham, 1908; Nepal
nigriceps Bingham, 1896; India
nigritus Gravenhorst, 1807; locality unknown
nigrohirtus Turner, 1918; Uganda, Kenya, Ruanda
    ssp. *calvus* Turner, 1918; Zambia
nitidus Magretti, 1884; Sudan
nobilis Kohl, 1891; sw. USSR
notatulus F. Smith, 1862; Celebes
    ssp. *concinnus* Bingham, 1896; Burma, N. Viet Nam, Malaya, Sumatra, Philippines
    ?*angustatus* Turner, 1919
    ssp. *javanus* van der Vecht, 1966; Java
    ssp. *wegneri* van der Vecht, 1966; Indonesia: Flores
nursei (Bingham), 1898 (*Trachypus*); Aden
occidentalis Strandtmann, 1946; U.S.: California

---

[1] Brèthes (1910, p. 277, footnote) proposed this new name for *Philanthus petiolatus* Saussure, but he cited Saussure's name as being published in Mitt. Schweizer. Ent. Ges. 8:261 (1891). However, *P. petiolatus* Saussure was published in Histoire Physique, Naturelle et Politique de Madagascar, vol. 20, p. 538 (1892). Under the name *petiolatus* Saussure referred back to p. 261 of his 1891 paper, and this is apparently where Brèthes was led astray. There is only one *Philanthus* described in the 1891 paper, *bimaculus*. This is species #12, the number to which Saussure (1892, p. 538) referred in his erroneous reference to *petiolatus*. *Philanthus bimaculus* Saussure, 1891, was based on a female. In 1892, p. 535, Saussure redescribed *bimaculus* but said it was known only by the male.

oraniensis Arnold, 1925; s. Africa
ordinarius Bingham, 1896; Burma
pacificus Cresson, 1879; U.S.: Nevada, new status by R. Bohart
    ssp. *arizonae* Dunning, 1898; U.S.: Wyoming to Arizona & California, new status by R. Bohart
      hirticeps Cameron, 1905, new synonymy by R. Bohart
      assimilis Banks, 1913, new synonymy by R. Bohart
pallidus Klug, 1845; Egypt to Ethiopia
pilifrons Cameron, 1908 (March); Uganda, Tanzania
    ugandicus Magretti, 1908
    ssp. *xanthogaster* Cameron, 1908; Kenya
politus Say, 1824; e. U.S., se. Canada
    dubius Cresson, 1865
    texanus Banks, 1913, new synonymy by R. Bohart
promontorii Arnold, 1925; S. Africa
psyche Dunning, 1896; N. America: Iowa to Alberta, Montana, Texas, Arizona, Chihuahua, new status by R. Bohart
    punctinudus Viereck and Cockerell, 1904
    hermosus Banks, 1913
pulchellus Spinola, 1842; sw. Europe (not n. nudum)
    sieboldti Dahlbom, 1845
    andalusiacus Kohl, 1888
pulcher Dalla Torre, 1897; w. U.S.; Canada: Alberta, new status by R. Bohart
    pulchellus Cresson, 1865, nec Spinola, 1842
    clarconis Viereck, 1906
pulcherrimus F. Smith, 1856; India
punjabensis Nurse, 1902; n. India
radamae Arnold, 1945; Madagascar
ramakrishnae Turner, 1918; s. India
raptor Lepeletier, 1845; nw. Africa
    marocanus Shestakov, in Nadig, 1933
    ssp. *siculus* Giordani Soika, 1944; Sicily
reinigi Bischoff, 1930; sw. USSR
†*rubidus* Arnold, 1946; Ethiopia, nec Jurine, 1807 (now in *Cerceris*)
    abyssinicus Arnold, 1932, nec Arnold, 1925
rubriventris Kazenas, 1970; sw. USSR, w. China
rugosifrons Arnold, 1949; Zambia
rugosus Kohl, 1891; S. Africa
rutilus Spinola, 1838; Algeria, Egypt
    ssp. *pachecoi* Giner Marí, 1945; Spanish Sahara
sanbornii Cresson, 1865; e. and centr. U.S.
    scutellaris Cresson, 1879
    eurynome W. Fox, 1890, new synonymy by R. Bohart
    trumani Dunning, 1897
    magdalenae "Viereck" Strandtmann, 1946
schulthessi Maidl, 1924; Sudan
    ssp. *nigrinus* Bytinski-Salz, 1959; Israel
schusteri R. Bohart, 1972; U.S.: California, Arizona
scrutator Nurse, 1902; w. India
serrulatae Dunning, 1898; sw. U.S.; Mexico: Sonora, new status by R. Bohart
sicarius F. Smith, 1856; Fernando Póo
siouxensis Mickel, 1916; w. U.S.; Mexico: Chihuahua, Coahuila, new status by R. Bohart

soikai Beaumont, 1961; n. Africa
solivagus Say, 1837; ne. U.S.; Canada: Quebec
    solidagus Howard, 1901, lapsus
sparsipunctatus Arnold, 1946; Zambia
?*spiniger* Thumberg, 1815; locality unknown
stecki Schulz, 1906; Fernando Póo
strigulosus Turner, 1918; S. Africa, Lesotho
stygius Gerstaecker, 1857; Mozambique
    ssp. *atronitens* Arnold, 1925; Rhodesia
subconcolor (Bingham), 1898 (*Trachypus*); Aden
sulphureus F. Smith, 1856; n. India
sumptuosus Turner, 1917; e. India
taantes Gribodo, 1895; Mozambique
tarsatus H. Smith, 1908; centr. U.S., new status by R. Bohart
tenellus Arnold, 1925; Rhodesia
triangulum (Fabricius), 1775 (*Vespa*); Europe
    fasciatus Fourcroy, 1785 (*Vespa*)
    maculatus Christ, 1791 (*Sphex*)
    limbatus Olivier, 1792 (*Vespa*)
    androgynus Rossi, 1792 (*Crabro*)
    pictus Panzer, 1797
    discolor Panzer, 1799
    apivorus Latreille, 1799
    allionii Dahlbom, 1845
    allwini Dalla Torre, 1897, lapsus for *allionii*
    ssp. *diadema* (Fabricius), 1781 (*Crabro*); Ethiopian Region
      frontalis Gerstaecker, 1857
      transversus Cameron, 1910
    ssp. *abdelcader* Lepeletier, 1845; Palearctic Africa
      abdelkader Lucas, 1880, emendation
    ssp. *bimaculatus* Magretti, 1908; Kenya
    ssp. *obliteratus* Pic, 1917; Algeria, Egypt
†*tricinctus* Gimmerthal, 1836; Russia, nec Spinola, 1805 (now in *Cerceris*)
tricolor Fairmaire, 1858; Guinea
turneri Arnold, 1925; S. Africa
variegatus Spinola, 1838; Egypt, Israel
    osbecki Dahlbom, 1845
    distinguendus Kohl, 1891
    palestinensis Balthasar, 1953
    ssp. *ecoronatus* Dufour, 1853; Morocco to Libya, Chad
      septralis Radoszkowski, 1888
    ssp. *nabateus* Bytinski-Salz, 1959; Israel, Saudi Arabia
variolosus Arnold, 1932; Ethiopia
ventilabris Fabricius, 1798; transcontinental in N. America
    vertilabris Fabricius, 1804, lapsus
    frontalis Cresson, 1865, nec Gerstaecker, 1857
    rugosus Provancher, 1895 (*Liris*), nec Kohl, 1891
    "*ventralis,* Fabr." of Ashmead, 1899:296, lapsus for *ventilabris*
    "*ventralis*" of Howard, 1901: pl. 3, fig. 33, lapsus for *ventilabris*
    completus Banks, 1915
ventralis (Mickel), 1918 (*Ococletes*); w. U.S., new status by R. Bohart
    strandtmanni Burks, 1951, unnecessary new name for

lapsus of Howard, 1901 (see synonymy of *ventilabris*)
venustus (Rossi), 1790 (*Crabro*); s. Europe to Israel, Mongolia
   melliniformis F. Smith, 1856
walteri Kohl, 1891; sw. USSR
werneri Maidl, 1933; Morocco
yerburyi Bingham, 1898; Aden
zebratus Cresson, 1879; w. U.S.

basilaris Cresson, 1879, new synonymy by R. Bohart
nitens Banks, 1913 (*Ococletes*), new synonymy by R. Bohart
illustris Mickel, 1918 (*Ococletes*)

Fossil *Philanthus*

annulatus Theobald, 1937; France: Oligocene
saxigenus Rohwer, 1909; Colorado: Tertiary

FIG. 187. *Trachypus mexicanus* Saussure, male.

Genus Trachypus Klug

Generic diagnosis: Characters of generic value are: last antennal article obliquely and sharply truncate, truncation flat and polished; subantennal sclerite present and defined by lines through tentorial pits to clypeus; pronotum raised and not closely appressed to scutum; hindwing media usually diverging in front of cu-a but sometimes at cu-a or a little beyond it; gastral segment I a slender peduncle or petiole which is more than twice as long as broad (fig. 187); female sternum VI not cleft apically.

Geographic range: The 19 listed species are all from the Neotropical Region. Three of these are recorded from Mexico, one from Puerto Rico, and one from Chile. Most of the others are from Brazil, Argentina, and Paraguay.

Systematics: The close relationship of *Trachypus* and *Philanthus* and the taxonomic significance of pedunculation of the gaster have been discussed in the systematics section of the latter genus. The truncate antenna with its end round and polished is probably the most critical recognition character for *Trachypus*. There are no adequate keys to species, the best being that of Brèthes (1910). The genus is being revised by E. Rubio.

We have been able to study most of the known species. Characters of special value seem to be the clypeal shape and dentition, nature of the female propleuron, punctation of the propodeal enclosure and first three terga, form of the trochanters (spinose in *patagonensis*), relative proportions of tergum I, pubescence of the sterna in males, degree of humeral prominence, and wing maculation (forewing apically spotted in male *elongatus*). The general habitus of a representative species is shown by fig. 187 and a typical male face by fig. 186 D.

Biology: Janvier (1928) has studied at length the nesting habits of *Trachypus denticollis* (as *Philanthus*) in Chile, Other reports on unidentified South American species were made by Bertoni (1911) and Callan (1954). Evans (1964e) briefly studied *T. mexicanus* in Mexico, where small aggregations were found nesting in a sandy strip and along the slopes and top of piles of ground limestone. A mound of soil surrounded each nest entrance. One nest with only a single cell was excavated. The cell was at a depth of 9 cm and contained six bees; three other bees were found in the nest burrow. The prey were of the genera *Augochlora*, *Halictus* (Halictidae), and *Exomalopsis* (Anthophoridae).

Janvier (1928) found *T. denticollis* nesting in colonies in compact, sandy soil, frequently along margins of roads, rivers, and sand dunes. Nests were dug to 30-40 cm and up to seven brood cells were constructed. Two to three days were required to construct and provision a cell. Prey consisted of *Halictus* (seven species), *Apis mellifera*, and *Melitoma chilensis*. The nest was closed between trips for prey. Janvier noted that even though a nest of *Halictus* existed next to one of *Trachypus*, the wasp did not attack *Halictus* within or near the bee nest, but instead, captured and stung *Halictus* at flowers. Janvier noted *Photopsis* (Mutillidae) as a parasite of *T. denticollis*. Janvier (1933) also found *Plumarius* (Plumariidae) in *denticollis* nests.

The most recent study of *Trachypus* biology was that of Evans and Matthews (1973a). They observed *petiolatus* nesting in firm, bare clay soil in Colombia and Argentina. The burrows were at first oblique, then horizontal, some attaining a length of nearly 2 m and containing up to 39 cells. Re-use of the same burrow by succeeding generations was indicated. In Colombia two females and one male were associated with a single burrow, thus suggesting communal use of burrows. Prey were mostly small bees but wasps were used also.

Checklist of *Trachypus*

annulatus Spinola, 1851; Brazil, Argentina
   ?*furcatus* Brèthes, 1910
basalis F. Smith, 1873, Brazil
batrachostomus Schrottky, 1909; Paraguay
cementarius (F. Smith), 1860 (*Philanthus*); Brazil
denticollis Spinola, 1851; Chile, Argentina
   incertus Spinola, 1851, new synonymy by E. Rubio
   rufipes Reed, 1894, new synonymy by E. Rubio
   chilensis "Spin." Brèthes, 1910, lapsus for *denticollis*?
disjunctus F. Smith, 1873; Brazil
elongatus (Fabricius), 1804 (*Zethus*); Guyana to Argentina
   gomesii Klug, 1810, new synonymy by E. Rubio
   apicalis F. Smith, 1856 (*Philanthus*), new synonymy by R. Bohart
   surinamensis Saussure, 1867, new synonymy by E. Rubio
   terminalis Taschenberg, 1875 (*Philanthus*), new synonymy by E. Rubio
   gomezi Strand, 1910 (*Philanthus*), emendation
   asuncionis Strand, 1910 (*Philanthus*), new synonymy by E. Rubio
flavidus (Taschenberg), 1875 (*Philanthus*); Brazil to Argentina
   ?*coriani* Schrottky, 1909
fulvipennis (Taschenberg), 1875 (*Philanthus*); Brazil to Argentina
   brevipetiolatus Schrottky, 1909, new synonymy by E. Rubio
gerstaeckeri Dewitz, 1881; Puerto Rico
gracilis (Cameron), 1890 (*Philanthocephalus*); Mexico to Costa Rica
   maculiceps Cameron, 1890 (*Philanthocephalus*), new synonymy by R. Bohart
hirticeps (Cameron), 1890 (*Philanthocephalus*); Mexico, Guatemala
mexicanus Saussure, 1867; U.S.: s. Texas; Mexico to Costa Rica
   mexicanus Cameron, 1890 (*Philanthocephalus*), nec Saussure, 1867
   annulitarsis Cameron, 1908, new synonymy by R. Bohart
miles Schrottky, 1909; Paraguay
patagonensis (Saussure), 1854 (*Philanthus*); Brazil to Argentina

egregius (Taschenberg), 1875 (*Philanthus*)
martialis Holmberg, 1903
magnificus Schrottky, 1909, new synonymy by E. Rubio
petiolatus (Spinola), 1841 (*Philanthus*); Panama to Argentina
    elegans F. Smith, 1856 (*Philanthus*), new synonymy by R. Bohart
    elegans Taschenberg, 1875 (*Philanthus*), nec F. Smith, 1856, new synonymy by E. Rubio
    punctifrons Cameron, 1890 (*Philanthocephalus*), new synonymy by R. Bohart
    mendozae Dalla Torre, 1897 (*Philanthus*), new name for *elegans* Taschenberg
    flavus Brèthes, 1910, new synonymy by E. Rubio
    ruficeps Brèthes, 1910, new synonymy by E. Rubio
    peruanus Giner Marí, 1944, new synonymy by E. Rubio
romandi (Saussure), 1854 (*Philanthus*); Brazil to Argentina
spegazzinii Brèthes, 1910; Peru, Bolivia, Argentina
    punctuosus Brèthes, 1910, new synonymy by E. Rubio
varius (Taschenberg), 1875 (*Philanthus*); Brazil to Argentina

Tribe Aphilanthopsini

The wasps of this tribe are medium-sized philanthines varying from about 7 to 15 mm in length. Many of the species are ornamented with red as well as with yellow or white. In all known cases ants are used as prey.

Systematics: The following characters seem to be of tribal importance: mandible edentate, antennal sockets located above clypeus and 1.0 to 2.5 diameters apart, no sharp interantennal carina and frons not strongly swollen above antennal sockets, lateral clypeal brushes present in male (fig. 186 A), palpal formula 6-4, episternal sulcus present or undefined, scrobal sulcus weak, midcoxal cavities separated but only moderately so, hindfemur simple at apex, plantulae usually present, marginal cell of forewing acutely narrowed toward apex, submarginal cell II not petiolate in front, first recurrent vein terminating on submarginal cell II, second recurrent on III, jugal lobe occupying more than half length of anal area (fig. 184 H), gaster sessile, male without cerci, female pygidial plate present (weak in *Philanthinus*), volsella differentiated into digitus and cuspis, white or yellow markings present.

Key to genera of Aphilanthopsini

1. Hindwing media diverging before cu-a (fig. 184 H), inner orbits of eyes bowed slightly away from each other, pronotal collar depressed below scutum and appressed to it, Mediterranean area (subtribe Philanthinina) ....
    .... *Philanthinus* Beaumont, p. 569
   Hindwing media diverging beyond cu-a, inner orbits of eyes bowed slightly toward each other (fig. 186 A,B), pronotal collar about as high as scutum, N. America (subtribe Aphilanthopsina) 2

2. Ocellocular distance much less than two hindocellus diameters; male sternum VIII with a broadly triangular apex in posterior view; male flagellomeres I to VI strongly compressed, VII to XI expanded into a club (fig. 186 A); female pygidial plate with an apicomedian ball-like projection .... *Listropygia* R. Bohart, p. 571
   Ocellocular distance about two hindocellus diameters or more, male sternum VIII thin apically, male flagellomeres not unusually flattened nor expanded into a club, female pygidial plate without an apicomedian ball-like projection 3

3. Female pygidial plate triangular, apex rounded; female sternum VI simple; female clypeus toothed toward apical middle (fig. 186 B); metanotum with a carina behind base of hindwing but no angular lamina overhanging lateral sinus on metanotum. *Aphilanthops* Patton, p. 570
   Female pygidium quadrate, surface concave; female sternum VI modified; female clypeus not toothed toward apical middle; metanotum with an angular lamina (behind base of hindwing) overhanging lateral sinus on metanotum .... *Clypeadon* Patton, p. 570

Subtribe Philanthinina

Genus Philanthinus Beaumont

Generic diagnosis: Clypeus of male apically tridentate to serrulate, clypeus of female ending in a lamella flanked by a tooth, antenna not clublike, antennal sockets more than two diameters apart, ocellocular distance not reduced, inner eye margins simple and bowed away from each other, no delimited subantennal sclerite, pronotal collar depressed and appressed to scutum, plantulae absent, hindwing media diverging before cu-a, female with a weakly defined pygidial plate, female sternum VI with a median groove and apically cleft, male sterna without fimbriae and VIII simple.

Geographic range: The four known species are all in the Palearctic Region: southwest USSR, Middle East, and north Africa.

Systematics: The exact phylogenetic position of *Philanthinus* is debatable. In some ways it seems related to *Philanthus,* but the inner eye margins are entire, and the face is not bulging. Since it has many features in common with *Aphilanthops*, we have placed it in the Aphilanthopsini. However, it has several distinctions from other members of the tribe, and it should probably be regarded as a rather generalized and relict genus in a separate subtribe. The depressed pronotal collar, which is appressed to the scutum, occurs in only one other genus, *Odontosphex*. Since there is little else to associate them, this is probably not a sign of close relationship. A similar situation exists with *Eremiasphecium* which shares with *Philanthinus* the bowing out of the inner eye margins. This, together with the rather basal divergence of the hindwing media, may simply be the unspecialized condition in the subfamily.

Useful references are Beaumont (1949a), Beaumont and Bytinski-Salz (1959), and Menke (1967d).

Biology: Unknown.

Checklist of *Philanthinus*

albiceps (Gussakovskij), 1952 (*Shestakoviella*); sw. USSR: Tadzhik S.S.R.

integer (Beaumont), 1949 (*Philanthus*); Morocco, Algeria, Egypt; Israel

quattuordecimpunctatus (F. Morawitz), 1888 (*Anthophilus*); sw. USSR: Kazakh S.S.R.

   elegans F. Morawitz, 1888 (*Anthophilus*), nec F. Smith, 1856, new synonymy by W. J. Pulawski

   eximius F. Morawitz, 1894 (*Philanthus*), new name for *elegans* (Morawitz)

theodori (Bytinski-Salz), 1959 (*Philanthus*); Israel, Syria

Subtribe Aphilanthopsina

Genus Aphilanthops Patton

Generic diagnosis: Clypeus of male apically tridentate, clypeus of female with three or more teeth (fig. 186 B), antenna not abruptly clublike, ocellocular distance not much reduced, inner eye margins simple and bowed toward each other, subantennal sclerite present (best seen in females) and defined by lines from antennal sockets through tentorial pits to clypeus, pronotal collar raised and distinct from scutum, no angular metanotal lamina at hindwing base, hindwing media diverging well beyond cu-a, female with a simple pygidial plate, female sternum VI not cleft apically, fimbriae present on male sterna IV-V (except in *foxi*), male sternum VIII simple.

Geographic range: The four known species are all Nearctic, two of them transcontinental, and the other two in southwestern U.S.

Systematics: Although clearly related to *Clypeadon* and *Listropygia*, there is no specialized ant clamp in *Aphilanthops*. This is presumably connected with its prey-gathering habits, the catching of queen ants during swarming rather than worker ants near the nest.

Best species characters are punctation, pubescence, and color pattern. Contours of a typical female face are shown in fig. 186 B.

Most useful references are Dunning (1898) and R. Bohart (1959, 1966).

Biology: The biology of only one species of *Aphilanthops* is well known. The nesting behavior of *frigidus* (F. Smith) was observed by Evans (1962), who also reviewed several earlier papers on this species. *A. frigidus* appears to have one generation per year, nesting from late June to mid August. Possibly only one nest per female is constructed during this time. Females nest gregariously in slightly sloping bare sand or gravel. Soil particles are loosened with the mandibles and then kicked 20 cm or more from the burrow, leaving no earth mound about the nest entrance. The main burrow descends obliquely at about 45 degrees to a single storage chamber at a depth of 12 to 25 cm beneath the soil surface. This cell is never closed and is used to store up to four prey members. Prey consists of winged queen ants of the genus *Formica*, which are captured and stung immediately upon landing after the nuptial flight. The ants are carried either venter up or sideways in the legs of the wasp while she flies to the nest. The ant may be taken directly into the burrow, but more commonly the wasp drops the ant just outside the burrow entrance, turns around, and drags it down the burrow by an antenna. During provisioning flights, the entrance to the nest may either be closed or open.

Four or more brood cells lie beneath the storage chamber at a depth of 25 to 45 cm from the soil surface. These chambers are provisioned with two or occasionally three de-alated ants. Evans discounted earlier ideas that *frigidus* larvae are fed progressively. Instead, according to Evans, several ants are first provisioned, and an egg is laid on the venter of the uppermost ant prior to plugging the brood chamber with soil.

A miltogrammine (Sarcophagidae) fly, *Senotainia trilineata* Wulp, has been reported attacking *A. frigidus* (Ristich, 1956). Evans (1962) has also seen the sarcophagids, *Metopia leucocephala* (Rossi) and *Euaraba tergata* (Coquillett) flying about a nesting area. *E. tergata* was recorded remaining within a nest for two minutes.

Checklist of *Aphilanthops*

foxi Dunning, 1898; U.S.: s. California

frigidus (F. Smith), 1856 (*Philanthus*); U.S.: transcontinental

   bakeri Dunning, 1896

   dawsoni Swenk, 1912 (*Nomada*)

hispidus W. Fox, 1894; sw. U.S.; Mexico: Baja California

subfrigidus Dunning, 1898; U.S.: transcontinental

   elsiae Dunning, 1898

Genus Clypeadon Patton

Generic diagnosis: Clypeus of male apically tridentate, clypeus of female ending in an edentate flange, antenna not abruptly clublike, ocellocular distance not much reduced, inner eye margins simple and bowed toward each other, subantennal sclerite defined by lines from antennal sockets through tentorial pits to clypeus, pronotal collar raised and distinct from scutum, an angular lamina present at posterior base of hindwing, hindwing media diverging well beyond cu-a, female with a concave and quadrate pygidial plate which with the cleft and otherwise modified sternum VI forms an ant clamp for carrying prey, male sterna without fimbriae and VIII simple.

Geographic range: The eight listed species are all from western United States and Mexico, with a concentration of species in southwestern desert areas of United States.

Systematics: The highly specialized gastral segment VI has evolved into an ant clamp for securing workers of *Pogonomyrmex* which serve as prey. This interesting structure, found in *Clypeadon* and *Listropygia*, is unique. Evans (1962) has illuminated the form and function concerned. There are specific differences in structure of the pygidial plate and sternum VI which constitute the clamp, and since the species of *Clypeadon* are apparently rather particular in their choice of prey, one wonders whether the adaptation has been carried to the level of ant specifics. So far, there is no evidence of this.

Clypeadon is closely related to *Aphilanthops* and was considered as a subgenus of it until recently. The entirely different female pygidial plate and the angular metanotal lamina in *Clypeadon* make distinction between them an easy matter. *Listropygia* is probably much more closely related to *Clypeadon,* and this is indicated by the similarity in prey. However, the subcapitate antennae and knobbed female pygidial plate of *Listropygia* set it apart.

A key to species has been given by R. Bohart (1966). Other useful references are Dunning (1898), R. Bohart (1959), and Menke (1967d). Specific characters stressed are the form of the female pygidial plate, development of shiny areas on the vertex, nature of pubescence on the male sterna, punctation of the scutum and terga, and color pattern of the gaster.

Biology: Evans (1962) has provided the most recent and complete review of the biology of *Clypeadon.* The nesting habits of *haigi, evansi,* and *laticinctus* have been well documented. Fragmentary records of hunting behavior have been given for *dreisbachi* and *taurulus.* In addition, prey records were given for *utahensis* (as *concinnula*) and *sculleni.*

In general, *Clypeadon* nest in sandy areas or clearings surrounded by sparse vegetation. Males have often been found on plants near the nesting site, but no mating activity has been seen. Burrowing is done by females which kick soil particles some distance from the nest entrance. Mounds are not usually formed, but in *laticinctus* a very low mound up to 10 cm long may occur. The nest may be excavated to a depth of about 25 cm. Brood cells are 8-12 mm in diameter and are almost spherical. Each cell may contain from 10-15 ants (*haigi, evansi*) or 15-26 ants (*laticinctus*) of the genus *Pogonomyrmex.*

Female *Clypeadon* appear to be species specific in their choice of prey, often ignoring nearby ants of the "wrong" species in favor of ants nesting farther away. Some female *Clypeadon* alight near an ant nest and remain motionless until an ant hurries by. The wasp then rushes the ant, stings it, and carries it away. Females of *dreisbachi* have been seen rushing at the ant nest opening, and then moving away with one or more ants in pursuit. At a chosen moment, after many pursuits, the wasp will turn about and sting an ant and then fly away with it. In flight, ants are carried venter up in the ant clamp (pygidial plate and hypopygium) which clasps the ant between its hindcoxae and midcoxae.

During prey capture the burrow entrance remains open. It may be closed from within at night. Ants are stored within the burrow at depths of from 6 to 15 cm beneath the soil surface until a brood cell is completed. An egg is laid on the venter of the last ant provisioned, and the cell is plugged.

One species of sarcophagid fly (*Senotainia trilineata* Wulp) has been reared from the brood cell of *Clypeadon evansi.* Miltogrammine flies (Sarcophagidae) have been noted to destroy over half of the brood cells excavated for study (Evans, 1962).

Checklist of *Clypeadon*

californicus (R. Bohart), 1959 (*Aphilanthops*); U.S.: California
dreisbachi (R. Bohart), 1959 (*Aphilanthops*); s. centr. U.S.; Mexico
evansi R. Bohart, 1966; sw. U.S.
haigi (R. Bohart), 1959 (*Aphilanthops*); sw. U.S.
laticinctus (Cresson), 1865 (*Philanthus*); w. U.S.
    quadrinotatus Ashmead, 1890 (*Aphilanthops*)
sculleni (R. Bohart), 1959 (*Aphilanthops*); sw. U.S., Mexico
taurulus (Cockerell), 1895 (*Aphilanthops*); sw. U.S.
    phoenix Pate, 1947 (*Aphilanthops*)
utahensis (Baker), 1895 (*Aphilanthops*); sw. U.S., Mexico
    concinnulus Cockerell, 1896 (*Aphilanthops*)

Genus Listropygia R. Bohart

Generic diagnosis: Clypeus of male apically bidentate, clypeus of female ending in an edentate flange, antenna with articles unusually flattened and apically expanded (especially in male) (fig. 186 A), ocellocular distance not more than 1.3 diameters, inner eye margins simple and bowed toward each other, subantennal sclerite present (obscured by hair in male) and defined by lines from antennal sockets through tentorial pits to clypeus, pronotal collar raised and distinct from scutum, episternal sulcus undefined, hindwing media diverging well beyond cu-a, female with a concave and quadrate pygidial plate terminating in a ball-like projection and forming an ant clamp with the cleft and otherwise modified sternum VI, male sterna without fimbriae, male sternum VIII broadly triangular in posterior view.

Geographic range: The unique species has been collected only in desert areas of southern California and western Arizona.

Systematics: The rather extraordinary, knobbed female pygidial plate of *Listropygia* is its most obvious feature. Rivaling it is the subcapitate male antenna which has the basal and medial flagellomeres slender and flattened, whereas the more apical ones are expanded (fig. 184 A). Also, it differs from its close relative *Clypeadon* in having a short nonangulate lamella on the metanotum at the hindwing base. The dense, brushlike silvery pubescence on the face of the male (fig. 184 A), as well as the expanded apex of its eighth tergum are remarkable. References are R. Bohart (1959, 1966) and Menke (1967d).

Biology: Little is known of the habits. One adult female was captured holding an ant worker (*Pogonomyrmex californicus* Buck) as prey (R. Bohart, 1959). Based on morphology, especially of the pygidial plate and hypopygium, Evans (1962) suggested that prey holding in *Listropygia* parallels that of *Clypeadon,* and that these two genera are unique among Hymenoptera in this respect.

Checklist of *Listropygia*

bechteli (R. Bohart), 1959 (*Aphilanthops*); sw. U.S.

Tribe Odontosphecini

These are small wasps, less than 10 mm long and resembling *Tachysphex* or small *Tachytes*. Nothing is known of the habits, and the group appears to be a relict one with disjunct distribution. Its occurrence in Bolivia, Argentina, northwest Africa, and Saudi Arabia may lend ammunition to the theory of continental drift.

Systematics: Characters of tribal significance are: mandible edentate and with socket not completely closed, upper margin of clypeus not unusually expanded upward medially, antennal sockets separated from clypeal margin and less than two diameters apart, a small tubercle between antennae, lateral ocelli vestigial and close to midocellus, face not unusually swollen above antennae, no lateral clypeal brushes in male, palpal formula 6-4, episternal and scrobal sulci absent, midcoxal cavities close together, hindfemur truncate apically, marginal cell of forewing narrowed acutely toward apex, submarginal cell II not petiolate in front, both recurrent veins terminating on II, hindwing media diverging before cu-a, jugal lobe occupying about four-fifths length of anal area (fig. 184 A), female with a pygidial plate, no cerci, volsella differentiated into digitus and cuspis, no white or yellow markings.

All of the above features are found in various combinations in other tribes of Philanthinae except for the unusually shaped clypeus, vestigial hindocelli, termination of both recurrent veins on submarginal cell II, and absence of white or yellow markings. A comparison of affinities between the monotypical genus *Odontosphex* and other philanthine genera is given under Systematics of the genus.

Genus Odontosphex Arnold

Generic diagnosis: Eyes in males nearly holoptic (fig. 188), in females strongly converging above, clypeus transverse and apicomedially with two to seven teeth, antennae short and with flagellomere I not longer than II, antennal sockets not more than two diameters apart, inner eye margins bowed away from each other, no subantennal sclerite defined by lines from antennal sockets through tentorial pits to clypeus, pronotum with collar thin, depressed and appressed to scutum, foretarsal rake of female made up of short spines, hindcoxa with lamella at inner apex, gaster sessile, female sternum VI not cleft apically, male sternum VIII somewhat narrowed and triangular toward apex, female tergum V covered with long and hairlike setae which project posteriorly.

Geographic range: One species from northwest Africa and Saudi Arabia and three from Argentina and Bolivia make up the disjunct distribution.

Systematics: The relationships of *Odontosphex* have been discussed by Arnold (1951) and more fully by

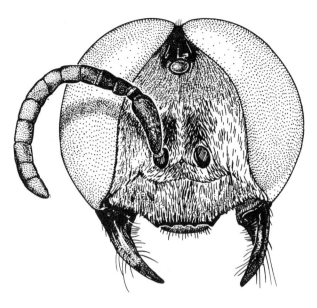

Odontosphex paradoxus

FIG. 188. Facial portrait of *Odontsphex paradoxus* male.

Menke (1967d). Arnold related the genus to the larrine genus *Tachysphex*, which has species strikingly similar in general appearance and in several morphological features of phylogenetic importance. Menke, on the other hand, pointed out the important similarities between *Odontosphex* and Philanthinae, particularly *Pseudoscolia*, showing at the same time that there were important structural reasons for excluding the genus from the Larrinae.

Essential morphological features held in common by *Odontosphex* and *Tachysphex* that might point to a larrine relationship are: (1) mandibular sockets not completely closed, (2) clypeus without an upwardly expanded basomedian lobe, (3) no delimited subantennal sclerite, (4) hindocelli vestigial and marked by swellings, (5) both recurrents terminating on submarginal cell II, (6) jugal lobe occupying most of length of anal area (fig. 184 A). Items 3 and 6 occur also in several philanthines; so 1, 2, 4, and 5 have the greatest weight. On the other hand, the close placement of the hindocelli to the midocellus and the absence of a circular elevation between them is atypical of the Larrini.

Characters of *Odontosphex* that point to a philanthine such as *Pseudoscolia* rather than to *Tachysphex* are: (1) mandibles edentate and without an externoventral notch, (2) antennal sockets placed somewhat above clypeus, (3) some male flagellomeres with flattened areas beneath, (4) episternal sulcus absent, (5) stigma of forewing tapering to an apex well beyond base of marginal cell, (6) hindwing media diverging well before cu-a, (7) nearly contiguous midcoxae, (8) hindfemur truncate at apex, (9) male sternum VIII triangular toward apex, and (10) volsella differentiated into digitus and cuspis.

The evidence seems clearly in favor of an alignment with Philanthinae along the *Pseudoscolia-Cerceris* axis. A very close relationship with *Pseudoscolia* is negated by the large jugal lobe, narrowly separated midcoxal cavities, transverse clypeus, partly open mandibular sockets, and vestigial hindocelli. The first three of these can be considered as generalized features which underwent phylogenetic change in the increasingly specialized genera, *Pseudoscolia, Cerceris,* and *Eucerceris* (fig. 183).

Specific differences within *Odontosphex*, as pointed out by Menke (1967d), are details of the male antenna, structure of the clypeal free margin in both sexes, and color pattern. Females are difficult to separate, but scutal punctation seems to be helpful.

Specimens of all four of the known species have been studied. A representative face is shown in fig. 188.

Biology: Unknown.

Checklist of *Odontosphex*

bidens Arnold, 1951: Mauritania (We have also seen a female from Saudi Arabia (Mus. London) that may well be this species.)
fritzi Menke, 1967; Argentina: Rio Negro
paradoxus Menke, 1967; Argentina: La Rioja; Bolivia: Cordillera Prov.
willinki Menke, 1967; Argentina: Rio Negro

Tribe Pseudoscoliini

Members of this tribe are small, colorful wasps which bear a superficial resemblance to species of *Clypeadon* and *Listropygia*. There is only a single genus, *Pseudoscolia*, that is restricted to the Old World.

Systematics: Tribal characters are: mandible edentate (fig. 189 A), antennal sockets located close to but distinctly above clypeus, and sockets separated by more than two diameters, no sharp interantennal carina, ocelli normal, face not swollen above antennae, lateral clypeal brushes in male present but weak, palpal formula 6-4, episternal and scrobal sulci absent, precoxal sulcus weak or absent, precoxal lobes thus not well defined, midcoxal cavities well separated, hindfemur truncate apically, plantulae present, submarginal cell II not petiolate in front, first recurrent vein terminating on submarginal cell II and second recurrent on III, marginal cell of forewing narrowed acutely toward apex, hindwing media diverging well before cu-a, jugal lobe occupying about half or less than half length of anal area (fig. 184 C), a pygidial plate present in both sexes, no cerci in male, volsella differentiated into digitus and cuspis, white or yellow markings present.

The tribe seems obviously related to the Cercerini with which it shares the following important characters: no episternal sulcus, hindfemur truncate, jugal lobe of hindwing reduced, midcoxal cavities well separated, both sexes with a pygidial plate, and inner eye margins bowed toward each other. In spite of this impressive list of similarities, there are many significant differences, all of which can be considered as the generalized rather than specialized condition. These are: volsella differentiated into digitus and cuspis, plantulae present, hindwing media diverging before cu-a (fig. 184 C), antennal sockets rather widely separated with no strong ridge between, and forewing marginal cell acutely narrowed distally. The absence of the scrobal sulcus and the weak or absent precoxal sulcus seem to be specializations. Additional characters that occur in Cercerini but which are not characteristic of that tribe are submarginal cell II sessile in front and mandible edentate. It is possible, then, to draw the conclusion that the Cercerini developed from a wasp very much like *Pseudoscolia*.

Genus Pseudoscolia Radoszkowski

Generic diagnosis: Ocellocular distance greatly reduced in some males, inner eye margins bowed toward each other, a special brush present near outer base of male mandible, no delimited subantennal sclerite, pronotum raised and distinct from scutum, gaster sessile, female sternum VI not apically cleft, male sternum VIII simple.

Geographic range: The 19 listed species are all Palearctic: Mongolia, southwest USSR, Middle East and north Africa.

Systematics: Beaumont (1949a) gave a key to species. Differentiating characters of importance are: clypeal dentition, male antennal structure, ocellocular distance, length of postocular area of head, punctation (especially on scutum, mesopleuron and terga), propodeal sculpture, and markings. Other references of note are Mochi (1939a) and Beaumont and Bytinski-Salz (1959). We have been able to study six species: *dewitzii, lyauteyi, pharaonum, sinaitica, tricolor,* and *theryi*. A representative face is shown in fig. 189 A.

Biology: According to Beaumont (1949a), he captured a female *Pseudoscolia tricolor* carrying a male *Halictus*.

Checklist of *Pseudoscolia*

angelae (Kohl), 1891 (*Philanthus*); Iraq, Israel
araxis (Kohl), 1891 (*Philanthus*); sw. USSR
berlandi (Beaumont), 1949 (*Philoponidea*); Algeria, Egypt; Israel
dewitzii (Kohl), 1889 (*Philoponus*); n. Africa; Israel
   minima Schulthess, 1923 (*Philanthus*)
diversicornis (F. Morawitz), 1894 (*Philanthus*); sw. USSR
efflatouni (Mochi), 1939 (*Philanthus*); Egypt
espanoli (Giner Marí), 1947 (*Philoponus*); nw. Africa
ferruginea Radoszkowski, 1880; sw USSR
lyauteyi (Schulthess), 1923 (*Philanthus*); nw. Africa
maculata Radoszkowski, 1876; sw. USSR
pharaonum (Kohl), 1898 (*Philoponus*); Egypt
shestakovi (Gussakovskij), 1952 (*Philoponidea*); sw. USSR
simplicicornis (F. Morawitz), 1894 (*Philanthus*); Mongolia, new name for *Philanthus variegatus* (Morawitz) (Art. 59c)
   variegatus F. Morawtiz, 1889 (*Anthophilus*), nec *Philanthus variegatus* Spinola, 1838
sinaitica (Mochi), 1939 (*Philanthus*); Egypt, Israel
soikae (Mochi), 1939 (*Philanthus*); Egypt
spinulicollis (Mochi), 1939 (*Philanthus*); Egypt

splendida (Giner Marí), 1945 (*Philoponus*); nw. Africa
theryi (Vachal), 1893 (*Aphilanthops*); Algeria
tricolor (Giner Marí), 1945 (*Philoponoides*); nw. Africa

Tribe Cercerini

These moderately small to large wasps are relatively common and widespread. They ordinarily have colorful markings and coarsely sculptured integument. They frequent flowers and no doubt contribute considerably toward pollination. Since the known prey habits of other Philanthinae tend exclusively toward adult Hymenoptera, it is interesting that Cercerini mainly provision with adult Coleoptera, although Hymenoptera are used by a few species.

Systematics: The following are tribal characteristics: mandible often dentate, antennal sockets located above clypeus but less than two diameters apart and divided by a strongly raised ridge or carina, lateral ocelli normal, inner eye margins not notched and somewhat bowed toward each other, lateral clypeal brushes in male well developed and often broad, palpal formula 6-4, episternal sulcus absent, scrobal sulcus a broad, deep horizontal depression, precoxal sulcus present, precoxal lobes well defined, midcoxal cavities widely separated, hindfemur truncate or flattened apically (fig. 190), plantulae absent, marginal cell of forewing broadly rounded distally, submarginal cell II sometimes petiolate in front, first recurrent vein terminating on submarginal cell II and second recurrent on III, hindwing media diverging well beyond cu-a, jugal lobe occupying less than half length of anal area (fig. 184 B, D), a well defined pygidial plate present in both sexes, male without cerci, volsella not differentiated into digitus and cuspis (fig. 6 B), white or yellow markings usually present.

The most significant of the above characters are the valleylike scrobal sulcus, absence of an episternal sulcus, truncate hindfemur, widely separated midcoxae, rounded apex of the marginal cell in the forewing, distal divergence of the hindwing media, and undifferentiated volsellae. The combination of any three of these will distinguish the tribe, but the broad, deep scrobal sulcus by itself is unique.

The truncation of the hind leg is similar to that of *Odontosphex* and *Pseudoscolia*. It consists basically of three parts: (1) apicodorsal plate on the hindfemur, (2) apicoventral plate on the hindfemur, and (3) basoposterior plate on the hindtibia. One or more of these may be somewhat reduced.

Key to genera of Cercerini

Outer veinlet of submarginal cell III meeting marginal cell before its outer third (fig. 184 B, D); terga without median or submedian transverse depressions; cosmopolitan .... *Cerceris* Latreille, p. 575
Outer veinlet of submarginal cell III meeting cell well beyond its outer third (fig. 184 E, F); terga with median or submedian transverse depressions (fig. 190); N. America .... *Eucerceris* Cresson, p. 589

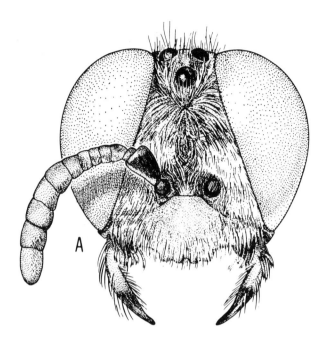

Pseudoscolia theryi

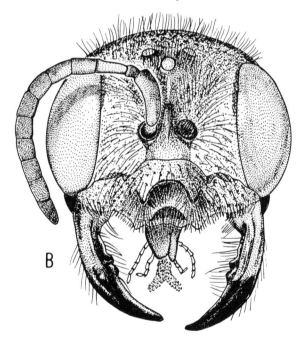

Cerceris bicornuta

FIG. 189. Facial portraits of females in the subfamily Philanthinae.

Genus Cerceris Latreille

Generic diagnosis: Clypeus of female often with distal teeth or other projections (fig. 189 B), ocellocular distance not reduced, subantennal sclerite nearly always defined by lines from antennal sockets through tentorial pits to clypeus, pronotum raised but often rather appressed to scutum, outer vein of submarginal cell III joining marginal cell at or before its outer third, submarginal cell II nearly always peiolate in front (fig. 184 B), first gastral segment usually forming a peduncle or sometimes a narrow petiole, terga without median or submedian transverse grooves, female sternum VI usually deeply cleft at apex; male pygidial plate not laterally denticulate, similar to that of female.

Geographic range: Cerceris, the largest genus in the family, is well represented on every continent and on many large islands. Over 200 species are Palearctic, nearly 200 each are Ethiopian and Oriental, over 100 are Neotropical, perhaps 75 are Nearctic, and more than 40 are Australasian. Together, with a few for which no locality is known, the total number of listed species is over 850.

Systematics: The generic names, *Nectanebus* Spinola and *Didesmus* Spinola, contain only a few species and seem to represent species groups at most. The trapezoidal submarginal cell II (fig. 184 D) and a few other features of *Nectanebus* are not of generic value in our opinion. Likewise, the somewhat bipetiolate gaster in *Didesmus* along with projections on the two basal sterna are remarkable but not generically significant. The use of *Apiraptrix* and *Apicerceris* as subgenera by some authors cannot be defended since both have the same type species as *Cerceris.* The customary petiolation of submarginal cell II has been used as a separation character from *Eucerceris,* but Scullen (1968) has shown that in the latter genus petiolation may often vary within a species according to sex. The nonpetiolate condition can probably be accepted as generalized. In this light, the specialized (petiolate) situation is more fully expressed in *Cerceris* than in *Eucerceris.*

An obvious close relationship exists between *Cerceris* and *Eucerceris,* but the formation of submarginal cell III in the latter, with its outer veinlet reaching the marginal cell near its end, is distinctive. In *Cerceris* this veinlet reaches the marginal cell somewhere near its middle but never beyond the distal third. Another separation character is the absence of transverse grooves near the middle of the terga in *Cerceris* and their presence in *Eucerceris.*

Brauns (1926) listed the important species characters among *Cerceris* of the Ethiopian Region as: shape of median part of clypeus, relative length of three basal flagellomeres, divergent or parallel inner orbits, ocellocular distance relationships, shape of prothorax, structure of mesopleuron and scutellum, shape and sculpture of propodeum, shape and sculpture of pygidial plate, presence and prominence of plate on sternum II, shape of gaster toward base, and coloration. He made four groups of the Ethiopian species on the basis of clypeal structure.

Tsuneki (1961) divided the *Cerceris* of northeastern Asia into two main groups by the presence or absence of the plate on sternum II. Females were subdivided primarily by clypeal characters, males mostly by sculpture, sternal pubescence, and antennal details. Van der Vecht (1964) divided the Javan species first by the presence or absence of a longitudinal carina on the inner side of the hindcoxa. Subdivisions utilized a variety of characters including, especially, the clypeus, structure of the first gastral segment and sternal pubescence. Scullen (1965) divided the North American species into five groups and a sixth miscellaneous lot. He emphasized clypeal characters in females and the relative width of the clypeal hair brushes in males. Also of importance were mandibular dentition, tegular form and sculpture, the mesopleural process, and pygidial shape. All authors agree that coloration is subject to great variation, more for some species than for others. A typical *Cerceris* face is shown in fig. 189 B.

The literature on *Cerceris* is extensive, as would be expected with such a large genus. A few of the more significant papers, with indication of the area covered, are: Schletterer (1887, 1889a, b) – Palearctic, Turner (1912a, c) – India, Australia, Brauns (1926) and Arnold (1931) – Ethiopian Region, Mochi (1939a) – Egypt, Beaumont (1951c) – north Africa, Giner Marí (1941) – Spain, Beaumont (1952c) – France, Tsuneki (1961, 1968b, 1970c) – eastern Asia, Tsuneki (1968a) – Australia, Krombein (1969) – Melanesia, Scullen (1965, 1972) – North America, van der Vecht (1964) – Java, Alayo (1968) – Cuba, and Fritz and Toro (1971) – South America. The papers by Tsuneki and Scullen are especially useful because they contain rather comprehensive and up-to-date bibliographies. Recent important contributions to synonymy have been made by Beaumont (various papers), Scullen (1965, 1972), van der Vecht (1961a), Empey (1969, 1971, 1972) and Fritz (1970b).

Biology: Scullen (1965) provided an extensive review of literature on habits of North American *Cerceris.* Scullen and Wold (1969) gave a complete listing of prey records in North America and a generalized biology for the genus. Linsley and MacSwain (1956) reviewed some prey habits for species occurring in Europe, Central and North America, and the Philippines. Evans (1971) made observations on nesting behavior of nine North American species. In Asia, extensive biological and review work has been done by Tsuneki (1965d). An additional source of Asian references is Yasumatsu and Watanabe (1964).

Workers in other regions include: South America: Janvier (1928); Malaya: Pagden (1934); Philippines: F. Williams (1919); Guyana: F. Williams (1928a); and Australia: Rayment (1947) and Evans and Matthews (1970).

It is feasible to give only a generalized account of *Cerceris* biology with unusual behavior being noted where appropriate. Females of *Cerceris* nest in aggregations in bare soil of a firm nature. Some nests have been reported from vertical sand banks (Krombein, 1959b). Most reported sites have been sand clay or gravel clay, generally on level ground, and often near dirt roads, paths, or structures such as brick walls or walls of buildings. Occasionally old nest entrances of *Cerceris* or other waps may be used to start a burrow. Under these circumstances females generally ignore the previous tunnels and prefer to make their own. In excavating a burrow on level ground the female forms a mound of loose soil about the entrance as she excavates the first several centimeters of the nest burrow. As burrowing progresses, some soil is not kicked to the surface but is pushed along the burrow to form an inner plug, the length of which must be breached whenever the female begins to construct brood cells. The burrow is generally a vertical tube from which horizontal cells are constructed or which, in itself, may become horizontal for some length. Excavated nests have ranged from 2.5 cm (Strandtmann, 1945) to an extreme depth of 1.3 meters (F. Williams, 1928a), but indications are that most nests average from 10 to 20 cm. It appears that after the main burrow is dug, female wasps begin to hunt prey which are then stored, mid-burrow, in the plug of soil remaining in the nest. After sufficient prey are collected, females dig and provision several cells and then repeat the collecting and storing process. Cells may be constructed from the nest entrance towards the interior (e.g., *hortivaga*) or from the end of the burrow towards the entrance (e.g., *arenaria*); (Tsuneki, 1965d). Cells usually number less than 10, and up to 15 or 20 prey may be stored per cell. Rayment (1947) recorded 40-60 prey per cell.

While the burrow entrance is not closed during hunting flights, the female generally closes the entrance from within as soon as she returns with prey. North American *Cerceris* provision exclusively with adult Coleoptera of the families Curculionidae, Buprestidae, Chrysomelidae, Tenebrionidae, and Bruchidae (Scullen and Wold, 1969). Prey in Australia was recorded by Evans and Matthews (1970) as Scarabaeidae and Chrysomelidae. Cerambycidae have been recorded from the Philippines (Williams, 1919). Coccinellidae and Curculionidae were given as prey in Chile by Jaffuel and Pirion (1926). In Europe and Japan, *C. rybyensis* is known to provision with various Apoidea, especially Halictidae (Tsuneki, 1965d). In Europe, a few other *Cerceris* are known to have this habit. According to Grandi (in Tsuneki, 1965d), *C. stratiotes* is believed to prey only upon the chalcid *Stilbula cynipiformis* (Rossi). Tsuneki has taken *Psen dahlbomi pacificus* from the brood cells of a *Cerceris*, which was provisioning with halictid and andrenid bees. He also reported a female of *C. pekingensis*, which took a species of *Pison* into her burrow. In Malaya, Pagden (1934) recorded a female of *C. langkasukae*, which was carrying a crabronine of the genus *Hingstoniola*, although her regular prey were buprestids.

Cerceris rubida is of interest because daughters from one nest may assist the mother in guarding the nest (Grandi, 1961).

Several dipterans have been recorded entering *Cerceris* burrows or chasing female wasps laden with prey. Noted flies were of the families Sarcophagidae and Anthomyiidae and included *Metopia leucocephala* (Rossi) (Hamm and Richards, 1930; Linsley and MacSwain, 1956; Byers, 1962), *M. campestris* (Fallén) (Byers, 1962), *Setulia grisea* Meigen and *Hammomyia grisea* Fallén (Hamm and Richards, 1930), and *Amobia floridensis* (Townsend) and *Senotainia trilineata* (Van der Wulp) (Linsley and MacSwain, 1956).

Hymenopterous parasites include the following Chrysididae: *Hedychrum nobile* Scopoli and *H. gerstaeckeri* Chevrier (Hamm and Richards, 1930; Móczár, 1967), *H. violaceum* Brullé (Byers, 1962), and *H. gerstaeckeri japonicum* Tsuneki (Tsuneki, 1965d). Two mutillids have been reared from the nests of *Cerceris: Dasymutilla coccineohirta* (Blake) (Linsley and MacSwain, 1956) and *Smicromyrme lewisi yanoi* Mickel (Tsuneki, 1965d).

Checklist of *Cerceris*

abacta Shestakov, 1915; Algeria
abdominalis (Fabricius), 1804 (*Philanthus*); nw. Africa; Spain
   hispanica Radoszkowski, 1869, nec Gmelin, 1790
   radoszkowskyi Schletterer, 1887
   ceballosi Giner Marí, 1941
abuensis Turner, 1912; w. India
acanthophila Cockerell, 1897; sw. U.S., Mexico
   cockerelli Viereck, 1903, new synonymy by R. Bohart
   huachuca Banks, 1947
acolhua Saussure, 1867; Mexico
acuta Radoszkowski, 1877; sw. USSR
adae Turner, 1936; w. Australia
adelpha Kohl, 1887; Korea
aemula Arnold, 1945; Madagascar
aemuloides Leclercq, 1962; Madagascar
aequalis Provancher, 1888; w. U.S.
   vicinoides Viereck and Cockerell, 1904
   psamathe Banks, 1912
   ssp. *bolingeri* Scullen, 1965; U.S.: Oregon, Nevada
   ssp. *idahoensis* Scullen, 1965; nw. U.S.
aerata Kazenas, 1972; sw. USSR
africanula Brauns, 1914; S. Africa
agnata Turner, 1912; Burma
alamos Scullen, 1972; nw. Mexico
alastoroides Turner, 1914; n. Australia
albicincta Klug, 1845; Sudan
albifrons F. Smith, 1856; "Africa"
albigena Cameron, 1905; S. Africa
albipicta F. Smith, 1873; India
albispinosa Arnold, 1952; Kenya
alboatra Walker, 1871; Egypt, Israel
albofasciata (Rossi), 1790 (*Vespa*); S. Palearctic Region
   albofasciata Thunberg, 1815 (*Philanthus*), nec Rossi, 1790
   luctuosa A. Costa, 1869
   navitatis F. Smith, 1873
   cribrata Mocsáry, 1879
   nativitatis Costa, F. Smith, 1856, emendation or lapsus
   ssp. *navigatrix* Strand, 1913; Taiwan
   ssp. *oasicola* Tsuneki, 1971; Mongolia
albolineata (Cameron), 1908 (*Trachypus*); Kenya
albopectoris Empey, 1973; s. Arabian penin.

alceste Mickel, 1918; U.S.: Nebraska
alcyone Arnold, 1951; Ethiopia
alexandrae F. Morawitz, 1889; Mongolia
amakosa Banks, 1926; Africa s. of equator
    vansoni Arnold, 1935
amathusia Beaumont, 1958; Cyprus
amatoria Arnold, 1931; Rhodesia
ameghinoi Brèthes, 1910; Argentina
andalgalensis Fritz & Toro, 1971; Argentina
andersoni Turner, 1918; Kenya
andina Brèthes, 1910; Argentina
andrei Gussakovskij, 1952; sw. USSR
angularis Cockerell, 1914; Philippines
angustata F. Morawitz, 1893; sw. USSR, Turkey
angustifrons Tsuneki, 1971; Mongolia
angustirostris Shestakov, 1922; sw. USSR
annulata (Rossi), 1790 (*Vespa*); Italy
    annulata Rossi, 1794 (*Crabro*), nec Fourcroy, 1785
annuligera Taschenberg, 1875; Brazil, Argentina
    diademata Holmberg, 1903
annulipes Brèthes, 1913; Argentina
ansa Shestakov, 1914; sw. USSR
    ssp. *manflava* Tsuneki, 1970; Mongolia
    flava Tsuneki, 1961, nec Mochi, 1939
antennata F. Morawitz, 1890; sw. USSR
antilope Tsuneki, 1971; Mongolia
antipodes F. Smith, 1856; Australia
apakensis Tsuneki, 1961; Mongolia
aquilina F. Smith, 1856; Brazil
arechavaletai Brèthes, 1909; Paraguay
arenaria (Linnaeus), 1758 (*Sphex*); Sweden
    xanthocephala J. Forster, 1771 (*Sphex*) (synonymy suggested by Morice, 1914)
    exulta Harris, 1776 (*Vespa*)
    petulans Harris, 1776 (*Vespa*)
    arenosa Gmelin, 1790 (*Vespa*)
    aurita Fabricius, 1794 (*Philanthus*)
    striolata Schletterer, 1887
    exultans Dalla Torre, 1897, lapsus
    ssp. *iberica* Schletterer, 1887; Spain
    ssp. *flavescens* Schletterer, 1889; sw. USSR
    ssp. *stecki* Schletterer, 1889; sw. USSR
    ssp. *nadigi* Shestakov, 1933; Morocco
    ssp. *schulzi* Beaumont, 1951; Tunisia
    ssp. *yanoi* Tsuneki, 1971; Japan, new name for *quinquecincta* Ashmead
    quinquecincta Ashmead, 1904, nec Fabricius, 1787
    ssp. *schnitnikovi* Kazenas, 1972; USSR: Kazakh S.S.R.
    ssp. *gotlandica* Eck, 1973; Sweden: Gotland
    ssp. *incognita* Eck, 1973; Germany, Poland
    ssp. *erlandssoni* Eck, 1973; Denmark, s. Sweden
argentifrons Guérin-Méneville, 1844; Madagascar
    ? *albotegulata* Arnold, 1945
argentina Brèthes, 1910; Argentina
argentosa Shestakov, 1912; sw. USSR
argia Mickel, 1916; sw. and centr. U.S., n. Mexico
ariadne Turner, 1912; Tibet
arizonella Banks, 1947; U.S.: Arizona
armata Beaumont, 1959; Israel

armaticeps Cameron, 1910; S. Africa
armigera Turner, 1917; Australia
    ssp. *rufofusca* Turner, 1936; Australia
arnoldi Brauns, 1926; Rhodesia
arrogans Arnold, 1931; S.-W. Africa
associa Kohl, 1898; s. Siberia
astarte Banks, 1913; ne. U.S.
aterrima Arnold, 1942; Urundi, Uganda, Tanzania
    flavobilineolata Giordani Soika, 1942
    rama Leclercq, 1955
    styrax Leclercq, 1955
atlacomulca Scullen, 1972; Mexico
atramontensis Banks, 1913; centr. and n.e. U.S.
    arbuscula Mickel, 1916
atriceps F. Smith, 1856; Brazil
augagneuri Arnold, 1945; Madagascar
aurantiaca F. Smith, 1873; Australia
australis Saussure, 1854; Tasmania
    nigrocincta F. Smith, 1856, nec Dufour, 1853
azteca Saussure, 1867; U.S.: Arizona; Mexico to Argentina
    seminigra Banks, 1947, nec Taschenberg, 1875
baluchistanensis Cameron, 1907; Pakistan
bambesae Empey, 1972; Zaire
banksi Scullen, 1965; e. U.S.
bannisteri Empey, 1971; S. Africa
bantamensis van der Vecht, 1964; Java
barnardi Brauns, 1926; S.-W. Africa
barrei Radoszkowski, 1893; sw. USSR
basalis F. Smith, 1856; Brazil
basimacula Cameron, 1907; Sikkim
basiornata Cameron, 1908; Kenya
beharensis Leclercq, 1967; Madagascar
bella Brèthes, 1910; Argentina
belli Turner, 1912; s. India
bellona Mercet, 1914; sw. Europe: Iberian Penin.
bequaerti Brauns, 1926; Zaire
    adulatrix Arnold, 1933
berenice Beaumont, 1966; Egypt
bicarinata Arnold, 1935; s. Africa
bicava Shestakov, 1917; Eritrea
bicincta Klug, 1835; s. Europe, n. Africa, Afghanistan, Mongolia
    interrupta Klug, 1835, nec Panzer, 1799
    sesquicincta Klug, 1835
    quadrimaculata Dufour, 1849
    variolosa A. Costa, 1869
    bimaculata Vogrin, 1954, nec Cameron, 1905
    ssp. *leucozonica* Schletterer, 1887; e. Europe, sw. USSR, Israel
    schulthessi Schletterer, 1889
bicornuta Guérin-Méneville, 1844; transcontinental U.S.; n. Mexico
    dufourii Guérin-Méneville, 1844
    venator Cresson, 1865
    curvicornis Cameron, 1890
    venatrix Schulz, 1906, emendation
    fidelis Viereck & Cockerell, 1904, new synonymy by R. Bohart

bicuspidata Arnold, 1931; S.-W. Africa
bidentula Maidl, 1926; Celebes
bifasciata Guérin-Méneville, 1835; India
bifurcata Empey, 1970; Zaire
bimaculata Cameron, 1905; e. India
    canaliculata Cameron, 1905, nec Pérez, 1895
    cameroni Schulz, 1906
binghami Turner, 1912; Burma
binodis Spinola, 1841; Centr. and S. America
    spinolae Dahlbom, 1844 (*Diamma*)
    viduata F. Smith, 1856
    singularis Brèthes, 1910
    guarani Strand, 1910
biplicatula Gussakovskij, 1938; e. China
bituberculata Tsuneki, 1963; Thailand
blakei Cresson, 1865; se. U.S.
    elegans F. Smith, 1856, nec Eversmann, 1849
    elegantissima Schletterer, 1887
boetica Pérez, 1913 (*Nectanebus*); Spain
boharti Scullen, 1965; sw. U.S.
bohartiana Fritz & Toro, 1971; Argentina
bolanica Turner, 1912; Pakistan
†*bolingeri* Scullen, 1972; sw. U.S., Mexico, nec Scullen, 1965
bonaerensis Holmberg, 1903; Argentina
borealis Mocsáry, 1901; s. Siberia
boschmai van der Vecht, 1964; Java
bothavillensis Brauns, 1926; S. Africa, Rhodesia, Lesotho
bothriophora Schletterer, 1887; Mexico
bougainvillensis Tsuneki, 1968; New Ireland, Bougainville
    ssp. *solomonis* Krombein, 1969; Solomons: Guadalcanal, Santa Ysabel
    ssp. *lavellensis* Krombein, 1969; Solomons: Vella Lavella, Giza
    ssp. *novogeorgica* Krombein, 1969; Solomons: New Georgia
boysi Turner, 1912; n. India
bracteata Eversmann, 1849; se. Europe
    penicillata Mocsáry, 1879
    mirabilis Shestakov, 1927
bradleyi Scullen, 1972; Mexico, Nicaragua
brandti Krombein, 1969; New Ireland
bredoi Arnold, 1958; Zambia
bridwelli Scullen, 1965; sw. U.S., nw. Mexico
brisbanensis Cockerell, 1930; Australia
bucculata A. Costa, 1860; s. Europe
    lunulata Thunberg, 1815 (*Philanthus*), nec Rossi, 1792
    bucculenta Ed. André, 1889, emendation or lapsus
bulawayoensis Brauns, 1926; Rhodesia
bulganensis Tsuneki, 1971; Mongolia
    fulviventris Tsuneki, 1971, nec Guérin-Méneville, 1844
bupresticida Dufour, 1841; Palearctic Region
    argentifrons Lepeletier, 1845, nec Guérin-Méneville, 1844
    frontalis F. Smith, 1856
    ? *brutia* A. Costa, 1869
    mixta Radoszkowski, 1877
    quadripunctata Radoszkowski, 1877
    ssp. *nigrina* Giner Marí, 1941; Spain
    ssp. *libyca* Beaumont, 1960; Libya
butleri Scullen, 1965; sw. U.S., nw. Mexico
cacaloapana Scullen, 1972; Mexico
caelebs Giner Marí, 1942; e. China
calida Turner, 1915; Australia
californica Cresson, 1865; w. and sw. U.S., nw. Mexico
    populorum Viereck and Cockerell, 1904
    garciana Viereck and Cockerell, 1904
    ferruginior Viereck and Cockerell, 1904
    argyrotricha Rohwer, 1908
    cognata Mickel, 1916
    denticularis Banks, 1917
    interjecta Banks, 1919
    arno Banks, 1947, new synonymy by R. Bohart
    calodera Banks, 1947
    illota Banks, 1947
    isolde Banks, 1947
callani Krombein, 1972; Trinidad
calochorti Rohwer, 1908; w. centr. U.S.
    ssp. *hidalgo* Scullen, 1972; sw. U.S., centr. Mexico
campestris Holmberg, 1903; Argentina
campicola Arnold, 1942; Zaire
campsomeroides Arnold, 1951; Ghana
canaliculata Pérez, 1895; n. Africa
    berlandi Giner Marí, 1941
    ssp. *tingitana* Beaumont, 1951; Morocco
    ssp. *palaestina* Beaumont, 1959; Israel, Syria
capito Lepeletier, 1845; nw. Africa
carinalis Pérez, 1905; Japan
carrizonensis Banks, 1915; U.S.: Texas
caucasica Shestakov, 1915; sw. USSR
cavagnaroi Scullen, 1972; El Salvador
celaeno Arnold, 1951; Ethiopia
celebensis Maidl, 1926; Celebes
cerussata Shestakov, 1918; sw. USSR
cerverae Giner Marí, 1941; Cuba
chacoana Brèthes, 1910; Argentina
changi Tsuneki, 1972; Taiwan
cheops Beaumont, 1951; n. Africa, Israel
cheskesiana Giner Marí, 1945; Cyprus, e. Mediterranean area
    cherkesiana Beaumont, 1947, emendation
chilensis Spinola, 1851; Chile
chilopsidis Viereck and Cockerell, 1904; sw. U.S.
chirindensis Arnold, 1932; Rhodesia, Zaire
chlorotica Spinola, 1838; Egypt, Israel
    lutea Taschenberg, 1875
    nilotica Schletterer, 1887
    ssp. *mateui* Giner Marí, 1945; Spanish Sahara
chromatica Schletterer, 1887; Egypt, Israel
    flava Mochi, 1939
chrysothemis Turner, 1912; Pakistan
circularis (Fabricius), 1804 (*Philanthus*); n. Africa
    clitellata Lepeletier, 1845
    elegans Dufour, 1853, nec Eversmann, 1849
    nobilis Radoszkowski, 1877

elegantula Shestakov, 1918
opulenta Morice, 1911
ssp. *dacica* Schletterer, 1887; s. Europe
ssp. *magnifica* Schletterer, 1887; e. Mediterranean area
ssp. *slovaca* Balthasar, 1953; Czechoslovakia
circumcincta Turner, 1912, n. India
citrinella F. Smith, 1856; Siberia
cleomae Rohwer, 1908; U.S.: Colorado
clypearis Saussure, 1887; Madagascar
clypeata Dahlbom, 1844; e. N. America
    imitator Cresson, 1865, nec F. Smith, 1856
    imitatoria Schletterer, 1887
    chryssipe Banks, 1912
    clymene Banks, 1912
    zobeide Brimley, 1929
    zosma Brimley, 1929
ssp. *tepaneca* Saussure, 1867; Mexico
    thermophila Schletterer, 1887
ssp. *prominens* Banks, 1912; e. U.S.
    alaope Banks, 1912
ssp. *gnarina* Banks, 1913; centr. U.S.
ssp. *dakotensis* Banks, 1915; n. centr. U.S.
clytia Beaumont, 1959; n. Africa, Israel
cochisi Scullen, 1965; sw. U.S., nw. Mexico
coelicola Giner Marí, 1942; e. China, Taiwan
colon (Thunberg), 1815 (*Philanthus*); no locality data
colorata Schletterer, 1889; sw. USSR
comberi Turner, 1912; Pakistan
compacta Cresson, 1865; e. and sw. U.S., Mexico
    aureofacialis Cameron, 1890
    solidaginis Rohwer, 1908
    belfragei Banks, 1917
compar Cresson, 1865; e. U.S., centr. Mexico
    jucunda Cresson, 1872
    carolina Banks, 1912
    catawba Banks, 1912
ssp. *geniculata* Cameron, 1890; sw. U.S. to Guatemala
    feralis Cameron, 1890
ssp. *orestes* Banks, 1947; sw. U.S., Mexico
ssp. *albinota* Scullen, 1972; sw. U.S., n. Mexico
completa Banks, 1919; U.S.: California, Oregon
    percna Scullen, 1965, new synonymy by R. Bohart
compta Turner, 1912; Pakistan
concinna Brullé, 1839; Canary Is.
confraga Shestakov, 1914; sw. USSR
confusa Giner Marí, 1942; e. China
congesta Giordani Soika, 1942; Somalia
conica Shestakov, 1922; sw. USSR
conifrons Mickel, 1916; w. U.S., n. Mexico
contigua (Villers), 1789 (*Sphex*); Spain, new name for *Sphex quinquecinctus* (Fabricius), (Art. 59c)
    quinquecinta Fabricius, 1787 (*Crabro*), nec *Sphex quinquecinctus* (Fabricius), 1775 (*Tiphia*), now in *Myzinum*
    cingulata Gmelin, 1790 (*Vespa*), new name for *Crabro quinquecinctus* Fabricius, 1787 in *Vespa*, nec *Vespa quinquecincta* Fabricius, 1787
    quinquelineata Turton, 1802 (*Vespa*), new name for *quinquecincta* Fabricius, 1787
    ? *quinquecincta* Schrank, 1802 (*Crabro*), nec Fabricius, 1787
convergens Viereck and Cockerell, 1904; w. U.S.
    rinconis Viereck and Cockerell, 1904
    hesperina Banks, 1917
    pudorosa Mickel, 1918
    snowi Banks, 1919
cooperi Scullen, 1972; s. Mexico to Costa Rica
cordillera Perez and Toro, 1973; Chile
coreensis Tsuneki, 1961; Korea
cornigera (Gmelin), 1790 (*Vespa*); India, new name for *Vespa cornuta* (Fabricius), (Art. 59c)
    cornuta Fabricius, 1787 (*Crabro*), nec *Vespa cornuta* Linnaeus, 1758, now in *Synagris*
    wroughtoni Cameron, 1890
costarica Scullen, 1972; s. Mexico to Brazil
ssp. *mitla* Scullen, 1972; s. Mexico
crandalli Scullen, 1965; sw. U.S., nw. Mexico
crassidens Cameron, 1902; Borneo
crepitans Leclercq, 1967; Madagascar
cribrosa Spinola, 1841; Mexico, Centr. and S. America
    subpetiolata Saussure, 1867
    pullata F. Smith, 1873
    albimana Taschenberg, 1875
cristovalensis Krombein, 1969; Solomons: San Cristoval
crotonella Viereck & Cockerell, 1904; sw. U.S.
cuernavaca Scullen, 1972; centr. Mexico
culcullata Bingham, 1912; n. Australia
cupes Shestakov, 1918; sw. USSR
curvitarsis Schletterer, 1887; S. Africa
    oneili Cameron, 1905
cuthbertsoni Arnold, 1946; Rhodesia
cyclops Krombein, 1969; w. New Guinea
darrensis Cockerell, 1930; Australia
decolorata Arnold, 1931; Kenya, Zaire
decorata Brèthes, 1910; Argentina
dedariensis Turner, 1936; Australia
dejecta Arnold, 1931; Kenya
dentata Cameron, 1890; e. India
denticulata Schletterer, 1889; sw. USSR
dentifrons Cresson, 1865; ne. U.S.
dentiventris Arnold, 1945; Madagascar
deserta Say, 1824; n. centr. and n.e. U.S.; Canada
    fulvipes Cresson, 1865, nec Eversmann, 1849
    fulvipediculata Schletterer, 1887
deserticola F. Morawitz, 1890; sw. USSR, Turkey
diabolica Giner Marí, 1942; e. China
†*dichroa* Brèthes, 1909; Uruguay, nec Dalla Torre, 1890 (now in *Eucerceris*)
dilatata Spinola, 1841; sw. U.S. to Argentina
    maximiliani Saussure, 1867
    contracta Taschenberg, 1875
    caridei Holmberg, 1903
    vigilii Brèthes, 1910
    divisa Brèthes, 1910
    olymponis Strand, 1910
    semiatra Banks, 1947
ssp. *chisosensis* Scullen, 1965; sw. U.S., n. centr. Mexico

diodonta Schletterer, 1887; S. Africa
   mitrata Bingham, 1902
   melanospila Cameron, 1905
   jansei Cameron, 1910
  ssp. *barbifera* Bischoff, 1911; Zaire, Uganda
   bagandarum Turner, 1918
  ssp. *sodalis* Turner, 1918; Kenya
dione Fritz, 1961; Argentina
discrepans Brauns, 1926; S. Africa
  ssp. *perplexa* Arnold, 1931; S.-W. Africa
dispar Dahlbom, 1845; se. Europe, e. Mediterranean area
   cypriaca Giner Marí, 1945
dissecta (Fabricius), 1798 (*Philanthus*); India
dissona Arnold, 1955; S.-W. Africa
distinguenda Shestakov, 1922; China
doderleini Schulz, 1905; n. Africa
dogonensis Krombein, 1974, Papua
dominicana Brauns, 1926; S. Africa
dondoensis Brauns, 1926; Mozambique, Rhodesia
   ephippiorhyncha Arnold, 1940
dorsalis Eversmann, 1849; sw. USSR
   excavata Schletterer, 1889
   caspica F. Morawitz, 1891, nec Gmelin, 1790
  ssp. *solskii* Radoszkowski, 1877; sw. USSR: Uzbek S.S.R.; Afghanistan, Mongolia, e. China
   solskyi Schletterer, 1887, emendation
   crassicollis Tsuneki, 1968
dowi Tsuneki, 1968; Philippines
downesivora Turner, 1912; Burma
dreisbachi Scullen, 1972; s. centr. Mexico
duchesnei Arnold, 1945; Madagascar
duisi Scullen, 1972; Mexico
duplicata Brèthes, 1910; Argentina
durango Scullen, 1972; centr. Mexico
dusmeti Giner Marí, 1941; Spain
eburneofasciata Brauns, 1926; S. Africa
echo Mickel, 1916; U.S. w. of Mississippi River, n. centr. Mexico
  ssp. *atrata* Scullen, 1965; e. U.S.
edolata Shestakov, 1912; sw. USSR
egena Arnold, 1931; Rhodesia, Malawi
  ssp. *mombasae* Arnold, 1942; Kenya
electra Arnold, 1951; Ghana
elegans Eversmann, 1849; sw. USSR
elizabethae Bingham, 1897; Burma
emeryana Gribodo, 1894; Mozambique
  ssp. *varilineata* Cameron, 1905; S. Africa
  ssp. *bechuana* Brauns, 1926; Botswana
  ssp. *multicolor* Arnold, 1946; Lesotho
enodens Brèthes, 1910; Argentina
ephippium Turner, 1912; Pakistan
errata Shestakov, 1918; sw. USSR
erronea Giner Marí, 1942; e. China
eryngii Marquet, 1875; s. Europe, e. Mediterranean area, Iran
   haueri Schletterer, 1887
   robusta Shestakov, 1915
  ssp. *ponantina* Beaumont, 1970; Morocco
   occidentalis Giner Marí, 1941, nec Saussure, 1867

erynnis Arnold, 1931; Malawi
erythrogaster Kazenas, 1972; USSR: Kazakh S.S.R.
erythropoda Cameron, 1890; Mexico
erythrosoma Schletterer, 1887; Mozambique, S. Africa
   pictiventris Gerstaecker, 1857, nec Dahlbom, 1845
   ornativentris Cameron, 1905
   rebaptizata Schulz, 1906
erythrospila Cameron, 1910; S. Africa
erythroura Cameron, 1908; Tanzania
escalerae Giner Marí, 1941; Morocco
eucharis Schletterer, 1887; Syria
euchroma Turner, 1910; Australia
eugenia Schletterer, 1887; sw. USSR, new name for *orientalis*
   orientalis Mocsáry, 1883, nec F. Smith, 1856
   mocsaryi Kohl, 1888
eulalia Brauns, 1926; S. Africa
  ssp. *transkeica* Empey, 1971; S. Africa
euryanthe Kohl, 1888; se. Europe, Turkey, sw. USSR
eustylicida Williams, 1928; Guyana
evansi Scullen, 1972; Mexico
eversmanni Schulz, 1912; Europe, sw. USSR
   cornuta Eversmann, 1849, nec Fabricius, 1787
  ssp. *clypeodentata* Tsuneki, 1971; Mongolia
†*excavata* Cameron, 1902; Borneo, nec Schletterer, 1889
expleta Brèthes, 1910; Argentina
expulsa Turner, 1920; e. India
faceta Arnold, 1951; Senegal
falcifera Tsuneki, 1961; e. China
farri Scullen, 1971; Jamaica
fastidiosa Turner, 1912; Pakistan
femurrubrum Viereck & Cockerell, 1904; sw. U.S., n. Mexico
   thione Banks, 1947, new synonymy by R. Bohart
   athene Banks, 1947, new synonymy by R. Bohart
   rossi Scullen, 1972, new synonymy by R. Bohart
ferocior Turner, 1912; Borneo
ferox F. Smith, 1856; Sumatra, Malaya
   annandali Bingham, 1903
ferruginea Brèthes, 1910; Argentina
festiva Cresson, 1865; Cuba
   gratiosa Schletterer, 1887
fimbriata (Rossi), 1790 (*Crabro*); Egypt, e. Mediterranean area to Afghanistan
   sexpunctata Fabricius, 1793 (*Philanthus*)
   scabra Fabricius, 1798 (*Mellinus*)
   signata Klug, 1835
  ssp. *pallidopicta* Radoszkowski, 1877; sw. USSR, Mongolia
   ?*polita* Schletterer, 1889
? ssp. *cogens* Kohl, 1915; sw. USSR
fingo Brauns, 1926 (Plate 43); S. Africa
finitima Cresson, 1865; U.S.: widespread; Mexico; Centr. America
   nigroris Banks, 1912
   vierecki Banks, 1947, new synonymy by R. Bohart
   citrina Scullen, 1965, new synonymy by R. Bohart
   morelos Scullen, 1972, new synonymy by R. Bohart
fischeri Spinola, 1838; n. Africa

histrio Dahlbom, 1845
? *contigua* Walker, 1871, nec Villers, 1789
fitzgeraldi Empey, 1973; Trucial Oman Coast, Oman
flavicornis Brullé, 1833; s. Europe, w. Asia
   conigera Dahlbom, 1845
   rostrata Marquet, 1875, nec F. Smith, 1873
   antoniae Fabre, 1879
flavida Cameron, 1890; Mexico
flavifrons F. Smith, 1856; w. Africa
flavilabris (Fabricius), 1793 (*Hylaeus*); s. Europe, e. Mediterranean area, nw. Africa
   aurita Latreille 1804-1805, ? nec Fabricius, 1794
   ferreri Vander Linden, 1829
   bidentata Lepeletier, 1845, nec Say, 1824
   laminata Eversmann, 1849
   propinqua A. Costa, 1860
   scutellaris A. Costa, 1869
   ferreroi Schulz, 1906, emendation
   ssp. *insularis* F. Smith, 1856; Sicily
flaviventris Vander Linden, 1829; Spain
   quilisi Giner Marí, 1941
   ssp. *pardoi* Giner Marí, 1941; Morocco
   ssp. *lusitana* Beaumont, 1953; Portugal
flavocostalis Cresson, 1865; Cuba
flavofasciata F. Smith, 1908; e. U.S.
   natallena Brimley, 1927
   ssp. *floridensis* Banks, 1915; se. U.S.
flavofemorata Arnold, 1942; Ethiopia, Kenya, Rhodesia, S. Africa
   ponderosa Arnold, 1931, nec Brèthes, 1920
   maia Arnold, 1951
flavomaculata Cameron, 1890; Costa Rica
flavonasuta Arnold, 1951; Ethiopia
flavopicta F. Smith, 1856; India
flavoplagiata Cameron, 1905; Himalaya Mts.
flavotrochanterica Rohwer, 1912; Mexico
fletcheri Turner, 1912; e. India
fluvialis F. Smith, 1873; Australia
fodiens Eversmann, 1849; sw. USSR
   charusini F. Morawitz, 1891
   ssp. *shur* Shestakov, 1922; sw. USSR
formicaria Eschscholtz, 1822; Philippines: Luzon
formidolosa Saussure, 1890 (pl. 19, fig, 9); Madagascar
formosa Dahlbom, 1845; S. Africa
   sumptuosa Arnold, 1931
   marleyi Arnold, 1949
   ssp. *nigrifemur* Arnold, 1955; S. Africa
   ssp. *hoeveldensis* Empey, 1971; S. Africa
formosana Strand, 1913; Taiwan
   ssp. *klapperichi* Giner Marí, 1942; China
   dubitabilis Giner Marí, 1942
forticula Arnold, 1955; Zambia
fortinata Cameron, 1902; e. India
   aureobarba Cameron, 1905
fortin Scullen, 1972; Mexico
fraterna Beaumont, 1959; Israel
freymuthi Radoszkowski, 1877; sw. USSR
   freimuthi Shestakov, 1918, emendation
frigida Mocsáry, 1901; s. Siberia

froggatti Turner, 1911; Australia
frontata Say, 1823; sw. and s. centr. U.S.; n. Mexico
   occidentalis Saussure, 1867
   texensis Saussure, 1867
   raui Rohwer, 1920, new synonymy by R. Bohart
fuliginosa F. Smith, 1856; Celebes
fulva Mocsáry, 1883; sw. USSR
fulvipes Eversmann, 1849; s. Siberia
   shaman Shestakov, 1922
fulviventris Guérin-Méneville, 1844; Senegal to Ethiopia
   bicolor F. Smith, 1856
   fossor F. Smith, 1856
fumipennis Say, 1837; centr. and e. U.S.
   cincta Dahlbom, 1844
   unicincta Taschenberg, 1875
fumosipennis Strand, 1910; Paraguay
furcata F. Morawitz, 1890; sw. USSR
furcifera Schletterer, 1887; Paraguay
fuscina Shestakov, 1914; sw. USSR
gaetula Beaumont, 1951; Morocco
galathea Beaumont, 1959; Israel, Syria
gallieni Arnold, 1945; Madagascar
   ssp. *bekiliensis* Arnold, 1945; Madagascar
gandarai Rohwer, 1912; sw. U.S., Mexico
garleppi Schrottky, 1911; Peru
gaudebunda Holmberg, 1903; Argentina
gaullei Brèthes, 1920; Argentina
gayi Spinola, 1851; Chile, Argentina
geboharti Tsuneki, 1969; Ryukyus: Okinawa
   boharti Tsuneki, 1968, nec Scullen, 1965
gemmina Shestakov, 1927; USSR: "Vladimirskoje"
geneana A. Costa, 1869; Sardinia
gibbosa Sickman, 1894; China
gigantea Gistl, 1857; Yugoslavia
gilberti Turner, 1916; Australia
gilesi Turner, 1910; Australia
gineri Beaumont, 1951; Morocco
? *globulosa* (Fourcroy), 1785 (*Vespa*); locality ?
goddardi Cockerell, 1930; Australia
gomphocarpi Brauns, 1926; S. Africa
grana Shestakov, 1922; China
grandis Banks, 1913; sw. U.S., nw. Mexico
graphica F. Smith, 1873; Mexico to centr. and sw. U.S.
   hebes Cameron, 1890
   macrosticta Viereck and Cockerell, 1904, new synonymy by R. Bohart
   ampla Banks, 1912
grata Arnold, 1931; S. W. Africa
greeni Turner, 1912; s. India
guichardi Beaumont, 1951; Algeria, Morocco
guigliae Giordani Soika, 1942; Eritrea
hackeriana Cockerell, 1930; Australia
haematina Kohl, 1915; locality unknown.
halone Banks, 1912; e. U.S.
   architis Mickel, 1916
   alacris Mickel, 1918
   salome Banks, 1923
   shermani Brimley, 1928
hamiltoni Arnold, 1931; Kenya

haramaiae Arnold, 1951; Ethiopia
harbinensis Tsuneki, 1961; ne. China
hatuey Alayo, 1968; Cuba
hausa Arnold, 1931; Nigeria
herbsti Empey, 1971; S.-W. Africa
heterospila Cameron, 1910; Rhodesia, S. Africa
    macololo Brauns, 1926
  ssp. *capensis* Arnold, 1931; S. Africa
    griqua Arnold, 1942
hexadonta Strand, 1913; Taiwan
hilaris F. Smith, 1856; n. India
    himalayensis Bingham, 1898
    simlaensis Cameron, 1905
hilbrandi van der Vecht, 1964; Java
hildebrandti Saussure, 1891; Madagascar
histerisnica (Spinola), 1838 (*Nectanebus*); Egypt, Israel, new combination by Bohart and Menke
    fischeri Spinola, 1838 (*Nectanebus*), nec Spinola, 1838
  ssp. *algeriensis* (Schulz), 1904 (*Nectanebus*); Algeria
histrionica Klug, 1845; Spanish Sahara to Egypt, Ethiopia, Israel
    syrkuti Dahlbom, 1845
    eatoni Morice, 1911
    fluxa Kohl, 1915
    honorei Mochi, 1939
    saharica Giner Marí, 1945
hohlbecki Shestakov, 1914; sw. USSR
    fragosa Kohl, 1915
hokkanzana Tsuneki, 1961; Korea
holconota Cameron, 1905; S. Africa
  ssp. *labiosa* Empey, 1971; S. Africa, Rhodesia
  ssp. *oculata* Empey, 1971; S. Africa
hortivaga Kohl, 1880; Palearctic Region
  ssp. *amamiensis* Tsuneki, 1961; Ryukyus
hora Arnold, 1931; S.-W. Africa
huastecae Saussure, 1867; Mexico
hurdi Scullen, 1972; sw. U.S., Mexico to Nicaragua
hypocritica Brauns, 1926; S. Africa
†*iberica* Schletterer, 1889; sw. Europe, nw. Africa, nec Schletterer, 1887
icta Shestakov, 1918; sw. USSR
iliensis Kazenas, 1972; USSR: Kazakh S.S.R.
illustris Arnold, 1931; Rhodesia
imitator F. Smith, 1856; Brazil
    imitatrix Schulz, 1906; emendation
impercepta Beaumont, 1950; Hungary
imperialis Saussure, 1867; s. Mexico to Costa Rica
    exsecta F. Smith, 1873
    pilosa Cameron, 1890
inara Beaumont, 1967; Turkey
inconspicua Arnold, 1931; Rhodesia, Malawi, S. Africa
indica (Thunberg), 1815 (*Philanthus*); "India Orientali"
inexorabilis Turner, 1912; Pakistan
inexpectata Turner, 1908; Australia
infumata Maidl, 1926; Celebes
iniqua Kohl, 1894; Sierra Leone
  ssp. *cratocephala* Cameron, 1908; Ethiopia to S. Africa
  ssp. *deceptrix* Brauns, 1926; S. Africa, Rhodesia

  ssp. *libitina* Brauns, 1926; S. Africa
  ssp. *arida* Arnold, 1931; Nigeria, Mauretania
insignita Arnold, 1951; Ethiopia
    opulenta Arnold, 1951, nec Morice, 1911
insolita Cresson, 1865; e. U.S., n. Mexico
    intractibilis Mickel, 1916
  ssp. *otomia* Saussure, 1867; s. Mexico to Argentina
    otomita Dalla Torre, 1897
  ssp. *chiriquensis* Cameron, 1890; s. Mexico to Ecuador
  ssp. *albida* Scullen, 1965; sw. U.S., s. Mexico
  ssp. *atrafemori* Scullen, 1965; sw. U.S.
  ssp. *cortezi* Scullen, 1972; s. centr. Mexico
  ssp. *panama* Scullen, 1972; Mexico to Panama
instabilis F. Smith, 1856; Oriental Region
    velox F. Smith, 1875
integra F. Morawitz, 1894; sw. USSR
interrupta (Panzer), 1799 (*Philanthus*); Europe, Asia
    labiata Fabricius, 1793 (*Crabro*), nec Olivier, 1792
    brevirostris Lepeletier, 1845
  ssp. *peninsularis* Mercet, 1903; Iberian penin., sw. France
interstincta (Fabricius), 1798 (*Philanthus*); Sri Lanka, India
    humbertiana Saussure, 1867
    emortualis Saussure, 1867
  ssp. *viscosa* F. Smith, 1875; e. India
    rufinodis F. Smith, 1875
    rufinodula Dalla Torre, 1897
intricata F. Smith, 1856; Mexico to Argentina
    simplex F. Smith, 1856
    vulpina F. Smith, 1856
    larvata Taschenberg, 1875
    affumata Schletterer, 1887
    melanogaster Holmberg, 1903
    elephantinops Holmberg, 1903
    dissita Holmberg, 1903
    catamarcensis Schrottky, 1909
    cisandina Brèthes, 1913
invalida Kohl, 1906; Aden
invita Turner, 1912; Sumatra
irene Banks, 1912; centr. U.S., ne. Mexico
irwini Scullen, 1972; Costa Rica, El Salvador
isis Arnold, 1931; Rhodesia
jackal Brauns, 1926; S. Africa
jakowleffii Kohl, 1898; s. Siberia
    jakovlevi Gussakovskij, 1938, emendation
japonica Ashmead, 1904; Japan
    harmandi Pérez, 1905
    interrupta Matsumura, 1912, nec Panzer, 1799
jatahyna Brèthes, 1920; Brazil
kalaharica Bischoff, 1913; S. Africa
  ssp. *ovambo* Empey, 1972; S.-W. Africa
kansuensis Gussakovskij, 1934; China
karimuiensis Krombein, 1969; ne. New Guinea
kasachstanica Kazenas, 1972; USSR: Kazakh S.S.R.
kaszabi Tsuneki, 1971; Mongolia
katangae Brauns, 1914; Zaire, Tanzania, Zambia
kazenasi Pulawski; USSR: Kazakh S.S.R., new name by W. Pulawski for *pulawskii* Kazenas

pulawskii Kazenas, 1972, nec Tsuneki, 1971
kedahae Pagden, 1934; Malaya, Sumatra, Java
kennicottii Cresson, 1865; U.S., Mexico
    eriogoni Viereck & Cockerell, 1904
  ssp. *zapoteca* Saussure, 1867; sw. U.S., Mexico, Centr. America
    montivaga Cameron, 1890
    beali Scullen, 1965
  ssp. *smithiana* Cameron, 1890; s. Mexico, Centr. America
  ssp. *bakeri* Cameron, 1904; Nicaragua, Guatemala
    chinandegaensis Cameron, 1904
    iresinides Rohwer, 1914
kilimandjaroensis Cameron, 1908; Ethiopia, Uganda, Tanzania
kirbyi Bingham, 1897; Burma
klugii F. Smith, 1856; Egypt
    annulata Klug, 1845, nec Rossi, 1790
    klugi Kirchner, 1867, nec F. Smith, 1856
    klugii Schletterer, 1887, nec F. Smith, 1856
koala Tsuneki, 1968; ne. Australia: Prince of Wales Island
kobrowi Brauns, 1926; S. Africa
kohlii Schletterer, 1887; sw. USSR
    perdita Kohl, 1898
kokuevi Shestakov, 1912; sw. USSR
    kokujevi Shestakov, 1918, emendation
koma Tsuneki, 1961; Korea
koryo Tsuneki, 1961; Korea
koshantshikovi Shestakov, 1914; sw. USSR
koulingensis Tsuneki, 1968; China
kozlovi Shestakov, 1922; China
krombeini Scullen, 1965; U.S.: Arizona
krugi Dewitz, 1881; Puerto Rico
kwangtsehiana Giner Marí, 1942; e. China, Taiwan
labeculata Turner, 1908; Australia
?*labiata* (Olivier), 1792 (*Crabro*); France
lacinia Tsuneki, 1961; ne. China
laeta (Fabricius), 1793 (*Philanthus*); Spain
laevigata F. Smith, 1856; Hispaniola
    levigata Dalla Torre, 1897, emendation
lama Turner, 1912; Tibet
lamarquensis Fritz & Toro, 1971; Argentina
langkasukae Pagden, 1934; Malaya
languida Cameron, 1905; S. Africa
lateridentata Arnold, 1945; Madagascar
laterifurcata Empey, 1973; Saudi Arabia, Aden, Yemen
laterimaculata Empey, 1973; Aden
lateriproducta Mochi, 1939; Egypt
latibalteata Cameron, 1904; n. India
latiberbis Tsuneki, 1968; ne. Australia: Prince of Wales Island
laticincta Lepeletier, 1845 (♀ only); nw. Africa
    atlantica Schletterer, 1887
latidens Cameron, 1902; Java, Borneo
latifrons Bingham, 1902; S. Africa
  ssp. *sedula* Arnold, 1940; Rhodesia
lativentris Gussakovskij, 1938; e. China

latro F. Smith, 1856; Brazil
laxata Shestakov, 1918; sw. USSR
lepcha Cameron, 1905; e. India
lepida Brullé, 1839; Canary Is.
leucochroa Schletterer, 1887; Sudan
lobaba W. F. Kirby, 1900; Socotra
longilabris Arnold, 1946; Zambia
longitudinalis Giordani Soika, 1942; Somalia
longiuscula Arnold, 1951; Ghana
luchti van der Vecht, 1964; Java
lunata A. Costa, 1869; s. Europe, sw. USSR
  ?*affinis* Rossi, 1792 (*Crabro*)
    dorsalis Dufour, 1849, nec Eversmann, 1849
    pyrenaica Schletterer, 1887
  ssp. *funerea* A. Costa, 1869; Sicily
  ssp. *albicolor* Shestakov, 1922; sw. USSR
  ssp. *tenebricosa* Giner Marí, 1941; Morocco
lunigera Dahlbom, 1845; S. Africa
    whiteana Cameron, 1905
    holconotula Brauns, 1926 (pl. 44)
†*lunulata* (Rossi), 1792 (*Crabro*); Italy, nec Fourcroy, 1785
lutzi Scullen, 1972; Panama
luxuriosa Dahlbom, 1845; Egypt, Ethiopia
    subimpressa Schletterer, 1887
luzonensis Crawford, 1910; Philippines
    spinigera Rohwer, 1919
  ssp. *fukaii* Rohwer, 1911; Taiwan
    superflua Strand, 1913
lynchii Brèthes, 1910; Argentina
lynx Brèthes, 1913; Argentina
macalanga Brauns, 1926; Rhodesia
macswaini Scullen, 1965; sw. U.S.
macula (Fabricius), 1804 (*Philanthus*); s. Africa
maculata Radoszkowski, 1877; sw. USSR
maculicrus Beaumont, 1967; Turkey, Syria
manchuriana Tsuneki, 1961; ne. China
mandibularis Patton, 1881; e. U.S.
manifesta Arnold, 1952; Kenya
maracandica Radoszkowski, 1877; sw. USSR
    ignaruris Kohl, 1915
marcia Nurse, 1903; n. India
margaretella Rohwer, 1915; Puerto Rico
margarita Beaumont, 1966; Egypt
marginata F. Smith, 1856; Brazil
marginula Dalla Torre, 1897; s. Mexico to Panama
    marginata Cameron, 1890, nec F. Smith, 1856
maritima Saussure, 1867; Mauritius
martialis Giner Marí, 1942; e. China
mastogaster F. Smith, 1856; India
mazimba Brauns, 1926; Mozambique, Rhodesia, Mali
media Klug, 1835; Europe, e. Mediterranean area
    capitata F. Smith, 1856
meditata Shestakov, 1918; sw. USSR
megacephala Brèthes, 1913; Argentina
melaina Turner, 1912; w. India
melanthe Banks, 1947; e. and sw. U.S.
    nitida Banks, 1913, nec Wesmael, 1852, new synonymy by H. Scullen

mellicula Turner, 1912; Pakistan
mendesensis Brèthes, 1920; Brazil
mendozana Brèthes, 1913; Argentina
    ssp. *melanopus* Brèthes, 1913; Argentina
menkei Fritz & Toro, 1971; Argentina
merope Arnold, 1951; Ghana
merredinensis Turner, 1936; Australia
mesopotamica Brèthes, 1913; Argentina
metatarsalis Turner, 1926; Thailand
mexicana Saussure, 1867; Mexico
micheneri Scullen, 1972; Mexico
micropunctata Shestakov, 1922; Siberia
militaris Dahlbom, 1844; S. Africa (possibly erroneous locality)
millironi Krombein, 1969; Solomons: Guadalcanal
    ssp. *tulagiensis* Krombein, 1969; Solomons: Florida
    ssp. *malaitensis* Krombein, 1969; Solomons: Malaita
mimica Cresson, 1872; s. U.S.: Transcontinental; Mexico
    esau Schletterer, 1887
    minima Schletterer, 1887, lapsus
    englehardti Banks, 1947
minax Mickel, 1918; w. U.S.
minuscula Turner, 1910; Australia
    ssp. *sculleniana* Krombein, 1969; New Guinea, Solomons, New Ireland
    ssp. *stanleyensis* Krombein, 1969; New Guinea
    ssp. *korovensis* Krombein, 1969; Solomons: Shortland
minutior Maidl, 1924; Sudan
minutissima Maidl, 1924; Sudan
misoolensis Krombein, 1969; W. New Guinea: Misool
mochiana Giordani Soika, 1942; Ethiopia
moczari Tsuneki, 1971; Mongolia
†*modesta* F. Smith, 1873; Brazil, nec F. Smith, 1856
moesta De Stefani, 1884; Sicily
    maesta Dalla Torre, 1897, emendation
moestissima Guiglia, 1941; Ethiopia
moggionis Arnold, 1951; Ethiopia
monocera Kohl, 1898; Senegal, Liberia
montealban Scullen, 1972; Mexico
montezuma Cameron, 1890; Mexico
monticola Arnold, 1931; Malawi, Rhodesia
morata Cresson, 1872; s. centr. U.S., Mexico
    nasica Viereck & Cockerell, 1904
morawitzi Mocsáry, 1883, sw. USSR
    ssp. *mongolensis* Tsuneki, 1971; Mongolia
    ssp. *changatai* Tsuneki, 1971; Mongolia
mordax Krombein, 1969; New Guinea: Papua
moroderi Giner Marí, 1941; Spain
morrae Strand, 1910; Paraguay, Argentina
    paraguayana Strand, 1910
        nigra Brèthes, 1910, nec Ashmead, 1900
    atra Scullen, 1965
morula Giordani Soika, 1942; Ethiopia
mossambica Gribodo, 1895; Mozambique, S. Africa
moyanoi Holmberg, 1903; Argentina
multiguttata Turner, 1908; Australia
multipicta F. Smith, 1873; Ethiopian Region
    simoni Buysson, 1897
    speculata Shestakov, 1917

murgabica Radoszkowski, 1893; sw. USSR
mutabilis Arnold, 1931; S. Africa, Lesotho
    empeyi Arnold, 1962
nagamasa Tsuneki, 1963; Thailand
nana Shestakov, 1918; India
nasidens Schletterer, 1887; S. Africa
    karooensis Brauns, 1914
    ssp. *obscura* Schletterer, 1887; S. Africa
        africana Cameron, 1905
natalensis Saussure, 1867; S. Africa
    morosula Brauns, 1914
neahminax Scullen, 1965; sw. U.S.
nebrascensis H. S. Smith, 1908; n. centr. U.S.
nebulosa Cameron, 1890; e. India
    erythropoda Cameron, 1902, nec Cameron, 1890
    himalayensis Cameron, 1905, nec Bingham, 1898
    intimella Cameron, 1905
    assamensis Cameron, 1905
neghelliensis Guiglia, 1939; Ethiopia
nenicra Saussure, 1887; Madagascar
    nenitra Saussure, 1892, emendation
nenitroidea Bischoff, 1913; S. Africa
neogenita Schulz, 1906; Argentina
    laevigata Holmberg, 1903, nec F. Smith, 1856
    holmbergi Brèthes, 1910
    antemissa Brèthes, 1910
    joergenseni Brèthes, 1913
nephthys Arnold, 1931; S. Africa
    ssp. *platyrhyncha* Arnold, 1942; Rhodesia
nigeriae Brauns, 1926; Nigeria
nigra Ashmead, 1900; W. Indies: St. Vincent
nigrescens F. Smith, 1856; U.S.: widespread
    arelate Banks, 1912, new synonymy by R. Bohart
    nigritula Banks, 1915
    munda Mickel, 1918, new synonymy by R. Bohart
    abbreviata Banks, 1919
    crawfordi Brimley, 1928
nigriceps F. Smith, 1873; Brazil
nigrifrons F. Smith, 1856; S. Africa, Mozambique
    trivialis Gerstaecker, 1857
    erythrospila Cameron, 1910
    ssp. *amaura* Kohl, 1891; Zaire, Liberia
        theryi Arnold, 1932
        ancilla Arnold, 1942
nigrostoma Brauns, 1926; S. Africa
nipponensis Tsuneki, 1961; Japan
nitrariae Morice, 1911; Algeria
nobilitata Cameron, 1905; S. Africa
    ssp. *sordidula* Arnold, 1931; S. Africa
    ssp. *walegae* Arnold, 1951; Ethiopia
nortinus Fritz and Toro, 1971; Chile
nugax Arnold, 1931; Mali, Nigeria, Somalia
    ssp. *dahlak* Menke; Dahlak Archipelago, new name for *insularis* Arnold
        insularis Arnold, 1951, nec F. Smith, 1856
nupta Shestakov, 1922; Mongolia
    ssp. *echingol* Tsuneki, 1971; Mongolia
nursei Turner, 1912; Pakistan

oaxaca Scullen, 1972; s. Mexico, Guatemala
obo Tsuneki, 1971; Mongolia
obregon Scullen, 1972; nw. Mexico
obsoleta Cameron, 1890; s. Mexico, Costa Rica
occipitomaculata Packard, 1866; sw. and centr. U.S.,
  n. Mexico
    fasciola Cresson, 1872
    novomexicana Viereck and Cockerell, 1904
oceania Beaumont, 1951; Morocco
odontophora Schletterer, 1887; se. Europe, Turkey, Iran
okumurai Tsuneki, 1968; Ryukyus
opposita F. Smith, 1873; Australia
orangiae Brauns, 1926; S. Africa
    korana Brauns, 1926
oraniensis Brauns, 1926; S. Africa
    lingnaui Arnold, 1933
orientalis F. Smith, 1856; India
osiris Arnold, 1931; S.-W. Africa
paleata Saussure, 1891; Madagascar
    carnaria Saussure, 1892
pallida Arnold, 1935; Mauretania
    lutulenta Arnold, 1951
pallidula Morice, 1897; Egypt, Israel
    ssp. *annexa* Kohl, 1898; Algeria
palmetorum Beaumont, 1951; n. Africa
papuensis Krombein, 1969; w. New Guinea
parkeri Scullen, 1972; sw. U.S., Mexico
paupercula Holmberg, 1903; Argentina
pauxilla Brèthes, 1913; Argentina
pearstonensis Cameron, 1905; S. Africa
    ssp. *bantu* Brauns, 1926; Rhodesia
pectinata Shestakov, 1922; Mongolia
pedestris Brèthes, 1910; Argentina
pedetes Kohl, 1887; e. Palearctic
    bicornuta F. Smith, 1856, nec Guérin-Méneville, 1844
    smithii Dalla Torre, 1890
pekingensis Tsuneki, 1961; e. China
    ssp. *alini* Tsuneki, 1961; ne. China
    ssp. *mongolica* Tsuneki, 1961; Mongolia
penai Fritz and Toro, 1971; Chile
pentadonta Cameron, 1890; e. India
perboscii Guérin-Méneville, 1844; Brazil
perfida Saussure, 1890 (pl. 19, fig. 10); Madagascar
    spinifrons Saussure, 1891
perfoveata Arnold, 1947; Madagascar
perkinsi Turner, 1910; Australia
perspicua Holmberg, 1903; Argentina
petiolata Saussure, 1891; Madagascar
pharaonum Kohl, 1898; Egypt, Israel, Iran
pharetrigera Shestakov, 1923; Mongolia, China
picta Dahlbom, 1844; Senegal (possibly erroneous
  locality)
pictifacies Brauns, 1926; Rhodesia, S. Africa
pictinoda Cameron, 1908; Tanzania
pictiventris Dahlbom, 1845; Java, Sumatra, Borneo
    malayana Cameron, 1903
    thaiana Tsuneki, 1963
    ssp. *praedata* F. Smith, 1861; n. Moluccas
    ssp. *immolator* F. Smith, 1864; New Guinea, Solomons
        immolatrix Schulz, 1906, emendation
        papuana Cameron, 1906
    ssp. *novarae* Saussure, 1867; Sri Lanka, India
        fervens F. Smith, 1873
    ssp. *formosicola* Strand, 1913; China, Taiwan
    ssp. *kawasei* Tsuneki, 1963; Malaya
picturata Taschenberg, 1875; Brazil
placida Arnold, 1931; Nigeria
placita Arnold, 1931; Kenya
    ssp. *arnoldiana* Guiglia, 1939; Ethiopia
pleuralis Tsuneki, 1968; Philippines
pleurispina Beaumont, 1959; Israel
poculum Scullen, 1965; sw. U.S.
podagrosa Kohl, 1906; Socotra
pollens Schletterer, 1887; Brazil, Paraguay
    tridentifera Brèthes, 1913 (*Paracerceris*)
polybioides Pendlebury, 1927; Malaya
polychroma Gribodo, 1895; S. Africa
ponderosa Brèthes, 1920; Brazil
posticata Banks, 1916; U.S.: New Mexico
potanini Shestakov, 1922; Mongolia
povolnyi Beaumont, 1970; Afghanistan
praedura Turner, 1908; Australia
priesneri Mochi, 1939; n. Africa
proboscidea Holmberg, 1903; Argentina
protea Turner, 1912; w. India
proteles Brauns, 1926; Zaire
pruinosa Morice, 1897; n. Africa, Israel
przewalskii Shestakov, 1922; Mongolia
pseudoerythrocephala Schulthess, 1926; ? S. Africa
pseudoflavescens Shestakov, 1925; sw. USSR
pseudoproteles Arnold, 1952; Kenya
pseudotridentata Maidl, 1926; Sumatra, Java
pucilii Radoszkowski, 1869; Siberia
    poutziloi Schulz, 1906, emendation
pulawskii Tsuneki, 1971; Mongolia
pulchella Klug, 1845; Libya, Egypt, Israel
    alfierii Mochi, 1939
    picta Mochi, 1939, nec Dahlbom, 1844
    ssp. *judaea* Beaumont, 1970; Israel, Syria
    scabra Beaumont, 1959, nec Fabricius, 1798
pulchra Cameron, 1890; e. India
puncticeps F. Morawitz, 1894; sw. USSR
purpurea Schletterer, 1887; Zaire, Senegal
†*pygmaea* Saussure, 1867; China, nec Thunberg, 1815
quadricincta (Panzer), 1799 (*Philanthus*); Palearctic
  Region
    fasciata Spinola, 1806
    ssp. *corsica* Beaumont, 1952; Corsica, Sardinia
    ssp. *segregata* Beaumont, 1970; nw. Africa
    divisa Giner Marí, 1947, nec Brèthes, 1910
quadricolor F. Morawitz, 1889; Mongolia
quadricornis Gussakovskij, 1938; e. China
quadridentata Arnold, 1931; S.-W. Africa
quadrifasciata (Panzer), 1799 (*Philanthus*); Europe,
  Turkey
    tricincta Thunberg, 1815 (*Philanthus*), nec Spinola,
      1805
    truncatula Dahlbom, 1844

bidentata Lepeletier, 1845
dufourii Lepeletier, 1845, nec Guérin-Méneville, 1844
nitida Wesmael, 1852
fargei F. Smith, 1856
spreta A. Costa, 1858
euphorbiae Marquet, 1875
queretaro Scullen, 1972; centr. Mexico
querula Kohl, 1906; Aden
quettaensis Cameron, 1907; Pakistan
quinquefasciata (Rossi), 1792 (*Crabro*); Palearctic Region
    nasuta Dahlbom, 1844, nec Latreille, 1809
    subdepressa Lepeletier, 1845
    boscae Giner Marí, 1941
  ssp. *consobrina* Kohl, 1898; Spain
  ssp. *seoulensis* Tsuneki, 1961; Korea
radjamandalae van der Vecht, 1964; Java
raptor F. Smith, 1856; w. Africa
    raptrix Schulz, 1906, emendation
rasoherinae Arnold, 1945; Madagascar
  ssp. *debilis* Arnold, 1945; Madagascar
raymenti Turner, 1936; Australia
reginula Brauns, 1926; S. Africa
reicula Krombein, 1969; New Guinea
rejecta Turner, 1917; n. India
renominata Turner, 1917; w. India
    opulenta Turner, 1912, nec Morice, 1911
repraesentans Turner, 1919; Kenya
reversa F. Smith, 1873; Brazil
rhinoceros Kohl, 1888; Egypt, Syria, Turkey
rhodesiae Brauns, 1926; Rhodesia
rhodesiensis Empey, 1971; Rhodesia
rhois Rohwer, 1908; sw. U.S.
rhynchophora Turner, 1912; Pakistan
rigida F. Smith, 1856; S. Africa, Mozambique
    charimorpha Brauns, 1926
rixosa F. Smith, 1856; Brazil
robertsonii W. Fox, 1893; e. U.S.
    austrina W. Fox, 1893
    pleuralis H. Smith, 1908
  ssp. *emmiltosa* Scullen, 1964; se. U.S.
  ssp. *bifida* Scullen, 1965; U.S.: N. Carolina
  ssp. *miltosa* Scullen, 1965; U.S.: Florida
roepkei Maidl, 1926; Sumatra, Java, Borneo
rossica Shestakov, 1915; sw. USSR
rostrata F. Smith, 1873; Mexico
rostrifera Brauns, 1926; S. Africa
rothi Giner Marí, 1942; e. China
rothneyi Cameron, 1890; e. India
rozeni Scullen, 1971; U.S.: Florida
rubida (Jurine), 1807 (*Philanthus*) s. Europe, w. Asia
    ornata Spinola, 1806, nec Fabricius, 1790
    modesta F. Smith, 1856
  ssp. *albonotata* Vander Linden, 1829; s. Europe
    julii Fabre, 1879
  ssp. *conjuncta* Schletterer, 1887; Iran, Afghanistan
  ssp. *pumilio* Giner Marí, 1945; Cyprus, Israel
rubrata R. Bohart and Menke; Florida, new name for *rufa* Scullen

rufa Scullen, 1965, nec Taschenberg, 1875
ruficapoides Strand, 1910; Paraguay
  ssp. *derufata* Strand, 1910; Paraguay
ruficauda Cameron, 1905; S. Africa, Lesotho
    ludibunda Arnold, 1931
  ssp. *lichtenburgensis* Brauns, 1926; S. Africa
  ssp. *barbara* Arnold, 1931; Kenya
ruficeps F. Smith, 1873; Brazil
ruficornis (Fabricius), 1793 (*Philanthus*); Europe, Turkey
    bidens Schrank, 1802 (*Crabro*)
    cunicularia Schrank, 1802 (*Crabro*)
    quadricincta Fabricius, 1804 (*Mellinus*), nec Panzer 1799
    trifida Fabricius, 1804 (*Philanthus*)
    ?*nasuta* Latreille, 1809
  ? ssp. *costai* Beaumont, 1950; Italy
  ssp. *laminifera* A. Costa, 1869; Italy
  ssp. *saghaliensis* Tsuneki, 1968; e. USSR: Sakhalin
rufifrons Arnold, 1932; Zaire
rufimana Taschenberg, 1875; Brazil
rufinoda Cresson, 1865; U.S.: widespread; n. Mexico
    crucis Viereck and Cockerell, 1904, new synonymy by R. Bohart
rufiscutis Cameron, 1908; Tanzania
  ssp. *conradsi* Giordani Soika, 1942; Tanzania
rufiventris Lepeletier, 1845; nw. Africa
rufocincta Gerstaecker, 1857; Mozambique, Rhodesia
    manicana Arnold, 1931
rufofacies Empey, 1973; Trucial Oman Coast
rufonigra Taschenberg, 1875; Brazil
  ssp. *turrialba* Scullen, 1972; Costa Rica to Colombia
rufopicta F. Smith, 1856; U.S.: Florida
rugosa F. Smith, 1856; Brazil
rugulosa Schrottky, 1909; Argentina
    polychroma Holmberg, 1903, nec Gribodo, 1895
  ssp. *dismorphia* Schrottky, 1909; Argentina
rustica Taschenberg, 1875; Brazil
    asuncionis Strand, 1910 (p. 136), nec Strand, 1910, p. 141, now in *Foxita*
rutila Spinola, 1838; Egypt, Tunisia
    excellens Klug, 1845
    rubecula Schletterer, 1889
    onophora Schletterer, 1889
  ssp. *lindenii* Lepeletier, 1845; Algeria
  ssp. *mavromoustakisi* Giner Marí, 1945; Cyprus, Israel, Turkey
rybyensis (Linnaeus), 1771 (*Sphex*); Palearctic Region
  ?*infundibuliformis* Fourcroy, 1785 (*Vespa*)
    ornata Fabricius, 1790 (*Philanthus*)
    apifalco Christ, 1791 (*Sphex*)
    semicincta Panzer, 1797 (*Philanthus*)
    hortorum Panzer, 1799 (*Philanthus*)
    variabilis Schrank, 1802 (*Crabro*)
    biguttata Thunberg, 1815 (*Philanthus*)
    rybiensis Ed. André, 1889, emendation
    kashmirensis Nurse, 1903
  ssp. *sicana* De Stefani, 1884; Sicily
  ssp. *dittrichi* Schulz, 1904; Turkey, Siberia

jacobsoni Shestakov, 1923
    ssp. *fertoni* Beaumont, 1952; Corsica
sabulosa (Panzer), 1799 (*Philanthus*); Palearctic Region
    emarginata Panzer, 1799 (*Philanthus*)
    ?*pygmaea* Thunberg, 1815 (*Philanthus*)
    superba Shestakov, 1923
    ssp. *algirica* (Thunberg), 1815 (*Philanthus*); n. Africa
        ariasi Giner Marí, 1941
    ssp. *minuta* Lepeletier, 1845; France
    ssp. *subgibbosa* Yasumatsu, 1935; China
    ssp. *dahlbomi* Beaumont, 1950; e. Europe
    ssp. *sinica* Tsuneki, 1961; e. China
    ssp. *lieftincki* Empey, 1969; Corsica
    ssp. *duplipunctata* Tsuneki, 1971; Mongolia
    ssp. *talynensis* Tsuneki, 1971; Mongolia
saegeri Empey, 1970; Zaire
saeva F. Smith, 1873; Australia
    ?*goodwini* Cockerell, 1930
        saeba Tsuneki, 1968; lapsus
saevissima F. Smith, 1856; Brazil
sahlbergi Shestakov, 1918; sw. USSR
saishuensis Tsuneki, 1968; South Korea
salai Giner Marí, 1945; w. India
samarensis Tsuneki, 1968; Philippines
sandiegensis Scullen, 1965; sw. U.S.
sanluis Fritz and Toro, 1971; Argentina
sareptana Schletterer, 1887; sw. USSR
    opalipennis Kohl, 1888
saussurei Radoszkowski, 1877; sw. USSR
schaeuffelei Beaumont, 1970; Iran
schariniensis Kazenas, 1972; USSR: Kazakh S.S.R.
schlettereri Radoszkowski, 1888; sw. USSR
schoutedeni Brauns, 1914; Zaire
sculleni Tsuneki, 1968; Australia
scutifera Shestakov, 1914; sw. USSR
semenovi Shestakov, 1914; sw. USSR
semilunata Radoszkowski, 1869; China, Siberia
seminigra Taschenberg, 1875; Sudan
semipetiolata Saussure, 1867; Mexico
sepulchralis F. Smith, 1857; Borneo
seraxensis Radoszkowski, 1893; sw. USSR
serripes (Fabricius), 1781 (*Vespa*); "N. Amer."
severini Kohl, 1913; Ethiopia to Mali, Zambia and Zaire
sexta Say, 1837; sw. and centr. U.S.
    biungulata Cresson, 1865
    orphne Banks, 1947
sextoides Banks, 1947; w. U.S.
    eurymele Banks, 1947
seyrigi Arnold, 1945; Madagascar
shelfordi Turner, 1912; Borneo
shestakovi Gussakovskij, 1952; sw. USSR
shestakoviana Gussakovskij, 1952; sw. USSR
shirozui Tsuneki, 1968; South Korea
sibirica F. Morawitz, 1892; Siberia
silvana Schletterer, 1887; Brazil
simulans Saussure, 1867; s. Mexico to Costa Rica
    scapularis Schletterer, 1887
    chrysogaster Schletterer, 1887
sinaitica Beaumont, 1951; Sinai, Israel

sinensis F. Smith, 1856; China, Taiwan
sirdariensis Radoszkowski, 1877; sw. USSR
smithii Dalla Torre, 1890; n. China
    bicornuta F. Smith, 1856, nec Guérin-Méneville, 1844
sobo Yasumatsu & Okabe, 1936; Japan
sokotrae Kohl, 1906; Socotra
solitaria Dahlbom, 1845; n. Africa
    nasuta Lepeletier, 1845, nec Latreille, 1809
    fasciata Lepeletier, 1845, nec Spinola, 1806
    erythrocephala Dahlbom, 1845
    rufa Taschenberg, 1875
    variegata Taschenberg, 1875
    selifera Schletterer, 1887
    algirica Schletterer, 1887, nec Thunberg, 1815
    delepineyi Arnold, 1935
    nemaensis Arnold, 1935
    gynochroma Mochi, 1939
somalica Arnold, 1940; Somalia
somotorensis Balthasar, 1956 (May); centr. and se. Europe
    beaumonti Bajári, 1956 (July)
sororcula Brèthes, 1913; Argentina
spathulifera Brèthes, 1913; Argentina
specifica Turner, 1912; Sri Lanka
spectabilis Radoszkowski, 1886; sw. USSR, Iran
spectrum Arnold, 1936; Rhodesia
    ssp. *multipictoides* Giordani Soika, 1942; Kenya
specularis A. Costa, 1869; s. Europe, sw. USSR, n. Africa
    punctuosa Schletterer, 1887
    schmiedeknechti Kohl, 1898
    octonotata Radoszkowski, 1877
    ssp. *punctosa* Schletterer, 1887; s. Europe
    ssp. *fergusoni* Beaumont, 1958; Cyprus
spinaea Beaumont, 1970, Israel
    spinipleuris Beaumont, 1959, nec Turner, 1918
spinicaudata Cameron, 1905; Ethiopian Region
spinipectus F. Smith, 1856; ne. Mediterranean area, sw. USSR, Afghanistan
    prisca Schletterer, 1887
    ssp. *spinolica* Schletterer, 1887; Libya, Egypt, Israel
        flaviventris Spinola, 1838, nec Vander Linden, 1829
    ssp. *teterrima* Gribodo, 1894, Algeria, Tunisia, Libya
        hartliebi Schulz, 1905
    ? ssp. *accola* Kohl, 1915; sw. USSR, Afghanistan
    ssp. *peloponesia* Beaumont, 1963; Greece
spinipleuris Turner, 1918; Australia
    varipes F. Smith, 1873, nec F. Smith, 1858
spiniventris Tsuneki, 1963; Thailand
spirans Saussure, 1892; Madagascar
splendidissima Giordani Soika, 1942; Tanzania
squamulifera Mickel, 1916; s. centr. U.S.
stefanii Ed. André, 1889; Sicily
stella Shestakov, 1914; sw. USSR
sternodonta Gussakovskij, 1938; e. China
    ?*fervida* F. Smith, 1856
    ssp. *fukiensis* Giner Marí, 1942; e. China
sterope Arnold, 1951; Ethiopia
stevensoni Brauns, 1926; Mozambique

stigmosalis Banks, 1916; sw. Canada, n. centr. U.S., n. Mexico
   fugatrix Mickel, 1918
   sayi Banks, 1923
   stevensi Banks, 1923
stockleini Giner Marí, 1942; e. China
straminea Dufour, 1853; n. Africa
   ?*waltlii* Spinola, 1838
   hirtiventris Morice, 1897
   morici Shestakov, 1918
   moricei Shestakov, 1928, emendation
  ssp. *komarovii* Radoszkowski, 1886; sw. USSR, Iran
   transversa Schletterer, 1889
   cavicornis F. Morawitz, 1890
   komarowii Ed. André, 1891, emendation
  ssp. *hebraea* Beaumont, 1959; Israel
strandi Giner Marí, 1943; Taiwan
stratiotes Schletterer, 1887; se. Europe, Turkey
striata F. Smith, 1873; Brazil
strigosa Cameron, 1890; Mexico
sulcipyga Mochi, 1939; Egypt
sulphurea Cameron, 1890; w. India
sungari Tsuneki, 1961; ne. China
supposita Kohl, 1915; sw. USSR
supraconica Tsuneki, 1961; Korea
synagroides Turner, 1912; Malawi
szechuana Tsuneki, 1968; centr. China
tadzhika Gussakovskij, 1952; sw. USSR
tango Shestakov, 1922; Mongolia
taygete Arnold, 1951; Ethiopia
tenuiventris Arnold, 1958; Rhodesia
tenuivittata Dufour, 1849; s. Europe, e. Mediterranean area
    rufipes F. Smith, 1856, nec Fabricius, 1787
     fuscipennis A. Costa, 1869
     melanothorax Schletterer, 1887
teranishii Sato, 1927; Japan
tetradonta Cameron, 1890; w. India
texana Scullen, 1965; U.S.: Texas
thermophila Schletterer, 1887; Mexico
tibialis Brèthes, 1910; Argentina
tienchiao Tsuneki, 1968; China
tiendang Tsuneki, 1961; n. China, e. Mongolia
   gegen Tsuneki, 1961
tinnula Gussakovskij, 1952; sw. USSR
tokunosimana Tsuneki, 1973; Ryukyus
tolteca Saussure, 1867; sw. U.S. to Costa Rica
   cosmiocephala Cameron, 1904
tonkinensis Turner, 1919; N. Viet Nam
townsendi Viereck and Cockerell, 1904; U.S.: New Mexico
toxopeusi Krombein, 1969; w. New Guinea
transversalis Brèthes, 1910; Argentina
triangulata Cresson, 1865; Cuba
   bilunata Cresson, 1865
trichionota Cameron, 1908; Kenya, Mozambique, S. Africa
   schultzei Bischoff, 1913
   transvaalicola Brauns, 1914
trichiosoma Cameron, 1890; Mexico
trichobunda Strand, 1913; Taiwan
tricincta (Spinola), 1805 (*Philanthus*); Italy
tricolor F. Smith, 1856; Brazil
tricolorata Spinola, 1838; Egypt, Saudi Arabia, Israel
   insignis Klug, 1845
   vidua Klug, 1845
   bulloni Giner Marí, 1945
tridentata Maidl, 1926; Sumatra, Java
trifasciata F. Smith, 1887; USSR
trimaculata Maidl, 1926; Java
trinitaria Alayo, 1968; Cuba
tristior Morice, 1911; Algeria
tristis Cameron, 1890; e. India
truncata Cameron, 1890; sw. U.S., Mexico to Costa Rica
tshontandae Arnold, 1946; Rhodesia
tuberculata (Villers), 1789 (*Sphex*); s. Europe, e. Mediterranean area, Iran
   rufipes Fabricius, 1787 (*Crabro*), nec *Vespa rufipes* Fabricius, 1775
   hispanica Gmelin, 1790 (p. 2764) (*Vespa*), new name for *Vespa rufipes* (Fabricius), 1787
   caspica Gmelin, 1790 (*Sphex*), see Bischoff, 1940
   ?*vespoides* Rossi, 1790 (*Bembex*)
   major Spinola, 1808
   dufouriana Fabre, 1854
   semirufa F. Smith, 1856
  ssp. *evecta* Shestakov, 1922; e. China, Mongolia
   draco Shestakov, 1927
   hunchuz Shestakov, 1927
   mickeli Giner Marí, 1942
  ssp. *cypria* Beaumont, 1958; Cyprus
  ssp. *ogotai* Tsuneki, 1971; Mongolia
tumulorum F. Smith, 1864; Indonesia: Halmahera
turbata Shestakov, 1918; sw. USSR
turkestanica Radoszkowski, 1893; e. Mediterranean area, sw. USSR, Iran, Afghanistan
   rufonoda Radoszkowski, 1877, nec Cresson, 1865 (Art. 58)
turneri Shestakov, 1918; Iran
tyrannica F. Smith, 1856; Gambia
ugandensis Arnold, 1931; Uganda
ulcerosa Arnold, 1936; Rhodesia
umbelliferarum Schrottky, 1911; Paraguay
umbinifera Maidl, 1926; Java, Sumatra
umhlangae Arnold, 1942; S. Africa
uncifera Arnold, 1931; Nigeria
uncta Arnold, 1931; Gambia
unidentata F. Morawitz, 1890; sw. USSR
unifasciata F. Smith, 1856; China
unispinosa Turner, 1917; Australia
vafra Bingham, 1895; Philippines
vagans Radoszkowski, 1877; sw. USSR
vagula Kohl, 1906; sw. Asia: Aden
vanduzeei Banks, 1917; U.S. and sw. Canada
   complanata Mickel, 1918
   eburnea Scullen, 1965, new synonymy by R. Bohart
varia Maidl, 1926; Sumatra, Java, Thailand
   obtusedentata Maidl, 1926

maculiceps Tsuneki, 1963
  ssp. *kalensis* Tsuneki, 1972; Taiwan
variaesimilis Maidl, 1926; e. China, Japan, Ryukyus, Taiwan, Thailand, Malaya, Java, Philippines
  spinicollis Giner Marí, 1942
  basiferruginea Tsuneki, 1963
varians Mickel, 1918; U.S.: California, Nevada; Mexico
varicincta Cameron, 1905; S. Africa
varipes F. Smith, 1858; Celebes
vechti Krombein, 1969; ne. New Guinea
vegeta Arnold, 1940; Rhodesia
vellensis Krombein, 1969; Solomons: Vella Lavella
  ssp. *obrieni* Krombein, 1969; Solomons: Santa Ysabel
  ssp. *fordi* Krombein, 1969; Solomons: Bougainville, Buka
  ssp. *segiensis* Krombein, 1969; Solomons: New Georgia
velutina Taschenberg, 1875; Brazil
ventripilosa Empey, 1972; S. Africa
  turneri Arnold, 1931; nec Shestakov, 1918
venusta F. Smith, 1873; Australia
  ssp. *keiensis* Strand, 1911; Indonesia: Aru, Ewab
  ssp. *oceanica* Brèthes, 1920; e. New Guinea, ne. Australia
    insulicola Tsuneki, 1968
  ssp. *atrescens* Krombein, 1969; w. New Guinea
veracruz Scullen, 1972; Mexico; U.S.: Texas
  ssp. *josei* Scullen, 1972; El Salvador
verecunda Arnold, 1958; Zambia
verhoeffi Tsuneki, 1961; ne. China
vernayi Arnold, 1935; Botswana
versicolor Schrottky, 1909; Argentina
verticalis F. Smith, 1856; s. centr. and se. U.S., ne Mexico
  gnara Cresson, 1872
  firma Cresson, 1872
vianai Fritz and Toro, 1971; Argentina
vicaria Shestakov, 1915; sw. USSR
vicina Cresson, 1865, centr. U.S.
  platyrhina Viereck and Cockerell, 1904
victrix Turner, 1910; Australia
vigilans F. Smith, 1856; s. India
  lanata Cameron, 1907
  ssp. *pervigilans* Turner, 1913; Kenya, S. Africa
    trilineata Bischoff, 1913
    fongosi Brauns, 1926
villersi Berland, 1950; Mali
violaceipennis Cameron, 1904; n. India
  rufoplagiata Cameron, 1905
virgina Shestakov, 1915; sw. USSR
vischnu Cameron, 1890; India
  dolosa Nurse, 1903
vittata Lepeletier, 1845; Algeria
  foveata Lepeletier, 1845
  nigrocincta Dufour, 1853
  ssp. *eurypyga* Kohl, 1898; Algeria
  ssp. *littorea* Beaumont, 1951; Morocco
vitticollis A. Morawitz, 1894; sw. USSR
vulcanica van der Vecht, 1964; Java
  ssp. *ardjunae* van der Vecht, 1964; Java
  ssp. *patuhana* van der Vecht, 1964; Java
  ssp. *merbabunda* van der Vecht, 1964; Java
vulpecula Empey, 1971; S.-W. Africa
  ssp. *fuscicauda* Empey, 1971; S. Africa
vulpinides Strand, 1910; Paraguay
waltoni Arnold, 1940; S. Africa
wickwari Turner, 1912; Sri Lanka
williamsi Scullen, 1972; centr. Mexico
willineri Fritz, 1960; Argentina
windorum Tsuneki, 1968; ne. Australia: Prince of Wales Island
wyomingensis Scullen, 1965; U.S.: Wyoming, N. Dakota
xanthogaster Arnold, 1942; Rhodesia
xanthostigma Arnold, 1945; Madagascar
xosa Brauns, 1926; Zaire to S. Africa
yalensis Turner, 1913; Kenya, Uganda, Ruanda, Burundi
  uxor Leclercq, 1955
yenpingensis Tsuneki, 1968; China
yngvei Cameron, 1908; Tanzania; Rhodesia
  massaica Cameron, 1908
  yungvei Arnold, 1931, emendation
  vumbui Arnold, 1931
yunnanensis Tsuneki, 1968; s. China
zacatecas Scullen, 1972; Mexico
zavattarii Guiglia, 1939; Ethiopia
zelica Banks, 1912; e. U.S.
zethiformis Giordani Soika, 1942; Sudan
ziegleri Rayment, 1947; Australia
zonalis F. Smith, 1852; China
zonata Cresson, 1865 (vol. 4); Cuba
  cubensis Cresson, 1865 (vol. 5)
zumpango Scullen, 1972; s. Mexico
zyx Leclercq, 1955; Burundi, Ruanda, Zambia, Zaire
  molesta Arnold, 1958

Fossil *Cerceris*

†*berlandi* Timon-David, 1944; Oligocene-France, nec Giner Marí, 1941

Nomen nudum in *Cerceris*

nivalis Tsuneki, 1971

Genus Eucerceris Cresson

Generic diagnosis: Typical body profile as in fig. 190. Clypeus of female often with discal teeth or other projections, ocellocular distance not reduced, subantennal sclerite defined by lines from antennal sockets through tentorial pits to clypeus, pronotum raised but often rather appressed to scutum, outer vein of submarginal cell III joining marginal cell beyond its outer third (fig. 184 E), submarginal cell II sometimes petiolate in front (fig. 184 F), first gastral segment usually somewhat narrowed and forming a peduncle, terga with median or submedian transverse grooves (graduli), female sternum VI deeply cleft at apex, male pygidial plate laterally denticulate.

Geographic range: Of the 38 known species, at least two-thirds may be considered Nearctic, most of these from southwestern United States. Approximately one-third are Mexican, and one is recorded from Panama.

Systematics: According to Scullen (1968), there are five main characteristics that separate *Eucerceris* from *Cerceris*. These are: submarginal cell III is more "inflated" in *Eucerceris* (fig. 184 E,F), submarginal cell II is less frequently petiolate, male clypeal brushes are less distinct and not "waxed," terga have transverse mesal depressions (fig. 190), and the male pygidial plate is laterally denticulate and unlike that of the female. Of these five characters the most obvious is the shape of submarginal cell III. It seems to us that the fact that it is "inflated" is less diagnostic than the termination of the outer veinlet far distad on the marginal cell.

Scullen (1939) subscribed to the view that *Cerceris* arose as an offshoot of *Eucerceris*, citing petiolate submarginal cell II in the former genus and its usually sessile condition in the latter. It seems to us that the genera must have arisen independently and both from ancestors with sessile submarginal cell II. Thus, *Cerceris histerisnica* (sometimes placed in the separate genus *Nectanebus*) would be closest to the ancestral type of that genus on the basis of the sessile second cell (fig. 184 D). In *Eucerceris* the more generalized types would be *lacunosa, brunnea, velutina, violaceipennis,* and *punctifrons* in which cell II is sessile in both sexes. This might be followed in phylogenetic sequence by the approximately two dozen species in which males have cell II sessile but the females have it petiolate (fig. 184 E,F). The final

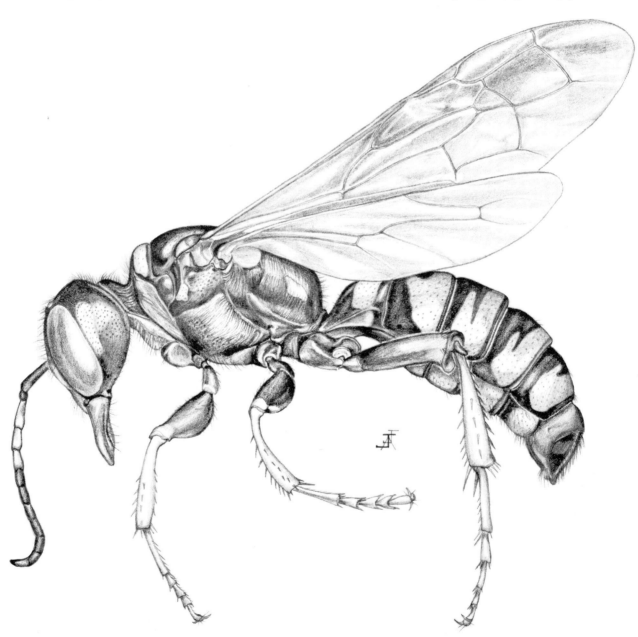

FIG. 190. *Eucerceris flavocincta* Cresson, female.

group in the series, containing *tricolor, angulata, baccharidis, vittatifrons, stangei* and *montana* have cell II petiolate in both sexes. In our opinion the unique submarginal cell III and transversely depressed terga of *Eucerceris* along with more typical cercerine features indicate that it is the most specialized genus in the subfamily. In the list of generalized versus advanced characters given under discussion of advancement within the subfamily, *Cerceris* has a rating of "25" and *Eucerceris* "27," whereas the next highest rating is *Listropygia* with "15."

Most important species characters, in addition to the shape of the submarginal cell II, are the form of the apical margin of the clypeus in the female, discal projections on one or more of the clypeal lobes in the female, fimbriae on various male sterna, and pygidial contours in both sexes. Definitive publications on taxonomy are those of Scullen (1939, 1948, 1968).

Biology: Nesting habits of three species of *Eucerceris* have been reported by Scullen (1939, 1968), Linsley and MacSwain (1954), Bohart and Powell (1956), and Krombein (1960b). In addition, prey records have been summarized by Scullen and Wold (1969).

Eucerceris pimarum and *ruficeps* appear to have similar nesting habits, while those of *flavocincta* are slightly different. *E. ruficeps* was found nesting in hard-packed sand and using the vertical shafts of old halictid bee burrows as a basis for its nests (Linsley and MacSwain, 1954). The halictid nest is modified by the addition of a partial, thin soil plug a little beneath the entrance, and by a complete soil plug 20 to 23 cm deep. From this point a lateral tunnel is excavated which may extend 29 to 42 cm beneath the soil surface. Another soil plug near the end of the lateral tunnel is used to store prey before brood cells are excavated. Prey consists of adult weevils of *Dysticheus rotundicollis* Van Dyke and *Sitona californicus* Fahrens. About four brood cells per nest are excavated at depths of 28 to 31 cm and are provisioned with about 15 weevils.

Work on *Eucerceris pimarum* has not been as complete as for *ruficeps*. Only one burrow of *pimarum* (as *triciliata*) has been recorded and it occurred near the basal slope of a roadside ditch (Krombein, 1960b). The site was composed of loose soil and gravel overlying a layer of hardened mud. Brood cells were found from 45 to 58 cm beneath the soil surface. Eight cells were found, two with empty cocoons, five with fresh cocoons, and one apparently being provisioned. The fragments of about 20 weevils (*Minyomerus languidus* Horn) were in each of two cells. Considering the habits of *pimarum* and *ruficeps*, Krombein (1960b) and Linsley and MacSwain (1954) have suggested that two generations per year are likely.

E. flavocincta excavates a shallow nest about 12 cm deep. Bohart and Powell (1956) reported four cells per nest with six to nine weevils per cell (undescribed genus near *Dyslobus*), but Scullen (1939) reported from six to 10 cells or more per nest with four weevils per cell (*Dyslobus lecontei* Casey). Scullen observed that it took about an hour to capture and store each weevil in the nest which was left open at all times.

Scullen (1968) recorded the finding by M. A. Cazier of weevil prey (*Peritaxia* sp.) for *rubripes* near Portal, Arizona; and the use of the weevil *Ophryastes sulcirostris* (Say) by *E. superba dichroa* (as *bicolor*) in Medicine Hat, Alberta, Canada. Evans (1970) reported that *cressoni* and *flavocincta* in Wyoming used weevils of several genera.

Linsley and McSwain (1954) found tachinid flies destroying cells of *E. ruficeps,* and Bohart and Powell (1956) observed *Hedychrum nigropilosum* Mocsáry frequenting the tunnels of *E. flavocincta.* Evans (1970) reported *Hedychrum parvum* Aaron in the tunnel of *E. cressoni* in Wyoming.

Checklist of *Eucerceris*

angulata Rohwer, 1912; sw. U.S., nw. Mexico
arenaria Scullen, 1948; sw. U.S.
atrata Scullen, 1968; Mexico
baccharidis Scullen, 1968; Mexico
baja Scullen, 1948; Mexico: Baja California
*bitruncata Scullen, 1939; U.S.: Texas, New Mexico, new status by R. Bohart
 ?*triciliata* Scullen, 1948
brunnea Scullen, 1948; Mexico
canaliculata (Say), 1823 (*Philanthus*); sw. and centr. U.S., Mexico
 bidentata Say, 1823 (*Cerceris*), nec Lepeletier, 1845
 cameroni Schulz, 1906, new name for *bidentata* Say
 atronitida Scullen, 1939, new synonymy by R. Bohart
 biconica Scullen, 1948
cerceriformis Cameron, 1891; Mexico
conata Scullen, 1939; w. U.S., new status by R. Bohart
 hespera Scullen, 1948
cressoni (Schletterer), 1887 (*Cerceris*); w. and centr. U.S., new name for *Cerceris fulvipes* (Cresson), (Art. 59c)
 fulvipes Cresson, 1865, nec *Cerceris fulvipes* Eversmann, 1849, now in *Cerceris*
 simulatrix Viereck and Cockerell, 1904
flavocincta Cresson, 1865; w. U.S. and Canada
 cingulatus Cresson, 1865
 striareata Viereck and Cockerell, 1904
 chapmanae Viereck and Cockerell, 1904
lacunosa Scullen, 1939; sw. U.S., nw. Mexico
 arizonensis Scullen, 1939
 ssp. *sabinasae* Scullen, 1968; U.S.: Arizona; Mexico
lapazensis Scullen, 1968; Mexico: Baja California
melanosa Scullen, 1948; Mexico
melanovittata Scullen, 1948; sw. U.S., Mexico
mellea Scullen, 1948; sw. U.S., Mexico
menkei Scullen, 1968; Mexico: Oaxaca
montana Cresson, 1882; sw. U.S.
 sonorensis Cameron, 1891 (*Cerceris*)
morula Scullen, 1968; U.S.: New Mexico; Mexico
 ssp. *albarenae* Scullen, 1968; sw. U.S.
nevadensis (Dalla Torre), 1890 (*Cerceris*); U.S.: Nevada, California; Baja California, new name for *Cerceris*

elegans (Cresson), (Art. 59c)
   elegans Cresson, 1879, nec *Cerceris elegans* Eversmann, 1849, now in *Cerceris*
   ssp. *ferruginosa* Scullen, 1939; U.S.: so. California, new status by R. Bohart
     mohavensis Scullen, 1968, new synonymy by R. Bohart
   ssp. *monoensis* Scullen, 1968; U.S.: e. centr. California, new combination by R. Bohart
pacifica Scullen, 1948; Mexico: Baja California
*pimarum Rohwer, 1908; sw. U.S.; Mexico
   apicata Banks, 1915, new synonymy by R. Bohart
provancheri (Dalla Torre), 1890 (*Cerceris*); w. U.S.: Nevada, California; Mexico: Baja California, new name for *Cerceris insignis* (Provancher), (Art. 59c)
   insignis Provancher, 1889, nec *Cerceris insignis* Klug, 1845, now in *Cerceris*
punctifrons (Cameron), 1890 (*Aphilanthops*); Mexico
   ssp. *cavagnaroi* Scullen, 1968; Mexico
rubripes Cresson, 1879; centr. U.S.
   unicornis Patton, 1879
   marginipennis Cameron, 1890 (*Aphilanthops*)
ruficeps Scullen, 1948; U.S.: Nevada, California
similis Cresson, 1879; w. U.S.
   barri Scullen, 1968, new synonymy by R. Bohart
sinuata Scullen, 1939; U.S.: Texas; Mexico
sonorae Scullen, 1968; Mexico
stangei Scullen, 1968; Mexico
superba Cresson, 1865; centr. U.S.
   fulviceps Cresson, 1879
   rhodops Viereck and Cockerell, 1904
   ssp. *dichroa* (Dalla Torre), 1890 (*Cerceris*); n. centr. U.S., new name for *Cerceris bicolor* (Cresson), (Art. 59c)
     bicolor Cresson, 1882, nec *Cerceris bicolor* F. Smith, 1856, now in *Cerceris*
tricolor Cockerell, 1897; sw. U.S., Mexico
velutina Scullen, 1948; Mexico
violaceipennis Scullen, 1939; Panama
vittatifrons Cresson, 1879; w. U.S.
zimapanensis Scullen, 1968; Mexico
zonata (Say), 1823 (*Philanthus*); n. centr. and ne. U.S.
   laticeps Cresson, 1865

LITERATURE CITED

Some of the following papers have been annotated. Data in parentheses include references to authors who have cleared up dates of publication. The complete citations of these special papers are also listed among the references. The names of most of the principal deceased authors in the Sphecidae are followed by references to obituary notices, biographies, portraits, and bibliographies. We have not given these references for older workers such as Linnaeus and Fabricius because their life work is well known. Biographies of these and other taxonomists can be found in M. M. Carpenter's (1945, 1953) "Bibliography of biographies of entomologists."

Adlerz, G.
  1903. Lefnadsförhallanden och instinkter inom familjerna Pompilidae och Sphegidae. I. Kungl. Svenska Vetenskap. -Akad. Handling. 42 (1): 1-48.
  1904. Lefnadsförhallanden och instinkter inom familjerna Pompilidae och Sphegidae. Kungl. Svenska Vetenskap. -Akad. Handling. (n.s.) 37(5):1-181.

Adriaanse, A.
  1948. *Ammophila campestris* Latr. und *Ammophila adriaansei* Wilcke, eind Beitrag zur vergleichenden Verhaltensforschung. Behaviour 1:1-35.

Agassiz, L.
  1847. Nomenclator Zoologiscus, fasc. 12, Indecim Universalem. viii + 393 pp. Jent and Gassmann, Soliduri (dating based on Bowley and Smith, 1968, and Kevan, 1970).

Ahrens, L. E.
  1925. Observations sur la vie de la guêpe *Palarus flavipes* Fabr. Bull. Inst. Sci. Lesshaft St. Petersbourg 11:57-68 (in Russian with French summary).
  1949. [On the biology and systematic position of *Nitela* Latreille and other members of the Miscophinae.] Dokl. Akad. Nauk S.S.S.R. 68 (2): 413-415 (in Russian).

Ahrens, L. E. and E. L. Ahrens
  1953. The behavior of the wasp *Stizoides tridentata* F. Ent. Obozr. (Moscow) 33:190-193 (in Russian).

Alayo, D.P.
  1968. Estudios sobre los Himenópteros de Cuba. I. Subfamilia Philanthinae. Poeyana (ser. A) 54:1-23.

Alcock, J. and A. F. Ryan
  1973. The behavior of *Microbembex nigrifrons*. Pan-Pac. Ent. 49:144-148.

Alfieri, A.
  1946. Les espèces égyptiennes du genre *Ammophila* Kirby. Bull. Soc. Fouad 1er Ent. 30:105-142.

Andrade, N. F. de
  1952. Sphecidae of Portugal. Genus *Miscophus* Jurine. Mem. Estud. Mus. Zool. Univ. Coimbra (211): 1-41.
  1953. *Miscophus* of Cyprus. Mem. Estud. Mus. Zool. Univ. Coimbra (216):1-40.
  1954. Palaearctic *Miscophus* of the *gallicus* group. Mem. Est. Mus. Zool. Univ. Coimbra (226):1-87.
  1956a. Note on the palaearctic *Miscophus* of the *soikai* group. Mem. Est. Mus. Zool. Univ. Coimbra (238):1-2.
  1956b. Western palaearctic *Miscophus* of the *handlirschii* group. Mem. Est. Mus. Zool. Univ. Coimbra (239):1-40.
  1960. Palaearctic *Miscophus*: *bicolor* group and isolated species. Mem. Est. Mus. Zool. Univ. Coimbra (262):1-136.

Araujo, R. L.
  1939. Contribuição para o conhecimento do gênero *Editha* Parker. Bol. Biol., São Paulo 4:505-511.

Arlé, R.
  1933. Sobre a nidificação, a biologia e os parasitos de *Sceliphron (Trigonopsis) abdominalis* Perty. Ann. Acad. Brasileira Sci. 5:205-211.

Armitage, K. B.
  1965. Notes on the biology of *Philanthus bicinctus*. J. Kansas Ent. Soc. 38:89-100.

Arnold, G.
1922-1931. The Sphegidae of South Africa, Parts I-XV. Ann. Transvaal Mus. 9:101-138 (= part I, 1922), 9:143-253 (= parts II-III, 1923a), 10:1-58 (= part IV, 1923b), 11:1-73 (= part V, 1924); 11: 137-175 (= part VI, 1925), 11:338-376 (= part VII, 1926), 12:55-131 (= part VIII, 1927), 12: 191-279, 338-375 (= parts IX-XI, 1928), 13: 217-418 (= parts XII-XIV, 1929), 14:135-220 (= part XV, 1931).
1932. New species of the Ethiopian Sphegidae. Occ. Pap. Rhodesian Mus. (1):1-31.
1936. New African Hymenoptera. No. 3. Occ. Pap. Rhodesian Mus. (5):1-38.
1940. New species of African Hymenoptera. No. 4. Ann. Transvaal Mus. 20:101-143.
1945. The Sphecidae of Madagascar. 193 pp. Cambridge Univ. Press, England.
1951. Sphecidae and Pompilidae collected by Mr. K. M. Guichard in west Africa and Ethiopia. Bull. Brit. Mus. (Nat. Hist.), Ent. 2:97-183.
1952. New species of African Hymenoptera. No. 10. Occ. Pap. Natl. Mus. Rhodesia (17):460-493.
1955. New species of African Hymenoptera. No. 11. Occ. Pap. Natl. Mus. So. Rhodesia (20):733-762.
1959. New species of African Hymenoptera. No. 14. Occ. Pap. Natl. Mus. So. Rhodesia (23B): 316-339.
Obit.: Stevenson, R. F., 1962. Occ. Pap. Natl. Mus. So. Rhodesia (26B):913-914, portrait. Pinhey, E., 1963. Nature 199:429-430. Stevenson and Pinhey, 1964. Proc. Trans. Rhodesia Sci. Assoc. 50: 9-11.

Ashmead, W. H.
1897. Nitelopterus, a new larrid genus. Ent. News 8:22-23.
1898. Two new genera of sand wasps. Ent. News 9:187-189.
1899. Classification of the entomophilous wasps, or the superfamily Sphegoidea. Canad. Ent. 31: 145-155, 161-174, 212-225, 238-251, 291-300, 322-330, 345-357.
1904. A new genus and some new species of Hymenoptera from the Philippine Islands. Canad. Ent. 36: 281-285.
Obit.: Entomological Society of Washington, 1909. Proc. Ent. Soc. Wash. 10:126-156, portrait and bibliography.

Audouin, J.
1822. *Ampulex*, p. 301. In: J. Bory de Saint-Vincent, Dictionnaire classique d'histoire naturelle, vol. 1, xvi + 604 pp. Paris.

Baerends, G. P.
1941. Fortpflanzungsverhalten und Orientierung der Grabwespe *Ammophila campestris* Jur. Tijdschr. Ent. 84:68-275.

Bajári, N.
1957. Hymenoptera III. Sphecoidea. Fauna Hungariae. 13(7):71-117.

Balduf, W. V.
1936. Observations on *Podalonia violaceipennis* (Lep.) and *Vespula maculata* (Linn.) Canad. Ent. 68: 137-139.

Baltazar, C. R.
1966. A catalogue of Philippine Hymenoptera. Pacific Insects Monogr. 8:1-488.

Balthasar, V.
1953. Ein Beitrage zur Kenntnis der Sphegiden und Chrysididen der Insel Cypern. Acta Ent. Mus. Natl. Pragae 28:39-56.
1972. Grabwespen – Sphecoidea. Fauna ČSSR 20: 1-471.

Banks, N.
1913. New American Philanthidae. Bull. Amer. Mus. Nat. Hist. 32:421-425.
1942. Notes on the United States species of *Tachytes*. Bull. Mus. Comp. Zool. 89:395-436.

Barber, H. S.
1915. *Macrosiagon flavipennis* in cocoon of *Bembix spinolae*. Proc. Ent. Soc. Washington 17: 187-188.

Barth, G. P.
1907a. Observations on the nesting habits of *Gorytes canaliculatus* Pack. Bull. Wisconsin Nat. Hist. Soc. (n.s.) 5:141-149.
1907b. On the nesting habits of *Psen barthi* Viereck. Bull. Wisconsin Nat. Hist. Soc. (n.s.) 5:251-257.
1909. The nesting of *Anacrabro ocellatus* Pack. Bull. Wisconsin Nat. Hist. Soc. (n.s.) 6:147-153.

Bates, H. W.
1876. The naturalist on the river Amazons. 394 pp. Roberts Brothers, Boston.

Beadnell, C. B.
1906. The sand wasp *Sphex lobatus*. J. Bombay Nat. Hist. Soc. 17:546.

Beaumont, J. de
1936a. Les *Tachytes* et les *Tachysphex* de la collection du général Radoszkowski. Rev. Suisse Zool. 43:597-621.
1936b. Les *Tachysphex* de la faune française. Ann. Soc. Ent. France 105:177-212.
1937. Les Psenini de la région paléarctique. Mitt. Schweiz. Ent. Ges. 17:33-93.
1940. Les *Tachysphex* de la faune égyptienne. Bull. Soc. Fouad 1er Ent. Egypte 24:153-179.
1943. Systématique et croissance dysharmonique. Mitt. Schweiz. Ent. Ges. 19:45-52.
1947a. Nouvelle étude des *Tachysphex* de la faune égyptienne. Bull. Soc. Fouad 1er Ent. 31:141-216.
1947b. Contribution à l'étude du genre *Tachysphex*. Mitt. Schweiz. Ent. Ges. 20:661-677.
1947c. Sphecidae de l'île de Chypre. Mitt. Schweiz. Ent. Ges. 20:381-402.
1949a. Les *Philanthus* et *Philoponidea* de l'Afrique du N.-O. Mitt. Schweiz. Ent. Ges. 22:173-216.
1949b. Contribution à l'étude du genre *Palarus* Latr. Rev. Suisse Zool. 56:627-673.
1950. Sphecidae récoltés en Algérie et au Maroc par Kenneth M. Guichard. Bull. Brit. Mus. (Nat. Hist.), Ent. 1:391-427.
1951a. Hyménoptères récoltés par une mission Suisse au Maroc (1947). Sphecidae. Bull. Soc. Sci. Nat. Phys. Maroc 29:259-284.
1951b. Les espèces européennes du genre *Philanthus*. Mitt. Schweiz. Ent. Ges. 24:299-315.
1951c. Contribution á l'étude des *Cerceris* nord-africains. Rev. Española Ent. 27:299-408.
1952a. Les *Hoplisoides* et les *Psammaecius* de la région paléarctique. Mitt. Schweiz. Ent. Ges. 25:211-

238.
1952b. La valeur systématique des caractères éthologiques. Rev. Suisse Zool. 59:306-313.
1952c. Les *Cerceris* de la faune française. Ann. Soc. Ent. France 119:23-80.
1953a. Le genre *Olgia* Radoszk. Rev. Suisse Zool. 60: 205-223.
1953b. Les *Gorytes* s.s. (= *Hoplisus*) de la région paléarctique. Mitt. Schweiz. Ent. Ges. 26:161-200.
1953c. Notes sur quelques types de Sphecidae décrits par A. G. Dahlbom. Opusc. Ent. 18:193-198.
1954a. Remarques sur la systématique des Nyssoninae paléarctiques. Rev. Suisse Zool. 61:283-322.
1954b. Les *Bembecinus* de la région paléarctique. Mitt. Schweiz. Ent. Ges. 27:241-276.
1954c. Sphecidae de l'Institut d'Entomologie de l'Université de Bologne. II. Larrinae. Boll. Ist. Ent. Univ. Bologna 20:53-64.
1954d. Notes sur le genre *Prosopigastra*. Mitt. Schweiz. Ent. Ges. 27:153-156.
1955a. Hyménoptères récoltés par une mission suisse an Maroc (1947). Sphecidae 3. Bull. Soc. Sci. Nat. Phys. Maroc 34:169-197.
1955b. La stylopisation chez les Sphecidae. Rev. Suisse Zool. 62(suppl.):51-72.
1956a. Sphecidae récoltés en Libyae et au Tibesti par M. Kenneth M. Guichard. Bull. Brit. Mus. (Nat. Hist.), Ent. 4:165-215.
1956b. Notes sur les *Stigmus* Panz. et *Spilomena* Shuck. de la Suisse. Mitt. Schweiz. Ent. Ges. 29:385-390.
1956c. Notes sur les *Lindenius* paléarctiques. Mitt. Schweiz. Ent. Ges. 29:145-185.
1958. La classification des *Ammophila* et la valeur taxonomique de l'armature génitale. Rev. Suisse Zool. 65:287-293.
1959. Le genre *Laphyragogus* Kohl. Rev. Suisse Zool. 66:723-734.
1960a. Sphecidae récoltés en Tripolitaine et en Cyrénaïque par M. Kenneth M. Guichard. Bull. Brit. Mus. (Nat. Hist.), Ent. 9:219-251.
1960b. Quelques *Ammophila* K. de la Zoologische Sammlung des Bayerischen Staates. Opusc. Zool. (52):1-5.
1960c. Le genre *Dinetus* Panz. Polskie Pismo Ent. 30: 251-271.
1960d. Sphecidae de l'île de Rhodes. Mitt. Schweiz. Ent. Ges. 33:1-26.
1961a. Notes sur les *Philanthus* paléarctiques. Mitt. Schweiz. Ent. Ges. 33:201-212.
1961b. Les espèces méditerranéennes du genre *Pison* Jur. Mitt. Schweiz. Ent. Ges. 34:53-56.
1961c. Les *Liris* F. du bassin méditerranéen. Mitt Schweiz. Ent. Ges. 34:213-252.
1962. Notes sur les *Sphex* paléarctiques du sous-genre *Chlorion* Latr. Boll. Ist. Ent. Univ. Bologna 26: 29-41.
1963. Les Ammophiles paléarctiques du groupe de *nasuta*. Rev. Suisse Zool. 70:1-24.
1964a. Notes sur les Sphecidae de la Suisse, deuxième série. Mitt. Schweiz. Ent. Ges. 36:289-302.
1964b. Insecta Helvetica Fauna 3. Hymenoptera: Sphecidae. 169 pp. Soc. Ent. Suisse, Lausanne.
1964c. Le genre *Solierella* Spinola en Europe et dans la Méditerranée orientale. Mitt. Schweiz. Ent. Ges. 37:49-68.
1965. Les Sphecidae de la Grèce. Mitt. Schweiz. Ent. Ges. 38:1-65.
1967a. Hymenoptera: Sphecidae. South African Animal Life 13:502-512.
1967b. Hymenoptera from Turkey, Sphecidae 1. Bull. Brit. Mus. (Nat. Hist.), Ent. 19:253-382.
1968a. Sphecidae des Îles Canaries. Bull. Brit. Mus. (Nat. Hist.), Ent. 21:245-278.
1968b. Sphecidae paléarctiques nouveaux ou peu connus. Mitt. Schweiz. Ent. Ges. 41:145-168.
1970. Noms nouveaux de divers Sphecidae. Mitt. Schweiz. Ent. Ges. 43:38.

Beaumont, J. de and H. Bytinski-Salz,
1959. The Sphecidae of Eretz Israel. II. Subfam.: Nyssoninae (tribes: Gorytini, Nyssonini, Alyssonini) and Philanthinae. Bull. Res. Council Israel (Sect. B, Zool.) 8:99-151.

Benno, P.
1957. Aantekeningen bij de rubicole Aculeaten-fauna in Nederland. Ent. Bericht. 17:143-146.
1966. Enige aantekeningen bij de fenologie van *Lestiphorus bicinctus* (Rossi) en zijn koekoekswesp, *Nysson trimaculatus* (Rossi). Ent. Bericht. 26:7-11.

Benoist, R.
1915. Sur l'*Entomognathus* brevis Lind., Hyménoptère chasseur d'Altises. Bull. Soc. Ent. France 20: 241-242.
1927. Sur la biologie des *Dolichurus*. Ann. Soc. Ent. France 96:111-112.
1942. Les Hyménoptères qui habitent les tiges de ronce aux environs de Quito (Équateur). Ann. Soc. Ent. France 111:75-90.

Benson, R. B., C. Ferrière, and O. W. Richards
1947a. Proposed emendation to *Nysson* of the name *Nysso* Latreille, 1796 (Z.N.(S.) 133). Bull. Zool. Nomencl. 1:214.
1947b. Proposed suspension of the règles for *Rhopalum* (Kirby ms.) Stephens, 1829 (Z.N.(S)133). Bull. Zool. Nomencl. 1:217.

Benz, G.
1959. Beobachtungen uber das brutbiologische Verhalten von *Sphex albisectus*, Lepeletier. Vierteljahrsschr. Nat. Ges. Zurich 104:307-319.

Bequaert, J.
1918. A revision of the Vespidae of the Belgian Congo based on the collection of the American Museum Congo expedition, with a list of Ethiopian diplopterous wasps. Bull. Amer. Mus. Nat. Hist. 39:1-384.
1933. The nesting habits of *Paranysson*, an African genus of fossorial wasps. Ent. News 44:36-39.
1937. The nest and prey of *Chlorion (Ammobia) caliginosum* in Colombia. Bull. Brooklyn Ent. Soc. 32:186.

Berland, L.
1923. Notes sur les Hyménoptères fouisseurs de France. II, synonymie de quelques noms employés par J.-H. Fabre. Bull. Soc. Ent. France 1923:171-174.
1925a. Notes sur les Hyménoptères fouisseurs de France. Ann. Soc. Ent. France 94:39-53.
1925b. Hyménoptères Vespiformes. I. In: Faune de

France 8:1-364.
1926. Les Sphegidae du Muséum National de Paris. Bull. Mus. Natl. Hist. Nat. 32:163-170, 200-206, 282-285.
1927. Les Sphegidae du Muséum National de Paris. Bull. Mus. Natl. Hist. Nat. 33:150-156.
1928. Les Sphegidae du Muséum National de Paris. Bull. Mus. Natl. Hist. Nat. 34:329-331.
1929. Les Sphegidae du Muséum National de Paris. Bull. Mus. Natl. Hist. Nat. (2)1:309-312.
1938. La proie et le terrier de *Sphex occitanicus* Lep. Rev. Française Ent. 5:195-197.
1956. Les *Sphex* africains. Bull. Ist. Francais Afrique Noire (A)18:1161-1181.
1958. Observations sur le comportement du *Sphex argyrius*. Bull. Soc. Ent. France 63:66-73.
1959. La nidification du *Sphex paludosus*. Bull. Soc. Ent. France 64:195-197.

Berland, L. and F. Bernard
1947. Les *Sphex* de France. Ann. Soc. Ent. France 116:1-16.

Bernard, F.
1934. Observations sur les proies de quelques Hyménoptères. Bull. Soc. Ent. France 39:247-250.

Bertoni, A.
1911. Contribución a la biologia de las avispas y abejas del Paraguay. Anal. Mus. Nac. Buenos Aires 22:97-146.

Billberg, G. J.
1820. Enumeratio Insectorum. ii ¡ 138 pp. Gadelianis, Holmiae.

Bingham, C. T.
1896. *Leianthrena* new genus, p. 381. In: Kohl, F. F., Die Gattungen der Sphegiden. Ann. Naturhist. Hofmus. Wien 11:233-516.
1897. The fauna of British India. Hymenoptera vol. 1, xiii and 579 pp. Taylor and Francis, London.
1900. Account of a remarkable swarming for breeding purposes of *Sphex umbrosus* Christ, with notes on the nests of two other species of *Sphex* and of certain of the Pompilidae. J. Bombay Nat. Hist. Soc. 13:177-180.

Bischoff, H.
1937. Uber das Vorkommen von *Stizus perrisi* Duf. in der Mark Brandenburg und seine sonstige Verbreitung. Markische Tierwelt (Berlin) 2:236-240.
1940. Deutung einiger von Gmelin 1790 beschriebenen Hymenopterenarten aus Südost-Russland. Sitzungb. Ges. Naturforsch. Freunde Berlin 1940:68-71.

Blackwelder, R. E.
1947. The dates and editions of Curtis' British entomology. Smithsonian Misc. Coll. 107(5):1-27.
1949. Studies on the dates of books on Coleoptera. I and III. Coleopterist's Bull. 3:42-46, 92-94.

Blanchard, E.
1846. Hyménoptères, pp. 113-227 (vol. 13), pls. 107-129 (vol. 14). In: G. Cuvier, Le Règne Animal etc. Fortin, Masson et Cie, Paris (dating after Sherborne, 1922a).

Blüthgen, P.
1949. Neues oder Wissenswertes über mitteleuropäische Aculeaten und Goldwespen. Beitr. Tax. Zool. 1:77-100.
1953. Alte und neue paläarktische *Spilomena*-Arten. Opuscula Ent. 18:160-179.
1954. Neues oder Wissenswertes ueber mitteleuropaeische Aculeata und Goldwespen III. Bonner zool. Beitr. 5:139-155.
1960. Zur Verbreitung und Lebenweise der europäischen *Spilomena*-Arten. Nachrichtenblatt 9(1):1-5.

Bodkin, G. E.
1917. "Cowfly tigers", an account of the hymenopterous family Bembecidae in British Guiana. J. Board Agr. British Buiana 10:119-125.

Bohart, G. E.
1951. Tribes Tachytini and Larrini. In: C. F. W. Muesebeck, *et al.*, Hymenoptera of America north of Mexico. U. S. Dept. Agr. Monograph 2:945-954.
1954. Honeybees attacked at their hive entrance by the wasp *Philanthus flavifrons* Cresson. Proc. Ent. Soc. Washington 56:26-27.

Bohart, G. E. and R. M. Bohart
1966. A revision of the genus *Larropsis* Patton. Trans. Amer. Ent. Soc. 92:653-685.

Bohart, G. E. and J. W. MacSwain
1939. The life history of the sand wasp, *Bembix occidentalis beutenmulleri* Fox and its parasites. Bull. So. Calif. Acad. Sci. 38:84-97.
1940. Notes on two chrysidids parasitic on western bembicid wasps. Pan-Pacific Ent. 16:92-93.

Bohart, R. M.
1941. A revision of the Strepsiptera with special reference to the species of North America. Univ. Calif. Pub. Ent. 7:91-160.
1958a. A North American species of the genus *Prosopigastra*. Proc. Ent. Soc. Washington 60:122-124.
1958b. A new *Prionyx* and a key to the North American species. Bull. Brooklyn Ent. Soc. 53:90-93.
1959. New species of *Aphilanthops* from western North America. Ann. Ent. Soc. Amer. 52:105-108.
1962a. New species of black *Tachysphex* from North America. Proc. Biol. Soc. Washington 75:33-40.
1962b. The *Tachytes pepticus* group in North America. Pan-Pacific Ent. 38:117-129.
1966. A review of *Aphilanthops* and related genera. Proc. Ent. Soc. Washington 68:158-167.
1967. New genera of Gorytini. Pan-Pacific Ent. 43:155-161.
1968a. New species of *Nysson* from southwestern United States. Pan-Pacific Ent. 43:315-325.
1968b. New *Hoplisoides* from the United States. Proc. Ent. Soc. Washington 70:287-292.
1968c. New Nyssoninae from North and South America. Pan-Pacific Ent. 44:228-236.
1968d. A synopsis of the American species of the genus *Oryttus*. Proc. Biol. Soc. Washington 81:431-438.
1968e. New species of *Pseudoplisus* from North and Central America. I. The *P. phaleratus* group. J. Kansas Ent. Soc. 41:494-501.
1969. New species of *Pseudoplisus* from North America, II. J. Kansas Ent. Soc. 42:392-405.
1970a. The genus *Ochleroptera* in New Guinea. Proc. Ent. Soc. Washington 72:386-387.

1970b. New species and lectotype designations in North American Bembicini. Pan-Pacific Ent. 46:201-207.
1971. New species of Gorytini from western North America. Proc. Biol. Soc. Washington 83: 445-454.
1972. New North American *Philanthus*. Proc. Ent. Soc. Washington 74:397-403.

Bohart, R. M and G. E. Bohart
1962. A revision of *Larropsis* subgenus *Ancistromma* Fox. Proc. Ent. Soc. Washington 64:21-37.

Bohart, R. M. and R. L. Brumley
1967. Two new species of *Hedychridium* from California. Pan-Pacific Ent. 43:232-235.

Bohart, R. M. and L. E. Campos
1960. A review of the genus *Omalus* Panzer in North America. Ann. Ent. Soc. Amer. 53:235-250.

Bohart, R. M. and E. E. Grissell
1972. Nesting habits and larva of *Pulverro monticola*. Pan-Pacific Ent. 48:145-149.

Bohart, R. M. and J. F. Holland
1965. Unpublished notes on nesting habits of Sphecidae in southern Oklahoma.
1966. Nesting habits of *Enchemicrum australe* Pate. Pan-Pacific Ent. 42:161.

Bohart, R. M. and D. S. Horning, Jr.
1971. California bembicine sand wasps. Bull. Calif. Insect Surv. 13:1-49.

Bohart, R. M., C. S. Lin, and J. F. Holland
1966. Bionomics of *Oxybelus sparideus* Cockerell at Lake Texoma, Oklahoma. Ann. Ent. Soc. Amer. 59:818-820.

Bohart, R. M. and P. M. Marsh
1960. Observations on the habits of *Oxybelus sericeum* Robertson. Pan-Pacific Ent. 36:115-118.

Bohart, R. M. and A. S. Menke
1961. A review of the genus *Palmodes* in North America. Proc. Ent. Soc. Washington 63:179-191.
1963. A reclassification of the Sphecinae with a revision of the nearctic species of the tribes Sceliphronini and Sphecini. Univ. Calif. Pub. Ent. 30:91-182.
1965 *Diodontus* Curtis, 1834: proposed designation of a type-species under the Plenary Powers. Z.N.(S.) 1711. Bull. Zool. Nomencl. 22:257-258.

Bohart, R. M. and J. A. Powell
1956. Observations on the nesting habits of *Eucerceris flavocinctus* Cresson. Pan-Pacific Ent. 32:143-144.

Bohart, R. M. and E. I. Schlinger
1956. An annotated synonymical list of North American *Oxybelus*. Pan-Pacific Ent. 32:157-165.
1957. California wasps of the genus *Oxybelus*. Bull. Calif. Insect Surv. 4:103-142.

Bondar, G.
1930. Vespas que protegem os animaes domesticos contra as muscas. Correio Agricola 8:179-181.

Bonelli, P.
1966. Osservazioni biologische sugli imenotteri melliferi e predatori della Val di Fiemme, XI. Contributo, *Gorytes (Harpactus) affinis* Spin. Stud. Trentini Sci. Nat. (b) 43:8-19.
1967. Osservazioni biologiche sugli Imenotteri melliferi e predatori della Val di Fiemme, XXV. Boll. Ist. Ent. Univ. Bologna 28:291-303.
1969. Osservazioni biologiche sugli Imenotteri melliferi e predatori della Val di Fiemme. Boll. Ist. Ent. Univ. Bologna 29:149-154, 165-172.

Bordage, E.
1912. Notes biologiques recueilies à l'Île de la Réunion. Bull. Sci. France Belgique (7)46:1-64.

Börner, C.
1919. Stammesgeschichte der Hautfluegler. Biol. Zentrablatt 39:145-186.

Bougy, E.
1935. Observations sur l'*Ammophila hirsuta* Scop. et sur *Hilarella stictica* Meig., son parasite. Rev. Française Ent. 2:19-27.

Bowden, J.
1964. Notes on the biology of two species of *Dasyproctus* Lep. and Br. in Uganda. J. Ent. Soc. S. Africa 26:425-437.

Bowley, D. R. and H. M. Smith
1968. The dates of publication of Louis Agassiz's Nomenclator zoologicus. J. Soc. Bibliog. Nat. Hist. 5:35-36.

Bradley, J. C.
1934. The status of the genus *Rhinopsis* with description of a new species from Texas. Ent. News 45:273-276.
1956. On the use of "Cresson" as a generic name. Ent. News 67:257.
1958. The phylogeny of the Hymenoptera. Proc. Tenth Internat. Congr. Ent. 1:265-269.

Brauns, H.
1896. [New genera on pp. 448-449]. In: Kohl, F. F., Die Gattungen der Sphegiden. Ann. Naturhist. Hofmus. Wien. 11:233-516.
1899. Zur Kenntnis der südafrikanischer Hymenopteren. Ann. Naturhist. Hofmus. Wien 13:382-423.
1906. Zur Kenntnis der Südafrikanischen Hymenopteren, II. Verhandl. zool.-bot. Ges. Wien 56:43-59.
1910. Neue Sphegiden aus Südafrika. Deutsche Ent. Zeitschr. 1910:666-670.
1911a. Biologisches über südafrikanische Hymenopteren. Zeitschr. Wiss. Insektenbiol. 7:117-120.
1911b. Ueber *Gorytes*-Arten aus Südafrika. Verhandl. zool.-bot. Ges. Wien 61:130-134.
1917. Notes and synonymy of Hymenoptera in the collection of the Transvaal Museum. Ann. Transvaal Mus. 5:238-245.
1926. The Ethiopian *Cerceris* species. Ann. Transvaal Mus. 11:268-337.

Brèthes, J.
1902. Notes biologiques sur trois Hyménoptères du Buenos Aires. Rev. Mus. La Plata 10:195-205.
1910. Himenópteros Argentinos. Anal. Mus. Nac. Hist. Nat. Buenos Aires 20:205-316.
1913. Himenópteros de la America Meridional. Anal. Mus. Nac. Hist. Nat. Buenos Aires 24:35-165.
1918. Un bembécido cazador de hemípteros. Physis 4:348-349.

Obit.: Dallas, E., 1928. Rev. Soc. Ent. Argentina 2:101-112, portrait and bibliography.

Bridwell, J. C.
- 1916. Notes and exhibitions. Proc. Hawaiian Ent. Soc. 3:275.
- 1919. Notes on *Nesomimesa antennata* (Smith). Proc. Hawaiian Ent. Soc. 4:40-41.
- 1920. Miscellaneous notes on Hymenoptera, 2nd paper, with descriptions of new species. Proc. Hawaiian Ent. Soc. 4:386-403.

Bristowe, W. S.
- 1948. Notes on the habits and prey of twenty species of British hunting wasps. Proc. Linn. Soc. London 160:12-37.
- 1971. The habits of a West Australian sphecid wasp. Entomologist 104:42-44.

Britton, W. E. and L. O. Howard
- 1921. William Hampton Patton. Ent. News 32:33-40.

Bruch, C.
- 1930. Nidificación de *Sceliphron figulus* (Dahlb.) D. T. y observaciones biológicas sobre esta especie. Anal. Soc. Cien. Argentina 110:367-386.

Brues, C. T. and A. L. Melander
- 1932. Classification of insects. Bull. Mus. Comp. Zool. 73:1-672.

Brullé, A.
- 1833. Mémoire sur un insecte Hyménoptère parasite et voisin du genre *Alyson*. Ann. Soc. Ent. France 2:403-410.

Burks, B. D.
- 1951. Tribe Bembicini. In: C. F. W. Muesebeck, et al., Hymenoptera of America north of Mexico. U. S. Dept. Agr. Monograph 2:995-1000.

Burmeister, H.
- 1874. Bembicidae Argentini. Bol. Acad. Nac. Cienc. Cordoba 1:97-129.

Byers, G. W.
- 1962. Observations at nest of *Cerceris halone* Banks, J. Kansas Ent. Soc. 35:317-321.

Callan, E. McC.
- 1942. A note on *Timulla* (*Timulla*) *eriphyla* Mickel, a parasite of *Tachysphex blatticidus* F. X. Williams, from Trinidad. Proc. Roy. Ent. Soc. London (A) 17:18.
- 1954. Observations on Vespoidea and Sphecoidea from the Paria Peninsula and Patos Island, Venezuela. Bol. Ent. Venezolana 9:13-27.

Cameron, P.
- 1888-1890. Insecta, Hymenoptera, vol. 2 (Fossores), xi +413 pp. In: F. D. Godman and D. Salvin, Biologia Centrali-Americana, Taylor and Francis, London (pp. 1-32 = 1888, 33-64 = 1889, 65-128 = 1890, 129-176 = 1891, Sphecidae = pp. 1-158).
- 1899. Description of a new genus and some new species of fossorial Hymenoptera from the Oriental Zoological Region. Ann. Mag. Nat. Hist. (7) 4:52-69.
- 1900. Descriptions of new genera and species of aculeate Hymenoptera from the Oriental Zoological Region. Ann. Mag. Nat. Hist. (7) 5:17-41.
- 1901. Descriptions of three new genera and seven new species of Hymenoptera from eastern Asia and Australia. Ann. Mag. Nat. Hist. (7) 8:116-122.
- 1902. Descriptions of new genera and species of Hymenoptera collected by Major C. S. Nurse at Deesa, Simla and Ferozepore. J. Bombay Nat. Hist. Soc. 14:267-293.
- 1904. Descriptions of new genera and species of Hymenoptera from Mexico. Trans. Amer. Ent. Soc. 30:251-267.
- 1905a. A new genus and species of Larridae from Central America. Entomologist 38:21-22.
- 1905b. Descriptions of new species of Neotropical Hymenoptera. Trans. Amer. Ent. Soc. 31:373-391.
- Obit.: Anonymous, 1912. News of the World, London, Dec. 8, 1912, p. 5. Morely, C., 1913. Entomologist 46:24. G. M. W., 1913, Ent. Mon. Mag. 49:20-21.

Carpenter, G.
- 1942. Note on the bionomics of the sphegid wasp *Dasyproctus bipunctatus* Lepeletier. Proc. Roy. Ent. Soc. London (A) 47:48.

Carpenter, M.
- 1945. Bibliography of biographies of entomologists. Amer. Midl. Nat. 33:1-116.
- 1953. Bibliography of biographies of entomologists (Supplement). Amer. Midl. Nat. 50:257-348.

Carrillo S., J. L.
- 1967. Larval stages in *Solierella blaisdelli* (Bridwell) and *S. peckhami* (Ashmead). Pan-Pacific Ent. 43:201-203.

Carrillo S., J. L. and L. E. Caltagirone
- 1970. Observations on the biology of *Solierella peckhami*, *S. blaisdelli* and two species of Chrysididae. Ann. Ent. Soc. Amer. 63:672-681.

Cazier, M. A. and M. A. Mortenson
- 1964. Studies on the bionomics of sphecoid wasps. I. *Moniaecera asperata*. Pan-Pacific Ent. 40:111-114.
- 1965a. Studies on the bionomics of sphecoid wasps. III. *Hapalomellinus albitomentosus* (Bradley). Wasmann J. Biol. 22:261-276.
- 1965b. Studies on the bionomics of sphecoid wasps. IV. *Nitelopterus laticeps* and *Nitelopterus texanus*. Pan-Pacific Ent. 41:21-29.
- 1965c. Studies on the bionomics of sphecoid wasps. V. *Bothynostethus distinctus* and *Entomognathus texana*. Pan-Pacific Ent. 41:30-33.
- 1965d. Studies on the bionomics of sphecoid wasps. VI. *Fernaldina lucae*. Pan-Pacific Ent. 41:34-43.

Ceballos, G.
- 1956. Catálogo de los Himenópteros de España. Trab. Inst. Español Ent. Madrid, 554 pp.
- 1959. Primer suplemento al catálogo de los Himenópteros de España. Rev. Española Ent. 35:215-242.
- 1964. Segundo suplemento al catálogo de los Himenópteros de España. Rev. Española Ent. 40:44-97.

Chandler, L. G.
- 1925. Habits of the sand wasp. Victorian Nat. 42:107-114.
- 1928. Notes on two grasshopper wasps. Victorian Nat. 45:176-181.

Cheesman, L. E.
- 1928. A contribution towards the insect fauna of French Oceania, part II. Ann. Mag. Nat. Hist. (10)1:169-194.

Clausen, C. P.
1940. Entomophagous insects. 688 pp. McGraw-Hill Book Co., New York.

Clausen, C. P., T. R. Gardner and K. Sato
1932. Biology of Japanese and Chosenese grub parasites. U.S.D.A. Tech. Bull. 308:1-26.

Cockerell, T. D. A.
1898. New North American insects. Ann. Mag. Nat. Hist. (7) 1:321-331.
1903. The habits of *Tachytes* and *Paranysson*. Entomologist 36:100.
1932. Some wasps of the genus *Zoyphium*. Mem. Queensland Mus. 10:117-118.

Conil, A.
1878. Études sur l'*Acridium paranense* Burm., ses variétés et plusiers insectes qui le détruisent. Periodico Zool. 3:177-257 (reprinted 1879, Bol. Acad. Nac. Cienc. Cordoba 3:386-472).

Copello, A.
1933. Biologia de *Hyperalonia morio*. Rev. Soc. Ent. Argentina 5:117-120.

Costa, A.
1858-1872. Immenotteri aculeati, famiglia degli Sfecidei. In: O. G. and A. Costa, 1829-1886, Fauna del Regno di Napoli, 11 vols., Napoli. Sphecidea: pp. 1-28 = 1858, pp. 29-36 = 1861, Nyssonidea: pp. 1-56 = 1859, Philanthidea: pp. 1-16 = 1860, pp. 17-38 = 1872, Bembicidea: pp. 1-8 = 1872, *Bembex*: pp. 1-8 = 1872, (for dating see Sherborne, 1937).
1867. Prospetto sistematico degli Imenotteri Italiani da servire di prodromo della Immenotterologia Italiana. Ann. Mus. Zool. Univ. Napoli 4:59-100.
1871. Prospetto sistematico degli Imenotteri Italiani da servire d'prodromo della Imenotterologia Italiana. Ann. Mus. Zool. Univ. Napoli 6:28-83.
1882. Notizie ed osservazioni sulla geo-fauna Sarda, memoria prima, resultamento di ricerche fatte in Sardegna nel Settembre 1881. Atti Roy. Acad. Sci. Fis. Mat. Napoli 9(11):1-41.
Obit.: Della Valle, A., 1899. Atti Accad. Pontaniana Naples 29 (Necrologia 4):1-6.

Court, H. K.
1961. Taxonomy and biology of the genus *Lindenius* in North America. Unpublished Master's thesis, University of California, Davis.

Court, H. K. and R. M. Bohart
1958. New species of *Lindenius* from western North America. Pan-Pacific Ent. 34:161-167.
1966. Systematic notes on crabronids with description of a new species. Pan-Pacific Ent. 42:329-332.

Court, H. and A. S. Menke
1968. *Solenius* Lepeletier and Brullé, 1834: proposed designation of a type species under the plenary powers. Z. N. (S) 1827. Bull. Zool. Nomencl. 24:357-358.

Cowan, C. F.
1971. On Guérin's Iconographie: Particularly the insects. J. Soc. Bibliog. Nat. Hist. 6:18-29.

Cowley, D. R.
1962. Aspects of the biology of the immature stages of *Pison spinolae* Shuckard. Trans. Roy. Soc. New Zealand 1:355-363.

Cresson, E. T.
1865a. Catalogue of Hymenoptera in the collection of the Entomological Society of Philadelphia, from the Colorado Territory. Proc. Ent. Soc. Philadelphia 4:242-313, 426-488.
1865b. Monograph of the Philanthidae of North America. Proc. Ent. Soc. Philadelphia 5:85-132.
1882. Descriptions of species belonging to the genus *Nysson* inhabiting North America. Trans. Amer. Ent. Soc. 9:273-284.
Obit.: Calvert, P. *et al.*, 1928. Trans. Amer. Ent. Soc. 52(Suppl.):i-lxiii, portrait.

Crèvecoeur, A.
1930. Remarques éthologiques sur quelques Hyménoptères, II. Bull. Ann. Soc. Ent. Belgique 69:358-366.

Cros, A.
1936. Étude biologique sur un Hyménoptère chasseur de mantes *Tachysphex fluctuatus*. Ann. Soc. Ent. France 105:355-368.

Curtis, J.
1829. A guide to an arrangement of British insects, etc. vi + 256 pp. F. Westley and A. Davis, London.
1834-1837. British entomology, etc. Vol. 11, plates 482-529 (= 1834); vol. 13, plates 578-625 (= 1836); vol. 14, plates 626-673 (= 1837). London (dating and editions based on Blackwelder, 1947).

Dahlbom, A. G.
1842. Dispositio methodica specierum Scandinavicarum ad familias Hymenopterorum naturales pertinentium. 16 pp. C. Berling, Lund.
1843-1845. Hymenoptera Europaea praecipue borealia etc. Vol. 1, xliv + 528 pp. Lundbergiana, Lund, pp. 1-172 = fasc. 1, 1843; pp. 173-352 = fasc. 2, 1844; pp. 353-528, i-xliv, tables = fasc. 3, 1845 (for dating see Menke, 1974b).
Obit.: Anonymous, 1859. Stett. Ent. Zeit. 20:337-340.

Dalla Torre, C. G. de (= K. W. or C. W. de)
1897. Catalogus Hymenopterorum hucusque descriptorum systematicus et synonymicus, vol. 8 Fossores. viii + 749 pp. G. Engelmann, Lipsiae.

Daly, H. V.
1964. Skeleto-muscular morphogenesis of the thorax and wings of the honeybee *Apis mellifera*. Univ. Calif. Pub. Ent. 39:1-77.
1965. Skeleto-muscular morphogenesis in the thorax of the Hymenoptera. Proc. 12th Internat. Congr. Ent., London, p. 151.

Dambach, C. A., and E. Good
1943. Life history and habits of the cicada killer (*Sphecius speciosus*) in Ohio. Ohio J. Sci. 43:32-41.

Danks, H. V.
1971. Biology of some stem-nesting aculeate Hymenoptera. Trans. Roy. Ent. Soc. London 122:323-399.

Davidson, M. D.
1895. Habits and parasites of *Stigmus inordinatus* Fox. Psyche 7:271-272.

Davidson, R. H. and B. J. Landis
1938. *Crabro davidsoni* Sandh., a wasp predacious on adult leafhoppers. Ann. Ent. Soc. Amer. 31:5-8.

Davis. W. T.
1920. Mating habits of *Sphecius speciosus*, the cicada-killing wasp. Bull. Brooklyn Ent. Soc. 15:128-129.

Deleurance, E. P.
1941. Contributions á l'étude biologique de la Camargue (1). Bull. Mus. Hist. Nat. Marseille 1:275-

1943. Notes sur la biologie de quelques prédateurs de la région de Montignac (Dordogne). Bull. Mus. Hist. Nat. Marseille 3:56-73.

1945a. Sur l'éthologie d'un *Tachytes* chasseur de mantes *Tachysphex costai* Dest. Bull. Mus. Hist. Nat. Marseille 5:25-29.

1945b. Note biologique sur le *Gorytes (Harpactus) concinnus* Rossi et sur son parasite le *Nysson trimaculatus* Rossi. Bull. Soc. Ent. France 50:122-126.

De Stefani, T.
1887. Un nuovo genere di Crabronidi ed altri imenotteri nuovi o poco cogniti raccolti in Sicilia. Nat. Siciliano 6:59-62, 85-90, 110-114, 143-147.

1896. Sulla nidificazione e biologia dello *Sphex paludosus* Rossi. Nat. Siciliano (n.s.) 1:131-136.

1901. Ulteriori osservazioni sulla nidificazione della *Sphex paludosus.* Monitore Zool. Italiano 12:222-223.

Diederichs, H.
1953. Sand Wespen. Kosmos 49:411-415.

Dow, R.
1932. Biological notes on Cuban wasps and their parasites. Psyche 39:8-19.

1941. A new *Stizus* from Utah, with notes on the other North American species. Psyche 48:171-181.

1942. The relation of the prey of *Sphecius speciosus* to the size and sex of the adult wasp. Ann. Ent. Soc. Amer. 34:310-317.

Drapiez, A.
1819. Description de huit espèces d'insectes nouveaux. Ann. Gen. Sci. Phys. 1:45-55.

Ducke, A.
1902. Neue Arten des Genus *Bothynostethus* Kohl. Verhandl. zool.-bot. Ges., Wien 52:575-580.

1903. Neue Grabwespen vom Gebiete des unteren Amazonas. Verhandl. zool.-bot. Ges., Wien 53:265-270.

1904. Zur Kenntnis der Sphegiden Nordbrasiliens. Zeitschr. Syst. Hymen. Dipt. 4:91-98. (Nachtrag, pp. 189-190).

1907a. Nouveau genre de Sphégides. Ann. Soc. Ent. France 76:28-30.

1907b. Contribution à la connaissance de la faune hyménoptérologique du nord-est du Brésil. Rev. Ent. (Caen) 26:73-96.

Obit.: Egler, W., 1963. Bol. Mus. Paraense Emilio Goeldi (n.s.), Botanica (18), 129 p., portrait. bibliography.

Dufour, L.
1841. Observations sur les métamorphoses du *Cerceris bupresticida*, et sur l'industrie et l'instinct entomologique de cet Hyménoptère. Ann. Sci. Nat., Zool. 15:353-370.

Duncan, C. D.
1939. A contribution to the biology of North American Vespine wasps. Stanford Univ. Press, Stanford Univ., Calif. 272 pp.

Dunning, S. N.
1898. Monograph of the species of *Aphilanthops* inhabiting boreal America. Trans. Amer. Ent. Soc. 25:19-26.

Dupont, F. and C. J. H. Franssen
1937. Aanteekeningen over de levenswijze van *Piagetia cavernicola* v.d. Vecht. Ent. Med. Nederland.-Indie 3:27-29.

Dusmet, J. M. and R. Garcia Mercet
1906. Los *Sphex* de España. Bol. Real Soc. Española Hist. Nat. 6:500-517.

Dutt, G. B.
1912. Life histories of Indian insects IV. Mem. Dept. Agric. India 4:183-267.

Eberhard, W. G.
1971. The predatory behavior of two wasps, *Agenoideus humilis* and *Sceliphron caementarium* on the orb weaving spider, *Araneus cornutus.* Psyche 77:243-251.

1972. Altruistic behavior in a sphecid wasp: support for kin-selection theory. Science 175:1390-1391.

1974. The natural history and behaviour of the wasp *Trigonopsis cameronii* Kohl. Trans. Roy. Ent. Soc. London 125:295-328.

Eickwort, G. C.
1967. Aspects of the biology of *Chilicola ashmeadi* in Costa Rica. J. Kansas Ent. Soc. 40:42-73.

1969. A comparative morphological study and generic revision of the augochlorine bees. Univ. Kansas Sci. Bull. 48:325-524.

Empey, H. N.
1969. Revision of the Ethiopian species of *Cerceris* Latreille, 1802. 1. A synonymical checklist of the described species. J. Ent. Soc. S. Africa 32:297-331.

1971. Notes and redescriptions of Dahlbom's *Cerceris* species from the Ethiopian Region, with a new subspecies of *Cerceris formosa* Dahlbom, 1845 from South Africa. J. Ent. Soc. S. Africa 34:231-249.

1972. New and little known species of *Cerceris* Latreille, 1802 from central and southern Africa. J. Ent. Soc. S. Africa 35:45-68.

Essig. E. O.
1942. College entomology. viii + 900 pp. Macmillan Co., N.Y.

Evans, H. E.
1955a. *Philanthus sanbornii* Cresson as a predator on honeybees. Bull. Brooklyn Ent. Soc. 50:47.

1955b. An ethological study of the digger wasp *Bembecinus neglectus*, with a review of the ethology of the genus. Behaviour 7:287-303.

1957a. Notes on a *Stictia* new to the United States. Ent. News 68:76-77.

1957b. Studies on the comparative ethology of digger wasps of the genus *Bembix*. 248 pp. Comstock Publishing Associates, Ithaca, N.Y.

1957c. Studies on the larvae of digger wasps. Part III: Philanthinae, Trypoxyloninae, and Crabroninae. Trans. Amer. Ent. Soc. 83:79-117.

1957d. Ethological studies on digger wasps of the genus *Astata.* J. New York Ent. Soc. 65:159-185.

1958a. Studies on the larvae of digger wasps. Part IV: Astatinae, Larrinae, Pemphredoninae. Trans. Amer. Ent. Soc. 84:109-139.

1958b. Observations on the nesting behavior of *Larropsis distincta* (Smith). Ent. News 69:197-200.

1958c. Studies on the nesting behavior of digger wasps of the tribe Sphecini. Part 1. Genus *Priononyx* Dahlbom. Ann. Ent. Soc. Amer. 51:177-186.

1959a. Studies on the larvae of digger wasps, part V:

conclusion. Trans. Amer. Ent. Soc. 85:137-191.
1959b. Observations on the nesting behavior of digger wasps of the genus *Ammophila*. Amer. Midl. Nat. 62:449-473.
1959c. *Isodontia*, the grass-carrying wasp. Nat. Mag. 52:237-239.
1960a. A study of *Bembix u-scripta*, a crepuscular digger wasp. Psyche 67:45-61.
1960b. Observations on the nesting behavior of three species of the genus *Crabro*. J. New York Ent. Soc. 68:123-134.
1961. Notes on the nesting behavior of *Plenoculus davisi* Fox. Ent. News 72:225-228.
1962. The evolution of prey-carrying mechanisms in wasps. Evolution 16:468-483.
1963a. Notes on the prey and nesting behavior of some solitary wasps of Jackson Hole, Wyoming. Ent. News 74:233-239.
1963b. Wasp Farm. viii + 178 pp. Natural History Press, Garden City.
1964a. The classification and evolution of digger wasps as suggested by larval characters. Ent. News 75:225-237.
1964b. Notes on the nesting behavior of *Philanthus lepidus* Cresson. Psyche 71:142-149.
1964c. Further studies on the larvae of digger wasps. Trans. Amer. Ent. Soc. 90:235-321.
1964d. Observations on the nesting behavior of *Moniaecera asperata* (Fox) with comments on communal nesting in solitary wasps. Insectes Sociaux 11:71-78.
1964e. Notes on the prey and nesting behavior of some solitary wasps of Mexico and southwestern United States. J. Kansas Ent. Soc. 37:302-307.
1965. Simultaneous care of more than one nest by *Ammophila azteca* Cameron. Psyche 72:8-23.
1966a. The comparative ethology and evolution of the sand wasps. xvi + 526 pp. Harvard Univ. Press, Cambridge.
1966b. The behavior patterns of solitary wasps. Ann. Rev. Ent. 11:123-154.
1966c. The accessory burrows of digger wasps. Science 152:465-471.
1968. Notes on some digger wasps that prey upon leafhoppers. Ann. Ent. Soc. Amer. 61:1343-1344.
1969a. Notes on the nesting behavior of *Pisonopsis clypeata* and *Belomicrus forbesii*. J. Kansas Ent. Soc. 42:117-125.
1969b. Three new Cretaceous aculeate wasps. Psyche 76:251-261.
1970. Ecological-behavioral studies of the wasps of Jackson Hole, Wyoming. Bull. Mus. Comp. Zool. 140:451-511.
1971. Observations on the nesting behavior of wasps of the tribe Cercerini. J. Kansas Ent. Soc. 44:500-523.
1972. The larva of *Heliocausus larroides*. Psyche 78:166-168.
1973a. Burrow sharing and nest transfer in the digger wasp *Philanthus gibbosus* (Fabricius). Anim. Behav. 21:302-308.
1973b. Cretaceous aculeate wasps from Taimyr, Siberia. Psyche 80:166-178.
1973c. Observations on the nests and prey of *Sphodrotes nemoralis* sp. n. J. Australian Ent. Soc. 12:311-314.

Evans. H. E. and M. Eberhard
1970. The wasps. vi + 265 pp. Univ. of Michigan Press, Ann. Arbor.

Evans, H. E. and J. E. Gillaspy
1964. Observations on the ethology of digger wasps of the genus *Steniolia*. Amer. Midl. Nat. 72:257-280.

Evans, H. E. and F. E. Kurczewski
1966. Observations on the nesting behavior of some species of *Tachytes*. J. Kansas Ent. Soc. 39:323-332.

Evans, H. E. and C. S. Lin
1956a. Studies on the larvae of digger wasps, part I: Sphecinae. Trans. Amer. Ent. Soc. 81:131-166.
1956b. Studies on the larvae of digger wasps, part II: Nyssoninae. Trans. Amer. Ent. Soc. 82:35-66.
1959. Biological observations on digger wasps of the genus *Philanthus*. Wasmann J. Biol. 17:115-132.

Evans, H. E., C. S. Lin and C. M. Yoshimoto
1954. Biological notes on *Psammaecius tricolor* (Cresson). Ent. News 65:6-11.

Evans, H. E. and E. G. Linsley
1960. Notes on a sleeping aggregation of solitary bees and wasps. Bull. So. Calif. Acad. Sci. 59:30-37.

Evans, H. E. and R. W. Matthews
1968a. The larva of *Microstigmus comes*, with comments on its relationship to other pemphredonine genera. Psyche 75:132-134.
1968b. North American *Bembix*, a revised key and suggested grouping. Ann. Ent. Soc. Amer. 61:1284-1299.
1970. Notes on the nests and prey of Australian wasps of the genus *Cerceris*. J. Australian Ent. Soc. 9:153-156.
1971a. Notes on the prey and nests of some Australian Crabronini. J. Australian Ent. Soc. 10:1-4.
1971b. Nesting behaviour and larval stages of some Australian nyssonine sand wasps. Austr. J. Zool. 19:293-310.
1973a. Observations on the nesting behavior of *Trachypus petiolatus* (Spinola) in Colombia and Argentina. J. Kansas Ent. Soc. 46:165-175.
1973b. Systematics and nesting behavior of Australian *Bembix* sand wasps. Mem. Amer. Ent. Inst., no. 20, iv + 387 pp.

Fabre, J. H.
1856. Étude sur l'instinct et les metamorphoses des Sphégiens Ann. Sci. Nat. (4) 6:137-183.
1915. The hunting wasps. viii + 427 pp. Dodd, Mead and Co., New York (translation by A. Teixiera de Mattos).

Fabricius, J. C.
1775. Systema entomolgiae, etc. xxviii + 832 pp. Kortii, Flensburgi et Lipsiae.
1776. Genera insectorum, etc., 310 pp. Bartschii, Chilonii.
1790. Nova insectorum genera. Skrivter Naturhist.-Selskabet Copenhagen 1:213-228.
1793. Entomologia systematica emendata et aucta, etc. Vol. 2, viii + 519 pp. C. G. Profit, Hafniae.
1804. Systema Piezatorum. xiv + 15-439 + 30 pp. C. Reichard, Brunsvigae (dating based on Hedicke, 1941, rather than Richards, 1935a).

Faester, K.
1949. Westeuropaische Spheciden. I. *Oxybelus* Latr.

Univ. Zool. Copenhagen, pp. 1-46.

Fahringer, J.
- 1922. Hymenopterologische Ergebnisse einer wissenschaftlichen Studienreise nach der Türkei und Kleinasien (mit Ausschluss des Amanusgebirges). Archiv Naturges. (A) 88:149-222.

Fain, A.
- 1973. Notes sur les hypopes des Saproglyphidae. III. Le genre *Crabrovidia* Zachvatkin, 1941. Description de 8 espèces nouvelles symphorétiques sur les Sphecidae. Bull. Ann. Soc. Roy. Belgique Ent. 109:153-189.

Fattig, P. W.
- 1936. Nest building of *Trypoxylon politiforme* Roh. Canad. Ent. 68:44-45.

Fernald, H. T.
- 1905. The type of the genus *Sphex*. Ent. News 16:163-166.
- 1906. The digger wasps of North America and the West Indies belonging to the subfamily Chlorioninae. Proc. U.S. Natl. Mus. 31:291-423.
- 1927. The digger wasps of North America of the genus *Podalonia (Psammophila)* Proc. U.S. Natl. Mus. 71:1-42.
- 1934. The North American and West Indian digger wasps of the genus *Sphex* (*Ammophila* auct.). 167 pp. E. O. Painter, Deland, Fla.
- 1945. A colony of solitary wasps. Ann. Ent. Soc. Amer. 38:458-460.
- Obit.: Alexander, C. P., 1953. J. Econ. Ent. 45:1111-1112, portrait.

Ferton, C.
- 1892. Un hyménoptère ravisseur de fourmis. Actes Soc. Linn. Bordeaux 44:341-346.
- 1894. Sur les moeurs du *Dolichurus haemorrhous*. Actes Soc. Linn. Bordeaux 47:215-221.
- 1896. Nouveaux hyménoptères fouisseurs et observations sur l'instinct de quelques espèces. Actes Soc. Linn. Bordeaux 48:261-272.
- 1901. Notes detachées sur l'instinct des hyménoptères mellifères et ravisseurs. Ann. Soc. Ent. France 70:83-148.
- 1902. Notes detachées sur l'instinct des hyménoptères mellifères et ravisuers. II. Ann. Soc. Ent. France 71:499-531.
- 1905. Notes sur l'instinct des hyménoptères mellifères et ravisseurs. III. Ann. Soc. Ent. France 74:56-103.
- 1908. Notes detachées sur l'instinct des hyménoptères mellifères et ravisseurs. IV. Ann. Ent. Soc. France 77:558.
- 1910. Notes detachées sur l'instinct des hyménoptères mellifères et ravisseurs. VI. Ann. Soc. Ent. France 79:145-178.
- 1912a. Notes detachées sur l'instinct des hyménoptères mellifères es ravisseurs. VII. Ann. Ent. Soc. France 80:351-412.
- 1912b. Hyménoptères nouveaux d'Algérie et observations sur l'instinct d'une espèce. Bull. Soc. Ent. France 1912:186-191.
- 1914. Notes detachées sur l'instinct des hyménoptères mellifères et ravisseurs. VIII. Ann. Ent. Soc. France 83:81-119.

Fox, W. J.
- 1891. On the species of *Trypoxylon* inhabiting America north of Mexico. Trans. Amer. Ent. Soc. 18:136-148.
- 1892a. Monograph of the North American species of *Tachytes*. Trans. Amer. Ent. Soc. 19:234-252.
- 1892b. Synopsis of the North American species of *Astatus,* Latr. Canad. Ent. 24:232-235.
- 1892c. The North American Pemphredonidae. Trans. Amer. Ent. Soc. 19:307-326.
- 1893a. A new genus of Larridae. Trans. Amer. Ent. Soc. 20:38.
- 1893b. The North American Larridae. Proc. Acad. Natl. Sci. Phila. 45:467-551.
- 1893c. Descriptions of new aculeate Hymenoptera. Psyche 6:553-556.
- 1894a. Studies among the fossorial Hymenoptera. I. Synopsis of the North American species of *Alyson*. Ent. News 5:86-89.
- 1894b. Studies among the fossorial Hymenoptera. II. Synopsis of the N. Am. species of the genus *Didineis* Wesmael. Ent. News 5:126-128.
- 1894c. Studies among the fossorial Hymenoptera. III. Synopsis of the N. Am. species of the genus *Mellinus* Fab. Ent. News 5:201-203.
- 1894d. A proposed classification of the fossorial Hymenoptera of North America. Proc. Acad. Nat. Sci. Phila. 46:292-307.
- 1895. The Crabronidae of boreal America. Trans. Amer. Ent. Soc. 22:129-226.
- Obit.: Anonymous, 1947. Frontiers 12(1):27.

Freeman, P.
- 1938. Notes on the nesting of five species of solitary wasps. Proc. Roy. Ent. Soc. London (Ser. A) 13:1-6.

Friederichs, K.
- 1918. Beobachtungen über einige solitäre Wespen in Madagaskar. Mitt. Zool. Mus. Berlin 9:27-45.

Frisch, J. S.
- 1937. The life-history and habits of the digger-wasp *Ammobia ichneumonea* (Linn.). Amer. Midl. Nat. 18:1043-1062.
- 1938. The life-history and habits of the digger-wasp, *Ammobia pennsylvanica* (Linn.). Amer. Midl. Nat. 19:673-677.
- 1940. Did the Peckhams witness the invention of a tool by *Ammophila urnaria*? Amer. Midl. Nat. 22:345-350.

Fritz, M. A.
- 1955. Nyssonini neotropicales. Rev. Soc. Ent. Argentina 18:11-16.
- 1957. Nyssonini neotropicales, II. Las especies Argentinas de *Metanysson* e *Idionysson*. Neotropica 2:91-96.
- 1959. Nyssonini neotropicales, III. Rev. Soc. Ent. Argentina 21:133-135.
- 1970a. Nyssonini Neotropicales, IV. Ann Mus. Hist. Nat. Argentino 1:143-159.
- 1970b. Los tipos de Cercerini en la colecion del Museo Argentino de Ciencias Naturales "Bernadino Rivadavia". Ann. Mus. Hist. Nat. Valparaiso (1):161-171.
- 1971. Notas sinonimicas sobre el genero *Cerceris*. Rev. Soc. Ent. Argentina 32(1-4):157-161.
- 1972. Nyssonini Neotropicales, V. El genero *Foxia* Ashmead en la Argentina y Bolivia. Acta Scientifica (ser. Entomología) (6):1-12.

Fritz, M. and H. Toro
1971. Contribución al estudio de los Cercerini Neotropicales. Ann. Mus. Hist. Nat. Valparaiso (2):139-166.

Froggatt, W. W.
1903. Cicadas ("locusts") and their habits. Agr. Gazette N.S. Wales 14:334-341, 418-425.

Fulcrand, J. and J. Gervet
1968. Données préliminaires sur le cycle nidificateur d'un Hyménoptère Sphégidé: *Podalonia hirsuta* Scopoli cycle moyen et variantes. Rev. Comport. Anim. 2:59-68.

Fye, R. E.
1965. Methods for placing wasp trap nests in elevated locations. J. Econ. Ent. 58:803-804.

Gemignani, E. V.
1933. Los caracteres diferenciales de tres especies del genero *Trypoxylon*. Rev. Soc. Ent. Argentina 5:309-313.
1940. Los tipos de las especies del genero *Trypoxylon* existentes en el Museo Argentino de Ciencias Naturales. Rev. Soc. Ent. Argentina 10:434-447.
1941. Una nueva especie del genero *Trypoxylon*. Rev. Soc. Ent. Argentina 11:42-44.

Gerstaecker, A.
1867. Die Arten der Gattung *Nysson* Latr. Abhandl. Naturforsch. Ges. Halle 10:71-122.

Gervet, J. and J. Fulcrand
1970. Le thème de piqûre dans la paralysation de sa proie par l'Ammophile *Podalonia hirsuta* Scopoli. Zeitschr. Tierpsychol. 27:82-97.

Gess, F. W.
1973. A new species of *Handlirschia* Kohl, a very poorly known genus from South Africa. Ann. Cape Prov. Mus., Nat. Hist. 9:103-107.

Gillaspy, J. E.
1951. Nesting habits of *Steniolia nigripes* Parker. Pan-Pacific Ent. 27:167-168.
1959. A new bembicine wasp related to *Stictiella tenuicornis* (Fox) with certain phylogenetic considerations. Pan-Pacific Ent. 35:187-194.
1963a. The genus *Stizoides* in North America, with notes on the Old World Fauna. Bull. Mus. Comp. Zool. 128:369-391.
1963b. Two new genera and a new species of Bembicini from North America, with a key to the genera having recessed ocelli. Ent. News 74:187-199.
1963c. The identity of *Stictiella corniculata* Mickel with a note on synonymy in *Stictiella*. Ent. News 74:251-252.
1964. A revisionary study of the genus *Steniolia*. Trans. Amer. Ent. Soc. 89:1-117.

Gillaspy, J. E., H. E. Evans, and C. S. Lin
1962. Observations on the behavior of digger wasps of the genus *Stictiella* with a partition of the genus. Ann. Ent. Soc. Amer. 55:559-566.

Gimmerthal, B. A.
1836. Beschreibung einiger neuen in Liefland ausgefundenen insecten. Bull. Soc. Imper. Nat. Moscou 9:429-449.

Giner Marí, J.
1941. Monographia de los *Cerceris* de España. Rev. Española Ent. 15:7-93.
1943a. *Ammoplanus* Palearcticos. III. Monographia de los *Ammoplanus* Gir. de España. Rev. Española Ent. 19:281-294.
1943b. Himenópteros de España, fam. Sphecidae. Trab. Inst. Español Ent., 270 pp.
1945. Resultados cientificos de un viaje entomologico al Sahara Español y zona Oriental del Marruecos Español, familias Sphecidae y Mutillidae. Rev. Española Ent. 20:351-385.

Giraud, J.
1869. Observations hyménoptèrologiques. Ann. Soc. Ent. France (4)9:469-473.

Gittins, A. R.
1958. Nesting habit and prey record of *Harpactostigma* (*Arcesilas*) *laminiferum* (Fox). Pan-Pacific Ent. 34:142.
1969. Revision of the Nearctic Psenini I. Redescriptions and keys to the genera and subgenera. Trans. Amer. Ent. Soc. 95:49-76.

Gmelin, J. E.
1790. Caroli a Linné, Systema Naturae per regna tria naturae, ed. 13, vol. 1, Regnum Animale, pt. 5, pp. 2225-3020. Lipsiae.

Grandi, G.
1925. Sull'istinto gregario della *Psammophila hirsuta* Scop. Riv. Sci. Nat. "Natura" 16:89-95
1926a. Contributi alla conoscenza degli Imenotteri melliferi el predatori. III. Boll. Lab. Zool. Portici 19:269-327.
1926b. Contributi alla conoscenza della biologia e della morfologia degli Imenotteri melliferi e predatori, IV. Mem. Soc. Ent. Italiana 5:187-213.
1928a. Contributi alla conoscenza biologica e morfologica degli Imenotteri melliferi e predatori, VI. Boll. Lab. Ent. R. Ist. Sup. Agr. Bologna 1:3-31.
1928b. Contributi alla conoscenza biologica e morfologica degli Imenotteri melliferi e predatori, VII. Boll. Lab. Ent. R. Ist. Sup. Agr. Bologna 1:259-326.
1929. Contributi alla conoscenza biologica e morfologica degli Imenotteri melliferi e predatori, XI. Boll. Lab. Ent. R. Ist. Sup. Agr. Bologna 2:255-290.
1930. Contributi alla conoscenza biologica e morfologica degli imenotteri melliferi e predatori, XI. Boll. Lab. Ent. R. Ist. Sup. Agr. Bologna 3:302-343.
1931. Contributi alla conoscenza degli Imenotteri melliferi e predatori, XII. Boll. Lab. Ent. R. Inst. Sup. Agr. Bologna 4:18-72.
1934. Contributi alla conoscenza degli Imenotteri melliferi e predatori, XIII. Boll. Lab. Ent. R. Ist. Sup. Agr. Bologna 7:1-144.
1954. Contributi alla conoscenza degli Imenotteri Aculeati, XXVI. Boll. Ist. Ent. Univ. Bologna 20:81-255.
1957. Contributi alla conoscenza degli Imenotteri Aculeati, XXVII. Boll. Ist. Ent. Univ. Bologna 22:307-398.
1961. Studi di un entomologo sugli Imenotteri superiori. Boll. Ist. Ent. Univ. Bologna 25:i-xvi, 1-659.
1962. Contributi alla conoscenza degli Imenotteri Aculeati, XXXI. Boll. Ist. Ent. Univ. Bologna 26:55-102.

Gribodo, G.
1883. Alcune nuove specie e nuovo genere di Imenot-

teri Aculeati. Ann. Mus. Civ. Stor. Nat. Genoa 18:261-268.
1884. Diagnosi de nuove specie di Imenotteri scavatori ed osservazioni sopra alcune specie poco conosciute. Boll. Soc. Ent. Italiana 16:257-284.

Griffin, F. J.
1937. A further note on "Palisot de Beauvois, Insectes Rec. Afr. Amér." 1805-1821. J. Soc. Bibliog. Nat. Hist. 1:121-122.
1938. On the date of publication of Latreille (in Sonnini's Buffon), an X (sic), Hist. Nat. Gen. Partic. Crust. Ins. 3. J. Soc. Bibliog. Nat. Hist. 1:157.

Guérin-Méneville, F. E.
1829-1844. Iconographie du règne animal de G. Cuvier, etc. 3 vols. (2 of plates, 1 of text), Fain and Thunot, Paris. Insectes, 576 pp. (vol. 3) = 1844; plates 70, 71 (vol. 2, livr. 39) = 1835 (dating based on Cowan, 1971).

Guiglia, D.
1948. I tipi di Imenotteri del Guérin esistenti nelle collezioni del Museo di Genova. Ann. Mus. Civ. Stor. Nat. Genova 63:175-191.
1953. Gli Oxybelini d'Italia. Ann. Mus. Civ. Storia Nat. Genova 66:55-158.

Gurney, A. B.
1951. The nesting habits of *Mimesa (Mimumesa) nigra* (Packard). Proc. Ent. Soc. Washington 53:280.

Gussakovskij, V. V.
1927. Les espèces paléarctiques du genre *Astatus* Latr. Ann. Mus. Zool. Acad. Sci. URSS 28:265-296.
1928a. Sphecidarum species novae. Bull. Inst. Zool. Appl. Phytopath. Leningrad 4:3-19.
1928b. Generis *Solierella* Spin. species palaearcticae. Rev. Russe Ent. 22:78-84.
1930a. Eine neue Sphecidengattung von Transcaspien. Rev. Española Ent. 6:275-286.
1930b. Species novae vel parum cognitae generum *Ammophila* Kby. et *Sphex* L. Rev. Russe Ent. 24:199-211.
1930c. Corrigenda et additamenta ad revisionem generis *Solierella* Spin. Rev. Russe Ent. 24:232-235.
1931a. Contributions à la connaissance des espèces paléartiques orientales du genre *Gastrosericus* Spin. Ann. Mus. Zool. Acad. Sci. URSS 31:449-457.
1931b. Revision der Gattung *Ammoplanus* Giraud und einigen verwandten Sphecidengattungen. Bol. Soc. Española Hist. Nat. 31:437-465.
1933a. Revisio generis *Prosopigastra* Costa. Rev. Ent. URSS 25:154-173.
1933b. Sphecidae et Psammocharidae, a cl. N. Zarudnyi in Persia orientali collectae. Trav. Inst. Zool. Acad. Sci. URSS 1:269-304, (pagination in error, should be 369-404).
1934a. Beitrag zur Kentniss der Pseninen und Pemphredoninen-Fauna Japans. Mushi 7:79-89.
1934b Schwedisch-chinesische wissenschaftliche Expedition nach den nordwestlichen Provinzen Chinas. Archiv Zool. 27A (21):1-15.
1935. Sphecodea und Vespodea von Tadjikistan. Trav. fil. Acad. Sci. URSS Tadjikistan 5:409-467.
1936. Les espèces paléarctiques du genre *Trypoxylon*. Trav. Inst. Zool. Acad. Sci. URSS 3:639-667.
1937. Espèces paléarctiques des genres *Didineis* Wesm., *Pison* Latr. et *Psen* Latr. Trav. Inst. Zool. Acad. Sci. URSS 4:599-698.
1952. [New and little known species of the Psammocharidae and Sphecidae of western Tadzhikistan] Trudy Zool. Inst. Akad. Nauk SSSR 10:199-288. (in Russian)
Obit.: Narzikulov, M., et al., 1970. Rev. Ent. URSS 49:502-507, bibliography (in Russian, English translation: 1971, Ent. Rev. 49:302-305).

Gutbier, A.
1915. Essai sur la classification et sur le développement des nids des guêpes et des abeilles. Horae Soc. Ent. Rossicae 41(7):1-56.

Hacker, H. and T. D. A. Cockerell
1922. Some Australian wasps of the genera *Zoyphium* and *Arpactus*. Mem. Queensland Mus. 7:283-290.

Hamm, A. H. and O. W. Richards
1926. The biology of the British Crabronidae. Trans. Ent. Soc. London 74:297-331.
1930. The biology of the British fossorial wasps of the families Mellinidae, Gorytidae, Philanthidae, Oxybelidae, and Trypoxylidae. Trans. Ent. Soc. London 78:95-131.

Handlirsch, A.
1887-1895. Monographie der mit *Nysson* und *Bembex* verwandten Grabwespen (parts I-VII and "Nachtrag und Schlusswort"). Sitzungber. Akad. Wiss. Wien. Math.-nat. Classe 95:246-421 (= 1887), 96:219-311 (= 1888a), 97:316-565 (= 1888b), 98:440-517 (= 1889), 99:77-166 (= 1890), 100:25-205 (=1892), 102:656-942 (= 1893), 104:801-1079 (= 1895).
1889. Über die Lebensweise von *Dolichurus corniculus* Spinola. Sitzungsber. zool.-bot. Ges. Wien 39:81-83.
1908. Kleiner Beitrag zur Kenntnis der Grabwespengattung *Stizus* Latr. Verhandl. zool.-bot. Ges. Wien 58:240-244.
1925. Familie Sphegidae, pp. 804-813. In: C. Schröder, Handbuch der Entomologie. Vol. 3. viii + 1202 pp. G. Fischer, Jena.
1933. Hymenoptera, pp. 895-1035. In: W. Kükenthal, Handbuch der Zoologie 4(2), Insecta part 2, section 1.
Obit.: Beier, M., 1935. Konowia 14:340-349, portrait and bibliography.

Hartman, C.
1905. Observations on the habits of some solitary wasps of Texas. Bull Univ. Texas (65):1-73. (also Trans. Texas Acad. Sci. 7:19-85).

Haskell, A. T.
1955. Further observations on the occurrence of *Sphex aegyptius* Lep. with swarms of the desert locust. Ent. Mon. Mag. 91:284-285.

Haupt, H.
1931. Ableitung und Benennung des Flügelgeäders bei den aculeaten Hymenopteren. Mitt. Deutsch. Ent. Ges. Berlin 2:118-126.

Hedicke, H.
1941. Über das Erscheinungsjahr von Fabricius' Systema Piezatorum. Mitt. Deutsch. Ent. Ges. Berlin 10:82-83.

Heinrich, G. H.
1969. The charcoal digger wasp, *Isodontia pelopaeform-*

Hellén, W.
- 1954. Die *Spilomena*-Arten Finnlands. Not. Ent. 34: 60-63.

Hemmingsen, A. M.
- 1960. Instincts of *Ammophila (Psammophila) tydei* Guillou. Ent. Medd. 29:325-328.

Herbst, P.
- 1920. Un nuevo jenero de avispas cavadoras. Bol. Mus. Nac. Hist. Nat. Santiago 11:217-220.
- 1921a. Ueber chilenische Hymenopteren welche Brèthes erwähnte. Stett. Ent. Zeit. 82:108-112.
- 1921b. Zur synonymie chilenischer Grabwespen. Stett. Ent. Zeit. 82:113-116.

Herrich-Schaeffer, G. A. V.
- 1840. Nomenclator Entomologicus, zweites Heft, viii + 40 + 244 pp. Regensburg.

Hicks, C. H.
- 1931. On the digger wasp, *Podalonia luctuosa* (F. Smith). Pan-Pacific Ent. 8:49-51.
- 1933. Note on the relationship of an ichneumonid to certain digger wasps. Pan-Pacific Ent. 9:49-52.
- 1934. Biological notes on *Sphex wrightii* (Cresson). Psyche 41:150-157.

Hine, J. S.
- 1906. A preliminary report on the horseflies of Louisiana with a discussion of remedies and natural enemies. Circ. State Crop Pest Commission Louisiana (6):1-43.

Hingston, R. W. G.
- 1925-1926. An oriental hunting wasp *Sphex lobatus*. J. Bombay Nat. Hist. Soc. 30:735-743, 31:147-157.

Holmberg, E. L.
- 1903. Delectus Hymenopterologicus Argentinus. Anal. Mus. Nac. Buenos Aires (3)2:377-517.

Honoré, A.-M.
- 1941. Contribution à la connaissance des espèces égyptiennes du genre *Palarus* Latr. Bull. Soc. Fouad 1er Ent. 25:191-202.
- 1942. Introduction à l'étude des Sphégides en Egypte. Bull. Soc. Fouad 1er Ent. 26:25-80.
- 1943. Nomenclature et espèces-types des genres de Sphégides Paléarctiques. Bull. Soc. Fouad 1er Ent. 27:29-56.
- 1944a. Revue des espèces égyptiennes du genre *Sphex* Linné, 1758. Bull. Soc. Fouad 1er Ent. 28:45-79.
- 1944b. Materiaux pour une monographie des *Miscophus* d'Egypte. Bull. Soc. Fouad 1er Ent. 28:119-143.

Horne, C.
- 1870. Notes on the habits of some hymenopterous insects from the north-west provinces of India. Trans. Zool. Soc. London 7:161-185.

Horn, W. and I. Kahle
- 1935-1937. Ueber entomologische Sammlungen, Teil I. Ent. Beihefte Berlin-Dahlem 2:1-160; Teil II, 3:161-296; Teil III, 4:297-536.

Horn, W. and S. Schenkling
- 1928. Index litteraturae entomologicae, serie I, vol. 3, pp. 705-1056, Berlin.

Howard, L. O.
- 1901. The insect book. xxx + 429 pp. Doubleday, Page and Co., New York.

Howard, L. O., H. G. Dyar, and F. Knob
- 1912. The mosquitoes of North and Central America and the West Indies. 1064 pp. Pub. 159, Carnegie Inst., Washington.

Howes, P. G.
- 1919. Insect behavior. 176 pp. R. G. Badger, Boston.

Huber, A.
- 1961. Zur Biologie von *Mellinus arvensis*. Zool. Jahrb. Abt. Syst. Oekol. Geog. Tiere 89:43-118.

Hungerford, H. B. and F. X. Williams
- 1912. Biological notes on some Kansas Hymenoptera. Ent. News 23:241-260.

Iida, T.
- 1967. A study on the larvae of the genus *Sphex* in Japan. Etizenia (19):1-8.
- 1969a. Contributions to the knowledge on the sphecid larvae in Japan, part 1. Kontyû 37:272-279.
- 1969b. Contributions to the knowledge on the sphecid larvae in Japan, part 2. Kontyû 37:280-289.

Illiger, J. C. (or K.) W.
- 1807a. Fauna Etrusca etc. Iterum edita et annotatis perpetuis aucta, vol. 2. vi + 511 pp. Fleckeisen, Helmstadii.
- 1807b. Vergleichung der Gattungen der Hautflügler. Illiger's Mag. Insektenkunde 6:189-199.

International Commission on Zoological Nomenclature
- 1939. Opinion 135: the suppression of the so-called "Erlangen List" of 1801. Opin. Rend. Internat. Comm. Zool. Nomencl. 2:7-12.
- 1943a. Opinion 139: the names *Cephus* Latreille (1802-1803) and *Astata* Latreille, 1796, in the Hymenoptera added to the Official List of Generic Names. Opin. Rend. Internat. Comm. Zool. Nomencl. 2:35-46.
- 1943b. Opinion 144: on the status of the names *Crabro* Geoffroy, 1762, *Crabro* Fabricius, 1775, and *Cimbex* Olivier, 1790. Opin. Rend. Internat. Comm. Zool. Nomencl. 2:89-98.
- 1946. Opinion 180: on the status of the names *Sphex* Linnaeus, 1758, and *Ammophila* Kirby, 1798. Opin. Decl. Rend. Internat. Comm. Zool. Nomencl. 2:569-585.
- 1963. Opinion 675: *Crabro bicinctus* Rossi, 1794: validated under the plenary powers. Bull. Zool. Nomencl. 20:331-332.
- 1968a. Opinion 884: *Diodontus* Curtis, 1834: designation of a type species under the plenary powers. Bull. Zool. Nomencl. 25:10-11.
- 1968b. Opinion 857: *Podalonia* Fernald, 1927: validation and designation of a type species under the plenary powers. Bull. Zool. Nomencl. 25:88-89.

Invrea, F.
- 1941. Brevi notizie ecologiche su alcuni Crisidi. Boll. Soc. Ent. Italiana 73:144-146.

Irving, W. G. and E. H. Hinman
- 1935. The blue mud-dauber as a predator of the black widow spider. Science 82:395-396.

Iwata, K.
- 1938a. On the habits of some Larridae in Japan. Kontyû 12:1-13 (in Japanese).
- 1938b. Habits of some Japanese pemphredonids and crabronids. Mushi 11:20-41.
- 1938c. Habits of a non-burrowing *Ammophila* from Japan. Mushi 11:70-74.
- 1939a. Habits of *Rhinonitela domestica* Williams, a minute booklouse-hunter in Formosa. Mushi

12:13-16.
1939b. Habits of *Sceliphron (Chalybion) inflexum* Sickmann. Mushi 12:92-101.
1939c. Habits of some solitary wasps in Formosa (IV). Trans. Nat. Hist. Soc. Formosa 29:161-178.
1941. Habits of some Japanese crabronids. Mushi 14:1-7.
1942. Comparative studies on the habits of solitary wasps. Tenthredo 6:1-146.
1963. Miscellaneous biological notes on aculeate Hymenoptera in Kagawa in the years of 1948 and 1949. Trans. Shikoku Ent. Soc. 7:114-118.
1964a. Bionomics of non-social wasps in Thailand. Nat. Life Southeast Asia 3:323-383.
1964b. Ethological notes on four Japanese species of *Pison*. Mushi 38:1-6.
1972. [Evolution of instinct-comparative studies of Hymenoptera behavior]. vi + 503 pp. Mano-Shoten, Tokyo (in Japanese).

Iwata, K. and M. Tanihata
1963. Biological observations on *Larra amplipennis* (Smith) in Kagawa, Japan. Trans. Shikoku Ent. Soc. 7:101-105.

Iwata, K. and K. Yoshikawa
1964. Biological records on two Saltatoria-hunters of the genera *Dicranorhina* and *Gastrosericus* in Thailand. Nat. Life Southeast Asia 3:385-390.

Jaffuel, P. P. F. and A. Pirión
1926. Himenópteros del valle de Marga-Marga. Rev. Chilena Hist. Nat. 30:362-383.

Janvier, H. (formerly known as Claude-Joseph)
1923. Observaciones entomolójicas. Instinto i costumbres del Celifrón (*Sceliphron vindex* Lep.). Anal. Univ. Chile (2) 1:81-115.
1926. Los esfexos de Chile. Anal. Univ. Chile (2) 4:5-69.
1928. Recherches biologiques sur les prédateurs du Chili. Ann. Sci. Nat. Zool. (10) 11:67-207.
1933. Étude biologique de quelques Hyménoptères du Chili. Ann. Sci. Nat. Zool. (10) 16:209-356.
1960. Recherches sur les Hyménoptères nidifiants aphidivores. Ann. Sci. Nat. Zool. (12) 2:281-321.
1961a. Recherches sur les Hyménoptères nidifiants aphidivores. II. Le genre *Pemphredon*. Ann. Sci. Nat. Zool. (12) 3:1-51.
1961b. Recherche sur les Hyménoptères nidifiants aphidivores. Ann. Sci. Nat. Zool. (12) 4:489-516. Ann. Sci. Nat. Zool. (12) 3:847-883.
1962. Recherches sur les Hyménoptères nidifiants aphidovores. Ann. Sci. Nat. Zool. (12) 4:489-516.

Jayakar, S. D. and R. S. Mangipudi
1965. Dormitories of *Chalybion bengalense* Dahlb. J. Bombay Nat. Hist. Soc. 61:708-711.

Jayakar, S. D. and H. Spurway
1963. Use of vertebrate faeces by the sphecoid wasp *Chalybion bengalense* Dahlb. J. Bombay Nat. Hist. Soc. 60:747-748.
1965a. Winter diapause in the squatter wasps *Antodynerus flavescens* (Fabr.) and *Chalybion bengalense* (Dahlb.). J. Bombay Nat. Hist. Soc. 61:662-667.
1965b. Variant behaviour of *Chalybion bengalense* Dahlb. J. Bombay Nat. Hist. Soc. 62:169-172.

Jurine, L.
1807. Nouvelle méthode de classer les Hyménoptères et les Diptères, Vol. 1, Hyménoptères. 319 + 4 pp. J. J. Paschoud, Geneva.

Katayama, H. and Y. Ikushima
1935. Some notes on the biology of *Sceliphron tubifex* Latreille. Trans. Kansai Ent. Soc. 6:41-54.

Kazenas, V. L.
1968. [Contributions to the biology of *Sphex mocsaryi* Kohl.]. Rev. Ent. URSS 47:806-808 (in Russian, English translation: 1969, Ent. Rev. 47:492-493).
1970. [On the biology of *Ammophila (Eremochares) dives* Brullé]. Rev. Ent. URSS 49:292-302. (in Russian, English translation: 1971, Ent. Rev. 49:172-180).

Kelner-Pillault, S.
1962. Un *Sphex* américain introduit dans le sud de la France *Sphex (Isodontia) harrisi* Fernald. Entomologiste 18:102-110.

Kevan, D. K. McE.
1970. Agassiz's Nomenclatoris zoologici index universalis — a correction. J. Soc. Bibliog. Nat. Hist. 5:286.

Kirby, W.
1798. *Ammophila*, a new genus of insects in the class Hymenoptera, including the *Sphex sabulosa* of Linnaeus. Trans. Linn. Soc. London 4:195-212.

Kirby, W. F.
1883. Notes on new or little known species of Hymenoptera, chiefly from New Zealand. Trans. Roy. Ent. Soc. London 1883:199-203.

Klug, F.
1801. Absonderung einiger Raupentödter und vereinigung derselben zu einer neuen Gattung *Sceliphron*. Neue Schrift. Ges. Naturforsch. Freunde Berlin 3:555-566.
1810. Einige neue Piezatengattungen. Mag. Ges Naturforsch. Freunde Berlin 4:31-45.
1845. Insecta, decas quinta (plates 41-50 and 41 unnumbered text pages). In: C. G. Ehrenberg, 1828-1845, Symbolae physicae, seu icones et descriptiones corporum novorum aut minus cognitorum, etc. 4 vol. (in 2), Berolini, Reimeri.

Kohl, F. F.
1878. Hymenopterologischer Beitrag. Verhandl. zool.-bot. Ges. Wien 27:701-710.
1880. Die Raubwespen Tirol's nach ihrer horizontalen und verticalen Verbreitung mit einem Anhange Biologischer und kirtischer Notizen. Zeitschr. Ferdinaneums Tirol Vorarlberg (3) 24:95-242.
1883a. Neue Hymenopteren in den Sammlungen des K.K. zoolog. Hof-Cabinetes zu Wien. Verhandl. zool.-bot. Ges. Wien 33:333-386.
1883b. Ueber neue Grabwespen des Mediterrangebietes. Deutsche Ent. Zeitschr. 27:161-186.
1883c. Die Fossorien der Schweiz. Mitt. Schweizer Ent. Ges. 6:647-684.
1884a. Beitrag zur Kenntniss der Hymenopteren-Gattung *Oxybelus* Latr. Termész. Füzetek 8:101-116.
1884b. Neue Hymenopteren in den Sammlungen des K. K. zoologischen Hof-Cabinetes zu Wien, II. Verhandl. zool.-bot. Ges. Wien 33:331-386 (publ. in March).
1885a. Die Gattungen und Arten der Larriden Autorum. Verhandl. zool.-bot. Ges. Wien 34:171-268, 327-454.

1885b. Die Gattungen der Sphecinen und die palearktischen *Sphex*-Arten. Termész. Füzetek 9: 154-207.
1889. Neue Gattungen aus der Hymenopteren-Familie der Sphegiden. Ann. Naturhist. Hofmus. Wien 4:188-196.
1890. Die Hymenopterengruppe der Sphecinen, I: Monographie der natürlichen Gattung *Sphex* Linné (sens. lat.). Abt. I-II. Ann. Naturhist. Hofmus. Wien 5:77-104, 317-461.
1891. Zur Kenntnis der Hymenopteren-Gattung *Philanthus* Fabr. (sens. lat.). Ann. Naturhist. Hofmus. Wien. 6:345-370.
1892. Neue Hymenopterenformen. Ann. Naturhist. Hofmus. Wien 8:197-234.
1893a. *Zoyphium*, eine neue Hymenopterengattung. Verhandl. zool.-bot. Ges. Wien 43:569-572.
1893b. Über *Ampulex* Jur. (s.l.) und die damit enger verwandten Hymenopteren Gattungen. Ann. Naturhist, Hofmus. Wien 8:455-516.
1895. Zur Monographie der natürlichen Gattung *Sphex* Linne. Ann. Naturhist. Hofmus. Wien 10:42-74.
1896. Die Gattungen der Sphegiden. Ann. Naturhist. Hofmus. Wien 11:233-516.
1897. *Eremiasphecium* Kohl. Eine neue Gattung der Hymenopteren aus der Familie der Sphegiden. Ann. Naturhist. Hofmus. Wien 12:67-70.
1899. Zur Kenntniss neuer gestachelter Hymenopteren. Ann. Naturhist. Hofmus. Wien 14:305-316.
1901. Zur Kenntnis der paläarktischen *Diodontus*-Arten. Verhandl. zool.-bot. Ges. Wien 51:120-135.
1902. Die Hymenopterengruppe der Sphecinen, II. Monographie der neotropischen Gattung *Podium*. Abhandl. zool.-bot. Ges. Wien 1:1-101.
1905. Hymenopterentypen aus der neotropischen Fauna. Verhandl. zool.-bot. Ges. Wien 60:338-366.
1906. Die Hymenopterengruppe der Sphecinen. III. Monographie der Gattung *Ammophila* W. Kirby. Abt. A. Die Ammophilinen der paläarktischen Region. Ann. Naturhist. Hofmus. Wien 21: 228-382.
1907. *Eparmostethus* novum genus Larridarum. Verhandl. zool.-bot. Ges. Wien 57:167-169.
1915. Die Crabronen der paläarktischen Region. Ann. Naturhist. Hofmus. Wien 29:1-288, 289-453.
1918. Die Hautflüglergruppe "Sphecinae". IV. Die natürliche Gattung *Sceliphron* Klug (*Pelopoeus* Kirby). Ann. Naturhist. Hofmus. Wien 32:1-171.
1923. Die Hymenopteren-Gattung *Belomicrus* A. Costa. Konowia 2:98-122, 180-202, 258-278.
Obit.: Maidl, F., 1925. Konowia 4:89-96, portrait and bibliography.

Kokujev, N.
1906. Les representants de la sous-famille Alysonini Dalla Torre dans la faune de la Russie. Horae Soc. Ent. Rossicae 37:209-219 (in Russian).

Kriechbaumer, J.
1874. Ueber die Gattung *Ampulex*, die 2 europäischen Arten derselben und eine neue Art aus Sikkim. Stett. Ent. Zeit. 35:51-56.

Krombein, K. V.
1936. Biological notes on some solitary wasps. Ent. News 47:93-99.
1938. Notes on the *Passaloecus* of New York State with descriptions of two new species. Bull. Brooklyn Ent. Soc. 33:122-127.
1939. Descriptions and records of new wasps from New York State. Bull. Brooklyn Ent. Soc. 34:135-144.
1943. A new Philippine *Nippononysson* with remarks on the affinities of the genus. Ann. Ent. Soc. Amer. 36:451-454.
1948. Synonymical notes on North American sphecoid wasps: I and II. Bull. Brooklyn Ent. Soc. 43: 18-21.
1949. The Aculeate Hymenoptera of Micronesia. I. Scollidae, Mutillidae, Pompilidae and Sphecidae. Proc. Hawaiian Ent. Soc. 13:367-410.
1950a. The Aculeate Hymenoptera of Micronesia. II. Colletidae, Halictidae, Megachilidae, and Apidae. Proc. Hawaiian Ent. Soc. 14:101-142.
1950b. Synonymical notes on North American sphecoid wasps: III. The Nearctic species of *Diodontus* Curtis. Bull. Brooklyn Ent. Soc. 45:35-40.
1950c. Taxonomic notes on the wasps of the subgenus *Pseneo* Malloch. Proc. Ent. Soc. Washington 52: 277-287.
1950d. A new *Nitela* from California. Pan-Pacific Ent. 26:130-131.
1951. Superfamily Sphecoidea, pp. 937-1034. In: Muesebeck *et al.*, Hymenoptera of America north of Mexico, synoptic catalog. U. S. Dept. Agr. Monogr. 2, 1420 pp.
1952. Preliminary annotated list of the wasps of Westmoreland State Park, Virginia, with notes on the genus *Thaumatodryinus*. Trans. Amer. Ent. Soc. 78:89-100.
1953. Biological and taxonomic observations on the wasps in a costal area of North Carolina. Wasmann J. Biol. 10:257-341.
1954a. Wasps collected at Lost River State Park, West Virginia in 1953. Bull. Brooklyn Ent. Soc. 49:1-7.
1954b. Taxonomic notes on some wasps from Florida with descriptions of new species and subspecies. Trans. Amer. Ent. Soc. 80:1-27.
1955a. Miscellaneous prey records of solitary wasps. I. Bull. Brooklyn Ent. Soc. 50:13-17.
1955b. Some notes on the wasps of Kill Devil Hills, North Carolina, 1954. Proc. Ent. Soc. Washington 57:145-160.
1956a. Miscellaneous prey records of solitary wasps. II. Bull. Brooklyn Ent. Soc. 51:42-44.
1956b. Biological and taxonomic notes on the wasps of Lost River State Park, West Virginia, with additions to the faunal list. Proc. Ent. Soc. Washington 58:153-161.
1956c. *Chrysis fuscipennis* Br., a recent adventive wasp in Washington, D. C., from the Old World. Proc. Ent. Soc. Washington 58:275.
1958a. Miscellaneous prey records of solitary wasps. III. Proc. Biol. Soc. Washington 71:21-26.
1958b. Additions during 1956 and 1957 to the wasp fauna of Lost River State Park, West Virginia, with biological notes and descriptions of new species. Proc. Ent. Soc. Washington 60:49-64.
1958c. Biological notes on some wasps from Kill

Devil Hills, North Carolina, and additions to the faunal list. Proc. Ent. Soc. Washington 60: 97-110.
1958d. Sphecoidea, pp. 186-204. In: Krombein, Hymenoptera of America North of Mexico, synoptic catalog. U.S. Dept. of Agr. Monogr. 2, first supplement, 305 pp.
1958e. *Pison (Paraceramius) koreense* (Rad.), a new adventive wasp in eastern United States. Ent. News 69:166-167.
1959a. A new genus and species of Psenini from the southwestern deserts. Bull. Brooklyn Ent. Soc. 54:18-21.
1959b. Biological notes on some ground-nesting wasps at Kill Devil Hills, North Carolina, 1958, and additions to the faunal list. Proc. Ent. Soc. Washington 61:193-199.
1959c. Natural history of Plummers Island, Maryland, XII. A biological note on *Trypoxylon richardsi* Sandhouse. Proc. Biol. Soc. Washington 72:101-102.
1960a. Biological notes on some Hymenoptera that nest in sumach pith. Ent. News 71:29-36, 63-69.
1960b. Biological notes on several southwestern ground-nesting wasps. Bull. Brooklyn Ent. Soc. 55:75-79.
1961a. Miscellaneous prey records of solitary wasps. IV. Bull. Brooklyn Ent. Soc. 56:62-65.
1961b. *Passaloecus turionum* Dahlbom, an adventive European wasp in the United States. Ent. News 72:258-259.
1962. Natural history of Plummers Island, Maryland. XIII. Descriptions of new wasps from Plummers Island, Maryland. Proc. Biol. Soc. Washington 75:1-17.
1963a. Natural history of Plummers Island, Maryland. XVII. Annotated list of the wasps. Proc. Biol. Soc. Washington 76:255-280.
1963b. Biological notes on *Cerceris blakei* Cresson. Bull. Brooklyn Ent. Soc. 43:72-79.
1963c. A new *Tachysphex* from southwestern United States. Ent. News 74:177-180.
1963d. Notes on the *Entomognathus* of eastern United States. Proc. Biol. Soc. Washington 76:247-254.
1964a. Miscellaneous prey records of solitary wasps. V. Bull. Brooklyn Ent. Soc. 58:118-120.
1964b. Results of the Archbold expeditions. No. 87. Biological notes on some Floridian wasps. Amer. Mus. Novitates (2201):1-27.
1964c. Natural history of Plummers Island, Maryland. XVIII. The hibiscus wasp, an abundant rarity, and its associates. Proc. Biol. Soc. Washington 77:73-112.
1967a. Superfamily Sphecoidea, pp. 386-421. In: Krombein and Burks, Hymenoptera of America north of Mexico, synoptic catalog. U.S. Dept. of Agr. Monogr. 2, second supplement, 584 pp.
1967b. Trap-nesting wasps and bees: life histories, nests, and associates. iii-vi + 570 pp. Smithsonian Press, Washington D.C.
1968. A fifth species of *Nitela* from North America. Nat. Canad. 95:699-702.
1969. A revision of the Melanesian wasps of the genus *Cerceris* Latreille, Smithsonian Contr. Zool. (22):1-36.
1970. Behavioral and life history notes on three Floridian solitary wasps. Smithsonian Contrib. Zool. (46):1-26.
1973. Notes on North American *Stigmus* Panzer. Proc. Biol. Soc. Washington 88:211-230.

Krombein, K. V. and H. E. Evans
1954. A list of wasps collected in Florida, March 29 to April 5, 1953, with biological annotations. Proc. Ent. Soc. Washington 56:225-236.
1955. An annotated list of wasps collected in Florida, March 20 to April 3, 1954. Proc. Ent. Soc. Washington 57:223-235.

Krombein, K. V. and F. E. Kurczewski
1963. Biological notes on three Floridian wasps. Proc. Biol. Soc. Washington 76:139-152.

Krombein, K. V. and L. M. Walkley
1962. Three hymenopterous parasites of an African mud-dauber wasp, *Sceliphron spirifex* (L.). Proc. Ent. Soc. Washington 64:78.

Krombein, K. V. and A. Willink
1951. The North American species of *Bembecinus*. Amer. Midl. Nat. 44:699-712.

Kurczewski, F. E.
1963. A first Florida record and note on the nesting of *Trypoxylon (Trypargilum) texense* Saussure. Florida Ent. 46:243-245.
1966a. Behavioral notes on two species of *Tachytes* that hunt pygmy mole-crickets. J. Kansas Ent. Soc. 39:147-155.
1966b. *Tachysphex terminatus* preying on Tettigoniidae — an unusual record. J. Kansas Ent. Soc. 39:317-322.
1966c. Comparative behavior of male digger wasps of the genus *Tachysphex*. J. Kansas Ent. Soc. 39:436-453.
1967a. A note on the nesting behavior of *Solierella inermis*. J. Kansas Ent. Soc. 40:203-208.
1967b. *Hedychridium fletcheri*, a probable parasite of *Tachysphex similis*. J. Kansas Ent. Soc. 40:278-284.
1968. Nesting behavior of *Plenoculus davisi*. J. Kansas Ent. Soc. 41:179-207.
1969. Comparative ethology of female digger wasps in the genera *Miscophus* and *Nitelopterus*. J. Kansas Ent. Soc. 42:470-509.
1971. A new *Tachysphex* from Florida, with keys to the males and females of the Florida species. Proc. Ent. Soc. Washington 73:111-116.
1972. Observations on the nesting behavior of *Diploplectron peglowi* Krombein. Proc. Ent. Soc. Wash. 74:385-397.

Kurczewski, F. E., N. A. Burdick and G. C. Baumier
1969. Observations on the nesting behavior of *Crossocerus (C.) maculiclypeus* Fox. J. New York Ent. Soc. 77:92-104.

Kurczewski, F. E. and H. E. Evans
1972. Nesting behavior and description of the larva of *Bothynostethus distinctus* Fox. Psyche 79:88-103.

Kurczewski, F. E. and S. E. Ginsberg
1971. Nesting behavior of *Tachytes (Tachyplena) validus*. J. Kansas Ent. Soc. 44:113-131.

Kurczewski, F. E. and B. J. Harris
1968. The relative abundance of two digger wasps, *Oxybelus bipunctatus* and *Tachysphex terminatus*, and their associates in a sand pit in cen-

tral New York. J. New York Ent. Soc. 76: 81-83.

Kurczewski, F. E. and D. J. Peckham
 1970. Nesting behavior of *Anacrabro ocellatus ocellatus*. Ann. Ent. Soc. Amer. 63:1419-1424.

Lanham, U. N.
 1952. Review of the wing venation of the higher Hymenoptera (suborder Clistogastra) and speculations on the phylogeny of the Hymenoptera. Ann. Ent. Soc. Amer. 44:614-628.
 1960. A neglected diagnostic character of the Apoidea. Ent. News 71:85-86.

La Rivers, I.
 1942. Notes on the bembicid, *Stictiella pulla* (Handlirsch). Pan-Pacific Ent. 18:4-8.
 1945. The wasp *Chlorion laeviventris* as a natural control of the mormon cricket. Amer. Midl. Nat. 33:743-763.

Latreille, P. A.
 1796. Précis des caractères génériques des insectes, disposés dans un ordre naturel. xiv + 208 pp. F. Bourdeaux, Paris.
 1802-1803. Histoire naturelle, générale et particulière des Crustacés et des insectes. Vol. 3. xii + 13 + 467 pp. F. Dufart, Paris (dating after Griffin, 1938).
 1804. Tableau méthodique des insectes, pp. 129-200. In: Nouveau dictionnaire d'histoire naturelle, etc. Vol. 24. 238 pp. Déterville, Paris.
 1809. Genera crustaceorum et insectorum. Vol. 4. 397 pp. A. Koenig, Parisiis et Argentorati.
 1810. Considérations générales dur l'ordre naturel des animaux composant les classes des Crustacés des Arachnides, et des Insectes. 444 pp. F. Schoell, Paris.
 1829. Les Crustacés, les Arachnides et les Insectes. Vol. 5. xxiv + 556 pp. In: G. C. Cuvier, Le règne animal, nouvelle édition, revue et augmentée. Déterville, Paris.

Leclercq, J.
 1939. La biologie des *Passaloecus*. Lambillionea 39: 59-62.
 1940. La biologie des *Passaloecus* (2e note). Lambillionea 40:49-52.
 1941. Notes écologiques sur les insectes de pays de Liège. Nat. Maandbl. 30:67-70.
 1950a. Description d'une espèce nouvelle d'*Entomocrabro* de la République d'Equateur. Bull. Soc. Ent. France 55:93-94.
 1950b. Sur les Crabroniens orientaux et australiens rangés par R. E. Turner (1912-1915) dans le genre *Crabro* (subgenus *Solenius*). Bull. Ann. Soc. Ent. Belgique 86:191-198.
 1950c. Notes systématiques sur les Crabroniens pédonculés. Bull. Inst. Roy. Sci. Nat. Belgique 26(15): 1-19.
 1950d. Les *Ectemnius* sud-américains du sous-genre *Apoctemnius*. Rev. Franciase Ent. 17:200-211.
 1951a. Notes systématiques sur quelques Crabroniens américains, orientaux et australiens. Bull. Ann. Soc. Ent. Belgique 87:31-56.
 1951b. Notes systématiques sur les *Anacrabro* (Packard, 1866). Bull. Soc. Ent. France 55:61-64.
 1951c. Sur quelques *Neodasyproctus* nouveaux ou peu connus. Rev. Zool. Bot. Afr. 44:333-337.
 1951d. La position générique du *Crabro tabanicida* Fischer. Bull. Soc. Ent. France 56:105-106.
 1953. Notes détachées sur les Hyménoptères Aculéates de Belgique. Bull. Ann. Soc. Ent. Belgique 89: 245-250.
 1954a. Monographie systématique, phylogénétique et zoogéographique des Hyménoptères Crabroniens. 371 pp. Lejeunia press, Liège.
 1954b. Notes détachées sur les Hyménoptères Aculéates de Belgique. Bull. Ann. Soc. Ent. Belgique 90:290-292.
 1955a. *Quexua* (Pate, 1942) nouveaux ou peu connus. Bull. Ann. Soc. Ent. Belgique 91:296-299.
 1955b. Sur les *Foxita* (Pate, 1942) des faunes brésilienne et bolivienne. Mem. Soc. Ent. Belgique 27:336-342.
 1955c. Hymenoptera Sphecoidea (Sphecidae I, subfam. Sphecinae). Parc. Natl. Upemba. Mission G. F. Witte, fasc. 34, 134 pp.
 1955d. Révision des *Podagritus* australiens. Bull. Ann. Soc. Ent. Belgique 91:300-304.
 1955e. Révision des *Rhopalum* (Kirby, 1829) néozélandais. Bull. Inst. Roy. Sci. Nat. Belgique 31(82):1-18.
 1956a. Le genre *Piyuma*, curiosité biogéographique. Bull. Inst. Belgique 32(16):1-4.
 1956b. Les *Dasyproctus* du sud-est asiatique et de l'Océanie. Bull. Ann. Soc. Ent. Belgique 92: 139-167.
 1956c. *Encopognathus* (*Florkinus*, subgen. nov.) *evolutionis* n. sp., Crabronien nouveau du Mexique. Notes sur sa signification phylogénétique, et remarques sur deux *Encopognathus* de l'Inde. Bull. Inst. Roy. Sci. Nat. Belgique 32(20):1-12.
 1956d. Contribution à l'étude des *Crossocerus* vivant au sud de l'Himalaya. Bull. Ann. Soc. Ent. Belgique 92:217-235.
 1957a. Recherches systématiques et taxonomiques sur le genre *Podagritus*. I. Sur onze espèces australiennes et une espèce des îles Fidji. Bull. Inst. Roy. Sci. Nat. Belgique 33(15):1-7.
 1957b. Sur le genre *Vechtia*, lignée orientale de Crabroniens évolués. Ent. Bericht. 17:106-107.
 1957c. Le genre *Rhopalum* en Australie. Bull. Ann. Soc. Ent. Belgique 93:177-232.
 1957d. Recherches systématiques et taxonomiques sur le genre *Podagritus*. II. Introduction à l'étude des espèces sud-amèricaines et révision des sousgenres *Echucoides* et *Echuca*. Bull. Inst. Roy. Sci. Nat. Belgique. 33(22):1-23.
 1957e. Sur les éléments du mésopectus des Hyménoptères Sphecides. Bull. Inst. Roy. Sci. Nat. Belgique 33(34):1-8.
 1957f. Recherches systématiques et taxonomiques sur le genre *Podagritus*. III. Révision des *Podagritus* subg. *Podagritus*. Bull. Inst. Roy. Sci. Nat. Belgique 33(46):1-18.
 1958a. Sphecoidea (Sphecidae II. subfam. Crabroninae). Parc. Natl. Upemba. Mission G. F. Witte, fasc. 45, 114 pp.
 1958b. Crabroniens du sud-est asiatique, nouveax ou peu connus. III. Genres *Encopognathus* et *Entomognathus*. Bull. Ann. Soc. Ent. Belgique 94: 99-101.
 1958c. Genre *Ectemnius*: tableau des sous-genres;

espèces appartenant aux sous-genres *Thyreocerus, Policrabro, Yanonius, Clytochrysus* et *Metacrabro*. Bull. Ann. Soc. Ent. Belgique 94:102-117.

1958d. Crabroniens du sud-est asiatique, nouveaux ou peu connus. V. Révision des *Ectemnius* subg. *Cameronitus* Leclercq. Bull. Ann. Soc. Ent. Belgique 94:134-155.

1959. Pemphredoninae. Parc. Natl. Upemba, I. Mission G. F. Witte, fasc. 53(2):17-62.

1961a. Psenini. Parc. Natl. Upemba. Mission G. F. Witte, fasc. 60 (3):13-36.

1961b. Sphecoidea: Sphecidae (subfam. Sphecinae, Pemphredoninae et Crabroninae). Parc. Natl. Garamba. Mission H. Saeger, fasc. 20(3):43-105.

1961c. Découverte d'une nouvelle Ammophile d'Espagne appartenant au groupe *fallax* Kohl. Rev. Española Ent. 37:211-214.

1961d. Hyménoptères Ampulcides et Sphecides récoltés par le Dr. Fred Keiser à Madagascar. Verhandl. Nat. Ges. Basel 72:100-119.

1961e. Sur les *Sphex* africains du groupe *haemorrhoidalis* Fabricius. Bull. Inst. Agron. Sta. Recher. Gembloux 29:323-327.

1961f. Diagnoses de quatre Crabroniens du sud-est Asiatique. Bull. Inst. Agron. Sta. Recher. Gembloux 29:71-78.

1963. Crabroniens d'Asie et des Philippines. Bull. Ann. Soc. Ent. Belgique 99:1-82.

1964. Notes detachées sur les Hyménoptères Aculéates de Belgique. Bull. Ann. Soc. Ent. Belgique 90:290-292.

1965. Sphecidae subfam. Trypoxyloninae. Parc. Natl. Garamba. Mission 4. Saeger, fasc. 46(5):67-153.

1966. *Chalybion bonneti* n. sp., Sphécide nouveau de Madagascar. Bull. Recher. Agron. Gembloux (n.s.) 1:55-59.

1968a. Sphecidae, subfam. Larrinae, genre *Paranysson*. Parc. Natl. Garamba. Mission H. Saeger, fasc. 53(5):79-95.

1968b. Les Crabroniens du genre *Ectemnius* en Amérique Latin. Ann. Soc. Ent. France (n.s.) 4:298-328.

1968c. Crabroniens des genres *Crossocerus* et *Enoplolindenius* trouvés en Amérique Latine. Bull. Soc. Roy. Sci. Liège 37:90-107.

1970. Crabroniens du genre *Rhopalum* trouvés en Amérique Latine. Bull. Soc. Roy. Sci. Liège 39:85-104.

1972a. Crabroniens du genre *Dasyproctus* trouvés en Asie et en Océanie. Bull. Soc. Roy. Sci. Liège 41:101-122.

1972b. Identification des Crabroniens du genre *Ectemnius* trouvé en Amérique Latine. Bull. Soc. Roy. Sci. Liège 41:195-226.

1973a. Un néotype pour *Blepharipus nigrita* Lepeletier et Brullé. Ent. Bericht. 33:52-53.

1973b. Crabroniens de Formose. Bull. Ann. Soc. Roy. Belge Ent. 109:285-304.

1973c. Espèces neotropicales d'*Anacrabro*. Acta Zool. Lilloana 30:37-51.

Lefroy, H. M.
1904. A note on the habits of *Chlorion (Sphex) lobatus*. J. Bombay Nat. Hist. Soc. 15:531-532.

Lepeletier de Saint-Fargeau, A.
1832. Mémoire sur G. *Gorytes* Latr. *Arpactus* Jur. Ann. Soc. Ent. France 1:52-79.

1845. Histoire naturelle des insectes. Hyménoptères. Vol. 3. 644 pp. Roret, Paris.
Obit.: Audinet-Serville, M., 1846. Ann. Soc. Ent. France (2) 4:193-200, bibliography.

Lepeletier de Saint-Fargeau, A. and A. Brullé
1834. Monographie du genre *Crabro*, de la famille des Hyménoptères fouisseurs. Ann. Soc. Ent. France 3:683-810.

Lichtenstein, A. A. H.
1796. Catalogus Musei zoologici ditissimi Hamburgi. Auctionis lege distrahendi. Sectio tertia: Insecta. 222 pp. Hamburg, G. F. Schniebes.

Liebermann, J.
1931. Esfégidos argentinos del género *Sphex*, con la descripción de una neuva especie de esfégido tucuricida. Anal. Soc. Cien. Argentina 112:5-26, 79-101.

Lin, C. S.
1966. Bionomics of *Isodontia mexicana* with a review of generic ethology. Wasmann J. Biol. 24:239-247.

1967. Nesting behavior of *Tachytes (Tachyplena) praedator* Fox, with a review of the biology of the genus. Amer. Midl. Nat. 77:241-245.

1969. Biology of *Trypargilum spinosum*, with notes on *T. texense* and *T. politum* at Lake Texoma. Wasmann J. Biol. 27:125-133.

1971. Bionomics of *Stictia carolina* at Lake Texoma, with notes on some Neotropical species. Texas J. Sci. 23:275-286.

Lin, N.
1963. Territorial behaviour in the cicada killer wasp, *Sphecius speciosus* (Drury). I. Behaviour 20:115-133.

1966. The use of emergence holes of the cicada killer as nest burrows by *Tachytes*. Bull. Brooklyn Ent. Soc. 59-60:82-84.

Lin, N., and C. D. Michener
1971. Evolution of sociality in insects. Quart. Rev. Biol. 47:131-159.

Linnaeus, C. von
1758. Systema naturae ... etc., 10th ed. Tomus 1. 823 pp. Laurentii Salvii, Holmiae.

Linsley, E. G. and J. W. MacSwain
1954. Observations on the habits and prey of *Eucerceris ruficeps* Scullen. Pan-Pacific Ent. 30:11-14.

1956. Some observations on the nesting habits and prey of *Cerceris californica* Cresson. Ann. Ent. Soc. Amer. 49:71-84.

Linsley, E. G., J. W. MacSwain, and R. F. Smith
1952. The bionomics of *Diadasia consociata* Timberlake and some biological relationships of emphorine and anthophorine bees. Univ. Calif. Pub. Ent. 9:267-290.

Llano, R. J.
1959. Observaciones biologicas de insectos bonarenses. Suplemento Revista Educación, Ministerio Educación Provincia Buenos Aires, La Plata, 136 pp.

Lohrmann, E.
1943. Die Grabwespengruppe der Stizinen. Mitt. Münchner Ent. Ges. 33:188-208.

1948. Die Grabwespengruppe der Bembicinen. Mitt. Münchner Ent. Ges. 34:420-471 (dated 1944, but published Dec. 1, 1948).

McCulloch, A. R.
1923. War in the garden. Australian Mus. Mag. 1: 209-212.

Maidl, F.
1914. Neue Sphegiden aus Westafrika. Boll. Lab. Zool. Gen. Agraria Univ. Portici 9:147-150.

Maidl, F. and A. Klima
1939. Sphecidae I (Astatinae-Nyssoninae). In: H. Hedicke, Hymenopterorum Catalogus. Verlag für Naturwiss. 150 pp. W. Junk, 's-Gravenhage.

Malloch, J. R.
1933. Review of the wasps of the subfamily Pseninae of North America. Proc. U.S. Natl. Mus 82(26): 1-60.

Malloch, J. R. and S. A. Rohwer
1930. New forms of sphecoid wasps of the genus *Didineis* Wesmael. Proc. U.S. Natl. Mus. 77(14):1-7.

Malyshev, S. I.
1968. Genesis of the Hymenoptera and the phases of their evolution, (translated from the Russian), edited by O. W. Richards and B. Uvarov. viii + 319 pp. Methuen and Co., London.

Maneval, H.
1929. Notes sur quelques Hyménoptères. Ann. Soc. Ent. France (5) 98:288-300.
1932. Notes recueillies sur les Hyménoptères. Ann. Soc. Ent. France (6) 101:85-110.
1939. Notes sur les Hyménoptères. Ann. Soc. Ent. France (6) 108:49-108.

Marchal, P.
1892. Notes sur la vie et les moeurs des insectes. Archiv. Zool. Expér. Gen. (2) 10:23-36.

Maréchal, P.
1936. Ethologie des *Trypoxylon* et observations sur *T. attenuatum* Sm. Bull. Ann. Soc. Ent. Belgique 76:373-396.
1938. Recherches sur les *Ammoplanus wesmaeli* et *perrisi* Gir. et description d'une nouvelle espèce d'*Ammoplanus*. Bull. Ann. Soc. Ent. Belgique 78:397-407.

Masuda, H.
1939. Biological notes on *Pison iwatai* Yasumatsu. Mushi 12:114-146.

Matsuda, R.
1971. Morphology and evolution of the insect-thorax. Mem. Ent. Soc. Canada (76):il-431.

Matthews, R. W.
1968a. *Microstigmus comes:* Sociality in a sphecid wasp. Science 160:787-788.
1968b. Nesting biology of the social wasp *Microstigmus comes*. Psyche 75:23-45.
1970. A new thrips-hunting *Microstigmus* from Costa Rica. Psyche 77:120-126.

Matthews, R. W. and J. R. Matthews
1968. A note on *Trypargilum arizonense* in trap nests from Arizona, with a review of prey preferences and cocoon structure in the genus. Psyche 75: 285-293.

Matthews, R. W. and H. E. Evans
1971. Biological notes on two species of *Sericophorus* from Australia. Psyche 77:413-429.

Mayer, J. and A. von Schulthess
1923. Beobachtungen an Nestern von geselligen und solitären Wespen. Mitt. Ent. Zurich umgebung (6):357-367.

Mazek-Fialla, K.
1936. Angaben zur Lebensweise von *Sceliphron destillatorius* Illig. am Neusiedlersee, mit besonderer Berücksichtigung des Nestbaues. Zeitschr. Wiss. Zool. (A) 148:298-308.

Medler, J. T.
1965. Biology of *Isodontia (Murrayella) mexicana* in trap nests in Wisconsin. Ann. Ent. Soc. Amer. 58:137-142.
1967. Biology of *Trypoxylon* in trap nests in Wisconsin. Amer. Midl. Nat. 78:344-358.

Menke, A. S.
1961. A review of the Nearctic species of *Chlorion* with a description of a new species from Baja California. Ann. Ent. Soc. Amer. 54:667-669.
1962a. Notes on two species of sphecine wasps described by H. T. Fernald from South America. Pan-Pacific Ent. 38:63.
1962b. A new genus of digger wasps from South America. Proc. Biol. Soc. Washington 75:303-306.
1963a. Dates of publication of Palisot de Beauvois' "Insectes recueilles en Afrique et en Amérique" Ann. Mag. Nat. Hist. (13) 5:701.
1963b. Notes and synonymy of some Neotropical *Sphex* and *Isodontia* described by E. Taschenberg and S. Rohwer. Pan-Pacific Ent. 39: 228-230.
1964a. New species of North American *Ammophila*. Acta Hymenopterologica 2:5-27.
1964b. Miscellanous notes on *Ammophila*. Ent. News 75:149-155.
1964c. A new subgenus of *Ammophila* from the Neotropical Region. Canad. Ent. 96:874-883.
1964d. Notes on two Sphecinae described by Fox. Pan-Pacific Ent. 40:238-240.
1965a. Synonymical notes on New World wasps of the subfamily Sphecinae. Tijdschr. Ent. 108:205-217.
1965b. The identity of some *Ammophila* observed by C. H. Hicks, H. E. Evans and others in connection with biological studies. Ent. News 66: 257-261.
1966a. The genera of the Ammophilini. Canad. Ent. 98:147-152.
1966b. New species of North American *Ammophila*, part II. Proc. Biol. Soc. Washington 79:25-40.
1966c. The identity of *Parapsammophila lutea* Taschenberg, 1869. Polskie Pismo Ent. 36:57-61.
1967a. New species of North American *Ammophila*, part III. Los Angeles Co. Mus. Contrib. Sci. (123):1-8.
1967b. New genera of Old World Sphecidae. Ent. News 78:29-35.
1967c. A new South American species of *Bothynostehus*. Pan-Pacific Ent. 43:140-141.
1967d. *Odontosphex* Arnold, a genus of the Philanthinae, with a key to the tribes and genera of the subfamily. Pan-Pacific Ent. 43:141-148.
1968a. New genera and species of wasps of the tribe Trypoxylonini from the Neotropical Region. Los Angeles Co. Mus. Contrib. Sci. (135):1-9.
1968b. New species of Neotropical Sphecidae. Pan-Pacific Ent. 43:309-314.
1968c. A review of the New World species of *Pison*.

1. The subgenus *Krombeiniellum.* Canad. Ent. 100:1100-1107.
1968d. New South American genera and species of the tribe Bothynostethini. Acta Zool. Lilloana 22: 89-99.
1968e. Two new species of *Nitela* with taxonomic notes on the taxa *Tenila, Rhinonitela* and *Nitela rugosa.* Mushi 42:133-139.
1969. A new *Nitela,* subgenus *Tenila,* with a key to the species of the subgenus. Proc. Ent. Soc. Washington 71:197-201.
1970. The genus *Ammophila* in the West Indies. Proc. Ent. Soc. Washington 72:236-239.
1974a. A preliminary review of the *agile* group of *Podium* Fabricius. Proc. Wash. Acad. Sci. 64: 147-153.
1974b. The dates of publication of A. G. Dahlbom's Hymenoptera Europaea, Vol. 1. Polskie Pismo Ent. 44:315-317.
1974c. *Hoplammophila* de Beaumont: a key to the species. Polskie Pismo Ent. 44: 581-583.

Menke, A. S., R. M. Bohart, and O. W. Richards
1974a. Proposed emendation to *Nysson* of the name *Nysso* Latreille, 1796 (Hymenoptera, Sphecidae) (Z.N.(S.)2055). Bull. Zool. Nomencl. 30:217-218.
1974b. *Euplilis* Risso, 1826 (Hymenoptera, Sphecidae): proposed suppression under the plenary powers in favor of *Rhopalum* Stephens, 1829 (Z.N.(S.)2056). Bull. Zool. Nomencl. 30: 219-220.

Menke, A. S., R. M. Bohart, and J. van der Vecht
1966. *Podalonia* Spinola, 1853: proposed suppression under plenary powers in favour of *Podalonia* Fernald, 1927, with *Ammophila violaceipennis* Lepeletier, 1845, as type-species, Z. N. (S.) 1735. Bull. Zool. Nomencl. 23:48-51.

Menke, A. S. and C. D. Michener
1973. *Sericogaster* Westwood, a senior synonym of *Holohesma* Michener. J. Australian Ent. Soc. 12: 173-174.

Menke, A. S. and A. Willink
1964. A survey of the Neotropical species of *Chlorion.* Ann. Ent. Soc. Amer. 57:548-552.

Mercet, R. G.
1906. Los "*Gorytes y Stizus*" de España. Mem. Real Soc. Española Hist. Nat. 4:111-158.
1907. El género *Prosopigastra.* Bol. Real Soc. Española Hist. Nat. 7:292-304.
1915. Una especie nueva de Crabrónido. Bol. Real Soc. Española Hist. Nat. 15:366-369.

Merisuo, A. K.
1932. Drei interessante *Crabro*-Funde. Notulae Ent. 12:89-92.
1972. Über die Tyloide der finnischen *Passaloecus*-Männchen. Ann. Ent. Fennici 38:203-207.
1973a. Zur Biologie von *Passaloecus clypealis* Faester. Ann. Ent. Fennici 39:103-108.
1973b. Beiträge zur Kenntnis der finnischen Arten der Gattung *Passaloecus* Shuckard. Ann. Ent. Fennici 39:108-119.

Merisuo, A. K. and E. Valkeila
1972. Beiträge zur Kenntnis der paläarktischen Arten der Gattung *Pemphredon* Latreille. Ann. Ent. Fennici 38:7-24.

Meunier, F.
1889. Description d'une nouvelle espèce ou peu connue de "Crabronides" de la tribu des Mellinites. Le Naturaliste (2) 3:24-25.

Micheli, L.
1933. Note biologiche e morfologiche sugli imenotteri (contributo 4). Mem. Soc. Ent. Italiana 12:5-15.

Michener, C. D.
1944. Comparative external morphology, phylogeny, and a classification of the bees. Bull. Amer. Mus. Nat. Hist. 82:151-326.
1965. A classification of the bees of the Australian and South Pacific Regions. Bull. Amer. Mus. Nat. Hist. 130:1-362.
1971. Notes on crabronine wasp nests. J. Kansas Ent. Soc. 44:405-407.

Mickel, C. E.
1916a. New species of Hymenoptera of the superfamily Sphecoidea. Trans. Amer. Ent. Soc. 42:399-434.
1916b. A review of the American species of *Xylocelia.* Ann. Ent. Soc. Amer. 9:344-352.

Miller, R. C. and F. E. Kurczewski
1972. A review of nesting behavior in the genus *Entomognathus* with notes on *E. memorialis* Banks. Psyche 79:61-78.

Mingo, P. E.
1964. Los Psenini de España. Bol. Real Soc. Española Hist. Nat. 62:155-173.

Minkiewicz, R.
1933. Nids et proies des Sphégiens de Pologne; troisième série. Polskie Pismo Ent. 12:181-261.

Mochi, A.
1939a. Revisione delle specie egiziane del genere *Cerceris* Latr. Bull. Soc. Fouad Ier Ent. 22:136-228.
1939b. Revisione delle specie egiziane dei generi *Philanthus* F. e *Nectanebus* Spin. Bull. Soc. Fouad Ier Ent. 23:86-136.
1939c. Revisione delle specie egiziane del genere *Stizus* Latr. Bull. Soc. Fouad Ier Ent. 23:183-237.
1940. *Ammoplanopterus* nov. gen. *sinaiticus* nov. spec. Bull. Soc. Ent. Egypt 24:27-30.

Móczár, L.
1952. Contribution à l'éthologie du *Palarus variegatus* F. Ann. Hist.-Nat. Mus. Natl. Hungarici 2: 119-124.
1958. Die Ungarischen Vertreter der Tribus Oxybelini unter Berücksichtigung der westpaläarktischen Arten. Ann. Hist.-Nat. Mus. Natl. Hungarici 50:281-299.
1959. Hymenoptera III. Specoidea II. Fauna Hungariae 44(8):1-87.
1967. Hymenoptera III. Chrysidoidea. Fauna Hungariae 86(2):1-118.

Morawitz, A.
1864. Verzeichniss der um St.-Petersburg aufgefundenen Crabroninen. Bull. Acad. Imp. Sci. St. Petersbourg 7:451-463.
1866. Einige Bemerkungen über die *Crabro*-Artigen Hymenopteren. Bull. Acad. Imp. Sci. St. Petersbourg 9:243-273.

Morice, F. D.
1897. New or little-known Sphegidae from Egypt. Trans. Ent. Soc. London 1897:301-316.
1911. Hymenoptera Aculeata collected in Algeria.

The Sphegidae. Trans. Ent. Soc. London 1911: 62-135.
1914. What was *Sphex xanthocephala* Forster (A British insect, but ignored in British lists)? Ent. Mon. Mag. (2)25:268-288.

Morice, F. D. and J. H. Durrant
1915. The authorship and first publication of the "Jurinean" genera of Hymenoptera. Trans. Ent. Soc. London 1914:339-436.

Muesebeck, C. F. W., et al.
1951. Hymenoptera of America north of Mexico. synoptic catalog. U.S. Dept. Agr. Monograph No. 2, 1420 pp.

Müller, M.
1911. Hymenopteren in *Lipara*-Gallen, mit besonderer Berücksichtigung der Raubwespe *Cemonus*. Ent. Rundschau 28:105-114.

Muma, M. H. and W. F. Jeffers
1945. Studies of the spider prey of several mud-dauber wasps. Ann. Ent. Soc. Amer. 38:245-255.

Murray, W. D.
1938. Some revisions in the genus *Sphex* with one new species, a subspecies, and a new name. Ann. Ent. Soc. Amer. 31:17-42.
1940. *Podalonia* of North and Central America. Entomologica Americana 20:1-82.

Musgrave. A.
1925. The sand wasp's burrow. Australian Mus. Mag. 2:243.

Myartseva, S. N.
1963. [On the biology and nesting behavior of the wasp *Ammophila ebenina* Spin.] Izvest. Akad. Nauk Turkmen SSR, Ser. Biol. Nauk 1963 (2):62-71 (in Russian).
1968. [Fossorial wasps of the genus *Sceliphron* in Turkmenia] Izvest. Akad. Nauk Turkmen SSR, Ser. Biol. Nauk 1968 (2):61-66 (in Russian).
1969. [Peculiarities of the biology and ethology of the sphecid tribe Ammophilini in Turkmenia] Izvest. Akad. Nauk Turkmen SSR, Ser. Biol. Nauk 1969(3):48-56 (in Russian).

Myers, J. G.
1934. Two Collembola-collecting crabronids in Trinidad. Trans. Roy. Ent. Soc. London 82:23-26.

Nagy, C. G.
1969. A new taxon of the family Heterogynidae Latreille. Ent. Mitt. Zool. Staatsinst. Zool. Mus. Hamburg 64:7-11.
1971. Studies on the Ampulicidae. Boll. Soc. Ent. Italiana 103:103-107.

Nambu T.
1966-1967. Studies on the biology of eight species of *Trypoxylon* occurring in Japan, with some notes on their parasites. Life Study (Fukui) 10:25-34 (1966), 11:6-16 (1967).
1970. On the habits of *Larra amplipennis* (Smith). Life Study (Fukui) 14:1-8.
1971a. Biology of *Trypoxylon responsum hatogayuum* Tsuneki in Saitama, Japan. Life Study (Fukui) 15:1-7.
1971b. Biological observation on *Trypoxylon pennsylvanicum japonense* Tsuneki in Saitama, Japan. Life Study (Fukui) 15:69-74.

Newton, R. C.
1956. Digger wasps, *Tachysphex* ssp., as predators of a range grasshopper in Idaho. J. Econ. Ent. 49:615-619.
1960. The nesting habits of the wasp, *Solierella* sp., and the location of its egg on the grasshopper, *Ageneotettix deorum*. J. Econ. Ent. 53:958-959.

Nicolas, M.
1894. Le *Sphex splendidulus* da Costa. Assoc. Française Avancement Sci., Compte Rendu 22 Sess., Besançon 1893 (2):636-647.

Nielsen, E. T.
1945. Moeurs des *Bembex*. Monographie biologique ave quelques considérations sur la variabilité des habitudes. Spolia Zool. Mus. Hauniensis 7:1-174.

Noskiewicz, J. and S. Chudoba
1949. Nouvelles espèces des Hyménoptères fouisseurs de la Silésie et les nouvelles stations des espèces plus rares. Polskie Pismo Ent. 18:406-431.
1950. Les espèces silésiennes due genre *Oxybelus* Latr. Polskie Pismo Ent. 19:299-323.

Noskiewicz, J. and W. Pulawski
1960. Klucze do oznaczania owadów Polski. Hymenoptera, Sphecidae. Polski Zwiazek Ent. 24(67):1-185.

Oehlke, J.
1965. Beitrag zur Systematik und Faunistik des Genus *Psen* Latr. Subgenus *Mimesa* Shuck. Acta Ent. Mus. Natl. Prague 36:339-347.
1969. Beiträge zur Insekten-Fauna der DDR: Hymenoptera – Bestimmungstabellan bis zu den Unterfamilien. Beitr. Ent. 19:753-801.
1970. Beiträge zur Insekten-Fauna der DDR: Hymenoptera – Sphecidae. Beitr. Ent. 20:615-812.

Oeser, R.
1971. Die Abdominal-Basis von *Ammophila sabulosa* (L., 1758). Mitt. Zool. Mus. Berlin 47:33-42.

Ohgushi, R.
1954. On the plasticity of the nesting habit of a hunting wasp, *Pemphredon lethifer fabricii* Müller. Mem. College Sci. Univ. Kyoto (B) 21:45-48.

Okumura, T.
1971. Biology of *Liris (Liris) aurulenta* Fabr. and *L. (Dociliris) deplanata binghami* Tsun. observed on the island of Amami-Ohshima, the Ryukyus. Life Study (Fukui) 15:22-26.

Olberg, G.
1952. Die Sandwespen. Die neue Brehm-Bücherei, Heft 68, 58 pp.
1953. Der Bienenfeind *Philanthus* (Bienenwolf). Die neue Brehm-Bücherei, Heft 94, 88 pp.
1959. Des Verhalten der solitären Wespen Mitteleuropas. xiv + 402 pp. Deut. Verlag Wiss., Berlin.
1966. Brutfürsorge und koloniales Verhalten der Heuschreckensandwespe *Sphex maxillosus*. Natur und Mus. 96:1-8.

Oniki, Y.
1970. Brazilian sphecid wasps in occupied hummingbird nests. J. Kansas Ent. Soc. 43:354-356.

Orfila, R. N. and J. Salellas
1929. Notas biologicas sobre *Sceliphron figulus* (Dahl.) D. T., Rev. Soc. Ent. Argentina 2:247-250.

Packard, A. S.
1866-1867. Revision of the fossorial Hymenoptera of North America I. Crabronidae and Nyssonidae.

Proc. Ent. Soc. Philadelphia 6:39-115 (1866), 353-444 (1867).
Obit.: Cockerell, T. D. A., 1920. Biogr. Mem. Natl. Acad. Sci. 9:181-236, portrait and bibliography.

Paetzel, M. M.
1973. Behavior of the male *Trypoxylon rubrocinctum*. Pan-Pac. Ent. 49:26-30.

Pagden, H. T.
1933. Two new Malayan sphecoids. Trans. Ent. Soc. London 81:93-101.
1934. Biological notes on some Malayan aculeate Hymenoptera. J. Fed. Malay States Mus. 17:458-486.

Palmer, J. A. and A. W. Stelfox
1931. On the habits of *Sphex affinis (lutaria)* in Ireland. Ent. Mon. Mag. 67:130-133.

Panzer, G. W. F.
1804. Faunae Insectorum Germanicae initiae oder Deutschlands Insekten. Heft 86, 24 plates (dating after Sherborn, 1923).
1805. Faunae Insectorum Germanicae initiae; oder Deutschlands Insekten. Heft 98, 24 plates (dating after Sherborn, 1923).
1806. Kritische Revision der Insektenfaune Deutschlands. Vol. 2. xii + 271 pp. Nürnberg.

Parker, F. D.
1960. A systematic study of the North American *Priononyx*. Pan-Pacific Ent. 36:205-208.
1962a. Two hosts of *Lomachaeta variegata* Mickel. Pan-Pacific Ent. 38:116.
1962b. A host of *Chrysis (Trichrysis) mucronata* Brullé and an additional host of *Chrysis (Chrysis) coerulans* Fabricius. Pan-Pacific Ent. 38:140.
1962c. On the subfamily Astatinae, with a systematic study of the genus *Astata* of America north of Mexico. Ann. Ent. Soc. Amer. 55:643-659.
1964. On the subfamily Astatinae, part II. A review of the genus *Astata* from Mexico and Central America. Ann. Ent. Soc. Amer. 57:552-559.
1966a. On the subfamily Astatinae. Part III. A new genus of Mexican Astatinae with notes on related genera. Ann. Ent. Soc. Amer. 59:765-767.
1966b. A review of the genus *Xenosphex* Williams with biological notes. Pan-Pacific Ent. 42:190-195.
1968a. On the subfamily Astatinae. Part V. Notes on the Caribbean species in the genus *Astata* Latreille. Pan-Pac. Ent. 43:328-330.
1968b. On the subfamily Astatinae. Part IV. The South American species in the genus *Astata* Latreille. Ann. Ent. Soc. Amer. 61:844-852.
1969. On the subfamily Astatinae. Part VI. The American species in the genus *Dryudella* Spinola. Ann. Ent. Soc. Amer. 62:963-976.
1972. On the subfamily Astatinae. Part VII. The genus *Diploplectron* Fox. Ann. Ent. Soc. Amer. 65:1192-1203.

Parker, F. D. and R. M. Bohart
1966. Host-parasite associations in some twig-nesting Hymenoptera of western North America. Pan-Pacific Ent. 42:91-98.
1968. Host-parasite associations in some twig-nesting Hymenoptera from western North America. Part II. Pan-Pacific Ent. 44:1-6.

Parker, J. B.
1910. Notes on the nesting habits of *Bembex nubilipennis*. Ohio Nat. 10:163-165.
1915. Notes on the nesting habits of some solitary wasps. Proc. Ent. Soc. Washington 17:70-77.
1917. A revision of the bembicine wasps of America north of Mexico. Proc. U.S. Natl. Mus. 52(2173):1-155.
1921. Notes on the nesting habits of *Tachytes*. Proc. Ent. Soc. Washington 23:103-107.
1925. Notes on the nesting habits of *Bembix comata* Parker. Proc. Ent. Soc. Washington 27:189-195.
1929. A generic revision of the fossorial wasps of the tribes Stizini and Bembicini with notes and descriptions of new species. Proc. U.S. Natl. Mus. 75(5):1-203.

Parker, J. R. and W. B. Mabee
1928. Montana insect pests for 1927 and 1928. Montana Agr. Exp. Sta. Bull. 216, 23 pp.

Pate, V. S. L.
1929. Studies on oxybeline wasps. I. *Enchemicrum*, an annectant between *Belomicrus* and *Oxybelus*. Ent. News 40:219-222.
1935. Synonymical notes on the fossorial wasps. Ent. News 46:244-250, 264-267.
1936a. Studies in the nyssonine wasps. I: the species of *Psammaletes*, a new subgenus of *Hoplisoides*. Trans. Amer. Ent. Soc. 42:49-56.
1936b. *Rhectognathus*, a new group in the *Lindenius* complex. Ent. News 47:147-153.
1937a. Studies in the nyssonine wasps. II. The subgenera of *Sphecius*. Bull. Brooklyn Ent. Soc. 31:198-200.
1937b. The third Nearctic species of *Nitela* with remarks on the genera *Tenila* Bréthes and *Rhinonitela* Williams. Bull. Brooklyn Ent. Soc. 32:5-7.
1937c. Studies in the pemphredonine wasps, I. New genera and species of the ammoplanoid complex. Trans. Amer. Ent. Soc. 63:89-125.
1937d. The generic names of the sphecoid wasps and their type species. Mem. Amer. Ent. Soc. 9:1-103.
1937e. The oxybeline wasps of the Philippines, with a synonymic catalogue of the Oriental species. Philippine J. Sci. 64:373-396.
1938a. Studies in the nyssonine wasps. III. A revision of the genus *Harpactostigma*. Trans. Amer. Ent. Soc. 64:57-77.
1938b. Studies in the nyssonine wasps. IV. New or redefined genera of the tribe Nyssonini, with descriptions of new species. Trans. Amer. Ent. Soc. 64:117-190.
1939. Studies in the pemphredonine wasps. II. Records and descriptions of new forms in the ammoplanoid complex from the southwestern United States. Trans. Amer. Ent. Soc. 64:373-420.
1940a. The taxonomy of the Oxybeline wasps. I. A review of the genera *Belomicroides*, *Brimocelus* and *Belomicrus* with particular reference to the Nearctic species. Trans. Amer. Ent. Soc. 46:1-99.
1940b. The taxonomy of the Oxybeline wasps. II. The classification of the genera *Belomicrus* and *Enchemicrum*. Trans. Amer. Ent. Soc. 46:209-264.
1940c. On two new genera of nyssonine wasps from the

Neotropical Region. Notulae Nat. (55):1-9.
1941a. A review of the genus *Entomocrabro*. Rev. Ent. 12:45-61.
1941b. On a new subgenus of pemphilidine wasps from Cuba. Ent. News 52:121-125.
1942a. On *Quexua*, a new genus of pemphilidine wasps from tropical America. Rev. Ent. 13:54-75.
1942b. A review of the myrmecotherous genus *Tracheliodes*. Lloydia 5:222-244.
1942c. The New World genera and species of the *Foxita* complex. Rev. Ent. 13:367-421.
1943a. Nearctic *Ammoplanus*. Bull. So. Calif. Acad. Sci. 41:141-163.
1943b. On the taxonomy of the genus *Encopognathus*. Lloydia 6:53-76.
1944a. The subgenera of *Crossocerus*, with a review of the Nearctic species of the subgenus *Blepharipus*. Lloydia 6:267-317.
1944b. Conspectus of the genera of pemphilidine wasps. Amer. Midl. Nat. 31:329-384.
1945. Notes on *Ammoplanus*. Pan-Pacific Ent. 21:81-88.
1946a. Prey records of gorytine wasps. Bull. Brooklyn Ent. Soc. 41:99.
1946b. On *Eupliloides*, an Oriental subgenus of *Crossocerus*. Proc. Ent. Soc. Washington 48:53-60.
1947a. On the genera of philanthine wasps, with the description of a new species from Arizona. Pan-Pacific Ent. 23:63-67.
1947b. A revision of the genus *Arigorytes*. Canad. Ent. 79:51-56.
1947c. On *Williamsita*, a genus of wasps from New Caledonia. Proc. Ent. Soc. Washington 49:107-112.
1947d. New pemphilidine wasps, with notes on previously described forms: II. Notulae Nat. (185):1-14.
1948a. A review of the genus *Moniaecera*. Trans. Amer. Ent. Soc. 74:41-60.
1948b. New pemphilidine wasps from southern Nigeria. Proc. U.S. Natl. Mus. 98:149-162.
Obit.: Kempf, W., 1961. Stud. Ent. 4:542-545. Krombein, K., 1961. Ent. News 72:1-5, portrait.

Patton, W. H.
1879a. The American Bembecidae: tribe Stizini. Bull. U.S. Geol. Geogr. Surv. 5:341-347.
1879b. List of a collection of aculeate Hymenoptera made by Mr. S. W. Williston in northwestern Kansas. Bull. U.S. Geol. Geogr. Surv. 5:349-370.
1881a. Some characters useful in the study of the Sphecidae. Proc. Boston Soc. Nat. Hist. 20:378-385 (dating after Britton and Howard, 1921).
1881b. List of the North American Larradae. Proc. Boston Soc. Nat. Hist. 20:385-397 (dating after Britton and Howard, 1921).
1881c. Notes on the Philanthinae. Proc. Boston Soc. Nat. Hist. 20:297-405 (dating after Britton and Howard, 1921).
1892. Notes upon Larradae. Ent. News 3:89-90.
1897. *Clypeadon*, new genus (*C. 4-notatus* Ashm.). Ent. News 8:13.
1909. A synonymical definition of *Nysson* and of *N. aurinotus*. Ohio Nat. 9:442-445.

Peckham, D. J., F. E. Kurczewski, and D. B. Peckham
1973. Nesting behavior of Nearctic species of *Oxybelus*. Ann. Ent. Soc. Amer. 66:647-661.

Peckham, G. W. and E. G. Peckham
1898. On the instincts and habits of the solitary wasps. Wisconsin Geol. Nat. Hist. Survey, Sci. Ser., Bull. no. 2, iv + 245 pp.
1900. Additional observations on the instincts and habits of the solitary wasps. Bull. Wisconsin Nat. Hist. Soc. 1:85-93.

Petzer, J.
1936. Der Bienenwolf *Philanthus triangulum*. Natur am Niederrhein 12:41-49.

Perkins, R. C. L.
1899. Hymenoptera Aculeata, Fossores, pp. 7-115. In: D. Sharp, Fauna Hawaiiensis. Vol. 1, part 1. 122 pp. J. Clay, Cambridge.
1902. On the generic characters of Hawaiian Crabronidae; four new genera characterized. Trans. Ent. Soc. London 1902:145-148.
1913. On the classification of British Crabronidae. Trans. Ent. Soc. London 1913:383-398.

Perty, M.
1833. Insecta Brasiliensia, fasc. 3, pp. 125-224, pls. 25-40. In: M. Perty, 1830-1833. Delectus animalium articulatorum, etc. 44 + 224 pp. Monachii (dating after Horn and Schenkling, 1928).

Peters, D. S.
1973. *Crossocerus dimidiatus* (Fabricius, 1781), eine weitere soziale Crabroninen-Art. Insectes Soc. 20:103-108.

Picard, F.
1903. Recherches sur l'éthologie du "*Sphex maxillosus*" F. Mem. Soc. Natl. Sci. Nat. Math. Cherbourg 33:97-130.

Piccioli, F.
1869-1870. Descrizione di un nuovo genere d'Imenotteri della famiglia degli Sfecidei spettante alla fauna Toscana. Boll. Soc. Ent. Italiana 1:282-285 (1869) (plate 1 was published in vol. 2, 1870).

Piel, R. P.
1933. Recherches biologiques sur les Hyménoptères du bas Yang-Tse (Chine). Ann. Soc. Ent. France 102:109-154.
1935a. Nidification de *Notogonidea japonica* Kohl. Notes Ent. Chinoise 2:165-173.
1935b. Recherches biologiques sur les Hyménoptères du Yang-Tse (Chine), étude sur les Sphégides. Ann. Soc. Ent. France. 104:273-306.

Poulton, E. B.
1917. Predaceous reduviid bugs and fossors, with their prey, from the S. Paulo district of southeast Brazil. Proc. Ent. Soc. London 1917:24-31.

Powell, J. A.
1963. Biology and behavior of Nearctic wasps of the genus *Xylocelia*, with special reference to *X. occidentalis* (Fox). Wasmann J. Biol. 21:155-176.
1964. Additions to the knowledge of the nesting behavior of North American *Ammophila*. J. Kansas Ent. Soc. 37:240-258.
1967. Behavior of ground nesting wasps of the genus *Nitelopterus*, particularly *N. californicus*. J.

Kansas Ent. Soc. 40:331-346.
Powell, J. A. and D. J. Burdick
1960. Observations on the nesting behavior of *Astata occidentalis* Cresson in central California. Pan-Pacific Ent. 36:25-30.
Powell, J. A. and J. A. Chemsak
1959a. Some biological observations on *Philanthus politus pacificus* Cresson. J. Kansas Ent. Soc. 32:115-120.
1959b. Biological observations on *Psammaecius adornatus* (Bradley). Pan-Pacific Ent. 35:195-201.
Priesner, H.
1958. The Egyptian species of the genus *Bembyx* F. Bull. Soc. Ent. Egypte 42:1-36.
Pulawski, W. J.
1955. Les espèces européennes du genre *Astata* Latr. Polskie Pismo Ent. 35:33-71.
1958a. Une espèce paléarctique du genre *Diploplectron* Fox. Bull. Soc. Ent. Egypte 42:473-476.
1958b. Sphecidae récoltés pendant un voyage en Bulgarie. Polskie Pismo Ent. 27:161-192.
1959. Espèces nouvelles on peu connues du genre *Astata* Latr. Polskie Pismo Ent. 39:359-416.
1961. Remarques sur les *Parapiagetia* Kohl d'Égypte. Polskie Pismo Ent. 31:85-92.
1962. Les *Tachytes* Panz. de la région paléarctique occidentale et centrale. Polskie Pismo Ent. 32:311-475.
1964. Études sur les Sphecidae d'Égypte. Polskie Pismo Ent. 34:64-155.
1965a. *Diploplectron asiaticum* sp.n. Polskie Pismo Ent. 35:221-224.
1965b. La structure du premier segment abdominal dans le genre *Ammophila* K. et ses conséquences systematiques. Polskie Pismo Ent. 35:259-262.
1965c. Sur la synonymie de certains Sphecidae paléarctiques. Polskie Pismo Ent. 35:563-578.
1967. Hymenoptera from Turkey. Sphecidae, II. (Genera *Astata* Latreille and *Tachysphex* Kohl) Bull. Brit. Mus. (Nat. Hist.) Ent. 19:385-410.
1971. Les *Tachysphex* de la région paléarctique occidentale et centrale. 464 pp. Zaklad Zool. Syst., Polskiej Akad. Nauk, Wroclaw.
1973. Les *Ammatomus* A. Costa de la région paléarctique occidentale et centrale. Polskie Pismo Ent. 43:273-288.
1974. A revision of the Neotropical *Tachysphex*. Polskie Pismo Ent. 44:3-80.
Radoszkowski, O.
1876. Matériaux pour servir à une faune hyménoptérologique de la Russie (suite). Horae Soc. Ent. Rossicae 12:82-150.
1877. Chrysidiformis, Mutillidae and Sphegidae (= section III of part 5), pp. i-ii, 1-42, 1-87 (= section on Sphecidae). In: A. Fedtschenko, Reise in Turkestan. Vol. 2 (Zool. Theil), part 5.
1887. Hyménoptères de Korée. Horae Soc. Ent. Rossicae 21:428-436.
1892. Essai sur une classification des Sphegides in sensu Linneano d'après la structure des armures copulatrices. Bull. Soc. Imp. Nat. Moscou (n.s.) 5:571-596.
Obit.: Porchinsky, I., 1896. Horae Soc. Ent. Rossicae 30:i-vi, portrait and bibliography.
Rafinesque-Schmaltz, C. S.
1815. Analyse de la nature ou tableau de l'univers et des corps organisés. 224 pp. Palerme.
Rasnitsyn, A. P.
1966. [Key to superfamilies and families of Hymenoptera]. Ent. Obozr. 45:599-611 (in Russian, English translation: 1967, Ent. Rev. 45:340-347).
Rathmayer, W.
1962. Paralysis caused by the digger wasp *Philanthus*. Nature 196:1148-1151.
Rau, P.
1915. The differentiation of the cocoons of *Pelopoeus caementarius* and *Chalybion caeruleum*. Psyche 22:62-63.
1922. Ecological and behaviour notes on Missouri insects. Trans. Acad. Sci. St. Louis 24(7):1-71.
1923. A note on the nesting habits of *Tachytes distinctus* Sm. Psyche 30:220-221.
1928a. The nesting habits of the wasp, *Chalybion caeruleum*. Ann. Ent. Soc. Amer. 21:25-35.
1928b. Field studies in the behavior of the non-social wasps. Trans. Acad. Sci. St. Louis 25:325-489.
1933. The jungle bees and wasps of Barro Colorado Island. 317 pp. Phil. Rau, Kirkwood, Mo.
1935a. Additional *Trypoxylon* names in "Jungle Bees and Wasps of Barro Colorado Island". Ent. News 46:188.
1935b. The wasp, *Chalybion cyaneum* Fab., preys upon the black widow spider, *Latrodectus mactans* Fab. Ent. News 46:259-260.
1937. A note on the nesting habits of the roach-hunting wasp, *Podium (Parapodium) carolina* Rohwer. Ent. News 48:91-94.
1938. Additional observations on the sleep of insects. Ann. Ent. Soc. Amer. 31:540-556.
1940. Some mud-daubing wasps of Mexico and their parasites. Ann. Ent. Soc. Amer. 33:590-595.
1943. The nesting habits of certain sphecid wasps of Mexico, with notes on their parasites. Ann. Ent. Soc. Amer. 36:647-653.
1944a. The prey and hunting habits of the wasp, *Trypoxylon politum* Say. Ent. News 55:9-10.
1944b. The nesting habits of the wasp, *Chlorion (Ammobia) pennsylvanicum* L. Ann. Ent. Soc. Amer. 37:439-440.
1946. Notes on the behavior of a few solitary wasps. Bull. Brooklyn Ent. Soc. 41:10-11.
1948. A note on the nesting habits of the wasp, *Pemphredon inornatus* Say. Ann. Ent. Soc. Amer. 41:326.
Rau, P. and N. Rau
1918. Wasp studies afield. xv + 372 pp. Princeton Univ. Press, Princeton.
Rayment, T.
1930. *Microglossa* and *Melitribus*, new genera of Australian bees. Proc. Roy. Soc. Victoria (n.s.) 42:211-220.
1935. A cluster of bees. 750 pp. Endeavour Press, Sydney.
1947. New bees and wasps — part IV. Victorian Nat. 63:256-260.
1953. New bees and wasps — part XXI. Parasites on sericophorine wasps. Victorian Nat. 70:123-127.

1955a. Taxonomy, morphology and biology of sericophorine wasps, with diagnoses of two new genera and descriptions of forty new species and six subspecies. Mem. Natl. Mus. Victoria (19):11-105.
1955b. Biology of two hunting wasps. Australian Zool. 12:132-141.
1959. Hyperparasitism by a minute fly and the specific description of a new species. Australian Zool. 12:330-333.

Reed, E. C.
1894. Entomolojía Chilena. Los fosores o avispas cavadoras. Anal. Univ. Repub. Chile 85: 599-653. (pp. 1-57 of separate).

Reinhard, E. G.
1924. The life history and habits of the solitary wasp, *Philanthus gibbosus*. Smithsonian Rep. 1922, pp. 363-376.
1925a. The wasp *Hoplisus costalis*, a hunter of treehoppers. J. Washington Acad. Sci. 15:107-110.
1925b. The wasp *Nysson hoplisivora*, a parasitic relative of *Hoplisus costalis*. J. Washington Acad. Sci. 15:122-177.
1929a. *Pemphredon* and her enemies. Nature Mag. 13:154-157.
1929b. The witchery of wasps. xxi + 291 pp. Century Co., New York.

Ribaut, H.
1952. Espèces françaises du genre *Passaloecus*. Bull. Soc. Ent. France 57:23-28.

Richards, O. W.
1930. A stylopized *Trypoxylon*. Ent. Mon. Mag. 66: 127-128.
1932. A note on the genus *Microstigmus* Ducke, with the description of a new species. Ann. Mag. Nat. Hist. (10) 9:372-377.
1934. The American species of the genus *Trypoxylon*. Trans. Roy. Ent. Soc. London 82:173-360.
1935a. Notes on the nomenclature of the aculeate Hymenoptera, with special reference to British genera and species. Trans. Roy. Ent. Soc. London 83:143-176.
1935b. Two new parasites of aculeate Hymenoptera from Trinidad. Stylops 4:131-133.
1936. Notes on American species of *Trypoxylon*. Ann. Mag. Nat. Hist. (10) 18:457-463.
1937. Results of the Oxford University expedition to British Guiana, 1929. Hymenoptera, Sphecidae and Bembecidae. Trans. Roy. Ent. Soc. London 86:101-118.
1956a. An interpretation of the ventral region of the hymenopterous thorax. Proc. Roy. Ent. Soc. Lond. (A) 31:99-104.
1956b. Hymenoptera. Introduction and keys to families. Handbooks for the identification of British insects 4(1):1-94. Roy. Ent. Soc. London.
1962. A revisional study of the masarid wasps. v + 294 pp. Wm. Clowes and Sons, London and Beccles.
1969. Species of *Trypoxylon* Latreille from Cuba. Proc. Roy. Ent. Soc. London (B) 38:121-124.
1971. The habits of *Editha magnifica* (Perty). Ent. Mon. Mag. 107:112-113.
1972a. The thoracic spiracles and some associated structures in the Hymenoptera and their significance in classification, especially of the Aculeata, pp. 1-13. In: Entomological essays to commemorate the retirement of Professor K. Yasumatsu. vi + 389 pp. Hokuryukan Publ. Co., Tokyo.
1972b. The species of the South American wasps of the genus *Microstigmus* Ducke. Trans. Roy. Ent. Soc. London 124:123-148.

Richardson, C. H.
1915. An observation on the breeding habits of *Stigmus conestogorum* Rohwer. Psyche 22:104-105.

Riek, E. F.
1955. Australian Ampulicidae. Australian J. Zool. 3:131-144.
1970. Hymenoptera, pp. 867-959. In: The Insects of Australia. xiv + 1029 pp. Melbourne Univ. Press, Carlton.

Riley, C. V.
1892. The larger digger wasp. Insect Life 4:248-252.

Risso, A.
1826. Histoire naturelle des principales productions de l'Europe méridonale et particulièrement de celles des environs de Nice et des Alpes maritimes. Vol. 5. viii + 403 pp. Levrault, Paris.

Ristich, S.
1953. A study of the prey, enemies and habits of the great golden digger wasp, *Chlorion ichneumoneum* (L.). Canad. Ent. 85:374-386.
1956. The host relationship of a miltogrammid fly *Senotainia trilineata* (VDW). Ohio J. Sci. 56: 271-274.

Ritsema, C.
1872. Description of a new genus and two new exotic species of the family Larridae. Ent. Mon. Mag. 9:121-123.
1884a. A new genus and species of the hymenopterous family Larridae. Notes Leyden Mus. 6:81-83.
1884b. [Untitled note on *Dalara*, new name]. Le Naturaliste 2:559-560.

Rohwer, S. A.
1909. The bembecid wasps of Boulder County, Colorado. Univ. Colorado Studies 6:243-248.
1911a. Two new names of sphecoid wasps. Psyche 18: 153-155.
1911b. A preoccupied name in Sphecoidea. Proc. Ent. Soc. Washington 13:234.
1911c. Descriptions of new species of wasps with notes on described species. Proc. U.S. Natl. Mus. 40: 551-587.
1912. Descriptions of new species of wasps in the collections of the United States National Museum. Proc. U.S. Natl. Mus. 41:447-478.
1915. Descriptions of new species of Hymenoptera. Proc. U.S. Natl. Mus. 49:205-249.
1916. Sphecoidea, pp. 645-697. In: H. L. Viereck, Guide to the insects of Connecticut, part III, the Hymenoptera, or wasp-like insects, of Connecticut. Conn. St. Geol. Nat. Hist. Surv. Bull. (22):1-824.
1917. The North American wasps of the subgenus *Pemphredon* Latreille. Bull. Brooklyn Ent. Soc. 12:97-102.
1919. Philippine wasp studies. Part 1. Descriptions of new species. Bull. Exp. Sta. Hawaiian Sugar Planters' Assoc. (Ent.). 14:1-18.
1921a. Some notes on wasps of the subfamily Nyssoni-

nae, with descriptions of new species. Proc. U.S. Natl. Mus. 59:403-413.
1921b. The Philippine wasps of the subfamily Sphecinae. Philippine J. Sci. 19:665-676.
Obit.: Muesebeck *et al.*, 1951. Proc. Ent. Soc. Washington 53:227-230, portrait.

Roman, A.
1912. Die Ichneumonidentypen C. P. Thunbergs. Zool. Bidrag. Uppsala 1:229-293.

Ross, H. H.
1936. The ancestry and wing venation of the Hymenoptera. Ann. Ent. Soc. Amer. 29:99-111.

Rotarides, M.
1934. Daten zur Biologie von *Sceliphron destillatorium* Illig. auf der Halbinsel Tihany. Arb. Ungar. Biol. Forschungsinst., Tihany 7:66-79.

Roth, P.
1925. Les *Sphex* de l'Afrique du Nord. Ann. Soc. Ent. France 94:365-404.
1928. Les Ammophiles de l'Afrique du Nord. Ann. Soc. Ent. France 97:153-240.
1929. Les Ammophiles de l'Espagne. Rev. Española Ent. 5:161-190.
1951. Les *Sphex* paléarctiques du sous-genre *Palmodes*. France 118:79-94.
1963. Les *Sphex* Paléarctiques du sous-genre *Palmodes*. Mem. Mus. Natl. Hist. Nat. (ns.) A, Zool.) 18:139-186.

Rudow, F.
1912. Lebensweise und Nestbau der Raub-, Mord-, und Grabwespen, Sphegidae und Crabronidae. Ent. Zeitschr. 26:30-32, 35-36, 39-40, 42-44, 46, 54-55, 59-60, 64, 66-67, 70-72, 75-76.

Sabrosky, C. W.
1974. *Sphex viatica* Linnaeus: a problem of types and revisers. Z.N. (S.) 2061 Bull. Zool. Nomencl. 31:159-163.

Sachtleben, H.
1961. Nachträge zu "Walther Horn und Ilse Kahle: Über entomologischen Sammlungen". Beitr. Ent. 11:481-540.

Sandhouse, G. A.
1940. A review of the Nearctic wasps of the genus *Trypoxylon*. Amer. Midl. Nat. 24:133-174.

Saunders, E.
1896. The Hymenoptera Aculeata of the British Islands. viii + 391 pp. L. Reeve & Co., London.

Saunders, S. S.
1873. On the habits and economy of certain hymenopterous insects which nidificate in briars; and their parasites. Trans. Roy. Ent. Soc. London 1873:407-414.

Saussure, H. de
1854. Mélanges Hyménoptérologiques, fasc. 1, 67 pp. Geneva (later published in: Mém. Soc. Phys. Hist. Nat. Genève 14:1-67, 1855).
1867. Hymenoptera, pp. 1-138. In: Reise der österreichischen Frigatte Novara, etc., Zoologischer Theil, vol. 2. 156 pp. Wien.
1887. Sur quelques Hyménoptères de Madagascar. Soc. Ent. Zurich 2:2-3, 9, 17-18, 25-26.
1890-1892. Histoire naturelle des Hyménoptères. Vol. 20. xxi + 590 pp. In: A. Grandidier, Histoire physique, naturelle et politique de Madagascar, Paris (only pages 177-590 and plates 21-27 date 1892, earlier pages and plates appeared in 1890).
Obit.: Bedot, M., 1906. Rev. Suisse Zool. 14:1-32, portrait and bibliography.

Say, T.
1823.
opterous insects. West. Quart. Rep. (Med., Surg.,
1837. vorzugsweiser Berücksichtigung der paläark-
Hymenopteren Studien. 147 pp. Engleman,

Schlette
1887. Grabwespen-Typen Tourniers, Brullés, Lepele- vorzugsweiser Berucksichtigung der paläarktischen Arten. Zool. Jahrb., Abt. Syst. Geol. Biol. 2:349-510.
1889a. Nachträgliches über die Hymenopteren-Gattung *Cerceris* Latr. Zool. Jahrb., Abt. Syst. Geol. Biol. 4:879-904.
1889b. Beitrag zur Kenntniss der Hymenopteren-Gattung *Cerceris* Latr. Zool. Jahrb., Abt. Syst. Geol. Biol. 4:1124-1131.

Schmidt, K.
1971. Die Grabwespen-Typen A. Schencks in der Sammlung C. L. Kirschbaum im Landesmuseum Wiesbaden. Beitr. Ent. 21:61-66.

Schmiedeknecht, O.
1930. Die Hymenopteren Nord- und Mitteleuropas. x + 1062 pp. G. Fischer, Jena.

Schrottky, C.
1911. Neue südamerikan Hymenoptera. Ent. Rundschau 28:2-3, 10-11, 19-20, 27-29, 38-39.

Schulz, W. A.
1904. Hymenopteren Amazoniens. Sitzber. Math, -Phys. Klasse Kgl. Bayer. Akad. Wiss. München 33:757-832.
1905. Hymenopteren Studien. 147 pp. Engleman, Leipzig.
1906. Spolia hymenopterologica. 355 pp. A. Pape, Paderborn.
1911a. Grabwespen-Typen Tourniers, Brullés, Lepeletiers und Schenks. Soc. Ent. 26:57-59, 63-64.
1911b. Zweihundert alte Hymenopteren. Zool. Ann. 4:1-220.
1912. Aelteste und alte Hymenopteren skandinavischer Autoren. Berlin Ent. Zeitschr. 57:52-102.

Scopoli, J.
1772. Observationes Zoologicae. Historico-Naturalis 5:70-128.

Scullen, H. A.
1939. A review of the genus *Eucerceris*. Oregon State Monogr. Studies Ent. 1, 80 pp.
1948. New species in the genus *Eucerceris* with notes on recorded species and a revised key to the genus. Pan-Pac. Ent. 24:155-180.
1965. Review of the genus *Cerceris* in America north of Mexico. Proc. U.S. Natl. Mus. 116:333-548.
1968. A revision of the genus *Eucerceris* Cresson. U.S. Natl. Mus. Bull. 268:1-97.
1972. Review of the genus *Cerceris* Latreille in Mexico and Central America. Smithsonian Contr. Zool. 110:1-121.

Scullen, H. A. and J. L. Wold
1969. Biology of wasps of the tribe Cercerini with a list of the Coleoptera used as prey. Ann. Ent.

Soc. Amer. 62:209-214.

Shafer, G. D.
1949. The ways of a mud-dauber. ix + 78 pp. Stanford Univ. Press, Stanford.

Sheldon, J. K.
1968. The nesting behavior and larval morphology of *Pison koreense* (Radoszkowski). Psyche 75:107-117.

Sherborn, C. D.
1922a. On the dates of Cuvier, 'Le régne Animal' etc. (disciples edition). Ann. Mag. Nat. Hist. (9) 10: 151-152.
1922b. Index animalium. Section 2, vol. 1. cxxi + 943 pp. Cambridge Univ. Press, London.
1923. On the dates of G. W. F. Panzer's 'Fauna Insect. German.', 1792-1844. Ann. Mag. Nat. Hist. (9) 11:566-568.
1937. On the dates of publication of Costa (O. G.) and (A.), Fauna del regno di Napoli, 1829-1886. J. Soc. Bibl. Nat. Hist. 1:35-47.

Sherborn, C. D. and B. B. Woodward
1901. Dates of publication of the zoological and botanical portions of some French voyages. Part II. Ann. Mag. Nat. Hist. (7) 8:161-164, 333-336.
1906. On the dates of publication of the natural history portions of the Encyclopédie Méthodique. Ann. Mag. Nat. Hist. (7) 17:577-582.

Shestakov, A.
1923. Revisio specierum palearcticis subgeneris *Apiraptrix* novi pertinentium. Sbornik Yaroslav. Gosudarstvennogo Univ. 2:101-115.

Shuckard, W. E.
1837. Essay on the indigenous fossorial Hymenoptera. xii + 259 pp. Shuckard, London (dated Jan. 2 according to Richards, 1935a:160).
1838. Descriptions of new exotic aculeate Hymenoptera. Trans. Ent. Soc. London 2 (1):68-82 (dating after Wheeler, 1912).
1840. In: W. Swainson and W. Shuckard, On the history and natural arrangement of insects, vol. 129 (of the Cabinet Cyclopedia of D. Lardner), iv + 406 pp. London.

Sickmann, F.
1893. Die Hymenopterenfauna von Iburg und seiner nächsten Umgebung, mit biologischen und kritschen Bemerkungen. Die Grabwespen. Jahresber. Nat. Ver. Osnabrück 9:39-112.

Smirnov, D.
1915. Sur les moeurs d'*Ammophila (Eremochares) dives* Brullé. Rev. Russ. Ent. 15:153-155.

Smith, C. E.
1935. *Larra analis* Fabricius, a parasite of the mole cricket *Gryllotalpa hexadactyla* Perty. Proc. Ent. Soc. Washington 37:65-82.

Smith, E. L.
1970. Evolutionary morphology of the external insect genitalia. 2. Hymenoptera. Ann. Ent. Soc. Amer. 63:1-27.

Smith, F.
1847. Description of a new species of hymenopterous insect belonging to the family Sphegidae. Ann. Mag. Nat. Hist. (1) 20:394-396.
1851. Descriptions of some new species of exotic Hymenoptera in the British Museum and other collections. Ann. Mag. Nat. Hist. (2) 7:28-33.
1856. Catalogue of hymenopterous insects in the collection of the British Museum, part IV, Sphecidae, Larridae and Crabronidae, pp. 207-497. London.
1858. Catalogue of the Hymenopterous insects collected at Sarawak, Borneo; Mt. Ophir, Malacca; and at Singapore, by A. R. Wallace. J. Proc. Linn. Soc. London. Zool. 2:42-130.
1864. Catalogue of hymenopterous insects collected by Mr. A. R. Wallace in the islands of Mysol, Ceram, Waigiou, Bouru and Timor. J. Proc. Linn. Soc. London. Zool. 7:6-48.
1869a. Descriptions of new genera and species of exotic Hymenoptera. Trans. Ent. Soc. London, 1869:301-311.
1869b. Descriptions of new species of the genus *Pison;* and a synonymic list of those previously described. Trans. Ent. Soc. London 1869:289-300.
1873. Descriptions of new species of fossorial Hymenoptera in the British Museum and of a species of the rare genus *Iswara* belonging to the family Dorylidae. Ann. Mag. Nat. Hist. (4) 12:253-260, 291-300, 402-415.

Obit.: Dunning, J. W., 1879. Entomologist 12:89-92, portrait.

Smith, H. S.
1908a. A new *Mellinus*. Ent. News 19:299-300.
1908b. The Sphegoidea of Nebraska. Univ. Nebraska, Univ. Studies 8:233-410.

Smith, M. R.
1923a. The life history and habits of *Bicyrtes quadrifasciata* Say. Ann. Ent. Soc. Amer. 16:238-246.
1923b. Unusual damage to the floors of a house by a species of pemphredonid wasp, *Stigmus fulvicornis* Rohwer. J. Econ. Ent. 16:553-554.

Smithers, C. N.
1958. Notes on the life history of *Sphex (Isodontia) pelopaeiformis* Dahlb. in Southern Rhodesia. Ent. Mon. Mag. (4) 19:193-197.

Snodgrass, R. E.
1910. The thorax of the Hymenoptera. Proc. U.S. Natl. Mus. 39:37-92.
1935. Principles of insect morphology. ix + 667 pp. McGraw-Hill Book Co., New York and London.
1956. Anatomy of the honeybee. 334 pp. Comstock Publ. Assn., Ithaca.
1960. Facts and theories concerning the insect head, Smithsonian Misc. Coll. 142(1):1-61.
1963. A contribution toward an encyclopedia of insect anatomy. Smithsonian Misc. Coll. 146 (2):1-48.

Soika, A. Giordani
1933. Sull 'etologia dell '*Ammophila tydei* Guill. Boll. Soc. Ent. Italiana 65:60-64.
1942. Sfegidi raccolti nell'Africa orientale dal Prof. Alberto Mochi e dal Dott. Marcello Mochi. Atti Soc. Italiana Sci. Nat. 81:196-209.

Soyer, B.
1947. Notes sur les Sphégiens et les Pompiles, VI. Bull. Soc. Linn. Lyon 16:117-121.

Spinola, M.
1806-1808. Insectorum Liguriae species novae aut rariores, etc. Vol. 1. xvii + 160 pp. (1806). Vol. 2, ii + 262 pp. (1808). A. Koenig, Parisiis et Argentorati.

1836. [Untitled note]. Ann. Soc. Ent. France 5:xxiii.
1838. Compte rendu des Hyménoptères recueillis par M. Fischer pendant son voyage en Égypte. Ann. Soc. Ent. France 7:437-546.
1843. Notes sur quelques Hyménoptères peu connus, recueillis en Espagne, pendant l'année 1842, par M. Victor Ghiliani, voyageur-naturaliste. Ann. Soc. Ent. France (2) 1:111-114.
1851. Himenopteros, pp. 153-569. In: C. Gay, Historia fisica y politica de Chile. Zoologia, vol. 6. 572 pp. Maulde and Renon, Paris.
1853. Compte rendu des Hyménoptères inédits provenants du voyage entomologique de M. Ghiliani dans le Parà en 1846. Mem. Accad. Sci. Torino 13:19-94.
Obit.: Gestro, R., 1915. Ann. Mus. Civ. Stor. Nat. Genoa 47:33-53, portrait and bibliography.

Spooner, G. M.
1948. The British species of psenine wasps. Trans. Roy. Ent. Soc. London 99:129-172.

Spurway, H., K. R. Dranamraju, and S. D. Jayakar
1964. One nest of *Sceliphron madraspatanum* (Fabr.). J. Bombay Nat. Hist. Soc. 61:1-26.

Stadelmann, H.
1897. Bemerkungen zu Dalla-Torre's Fossorienkatalog. Ent. Nachr. 23:254-255.

Statz, G.
1936. Ueber alte und neue fossile Hymenopterenfunde aus den tertiären Ablagerungen von Rott am Siebengebirge. Decheniana 93:256-312.

Stearn, W. T.
1937. On the dates of publication of Webb and Berthelot's "Histoire naturelle des Îles Canaries". J. Soc. Bibl. Nat. Hist. 1:49-63.

Steiner, A. L.
1962. Étude du comportement prédateur d'un Hyménoptère sphégien: *Liris nigra* V. de L. (= *Notogonia pompiliformis* Pz.). Ann. Sci. Nat. (Zool.) France (12) 4:1-126.
1965. Mise au point d'une technique d'élevage d'Hyménoptères fouisseurs en laboratorie. Bull. Soc. Ent. France 70:12-18.
1968. Behavioral interactions between *Liris nigra* Van der Linden and *Gryllulus domesticus* L. Psyche 75:256-273.
1971. Behavior of the hunting wasp *Liris nigra* V.d.L. in particular or in unusual situations. Canad. J. Zool. 49:1401-1415.

Stephen, W. P., G. E. Bohart and P. F. Torchio
1969. The biology and external morphology of bees, with a synopsis of the genera of northwestern America. 140 pp. Oregon State Agr. Exp. Sta., Corvallis.

Stephens, J. F.
1829a. The nomenclature of British insects, etc. 68 pp. Baldwin and Cradock, London.
1829b. A systematic catalogue of British insects, etc. xxxiv + 416 + 388 pp. Baldwin and Cradock, London (published July 15 according to Blackwelder, 1949).

Stone, A., *et al.*
1965. A catalog of the Diptera of America north of Mexico. U.S. Dept. Agric., Agric. Handbook 276, 1696 pp.

Strandtmann, R. W.
1945. Observations on the habits of some digger wasps. Ann. Ent. Soc. Amer. 38:305-313.
1946. A review of the North American species of *Philanthus*, north of Mexico. 126 pp. Ohio State Univ. Press, Columbus.
1953. Notes on the nesting habits of some digger wasps. J. Kansas Ent. Soc. 26:45-52.

Taschenberg, E.
1869. Die Sphegidae des Zoologische Museums der Universität in Halle. Zeitschr. Ges. Naturwiss. Berlin 34:407-435.
Obit.: Anonymous, 1898. Nature 57:300-301.

Telford, A. D.
1964. The Nearctic *Parnopes* with an analysis of the male genitalia in the genus. Univ. Calif. Pub. Ent. 36:1-42.

Teschner, W.
1959. Starrheit und Variabilität in Verhalten von Sandwespen. Zool. Beitr. (n.s.) 4:411-472.

Thomas, S. J.
1962. A new host record of the chrysidid, *Omalus auratus* Linnaeus. Ent. News 73:217-218.

Thomson, C. G.
1874. Skandinaviens Hymenoptera, vol. 3, part 2, *Mutilla* och *Sphex* Lin., pp. 99-295.

Thunberg, C. P.
1791. Donation. Thunbergianae. Append. II. Mus. Nat. Acad. Lipsaliensis, pp. 123-129.

Timberlake, P. H.
1920. Notes and exhibitions. Proc. Hawaiian Ent. Soc. 4:330.

Tinbergen, N.
1932. Über die Orientierung des Bienenwolfes. I. Zeitschr. Vergleich. Physiol. 16:305-334.
1935. Über die Orientierung des Bienenwolfes. II. Zeitschr. Vergleich. Physiol. 21:699-716.
1958. Curious naturalists. 280 pp. Country Life Limited, London.

Tinbergen, N. and W. Kruyt
1938. Über die Orientierung des Bienenwolfes. III. Zeitschr. Vergleich. Physiol. 25:292-334.

Tinbergen, N. and R. J. van der Linde
1938. Über die Orientierung des Bienenwolfes. IV. Biol. Zentr. 58:425-35.

Townes, H.
1973. The type of *Sphex viatica* Linnaeus. Polskie Pismo Ent. 43:91-96.

Tsuneki, K.
1947a. Nesting habits of *Astata boops* (Schrank). Mushi 17:103-111.
1947b. On the wasps of the genus *Crabro* s.l. from Hokkaido, with descriptions of new species and subspecies. J. Fac. Sci. Hokkaido Univ. (6) 9:397-435.
1951. The genus *Pemphredon* Latreille of Japan and the adjacent regions. J. Fac. Sci. Hokkaido Univ. (6) 10:163-208.
1952a. Ethological studies on the Japanese species of *Pemphredon*, with notes on their parasites, *Ellampus* spp. J. Fac. Sci. Hokkaido Univ. (6) 11:57-75.
1952b. The genus *Rhopalum* Kirby (1829) of Japan, Korea, Saghalien and the Kuriles, with a suggested reclassification of the subgenera and descriptions of four new species. J. Fac. Sci.

Hokkaido Univ. (6) 11:110-125.
1954a. The genus *Stigmus* Panzer of Europe and Asia with descriptions of eight new species. Mem. Fac. Lib. Arts, Fukui Univ. (2, Nat. Sci.) (3): 1-38.
1954b. The genus *Crossocerus* Lepeletier et Brullé (1834) of Japan, Korea, Saghalien and the Kuriles. Mem. Fac. Lib. Arts, Fukui Univ. (2, Nat Sci.) (3):57-58.
1954c. Descriptions and records of wasps of the families Chrysididae and Sphecidae of Japan. Mem. Fac. Lib. Arts, Fukui Univ. (2, Nat. Sci.) (4): 37-54.
1955. The genus *Passaloecus* Shuckard of Japan, with ethological observations on some species. Mem. Fac. Lib. Arts, Fukui Univ. (2, Nat. Sci.) (5):1-21.
1956a. Two new species of the genus *Nitela* in Japan. Akitu 5:33-35.
1956b. The occurrence of the genus *Spilomena* Shuckard in Japan, with descriptions of two new species. Kontyû 24:73-76.
1956c. On the taxonomical position, curious distribution and male polymorphism of *Ectemnius (Yanonius* nov.*) martjanowii* F. Morawitz, 1892. Kontyû 24:128-132.
1956d. Die Trypoxylonen der nordoestlichen Gebiete Asiens. Mem. Fac. Lib. Arts, Fukui Univ. (2, Nat. Sci.) (6):1-42.
1958. Ethological studies on *Bembix niponica* Smith, with emphasis on the psychobiological analysis of behaviour inside the nest. III. Conclusive Part. Mem. Fac. Lib. Arts, Fukui Univ. (2, Nat. Sci.) (8):1-78.
1959a. Contributions to the knowledge of the Cleptinae and Pseninae Faunae of Japan and Korea. Mem. Fac. Lib. Arts, Fukui Univ. (2, Nat. Sci.) (9): 1-78.
1959b. Notes on some synonymy of the Japanese Crabroninae, with the erection of a new subgenus of *Ectemnius*. Akitu 8:7-8.
1959c. Une nouvelle espèce du Crabronien du Japon. Akitu 8:83-84.
1960. Biology of the Japanese Crabroninae. Mem. Fac. Lib. Arts, Fukui Univ. (2, Nat. Sci.) (10):1-53.
1961. Studies on *Cerceris* of north eastern Asia. Mem. Fac. Lib. Arts, Fukui Univ. (2, Nat. Sci.) (11): 1-72.
1962. Taxonomic notes on *Ammophila* Kirby of Japan and Korea. Life Study (Fukui) 6:24-29.
1963a. The tribe Gorytini of Japan and Korea. Etizenia (1):1-20.
1963b. A new study on the nesting biology of the tube renting *Ammophila, A. aemulans* Kohl. Life Study (Fukui) 7:44-48.
1963c. Comparative studies on the nesting biology of the genus *Sphex* (s.l.) in East Asia. Mem. Fac. Lib. Arts, Fukui Univ. (2, Nat. Sci.) (13):13-78.
1963d. Chrysididae and Sphecidae from Thailand. Etizenia (4):1-50.
1963e. *Crossocerus leclercqi* is a species of *Piyumoides*. Ins. Matsumurana 26:103.
1964a. Notes on *Pison* of Japan. Akitu 11:43.
1964b. An interesting new species of the genus *Psen* captured in Japan with the erection of a new subgenus. Ins. Matsumurana 27:12-15.
1964c. The genus *Tachytes* Panzer of Japan and Korea. Etizenia (5):1-11.
1964d. Supplementary notes on the nesting biology of three species of *Sphex (Isodontia)* occurring in Japan. Etizenia (7):1-14.
1965a. Variation in characters of *Bembecinus hungaricus* Frivaldzky occurring in East Asia, with taxonomic notes on hitherto known species. Etizenia. (8):1-17.
1965b. The nesting biology of *Stizus pulcherrimus* F. Smith with special reference to the geographical variation. Etizenia (10):1-21.
1965c. Nesting biology of *Argogorytes mystaceus grandis* Gussakovskij. Life Study (Fukui) 9:41-42.
1965d. The biology of east-Asiatic *Cerceris* with special reference to the peculiar social relationships and return to the nest in *Cerceris hortivaga* Kohl. Etizenia (9):1-46.
1966a. Taxonomic notes on *Trypoxylon* of Formosa and the Ryukyus with descriptions of new species and subspecies. Etizenia (13):1-19.
1966b. Contribution to the knowledge of the Pemphredoninae fauna of Formosa and the Ryukyus. Etizenia (14):1-21.
1966c. Contribution to the knowledge of Crabronine fauna of Formosa and the Ryukyus. (15):1-21.
1967a. Studies on the Formosan Sphecidae (I). The subfamily Larrinae. Etizenia (20):1-60.
1967b. On some Ampulicidae from Formosa. Etizenia (21):1-13.
1967c. Studies on the Formosan Sphecidae (II). The subfamily Trypoxyloninae. Etizenia (22):1-21.
1967d. Studies on the Formosan Sphecidae IV, the subfamily Sphecinae with special reference to the genus *Ammophila* in eastern Asia. Etizenia (26): 1-24.
1968a. On some *Cerceris* from Australia, with a tentative key to the Australian species. Etizenia (28):1-32.
1968b. *Cerceris* from the western Pacific areas. Etizenia (29):1-36.
1968c. Studies on the Formosan Sphecidae (V). The subfamily Crabroninae with a key to the species of Crabronini occurring in Formosa and the Ryukyus. Etizenia (30):1-34.
1968d. The biology of *Ammophila* in East Asia. Etizenia (33):1-64.
1969a. Gleanings on the bionomics of the east-Asiatic non-social wasps. II. Some species of Tachytini, Larrini and Palarini. Etizenia (39):1-22.
1969b. Gleanings on the bionomics of the east-Asiatic non-social wasps. III. *Astata boops* Schrank in Korea, Etizenia (40):1-12.
1970a. Gleanings on the bionomics of the east-Asiatic non-social wasps. V. Some species of Pemphredoninae. Etizenia (42):1-20.
1970b. Gleanings on the bionomics of the east-Asiatic non-social wasps. VI. Some species of Trypoxyloninae. Etizenia (45):1-20.
1970c. Studies on the Formosan Sphecidae (VII). The subfamily Philanthinae. Etizenia (44):1-24.
1970d. Change of the taxonomic position of three species of Crabronini occurring in Japan with notes on some species. Etizenia (50):1-8.
1971a. Studies on the Formosan Sphecidae (VIII). A

supplement to the subfamily Crabroninae. Etizenia (51):1-29.
1971b. Revision der *Spilomena*-Arten Japans, mit den Beschreibungen von drei neuen Arten. Life Study (Fukui) 15:8-18.
1971c. Studies on the Formosan Sphecidae (XIII). A supplement to the subfamily Pemphredoninae with a key to the Formosan species. Etizenia (57):1-21.
1971d. Ergebnisse der zoologischen Forschungen von Dr. Z. Kaszab in der Mongolei. Sphecidae I-II. Acta. Zool. Acad. Sci. Hungaricae 17:139-217.
1971e. A guide to the study of the Japanese Hymenoptera, VII. The genus *Passaloecus* Shuckard, Life Study (Fukui) 15:85-89.
1971f. Studies on the Formosan Sphecidae (X). Revision of and supplement to the subfamily Trypoxyloninae. Etizenia (54):1-19.
1972a. Ergebnisse der Zoologischen Forschungen von Dr. Z. Kaszab in der Mongolei. Sphecidae, IV-V. Acta Zool. Acad. Sci. Hungaricae 18:147-232.
1972b. On some species of the Japanese Sphecidae, notes and descriptions. Etizenia (59):1-20.
1972c. Studies on the Formosan Sphecidae (XIV). Notes on some specimens newly examined, with a description of a related Japanese subspecies. Etizenia (60):1-13.
1973a. New and first recorded species and subspecies of Sphecidae and Mutillidae from Japan, with taxonomic notes on some species. Etizenia (65):1-28.
1973b. The biology of some pith burrowing silver mouth wasps. Life Study (Fukui) 17:14-20.

Tsuneki, K. and T. Iida
1969. The biology of some species of the Formosan Sphecidae, with descriptions of their larvae. Etizenia (37):1-21.

Tulloch, G. S.
1929. The proper use of the terms parapsides and parapsidal furrows. Psyche 36:376-381.

Turner, R. E.
1907. New species of Sphegidae from Australia. Ann. Mag. Nat. Hist. (7) 19:268-276.
1910. Additions to our knowledge of the fossorial wasps of Australia. Proc. Zool. Soc. London 1910:253-356.
1911. Notes on fossorial Hymenoptera. IV. Ann. Mag. Nat. Hist. (8) 7:479-485.
1912a. A monograph of the wasps of the genus *Cerceris* inhabiting British India. J. Bombay Nat. Hist. Soc. 21:476-516.
1912b. Notes on fossorial Hymenoptera, VI, on the species collected in New Guinea by the expedition of the British Ornithologists Union. Ann. Mag. Nat. Hist. (8) 9:194-202.
1912c. Notes on fossorial Hymenoptera, IX. Ann. Mag. Nat. Hist. (8) 10:48-63.
1912d. A revision of the Australian species of the genus *Cerceris*. Proc. Linn. Soc. N. S. Wales 36:644-678.
1914. Notes on fossorial Hymenoptera. XIII. A revision of the Paranyssoninae. Ann. Mag. Nat. Hist. (8) 14:337-359.
1915a. Notes on fossorial Hymenoptera. XV. New Australian Crabronidae. Ann. Mag. Nat. Hist. (8) 15:62-96.
1915b. Notes on fossorial Hymenoptera. XVIII. On the Australian species of *Bembex*. Ann. Mag. Nat. Hist. (8) 16:434-447.
1916a. Notes on fossorial Hymenoptera. XIX. On new species from Australia. Ann. Mag. Nat. Hist. (8) 17:116-136.
1916b. Notes on fossorial Hymenoptera XX. On some Larrinae in the British Museum. Ann. Mag. Nat. Hist. (8) 17:248-259.
1916c. Notes on fossorial Hymenoptera, XXI. On the Australian Larrinae of the genus *Tachytes*. Ann. Mag. Nat. Hist. (8) 17:299-306.
1916d. Notes on fossorial Hymenoptera, XXIII. Ann. Mag. Nat. Hist. (8) 18:277-288.
1916e. Notes on the wasps of the genus *Pison* and some allied genera. Proc. Zool. Soc. London 1916:591-629.
1917. A revision of the wasps of the genus *Tachytes* inhabiting the Ethiopian Region. Ann. Mag. Nat. Hist. (8) 20:1-43.
1918. Notes on fossorial Hymenoptera. XXXV. On new Sphecoidea in the British Museum. Ann. Mag. Nat. Hist. (9) 1:356-364.
1926. New Sphegidae from the Malay Peninsula. J. Fed. Malay St. Mus. 13:199-202.
1929. A new species of *Microstigmus*. Bull. Ent. Res. 20:407-408.
Obit.: Benson, R., 1946. Ent. Mon. Mag. 82:47.

Turner, R. E. and J. Waterston
1926. On a new subgenus of *Crabro*. Ann. Mag. Nat. Hist. (9) 17:189-191.

Turton, W.
1802. System of nature through the three grand kingdoms of animals, vegetables, and minerals, etc. Vol. 3. 784 pp. Lackington, Allen and Co., London.

Ulrich, W.
1924. Die Mundwerkzeuge der Spheciden. Zeitschr. Morph. Ökol. Tiere 1:540-636.

Vachal, J.
1893. Une nouvelle espèce d'Hyménoptères, de la sous-famille des Philanthinae, de l'Algérie. Ann. Soc. Ent. France 62:cclxiv-cclxv.
1899. Contributions hyménoptériques. Ann. Soc. Ent. France 68:534-539, errata, p. 843.
1900. Rectification d'un nom de genre préoccupé. Bull. Soc. Ent. France 1900:233.

Valdeyron-Fabre, L.
1952. Sur le parasitisme de *Brachytrypes megacephalus* Lef. par un *Sphex*, *S. xanthocerus* Guérin. Bull. Soc. Sci. Nat. Tunisie 5:107-108.
1955. Observations sur la biologie de *Brachytrypes megacephalus* Lef. en Tunisie. Rev. Path Vég. Ent. Agr. France 34:136-158.

Valkeila, E.
1955. Observations on the biology and development of *Nitela spinolae* Dhlb. Ann. Ent. Fennici 21:54-57.
1956. A note on the taxonomy and nomenclature of two European species of the genus *Stigmus* Panzer. Ann. Ent. Fennici 22:165-166.
1957. Mitteilungen über die nordeuropäischen *Spilomena*-Arten. Ann. Ent. Fennici 23:163-178.

1961. Beiträge zur Kenntnis der nordeuropäischen Raubwespen. Ann. Ent. Fennici 27:141-146.
1963. Zur Lebensweise von *Belomicrus borealis* Fors. Ann. Ent. Fennici 29:231-236.

Valkeila, E. and J. Leclercq
1972. Données pour un atlas des Hyménoptères de l'Europe occidentale. XI, *Pemphredon* from Belgium and elsewhere. Bull. Recherch. Agron. Gembloux (n.s.) 5:695-708. (dated 1970 but published Oct. 20, 1972).

Vander Linden, P. L.
1827. Observations sur les Hyménoptères de la famille des fouisseurs, première partie. Nouv. Mem. Acad. R. Sci. Belle-Lettres Bruxelles 4:271-367 (pp. 1-97 of authors separate).

van der Vecht, J.
1937. On a new *Piagetia*, with notes on other species. Ent. Med. Nederland.-Indië 3:21-26.
1939. Introduction to the study of the Sphecidae of Java, with a key to the genera. Ent. Med. Nederland.-Indië 5:72-86.
1949. On Indo-Australian *Bembecinus*, with special reference to the species occurring in Java. Treubia 20:289-307.
1950. A note on the larrid wasps *Dalara schlegelii* (Rits.) and *Hyloliris mandibularis* Will. Idea 8:14-16.
1951. Over nestbouw en prooi van *Dasyproctus ceylonicus* (Sauss.). Idea 8:103-104.
1957. The Sphecoidea of the Lesser Sunda Islands, 1. Sphecinae. Verhandl. Naturf. Gcs. Basel 68: 358-372.
1958. The identity of *Sphex viatica* Linné, 1758. Ent. Bericht. 18:47-48.
1959. Notes on aculeate Hymenoptera described in the period 1758-1810. Ent. Bericht. 19:211-215.
1961a. Hymenoptera Sphecoidea Fabriciana. Zool. Verhandl. (48):1-85.
1961b. Über Taxonomie und Evolution der Grabwespengattung *Sceliphron* Klug. Verhandl. XI. Internatl. Kong. Ent. Wien 1:251-256.
1964. The *Cerceris* species of Java. Zool. Meded. 39:348-368.
1966. Notes on some Oriental *Philanthus* species. Proc. Kon. Nederl. Akad. Wetenschappen, (c) 69:420-431.
1973. Contribution to the taxonomy of the Oriental and Australian Sphecini. Proc. Kon. Nederl. Akad. Wetenschappen (c) 75:341-353.

van der Vecht, J. and F. M. A. van Breugel
1968. Revision of the nominate subgenus *Sceliphron* Latreille. Tijdschr. Ent. 111:185-255.

van der Vecht, J. and K. V. Krombein
1955. The subspecies of *Sphex sericeus* (Fabr.) (= *S. aurulentus* auct., nec Fabr. 1787). Idea 10:33-43.

van Lith, J. P.
1949. Le sous-genre *Psen Mimumesa* Malloch (avec une liste des Psenini capturés aux Pays-Bas). Tijdschr. Ent. 91:135-148.
1951. Over de biologie van het geslacht *Psenulus* Kohl. Ent. Bericht. 13:211-217.
1955. De Nederlandse *Spilomena*-soorten. Ent. Bericht. 15:525-527.
1959. Contribution to the knowledge of the Indo-Australian Pseninae. Zool. Verhandl. (39): 1-69.
1962. Contribution to the knowledge of the Indo-Australian Pseninae. Part II. *Psenulus* Kohl, 1896. Zool. Verhandl. (52):1-118.
1965. Contribution to the knowledge of the Indo-Australian Psenini III. Zool. Verhandl. (73): 1-80.
1968. Contribution to the knowledge of Indo-Australian, South Pacific and East Asiatic Psenini. Genus *Psen* Latreille. Tijdschr. Ent. 111:89-135.
1972. Contribution to the knowledge of Oriental *Psenulus*. Tijdschr. Ent. 115:153-203.

Verhoeff, C.
1890. Zusätze zu den in no. 21 beschriebenen Hymenopteren. Ent. Nachr. 16:382-386.
1892. Ueber einige neue und seltene Fossorien. Ent. Nachr. 18:65-72.

Verhoeff, P. M. F.
1948. Systematisches Verzeichnis der niederländischen *Oxybelus*-Arten. Tijdschr. Ent. 89:158-208.
1951. Notes on *Astata* Latreille. Zool. Meded. 31: 149-164.
1955. Zur Taxonomie der palaearktischen *Miscophus*. Ent. Bericht. 15:374-382.

Verlaine, L.
1925. L'instinct et l'intelligence chez les Hyménoptères II. L'instinct de nidification chez le *Pelopaeus clypeatus* Kohl du Congo Belge. Ann. Soc. Ent. Belgique 64:197-237.
1926. Les moeurs du *Pelopaeus clypeatus*, guêpe maçonne du Congo. Rev. Zool. Bot. Africaine 13:174-180.

Vesey-Fitzgerald, D.
1936. Nesting habits of *Trypoxylon* from Trinidad. Proc. Roy. Ent. Soc. London (A) 11:111-114.
1940. Notes on Bembicidae and allied wasps from Trinidad. Proc. Roy. Ent. Soc. London (A) 15:37-39.
1950. Nesting habits of some aculeate Hymenoptera in the Seychelles. Proc. Roy. Ent. Soc. London (A) 25:75-80.

Viereck, H. L.
1901. New species of the subfamily Pseninae. Trans. Amer. Ent. Soc. 27:338-342.
1916. Guide to the insects of Connecticut III. The Hymenoptera or wasp-like insects of Connecticut. Introduction. Bull. Conn. St. Geol. Nat. Hist. Surv. 22:9-24.

Villers, C.
1789. Caroli Linnaei entomologia etc. Vol. 3. 657 pp. Piestre et Delamolliere, Lugduni.

Vincens, F.
1910. Observations sur les moeurs et l'instinct d'un insecte hyménoptère, le *Nitela spinolae* Latr. Bull. Soc. Hist. Nat. Toulouse 43:11-18.

Wagner, A. C. W.
1931. Einige Bemerkungen über *Cemonus* Jur. (*Diphlebus* Westw.). Mitt. Deutsche Ent. Ges. 2:13.

Wagner, H. O.
1955. Nestbaurhythmus einer mexikanischen Töpferwespe aus der Gattung *Trypoxylon*. Zeitschr. Tierpsychol. 11:353-357.

Ward, G.

1970. The occurrence of *Chalybion zimmermanni* Dahlbom in Indiana. Proc. Indiana Acad. Sci. 79:231-233.
1971. Nest site preference of *Chalybion zimmermanni* Dahlbom. Proc. Indiana Acad. Sci. 80:264-266.
1972. Aggregations of *Chalybion californicum* (Saussure) near Centerville, Wayne County, Indiana. Proc. Indiana Acad. Sci. 81:177-181.

Wasbauer, M. S. and W. E. Simonds
1964. A note on the prey and nest structure of *Stigmus inordinatus inordinatus*. Pan-Pacific Ent. 40:114-116.

Weber, P. W.
1950. Notes and exhibitions. Proc. Hawaiian Ent. Soc. 14:14-16.

Wesmael, M.
1851-1852. Revue critique des Hyménoptères fouisseurs de Belgique. Bull. Acad. Roy. Sci. Lettres Beaux-Arts Belgique. 18:362-384, 451-494 (1851); 19:82-110, 261-286, 589-635 (1852).

Westwood, J. O.
1834. Insectorum arachnoidumque novorum decades duo. Zool. J. 5:440-453.
1835. (New hymenopterous insects from the collections of Rev. F. W. Hope and Mr. Westwood). Proc. Zool. Soc. London 3:68-72.
1837. On generic nomenclature. Mag. Nat. Hist. (n.s.) 1:169-173.
1839-1840. Synopsis of the genera of British insects, pp. 1-158 (pp. 65-80 = 1839; pp. 81-96 = 1840). In: An introduction to the modern classification of insects, vol. 2. London (dating after Blackwelder, 1949).
1841a. (April) (Descriptions of exotic hymenopterous insects belonging to the family Sphecidae). In: Proc. Meetings Ent. Soc. London. Ann. Mag. Nat. Hist. (1) 7:151-152 (dating after Wheeler, 1912).
1841b. (June) Descriptions of exotic hymenopterous insects belonging to the family Sphegidae. Jour. Proc. Ent. Soc. London 1:16 (dating after Wheeler, 1912).
1842. Descriptions of some new species of exotic hymenopterous insects. Trans. Ent. Soc. London (1) 3:223-231.
1844. Part XVII, pp. 65-80. In: Arcana entomologica. Vol. 2. iv + 192 pp., W. Smith, London.

Wheeler, G.
1912. On the dates of the publications of the Entomological Society of London. Trans. Ent. Soc. London 1911:750-767.

Wheeler, W. M. and R. Dow
1933. Unusual prey of *Bembix*. Psyche 40:57-59.

White, E.
1962. Nest building and provisioning in relation to sex in *Sceliphron spirifex* L., J. Anim. Ecol. 31:317-329.

Williams, C. B.
1933. Observations on the desert locust in East Africa from July, 1928 to April, 1929. Ann. Appl. Biol. 20:463-497.

Williams, F. X.
1914a. The Larridae of Kansas. Univ. Kansas Sci. Bull. 18:121-213.
1914b. Notes on the habits of some solitary wasps that occur in Kansas, with description of a new species. Univ. Kansas Sci. Bull. 18:223-230.
1919. Philippine wasp studies. Bull. Exp. Sta. Hawaiian Sugar Planters' Assoc. (Ent.) (14):1-186.
1927. Notes on the habits of bees and wasps of the Hawaiian Islands. Proc. Hawaiian Ent. Soc. 6:425-464.
1928a. Studies in tropical wasps — their hosts and associates (with descriptions of new species). Bull. Exp. Sta. Hawaiian Sugar Planters' Assoc. (Ent.) (19):1-179.
1928b. The sphecid wasp, *Podalonia violaceipennis* (Lep.) Proc. Hawaiian Ent. Soc. 7:163.
1929. Notes on the habits of the cockroach-hunting wasps of the genus *Ampulex* sens. lat., with particular reference to *Ampulex (Rhinopsis) canaliculatus* Say. Proc. Hawaiian Ent. Soc. 7:315-329.
1936. Notes on two oxybelid wasps in San Francisco, California. Pan-Pac. Ent. 12:1-8.
1942. *Ampulex compressa* (Fabr.), a cockroach-hunting wasp introduced from New Caledonia into Hawaii. Proc. Hawaiian Ent. Soc. 11:221-233.
1945. The aculeate wasps of New Caledonia with natural history notes. Proc. Hawaiian Ent. Soc. 12:407-452.
1946. Two new species of Astatinae, with notes on the habits of the group. Proc. Hawaiian Ent. Soc. 12:641-650.
1950. The wasps of the genus *Solierella* in California. Proc. Calif. Acad. Sci. (4) 26:355-417.
1954a. *Xenosphex xerophila*, an apparently new genus and species of wasp from southern California. Wasmann J. Biol. 12:97-103.
1954b. The wasps of the genus *Pisonopsis* Fox. Pan-Pacific Ent. 30:235-246.
1955. A correction: a second species of *Xenosphex*. Wasmann J. Biol. 13:313-315.
1960a. Three apparently rare sphecoid wasps from southern California, including a new subgenus and species of *Dolichurus*. Wasmann J. Biol. 17:299-306.
1960b. The wasps of the genus *Plenoculus*. Proc. Calif. Acad. Sci. (4) 31:1-49.
1962. A new species of wasp of the genus *Dolichurus* from southern Mexico. Proc. Ent. Soc. Washington 63:290-293.
Obit.: Zimmerman, E. C., 1969. Pan-Pac. Ent. 45:135-146, portrait. Arnaud, P., 1970. Occ. Pap. Calif. Acad. Sci. (80):1-33, portrait and bibliography.

Willink, A.
1947. Las especias Argentinas de "Bembicini". Acta Zool. Lilloana 4:509-651.
1949. Las especies neotropicales de "*Bembecinus*". Acta Zool. Lilloana 7:81-112.
1951. Las especies Argentinas y Chilenas de Chlorionini. Acta Zool. Lilloana 11:53-225.
1958. Descripcion de un nuevo genera y especie de Bembicini y observaciones sobre otra. Acta Zool. Lilloana 16:47-54.

Wissmann, O.
1849. Verzeichniss der im Königreich Hannover, zu-

mal in südlichen Theile und am Harze, bisher aufgefunden Mordwespen. Stett. Ent. Zeit. 10:8-17.

Wolcott, G. N.
1938. The introduction into Puerto Rico of *Larra americana* Saussure, a specific parasite of the "changa" or Puerto Rican mole-cricket *Scapteriscus vicinus* Scudder. J. Agr. Univ. Puerto Rico 22:193-218.

Yamamoto, D.
1942. Habits of *Sceliphron (Chalybion) inflexum* Sickmann. Kontyû 16:69-75.
1958. Habits of *Chalybion inflexum* Sickmann, II. Bull. Fac. Agr. Meiji Univ. (7):63-72.

Yarrow, I. H. H.
1969. Some additional and little known British species of the solitary wasp genus *Spilomena* Shuckard. Ent. Gazette 10:97-104.
1970. Some nomenclatorial problems in the genus *Passaloecus* Shuckard and two species not before recognized as British. Ent. Gazette 21:167-189.

Yasumatsu, K.
1934. Notes on the genus *Passaloecus* Shuckard. Mushi 7:109-114.
1935a. The Oxybelidae of Japan and Korea. Trans. Sapporo Nat. Hist. Soc. 14:38-41.
1935b. The genus *Pison* Spinola of the Japanese Empire. Annot. Zool. Japon. 15:227-238.
1936. Ampulicidae of the Japanese Empire. Tenthredo 1:165-232.
1938a. A revision of the genus *Sphex* Linné of the Japanese Empire. Tenthredo 2:44-135.
1938b. Two new wasps from Japan. Mushi 11:83-86.
1939. Notes supplémentaires sur le genre *Pison* Spinola du Japon. Festschrift zum 60. Geburtstage von Professor Dr. Embrik Strand. Riga 5:81-84.
1953. Sphecoidea of Micronesia. 4. Revision of the genus *Pison* Spinola. Part 1. J. Fac. Agr. Kyushu Univ. 10:133-150.

Yasumatsu, K. and F. Maidl
1936. A new genus and species of the family Nyssonidae sensu Kohl from the Far East. Festschrift zum 60. Geburtstage von Professor Dr. Embrik Strand. Riga 1:501-504.

Yasumatsu K. and H. Masuda
1932. On a new hunting wasp from Japan. Fukuoka Hakubutsugaku Zasshi 1:53-64.

Yasumatsu, K. and C. Watanabe
1964. A tentative catalogue of insect natural enemies of injurious insects in Japan. Part 1. Parasite – predator host catalogue. 166 pp. Ent. Lab. Fac. Agr. Kyushu Univ., Fukuoka, Japan.
1965. A tentative catalogue of insect natural enemies of injurious insects in Japan. Part 2. Host parasite-predator catalogue. 116 pp. Ent. Lab. Fac. Agr. Kyushu Univ. Fukuoka, Japan.

Yoshikawa, K.
1964. Predatory hunting wasps as the natural enemies of insect pests in Thailand. In: T. Kira and T. Umesao, Nature and Life in Southeast Asia. 3:391-397.

Yoshimoto, C. M.
1958. Revision of the Hawaiian Pemphredoninae. Proc. Hawaiian Ent. Soc. 17:128-149.
1960. Revision of Hawaiian Crabroninae with synopsis of Hawaiian Sphecidae. Pacific Insects 2:301-337.
1964. Nesting activity and larval description of *Trypoxylon (Trypoxylon) bicolor* Smith in Hawaii. Pacific Insects 6:517-521.
1965. Nesting activity of the mud-daubing wasp, *Pison argentatum* Shuckard in Hawaii. Pacific Insects 7:291-294.

Zavadil, V.
1948. Oddil II, pp. 101-179. In: V. Zavadil and J. Šnoflák, Kutilky (Sphecidae) Československé Republiky. Ent. Prírucky Ent. Listů (13):1-179.

Zavadil, V. and J. Šnoflák
1948. Kutilky (Sphecidae) Československé Republiky. Ent. Prírucky Ent. Listů (13):1-179.

Zetterstedt, J. W.
1838. Insecta Lapponica. Heft 2, pp. 257-476. L. Voss, Lipsiae (dating after Stone *et al.*, 1965).

Zimsen, E.
1964. The type material of J. C. Fabricius. 656 pp. Munkogaard, Copenhagen.

ADDENDUM

Since completion of the manuscript and reading of galley proof a number of important taxonomic matters have come to our attention. These include correction of errors, changes in synonymy, new taxa, various nomenclatorial problems, significant recent literature, etc. These are noted below. Last entries were made on June 6, 1975. This information should make our checklists accurate through December 31, 1974.

According to Leclercq (in litt.), *Lamocrabro* should be treated as a subgenus of *Pae*. Also *Taruma* may prove to be a synonym of *Foxita*. Leclercq has examined the types of *Taruma bara, Foxita galibi* and *F. tarumoides* and found them to be very similar.

Gmelin (1790) described a number of new species in *Vespa*, section "Crabrones Fabricii", that may prove to be crabronines or at least sphecids. All were described from "Europa" except one without locality noted below; *argyrostoma* (Turton, 1802, spelled it *Vespa argurostoma*), *braccata, canescens, chrysogona, chrysopus, chrysostoma, clavata, cylindrica, dichroa, ferruginea, flavicans, flavipes* (Kohl, 1915, said "kein *Crabro*"), *immaculata* (Kohl, 1915, spelled it *immacula*), *leskii* (Spinola, 1806:103, tentatively associated the name with species now placed in *Ectemnius*), *lutea, marginata, melanostoma* (Turton, 1802, spelled it *melanops*), *nigerrima* (no locality and description very brief), *ochropus, punctulata, tibialis, unifasciata*.

The type material (female holotype) of *Ectemnius rufipes* (Lepeletier and Brullé), 1834:741 (*Ceratocolus*) from "Caroline" has apparently been lost. The description closely fits the eastern U. S. species *E. texanus* (Cresson, 1872). To clarify the name *rufipes* and stabilize the nomenclature a neotype for *rufipes* is here designated: female, Decatur, Alabama, August, 1944 (G. E. Bohart, Univ. California Davis Entomology Museum).

The publication dates of two Fox papers have been found to differ from those used by us. Fox (1893b – see literature cited) was published in March of 1894. This affects the dates of two generic names, *Ancistromma* and *Dienoplus*, as well as many specific names, especially in *Tachysphex*. Fox's "Synopsis of the North American species of *Gorytes* Latr." (Proc. Acad. Nat. Sci. Philadelphia, 1895:517-539) appeared early in 1896. This affects 11 species names in the Gorytini which we cited as 1895.

Nomina oblita: Evans and Matthews (1968b) proposed two nomina oblita in *Bembix (dentilabris* and *pallidipicta)*. However, they did not fulfill the requirements of Art. 23(b)i of the pre Monaco Code, and we do not feel bound to accept their nomina oblita. Consequently, in our checklist we followed the law of priority and used *dentilabris* in place of *u-scripta* and *pallidipicta* in place of *pruinosa*.

As emended by the Monaco Congress, Art. 32(a)ii of the Code is very clear in stating that the original spelling of a name must be used unless "there is in the original publication itself, without recourse to any external source of information, clear evidence of an inadvertent error," . . . etc. Hence most emendations proposed in the past, even by the original author, are invalid. A few emendations involving patronyms are incorrect for another reason. Early authors occasionally latinized proper names by adding -ius before forming the genitive ending (i.e., Kirby to Kirbius, Brullé to Brullius). In the genitive case such names become *kirbii* and *brullii*. Some authors have emended such names (*kirbii* to *kirbyi* for example). A third situation involves the use of the suffix -e for masculine patronyms. The correct Latin ending for patronyms based on names ending in -a is -e, whether the person is a man or woman. Hence the names *Nitela spinolae* (after Maximilian Spinola) and *Tachysphex costae* (after Achille Costa) are correct. The automatic changing of these masculine patronyms to *spinolai* and *costai* is no longer necessary because Art. 31A makes such action a recommendation only. Furthermore, Art. 32 is pertinent. Listed below are the correct spellings of names improperly cited by us in the text:

Prionyx kirbii (Vander Linden), not *kirbyi*, the latter emendation apparently first proposed by Dours, 1874.

Pemphredon grinnelli (Rohwer), not *gennelli* (although the latter spelling was the original orthography, we believe *grinnelli* was intended because Grinnell's name was mentioned as the collector.) Rohwer

first used the spelling *grinnelli* in 1911.
Crossocerus varus Lepeletier and Brullé, not *varius* (because of homonymy with *varus* (Panzer), this species must be called *pusillus* Lepeletier and Brullé).

The following changes should be made in our checklists:

The subspecies *infesta* Smith under *Ammophila sabulosa* (Linnaeus) should be raised to species.

The name *sicula* (Kohl) in the synonymy of *Astata boops* (Schrank) should be raised to species, distribution: Sicily.

The name *massiliensis* (Morice) in the synonymy of *Astata rufipes* Mocsáry is a subspecies, distribution: Iberian penin., s. France.

Bembix julii Fabre and its synonym *occitanica* are synonyms of *B. sinuata* Panzer.

Carinostigmus major (Tsuneki), 1954 (*Stigmus*) should be added to synonymy of *C. maior* as an emendation.

Carinostigmus monstrosus (Tsuneki) is a synonym of *C. maior* (Maidl).

Carinostigmus thailandinus (Tsuneki) is a synonym of *C. iwatai* (Tsuneki).

Cerceris basiornata Cameron is a subspecies of *C. rufocincta* Gerstaecker.

Cerceris bredoi Arnold is a synonym of *C. rufocincta* ssp. *basiornata* Cameron.

Cerceris heterospila Cameron is a subspecies of *C. kilimandjaroensis* Cameron; *macololo* is a synonym.

Cerceris heterospila ssp. *capensis* Arnold is a subspecies of *C. kilimandjaroensis* Cameron; *griqua* is a synonym.

Cerceris jackal Brauns is a synonym of *C. rufocincta* ssp. *polychroma* Gribodo.

Cerceris kalaharica Bischoff is a subspecies of *C. trichionota* Cameron.

Cerceris kalaharica ssp. *ovambo* Empey is a subspecies of *C. trichionota* Cameron.

Cerceris mochiana Soika is a synonym of *C. rufiscutis* Cameron, s.s.

Cerceris pictinoda Cameron is a senior synonym of *C. ruficauda* ssp. *barbara* Arnold.

Cerceris polychroma Gribodo is a subspecies of *C. rufocincta* Gerstaecker.

Cerceris sokotrae Kohl is a subspecies of *C. rufocincta* Gerstaecker.

Cerceris sterope Arnold is a synonym of *C. kilimandjaroensis* Cameron, s.s.

Cerceris stevensoni Brauns is a synonym of *C. kilimandjaroensis* Cameron, s.s.

Cerceris zyx Leclercq and its synonym *molesta* are synonyms of *C. kilimandjaroensis* Cameron, s.s.

The name *schultzei* Bischoff in the synonymy of *Cerceris trichionota* Cameron is a valid species; *transvaalicola* is a synonym; S. Africa, Mozambique.

Crabro brachycarpae Rohwer and its synonym *gillettei* Rohwer are synonyms of *Crabro vernalis* (Packard). New synonymy by R. Bohart.

Crabro bruneri in the synonymy of *Crabro vernalis* is a valid species; U.S.: Nebraska and Oregon. New status by R. Bohart.

Crabro chalybaeus Kohl, 1915 should be added to synonymy of *C. chalybeus* as an alternate original spelling.

The name *clarconis* Viereck in the synonymy of *Crabro hilaris* Smith should be transferred to the synonymy of *Crabro cingulatus* (Packard). New synonymy by R. C. Miller.

Crossocerus asiaticus Tsuneki is a synonym of *Cr. pseudopalmarius* (Gussakovskij)

Crossocerus hedgreni (Kjellander) is a synonym of *Cr. heydeni* (Kohl)

Crossocerus sulcus (Fox) and its synonym *plesius* (Rohwer) are synonyms of *Crossocerus elongatulus* (Vander Linden). The range of the last should be emended to include e. U.S.

Crossocerus pusillus Lepeletier and Brullé is the proper name for *Crossocerus varius* Lepeletier and Brullé (see discussion above on emendations).

"New status by F. D. Parker" should have been indicated after the following species names in *Dryudella*: *ammochtona*, *anatolica*, *elbae*, *eurygnatha*, and *modesta*

Ectemnius hispanicus (Kohl) is a synonym of *E. continuus* (Fabricius)

The name *interruptefasciatus* Retzius in the synonymy of *Ectemnius cavifrons* (Thomson) should be transferred to the synonymy of *Ectemnius sexcinctus* (Fabricius) and the "?" eliminated.

The following name was omitted from the synonymy of *Gorytes angustus* (Provancher): *angustatus* Ashmead, 1899 (*Hoplisus*)

Lestica ochotica A. Morawitz is a subspecies of *L. subterranea* (Fabricius).

The name *foersteri* Fritz in the synonymy of *Metanysson alfkeni* (Ducke) is a valid species, distribution: Argentina.

Miscophus reptans Arnold is a synonym of *Saliostethus saltator* Arnold, new synonymy by O. Lomholdt.

Philanthus fomosanus Tsuneki is a subspecies of *P. notatulus* Smith.

The name *longiventris* (Cameron) in the synonymy of *Pseneo montezuma* should be raised to species, and *montezuma* placed under it as a synonym.

Pseneo montivagus (Dalla Torre) should be transferred to *Psen*; *Mimesa cameroni* Ashmead, 1899, which we overlooked, should be added as a synonym.

Pseneo spicatus (Malloch) is a synonym of *P. longiventris*

Psenulus brevitarsis Merisuo is a synonym of *P. pallipes* (Panzer)

Sericophorus rufonigrus should read *rufoniger*

Subspecies *yaeyamanus* Tsuneki under *Tachysphex bengalensis* should be transferred to *Tx. tinctipennis* Cameron.

Subspecies *major* Kohl and *minor* Beaumont under *Trypoxylon figulus* are synonyms of *figulus* s.s.

Trypoxylon striatum and its synonyms are synonyms of *T. lactitarse* Saussure

The following new taxa have been published since completion of the manuscript:

New genera:
Corenocrabro Tsuneki, 1974. Type species: *Corenocra-*

bro ectemiformis Tsuneki, 1974, original designation.
Cretosphex Rasnitsyn, 1975. Type species: Cretosphex incertus Rasnitsyn, 1975, original designation (Lower Cretaceous fossil).

New subgenera:
Diodontus (Corenius) Tsuneki, 1974. Type species: Diodontus chosenensis Tsuneki, 1974, original designation.
Liris (Colloliris) Tsuneki, 1974. Type species Notogonidea negrosensis Williams, 1928, original designation. (a synonym of Liris subgenus Leptolarra, new synonymy by A. Menke).

New species:
Ammatomus tanoi Tsuneki, 1974; Thailand
Antomartinezius boharti Fritz, 1973; Argentina
Bembecinus nyamadanus Tsuneki, 1974; Thailand
Bembix borrei ssp. bariensis Tsuneki, 1974; Indonesia:Bari
Bicyrtes colombica Fritz, 1974; Colombia
Carinostigmus borneanus (Tsuneki), 1974 (Stigmus); Borneo (new combination by Menke)
Cretosphex incertus Rasnitsyn, 1975; Lower Cretaceous fossil
Cerceris armaticeps ssp. caffrariae Empey, 1974; S. Africa: Cape Prov.
Cerceris carrascoi Fritz and Toro, 1973; Peru
Cerceris crenulifera Kazenas, 1974; sw. USSR: Kazakh S.S.R.
†Cerceris geniculata Kazenas, 1974; sw. USSR: Kazakh S.S.R., nec Cameron, 1890
Cerceris languida ssp. agulhas Empey, 1974; S. Africa: Cape Prov.
Cerceris languida ssp. shangaani Empey, 1974; Rhodesia, S. Africa
Cerceris languida ssp. tanzana Empey, 1974; Tanzania: Tanganyika
Cerceris pictiventris ssp. bariana Tsuneki, 1974; Indonesia: Bari
Cerceris ruficauda ssp. senegalensis Empey, 1974; Senegal
Cerceris rufiscutis ssp. aethiopica Empey, 1974; Ethiopia (also spelled aethopica)
Cerceris rufiscutis ssp. beniensis Empey, 1974; Zaire
Cerceris rufiscutis ssp. kenyae Empey, 1974; Kenya
Cerceris rufiscutis ssp. matabele Empey, 1974; Rhodesia
Cerceris rufiscutis ssp. umtaliensis Empey, 1974; Rhodesia, Zaire
Cerceris rufocincta ssp. voltaica Empey, 1974; Upper Volta
Cerceris spinicaudata ssp. maliensis Empey, 1974; Mali
Cerceris spinifera Kazenas, 1974; sw. USSR: Kazakh S.S.R.
Cerceris trichionota ssp. somereni Empey, 1974; Kenya
Cerceris tucuman Fritz and Toro, 1974; Argentina
Corenocrabro ectemiformis Tsuneki, 1974; Korea
Crossocerus assimilis ssp. collaris Tsuneki, 1974; Korea
Crossocerus podagricus ssp. hokusenensis Tsuneki, 1974; Korea
Crossocerus (Eupliloides) tanoi (Tsuneki), 1974 (Eupliloides); Thailand (new combination by Bohart and Menke, we treat Eupliloides as a subgenus of Crossocerus, but Tsuneki elevates it to genus without giving his reasons for doing so.)

Crossocerus wesmaeli ssp. parvicorpus Tsuneki, 1974; Korea
Diodontus chosenensis Tsuneki, 1974; Korea
Diodontus minutus ssp. orientalis Tsuneki, 1974; Korea
Diodontus wahisi Leclercq, 1974; France
Diploplectron pulawskii Kazenas, 1975; sw. USSR: Kazakh S.S.R.
Ectemnius chrysites ssp. chosenensis Tsuneki, 1974; Korea
Ectemnius horvatovichi Tsuneki, 1974; Korea
Encopognathus kinabalensis Tsuneki, 1974; Borneo
Eremiasphecium dzhanokmenae Kazenas, 1974; USSR: Kazakh SSR
Eremiasphecium gussakovskii Kazenas, 1974; USSR: Kazakh SSR
Gastrosericus siamensis Tsuneki, 1974; Thailand
Gastrosericus thailanditus Tsuneki, 1974; Thailand
Liris borneana Tsuneki, 1974; Borneo
Liris clypeopunctata Tsuneki, 1974; Thailand
Liris punctata Tsuneki, 1974; Thailand
Liris tanoi Tsuneki, 1974; Thailand
†Liris trifasciata Tsuneki, 1974; Thailand, nec Smith, 1856
Losada penai Fritz, 1973; Paraguay
Mellinus satanicus Siri and Bohart, 1974; s. Mexico
Metanysson cicheroi Fritz, 1973; Argentina
Metanysson horacioi Fritz, 1973; Bolivia
Metanysson toba Fritz, 1973; Argentina
Mimesa empeyi (van Lith), 1974 (Psen); S. Africa (new combination by Bohart and Menke)
Nitela borealis Valkeila, 1974; n. Europe
Nitela spinolai Valkeila, 1974, emendation of spinolae Latreille, 1809
Notocrabro micheneri Leclercq, 1974; Australia: Queensland
Oxybelus ayuttayanus Tsuneki, 1974; Thailand
Oxybelus koreanus Tsuneki, 1974; Korea
Oxybelus thailanditus Tsuneki, 1974; Thailand (new name and new status for thaianus Tsuneki, 1963)
Passaloecus annulatus ssp. koreanus Tsuneki, 1974; Korea
Passaloecus ribauti Merisuo, 1974; s. France
Prosopigastra numida Pulawski, 1975; Tunisia
Psen leclercqi van Lith, 1974; Madagascar
Psenulus alveolatus van Lith, 1974; Rhodesia
Psenulus aurifasciatus van Lith, 1974; Sierra Leone
Psenulus bidentatus ssp. pallidus van Lith, 1974; Zaire, Malawi, Rhodesia, Guinea
Psenulus freetownensis van Lith, 1974; Sierra Leone, Guinea
Psenulus jacoti van Lith, 1974; S. Africa
Psenulus leoninus van Lith, 1974; Sierra Leone
Psenulus oweni van Lith, 1974; Sierra Leone
Psenulus rugifrons van Lith, 1974; Ethiopia
Psenulus sapobaensis van Lith, 1974; Nigeria
Psenulus thaianus Tsuneki, 1974; Thailand
Psenulus uelleburgi van Lith, 1974; Guinea
Pseudoturneria couloni Leclercq, 1974; Australia: Victoria
Pseudoturneria territorialis Leclercq, 1974; Australia: ACT

Rhopalum atlanticum R. Bohart, 1974; U.S.: Atlantic States
Rhopalum coarctatum ssp. *koreense* Tsuneki, 1974; Korea
Rhopalum nipponicum ssp. *chosenense* Tsuneki, 1974; Korea
Rhopalum pacificum R. Bohart, 1974; U.S.: Calif., Ore., Nevada
Sphex tanoi Tsuneki, 1974; Thailand
Tachytes codonocarpi Pulawski, 1975; Australia: S. Australia
Tachytes dubiosus Tsuneki, 1974; Thailand
Tachytes fruticis ssp. *taianus* Tsuneki, 1974; Thailand
Trypoxylon appendiculatum Tsuneki, 1974; Borneo
Trypoxylon monstruosum Tsuneki, 1974; Thailand
Trypoxylon pappi Tsuneki, 1974; Korea
Trypoxylon varipes ssp. *nasutum* Tsuneki, 1974; Korea
Williamsita bushiella Leclercq, 1974; Australia: S. Australia
Williamsita riekiella Leclercq, 1974; Australia: N.S. Wales
Williamsita vedetta Leclercq, 1974; Australia: W. Australia

Literature (partially annotated):

Alayo, D. P., 1973. Catálogo de los Himenópteros de Cuba. 218 pp. Instituto Cubano del Libro, Havana. (erroneously dated Nov. 1970 on title page).

Alcock, J., 1973. Notes on a nesting aggregation of digger wasps in Seattle, Washington. Wasmann J. Biol. 31:323-336.

———, 1974. The behavior of *Philanthus crabroniformis*. J. Zool. 173:223-246.

———, 1974. The nesting behavior of *Cerceris simplex macrosticta*. J. Nat. Hist. 8:645-652.

———, 1975. The nesting behavior of *Philanthus multimaculatus* Cameron. Amer. Midl. Nat. 93:222-226.

Bohart, R. M., 1974. A review of the genus *Rhopalum* in America north of Mexico. J. Georgia Ent. Soc. 9:252-260. (key to species)

Eady, R. D., 1974. The present state of nomenclature of wing venation in the Braconidae (Hymenoptera); its origins and comparison with related groups. J. Ent. (B) 43:63-72. (excellent review of wing terminology in Hymenoptera, proposed nomenclature differs little from that used in this book).

Eck, R., 1971. Zur bionomie von *Cerceris arenaria* (L.). Ent. Abhandl. Staatl. Mus. Tierkunde Dresden 37: 337-361.

Else, G. R., 1974. *Ectemnius nigrinus* (Herrich-Schaeffer) a crabronine wasp new to Britain, with a key to the British species of *Ectemnius* Dahlbom. Ent. Gazette 25:203-211.

Empey, H. N., 1974. The status, sex and synonymy of Cameron's types of *Cerceris* Latreille from the Ethiopian Region. Ann. Transvaal Mus. 29:99-134.

Evans, H. E. and R. W. Matthews, 1974. Observations on the nesting behavior of South American sand wasps. Biotropica 6:130-134.

Evans, H. E., R. W. Matthews and E. Callan, 1974. Observations of the nesting behavior of *Rubrica surinamensis* (De Geer). Psyche 81:334-352.

Fritz, M. A., 1973. Nyssonini Neotropicales VI (Hym. Sphecidae: Nyssoninae). Anal. Mus. Hist. Nat. Valparaiso 6:191-202. (key to Neotropical species of *Metanysson*)

International Commission on Zoological Nomenclature, 1974. Opinion 1015: *Solenius* Lepeletier and Brullé, 1834 (Insecta, Hymenoptera): designation of a type-species under the plenary powers. Bull. Zool. Nomencl. 31:16-18. (*Solenius interruptus* Lepeletier and Brullé designated as type species).

Janvier, H., 1974. Una colonia de *Argogorytes hispanicus* (Merc., 1906) en Malaga. Graellsia 27:67-77.

Kazenas, V. L., 1972. Sphecidae of the Southeast Kazakhstan. Horae Soc. Ent. Unionis Soveticae 55:93-186.

Kurczewski, F. E. and C. J. Lane, 1974. Observations on the nesting behavior of *Mimesa (Mimesa) basirufa* Packard and *M. (M.) cressonii* Packard. Proc. Ent. Soc. Wash. 76:375-384.

Leclercq, J., 1974. Crabroniens d'Australie. Bull. Ann. Soc. Roy. Belge Ent. 110:37-57. (Key to species of *Williamsita*; subgenus *Androcrabro* placed in synonymy; notes on species of *Chimiloides*).

———, 1974. Données pour un atlas des Hyménoptères das Pemphredoninae (sauf *Pemphredon*). Bull. Recher. Agron. Gembloux 7:191-222. (key to *Diodontus* of Western Europe).

———, 1974. Noms, types et néotypes d'une trentaine de crabroniens européens. Bull. Ann. Soc. Roy. Belge Ent. 110:258-286. (considerable new synonymy of old names)

Lomholdt, O., 1973. Biological observations on the digger-wasp *Passaloecus eremita* Kohl. Vidensk. Medd. Dansk Naturhist. For. 136:29-41.

Marshakov, V. G., 1975. A review of the genera of the tribe Crabronini from the USSR. The genus *Lestica* Billberg, 1820. Rev. Ent. URSS 54:151-163. (key to genera of USSR and species of *Lestica*).

Matthews, R. W. and H. E. Evans, 1974. Notes on the behavior of three species of *Microbembex* in South America. J. Georgia Ent. Soc. 9:79-85.

Menke, A. S., 1974. Some sphecid workers past and present. 3 unnumbered pages. Agricultural Research Service, USDA, unnumbered publication. (consists of photographs).

———, 1974. The correct name for a common North American *Trypoxylon* wasp. Proc. Ent. Soc. Wash. 76:418.

Merisuo, A. K., 1974. Zur Kenntnis der europäischen Arten der Gattung *Passaloecus* Shuckard. Ann. Ent. Fennici 40:10-15 (key to species groups).

Miller, R. C. and F. E. Kurczewski, 1973. Ecology of digger wasps, pages 204-217 in: Dindal, D. L., Soil microcommunities, I. Symbiotic relationships of soil invertebrates. Proc. First Soil Micrcommunities Confer. USAEC, Off. of Inform. Serv. CONF-711076. Natl. Tech. Inform. Serv., USDC, Springfield, Va.

___, 1973. Intraspecific interactions in aggregations of *Lindenius*. Insects Soc. 20:365-378.

Nambu, T., 1973. Biology of *Crossocerus (Towada) flavitarsus* Tsuneki, using resin to close the nest entrance. Life study (Fukui) 17:55-60.

Powell, J., 1974. On the nesting behavior of the sand-wasp *Gorytes canaliculatus* (Packard) in California. J. Kansas Ent. Soc. 47:1-7.

Pulawski, W. J., 1974. Notes sur la biologie de deux *Tachysphex* rares: *T. rugosus* Guss. et *T. plicosus* Costa. Polskie Pismo Ent. 44:715-718.

___, 1975. Synonymical notes on Larrinae and Astatinae. J. Wash. Acad. Sci. 64:308-323 (erroneously dated 1974)

Rasnitsyn, A. P., 1975. Hymenoptera Apocrita of Mesozoic. Trans. Palaeontological Inst. 147:1-134 (Review of all Mesozoic fossils of the Apocrita with keys to taxa.)

Siri, M. L. and R. M. Bohart, 1974. A review of the genus *Mellinus*. Pan Pac. Ent. 50:169-176. (key to species of the world.)

Steiner, A. L., 1974. Unusual caterpillar-prey records and hunting behavior for a *Podalonia* digger wasp: *Podalonia valida* (Cresson). Pan-Pacific Ent. 50:73-77.

Tsuneki, K., 1973. Nests of some pemphredonine wasps in the pith of *Miscanthus*. Life Study (Fukui) 17: 63-73.

___, 1974. A contribution to the knowledge of Sphecidae occurring in Southeast Asia. Polskie Pismo Ent. 44:585-660.

___, 1974. Sphecidae from Korea. Ann. Hist.-Nat. Mus. Hungarici 66:359-387.

Valkeila, E., 1974. *Nitela spinolai* Latr. s. auct.: a confusion of two European species. Ann. Ent. Fennici 40:75-85. (key to European species and important description of larva).

van der Vecht, J., 1975. The date of publication of M. Spinola's paper on the Hymenoptera collected by V. Ghiliani in Para, with notes on the Eumenidae described in this work. Ent. Bericht. 35:60-63. (paper cited by us as Spinola, 1853, was first published in 1851 with different pagination.)

van Lith, J. P., 1974. Revision of the Psenini of the Ethiopian Region, including Malagasy. Tijdschr. Ent. 117:39-101. (keys to species).

Ward, G. L., 1973. Growth of *Chalybion zimmermanni* Dahlbom in captivity. Proc. Indiana Acad. Sci. 82: 231-233.

INDEX

Names are Indexed as follows:
CAPITALS: All names for taxa above the generic level. The first letter of each generic or subgeneric name is capitalized.
Boldface: Valid generic and subgeneric names.
Roman: Specific and subspecific names of Sphecidae, followed by the author of the name and the genus in which the name is now placed. The author's name is in parentheses if he used a different genus originally. Endings of species-group names have been corrected where necessary to agree in gender with the genus to which they are presently referred.

abacta Shestakov, Cerceris, 576
abata J. Parker, Bembix, 545
abax Leclercq, Dasyproctus, 420
abbottii W. Fox, Sphex, 114
abbottii Westwood, Ampulex, 77
abbreviata Banks, Cerceris, 584
abbreviata (Fabricius), Eremnophila, 147
abbreviata (Turner), Liris, 243
abbreviatus Strand, Dolichurus, 69
abdelcader Lepeletier, Philanthus, 566
abdelkader Lucas, Philanthus, 566
abdita (Kohl), Isodontia, 123
abditum Arnold, Trypoxylon, 345
abditus (Kohl), Lindenius, 384
abditus Kohl, Tachysphex, 275
abdominalis Ashmead, Plenoculus, 311
abdominalis Baker, Oxybelus, 366
abdominalis Cresson, Mellinus, 449
abdominalis (Cresson), Palmodes, 127
abdominalis (Cresson), Pseudoplisus, 503
abdominalis (Dahlbom), Stizoides, 528
abdominalis F. Smith, Dolichurus, 69
abdominalis (Fabricius), Cerceris, 576
abdominalis Guérin-Méneville, Larra, 237
abdominalis (Guérin-Méneville), Paranysson, **44, 308**
abdominalis (Panzer), Astata, 212
abdominalis Perty, Trigonopsis, 98
abdominalis (Say), Tachytes, 263
abdominalis (W. Fox), Moniaecra, 395
abdominalis Williams, Solierella, 313
abeillei (Marquet), Podalonia, 144
abercornensis Arnold, Bembix, 545
abercornensis Arnold, Tachytes, 263
aberrans Ducke, Bothynostethus, 352
aberrans (Eversmann), Sphecius, 511
aberrans (Gussakovskij), Lestica, 431
aberrans Turner, Pison, 335
aberti Haldeman, Ammophila, 150
ablatus Nurse, Tachysphex, 275
Ablepharipus Perkins, 48, 398
abjectus Kohl, Tachysphex, 273

abnormis (Blackburn), Ectemnius, 424
abnormis (Kohl), Polemistus, 185
abnormis Tsuenki, Odontocrabro, 418
abnormis (Turner), Larrisson, 44, **305**
aborensis Nurse, Ampulex, 77
abragensis Priesner, Bembix, 545
abscondicus Andrade, Miscophus, 317
absoluta Nurse, Astata, 212
abyssinica (Arnold), Liris, 243
abyssinicus (Arnold), Ectemnius, 424
abyssinicus Arnold, Philanthus, 564, 566
abyssinicus (Arnold), Sphex, 114
Acanthocrabro Perkins, 48, 398
acanthomerus (Morice), Bembecinus, 530
acanthomerus Pate, Encopognathus, 380
acanthophila Cockerell, Cerceris, 576
acanthophora Gussakovskij, Prosopigastra, 285
acanthophorus (Kohl), Crossocerus, 400
Acanthostethus F. Smith, 51, **473**
acavai Pate, Foxita, 416
accepta Leclercq, Piyuma, 410
accline (Kohl), Chalybion, 102
accola Kohl, Cerceris, 587
accumulator F. Smith, Trypoxylon, 345
acephalus Leclercq, Crossocerus, 400
aciculata (Cameron), Liris, 244
aciculatus (Provancher), Ectemnius, 425
acolhua Saussure, Cerceris, 576
acollae Leclercq, Podagritus, 392
Acolpus Vachal, 55
acrobates (Kohl), Tachysphex, 273
actaeon Beaumont, Tachysphex, 276
aculeata (Fernald), Ammophila, 151
aculeatus Thomson, Oxybelus, 367
acuta (Fernald), Ammophila, 150
acuta Radoszkowski, Cerceris, 576
acutangulum Arnold, Trypoxylon, 345
acutemarginatus Strand, Tachysphex, 272
acuticornis F. Morawitz, Oxybelus, 368
acutissimus Bischoff, Oxybelus, 366

acutus Baker, Oxybelus, 369
acutus (Patton), Tachysphex, 272
adae Turner, Cerceris, 576
adamsi Titus, Diodontus, 179
adamsoni Arnold, Philanthus, 564
adelaidae Turner, Tachysphex, 272
adelpha Kohl, Ammophila, 150
adelpha Kohl, Cerceris, 576
adelphiae Sandhouse, Trypoxylon, 346
adelphus Richards, Microstigmus, 192
adelphus W. F. Kirby, Stizus, 526
adhaesus (Kohl), Crossocerus, 400
adiaphilus Krombein, Nippononysson, 467
adjunctus Kohl, Tachysphex, 272
admirabilis Radoszkowski, Bembix, 545
admirabilis Turner, Tachytes, 263
adonis (Handlirsch), Editha, 542
adornatus (Bradley), Hoplisoides, 521
adriaansei Wilcke, Ammophila, 153
adspectans (Blackburn), Ectemnius, 424
adulatrix Arnold, Cerceris, 577
adumbrata (Handlirsch), Rubrica, 540
advena F. Smith, Crabro, 407
adventicus (Beaumont), Dienoplus, 496
advenus Pulawski, Tachytes, 272
aedilis (F. Smith), Liris, 244
aegyptia Kohl, Solierella, 313
aegyptiaca Beaumont, Liris, 246
aegyptiaca Radoszkowski, Mimesa, 161
aegyptiacum Gussakovskij, Trypoxylon, 345
aegyptiacum Klug, Sceliphron, 106
aegyptiacus Morice, Tachysphex, 274
aegyptiacus Pulawski, Tachytes, 265
aegyptiacus (Schulz), Dienoplus, 496
aegyptium Kohl, Trypoxylon, 345
aegyptium (Linnaeus), Sceliphron, 106
aegyptius (Kohl), Lindenius, 384
aegyptius (Lepeletier), Prionyx, 133
aegyptius Lepeletier, Stizus, 526

633

aegyptius Morice, Miscophus, 317
aellos Menke, Ammophila, 150
aemula Arnold, Cerceris, 576
aemulans (Kohl), Hoplammophila, 141
aemulans (Kohl), Podagritus, 392
aemuloides Leclercq, Cerceris, 576
aemulum Kohl, Sceliphron, 106
aemulus (Handlirsch), Bembecinus, 530
aemulus Kohl, Tachysphex, 272
aenea (Fabricius), Ampulex, 77
aenea (Handlirsch), Ochleroptera, 490
aenea Spinola, Ampulex, 77
aenescens (Bridwell), Miscophus, 317
aenescens (Dahlbom), Lindenius, 384
aeneus E. Saunders, Tachytes, 263
aeneus Turner, Auchenophorus, 327
aeneipenne Arnold, Trypoxylon, 345
aenigma Honoré, Miscophus, 318
aequadoricus (Strand), Podagritus, 392
aequalis (Handlirsch), Pseudoplisus, 503
aequalis (Patton), Synnevrus, 470
aequalis Provancher, Cerceris, 576
aequalis W. Fox, Crabro, 407
aequalis (W. Fox), Liris, 244
aequalis W. Fox, Tachysphex, 272
aequipunctatus Pate, Oxybelus, 368
aerarium Patton, Chlorion, 89
aerata Kazenas, Cerceris, 576
aereola Bradley, Ampulex, 77
aeripilosa Tsuneki, Larra, 237
aerofacies (Malloch), Pluto, 171
aestivale Richards, Trypoxylon, 348
aestivalis Mercet, Stizus, 526
aestuans Turner, Tachytes, 263
aestuosus Bingham, Oxybelus, 366
aeta Pate, Crossocerus, 400
aethiopica Empey, Cerceris, 629
aethiopica Kohl, Lyroda, 299
aethiopicus Arnold, Tachysphex, 272
aethiopicus Cameron, Oxybelus, 366
aethiops (Cresson), Tachysphex, 272
aethiops (F. Smith), Larra, 237
aethiops (Handlirsch), Handlirschia, 52, **508**
afer Lepeletier, Sphex, 114
afer Morice, Diodontus, 179
affine. F. Smith, Chlorion, 90
affine (Fabricius), Sceliphron, 105
affine (Maindron), Sceliphron, 106
affinis (Cameron), Bicyrtes, 537
affinis Dahlbom, Bembix, 545
affinis F. Morawitz, Palarus, 291
affinis (F. Smith), Ectemnius, 426
affinis F. Smith, Trigonopsis, 98
affinis Gussákovskij, Belomicrus, 363
affinis Gussakovskij, Psen, 166
affinis Gussakovskij, Solierella, 313
affinis (Hacker & Cockerell), Sericophorus, 302
affinis Kazenas, Lindenius, 384
affinis Lepeletier and Brullé, Crossocerus, 401
affinis (Lucas), Isodontia, 124
affinis Marquet, Oxybelus, 368
affinis Pulawski, Miscophus, 318
affinis (Rohwer), Solierella, 313
affinis (Rossi), Cerceris 583
affinis (Spinola), Dienoplus, 496
affinis Turner, Aphelotoma, 70
affinis Vander Linden, Astata, 212
affinis (W. Kirby), Podalonia, 143
affinis (Wesmael), Crossocerus, 400

afghanica Balthasar, Ammophila, 150
afghanica Balthasar, Podalonia, 143
afghaniensis (Beaumont), Prionyx, 131
afra Handlirsch, Bembix, 545
africana (Arnold), Dryudella, 214
africana Beaumont, Prosopigastra, 286
africana Cameron, Ampulex, 77
africana Cameron, Cerceris, 584
africana (Leclercq), Xysma, 194
africana Turner, Liris, 244
africanula Brauns, Cerceris, 576
africanus Beaumont. Palarus, 291
africanus Kohl, Oxybelus, 367
africanus Leclercq, Encopognathus, 380
africanus (Maidl), Gastrosericus, 256
africanus Mercet, Gorytes, 501
africanus Pulawski, Tachysphex, 274
africanus (Radoszkowski), Pseudoplisus, 503
africanus Turner, Ammatomus, 513
Afrogorytes Menke, 53, **522**
affumata Schletterer, Cerceris, 582
agadiriensis Andrade, Miscophus, 318
agadiriensis Nadig, Tachytes, 263
agalena Gittins, Mimesa, 161
agamemnon Richards, Trypoxylon, 348
agile (F. Smith), Pison, 335
agile Kohl, Podium, 96
agilis F. Smith, Astata, 212
agilis (F. Smith), Bembecinus, 530
agilis (F. Smith), Dasyproctus, 420
agilis (F. Smith), Liris, 244
agilis F. Smith, Oxybelus, 366
agilis (F. Smith), Tachypshex, 272
agitata (Turner), Liris, 244
aglaia (Handlirsch), Hoplisoides, 520
agnata Turner, Cerceris, 576
agnitus Brèthes, Oxybelus, 366
Agnosicrabro Pate, 49
agnus Pulawski, Tachysphex, 272
Agraptus Wesmael, 52
agrestis J. Parker, Bembix, 545
agulhas Empey, Cerceris, 629
agycus (Cameron), Ectemnius, 424
ahasverus Kohl, Sphex, 114
aino (Tsuneki), Crossocerus, 400
aino Tsuneki, Gorytes, 501
Ainocrabro Tsuneki, 48, **398**
airensis Berland, Ammophila, 150
ais Pate, Ectemnius, 428
aiurnnensis Giner Marí, Gastrosericus, 256
ajax (Rohwer), Psenulus, 173
ajaxellus (Rohwer), Psenulus, 172
ajjer Beaumont, Laphyragogus, 220
akrofisianus Balthasar, Miscophus, 318
alacer (Bingham), Lestica, 430
alacer Kohl, Sphex, 114
alacris Arnold, Tachytes, 263
alacris Mickel, Cerceris, 581
alacris J. Parker, Bembix, 545
alamos Scullen, Cerceris, 576
alaope Banks, Cerceris, 579
alaris Brauns, Kohliella, 44 **286**
alaris (Saussure), Liris, 244
alastoroides Turner, Cerceris, 576
alata (Panzer), Lestica, 430
alatulus (Dahlbom), Ectemnius, 426
alaya (Pate), Hoplisoides, 520
alayoi Pulawski, Tachysphex, 272
albarenae Scullen, Eucerceris, 591
alberti (Arnold), Liris, **244**
albertus (Carter), Crossocerus, 402

albibarbis (Arnold), Hovanysson, 472
albicantia J. Parker, Stenolia, 553
albicapilla Arnold, Bembix, 545
albiceps (Gussakovskij), Philanthinus, 570
albicera (C. Fox), Glenostictia, 552
albicincta Klug, Cerceris, 576
albicolor Shestakov, Cerceris, 583
albida Scullen, Cerceris, 582
albidula Turner, Bembix, 545
albidulus (Lepeletier), Gorytes, 501
albifacies (Malloch), Pluto, 171
albifrons Cresson, Philanthus, 564
albifrons F. Smith, Cerceris, 576
albifrons Fabricius, Sphex, 114
albifrons Villers, Sphex, 117
albigena Cameron, Cerceris, 576
albilabris (Fabricius), Lindenius, 384
albilabris (Lepeletier), Gorytes, 501
albimaculatus Kazenas, Oxybelus, 366
albimana Taschenberg, Cerceris, 579
albinota Scullen, Cerceris, 579
albipes (Ashmead), Solierella, 313
albipes F. Morawitz, Oxybelus, 366
albipes (F. Smith), Podagritus, 392
albipes F. Smith, Trypoxylon, 345
albipicta F. Smith, Cerceris, 576
albisectus (Lepeletier and Serville), Prionyx, 133
albispinosa Arnold, Cerceris, 576
albitarse Fabricius, Trypoxylon, 348
albitarse of Richards, Trypoxylon, 349
albitomentosus (Bradley), Hapalomellinus, 52, **497**
alboatra Walker, Cerceris, 576
albocinctus (Lucas), Tachysphex, 272
alboclypeata Bradley, Spilomena, 193
albocollaris (Ashmead), Crossocerus, 400
albofasciata F. Smith, Bembix, 545
albofasciata (Rossi), Cerceris, 576
albofasciata (Thunberg), Cerceris, 576
albohirsuta (Tsuneki), Podalonia, 144
albohirta (Turner), Isodontia, 123
albolineata (Cameron), Cerceris, 576
albomaculatus Andrade, Miscophus, 318
albomaculatus (Mickel), Oxybelus, 369
albomaculatus Tsuneki, Ectemnius, 424
albomarginatus (Cresson), Epinysson, 472
albonigrum Richards, Trypoxylon, 348
albonotata Vander Linden, Cerceris, 586
albonotatus, Walker, Tachytes, 263
albopectinatus (Taschenberg), Prionyx, 133
albopectoris Empey, Cerceris, 576
albopictus Radoszkowski, Oxybelus, 366
albopictus Taschenberg, Philanthus, 564
albopilosa Arnold, Bembix, 545
albopilosa Tsuneki, Liris, 244
albopilosa (Tsuneki), Mimesa, 161
albopilosella Cameron, Astata, 212
albopilosum W. Fox, Trypoxylon, 349
albopilosus Cresson, Philanthus, 564
alboscutellatus Arnold, Psenulus, 172
albosignatus H. Smith, Oxybelus, 369
albosignatus W. Fox, Gorytes, 501
albospiniferus (Reed), Prionyx, 134
albotegulata Arnold, Cerceris, 577
albotomentosa Morice, Ammophila, 150
albovillosa Cameron, Astata, 212
albovillosulus (Giordani Soika), Prionyx, 134
albovillosum (Cameron), Penepodium, 92
albufeirae Andrade, Miscophus, 318

alcatae Leclercq, Encopognathus, 380
alceste Mickel, Cerceris, 577
alcyone Arnold, Cerceris, 577
aldabra J. Parker, Bembix, 545
aldrichi Sandhouse, Trypoxylon, 345
alecto (F. Smith), Larra, 237
alemon van der Vecht, Isodontia, 123
Alepidaspis A. Costa, 46
alevinus Leclercq, Podagritus, 392
alexandrae F. Morawitz, Cerceris, 577
alfierii Honoré, Miscophus, 318
alfierii Mochi, Cerceris, 585
alfierii Priesner, Bembix, 545
alfierii Pulawski, Tachytes, 263
alfkeni (Ducke), Metanysson, 481
algeriensis Lohrmann, Bembix, 546
algeriensis Schulz, Bembix, 548
algeriensis (Schulz), Cerceris, 582
algira (Kohl), Parapsammophila, 139
algirica Schletterer, Cerceris, 587
algirica (Thunberg), Cerceris, 587
algiricum Kohl, Pison, 335
algirus (Kohl), Lindenius, 384
algirus Kohl, Tachysphex, 275
algoense Arnold, Trypoxylon, 345
algoensis Arnold, Bembix, 547
alicantina (Mercet), Dienoplus, 496
alicantinus Mercet Nysson, 469
aliceae Turner, Sericophorus, 302
aliciae (Turner), Podagritus, 392
Alicrabro Tsuneki, 48, **397**
alienus (Krombein), Psenulus, 172
alini Tsuneki, Cerceris, 585
alipes (Bingham), Ammatomus, 513
alisana Tsuneki, Ampulex, 77
alishanus Tsuneki, Ectemnius, 424
alishanus (Tsuneki), Polemistus, 185
alishanus Tsuneki, Psen, 166
alishanus Tsuneki, Stigmus, 189
alleni Richards, Trypoxylon, 348
Alliognathus Ashmead, 47
allionii Dahlbom, Philanthus, 566
allonymum Schulz, Pison, 335
alluaudi (Berland), Prionyx, 133
allunga Evans and Matthews, Bembix, 545
allwini Dalla Torre, Philanthus, 566
almagrus Leclercq, Podagritus, 393
alorus Nagy, Dolichurus, 69
alpaca F. Parker, Astata, 212
alpestris Cameron, Astata, 212
alpestris Cameron, Crabro, 407
alpestris Cameron, Mellinus, 449
alpestris (Cameron), Podalonia, 144
alpestris (Cameron), Pseudoplisus, 503
alpestris Rohwer, Tachysphex, 272
alpheus Pate, Ectemnius, 424
alpicola Beaumont, Ammophila, 153
alpina (Kohl), Podalonia, 144
alpinus Handlirsch, Mellinus, 449
alpinus Imhoff, Crabro, 407
Alpiothyreopus Noskiewicz and Chudoba, 37
altaicus F. Morawitz, Crabro, 407
altaiensis (Tsuneki), Podalonia, 144
altamazonica Williams, Larra, 237
alternatus van der Vecht, Bembecinus, 530
altibia (Strand), Prionyx, 134
alticola Cameron, Ammophila, 153
alticola Cameron, Crabro, 407
alticola (Cameron), Pseudoplisus, 503
alticola F. Morawitz, Crabro, 407

alticola Tsuneki, Crossocerus, 400
alticola (Viereck), Mimumesa, 164
alticollis Beaumont, Lindenius, 384
altigena Dalla Torre, Crabro, 407
altigena Gussakovskij, Ammophila, 150
alutaceum Richards, Trypoxylon, 349
alutaceus Leclercq, Podagritus, 393
alveolatus van Lith, Psenulus, 629
Alyson Jurine, 51
Alysson Panzer, 51, **457**
ALYSSONINI, 59, **453**, 456
amabilis Arnold, Philanthus, 564
amahuaca Pate, Entomocrabro, 377
amakosa Banks, Cerceris, 577
amamiensis Tsuneki and Iida, Dolichurus, 69
amamiensis Tsuneki, Bembecinus, 531
amamiensis Tsuneki, Cerceris, 582
Amaripa Pate, 47
amaripa Pate, Pae, 413
amathusia Beaumont, Cerceris, 577
amator F. Smith, Sphex, 114
amatoria Arnold, Cerceris, 577
amaura Kohl, Cerceris, 584
amazonica Ducke, Nitela, 45, **325**
amazonica Ducke, Solierella, 313
amazonus F. Smith, Tachytes, 263
ambidens Kohl, Tachytes, 263
ambiguum Tsuneki, Trypoxylon, 345
ambiguus Arnold, Tachysphex, 272
ambiguus (Dahlbom), Crossocerus, 400
ambiguus Gerstaecker, Oxybelus, 368
ambiguus Handlirsch, Gorytes, 501
ambiguus (Schenck), Psenulus, 172
amboinensis van Lith, Psen, 166
ambositrae Leclercq, Tachysphex, 272
ambustus Klug, Palarus, 290
ameghinoi Bréthes, Cerceris, 577
amenartais (Pulawski), Dryudella, 213
americana (Cresson), Larra, 237
americana Fabricius, Bembix, 545
americana (Saussure), Larra, 237
americanus Packard, Diodontus, 179
americanus Packard, Stigmus, 188
americanus (Robertson), Oxybelus, 367
americanus Spinola, Oxybelus, 366
americanus W. Fox, Miscophus, 318
ametinus Cameron, Tachytes, 265
ami Tsuneki, Stigmus, 189
amiculus Banks, Tachytes, 265
ammanitus Leclercq, Ectemnius, 424
Ammatomus A. Costa, 53, **512**
ammatticum Leclercq, Rhopalum, 389
Ammobia Billberg, 39
ammochrysus Schulz, Philanthus, 564
Ammophila W. Kirby, 39, **147**
AMMOPHILINI, 58, **134**, 137
ammophiloides A. Costa, Trypoxylon, 345
Ammophilus Latreille, 40
Ammophilus Perty, 38
Ammophylus Latreille, 40
Ammoplanellus Gussakovskij, 42, **198**
Ammoplanops Gussakovskij, 42, **197**
Ammoplanopterus Mochi, 42
Ammoplanus Gussakovskij, 42, **197**
Ammopsen Krombein, 40, **159**
Ammosphecidium Kohl, 45
amochtona (Pulawski), Dryudella, 213
amoena Handlirsch, Bembix, 545
amoena Stal, Ampulex, 77
amoenus (F. Smith), Stizoides, 528

ampla Banks, Cerceris, 581
ampliceps Krombein, Spilomena, 193
amplipennis (F. Smith), Larra, 237
amplus W. Fox, Tachysphex, 272
Ampulex Jurine, 38, **74**
AMPULICINI, 58, **74**
AMPULICINAE, 57, **63**, 66
amu Pate, Tracheliodes, 405
amurensis (Kohl), Crossocerus, 400
Anacrabro Packard, 46, **377**
Anacrucis Rayment, 44
anaetis (Pate), Xerogorytes, 53, **517**
analis Arnold, Tachysphex, 273
analis Cresson, Oxybelus, 366
analis Fabricius, Larra, 237
analis Gerstaecker, Oxybelus, 370
anamiense Tsuneki, Trypoxylon, 346
anathema (Rossi), Larra, 42, **237**
anatolica (Beaumont), Hoplammophila, 141
anatolica (Pulawski), Dryudella, 213
anatolicus Beaumont, Bembecinus, 532
anatolicus Beaumont, Lindenius, 384
anatolicus Beaumont, Miscophus, 318
anatolicus (Kohl), Palmodes, 127
ancara (Saussure), Liris, 244
anceps Arnold, Tachysphex, 272
anceps (Mocsáry), Gorytes, 501
anchoratus Mantero, Stizus, 526
anchorites Turner, Stizus, 526
ancilla Arnold, Cerceris, 584
Ancistromma W. Fox, 43, **259**
andalgalensis Fritz and Toro, Cerceris, 577
andalsiacus Tsuneki, Oxybelus, 366
andalusiacus Kohl, Philanthus, 566
andalusiacus Spinola, Oxybelus, 366
andalusiaticus Móczár, Oxybelus, 366
andamanicum Kohl, Sceliphron, 106
andersoni Turner, Cerceris, 577
andina Bréthes, Cerceris, 577
andinus Bréthes, Oxybelus, 366
andradei Beaumont, Solierella, 313
andradei P. Verhoeff, Miscophus, 318
andrei (F. Morawitz), Podalonia, 144
andrei Gussakovskij, Cerceris, 577
andrei (Handlirsch), Stictia, 542
andreniformis Cameron, Tachytes, 263, 265
Androcrabro Leclercq, 49, **422**
androgynus (Rossi), Philanthus, 566
anerus Leclercq, Podagritus, 393
angelae (Kohl), Pseudoscolia, 573
angelicus (Kincaid), Crossocerus, 400
angularis Cockerell, Cerceris, 577
angularis Gussakovskij, Ammoplanus, 198
angularis (H. Smith), Synnevrus, 470
angularis Mickel, Tachysphex, 272
angularis W. Fox, Pemphredon, 182
angulata (F. Smith), Bicyrtes, 537
angulata Rohwer, Eucerceris, 591
angulatus (Malloch), Pseneo, 165
angulicolle Cameron, Rhopalum, 389
angulicollis (Arnold), Dasyproctus, 420
angulicollis (Tsuneki), Mimesa, 161
angulicornis (Malloch), Pluto, 171
angulifera (Strand), Bicyrtes, 537
angulifrons van Lith, Psen, 166
angustata F. Morawitz, Cerceris, 577
angustata (Taschenberg), Liris, 245
angustatus (Ashmead), Gorytes, 628
angustatus Pulawski, Tachypshex, 272

angustatus Turner, Philanthus, 565
angusticollis Spinola, Ampulex, 77
angustifrons (Arnold), Bembecinus, 530
angustifrons (Arnold), Dasyproctus, 420
angustifrons Kohl, Larra, 237
angustifrons Kohl, Philanthus, 564
angustifrons Kohl, Podium, 96
angustifrons Schulthess, Prosopigastra, 285
angustifrons Tsuneki, Cerceris, 577
angustipetiolatum Tsuneki, Rhopalum, 389
angustirostris Shestakov, Cerceris, 577
angustiventris (Arnold), Liris, 244
angustus Arnold, Tachysphex, 274
angustus Gussakovskij, Passaloecus, 184
angustus J. Parker, Selman, 541
angustus (Provancher), Gorytes, 501
angustus Saussure, Oxybelus, 366
anilis (Handlirsch), Microbembex, 539
anisitsi (Strand), Bicyrtes, 537
anisitsi Strand, Tachytes, 263
anna Dunning, Philanthus, 565
annae Dunning, Philanthus, 565
annamensis van Lith, Psenulus, 172
annandali Bingham, Cerceris, 580
annandali (Bingham), Crossocerus, 400
annexa Kohl, Cerceris, 585
annosa Heer, Ammophila, 31, 154
annulare Dahlbom, Trypoxylon, 348
annularis (Christ), Mellinus, 449
annularis Poda, Sphex, 117
annulata J. Parker, Bicyrtes, 537
annulata Klug, Cerceris, 583
annulata (Rossi), Cerceris, 577
annulatum Brèthes, Trypoxylon, 348
annulatum (Cresson) Sceliphron, 105
annulatum Richards, Trypoxylon, 349
annulatus (Eversmann), Dienoplus, 496
annulatus (Gimmerthal), Mellinus, 449
annulatus (Klug), Stizus, 526
annulatus Lepeletier and Brullé, Crossocerus, 401
annulatus Lichtenstein, Sphex, 117
annulatus (Say), Passaloecus, 184
annulatus Spinola, Trachypus, 568
annulatus Theobald, Philanthus, 31, 567
annulatus (Walker), Palarus, 291
annulicornis (Tsuneki), Polemistus, 185
annuligera Taschenberg, Cerceris, 577
annulipes Brèthes, Cerceris, 577
annulipes Cameron, Alysson, 458
annulipes (Cameron), Pluto, 171
annulipes Cameron, Trypoxylon, 346
annulipes (Lepeletier and Brullé), Crossocerus, 400
annulipes Motschulsky, Ampulex, 77
annulipes Taschenberg, Trypoxylon, 349
annulitarsis Cameron, Trachypus, 568
annulus (Dahlbom), Crossocerus, 400
anomala Mantero, Pisonopsis, 332
anomala Taschenberg, Ammophila, 153
anomalipes Arnold, Bembix, 545
Anomiopteryx Gussakovskij, 42, **196**
anomoneurae (Yasumatsu), Psenulus, 172
anonymus Leclercq, Sphex, 115
Anothyreus Dahlbom, 48
Anoxybelus Kohl, 46
ansa Shestakov, Cerceris, 577
antaca (Saussure), Liris, 244
antaeus Holthusius, Sphex, 117

antaeus Lichtenstein, Sphex, 117
antaka (Saussure), Liris, 244
antarcticus Linnaeus, Sphex, 117
antares Andrade, Miscophus, 318
antemissa Brèthes, Cerceris, 584
antennalis Beaumont, Solierella, 313
antennalis Kohl, Belomicrus, 363
antennata Ducke, Solierella, 313
antennata F. Morawitz, Cerceris, 577
antennata (F. Smith), Nesomimesa, 167
antennatus F. Smith, Sphex, 115
antennatus (Klug), Sphecius, 510
antennatus (Mickel), Diodontus, 179
antennatus (Rohwer), Psenulus, 173
antennatus W. Fox, Tachysphex, 272
anteum Leclercq, Rhopalum, 389
Anthophilus Dahlbom, 54
anthracina (Cameron), Liris, 244
anthracina Kohl, Liris, 244
anthracinus (A. Costa), Prionyx, 134
anthracinus F. Smith, Pemphredon, 182
anthracinus (Handlirsch), Bembecinus, 530
anthracipenellus (Taschenberg), Sagenista, 522
anthreniformis Dalla Torre, Tachytes, 263
anthrisci Wolf, Passaloecus, 184
antica (F. Smith), Liris, 244
antigae Antiga and Bofill, Oxybelus, 370
antigae Antiga and Bofill, Tachysphex, 277
antillarum Cameron, Tachytes, 263
antillarum Leclercq, Rhopalum, 389
antillarum Pulawski, Tachypshex, 272
antillarum (Saussure), Prionyx, 134
antilleana Evans and Matthews, 545
antilles (Krombein), Liris, 244
antilope Tsuneki, Cerceris, 577
antiopa (Handlirsch), Stictia, 542
antipodes F. Smith, Cerceris, 577
antipodum (Handlirsch), Bembecinus, 530
antiquella Cockerell, Ammophila, 33, 154
Antomartinezius Fritz, 51, **476**
antoninae Fabre, Cerceris, 581
Antronius Zetterstedt, 41
anubis Beaumont, Liris, 247
anubis Pulawski, Tachypshex, 272
anxius (Wesmael), Crossocerus, 402
apache Banks, Tachytes, 264
apache (Pate), Ammoplanellus, 199
apache Pate, Belomicrus, 363
apakaensis Tsuneki, Tachysphex, 272
apakensis Tsuneki, Cerceris, 577
apakensis (Tsuneki), Podalonia, 145
apakensis (Tsuneki), Prionyx, 133
Aphelotoma Westwood, 38, **70**
aphidiperda Rohwer, Stigmus, 189
aphidium Meunier, Mellinusterius, 38
aphidum Lepeletier and Brullé, Crossocerus, 401
Aphilanthops Patton, 55, **570**
APHILANTHOPSINI, 60, **569**
aphrodite Menke, Ammophila, 150
apicale (Guérin-Méneville), Chlorion, 90
apicale W. Fox, Trypoxylon, 346
apicalis Brullé, Ammophila, 153
apicalis F. Smith, Ampulex, 77
apicalis (F. Smith), Isodontia, 123
apicalis (F. Smith), Pseudoplisus, 503
apicalis (F. Smith), Trachypus, 568

apicalis Guérin-Méneville, Ammophila, 150
apicalis Guérin-Méneville, Stizus, 526
apicalis (Harris), Isodontia, 123
apicalis (Klug), Stizoides, 528
apicalis (Lepeletier and Brullé), Entomognathus, 382
apicalis (Saussure), Isodontia, 123
apicalis (W. Fox), Liris, 244
apicalis W. Fox, Tachysphex, 272
apicalis Williams, Plenoculus, 311
apicata Banks, Eucerceris, 592
apicata (Bingham), Isodontia, 123
apicatus H. Smith, Oxybelus, 369
apicepennis Cameron, Larra, 237
Apicerceris Pate, 55
apicipenne Cameron, Trypoxylon, 348
apicipennis Cameron, Astata, 213
apicipennis (Cameron), Liris, 244
apifalco (Christ), Cerceris, 586
apiformis (Arnold), Entomognathus, 382
apiformis F. Smith, Tachytes, 263, 265
apilinguaria (Christ), Bembix, 548
Apiraptrix, Shestakov, 55
Apiratryx Balthasar, 55
Apius Jurine, 45
Apius Panzer, 45
apivorus Latreille, Philanthus, 566
Apobembex Pate, 54
Apocrabro Pate, 48, **398**
Apoctemnius Leclercq, 50
apoctenus Pulawski, Tachysphex, 272
Apoides Tsuneki, 48, **398**
Aporhopalum Leclercq, 47, **388**
Aporia Wesmael, 40
Aporina Gussakovskij, 40
apostata Mercet, Astata, 212
apostoli Beaumont, Gastrosericus, 256
appendiculata (Cameron), Liris, 244, 247
appendiculata (Taschenberg), Liris, 244
appendiculatum Cameron, Pison, 336
appendiculatum Tsuneki, Trypoxylon, 630
approximata Turner, Ampulex, 77
approximatus (Turner), Arpactophilus, 186
approximatus Turner, Tachytes, 263
aprunatus Leclercq, Ectemnius, 424
apterinus (Leclercq), Polemistus, 185
aptus Marshakov, Lindenius, 384
Apycnemia Leclercq, 40
aquilina F. Smith, Cerceris, 577
aquitana (Pulawski), Dryudella, 213
arabica Lohrmann, Bembix, 546
arabica W. F. Kirby, Ammophila, 150
arabicum Gussakovskij, Trypoxylon, 345
araboides Leclercq, Dasyproctus, 420
arabs (Kohl), Dasyproctus, 420
arabs Lepeletier, Oxybelus, 366
arabs (Lepeletier), Sceliphron, 105
arania Leclercq, Spilomena, 193
arapaho (Pate), Oryttus, 507
arapaho (Pate), Rhopalum, 390
arasianum Tsuneki, Rhopalum, 390
arator Turner, Arpactophilus, 186
araucanus Pulawski, Tachypshex, 272
araucarius van Lith, Psenulus, 172
araxis (Kohl), Pseudoscolia, 573
arbuscula Mickel, Cerceris, 577
Arcesilas Pate, 52
archaeophilus Pulawski, Tachytes, 263
Archarpactus Pate, 52

INDEX 637

archboldi Krombein, Trypoxylon, 349
Archisphex Evans, 31
architectum (Lepeletier), Sceliphron, 105
architis Mickel, Cerceris, 581
archoryctes Cockerell, Hoplisus, 31
arcuata (F. Smith), Lyroda, 299
arcuata J. Parker, Bembix, 546
arcuata (Taschenberg), Stictia, 542
arcuata Williams, Solierella, 313
arcuatus (Say), Ectemnius, 424
ardens (Cameron), Crossocerus, 400
ardens F. Smith, Ammophila, 150
ardjunae van der Vecht, Cerceris, 589
areatus (Taschenberg), Argogorytes, 492
arechavaletai Brèthes, Ammophila, 152
arechavaletai (Brèthes), Bembecinus, 530
archavaletai Brèthes, Cerceris, 577
arechavaletai (Brèthes), Podagritus, 393
Arecuna Pate, 47
arelate Banks, Cerceris, 584
arenaria (Fabricius), Podalonia, 144
arenaria Handlirsch, Bembix, 545
arenaria Krombein, Solierella, 314
arenaria (Linnaeus), Cerceris, 577
arenaria (Luderwaldt), Podalonia, 144
arenaria Scullen, Eucerceris, 591
arenarius Arnold, Belomicroides, 363
arenarius Arnold, Tachysphex, 272
arenarius Kazenas, Dinetus, 216
arenarius (Panzer), Gorytes, 501
arenarum (Beaumont), Dienoplus, 496
arenarum Cockerell, Miscophus, 318
arenarum Handlirsch, Stizus, 526
arenivagus Krombein, Entomognathus, 382
arenivagus Krombein, Pluto, 171
arenosa (Gmelin), Cerceris, 577
arenosa (Gmelin), Podalonia, 144
arentis R. Bohart, Epinysson, 472
areolata (Spinola), Pisonopsis, 332
areolata Walker, Ammophila, 150
argentarius Mickel, Oxybelus, 367
argentata (C. Fox), Glenostictia, 552
argentata Hart, Ammophila, 152
argentata (Lepeletier), Podalonia, 145
argentata (Palisot de Beauvois), Liris, 244
argentatum Schuckard, Pison, 335
argentatus Brullé, Tachytes, 265
argentatus Curtis, Oxybelus, 366
argentatus Fabricius, Sphex, 114
argentatus Gussakovskij, Tachypshex, 272
argentatus (Lepeletier and Brullé), Encopognathus, 380
argentatus (Mocsáry), Prionyx, 134
argentea (W. Kirby), Podalonia, 144
argenteofacialis (Cameron), Lyroda, 299
argenteofasciatus Radoszkowski, Nysson, 469
argenteolineatus (Cameron), Encopognathus, 380
argenteopilosus Cameron, Oxybelus, 367
argenteopilosus Rohwer, Trichogorytes, 52, **498**
argenteovestitus Cameron, Tachytes, 263
argenteum (Ashmead), Pison, 335
argenteus Gussakovskij, Tachytes, 263
argenteus (Schenck), Ectemnius, 427
argenteus (Taschenberg), Stizus, 527
argenteus Turton, Sphex, 114
argenticauda (Cameron), Liris, 244
argenticeps Arnold, Tachysphex, 272
argenticrus Dalla Torre, Tachytes, 264

argenticus R. Bohart, Nysson, 469
argentiferus Walker, Sphex, 114
argentifrons Arnold, Tachysphex, 272
argentifrons (Cameron), Liris, 244
argentifrons (Cresson), Microbembex, 539
argentifrons (Cresson), Mimesa, 162
argentifrons (Cresson), Pluto, 171
argentifrons (Cresson), Podalonia, 144
argentifrons (Cresson), Sceliphron, 105
argentifrons (F. Smith), Bembecinus, 530
argentifrons F. Smith, Sphex, 116
argentifrons Guérin-Méneville, Cerceris, 577
argentifrons Lepeletier, Cerceris, 578
argentifrons Lepeletier, Sphex, 114
argentina Brèthes, Cerceris, 577
argentina Brèthes, Microbembex, 539
argentina (Brèthes), Pseneo, 165
argentina Gussakovskij, Ammophila, 153
argentina Menke, Willinkiella, 352
argentina (Schrottky), Willinkiella, 46, **352**
argentinae Rohwer, Diodontus, 179
argentinum Brèthes, Trypoxylon, 349
argentinus Brèthes, Anacrabro, 379
argentinus Brèthes, Heliocausus, 453
argentinus Brèthes, Oxybelus, 367
argentinus (Brèthes), Zanysson, 475
argentinus Taschenberg, Sphex, 114
argentipes Cameron, Tachytes, 264
argentipes F. Smith, Tachytes, 264
argentipilis (Provancher), Podalonia, 144
argentosa Shestakov, Cerceris, 577
argia Mickel, Cerceris, 577
Argogorytes Ashmead, 52, **491**
augurostoma Turton, Vespa, 627
argus (Christ), Crabro, 408
argus Harris, Crabro, 407
argus Packard, Crabro, 407
argusinus R. Bohart, Crabro, 407
argypheus R. Bohart and Schlinger, Oxybelus, 367
Argyrammophila Gussakovskij, 40
argyreus (F. Smith), Tachytes, 263
argyreus (Hacker & Cockerell), Sericophorus, 302
argyrius (Brullé), Chilosphex, 39, **128**
argyrius Gussakovskij, Tachysphex, 272
argyrocephala Arnold, Ammophila, 150
argyrofacies Strand, Tachytes, 263
argyropis Saussure, Tachytes, 263
argyropleura R. Bohart, Microbembex, 539
argyropleura van der Vecht, Ammophila, 152
argyropyga (A. Costa), Liris, 246
argyrosticta Lichtenstein, Sphex, 117
argyrostoma Gmelin, Vespa, 627
argyrostoma Lichtenstein, Sphex, 117
argyrotricha Rohwer, Cerceris, 578
argyrotrichus Rohwer, Tachysphex, 275
ariadne Pulawski, Astata, 212
ariadne Turner, Cerceris, 577
ariasi Giner Marí, Cerceris, 587
ariasi (Mercet), Podalonia, 143
aricae Leclercq, Podagritus, 393
aricana Lohrmann, Stictia, 542
arida Arnold, Cerceris, 582
aridulus R. Bohart, Nysson, 469
ariel (Cameron), Ectemnius, 427
ariellus (Banks), Tachytes, 266

Arigorytes Rohwer, 53, **516**
arivaipa Pate, Metanysson, 481
arizonae Dunning, Philanthus, 566
arizonella Banks, Cerceris, 577
arizonense W. Fox, Trypoxylon, 348
arizonensis Baker, Hoplisoides, 521
arizonensis G. and R. Bohart, Larropsis, 258
arizonensis (Malloch), Mimesa, 161
arizonensis Scullen, Eucerceris, 591
arizonicus Banks, Tachytes, 266
arizonicus R. Bohart, Philanthus, 564
arlei Richards, Microstigmus, 192
arlettae Beaumont, Bembix, 545
armata Beaumont, Cerceris, 577
armata (Cresson), Microbembex, 539
armata (Illiger), Hoplammophila, 39, **141**
armata (Sulzer), Bembix, 548
armaticeps Cameron, Cerceris, 577
armaticeps (W. Fox), Lindenius, 384
armatum Taschenberg, Trypoxylon, 348
armatus (Cresson), Zanysson, 475
armatus (Provancher), Hoplisoides, 521
armatus (Vander Linden), Lindenius, 384
armeniacae Cockerell and W. Fox, Passaloecus, 184
armiger Olivier, Oxybelus, 368
armigera Turner, Cerceris, 577
armipes (Siebold), Crossocerus, 401
arno Banks, Cerceris, 578
arnoldi Benoit, Brachystegus, 473
arnoldi Benoit, Liris, 247
arnoldi (Benoit), Oryttus, 507
arnoldi Benoit, Oxybelus, 367
arnoldi Berland, Philanthus, 564
arnoldi Brauns, Ampulex, 77
arnoldi Brauns, Cerceris, 577
arnoldi Leclercq, Diodontus, 179
arnoldi Lohrmann, Stizus, 527
arnoldi Menke, Trypoxylon, 345
arnoldi Mochi, Stizus, 526
arnoldi R. Bohart and Menke, Entomognathus, 382
arnoldi Turner, Miscophus, 318
arnoldiana Guiglia, Cerceris, 585
Arnoldita Pate, 49, **418**
Arpactophilus F. Smith, 41, **186**
arpactus (Fabricius), Argogorytes, 492
Arpactus Jurine, 52
Arpactus Panzer, 52
arreptus (Kohl), Ectemnius, 427
arrogans Arnold, Cerceris, 577
arroyense Richards, Trypoxylon, 345
arthriticus (Rossi), Mellinus, 449
artisanus Leclercq, Dasyproctus, 420
arudum Leclercq, Trypoxylon, 345
arvensis (Dahlbom), Ammophila, 153
arvensis Lepeletier, Ammophila, 150
arvensis (Linnaeus), Mellinus, 449
Aryana Pate, 46
Asaconoton Arnold, 46
aschabadensis Radoszkowski, Astata, 212
aselenos Lichtenstein, Sphex, 117
ashmeadi Baltazar, Trypoxylon, 345
ashmeadi (Fernald), Sphex, 114
ashmeadi Pate, Ammoplanops, 197
ashmeadi Rohwer, Diplectron, 211
ashmeadi Turner, Pison, 335
ashmeadii W. Fox, Tachysphex, 272
asiagenes Pulawski, Tachytes, 263
asiatica Gussakovskij, Ancistromma, 259
asiatica Radoszkowski, Bembix, 547

asiatica Tsuneki, Ammophila, 150
asiaticum (Linnaeus), Sceliphron, 105
asiaticum Pulawski, Diploplectron, 211
asiaticus Gussakovskij, Bembecinus, 530
asiaticus Leclercq, Ectemnius, 424
asiaticus (Radoszkowski), Ammatomus, 513
asiaticus Tsuneki, Crossocerus, 400
asiaticus Tsuneki, Diodontus, 179
asinum Leclercq, Trypoxylon, 345
asinus Arnold, Tachysphex, 272
asmarensis Arnold, Philanthus, 564
asmarensis Giordani Soika, Philanthus, 564
asper (W. Fox), Dryudella, 215
aspera (Christ), Podalonia, 144
aspera (Dalla Torre), Dryudella, 215
asperata (W. Fox), Eremnophila, 147
asperata (W. Fox), Moniaecra, 395
asperatus W. Fox, Gorytes, 501
asperatus W. Fox, Tachysphex, 272
asperiformis (Rohwer), Dryudella, 215
asperithorax (Rayment), Sericophorus, 302
asphaltites Beaumont, Bembecinus, 530
aspidiphorus (Schrank), Ectemnius, 426
assamensis Cameron, Ampulex, 77
assamensis Cameron, Cerceris, 584
assamensis (Cameron), Crossocerus, 400
assamensis Cameron, Tachytes, 263
assamensis van Lith, Psen, 166
assentator (Arnold), Bembecinus, 530
assimile (Dahlbom), Sceliphron, 105
assimile Sickmann, Pison, 335
assimilis Banks, Philanthus, 564, 566
assimilis (Bingham), Paranysson, 308
assimilis (F. Smith), Crossocerus, 400
assimilis (Fabricius), Stizoides, 53, **528**
assimilis Kohl, Ammophila, 150
assimilis Kohl, Ampulex, 77
associa Kohl, Cerceris, 577
associatus Turner, Tachytes, 263
astarte Banks, Cerceris, 577
Astata Latreille, 42, **211**
astatiformis Tsuneki, Tachytes, 263
ASTATINAE, 57, **203**, 205
ASTATINI, 59, **205**
Astatus Latreille, 42
Astaurus Rayment, 51
astutus Nurse, Tachytes, 263
asuncicola Strand, Trypoxylon, 345
asuncionis (Strand), Bembecinus, 530
asuncionis Strand, Cerceris, 586
asuncionis (Strand), Foxita, 416
asuncionis (Strand), Hoplisoides, 520
asuncionis Strand, Tachytes, 265
asuncionis (Strand), Trachypus, 568
aswad (Nurse), Crossocerus, 400
Atelosphex Arnold, 44
ater (Cresson), Crossocerus, 404
ater (Fabricius), Psen, 166
ater (Gmelin), Hoplisoides, 520
ater Jurine, Stigmus, 189
ater Lepeletier, Miscophus, 318
ater (Mickel), Diodontus, 179
ater (Olivier), Psen, 40, **166**
aterrima Arnold, Cerceris, 577
aterrima (F. Smith), Liris, 244
aterrima (Turner), Aphelotoma, 38, 73
aterrimus Arnold, Tachysphex, 272
aterrimus Turner, Carinostigmus, 191
athene Banks, Cerceris, 580
atitlanae (Cameron), Ectemnius, 428

atkinsoni Richards, Trypoxylon, 348
atlacomulca Scullen, Cerceris, 577
atlanteus Andrade, Miscophus, 318
atlanteus Beaumont, Lindenius, 384
atlanteus Beaumont, Tachysphex, 272
atlantica Roth, Ammophila, 151
atlantica Schletterer, Cerceris, 583
atlanticum R. Bohart, Rhopalum, 630
atlanticus Berland, Tachytes, 264
atlanticus Viereck, Plenoculus, 311
atomus Banks, Tachytes, 264
Atopostigmus Krombein, 41, 188
atorai Pate, Foxita, 416
atra Reed, Solierella, 313
atra Scullen, Cerceris, 584
atra Williams, Larropsis, 258
atrafemori Scullen, Cerceris, 582
atramontensis Banks, Cerceris, 577
atrata Scullen, Cerceris, 580
atrata Scullen, Eucerceris, 591
atrata (Spinola), Liris, 244
atratina (F. Morawitz), Mimumesa, 164
atratulus Taschenberg, Diodontus, 179
atratus (Arnold), Bembecinus, 530
atratus (Fabricius), Psenulus, 173
atratus Jurine, Psen, 166
atratus (Lepeletier), Prionyx, 131
atratus (Lepeletier), Tachysphex, 272
atrescens Krombein, Cerceris, 589
atriceps (Cresson), Ectemnius, 424
atriceps F. Smith, Cerceris, 577
atriceps (F. Smith), Podalonia, 144
Atrichothyreopus Noskiewicz and Chudoba, 37
atricornis (Malloch), Pluto, 171
atricornis Packard, Gorytes, 501
atrifrons (Cresson), Lestica, 431
atrifrons F. Smith, Bembix, 545
atrifrons W. Fox, Gorytes, 501
atripenne Gussakovskij, Pison, 335
atripennis (Perkins), Ectemnius, 424
atripes (F. Morawitz), Sceliphron, 106
atripes F. Smith, Ammophila, 151
atriventris (Malloch), Mimesa, 162
atrocyanea (Eversmann), Podalonia, 144
atrohirta Turner, Ampulex, 77
atrohirtus Kohl, Sphex, 114
atronitens Arnold, Philanthus, 566
atronitida Scullen, Eucerceris, 591
atropilosus Kohl, Sphex, 114
atropos Gribodo, Liris, 244
atrospinosa Turner, Bembix, 545
atrox Arnold, Liris, 244
atrox (F. Smith), Stizus, 526
atrum (Scopoli), Sceliphron, 106
atrum (Spinola), Pison, 45, **335**
attenuata (Christ), Ammophila, 152
attenuatum F. Smith, Trypoxylon, 345
attenuatus Turner, Gastrosericus, 256
aucella Menke, Ammophila, 151
Auchenophorus Turner, 45, **325**
aucklandi Leclercq, Rhopalum, 389
augagneuri Arnold, Cerceris, 577
Aulacophilus F. Smith, 45, **337**
aurantiaca F. Smith, Cerceris, 577
aurantiaca (Kohl), Lestica, 430
aurantica (W. Fox), Ancistromma, 259
aurantiacus Arnold, Sphecius, 511
aurantiacus Mocsáry, Oxybelus, 367
aurantiacus Radoszkowski, Palarus, 290
aurantiacus (Turner), Austrogorytes, 498

auranticus Arnold, Brachystegus, 473
aurarius (Matsumura), Ectemnius, 424
aurata (Fabricius), Liris, 244
aurata J. Parker, Microbembex, 539
auratum Shuckard, Pison, 335
auratus (F. Smith), Ectemnius, 426
auratus (Schummel), Brachystegus, 473
auratus (van Lith), Pseneo, 165
aureobalteatus (Cameron), Zanysson, 475
aureobarba Cameron, Cerceris, 581
aureocinctus Cameron, Tachytes, 264
aureofaciale Strand, Pison, 337
aureofacialis Cameron, Cerceris, 579
aureofasciata Turner, Bembix, 545
aureohirtus Rohwer, Psen, 166
aureolum Rohwer, Trypoxylon, 349
aureonitens Lichtenstein, Sphex, 117
aureonotata (Cameron), Eremnophila, 147
aureopilosa Brauns, Tachytella, 257
aureopilosum Cameron, Pison, 336
aureopilosus Berland, Sphex, 114
aureopilosus Tsuneki, Tachysphex, 272
aureosericea (Cameron), Liris, 244
aureosericeum Kohl, Podium, 96
aureosericeum Rohwer, Pison, 335
aureosericeum Schrottky, Trypoxylon, 349
aureovestitum Taschenberg, Trypoxylon, 348
aureus (Beaumont), Dienoplus, 496
auriceps Cameron, Tachysphex, 272
auriceps (Cresson), Ectemnius, 424
aurichalceus Kohl, Tachytes, 264
auricollis van der Vecht, Ammophila, 152
auricomans Bradley, Tachytes, 264
auricomus (Bingham), Ectemnius, 425
auricomus van Lith, Psen, 166
auricula Riek, Aphelotoma, 70
aurifasciatus van Lith, Psenulus, 629
aurifera R. Turner, Ammophila, 151
aurifex F. Smith, Pison, 335
aurifex F. Smith, Sphex, 116
aurifex F. Smith, Tachytes, 264
aurifluus Perty, Sphex, 115
aurifluus Turner, Stizus, 526
aurifrons Cameron, Oxybelus, 369
aurifrons (Cameron), Tachysphex, 275
aurifrons (F. Smith), Ectemnius, 425
aurifrons (F. Smith), Isodontia, 123
aurifrons (F. Smith), Liris, 244
aurifrons F. Smith, Oxybelus, 367
aurifrons (Lucas), Tachysphex, 275
aurifrons Shuckard, Trypoxylon, 348
aurifrons (Taschenberg), Pseneo, 165
aurifrons Tsuneki, Psen, 166
auriginosus Eversmann, Palarus, 291
aurilabris (Herrich-Schaeffer), Ectemnius, 427
aurinotus (Say), Synnevrus, 470
auripes Berland, Tachytes, 264
auripes (Fernald), Isodontia, 123
auripygata (Strand), Isodontia, 123
aurita (Fabricius), Cerceris, 577
aurita Latreille, Cerceris, 581
auriventre Turner, Pison, 335
auriventris (Guérin-Méneville), Liris, 245
auriventris Turner, Aphelotoma, 70
aurocapillus Templeton, Sphex, 115
auromaculata (Peréz), Eremnophila, 147
auropilosa (Rohwer), Liris, 244

INDEX 639

auropilosellus (Cameron), Hoplisoides, 521
auropilosus Turner, Tachysphex, 272
aurovestitus F. Smith, Tachytes, 264
aurovestitus Turner, Dasyproctus, 420
aurulenta (Fabricius), Liris, 43, **244**
aurulenta W. Fox, Ancistromma, 259
aurulentus Fabricius, Sphex, 116
aurulentus (Fabricius), Tachytes, 264
aurulentus Gistel, Sphex, 115
aurulentus Guérin-Méneville, Sphex, 115
ausiana Leclercq, Spilomena, 193
aussi (Leclercq), Ammoplanellus, 199
austeni Turner, Psammaecius, 516
austerus (Handlirsch), Sagenista, 522
austerus Mickel, Tachytes, 265
austragilis Leclercq, Dasyproctus, 420
australasiae Shuckard, Astata, 212
australe Pate, Enchemicrum, 46, **364**
australe Saussure, Pison, 336
australiae Leclercq, Rhopalum, 389
australiensis Schulz, Nitela, 325
australis Fritz, Pisonopsis, 332
australis Saussure, Cerceris, 577
australis (Saussure), Liris, 244
australis (Saussure), Palmodes, 127
australis (Saussure), Tachysphex, 272
australis Turner, Spilomena, 193
australis Williams, Solierella, 313
austriacum (Kohl), Rhopalum, 389
austriacus (Kohl), Pemphredon, 180
austriacus Kohl, Tachysphex, 276
austriacus Schrank, Sphex, 117
austrina W. Fox, Cerceris, 586
austrinus Banks, Tachytes, 264
austrinus (Bingham), Ammatomus, 513
austrocaledonica Williams, Nitela, 325
Austrogorytes R. Bohart, 52, **498**
Austrostigmus Turner, 41,
Austrotoma Riek, 38, **70**
avellanipes (Saussure), Liris, 244
avernus Leclercq, Psenulus, 172
avexum Leclercq, Rhopalum, 389
avidus Bingham, Philanthus, 564
avius (Arnold), Dasyproctus, 420
aymara Pate, Enoplolindenius, 415
ayuttayanus Tsuneki, Oxybelus 629
azteca Cameron, Ammophila, 151
azteca Saussure, Cerceris, 577
azteca (Saussure), Isodontia, 123
aztecum (Saussure), Chalybion, 103
aztecum Saussure, Trypoxylon, 349
aztecus Cameron, Oxybelus, 367
aztecus (Cresson), Epinysson, 472
aztecus Leclercq, Ectemnius, 424
azureum F. Parker, Uniplectron, 42, **208**
azureum Lepeletier & Serville, Chlorion, 90
azuzium Blanchard, Chlorion, 90

baal Pulawski, Tachysphex, 274
babaulti Beaumont, Solierella, 313
baccharidis Scullen, Eucerceris, 591
backi Sandhouse, Trypoxylon, 345
bactriana Gussakovskij, Alysson, 458
bactriana Gussakovskij, Solierella, 313
badius Banks, Tachytes, 264
badurensis van Lith, Psen, 166
bagandarum Turner, Cerceris, 580
baguenai (Giner Marí), Liris, 246
bahamas Krombein, Sphecius, 511

bahiacus Leclercq, Ectemnius, 424
bahiae J. Parker, Bembix, 545
bahiae Saussure, Trypoxylon, 349
baja Scullen, Eucerceris, 591
bajanica (Tsuneki), Dryudella, 215
bakeri Cameron, Cerceris, 583
bakeri Dunning, Aphilanthops, 570
bakeri F. Parker, Astata, 212
bakeri R. Bohart, Nysson, 469
bakeri (Rohwer), Parapsammophila, 139
bakeri Rohwer, Psen, 166
bakeri (Rohwer), Psenulus, 172
bakeri (Williams), Liris, 244
bakeri Williams, Tachytes, 264
baliensis van Lith, Psenulus, 173
baltazarae van Lith, Psenulus, 172
balteatum Cameron, Trypoxylon, 345
balteatus (Cameron), Neoplisus, 505
balticus Merisuo, Pemphredon, 180
balucha (Nurse), Ectemnius, 425, 426
baluchistanensis Cameron, Cerceris, 577
bambesae Empey, Cerceris, 577
bambosicola Tsuneki, Crossocerus, 401
bamendae (Leclercq), Rhopalum, 385
banabacoa Alayo, Philanthus, 564
bandraensis (Giner Marí), Hoplisoides, 520
bandraensis (Giner Marí), Polemistus, 185
banksi Ashmead, Oxybelus, 368
banksi (Rohwer), Ectemnius, 426
banksi Scullen, Cerceris, 577
bannisteri Empey, Cerceris, 577
bannitus Kohl, Sphex, 115
banoensis Rohwer, Tachytes, 264
bantamensis van der Vecht, Cerceris, 577
bantu Brauns, Cerceris, 585
bara Pate, Taruma, 417, 627
barabbas (Pagden), Polemistus, 185
barbara Arnold, Cerceris, 586
barbara Handlirsch, Bembix, 545
barbara (Lepeletier), Ammophila, 151
barbarorum Arnold, Ammophila, 151
barbarum Beaumont, Trypoxylon, 346
barbarus (Beaumont), Bembecinus, 530
barbarus Beaumont, Gorytes, 501
barbarus (Roth), Palmodes, 127
barbata F. Smith, Ammophila, 153
barbatulus (Handlirsch), Hoplisoides, 521
barbatus (Arnold), Carinostigmus, 191
barbatus F. Smith, Philanthus, 564
barberi (Krombein), Miscophus, 319
barberi Krombein, Spilomena, 193
barberi Richards, Trypoxylon, 345
barberi (Rohwer), Epinysson, 472
barbieri Beaumont, Didineis, 458
barbifera Bischoff, Cerceris, 580
barbiger Mickel, Philanthus, 564
barbipes (Dahlbom), Crossocerus, 400
barbiventris F. Morawitz, Bembix, 545
bareii Radoszkowski, Oxybelus, 367
bariana Tsuneki, Cerceris, 629
bariensis Tsuneki, Bembix, 629
baringa Evans and Matthews, Bembix, 545
barkeri (Arnold), Bembecinus, 530
barkeri (Arnold), Dasyproctus, 420
barkeri Arnold, Tachysphex, 272
barnardi Brauns, Cerceris, 577
barrei (Radoszkowski), Bembecinus, 531
barrei Radoszkowski, Cerceris, 577
barrei Radoszkowski, Nysson, 469
barri Gittins, Mimesa, 161
barri Scullen, Eucerceris, 592

barthi Viereck, Psen, 166
barticense Richards, Trypoxylon, 348
basale F. Smith, Pison, 335
basale Rohwer, Trypoxylon, 349
basalis Cameron, Tachytes, 266
basalis Dahlbom, Bembix, 547
basalis F. Smith, Acanthostethus, 474
basalis F. Smith, Ammophila, 151
basalis (F. Smith), Argogorytes, 492
basalis F. Smith, Cerceris, 577
basalis (F. Smith), Lestica, 430
basalis F. Smith, Nysson, 469
basalis F. Smith, Philanthus, 564
basalis F. Smith, Trachypus, 568
basalis Guérin-Méneville, Stizus, 526
basalis (Stephens), Mimesa, 161
basifasciatus (Arnold), Dasyproctus, 420
basiferruginea Tsuneki, Cerceris, 589
basiflavus (Brèthes), Ectemnius, 424
basilanensis (Rohwer), Psenulus, 172
basilaris (Cresson), Epinysson, 472
basilewskyi Leclercq, Philanthus, 564
basilewskyi Leclercq, Psenulus, 172
basilicus (Guérin-Méneville), Tachytes, 264
basilicus (R. Turner), Sphex, 114
basiliris Cresson, Philanthus, 567
basilissa (Turner), Liris, 244
basimacula Cameron, Cerceris, 577
basiornata Cameron, Cerceris, 577
basirufa Packard, Mimesa, 161
basirufus (Brèthes), Perisson, 476
basirufus (Rohwer), Epinysson, 472
basirufus Rohwer, Tachytes, 267
bastiniana Richards, Isodontia, 123
basuto Arnold, Dolichurus, 69
basuto Pate, Taruma, 417, 627
basuto (Arnold), Sphex, 115
basutorum (Arnold), Hoplisoides, 520
basutorum (Turner), Neodasyproctus, 419
bataviana Strand, Bembix, 545
batesianum Schulz, Podium, 96
batrachostomus Schrottky, Trachypus, 568
baumanni Handlirsch, Bembix, 545
baumanni Handlirsch, Stizus, 526
bazilanensis Yasumatsu, Bembix, 545
beali Scullen, Cerceris, 583
beata Blüthgen, Spilomena, 193
beata (Cameron), Liris, 244
beatus Cameron, Sphex, 115
beaumonti Andrade, Plenoculus, 311
beaumonti Bajári, Cerceris, 587
beaumonti Balthasar, Miscophus, 318
beaumonti Bytinski-Salz, Palarus, 290
beaumonti Hellén, Pemphredon, 180
beaumonti Móczár, Rhopalum, 389
beaumonti (Pulawski), Dryudella, 214
beaumonti Pulawski, Tachysphex, 272
beaumonti (van Lith), Mimumesa, 164
beccarii Mantero, Stizus, 526
beccum F. Parker, Diploplectron, 211
bechteli F. Parker, Astata, 212
bechteli (R. Bohart), Listropygia, 55, **572**
bechuana Brauns, Cerceris, 580
bechuana (Turner), Ammophila, 151
bechuanae Arnold, Oxybelus, 367
beckeri Tournier, Mimesa, 161
beharensis Leclercq, Cerceris, 577
behni (Dahlbom), Hoplisoides, 520
beidzmiao Tsuneki, Tachysphex, 272
beieri Leclercq, Foxita, 416

bekiliensis Arnold, Cerceris, 581
bekiliensis (Arnold), Crossocerus, 400
belfragei Banks, Cerceris, 579
belfragei Banks, Tachytes, 267
belfragei Cresson, Bembix, 545
belfragei Cresson, Sphex, 117
belfragei (Cresson), Tachysphex, 272
belfragei (Cresson), Tanyoprymnus, 512
belgarum Leclercq, Rhopalum, 389
belgica (Bondroit), Mimumesa, 164
belgicus (Wesmael), Dienoplus, 496
belizensis (Cameron), Stictia, 542
bella Brèthes, Cerceris, 577
bella (Cresson), Dryudella, 214
bella (Cresson), Lestica, 431
bella (Lepeletier), Liris, 244
bella Menke, Ammophila, 151
bella (Rohwer), Liris, 244
bellardi Richards, Trypoxylon, 345
bellatrix J. Parker, Bembix, 545
belli Turner, Cerceris, 577
bellicosus (F. Smith), Austrogorytes, 52, **498**
bellicosus Olivier, Oxybelus, 368
bellona Mercet, Cerceris, 577
bellula Menke, Ammophila, 151
bellulus Arnold, Miscophus, 318
bellulus (Schulz), Encopognathus, 380
bellum (Cameron), Penepodium, 92
bellus (Cameron), Encopognathus, 380
bellus (Cresson), Epinysson, 472
bellus Dahlbom, Oxybelus, 368
bellus van Lith, Psen, 166
Belomicroides Kohl, 46, **360**
Belomicrus A. Costa, 46, **360**
beludzhistanicus Gussakovskij, Tachytes, 263, 264
belveriensis Andrade, Miscophus, 318
Bembecinus A. Costa, 53, **529**
bembesiana (Bischoff), Liris, 244
Bembex Fabricius, 54
BEMBICINI, 59, **532**, 534
Bembidula Burmeister, 53
Bembix Fabricius, 54, **543**
Bembyx Fabricius, 54
bengalense (Dahlbom), Chalybion, 102
bengalensis (Cameron), Liris, 244
bengalensis Cameron, Tachysphex, 272
bengalensis van Lith, Psenulus, 172
benidormicus P. Verhoeff, Miscophus, 318
beniensis Empey, Cerceris, 629
benignum (F. Smith), Sceliphron, 105
beniniensis (Palisot de Beauvois), Ammophila, 151
benoisti Leclercq, Enoplolindenius, 415
benoistianus Leclercq, Anacrabro, 379
benoiti Leclercq, Chalybion, 102
benoiti Leclercq, Dasyproctus, 420
benoiti Leclercq, Psenulus, 172
bensoni Arnold, Stizus, 526
bensoni (Beaumont), Olgia, 491
bensoni Richards, Trypoxylon, 348
bequaerti Arnold, Bembix, 545
bequaerti (Arnold), Liris, 244
bequaerti (Arnold), Polemistus, 185
bequaerti Brauns, Cerceris, 577
bequaerti Pate, Entomocrabro, 377
becquarti (Yasumatsu), Lestiphorus, 506
berenice Beaumont, Cerceris, 577
bergii (Handlirsch), Neoplisus, 505

berissus Leclercq, Ectemnius, 424
berlandi Andrade, Miscophus, 318
berlandi Arnold, Stizus, 526
berlandi Beaumont, Psenulus, 172
berlandi (Beaumont), Pseudoscolia, 573
berlandi Giner Marí, Cerceris, 578
berlandi (Richards), Crossocerus, 401
berlandi Timon-David, Cerceris, 31, 589
berlandi Willink, Bembecinus, 530
bermudensis (Malloch), Mimumesa, 164
bernardi Beaumont, Bembecinus, 530
bernardi Beaumont, Palarus, 290
berontha Evans and Matthews, Bembix, 545
besseyae (Rohwer), Ectemnius, 424
betremi van Lith, Psen, 166
betsilea Saussure, Larra, 237
beulahensis (Rohwer), Diodontus, 179
beutenmuelleri W. Fox, Bembix, 547
bezicus (Leclercq), Ammoplanellus, 199
biarmatus Mocsáry, Bembecinus, 531
biarti Richards, Trypoxylon, 345
bibundica Leclercq, Lestica, 430
bicalcaratum Richards, Trypoxylon, 348
bicarinata Arnold, Cerceris, 577
bicava Shestakov, Cerceris, 577
bicellulalis Strand, Ammophila, 152
bicincta Klug, Cerceris, 577
bicinctum van der Vecht, Sceliphron, 106
bicinctus (Fabricius), Argogorytes, 492
bicinctus (Mickel), Philanthus, 564
bicinctus (Rossi), Lestiphorus, 52, **506**
bicinctus (Taschenberg), Bebecinus, 530
bicinctus Turner, Psenulus, 172
biclipeatus (Christ), Stizus, 526
bicolor (Brullé), Tachysphex, 272
bicolor Cresson, Eucerceris, 592
bicolor Cresson, Hyponysson, 467
bicolor Dahlbom, Sphex, 114
bicolor F. Smith, Arpactophilus, 41, **186**
bicolor F. Smith, Cerceris, 581
bicolor F. Smith, Sericophorus, 302
bicolor F. Smith, Trypoxylon, 345
bicolor Fabricius, Larra, 237
bicolor (Gussakovskij), Eremiasphecium, 561
bicolor (Jurine), Mimesa, 161
bicolor Jurine, Miscophus 45, **318**
bicolor Lepeletier, Dolichurus, 69
bicolor of authors, Mimesa, 162
bicolor Radoszkowski, Bembix, 545
bicolor Richards, Microstigmus, 192
bicolor Saussure, Chlorion, 89
bicolor Say, Astata, 212
bicolor Schilling, Oxybelus, 368
bicolor (Taschenberg), Stizoides, 528
bicolor (W. F. Kirby), Liris, 244
bicolor Walker, Chlorion, 90
bicolor Williams, Solierella, 313
bicolorata Cameron, Larra, 237
bicolorisquama Strand, Oxybelus, 368
biconica Scullen, Eucerceris, 591
bicornis Williams, Nitela, 325
bicornuta F. Smith, Cerceris, 585, 587
bicornuta Guérin-Méneville, Cerceris, 577
bicornutus Arnold, Belomicrus, 363
bicornutus Arnold, Oxybelus, 367
bicuspidata Arnold, Cerceris, 578
bicuspidatus (Arnold), Dasyproctus, 420
Bicyrtes Lepeletier, 53, **535**

bidecoratus (Perkins), Ectemnius, 424
bidens (Arnold), Bembecinus, 530
bidens Arnold, Odontosphex, 55, **573**
bidens Gussakovskij, Tachytes, 264
bidens (Haliday), Crossocerus, 402
bidens J. Parker, Microbembex, 539
bidens (Linnaeus), Nysson, 469
bidens (Schrank), Cerceris, 586
bidentata (Arnold), Liris, 244
bidentata (Gussakovskij), Mimesa, 162
bidentata Lepeletier, Cerceris, 581, 586
bidentata (Say), Eucerceris, 591
bidentata Vander Linden, Bembix, 545
bidentatiforme Rohwer, Diploplectron, 211
bidentatum Ashmead, Diploplectron, 211
bidentatum W. Fox, Trypoxylon, 345
bidentatus (Arnold), Entomognathus, 382
bidentatus Arnold, Gastrosericus, 256
bidentatus (Cameron), Psenulus, 172
bidentatus Rohwer, Diodontus, 179
bidentula Maidl, Cerceris, 578
bifasciata Guérin-Méneville, Cerceris, 578
bifasciata (Schulthess), Dryudella, 214
bifasciatus (Brèthes), Epinysson, 472
bifasciatus (Brèthes), Hoplisoides, 520
bifasciatus (Fabricius), Stizoides, 529
bifasciatus O. Müller, Sphex, 117
bifida Menke, Nitela, 325
bifida Scullen, Cerceris, 586
bifidus Leclercq, Crossocerus, 400
bifoveolatus (Taschenberg), Prionyx, 133
bifurcata (C. Fox), Glenostictia, 552
bifurcata Empey, Cerceris, 578
bigeloviae Cockerell and W. Fox, Astata, 212
bigeloviae (Cockerell and W. Fox), Psammaletes, 52, **50**
bigeminus (Patton), Ectemnius, 425
biguttata (Thunberg), Cerceris, 586
biguttatum (Taschenberg), Podium, 96
biguttatus Arnold, Ammatomus, 513
biguttatus F. Morawitz, Crabro, 408
bilbaoensis Leclercq, Crabro, 408
bilineatum (F. Smith), Sceliphron, 106
bilineatus Gravenhorst, Philanthus, 564
bilobatum Tsuneki, Trypoxylon, 345
bilobatus Kohl, Sphex, 114
bilunata Cresson, Cerceris, 588
bilunatus Cresson, Philanthus, 564
bilunulatus A. Costa, Lestiphorus, 506
bimaculata Cameron, Cerceris, 578
bimaculata (Rayment), Spilomena, 193
bimaculata Vogrin, Cerceris, 577
bimaculatum (Lepeletier), Sceliphron, 105
bimaculatus Arnold, Dolichurus, 69
bimaculatus Lepeletier and Brullé, Crossocerus, 403
bimaculatus Magretti, Philanthus, 566
bimaculatus (Matsumura and Uchida), Bembecinus, 530
bimaculatus Packard, Mellinus, 449
bimaculatus (Panzer), Alysson, 458
bimaculigera (Strand), Eremnophila, 147
bimaculus Saussure, Philanthus, 564
bimetallicus Turner, Tachytes, 264
bimini Krombein, Oxybelus, 366
bimini Krombein, Tachytes, 264

binghami Cameron, Gastrosericus, 256
binghami Tsuneki, Liris, 245
binghami Turner, Cerceris, 578
binodis (Fabricius), Eremnophila, 147
binodis Spinola, Cerceris, 578
binotatus Lepeletier and Brullé, Crossocerus, 400
bipartior W. Fox, Diodontus, 180
bipartitus (Handlirsch), Pseudoplisus, 503
biplicatula Gussakovskij, Cerceris, 578
bipunctata Dufour, Bembix, 549
bipunctata Radoszkowski, Bembix, 545
bipunctata Rohwer, Isodontia, 123
bipunctata (Say), Ochleroptera, 490
bipunctatus Bingham, Dolichurus, 69
bipunctatus F. Morawitz, Tachysphex, 273
bipunctatus (F. Smith), Stizus, 526
bipunctatus (Fabricius), Mellinus, 449
bipunctatus Lepeletier and Brullé, Dasyproctus, 420
bipunctatus Olivier, Oxybelus, 367
bipunctatus (Zetterstedt), Ectemnius, 424
bipustulatus (Arnold), Hoplisoides, 520
bipustulosus Arnold, Tachysphex, 272
birganjensis van Lith, Psenulus, 172
birkmanni Baker, Hoplisoides, 521
birkmanni Rohwer, Pisonopsis, 332
birkmanni Rohwer, Tachytes, 264
birmanicus van Lith, Psen, 167
bischoffi Maréchal, Ammoplanus, 198
bischoffi Zeuner, Sphex, 33, 117
bisexmaculatus (Viereck), Ectemnius, 425
bishopi van Lith, Psen, 166
bishoppi Krombein and Willink, Bembecinus, 530
bisicatus van Lith, Psenulus, 172
bisignatus F. Morawitz, Palarus, 290
biskrensis E. Saunders, Tachytes, 264
biskrensis (Roth), Prionyx, 134
bismarkensis van Lith, Psenulus, 172
bison (A. Costa), Crossocerus, 402
bispinosa (Reich), Ampulex, 77
bispinosus Beaumont, Crossocerus, 400
bistillatus Balthasar, Oxybelus, 368
bitruncata Scullen, Eucerceris, 591
bituberculata (Bytinski-Salz), Parapsammophila, 139
bituberculata (J. Parker), Glenostictia, 552
bituberculata Tsuneki, Cerceris, 578
bituberculatus Arnold, Tachysphex, 272
bituberculatus Cameron, Tachysphex, 272
biungulata Cresson, Cerceris, 587
bivittata (Turner), Williamsita, 442
bizonatus Spinola, Stizus, 526
blaisdelli (Bridwell), Solierella, 313
blakei Cresson, Cerceris, 578
blanda Saussure, Astata, 212
blandinus (F. Smith), Stizoides, 528
blatticidus Williams, Tachysphex, 274
blattivorus Gussakovskij, Tachysphex, 272
Blepharipus Lepeletier and Brullé, 47, **398**
bnun Tsuneki, Crossocerus, 400
bnun Tsuneki, Psen, 166
Bnunius Tsuneki, 48, **398**
boarliri Evans and Matthews, Bembix, 545
bocandei (Spinola), Chalybion, 102
bodkini Richards, Trypoxylon, 347

boer Arnold, Tachysphex, 272
boer (Handlirsch), Bembecinus, 530
boerhaviae Cockerell, Anacrabro, 379
boetica (Pérez), Cerceris, 578
bogorensis Leclercq, Ectemnius, 424
bogotense Richards, Trypoxylon, 348
Bohartella Menke, 46, **355**
boharti F. Parker, Astata, 212
boharti F. Parker, Xenosphex, 439
bohart Fritz, Antomartinezius, 629
boharti Gillaspy, Xerostictia, 552
boharti Krombein, Tachysphex, 272
boharti Menke, Ammophila, 151
boharti Menke, Chlorion, 89
boharti Scullen, Cerceris, 578
boharti Tsuneki, Cerceris, 581
boharti Williams, Plenoculus, 311
boharti Williams, Solierella, 313
bohartiana Fritz and Toro, Cerceris, 578
bohartorum Tsuneki, Rhopalum, 389
bohartum Tsuneki, Rhopalum, 389
bohemanni Dahlbom, Sphex, 114
bohemicus Zavadil, Nysson, 469
boholensis van Lith, Psenulus, 172
bojus (Schrank), Crossocerus, 403
bola F. Parker, Astata, 212
bolanica (Nurse), Podalonia, 144
bolanica Turner, Cerceris, 578
boletus Leclercq, Ectemnius, 424
bolingeri Scullen, Cerceris, 576, 578
bolivari (Handlirsch), Bembecinus, 530
bolivari Handlirsch, Bembix, 546
bolivari Mercet, Philanthus, 565
boliviae Leclercq, Foxita, 416
boliviense Richards, Trypoxylon, 347
bollii (Cresson), Argogorytes, 492
bonaerensis Holmberg, Cerceris, 578
bonaespei Lepeletier, Ammophila, 151
bonifaciensis Ferton, Miscophus, 318
boninense Tsuneki, Pison, 336
boninensis (Tsuneki), Isodontia, 123
bonneti Leclercq, Chalybion, 102
boonamin Evans and Matthews, Bembix, 545
boops Gussakovskij, Prosopigastra, 285
boops (Schrank), Astata, 42, **212**
boranensis (Guiglia), Hoplisoides, 520
bordai Fritz, Podagritus, 393
borealis (Dahlbom), Ectemnius, 426
borealis (Dahlbom), Mimumesa, 164
borealis Dahlbom, Passaloecus, 184
borealis (F. Smith), Mimumesa, 164
borealis Forsius, Belomicrus, 363
borealis Mocsáry, Cerceris, 578
borealis Pulawski, Tachysphex, 275
borealis Valkeila, Nitela, 629
borealis (Zetterstedt), Ectemnius, 424
boregensis Williams, Plenoculus, 311
boregensis Williams, Solierella, 313
borero Pate, Idionysson, 480
borinquinensis (Pate), Epinysson, 472
borneana Cameron, Bembix, 545
borneana Tsuneki, Liris, 629
borneanus (Cameron), Bembecinus, 530
borneanus Cameron, Tachytes, 264
borneanus (Tsuneki), Carinostigmus, 629
borneensis (Rohwer), Psenulus, 174
bornicus Leclercq, Ectemnius, 427
borrei Handlirsch, Bembix, 545
boscae Giner Marí, Cerceris, 586
boschmai van der Vecht, Cerceris, 578

bothavillensis Brauns, Cerceris, 578
bothriophora Schletterer, Cerceris, 578
BOTHYNOSTETHINI, 59, **349**, 351
Bothynostethus Kohl, 46, **351**
botscharnikovi Gussakovskij, Didineis, 458
bougainvillae Pate, Crossocerus, 400
bougainvillensis Tsuneki, Cerceris, 578
boulderensis (Rohwer), Crossocerus, 400
bourgeoisi Strand, Trypoxylon, 345
bouwmani P. Verhoeff, Oxybelus, 367
boyaca Pate, Enoplolindenius, 415
boysi Turner, Cerceris, 578
braccata Gmelin, Vespa, 627
brachycarpae Rohwer, Crabro, 408
brachycerus Arnold, Tachysphex, 272
brachycerus Kohl, Diodontus, 179
Brachymerus Dahlbom, 48
brachyptera Arnold, Bembix, 545
brachyptera Lichtenstein, Sphex, 117
Brachystegus A. Costa, 51, **472**
brachystomus Kohl, Sphex, 114
brachystylus (Kohl), Chalybion, 103
bracteata Eversmann, Cerceris, 578
bradleyi (G. and R. Bohart), Ancistromma, 259
bradleyi J. Parker, Bembecinus, 537
bradleyi (Maidl), Liris, 244
bradleyi Scullen, Cerceris, 578
brahmanus Leclercq, Crossocerus, 400
brandti Krombein, Cerceris, 578
brasiliana (Brèthes), Eremnophila, 147
brasilianum Saussure, Trypoxylon, 345
brasilianus Handlirsch, Scapheutes, 355
brasilianus Pulawski, Tachysphex, 272
brasilianus Saussure, Sphex, 114
brasiliensis (Shuckard), Sagenista, 522
braueri (Handlirsch), Brachystegus, 473
braueri (Kohl), Encopognathus, 380
braueri Kohl, Liris, 244
braunsi Arnold, Gastrosericus, 256
braunsi (Arnold), Liris, 244
braunsi Arnold, Oxybelus, 367
braunsi Arnold, Tachysphex, 272
braunsi Arnold, Trypoxylon, 345
braunsi Mercet, Encopognathus, 380
braunsi (Turner), Ammophila, 151
braunsi Turner, Tachytes, 264
braunsii Arnold, Nitela, 325
braunsii (Handlirsch), Bembecinus, 530
braunsii Handlirsch, Bembix, 545
braunsii (Handlirsch), Hoplisoides, 520
braunsii Handlirsch, Nysson, 469
braunsii Kohl, Belomicrus, 363
braunsii (Kohl), Dasyproctus, 420
braunsii Kohl, Larra, 237
braunsii (Kohl), Polemistus, 185
braziliense (Schrottky), Penepodium, 92
bredoi Arnold, Ampulex, 77
bredoi (Arnold), Bembecinus, 530
bredoi Arnold, Cerceris, 578
bredoi (Arnold), Dasyproctus, 420
bredoi Arnold, Philanthus, 564
bredoi Arnold, Tachytes, 264
bredoi Benoit, Liris, 244
breijeri (Arnold), Tachysphex, 272
brendeli (Taschenberg), Stizus, 526
brethesi Gemignani, Trypoxylon, 348
brethesi Leclercq, Podagritus, 393
brevicarinata (Cameron), Liris, 244
brevicarinatum Cameron, Trypoxylon, 346

breviceps F. Smith, Ammophila, 151
breviceps Pulawski, Tachysphex, 272
brevicolle Kohl, Podium, 96
brevicornis A. Morawitz, Passaloecus, 184
brevicornis Arnold, Ammoplanus, 198
brevicornis Cameron, Ampulex, 78
brevilabris Beaumont, Diodontus, 179
brevilabris Wolf, Passaloecus, 184
brevinodum (Spinola), Rhopalum, 389
brevior Cockerell, Sceliphron, 31, 106
brevipecten Beaumont, Tachysphex, 272
brevipenne Saussure, Trypoxylon, 345
brevipennis Bingham, Ammophila, 151
brevipennis Cameron, Tachytes, 264
brevipennis Mercet, Tachysphex, 272
brevipennis Walsh, Stizus, 526
brevipes Pulawski, Tachysphex, 272
brevipetiolatus (Rohwer) Pluto, 171
brevipetiolatus Schrottky, Trachypus, 568
brevipetiolatus Wagner, Pemphredon, 182
brevipilis Banks, Tachytes, 266
brevirostris Lepeletier, Cerceris, 582
brevis (Eversmann), Crossocerus, 401
brevis Lohrmann, Bembix, 546
brevis Maidl, Mimesa, 162
brevis (Vander Linden), Entomognathus, 382
brevis Walker, Tachytes, 264
brevisericea Murray, Ammophila, 151
brevispinosus Arnold, Paranysson, 308
brevitarsis Kohl, Tachysphex, 272
brevitarsis Merisuo, Psenulus, 172
brevitarsis Pulawski, Astata, 212
breviusculus (Gussakovskij), Entomognathus, 382
breviventre Arnold, Trypoxylon, 345
breviventris F. Morawitz, Mimesa, 162
breviventris W. Fox, Tachytes, 267
breyeri Arnold, Stizus, 526
bridarollii Willink, Bembecinus, 530
bridwelli Fernald, Sphex, 115
bridwelli Pate, Belomicrus, 363
bridwelli Pate, Encopognathus, 380
bridwelli Sandhouse, Trypoxylon, 345
bridwelli Scullen, Cerceris, 578
bridwelli Williams, Solierella, 313
Brimocelus Arnold, 46, **364**
brinchangensis van Lith, Psen, 166
brinckerae Turner, Tachysphex, 272
brisbanensis Cockerell, Cerceris, 578
brisbanensis Rayment, Sericophorus, 303
brisbanensis (Turner), Acanthostethus, 474
brodiei Provancher, Oxybelus, 370
brookii (Bingham), Dasyproctus, 420
brownei Turner, Encopognathus, 380
browni (Ashmead), Pison, 337
browni (Turner), Austrogorytes, 498
bruchi (Schrottky), Neoplisus, 505
bruchi Schrottky, Rhopalum, 389
bruesi Rohwer, Tachysphex, 272
bruinjnii (Maindron), Sceliphron, 106
brullei Guérin-Méneville, Bembix, 545
brullii (F. Smith), Tachysphex, 272
bruneiceps Arnold, Tachysphex, 273
bruneri (Fernald), Isodontia, 123
bruneri (H. Smith), Larropsis, 258
bruneri Mickel, Crabro, 373
brunnea Scullen, Eucerceris, 591
brunneicornis Viereck, Diodontus, 179
brunneipes (Cresson), Diploplectron, 42, **211**

brunneipes (Packard), Ectemnius, 424
brunneofasciata Giordani Soika, Ampulex, 77
brunneomarginatus Arnold, Tachytes, 264
brunneri Handlirsch, Bembix, 545
brunneri J. Parker, Trichostictia, 549
brunnescens Honoré, Miscophus, 319
brunneus Pulawski, Tachytes, 264
brunnipennis Arnold, Liris, 244
brunnipes (Cresson), Prionyx, 133
brunniventris (Arnold), Crossocerus, 400
brunniventris Rohwer, Microstigmus, 192
brutia A. Costa, Cerceris, 578
bruxellensis Bondroit, Mimesa, 162
bruynii Cameron, Sceliphron, 106
bryani Perkins and Cheesman, Psen, 166
bryanti Richards, Trypoxylon, 348
bryanti Turner, Ampulex, 77
bubalus Handlirsch, Bembix, 545
buccadentis Mickel, Lindenius, 384
bucculata A. Costa, Cerceris, 578
bucculenta Ed. André, Cerceris, 578
bucephalus (Christ), Ectemnius, 426
bucephalus F. Smith, Philanthus, 564
bucephalus (F. Smith), Tracheliodes, 405
bucephalus (F. Smith), Vechtia, 417
Bucerceris Minkiewicz, 55
bucharica Gussakovskij, Didineis, 458
bucharicus Gussakovskij, Pemphredon, 180
buchwaldi Richards, Trypoxylon, 348
buddha Cameron, Ammophila, 151
buddha (Cameron), Dasyproctus, 420
buddha Cameron, Trypoxylon, 345
buddha Dalla Torre, Bembix, 545
budha Handlirsch, Bembix, 545
bugabense Cameron, Podium, 96
bugabensis Cameron, Oxybelus, 367
bugobensis Dalla Torre, Oxybelus, 367
bulawayoensis (Arnold), Crossocerus, 400
bulawayoensis Bischoff, Larra, 237
bulawayoensis Bischoff, Tachytes, 264
bulawayoensis Brauns, Cerceris, 578
bulganensis Tsuneki, Cerceris, 578
bulganicus Tsuneki, Miscophus, 318
bulgarica Pulawski, Prosopigastra, 285
bulgaricus Balthasar and Hrubant, Ectemnius, 427
bulloni Giner Marí, Bembix, 545
bulloni Giner Marí, Cerceris, 588
bulsanensis (Kohl), Ectemnius, 428
bumbanus Leclercq, Paranysson, 308
buntor Evans and Matthews, Bembix, 545
burando Evans and Matthews, Bembix, 545
burgdorfi Leclercq, Ectemnius, 424
burmeisteri (Burmeister), Dynatus, 94
burmeisteri (Handlirsch), Bicyrtes, 537
burmeisteri Willink, Hemidula, 540
burmeisterii (E. Lynch Arribalzaga), Larra, 237
burnettianus Turner, Dasyproctus, 420
burnsi Leclercq, Podagritus, 393
burnsiellus Rayment, Sericophorus, 302
burraburra Evans and Matthews, Bembix, 545
burungaensis (Arnold), Crossocerus, 400
busckii Richards, Trypoxylon, 348
bushiella Leclercq, Williamsita, 630
butleri R. Bohart, Pseudoplisus, 503
butleri Scullen, Cerceris, 578
butuana Leclercq, Piyuma, 410

butuanensis van Lith, Psenulus, 172
butuani Leclercq, Ectemnius, 425
buyssoni (Arnold), Bembecinus, 530
buyssoni Morice, Tachysphex, 273
bytinskii Beaumont, Bembecinus, 530
bytinskii P. Verhoeff, Miscophus, 318

cabrerae Leclercq, Lindenius, 384
cabrerai Mercet, Tachysphex, 273
cacaloapana Scullen, Cerceris, 578
caelebs Giner Marí, Cerceris, 578
caelebs (Kohl), Parapsammophila, 139
caelebs Patton, Tachytes, 265
caelestina Dalla Torre, Larra, 237
caementarium (Drury), Sceliphron, 105
Caenolarra Cameron, 43
Caenopsen Cameron, 40
caerulea (Cresson), Dryudella, 214
caerulea Murray, Podalonia, 144
caerulea Westwood, Trirogma, 38, **73**
caeruleanus Drury, Sphex, 114
caeruleipennis (Maidl), Liris, 244
caerulescens Le Guillou, Sphex, 115
caerulescens (Turner), Argogorytes, 492
caeruleum (Linnaeus), Chalybion, 102
caesarea (Handlirsch), Rubrica, 541
caesariensis Pate, Belomicrus, 363
caesius (Compte Sart), Bembicinus, 532
caffer (Saussure), Bembecinus, 530
caffrariae Empey, Cerceris, 629
cahuilla Pate, Belomicrus, 363
cahuilla Pate, Timberlakena, 201
calapensis van Lith, Psenulus, 173
calcarale Strand, Trypoxylon, 346
calcaratiformis Rohwer, Tachytes, 267
calcaratus W. Fox, Tachytes, 267
calcarina Handlirsch, Bembix, 546
calceatum (Tsuneki), Rhopalum, 390
calceatus (Rossi), Gorytes, 501
Calceorhopalum Tsuneki, 47, **388**
calderoni Leclercq, Rhopalum, 389
caldesianum Richards, Trypoxylon, 345
caliban Arnold, Tachysphex, 275
calida Turner, Cerceris, 578
californica Cresson, Cerceris, 578
californica Menke, Ammophila, 151
californica R. Bohart, Microbembex, 539
californica Williams, Psionopsis, 332
californica Williams, Solierella, 313
californicum F. Parker, Diploplectron, 211
californicum (Saussure), Chalybion, 38, **102**
californicum Saussure, Trypoxylon, 348
californicus (Ashmead), Miscophus, 318
californicus Bohart and Menke, Palmodes, 127
californicus Court and R. Bohart, Lindenius, 384
californicus Cresson, Philanthus, 564
californicus Eighme, Pulverro, 196
californicus (R. Bohart), Clypeadon, 571
californicus R. Bohart and Schlinger, Oxybelus, 367
californicus R. Bohart, Pseudoplisus, 503
californicus R. Bohart, Tachytes, 264
californicus (Williams), Paradolichurus, 38, **70**
californiensis Gillaspy, Steniolia, 553
caliginosum Turner, Pison, 335
caliginosus Erichson, Sphex, 114
caliptera (Say), Lyroda, 299
calixtum Leclercq, Rhopalum, 389

INDEX 643

callangae Leclercq, Enoplolindenius, 415
callani Krombein, Cerceris, 578
callani Pate, Crossocerus, 400
callani Pate, Oxybelus, 367
calligaster (Viereck), Oxybelus, 366
calliope (Arnold), Hoplisoides, 520
callista J. Parker, Stictiella, 551
callosus Kohl, Tachytes, 266
calochorti Rohwer, Cerceris, 578
calodera Banks, Cerceris, 578
calopterus Kohl, Sphex, 116
calopteryx (Gussakovskij), Dryudella, 215
calopteryx Gussakovskij, Tachysphex, 276
calopteryx (Handlirsch), Stizoides, 528
Calosphex Kohl, 39
Calotachytes Turner, 44
calva (Arnold), Ammophila, 151
calverti (Pate), Rhopalum, 389
calvus Turner, Philanthus, 565
calvus Turner, Tachytes, 266
campbellii (W. Saunders), Chlorion, 90
camelus (Arnold), Hovanysson, 472
camelus (Eversmann), Lestica, 430
cameroni (Ashmead), Psen, 628
cameroni Rohwer, Bembix, 545
cameroni Schulz, Cerceris, 578
cameroni Schulz, Eucerceris, 591
cameroni Yasumatsu and Masuda, Alysson, 458
cameronianus Morice, Tachytes, 264
cameronii Dalla Torre, Bembix, 545
cameronii Dalla Torre, Trypoxylon, 345
cameronii Kohl, Pison, 335
cameronii (Kohl), Trigonopsis, 98
cameronis Handlirsch, Bembix, 545
cameronis (Handlirsch), Pseudoplisus, 503
Cameronitus Leclercq, 50
camerunensis Tullgren, Philanthus, 564
camerunicus Strand, Sphex, 115
campaspe Richards, Trypoxylon, 348
campestris (F. Smith), Liris, 244
campestris Holmberg, Cerceris, 578
campestris Latreille, Ammophila, 151
campestris (Linnaeus), Argogorytes, 492
campicola Arnold, Cerceris, 578
camposi Campos, Sphex, 114
campsomeroides Arnold, Cerceris, 578
canadense (F. Smith), Sceliphron, 105
canadensis Dalla Torre, Crabro, 408
canadensis (Malloch), Mimumesa, 164
canadensis (Provancher), Prionyx, 133
canaliculata Cameron, Cerceris, 578
canaliculata Pérez, Cerceris, 578
canaliculata Say, Ampulex, 77
canaliculata (Say), Eucerceris, 591
canaliculatum Cameron, Trypoxylon, 345
canaliculatus Packard, Gorytes, 501
canaliculatus Radoszkowski, Oxybelus, 367
canalifera (Arnold), Arnoldita, 418
canalis R. Bohart and Schlinger, Oxybelus, 367
canariensis Beaumont, Miscophus, 318
canariensis Beaumont, Tachysphex, 273
canariensis Bischoff, Spilomena, 193
canariensis E. Saunders, Solierella, 313
canariensis Pulawski, Astata, 212
candiotes Schulz, Bembix, 547
canescens Cameron, Oxybelus, 367

canescens (Dahlbom), Podalonia, 144
canescens (Dahlbom), Prionyx, 134
canescens (F. Smith), Larra, 237
canescens F. Smith, Sphex, 114
canescens (Gmelin), Bembix, 545
canescens Gmelin, Vespa, 627
caninus Andrade, Miscophus, 318
canlaoensis van Lith, Psenulus, 172
canlaoni Leclercq, Rhopalum, 389
canonicola (Viereck), Crossocerus, 402
caocinnus Tsuneki, Psen, 166
capax W. Fox, Ancistromma, 259
capense Cameron, Trypoxylon, 345
capensis Arnold, Ammoplanus, 198
capensis Arnold, Cerceris, 582
capensis (Brauns), Gastrosericus, 256
capensis Brauns, Parapiagetia, 281
capensis Brauns, Prospigastra, 285
capensis Brauns, Psenulus, 172
capensis Cameron, Ampulex, 77
capensis Cameron, Oxybelus, 369
capensis Dahlbom, Philanthus, 564
capensis (Handlirsch), Brachystegus, 473
capensis Lepeletier, Bembix, 545
capensis (Lepeletier), Podalonia, 144
capensis (Saussure), Tachysphex, 273
capicola Arnold, Saliostethus, 320
capicola Brauns, Nitela, 325
capicola Handlirsch, Bembix, 545
capistratus (Schrank), Mellinus, 449
capitale Richards, Trypoxylon, 345
capitalis (E. Saunders), Parapiagetia, 281
capitalis Leclercq, Crossocerus, 400
capitalis (Radoszkowski), Liris, 245
capitata F. Smith, Cerceris, 583
capitata Gussakovskij, Spilomena, 193
capitatus (Nurse), Hoplisoides, 521
capito (Dahlbom), Crossocerus, 400
capito Lepeletier, Cerceris, 578
capitosus (Shuckard), Crossocerus, 400
capnoptera (Handlirsch), Bicyrtes, 537
cappadocicus Beaumont, Miscophus, 318
capparidis Gussakovskij, Solierella, 313
caprella Arnold, Ammophila, 151
capricornicus Leclercq, Ectemnius, 424
capuccina (A. Costa), Podalonia, 145
Carabro Say, 48
caramuru Holmberg, Crabro, 409
carbo Bohart and Menke, Palmodes, 127
carbonaria A. Costa, Mimesa, 162
carbonaria (Burmeister), Stictia, 542
carbonaria (F. Smith), Larra, 237
carbonaria Kohl, Astata, 212
carbonaria Saussure, Larra, 238
carbonaria (Tournier), Mimumesa, 164
carbonarius (Dahlbom), Crossocerus, 402
carbonarius (F. Smith), Argogorytes, 52, 492
carbonarius F. Smith, Dolichurus, 69
carbonarius (F. Smith), Podagritus, 393
carbonarius F. Smith, Sphex, 114
carbonarius (F. Smith), Psen, 166
carbonicolor (Dalla Torre), Podagritus, 393
carbonicolor van der Vecht, Sphex, 114
carbunculus Dalla Torre, Larra, 238
carcavalloi Fritz, Metanysson, 481
carceli (Lepeletier), Dienoplus, 496
carcinum Leclercq, Trypoxylon, 345
cardinalis Gussakovskij, Nysson, 469
caribana Fritz, Stictia, 542
caridei Holmberg, Cerceris, 579
caridei (Lieberman), Prionyx, 133

carinalis Pérez, Cerceris, 578
carinata Arnold, Prosopigastra, 285
carinata Cameron, Nursea, 465
carinata F. Smith, Bembix, 545
carinata (Gussakovskij), Mimesa, 162
carinatum Say, Trypoxylon, 345
carinatum Turner, Pison, 335
carinatus Berland, Tachytes, 264
carinatus (F. Smith), Ectemnius, 424
carinatus Gussakovskij, Ammoplanops, 42, **197**
carinatus Gussakovskij, Oxybelus, 367
carinatus Lohrmann, Bembecinus, 530
carinatus R. Bohart, Hoplisoides, 520
carinatus Rayment, Sericophorus, 302
carinatus Thomson, Pemphredon, 182
cariniceps Richards, Trypoxylon, 348
carinifrons Cameron, Ampulex, 77
carinifrons (Cameron), Psenulus, 172
carinifrons Cameron, Trypoxylon, 348
carinifrons Gussakovskij, Trypoxylon, 346
carinifrons Menke, Nitela, 325
carinifrons W. Fox, Trypoxylon, 346
Carinostigmus Tsuneki, 41, **189**
cariosum Arnold, Trypoxylon, 345
carli Beaumont, Tachysphex, 273
Carlobembix Willink, 53, **539**
carnaria Saussure, Cerceris, 585
caroli (Pérez), Podalonia, 144
carolina Banks, Cerceris, 579
carolina Banks, Rhopalum, 389
carolina (Fabricius), Stictia, 542
carolina Rohwer, Podium, 96
carolina (Rohwer), Pseneo, 165
carolinensis Banks, Philanthus, 565
carolinensis Yasumatsu, Liris, 244
carolinus Banks, Oxybelus, 369
carolus Leclercq, Podagritus, 393
carpenteri Richards, Trypoxylon, 345
carpetanus (Mercet), Bembecinus, 530
carrascoi Fritz, Podagritus, 393
carrascoi Fritz and Toro, Cerceris, 629
carripan Evans and Matthews, Bembix, 545
carrizonensis Banks, Cerceris, 578
casali (Fritz), Epinysson, 472
cashibo Pate, Quexua, 387
caspica F. Morawitz, Cerceris, 580
caspica (Gmelin), Cerceris, 588
caspica (Gussakovskij), Parapsammophila, 139
caspicus (F. Morawitz), Crabro, 408
cassandra Schrottky, Larra, 237
cassus Leclercq, Ectemnius, 424
castaneipes Dahlbom, Sphex, 114
castaneus Radoszkowski, Nysson, 469
castaneus Rayment, Sericophorus, 302
castellanus Mercet, Nysson, 469
castor (Handlirsch), Dienoplus, 496
castoris Leclercq, Trypoxylon, 345
catalinae R. Bohart, Pseudoplisus, 503
catamarcensis Schrottky, Cerceris, 582
catamarcensis (Schrottky), Eremnophila, 147
catamarcensis (Schrottky), Metanysson, 481
catamarcensis (Schrottky), Oxybelus, 367
cataractae Arnold, Brachystegus, 473
cataractae Arnold, Tachytes, 264
cataractae Arnold, Trypoxylon, 345
catarinae R. Bohart, Liogorytes, 516
catawba Banks, Cerceris, 579
catherinae Fritz, Podagritus, 393

catherinae Pulawski, Tachysphex, 275
catinum Leclercq, Trypoxylon, 345
caucasica Maidl, Mimesa, 162
caucasica (Moscáry), Podalonia, 144
caucasica Shestakov, Cerceris, 578
caucasicum (Ed. André), Sceliphron, 105
caucasicus (Radoszkowski), Dienoplus, 496
caucasicus Radoszkowski, Oxybelus, 367
caucasicus (Radoszkowski), Tachysphex, 273
caucasicus Radoszkowski, Tachytes, 263
cavagnaroi Scullen, Cerceris, 578
cavagnaroi Scullen, Eucerceris, 592
cavallum Leclercq, Trypoxylon, 345
cavernicola (van der Vecht), Dicranorhina, 251
cavicornis F. Morawitz, Cerceris, 588
cavifrons (Thomson), Ectemnius, 424
cavifrons van Lith, Psenulus, 172
cayennensis (Spinola), Sagenista, 522
cayerae Leclercq, Ectemnius, 425
cazieri R. Bohart, Hoplisoides, 520
ceanothae (Viereck), Xysma, 41, **194**
cearensis Ducke, Dolichurus, 69
cearensis (Ducke), Neoplisus, 505
ceballosi Andrade, Miscophus, 318
ceballosi Giner Marí, Ammoplanus, 198
ceballosi Giner Marí, Bembix, 547
ceballosi Giner Marí, Cerceris, 576
ceballosi Leclercq, Lindenius, 384
Ceballosia Giner Marí, 42
celaeno Arnold, Cerceris, 578
celebensis Maidl, Cerceris, 578
celebesianus Strand, Sphex, 116
Celia Shuckard, 41
cellularis Gussakovskij, Ammophila, 151
celsissimus Turner, Tachytes, 264
celtica (Spooner), Mimumesa, 164
cementaria (F. Smith), Podalonia, 145
cementarius (F. Smith), Trachypus, 568
Cemonus Jurine, 41
Cemonus Panzer, 41
Cenomus Gimmerthal, 41
centrale (Cameron), Megistommum, 504
centrale W. Fox, Trypoxylon, 345
centralis Berland, Bembix, 546
centralis Cameron, Ammophila, 151
centralis (Cameron), Ectemnius, 424
centralis van Lith, Psenulus, 174
cephalotes Handlirsch, Kohlia, 53, **513**
cephalotes (Olivier), Ectemnius, 424
cephalotes (Panzer), Ectemnius, 424
cephalotes Walker, Tachytes, 264
cerasicola Pate, Nitela, 325
Ceratocolus Lepeletier and Brullé, 50
Ceratocrabro Tsuneki, 50
ceratophorus Pulawski, Tachytes, 264
Ceratophorus Shuckard, 41
Ceratosphex Rohwer, 39
Ceratostizus Rohwer, 53
ceraunius (Rossi), Dinetus, 216
cercericus (Leclercq), Echucoides, 394
cerceriformis Cameron, Eucerceris, 591
CERCERINI, 59, **574**
Cerceris Latreille, 55, **575**
cereolus Morice, Dinetus, 216
ceres Cameron, Ammophila, 153
cerussata Shestakov, Cerceris, 578
cerverae Giner Marí, Cerceris, 578
cervinicornis Tsuneki, Diodontus, 179
cetonicus Leclercq, Ectemnius, 427

cetřatus (Shuckard), Crossocerus, 400
cevirus Leclercq, Dasyproctus, 420
ceylonicus Cameron, Oxybelus, 367, 369
ceylonicus (Cameron), Tachysphex, 275
ceylonicus Saussure, Dasyproctus, 420
ceylonicus van Lith, Psenulus, 172
chacoana Brèthes, Cerceris, 578
chagrinatus Leclercq, Ectemnius, 425
chahariana (Tsuneki), Podalonia, 144
chalcifrons (Packard), Mimumesa, 164
chalcithorax Arnold, Gastrosericus, 256
Chalcolampus Wesmael, 46
chalybaeum Kohl, Podium, 96
chalybaeus Kohl, Crabro, 628
chalybea F. Smith, Ampulex, 77
chalybea (Kohl), Podalonia, 144
chalybeum (F. Smith), Chalybion, 103
chalybeum (Vander Linden), Chalybion, 103
chalybeus F. Smith, Sericophorus, 44, **302**
chalybeus Kohl, Crabro, 408
Chalybion Dahlbom, 38, **98**, 101
Chalybium Agassiz, 38
Chalybium Schulz, 38
championi Cameron, Ammophila, 153
championi (Cameron), Ectemnius, 428
championi (Cameron), Liris, 244
championi (Cameron), Ochleroptera, 490
championi Turner, Protostigmus, 42, **196**
changaiensis Tsuneki, Diodontus, 179
changatai Tsuneki, Cerceris, 584
changi Tsuneki, Cerceris, 578
changi Tsuneki, Ectemnius, 427
changi Tsuneki, Rhopalum, 389
changi Tsuneki, Tachysphex, 273
changuina Pate, Zanysson, 475
chapmanae Viereck and Cockerell, Eucerceris, 591
chapmani (Cameron), Liris, 246
chapraensis (Turner), Encopognathus, 380
chariis van Lith, Psenulus, 172
charimorpha Brauns, Cerceris, 586
charusini F. Morawitz, Cerceris, 581
cheesmanae Krombein, Psen, 166
cheesmanae Menke, Liris, 244
cheesmani Leclercq, Crossocerus, 400
Cheilopogonus Westwood, 54
chelatus R. Bohart, Tachytes, 264
chemehuevi Pate, Ammoplanus, 198
cheops Beaumont, Cerceris, 578
cheops Beaumont, Tachysphex, 273
cheops Bingham, Stizus, 527
cheops Morice, Stizus, 527
chephren Beaumont, Tachysphex, 273
cherkesiana Beaumont, Cerceris, 578
cheskesiana Giner Marí, Cerceris, 578
chevrieri Kohl, Nysson, 469
chevrieri (Tournier), Psenulus, 173
Chevrieria Kohl, 41
chibcha Pate, Enoplolindenius, 415
chichidzimaense Tsuneki, Trypoxylon, 345
chichimeca Pate, Zanysson, 475
chichimecum Saussure, Trypoxylon, 345
chichimecus (Saussure), Bembecinus, 531
chichimecus Saussure, Sphex, 116
chiengmaiensis Tsuneki, Nysson, 469
chihpense Tsuneki, Trypoxylon, 345
chilense Spinola, Pison, 335
chilense (Spinola), Sceliphron, 105
chilensis (Eschscholz), Zyzzyx, 54, **550**

chilensis Herbst, Spilomena, 193
chilensis Kohl, Solierella, 313
chilensis Reed, Ammophila, 152
chilensis Reed, Oxybelus, 367
chilensis (Reed), Rhopalum, 389
chilensis Saussure, Astata, 212
chilensis (Saussure), Clitemnestra, 489
chilensis "Spin." Brèthes, Trachypus, 568
chilensis Spinola, Cerceris, 578
chilensis (Spinola), Prionyx, 134
chilensis (Spinola), Tachytes, 264
chiliensis Lepeletier, Sphex, 115
chillcotti van Lith, Psenulus, 172
chilon Lichtenstein, Sphex, 119
Chilopogon Kohl, 54
chilopsidis (Cockerell and W. Fox), Larropsis, 258
chilopsidis Cockerell, Philanthus, 565
chilopsidis Viereck and Cockerell, Cerceris, 578
Chilosphex Menke, 39, **127**
Chimila Pate, 49, **412**
Chimiloides Leclercq, 49, **413**
chinandegaensis Cameron, Cerceris, 583
chinense Gussakovskij, Trypoxylon, 345
chinense van Breugel, Sceliphron, 106
chinensis (Gussakovskij), Crossocerus, 403
chinensis (Gussakovskij), Mimesa, 162
chinensis (Sickmann), Ectemnius, 428
chingi Tsuneki, Trypoxylon, 345
chipsanii (Matsumura), Ectemnius, 427
chirindense Arnold, Trypoxylon, 345
chirindensis Arnold, Cerceris, 578
chirindensis (Arnold), Encopognathus, 380
chiriquensis Cameron, Ammophila, 153
chiriquensis Cameron, Cerceris, 582
chisosensis Scullen, Cerceris, 579
chivensis Gussakovskij, Solierella, 313
chivensis Pulawski, Tachytes, 264
Chlorampulex Saussure, 38
chlorargyricus A. Costa, Sphex, 114
Chlorion Latreille, 38, **88**
Chlorium Schulz, 38
chlorotica Spinola, Bembix, 545
chlorotica Spinola, Cerceris, 578
chobauti (Roth), Prionyx, 133
chongar Tsuneki, Trypoxylon, 346
chontale (Pate), Huavea, 396
chopardi Berland, Bembix, 545
chorasmius (Gussakovskij), Ammoplanellus, 42, **199**
chosenense Tsuneki, Rhopalum, 630
chosenense Tsuenki, Trypoxylon, 345
chosenensis Tsuneki, Crossocerus, 402
chosenensis Tsuneki, Diodontus, 629
chosenensis Tsuneki, Ectemnius, 629
chromatica Schletterer, Cerceris, 578
chromatipus Pate, Crossocerus, 400
chrysarginus (Lepeletier), Ectemnius, 425
chrysargurus (Dahlbom), Ectemnius, 425
chrysargynus (Mickel), Ectemnius, 424
chrysargyrus (Lepeletier and Brullé), Ectemnius, 425
chrysis (Christ), Chlorion, 90
chrysis Kohl, Miscophus, 318
chrysis (Lepeletier and Brullé), Enoplolindenius, 415
chrysites (Kohl), Ectemnius, 425
chrysobapta (F. Smith), Isodontia, 123
chrysobapta (F. Smith), Liris, 244

INDEX 645

chrysocercus Rohwer, Tachytes, 264
chrysogaster Schletterer, Cerceris, 587
chrysogona Gmelin, Vespa, 627
Chrysolarra Cameron, 43
chrysomaila (van Lith), Pseneo, 165
chrysonota (F. Smith), Liris, 244
chrysophorus Kohl, Sphex, 115
chrysopterus (Ruthe and Stein), Prionyx, 134
chrysopus Gmelin, Vespa, 627
chrysopyga (Spinola), Tachytes, 264
chrysorrhoea (Kohl), Isodontia, 123
chrysorrhoeus Handlirsch, Stizus, 526
chrysosticta Lichtenstein, Sphex, 119
chrysostoma Gmelin, Vespa, 627
chrysostoma Lichtenstein, Sphex, 119
chrysostomus (Lepeletier and Brullé), Ectemnius, 426
chrysothemis Turner, Cerceris, 578
chrysozonus (Gerstaecker), Idionysson, 480
chrysozonus (Turner), Austrogorytes, 498
chryssipe Banks, Cerceris, 579
chrysura (Cameron), Liris, 244
chudeaui (Berland), Prionyx, 133
chudeaui Berland, Tachytes, 264
chumash Pate, Nysson, 469
chumashano Pate, Pulverro, 196
cicheroi Fritz, Metanysson, 629
ciliata (F. Smith), Liris, 244
ciliata (Fabricius), Microbembex, 539
ciliatum (Fabricius), Chlorion, 90
ciliatus (Handlirsch), Austrogorytes, 498
ciliciensis Beaumont, Bembix, 545
cimiciraptor Williams, Anacrabro, 379
cimicivora (Ferton), Prosopigastra, 285
cincta Dahlbom, Cerceris, 581
cincta (Spinola), Dryudella, 215
cinctella Handlirsch, Bembix, 545
cinctella (W. Fox), Lestica, 430
cinctellum Richards, Trypoxylon, 345
cinctibella (Viereck), Lestica, 431
cinctipes (Provancher), Crossocerus, 400
cinctitarsis (Ashmead), Crossocerus, 400
cinctus Fabricius, Sphex, 117
cinctus Latreille, Gorytes, 501
cinctus (Rossi), Bembecinis, 532
cinctus (Spinola), Ectemnius, 426
cinctus (Turner), Ectemnius, 427
cinerascens Arnold, Tachytes, 264
cinerascens Dahlbom, Sphex, 115
cinerea Fernald, Isodontia, 123
cinerea Handlirsch, Bembix, 545
cinereohirtum Cameron, Trypoxylon, 349
cinereorufocinctus Dahlbom, Sphex, 114
cinereum Cameron, Trypoxylon, 349
cingulata (Burmeister), Bicyrtes, 537
cingulata (Gmelin), Cerceris, 579
cingulata Lichtenstein, Bembix, 549
cingulata Packard, Mimesa, 162
cingulatus Cresson, Eucerceris, 591
cingulatus (F. Smith), Bembecinus, 530
cingulatus (Fabricius), Stizus, 526
cingulatus (Packard), Crabro, 408
cingulatus (Rayment), Sericophorus, 302
cinguliger (F. Smith), Bembecinus, 530
cinxius (Dahlbom), Crossocerus, 400
circularis (Fabricius), Cerceris, 578
circumcincta Turner, Cerceris, 579
circumscriptus (Kohl), Enoplolindenius, 415

cisandina Brèthes, Cerceris, 582
cisandina Pate, Bicyrtes, 537
citipes Krombein, Dienoplus, 496
citrina Mercet, Bembix, 546
citrina Scullen, Cerceris, 580
citrinella F. Smith, Cerceris, 579
citrinulus Gussakovskij, Oxybelus, 367
citrinus Arnold, Sphecius, 511
citrinus (Klug), Stizoides, 528
citrinus Radoszkowski, Oxybelus, 367
citripes Taschenberg, Bembix, 545
cladothricis (Cockerell), Belomicrus, 363
clandestinus Kohl, Oxybelus, 367
clarconis Viereck, Crabro, 408
clarconis (Viereck), Epinysson, 472
clarconis Viereck, Philanthus, 566
clarconis Viereck, Tachysphex, 273
claripennis (Arnold), Ammoplanellus, 199
claripennis Arnold, Tachysphex, 273
claripennis Morice, Sphecius, 511
claripennis R. Bohart, Pseudoplisus, 503
claripes Arnold, Tachysphex, 273
clarkei Krombein, Trypoxylon, 346
clarus Cockerell, Philoponites, 31
clarus Pulawski, Diodontus, 179
claudii (Janvier), Rhopalum, 389
clavata Gmelin, Vespa, 627
clavatum Say, Trypoxylon, 348
clavatus (Baker), Arigorytes, 517
clavatus Cameron, Psen, 166
clavatus (Retzius), Mellinus, 449
clavicerum Lepeletier and Serville, Trypoxylon, 346
clavicornis (Handlirsch), Bembecinus, 530
clavicornis (Malloch), Pluto, 171
claviger (Kohl), Sericophorus, 302
clavigera (F. Smith), Isodontia, 123
clavimana Gussakovskij, Didineis, 458
clavipes Cameron, Dolichurus, 69
clavipes (Dahlbom), Dolichurus, 69
clavipes Kohl, Sphex, 114
clavipes (Linnaeus), Rhopalum, 389
claviventre (Cresson), Rhopalum, 389
claviventris (Cameron), Pseneo, 165
clavus (Fabricius), Ammophila, 151
clearei Leclercq, Ectemnius, 425
clemente Menke, Ammophila, 151
cleomae Dunning, Philanthus, 565
cleomae Rohwer, Cerceris, 579
cleopatra Beaumont, Liris, 244
cleopatra Menke, Ammophila, 151
cleopatra Pulawski, Astata, 212
cliffordi Rayment, Sericophorus, 302
clitellata Lepeletier, Cerceris, 578
Clitemnestra Spinola, 51, **485**
cloevorax (Nielson), Crossocerus, 404
clymene Banks, Cerceris, 579
Clypeadon Patton, 55, **570**
clypealis Faester, Pasaloecus, 184
clypealis Thomson, Pemphredon, 180
clypealis (Tsuneki), Ammoplanellus, 199
clypearia (Schreber), Lestica, 430
clypearis Ducke, Bothynostethus, 352
clypearis Honoré, Miscophus, 318
clypearis Saussure, Cerceris, 579
clypearis (Schenck), Crossocerus, 404
clypeata (Christ), Bembix, 548
clypeata Dahlbom, Cerceris, 579
clypeata F. Parker, Astata, 212
clypeata (F. Smith), Liris, 244

clypeata (Gillaspy), Glenostictia, 552
clypeata Malloch and Rohwer, Didineis, 459
clypeata (Mocsáry), Hoplammophila, 141
clypeata Murray, Podalonia, 144
clypeata (Schreber), Lestica, 430
clypeata (W. Fox), Mimumesa, 164
clypeata W. Fox, Pisonopsis, 45, **332**
clypeata Williams, Solierella, 313
clypeatum Cameron, Pison, 336
clypeatum (Fairmaire), Chalybion, 103
clypeatus Arnold, Tachysphex, 273
clypeatus F. Smith, Sphex, 115
clypeatus (Rayment), Sericophorus, 302
clypeatus (Rohwer), Enoplolindenius, 415
clypeatus Taschenberg, Tachytes, 263
clypeatus Tsuneki, Philanthus, 564
Clypeocrabro Richards, 50
clypeodentata Tsuneki, Cerceris, 580
clypeopunctata Tsuneki, Liris, 629
Clytaemnestra Handlirsch, 51
Clytemnestra Saussure, 51
clytia Beaumont, Cerceris, 579
Clytochrysus A. Morawitz, 49
cnemophila (Cameron), Liris, 247
coachella Menke, Ammophila, 151
coachellae R. Bohart, Arigorytes, 516
coahuila Pate, Metanysson, 481
coarctatum (Scopoli), Rhopalum, 389
coarctatus (Spinola), Ammatomus, 53, **513**
coarctatus Spinola, Philanthus, 564
cobosi Giner Marí, Astata, 212
cocacolae P. Verhoeff, Oxybelus, 367
coccineus (Balthasar), Dienoplus, 496
coccineus Gmelin, Sphex, 117
cochise Pate, Oxybelus, 367
cochisensis R. Bohart, Gorytes, 501
cochisi Scullen, Cerceris, 579
cockerellae Rohwer, Pison, 33, **337**
cockerellae Rohwer, Tachysphex, 273
cockerellae Rohwer, Trypoxylon, 348
cockerelli (Ashmead), Ammoplanops, 197
cockerelli (Ashmead), Trichogorytes, 498
cockerelli Dunning, Philanthus, 565
cockerelli Menke, Sericophorus, 302
cockerelli (Rohwer), Crossocerus, 402
cockerelli Rohwer, Diodontus, 179
cockerelli (Rohwer), Lestiphorus, 506
cockerelli Rohwer, Lyroda, 299
cockerelli Rohwer, Pemphredon, 182
cockerelli Viereck, Cerceris, 576
cockerellii Baker, Oxybelus, 369
cockerellii W. Fox, Plenoculus, 311
cocopa Pate, Oxybelus, 367
cocorite Richards, Trypoxylon, 346
codonocarpi Pulawski, Tachytes, 630
coelebs Dalla Torre, Tachytes, 265
coelestina (F. Smith), Larra, 237
coelicola Giner Marí, Cerceris, 579
Coelocrabro Thomson, 47
Coeloecus C. Verhoeff, 41
coeruleornata Cameron, Ammophila, 151
coerulescens Arnold, Miscophus, 318
cogens Kohl, Cerceris, 580
cogens (Kohl), Lindenius, 384
cognata Kohl, Ampulex, 77
cognata Mickel, Cerceris, 578
cognata Perkins, Deinomimesa, 169
cognatum Arnold, Trypoxylon, 345

cognatum Cameron, Trypoxylon, 346
cognatus F. Smith, Sphex, 114
cognatus W. Fox, Crabro, 408
collare Kohl, Pison, 335
collaris (Arnold), Dasyproctus, 420
collaris Cresson, Ammophila, 152
collaris Ducke, Bothynostethus, 352
collaris (Hacker & Cockerell), Sericophorus, 302
collaris Honoré, Miscophus, 318
collaris Kohl, Oxybelus, 367
collaris Kohl, Tachysphex, 274
collaris Linnaeus, Sphex, 117
collaris (Matsumura), Lestica, 431
collaris Tsuneki, Crossocerus, 629
collaris Tsuneki, Diodontus, 179
collaris Turner, Nitela, 325
collarti Leclercq, Entomognathus, 382
collator Bradley, Ampulex, 77
collinum F. Smith, Trypoxylon, 348
collinus (F. Smith), Ectemnius, 425
Colloliris Tsuneki, 629
colombica Fritz, Bicyrtes, 629
colon (Thunberg), Cerceris, 579
Coloptera Lepeletier, 40
coloradensis Baker, Oxybelus, 369
coloradensis Banks, Tachytes, 267
coloradensis (Packard), Crabro, 408
colorado Pate, Pulberro, 196
coloradoensis (Cameron), Mimumesa, 164
coloradoensis Rohwer, Stigmus, 188
colorata Leclercq, Lestica, 431
colorata Schletterer, Cerceris, 579
colorata (W. Fox), Ochleroptera, 490
coloratum F. Smith, Trypoxylon, 346
coloratum Taschenberg, Trypoxylon, 349
coloratus Baker, Belomicrus, 363
coloratus Nurse, Stizus, 526
coloripes (Arnold), Liris, 246
columbiae (Bradley), Crossocerus, 400
columbiae W. Fox, Tachytes, 264
columbianum Gribodo, Chlorion, 90
columbianum Saussure, Trypoxylon, 348
columbianus (Kohl), Belomicrus, 363
columbianus (Kohl), Lindenius, 384
columbianus (Kohl), Pulverro, 196
columbianus Saussure, Tachytes, 264
comanche Banks, Tachytes, 265
comanche Cameron, Ammophila, 154
comata J. Parker, Bembix, 545
comatus Reed, Oxybelus, 367
comberi Leclercq, Crabro, 408
comberi (Turner), Bembecinus, 530
comberi Turner, Cerceris, 579
comberi Turner, Palarus, 291
comberi Turner, Tachytes, 264
combinata Leclercq, Lestica, 431
combustus (F. Smith), Stizus, 526
comechingon Willink, Bembecinus, 530
comes Krombein, Microstigmus, 192
commixta (Turner), Liris, 244
communis (Cresson), Podalonia, 144
compacta Cresson, Cerceris, 579
compacta Fernald, Podalonia, 144
compacta (Kohl), Lestica, 431
compactus Arnold, Tachytes, 264
compactus Cresson, Nysson, 469
compactus Handlirsch, Mellinus, 449
compactus (Perkins), Ectemnius, 425
compactus van Lith, Psenulus, 172
compactus W. Fox, Tachysphex, 275
compar Cresson, Cerceris, 579

compar Handlirsch, Stizus, 527
compedita (Piccioli), Solierella, 313
compedita Turner, Bembix, 546
complanata Mickel, Cerceris, 588
complanatum (F. Smith), Penepodium, 92
completa Banks, Cerceris, 579
complex Kohl, Sceliphron, 105
compressa (Fabricius), Ampulex, 38, 77
compressicornis (Fabricius), Psen, 166
compressifemur (Giner Marí), Liris, 427
compressiventris Guérin-Méneville, Ampulex, 77
compta Nurse, Astata, 212
compta Turner, Cerceris, 579
comptus (Lepeletier and Brullé), Ectemnius, 426
conata Scullen, Eucerceris, 591
conator (Turner), Dasyproctus, 420
conceptus Pulawski, Tachysphex, 273
concinna Brullé, Cerceris, 579
concinna (F. Smith), Lyroda, 299
concinnulus (Cockerell), Clypeadon, 571
concinnus Bingham, Philanthus, 565
concinnus (Dahlbom), Entomosericus, 434
concinnus F. Smith, Tachytes, 264
concinnus (Rossi), Oryttus, 52, **507**
conclusus Nurse, Tachysphex, 273
concolor (Brullé), Podalonia, 143
concolor Dahlbom, Miscophus, 318
concolor (Dahlbom), Psenulus, 172
concolor (Eversmann), Stizoides, 529
concolor Provancher, Pemphredon, 181
concolor (Radoszkowski), Mimumesa, 164
concolor Say, Pemphredon, 181
concordius Leclercq, Podagritus, 393
concors (Gussakovskij), Mimesa, 162
conditor F. Smith, Ammophila, 151
conestogorum Rohwer, Stigmus, 189
confalonierii (Guiglia), Podalonia, 144
conferta (W. Fox), Larropsis, 258
confertim W. Fox, Pemphredon, 181
confertus (Turner), Acanthostethus, 474
confertus (W. Fox), Crossocerus, 402
confertus (W. Fox), Hoplisoides, 520
confinis (Dahlbom), Palmodes, 127
conicus Provancher, Alysson, 458
coniferus Arnold, Oxybelus, 367
confinis Gussakovskij, Tachysphex, 275
confinis (Walker), Ectemnius, 425
confluens (LeConte), Lestica, 431
confluens Richards, Trypoxylon, 349
confluenta (Say), Lestica, 431
conforme F. Smith, Pison, 335
confraga Shestakov, Cerceris, 579
confrater Kohl, Sphex, 114
confrater Kohl, Trypoxylon, 346
confrater Pulawski, Tachysphex, 273
confusa A. Costa, Ammophila, 151
confusa Giner Marí, Cerceris, 579
confusa (Radoszkowski), Liris, 246
confusus (Alayo), Hoplisoides, 520
confusus Arnold, Tachytes, 264
confusus (Dutt), Hoplisoides, 520
confusus Schulz, Crossocerus, 400
confusus Turner, Palarus, 290
confusus Wagner, Pemphredon, 182
congener (Dahlbom), Crossocerus, 400
congener Kohl, Sphex, 116
congenerum Turner, Pison, 335
congesta Giordani Soika, Cerceris, 579

conglobatus (Turner), Ectemnius, 425
congoensis (Arnold), Liris, 245
congoensis Arnold, Paranysson, 308
congoensis Arnold, Tachysphex, 272
congoensis (Berland), Prionyx, 133
congoensis Mantero, Stizus, 526
congolus Leclercq, Psen, 166
congophilus Benoit, Oxybelus, 367
congruus (Walker), Carinostigmus, 41, **191**
conica H. Smith, Mimesa, 162
conica Shestakov, Cerceris, 579
conicus (Germar), Sphecius, 511
conicus Radoszkowski, Sphex, 117
conicus Villers, Sphex, 117
conifera (Arnold), Ammophila, 151
conifrons Mickel, Cerceris, 579
conigera Dahlbom, Cerceris, 581
conigera Kohl, Ampulex, 77
conjuncta Schletterer, Cerceris, 586
conjungens (Walker), Liris, 244
connexa W. Fox, Bembix, 545
connexum Turner, Trypoxylon, 346
Conocercus Shuckard, 38
conradsi Giordani Soika, Cerceris, 586
conradti Berland, Sphex, 115
consanguineum (F. Smith), Penepodium, 92
consanguineum (Kohl), Chlorion, 89
consanguineus Arnold, Tachysphex, 273
consanguineus (Handlirsch), Dienoplus, 496
conscriptus (Nurse), Stizoides, 528
consimilis Banks, Philanthus, 564
consimilis Kohl, Ampulex, 77
consimilis Kohl, Philanthus, 565
consimilis (W. Fox), Larropsis, 258
consimilis W. Fox, Tachysphex, 275
consimiloides Williams, Tachysphex, 274
consobrina Arnold, Liris, 245
consobrina (Arnold), Parapsammophila, 139
consobrina Kohl, Cerceris, 586
consobrinus (Arnold), Ammoplanellus, 199
consobrinus (Handlirsch), Bembecinus, 530
consocius Kohl, Tachysphex, 273
consolator Leclercq, Ectemnius, 431
consors Cameron, Ammophila, 151
conspiciendus (Mickel), Ectemnius, 428
conspicillatum (A. Costa), Sceliphron, 106
conspicua (F. Smith), Liris, 245
conspicuus Arnold, Stizus, 526
conspicuus Cresson, Crabro, 408
constanceae (Cameron), Ampulex, 77
constanceae (Cameron), Lestica, 431
constricta Krombein, Lestica, 431
constrictum Turner, Pison, 336
constrictus (F. Smith), Argogorytes, 492
constrictus (Provancher), Pulverro, 196
consuetipes (Turner), Austrogorytes, 498
contigua (Villers), Cerceris, 579
contigua Walker, Cerceris, 581
contiguus (Cresson), Ectemnius, 427
continentis van Lith, Psenulus, 172
continua (Fabricius), Rubrica, 541
continuus Dahlbom, Oxybelus, 367
continuus (Fabricius), Ectemnius, 425
continuus (Klug), Stizus, 526

contracta Taschenberg, Cerceris, 579
contractus Arnold, Dolichurus, 69
contractus W. Fox, Tachytes, 264
contractus Walker, Tachytes, 264
convallis Patton, Sphecius, 511
convergens Tsuneki, Stigmus, 189
convergens Viereck and Cockerell, Cerceris, 579
convexifrons Taschenberg, Pison, 337
convexum (F. Smith), Chalybion, 102
convexus Pulawski, Tachysphex, 273
cooba Evans and Matthews, Bembix, 546
cookii Baker, Belomicrus, 363
coonundura Evans and Matthews, Bembix, 546
cooperi Beaumont, Liris, 244
cooperi Scullen, Cerceris, 579
copiosus Arnold, Tachytes, 264
coquilletti (Rohwer), Mimesa, 162
coquilletti Rohwer, Stigmus, 189
coquilletti Rohwer, Tachysphex, 273
coquillettii (W. Fox), Arigorytes, 53, **516**
cora Cameron, Ammophila, 151
cora (Cameron), Podagritus, 393
coracinus Valkeila, Pemphredon, 181
coraxus van der Vecht, Sphex, 117
cordatus Spinola, Oxybelus, 367
cordialis Fritz, Idionysson, 480
cordiformis Gussakovskij, Oxybelus, 367
cordillera Perez and Toro, Cerceris, 579
cordillierae Leclercq, Ectemnius, 426
cordobensis (Fritz), Liogorytes, 516
coreensis Tsuneki, Cerceris, 579
Corenius Tsuneki, 629
Corenocrabro Tsuneki, 628
coriaceus (A. Costa), Tachysphex, 273
coriaceus van Lith, Psen, 166
coriani Schrottky, Trachypus, 568
coriolum Leclercq, Rhopalum, 389
corizi Williams, Solierella, 313
corniculata Mickel, Stictiella, 551
corniculus (Spinola), Dolichurus, 69
corniger Gussakovskij, Tachytes, 264
corniger Shuckard, Passaloecus, 184
corniger Williams, Liris, 245
cornigera (Gmelin), Cerceris, 579
cornigera Gussakovskij, Tachytes, 264
cornigerum Cameron, Trypoxylon, 346
cornuta Eversmann, Cerceris, 580
cornuta (Fabricius), Cerceris, 579
cornutum Gussakovskij, Trypoxylon, 346
cornutus (F. Smith), Stizoides, 528
cornutus Robertson, Oxybelus, 367
coromandelicum (Lepeletier), Sceliphron, 106
coronalis (F. Smith), Liris, 245
coronata A. Costa, Ammophila, 151
coronatus Fabricius, Philanthus, 54, **564**
coronatus (Thunberg), Philanthus, 564
corporaali Leclercq, Ectemnius, 425
corporali van Lith, Psenulus, 172
corpulentus (Arnold), Bembecinus, 530
correntium Brèthes, Trypoxylon, 348
corrugata (G. and R. Bohart), Ancistromma, 259
corrugata Turner, Larra, 238
corrugatus (Packard), Ectemnius, 425
corrugis (Mickel), Gorytes, 501
corsa Beaumont, Solierella, 314
corsica Beaumont, Cerceris, 585
corsicus Andrade, Miscophus, 318
cortezi Scullen, Cerceris, 582

coruleter Pate, Anacrabro, 379
corusanigrens (Rohwer), Psenulus, 174
coruscans Turner, Auchenophorus, 45, **327**
corvidus Leclercq, Ectemnius, 425
corynetes Lichtenstein, Sphex, 119
Corynopus Lepeletier and Brullé, 47, 388
cosmiocephala Cameron, Cerceris, 588
cosquin F. Parker, Astata, 212
costae A. Costa, Astata, 212
costae (De Stefani), Tachysphex, 273
costae Ed. André, Mimesa, 162
costae (Handlirsch), Synnevrus, 470
costae (Magretti), Liris, 244
costae Piccioli, Astata, 212
costai Beaumont, Alysson, 458
costai Beaumont, Cerceris, 586
costalis (Cresson), Hoplisoides, 520
costalis Taschenberg, Tachytes, 264
costano Pate, Pulverro, 196
costarica Scullen, Cerceris, 579
costaricensis Brauns, Nitela, 325
costaricensis Cameron, Crabro, 408
costipennis (Spinola), Isodontia, 123
couloni Leclercq, Pseudoturneria, 629
cowani (W. F. Kirby), Liris, 245
coxalis Morice, Kohlia, 513
coxalis Patton, Tachytes, 265
coyotero (Pate), Synnevrus, 470
Crabro Fabricius, 48, **407**
crabroneus (Thunberg), Mellinus, 449
crabroniformis F. Smith, Bembix, 549
crabroniformis F. Smith, Philanthus, 564
crabroniformis (F. Smith), Psenulus, 172
CRABRONINAE, 58, **357**, 359
CRABRONINI, 60, **370**, 372
cradockensis Arnold, Ammoplanus, 198
craesus (Lepeletier and Brullé) Ectemnius, 425
crandalli R. Bohart and Schlinger, Oxybelus, 367
crandalli Scullen, Cerceris, 579
craspedota (Fernald), Ammophila, 152
crassiceps (Mickel), Ectemnius, 425
crassicollis Tsuneki, Cerceris, 580
crassicorne Cameron, Pison, 336
crassicorne (Gussakovskij), Eremiasphecium, 561
crassicornis (A. Costa), Hoplisoides, 521
crassicornis Arnold, Tachysphex, 273
crassicornis (Cockerell), Sericophorus, 302
crassicornis (Fabricius), Stizoides, 528
crassicornis Gribodo, Diodontus, 179
crassicornis Handlirsch, Didineis, 458
crassicornis Kohl, Ampulex, 77
crassicornis (Maidl), Liris, 245
crassicornis (Mantero), Ammatomus, 513
crassicornis (Spinola), Ectemnius, 425
crassicornis Viereck, Didineis, 461
crassicornus Viereck, Diodontus, 179
crassidens Cameron, Cerceris, 579
crassifemoralis (Turner), Ammophila, 151
crassiformis Viereck, Tachysphex, 273
crassifrons Tsuneki, Trypoxylon, 346
crassinodum (Spinola), Rhopalum, 389
crassipes A. Costa, Mimesa, 162
crassipes Arnold, Tachysphex, 273
crassipes (Cameron), Dynatus, 94
crassipes (F. Smith), Liris, 244
crassipes (Fabricius), Rhopalum, 389

crassipes (Handlirsch), Bembecinus, 530
crassipes (Walckenaer), Oxybelus, 369
crassipunctatum Arnold, Trypoxylon, 346
crassus Arnold, Belomicrus, 363
crassus Patton, Tachytes, 264
cratocephala Cameron, Cerceris, 582
Cratolarra Cameron, 42, 235
craverii (A. Costa), Hoplisoides, 521
crawfordi Brimley, Cerceris, 584
crawfordi (Rohwer), Liris, 246
crawshayi Turner, Ampulex, 77
crenicornis Marshakov, Lindenius, 384
crenulatus W. Fox, Tachysphex, 273
crenulifer (Kohl), Lindenius, 384
crenulifera Kazenas, Cerceris, 629
crenuloides Williams, Tachysphex, 276
creon (Nurse), Prosopigastra, 285
crepitans Leclercq, Cerceris, 579
crepitans Leclercq, Liris, 245
Cresson Pate, 51, **476**
cressoni Banks, Tachytes, 266
cressoni (Cameron), Bembecinus, 532
cressoni (H. Smith), Ammophila, 152
cressoni Pate, Ammoplanops, 197
cressoni Rohwer, Diploplectron, 211
cressoni Rohwer, Pison, 335
cressoni (Schletterer), Eucerceris, 55, **591**
cressonii Dalla Torre, Bembix, 545
cressonii Dalla Torre, Pemphredon, 181
cressonii Packard, Mimesa, 162
cressonii Robertson, Oxybelus, 367
cressonii W. Fox, Larra, 237
cressonis Handlirsch, Bembix, 545
Cressonius Bradley, 51
cretica Beaumont, Solierella, 313
creticus Beaumont, Bembecinus, 531
creticus (Beaumont), Dienoplus, 496
creticus Beaumont, Sphecius, 511
Cretosphex Rasnitsyn, 628
cribrarius (Linnaeus), Crabro, 408
cribrata Kohl, Ampulex, 78
cribrata Mocsáry, Cerceris, 576
cribratus Eversmann, Crabro, 408
cribratus F. Morawitz, Gorytes, 501
cribrellifer (Packard), Crabro, 408
cribrosa Spinola, Cerceris, 579
cricetum Leclercq, Trypoxylon, 346
crinita Saussure, Bembix, 547
crippsi (Arnold), Ectemnius, 425
crispus Arnold, Miscophus, 318
cristatus Mocsáry, Bembix, 548
cristatus (Packard), Ectemnius, 426
cristatus Saussure, Oxybelus, 367
cristovalensis Krombein, Cerceris, 579
croaticus (Vogrin), Dienoplus, 496
croceata (Lepeletier), Stictia, 542
croceipes (Eversmann), Argogorytes, 492
croceosignatus (Arnold), Dasyproctus, 420
crocidilus Pulawski, Tachysphex, 273
croesus (F. Smith), Liris, 245
croesus (Handlirsch), Zanysson, 475
croesus Lepeletier, Sphex, 115
crosi Ferton, Prosopigastra, 285
Crossocerus Lepeletier and Brullé, 47, **397**
crotonella Viereck and Cockerell, Cerceris, 579
crotoniphilus Viereck and Cockerell,

Philanthus, 564
crowsoni Evans, Archisphex, 31
crugiera (Hacker and Cockerell), Argogorytes, 492
crucis (Cockerell and W. Fox), Psammaletes, 508
crucis (Fabricius), Prionyx, 134
crucis Viereck and Cockerell, Cerceris, 586
crudator Leclercq, Ectemnius, 425
crudele Richards, Trypoxylon, 346
crudelis Bingham, Ampulex, 77
crudelis (F. Smith), Prionyx, 133
crudelis (Saussure), Dasyproctus, 420
cruentatum (Arnold), Rhopalum, 389
cruentatus (Latreille), Dienoplus, 496
cruentus (Fabricius), Dienoplus, 496
ctenophorus Pulawski, Tachysphex, 273
ctenopus (Arnold), Stizoides, 528
ctenopus (Cameron), Ectemnius, 426
ctenopus Kohl, Miscophus, 318
cubana R. Bohart, Microbembex, 539
cubanus Pulawski, Tachysphex, 273
cubense Richards, Trypoxylon, 348
cubensis Cresson, Cerceris, 589
cubensis (Cresson), Lestica, 431
cubensis Cresson, Tachytes, 267
cubensis (Fernald), Sphex, 114
cubiceps (Packard), Ectemnius, 426
cubiceps (Taschenberg), Ectemnius, 427
cubitalis (Saussure), Liris, 245
cubitaloide (Strand), Chalybion, 103
cucamonga Pate, Belomicrus, 363
cuculus Dudgeon, Stigmus, 189
cuernavaca Scullen, Cerceris, 579
cuernosi Leclercq, Ectemnius, 425
culcullata Bingham, Cerceris, 579
cultrifera Arnold, Bembix, 546
cuna Pate, Foxia, 478
cuneatus Williams, Plenoculus, 311
cunicularia (Schrank), Cerceris, 586
cupes Shestakov, Cerceris, 579
Cuphopterus A. Morawitz, 48, 398
cuprea F. Smith, Ampulex, 77
curictensis (Mader), Ectemnius, 425
curiosus Gussakovskij, Tachytes, 266
curruca (Dahlbom), Spilomena, 193
cursitans Handlirsch, Bembix, 546
curtipes (Perkins), Ectemnius, 425
curtulus (A. Costa), Hoplisoides, 521
curtulus F. Morawtiz, Nysson, 469
curtus Lepeletier and Brullé, Lindenius, 384
curvatum (F. Smith), Sceliphron, 106
curvatum Ritsema, Chalybion, 103
curvicollis (Cameron), Foxita, 416
curvicornis Cameron, Cerceris, 577
curvidens Tsuneki, Ammoplanus, 198
curvilineatus Cameron, Palarus, 291
curvilineatus (Cameron), Prionyx, 133
curvimaculatus (Cameron), Philanthus, 564
curvipilosus van Lith, Psen, 166
curviscutis Arnold, Oxybelus, 367
curvistriata Cameron, Ammophila, 151
curvitarsis (Herrich-Schaeffer), Tracheliodes, 405
curvitarsis Schletterer, Cerceris, 579
cuspidatus F. Smith, Passaloecus, 184
cuthbertsoni Arnold, Cerceris, 579
cyanator (Thunberg), Ampulex, 77
cyanauges Pate, Ectemnius, 425

cyaneipes Dalla Torre, Ampulex, 77
cyanescens Dahlbom, Ammophila, 153
cyanescens (Radoszkowski), Bembecinus, 531
cyanescens (Radoszkowski), Chlorion, 90
cyanescens Turner, Miscophus, 318
cyaneum (Cameron), Chalybion, 103
cyaneum Dahlbom, Chlorion, 89
cyaneum (Fabricius), Chalybion, 102
cyaneus (Saussure), Sericophorus, 302
cyanipennis (Fabricius), Isodontia, 123
cyanipennis (Lepeletier), Parapsammophila, 39, **139**
cyanipennis (Saussure), Stizoides, 528
cyanipes (Westwood), Ampulex, 77
cyaniventris Guérin-Méneville, Ammophila, 151
cyaniventris (Guérin-Méneville), Stangeella, 87
cyanophilus Rayment, Sericophorus, 302
cyanopterus (Gussakovskij), Stizoides, 528
cyanura Kohl, Ampulex, 77
cyanurus (Rohwer), Miscophus, 318
cybele Menke, Ammophila, 151
cyclocephalus F. Smith, Trigonopsis, 98
cyclops Krombein, Cerceris, 579
cyclostoma Gribodo, Ampulex, 77
cygnorum (Turner), Austrogorytes, 498
cygnorum (Turner), Podagritus, 393
cygnorum Turner, Sceliphron, 106
cygnorum Turner, Sphodrotes, 304
cylindrica Gmelin, Vespa, 627
cylindrica (W. Fox), Mimumesa, 164
cyphononyx (Kohl), Liris, 247
cypria Beaumont, Cerceris, 588
cypriaca Beaumont, Liris, 247
cypriaca Beaumont, Prosopigastra, 285
cypriaca Beaumont, Solierella, 313
cypriaca Giner Marí, Cerceris, 580
cypriacus Andrade, Miscophus, 318
cypriacus van Lith, Psenulus, 172
cyprium Gussakovskij, Pison, 335
cyprius Beaumont, Bembecinus, 531
cyprius Pulawski, Tachysphex, 275
cyrenaicus (Beaumont), Dienoplus, 496
cyrenaicus Beaumont, Tachysphex, 274
cyrenaicus (Gribodo), Palmodes, 127

dacica Schletterer, Cerceris, 579
daeckei (Rohwer), Crossocerus, 402
daeckei Viereck, Nysson, 469
daemonius (Perkins), Ectemnius, 424
daggyi (Murray), Palmodes, 127
dahlak Menke, Cerceris, 584
dahlbomi A. Morawitz, Diodontus, 179
dahlbomi Beaumont, Cerceris, 487
dahlbomi (Cresson), Liris, 245
dahlbomi Schluz, Oxybelus, 367
dahlbomi Sparre-Schneider, Passaloecus, 184
dahlbomi Thomson, Diodontus, 179
dahlbomi Tischbein, Dolichurus, 69
dahlbomi (Wesmael), Mimumesa, 164
Dahlbomia Wissman, 40
dahlbomii Handlirsch, Bembix, 546
dahlbomii Kohl, Ampulex, 77
dakotensis Banks, Cerceris, 579
dakotensis (Rohwer), Epinysson, 472
dakotensis Rohwer, Tachysphex, 273
Dalara Ritsema, 43, **248**
dallatorreanus (Kohl), Ectemnius, 427
damascenus (Beaumont), Prionyx, 133

dalmatica Kriechbaumer, Bembix, 545
danae Arnold, Tachytes, 264
dantani Roth, Ammophila, 151
dapitanensis (Rohwer), Psenulus, 173
Darala Ritsema, 43
darium Leclercq, Trypoxylon, 346
darrensis Cockerell, Cerceris, 579
dartanus Leclercq, Ectemnius, 425
darwini Turner, Nitela, 325
darwiniensis R. Turner, Sphex, 114
dasycheira (Dalla Torre), Liris, 245
dasymera Pate, Lestica, 431
Dasyproctus Lepeletier and Brullé, 49, **419**
davanus (Rohwer), Psenulus, 173
davidsoni (Sandhouse), Crossocerus, 400
davisi (Fernald), Sphex, 114
davisi W. Fox, Plenoculus, 44, **311**
dawsoni Mickel, Mimesa, 162
dawsoni (Swenk), Aphilanthops, 570
debeaumonti P. Verhoeff, Oxybelus, 367
debilis Arnold, Cerceris, 586
debilis Arnold, Tachysphex, 272
debilis F. Morawitz, Ammophila, 152
debilis (Pérez), Tachytes, 263
debilis Turner, Tachysphex, 277
decavum Leclercq, Rhopalum, 389
decemguttatus (Dahlbom), Synnevrus, 470
decemmaculata (Packard), Stictia, 542
decemmaculatus Eversmann, Philanthus, 564
decemmaculatus (Say), Ectemnius, 425
decemmaculatus (Spinola), Synnevrus, 470
decemnotatus A. Costa, Nysson, 469
decemnotatus (F. Morawitz), Brachystegus, 473
deceptor Krombein, Gorytes, 501
deceptor (Turner), Liris, 245
deceptrix Brauns, Cerceris, 582
decimmaculatus (Donovan), Oxybelus, 367, 370
decipiens (Arnold), Brachystegus, 473
decipiens (Arnold), Dienoplus, 496
decipiens Arnold, Gastrosericus, 256
decipiens Arnold, Oxybelus, 368
decipiens Arnold, Sceliphron, 105
decipiens Brèthes, Oxybelus, 367
decipiens F. Smith, Pison, 335
decipiens Honoré, Palarus, 291
decipiens Kohl, Sphex, 114
decipiens Priesner, Bembix, 546
decolorata Arnold, Cerceris, 579
decorata Brèthes, Cerceris, 579
decorata (F. Smith), Liris, 245
decorata (Taschenberg), Stictia, 542
decoratus (F. Smith), Crossocerus, 402
decoratus F. Smith, Sphex, 114
decoratus (Handlirsch), Ammatomus, 513
decoratus (Turner), Brachystegus, 473
decoratus van Lith, Psenulus, 174
decoratus Walker, Tachytes, 264
decorosus (Mickel), Oxybelus, 367
decorsei Berland, Tachytes, 264
decorus (W. Fox), Crossocerus, 400
decorus W. Fox, Gorytes, 501
decorus W. Fox, Tachysphex, 273
dedariensis Turner, Cerceris, 579
dedarum Leclercq, Rhopalum, 389
dedzcli Tsuneki, Spilomena, 193
defecta (Brèthes), Bicyrtes, 538
defectus Cockerell and Baker,

Oxybelus, 369
difiguratus Leclercq, Ectemnius, 425
deforme (F. Smith), Sceliphron, 106
degenerans (Kohl), Chalybion, 103
Deinomimesa Perkins, 40, **167**
dejecta Arnold, Cerceris, 579
dejecta Arnold, Liris, 245
dejecta Cameron, Ammophila, 151
delepineyi Arnold, Cerceris, 587
delessertii Guérin-Méneville, Stizus, 526
delicatulus (Morice), Dienoplus, 496
delicatus Mickel, Oxybelus, 369
deltaensis Johnson and Rohwer, Microbembex, 539
dendrophilum Arnold, Trypoxylon, 346
denisi Beaumont, Tachysphex, 273
denningi Murray, Ammophila, 153
densepunctatus (Arnold), Neodasyproctus, 419
densipunctatus (Yasumatsu), Lestiphrous, 706
dentata Cameron, Cerceris, 579
dentata Matsumura and Uchida, Ampulex, 77
dentatus Cameron, Philanthus, 564
dentatus (F. Smith), Lamocrabro, 413
dentatus Puton, Pemphredon, 181
dentatus Snoflak, Ammoplanus, 198
dentatus van Lith, Psenulus, 172
dentatus (W. Fox), Hoplisoides, 521
dentatus Williams, Tachysphex, 273
denticauda Arnold, Bembix, 546
denticeps Cameron, Pison, 335
denticollis (Cameron), Ampulex, 77
denticollis Spinola, Trachypus, 568
denticollis Tsuneki, Ampulex, 77
denticollis Tsuneki, Diodontus, 179
denticornis (F. Smith), Ectemnius, 427
denticornis (Gussakovskij), Crossocerus, 403
denticornis (Handlirsch), Rubrica, 541
denticoxa (Bischoff), Crossocerus, 400
denticrus Herrich-Shcaeffer, Crossocerus, 401
denticularis Banks, Cerceris, 578
denticulata (Morice), Parapiagetia, 281
denticulata Packard, Mimesa, 162
denticulata Schletterer, Cerceris, 579
denticulata (Turner), Liris, 245
denticulatum F. Smith, Podium, 96
denticulatus (Packard), Ectemnius, 426
denticulatus (Packard), Hoplisoides, 521
dentifer (Noskiewicz), Entomognathus, 382
dentifrons Cresson, Cerceris, 579
dentigera Gussakovskij, Ammophila, 151
dentilabris Handlirsch, Bembix, 546, 627
dentipes E. Saunders, Dinetus, 216
dentipes (Gussakovskij), Bembecinus, 531
dentipes Panzer, Crabro, 408
dentipes (Turner), Liris, 245
dentipleuris (Cameron), Piyuma, 410
dentiventris Arnold, Cerceris, 579
dentiventris (Handlirsch), Bembecinus, 531
denverensis (Cameron), Glenostictia, 552
denverensis Rohwer, Oxybelus, 369
deperditum Turner, Pison, 335
depilosellus Turner, Tachysphex, 273
deplanata (Kohl), Liris, 245
deplanatus Kohl, Sphex, 114
depredator F. Smith, Philanthus, 564

depressiventris Turner, Tachysphex, 273
depressus (Saussure), Tachysphex, 275
derivus Leclercq, Crossocerus, 401
derufata Strand, Cerceris, 586
descendentis Mercet, Tachysphex, 273
deserta G. and R. Bohart, Larropsis, 258
deserta Say, Cerceris, 579
deserti Berland, Miscophus, 318
deserti Williams, Plenoculus, 311
deserticola Beaumont, Oxybelus, 369
deserticola Beaumont, Tachysphex, 273
deserticola F. Morawitz, Cerceris, 579
deserticola Fritz, Foxia, 478
deserticola Tsuneki, Ammophila, 151
deserticola Tsuneki, Oxybelus, 367
deserticola Turner, Miscophus, 318
deserticolus Turner, Arpactophilus, 186
deserticus Giner Marí, Stizus, 526
desertorum (Eversmann), Prionyx, 134
desertorum F. Morawitz, Philanthus, 564
desertorum F. Morawitz, Tachysphex, 273
desertorum (Gussakovskij), Eremiasphecium, 561
desertorum Gussakovskij, Prospigastra, 285
desertus R. Bohart, Epinysson, 472
destillatorium (Illiger), Sceliphron, 105
destructus Cockerell, Prophilanthus, 33
detritus Arnold, Tachysphex, 276
deuterium Leclercq, Trypoxylon, 346
dewitzii Handlirsch, Stizus, 526
dewitzii (Kohl), Pseudoscolia, 573
diablense Williams, Diploplectron, 211
diabolica Balthasar, Astata, 212
diabolica (F. Smith), Liris, 245
diabolica Giner Marí, Cerceris, 579
diabolicus Arnold Tachysphex, 273
diabolicus F. Smith, Sphex, 114
diacanthus Beaumont, Bembecinus, 531
diacanthus (Gussakovskij), Crossocerus, 401
diadema (Fabricius), Philanthus, 566
diademata Holmberg, Cerceris, 577
Diamma Dahlbom, 55
diana (Handlirsch), Editha, 543
dichroa Brèthes, Cerceris, 579
dichroa (Dalla Torre), Eucerceris, 592
dichroa Gmelin, Vespa, 627
dichrous F. Smith, Tachytes, 264
dichrous Herbst, Tachytes, 264
dichrous Kohl, Philanthus, 564
dichrous Mercet, Gorytes, 501
dicksoni Arnold, Tachysphex, 273
Dicranorhina Shuckard, 43, **250**
Didesmus Dahlbom, 55
Didineis Wesmael, 51, **458**
Dienoplus W. Fox, 52, **495**
diervillae Iwata, Pemphredon, 181
diezguitas Fritz, Metanysson, 481
differens Blüthgen, Spilomena, 193
differens Turner, Pison, 337
differentia Murray, Podalonia, 145
difficile Turner, Pison, 337
difficillimus (Kohl), Lindenius, 384
difficilis Nurse, Miscophus, 318
difficilis Spinola, Sphex, 115
difficilis Strand, Ampulex, 77
difformis Handlirsch, Bembix, 546
difformis (Handlirsch), Microbembex, 539
digenus Dalla Torre, Tachysphex, 273

digitatum (Gussakovskij), Eremiasphecium, 561
dignus Kohl, Tachysphex, 273
digueti (Berland), Isodontia, 124
dilata Malloch and Rohwer, Didineis, 459
dilata Radoszkowski, Bembix, 546
dilata Spinola, Cerceris, 579
dilatatus Herrich-Schaeffer, Crossocerus, 400
dilaticornis (F. Morawitz), Ectemnius, 425
dilaticornis Turner, Tachytes, 264
dilectiformis (Rohwer), Ectemnius, 425
dilectus (Cresson), Ectemnius, 425
dilectus (Saussure), Psenulus, 172
dilutus Baker, Oxybelus, 367
dilwara Nurse, Tachytes, 266
dimidiata A. Costa, Dryudella, 215
dimidiata (Christ), Ammophila, 153
dimidiata F. Smith, Ammophila, 151
dimidiata Taschenberg, Astata, 212
dimidiatum F. Smith, Pison, 335
dimidiatus (De Geer), Palmodes, 127
dimidiatus (Fabricius), Crossocerus, 401
dimidiatus Jurine, Nysson, 469
dimidiatus Klug, Philanthus, 564
dimidiatus Lepeletier, Sphex, 115
dimidiatus (Panzer), Tachysphex, 276
dimidiatus Saussure, Tachysphex, 273
dimidiatus (Taschenberg), Stizus, 527
Dimorpha Panzer, 42
DINETINI, 59, **215**
Dinetomorpha Pate, 43
Dinetus Jurine, 42
Dinetus Panzer, 42, **216**
dineurum Leclercq, Rhopalum, 389
Dineurus Westwood, 41
diodon (Kohl), Isodontia, 123
diodonta (Handlirsch), Bicyrtes, 537
diodonta Schletterer, Cerceris, 580
Diodontus Curtis, 40, **176**
diodontus Pulawski, Tachytes, 264
dione Fritz, Cerceris, 580
diopura (Pate), Rhopalum, 389
Diphlebus Westwood, 41
diphyllus (A. Costa), Oxybelus, 367
Diploplectron W. Fox, 42, **208**
dipteroides Turner, Sericophorus, 302
dira Arnold, Bembix, 545
directa (C. Fox), Glenostictia, 552
discifer (Packard), Crabro, 408
discisa (Taschenberg), Bicyrtes, 537
discolor (Frivaldsky), Tachysphex, 275
discolor (Handlirsch), Bembecinus, 431
discolor Panzer, Philanthus, 566
discrepans Arnold, Tachytes, 264
discrepans Brauns, Cerceris, 580
discrepans (Giffard), Ectemnius, 425
discrepans Turner, Tachysphex, 273
discreta (W. Fox), Larropsis, 258
discretus W. Fox, Crabro, 407
disjunctus F. Smith, Trachypus, 568
dismorphia Schrottky, Cerceris, 586
dispar Dahlbom, Cerceris, 580
dispar F. Morawtiz, Stizus, 526
dispar Gussakovskij, Mimesa, 162
dispar Pulawski, Solierella, 313
dispar Reed, Astata, 212
dispar (Taschenberg), Podalonia, 144
dispar Valkeila, Pemphredon, 181
dispar W. Fox, Passaloecus, 184
dispersus Andrade, Miscophus, 318
dispersus Turner, Tachytes, 264

disputabilis Morice, Palarus, 291
disputabilis Turner, Tachytes, 264
dissecta Dahlbom, Bembix, 548
dissecta (Fabricius), Cerceris, 580
dissector (Thunberg), Ampulex, 77
dissectus Dahlbom, Oxybelus, 367
dissectus (Panzer), Gorytes, 501
dissectus (Panzer), Nysson, 469
dissimilis Arnold, Oxybelus, 367
dissimilis C. Fox, Steniolia, 553
dissimilis W. F. Kirby, Bembix, 546
dissita Holmberg, Cerceris, 582
dissona Arnold, Cerceris, 580
distanti Turner, Tachytes, 264
distincta (F. Smith), Ancistromma, 259
distinctus (Arnold), Bembecinus, 531
distinctus Arnold, Philanthus, 565
distinctus (Chevrier), Psenulus, 173
distinctus (F. Smith), Ectemnius, 425
distinctus F. Smith, Tachytes, 264
distinctus of authors, Tachytes, 264
distinctus W. Fox, Bothynostethus, 352
distinctus W. Fox, Passaloecus, 184
distinguenda Kohl, Ampulex, 77
distinguenda Shestakov, Cerceris, 580
distinguenda (Spinola), Liris, 245
distinguendum (Kohl), Penepodium, 92
distinguenda (A. Morawitz), Crossocerus, 401
distinguendus Chevrier, Nysson, 469
distinguendus Handlirsch, Stizus, 527
distinguendus Kohl, Philanthus, 566
distinguendus (Yasumatsu), Hoplisoides, 521
distortus Leclercq, Crossocerus, 401
ditior (Morice), Synnevrus, 470
dittrichi Schulz, Cerceris, 586
divaricatus Pate, Oxybelus, 368
divergens Arnold, Gastrosericus, 256
divergens (Ducke), Foxia, 478
divergens J. Parker, Stictiella, 551
diversa (Walker), Larra, 238
diversicornis (Arnold), Entomognathus, 382
diversicornis (F. Morawitz), Pseudoscolia, 573
diversicornis Turner, Tachytes, 264
diversidens Arnold, Bembix, 546
diversilabris Arnold, Tachysphex, 273
diversipennis F. Smith, Bembix, 546
diversipes F. Morawitz, Bembix, 546
diversipes Gussakovskij, Ammoplanus, 198
diversipes Herrich-Schaeffer, Crossocerus, 402
diversipes Pulawski, Astata, 212
diversus (Olivier), Mellinus, 449
diversus van Lith, Psenulus, 172
diversus (W. Fox), Hoplisoides, 521
dives (Brullé), Eremochares, 39, **146**
dives (Handlirsch), Stictia, 542
dives (Handlirsch), Zanysson, 475
dives Holmberg, Tachytes, 263
dives Kohl, Ampulex, 77
dives (Lepeletier and Brullé), Ectemnius, 425
dives (Lepeletier), Liris, 245
dives (Lepeletier), Prionyx, 133
dives Turner, Pison, 335
divina Kohl, Ammophila, 153
divisa Brèthes, Cerceris, 579
divisa Giner Marí, Cerceris, 585

divisa (Patton), Larropsis, 259
divisus (F. Smith), Pseudoplisus, 503
divitoides (C. Verhoeff), Ectemnius, 426
dixeyi (Bingham), Liris, 247
dizonus (Handlirsch), Austrogorytes, 448
dizoster Pate, Ectemnius, 426
djaouak Beaumont, Ammophila, 151
Dociliris Tsuneki, 43
docilis (F. Smith), Larra, 237
docilis (F. Smith), Liris, 245
doddi (Turner), Sericophorus, 302
doddii (Turner), Chimiloides, 414
doderleini Schulz, Cerceris, 580
dogonensis Krombein, Cerceris, 580
dolichocephala Cameron, Ammophila, 151
dolichocerus Kohl, Sphex, 115
Dolichocrabro Ashmead, 48
dolichodera Kohl, Ammophila, 151
dolichoderus (Kohl), Prionyx, 134
dolichostoma (Kohl), Parapsammophila, 139
dolichothorax (Kohl), Chalybion, 102
DOLICHURINI, 58, **66**
Dolichurus Latreille, 38, **66**
dolosa (Kohl), Isodontia, 123
dolosa Nurse, Cerceris, 589
dolosa W. Fox, Ancistromma, 259
dolosana (Rohwer), Ancistromma, 259
dolosus Arnold, Tachysphex, 275
dolosus Kohl, Philanthus, 564
domestica (Williams), Nitela, 325
domesticum Williams, Rhopalum, 389
domicola Tsuneki, Crossocerus, 401
domingana (Strand), Liris, 245
domingensis Leclercq, Ectemnius, 426
dominica Pate, Astata, 212
dominicana Brauns, Cerceris, 580
dominicana Evans, Liris, 245
dominicanus Evans, Ectemnius, 426
dondoensis Brauns, Cerceris, 580
dongalensis Klug, Palarus, 291
doreeni Leclercq, Podagritus, 393
doriae Gribodo, Eremochares, 146
doriae Gribodo, Nysson, 469
doriae Magretti, Bembix, 546
dorothyae Krombein, Gorytes, 501
dorsalis Dufour, Cerceris, 583
dorsalis Eversmann, Cerceris, 580
dorsalis (Kohl), Polemistus, 185
dorsalis Lepeletier, Sphex, 114
dorycus Guérin-Méneville, Sphex, 114
doumerci (Lepeletier), Prionyx, 133
dowi Tsuneki, Cerceris, 580
downesivora Turner, Cerceris, 580
draco Shestakov, Cerceris, 588
dreisbachi (R. Bohart), Clypeadon, 571
dreisbachi Scullen, Cerceris, 580
drewseni (Dahlbom), Gastrosericus, 256
dromedarius Nagy, Dolichurus, 69
droserum (Leclercq), Rhopalum, 389
drymocallidis (Rohwer), Ectemnius, 425
Dryphus Herrich-Schaeffer, 47
Dryudella Spinola, 42, **213**
dubia (F. Smith), Lestica, 431
dubia (Fernald), Ammophila, 153
dubia Gussakovskij, Bembix, 546
dubia Kohl, Ammophila, 151
dubia Kohl, Ampulex, 77
dubiosum Tsuneki, Trypoxylon, 346
dubiosus (Arnold), Dasyproctus, 420
dubiosus (Ashmead), Ectemnius, 427
dubiosus Dalla Torre, Tachysphex, 277

dubiosus (Guiglia), Bembecinus, 532
dubiosus Tsuneki, Tachytes, 630
dubitabilis Giner Marí, Cerceris, 581
dubitatus Cresson, Sphex, 114
dubitatus Rohwer, Tachytes, 265
dubitatus (Turner), Brachystegus, 473
dubium (Taschenberg), Penepodium, 92
dubium (W. F. Kirby), Pison, 336
dubius A. Costa, Nysson, 469
dubius Cresson, Philanthus, 566
dubius Dahlbom, Oxybelus, 370
dubius Ducke, Bothynostethus, 352
dubius (F. Smith), Bembecinus, 530
dubius Gussakovskij, Gastrosericus, 256
dubius Kohl, Heliocausus, 453
dubius Olivier, Nysson, 469
dubius (Panzer), Miscophus, 318
dubius (Radoszkowski), Tachysphex, 273
dubius Tsuneki, Passaloecus, 184
dubius (Turner), Arpactophilus, 186
dubius van Lith, Psenulus, 173
dubius W. Fox, Tachysphex, 277
duboulayi (Turner), Clitemnestra, 489
ducalis (F. Smith), Liris, 245
duchesnei Arnold, Cerceris, 580
duckei (Kohl), Entomocrabro, 377
duckei Menke, Bothynostethus, 352
duckei Menke, Pison, 337
duckei Richards, Trypoxylon, 346
dudgeoni (Nurse), Polemistus, 185
dufouri (Dahlbom), Brachystegus, 473
dufouri (Dahlbom), Psenulus, 172
dufouri Marquet, Oxybelus, 367
dufouriana Fabre, Cerceris, 588
dufourianum (Blanchard), Brachystegus, 473
dufourii Guérin-Méneville, Cerceris, 577
dufourii (Lepeletier), Brachystegus, 473
dufourii Lepeletier, Cerceris, 586
dufourii Lucas, Philanthus, 564
dugesianus Leclercq, Lindenius, 384
duisi Scullen, Cerceris, 580
Dumonela Reed, 53
dunbrodyensis Cameron, Ammophila, 151
dunensis Beaumont, Belomicrus, 363
dungensis Leclercq, Ectemnius, 426
duplicata Brèthes, Cerceris, 580
duplicata Provancher, Steniolia, 553
duplicata (Turner and Waterston), Hingstoniola, 417
duplicatus Rohwer, Tachytes, 264
duplipunctata Tsuneki, Cerceris, 587
durango Scullen, Cerceris, 580
durga (Nurse), Podalonia, 145
dusmeti E. Pérez, Oxybelus, 367
dusmeti Giner Marí, Ammoplanus, 198
dusmeti Giner Marí, Cerceris, 580
dusmeti Giner Marí, Tachysphex, 272
dusmeti Mercet, Nysson, 469
dutti Turner, Nysson, 469
dux (Kohl), Larra, 238
dyeri Richards, Trypoxylon, 346
Dynatus Lepeletier, 38, **92**
dyscheira (Saussure), Liris, 245
Dyscolocrabro Pate, 49
dysmica Menke, Ammophila, 151
dzhanokmenae Kazenas, Eremiasphecium, 629
dzimm Tsuneki, Psen, 166
dzinghis Tsuneki, Dienoplus, 496
dzinghis Tsuneki, Tachysphex, 273
dzingis (Tsuneki), Mimesa, 162

dziuroo Tsuneki, Diodontus, 179

ealae Leclercq, Psenulus, 172
ealaenis Leclercq, Neodasyproctus, 419
eatoni E. Saunders, Miscophus, 318
eatoni (E. Saunders), Prionyx, 133
eatoni E. Saunders, Tachysphex, 275
eatoni Morice, Cerceris, 582
ebenina (Lepeletier), Podalonia, 144
ebenina (Spinola), Podalonia, 144
ebeninus Arnold, Tachysphex, 273
eberhardi Richards, Microstigmus, 192
ebetsuense Tsuneki, Rhopalum, 389
ebneri Maidl, Trypoxylon, 346
eburnea Radoszkowski, Bembix, 546
eburnea Scullen, Cerceris, 588
eburnea (Taschenberg), Lestica, 431
eburneofasciata Brauns, Cerceris, 580
eburneofasciatus Dahlbom, Oxybelus, 367
eburnoeguttatus Arnold, Oxybelus, 367
eburneopictus (Arnold), Neodasyproctus, 419
eburneus (Chevrier), Gorytes, 501
eburneus Dours, Oxybelus, 370
eburneus Radoszkowski, Oxybelus, 367
eburneus van Lith, Psenulus, 173
ebusiana Giner Marí, Bembix, 547
echingol Tsuneki, Astata, 213
echingol Tsuneki, Cerceris, 584
echo Mickel, Cerceris, 580
Echuca Pate, 47, **392**
Echucoides Leclercq, 47, **394**
eckloni (Dahlbom), Chalybion, 103
economicum (Curtiss), Sceliphron, 105
ecoronatus Dufour, Philanthus, 566
ectemiformis Tsuneki, Corenocrabro, 629
Ectemnius Dahlbom, 49, **422**
ecuadorae R. Bohart, Neogorytes, 52, 494
ecuadorium Richards, Trypoxylon, 346
edax (Bingham), Isodontia, 123
edentata (Malloch), Mimesa, 162
edgarus Leclercq, Podagritus, 393
Editha J. Parker, 54, **542**
edolata Shestakov, Cerceris, 580
eduardi Morice, Tachysphex, 274
edwardsi (Cameron), Prionyx, 134
edwardsii Patton, Steniolia, 553
efflatouni (Mochi), Pseudoscolia, 573
effossus (Packard), Ectemnius, 426
effrenus (Kohl), Lindenius, 384
effugiens (Brauns), Pseudoplisus, 503
eganus Leclercq, Anacrabro, 379
egena Arnold, Cerceris, 580
egens (Handlirsch), Bembecinus, 531
egens Handlirsch, Bembix, 546
egens (Kohl), Isodontia, 123
egregia (Arnold), Liris, 248
egregia Mocsáry, Ammophila, 152
egregium Arnold, Trypoxylon, 346
egregium (Saussure), Penepodium, 92
egregius Arnold, Ammoplanus, 198
egregius (Arnold), Encopognathus, 380
egregius Arnold, Tachysphex, 273
egregius (Gussakovskij), Stizoides, 528
egregius (Handlirsch), Lestiphorus, 706
egregius (Taschenberg), Trachypus, 569
elbae (Pulawski), Dryudella, 214

eldoradensis Rohwer, Tachysphex, 273
electa Kohl, Ammophila, 154
electra Arnold, Cerceris, 580
electus Nurse, Gastrosericus, 256
eleebana Evans and Matthews, Bembix, 546
elegans Andrade, Miscophus, 318
elegans (Bingham), Liris, 245
elegans (Cresson), Dryudella, 214
elegans Cresson, Eucerceris, 592
elegans Dahlbom, Stizus, 526
elegans Dufour, Cerceris, 578
elegans Eversmann, Cerceris, 580
elegans (F. Morawitz), Philanthinus, 570
elegans F. Smith, Cerceris, 578
elegans (F. Smith), Eremochares, 146
elegans F. Smith, Gorytes, 501
elegans (F. Smith), Isodontia, 123
elegans (F. Smith), Megistommum, 504
elegans F. Smith, Philanthus, 565
elegans (F. Smith), Trachypus, 569
elegans G. and R. Bohart, Larropsis, 259
elegans J. Parker, Steniolia, 553
elegans (Lepeletier), Dienoplus, 496
elegans Mocsáry, Oxybelus, 367
elegans (Provancher), Crabro, 407
elegans (Taschenberg), Trachypus, 569
elegans van Lith, Psenulus, 172
elegantior Rayment, Sericophorus, 302
elegantissima Schletterer, Cerceris, 578
elegantissimus Dalla Torre, Philanthus, 565
elegantula Kohl, Ampulex, 77
elegantula Shestakov, Cerceris, 579
elegantula Turner, Spilomena, 193
elegantulum F. Smith, Trypoxylon, 346
elegantulus Gerstaecker, Oxybelus, 370
elegantulus (H. Smith), Gorytes, 501
elegantulus (Turner), Prionyx, 133
elephantinops Holmberg, Cerceris, 582
elizabethae Bingham, Cerceris, 580
ellenbergeri Arnold, Stizus, 526
elongata Fischer-Waldheim, Ammophila, 151
elongata (Packard), Mimumesa, 164
elongatulus Turner, Ammatomus, 513
elongatulus (Vander Linden), Crossocerus, 401
elongatum Ashmead, Trypoxylon, 345
elongatum F. Smith, Trypoxylon, 346
elongatus Cresson, Tachytes, 264
elongatus (Dudgeon), Crossocerus, 400
elongatus (Fabricius), Trachypus, 54, **568**
elongatus Lepeletier, Crossocerus, 401
elongatus (Provancher), Crabro, 408
elongatus Radoszkowski, Oxybelus, 367
elongatus Villers, Sphex, 117
elsiae Dunning, Aphilanthops, 570
elvinus (Cameron), Ectemnius, 427
emarginata (Cresson), Stictiella, 551
emarginata (Panzer), Cerceris, 587
emarginata Sichel, Bembix, 545
emarginatus (Brullé), Chilosphex, 128
emarginatus (Hacker & Cockerell), Sericophorus, 303
emarginatus (Kohl), Crossocerus, 401
emarginatus Say, Oxybelus, 367
emarginatus van Lith, Psen, 166
embeliae Leclercq, Ectemnius, 426
emdeni Richards, Trypoxylon, 348

emeritum Leclercq, Trypoxylon, 346
emeryana A. Costa, Dryudella, 215
emeryana Gribodo, Cerceris, 580
emeryi (Gribodo), Hoplisoides, 521
emir Handlirsch, Stizus, 526
emirus (Leclercq), Carinostigmus, 191
emmiltosa Scullen, Cerceris, 586
emortualis Saussure, Cerceris, 582
empeyi Arnold, Cerceris, 584
empeyi (van Lith), Mimesa, 629
Enchemicrum Pate, 46, **364**
Encopognathus Kohl, 46, **379**
englebergi (Brauns), Prionyx, 133
englehardti Banks, Cerceris, 584
enodens Brèthes, Cerceris, 580
Enodia Dahlbom, 39
Enoplolindenius Rohwer, 49, **414**
enslini Blüthgen, Spilomena, 193
enslini Bytinski-Salz, Bembix, 545
enslini Maidl, Astata, 212
enslini Wagner, Pemphredon, 181
ENTOMOSERICINI, 59, **434**
Entomocrabro Kohl, 46, **375**
Entomognathus Dahlbom, 46, **381**
Entomopison Menke, 45, 334
ENTOMOSERICINAE, 57, **433**
Entomosericus Dahlbom, 51, **434**
Eogorytes R. Bohart, 52, **505**
Eopsenulus Gussakovskij, 40
eous Gussakovskij, Gorytes, 501
Eparmostethus Kohl, 43
epeoliformis (F. Smith), Synnevrus, 470
ephippiatus Packard, Gorytes, 501
ephippiorhyncha Arnold, Cerceris, 580
ephippium F. Smith, Sphex, 114
ephippium Turner, Cerceris, 580
Epibembex Pate, 54
Epicrossocerus Ashmead, 48, 398
Epinysson Pate, 51, **471**
Epiphilanthus Ashmead, 54
equalis J. Parker, Microbembex, 539
equalis Viereck, Passaloecus, 184
eques (Nurse), Crossocerus, 401
equestris (Fabricius), Mimesa, 40, **162**
equestris of authors, Mimesa, 162
erberi Mocsáry, Bembecinus, 531
erebus (F. Smith), Larra, 237
erebus Leclercq, Ectemnius, 426
erebus W. F. Kirby, Sphex, 114
EREMIASPHECIINI, 60, **558**
Eremiasphecium Kohl, 54, **560**
eremica Gillaspy, Steniolia, 553
eremicus Bytinski-Salz, Stizus, 527
eremita Kohl, Passaloecus, 184
eremita Pulawski, Astata, 212
eremnon Menke, Pison, 335
Eremnophila Menke, 39, **146**
Eremochares Gribodo, 39, **145**
eremophila (Turner), Parapsammophila, 139
eremorum Beaumont, Gastrosericus, 256
eremorum Beaumont, Psammaecius, 516
erfilai Willink, Bicyrtes, 538
eriogoni Pate, Belomicrus, 363
eriogoni (Rohwer), Crossocerus, 403
eriogoni (Rohwer), Pulvero, 196
eriogoni Viereck and Cockerell, Cerceris, 583
eritreanus Bischoff, Philanthus, 565
erlandssoni Eck, Cerceris, 577
erminea Kohl, Ammophila, 153
ermineus Banks, Tachytes, 265

ermineus Kohl, Sphex, 114
errabunda (Kohl), Parapsammophila, 139
errabunda (Mercet), Podalonia, 145
errabundus Kohl, Sphex, 114
errans (Beaumont), Bembecinus, 532
errans (W. Fox), Lindenius, 384
errans Rohwer, Pemphredon, 181
errans Saussure, Trypoxylon, 346
errans Turner, Gastrosericus, 256
errata Shestakov, Cerceris, 580
erratica Bingham, Larra, 238
erraticus F. Smith, Psen, 166
erraticus (F. Smith), Psenulus, 172
erraticum Tsuneki, Rhopalum, 389
erronea Giner Marí, Cerceris, 580
erubescens Morice, Nysson, 469
erugatus R. Bohart, Pseudoplisus, 503
erusus Leclercq, Psenulus, 172
eryngii Marquet, Cerceris, 580
erynnis Arnold, Cerceris, 580
erynnis Turner, Tachytes, 265
erythopoda Rohwer, Psen, 166
erythraeensis Mantero, Stizus, 526
erythraeus Mickel, Tachysphex, 273
erythraspis Cameron, Stizus, 526
erythrinus Guiglia, Sphex, 114
erythrocephala Dahlbom, Cerceris, 587
erythrocephala (Taschenberg), Stizus, 526
erythrocephalus (Fabricius), Parapsammophila, 139
erythrocerum Kohl, Pison, 335
erythrogaster (A. Costa), Tachysphex, 273
erythrogaster Cameron, Tachysphex, 273
erythrogaster Kazenas, Cerceris, 580
erythrogaster (Turner), Chimiloides, 414
erythrogaster (Turner), Stizoides, 528
erythrogastra (Rohwer), Prionyx, 133
erythrogastrum Rohwer, Pison, 335
erythropa (Dalla Torre), Parapiagetia, 281
erythrophorus Dalla Torre, Tachysphex, 273
erythropoda Cameron, Cerceris, 580, 584
erythropoda (Cameron), Liris, 245
erythropoda Cameron, Nysson, 469
erythropoda (Cameron), Parapiagetia, 281
erythropoda Cameron, Sphex, 117
erythropoda Malloch, Psen, 166
erythropterus Cameron, Sphex, 114
erythropus (Brèthes), Podagritus, 393
erythropus (F. Smith), Podalonia, 144
erythropus Herbst, Tachysphex, 276
erythropus Kohl, Ampulex, 77
erythropus Kohl, Pison, 335
erythropus (Spinola), Tachysphex, 273
erythropus Taschenberg, Ammophila, 151
erythropyga (Arnold), Liris, 245
erythropyga Turner, Larra, 230
erythrosoma (Pulawski), Dryudella, 214
erythrosoma Schletterer, Cerceris, 580
erythrosoma (Turner), Sericophorus, 303
erythrospila Cameron, Ammophila, 151
erythrospila Cameron, Cerceris, 580, 584
erythrothorax Cameron, Alysson, 458
erythrotoma (Cameron), Dasyproctus, 420
erythrotoma (Cameron), Liris, 245
erythroura Cameron, Cerceris, 580
erythrozonatum Cameron, Trypoxylon, 346
erytrhinus Magretti, Sphex, 114
esakii Yasumatsu, Ampulex, 77
esakii (Yasumatsu), Crossocerus, 403

esakii Yasumatsu, Liris, 245
esakii Yasumatsu, Pison, 335
esau Beaumont, Crossocerus, 401
esau Schletterer, Cerceris, 584
escalerae Giner Marí, Cerceris, 580
escalerae (Turner), Bembecinus, 531
escalarae (Turner), Dienoplus, 496
escalerai Giner Marí, Astata, 212
esoterus Leclercq, Encopognathus, 380
espanoli (Giner Marí), Pseudoscolia, 573
essequibo Pate, Quexua, 387
esterensis Leclercq, Ectemnius, 424, 426
esuchus (Rohwer), Psenulus, 172
etruscus (Rossi), Tachytes, 265
ettingol (Tsuneki), Prionyx, 133
eucalypti Turner, Rhopalum, 389
Eucerceris Cresson, 55, **589**
eucharis Schletterer, Cerceris, 580
euchroma Turner, Cerceris, 580
euchromus Handlirsch, Stizus, 526
eugenia (F. Smith), Eremnophila, 147
eugenia Schletterer, Cerceris, 580
eugeniae Gussakovskij, Stizus, 526
eugeniae Gussakovskij, Trypoyxlon, 346
eulalia Brauns, Cerceris, 580
eumenoides Ducke, Aulacophilus, 338
Euoxybelus Noskiewicz and Chudoba, 37
euphorbiae Marquet, Cerceris, 586
euphorbiae R. Bohart, Nysson, 469
euphrosyne (Handlirsch), Hoplisoides, 520
Euplilis Risso, 47
Eupliloides Pate, 48, **398**
europaea Giraud, Ampulex, 77
europaea Mercet, Ancistromma, 259
europaeus Kohl, Tachytes, 265
europaeus Tsuneki, Stigmus, 189
eurous Pulawski, Tachytes, 265
euryanthe Kohl, Cerceris, 580
eurygnatha (Pulawski), Dryudella, 214
eurymele Banks, Cerceris, 587
eurynome (Arnold), Hoplisoides, 521
eurynome W. Fox, Philanthus, 566
euryops (Kohl), Entomognathus, 382
eurypus (Kohl), Lestica, 431
eurypyga Kohl, Cerceris, 589
eurypygus van Lith, Psen, 166
Euspongus Lepeletier, 52
eustylicida Williams, Cerceris, 580
euteles (Handlirsch), Bembecinus, 531
euxinus Pulawski, Tachysphex, 273
Euzonia Stephens, 52
evansi Gillaspy, Stictiella, 551
evansi (Krombein), Miscophus, 318
evansi Menke, Ammophila, 151
evansi Pate, Moniacera, 395
evansi R. Bohart, Clypeadon, 571
evansi Scullen, Cerceris, 580
evecta Shestakov, Cerceris, 588
eversmanni (Ed. André), Isodontia, 123
eversmanni Radoszkowski, Stizus, 526
eversmanni Schulz, Cerceris, 580
evolutionis (Leclercq), Entomognathus, 382
exaratus (Eversmann), Psen, 166
excavata Cameron, Cerceris, 580
excavata Schletterer, Cerceris, 580
excavatum F. Smith, Trypoxylon, 348
excavatus (Turner), Synnevrus, 470
excavatus (W. Fox), Ectemnius, 426
excellens Cameron, Tachytes, 265
excellens Klug, Cerceris, 586
excellens Lohrmann, Stizus, 526

excellens Strnad, Trypoxylon, 346
excelsus Turner, Tachysphex, 273
excerptus Turner, Tachysphex, 273
excisus Arnold, Tachysphex, 273
excisus Gussakovskij, Belomicrus, 363
excisus (Handlirsch), Bembecinus, 530
excisus (Kohl), Prionyx, 133
exclamans Viereck, Oxybelus, 367
exclusum Turner, Pison, 335
exclusus Turner, Tachytes, 265
exegetus Leclercq, Podagritus, 393
Exeirus Shuckard, 52, **494**
exigua (W. Fox), Stictiella, 551
exiguum Tsuneki, Trypoxylon, 346
exiguus (Handlirsch), Dienoplus, 496
exiguus (Vander Linden), Crossocerus, 401, 402
exilipes (F. Smith), Liris, 248
eximia (Lepeletier), Eremnophila, 147
eximium F. Smith, Trypoxylon, 346
eximium (Kohl), Chlorion, 89
eximium (Strand), Chlorion, 89
eximius Arnold, Stizus, 526
eximius (F. Morawitz), Philanthinus, 570
eximius F. Morawitz, Stizus, 526
eximius (F. Smith), Austrogorytes, 498
eximius Gussakovskij, Miscophus, 318
eximius Honoré, Palarus, 291
eximius Lepeletier, Sphex, 116
eximius (Provancher), Hapalomellinus, 497
eximius Pulawski, Tachysphex, 273
eximius Sickmann, Oxybelus, 367
Exirus Schulz, 52
exornata Fernald, Isodontia, 123
exornatum Turner, Pison, 335
exornatus W. Fox, Tachytes, 265
exoticus Taschenberg, Miscophus, 318
expansa Gribodo, Bembix, 546
expectata Valkeila, Spilomena, 193
expectatus Turner, Dasyproctus, 420
expedita (Kohl), Liris, 248
expeditionis Leclercq, Rhopalum, 389
expleta Brèthes, Cerceris, 580
expulsa Turner, Cerceris, 580
exsecta F. Smith, Cerceris, 582
exsecta Kohl, Ammophila, 151
exsectus W. Fox, Tachysphex, 273
extensa (Walker), Liris, 245
extrema Dahlbom, Larra, 238
extremitata Cresson, Ammophila, 151
extremus van Lith, Psenulus, 172
exul (Turner), Polemistus, 185
exulta (Harris), Cerceris, 577
exultans Dalla Torre, Cerceris, 577
exultans Turner, Pison, 335
exultatum Leclercq, Rhopalum, 389
eyeni Leclercq, Trypoxylon, 346
eyrensis R. Turner, Ammophila, 151
ezra (Pate), Mimesa, 162
ezrae (Cameron), Crossocerus, 401

fabius Nurse, Palarus, 291
fabricator (F. Smith), Chalybion, 102
fabricator F. Smith, Pison, 336
fabricator F. Smith, Trypoxylon, 346
fabricii Dahlbom, Sphex, 116
fabricii (Müller), Pemphredon, 182
faceta Arnold, Cerceris, 580
faceta Bingham, Paraliris, 250
facialus (Handlirsch), Bembecinus, 531
facilis (F. Smith), Liris, 245

INDEX 653

facilis (F. Smith), Neoplisus, 505
fairmairei (Handlirsch), Argogorytes, 492
fairmairei Kohl, Heliocausus, 453
falcifera Tsuneki, Cerceris, 580
flaciger Arnold, Tachytes, 265
flacigera Arnold, Tachytes, 265
fallax Arnold, Ammatomus, 513
fallax Arnold, TAchysphex, 273
fallax F. Morawitz, Mimesa, 162
fallax Gerstaecker, Oxybelus, 370
fallax Handlirsch, Gorytes, 501
fallax Kohl, Ammophila, 152
fallax Kohl, Nitela, 325
fallax (Kohl), Penepodium, 92
fallax Kohl, Sceliphron, 106
fallax Mercet, Bembix, 546
fallax W. Fox, Trypoxylon, 349
falsa (F. Morawitz), Prospigastra, 285
famelicus Pulawski, Tachytes, 265
familiaris (F. Smith), Piyuma, 410
fantiorum Arnold, Bembix, 546
fanuiensis Cheesman, Tachysphex, 273
fargei F. Smith, Cerceris, 586
fargeii (F. Smith), Ectemnius, 425
fargeii (Shuckard), Argogorytes, 492
farri Scullen, Cerceris, 580
fasciata (F. Smith), Liris, 245
fasciata (F. Smith), Lyroda, 299
fasciata Fabricius, Bembix, 548
fasicata Jurine, Ampulex, 77
fasciata Lepeletier, Cerceris, 587
fasciata Spinola, Cerceris, 585
fasciatiipennis (Cameron), Dicranorhina, 251
fasciatipennis (Cameron), Hoplisoides, 521
fasciatiipennis Kohl, Pison, 335
fasciatum (Lepeletier), Sceliphron, 105
fasciatum (Radoszkowski), Pison, 335
fasciatus (A. Costa), Crossocerus, 403
fasciatus Dahlbom, Oxybelus, 370
fasciatus (Fabricius), Stizoides, 528
fasciatus (Fabricius), Stizus, 526
fasciatus (Fourcroy), Philanthus, 566
fasciatus (Lepeletier and Brullé), Ectemnius, 426
fasciatus Morice, Tachysphex, 273
fasciatus (Olivier), Zanysson, 475
fasciatus (Radoszkowski), Belomicrus, 364
fasciatus Rohwer, Passaloecus, 33, 184
fasciatus Taschenberg, Oxybelus, 370
fasciatus (W. Fox), Pseudoplisus, 503
fasciatus Westwood, Sericogaster, 38
fasciipennis Gussakovskij, Tachysphex, 276
fasciola Cresson, Cerceris, 585
fastidosa Turner, Cerceris, 580
fastidiosus Beaumont, Lindenius, 384
fastigatus Mickel, Oxybelus, 369
fastigium W. Fox, Trypoxylon, 346
fatalis Turner, Tachytes, 265
faunus (Arnold), Entomognathus, 382
feae (Handlirsch), Hoplisoides, 521
federationis Leclercq, Crossocerus, 401
fedorensis Rohwer, Tachysphex, 276
fedtschenkoi Radoszkowski, Oxybelus, 367
fedtschenkoi Radoszkowski, Stizus, 526
felina (Arnold), Liris, 244
femoralis Arnold, Philanthus, 565
femoralis F. Morawitz, Crabro, 408
femoralis Kohl, Belomicrus, 363
femoralis (Mocsáry), Dryudella, 214
femoralis Radoszkowski, Bembix, 545
femoralis (Saussure), Liris, 247

femorata (Cameron), Larra, 238
femorata F. Parker, Astata, 212
femorata (Saussure), Larra, 238
femorata (W. Fox), Stictiella, 551
femoratum (Fabricius), Chalybion, 103
femoratus (Bradley), Oryttus, 508
femurrubra W. Fox, Ammophila, 151
femurrubrum Viereck and Cockerell, Cerceris, 580
fenchihuense Tsuneki, Trypoxylon, 346
fenchihuensis Tsuneki, Larra, 238
fenestratum F. Smith, Pison, 335
fenestratus (F. Smith), Stizoides, 528
fenimorum Leclercq, Rhopalum, 389
fennicus Merisuo, Pemphredon, 181
fera (Lepeletier), Podalonia, 144
feralis Cameron, Cerceris, 579
ferghanica (Gussakovskij), Eremochares, 146
fergusoni Beaumont, Belomicroides, 363
fergusoni Beaumont, Cerceris, 587
fergusoni Pate, Crossocerus, 401
ferocior Turner, Cerceris, 580
ferocior van der Vecht and Krombien, Sphex, 116
ferox Arnold, Parapiagetia, 281
ferox F. Smith, Cerceris, 580
ferox F. Smith, Sphex, 116
ferox (F. Smith), Tachytes, 264
ferox F. Smith, Trypoxylon, 346
ferox Perkins, Deinomimesa, 40, **169**
ferox (Saussure), Dasyproctus, 420
ferox Shuckard, Oxybelus, 367
ferox (Westwood), Chalybion, 102
ferox Westwood, Chlorion, 90
fernaldi (Murray), Ammophila, 151
Fernaldina R. Bohart and Menke, 39, 108
ferrandii Magretti, Stizus, 526
ferreri Vander Linden, Cerceris, 581
ferreroi Schulz, Cerceris, 581
ferrierei Leclercq, Neodasyproctus, 419
ferrieri Kohl, Belomicrus, 363
ferrugatus (Gussakovskij), Dienoplus, 496
ferruginea Bradley, Ampulex, 77
ferruginea Brèthes, Cerceris, 580
ferruginea Gmelin, Vespa, 627
ferruginea Radoszkowski, Pseudoscolia, 573
ferrugineipes (Arnold), Polemistus, 185
ferrugineipes Cameron, Tachytes, 265
ferrugineipes Lepeletier, Ammophila, 151
ferrugineipes (Lepeletier), Liris, 245
ferrugineipes (Rohwer), Ectemnius, 428
ferrugineipes W. Fox, Sphex, 114
ferrugineum Ashmead, Diploplectron, 211
ferrugineus (F. Smith), Stizus, 526
ferrugineus Lepeletier, Sphex, 116
ferrugineus Pulawski, Tachysphex, 273
ferrugineus (Rayment), Sericophorus, 303
ferrugineus (Spinola), Hoplisoides, 521
ferrugineus (Viereck), Pseneo, 165
ferrugineus (W. Fox), Prionyx, 133
ferruginior Viereck and Cockerell, Cerceris, 578
ferruginosa Cresson, Ammophila, 152
ferruginosa Scullen, Eucerceris, 592
fertoni Beaumont, Cerceris, 587
fertoni (Handlirsch), Bembecinus, 531
fertoni (Handlirsch), Dienoplus, 496
fertoni Pulawski, Tachysphex, 273
Fertonius Pérez, 48
ferum (Drury), Chalybion, 102
ferum (Drury), Chlorion, 90

ferus (Dahlbom), Palmodes, 127
fervens F. Smith, Cerceris, 585
fervens (F. Smith), Sceliphron, 106
fervens F. Smith, Tachytes, 263, 265
fervens (Linnaeus), Prionyx, 133
fervida F. Smith, Cerceris, 587
fervidus (Cresson), Sphecius, 511
fervidus F. Smith, Tachytes, 265
festinans (F. Smith), Liris, 245
festiva Cresson, Cerceris, 580
festiva (F. Smith), Eremochares, 146
festiva J. Parker, Bembix, 545
festivum Brèthes, Trypoxylon, 349
festivum F. Smith, Pison, 335
festivum Mocsáry, Alysson, 458
festivus (Marquet), Tracheliodes, 405
fidelis Cresson, Nysson, 469
fidelis Pulawski, Tachytes, 265
fidelis Viereck & Cockerell, Cerceris, 577
fiebrigi (Brèthes), Heliocausus, 453
fiebrigi Brèthes, Sagenista, 522
fiebrigi Brèthes, Tachytes, 263
fiebrigi Leclercq, Enoplolindenius, 415
fiebrigi Richards, Trypoxylon, 346
fieuzeti Giner Marí, Trypoxylon, 346
figulum (Dahlbom), Sceliphron, 105
figulus (Linnaeus), Trypoxylon, 45, **346**
filata Walker, Ammophila, 152
filicornis Kohl, Tachysphex, 273
filicornis Rohwer, Larropsis, 259
filicornis (Rohwer), Psenulus, 172
filiformis Radoszkowski, Crabro, 408
filigranus (A. Costa), Tracheliodes, 405
filipina Lohrmann, Bembix, 546
filippovi (Gussakovskij), Carinostigmus, 191
filippovi (Gussakovskij), Mimesa, 162
fimbriata (Pagden), Hingstoniola, 418
fimbriata (Rossi), Cerceris, 580
fingo Brauns, Cerceris, 580
finitima Cresson, Cerceris, 580
finitimus (F. Morawitz), Ectemnius, 427
finschii Handlirsch, Bembix, 546
finschii Kohl, Sphex, 114
firma Cresson, Cerceris, 589
fischeri Spinola, Bembix, 546
fischeri Spinola, Cerceris, 580
fischeri (Spinola), Cerceris, 582
fischeri Spinola, Oxybelus, 367
fischeroides Magretti, Bembix, 546
fissus Lepeletier, Oxybelus, 367
fistularium (Dahlbom), Sceliphron, 105
fitzgeraldi Empey, Cerceris, 581
fitzgeraldi Richards, Trypoxylon, 346
flagellarius (F. Morawitz), Ectemnius, 426
flagellarius Turner, Philanthus, 565
flagellatus Nurse, Tachytes, 265
flagellatus R. Bohart, Gorytes, 501
flammitrichus Strand, Sphex, 114
flava Mochi, Cerceris, 578
flava Tsuneki, Cerceris, 577
flavescens F. Smith, Bembix, 546
flavescens Schletterer, Cerceris, 577
flavicans Gmelin, Vespa, 627
flavicinctus (Donovan), Argogorytes, 492
flaviclypeus (W. Fox), Lindenius, 384
flavicollis (F. Morawitz), Ectemnius, 425
flavicornis Arnold, Oxybelus, 367
flavicornis (Baker), Oryttus, 508
flavicornis Brullé, Cerceris, 581
flavicornis Gussakovskij, Gastrosericus,

256
flavicornis Gussakovskij, Solierella, 313
flavicornis (Harris), Pseudoplisus, 503
flavicornis (Packard), Pseudoplisus, 503
flavicornis Tsuneki, Stigmus, 189
flavicornis van Lith, Psenulus, 173
flavicornis Villers, Sphex, 117
flavida Cameron, Cerceris, 581
flavida (Kohl), Podalonia, 144
flavidulus R. Bohart, Pseudoplisus, 503
flavidus (Taschenberg), Trachypus, 568
flavifrons Cresson, Philanthus, 564
flavifrons F. Smith, Bembix, 546
flavifrons F. Smith, Cerceris, 581
flavilabris F. Smith, Bembix, 545
flavilabris (Fabricius), Cerceris, 581
flavimanum Arnold, Trypoxylon, 346
flavinerva (Cameron), Liris, 245
flavipenne (Latreille), Penepodium, 92
flavipennis Fabricius, Sphex, 39, **114**
flavipennis (Lepeletier and Brullé), Ectemnius, 426
flavipennis (Turner), Astata, 212
flavipennis (Williams), Liris, 245, 247
flavipes Arnold, Philanthus, 565
flavipes Cameron, Oxybelus, 368
flavipes (Christ), Sceliphron, 105, 106
flavipes (F. Smith), Bembecinus, 531
flavipes F. Smith, Bembix, 546
flavipes F. Smith, Sphex, 114
flavipes (Fabricius), Palarus, 291
flavipes (Fabricius), Sceliphron, 105
glavipes Gmelin, Vespa, 627
flavipes Kazenas, Oxybelus, 368
flavipes (Lepeletier and Brullé), Crabro, 408
flavipes (Lepeletier and Brullé), Ectemnius, 428
flavipunctatum (Christ), Sceliphron, 105
flavissimus Leclercq, Crossocerus, 48, 401
flavistigma Thomson, Pemphredon, 181
flavitarsis (Arnold), Bembecinus, 530
flavitarsis (Fernald), Sphex, 114
flavitarsis W. Fox, Diodontus, 179
flavitarsus (Tsuneki), Towada, 412
flavitibialis Tsuneki, Oxybelus, 368
flavitincta (Arnold), Liris, 245
flavitrochantericus (Viereck), Crossocerus, 403
flavivena Arnold, Liris, 244
flaviventris Arnold, Oxybelus, 368
flaviventris F. Morawitz, Gorytes, 501
flaviventris F. Smith, Bembix, 546
flaviventris Spinola, Cerceris, 587
flaviventris Vander Linden, Cerceris, 581
flavobilineolata Giordani Soika, Cerceris, 577
flavocincta Cresson, Eucerceris, 591
flavocincta Turner, Bembix, 546
flavocinereus Arnold, Tachytes, 265
flavocostalis Cresson, Cerceris, 581
flavofasciata F. Smith, Cerceris, 581
flavofasciatum (F. Smith), Sceliphron, 106
flavofasciatus (Turner), Sericophorus, 303
flavofemorata Arnold, Cerceris, 581
flavofimbriatus Arnold, Tachysphex, 276
flavogeniculatus (Taschenberg), Tachysphex, 273
flavohirtus Tsuneki, Ectemnius, 426
flavolatera J. Parker, Bembix, 547
flavolimbatum Turner, Pison, 335

flavolineatus Cameron, Philanthus, 565
flavomaculata Cameron, Cerceris, 581
flavomaculatum (De Geer), Sceliphron, 105
flavomaculatus Cameron, Alysson, 458
flavomaculatus (F. Smith), Stizus, 526
flavonasuta Arnold, Cerceris, 581
flavoniger Dutt, Crabro, 408
flavopicta Arnold, Bembix, 547
flavopicta F. Smith, Cerceris, 581
flavopictus (Arnold), Bembecinus, 531
flavopictus (F. Smith), Crossocerus, 401
flavopictus (F. Smith), Scapheutes, 355
flavopictus Pulawski, Miscophus, 318
flavoplagiata Cameron, Cerceris, 581
flavotrochanterica Rohwer, Cerceris, 581
flavoundata (Arnold), Dryudella, 214
flavovestitus F. Smith, Sphex, 114
flavus Brèthes, Trachypus, 569
flavus (Cameron), Bembecinus, 532
flebile (Lepeletier), Chalybion, 103
fletcheri (Turner), Astata, 213
fletcheri Turner, Cerceris, 581
fletcheri Turner, Diodontus, 179
fletcheri Turner, Trypoxylon, 346
flexuosa (Taschenberg), Stictia, 542
flexuosefasciatus (Mantero), Bembecinus, 531
fliranus Leclercq, Ectemnius, 426
florale Richards, Trypoxylon, 346
floralis (Olivier), Ectemnius, 424
floridana Pate, Nitela, 325
floridana (Rohwer), Mimumesa, 164
floridanus Krombein and Willink, Bembecinus, 531
floridanus Robertson, Oxybelus, 368
floridanus Rohwer, Tachytes, 265
floridanus (W. Fox), Pseudoplisus, 503
floridensis Banks Cerceris, 581
floridensis (Fernald), Ammophila, 154
floridicus R. Bohart, Hoplisoides, 521
florissantensis Rohwer, Crabro, 408
florissantensis Rohwer, Diodontus, 179
florissantensis (Rohwer), Dryudella, 215
florkini Leclercq, Lestica, 431
Florkinus Leclercq, 46
fluctuatus Gerstaecker, Tachysphex, 273
fluctuatus (Gerstaecker), Tachysphex, 276
fluvialis F. Smith, Cerceris, 581
fluviatilis Arnold, Gastrosericus, 256
fluxa Kohl, Cerceris, 582
fodiens (Handlirsch), Bicyrtes, 537
fodiens Eversmann, Cerceris, 581
foeniforme (Perty), Penepodium, 92
foersteri Fritz, Metanysson, 481
foleyi (Beaumont), Parapsammophila, 139
fongosi Brauns, Cerceris, 589
fonti Mercet, Bembix, 546
forbesii (Robertson), Belomicrus, 363
forcipata Handlirsch, Bembix, 546
fordi Krombein, Cerceris, 589
forestus Leclercq, Ectemnius, 424
forficula Saussure, Chlorion, 90
formicaria Eschscholtz, Cerceris, 581
formicarius (Ferton), Tracheliodes, 405
formicaroides Gistel, Sphex, 117
formicoides Menke, Ammophila, 152
formicoides Turner, Ampulex, 77
formidolosa Saussure, Cerceris, 581
formosa (Cresson), Stictiella, 54, **551**

formosa Dahlbom, Cerceris, 581
formosa (F. Smith), Lyroda, 299
formosa Kohl, Ampulex, 77
formosana Bischoff, Bembix, 546
formosana Strand, Ammophila, 151
formosana Strand, Cerceris, 581
formosana Tsuneki, Ammophila, 152
formosana Tsuneki, Leclerqia, 411
formosana Tsuneki, Liris, 245
formosana (Tsuneki), Spilomena, 193
formosanum Tsuneki, Rhopalum, 389
formosanum van der Vecht, Sceliphron, 106
formosanus (Sonan), Bembecinus, 531
formosanus Tsuneki, Alysson, 458
formosanus (Tsuneki), Carinostigmus, 191
formosanus Tsuneki, Dasyproctus, 421
formosanus Tsuneki, Dolichurus, 69
formosanus Tsuneki, Ectemnius, 426
formosanus Tsuneki, Philanthus, 565
formosanus Tsuneki, Psenulus, 173
formosanus Tsuneki, Tachysphex, 273
formosanus Tsuneki, Tachytes, 266
formosellus van der Vecht, Sphex, 114
formosensis Tsuneki, Ammophila, 152
formisicola Strand, Cerceris, 585
formosicola (Strand), Isodontia, 123
formosicola Strand, Psenulus, 172
formosicola Strand, Trypoxylon, 346
formosissimus Turner, Tachytes, 265
formosum (F. Smith), Sceliphron, 106
formosus Arnold, Miscophoides, 45, **322**
formosus Arnold, Tachytes, 267
formosus F. Smith, Philanthus, 565
formosus F. Smith, Sphex, 114
formosus (Jurine), Dienoplus, 496
formosus Tsuneki, Oxybelus, 369
formosus (Tsuneki), Polemistus, 185
forticarinata Leclercq, Vechtia, 417
forticarinatus Cameron, Oxybelus, 368
forticula Arnold, Cerceris, 581
fortin Scullen, Cerceris, 581
fortinata Cameron, Cerceris, 581
fortior Turner, Tachysphex, 273
fortistriolatus Cameron, Palarus, 291
fortuitus (Kohl), Entomognathus, 382
fortunatus Beaumont, Dasyproctus, 420
fortunatus Beaumont, Tachysphex, 275
fossor F. Smith, Cerceris, 581
fossor Rohwer and Cockerell, Oxybelus, 368
fossor Rohwer, Diploplectron, 211
fossor (Rohwer), Solierella, 313
fossoria F. Smith, Bembix, 546
fossorius (Linnaeus), Ectemnius, 426
fossulatus Turner, Philanthus, 565
fossuleus Leclercq, Crossocerus, 401
fossuliferum (Gribodo), Sceliphron, 105
foucauldi Beaumont, Tachysphex, 277
foveata Lepeletier, Cerceris, 589
foveatum Cameron, Holcorhopalum, 385
foveatum Cameron, Trypoxylon, 346
foveatus Arnold, Philanthus, 565
foveifrons Cameron, Ampulex, 77
foveiscutis (Cameron), Liris, 245
foveiscutis (Gerstaecker), Zanysson, 475
foveolatus Handlirsch, Gorytes, 501
foveolatus (Holmberg), Crossocerus, 401
foveolatus (W. Fox), Pseudoplisus, 503
foveolineatus (Viereck), Tracheliodes, 405
foxi Banks, Tachytes, 265
foxi Bohart and Menke, Prionyx, 133

foxi Dunning, Aphilanthops, 570
foxi Gillaspy, Stizoides, 528
foxi J. Parker, Bembix, 545
foxi Pate, Ammoplanops, 197
Foxia Ashmead, 51, **478**
foxiana Pate, Moniacera, 395
foxii Ashmead, Diploplectron, 211
foxii (Ashmead), Paranysson, 308
foxii Cockerell, Spilomena, 193
foxii (Handlirsch), Neoplisus, 505
foxii (Kincaid), Ectemnius, 424
foxii Kohl, Podium, 96
foxii (Rohwer), Epinysson, 472
foxii Rohwer, Pemphredon, 181
foxii Rohwer, Tachysphex, 274
foxii (Viereck), Solierella, 313
Foxita Pate, 49, **415**, 627
fractum Richards, Trypoxylon, 348
fragilis F. Smith, Ammophila, 152
fragilis (Nurse), Prionyx, 133
fragilis of authors, Ammophila, 153
fragosa Kohl, Cerceris, 582
franciscus Pate, Belomicrus, 363
franclemonti (Krombein), Diodontus, 179
franzi (Cameron), Isodontia, 123
franzi Turner, Stizus, 526
fratellus Kohl, Crabro, 408
frater (Dahlbom), Dasyproctus, 420
frater Honoré, Miscophus, 319
fraterculum Turner, Pison, 335
fraterculus (Arnold), Bembecinus, 531
fraterculus (Gussakovskij), Brachystegus, 473
fraterna Beaumont, Cerceris, 581
fraternus (Arnold), Entomognathus, 382
fraternus Brèthes, Heliocausus, 453
fraternus Fritz, Metanysson, 481
fraternus Mercet, Gorytes, 501
fraternus Mercet, Nysson, 469
fraternus Rohwer, Diodontus, 179
fraternus Say, Stigmus, 189
fraternus Taschenberg, Tachytes, 265
fraudulenta Arnold, Bembix, 546
fraudulenta (Kohl), Liris, 246
fraudulentus (Arnold), Bembecinus, 531
fraudulentus Arnold, Oxybelus, 368
fredericismithi (Schulz), Ectemnius, 426
freetownensis van Lith, Psenulus 629
freimuthi Shestakov, Cerceris, 581
frenchii (Turner), Austrogorytes, 498
frenchii (Turner), Rhopalum, 389
freygessneri (Carl), Dryudella, 214
freygessneri (Handlirsch), Synnevrus, 470
freygessneri Kohl, Tachytes, 265
freygessneri Morice, Bembix, 546
freyi Bischoff, Diodontus, 179
freyi Handlirsch, Sphecius, 511
freymuthi Radoszkowski, Cerceris, 581
friburgensis Brèthes, Scapheutes, 355
friedrichi (Schrottky), Eremnophila, 147
friesei Handlirsch, Nysson, 470
friesei Kohl, Diodontus, 179
friesei Kohl, Podium, 96
frigida Mocsáry, Cerceris, 581
frigidum F. Smith, Trypoxylon, 346
frigidus (F. Smith), Aphilanthops, 55, **570**
frigidus (F. Smith), Ectemnius, 427
frioense Richards, Trypoxylon, 346
frioensis J. Parker, Bembix, 547
frischii (Fourcroy), Ammophila, 153

fritzi Leclercq, Anacrabro, 379
fritzi Menke, Odontosphex, 573
fritzi R. Bohart, Antomartinezius, 478
frivaldszkyi Mocsáry, Trigonopsis, 98
froggati Turner, Cerceris, 581
froggatti Rayment, Sericophorus, 302
froggatti Turner, Cerceris, 581
frommeri R. Bohart, Bembix, 546
frondigera (A. Costa), Oxybelus, 366
frontale F. Smith, Trypoxylon, 346
frontale (Kohl), Chalybion, 102
frontalis Cresson, Philanthus, 566
frontalis F. Smith, Cerceris, 578
frontalis F. Smith, Tachytes, 265
frontalis Gerstaecker, Philanthus, 566
frontalis Olivier, Bembix, 546
frontalis (Panzer), Mellinus, 449
frontalis (Radoszkowski), Dryudella, 214
frontalis Robertson, Oxybelus, 368
frontalis (Turner), Sericophorus, 303
frontalis (W. Fox), Psenulus, 172
frontata Say, Cerceris, 581
fronticorne Gussakovskij, Trypoxylon, 346
fruticis Tsuneki, Tachytes, 265
fucatus Arnold, Tachytes, 265
fucosa J. Parker, Bembix, 546
fugatrix Mickel, Cerceris, 588
fugax Fabricius, Trypoxylon, 348
fugax (Radoszkowski), Tachysphex, 273
fugax (Turner), Pseudoplisus, 503
fukaii Rohwer, Cerceris, 583
fukiensis Giner Marí, Cerceris, 587
fukuiensis Tsuneki, Crossocerus, 401
fukuiensis Tsuneki, Sphex, 115
fulgens Beaumont, Ampulex, 77
fulgidus Arnold, Psenulus, 172
fulgidus Arnold, Tachysphex, 273
fulginosa (Dahlbom), Liris, 245
fulginosa F. Smith, Cerceris, 581
fulginosum (Scopoli), Trypoxylon, 346
fulginosus Dahlbom, Sphex, 116
fulginosus (Klug), Stizus, 526
fuliginosus Tsuneki, Argogorytes, 492
fulleri Rayment, Sericophorus, 302
fulva Mocsáry, Cerceris, 581
fulvicaudis Cameron, Oxybelus, 368
fulviceps Cresson, Eucerceris, 592
fulvicornis Fabricius, Mellinus, 449
fulvicornis Gussakovskij, Spilomena, 193
fulvicornis Rohwer, Stigmus, 189
fulvicornis (Schenck), Psenulus, 172
fulvicornis Tsuneki, Tachysphex, 273
fulvicornis Turner, Auchenophorus, 327
fulvicornis Turner, Ectemnius, 426
fulvicornis Turner, Tachysphex, 273
fulvicrus (Perkins), Dasyproctus, 420
fulvipediculata Schletterer, Cerceris, 579
fulvipennis (F. Smith), Pseudoplisus, 503
fulvipennis (Taschenberg), Trachypus, 568
fulvipes A. Costa, Nysson, 469
fulvipes Cameron, Trypoxylon, 349
fulvipes Cresson, Cerceris, 579
fulvipes Cresson, Eucerceris, 591
fulvipes Cresson, Podium, 96
fulvipes Eversmann, Cerceris, 581
fulvipes (Eversmann), Stizoides, 528
fulvipes (F. Smith), Tachytes, 264
fulvipes Fabricius, Liris, 245
fulvipes Gerstaecker, Chlorion, 90
fulvipes (Malloch), Pseneo, 165
fulvipes Robertson, Oxybelus, 368
fulvipes Schrottky, Tachytes, 265

fulvipes W. Fox, Stigmus, 189
fulvispine Cameron, Trypoxylon, 349
fulvitarsis (A. Costa), Tachysphex, 273
fulvitarsis (Gussakovskij), Mimumesa, 164
fulviventris (Antiga and Bofill), Lindenius, 384
fulviventris Cresson, Tachytes, 265
fulviventris Guérin-Méneville, Cerceris, 581
fulviventris (Guérin-Méneville), Liris, 245
fulviventris Kohl, Sphex, 115
fulviventris Latreille, Palarus, 291
fulviventris Tsuneki, Cerceris, 578
fulviventris Tsuneki, Nysson, 469
fulvocollare Cameron, Trypoxylon, 346
fulvohirtum Arnold, Sceliphron, 106
fulvohirtus Bingham, Sphex, 114
fulvohirtus (Cameron), Podagritus, 393
fulvohirtus (Tsuneki), Eogorytes, 52, **505**
fulvopilosellus (Cameron), Ectemnius, 426
fulvopilosus Cameron, Oxybelus, 368
fulvopilosus Cameron, Tachytes, 265
fulvopilosus (Meade-Waldo), Ectemnius, 426
fulvovestitus Cameron, Tachytes, 265
fumicatus Christ, Sphex, 114
fumida J. Parker, Bembix, 546
fumigatum (Perty), Podium, 96
fumipenne (Taschenberg), Penepodium, 92
fumipennis E. Saunders, Astata, 212
fumipennis (F. Smith), Ectemnius, 426
fumipennis (F. Smith), Neoplisus, 505
fumipennis F. Smith, Sphex, 115
fumipennis Gussakovskij, Prosopigastra, 285
fumipennis Say, Cerceris, 581
fumipennis W. Fox, Tachysphex, 272
fumosipennis Strand, Cerceris, 581
fumosus (Brèthes), Ectemnius, 428
fumosus Kohl, Sphex, 114
funatui Tsuneki, Trypoxylon, 346
fundatus Rohwer, Tachytes, 267
funebris (Berland), Prionyx, 133
funebris (Handlirsch), Stizoides, 528
funebris Honoré, Miscophus, 318
funebris (Radoszkowski), Liris, 246
funebris Turner, Bembix, 545
funebris Turner, Sericophorus, 302
funerarius F. Morawitz, Palarus, 291
funerarius Gussakovskij, Sphex, 115
funerea A. Costa, Cerceris, 583
funerea (F. Smith), Liris, 245
funerea (Nurse), Parapsammophila, 139
funereum Gribodo, Chlorion, 89
funereus (Arnold), Dasyproctus, 420
funereus (Giner Marí), Dienoplus, 496
funereus Gussakovskij, Gastrosericus, 256
funestum Kohl, Sceliphron, 106
funestus Arnold, Belomicrus, 363
funestus Kohl, Crabro, 408
funestus Kohl, Sphex, 115
funestus Turner, Daysproctus, 420
furcata Erichson, Bembix, 546
furcata F. Morawitz, Cerceris, 581
furcatus Brèthes, Trachypus, 568
furcatus Lepeletier, Oxybelus, 369
furcifer Turner, Oxybelus, 368
furcifera Schletterer, Cerceris, 581
furuichii (Iwata), Ectemnius, 426
fuscata (Dahlbom), Isodontia, 123
fuscata Riek, Aphelotoma, 70
fuscata Tsuneki, Liris, 245

fuscicauda Empey, Cerceris, 589
fuscatus Bingham, Stizus, 526
fuscatus (Fabricius), Alysson, 458
fuscatus (Wagner), Pemphredon, 181
fuscilabris Mocsáry, Bembix, 547
fuscina Shestakov, Cerceris, 581
fuscinerva Cameron, Larra, 238
fuscinerva (Cameron), Liris, 245
fuscinervis (Cameron), Psen, 166
fuscipalpe Cameron, Pison, 335
fuscipenne F. Smith, Pison, 335
fuscipenne (F. Smith), Chalybion, 102
fuscipenne Fabricius, Trypoxylon, 348
fuscipennis A. Costa, Cerceris, 588
fuscipennis Cameron, Larra, 238
fuscipennis (Cameron), Neoplisus, 505
fuscipennis (Cameron), Pemphredon, 181
fuscipennis (Dahlbom), Psenulus, 40, **172**
fuscipennis (F. Smith), Parapsammophila, 139
fuscipennis (F. Smith), Stizus, 526
fuscipennis (Fabricius), Isodontia, 123
fuscipennis Guérin-Méneville, Philanthus, 565
fuscipennis (Hacker & Cockerell), Sericophorus, 303
fuscipennis (Lepeletier and Brullé), Ectemnius, 426
fuscipennis Lepeletier, Bembix, 546
fuscipennis (Lepeletier), Editha, 543
fuscipennis (Radoszkowski), Mimumesa, 164
fuscipes (Arnold), Ammatomus, 513
fuscipes (Cresson), Zanysson, 475
fuscipes (Packard), Mimumesa, 164
fuscipes Tsuneki, Psenulus, 173
fuscispina Pulawski, Tachysphex, 274
fuscistigma Cameron, Astata, 212
fuscistigma (Cameron), Liris, 245
fuscitarsis (Herrich-Schaeffer), Ectemnius, 425
fuscitarsus (Schenck), Ectemnius, 425
fusciventre Cameron, Trypoxylon, 348
fusciventris Gussakovskij, Solierella, 313
fuscohirtus Gussakovskij, Oxybelus, 368
fuscum Klug, Sceliphron, 105
fuscum (Lepeletier), Chalybion, 103
fuscus Lepeletier, Sphex, 114
fuscus (Taschenberg), Hoplisoides, 521
fuscus W. Fox, Tachysphex, 274

gaditana (Gmelin), Stizus, 527
gaetula Beaumont, Cerceris, 581
gaetula Beaumont, Prosopigastra, 285
gaetulus (Roth), Palmodes, 127
gagates Arnold, Tachysphex, 274
gagates Bradley, Nysson, 469
galactina Dufour, Bembix, 546
galapagensis Rohwer, Tachysphex, 274
galathea Beaumont, Cerceris, 581
galei (Rohwer), Miscophus, 318
galibi Pate, Foxita, 416, 627
galileus Beaumont, Tachysphex, 273
gallica Beaumont, Astata, 212
gallica Mocsáry, Bembix, 548
gallicus Kohl, Miscophus, 318
gallicus Kohl, Tachysphex, 276
gallieni, Arnold, Cerceris, 581
ganahlii (Dalla Torre), Liris, 245
gandarai Rohwer, Cerceris, 581
gandarai Rohwer, Trypoxylon, 348
ganglbaueri Handlirsch, Bembix, 546
ganglebaueri Kohl, Nysson, 469
garamantis (Roth) Palmodes, 127

garambae Leclercq, Psenulus, 173
garciai Fritz, Foxia, 478
garciana Viereck and Cockerell, Cerceris, 578
gardneri Cameron, Trypoxylon, 346
garianensis Andrade, Miscophus, 318
garleppi Schrottky, Cerceris, 581
Gasterosericus Dahlbom, 43
Gastrargyron Pate, 43
gastrica (Taschenberg), Larra, 238
gastrifera (Strand), Liris, 245
Gastrosericus Spinola, 43, **252**
Gastrosphaeria A. Costa, 39
gaucho Richards, Trypoxylon, 349
gaudebunda Holmberg, Cerceris, 581
gaullei Berland, Sphex, 115
gaullei Brèthes, Cerceris, 581
gaumeri Cameron, Ammophila, 152
gayi (Berland), Prionyx, 133
gayi Spinola, Cerceris, 581
gayi (Spinola), Clitemnestra, 51, **489**
gayi Spinola, Podagritus, 393
gayi (Spinola), Tachysphex, 276
gayi (Spinola), Tachytes, 264
gayi (Spinola), Zanysson, 475
gazagnairei (Handlirsch), Bembecinus, 531
gazagnairei (Handlirsch), Hoplisoides, 521
geboharti Tsuneki, Cerceris, 581
gedehensis Leclercq, Ectemnius, 424
gegen Tsuneki, Ammophila, 152
gegen Tsuneki, Ammoplanus, 198
gegen Tsuneki, Bembix, 547
gegen Tsuneki, Cerceris, 588
gegen Tsuneki, Diodontus, 179
gegen Tsuneki, Tachysphex, 274
gegensumus Tsuneki, Miscophus, 318
gelane Evans and Matthews, Bembix, 546
gemblancencis Leclercq, Crossocerus, 401
gemellus (Arnold), Ammatomus, 513
geminatus (A. Costa), Gorytes, 501
geminus Handlirsch, Gorytes, 501
geminus Valkeila, Pemphredon, 181
gemmina Shestakov, Cerceris, 581
genalis Kohl, Philanthus, 565
geneana A. Costa, Bembix, 546
geneana A. Costa, Cerceris, 581
generosa J. Parker, Bembix, 546
genicularis (F. Morawitz), Parapiagetia, 281
geniculata Cameron, Cerceris, 579
geniculata Kazenas, Cerceris, 629
geniculatum Cameron, Trypoxylon, 346, 349
geniculatus Cameron, Diodontus, 179
geniculatus Fabricius, Crabro, 409
geniculatus Lepeletier, Nysson, 470
geniculatus (Olivier), Ectemnius, 424
geniculatus Olivier, Nysson, 470
geniculatus (Shuckard), Crossocerus, 404
geniculatus (Spinola), Tachysphex, 274
gennelli (Rohwer), Pemphredon, 181
geometricus Leclercq, Entomognathus, 382
georgia Pate, Enoplolindenius, 415
georgii Arnold, Tachysphex, 274
geraesae Leclercq, Podagritus, 393
gerardi Leclercq, Crossocerus, 401
germabenis (Radoszkowski), Liris, 246
gerstaeckeri Dewitz, Trachypus, 568
gerstaeckeri Handlirsch, Nysson, 469
gerstaeckeri P. Verhoeff, Oxybelus, 367
gertrudis Krombein, Passaloecus, 184

getricus Leclercq, Podagritus, 393
ghesquierei Leclercq, Psenulus, 173
ghesquieri (Leclercq), Polemistus, 185
ghilianii Spinola, Dryudella, 215
giacomellii (Schrottky), Eremnophila, 147
gibba (Alfieri), Parapsammophila, 139
gibbicollis Giner Marí, Miscophus, 318
gibbosa Kohl, Liris, 245
gibbosa Sickmann, Cerceris, 581
gibbosus (Arnold), Afrogorytes, 523
gibbosus (Dahlbom), Philanthus, 565
gibbosus (Fabricius), Philanthus, 565
gibbosus Rossi, Sphex, 117
gibbus Kohl, Tachysphex, 275
gibbus Pulawski, Tachysphex, 274
gibbus (Villers), Mellinus, 449
giffardi (Rohwer), Ectemnius, 425
giffardi (Rohwer), Pemphredon, 181
giffardi Rohwer, Tachysphex, 276
giffardi Yoshimoto, Ectemnius, 428
gigantea Gistl, Cerceris, 581
gigantea Gussakovskij, Prosopigastra, 285
gigantea (Kohl), Parapsammophila, 140
gigantea Schöberlin, Ammophila, 31
giganteus (Erichson), Dynatus, 94
giganteus Heer, Sphex, 31, 117
gigas Bischoff, Tachytes, 266
gigas Giner Marí, Miscophus, 318
gigas Lichtenstein, Sphex, 119
gigas Taschenberg, Astata, 212
gigas Walsh and Riley, Stizus, 527
gilberti (Turner), Acanthostethus, 474
gilberti Turner, Cerceris, 581
gilberti Turner, Dolichurus, 69
gilesi Turner, Cerceris, 581
gillaspyi Evans and Matthews, Bembix, 546
gillaspyi Krombein, Plenoculus, 311
gillettei Rohwer, Crabro, 408
gillettei Rohwer, Tachysphex, 272
gillettei W. Fox, Diodontus, 179
gilva Gillaspy, Glenostictia, 552
gineri Beaumont, Cerceris, 581
gineri P. Verhoeff, Miscophus, 318
ginjulla Evans and Matthews, Bembix, 546
gisteli Strand, Sphex, 115
glaber Kohl, Philanthus, 565
glaber Kohl, Tachysphex, 274
glabicornis (Arnold), Crossocerus, 401
glabra Kohl, Astata, 212
glabrata (F. Smith), Larra, 238
glabratus Kohl, Stigmus, 189
glabratus R. Bohart, Hoplisoides, 521
glabrellus (Turner), Arpactophilus, 186
glabrior Williams, Tachysphex, 275
glabriusculus Arnold, Tachysphex, 266
glabrum Kohl, Pison, 335
glasunowi F. Morawitz, Oxybelus, 368
glauca Dahlbom, Bembix, 546
glauca Fabricius, Bembix, 546
glenensis H. Smith, Oxybelus, 369
Glenostictia Gillaspy, 54, **551**
globiceps (F. Morawitz), Prosopigastra, 285
globosus (F. Smith), Prionyx, 133
globulosa (Fourcroy), Cerceris, 581
globulosus (Fourcroy), Mellinus, 449
gloriosus Cresson, Philanthus, 565
gnara Cresson, Cerceris, 589
gnarina Banks, Cerceris, 579
gnatho Lichtenstein, Sphex, 119
gnavum (Kohl), Chalybion, 102
globiensis Tsuneki, Bembix, 546
gobiensis Tsuneki, Diodontus, 179

gobiensis Tsuneki, Lindenius, 384
gobiensis Tsuneki, Miscophus, 318
gobiensis Tsuneki, Oxybelus, 368
gobiensis (Tsuneki), Podalonia, 144
gobiensis (Tsuneki), Prionyx, 133
gobiensis Tsuneki, Stizus, 526
godavariensis van Lith, Psenulus, 173
goddardi Cockerell, Cerceris, 581
godeffroyi Saussure, Sphex, 116
godmani (Cameron), Bembecinus, 532
godmani Cameron, Larra, 238
golbachi Leclercq, Anacrabro, 379
gombaki Leclercq, Encopognathus, 380
gomesii Klug, Trachypus, 568
gomezi (Strand), Trachypus, 568
gomphocarpi Brauns, Cerceris, 581
gonager Lepeletier and Brullé, Crossocerus, 400
gondarabai (Guiglia), Tachytes, 265
Gonioxybelus Pate, 46
Gonius Jurine, 44
Gonius Panzer, 44
Gonostigmus Rohwer, 41
goodwini Cockerell, Cerceris, 587
gorgon Kohl, Sphex, 116
gorongozae (Arnold), Rhopalum, 389
goryanum (Lepeletier), Penepodium, 92
Gorystizus Pate, 53
Gorytes Latreille, 52, **500**
GORYTINI, 59, **481**, 483
gorytoides (Handlirsch), Bembecinus, 531
gotlandica Eck, Cerceris, 577
gounellei Richards, Trypoxylon, 346
gowdeyi (Turner), Liris, 245
goyarra Evans and Matthews, Bembix, 546
gracile Taschenberg, Trypoxylon, 346
gracile Wesmael, Rhopalum, 389
gracilens J. Parker, Bembix, 546
gracilescens F. Smith, Trypoxylon, 346
gracilicorne Arnold, Trypoxylon, 346
gracilicornis Arnold, Astata, 212
gracilicornis (Arnold), Liris, 245
gracilicornis Arnold, Tachytes, 265
gracilicornis (Handlirsch), Bembecinus, 531
gracilicornis Mercet, Tachysphex, 274
gracilipes E. Saunders, Diodontus, 179
gracilipes Handlirsch, Stizus, 526
gracilis (Cameron), Trachypus, 568
gracilis (Curtis), Passaloecus, 184
gracilis (Eversmann), Crossocerus, 402
gracilis Giner Marí, Dinetus, 216
gracilis Gussakovskij, Astata, 212
gracilis Handlirsch, Bembix, 546
gracilis J. Parker, Bicyrtes, 538
gracilis Lepeletier, Ammophila, 152
gracilis Kohl, Palarus, 291
gracilis of authors, Passaloecus, 184
gracilis (Patton), Oryttus, 507
gracilis Pulawski, Tachysphex, 276
gracilis Rayment, Sericophorus, 302
gracilis Rohwer, Larropsis, 258
gracilis (Schulthess), Bembecinus, 531
gracilis Williams, Plenoculus, 311
gracilissimus (Packard), Ectemnius, 426
gracilitarsus Morice, Tachysphex, 274
gracillima Taschenberg, Ammophila, 152
gracillimum F. Smith, Trypoxylon, 346
gradilis J. Parker, Bembix, 545
graeca Beaumont, Astata, 212
graecus Kohl, Tachysphex, 274

graecus (Mocsáry), Prionyx, 133
grahami Leclercq, Rhopalum, 389
grahami van Lith, Psen, 166
grana Shestakov, Cerceris, 581
grandidieri (Saussure), Sphecius, 511
grandii Beaumont, Tachysphex, 274
grandii Maidl, Mimesa, 162
grandis Banks, Cerceris, 581
grandis (Chevrier), Larra, 237
grandis Gistel, Ammophila, 152
grandis (Gussakovskij), Argogorytes, 492
grandis (Lepeletier and Brullé), Ectemnius, 426
grandis Lepeletier, Stizus, 527
grandis Radoszkowski, Prionyx, 133
grandis (Say), Sphecius, 511
grandissimus Gussakovskij, Tachysphex, 274
grandissimus (Radoszkowski), Synnevrus, 470
grangeri Beaumont, Miscophus, 318
granti (W. F. Kirby), Prionyx, 134
granulatus (Arnold), Encopognathus, 380
granulatus (Walker), Ectemnius, 425
granulosa (G. and R. Bohart), Ancistromma, 259
granulosa (W. Fox), Mimesa, 162
granulosus Mickel, Tachysphex, 277
graphica F. Smith, Cerceris, 581
grassator Bingham, Crabro, 409
grata Arnold, Cerceris, 581
gratiosa (F. Smith), Microbembex, 539
gratiosa Kohl, Ampulex, 77
gratiosa Schletterer, Cerceris, 580
gratiosissimus Dalla Torre, Sphex, 116
gratiosus Andrade, Miscophus, 318
gratiosus F. Smith, Sphex, 115, 116
gratuitus Andrade, Miscophus, 318
gravida (Handlirsch), Rubrica, 53, **541**
gredleri (Kohl), Chalybion, 102
gredleri (Kohl), Ectemnius, 424
greenei Rohwer, Dolichurus, 69
greenei Rohwer, Larropsis, 259
greeni Turner, Cerceris, 581
greenii (Bingham), Lestiphorus, 506
gregaria (W. Fox), Mimesa, 162
gregarius Scopoli, Sphex, 117
gregoryi (Turner), Brachystegus, 473
grenadense Richards, Trypoxylon, 346
grenadinum (Pate), Rhopalum, 389
gressitti Tsuneki, Psen, 166
grinnelli (Rohwer), Pemphredon, 181
griqua Arnold, Cerceris, 582
griseola (Arnold), Liris, 244
griseolus Arnold, Tachytes, 265
grisescens Dahlbom, Bembix, 546
grossa (Cresson), Podalonia, 145
gryllicida Evans, Liris, 245
grylloctonus Richards, Trigonopsis, 98
gryphus F. Smith, Ammophila, 153
gryphus Harris, Crabro, 408
guadalajarae R. Bohart, Pseudoplisus, 503
guadalupensis Leclercq, Ectemnius, 426
guarani Strand, Cerceris, 578
guaranitica Willink, Isodontia, 123
guatemalensis Cameron, Sphex, 115
guatemalensis Cameron, Tachytes, 265
guatemalensis (Rohwer), Epinysson, 472
guatemalensis (Rohwer), Steniolia, 553
gueinzius (Turner), Carinostigmus, 191
guerini Dahlbom, Ampulex, 77
guerinii Dalla Torre, Ammophila, 150

guerrerensis (Cameron), Crossocerus, 401
guiana (Cameron), Bicyrtes, 538
guiana (Cameron), Eremnophila, 147
guiana Cameron, Larra, 238
guiana (Williams), Nitela, 325
guianensis Rohwer, Microstigmus, 192
guichardi Arnold, Alysson, 458
guichardi Arnold, Tachytes, 264
guichardi Beaumont, Ammophila, 152
guichardi Beaumont, Cerceris, 581
guichardi Leclercq, Crossocerus, 401
guichardi Beaumont, Miscophus, 318
guichardi Beaumont, Oxybelus, 368
guichardi (Beaumont), Prionyx, 133
guichardi Beaumont, Solierella, 313
guichardi (Beaumont), Synnevrus, 470
guigliaae Andrade, Miscophus, 318
guigliae Arnold, Bembix, 546
guigliae Beaumont, Gastrosericus, 256
guigliae Giordani Soika, Cerceris, 581
guigliae Gussakovskij, Miscophus, 319
guignardi Provancher, Alysson, 458
guillarmodi Arnold, Alysson, 458
guillarmodi (Arnold), Carinostigmus, 191
guillarmodi Arnold, Dolichurus, 69
guineensis Ritsema, Ammophila, 151
gujarticus Nurse, Tachysphex, 274
gulielmi (Viereck), Hoplisoides, 521
gulmargensis Nurse, Crabro, 408
gulussa (Morice), Podalonia, 144
gunamarra Evans and Matthews, Bembix, 546
gusenleitneri Beaumont, Bembecinus, 531
gussachaos Menke, Trypoxylon, 346
gussakovskii Kazenas, Eremiasphecium 629
gussakovskii Marshakov, Lindenius, 384
gussakovskii Pulawski, Tachysphex, 274
gussakovskij van Lith, Psenulus, 173
gussakovskiji Andrade, Miscophus, 318
gussakovskiji (Beaumont), Mimesa, 162
gussakovskiji Menke, Solierella, 313
gussakovskiji R. Bohart, Crabro, 408
gussakovskiji Tsuneki, Trypoxylon, 346
gustatum Leclercq, Trypoxylon, 346
guttata Lichtenstein, Bembix, 549
guttata (Taschenberg), Trichostictia, 549
guttatula (Turner), Clitemnestra, 489
guttulatus (Handlirsch), Bembecinus, 531
guttatum Tsuneki, Rhopalum, 389
guttatus (Fabricius), Dinetus, 216
guttatus Gmelin, Sphex, 117
guttatus Olivier, Nysson, 469
guttatus (Vander Linden), Ectemnius, 426
gwaaiensis Arnold, Philanthus, 565
gynandromorphus (Handlirsch), Bembecinus, 531
gynochroma Mochi, Ceraceris, 587
gyponacinus Rohwer, Gorytes, 501
gyponae (Williams), Dienoplus, 496

habenus Say, Sphex, 115
haberhaueri (Radoszkowski), Prionyx, 133
habilis Turner, Tachytes, 265
hackeri (Cockerell), Sericophorus, 303
hackeri Rayment, Sericophorus, 302
hackeriana Cockerell, Ceraceris, 581
hades Schrottky, Tachytes, 265
hadrus R. Bohart, Pseudoplisus, 503
haematogastrum (Spinola), Penepodium, 92
haemitina Kohl, Ceraceris, 581

haemodes (Kohl), Lindenius, 384
haemorrhoidalis (A. Costa), Psenulus, 173
haemorrhoidalis (Berland), Psenulus, 172
haemorrhoidalis F. Smith, Trigonopsis, 98
haemorrhoidalis (Fabricius), Liris, 245
haemorrhoidalis Fabricius, Sphex, 115
haemorrhoidalis (Handlirsch), Bembecinus, 531
haemorrhoidalis Magretti, Sphex, 116
haemorrhoidalis of authors, Sphex, 117
haemorrhoidalis Olivier, Oxybelus, 368
haemorrhous A. Costa, Dolichurus, 69
haigi (R. Bohart), Clypeadon, 571
haimatosoma Kohl, Ammophila, 152
hakodatense Tsuneki, Rhopalum, 389
hakusanicus Tsuneki, Passaloecus, 184
hakusanus Tsuneki, Crossocerus, 401
hakusanus Tsuneki, Psen, 166
hakutozanus Tsuneki, Gorytes, 501
haleakalae Perkins, Deinomimesa, 169
haleakalae (Perkins), Ectemnius, 426
halictiformis Arnold, Tachysphex, 274
halone Banks, Cerceris, 581
halophilus Arnold, Tachysphex, 275
hamata C. Fox, Bembix, 545
hamatus Arnold, Stizus, 526
hamatus (Handlirsch), Hoplisoides, 521
hamiger (Kohl), Lindenius, 384
hamilcar (Kohl), Lindenius, 384
hamiltoni Arnold, Cerceris, 581
hamiltoni Turner, Tachytes, 265
handlirschella Ferton, Bembix, 546
handlirschi Brauns, Palarus, 291
handlirschi Cameron, Bembix, 546
handlirschi (Cameron), Pseudoplisus, 503
handlirschi Ferton, Bembix, 546
handlirschi Gussakovskij, Ammoplanus, 198
handlirschi Hellén, Nysson, 469
handlirschi Mercet, Bembix, 547
handlirschi Morice, Prosopigastra, 285
handlirschi Radoszkowski, Stizus, 526
handlirschi Rohwer, Mellinus, 33, 449
handlirschi (Schmiedeknecht), Synnevrus, 470
Handlirschia Kohl, 52, **508**
handlirschii Brauns, Miscophoides, 45, **322**
handlirschii (F. Morawitz), Ammatomus, 513
handlirschii Kohl, Diodontus, 179
handlirschii Kohl, Miscophus, 318
hanedai Tsuneki, Rhopalum, 389
hanedai Tsuneki, Liris, 245
hanedai (Tsuneki), Odontopsen, 163
hanedai Tsuneki, Pison, 335
hannibal (Kohl), Lindenius, 384
hannibalis Gribodo, Trypoxylon, 345
hannonis (Gribodo), Ectemnius, 425
Hapalomellinus Ashmead, 52, **496**
haplocerus (Handlirsch), Bembecinus, 531
Haplognatha Gussakovskij, 44
haramaiae Arnold, Cerceris, 582
harbecki Rohwer, Diodontus, 180
harbinensis Tsuneki, Alysson, 458
harbinensis Tsuneki, Cerceris, 582
harbinensis Tsuneki, Gorytes, 501
harbinensis Tsuneki, Lindenius, 384
harenarum Arnold, Bembix, 546
harmandi Pérez, Cerceris, 582
harmandi (Pérez), Isodontia, 123
Harpactes Dahlbom, 52

harpactoides (F. Smith), Lyroda, 299
Harpactophilus Kohl, 41
Harpactopus F. Smith, 39
Harpactostigma Ashmead, 52
Harpactus Shuckard, 52
harpax Arnold, Tachysphex, 274
harpax (Kohl), Prionyx, 134
harpax Leclercq, Rhopalum, 389
harpax Patton, Tachytes, 265
harraricus Arnold, Oxybelus, 368
harringtonii (Provancher), Hoplisoides, 521
harringtonii (W. Fox), Crossocerus, 401
harrisi (Fernald), Isodontia, 123
harrisii (Packard), Crossocerus, 401
harti (Fernald), Ammophila, 152
hartliebi Schulz, Cerceris, 587
harudus (Leclercq), Carinostigmus, 191
harveyi (Beaumont), Podalonia, 144
harveyi (Beaumont), Synnevrus, 470
hasdrubal Beaumont, Lindenius, 384
hastatifrons Turner, Palarus, 291
hastatus Fabricius, Oxybelus, 368
hatogayuum Tsuneki, Trypoxylon, 347
hatuey Alayo, Cerceris, 582
haueri Schletterer, Cerceris, 580
hausa Arnold, Cerceris, 582
hawaiiensis Perkins, Deinomimesa, 169
hawaiiensis (Perkins), Ectemnius, 426
hawaiiensis Perkins, Nesomimesa, 40, **167**
hebes Cameron, Cerceris, 581
hebetescens (Turner), Ectemnius, 426
hebraea Beaumont, Cerceris, 588
hebraeus Andrade, Miscophus, 318
hebraeus Balthasar, Stizus, 526
hebraeus Beaumont, Bembecinus, 531
hebraeus Beaumont, Gorytes, 501
hector (Cameron), Ectemnius, 427
hedgreni (Kjellander), Crossocerus, 401
hedickei Giner Marí, Bembix, 546
heidenii (Saussure), Larra, 238
heinii (Kohl), Chalybion, 102
heinrichi van Lith, Psen, 166
helena Beaumont, Olgia, 491
helianthi (Rohwer), Hoplisoides, 521
helianthi Rohwer, Tachysphex, 274
helianthopolis J. Parker, Bembix, 549
HELIOCAUSINI, 59, **449**
Heliocausus Kohl, 51, **451**
heliophilus Nurse, Tachysphex, 276
heliophilus Pulawski, Miscophus, 318
heliopolites Morice, Tachysphex, 272, 273
Helioryctes F. Smith, 44
helioryctoides (Turner), Paranysson, 308
helleri (Kohl), Lindenius, 384
helleri (Kohl), Solierella, 313
hellmanni (Eversmann), Philanthus, 565
hellmayri Schulz, Ampulex, 77
helmsii (Cameron), Tachysphex, 275
helvetica Tournier, Mimesa, 162
helveticus Kohl, Miscophus, 318
helveticus Kohl, Tachysphex, 274
Hemichalybion Kohl, 39, **101**
hemicyclius van Lith, Psenulus, 173
Hemidula Burmeister, 53, **539**
hemilauta Kohl, Ammophila, 152
hemiprasinum (Sichel), Chlorion, 89
hemipterum (Fabricius), Sceliphron, 105
hemipyrrhum (Sichel), Chlorion, 89
Hemithyreopus Pate, 49
hemixanthopterus Morice, Sphecius, 511

hengchunensis Tsuneki, Tachytes, 265
henrici Krombein, Crabro, 408
henrici Turner, Nitela, 325
henricus Dunning, Philanthus, 564
hentyi (Rayment), Acanthostethus, 474
heptapotamiensis Kazenas, Gorytes, 501
heraclei (Rohwer), Ectemnius, 426
herbsti (Arnold), Bembecinus, 531
herbsti Empey, Cerceris, 582
herbsti R. Bohart, Neonysson, 471
herbstii (Kohl), Rhopalum, 389
herbstii Kohl, Tachysphex, 277
herero (Arnold), Parapsammophila, 139
herinaceus (Gribodo), Tracheliodes, 405
hermia Arnold, Tachysphex, 274
hermosa Menke, Ammophila, 152
hermosus Banks, Philanthus, 566
Heroecus C. Verhoeff, 41
heros (Fabricius), Stictia, 542
heros (Kohl), Lestica, 431
herrerai (Brèthes), Prionyx, 133
hesione Bingham, Bembix, 546
hespera Scullen, Eucerceris, 591
hesperina Banks, Cerceris, 579
hesperus Banks, Tachytes, 266
hesperus Bohart and Menke, Palmodes, 127
hesperus (Pate), Psenulus, 172
hesperus R. Bohart, Nysson, 469
hessei (Arnold), Encopognathus, 380
hessei Arnold, Oxybelus, 368
heteracantha Gussakovskij, Bembix, 546
heterocera Beaumont, Solierella, 313
heterocerum (Mantero), Rhopalum, 389
heterochromus Beaumont, Tachysphex, 274
heterolepis Cockerell and Baker, Oxybelus, 369
heterospila Cameron, Cerceris, 582
hewitti (Cameron), Isodontia, 123
hewitti (Cameron), Piyumoides, 411
hewitti (Cameron), Crossocerus, 401
hexadonta Strand, Cerceris, 582
hexagonalis W. Fox, Stigmus, 189
hexaspila J. Parker, Bembix, 546
heydeni Dahlbom, Ammophila, 152
heydeni Kohl, Crossocerus, 401
heydenii Saussure, Larra, 238
hicksi Sandhouse, Tracheliodes, 405
hidalgo Scullen, Cerceris, 578
hilaris F. Smith, Cerceris, 582
hilaris F. Smith, Crabro, 408
hilbrandi van der Vecht, Cerceris, 582
hildebrandti Saussure, Cerceris, 582
hillorum Leclercq, Rhopalum, 389
hilota Leclercq, Enoplolindenius, 415
himalayensis Bingham, Cerceris, 582
himalayensis Cameron, Ampulex, 77
himalayensis Cameron, Cerceris, 584
himalayensis (Cameron), Ectemnius, 428
himalayensis van Lith, Psen, 166
hinei J. Parker, Bembix, 546
hingstoni Leclercq, Crossocerus, 401
hingstoni Richards, Microstigmus, 192
Hingstoniola Turner and Waterston, 49, **417**
hippolyta Arnold, Tachysphex, 274
hirashimai Tsuneki, Crossocerus, 401
hirashimai Tsuneki, Psen, 166
hirsuta J. Parker, Microbembex, 539
hirsuta (Scopoli), Podalonia, 144
hirsutaffinis (Tsuneki), Podalonia, 144

INDEX 659

hirsuitfrons Banks, Tachytes, 264
hirsutula Gussakovskij, Astata, 213
hirsutus Baker, Oxybelus, 369
hirsutus F. Smith, Tachytes, 265
hirsutus Saussure, Sphex, 114
hirsutus (Sonan), Bembecinus, 531
hirta (Handlirsch), Ochleroptera, 490
hirticeps Cameron, Philanthus, 566
hirticeps (Cameron), Podalonia, 144
hirticeps (Cameron), Trachypus, 568
hirticeps (F. Morawitz), Podalonia, 145
hirticulus (Mickel), Philanthus, 564
hirtipes (A. Morawitz), Crossocerus, 400
hirtipes (Fabricius), Prionyx, 133
hirtipes Fabricius, Sphex, 115
hirtitibia (Arnold), Crossocerus, 401
hirtiusculus (Arnold), Bembecinus, 531
hirtiventris Morice, Cerceris, 588
hirtulus (F. Smith), Bembecinus, 531
hirtum (Kohl), Chlorion, 90
hispanica (Gmelin), Cerceris, 588
hispanica Mercet, Bembix, 547
hispanica Mocsáry, Ammophila, 152
hispanica Radoszkowski, Cerceris, 576
hispanicum Richards, Trypoxylon, 349
hispanicus Ceballos, Miscophus, 318
hispanicus Giner Marí, Oxybelus, 370
hispanicus (Kohl), Ectemnius, 426
hispanicus (Mercet), Argogorytes, 492
hispanicus Mocsáry, Stizus, 526
hispidus (F. Morawitz), Prionyx, 134
hispidus W. Fox, Aphilanthops, 570
hispidus W. Fox, Crabro, 408
hissarica (Gussakovskij), Mimesa, 162
hissaricum Gussakovskij, Pison, 336
hissaricus Gussakovskij, Ammoplanus, 198
hissaricus (Gussakovskij), Dienoplus, 496
hissaricus Gussakovskij, Miscophus, 318
histerisnica (Spinola), Cerceris, 582
histrio Dahlbom, Cerceris, 581
histrio Dahlbom, Palarus, 291
histrio F. Morawitz, Stizus, 526
histrio Fabricius, Philanthus, 565
histrio Lepeletier, Astata, 212
histrio (Lepeletier), Palarus, 290
histrio (Lepeletier), Sceliphron, 105
histrio (Saussure), Dienoplus, 496
histrio Spinola, Palarus, 291
histrionica Klug, Cerceris, 582
histronicus Balthasar, Miscophus, 318
hitei Rohwer, Tachysphex, 277
hiurai Tsuneki, Crossocerus, 401
hobartia Turner, Spilomena, 193
hoevaldensis Empey, Cerceris, 581
hofferi Snoflak, Ammoplanus, 198
Hogardia Lepeletier, 53
hogardii (Latreille), Sphecius, 511
hohlbecki Shestakov, Cerceris, 582
hokana Evans and Matthews, Bembix, 546
hokkaidense Tsuneki, Rhopalum, 389
hokkaidensis Tsuneki, Crossocerus, 400
hokkanza Tsuneki, Cerceris, 582
hokusenensis Tsuneki, Crossocerus 629
holconota Cameron, Cerceris, 582
holconotula Brauns, Cerceris, 583
Holcorhopalum Cameron, 47, **385**
holmbergi Brèthes, Cerceris, 584
holmbergii Dalla Torre, Tachytes, 263
Hologambrus Morice, 44
holognathus (Morice), Holotachysphex, 44, **282**

holoneurum Schrottky, Trypoxylon, 349
holoni Bytinski-Salz, Bembix, 546
holosericea (Fabricius), Ammophila, 152
Holotachysphex Beaumont, 44, **282**
Holotachytes Turner, 43
holtensis Leclercq, Crossocerus, 400
holtmanni van Lith, Psenulus, 173
hombceanum Tsuneki, Rhopalum, 389
Homogambrus Kohl, 44
homogenea (Mercet), Podalonia, 145
homonymus (Schulz), Hoplisoides, 521
honesta Kohl, Ampulex, 77
honestus (Cresson), Ectemnius, 424
honorei Alfieri, Ammophila, 152
honorei Balthasar, Miscophus, 318
honorei Mochi, Cerceris, 582
hoozanius van Lith, Psenulus, 173
Hoplammophila Beaumont, 39, **140**
Hoplisidea Cockerell, 33
hoplisivora (Rohwer), Epinysson, 472
Hoplisoides Gribodo, 53, **517**
Hoplisus Lepeletier, 52
hoplites (Handlirsch), Bembecinus, 531
Hoplocrabro Thomson, 48, **397**
Hoplocrabron De Stefani, 42
hora Arnold, Cerceris, 582
horacioi Fritz, Metanysson, 629
horni Schulthess, Ammophila, 152
horni (Strand), Chalybion, 103
horni Strand, Nysson, 469
hortensis (Poda), Ammophila, 153
hortivaga Kohl, Cerceris, 582
hortivagans (Strand), Penepodium, 92
hortorum (Panzer), Cerceris, 586
horus Beaumont, Tachysphex, 274
horvatovichi Tsuneki, Ectemnius, 629
hospes Bingham, Tachytes, 264
hospes F. Smith, Ampulex, 77
hospes F. Smith, Pison, 336
hospitalis Leclercq, Crossocerus, 401
hostilis Kohl, Tachysphex, 274
houskai (Balthasar), Bembecinus, 532
houskai Balthasar, Bembix, 546
houskai (Balthasar), Oryttus, 508
hova Arnold, Astata, 213
hova Arnold, Liris, 247
hova Saussure, Bembix, 546
hova Saussure, Trypoxylon, 346
Hovanysson Arnold, 51, **472**
howardi (Ashmead), Dienoplus, 496
hrubanti Balthasar, Nysson, 469
huachuca Banks, Cerceris, 576
Huachuca Pate, 51
hualga Pate, Timberlakena, 201
huastecae Saussure, Cerceris, 582
Huavea Pate, 47, **395**
hubbardi Rohwer, Stigmus, 189
hugeli Handlirsch, Stizus, 526
humahuaca Pate, Enoplolindenius, 415
humbertana Saussure, Ammophila, 151
humbertiana Saussure, Cerceris, 582
humeralis Dufour, Palarus, 291
humile Arnold, Pison, 336
humilicollis Beaumont, Lindenius, 384
humilis (Cockerell), Sericophorus, 303
humilis Handlirsch, Nysson, 469
hunchuz Shestakov, Cerceris, 588
hungarica Mocsáry, Ammophila, 152
hungarica Pulawski, Astata, 213
hungaricus (Frivaldsky), Bembecinus, 531
hurdi (G. and R. Bohart), Ancistromma, 259

hurdi Menke, Ammophila, 152
hurdi R. Bohart and Schlinger, Oxybelus, 368
hurdi R. Bohart, Tachysphex, 274
hurdi Scullen, Cerceris, 582
hurdi Williams, Plenoculus, 311
hyalinus (Shuckard), Crossocerus, 401
hyalipennis Handlirsch, Stizus, 526
hyalipennis Kohl, Diodontus, 179
hyalipennis (Kohl), Prionyx, 134
hybridus (Eversmann), Ectemnius, 425
hybridus (Leclercq), Carinostigmus, 191
hylaeioides (Rayment), Clitemnestra, 489
Hylocrabro Perkins, 50
Hyloliris Williams, 43
hypenetes (Handlirsch), Hoplisoides, 521
hyperocrus (Arnold), Bembecinus, 531
hyperorientale Strand, Trypoxylon, 347
Hypocrabro Ashmead, 49
hypocritica Brauns, Cerceris, 582
hypoleius (F. Smith), Tachysphex, 274
Hypomellinus Ashmead, 52
Hypomiscophus Cockerell, 45
Hyponysson Cresson, 51, **467**
hypotheticus Kokujev, Crabro, 408
Hypothyreus Ashmead, 50
hypsae (De Stefani), Ectemnius, 426

iberica Ed. André, Ammophila, 152
iberica Schletterer, Cerceris, 577, 582
ibericus Beaumont, Philanthus, 564
ibericus Beaumont, Stizus, 527
ibericus Dusmet and Alonso, Mellinus, 449
ibericus Handlirsch, Nysson, 469
ibericus (Kohl), Lindenius, 384
ibericus (Mercet), Hoplisoides, 521
ibericus (Roth), Palmodes, 127
ibericus (Saussure), Tachysphex, 275
ibex Kohl, Lindenius, 384
icariiformis (Bingham), Lestiphorus, 706
icarioides (Turner), Ammatomus, 513
icarioides Turner, Pison, 337
ichneumoneus (Linnaeus), Sphex, 115
ichneumoniforme (Arnold), Rhopalum, 389
ichneumoniformis Fabricius, Larra, 237
ichneumonoides Arnold, Miscophus, 318
icta Shestakov, Cerceris, 582
Icuma Cameron, 53
idahoensis Scullen, Cerceris, 576
Idionysson Pate, 51, **480**
idoneus (Turner), Notocrabro, 394
idrieus (Cameron), Dasyproctus, 420
idzekii Tsuneki, Tachysphex, 274
ignaruris Kohl, Cerceris, 583
ignavum Turner, Pison, 336
ignavus (Arnold), Entomognathus, 382
ignipennis (F. Smith), Liris, 246
ignitus F. Smith, Dolichurus, 69
ignotus Strand, Sphex, 115
iheringii (Handlirsch), Hoplisoides, 521
iheringii Kohl, Sphex, 116
iliensis Kazenas, Cerceris, 582
illebefactus Pulawski, Tachytes, 265
illota Banks, Cerceris, 578
illudens (Lepeletier), Liris, 245
illustris Arnold, Cerceris, 582
illustris Cresson, Sphex, 115
illustris (Mickel), Philanthus, 567
imayoshii Yasumatsu, Trypoxylon, 346
imbelle (Turner), Podagritus, 393

imbellis Turner, Tachysphex, 274
imbutus (W. Fox), Ectemnius, 428
imerinae Saussure, Ammophila, 151
imitans Giner Marí, Miscophus, 318
imitans (Kohl), Crossocerus, 401
imitator Cresson, Cerceris, 579
imitator F. Smith, Cerceris, 582
imitator (Handlirsch), Neoplisus, 505
imitator Menke, Ammophila, 152
imitatoria Schletterer, Cerceris, 579
imitatrix Schulz, Cerceris, 582
immacula Kohl, Vespa, 627
immaculata Gmelin, Vespa, 627
immaculatus Guiglia, Oxybelus, 368
immaculatus Krombein, Dasyproctus, 421
immanis (Saussure), Dasyproctus, 420
immigrans (Williams), Dryudella, 214
immitis (Saussure), Dasyproctus, 420
immitis Saussure, Philanthus, 564
immolator F. Smith, Cerceris, 585
immolatrix Schulz, Cerceris, 585
impatiens F. Smith, Ammophila, 152
impatiens F. Smith, Oxybelus, 370
impatiens Kohl, Philanthus, 565
imperator Nurse, Stizus, 526
impercepta Beaumont, Cerceris, 582
imperfectus Beaumont, Tachysphex, 273
imperialensis (Bradley), Hoplisoides, 520
imperialis (Handlirsch), Bembecinus, 531
imperialis Gerstaecker, Oxybelus, 368
imperialis Handlirsch, Stizus, 526
imperialis Kohl, Sphex, 117
imperialis R. Bohart, Mellinus, 449
imperialis R. Bohart, Pseudoplisus, 503
imperialis Saussure, Cerceris, 582
imperialis Saussure, Tachytes, 265
impetuosus (Cameron), Dasyproctus, 420
impiger (Bingham), Hoplisoides, 521
impressifrons Arnold, Larra, 238
impressifrons (F. Smith), Crossocerus, 401
impressifrons (Malloch), Mimesa, 162
impressus (F. Smith), Ectemnius, 425, 427
impressus Kokujev, Sphecius, 511
impudens Andrade, Miscophus, 318
impudens (Nurse), Dienoplus, 496
impunctatum Turner, Pison, 336
imsganensis (Nadig), Hoplisoides, 521
inaequale Turner, Pison, 336
inara Beaumont, Cerceris, 582
inca Leclercq, Quexua, 387
incana (Dahlbom), Podalonia, 144
incanus Beaumont, Tachysphex, 274
incavus (W. Fox), Crossocerus, 402
incerta Arnold, Liris, 246
incertus Radoszkowski, Alysson, 458
incertus (Radoszkowski), Brachystegus, 473
incertus Radoszkowski, Palarus, 291
incertus (Radoszkowski), Tachysphex, 274
incertus Rasnitsyn, Cretosphex, 629
incertus Spinola, Trachypus, 568
incertus W. Fox, Crabro, 408
incisura (Mickel), Oxybelus, 369
incognita Eck, Cerceris, 577
incognita J. Parker, Bembix, 546
incognitum Brèthes, Trypoxylon, 349
incognitus Pulawski, Tachytes, 266
incomptus Gerstaecker, Oxybelus, 370
incomptus Gerstaecker, Sphex, 115
inconspicua Arnold, Cerceris, 582
inconspicuum Turner, Pison, 336
inconspicuus Andrade, Miscophus, 318
inconspicuus (Ducke), Epinysson, 472

inconspicuus (W. F. Kirby), Tachysphex, 274
inconstans Arnold, Trypoxylon, 346
incripta Dahlbom, Bembix, 546
inda (Linnaeus), Prionyx, 133
indica Arnold, Liris, 247
indica (Cameron), Liris, 246
indica (Dalla Torre), Parapsammophila, 139
indica Handlirsch, Bembix, 546
indica (Thunberg), Cerceris, 582
indicum Menke, Trypoxylon, 346
indicus Dalla Torre, Tachytes, 265
indicus (Gmelin), Palarus, 291
indicus Nurse, Palarus, 291
indicus (Saussure), Dasyproctus, 420
indicus van Lith, Psen, 166
indifferens Arnold, Tachytes, 265
indonesiae Leclercq, Crossocerus, 401
indonesica Leclercq, Lestica, 431
indostana (Linnaeus), Prionyx, 133
indostani Turner, Spilomena, 193
indulgens Cheesman, Tachytes, 265
induta Kohl, Ammophila, 152
inepta Cresson, Ammophila, 154
inermis Arnold, Tachysphex, 275
inermis (Cresson), Solierella, 313
inermis (Handlirsch), Bembecinus, 531
inermis (Handlirsch), Glenostictia, 552
inermis Leclercq, Paranysson, 308
inermis (Thomson), Crossocerus, 402
inexorabilis Turner, Cerceris, 582
inexorabilis Turner, Tachytes, 265
inexpectata Turner, Cerceris, 582
inexspectatus Beaumont, Nippononysson, 467
inextracabilis Pulawski, Tachysphex, 274
infantulus (Kohl), Dasyproctus, 420
inferioris Banks, Tachytes, 266
inferna Heer, Ammophila, 31, 154
infernalis Arnold, Miscophus, 318
infernalis (Handlirsch), Oryttus, 508
infesta F. Smith, Ammophila, 153
infimum Arnold, Trypoxylon, 346
inflata (van Lith), Mimesa, 162
inflexum (Sickmann), Chalybion, 103
infracta J. Parker, Stictia, 542
infranensis (Nadig), Dienoplus, 496
infrarugosus (Arnold), Dasyproctus, 421
infumata Handlirsch, Bembix, 546
infumata Maidl, Cerceris, 582
infumatum Turner, Pison, 336
infumatus Mickel, Pseudoplisus, 503
infundibuliformis (Fourcroy), Cerceris, 586
infundibuliformis (Fourcroy), Mellinus, 449
infuscatus (Klug), Stizus, 526
ingens (F. Smith), Dynatus, 94
ingens F. Smith, Sphex, 115
ingricus (F. Morawitz), Crabro, 408
inimica Beaumont, Bembix, 546
inimicus (M. Harris), Argogorytes, 492
iniqua Kohl, Cerceris, 582
innocens Beaumont, Bembecinus, 531
innominatus Bingham, Philanthus, 565
inopides Strand, Bembix, 545
inopinata Beaumont, Liris, 246
inops Handlirsch, Bembix, 545
inordinatus W. Fox, Stigmus, 189
inornatum Matsumura, Trypoxylon, 346
inornatus Beaumont, Nysson, 469

inornatus (Matsumura), Crossocerus, 400
inornatus Mocsáry, Crabro, 408
inornatus Rayment, Sericophorus, 302
inornatus (Robertson), Oxybelus, 368
inornatus Say, Pemphredon, 181
insidiatrix (Handlirsch), Bicyrtes, 537
insidiosa Beaumont, Solierella, 313
insidiosus Spooner, Diodontus, 179
insignatus Banks, Philanthus, 565
insigne Sickmann, Pison, 336
insignis E. Saunders, Prospigastra, 285
insignis F. Smith, Ammophila, 152
insignis (F. Smith), Ectemnius, 426
insignis Handlirsch, Bembix, 545
insignis Klug, Cerceris, 588
insignis (Kohl), Prionyx, 133
insignis of authors, Passaloecus, 184
insignis Provancher, Eucerceris, 592
insignis (Vander Linden), Passaloecus, 41, **184**
insignita Arnold, Cerceris, 582
insobricus A. Costa, Miscophus, 318
insolens (W. Fox), Crossocerus, 401
insolita Cresson, Cerceris, 582
insolita F. Smith, Ammophila, 152
insolitum W. Fox, Trypoxylon, 346
insolitus Andrade, Miscophus, 318
insolitus (W. Fox), Arigorytes, 517
instabile (F. Smith), Chlorion, 90
instabilis F. Smith, Ammophila, 152
instabilis F. Smith, Cerceris, 582
instabilis (F. Smith), Isodontia, 123
instabilis Turner, Tachytes, 265
instructus Nurse, Tachysphex, 274
insulana Lohrmann, Stictia, 542
insulanus Beaumont, Bembecinus, 532
insulare F. Smith, Pison, 336
insularis Arnold, Cerceris, 584
insularis Bohart and Menke, Palmodes, 127
insularis (Cameron), Isodontia, 123
insularis (Cameron), Liris, 248
insularis Cresson, Astata, 213
insularis (Cresson), Hoplisoides, 521
insularis Cresson, Tachytes, 264
insularis (Dahlbom), Stictia, 542
insularis F. Smith, Ampulex, 78
insularis F. Smith, Cerceris, 581
insularis Giner Marí, Ammoplanus, 198
insularis (Handlirsch), Bembecinus, 531
insularis Kohl, Oxybelus, 368
insularis (Saussure), Liris, 246
insulicola Balthasar, Miscophus, 318
insulicola Tsuneki, Cerceris, 589
insulicola Tsuneki, Ectemnius, 427
insulsus Arnold, Tachysphex, 274
insulum Arnold, Trypoxylon, 347
intaminata (Turner), Dicranorhina, 251
integer (Beaumont), Philanthinus, 55, **570**
integer (Fabricius), Stizus, 526
integer Gussakovskij, Tachytes, 265
integer (Morice), Holotachysphex, 282
integra (Burmeister), Editha, 543
integra F. Morawitz, Cerceris, 582
integra (Kohl), Parapiagetia, 281
integrus (Arnold), Sphex, 116
intercedens Handlirsch, Gorytes, 501
intercepta Lepeletier, Ammophila, 152
interjecta Banks, Cerceris, 578
intermedia (Arnold), Liris, 247
intermedia (Cameron), Liris, 246

intermedia Dahlbom, Bembix, 548
intermedia (Dahlbom), Dryudella, 215
intermedia Murray, Podalonia, 144
intermedia Saussure, Trigonopsis, 98
intermedia Tsuneki, Liris, 246
intermedia Williams, Liris, 246
intermedius A. Morawitz, Ectemnius, 427
intermedius Baker, Oxybelus, 369
intermedius Handlirsch, Sphecius, 511
intermedius Tsuneki, Pemphredon, 182
intermedius Tsuneki, Psen, 166
intermedius (Schenck), Psenulus, 172
intermedius (Viereck), Synnevrus, 470
intermedius (Viereck), Tachytes, 265
intermissum Kohl, Podium, 96
interocularis G. and R. Bohart, Larropsis, 259
interrupta Klug, Cerceris, 577
interrupta (Lepeletier and Brullé), Lestica, 431
interrupta Matsumura, Cerceris, 582
interrupta (Panzer), Cerceris, 582
interruptefasciatus (Retzius), Ectemnius, 424
interruptulus Dalla Torre, Crabro, 407
interruptum (Palisot de Beauvois), Sceliphron, 106
interruptus Brèthes, Oxybelus, 368
interruptus Cresson, Oxybelus, 370
interruptus (Dahlbom), Ectemnius, 425
interruptus (Fabricius), Nysson, 469
interruptus (Fabricius), Palarus, 291
interruptus (Latreille), Brachystegus, 473
interruptus (Lepeletier and Brullé), Crabro, 407
interstincta (Fabricius), Cerceris, 582
interstinctus (F. Smith), Ectemnius, 426
interstitialis Cameron, Alysson, 458
interstitialis Cameron, Ampulex, 78
interstitialis (Cameron), Dryudella, 215
interstitialis (Cameron), Mimumesa, 164
interstitialis Cameron, Psenulus, 173
interstitialis Cameron, Tachytes, 265
intimella Cameron, Cerceris, 584
intractibilis Mickel, Cerceris, 582
intricans Gribodo, Hoplisoides, 53, **521**
intricata F. Smith, Cerceris, 582
intricatus (F. Smith), Crossocerus, 404
intrudens (F. Smith), Sceliphron, 105
intrudens F. Smith, Trypoxylon, 346
intrudens Nurse, Gorytes, 501
inusitatus W. Fox, Tachysphex, 275
inusitatus Yasumatsu, Sphex, 115
invalida Kohl, Cerceris, 582
invalidus Leclercq, Ectemnius, 426
inventus Nurse, Tachysphex, 273
inversus Patton, Philanthus, 565
invita Turner, Cerceris, 582
inyoensis Court and R. Bohart, Lindenius, 384
irene Banks, Cerceris, 582
iresinides Rohwer, Cerceris, 583
iresinides (Rohwer), Solierella, 313
iridescens Kohl, Podium, 96
iridescens Turner, Rhopalum, 389
iridescens Turner, Spilomena, 193
iridifrons (Pérez), Ectemnius, 426
iridipenne Cameron, Pison, 335
iridipenne F. Smith, Pison, 336
iridipennis (Cameron), Eremnophila, 147
iridipennis (Cameron), Liris, 246
iridipennis (F. Smith), Hoplisoides, 521

iridipennis (F. Smith), Tachysphex, 274
iridipennis (Maidl), Liris, 246
iridipennis (Turner), Sericophorus, 303
iridis Dow, Stizus, 526
iriomotoensis Tsuneki, Liris, 246
irrequietus (Kohl), Lindenius, 384
irreverens Cheessman, Tachytes, 265
irritabilis Turner, Tachytes, 265
irritata Nurse, Bembix, 546
irrorata (F. Smith), Liris, 246
irwini F. Parker, Diploplectron, 211
irwini R. Bohart and Grissell, Pseneo, 165
irwini Scullen, Cerceris, 582
isabellae Beaumont, Bembix, 546
Ischnolynthus Holmberg, 47
ishigakiensis Tsuneki, Ectemnius, 428
isigakiense Tsuneki, Trypoxylon, 346
isis Arnold, Cerceris, 582
isis Beaumont, Tachysphex, 274
Iskutana Pate, **49**, **415**
Isodontia Patton, **39**, **119**
isolatus (Turner), Encopognathus, 380
isolde Banks, Cerceris, 578
isolitum Turner, Pison, 336
isselii (Gribodo), Prionyx, 133
istam Pate, Belomicrus, 363
italicus A. Costa, Belomicrus, 46, **363**
italicus A. Costa, Miscophus, 318
italicus Beaumont, Crossocerus, 401
ithacae Krombein, Passaloecus, 184
ituriensis Benoit, Liris, 246
iwatai Gussakovskij, Psenulus, 172
iwatai (Tsuneki), Carinostigmus, 191
iwatai Yasumatsu, Pison, 336
iwatai (Yasumatsu), Piyuma, 410
Iwataia Tsuneki, **50**

jackal Brauns, Cerceris, 582
jacksoni (Arnold), Bembecinus, 531
jacobsoni (Gussakovskij), Mimesa, 162
jacobsoni (Kohl), Dasyproctus, 421
jacobsoni Maidl, Spilomena, 193
jacobsoni Shestakov, Cerceris, 587
jacoti van Lith, Psenulus, 629
jaculator (F. Smith), Dryudella, 215
jaculator (F. Smith), Isodontia, 123
jaculator (F. Smith), Liris, 246
jaffueli (Herbst), Solierella, 313
jakovlevi Gussakovskij, Cerceris, 582
jakowleffii Kohl, Cerceris, 582
jakowlewi (F. Morawitz), Ectemnius, 428
jalapensis R. Bohart and Grissell, Psenulus, 173
jamaica Christ, Sphex, 115
jamaica Pate, Ochleroptera, 490
jamaicense (Fabricius), Sceliphron, 105
jamacensis (Drury), Sphex, 115
jansei Cameron, Ampulex, 78
jansei Cameron, Cerceris, 580
jansei Cameron, Sphex, 115
jansoni Turner, Aulacophilus, 338
japonense Tsuneki, Trypoxylon, 347
japonensis Tsuneki, Dienoplus, 496
japonica Ashmead, Cerceris, 582
japonica (Cameron), Liris, 246
japonica Iwata, Lyroda, 299
japonica Kohl, Ammophila, 151
japonica Kohl, Ampulex, 77
japonica (Kohl), Liris, 245
japonica Pérez, Mimesa, 162
japonica (Schulz), Lestica, 430

japonica Tsuneki, Spilomena, 193
japonicum (Gribodo), Chalybion, 102
janponicum Gussakovskij, Pison, 336
japonicum Pérez, Chalybion, 103
japonicus Iwata, Tachysphex, 275
japonicus Kohl, Tachytes, 267
japonicus Matsumura, Pemphredon, 181
japonicus (Sonan), Bembecinus, 531
japonicus Tsuneki, Nysson, 470
japonicus Tsuneki, Psenulus, 173
japonicus Tsuneki, Stigmus, 189
jaragua Pate, Enoplolindenius, 415
jaroschewskyi F. Morawitz, Crabro, 408
jaroslavensis Kokujev, Alysson, 458
jason (Cameron), Crossocerus, 401
jason (Cameron), Podalonia, 144
jatahyna Brèthes, Cerceris, 582
jaumei (Alayo), Hoplisoides, 521
javana Cameron, Ampulex, 78
javanensis van Lith, Psenulus, 173
javanum Cameron, Pison, 336
javanum (Lepeletier), Sceliphron, 105
javanum Taschenberg, Trypoxylon, 346
javanus (Handlirsch), Bembecinus, 531
javanus Leclercq, Dasyproctus, 421
javanus van der Vecht, Philanthus, 565
jentinki (Handlirsch), Hoplisoides, 521
jerichoensis Pulawski, Astata, 212
jessonicum (Bischoff), Rhopalum, 389
jibacoa (Alayo), Hoplisoides, 521
jocositarsa Saussure, Liris, 245
jocosum (Cameron), Podagritus, 393
joeli Bytinski-Salz, Bembix, 546
joergenseni Brèthes, Cerceris, 584
joergenseni Brèthes, Heliocausus, 453
joergenseni (Brèthes), Liogorytes, 516
joergenseni Brèthes, Oxybelus, 368
joergenseni (Brèthes), Parapiagetia, 281
joergenseni (Brèthes), Podagritus, 393
joergenseni (Brèthes), Pseneo, 165
joergenseni Brèthes, Sphex, 116
joergenseni Brèthes, Trypoxylon, 348
johannis (Arnold), Carinostigmus, 191
johannis (Cameron), Bembecinus, 531
johannis (Fabricius), Prionyx, 133
johannis Richards, Trypoxylon, 348
johnsoni Rohwer, Tachysphex, 272
johnsoni (Viereck) Mimumesa, 164
johnsoni W. Fox, Trypoxylon, 346
johnstoni Turner, Bembix, 546
jokischianus (Panzer), Tachysphex, 276
jonesi (Turner), Gorytes, 501
jordani (Handlirsch), Hoplisoides, 521
jordanicus Lohrmann, Bembix, 546
jordanicus Lohrmann, Stizus, 527
jozankeanus (Matsumura), Ectemnius, 428
josei Scullen, Cerceris, 589
jubilans (Kohl), Crossocerus, 401
jucunda Arnold, Liris, 246
jucunda Cresson, Cerceris, 579
jucunda Pulawski, Astata, 212
jucundus Arnold, Ammoplanus, 198
jucundus (Arnold), Dasyproctus, 420
jucundus F. Smith, Tachytes, 265
judea Beaumont, Cerceris, 585
judaeorum Kohl, Ammophila, 152
judaeus (Beaumont), Prionyx, 133
judicum Leclercq, Trypoxylon, 346
jugurthae (Gribodo), Liris, 246
jujuyensis Brèthes, Tachysphex, 274
julii Fabre, Ammophila, 153
julii Fabre, Bembix, 546

julii Fabre, Cerceris, 586
julliani Kohl, Tachysphex, 274
juncea Cresson, Ammophila, 152
juncea of authors, Ammophila, 151
jungi (Maa), Dasyproctus, 421
juniatae Krombein, Crabro, 408
junodi Arnold, Bembix, 546
junonium (Schrottky), Penepodium, 92
jurinei (Vander Linden), Tachysphex, 277
jurini Spinola, Pison, 335
jurinii (Drapiez), Tachysphex, 277
jurinii Stephens, Pemphredon, 182
jurumpa Pate, Belomicrus, 363

kabeyae Leclercq, Trypoxylon, 346
kachelibae (Arnold), Bembecinus, 531
kalaharica (Arnold), Ammophila, 152
kalaharica Bischoff, Cerceris, 582
kalaharicus Arnold, Tachysphex, 276
kalaharicus Bischoff, Oxybelus, 368
kalensis Tsuneki, Cerceris, 589
kalimantan Menke, Trypoxylon, 346
kallipygus Pulawski, Tachysphex, 274
kamateensis Tsuneki, Crossocerus, 401
kamtschatica Gussakovskij, Ammophila, 153
kamulla Evans and Matthews, Bembix, 546
kankauense Strand, Trypoxylon, 346
kankauensis Strand, Oxybelus, 368
kankauensis Strand, Psenulus, 173
kansensis (Slansky), Miscophus, 318
kansensis (Williams), Solierella, 313
kansitakuanus Tsuneki, Crossocerus, 401
kansitakuanus Tsuneki, Stigmus, 189
kansitakum Tsuneki, Trypoxylon, 346
kansuensis Gussakovskij, Cerceris, 582
kantsi Pate, Diploplectron, 211
kapiricum Leclercq, Trypoxylon, 346
karenae Menke, Ammophila, 152
karimuiensis Krombein, Cerceris, 582
karnyi (Maidl), Liris, 246
karooensis Brauns, Cerceris, 584
karooensis (Brauns), Dienoplus, 496
karooensis (Brauns), Gastrosericus, 256
Karossia Arnold, 46
karrooense Arnold, Pison, 335
karrooensis (Arnold), Bembecinus, 531
karrooensis Arnold, Miscophus, 318
karrooensis Arnold, Tachysphex, 274
karrooensis Arnold, Tachytes, 267
karschii Handlirsch, Bembix, 545
kasachstanica Kazenas, Cerceris, 582
kashmirensis Nurse, Astata, 212
kashmirensis Nurse, Cerceris, 586
kashmirensis (Nurse), Mimumesa, 164
kaskaskia Pate, Nysson, 469
kaszabi Tsuneki, Ammoplanus, 198
kaszabi Tsuneki, Belomicrus, 363
kaszabi Tsuneki, Cerceris, 582
kaszabi Tsuneki, Diodontus, 179
kaszabi (Tsuneki), Dryudella, 214
kaszabi Tsuneki, Lindenius, 384
kaszabi (Tsuneki), Mimesa, 162
kaszabi Tsuneki, Oryttus, 508
kaszabi Tsuneki, Parapiagetia, 281
kaszabi Tsuneki, Philanthus, 565
kaszabi (Tsuneki), Podalonia, 144
kaszabi Tsuneki, Spilomena, 193
kaszabi Tsuneki, Stizus, 527
kaszabi Tsuneki, Tachysphex, 274
kaszabi Tsuneki, Trypoxylon, 346
katangae Brauns, Cerceris, 582

katangae Leclercq, Trypoxylon, 346
kathovi Kokujev, Alysson, 458
kauaiensis Perkins, Nesomimesa, 167
kaufmanni Radoszkowski, Entomosericus 434
Kaufmannia Radoszkowski, 52
kawasei Tsuneki, Cerceris, 585
kazenasi Pulawski, Cerceris, 582
kedahae Pagden, Cerceris, 583
kedahense Leclercq, Rhopalum, 389
keiensis Strand, Cerceris, 589
keiseri (Leclercq), Liris, 246
kennedii Curtis, Didineis, 459
kennedyi (Murray), Ammophila, 152
kennicottii Cresson, Cerceris, 583
kenyae Empey, Cerceris, 629
kerangi Leclercq, Rhopalum, 389
khasiae Cameron, Trypoxylon, 346
khasiana Cameron, Ampulex, 78
khasiana Cameron, Bembix, 547
khasiana (Cameron), Liris, 246
khasianus (Cameron), Ectemnius, 425
kiatae Leclercq, Podagritus, 393
kibonotensis Cameron, Dasyproctus, 421
kiesenwetteri (A. Morawitz), Rhopalum, 389
kigonseranum (Strand), Chlorion, 90
kilimandjaroensis Cameron, Cerceris, 583
kilimandjaroensis Cameron, Sphex, 114
kiloense (Leclercq), Chalybion, 103
kinabalensis Tsuneki, Encopognathus, 629
kinkadzanense Tsuneki, Trypoxylon, 347
kirbii (Vander Linden), Prionyx, 627
kirbyi Bingham, Cerceris, 583
kirbyi (Dours), Prionyx, 627
kirbyi (Vander Linden), Prionyx, 39, **133**
kirgisica F. Morawitz, Ammophila, 154
kirgisica F. Morawitz, Bembix, 546
kirgisicus Radoszkowski, Oxybelus, 368
kiseritzkii Gussakovskij, Stizus, 527
kisilkumii Radoszkowski, Stizus, 526
kittyae Beaumont, Bembix, 546
kizanensis Tsuneki, Ectemnius, 424
kizilkumii Radoszkowski, Oxybelus, 368
kizilkumii Radoszkowski, Philanthus, 565
kizilkumii (Radoszkowski), Prosopigastra, 285
klapperichi (Balthasar), Chalybion, 103
klapperichi Balthasar, Trypoxylon, 346
klapperichi Beaumont, Crossocerus, 401
klapperichi Giner Marí, Cerceris, 581
klugi Kirchner, Cerceris, 583
klugi Menke, Palarus, 291
klugii F. Smith, Cerceris, 583
klugii (F. Smith), Stizoides, 528
klugii (Lepeletier), Podalonia, 145
klugii Schletterer, Cerceris, 583
knabi (Rohwer), Hoplisoides, 521
knoxensis (Mickel), Crabro, 407
koala Tsuneki, Cerceris, 583
kobrowi (Arnold), Bembecinus, 531
kobrowi Brauns, Cerceris, 583
kockensis Leclercq, Crossocerus, 401
kodairai Tsuneki, Tachysphex, 274
kodamanum Tsuneki, Trypoxylon, 347
koenigi F. Morawitz, Stizus, 527
kohli Arnold, Dolichurus, 69
kohli (Arnold), Neodasyproctus, 419
kohli Arnold, Psenulus, 173
kohli Arnold, Trypoxylon, 347

kohli (Bischoff), Crossocerus, 401
kohli (Brauns), Dicranorhina, 251
kohli Brauns, Miscophus, 318
kohli (Cameron), Dryudella, 215
kohli (Ed. André), Chlorion, 89
kohli Gussakovskij, Psen, 166
kohli Mercet, Prosopigastra, 285
kohli Sickmann, Sceliphron, 106
kohli Tsuneki, Diodontus, 179
Kohlia Handlirsch, 53, **513**
kohliana Cockerell, Hoplisoidea, 33
kohlianus Schulthess, Belomicrus, 363
kohlianus Strand, Sphex, 116
Kohliella Brauns, 44, **286**
kohlii (Bingham), Laphyragogus, 220
kohlii Bingham, Pison, 336
kohlii (Brauns), Belomicrus, 363
kohlii Ducke, Bothynostethus, 352
kohlii F. Morawitz, Philanthus, 565
kohlii (Gussakovskij), Eremochares, 146
kohlii (Handlirsch), Pseudoplisus, 503
kohlii Mocsáry, Stizus, 527
kohlii Morice, Bembix, 547
kohlii Schletterer, Cerceris, 583
kohlii Schmiedeknecht, Ammoplanus, 198
kohlii Turner, Arpactophilus, 186
kohlii Turner, Bembix, 546
kohlii (Turner), Sericophorus, 303
kohlii (W. Fox), Pseneo, 40, **165**
kohlii Zavattari, Podium, 96
koikense Tsuneki, Trypoxylon, 347
koikensis Tsuneki, Spilomena, 193
kokandicus Radoszkowski, Philanthus, 565
kokuevi Shestakov, Cerceris, 583
kolaensis Turner, Tachytes, 266
kolazyi Handlirsch, Nysson, 469
kolazyi Kohl, Trypoxylon, 347
kollari (Dahlbom), Ectemnius, 426
kolthoffi Gussakovskij, Sphex, 115
kolthoffi Gussakovskij, Trypoxylon, 347
koma Tsuneki, Cerceris, 583
koma Tsuneki, Trypoxylon, 346
komarovi Radoszkowski, Stizus, 527
komarovii Radoszkowski, Cerceris, 588
komarowi F. Morawitz, Philanthus, 565
komarowii Ed. André, Cerceris, 588
konosuense Tsuneki, Trypoxylon, 347
konowi (Kohl), Ectemnius, 426
konowi Mercet, Nysson, 469
koppenfelsii Taschenberg, Ammophila 152
kora Evans and Matthews, Bembix, 547
korana Brauns, Cerceris, 585
korbi (Kohl), Crabro, 408
koreanum Tsuneki, Trypoxylon, 347
koreanum Uchida, Sceliphron, 106
koreanus Handlirsch, Gorytes, 501
koreanus Tsuneki, Crabro, 408
koreanus Tsuneki, Crossocerus, 403
koreanus Tsuneki, Oxybelus, 629
koreanus Tsuneki, Passaloecus, 629
koreanus Tsuneki, Pemphredon, 181
koreanus Tsuneki, Psen, 166
koreense (Radoszkowski), Pison, 45, **337**
koreense Tsuneki, Rhopalum, 630
korovensis Krombein, Cerceris, 584
korrorense Yasumatsu, Pison, 336
koryo Tsuneki, Cerceris, 583
kosemponis Strand, Bembix, 548
koshantshikovi Shestakov, Cerceris, 583
koshevnikovi Kokujev, Didineis, 459

koshunicon Strand, Trypoxylon, 347
kotoshonus (Sonan), Bembecinus, 531
kotschyi (Handlirsch), Bembecinus, 531
koulingensis Tsuneki, Cerceris, 583
kowbrowi (Arnold), Sphex, 115
Koxinga Pate, 46, **381**
koxinga Pate, Piyuma, 410
kozlovi Shestakov, Cerceris, 583
kozlovii (Kohl), Podalonia, 144
kraepelini (Brauns), Oryttus, 508
kratochvili Snoflak, Lindenius, 384
kriechbaumeri Brauns, Miscophus, 318
kriechbaumeri Handlirsch, Bembix, 547
kriechbaumeri (Kohl), Ectemnius, 426
kriechbaumeri Kohl, Paraliris, 250
kriegeri Brauns, Diploplectron, 211
kristenseni Turner, Ampulex, 78
kristenseni Turner, Tachytes, 265
krombeini Bohart & Menke, Podium, 96
krombeini Kurczewski, Tachysphex, 274
krombeini Leclercq, Podagritus, 393
krombeini Menke, Liris, 246
krombeini Menke, Pison, 337
krombeini R. Bohart and Schlinger, Oxybelus, 368
krombeini Scullen, Cerceris, 583
krombeini Scullen, Trypoxylon, 348
krombeini Tsuneki, Pemphredon, 181
krombeini van Lith, Psen, 166
Krombeiniellum Richards, **45**, **334**
krugeri Schulthess, Philanthus, 564
krugi Dewitz, Cerceris, 583
krusemani Leclercq, Crossocerus, 401
krusemani (Leclercq), Ectemnius, 426
kuchingensis (Cameron), Liris, 246
kuehlhorni Leclercq, Rhopalum, 389
kuehlhorni Willink, Bembecinus, 531
kulingensis van Lith, Psen, 166
kulingensis (Yasumatsu), Gorytes, 501
kumaso Tsuneki, Trypoxylon, 347
kununurra Evans and Matthews, Bembix, 547
kuramaensis (Iwata), Lestica, 431
kuramensis (Tsuneki), Lestica, 431
kurandae Turner, Nitela, 325
kurarensis Yasumatsu, Ampulex, 78
kurdistanicus (Balthasar), Prionyx, 133
kuroo Tsuneki, Diodontus, 179
kutui Leclercq, Dasyproctus, 421
kuwayamai Tsuneki, Rhopalum, 389
kuznetzovi Gussakovskij, Belomicrus, 363
kwangtsehiana Giner Marí, Cerceris, 583
kyotoense Tsuneki, Trypoxylon, 347

labeculata Turner, Cerceris, 583
labiata Fabricius, Bembix, 547
labiata (Fabricius), Cerceris, 582
labiata (Fabricius), Liris, 246
labiata (Olivier), Cerceris, 583
labiatus (Olivier), Mellinus, 449
labidura Handlirsch, Bembix, 547
labilis Turner, Tachytes, 265
labiosa Empey, Cerceris, 582
laboriosa (F. Smith), Liris, 246
laboriosum (F. Smith), Sceliphron, 105
labrosus (Harris), Prionyx, 131
lacertosus Arnold, Tachysphex, 274
lachesis Turner, Tachytes, 265
lacinia Tsuneki, Cerceris, 583
lacordairei (Lepeletier), Gorytes, 501
lactarius (Chevrier), Crabro, 407
lactea Cameron, Bembix, 547
lacteipennis Mocsáry, Stizus, 527

lacteipennis Rohwer, Crabro, 408
lactitarse Saussure, Trypoxylon, 348
lacunosa Scullen, Eucerceris, 591
lacustre Arnold, Trypoxylon, 347
lacustris Arnold, Liris, 246
laerma (Cameron), Prionyx, 133
laeta (Bingham), Podalonia, 145
laeta E. Saunders, Astata, 212
laeta (Fabricius), Cerceris, 583
laeta J. Parker, Bembix, 547
laetus (Antiga and Bofill), Ectemnius, 426
laetum (F. Smith), Sceliphron, 106
laetus (F. Smith), Scapheutes, 355
laetus Gribodo, Philanthus, 565
laetus Klug, Palarus, 291
laetus Say, Oxybelus, 368
laeve F. Smith, Pison, 336
laeve W. Fox, Trypoxylon, 348
laeviceps F. Smith, Ammophila, 152
laeviceps (F. Smith), Crossocerus, 402
laeviceps Gussakovskij, Pemphredon, 182
laeviceps Tsuneki, Spilomena, 193
laevicollis Ed. André, Ammophila, 152
laevidorsis (Pérez), Ancistromma, 260
laevifrons (F. Smith), Tachysphex, 274
laevifrons F. Smith, Trypoxylon, 348
laevigata F. Smith, Ammophila, 152
laevigata F. Smith, Cerceris, 583
laevigata Holmberg, Cerceris, 584
laevigata Kohl, Ampulex, 78
laevigatum (Kohl), Chalybion, 103
laevigatus Arnold, Sphex, 116
laevigatus De Stefani, Ectemnius, 425
laevigatus (Kohl), Gorytes, 502
laevigatus (Rayment), Sericophorus, 303
laevigatus (Schenck), Psenulus, 173
laevigatus Schilling, Oxybelus, 367
Laevigorytes Zavadil, 52
laevior Arnold, Psenulus, 172
laevior Morice, Prosopigastra, 285
laevipes Lepeletier and Brullé, Crossocerus, 403
laevipes (W. Fox), Isodontia, 123
laevis A. Costa, Lindenius, 384
laevis F. Smith, Dolichurus, 69
laevis Gussakovskij, Psenulus, 173
laevis (Latreille), Dienoplus, 496
laevis (Provancher), Pulverro, 196
laevis Pulawski, Nysson, 469
laeviventris (Cresson), Palmodes, 127
lagunae Ashmead, Pison, 336
lagunensis (Williams), Solierella, 313
lama Tsuneki, Bembix, 549
lama Turner, Cerceris, 583
lamarquensis Fritz and Toro, Cerceris, 583
lambertoni Leclercq, Dasyproctus, 421
lamborni Arnold, Tachytes, 265
lamellata Handlirsch, Bembix, 547
lamellatus Olivier, Oxybelus, 368
lamellatus Turner, Gastrosericus, 256
lamentabilis Arnold, Tachytes, 265
laminata Eversmann, Cerceris, 581
laminifera A. Costa, Cerceris, 586
laminiferus (W. Fox), Oryttus, 508
Lamocrabro Leclercq, 49, **413**, 627
lampei Strand, Ammophila, 152
lamprus van Lith, Psenulus, 173
lanaiensis (Perkins), Ectemnius, 425
lanata Cameron, Cerceris, 589
lanatus Arnold, Tachytes, 274
lanatus Mocsáry, Sphex, 115

lanceiventris Vachal, Sphex, 116
lanceolatus Gerstaecker, Oxybelus, 368
lancifer Olivier, Oxybelus, 368
lanciger Kohl, Sphex, 115
langkasukae Pagden, Cerceris, 583
languida Cameron, Cerceris, 583
lanierii Guérin-Méneville, Sphex, 115
laniger Pulawski, Tachysphex, 274
lanio (Stal), Exeirus, 495
lanuginosa (Marquet), Podalonia, 145
lanuginosus Arnold, Gastrosericus, 256
lanuginosus Pulawski, Tachytes, 265
lapazae R. Bohart, Oryttus, 508
lapazensis Scullen, Eucerceris, 591
LAPHYRAGOGINAE, 57, **217**
LAPHYRAGOGINI, 59, **219**
Laphyragogus Kohl, 42, **219**
lapidaria (Fabricius), Lestica, 430
lapidarius (Panzer), Ectemnius, 426
lapillus Beaumont, Nysson, 469
laportaei (Lepeletier), Ectemnius, 426
laportei (Lepeletier and Brullé), Ectemnius, 426
lapponicus Zetterstedt, Crabro, 408
Lara Drapiez, 42
largior W. Fox, Crabro, 408
larimerensis Rohwer, Belomicrus, 363
Larra Fabricius, 42, 53, **233**
Larrada F. Smith, 42
Larrana Rafinesque-Schmaltz, 42
Larraxena F. Smith, 42
larriformis (Williams), Liris, 246
LARRINAE, 58, **221**, 226
LARRINI, 59, **226**, 229
Larrisson Menke, 44, **304**
larroides (Spinola), Heliocausus, 453
larroides (Williams), Liris, 245, 246
Larrophanes Handlirsch, 33
Larropsis Patton, 43, **257**
larutae Leclercq, Crossocerus, 401
larvata Taschenberg, Cerceris, 582
larvatus (Wesmael), Ectemnius, 427
lasseni Williams, Solierella, 313
latebrosa Kohl, Bembix, 547
latebrosus (Kohl), Lindenius, 384
lateralis (Bingham), Bembecinus, 531
lateralis Packard, Nysson, 469
latericaudatus (Arnold), Bembecinus, 531
lateridentata Arnold, Cerceris, 583
laterifurcata Empey, Cerceris, 583
laterimacula (Handlirsch), Bembecinus, 531
laterimaculata Empey, Cerceris, 583
lateriproducta Mochi, Cerceris, 583
laterisetosa (Spinola), Liris, 246
lateritia Taschenberg, Parapsammophila, 139
lateritius (Handlirsch), Dienoplus, 496
lateritius Shuckard, Exeirus, 52, 495
latiannulatus Cameron, Psenulus, 172
latibalteata Cameron, Cerceris, 583
latiberbis Tsuneki, Cerceris, 583
laticauda Gussakovskij, Tachysphex, 274
laticeps (Arnold), Ammophila, 152
laticeps Arnold, Gastrosericus, 256
laticeps Arnold, Philanthus, 565
laticeps Arnold, Tachysphex, 274
laticeps (Ashmead), Miscophus, 318
laticeps Cresson, Eucerceris, 592
laticeps Gussakovskij, Ammoplanops, 197
laticincta Lepeletier, Cerceris, 583
laticinctus (Arnold), Bembecinus, 531
laticinctus (Cresson), Clypeadon, 55, **571**
laticinctus (Lepeletier), Gorytes, 501

laticinctus Provancher, Gorytes, 501
laticorne (Tsuneki), Rhopalum, 389
latidens Cameron, Cerceris, 583
latidens Gerstaecker, Oxybelus, 368
latifascia (Walker), Bembecinus, 531
latifasciata Turner, Bembix, 547
latifrons Bingham, Cerceris, 583
latifrons Gussakovskij, Prosopigastra, 285
latifrons J. Parker, Bembix, 548
latifrons Kohl, Ampulex, 78
latifrons Kohl, Oxybelus, 368
latifrons Kohl, Palarus, 291
latifrons Kohl, Tachysphex, 274
latifrons (Spinola), Hoplisoides, 521
latifrons Tsuneki, Tachytes, 265
latifrons (W. Fox), Lindenius, 384
latigenata Willink, Bembix, 547
latilineatus Cameron, Oxybelus, 368
latimana Malloch and Rohwer, Didineis, 459
latiornatum Cameron, Trypoxylon, 348
latipes F. Smith, Crabro, 408
latiscapus Leclercq, Ammoplanus, 198
latiscutatum Arnold, Trypoxylon, 347
latissimus Turner, Tachysphex, 273
latitarsis Handlirsch, Bembix, 547
latitarsis Marshakov, Lindenius, 384
lativalvis Gussakovskij, Ammophila, 152
lativalvis Kohl, Larra, 238
lativalvis (Thomson), Tachysphex, 275
lativentris Gussakovskij, Cerceris, 583
latreillei Lepeletier, Bembix, 547
latreillei Lepeletier, Sphex, 115
latreillei (Spinola), Penepodium, 92
latro (Beaumont), Didineis, 459
latro Erichson, Sphex, 115
latro F. Smith, Cerceris, 583
latro (Kohl), Penepodium, 92
latro Olivier, Oxybelus, 368
latronum (Kohl), Rhopalum, 389
Latrorhopalum Tsuneki, 47, **388**
Latroxybelus Noskiewicz and Chudoba, 46
latus Arnold, Ammatomus, 513
laufferi Mercet, Nysson, 469
Lautara Herbst, 45
lautus Cresson, Sphex, 115
lavellensis Krombein, Cerceris, 578
Lavia Rayment, 53
lawuensis van Lith, Psen, 167
laxata Shestakov, Cerceris, 583
layano Pate, Metanysson, 481
layouanum Evans, Trypoxylon, 347
lazulina Kohl, Ampulex, 78
lebedevi Gussakovskij, Tachysphex, 277
leclercqi Beaumont, Lindenius, 384
leclercqi Menke, Ammophila, 152
leclercqi (Tsuneki), Towada, 412
leclercqi van Lith, Psen, 629
Leclercqia Tsuneki, 49, **411**
leensis Rohwer, Tachysphex, 274
leeuwinensis Turner, Bembix, 547
lefebvrei Lepeletier and Brullé, Crossocerus, 403
leguminiferus Cockerell and W. Fox, Diodontus, 179
Leianthrena Bingham, 42
leila Pulawski, Astata, 212
leioceps Strand, Dolichurus, 69
lenape (Pate), Ammoplanellus, 199
lenapeorum Viereck, Entomognathus, 382
lenis (Nurse), Ammatomus, 513

lentifrons Brauns, Saliostethus, 45, **320**
lentus (W. Fox), Crossocerus, 401
Leocrabro Leclercq, 50
leonesus Leclercq, Ectemnius, 426
leoni Krombein, Nitela, 325
leonina (Saussure), Isodontia, 123
leoninus van Lith, Psenulus, 629
leontopolites Pate, Crossocerus, 401
Leontosphex Arnold, 39
leoparda (Fernald), Ammophila, 152
lepcha Cameron, Cerceris, 583
lepeletieri (F. Smith), Crossocerus, 404
lepeletierii Saussure, Sphex, 116
lepida Brullé, Cerceris, 583
lepidum (Strand), Chlorion, 90
lepidus (Arnold), Dienoplus, 496
lepidus Arnold, Tachytes, 265
lepidus Cresson, Philanthus, 565
lepidus Klug, Palarus, 291
lepidus (Klug), Stizus, 527
leprieurii (Spinola), Tachytes, 265
leptogaster Kohl, Trypoxylon, 347
leptogaster Cameron, Sceliphron, 106
Leptolarra Cameron, 43, **241**
leptospermi (Turner), Podagritus, 393
lepturus Arnold, Oxybelus, 368
leskii Gmelin, Vespa, 627
Lestica Billberg, 50, **428**
lesticoides Leclercq, Ectemnius, 426
Lestiphorus Lepeletier, 52, **505**
Lestopodius Agassiz, 52
lethifer (Shuckard), Pemphredon, 181
lethifer Thomson, Pemphredon, 182
letiferum Arnold, Trypoxylon, 347
leucarthrum Richards, Trypoxylon, 347
leucochroa Schletterer, Cerceris, 583
leucognathus (Perkins), Ectemnius, 428
leuconotus Brullé, Sphex, 116
leuconotus (F. Morawitz), Prionyx, 133
leucopterus Pallas, Sphex, 117
leucosoma (Kohl), Prionyx, 134
leucostoma (Linnaeus), Crossocerus, 401
leucostomoides (Richards), Crossocerus, 402
leucostomus of authors, Crossocerus, 402
leucotrichium Rohwer, Trypoxylon, 348
leucozonica Schletterer, Cerceris, 577
leucurus (A. Costa), Dienoplus, 496
leuthstromi Ashmead, Astata, 212
levantinus Balthasar, Miscophus, 318
levantinus Pulawski, Gorytes, 501
levantinus Pulawski, Tachytes, 265
leviceps (Dalla Torre), Crossocerus, 402
levifrons (Arnold), Carinostigmus, 191
levifrons Dalla Torre, Tachysphex, 274
levigata Dalla Torre, Ampulex, 78
levigata Dalla Torre, Cerceris, 583
levigatum (Dalla Torre), Chalybion, 103
levilabris (Cameron), Chlorion, 90
levini R. Bohart, Philanthus, 565
levinotus Merisuo, Pemphredon, 182
levipes (Bingham), Polemistus, 185
levipes (Vander Linden), Crossocerus, 403
levis (Dalla Torre), Dienoplus, 496
levis Dalla Torre, Dolichurus, 69
levis (Dalla Torre), Lindenius, 384
levis J. Parker, Bembix, 547
levis Williams, Solierella, 313
levius Richards, Trypoxylon, 345
lewisi Cameron, Oxybelus, 368
leytensis R. Bohart and Grissell, Pseneo, 165

Lianthrena Bingham, 42
libanonis (Kohl), Entomognathus, 382
liberiae Leclercq, Dasyproctus, 421
liberiensis (Arnold), Hoplisoides, 521
liberiensis J. Parker, Bembix, 547
libertinus (Arnold), Neodasyproctus, 419
libitina Brauns, Cerceris, 582
libyca Beaumont, Cerceris, 578
libycus Beaumont, Sphex, 115
lichtenburgensis (Arnold), Dasyproctus, 420
lichtenburgensis Arnold, Tachytes, 266
lichtenburgensis Brauns, Cerceris, 586
lichtensteini Mocsáry, Bembix, 549
lieftincki Empey, Cerceris, 587
lieftincki Leclercq, Lestica, 431
lieftincki van Lith, Psen, 166
lignarius (F. Smith), Dasyproctus, 421
ligulata (Williams), Liris, 246
lihyuetanus Tsuneki, Tachysphex, 274
lilacinus Rayment, Sericophorus, 302
lilliputianus Turner, Tachysphex, 275
lilloi Willink, Bicyrtes, 537
limatus Arnold, Oxybelus, 368
limatus Arnold, Tachysphex, 274
limatus Bingham, Philanthus, 565
limbata (Kriechbaumer), Eremochares, 146
limbatus (Olivier), Philanthus, 566
limbellus R. Bohart, Gorytes, 501
limpidipennis (F. Smith), Liris, 246
lindbergi (Beaumont), Crossocerus, 402
lindbergi Beaumont, Tachysphex, 274
lindenii (Dalla Torre), Ectemnius, 425
lindenii Lepeletier, Cerceris, 586
Lindenius Lepeletier and Brullé, 46, **382**
lindenius (Shuckard), Ectemnius, 425
lindensis (Inchbald), Ectemnius, 425
lineata Fabricius, Stictia, 542
lineata Mocsáry, Dryudella, 215
lindneri (Schulthess), Dynatus, 94
lineatifrons Cameron, Bembix, 547
lineatifrons Cameron, Palarus, 291
lineatipes Cameron, Sceliphron, 106
lineatotarsis (Matsumura), Ectemnius, 427
lineatus (Cameron), Bembecinus, 532
lineatus (Fabricius), Oxybelus, 368
lineatus (Fabricius), Stizus, 527
lineatus Lichtenstein, Sphex, 119
lineatus Villers, Sphex, 117
lineolatus Schenck, Nysson, 469
lineolus Lepeletier, Sphex, 116
lingnaui Arnold, Cerceris, 585
lingnaui Arnold, Tachytes, 265
linguiferus Turner, Oxybelus, 368
lingula Gerstaecker, Oxybelus, 368
lingyuanensis Yasumatsu, Philanthus, 565
linsleyi R. Bohart and Schlinger, Oxybelus, 368
linsleyi R. Bohart, Tachysphex, 274
Liogorytes R. Bohart, 53, **516**
lipan Pate, Metanysson, 481
lipatus Leclercq, Crossocerus, 402
liriformis Pulawski, Tachysphex, 274
liriformis (Williams), Liris, 246
Liris Fabricius, 43, **238**
Lirisis Rafinesque-Schmaltz, 43
liroides (Turner), Liris, 245
liroides (Williams), Liris, 248
Lirosphex Brèthes, 44
Lisponema Evans, 31, **176**
lissipes Pulawski, Prosopigastra, 285

lissonotum Cameron, Trypoxylon, 347
lissus Bohart and Menke, Palmodes, 127
Listropygia R. Bohart, 55, **571**
litigiosa (Kohl), Parapsammophila, 140
litoralis (Arnold), Ammophila, 152
litoralis Arnold, Bembix, 547
litoralis Tsuneki, Liris, 246
littorale Turner, Rhopalum, 389
littoralis (Bondroit), Mimumesa, 164
littoralis (Malloch), Pluto, 171
littoralis Rayment, Sericophorus, 302
littoralis Turner, Bembix, 547
littoralis van der Vecht, Bembecinus, 531
littoralis (Wagner), Pemphredon, 181
littorea Beaumont, Cerceris, 589
littoreus Andrade, Miscophus, 318
liturata Turner, Bembix, 547
lituratus (Panzer), Ectemnius, 426
liventis J. Parker, Bembix, 547
lividocinctus (A. Costa), Prionyx, 133
lixivia Tournier, Mimesa, 162
llameo Pate, Quexua, 387
llanoi (Fritz), Liogorytes, 516
loa Pate, Crossocerus, 400
lobaba W. F. Kirby, Cerceris, 583
lobatifrons Turner, Bembix, 547
lobatum (Fabricius), Chlorion, 38, **90**
lobicollis (Cameron), Eremnophila, 147
lobicornis van Lith, Psen, 166
lobiferum Arnold, Pison, 336
lobifex Richards, Microstigmus, 192
lobimana Handlirsch, Bembix, 547
localis Leclercq, Dasyproctus, 421
loeflingi Dahlbom, Philanthus, 565
loeuvi Thomson, Crabro, 408
loewi Dahlbom, Crabro, 408
lohrmanni R. Bohart, Stizus, 527
lomii (Guiglia), Bembecinus, 531
longiceps (Gussakovskij), Eremiasphecium, 561
longiceps Turner, Spilomena, 193
longicollis Cameron, Ampulex, 78
longicollis Kohl, Ammophila, 152
longicornis Arnold, Parapiagetia, 281
longicornis Beaumont, Diodontus, 179
longicornis (Cameron), Liris, 246
longicornis Handlirsch, Gorytes, 501
longicornis Pulawski, Solierella, 313
longicornis (Rossi), Argogorytes, 492
longicornis Tsuneki, Psen, 167
longicornis (W. Fox), Mimumesa, 164
longifrons (Rayment), Spilomena, 193
longilabris Arnold, Cerceris, 583
longilabris Gillaspy, Xerostictia, 54, **552**
longinodum (Spinola), Rhopalum, 389
longipalpis Beaumont, Tachysphex, 274
longipalpis (C. Verhoeff), Ectemnius, 427
longipennis J. Parker, Bembix, 547
longipes Merisuo, Passaloecus, 184
longipilosella (Cameron), Podalonia, 144
longipilosellum Cameron, Podium, 96
longirostra (Say), Steniolia, 54, **553**
longirostris Arnold, Tachytes, 265
longirostris Handlirsch, Steniolia, 553
longispina Cameron, Oxybelus, 367
longispinis (Cameron), Zanysson, 475
longitarsis (Cameron), Liris, 246
longitarsis Cameron, Tanyoprymnus, 512
longitudinalis Giordani Soika, Cerceris, 583
longiuscula Arnold, Cerceris, 583
longiventris (Cameron), Liris, 245

longiventris (Cameron), Pseneo, 165
longiventris (Malloch), Pluto, 171
longiventris Saussure, Ammophila, 151
longiventris (Saussure), Isodontia, 123
longoevus Cockerell, Crabro, 33
longoevus (Cockerell), Ectemnius, 428
longula (Gussakovskij), Mimumesa, 164
longulus (Tournier), Psenulus, 173
longus (Christ), Crabro, 408
loorea Evans and Matthews, Bembix, 547
Lophocrabro Rohwer, 50
lorentzi Cameron, Sceliphron, 105
loriculatus (F. Smith), Bembecinus, 531
Lorrheum Schuckard, 38
Losada Pate, 51, **479**
loti Pate, Ammoplanus, 198
lotus Andrade, Miscophus, 319
lounsburyi (Bridwell), Saliostethus, 320
loupata J. Parker, Bembix, 547
lowei Dalla Torre, Crabro, 408
lubricata Nurse, Astata, 212
lubricus Beaumont, Oxybelus, 368
lubricus (Pérez), Psenulus, 173
lubutana Arnold, Nitela, 325
lucae (Saussure), Sceliphron, 105
lucae Saussure, Sphex, 117
luchti van der Vecht, Cerceris, 583
luciati Brèthes, Sphex, 116
lucida C. Fox, Bembix, 545
lucida G. and R. Bohart, Larropsis, 259
lucida (Rohwer), Solierella, 313
lucidula (Turner), Clitemnestra, 489
lucidum Arnold, Trypoxylon, 347
lucidum Rohwer, Rhopalum, 390
lucidus Rohwer, Stigmus, 188
lucidus Villers, Sphex, 117
lucillus Pulawski, Tachysphex, 274
lucinda Nurse, Astata, 212
luctuosa A. Costa, Cerceris, 576
luctuosa (F. Smith), Liris, 246
luctuosa (F. Smith), Podalonia, 144
luctuosum F. Smith, Podium, 96
luctuosus, Andrade, Miscophus, 318
luctuosus Arnold, Psenulus, 173
luctuosus Arnold, Tachysphex, 276
luctuosus F. Smith, Sphex, 115
luctuosus Shuckard, Pemphredon, 182
ludibunda Arnold, Cerceris, 586
ludovici (Turner), Brachystegus, 473
ludovicus (F. Smith), Parapsammophila, 140
luederwaldti Richards, Microstigmus, 192
luffi (E. Saunders), Podalonia, 144
lugens Dahlbom, Pemphredon, 182
lugens (Kohl), Prionyx, 133
lugens Taschenberg, Astata, 212
lughensis Magretti, Stizus, 527
lugubre (Christ), Sceliphron, 106
lugubris Arnold, Ampulex, 78
lugubris (Arnold), Dasyproctus, 420
lugubris Arnold, Miscophus, 318
lugubris (Arnold), Tachysphex, 274
lugubris (Beaumont), Dienoplus, 496
lugubris (Fabricius), Pemphredon, 40, **182**
lugubris Gerstaecker, Ammophila, 151
lugubris Villers, Sphex, 117
lugubris Walker, Tachytes, 265
lukombensis Cameron, Ammophila, 151
luluana Leclercq, Ampulex, 78
lumpuri Leclercq, Encopognathus, 380
lunata A. Costa, Cerceris, 583
lunata Fabricius, Bembix, 547

lunatum (Fabricius), Sceliphron, 105
lunatus (Christ), Crabro, 408
lunatus (Dahlbom), Dienoplus, 496
lunatus Schrank, Crabro, 409
lundbladi (Kjellander), Crossocerus, 402
lunicornis (Fabricius), Didineis, 459
luniger (Eversmann), Sphecius, 511
lunigera Dahlbom, Cerceris, 583
lunulata (Rossi), Cerceris, 583
lunulata (Thunberg), Cerceris, 578
luperus Shuckard, Diodontus, 179
lusca Spinola, Bembix, 547
lusingae Leclercq, Psenulus, 173
lusingi (Leclercq), Chalybion, 103
lusingum Leclercq, Trypoxylon, 347
lusitana Beaumont, Cerceris, 581
lusitanica Pulawski, Astata, 212
lusitanicus Andrade, Miscophus, 318
lutaria (Fabricius), Mimesa, 162
lutaria of authors, Podalonia, 143
lutasator (Williams), Liris, 246
lutea Gmelin, Vespa, 627
lutea of authors, Parapsammophila, 140
lutea Sonan, Bembix, 548
lutea Taschenberg, Cerceris, 578
lutea (Taschenberg), Eremochares, 146
luteicollis (Lepeletier and Brullé), Tracheliodes, 405
luteifrons Radoszkowski, Sphex, 117
luteipalpis Lepeletier and Brullé, Crossocerus, 401
luteipenne (Fabricius), Penepodium, 38, **92**
luteipennis (Cameron), Liris, 245
luteipennis (Cresson), Liris, 246
luteipennis (Gerstaecker), Zanysson, 475
luteipennis Mocsáry, Sphex, 116
luteipes F. Smith, Crabro, 408
luteitarse Saussure, Trypoxylon, 349
luteiventris (A. Morawitz), Lindenius, 384
luteiventris Turner, Spilomena, 193
luteopictus (Rohwer), Psenulus, 173
luteosignatum Arnold, Trypoxylon, 347
luteotaeniatus Gussakovskij, Stizus, 527
lutescens Radoszkowski, Bembix, 547
lutescens (Radoszkowski), Sphecius, 511
lutescens Radoszkowski, Stizus, 527
lutescens Turner, Pison, 336
lutescens (Turner), Psenulus, 173
luteus van Lith, Psenulus, 172
lutulenta Arnold, Cerceris, 585
lutzi Scullen, Cerceris, 583
luxuriosa Dahlbom, Cerceris, 583
luxuriosa A. Costa, Crabro, 409
luxuriosus Morice, Tachysphex, 274
luxuriosus (Radoszkowski), Psammaecius, 516
luxuriosus (Schrottky), Zanysson, 475
luzonense Rohwer, Sceliphron, 106
luzonensis Crawford, Cerceris, 583
luzonensis J. Parker, Bembix, 547
luzonensis Rohwer, Dicranorhina, 252
luzonensis Rohwer, Larra, 238
luzonensis (Rohwer), Liris, 248
luzonensis (Rohwer), Polemistus, 185
luzonensis (Rohwer), Psenulus, 173
luzonensis Williams, Nitela, 325
luzonia Leclercq, Lestica, 431
lyauteyi (Schulthess), Pseudoscolia, 537
lynchi Brèthes, Trypoxylon, 349
lynchii Brèthes, Cerceris, 583
lynchii (Holmberg), Podagritus, 393

lynx Brèthes, Cerceris, 583
Lyroda Say, 44, **295**
Lyrodon Howard, 44
Lyrops Dahlbom, 42
Lyrops Illiger, 43
lysias (Cameron), Ectemnius, 427

maai van Lith, Psenulus, 173
mabwense Leclercq, Trypoxylon, 347
macalanga Brauns, Cerceris, 583
maccus (Handlirsch), Stictia, 542
macilentis Saussure, Polemistus, 41, **185**
mackayensis Turner, Bembix, 547
mackayensis (Turner), Ectemnius, 427
mackayensis Turner, Tachysphex, 274
macololo Brauns, Cerceris, 582
macra Cresson, Ammophila, 152
macrocephala (W. Fox), Isodontia, 124
macrocephalum Turner, Rhopalum, 389
macrocola Kohl, Ammophila, 151
macrodentatus van Lith, Psenulus, 173
macrogaster (Dahlbom), Podalonia, 144
macrosticta Viereck and Cockerell, Cerceris, 581
macswaini R. Bohart and Schlinger, Oxybelus, 368
macswaini Scullen, Cerceris, 583
macula (Fabricius), Cerceris, 583
macula (Fabricius), Prionyx, 133
macularis (Gussakovskij), Parapsammophila, 139
maculata (Fabricius), Stictia, 542
maculata Radoszkowski, Astata, 212
maculata Radoszkowski, Cerceris, 583
maculata Radoszkowski, Pseudoscolia, 55, **573**
maculatus (Christ), Philanthus, 566
maculatus Dahlbom, Palarus, 291
maculatus (Fabricius), Ectemnius, 427
maculatus (Fabricius), Nysson, 469
maculatus (Handlirsch), Hoplisoides, 521
maculatus (Lepeletier and Brullé), Crossocerus, 402
maculatus Reed, Heliocausus, 453
maculatus van Lith, Psenulus, 173
maculatus of Rossi, Crossocerus, 401
maculiceps (Cameron), Trachypus, 568
maculiceps Tsuneki, Cerceris, 589
maculiclypeus (W. Fox), Crossocerus, 402
maculicollis Dalla Torre, Ampulex, 78
maculicollis Tsuneki, Dolichurus, 69
maculicornis (Cameron), Ampulex, 78
maculicornis E. Saunders, Tachytes, 265
maculicornis (F. Morawitz), Gorytes, 501
maculicornis (Taschenberg), Ectemnius, 424
maculicrus Beaumont, Cerceris, 583
maculifrons Cameron, Ammophila, 151
maculifrons (Cameron), Dryudella, 214
maculifrons Cameron, Philanthus, 565
maculipenne F. Smith, Pison, 336
maculipennis (Cameron), Hoplisoides, 521
maculipennis Cameron, Tachytes, 266
maculipennis (Giraud), Hoplisoides, 521
maculipennis F. Smith, Crossocerus, 402
maculipes Arnold, Miscophus, 318
maculipes F. Smith, Oxybelus, 368
maculipes (Mickel), Nysson, 469
maculipes Tsuneki, Psenulus, 173
maculipes (W. Fox), Mimesa, 162
maculitarsis (Cameron), Crossocerus, 402

maculitarsis Cameron, Tachytes, 264
maculitarsis Schrottky, Stictia, 542
maculiventris (Cameron), Philanthus, 565
maculiventris Tournier, Oxybelus, 368
maculosus (Gmelin), Ectemnius, 427
maculosus (Gmelin), Nysson, 469
macuxi Pate, Zanysson, 475
madagascariensis Arnold, Solierella, 313
madagascariensis Brèthes, Philanthus, 565
madasummae van der Vecht, Sphex, 115
madecassa Arnold, Lyroda, 299
madecassa (Kohl), Podalonia, 144
madecassa Saussure, Bembix, 547
madecassa Saussure, Larra, 238
madecassum Arnold, Trypoxylon, 345
madecassum (Gribodo), Chalybion, 103
madecassus Arnold, Alysson, 458
madecassus Arnold, Psen, 166
madecassus (Kohl), Ammoplanellus, 200
madecassus (Kohl), Gastrosericus, 256
madecassus (Schulthess), Ammatomus, 513
madeirae (Dahlbom), Podalonia, 145
maderae R. Bohart, Synnevrus, 470
maderospatanum (Gmelin), Sceliphron, 106
madrasiensis van Lith, Psen, 166
madraspatanum (Fabricius), Sceliphron, 106
maeklini A. Morawitz, Crabro, 408
maerens Turner, Tachytes, 265
maesta Dalla Torre, Cerceris, 584
maestus (Mickel), Diodontus, 179
maestus Mickel, Tachytes, 265
magarra Evans and Matthews, Bembix, 547
magdalena C. Fox, Bembix, 547
magdalenae "Viereck" Strandtmann, Philanthus, 566
magellanicus Williams, Tachytes, 265
magellanum (Leclercq), Rhompalum, 389
magnifica Kohl, Liris, 245
magnifica (Perty), Editha, 54, **543**
magnifica Schletterer, Cerceris, 579
magnificum F. Morawitz, Chlorion, 90
magnificus (F. Smith), Stizus, 527
magnificus (Provancher), Philanthus, 564
magnificus Schrottky, Trachypus, 569
magrettii Gribodo, Sphex, 114
magrettii Gribodo, Trypoxylon, 347
magrettii (Handlirsch), Bembecinus, 532
magrettii (Mantero), Brachystegus, 473
mahatma (Turner), Podalonia, 144
mahensis (Cameron), Liris, 246
maia Arnold, Cerceris, 581
maia (Bingham), Isodontia, 123
maiae (Pate), Epinysson, 472
maidli (Arnold), Liris, 246
maidli Beaumont, Tachysphex, 274
maidli Gussakovskij, Ammoplanus, 198
maidli Richards, Trypoxylon, 347
maidli Schulthess, Bembix, 547
maidli (Yasumatsu), Isodontia, 123
maidlii Kohl, Oxybelus, 368
maindroni Berland, Tachytes, 265
maindroni van der Vecht, Sceliphron, 106
maior (Handlirsch), Bembecinus, 531
maior (Maidl), Carinostigmus, 191
major Beaumont, Lindenius, 384
major (Dalla Torre), Bembecinus, 531
major (F. Morawitz), Prosopigastra, 285

major (Gronblom), Dryudella, 215
major Kohl, Ampulex, 78
major Kohl, Diodontus, 179
major Kohl, Philanthus, 565
major Kohl, Trypoxylon, 346
major Mickel, Oxybelus, 368
major (Rohwer), Solierella, 313
major Spinola Cerceris, 588
major (Tsuneki), Carinostigmus, 628
majus Richards, Trypoxylon, 349
majuscula (Kohl), Crossocerus, 401
makilingi (Williams), Crabro, 410
malagassus Saussure, Sphex, 115
malaisei (Gussakovskij), Crossocerus, 402
malaisei Gussakovskij, Nysson, 470
malaisei Gussakovskij, Trypoxylon, 347
malaitensis Krombein, Cerceris, 584
malayana Cameron, Cerceris, 585
malayana (Cameron), Isodontia, 123
malayanus Handlirsch, Sphecius, 511
malayanus van Lith, Psenulus, 172
maldiviensis Cameron, Bembix, 547
maliensis Empey, Cerceris, 629
maligna Mercet, Ancistromma, 259
malignum (Kohl), Chalybion, 103
maliki Evans and Matthews, Bembix, 547
malkovskii Pulawski, Tachysphex, 274
malyshevi Ahrens, Crabro, 408
mamillata (Handlirsch), Stictiella, 551
manchuriana Tsuneki, Cerceris, 583
manchurianus Tsuneki, Lindenius, 384
mandarina (F. Smith), Palmodes, 127
mandibulare Fabricius, Chlorion, 90
mandibularis Beaumont, Solierella, 313
mandibularis Cameron, Ammoplanus, 198
mandibularis (Cresson), Passaloecus, 184
mandibularis Cresson, Sphex, 115
mandibularis Dahlbom, Oxybelus, 368
mandibularis (F. Smith), Ectemnius, 427
mandibularis (H. Smith), Mimumesa, 164
mandibularis Menke, Liris, 246
mandibularis Patton, Cerceris, 583
mandibularis Patton, Tachytes, 264
mandibularis Tsuneki, Pemphredon, 181
mandibularis Tsuneki, Psen, 167
mandibularis Tsuneki, Psenulus, 172
mandibularis (Williams), Dalara, 250
mandibularis Williams, Solierella, 314
mandibulata (W. F. Kirby), Podalonia, 144
mandibulatum Richards, Trypoxylon, 347
mandibulatum Turner, Pison, 336
manduco Lichtenstein, Sphex, 119
maneei Banks, Tachysphex, 276
manicana Arnold, Cerceris, 586
manicatus Lichtenstein, Sphex, 119
manifesta Arnold, Cerceris, 583
manifestata (Turner), Williamsita, 422
manflava Tsuneki, Cerceris, 577
manilae (Ashmead), Liris, 245
manilensis (Rohwer), Liris, 245
manjikuli Tsuneki, Hoplisoides, 521
manjikuli Tsuneki, Tachytes, 265
manni Richards, Trypoxylon, 347
manni Rohwer, Oxybelus, 370
mansueta (F. Smith), Larra, 238
mantiraptor Ferton, Tachysphex, 272
mantivorus Beaumont, Tachysphex, 273
manzonii Gribodo, Miscophus, 318
maracandensis Radoszkowski, Alysson, 458
maracandica Radoszkowski, Cerceris, 583
maracandica (Radoszkowski), Olgia, 491

maracandicus Radoszkowski, Gastrosericus, 256
maracandicus (Radoszkowski), Prionyx, 133
maracandicus (Radoszkowski), Tachysphex, 273
maranhensis (Ducke), Paradolichurus, 70
marathroicus (De Stefani), Ammoplanus, 198
marcellus Leclercq, Podagritus, 393
marcia Nurse, Cerceris, 583
mareeba Evans and Matthews, Bembix, 547
margaretella Rohwer, Cerceris, 583
margarita Beaumont, Cerceris, 583
marginalis Gussakovskij, Gastrosericus, 256
marginalis Pérez, Ammophila, 153
marginalis Turner, Sphodrotes, 304
marginata Cameron, Cerceris, 583
marginata F. Smith, Cerceris, 583
marginata Gmelin, Vespa, 627
marginatum Cameron, Trypoxylon, 347
marginatum F. Smith, Pison, 336
marginatus F. Smith, Oxybelus, 368
marginatus (F. Smith), Prionyx, 133
marginatus (Malloch), Pluto, 171
marginatus (Say), Passaloecus, 184
marginatus (Spinola), Zanysson, 475
marginellus Spinola, Oxybelus, 368
marginicollis (Cameron), Stigmus, 189
marginicollis Gussakovskij, Oxybelus, 368
marginifrons Cameron, Trypoxylon, 347
marginipennis (Cameron), Eucerceris, 592
marginula Dalla Torre, Cerceris, 583
marhra Evans and Matthews, Bembix, 547
mariannense Yasumatsu, Pison, 336
maricandicus Radoszkowski, Oxybelus, 368
maricopa Pate, Belomicrus, 363
marismortui (Bytinski-Salz), Podalonia, 144
maritima Saussure, Cerceris, 583
maritimus F. Smith, Miscophus, 318
maritimus Marquet, Oxybelus, 369
marjoriae van Lith, Psen, 166
marlatti (Rohwer), Synnevrus, 470
marleyi Arnold, Belomicroides, 363
marleyi Arnold, Cerceris, 581
marnonis Handlirsch, Stizus, 527
marocanus Shestakov, Philanthus, 566
maroccana (Giner Marí), Dryudella, 214
maroccana Mercet, Bembix, 548
maroccanus Beaumont, Belomicrus, 364
maroccanus Beaumont, Tachysphex, 275
maroccanus (Dusmet), Hoplisoides, 521
maroccanus Pulawski, Tachytes, 265
marocensis (Tsuneki), Carinostigmus, 191
marshalli (Turner), Hoplisoides, 521
marshalli Turner, Stizus, 527
marshalli Turner, Tachysphex, 274
marshalli Turner, Tachytes, 265
marshi Menke, Ammophila, 152
marsupiata Handlirsch, Bembix, 547
marthae Handlirsch, Stizus, 527
marthae Willink, Carlobembix, 53, **539**
martialis Giner Marí, Cerceris, 583
martialis Holmberg, Trachypus, 569
martinezi Fritz, Foxia, 479
martjanowii (F. Morawitz), Ectemnius, 427
marunum Leclercq, Isorhopalum, 390
marusius Pate, Belomicroides, 363
masaicum Turner, Sceliphron, 105

mashona Arnold, Tachysphex, 275
Mashona Pate, 46, 381
masinissa (Morice), Podalonia, 144
masoni Krombein, Ammopsen, 40, **161**
masoni Williams, Solierella, 313
massaica Cameron, Ammophila, 151
massaica Cameron, Bembix, 546
massaica Cameron, Cerceris, 589
massaica (Cameron), Liris, 244
massaicum (Cameron), Chlorion, 90
massaicum Cameron, Trypoxylon, 347
massaicus Cameron, Dasyproctus, 420
massiliensis (Kohl), Ectemnius, 427
massiliensis Morice, Astata, 213
massinissa of authors, Podalonia, 144
mastogaster F. Smith, Cerceris, 583
matabele Arnold, Oxybelus, 368
matabele Empey, Cerceris, 629
matalensis Turner, Psen, 166
matangensis (Turner), Argogorytes, 492
mateui Giner Marí, Cerceris, 578
matinecoc Pate, Zanysson, 475
matronalis Dahlbom, Tachytes, 265
mattheyi (Beaumont), Bembecinus, 531
mauiensis (Blackburn), Ectemnius, 427
maura (Fabricius), Larra, 238
mauretanica Beaumont, Bembix, 548
mauretanicus Handlirsch, Gorytes, 501
mauretanus Pulawski, Tachysphex, 274
mauritanica (Mercet), Podalonia, 144
mauritanicus Linnaeus, Sphex, 117
mauritianus Christ, Sphex, 117
mauritii van Lith, Psenulus, 173
maurus F. Smith, Sphex, 115
maurus (Lepeletier and Brullé), Crossocerus, 404
maurus (Rohwer), Miscophus, 318
maurus (Rohwer), Psenulus, 173
maurus Rohwer, Tachysphex, 274
maurusius Beaumont, Belomicrus, 363
maurusius Valkeila, Pemphredon, 182
mavromoustakisi Andrade, Miscophus, 318
mavromoustakisi (Balthasar), Ammatomus, 513
mavromoustakisi Beaumont, Sphex, 116
mavromoustakisi Giner Marí, Cerceris, 586
maxillare (Palisot de Beauvois), Chlorion, 90
maxillosum (Poiret), Chlorion, 90
maxillosus Fabricius, Sphex, 116
maxima Maidl, Nitela, 325
maximiliani Kohl, Sphex, 115
maximiliani Saussure, Cerceris, 579
mayeri (De Witz), Ectemnius, 427
mayoni Leclercq, Isorhopalum, 390
mayri Brauns, Mesopalarus, 44, **306**
mayri (Handlirsch), Bembecinus, 531
mayri Richards, Trypoxylon, 349
mazaruni Richards, Trypoxylon, 349
mazimba Brauns, Cerceris, 583
mcateei Krombein, Gorytes, 501
mcclayi Menke, Ammophila, 152
medea (Handlirsch), Stictia, 542
media Klug, Cerceris, 583
medianum W. Fox, Trypoxylon, 348
mediata Cresson, Ammophila, 152
mediator Nurse, Trypoxylon, 347
mediatus Fabricius, Crabro, 408
medicus Leclercq, Dasyproctus, 421
meditata Shestakov, Cerceris, 583
mediterranea Gistl, Larra, 238

mediterranea Handlirsch, Bembix, 548
mediterraneus (Kohl), Lindenius, 384
mediterraneus Kohl, Tachysphex, 274
medium Beaumont, Trypoxylon, 346
medius Dahlbom, Diodontus, 179
medius (F. Smith), Pluto, 171
medius W. Fox, Crabro, 408
megacephala Brèthes, Cerceris, 583
megacephala (F. Smith), Foxita, 416
megacephalus (Rohwer), Ectemnius, 425
megacephalus (Rossi), Crossocerus, 402
megacera (J. Parker), Glenostictia, 552
megadonta Cameron, Bembix, 547
megaerus Turner, Tachytes, 265
Megalomma F. Smith, 52
megalommiforme (Strand), Megistommum, 503
Megalommus Shuckard, 52
megalophthalma (Handlirsch), Clitemnestra, 489
Megistommum Schulz, 52, **503**
Megalostizus Schulz, 53
Megapodium Dahlbom, 49
Megastizus Patton, 53
megerlei Dahlbom, Bembex, 547
megerlei (Dahlbom), Tracheliodes, 405
Megistommum Schulz, 52, 503
melaena F. Smith, Bembix, 549
melaena Murray, Podalonia, 144
melaenus (Spinola), Prionyx, 133
melaina Turner, Cerceris, 583
melampous J. Parker, Stictiella, 551
melanaria Cameron, Astata, 212
melanaria (Dahlbom), Eremnophila, 147
melanaria (Kohl), Ancistromma, 260
melanaria Kohl, Larra, 237
melanarius Arnold, Oxybelus, 369
melanarius (Mocsáry), Palmodes, 127
melanarius (Wesmael), Crossocerus, 402
melanaspis (Cameron), Philanthus, 565
melanaspis J. Parker, Bembix, 547
melancholica F. Smith, Bembix, 547
melancholicus Arnold, Tachytes, 265
melancholicus Chevrier, Oxybelus, 370
melanderi Arnold, Philanthus, 565
melania Turner, Liris, 246
melanica (Giner Marí), Dryudella, 214
melanius (Rohwer), Crossocerus, 402
melanocephalum Turner, Pison, 336
melanocera Cameron, Ampulex, 78
melanochilos Pate, Crossocerus, 402
melanochlorus Gmelin, Sphex, 117
melanocnemis Kohl, Sphex, 117
melanocnemis Turner, Larra, 238
melanocorne Strand, Trypoxylon, 347
Melanocrabro Perkins, 50
melanocrus Rohwer, Passaloecus, 184
melanogaster (Brèthes), Prionyx, 133
melanogaster Holmberg, Cerceris, 582
melanogaster (Kohl), Crossocerus, 402
melanogaster Riek, Aphelotoma, 70
melanogaster (Schrottky), Megistommum, 503
melanognathus Rohwer, Passaloecus, 184
melanoleucum Richards, Trypoxylon, 349
melanonotus van Lith, Psenulus, 173
melanopa Handlirsch, Bembix, 547
melanoda Strand, Sphex, 116
melanops Turton, Vespa, 627
melanoptera Kohl, Larra, 238
melanopteron Richards, Trypoxylon, 348

melanopterus (Dahlbom), Stizoides, 529
melanopus Brèthes, Cerceris, 584
melanopus Dahlbom, Sphex, 115
melanopus (Lucas), Eremochares, 146
mealnopus Pate, Nysson, 470
melanopygus A. Costa, Tachytes, 263
melanopyrus (F. Smith), Paranysson, 308
melanosa Scullen, Eucerceris, 591
melanosoma F. Smith, Chlorion, 90
melanosoma Gribodo, Bembix, 547
melanosoma Rohwer, Psen, 166
melanospila Cameron, Cerceris, 580
melanosterna J. Parker, Stictiella, 551
melanosticta (Gmelin), Mellinus, 449
melanostoma A. Costa, Bembix, 546
melanostoma Gmelin, Vespa, 627
melanotarsis (Cameron), Ectemnius, 427
melanotica Pulawski, Astata, 212
melanthe Banks, Cerceris, 583
melanothorax Schletterer, Cerceris, 588
melanotica Pulawski, Astata, 212
melanotus (F. Morawitz), Prionyx, 133
melanovittata Scullen, Eucerceris, 591
melanoxanthum (Schrottky), Megistommum, 504
melanoxanthus (F. Smith), Stizus, 527
melanura F. Morawitz, Bembix, 547
melanurum Cameron, Trypoxylon, 347
melanurus Handlirsch, Stizus, 527
melas Gussakovskij, Sphex, 115
melas Kohl, Tachysphex, 275
melas M. Müller, Nysson, 470
meliloti (Johnson and Rohwer), Bicyrtes, 538
melinopus (Kohl), Lindenius, 384
mellea Scullen, Eucerceris, 591
melleus (F. Smith), Stizus, 527
melleus Say, Alysson, 458
mellicula Turner, Cerceris, 584
melliniformis F. Smith, Philanthus, 567
MELLININI, 59, **445**
Mellinogastra Ashmead, 52
mellinoides (W. Fox), Lestiphorus, 506
Mellinus Fabricius, 51, **447**
Mellinusterius Meunier, 37
mellipes (Cresson), Epinysson, 472
mellipes (Say), Mimumesa, 164
melpomene (Arnold), Gorytes, 501
memnon Turner, Tachytes, 265
memnonia (F. Smith), Liris, 246
memorialis Banks, Entomognathus, 382
mendax (F. Smith), Larra, 238
mendesensis Brèthes, Cerceris, 584
mendica (Handlirsch), Bicyrtes, 537
mendicus (Handlirsch), Dienoplus, 496
mendozae (Dalla Torre), Trachypus, 569
mendozana Brèthes, Cerceris, 584
mendozanum Richards, Trypoxylon, 349
mendozanus Brèthes, Heliocausus, 453
mendozanus (Brèthes), Hoplisoides, 521
mendozanus Brèthes, Sphex, 115
mendozanus Brèthes, Tachysphex, 275
menelaus (Nurse), Prosopigastra, 285
meneliki Arnold, Philanthus, 565
menkei Fritz and Toro, Cerceris, 584
menkei Scullen, Eucerceris, 591
menyllus (Cameron), Ectemnius, 427
meraukensis Cameron, Tachytes, 265
merbabunda van der Vecht, Cerceris, 589
merceti Andrade, Miscophus, 318
merceti Arnold, Spilomena, 193
merceti (Beaumont), Hoplisoides, 521

merceti Beaumont, Stizus, 526
merceti Brauns, Nitela, 325
merceti J. Parker, Bembix, 547
merceti (Kohl), Lindenius, 384
merceti (Kohl), Podalonia, 144
mergus W. Fox, Tachytes, 265
meridianus van Lith, Psen, 167
meridionale Gussakovskij, Trypoxylon, 349
meridionale Richards, Trypoxylon, 346
meridionale Turner, Pison, 336
meridionalis A. Costa, Bembecinus, 53, **531**
meridionalis (A. Costa), Ectemnius, 427
meridionalis (Arnold), Sphex, 114
meridionalis Beaumont, Psenulus, 173
meridionalis C. Fox, Steniolia, 553
meridionalis Ducke, Anacrabro, 379
meridionalis Mocsáry, Oxybelus, 368
meridionalis (Tsuneki), Mimumesa, 164
merope Arnold, Cerceris, 584
Merospis Pate, 50
merredinensis Turner, Cerceris, 584
mertoni Strand, Sphex, 116
meruensis Cameron, Ammophila, 151
meruensis (Cameron), Isodontia, 123
mervensis Dalla Torre, Oxybelus, 368
mervensis Radoszkowski, Bembix, 548
mervensis (Radoszkowski), Podalonia, 144
merwensis Radoszkowski, Oxybelus, 368
mescalero Menke, Ammophila, 152
mescalero Pate, Belomicrus, 363
mescalero (Pate), Liris, 246
mescalero Pate, Pulverro, 42, **197**
mesillensis Cockerell, Bicyrtes, 537
Mesocrabro C. Verhoeff, 49
mesomelaena Lichtenstein, Sphex, 119
Mesopalarus Brauns, 44, **305**
mesopleuralis (F. Morawitz), Lindenius, 384
Mesopora Wesmael, 40
mesopotamica Brèthes, Cerceris, 584
mesostenus (Handlirsch), Ammatomus, 513
Metacrabro Ashmead, 50
metallescens Rayment, Sericophorus, 302
metallica Kohl, Ampulex, 78
metallicum Taschenberg, Chlorion, 89
metallicus C. Verhoeff, Miscophus, 318
metallicus Taschenberg, Sphex, 114
metamelanica Strand, Bembix, 546
Metanysson Ashmead, 51, **480**
metatarsalis Gussakovskij, Ammoplanus, 198
metatarsalis Turner, Cerceris, 584
metathoracicus (H. Smith), Epinysson, 472
metathoracicus (Mickel), Diodontus, 179
metopias Kohl, Oxybelus, 368
mexicana Cameron, Mimesa, 162
mexicana Cresson, Astata, 213
mexicana Gillaspy, Steniolia, 553
mexicana (Handlirsch), Stictia, 542
mexicana (Rohwer), Solierella, 313
mexicana Saussure, Cerceris, 584
mexicana (Saussure), Isodontia, 123
mexicana (Saussure), Podalonia, 144
mexicana Saussure, Trypoxylon, 349
mexicanum Saussure, Trypoxylon, 349
mexicanus (Cameron), Enoplolindenius, 415
mexicanus Cameron, Entomognathus, 382

mexicanus (Cameron), Psammaletes, 508
mexicanus (Cameron), Trachypus, 568
mexicanus (Cresson), Zanysson, 475
mexicanus (Handlirsch), Bembecinus, 531
mexicanus Robertson, Oxybelus, 368
mexicanus (Saussure), Prionyx, 134
mexicanus Saussure, Trachypus, 568
mexicanus Taschenberg, Sphex, 117
mexicanus Williams, Plenoculus, 311
meyeri Kohl, Belomicrus, 363
mianga Evans and Matthews, Bembix, 547
micado Cameron, Ampulex, 78
Micadophila Tsuneki, 39
micans Cameron, Ammophila, 152
micans (Eversmann), Prionyx, 134
micans (Handlirsch), Bicyrtes, 537
micans Kohl, Ampulex, 78
micans (Radoszkowski), Tachysphex, 275
micans (Spinola), Liris, 244
micans Taschenberg, Sphex, 114
micantipygus Strand, Tachytes, 266
michaelsensi Schulz, Lyroda, 299
michelbacheri R. Bohart, Philanthus, 565
micheneri Leclercq, Anacrabro, 379
micheneri Leclercq, Notocrabro 629
micheneri Scullen, Cerceris, 584
micipsa (Morice), Podalonia, 144
mickeli Giner Marí, Cerceris, 588
mickeli Murray, Podalonia, 144
miconiae Richards, Microstigmus, 192
Microbembex Patton, 53, **538**
microcephalus (Handlirsch), Hoplisoides, 521
Microcrabro Saussure, 48, **397**
Microglosella Rayment, 41
Microglossa Rayment, 41
micromegas (Saussure), Crossocerus, 402
micromegas Saussure, Tachysphex, 275
micropunctata Shestakov, Cerceris, 584
Microstictia Gillaspy, 54
microstictus (Herrich-Schaeffer), Ectemnius, 427
Microstigmus Ducke, 41, **191**
midas Arnold, Tachytes, 266
midus (Arnold), Entomognathus, 382
miegi Mercet, Nysson, 469
miegii Dufour, Astata, 212
mielatti Leclercq, Crossocerus, 402
migiurtinicum (Giordani Soika), Chlorion, 90
mildei Dahlbom, Bembix, 547
miles Schrottky, Trachypus, 568
miles Taschenberg, Parapsammophila, 139
miliaris (Cameron), Eremnophila, 147
militaris Dahlbom, Cerceris, 584
militaris (Gerstaecker), Synnevrus, 470
militaris O. Müller, Sphex, 117
militaris Saussure, Bembix, 547
militaris (Turner), Podagritus, 393
milleri Turner, Sphecius, 511
Millimus Gimmerthal, 51
millironi Krombein, Cerceris, 584
millsi Cockerell, Dryudella, 215
miltoni (van Lith), Pseneo, 165
miltosa Scullen, Cerceris, 586
mima Handlirsch, Bembix, 547
Mimesa Shuckard, 40, **161**
mimetes (Handlirsch), Megistommum, 503
mimetica (Cockerell), Clitemnestra, 489
mimeticus Honoré, Miscophus, 318

mimeticus Schrottky, Tachytes, 263
mimica Cresson, Cerceris, 584
mimica Menke, Ammophila, 152
mimicus (Arnold), Entomognathus, 382
mimicus (Arnold), Lestiphorus, 506
mimulus Valkeila, Nysson, 469
mimulus R. Turner, Sphex, 115
mimicus (Bridwell), Saliostethus, 320
Mimumesa Malloch, 40, **163**
minamikawai Tsuneki, Crossocerus, 402
minarum Ducke, Solierella, 313
minax Arnold, Astata, 213
minax (Kohl), Podalonia, 144
minax Mickel, Cerceris, 584
mindanao Menke, Liris, 246
mindanaoensis (Rohwer), Psenulus, 173
mindanaoensis Williams, Liris, 246
mindanaoensis (Williams), Liris, 246
mindoroensis (van Lith), Pseneo, 165
mindorensis Williams, Tachysphex, 276
miniatulus Arnold, Tachysphex, 275
minicum Arnold, Pison, 336
minidoka Pate, Belomicrus, 363
minima Beaumont, Prosopigastra, 285
minima (Arnold), Liris, 246
minima (Maidl), Liris, 246
minima Schletterer, Cerceris, 584
minima Schöberlin, Ammophila, 31
minima (Schulthess), Pseudoscolia, 573
minima Turner, Lyroda, 299
minimus (Gussakovskij, Belomicrus, 363
minimus (Packard), Crossocerus, 402
minimus Rohwer, Nysson, 469
minimus (Turner), Acanthostethus, 474
minimus (W. Fox), Tachysphex, 272
minitulus (Arnold), Crossocerus, 402
minor (Arnold), Bembecinus, 531
minor Beaumont, Trypoxylon, 346
minor (F. Morawitz), Palmodes, 127
minor Gussakovskij, Pemphredon, 182
minor Kohl, Ampulex, 78
minor Kohl, Astata, 213
minor Kohl, Philanthus, 565
minor Lepeletier, Ammophila, 152
minor Maréchal, Ammoplanus, 198
minor (Mickel), Oxybelus, 367
minor (Provancher), Tachysphex, 277
minor, Rohwer, Tachytes, 266
minor van Lith, Psen, 166
minor Williams, Larropsis, 258
minos (Beaumont), Chalybion, 103
minos Schrottky, Tachytes, 265
minuscula Turner, Cerceris, 584
minusculum Leclercq, Rhopalum, 389
minuta Frivaldsky, Ammophila, 153
minuta Lepeletier, Cerceris, 587
minutior Maidl, Cerceris, 584
minutissima Maidl, Cerceris, 584
minutissimus Arnold, Belomicrus, 363
minutissimus F. Morawitz, Oxybelus, 368
minutissimus (Radoszkowski), Spilomena, 193
minutula (Handlirsch), Stictiella, 551
minutulus Arnold, Tachysphex, 273, 275
minutus Andrade, Miscophus, 318
minutus Arnold, Miscophoides, 322
minutus Arnold, Nysson, 469
minutus (Fabricius), Diodontus, 179
minutus (Malloch), Pluto, 171
minutus (Mocsáry), Hoplisoides, 521
minutus Nurse, Tachysphex, 275
minutus Rayment, Sericophorus, 302

minutus Rohwer, Tachytes, 265
minutus (Tournier), Psenulus, 173
minutus (Wagner), Pemphredon, 182
minya Evans and Matthews, Bembix, 547
mionii Guérin-Méneville, Stizoides, 529
mirabilis Berland, Ampulex, 78
mirabilis (Gussakovskij), Eremochares, 146
mirabilis Shestakov, Cerceris, 578
mirandum (Kohl), Chlorion, 90
mirandus (W. Fox), Oryttus, 508
mirifica Pate, Solierella, 313
mirificus Kohl, Belomicrus, 363
mirus (Arnold), Bembecinus, 531
mirus Kohl, Tachytes, 266
miscella Mercet, Bembix, 545
MISCOPHINI, 60, **291**, 294
Miscophinus Ashmead, 45
Miscophoidellus Menke, 45, **322**
miscophoides (Arnold), Liris, 246
miscophoides (Arnold), TAchysphex, 275
Miscophoides Brauns, 45, **322**
miscophoides Spinola, Solierella, 45, **313**
Miscophus Jurine, 45, **314**
Miscothyris F. Smith, 51
Miscothyris Shuckard, 51
Miscus Jurine, 40
miserabilis J. Parker, Bembix, 547
misoolensis Krombein, Cerceris, 584
mitis Turner, Tachytes, 266
mitjaevi (Kazenas), Lestiphorus, 506
mitla Scullen, Cerceris, 579
mitlaensis Alfieri, Ammophila, 152
mitrata Bingham, Cerceris, 580
mitulus (Arnold), Bembecinus, 531
mixta Radoszkowski, Cerceris, 578
mixta (W. Fox), Mimumesa, 164
mixtus Fabricius, Sphex, 117
mixtus Snoflak, Lindenius, 384
miyanoi Holmberg, Cerceris, 584
mizubo Leclercq, Ectemnius, 427
mizuho (Tsuneki), Ectemnius, 427
mizuko Leclercq, Ectemnius, 427
mocanus Leclercq, Anacrabro, 379
mochiana Giordani Soika, Cerceris, 584
mochii Arnold, Miscophus, 318
mochii Arnold, Nitela, 325
mochii (Beaumont), Holotachysphex, 282
mochii Giordani Soika, Sphex, 115
mocsaryi Brauns, Prosopigastra, 285
mocsaryi Dalla Torre, Sceliphron, 106
mocsaryi Frivaldsky, Ammophila, 153
mocsaryi Handlirsch, Scapheutes, 46, **355**
mocsaryi Handlirsch, Stizus, 527
mocsaryi Kohl, Ampulex, 78
mocsaryi Kohl, Cerceris, 580
mocsaryi Kohl, Crabro, 408
mocsaryi (Kohl), Penepodium, 92
mocsaryi (Kohl), Prionyx, 134
mocsaryi Kohl, Spilomena, 193
mocsaryi Kohl, Tachysphex, 275
moczari Tsuneki, Cerceris, 584
moczari Tsuneki, Oxybelus, 368
moczari (Tsuneki), Podalonia, 144
moczari Tsuneki, Tachysphex, 275
modesta F. Smith, Cerceris, 584, 586
modesta (F. Smith), Liris, 246
modesta Handlirsch, Bembix, 547
modesta Mocsáry, Ammophila, 152
modesta Mocsáry, Dryudella, 215
modesta Radoszkowski, Olgia, 52, **491**
modesta (Rohwer), Mimumesa, 164
modesta (Rohwer), Solierella, 313

modestum Rohwer, Rhopalum, 389
modestus Arnold, Gastrosericus, 256
modestus Arnold, Miscophus, 318
modestus Arnold, Tachysphex, 275
modestus Brèthes, Oxybelus, 368
modestus (Cresson), Pseudoplisus, 503
modestus (F. Smith), Bembecinus, 531
modestus F. Smith, Sphex, 115
modestus F. Smith, Tachytes, 266
modestus (Kohl), Belomicrus, 363
moebii Handlirsch, Bembix, 547
moebii Kohl, Ampulex, 78
moenkopi Menke, Ammophila, 152
moenkopi Pate, Ammoplanops. 197
moerens Arnold, Tachytes, 265
moerens (Turner), Acanthostethus, 474
moesta De Stefani, Cerceris, 584
moestissima Guiglia, Cerceris, 584
moestus (Cresson), Epinysson, 472
moggionis Arnold, Cerceris, 584
Mohavena Pate, 42
mojavensis Scullen, Eucerceris, 592
mojavensis Williams, Plenoculus, 311
mokari Evans and Matthews, Bembix, 547
moksari Marquet, Ammophila, 154
molesta Arnold, Cerceris, 589
molokaiensis (Perkins), Ectemnius, 427
molucca Leclercq, Lestica, 431
moma Evans and Matthews, Bembix, 547
mombasae Arnold, Cerceris, 580
monachi Menke, Ammophila, 152
monachus Gerstaecker, Oxybelus, 367
monachus Mercet, Nysson, 469
monedula Handlirsch, Bembix, 547
Monedula Latreille, 54
moneduloides (F. Smith), Bembecinus, 531
moneduloides (Packard), Tanyoprymnus, 53, **512**
moneta F. Smith, Ammophila, 152
monetarius F. Smith, Tachytes, 266
mongolensis Tsuneki, Ammophila, 152
mongolensis Tsuneki, Ammoplanus, 198
mongolensis Tsuneki, Argogorytes, 492
mongolensis Tsuneki, Cerceris, 584
mongolensis Tsuneki, Crossocerus, 404
mongolensis (Tsuneki), Synnevrus, 470
Mongolia Tsuneki, 54
mongolica F. Morawitz, Mimesa, 162
mongolica (F. Morawitz), Parapiagetia, 281
mongolica Tsuneki, Bembix, 547
mongolica Tsuneki, Cerceris, 585
mongolica (Tsuneki), Dryudella, 215
mongolica Tsuneki, Spilomena, 193
mongolicus F. Morawitz, Philanthus, 565
mongolicus Gussakovskij, Gastrosericus, 256
mongolicus Kohl, Tachysphex, 275
mongolicus Tsuneki, Ammoplanops, 197
mongolicus Tsuneki, Belomicrus, 363
mongolicus Tsuneki, Bembecinus, 532
mongolicus Tsuneki, Crabro, 408
mongolicus Tsuneki, Diodontus, 179
mongolicus Tsuneki, Gorytes, 501
mongolicus Tsuneki, Lindenius, 384
mongolicus Tsuneki, Miscophus, 318
mongolicus Tsuneki, Oxybelus, 367
mongolicus Tsuneki, Passaloecus, 184
mongolicus Tsuneki, Stizus, 527
monglicus Tsuneki, Tachytes, 266

Moniaecera Ashmead, 47, **394**
monilicornis Dahlbom, Passaloecus, 184
monilicornis Morice, Parapsammophila, 139
mono Pate, Belomicrus, 363
monocera Kohl, Cerceris, 584
monochroma (Pulawski), Dryudella, 215
monodi (Berland), Bembecinus, 531
monodon (Handlirsch), Bembecinus, 531
monodonta (Say), Microbembex, 53, **539**
monoensis Scullen, Eucerceris, 592
Monomatium Shuckard, 42
monozonus (Cameron), Ectemnius, 427
monstrificum Kohl, Trypoxylon, 349
monstrosum (Kohl), Chalybion, 103
monstrosus (Dahlbom), Crossocerus, 400
monstrosus (Handlirsch), Afrogorytes, 53, **523**
monstrosus of authors, Crossocerus, 403
monstrosus (Tsuneki), Carinostigmus, 191
monstruosum Tsuneki, Trypoxylon, 630
montana Cameron, Ampulex, 78
montana (Cameron), Podalonia, 144
montana (Cresson), Dryudella, 215
montana Cresson, Eucerceris, 591
montana (Leclercq), Dicranorhina, 252
montanum Cameron, Pison, 336
montanum Schulz, Trypoxylon, 347
montanus (A. Costa), Psenulus, 173
montanus (Cameron), Psenulus, 173
montanus (Cameron), Pseudoplisus, 503
montanus (Cresson), Ectemnius, 426
montanus (Cresson), Tachysphex, 275
montanus Dahlbom, Pemphredon, 182
montanus (F. Morawitz), Palmodes, 127
montanus (Gistl), Ectemnius, 426
montanus Gussakovskij, Miscophus, 318
montanus (Radoszkowski), Tachysphex, 272
montanus Robertson, Oxybelus, 370
montealban Scullen, Cerceris, 584
montezuma Cameron, Ammophila, 152
montezuma Cameron, Cerceris, 584
montezuma (Cameron), Lindenius, 384
montezuma (Cameron), Liris, 244
montezuma (Cameron), Pseneo, 165
montezuma F. Smith, Steniolia, 553
monticola Arnold, Ampulex, 77
monticola Arnold, Cerceris, 584
monticola (Cameron), Pseneo, 165
monticola Dalla Torre, Tachysphex, 272
monticola Eighme, Pulverro, 197
monticola (Giner Marí), Dryudella, 215
monticola Gussakovskij, Ammoplanus, 198
monticola (Packard), Crabro, 408
monticola (Packard), Psen, 166
monticola (Perkins), Ectemnius, 427
monticola Tsuneki, Diodontus, 179
monticola Tsuneki, Trypoxylon, 347
montivaga Cameron, Cerceris, 583
montivaga (Cameron), Liris, 246
montivagans (Strand), Ectemnius, 426
montivagum Arnold, Trypoxylon, 347
montivagus (Arnold), Hoplisoides, 521
montivagus Arnold, Tachysphex, 274
montivagus (Cameron), Ectemnius, 428
montivagus Cameron, Stigmus, 189
montivagus (Dalla Torre), Pseneo, 165
montivagus (Mocsáry), Gorytes, 501
mookoensis Tsuneki, Diodontus, 179

mookosis Tsuneki, Tachysphex, 273
moonga Evans and Matthews, Bembix, 547
mopsus (Handlirsch), Synnevrus, 470
moraballi Richards, Trigonopsis, 98
moraballi Richards, Trypoxylon, 347
moralesi Giner Marí, Astata, 213
morata Cresson, Cerceris, 584
moravicus Balthasar, Miscophus, 318
moravicus (Snoflak), Dienoplus, 496
morawitzi (Ed. André), Podalonia, 144
morawitzi (Gussakovskij), Crossocerus, 402
morawitzi Mocsáry, Cerceris, 584
morawitzi Pulawski, Tachysphex, 275
morawitzi (Radoszkowski), Dienoplus, 496
mordax Kohl, Liris, 246
mordax Krombein, Cerceris, 584
morelensis (Williams), Paradolichurus, 70
morelos Scullen, Cerceris, 580
moricei Beaumont, Oxybelus, 368
moricei E. Saunders, Gastrosericus, 256
moricei Kohl, Belomicrus, 363
moricei Kohl, Diodontus, 179
moricei Mercet, Prospigastra, 285
moricei Shestakov, Cerceris, 588
morici Shestakov, Cerceris, 588
morio Cresson, Diodontus, 181
morio Kohl, Liris, 246
morio (Kohl), Palmodes, 127
morio Lepeletier and Brullé, Crossocerus, 401
morio Vander Linden, Pemphredon, 182
moroderi Giner Marí, Cerceris, 584
morosa (F. Smith), Isodontia, 123
morosula Brauns, Cerceris, 584
morosum F. Smith, Pison, 336
morosus Chevrier, Oxybelus, 367
morosus (F. Smith), Tachysphex, 275
Morphota F. Smith, 44
morrae Strand, Cerceris, 584
morrae (Strand), Liris, 246
morrensis (Strand), Hoplisoides, 521
morrisoni (Cameron), Podalonia, 144
morrisoni Richards, Trypoxylon, 345
mortifer Valkeila, Pemphredon, 182
mortuellus Cockerell, Tracheliodes, 31, 405
mortuum Cockerell, Chalybion, 33, 103
morula Giordani Soika, Cerceris, 584
morula Scullen, Eucerceris, 591
mossambica Gribodo, Cerceris, 584
Motes Kohl, 43, 241
mottensis Mickel, Oxybelus, 369
moultoni Turner, Ampulex, 78
mozambicus (Arnold), Ectemnius, 425
mucronata (Jurine), Ammophila, 153
mucronatus (Fabricius), Oxybelus, 368
mucronatus Packard, Oxybelus, 369
mucronatus (Thomson), Crossocerus, 401
muesebecki (Krombein), Liris, 246
muiri Turner, Dasyproctus, 421
mukodzimaensis Tsuneki, Bembecinus, 530
mullewanus Leclercq, Podagritus, 393
multiannulatus Dalla Torre, Philanthus, 565
multicolor Arnold, Cerceris, 580
multicolor Turner, Stizus, 527
multifasciatus Tsuneki, Belomicrus, 363

multiguttata Turner, Cerceris, 584
multiguttatus (Arnold), Bembecinus, 531
multiguttatus Arnold, Palarus, 291
multimaculatus Cameron, Philanthus, 565
multipicta F. Smith, Bembix, 547
multipicta F. Smith, Cerceris, 584
multipictoides Giordani Soika, Cerceris, 587
multipictus (Rohwer), Psenulus, 173
multipunctatus (van Lith), Pseneo, 165
multistrigatum Turner, Pison, 336
multistrigosa Reed, Clitemnestra, 489
munakatai (Tsuneki), Ectemnius, 427
munakatai Tsuneki, Stigmus, 189
munda Mickel, Cerceris, 584
mundula Kohl, Larra, 238
mundurra Evans and Matthews, Bembix, 547
mundus (Beaumont), Dienoplus, 496
mundus W. Fox, Tachysphex, 275
munta Evans and Matthews, Bembix, 547
murarium (F. Smith), Sceliphron, 106
murgabensis (Gussakovskij), Plenoculus, 311
murgabica Radoszkowski, Cerceris, 584
murina (Dahlbom), Liris, 246
murorum (Latreille), Crossocerus, 403
murotai Tsuneki, Ampulex, 78
murotai Tsuneki, Rhopalum, 389
murotai Tsuneki, Trypoxylon, 347
Murrayella R. Bohart and Menke, 39
murrayi Menke, Ammophila, 152
musaeus (Cameron), Dasyproctus, 420
musca Handlirsch, Bembix, 547
muscicapa Handlirsch, Bembix, 545
mushaense Tsuneki, Rhopalum, 389
muspa (Pate), Liris, 248
mutabilis (Arnold), Bembecinus, 531
mutabilis Arnold, Cerceris, 584
mutatum Kohl, Trypoxylon, 349
mutica Dahlbom, Ammophila, 152
muticus Kohl, Sphex, 115
mutilatum Richards, Trypoxylon, 347
mutilloides (Ducke), Losada, 480
mutilloides Kohl, Ampulex, 78
mutilloides Walker, Tachytes, 266
Mutillonitela Bridwell, 45
mweruensis (Arnold), Hoplisoides, 521
mweruensis (Arnold), Sphex, 115
mycerinus Beaumont, Tachysphex, 275
myersi Turner, Microstigmus, 192
myersianus (Rohwer), Psen, 166
mystaceus (Linnaeus), Argogorytes, 492
mysticus (Gerstaecker), Acanthostethus, 474
mysticus Pulawski, Tachysphex, 273

nabataeus Beaumont, Dinetus, 216
nabateus Bytinski-Salsz, Philanthus, 566
nadigi Roth, Ammophila, 153
nadigi Roth, Stizus, 527
nadigi Shestakov, Cerceris, 577
naefi (Beaumont), Bembecinus, 531
nagamasa Tsuneki, Cerceris, 584
nagamasae Tsuneki, Trypoxylon, 347
nalandicum Strand, Sceliphron, 105
namana (Bischoff), Liris, 246
namaqua Arnold, Philanthus, 565
nambui Tsuneki, Bembecinus, 531
nambui Tsuneki, Tachysphex, 275
nambui Tsuneki, Trypoxylon, 347
namkumiensis (Laidlaw), Prionyx, 134
nana Arnold, Astata, 212

INDEX 671

nana Arnold, Tachytella, 257
nana (Bingham), Liris, 246
nana Shestakov, Cerceris, 584
nannophyes Merisuo, Pemphredon, 182
nanulus Strand, Sphex, 114
nanus Arnold, Tachysphex, 275
nanus Bingham, Oxybelus, 369
nanus (Cameron), Entomognathus, 382
nanus (Handlirsch), Bembecinus, 531
nanus Handlirsch, Nysson, 469
naranhun Tsuneki, Tachysphex, 275
narcissus Leclercq, Piyumoides, 411
narratus Leclercq, Entomognathus, 382
nasalis F. Morawitz, Tachysphex, 275
nasalis Provancher, Ammophila, 152
nasalis R. Bohart, Philanthus, 565
nasica Viereck & Cockerell, Cerceris, 584
nasicornis (F. Smith), Lamocrabro, 413
nasicornis Gussakovskij, Tachytes, 266
nasicornis van Lith, Psenulus, 173
nasidens Schletterer, Cerceris, 584
nasuta (Christ), Rubrica, 541
nasuta Dahlbom, Cerceris, 586
nasuta Er. André, Ampulex, 78
nasuta J. Parker, Microbembex, 539
nasuta Latreille, Cerceris, 586
nasuta Lepeletier, Ammophila, 152
nasuta Lepeletier, Cerceris, 587
nasuta Morice, Bembix, 547
nasutum Tsuneki, Trypoxylon, 630
nasutus Bischoff, Oxybelus, 369
nasutus (Cameron), Brachystegus, 473
nasutus Gribodo, Lindenius, 384
nasutus Tsuneki, Ammoplanus, 198
natalensis Arnold, Oxybelus, 369
natalensis Arnold, Philanthus, 565
natalensis (F. Smith), Pseudoplisus, 503
natalensis Saussure, Cerceris, 584
natalensis Saussure, Tachytes, 266
natalis Dahlbom, Bembix, 545
natalis J. Parker, Microbembex, 539
natallena Brimley, Cerceris, 581
nathani Leclercq, Entomognathus, 382
nativitatis Costa, Cerceris, 576
nattereri Kohl, TAchysphex, 274
nattereri Richards, Trypoxylon, 349
navajo Pate, Foxia, 479
naviculatus Villers, Sphex, 117
navigatrix Strand, Cerceris, 576
navitatis F. Smith, Cerceris, 576
neahminax Scullen, Cerceris, 584
neara Nurse, Larra, 238
nearctica Kohl, Ammophila, 152
nearctica R. Bohart, Prosopigastra, 285
nearcticum (Kohl), Chlorion, 89
nearcticus Kohl, Pemphredon, 182
neavei (Arnold), Sphex, 115
neavei Kohl, Stizus, 526
neavei Turner, Gastrosericus, 256
neavei (Turner), Liris, 246
neavei Turner, Prosopigastra, 285
neavei Turner, Stizus, 527
neavei Turner, Tachytes, 267
neboissi Leclercq, Rhopalum, 389
nebrascensis H. Smith, Mimesa, 161
nebrascensis H. S. Smith, Cerceris, 584
nebulosa Cameron, Cerceris, 584
nebulosa F. Smith, Ampulex, 78
nebulosus (Packard), Hoplisoides, 521
Nectanebus Spinola, 55
nectarinoides (Ducke), Bembecinus, 530

nefertiti Menke, Ammophila, 152
neghelliensis Guiglia, Cerceris, 584
neglecta Dahlbom, Bembix, 547
neglecta (F. Smith), Williamsita, 422
neglectum Kohl, Trypoxylon, 349
neglectus (Cresson), Bembecinus, 531
neglectus Handlirsch, Gorytes, 501
neglectus (Rohwer), Argogorytes, 492
neglectus Turner, TAchytes, 266
neglectus (Wagner), Pemphredon, 181
negrosensis (Williams), Liris, 246
nelli (Viereck), Crossocerus, 402
nemaensis Arnold, Cerceris, 587
nemoralis (Arnold), Bembecinus, 531
nemoralis (Arnold), Crossocerus, 403
nemoralis Evans, Sphodrotes, 304
nenicra Saussure, Cerceris, 584
nenitra Saussure, Cerceris, 584
nenitroidea Bischoff, Cerceris, 584
Neoblepharipus Leclercq, 48, **398**
Neodasyproctus Arnold, 49, **418**
Neodiodontus Tsuneki, 40, **178**
Neofoxia Viereck, 40
neogenita Schulz, Cerceris, 584
Neogorytes R. Bohart, 52, **492**
neomexicna Johnson and Rohwer, Microbembex, 539
neomexicanus Court and R. Bohart, Lindenius, 384
neomexicanus Rohwer, Diodontus, 179
neomexicanus Rohwer, Tachysphex, 272
neomexicanus Strandtmann, Philanthus, 565
Neomimesa Perkins, 40, **167**
Neonysson R. Bohart, 51, **470**
Neoplisus R. Bohart, 52, **504**
neorusticus R. Bohart, Nysson, 469
Neosphex Reed, 39
neotropica Kohl, Ampulex, 78
neotropicum F. Parker, Diploplectron, 211
neotropicum Menke, Pison, 337
neotropicus Kohl, Sphex, 115
neotropicus Kohl, Stigmus, 189
neoxena F. Smith, Ammophila, 151
neoxenus (Kohl), Prionyx, 133
nepalensis Bingham, Philanthus, 565
nepalensis van Lith, Psen, 166
nepalensis (Zavattari), Podalonia, 144
nephertiti (Pulawski), Dryudella, 215
Nephridia Brullé, 45
nephthys Arnold, Cerceris, 584
neptunus Leclercq, Ectemnius, 427
neptunus van Lith, Psenulus, 173
nescius Merisuo, Pemphredon, 182
nesiotes (Pate), Ectemnius, 428
Nesocrabro Perkins, 50
neuqueni Leclercq, Podagritus, 393
neuvillei Magretti, Oxybelus, 369
nevadensis (Cresson), Sphecius, 511
nevadensis (Dalla Torre), Eucerceris, 591
nevadensis R. Bohart, Tachytes, 266
nevadensis Rodeck, Bembix, 545
nevadensis W. Fox, Gorytes, 501
nevadica Cresson, Astata, 213
nevesi Andrade, Miscophus, 319
nicaraguaense Cameron, Rhopalum, 389
nicaraguanum Kohl, Sceliphron, 105
nicholi (Carter), Podalonia, 144
nicolai Cockerell, Bembix, 545
nicolai Ferton, Miscophus, 319
nielseni (Kohl), Ectemnius, 427
nietneri van Lith, Psenulus, 173

nigella (F. Smith), Isodontia, 123
nigelloides (Strand), Isodontia, 123
nigellum Krombein, Pison, 336
nigellus Lichtenstein, Sphex, 119
niger (A. Costa), Dienoplus, 496
niger Beaumont, Palarus, 291
niger Berland, Tachytes, 264
niger Chevrier, Nysson, 469
niger Dahlbom, Miscophus, 319
niger (Dufour), Gastrosericus, 256
niger (Hacker & Cockerell), Sericophorus, 303
niger Lepeletier and Brullé, Crossocerus, 402
niger (Motschulsky), Carinostigmus, 191
niger O. Müller, Sphex, 117
niger (Provancher), Crossocerus, 400
niger (Radoszkowski), Stizoides, 529
niger Robertson, Oxybelus, 369
niger (Vachal), Stizoides, 529
nigeriae Brauns, Cerceris, 584
nigeriae Leclercq, Psenulus, 173
nigerrima Gmelin, Vespa, 627
nigerrimus A. Costa, Sphex, 115
nigerrimus (F. Smith), Tachysphex, 275
nigerrimus Schrank, Sphex, 117
nigerrimus (White), Tachysphex, 274
Nigliris Tsuneki, 43
nigra Ashmead, Cerceris, 584
nigra (Ashmead), Solierella, 313
nigra Brèthes, Cerceris, 584
nigra (Brullé), Podalonia, 143
nigra (Cameron), Lyroda, 299
nigra Cameron, Trirogma, 73
nigra F. Smith, Astata, 213
nigra (Fabricius), Liris, 246
nigra (Maidl), Dicranorhina, 252
nigra of (Vander Linden), Liris, 246
nigra (Packard), Mimumesa, 40, **164**
nigra (Rohwer), Solierella, 313
nigrans Krombein, Solierella, 313
nigratus (Brischke), Psenulus, 173
nigrellum Rohwer, Trypoxylon, 346
nigrescens (Arnold), Ectemnius, 428
nigrescens Berland, Tachytes, 266
nigrescens F. Smith, Cerceris, 584
nigrescens (Rohwer), Mimesa, 162
nigrescens (Rohwer), Miscophus, 319
nigrescens Rohwer, Tachysphex, 275
nigrescens van der Vecht and Krombein, Sphex, 116
nigricans Cameron, Ampulex, 78
nigricans Cameron, Astata, 213
nigricans Cameron, Miscophus, 319
nigricans Cameron, Trypoxylon, 347
nigricans Dahlbom, Ammophila, 152
nigricans (F. Smith), TAchysphex, 275
nigricans Morice, Didineis, 459
nigricans Turner, Nitela, 325
nigricans (Walker), Liris, 246
nigricapillus (Berland), Pionyx, 133
nigricaudus Brèthes, Tachytes, 263
nigriceps Bingham, Philanthus, 565
nigriceps F. Smith, Cerceris, 584
nigriceps (F. Smith), Megistommum, 503
nigriceps (Rohwer), Miscophus, 319
nigriclypeus (Sonan), Bembecinus, 531
nigricollis Arnold, Ammophila, 152
nigricolor (Dalla Torre), Tachysphex, 275
nigricomus R. Bohart, Pseudoplisus, 503
nigricornis (Dufour), Sphecius, 511
nigricornis (Provancher), Crossocerus, 402

672 SPHECID WASPS

nigricornis Rayment, Sericophorus, 302
nigricornis Shuckard, Oxybelus, 367
nigricornis (Tournier), Psenulus, 173
nigricoxis Strand, Stigmus, 189
nigricula Riek, Aphelotoma, 70
nigridens (Herrich-Schaeffer), Lestica, 430
nigridorsum Pulawski, Solierella, 313
nigrifacies (Mocsáry), Gorytes, 501
nigrifemur Arnold, Cerceris, 581
nigrifemur Beaumont, Miscophus, 318
nigrifex Richards, Microstigmus, 192
nigrifrons (Cresson), Ectemnius, 427
nigrifrons (F. Smith), Argogorytes, 492
nigrifrons F. Smith, Cerceris, 584
nigrifrons (Provancher), Microbembex, 539
nigrina F. Morawitz, Ammophila, 151
nigrina Giner Marí, Cerceris, 578
nigrinervis Cameron, Psen, 166
nigrinum Kiesenwetter, Rhopalum, 389
nigrinus Bytinski-Salz, Philanthus, 566
nigrinus (Herrich-Schaeffer), Ectemnius, 424
nigrior Arnold, Palarus, 291
nigrior (W. Fox), Crossocerus, 400
nigrior W. Fox, Tachysphex, 275
nigripennata (Dalla Torre), Liris, 244
nigripennis Cameron, Liris, 245
nigripennis (Spinola), Tachysphex, 276
nigripennis Tsuneki, Psen, 166
nigripennis (W. Fox), Liris, 244
nigripes F. Smith, Ammophila, 151
nigripes F. Smith, Sphex, 117
nigripes (Fabricius), Larra, 238
nigripes (Guérin-Méneville), Chlorion, 90
nigripes Holthusius, Sphex, 119
nigripes Honoré, Miscophus, 319
nigripes J. Parker, Steniolia, 553
nigripes Lichtenstein, Sphex, 119
nigripes Olivier, Oxybelus, 370
nigripes Provancher, Nysson, 470
nigripes Pulawski, Tachysphex, 275
nigripes Reed, Ammophila, 152
nigripes (Saussure), Liris, 247
nigripes Spinola, Nysson, 470
nigripes Tsuneki, Tachysphex, 274
nigripes (Westwood), Dynatus, 94
nigrispine Cameron, Trypoxylon, 347
nigristoma Brauns, Cerceris, 584
nigrita Eversmann, Mimesa, 162
nigrita (Lepeletier), Liris, 244
nigritaria (Walker), Eremochares, 146
nigritarsus (Herrich-Schaeffer), Ectemnius, 427
nigrithorax (Benoit), Chalybion, 103
nigritula Banks, Cerceris, 584
nigritulus Turner, Oxybelus, 369
nigritus (Gimmerthal), Crossocerus, 400
nigritus Gravenhorst, Philanthus, 565
nigritus (Lepeletier and Brullé), Crossocerus, 402
nigritus (Lucas), Prionyx, 134
nigritus W. Fox, Diodontus, 179
nigriventre (A. Costa), Sceliphron, 105
nigriventris Arnold, Stizus, 527
nigriventris (Brèthes), Podagritus, 393
nigriventris (Cameron), Liris, 246
nigriventris (Gussakovskij), Podalonia, 144
nigriventris Tournier, Oxybelus, 368
nigriventris Tsuneki, Oxybelus, 369
nigriventris van Lith, Psen, 166
nigroaeneus Shuckard, Oxybelus, 367

nigroannulatus Bischoff, Tachytes, 266
nigrocaerulea Cameron, Ammophila, 151
nigrocaerulea Saussure, Ampulex, 78
nigrocaudatus Williams, Tachysphex, 275
nigrocincta Dufour, Cerceris, 589
nigrocincta F. Smith, Cerceris, 577
nigrocincta (Fernald), Eremnophila, 147
nigrocoerulea (Taschenberg), Isodontia, 123
nigrocornuta J. Parker, Bembix, 547
nigrohirta (Kohl), Podalonia, 144
nigrohirtus Kohl, Sphex, 115
nigrohirtus Turner, Philanthus, 565
nigrolineatus (Cameron), Psenulus, 173
nigromaculatus (Cameron), Psenulus, 173
nigromaculatus (F. Smith), Chimiloides, 414
nigropectinata Turner, Bembix, 547
nigropectinatus (Taschenberg), Prionyx, 133
nigropilosa Cresson, Astata, 213
nigropilosa (Rohwer), Ammophila, 153
nigropilosellus (Cameron), Tachytes, 266
nigror Rayment, Sericophorus, 302
nigror (W. Fox), Crossocerus, 400
nigroris Banks, Cerceris, 580
nigrum (Latreille), Pison, 335
nikkoense Tsuneki, Rhopalum, 389
nikkoensis Tsuneki and Tanaka, Crossocerus, 402
nikkoensis Tsuneki, Psenulus, 173
nikkoensis Tsuneki, Spilomena, 193
nikolajevskii (Gussakovskij), Ammatomus, 513
nilotica Priesner, Bembix, 547
nilotica Schletterer, Cerceris, 578
niloticus F. Smith, Philanthus, 564
niloticus Handlirsch, Stizus, 527
niloticus Honoré, Miscophus, 319
niloticus Pulawski, Tachysphex, 275
niloticus Turner, Tachytes, 266
niponica F. Smith, Bembix, 547
nipponensis Tsuneki, Cerceris, 584
nipponensis Tsuneki, Ectemnius, 424
nipponensis Yasumatsu, Psenulus, 173
nipponica Tsuneki, Ammophila, 153
nipponica Tsuneki, Didineis, 459
nipponicola Tsuneki, Passaloecus, 184
nipponicum (Kohl), Rhopalum, 389
nipponicum Tsuneki, Sceliphron, 106
nipponicum Tsuneki, Trypoxylon, 347
nipponicus Tsuneki, Oxybelus, 369
nipponicus Tsuneki, Tachytes, 266
nipponis Tsuneki, Argogorytes, 492
nipponis Tsuneki, Ectemnius, 427
nipponius Tsuneki, Crossocerus, 401
Nippononysson Yasumatsu and Maidl, 51, **465**
Nipponopsen Yasumatsu, 40
nirginus Kazenas, Belomicrus, 363
nisera Leclercq, Enoplolindenius, 415
Nitela Latreille, 45, **322**
Niteliopsis S. Saunders, 45
nitelopteroides Williams, Tachysphex, 275
Nitelopterus Ashmead, 45, 315
nitens (Arnold), Liris, 247
nitens (Banks), Philanthus, 567
nitens Handlirsch, Bothynostethus, 352
nitens Williams, Solierella, 313
nitida Banks, Cerceris, 583
nitida Cameron, Liris, 244

nitida (Cheesman), Liris, 244
nitida Fischer-Waldheim, Ammophila, 152
nitida Gussakovskij, Solierella, 313
nitida Perkins, Nesomimesa, 167
nitida Wesmael, Cerceris, 586
nitidicollis Turner, Ampulex, 78
nitidicorpus Tsuneki, Crossocerus, 402
nitidior Beaumont, Miscophus, 319
nitidior Beaumont, Tachysphex, 275
nitidissimum Richards, Trypoxylon, 347
nitidissimus Beaumont, Tachysphex, 275
nitidiusculus (F. Smith), Tachysphex, 275
nitidiusculus (F. Smith), Tachytes, 266
nitidiventris F. Smith, Sphex, 116
nitidiventris Spinola, Sphex, 115
nitidiventris (W. Fox), Crossocerus, 402
nitidulum (Christ), Chalybion, 102
nitidulus (Fabricius), Tachytes, 266
nitidum F. Smith, Pison, 335, 336
nitidum F. Smith, Trypoxylon, 46, **349**
nitidum (Spinola), Penepodium, 92
nitidus Andrade, Miscophus, 319
nitidus Arnold, Diodontus, 179
nitidus Beaumont, Lindenius, 384
nitidus (Cameron), Entomognathus, 382
nitidus Magretti, Philanthus, 565
nitidus of authors, Tachysphex, 277
nitidus (Spinola), Tachysphex, 275
nitidus van Lith, Psen, 166
nitobei (Matsumura), Lestica, 431
nitraria Pulawski, Solierella, 313
nitrariae Morice, Cerceris, 584
nivalis Tsuneki, Cerceris, 589
niveatus (Dufour), Prionyx, 133
niveifrons Rayment, Sericophorus, 302
niveitarse Saussure, Trypoxylon, 349
nivosus (F. Smith), Prionyx, 133
nobilis Andrade, Miscophus, 319
nobilis Brèthes, Sphecius, 511
nobilis Kohl, Philanthus, 565
nobilis O. Müller, Sphex, 117
nobilis Radoszkowski, Cerceris, 578
nobilitata Cameron, Cerceris, 584
nobilitatum Taschenberg, Chlorion, 89
noctulum Turner, Pison, 336
nodosa W. Fox, Didineis, 459
nodosicorne Turner, Trypoxylon, 347
nodosum Arnold, Trypoxylon, 347
nodosus Klug, Philanthus, 565
nokomis (Rohwer), Ectemnius, 424
nolcha Pate, Timberlakena, 42, **201**
nomadus Andrade, Miscophus, 318
nomarches Pulawski, Tachytes, 266
nonakai Tsuneki, Tachysphex, 275
noonadanius van Lith, Psenulus, 173
nordicus Leclercq, Philanthus, 564
nortinus Fritz and Toro, Cerceris, 584
Norumbega Pate, 49
notabilis Arnold, Bembix, 547
notabilis (Handlirsch), Neoplisus, 52, **505**
notabilis (Handlirsch), Synnevrus, 470
notabilis Turner, Tachytes, 266
notata Dahlbom, Bembix, 548
notatulus F. Smith, Philanthus, 565
notatus (Illiger), Crossocerus, 401
notatus (Taschenberg), Selman, 54, **541**
Nothocrabro Pate, 48, 398
Nothosphecius Pate, 53, 510
notipilis R. Bohart, Pseudoplisus, 503
Notocrabro Leclercq, 47, **394**
notogeum Leclercq, Rhopalum, 389
Notoglossa Dahlbom, 46

INDEX 673

Notogonia A. Costa, 43
notogoniaeformis Nadig, Tachysphex, 275
Notogonidea Rohwer, 43
Notogonius Howard, 43
notonitidus (Willink), Prionyx, 133
notostictus (Perkins), Ectemnius, 425
Nototis Arnold, 46
novaguineensis R. Bohart, Ochleroptera, 490
novahibernicus van Lith, Psen, 166
novanus (Rohwer), Ectemnius, 425
novarae Saussure, Ampulex, 77
novarae Saussure, Cerceris, 585
novarae (Saussure), Tachysphex, 275
noverca (Kaye), Prionyx, 134
novita (Fernald), Ammophila, 152
novocaledonica (Williams), Williamsita, 422
novocaledonicum Krombein, Pison, 336
novogeorgica Krombein, Cerceris, 578
novomexicana Viereck and Cockerell, Cerceris, 585
nubecula Cresson, Astata, 213
nubigera Gussakovskij, Prosopigastra, 285
nubilipennis (Arnold), Carinostigmus, 191
nubilipennis Beaumont, Tachysphex, 275
nubilipennis Cameron, Bembix, 546
nubilipennis Cresson, Bembix, 547
nubilipennis (F. Smith), Stizus, 526
nubilis Beaumont, Sphex, 116
nubilosa Gillaspy, Stictiella, 551
nubilosa J. Parker, Bembix, 547
nuda (Murray), Ammophila, 151
nuda (Nurse), Prosopigastra, 285
nuda (Taschenberg), Liris, 244
nudatus (Kohl), Prionyx, 133
nudiventris A. Costa, Larra, 237
nudiventris (Turner), Acanthostehus, 474
nudiventris Turner, Tachytes, 266
nudus Fernald, Sphex, 115
nugax Arnold, Cerceris, 584
nugax (Kohl), Liris, 247
nugenti (Turner), Isodontia, 123
numida Pulawski, Prosopigastra, 629
numidicus Gribodo, Crabro, 408
numidus Beaumont, Miscophus, 319
numidus (Schulz), Stizoides, 529
nupera J. Parker, Bembix, 547
nupta Shestakov, Cerceris, 584
nuristanica (Balthasar), Dryudella, 215
Nursea Cameron, 51, 465
nursei (Bingham), Philanthus, 565
nursei (Kohl), Ectemnius, 428
nursei Turner, Cerceris, 584
nursei Turner, Palarus, 291
nyamadanus Tsuneki, Bembecinus, 629
nyasae Arnold, Nitela, 325
nyasae (Turner), Bembecinus, 531
nyasicus (Turner), Pseudoplisus, 503
nyanzae (R. Turner), Sphex, 115
nyassae Arnold, Oxybelus, 368
nyholmi (Arnold), Dasyproctus, 420
nylanderi (Dahlbom), Psenulus, 172
nyssae Arnold, Liris, 248
Nysso Latreille, 51
Nysson Latreille, 51, 467
NYSSONINAE, 57, 441, 445
NYSSONINI, 59, 461, 463, 465
Nyssonus Rafinesque-Schmaltz, 51

oakleyi Krombein, Pison, 336
oasicola Tsuneki, Cerceris, 576
oasicola Tsuneki, Diodontus, 179
oasicola Tsuneki, Oxybelus, 369
oasicola Tsuneki, Prosopigastra, 286
oasis (Tsuneki), Prionyx, 133
oaxaca Scullen, Cerceris, 585
obductus W. Fox, Tachytes, 266
oberon Arnold, Tachysphex, 275
obesum Arnold, Pison, 336
obesus Arnold, Palarus, 291
obesus Kohl, Tachytes, 265
obesus (Turner), Austrogorytes, 498
obidense Richards, Trypoxylon, 349
obidensis (Ducke), Paradolichurus, 70
obliqua (Cresson), Steniolia, 553
obliqua (F. Smith), Ancistromma, 260
obliquestriatus (Mocsáry), Prionyx, 133
obliquetruncata (Turner), Liris, 247
obliquum F. Smith, Pison, 337
obliquus (Shuckard), Crossocerus, 401
oblita Holmberg, Ochleroptera, 52, 490
obliterata Turner, Spilomena, 193
obliteratum F. Smith, Pison, 45, 337
obliteratus Pic, Philanthus, 566
obliteratus (Turner), Acanthostethus, 474
oblongus (Packard), Ectemnius, 427
obo Tsuneki, Cerceris, 585
obo Tsuneki, Diodontus, 179
obo (Tsuneki), Dryudella, 215
obo (Tsuneki), Podalonia, 144
obockensis Leclercq, Dasyproctus, 420
obregon Scullen, Cerceris, 585
obrieni Krombein, Cerceris, 589
obscura Arnold, Astata, 213
obscura Beaumont, Solierella, 313
obscura Bischoff, Ammophila, 152
obscura (Magretti), Liris, 246
obscura Schletterer, Cerceris, 584
obscura Sickmann, Larra, 238
obscuranus Rohwer, Tachytes, 264, 266
obscurella (F. Smith), Isodontia, 123
obscurior Dalla Torre, Larra, 238
obscurior Gussakovskij, Spilomena, 193
obscuripennis (Schenck), Tachysphex, 275
obscuritarsis Pulawski, Miscophus, 319
obscurum Shuckard, Pison, 337
obscurus Andrade, Miscophus, 318
obscurus (Beaumont), Dienoplus, 496
obscurus Cresson, Tachytes, 264
obscurus (F. Smith), Ectemnius, 426
obscurus (Fabricius), Sphex, 115
obscurus Fischer-Waldheim, Sphex, 116
obscurus Handlirsch, Mellinus, 449
obscurus (Kohl), Belomicrus, 363
obscurus Pulawski, Tachysphex, 275
obscurus Reed, Heliocausus, 453
obscurus Statz, Sphex, 31, 117
observabilis Kohl, Tachytes, 266
observabilis (R. Turner), Sphex, 115
obsoleta Cameron, Cerceris, 585
obsoleta Howard, Bembix, 547
obsoletus (Rossi), Tachytes, 266
obsonator F. Smith, Trypoxylon, 347
obstrictus Gussakovskij, Ectemnius, 428
obtusa Turner, Bembix, 547
obtusedentata Maidl, Cerceris, 588
obtusedentata (Maidl), Liris, 247
occidentale (Beaumont), Chlorion, 90
occidentale (W. Fox), Rhopalum, 389
occidentalis Andrade, Miscophus, 319

occidentalis (Arnold), Dasyproctus, 421
occidentalis Arnold, Larra, 237
occidentalis Arnold, Palarus, 291
occidentalis Beaumont, Philanthus, 564
occidentalis Cresson, Astata, 213
occidentalis Giner Marí, Cerceris, 580
occidentalis J. Parker, Stizus, 527
occidentalis Johnson and Rohwer, Microbembex, 539
occidentalis (Malloch), Psenulus, 172
occidentalis Murray, Podalonia, 144
occidentalis Pulawski, Tachytes, 266
occidentalis Rayment, Sericophorus, 302
occidentalis Saussure, Cerceris, 581
occidentalis Strandtmann, Philanthus, 565
occidentalis (Viereck), Liris, 244
occidentalis W. Fox, Bembix, 547
occidentalis W. Fox, Diodontus, 179
occidentalis Williams, Pisonopsis, 332
occipitalis Arnold, Ampulex, 78
occipitalis F. Morawitz, Ammophila, 153
occipitomaculata Packard, Cerceris, 585
occitanica Mocsáry, Bembix, 546
occitanicus (Lepeletier and Serville), Palmodes, 39, 127
occitanicus Marquet, Oxybelus, 369
occulata (Jurine), Astata, 212
occultum Kohl, Chlorion, 89
occultus Fabricius, Crabro, 408
oceania Beaumont, Cerceris, 585
oceanica Brèthes, Cerceris, 589
oceanicum (Turner), Rhopalum, 390
oceanicus Beaumont, Belomicrus, 363
oceanicus Beaumont, Palarus, 291
oceanicus (Schulz), Podagritoides, 394
ocellare Kohl, Sceliphron, 106
ocellaris Gimmerthal, Pemphredon, 182
ocellatus Beaumont, Alysson, 458
ocellatus Gistel, Stizus, 527
ocellatus Packard, Anacrabro, 379
ocellatus R. Bohart, Pseudoplisus, 503
ocha Pate, Timberlakena, 201
Ochleroptera Holmberg, 52, 489
ochotica (A. Morawitz), Lestica, 431
ochracea Handlirsch, Bembix, 547
ochripes Arnold, Nitela, 325
ochroptera A. Costa, Mimesa, 162
ochroptera (Kohl), Isodontia, 123
ochropus Gmelin, Vespa, 627
ocliterius (F. Morawitz), Lindenius, 384
Oclocletes Banks, 54
Ococletes Mickel, 54
octavonotatus (Lepeletier), Ectemnius, 426
octodentatus Arnold, Tachysphex, 275
octomaculatus (Preyssler), Ectemnius, 428
octomaculatus (Schrank), Ectemnius, 428
octonotata Radoszkowski, Cerceris, 587
octonotatus (Lepeletier and Brullé), Ectemnius, 425
octopunctata (Donovan), Bembix, 548
octosetosa Lohrmann, Bembix, 547
oculare Menke, Trypoxylon, 347
oculata Empey, Cerceris, 582
oculata Panzer, Bembix, 547
oculata Panzer of Latreille, Bembix, 547
Odontocrabro Tsuneki, 49, 418
Odontolarra Cameron, 44
odontomerus (Handlirsch), Zyzzyx, 550
odontophora (Handlirsch), Bicyrtes, 537
odontophora (Kohl), Liris, 247
odontophora Schletterer, Cerceris, 585
odontophorus (Cameron), Crossocerus, 402

674 SPHECID WASPS

odontophorus (Kohl), Belomicrus, 363
Odontopsen Tsuneki, 40, **162**
odontopyga Turner, Bembix, 548
ODONTOSPHECINI, 59, **572**
Odontosphex Arnold, 55, **572**
odontostoma (Kohl), Parapiagetia, 44, **281**
odyneroides (Cresson), Ectemnius, 427
oedignathus (Arnold), Dasyproctus, 421
ogasawaraensis Tsuneki, Bembecinus, 530
ogloblini Gussakovskij, Didineis, 459
ogotai Tsuneki, Cerceris, 588
ohgushii Tsuneki, Nitela, 325
ohnonis Tsuneki, Psen, 166
okabei Yasumatsu, Crabro, 408
okiellus Rayment, Sericophorus, 302
okinawanum Tsuneki, Trypoxylon, 347
okinawanus (Sonan), Bembecinus, 530
okumurai Tsuneki, Cerceris, 585
olamentke (Pate), Ammoplanellus, 200
olba Evans and Matthews, Bembix, 548
olfersi Richards, Trypoxylon, 349
Olgia Radoszkowski, 52, **491**
oligocaenum Cockerell, Pison, 31, 337
oligocenum Cockerell, Pison, 31, 337
olivacea Cyrillo, Bembix, 548
olivacea Fabricius, Bembix, 548
olivata Dahlbom, Bembix, 548
olymponis Strand, Cerceris, 579
ombrodes Nagy, Dolichurus, 69
omissus Dahlbom, Nysson, 469
omissus (Kohl), Prionyx, 133
ommissum (Kohl), Chalybion, 103
omoi Guiglia, Tachysphex, 275
Omphalius Vachal, 53
oneili Brauns, Palarus, 291
oneili Cameron, Cerceris, 579
onoi (Yasumatsu), Crossocerus, 402
onophora Schletterer, Cerceris, 586
oomborra Evans and Matthews, Bembix, 548
opaca (Pulawski), Dryudella, 215
opacifrons (Tsuneki), Crossocerus, 402
opacum Brèthes, Trypoxylon, 349
opacum Rohwer, Rhopalum, 389
opacus Dahlbom, Sphex, 116
opacus F. Morawitz, Tachysphex, 275
opacus Tournier, Oxybelus, 368
opacus van Lith, Psen, 166
opalinum F. Smith, Podium, 96
opalipennis Kohl, Cerceris, 587
opalipennis (Kohl), Liris, 247
operus (Rohwer), Ectemnius, 425
ophthalmicus Handlirsch, Larrophanes, 33
opifex (Bingham), Dasyproctus, 421
opima Turner, Bembix, 548
opinabilis J. Parker, Bembix, 548
opipara (Kohl), Liris, 245
oppidanus Leclercq, Dasyproctus, 421
opposita F. Smith, Cerceris, 585
oppositus Say, Alysson, 458
oppositus Turner, Tachytes, 266
optimum Richards, Trypoxylon, 349
optimus F. Smith, Sphex, 116
opulenta Arnold, Cerceris, 582
opulenta (Guérin-Méneville), Eremophila, 39, **147**
opulenta (Lepeletier), Liris, 244
opulenta Morice, Cerceris, 579
opulenta Turner, Cerceris, 586
opulentus (F. Smith), Ectemnius, 428
opulentus F. Smith, Sphex, 114

opulentus (Gerstaecker), Epinysson, 472
opulentus Nurse, Tachytes, 266
opuntiae (Rohwer), Palmodes, 127
opwana (Rohwer), Lestica, 431
opwanus Rohwer, Tachysphex, 272
oraclensis R. Bohart, Pseudoplisus, 503
orangiae Brauns, Cerceris, 585
oraniensis Arnold, Philanthus, 566
oraniensis Brauns, Cerceris, 585
oraniensis (Brauns), Gastrosericus, 256
oraniensis Brauns, Miscophus, 319
oraniensis (Lepeletier), Diodontus, 179
oraniensis (Lepeletier), Tachysphex, 275
oraniensis Roth, Ammophila, 153
orbiculatus (Tsuneki), Ammoplanellus, 200
orbignyi (Guérin-Méneville), Zyzzyx, 550
ordinaria (Arnold), Liris, 247
ordinaria (Turner), Williamsita, 422
ordinarium Richards, Trypoxylon, 347
ordinarius Bingham, Philanthus, 566
oreades Valkeila, Pemphredon, 182
Oreocrabro Perkins, 50
oreophilus (Kuznetzov-Ugamskij), Lestiphorus, 506
orestes Banks, Cerceris, 579
orfilai Willink, Bicyrtes, 538
oribatea Pate, Bicyrtes, 537
orichalcea Dahlbom, Liris, 245
orichalceomicans (Strand), Eremochares, 146
orientale Cameron, Pison, 336
orientale Cameron, Trypoxylon, 347
orientalis (Alayo), Epinysson, 472
orientalis (Arnold), Ammoplanellus, 200
orientalis Beaumont, Laphyragogus, 220
orientalis Beaumont, Miscophus, 318
orientalis Beaumont, Prosopigastra, 286
orientalis Bytinski-Salz, Philanthus, 564
orientalis Cameron, Ammophila, 151
orientalis (Cameron), Dasyproctus, 420
orientalis Cameron, Didineis, 459
orientalis Cameron, Psen, 166
orientalis Cameron, Stizus, 527
orientalis F. Smith, Cerceris, 585
orientalis (F. Smith), Dryudella, 215
orientalis Handlirsch, Bembix, 548
orientalis (Handlirsch), Hoplisoides, 521
orientalis Gussakovskij, Psen, 166
orientalis (Gussakovskij), Psen, 167
orientalis Kohl, Palarus, 291
orientalis Mocsáry, Cerceris, 580
orientalis (Mocsáry), Palmodes, 127
orientalis Tsuneki, Diodontus, 629
orientalis Valkeila, Pemphredon, 182
orientinum Richards, Trypoxylon, 347
orientis Pulawski, Tachytes, 265
orinus van Lith, Psenulus, 173
oriolum Leclercq, Rhopalum, 389
orius Leclercq, Ectemnius, 427
orius van der Vecht, Sphex, 117
orizabense F. Parker, Diploplectron, 211
orizabense Richards, Trypoxylon, 349
orizabinus Leclercq, Ectemnius, 427
ornata (Fabricius), Cerceris, 586
ornata Spinola, Cerceris, 586
ornaticauda (Cameron), Bembecinus, 530
ornaticuada (Cameron), Bicyrtes, 537
ornatilabiata Cameron, Bembix, 548
ornatipennis Gussakovskij, Tachysphex, 276

ornatipes Cameron, Tachytes, 265, 266
ornatipes Cameron, Trypoxylon, 346
ornatipes (Turner), Crossocerus, 402
ornatipes W. Fox, Trypoxylon, 346
ornatitarsis (Cameron), Liris, 247
ornativentris Cameron, Cerceris, 580
ornatum F. Smith, Trypoxylon, 348
ornatum (Gussakovskij), Eremiasphecium, 561
ornatus Dahlbom, Stizus, 527
ornatus F. Smith, Ammatomus, 513
ornatus (F. Smith), Dienoplus, 496
ornatus (Hacker & Cockerell), Sericophorus, 303
ornatus Lepeletier and Brullé, Crossocerus, 402
ornatus (Lepeletier and Brullé), Ectemnius, 424
ornatus Lepeletier, Sphex, 115
ornatus (Ritsema), Psenulus, 173
Ornicrabro Leclercq, 48
orotina Pate, Enoplolindenius, 415
orphne Banks, Cerceris, 587
Orthoxybelus Pate, 46
Oryttus Spinola, 52, **506**
osbecki Dahlbom, Philanthus, 566
oscari Turner, Paranysson, 308
osdroene (Beaumont), Dienoplus, 496
osiriaca (Pulawski), Dryudella, 215
osiris Arnold, Cerceris, 585
osiris Beaumont, Tachysphex, 275
othello Balthasar, Miscophus, 319
otomia Saussure, Cerceris, 582
otomita Dalla Torre, Cerceris, 582
ottomanus (Mocsáry), Hoplisoides, 521
ourapilla Evans and Matthews, Bembix, 548
outeniqua Arnold, Larra, 238
ovalis Lepeletier and Brullé, Crossocerus, 402
ovambo Empey, Cerceris, 582
ovans Bingham, Bembix, 548
ovata (Christ), Lestica, 430
ovatus (Schulz), Crossocerus, 402
overlaeti Leclercq, Ampulex, 78
oviventris Saussure, Tachytes, 266
oweni van Lith, Psenulus, 629
oxanus Nagy, Dolichurus, 69
oxiana Gussakovskij, Nitela, 325
oxianus Gussakovskij, Sphex, 116
OXYBELINI, 60, **359**
oxybeloides Pate, Encopognathus, 380
Oxybeloides Radoszkowski, 46
Oxybelomorpha Brauns, 46
Oxybelus Latreille, 46, **364**
Oxycrabro Leclercq, 48, **398**
oxydorcus (Handlirsch), Bembecinus, 531
oxystoma Cameron, Eremnophila, 147

pachappa Pate, Belomicrus, 363
pachecoi Giner Marí, Philanthus, 566
pachydermus Strand, Sphex, 115
pachygaster Richards, Trypoxylon, 347
pachysoma Kohl, Sphex, 114
pacifica Ashmead, Foxia, 479
pacifica (Melander and Brues), Podalonia, 144
pacifica Scullen, Eucerceris, 592
pacifica (Tsuneki), Mimumesa, 164
pacifica Williams, Larra, 238
pacificatrix (Turner), Liris, 247
pacificum Gussakovskij, Trypoxylon, 347
pacificum R. Bohart, Rhopalum, 630

pacificus Bohart and Menke, Palmodes, 127
pacificus Cresson, Philanthus, 566
pacificus (Gussakovskij), Crossocerus, 401
pacificus (Gussakovskij), Lestiphorus, 506
pacificus Gussakovskij, Pemphredon, 182
pacificus (Rohwer), Epinysson, 472
pacificus (Rohwer), Oxybelus, 367
pacificus Tsuneki, Bembecinus, 531
pacificus (Turner), Bembecinus, 531
pacificus Turner, Tachysphex, 275
packardi Dalla Torre, Oxybelus, 369
packardii (Cresson), Ectemnius, 424
packardii Robertson, Oxybelus, 369
pacuarus Leclercq, Ectemnius, 427
Pae Pate, 49, **412**, 627
pae Pate, Chimila, 412
paenemarginatus (Viereck), Oxybelus, 369
paenerugosa (Viereck), Larropsis, 258
pagana (Dahlbom), Larra, 238
pagdeni Leclercq, Hingstoniola, 418
pagdeni van Lith, Psenulus, 173
pahangensis van Lith, Psen, 166
pahangi Leclercq, Ectemnius, 427
palaestina Beaumont, Cerceris, 578
palaestinensis Lohrmann, Bembix, 548
palamosi Leclercq, Ectemnius, 427
PALARINI, 60, **286**
Palarus Latreille, 44, **288**
palauense Yasumatsu, Pison, 336
palawanensis Williams, Dicranorhina, 252
palearcticum Pulawski, Diploplectron, 211
paleata Saussure, Cerceris, 585
paleotropicum Schulz, Trypoxylon, 348
palestinensis Balthasar, Philanthus, 566
palitans (Bingham), Ectemnius, 427
palitoides Leclercq, Ectemnius, 427
pallescens (Giorna), Bembix, 547
pallescens Priesner, Bembix, 548
pallida Arnold, Cerceris, 585
pallida Radoszkowski, Bembix, 548
pallidehirtus Kohl, Sphex, 116
pallidicinctus van der Vecht, Bembecinus, 531
pallidicollis van Lith, Psenulus, 173
pallidicornis (F. Morawitz), Lindenius, 384
pallidipalpe F. Smith, Pison, 336
pallidipalpis Lepeletier and Brullé, Crossocerus, 40
pallidipenne Taschenberg, Chlorion, 89
pallidipes Arnold, Solierella, 313
pallidipes (Dalla Torre), Liris, 245
pallidipes (Dalla Torre), Rhopalum, 389
pallidipicta F. Smith, Bembix, 548, 627
pallidistigma (Malloch), Pluto, 171
palliditarse Cameron, Trypoxylon, 345
palliditarse of Richards, Trypoxylon, 348
palliditarse Saussure, Trypoxylon, 348
palliditarsis (E. Saunders), Mimumesa, 164
pallidiventris Arnold, Tachytes, 266
pallidopicta Radoszkowski, Cerceris, 580
pallidula Morice, Cerceris, 585
pallidus Arnold, Oxybelus, 369
pallidus Klug, Philanthus, 566
pallidus Richards, Microstigmus, 192
pallidus van Lith, Psenulus, 629
pallidus W. Fox, Crabro, 408
pallipes (F. Smith), Liris, 245
pallipes (Lepeletier and Brullé), Rhopalum, 389
pallipes (Panzer), Psenulus, 173
palmarius (Schreber), Crossocerus, 402
palmarum Williams, Plenoculus, 311

palmata F. Smith, Bembix, 548
palmatus (De Stefani), Crossocerus, 403
palmatus Panzer, Crabro, 408
palmerii Leclercq, Dasyproctus, 420
palmetorum Beaumont, Cerceris, 585
palmetorum Beaumont, Oxybelus, 369
palmetorum (Roth), Palmodes, 127
palmipes auctt, Crossocerus, 403
palmipes (Linnaeus), Crossocerus, 402
Palmodes Kohl, 39, **124**
palona Evans and Matthews, Bembix, 548
palopterus (Dahlbom), Tachysphex, 275
paludosa (Rossi), Isodontia, 123
palumbula (Kohl), Liris, 246
pammelas Pate, Crossocerus, 404
pamparum Brèthes, Oxybelus, 369
pamparum (Brèthes), Podagritus, 393
pan Arnold, Trypoxylon, 347
pan Beaumont, Psenulus, 173
panama Scullen, Cerceris, 582
panamensis (Cameron), Liris, 247
panamensis (Maidl and Klima), Hoplisoides, 521
paniquita Pate, Pae, 413
pannonica Handlirsch, Didineis, 459
pannonica Maidl, Mimesa, 162
pannonica Mocsáry, Bembix, 547
pano Pate, Quexua, 387
pantherina (Handlirsch), Stictia, 542
panurgoides (Viereck), Entomognathus, 382
panzeri Handlirsch, Bembix, 547
panzeri Lepeletier, Nysson, 469
panzeri (Vander Linden), Lindenius, 384
panzeri (Vander Linden), Tachysphex, 275
papagorum (Viereck), Ectemnius, 426
papagorum (Viereck), Pseudoplisus, 503
pappi Tsuneki, Trypoxylon, 630
papua Handlirsch, Bembix, 548
papuana Cameron, Cerceris, 585
papuanum Cameron, Sceliphron, 106
papuanum Schulz, Pison, 336
papuanus (Cameron), Bembecinus, 531
papuensis Krombein, Cerceris, 585
papyrus Andrade, Miscophus, 319
Paraceramius Radoszkowski, 45
Paracerceris Brèthes, 55
paracochise R. Bohart and Schlinger, Oxybelus, 369
parcorum Leclercq, Crossocerus, 402
Paracrabro Turner, 41, **186**
paradisiacus (Beaumont), Oryttus, 507
Paradolichurus Williams, 38, **69**
paradoxa Giner Marí, Bembix, 548
paradoxa Gussakovskij, Anomiopteryx, 42, **196**
paradoxa Gussakovskij, Solierella, 314
paradoxus Menke, Odontosphex, 573
paraensis (Spinola), Bothynostethus, 352
paraguayana (Strand), Bicyrtes, 537
paraguayana Strand, Cerceris, 584
paraguayana Strand, Larra, 238
paraguayensis Brèthes, Oxybelus, 369
Paraliris Kohl, 43, **250**
parallela Arnold, Nitela, 325
parallela Murray, Podalonia, 144
Paralellopsis Maidl, 43
Parallelopsis Pate, 43
parallelus Say, Stigmus, 189
Paramellinus Rohwer, 52
Parammoplanus Pate, 42, **176**

paranensis (Berland), Isodontia, 124
Paranothyreus Ashmead, 48
Paranysson Guérin-Méneville, 44, **306**
Parapiagetia Kohl, 44, **277**
Parapison F. Smith, 45
Parapodium Taschenberg, 38
parapolita (Fernald), Ammophila, 153
Parapsammophila Taschenberg, 39, **137**
parasinus (Kittel), Oxybelus, 369
Parasphex F. Smith, 39
parata (Provancher), Bicyrtes, 538
Parathyreopus Pate, 49
parcimonium Leclercq, Rhopalum, 389
pardoi Andrade, Miscophus, 319
pardoi Giner Marí, Bembix, 548
pardoi Giner Marí, Cerceris, 581
Parechuca Leclercq, 47, **392**
parenosas (Pate), Psenulus, 173
paria Pate, Enoplolindenius, 415
paria Pate, Losada, 480
parietalis Mercet, Nysson, 469
parkanensis Zavadil, Lindenius, 384
parkeri (Banks), Crossocerus, 400
parkeri Bohart and Menke, Prionyx, 133
parkeri Lohrmann, Bembix, 546, 548
parkeri Menke, Ammophila, 153
parkeri Scullen, Cerceris, 585
Paroxycrabro Leclercq, 48, **398**
parrotti (Leclercq), Podagritus, 393
partamona Pate, Enoplolindenius, 415
partamona (Pate), Epinysson, 472
parthenia (A. Costa), Isodontia, 123
partitus (W. Fox), Neoplisus, 505
parva (Cameron), Liris, 247
parva (Rohwer), Solierella, 314
parvicorpus Tsuneki, Crossocerus, 629
parvidentatus van Lith, Psenulus, 173
parviornata (Leclercq), Hingstoniola, 418
parvispinosus (Reed), Cresson, 476
parvula F. Morawitz, Bembix, 548
parvula Gussakovskij, Astata, 212
parvula (Handlirsch), Ochleroptera, 490
parvula Schrottky, Larra, 238
parvulus Beaumont, Palarus, 291
parvulus (Cresson), Tachysphex, 275
parvulus (Herrich-Schaeffer), Ectemnius, 426
parvulus Packard, Ectemnius, 427
parvulus (Radoszkowski), Belomicrus, 363
parvulus (Radoszkowski), Diodontus, 179
parvulus (Roth), Palmodes, 127
parvum Schrottky, Trypoxylon, 347
parvus Cresson, Oxybelus, 369
parvus W. Fox, Plenoculus, 311
parvus W. Fox, Tachytes, 266
Passaloecus Shuckard, 41, **182**
pasteelsi Leclercq, Pison, 336
patagonensis (Saussure), Trachypus, 568
patagonica (Brèthes), Microbembex, 539
patagonicus (Fritz), Liogorytes, 516
patagonicus (Holmberg), Podagritus, 393
patagonicus Mantero, Stigmus, 189
patei Beaumont, Belomicrus, 363
patei (Fritz), Antomartinezius, 478
patei Leclercq, Foxita, 416
patei Willink, Bicyrtes, 537
patellarius (Schreber), Crabro, 408
patellatus Arnold, Psen, 166
patellatus Panzer, Crabro, 408
patongensis Rayment, Sericophorus, 302

patricius (Arnold), Entomognathus, 382
patrizii Guiglia, Tachytes, 264
patruele Arnold, Trypoxylon, 347
pattoni Banks, Tachytes, 266
patuhana van der Vecht, Cerceris, 589
paucimaculatus (Packard), Ectemnius, 427
paucipunctatus Arnold, Oxybelus, 369
pauli Mantero, Stizus, 527
paulinierii Guérin-Méneville, Sphex, 116
paulisae Leclercq, Psenulus, 173
paulisum Leclercq, Trypoxylon, 347
pauloense (Schrottky), Penepodium, 92
paulus van Lith, Psen, 166
pauper Evans, Pittoecus, 31
pauper (Packard), Ectemnius, 425
pauper Packard, Mimesa, 162
pauperatus (Lepeletier and Brullé), Crossocerus, 401
paupercula Holmberg, Cerceris, 585
pauxilla Brèthes, Cerceris, 585
pauxillus Arnold, Psenulus, 172
pauxillus (Gussakovskij), Crossocerus, 402
pauxillus W. Fox, Tachysphex, 275
Pavlovskia Gussakovskij, 45
pavlovskii (Gussakovskij), Crossocerus, 402
pavlovskii (Gussakovskij), Kohlia, 513
paxinus Leclercq, Ectemnius, 427
pearstonensis Cameron, Cerceris, 585
pechumani Krombein, Tachysphex, 275
pechumani Pate, Psammaletes, 508
peckhami (Ashmead), Solierella, 313
peckhami (Fernald), Ammophila, 153
pectinata Pulawski, Solierella, 314
pectinata Shestakov, Cerceris, 585
pectinatus Beaumont, Laphyragogus, 220
pectinatus Pate, Encopognathus, 380
pectinatus Pulawski, Tachysphex, 275
pectinipes Handlirsch, Bembix, 548
pectinipes of authors, Tachysphex, 276
pectorale (Dahlbom), Sceliphron, 106
pectorale Richards, Trypoxylon, 349
pectoralis Dahlbom, Bembix, 547
pectoralis F. Morawitz, Oxybelus, 369
pectoralis (F. Smith), Sphecius, 511
pectoralis Pulawski, Tachysphex, 273
pectorosus Mickel, Oxybelus, 367
peculator Nurse, Tachysphex, 272
peculiaris W. Fox, Didineis, 459
pedestris Brèthes, Cerceris, 585
pedestris Saussure, Liris, 245
pedetes Kohl, Cerceris, 585
pedibusnigris Zanon, Sphex, 116
pedicellaris (F. Morawitz), Ectemnius, 425
pedicellatum Packard, Rhopalum, 389
pegasus Harris, Crabro, 407
pegasus (Packard), Crabro, 407
peglowi Krombein, Diploplectron, 211
pekingensis Tsuneki, Cerceris, 585
pekingensis (Tsuneki), Mimesa, 162
pekingensis Tsuneki, Nysson, 469
pekingensis Tsuneki, Tachysphex, 275
pelas Pate, Crossocerus, 402
peletieri Le Guillou, Pison, 336
pelletieri Le Guillou, Pison, 336
Pelopaeus Latreille, 39
pelopoeiformis (Dahlbom), Isodontia, 124
peloponesia Beaumont, Cerceris, 587
pelops Morice, Astata, 212
pelotarum Leclercq, Ectemnius, 427
peltarius (Schreber), Crabro, 408
peltatus Fabricius, Crabro, 408
peltista Kohl, Crabro, 408

peltopse Kohl, Trypoxylon, 347
Pemphilis Pate, 48
Pemphredon Latreille, 40, **179**
PEMPHREDONINAE, 57, 58, **155**, 158
PEMPHREDONINI, 59, **174**, 175
pempuchi Tsuneki, Ectemnius, 427
pempuchi (Tsuneki), Isodontia, 124
pempuchiensis Tsuneki, Dolichurus, 69
pempuchiensis Tsuneki, Psenulus, 173
penai Fritz, Losada, 629
penai Fritz and Toro, Cerceris, 585
penangensis (Rohwer), Psenulus, 173
pendleburyi Leclercq, Ectemnius, 427
pendleburyi Turner, Nitela, 325
pendleburyi van Lith, Psenulus, 173
pendulus Panzer, Stigmus, 41, **189**
Penepodium Menke, 38, **90**
penicillata Mocsáry, Cerceris, 578
peninsularis (Kohl), Lindenius, 384
peninsularis Mercet, Cerceris, 582
peninsularum Bohart and Menke, Chalybion, 103
pennsylvanica (F. Smith), Liris, 244
pennsylvanicum Saussure, Trypoxylon, 347
pennsylvanicus Banks, Tachytes, 266
penpuchiensis Tsuneki, Bembecinus, 531
pensile (Illiger), Sceliphron, 105
pensylvanica Haldeman, Ampulex, 77
pensylvanica (Palisot de Beauvois), Liris, 244
pensylvanicus Linnaeus, Sphex, 116
pentadonta Cameron, Cerceris, 585
pentapolitanus (Beaumont), Holotachysphex, 282
pentheres Handlirsch, Stizus, 526
pentheri Brauns, Palarus, 291
pentheri Cameron, Tachysphex, 275
pentheri Leclercq, Dasyproctus, 421
pentheri Richards, Trypoxylon, 347
penuti Pate, Belomicrus, 363
pepticus (Say), Tachytes, 266
peraki Leclercq, Lestica, 431
perarmata (Arnold), Arnoldita, 418
peratus Leclercq, Podagritus, 393
perboscii Guérin-Méneville, Cerceris, 585
percna Scullen, Cerceris, 579
percussor Handlirsch, Sphecius, 511
perdita Kohl, Cerceris, 583
peregrinus (F. Smith), Bembecinus, 531
peregrinus (Yasumatsu), Lestiphorus, 506
perezi (Berland), Prionyx, 134
perezi Ferton, Dinetus, 216
perfida Saussure, Cerceris, 585
perforator F. Smith, Rhopalum, 390
perfoveata Arnold, Cerceris, 585
pergandei (Handlirsch), Hoplisoides, 521
Pericrabro Leclercq, 48, **398**
peringueyi (Arnold), Ammophila, 153
peringueyi Saussure, Oxybelus, 369
Perisson Pate, 51, **475**
perkinsi (Turner), Austrogorytes, 498
perkinsi Turner, Cerceris, 585
perkinsi Yoshimoto, Nesomimesa, 167
perlucida (Turner), Clitemnestra, 489
perlucida (Turner), Pseudoturneria, 409
permagnus (Willink), Sphex, 116
permixtus Nouvel and Ribant, Entomognathus, 382
permutans (Turner), Isodontia, 124
pernniger Arnold, Tachysphex, 275
perniger Williams, Plenoculus, 311

pernix F. Parker, Dryudella, 215
perornatus Arnold, Oxybelus, 369
perornatus Turner, Tachytes, 266
perplexa Arnold, Cerceris, 580
perplexa (Rohwer), Mimesa, 162
perplexum F. Smith, Pison, 336
perplexus (F. Smith), Palmodes, 127
perpulchrum (Arnold), Chalybion, 103
perpusillus (Walker), Crossocerus, 402
perrisi Giraud, Ammoplanus, 42, **198**
perrisii Dufour, Stizus, 527
persa Gussakovskij, Belomicrus, 363
persa Gussakovskij, Sphecius, 511
persa Gussakovskij, Tachysphex, 275
persa Schulthess, Bembix, 548
persicus (Kohl), Ectemnius, 427
persicus (Mocsáry), Prionyx, 134
persimilis Turner, Bembix, 548
persimilis (Turner), Lestiphorus, 506
persimilis (Turner), Stizoides, 529
persistans Turner, Tachysphex, 275
personata (F. Smith), Larra, 238
personatus W. Fox, Mellinus, 449
perspicua Holmberg, Cerceris, 585
pertheesi Gorski, Alysson, 458
pertinax Turner, Pison, 336
peruana (Brèthes), Liris, 247
peruanus Giner Marí, Trachypus, 569
peruanus Kohl, Sphex, 116
peruviana (Rohwer), Ammophila, 152
peruvianus (Guérin-Méneville), Zyzzyx, 550
pervigilans Turner, Cerceris, 589
pescotti Rayment, Sericophorus, 302
peterseni van Lith, Psenulus, 173
petiolare Kohl, Sceliphron, 106
petiolata (Drury), Isodontia, 124
petiolata (F. Smith), Isodontia, 123
petiolata (F. Smith), Mimumesa, 164
petiolata Saussure, Cerceris, 585
petiolatum F. Smith, Trypoxylon, 347
petiolatum (Nurse), Rhopalum, 390
petiolatum (Taschenberg), Megistrommum, 503
petiolatus Cameron, Alysson, 458
petiolatus (Fourcroy), Mellinus, 449
petiolatus (Lepeletier and Brullé), Ectemnius, 426
petiolatus (Panzer), Mellinus, 449
petiolatus Saussure, Philanthus, 565
petiolatus (Spinola), Trachypus, 569
petioloides Strand, Trypoxylon, 346
petrosus Eversmann, Crabro, 408
petulans (Harris), Cerceris, 577
petularum Leclercq, Pison, 336
pexa J. Parker, Bicyrtes, 538
phaeochilos Pate, Crossocerus, 402
phaleratus (Say), Pseudoplisus, 503
pharao Kohl, Oxybelus, 367
pharaonis Arnold, Miscophus, 319
pharaonum Kohl, Cerceris, 585
pharaonum (Kohl), Liris, 244
pharaonum (Kohl), Pseudoscolia, 573
pharetrigera Shestakov, Cerceris, 585
philadelphica (Lepeletier), Isodontia, 39, **124**
PHILANTHINAE, 58, **555**, 558
PHILANTHINI, 59, **561**
Philanthinus Beaumont, 55, **569**
Philanthocephalus Cameron, 54
philanthoides (Panzer), Lestica, 431
Philanthus Fabricius, 54, **561**

philippensis (Rohwer), Isodontia, 123
philippi E. Saunders, Tachysphex, 275
philippinense Ashmead, Trypoxylon, 347
philippinensis (Ashmead), Dasyproctus, 420
philippinensis Pate, Oxybelus, 369
philippinensis (Rohwer), Psenulus, 173
Phillanthus Guérin-Méneville, 54
philomela Nurse, Ammophila, 153
Philoponidea Pate, 55
Philoponites Cockerell, 31
Philoponoides Giner Marí, 55
Philoponus Kohl, 55
phoenix (Pate), Clypeadon, 571
phyllophorus Kohl, Oxybelus, 369
physopoda Handlirsch, Bembix, 548
Physoscelis Westwood, 47
Physoscelus Lepeletier and Brullé, 47
Phytosphex Arnold, 44
Piagetia Ritsema, 43
piagetiodes (E. Saunders), Parapiagetia, 281
picea A. Costa, Astata, 212
piceiventris (Cameron), Podalonia, 144
piceus (Handlirsch), Lestiphorus, 506
picicornis (F. Morawitz), Palmodes, 127
picicornis (F. Morawitz), Psen, 166
picina (Mickel), Ancistromma, 259
picipes Cameron, Ammophila, 153
picipes (Cameron), Liris, 247
picta Dahlbom, Cerceris, 585
picta (Kohl), Dryudella, 215
picta Mochi, Cerceris, 585
pictetii Handlirsch, Alysson, 458
picticollis F. Morawitz, Bembix, 547
picticornis Arnold, Tachytes, 266
picticornis F. Morawitz, Stizus, 526
picticornis (Gussakovskij), Dryudella, 215
picticornis (Vogrin), Dienoplus, 496
pictifacies Brauns, Cerceris, 585
pictifrons (F. Smith), Glenostictia, 552
pictifrons W. Fox, Dienoplus, 52, **496**
pictinoda Cameron, Cerceris, 585
pictipennis (Maidl), Liris, 247
pictipennis Walsh, Ammophila, 153
pictipes (Arnold), Ammatomus, 513
pictipes F. Morawitz, Oxybelus, 368
pictipes (Herrich-Schaeffer), Ectemnius, 426
pictipes (W. Fox), Crossocerus, 400
pictiscutis Turner, Oxybelus, 369
pictisentis Cameron, Oxybelus, 369
pictithorax (Strand), Stictia, 542
pictiventris Dahlbom, Cerceris, 585
pictiventris F. Morawitz, Palarus, 291
pictiventris Gerstaecker, Cerceris, 580
pictum (F. Smith), Sceliphron, 106
picturata Bytinski-Salz, Bembix, 546
picturata Taschenberg, Cerceris, 585
pictus Arnold, Belomicroides, 363
pictus Arnold, Oxybelus, 369
pictus Dahlbom, Stizus, 527
pictus (F. Smith), Crossocerus, 402
pictus (F. Smith), Hoplisoides, 521
pictus (Fabricius), Dinetus, 42, **216**
pictus Kohl, Laphyragogus, 42, **220**
pictus Panzer, Philanthus, 566
pictus Ribaut, Passaloecus, 184
pictus (Schenck), Ectemnius, 426
pieli (Yasumatsu), Brachystegus, 473
pieli Yasumatsu, Gorytes, 501
pietschmanni Kohl, Sceliphron, 106

pignatus Leclercq, Crossocerus, 402
pikati Evans and Matthews, Bembix, 548
pileatum F. Smith, Trypoxylon, 347
piliferus Leclercq, Chimiloides, 414
pilifrons Cameron, Philanthus, 566
pilimarginata Cameron, Ammophila, 153
pilipes Kohl, Ampulex, 78
piliventris (Cameron), Liris, 247
pillara Evans and Matthews, Bembix, 548
pilosa Arnold, Prosopigastra, 286
pilosa Cameron, Ampulex, 78
pilosa Cameron, Cerceris, 582
pilosa (Cameron), Liris, 247
pilosa (Fernald), Ammophila, 151
pilosellus Pulawski, Tachysphex, 276
pilosellus Turner, Sphodrotes, 304
pilosifrons (Turner), Liris, 247
pilosulus Turner, Tachysphex, 276
pilosum F. Smith, Pison, 337
pilosus Arnold, Oxybelus, 369
pilosus (F. Smith), Zanysson, 475
pilosus (Gimmerthal), Pemphredon, 182
pilosus van Lith, Psen, 166
pima Court and R. Bohart, Huavea, 396
pimarum Rohwer, Eucerceris, 592
pinal Pate, Moniaecera, 395
pinguis (Dahlbom), Dryudella, 215
pinguis Handlirsch, Bembix, 548
pinguis (W. Fox), Lindenius, 384
pinguis (Zetterstedt), Dryudella, 215
pinnatus Saussure, Oxybelus, 368
piraporae J. Parker, Bembix, 548
piratus (Leclercq), Echucoides, 394
pirita Fritz, Foxia, 479
Pison Jurine, 45, **332**
Pisonitus Shuckard, 45
Pisonoides F. Smith, 45, **335**
pisonoides (Reed), Tachysphex, 276
pisonoides (S. Saunders), Solierella, 314
pisonopsis Pulawski, Tachysphex, 276
Pisonopsis W. Fox, 45, **329**
Pisoxylon Menke, 45, **338**
Pisum Agassiz, 45
Pisum Schulz, 45
pitamawa (Rohwer), Liris, 247
pitanta Pate, Oxybelus, 369
Pittoecus Evans, 31
pius (Strand), Podagritus, 393
Piyuma Pate, 49, **409**
Piyumoides Leclercq, 49, **410**
pizarrus Leclercq, Podagritus, 393
placida Arnold, Cerceris, 585
placida F. Smith, Ammophila, 153
placida F. Smith, Bembix, 548
placidum Cameron, Trypoxylon, 347
placidum F. Smith, Trypoxylon, 347
placidus F. Smith, Hoplisoides, 521
placita Arnold, Cerceris, 585
plagiatus (Cresson), Synnevrus, 470
plagiatus Walker, Tachytes, 266
plana (W. Fox), Stictiella, 551
planaris (Mickel), Lestica, 431
planata Tsuneki, Liris, 247
planatum (Arnold), Chalybion, 102
planifemur Krombein, Crossocerus, 402
planifrons F. Morawitz, Bembix, 548
planifrons (Thomson), Ectemnius, 427
planifrons (Wesmael), Gorytes, 501
planipes (W. Fox), Crossocerus, 402
planoense Rohwer, Trypoxylon, 349
platense Brèthes, Trypoxylon, 349
platensis Brèthes, Ammophila, 153

platensis Brèthes, Oxybelus, 369
platensis (Brèthes), Prionyx, 134
platensis Brèthes, Solierella, 314
platycerus Menke, Plenoculus, 311
platynota (G. and R. Bohart), Ancistromma, 260
platynotus (Matsumura), Prionyx, 134
platyrhina Viereck and Cockerell, Cerceris, 589
platyrhyncha Arnold, Cerceris, 584
platytarsus Gussakovskij, Ammoplanus, 198
platyurus Gussakovskij, Pemphredon, 182
plaumanni Leclercq, Rhopalum, 390
plaumanni Menke, Pison, 337
plebeja (Taschenberg), Liris, 247
plenoculiformis Williams, Tachysphex, 276
plenoculoides (W. Fox), Solierella, 314
Plenoculus W. Fox, 44, **308**
plesia (Rohwer), Zanysson, 475
plesiosaurus (F. Smith), Podium, 96
plesium Rohwer, Trypoxylon, 346
plesius (Rohwer), Crossocerus, 403
plesius Rohwer, Tachysphex, 276
pleuracutus Leclercq, Crossocerus, 402
pleuralis H. Smith, Cerceris, 586
pleuralis Tsuneki, Cerceris, 585
pleuralis W. Fox, Crabro, 408
pleuripunctatus (A. Costa), Gorytes, 501
pleurispina Beuamont, Cerceris, 585

plicosus (A. Costa), Tachysphex, 276
plomleyi (Turner), Clitemnestra, 489
plumata Leclercq, Lestica, 431
plumiferus A. Costa, Sphex, 114
plumipes Radoszkowski, Sphex, 114
pluridentata Arnold, Parapiagetia, 281
pluschtschevskii Radoszkowski, Stizus, 527
pluschtschevskyi (F. Morawitz), Lestica, 431
pluto (F. Smith), Liris, 247
Pluto Pate, 40, **169**
plutocraticus Turner, Tachytes, 266
plutonius Leclercq, Ectemnius, 427
poculum Scullen, Cerceris, 585
podager (Beaumont), Bembecinus, 531
podagrica Chevrier, Pemphredon, 182
podagricus Kohl, Stigmus, 189
podagricus (Vander Linden), Crossocerus, 402
Podagritoides Leclercq, 47, **393**
Podagritus Spinola, 47, **390**
podagrosa Kohl, Cerceris, 585
Podalonia Fernald, 39, **141**
Podalonia Spinola, 39
Podium Fabricius, 38, **94**
poecilocnemis Morice, Ammophila, 153
poecilopterus (Handlirsch), Stizoides, 529
poeyi Pate, Isodontia, 124
pogonodes (Bradley), Hoplisoides, 521
Polemistus Saussure, 41, **184**
Policrabro Leclercq, 50
polita Cresson, Ammophila, 153
polita (F. Smith), Larra, 238
polita (Malloch), Mimesa, 162
polita (Mocsáry), Podalonia, 144
polita Schletterer, Cerceris, 580
polita (Taschenberg), Liris, 247
politica (Dalla Torre), Liris, 247
politiforme Rohwer, Trypoxylon, 349
politiventris Rohwer, Psen, 166

politum (F. Smith), Megistommum, 503
politum Say, Trypoxylon, 349
politus (A. Costa), Tacheliodes, 405
politus (Bingham), Hoplisoides, 521
politus Honoré, Miscophus, 317
politus Say, Philanthus, 566
politus Turner, Dinetus, 216
pollens (Kohl), Prionyx, 134
pollens Schletterer, Cerceris, 585
pollux (Handlirsch), Dienoplus, 496
pollux (Nurse), Holotachysphex, 282
polyacanthus A. Costa, Oxybelus, 369
polybia (Handlirsch), Liogorytes, 53, **516**
polybia Schrottky, Podagritus, 393
polybioides Pendlebury, Cerceris, 585
polyceros Pate, Oxybelus, 369
polychroma Gribodo, Cerceris, 585
polychroma Holmberg, Cerceris, 586
polychromus (Handlirsch), Bembecinus, 531
polygoni Rohwer, Oxybelus, 367
polynesialis (Cameron), Ectemnius, 427
pompiliformis (Panzer), Astata, 212
pompiliformis (Panzer), Liris, 246
pompiliformis (Panzer), Tachysphex, 276
ponantina Beaumont, Cerceris, 580
ponape Krombein, Pison, 336
ponca Pate, Enoplolindenius, 415
ponderosa Arnold, Cerceris, 581
ponderosa Brèthes, Cerceris, 585
ponderosa (Gerstaecker), Parapsammophila, 140
pondola Leclercq, Spilomena, 193
pontica Pulawski, Astata, 213
ponticus Beuamont, Gorytes, 501
ponticus Beaumont, Lindenius, 384
pontilis van Lith, Psenulus, 174
popovi Pulawski, Tachytes, 266
populorum Viereck and Cockerell, Cerceris, 578
porexus Leclercq, Crossocerus, 403
porteri (Ruiz), Neonysson, 471
portiana (Rohwer), Ancistromma, 260
portlandensis (Rayment), Acanthostethus, 474
portoi Andrade, Miscophus, 319
portschinskii Radoszkowski, Bembix, 548
posterorubrum Richards, Trypoxylon, 347
posterus (Sonan), Bembecinus, 531
posterus W. Fox, Tachysphex, 272
posticata Banks, Cerceris, 585
postumus Bischoff, Miscophus, 319
potanini Shestakov, Cerceris, 585
potawatomi Pate, Belomicrus, 363
potosium Leclercq, Rhopalum, 390
potosus Leclercq, Crossocerus, 403
poultoni Richards, Trypoxylon, 349
poutziloi Schulz, Cerceris, 585
povolnyi Beaumont, Cerceris, 585
powelli Gillaspy, Steniolia, 553
powelli R. Bohart, Tachysphex, 276
praeclarus (Arnold), Ectemnius, 427
praedata F. Smith, Cerceris, 585
praedator F. Smith, Sphex, 116
praedator W. Fox, Tachytes, 266
praedatrix (Strand), Liris, 247
praedura Turner, Cerceris, 585
praestabilis Turner, Tachytes, 266
praestans F. Morawitz, Stizus, 527
praestans (Kohl), Palmodes, 127

praetermissa (Richards), Liris, 247
praetextus F. Smith, Sphex, 117
praevius (Kohl), Ectemnius, 427
pragensis Snoflak, Ammoplanus, 198
praslinia (Guérin-Méneville), Isodontia, 124
pratensis Arnold, Gastrosericus, 256
pratensis Jurine, Mellinus, 449
pratensis Mercet, Nysson, 469
pratus Carter, Crabro, 408
pregustum Leclercq, Pison, 336
premunitum Leclercq, Pison, 336
prerugosa Leclercq, Vechtia, 417
pretiosum Taschenberg, Chlorion, 89
pretiosus Cameron, Tachytes, 266
pretiosus Kohl, Miscophus, 319
pretiosus (Turner), Austrogorytes, 498
pretoriaensis (Cameron), Liris, 244
pretoriensis (Arnold), Ammatomus, 513
priesneri Beaumont, Bembix, 548
priesneri Beaumont, Tachysphex, 276
priesneri Mochi, Cerceris, 585
priesneri Pulawski, Tachytes, 266
primania (Kohl), Liris, 247
primitiva Leclercq, Lestica, 431
primogeniti Andrade, Miscophus, 319
princeps (F. Smith), Larra, 238
princeps (Kohl), Penepodium, 92
princeps Kohl, Sphex, 115
princesa Pate, Crossocerus, 400
principale (Strand), Chlorion, 89
Priononyx Dahlbom, 39
Prionyx Vander Linden, 39, **128**
prisca Schletterer, Cerceris, 587
priscum Turner, Pison, 336
prismatica F. Smith, Trirogma, 74
prismatica Saussure, Larra, 238
prismaticus (F. Smith), Bembecinus, 531
prisonium Leclercq, Rhopalum, 390
pristinus Evans, Taimyrisphex, 31
proboscidea Holmberg, Cerceris, 585
procera Dahlbom, Ammophila, 153
procera Lepeletier, Ammophila, 153
procerulides (Strand), Megistommum, 503
procerus (A. Costa), Psenulus, 173
procerus A. Costa, Synnevrus, 470
procerus A. Costa, Tachytes, 266
procerus (Handlirsch), Megistommum, 52, **504**
procrustes Handlirsch, Gorytes, 501
proditor Kohl, Larra, 237
proditor (Lepeletier), Palmodes, 127
producticollis Morice, Ammophila, 152, 153
producticollis (Packard), Lestica, 431
productus (W. Fox), Ectemnius, 427
projectum W. Fox, Trypoxylon, 349
projectus Nurse, Tachysphex, 276
projectus R. Bohart, Hoplisoides, 521
projectus van Lith, Psenulus, 173
proletarius (Mickel), Ectemnius, 427
prominens Banks, Cerceris, 579
promontorii (Arnold), Ammophila, 152
promontorii Arnold, Philanthus, 566
promontorii Brauns, Nitela, 325
promontorii Lohrmann, Bembix, 548
Pronoeus Latreille, 38
Prophilanthus Cockerell, 33
propinqua A. Costa, Cerceris, 581
propinqua (Kincaid), Mimumesa, 164
propinqua Taschenberg, Ammophila, 153
propinqus Arnold, Oxybelus, 366

propinquus (Cresson), Pseudoplisus, 503
propinquus Rohwer, Tachytes, 264
propinquus (Shuckard), Crossocerus, 401
propinquus Viereck, Tachysphex, 276
propinquus (W. Fox), Crossocerus, 402
proinquus W. Fox, Plenoculus, 311
propodealis Bischoff, Stizus, 526
Prosceliphron van der Vecht, 39, 106
proserpina (Handlirsch), Stictia, 542
prosopidis Pate, Belomicrus, 363
prosopidis Williams, Solierella, 314
prospiformis (Nurse), Lindenius, 384
Prosopigastra A. Costa, 44, **282**
prosopigastroides Bischoff, Tachysphex, 276
prosopigastroides (Gussakovskij), Holotachysphex, 282
prosopis R. Bohart, Gorytes, 501
prosopoides (Turner), Piyuma, 410
prosper Kohl, Sphex, 116
protea Turner, Cerceris, 585
proteles Brauns, Cerceris, 585
protensus (Arnold), Neodasyproctus, 419
Proterosphex Fernald, 39
proteus (Arnold), Bembecinus, 532
Protoctemnius Leclercq, 50
Protostigmus Turner, 42, **194**
Protothyreopus Ashmead, 50
provancheri (Dalla Torre), Eucerceris, 592
provancheri Dalla Torre, Pemphredon, 181
provancheri Handlirsch, Gorytes, 501
provancheri W. Fox, Crabro, 408
providum F. Smith, Trypoxylon, 347
provincialis Richards, Astata, 212
proxima Cresson, Mimesa, 162
proxima (F. Smith), Ammophila, 153
proximus F. Smith, Sphex, 115
proximus (Handlirsch), Bembecinus, 532
proximus Handlirsch, Gorytes, 501
proximus Nurse, Tachytes, 267
proximus (Shuckard), Crossocerus, 401
primaaestate Johnson and Rohwer, Bembix, 545
pruinosa (Cameron), Liris, 247
pruinosa Cresson, Ammophila, 153
pruinosa (F. Smith), Liris, 247
pruinosa Morice, Cerceris, 585
pruinosa W. Fox, Bembix, 548, 627
pruinosum (Cameron), Pison, 336
pruinosus Baker, Hoplisoides, 521
pruinosus (Cameron), Bembecinus, 530
pruinosus Germar, Sphex, 116
przewalskii Shestakov, Cerceris, 585
przewalskyi Kazenas, Argogorytes, 492
psamathe Banks, Cerceris, 576
Psammaecius Lepeletier, 53, **515**
Psammaletes Pate, 52, **508**
psammobius Kohl, Oxybelus, 368
psammobius (Kohl), Tachysphex, 276
psammodes (Lepeletier), Podalnia, 145
Psammophila Dahlbom, 39
psammophilus Bytinski-Salz, Philanthus, 564
Psammosphex Gussakovskij, 44
Psen Latreille, 40, **165**
Pseneo Malloch, 40, **164**
Psenia Malloch, 40
Psenia Stephens, 40
PSENINI, 58, **158**, 159
Psenulus Kohl, 40, **171**
pseudajax van Lith, Psenulus, 173
pseudanathema Kohl, Larra, 237
Pseudanthophilus Ashmead, 54

pseudoargyrius (Roth), Chilosphex, 128
pseudocaucasica Balthasar, Podalonia, 144
Pseudocrabro Ashmead, 50
pseudocroesus (Arnold), Liris, 244
pseudoerythrocephala Schulthess, Cerceris, 585
pseudoflavescens Shestakov, Cerceris, 585
Pseudohelioryctes Ashmead, 44
Pseudolarra Reed, 51
pseudolineatus van Lith, Psenulus, 173
pseudoliris (Turner), Liris, 245
pseudomimeticus Andrade, Miscophus, 319
pseudonasuta Bytinski-Salz, Ammophila, 153
pseudonotogonia Brauns, Miscophus, 319
Pseudonysson Radoszkowski, 45
pseudopalmarius (Gussakovskij), Crossocerus, 403
pseudopanzeri Beaumont, Tachysphex, 276
Pseudoplisus Ashmead, 52, **502**
pseudoproteles Arnold, Cerceris, 585
Pseudoscolia Radoszkowski, 55, **573**
PSEUDOSCOLIINI, 59, **573**
pseudoscutus (Leclercq), Carinostigmus, 191
Pseudosphex Taschenberg, 39
pseudostriatus (Giner Marí), Prionyx, 134
pseudotridentata Maidl, Cerceris, 585
Pseudoturneria Leclercq, 49, **409**
Pseudoxybelus Gussakovskij, 46
psilocera Kohl, Larra, 238
psilocera (Kohl), Podalonia, 144
psilocerus Kohl, Tachysphex, 276
psilopa Kohl, Ampulex, 78
psilopus Kohl, Tachysphex, 276
psyche Dunning, Philanthus, 566
psychra (Pate), Mimumesa, 164
ptah Pulawski, Tachysphex, 276
pterotus Panzer, Crabro, 408
Pterygorytes R. Bohart, 53, **513**
Ptygosphex Gussakovskij, 44
Ptyx Pate, 50
pubescens Arnold, Stizus, 527
pubescens Curtis, Ammophila, 153
pubescens (Fabricius), Prionyx, 134
pubescens (Gmelin), Crabro, 408
pubescens (Klug), Stizus, 527
pubescens Murray, Podalonia, 145
pubescens (Shuckard), Crossocerus, 402
pubidorsum (A. Costa), Prionyx, 134
pucarense Leclercq, Rhopalum, 389
pucilii Radoszkowski, Cerceris, 585
pudorosa Mickel, Cerceris, 579
puente Pate, Oxybelus, 368
pugillator A. Costa, Crabro, 408
pugillatrix Handlirsch, Bembix, 548
pugnans (F. Smith), Enoplolindenius, 415
pugnator Turner, Tachysphex, 276
pugnax Olivier, Oxybelus, 368
pulawskii Beaumont, Dinetus, 216
pulawskii Kazenas, Cerceris, 583
pulawskii Kazenas, Diploplectron, 629
pulawskii Tsuneki, Ammophila, 153
pulawskii Tsuneki, Cerceris, 585
pulawskii Tsuneki, Oxybelus, 369
pulawskii Tsuneki, Trypoxylon, 347
pulchella (Cresson), Stictiella, 551
pulchella F. Smith, Ammophila, 151
pulchella Klug, Cerceris, 585
pulchellum Gussakovskij, Sceliphron, 106
pulchellus (A. Costa), Dienoplus, 496
pulchellus Arnold, Gastrosericus, 256
pulchellus Cresson, Philanthus, 566
pulchellus Gerstaecker, Oxybelus, 370
pulchellus (Mercet), Bembecinus, 532
pulchellus Spinola, Philanthus, 566
pulchellus (Wesmael), Hoplisoides, 521
pulcher Andrade, Miscophus, 319
pulcher (Cameron), Pseneo, 165
pulcher Dalla Torre, Philanthus, 566
pulcher Pulawski, Tachysphex, 276
pulcherrima (Cameron) Liris, 247
pulcherrima J. Parker, Editha, 543
pulcherrimus (Bingham), Psenulus, 173
pulcherrimus F. Smith, Philanthus, 566
pulcherrimus (F. Smith), Stizus, 527
pulchra Cameron, Cerceris, 585
pulchriceps Cameron, Ampulex, 78
pulchricollis Cameron, Ammophila, 151
pulchricornis Turner, Tachytes, 266
pulchrinum Turner, Pison, 336
pulchrior Maréchal, Ammoplanus, 198
pulchripennis (Cameron), Liris, 245
pulchripennis Mocsáry, Sphex, 114
pulchritinctus (Cameron), Bembecinus, 530
pulchrivestitus Cameron, Tachytes, 266
pulchrum (Lepeletier), Chlorion, 90
pulicus Leclercq, Entomognathus, 382
puliense Tsuneki, Trypoxylon, 347
puliensis Tsuneki, Doichurus, 69
pulka Evans and Matthews, Bembix, 548
pulla (Handlirsch), Glenostictia, 54, **552**
pullata F. Smith, Cerceris, 579
pullata J. Parker, Bicyrtes, 538
pullatus (Arnold), Dasyproctus, 421
pullulus (A. Morawitz), Crossocerus, 403
pullulus (Strand), Tachysphex, 276
pulveris (Nurse), Dasyproctus, 421
pulverosus (Radoszkowski), Tachysphex, 275
Pulverro Pate, 42, **196**
pulvillata Sowerby, Ammophila, 153
pulvis R. Bohart, Hapalomellinus, 497
pumila Lichtenstein, Sphex, 119
pumilio Arnold, Trypoxylon, 347
pumilio Giner Marí, Cerceris, 586
pumilio (Taschenberg), Prionyx, 134
pumilus A. Costa, Ectemnius, 427
pumilus Cresson, Nysson, 469
punae Perkins, Deinomimesa, 169
puncta Murray, Podalonia, 145
puncta F. Smith, Ammophila, 153
punctata (Fabricius), Stictia, 542
punctata Tsuneki, Liris, 629
puncticeps (Arnold), Ammophila, 153
punctatiformis Arnold, Tachysphex, 276
punctatissima A. Costa, Prospigastra, 44, **286**
punctatissima Blüthgen, Spilomena, 193
punctatissimum Arnold, Trypoxylon, 347
punctatissimus (Turner), Acanthostethus, 474
punctatum (Kohl), Chalybion, 103
punctatus Ashmead, Nysson, 469
punctatus Baker, Oxybelus, 369
punctatus (F. Smith), Tachysphex, 276
punctatus (Fabricius), Oxybelus, 370
punctatus (Kirschbaum), Hoplisoides, 521
punctatus (Lepeletier and Brullé), Ectemnius, 425
punctatus Say, Philanthus, 565
punctatus (Snoflak), Crossocerus, 403
punctatus (W. Fox), Pseneo, 165
punctatus (Walckenaer), Oxybelus, 369
puncticeps (Cameron), Psenulus, 173
puncticeps Cameron, Tachysphex, 276
puncticeps F. Morawitz, Cerceris, 585
puncticeps Gussakovskij, Diodontus, 179
puncticeps Gussakovskij, Psenulus, 173
puncticeps H. Smith, Tachysphex, 276
puncticollis (Kohl), Palmodes, 127
punctifer Merisuo, Pemphredon, 182
punctifrons (Cameron), Eucerceris, 592
punctifrons (Cameron), Hoplisoides, 521
punctifrons (Cameron), Trachypus, 569
punctifrons (Malloch), Mimesa, 162
punctifrons Shuckard, Pison, 336
punctifrons (W. Fox), Tachysphex, 276
punctiger (Westwood), Philanthus, 565
punctinudus Viereck and Cockerell, Philanthus, 566
punctipes Pulawski, Tachysphex, 276
punctipleura (Cameron), Liris, 246
punctipleuris (Gussakovskij), Mimesa, 162
Punctipsen van Lith, 40
punctiventris Arnold, Tachysphex, 276
punctivertex Richards, Trypoxylon, 347
punctosa Schletterer, Cerceris, 587
punctulata Gmelin, Vespa, 627
punctulata (Kohl), Ancistromma, 260
punctulatum Ashmead, Pison, 335
punctulatum Kohl, Pison, 336
punctulatum Taschenberg, Trypoxylon, 349
punctulatus De Stefani, Ectemnius, 425
punctulatus H. Smith, Tachysphex, 276
punctulatus (Vander Linden), Psammaecius, 53, 516
punctuosa Schletterer, Cerceris, 587
punctuosus Arnold, Tachytes, 266
punctuosus Brèthes, Trachypus, 569
punctuosus (Eversmann), Hoplisoides, 521
punctuosus Kohl, Sphodrotes, 44, **304**
punctuosus (Rayment), Sericophorus, 303
punctus (Zetterstedt), Crossocerus, 402
pungens (Kohl), Podalonia, 145
punicus Andrade, Miscophus, 319
punicus Ed André, Diodontus, 179
punicus Gribodo, Diodontus, 179
punjabensis Nurse, Philanthus, 566
purpurea Schletterer, Cerceris, 585
purpurea (Westwood), Ampulex, 78
purpureipennis (Matsumura and Uchida), Liris, 244
purpurescens (Pérez), Chalybion, 103
pusanoides Leclercq, Crossocerus, 403
pusanus Leclercq, Crossocerus, 403
pusilla Arnold, Larra, 238
pusilla (Say), Spilomena, 193
pusillus Beaumont, Nysson, 469
pusillus (Gussakovskij), Palmodes, 127
pusillus (Hacker & Cockerell), Sericophorus, 303
pusillus (Handlirsch), Bembecinus, 532
pusillus (Harris), Crossocerus, 401
pusillus Lepeletier and Brullé, Crossocerus, 404, 628
pusillus Saussure, Polemistus, 185
pusulosus Beaumont, Tachysphex, 276
puttalamum Strand, Trypoxylon, 347
pygialis (Pérez), Lestica, 431
pygidiale R. Bohart, Rhopalum, 390

pygidialis Cameron, Larra, 238
pygidialis Gussakovskij, Oxybelus, 369
pygidialis (Handlirsch), Microbembex, 539
pygidialis Handlirsch, Stizus, 527
pygidialis Kohl, Tachysphex, 274
pygidialis (Malloch), Mimesa, 162
pygidialis W. Fox, Astata, 212
pygidialis (W. Fox), Hoplisoides, 520
pygmaea (Brèthes), Ochleroptera, 490
pygmaea (Cameron), Liris, 247
pygmaea Saussure, Cerceris, 585
pygmaea (Thunberg), Cerceris, 587
pygmaeum Cameron, Trypoxylon, 347
pygmaeum Gussakovskij, Trypoxylon, 346
pygmaeus Handlirsch, Mellinus, 449
pygmaeus Kohl, Tachytes, 263, 266
pygmaeus Olivier, Oxybelus, 370
pygmaeus (Rossi), Lindenius, 384
pygmaeus Schrank, Sphex, 117
pygmaeus (Tournier), Psenulus, 173
pygmaeus Tsuneki, Diodontus, 179
pyrenaica Schletterer, Cerceris, 583
pyrrhobasis (Morice), Dienoplus, 496
pyrrhus Leclercq, Crossocerus, 403
pyrura (Rohwer), Oxybelus, 369

quabajai Pate, Ammoplanus, 198
quadrangularis (Packard), Ectemnius, 427
quadrangulus E. T. Cresson, Jr., Ectemnius, 427
quadraticollis A. Costa, 153
quadriceps (Bingham), Lestica, 431
quadriceps Tsuneki, Stigmus, 189
quadriceps Tsuneki, Trypoxylon, 347
quadricincta (Fabricius), Cerceris, 586
quadricincta (Panzer), Cerceris, 585
quadricinctus (Fabricius), Ectemnius, 428
quadricolor (Arnold), Synnevrus, 470
quadricolor Cockerell and Baker, Oxybelus, 367
quadricolor F. Morawitz, Cerceris, 585
quadricolor (Gerstaecker), Tachysphex, 276
quadricolor (W. F. Kirby), Dasyproctus, 421
quadricornis Gussakovskij, Cerceris, 585
quadridentata Arnold, Cerceris, 585
quadridentata (Cameron), Podalonia, 144
quadridentatus Arnold, Dolichurus, 69
quadridentatus (Cameron), Paranysson, 308
quadridentatus van Lith, Psenulus, 173
quadrifasciata (F. Smith), Liris, 247
quadrifasciata (Panzer), Cerceris, 585
quadrifasciata (Say), Bicyrtes, 538
quadrifasciatus Dreisbach, Tachytes, 267
quadrifasciatus (Fabricius), Gorytes, 501
quadrifasciatus O. Müller, Sphex, 117
quadrifasciatus Pulawski, Tachysphex, 274
quadrifasciatus Villers, Sphex, 117
quadrifurci Pulawski, Tachysphex, 276
quadriguttatus Olivier, Nysson, 469
quadriguttatus Spinola, Nysson, 469
quadrimaculata Dufour, Cerceris, 577
quadrimaculata (Provancher), Lestica, 431
quadrimaculata Taschenberg, Bembix, 548
quadrimaculatus (Fabricius), Crossocerus, 403
quadrimaculatus (Sonan), Bembecinus, 531
quadrinotatus A. Costa, Oxybelus, 369
quadrinotatus (Ashmead), Clypeadon, 571

quadrinotatus Say, Oxybelus, 370
quadripunctata (Provancher), Lestica, 431
quadripunctata Raodszkowski, Cerceris, 578
quadripunctata (Radoszkowski), Dryudella, 215
quadripunctatus (Fabricius), Crossocerus, 403
quadrisignata (Arnold), Dryudella, 214
quadrisignatus (Palma), Dienoplus, 496
quadristrigatus (Arnold), Bembecinus, 532
quartinae of authors, Sceliphron, 105
quartinae (Gribodo), Sphecius, 511
quartinae (Gribodo), Stizus, 527
quartinae Gribodo, Trypoxylon, 347
quattuordecimguttatus Shuckard, Oxybelus, 369
quattuordecimnotatus Dahlbom, Oxybelus, 367
quattuordecimnotatus Jurine, Oxybelus, 369
quattuordecimpunctatus (F. Morawitz), Philanthinus, 470
quatuorcinctus (Christ), Ectemnius, 428
quatuordecimmaculatus (Packard), Ectemnius, 427
quatuormaculatus (Christ), Crabro, 408
quebecensis (Provancher), Tachysphex, 276
quedenfeldti (Handlirsch), Hoplisoides, 521
queenslandensis (Turner), Arpactophilus, 186
queenslandensis Turner, Lyroda, 299
quemaya Pate, Belomicrus, 363
querecho Pate, Belomicrus, 364
queretaro Scullen, Cerceris, 586
querula Kohl, Cerceris, 586
quettae Nurse, Astata, 213
quettaensis Cameron, Cerceris, 586
quettaensis Nurse, Miscophus, 319
Quexua Pate, 47, **385**
quiescens Nurse, Palarus, 291
quilisi Giner Marí, Cerceris, 581
quinquecincta Ashmead, Cerceris, 577
quinquecincta (Fabricius), Cerceris, 579
quinquecincta (Schrank), Cerceris, 579
quinquecinctus (Fabricius), Gorytes, 52, **501**
quinquecinctus Schrank, Crabro, 409
quinquedentatus Tsuneki, Crossocerus, 403
quinquefasciata (Rossi), Cerceris, 586
quinquefasciatus (Panzer), Gorytes, 501
quinqueguttatus Lichtenstein, Sphex, 119
quinquelineata (Turon), Cerceris, 579
quinquemaculata J. Parker, Bicyrtes, 538
quinquemaculatus (Fabricius), Mellinus, 449
quinquemaculatus (Lepeletier and Brullé), Crossocerus, 403
quinquemaculatus (Maa), Dasyproctus, 421
quinquemaculatus (Sonan), Bembecinus, 531
quinquemacutatus (Dahlbom), Crossocerus, 403
quinquenotatus (Jurine), Tracheliodes, 40
quinquesignatus (Bignell), Tracheliodes, 405
quinquespinosa J. Parker, Bembix, 548
quinquespinosus (Say), Bembecinus, 532

quintile Viereck, Trypoxylon, 348
quitense (Benoist), Rhopalum, 390
quitensis (Benoist), Solierella, 314
quodi Vachal, Sceliphron, 105

rabiosus (Kohl), Dasyproctus, 420
rachiticus (Rossi), Mellinus, 449
radamae Arnold, Philanthus, 566
radamae (Saussure), Liris, 247
raddei Handlirsch, Stizus, 527
radialis E. Saunders, Astata, 213
radialis of Arnold, Liris, 246
radialis (Saussure), Liris, 247
radiatus (Arnold), Brimocelus, 46, **364**
radiatus Gussakovskij, Tachysphex, 276
radiatus (Pérez), Ectemnius, 427
radiatus W. Fox, Alysson, 458
radjamandalae van der Vecht, Cerceris, 586
radoszkowskii Beaumont, Tachytes, 263
radoszkowskii Dalla Torre, Bembix, 548
radoszkowskii Pulawski, Astata, 212
radoszkowskyi (Dalla Torre), Belomicrus, 364
radoszkowskyi F. Morawitz, Tachysphex, 276
radoszkowskyi Handlirsch, Bembix, 548
radoszkowskyi Handlirsch, Gorytes, 501
radoszkowskyi (Kohl), Prionyx, 134
radoszkowskyi Schletterer, Cerceris, 576
radulina Evans, Liris, 247
ralumus Leclercq, Dasyproctus, 421
rama Leclercq, Cerceris, 575
ramakrishnae Turner, Philanthus, 566
ramses Pulawski, Tachysphex, 276
ranavalonae Arnold, Philanthus, 565
rancocas Pate, Oxybelus, 369
ranosahae (Arnold), Pseudoplisus, 503
rapax Handlirsch, Stizus, 527
raptor F. Smith, Ampulex, 78
raptor F. Smith, Bembix, 549
raptor F. Smith, Cerceris, 586
raptor Lepeletier, Oxybelus, 369
raptor Lepeletier, Philanthus, 566
raptor Handlirsch, Sphecius, 511
raptrix Schulz, Cerceris, 586
rarus Arnold, Tachytes, 266
rasoherinae Arnold, Cerceris, 586
ratzeburgi Dahlbom, Alysson, 458
raui Rohwer, Cerceris, 581
raui Rohwer, Crossocerus, 403
raui (Rohwer), Hyponysson, 467
raui Rohwer, Stigmus, 189
rava Arnold, Bembix, 548
ravinus Leclercq, Ectemnius, 426
raymenti Menke, Sericophorus, 302
raymenti Turner, Cerceris, 586
rebaptizata Schulz, Cerceris, 580
rebellus Leclercq, Dasyproctus, 421
rechingeri Kohl, Larra, 238
rechingeri Kohl, Pison, 336
recondita (Turner), Liris, 247
rectangularis (Gussakovskij), Crossocerus, 403
recticornis Bradley, Nysson, 469
rectilateralis (Arnold), Bembecinus, 532
rectirugosum Arnold, Trypoxylon, 347
rectum Kohl, Sceliphron, 106
recuperatus Leclercq, Ectemnius, 427
recurva J. Parker, Bembix, 548
redivivus Kohl, Tachysphex, 277
reductus Banks, Philanthus, 565

reedi Menke, Tachysphex, 276
refractus Nurse, Psen, 166
refuscata J. Parker, Bembix, 548
regale F. Smith, Chlorion, 90
regale F. Smith, Pison, 336
regalis F. Smith, Ampulex, 78
regalis van Lith, Psen, 167
regia J. Parker, Bembix, 548
regina Menke, Ammophila, 153
regina (Turner), Liris, 247
reginellus Leclercq, Ectemnius, 427
reginula Brauns, Cerceris, 586
regium Gussakovskij, Trypoxylon, 347
regium Richards, Trypoxylon, 349
regnata J. Parker, Bembix, 548
regulare Viereck, Trypoxylon, 347
regularis (W. Fox), Mimumesa, 164
reicula Krombein, Cerceris, 586
reinigi Bischoff, Philanthus, 566
reiseri Kohl, Tachysphex, 273
reiteri (Kohl), Lestica, 431
rejecta Turner, Cerceris, 586
rejector F. Smith, Trypoxylon, 347
rejectus Baker, Oxybelus, 369
relativum Rohwer, Trypoxylon, 349
relativum Rohwer, Diploplectron, 211
relativus W. Fox, Passaloecus, 184
relegata Turner, Bembix, 548
relicta (Leclercq), Lestica, 431
relucens F. Smith, Sericophorus, 302
relucens Turner, Tachytes, 266
remotus Pulawski, Tachysphex, 276
remotus (Turner), Hoplisoides, 521
renicinctus (Say), Stizoides, 529
renominata Turner, Cerceris, 586
repanda Fabricius, Bembix, 549
repandus (Fabricius), Tachytes, 266
repandus (Panzer), Bembecinus, 532
repentinum Arnold, Pison, 336
repositus (Arnold), Crossocerus, 403
repraesentans Turner, Cerceris, 586
reptans Arnold, Miscophus, 319
republicus Leclercq, Crossocerus, 403
residua J. Parker, Bembix, 548
resinipes (Fernald), Sphex, 116
resoluta Nurse, Astata, 213
resplendens Kohl, Sphex, 116
resplendens (Kohl), Trigonopsis, 98
responsum Nurse, Trypoxylon, 347
retiaria (Turner), Liris, 245
reticollis (A. Costa), Parapsammophila, 139
reticulata (Cameron), Liris, 246
reticulata Ducke, Nitela, 325
reticulata (Malloch), Mimumesa, 164
reticulata (Saussure), Liris, 246
reticulata Turner, Nitela, 325
reticulatum Williams, Diploplectron, 211
reticulatus Antiga and Bofill, Tachysphex, 277
reticulatus Cameron, Diodontus, 179
reticulatus Cameron, Dolichurus, 69
reticulatus Cameron, Philanthus, 565
reticulatus (Cameron), Polemistus, 185
reticulatus Cameron, Psen, 166
reticulatus (Lepeletier and Brullé), Crabro, 408
reticulatus (Lepeletier and Brullé), Ectemnius, 426
reticulatus Mickel, Stigmus, 189
reticulatus Pate, Oxybelus, 369

reticulatus (Turner), Arpactophilus, 186
reticuloides (Richards), Liris, 246
reticulosus Arnold, Psenulus, 173
retifera Arnold, Nitela, 325
retowskii (Konow), Eremochares, 146
retractus Nurse, Sphex, 116
retusa Gistel, Ammophila, 151
revelatus (Cameron), Dasyproctus, 420
reversa F. Smith, Cerceris, 586
reversus Arnold, Gastrosericus, 256
reversus (F. Smith), Bembecinus, 532
revindicatus (Schulz), Bembecinus, 532
reymondi (Roth), Prionyx, 134
rhaetica Kohl, Ammophila, 153
rhaeticus Aichinger and Kriechbaumer, Crabro, 408
rhaibopus Kohl, Lindenius, 384
Rhectognathus Pate, 46
rhimpa F. Parker, Dryudella, 215
rhinoceros Gussakovskij, Tachytes, 266
rhinoceros Kohl, Cerceris, 586
rhinoceros (Strand), Hoplammophila, 141
Rhinonitela Williams, 45
Rhinopsis Westwood, 38
rhodesiae (Arnold), Brachystegus, 473
rhodesiae Arnold, Nitela, 325
rhodesiae Brauns, Cerceris, 586
rhodesiana Arnold, Ampulex, 77
rhodesiana Arnold, Parapiagetia, 281
rhodesiana Arnold, Solierella, 314
rhodesianum Bischoff, Pison, 335
rhodesianus (Arnold), Ammoplanellus, 200
rhodesianus Arnold, Belomicrus, 364
rhodesianus (Arnold), Encopognathus, 380
rhodesianus (Arnold), Prionyx, 134
rhodesianus Arnold, Saliostethus, 320
rhodesianus Bischoff, Tachysphex, 276
rhodesianus Bischoff, Tachytes, 266
rhodesianus Turner, Miscophus, 319
rhodesiensis (Arnold), Dasyproctus, 421
rhodesiensis Empey, Cerceris, 586
rhodius Beaumont, Bembecinus, 531
rhodius Beaumont, Tachysphex, 276
rhododactylus Taschenberg, Tachytes, 266
rhodops Viereck and Cockerell, Eucerceris, 592
rhodosoma (R. Turner), Sphex, 116
rhois Rohwer, Cerceris, 586
rhopaloceroides (Arnold), Bembecinus, 532
rhopalocerus (Handlirsch), Ammatomus, 513
rhopalocerus (Handlirsch), Bembecinus, 532
rhopaloides Leclercq, Podagritus, 393
Rhopalum Stephens, 47, **387**
rhynchophora Turner, Cerceris, 586
ribauti Merisuo, Passaloecus, 629
ricata Leclercq, Quexua, 387
richardsi Arnold, Stizus, 527
richardsi (Beaumont), Crossocerus, 403
richardsi Pate, Entomocrabro, 377
richardsi Sandhouse, Trypoxylon, 347
richardsi Tsuneki, Psen, 167
richteri Beaumont, Parapiagetia, 281
rieki Leclercq, Podagritus, 393
riekiella Leclercq, Williamsita, 630
rigida F. Smith, Cerceris, 586
rileyi W. Fox, Pemphredon, 182
rimatus Leclercq, Crossocerus, 403
rinconis Viereck and Cockerell, Cerceris, 579
riojacus Leclercq, Ectemnius, 424
riosorum Leclercq, Ectemnius, 427

Riparena Pate, 42
riparia Gussakovskij, Prosopigastra, 286
riparium (Arnold), Crossocerus, 403
ritsemae (Dalla Torre), Chalybion, 103
ritsemae Handlirsch, Stizus, 526
ritsemae (Ritsema), Dicranorhina, 252
rivertonensis Viereck, Passaloeucs, 184
riveti (Strand), Podagritus, 393
rixosa F. Smith, Cerceris, 586
rjabovi Gussakovskij, Ammoplanus, 198
robertsoni Rohwer, Anacrabro, 379
robertsoni (Rohwer), Enoplolindenius, 415
robertsonii Baker, Oxybelus, 369
robertsonii W. Fox, Cerceris, 586
roborovskyi Kohl, Ammophila, 153
robusta Arnold, Spilomena, 193
robusta (Cameron), Isodontia, 123
robusta (Cresson), Podalonia, 145
robusta Lohrman, Bembix, 548
robusta Shestakov, Cerceris, 580
robusta (Williams), Liris, 247
robustior Williams, Tachysphex, 276
robustisoma Strand, Sphex, 116
robustoides (Williams), Liris, 247
robustus (Arnold), Bebecinus, 531
robustus Cameron, Oxybelus, 369
robustus (Handlirsch), Hoplisoides, 521
roddi Rayment, Sericophorus, 302
roepkei Maidl, Cerceris, 586
roettgeni C. Verhoeff, Passaloecus, 184
rogenhoferi (Handlirsch), Ammatomus, 513
rogenhoferi Kohl, Trypoxylon, 349
rohweri (Bridwell), Solierella, 313
rohweri Lohrmann, Bembix, 545
rohweri Richards, Trypoxylon, 349
rohweri van Lith, Psenulus, 172
rohweri (Williams), Liris, 247
romandi (Saussure), Trachypus, 569
romandinum (Saussure), Penepodium, 92
rondani (A. Costa), Tracheliodes, 405
roratus Kohl, Sphex, 115
roseiventris (F. Morawitz), Prosopigastra, 285
roshanica Gussakovskij, Spilomena, 193
rossi R. Bohart, Tachytes, 265
rossi Scullen, Cerceris, 580
rossica (Gussakovskij), Mimesa, 162
rossica Shestakov, Cerceris, 586
rossii Dahlbom, Bembix, 547
rostrata F. Smith, Cerceris, 586
rostrata (Gmelin), Bembix, 548
rostrata (Linnaeus), Bembix, 54, **548**
rostrata Marquet, Cerceris, 581
rostratum Taschenberg, Trypoxylon, 348
rostratus Berland, Tachytes, 266
rostrifera Brauns, Cerceris, 586
rotaense Tsuneki, Pison, 336
rothi (Beaumont), Podalonia, 145
rothi Giner Marí, Cerceris, 586
rothneyi Bingham, Miscophus, 319
rothneyi Cameron, Ampulex, 78
rothneyi Cameron, Cerceris, 586
rothneyi Cameron, Gastrosericus, 256
rothneyi Cameron, Pison, 336
rothneyi Cameron, Sphex, 116
rothneyi Cameron, Tachytes, 267
rothschildi Magretti, Palarus, 291
rotolum Leclercq, Rhopalum, 390
rottensis Meunier, Nysson, 31
rotundarius (Dahlbom), Crossocerus, 403
rotundicollis Arnold, Philanthus, 564

roubali Zavadil, Nysson, 469
rouxi Schulthess, Sphex, 115
rozeni Scullen, Cerceris, 586
ruandanus Bischoff, Philanthus, 565
ruandensis (Arnold), Crossocerus, 403
rubecula Schletterer, Cerceris, 586
rubella (F. Smith), Liris, 247
rubellum Richards, Trypoxylon, 347
rubellus Turner, Stizus, 527
rubellus Turner, Tachytes, 266
rubens (Mickel), Ancistromma, 260
rubi Wolf, Trypoxylon, 346
rubicola (Dufour and Perris), Ectemnius, 427
rubicola Harttig, Psenulus, 173
rubicundis van Lith, Psen, 167
rubicundus Arnold, Ammatomus, 513
rubicundus Pulawski, Tachysphex, 276
rubida (Jurine), Cerceris, 586
rubidus Arnold, Philanthus, 566
rubiginosa Lepeletier, Ammophila, 153
rubiginosum Gussakovskij, Trypoxylon, 347
rubiginosus (Handlirsch), Pseudoplisus, 503
rubra Radoszkowski, Ammophila, 152
rubrata R. Bohart and Menke, Cerceris, 586
Rubrica J. Parker, 53, **540**
rubricans Pérez, Liris, 245
rubricata (F. Smith), Liris, 247
rubricatus Turner, Sphodrotes, 304
rubriceps Taschenberg, Ammophila, 151
rubrifemoratum Richards, Trypoxylon, 347
rubripes Cresson, Eucerceris, 592
rubripes Spinola, Ammophila, 153
rubripyx Arnold, Dolichurus, 69
rubriventris A. Costa, Ammophila, 152
rubriventris Ferton, Miscophus, 318
rubriventris Honoré, Miscophus, 318
rubriventris Kazenas, Philanthus, 566
rubrocaudatus Arnold, Oxybelus, 369
rubrocaudatus (Blackburn), Ectemnius, 427
rubrocaudatus Turner, Psenulus, 172
rubrocinctulus Strand, Gorytes, 501
rubrocinctum Packard, Trypoxylon, 348
rubrocinctum Peckham and Peckham, Rhopalum, 389
rubroflavus Turner, Stizus, 527
rubropictus (Matsumura), Ectemnius, 426
rubrosignatus (Turner), Argogorytes, 492
rudesculpta Gussakovskij, Spilomena, 193
rufa Arnold, Larra, 237
rufa (Panzer), Mimesa, 162
rufa Scullen, Cerceris, 586
rufa Taschenberg, Cerceris, 587
rufacapoides Strand, Cerceris, 586
rufescens Andrade, Miscophus, 319
rufescens Cockerell, Plenoculus, 311
rufescens (F. Smith), Stizus, 527
rufescens (Gussakovskij), Parapiagetia, 281
rufescens Krombein, Ectemnius, 428
rufescens (Lepeletier), Sphecius, 511
rufescens Strand, Sceliphron, 105
rufibasis (Banks), Crabro, 408
rufibasis (Malloch), Pluto, 171
ruficauda Cameron, Cerceris, 586
ruficaudata (Turner), Astata, 213
ruficaudatus (Arnold), Entomognathus, 382

ruficaudis (Arnold), Dasyproctus, 421
ruficaudis Cameron, Oxybelus, 369
ruficaudis (Taschenberg), Tachysphex, 276
ruficaudus Taschenberg, Sphex, 115
ruficeps Brèthes, Trachypus, 569
ruficeps F. Smith, Cerceris, 586
ruficeps Scullen, Eucerceris, 592
ruficollis A. Morawitz, Ammophila, 153
ruficollis Cameron, Alysson, 458
ruficollis Cameron, Ampulex, 78
ruficollis (Cameron), Dicranorhina, 252
ruficollis (Fabricius), Dienoplus, 496
ruficollis Lichtenstein, Sphex, 119
ruficollis Reed, Ammophila, 152
ruficollis (Turner), Arpactophilus, 186
ruficorne (F. Smith), Pison, 335, 336
ruficornis (Cameron), Ampulex, 78
ruficornis (Cameron), Dicranorhina, 252
ruficornis F. Morawitz, Diodontus, 179
ruficornis F. Smith, Oxybelus, 369
ruficornis (Fabricius), Cerceris, 586
ruficornis Fabricius, Mellinus, 449
ruficornis (Fabricius), Stizus, 526, 527
ruficornis (J. Forster), Stizus, 527
ruficornis Latreille, Gorytes, 501
ruficornis (Provancher), Oryttus, 508
ruficornis Rayment, Sericophorus, 302
ruficornis (Villers), Mellinus, 449
ruficornis (Zetterstedt), Ectemnius, 427
ruficosta Spinola, Ammophila, 153
ruficoxa van der Vecht, Ammophila, 152
ruficoxis Cameron, Ampulex, 78
ruficrus (Dufour), Tachysphex, 272
ruficrus R. Bohart, Arigorytes, 517
ruficrus van Lith, Psen, 167
rufidens Cameron, Trypoxylon, 347
rufifemur (Packard), Ectemnius, 427
rufifrons Arnold, Cerceris, 586
rufigaster Packard, Rhopalum, 390
rufilumbis Lichtenstein, Sphex, 119
rufimana Taschenberg, Cerceris, 586
rufimanum Spinola, Trypoxylon, 347
rufinervis Pérez, Sphex, 116
rufinoda Cresson, Cerceris, 586
rufinodis F. Smith, Cerceris, 582
rufinodula Dalla Torre, Cerceris, 582
rufinodus Cresson, Mellinus, 449
rufipalpis Cameron, Tachytes, 266
rufipennis Fabricius, Liris, 247
rufipennis (Fabricius), Prionyx, 133
rufipes (Aichinger), Tachysphex, 272
rufipes (Arnold), Bembecinus, 530
rufipes Berland, Tachytes, 266
rufipes (Brullé), Ectemnius, 425
rufipes F. Smith, Cerceris, 588
rufipes (F. Smith), Hoplisoides, 521
rufipes (F. Smith), Larra, 238
rufipes (F. Smith), Pison, 335
rufipes (Fabricius), Cerceris, 588
rufipes Fabricius, Podium, 38, **96**
rufipes (Fabricius), Stizus, 527
rufipes (Guérin-Méneville), Ammophila, 153
rufipes (Guérin-Méneville), Chlorion, 90
rufipes Latreille, Palarus, 44, **291**
rufipes Lepeletier and Brullé, Crossocerus, 402
rufipes (Lepeletier and Brullé), Ectemnius, 427, 627
rufipes (Lepeletier), Podalonia, 144
rufipes Lepeletier, Sphex, 116

rufipes Mocsáry, Astata, 213
rufipes (Mocsáry), Sceliphron, 106
rufipes (Olivier), Brachystegus, 473
rufipes (Olivier), Stizoides, 528
rufipes (Provancher), Tachysphex, 272
rufipes Rayment, Sericophorus, 302
rufipes (Reed), Tachysphex, 276
rufipes Reed, Trachypus, 568
rufipes (Rohwer), Sericophorus, 302
rufipes (Saussure), Liris, 245
rufipes Shuckard, Pison, 336
rufipes Spinola, Palarus, 291
rufipes Taschenberg, Oxybelus, 369
rufiscutis Cameron, Cerceris, 586
rufiscutis Turner, Tachytes, 266
rufiscutus (R. Turner), Sphex, 116
rufitarse Arnold, Pison, 336
rufitarsis Cameron, Gastrosericus, 256
rufitarsis (Cameron), Liris, 247
rufitarsis (Cameron), Tachysphex, 272
rufitarsis (Dalla Torre), Ectemnius, 425
rufitarsis F. Morawitz, Oxybelus, 367
rufitarsis F. Smith, Astata, 213
rufitarsis (Spinola), Tachysphex, 276
rufitarsus (Rayment), Spilomena, 193
rufithorax Arnold, Ampulex, 77
rufithorax (Brauns), Dienoplus, 496
rufithorax (Ducke), Liris, 245
rufitibialis Arnold, Tachytes, 265
rufiventralis Ferton, Tachysphex, 274
rufiventre (Panzer), Rhopalum, 389
rufiventris Arnold, Belomicrus, 364
rufiventris (Cameron), Lyroda, 299
rufiventris Cameron, Psen, 167
rufiventris Cresson, Astata, 213
rufiventris Cresson, Nysson, 469
rufiventris (Cresson), Palmodes, 127
rufiventris F. Morawitz, Gastrosericus, 256
rufiventris (Fabricius), Trigonopsis, 38, **98**
rufiventris Gussakovskij, Prosopigastra, 286
rufiventris Lepeletier, Cerceris, 586
rufiventris Priesner, Bembix, 548
rufiventris R. Bohart, Microbembex, 539
rufiventris Radoszkowski, Stizus, 527
rufiventris (Reed), Tachysphex, 276
rufiventris (Spinola), Tachysphex, 275
rufiventris Timberlake, Encopognathus, 380
rufiventris Tsuneki, Crossocerus, 403
rufiventris Tsuneki, Miscophus, 319
rufiventris Turner, Aphelotoma, 70
rufiventris Turner, Nitela, 325
rufoannulatus Cameron, Psen, 167
rufoannulatus Strand, Tachytes, 263
rufoantennatum Rohwer, Diploplectron, 211
rufobalteatus (Cameron), Psenulus, 173
rufobasalis Rayment, Sericophorus, 302
rufocaudatus (Mickel), Hoplisoides, 521
rufocincta Gerstaecker, Cerceris, 586
rufocinctus Brullé, Sphex, 116
rufocinctus Dahlbom, Stizus, 527
rufocinctus (W. Fox), Lestiphorus, 506
rufodorsatus De Stefani, Sphex, 114
rufofacies Empey, Cerceris, 586
rufofasciatus Cresson, Tachytes, 263
rufofemorata Arnold, Astata, 213
rufofemorata Cameron, Ampulex, 78
rufofemorata (Cameron), Liris, 247
rufoflavus R. Bohart, Nysson, 469
rufofusca Turner, Cerceris, 577
rufogeniculata (Cameron), Liris, 244

rufoluteus (Packard), Pseudoplisus, 503
rufomaculatus (Cameron), Tachytes, 263
rufomaculatus (W. Fox), Pseudoplisus, 503
rufomarginatus Arnold, Tachytes, 266
rufomixtus (Turner), Argogorytes, 492
rufoniger Bingham, Tachysphex, 276
rufoniger Mochi, Stizus, 527
rufoniger (Turner), Sericophorus, 303, 628
rufoniger (Turner), Synnevrus, 470
rufonigra Taschenberg, Cerceris, 586
rufonoda Radoszkowski, Cerceris, 588
rufonodis (Radoszkowski), Ammatomus, 513
rufopetiolatus van Lith, Psen, 166
rufopicta F. Smith, Cerceris, 586
rufopictum (F. Smith), Sceliphron, 106
rufopictum (Magretti), Chalybion, 103
rufopictus Arnold, Tachysphex, 272
rufopictus F. Morawitz, Oxybelus, 369
rufopictus (F. Smith), Synnevrus, 470
rufopictus Yasumatsu and Maidl, Nippononysson, 467
rufoplagiata Cameron, Cerceris, 589
rufoscapa (Cameron), Liris, 247
rufosignatum Taschenberg, Trypoxylon, 349
rufotaeniatus (Kohl), Podagritus, 393
rufotibialis Rayment, Sericophorus, 302
rufozonale W. Fox, Trypoxylon, 348
rufulicornis (Maidl and Klima), Oryttus, 508
rufus Giner Marí, Miscophus, 318
rufus Handlirsch, Nysson, 469
rugiceps Dalla Torre, Trypoxylon, 347
rugicollis Gussakovskij, Ammophila, 153
rugicollis Lepeletier, Ammophila, 151
rugicollis (Viereck), Crabro, 407
rugidorsatus Turner, Tachysphex, 276
rugifer (Dahlbom), Ectemnius, 428
rugifer Dahlbom, Pemphredon, 182
rugifer Kohl, Sphex, 116
rugifera (Turner), Liris, 247
rugifrons Arnold, Astata, 213
rugifrons Cameron, Trypoxylon, 347
rugifrons F. Smith, Trypoxylon, 347
rugifrons van Lith, Psenulus, 629
rugosa F. Smith, Cerceris, 586
rugosa (F. Smith), Vechtia, 417
rugosa J. Parker, Bembix, 548
rugosa (W. Fox), Larropsis, 259
rugosa Williams, Nitela, 325
rugosifrons (Arnold), Carinostigmus, 191
rugosifrons Arnold, Philanthus, 566
rugosissima Arnold, Nitela, 325
rugosissimus Turner, Entomognathus, 382
rugosopunctatus (Provancher), Anacrabro, 379
rugosopunctatus (Taschenberg), Ectemnius, 425
rugospunctatus Turner, Encopognathus, 380
rugosulopunctatus (Dalla Torre), Anacrabro, Anacrabro, 379
rugosum F. Smith, Chlorion, 90
rugosum F. Smith, Pison, 336
rugosum Menke, Trypoxylon, 349
rugosus Cameron, Nysson, 469
rugosus Gussakovskij, Tachysphex, 276
rugosus Herrich-Schaeffer, Crossocerus, 402
rugosus Kohl, Philanthus, 566
rugosus Matsumura, Sphex, 116
rugosus Packard, Gorytes, 501

rugosus (Provancher), Philanthus, 566
rugosus Rayment, Sericophorus, 302
rugosus van Lith, Psenulus, 173
rugosus W. Fox, Diodontus, 179
rugulosa Schrottky, Cerceris, 586
rugolosus Móczár, Oxybelus, 368
rukwaensis (Arnold), Prionyx, 134
rumipámbae Leclercq, Rhopalum, 390
rumipambensis Benoist, Stigmus, 189
ruspolii Schulthess, Nysson, 469
russeolus Krombein, Gorytes, 501
rustica Taschenberg, Cerceris, 586
rusticus Cresson, Nysson, 469
rusticus Nurse, Diodontus, 179
ruthenica Gussakovskij, Didineis, 459
ruthenicus Birula, Nysson, 469
ruthenicus (F. Morawitz), Ectemnius, 425
rutila Spinola, Cerceris, 586
rutilus (Pate), Oryttus, 507
rutilus Spinola, Philanthus, 566
ruwenzoriensis (Arnold), Crossocerus, 403
rybiensis Ed. André, Cerceris, 586
rybyensis (Linnaeus), Cerceris, 55, **586**
ryukyuense Tsuneki, Trypoxylon, 347
ryukyuensis Tsuneki, Bembecinus, 530

sabina Gittins, Mimesa, 162
sabinasae Scullen, Eucerceris, 591
sabrina (Leclercq), Liris, 247
sabulosa Bytinski-Salz, Bembix, 546
sabulosa (F. Smith), Liris, 247
sabulosa Gussakovskij, Astata, 213
sabulosa (Linnaeus), Ammophila, 39, **153**
sabulosa (Panzer), Cerceris, 587
sabulosus F. Smith, Oxybelus, 366
sabulosus (Fabricius), Mellinus, 449
sabulosus (Olivier), Mellinus, 449
sacalava (Arnold), Holotachysphex, 282
sacra (Bytinski-Salz), Parapsammophila, 139
sacricola Pulawski, Tachytes, 266
sacuya Pate, Entomocrabro, 377
saeba Tsuneki, Cerceris, 587
saegeri Empey, Cerceris, 587
saegeri Leclercq, Diodontus, 179
saeva F. Smith, Ammophila, 153
saeva F. Smith, Cerceris, 587
saevissima F. Smith, Cerceris, 587
saevus Arnold, Tachysphex, 276
saevus (F. Smith), Prionyx, 134
saevus (Saussure), Dasyproctus, 421
sagakuchii Leclercq, Ectemnius, 428
sagani Guiglia, Tachytes, 266
sagax Kohl, Ampulex, 78
sagax Kohl, Liris, 247
sagax (Kohl), Palmodes, 127
sagittatus (Dahlbom), Oxybelus, 368
Sagenista R. Bohart, 53, **522**
saghaliensis Tsuneki, Cerceris, 586
sagittatus (Dahlbom), Oxybelus, 368
saharae Giner Marí, Bembix, 548
saharae (Giner Marí), Dienoplus, 496
saharae (Giner Marí), Podalonia, 145
saharae (Handlirsch), Ammatomus, 513
saharae Roth, Stizus, 527
saharica Beaumont, Parapiagetia, 281
saharica Giner Marí, Cerceris, 582
saharicus Beaumont, Belomicrus, 363
saharicus Pulawski, Tachytes, 266
sahlbergi (A. Morawitz), Entomognathus, 382
sahlbergi Shestakov, Cerceris, 587
saiguesei (Tsuneki), Carinostigmus, 191

saishiuensis Okamoto, Palarus, 291
saishuensis Tsuneki, Cerceris, 587
saitamaense Tsuneki, Trypoxylon, 347
sakaguchii (Matsumura and Uchida), Ectemnius, 428
sakalavus Leclercq, Dasyproctus, 421
sakuranus Tsuneki, Oxybelus, 368
salai Giner Marí, Cerceris, 587
salai Giner Marí, Lyroda, 299
salicis (Cockerell), Lindenius, 384
salicis Rohwer, Diodontus, 179
salina Lohrmann, Bembix, 548
Saliostethoides Arnold, 45, **320**
Saliostethus Brauns, 45, **319**
sallei Guérin-Méneville, Bicyrtes, 538
sallitus Andrade, Miscophus, 319
salome Banks, Cerceris, 581
saltator Arnold, Saliostethoides, 45, **320**
salti Richards, Trypoxylon, 349
saltitans Arnold, Psenulus, 173
salvadorius Leclercq, Anacrabro, 379
salvus Kohl, Tachytes, 266
samarensis Tsuneki, Cerceris, 587
sambucicola (C. Verhoeff), Crossocerus, 402
sameshimai (Yasumatus), Mimumesa, 164
samiatus R. Bohart, Pseudoplisus, 503
samoa Menke, Liris, 247
samoensis Williams, Liris, 247
samoensis (Williams), Liris, 247
sanbornii Cresson, Philanthus, 566
sanctae-rosae Cockerell, Bembix, 545
sanctum Richards, Trypoxylon, 347
sanctus Pulawski, Gastrosericus, 256
sandakaensis (Rohwer), Psenulus, 173
sandakanus Leclercq, Dasyproctus, 421
sandiegensis Scullen, Cerceris, 587
sanguinans (Dominique), Sagenista, 522
sanguinea Williams, Larra, 237
sanguinicollis Brauns, Ampulex, 78
sanguinolenta (Turner), Clitemnestra, 489
sanguinosus Mickel, Tachysphex, 276
sanluis Fritz and Toro, Cerceris, 587
santacrucae Bradley, Didineis, 459
santamartae Richards, Trypoxylon, 348
santoro Yasumatsu, Psen, 166
santschii (Schulthess), Belomicroides, 363
sapellonis (Baker), Argogorytes, 492
sapobaensis van Lith, Psenulus, 629
sapporoense Tsuneki, Trypoxylon, 347
sapporoensis (Kohl), Crossocerus, 401
sapporensis (Matsumura), Lestica, 430
sarafschani Radoszkowski, Bembix, 547
sarafschani Radoszkowski, Oxybelus, 369
sarawakense Cameron, Pison, 336
sarawakensis J. Parker, Bembecinus, 531
sarda Kohl, Ammophila, 152
sardonium (Lepeletier), Sceliphron, 105
sardous (Carrucio), Sceliphron, 105
sarekandana Balthasar, Ammophila, 153
sareptana Gussakovskij, Astata, 213
sareptana Kohl, Ammophila, 153
sareptana Schletterer, Cerceris, 587
sareptanus (Handlirsch), Dienoplus, 496
sareptanus Pulawski, Tachysphex, 275
sarmaticus F. Morawitz, Stizus, 527
satan Pate, Ectemnius, 428
satanas Kohl, Sphex, 116
satanas Pulawski, Tachysphex, 276
satanicus Siri and Bohart, Mellinus, 629
satoi Yasumatus, Ampulex, 78

satschouanus (Kohl), Lindenius, 384
satsumanus (Sonan), Bembecinus, 532
saturnus Arnold, Tachysphex, 276
saundersi Mercet, Tachysphex, 276
saundersi Morice, Palarus, 291
saundersi (Perkins), Ectemnius, 428
saundersi Richards, Trypoxylon, 349
saundersii Bingham, Tachytes, 266
saussurei (du Buysson), Ammophila, 153
saussurei (Fernald), Sphex, 114
saussurei (Handlirsch), Acanthostethus, 474
saussurei Handlirsch, Bembix, 546
saussurei Handlirsch, Stizus, 527
saussurei Kohl, Bothynostethus, 46, **352**
saussurei (Kohl), Chalybion, 103
saussurei (Kohl), Dasyproctus, 421
saussurei Kohl, Larra, 238
saussurei (Kohl), Parapiagetia, 281
saussurei Radoszkowski, Cerceris, 587
saussurei Reed, Tachytes, 265
saussurei Rohwer, Trypoxylon, 349
sauteri van Lith, Psen, 167
savignyi (Spinola), Liris, 245
savignyi Spinola, Oxybelus, 368
savignyi Spinola, Stizus, 527
saxatilis (Cameron), Ectemnius, 428
saxigenus Rohwer, Philanthus, 33, 567
sayi Banks, Cerceris, 588
sayi Banks, Tachytes, 266
sayi (Cockerell), Ectemnius, 425
sayi Cresson, Bembix, 548
sayi (Rohwer), Pluto, 171
sayi (Rohwer), Solierella, 314
sayi W. Fox, Astata, 213
scaber (Lepeletier and Brullé), Ectemnius, 428
scabra Beaumont, Cerceris, 585
scabra (Fabricius), Cerceris, 580
scabriuscula Arnold, Liris, 247
scabrosus Arnold, Tachysphex, 276
scabrum R. Turner, Pison, 336
scalaris (Illiger), Brachystegus, 473
scalaris Taschenberg, Tachytes, 263
Scapheutes Handlirsch, 46, **355**
SCAPHEUTINI, 59, **352**, 355
scapheutoides Menke, Bohartella, 46, **355**
scaposus "Zetterstedt", Crossocerus, 404
scapteriscica Williams, Larra, 238
scapularis Kohl, Astata, 212
scapularis Schletterer, Cerceris, 587
scaura Arnold, Bembix, 548
scaurus Arnold, Tachysphex, 276
scelesta Turner, Larra, 238
scelestus Cresson, Philanthus, 564
Sceliphron Klug, 39, **103**
SCELIPHRONINI, 58, **82**, 86
Sceliphrum Schulz, 39
schaeuffelei Beaumont, Cerceris, 587
schariniensis Kazenas, Cerceris, 587
schencki (Tournier), Psenulus, 173
schenckii (Cockerell), Ectemnius, 426
Schistosphex Arnold, 44
schlegelii (Ritsema), Larra, 250
schlettereri (Kohl), Ectemnius, 428
schlettereri Handlirsch, Gorytes, 501
schlettereri Radoszkowski, Cerceris, 587
schlingeri R. Bohart, Nysson, 469
schlingeri R. Bohart, Tachysphex, 276
schmidti Beaumont, Entomognathus, 382
schmidti Brauns, Nitela, 325
schmidti Richards, Trypoxylon, 347
schmiedeknechti (Handlirsch), Bembecinus, 532
schmiedeknechti Handlirsch, Stizus, 527
schmiedeknechti Kohl, Cerceris, 587
schmiedeknechti Kohl, Diodontus, 179
schmiedeknechti Kohl, Tachysphex, 276
schmiedeknechtii Handlirsch, Gorytes, 501
schmiedeknechtii Handlirsch, Nysson, 469
schmiedeknechtii Kohl, Belomicroides, 46, **363**
schmiedeknechtii (kohl), Entomognathus, 382
schmiedeknechtii Kohl, Eremiasphecium, 54, **561**
schmiedeknechtii (Kohl), Podalonia, 145
schmiedeknechtii Kohl, Trypoxylon, 347
schnitnikovi Kazenas, Cerceris, 577
schnusei Richards, Trypoxylon, 349
schoenherri Dahlbom Philanthus, 565
schoenlandi Cameron, Dasyproctus, 421
schoenlandi Cameron, Tachysphex, 276
schoutedeni Brauns, Cerceris, 587
schoutedeni Kohl, Sphex, 116
schoutedeni (Leclercq), Polemistus, 185
schrottkyi (Bertoni), Sphex, 116
schrottkyi (Fritz), Neoplisus, 505
schrottkyi Willink, Microbembex, 539
schubotzii (Arnold), Hoplisoides, 521
schulthessi (Giner Marí), Liris, 247
schulthessi Maidl, Philanthus, 566
schulthessi Richards, Trypoxylon, 349
schulthessi Roth, Sphecius, 511
schulthessi Schletterer, Cerceris, 577
schulthessii Kohl, Belomicrus, 364
schulthessirechbergi (Kohl), Chalybion, 103
schultzei Bischoff, Cerceris, 588
schulzi Beaumont, Cerceris, 577
schusteri R. Bohart, Philanthus, 566
schwarzi Beaumont, Bembecinus, 532
schwarzi (Beaumont), Dienoplus, 496
schwarzi (Pulawski), Holotachysphex, 282
schwarzi (Rohwer), Ectemnius, 428
sciaphila (Arcidiacono), Mimesa, 162
sciaphillus Leclercq, Crossocerus, 403
scintillans Andrade, Miscophus, 319
scioensis Gribodo, Sphex, 116
sciophanes (Nagy), Ampulex, 78
sciopteryx Perkins, Nesomimesa, 167
scitula Arnold, Bembix, 547
scitula (W. Fox), Glenostictia, 552
scitulus (Cresson), Hoplisoides, 520
scoliaeformis Arnold, Stizus, 527
scolioides Kokujev, Stizus, 527
scolopacea Handlirsch, Steniolia, 553
scoticus Perkins, Pemphredon, 182
Scotomphales Vachal, 53, **528**
scotti Turner, Bembix, 548
scotti (Turner), Dasyproctus, 421
scrobiculata Arnold, Solierella, 314
scrobiferum Richards, Trypoxylon, 349
scruposum Arnold, Pison, 336
scrutator Nurse, Philanthus, 566
scudderi Cockerell, Passaloecus, 33, 184
sculleni (R. Bohart), Clypeadon, 571
sculleni R. Bohart, Tachytes, 266
sculleni Sandhouse, Trypoxylon, 347
sculleni Tsuneki, Cerceris, 587
sculleniana Krombein, Cerceris, 584
sculptilis W. Fox, Tachysphex, 276
sculptiloides Williams, Tachysphex, 276
sculpturata (F. Smith), Lestica, 431
sculpturata (Kohl), Liris, 246
sculpturata Turner, Nitela, 325
sculpturatus Rayment, Sericophorus, 302
scutatum Chevrier, Trypoxylon, 347
scutatus (Fabricius), Crossocerus, 402
scutatus (Rohwer), Psenulus, 173
scutellaris A. Costa, Cerceris, 581
scutellaris A. Costa, Oxybelus, 368
scutellaris Cresson, Philanthus, 566
scutellaris (F. Smith), Crossocerus, 401
scutellaris (Gimmerthal), Crossocerus, 402
scutellaris (Spinola), Sagenista, 53, **522**
scutellaris W. F. Kirby, Stizus, 527
scutellatus (Packard), Crossocerus, 401
scutellatus (Say), Crossocerus, 403
scutellatus (Scheven), Crabro, 408
scutellatus Turner, Psenulus, 172, 174
scutellifer (Dalla Torre), Crossocerus, 403
scutifera Shestakov, Cerceris, 587
scutiferum Taschenberg, Trypoxylon, 347
scutifrons Saussure, Trypoxylon, 347
scutigerum Taschenberg, Trypoxylon, 347
scutularius (Schreber), Crabro, 408
scythicus Valkeila, Pemphredon, 182
seabrai Andrade, Solierella, 314
seamansi (Carter), Lestica, 431
secernendus (Turner), Argogorytes, 492
sechi Pate, Ammoplanus, 198
sechi Pate, Belomicrus, 364
secoense F. Parker, Diploplectron, 211
seculata J. Parker, Bembix, 548
secunda (Rohwer), Foxia, 479
secundus Saussure, Dolichurus, 69
sedlaceki van Lith, Psen, 167
sedula Arnold, Cerceris, 583
sedulus F. Smith, Tachytes, 266
sedulus Merisuo, Pemphredon, 182
segiensis Krombein, Cerceris, 589
segregata Beaumont, Cerceris, 585
segregatum Richards, Trypoxylon, 347
segregatus Leclercq, Crossocerus, 403
segrex van Lith, Psenulus, 174
seitzii Kohl, Ampulex, 78
selangori Leclercq, Piyuma, 410
selecta Nurse, Astata, 213
selectus Nurse, Diodontus, 179
selectus Nurse, Tachysphex, 276
selifera Schletterer, Cerceris, 587
sellae Gribodo, Sphex, 114
Selman J. Parker, 54, **541**
semenovi Gussakovskij, Tachysphex, 274
semenovi Shestakov, Cerceris, 587
semenowi F. Morawitz, Chlorion, 90
semiappendiculata (Cameron), Liris, 247
semiargentia (Taschenberg), Liris, 246
semiatra Banks, Cerceris, 579
semicincta (Panzer), Cerceris, 586
semicinctus Villers, Sphex, 117
semiflavum Richards, Trypoxylon, 349
semifossulatus van der Vecht, Sphex, 116
semilunata Radoszkowski, Cerceris, 587
seminiger (Dahlbom), Sagenista, 522
seminigra Banks, Cerceris, 577
seminigra F. Morawitz, Bembix, 548
seminigra Taschenberg, Cerceris, 587
seminitidus van Lith, Psen, 166
seminole Banks, Tachytes, 264
seminole (Bradley), Epinysson, 472
seminudus Arnold, Tachytes, 266

semipetiolata Saussure, Cerceris, 587
semipunctatus (Lepeletier and Brullé), Ectemnius, 428
semipunctatus (Taschenberg), Hoplisoides, 421
semirubra Bischoff, Larra, 237
semirufa (Banks), Ancistromma, 259
semirufa F. Smith, Cerceris, 588
semirufus (Cresson), Tachysphex, 276
semirugosa Williams, Solierella, 314
semistriatus Schmiedeknecht, Lestiphorus, 506
semoni Cameron, Bembix, 548
semota Beaumont, Ammophila, 151
semperi (Handlirsch), Bembecinus, 532
senci Pate, Foxita, 416
senegalensis Arnold, Brachystegus, 473
senegalensis Arnold, Gastrosericus, 256
senegalensis (Arnold), Prionyx, 134
senegalensis Berland, Tachytes, 266
senegalensis Empey, Cerceris, 629
senegambicum Kohl, Trypoxylon, 348
senex (Arnold), Arnoldita, 418
senex Bischoff, Ampulex, 78
senilis (Dahlbom), Podalonia, 145
senilis Fabricius, Bembix, 548
senilis (Morice), Prionyx, 134
sennacus Leclercq, Ectemnius, 428
sennae (Mantero), Prionyx, 134
senonus Leclercq, Crossocerus, 403
seoulensis Tsuneki, Cerceris, 586
separabilis Turner, Tachytes, 266
separanda F. Morawitz, Ammophila, 151
separanda Handlirsch, Bembix, 545
separatum F. Smith, Pison, 336
separatum F. Smith, Sceliphron, 106
separatus van Lith, Psenulus, 174
sepicola (F. Smith), Isodontia, 124
septentrionalis (Packard), Ectemnius, 427
septralis Radoszkowski, Philanthus, 566
sepulchralis (Beaumont), Dryudella, 215
sepulchralis F. Smith, Cerceris, 587
sepulchralis (Gerstaecker), Liris, 247
sepulchralis (Handlirsch), Sagenista, 522
sepulchralis of (Arnold), Liris, 246
sepulcralis Williams, Tachysphex, 276
sepultus (Cockerell), Gorytes, 502
sepultus Cockerell, Hoplisus, 31
seraksensis Radoszkowski, Oxybelus, 367
serapis Pulawski, Tachytes, 265
seraxensis Radoszkowski, Cerceris, 587
seraxensis Radoszkowski, Palarus, 291
serena (Turner), Liris, 247
serena (Turner), Williamsita, 422
sericans Gussakovskij, Tachysphex, 276
sericans (Morice), Prosopigastra, 285
sericatus Cresson, Tachytes, 266
sericatus (F. Smith), Lestiphorus, 506
sericatus Gerstaecker, Oxybelus, 368
sericea G. and R. Bohart, Larropsis, 259
sericea Lepeletier and Serville, Ammophila, 152
sericea Murray, Podalonia, 145
sericea (Spinola), Bicyrtes, 538
sericeum Kohl, Pison, 336
sericeus (Cameron), Hoplisoides, 521
sericeus (F. Smith), Tachysphex, 276
sericeus (Fabricius), Sphex, 116
sericeus (Kohl), Sericophorus, 44, **303**
sericeus Radoszkowski, Miscophus, 319
sericeus Robertson, Oxybelus, 369

sericifrons H. Smith, Ancistromma, 260
Sericogaster Westwood, 38
Sericophorus F. Smith, 44, **299**
Sericophorus Shuckard, 44
sericops (F. Smith), Tachysphex, 275
sericosoma (Turner), Liris, 247
serotinus De Stefani, Ectemnius, 426
serotinus O. Muller, Sphex, 117
serpentinus Lichtenstein, Sphex, 119
serrano Pate, Belomicrus, 364
serrano Pate, Pulverro, 196
serrata (Handlirsch), Stictiella, 551
serratocornis Jurine, Psen, 166
serratus Tsuneki, Ammoplanus, 198
serrei Leclercq, Enoplolindenius, 415
serripes (Fabricius), Cerceris, 587
serripes (Panzer), Crossocerus, 401
serrulatae Dunning, Philanthus, 566
servillei (Lepeletier), Sceliphron, 105
servillei Lepeletier, Sphex, 116
servillii Lepeletier, Bicyrtes, 538
servitorius Leclercq, Ectemnius, 428
servus Arnold, Paranysson, 308
servus Dalla Torre, Crossocerus, 400
sesquicincta Klug, Cerceris, 577
seth Pulawski, Tachysphex, 274
setigera (Arnold), Liris, 246
setigera Kohl, Tachytes, 266
setosus Taschenberg, Tachytes, 266
severa F. Smith, Bembix, 548
severini Kohl, Cerceris, 587
severini (Kohl), Isodontia, 123
sexcinctus (Fabricius), Ectemnius, 428
sexdentatum Taschenberg, Podium, 96
sexfasciatus (Fabricius), Stizus, 527
sexguttatus Gussakovskij, Nysson, 469
sexinus Leclercq, Tachysphex, 276
sexmaculatus (Olivier), Crossocerus, 401
sexmaculatus (Say), Ectemnius, 425
sexpunctata (Fabricius), Cerceris, 580
sexta Say, Cerceris, 587
sextoides Banks, Cerceris, 587
seychellense Turner, Rhopalum, 390
seychellensis (Cameron), Liris, 247
seyrigi Arnold, Alysson, 458
seyrigi Arnold, Ammatomus, 513
seyrigi Arnold, Cerceris, 587
seyrigi (Arnold), Ectemnius, 428
seyrigi Arnold, Miscophus, 319
seyrigi Arnold, Pison, 336
seyrigi Arnold, Spilomena, 193
seyrigi Arnold, Tachysphex, 276
seyrigi Arnold, Trypoxylon, 348
shaman Shestakov, Cerceris, 581
shangaani Empey, Cerceris, 629
shannoni Richards, Trypoxylon, 348
shappirioi (G. and R. Bohart), Ancistromma, 260
shawi Bradley, Alysson, 458
shawii Rohwer, Pemphredon, 181
sheffieldi (R. Turner), Podalonia, 145
shelfordi Turner, Cerceris, 587
sheppardi Arnold, Tachytes, 267
shermani Brimley, Cerceris, 581
shestakovi Gussakovskij, Ammoplanus, 198
shestakovi Gussakovskij, Ancistromma, 259
shestakovi Gussakovskij, Cerceris, 587
shestakovi Gussakovskij, Gastrosericus, 256
shestakovi (Gussakovskij), Mimesa, 162
shestakovi (Gussakovskij), Pseudoscolia, 573

shestakovi Gussakovskij, Sceliphron, 106
Shestakovia Gussakovskij, 54
shestakoviana Gussakovskij, Cerceris, 587
Shestakoviella Gussakovskij, 55
shibuyai (Iwata), Crossocerus, 403
shimoyamai Tsuneki, Ectemnius, 428
shimoyamai Tsuneki, Trypoxylon, 348
shirozui Tsuneki, Cerceris, 587
shirozui Tsuneki, Dolichurus, 69
shirozui Tsuneki, Liris, 248
shirozui Tsuneki, Pemphredon, 182
shirozui Tsuneki, Psen, 167
shirozui Tsuneki, Rhopalum, 390
shirozui Tsuneki, Stigmus, 189
shirozui Tsuneki, Tachytes, 263
shirozui Tsuneki, Trypoxylon, 348
shiva Nurse, Tachytes, 266
shoshone Menke, Ammophila, 153
shuckardi (A. Morawitz), Pemphredon, 181
shuckardi (Dahlbom), Ectemnius, 425
shuckardi (F. Smith), Crossocerus, 402
shuckardi Wesmael, Mimesa, 162
shuckardi Wesmael, Nysson, 469
shuckardi Yasumatsu, Passaloecus, 184
shukuzanus Tsuneki, Psen, 167
shur Shestakov, Cerceris, 581
siamensis (Cockerell), Polemistus, 185
siamensis Taschenberg, Sphex, 117
siamensis Tsuneki, Gastrosericus, 629
sibilans Handlirsch, Bembix, 548
sibirica (Beaumont), Mimesa, 162
sibirica F. Morawitz, Cerceris, 587
sibirica (Fabricius), Ampulex, 77
sibirica Gussakovskij, Didineis, 459
sibirica (Gussakovskij), Mimumesa, 164
sibiricana R. Bohart, Mimumesa, 164
sibiricus A. Morawitz, Crabro, 408
sibiricus Gussakovskij, Ammoplanus, 198
sibiricus Gussakovskij, Tachytes, 267
sibiricus (Mocsáry), Bembecinus, 531
sibiricus Radoszkowski, Philanthus, 565
siblina Leclercq, Lestica, 431
sibuyanensis van Lith, Psenulus, 173
sicana De Stefani, Cerceris, 586
sicarius F. Smith, Philanthus, 566
sickmanni Kohl, Ammophila, 153
sicula Kohl, Astata, 212
siculus (Beaumont), Dienoplus, 496
siculus De Stefani, Ectemnius, 425
siculus Giordani Soika, Philanthus, 566
sieberti Strand, Sphex, 114
sieboldti Dahlbom Philanthus, 566
sierrense F. Parker, Diploplectron, 211
signata Klug, Cerceris, 580
signata (Linnaeus), Stictia, 54, **542**
signaticrus F. Morawitz, Crabro, 408
signatus (Handlirsch), Bembecinus, 532
signatus (Olivier), Crossocerus, 401
signatus (Panzer), Crossocerus, 400
signifer (Packard), Crabro, 407
sigua (Pate), Epinysson, 472
siitanus Tsuneki, Tachysphex, 276
sikkimensis (Kriechbaumer), Ampulex, 78
sikorae Saussure, Tachysphex, 275
Silaon Kohl, 45
Silaon Piccioli, 45
silvana Schletterer, Cerceris, 587
silvaticus Arnold, Psen, 167
silverlocki (Turner), Afrogorytes, 523
silverlocki Turner, Gastrosericus, 256
silverlocki Turner, Tachytes, 267

silvestre Richards, Trypoxylon, 348
silvestrii Maidl, Bembix, 548
silvicola (Williams), Liris, 247
silvicola Williams, Tachytes, 267
silvicoloides Williams, Tachytes, 267
simalurense (Maidl), Rhopalum, 390
Simblephilus Dahlbom, 54
Simblephilus Jurine, 54
similans Rohwer, Tachysphex, 276
similans W. Fox, Bembix, 545
similicolor (Dow), Argogorytes, 492
similis (Bridwell), Solierella, 314
similis Cresson, Eucerceris, 592
similis Cresson, Oxybelus, 369
similis F. Morawitz, Miscophus, 319
similis (Gussakovskij), Dryudella, 215
similis (Mocsáry), Larra, 238
similis (Rohwer), Mimumesa, 164
similis Rohwer, Tachysphex, 276
similis (W. Fox), Crossocerus, 403
simillima F. Smith, Ammophila, 151
simillima (F. Smith), Bicyrtes, 538
simillima (F. Smith), Larra, 238
simillimum F. Smith, Pison, 336
simillimus Cresson, Philanthus, 564
simillimus (F. Smith), Bembecinus, 532
simillimus (F. Smith), Dasyproctus, 421
simillimus F. Smith, Gorytes, 501
simillimus (Fernald), Prionyx, 134
simillimus Gussakovskij, Tachysphex, 274
simillimus Schulthess, Tachytes, 267
simlaense Richards, Trypoxylon, 349
simlaensis Cameron, Cerceris, 582
simlaensis (Nurse), Crossocerus, 403
simlensis van Lith, Psen, 167
simoni Buysson, Cerceris, 584
simoni (Buysson), Isodontia, 124
simonyi Kohl, Tachysphex, 277
simplex Arnold, Gastrosericus, 256
simplex Dahlbom, Oxybelus, 370
simplex F. Smith, Cerceris, 582
simplex Gussakovskij, Ammoplanus, 198
simplex (Kohl), Isodontia, 124
simplex (Malloch), Mimesa, 162
simplex Pulawski, Tachysphex, 274
simplex (Tournier), Psenulus, 173
simplicicornis (F. Morawitz), Pseudoscolia, 573
simplicicornis W. Fox, Nysson, 469
simplicicornis (W. Fox), Pseneo, 165
simplicipes E. Saunders, Dinetus, 216
simplicipes (F. Morawitz), Rhopalum, 389
simpsoni (Turner), Stizoides, 529
simulans F. Smith, Tachytes, 267
simulans Saussure, Cerceris, 587
simulans Turner, Pison, 336
simulatrix (Arnold), Liris, 247
simulatrix, Turner, Tachytes, 267
simulatrix Viereck and Cockerell, Eucerceris, 591
sinaitica Alfierei, Ammophila, 152
sinaitica Beaumont, Cerceris, 587
sinaitica (Mochi), Pseudoscolia, 573
sinaiticus (Mochi), Protostigmus, 196
sinaiticus Pulawski, Tachysphex, 276
sinclairi Lal, Psenulus, 174
sinensis F. Smith, Cerceris, 587
sinensis F. Smith, Tachytes, 267
sinensis Lichtenstein, Sphex, 119
sinensis (Mocsáry), Larra, 238
sinensis Saussure, Ampulex, 77
sinensis Sickmann, Ammophila, 153

sinensis (Yasumatsu), Ammatomus, 513
singarae Leclercq, Entomognathus, 382
singularis Brèthes, Cerceris, 578
singularis Dahlbom, Passaloecus, 184
singularis Evans, Lisponema, 31
singularis (F. Smith), Ectemnius, 427
singularis F. Smith, Sphex, 114
singularis (Taschenberg), Hemidula, 53, **540**
singularis van Lith, Psenulus, 174
sinica Tsuneki, Cerceris, 587
sinicus Leclercq, Crossocerus, 403
sintangense Strand, Sceliphron, 105
sinuata Panzer, Bembix, 548
sinuata Panzer of Latreille, Bembix, 548
sinuata Scullen, Eucerceris, 592
sinuatus (A. Costa), Gorytes, 501
sinuatus (Fabricius), Ectemnius, 426
sinuatus (Latreille), Bembecinus, 532
sinuatus (Provancher), Crabro, 408
sinuatus Pulawski, Tachytes, 267
sinuatus Williams, Plenoculus, 311
sinuosiscutis Arnold, Trypoxylon, 348
siouxensis (Mickel), Diodontus, 179
siouxensis Mickel, Philanthus, 566
sipapomae (Arnold), Bembecinus, 532
sipapome Arnold, Tachysphex, 276
sirdariensis Radoszkowski, Cerceris, 587
sirdariensis (Radoszkowski), Prionyx, 134
sirius Andrade, Miscophus, 319
sjoestedti Cameron, Ammophila, 151
sjoestedti Cameron, Dasyproctus, 420
sjoestedti (Cameron), Prionyx, 133
sjoestedti Cameron, Tachytes, 267
sjoestedti Gussakovskij, Ammophila, 153
slateri (Arnold), Ectemnius, 428
slimmatus Leclercq, Crossocerus, 403
slossonae (Ashmead), Ectemnius, 425
slossonae (Ashmead), Miscophus, 319
slovaca Balthasar, Cerceris, 579
slovaca Zavadil, Ammophila, 153
smaragdina F. Smith, Ampulex, 78
smaragdinum (Christ), Cahlybion, 102
smaragdinum (Christ), Chlorion, 90
smithiana Cameron, Cerceris, 583
smithiensis Leclercq, Williamsita, 422
smithii Ashmead, Stigmus, 189
smithii (Cresson), Pseudoplisus, 52, **503**
smithii Dalla Torre, Cerceris, 585, 587
smithii Handlirsch, Bembix, 548
smithii F. Smith, Ammophila, 153
smithii (W. Fox), Pluto, 171
smohalla Pate, Arigorytes, 517
snoflaki (Zavadil), Crossocerus, 402
snowi Banks, Cerceris, 579
snowi G. and R. Bohart, Larropsis, 259
snowii W. Fox, Crabro, 408
sobo Yasumatsu and Okabe, Cerceris, 587
socia (Kohl), Ancistromma, 260
sociabilis (Arnold), Crossocerus, 403
socius (Handlirsch), Bembecinus, 531
socius Thompson, Crossocerus, 400
sodalicia Kohl, Ampulex, 78
sodalis (Bingham), Ectemnius, 428
sodalis Turner, Cerceris, 580
sogatophagus Pagden, Psenulus, 174
sogdianum Gussakovskij, Pison, 336
sogdianum Gussakovskij, Trypoxylon, 348
sogdianus Gussakovskij, Gorytes, 502
soikai Beaumont, Miscophus, 319

soikai Beaumont, Philanthus, 566
soikae (Mochi), Pseudoscolia, 573
sokotrae Kohl, Cerceris, 587
sola J. Parker, Bicyrtes, 538
solani (Cockerell), Metanysson, 51, **481**
Solenius Lepeletier and Brullé, 50
solidaginis Rohwer, Cerceris, 579
solidagus Howard, Philanthus, 566
solidescens Scudder, Didineis, 31
Solierella Spinola, 45, **311**
solieri (Lepeletier), Palmodes, 127
solieri (Lepeletier), Sceliphron, 105
solitaria Dahlbom, Cerceris, 587
solitarius (Arnold), Bembecinus, 532
solitarius Arnold, Oxybelus, 369
solitarius (F. Smith), Dasyproctus, 421
solivagum Arnold, Trypoxylon, 348
solivagus (Bondroit), Pemphredon, 182
solivagus Say, Philanthus, 566
solomonensis van Lith, Psenulus, 173
solomonis Krombein, Cerceris, 578
solowiyofkae Matsumura, Ammophila, 153
solskii Radoszkowski, Cerceris, 580
solskii Radoszkowski, Oxybelus, 368
solskyi A. Morawitz, Stigmus, 189
solskyi Schletterer, Cerceris, 580
solstitialis (F. Smith), Liris, 247
somalica Arnold, Cerceris, 587
somalicus (Arnold), Bembecinus, 532
sombrana J. Parker, Stictia, 542
sombratus Leclercq, Podagritus, 393
somereni Empey, Cerceris, 629
sommereni (R. Turner), Cahlybion, 103
somotorensis Balthasar, Cerceris, 587
sonani Yasumatsu, Ampulex, 78
sonani (Yasumatsu), Isodontia, 124
songaricus (Eversmann), Prionyx, 134
sonnerati Kohl, Ampulex, 78
sonorae Scullen, Eucerceris, 592
sonorae Williams, Solierella, 314
sonorense Cameron, Trypoxylon, 349
sonorensis (Cameron), Ectemnius, 428
sonorensis (Cameron), Eucerceris, 591
sonorensis (Cameron), Podalonia, 145
sonorensis (Cameron), Tachysphex, 276
sophiae Evans, Liris, 247
sorbicus Leclercq, Podagritus, 393
sordidatus Arnold, Miscophus, 319
sordidula Arnold, Cerceris, 584
sordidus Arnold, Belomicrus, 364
sordidus Dahlbom, Sphex, 114
sordidus Dahlbom, Tachysphex, 276
soror Dahlbom, Bembix, 547
soror (Dahlbom), Prionyx, 134
soror Mocsáry, Trigonopsis, 98
soror Richards, Microstigmus, 192
sororcula Brèthes, Cerceris, 587
sotirus Leclercq, Crossocerus, 403
Spalagia Shuckard, 51
spangleri Krombein, Crossocerus, 403
Spanolarra Cameron, 43
sparideus Cockerell, Oxybelus, 369
sparsa G. and R. Bohart, Larropsis, 259
sparsa Rohwer, Larra, 237
sparsipunctatus Arnold, Philanthus, 566
spathulifera Brèthes, Cerceris, 587
spathulifer (Turner), Liris, 247
spatulata J. Parker, Bembix, 548
spatulatum Richards, Trypoxylon, 349
spatulatus W. Fox, Tachytes, 267

specialis (F. Smith), Neoplisus, 505
specifica Turner, Cerceris, 587
speciosa Arnold, Bembix, 548
speciosa (Cresson), Stictiella, 551
speciosissimus Morice, Tachysphex, 277
speciosus Dahlbom, Sphecius, 511
speciosus (Drury), Sphecius, 52, **511**
spectabilis Gerstaecker, Oxybelus, 369
spectabilis Kohl, Ampulex, 78
spectabilis Radoszkowski, Cerceris, 587
spectabilis (Taschenberg), Sphecius, 511
spectrum Arnold, Cerceris, 587
spectrum Handlirsch, Stizus, 527
speculare Turner, Pison, 336
specularis A. Costa, Cerceris, 587
specularis Andrade, Miscophus, 319
speculata Shestakov, Cerceris, 584
spegazzinii (Brèthes), Bembecinus, 530
spegazzinii Brèthes, Trachypus, 569
sphaerosoma (Handlirsch), Ochleroptera, 490
Sphaex Scopoli, 39
Sphecienus Patton, 53, **509**
SPHECINAE, 57, **79**, 82
SPHECINI, 58, **106**, 108
Sphecius Dahlbom, 52, **509**
sphecodoides Bradley, Nysson, 469
sphecodoides Rohwer, Tachysphex, 276
sphegiformis (Klug), Stizus, 526
Sphex Linnaeus, 39, **109**
Sphodrotes Kohl, 44, **304**
spicatus (Malloch), Pseneo, 165
spilaspis (Cameron), Crossocerus, 403
spilaspis Cameron, Philanthus, 564
spilographus (Handlirsch), Hoplisoides, 521
Spilomena Shuckard, 41, **192**
spiloptera Cameron, Ampulex, 78
spilopterus (Handlirsch), Hoplisoides, 521
spilostomus (Kohl), Lindenius, 384
spinaea Beaumont, Cerceris, 587
spinibuccus (Viereck), Crossocerus, 400
spinicaudata Cameron, Cerceris, 587
spinicollis Giner Marí, Cerceris, 589
spinicollis Gussakovskij, Diodontus, 179
spinicollis (Herrich-Schaeffer), Ectemnius, 426
spinicollum Tsuneki, Rhopalum, 390
spinicornis (Saussure), Bembecinus, 532
spinicornis (Turner), Austrogorytes, 498
spinifemur (Beaumont), Bembecinus, 532
spinifer Blanchard, Sceliphron, 106
spinifera Kazenas, Cerceris, 629
spinifera (Mickel), Stictiella, 551
spiniferus Cameron, Oxybelus, 368
spiniferus (du Buysson), Ammatomus, 513
spiniferus (Mickel), Diodontus, 179
spiniferus (W. Fox), Ectemnius, 428
spinifrons (Bingham), Vechtia, 417
spinifrons Saussure, Cerceris, 585
spiniger Kohl, Sphex, 114
spiniger Thunberg, Philanthus, 566
spiniger (Turner), Acanthostethus, 474
spinigera Rohwer, Cerceris, 583
spinigerus (Cameron), Crossocerus, 403
spinipectus F. Smith, Cerceris, 587
spinipectus (Shuckard), Crossocerus, 404
spinipes (A. Morawitz), Ectemnius, 428
spinipes (F. Smith), Podalonia, 145
spinipes Gmelin, Sphex, 117
spinipleuris Beaumont, Cerceris, 587
spinipleuris Turner, Cerceris, 587

spiniventris Tsuneki, Cerceris, 587
spinolae A. Costa, Dryudella, 215
spinolae (Dahlbom), Cerceris, 578
spinolae (F. Smith), Prionyx, 134
spinolae Kohl, Solierella, 313
spinolae Latreille, Nitela, 45, **325**
spinolae Lepeletier, Bembix, 545
spinolae (Lepeletier), Chalybion, 103
spinolae Lepeletier, Dynatus, 38, **94**
spinolae Reed, Tachytes, 264
spinolae Saussure, Astata, 213
spinolae Saussure, Palarus, 291
spinolae Shuckard, Pison, 336
spinolai Valkeila, Nitela, 629
spinolica Schletterer, Cerceris, 587
spinosa F. Smith, Ammophila, 151
spinosa (Fabricius), Bicyrtes, 538
spinosum Cameron, Trypoxylon, 349
spinosus (Fabricius), Nysson, 469
spinosus (J. Forster), Nysson, 469
spinosus (Panzer), Alysson, 458
spinosus Pulawski, Tachysphex, 277
spinosus W. Fox, Tachysphex, 272
spinulicollis (Mochi), Pseudoscolia, 573
spinulifer (Turner), Notocrabro, 394
spinulosa Beaumont, Olgia, 491
spinulosus Gussakovskij, Oxybelus, 369
spinulosus Pulawski, Tachysphex, 277
spinulosus Radoszkowski, Stizus, 527
spirans Saussure, Cerceris, 587
spirifex (Linnaeus), Sceliphron, 39, **106**
spiritalis J. Parker, Bembix, 548
spissatus W. Fox, Tachysphex, 272
splendens (Ashmead), Liris, 247
splendida Arnold, Bembix, 548
splendida (Giner Marí), Pseudoscolia, 574
splendidissima Giordani Soika, Cerceris, 587
splendidula (A. Costa), Isodontia, 124
splendidula Kohl, Ampulex, 78
splendidulus (Bradley), Hoplisoides, 521
splendidulus F. Morawitz, Tachysphex, 277
splendidum Fabricius, Chlorion, 90
splendidum (Handlirsch), Megistommum, 504
splendidus Berland, Sphex, 115
splendidus (Hacker & Cockerell), Sericophorus, 303
splendidus O. Müller, Sphex, 117
splendidus Reich, Sphex, 117
spoliatus (Girard), Tachysphex, 272
spooneri (Richards), Mimumesa, 164
spreta A. Costa, Cerceris, 586
spretum (Kohl), Penepodium, 92
spretus Kohl, Tachysphex, 277
spryi Rayment, Sericophorus, 302
spryi (Turner), Austrogorytes, 498
spurius (Dahlbom), Miscophus, 319
squamosus F. Smith, Oxybelus, 368
squamulifera Mickel, Cerceris, 587
stachi Beaumont, Tachysphex, 277
stadelmanni Handlirsch, Bembix, 548
stadelmanni Kohl, Sphex, 116
stadelmannii Dalla Torre, Bembix, 548
staegeri Dahlbom, Tachytes, 264
Stangeella Menke, 38, **87**
stangei Menke, Ammophila, 153
stangei Menke, Pison, 337
stangei Menke, Solierella, 314
stangei Scullen, Eucerceris, 592
stanleyensis Krombein, Cerceris, 584

stanleyi (Kohl), Isodontia, 124
stantoni (Ashmead), Dolichurus, 38, **69**
staudingeri Gribodo, Sphex, 116
staudingeri Richards, Trypoxylon, 348
stecki Beaumont, Astata, 212
stecki Schletterer, Cerceris, 577
stecki Schulz, Philanthus, 566
steckii Kohl, Belomicrus, 364
stefanii Ed. André, Cerceris, 587
steidachneri Kohl, Arpactophilus, 186
stella Shestakov, Cerceris, 587
stenaspis J. Parker, Bembecinus, 531
Steniolia Say, 54, **552**
stenobdoma J. Parker, Bembix, 548
Stenocrabro Ashmead, 47
Stenogorytes, Schrottky, 52
Stenomellinus Schulz, 40
stenopus (Arnold), Stizoides, 529
stenopygus (Handlirsch), Argogorytes, 492
steppicola (Tsuneki), Eremiasphecium, 561
Stercobata Gussakovskij, 55
sternalis (Rohwer), Liris, 248
sternodonta Gussakovskij, Cerceris, 587
sterope Arnold, Cerceris, 587
Stethorectus F. Smith, 38
stevensi Banks, Cerceris, 588
stevensi Rohwer, Didineis, 459
stevensoni (Arnold), Ammatomus, 513
stevensoni Arnold, Astata, 213
stevensoni Arnold, Bembix, 549
stevensoni (Arnold), Dasyproctus, 421
stevensoni (Arnold), Entomognathus, 382
stevensoni Arnold, Kohliella, 286
stevensoni Arnold, Miscophus, 319
stevensoni Arnold, Oxybelus, 369
stevensoni Arnold, Spilomena, 193
stevensoni Arnold, Stizus, 527
stevensoni Arnold, Tachysphex, 277
stevensoni Arnold, Trypoxylon, 348
stevensoni Brauns, Cerceris, 587
stevensoni J. Parker, Bembix, 548
stevensonianus (Arnold), Dasyproctus, 421
Stictia Illiger, 54, **541**
Stictiella J. Parker, 54, **550**
stictochilos Pate, Crossocerus, 403
Stictoptila Pate, 48, **398**
stieglmayri (Kohl), Polemistus, 185
stieglmayri Richards, Trypoxylon, 348
stigma Linnaeus, Sphex, 117
stigma (Panzer), Dryudella, 215
stigmosalis Banks, Cerceris, 588
Stigmus Panzer, 41, **188**
stimulator Turner, Tachysphex, 277
stirocephalus (Cameron), Enoplolindenius, 415
stirpicola (Packard), Ectemnius, 428
STIZINI, 59, **523**, 525
Stizobembex Gussakovskij, 53
Stizoides Guérin-Méneville, 53, **527**
Stizolarra Saussure, 53
Stizomorphus, A. Costa, 53
Stizus Latreille, 53, **525**
stockleini Giner Marí, Cerceris, 588
storeyi Turner, Stizus, 527
straboni (Berland), Palmodes, 127
straminea Dufour, Cerceris, 588
stramineipes (Arnold), Rhopalum, 389
strandi Giner Marí, Cerceris, 588
strandi Willink, Chlorion, 90
strandi Yasumatsu, Oxybelus, 369

strandi Yasumatsu, Pison, 336
strandtmanni Burks, Philanthus, 566
strangulatus (Bischoff), Crossocerus, 403
stratiotes Schletterer, Cerceris, 588
strenua (Cameron), Liris, 248
strenua Cresson, Ammophila, 153
strenua (Walker), Podalonia, 144
strenuum Turner, Pison, 336
strenuus Mickel, Bembecinus, 531
striareata Viereck and Cockerell, Eucerceris, 591
striata (Cameron), Dicranorhina, 252
striata F. Smith, Cerceris, 588
striata (Fabricius), Rubrica, 541
striata G. and R. Bohart, Larropsis, 259
striata Mocsáry, Ammophila, 153
striata (Viereck), Mimumesa, 164
striaticollis (Cameron), Liris, 248
striaticollis F. Morawitz, Ammophila, 151
striaticollis Turner, Aphelotoma, 70
straitifrons Cameron, Ampulex, 78
straitfrons (Cameron), Polemistus, 185
striatifrons (Mickel), Oxybelus, 369
striatipes (Ashmead), Solierella, 314
striatiscutis Cameron, Oxybelus, 368
striatulus (Brèthes), Prionyx, 133
striatulus Lepeletier and Brullé, Crossocerus, 404
striatulus (Lepeletier and Brullé), Ectemnius, 424, 425
striatulus (Rayment), Sericophorus, 303
striatum Provancher, Trypoxylon, 349
striatus Baker, Oxybelus, 369
striatus (F. Smith), Prionyx, 133
striatus Lepeltier, Crossocerus, 404
striatus (Lepeletier and Brullé), Ectemnius, 424
striatus (Mickel), Diodontus, 179
striatus W. Fox, Alysson, 458
stricklandi Pate, Crossocerus, 403
strictifrons Vachal, Pison, 336
stridulans (Strand), Editha, 542
strigatus (Chevrier), Pemphredon, 181
strigatus Mochi, Stizus, 527
strigatus Turner, Tachysphex, 277
strigosa Cameron, Cerceris, 588
strigosa Kohl, Astata, 212
strigosus (Mocsáry), Tachysphex, 273
strigulosum Turner, Pison, 336
strigulosus (A. Costa), Palmodes, 127
strigulosus Turner, Philanthus, 566
striolata Cameron, Ammophila, 153
striolata (Saussure), Ampulex, 77
striolata Schletterer, Cerceris, 577
striolatum Cameron, Pison, 336
striolatus (Arnold), Neodasyproctus, 419
striolatus Cameron, Diodontus, 179
striolatus (Cameron), Psen, 167
striolatus Cameron, Tachysphex, 274, 276
stroudi Gribodo, Trypoxylon, 348
strumosa Kohl, Ammophila, 153
stschurowskii (Radoszkowski), Prionyx, 134
stuckenbergi Arnold, Psenulus, 174
stueberi van der Vecht and Krombein, Sphex, 116
stygicus Bohart and Menke, Palmodes, 127
stygius (Blackburn and Kirby), Ectemnius, 428
stygius Gerstaecker, Philanthus, 566
stygius (Mickel), Crossocerus, 400

stygius Williams, Plenoculus, 311
styrax Leclercq, Cerceris, 577
styrius (Kohl), Crossocerus, 403
suada (F. Smith), Bembecinus, 532
suavis Arnold, Tachysphex, 277
suavis Burmeister, Ammophila, 152
suavis (F. Morawitz), Prionyx, 133
subaeneus Lepeletier and Brullé, Lindenius, 385
subaequalis (Rohwer), Liris, 244
subalpinus (Cockerell), Bembecinus, 532
subandinus Pulawski, Tachysphex, 277
subapicalis (Walker), Stizus, 526
subassimilis Strand, Ammophila, 153
subatratus (R. Bohart), Prionyx, 134
subaustralis (Viereck), Pseudoplisus, 503
subcitripes Willink, Bembix, 546
subconcolor (Bingham), Philanthus, 566
subcoriaceus Arnold, Tachysphex, 277
subcornutus Cockerell, Oxybelus, 369
subcristatus Saussure, Oxybelus, 369
subcyaneum Gerstaecker, Chlorion, 90
subdentatus F. Morawitz, Tachysphex, 277
subdepressa Lepeletier, Cerceris, 586
subeburnea Tsuneki, Bembix, 548
subeditus Leclercq, Tachysphex, 277
subexcisus (Brèthes), Prionyx, 133
subfasciata (Walker), Liris, 248
subfimbriatus Arnold, Tachysphex, 277
subfrigidum Rohwer, Trypoxylon, 346
subfrigidus Dunning, Aphilanthops, 570
subfuscatus (Dahlbom), Prionyx, 134
subfuscatus (Eversmann), Prionyx, 134
subfuscatus Turner, Tachysphex, 277
subgibbosa Yasumatsu, Cerceris, 587
subgratiosa (Strand), Microbembex, 539
subhyalinus W. Fox, Sphex, 116
subimpressa Schletterer, Cerceris, 583
subimpressum F. Smith, Trypoxylon, 349
subita (Say), Lyroda, 44, 299
sublaevis (Beaumont), Mimesa, 162
sublimis Cresson, Philanthus, 564
submellipes (Viereck), Epinysson, 472
subnasutus (Arnold), Entomognathus, 382
subopacus Turner, Tachysphex, 277
subpetiolata (Brèthes), Parapiagetia, 281
subpetiolata (F. Smith), Liris, 248
subpetiolata Saussure, Cerceris, 579
subpileatum Strand, Trypoxylon, 348
subpunctatus (Illiger), Crossocerus, 401
subpunctatus (Rossi), Crossocerus, 401
subspinosus Klug, Oxybelus, 369
substriatula (Turner), Prapiagetia, 281
subterranea (Fabricius), Lestica, 431
subtessellata (F. Smith), Liris, 248
subtilis (Antiga and Bofill), Ectemnius, 425
subtilis Gussakovskij, Oxybelus, 369
subtilis van Lith, Psen, 166
subtilis W. Fox, Nysson, 470
subtilus (Handlirsch), Ochleroptera, 490
subtruncatus Dahlbom, Sphex, 116
subulatus (Dahlbom), Crossocerus, 403
subulatus Robertson, Oxybelus, 369
subversus Banks, Philanthus, 565
subviridis Rayment, Sericophorus, 302
succincta (Cockerell), Tracheliodes, 31, 405
succinctum Cresson, Trypoxylon, 348
succinctus Cresson, Crabro, 407
succinctus (Olivier), Mellinus, 449
succineicollarum Tsuneki, Rhopalum, 390

succineus (Klug), Stizus, 527
sucrensis Leclercq, Enoplolindenius, 415
sudai Tsuneki, Pemphredon, 182
sudanensis (Arnold), Bembecinus, 532
suffusus (W. Fox), Pluto, 171
sugihari (Iwata), Crossocerus, 403
suifuensis van Lith, Psenulus, 174
sulcatoides Richards, Trypoxylon, 348
sulcatus (Malloch), Psenulus, 174
sulcatus Turner, Arpactophilus, 186
sulcatus Turner, Tachytes, 267
sulcidorsum Beaumont, Tachysphex, 277
sulcifrons (A. Costa), Gorytes, 502
sulcifrons (Cameron), Liris, 248
sulcifrons Gussakovskij, Trypoxylon, 348
sulcipyga Mochi, Cerceris, 588
sulciscuta Gribodo, Sphex, 117
sulcus La Munyon, Trypoxylon, 349
sulcus (W. Fox), Crossocerus, 403
sulphurata (Dalla Torre), Lestica, 431
sulphurea Cameron, Cerceris, 588
sulphurea (Spinola), Microbembex, 539
sulfurea W. Fox, Steniolia, 553
sulphureipes (F. Smith), Ectemnius, 425
sulphurescens Dahlbom, Bembix, 548
sulphureus F. Smith, Philanthus, 566
sulphureus van Lith, Psenulus, 173
suluensis Pate, Oxybelus, 369
suluensis van Lith, Psenulus, 174
suluensis Williams, Tachytes, 266
sumatrae Leclercq, Rhopalum, 390
sumatrana Kohl, Larra, 238
sumatranum (Kohl), Chalybion, 103
sumatranus (Ritsema), Psenulus, 172
sumatrensis (Maidl), Liris, 248
sumatrensis (Maidl), Polemistus, 185
sumptuosa Arnold, Cerceris, 581
sumptuosus A. Costa, Sphex, 115
sumptuosus Turner, Philanthus, 566
sundewalli (Dahlbom), Prionyx, 134
sungari Tsuneki, Cerceris, 588
superba Cresson, Eucerceris, 592
superba Shestakov, Cerceris, 587
superbiens Morice, Tachytes, 264
superbum F. Smith, Trypoxylon, 349
superbum Radoszkowski, Chlorion, 90
superbus (Harris), Mellinus, 449
superbus (Tournier), Psen, 166
superciliaris Saussure, Ammophila, 151
superciliosum Richards, Trypoxylon, 349
superflua Strand, Cerceris, 583
supposita Kohl, Cerceris, 588
supraconica Tsuneki, Cerceris, 588
surgicus Leclercq, Entomognathus, 382
surigensis Williams, Tachytes, 267
surinamense Richards, Trypoxylon, 349
surinamensis (De Geer), Rubrica, 541
surinamensis (Retzius), Sphex, 115
surinamensis Saussure, Ampulex, 78
surinamensis Saussure, Trachypus, 568
surusumi Tsuneki, Crossocerus, 403
surusumi Tsuneki, Liris, 248
susanae Chessman, Pison, 336
Susanowo Tsuneki, 41
suspicax Kokujev, Pison, 336
suspiciosa (F. Smith), Podalonia, 145
suspiciosum F. Smith, Pison, 336
susterai Balthasar, Miscophus, 319
susterai Snoflak, Ammophila, 153
susterai Zavadil, Nysson, 470
sutshanicus (Gussakovskij), Crossocerus, 403

sutteri van der Vecht, Sceliphron, 106
suzukii (Matsumura), Crossocerus, 403
svetlanae Pulawski, Tachysphex, 277
swalei (Leclercq), Podagritus, 393
swalei Turner, Gastrosericus, 256
swellendamensis (Arnold), Entomognathus, 382
sybarita Kohl, Ampulex, 78
sycorax Arnold, Tachysphex, 275
sycorax (F. Smith), Paraliris, 250
sydneyi Rayment, Sericophorus, 302
Sylaon Piccioli, 45
sylvatica (Arnold), Lestica, 431
Symblephilus Panzer, 54
synagroides Turner, Cerceris, 588
Synneurus Gerstaecker, 51
Synnevrus A. Costa 51, **470**
synoecoides "Perez", Campos, Sphex, 117
Synorhopalum Ashmead, 47
Synothyreopus Ashmead, 48
syriaca Beaumont, Solierella, 314
syriaca Mocsáry, Ammophila, 153
syriacum Mercet, Trypoxylon, 348
syriacus Andrade, Miscophus, 319
syriacus (Klug), Sphecius, 511
syriacus (Kohl), Lindenius, 384
syriacus Kohl, Tachysphex, 272
syriacus (Mocsáry), Palmodes, 127
syrittus Leclercq, Entomognathus, 382
syrkuti Dahlbom, Cerceris, 582
szechuana Tsuneki, Cerceris, 588

taantes Gribodo, Philanthus, 566
tabanicida (Fisher), Ectemnius, 428
tabascense Strand, Trypoxylon, 345
tabrobanae Cameron, Tachytes, 267
Tachybulus Latreille, 45
Tachynana Banks, 44
Tachyoides Banks, 44
Tachyplena Banks, 44
Tachyptera Dahlbom, 43
Tachyrrhostus Saussure, 44
tachyrrostus Saussure, Tachytes, 267
tachysphecoides (Viereck), Larropsis, 258
Tachysphex Kohl, 44, **267**
Tachystizus Pate, 53, **528**
Tachytella Brauns, 43, **256**
Tachytes Panzer, 43, **260**
tachytoides Tsuneki, Liris, 248
tadzhicus Gussakovskij, Tachysphex, 273
tadzhika Gussakovskij, Cerceris, 588
tadzhikus (Gussakovskij), Plenoculus, 311
taeniatum Leclercq, Rhopalum, 390
taenigaster (Viereck), Oxybelus, 369
tages W. F. Kirby, Stizus, 527
tagiurae Andrade, Miscophus, 319
tahitense Saussure, Pison, 336
tahitense (Saussure), Sceliphron, 105
Taialia Tsuneki, 41, 193
taianus Tsuneki, Tachytes, 630
taihorinus Strand, Psenulus, 174
Taimyrisphex Evans, 31
tainanense Strand, Trypoxylon, 348
taino Pate, Ectemnius, 428
taipingshanum Tsuneki, Rhopalum, 390
taiwana Bischoff, Bembix, 548
taiwana Tsuneki, Ammophila, 151
taiwana Tsuneki, Lyroda, 299
taiwanense of Tsuneki, Trypoxylon, 348
taiwanum Tsuneki, Rhopalum, 390
taiwanum Tsuneki, Sceliphron, 106

taiwanum Tsuneki, Trypoxylon, 347
taiwanus Sonan, Alysson, 458
taiwanus (Tsuneki), Carinostigmus, 191
taiwanus Tsuneki, Crossocerus, 403
taiwanus Tsuneki, Ectemnius, 427
taiwanus (Tsuneki), Eogorytes, 505
taiwanus (Tsuneki), Liris, 246
taiwanus Tsuneki, Nysson, 469
taiwanus Tsuneki, Passaloecus, 184
taiwanus Tsuneki, Psen, 166
takasago Tsuneki, Crossocerus, 403
takasago Tsuneki, Lyroda, 299
takasago Tsuneki, Psen, 166
takasago Tsuneki, Trypoxylon, 348
takeuchii Tsuneki, Gorytes, 502
takeuchii Yasumatsu, Ampulex, 78
tallongensis Rayment, Sericophorus, 302
Talthybius Rafinesque-Schmaltz, 38
talynensis Tsuneki, Cerceris, 587
tanakai Tsuneki, Crossocerus, 403
tanakai Tsuneki, Psenulus, 174
tanganyikae Arnold, Liris, 245
tango Shestakov, Cerceris, 588
tanoi Tsuneki, Ammatomus, 629
tanoi Tsuneki, Bembecinus, 532
tanoi Tsuneki, Crossocerus, 403
tanoi (Tsuneki), Crossocerus, 629
tanoi Tsuneki, Liris, 629
tanoi Tsuneki, Psen, 167
tanoi Tsuneki, Sphex, 630
tanoi Tsuneki, Trypoxylon, 348
Tanyoprymnus Cameron, 53, **511**
tanythrix R. Bohart, Pseudoplisus, 503
tanzana Empey, Cerceris, 629
taprobanae F. Smith, Dolichurus, 69
taprobane (Cameron), Dasyproctus, 420
taprobanense (Strand), Chalybion, 103
taprobanensis Pate, Oxybelus, 369
tara Beaumont, Crossocerus, 403
Taranga W. F. Kirby, 45
tarda Cameron, Astata, 213
targionii (Caruccio), Chalybion, 103
tarijensis Brèthes, Oxybelus, 370
tarsalis (Dalla Torre), Tachytes, 267
tarsalis Krombein, Stigmus, 189
tarsalis Rohwer, Microbembex, 539
tarsalis (Spinola), Tachytes, 267
tarsalis (W. Fox), Crossocerus, 403
tarsata F. Smith, Ammophila, 150
tarsata (F. Smith), Larra, 238
tarsata Latreille, Bembix, 548
tarsatus (F. Smith), Austrogorytes, 498
tarsatus F. Smith, Tachytes, 267
tarsatus (F. Smith), Tachytes, 267
tarsatus H. Smith, Philanthus, 566
tarsatus (Say), Tachysphex, 277
tarsatus (Shuckard), Crossocerus, 403
tarsinus (Lepeletier), Tachysphex, 277
Taruma Pate, 49, **416**, 627
tarumoides Leclercq, Foxita, 416, 627
taschenbergi Cameron, Ammophila, 151
taschenbergi Kohl, Oxybelus, 370
taschenbergi (Kohl), Penepodium, 92
taschenbergi Magretti, Sphex, 117
taschenbergii Handlirsch, Bembix, 548
tasmanica (F. Smith), Williamsita, 422
tasmanica Turner, Lyroda, 299
tasmanica Westwood, Aphelotoma, 38, **70**
tasmanicum F. Smith, Pison, 336
tassilicus Pulawski, Tachytes, 267
tasmanicus (Turner), Acanthostethus, 474
tau (Palisot de Beauvois), Podium, 96

tauricus (Radoszkowski), Dienoplus, 496
taurulus (Cockerell), Clypeadon, 571
taxus Leclercq, Crossocerus, 403
tayalum Tsuncki, Rhopalum, 390
taygete Arnold, Cerceris, 588
tecuya Pate, Lindenius, 385
tegularis (Cameron), Liris, 248
tegularis E. Saunders, Oxybelus, 367
tegularis (F. Morawitz), Dryudella, 215
tekkensis Gussakovskij, Ammophila, 153
teleges Pate, Ectemnius, 428
Telexysma Leclercq, 41
teliferopodus Rayment, Sericophorus, 302
temerarius Kohl, Philanthus, 565
temperatus Balthasar, Miscophus, 319
temporalis Beaumont, Gastrosericus, 256
temporalis (Gussakovskij), Crossocerus, 403
temporalis Kohl, Diodontus, 179
temporalis Kohl, Stigmus, 189
temporalis Leclercq, Dasyproctus, 421
tenax Pulawski, Tachysphex, 274
tenax W. Fox, Pemphredon, 181
tenebricosa Giner Marí, Cerceris, 583
tenebrosa (F. Smith), Liris, 248
tenebrosa J. Parker, Bembix, 546
tenebrosum Turner, Pison, 336
tenebrosus (Kohl), Lindenius, 384
tenellus (Arnold), Carinostigmus, 191
tenellus Arnold, Philanthus, 566
tenellus (Klug), Bembecinus, 532
tener Valkeila, Pemphredon, 182
tenera (Handlirsch), Ochleroptera, 490
tenggarae van der Vecht, Sceliphron, 106
tengu Tsuneki, Oxybelus, 370
Tenila Brèthes, 45, 325
tenoctitlan Richards, Trypoxylon, 349
tenthredinoides Scopoli, Sphex, 117
tenuicornis (F. Morawitz), Prionyx, 134
tenuicornis (F. Smith), Larropsis, 259
tenuicornis (F. Smith), Stizus, 527
tenuicornis (W. Fox), Glenostictia, 552
tenuifasciata J. Parker, Bembix, 548
tenuiglossa Packard, Crabro, 408
tenuipunctus W. Fox, Tachysphex, 277
tenuis A. Morawitz, Passaloecus, 184
tenuis Nurse, Diodontus, 179
tenuis (Oehlke), Mimesa, 162
tenuis (Palisot de Beauvois), Ammophila, 151
tenuis Turner, Tachysphex, 277
tenuis W. Fox, Crabro, 408
tenuiventre (Turner), Rhopalum, 390
tenuiventris Arnold, Cerceris, 588
tenuivittata Dufour, Cerceris, 588
tepaneca Saussure, Cerceris, 579
tepanecus Saussure, Sphex, 117
tepicum Leclercq, Rhopalum, 390
tequesta Pate, Ectemnius, 425
teranishii Sato, Cerceris, 588
tercera Fritz, Foxia, 479
teren Pate, Hapalomellinus, 497
terlinguae (C. Fox), Glenostictia, 552
terminalis Dahlbom, Stizus, 526
terminalis (Taschenberg), Trachypus, 568
terminata Cresson, Astata, 212
terminata F. Smith, Ammophila, 153
terminatus (F. Smith), Tachysphex, 279
terminus (W. F. Kirby), Sphecius, 511
terpenus Leclercq, Podagritus, 393
terricola Leclercq, Entomocrabro, 377

terrificus Arnold, Tachysphex, 274
terrigenus van Lith, Psen, 167
territorialis Leclercq, Pseudoturneria, 629
tersus Kazenas, Crossocerus, 403
tertiarium Meunier, Sceliphron, 31, 206
tertius Saussure, Dolichurus, 69
tervureni Leclercq, Dasyproctus, 420
tessellatus Dahlbom, Tachysphex, 277
testacea G. and R. Bohart, Larropsis, 259
testaceicauda Handlirsch, Bembix, 547
testaceicorne Cameron, Trypoxylon, 348
testaceicornis (Cameron), Liris, 248
testaceinerva Cameron, Tachytes, 267
testaceipalpis Cameron, Dasyproctus, 420
testaceipes Arnold, Trypoxylon, 348
testaceipes Bingham, Tachysphex, 277
testaceipes (R. Turner), Parapsammophila, 140
testaceipes Turner, Pison, 336
testaceitarsis Cameron, Alysson, 458
testaceum Turner, Rhopalum, 390
testaceus Gmelin, Sphex, 117
teterrima Gribodo, Cerceris, 587
tetli Pate, Ammoplanus, 198
tetracanthus (Pérez), Ectemnius, 428
tetradonta Cameron, Cerceris, 588
tetraedus (Blanchard), Ectemnius, 428
tetraedrus (Dahlbom), Ectemnius, 426
teutona (Fabricius), Larra, 237
texana Cresson, Bembix, 548
texana (Cresson), Didineis, 459
texana Scullen, Cerceris, 588
texanum (Cresson), Chalybion, 103
texanus (Ashmead), Miscophus, 319
texanus Banks, Philanthus, 566
texanus (Cresson), Ectemnius, 428, 627
texanus (Cresson), Entomognathus, 382
texanus Cresson, Sphex, 117
texanus Cresson, Stizus, 527
texanus (Cresson), Tachysphex, 277
texanus Cresson, Tachytes, 264
texanus (Cresson), Zanysson, 475
texanus (Malloch), Pluto, 171
texanus Robertson, Oxybelus, 369
texense Saussure, Trypoxylon, 349
texensis G. and R. Bohart, Larropsis, 259
texensis Saussure, Cerceris, 581
thaiana Tsuneki, Bembix, 545
thaiana Tsuneki, Cerceris, 585
thaiana Tsuneki, Liris, 248
thaianum Tsuneki, Trypoxylon, 348
thaianus Tsuneki, Ammatomus, 513
thaianus Tsuneki, Encopognathus, 380
thaianus Tsuneki, Oxybelus, 370
thaianus Tsuneki, Psenulus, 629
thailandinus (Tsuneki), Carinostigmus, 191
thailanditus Tsuneki, Gastrosericus, 629
thailanditus Tsuneki, Oxybelus, 629
thalassina Gussakovskij, Prosopigastra, 286
thalassinus Pulawski, Tachytes, 263
thalia (Handlirsch), Hoplisoides, 521
thauma (Pate), Holcorhopalum, 385
theodori (Bytinski-Salz), Philanthinus, 570
Therapon J. Parker, 54
theridii Ducke, Microstigmus, 41, **192**
thermophila Schletterer, Cerceris, 579, 588

thermophilus Beaumont, Oxybelus, 367
theryi Arnold, Cerceris, 584
theryi (Gribodo), Ammophila, 154
theryi (Vachal), Pseudoscolia, 574
theseus Arnold, Tachysphex, 277
thione Banks, Cerceris, 580
thomae (Fabricius), Prionyx, 134
thooma Evans and Matthews, Bembix, 548
thoracica F. Smith, Ampulex, 78
thoracica (F. Smith), Clitemnestra, 489
thoracicus Ashmead, Stigmus, 189
thracius Pulawski, Miscophus, 318
thripoctenus Richards, Microstigmus, 192
thunbergi Lepeletier, Sphex, 115
Thyreocerus A. Costa, 49
Thyreocnemus A. Costa, 48
thyreophorus Kohl, Crabro, 409
Thyreopus Lepeletier and Brullé, 48
Thyreosphex Ashmead, 38
Thyreus Lepeletier and Brullé, 50
thysanomera (Kohl), Liris, 248
tibesti Beaumont, Bembix, 546
tibeticus Leclercq, Ectemnius, 427
tibiale Cameron, Sceliphron, 106
tibiale F. Smith, Pison, 336
tibiale (Fabricius), Chalybion, 103
tibiale (Strand), Chlorion, 90
tibiale Zetterstedt, Trypoxylon, 346
tibialis (Arnold), Liris, 246
tibialis Beaumont, Liris, 247
tibialis Brèthes, Cerceris, 588
tibialis (Cresson), Pluto, 40, **171**
tibialis (Fabricius), Rhopalum, 389
tibialis Gmelin, Vespa, 627
tibialis Gussakovskij, Astata, 213
tibialis Handlirsch, Steniolia, 553
tibialis (Lepeletier), Isodontia, 123
tibialis (Olivier), Ectemnius, 424, 428
tibialis (Say), Crossocerus, 401
tibooburra Evans and Matthews, Bembix, 548
tiemudzin Tsuneki, Diodontus, 179
tienchiao Tsuneki, Cerceris, 588
tiendang Tsuneki, Cerceris, 588
timberlakei (Bridwell), Miscophus, 319
timberlakei Pate, Ammoplanops, 197
timberlakei Pate, Belomicrus, 364
timberlakei R. Bohart and Schlinger, Oxybelus, 370
timberlakei R. Bohart, Nysson, 470
timberlakei Sandhouse, Trypoxylon, 348
timberlakei Williams, Plenoculus, 311
timberlakei Williams, Solierella, 314
timberlakei Williams, Xenosphex, 439
Timberlakena Pate, 42, **200**
timidus Chevrier, Oxybelus, 370
timorense van der Vecht, Sceliphron, 106
tinctipenne Cameron, Trypoxylon, 348
tinctipennis Cameron, Astata, 213
tinctipennis (Cameron), Liris, 248
tinctipennis Cameron, Pemphredon, 182
tinctipennis Cameron, Sphex, 117
tinctipennis Cameron, Tachysphex, 277
tinctipennis Cameron, Tachytes, 266
tinctus Andrade, Miscophus, 319
tingitana Beaumont, Cerceris, 578
tingitanus Beaumont, Oxybelus, 367
tingitanus Pate, Tachytes, 266
tinnula Gussakovskij, Cerceris, 588
tiphia Gistel, Sphex, 117
tirolensis Kohl, Crossocerus, 400
tirolensis (Kohl), Gorytes, 501
tischbeinii (Dahlbom), Crossocerus, 402

tisiphone (F. Smith), Liris, 245, 248
tisiphonoides (Dalla Torre), Liris, 245
titania Arnold, Tachysphex, 277
toba Fritz, Metanysson, 629
tokunosimana Tsuneki, Cerceris, 588
toledensis Leclercq, Crossocerus, 403
tolteca Saussure, Cerceris, 588
toltecum Saussure, Trypoxylon, 348
tomentosa (Arnold), Ammophila, 151
tomentosa Tsuneki, Ammophila, 154
tomentosa Fabricius, Sphex, 117
tomentosus Gmelin, Sphex, 117
tomentosus (Handlirsch), Epinysson, 472
tomentosus Kohl, Tachytes, 267
Toncahua Pate, 46, **381**
tongyaii Tsuneki, Rhopalum, 390
tonkinensis Turner, Cerceris, 588
tonkinensis (Yasumatsu), Argogorytes, 492
tonto Pate, Zanysson, 475
tonus (Bondriot), Argogorytes, 492
tornquisti (Cockerell), Tracheliodes, 31, 404
toroensis Turner, Ampulex, 78
torosa J. Parker, Bembix, 548
torridus F. Smith, Sphex, 117
torridus R. Bohart, Epinysson, 472
tosawai Yasumatsu, Pison, 336
touareg Beaumont, Bembecinus, 532
touareg Ed. André, Ammophila, 153
Towada Tsuneki, 49, **411**
townesi (Krombein), Liris, 248
townesi Leclercq, Dasyproctus, 421
townesi (van Lith), Pseneo, 165
townesorum Krombein, Nitela, 325
townsendi (Cockerell), Pluto, 171
townsendi (Rohwer), Lestica, 431
townsendi Rohwer and Cockerell, Oxybelus, 367
townsendi Viereck and Cockerell, Cerceris, 588
toxopeusi Krombein, Cerceris, 588
toxopeusi van Lith, Psen, 167
toxopterus Leclercq, Dasyproctus, 421
toyensis Tsuneki, Tachytes, 267
Tracheliodes A. Morawitz, 48, **404**
Trachelosimus A. Morawitz, 47
Trachypus Klug, 54, **568**
traductor (Nurse), Crossocerus, 403
tramosericus (Viereck), Epinysson, 472
tranquebarica (Gmelin), Bembix, 548
transandina Williams, Larra, 238
transcaspica F. Morawitz, Larra, 238
transcaspica Radoszkowski, Bembix, 549
transcaspicum (Radoszkowski), Sceliphron, 106
transcaspicus Andrade, Miscophus, 319
transcaspicus Gussakovskij, Ammoplanus, 198
transcaspicus (Kokujev), Dienoplus, 496
transcaspicus (Radoszkowski), Brachystegus, 473
transcaspicus Radoszkowski, Oxybelus, 370
transcaspicus Radoszkowski, Stizus, 527
transiens (A. Costa), Dienoplus, 496
transiens (Kohl), Ectemnius, 427
transiens Turner, Oxybelus, 370
transiens (Turner), Rhopalum, 390
transkeica Empey, Cerceris, 580
transvaalense Cameron, Pison, 336
transvaalensis (Arnold), Holotachysphex, 282

transvaalensis Brauns, Nitela, 325
trasvaalensis Cameron, Ammophila, 152
transvaalensis (Cameron), Hoplisoides, 521
transvaalensis (Cameron), Liris, 244
transvaalensis Cameron, Sphex, 114
transvaalensis Cameron, Tachytes, 267
transvaalicola Brauns, Cerceris, 588
transversa Cheesman, Liris, 248
transversa Schletterer, Cerceris, 588
transversalis Brèthes, Cerceris, 588
transversalis (Shuckard), Crossocerus, 401
transversistriatum Strand, Trypoxylon, 349
transversus Cameron, Philanthus, 566
transversus (Fernald), Ammophila, 150
transversus Williams, Plenoculus, 311
trapezoideus (Packard), Ectemnius, 427
treforti Sajo, Oxybelus, 367
tremulum Arnold, Trypoxylon, 348
trepanda Dahlbom, Bembix, 549
trepida Handlirsch, Bembix, 549
trevirus Leclercq, Psenulus, 174
triangularis Ashmead, Pisonopsis, 332
triangularis (F. Smith), Pterygorytes, 515
triangularis (Turner), Acanthostethus, 474
triangulata Cresson, Cerceris, 588
triangulatus van Lith, Psen, 167
triangulifer Provancher, Alysson, 458
triangulifera Arnold, Bembix, 549
triangulum (Fabricius), Philanthus, 566
triangulus Brullé, Sphex, 116
triangulus Villers, Sphex, 117
tricarinata Cameron, Ampulex, 78
trichargyrus (Spinola), Prionyx, 134
trichinotus (Cameron), Bembecinus, 532
trichiocephalus Cameron, Philanthus, 565
trichionota Cameron, Cerceris, 588
trichionota (Cameron), Isodontia, 124
trichiosoma Cameron, Ammophila, 154
trichiosoma Cameron, Ampulex, 78
trichiosoma (Cameron), Argogorytes, 492
trichiosoma Cameron, Cerceris, 588
trichiosomus (Cameron), Ectemnius, 428
trichobunda Strand, Cerceris, 588
trichogastor Valkeila, Pemphredon, 182
Trichogorytes Rohwer, 52, **497**
trichopus Pulawski, Tachytes, 264
trichopygus (Beaumont), Synnevrus, 470
trichopygus Pulawski, Tachytes, 267
Trichostictia J. Parker, 54, **549**
Trichothyreopus Noskiewicz and Chudoba, 37
trichrus (Mickel), Nysson, 470
triciliata Scullen, Eucerceris, 591
tricincta (Spinola), Cerceris, 588
tricincta (Thunberg), Cerceris, 585
tricinctus Dahlbom, Tachytes, 267
tricinctus (Fabricius), Hoplisoides, 520
tricinctus (Fabricius), Nysson, 470
tricinctus (Fabricius), Sphecius, 511
tricinctus (Fabricius), Tachytes, 267
tricinctus Gimmerthal, Philanthus, 566
tricinctus (Mickel), Psammaletes, 508
tricinctus (Pérez), Gorytes, 502
tricinctus Pérez, Tachytes, 264
tricinctus (Schrank), Mellinus, 449
tricolor Cockerell, Eucerceris, 592
tricolor (Cresson), Hoplisoides, 521
tricolor Dahlbom, Bembix, 549
tricolor F. Smith, Cerceris, 588
tricolor (F. Smith), Hoplisoides, 521
tricolor (F. Smith), Podagritus, 393

tricolor (Fabricius), Tachytes, 266
tricolor Fairmaire, Philanthus, 566
tricolor Gess, Handlirschia, 509
tricolor (Giner Marí), Pseudoscolia, 574
tricolor Gussakovskij, Crabro, 408
tricolor Gussakovskij, Oxybelus, 370
tricolor Handlirsch, Stizus, 527
tricolor Lepeletier and Serville, Alysson, 458
tricolor of Panzer, Tachytes, 265
tricolor Reich, Sphex, 117
tricolor Sickmann, Trypoxylon, 348
tricolor Turner, Arpactophilus, 186
tricolor (Vander Linden), Dryudella, 42, **215**
tricolorata J. Parker, Bicyrtes, 538
tricolorata Spinola, Cerceris, 588
tricoloratus (Turton), Tachytes, 266
tricoloripes (Arnold), Entomognathus, 382
tricosa J. Parker, Microbembex, 539
tricuspis Schrank, Crabro, 409
tridens Arnold, Entomognathus, 382
tridens Brèthes, Heliocausus, 453
tridens (Fabricius), Bembecinus, 532
tridens (Fabricius), Oxybelus, 370
tridens Gerstaecker, Nysson, 470
tridens (Taschenberg), Lyroda, 299
tridentata (F. Smith), Williamsita, 422
tridentata Maidl, Cerceris, 588
tridentata (Strand), Bembecinus, 537
tridentata Tsuneki, Parapiagetia, 281
tridentatum Packard, Trypoxylon, 349
tridentatus Arnold, Tachysphex, 277
tridentatus F. Smith, Oxybelus, 370
tridentatus (Fabricius), Stizoides, 529
tridentatus Gussakovskij, Pemphredon, 182
tridentatus (Rohwer), Crossocerus, 401
tridentatus (van Lith), Pseneo, 165
tridentifera (Brèthes), Cerceris, 585
tridentifera F. Smith, Bembix, 548
trifasciata F. Smith, Cerceris, 588
trifasciata (F. Smith), Liris, 248
trifasciata J. Parker, Stictia, 542
trifasciata Tsuneki, Liris, 629
trifasciatus Arnold, Tachysphex, 273
trifasciatus O. Müller, Sphex, 117
trifasciatus (Say), Ectemnius, 428
trifida (Fabricius), Cerceris, 586
trifidis Cockerell and Baker, Oxybelus, 367
trigeminum Richards, Trypoxylon, 348
trigona Cameron, Ampulex, 77
trigonalis Saussure, Tachytes, 267
trigonopsis F. Smith, Ampulex, 78
Trigonopsis Perty, 38, **96**
trigonopsoides Menke, Podium, 96
trilineata Bischoff, Cerceris, 589
trilineatus (Turton), Nysson, 470
triloba (Say), Lyroda, 299
trimaculata Maidl, Cerceris, 588
trimaculatus (Rossi), Nysson, 470
trimaculatus van Lith, Psenulus, 174
trimaculigera (Strand), Eremnophila, 147
trimarginatus O. Müller, Sphex, 117
trinacriense (De Stefani), Sceliphron, 105
trinacrius Beaumont, Crossocerus, 401
trinidadense Richards, Trypoxylon, 348
trinidadianum Richards, Trypoxylon, 347
trinitaria Alayo, Cerceris, 588

trinotatus (A. Costa), Ectemnius, 427
triodon (Kohl), Isodontia, 123
triodon Richards, Trypoxylon, 349
Tripoxilon Spinola, 45
triquetrus W. Fox, Tachysphex, 277
Trirhogma Agassiz, 38
Trirogma Westwood, 38, **73**
trispinosus (Fabricius), Oxybelus, 370
tristani (Pate), Rhopalum, 390
triste Arnold, Trypoxylon, 348
triste (Kohl), Penepodium, 92
tristior Morice, Cerceris, 588
tristis Cameron, Cerceris, 588
tristis C. Fox, Bicyrtes, 537
tristis Cresson, Nysson, 470
tristis Dahlbom, Diodontus, 179
tristis (F. Smith), Liris, 248
tristis Kohl, Sphex, 114
tristis Pérez, Mellinus, 449
tristis (Vander Linden), Diodontus, 40, **179**
tristrigatus (Fabricius), Hoplisoides, 520
trisulcus (W. Fox), Psenulus, 174
tritis van Lith, Psenulus, 173
tritospilus R. Bohart, Pseudoplisus, 503
trivialis Gerstaecker, Cerceris, 584
trivittata (W. F. Kirby), Liris, 248
trochanteratum Cameron, Trypoxylon, 348
trochanterica Beaumont, Astata, 213
trochantericus (Herrich-Schaeffer), Tracheliodes, 405
troglodytes Handlirsch, Bembix, 549
troglodytes (Vander Linden), Spilomena, 41, **193**
tropicale Tsuneki, Trypoxylon, 347
tropicalis Arnold, Astata, 213
tropicalis (Arnold), Crossocerus, 403
tropicalis Arnold, Liris, 248
tropicalis Fritz, Metanysson, 481
trukense Yasumatsu, Pison, 336
trumani Dunning, Philanthus, 566
truncata Cameron, Cerceris, 588
truncata (Cameron), Liris, 248
truncata (F. Smith), Liris, 248
truncata Handlirsch, Bembix, 549
truncatifrons Turner, Tachysphex, 277
truncatula Dahlbom, Cerceris, 585
truncatula (Dalla Torre), Liris, 248
Trypargilum Richards, 46, 343
Trypoxilon Jurine, 46
Trypoxylon Latreille, 45, **339**
TRYPOXYLONINI, 59, **327**, 328
Trypoxylum Agassiz, 46
Trypoxylum Schulz, 46
Tsaisuma Pate, 46
tshontandae Arnold, Cerceris, 588
tsingtauensis Strand, Sphex, 117
tsuifengensis Tsuneki, Crossocerus, 403
tsuifenicum Tsuneki, Rhopalum, 390
tsuifenicus Tsuneki, Ectemnius, 428
tsunekiense Leclercq, Rhopalum, 390
tsunekii Andrade, Miscophus, 319
tsunekii Menke, Ammophila, 154
tsunekii van Lith, Psen, 167
tubarum Leclercq, Rhopalum, 390
tuberculata (Villers), Cerceris, 588
tuberculata (W. Fox), Stictiella, 551
tuberculatum F. Smith, Pison, 336
tuberculatus F. Smith, Sphex, 117
tuberculatus Giner Marí, Tachytes, 267
tuberculatus (Handlirsch), Epinysson, 472

tuberculatus (Turner), Sericophorus, 303
tuberculicorne Turner, Rhopalum, 390
tuberculifer Gussakovskij, Ammoplanops, 197
tuberculifrons Arnold, Trypoxylon, 348
tuberculifrons (Rohwer), Psenulus, 174
tuberculiger Kohl, Crabro, 409
tuberculiscutis (Turner), Ammophila, 154
tuberculiventris Turner, Bembix, 549
tuberculiventris (Turner), Stizoides, 529
tubifex (Latreille), Sceliphron, 106
tubulentum Arnold, Trypoxylon, 348
tuckeri Arnold, Tachysphex, 277
tucumanensis (Strand), Prionyx, 134
tucuman Fritz, and Toro, Cerceris, 629
tucumanum Brèthes, Trypoxylon, 348
tuktum Pate, Belomicrus, 364
tulagiensis Krombein, Cerceris, 584
tulbaghensis Arnold, Stizus, 526
tumidoventris (Perkins), Ectemnius, 428
tumidus (Packard), Crabro, 409
tumidus (Panzer), Dienoplus, 496
tumulorum F. Smith, Cerceris, 588
tunetana Gussakovskij, Prospigastra, 285
tunetanus A. Costa, Stizus, 527
turanica F. Morawitz, Parapsammophila, 140
turanica Gussakovskij, Didineis, 459
turanicum (Gussakovskij), Chalybion, 103
turanicus Gussakovskij, Dolichurus, 69
turanicus Gussakovskij, Laphyragogus, 220
turanicus Roth, Sphecius, 511
turbata Shestakov, Cerceris, 588
turca Dahlbom, Bembix, 549
turcestanica (Dalla Torre), Podalonia, 145
turcestanica Kohl, Ammophila, 154
turcica Mocsáry, Ammophila, 152
turcmenicus (Radoszkowski), Dienoplus, 496
turcomanica Gussakovskij, Prosopigastra, 286
turcomanica Radoszkowski, Tachytes, 267
turcomanicus (Radoszkowski), Prionyx, 133
turionum Dahlbom, Passaloecus, 184
turkestana Kohl, Ammophila, 154
turkestanica Radoszkowski, Cerceris, 588
turkestanicum Gussakovskij, Trypoxylon, 348
turkestanicus Mocsáry, Oxybelus, 367
turneri Arnold, Belomicrus, 364
turneri Arnold, Cerceris, 589
turneri (Arnold), Crossocerus, 403
turneri Arnold, Gastrosericus, 256
turneri (Arnold), Holotachysphex, 282
turneri Arnold, Mesopalarus, 306
turneri Arnold, Philanthus, 566
turneri Arnold, Psenulus, 174
turneri Arnold, Spilomena, 193
turneri Arnold, Tachytes, 267
turneri Brauns, Palarus, 291
turneri Dutt, Solierella, 314
turneri (Froggatt), Bembecinus, 532
turneri Shestakov, Cerceris, 588
Turneriola Leclercq, 49
turrialba Scullen, Cerceris, 586
tydei (Le Guillou), Podalonia, 145
tylotis Court and R. Bohart, Lindenius, 385
typicus (Rohwer), Stigmus, 189
tyrannica Cameron, Ammophila, 154
tyrannica F. Smith, Cerceris, 588

tyrannicus F. Smith, Sphex, 116
tyrannus (F. Smith), Prionyx, 133
tyronus Leclercq, Dasyproctus, 421
tyunzendzianus Tsuneki, Crossocerus, 403

uchidai (Tsuneki), Crossocerus, 403
uelensis Benoit, Liris, 248
uelleburgi van Lith, Psenulus, 629
u-flavum (Panzer), Mellinus, 449
ugandense Arnold, Pison, 336
ugandensis Arnold, Cerceris, 588
ugandensis Turner, Bembix, 549
ugandensis Turner, Tachytes, 267
ugandicus (Leclercq), Carinostigmus,, 191
ugandicus Magretti, Philanthus, 566
ulanbaatorensis (Tsuneki), Podalonia, 143
ulcerosa Arnold, Cerceris, 588
uljanini Radoszkowski, Crabro, 409
uljanini (Radoszkowski), Sphecius, 511
uloola Evans and Matthews, Bembix, 549
ulula Arnold, Bembix, 549
umatilla (Pate), Ammoplanellus, 200
umatillae R. Bohart, Gorytes, 502
umbelliferarum Schrottky, Cerceris, 588
umbinifera Maidl, Cerceris, 588
umbonatus (Baker), Oryttus, 508
umbonicida Pate, Hoplisoides, 521
umbripennis (Cameron), Liris, 248
umbrosus Christ, Sphex, 114
umbrosus Mickel, Oxybelus, 368
umbrosus of authors, Sphex, 115
umbrosus (Schrottky), Ectemnius, 424
umhlangae Arnold, Cerceris, 588
umtalicus (Arnold), Hoplisoides, 521
umtalicus Strand, Sphex, 115
umtaliensis Empey, Cerceris, 629
unami Pate, Ammoplanus, 198
uncifera Arnold, Cerceris, 588
uncta Arnold, Cerceris, 588
undata Dahlbom, Bembix, 547
undatus (F. Smith), Tachysphex, 277
undeneya Evans and Matthews, Bembix, 549
undulata Dahlbom, Bembix, 545
undulata Spinola, Bembix, 549
undussumae Stadelmann, Philanthus, 565
unguicorne Richards, Trypoxylon, 348
unguicularis (Kohl), Parapsammophila, 140
unguiculatus Arnold, Tachysphex, 277
unicincta Cresson, Mimesa, 162
unicincta Taschenberg, Cerceris, 581
unicinctus (Brèthes), Liogorytes, 516
unicinctus (Say), Stizoides, 529
unicolor Arnold, Gastrosericus, 256
unicolor (F. Smith), Ectemnius, 428
unicolor Fabricius, Sphex, 114
unicolor Lepeltier, Astata, 213
unicolor of (Panzer), Psen, 166
unicolor (Panzer), Pemphredon, 182
unicolor (Panzer), Tachysphex, 277
unicolor Saussure, Chlorion, 90
unicolor Say, Astata, 213
unicolor Schummel, Miscophus, 319
unicolor (Vander Linden), Mimumesa, 164
unicornis Patton, Eucerceris, 592
unicus Mickel, Oxybelus, 369
unicus (Patton), Crossocerus, 402
unidentata F. Morawitz, Cerceris, 588
unifasciata F. Smith, Cerceris, 588
unifasciata Gmelin, Vespa, 627

unifasciatum (F. Smith), Sceliphron, 106
unifasciatus (Radoszkowski), Stizoides, 529
unifasciculatus Malloch, Psen, 167
uniformis G. and R. Bohart, Larropsis, 259
uniformis (Pérez), Synnevrus, 470
unigena Balthasar, Miscophus, 319
uniglumis (Linnaeus), Oxybelus, 370
uniglummis (Christ), Oxybelus, 370
uniguttatus (Arnold), Dasyproctus, 421
uniocellatus (Dufour), Liris, 246
Uniplectron Parker, 42, **208**
unispinosa Turner, Cerceris, 588
unita Menke, Ammophila, 154
universitatis (Rohwer), Crossocerus, 403
universitatis Rohwer, Stigmus, 189
urichi Richards, Trypoxylon, 348
urnaria Dahlbom, Ammophila, 154
urnaria Lepeletier, Ammophila, 150
urophori Radoszkowski, Crabro, 409
ursidus Leclercq, Crossocerus, 403
uruensis Leclercq, Dasyproctus, 421
uruguayensis (Holmberg), Microbembex, 539
usambaraensis (Cameron), Liris, 248
usambarensis Stadelmann, Philanthus, 565
u-scripta W. Fox, Bembix, 546, 627
usheri Arnold, Bembix, 549
usitata (W. Fox), Glenostictia, 552
ussuriense Gussakovskij, Pison, 336
ussuriensis Gussakovskij, Crabro, 409
ussuriensis van Lith, Psen, 167
ustulata (Kohl), Isodontia, 124
usurpator Schulz, Gorytes, 501
utahensis (Baker), Clypeadon, 571
utahensis Banks, Tachytes, 267
utahensis (Rohwer), Pemphredon, 181
utensis (Mickel), Crossocerus, 400
utopica (Leclercq), Liris, 248
uturoae Cheesman, Oxybelus, 370
uxor Leclercq, Cerceris, 589

vacus (Rossi), Mellinus, 449
vadosus van Lith, Psen, 167
vafra Bingham, Cerceris, 588
vaga (Christ), Isodontia, 124
vagabunda F. Smith, Ammophila, 153
vagabundus (Panzer), Crossocerus, 403
vagans (Arnold), Liris, 248
vagans Blüthgen, Spilomena, 193
vagans (Fokker), Ectemnius, 425
vagans Radoszkowski, Cerceris, 588
vagatus (F. Smith), Ectemnius, 425
vagula Kohl, Cerceris, 588
vagulum Richards, Trypoxylon, 349
vagum F. Smith, Trypoxylon, 349
vagus (F. Smith), Argogorytes, 492
vagus (Linnaeus), Mellinus, 449
vagus of authors, Ectemnius, 425
vagus (Radoszkowski), Prionyx, 133
vagus Radoszkowski, Tachytes, 267
valdiviae Leclercq, Podagritus, 393
valenciai Fritz, Podagritus, 393
valens (W. Fox), Pterygorytes, 53, **515**
valida (Cresson), Podalonia, 145
validior Gussakovskij, Bembecinus, 532
validus Cresson, Tachytes, 267
validus (De Stefani), Ectemnius, 425
vallicola Tsuneki, Trypoxylon, 348
vallicollae Rohwer, Diodontus, 179

INDEX 693

vandeli Ribaut, Passaloecus, 184
vanderlindenii (Dahlbom), Crossocerus, 400
vanderlindenii Robert, Astata, 213
vanduzeei Banks, Cerceris, 588
vanduzeei Gillaspy, Steniolia, 553
vandykei Williams, Solierella, 314
vanlithi (Tsuneki), Mimumesa, 164
vanrhynsi Arnold, Tachysphex, 277
vansoni Arnold, Cerceris, 577
vanyume Pate, Belomicrus, 364
vanyumi Pate, Ammoplanus, 198
vaporus Leclercq, Dasyproctus, 421
vaquero F. Parker, Astata, 213
vaqueroi (Giner Marí), Prionyx, 134
varelai Mercet, Nysson, 470
varentzowi (F. Morawitz), Ectemnius, 428
varia Maidl, Cerceris, 588
variabilis Chevrier, Nysson, 470
variabilis F. Smith, Bembix, 549
variabilis (Schrank), Cerceris, 586
variaesimilis Maidl, Cerceris, 589
varians (Arnold), Bembecinus, 532
varians F. Morawitz, Oxybelus, 370
varians (Fabricius), Tachytes, 267
varians Mickel, Cerceris, 589
varicincta Cameron, Cerceris, 589
varicolor Turner, Ampulex, 78
varicoloratus Baker, Oxybelus, 369
varicornis (Cameron), Dicranorhina, 252
varicornis Fabricius, Crabro, 409
variegata (Olivier), Bicyrtes, 538
variegata Taschenberg, Cerceris, 587
variegatus (F. Morawitz), Pseudoscolia, 573
variegatus (Fabricius), Palarus, 291
variegatus Spinola, Philanthus, 566
variegatus Wesmael, Oxybelus, 370
varihirtus Cameron, Tachysphex, 276
variicornis (Reed), Pisonopsis, 332
variipes Saussure, Larra, 238
variitarse Turner, Rhopalum, 390
varilineata Cameron, Cerceris, 580
variolatus (A. Costa), Synnevrus, 470
variolosa A. Costa, Cerceris, 577
variolosa Giner Marí, Ammophila, 153
variolosus Arnold, Philanthus, 566
varipenne Reiche and Fairmaire, Chlorion, 90
varipes Cresson, Ammophila, 154
varipes F. Smith, Cerceris, 587, 589
varipes Lepeletier and Brullé, Crossocerus, 401
varipes Pérez, Trypoxylon, 348
varipes Tsuneki, Dienoplus, 496
varipilosa (Cameron), Liris, 248
varipilosellus (Cameron), Zanysson, 475
varipilosum Cameron, Trypoxylon, 348
varipilosus Cameron, Tachytes, 266
varipunctus H. Smith, Pseudoplisus, 503
varius Lepeletier, Crossocerus, 404
varius Lepeletier and Brullé, Crossocerus, 403, 628
varius Sickmann, Palarus, 291
varius (Taschenberg), Trachypus, 569
varius van Lith, Psenulus, 174
varus Lepeletier and Brullé, Crossocerus, 403, 628
varus (Panzer), Crossocerus, 404
vasta Lohrmann, Bembix, 549
vechti Krombein, Cerceris, 589
vechti van Lith, Psen, 167
Vechtia Pate, 49, **417**

vechtinus Leclercq, Dasyproctus, 421
vedetta Leclercq, Williamsita, 630
vegeta Arnold, Cerceris, 589
vegeta (W. Fox), Larropsis, 259
vegetoides (Viereck), Larropsis, 258
veitchi (Turner), Neodasyproctus, 419
veles Carter, Crabro, 409
vellensis Krombein, Cerceris, 589
velox F. Smith, Cerceris, 582
velox F. Smith, Tachytes, 267
velox Handlirsch, Bembix, 549
velutina (Schrottky), Eremnophila, 147
velutina Scullen, Eucerceris, 592
velutina Taschenberg, Cerceris, 589
velutinus (Spinola), Oryttus, 508
venans (Kohl), Dasyproctus, 421
venator Arnold, Dolichurus, 69
venator Arnold, Tachysphex, 277
venator Cresson, Cerceris, 577
venator F. Smith, Bembix, 545
venator (Rohwer), Crabro, 409
venatrix Schulz, Cerceris, 577
venetianus Gussakovskij, Miscophus, 319
venetus Pate, Psen, 167
veniperdus (Lohrmann), Bembecinus, 532
ventilabris Fabricius, Philanthus, 566
ventralis Dahlbom, Bembix, 545
"ventralis, Fabr.", Philanthus, 566
ventralis Fox, Oxybelus, 370
ventralis Lepeletier, Bembix, 545
ventralis (Mickel), Philanthus, 566
"ventralis" of Howard, Philanthus, 566
ventralis (Say), Bicyrtes, 53, **538**
ventralis (W. Fox), Crossocerus, 402
ventripilosa Empey, Cerceris, 589
venturii (Schrottky), Podagritus, 393
venusta Arnold, Bembix, 549
venusta Bingham, Lyroda, 299
venusta F. Smith, Cerceris, 589
venusta Stal, Ampulex, 78
venustiformis (Rohwer), Pseudoplisus, 503
venustum Tsuneki, Rhopalum, 390
venustus Beaumont, Dinetus, 216
venustus Beaumont, Miscophus, 319
venustus (Cresson), Pseudoplisus, 503
venustus Lepeletier and Brullé, Lindenius, 384
venustus (Mickel), Psammaletes, 508
venustus (Rossi), Philanthus, 567
venustus Sickmann, Oxybelus, 370
veracruz Scullen, Cerceris, 589
verecunda Arnold, Cerceris, 589
verecundus (Arnold), Entomognathus, 382
verecundus Arnold, Miscophus, 319
verhoeffi Andrade, Miscophus, 319
verhoeffi Beaumont, Oxybelus, 370
verhoeffi Beaumont, Solierella, 314
verhoeffi Bytinski-Salz, Stizoides, 529
verhoeffi Pulawski, Tachysphex, 277
verhoeffi Tsuneki, Alysson, 458
verhoeffi Tsuneki, Bembecinus, 531
verhoeffi Tsuneki, Cerceris, 589
verhoeffi Tsuneki, Crossocerus, 404
verhoeffi Tsuneki, Gorytes, 502
verhoeffi Tsuneki, Stigmus, 189
verhoeffii (Dalla Torre), Ectemnius, 426
vernalis Brauns, Parapiagetia, 281
vernalis (Packard), Crabro, 409
vernayi Arnold, Cerceris, 589
versatilis Turner, Tachytes, 267
versicolor Beaumont, Psammaecius, 516
versicolor (Handlirsch), Bembecinus, 532

versicolor Schrottky, Cerceris, 589
versuta Arnold, Bembix, 549
verticalis F. Smith, Cerceris, 589
verticalis (F. Smith), Quexua, 387
vertilabris Fabricius, Philanthus, 566
verutus Rayment, Dasyproctus, 421
vespiformis (De Geer), Stictia, 542
vespiformis F. Smith, Bembix, 549
vespiformis (Fabricius), Stizus, 527
vespiformis (Latreille), Sphecius, 511
vespiformis (Panzer), Ectemnius, 428
vespiformis Schrank, Sphex, 117
vespoides F. Smith, Aulacophilus, 45, **338**
vespoides (F. Smith), Hoplisoides, 521
vespoides (Rossi), Cerceris, 588
vespoides (Walker), Stizus, 527
vestitum F. Smith, Pison, 337
vestitus (F. Smith), Ectemnius, 427
vestitus F. Smith, Sphex, 117
vestitus Kohl, Tachysphex, 277
vestitus (F. Smith), Tachytes, 267
vestor (Ashmead), Ectemnius, 427
vexillata (Panzer), Lestica, 430
vianai Fritz and Toro, Cerceris, 589
viarius Arnold, Tachysphex, 277
viatica of authors, Podalonia, 144
viator Andrade, Miscophus, 318
vicaria Shestakov, Cerceris, 589
vicarius (Handlirsch), Dienoplus, 496
vicina Cresson, Cerceris, 589
vicina (Dalla Torre), Trigonopsis, 98
viciniformis Viereck, Crabro, 408
vicinoides Viereck and Cockerell, Cerceris, 576
vicinus Cameron, Tachytes, 267
vicinus Cresson, Crabro, 408
vicinus Dahlbom, Crossocerus, 402
vicinus Handlirsch, Gorytes, 502
vicinus (Lepeletier), Gorytes, 501
vicinus Lepeletier, Sphex, 116
victor Curtis, Astata, 212
victor Lepeletier, Oxybelus, 370
victoriensis Lohrmann, Bembix, 549
victoriensis Rayment, Sericophorus, 302
victrix Turner, Cerceris, 589
vidua (F. Smith), Isodontia, 124
vidua Klug, Cerceris, 588
vidua Lepeletier, Bembix, 548
viduata F. Smith, Cerceris, 578
viduata (Handlirsch), Bicyrtes, 538
viduatum (Kohl), Penepodium, 92
viduatus (Christ), Prionyx, 134
viduum Arnold, Trypoxylon, 348
viennensis Leclercq, Crossocerus, 404
viereki Banks, Cerceris, 580
viereki (H. Smith), Crossocerus, 400
viereki Pate, Ammoplanops, 197
viereki Pate, Belomicrus, 364
viereki Pate, Diploplectron, 211
viereki (Rohwer), Crabro, 408
viereki Rohwer, Didineis, 459
viereki (Rohwer), Solierella, 314
vigilans F. Smith, Cerceris, 589
vigilans (F. Smith), Liris, 248
vigilii Brèthes, Cerceris, 579
vilarrubiai (Giner Marí), Prionyx, 133
villersi Berland, Cerceris, 589
villosa (W. Fox), Glenostictia, 552
villosifrons (Packard), Ectemnius, 424
villosus A. Costa, Stizus, 526
villosus Arnold, Tachysphex, 277
villosus W. Fox, Crabro, 409
vindex (F. Smith), Liris, 246

694 SPHECID WASPS

vindex (Lepeletier), Sceliphron, 105
vindobonensis Maidl, Mimesa, 162
vinulenta (Cresson), Liris, 248
violacea F. Smith, Trigonopsis, 98
violacea (Handlirsch), Ochleroptera, 490
violaceipennis (Cameron), Brachystegus, 473
violaceipennis Cameron, Cerceris, 589
violaceipennis (Cameron), Ectemnius, 428
violaceipennis Cameron, Liris, 245, 248
violaceipennis (Cameron), Liris, 248
violaceipennis (Cameron), Parapsammophila, 139
violaceipennis (Lepeletier), Palmodes, 127
violaceipennis (Lepeletier), Podalonia, 39, **145**
violaceipennis Scullen, Eucerceris, 592
violaceum (Fabricius), Chalybion, 102
violaceus Rayment, Sericophorus, 302
violascens (Dalla Torre), Trigonopsis, 98
virchu Cameron, Tachytes, 267
virgatus W. Fox, Crabro, 409
virgina Shestakov, Cerceris, 589
virginianus (Rohwer), Diodontus, 179
virginianus Rohwer, Pemphredon, 182
virginiensis Rohwer, Nitela, 325
viridescens Arnold, Ampulex, 78
viridicoeruleum Lepeletier & Serville, Chlorion, 90
virdis (Barbut), Chlorion, 90
viridis (Saussure), Sericophorus, 302
virosum Turner, Pison, 337
vischnu Cameron, Cerceris, 589
vischnu Cameron, Tachytes, 267
vischnu Schulz, Tachytes, 267
vischu Cameron, Ammophila, 153
vischu Dalla Torre, Tachytes, 267
viscosa F. Smith, Cerceris, 582
vishnu Bingham, Tachytes, 267
visnagae Beaumont, Laphyragogus, 220
visseri Willink, Isodontia, 124
vitiensis Williams, Tachysphex, 277
vittata Lepeletier, Cerceris, 589
vittatifrons Cresson, Eucerceris, 592
vittatus (Kohl), Prionyx, 134
vitticollis A. Morawitz, Cerceris, 589
vivax (Cameron), Liris, 248
vivida (Handlirsch), Stictia, 542
vividus (Turner), Dienoplus, 496
voeltzkovii Kohl, Sceliphron, 105
voeltzkowii (Kohl), Liris, 248
voeltzkowii Kohl, Sceliphron, 105
voeltzkowii Kohl, Sphex, 115
volatilis (F. Smith), Isodontia, 123
volcanica Cameron, Ammophila, 153
volitans Arnold, Trypoxylon, 348
vollenhovenia (Schulz), Liris, 248
vollenhovia (Ritsema), Liris, 248
voltaica Empey, Cerceris, 629
volubilis Kohl, Sphex, 115
volubilis Turner, Tachytes, 267
volucris (Handlirsch), Stictia, 542
vortex Tsuneki, Liris, 245
vulcania du Buysson, Ammophila, 154
vulcanica van der Vecht, Cerceris, 589
vulgaris Cresson, Ammophila, 152
vulgaris W. Kirby, Ammophila, 153
vulneratus Turner, Tachysphex, 277
vulpecula Empey, Cerceris, 589
vulpina F. Smith, Cerceris, 582
vulpina (Handlirsch), Trichostictia, 54, **549**

vulpinides Strand, Cerceris, 589
vumbui Arnold, Cerceris, 589
vumbuiensis (Arnold), Dasyproctus, 420

Waagenia Kriechbaumer, 38
wadamiri Evans and Matthews, Bembix, 549
wagleri Gistel, Bembix, 549
wagneri Arnold, Pison, 337
wagneri (Berland), Prionyx, 133
wagneri du Buysson, Microstigmus, 192
wahisi Leclercq, Diodontus, 629
wahlbergi Dahlbom, Ammophila, 154
walegae Arnold, Cerceris, 584
walkeri (Dalla Torre), Podalonia, 144
walkeri (Handlirsch), Palarus, 291
walkeri (Shuckard), Crossocerus, 404
walkeri Turner, Tachysphex, 277
wallacei R. Turner, Sphex, 116
walteri (Handlirsch), Dienoplus, 496
walteri (Kohl), Chalybion, 103
walteri (Kohl), Ectemnius, 428
walteri Kohl, Philanthus, 567
waltlii Spinola, Cerceris, 588
waltlii Spinola, Gastrosericus, 256
waltoni Arnold, Cerceris, 589
waltoni Arnold, Tachysphex, 277
wangoola Evans and Matthews, Bembix, 549
wanna Evans and Matthews, Bembix, 549
warawara Evans and Matthews, Bembix, 549
washingtoni Rohwer, Tachysphex, 272
washoensis G. and R. Bohart, Larropsis, 259
watanabei Tsuneki, Rhopalum, 390
waterstonii Kohl, Belomicrus, 364
wattanapongsiri Tsuneki, Ectemnius, 426
weberi Handlirsch, Bembix, 549
weberi Williams, Solierella, 314
weberi Yoshimoto, Ectemnius, 428
weema Evans and Matthews, Bembix, 549
wegneri van der Vecht, Philanthus, 565
wegneri van der Vecht and Krombein, Sphex, 116
wenonah (Banks), Encopognathus, 380
werestchagini Gussakovskij, Crabro, 409
werneri Maidl, Philanthus, 567
werneri (Maidl), Prospigastra, 286
werneri R. Bohart, Pseudoplisus, 503
wesmaeli (A. Morawitz), Pemphredon, 182
wesmaeli Giraud, Ammoplanus, 198
wesmaeli Lepeletier, Nysson, 469
wesmaeli (Vander Linden), Crossocerus, 404
westermanni (Dahlbom), Dasyproctus, 421
westermanni Spinola, Bembix, 549
westoni Bingham, Bembix, 549
westwoodii Shuckard, Pison, 337
wheeleri Arnold, Liris, 248
wheeleri Arnold, Stizus, 527
wheeleri Krombein and Willink, Bembecinus, 532
wheeleri Richards, Trypoxylon, 348
wheeleri Rohwer, Tachysphex, 277
whiteana Cameron, Cerceris, 583
whitei (Cameron), Hoplisoides, 521
wickhami (Ashmead), Crossocerus, 404
wickhamii (Ashmead), Crossocerus, 404
wickwari Turner, Cerceris, 589
wickwari (Turner), Ectemnius, 428
wickwari Turner, Parapiagetia, 281

wilcannia Evans and Matthews, Bembix, 549
williamsi F. Parker, Astata, 213
williamsi Krombein, Liris, 246
williamsi Menke, Nitela, 325
williamsi (Mickel), Lestiphorus, 506
williamsi R. Bohart, Tachysphex, 277
williamsi (Rohwer), Liris, 245
williamsi Scullen, Cerceris, 589
Williamsita Pate, 49, **421**
willineri Fritz, Cerceris, 589
willinki Leclercq, Podagritus, 393
willinki (Menke), Eremnophila, 147
willinki Menke, Odontosphex, 573
Willinkiella Menke, 46, **352**
willistoni (Fernald), Ammophila, 153
willowmorensis (Arnold), Bembecinus, 530
willowmorensis Arnold, Oxybelus, 370
willowmorensis Arnold, Tachysphex, 277
willowmorensis Brauns, Nysson, 470
wiluna Evans and Matthews, Bembix, 549
windorum Tsuneki, Cerceris, 589
winthemi Richards, Trypoxylon, 348
witoto Pate, Quexua, 387
wittei Leclercq, Crossocerus, 400
wittei Leclercq, Spilomena, 193
witzenbergensis (Arnold), Bembecinus, 532
woerdeni (Ritsema), Dicranorhina, 252
wolcotti H. S. Smith, Mellinus, 449
wollastoni Turner, Dicranorhina, 252
wollastoni Turner, Pison, 337
wollmanni (Kohl), Lestica, 431
wollowra Evans and Matthews, Bembix, 549
wolpa Evans and Matthews, Bembix, 549
woosnami Arnold, Oxybelus, 370
wouroukatte Beaumont, Belomicrus, 364
wowine Evans and Matthews, Bembix, 549
woyowai Pate, Foxita, 416
wrightii (Cresson), Ammophila, 154
wroughtoni Cameron, Cerceris, 579
wroughtonii Cameron, Gastrosericus, 256
wuestneii (Faester), Mimumesa, 164
wuestneii Handlirsch, Didineis, 461
wusheense Tsuneki, Rhopalum, 390
wusheensis Tsuneki, Ammophila, 153
wyomingensis Scullen, Cerceris, 589

xambeui (Er. André), Solierella, 313
xanthandrum Richards, Trypoxylon, 349
xanthocephala (J. Forster), Cerceris, 577
xanthocera Gussakovskij, Solierella, 314
xanthocera (Pulawski), Dryudella, 215
xanthoceros (Illiger), Chlorion, 90
xanthochilos Pate, Crossocerus, 404
xanthochrous (Handlirsch), Bembecinus, 531
xanthogaster Arnold, Cerceris, 589
xanthogaster Cameron, Philanthus, 566
xanthogaster Pate, Oxybelus, 370
xanthognatha (Pérez), Isodontia, 123
xanthognathus (Rohwer), Crossocerus, 404
xanthognathus Rohwer, Psenulus, 174
xanthonotus van Lith, Psenulus, 174
xanthoptera Arnold, Liris, 248
xanthoptera Cameron, Ammophila, 154
xanthopterus Cameron, Sphex, 115
xanthopus (Brullé), Pison, 337
xanthopus Turner, Pison, 335

xanthosoma Menke, Pisoxylon, 45, **338**
xanthostigma Arnold, Cerceris, 589
xanthostigma Cameron, Philanthus, 565
xantianum Saussure, Trypoxylon, 349
Xenocrabro Perkins, 50
xenoferus Rohwer, Tachytes, 267
xenophon Richards, Trypoxylon, 348
XENOSPHECINAE, 57, **437**
XENOSPHECINI, 59, 437
Xenosphex Williams, 51, **439**
xenum Leclercq, Rhopalum, 390
Xerogorytes R. Bohart, 53, **517**
xerophilus R. Bohart and Schlinger, Oxybelus, 370
xerophilus Williams, Xenosphex, 439
Xerostictia Gillaspy, 54, **552**
Xestocrabro Ashmead, 50
xila (Pate), Ammoplanellus, 200
xosa Brauns, Cerceris, 589
xuthus van der Vecht, Sphex, 117
Xylocelia Rohwer, 40
Xylocrabro Ashmead, 50
Xyloecus Shuckard, 41
xylurgus (Shuckard), Ectemnius, 426
Xysma Pate, 41, **193**

yaeyamanus Tsuneki, Tachysphex, 272
yaeyamanus Tsuneki, Tachytes, 267
yakima Banks, Philanthus, 565
yalensis Turner, Cerceris, 589
yalta Evans and Matthews, Bembix, 549
yamato Tsuneki, Crossocerus, 400
yamato Tsuneki, Passaloecus, 184
yamatonicus (Tsuneki), Crossocerus, 403
yamatonis Tsuneki, Lestiphorus, 506
yamatonis Tsuneki, Psenulus, 173
yanoi Tsuneki, Cerceris, 577
yanoi (Tsuneki), Crossocerus, 404
Yanonius Tsuneki, 50
yarrowi Beaumont, Tachysphex, 277
yarrowi Cresson, Ammophila, 150
yarrowi Leclercq, Ammophila, 152
yarrowi Leclercq, Podagritus, 393
yasumatsui Gussakovskij, Psen, 167
yasumatsui (Tsuneki), Crossocerus, 404

yasumatsui Tsuneki, Nitela, 325
yatesi Mickel, Larropsis, 259
yavapai Pate, Metanysson, 481
yenpingensis Tsuneki, Cerceris, 589
yerburii (Cameron), Crossocerus, 404
yerburyi Bingham, Philanthus, 567
yerburyi Bingham, Tachytes, 267
yercaudi Leclercq, Rhopalum, 390
yermasoyensis Balthasar, Miscophus, 319
yeto Tsuneki, Crossocerus, 400
yezo Tsuneki, Trypoxylon, 346
yngvei Cameron, Cerceris, 589
yomasanus van Lith, Psen, 167
yorki Leclercq, Dasyproctus, 421
yorkoides Leclercq, Dasyproctus, 421
yosemite Pate, Ectemnius, 428
yoshikawai Tsuneki, Ammatomus, 513
yoshikawai (Tsuneki), Polemistus, 185
yoshimotoi R. Bohart, Ectemnius, 428
yoshimotoi van Lith, Psenulus, 174
ypsilon Arnold, Bembix, 546
yucaipa Pate, Timberlakena, 201
yucatanensis (Cameron), Enoplolindenius, 415
yucatanensis Cameron, Tachytes, 264
Yuchiha Pate, 48
yumae R. Bohart, Oryttus, 508
yungvei Arnold, Cerceris, 589
yunkara Evans and Matthews, Bembix, 549
yunnanensis Tsuneki, Cerceris, 589

zacatecas Scullen, Cerceris, 589
zaidamensis (Radoszkowski), Crossocerus, 402
sakakiensis Balthasar, Miscophus, 319
zalinda Beaumont, Prosopigastra, 286
zanoni (Gribodo), Prionyx, 134
zanthoptera Cameron, Ammophila, 154
Zanysson Rohwer, 51, **475**
zapoteca Saussure, Cerceris, 583
zapotecus (Cresson), Epinysson, 472
zarcoi Giner Marí, Ammoplanus, 198
zarudniana Gussakovskij, Larra, 238
zarudnyi Gussakovskij, Bembix, 549
zarudnyi (Gussakovskij), Prionyx, 134

zarudnyi Gussakovskij, Tachysphex, 273
zavadili Snoflak, Spilomena, 193
zavattari Guiglia, Oxybelus, 370
zavattarii Guiglia, Cerceris, 589
zavattarii Guiglia, Tachysphex, 275
zealandum Leclercq, Rhopalum, 390
zebratus Cresson, Philanthus, 567
zelica Banks Cerceris, 589
zellus (Rohwer), Lindenius, 384
Zelorhopalum Leclercq, 47, 388
zerbeii (Viereck), Larropsis, 258
zethiformis Giordani Soika, Cerceris, 589
zetterstedti (Dahlbom), Editha, 543
zhelochovtzevi Gussakovskij, Stizus, 527
zibanensis (Morice), Bembecinus, 532
ziegleri Lepeletier and Brullé, Crossocerus, 404
ziegleri Rayment, Cerceris, 589
zikae Arnold, Trypoxylon, 348
zimini (Gussakovskij), Belomicroides, 363
zimini Gussakovskij, Didineis, 461
zimini Gussakovskij, Solierella, 314
zimini Gussakovskij, Stizus, 527
zimmeri Mickel, Tachysphex, 277
zimmermanni Dahlbom, Chalybion, 103
zimpanensis Scullen, Eucerceris, 592
zobeide Brimley, Cerceris, 579
zonalis F. Smith, Cerceris, 589
zonata Cresson, Cerceris, 589
zonata Klug, Bembix, 549
zonata (Say), Eucerceris, 592
zonatum Saussure, Chlorion, 90
zonatus (Klug), Stizus, 527
zonatus (Panzer), Ectemnius, 428
zonosoma Handlirsch, Stizus, 527
zorah Beaumont Parapiagetia, 281
zosma Brimely, Cerceris, 579
Zoyphidium Pate, 44, 301
Zoyphium Kohl, 44
zumpango Scullen, Cerceris, 589
zurki Leclercq, Trypoxylon, 348
zyx Leclercq, Cerceris, 589
Zyzzyx Pate, 54, **549**